FOUNDATIONS OF AUGMENTED COGNITION

VOLUME 11

FOUNDATIONS OF AUGMENTED COGNITION

VOLUME 11

Edited By

Dylan D. Schmorrow

CRC Press
Taylor & Francis Group
Boca Raton London New York

CRC Press is an imprint of the
Taylor & Francis Group, an **informa** business

CRC Press
Taylor & Francis Group
6000 Broken Sound Parkway NW, Suite 300
Boca Raton, FL 33487-2742

First issued in paperback 2019

ISBN-13: 978-0-8058-5806-8 (hbk)
ISBN-13: 978-0-367-39186-7 (pbk)

Visit the Taylor & Francis Web site at
http://www.taylorandfrancis.com

and the CRC Press Web site at
http://www.crcpress.com

HCI International 2007

The 12th International Conference on Human-Computer Interaction, HCI International 2007, will take place jointly with:

- Symposium on Human Interface (Japan) 2007
- 7th International Conference on Engineering Psychology and Cognitive Ergonomics
- 4th International Conference on Universal Access in Human-Computer Interaction
- 2nd International Conference on Virtual Reality
- 2nd International Conference on Usability and Internationalization
- 2nd International Conference on Online Communities and Social Computing
- 3rd International Conference on Augmented Cognition
- 1st International Conference on Digital Human Modeling

The Conference will be held in Beijing, P.R. China, 22-27 July, 2007, and will cover a broad spectrum of HCI-related themes on theoretical, generic and applied areas of HCI through the following modes of communication: plenary / keynote presentation(s), parallel sessions, poster sessions and tutorials. For more information, please visit the Conference website: www.hci-international.org

General Chair:

Constantine Stephanidis
University of Crete and FORTH-ICS
Greece
cs@ics.forth.gr

Parallel paper presentations as well as abstracts of posters will be included in the Conference Proceedings, to be published in CD-ROM by Lawrence Erlbaum Associates, Inc (LEA). In addition, the Conference Proceedings will also be published on demand by LEA in a multi-volume book edition.

Table of Contents

Chapter 9: Transforming Sensors into Cognitive State Gauges

Section 3: Augmented Cognition Technology

Chapter 10: Fundamentals of Augmented Cognition

Chapter 11: Context Modeling for Augmented Cognition

Chapter 12: Issues of Trust in Adaptive Systems

Chapter 13: Closed Loop Systems - Stability and Predictability

Section 4: Augmented Cognition and Advanced Computing

Chapter 14: Stress in the Computing Environment

Chapter 18: Affective Computing

Section 5: AugCog New Directions

Chapter 19: Augmented Cognition for Training Superiority

Chapter 23: Augmented Cognition and its Influence on Decision Making

Chapter 24: Team Cognition

Preface

This volume brings together a comprehensive and diverse collection of research, theory and thought that serves as the basis for the new field of Augmented Cognition research and development.

The goal of Augmented Cognition research is to create revolutionary human-computer interactions that capitalize on recent advances in the fields of neuroscience, cognitive science and computer science. Augmented Cognition can be distinguished from its predecessors by the focus on the real-time cognitive state of the user, as assessed through modern neuroscientific tools. At its core, an Augmented Cognition system is a 'closed loop' in which the cognitive state of the operator is detected in real time with a resulting compensatory adaptation in the computational system, as appropriate (Kruse & Schmorrow, Chapter 10).

Augmented Cognition research and development became a possibility through the intersection of two separate scientific revolutions that were culminating at the start of the 21st century. The information technology revolution began with the birth of modern computer science and the dramatic advances that followed in information processing capabilities. This revolution gave rise to ubiquitous and pervasive information processing tools and provided scientists, engineers and human systems integration practitioners with a powerful capability to sort, process and manage vast amounts of data at incredible speeds. These speeds have allowed us to have near instantaneous access to information and analysis; quantitatively based, instant answers were now possible within the human decision time frame. Tools from the information technology revolution enabled another revolution, this one involving the human brain. At the close of the 20th century, basic science understanding of the human brain began to provide quantitative insight on how the human brain functioned at the most basic elemental levels and how human thought and performance was associated with brain functioning. The human brain revolution provided the beginnings of a detailed knowledge of human thought processing, functioning and performance. Augmented Cognition researchers and developers seized this opportunity and used the emergent technologies and tools to advance and build computational systems that were able to dynamically adapt to real time human thought, supporting orders of magnitude gains in performance.

The precursors for Augmented Cognition lie in the worlds of Psychology, Neuroscience, Mathematics, and Computer Science. Psychology has given us the questions to ask regarding the ways in which people perceive and process information. Neuroscience, and in large part, the human brain revolution, has given us the understanding of the brain's components, as well as the sensing technologies to begin to quantitatively understand the brain. Advanced Mathematics has given us the ability to parse through the information from these sensors to draw conclusions. And, were it not for the huge leaps in processing power through Computer Science, it would not be possible to process this information in real time, as is required for the dynamically adaptable systems that comprise Augmented Cognition.

Today, we can make what was once imagined as impossible, possible. With technology and computing power continuing to enable significant advances in science, Augmented Cognition has not only contributed to scientific advances, but is rapidly turning these advances into practice.

Thanks in large part to the Improving Warfighter Information Intake Under Stress program at the Defense Advanced Research Projects Agency (DARPA), major developments have been made in sensing technologies (real time spatial and temporal imaging of brain activity), cognitive state detection (real time processing and characterization of brain function), and information presentation (improved encoding, storage, and retrieval of information through cognitively designed information system). This program is now applying developed technologies to various platforms, to engender interest and support by future customers.

DARPA's program, however, is only one part of this new and growing field of research. As demonstrated by the contributors to this volume, there is a wide and varied field of scientists investigating Augmented Cognition. There is also a growing collection of application areas for this technology. Augmented Cognition technologies are particularly well-suited for the design and evaluation of Human Computer Interfaces (we can more accurately detect aspects that cause a high or low workload in users), in training/education applications (we can get individualized, quantitative measures of how well a student understands material, to help speed the progression from novice to expert), and for military systems (many environments bombard today's warfighter with so much information that they absolutely require new strategies for sorting through everything).

This volume will provide you with insight into the field of Augmented Cognition: where it has been, and – more importantly – where it is going. The chapters of papers are grouped into sections: Human Information Processing, Cognitive State Sensors, Augmented Cognition Technology, Augmented Cognition and Advanced Computing, and Augmented Cognition New Directions. These themes represent distinct aspects of the field of Augmented Cognition, and run the gamut between the information processing capabilities of humans, to the detection of that processing, to the systems that are being developed to advance and exploit Augmented Cognition technologies, to the aspects and requirements of more advanced technology applications, and even to the future of the field.

The work of Augmented Cognition academics and practitioners has begun to be disseminated and exchanged and this scientific information and understanding will continue to drive further technological advances. Humans can no longer keep pace with the overwhelming amount of information and demanding decisions that must constantly be made. The need for computational systems to know what is going on in both the environment and in the human brain has only begun. What follows in these pages serves as the foundation of this new field of research and development. In future years, the information in this volume will likely be seen as immature and suffering from gaps in knowledge. I choose to see this as an opportunity. I choose to see this as the beginning of a golden age of exploration. I choose to see this as the moment when computational systems will truly begin to adapt to humans. These foundational papers are "diamonds-in-the-rough." It is my hope that the authors of the following chapters, along with others who will join the ranks of these pioneers, will bring about the true revolution in Human-Computer Interaction.

-- Dylan Schmorrow

The papers contained in this volume were presented at the 1st International Conference on Augmented Cognition held in Las Vegas, Nevada on July 22-27, 2005. This conference was held in conjunction with the 11th International Conference on Human-Computer Interaction, the Symposium on Human Interface (Japan) 2005, the 6th International Conference on Engineering Psychology & Cognitive Ergonomics, the 3rd International Conference on Universal Access in Human-Computer Interaction, the 1st International Conference on Virtual Reality, the 1st International Conference on Usability and Internationalization, and the 1st International Conference on Online Communities and Social Computing. This unique cross-sectional gathering provided a forum for Augmented Cognition researchers to discuss not only the advances in emerging scientific research and associated technologies, but also an opportunity to share practical application advances.

Foreword

With the rapid introduction of highly sophisticated computers, (tele) communication, service, and manufacturing systems, a major shift has occurred in the way people use technology and work with it. The objective of this book series on Human Factors and Ergonomics is to provide researchers and practitioners a platform where important issues related to these changes can be discussed, and methods and recommendations can be presented for ensuring that emerging technologies provide increased productivity, quality, satisfaction, safety, and health in the new workplace and the Information Society.

The present volume is published at a very opportune time, when the Information Society Technologies are emerging as a dominant force, both in the workplace, and in everyday life activities. In order for these new technologies to be truly effective, they must provide communication modes and interaction modalities across different languages and cultures, and should accommodate the diversity of requirements of the user population at large, including disabled and elderly people, thus making the Information Society universally accessible, to the benefit of mankind.

Augmented Cognition is a rapidly evolving disciple being the brain child of Dylan D. Schmorrow, a Program Manager at DARPA and ONR, and Commander in the U.S. Navy. This book is the proceedings of the 1st International Conference on Augmented Cognition jointly held with HCI International 2005 Conference in Las Vegas, July 22–27, 2005. The 2nd International Conference on Augmented Cognition will be held in conjunction with the Human Factors and Ergonomics Society 50th Annual Meeting, October 15–16, 2006 at the San Francisco Hilton in San Francisco, CA. The 3rd International Conference on Augmented Cognition will be held jointly with HCI International 2007 in July 22–27, 2007 in Beijing, P.R. China (http://www.hcii2007.org/).

The book's five sections and 24 chapters present the theoretical foundation and operational models for the understanding, design and operation of engineering systems requiring increased interaction with humans. The effectiveness of the operation of such systems is significantly unchanged through augmented cognition integration into the human information processing utilized for decisions and actions.

The book would be of special value to individuals working in human-computer interaction, human factors and ergonomics and cognitive science who are interested to learn today where the future of the discipline will lead.

—Gavriel Salvendy
Purdue University, USA
Tshighua University, P.R. China

Section 1
Human Information Processing

Chapter 1

Multimodal Interfaces in Human Computer Interaction

Implications of Compatibility and Cuing Effects for Multimodal Interfaces

Robert W. Proctor

Purdue University
Dept. of Psychological Sciences
703 Third St.
West Lafayette, IN 47907
proctor@psych.purdue.edu

Hong Z. Tan

Purdue University
School of Electrical and Computer Engineering
465 Northwestern Avenue
West Lafayette, IN 47907
hongtan@purdue.edu

Kim-Phuong L. Vu

California State University, Northridge
Dept. of Psychology
18111 Nordhoff St.
Northridge, CA 91325
kimvu@csun.edu

Rob Gray

Arizona State University East
Dept. of Applied Psychology
7001 E Williams Field Road
Mesa, AZ 85212
robgray@asu.edu

Charles Spence

Oxford University
Dept. of Experimental Psychology
South Parks Road
Oxford, OX1 3UD
charles.spence@psy.ox.ac.uk

Abstract

Studies of stimulus-response compatibility show that certain stimulus modalities go more naturally with particular response modalities than others. For example, it is more natural to say the word "left" to a spoken word LEFT than to move a control left, whereas it is more natural to move a control left to a light onset to the left than it is to say the word "left". Such modality relations determine not only how quickly a person can respond to relevant information but also the amount of interference that they may get from irrelevant distracting information. Cues that signal the spatial location of an upcoming target stimulus facilitate processing not only when the two stimuli are presented in the same modality but also when they occur in different modalities. In general, these crossmodal cuing effects occur when the cue and target stimulus are spatially co-located. In this paper, we review the findings on compatibility and cuing studies across stimulus modalities and discuss their implications for multimodal interface design.

1   Introduction

Multimodal interfaces are increasingly being used for a variety of purposes because they have the potential to facilitate effective and efficient interactions between humans and computers (Hempel & Altınsoy, 2005; Sorkin, 1987). The use of multiple display and control modalities enables different ways of presenting and responding to information, the incorporation of redundancy into displays, and emulation of real-life environments. Multimodal interfaces can reduce mental workload and make human-computer interactions more naturalistic. However, designing effective multimodal interfaces is a challenge because many interactive effects between different modalities may arise. These effects must be taken into account if the full benefits of multimodal interfaces are to be realized.

The present paper reviews findings from two areas of research that have direct implications for multimodal interface design. The first area, that of compatibility effects, examines performance as a function of the modalities of displays and controls and the individual mappings of display elements to control elements. The second area, that of cuing effects, examines interactions of spatial location information across different sensory modalities.

2   Compatibility Effects

In many applied contexts, people are required to make rapid and accurate controlling actions in response to different displayed events. Performance on these tasks can vary in efficiency as a function of the extent to which the required action to the event is consistent with the operator's natural response tendencies, or, how compatible the stimuli are with the responses. Compatibility is particularly important to take into account for multimodal interfaces because

3

compatibility effects occur for a variety of stimulus and response modes, are stronger for some modes than for others, and may have interactive effects across modes and tasks.

People's natural response tendencies for different display-control configurations have been documented in the literatures on *population stereotypes* for control-display mappings (Hoffmann, 1997) and *stimulus-response compatibility effects* (Hommel & Prinz, 1997). Several compatibility principles that have emerged from this research are summarized in Table 1.

Table 1: Compatibility Principles

- **Spatial Compatibility**
  - o Compatible mappings of stimuli assigned to their spatially corresponding responses typically yield better performance.
  - o Better performance occurs when the mapping of stimuli to responses can be characterized by a rule or relation than when it is random.
- **Movement Compatibility**
  - o The motion of the display should move in the same direction as the motion of the control.
  - o Clockwise movement is used to indicate upward movement or an increase in magnitude of the display.
- **Proximity Compatibility**
  - o Controls should be placed closest to the display they are controlling.
  - o Controls and displays should be arranged in functionally corresponding groups.
  - o Control and displays should be sequentially arranged.
- **Mode Compatibility**
  - o Better performance occurs when there is a match between display and control modes (visuospatial-manual and verbal-vocal) than when there is not.
  - o Less interference from irrelevant information when it is conveyed by a different stimulus mode than the relevant information.
- **Others**
  - o The up-right/down-left mapping is often better than the up-left/down-right mapping.
  - o Pure tasks of a single stimulus-response mapping produce better performance than mixed tasks with multiple mappings.

Much of the research on stimulus-response compatibility effects has examined differences in performance for different mappings of stimuli to responses. In the simplest type of choice reaction task, involving two alternatives, a left or right keypress is made to a visual stimulus appearing on the left or right. The spatially corresponding mapping of left stimuli to left responses and right stimuli to right responses yields better performance (i.e., faster reactions and fewer errors) than the spatially non-corresponding mapping of left stimuli to right responses and right stimuli to left responses (e.g., Vu & Proctor, 2004). Mapping effects such as these, in which performance varies as a function of the mappings of individual stimulus and response elements within the same stimulus and response sets, are sometimes called element-level compatibility effects (Kornblum, Hasbroucq, & Osman, 1990). An interesting variant of spatial compatibility is the Simon effect, in which, when stimulus location is irrelevant and another dimension such as color is relevant, performance is still better when stimulus and response locations correspond than when they do not (see Simon, 1990, for a comprehensive review). Both stimulus-response compatibility proper and the Simon effect also occur for auditory (e.g., Roswarski & Proctor, 2000; Simon, 1990) and vibrotactile stimuli (Hasbroucq, Guiard, & Kornblum, 1989), as well as when the responses consist of left versus right footpedal responses (e.g., Vu & Proctor, 2001), movements of a joystick or switch to the left or right (e.g., Vu & Proctor, 2001), aimed movements of a finger to a left or right location (Wang & Proctor, 1996), or turns of a steering wheel clockwise (right) or counterclockwise (left; Proctor, Wang, & Pick, 2004).

Element-level compatibility effects occur not only when the stimuli differ in physical location but also when the spatial information is conveyed symbolically (e.g., left or right pointing arrow) or verbally (e.g., the word left or right; Vu & Proctor, 2004). Moreover, the responses can also be spoken directional words, such as "left" or "right" (Vu & Proctor, 2004). These outcomes indicate that compatibility effects occur both when there is physical similarity between the stimulus and response dimensions and when there is only conceptual similarity. Although a match between stimulus and response dimensions at the conceptual level produces a compatibility effect, the effect is typically smaller than when the dimensions also match at a physical, or mode, level as well. With visual stimuli, performance is better when the stimulus and response modes match (i.e., spoken responses are made to written location words or manual responses to visuospatial stimuli) than when they do not (i.e., spoken responses are made to visuospatial stimuli or manual responses to written location words; Wang & Proctor, 1996; see also Wickens, 1992). Such differences in mode relations are a type of set-level compatibility effect (Kornblum et al., 1990), that is, a difference in the compatibility of the overall stimulus set with the response set. According to Kornblum et al.'s dimensional overlap model, which attributes the set-level compatibility effects to stronger automatic activation of the corresponding response when set-level compatibility is high than when it is low, the high compatibility sets should yield both faster responses with a compatible mapping and slower responses with an incompatible mapping than the low compatibility sets. However, although the match between stimulus and response modes facilitates responding when the mapping is compatible (Kornblum & Lee, 1995; Proctor, Wang, & Vu, 2002), a mode mismatch does not necessarily cause interference when the mapping is incompatible (see Proctor et al., 2002). Greenwald (1970) proposed that ideomotor compatible stimuli and responses, that is, those for which the stimulus is similar to the sensory feedback provided by the response (e.g., saying the word "ten" in response to the spoken word "ten"), have a special advantage. According to Greenwald, response selection is bypassed when ideomotor compatible responses to stimuli are required because action is mediated by an image of the sensory feedback in this case.

Physical similarity is important in determining the amount of interference produced by irrelevant stimulus information. For example, Baldo, Shimamura, and Prinzmetal (1998) used left or right pointing arrows and the words left and right as stimuli. Robust Simon effects were obtained when the participants responded manually to the words, with the arrows being irrelevant, and when they responded vocally (saying "left" or "right") to the arrows while ignoring the irrelevant words. The effects of irrelevant information on performance were reduced substantially when the response mode was physically similar to that of the relevant stimulus dimension, that is, when responding manually to arrows and vocally to words. Lu and Proctor (2001) obtained similar results using stimuli for which a location word (left or right) was embedded inside of an outline arrow pointing to the left or right. When arrow direction was relevant, keypress responses showed little influence of the irrelevant location word, but when location word was relevant, keypress responses showed a substantial correspondence effect for irrelevant arrow direction.

Compatibility is a significant factor in dual-task performance. When the stimuli for two reaction tasks are presented close together in time, reaction time to the second task is typically slowed. This phenomenon is called the *psychological refractory period* (PRP) effect. Greenwald and Shulman (1973) provided evidence suggesting that no PRP effect occurs when both tasks are ideomotor compatible (e.g., moving a joystick left or right in response to a left or right arrow and saying "A" of "B" to the spoken letter A or B). Although the elimination of PRP effects has not been replicated in several recent experiments (Lien, Proctor, & Allen, 2002; though see also Greenwald, 2003; Lien, Proctor, & Ruthruff, 2003), the effect size is clearly much smaller than that obtained in most situations (see also Schumacher et al., 2001). Another finding of importance in the PRP literature is that crosstalk effects across tasks (similar to the Simon effect within a single task) occur in many situations, with a stimulus for one task tending to activate its corresponding response for the other task (e.g., Hommel, 1998). Interestingly, such crosstalk effects are largely absent when the two tasks are ideomotor compatible (Lien, McCann, Ruthruff, & Proctor, 2005). Discussion of a broader range of compatibility issues in dual-task performance can be found in Lien and Proctor (2002).

In addition to physical and conceptual similarity contributing to compatibility effects, it is generally accepted that what is called structural similarity contributes to performance as well (e.g., Cho & Proctor, 2003; Kornblum & Lee, 1995; Reeve & Proctor, 1984). That is, performance benefits when correspondence in the structure of the stimulus and response sets is maintained, even in the absence of physical or conceptual similarity. For example, performance is better when 10 digits are mapped to the 10 fingers of the hands in a left-to-right order than when they are randomly assigned. When a symbolic stimulus set composed from two values on each of two dimensions (e.g., letter identity and size) is mapped to a row of four response keys, performance is best when the most salient dimension corresponds with the salient distinction between the two leftmost and two rightmost responses (Proctor & Reeve, 1985). Additionally, when stimuli and responses vary along orthogonal spatial dimensions, the mapping of an upper stimulus location to a right response and lower stimulus location to a left response often produces better performance than the alternative mapping because it maintains correspondence between the positive and negative alternatives of the two dimensions (Cho & Proctor, 2003). The point is that in mapping stimuli from different modes to responses, one has to accommodate the properties of the entire stimulus and response sets, as well as those of the elements from which they are composed.

When the elements of the display and control configurations can be coded along two spatial dimensions simultaneously (i.e., when they are arrayed along a diagonal), compatibility effects occur for both dimensions. However, depending upon which dimension is made more salient by the stimulus-response configuration, the compatibility effect can be larger for one dimension than for the other. This prevalence of one dimension can occur regardless of whether the stimuli are presented visually or auditorily (Nicoletti & Umiltà, 1984, 1985), and no matter whether the responses are executed with both hands and feet, a single hand and foot (Rubichi, Nicoletti, Pelosi, & Umiltà, 2004), or with unimanual joystick movements (Vu & Proctor, 2001). Thus, the manner in which the display and control elements are configured can provide the basis for coding along the salient dimension when more than one spatial reference frame is provided. More generally, when stimulus and response sets can be coded with respect to more than one frame of reference, which pairings of stimuli and responses are most compatible is dependent upon the frames on which coding is based (Proctor et al., 2004).

Although compatibility effects are robust, occurring in essentially all stimulus and response modalities and even after extended practice, there have been several demonstrations that the benefit for spatial compatibility can be eliminated through the influence of other associations defined for a task performed prior to, or concurrent with, a spatially compatible task (e.g., see Proctor & Lu, 1999; Proctor, Vu, & Marble, 2003; Tagliabue, Zorzi, & Umiltà, 2002). Vu and Proctor (2004) showed that when set-level compatibility is high (i.e., the stimulus-response sets are visuospatial-manual or verbal-vocal), mixing compatible and incompatible mappings within a block of trials decreases overall task performance in comparison to when each mapping is performed in isolation, and, more important, eliminates the benefit for the compatible mapping. The cost of mixing on the compatibility effect can be reduced, though, by presenting the stimuli for each mapping in different stimulus modes (e.g., location words as stimuli for one mapping and physical locations as stimuli for the other; Proctor, Marble, & Vu, 2000; Proctor & Vu, 2002).

Although our discussion of compatibility effects has focused on stimuli and responses with spatial or directional properties, it is important to emphasize that compatibility effects for both relevant and irrelevant stimulus dimensions occur whenever there is any similarity, or overlap, between stimulus and response dimensions. For example, a Simon effect has been shown to occur on the basis of the irrelevant positive or negative affective content of words to which a vocal response "positive" or "negative" is to be made based on another stimulus attribute such as whether the word is a noun or verb (e.g., De Houwer, Crombez, Baeyens, & Hermans, 2001). A Simon effect has also been demonstrated to occur when responses of long and short durations must be made to stimuli that vary on an irrelevant dimension of duration (i.e., long vs. short; Kunde & Stöcker, 2002).

3   Cuing Effects

Like many studies on stimulus-response compatibility, studies on cuing effects typically also use reaction time and error rate as performance metrics (though see Prinzmetal, McCool, & Park, 2005). In a typical experiment using the *orthogonal cuing paradigm* (see Spence, McDonald, & Driver, 2004, and Driver & Spence, 2004, for recent reviews), a participant receives vibrotactile stimulation to the left or right hand (the cue) followed by the illumination of one of two visual LEDs held by the left or right hand (the target), and is asked to make a speeded response to indicate whether an upper or lower LED was illuminated by pressing one of two footpedals. Cuing effects are measured in terms of the difference in reaction times between the valid (when the cue and target occur on the same side) and invalid cuing conditions (when the cue and target occur on different sides). This difference in performance between valid and invalid trials has been taken as providing a measure of the extent to which the presentation of stimuli in one sensory modality can direct, or capture, *spatial* attention in another modality (e.g., Spence, 2001).

Auditory, visual, as well as vibrotactile stimuli have been examined in spatial cuing experiments. It has been demonstrated that the speeded detection of a visual target is faster and tends to be more accurate following the presentation of a spatially-noninformative peripheral auditory cue presented on the same side of the visual target rather than on the opposite side (e.g., Bolognini et al., 2005; Spence & Driver, 1997; see also Prinzmetal, Park, &

Garrett, in press). By contrast, speeded discrimination responses for auditory targets are only affected by the prior presentation of spatially noninformative visual cues under certain situations, but not others (see Ward, McDonald, & Lin, 2000; McDonald, Teder-Sälejärvi, Heraldez, & Hillyard, 2001; and see Spence et al., 2004, for a recent review). As far as the crossmodal pairing of visual and tactile stimuli goes, there is evidence that visual target judgments can be significantly affected by spatially non-predictive tactile cues, and vice versa (Spence, Nicholls, Gillespie, & Driver, 1998; Kennett, Eimer, Spence, & Driver, 2001; Kennett, Spence, & Driver, 2002; Gray & Tan, 2002; Tan, Gray, Young, & Traylor, 2003). Finally, spatially non-predictive tactile cues can also lead to significant crossmodal spatial-cuing effects upon auditory target judgments, and vice versa (Spence et al., 1998).

Most of the crossmodal (among vision, audition and touch) spatial-cuing effects discussed thus far fall into the category of *exogenous* (or involuntary) cuing, where attention to a spatial location is automatically elicited by the presentation of spatially non-informative peripheral cues at the same time as, or shortly before the onset of, the target stimuli. The orthogonal cuing paradigm is effective at eliciting exogenous cuing when the cues (typically presented on the left or right) do not predict the likely target location (e.g., up and down). The other type, *endogenous* orienting, is elicited if the participants are informed that the targets are more likely to occur at the cued location than at the non-cued location, or when an informative central arrow is used to indicate the likely site of target stimuli (see Driver & Spence, 2004, for a detailed review; see also Chambers, Stokes, & Mattingley, 2004). This is a voluntary form of directing one's spatial attention to the expected location. In addition to the simple target detection and discrimination tasks described above, many studies of crossmodal links in spatial attention have examined unspeeded temporal order judgments (e.g., Spence, Shore & Klein, 2001; Zampini, Shore, & Spence, in press), as well as perceptual sensitivity (e.g., Bolognini et al., 2005; Ho & Spence, in press; McDonald, Teder-Sälejärvi, & Hillyard, 2000; Prinzmetal et al., in press). Neuroimaging techniques, such as event-related potentials (ERPs) and functional magnetic resonance imaging (fMRI) have also been used to investigate the neural underpinnings of spatial attentional orienting (e.g., see chapters in Spence & Driver, 2004). We have chosen to focus our discussion on early (i.e., short SOA) exogenous crossmodal spatial-attention cuing using detection and discrimination latency tasks because of our interest in exploiting these results for the design of effective multimodal warning systems that can automatically capture a user's attention.

The issue of collocation is an important one in studies of crossmodal spatial attention. In general, performance is enhanced if information coming from more than one sensory modality is presented from approximately the same external location; and conversely, it is easier to reject distracting information presented at a different spatial location. Even when auditory and visual tasks are entirely unrelated, actively performing them together can be more efficient when the visual and auditory presentations originate from a common external spatial source, rather than from different locations. For example, Spence and Read (2003) reported that participants in a driving simulator found it easier to shadow (repeat) speech presented from the front of the vehicle (where visual attention is typically focused upon during driving; e.g., Lansdown, 2002) than that presented from the side. It has been suggested that some of the null crossmodal results found in earlier studies (e.g., Tassinari & Campara, 1996, using a tap on the shoulder and the illumination of a square on a screen) might have been due to the fact that the cue and target stimuli were presented from very different spatial locations even when they were on the same side with respect to the participant's torso (cf. Spence et al., 1998; although see later experiments by Tan et al., 2003, discussed below). The importance of collocation is further underscored by a recent study using moving cues. Gray and Tan (2002) suggested dynamic and predictive spatial links between touch and vision by demonstrating faster visual discrimination performance at the final tactile pulse location derived from moving tactile cues, and by demonstrating the dependence of tactile discrimination performance on the visual cuing object's time to contact.

An exception to the cue-target collocation rule is provided by a study of Tan et al. (2003) in which participants received vibrotactile cues on the four corners of their backs prior to searching for a visual change on a computer monitor. Participants were informed that the location of the tactile cue predicted the quadrant of the visual change on 50% of the trials, hence the task elicited both endogenous and exogenous spatial attention cuing. Tan et al. (2003) reported that visual detection time decreased significantly when the location of the tactile cue was in the same quadrant as that of the visual change, and that detection time increased significantly when the tactile cue and the visual target occurred in different quadrants. Another recent study by Ho, Tan, and Spence (2004; submitted) confirmed that crossmodal attention cuing effects can be elicited when the (tactile) cue and the (visual) target are presented from very different locations (so long as the direction in which the stimuli are presented was matched). In a simulated driving environment, participants felt a vibrotactile stimulus presented on the front or back around the waist, and were required to brake, accelerate or maintain constant speed by checking the front or the rearview mirror for a potential emergency driving situation (i.e., the rapid approach of a car from either in front or behind). It was found that participants responded significantly more rapidly following valid vibrotactile cues than following invalid cues. A further twist to the spatial setup of this experiment was that when prompted by a vibrotactile cue to the *back*, participants were able to look at the rearview mirror in the *front* in order to check the traffic condition *behind* the vehicle. Therefore, it would appear that the cue-target collocation rule can be relaxed when a tactile cue is involved, and when the spatial mapping between the cue and target is overlearned (such as in driving). This is a useful result that should be explored in designing multimodal warning systems. Whereas it is generally desirable to match the cue and target stimuli locations to maximize spatial cuing effects, tactile cues may be effectively deployed when it is not feasible to place warning signals at exactly the same location as that of dangerous events.

The relative effectiveness of vision and kinesthesis for crossmodal spatial attention cuing was examined by Klein (1977). He showed that whereas it takes the same amount of time to switch attention *from* vision and *from* kinesthesis, people can more rapidly switch their attention *to* kinesthesis than *to* vision. In general, kinesthetic stimuli seem to be superior to visual stimuli in alerting attention, and kinesthetic responses are generally faster (albeit less accurate) than visual or auditory responses (Robinson, 1934, Table 7). Klein (1977) speculated that our attentional bias to vision may stem from the relatively poor alerting capability of visual stimuli. Numerous studies have now shown that spatially non-informative tactile cues can effectively elicit an automatic exogenous shift of attention that facilitates subsequent responses to visual, audio and tactile stimuli (Spence et al., 1998; Spence & McGlone, 2001; Kennett et al., 2001; Kennett et al., 2002). Therefore, touch is an extremely effective modality for alerting. Touch stimuli with spatial information can potentially speed up visual response to pending hazardous situations. Given the effectiveness of exogenous cuing, there seems to be no need for extensive user training in order for a multimodal warning system to be highly effective.

4   Summary

Human factors specialists have recognized the importance of compatibility effects in designing interfaces since the earliest days of research in the field. However, the robustness of such effects and the fact that there are many aspects of compatibility that must be considered when designing multimodal displays have not been fully appreciated. Element-level compatibility effects, that is, differences in performance as a function of stimulus-response mapping, occur for a variety of situations in which the stimulus and response dimensions have some similarity. This similarity need not be spatial, nor does it need to be a physical, perceptual property. Compatibility effects can occur solely as a function of conceptual or structural similarity between stimulus and response sets, and thus will be evident across stimulus and response modalities as well as within them. It is important for designers to realize, though, that the set-level compatibility for certain combinations of stimulus and response modalities is higher than for others. Those combinations for which set-level compatibility is high, such as physical stimulus location and physical response location, will yield the best performance when the modality relation is relevant and the mapping of display and control elements is also compatible. However, when the modality relation is irrelevant, as when a relevant stimulus dimension such as color is responded to with a keypress and stimulus location is irrelevant, the most intrusion will tend to be observed. When compatible and incompatible mappings are mixed or multiple tasks must be performed, interactions among mappings and tasks often occur. In particular, the performance benefit of a compatible mapping is often drastically reduced. For multiple tasks, crosstalk between tasks can occur, and certain combinations of tasks are easier to perform together than others.

In recent years, numerous studies on cuing effects have suggested substantial spatial attentional links between vision, audition and touch. The consensus seems to be that auditory and tactile cues are more effective at directing visual attention than vice versa, although vision is still the preferred modality for detailed information processing. Auditory and tactile stimuli may be more automatically alerting than visual stimuli (e.g., see Posner et al., 1976; Klein, 1977). Spatially-informative as well as spatially-noninformative cues can effectively elicit shifts in spatial attention both in the same sensory modality and across modalities. Stronger cuing effects are achieved when the cue and target stimuli occur at the same spatial location, although non-collocated haptic cues can be just as effective when there is a logical mapping between the cue and target locations. In the context of designing multimodal warning signals, a recommended approach is to use spatial sounds or haptic cues to direct an operator's visual attention towards a critical location (e.g., the side of the vehicle with an impending collision, or an area on a large display that demands immediate action). Given that it is not always practical to deliver cues at the same location as that of an upcoming event, future research should focus on the conditions under which cues presented in the peripersonal space can be effectively used to elicit attention shift in the external space.

In the typical cuing task, the concern is primarily with directing attention to the location at which the target stimulus is expected to appear. Better performance for stimuli occurring at the cued location than at uncued locations is attributed to attention. Because the main purpose of cuing studies is to examine the facilitatory effects of spatial attention on stimulus identification, little consideration is typically given to which action is made in response to the stimulus. In contrast, in most compatibility studies, the concern is with the mapping or spatial correspondence between individual members of the stimulus set and individual members of the response set. Because of expected interactions between spatial cuing and compatibility effects, a more comprehensive approach to these effects would be to examine both together. As one example, the use of orthogonal cuing and stimulus (response) dimensions in cuing studies was intended to preclude response priming as a factor. However, stimulus-response compatibility effects can occur for orthogonal stimulus-response mappings (Cho & Proctor, 2003), even when stimulus location is irrelevant. Whether differential priming of responses occurs in the typical cuing task can be evaluated by examining correspondence effects between the cued location and the response location assigned to the target stimulus.

Interactions of other types between cuing and stimulus-response compatibility need to be examined as well. A tactual or auditory cue might be used to direct visual attention to a desired location, but it could have an inadvertent effect of priming the person to make a response corresponding to that location. Similarly, it may seem intuitive to use a warning signal to direct a person's attention toward the location of a potentially critical event, such as an impending collision with another vehicle. However, because this location is opposite to the direction in which the action should be taken to avoid the collision, any tendency that the warning induces to respond in the corresponding direction would be undesirable. Issues such as these are critical for the design of any multimodal interface that uses spatial information.

References

Baldo, J. V., Shimamura, A. P., & Prinzmetal, W. (1998). Mapping symbols to response modalities: Interference effects on Stroop-like tasks. *Perception & Psychophysics, 60,* 427-437.

Bolognini, N., Frassinetti, F., Serino, A., & Làdavas, E. (2005). "Acoustical vision" of below threshold stimuli: Interaction among spatially converging audiovisual inputs. *Experimental Brain Research, 160,* 273-282.

Chambers, C. D., Stokes, M. G., & Mattingley, J. B. (2004). Modality-specific control of strategic spatial attention in parietal cortex. *Neuron, 44,* 925-930.

Cho, Y. S., & Proctor, R. W. (2003). Stimulus and response representations underlying orthogonal stimulus-response compatibility effects. *Psychonomic Bulletin & Review, 10,* 45-73.

De Houwer, J., Crombez, G., Baeyens, F., & Hermans, D. (2001). On the generality of the affective Simon effect. *Cognition & Emotion, 15,* 189-206.

Driver, J., & Spence, C. (2004). Crossmodal spatial attention: Evidence from human performance. In C. Spence & J. Driver (Eds.), *Crossmodal space and crossmodal attention* (pp. 179-220). Oxford, UK: Oxford University Press.

Gray, R., & Tan, H. Z. (2002). Dynamic and predictive links between touch and vision. *Experimental Brain Research, 145,* 50-55.

Greenwald, A. G. (1970). A choice reaction time test of ideomotor theory. *Journal of Experimental Psychology, 86,* 20-25.

Greenwald, A. G. (2003). On doing two things at once: III. Confirmation of perfect timesharing when simultaneous tasks are ideomotor compatible. *Journal of Experimental Psychology: Human Perception & Performance, 29,* 859-868.

Greenwald, A. G, & Shulman, H. G. (1973). On doing two things at once: II. Elimination of the psychological refractory period effect. *Journal of Experimental Psychology, 101,* 70-76.

Hasbroucq, T., Guiard, Y., & Kornblum, S. (1989). The additivity of stimulus-response compatibility with the effects of sensory and motor factors in a tactile choice reaction time task. *Acta Psychologica, 72,* 139-144.

Hempel, T., & Altınsoy, E. (2005). Multimodal user interfaces: Designing media for the auditory and the tactile channel. In R. W. Proctor & K.-P. L. Vu (Eds.), *Handbook of human factors in Web design* (pp. 134-155). Mahwah, NJ: Erlbaum.

Ho, C., & Spence, C. (in press). Verbal interface design: Do verbal directional cues automatically orient visual spatial attention? *Computers in Human Behaviour.*

Ho, C., Tan, H. Z., & Spence, C. (2004; submitted). Using spatial vibrotactile cues to direct a driver's visual attention. *Transportation Research Part F: Traffic Psychology and Behavior.*

Hoffmann, E. R. (1997). Strength of component principles determining direction of turn stereotypes--Linear displays with rotary controls. *Ergonomics, 40,* 199-222.

Hommel, B. (1998) Automatic stimulus-response translation in dual-task performance. *Journal of Experimental Psychology: Human Perception and Performance, 24,* 1368-1384.

Hommel, B., & Prinz, W. (Ed.) (1997). *Theoretical issues in stimulus-response compatibility.* Amsterdam: North-Holland.

Kennett, S., Eimer, M., Spence, C., & Driver, J. (2001). Tactile-visual links in exogenous spatial attention under different postures: Convergent evidence from psychophysics and ERPs. *Journal of Cognitive Neuroscience, 13,* 462-478.

Kennett, S., Spence, C., & Driver, J. (2002). Visuo-tactile links in covert exogenous spatial attention remap across changes in unseen hand posture. *Perception & Psychophysics, 64,* 1083-1094.

Klein, R. M. (1977). Attention and visual dominance: A chronometric analysis. *Journal of Experimental Psychology: Human Perception and Performance, 3,* 365-378.

Kornblum, S., Hasbroucq, T., & Osman, A. (1990). Dimensional overlap: Cognitive basis for stimulus-response compatibility: A model and taxonomy. *Psychological Review, 97,* 253-270.

Kornblum, S., & Lee, J.-W. (1995). Stimulus-response compatibility with relevant and irrelevant stimulus dimensions that do and do not overlap with the response. *Journal of Experimental Psychology: Human Perception & Performance, 21,* 855-875.

Kunde, W., & Stöcker, C. (2002). A Simon effect for stimulus-response duration. *Quarterly Journal of Experimental Psychology, 55A,* 581-592.

Lansdown, T. C. (2002). Individual differences during driver secondary task performance: Verbal protocol and visual allocation findings. *Accident Analysis & Prevention, 34,* 655-662.

Lien, M.-C., & Proctor, R. W. (2002). Stimulus-response compatibility and psychological refractory period effects: Implications for response selection. *Psychonomic Bulletin & Review, 9,* 212-238.

Lien, M.-C., Proctor, R. W., & Allen, P. A. (2002). Ideomotor compatibility in the psychological refractory period effect: 29 years of oversimplification. *Journal of Experimental Psychology: Human Perception and Performance, 28,* 396-409.

Lien, M. -C., Proctor, R. W., & Ruthruff, E. (2003). Still no evidence for perfect timesharing with two ideomotor-compatible tasks: A reply to Greenwald (2003). *Journal of Experimental Psychology: Human Perception and Performance, 29,* 1267-1272.

Lien, M.-C., McCann, R. S., Ruthruff, E., & Proctor, R. W. (2005). Dual-task performance with ideomotor compatible tasks: Is the central processing bottleneck intact, bypassed, or shifted in locus? *Journal of Experimental Psychology: Human Perception and Performance, 31,* 122-144.

Lu, C.-H., & Proctor, R. W. (2001). Influence of irrelevant information on human performance: Effects of S-R association strength and relative timing. *Quarterly Journal of Experimental Psychology, 54A,* 95-136.

McDonald, J. J., Teder-Sälejärvi, W. A., Heraldez, D., & Hillyard, S. A. (2001). Electrophysiological evidence for the "missing link" in crossmodal attention. *Canadian Journal of Experimental Psychology, 55,* 141-149.

McDonald, J. J., Teder-Sälejärvi, W. A., & Hillyard, S. A. (2000). Involuntary orienting to sound improves visual perception. *Nature, 407,* 906-908.

Nicoletti, R., & Umiltà, C. (1984). Right-left prevalence in spatial compatibility. *Perception & Psychophysics, 35,* 333-343.

Nicoletti, R., & Umiltà, C. (1985). Responding with hand and foot: The right-left prevalence in spatial compatibility is still present. *Perception & Psychophysics, 38,* 211-216.

Posner, M. I., Nissen, M. J., & Klein, R. M. (1976). Visual dominance: An information-processing account of its origins and significance. *Psychological Review, 83,* 157-171.

Prinzmetal, W., McCool, C., & Park, S. (2005). Attention: Reaction time and accuracy reveal different mechanisms. *Journal of Experimental Psychology: General, 134,* 73-92.

Prinzmetal, W., Park, S., & Garrett, R. (in press). Involuntary attention and identification accuracy. *Perception & Psychophysics.*

Proctor, R. W., & Lu, C.-H. (1999). Processing irrelevant information: Practice and transfer effects in choice-reaction tasks. *Memory & Cognition, 27,* 63-77.

Proctor, R. W., Marble, J., & Vu, K.-P. (2000). Mixing incompatibly mapped location-relevant trials with location-irrelevant trials: Effects of stimulus mode on performance. *Psychological Research/Psychologische Forschung, 64*, 11-24.

Proctor, R. W., & Reeve, T. G. (1985). Compatibility effects in the assignment of symbolic stimuli to discrete finger responses. *Journal of Experimental Psychology: Human Perception and Performance, 11*, 623-639.

Proctor, R. W., & Vu, K.-P. L. (2002). Mixing location-irrelevant and location-relevant trials: Influence of stimulus mode on spatial compatibility effects. *Memory & Cognition, 30*, 281-293.

Proctor, R. W., Vu, K.-P. L., & Marble, J. G. (2003). Spatial compatibility effects are eliminated when intermixed location-irrelevant trials produce the same spatial codes. *Visual Cognition, 10*, 15-50.

Proctor, R. W., Wang, D.-Y. D., & Pick, D. F. (2004). Stimulus-response compatibility with wheel-rotation responses: Will an incompatible response coding be used when a compatible coding is possible? *Psychonomic Bulletin & Review, 11*, 841-847

Proctor, R. W., Wang, H., & Vu, K.-P. L. (2002) Influences of conceptual, physical, and structural similarity on stimulus-response compatibility. *Quarterly Journal of Experimental Psychology, 55A*, 59-74.

Reeve, T. G., & Proctor, R. W. (1984). On the advance preparation of discrete finger responses. *Journal of Experimental Psychology: Human Perception and Performance, 10*, 541-553.

Robinson, E. S. (1934). Work on the integrated organism. In C. Murchinson (Ed.), *A handbook of general experimental psychology* (pp. 571-650). Worcester, MA: Clark University Press.

Rorden, C., Greene, K., Sasine, G. M., & Baylis, G. C. (2002). Enhanced tactile performance at the destination of an upcoming saccade. *Current Biology, 12*, 1429-1434.

Roswarski, T. E., & Proctor, R. W. (2000). Auditory stimulus-response compatibility: Is there a contribution of stimulus-hand correspondence? *Psychological Research/Psychologische Forschung, 63*, 148-158.

Rubichi, S., Nicoletti, R., Pelosi, A., & Umiltà, C. (2004). Right-left prevalence effect with horizontal and vertical effectors. *Perception & Psychophysics, 66*, 255-263.

Schumacher, E. H., Seymour, T. L, Glass, J. M., Fencsik, D. E., Lauber, E. J., Kieras, D. E., & Meyer, D. E. (2001). Virtually perfect time sharing in dual-task performance: Uncorking the central cognitive bottleneck. *Psychological Science, 12*, 101-108.

Selcon, S. J., Taylor, R. M., & McKenna, F. P. (1995). Integrating multiple information sources: Using redundancy in the design of warnings. *Ergonomics, 38*, 2362-2370.

Simon, J. R. (1990). The effects of an irrelevant directional cue on human information processing. In R. W. Proctor & T. G. Reeve (Eds.), *Stimulus–response compatibility: An integrated perspective* (pp. 31–86). Amsterdam: North-Holland.

Sorkin, R. D. (1987). Design of auditory and tactile displays. In G. Salvendy (Ed.), *Handbook of human factors* (pp. 549-576). New York: Wiley.

Spence, C. (2001). Crossmodal attentional capture: A controversy resolved? In C. Folk & B. Gibson (Eds.), *Attention, distraction and action: Multiple perspectives on attentional capture* (pp. 231-262). Amsterdam: Elsevier Science BV.

Spence, C., & Driver, J. (1997). Audiovisual links in exogenous covert spatial orienting. *Perception & Psychophysics, 59*, 1-22.

Spence, C., & Driver, J. (Eds.). (2004). *Crossmodal space and crossmodal attention.* Oxford: Oxford University Press.

Spence, C., McDonald, J., & Driver, J. (2004). Exogenous spatial cuing studies of human crossmodal attention and multisensory integration. In C. Spence & J. Driver (Eds.), *Crossmodal space and crossmodal attention* (pp. 277-320). Oxford, UK: Oxford University Press.

Spence, C., & McGlone, F. P. (2001). Reflexive spatial orienting of tactile attention. *Experimental Brain Research, 141*, 324-330.

Spence, C., & Read, L. (2003). Speech shadowing while driving: On the difficulty of splitting attention between eye and ear. *Psychological Science, 14*, 251-256.

Spence, C., Nicholls, M. E. R., Gillespie, N., & Driver, J. (1998). Cross-modal links in exogenous covert spatial orienting between touch, audition, and vision. *Perception & Psychophysics, 60*, 544-557.

Spence, C., Shore, D. I., & Klein, R. M. (2001). Multisensory prior entry. *Journal of Experimental Psychology: General, 130*, 799-832.

Tagliabue, M., Zorzi, M., & Umiltà, C. (2002). Cross-modal re-mapping influences the Simon effect. *Memory & Cognition, 30*, 18-23.

Tan, H. Z., Gray, R., Young, J. J., & Traylor, R. (2003). A haptic back display for attentional and directional cueing. *Haptics-e: The Electronic Journal of Haptics Research, 3*, 20 pp.

Tassinari, G., & Campara, D. (1996). Consequences of covert orienting to non-informative stimuli of different modalities: A unitary mechanism? *Neuropsychologia, 34*, 235-245.

Vu, K.-P. L., & Proctor, R. W. (2001). Determinants of right-left and top-bottom prevalence for two-dimensional spatial compatibility. *Journal of Experimental Psychology: Human Perception & Performance, 27*, 813-828.

Vu, K.-P. L., & Proctor, R. W. (2004). Mixing compatible and incompatible mappings: Elimination, reduction, and enhancement of spatial compatibility effects. *Quarterly Journal of Experimental Psychology, 57A*, 539-556.

Wang, H., & Proctor, R. W. (1996). Stimulus-response compatibility as a function of stimulus code and response modality. *Journal of Experimental Psychology: Human Perception & Performance, 22*, 1201-1217.

Ward, L. M., McDonald, J. J., & Lin, D. (2000). On asymmetries in cross-modal spatial attention orienting. *Perception & Psychophysics, 62*, 1258-1264.

Wickens, C. D. (1992). *Engineering psychology and human performance* (2nd. Edition). NY: HarperCollins.

Zampini, M., Shore, D. I., & Spence, C. (in press). Audiovisual prior entry. *Neuroscience Letters.*

Multimodal Interfaces Improve Memory

Jeanine K. Stefanucci

University of Virginia
102 Gilmer Hall, Box 400400
Charlottesville, VA 22904
jks8s@virginia.edu

Dennis R. Proffitt

University of Virginia
102 Gilmer Hall, Box 400400
Charlottesville, VA 22904
drp@virginia.edu

Abstract

We investigated whether multimodal interfaces improve memory. In previous studies, we showed that people who learned information on a multimodal display system were more likely to remember it later compared to people who learned information on a standard desktop computer (Tan, Stefanucci, Proffitt, & Pausch, 2001). We also found that brain areas involved in processing this multimodal information during encoding were active during retrieval of the information (Stefanucci, Downs, Snyder, Downs, & Proffitt, 2002). We conducted several follow-up studies aimed at determining which aspect of the display environment was responsible for the observed memory advantage. We found that multimodal environments were necessary to improve memory; unimodal environments were not sufficient (Stefanucci & Proffitt, 2002; Stefanucci & Proffitt, 2005). Finally, we added cues to the multimodal interface to maximize the memory benefit. We also reinstated these cues at retrieval to assess whether they increased the memory benefit as well (Stefanucci, O'Hargan, & Proffitt, 2005). These findings suggest that multimodal stimulation is an important mediator in remembering information.

1 Introduction

This paper will discuss research, which combines findings from two major fields: cognitive psychology and human-computer interface design. Specifically, we will show that human- computer interfaces can be designed to support memory by applying past findings in cognitive psychology. In particular, our data indicate that the inclusion of multiple modalities is a key element effective human computer interfaces. Before we discuss our findings, we will review relevant findings in memory research and human-computer interface design that motivated our work.

1.1 Human Memory Research

Memory for information is often tied to the place in which it was learned. For instance, if you are trying to remember a conversation you had with a friend last week, you might consider visualizing the place where you had the conversation. Research in cognitive psychology has amassed evidence in support of the claim that ambient cues present in an environment at the time of memory encoding may aid subsequent recall (for review, see Smith & Vela, 2001). These studies have shown that memory is improved when distinctive environments are present at both encoding and retrieval. However these environments need not be related to the to-be-remembered information. For example, Godden and Baddeley (1975) found that people who learned lists of words underwater were better able to recall that information when they were underwater rather than on land. This finding was extended to more typical environments by Smith, Glenberg, and Bjork (1978). They found that people who learned information and recalled it in the same laboratory setting showed better memory for the material than people who had to recall the information in a different laboratory room. Numerous other studies have examined the effects of environmental congruency on the ability to retrieve information (Smith, 1979; Eich, 1985; Dolinsky & Zabrucky, 1983). Although the generality of these findings has been contested by some, (Fernandez & Glenberg, 1985; Nixon & Kanak, 1981) we believe that they nonetheless provide a useful base for our research.

1.2 Human-Computer Interface Research

Traditionally, research on user interface design has focused on making systems more usable for the average person (Card, Moran, & Newell, 1983). Specifically, interfaces have been evaluated on the time it takes a user to complete a task and the errors committed while doing the task. In order to make systems more usable, researchers have made the interfaces consistent so as not to distract user's attention from the task they are performing. Consistency in the interface also decreases the time taken to learn the interface and increases proficiency across interfaces with similar designs. While these design principles enhance usability, they do not increase the user's ability to remember information, nor do they take advantage of the user's environment.

Utilizing principles sometimes counter to traditional user interface designs, the use of ambient environments to create user interfaces has recently increased in the domain of human-computer interface design. For example, Ishii and Ullmer (1997) argued that current interfaces were too confined to the conventional GUI (Graphical User Interface) that includes only a keyboard, monitor and mouse. They claimed that ambient displays could be constructed that took advantage of more of the user's physical environment. Following this argument, Wisneski et al. (1998) constructed an "ambientROOM" that had ambient sounds to alert the user of global events happening in central workspaces, and local events such as the quantity of unread emails. However, most of these implementations emphasized awareness of surroundings and alert systems rather than memory for the information presented upon them.

More relevant for the current paper is research done by Davis, Scott, Pair, Hodges and Oliverio (1999), who showed that the presentation of ambient environmental auditory cues increased users' abilities to remember information. Using a head-mounted display, they had participants enter virtual environments that contained objects such as bookshelves, a rug, some furniture and a window. In the room, participants could view the outdoors through the window and often there was something interesting to watch (e.g., a thunderstorm). Participants heard either high fidelity sounds that were congruent with the outdoor scene (thunder and rain), low fidelity sounds that were congruent with the scene (AM radio quality) or no sounds at all. Each participant was asked to examine objects that were placed on the bookshelves in the rooms. At a later testing session in a different environment, those participants who received the high fidelity sounds recognized more of the objects located on the shelves and correctly attributed them to the appropriate rooms.

2 Can multimodal interfaces improve memory?
Our laboratory has recently started a research program, which integrates the psychological research on context-dependent memory and the computer science research on human-computer interfaces mentioned above. Our goal is to show that interfaces can be built to improve memory for the users. In the following studies we describe iterative research, which shows that multimodal interfaces provide the most robust context-dependent memory effects.

2.1 Implementing a multimodal interface for memory enhancement
Tan, Stefanucci, Proffitt & Pausch (2001) built a multimodal large-scale display system, termed the InfoCockpit, to test whether user interfaces could increase memory for the information presented on them (see Figure 1).

Figure 1: The InfoCockpit - a multimodal display system that includes multiple monitors, ambient visuals, and ambient audio.

The motivation for building the system came from the fact that user interfaces are typically designed for usability and consistency, which does not make the interfaces (or the information presented on them) memorable. Borrowing from the context-dependent memory literature discussed earlier, we constructed a multimodal interface that included three flat-panel monitors and a large projection screen, as well as a three-dimensional surround sound system. We asked people to learn lists of words on this interface or on a standard desktop computer with one monitor and no added cues. Those participants that were placed in the multimodal condition learned each word list on a separate monitor (spatial cue) with different contextual environments displayed while the lists were being learned (visual and sound cues). The environments included projected 360° panoramic images of a place (e.g., the Lawn at the University of Virginia) and congruent three-dimensional surround sounds (dogs barking, birds chirping, cicadas). In addition, the word lists interfered with one another because the first word of the pairs in each of the lists was the same across lists. For example, participants learned that "plate" went with "scientist" in one list, but "plate" was paired with "string" in another list. After learning all of the word lists, participants were asked to return the following day. An unexpected memory test was then given and participants tried to recall as many of the words from the previous day as they could. None of the added multimodal cues were present during testing. The added spatial and environmental memory cues in the multimodal encoding condition improved memory performance by 56%. Participants who had multimodal environments present during learning showed better retrieval of the information following a 24-hour delay.

2.2   Brain activations correlated with multimodal memories

Following the initial study by Tan et al. (2001), we were interested in whether brain areas that were active in the multimodal group during encoding (likely visual and spatial areas associated with processing the added cues) would also be active at retrieval. It was our belief that if participants were indeed using the added multimodal cues present at encoding to facilitate their later recall, the same areas would be active. This argument is not without support because many researchers in cognitive neuroscience have found that regions of the sensory cortex are reactivated when sensory-specific encoded material is retrieved (Wheeler, Petersen, & Buckner, 2000).

Many functional imaging studies have localized brain areas active during episodic memory retrieval. However, little research has focused on the effect that environmental context has on these activations. Based on behavioral research, we would expect differential activations at retrieval due to differences in the contextual environment present when episodic memories are encoded. In the current study, we used functional magnetic resonance imaging (fMRI) to measure differences in brain activations at retrieval associated with changes in the multimodal environment present at encoding.

Similar to the study previously, participants in the multimodal condition learned word lists on a multiple monitor computer, encompassed by an immersive, projected context. Participants in the standard desktop condition learned the word pairs on a single-monitor desktop computer with no added spatial, visual or auditory cues. fMRI was used to compare brain activations in the two conditions while participants retrieved the previously-learned information during testing. Behavioral data for the testing phase was collected on a laptop immediately after scanning. Participants who learned semantic information in the multimodal condition showed improved memory performance compared to participants who learned on the standard desktop computer, replicating the findings of Tan et al. (2001). Those participants who encoded in a multimodal environment, with added spatial, visual and auditory cues, showed more activation in the prefrontal cortex, the parietal lobe, and the occipital lobe (see Figure 2).

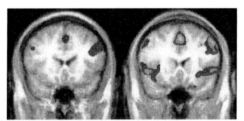

Figure 2: The desktop condition (left) showed fewer activations in the occipital, parietal and frontal areas than the multimodal (InfoCockpit) condition (right).

We argue that learning information in a spatially distributed, multimodal context facilitates later recall of information by involving a larger network of brain areas associated with episodic memory retrieval. Consistent with the behavioral literature on context-dependent memory, these findings suggest that brain areas activated during episodic memory retrieval include areas associated with retrieving information about the environmental conditions present at encoding.

2.3   Which aspects of the multimodal interface improve memory?

Because the research discussed above showed that a display environment with both multiple monitors and projected contextual environments increased memory, we became interested in which aspect of the system was the most responsible for the memory benefit. For example, was the combination of the spatial, visual and auditory cues necessary to support memory or was just one of the cues sufficient to improve memory?

Stefanucci and Proffitt (2002) systematically isolated the components of the InfoCockpit included in the previous studies to test whether they were necessary to provide a memory advantage. Participants in this study were placed in one of four conditions, defined by the computer system on which they learned the word lists. The desktop condition learned the information on a single monitor with no added cues, as in the previous experiments. The InfoCockpit condition contained spatial cues (each word list was learned on a separate monitor) and multimodal environmental context (each word list was associated with projected images of a place and congruent 3D sound environments). The spatial condition learned the word lists each on a different monitor, but did not have added projected environments. The context condition learned the word lists on a single monitor, but different multimodal environments were presented with each list. All other aspects of the learning and testing phases were the same as in the Tan et al. (2001) study.

Interestingly, participants who learned information with only the multimodal environmental contexts present remembered more information than the InfoCockpit, spatial, and standard desktop conditions. However, the analyses also revealed that the InfoCockpit group remembered more than the standard desktop group, replicating our preliminary results. Our assumption was that the InfoCockpit would improve memory relative to the standard desktop condition and the conditions that only included individual components of the InfoCockpit. However, the multimodal, projected environment without the added spatial cue from multiple monitors was better for improving memory than the ensemble of cues in the InfoCockpit (see Figure 3).

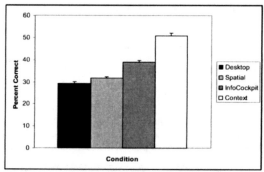

Figure 3. Participants that had only multimodal, ambient environments present during learning, showed superior performance on the memory test.

This counterintuitive finding may make more sense upon closer inspection. We believe that the participants in the InfoCockpit condition had two strategies available to them at retrieval, one involving the location where they learned the information and the other involving the environment present while they learned the information. The lower relative performance of the InfoCockpit condition may be attributed to a competition between these two retrieval strategies. Another alternative is that these participants relied on the less effective spatial cue, whereas participants in the context condition had only the best cue available at retrieval. When the cues were systematically tested against one another in this task, we found that the multimodal information (the context) was a more reliable cue for later recall and those participants in the context condition exploited it to the fullest.

2.3.1   Which aspects of the multimodal contextual environment improve memory?
The motivation of the next study was to assess whether the multimodal ambient environments that proved to be the most beneficial memory cue could improve memory relative to the unimodal cues that they included. For instance, does a unimodal ambient environment provide the same memory benefit as the multimodal contextual environments of the previous studies? When examining the effect of a multimodal environment on memory, one is always concerned that the effect may be driven by only one parameter of the multimodal environment. Therefore, in this experiment we systematically isolated and combined the two ambient cues – sights and sounds – that made up the multimodal contextual environments used in the previous studies to assess whether these cues reliably produce a memory effect alone or in combination.

Stefanucci and Proffitt (2005) did an experiment to test whether the combination of the sights and sounds that made up the multimodal, contextual environments in the previous experiments were necessary to produce a memory advantage. Our hypothesis was that people may be recalling only the visuals of the projected environments at retrieval, because it would be hard to engage in auditory imagery to recreate the sound environments present at encoding. Therefore, it is possible that only the visuals are needed at encoding to produce a memory effect at retrieval.

Contrary to this prediction, we found that people who had sounds and visuals present at encoding, which created an ambient multimodal environment rather than a unimodal environment, performed better on the memory test than participants who had only one of the cues present during encoding. As in the previous experiment, participants were placed in one of four conditions: multimodal (sights and sounds), unimodal sights, unimodal sounds or the standard desktop condition. The procedure for learning the lists was the same as in the previous studies and the testing phase occurred one day after the learning phase. None of the ambient cues were reinstated during testing.

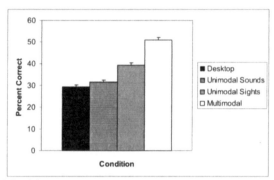

Figure 4: Participants that had the multimodal, ambient environments present during learning performed better on the memory test than participants that only had unimodal environments.

Those participants that learned information with the multimodal ambient environment present at encoding remembered more words than the unimodal conditions and the desktop condition (see Figure 4). In fact, the participants who received visuals and sounds at encoding remembered 73% more than participants in the desktop condition. We concluded that the combination of sounds and images was important in producing a memory advantage. We believe that the participants in the multimodal condition received a benefit from the sounds at encoding because the sounds kept the participant immersed in the surrounding "environment" while they learned the information on the monitor. The projected (and unchanging) contextual images were easy for the participants to ignore while they trained on the word lists. However, the dynamic sound environment was not easy to ignore while learning the lists. The sounds may have served as a constant reminder of the images that were being projected while participants learned the lists. Intuitively, it seems more likely that the participants recalled the visuals of the environment when trying to retrieve the learned information, but the sounds may have provided a stronger cue for binding the learned information to the visuals during encoding.

3 How much can multimodal interfaces improve memory?

Finally, we conducted an experiment to show that the memory benefits of the previous experiments could be even bigger with more environmental cues present during encoding and with the cues available at retrieval. Stefanucci, O'Hargan, and Proffitt (2005) implemented a "new" InfoCockpit that had additional memory cues, for the purpose of increasing memory performance as much as possible. Specifically, Smith and Vela (2001) claimed that the experimenter was actually a valuable part of the contexts used in previous studies. For instance, if the experimenter stayed the same across learning and testing sessions, this helped people remember more from the learning session than if the experimenter was different across these sessions. Therefore, we added "experimenters", or videos of people who served as the experimenter for each list, to see whether having a person associated with the contexts in the experiment would further increase memory performance. These experimenters gave all instructions to the participants for the training and testing sessions. In addition, we decided to test whether a large memory benefit could be obtained by reinstating all of the cues present during encoding at retrieval (experimenters, sounds, visuals).

In laboratory settings, the memory benefits of context-dependent memory experiments have been fairly small. In contrast, we believed that when the combination of the cues mentioned above (faces, voices, sounds, reinstatement of cues at retrieval) was experienced in real world, office environments, large memory benefits could be observed.

Figure 5: Faces and voices were added to the ambient, multimodal environments for each of the lists.

Participants in this experiment were placed into the InfoCockpit group or the standard desktop group, as in previous studies; however the InfoCockpit condition included the new memory cues discussed above (see Figure 5). Participants in both conditions were led to their display system by an experimenter, but then were told that all instructions and information would be presented to them on the computer. The InfoCockpit condition was given instructions from a video taped experimenter who was a different person for each of the word lists. The desktop condition read the exact same text that was presented to the InfoCockpit group, but there was no videotape to accompany the text. It is important to note that in both cases, a "person" was associated with each of the word lists, but for the control condition, this person was abstract and did not have a distinct face or voice.

The learning sessions proceeded as in the previous experiments described. However, all participants were told that they should try to remember the word pairs because there would be a test the following day. Participants then returned one day later for testing, but were placed back in the environments they were in for training. Therefore, context was reinstated for both groups.

As expected, the InfoCockpit condition recalled more of the words than the standard desktop condition. However, the memory benefit for the InfoCockpit group was much larger than in previous experiments. When presented with a combination of distinct images, sounds and faces participants in this study remembered 131% more than those who had no unique contexts at encoding.

4   Conclusions

A large literature in the cognitive psychology of memory has documented that people recall more information if they are in the same place at retrieval as they were at encoding (for review, see Smith & Vela, 2001). The problem with this research is that the effects are small and sometimes unreliable. Also, "context" is a broad, often ill-defined concept and many researchers have studied it in very different ways.

We believe that some of these problems arise from the fact that the research is being done in laboratory settings that do not provide rich, real-world contextual settings. In this paper, we described a new multimodal display system that simulates real-world environments. The system displays images of distinct places accompanied by congruent three-dimensional surround sounds. These environments were used as contextual settings for context-dependent memory studies. We asked two main questions in the studies outlined: 1) would these environments evoke reliable and large context-dependent memory effects? 2) which aspects of the display system were responsible for the memory effects?

Our findings suggest that multimodal interfaces can improve memory on a larger scale than indicated in previous laboratory studies. In addition, sound played an important role in the observed effects when combined with congruent images. These results suggest that office display systems of the future should incorporate new principles of design, in order to include ambient multimodal environments to enhance both user productivity and memory performance.

5   References

Card, S. K., Moran, T. P., & Newell, A. (1983). *The Psychology of Human-Computer Interaction.* Hillsdale, New Jersey: Lawrence Erlbaum and Associates.

Davis, E. T., Scott, K., Pair, J., Hodges, L. F., Oliverio, J. (1999).   Can audio enhance visual perception and performance in a virtual environment? *Proceedings Of The Human Factors And Ergonomics Society 43rd Annual Meeting*: 1197-1201.

Dolinsky, R., & Zabrucky, K. M. (1983). Effects of environmental context changes on memory. *Bulletin of the Psychonomic Society, 21* (6), 423-426.

Eich, E. (1985). Context, memory, and integrated item/context imagery. *Journal of Experimental Psychology: Learning, Memory & Cognition, 11*, 764-770.

Fernandez, A., & Glenberg, A. M. (1985).   Changing environmental context does not reliably affect memory. *Memory & Cognition, 13*(4), 333-345.

Godden, D. R., & Baddeley, A. D. (1975). Context-dependent memory in two natural environments: On land and underwater. *British Journal of Psychology*, 66(3), 325-331.

Ishii, H., & Ullmer, B. (1997). Tangible Bits: Towards Seamless Interfaces between People, Bits and Atoms. *Proceedings of CHI '97*, ACM Press, 234-241.

Nixon, S. J., & Kanak, N. J. (1981). The interactive effects of instructional set and environmental context changes on the serial position effect. *Bulletin of the Psychonomic Society, 18*(5), 237-240.

Smith, S. M. (1979). Remembering in and out of context. *Journal of Experimental Psychology: Human Learning and Memory, 5*, 460-471.

Smith, S. M., & Vela, E. (2001). Environmental context-dependent memory: A review and meta-analysis. *Psychonomic Bulletin & Review, 8*, 203-220.

Smith, S. M., Glenberg, A., & Bjork, R. A. (1978). Environmental context and human memory. *Memory & Cognition, 6*, 342-353.

Stefanucci, J. K., Downs, T. H., Snyder, A. P., Downs, J. H., & Proffitt, D. R. (2002). Context-dependent memory engages a frontal-parietal-occipital network. *Poster presented at the Annual Meeting of the Cognitive Neuroscience Society*, San Francisco, CA.

Stefanucci, J. K., O'Hargan, S. P., & Proffitt, D. R. (2005). *Augmenting context-dependent memory*. Manuscript submitted for publication.

Stefanucci, J. K., & Proffitt, D. R. (2005). *Which aspects of contextual environments bind to memories?* Manuscript submitted for publication.

Stefanucci, J. K., & Proffitt, D. R. (2002). Providing distinctive cues to augment human memory. In W. Gray & C. Schunn (Eds.), *Proceedings of the Twenty-fourth Annual Conference of the Cognitive Science Society* (p. 840). Mahwah, NJ: Erlbaum.

Tan, D.S., Stefanucci, J.K., Proffitt, D.R., Pausch, R. (2001). The Infocockpit: Providing Location and Place to Aid Human Memory. *Workshop on Perceptive User Interfaces 2001*, Orlando, Florida.

Wheeler, M. E., Petersen, S. E., Buckner, R. L. (2000). Memory's echo: Vivid remembering reactivates sensory-specific cortex. *Proceedings of the National Academy of Sciences of the United States of America* 97: 11125-11129.

Wisneski, C., Ishii, H., Dahley, A., Gorbet, M., Brave, S., Ullmer, B., & Yarin, P. (1998). Ambient displays: Turning architectural space into an interface between people and digital information. *Proceedings of the First International Workshop on Cooperative Buildings (CoBuild '98)*, 22-32.

Multi-Modal Interfaces for Future Applications of Augmented Cognition

Jack Vice
Anna Lockerd
Corinna Lathan

AnthroTronix, Inc.
8737 Colesville Rd., 10th Floor
Silver Spring, MD 20910
info@atinc.com

Abstract

As computational devices have become prevalent in military applications and rather complex with regard to the volume and type of information presented, the design considerations for the associated human machine interfaces have increased in importance. In order to maximize performance, information must be provided in a format that enables operators to quickly and easily interpret and appropriately react to the data. Ideally, information is provided in several modalities: visual, auditory, tactile, and/or olfactory, to facilitate redundancy and accommodate the modality that the operator can best attend to at any given time (see, for example, Miller, 1982). Additionally, alternative user input modalities such as applied force, speech, and gesture can improve the human computer interaction.

AnthroTronix, Inc. (ATinc), a research and development engineering firm specializing in advanced human machine interface devices, has extensive experience developing multi-modal interfaces for communication and command/control of computer-based systems such as wearable computers and robotic platforms. Technologies developed by ATinc include instrumented gloves, wearable and weapon-mounted robotic control devices and wearable, haptic, and olfactory feedback devices. This paper will highlight some of the technologies developed by ATinc, as well as explore the intended and potential alternative applications for such devices in both military and commercial environments. Examples of the integration of these technologies into future applications in augmented cognition will be discussed.

A primary application for such multi-modal interface technologies includes use by the dismounted warfighter. Intelligent wearable computing devices allow warfighters to communicate with each other, obtain information, and control remote devices without impeding their ability to perform tasks in a field environment.

A potential near-term, application for these technologies is in virtual reality training environments. An effective interface device can be used within a simulated environment for command and control field tasks such as tele-operation of remote robotic platforms. This allows warfighters to perform much-needed training exercises in a controlled environment while providing a test bed for the technology until it has reached a fieldable technology readiness level. Additionally, this allows the design team to receive feedback from the target users with respect to the usability and effectiveness of the device for future design iterations. This advises the design process, resulting in an effectual transition to field use.

1   Introduction

Research in cognitive psychology has shown that multi-modal data representation can be used to maximize the amount of information that can be effectively processed by a human operator performing a task. In 1982, Miller found that multi-modal presentation of redundant signals resulted in increased reaction time. Research by Sulzen (2001) demonstrated improved memory in cases where information was distributed across multiple modalities. Therefore, by presenting task information in multiple modalities, performance can be improved. Additionally, by tapping different cognitive demands, individuals are able to share tasks (Wickens, 2002). Augmented Cognition is an emerging field of research seeking to enhance a user's abilities via computational technologies that are explicitly designed to address bottlenecks, limitations, and biases in cognition, specifically those present in the human-computer interaction.

AnthroTronix, Inc. (ATinc) has extensive experience in the design and development of multi-modal human machine interface systems for optimization of user performance, specifically in the military domain. Under the DARPA-funded Augmented Cognition program, ATinc researched methods of assessing a military operator's cognitive state and mitigating workload during periods of cognitive overload. Algorithms developed by ATinc and other researchers can now be integrated into closed-loop systems for performance enhancement. One primary means of increasing operator performance is to employ multiple modalities for data presentation and entry.

A primary application for such multi-modal interface technologies includes use by the dismounted warfighter. The dismounted warfighter has the special situation in which visual and auditory data must be presented to the warfighter in such a way that does not adversely affect the warfighter's own situational awareness. The challenge is not only to increase the warfighter's net cognitive abilities but also to do it in such a way that does not hinder the warfighter's ability to engage the enemy when the situation necessitates. This will require the software to not only be aware of the warfighter's current cognitive state, but also be aware of the warfighter's current tactical situation. For example, if the system is providing data to the warfighter in a visual modality, and the warfighter is approaching a potential enemy position, the system might start to migrate the data representation from visual representations to auditory. As the warfighter reaches the potential enemy position, the system might migrate the data from auditory to tactile representation, thus freeing up the warfighter's auditory modality for his or her own situational awareness when most necessary.

This paper will discuss a number of human input and output modalities as well as specific technological solutions developed to facilitate the use of those modalities. The input modalities discussed will be applied force, speech and gesture. The output modalities discussed will be visual, auditory, tactile and olfactory.

2   Input Modalities

One challenge that is integral to the dismounted warfighter domain is that of the human computer interface. Historically, the dismounted warfighter has not had the comprehensive visual and auditory modalities, or the typical input modalities such as keyboard and mouse, commonly available to a workstation operator. This is primarily because of the mobility of the dismounted warfighter as well as the tactical and situational awareness impact of a given input or output modality.

Providing users with a variety of input modalities enables them to use whichever method is most efficient and effective for a given task in a given situation. For example, simple, discrete inputs, such as making a selection from a list of options, can be facilitated by binary pushbuttons and switches. When high-fidelity input is necessary, an analog or proportional-type input such as mouse or applied force joystick, might be most effective. For situations in which users' hands are otherwise engaged, speech might be a preferable input. Additionally, in tactical situations such as that of the dismounted warfighter where noise discipline is required, hand and arm gesture recognition could be used for discrete inputs. ATinc has designed and fabricated a number of interfaces that support these primary means of data input: applied force, speech, and gesture.

2.1   Applied Force

Applied force input modalities, such as pushbutton, slide, rotational, and pressure sensors, can provide an intuitive and robust means of both binary and proportional inputs. Such inputs are used in everyday life, and are therefore intuitive. Pushbuttons and switches are ideal for discrete inputs, while analog force sensors, such as joysticks, provide an effective interface for proportional inputs, including mouse cursor control or tele-operation of an unmanned vehicle.

ATinc is developing an Isometric Controller Grip (ICG) to allow the warfighter to maintain an immediate enemy engagement posture while providing high fidelity input to the system. For example, the warfighter could be operating and getting feedback from an unmanned ground vehicle without sacrificing the ability to immediately engage the enemy, if necessary. The ICG, shown in Figure 1, consists of a force/torque sensor embedded within an aluminum grip, which acts as an isometric joystick that mounts to the front of an M4 rifle and can be used as a front vertical grip when firing the weapon. The ICG detects forces and torques applied to the grip and translates applied forces/torques into proportional control inputs. In addition to maintaining warfighter lethality when mounted to the weapon, the isometric design provides for a very robust (no moving parts), ruggedizable input device.

Fig 1: Isometric Controller Grip

2.2   Speech

Speech as an input modality has significant potential for the dismounted warfighter for several reasons. First, speech is an inherently intuitive form of communication. Speech is hands free, it requires very little additional hardware (with the exception of a microphone), and is extremely comprehensive in communication capability. Because speech is a natural form of human communication, no adaptation or training would be required of the warfighter to utilize speech as an input modality. ATinc is working with a natural dialog system called Listen, Communicate, Show (LCS) from Lockheed Martin's Advanced Technology Lab (ATL). The LCS system was used for Augmented Cognition research as a speech input and audio output modality. LCS combines speech recognition, natural language processing, context tracking and voice synthesis to facilitate natural conversation with the system. Because speech is a hands-free form of data entry, it is an optimal modality for combination with other input and output modalities, particularly those that are spatial, as speech is almost exclusively verbal in nature. For example, an operator could use an applied force sensor to control a unmanned vehicle (acting as a point or scout) while viewing video feedback from the remote device, and simultaneously communicate information to other dismounted warfighters regarding what is being seen and how to respond.

2.3   Gesture

One primary form of communication between dismounted warfighters, while conducting combat maneuvers is hand and arm signals. Dismounted warfighters in the field often utilize an established set of hand signals in order to communicate with others while maintaining noise discipline. ATinc has designed and developed an Instrumented Glove (iGlove), shown in Figure 2, which contains embedded analog sensors capable of detecting hand and finger orientation and position. When combined with custom pattern-matching software algorithms developed by ATinc, the iGlove is capable of recognizing hand signals, such as those commonly used by the warfighter, as discrete inputs to the system. Hand signals can thus be electronically communicated as commands to other team members in the form of audio or text, or in the domain of unmanned vehicle tele-operation, as commands to the remote vehicle. The advantage of this modality is that communication and control inputs can be conveyed without violating noise discipline and without being within line of sight of the other team members.

23

Fig 2: iGlove Interface

In addition to detecting discrete gestures, the iGlove can also be used in an "air joystick" modality for proportional input or control. In this modality, the operator makes a fist, as if holding a joystick, and uses roll and pitch of the hand to control two proportional input signals. This modality, like the analog applied force modality, is ideal for proportional control of a mouse cursor or directional control of an unmanned vehicle. The advantage of this form factor, is that the operator is not required to hold an actual sensor, and therefore, can switch back and forth from this control modality to another task which might require the hand, without having to stow a piece of equipment. During the Digital Military Police Program sponsored by the U.S. Army Natick Soldier Systems Center, ATinc integrated an iGlove-based gesture to voice and text system into a wearable communication. Currently, ATinc is working with the U.S. Army Research Laboratory to implement iGlove control of unmanned ground vehicle platforms.

3   Output/Feedback Modalities

The way in which information is conveyed to a human operator greatly affects the operator's ability to process the information. Multi-modal feedback can be used to decrease cognitive load and enhance user performance by allowing users to complete more tasks in a shorter period of time, while reducing errors. Output modality switching as a mitigation strategy for cognitive overloading has proven to be effective for augmenting cognition (Miller, 1982). Additionally, information presentation redundancy can improve performance by decreasing reaction time (Miller, 1982). ATinc's multi-modal research efforts have primarily focused on two domains: the dismounted warfighter who at times cannot have visual and auditory modalities obstructed by the system and the workstation operator who can take advantage of modalities in addition to the visual modality typical to human-computer interaction.

3.1   Visual

Most computational devices use visual as the primary output modality. With respect to perception, humans are primarily visual creatures and can process several pieces of information simultaneously in the visual domain. For example, while driving, vehicles in the periphery can be detected even when attention is focused on the road surface directly ahead. Visual information presentation can be further broken down into verbal and spatial: verbal consisting of actual text or words, and spatial consisting of images and placement of those images within a display. Computational devices can be used to send visual information via radio frequency; this can be used for communication between multiple dismounted warfighters, between individual dismounted warfighters and a command post, or between warfighters and remote robotic platforms such as unmanned vehicles. Live video, on-screen maps, graphics, and text can be displayed on a variety of wearable computing monitors or displays.

One of the primary drawbacks of the visual feedback/output modality for the dismounted warfighter domain is that it requires that the operator look at a screen or display monitor, diverting his/her visual attention from the surrounding environment, potentially degrading the warfighter's situational awareness. When in the field, it is often crucial that dismounted warfighters direct some or all of their visual assets towards viewing/assessing their surroundings for navigation, inter-team communication, or engaging the enemy. A monocular display can allow users to divide their fields of view partially, without total visual obstruction. Several form factors incorporating visual feedback have been designed and/or fabricated during past and current research efforts at ATinc. These field-portable input/output devices include a Wrist-Mounted Operator Control Unit (OCU), Dual Joystick Controller, Monocular OCU (MOCU), and Immersive OCU (IOCU), which uses internal sensors to track absolute and relative head movement of the operator, and stereo video displays, which provide immersive remote situational awareness (See Table 1).

Table 1: Controller Concept Designs

Controller Description	Concept Photo/Graphic
Wrist-Mounted Operator Control Unit - Ruggedized wearable handheld computer with joystick, analog to digital converter, battery pack, and network communications.	
Dual Joystick Controller - Modeled after input controllers commonly used in the video game industry. This type of controller is very familiar and intuitive due to the popularity of console games among Soldiers and Marines.	
Monocular Operator Control Unit (MOCU) - Inspired by the 'camcorder' style user interface. Contains CPU, radio, power supply, audio/video, and control inputs. The advantage of the MOCU is that it provides high fidelity immersive optics while not totally obstructing the operator's vision and hearing.	
Immersive Operator Control Unit (IOCU) - Similar functionality as the MOCU. Internal sensors track absolute and relative head movement of the operator. Because of the stereo audio and video displays, it provides immersive remote situational awareness.	

3.2   Auditory

At times when the operator cannot divert his or her visual attention from combat tasks, auditory feedback can be largely beneficial for conveying information. Additionally, some types of information, such as verbal messaging, are most intuitively represented to the user via an auditory modality. Because of the human brain's natural ability to localize audio sources, spatial data can be represented via audio, providing directional information in addition to the data content. During the Digital MP program, ATinc implemented a gesture to voice interface that translated a hand signal input from one warfighter to a voice output, heard by another team member, not within line-of sight. An ear bud or bone phone can be used to convey information to a team member without obstructing the reception of local audio, and while maintaining noise discipline. Combining verbal auditory tasks, such as communication or information retrieval, with spatial visual tasks, such as looking for enemy positions, tele-operation of unmanned vehicles, or room clearing, has the potential to maximize warfighter performance capabilities.

3.2   Tactile

For situations in which data representation redundancy is desired, or when the warfighter's vision and hearing must not be obstructed, a tactile feedback modality can be largely beneficial. Tactile interfaces can be used to reinforce visual or auditory information, or can be used as an additional means of data presentation. Burke, Gilson, and Jagacinski demonstrated improved operator performance in cases in which a kinesthetic-tactual display was used in conjunction with a visual display, as compared to performance with two visual displays (Burke, Gilson, & Jagacinski, 1980). Additionally, tactile displays can be placed in a number of sites on the body, which allows for a great deal of flexibility in terms of an intuitive interface for the given application or form of information, as well as not being restrictive or conflicting with other pieces of equipment.

One such form factor for vibrotactile feedback to a user, developed as part of a collaborative effort between ATinc and George Washington University, is a belt with an embedded array of vibrotactile motors. Of the many methods for tactile data representation, the simplest is alerting the warfighter. Additional applications include navigation, obstacle avoidance and complex messaging. Using GPS and compass data sent from a wearable computing system, the tactor-belt can indicate to the warfighter, the direction to move or avoid. As shown in previous research, a tactor array can be used to communicate to the operator the location of obstacles near an unmanned ground vehicle in order to assist in vehicle tele-operation (Lathan & Tracey, 2002). Complex tactile messaging can be used to provide richer information through the tactile modality. Similar to audio communication of Morse code, a library of tactile messages defined by a series of tactile motor activations at various frequencies over a short period of time, is being developed to provide a redundant or additional modality of data communication to field users who may not be within line of sight, without detracting from auditory or visual tasks being performed, and without compromising noise discipline.

3.3   Olfactory

Olfactory stimulation as an output modality has potential in training, as well as operational domains. Though olfaction lacks the temporal resolution and richness of information when compared with other modalities, the close coupling between olfaction and memory has the potential to increase performance (Herz & Engen, 1996). Through olfactory conditioning, the warfighter can be conditioned to associate specific odors with specific operational information. For example, an unmanned vehicle operator could be conditioned to associate specific scents with unsafe engine temperatures; further, increases in engine temperature could be conveyed by an increase in the odorant intensity. A 'burning' scent would likely be most intuitive for this application.

ATinc, in collaboration with University of Southern California's Institute for Creative Technologies (ICT), has developed a 'Scent Collar'. This device, shown in Figure 3, is worn around the neck of the user, and houses self-contained modular scent delivery cartridges, which are controlled wirelessly by a host computer. In addition to its use as an output modality, the olfactory stimulation can be used for realism in VR environments as well as training for the detection of specific scents in the field. The army is currently using the scent collar to enhance the immersive nature of the VR environment.

Fig 3: Scent Collar

4   Integration with Future Applications

Integration of novel input and output modalities into the development platforms of dismount soldier programs such as the Future Force Warrior (FFW) and Future Combat Systems (FCS) will require a combination of efforts. Integration of software algorithms into the existing or developing software system would provide warfighter improvements without requiring hardware changes. The new algorithms would provide a means to enhance the displaying of information through the two existing modalities: visual and auditory. The hardware and textile integration of vibrotactile actuators would add a third data presentation modality.

With the delivery of a vibrotactile belt for Honeywell's Augmented Cognition effort and the delivery and support of vibrotactile armbands to Micro Analysis and Design for their Future Force warrior effort, ATinc is actively pursuing and supporting field tests for tactile feedback. ATinc has been communicating with FFW systems engineers at Natick Soldier Systems in order to ensure that all dismounted warfighter input and output modalities currently being developed are technologically compatible with the FFW system. Collaboration with the U.S. Army Research Laboratory will facilitate additional field testing and technology refinement of these modalities.

5   Conclusion

As the Augmented Cognition program progresses forward, performance of the human-computer system will continue to be optimized. One aspect of this optimization will be the use of multiple modalities for data input and output. Combining visual, auditory, tactile and olfactory modalities, multi-modal interfaces can increase performance by providing redundancy of data representation, by allowing modality switching as a mitigation strategy, and by representing specific types of data via the modality most intuitive to the data type. For the dismounted warfighter, the challenge is magnified by the specific domain, which lacks typical human-computer modalities. Intuitive, unobtrusive input modalities such as speech, applied force and gesture activation, can improve warfighter performance by facilitating system input control without compromising warfighter lethality or combat sustainability.

6   Acknowledgements

AnthroTronix would like to thank the many partners on past and present development efforts including the U.S. Army Research Laboratory in Adelphi, MD, U.S. Army Natick Soldier Systems, Lockheed Martin Advanced Technology Laboratories, George Washington University, the Institute for Creative Technology at the University of Southern California, and the DARPA Augmented Cognition Program and its many team players.

7   References

Burke, M. W., Gilson, R. D., & Jagacinski, R. (1980). Multi-modal information processing for visual workload relief. *Ergonomics, 23,* 961- 975.

Herz, R. S. & Engen T. (1996). Odor memory: review and analysis. *Psychonomic Bulletin and Review 3,* (3), 300-313.

Lathan, C. and Tracey, M. (2002). The effects of operator spatial perception and sensory feedback on human-robot teleoperation performance. Presence: Teleoperators and Virtual Environments, MIT Press, 11 (4):368-377.

Miller, J. O. (1982). Divided Attention: Evidence for Coactivation with Redundant Signals. *Cognitive Psychology, 14,* 247-279.

Sulzen, J. (2001). Modality based working memory. School of Education, Stanford University. Retrieved, February 5 2003, from http://ldt.stanford.edu/~jsulzen/james-sulzen-portfolio/classes/PSY205/modality-project/paper/modality-expt-paper.PDF.

Wickens, C. (2002). Multiple resources and performance prediction. *Theoretical Issues in Ergonomics Science, 3*(2), 150-177.

Monitoring of Dual Coded Information:
The Impact of Task Switching

David A. Kobus and Christine M. Brown

Pacific Science & Engineering Group
9180 Brown Deer Road
San Diego, CA 92121
dakobus@pacific-science.com

Abstract

Thirty participants completed two experimental conditions investigating the effects of bisensory facilitation upon reaction time. Participants first completed a Simple Reaction Time task by responding to all stimuli (auditory, visual or both stimuli occurring simultaneously) that were presented using the same response key. They then participated in a Choice Decision-Making task designed to explore whether facilitory effects were sustained during a four choice decision-making task. Participants were to respond to the correct quadrant in which the stimulus occurred. Participants completed two forms of the choice reaction time task. In one condition, called Unambiguous tasks, participants were previously informed which modality to monitor. During a second condition, called Ambiguous task, participants did not know on any given trial which modality would be receiving the target information. Regardless, bisensory facilitation occurred during Choice Decision-Making. This effect occurs even though the addition of auditory stimulus provides only partially redundant information. However, during Ambiguous trials, responses to auditory stimuli were significantly slowed. This may be due to a "wait and see phenomenon" for confirmatory visual information during auditory trials. Results may have implications for designers considering modality switching as a mitigation strategy in closed-loop systems. Further investigation into effects of alternating between uni-modal and multi-modal stimuli environments may provide insight for how to enhance response time during simple and complex decision-making tasks.

1 Introduction

Evolving technology has changed our environment. We receive a plethora of visual and auditory signals on a regular basis via computers, cellular telephones, and automobile navigational systems, to name a few. These signals may be sensed independently or arrive simultaneously at sensory systems providing redundant coding of a single piece of information. Due to these changes in our sensory environment and the prominent use of bisensory coding, a better understanding of the nature of bisensory processing is needed.

Several studies have shown a performance advantage when stimuli carrying redundant information are presented in two modalities simultaneously (Doyle & Snowden, 2001). This advantage may be speed or accuracy related. Miller (1991) found that trials with redundant auditory and visual targets also produced faster response times than trials that were presented to either modality alone. This is often referred to as the *redundant-signals effect*. Lewandowski and Kobus, (1989), found that simultaneous presentation of redundant auditory and visual stimuli significantly lowered target detection threshold and recognition accuracy over trials in which each modality was presented stimuli independently. Facilitation of recall performance has also been shown when redundant signals are presented (Kobus, Moses & Bloom, 1994). The present study further investigated the nature of the *redundant-signals effect* during a four-choice decision-making task. In addition, the effects of modality switching were also investigated.

2 Method

2.1 Participants

Thirty-one undergraduates volunteered to participate in exchange for research credit in their psychology course. One participant was dropped from the analyses due to an excessive error rate and exceptionally high response times denoting that they failed to understand the instructions. The remaining participants consisted of eight males and 22 females ranging in age from 17 to 25 ($M = 18.52$, $SD = 1.59$). All participants had normal or corrected to normal vision and hearing.

2.2 Apparatus

Participants were seated comfortably in front of a standard computer color monitor (19 inch) and were provided access to four number keys on the number pad (1, 2, 4, and 5). Stimuli were presented to the auditory or visual modalities. Visual stimuli consisted of a one-inch black letter "O" or a 1 inch black dot. The auditory stimulus was an 800 Hz tone, presented in stereo at 83 db, on Altec Lansing speakers. A Gateway Pentium III desktop computer controlled (using laboratory developed Visual Basic software) the presentation of all stimuli and ensured a random inter-stimulus interval between one and five seconds.

2.3 Procedure

This experiment was comprised of two conditions, and all participants completed both. Table 1 provides detailed descriptions of stimuli used for each of the tasks and number of trials each participant completed.

Table 1. Overview of the stimulus conditions for each experimental task.

	Visual Stimulus	Auditory Stimulus	Bisensory Stimulus	Total Trials
Simple Reaction Time	1 inch black "O", in screen center (25 trials)	800 hz tone, 83 db (25 trials)	Both auditory & visual stimulus simultaneously (25 trials)	75
Choice-Decision Making				
Unambiguous Visual	Solid 1 inch black dot – (20 trials each quadrant)			80
Unambiguous Bisensory			Both auditory and visual stimulus presented simultaneously (20 trials each quadrant)	80
Ambiguous	Solid 1 inch black dot – (20 trials each quadrant)	800 hz tone, 83 db (20 trials)	Both auditory and visual stimulus presented simultaneously (20 trials each quadrant)	180

The first condition, *Simple Reaction Time*, was designed to replicate previous research findings regarding bisensory facilitation of reaction time. All participants completed the Simple Reaction Time condition first. Participants responded by using the mouse to click on a response bar as soon as they detected a stimulus. Immediately upon responding the next trial was automatically initiated. The second condition, *Choice Decision-Making*, was designed to explore whether facility effects occur during a higher-level four-choice decision-making task. The Choice Decision-Making condition was comprised of three subtasks. Two of the subtasks were referred to as Unambiguous tasks during which the participant knew what type of stimulus to expect on every trial. During the third subtask, referred to as the Ambiguous task, stimuli were randomly presented via the auditory, visual or a combination of both modalities. Participants were unaware of what type of stimulus to expect on a given trial. Participants were asked to use the index finger of their preferred hand to respond to all stimuli using the 1, 2, 4 and 5 keys of the number pad (see Figure 1) that corresponded to locations within the quadrant displayed on the monitor. For trials of a tone alone, participants were instructed that they would be allowed to press any of the response keys to provide the fastest response immediately upon stimulus detection.

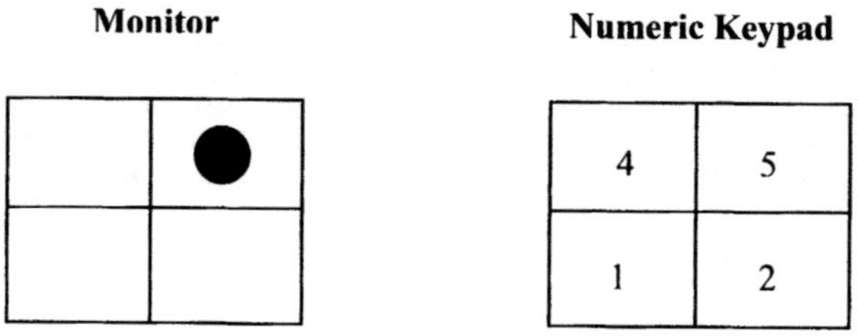

Monitor   **Numeric Keypad**

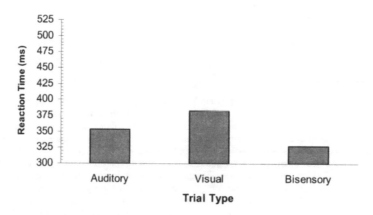

Figure 1. Example of quadrant displayed on the monitor (left) and the visual stimulus presented in one of the four locations. Four keys of the numeric keypad (right) corresponded to the spatial locations of the four quadrants on the monitor. In the example shown participants would respond by using the "5" key.

3   Results

Separate one-way repeated measures ANOVAs were conducted for each condition using SPSS (v11.5). Median reaction times and number of errors were calculated for each participant and condition. Data for one participant was excluded due to an excessive error rate and exceptionally high response times possibly related to misunderstanding instructions. Results show that for the *Simple Reaction Time* condition, mean reaction time differences between the three presentation modes were statistically significant, $F(2, 58) = 51.34$, $p < .05$. These results are graphically displayed in Figure 2. Pairwise comparisons (Bonferroni) revealed that responses to bisensory stimuli were significantly faster than trials in which stimuli were presented to either the auditory or visual modalities alone. Responses to auditory stimuli were significantly faster than to trials displayed to the visual modality.

Figure 2. Mean response time for the Simple Reaction Time condition for each trial type.

A one-way repeated measures ANOVA was also completed for the Choice Decision-Making condition of the Ambiguous task. Results indicate a statistically significant reaction time difference between presentation modes $F(2, 58) = 77.96$, $p < .05$. All pairwise comparisons (Bonferroni) for the three presentation modes were statistically different. Figure 3 represents these differences (Ambiguous task). When the modality of stimulus presentation was ambiguous, responses were fastest to Bisensory stimuli followed by the Visual Choice trials. The responses during the Auditory trials were by far the slowest.

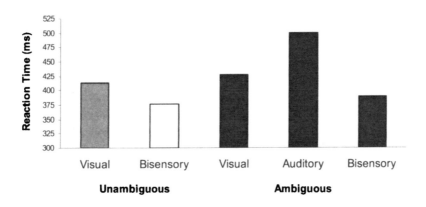

Figure 3. Mean response time for the Choice Decision-Making condition for each trial type.

A further one-way ANOVA was used to compare response times between the Ambiguous and Unambiguous bisensory choice conditions. The mean reaction time of the Unambiguous Bisensory Choice task was significantly faster than the mean reaction time during the Ambiguous Bisensory Choice trials, $F(1, 29) = 6.043$, $p < .05$. This difference is shown in the graph in Figure 3.

A one-way ANOVA was also conducted to compare the Ambiguous and Unambiguous visual choice conditions. The reaction time of the Unambiguous Visual Choice task was significantly faster than the reaction time of the Ambiguous Visual Choice trials when performed within the Ambiguous task, $F(1, 29) = 5.751$, $p < .05$. Figure 3 also graphically illustrates this difference.

4   Discussion

The findings of the Simple Reaction Time condition are in concert with previous research, indicating that the use of redundant information presented to both the auditory and visual modalities simultaneously significantly lowers reaction time (Miller, 1991). This study extends these findings by demonstrating bisensory facilitation during a four-choice decision-making task. The results show that facilitation occurs even when the auditory stimulus provides only partially redundant information. The auditory stimulus assisted in stimulus detection; however it did not provide redundant information to the participant regarding the location of the stimulus so that the correct response key could be chosen. Interestingly, in the Simple Reaction Time condition, the reaction time to auditory stimuli was significantly faster than visual stimuli as one would predict based on physiology; however during the Ambiguous task of the Choice Decision-Making condition, reaction time to auditory alone trials were significantly _slower_ than all other tasks performed. Overall mean reaction times during the Ambiguous task are significantly slower than their counterpart condition during the Unambiguous tasks. However the change in reaction time during the Auditory trials makes it by far the slowest mode of presentation, which was not expected. We hypothesize that this may be due to a "wait and see" strategy adopted by participants for visual confirming information.

One might wonder why reaction time during the Visual Choice trials is attenuated during the Ambiguous task. Here it may be that participants wait momentarily for the confirming information from the other modality. Perhaps in an environment where information may vary between bisensory and uni-sensory presentation, sensory integration leads one to be biased toward information redundancy on all trials. When this does not occur on a given trial, such as during uni-modal trials, response time is slowed while the participant waits to verify information from the second channel. Further investigation into the effect of alternating between a uni-modal and multi-modal stimulus environment may provide insight for how to enhance response time during simple and complex decision-making tasks. Such findings may be valuable for alerting systems in healthcare, automobiles, and other environments where decisions must be made quickly and acted upon swiftly.

References

Doyle, M., & Snowden, R. J., (2001). Identification of visual stimuli is improved by accompanying auditory stimuli: The role of eye movements and sound location. *Perception, 30*, 795-810.

Kobus, D.A., Moses, J. D. & Bloom, F. A. (1994). Effect of multimodal stimulus presentation of recall. *Perceptual & Motor Skills, 78*, 320-322.

Lewandowski, L. J., & Kobus, D. A. (1989). Bimodal information processing in sonar performance. *Human Performance, 2*(1), 73-84.

Miller, J. (1991). Channel interaction and the redundant-target effect in bimodal divided attention. *Journal of Experimental Psychology: Human Performance and Perception, 17*(1), 160-169.

Assessing Influences of Verbal and Spatial Ability on Multimodal C² Task Performance

Leah M. Reeves, Ali Ahmad, Kay M. Stanney

Department of Industrial Engineering & Mgmt. Systems
University of Central Florida
4000 Central Florida Blvd.
Orlando, FL USA 32816-2450
leah@intouchcomputing.com

Abstract

Due to the current lack of empirical evidence and principle-driven guidelines, designers often encounter difficulties when choosing the most appropriate modal display and interaction techniques for given users, applications, or specific military command and control (C^2) tasks within C4ISR systems. The development of multimodal design guidelines from both a user and task domain perspective is thus critical to the achievement of successful Human Systems Integration (HSI) within military environments such as C^2 systems. The present study focused on preliminary evidence indicating that how well a person processes spatial and/or verbal information may be a significant factor in determining how they will perform in multimodal, multi-task situations. The current results provide initial empirical support in identifying user attributes, such as spatial ability ($p < 0.02$) and learning style ($p < 0.03$), which may aid in developing principle-driven guidelines for how and when to effectively present task-specific modal information to improve C^2 operators' performance. Future research will examine more spatial and verbal ability tests that may be significant predictors of performance and thus likely candidates for incorporation into a Tool for Information Processing Capacity Assessment (TIPCA) currently under development (Stanney, Reeves, Hale, Samman & Buff, 2003).

1 Introduction

Military operations and friendly fire mishaps over the last decade have demonstrated that Command, Control, Communications, Computers, Intelligence, Surveillance, and Reconnaissance (C4ISR) systems may often lack the ability to efficiently and effectively support operations in complex, time critical environments and that certain users may not be able to effectively process and successfully react to the immense amount of information that is being pushed on them. With the vast increase in the amount and type of information available, the challenge to today's military system designers is to create interfaces that allow operators to proficiently process the optimal amount of mission essential data (Stanney, Reeves, Hale, Samman & Buff, 2003). To meet this challenge, multimodal system technology is showing great promise because, as the technology that supports C4ISR systems advances, the possibility of leveraging all of the human sensory systems becomes possible (Stanney, Reeves, Hale, Samman & Buff, 2003). The implication is that by facilitating the efficient use of a C4ISR operator's multiple information processing resources, substantial gains in the information management capacity of the operator-computer integral may be realized (Stanney, Samman, Reeves, et al., 2004). The assumption that all operators will be equally effective in interacting with newly implemented multimodal technologies, however, may be premature, as not all operators may be able to efficiently process various combinations of modalities (e.g., visual, auditory, haptic) and information formats (e.g., verbal, spatial) simultaneously.

Consequently, and despite its great promise, the potential of multimodal technology as a tool for streamlining interaction within military C4ISR environments may not be fully realized until guiding principles such as the following are identified:
- how to combine multisensory display techniques for given tasks, problem domains, and user capabilities;
- how to design task information presentation (e.g., via which modality(s), in a verbal and/or spatial format) to capitalize on individual operator's information processing capabilities, and;

- how to identify operators who have the potential capacity for successfully interacting with enormously increasing amounts and varied forms of information prevalent within today's military.

The present study focused on the latter issue based on preliminary evidence indicating that how well a person processes spatial and/or verbal information may be a significant factor in determining how they will perform in a multimodal, multi-task environment such as a military command and control (C^2) system (Stanney, Reeves, Hale, Samman & Buff, 2003). Recent findings suggest that the assessed spatial ability of a particular operator, as reflected by spatial visualization, spatial recognition, and spatial manipulation from standardized paper/pencil based tests, shows significant correlations to general multimodal tasks (Mayer & Massa, 2003) and C4ISR task performance (Doane, 2002; Lathan & Tracy, 2004). The present study aims to establish preliminary evidence for verbal and spatial ability predictors of performance on tasks that emulate the time-critical, externally paced mission situations often encountered by C^2 operators.

2   Method

A lab-controlled experiment was designed to examine how well particular standardized spatial and verbal ability tests predicted performance of single- and dual-tasks, which incorporated basic working memory (WM) constructs associated with most C^2 system tasks, namely visual-spatial, visual-verbal, and auditory-verbal components. The task scenarios were designed to capture a user's ability to perform the WM-specific tasks individually (uni-modally) and in a more difficult, combined situation (multimodally), which could be more representative of a C^2 task environment.

2.1   Participants

Eight college graduates (7 males, 1 female) were recruited to participate in this study. Participants had a mean age of 30.4 years (s.d. = 3.8) with a range of 26-37 years. Six participants were right-handed and two considered left-handed but stated they use their right hand for many daily tasks and their computer mouse interaction. All participants use a computer for daily work, with four being proficient computer programmers and four using the computer for basic office and multimedia applications. Each had more than 10 years experience using computers.

2.2   Apparatus

Tasks were performed on a 3.0 GHz Intel P4 processor computer with an MSI K7N2G-ILSR NF2 AGP 8X motherboard, GEFORCE-4 TI 4600 8x AGP video card, two CORSAIR 512 Mb PC3200 PC400 DDR memory chips, and a Creative SB Audigy 2 Platinum 6.1 sound card. The interface was presented on a 19" Viewsonic 0.22 dot pitch flat screen monitor at 85 Hz refresh rate and 1024x768 screen resolution. Audio was presented through 2 Creative THX 550 speakers. All user input was done with a standard keyboard and mouse.

2.3   Tasks

2.3.1   Verbal-Auditory (VA) Task – "Listen for Missile IDs"

In this task participants were auditorally via synthesized speech presented with and asked to remember an alphanumeric list of Missile IDs (c.g., 56P, 63U) while looking at the visual screen (see Fig. 1). As the visual Remember prompt disappeared (see Fig. 2), a "Listen" auditory cue was given followed by a spoken missile ID. Participants were instructed to with their left hand, press the keyboard's left (<= Yes) or right (=>No) arrow keys to answer whether or not the *spoken* Missile ID was in the initially *spoken* "Remember" list. Pressing an arrow key to answer also prompted the next Missile ID "Listen" cue to be provided.

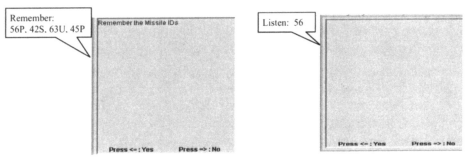

Figure 1. Remember Missile ID list.   Figure 2. Missile ID recognition/response.

2.3.2   Verbal-Visual (VV) Task – "Mental Arithmetic"

In this task participants were visually presented with a single number on the center of the display and instructed to click the "next" button with the mouse and their right hand in order to have the next number presented (see Fig. 3). Participants were instructed to mentally add the series of numbers at their own pace but to work as quickly and accurately as possible because their score depended on both the amount of numbers they could add in the given time frame and if they got the total correct. At the end of a series of numbers, participants were prompted to enter the sum (see Fig. 4) via the keyboard with their right hand and use the keyboard's "enter" key. They had only 10 seconds to enter the sum before the next series of numbers automatically began. Right-handed responses for this task were necessary to limit conflicts when the later dual-task condition was performed.

   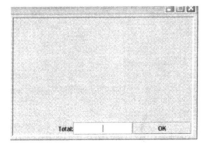

Figure 3. Visually presented number for   Figure 4. Enter total from mental arithmetic.
mental addition.

2.3.3   Spatial-Visual (SV) Task – "Remember Missile Locations"

In this task participants were visually presented with a 5 x 5 grid for 20 seconds that contained from three to ten missiles (see Fig. 5). Then the grid was removed for 20 seconds for users to remember the locations of the missiles. The grid reappeared with a subset of the originally shown missiles, and participants were instructed to use their right hand and mouse to click the grid space(s) of the missing missiles. Participants had approximately 10 seconds to indicate locations of missing missiles before the grid disappeared and the next trial began.

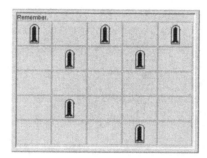

Figure 5. Remember Missile locations.

2.3.4 VA Task Combined with VV Task

For this task condition, participants performed both the Listen for Missile IDs and Mental Arithmetic tasks at the same time. Each task was performed with the same respective instructions and conditions as when it was performed alone. The only difference was that each task's respective visual screens (see Figs. 1 - 4) were side by side, with the VA task's screens on the left and the VV task's screens on the right.

2.3.5 VA Task Combined with SV Task

For this condition, participants performed both the Listen for Missile IDs and Remember Missile Locations tasks at the same time. Each task was performed with the same respective instructions and conditions as when it was performed alone. The only difference was that each task's respective visual screens (see Figs. 1-2 and 5) were side by side, with the VA task's screens on the left and the SV task's screens on the right.

2.4 Procedure

Before the start of the experiment, participants completed an informed consent and demographics questionnaire. Each participant then completed a series of standardized spatial and verbal ability tests and a verbal/visual preference questionnaire:

- Surface Development (ETS, 1976a) and Map Planning (ETS, 1976b) tests from the Kit of Factor-Referenced Cognitive Tests assessed spatial abilities;
- Auditory Number Span Test (ETS, 1976c) assessed auditory verbal working memory abilities, and;
- Visualizer/Verbalizer Questionnaire (VVQ) (Richardson, 1977) captured participants' self-reporting as to whether they considered themselves more visual or verbal in nature.

Participants were then randomly assigned to a particular experimental condition to ensure all conditions were properly counter-balanced. Regardless of task scenario order, single tasks were always performed before dual task conditions. Participants read written instructions for each task scenario prior to beginning. To further ensure each participant was trained on a task to a level of approximately 90% performance to avoid practice and learning confounds, each person was given two to three practice sessions as required of a task scenario prior to commencing the actual test scenario.

2.5 Experimental Design

In order to evaluate an individual's performance with different combinations of information format and presentation modality in both single- and dual-task conditions for later comparison to individual ability scores, a 3x5 (info format/presentation modality x task scenario) within-subjects design was implemented. The three info-format/presentation modality combinations were VV, VA, and SV. The five task scenarios were: VV alone, VA alone; SV alone; VA and VV together, and; VA and SV together. Participants' scores on each of the standardized spatial and verbal ability tests and VVQ were assessed along with their response times and accuracy scores on the single- and dual-task conditions. To determine whether standardized ability tests and/or the VVQ were significant predictors of performance in any of the task conditions, particularly the multimodal dual-task conditions, multiple linear regressions were performed with the ability tests and VVQ as independent variables and task performance scores as dependent variables.

3 Results

The response and accuracy rates for single- and dual-task performance were compared to the different individual tests using simple linear regression. Simple linear regression shows whether there is a significant correlation between the participants' task performance (dependent variable) and their spatial or verbal ability scores (independent variable). The R^2 is a good measure for the strength of the linear relationship between the response and the independent variable. Regression results are presented in Table 1. The shaded cells represent values that are significant at $\alpha=0.05$. From the table, The Surface Development Test shows significance for predicting all dual task performance conditions: Task 1 (Listen for Missile IDs, VA) response score when it is performed with Task 2 (Mental Arithmetic, VV); Task 1 accuracy when it is performed with either Task 2 or Task 3 (Remember Missile Locations, SV). The Map Planning Test shows significance for predicting Task 2 response when performed with Task 1, as well as for Task 2's single task accuracy. The VVQ showed significance for single Task 3 accuracy performance, while the Auditory Number Span Test showed no significance for any of the task conditions.

Table 1. Multiple Linear Regression Results

			Auditory Num. Span		Visualizer/ Verbalizer		Surface Develop.		Map Planning	
			p	R^2	p	R^2	p	R^2	p	R^2
Response	Single	Task1	0.729	4.6%	0.448	0.0%	0.554	0.0%	0.498	0.0%
		Task2	0.965	0.0%	0.754	0.0%	0.215	26.8%	0.067	63.5%
		Task3	*	*	*	*	*	*	*	*
	Dual 1-2	Task1	0.777	0.0%	0.344	6.0%	0.002	95.7%	0.397	0.0%
		Task2	0.339	6.6%	0.417	0.0%	0.227	24.5%	0.047	70.8%
	Dual 1-3	Task1	0.534	0.0%	0.285	14.6%	0.080	59.0%	0.515	0.0%
		Task3	*	*	*	*	*	*	*	*
Accuracy	Single	Task1	0.951	0.1%	0.946	0.0%	0.100	53.1%	0.974	0.0%
		Task2	0.974	0.0%	0.381	1.2%	0.071	61.9%	0.052	68.8%
		Task3	0.908	0.0%	0.037	74.7%	0.112	49.9%	0.203	29.0%
	Dual 1-2	Task1	0.737	0.0%	0.769	0.0%	0.028	79.1%	0.545	0.0%
		Task2	0.288	14.2%	0.970	0.0%	0.301	12.2%	0.903	0.0%
	Dual 1-3	Task1	0.778	0.0%	0.723	0.0%	0.047	70.8%	0.767	0.0%
		Task3	0.433	0.0%	0.363	3.5%	0.949	0.0%	0.303	11.8%

Note (1): Task 1 refers to *Listen for Missile IDs"* task (VA); Task 2 refers to the *Mental Arithmetic* task (VV), and; Task 3 refers to the *Remember Missile Locations* task (SV).
Note (2): The presence of the (*) in Task 3's response in both single and dual conditions is due to the fact that performance on this task did not depend on response rate, as all participants had to respond a fixed number of times and were only scored for accuracy on this task.

The results of multiple linear regression of the single- and dual-task accuracy and response performance on Surface Development and Map Planning Tests show significance (α=0.05) only for single Task 3 accuracy. The p-values for Surface Development, Map Planning, and Interaction are 0.021, 0.019 and 0.037, respectively. From the resulting regression model, the accuracy on task 3 could be predicted by:

STask3A = 0.576 + 3.75 SurfDev + 4.68 MapPlan - 5.22 SurfDev*MapPlan.

In addition, multiple linear regression runs showed the Surface Development Test is significant at α=0.1 for predicting the dual task performance of Task 3 response rate when performed concurrently with Task 1 (p = 0.077).

As an aside and to check for whether Wickens' (1984; 1992) Multiple Resource Theory (MRT) model and dual-task predictions (Wickens & Liu, 1988) held true, paired t comparisons for dual-task performance were done on task scenarios 1-2 and 1-3 to assess potential task interference issues when tasks tap resources from within a single WM mode (e.g., verbal, spatial) versus when divided across modes. The performance on task 1 when performed concurrently with tasks 2 and 3 was significantly different in terms of both response rate (t = -4.85, p = 0.002) and accuracy (t = -5.41, p = 0.001). As expected from MRT and dual-task predictions, both the response rate (1.41x) and accuracy (1.74x) are better when task 1 is performed with task 3 versus when performed with Task 2, as 1 and 2 require the same verbal WM channels but 1 and 3 are using verbal and spatial channels, respectively.

4 Discussion

The above results suggest some of the standardized tests used have potential predictive validity, namely the Surface Development and Map Planning Tests, but clearly indicate more empirical results are needed to improve significance and R^2 values. Using simple linear regression to identify significant predictors may have resulted in a loss of important information regarding combined performance on tests and potential interactions. Further, the participant size should also be increased in future investigations when trying to determine predictive power and/or for fitting a valid regression model of spatial and/or verbal ability tests that could reliably predict single- and dual-task performance. Future research will focus on such issues, as well as identifying other potential standardized verbal or spatial ability tests to evaluate.

The fact that the VVQ was only significant for predicting the accuracy on the single Remember Missile Locations task is not too surprising, as Mayer and Massa's (2003) latest assessment of the VVQ indicates it is still an inconclusive measure for predicting actual effects on visual and/or auditory task performance. Their current research is investigating the development and validation of a new visual-verbal questionnaire, which may be evaluated in follow-on studies to assess whether it warrants inclusion in a future spatial/verbal ability test battery.

The significance of The Surface Development Test for predicting performance in most task conditions indicates it as a likely candidate for incorporation into a test battery that assesses general spatial ability as a predictor on multimodal performance. It is interesting to note that the Map Planning Test showed significance for predicting the ability to perform mental arithmetic alone as well as when performed concurrently with an auditory verbal task. An implication may be that the Map Planning Test could be a likely candidate for assessing a person's ability to perform multiple tasks at a time, whether such tasks are designed to facilitate effective MRT mappings or not, and thus reveal such a person as a potential high information processor. More research is needed to determine if such an assumption is founded or not. For, if such potential could be assessed a priori, the ability to determine optimum C4ISR operator selection assignments could be greatly improved.

5 Conclusions and Future Research

The preliminary findings from this study indicate that how well a person processes spatial and/or verbal information may be a significant factor in determining how they will perform in multimodal, multi-task situations. The current results provide initial empirical support in identifying user attributes, such as spatial ability and learning style, which may aid in developing tools for identifying high multimodal information processors and predicting their performance in operational settings. Such tools could also aid in determining principle-driven guidelines for how and when to effectively present task-specific modal information to improve C^2 operators' performance when their capacity for processing such information may not be at an optimum level and/or when they may have an affinity for processing certain combinations of information formats (e.g., verbal, spatial) per modality (e.g., visual, auditory, haptic).

Future research is focused on identifying other standardized spatial/verbal aptitude tests that may lead to improved validity in predicting performance on complex, operationally-relevant multimodal C^2 tasks. This research will also investigate other potential performance predictors, such as multimodal working memory capabilities (i.e., both individual and combined information processing channel capacity at various workload levels (intra- and inter-modally) when information is presented serially and simultaneously) and operators' ability to interpret modal mappings to various types of task information based on speed, accuracy, and signal discrimination sensitivity. The overall goal is to use visual, auditory and haptic displays for incorporating the most valid tests into a Tool for Information Processing Capacity Assessment (TIPCA), which is currently under development (Stanney, Reeves, Hale, Samman & Buff, 2003), that can assess an individual's ability to process multimodal information within a general complex, multi-display environment and thus predict future task performance. Although initial application of such a tool is focused on C^2 task domains, there is great potential for its ability to predict performance on general multimodal computer- supported tasks.

6   Acknowledgements

This work has been sponsored in part by the Office of Naval Research, as preliminary dissertation fellowship research for Ph.D. candidate and DARPA Augmented Cognition Lockheed-Martin-ATL sub-contractor, Leah M. Reeves. We would like to thank Polly Tremoulet, Helen Hastie, and Michael Thomas, of LM-ATL, for their testbed support. Any opinions, findings, and conclusions or recommendations expressed in this material are those of the authors and do not necessarily reflect the views or endorsement of Lockheed-Martin-ATL, DARPA or the Office of Naval Research.

References

Doane, S. (2002). *New Measures of Complex Spatial Processing Abilities: Relating Spatial Abilities to Learning and Performance.* Mississippi State, MS: Mississippi State University Department of Psychology. (DTIC No. ADA413799, found on-line at: http://handle.dtic.mil/100.2/ADA413799)

ETS (1976a). Surface Development Test. *Kit of Factor-Referenced Cognitive Tests.* Princeton, NJ: Educational Testing Service.

ETS (1976b). Map Planning Test. *Kit of Factor-Referenced Cognitive Tests.* Princeton, NJ: Educational Testing Service.

ETS (1976c). Auditory Number Span Task. *Kit of Factor-Referenced Cognitive Tests.* Princeton, NJ: Educational Testing Service.

Lathan, C. E. & Tracy, M. R. (2002). The effects of operator spatial perception and sensory feedback on human-robot teleoperation performance. *Presence: Teleoperators and Virtual Environments,* 11(4), 368-377.

Mayer, R. E., & Massa, L. J. (2003). Three facets of visual and verbal learners: Cognitive ability, cognitive style, and learning preference. *Journal of Educational Psychology,* 95(4), 833-846.

Richardson, A. (1977). Verbalizer-visualizer: A cognitive style dimension. *Journal of Mental Imagery, 1,* 109-126.

Stanney, K., Reeves, L., Hale, K., Samman, S., Buff, W. (2003). OSD02-CR14: Multimodal Information Perceptual-ization (MIP) for C4ISR Systems. Navy Technical Report TR-203-002.

Stanney, K., Samman, S., Reeves, L., Hale, K.S., Buff, W., Bowers, C., Goldiez, B., Nicholson, D., & Lackey, S. (2004). A paradigm shift in interactive computing: Deriving multimodal design principles from behavioral and neurological foundations, *International Journal of Human-Computer Interaction, 17*(2), p.229-257.

Wickens, C. D. (1984). Engineering Psychology and Human Performance. Columbus, OH: Charles E. Merrill Publishing Co.

Wickens, C. D. (1992). Engineering psychology and human performance (2nd ed.). New York: Harper Collins Publishers.

Wickens, C. D., and Liu, Y. (1988). Codes and modalities in multiple resources: A success and a qualification. *Human Factors, 30,* 599- 616.

Section 1
Human Information Processing

Chapter 2

Cognitive Foundations of Human Information Processing

The Cognitive Pyramid - Fundamental Components that Influence Human Information Processing

James Patrey

NAVAIR
Patuxent River, MD
james.patrey@navy.mil

Amy A. Kruse

DARPA
Arlington, VA
akruse@darpa.mil

Abstract

Over the last decade, significant advances in the fields of neuroscience, cognitive psychology, and psychophysiology have occurred. These developments have revolutionized the way we understand cognitive functioning at a basic level, as well as the more abstract features of human information processing (situational awareness, decision making).

The Cognitive Pyramid is comprised of four hierarchical, unidirectional components: cognitive requirements, cognitive mediators, cognitive functional components, and integrated cognition. Cognitive requirements are biologically based, and address the presence of consciousness - mitigating factors relate primarily to the presence of glucose, oxygen and the absence of trauma. Cognitive mediators are primarily physiological factors that can impact cognition and include fatigue, hydration, noise, vibration, thermal stress, and illness. In and of themselves, they are wholly non-cognitive, but their influence on cognition can be profound. The cognitive functional components include the traditional cognitive constructs of memory, attention, and decision-making processes. Lastly, integrated cognition is perhaps best captured by the construct of situational awareness - the coherent, unitary picture emergent from the cognitive components. The Cognitive Pyramid approach will permit the range of cognitive issues to be investigated in both laboratory and applied settings. This paper will discuss the utility of sensing these components in the laboratory and operational environments, as well as future directions for the development of this approach.

1 Cognitive Pyramid

The cognitive performance of our warfighters is impacted by wide variety of factors. In understanding cognitive performance in military operations, it is essential to consider the full range of influences. Existing relevant cognitive models, such as Hancock & Warm (1989) and Staal (2004), superbly detail the range of task driven cognitive factors, such as fatigue, overload and intrinsic constraints, such as sleep deprivation and anxiety. However, the comprehensive range of cognitive mediators and moderators present in military operations are often unaddressed in these models. During military operations, there is a high likelihood of the presence of environmental factors that can cause physiological and cognitive impairment. It is thereby crucial that these are fully incorporated when understanding cognitive performance of the warfighter.

The North Atlantic Treaty Organization's Human Factors in Medicine (NATO, 2004) assembled a team of internationally-recognized experts to address a concept called "Operator Functional State" (OFS) to review the greater breadth required of models designed to address human performance across a variety of domains often unique to the warfighter. In addition to addressing cognitive factors driven by task characteristics, such as cognitive load, mental fatigue, and situational awareness, they also included a range of environmental and individual state factors as follows:

- *Noise*
- *Hypoxia*
- *Thermal stress*
- *Vibration*
- *Hyperbaric*
- *Hypoxic*

43

- *Sustained Acceleration*
- *Circadian Rhythms*
- *Hydration*
- *Fatigue*
- *Sleep loss*

The OFS model is a significant step in moving models of cognition towards greater applicability in applied settings and developing a hierarchical conception of cognition. In this regard, higher order cognition (such as cognitive load, mental fatigue, and situational awareness addressed in OFS) is subject to variance and, to some extent, dependencies upon non-cognitive components (oxygen levels, thermal stress, etc.). There have been addressed in parts elsewhere (e.g., Hancock & Vasmatzidis, 1998, Calloway & Dembo, 1958) but is arguably currently incomplete for the purposes of understanding warfighter performance. Future models must address foundational components, such as hypoxia, sustained acceleration, and thermal stress, and mediating components, such as vibration, fatigue, and sleep loss, in order to sufficiently understand cognition in the applied setting of the warfighter.

References

Calloway, E. & Dembo, D. (1958). *Narrowed attention: A psychological phenomenon that accompanies a certain physiological change.* Archives of Neurology and Psychiatry, 79, 74-90.

Hancock, P.A., & Vasmatzidis, I. (1998). *Human occupational and performance limits under stress: the thermal environment as a prototypical example.* Ergonomics, 41 (8), 1169-1191.

Hancock, P. A., & Warm, J. S. (1989). *A dynamic model of stress and sustained attention.* Human Factors, 31, 519-537.

Staal, M. (2004). *Stress, cognition, and human performance: a literature review and conceptual framework.* NASA/TM – 2004-212824.

Wilson, G. F., & Schlegel, R. E. (Eds.) (2004). *Operator Functional State Assessment,* NATO RTO Publication RTO-TR-HFM-104. Neuilly sur Seine, France: NATO Research and Technology Organization.

Computation of the Correlation between Consciousness and Physiology in Stress Environments

Tadashi Kitamura   *Akihiro Nanjoh*   *Kohji Hirakoba*

Kyushu Institute of Technology

Department of Mechanical System Engineering   Department of Biological Functions and Systems Engineering

kita@mse.kyutech.ac.p   jo@mse.kyutech.ac.p   hirakoba@lai.kyutech.ac.jp

Abstract

The purpose of this study is to report a simulator of the mind and body correlation of an agent experiencing two short term stress conditions: constraint and exercise. A large scale model of circulatory physiology and a biologically-inspired robot control architecture are combined to evaluate stress, behavior and emotions. The circulatory model was built to control artificial heart, and the robot control architecture, named Consciousness-based Architecture (CBA), was designed to mimic human/animal behavior, emotion and consciousness. The whole linked model is tuned using rat's constraint stress condition data and human exercise stress data. Using the tuned model, an autonomous process of exercise and rest in a loose constraint environment is simulated. The linked model provides a psycophysiological basis for displaying the agent's subjectivity of need for rest, fatigue feeling, need for escaping from constraint .

1 Introduction

Since H.Selye proposed a concept of stresss as physiological response to invasion (Selye, 1946), and 'stressor' (Selye, 1955), as a stimulus resulting in stress, scientific efforts have been made to investigate and elucidate the relationship between stressor and stress. Descriptive and schematic explanations about stress production mechanisms were proposed including internal secretions and autonomic nerve activities (Long, 1947, Stratakis, 1995). Concept stressor has been applied in a variety of life events such as move, family death and work change, and analysed to evaluate stresses of subjects (Holmes, et al., 1967).

Studies on stress in a more specific environment have been made such as stress in a human-machine interface from psyco-physiological view. Cardiovascular autonomic response is evaluated to phasic mental work using indices built on the basis of a simple circulatory physiology model (Ohsuga, 2001). Most psycophysiological studies of the stressor-physiology relationship, however, have been made in the fashion that physiological outputs are statistically or descriptively analysed to applied physical or mental stressors (for example, Salvador, 2001; Hinz, 2001 and Dishman, 2000). In general, few mathematical model-based approaches of mind-body correlations respecting stress have been taken, nor is model building of such relationship focused on in this field. Such model building including stress modeling is obviously important and useful for cognitive modelling and human-like robotic design as well as psychophysiological data analysis.

This paper reports combination of a large scale model of circulatory physiology and a biologically-inspired robot control architecture to simulate stress, behavior and emotions under exercise and constraint conditions. Originally, the circulatory model was built for controlling artificial hearts (Kitamura, 1999), and the robot control architecture, named Consciousness-based Architecture (CBA), was built to mimic human/animal behavior, emotion and consciousness (Kitamura, 2000, 2001, 2001). In the following sections, the models are described, linked, tuned on the basis of rat's constraint stress condition data and human exercise stress data. Using the tuned combined model, an autonomous process of agent's exercise and rest in a loose constraint environment is computed and discussed.

2 Review of CBA's Self-Adaptability

2.1 Basic Architecture

CBA, *Consciousness-based Architecture*, is a biologically-inspired six-layered developmental hierarchy model of the relationship between consciousness and behaviour as shown in Figure 1 (Kitamura, 2000). Level 0 represents sleep or rest, and level 5 is symbolic manipulation without corresponding behavior. Levels from 1 to 4 make self-adaptive behaviors possible without learning. A set of advanced behaviors is piled on top of each level up through the whole hierarchy of CBA between level 1 and 4. B_{ij}, jth behavior at level i, is a predetermined functional of behaviors of level i-1, (i=2,3,4), that is, $B_{ij} = B_{ij} (B_{i-1,\bullet})$ for $B_{i-1,\bullet} \in \{B_{ik}, (k=1,...j_{i-1})\}$. Thus, each behavior is decoded into a functional of four reflective actions at level 1, *move forward*, *move backward*, *turn right* and *turn left*.

"*Move*" at level 2 is a linear motion directing to an invisible, known object: It is performed by repetition of reflective actions, "*move forward*" or "*move backward*". Three motions to an only visible object at level 3, "*approach*", "*escape*" and "*avoid*", are conducted by a sequence of Move toward a visible object, Move away from it, and turn left or right around it respectively. The direction of *avoid* is selected at random.

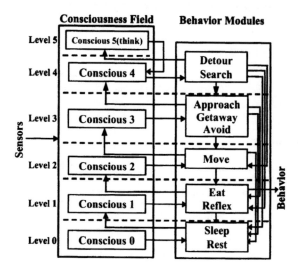

Figure 1 CBA (Consciousness-based Architecture)

While every motion at level 3 makes sense in sight of an object, behaviors at level 4 such as "*search*" and "*detour*" are performed to an object out of sight. *Detour* is used evaluating *persistence* in a target object if it disappears behind an obstacle. *Detour* is successful only if the obstacle is straight and short in width enough for the agent to be able to keep *persistent* in the target. If the obstacle is too long for a persistent agent, *detour* is obstructed into level 5. The magnitude of *persistence* depends on a variable representing *anticipation* as mentioned in 2.2.

Think at level 5, although it has no corresponding set of behaviors at this level, enables a robot to perform high level detour on a rule-base for controlling behaviors at level 4.

2.2 Behavior Selection on CBA

Behavior selection on CBA, vertical or horizontal, is conducted on the basis of winner-take-all arbitration. It depends on *emotional* two criteria, $C_k(t)$, a *degree of consciousness* for a level selection at level k at t=t, and I_i representing *a degree of pleasure* for horizontal selection at a selected level i. I_i is a functional of a behavior as defined later.

Suppose a robot is conducting a behavior at level p: It takes one of the three loops according to the evaluation of $C_k(t)$, (k=1,2,..,p) and I_p once within one time step. The first, major loop takes place with "*behavior obstruction*", at which the level of the consciousness goes up to level $p+1$ if I_p is not increased for any behavior at level p. In the second loop, the robot stays at the same level p if there is a behavior which improves I_p by a behavior at level p, and in the third, the level goes down to $j<p$ if $C_j(t)$ is maximal at $i=j$ for i=1,2,..,p-1. This last loop may be the case where the robot finds a favorite object nearer than the target it has been running after, for instance.

$C_i(t)$ is defined as,

$$C_i(t) = \sum_{j=1}^{N_E} \left| b_{ij}(t) \right| + \sum_{j=1}^{N_I} \left| g_{ij}(t) \right| + \sum_{j=1}^{N_E} \left| a_{ij}(t) \right| \quad (1)$$

where β_{ij} and γ_{ij} and a_{ij} are the intensities of external stimuli at level i, those of jth internal perceptions, such as *hunger* or *energy level and fatigue*, and *anticipations* depending on *emotion* about an external perception β_{ij} respectively, and where N_E and N_I are the numbers of external objects and internal perceptions respectively. β_{ij} and γ_{ij} are defined as normalized between 1 and -1.

$\beta_{ij}>0$ for an object j which gives a positive perception is defined as a decreasing function g_j of two variables: distance d_j between the robot and object j and deflection angle θ_j of the object's direction relative to the robot's front.

$$g_j(d_j,q_j) = \begin{cases} k_j(1-|q_j|/p)(1-d_j/d_{j0}) & \text{for } 0 \pounds d_j \pounds d_{j0} \text{ and } |q_j|<p \\ 0, & \text{else} \end{cases} \quad (2)$$

where $k_j>0$ and $d_{j0}>0$ are predetermined depending on the object. For a negative perception $\beta_{ij}<0$, $k_j<0$ is used in g_j instead of $k_j>0$.

Criterion $I_i(t)$ is defined as,

$$I_i(t) = \overset{N_E}{\underset{j=1}{\mathring{a}}} b_{ij} + \overset{N_I}{\underset{j=1}{\mathring{a}}} g_{ij} + \overset{N_E}{\underset{j=1}{\mathring{a}}} a_{ij} \quad (3)$$

This criterion for behavior selection represents a summation of emotion values at level i, meaning *pleasure* for positive emotion and *displeasure* for negative one. From the definitions of I_i and β_{ij}, I_i is a functional of behavior conducted at level i, i.e., $I_i(t), B_{ij})$, because the distance and deflection angle respecting object j depends on the behavior. This is because, as mentioned in 2.1, each behavior at any level is decoded into some of the four reflective actions at level 1, *move forward, move backward, turn right* and *turn left*. Therefore, a better behavior B_{ij} belonging to a set of behaviors at level i can be estimated in the environment fixed at t=t so that $I_i(t, B_{ij})>I_i(t)$. I_i therefore can be also maximized respecting the behavior.

Anticipation a_{ij} is defined as functions of β_{ij} for i=4 and 5. It basically encourages a seemingly achievable behavior by increasing I_i, and discourages an unachievable one by decreasing it. It is defined as follows.

$$a_{ij} = \begin{cases} 0 & \text{for } i \pounds 2 \\[2mm] f_3(b_{ij}) = \begin{cases} a_3 b_{ij}, & \text{if } |b_{ij}(t)|>|b_{ij}(t-1)| \\ -a_3 b_{ij}, & \text{else} \end{cases} & \text{for } i=3 \\[4mm] f_4(b_{ij}) = \begin{cases} a_4 b_{ij}(t_{invis})2^{-t?T}, & \text{if } T_0>t>0 \\ -a_4 b_{ij}(t_{invis})2^{-t?T}, & \text{else} \end{cases} & \text{for } i=4,5 \end{cases} \quad (4)$$

where $a_i>0$, (i=3,4) are predetermined, T and $T_0<T$ are appropriate time constants predetermined and t_{invis} is the time when the object became invisible.

Anticipation at level 3 as defined in Equation 3 makes a robot continue to run after the target if the distance to it becomes shorter; otherwise, the agent gives it up. Thus, anticipation is the intensity of persistence in a target. *Anticipation* at level 4 represents persistence in the memory of an invisible object the robot has been searching for. Therefore it includes a forgetting factor, but cannot include the distance to the invisible object at level 4. Thus, it can be said that positive f_i gives optimism about an object, and conversely, the negative one means pessimism.

All the initial information given to CBA is hungry parameter γ_{il}, location of Goal, and functions g_j defined by (2) for β_{ij} representing the goal and an unknown obstacle to be encountered. In this study, β functional for Goal is defined only for level 2, that is, $\beta_{2,Goal}>0$, and $\beta_{i,Goal}=0$ for i=1,3,4,5. β functionals for obstacles are defined as identical for all levels. γ_{il} (i=1,2,...,5) is a negative decreasing function of time defined as,

$$g_{i1}(t) = \begin{cases} g_0 t^2, & \text{if } t_1 \,^3 \, t \,^3 \, 0 \\ g_0 t_1^2, & \text{if } t>t_1 \end{cases} \quad (5)$$

where $-1/t_1^2<\gamma_0<0$ is predetermined. It triggers behavior at each level as long as it is negative; otherwise, no behavior is activated at any level, i.e., the robot is static. All other necessary parameters for f_j and g_j are determined according to the purpose of experiments.

2.3 *Think* at Level 5

In this module, the following rule-based function to control a symbolic behavior, 'high-level detour', is installed at the top level of CBA.

(R-D1) Set as the subgoal the intersecting point between the obstacle and the circle of the vision field boundary.

This rule creates a positive emotion value β_y as a sugoal to the intersecting point on the obstacle, resulting in driving 'Search' for the subgoal at level 4 toward the intersecting point until finding an edge of the obstacle. If a behavior is obstructed at level 4, and 'high-level detour' is reentered, uniquely chosen and then activated, then this function finds a subgoal to move to, which gives a positive value to the corresponding 'anticipation' to the subgoal.

3 Model of Circulatory Physiology

This circulatory model (Kitamura, 1999), modified from Human model developed by Coleman (Coleman, 1981), simulates physiological phenomena on the longer time scale than one heart cycle. It consists of over 25 physiological modules of major circulatory organs and other related functional units with over 20 variables and 60 parameters. As Figure 2 shows, the major 25 modules include Control of Exercise, Pharmacology, Reflexes, Cardiac Function, General Circulation, Gas Exchange, Basic Renal Hormones, Status of Kidney, Water Balance, Sodium Balance, Acid/Base Balance, Potassium Balance, Protein Balance, Temperature Regulation.

CARDFUNC and CIRC in this circulatory model are linked to a lumped parameter hemodynamic model computed on the beat-by-beat basis. The lumped parameter hemodynamic model includes two cardiac functions in terms of elastance with four cardiac valves modeled as diodes and Windkessel models for the two sides of circulation.

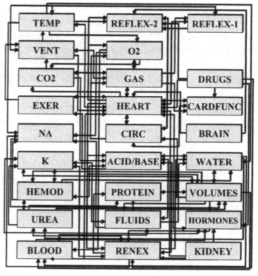

Figure 2 Model of Circulatory Physiology

4 Linking Two Models

In bilaterally linking the two models of CBA and circulatory physiology, there are two basic assumptions as shown in Figure 3: (1) motion and internal perceptions in CBA are influenced by changes in the circulatory physiology, and (2) behavior obstruction and/or excess of exercise produce a stressor in CBA which causes a stress in circulatory physiology. On the first assumption, one of γ_{i2} is proportional to blood lactate computed in ACID/BASE, whose accumulation causes agent's fatigue making the agent conscious of fatigue. At the same time, the increase in blood lactate reduces motion speed of the agent. This will be used for exercise stress which makes an agent take a rest.

On the second one, a stressor module and module Brain (Figure 2) are made in CBA and the model of circulatory physiology respectively. A stressor takes place when a behavior is obstructed, and then is felt as a stress. Stressor is considered to represent the magnitude of behavior obstruction or excess of exercise. It is given by

$$Stressor = K_{stress}\, e^{-t/T_{imen}}\,\text{ [intensity of stress duration]} \qquad (6)$$

where intensity of stress duration is defined as integral of the absolute value of negative anticipation term to the constraining obstacle in CBA for constraint stress, and as exercise duration time for exercise stress respectively. In REFLEX-2 of the physiological model in Figure 2, noradrenaline as a transmitter in the brain and a hormone in the body are computed.

For constraint stress, module Brain detects a stress with a threshold in the amygdala, and computes the magnitude of stress and the released volume of noradrenaline as a transmitter in the hypothalamus as follows.

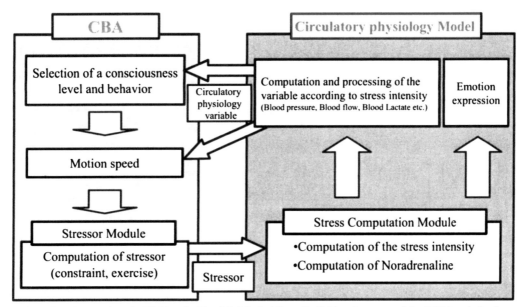

Fig. 3 Two Assumptions for Linking the Two Models: In bilaterally linking the two models of CBA and circulatory physiology, there are two basic assumptions as shown in Figure 3: (1) motion and internal perceptions in CBA are influenced by changes in the circulatory physiology, and (2) behavior obstruction produces a stressor in CBA which causes a stress in circulatory physiology.

$$\frac{dStress}{dt} = \frac{1}{T_s}(-\,Stress + k'\,Stressor) \tag{7}$$

$$\frac{dNorad_{BRAIN}}{dt} = \frac{1}{T_{BNORAD}}(-\,Norad_{BRAIN} + k_{BNORAD}'\,Stress) \tag{8}$$

where T_s, k, T_{BNORAD} and k_{BNORAD} are tuned using rat' data.

For exercise stress, the same 1st order delay system as Equation (7) is made for stress response of the hypothalamus-adrenal cortex system in Brain to a stressor input. Stress is detected as a stress if it is over a threshold corresponding a lactate threshold, about 2mmol/L. Stress is an input to the same 1st order delay system as Equation (8) to compute noradrenal secretion in the body.

5   Simulations for Stress Conditions and Discussion

5.1   Parameter Tuning

Parameter tuning for Equations 6, 7 and 8 in Brain of the combined model is done with the two stress conditions: constraint stress and exercise stress. For the constraint stress as shown in Figure 4, the simulation run is that an agent with the linked model of consciousness and physiology is constrained in a cell of its body size for 20 minutes, and is re-

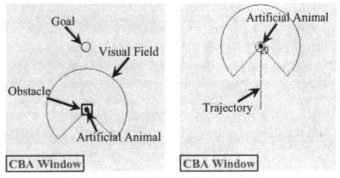

Figure 4 Agent's Environment for Constraint Stress

leased in an open field. Figure 5 shows that tuning results in a good agreement of the released noradorenaline in the hypothalamus of the rat data (Yokoo, 1990) and simulation.

To tune parameters of Equations (6), (7) and (8) for exercise stress, an exercise is imposed on an agent so that that it makes straight for a given goal. This exercise of the agent is compared with a human exercise on an ergometer for 15 minutes with 88% of %VO2max (i.e., equivalent to maximum O2 intake) (Marliss, 2000). As a result of parameter tuning, Figures 6 and 7 show good agreements of the simulation and human data for blood noradrenaline and blood lactate respectively.

In this simulation, for CBA, two internal perceptions are given, hunger for γ_{i1} as a negative decreasing time-function, and fatigue for $\gamma_{i2}=k_{LAC}*$BLAC where BLAC is blood lactate in the body and $k_{LAC}<0$. There is only one β which is for the surrounding obstacle.

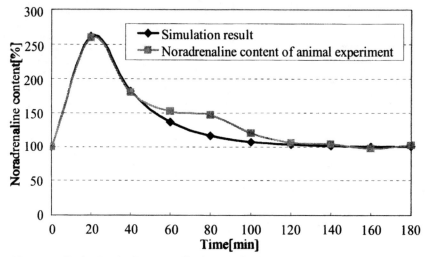

Figure 5 Parameter Tuning Result of Noradrenaline in Brain for Rat Data and Simulation for Constraint Stress

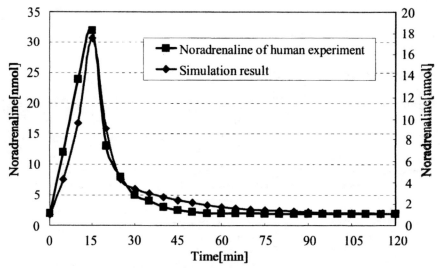

Figure 6 Parameter Tuning Result of Noradrenaline in Body for Human Data and Simulation for Exercise Stress

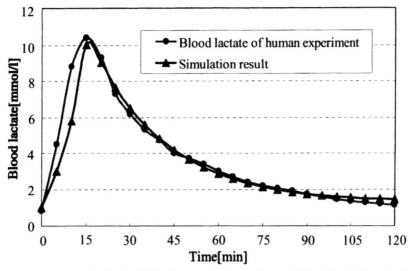

Figure 7 Parameter Tuning Result of Blood Lactate for Human Data and Simulation for Exercise Stress

5.2   Exercise and Rest in Closed Field

The purpose of the simulation is to compute an autonomous process of exercise and rest due to fatigue and constraint by increase in blood lactate under a loosely constraining environment. The simulation setup as shown in Figure 8 is that the agent at the beginning placed in the field surrounded by an obstacle directs to a given goal outside the closed field. Figure 9 shows a behavior track under the stress conditions.

Figure 8 Agent's Environment for Exercise and Rest

Figure 10 shows time courses of %VO2max, blood lactate, agent's moving speed, and noradrenalines in the body and brain. Times T1 and T3 in Figures 9 and 10 mean the starts of speed reduction, and times T2 and T4, the starts of rest due to maximal fatigue. At T1 and T3, blood lactate is around LT (about 2mmol/l), lactate threshold where the agent begins to feel fatigue as a stress. LT corresponds to 60% of exercise intensity (Pederson, 2004) At T2 and T3, blood lactate reaches OBLA (about 4mmol/l), onset of blood lactate accumulation which causes the agent to stop in spite of the continuation of brain command to move.

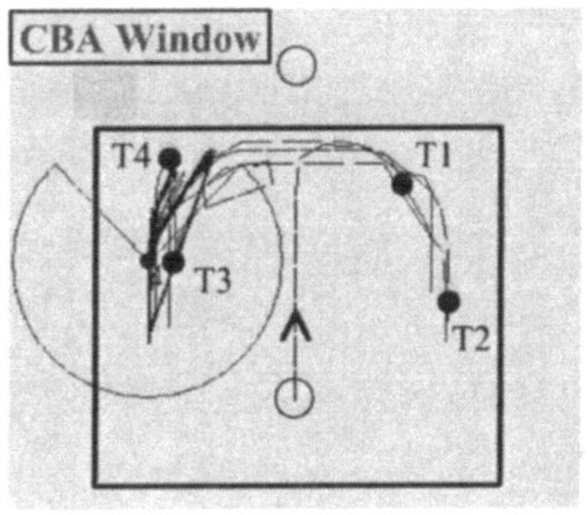

This simulation is successful because the two stresses increase according to the increases in noradrenalines in the body and brain. Also because the agent repeats autonomous exercise and rest as two peaks of the five variables for 60 minutes as shown in Figure 9.

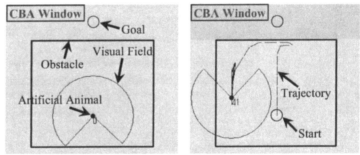

Figure 9 Behavior Track for Exercise and Rest

51

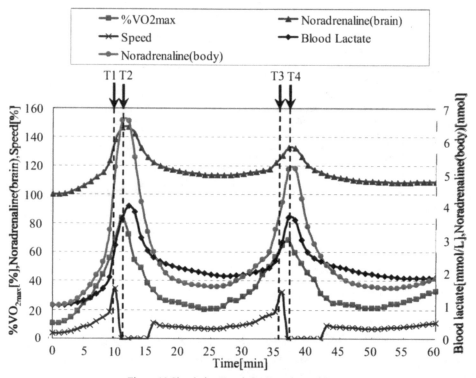

Figure 10 Simulation Result for Exercise and Rest

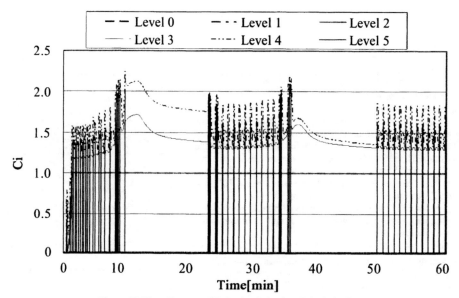

Figure 11 Time Courses of Behavior Selection Criteria in CBA

Figure 11 shows time courses of the behavior selection criteria, $C_i(t)$ and $,I_i(t)$. for 60 minutes corresponding to the above simulation of the autonomous process. It clearly shows the events of rests at level 4 and intermittent reentering in level 5. These rests are taken due to muscular fatigue, not the ones at level 0.

5.3 Limitations of Models

CBA was first built to produce animal-like behaviors with a model of the hierarchical relationship between consciousness and behavior (Kitamura, 2001, 2001, 2000). It was shown that it linked low level behaviors to autonomous symbolic behaviors including enclosure in computer simulation, and ambush and pursuit in experiments by Khepera robots. A learning algorithm was also successfully simulated, and shows much more efficient than Q-learning (Kitamura, 2002). These symbolic behaviors were controlled on the rule base at level 5. They should be integrated into a whole 'Think' model so that CBA can be more adaptive to environmental changes.

Although qualitative discrepancy between the model of circulatory physiology and animal data can be minimized by adjusting its parameters, the circulatory model showed limited quantitative accuracy for both short and long term simulations (Kitamura, 1999). The above limitations of CBA and the circulatory model are taken over to the proposed whole model.

In addition, in the module Brain built in the circulatory model, a mechanism of positive emotions due to released transmitters other than catecholamine should be embedded. Agent's motion by CBA should be linked to its limb muscular O2 consumption computed in the circulation, while the total O2 consumption of the whole body is linked to the agent's motion in the present simulation.

6 Conclusion

Mathematical model building including stress models is important and useful for cognitive modelling and human-like robotics as well as psycophysiological data analysis. This paper reports combination of a large scale model of circulatory physiology and a biologically-inspired robot control architecture to simulate stress, behavior and circulatory physiology. The two models are linked and tuned on the basis of rat's constraint stress condition data and human exercise stress data. Using the tuned combined model, an autonomous process of exercise and rest in a loosely constraining environment is simulated. The linked model provides a psychophysiological basis for displaying the agent's subjectivity of need for rest, fatigue feeling, need for escaping from a stressor as well as the existing emotions about objects and internal perception of hunger. The linked model will be useful to design a human interface of a home robot expressing its emotions and feelings depending on the given environment. For future work, brain transmitters producing positive emotions such as dopamin should be installed in the model.

References

Coleman, T.G. (1981). Human Physiological Basis of a Comprehensive Model for Body Function, University of Mississippi Medical Center.

Dishman, R.K., Nakamura, Y., Garcia, M.E., Thompson, R.W., Dunn, A.L., and Blair, S.N. (2000). Heart rate variability, trait anxiety, and perceived stress among physically fit men and women,vol.37, no.2, 2000, Pages 121-133

Hinz, A., Seibt,R., Scheuch, K. (2001). Covariation and Temporal Stability of Peripheral and Brachial Blood Pressure Responses to Mental and Static Stress, Journal of Psychophysiology, Vol. 15, No. 3

Holmes,T.H., Rahe,R.H. (1967). The social readjustment rating scale, J.Psycosom Res., 11, pp.213-218.

Kitamura.T (1999). A Computer-aided Approach to the Structural Analysis and Modification of a Large Circulatory System Model, IEEE Trans. BME, Vol.46, No.5, pp.485-493.

Kitamura.T. Tahara, T. Asami, K. (2000). How can a robot have consciousness? Advanced Robotics, vol. 14, no.4, pp. 263—277.

Kitamura.T and Ono, K. (2001). An Architecture of Emotion-Based Behavior Selection for Mobile Robots, Cybernetics and Systems, vol.32, no.6, pp.671-690.

Kitamura.T (2001) An Architecture for Animal-like Behavior Selection, in What Should be Computed to Understand and Make Brain Function?-From Robotics, Soft Computing, Biology and Neuroscience to Cognitive Philosophy-FLSI Series, vol.3, pp.23-42, (Ed. T.Kitamura), World Scientific Publishing Company.

Kitamura.T, Otsubo, J. and Abe, M. (2002). Emotional Intelligence for Linking Symbolic Behaviors, IEEE Proc. ICRA, Washington DC, vol.1, pp.1001-1006.

Marliss, E.B., Kreisman, S.H., Manzon, A., Halter, J.B., Vranic, M. and SNessim, S.J., et al. (2000) Gender differences in glucoregulatory responses to intense exercise. J.Appl. Physiol., 88, pp.457-466

Long, C.N.H. (1947). Conditions Associated with Secretion of Adrenal Cortex, Federation Proc., 6, 461.

Ohsuga, M., Shimono, F, and Genno, H. (2001) Assessment of phasic work stress using autonomic indices, International Jounal of Psychophysiology, Vol.40, No.3, pp.211-220.

Pedersen, P. K., Sorensen, J. B., Jensen, K., Johansen, L., and Levin, K. (2002). Medicine and Science in Sports and Exercise, Vol. 34, No. 4, pp. 655-661.

Salvador, A., Ricarte, J., González-Bono, E., Moya-Albiol, L. (2001). Effects of Physical Training on Endocrine and Autonomic Response to Acute Stress, Journal of Psychophysiology, Vol. 15, No. 2.

Selye, H. (1946) The general adaptation syndrome and the disease of adaptation, J. Clin. Endocrinol., 6, pp.117-230,.

Selye, H., et al. (1955/56). 5th annual report on stress, MDPublication Inc., New York.

Stratakis, C. A., et al. (1995) Neuroendocrinology and pathophysiology of the stress system, Ann. N.Y. Acad. Sci., 771, pp.1-18.

Yokoo, H., Tanaka, M., Tanaka, K., et al. (1990) Stress-induced increases in noradrenaline release in the rat hypothalamus assessed by intracranial microdialysis, Experimentia, 46, pp.290-292.

Training Effects on the Neural Circuits that Support Visual Attention and Multi-Task Performance

Paige E. Scalf, Kirk I. Erickson, Stanley J. Colcombe, Jenny S. Kim, Maritza Alvadro, Ruchika Wadhwa and Arthur F. Kramer
University of Illinois at Urbana-Champaign, Beckman Institute and Department of Psychology

Introduction

A critical role of attention is to identify those items in the environment that require a response. Because the amount of information in the environment is practically unlimited, all models of attention share the idea that attention is limited in capacity. The experiments discussed in this paper address the neural underpinnings of training-induced increases in attentional capacity. Because not all tasks that are attentionally demanding necessarily rely on the same attentional resources, we examined training-related changes in the performance of two task paradigms believed to tap different types of attentional capacity. We used fMRI to assess the neural impact of cognitive training. In order to isolate changes that occur as a result of training from those that occur as a result of increased task familiarity, we took the unusual step (within the functional neuroimaging field) of making longitudinal measurements on both trained and untrained individuals.

The first of these tasks employed the Functional Field of View (FFOV) paradigm, which is believed to index the spatial area over which visual attention may be allocated. The FFOV is typically thought of as the maximum distance a stimulus can be moved from the central focus of attention and still be reliably detected or located (Mackworth, 1965; Mackworth, 1976). One of the most interesting features of the FFOV task is that it shrinks with increasing cognitive load. Although an individual's FFOV might be quite large during the performance of a simple detection task, it may grow smaller if distracting items are added to the display or if an additional stimulus requires a response. Neuroimaging data from our lab (Scalf et al., in preparation) demonstrates that reduction of the FFOV through the inclusion of an additional task is associated with decreased response of the visual system to task-relevant information. We interpret these data as suggesting that increases in task demand reduce attention to task relevant items and thereby decrease the strength with which the visual system will represent those items. In consequence, these representations are less able to influence higher-level neural systems responsible for generating task appropriate responses. Consistent with this idea, we find that inferior prefrontal regions associated with response selection are less active during task conditions that produce a reduced FFOV.

Our second task employed the Psychological Refractory Period (PRP) paradigm, which indexes a temporal limitation in individuals' ability to respond to multiple stimuli. Specifically, making a speeded response to an initial stimulus item delays participants' ability to make a response to a second target item (Pashler, 1984). This effect is observed even when stimuli are presented to different sensory modalities and responded to via different output modalities. The PRP is thought to reflect a bottleneck in central cognitive operations such as response selection, response preparation, mental rotation, or memory retrieval (Pashler, 1984). Processing of the second item is delayed during the time that the first item is undergoing these central operations. When processing of the initial task is completed, processing of the second task may proceed. Neuroimaging data from our lab (Erickson et al., submitted) indicate that responding to multiple tasks increases activation in parietal and prefrontal regions relative to responding to single tasks, suggesting that these regions may be called upon to help coordinate the performance of multiple tasks.

Performance in both FFOV and PRP tasks is highly amenable to training. The loci of training-related improvements for the two tasks are different, however, despite the fact that the most demanding condition of both paradigms requires the performance of multiple tasks. For example, training-related improvements in paradigms similar to the FFOV may be entirely accounted for by improvements in participants' ability to represent and respond to stimuli under single task conditions (Ahissar, Laiwand, & Hochstein, 2001). Training in the FFOV paradigm, then, appears to improve performance primarily by improving participants' ability to represent and respond to stimuli throughout the visual environment, rather than by improving their ability to coordinate and control the component tasks' simultaneous performance. Training-related improvements in tasks similar to the PRP, however, are proportionally greater for dual-task than for single-task conditions (Kramer et al., 1999; Bherer et al., submitted). Training in the

PRP paradigm, then, primarily affects participants' ability to coordinate and control simultaneous performance of the component tasks rather than their ability to perform the component tasks individually.

fMRI allows us to asses the neural impact of cognitive training. Changes in behavior can occur for a number of reasons; the participant may become better at directing attention to the task at hand, may be able to use attention more efficiently or may adopt a different task strategy. We are beginning to use neuroimaging data to discriminate between these different explanations. Training-related changes in neural activity may cause increases in activation magnitude, decreases in activation magnitude or shifts in activation location. Training related increases in activation are typically considered to reflect improvements in the recruitment of neural networks that support task performance; for example, the amount of tissue recruited to support a particular function may increase, or the cognitive function performed by that tissue may become more homogenous (Furmanski et al., 2004; Nyberg et al., 2003; Oleson et al., 2004; Tracey et al., 2001). Task-related decreases in activation are usually thought to reflect increased neural network efficiency; practice with a task may increase the selectivity and specificity of the tissue that responds to task demands, resulting in a net decrease in the amount of tissue that must be recruited to support successful task performance (Callan et al., 2003; Garavan et al., 2000; Landau et al., 2004; Kassubeck et al., 2001; Poldrack et al., 2001). Shifts in the location of activation are thought to reflect changes in task or neural strategy; specifically, either the cognitive strategy employed by participants or the neural systems used to support a cognitive operation may change with training (Fletcher et al., 1999; Poldrack et al., 1998; Staines et al., 2002).

One characteristic of our study that is unique in the fMRI literature is the inclusion of untrained control groups. The decision to abandon the practice of including a control group is usually dictated by practical considerations; simply the expense of collecting neuroimaging data discourages most researchers from testing participants in whom they do not expect to observe an effect. The justification for eliminating the control group is that each trained participant may serve as his/her own control; any within subject changes must be due to exposure to the task. The difficulty in interpreting such changes in the absence of a control group, however, is that the "task" begins to encompass many factors in which the experimenter may not be particularly interested, such as exposure to the neuroimaging environment, adaptation to stimulus presentation and response collection equipment, comprehension of task instructions or fatigue and boredom. In short, many training studies may index the neural changes due to exposure to the task environment rather than those due to training in the specific cognitive components of a task. It is important to distinguish between these possibilities if we wish to discriminate neural changes that reflect increased task or skill proficiency from those that reflect task familiarity.

Methods

Because behavioral data indicates that cognitive loci of training related changes in FFOV and PRP task performance are different, we suspected that the neural loci of training related changes might also differ between the two paradigms. In the current study, we examined how cognitive training in both a PRP task and an FFOV task altered neural function to improve attentional efficiency. We tested over 60 young adults; half of these were randomly assigned to the FFOV paradigm and the remaining half were assigned to the PRP paradigm. All participants, whether assigned to training or control groups, performed both neuroimaging and behavioral pretesting and posttesting sessions.

During neuroimaging testing, participants in the FFOV condition were asked to determine whether a "T" placed 5 degree from fixation and a "/" placed 7.5 or 15 degrees from fixation occurred on the right or left side of the screen. Participants indicated the location of the "T" by using their right or left index fingers on a four-button response pad. Participants indicated the location of the "/" by using their right or left middle fingers on the same response pad. Participants performed each of these tasks in isolation before performing them simultaneously. Stimulation remained constant across task conditions. During behavioral testing and training, participants performed a more difficult version of this task in which the "T" could be located in one of four locations 5 degrees from fixation and the "/" could be located in one of six locations 7.5 or 15 degrees from fixation. Participants indicated the exact location of the "T" and "/" via mouse click. Although the behavioral version of our FFOV task is more similar to classical versions of the FFOV paradigm, the large number of responses it required made it impossible to implement in the MRI environment.

During both behavioral and neuroimaging testing, participants in the PRP condition determined whether the color of an "X" located directly above fixation was yellow or green and whether the identity of a letter located directly below fixation was "B" or "C". Participants indicated their decision via button press; the exact button and hand with which they responded to each item was counterbalanced across participants. Participants performed both of these tasks in isolation as well as simultaneously. Only task-relevant stimuli appeared on any given trial; participants saw only a single stimulus when performing tasks individually. Behavioral training and testing procedures were identical to those used during neuroimaging except for the addition of a placeholder star (*) during single-trials in the fMRI sessions that served to equate the visual stimulation between single and dual task condition. .

Results and Discussion

Consistent with the behavioral training results observed in other studies, we found that training related changes in neural activation were specific to single task conditions in the FFOV task and were greater for dual task conditions in the PRP task. The effects reported in the following paragraph reflect differences between pretest and posttest activation that were significantly larger in trained participants than in control participants. Participants trained in the FFOV showed decreases in activation to single task conditions in posterior ventral regions associated with the representation of visual items (Corbetta et al., 1991) and dorsal prefrontal regions associated with increasing the relative influence of task-relevant stimuli on response selection and execution processes (Banich et al., 2000). Relative to control participants, participants trained in the PRP paradigm showed decreases in activation that were specific to the dual task condition in parietal regions typically associated with dividing attention between multiple task-relevant stimuli (Corbetta & Schulman, 2002) as well as in medial prefrontal regions typically associated with mediating response conflict (Milham et al., 2003). Participants trained in the PRP condition also showed dual-task specific activation increases in dorsal prefrontal regions associated with increasing the relative influence of task-relevant stimuli on response selection and execution processes. Our neuroimaging data, then, indicates that the neural loci of training-related changes to the attentional system are determined by the specific characteristics of the task with which participants are trained.

Notably, we found that training in both the FFOV and PRP tasks leads primarily to reductions in task-related activation, suggesting that cognitive training in both tasks improves the efficiency of neural systems that responds to them. This improved efficiency may reflect increases in the selectivity and specificity of the neural populations recruited to support task performance. Such improved efficiency occurred during single task conditions of the FFOV task in the visual regions that support the representation of task-relevant visual items and the prefrontal regions that influence response selection and execution processes. Training in the PRP task improved neural efficiency in those regions that mediate response conflict and those regions that allow attention to be divided among task relevant items. Viewed in this light, training-related changes in neural function may increase attentional capacity primarily by reducing the number of neural resources dedicated to the performance of a particular cognitive operation.

Our data from control participants in the FFOV paradigm showed clear evidence of the changes in neural activation that can occur as a result of increased task familiarity rather than as a result of cognitive training per se. We found that untrained participants showed increased activation to single-task conditions during the post-testing session relative to the pre-testing session. These increases in activation occurred in the same dorsal prefrontal regions associated with increasing the relative influence of task-relevant stimuli on response selection and execution processes (although were restricted to the left hemisphere), as well as in medial prefrontal structures associated with mediating response conflict. Note that these medial prefrontal regions are identical to those in which training in the PRP paradigm caused a decrease in task-related activation, while the dorsal prefrontal regions are identical to those in which training-related increases were correlated with improved PRP performance. As we said earlier, increases in task-related activation are typically associated with better recruitment of the neural networks that support task performance. Because our control group received some practice with the task during behavioral testing and showed significant improvement in behavioral performance, we interpret these findings as reflecting an intermediate stage of training-related improvement in which prefrontal regions that support the distribution of attention to visual items and mediate response conflict are better recruited to support task performance, but are not yet improved with regard to the selectivity and specificity with which they respond to stimulus items. Our data, therefore, highlight the importance of including control groups in the design of neuroimaging studies that investigate cognitive training if training-related functional plasticity is to be isolated from familiarity driven changes in neural response.

References

Ahissar, M., Laiwand, R., & Hochstein, S. (2001). Attentional demands following perceptual skill training. Psychological Science, 12(1), 56-62.

Banich, M. T., Milham, M. P., Atchley, R. A., Cohen, N. J., Webb, A., Wszalek, T., et al. (2000). Prefrontal regions play a predominant role in imposing an attentional 'set': evidence from fMRI. Brain research. Cognitive brain research., 10(1-2), 1-9.

Brass, M., Von Cramon, D.Y., 2004. Decomposing components of task preparation with functional magnetic resonance imaging. Journal of Cognitive Neuroscience. 16(4), 609-620.

Callan, D.E., Tajima, K., Callan, A.M., Kubo, R., Masaki, S., Akahane-Yamada, R., 2003. Learning-induced neural plasticity associated with improved identification performance after training of a difficult second-language phonetic contrast. Neuroimage. 19, 113-124.

Corbetta, M., F. M. Miezin, et al. (1991). "Selective and divided attention during visual discriminations of shape, color, and speed: functional anatomy by positron emission tomography." The Journal of neuroscience : the official journal of the Society for Neuroscience. 11(8): 2383-402.

Corbetta, M., Shulman, G.L., 2002. Control of goal-directed and stimulus-driven attention in the brain [Review]. [Review] Nature Reviews Neuroscience. 3(3), 201-215.

Erickson, K. I., M. P. Milham, et al. (2004). "Behavioral conflict, anterior cingulate cortex, and experiment duration: implications of diverging data." Human brain mapping. 21(2): 98-107.

Fletcher, P., Buchel, C., Josephs, O., Frsiton, K., Dolan, R., 1999. Learning-related neuronal responses in prefrontal cortex studied with functional neuroimaging. Cerebral Cortex. 9, 168-178.

Furmanski, C.S., Schluppeck, D., Engel, S.A., 2004. Learning strengthens the response of primary visual cortex to simple patterns. Current Biology. 14, 573-578.

Garavan, H., Kelley, D., Rosen, A., Rao, S.M., Stein, E.A., 2000. Practice-related functional activation changes in a working memory task. Microscopy Research and Technique. 51, 54-63.

Kassubek, J., Schmidtke, K., Kimming, H., Lucking, C.H., Greenlee, M.W., 2001. Changes in cortical activation during mirror reading before and after training: An fMRI study of procedural learning. Brain Research and Cognition. 10, 207-217.

Kramer, A.F., Humphrey, D.G., Larish, J.F., Logan, G.D., Strayer, D.L., 1994. Aging and Inhibition – Beyond a unitary view of inhibitory processing in attention. Psychology & Aging. 9(4), 491-512.

Landau, S. M., E. H. Schumacher, et al. (2004). "A functional MRI study of the influence of practice on component processes of working memory." Neuroimage 22(1): 211-21.

Mackworth, N. H. (1965). "Visual noise causes tunnel vision." Psychonomic Science 3(2): 67-68.

Mackworth, N. H. (1976). Stimulus density limits the useful field of view. Eye movements and psychological processes. R. A. Monty and J. W. Senders. Oxford, England, Lawrence Erlbaum: x, 550.

Milham, M.P., Banich, M.T., Claus, E.D., Cohen, N.J., 2003. Practice-related effects demonstrate complementary roles of anterior cingulated and prefrontal cortices in attentional control small star filled. NeuroImage. 18, 483-493.

Milham, M. P., Banich, M. T., Barad, V., Cohen, N.J., Wszalek, T., & Kramer A.F. (2003). Competition for priority in processing increases prefrontal cortex's involvement in top-down control: an event-related fMRI study of the stroop task. Cognitive brain research., 17(2), 212-222.

Nyberg, L., J. Sandblom, et al. (2003). "Neural correlates of training-related memory improvement in adulthood and aging." Proc Natl Acad Sci U S A 100(23): 13728-33.

Olesen, P. J., H. Westerberg, et al. (2004). "Increased prefrontal and parietal activity after training of working memory." Nat Neurosci 7(1): 75-9.

Pashler, H., 1984. Processing stages in overlapping tasks: Evidence for a central bottleneck. Journal of Experimental Psychology: Human Perception and Performance. 10, 358–377.

Poldrack, R.A., Desmond, J.E., Glover, G.H., Gabrieli, J.D.E., 1998. The neural basis of visual skill learning: An fMRI study of mirror reading. Cerebral Cortex. 8, 1-10.

Poldrack, R.A., Gabrieli, J.D.E., 2001. Characterizing the neural mechanisms of skill learning and repetition priming: Evidence from mirror reading. Brain. 124, 67-82.

Staines, W.R., Padilla, M., Knight, R.T., 2002. Frontal-parietal event-related potential changes associated with practicing a novel visuomotor tasl. Brain Research and Cognition. 13, 195-202.

Tracy, J.I., Faro, S.S., Mohammed, F., Pinus, A., Christensen, H., Burkland, D., 2001. A comparison of 'early' and 'late' stage brain activation during brief practice of a simple motor task. Cognitive Brain Research. 10, 303-316.

Weissman, D.H., Woldorff, M.G., Hazlett, C.J., Mangun, G.R., 2002. Effects of practice on executive control investigated with fMRI. Cognitive Brain Research. 15, 47-60.

Nonlinear Dynamics Model and Simulation of Spontaneous Perception Switching with Ambiguous Visual Stimuli

Norbert Fürstenau

German Aerospace Center, Institute for Flight Guidance
Lilienthalplatz 7, D-38022 Braunschweig, Germany
norbert.fuerstenau@dlr.de

Abstract

A nonlinear dynamics recursive coupled attention – perception model of cognitive multistability is described and numerical simulations of the quasiperiodic reversals between perception states of ambiguous visual stimuli are presented. The perception state is treated as a macroscopic order parameter which is formalized as the phase variable of a recursive cosinuidal map with two control parameters µ = difference of meaning and G ~ attention. The mapping function is closely related to the neuronal mean field phase oscillator theory of temporal binding. Mean field interference with delayed phase feedback with gain ~ G, delay T, and damping time τ enables transitions between chaotic and limit cycle attractors representing the perception states. Quasiperiodic perceptual reversals are induced by attention satiation (fatigue) G(t) with time constant γ. For $\tau/T \ll 1$ the coupled attention – perception dynamics reproduces the experimentally observed Γ-distribution of the reversal time statistics if a small noise term is added to the attention equation. Mean reversal times of typically 3 – 5 s as reported in the literature, are correctly predicted if the delay time T is associated with the delay of 40 ms between stimulus onset and primary visual cortex (V1) response. Numerically determined perceptual transition times of 4 – 5 T are in reasonable agreement with stimulus – conscious perception delay of 150 – 200 ms. Eigenfrequencies of the limit cycle oscillations as obtained from a linear stability analysis, are in the range of 10 – 100 Hz, in agreement with typical EEG frequencies.

1 Introduction

From the use of (transparent) monocular head – mounted displays (HMD) in military helicopters it is well known that different visual input into both eyes may involve adverse perceptual and attentional effects like binocular rivalry which may lead to significant reduction of reaction times (Peli, 1990). Binocular rivalry is the spontaneous switching of conscious awareness between the different percepts corresponding to the different stimuli of both eyes (Engel et.al., 1999)(Blake & Logothetis, 2002). Laramee et.al. (Laramee & Ware, 2002) determined more than 100% response time increase in a HMD based visual search task due to binocular rivalry and visual interference effects.

Binocular rivalry belongs to a larger class of cognitive multistability effects observed with ambiguous visual stimuli which open a unique perspective in consciousness research (e.g. (Engel et.al., 1999), (Leopold & Logothetis, 1999). Attneave (1971) presented an overview of different types of perceptual multistability such as spontaneous perspective reversal (e.g. the perspective switching of the Necker-cube), figure ground reversal and rival – schemata reversal, and he proposed the existence of a cognitive positive feedback loop with locking into the alternative schemata and fatigue of associated different neural structures. Attneave (1971) and Lehky (1988) pointed out a similarity of cognitive bistability and electronic multivibrator circuits and suggested the possibility of analoguous neural structures. Natsuki et.al. (Natsuki, Nishimura, & Matsui, 2000) showed that a neural network based deterministic chaos model of multistable perception yields a Γ- distribution of persistance durations. In contrast to this microscopic neural network approach, however, the present formal model is based on the dynamic coupling of the macroscopic behavioral variables perception and attention with attentional fatigue (satiation, (Orbach et.al. 1963)) as origin of the quasiperiodic perceptual reversals.

Engel et.al. (Engel et.al., 1999) use experimental results of binocular rivalry for discussing neural coherence and synchronuous firing as fundmental aspect of temporal binding. A recent review of Blake et.al. (Blake & Logothetis, 2002) focuses on neurophysiological research including fMRI and concludes that despite remaining controversial issues the present results indicate rivalry to be based on multiple distributed processes instead of a single neural mechanism. Leopold & Logothetis (1999) claim that experimental evidence suggests perceptual reversals to be more closely related to the expression of behavior than to passive sensory responses. This provides a motivation for the present macroscopic coupled attention – perception model. It is closely related to the neural mean field phase oscillator theory (Schuster & Wagner, 1990) which was used for modeling the proposed synchronization of neural oscillations as the physiological basis of dynamic temporal binding (Engel et.al., 2001). It takes into account the recursive character of the neural processes between stimulus onset and conscious perception with the delay of T = 40 ms between stimulus and primary visual cortex (V1) response, and delay of $T_c \approx 5$ T between stimulus and creation of the conscious percept (Lamme, 2003). Concepts of mean field coherence and interference as the basis of the present model are used in an analoguous manner by von der Malsburg (1997) with respect to consciousness.

Atmanspacher et.al. (Atmanspacher, Filk & Römer, 2004) recently treated the bistable perception in terms of the evolution of an unstable two-state system within a generalized quantum theoretical framework. The time dependent probability of the system to spontaneously switch from one state to the other yields a $\cos^2(kt)$–characteristic (k = coupling constant) which is typical for interference effects. A comparable cosinuidal transfer characteristic is used in the present recursive model, however, in contrast to Atmanspacher et.al. based on classical mean field interference with delayed feedback, which is formally analoguous to multistable interference effects in nonlinear optical systems (Watts & Fürstenau 1989) (Fürstenau 1991).

Based on the previously outlined nonlinear dynamics model of cognitive multistability (Fürstenau, 2003, 2004), I present new numerical results obtained by integrating the coupled differential – delay equations by means of the dynamical systems tool Matlab –Simulink. The simulations support the earlier approximate results that the quasiperiodic reversals between two perception states of an ambiguous stimulus can be represented by self - oscillations between limit cycle and chaotic attractors in perception delay space. The perception state variable is represented by the phase v(t) of a recursive cosinuidal mapping function with feedback delay time T and gain g. According to (Hillyard et.al.,1999) feedback gain g as a neurophysiological quantity is associated with attention. Like in the multistability model of Ditzinger & Haken (e.g. Haken, 1996, pp. 247 - 257) the perception variable (phase) is treated as order parameter within the formal framework of Synergetics (Haken 1978), with the slowly time varying attention parameter g(t) modelling satiation (attentional fatigue). The new aspect of the present approach in contrast to the Ditzinger&Haken model is the neurophysiologically motivated delay T, which is associated with the stimulus onset – primary visual cortex (V1) response delay of T = 40 ms. It gives rise to the existence of limit cycle and chaotic attractors. The attractors are interpreted as perception states in agreement with Freeman et.al. (e.g. (Freeman, 2000)). A small random attention noise component is added to the deterministic equations for taking account of the dissipative processes of realistic dynamical systems. The statistical analysis of simulated time series predicts gamma distributions of the perceptual reversal times with mean values and variance in reasonable agreement with experimental results of Borsellino et.al. (1972). The contributions of chaotic dynamics as well as random noise supports the finding of (Richards, Wilson & Somme, 1994) who separated the chaotic and random contributions of perception state time series of quite different visual perception phenomena including multistability, via determination of the correlation dimension (Grassberger, 1983).

In the following section 2 I describe the nonlinear dynamics model, followed by an analysis of the stationary solutions of the differential – delay perception equation in section 3. Section 4 provides a linear stability analysis as well as an evaluation of the Lyapunov coefficient. Simulated time series as obtained by numerical integration with Matlab - Simulink are shown in section 5. A statistical analysis of the reversal time intervals in section 6 proves the good agreement with published experimental data. A summary and conclusion follows in section 7.

2 The Recursive Mean Field Interference Model

The present macroscopic approach is related to the mesoscopic mean field phase oscillator theory of coupled neuronal columns in the visual cortex (Schuster & Wagner, 1990). It was used for modeling the synchronization of neuronal oscillations as the physiological basis of dynamic temporal binding which in turn is thought to be crucial for the selection of perceptualy or behavioraly relevant information (Engel et.al., 2001) (Engel et.al., 1999). Self oscillation of neuronal groups within columns and coupling between columns is excited when the external stimulus exceeds a certain threshold (Schuster & Wagner, 1990). Single columns exhibit multistable characteristics of the neuronal mean field as function of the stimulus, similar to the present model. Within the phase synchronization theory phase locking between two different groups of neurons is described by means of the circle map which in a discrete version is written as (Anishenko et.al., 2002, pp. 88 - 91):

$$x_{k+1} = x_k + \delta - \frac{K}{2\pi}\sin(2\pi x_k), \text{ mod } 1 \qquad (1)$$

with x_k = phase difference between two oscillators. For K=0 the parameter δ represents the winding number, which characterizes the ratio of two frequencies of uncoupled oscillators. If $0<K<1$, δ is the frequency detuning. For $K > 1$ the map demonstrates scenarios of torus breakdown and appearance of chaos.

As a kind of minimum architecture allowing for multistability I propose coupling of the attention and perception dynamics via delayed phase feedback interference, and attention satiation which gives rise to discontinuous state transitions between different perception states. The model is analoguous to the well known multistability phenomena in nonlinear optical systems with feedback (e.g. (Gibbs, 1985)), and detailed examples of corresponding interferometric circuits were described e.g. in (Watts & Fürstenau, 1989) (Fürstenau, 1991). Interference is the superposition of (electromagnetic) fields, yielding extinction or amplification of each other, depending on the relative phase shift $\Delta\Phi$. It may be compared with the phase shift between the coupled self - oscillating neuronal columns of the mean field theory. A simplified block diagram is shown in Figure 1. It depicts interference between two phase shifted fields representing the two possible interpretations of the stimulus. Interference creates the typical cosinuidal dependence of the output U_t (= squared modulus of the sum of the field amplidudes) on the phase difference $\Delta\Phi$ between the two superimposed waves. The normalized output $v_t = U_t / U_\pi$ of the feedback interference circuit with phase – voltage modulation factor $d\Phi/dU = \pi/U_\pi$ defines the perception state as synergetic order parameter.

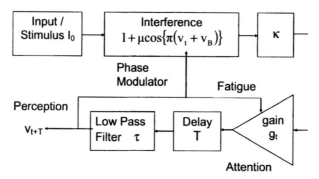

Figure 1: Nonlinear delayed phase feedback circuit. Interference = superposition of separate phase shifted fields yielding fringe contrast (~difference of meaning) μ, excited by stimulus I_0. v_t = perception state (phase variable), g_t ~ attention (gain control parameter), κ = conversion factor, τ = damping time.

Phase-feedback is realized by a suitable phase modulation mechanism. One possibility is frequency modulation of the input field (Fürstenau 1991), corresponding to the stimulus induced modulation of the neuronal mean field limit cycle oscillations (Schuster & Wagner 19990). According to Hillyard et.al. (1999) stimulus-evoked neural activity can be modified by an attentional induced additive bias or by a true gain modulation. Increase of gain is correlated with increased blood flow through the respective cortical areas. Accordingly the feedback gain g serves as a control parameter which induces the quasi - discontinuous transitions between different stationary states through attention satiation or fatigue (Orbach et.al., 1963). The relevance of feedback gain control in cognitive models was shown by Grossberg et. al. (e.g. (Grossberg, 1995)) within the context of adaptive resonance theory (ART) neural circuits, e.g. for describing the matching process of patterns arriving in the short term memory. Freeman et.al. (e.g. Freeman (2000)) developed a mesoscopic nonlinear dynamics model of olfactory perception with nonlinear gains and delays coupling the different structures (groups of neurons) of the olfactory system. Corresponding to the present macroscopic cognitive model, it predicts chaotic attractor states representing the olfactory percepts, in agreement with EEG measurements. Feedback delay T is introduced according to the recursive character of the perception process (Lamme, 2003) and it is defined as delay between stimulus onset and primary visual cortex response (T = 40 ms).

The two parameters characterizing the visual input are stimulus strength I_0 and interference contrast μ. In psychological terms an interpretation of μ as difference of meaning of the corresponding two percepts is suggested. A strongly damped (overdamped) feedback system is assumed with time constant $\tau \gg$ coefficient of d^2v/dt^2 which is neglected. For modeling the slow (compared to T) rivalry self – oscillation as spontaneous switching between percept 1 (=P1) and percept 2 (P2), like in Haken's approach (e.g. (Haken, 1996) the concept of attention satiation is included into the model. Formally this is realized by a slow time dependence of g (satiation time constant $\gamma \gg \tau$, T) which leads to a coupled system of two first order nonlinear differential delay equations for v(t) and g(t). In a first approach to model the random disturbances a δ-correlated stochastic force L(t) (e.g (Risken, 1996)) with random amplitude r s_j, $-1 \leq s_j \leq 1$ is added to the attention equation g(t), similar to Haken (Haken, 1996) and Lehky (Lehky, 1988). It accounts for random noise unavoidable in any realistic dynamical system due to dissipative processes, and it adds pushes of random amplitude to the attention parameter at discrete times $t_j = j$ T.:

$$\tau \dot{v}_{t+T} + v_{t+T} = G[1 + \mu \cos(\pi(v_t + v_B))] \tag{2a}$$

$$\dot{G}_t = G_0(v_b - v_t)/\gamma - G_t(v_b - v_t)/\tau_G + L_t \tag{2b}$$

The rhs. of equ. (2a) describes the conventional interference between two coherent fields. The interference contrast μ $(0 \leq \mu \leq 1)$ depends on the coherence of the fields. In what follows I assume the phase bias $v_B = 0$ mod 2. The gain parameter G(t) = $\kappa I_0 g(t)/ U_\pi$ with phase – voltage modulation factor $d\Phi/dU = \pi/U_\pi$ is the product of feedback gain g(t) and input (stimulus) I_0 with satiation speed G_0/γ and attention decay time τ_G. In contrast to (Fürstenau, 2004) like G_0/γ also τ_G is coupled to the perception state for symmetry reasons. Time scale is in units of T_S (= integration time interval) which in the simulations is choosen as $T_S = T/2$.

3   Stationary Solutions of the Recursive Interference Equations

Two types of instabilities are observed with systems described by equation (2a): period doubling and node bifurcation. They are visualised in Figure 2 by plotting the stationary solutions including period doubling up to period 8, $v_{t+iT} = v_t = v^*$, i = 1, 2, 4, 8.

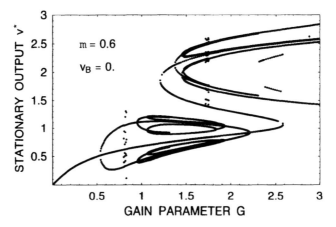

Figure 2: Stationary solutions of equation (2a) with period doubling bifurcations.

For this analysis the FindRoot procedure of the software tool Mathematica (vers.5) was used (Wolfram1999). Like in the case of the quadratic map (Schuster1984) period doubling pitchfork bifurcations are observed on both positive slope regions. The graph yields the control parameter values of the first three bifurcation points providing a first approximation to the Feigenbaum constant $\delta_\infty = 4.6692$. On the first stationary level (P1): $G_1 = 0.535$, $G_2 = 0.947$, $G_3 = 1.058$ yielding $\delta_\infty^1 \approx \delta_1 = (G_2 - G_1)/(G_3 - G_2) = 3.71$ and on the second level (P2): $G_1 = 1.3645$, $G_2 = 1.1.4445$, $G_3 = 1.4625$ yielding: $\delta_\infty^2 \approx \delta_2 = (G_2 - G_1)/(G_3 - G_2) = 4.44$. An estimate for the chaotic boundaries is obtained from $\delta_\infty \approx (G_3 - G_2)/(G_\infty - G_3)$ as $G_\infty^1 \approx 1.08$ and $G_\infty^2 \approx 1.47$ for P1 and P2 respectively.

Node bifurcation is required for explaining the existence of ambiguous perception in the present model. In (Fürstenau, 2003) I have shown a graph depicting the dependence of the stationary solutions on the contrast parameter μ (difference of meaning between the two percepts). For maximum contrast $\mu = 1$ the horizontal slope $(dG / dv)^{-1} = 0$ yields $v_u^\infty = 2i - 1$, $i = 1,2,3,\ldots$ as stable belief / perception levels in the limit G -> ∞. At a critical value of the contrast parameter $\mu_n = 0.18$ (node bifurcation) the slope of the stationary system state v^* as function of G becomes infinite. For $\mu > \mu_n$ the stationary solution $v^*(G)$ becomes multivalued. For $\mu < \mu_n$ both percepts are fused into a single meaning. Under increasing input I_0 or feedback gain g the stationary (1st order) perception state v^* jumps discontinuously from P1 to P2 at the turning points of the S-shaped hysteresis curve (= extrema of the inverse curve $G(v^*)$). The transition of P2 back to P1 occurs at a lower stimulus or gain parameter (~attention) value due to the hysteresis. The width of the instable negative slope section and the multivalued G – range is controlled by μ. A similar hysteresis is observed for the coupling constants of columns of the visual cortex within the neuronal mean field theory (Schuster & Wagner, 1990). Due to the periodic nature of the cos-function, in principle an infinite number of discontinuous transitions between stationary levels (positive slope regions) exists, with the stability determined by the damping time τ (Watts & Fürstenau, 1989).

4   Stability Analysis

A linear stability analysis of equation (2a) yields time constant values τ/T for the onset of limit cycle and chaotic oscillations as well as eigenfrequencies of the system. It is performed by investigating small deviations $x = v - v^*$ from the fixed points v^*. Using the ansatz $x = \exp\{(\alpha + i\beta)t\}$ the boundary of instability ($\alpha = 0$) is obtained as:

$$v^* = \frac{1}{\pi}\arcsin\left\{\sqrt{(\beta\tau)^2 + 1 - \tau/T}\Big/2\mu\pi G\right\} - v_B \tag{3}$$

At the boundary to the instable areas the eigenfrequencies $\beta = 2\pi f$ are obtained via $\beta\tau = -\tan(\beta T)$. The analytical approximation for $\tau \ll T$ yields $f \approx f_1\, i / (1 - \tau/T)$, $i = 0, 1, 2, \ldots$ with $f_1 = 1 / 2T$, i.e. half of the inverse feedback delay time. With T = 40 ms (delay between stimulus onset and V1 - response, see above) we obtain $f_i \approx i\, f_1 = i\ 12.5$ Hz, $i = 1,2,3,\ldots$. With large damping time τ of the order of T, period doubling oscillations are suppressed and multistability, i.e. a large number of stable stationary states is obtained (Watts & Fürstenau, 1989). Figure 3 shows

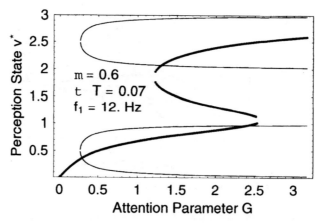

Figure 3: Stationary fixed points v^* of cosinuidal map with boundaries of instable areas extending to the right for first Eigenfrequency $f_1 = 12$ Hz.

the first order stationary fixed points of (2a) for low damping $\tau/T = 0.07$ with $\mu = 0.6$, together with the boundaries of the instable areas as calculated with equation (3) using the lowest eigenfrequency $f_1 = 11.7$ Hz (T = 40 ms).

For higher eigenfrequencies (like for higher damping) these instability boundaries (horizontal, parabola like curves) shift to the right, indicating suppression of oscillations superimposed to the first order stationary solutions (hysteresis curve) on the left side of the intersections between the hysteresis curve and the instability boundaries. The instable areas extend to the right of the instability boundaries. With the selected τ both stationary levels (P1, P2) of the fixed point curve exhibit intersections with the boundaries, which allows oscillations for G > G(intersection), i.e. for sufficiently large feedback gain g or input I_0. At the turning points of the hysteresis curve quasi discontinuous transitions occur between the stationary levels P1 and P2, ending in an oscillatiory state due to the instability boundary intersecting the first order fixed point curve on the left side of the transition end point.

The period doubling behavior of the bifurcations shown in Fig.2 proves that within certain parameter ranges (μ, τ) any system noise has chaotic contributions. This is confirmed by numerical evaluation of the Lyapunov coefficient λ (Schuster, 1984) which for the discrete approximation of equ.(2a) (Fürstenau2003, 2004) is given by

$$\lambda = \lim_{N \to \infty} \sum_{i=0}^{N-1} \ln\left(\left|f'(v_i)\right|\right) \qquad (4)$$

with N = number of iteration steps and $f'(v_i)$ = derivative of the rhs. of equation (2a):

$$f'(v) = -\frac{G\mu\mu}{1+\tau}\sin(\pi(v - v_B)) + \frac{\tau}{1+\tau} \qquad (5)$$

λ measures the exponential separation of neighbouring initial states v_0 during the iteration process under the action of the mapping. It corresponds to the average loss of information per iteration. Figure 4 shows the Lyapunov exponent for zero damping $(\tau = 0)$ with N=1000 iterates as plotted versus the attention parameter G.

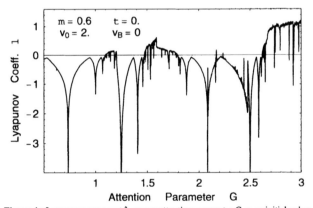

Figure 4 : Lyapunov exponent λ versus attention parameter G. v_0 = initial value.

Within the regions $1.1 < G < 1.2$ and $1.5 < G < 1.8$, corresponding to the bifurcation graphs on the perception levels P1 and P2 of Figure 2, chaotic behavior is observed $(\lambda > 0)$. The bifurcation points $(\lambda = 0)$ indicate the period doubling route to chaos (Feigenbaum, 1979). With increasing damping $(\tau > 0)$ the chaotic and limit cycle oscillations as indicated by $\lambda \geq 0$ are suppressed. E.g. $\lambda > 0$ is no longer observed for $G < 1.6$ if $\tau = 0.03$.

5 Simulated Perception – Attention Dynamics

In (Fürstenau, 2003) (Fürstenau, 2004) I have shown that like with the synergetic model of cognitive multistability (e.g. Haken, 1996, pp. 247-257) (quasi-) periodic transitions between the two positive slope sections of the hysteresis curve are induced via attention satiation or adaptation. It is realized by means of the perception (v) – attention (G) coupling, due to the slow time variation of the feedback gain g(t) with fatigue time constant γ (satiation speed G_0/γ, quantifying the degree of coupling). For large damping $\tau > T$ the limit cycle (period doubling) and chaotic oscillations are suppressed and a continuous approximation to equations (2) exhibits the dependence of dominance and suppression periods of the percepts on the perception bias v_b (Fürstenau, 2003). For $\tau \ll T$ a recursive approximation to (2) was used, with approximate numerical results (Fürstenau, 2003, 2004) yielding quasiperiodic transitions between limit cycle and chaotic oscillations corresponding to the bifurcations on the two stationary levels of the hysteresis curve (Figure 2). In this section I present numerical results of simulations for the exact coupled differential – delay equations (2a, b) as obtained via numerical integration with the dynamical systems tool Matlab – Simulink. Figure 5 shows the numerical evaluation of equations (2) for $\mu = 0.6$, $\tau/T = 0.07$, $v_b = 1.5$, $\gamma = 60$, $\tau_G = 1400$, a random noise amplitude $r = 0.03$, $G_0 = 1$, and time scale in units of the integration intervall $T_S = T/2 = 20$ ms (with $T :=$ stimulus – V1-response delay = 40 ms).

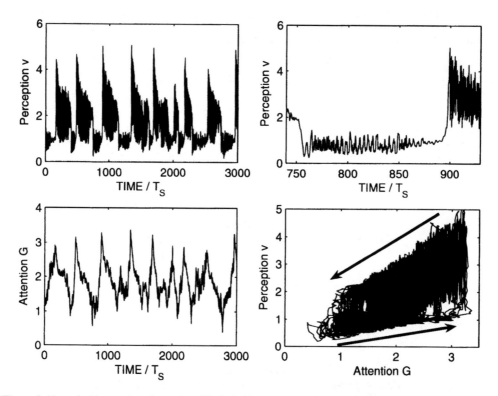

Figure 5: Numerical integration of equations (2). Left side, top: perception state v_t time series during time interval 3000 T_S; bottom: attention parameter $G(t)$ (~ feedback gain). Right side, top: v_t during ca. 0.5 rival period time interval (200T_S) exhibiting limit cycle and chaotic oscillations, and transition times $\Delta_t \approx 10$ T_S;

The time series of the perception state shows the spontaneous transitions between stationary states near $v^* \approx 1$ (P1) and $v^* \approx 2.5$ (P2) with the expected superimposed limit cycle and chaotic oscillations. From the 250 T_S - time section shown in the top row of Fig. 5 a transition time between P1 and P2 of the order of 10 $T_S \approx 150$ - 200 ms is obtained. It is in reasonable agreement with the time interval between (visual) stimulus onset and the beginning of conscious perception (Lamme, 2003). A difference as compared to previous results (Fürstenau, 2004) is the stronger damping and decay of the oscillation within the lower level (P1), which is due to the perceptional coupling of the attentional damping term G_t/τ_G via $(v_b - v_t)$. The lower left graph depicts the satiation / adaptation process of the attention (gain) parameter G. It exhibits gradual decrease after $v(t)$ has jumped to the upper level P2 (because $v_b - v(t) < 0$) and gradual increase after transition of v to P1 (here $v_b - v(t) > 0$). The reversal time period is determined by the satiation time constant γ of the attention control parameter G(t). The right graph in the lower row depicts the satiation / fatigue induced v - G dynamics in perception – attention phase space. The dense filling of trajectories in phase space indicates a significant degree of chaotic dynamics. It appears only weakly expressed within the reversal time statistics (see below)and is mainly a characteristic of the individual perception state dynamics.

6   Reversal Time Statistics

If $\mu > \mu_n = 0.18$ the stationary first order solution of equation (2a) becomes multivalued (Figure 2: S – shaped hysteresis curve) and the coupled attention – perception equations (2a,b) allow for self oscillations between the positive slope regions (percept P1 and P2). The relative durations of the dominant and suppressed phase of a percept is determined by the bias parameter v_b (Fürstenau, 2003) which qualitatively models experimental results reviewed in (Engel et.al., 1999). Figure 6 depicts the relative frequencies of the perceptual duration times as obtained by averaging 10 time series consisting of N = 50000 iterations each.

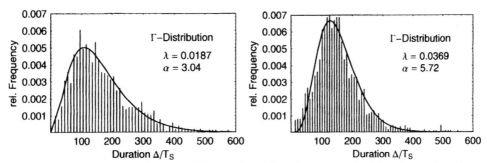

Figure 6: Relative frequencies of perceptual duration time Δ, in units T_S, as obtained from time series with μ = 0.6, τ /T = 0.07, γ = 60, τ_G = 1400, r = 0.03. Fit with Γ-distribution (probability density, solid line). Left: Percept P1 (first stationary level), right: Percept P2 (second stationary level).

Plotted are the two distributions of the perceptual durations Δ of percept 1 on the lower level Δ(P1) and percept 2 on the higher level Δ(P2). The relative frequencies are fitted by a Γ – distribution as probability density with shape parameter α and scale parameter λ, according to (Borsellino et.al., 1972). Mean and variance are given by $\Delta_m = \alpha/\lambda$ and $\sigma^2 = \alpha / \lambda^2$ respectively. The Γ-distribution yields the best fit parameters as compared with log-normal and normal distributions. For P1 mean Δ_m = 162 T_S, standard deviation σ = 93 T_S; for percept P2 mean Δ_m = 155 T_S, standard deviation σ = 65 T_S. The ratios $\sigma/\Delta_m \approx 0.4 - 0.6$ are in good agreement with experimental findings reported in the literature. In contrast to (Fürstenau, 2003) with purely deterministic time series, and in agreement with (Fürstenau, 2004) the addition of the small random attention noise L(t) in equation (2b) leads to a significant increase of the variance, whereas the mean values remain roughly the same, indicating the dominant influence of the deterministic dynamics on Δ_m. A preliminary analysis of the present simulations however, shows the chaotic contribution to the total variance significantly smaller as compared to the recursive approximation (Fürstenau, 2004). This is in agreement with findings of (Lehky, 1995) and (Merk & Schnakenberg, 2002) (Gao et.al. 2003) who find no significant chaotic contribution in the reversal time distributions. It appears that the limit cycle and chaotic oscillations of the perception states are only weakly if at all reflected in the statistics of the perceptual duration times.

With feedback delay time T identified with the stimulus onset - primary visual cortex (V1) response delay of ~ 40 ms (Lamme, 2003) absolute values of mean perceptual duration times of 3.1 ~ 3.2 s are obtained, in good agreement with published experimental results (Borsellino et.al., 1972), (Lehky, 1988), (Merk & Schnakenberg, 2002). The details of course depend on the choice of the model parameter values μ, v_b, τ, γ, τ_G, r, which may be tuned to match the inter – subject variations of Δ_m (via γ), the relative duration of dominance vs. suppression times (via v_b) and the strength of limit cycle and chaotic oscillations (via τ/T).

7 Conclusion

New numerical results of the recursive interference attention-perception model of cognitive multistability are presented, with the coupled delay–differential equations integrated by means of the dynamical systems software tool Matlab – Simulink. The model is related to the mean field phase oscillator theory of coupled neuronal groups in the visual cortex (Schuster & Wagner, 1990) which was used for modeling the synchronization of neural oscillations as the physiological basis of dynamic temporal binding (Engel et.al., 2001). Numerical analysis for small damping exhibits quasiperiodic perceptual reversals as transitions between limit cycle and chaotic attractors which qualitatively agrees with findings of Freeman et.al. (e.g. (Freeman, 2001)). The difference of meaning μ ($=$ interference contrast) between the two percepts allows for modelling the transition from unambiguous perception (small μ) to rivalry (large $\mu > 0.18$, rival percepts) via a node bifurcation.

Experimentally observed Γ- or log-normal probability densities and stochastic independence of dominance and supression periods were modeled earlier by a microscopic neural multivibrator model (Lehky, 1988) as well as by a macroscopic synergetic model (Haken, 1996). Like with the present model, both approaches add stochastic terms to the dynamical equations and yield reasonable qualitative, partly even quantitative agreement with the experimentally observed relative frequencies of rivalry dominance durations (e.g. (Borsellino, 1972). They do not, however, yield absolute values of mean reversal and transition times which in the present model is achieved by means of the feedback delay T which is associated with the primary visual cortex (V1) response delay of ≈ 40 ms (Lamme, 2003). The recursive dynamics with delay T also explains the contribution of deterministic chaos to the measured perceptual duration time series of certain types of multistabel visual perception (Richards et.al., 1994).

The analysis of simulated perception state time series yields the relative frequencies of the perceptual reversal times. They are fitted by Γ-distributions with mean of $3 - 4$ s and variance in agreement with experimental results of Borsellino et.al. (1972). The simulated time series exhibit transition times between the two alternative percepts of ca.150 - 200 ms which is of the same order as the delay of conscious perception, after presenting a new stimulus (Lamme, 2003). Limit cycle oscillations of the perception states with frequencies in the $10 - 100$ Hz range are in agreement with typical EEG – frequencies. Although perception state oscillations exhibit significant chaotic contributions the reversal time statistics is dominated by stochastic noise, in agreement with results reported in the literature (Lehky, 1995)(Merk & Schnakenberg, 2002).

Further research will focus on the comparison of numerical simulations and experimental results with regard to periodically interrupted stimulus for testing the satiation hypothesis, and on a more detailed analysis of the influence of the different parameters and time constants on the statistical properties of the time series.

Acknowledgement

I am indebted to Monika Mittendorf for help in writing the Mathematica and Matlab code of the numerical analysis and simulation software and for support in performing the statistical analysis.

References

Anishchenko, V.S., Astakhov, V.V., Neiman, A.B., Vadivasova, T.E., Schimanski-Geier, L., (2002). Nonlinear Dynamics of Chaotic and Stochastic Systems, Heidelberg: Springer.

Atmanspacher, H. Filk T., Römer, H. (2004): Quantum Zeno Features of Bistable Perception, *Biol. Cybern,* .vol. 90, pp. 33 – 40

Attneave, F. (1971). Multistability in perception, *Scientific American* 225, 63 - 71

Blake, R., Logothetis, N.K. (2002). Visual competition, *Nature Reviews / Neuroscience* 3, 1 - 11

Borsellino, A., de Marco, A., Allazetta, A., Rinesi, S., Bartolini, B. (1972). Reversal time distribution in the perception of visual ambiguous stimuli, *Kybernetik,* 10, 139 - 144

Engel, A.K., Fries, P, König, P., Brecht, M., Singer, W. (1999). Temporal binding, binocular rivalry, and consciousness, *Consciousness and Cognition* 8, 128 - 151

Engel A.K., Fries P., Singer W. (2001). Dynamic Predictions: Oscillations and Synchrony in Top-Down Processing. *Nature Reviews Neuroscience,* 2, 704 – 718

Feigenbaum, M.J. (1979). The universal metric properties of nonlinear transformations, *J. Statistical Physics* 21, 669 - 706

Freeman, W.J.: *Neurodynamics*, Springer, London 2000.

Fürstenau, N. (1991). Bistable fiber-optic Michelson interferometer that uses wavelength control, *Optics Letters* 16, 1896 – 1898

Fürstenau N. (2003): Nonlinear dynamics model of cognitive multistability and binocular rivalry, *Proceedings of the IEEE 2003 Int. Conf. on Systems, Man and Cybernetics*, (Ed.), ISBN 0-7803-7953-5, IEEE cat. no. 03CH37483C,. pp. 1081-1088

Fürstenau N. (2004): A chaotic attractor model of cognitive multistability, *Proceedings of the IEEE 2004 Int. Conf. on Systems, Man and Cybernetics*, (Ed.), ISBN 0-7803-7953-5, IEEE cat. no. 03CH37483C,. pp. 1081-1088

Gibbs, H.M. (1985). *Optical bistability: controlling light with light*. London: Academic Press

Grassberger, P. , Procaccia, I.. (1983)"Measuring the strangeness of strange attractors", Physica 9D 189 - 208

Grossberg, S. (1995). Neural dynamics of motion perception, recognition learning, and spatial attention. In R.F. Port, T. van Gelder (Eds.), *Mind as Motion*. Cambridge/MA: The MIT Press

Haken, H. (1978). *Synergetics*, Berlin: Springer

Haken, H. (1996). *Principles of brain functioning*, Berlin: Springer

Hillyard, S.A., Vogel, E.K. Luck, S.J.(1999). Sensory gain control (amplification) as a mechanism of selective attention: electrophysiological and neuroimaging evidence. in: G.W. Humphreys, J. Duncan, A. Treisman (eds.), *Attention, Space, and Action*, Oxford University press

Lamme, V.A.F. (2003): Why visual attention and awareness are different. *Trends in cognitive Sciences, 7*, 12 – 18

Laramee R.S., Ware C., *Rivalry and Interference with a Head Mounted Display*, Technical Report TR-VRVis-2002-005, and: *ACM Transactions on Computer-Human Interactions* 9, 238-251

Lehky S. R. (1988). An astable multivibrator model of binocular rivalry, *Perception* 17, 215-228

Lehky, S. R. (1995). Binocular rivalry is not chaotic, *Proc. R. Soc. Lond.* B 259, 71 - 76

Leopold, D.A. & Logothetis, N.K. (1999). Multistable phenomena: changing views in perception, *Trends in Cognitive Sciences* 3, 254 – 264

Merk, I., Schnakenberg, J. (2002), A stochastic model of multistable visual perception, *Biol. Cybern.* 86, 111-116

Natsuki, N., Nishimura, H., Matsui, N. (2000). A neural chaos model of multistable perception, *Neural Processing Letters* 12, 267 – 276

Orbach, J, Ehrlich, D. Heath, H.A. (1963). Reversibility of the Necker Cube: An examination of the concept of satiation of orientation, *Perceptual and Motor Skills vol.* 17, 439 - 458

Peli E. (1990). Visual issues in the use of a head-mounted monocular display. *Optical Engineering 29*, 883-892

Richards W., Wilson H.R., Sommer M.A. (1994). Chaos in percepts. *Biol. Cybern.* 70, 345-349

Risken, H. (1996). *The Fokker – Planck Equation*. Berlin: Springer

Schuster, H.G. (1984). Deterministic Chaos. Weinheim: Physik - Verlag

Schuster, H.G., Wagner, P. (1990). A Model for Neural Oscillations in the Visual Cortex, Biol. Cybernetics, 64, 77 - 82

von der Malsburg, C. (1997). The coherence definition of consciousness. In M. Ho, Y. Miyashita, E.T. Rolls (Eds.). *Cognition, computation and consciousness*, (193 – 204). Oxford University Press

Watts, C., Fürstenau, N. (1989). Multistable fiber-optic Michelson Interferometer exhibiting 95 stable states, *IEEE J. Quantum Electron.* 25, 1 – 5

Wolfram, S. (1999). *The Mathematica Book – Mathematica version 4* (4th ed.). Cambridge University Press

Augmented Cognition and Time Perception

G.S. Bahr[*], B. Wheeler Atkinson[§], M. Walwanis Nelson[*], & E. Stewart[§]
[*]NAVAIR Training Systems Division, [§]JHT, Inc.
12350 Research Parkway, Orlando, FL 32826
gisela.bahr@navy.mil, batkinson@jht.com, melissa.walwanis@navy.mil,
Eric.Stewart@navy.mil

Abstract

Augmented Cognition Technology appears to provide an opportunity for cognitive scientists to conduct non-invasive experimentation involving human subjects. Within an Augmented Cognition framework, we suggest that the study of human time perception may provide insight into the neurological underpinnings of time perception, their manipulation and calibration, as well as the impact of environmental and cognitive factors, such as varying levels of workload, fatigue, expertise, and individual differences. Such findings are likely to impact the way we understand and design performance support systems across multiple domains.

1 A Cognitive Moment

The notion that our perception of time results from the continuous flow of processing has long been questioned by psychologists. For instance, William James in his "principles of psychology" speaks to the variable, uncertain frequency of our perceptual scanning rate and how its graininess ultimately shapes our experience and awareness of the environment (1890, 1950). Similarly, in 1955, Stroud suggested that time perception emerges from a series of perceptual blocks. He defined the "Perceptual Moment" as our basic unit of time. His operational definition of the Perceptual Moment is based on event parsing: events that appear to occur simultaneously are perceived within the same Perceptual Moment. On the other hand, events that appear to occur successively, perceptually reside in separate Perceptual Moments. Ergo, knowledge of sequential order is not available within a single perceptual moment. As a result, time progression appears to be an artifact of strings of events residing in discrete perceptual moments.

Stroud's theory has primarily been tested on lower level visual processing to investigate the possibility of a fixed perceptual interval similar to a refractory period (for review see Patterson, 1990). Researchers were perplexed to discover a relatively consistent range of latencies rather than a single cycle frequency. Studies suggest that latencies are sensitive to context, i.e., their variability appears related to stimuli characteristics.

Unlike the view from a "bottom-up" visual perception approach, from a "top-down" cognitive perspective the notion that event parsing is context sensitive and moderated by environmental and cognitive factors appears reasonable. For instance, one would expect that prior experience, knowledge, expectations, attention, task complexity, workload, expertise and fatigue influence the parsing of events in more complex, real-world environments. For the cognition based investigation of event ordering on a temporal axis, we propose the construct of the cognitive moment for three reasons:

- elimination of fixed interval assumption
- assumption of context sensitivity
- assumption of moderation by cognitive and environmental factors

2 More Efficient Use of Processing Time?

Empirical studies on duration estimation have reliably demonstrated that expectations and attention influence our perception of time. For instance the watched pot paradigm provides empirical validation of our personal experience that maintaining attention and focus on a particular event appears to "slow down time" (Block, George & Reed, 1980). Further anecdotal evidence indicates that highly focused individuals report an expansion of time available for observation (Sacks, 2004): for instance, athletes describe how their surroundings prior to a competition appear in slow-motion; similarly, individuals involved in highly threatening or dangerous situations describe a slowing down of time. Such individuals report the perception of more time available for observation and decision making. This phenomenon, i.e., the possible human capacity for an enhanced, fine grained information intake under stress, is particularly of operational interest because it may afford a) enhanced cognitive functioning during critical events and b) enhanced observation opportunity facilitating situation awareness (Endsley & Garland, 2000).

3   Gaining Control of Time: Empirical Questions

The lack of integration among cognitive and physiological models of time reflects the fact that psychologists and neuroscientists do not yet understand, or control human time perception. In line with Stroud's seminal work, it is possible that the perceived expansion of time is due to increased event parsing and an increase of a cognitive rather than perceptual scanning rate. The empirical question arises whether individuals who report an expansion of time exhibit a more fine-grained perception of their environment, and, whether the theoretical construct of the Cognitive Moment and its proposed moderators can account for the variability. An investigation may also shed light on expert - novice distinctions: for instance, many novices report a sense of being overwhelmed when first entering a complex environment and describe an onslaught of information seemingly to occur "simultaneously". From a time perception perspective, this phenomenon may imply that cognitive moment latency is inversely related to domain knowledge and expertise, workload and situational awareness. Identification of cognitive moment variables and comparing expert to novice processing may enhance our understanding and facilitate the development of training tools for construction and transfer of knowledge.

4   The Potential of Non-Invasive Augmented Cognition Technology

For the investigation of the manipulation and control of time perception, we propose a reverse inference paradigm based on the assumption that activation patterns in particular brain regions serve as markers for the engagement of respective localized cognitive processes (Poldrack & Wagner, 2004). It appears that the sensor-suite technology developed within the Augmented Cognition Project (i.e., the real time measurement of cognitive activity driving instantaneous environmental mitigations) can provide the technological concept for non-invasive study of time perception involving human participants. Using an "AugCog" apparatus, the systematic investigation of the factors moderating cognitive moment duration may lead us to the detection and the enhanced understanding of the implicated patterns and structures, as well as their interaction with environmental variables and individual differences.

For instance, the suprachiasmatic nucleus (SCN) of the mammalian hypothalamus has been referred to as the master circadian pacemaker also known as *Zeitgeber* (e.g., Abe et al., 2002). In addition to the SCN, recent evidence suggests that a multitude of brain structures appear to be involved in exhibiting oscillations in line with Zeitgeber concept (Abe et al., 2002): Primarily structures in the diencephalon, in particular hypothalamic and thalamic nuclei, as well as paraneural structures (pineal and pituitary glands) display rhythmic activity that is likely to be implicated in the perception of time. However, questions regarding (a) how these structures integrate, (b) which factors affect oscillation pattern, and (c) whether and how variations in task complexity, fatigue, expertise and workload, are expressed by these structures, remain empirical. Moreover, the possibility that such factors and their impact may be mitigated through the real-time assessment of operator state remains unexplored (Miro, Cano, Espinosa-Fernandez, & Buela-Casal, 2003).

5   Conclusion

In summary, Augmented Cognition Technology appears to provide an opportunity for cognitive scientists to conduct non-invasive experimentation involving human subjects. Within an Augmented Cognition framework, we suggested that the study of human time perception may provide insight into the neurological underpinnings of time perception, their manipulation and calibration, as well as the impact of environmental and cognitive factors, such as varying levels of workload, fatigue, expertise, and individual differences. Gaining insight, and thus acquiring control of human time perception, is likely to impact the way we understand and design performance support systems across multiple domains.

References

Abe, M., Herzog, E. D., Yamazaki, S., Straume, M., Tei, H., Sakaki., Y., Menaker, M. & Block, G. D. (2002). Circadian rhythms in isolated brain regions. *Journal of Neuroscience, 22*, 350-356.

Block, R. A., George, E. J., & Reed, M. A. (1980). A watched pot sometimes boils: A study of duration experience. *Acta Psychologica, 46*, 81-94.

Endsley, M. R., & Garland, D. J. (2000). *Situation awareness analysis and measurement.* Hillsdale, NJ: Lawrence Erlbaum Associates.

James, W. (1890, 1950). *The principles of psychology* (Vol. 1). Dover, New York, 1950.

Patterson, R. (1990). Perceptual moment models revisited. In R. A. Block (Ed.), *Cogntive Models of Psychological Time* (pp. 85-101). Hillsdale, NJ: Lawrence Erlbaum Associates.

Poldrack, R. A., & Wagner, A. D. (2004). What can neuroimaging tell us about the mind? *Current Directions in Psychological Science, 13*, 177-181.

Miro, E., Cano, M. C., Espinosa-Fernandez, L. & Buela-Casal, G. (2003). Time estimation during prolonged sleep deprivation and its relationships with activation measures. *Human Factors, 45*, 148-159.

Sacks, O. (2004, August 23). Speed: Aberrations of time and movement. *New Yorker*, pp. 60-69.

Stroud, J. M. (1955). The fine structure of psychological time. In H. Quastler (Ed.), *Information theory in psychology: Problems and methods* (pp. 174-207). Glencoe, IL: Free Press.

Change Blindness: What you don't see is what you don't get

Paula J. Durlach

U.S. Army Research Institute for the Behavioral and Social Sciences
12350 Research Parkway, Orlando, FL, 32826
Paula.Durlach@peostri.army.mil

Abstract

Change blindness refers to a failure to detect what should be an obvious visual change. It is likely to occur during human-computer interaction when more than one change occurs on the display at a time. It is also likely to occur as a result of a distraction that takes the user's attention away from the display. Experiments documenting the occurrence of change blindness during the use of a military digital system are presented. Using that system, a distraction such as the closing of a task window decreased the detection of an icon change by about 50%, compared with no distraction. The difficulty people had detecting simultaneous icon changes is also described. Data are presented suggesting that people learn strategies to aid change detection in a specific context; but that these strategies may not generalize to atypical conditions. Rather than relying on the user to detect changes in a visual display, designers of complex monitoring and control systems should represent change information explicitly.

1   What is change blindness?

Change blindness refers to the failure to detect what should be an obvious visual change. It tends to occur when there is a distraction that draws visual attention away from the location of the change. Even an eye blink can be a sufficient enough distraction to cause people to miss a change (O'Regan, Deubel, Clark, & Rensink, 2000). Normally, when a visual stimulus changes, the change produces local "visual transients" that cause an orienting response to the location of the change. This process usually ensures that the change is noticed; however, those local visual transients may fail to draw attention, either because they are not visible (e.g., if eye is closed or the change is occluded) or because of competing attention-attracting events elsewhere (O'Regan, Rensink, & Clark, 1999). After the change has occurred, and is clearly visible, there is no certainty that it will be noticed. In fact, most people are unaware of just how easily visual changes can be missed and tend to overestimate their ability to detect changes (Levin, Momen, & Drivdahl, & Simons, 2000; Levin, 2002).

Cognitive psychologists have been interested in the phenomenon of change blindness because its occurrence implies more fleeting and less complete visual representations than has been traditionally hypothesized (e.g., Noë, Pessoa, & Thompson, 2000; Rensink, 2002; Simons, 1996). If a change is clearly visible and yet still not detected, it implies that the original state may not have been represented in memory. The phenomenon has practical implications for human-computer interaction, as well as for our understanding of visual representation. This is the topic of concern of this paper. As computer-based systems become more complex and multifunctional, and their display capabilities more sophisticated, there has been a tendency to put more and more information onto visual displays, without consideration of how this will affect information processing by the user.

The pop-up advertisement is a prime example of a display feature likely to cause change blindness. It may compete for attention and even occlude an important change. Perhaps pop-up windows aren't that much of a concern if you are searching the Internet for a new rum cake recipe or the cheapest tickets to New Orleans. On the other hand, if you are controlling air traffic, monitoring the processes of an oil refinery, or commanding a military operation, you probably don't need pop-up messages hindering your ability to detect changes in your display. Yet, as the ability to digitize information has increased, so has the tendency to make it visually available and communicable over networks. The Army, in particular, is planning to digitize more units, and eventually change the nature of its operations, largely based on the premise that remote sensors and networking will bestow it with information and co-ordination superiority (Objective Force Task Force, 2002). In the future, intelligent agents may be able to analyze and fuse incoming data in order to present human observers and decision makers with a holistic, integrated information picture; however, in the nearer term, interpretation of incoming information will require humans to sift through it, interpret it, and act on it. To avoid overwhelming the humans in the loop, and ensure that critical changes are detected, an understanding of human attention and memory capabilities ought to drive design of human-computer interfaces intended for command and control purposes. In particular, the need to remember details of past states in order to extract the implications of current states (what changed and does it matter?) should be minimized. If people are prone to miss changes, why not represent change in some explicit way, instead of relying on the user to detect it (Durlach, 2004b; Durlach & Meliza, 2004; Smallman & St. John, 2003)?

2 Experiments using a fielded military system

To determine whether change blindness is truly a problem for human-computer interaction, we wanted to establish the extent to which it might occur during the use of an actual system, where failure to detect changes might pose serious consequences. The system we used was a modified version of Force XXI Battle Command Brigade and Below (FBCB2). This is a fielded Army system that is situated in the vehicles of digitized units, and is also used in tactical operating centers to coordinate the information that is networked to the digitized units. FBCB2 includes a terrain map, on which icons representing friendly, enemy, and other elements can be displayed. Icons are coded according to Mil-Standard 2525B (1999), and represent affiliation and unit type. Affiliation is coded redundantly by color and shape (e.g., blue circle is friendly, red diamond is enemy). A pattern within an icon indicates platform (more detailed information about the nature of the unit). Our system displayed the map on a PC; but otherwise resembled the display used in actual operations. The map was 7 by 5.75 inches (17.9 by 14.3 cm), and the map icons were approximately .25 by .25 inches (.7 by .7 cm).

Besides the map, FBCB2 includes various status indicators and task buttons. The task buttons allow users to conduct tasks, such as send messages, keep records, or perform terrain analysis. When one of these buttons is pressed, a window or series of windows appears superimposed on the map. The user must supply requested information in these windows in order to perform the desired task.

Our version of FBCB2 included features that allowed for the scheduling of icon changes and the collection of change detection data. When they saw a change, participants were instructed to use the mouse to click on a response bar located below the normal FBCB2 display. This paused the experimental session and opened a choice menu. Participants selected the type of change they saw from the menu, and then clicked a command button to return to the regular display and restart the session. Menu choices were grouped by type (e.g., new unit appeared, unit disappeared), and listed by color within type (e.g., new blue unit appeared, new red unit appeared, etc.). Icon changes included appearance, disappearance, position change, color change, and platform change.

Our participants were university students with normal color vision. They had no experience with FBCB2 prior to the experiments, and little understanding of the significance of the changes they observed. Accordingly, our research informs on people's ability to detect changes based on the physical properties of the stimuli, divorced from any meaning the changes might signify. As meaningful changes are better detected than non-meaningful changes (Rensink, 2002), we cannot claim that our data represent the behavior that might be expected from experienced tacticians using this system. Our participants were given a brief tutorial about the system prior to testing, and were trained to perform all the tasks they might be asked to conduct during the experiment.

We were primarily interested in examining change detection during monitoring the FBCB2 screen, and change blindness vulnerabilities that were inherent to the system. As such, there were no external distractions as might be found in realistic use of the system. In an operational setting, users typically have other tasks to perform besides monitoring FBCB2; these distractions and interruptions would be expected to worsen change detection. Our sessions were essentially vigilance experiments, in which participants were asked to monitor the map for changes. Sessions lasted between 45 and 90 minutes, depending on the experiment. Once a change occurred, participants were free to report it any time afterwards, up until the point at which the same event was scheduled to occur again. This was never less than 5 seconds; but typically was quite a bit longer. At that time, the change was logged as missed.

In some experiments, participants were also asked to perform some of the FBCB2 tasks such as send a message to another unit, or set a periodic reminder. This was done via a pop-up window on the screen. The message in the window instructed the participant to conduct a task, and provided the required details, such as 'set a periodic reminder to have lunch with your commanding officer every Wednesday at noon.' These instructions followed an icon change by at least 5 seconds. Participants clicked OK on the message and then performed the task. On certain occasions, completion of the task (clicking on the close button of the last task window) coincided with an icon change. We were interested in how this "distraction" of the window closing affected people's ability to detect the icon change. Figure 1 illustrates schematically an icon position change with or without such distraction. It's been shown in a lab setting that color patches flashed on a display screen at the same time as a change in a visual scene interfere with the detection of the change, even when the patches don't occlude the change (O'Regan, Rensink, & Clark, 1999). Our manipulation was similar but involved only the disappearance of the task window (as opposed to appearance and disappearance).

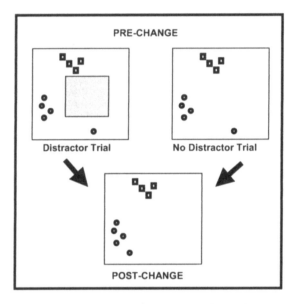

Figure 1. Schematic illustration of an icon position change with a closing-window distractor (left), or no distractor (right).

We ran two experiments examining the effect of a closing-window distractor on icon change. In the first experiment (Durlach & Chen, 2003), the map only contained two icons, and only one of these ever changed. Changes could include icon appearance, disappearance, color, position and platform. In the second experiment (Durlach, 2004a), the map was busier, containing between 9 and 11 icons, and while several types of changes occurred during the study, only position changes ever occurred coincident with the closing of a task window. Detection of these was compared (within-participant) to comparable position changes without the distraction. The results from both studies are shown in Figure 2. The figure allows comparison of the distributions for percent changes detected, for the specific changes of interest. As can be seen in the figure, in both studies, there was a dramatic effect of distraction, such that detection rate fell approximately to half its rate when there was a closing-window distractor, compared to when there was no distractor.

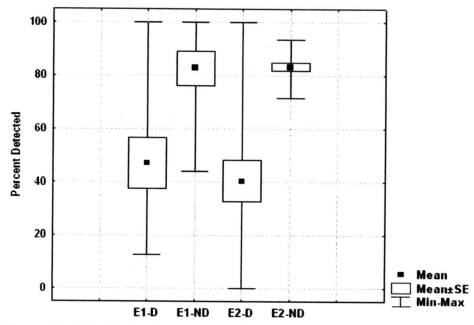

Figure 2. Response distributions for detected changes in Experiments 1 (E1) and 2 (E2), depending if the change was accompanied by a closing-window distractor (D) or not (ND). SE = standard error of the mean.

Another system-inherent source of change blindness we investigated was simultaneous icon changes (Durlach, Kring & Bowens, submitted for publication). During operations, FBCB2's map display is updated and redrawn at specific time intervals (not in real time). New information sent by any of the networked units between the refresh times will appear simultaneously the next time there is a display refresh. If changes occur simultaneously, there is likely to be a competition for attention among them. Consequently, it is possible that change detection will suffer accordingly. To investigate this issue, the changes we used were icon appearance and disappearance. Icons appeared on the map in color groups (red, yellow, and blue), with 6-9 icons in each group at any given time. As in the previous experiments, participants were asked to monitor the map and report observed changes. They were instructed that more than one change might occur at a time and that multiple choices could be made from the menu describing the changes. Icons appeared or disappeared one, two, or three at a time. Only one icon from each color could change at a time. Thus, for 3 simultaneous changes, the changes always involved one red, one yellow, and one blue icon.

As predicted, a change was less likely to be detected if it was accompanied by one or two simultaneous changes, compared with if it occurred alone. One change was detected 79.7% of the time when it occurred alone, 75.6% of the time when it was accompanied by one other change (regardless of whether that change was detected), and 69.6% of the time when accompanied by two other changes (regardless of whether these were detected). When 2 changes occurred simultaneously, people detected both of them only 59% of the time. When 3 changes occurred simultaneously, people detected all three of them only 37% of the time. Interestingly, people were better at detecting exactly one change when more than one occurred at the same time, compared to when only one change occurred. People failed to detect any change 20.4% of the time for a single change, but only 8.1% of the time for two changes, and only 5.8% of the time for 3 changes. The most plausible explanation is that multiple changes made it more likely that people were looking at the location of a change when it occurred, and therefore detected it. Unfortunately, they were not so good at detecting the other simultaneous changes.

Besides demonstrating the impact of simultaneity on detection, the results of this study indicated that detection of icon appearance was superior to detection of icon disappearance. Overall, the difference in mean percent detected was not large, but it was statistically significant. It was 76.5% for appearance and 74% for disappearance. When we examined the false alarms participants made—reporting a change that did not actually occur—we found that people made significantly more disappear false alarms than appear false alarms. We tried to take this into account by creating a detection index for appearance and disappearance, as follows.

$$\frac{100 * (\# \text{ correct detections} - \# \text{ false alarms})}{\text{Total } \# \text{ of changes}}$$

For people who made no false alarms, the index was the same as percent correct. The resulting mean detection index was 74.1 for appearance, and 60.5 for disappearance, a difference that was highly significant, $F(1, 34) = 126.5; p < .001$.

Within the attention literature, it has been suggested that object onset (or appearance) has a privileged power (compared to other changes) to capture attention (e.g., Cole, Kentridge, & Heywood, 2004; Cole, Kentridge, Gellatly, & Heywood, 2003); however, it has also been suggested this is only the case when people expect onset to be a target change (Folk, Remington, & Johnston, 1992). As discussed by Most, Scholl, Clifford, & Simons (2005), although the effect of onset might not be purely automatic, it nevertheless is a change type that draws attention very reliably. Here we have shown that onset is more likely to draw attention than offset.

Our experiments have shown that change blindness can occur during the use of current systems. It is possible that more experienced users would be less prone to change blindness, and if this is the case, it might be beneficial to understand why. During our appearance/disappearance study, we did find that change detection improved as a function of time into the session. We suspect that this may have resulted from the adoption of pneumonic strategies. For example, people could have learned to count the number of icons of each color periodically in order to detect changes. In a survey after the session, 17 (of 36) participants said that they used some form of counting as their primary strategy. In order to better understand how experience detecting changes might lead to better detection, we conducted a series of studies examining how practice affected detection.

3   Can change detection improve with training?

For these experiments, we used a research paradigm called the flicker procedure (Rensink, O'Regan, & Clark, 2000). The advantage of this procedure over the FBCB2 set-up was the ability for better experimental control. In this paradigm, each trial consists of a repeating series of images: X, A, X, B, X, A, X, B, etc., where X is a uniform grey screen. The participant's task is to determine whether A and B are the same or different pictures. This can be challenging because presentation of X essentially masks local visual transients that would normally signal a difference between A and B, were they to follow one another immediately. In essence, X simulates the effect of a distraction. In the study described below, A and B were presented for 250 ms each time they appeared, whereas X was presented for 750 ms. Participants watched the repeating sequence until they made their decision. Once they made their decision they clicked one of two command buttons below the images to indicate their choice. They were given feedback as to whether their response was correct or incorrect.

A and B consisted of arrays of 10 icons, semi-randomly arranged on the screen. In one study (Neumann & Durlach, 2004) we used icons similar to those used by FBCB2 in that they were 3-dimensional: color, outer shape, and inner shape. There were four values used for each dimension (e.g., red, blue, green, and yellow for color), thus making 64 possible stimuli. On any trial, the arrays shown in A and B were the same, or differed by a change in one icon, on one dimension. We were interested in whether repeated practice detecting the same change (over different arrays) would lead to improvement in detecting that change, and to what extent that improvement would generalize to novel changes. For example, given practice on red/blue changes, how would that affect performance not only for red/blue changes, but also on green/yellow changes (intradimensional transfer) and on circle/diamond changes (extradimensional transfer). Each participant was given practice with 2 changes. Each involved one pair of values from two different dimensions (e.g., red/blue and circle/diamond). These were given at different relative frequencies such that the frequent change occurred five times as often as the rare change. No changes ever occurred in the third dimension. Trials without a change were also intermixed. Assignment of dimension to change frequently, rarely, or not at all was counterbalanced across participants. We predicted that greater improvement would be seen for the more frequent change; but we were uncertain whether any transfer to novel changes would occur.

There is evidence that change detection ability can be altered by experience. For example, expert players are better at detecting changes in chess positions compared to novice players (Kämpf & Stobel, 1998); but only a few studies have examined the development of improved detection with systematically manipulated experience (Austen & Enns, 2000; Williams & Simons, 2000). In particular, Williams and Simons (2000) examined the ability of people to detect changes in artificial beings called fribbles. Different features were associated with 4 different "species" of fribbles. Changes occurred while fribbles moved behind an occluding object or briefly disappeared and then reappeared. In the four studies reported in the paper, 59 out of 84 participants showed an increase in sensitivity to changes over the course of the change detection session (as determined by a nonparametric signal detection analysis). One study was designed to determine if this increase in sensitivity was due to a general practice effect (e.g., familiarity with the change detection procedure), or rather was specific to the types of changes experienced. Results favored the latter interpretation. Practice detecting changes between species A and B led to enhanced sensitivity in detecting novel changes between species A and B, but not species C and D. It is important to note that enhanced change sensitivity was observed for *novel* A/B pairs; i.e., feature-pairs that had not been scheduled for change before. Therefore, the effect of practice generalized within, but not across species. Based on these results, we might expect to find intradimensional, but not extradimensional transfer in our study.

To test this, following practice, participants in our flicker study were given change trials involving both the changes experienced in practice, and those stimulus-pairs never changed during practice, plus no-change trials, all in equal proportion. For the trained changes, we did find greater detection of frequent vs. rare changes; however, the size of this difference depended on which stimuli played the role of the frequent and rare changes. In general, people were better able to detect color than shape changes, so detection of color changes was high regardless of whether a color change had been practiced frequently or not.

Figure 3 compares the results from training and testing, collapsed across frequent and rare dimensions, for simplicity. Mean sensitivity, calculated according to a nonparametric signal detection analysis, is shown on the Y-axis. The two left bars show sensitivity at the beginning of training (first eight change trials), and at the end of training. Clearly, there was an increase in sensitivity in the practiced changes as a result of training. The two right bars show sensitivity on the intra- and extra-dimensional transfer trials. Comparing these to detection sensitivity at the start of training, there appears to be no transfer either within the training dimensions (intra-) or for a previously unchanged dimension (extra-).

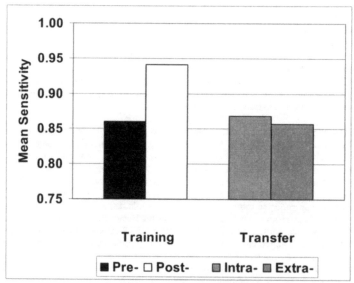

Figure 3. Mean change detection sensitivity at the beginning of training (Pre-) and the end of training (Post-), and to novel changes not seen during training (Intra- and Extra-), for the flicker experiment.

In other experiments with only 2-dimensional stimuli, we have found evidence of both improved detection sensitivity with practice and transfer to novel changes. As yet, we don't understand the factors that determine whether transfer will occur or not. We suspect that in the experiment described above, part of the improvement observed as a result of training was due to participants learning to ignore the non-changing features while they searched the screen for a change. Such learning to ignore would inhibit detecting changes involving these features when they were actually changed during the test. Attempts to improve change detection through training may result in specific improvements determined by the training regime. It seems that people will learn strategies to improve their performance that depend on the specific training conditions. If those conditions change, the learned strategies may no longer be beneficial. Unless a training regime can be found that produces general improvement in change detection, it is not clear that training is a viable approach to dealing with the problem of change blindness in the context of human-computer interaction.

4   Implications for human-computer interface design.

The implication of the work described above is that people are likely to miss changes in visual displays if the change occurs at the same time as another visual event. Change blindness will be exacerbated in situations in which people have multiple tasks to perform, in which visual attention has to be divided, and several things happen at the same time. Practice with a particular system may lead to the adoption of strategies that improve change detection of common changes; but this is not necessarily good for detection of rare or unexpected changes. Because we know that people are apt to miss changes, human-computer interfaces should represent change explicitly, instead of relying on the user to detect it.

As discussed by Durlach (2004b) and Durlach & Meliza (2004), there are several ways that changes could be represented explicitly. One important issue in designing a system that explicitly represents change is determining which changes to represent. If all changes are represented, this could itself be an overwhelming amount of information. On the other hand, if only certain changes are represented, what criteria should be used for the selection process? A second important issue is how to notify the user that a change has occurred. At one extreme, all changes could trigger some form of alert; but this is likely to be too intrusive and be more of an annoyance than an aid. At the other extreme, changes could be passively logged without any notification to the user. It would then be left to the initiative of the user to inspect the log.

A compromise between the extremes laid out above is most likely a good solution. All changes could be logged; but (if a table) this log should be sortable and searchable so that the operator can use it to satisfy specific goals. Various ways to access the log data should be available with cross referencing across different relevant forms of information representation. For example, if an operator wanted to know about changes to a specific icon, selecting the icon could bring up its logged history. Conversely, selecting an item in the change log should highlight the relevant icon on the map (see Smallman & St. John, 2003). Whether in table form or "instant replay," the operator should have some way to recover the relatively recent history of important elements represented in their system.

Some selection process needs to be applied in order to determine which changes should trigger alerts without waiting for the operator to ask for change information. A set of default change conditions, based on expert knowledge of the operational context is advisable. For example, Durlach and Meliza (2004) described a system for integration with FBCB2 that would help prevent fratricide. It would send an alert (to both parties) if a friendly unit entered the firing zone of another friendly unit. In addition to defaults, there will likely be some alerts that an operator will want (or not want), depending on the specific operating context at a given time. An authoring tool that allowed the operator (or supervisor) to write and activate (or deactivate) customized alerts should augment the default alerting rules. An editable alerting system would also be of great benefit in training, acting like a tutor to instruct the trainee about what to attend to.

Some selection process also needs to be applied to determine how an alert is delivered. Doing justice to the large literature on the design of alerts is not possible in this paper (e.g., Mumaw, Roth, Vicente, & Burns, 2000; Wogalter, Conzola, & Smith-Jackson, 2002); however, considerations include aspects such as modality, information content, requirement for acknowledgement, persistence, and data logging of user reactions. The intrusiveness of the alert should depend on its priority. Allowing the operator to choose the characteristics of alerts could be a feature integrated into an alert-editing tool.

The objective of the paper was to raise awareness about change blindness and its implications for human-computer interaction. Detection of changes requires both attention and memory. Rather than relying on the user to detect changes in a visual display, designers of complex monitoring and control systems should represent change information explicitly.

References

Austen, E. & Enns, J. T. (2000). Change detection: Paying attention to detail. *Psyche*, 6. Retrieved on March 28, 2002 from http://psyche.cs.monas.edu.au/v6/psyche-6-11-auten.html.

Cole, G. G., Kentridge, R., Gellatly, A. R. H., & Heywood, C. (2003). Detectability of onsets versus offsets in the change detection paradigm. *Journal of Vision*, 3, 22-31.

Cole, G. G., Kentridge, R., & Heywood, C. (2004). Visual salience in the change detection paradigm: The special role of oject onset. *Journal of Experimental Psychology:Human Perception and Performance*, 30, 464-477.

Durlach, P.J. (2004a). Army digital systems and vulnerability to change blindness. 24th Army Science Conference, November, Orlando, FL.

Durlach, P.J. (2004b). Change blindness and its implications for complex monitoring and control systems design and operator training. *Human-Computer Interaction*, 19, 423-451.

Durlach, P.J. and Chen, J.Y.C. (2003). Visual change detection in military visual displays. I/ITSEC, December, Orlando, FL.

Durlach, P.J., Kring, J.P, & Bowens, L.D. (submitted for publication). *Journal of Experimental Psychology: Applied.*

Durlach, P.J. & Meliza, L.L. (2004). The need for intelligent change alerting in complex monitoring and control systems. *Interaction between humans and autonomous systems over extended operation*. AAAI Technical Report SS-04-03, 93-97.

Folk, C. L., Remington, R. W., & Johnston, J. C. (1992). Involuntary covert orienting is contingent on attentional control settings. *Journal of Experimental Psychology: Human Perception and Performance*, 18, 1030-1044.

Kämpf, U. & Strobel, R. (1998). "Automatic" position evaluation in "controlled" change detection: Data-driven vs. concept-guided encoding and retrieval strategy components in chess players with varying degrees of expertise. *Zeitschrift für Psychologie*, 206, 23-46.

Levin, D. (2002). Change blindness blindness: as visual metacognition. *Journal of Consciousness Studies*, 9, 111-130.

Levin, D. T., Momen, N., Drivdahl, S. B., & Simons, D. J. (2000). Change blindness: the metacognitive error of overestimating change-detection ability. *Visual Cognition*, 7, 397-412.

Mil-Std 2525B: Department of Defense Interface Standard: Common Warfighting Symbology (1999). Retrieved May 28, 2003, from the General Dennis J. Reimer Training and Doctrine Digital Library website: http://www.adtdl.army.mil/cgi-bin/atdl.dll/fm/3-25.26/toc.htm

Most, S. B., Scholl, B. J., Clifford, E. R., & Simons, D. J. (2005). What you see is what you set: Sustained inattentional blindness and the capture of awareness. *Psychological Review*, 112, 217-242.

Mumaw, R., Roth, E.M., Vicente, K.J., & Burns, C.M. (2000). There is more to monitoring a nuclear power plant than meets the eye. *Human Factors*, 42, 36-55.

Neumann, J.L. & Durlach, P.J. (2004). The effect of practice on visual change detection in computer displays. 48th Annual Meeting of the Human Factors and Ergonomics Society. New Orleans, LA.

Noe, A., Pessoa, L., & Thompson, E. (2000). Beyond the grand illusion: What change blindness really teaches us about vision. *Visual Cognition*, 7, 93-106.

Objective Force Task Force. 2002. The Objective Force in 2015 White Paper. Retrieved December 15, 2002 from http://www.objectiveforce.army.mil/pages/ OF%20in%202015%20White%20Paper%20(final).pdf

O'Regan, J. K., Deubel, H., Clark, J. J., & Rensink, R. A. (2000). Picture changes during blinks: Looking without seeing and seeing without looking. *Visual Cognition*, 7, 191-211.

O'Regan, J. K., Rensink, R. A., & Clark, J. J. (1999). Change-blindness as a result of "mudsplashes." *Nature*, 398, 34.

Rensink, R. A., O'Regan, J. K., & Clark, J. J. (2000). On the failure to detect changes in scenes across brief interruptions. *Visual Cognition*, 7, 127-145.

Rensink, R. A. (2002). Change Detection. *Annual Review of Psychology*, 53, 245-277.

Simons, D. J. (1996). In sight, out of mind: When object representations fail. *Psychological Science*, 7, 301-305.

Smallman, H.S. & St. John, M. (2003). CHEX (change history explicit): New HCI concepts for change awareness. 47th Annual Meeting of the Human Factors and Ergonomics Society. Denver, CO: Human Factors and Ergonomics Society.

Williams, P. & Simons, D. J. (2000). Detection changes in novel, complex, three-dimensional objects. *Visual Cognition*, 7, 297-322.

Wogalter, M.S., Conzola, V.C., & Smith-Jackson, T.L. (2002). Research-based guidelines for warning design and evaluation. *Applied Ergonomics*, 33, 219-230.

Stress, Cognition, and Human Performance: A Conceptual Framework

Mark A. Staal

Multi-National Force-Iraq HQ
mark.staal@iraq.centcom.mil

Abstract

The following paper addresses the effects of various stressors on cognition. While attempting to be as inclusive as possible, the review focuses its examination on the relationships between cognitive appraisal, attention, memory, and stress as they relate to information processing and human performance. In summary of this review, a conceptual framework for cognitive process under stress has been assembled. The research literature that addresses stress, theories governing its effects on human performance, and experimental evidence that supports these notions is large and diverse. Accordingly, the present organization and synthesize of this body of work has relied upon several earlier efforts (Bourne & Yaroush, 2003; Driskell, Mullen, Johnson, Hughes, & Batchelor, 1992; Driskell & Salas, 1996; Handcock & Desmond, 2001; Stokes & Kite, 1994). However, each of these previous efforts either simply reported general findings, without sufficient experimental illustration, or narrowed their scope of investigation to the extent that the breadth of such findings remained hidden from the reader. Moreover, none of these examinations yielded an architecture that adequately describes or explains the inter-relations between information processing elements under stress conditions. The present review provides an initial step toward this end. Limitations and key concerns that remain unanswered by the research community are also discussed.

1 Stress, Cognition, and Human Performance: A Conceptual Framework

1.1 What is stress?

Stress is a concept that we as a scientific community continue to struggle to define. With broad variations in descriptions the research literature in this area is often confusing and contradictory. For the better part of the last century research has relied on various theories to help explain the relationship between stress and performance. These include arousal, activation, energetical, and resource models. Each has struggled with its own limitations and the result has been a research community that is factional and disconnected. The lack of agreement between researchers has presented additional difficulties in unifying the literature. Furthermore, without an overarching theory to draw these elements together, there can be neither hope for a unitary explanation of mechanisms nor a sense of cohesion among concepts. This review reflects these disconnects within the research community while attempting to coordinate the extant body of diverse material. For the sake of simplicity and coherence, I have selected a definition proposed by McGrath (1976) that seems to be broad enough to incorporate most of the current assumptions about what stress is and is not, yet focused enough to be meaningful. McGrath conceptualized stress as the interaction between three elements: perceived demand, perceived ability to cope, and the perception of the importance of being able to cope with the demand. Unlike many previous definitions of stress, this formulation distinctly incorporates the transactional process believed to be central to current cognitive appraisal theories.

1.2 In the eyes of the beholder

There is growing consensus that a transactional model of cognitive appraisal, one that views stress as a dynamic relationship between demands, resources, and cognition, provides an accurate account of how individuals react to stressors. Moreover, the research literature provides strong empirical support for the notion that evaluations of threat and/or control are clearly related to the experience of subjective stress (or its reversal). These evaluations are also directly related to improvements or decrements in performance in some instances. Much of the research points to a rapid initial evaluation followed by a more involved higher-order cognitive process (i.e., a cognitive appraisal system). Integrally related to cognitive appraisal and its effect on human performance under stress, is the predictability or perceived controllability of the stressor. These factors are part of the appraisal system and have been clearly shown to influence the human stress response.

1.3   The direct and indirect effects of putative stressors

An extension of this discussion can be found in the debate over direct and indirect stress effects. It is proposed that direct stress effects are those incurred by the taskload alone irrespective of any psychological stress that may also be generated. Accordingly, indirect stress effects are those that evolve out of psychological factors associated with the taskload demands. There is a fine line that separates these two, and they can be indistinguishable at times. This fact has made their separation and measurement particularly difficult. At the heart of the debate between these two effects are several challenges. For example, is the application of some task demand (i.e., workload or time pressure) an application of stress? Many would argue that it is while others would contend the contrary. Proponents of the former typically offer one of two arguments. The first states that stress is a term that can be applied to any demand on a system. Therefore, any task that requires mental resources to accomplish qualifies as a stressor – it places a demand or stress on the system. This argument is one of pure semantics. The second argument proposes that most significant demands incur a psychological cost as well. Through frustration, anxiety, psychological discomfort, or some other negative affective state, these demands trigger a psychological response. This response often contains both physiological and mental components that vie for resources. In this way, stress acts as a secondary workload. However, a compelling argument can be made that workload is simply a demand, one that does not require, nor regularly incur, a secondary psychological cost.

If we agree that subjective experience and specifically cognitive appraisal (a transactional model assumption) is elemental in defining stress, then one must assume it plays a significant role in answering questions about whether workload, time pressure, or other putative stressors carry both direct and indirect effects. Does this suggest that when a demand is deemed stressful or upsetting it is necessarily a stressor, regardless of the objective outcome? If an increase in workload does not impair performance yet is viewed as stressful by the operator, does this indicate that it should be considered a stressor? Reasonable arguments can be made to support both positions, and the research literature, in its current state, is a reflection of this fact.

Several researchers have attempted to side-step this issue by relying on descriptions of taskload alone, ignoring the potential psychological stress consequences related to this taskload. In doing so, they have circumvented a direct discussion of stress and its role in performance degradation or enhancement. However, in leaving this issue unaddressed, these authors have left the reader to infer a stress effect, correctly or not. I have not attempted to resolve this issue but to make the reader aware of it. At the end of this review I describe a conceptual framework that helps organize data and concepts, and attempts to provide more coherence than is apparent in the literature.

1.4   Moderating variables

A growing body of work implicates numerous intervening variables between demands, resources, and cognition. Such moderators appear to influence the stress-performance process at its root, the initial evaluation and appraisal system, as well as during other points in the process. For instance, various experiences, motivations, and personality characteristics have been demonstrated to facilitate performance under certain circumstances and at other times degrade it. Related to this discussion of individual differences is the finding that the selective nature of attentive processes can be connected to affective states. Specifically, a large body of literature indicates that high-trait (a stable characteristic of the individual) and high-state (a temporary fluctuation) anxious individuals demonstrate an attentional bias toward threatening stimuli. It is believed that such biases modulate the effect of stress on performance outcome by diverting mental resources toward task-irrelevant processes.

1.5   The effects of stress on attention

The research literature concerning stress' effects on attentional processes is relatively clear. Psychological stress along with various forms of workload tend to tunnel attention, reducing focus on peripheral information and tasks and centralizing focus on main tasks. What distinguishes a main task from a peripheral task appears to depend on whichever stimulus is perceived to be of greatest importance to the individual or that which is perceived as most salient. Threat-relevance is strongly associated with salience. Therefore, when environmental stimuli are threat-related, such stimuli are often considered to be most salient by the individual. This tunneling of attention can result in either enhanced performance or reduced performance, depending on the nature of the task and the situation. For instance, when peripheral cues are irrelevant to task completion the ability to tune them out is likely to improve performance. On the other hand, when these peripheral cues are related to the task and their incorporation would otherwise facilitate success on the task, performance suffers when they are unattended. This finding may apply to both visual attention and auditory attention; however, auditory attention has received little study. Experimental designs that incorporate a stress manipulation check (assessing the effectiveness of the supposed stress manipulation) are unfortunately not common. Researchers often only assume that their manipulations (e.g., increased workload, time pressure, physical or emotional threat, etc.) function as psychological stressors. Further, most of these studies fail to distinguish possible direct effects of manipulations from indirect effects.

1.6   Dual-task management

We live in a dynamic and complex environment which often demands the concurrent management of multiple tasks. Although portions of the research examining multi-task performance are covered in the attention and memory sections of this review, the dual and multi-task performance literature is significant and deserves individual attention as well. As one might imagine, concurrent task management results in degraded performance on either the primary or secondary task. Typically, research on concurrent task management employs only two tasks (for methodological purposes one is designated as the primary and the other, the secondary); however, the reader will note that naturalistic settings frequently require attention to more than two tasks concurrently. There are several explanations for the finding that one of these tasks (typically the target task or primary) is degraded. This will be discussed in greater detail in the review; however, the concept of capacity and the presumption of limited resources has been the most popular explanation of dual-task performance decrements.

1.7   The effects of stress on memory

The research literature concerning the effects of stress on memory consistently demonstrates that elements of working memory are impaired. Although the mechanisms behind these effects are poorly understood, it seems likely that encoding and maintenance processes are the most affected. Some have concluded that this reflects a reduction in resource capacity. Resources may be eliminated in some way, the span of time in which they can be accessed may be reduced, or these resources may be drawn away as a result of resource sharing (the absorption of resources by competing demands). Furthermore, little is known about what stage in the process this depletion or occupation takes place. It may be that resources or capacity are reduced at several points in the process (i.e., encoding, rehearsal, or retrieval). Few, if any, studies have attempted to separate these dimensions within memory processes while under stress conditions.

There are a variety of tasks and putative stressors under which memory has been measured. Anxiety is perhaps the most common stress condition under which researchers have examined memory performance. This research has generally directed the field toward resource-depletion models. These assert that worry and intrusive thoughts compete for a limited pool of resources. This competition necessarily results in fewer available resources that can be devoted to the primary task. A complementary view contends that attention may reduce the bandwidth of perceptual cues thus reducing the number or scope of attended stimuli (following Easterbrook's hypothesis). The reduction in cue sampling may in turn result in a reduction in the number of items encoded into memory, thus further reducing the number or range of possible target items available for later recollection.

The effects of divided attention on memory performance may be significantly related to this discussion. Analogous to the role of anxiety described previously, secondary tasks require and draw away resources and attention from the primary task. Thus, dividing attention between tasks reduces the attentional resources available to apply to either task. In such cases when the recall or recognition of information is required, this division often results in a decreased capacity to recall or recognize information. It has been demonstrated that dividing attentional resources has a direct negative effect on the encoding of information, although research suggests that several mechanisms may be at work. For example, divided attention may lead to a reduction in the time available to process incoming stimuli (due to time devoted to a secondary task) or it may result in reduced depth of processing and less elaborative coding.

Several consistent observations have been made concerning memory for emotional events. First, memory tends to be impaired temporarily when recalling information prior to or following an emotional event. Second, memory for a targeted emotional event may or may not be impaired under stressful conditions; however, there is a tendency for improved recall of central features when such events are emotional as compared to neutral. These "tunnel" memories resemble what has been observed in attentional processing (Easterbrook's hypothesis). It has been argued that such memories result from a combination of selective attention, preattentive bias, and post-event elaboration. Third, peripheral details are less often remembered when the main events witnessed are emotional in nature. Fourth, memory tends to be impacted by context effects. Specifically, memory improves when retrieval conditions are congruent with encoding conditions (i.e., mood-congruency effects). Finally, research points to the notion that individuals may be predisposed or primed toward emotionally valent information.

Recent research has made the connection between hippocampal function and memory. Damage to the hippocampus often leads to impairment in learning and memory. The hippocampus is also implicated in the human stress response and the activation of glucocorticoids. Moreover, exposure to high doses of cortisol (a known marker of the human stress response) has also been found to block hippocampal potentiation. Thus, the hormonal stress response may cause direct effects on the brain structure mediating some memory functions. These neurophysiological and electrophysiological relationships appear to be the most promising link to an underlying biological mechanism and process at this time.

1.8   The effects of stress on judgment and decision making

The research literature concerning the effects of stress on judgment and decision making demonstrates that both individual and team processes are degraded. Many putative stressors (e.g., noise, fatigue, physical or emotional threat, etc.) have been shown to negatively affect these processes. The ways in which this occurs are diverse. Stress can lead to hypervigilance, a state of disorganized and somewhat haphazard attentional processing. This condition often results in a frantic search, rapid attentional shifting and a reduction in the number and quality of alternatives considered. As fewer alternatives are considered, there is a recursion to previously sampled possibilities. Individuals tend to rely on these previous responses regardless of their previous response success. Thus, in addition to experiencing greater rigidity, individuals may tend to persist with a method or strategy even after it has ceased to be helpful. This assumes that the previous strategy or approach is well-learned.

Affect or mood has also been implicated in judgment and decision making. Individuals experiencing negative affect tend to access more information in general as well as more helpful information in their assessment of situations. What has been termed "depressive realism" in the past has found some support in this regard. The inverse of this assertion appears also to be true. Those experiencing positive mood states tend to approach tasks with less effort and less time. This trend appears most characteristic of decisions or judgments that are made when the task or conditions are ambiguous. Thus, individuals in positive moods may be less likely to use systematic and detail-oriented processing strategies in their decision as compared to individuals who are experiencing a negative mood.

Effective or adaptive teams tend to shift strategies under stress. This often takes the form of a shift from explicit coordination toward implicit coordination, subsequently enhancing performance. This strategy has been suggested to reduce coordination overhead or the typical costs in time, resources, and effort that teams using explicit strategies alone incur. The opposite is true for less adaptive teams under stress. These groups tend to lose implicit coordination and fall back to explicit, on-line control strategies. The result can be a heavy cost in resources and ultimately decrements in performance. This phenomenon has also been described as a loss in team perspective. In this regard, teams under stress may lose their shared mental models and their collective comprehension (awareness of each other's efforts), shifting to an individualistic self-focus. Although it has not been definitively determined, team's that share common mental models are believed to be those that are able to shift from explicit to implicit coordination.

In addition to strategy shifting, various research results point to task or load shedding as another adaptive strategy. This form of task simplification has been studied in a variety of contexts and has been characterized as economizing workload with a shift in strategy or method that reduces any redundant information or non-essential information from being processed. This type of resource management seems to happen logically and/or systematically at first (paring tasks appropriately) but may result in less-organized and less-reasoned shedding as workload and stress increase to dramatic levels.

1.9 Unanswered questions

Finally, perhaps the most important section of the review highlights its limitations and the many questions it has unearthed that remain unanswered. The majority of this review is based on a synthesis of previous reviews and as such is a reflection of what these reviews have addressed as well as what they have neglected. There are many questions that have been left unanswered (or inadequately answered) by the research community. As in most areas of science, it seems that the more one learns about a complex human system the more one realizes how little is really understood about that system. Furthermore, such inquiry seems to frequently result in more questions being raised than answered. Many of these are questions that our field is currently ill-equipped to answer.

- What is stress and can we measure it? Do all stressors create the same physiological pattern? If not, why not, and in what way do they differ? What is the mechanism that facilitates these differences? Is there a unitary mechanism that underlies the human stress response such as arousal? Are there several different systems?
- What causes cognitive or information processing decrements? Are these the result of the direct effects of arousal or some other physiological system? What are the boundaries of impairment for each cognitive process and each stressor?
- Why does positive appraisal improve performance? Are the positive effects of appraisal related to effort and mobilization? What mechanism is activated differently when viewing a stressor as a challenge as opposed to an overwhelming threat?
- Why do various external stressors (heat, cold, noise, fatigue, etc.) cause decrements? Are these due to direct or indirect effects, or both? For example, are thermals (heat and cold) merely an irritant that plays into focus and motivation, or do they operate on physiological or thermodynamic principles to degrade performance? How can we systematically separate and measure these two types of effects to better understand their relative contributions?
- Where do psychological resources come from? Are they a static pool upon which we draw? Are they called up via physiological responses (arousal/activation hypothesis) as described in the traditional fight/flight models? Are they limited in capacity? Is this just a regulation-of-attention problem? Are the cognitive structures associated with information processing the resources described by others, or do they simple require resources to process?
- To what degree are top-down processes engaged in information processing and to what degree are bottom-up processes involved? In other words, are resources drawn and pulled by stimuli or directed by executive functions? Does this depend on the process, the task, or both? What is the nature of the central executive or homunculus function in the allocation or resources and other processes?
- Is attention the primary gatekeeper for all other information processing decrements? For example, can working memory resources or capacity be diminished directly or are such deficits the result of earlier effects on attention? Similarly, does psychological stress inhibit attention, and thus cause memory to be degraded (encoding is disrupted), does it interfere with the quality of what is encoded (bad in = bad out), does it disrupt maintenance functions (i.e., rehearsal), or simply make retrieval more difficult in some other way?
- Are biases in attention toward threat-related cues a function of trait, state, or a different underlying mechanism? Do these biases result in preferential orienting, difficulties disengaging, more depthful processing, or a combination of effects?
- Are performance decrements that result from stress and/or workload catastrophic or gradual? Does this correspond to physiological changes that are catastrophic or gradual? Does this depend on the task? Does this depend on the source of the stress? When task-shedding occurs, which tasks are abandoned? At what point are they resumed? Why are some tasks protected and others shed?
- Can we map the cognitive architecture of the human stress response to a neuro-biological basis? Are there corresponding neuro-anatomical structures that support our divisions of labor and projections of relationship? Will this provide us a definitive answer as to what resources really are? Are neuroendocrine and biochemical correlates causal agents or just transmission agents?
- Well-learned tasks are generally resistant to stressors as compared to newly learned tasks (this is true of implicit material as compared to explicit). Why? Does it depend upon which type of task or which type of stressor?

- What are the protective factors against stress and how do they work? What are the underlying mechanisms that explain how they operate?
- Why do various moderating variables change the way putative stressors affect performance and cognition? Do they work off of a common mechanism such as effort or motivation or are there different mechanisms that explain their effects?

2   The Conceptual Framework of Information Processing under Stress

I present here, and in appendices A-C, a brief sketch of a conceptual framework of information processing under stress (Staal, Nowinski, Holbrook, & Dismukes, in preparation). As with any such model its benefits are descriptive and not predictive. This framework is an instantiation of the transactional perspective and is generally consistent with previous models (e.g. Hancock & Warm, 1989; Matthews, 2002; Park & Folkman, 1997). It is also grounded in well-established and accepted cognitive architectural principles (Anderson, 2000; Lebiere & Anderson, 2001). However, while attempting to accommodate previous frameworks, it also extends beyond these in attempting to integrate various perspectives and bodies of information.

This framework relies on several assumptions that have been drawn from the existing cognitive science literature. It does not assert a unique mechanism or explanation for the nature of resources nor does it attempt to displace existing explanations beyond that which has been provided by the literature review, concerning arousal, effort, or other activation-based theories of energy mobilization. The model does assume that energy mobilizations typically occur under one of two conditions: 1) task-induced situations (activation results from the stimulation of the task or environment itself), and 2) internally-guided mental effort (a voluntary mobilization). This set of assumptions is somewhat distinctive from traditional resource theories (Kahneman, 1973; Wickens, 1992), as it emphasizes regulatory processes and not the availability of supplies. Accordingly, the framework contends that activation and energetical processes can be allocated, controlled, and subject to resource management decisions. Moreover, this model assumes that behavior is largely goal-directed and self-regulated and that this regulation incurs costs to various portions of the processing system (i.e., further resources can be acquired at the expense of increased effort). These additional assumptions rest predominantly in the work of Hockey (1997) and Gaillard (2001). It should be noted; however, that these propositions are not in conflict with a strictly bottom-up-driven processing model either, and it is believed that both processes play a role in information processing under stress.

This construction assumes a transactional model of cognitive appraisal. As the reader may recall, transactional models view stress as the interaction between the environment and individual, emphasizing the role of appraisal in designing a response to stress (Lazarus, 1966; Lazarus & Folkman, 1984; McGrath, 1976). Thus, to a certain degree, cognitive mediation is required for perceptions of threat, fear, or anxiety, particularly in the later stages of the initial stress response. Prior to such processing, this model also assumes that an early, likely "hard-wired" and subcortical evaluation occurs after orienting to a given stimulus (Crawford & Cacioppo, 2002; Duckworth, Bargh, Garcia, Chaiken, 2002; Rohrbaugh, 1984). This initial evaluation is believed to be rooted in bio-evolutionary mechanisms and as a result occurs prior to conscious awareness (perhaps arguably at a pre-cognitive level).

Following this initial stage of recognition, evaluation, and higher-order appraisal, information regarding the stimulus and one's response feeds back into the system. As a result, there is an activation and mobilization of resources. These resources are both drawn and directed toward further processing of the stimuli. This process is moderated by various influences such as goal structure, motivation, and effort as well as individual difference characteristics and previous experience and learning. Continued information processing is accompanied by further appraisal and evaluation that periodically feeds back into the process. Such continuous input and situation assessment updates provide adjustments to resource allocation and processing accordingly. This process is believed to be similar to what has been described in the situation models literature (Zwaan & Madden, 2004; O'Brien, Cook, & Peracchi, 2004), and that encapsulated by research on situation assessment and awareness (McCarley, Wickens, Goh, & Horrey, 2002; Uhlarik & Comerford, 2002; Wickens, 2002). As this processing directs and draws attention, the effects of stress and workload impinge on the resources available for processing. This can result from a number of factors including resource-depletion and re-allocation. For example, emotion or threat-based cognitions (i.e., anxious thoughts) may re-distribute resources away from a task-relevant response or may simply consume or occupy them, resulting in fewer available resources to devote elsewhere (Ashcraft, 2002; Ashcraft & Kirk, 2001; Eysenck & Calvo, 1992).

The general effect of both psychological stress as well as high degrees of workload on attentional processing tends to be a reduction in the processing of peripheral information and an enhanced focus on centralized cues (Chajut & Algom, 2003; Easterbrook, 1959). The determinant between central and peripheral stimuli seems to be based on which stimulus is perceived to be of greatest importance to the individual or that which is perceived as most salient. Salience, from this perspective, may relate to distinctiveness as well as emotional valence (i.e., threat relevance). In some circumstances, filtering out peripheral cues is beneficial to performance and in other instances it is detrimental. From the perspective of this framework, attentional processing has been positioned as a conduit between cognitive appraisal and the direct effects of stress on information processing and later processes such as memory and decision making.

The basic mechanism by which stress affects memory performance is posited as the siphoning of attentional resources (Berntsen, 2002; Christianson, 1992). The extent to which a given memory process is affected is determined by the extent to which that memory process is attention-demanding (and the extent to which a stressor is attention-demanding). Encoding, rehearsal, and effortful retrieval of memory are all impaired under conditions of stress while well-learned behaviors and retrieval of information from long-term memory remain relatively intact.

It seems obvious that all information processing requires some amount of mental resources to occur. This allocation and use of resources begins at the appraisal process itself. Once a situation has been identified as threatening, rumination may begin to intrude upon an individual's thoughts. These worries occupy attentional resources that are necessary for other cognitive processing (Metzger, Miller, Cohen, & Sofka, 1990). In particular, it seems that worry intrudes on processing of verbal information (Ikeda, Iwanaga, & Seiwa, 1996; Rapee, 1993). Further, such negative thoughts perpetuate themselves by supporting involuntary retrieval of other anxious thoughts through their association in memory (Bower, 1981). Some individuals in certain situations may feel compelled to suppress an expression of their emotional response to stress. This activity, like rumination, requires the involvement of cognitive resources, leaving fewer resources available for other information processing (Richards & Gross, 2000; Richards, Butler & Gross, 2003). The amount of resources pre-empted by each action is believed to be quite variable. For instance, proceduralized tasks require few limited resources while novel or complex tasks require many more (Beilock, Carr, MacMahon, & Starkes, 2002).

The allocation of remaining available resources means that there are fewer left to devote to the processing of other stimuli. Working memory capacity is therefore effectively reduced (Burrows, 2002). Attentional resources are required to maintain information in working memory through rehearsal. As mentioned above, it seems to be the case that rumination interferes with phonological rehearsal of information in particular, leaving verbal information particularly vulnerable. Encoding and rehearsal of long-term memory are also negatively affected, as both are resource-demanding activities. Thus, information, other than that considered most salient, is not encoded as deeply or rehearsed as frequently. Subsequently, its recollection and recognition is likely to be reduced in both quantity and quality.

Retrieval from long-term memory should be relatively unaffected as it is not generally resource-demanding. Well-learned information, such as that associated with procedures or skills, is retrieved relatively automatically. However, several researchers have made a distinction between the processes involved in familiarity and recollection (Jacoby, 1991; Kensinger & Corkin, 2003). Familiarity occurs automatically, while recollection involves a more effortful process. Retrieval of ingrained knowledge from long-term memory should not be diminished under conditions of stress. However, more involved retrieval from long-term memory, that which requires an active search of memory, should show deficits in stressful situations.

Once accounting for diminished capacity, the direct effects of stress on attentional processing, and the subsequent effect on working memory, it is not difficult to understand why decision making is often compromised and concurrent task management degraded. As resources are ultimately used to capacity individuals must voluntarily or involuntarily shift strategies. Such shifts include the shedding of tasks as well as their simplification.

Appendices A through C provide a schematic representation of each component part among the three phases of processing, sense making, resource management, and performance. Appendix A presents an overview of cognitive appraisal processing elements while appendix B provides a review of the processing framework associated with attention and memory. Appendix C integrates all of the framework's component parts.

Appendix A: Cognitive Appraisal Processing Elements

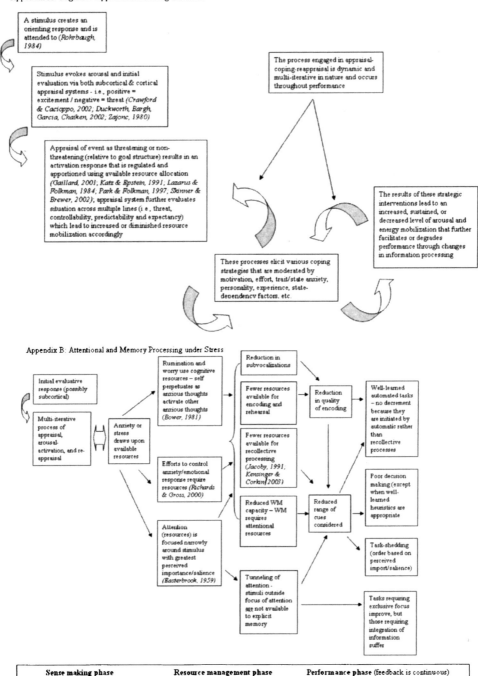

A stimulus creates an orienting response and is attended to (*Rohrbaugh, 1984*)

Stimulus evokes arousal and initial evaluation via both subcortical & cortical appraisal systems - i.e., positive = excitement / negative = threat (*Crawford & Cacioppo, 2002; Duckworth, Bargh, Garcia, Chaiken, 2002; Zajonc, 1980*)

Appraisal of event as threatening or non-threatening (relative to goal structure) results in an activation response that is regulated and apportioned using available resource allocation (*Gaillard, 2001; Katz & Epstein, 1991; Lazarus & Folkman, 1984; Park & Folkman, 1997; Skinner & Brewer, 2002*); appraisal system further evaluates situation across multiple lines (i.e., threat, controllability, predictability and expectancy) which lead to increased or diminished resource mobilization accordingly

The process engaged in appraisal-coping-reappraisal is dynamic and multi-iterative in nature and occurs throughout performance

The results of these strategic interventions lead to an increased, sustained, or decreased level of arousal and energy mobilization that further facilitates or degrades performance through changes in information processing

These processes elicit various coping strategies that are moderated by motivation, effort, trait/state anxiety, personality, experience, state-dependency factors, etc.

Appendix B: Attentional and Memory Processing under Stress

Initial evaluative response (possibly subcortical)

Multi-iterative process of appraisal, arousal-activation, and re-appraisal

Anxiety or stress draws upon available resources

Rumination and worry use cognitive resources – self perpetuates as anxious thoughts activate other anxious thoughts (*Bower, 1981*)

Efforts to control anxiety/emotional response require resources (*Richards & Gross, 2000*)

Attention (resources) is focused narrowly around stimulus with greatest perceived importance/salience (*Easterbrook, 1959*)

Reduction in subvocalizations

Fewer resources available for encoding and rehearsal

Fewer resources available for recollective processing (*Jacoby, 1991; Kensinger & Corkin, 2003*)

Reduced WM capacity – WM requires attentional resources

Tunneling of attention - stimuli outside focus of attention are not available to explicit memory

Reduction in quality of encoding

Reduced range of cues considered

Well-learned automated tasks – no decrement because they are initiated by automatic rather than recollective processes

Poor decision making (except when well-learned heuristics are appropriate)

Task-shedding (order based on perceived import/salience)

Tasks requiring exclusive focus improve, but those requiring integration of information suffer

Sense making phase	Resource management phase	Performance phase (feedback is continuous)

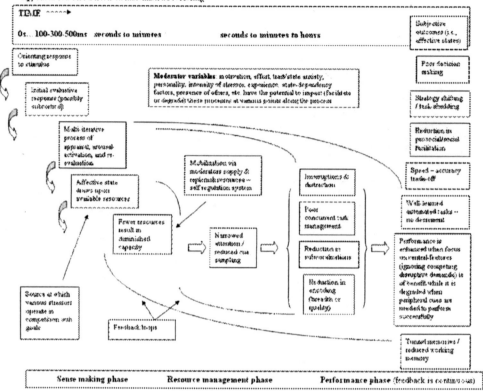

Section 1
Human Information Processing

Chapter 3

Modulators of Human Information Processing in Realistic Environments

Biocybernetic Systems:
Information Processing Challenges that Lie Ahead

Mark W. Scerbo

Old Dominion University
Norfolk, VA 23529 USA
mscerbo@odu.edu

Abstract

The idea of using psychophysiological measures to improve human performance through a tighter coupling of operators and their systems has come a long way since initial work in the 1990s. Research sponsored by DARPA's Augmented Cognition program has greatly expanded efforts to develop systems that can detect an individual's cognitive state and manipulate task parameters to overcome perceptual, attentional, and working memory bottlenecks. The goal of the present paper is to discuss several hurdles that must be overcome if the biocybernetic approach is to move from laboratory demonstrations to the field. These include efforts to validate the approach, establish the magnitude of effects, gain a better understanding of underlying constructs, and the need to distinguish between task-related and task-unrelated thoughts. Finally, there is the notion that augmentation is good. However, research that addresses human interaction with highly automated systems has shown that the benefits of automation are often accompanied by novel, unexpected, and undesirable consequences. Thus, a system that augments cognitive functioning must be evaluated within the total context it is used and its potential advantages must be weighed against its potential liabilities, both anticipated and unanticipated.

1   Introduction

In the popular television series, *Star Trek: The Next Generation*, the crew encounters an alien race called the Borg. The Borg is comprised of individuals who are part human and part machine and are connected through a collective network such that each individual contributes to and is a recipient of a shared consciousness. The Borg exist to assimilate new life forms and technology and therefore threaten any life form that wishes to retain its individuality.

The idea of cyborgs, humans that are part organic and part machine, has been popular in science fiction for years and tends to conjure up feelings of fear in readers who envision a loss of control over portions of their bodies or their minds. However, that connotation is at odds with most examples of machines that augment and benefit humans. For example, prosthetic legs restore the freedom of mobility to individuals who have had one or both legs amputated. In fact, according to Clark (2003), there may be nothing unusual at all about cyborgs. Clark argues that what makes human brains and human intelligence special is the ability to engage in complex relationships with nonbiological devices. For instance, he suggests that using a pen and paper to store information while multiplying large numbers is a common example of merging our mental abilities with external, nonbiological resources. He argues further that it is human nature to use and develop tools. Whether those tools are physically attached or embedded in us is not the distinguishing characteristic of "natural-born cyborgs". Instead, it is the ability of humans to dovetail their minds to their tools that makes them cyborgs. In fact, Clark argues that the most seamless integrations of man and machine are those that operate deep beneath the level of conscious awareness. Further, he contends that, "New waves of almost invisible, user-sensitive, semi-intelligent, knowledge-based electronics and software are perfectly posed to merge seamlessly with individual biological brains. In so doing they will ultimately blur the boundary between the user and her knowledge-rich, responsive, unconsciously operating environments" (p. 34). Thus, what could be so unnatural about biocybernetic systems that extend our cognitive abilities?

2   History

Interest in biocybernetic control of systems can be traced to the late 1970s and early 1980s. Researchers working in the area of artificial intelligence began developing adaptive aids to facilitate decisions about allocating tasks between a human and computer (Rouse, 1976). For example, Geddes (1985) and his colleagues (Rouse, Geddes, & Curry, 1987, 1988) argued that automation could be invoked using information about the current state of a system, external events, and expected operator actions. The operator's intentions could be predicted from patterns of his or her activity. Likewise, Hancock and Chignell (1987, 1988) suggested that tasks or subtasks be allocated to the human or the system based upon both current and future levels of operator workload that could be determined by deviations from an ideal state. In 1981, Gomer suggested that the distribution of responsibilities between an operator and computer could be modified in real time and that psychophysiological indices could be used to trigger those changes.

The first comprehensive discussion of adaptive automation from a human performance perspective was offered by Parasuraman, Bahri, Deaton, Morrison, and Barnes (1992). The authors discussed the benefits and problems associated with automated systems in aviation and suggested that if automation could be adapted to the operator's needs in real time, it would provide a much tighter coupling between the demands on the operator and system performance. They discussed several mechanisms for triggering changes among states of automation including the viability of several psychophysiological indices. Later, Byrne and Parasuraman (1996) presented a more detailed discussion of the viability of psychophysiological measures in adaptive systems. They argued that psychophysiological measures could provide a continuous index of the operator's mental activity that would be particularly important during periods when the operator makes few or no overt responses. Psychophysiological measures could provide additional information when coupled with behavioral measures. Further, psychophysiological measures could also reveal *which* brain networks were active or inactive.

In 1995, Pope, Bogart and Bartolome described the first biocybernetic adaptive system. Their system used an index of "task engagement" based upon ratios of EEG power bands (alpha, beta, theta, etc.). Pope and his colleagues chose to use power band ratios based on research showing that the collective activity among multiple power bands was more diagnostic in distinguishing among states of wakefulness and vigilance than any single power band alone. In their system, EEG signals are recorded from various locations on the scalp and are sent to a LabView Virtual Instrument (VI) that determines the power in each band for each recording site. The VI then calculates the engagement index used to change the operator's task between automatic and manual modes. The system recalculates the engagement index every two seconds using a moving window procedure and re-evaluates the need to change task modes with each successive update. Pope and his colleagues studied several engagement indices comprised of different power band ratios. In addition, they studied system operation under both negative and positive feedback contingencies. They argued that under negative feedback, the system should switch modes more frequently in order to maintain a stable level of engagement. By contrast, under positive feedback the system should be driven to extreme levels, remain at those levels for longer durations, and therefore switch modes less often. Pope and his colleagues examined differences in the frequency of task mode switches under each feedback contingency as a means to assess the sensitivity of the various engagement indices. They observed that the index based on the ratio of beta/(alpha +theta) was most sensitive to differences between positive and negative feedback.

Later, Freeman, Milkulka, Prinzel and Scerbo (1999) expanded upon the work of Pope et al. (1995) and studied the biocybernetic system in an adaptive context. They had operators perform compensatory tracking, resource management, and system monitoring tasks simultaneously. In their study, all tasks remained in automatic mode except the tracking task that shifted between automatic and manual modes. Like Pope et al., they examined performance under both negative and positive feedback conditions. Under negative feedback, the tracking task was switched to or maintained in automatic mode when the index reflected high engagement (i.e., an increase above a pre-established baseline level). On the other hand, the tracking task was switched to or maintained in manual mode when the index reflected low engagement (i.e., a decrease below the baseline level). The opposite schedule of task changes occurred under the positive feedback condition. Freeman and his colleagues also analyzed performance on the tracking task. They argued that if the system was effective at moderating workload, better tracking performance should be observed under negative as compared to positive feedback conditions. Their results confirmed this prediction. Regarding system operation under negative feedback, when the value of the index was high (i.e., reflecting higher engagement) the task did indeed switch to automatic mode and when the value was low (i.e., reflecting lower engagement) the task switched to manual mode. The opposite pattern of switches occurred under positive feedback. Further, they found that tracking performance improved under the negative as compared to the positive feedback condition. In subsequent studies, similar findings were found when individuals performed the task over much longer intervals and under conditions of high and low task load (see Scerbo et al., 2003).

More recently, St. John, Kobus, Morrison, and Schmorrow (2004) have described a new DARPA program aimed at enhancing an operator's effectiveness by managing the presentation of information and cognitive processing capacity through cognitive augmentation driven by psychophysiological measures. The goal of the program is to develop systems that can detect an individual's cognitive state and then manipulate task parameters to overcome four primary "bottlenecks" in cognitive performance: attention, executive functioning, sensory input, and working memory. Efforts associated with the augmented cognition program expand upon the work of Pope and his colleagues (1995). The system developed by Pope et al. relied on a single psychophysiological measure. By contrast, the augmented cognition systems use multiple measures including near infrared spectrometry (NIRS), galvanic skin response (GSR), body posture, as well as EEG. The physiological measures are integrated to form "gauges" that reflect constructs such as effort, arousal, attention, and workload. Performance thresholds are established for each gauge and are used to trigger mitigation strategies for modifying the task (e.g., switching between verbal and spatial information formats, reprioritizing or rescheduling tasks, or changing the level of display detail). At present, efforts are underway to address:

- working memory bottlenecks in a task requiring threat detection and assessment for an Aegis system operator. The object is to identify areas of the brain that are overloaded with either verbal or spatial activities and to balance the information presentation formats across brain centers.

- high workload in an automobile that combines physiological data, inputs from the automobile's sensors, and models of driver behavior. The goal is to create a system that will monitor driver workload and use that information to manage primary and secondary automotive systems.

- attentional bottlenecks in a mobile system for the dismounted soldier. The goal is to record EEG, ECG, and EDR data to support five gauges addressing stress, arousal, P300 novelty, executive load, and task engagement. The values of the gauges are then used to change the priority, timing, and presentation format of messages in a communications system.

- executive functioning in a task where the goal is to control the simultaneous activities of four unmanned aerial vehicles. The system has four gauges addressing global cognitive workload, spatial workload, verbal working memory, and executive functioning. Data from the sensors are analyzed by a neural network to detect cognitive states and to balance verbal and spatial presentation formats.

3.   Challenges

Interest in biocybernetic systems has grown dramatically over the last 20 years. Researchers have produced working examples and prototypes of systems designed moderate workload and to facilitate cognitive functioning. Despite these initial successes, many significant challenges must be overcome if this technology is to move out of the laboratory. Although there are a number of obvious challenges (e.g., improving the signal-to-noise ratio of psychophysiological measures, the need to extract useful signals from within dynamic environments, etc.), in the following sections some less obvious, but nonetheless important challenges are presented. These topics are not intended to be exhaustive, but to highlight some of the most important work that still needs to be done.

3.1   Validation

One of the first challenges for biocybernetic and augmented cognition systems is to establish the validity of the approach. Obviously, this can be achieved through systematic controlled tests and replication; however, it may also require the use of convergent and divergent tests and an examination of the magnitude of effects. Regarding convergent and divergent tests, the research by Pope et al. (1995) and Freeman et al. (1999) described above used both negative and positive feedback contingencies as a means to determine the sensitivity among candidate physiological measures. Specifically, a negative feedback relationship was used to demonstrate stable system performance and a positive feedback was used to drive system performance to extreme levels. In this regard, one can have greater confidence in the augmented cognition approach if one can consistently demonstrate that the system performs according to expectations under a variety of test conditions that include appropriate controls. On the other hand, although consistent findings are necessary to achieve construct validation they do not address the magnitude of performance effects. A growing number of researchers in the behavioral sciences and engineering communities are beginning to measure importance and success not by statistical significance, but by the magnitude of effects (Shadish, Cook, & Campbell, 2002; Snow, Reising, Barry, & Hartsock, 1999). In this regard, demonstrating that these systems work is a necessary first step; however, the next critical step will be to show that the performance gains obtained with these systems are of practical importance.

3.2   The underlying constructs

There seems to be a general consensus among researchers and developers of biocyberetic and augmented cognition systems that the constructs under consideration (e.g., workload, stress, arousal, etc.) are quantitative in nature, i.e., they only change in magnitude. If one adopts the inverted U-shape function to describe the relationship between any of these constructs and performance, one would expect to see an initial increase and then subsequent decrease in performance with increases in the constructs. However, that notion rests on two assumptions: 1) that increases in magnitude are not accompanied by any qualitative changes in the underlying construct, and 2) performance differences only vary in a unidimensional manner. There is much evidence to suggest that both of these assumptions are false. The patterns of errors observed and coping mechanisms adopted under very high and low levels of stress, workload, arousal, tend to be different. For instance, Hockey and Hamilton (1983) have argued that the effects of stress on performance cannot be described by a single underlying dimension. Instead, they are best understood by patterns or states of multiple underlying processes, some of which have opposite effects on performance with increases in the underlying construct. Further, Mathews (2001) argues that the effects of stress cannot be considered without an understanding of one's cognitive appraisal of a given situation. Thus, significant benefits of a cognitive augmentation system may come from treating high and low levels of the underlying construct differently.

On a related note, many researchers and developers of biocybernetic and augmented cognition systems have chosen to address what could be considered moderate levels of the underlying constructs. On the one hand, working with moderate levels of stress, workload, or arousal may make it easier to maintain control over aspects of internal validity through traditional techniques of experimental design. Indeed, the necessity for tight experimental controls is a requirement for establishing the validity of the biocybernetic or augmented cognition approach (see above). On the other hand, however, focusing solely on moderate levels of stress, workload, or arousal might mask the most promising benefits of this technology. That is, by pursuing a broader range of the construct including high and low levels would likely produce systems that have greater external validity. For example, the system developed by Pope et al. (1995) evolved out of the need to address extended periods of underload often experienced by commercial pilots on long duration flights. The power band ratios were conceived as a means to "re-engage" someone in a task who might otherwise drift away. It could be argued that developing augmented cognition systems that specifically address periods of high or even excessive workload might ultimately prove to be most beneficial.

3.3   Task-related and task-unrelated thoughts

Recently, Scerbo and his colleagues (in press) discussed the nature of task-related and task-unrelated thoughts in information processing tasks. They argued that a normal conscious individual has an active mind and that at any moment, his or her thoughts may or may not be related to the task at hand. Further, task-unrelated thoughts compete for attention, can lead to hazardous states of awareness, and can ultimately have a negative effect on performance. Task-unrelated thoughts become more prevalent when individuals must perform an easy task over extended periods of time. Although one's level of effort and motivation can influence task-unrelated thoughts, they are not completely under one's control and attempts to manage them can contribute to sources of workload and stress. Consequently, any serious attempt to manage tasks based upon physiological indices of cognitive activity must be able to distinguish between task-related and task-unrelated thoughts.

3.4   Reactive vs. proactive systems

Another major problem for biocybernetic and augmented cognition systems is that most are reactive in nature. That is, brain activity must be recorded and analyzed before instructions can be sent to modify the system or interface. This takes time and even with short delays, the system must still wait for changes to occur in order to react.

As noted above, early work in adaptive interfaces drew upon models of operator workload and expected operator actions to guide the allocation of tasks between the system and the user (Hancock & Chignell, 1987, 1988; Rouse, Geddes, & Curry, 1987, 1988). Subsequent efforts in developing adaptive aids were geared toward creating associate systems that could function as a junior crew member and provide the operator with the right information, in the right format, at the right time. For example, the Rotorcraft Pilot's Associate system was an intelligent, adaptive cockpit for the next generation attack helicopter (Miller & Hannen, 1999). The system was designed to detect and organize incoming data and assess internal information regarding the status of the aircraft as well as external information about targets and the status of the mission. In addition, the system could also make inferences about current and impending activities for the crew and allocate tasks among crew members as well as the aircraft, prioritize information to be presented on limited display spaces, and adjust the amount of detail to be presented in displays. Miller and Hannen argued that the key to the Rotorcraft Pilot's Associate was its ability to *infer* operator intentions and communicate about them. Consequently, their experiences underscore the need for adaptive systems to be proactive.

Recently, Forsythe (in press) described an augmented cognition system being developed by DaimlerChrysler to support driver behavior that incorporates a cognitive model of the operator. Information obtained from the automobile (e.g., steering wheel angle, lateral acceleration) is combined with measures of operator behavior (e.g., head turning, postural adjustments, and vocalizations) and psychophysiological measures to generate inferences about workload levels corresponding to different driving situations. Thus, by incorporating models of operator behavior and/or intentions, the system can be more proactive than current biocybernetic systems that rely solely on psychophysiological measures.

4.   What good is augmentation?

Researchers and developers of cognitive augmentation systems seem to be operating on the premise that augmentation is good. However, that may not necessarily be true. In fact, one could argue that augmentation is just different. Consider the simple example of using binoculars that are used to "augment" vision. Binoculars augment vision by distorting the field of view such that distant objects appear closer and the relative distances among objects appear compressed. The advantage of having a better view of more distant objects is accompanied by the disadvantage of having a narrower field of view and increased difficulty scanning the visual field.

The same would be true of any system designed to augment cognitive functioning. The benefits for certain activities will undoubtedly come at the expense of others. For instance, consider the strategy of balancing information processing demands by switching between verbal and spatial presentation formats. If one assumes that this strategy does indeed lower overall workload for the task at hand, what is the impact on situation awareness or on other tasks that must be performed? Recently, Risser and his colleagues (Risser, Scerbo, Baldwin, & McNamara, 2003; Scerbo, Risser, Baldwin, & McNamara, 2003) studied the effects of different information presentation formats on an air traffic control communications task. They found that merely changing the message format from radio (speech) to data link (text) rendered performance susceptible to sources of interference from other tasks performed simultaneously *or* subsequently to the primary task.

Further, recent research addressing the nature of human interaction with automation has shown that the benefits of highly automated systems are often accompanied by novel, unexpected, and undesirable consequences (Woods, 1996). Adaptive interfaces are no exception. They can increase the workload associated with maintaining "system mode" awareness and often introduce a higher incidence of mode errors. Moreover, when they fail they do so in new ways and the source of errors is often difficult to detect. Ultimately, a system that augments cognitive functioning must be evaluated in the total context in which it is used and its potential advantages must be weighed against its potential liabilities, both anticipated and unanticipated.

5.   Conclusions

The evolution of biocybernetic systems that augment cognitive abilities is still in its infancy. The challenges outlined above are not exhaustive, but are a representation of technological hurdles that must be overcome for this technology to mature. Perhaps, in 20 years biocybernetic, augmented cognitive systems will be so common that they will be perceived as perfectly natural like Clark (2003) suggests.

However, lessons learned from other forms of complex technology suggest that the benefits cannot be championed without serious consideration of potential consequences. Adaptive automation systems require that users give up some degree of control (see Scerbo, in press). We would be well advised to fully understand the effects of each developmental step in the evolution of biocybernetic systems that usurp greater user authority, particularly those that adapt to our thoughts and intentions. It is quite possible that we may end up with systems that are a natural extension of our minds as Clark suggests. On the other hand, we might also find that we are unable to prevent ourselves from being permanently and irrevocably assimilated into the collective mind of some future Borg.

6.   References

Byrne, E.A., & Parasuraman, R. (1996). Psychophysiology and adaptive automation. *Biological Psychology, 42*, 249-268.

Clark, A. (2003). *Natural-Born Cyborgs: Minds, Technologies and the Future of Human Intelligence.* Oxford: University Press.

Freeman, F.G., Mikulka, P.J., Prinzel, L.J., & Scerbo, M.W. (1999). Evaluation of an adaptive automation system using three EEG indices with a visual tracking task. *Biological Psychology, 50,* 61-76.

Forsythe, C. (in press). Augmented cognition. In C. Forsythe (Ed.), *Cognitive systems: Human cognitive models in system design.*

Geddes, N.D. (1985). Intent inferencing using scripts and plans. In *Proceedings of the First Annual Aerospace Applications of Artificial Intelligence Conference* (pp.160-172). Wright-Patterson Air Force Base, OH: U.S. Air Force.

Gomer, F. E. (1981). Physiological monitoring and the concept of adaptive systems. In J. Moraal & K. F. Kraiss (Eds.), *Manned systems design: Methods, equipment, and applications* (pp. 271-287). New York: Plenum.

Hancock, P.A., & Chignell, M.H. (1987). Adaptive control in human-machine systems. In P.A. Hancock (Ed.), *Human factors psychology* (pp. 305-345). North Holland: Elsevier Science Publishers.

Hancock, P.A., & Chignell, M.H. (1988). Mental workload dynamics in adaptive interface design. *IEEE Transactions on Systems, Man, and Cybernetics, 18*, 647-658.

Hockey, R., & Hamilton, P. (1983). The cognitive patterning of stress states. In G. R. J. Hockey (Ed.), *Stress and fatigue in human performance* (pp. 331-362). Chichester: Wiley.

Matthews, G. (2001). Levels of transaction: A cognitive science framework for operator stress. In P.A. Hancock & P.A. Desmond (Eds.), *Stress, workload, and fatigue* (pp. 5-33). Mahwah, NJ: Erlbaum.

Miller, C. A., & Hannen, M. D. (1999). The Rotorcraft Pilot's Associate: design and evaluation of an intelligent user interface for cockpit information management. *Knowledge-Based Systems, 12*, 443-456.

Parasuraman, R., Bahri, T., Deaton, J.E., Morrison, J.G., & Barnes, M. (1992). *Theory and design of adaptive automation in aviation systems* (Progress Report No. NAWCADWAR-92033-60). Warminster, PA: Naval Air Warfare Center, Aircraft Division.

Pope, A. T., Bogart, E. H., & Bartolome, D. (1995). Biocybernetic system evaluates indices of operator engagement. *Biological Psychology, 40*, 187-196.

Risser, M.R., Scerbo, M.W., Baldwin, C.L., & McNamara, D. S. (2003). ATC Commands executed in speech and text formats: Effects of task interference. *Proceedings of the Twelfth International Symposium on Aviation Psychology* (pp. 999-1004), Wright State University.

Rouse, W.B. (1976). Adaptive allocation of decision making responsibility between supervisor and computer. In T.B. Sheridan & G. Johannsen (Eds.), *Monitoring behavior and supervisory control* (pp. 295-306). New York: Plenum Press.

Rouse, W.B., Geddes, N.D., & Curry, R.E. (1987, 1988). An architecture for intelligent interfaces: Outline of an approach to supporting operators of complex systems. *Human-Computer Interaction, 3*, 87-122.

Scerbo, M. W. (in press). Adaptive automation. In R. Parasuraman & M. Rizzo (Eds.), *Neuroergonmincs: The brain at work.* Oxford: Oxford University Press.

Scerbo, M. W., Bliss, J. P., Freeman, F.G., Mikulka, P.J., & Schultz Robinson, S. (in press). Measuring task related and unrelated thoughts. In D. K. McBride (Ed.), *Quantifying human information processing (QHIP) study report.* Lanham, MD: Lexington Books.

Scerbo, M. W., Freeman, F. G., & Mikulka, P. J. (2003). A brain-based system for adaptive automation. *Theoretical Issues in Ergonomic Science, 4,* 200-219.

Scerbo, M.W., Risser, M. R., Baldwin, C. L., & McNamara, D.S. (2003). The Effects of Task Interference and Message Length on Implementing Speech and Simulated Data Link Commands. *Proceedings of the Human Factors & Ergonomics Society 47th Annual Meeting* (pp. 95-99). Santa Monica, CA: Human Factors & Ergonomics Society.

Shadish, W. R., Cook, T. D., & Campbell, D. T. (2002). *Experimental and quasi-experimental designs for generalized causal inference.* Boston: Houghton-Mifflin.

Snow, M. P., Reising, J., Barry, T. P., & Hartsock, D. C. (1999). Comparing new designs with baselines. *Ergonomics in Design, 7,* (4), 28-33.

Woods, D. D. (1996). Decomposing automation: Apparent simplicity, real complexity. In R. Parasuraman & M. Mouloua (Eds.), *Automation and human performance: Theory and applications* (pp. 3-17). Mawhaw, NJ: Lawrence Erlbaum Associates.

Using Modes of Cardiac Autonomic Control to Assess Demands upon Processing Resources During Driving

Richard W. Backs, Jonathon Shelley, and John K. Lenneman
Department of Psychology
Central Michigan University
Mount Pleasant, MI 48859
backs1rw@cmich.edu

Abstract

We have used a bivariate model of autonomic nervous system (ANS) innervation of cardiac activity to successfully differentiate demands upon perceptual/central and manual response attention processing resources during single and dual task manual tracking in laboratory studies. Recently, we have attempted to apply the model to the driving domain in simulated driving studies. In the following sections of the paper we will discuss the physiological model, how it has been tested in laboratory studies, how it is been used in driving simulation, and the challenges of extending it to on-road driving.

1   The Autonomic Space Model

Bernston, Cacioppo, and Quigley (1991; 1993) have developed a model of autonomic determinism that states that the parasympathetic (PNS) and sympathetic (SNS) branches of the ANS have multiple modes of control over effector organs such as the heart (Figure 1). In the classic view of autonomic function, the PNS an SNS are reciprocally-coupled and respond to physiological and psychological demands in an antagonistic fashion, where an increase in one branch occurs concomitantly with a decrease in the other branch. As shown in Table 1, the PNS and SNS may be reciprocally-coupled as in the classic model, and also non-reciprocally-coupled (where the two branches increase or decrease together) and uncoupled (where each branch may increase or decrease without change in the other) in the Berntson et al. model. Table 1 presents the eight different modes of autonomic control for heart period. Heart period (i.e., the time in ms between successive heart beats) is preferred over heart rate (i.e., the number of heart beats in a minute) because of its superior biometric properties (Berntson, Cacioppo, & Quigley, 1995). Faster heart rate corresponds to shorter heart period, and slower heart rate corresponds to longer heart period.

Table 1: Modes of Cardiac Control for Heart Period (after Berntson et al., 1991)

Control Mode	Sympathetic Response	Parasympathetic Response	Heart Period Response
Reciprocally-Coupled Modes			
Sympathetic Activation/Parasympathetic Inhibition	Increase	Decrease	Shorter
Parasympathetic Activation/Sympathetic Inhibition	Decrease	Increase	Longer
Nonreciprocally-Coupled Modes			
Coactivation	Increase	Increase	Shorter, Longer, No Change
Coinhibition	Decrease	Decrease	Shorter, Longer, No Change
Uncoupled Modes			
Sympathetic Activation	Increase	---	Shorter
Sympathetic Inhibition	Decrease	---	Longer
Parasympathetic Activation	----	Increase	Longer
Parasympathetic Inhibition	----	Decrease	Shorter

Although most physiological demands (such as exercise) elicit reciprocally-coupled modes of autonomic control, task demands upon information processing resources are thought to elicit multiple modes of control that vary with the type of processing (Backs, 2001). Therefore, an end organ response such as heart period will be relatively uninformative about the precise nature of the task demands because heart period may shorten (e.g., due to reciprocally-coupled sympathetic activation/parasympathetic inhibition, uncoupled sympathetic activation, or uncoupled parasympathetic inhibition), lengthen (e.g., due to reciprocally-coupled parasympathetic activation/sympathetic inhibition, uncoupled sympathetic inhibition, or uncoupled parasympathetic activation), or even not change at all (e.g., due to coactivation or coinhibition) depending upon the underlying mode of autonomic control elicited by the demand. Further, because the same heart period response can be elicited by multiple modes of autonomic control (see in Figure 1), heart period alone is uninformative about the psychological processes responsible for the response. Knowledge of the autonomic mode of control will provide both greater sensitivity and diagnosticity for task demands upon psychological processes than heart period or any other ANS response. For example, sensitivity will be improved because for non-reciprocally coupled modes of control the PNS and SNS activity may not result in a change in heart period. Diagnosticity of information processing resources will be improved because there are eight modes of autonomic control which may have a more direct linkage to the brain structures that are involved in information processing.

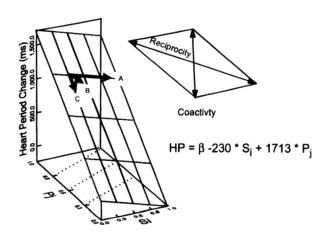

$$HP = \beta -230 * S_i + 1713 * P_j$$

Figure 1: Human autonomic space for heart period obtained from the equation in the figure. P_j and S_i are, respectively, the normalized activations of parasympathetic and sympathetic branch inputs to the heart at point ij. P_j and S_i are scaled to be proportional to the dynamic range for each branch. Heart period is shown as the relative change from baseline (β) in ms. Uncoupled control modes are represented by change along a single axis, whereas reciprocal and coactivity control modes are represented by change along a directional vector as indicated by the on the P by S plane on the right. Dotted lines on the P by S plane on the left represent isoresponse contours projected from the surface to the autonomic plane. Heart period does not change along an isoresponse contour even though autonomic input changes. Example autonomic control mode vectors for uncoupled sympathetic activation (A), reciprocally-coupled sympathetic activation and parasympathetic inhibition (B), and uncoupled parasympathetic inhibition (C). All vectors were anchored to an arbitrary point in the space, and the length of the vectors was computed using the equation so that each vector represents the same 100 ms decrease in heart period from the anchor point. (Adapted with permission from "Cardiac psychophysiology and autonomic space in humans: Empirical perspectives and conceptual implications" by G. G. Berntson, J. T. Cacioppo, & K. S. Quigley, 1993, *Psychological Bulletin, 114*, 296-322. Copyright 1993 by the American Psychological Association, Inc.)

2   Laboratory Studies

We use electro- and impedance cardiography to obtain noninvasive indices of the underlying parasympathetic and sympathetic ANS innervation of the heart in our studies. The parasympathetic index is respiratory sinus arrhythmia (RSA) measured using high frequency heart rate variability (Berntson et al., 1997), and the sympathetic index is pre-ejection period (PEP; Sherwood et al., 1990). RSA and PEP change from resting baseline conditions have been validated as indices of PNS and SNS innervation of the heart using pharmacological manipulations (Berntson et al., 1994; Cacioppo et al., 1994). We typically standardize RSA and PEP changes from baseline within a study by converting the difference scores to standard deviation units so that changes in autonomic space can be examined using a common metric for RSA and PEP. How manipulations of task demands elicit change in autonomic space can then be plotted on the P_j by S_i plane as in Figure 2.

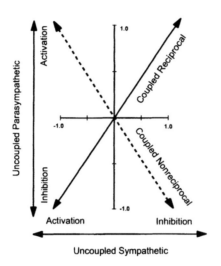

Figure 2: P by S plane used in our research to represent changes in autonomic space from resting baseline that are elicited by task demand manipulations. Parasympathetic innervation measured by RSA is represented on the ordinate, sympathetic innervation measured by PEP on the abscissa, and the origin represents baseline. Vectors on the positive diagonal indicate coupled reciprocal modes of autonomic control, whereas vectors on the negative diagonal indicate coupled nonreciprocal modes of control. Vectors parallel to one axis indicate uncoupled modes of control. Figure modified from G. G. Berntson, J. T. Cacioppo, & K. S. Quigley (1991). Autonomic determinism: The modes of autonomic control, the doctrine of autonomic space, and the laws of autonomic constraint, *Psychological Review, 98,* 459-487 with permission from the authors.

We measured RSA and PEP in a series of laboratory studies that manipulated the perceptual/central, and manual response processing (Wickens, 1991) demands of continuous visual-manual compensatory tracking tasks. In some studies, the tracking task was performed alone (e.g., Lenneman & Backs, 2000), whereas in others the tracking task was performed as part of a dual task in combination with other tasks such as visual and auditory monitoring (e.g., Backs, Rohdy, & Barnard, 2005). The basic findings of these studies for cardiac autonomic modes of control can be summarized as follows. Task demands upon manual processing resources such as increasing the bandwidth of the disturbance to be nulled elicit an uncoupled parasympathetic inhibition mode of control (lower panel of Figure 3). Demands upon perceptual (such as the increasing the number of items to be monitored) or manual (such as adding the tracking task to a dual task) processing resources also elicit an uncoupled parasympathetic inhibition mode of control (Figure 4). Task demands upon perceptual/central processing resources such as going from velocity to acceleration order-of-control in tracking (upper panel of Figure 3) or for perceptual monitoring (Figure 4) elicit a reciprocally-coupled sympathetic activation and parasympathetic inhibition mode of control.

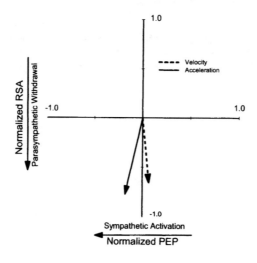

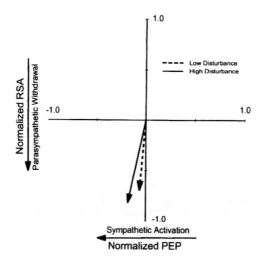

Figure 3: Upper panel: velocity tracking elicits uncoupled parasympathetic inhibition, whereas going from velocity to acceleration order-of-control elicits reciprocally-coupled sympathetic activation and parasympathetic inhibition. Lower panel: going from tracking low to high disturbance bandwidth perturbations elicits more intense uncoupled parasympathetic inhibition. Reproduced with permission from Lenneman, J. K., & Backs, R. W. (2000). The validity of factor analytically derived cardiac autonomic components for mental workload assessment. In Backs, R. W., & Boucsein, W. (Eds.), *Engineering Psychophysiology: Issues and Applications* (pp. 161-174). Mahwah, NJ: Lawrence Erlbaum.

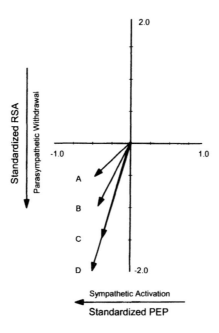

Figure 4: Each vector is a monitoring task: A) low difficulty single-task; B) high difficulty single-task; C) low difficulty dual-task; D) high difficulty dual-task. The autonomic mode of control for the monitoring task is reciprocally-coupled sympathetic activation and parasympathetic inhibition. The mode of control for increasing the perceptual demands of the monitoring task (compare A and B) is uncoupled parasympathetic inhibition. Adding the perceptual/manual demands of the tracking task to the monitoring task (compare A to C and B to D) also elicits uncoupled parasympathetic inhibition. From Backs, R. W., Rohdy, J., & Barnard, J. (2005). Cardiac control during dual-task performance of visual or auditory monitoring with visual-manual tracking. Manuscript submitted for publication.

3   Driving Simulation Studies

We wanted to see if the autonomic control modes for task demand upon attention processing resources identified from RSA and PEP in the laboratory studies generalized to the driving domain. Our initial driving simulation studies have examined driving only (Backs, Lenneman, Wetzel, & Green, 2003) and driving while using a simulated automated telematic device (Wetzel, Sheffert, & Backs, 2004). Data from Backs et al. (2003) are reproduced in Figure 5 showing the modes of autonomic control while drivers negotiated curves that varied in degree of curvature. The participants drove a course through a series of straightaways and curves at a fixed (cruise controlled) speed of 72.4 km/hr. Curves of three radii were driven, 582, 291 and 194 m (3, 6, and 9 degrees-of-curvature, respectively), and each curve lasted for 30 s so impedance cardiographic data could be obtained. The driving environment depicted in the simulation was a two-lane winding road with no traffic ahead and stationary oncoming cars, traffic signs, and road edge posts. The perceptual-manual demands of simulated driving elicited uncoupled parasympathetic inhibition in the 582 m and 194 m curves, but reciprocally coupled sympathetic activation and parasympathetic withdrawal in the 291 m curve.

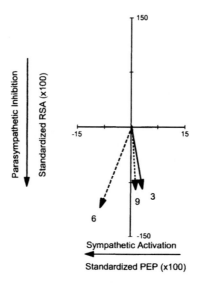

Figure 5: Autonomic space for driving through curves with 3, 6, and 9 degrees-of-curvature. The mode of autonomic control was uncoupled parasympathetic inhibition for the 3 and 9 degree curves, whereas it was reciprocally-coupled sympathetic activation and parasympathetic inhibition for the 6 degree curve. Reproduced with permission from Backs, R. W., Lenneman, J. K., Wetzel, J. M., & Green, P. (2003). Cardiac measures of driver workload during simulated driving with and without visual occlusion. *Human Factors, 45,* 525-538.

4 Modulators of Impedance Cardiographic Assessment of Information Processing During On-Road Driving

To take the next step and test whether changes in information processing demands during on-road driving elicit the cardiac modes of control identified in the laboratory will require an ambulatory impedance cardiograph that can produce the same quality data as can be obtained in the laboratory and driving simulator. The development of ambulatory instruments has rapidly advanced in the last few years and new instruments are under development that show much promise for achieving this goal. Listed below are some technological, physiological, psychological, and environmental modulators that we believe will be important for moving impedance cardiography from the laboratory to on-road driving (see also Fahrenberg & Wientjes, 2000). Not all modulators fit neatly into a single category, and many modulators will interact across categories, but we believe that this taxonomy is a useful way to think about the potential constraints upon using impedance cardiography to make inferences about information processing during on-road driving.

4.1   Technological Modulators

Technological factors are perhaps the easiest to address, and new instruments (e.g., Mindware Technologies, Inc.) have adequately addressed many of these (or soon will). Self-contained systems that offer the greatest personal freedom (as opposed to a driver in an instrumented research vehicle) will need to incorporate the physiological transducer and the bioamplification and filtering, data acquisition, and data storage stages. If the system is needed to do more than merely collect the electro- and impedance cardiograms for later off-line analyses, then data processing, including artifact handling, and output of appropriate cardiovascular parameters will also need to be incorporated into the system. Factors such as processing speed, data storage, and power requirements of a personal system are less of a concern as newer systems are built around powerful personal digital assistants (PDA). The amplification and acquisition stages can be based on a PDA flash card to have a system that is small, lightweight, and have a fairly long life with an external battery. PDA-based systems are also durable and reliable and are easy to synchronize with events of interest or other devices, especially if they are capable of wireless communication. One limitation that remains to be addressed is at the transducer stage: current ambulatory electro- and impedance cardiographs still rely on traditional band or spot electrodes attached to the driver. Future systems may use accelerometers or vibrometers built into the vehicle seat to detect cardiac responses, but of course these will introduce a new set of problems in isolating the biometric information from the background environment of the vehicle.

4.2   Physiological Modulators

The first concern with an ambulatory system is whether the cardiovascular parameters derived with the system are valid and reliable compared to the same parameters obtained in the laboratory. Fortunately, recent research indicates that the parameters are valid and reliable under both controlled (Nakonezny et al., 2001) and field (Riese et al., 2003) conditions. Ambulatory systems for use during on-road driving will need to handle a wide range of driver characteristics. For example, individual differences in body composition, physical fitness, age, and cardiovascular health status affect the quality of the impedance cardiogram and the ease with which the relevant parameters such as PEP can be extracted. Smoking and ingestion of prescription or over-the-counter medications, food, caffeine, and alcohol all have transient effects upon the cardiovascular system that may last for hours (Pickering, 1989). Other factors that also affect the physiological background against which the information processing demand of driving will be extracted are changes in posture and the circadian and ultradian rhythms inherent in most cardiovascular parameters. Finally, information processing demands are typically evaluated compared to baseline conditions in the laboratory and driving simulator, but it is not at all clear what the appropriate baseline should be (e.g., resting, simulator, or other driving conditions) during on-road driving.

4.3   Psychological Modulators

Drivers must accept and trust the ambulatory impedance system and comply with any additional task demands that it may impose (Fahrenberg & Wientjes, 2000). For example, drivers will have to accept wearing electrodes or other transducers for an extended period, and trust that the ambulatory system will function in the way that they have been instructed that it will. Drivers may have to comply with the need to activate the system at certain times or to enter event information, or perhaps even to troubleshoot the system if it malfunctions. Further, reactivity to the system is also a concern; drivers will clearly know that they are being monitored and may behave differently as a result. For example, drivers may avoid situations or engaging in certain behaviors (e.g., risky driving) that they believe would reflect poorly upon themselves.

The extent to which laboratory and simulator findings for the mappings between task demands upon information processing resources generalize to on-road driving may be limited by numerous other psychological modulators. The richness and unpredictability of the driving environment may influence perceptual and emotional processes in ways that are difficult to investigate in simulated driving compared to cognitive processes. For example, the feedback that the driver gets from the complex interaction of visual, auditory, tactile, and egomotion cues during driving is difficult to simulate even with the highest fidelity motion-based simulators. Perhaps of greater concern is that the effect of emotional stress on information processing during driving cannot be adequately understood in the laboratory, because although the risks in driving can be simulated (e.g., unexpected vehicle incursion), the consequences associated with those risks cannot. Similarly, social interactions with passengers or during a cellular telephone conversation may affect the ability to detect information processing changes during driving. Although talking does not greatly interfere with the ability to extract relevant cardiovascular parameters from the impedance cardiogram, the emotional content of the social interaction and the potential distraction caused by conversation may increase the difficulty of isolating the information processing effects upon the parameters. Other psychological states such as fatigue and boredom may also interact with information processing effects upon the cardiovascular parameters, and to our knowledge these have not been examined in driving simulation. Finally, like with the physiological modulators, individual differences are undoubtedly important: drivers vary widely in experience and ability, age, and sensory and perceptual efficiency that affect their information processing capacities.

4.4 Environmental Modulators

The potential impact of numerous modulators associated with the external environment and with the internal environment of the vehicle upon the ability to use impedance cardiography to assess information processing during driving is generally unknown. External factors such as weather and traffic can be simulated, but with only limited fidelity. As mentioned above, although the perceptual aspects of a factor such as poor weather can be reproduced in the simulator, the feedback associated with driving on a slippery road is more difficult to reproduce and the risks and consequences associated with driving in poor weather are generally absent. Also, the random variability of simulated traffic may not be a veridical representation of human behavior. Finally, internal factors such as ambient temperature, auditory noise, and vibration would generally not be concerns in passenger vehicles, but may be challenges to using impedance cardiography in commercial or military vehicles where augmented cognition systems may be most useful.

5 Conclusions

We believe that knowledge of the modes of autonomic control of the heart may be useful for augmented cognition applications in surface transportation vehicles in several ways. For example, Backs, Lenneman, and Sicard (1999) have proposed a tentative mental workload hierarchy for flight simulation based upon the autonomic control modes for the heart: coupled reciprocal control during low workload; uncoupled sympathetic activation with high but manageable workload; and coactivation to situations having critical consequences that must be immediately addressed. Our initial driving simulation study suggested that this tentative hierarchy may need modification for extension to simulated driving, where uncoupled parasympathetic inhibition may be a lower level of the hierarchy, one that was not observed during simulated flight (Backs et al., 2003). It may also be that our approach of using autonomic control modes of the heart to make inferences about information processing resources can be extended to provide information about the emotional state of the driver that may not be evident from other classes of measures of human performance.

6 References

Backs, R. W. (2001). An autonomic space approach to the psychophysiological assessment of mental workload. In P. A. Hancock, & P. A. Desmond (Eds.), *Stress Workload, and Fatigue* (pp.279-289). Mahwah, NJ: Lawrence Erlbaum and Associates.

Backs, R. W., Lenneman, J. K., & Sicard, J. L. (1999). The use of autonomic components to improve cardiovascular assessment of mental workload in flight simulation. *The International Journal of Aviation Psychology, 9*, 33-47.

Backs, R. W., Lenneman, J. K., Wetzel, J. M., & Green, P. (2003). Cardiac measures of driver workload during simulated driving with and without visual occlusion. *Human Factors, 45*, 525-538.

Backs, R. W., Rohdy, J., & Barnard, J. (2005). Cardiac control during dual-task performance of visual or auditory monitoring with visual-manual tracking. Manuscript submitted for publication.

Berntson, G. G., Bigger, J. T., Jr., Eckberg, D. L., Grossman, P., Kaufman, P. G., Malik, M., Nagaraja, H. N., Porges, S. W., Saul, J. P., Stone, P. H., & van der Molen, M. W. (1997). Heart rate variability: Origins, methods, and interpretive caveats. *Psychophysiology, 34*, 623-648.

Berntson, G. G., Cacioppo, J. T. & Quigley, K. S. (1991). Autonomic determinism: The modes of autonomic control, the doctrine of autonomic space, and the laws of autonomic constraint. *Psychological Review, 98*, 459-487.

Berntson, G. G., Cacioppo, J. T. & Quigley, K. S. (1993). Cardiac psychophysiology and autonomic space in humans: Empirical perspectives and conceptual implications. *Psychological Bulletin, 114*, 296-322.

Berntson, G.G., Cacioppo, J.T., Quigley, K.S. (1995). The metrics of cardiac chronotropism: Biometric perspectives. *Psychophysiology, 32*, 162-171.

Berntson, G.G., Cacioppo, J.T., Binkley, P.F., Uchino, B.N., Quigley, K.S., Fieldstone, A. (1994). Autonomic cardiac control. III. Psychological stress and cardiac response in autonomic space as revealed by pharmacological blockades. *Psychophysiology, 31*, 599-608.

Cacioppo, J. T., Berntson, G.G., Blinkley, P.F., Quigley. K.S., Uchino, B. N., & Fieldstone, A. (1994). Autonomic cardiac control. II. Noninvasive indices and basal response as revealed by autonomic blockade. *Psychophysiology, 31*, 586-598.

Fahrenberg, J., & Wientjes, C. J. E. (2000). Recording methods in applied environments. In R.W. Backs & W. Boucsein (Eds.), *Engineering Psychophysiology: Issues and Applications* (pp. 111-136). Mahwah, NJ: Erlbaum.

Lenneman, J. K., & Backs, R. W. (2000). The validity of factor analytically derived cardiac autonomic components for mental workload assessment. In Backs, R. W., & Boucsein, W. (Eds.), *Engineering Psychophysiology: Issues and Applications* (pp. 161-174). Mahwah, NJ: Lawrence Erlbaum.

Mindware Technologies, Inc. URL http://www.mindwaretech.com/products_Imp.htm

Nakonezny, P. A., Kowalewski, R. B., Ernst, J. M., Hawkley, L. C., Lozano, D. L., Litvack, D. A., Berntson, G. G., Sollers, III, J. L., Kizakevich, P., Cacioppo, J. T., & Lovallo, W. R. (2001). New ambulatory cardiograph validated against the Minnesota impedance cardiograph. *Psychophysiology, 38*, 465-473.

Pickering, T. G. (1989). Ambulatory monitoring: Applications and limitations. In N. Schneiderman, S. M. Weiss, & P. G. Kaufman (Eds.), *Handbook of Research Methods in Cardiovascular Behavioral Medicine* (pp. 261-291). New York: Plenum Press.

Riese, H., Groot, P. F. C., van den Berg, M., Kupper, N. H. M., Magnee, E. H. B., Rohaan, E. J., Vrijkotte, T. G. M., Willemsen, G., & de Geus, E. J. C. (2003). Large-scale ensemble averaging of ambulatory impedance cardiograms. *Behavior Research Methods, Instruments, & Computers, 35*, 467-477.

Sherwood, A., Allen, M. T., Fahrenberg, J., Kelsey, R. M., Lovallo, W. R., & van Dooran, L. J. P. (1990). Methodological guidelines for impedance cardiography. *Psychophysiology, 27*, 1-23.

Wetzel, J. M., Sheffert, S. M., & Backs, R. W. (2004). Driver trust, annoyance, and acceptance of an automated calendar system. *Proceedings of the Human Factors and Ergonomics Society 48th Annual Meeting* (pp. 2335-2339). Santa Monica, CA: Human Factors and Ergonomics Society.

Wickens, C. D. (1991). Processing resources and attention. In D. L. Damos (Ed.), *Multiple-task performance* (pp. 3-34). London: Taylor & Francis.

Sensor-Based Cognitive State Assessment in a Mobile Environment

Santosh Mathan, Natalia Mazaeva,
Stephen Whitlow

Honeywell Laboratories
3660 Technology Dr,
Minneapolis, MN 55418

Andre Adami, Deniz Erdogmus, Tian Lan,
Misha Pavel

Bio Medical Engineering Department
Oregon Health Sciences University,
Beaverton, OR 97006

Abstract

Inferring cognitive state from non invasive neurophysiological sensors is a challenging task even in pristine laboratory environments. Artifacts ranging from eye blinks, to muscle artifacts and electrical line noise can mask electrical signals associated with cognitive functions. These concerns are particularly pronounced in the context of the Honeywell team's ongoing efforts to realize neurophysiologically driven adaptive automation for the dismounted ambulatory soldier. Besides the typical sources of signal contamination, the Honeywell team has to deal with the effects of artifacts induced by shock, rubbing cables and gross muscle movement. This paper presents the Honeywell team's efforts to make reliable sensor based cognitive state assessments given the constraints just cited. Cognitive state classification results suggest that it is feasible to classify cognitive state in ambulatory, military-relevant task contexts.

1 Introduction

DARPA's (Defense Advanced Research Projects Agency) Augmented Cognition program is an effort aimed at tailoring computer based assistance to a user's cognitive state. Technologies that have matured under this program promise to foster a fundamentally new type of human computer interaction in complex task domains. Currently, computer based assistance in many challenging contexts takes the form of rigidly automated systems (Sarter, Woods, & Billings, 1997). These systems largely relegate the human to the role of a passive observer and occasionally force the human to take over in extremely demanding task conditions. Computer based assistance conceived by the AugCog program is of a more compliant nature, where the human is actively engaged in the task at all times. Automation merely serves to help users cope with the most difficult of circumstances (c.f. Norman, 1990). This type of mixed initiative interaction, offers the promise of realizing the best attributes of both humans and machines in the service of performing complex tasks.

This paper describes Honeywell's efforts in conjunction with the Augmented Cognition program. The Honeywell Augmented Cognition team focuses on the dismounted Future Force Warrior (FFW). FFW is a component of the US Army's Advanced Technology Development (ATD) program. A critical element of the FFW program is a reliance on networked communications and high density information exchange. Such an infrastructure is expected to increase situational awareness at every level of the operational hierarchy. It is hoped that an information technology based transformation of the military will facilitate better individual and collaborative decision making at every level. However, effective decision making on the basis of broad access to mission relevant information is constrained by the limits of the human information processing system. The goal of the Honeywell Augmented Cognition team is to use physiological and neurophysiological sensors to detect occasions when cognitive resources may be inadequate to cope with mission relevant demands. Efforts of the Honeywell team focus on ways to leverage automation to effectively manage information under difficult task conditions. To the best of our knowledge, the Honeywell team's efforts represent one of the first attempts to create a wearable cognitive state classification system in the context of a fully ambulatory individual. A major factor limiting such applications is the potential for artifacts induced by gross body movements to overwhelm task related neurophysiological signals.

Realizing the vision of the AugCog program in the context of an ambulatory soldier is constrained by several challenges. First, as Schmorrow and Kruse (2002) have noted, processing and analysis of neurophysiological data is largely conducted off-line by researchers and practitioners. However, in order for Augmented Cognition technologies to work in practical settings, effective and computationally efficient artifact reduction and signal processing solutions are necessary. Second, inferring the cognitive state of users demands pattern recognition solutions that are robust to noise and the inherent non stationarity in neurophysiological signals. Third, it requires the development of means to collect reliable neurophysiological data outside the laboratory. Hence, compact and robust form factors associated with neurophysiological sensors and processors are a matter of critical concern. Users should be able to move around freely.

In the following pages we describe a system designed to facilitate cognitive state classification in mobile environments. We describe a hardware configuration that allows neurophysiological data to be collected and processed in a body-worn wireless platform. We provide an overview of software components used for signal processing and artifact reduction. We highlight our classification approach. Additionally, we present results that show it is feasible to discriminate among workload levels on the basis of neurophysiological sensors in ambulatory contexts.

2   Hardware Configuration

The wireless sensor suite employed by Honeywell is assembled using a variety of off-the-shelf components. EEG data is collected using the BioSemi Active Two system using a 32 channel EEG cap and eye electrodes. This system integrates an amplifier with an Ag-AgCl electrode – this affords extremely low noise measurements without any skin preparation. The system also incorporates a wearable Arousal Meter. The Arousal Meter, developed by Clemson University, senses a subject's ECG signals and outputs inter-beat interval data in conjunction with a derived measure of a subject's cognitive arousal. Information regarding physical context is obtained using a combination of a Dead Reckoning Module (DRM) manufactured by Point Park Research and an Inertia Cube manufactured by InterSense. The DRM unit is a self contained navigation component that fuses information from several internal sensors to determine displacement from a specific geographical position. The internal sensors consist of a thermometer, barometer, magnetometer, accelerometer, gyroscope, and GPS receiver. The system is specifically designed to work with intermittent GPS signals. The InertiaCube provides information about head orientation about the head's pitch, roll, and yaw axes.

Figure 0: Body worn sensor suite and signal processing system

Information from the sensors described above is processed on a body worn laptop. The sensors are connected to the laptop via a combination USB, serial port and Bluetooth interfaces. The sensor electronics and the laptop are mounted in a backpack worn by the subject (Figure 1). Sensor data is collected and processed on the laptop computer during the experiment. A base station computer controls the experiment and communicates with the body-worn laptop computer via an 802.11 wireless network.

3   Signal Processing Software

The cognitive state classification efforts reported here rely primarily on EEG data. As mentioned earlier, the sensor monitoring equipment consists of a BioSemi Active Two EEG system with 32 electrodes. Vertical and horizontal eye movements and blinks are recorded with electrodes below and lateral to the left eye. All channels reference the right mastoid. EEG is sampled at 256Hz from 7 channels (CZ, P3, P4, PZ, O2, P04, F7) while the subject is performing tasks. These sites were selected based on a saliency analysis on EEG collected from various subjects performing cognitive test battery tasks (Russell & Gustafson, 2001). EEG signals are pre-processed to remove eye blinks using an adaptive linear filter based on the Widrow-Hoff training rule (Widrow & Hoff, 1960). Information from the VEOGLB ocular reference channel was used as the noise reference source for the adaptive ocular filter. DC drifts were removed using high pass

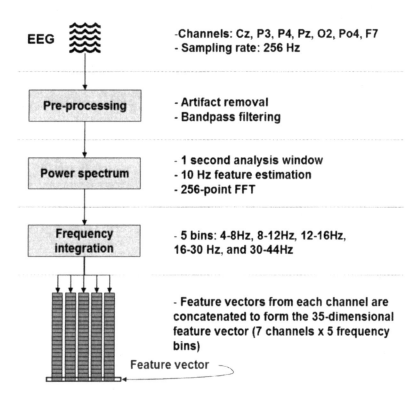

EEG
-Channels: Cz, P3, P4, Pz, O2, Po4, F7
- Sampling rate: 256 Hz

Pre-processing
- Artifact removal
- Bandpass filtering

Power spectrum
- 1 second analysis window
- 10 Hz feature estimation
- 256-point FFT

Frequency integration
- 5 bins: 4-8Hz, 8-12Hz, 12-16Hz, 16-30 Hz, and 30-44Hz

- Feature vectors from each channel are concatenated to form the 35-dimensional feature vector (7 channels x 5 frequency bins)

Feature vector

Figure 0: Signal processing system

filters (0.5Hz cut-off). A band pass filter (between 2Hz and 50Hz) is also employed, as this interval is generally associated with cognitive activity. The power spectral density (PSD) of the EEG signals is estimated using the Welch method (Welch, 1967). The PSD process uses 1-second sliding windows with 50% overlap. PSD estimates are integrated over five frequency bands:

4-8Hz (theta), 8-12Hz (alpha), 12-16Hz (low beta), 16-30Hz (high beta), 30-44Hz (gamma). These bands, sampled every 0.1 seconds, are used as the basic input features for cognitive classification. The particular selection of the frequency bands is based on well-established interpretations of EEG signals in prior cognitive and clinical (e.g. Gevins, Smith, McEvoy & Yu, 1997) contexts. The overall schematic diagram of the signal processing system is shown in Figure 2.

4   Cognitive State Classification System

Estimates of spectral power form the input features to a pattern classification system. The classification system uses parametric and non parametric techniques to asses the likely cognitive state on the basis of spectral features; i.e. estimate p(cognitive state | spectral features). The classification process relies on probability density estimates derived from a set of spectral samples. These spectral samples are gathered in conjunction with tasks representative of the eventual task environment. It is assumed that these sample patterns are representative of the population of spectral patterns one would expect in the performance environment. The classification system uses three distinct classification approaches: K nearest neighbor (KNN), Parzen Windows, and Gaussian Mixture Models (Figure 3). We describe each of these components next.

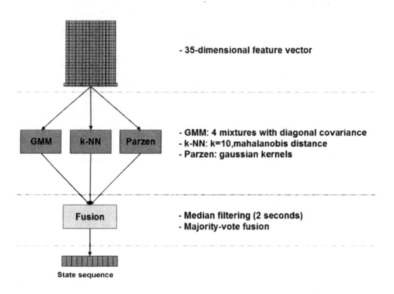

- 35-dimensional feature vector

- GMM: 4 mixtures with diagonal covariance
- k-NN: k=10,mahalanobis distance
- Parzen: gaussian kernels

- Median filtering (2 seconds)
- Majority-vote fusion

State sequence

Figure 0: Classification system

4.1 Gaussian Mixture Models

Gaussian Mixture models provide a way to model the probability density functions of spectral features associated with each cognitive state. This is accomplished using a superposition of Gaussian kernels. The unknown probability density associated with each class or cognitive state is approximated by a weighted linear combination of Gaussian density components. Given, an appropriate number of Gaussian components, and appropriately chosen component parameters (mean and covariance matrix associated with each component), a Gaussian mixture model can model any probability density to an arbitrary degree of precision.

The parameters associated with component Gaussians are iteratively determined using the Expectation Maximization algorithm (Dempster, Laird, and Rubin, 1977). Once the Gaussian parameters have been initialized, the system iterates through a two step procedure for each sample associated with each class. In the first step (expectation step), the system computes the probability of a particular training sample belonging to a particular class based on current model

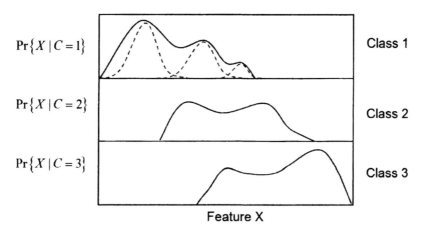

Figure 4: Gaussian mixture models

parameters (posteriori probability). In the maximization step, the model parameters are adjusted in the direction of increasing the class membership likelihood.

Once probability density functions associated with each cognitive state have been generated, it becomes possible to classify individual spectral samples. Each spectral vector is attributed to a class that has the highest posterior probability of representing it. Posterior probabilities are computed using Bayes rule. For example, Figure 4 shows the probability density functions associated with three distinct classes. These probability densities are estimated using three Gaussians. For example, very high values of the data point x are most likely to come from class 3, very low values of x are most likely to come from Class 1.

4.2   K Nearest Neighbor

The K-nearest neighbor approach is a non parametric technique that makes no assumption about the form of the probability densities underlying a particular set of data. Given a particular sample

Figure 5: K nearest neighbor

x, the classification process identifies k samples whose features come closest (as assessed by Euclidian or Mahalanobis distance metrics) to the features represented in x. The sample x would be assigned the modal class of the nearest k neighbors. For example, consider the data point represented by the question mark in Figure 5. Based on k = 5, it would be assigned the label associated with the most common class category of it's 5 nearest neighbors: 1. It can be shown that if k is large, but the overall cell small, that the classifier will approach the best possible classification (Bayes rate) (Duda, Hart, & Stork, 2000)

4.3   Parzen Windows

Parzen windows (Parzen, 1967) are a generalization of the *k*-nearest neighbor technique. Instead of choosing the nearest neighbors and assigning a sample x with the label associated with the modal class of its neighbors, one can weight each vote by using a kernel function. With Gaussian kernels, the weight decreases exponentially with the square of the distance. As a consequence, far away points become insignificant. Kernel volumes constrain the region within which neighbors are considered. Consequently, Parzen windows may be a better choice when there are large differences in the variability associated with each class. The data point shown in Figure 4, will be assigned to the dominant class in it's immediate vicinity.

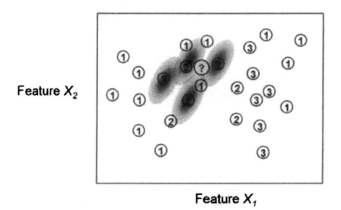

Feature X_2

Feature X_1

Figure 6: Parzen windows

4.4   Composite Classifier

These statistical classification techniques were chosen over multi-layer neural networks because they require minimal training time. KNN and Parzen Windows require no training, whereas the EM algorithm used to generate GMMs, converges relatively quickly. KNN and Parzen Window approaches require all training patterns to be held in memory. Every new feature vector has to be compared to each of these patterns. However, despite the computational cost of these comparisons at run time, the system was able to output classification decisions well within real-time constraints.

The composite classification system regards the output from each classifier as a vote for the likely cognitive state. The majority vote of the three component classifiers forms the output of the composite classifier. When there is no majority agreement, the Parzen window decision is selected. A classification decision is output at a rate of 10Hz. Outputs from the composite classifier are passed through a modal filter before an assessment of cognitive state is output by the classification system. Modal filtering serves to make the cognitive state assessment process more robust to undesirable fluctuations in the underlying EEG signal. Modal filtering is done over a sliding 2 second window with the assumption that cognitive state remains stable over that period of time.

5   Results

The system described here was empirically assessed. The experiment compared classification accuracy across three workload levels in two mobility conditions: stationary and walking. The tasks in the stationary case were: *relaxed* (waiting for orders), *communicate* (getting orders from base via radio communication), and *count* (starting from 100 and decreasing by 7). Tasks in the mobile case were: *navigate* (walking to a designated target), *navigate and visual search* (walking while looking for snipers), and *navigate and communicate* (receiving and giving mission status reports). The subject wore the sensor suite described earlier in this paper in both mobility conditions. EEG was collected as subjects performed each of the tasks mentioned above.

After the preprocessing and PSD feature extraction stages, approximately 3000 samples were obtained. One third of this data was used for training the classifiers, and the remaining two-thirds were used for testing. Classification results for both stationary and mobile cases are presented in the confusion matrix shown in Figure-7. As the diagonals associated with each confusion matrix indicate, classification accuracy was well over 90%. The results presented here are representative of outcomes replicated with a large number of independent data sets and cognitive tasks.

The cognitive state estimator described here was assessed in the context of a real-time, closed-loop, adaptive performance enhancement system. The system optimally scheduled communication traffic to the subject based on cognitive state assessments. Experiments conducted demonstrate that the assistance offered by this interface improves task-related performance greatly. For instance, the scheduling of communication based on the cognitive load assessment resulted in 100% improvement in message comprehension and 125% improvement in situation awareness.

STATIONARY   MOBILE

	classified as		
	Relaxed	Communicate	Count
Relaxed	1.000	0.021	0.000
Communicate	0.000	0.979	0.098
Count	0.000	0.000	0.902

true class

	classified as		
	Navigate	Search	Nav & Comm
Navigate	0.959	0.019	0.000
Search	0.003	0.981	0.047
Nav & Comm	0.038	0.000	0.953

true class

Figure 7: Probability of classifying test patterns correctly. Higher numbers on the diagonal of each matrix correspond to better performance.

6   Conclusion

The ability to detect and to classify the cognitive state of the operator is a prerequisite to successful augmentation of a user's task performance. However, there are numerous technical challenges that limit classification accuracy in the context of an ambulatory soldier. This paper describes hardware and software components that were used to gather, filter, and classify neurophysiological signals. Classification results show that it is indeed feasible to accurately classify the cognitive state of an ambulatory individual.

While these results are encouraging, it is important to emphasize that these results were obtained in conjunction with training and testing data obtained from the same experimental session. We have observed classification accuracy in the 60% to 70% range when training and testing data are drawn from different experimental sessions. Long term non stationarity in EEG limits classification accuracy across experimental sessions. However, it is important to note that EEG is only one component in the cognitive state assessment suite that the Honeywell team is developing. We expect information from sources such as fNIR (functional near infra red imaging), accelerometers, and context modeling to complement EEG based cognitive state assessments. We expect the complementary use of these components to compensate for the variability and unpredictability in operational environments.

Acknowledgments

The authors would like to acknowledge the technical contributions of Jim Carciofini, Michael Dorneich, and Jeff Rye. This paper/research was supported by contract number DAAD-16-03-C-0054 funded through DARPA and the U.S. Army Natick Soldier Center. LCDR Dylan Schmorrow serves as the Program Manager of the DARPA Augmented Cognition program and Mr. Henry Girolamo is the DARPA Agent. The opinions expressed herein are those of the authors and do not necessarily reflect the views of DARPA or the U.S. Army Natick Soldier Center.

References:

C.A. Russell, S.G. Gustafson, "Selecting Salient Features of Psychophysiological Measures," Air Force Research Laboratory Technical Report (AFRL-HE-WP-TR-2001-0136), 2001.

A.P. Dempster, N.M. Laird, D.B. Rubin, "Maximum Likelihood from Incomplete Data via the EM Algorithm," Journal of the Royal Statistical Society, vol. 39, pp. 1-38, 1977.

R.O. Duda, P.E. Hart, D.G. Stork, Pattern Classification, 2nd ed., Wiley, 2000.

A. Gevins, M.E. Smith, L.McEvoy, D. Yu, "High Resolution EEG Mapping of Cortical Activation Related to Working Memory: Effects of Task Difficulty, Type of Processing, and Practice," Cerebral Cortex, vol. 7, pp. 374-385, 1997.

Norman DA. The problem with "automation": inappropriate feedback and interaction, not "over-automation". Philosophical Transaction of the Royal Society of London, 1990; B327: 585-593.

E. Parzen, "On Estimation of a Probability Density Function and Mode", in Time Series Analysis Papers, Holden-Day, Inc., San Diego, California, 1967.

Sarter, N.B., Woods, D.D., and Billings, C.E. (1997). Automation Surprises. In G. Salvendy (Ed.), Handbook of Human Factors and Ergonomics (2nd edition) (pp. 1926-1943). New York, NY: Wiley.

Schmorrow, D.D., & Kruse, A.A., 2002. Improving human performance throughadvanced cognitive system technology. In: Proceedings of the Interservice/Industry Training, Simulation and Education Conference (I/ITSEC'02), Orlando, FL.

B. Widrow and M. E. Hoff, "Adaptive switching circuits," in IRE WESCON Convention Record, 1960, pp. 96--104.

P. Welch, "The Use of Fast Fourier Transform for the Estimation of Power Spectra: A Method Based on Time Averaging Over Short Modified Periodograms," IEEE Transactions on Audio and Electroacoustics, vol. 15, no. 2, pp. 70-73, 1967.

Augmentation of Cognition and Perception Through Advanced Synthetic Vision Technology

Lawrence J. Prinzel III[1], Lynda J. Kramer, Randall E. Bailey,
Jarvis J. Arthur, Steve P. Williams, Jennifer McNabb

MS 152
NASA Langley Research Center
Hampton, VA 23681
[1]l.j.prinzel@larc.nasa.gov

Abstract

Synthetic Vision System technology augments reality and creates a virtual visual meteorological condition that extends a pilot's cognitive and perceptual capabilities during flight operations when outside visibility is restricted. The paper describes the NASA Synthetic Vision System for commercial aviation with an emphasis on how the technology achieves Augmented Cognition objectives.

1. Introduction

Augmented Cognition (AugCog) seeks to extend an operator's abilities via computational technologies that are explicitly designed to address information processing bottlenecks, limitations, and biases. Synthetic vision can be described as an emerging AugCog technology that increases a pilot's perceptual and cognitive capabilities, resulting in enhanced situation awareness, visualization capabilities, and attention to help meet national aviation goals of reducing the fatal accident rate and improving National Airspace System efficiency and throughput.

Synthetic vision serves to augment reality through creation of a "virtual visual meteorological condition" that emulates a visual flight rules environment regardless of actual weather or visibility conditions. Synthetic vision is not merely an ersatz for visual perception but instead, represents a suite of technologies that together meet, or extend, a pilot's natural capabilities in visual flight rules operations.

There are many conceptualizations of synthetic vision that range from relatively rudimentary 3-D displays of terrain information to more sophisticated, integrated synthetic vision systems. The NASA Synthetic Vision System (SVS) concept extends the basic concept beyond a depiction of how the outside world would look to the pilot if he or she could see outside the cockpit window, but additionally incorporates other innovative safety and situation awareness features that together significantly increase aviation safety and pilot efficiency.

The purpose of the paper is to discuss the effects of synthetic vision as a modulator of human information processing in realistic environments. First, synthetic vision is defined and the need of synthetic vision technology is explored. Next, the NASA Synthetic Vision System is described, followed by discussion of how the system supports pilot augmented cognition and perception. Finally, a summary of flight test research and future directions is provided that demonstrates the potential of Synthetic Vision Systems to substantially augment human information cognition and processing when applied to commercial aviation operations.

1.1. Synthetic Vision

A "synthetic vision system" is an electronic means of displaying the pertinent and critical features of the environment external to the aircraft through a computer-generated image of the external scene topography using on-board databases (e.g., terrain, obstacles, cultural features), precise positioning information, and flight display symbologies that may be combined with information derived from a weather-penetrating sensor (e.g., runway edge detection, object detection algorithms) or with actual imagery from enhanced vision sensors (EVS). What characterizes the Synthetic Vision Systems technology is the intuitive representation of visual information and cues in a manner analogous to what the pilot or flight crews would normally have in day, visual meteorological conditions.

1.2. The Need for Synthetic Vision

Humans have always had an enduring fascination with the miracle of flight. From the Greek mythological story of Icarus and Daedalus to the enthusiastic 100-year celebration of the Wright Brothers inaugural flight, our dream of flying up high like a bird inspires the young and old alike. Rapid advancement in technology has compensated for our avian inadequacies. In the relatively short span of 100 years, we have progressed from flights of a few hundred feet to routine trips over oceans to distant parts of the world. Speed, altitude, and range all have increased a thousand-fold.

Aviation has also been witnessed the introduction of technologies to improve aviation safety as our thirst for efficiency and convenience have increased. The development of attitude indicators, flight management systems, instrument landings systems, and radio navigation aids have all extended aircraft operations into weather conditions in which forward visibility is restricted. Before these technologies were introduced, pilots often avoided flying in bad weather or did so at great personal risk. Today, commercial aviation is among the safest modes of transportation. The thousands of daily commercial operations without incident serve as testimony to the remarkable achievements made in aviation safety. But while standard instrumentation has served aviation well, trends in operations and accident statistics are showing disturbing forecasts. Human flight is far from the ancient ideal of visual, bird-like flight and aircraft accidents serve as powerful reminders of the dangers lurking whenever humans take to the skies.

The problems confronting modern aviation still involve limited visibility as a causal factor. For example, 30% of commercial aviation and 50% of all aviation fatalities are categorized as controlled-flight-into-terrain accidents. In general aviation (GA), almost three times more GA fatalities occurred in instrument meteorological conditions. Limited visibility also increases the potential for runway incursions that, from 2000 to 2003, averaged 5.6 runway incursions per million aircraft operations or 1,474 runway incursions out of 262 million aircraft operations. Finally, the most significant problem causing airport delays are limited runway capacity and the increased air traffic separation required when weather conditions fall below visual flight rules operations. Many of these visibility problems have much to do with how cognitively complex flying has become owing largely to the evolution of cockpit displays design, which require the pilot to extract and integrate information from multiple display sources to form a mental model. As a consequence, significant increases in aviation safety are unlikely to come from continued extrapolation from what exists today. As Theunissen (1997) observed, "new functionality and new technology cannot simply be layered onto previous design concepts because the current system complexities are already too high. Better human-machine interfaces require a fundamentally new approach" (p.7). Synthetic vision is one such new approach; it is a visibility solution to these visibility problems.

2. NASA Synthetic Vision System

The National Aeronautics and Space Administration (NASA) Aviation Safety and Security Program (AvSSP), Synthetic Vision Systems project is developing technologies that will mitigate low visibility as a causal factor to civil aircraft accidents while replicating operational benefits in unlimited ceiling and visibility day conditions, regardless of actual outside weather or visibility conditions. The goal is to augment pilot cognition and perception by creating a "virtual visual meteorological condition." To achieve this, synthetic vision must encompass more than visual representation of the outside world presented on a two-dimensional cockpit display. Synthetic vision must be, instead, a system that incorporates four complementary elements that together extend pilot cognitive abilities, in any weather or visibility condition, which meet or exceed those during visual flight rules flight without the Synthetic Vision System. These four elements are an enhanced intuitive view, hazard detection and display, integrity monitoring and alerting, and precision navigation guidance.

2.1. Enhanced Intuitive View

Synthetic vision systems present an enhanced intuitive view by display of pertinent and critical features of the environment external to the aircraft through computer-generated imagery irrespective of weather conditions that may prevent a pilot from effectively seeing these factors through the cockpit window.

One method for providing an enhanced view is by intuitive design where the display format presents these "flight-critical" data in the way that the pilot normally would see in day visual meteorological conditions. This design format is used predominately within the tactical flight displays - the head-down primary flight display (PFD) and on the Head-Up Display (HUD) (Figure 1). In particular, since the HUD depiction is always conformal, the SVS provides an augmented reality display within the HUD field-of-regard, when forward outside visibility may be limited, but still operating under Visual Flight Rules. Further, SVS depicted on both the PFD and HUD provide for emergent details for more rapid and positive recognition and awareness, as visibility transitions from instrument to visual conditions.

121

The displays also include advanced symbology and guidance features that reduce flight technical error and foster instant recognition and awareness for the aircraft planned flight path. This information is generated by highway-in-the-sky guidance information on the tactical displays and three-dimensional route depictions on the strategic displays. Enhanced intuitive view further is achieved through advanced surface guidance map displays, which uplink ATC taxi clearances and graphically present the route and ATC instructions on a moving map display of the airport environment (Figure 1).

Figure 1. Synthetic Vision System Displays

The Synthetic Vision System is also designed to support enhanced intuitive strategic views by understanding how a pilot's informational needs differ from tactical flight operations. For example, synthetic vision navigation displays have significant potential to help a pilot's cognitive understanding of map and ownship positioning and that of traffic, terrain, and obstacle hazards (Figure 1). An innovative feature for the Synthetic Vision System navigation display has been developed to include a multi-mode navigation function that allows pilots to select between 2-D and 3-D exocentric views (Figure 2). Pilots normally use the 2-D synthetic vision co-planar navigation displays. With SVS displays, the pilot can initiate a "situation awareness" mode that presents several 3-D exocentric perspectives that time-out back to the normal 2-D co-planar SVS navigation display. The "situation awareness" mode also provides a dynamic "rehearsal" tool that pilots can use to step through and rehearse complex or unfamiliar airport approaches and non-normal procedures prior to initial descent or departure during a low workload phase of flight (Figure 3).

The culmination of SVS display technology is not just a visual cue analog of visual flight rules operations but, with these mentioned benefits, SVS will provide greater situational awareness than an aircraft operating in actual VMC.

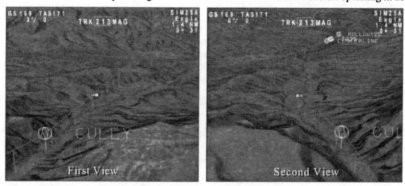

Figure 2. 3-D Exocentric Display Modes

2.2. Hazard Detection and Display

Terrain, cultural, traffic, obstacles, and other hazards are graphically represented to the pilot to maintain the pilot's situation awareness and proactively ensure terrain and hazard separation. The Synthetic Vision System provides for improved pilot detection, identification, geometry awareness, prioritization, action decision and assessment, and overall situation awareness not afforded by today's avionics. This allows the pilot to be proactive in avoiding hazardous conditions instead of reactive to alert cautions and warnings with traditional cockpit displays.

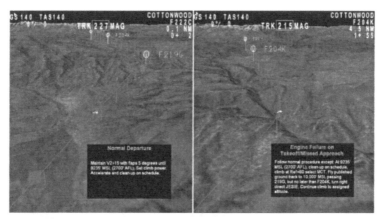

Figure 3. Situation Awareness Rehearsal Tool (shown here for rehearsal of missed approach)

Synthetic vision systems technology utilizes available sensor data to extract (automatically) traffic, terrain, and obstacle hazards and present these data – clearly and obviously – through icons or symbols to the flight crew. Currently, this processing is restricted to "known" hazards available in a database or through Notice-to-Airmen (NOTAM) information or transponding aircraft/vehicles. Eventually, this capability will grow to real-time imaging and non-imaging sensor processing technology whereby the information provided by infrared, milli-meter wave radar, and enhanced weather radars will be automatically processed and traffic, terrain, and obstacle hazards, not already contained in the SVS database, will be identified and shown on the displays. By using sensor image processing, the ability of the pilot to use, interpret, and understand sensor information, particularly sensor information outside of the normal visual wavelengths, would no longer be a determining factor for aviation safety using enhanced vision. The sensor/image processing will contain the necessary knowledge and information to perform this task. It will be unnecessary to conduct significant (and costly) pilot training to understand and use "enhanced vision" sensors.

Additionally, a runway incursion prevention system (RIPS) subsystem is operating as part of the SVS to alert the pilot to potential airborne or ground incursion aircraft or ground vehicles (Figure 4), as well as providing surface guidance and alerting to off-nominal conditions. In essence, RIPS provides an independent monitoring function for operations near and on the airport surface, to reduce the possibility of incidents or accidents caused by pilot errors or mistakes.

2.3. Integrity Monitoring and Alerting

Some level of integrity monitoring and alerting is required in all SVS applications because pilots must trust that the synthetic vision system provides an accurate portrayal (i.e., not hazardously misleading information). A flight-critical level of integrity, redundancy, and the inclusion of reversionary modes are needed to achieve the ultimate potential for a Synthetic Vision System. The Synthetic Vision System has independent sources to verify and validate the synthetic vision presentation (e.g., radar altimeters, enhanced vision sensors, TAWS, weather radar) fashioned to create integrity monitoring functions. If the integrity monitoring discovers a mismatch, the displays degrade gracefully to reversionary modes and trigger an alert to the pilot that synthetic vision is no longer available or reliable (Figure 5). The system effectively prevents a pilot from using erroneous or misleading synthetic vision information.

2.4. Precision Navigation Guidance

Synthetic Vision System features (e.g., surface guidance, taxi maps, tunnels/pathways/highways-in-the-sky, velocity vectors, command guidance cues) allow pilots to rapidly and accurately correlate ownship position to relevant terrain, desired flight paths/plans, cultural features, and obstacles. These elements enable the pilot to monitor navigation precision to meet Required Navigation Performance (RNP) criteria and compliance with complex approach and departure procedures (RNAV, GLS, curved, step-down, noise abatement) without the need for land-based navigation aids (e.g., ILS, VOR, DME, ADF, NDB, LORAN) that are expensive to install and maintain.

Figure 4. Runway Incursion Prevention System Alerting

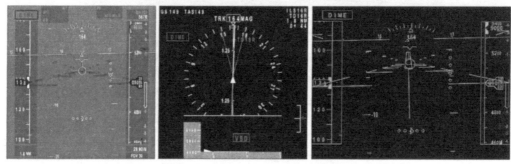

Figure 5. Database Integrity Monitoring Alert

There are several unique and innovative features of the Synthetic Vision System that support precision navigation guidance. First, the system uses a pathway-in-the-sky tunnel format that presents dynamic cues for instantaneous recognition of ownship positioning relative to the flight path. The tunnel depiction modulates based on flight path error in response to where the aircraft is positioned inside the flight path boundaries (Figure 6). If the pilot flies the aircraft outside these boundaries, tunnel cues help guide the pilot back into the tunnel. The tunnel is partnered with a flight path marker and (pursuit) guidance cue that enable the pilot to easily achieve RNP standards while significantly reducing workload and enhancing situation awareness. The pathway concept also provides 4-D visual guidance (speed channel symbology) to meet required time to arrival which help the pilot manage aircraft power inputs to more precisely arrive on time (Figure 7). Pathway guidance is also provided on the ground with symbology that helps the pilot manage ground speed, runway exit speed, braking, etc. (Figure 7).

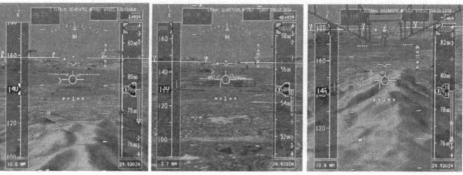

Figure 6. Dynamic Pathway-In-The-Sky Tunnel

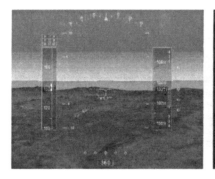

Figure 7. 4-D Tunnel (left figure) and Roll-Out And Take-Off Symbology (right figure)

3. Augmented Cognition through Synthetic Vision System Technology

The four elements of enhanced intuitive view, hazardous detection and display, integrity monitoring and alerting, and precision navigation guidance provide the necessary ingredients to augment pilot information processing.

Information processing can be represented simply as a four-stage sequential model. The first stage, or information acquisition, describes the acquisition and registration of sensory and perceptual inputs involving activation of sensory receptors, sensory processing, pre-processing of data prior to full perception, and orientation and selection of attended stimuli. The second stage, or information analysis, refers to active perception and information retrieval, encoding, and manipulation using working memory. Cognitive operations are active in this stage but occur prior to the point of decision, and involve processes of rehearsal, integration, and inference. The third stage, or decision and action selection, is where decisions are made based on cognitive processing. Finally, the fourth stage, or action implementation, occurs with the response or action consistent with the decision.

The four-stage model is a simplification of the richness of human information processing, but it does allow adoption for describing human-interaction with system functions (cf., Parasuraman, Sheridan, and Wickens, 2000) and helps to show how synthetic vision helps to achieve Augmented Cognition objectives. Below are examples of how the four Synthetic Vision System element technologies support each stage of human information processing to eliminate pilot information processing bottlenecks, limitations, and biases.

3.1. Information Acquisition

Information acquisition refers to the sensing and registration of inputs to the human sensory system. As stated in AFMAN 11-217, "Vision is by far the most important sensory system providing spatial orientation during flight." Synthetic vision, with its enhanced intuitive view, strives to mitigate the lack of outside visibility through the intuitive presentation of references that orient the pilot to the outside world and the hazards to be avoided. These references provide both spatial orientation and geographic orientation information, both critical ingredients to ensure avoidance of Controlled-Flight-Into-Terrain.

3.2. Information Analysis

Information analysis involves higher cognitive functions that integrate incoming data to predict future states. In an aviation context, integration and prediction of the vast array of data is not often easy. Pilots often speak of the need to "stay ahead of the aircraft" that is necessary to achieve higher levels of situation awareness. During IMC, this requires the pilot to integrate symbolic and alphanumeric data information from separate cockpit displays to create a mental image or model of that world beyond his or her visual range. Synthetic vision, however, integrates multiple sources of data and presents this flight information in an intuitive visual way, greatly adding to the pilot's information analysis task.

The Synthetic Vision System technology embraces specific features to enhance a pilot's information analysis by including, for example, intuitive planned flight path displays using pathway-in-the-sky or tunnel displays, geographical and conformal flight hazard information using synthetic terrain and enhanced / synthetic vision object fusion, and future path/prediction information via multiple-mode exocentric 2-D and 3-D navigation display capability, trend predictors, runway exit speed guidance, etc.

125

3.3. Decision and Action Selection

Decision and action selection refers to a selection among decision alternatives based on information analysis of input data. For cockpit display systems, decision and action selection would be supported through display alerting that provides a solution set of alternative actions for the pilot to choose from. Synthetic vision technology creates decision and action alternatives where none existed previously. When an enhanced ground proximity warning system alert occurs, in the absence of synthetic vision, the pilot's action is constrained to vertical maneuver only, because insufficient information is provided for lateral maneuvering. Information on whether this action is the best or proper decision is non-existent. The Synthetic Vision System provides the pilot with the information necessary for prioritization of significant threats and determination of immediacy and proximity of closed point of approach through integration with associated aircraft systems and graphical presentation. The system also supports decision-making to which action, if any, is appropriate with regard to aircraft maneuvering or communication to maintain safe separation from hazards.

Another example involves real-time detection of hazards, obstacles, runway, other aircraft, etc. and represents that information to the pilot. If a conflict exists, the system alerts the pilot and orients them to the hazard with specific information as to nature of the threat, proximity, and threat significance. The information is displayed in real-time so closure rates and proximity can be assessed for the pilot to select the best course of action, given their experience, understanding of the aircraft systems state and performance, and training, rather than rote procedures.

3.4. Action Implementation

Action implementation applies to actual execution of the action selected. An example would be the automatic ground collision avoidance system (auto-GCAS) that could automatically maneuver an aircraft for terrain avoidance.

The Synthetic Vision System was designed principally to support manual flight operations, and to develop provisions which may complement and do not restrict automatic flight. In this sense, SVS technology provides the pilot with sufficient information relating to relative success of maneuver decided upon for achieving design goal of maintaining safe operation. This includes further iteration of action decision, action implementation, and feedback.

4. Synthetic Vision System as Modulator of Information Processing in Real-World Environments

The Synthetic Vision System shares many commonalities with other AugCog technologies. The present paper has described how the technology supports pilot information processing through system elements of enhanced intuitive view, hazardous detection and display, integrity monitoring and alerting, and precision navigation guidance. These elements provide for global situation awareness of threats and future projection of path control decisions through display presentation, integrated hazard alerting, dedicated image object detection and fusion computation equipment, traffic and obstacle presentation, integrity monitoring, and other features of the Synthetic Vision System. The enhancement of local guidance and global situation awareness supported through augmentation of human information processing has significant safety and operational benefits.

The following sections present an overview of selected flight test research that illustrates the Synthetic Vision System as modulator of information processing in real-world environments. The pay-off from improved information processing by SVS for commercial aviation consists of safety and operational benefits that are described below. (Synthetic Vision System technology has also been employed for general aviation use, and demonstrated the efficacy of the technology to significantly extend pilots capabilities and mitigate low visibility as a causal factor in general aviation accidents (cf., Prinzel et al., 2004). These works are not, however, discussed here.)

4.1. Terrain-Challenged Commercial Aviation Operations

Flight test evaluations were conducted to test the efficacy of the Synthetic Vision System in and around terrain-challenged airport environments. These locations were chosen because of the significant high terrain that surrounds each of these airports and the complex operational procedures needed to ensure terrain separation. A primary goal of these flight tests was to demonstrate that manually flown approaches into these airports could be done safer and more efficiently than with today's instrumentation.

4.1.1. Eagle, Co. Regional Airport (EGE) Flight Test (2001)

The objective of the Eagle-Vail, CO (EGE) flight test was to demonstrate the operational and safety benefits of synthetic vision display technology in a terrain-challenged operating environment. Flight test evaluations of tactical display concepts were conducted over a three-week period. Seven evaluation pilots, representing Boeing, Delta, United, American, the FAA and NASA, participated in testing to evaluate pilot acceptability/usability and terrain awareness benefits of head-down and head-up primary flight displays. Predominately using existing approach and departure procedures (Figure 8), synthetic vision display configurations were evaluated consisting of several NASA terrain texturing and display size concepts. In total, 106 approaches and departures were conducted for data. The details of the EGE flight test are described in Kramer et al. (2004).

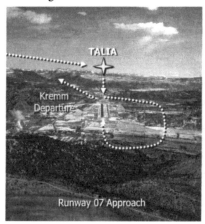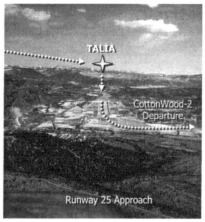

Figure 8. EGE Approach and Departure Experimental Tasks

4.1.2. Reno/Tahoe International Airport (RNO) / Wallops Flight Facility (WAL) (2004)

The objective of the Reno, NV and Wallops, VA flight test, known as GVSITE, was to evaluate the performance, usability, and acceptance of an integrated synthetic vision concept, which included advanced Synthetic Vision display concepts (Figures 9) for a transport aircraft flight deck, a Runway Incursion Prevention System (RIPS), Synthetic Vision and Enhanced Vision sensors, and real-time Database Integrity Monitoring Equipment (DIME). Fifty-nine flights were flown over a 66-day period at the Reno-Tahoe International Airport (RNO) and at NASA's Wallops Flight Facility (WAL). Over 129 flight hours were flown with 166 landings and 186 low approaches, in daylight and night conditions. The flights included Runway Incursion scenarios using NASA's Be-200 aircraft and SV-Sensors Radar Van, used to test the integration of Synthetic Vision System with the RIPS. The flight test also included the evaluation of various DIME operational scenarios and SV-Sensors performance on the Gulfstream-V aircraft. The flight test marked the first time NASA's technologies have been integrated as a complete system incorporating advanced weather radar object detection, synthetic vision database integrity monitoring, refined dynamic tunnel and guidance concepts, and RIPS. The details of the RNO/WAL flight test are described in Kramer et al. (in press).

Figure 9. Flight Testing Integrated Synthetic Vision System

127

The flight test results showed statistically significant improvements in pilot's situation awareness, lower workload, and reduced flight technical error when using the synthetic vision display concepts compared to approaches with current instrumentation. In addition, pilots were able to fly a complex visual arrival and non-normal departure procedures using the SVS in simulated IMC.

In terms of safety benefits, these flight tests showed that synthetic vision may help to reduce many accident precursors including:

- Loss of vertical and lateral path, terrain and traffic awareness
- Unclear escape or go-around path even after recognition of problem
- Loss of altitude awareness
- Loss of situation awareness relating to the runway environment and incursions
- Unclear path guidance on the surface
- Transition from instruments to visual flight and spatial disorientation

These safety benefits are particularly evident during non-normal and emergency situations. In these non-normal events, mental workload and tasking/attentional demands placed on the pilot are high and synthetic vision systems, through their intuitive display and presentation methods, off-load the pilots from basic spatial awareness tasking (to avoid terrain, traffic, and obstacles), increase their speed of situation recognition, and provide resolution recovery guidance.

4.2. Operationally-Complex Commercial Aviation Operations

The aviation safety benefits alone may be reason enough to pursue the technology, but operational and economic benefits must be considered for Part 121 and 135 operations because of the costs associated with implementation of these systems. A NASA-sponsored cost-benefit analysis of 10 major US airports calculated the average cost savings to airlines for the years 2006 to 2015 to be $2.25 Billion. While these savings are predicated on several technology developments and successful implementation/certification, there is a potential order of magnitude in economic savings and operational efficiency that can be achieved. These benefits, however, must be shown through operational demonstration that pilot information processing can best be supported using this technology because it allows for flight operations that otherwise would not be possible with traditional instrumentation.

5.2.1. Dallas Fort-Worth (DFW) Airport (2000)

The objectives of the Dallas Fort-Worth (DFW) flight test were: (1) to evaluate to the potential of retrofitting SVS technology into current flight decks and as a forward-fit technology into future flight deck concepts and (2) to evaluate an aircraft-based RIPS integrated with a FAA Runway Incursion Reduction Program surface infrastructure (Figure 10). The flight test evaluated the technical and operational system performance of synthetic vision displays, airport surface databases and runway incursion warning systems suitable for retrofit into current technology cockpits using complex, side-step runway maneuvers during nighttime operations. Ten evaluation pilots performed 122 test runs (approaches and surface operations) using NASA's Boeing 757 research aircraft at night in an operationally challenged, international airport environment. The details of the DFW flight test are described in Prinzel et al. (2004).

Figure 10. Dallas Fort-Worth Nighttime Operations

In terms of operational benefits, this flight test and others showed synthetic vision could serve to increase National Airspace System capacity by providing the potential for increased visual-like operations gate-to-gate even under extreme visibility restricted weather conditions (e.g., Category IIIb minimums), including:

- Intuitive depiction of ATC cleared flight paths and taxi clearances
- Enhanced surface operations (e.g., rollout, turn off and hold short, taxi)
- Reduced runway occupancy time in low visibility
- Reduced departure and arrival minimums
- Better allow for converging and circling approaches, especially for dual and triple runway configurations
- Provide for independent operations on closely-spaced parallel runways
- Provide for precise noise abatement operations
- Required Navigation Performance adherence and 4D navigation potential
- Enhanced path guidance, compliance monitoring, and alerting, and enhanced flight management
- Depiction of terminal, restricted and special use airspace
- Depiction of traffic and weather hazards and resolutions
- Approach operations to Type I and non-ILS runways
- Piloting aid support (e.g., flare guidance, runway remaining, navigation guidance)

6. Conclusions

Augmented Cognition refers to computational technologies designed to explicitly address human information processing bottlenecks, limitations, and biases. Synthetic vision can be considered an emerging AugCog technology that increases a pilot's perceptual and cognitive capabilities, resulting in enhanced situation awareness, visualization capabilities, and attention to help meet national aviation goals of reducing the fatal accident rate and improving National Airspace System efficiency and throughput. The paper described the four elements of the Synthetic Vision System and discussed how the technology supports AugCog objectives demonstrated through real-world flight test evaluations. New directions will develop prevention, intervention, and mitigation technologies integrated with the Synthetic Vision System to help address the safety and security needs of the National Airspace System. The future system will further embed AugCog-type technologies and continue to support the human user and enhance human performance, interaction and reliability in the use and design of these complex aerospace systems.

7. References

Kramer, L., Bailey, R., Arthur, J., & Prinzel, L. (in press). Flight testing an integrated synthetic vision system. In J. Verly (Ed.), SPIE, Enhanced and Synthetic Vision. Orlando, FL: Society for Optical Engineering

Kramer, L, Prinzel, L, Bailey, R, Arthur, J, & Parrish, R (2004). Flight test evaluation of synthetic vision concepts at a terrain challenged airport. NASA-TP-2004-212997.

Parasuraman, R., Sheridan, T. & Wickens, C. (2000). A model for types and levels of human interaction with automation. IEEE Transactions on Systems, Man, and Cybernetics, 30, 286-297.

Prinzel, L.J., Comstock, J.R., Glabb, L., Kramer, L., Arthur, T., & Barry, J. (2004). The Efficacy of Head-down and head-up synthetic vision display concepts for retro- and forward-fit aircraft. International Journal of Aviation Psychology. 14, 53-77.

Prinzel, L., Hughes, M, Kramer, L, Arthur, J (2004). Aviation safety benefits of NASA synthetic vision: Low visibility loss-of-control, runway incursion detection, and CFIT experiments. Proceedings of the Human Performance, Situation Awareness, and Automation Technology Conference, 5. Embry-Riddle: Daytona Beach, FL.

Theunissen, E. (1997). Integrated design of a man-machine interface for 4-D navigation. Netherlands: Delft University Press.

Cellulars, Cars, and Cortex – A Neurophysiological Study of Multi-task Performance Degradation

Curtis Ponton

Neuroscan Labs
5700 Cromo Drive
El Paso, TX 79912
cponton@neuroscan.com

Abstract

Due to the rapid evolution of technology, complex behaviors, such as listening to conversation on a cellular phone while driving, are often combined. The technological advances that allow multi-tasking at times outstrip the human brain's capacity to manage the simultaneous performance of complex behaviors proficiently. When, for example, cellular phone used is combined with driving, the deterioration in driving performance can have serious and potentially fatal consequences. Previous behavioral studies have clearly demonstrated that even under so-called "hands-free" driving conditions, cell phone use adversely impacts driving performance. Investigators have gone so far as to conclude that there is virtually no safety advantage conferred by hands-free cell phone operation while driving. This suggests that the additional attentional load from actively engaging in a conversation on a cell phone is a critically important factor in degraded driver performance. Previous studies suggest that divided attention produces significant changes in neurophysiological brain activity that can be measured with EEG. To investigate the possible neurophysiological bases of this degraded performance, a series of studies were performed in which subjects were required to monitor and verbally recall seven element digit span list (simulating recall of a phone number) presented via headphones while performing the driving simulation. Brain activity was monitored by recording electroencephalographic activity from 64 scalp electrode locations under a number of conditions in which the driving simulation and the digit span recall tasks were performed separately, and concurrently, with or without music presented at the same time. Analysis of the behavioral data showed that when the primary tasks were performed simultaneously, there was a significant adverse impact on both simulated driving performance and digit span recall. There was a significant increase in the number of upper speed limit violations as well as an impaired ability to recall the seven element digit span correctly, particularly for numbers from the middle of the list. In contrast, passively listening to music had only a minimal impact on driving performance and digit span recall performed in isolation or concurrently. When the digit span task was performed in isolation, brain activity was characterized by high levels of alpha oscillations (8-12 Hz) over areas of association cortex including lateral posterior temporal and parietal cortex. This focused-attention alpha, described in several previous studies, was not attenuated by near-masking level of music. However, this focus-attention alpha induced by performance of the digit span task appeared to be almost entirely suppressed when simultaneously completing the driving simulation. In addition, digit-evoked phase-locked brain activity over left lateral frontal cortex, often associated with working memory, was clearly suppressed while performing the driving task concurrently. When completing the driving simulation in isolation, brain activity was characterized by high beta and gamma band oscillations (13-29 Hz and 30-50 Hz, respectively) over most of frontal cortex. Performing the driving simulation while listening to music reduced levels beta and gamma band activity, but the suppression was much greater while performing the digit span task concurrently. The results of this investigation provide clear neurophysiological evidence that patterns of brain activity associated with optimal, isolated completion of a memory recall task and a driving simulation are adversely impacted by concurrent execution of these tasks. These data are consistent with those of previous studies showing degraded driving performance with concurrent cellular phone use. Further, this research clearly demonstrates that when complete simultaneously, there is widespread suppression of brain activity associated with optimal simulated driving and memory recall performance.

The Communications Scheduler: A Task Scheduling Mitigation For A Closed Loop Adaptive System

Michael C. Dorneich, Stephen D. Whitlow, Santosh Mathan, Jim Carciofini, Patricia May Ververs

Honeywell Laboratories
3660 Technology Drive; Minneapolis, MN 55418
michael.dorneich@honeywell.com

Abstract

This paper describes the Communications Scheduler, an adaptive system mitigation designed to scheduler incoming messages to improve message comprehension and situation awareness. The U.S. Army is currently defining the roles of the 2010-era Future Force Warrior (FFW) program, which seeks to push information exchange requirements to the lowest levels and posits that with enhanced capabilities a squad can cover the battlefield in the same way that a platoon now does. Among other capabilities, the application of a full range of netted communications and collaborative situational awareness will afford the FFW unparalleled knowledge and expand the effect of the Future Force three dimensionally. It is expected that in a highly networked environment the sheer magnitude of communication traffic could overwhelm the individual soldier. The Honeywell Augmented Cognition team has developed the Communications Scheduler, a task scheduling mitigation driven by real-time neurophysiological and physiological measurements of human cognitive state, which is used to augment the work environment to improve human-automation joint performance. Evaluations have shown over at least a 100% improvement in message comprehension an situation awareness.

1   Introduction

Task analysis interviews with existing military operations identified factors that negatively impact communications efficacy. In one example, in the first few minutes of any intense mission, radio communications are a suboptimal method of communications because everybody is intensely focused on their tasks at hand. In one famous raid, for example, the commander did not hear the radio communications informing him that the plan had changed until he was physically grabbed by the ground force commander and given this critical information. The commander responded by radioing his own troops, who also did not respond. The implications of these kinds of situations are many, but, first and foremost, mission critical information must be reliably communicated. What aspects of the communication method can be altered to improve the chances that a message will be received and understood? Does it require a multi-modal, physical alert? Should communications be limited to only critical messages during high workload situations?

The U.S. Army is currently defining the roles of the 2010-era Future Force Warrior (FFW). The FFW program seeks to push information exchange requirements to the lowest levels and posits that with enhanced capabilities a squad can cover the battlefield in the same way that a platoon now does. Among other capabilities, the application of a full range of netted communications and collaborative situational awareness will afford the FFW unparalleled knowledge and expand the effect of the Future Force three dimensionally.

An approach was adopted that considers the joint human-computer system when identifying bottlenecks to improve system performance. The Honeywell team is focusing primarily on the Attention Bottleneck. The appropriate allocation of attention is important to FFW because it directly affects two cornerstone technology thrusts within the FFW program: netted communications and collaborative situation awareness. The Honeywell team has developed a set of cognitive gauges based on real-time neurophysiological and physiological measurements of the human operator. The capability to assess cognitive state to determine allocation of attention provides the opportunity to adapt the soldier's current task environment. Cognitive assessment can drive adaptive strategies to mitigate the specific information processing bottlenecks.

With the aid of our proposed adaptive system we hope to increase the soldier's situation awareness, survivability, performance, and information intake by improving their ability to comprehend and act on available information. The results of this work will benefit soldiers by creating a system that alters task presentation based on an analysis of that soldier's cognitive state. It is hypothesized that this adaptation of the soldier's workspace will lead to greater joint human-automation performance in dismounted soldier operations. Honeywell has developed the Communications Scheduler to manage the incoming information by scheduling the communications to be received by the soldier at the most optimal period, offloading tasks or portions of tasks to automation when the soldier is overwhelmed, and providing information in multiple modalities (audio, visual, tactile) to ensure comprehension. A high task load condition will prompt the automation to defer all but the highest priority messages, offload tasks, or change the modality of information presentation; a low load condition would indicate an appropriate time for interruption and higher levels of soldier participation in on-going tasks. Without these augmentations the soldier can become overloaded with information having to decide when and where to focus this attention among the myriad of high priority communications and high priority tasks.

Section 2 describes the architecture of the Augmented Cognition Systems. Section 3 describes the Communications Scheduler in detail. Section 4 describes the results of two evaluations to test aspects of the Communications Scheduler's logic and ability to improve performance when driven correctly from the cognitive gauges.

2 Augmented Cognition System Design

2.1 Architecture

This section briefly describes the system architecture of the Honeywell Closed Loop Integrated Prototype (CLIP). (see Figure 1). The architecture is made up of the following components:

- *Cognitive State Assessor* (CSA) combines measures of cognitive state to produces the cognitive state profile (CSP).
- *Augmentation Manager* (AM), adapts the work environment to optimize joint human-automation cognitive abilities for specific domain tasks. The AM is comprised of three components:
  - o *Interface Manager*, responsible for realizing a dynamic interaction design in the HMI
  - o *Automation Manager*, responsible for the level and type of automation
  - o *Context Manager*, responsible for tracking tasks, goals, and performance
- *Human-Machine Interface*, where the human interacts with the system.
- *Automation*, where tasks can be partially or wholly automated.
- *Virtual Environment* (not shown) is a simulated approximation of the real world.
- *Experimenter's Interface* give the experimenter both insight and control over events within the system.

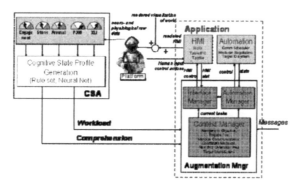

Figure 1. Honeywell Augmented Cognition system architecture.

2.2 Cognitive State Assessor

The Honeywell AugCog team has developed a comprehensive suite of sensors (including EEG, pupilometry, physiological measures such as EDR, and ECG) that feed a set of "cognitive gauges" that make up the Cognitive State Assessor (CSA). These include an engagement index, a stress gauge, an arousal meter, an executive load index, and a P300-driven novelty gauge.

2.2.1 Gauges

Engagement Index. The engagement index is a ratio of EEG power bands (beta/(alpha + theta)). The engagement index, as described by Freeman et al. (1999) is a measurement of how cognitively engaged a person is in a task, or their level of alertness. Adaptive systems have used this index to successfully control an automation system for tracking performance and a vigilance task (Freeman et al., 1999; Mikulka, Scerbo, & Freeman, 2002; Pope, Bogart, & Bartolome, 1995). Consistent with Freeman et al.'s work, EEG data is recorded from sites Cz, Pz, P3, and P4 with a ground site midway between Fpz and Fz. The Engagement Index is calculated from a running average of powers for different EEG frequency bands (Prinzel, et al., 1999). Prinzel, et al., (1999) reported that adaptive task allocation may be best reserved for the endpoints of the task engagement continuum. Therefore, two levels of engagement (low, high) were measured in this study. The engagement index reflects the selection and focus on some aspect at the expense of the other competing demands, thus it is a measure of focused attention. High levels of engagement reflect selection and attentional focus whereas lower levels of engagement indicate that the subject is not actively engaged with some aspect of the environment.

Arousal Meter. Clemson University's Arousal Meter (Hoover & Muth, 2003) derives autonomic arousal from the cardiac inter-beat interval (IBI), derived from the Electrocardiogram (ECG) at one ms accuracy. A three lead ECG is used to detect R-spikes and derive ms resolution IBIs that are then re-sampled at 4 Hz. A Fast Fourier transform (FFT) is computed for 16 s, 32 s, or 64 s worth of IBIs. A sliding window is established such that a new FFT is computed every .25 s. When the FFT is computed, the high frequency peak (max power between 9 and 30 cycles per minute) is identified and the power at that peak, termed respiratory sinus arrhythmia (RSA), is stored. Once one minute's worth of FFT results are stored, the arousal meter begins to generate a standardized arousal which is computed every 0.25 s using a z- log-normal score standardization and the running mean and standard deviation of the RSA values. The gauge has 3 levels (low, medium and high). Increases in this score are associated with increased autonomic arousal.

Stress Gauge. The Institute of Human and Machine Cognition developed a composite Stress Gauge (Raj et al., 2003; Kass et al., 2003). The gauge uses a weighted average of the four inputs (Video Pupilometry (VOG), High Frequency Electrocardiogram (HFQRS ECG), Electrodermal Response (EDR), and Electromyogram (EMG) from the left trapezius muscle to detect the subject's response to changes in cognitive load. The gauge was used to detect cognitive stress related to managing multiple competing tasks on a moment-to-moment basis.

P300 Novelty Detector. The EEG Auditory P300 reflects a central nervous system response to behaviorally relevant infrequent sounds. Previous literature (Wickens, Heffley, Kramer, & Donchin, 1980) suggests that P300 amplitude in response to a task relevant infrequent auditory stimulus is modulated by attentional resources: if the subject is very focused on a primary task the auditory stimulus will be missed and the corresponding P300 diminished. Columbia University and the City College of New York have created a gauge called the P300-novelty detector gauge (Sjada, Gerson, & Parra, 2003), that spatially integrates signals from sensors distributed across the scalp, learning a high dimensional hyperplane for discriminating between task-relevant (incoming message auditory alert) and task irrelevant responses. In the current task environment a tone is played before an auditory message to evoke a P300 activity. Mitigation strategies are based on the assumption that the presence of a strong evoked response indicates that subjects have sufficient attentional resources to process the incoming message. The gauge includes frontal and parietal electrodes.

Executive Load Index. Human Bionics developed the eXecutive Load Index (XLI) (DuRousseau, 2004) to measure cognitive state. It operates by measuring power in the EEG at frontal (FCZ) and central midline (CPZ) sites. The algorithm uses a weighted ratio of delta + theta/alpha bands calculated during a moving 2-second window. The current reading is compared to the previous 20-second running average to determine if the executive load is increasing, decreasing or staying the same. The index was designed to measure real-time changes in cognitive load related to the processing of messages. This gauge was previously validated to discern trial difficulty in a continuous performance high-order cognitive task battery.

2.2.2 Cognitive State Profile

The CSA outputs a Cognitive State Profile comprised of two decision state variables: *workload* and *comprehension*. The CSP drives the mitigations of the Augmentation Manager. Currently a simple set of rules is used to derive the assessment of Workload and Comprehension, although work is underway to define this step with neural networks, in order to better account for individual subject differences (see companion paper in these proceedings: Mathan, Mazaeva, & Whitlow, 2005). For this CVE, *workload* was considered high is any of the three gauges, Engagement, Arousal, or Stress, registered high. This was done to allow for the fact that on any given subject, only a subset of the gauges may be able to discriminate differing levels of workload, based on individual differences in subject's cognitive response to the scenario workload manipulation. Likewise, in order to bias *comprehension* towards false positives, both the P300 Novelty Detector Gauge and the XLI gauge had to be high (i.e. reporting a yes that the subject comprehended the message) for the *comprehension* variable to be set to "True".

2.3   Augmentation Manager

The Augmentation Manager (AM) is the primary reasoning component of Application, tasked with determining what and how information will be displayed to the human, the allocation of tasks between the human and automation, and the management of the automation and the interaction with the human. The goal of the Automation Manager is to adapt the human's work environment to the current state of the user, the current state of the world, and the current state of the tasks the joint human-automation system is performing. As such, when the CSA determines that the human is overloaded, stressed, or no longer able to handle the task load, the system employs a mitigation strategy to adapt the system to maintain or enhance performance. There are four broad categories of possible mitigations in an Augmented Cognition system:

- Task/Information Management
- Modality Management
- Task Offloading
- Task Sharing

The AM's major functions are described in Table 1 below:

Table 1. The major functions of the Augmentation Manager.

AM Function	Component	Requirements
Information Management	Interface Manager	• Decide information content • Decide information abstraction needed to support the human tasks. • Decide information presentation modality.
Task offloading.	Automation Manager	• Decide which functions allocated to human, which to automation. • Function allocation that is dynamic, responsive to changing context.
Task Sharing.	Automation Manager	• Decide which tasks can be shared • Determine effects of function allocation on task-subtask interactions, workload, mission management, and performance 7under stress.
Task Scheduling and Management.	Automation Manager	• Prioritize tasks to minimize multi-tasking costs. • Model the current and projected tasks, • Schedule tasks, and manage human's "automation awareness."
Task Tracking	Context Manager	• model dynamic multitasking • model task objectives
Performance tracking	Context Manager	• Model current performance, • predict performance

Honeywell has developed the Communications Scheduler to mitigate situations where the user is forced to divide their attention between multiple tasks and performance breaks down due to information overload. Of particular importance is the soldier's ability to handle continuous inflow of netted communications and directing his or her attention to the highest priority task to complete his/her mission in this highly dynamic environment. This is crucial to not only the soldier's own survival but that of his or her fellow soldiers (Dorneich et al, 2004). The Communications Scheduler mitigates divided attention demands via task-based management and modality-appropriate information presentation strategies. The Communications Scheduler is described in detail in Section 3.

2.4   Human-Machine Interface

The Human-Machine Interface (HMI) realizes the interaction design of the Interface Manager outputs by configuring a multi-modal user interface to optimally support cognitive throughput of the next generation warfighter. In the system configuration described in this paper, the HMI consists of four components:

- Head-Up Display (HUD). The subject will see various icons in his or her field of view. There is also a limited ability to display a text message. In addition, in the upper left corner a ego-centric compass displays the current heading.
- TabletPC Display Device. This device contains a messaging application that allows the soldier to interact with deferred messages.
- Tactile Display. This system was used to provide tactile navigation cues via 24 tactors (2 rows of 12 columns) worn in a belt about the upper abdomen (for a description of its use, see companion paper: Dorneich et al 2005).
- Aural Alerting System. Prior to presenting an incoming radio messages, the HMI had the ability to insert a tone to alert the user of the priority of the message.

2.5   Virtual Environment

The Honeywell Augmented Cognition program required a Virtual Environments (VE) with a with appropriate fidelity to support sensor-suite validation, and concept validation. The VE provides a realistic, tactically correct MOUT battlefield environment, with opposing forces (OPFOR) and friendly forces (BLUFOR) that can be controlled either by "botAI" (automated behavior scripts) or additional human operators. The VE is of a sufficient fidelity to represent the visual complexity of a MOUT environment in order to appropriately tax a subject's workload when interacting with the VE, with the following properties to add to the realism and immersiveness of the environment:

- Visually complex MOUT world
- Building interactivity to allow subject to enter buildings
- Three-dimensional world for mobility in the lateral and vertical directions
- Several subject behavior characteristics including: crouching, running, walking, jumping, firing weapon, climbing stairs, depreciated health upon sustaining enemy attack.
- Team members (BLUFOR) with the following characteristics: ability to fire at enemy, defend a position (or objective), move realistically, ability to follow the subject, ability to navigate to objective, ability to be tasked by subject, depreciate in health upon sustaining enemy hit.
- Audio to provide environmental sounds, weapons, and ability to insert audio messages from external Communications Scheduler

Evaluations of the Communications Scheduler (described in Section 4) utilized a desktop-PC Virtual Environment based on a modified Quake3 TeamArena game engine. The VE, illustrated in Figure 2, depicts a small area of a city, with realistic textures, and detailed models, but little interactivity (doors do not open, crates do not move, etc.). The city was comprised of narrow streets, surrounded by two and three story buildings. The environment had an industrial appearance. The subject is entered into the environment in one of many predetermined locations in the map. In addition to the subject there are some number of simulated players (bots), some opposing forces and some blue forces. These forces were presented both at street level and above as snipers. The specific numbers of OPFOR and BLUFOR were adjusted at runtime. Each bot was assigned a skill level between 1 (easy) and 5 (hard), therefore workload was adjusted easily. Each player (subject or bot) had a realistic visual representation, with subtle details (primarily color and pattern of uniform) distinguishing blue forces from opposing forces. The subject performed tasks in the environment using a combination of keyboard and mouse controls. The controls allowed the subject to look around the world. They also enabled the subject to move (walking forward or backward, sidestepping left or right, jumping, and crouching).

Figure 2. The Honeywell FFW Virtual Envuironment.

3 Communications Scheduler

3.1 Overview

The Communications Scheduler mitigates the attention bottleneck via task scheduling and modality management of incoming communications. The system is tasked with determining when and how information is displayed to the soldier. The Communications Scheduler schedules and presents messages to the soldier based on the cognitive state profile, the message characteristics, and the current context (tasks). Based on these inputs, the Communications Scheduler can pass through messages immediately, defer and schedule non-relevant or lower priority messages, escalate higher priority messages that were not attended to, divert attention to incoming higher-priority messages, change the modality of message presentation, or delete expired or obsolete messages.

3.2 Message Characteristics

All messages have a priority associated with them, depending on how critical they are. High priority are ,ission-critical and time-critical, medium priority messages are mission-critical only, and low priority messages are not critical (although they may still be important).

When the augmentation is in effect, messages will be scheduled according to certain rules. High priority messages are mission critical and time critical, which means they must be heard and understood as soon as they arrive. Medium priority messages are mission critical, but have a larger time window to work with. A medium priority message could potentially be deferred if the system finds that you are highly engaged in another task. All medium priority messages will be played before the end of the mission. Low priority messages are not mission critical or time critical. They will be presented if the subject is not engaged in another task. If the system finds that the subject is engaged in another task, the low priority message will be presented in text format in the message window.

3.3 Message Alert Modes

High priority messages have a tone played before they are presented. If the system finds that the subject is highly engaged in a task it will play the same tone ,but louder and more saliently. If the system finds that the subject missed a high priority message after it has been presented, it will repeat the message once using the same tone, but louder again. In summary, there are three versions of the same tones associated with high priority messages. Medium priority messages will also have a tone played before they are presented. It is recognizably different than the high priority tone. Medium priority messages will also be repeated once if the system finds that the subject missed a message, however, the tone will remain the same. It will not change in loudness like the high priority tone does. Low priority messages do not have a tone associated with them. Low priority messages will not be repeated.

3.4 System Logic

The Cognitive State Assessor (CSA) determines two cognitive state profile decision variables: *workload* and *comprehension* (see section 2.2.2). The Communications Scheduler determined the initial message presentation based on a user's current *workload*. The Communications Scheduler performed one of three actions when deciding how to first present the message:

- Present the message immediately in the audio modality with the appropriate "normal" tone proceeding it.
- Present the message immediately in the audio modality preceded by the appropriate "higher saliency" tone.
- Present the message immediately in the text modality on the subject's TabletPC.

After the first presentation of a message to the user (in audio modality), the Communications Scheduler determined whether to take further action on a message depending on the CSA's assessment of *comprehension*. Comprehension is an assessment of whether the subject had the attentional resources at the moment of message presentation to properly attend to and understand the message. Based on Comprehension, it performed one of four actions:

- Replay the message immediately in the audio modality preceded by the same tone used previously.
- Replay the message immediately in the audio modality preceded by a higher, more salient tone than used previously. Note if first presentation was the "higher" tone, this replay would use the "highest" tone.
- Do nothing as the gauges have sensed that the subject comprehended the message.
- Not Applicable – the "before" decision precludes any need to make an "after" decision.

The decision logic of the Communications Scheduler is summarized in Table 2. Each Workload cell has a rule P(modality, saliency) where P = Play, modality = audio or text, and saliency = normal, higher, highest. Each Comprehension cell has a either rule Replay(saliency) where saliency = up (i.e. the previous salience was higher, escalate to highest) or same, or Done or NA = not applicable.

Table 2. Communications Scheduler decision rule set, where each rule is of the form Play(modality, saliency).

CSP Variable -Priority	Before first message presentation			After first message presentation		
	High	Med	Low	High	Med	Low
Workload High	P(audio,higher)	P(text,normal)	P(text,normal)			
Workload Low	P(audio,normal)	P(audio,normal)	P(default,normal)			
Workload Low after High	P(audio,higher)	P(text,normal)	P(text,normal)			
Workload Unknown	P(audio,normal)	P(audio,normal)	P(audio,normal)			
Comprehension High				Done	Done	N/A
Comprehension Low				Replay(up)	Replay(same)	N/A
Comprehension Unknown				Done	Done	N/A

3.5   System Behavior

High priority messages are mission critical and time critical, which means they must be heard and understood as soon as they arrive. Thus the Communications Scheduler took the following actions on high priority messages:

- High priority messages were preceded by a tone (normal or escalated)
- A visual icon reminded the subject to pay attention (see left side of Figure 3)
- High priority messages that required an overt response were accompanied by a visual summary
- Message may have been repeated

Medium priority messages are mission critical, but have a larger time window to work with. A medium priority message may have been deferred if the system found that the subject was highly engaged in another task. All medium priority messages were played before the end of the mission. Low priority messages are not mission critical or time critical. They were presented if the subject is not engaged in another task. If the system found that the subject was engaged in another task, the low priority message were presented in text format in the message window. Specifically, low and medium priority messages were deferred to the Tablet PC application, and a visual icon appear on the HUD to alert to the action the scheduler has taken (see right half of Figure 3).

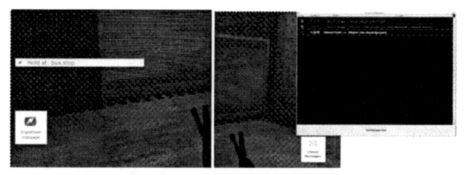

Figure 3. High priority messages are alerted by an icon and (possibly) a text summary on the HUD (right). Deferred messages appear on the Tablet PC with an icon on the HUD (left).

3.6 Automation Etiquette

Poorly designed automation can be dangerous. Research shows that unless users are able to predict clearly how an automated system is likely perform, automation may introduce more problems than it solves (Sarter, Woods, & Billings, 1997). The mitigations strategies described here have very clear rules to eliminate uncertainty and unpredictability. The Communications Scheduler benefits uses by allowing the user to defer responses to messages under conditions when attention has to be split between competing tasks, thus allowing users to focus on higher priority tasks first. However, this kind of automated system behavior has negative side effects: Loss of momentary situational awareness, and lags in responses could break coordination among teams and introduce inefficiencies in the mission. Thus it is important that the Communications Scheduler be invoked only when the benefits of its use outweigh its costs. For that reason the Communications Scheduler would not be used continuously, but rather only in times of high cognitive stress of the user, when faced with competing tasks that overload his or her ability to comprehend and process all incoming information.

Since the Communications Scheduler will not be used continuously, the issue of automation etiquette becomes important. The Communications Scheduler should be invoked (and should cease) in such a manner that does not exacerbate confusion. The Communications Scheduler mitigation was invoked when Workload was high – for instance low priority messages were deferred to the TabletPC. However, when workload lowered below the threshold used to trigger the message deferral, the Communications Scheduler continued to defer messages. This is due to the fact that deferring communications on the basis of moment to moment fluctuations in gauge values can be confusing. Messages could be misinterpreted without surrounding context if they were to be played in audio modality after its predecessor messages have been deferred to the TabletPC (and remain unread for a period of time). If expected messages were not heard, it may have been hard to disambiguate whether this is because of the communications scheduler or some mission related cause. To avoid confusion, once communications scheduling was activated, all low and medium priority messages were deferred to tablet PC until user has caught up on all messages and clicked a 'messages read' button. The subject of automation etiquette is discussed in more depth in a companion paper in these proceedings (Mathan, Dorneich, & Whitlow 2005).

4 Evaluation

Results from an evaluation on the first iteration of the Communications Scheduler (Dorneich et al 2004) indicated that augmentations benefited subjects under high workload conditions. Under high workload conditions, the communications scheduler produced better situation awareness and improved the ability of participants to correctly identify and engage the foes. These results are consistent with many program findings (e.g., Prinzel et al., 2003) that show the benefit of augmentation at the extreme ends of the workload space. In addition, the analysis of the behavior of the system with regards to subject acknowledgement and comprehension of messages was compelling. Based on cognitive state, the system was able to infer a subject's message comprehension and repeat unattended messages in the majority (71.6%) of cases, with a false alarm rate (23.7%) that can be partially attributed to the subjects' automatic acknowledgment of a truly unattended message. It is important to remember that these mitigations were driven solely by the cognitive state of the subject, as measured in real time by five cognitive gauges. These findings were used to guide the development of the second iteration of the Communications Scheduler, described in this paper.

The evaluation of the second iteration is detailed in a companion paper in these proceedings (Dorneich et al 2005), and briefly described below. Concurrent with this evaluation, another version of the Communications Scheduler was evaluated in a more mobile setting, as described in a companion paper in these proceedings (Whitlow and Ververs 2005).

The Concept Evaluation Experiment (CVE) involved 14 participants (M_{age} = 25.4 years) with an average education level of 15 years. To reduce the effect of learning (of the VE) for this experiment, participants were chosen who rated their skill level at first person shooter games as average to above average. The average skill rating was 3.4/5 (Range = 2-4), with only 1 person rating himself as a 2/5. Overall, participants' average time playing was 5.7 hours per week. There were two independent variables in this study: 1) Mitigation (on/off), and 2) Scenario (three, which vary by attention type). The study consisted of three two-factor experiments. Each experiment compared performance under gauge driven mitigation with performance without mitigation. Scenario 1 utilized the Communications Scheduler as the mitigation, and will be described here.

This scenario focused on divided attention bottleneck in multi-tasking, and consisted of the subject performing three tasks: 1) Navigate to Objective, 2) Engage Foes, and 3) Manage Communications. The subject was a platoon leader, whose goals were to lead the platoon along a known route through a hostile urban environment to the objective, while being careful to engage enemy soldiers. Participants also received incoming communications throughout the scenarios, with some messages requiring an overt, behavioral response. Participants received status updates, mission updates, requests for information, and reports; these incoming communications are a primary source of their situation awareness. Radio communications volume was extremely high. The scenario only included two or three high priority messages, which told the soldier to hold at certain locations for a specified amount of time, or that the objective location had changed. Failure to heed these high priority messages caused the subject to encounter an ambush.

Scenario 1 was designed to put subjects into distinct periods or extreme high and extreme low workload periods. The high workload times included a high volume of communications traffic to the subject, just at the time when their workload was high due to being targeted by foes. The principal metrics were

- *Message Comprehension* inferred from correct change in overt behavior due to message instruction, and correct responses to message queries.
- *Situation Awareness* inferred from the number of correct post-trial questions to ascertain if they could recall mission-critical information relayed through the communications.

The goal was to improved performance on these tasks while not degrading performance on the Navigate to Objective and Engage Foes tasks, and no increase in workload:

- *Message Comprehension.* Subjects in the unmitigated condition correctly responded to 57 of 143 possible messages (39.9%). Subjects in the mitigated condition correctly responded to 114 of 143 messages (79.7%). The mitigated condition shows a significant ($p<0.0001$) performance increase of 100%.
- *Situation Awareness.* Subjects in the unmitigated condition correctly responded to 22 of 84 situation awareness questions (26.2%). Subjects in the mitigated condition correctly responded to 49 of 84 situation awareness questions (58.9%). The mitigated condition shows a significant ($p=0.009$) performance increase of 125%.
- No negative affect on ability to engage foes.
- No negative affect on workload.
- 85% of subjects felt communication easier with augmentation

5   Discussion

The Communications Scheduler has been developed to help the Future Force Warrior manage the high volume of incoming information in a netted communications and collaborative situation awareness environment. With the aid of our proposed adaptive system we aim to decrease the soldier's risk by improving their ability to comprehend and act on available information. The Communications Scheduler was designed to improve message comprehension and situation awareness while not decrementing performance on concurrent tasks under high workload. Evaluations of the Communications Scheduler have shown over 100% improvement in message comprehension and situation Awareness. Situation Awareness is key to the ability to effectively manage mission priorities and coordinate with team members. Performance in this area is particularly difficult in high workload periods, as evidenced by the low overall scores in the evaluation. Even with the dramatic improvement as a result of the mitigation strategy, there is an opportunity here for further improvement.

6 [x] Acknowledgements

This paper was supported by a contract with DARPA and the U.S. Army Natick Soldier Center, DAAD-16-03-C-0054, for which LCDR Dylan Schmorrow serves as the Program Manager of the DARPA Augmented Cognition program and Mr. Henry Girolamo is the DARPA Agent. The opinions expressed herein are those of the authors and do not necessarily reflect the views of DARPA or the U.S. Army Natick Soldier Center.

The Honeywell team would like to acknowledge the efforts of Mr. Trent Reusser, Ms. Janet Creaser, Mr. Jeff Rye, and Ms. Danni Bayn for work on the pilot studies, scenario definition, and data analyses.

References

Dorneich, M.C., Mathan, S., Creaser, J., Whitlow, S.D., and Ververs, P.M. (2005). "Enabling Improved Performance though a Closed-Loop Adaptive System Driven by Real-Time Assessment of Cognitive State." *Proceedings of the 11th International Conference on Human-Computer Interaction (Augmented Cognition International)*, Las Vegas, Jul 22-27, 2005

Dorneich, M., Whitlow, S., Ververs, P.M., Carciofini, J., and Creaser, J. (2004) "Closing the Loop of an Adaptive System with Cognitive State," *Proceedings of the Human Factors and Ergonomics Society Conference 2004*, New Orleans, September.

Dorneich, Michael C., Whitlow, Stephen D., Ververs, Patricia May, Rogers, William H. (2003). "Mitigating Cognitive Bottlenecks via an Augmented Cognition Adaptive System," *Proceedings of the 2003 IEEE International Conference on Systems, Man, and Cybernetics*, Washington DC, October 5-8, 2003. (Invited)

DuRousseau, D.R., (2004) *Multimodal Cognitive Assessment System*. Final Technical Report, DARPA, DAAH01-03-C-R232.

Freeman, F.G., Mikulka, P.J., Prinzel, L.J., & Scerbo, M.W. (1999). Evaluation of an adaptive automation system using three EEG indices with a visual tracking system. *Biological Psychology, 50*, 61-76.

Hoover, A. & Muth, E.(in press). A Real-Time Index of Vagal Activity. *International Journal of Human-Computer Interaction*.

Kass, S J, Doyle, M Raj, AK, Andrasik, F, & Higgins, J (2003, April). Intelligent adaptive automation for safer work environments. In J.C. Wallace & G. Chen, *Occupational health and safety: Encompassing personality, emotion, teams, and automation*. Symposium conducted at the Society for Industrial and Organizational Psychology 18th Annual Conference, Orlando, FL.

Mathan, S., Mazaeva, N., & Whitlow, S. (2005). Sensor-based cognitive state assessment in a mobile environment. *Proceedings of the 11th International Conference on Human-Computer Interaction (Augmented Cognition International)*, Las Vegas, Jul 22-27, 2005.

Mathan, S., Dorneich, M., & Whitlow, S. (2005). Automation Etiquette in the Augmented Cognition Context. *Proceedings of the 11th International Conference on Human-Computer Interaction (Augmented Cognition International)*, Las Vegas, Jul 22-27, 2005.

Mikulka, P. J., Scerbo, M. W., & Freeman, F. G. (2002). Effects of a Biocybernetic System on Vigilance Performance. *Human Factors, 44*(4), 654-664.

Pope, A.T., Bogart, E.H., & Bartolome, D.S (1995). Biocybernetic system validates index of operator engagement in automated task. *Biological Psychology, 40*, 187-195.

Prinzel, III, L.J., Hadley, G., Freeman, F.G., & Mikulka, P.J. (1999). Behavioral, subjective, and psychophysiological correlates of various schedules of short-cycle automation. In M. Scerbo & K. Krahl (Eds.), *Automation technology and human performance: Current research and trends*. Mahwah, NJ: Lawrence Erlbaum Associates.

Raj A.K., Perry, J.F., Abraham, L.J., Rupert A.H. (2003) Tactile interfaces for decision making support under high workload conditions, *Aerospace Medical Association 74th Annual Scientific Meeting*, San Antonio, TX..

Sajda, P., Gerson, A., & Parra, L. (2003). Spatial Signatures of Visual Object Recognition Events Learned from Single-trial Analysis of EEG, *IEEE Engineering in Medicine and Biology Annual Meeting*, Cancun, Mexico.

Sarter, N. B., Woods, D. D., Billings, C. E., "Automation surprises," Handbook of Human Factors and Ergonomics, 2nd Edition, Salvendy, Gavriel (Ed)., Wiley and Sons, 1997.

Whitlow, S.D., and Ververs, P.M. (2005). "Scheduling Communications with an Adaptive System Driven by Real-Time Assessment of Cognitive State." Accepted to the *11th International Conference on Human-Computer Interaction (Augmented Cognition International)*, Las Vegas, Jul 22-27, 2005

Wickens, C.D., Heffley, E., Kramer, A. & Donchin, E. (1980) The event-related brain potential as an index of attention allocation in complex displays. *Proceedings of the 24th Annual Meeting of the Human Factors Society*. Santa Monica, CA, 1980.

Some Implications and Challenges for "Augmented Cognition" from the Perspective of The Theory of Complex and Cognitive Systems

Robert R. Hoffman

Institute for Human and Machine Cognition
40 S. Alcaniz St.
Pensacola, FL 32502
rhoffman@ihmc.us

Abstract

A number of cautionary tales for the Augmented Cognition Program can be derived from:
- The Theory of Complex and Cognitive Systems (TOCCS) (Hoffman & Woods, 2005; Woods and Hoffman, forthcoming) and
- The principles of Human-Centered Computing (HCC) (Endsley & Hoffman, 2002; Hoffman & Hayes, 2004; Hoffman, Hayes, & Ford, 2001; Hoffman, Lintern & Eitelman, 2004; Hoffman, Roesler, & Moon, 2004). These challenges stem from the assumptions and goals of the program, taken in light of what is known about cognitive work as it occurs in 'real-world' contexts.

This article invites reflection on the assumptions and goals of the program, with an eye toward anticipating and coping with potential difficulties and roadblocks that may occur as the program work progresses, especially the "reductive tendency" and the driving notion of bias mitigation.

1. Assumption One

Augmented cognition technologies will help humans conduct cognitive work in realistic contexts.

In general, contexts for cognitive work in complex sociotechnical systems are typified by dynamics, interactions, and context-dependencies—many of which cannot be anticipated as events unfold. In such domains, people do not conduct tasks. They engage in context-sensitive, knowledge-driven choice among action sequence alternatives. It only appears as if people conduct tasks when the case at hand is routine, or when a stepwise procedure is mandated. Kludges, work-arounds and local adaptations are an inevitable feature of human interaction with complex information processing systems (Koopman & Hoffman, 2003).

Designs for new decision-aiding systems are hypotheses about how the technologies will change the work in the "envisioned world" (Woods & Dekker, 2004). The introduction of new technology, including appropriately human-centered technology, will bring about changes in environmental constraints (i.e., features of the sociotechnical system, or the context of practice). Even though the domain constraints may remain unchanged, and even if cognitive constraints are leveraged and amplified, changes to the environmental constraints may be negative (Dekker, Nyce, & Hoffman, 2004, p. 74).

Furthermore, the sociotechnical workplace is always a moving target. By the time one has approached the point of testing an envisioned world hypothesis, the sociotechnical workplace in which the work will be carried out will *already* have changed.

As cognitive engineering research in aviation accidents and industrial process control accidents has shown, realistic contexts are also typified by "automation surprises," in which machine actions are not transparent to the human and sometimes are at odds with the human's goals. It is predictable that even though Augmented Cognition technologies might actively influence human cognition, they may do so in a way that is not transparent to the human. This will cause automation surprises, and disruptions will result. To be "team players," humans and machines must achieve and maintain a common ground, and the machines need to make their intent and stance always clear to the humans (Christofferson & Woods, 2002; Klein, et al., 2004).

An implication of these considerations is that micro-scale models of cognition can be brittle and transient and of limited help in evolutionary design (as opposed to the re-design of legacy or mandated systems). It is proving of value to describe cognitive work at a macro-level, involving such activities as problem detection, sensemaking, replanning, coordinating, etc., and these activities are always parallel and always strongly interactive (Klein, et al., 2003).

2. Assumption Two

Augmented cognition technologies are needed because human reasoning is typified by biases and limitations.

It can be argued that many of the so-called reasoning biases are not really biases at all (Hoffman, 2004). Humans are "limited" only from the perspective of a hypothetical agent that has capabilities that exceed those of the human (Flach & Hoffman, 2003). Historically the comparison has been to machines, which are praised for their logic, precision, reliability, and other features that are counterpoint to human irrationality, emotionality. This "machine-centered" approach is contrasted with a "human-centered" approach, which emphasizes the strengths of humans and the limitations of machines. Humans are the only "machine" capable of reasoning at the knowledge level and in a way that is sensitive to context (Hoffman, Hayes, & Ford, 2001; Hoffman, Hayes, Ford, & Hancock, 2002). It is always humans who repair the problems caused by machines that act reliably, but incorrectly due to their literal-mindedness and insensitivity to context.

In many current venues for the development of advanced decision architectures, the assumption being made is that human reasoning is characterized by bias and limitations, and therefore technologies are needed to compensate, and can compensate. This assumption permits the creation of certain kinds of technologies (interfaces, etc.) without a clear understanding of the nature of the cognitive work that actually has to be accomplished, and the extent to which the true work depends most critically not on overcoming human biases but on supporting cognitive work at a knowledge level and in a way that is sensitive to context (Dekker, Nyce, & Hoffman, 2003).

3. The Reductive Tendency

A theme to the challenges that TOCCS and HCC present to Augmented Cognition involves an important way in which human capabilities affect complex cognitive systems. This is the "reductive tendency" (Feltovich, Hoffman, & Woods, 2004). Research on the psychology of expertise has revealed a number of "dimensions of difficulty" that make complex problems difficult for humans. Examples are: Continuous events tend to be regarded as discrete; interactive processes tend to be regarded as separate; nonlinear relationships tend to be regarded as being linear; dependent events tend to be seen as independent, and so on. Humans tend to create simplified understandings of complex problems. (This is not a bias or a "limitation," but is a necessary consequence of human learning—at any point in time anyone's knowledge of something is bound to be incomplete.) In areas of complex cognition, the reductive tendency can lead to significant misconceptions and error-ridden performance.

In addition, the misconceptions often resist change. When learners are confronted with evidence contrary to their mental models, they perform reasoning maneuvers to rationalize their faulty beliefs without fundamentally altering them. These protective operations are called knowledge shields, and researchers have identified 23 of them (Feltovich, Coulson, & Spior, 2001; Feltovich, Spiro, & Coulson, 1997;Spiro, et al., 1988, 1989).

An example of how the reductive tendency affects the design of CSSs is the "substitution myth." This is the assumption that when a work system component is replaced by some new device or tool, that the work's fundamental nature will remain essentially the same. That is, the replacement device (or human-augmentation diad) will function essentially much the same way that the original component did (although it might be faster, more efficient, and so on). This is almost never the case; the work *system* changes dramatically, often in unintended, unanticipated, and undesirable ways.

The reductive tendency is directly pertinent to Augmented Cognition because the humans who are designing Augmented Cognition technologies are themselves subject to the reductive tendency. Collaborative research aimed at creating Augmented Cognition technologies is *itself* a complex cognitive system—a workplace in which individuals act as collectives with the support of information technology, to conduct cognitive work.

- The cognitive engineer might be prone to think of continuous processes in terms of discernable steps.
- The cognitive engineer might fail to appreciate the widespread interdependency of effects across workplace components.
- The cognitive engineer might think of the work practice as a linear set of workplace steps, as in an assembly line, rather than a matter of interactiveness and simultaneity.
- The cognitive engineer might regard the envisioned CSS as a set of low-level, direct causes and effects.
- The cognitive engineer might assume that changes, effects of interventions, and perturbations of various kinds to the CSS will have incremental, manageable consequences.
- The cognitive engineer might assume that a design principle has the same applicability and effects throughout the many different and changing contexts of work and practice.

These kinds of reductive assumptions can, and usually do, result in technologies that impose difficulties in coordination and communication, and in building and maintaining common ground. This arises especially when situations are challenging or novel. While designers might be confident that demands for coordination have been reduced, when new technologies are in use, the practitioners experience automation surprises. Designs for systems based on knowledge of how macrocognitive functions operate during difficult or tough cases will tend to be more robust than systems designed around models stemming from the analysis of routine cases.

4. Follow-on

On the basis of the laws of cognitive work that form the core of TOCCS and express the philosophy of HCC, it would be predicted that the issues and challenges raised here (and some additional ones) have *already* arisen in the Augmented Cognition program efforts. Those same laws, however, if taken as design guidance, may help the program anticipate and cope with the challenges. Some laws are described in Table 1, below.

Table 1. Some Laws of Complex and Cognitive Systems that might serve as design guidance.

Law	Departures from human-centeredness that often occur when the law is not followed.
The Triples Rule The unit of analysis for complex cognitive systems is the triple of people-machines-contexts (Hoffman, Hayes, Ford, & Hancock, 2002).	• Automation of everything that can possibly be automated. • The assumption that complex cognitive systems can be designed and used in a context-independent manner.
Aretha Franklin Law Do not devalue the human in order to justify the machine. Do not criticize the machine in order to rationalize the human. Advocate the human-machine system in order to amplify both (Flach & Hoffman, 2003).	• The assumption that the purpose of the technology is to overcome human limitations rather than serve to amplify and extend human abilities to perceive, know, reason, and collaborate. • Over-applying technology, and applying it inappropriately.
Sacagawea Law Human-centered computational tools need to support active organization of information, active search for information, active exploration of information, reflection on the meaning of information, and evaluation and choice among action sequence alternatives (Endsley & Hoffman, 2002).	• Failure to fully involve the users throughout the design and development process. Failure of the technologists to "live with" the users. • Systems that cause automation surprises and misfits between the enforced procedures and contextual circumstances.

(Table 1 continues)

(Table 1, continued)

The Lewis and Clark Law The human user of the guidance needs to be shown the guidance in a way that is organized in terms of their major goals. Information needed for each particular goal should be shown in a meaningful form, and should allow the human to directly comprehend the major decisions associated with each goal (Endsley & Hoffman, 2002).	• Imposition on the user of the designer's model of the task. • Systems that cause automation surprises and misfits between the enforced procedures and contextual circumstances.
The Pleasure Law Human-centered tools provide a feeling of direct engagement. They simultaneously provide a feeling of flow and challenge.	• Tools that force the human to adapt. • Effort is required to work the technology rather than work on the problem at hand.
The Janus Law Human-centered systems do not force a separation between learning and performance. They integrate them. Training always involves a performance or practice component, but cognitive work in context always requires continual learning.	• Empirically false and potentially dangerous distinction between learning aids and performance support systems. • Systems that do not permit the human to continually expand their knowledge and reasoning skills.
The Mirror-Mirror Law Every participant in a cognitive complex system will form a model, a potentially incomplete or incorrect model, of the other participant agents as well as a model of the controlled process and its environment.	• Systems that do not support direct perception or that fail to induce in the human a veridical model of the system that is being manipulated or controlled.
The Moving Target Law The sociotechnical workplace is constantly changing, and constant change in environmental constraints may entail constant change in cognitive constraints, even if domain constraints remain constant (Dekker, Nyce, & Hoffman, 2003).	• Failure to include capability for future expansion of deployed systems. • Failure to anticipate a need to incorporate the user's desirements. • Failure to see requirements specification as a process.
The Fort Knox Law The knowledge and skills of proficient workers is gold. It must be elicited and preserved, but the gold must not simply be stored and safeguarded. It must be disseminated and utilized within the organization when needed (Hoffman & Hanes, 2003).	• Failure to involve domain experts ("users") in the system design and development process, from the beginning. • Failure to fold on-going knowledge management processes into CCS operations and activities. The human-centered tool makes the true work easier, and at the same time enables knowledge capture and knowledge sharing as a "freebie."

5. References

Christoffersen, K., & Woods, D. D. (2002). How to make automated systems team players. *Advances in Human Performance and Cognitive Engineering Research, 2,* 1-12.

Dekker, S. W. A., Nyce, J. M., & Hoffman, R. R. (March-April 2003). From contextual inquiry to designable futures: What do we need to get there? *IEEE Intelligent Systems,* pp. 74-77.

Endsley, A., & Hoffman, R. (November/December 2002) The Sacagawea Principle. *IEEE Intelligent Systems,* pp. 80-85.

Feltovich, P. J., Coulson, R. L., Spiro, R. J (2001). Learners' (mis)understanding of important and difficult concepts: A challenge for smart machines in education. In K. D. Forbus & P. J. Feltovich (Eds.), *Smart machines in education: The coming revolution in educational technology* (pp. 349-37). Menlo Park, CA: AAAI/MIT Press.

Feltovich, P.J., Hoffman, R.R., & Woods, D. (May/June 2004). Keeping it too simple: How the reductive tendency affects cognitive engineering. *IEEE Intelligent Systems,* pp. 90-95.

Feltovich, P. J., Spiro, R. J., & Coulson, R. L (1997). Issues of expert flexibility in contexts characterized by complexity and change. In P. J. Feltovich, K. M. Ford, & R. R. Hoffman (Eds.), *Expertise in context: Human and machine.* (pp. 125-146). Cambridge: AAAI/MIT Press.

Flach, J., & Hoffman, R. R. (January-February 2003). The limitations of limitations. *IEEE Intelligent Systems.* (Pp. 94-97).

Hoffman, R. R. (2004). Biased About Biases: The "Theory of the Handicapped Mind" in the Psychology of Intelligence Analysis. In *Proceedings of the 48th Annual Meeting of the Human Factors and Ergonomics Society* (pp. 406-410). Santa Monica, CA: Human Factors and Ergonomics Society.

Hoffman, R. R., & Hanes, L. F. (July-August 2003). The boiled frog problem. *IEEE Intelligent Systems,* pp. 68-71.

Hoffman, R. R., & Hayes, P. J., (January/February 2004). The Pleasure Principle. *IEEE: Intelligent Systems,* pp. 86-89.

Hoffman, R. R., Hayes, P. J., & Ford, K. M. (September-October 2001). Human-Centered computing: Thinking in and outside the box. *IEEE: Intelligent Systems,* pp. 76-78.

Hoffman, R. R., Hayes, P., Ford, K. M., & Hancock, P. A. (May/June 2002). The Triples Rule. *IEEE: Intelligent Systems,* 62-65.

Hoffman, R. R., Lintern, G., & Eitelman, S. (March/April 2004). The Janus Principle. *IEEE: Intelligent Systems,* pp. 78-80.

Hoffman, R. R., Roesler, A., & Moon, B. M. (July/August 2004). What is design in the context of Human-Centered Computing? *IEEE: Intelligent Systems,* pp. 89-95.

Hoffman, R. R., & Woods, D. D. (January/February, 2005). Steps toward a theory of complex and cognitive systems. *IEEE: Intelligent Systems,* pp. 76-79.

Klein, G., Ross, K. G., Moon, B. M., Klein, D. E., Hoffman, R. R., & Hollnagel, E. (May/June 2003). Macrocognition. *IEEE Intelligent Systems,* pp. 81-85.

Klein, G., Woods, D. D., Bradshaw, J. D., Hoffman, R. R., & Feltovich, P. J. (November/December 2004). Ten Challenges for Making automation a "Team player" in joint human-agent activity. *IEEE: Intelligent Systems,* (pp. 91-95).

Koopman, P., & Hoffman, R. R., (November/December 2003). Work-arounds, make-work, and kludges. *IEEE: Intelligent Systems,* pp. 70-75.

Spiro, R. J., Coulson, R. L., Feltovich, P. J., and Anderson, D. K. (1988). Cognitive flexibility theory: Advanced knowledge acquisition in ill-structured domains. *Proceeding of the 10th Annual Conference of the Cognitive Science Society* (pp. 375-383). Hillsdale, NJ: Lawrence Erlbaum Associates.

Spiro, R. J., Feltovich, P. J., Coulson, R. L., & Anderson, D. K (1989). Multiple analogies for complex concepts: Antidotes for analogy-induced misconception in advanced knowledge acquisition. In S. Vosniadou & A. Ortony, (Eds.), *Similarity and analogical reasoning* (pp. 498-531). Cambridge: Cambridge University Press.

Woods, D. D., & Dekker, S. W. A. (2001). Anticipating the effects of technology change: A new era of dynamics for human factors," *Theoretical Issues in Ergonomics Science, 1,* 272–282.

4. H. Beyer and K. Holtzblatt,

Woods, D. D., & Hoffman, R. R. (forthcoming). *The theory of complex and cognitive systems.*

Section 1
Human Information Processing

Chapter 4

Task Specific Information Processing in Operational and Virtual Environments

Task Specific Information Processing in Operational and Virtual Environments

Klaus Mathiak

Psychiatry & Psychotherapie, RWTH Aachen
Pauwelsstr. 30, 52074 Aachen, Germany
KMathiak@UKAachen.de

Abstract

Virtual reality (VR) is a continuously evolving field. By understanding the basics of crossmodal perception in virtual environments, a more and more natural percept can be achieved under laboratory conditions. Based on such models, minute research on human behavior and physiology during almost unrestricted behavior can be conducted. Here we will present latest advances in the realization of multimodal environments (Berger, MPI Tübingen, Germany) and discuss spatial frameworks to implement augmented and virtual reality (Biocca, Michigan State University, East Lansing, MI). Then Russo (Fort Rucker, Alabama) will present data on task performance during highly complex behavior in VR and Rani (Vanderbilt University, Nashville, TN). Finally, we will discuss if brain physiology can be monitored in this situation. Computer games simulate social behavior and allow for the observation of cognitive and emotive brain activity during interaction (Mathiak, RWTH Aachen, Germany). Driving simulation can be as well studied by means of functional brain imaging; moreover, Kincses (DaimlerChrysler AG, Stuttgart, Germany) attempted to relate the findings to brain activity during real driving. VR and the observation of human physiology may narrow the gap between highly controlled lab experimentation and free field behavior of the human.

1 Overview

Traditionally, experimental psychology attempts to control the independent variables as much as possibly. The thought behind this is that human behavior can be described by a black-box approach. This indeed has proven to be very successful with regards to certain fundamental principles. The Weber law, for instance, allows estimating the psychological effect of changes of the physical properties of stimulus materials. However, effects of interaction are neglected. A first approximation of describing active contributions of the subject to sensory processing is named *attention*. We know that by directing selective attention to a specific point in space stimuli originating from this place are processed in general faster and with higher proficiency as compared to other locations. This mechanism however acts also without voluntary directing attention and seems to be a rather general mechanism of crossmodal integration (e.g., Mathiak, Hertrich, et al. 2005).

We must assume that the central nervous system is optimized to navigate the body through complex environments in an optimized fashion. Crossmodal interactions and the integration of predictive and contextual information must be considered the rule rather than exception. This thought might lead directly to a paradigmatic shift in experimental psychology. More likely, however, the improvements in computer technique and increased processing speeds favor this direction. Nowadays, rather standard personal computers are capable to simulate complex visual scenes which allow for interactions of the user. The video game development is maybe the largest driving force in that direction. In our study on virtual social interactions (Mathiak & Weber, 2005), we took profit from this situation and could simulate social interactions without investing into computers or software. Moreover, we could refer to a large pool of potential volunteers that were highly skilled in this environment and spend a rather high proportion of their lives in VR. This in turn reduced the remaining interferences from suboptimal sensory input. Clearly, visual field and auditory perception is limited; somatosensory input is hardly related to the virtual action; and vestibular input is absent (during functional imaging the brain must not move). To achieve maximal natural percepts for general audience many issues still need to be resolved.

Flight simulators are very close to the real experience and very expensive. Russo et al. (2005) investigated in such an environment effects of sleep suppression on behavioral and physiological measures. They recruited eight military pilots which in turn are also experienced with flight simulators. Oculomotor deficits served as indicator for risk of errors in behavior. The gain of physiological indicators is twofold: first, the measure might be applied as well in operational environments without interference with the behavior and, second, insight into the underlying physiology and brain mechanisms can be obtained. VR can be used to introduce additional tasks to study involvement and distraction with standardized stimuli. That is the classical psychophysics can be added to the complex situation which would be in real situations forbidden, e.g., during flying or driving. The behavior can now also be correlated with physiological measures; Russo et al. used oculomotor signals which can be obtained as well in most operational environments.

The connection between physiological measures and neuronal mechanisms is context dependent. One way to address this problem is to keep the physiology constant. In this direction, Rani et al. (2005) introduced a feedback that controlled the VR environment according to the physiological signal. Indeed this further enhances the level of interactivity with an environment. Common social interactions are based on direct interactions also on an unconscious level. Physiological parameters – such as a rush in the face, small changes in the mimic or gesture, or the diameter of the pupils – are detected and modify the behavior of the communication partner immediately. This occurs mostly unnoted but will be required to simulate social interactions fully. In these terms, another level of interactivity is opened up here.

Simulated spaces and VR conceivably will never achieve the complete natural resemblance. To detect and bridge the differences, brain measures can be compared between real and virtual situations. Kincses (2005a, 2005b) follows this line by using driving simulators on the one hand and standard car driving on the other hand. During car driving EEG can be recorded, and the cortical processing of specific interfering and informing stimuli is reflected by the evoked responses. As a general finding, we expect a higher variability because more and faster changes in the cognitive state occur.

Real and virtual environments can be directly merged. To study the correlation between existing and simulated environments, augmented reality (AR) attempt to integrate both frameworks. Biocca et al. (2005) formalize the specific cognitive sets of space representation in both modalities. This gives us a twofold benefit at hand again: first, virtual stimuli can be compared with effects of real objects and, second, this integration may be the ultimate aim to improve human information processing. So far multisensory stimulus application (e.g. Mathiak, Hertrich, et al. 2005) aims at improving responses to the stimuli itself; more important, however, is to improve responses to real objects such as to a merely visible pedestrian in the street during car driving. If this merging of virtual and real worlds proves to be successful, moreover, the responses to the virtual stimuli can be considered more valid than if considered independent from the outside environment.

2   Conclusion

The advent of improved computer and hardware is to change the scientific approach in experimental psychology. Without loosing objectivity, complex scenes can be investigated and, moreover, the interactivity of the subject can be considered. However, building up experimental designs and theoretical frameworks, we notice that we are just at the beginning: (1) presentation of specific modalities and their interactions still need to be optimized for VR; (2) some modalities still are neglected for interactivity; (3) physiological and brain imaging during VR has to be further developed; (4) VR has to be validated against operational environments; (5) the integration within AR stands at its beginning. This symposium aims at bringing together interdisciplinary groups to advance the current developments of an integrated view with advanced VR and physiological understanding of the neural mechanisms underlying complex interactive behavior.

References

Biocca, F., Tang, A., Owen, C., Mou, W., Xiao, F. (2005) Mobile Infospaces: Personal and Egocentric Space as Psychological Frames for Information Organization in Augmented Reality Environments. Proceedings of the HCI International 2005. Lawrence Erlbaum Associates, Mahwah, NJ.

Kincses, W. E. (2005a). Neuronal correlates of Driving. Proceedings of the HCI International 2005. Lawrence Erlbaum Associates, Mahwah, NJ.

Kincses, W. E. (2005b). Neural Correlates and Cognitive States of Driving. Proceedings of the HCI International 2005. Lawrence Erlbaum Associates, Mahwah, NJ.

Mathiak, K., Hertrich, I., Swirszcz, K., Zvyagintsev, M., Lutzenberger, W., Ackermann, H. (2005). Preattentive crossmodal information processing under distraction and fatigue. Proceedings of the HCI International 2005. Lawrence Erlbaum Associates, Mahwah, NJ.

Mathiak, K., Weber, R. (2005). fMRI of virtual social behavior: brain signals in virtual reality and operational environments. Proceedings of the HCI International 2005. Lawrence Erlbaum Associates, Mahwah, NJ.

Rani, P., Sarkar, N. (2005) Maintaining Optimal Challenge in Computer Games through Real-Time Physiological Feedback. Proceedings of the HCI International 2005. Lawrence Erlbaum Associates, Mahwah, NJ.

Russo, M., Sing, H., Kendall, A., Johnson, D., Santiago, S., Escolas, S., Holland, D., Thorne, D., Hall, S., Redmond, D. (2005). Correlations among visual perception, flight performance, and reaction time impairments in military pilots during 26 hours of continuous wake: Implications for automated workload control systems. Proceedings of the HCI International 2005. Lawrence Erlbaum Associates, Mahwah, NJ.

Mobile Infospaces: Personal and Egocentric Space as Psychological Frames for Information Organization in Augmented Reality Environments

Frank Biocca[†], Arthur Tang[†], Charles Owen[*], Weimin Mou[§], Xiao Fan[*]*

†Media Interface and Network Design (M.I.N.D.) Laboratories
*Media and Entertainment Technologies Laboratory
Michigan State University, East Lansing, MI 48824
Web-page: http://mindlab.org
E-mail: {biocca, tangkwo1, cbowen, xiaofan}@msu.edu

§Institute of Psychology
Chinese Academy of Sciences, Beijing, China
Web-page: http://www.psych.ac.cn
E-mail: mouw@psych.ac.cn

Abstract

How should designers of augmented reality (AR) interfaces organize the wide range of menus, messages, and data objects in the large working volume to support AR users in a wide-range of tasks? This is common challenge for designers of AR interfaces and there is very little research on optimizing the volume and continuity of information access in volumetric immersive AR systems. The Mobile Infospaces Project seeks ways to map and organize large volumes of mobile information by leveraging the human brain's prodigious capability for spatial cognition. Such a mapping may offer a potential route to biologically and psychologically driven models of interface design and potential guidelines for the design of mobile, AR interfaces. This paper introduces the overall framework for mobile infospaces, basing these on a body of research in neuropsychology, spatial cognition, and behavioral sciences. It then focuses on two sets of egocentric spaces, the personal and peripersonal spaces, and reviews mobile infospaces experiments suggesting that (1) users can adapt to experiencing and mentally updating fields of virtual objects attached to the body, and (2) there are perceptual, motor, and semantic asymmetries in the processing of icons and other data objects organized in mobile egocentric (body-centered) augmented reality spaces.

Keywords: Augmented Reality, Spatial Cognition, Human Computer Interaction, Spatial Interface Guideline, Augmented Cognition.

1   Introduction

Alan Kay described the personal computer as the first metamedium – an electronic medium which can be used to store, manipulate and access all media form such as text, images, audio, video, and 3D objects (Kay, 1984). The emergence of the World Wide Web in the last decade brought into existence the "global village interconnected by an electronic nervous system" envisioned by Marshall McLuhan (McLuhan, 1967). During this era, the computer has evolved into an information portal to databases in different media forms and a communication portal of different social activities. An unprecedented rate of information and activities can be transmitted continuously through this portal to the user. The *user interface* is analogous to a gate for this communication and information portal. It manages, and often limits, the information and commands the user is able to absorb from and deploy to the computer system.

2   Using Space as a Medium for Thought

Every medium, from traditional printed media to modern computer-mediated interactive presentations, uses spatial arrangement in some way to organize information. The standardized computer user interface for the last 25 years, the traditional WIMP (windows, icon, menu, pointer) direct manipulation interface (Shneiderman, 1983) is a 2D arrangement of icons with overlapping windows suggesting "depth" and containers, folders, that are "opened" to reveal arrays of icons. Motor interaction is also spatial, as the system is controlled by a virtual hand or pointer connected to a 2D spatial input device, the mouse. With the advent of body motion tracking systems and low-cost high-performance graphics workstations in recent years, novel highly spatial augmented reality (AR) interfaces visualized in Hollywood movies, video games and science fiction are getting closer to reality. These interfaces tightly couple spatial 3D stimuli to movement of the user's body in which the sensors and effectors of the computer system are mapped to the user's body schema (Biocca, 1997). Making use of the greater range of human sensorimotor capability, volumetric AR interfaces promise high communication bandwidth between the user and the computer by getting out of the shadow of its technological ancestor – the typewriter.

Volumetric AR interfaces have very unique characteristics as compared to other media and computer interfaces. The user interacts with the computer system through body motion in a volumetric space, instead of a two-dimensional screen surface. This is very different from other three dimensional screen-based interaction such as the Data Mountain (Robertson et al., 1998) and fish-tank VR (Ware, Arthur & Booth 1993 #18). These screen-based three dimensional interfaces are analogous to a window into cyberspace through which users are peering. AR is a truly immersive spatial electronic medium in which the user's perception is egocentrically immersed into the medium that arrays 2D and 3D information around the user. This unique spatial arrangement allows the display of large volumes of data, and designers are exploring ways to organize this information. In this article we examine ways to systematically organize information, focusing primarily on arrays of information around the moving body.

2.1   How Spatial Representations Leverage Spatial Cognition for Thinking

In the everyday world, humans organize and manipulate objects in space to facilitate thinking. Kirsh asserted that humans are constantly, whether consciously or subconsciously, organizing and reorganizing space in everyday life to enhance performance (Kirsh, 1995). He further categorized three epistemic actions in spatial problem solving strategies: spatial arrangements that simplify choices, spatial arrangements that simplify perception, and spatial dynamics that simplify internal computation.

Spatial schema and spatial reasoning is not just about space, but may be implicated in abstract reasoning. There is ample evidence in the fields of psychology and neuroscience that spatial cognition plays an important role in mathematical reasoning, modeling of time, language organization, and memory organization (Grabowska & Nowicka, 1996; Bryant, 1992; Bryant, Tversky & Franklin 1992; Gardner, 1983; Kirsh & Maglio, 1992). The use of spatial representation and organization to enhance human cognition has been a successful strategy since the effective mnemonic strategies of ancient Greece. Demosthenes, a Greek orator born around 384 B.C., used a strategy known as "Method of Loci" to memorize long speeches by mentally walking through his house, associating each element in the speech with different spots or objects in the house.

How information is spatially represented can facilitate cognition. Different spatial arrangement of physical objects, for example, can dramatically affect how people solve a problem. Zhang and Norman reported an experiment showing that subjects' performance when solving the Tower of Hanoi problem was drastically affected by the spatial placement of the problem pieces (Zhang & Norman, 1994). Much of the problem representation of the Tower of Hanoi problem can be offloaded to the external spatial representation of the problem pieces, and as a result, the load of internal working memory can be reduced and more working memory capacity can be used for problem solving.

There is some historical evidence that the arrival of new ways to visualize information has had dramatic impact on engineering and the sciences. Virtual environments and visualizations represent information spatially through proximity, color gradation, or spatial arrays to allow users to immediately grasp large amounts of quantitative data and complex mathematical relationships (Ware, 2000; Card, Mackinlay, & Shneiderman, 1999). Spatial arrays can be intuitive for even novel users. For example, Merickel found that virtual reality enhanced children's ability to solve spatially related problems (Merickel, 1992).

2.2   Spatial Cognition and Augmented Reality Space

This brings us to one of the most spatial of all media, mobile augmented reality systems. Wearable and mobile AR systems have the potential to provide continuous support for virtual space, continuous visualized information arrays, integrating, annotating, and interaction with physical space. They can be powerful "cognitive artifacts" (Norman, 1993) or "intelligence amplifying systems" (Brooks, 1996) to enhance human cognitive activities, such as attention, planning, decision making, and procedural and semantic memory.

But AR information objects have different spatial properties. Because of the nature of gravity, traditional information objects have to be physically attached to the body. In an AR computing environment, tools and information objects can remain stationary in space, and be stabilized with respect to the world and/or to user's body parts such as the head and the torso or other potential body-centric frames (Owen et al., 2005). Our space to organize information objects is increased by extending the working volume from the surface of our body to a peripersonal volume in the volumetric AR computing environment, and users are able to manipulate and access multiple information objects concurrently.

Understanding the cognitive properties of spatial frameworks is important for information organization in peripersonal space. In an AR computing environment, tools and information objects can remain stationary in space and be attached to different reference frames. Since it is impossible to generate this kind of "anti-gravity" feature in the physical environment, precious little is known about how human organize information objects in an egocentric and allocentric "weightless" environment. How might users manage and organize different information field around different frame of reference in this new environment? The current research project aims to provide a theory driven basis for AR interface guideline by mapping human spatial cognition properties to AR spaces.

3   Spatial Framework Theory: Conceptualizing the Organization of Information in Mobile Augmented Reality Spaces

So, given an environment where any kind of information can be placed anywhere in space around the user, how do we organize and separate information objects in that space? A biological or psychologically based approach might be guided by the way the brain partitions and organizes space.

Much of the human brain is allocated to track the location of the body and objects in space, especially in the planning of motor actions. Objects in the environment appear to be modelled in the brain using interrelated spatial coordinate frameworks organized around the body, objects, and the larger environment (Bryant, 1992; Previc, 1998). Neuropsychological data suggest that egocentric peripersonal space, organized around the body and especially the head, may be the primitive personal space that is most accessible (Previc, 1998).

Theory driven human-computer interaction design is necessary to develop a high performance AR interface. Kirsh argued that "methods used to manage our space are key to organization of our thought patterns and behavior" (Kirsh, 1995). If users of mobile AR systems will be accessing, organizing, and deploying large volumes of information in space, then an understanding of how the brain accesses and organizes spatial information is a sound, human factors basis for interface guidelines and research.

From the biological and psychological view the augmented reality space is not a continuous field of Cartesian space. Research in spatial cognition indicates that objects in space are represented using different spatial coordinate systems. The physical and virtual environments appear to be modelled in the brain using a set of interrelated spatial coordinate frameworks roughly organized around the body, objects, and the larger environment (Bryant, 1992; Previc, 1998; Cutting & Vishton, 1995). According to current research into brain models, we can divide spatial frameworks into the following spaces which we call *infospaces* to emphasize that they are designable frameworks: (1) personal infospace, composed primarily of the psychological perception of the volume and immediate periphery of the body; (2) peripersonal infospace (Bryant, 1992; Previc, 1990; Previc, 1998), a volume of space immediately around the body, primarily engaged during the processing and manipulation of information within the lower immediate frontal area immediately around the central axis of the body; and (2) extrapersonal (allocentric or environmental) infospace (Cutting & Vishton, 1995; Previc, 1990; Previc, 1998), the perception and modelling the information objects and environment in the upper visual field, further away and surrounding the body. Figure 1 illustrates the basic divisions.

With motion tracking technologies, AR systems afford many options for information placement relative to any of these information spaces: the environment, objects, and user's body. Figure 2 illustrates a prototype AR interface with information attached to different reference frames. Different reference frames in an AR space have different intrinsic properties.

In the section that follows we review some current research on how psychological spaces (spatial frames) might be used to organize information as spaces are designed into mobile infospaces.

156

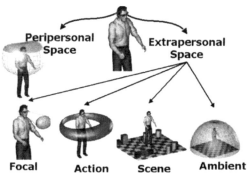

Figure 1. Representation of spatial frameworks in 3D space.

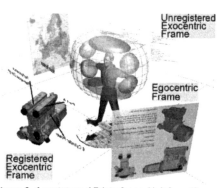

Figure 2. A prototype AR interface with information objects placed in different reference frame

4 Egocentric Reference Frames: Organizing AR Information around the Body

Egocentric reference frames are reference frames attached to the observer. Information objects can be attached to any body parts tracked by the motion tracking system. Examples include (but are not limited to) the head and body coordinate systems. Information objects attached to these reference frames have limited volume, and the mental representation of object spatial locations is tightly coupled with proprioceptive memory and action memory.

4.1 Personal (Body) Space

The clearest psychological boundary is defined by personal space, the psychological space that defines the boundary between the body, the proximal "me," and the space "outside This space not just as proprioceptive information about limb and body position but also where pericutaneous (tactile surfaces) and buccal (oral) interactions occur.

The psychological body space is not commonly seen as a possible digital information space for mobile computing. But there is evidence that personal space, defined as the shape and extent of body schema, expands to incorporate tools attached to the body. Although viewed by an objective observer, the boundary of the body appears physical and fixed, there is evidence that psychologically the sense of the boundary of the body is plastic.

Philosophers and psychologists (e.g., Heidegger, 1968; Bateson, 1972) have long speculated that the psychological boundary of the body sometimes expands so that tools near the body are integrated into the personal body space. Neuroscience studies by Maravita and Iriki (Maravita & Iriki, 2004) on neuronal motor responses during tool use by monkeys suggest that the body schema, defined as receptive fields of neurons to perceived body parts, expand to incorporated tools such as sticks and rakes after extended use. Furthermore, they show that this extension of the receptive fields extends to video representations of the monkey's body, so that the neurons respond to a displaced virtual hand. This suggests that at some level tools can appear to be incorporated into the body schema.

Another line of research that suggests how media tools can restructure body schema is the work on visual-motor adaptation in space perception. In these studies a technology is used to alter visual perception though the use of sensory prosthesis such as a prismatic lens, and adaptation to the sensory change, subsequent errors, and readaptation after the alteration is removed are observed. In studies on visual and motor hand adaptation in virtual environments, we found that augmented reality systems can remap the perceived location of the hands (motor space) relative to visual space (Biocca, 1998).

4.2 Mapping Digital Information to Personal Space in Augmented Reality Systems

In our mobile infospaces research, we have looked at ways in which there might be psychological advantages to placing virtual tools such as icons, buttons, and other digital objects on or very near to the body surface. There is

157

some neuroscience data, again using animals, that suggests that neuronal responses to body space may extend slightly beyond the skin surface:

> *That is, the visual space near the animal is represented as if it were a gelatinous medium surrounding the body that deforms whenever the head rotates or the limbs move. Such a map would divide the location of the visual stimulus with respect to the body surface, in somatotopic coordinates (Graziano, 1995/p. 1031).*

In the mobile infospaces project we consider the tracking of individual body parts to see how digital information about the user can be organized around the moving body (Owen et al., 2005).

4.2.1 Head Stabilized (Craniocentric) Information Spaces

It has become a common practice to attach digital information to the head-centered (craniocentric) reference frame of personal space. There is some evidence of a craniocentric reference frame, a neuronal spatial representation of both the location of the head but of the immediate space around the head (Graziano, 1995).

In any head mounted AR system, it is not unusual to see "virtual mask" views, that is information like status data (e.g., time, orientation, battery power) to a 2D plane in front of the field of view, sometimes in a location perceptually close to head. The information is attached to craniocentric space, in that it is fixed relative to the cranium and moves as the head moves. Information placed in the head-stabilized frame is stationary with respect to the user's viewpoint and is always visible to the user. This reference frame is being extensively used by military pilot Head-Up Displays (HUD). Although the visual region of this volume can be as wide as the field-of-view (approximately 220° by 160°), in AR and VR systems the working volume of this frame is more often very limited, due to the field-of-view of existing display technologies. Only a few information objects can be placed in the head stabilized frame.

There is little research or guidelines on the relative advantages and ideal placement of such heads-up information. The head-stabilized frame can be used for system or alert messages that need the immediate attention. However, studies related to automobile HUD design suggested that symbology place within a 5° radius of the fovea is annoying to the driver (Sojourner & Antin, 1990). Furthermore, although the fovea is not always fixed on objects in the center of display, we suggest that the center portion of the head-stabilized frame remained unused to reduce visual clutter of the real environment, and that the lower region in mobile systems be sparingly used so as to not interfere with walking and object manipulation. Interface designers also need to be aware that visual stimulus presented in this frame will not be perceived by the user after a period of time. A simple illustration of this is if a piece of dust is attached to glasses, the observer would not perceive its existence after a while.

4.2.2 Limb Surface and Stabilized Information Organization

The surfaces of the body are also information spaces. Of course, displays such as clothing, tattoos, woman's makeup, and other body-attached items are not used for communication. But these are representations used for signalling information such as status, sexual availability, and other social information to other observers. The use of the body surface in augmented reality systems to hold and display private information for the user is still rare.

There are cases in current VR and AR practice where information is attached to limb stabilized space, usually for viewing by both the user and others, for example, hands interaction is a main motor interaction for many VR and AR interfaces. In VR interfaces, hand stabilized information systems often present relevant information such as a crude cursor, representations of the hand, or virtual tools attached to the hand. Information objects attached to the hand-stabilized reference frame should be action orientated. For example, tools selected for the current action, menus, and selection trays can be attached to the non-dominant hand, and the dominant hand can be used for selection and action. But this is only a limited exploration of the use of body space for carrying information for the user.

In a sub-project affiliated with the mobile infospaces, the Digital Tattoo project, we are exploring the performance advantages to the user of attaching personalized digital information directly on the body surface, specifically to the hand and arm coordinate space. Design and data collection are currently underway, but we believe that adding proprioceptive cues can increase tool and object manipulation accuracy to the point that Fitt's Law will have different properties in personal space and that memory for the location of tools arrays can be increased.

5 Organizing Information around the Immediate Peripersonal Space of the Mobile Augmented Reality User

Another key egocentric space motivated by neuroscience research on spatial references frames is the peripersonal space (Previc, 1998). Located immediately in front on the body occupying the central 60° in the lower visual field and with a radial extension of 0-2 m, it overlaps considerably with the ergonomic space known as the **reach envelop** (Proctor and Van Zandt, 1994). Peripersonal space appears to be functionally organized for binocular object inspection, hand motion, and manipulations such as directly reaching and handling objects. This interpretation is supported by behavioural evidence, in that information and objects in this area are found and manipulated the fastest.

So, how can this psychological space becomes an information space? Motion tracking technologies open up the egocentric space as a user interface modality through the use of body-stabilized frames, i.e. displays that present 3D objects in a frame aligned with the user's torso. Information objects placed in this frame remain arm-reachable regardless of the user's position and orientation in the world, providing quick and easy access to objects placed in the frame. This reference frame is ideal for the placement of information objects that assist working memory (e.g. instructional procedures) and personal information objects that require continuous access.

5.1 Can Users Adapt to Automatically Tracking Large Amounts of Information Organized Around Their Body (Egocentric Reference Frame) in AR Space?

There has been limited research into information organization in an egocentric peripersonal "weightless" environment. A question that guided the mobile infospaces research project was: can naïve users perceive and memorize information attach to their own body? A study, conducted by Mou et al. (Mou et al., 2004), suggests that users with no prior experience with mobile AR systems tend to use the allocentric reference frame to access information arrays presented in AR environments. In other words, people expect the information arrays of virtual objects in AR environments to behave like arrays of objects in physical environments (i.e., when they rotate their body, objects stay in their location relative to the physical environment). But experiments in this study suggest that users who briefly experienced egocentrically centered displays of virtual objects, or those who are instructed that the display is egocentrically centered, are able to quickly adopt a body-stabilized frame of reference to code and access the locations of virtual objects in the physical environment.

6 When Information around the Body Has Different Psychological Properties: Perceptual Asymmetries in Head-centered and Body-centered Egocentric Space

The asymmetrical organization of the human brain creates significant asymmetries in human perception, cognition, and body morphology. Human handedness is a key example of this asymmetry. But it is the effect of brain asymmetry on human cognition, information processing, and representation that is of interest here although both the perceptual and motor asymmetries affect processing and perception of AR information objects organized around the moving body.

Because of this natural body asymmetry, a question emerges for AR interfaces that deploy information around egocentric space: Will the processing of icons, menus, and other information objects in body stabilized egocentric space show evidence of asymmetries in hemispheric processing in visual, auditory, and motor space? Asymmetries in motor control are certain and need little research exploration as they can be observed everyday in tool use: tools located on the side of the dominant hand will be more quickly accessed and be manipulated with greater motor control. But what about other aspects of information processing such as attention, decision making, and memory? Understanding these asymmetries might provide insight into how to optimize the placement of different classes of information so that they are suitably mapped to the strengths of asymmetrical processing.

6.1 Bilateral Asymmetry in the Visual Field

The concept of contralaterality (the difference in information processing between the two cerebral hemispheres) was documented as early as 2500 B.C. by the ancient Egyptian. Visual processing consistently shows a left-right bias in visual perception and information processing. A source of the hemispheric asymmetry is the organization of the retinas. Visual information of the left visual field of both eyes is received by the right side of the retinas, and then connected to the visual pathway leading to the visual cortex of the right hemisphere. Similarly, visual information of the right visual field of both eyes is received by the left side of the retinas, and then connected to the visual pathway leading to the visual cortex of the left hemisphere. Figure 3 illustrates these visual pathways. Hence, visual information from the left and right visual fields projects exclusively to the contralateral cerebral hemispheres (Bryden, 1982). Researchers in visual perception often use the tachistoscopic technique to test different hypotheses in perceptual bilateral asymmetry effects. A subject is asked to fixate his/her eyes on the center of a screen, and stimulus materials are flashed to either the left or right side of the visual field. Reaction time and/or task accuracy are measured. The differential reaction time measured in this experimental technique is very short, and typically 50 milliseconds is considered a large difference. From the experiments undertaken by various researchers using the tachistoscopic technique and lesion studies, the left hemisphere (right visual field) is found to be biased towards letters and words, functional or symbolic meaning, verbal memory, local patterns (Yovel, Yovel & Levy 2001; Robertson & Lamb, 1991), higher spatial frequencies (Sergent, 1983; Sergent, 1987), categorical spatial relationships (Kosslyn, 1987), and time; while the right hemisphere (left visual field) is found to be biased towards geometric patterns, visual appearance, visual memory, global patterns, lower spatial frequency, coordinate spatial relation, emotion (Dimond & Farrington, 1977), face recognition, and sustain attention (Whitehead, 1991). Some of these experimental results are summarized in Table 1.

6.2 Do differences in Visual Processing Affect Processing of Objects in Augmented Reality Space?

The bilateral asymmetry properties of the visual system are usually investigated in highly constrained laboratory environments. Also, the effects of bilateral asymmetry apply to the visual field of the retinal image only. In real life scenarios, images initially projected to one side of the visual field on the retina may move to another side due to body and eye movement. This issue is especially significant when the user is attempting to focus attention and fixate the eyes on the object (e.g. when reading text). Hence, simply placing information items to one side of the environment does not imply the image of the information will be projected to that visual field exclusively or even dominantly.

Augmented reality systems allow for the exploration of the possible real world effects of hemispheric asymmetries in visual motor and auditory processing. Head stabilized and body stabilized information displays guarantee that visual icons and data objects will rest on one side of the body *all the time.* This asymmetry of information organization relative to the moving body rarely occurs in physical environments. In the case of visual field asymmetry, objects on one side of egocentric space will predominately, if not always, fall one on side of the visual field. Will we find evidence of effects of visual field or other left-right based asymmetries when information is placed on one side of the user because the image is predominantly falling on one visual field and therefore predominantly processed by one hemisphere? A more robust evaluation is needed to investigate the strength and application of these effects on information display.

6.3 Do users have differential reaction times to cues presented in left or right egocentric AR space?

We conducted a study to evaluate the generalizability of perceptual asymmetry properties to information display. In the first experiment, reaction time of visual change detection in different spatial location in the head stabilized reference frame was compared. No statistically significant difference was observed between reaction times to stimuli presented in different spatial locations. In the second experiment, diagram or text of a short assembly instruction were placed either on the left or on the right side of the head stabilized reference frame, and time of completion of the assembly task were compared between conditions. No statistical significant difference was observed in terms of time of completion between the positions of the instruction. Based on our evaluations, we concluded that the perceptual effects of bilateral asymmetry in reaction time are not robust, and may have little, if any, practical effects

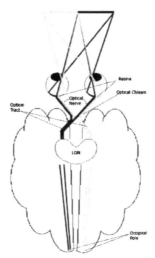

Figure 3. The visual pathway of left and right visual field. Retinal signal from the left and right visual fields projects exclusively to the contralateral cerebral hemispheres.

Left hemisphere (Right visual field)	Right hemisphere (Left visual field)
Letter and words	Geometric patterns
Function or symbolic meaning	Visual appearance
Verbal memory	Visual memory
Local patterns	Global patterns
High spatial frequency	Low spatial frequency
Categorical spatial relation	Coordinate spatial relation
Time	Emotion
	Face recognition
	Sustain attention

Table 1. Summary of cerebral hemispheric specializations.

on reaction to stimuli in AR and other information displays. Perceptual effects of bilateral asymmetry in explored in laboratory environments are usually in the scale of milliseconds. The differential effects on perceptual processing may not be significant when embedded within the longer time and other variables affecting the completion of a task (e.g. one assembly step usually takes at least a few seconds).

6.4 Do User's Respond differently to Virtual Objects and Faces Presented in Left or Right Egocentric Space or Within or Outside Peripersonal Space?

Research on the neuropsychology of 3D space and communication research on proxemics suggests that locations in space may differ in how they are processed by the brain, and that such differences affect information processing. We conducted an experiment to explore the semantic effect of egocentric spatial location of an agent (a virtual head) and a neutral object (a sphere). Results indicated that perceived urgency, aggressiveness and superiority systematically varied with position of the stimulus (Biocca, David, Tang & Lim 2004). When objects were within peripersonal space, within arm's reach and potentially threatening, all participants sensed increased urgency, aggressiveness, relevance, superiority of the all objects, and especially of the potential agent, represented by the 3D head model.

In another set of studies we examined differential memory of anthropomorphic icons and object based icons (representations) organized around the body of the user in an augmented reality space. Users had better memory for anthropomorphic and object based icons, as measured by speed to point to the item. There was no left-right asymmetry in the response. The findings suggest that spatial location of virtual objects or anthropomorphic agents (faces, anthropomorphic representations, or agents) around the body potentially influences the perceived meaning and that this effect is strongest within peripersonal space, and shows a weak but significant left-right hemisphere effect for semantic connotations. In these studies we only look into variation in connotations with shifts in spatial location. But spatial location may also interact with many other cognitive processes such as attention and memory.

7   Conclusion

The design of AR interfaces prompts a significant human factors challenge of mapping different metaphors, information, and functions of computer usage into the human cognitive system. This paper reviews some recent research in the mobile infospaces project, and a body of research in neuropsychology, spatial cognition, and behavioural sciences for the development of a theory-driven information display and organization in volumetric AR environments by mapping human spatial cognition properties to AR space. A series of major research questions in cognitive psychology tightly related to the design of AR interfaces is raised. The literature provides grounding for theory driven human-computer interaction design for the development of high performance AR interfaces, mobile infospaces potentially tailored to human spatial cognition.

8   Authors' Note

This project is based upon works supported by the National Science Foundation under Grand Number IIS 00-28743ITR. Any opinions, findings, and conclusions or recommendations expressed in this material are those of the author(s) and do not necessarily reflect the views of the National Science Foundation.

References

Bateson, G. (1972). Steps to an ecology of mind. New York, Ballantine Books.

Biocca, F. (1997). "The Cyborg's Dilemma: Progressive Embodiment in Virtual Environments." Journal of Computer-Mediated Communication 3(2).

Biocca, F., David, P., Tang, A., and Lim, L. (2004). Does virtual space come precoded with meaning? Location around the body in virtual space affects the meaning of objects and agents. Paper presented at the 54th Annual Conference of the International Communication Association, May 27 – 31, 2004, New Orleans, LA.

Brooks, F. (1996). "The Computer Scientist as Toolsmith II." Communications of the ACM 39(3): 61-68.

Bryant, D. J. (1992). "A Spatial Representation System in Humans." Psycholoquy 3(16).

Bryant, D. J., B. Tversky, et al. (1992). "Internal and External Spatial Frameworks for Representing Described Scenes." Journal of Memory and Language 31: 74-98.

Bryden, M. P. (1982). Laterality: Functional Asymmetry in the Intact Brain. New York, NY, Academic.

Card, S. K., J. D. Mackinlay, et al. (1999). Readings in information visualization: Using vision to think. San Francisco, Morgan Kaufmann.

Cutting, J. E. and P. M. Vishton (1995). Perceiving layout and knowing distances: The integration, relative potency, and contextual use of different information about depth. Perception of space and motion. W. Epstein and S. Rogers. San Diego, CA, Academic Press. 5: 69-117.

Dimond, S. J. and L. Farrington (1977). "Emotional Response to Films Shown to the Right or Left Hemisphere Measured by Heart Rate." Acta Psychologica 41: 259.

Gardner, H. (1983). Frames of Mind: The Theory of Multiple Intelligences. New York, NY, Basic Books.

Grabowska, A. and A. Nowicka (1996). "Visual-spatial-frequency Model of Cerebral Asymmetry: A Critical Survey of Behavioral and Electrophysiological Studies." Psychological Bulletin 120: 434-449.

Graziano, M. S. and C. Gross (1995). The representation of extrapersonal space: A possible role for bimodal visual-tacile neurons. The cognitive neurosciences. M. Gazzaniga. Cambridge, M.I.T. Press: 1021-1034.

Kay, A. (1984). "Computer Software." Scientific America **251**(3): 40-47.

Kirsh, D. and P. Maglio (1992). Some Epistemic Benefits of Action: Tetris, a Case Study. Fourteenth Annual Conference of the Cognitive Science Society, Hillsdale, NJ.

Kirsh, D. (1995). "The Intelligent Use of Space." Artificial Intelligence **73**(1-2): 31-68.

Heidegger, M. (1968). What is a thing? Chicago, H. Regnery Co.

Kosslyn, S. M. (1987). "Seeing and Imagining in the Cerebral Hemispheres: A Computational Approach." Psychological Review **94**: 148-175.

Maravita, A. and A. Iriki (2004). "Tools for the body (schema)." Trends in Cognitive Science **8**(2): 79-86.

McLuhan, M. (1967). Gutenberg Galaxy. Toronto, Canada, University of Toronto Press.

Mou, W., Biocca, F., Owen, C., Tang, A., Xiao, F., and Lim, L. (2004). **Frames of reference in mobile augmented reality displays**. Journal of Experimental Psychology: Appllied, Vol. 10, No. 4, pp. 238-244.ISSN: 1076-898X.

Norman, D. (1993). Things That Make us Smart: Defedning Human Attributes in the Age of the Machine. Menlo Park, CA, Addison-Wesley Publishing Co.

Owen, C., Xiao, F., Biocca, F., Tang, A., Mou, W., and Lim, L. (in press). **Information reference frames in mobile augmented reality interfaces. In** Proceedings of Human Computer Interaction International 2005, 11th International Conference on Human-Computer Interaction, July 22 – 27, 2005, Las Vegas, NV.

Previc, F. H. (1990). "Functional Specialization in the Lower and Upper Visual Fields in Humans: Its Ecological Origins and Neurophysiological Implications." Behavioral and Brain Sciences **13**: 519-542.

Previc, F. H. (1998). "The Neuropsychology of 3-D space." Psychological Bulletin **124**: 123-164.

Proctor, R. and T. Van Zandt (1994). Human factors in simple and complex systems. Boston, Allyn and Bacon.

Robertson, G., M. Czerwinski, et al. (1998). Data Mountain: Using Spatial Memory for Document Management. ACM UIST '98 Symposium on User Interface Software & Technology, San Francisco, CA.

Shneiderman, B. (1983). "Direct Manipulation: A Step Beyond Programming Languages." IEEE Computer **16**(8): 57-69.

Sergent, J. (1983). "Role of the Input in Visual Hemispheric Asymmetries." Psychological Bulletin **93**: 481-512.

Sergent, J. (1987). "Failures to Confirm the Spatial-frequency Hypothesis: Fatal Blow or Healthy Complication?" Canadian Journal of Psychology **41**: 412-428.

Sojourner, R. and J. Antin (1990). "The Effects of a Simulated Head-up Display Speedometer on Perceptual Task Performance." Human Factors **32**(3): 329-339.

Ware, C., K. Arthur, et al. (1993). Fish Tank Virtual Reality. SIGCHI conference on Human factors in computing systems, Amsterdam, The Netherlands.

Ware, C. (2000). Information visualization. San Francisco, Morgan Kaufmann.

Whitehead, R. (1991). "Right Hemisphere Superiority during Sustained Visual Attention." Journal of Cognitive Neuroscience **3**: 329-334.

Yovel, G., I. Yovel, et al. (2001). "Hemispheric Asymmetries for Global and Local Visual Perception Effects of Stimulus and Task Factors." Journal of Experimental Psychology: Human Perception and Performance **27**: 1369-1385.

Zhang, J. and D. Norman (1994). "Representations in Distributed Cognitive Tasks." Cognitive Science **18**(1): 87-122.

Cognitive Influences on Self-Rotation Perception

Daniel R. Berger

Max Planck Institute
for Biological Cybernetics
Spemannstr. 38
72076 Tübingen, Germany
daniel.berger@tuebingen.mpg.de

Markus von der Heyde

Bauhaus-Universität Weimar
Steubenstraße 6a
99421 Weimar, Germany
markus.von.der.heyde
@scc.uni-weimar.de

Heinrich H. Bülthoff

Max Planck Institute
for Biological Cybernetics
Spemannstr. 38
72076 Tübingen, Germany
heinrich.buelthoff@tuebingen.mpg.de

Abstract

In this study we examined the types of information that can influence the perception of upright (yaw) rotations. Specifically, we examined the influence of stimulus magnitude, task-induced attention and awareness of inter-sensory conflicts on the weights of visual and body cues.

Participants had to reproduce rotations that were presented as simultaneous physical body turns (via a motion platform) and visual turns displayed as a rotating scene. During the active reproduction stage, conflicts between the body and visual rotations were introduced by means of gain factors. Participants were instructed to reproduce either the visual scene rotation or the body rotation. After each trial participants reported whether or not they had perceived a conflict.

We found significant influences of the magnitude of the rotation, attention condition (instruction to reproduce platform or scene rotation), and reported awareness of a sensory conflict during the reproduction phase. Attention had a larger influence on the response of the participants when they noticed a conflict compared to when they did not perceive a conflict. Attention biased their response towards the attended modality.

Our results suggest that not only the stimulus characteristics, but also cognitive factors play a role in the estimation of the size of a rotation in an active turn reproduction task.

1 Introduction

When moving in the world, we need to generate reliable estimates of our own location and movement in space from the available sensory signals. The amount of influence each sensory modality has on the resulting estimate can depend on the characteristics of the stimulus itself, but also on cognitive factors such as attention, task, and awareness of conflicts between different sensory signals.

The different senses provide signals with different reliabilities depending on the velocity and acceleration range of self-motion. Very slow or long-lasting movements cannot be sensed accurately by the vestibular system because of its high-pass transduction characteristics. The visual sense, on the other hand, may be more accurate for small and slow movements than for fast and large movements. For an optimal estimation of the movement, given the available sensory inputs, the brain should weight the different modalities according to their reliabilities, giving more weight to more reliable cues (Ernst & Bülthoff, 2004). Effects of sensory reliability on the weights have for example been found in the integration of visual and haptic modalities (Ernst, Banks & Bülthoff, 2000; Ernst & Banks 2002; Gepshtein & Banks, 2003). Current models of the perception of self-motion contain filters to model the dynamic response of the different sensors. By such filtering the amount of influence each cue has on the resulting percept in the model is also dependent on movement magnitude (Mergener, Schweigart, Kolev, Hlavacka & Becker, 1995; Zupan, Merfeld & Darlot, 2002).

While the influence of stimulus characteristics on perception has been addressed in many multisensory integration studies, few have looked at the influence of cognitive factors such as attention (Calvert & Thesen, 2004). The normal psychophysical procedure is to limit the influence of cognitive factors in careful experimental designs, instead of using them as an experimental variable. A few studies have studied cognitive effects on the perception of self-motion. Kitazaki and Sato (2003) investigated the influence of guided attention on the illusory perception of self-motion induced by visual stimuli (vection). A study by Lambrey & Berthoz (2004) recently addressed the influence of conflict awareness on multisensory integration in self-motion in a navigation task.

Figure 1: Left: Schematic view of the motion platform with projection screen. In the experiments described in this paper, a flat projection screen was used. Right: Visual pattern used in the experiments. The scene was viewed through red-cyan glasses. The motion of the scene was displayed with moving triangles with limited lifetime to avoid tracking of individual elements.

The aim of this study was to examine the influence of stimulus-related and cognitive factors on the integration of visual and body cues of self-motion in the perception of upright (yaw) rotations. We used different rotation magnitudes as a means to manipulate the reliabilities of visual and body cues for self-motion in a natural way. To investigate the influence of task-related attention, we used two different tasks, to "turn back the visual scene" and to "turn back the platform", which guided the attention of the participant to either one of the modalities. After each turn, the participant indicated whether a conflict had been perceived or not. This helped us to examine the influence of conflict perception on the responses.

2   Materials and Methods

Thirteen participants took part in this experiment (7 female and 6 male), randomly drawn from the MPI subject database. They gave their informed consent and were paid for participation.

Participants were seated on a Stewart platform equipped with a projection screen (see Figure 1, left), which had a projected visual field of 86° horizontal × 63° vertical. They viewed a field of random triangles with limited lifetime of 1.2 seconds, presented in 3D by using red-cyan colour anaglyphic viewers (Figure 1, right). The colours of background and triangles were adjusted to minimize inter-ocular crosstalk. The 3D triangles appeared further away than the screen (up to 2-3 meters). Triangle brightness was attenuated with distance to enhance the impression of depth. A fixation cross at eye height was drawn approximately 10 cm in front of the screen. Participants were told to remain fixated during all turns.
Physical body rotations were performed by rotating the motion platform. To rotate the observer in the visual scene, the virtual camera was rotated in this random triangle field in the direction opposite to the platform rotation. The screen then appeared as a window which turned with the observer, and through which a non-rotating outside scene could be observed. The rotation did not introduce any parallax motion in the image.
Previous studies have shown that the impression of self-motion is enhanced by both stereoscopic cues (Palmisano, 2002) and the fact that the moving scene is perceived as further away than fixation point and projection frame (Kitazaki & Sato, 2003). We chose to use random triangles instead of random dots to reduce the correspondence problems when viewing a random dot field in stereo with dots that all look the same.

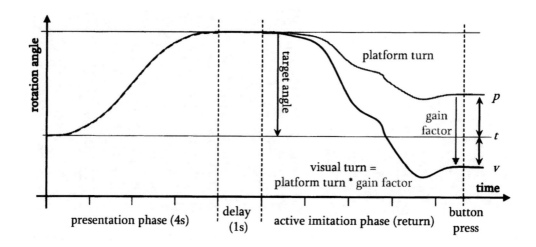

Figure 2: Example rotation trajectory of one trial. Red line is platform rotation, blue line visual scene rotation. During the presentation of a rotation (left), platform and visual scene turn equally far, using the same motion profile. During the active return (right), a gain factor causes the visual scene to turn more or less than the platform. The visual turn profile is equal to the platform turn profile multiplied by a constant gain value.
From the size of the active rotation, the weights of the visual cue and the body cues can be computed:
$visual\_weight = (t - p) / (v - p)$, $body\_cue\_weight = 1 - visual\_weight$.

Every trial of the experiment began with the presentation of a yaw rotation. The participant was rotated passively both on the platform and within the visual scene, by 10°, 12.5°, 15°, 20°, 25° or 30° within 3 seconds. The visual and body turns had consistent profiles (same speed and final angle). We used a cosine velocity profile, which ensured smooth accelerations and decelerations (Figure 2, *presentation phase*). The duration of each turn was kept constant to prevent subjects from estimating the size of a rotation from its duration. Thus, larger rotations also involved higher velocities and accelerations.

After the presentation phase, a short delay of 1 second was inserted before the participant was allowed to turn back. During the delay the instruction to "turn back the visual scene" or "turn back the platform" was given aurally to remind the participant of the current task condition. We kept the delay short to prevent decay of the memory of the presented rotation.

After the presentation phase the participant was given control of the movement of platform and scene via a joystick. The task was to turn the platform (and/or scene, see below) for the same angle in the opposite direction of the previously presented rotation (Figure 2, *active imitation phase*). After completing the rotation, the participant pressed one of three joystick buttons (see below). The next rotation was then presented.

The experiment consisted of two blocks of 108 trials (six rotation angles × nine gain factors (see below), each condition measured twice). In one block participants were told to return the visual scene ("RVis" condition), in the other to return the platform ("RBody" condition). These instructions guided attention to either the visual cue or the (inertial) body cues of self-motion. Note that the only difference between the RVis and RBody conditions was the focus of attention. The presented stimuli were the same in both conditions.

During the active return, a gain factor (factorial difference in rotation velocity) was introduced between the visual rotation and the body rotation. This factor was chosen from 0.35, 0.5, 0.71, 0.84, 1.0, 1.19, 1.41, 2.0, or 2.83. Every combination of a presented rotation angle and a gain factor during reproduction was tested twice. Participants were explicitly told about the gain factors. After each return movement, the participant had to press one of three joystick buttons, depending on whether the visual scene had been perceived as faster, the body rotation had been perceived as faster, or no difference between the two had been noticed. This allowed analysis of the influence of perceiving a conflict on the responses. As participants had to compare the rotation magnitudes of both modalities, they could not completely ignore the unattended modality.

The yaw rotation range of the motion platform was limited to 100°. Since we did not want to reposition the platform between trials, an online trial selection algorithm was used. Depending on the current position of the platform, one trial was selected from the subset of feasible trials, which maximized the balance of trials presented so far, balanced over rotation angles and gain factors. If several trials were equally good, the choice among them was made randomly. Only if no possible trial was left, the platform was repositioned. This happened approximately one or two times per experimental block of 108 trials, depending on the participant's behavior. Repositioning was for example necessary if a participant had turned way too far and the platform had reached the limit of its motion range.

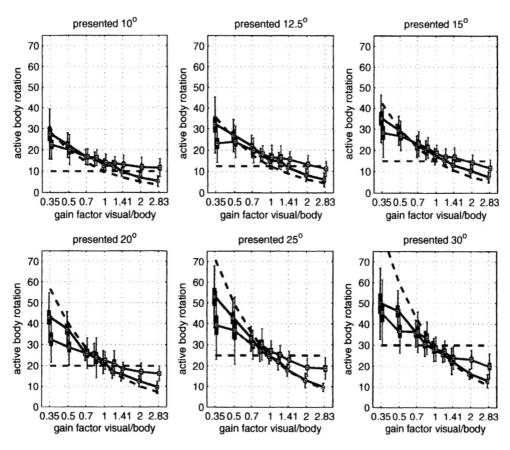

Figure 3: Response curves for RBody (red) and RVis (blue) attention conditions, for all 13 participants. Responses are plotted in active platform rotation angles over gain factors. Dashed red (horizontal) line: target angle when correctly turning back the platform; dashed blue (curved) line: target angle when correctly turning back the visual scene. For a gain factor of 0.35, for example, the platform has to be over-turned by a factor of 2.83 to return the visual scene accurately.

167

3 Results

For gain factors of 0.5 or less, or 2.0 or more, the responses in the RVis and RBody conditions differed from each other (Figure 3). Sensory conflicts caused by those gain factors were also often recognized by the participants. Three-way analysis of variance (ANOVA) calculated on the raw data with attention condition (RVis or RBody), presented rotation angle, and gain factor as within-subject factors revealed a significant interaction of attention condition with the gain factors ($F(8,96) = 10.69$, $p <.001***$). Thus attention to one or the other cue – expressed in the task to either "turn back the platform" or "turn back within the visual scene" – has a significant influence on the response. In Figure 3, this shows in the difference of slopes of the red and blue curves. The ANOVA also reported significant main effects for angle and gain – responses were significantly different for different presented angles, and for different gain factors. The interactions between angle and gain and the three-way interaction of block, angle and gain were also significant. For the complete ANOVA results, see Table 1.

Table 1: 3-way ANOVA results on raw responses. 'Attention' denotes the two attention conditions (RVis coded as 1, RBody coded as 2 for the ANOVA). 'Gain' is the set of gain factors, 'Angle' the set of target angles.

Condition	SS	df	MS	F	p
Attention	175.2	1, 12	175.2	0.49	.498
Angle	40437.6	5, 60	194.0	193.96	<.001***
Attention, Angle	60.5	5, 60	12.1	0.45	.813
Gain	79212.2	8, 96	9901.5	88.37	<.001***
Attention, Gain	7036.2	8, 96	879.5	10.69	<.001***
Angle, Gain	4705.0	40, 480	117.6	5.82	<.001***
Attention, Angle, Gain	1093.1	40, 480	20.5	1.33	.089

3.1 Analysis of Visual Weight and Offset

To assess the relative weights of visual cue and body cues of self-motion in the different conditions, linear functions over gain factors were fitted to the responses. By this method we calculated cue weights and response offsets for the different participants, rotation angles and attention conditions separately. Two free parameters were fitted by using least-squares fitting (via Matlab *fminsearch*): a cue weighting parameter w_v (visual weight) to interpolate linearly between the RVis and the Rbody target curves, and a constant offset c. For a given presented target angle α, visual target rotation angle t_v, body rotation angle t_b, gain factors g_i and response angle r, the visual weight w_v and the offset c were derived by minimizing

$$\sum_i ((w_v \cdot t_v(\alpha, g_i) + (1 - w_v) \cdot t_b(\alpha, g_i) + c) - r(\alpha, g_i))^2$$

Figure 4 shows the resulting best-fitting parameters in different conditions for individual participants.

Both for the visual weight and for the offset 2-way ANOVA analyses were computed, with attention condition (RBody, RVis) and presented rotation angle as within-subject factors.
We found a mean visual weight for the RVis condition of 0.73, whereas it was 0.39 for the RBody condition, and this difference was highly significant ($F(1,12)=15.24$, $p=.002**$). The visual weight also depended significantly on stimulus magnitude (target angle) ($F(5,60)=6.03$, $p<.001***$), with a higher visual weight for small rotations (Figure 4, left). In the RVis condition, the mean visual weight over all participants dropped from 0.89 for rotations of 10° to 0.51 for rotations of 30°, and in the RBody condition, it dropped from 0.47 for rotations of 10° to 0.33 for rotations of 30°.
The offset also depended significantly on stimulus magnitude ($F(5,60)=45.61$, $p<.001***$). Small rotations were over-estimated and large rotations under-estimated (Figure 4, right). For rotations of 10°, participants turned on average too far by 2.5° in the RVis and by 4.6° in the RBody condition. For rotations of 30°, they did not turn far enough (-3.5° in the RVis and -1.5° in the RBody condition). Complete ANOVA results are shown in Table 2.

Table 2: 2-way ANOVA results of model parameters 'visual weight' and 'offset', in dependency of stimulus parameters block and angle. 'Attention' denotes the two attention conditions (RVis coded as 1, RBody coded as 2 for the ANOVA). 'Angle' is the set of target angles.

variable	condition	SS	df	MS	F	p
Visual weight	Attention	4.64	1, 12	4.64	15.24	.002**
	Angle	1.11	5, 60	0.22	6.03	<.001***
	Attention, Angle	0.31	5, 60	0.06	1.72	.144
offset	Attention	144.2	1, 12	144.2	4.89	.047*
	Angle	725.1	5, 60	145.0	45.61	<.001***
	Attention, Angle	5.44	5, 60	1.09	0.67	.648

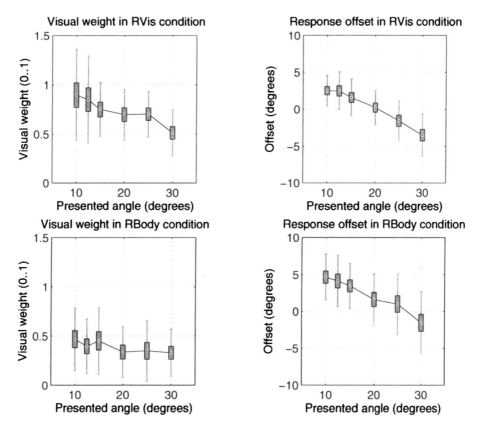

Figure 4: Offsets and visual weights of best fit response functions for all thirteen participants. The upper two subfigures show the results of the "turn back within the visual scene" condition, the lower two show the results of the "turn back platform" condition. A visual weight larger than 1 can result if the response over gain factors is steeper than the visual target curve (blue dashed curve in Figure 3). Visual weights larger than 1 have been measured for some participants particularly when they were instructed to turn back in the visual scene (RVis condition), and might happen if participants try to over-compensate for the influence of the body rotation on their response.

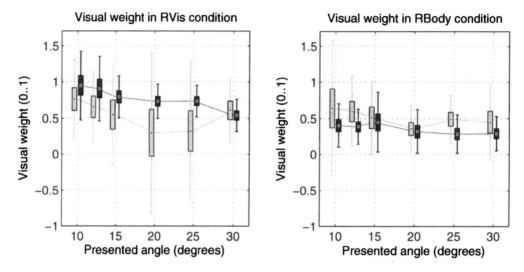

Figure 5: Visual weights of the participants' responses for different presented angles in the two attention conditions, and whether or not they noticed a conflict. Light cyan: no conflict detected, Dark magenta: conflict detected.

3.2 Analysis of the Influence of Conflict Awareness

After each trial, the participants had to press one of three joystick buttons, depending on whether during active return the rotated angle of the ignored modality (the visual rotation in the RBody condition, and the body rotation in the RVis condition) had appeared to them smaller, equal (no conflict) or larger than the angle of the returned modality.

To analyze the influence of the awareness of a conflict on the responses, we split the data in two sets, depending on whether or not the participant had perceived a conflict between visual and body rotation during active return.

The splitting causes an accumulation of "conflict detected" trials for large or small gain factors, and an accumulation of "no conflict detected" trials for gain factors near 1, because large conflicts are easier to detect than small conflicts. This makes statistical comparisons between the two sets difficult. Therefore we again fitted linear functions over gain factors, this time for "conflict detected" and "no conflict detected" trial subsets individually, and retrieved visual weights and offset values. We then used three-way ANOVA with attention condition (RBody, RVis), presented rotation angle, and detection of a conflict as within-subject factors.

One participant had to be excluded from this analysis because she had a strong bias to respond "conflict detected", which made curve fitting impossible in some conditions.

We found a significant influence of conflict awareness on the cue weights. If participants were not aware of the conflict, the visual weights were close to 0.5 in both RBody and RVis conditions (0.51 in RBody and 0.53 in RVis), whereas they were clearly separated as soon as a conflict was noticed (0.36 in RBody and 0.78 in RVis condition). This effect is significant in the ANOVA as interaction between attention condition and perception of a conflict $(F(1,11) = 5.6, p=.037^*)$. Figure 5 shows the visual weight response distributions for different presented angles. We did not find any significant effects of conflict perception on the response offset.

4 Discussion

This experiment showed that the weights of visual and vestibular/proprioceptive modalities in the sensor integration process for the perception of self-rotation were influenced both by stimulus characteristics (magnitude of the rotation) and by cognitive factors (task-induced attention on one modality and awareness of conflicts).

A dependency of the visual weight on the rotation magnitude – higher visual weight for small rotations – is expected by a sensor integration model in which the weight of each cue depends on its reliability (Ernst & Bülthoff, 2004), as the reliability of the vestibular sense is lower for small and slow rotations close to threshold than for rotations of larger magnitude. The smallest of our rotations (3.3°/sec average) were quite close to vestibular threshold compared to the largest rotations (10°/sec average). The threshold for such cosine yaw rotations of about 3 seconds duration is around 1.5°/sec in darkness and 0.55°/sec in the presence of a visual target (Benson, Hutt & Brown, 1989), but the variance between participants is high. The threshold for visual motion is lower – it is in the range of 0.3°/s (as used in the model by (Mergener et al., 1995)).
Attention is thought to bias the competition of different sensory inputs to reach higher brain areas (Desimone, 1998), and may thus also gate the inputs from lower brain areas processing single modalities to higher multimodal areas. It is therefore plausible that guiding attention to one of several modalities could have an effect on the influence this modality has on an area which integrates signals from several modalities. The brain could even use this attentional gating for the selection of appropriate inputs to perform a certain task. Our findings support this hypothesis, as attending to one cue significantly increased the weight of that cue in the resulting percept.

One important issue in multisensory integration is the identification of corresponding information. Only stimuli which are representing the same variable should be integrated. For example, the decision whether a given sound and a given visual stimulus belong together can often only be made if the stimulus has been identified to belong to a certain object category. Studies on multisensory integration in the superior colliculus provide evidence that the decision to integrate or not to integrate signals of different modalities is controlled by inputs from multimodal cortical areas (Wallace, 2004). These signals could be related to top-down attention, which would depend on the identification of the stimulus identity in higher cortical areas which are also involved in awareness. If a conflict is detected, top-down attention could be responsible for selecting the signals appropriate for the current task. Discordant stimuli may give rise to conflicting neural representations, inducing competition. It is known that attention has a large effect in biasing the competition (Desimone, 1998). Such processes would explain why we found a stronger integration of the multimodal signals when no conflict was detected, compared to trials in which participants were aware of a conflict between the two cues.

The effect of the awareness of a conflict on the response could also be interpreted differently. It could be that detection of a conflict and amount of multisensory integration are correlated because they depend on a common underlying mechanism. A possible model is shown in Figure 6. Assume that the perceived rotation angle in each modality is taken from a Gaussian distribution around the actual angle. Then, in trials in which the angles in the two modalities are perceived more alike (dotted lines), both the reproduced angle will be more towards the centre of the two distributions, and the conflict is detected less often, because it is very small. In trials in which the angles in the two modalities are perceived as very different (dashed lines), the reproduced angle would be more biased to one side (possibly determined by attention) and at the same time the difference between the two perceived angles would be large and typically above threshold for conflict awareness.

Because of this alternative interpretation we can not count our results as unequivocal evidence for a direct influence of conflict awareness on multisensory integration. Further experiments are needed to address this issue.

In conclusion, the results of the self-rotation perception experiment support models of multisensory integration in which the weight of each modality depends on its reliability. The experiment also showed that the processes of sensory integration for the perception of self-rotation in the human brain are not only governed by the stimulus characteristics alone, but also by task-related (top-down) attention, which can bias the influences of the relevant modalities. We found that the amount by which the influence of the modalities on the response is modulated by attention is correlated with the awareness of a conflict. The results are consistent with the idea that becoming aware of a conflict might trigger sensory biasing by means of top-down attention to generate robust behaviour under conflicts.

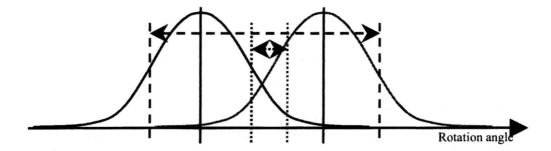

Figure 6: Hypothetical process for responses and awareness of conflicts. Blue: visual rotations, Red: body rotations. For two concurrently presented rotations with different angles (blue and red solid lines) the participant perceives rotation angles in the vicinity, presumably with a certain probability distribution. For a case in which the perceived rotation angles are similar (dotted lines), one would expect both the conflict to be less perceptible, and the response to be closer to the centre of the two distributions. If the perceived rotation angles are further apart (dashed lines), both the perceived conflict would be larger, and easier to detect, and the response would be biased to one of the sides. Attention could determine which side is selected.

5   References

Benson, A. J., Hutt, E. C. B. & Brown, S. F. (1989). Thresholds for the perception of whole body angular movement about a vertical axis. *Aviation, Space and Environmental Medicine*, 60, 205-213.

Calvert, G. A. & Thesen, T. (2004). Multisensory integration: methodological approaches and emerging principles in the human brain. *Journal of Physiology – Paris*, 98:191-205.

Desimone, R. (1998). Visual attention mediated by biased competition in extrastriate visual cortex. *Phil. Trans. R. Soc. Lond. B* 353:1245-1255.

Ernst, M. O. & Bülthoff, H. H. (2004). Merging the senses into a robust percept. *Trends in Cognitive Sciences* 8(4):162-169.

Ernst, M.O. & Banks, M. S. (2002). Humans Integrate Visual and Haptic Information in a Statistically Optimal Fashion. Nature 415, 429-433.

Ernst, M.O., Banks M. S. & Bülthoff H. H. (2000). Touch can change visual slant perception. Nature Neuroscience 3 (1), 69-73.

Gepshtein, S. & Banks, M. S. (2003). Viewing geometry determines how vision and haptics combine in size perception. *Current Biology*, 13:483-488.

Kitazaki, M. & Sato, T. (2003). Attentional modulation of self-motion perception. *Perception*, 32:475-484.

Lambrey, S. & Berthoz, A. (2003). Combination of conflicting visual and non-visual information for estimating actively performed body turns in virtual reality. *International Journal of Psychophysiology*, 50:101-115.

Mergener, T., Schweigart, G., Kolev, O., Hlavacka, F., & Becker, W. (1995). Visual-vestibular interaction for human ego-motion perception. In: T. Mergener and F. Hlavacka (Eds.), *Multisensory control of Posture*, Plenum Press, New York.

Palmisano, S. (2002). Consistent stereoscopic information increases the perceived speed of vection in depth. *Perception*, 31:463-480.

Wallace, M. T. (2004). The development of multisensory processes. *Cognitive Processes*, 5: 69-83.

Zupan, L. H., Merfeld, D. M., & Darlot, C. Using sensory weighting to model the influence of canal, otolith and visual cues on spatial orientation and eye movements. *Biological Cybernetics*, 86:209-230.

Acknowledgements
We would like to thank Ian Thornton, Manuel Vidal, Marc Ernst, Astros Chatziastros and Bernhard Riecke for valuable comments on the experiment, data analysis and an earlier draft of the manuscript. This work has been supported by the Max Planck Society and the Deutsche Forschungsgemeinschaft (Sonderforschungsbereich 550).

Relationships among visual perception, psychomotor performance, and complex motor performance in military pilots during an overnight air-refueling simulated flight: Implications for automated cognitive workload reduction systems

M. Russo MD*, A Kendall, D Johnson, H Sing, MS, S Escolas PhD, S Santiago, D Holland PhD, S Hall MS, D Redmond MD

* United States Army Aeromedical Research Laboratory, Fort Rucker, Alabama, 36362   334 255-6874
Michael.Russo@US.ARMY.MIL

1. Introduction

This paper explores the relationships between visual perceptual and complex motor impairments under conditions of 26.5 hours of continuous wakefulness and discusses the role of these relationships in the development of automated cognitive workload reduction systems. The occurrence and time of onset of performance decrements during visual perceptual, complex motor, and simple motor tasks are considered for cognitive monitoring and predictive mathematical modeling efforts. Visual perception is studied for its sensitivity as a surrogate or marker of potential operational performance. If perceptual impairments are found to occur prior to the onset of motor impairments, potentially, interventions can be made to prevent the deterioration of some motor functions. Among these interventions may be the activation of automated cognitive workload reduction systems. Automated workload reduction systems may be activated by threshold-based outputs from mathematical performance assessment and prediction models. The inputs to these performance models would be the neurophysiologic data indicating fatigue and cognitive degradation.

In an earlier paper (39) we described the visual perceptual impairment as the occurrence and pattern of significant neglect of visual stimuli presented through the far left to far right horizontal visual fields. This visual perceptual impairment began in the non-instructor pilot group at 19 hours of continuous wakefulness in a simulated overnight fixed wing aircraft flight and correlated with the number of simulator shutdowns due to exceeding the simulator tolerances. In this current follow-on paper, we proceed with comparisons of the entire pilot group's performance in the same visual perceptual task, a second complex motor task (flight heading), and psychomotor vigilance task time (simple reaction time) for the purpose of improving our overall understanding of the relationships among perceptual, complex motor, and simple motor functions.

Visual neglect is a person's unconscious inability to recognize or acknowledge some visual information despite a structurally intact visual system. In a normal visual system, the eyes, optic nerves, optic tracts, and primary visual cortices in the occipital lobes register the visual stimuli and forward the information to the higher order visual associational areas in the parietal brain regions. From the parietal regions, the processed visual information is forwarded to thalamic nuclei and then to the prefrontal regions where the visual image may be brought to awareness. Visual neglect occurs if the visual stimuli are not fully or correctly processed by the parietal lobes, not forwarded to or processed by the thalamic relays, or not forwarded to or attended to by the prefrontal regions.

Based upon clinical cases in patients with brain damage, parietal lobe dysfunction can result in patterns of impairment where primarily peripherally located visual stimuli are neglected, and one or more of several simultaneously presented visual stimuli are neglected (2, 37). Due to the attentional functions of the prefrontal regions, prefrontal region dysfunction may result in a pattern of neglect to stimuli from all fields of visual regard.

The finding of visual neglect is consistent with the findings of Thomas et al. (44, 45) who found hypometabolism in the parietal regions in a positron emission tomography study of the effects of total sleep deprivation. Thomas et al. showed significant decreases in cerebral glucose utilization in brain regions responsible for visual attention and visual cognition in awake volunteers at 24 hours of continuous wakefulness. The prefrontal regions are involved in executive functions, such as prioritizing and selecting tasks, and maintaining attention on a selected task. The bilateral parietal regions are responsible for processing and interpreting visual information, developing complex visual-spatial relationships, appreciating multiple visual images simultaneously, and with aspects of eye movement.

Multiple studies investigated the deterioration of perceptual and oculomotor function under conditions of fatigue or extended wake (3, 10, 11, 16, 19, 22, 23, 25, 32, 35, 37, 38, 40, 45, 49). Studies that examined perceptual function reported a decrease in visual vigilance, visual acuity, visual detection and visual scanning (3, 19, 22, 25, 35, 37), while studies that examined oculomotor function reported sympathetic-parasympathetic mediated fluctuations in pupil size, saccadic velocity, oculomotor fixation, and initial pupil diameter (10, 11, 16, 23, 32, 38, 40, 45, 49). McLaren et al. reported decreased sleep latency and correlations with pupilometric variables (24).

Several authors discussed relationships between impairments in visual perception and attentional processes in sustained task performance and prepared the foundation for exploration of visual perceptual impairments in sustained wake (26, 27, 33, 48). Roge at al. (34) described a tunnel-vision phenomenon during total sleep deprivation in an automotive simulator. We recently noted the occurrence and patterns of visual neglect beginning at 19 hours of continuous wake (37, 39).

Studies conducted at Walter Reed Army Institute of Research (46, 47), using a choice visual divided attention task during a driving simulator (42), reported a relationship between visual impairments and operational performance failure, with declines in recognition of the visual stimuli, and lapses and errors of commission beginning at 20 hours of continuous wake. While microsleep events (3 sec of Stage 1 theta) increased with sleep deprivation, the lapses rarely temporally correlated with operational performance (43), suggesting that the impairments observed were resulting from functional neural changes other than those attributable to frank sleep. These Walter Reed Studies lead to the design of our choice visual perception task to assess for possible visual neglect due to acute sleep deprivation.

Stern's (41) study of visual attentiveness in a truck simulator, using commercial drivers, showed eye-gaze shift impairments. The study examined divided attention in a moderate cognitive loading environment influenced by sleep deprivation. Results showed that sleep deprivation decreased the frequency of eye movements towards simulated side-view mirrors.

Studies using a restricted sleep design to assess vigilance and operational performance reported significant increases in reaction time and lapsing (4, 13) on a psychomotor vigilance task. In a sleep-dose response study where subjects received 3, 5, 7 or 9 hrs of sleep a night for 7 nights, Belenky et al. (4) reported a decrease in response speed for the 3, 5 and 7 hr groups and an increase in lapses in the 3 and 5 hr groups. After 3 nights of recovery sleep, the 3, 5 and 7 hr groups did not return to baseline performance (4, 20, 35). In a 7-day sleep restriction study where sleep was restricted to 4-5 hrs per night, Dinges et al. (13) reported that, by the second day, response time and lapses were significantly increased from baseline. Another study, conducted by Doran et al. (14), compared performance on a vigilance task between an 88 hour total sleep deprivation group, and a group that received a 2 hour nap. Results showed that the 2-hour nap group sustained vigilance at baseline levels whereas the total sleep deprivation group had an increase in reaction time and errors.

In the field of human factors-related aviation research, performance studies focused primarily on elucidating pilot flight performance decrements. Possible reasons why real-world flight performance decrements occur are less-than-optimal controls and displays, pilot over- and under-stimulation, loss of situational awareness, operations tempo- or workload-related fatigue, continuous wakefulness, time-zone crossings, and shift changes. Several authors have reported on changes in visual motor systems but minimal discussion related these changes to performance impairments. Some authors studied the effects of pharmacological aides on reversing sleep-deprivation-induced pilot performance decrements. Caldwell et al. (5, 7) found that both modafinil and dextroamphetamine maintained performance and alertness in fatigued rotary wing pilots in simulated flights, even during the circadian trough. In a study of F-117 Stealth Fighter pilots in simulators, Caldwell et al. (8) showed that modafinil improved mood and flying performance on some tasks as early as 20 hours of continuous wake, with greater gains seem from the stimulant with increasing periods of wake. A study conducted by Neri et al. (29) investigated controlled waking rest-breaks as a countermeasure for the effects of fatigue. The treatment group exhibited decreased slow eye movement 15 minutes after break with no significant difference on objective vigilance when compared to the control group.

Paul et al. (31) reported on the fatigue effects of pilots flying transatlantic flight re-supply missions. Administration of neuropsychological tests occurred during the transatlantic sortie. Results showed that performance decrements occurred over the three days of the transatlantic re-supply missions. Prior to the transatlantic flight, pilots reported an average of 8 hours 40 minutes of sleep, which decreased to 6 hours and 40 minutes of sleep during the flight sortie. Subjective measures of sleepiness also predicted flight performance decrements in fatigued pilots (17, 30).

In a case study of how fatigue from the demands of the operational world can affect pilots' performance, Armentrout and Holland (1) conducted multi-discipline investigations into a USAF C-5 Galaxy mishap where the pilots were so tired that they fully stalled a massive C-5 Galaxy aircraft on a dark, quiet night approach to an island. The pilots fell over 4,000 feet in their stalled and fully functional C-5 aircraft during a mixed visual-instrument approach. Holland and Armentrout utilized information from a variety of sources (including pilot recollections) to show that the aircrew experienced a loss of situation awareness, channelized attention, and spatial disorientation with fatigue from days of sleep interruption as a critical over-arching enabler. The fatigued pilots did not fully appreciate the relevance of visual instrument panel indications of slowly deteriorating airspeed, vertical velocity increases, and an increasing angle of attack that lead to the fully stalled condition. The pilots barely recovered the aircraft just 773 feet above the water's surface after passing through more than 90 degrees of wing bank twice during the free-fall of the stalled condition. Thus, this case study highlighted that fatigue can directly affect not only the ability to adequately process and integrate a wide variety of stimuli, including visually-presented information, but fatigue increases the chance that reductions in situation awareness may occur. This reduction can impede flying safety under some scenarios due to the brain's inability to process and integrate information from disparate sources into a meaningful whole with regard to the state of the aircraft.

Regarding perceptual and operational performance in aviation research, Behar (3) and Morris and Miller (28) noted different results in visual impairment and its relation to flight performance. Behar reported that although oculomotor decrements occurred in the pilots, their ability to sustain flight performance remained unaffected by fatigue. Morris and Miller, however, showed that flight performance impairments correlated with long eyelid closures and increases in blink amplitude in a simulator study of military pilots under conditions of extended wake. Differences in results could be due to length of sleep debt period and familiarity with the simulator. This leads to the question of visual impairment being a possible predictor for flight performance decrements.

In the aforementioned studies, the visual measurements used did not measure fixed visual deficits, but rather transient attentional deficits under a moderate cognitive loading environment. We demonstrated visual neglect in U.S. Air Force pilots beginning at 19 hours of continuous wake. The visual neglect preceded and correlated with shutdowns of the fixed wing simulator due to erratic flying. Impaired flight performance began at 21.5 hours of continuous wake. This paper is an extension of that Russo et al. paper (39) and reports on the relationship between visual perception and motor performance (complex: deviations from an established course heading; and simple: psychomotor vigilance) under conditions of high cognitive loading and modest sleep deprivation (across 26.5 hours of continuous wakefulness). The high cognitive loading environment is created by the complexity of the primary task: flying a simulated C-141 aircraft at approximately 275 knots and within approximately 50 feet of an air tanker. Currently, there is no research in the field of aviation reporting possible relationships among these variables using a sleep deprivation paradigm.

2. Methods

2.1 Task and Subject Selection
This study evaluated visual perception and motor performance in U.S. Air Force pilots in a simulated overnight aerial refueling flight, following a day of continuous wake. Although pilots do not normally fly long missions after a day-long period of wake, the operational tempo of today's military sometimes may not permit adherence to crew-rest recommendations. This design was selected to reflect a possible course of action available to military commanders and planners in times of high operational tempo. Because pilots often have only seconds to interpret and act upon visual information, small visual errors occurring over brief periods of time could possibly result in catastrophic mission failures. A choice visual perception task (CVPT) was designed and integrated into a U.S Air Force C-141 simulator cockpit to assess for possible impairments in a high cognitive loading environment. The simulator used in this study was an Air Refueling Part Task Trainer (ARPTT). The non-digital visual representation of a high-fidelity dynamic motion KC-135 tanker model is reported to be approximately as realistic as flying behind a real tanker under dusk and nighttime simulator conditions. Flying in formation behind this extremely realistic looking aircraft in the simulator is very compelling, and about as difficult as the real world exercise, according to pilots who use this simulator routinely to train for actual aerial refueling operations.

2.2 Subjects

Eight right-handed male pilots between the ages of 31 and 52 years old (mean age 37 years) from the 305th and 732nd Air Mobility Command Wings, McGuire Air Force Base, N.J, participated in this study. All volunteers were qualified to fly the C-141 airframe and the ARPTT. All were currently on flight status, indicating recent verification of good health, including visual and color acuity in accordance with Air Force Pamphlet 48-133, 1 June 2000, Chapter 5 (Visual Acuity Testing) and AFI 48-123, Attachment 7 (medical standards for flying duty) paragraph 7. Vision was tested using the Optec 2300 Vision Tester that assesses far vertical and far lateral phoria, distant visual acuity, fusion and depth perception. The pseudoisochromatic plate set (PIP) was used to establish normal color vision. Air Force pilots qualified on both the C-141 airframe and the ARPTT represented a small and geographically dispersed group of individuals. To increase the probability of obtaining qualified pilots, and to more closely reflect realistic operations, the investigators permitted the use of caffeine (<400mg/day) and nicotine in moderation. Thus, the risk of possibly introducing an uncontrolled variable was acknowledged, but pilots in the operational world often use caffeine or other stimulants on occasion, so this moderate caffeine use is rather generalizable to the real operational environment. Those on daily medications were automatically excluded.

Three of the pilots were active duty U.S. Air Force personnel, four were reserve U.S. Air Force personnel who primarily flew transcontinental and transoceanic flights for commercial airlines (United, US Airways, and American Airlines) on Boeing 747s, Boeing 777s, and Air France Airbus, and one was a reserve commissioned pilot on full-time active duty. Two of the commercial airline pilots and one of the active duty pilots were air-refueling instructor pilots. The eight pilots reported a mean of 7.75 actual refueling aircraft hours in the three months prior to the study with ranges from 0.5 to 30 hours. The eight pilots also reported a mean of 63.75 total simulator hours with ranges from 40 to 100 hours. The pilot volunteers were paid for their participation.

2.3 Measures

Complex motor performance was measured using an Air Refueling Partial Task Trainer (ARPTT). The ARPTT is a U.S. Air Force high fidelity optical simulator configured to represent the cockpit of the C-141B (cargo) airframe and is used to train pilots on the task of air refueling. The reference "Part Task" indicated that the trainer simulated only the refueling component of an actual air-refueling mission. In order to refuel, the C-141B pilot was expected to maneuver into a proscribed position behind and below a KC-135 tanker aircraft. In a subsequent maneuver, the receiver continued closure at a rate of 1 to 2 feet per second until the pre-contact position (50 feet aft, slightly below and in trail of the tanker) was achieved. The stabilization of the aircraft in the pre-contact location was somewhat difficult to maintain because of the effects of the receiver's bow wave on the tanker's aft fuselage and horizontal stabilizer. In this paper, pre-contact position data (i.e., azimuth deviations, see below) is presented.

Operational measures available through the ARPTT included simulator shutdowns, which occurred when the threshold parameters of the simulator have been exceeded. The simulator shut down (froze) when the pilots exceeded the following parameters: From the pre-contact position, the simulator froze for 28 degrees of azimuth deviation or 20 degrees of nose elevation. From the contact position, the preceding thresholds applied, and in addition, an advance towards the tanker of 10 feet would freeze the simulator. These thresholds were determined primarily by the simulator's optical and mechanical limitations. The parameters for freezing the simulator exceeded the normal operational thresholds.

Azimuth Deviation (drift from established course heading) also was recorded via the ARPTT, and deviations from the established azimuth were taken from the data output range from 0 to +/-30 off primary heading. Azimuth Deviations were scored as the total number of times the aircraft crossed the 0 degree heading and at each additional 5 degree crossing, that is at 5, 10, 15, 20, 25, and 30 degrees. The mean of the Azimuth Deviations during the 20-minute visual perception task was used for data analysis.

The Choice Visual Perception Task (CVPT) was designed and installed onto the instrument panel of the ARPTT cockpit to assess central and peripheral visual field awareness along the horizontal meridian in a high cognitive loading, multi-task environment. The display was a semicircular, small-caliber (approximately 1.5cm diameter) high-grade aluminum tube with light stimuli 18 inches from the center of the pilot's head just below eye-level. During this 20-minute task, 150 sequential single- or double-light stimuli were presented, and each stimulus lasted 0.25-seconds. The pilots divided their attention between performance of the task and active flight. (Stimuli spanned 75-degrees left to 75-degrees right of center at 15-degree intervals just below the simulated horizon and against the instrument panel's black background. At each of the 11 intervals (L75 degrees, L60, L45, L30, L15, Center, R15,

R30, R45, R60, and R75), a single stimulus was presented 10 times; double stimuli were presented 10 times each at positions L75 and Center, L60 and Center, R75 and Center, and R60 and Center. The presentation sequence of all stimuli was quasi-randomized (which allowed for each of the 15 types of stimulus to be presented 10 times). Inter-stimulus intervals were randomized between the range of 3 to 15 seconds. The presentation of two simultaneous stimuli allowed for the possibility that the pilot would see and respond to one, both, or neither of the visual stimuli.

The volunteers flew in the ARPTT and performed the pre-contact position maneuver each time they completed a CVPT. This was done 15 times on Day 2 over the 24-hour testing period at 0931, 1031, 1131, 1631, 1731, 2010, 2100, 2330, 0030, 0100, 0330, 0430, 0500, 0700, and 0800. Each session provided a unique combination of stimuli sequences and inter-stimulus intervals. All visual perception task iterations, sequences, and inter-stimulus intervals were equivalent for each pilot. The pilot was asked to respond verbally with "left," "right," "center," "center-left," or "center-right" because both hands and feet were fully engaged in controlling the aircraft. One infrared cockpit camera recorded pilot voice and facial movement, a second recorded the view appreciated by the pilot of the air tanker, and a VCR recorded flight parameters.

The CVPT occurred while the pilot flew in the pre-contact position. For the primary task, the pilots were instructed to fly safely at the pre-contact position, and for the secondary task to verbally acknowledge the light stimulus.

The Psychomotor Vigilance Task (PVT) was performed 16 times at 0817, 0902, 1002, 1102, 1617, 1702, 1802, 2132, 2302, 0132, 0202, 0302, 0533, 0602, 0632, and 0833. Past research determined that the PVT is sensitive to sleep deprivation (4, 12, 13). Simple reaction time tasks, as opposed to disjunctive or choice, require responding as quickly as possible to the occurrence of a single stimulus. Such tasks can assess motor speed relatively isolated from higher cognitive functions, requiring only the detection of stimulus presence or absence without further discrimination. If the inter-stimulus intervals are long and/or variable, or the task duration is long then such tasks may also assess attention and vigilance.

For this study, each PVT lasted 10 minutes. The metrics from the PVT were Speed (1/Reaction Time), Lapses, False Starts, and Anticipations. Speed is the reciprocal of reaction time or response latency (the length of time in milliseconds for the subject to respond to the single stimulus on the LED display) and was calculated as the number of responses per second. Lapses were defined as the number of reaction times exceeding 500 milliseconds. False Starts were the number of times when the subject responded without a stimulus presentation, and Anticipations were the number of responses less than 0.1 second and considered to be due to chance.

The Advanced Tri-Mode Actigraph (Precision Control Design, Inc, Fort Walton Beach, Florida) utilized in this study used a linear piezoelectric accelerometer to record human movement in 1-minute epochs for distinguishing wake from sleep (18). The wrist-worn device used for this study was lightweight (1.2 oz) and small (1.82" x 1.34" x .45"), with 128K memory, and collected data in the Tri-Mode setting (Zero Crossings, Time Above Threshold, and Proportional Integrating Measure simultaneously). The wrist-mounted actigraph operated within a bandwidth of 2-3Hz with a .002g at 1Hz sensitivity. Sleep was scored using the Cole-Kripke algorithm (9, 18, 21). Wrist-mounted actigraphy has become an acceptable method for discriminating wake from sleep, with correlations between polysomnographically and actigraphically measured sleep as high as .98 (9).

2.4 Procedure

The pilot volunteers flew the experimental conditions individually. Volunteers arrived at the ARPTT trainer, McGuire Air Force Base for training on Study Day One. The investigators explained the study in detail, reviewed the informed consent, and answered questions. The subjects provided their informed consent during the enrollment phase of recruitment. The Walter Reed Army Institute of Research Human Use Review Committee approved the protocol for this study.

The volunteers placed the actigraph watch on their non-dominant wrist to assess sleep and wake periods. Subjects learned the PVT and the visual perception task, and re-familiarized themselves with the ARPTT. After training and re-familiarization, the pilots returned home, went to bed at midnight, awoke at 0600 on Study Day Two, and reported to the ARPTT building by 0800.

The pilot volunteers complied with the requested sleep schedule (confirmed by the wrist-mounted actigraph) by sleeping an average of 5.8 hours with a range of 5.62 – 6.15 hours of sleep. One pilot reported moderate use of tobacco. This volunteer smoked outside the building during breaks between testing and flying. Caffeine containing products up to 400mg was permitted on Study Day One (training) and up to 12 noon on Study Day Two. Compliance was determined by observation and self-report. Over the counter analgesics (e.g., acetaminophen for headache) were available through the medical monitor. None of the volunteers requested medication.

On Study Day Two, volunteers arrived at the simulator building at 0800, began testing, and pursued ad lib activities. Continuous wake throughout the day was verified using the actigraph. The volunteers ate dinner between 1815 and 1930. From 1930 until 2000, they tested and received a pre-flight and mission brief. From 2000 on Study Day Two to 0830 on Study Day Three, the volunteers tested and flew the mission scenario. The pilots tested on the PVT in a quiet room adjacent to the simulator during break periods, before and after simulated flight.

In an actual flight operation, a co-pilot would be present. In order to simulate a co-pilot sharing flight operation responsibilities, each volunteer was permitted three scheduled breaks: 2130-2320, 0130-0320, and 0530-0650. In a departure from normal operational procedures, the subjects did not nap during these scheduled breaks. Actual flying time for each volunteer was 7.5 hours. The pilots performed four 15-minute fueling sequences during flight at 2040, midnight, 0400, and 0730.

The scenario of an overnight flight after a day without sleep was intended to simulate a possible scenario that may exist in a high operational tempo environment, where there may be limited opportunity for regular crew-rest.

2.5 Analyses
Paired t-tests were performed between trials 2 and 7 (4 vs 15 hours awake for decrements occurring by 15 hrs), and between trials 7 and 10 (15 vs 19 hours awake) for decrements occurring by 19 hrs) to determine whether significant differences began by the 19th hour of continuous wake for the 8-subject group, independent of instructor status. The trial comparisons were made for averaged CVPT Response Omissions, ARPTT Azimuth Deviations, and PVT Speed (1/RT), and averaged number of Lapses, False Starts, and Anticipations per trial. Since number of CVPT response omissions, and PVT Lapses, False Starts, and Anticipations are not normally distributed measures, transformation of these data was necessary to achieve a normal distribution. Therefore, 1 was added to each datum followed by log transformation. Statistical significance was set at $p < 0.05$ (two-tailed).

Pearson-product moment correlation analyses were performed between measures significantly affected by sleep deprivation: CVPT Response Omissions and ARPTT Azimuth Deviations, and CVPT Response Omissions and PVT Speed and Lapses; and ARPTT Azimuth Deviations and PVT Speed and Lapses. Each of the correlation analyses included all 15 data collection time points for the CVPT Response Omissions, ARPTT Azimuth Deviations that coincided with CVPT administrations, and 15 of the 16 data collection time points for the PVT (the first PVT data point was not used to provide an equal number of data points in the analysis). The means for each hour computed across all subjects for each of the metrics were used in separate analyses. Since each hour represented a different treatment or population, i.e., 3 hours awake, 4 h awake, 5 h awake and so forth—rather than a random sample from a single population at each of the hourly time points—the means across subjects for each hour were used, rather than individual subject's hourly means. Statistical significant was set at $p < 0.05$ (two-tailed)

3. Results
For the CVPT metric, 2250 stimuli were presented to each volunteer over the course of the study. Total errors included both Errors of Omission and Wrong Responses. Omission of a single stimulus occurred when the subject failed to respond to the single light. Omission of a double stimulus occurred when the subject failed to respond to one or both of the presented lights. Total Errors of Omissions were 963, with 760 to single stimuli and 203 to double stimuli. The average (mean) errors of omission for the 8 pilots are used for comparisons.

ARPTT Azimuth Deviation was measured by the number of crossings at zero, 5 degrees, 10 degrees, 15 degrees, 20 degrees, 25 degrees, and 30 degrees. The number of zero crossings was 13,023, the number of 5 degree crossings was 2356, 10 degree crossings was 438, 15 degree crossings was 177, 20 degree crossings was 78, 25 degree crossings was 33, and 30 degree crossings was 27. The total number of Azimuth Deviations was 16,132.

As previously found by Russo et al. (39) for the non-instructor pilots (5 total), the current analysis based on all pilots also showed that declining performance for CVPT Response Omissions occurred at 19 hours of continuous wake (p = 0.018) (Figure 1a). Declining performance at 19 hours of wake was shown as well for ARPTT Azimuth

Deviations ($p = 0.046$) (Figure 1b), and for PVT Speed ($p = 0.019$) (Figure 1c) and Lapses ($p = 0.023$) (Figure 1d). Significant differences were not found for mean number of PVT False Starts and Anticipations by 19 hours awake.

Since significant increases in omissions to visual stimuli, increases in course deviations from an established azimuth, and decreases in simple reaction time speed and increases in lapses were observed, follow-on correlation analyses determined if significant relationships existed among these outcomes.

CVPT Response Omissions and ARPTT Azimuth Deviations were significantly positively correlated ($r = 0.97, p \leq 0.0000$).

CVPT Response Omissions were also significantly associated with the PVT measurements as evidenced by a negative correlation with Speed ($r = -0.92$ $p = 0.0000$) and a positive correlation with Lapses ($r = 0.90, p = 0.0000$).

Because visual perceptual impairment correlated with both complex motor performance and psychomotor vigilance performance impairments, subsequent correlation analyses were conducted to determine the relationship between the latter two outcomes. Analyses confirmed a significant negative correlation between ARPTT Azimuth Deviations and PVT Speed ($r = -0.92, p = 0.0000$) and a significant positive correlation between ARPTT Azimuth Deviations and PVT Lapses ($r = 0.91, p = 0.0000$).

4. Discussion
Visual perceptual, complex motor, and vigilance decrements all occurred with less than 24 hours of continuous wake. These findings are consistent with findings from earlier studies and reaffirm that acute sleep deprivation impairs multiple aspects of performance. Analyses performed individually for each metric showed significant changes at 19 hours of wakefulness. There was no difference in time of onset of impairments among the three metrics (CVPT - visual perception, ARPTT Azimuth Deviations - complex motor performance, and PVT - psychomotor vigilance).

An important finding in this study is that visual perceptual impairment and complex motor performance are more strongly correlated with each other than either is to simple reaction time. The choice visual perception task simulated the visual input a pilot might experience during normal flight operations – that is the random appearance of a light stimulus along the instrument panel. The task required active visual perception, that is, both recognition of the stimulus and a relatively rapid acknowledgement of its location. Although in itself relatively simple, the CVPT required utilizing prefrontal regions for attentional focus on the occurrence of the stimuli, parietal regions for visual recognition and localization, and fronto-temporal regions for the verbal response. Multiple thalamic nuclei relayed the afferent signals and modulated the efferent responses. Thomas et al. (44, 45) found cerebral deactivation in the prefrontal, parietal, and thalamic regions in a positron emission tomography study beginning at 24 hours of total sleep deprivation. These brain regions are primarily responsible for visual attention and visual cognition. The complex motor performance task of flying in the pre-contact position behind the KC-135 tanker clearly required far more involved neural circuitry. Pilots perceive the relative location of their aircraft using multiple visual cues, such as the signals presented by the boomer and the relative wing position of the tanker. Pilots perceive acceleration and deceleration through proprioceptive and vestibular systems. The cerebellar system provides integration between proprioception and fine motor coordination. All sensory inputs relayed through, and were interpreted by, the thalamus. The pilot responded accordingly through the frontal motor systems with actions by both hands and feet. During a complex maneuver, such as is represented by the precontact and refueling activities, a pilot might often remain silent in order to conserve and focus cognitive abilities on the delicate and demanding flight task.

As continuous wake similarly impaired responses to both the sensory perception task and the complex motor task, a relatively close neurophysiological relationship may be hypothesized. The close relationship between decrementing sensory perception and decrementing motor performance would lead one to consider a causal relationship wherein the impaired visual perception was contributing to the impaired motor responses. The pilot attempting to maintain a steady precontact or contact position relies heavily, although not exclusively, on visual cues. Once perception of visual cues degrades, motor performance dependent upon these cues would logically degrade as well. Alternately, sleep deprivation may be decrementing independently both sensory and motor systems, and the impairments may be representing a simultaneously occurring global impairment of higher order cognitive functions. The positive correlation between response time and azimuth deviation clearly demonstrated a relationship between prefrontal modulated simple reaction time and frontal motor azimuth deviation.

In this study there were no differences found based upon age. A larger sample size and increased the age range could increase the power of the statistical findings, and possibly reveal age-related differences. Rogé et al. (34) measured the useful visual field (area around the fixation point, inside which information can be quickly found and extracted during a visual task) of 18-51 year olds on a driving simulator under the condition of sleep deprivation. Results showed that under the condition of sleep deprivation the useful visual field began to deteriorate in the older drivers. Other studies in general indicate that older pilots often perform better due to critical skills refinement gained from experience, thereby having more reserve cognitive capacity with which to engage in the flying tasks as the workload demands change. In regards to gender differences in aviation research, Caldwell (6) measured the differences of performance; mood and recovery sleep in male and female pilots over 40 hours of sustained wakefulness. Even though females reported feeling less tense and more energetic than the men, there were no interactions between sleep deprivation and gender on either the flight performance or psychological mood. In this study, all volunteers were male.

Impairments of visual perception may also be a contributing cause of ground vehicular accidents in awake but sleepy drivers. As visual perception impairments develop, the ability to appreciate changes in the road (curves, narrows or intersections) and to fully appreciate and integrate that information into situational awareness, could result in failure to properly adjust speed and direction in a timely manner, even in the presence of intact motor system. In pilots, impairments of visual attention resulting from total sleep deprivation may contribute to the occurrence of accidents during maneuvers that require tight formations, fast response times, and quick decision-making. Pilots may not be as attentive to the relative location of their wingtips as they would be when fully rested because they are not able to fully "capture" all of the relevant inputs needed in time to avoid difficulties. Similar comments could be made with regard to flying instrument or complex visual approaches in a wide variety of aircraft.

As fatigue increases, it is reasonable to speculate that the odds of having problems with spatial disorientation and of improperly handling visual illusions would increase as well. Fatigue affects situation awareness levels in general, because one critical component of situation awareness is the ability to project current aircraft states into the near future (15). Extreme fatigue short-circuits this process because the higher integrative functions of the brain do not work as efficiently to process disparate pieces of information into a unified whole, as is required for a global assessment of the operational situation. Possible future research could address the effect on pilot performance by incorporating crew-rest condition variables to assess the impact of restricted and shifted sleep schedules on visual perception, complex motor, and simple reaction time performance.

For the purposes of cognitive monitoring, assessment of visual perception would have to occur unobtrusively, that is, without a dedicated stimulus-response requirement. Measurement of visual perception as an indicator of visual awareness and cognitive performance may be accomplished through two mechanisms. First may be through the use of visual evoked potentials or occipital-parietal electrocortical signals. With the advent of dry-application high-impedance electrodes, evoked and electroencephalographic indices may be captured unobtrusively in pilots and soldiers. The electroencephalographic information may be able to provide evidence of failing cognition (see sister article in this supplement by Sing et al). A second technique for capturing visual perceptual information may be through changes in pupillary hippus. Pupil size changes appear to reflect recognition and cognitive processing of visual information. Measurements of these pupil oscillations are unobtrusive in that eye tracking devices may be mounted into instrument panels, helmet displays, or eye-glass frames.

More likely is that future neurophysiological monitoring systems will integrate electrocortical information with pupil and oculomotor information into algorithms that interpret visual perception as an input for mathematical models that assess and predict cognitive performance. The output of the performance assessment and prediction models may be an input for automated workload-assist computer programs. A workload-assist program would seek to balance the status of the human with the control requirements of the mobile platform and with the objectives at a specific point in a specific mission.

In a military aviation example, a bombing mission may be composed of a 3 hour flight to target area, a 5 minute target acquisition, confirmation, and armament launch period, followed by a 3 hour return flight. The adrenaline and possibly the pharmacologic countermeasures that provided alertness and cognitive support to successfully accomplish the primary objective (target destruction) may wane in effectiveness as the return flight begins. Unobtrusive capture of oculo-motor, electroencephalographic, and other neurophysiologic indices would detect decreasing alertness and attention. A cognitive performance assessment algorithm could predict increasing risk of fatigue-related performance failure if the pilot maintained the current workload. An automated workload-reduction system could at this point indicate to the pilot that unless countermanded, course, altitude, and airspeed would be maintained on preprogrammed parameters without human engagement. Pilot cognitive resources could then be conserved until needed for manual initiation of landing procedures.

In summary, significant visual perceptual, complex motor, and simple reaction time impairments began in the 19th hour of continuous wake. Visual perceptual impairment and complex motor performance decrements strongly correlated with each other. As such, visual perception is a sensitive indicator of operational performance during complex motor tasks under conditions of sleep deprivation. This research supports the use of visual perceptual measures in place of complex motor performance measures, in situations where the individual is dependent on primarily visual information. These findings support the utilization of oculomotor indices as potential components of unobtrusive neurophysiologic monitoring arrays. This research supports the development of automated workload reduction systems based upon mathematical cognitive assessment and performance prediction models.

DEPARTMENT OF DEFENSE DISCLAIMER

Human volunteers participated in these DoD studies after giving their free and informed consent. Protocols for these studies were approved by the Walter Reed Army Institute of Research Human Use Review Committee. Investigators adhered to AR 70-25 on the use of volunteers in research. Citations of commercial organizations and trade names in this report do not constitute an official Department of the Army endorsement or approval of the products or services of these organizations. The views expressed in this paper are those of the authors and do not reflect the official policy or position of the Department of the Army, Department of Defense, or the U.S. Government. Support for this work was provided by the Walter Reed Army Institute of Research through an In-Laboratory Innovative Research grant (ILIR), the U.S. Air Force, L3 Com, Inc. and LB&B Inc.

References

1. Armentrout J, Holland D. Analysis of crew fatigue factors in a C-5 loss of control mishap. Aviat Space Environ Med 2004; 75(4, Sec II), B81.
2. Balint R. Psychic paralysis of gaze, optic ataxia, and spatial disorder of attention. Cognitive Neuropsychology 1995; 12(3): 265-281.
3. Behar I, Kimball KA, Anderson DA. Dynamic visual acuity in fatigued pilots. Fort Rucker Al: Bio-Optics Division, US Army Aeromedical Research Laboratory; 1976 Jun. Report No. 76-24.
4. Belenky G, Wesensten NJ, Thorne DR, Thomas ML, Sing HC, Redmond DP, Russo MB, Balkin TJ. Patterns of performance degradation and restoration during sleep restriction and subsequent recovery: A sleep dose-response study. J Sleep Res 2003 Mar; 12(1): 1-12.
5. Caldwell JA, Caldwell JL. An in flight investigation of the efficacy of dextroamphetamine for sustaining helicopter pilot performance. Aviat Space Environ Med 1997 Dec; 68(12): 1073-80.
6. Caldwell JA Jr., LeDuc PA. Gender influences on performance, mood and recovery sleep in fatigued aviators. Ergonomics 1998; 41(12): 1757-1770.
7. Caldwell JA, Caldwell JL, Smyth III NK, Hall KK. A double-blind, placebo controlled investigation of the efficacy of modafinil for sustaining the alertness and performance of aviator: a helicopter simulator study. Psychopharmacology 2000 Mar; 150: 272-82.
8. Caldwell J, Caldwell JL, Smith J, Brown D. Modafil's effects on simulator performance and mood in pilots during 37 h without sleep. Aviat Space Environ Med 2004; 75(9): 777-784.
9. Cole RJ, Kripke DF, Gruen W, Mullaney DJ, Gillen JC. Automatic sleep/wake identification from wrist actigraphy. Sleep 1992; 15(5): 461-9.
10. De Gennaro L, Ferrara M, Urbani L, Bertini M. Oculomotor impairment after 1 night of total sleep deprivation: a dissociation between measures of speed and accuracy. Clin Neurophysiol 2000; 111:1771-8.
11. De Gennaro L, Ferrara M, Curcio G, Bertini M. Visual search performance across 40 h of continuous wakefulness: Measures of speed and accuracy and relation with oculomotor performance. Physiology and Behavior 2001; 74: 197-204.
12. Dinges DF, Powell JW. Microcomputer analyses of performance on a portable, simple visual RT task during sustained operations. Behav Res Methods Instrum Comput 1985; 17:652-5.
13. Dinges DF, Pack F, Williams K, Gillen KA, Powell JW, Ott GE, Aptowicz C, Pack AI. Cumulative sleepiness, mood disturbance, and psychomotor vigilance performance decrements during a week of sleep restricted to 4-5 hours per night. Sleep 1997; 20(4): 267-77.
14. Doran SM, Van Dongen HPA, Dinges DF. Sustained attention performance during sleep deprivation: evidence of state instability. Arch Ital Biol 2001; 139: 253-67.
15. Endsley M. Toward a theory of situation awareness in dynamic systems. Human Factors 1995; 37(1): 32-64.
16. Ferrara M, De Gennaro L, Bertini M. Voluntary oculomotor performance upon awakening after total sleep deprivation. Sleep 2000; 23(6): 801-11.
17. French J, Bisson RU, Neville KJ, Mitcha J, Storm WF. Crew fatigue during simulated, long duration B-1B bomber missions. Aviat Space Environ Med 1994 May; 65(5, Suppl): A1-6.
18. Girardin J-L, Kripke DF, Cole RJ, Assmus JD, Langer RD. Sleep detection with an accelerometer actigraph: comparisons with polysomnography. Physiology and Behavior 2001; 72:21-8.
19. Hatfield J. The effects of sleep deprivation on eye movements. Boston, MA: Walter Fernald State School Boston MA; 1971 Jul.
20. Johnson D, Thorne D, Rowland L, Balkin T, Sing H, Thomas M, Wesensten N, Redmond D, Russo M, Welsh A, Aladdin R, Cephus R, Hall S, Powel J, Dinges D, Belenky G. The effects of partial sleep deprivation psychomotor vigilance (abstract). Sleep 1998; 21: 137.

21. Kripke DF, Mullaney DJ, Messin S, Wyborney VG. Wrist actigraphy measures of sleep and rhythms. Electroencephalography and Clinical Neurophysiology 1978;44:674-6.
22. Lieberman HR, Coffey B, Kobrick J. A vigilance task sensitive to the effects of stimulants, hypnotics, and environmental stress: The scanning visual vigilance test. Natick, MA: US Army Research Institute of Environmental Medicine, Military Nutrition and Biochemistry Division; 1998 Aug. Report No. XA-USARIEM.
23. Lowenstein O, Loewenfeld I. Types of central autonomic innervation and fatigue. Archives of Neurology and Psychiatry 1951; 66:581-599.
24. McLaren JW, Hauri PJ, Lin SC, Harris CD. Pupillometry in clinical sleepy patients. Sleep Med 2002; 3: 347-52.
25. Meeteren AV. Het Effekt van Slaaponthouding op een Eenvoudige Visuele Detektietaak (Dutch) The effect of sleep deprivation upon a simple visual detection task. Netherlands: Institute for Perception RVO-TNO Soestrerberg; 1983 Jul. Report No(s). IZF-1983-11 and TDCK-78191.
26. Miura T. Coping with situational demands: A study of eye-movements and peripheral vision performance. In: Gale AG, Freeman MH, Haslegrave CM, Smith P, Taylor SP, eds. Vision in vehicles. Amsterdam: North-Holland; 1986: 126-37.
27. Miura T. Active function of eye-movement and useful field view in a realistic setting. In. Groner R, d'Ydewalle G, Parham R, eds. From eye to mind. Information acquisition in perception, search, and reading. Amsterdam: North-Holland; 1990: 119-27.
28. Morris TL, Miller JC. Electrooculographic and performance indices of fatigue during simulated flight. Biol Psych 1996; 42: 343-60.
29. Neri DF, Oyung RL, Colletti LM, Mallis MM, Tam PY, Dinges DF. Controlled breaks as a fatigue countermeasure on the flight deck. Aviat Space Environ Med 2002 Jul; 73(7): 654-64.
30. Neville KJ, Bisson RU, French J, Boll PA, Storm WF. Subjective fatigue of C-141 aircrews during Operation Desert Storm. Hum Factors 1994 Jun; 36(2): 339-49.
31. Paul, MA, Pigeau, RA, Weinberg, H. CC-130 pilot fatigue during re-supply missions to former Yugoslavia. Aviat Space Environ Med 2001; 72: 965-73.
32. Polakoff R, Hirshkowitz RS, Herman J. Deterioration in oculomotor fixation causes operator failure in a visual paradigm (abstract). Sleep Research 1997; 26:65.
33. Recarte MA, Nunes LM. Effects of verbal and spatial- imagery tasks on eye fixations while driving. J Exp Psychol Appl 2000 Mar; 6(1): 31-43.
34. Rogé J, Pébayle T, El Hannachi S, Muzet A. Effect of sleep deprivation and driving duration on the useful visual field in younger and older subjects during simulator driving. Vision Res 2003; 43: 1465-72.
35. Rosenberger PB. Concurrent schedule control of human eye movement behavior. Boston MA: Walter E Fernald State School Waverly Mass Eunice Kennedy Shriver Center; 1971 Sep. Report No. 1.
36. Rowland L, Thorne D, Balkin T, Sing H, Wesensten N, Redmond D, Johnson D, Anderson A, Cephus R, Hall S, Thomas M, Powell J, Dinges DF, Belenky G. The effects of four different sleep-wake cycles on psychomotor vigilance (abstract). Sleep Research 1997; 26:627.
37. Russo M, Thorne D, Thomas M, Sing H, Redmond D, Balkin T, Wesensten N, Welsh A, Rowland L, Johnson D, Cephus R, Hall S, Belenky G. Sleep deprivation induced balint's syndrome (peripheral visual field neglect): a hypothesis for explaining driving simulator accidents in awake but sleepy drivers. Sleep 1999; 22(1): 327.
38. Russo MB, Thomas M, Thorne D, Sing H, Redmond D, Rowland L, Johnson D, Hall S, Krichmar J, Balkin T. Oculomotor impairment during chronic partial sleep deprivation. Clin Neurophys 2003 114:723-736.
39. Russo MB, Sing H, Santiago S, Kendall AP, Johnson D, Thorne D, Escolas SM, Holland D, Hall S, Redmond D. Visual Neglect: Occurrence and patterns in pilots in simulated overnight flight. Aviat Space Environ Med 2004; 75, 4:323-332.
40. Schmidt D, Abel LA, Dell'Osso F, Daroff RB. Saccadic velocity characteristics: Intrinsic variability and fatigue. Aviat Space Environ Med 1979; 50:393-5.
41. Stern, J. Eye Activity Measures of Fatigue, and Napping as a Countermeasure. Alexandria, VA: Trucking Research Institute, Federal Highway Administration, Washington Office of Motor Carrier Research and Standards; 1998 Feb. Report No: FHWA-MC-99-028.
42. Systems Technology, Inc. Simulator, Systems Technology, Inc., Hawthorne, CA.
43. Thomas M, Thorne D, Sing H, Redmond D, Balkin T, Wesensten N, Russo M, Welsh A, Rowland L, Johnson D, Aladdin R, Cephus R, Hall S, Belenky G. The relationship between driving accidents and microsleep during cumulative partial sleep deprivation. J Sleep Res 1998; 7(2, Suppl):275.
44. Thomas M, Sing H, Belenky G, Holcomb H, Mayberg H, Dannals R, Wagner H, Thorne D, Popp K, Rowland L, Welsh A, Balwinski S, Redmond D. Neural basis of alertness and cognitive performance impairments during sleepiness. I. Effects of 24 hours of sleep deprivation on waking human regional brain activity. J Sleep Res 2000; 9(4): 335-352.
45. Thomas M, Sing H, Belenky G, Holcomb H, Mayberg H, Dannals R, Wagner H, Thorne D, Popp K, Rowland L, Welsh A, Balwinski S, Redmond D. Neural basis of alertness and cognitive performance impairments during sleepiness. II. Effects of 48 and 72 h of sleep deprivation on waking human regional brain activity. Thalamus and Related Systems 2003 Aug; 2(3): 199-229.
46. Thorne D, Thomas M, Sing H, Peters R, Kloeppel F, Belenky G. Accidents, attention, and performance in a driving simulator during 64 hours of progressive sleep deprivation. (abstract). Sleep Research 1997; 26: 634.
47. Thorne DR, Thomas ML, Russo MB, Sing HC, Balkin TJ, Wesensten NJ, Redmond DP, Johnson DE, Welsh A, Rowland L, Cephus R, Hall SW, Belenky G. Performance on a driving simulator divided-attention task during one week of restricted nightly sleep (abstract). Sleep 1999; 22: S301.
48. Underwood G, Radach R. Eye guidance and human information processing: Reading, visual search, picture, perception, and driving. In. Underwood G, ed. Eye guidance in reading and scene perception. Oxford, England: Elsevier; 1998: 1-28.
49. Yoss R, Moyer N, Hollenhorst R. Pupil size and spontaneous pupillary waves associated with alertness, drowsiness, and sleep. Neurology 1970; 20:545-54.

Maintaining Optimal Challenge in Computer Games through Real-Time Physiological Feedback

Pramila Rani
Electrical Engineering
Vanderbilt University
pramila.rani@vanderbilt.edu

Nilanjan Sarkar
Mechanical Engineering
Vanderbilt University
nilanjan.sarkar@vanderbilt.edu

Changchun Liu
Electrical Engineering
Vanderbilt University
changchun.liu@vanderbilt.edu

Abstract

Computer based games are steadily becoming a powerful tool for entertainment and education. When the aim of the game is to keep the player maximally involved, it would be useful to continuously alter the game difficulty level in order to maintain a high level of challenge. This paper presents a framework for providing real-time feedback regarding the affective state (anxiety level) of the player to modify game difficulty. Peripheral physiological signals were measured through wearable biofeedback sensors and an anxiety-based difficulty altering methodology for a Pong game was designed and implemented. The results from affect-elicitation tasks for human participants showed that it is possible to detect affective states of anxiety, engagement, boredom and frustration through physiological sensing in real-time. Performance feedback-based and anxiety-feedback based Pong experiments were then conducted to demonstrate that the latter resulted in higher performance improvement and greater perceived challenge by the participants under lower anxiety.

1. Introduction

Computer games play an important role in modern society. Although its primarily providing entertainment, it can be effectively used for education. It is well known that often the same children who are unable to focus in classes at school can exhibit extraordinary determination and engagement in video/computer games. They can not only remain focused for longer hours, they are also highly motivated to improve their skill level in these games. Thus, a well designed computer game that can provide optimal challenge level to the user and adapt dynamically can be used as an effective education delivery system.

Csikszentmihalyi [1] calls the feeling of deep engagement a state of "flow": "Flow tends to occur when a person's skills are fully involved in overcoming a challenge that is just about manageable. ... When goals are clear, feedback relevant, and challenges and skills are in balance, attention becomes ordered and fully invested." In many research works investigating computer games, it has been found that varying the challenge level is an important characteristic of these games that makes them so appealing [2]. There are two aspects of maintaining optimal challenge – at each stage of the game the player should be given, "a challenge that is just about manageable," and the level of challenge should be continuously adjusted according to the display of skill and involvement by the player. If the challenge is sub-optimal, the players will become bored; while impossible challenges will frustrate and dishearten them. The same principle can also be suitably used in educational settings, where computers can teach in the zone of optimal development for each student, motivating him/her with just the exact level of challenge and incrementing the challenge level to elevate the student to the next level of skill.

In most existing games, the level of difficulty is altered to increase or decrease the level of challenge. This alteration can be manual (that is the player explicitly chooses to play at a particular level) or automatic (where the performance of the player is assessed to automatically increase/decrease the difficulty level of the game).

In this paper, we propose affect-based modification of game difficulty to achieve higher challenge. Here, the anxiety of the player was detected in real-time and employed to alter game difficulty to keep the player maximally involved. Anxiety was chosen to be an important component to the overall challenge that a person faces. It was detected from physiological signals using wearable and affective computing techniques. The player's other discrete affective states of boredom, engagement, and frustration were also detected for post-game analysis.

Previous work done in this area consists of computer-based interactive learning environments that assess a learner's affective state during automated tutoring sessions [3]-[5]. In [3] the focus is on improving user interaction in educational computer games by achieving a tradeoff between engagement and learning. In [4] AutoTutor - a computerized tutor has been developed which serves as a learning scaffold to assists students by simulating the discourse patterns and pedagogical strategies of a human tutor. Mota et al in [5] present preliminary work done in the area of developing a Learning Companion, a computer-based system that is responsive to the affective aspects of learning.

3. Methodology

3.1. Task Design

The experiment was conducted in two phases- Phase I and Phase II. During Phase I which was essentially the modeling phase, physiological and task-related data was collected to learn the physiological patterns for various affective states of participants. In Phase II, the models built in Phase I were used to assess the affective state of the participant in real-time and relevant feedback was provided.

To obtain training data, human subjects were engaged in the computer game (a variant of the early, classic video game "Pong") and an anagram solving task that elicited various affective states. Fifteen individuals (eight female, and seven male) participated in the experiment. Their ages ranged from 18 to 54 years of age. The design was a fully within-participants design in which each participant engaged in six experimental sessions (one hour each) over the course of six days. During each session the participants were presented with a series of trials in which the game/task difficulty was systematically varied in order to produce various affective states such as anxiety, engagement, anger, frustration, and boredom. The anagram solving task has been previously employed to explore relationships between both electrodermal and cardiovascular activity with mental anxiety [6]. Emotional responses were manipulated in this task by presenting the participant with anagrams of varying difficulty levels, as established through pilot work. Pong game has been used in the past by researchers to study anxiety, performance, and gender differences [7]. Various parameters of the game were manipulated to elicit the required affective responses. These included: ball speed and size, paddle speed and size, sluggish or over-responsive keyboard and random keyboard response. The relative difficulties of various trial configurations were established through pilot work. During the tasks, the participant's physiology was monitored with the help of wearable biofeedback sensors. This sensor data indicating physiological response of the subject along with the self-reported affective states of the participants were used to train the inference system. After each trial the participants reported their perceived subjective emotional states. This information was collected using a battery of five self-report questions rated on nine-point Likert scales. Self-reports were used as reference points to link the objective physiological data to participants' subjective anxiety levels.

Phase II was the verification phase during which it was investigated which methodology of difficulty variation resulted in a better game - a performance-based design or an affective–state based design. By a "better" game we mean one in which the participant reported higher level of challenge, while exhibiting a greater improvement in performance and lower anxiety. In the performance-based design the level of difficulty of the Pong game was varied based on the performance of the player while in the affect-based design, the anxiety level of the player (which was detected in real-time using physiological signals) was employed to switch between difficulty levels of the game.

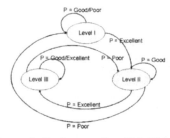

Figure 1. Stateflow Diagram for Performance-Based Modification of Game Difficulty

To keep the game design simple, three levels of difficult were designed - level I (easy), level II (moderately difficult) and level III (very difficult). Furthermore, three levels of performance were identified – poor, good and excellent and three levels of anxiety were defined – low, medium and high. Figure 1 and Figure 2 show the stateflow models that were utilized to switch difficulty based on performance (P) and anxiety (A). During the Phase II, the same participants who took part in Phase I were required to play two sessions of the game.

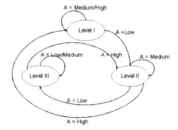

Figure 2. Stateflow Diagram for Anxiety-Based Modification of Game Difficulty

3.2 Experimental Set-up

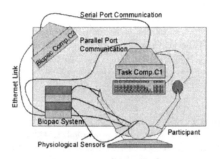

Figure 3. Experimental Set-up

Figure 3 shows the set-up for the experiment. The participant played the game on computer C1 while his/her physiological data was acquired via the Biopac system on C2. Physiological signals were transferred from the Biopac transducers to C2 through an ethernet link at 1000 Hz. after being amplified and digitized. C1 and C2 conversed with each other by means of serial communication between them. C1 was also connected to the Biopac system via a parallel port, through which the game related markers were recorded along with the physiological data in a time-synchronized manner.

The physiological signals were acquired and processed in Matlab environment on C2. Customized algorithms implemented in Matlab code were used to condition the signals and extract relevant features in real-time. The Pong game was implemented in Java and was presented to the participants on C1.

3.3 Physiological Indices for Affect Detection

Affect detection as implied in this paper involves perceiving or sensing of emotional states in humans. Affective states have potentially observable effects over a wide range of response systems, including facial expressions, vocal intonation, gestures, and physiological responses (such as cardiovascular activity, electrodermal responses, muscle tension, respiratory rate etc.) [8]. However, in our work we have chosen to determine a person's underlying affective state through the use of physiological signals for various reasons. Physical expressions (facial expressions, vocal intonation) are culture, gender, and age dependent thus complicating their analysis. On the other hand physiological signals are usually involuntary and tend to represent objective data points. Moreover, they offer an avenue for recognizing affect that may be less obvious to humans but more suitable for computers. Another important reason for choosing physiology is that our aim is to detect affective states of people engaged in real-life activities, such as working on their computers, controlling a robot, or operating a machine. In most of these cases, even if a person does not overtly express his/her emotion through speech, gestures or facial expression, a change in the physiology pattern is inevitable and detectable.

There is a rich history in the Human Factors and Psychophysiology literature of understanding occupational stress [9], operator workload [10], mental effort [11] and other similar measurements based on physiological measures such as Electromyogram (EMG), Electroencephalogram (EEG), and heart rate variability (HRV).

The physiological signals that were initially examined were ECG (Electrocardiogram), ICG (Impedance Cardiogram), PPG (Photoplethysmogram), Heart Sound, GSR (Galvanic Skin Response), peripheral temperature, and EMG. These signals were selected because they can be measured non-invasively and are relatively resistant to movement artifacts

Multiple features were derived for each physiological measure. $Power_{sym}$ is the power associated with the sympathetic nervous system activity of the heart (in the frequency band 0.04-0.15 Hz.). $Power_{parasym}$ is the power associated with the heart's parasympathetic nervous system activity (in the frequency band 0.15-0.4 Hz.). InterBeat Interval (IBI) is the time interval in milliseconds between two "R" waves in the ECG waveform in millisecond. IBI_{mean} and IBI_{std} are the mean and standard deviation of the IBI. Photoplethysmograph signal (PPG) measures changes in the volume of blood in the finger tip associated with the pulse cycle. Pulse transit time (PTT) is the time it takes for the pulse pressure wave to travel from the heart to the periphery, and it is estimated by computing the time between systole at the heart (as indicated by the R-wave of the ECG) and the peak of the pulse wave reaching the peripheral site where PPG is being measured. Heart Sound signal measures sounds generated during each heartbeat. These sounds are produced by blood turbulence due primarily to the closing of the valves within the heart. The features extracted from the heart sound signal consisted of the mean and standard deviation of the 3rd, 4th, and 5th level coefficients of the Daubechies wavelet transform. Bioelectrical impedance analysis (BIA) measures the impedance or opposition to the flow of an electric current through the body fluids contained mainly in the lean and fat tissue. A common variable in recent psychophysiology research, pre-ejection period (PEP) derived from bioimpedance and ECG measures the latency between the onset of electromechanical systole, and the onset of left-ventricular ejection and is most heavily influenced by sympathetic innervation of the heart. Electrodermal activity consists of two main components – Tonic response and Phasic response. Tonic skin conductance refers to the ongoing or the baseline level of skin conductance in the absence of any particular discrete environmental events. Phasic skin conductance refers to the event related changes that occur, caused by a momentary increase in skin conductance (resembling a peak). The EMG signal from corrugator supercilii muscle (eyebrow) captures a person's frown and detects the tension in that region. It is also a valuable source of blink information and helps us determine the blink rate. The EMG signal from the zygomaticus major muscle captures the muscle movements while smiling. Upper trapezius muscle activity measures the tension in the shoulders, one of the most common sites in the body for developing stress. The useful features derived from EMG activity were: mean, slope, standard deviation, mean frequency and median frequency. Blink movement could be detected from the corrugator supercilii activity. Mean amplitude of blink activity and mean interblink interval were also calculated from corrugator EMG. A detailed description of all these measures can be found in our previous work [12].

3.4 Regression Tree for Anxiety Prediction

Various machine learning and pattern recognition methods have been applied for determining the underlying affective state from cues such as facial expressions, vocal intonations, and physiology. Fuzzy logic [14], neural network [15], k-nearest neighbors algorithm [16], linear and nonlinear regression analysis [17], discriminant function analysis [18], a combination of Sequential Floating Forward Search and Fisher Projection methods [19], Bayesian classification [20], Naïve Bayes classifier [21] and Hidden Markov Model [22] are some of the methods that have been investigated in the past.

In this paper we have used regression trees (also known as decision trees) to determine the affective state from a set of features derived from physiological signals. This method has not been employed before for affect detection and recognition. Determining a person's probable affective state or the level of arousal for a given affective state from the physiological response resembles a classification problem where the attributes are the physiological features. An essential criterion for a good classification method is that it not only generates accurate classifiers but also offers insight and understanding into the predictive structure of the data. Classification And Regression Trees (CARTs) have been extensively applied in the medical field [23], speech recognition [24], and gait measurement. Regression tree learning is a frequently used inductive inference method.

Regression trees approximate discrete valued functions that adapt well to noisy data and are capable of learning disjunctive expressions. A regression tree takes as input a situation or an object characterized by a set of properties and outputs a decision [25]. It consists of several nodes, each representing a question that determines if a predictor satisfies a given condition. The system proceeds to the next question or arrives at a fitted response value dependent upon the provided answer. The regression tree creation began by choosing the best attribute to split the examples. The best attribute is the one that changes the classification the most. Once the examples were split, each outcome represented a new regression tree-learning problem containing fewer examples. Proceeding in this manner, the shortest regression tree that generalized the training data examples was constructed.

4. Results

4.1. Offline Analysis – Phase I

During Phase I, fifteen data sets were collected (one for each participant.) Each data set contained six hours of physiological signal recording. This was in addition to the ten minutes of base line data on each day. Each participant completed approximately 100 task epochs. These sessions spanned the anagram and Pong tasks. Wearable sensors were used to continuously monitor the person's physiological activities, and the physiological features as mentioned in Section 3.3 were calculated using customized algorithms. The self-reports of the participants indicated the underlying affective state of the person at various times while performing the tasks.

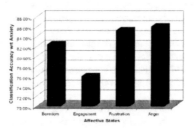

Figure 3 Classification accuracy of anxiety
with regard to other affective states

There were significant correlations observed between the physiology of the participants and their self-reported anxiety levels. This was in accordance with the claim of psychophysiologists that there is a distinct relationship between physiology and underlying affective states of a person. Due to the phenomena of person stereotypy [13] no

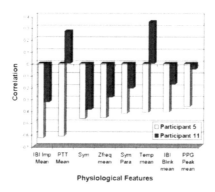

Figure 3 Person Stereotypy for the affective
state of anxiety

two participants had exactly the same set of useful (highly correlated) features. Figure 3 shows the physiological features that were highly correlated with the state of anxiety for participant 5 and the corresponding correlation of the same features with the state of anxiety for participant 11. An absolute correlation greater than equal to 0.3 is considered significant. It can be seen from Figure 3 that two features – mean of pulse transit time (PTT_{mean}) and mean of temperature ($Temp_{mean}$) are correlated differently for the two participants. While both are correlated positively with anxiety for participant 11, they are negatively correlated for participant 5. However, features like mean interbeat interval of impedance (IBI Imp_{mean}), sympathetic activity power (Sym) and mean frequency of EMG activity from zygomaticus major ($Zfreq_{mean}$) are similarly related for both participants. Using regression tree, varying levels of anxiety could be detected with a mean accuracy of 88.54% across all fifteen participants.

It was also observed that each affective state for any participant had a unique set of feature correlates – that is the set of features correlated with anxiety were distinct from those correlated with boredom or engagement. Since the signature of each affective state was different, it was expected that a distinction between anxiety and boredom/engagement/anger/frustration could be made based on the physiological features alone. Figure 3 shows the percentage accuracy in distinction between anxiety and other states across the fifteen participants.

It can be seen that on the basis of physiology alone, a state of anxiety could be distinguished from a state of boredom 82% of the times, from state of engagement 76% of the times and from states of frustration and anger 85% and 86% of the times respectively.

4.2. Online Results – Phase II

In Phase II, two variations of Pong game were designed. One in which the game difficulty was adapted based on player's performance in order to increase the level of challenge, and another in which a measure of the player's anxiety was employed to alter game difficulty. During each variation of the game, the player's improvement in skill, and overall anxiety during game were measured. A self-report was obtained at the end of the session in which the participants reported the challenge, anxiety, boredom, anger, frustration, and engagement experienced during the session. The results are based on the validation sessions with two participants. Figure 4 shows the improvement in performance after the performance-based and affect-based sessions. Both the participants showed a greater improvement in performance after the affect-based session. Participant 6 showed 6.7% improvement in performance in the first session and 13% improvement in the second session. Participant 8 showed 1.3% improvement in performance in the first session and 6.7% improvement in the second session.

189

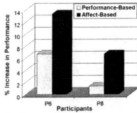

Figure 4. Percent Increase in Performance in Performance-Based and Affect-Based Games

During each session, anxiety was calculated using physiological signals and it was determined that both the participants showed higher levels of anxiety while engaged in the performance-based task (Figure 5). At the end of each session, the participants had reported the level of challenge that they had experienced and from this self-report, it was seen that they both perceived the affect-based task to be more challenging than the performance-based one (Figure 5). Their reports also indicated they were equally engaged in both the sessions. This could have possibly been because the sessions lasted only 30 minutes and the game difficulty was constantly varied, hence the chances of playing at the same level and getting bored were very less.

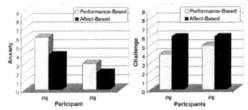

Figure 5 Session Anxiety and Perceived Challenge in Performance-Based and Affect-Based Games

4.3 Discussion

The above experimental results demonstrate two important points. 1) Physiological signals are a powerful indicator of underlying affective states and can play an important role in relaying implicit communication regarding a human to a machine. 2) In computer-based games, user affective feedback can be possibly utilized to make the game more challenging, while inducing the player to perform better under lower anxiety.

We conducted experiments that elicited various affective states in participants engaged in computer games. We also analyzed their physiological data offline in light of their self-reports to determine distinct patterns for anxiety, engagement, boredom, and frustration. Our results confirmed the occurrence of person-stereotypy and context-stereotypy in expressing emotions and we circumvented this problem by adopting a person-specific and context specific approach to affect recognition. Validation experiments are being run wherein affective feedback from the player is used to improve the challenge level of the game. The results of two participants have been presented in this paper. The outcome is promising and it seems that it may be feasible to successfully use affective feedback, specifically anxiety to adapt the challenge in computer games in real-time.

Among the related research works in the area of affective computing, most focus on methods for recognizing a person's discrete emotional states while deliberately expressing pure emotions such as joy, grief, anger, etc. The other works that detect affective states of people engaged in real-life tasks, most use overt signals such as facial expressions, voice or speech. Even those that use physiology have not used context and person specific techniques for learning physiological patterns. Also, there is no known human-computer interaction framework functional in this context that uses such a comprehensive set of physiological features for real-time affect detection. In this we work we detect the anxiety level purely on the basis of physiological signals.

5. Conclusions and Future Work

An approach to maintaining optimal challenge in computer games using physiological feedback from player is presented. In this work we focus on anxiety as the target affective state. A set of physiological indices have been presented that have been shown to have significant correlation with anxiety. The affect recognition technique infers the underlying affective state of the human from peripheral physiological signals using regression theoretic methodology.

In the Phase I, two separate cognitive tasks were designed to elicit affective states of anxiety, engagement, boredom, and frustration. Fifteen human participants took part in this study where each participant engaged in the cognitive tasks for six hours. Phase II experiments were conducted to verify if continuous feedback regarding the anxiety level of the player could be useful in modifying the difficulty level of game. The results from two participants indicated that physiological feedback was more effective than performance feedback in providing greater challenge to the players, lowering their anxiety and improving their performance.

Future work will involve completing the Phase II experiments with the remaining participants. We would also like to study the relationship between player's perceived challenge and his/her affective states of engagement, boredom and frustration.

References

[1]   M. Csikszentmihalyi, *Finding flow: The psychology of engagement with everyday life,* BasicBooks (New York), 1st edition , pp. 30-31, 1997.

[2]   T. W. Malone. "Towards a theory of intrinsically motivating instruction" in Cognitive Science, vol. 4, pp: 333-369, 1981.

[3]   C. Conati Probabilistic Assessment of User's Emotions in Educational Games . Journal of Applied Artificial Intelligence, special issue on " Merging Cognition and Affect in HCI", vol. 16 (7-8), pp. 555-575, 2002

[4]   A. C. Graesser, K. Wiemer-Hastings, P. Wiemer-Hastings, R. Kreuz,, AutoTutor: A simulation of a human tutor. Journal of Cognitive Systems Research, vol. 1, pp: 35-51, 1999

[5]   A. Kapoor, S. Mota, and R. Picard, "Towards a Learning Companion that Recognizes Affect", AAAI Fall Symposium, 2001.

[6]   G. Weidner, C. Kohlmann, M. Horsten, S. P. Wamala, K. Schenck-Gustafsson, M. Högbom, and K. Orth-Gomer, "Cardiovascular Reactivity to Mental Stress in the Stockholm Female Coronary Risk Study," *Psychosomatic Medicine,* vol. 63, pp: 917-924, 2001

[7]   R. M. Brown, L. R. Hall, R. Holtzer, S. L. Brown, N. L. Brown, "Gender and Video Game Performance, Sex Roles," vol. 36 (11-12), pp. 793 – 812, 1997

[8]   R. Picard, *Affective Computing,* The MIT Press, Cambridge, 1997

[9]   F.E. Gomer, L.D. Silverstein, W.K. Berg, and D.L. Lassiter. "Changes in electromyographic activity associated with occupational stress and poor performance in the workplace." Human Factors, vol. 29(2), pp: 131-143, 1987.

[10] A.F. Kramer, E.J. Sirevaag, and R. Braune. "A Psychophysiological assessment of operator workload during simulated flight missions." Human Factors, vol. 29(2), pp: 145-160, 1987.

[11] K.J. Vicente, D.C. Thornton, and N. Moray. "Spectral analysis of sinus arrhythmia: a measure of mental effort." Human Factors, vol. 29(2), pp: 171-182, 1987.

[12] (Under Review) P. Rani, N. Sarkar, "An Approach to Human-Robot Interaction Using Affective Cues," IEEE Transactions on Robotics and Automation.

[13] J. L. Lacey and B. C. Lacey, "Verification and extension of the principle of autonomic response-stereotypy," *American Journal of Psychology,* vol. 71(1), pp: 50-73, 1958.

[14] N. Tsapatsoulis, K. Karpouzis, G. Stamou, F. Piat and S. Kollias, "A Fuzzy System for Emotion Classification based on the MPEG-4 Facial Definition Parameter Set," Proc. EUSIPCO-2000, Tampere, Finland, 2000.

[15] V.A. Petrushin. "Emotion recognition in speech signal: experimental study, development and application." Proc. 6th International Conference on Spoken Language processing, pp: 454-457, 2000.

[16] K. R. Scherer, "Studying the emotion-antecedent appraisal process: An expert system approach," Cognition and Emotion, vol. 7, pp:325-355, 1993

[17] T. Moriyama, H. Saito, and S. Ozawa. "Evaluation of the relation between emotional concepts and emotional parameters on speech." IEICE Journal, J82-DII(10), pp: 1710-1720, 1999.

[18] W. Ark, D. Dryer, and D. Lu, "The Emotion Mouse." Human-Computer Interaction: Ergonomics and User Interfaces 1, Bullinger. H. J. and J. Ziegler (Eds.), Lawrence Erlbaum Assoc., London. pp. 818-823, 1999.

[19] E. Vyzas and R. W. Picard, "Affective Pattern Classification", 1998 AAAI Fall Symposium Series: Emotional and Intelligent: The Tangled Knot of Cognition, October 23-25, 1998, Orlando, Florida, 1998

[20] Y. Qi and R. W. Picard. "Context-sensitive Bayesian Classifiers and Application to Mouse Pressure Pattern Classification." Proc. International Conference on Pattern Recognition, Quebec City, Canada, 2002.

[21] N. Sebe, I. Cohen, A. Garg, and T. S. Huang. "Emotion Recognition using a Cauchy Naive Bayes Classifier." Proceedings of International Conference on Pattern Recognition, pp: 10017-10021, 2002.

[22] I. Cohen, A. Garg, and T.S. Huang. "Emotion recognition using multilevel HMM." Proc. NIPS Workshop on Affective Computing, Colorado, 2000.

[23] P. Kokol, M. Mernik, J. Završnik, K. Kancler and I. Malèiæ, "Decision Trees and Automatic Learning and Their Use in Cardiology," Journal of Medical Systems, vol. 9(4): pp: 201-206, 1994.

[24] S. Downey and M. J, Russell, "A Decision Tree Approach to Task Independent Speech Recognition," Proc. Inst Acoustics Autumn Conf on Speech and Hearing, vol. 14(6), pp: 181-188, 1992.

[25] L. Breiman, J. H. Friedman, R. A. Olshen, and C. J. Stone, Classification and Regression Trees, Wadsworth & Brooks/Cole Advanced Books & Software, Pacific Grove, CA., 1984.

fMRI of virtual social behavior: brain signals in virtual reality and operational environments

Klaus Mathiak

Psychiatry & Psychotherapie, RWTH Aachen
Pauwelsstr. 30, 52074 Aachen, Germany
KMathiak@UKAachen.de

René Weber

Dept. Communication, Michigan State University
East Lansing, Michigan 48824
renew@usc.edu

Abstract

Recent studies as well as meta-analyses demonstrate that exposure to violent video games is significantly linked to increases in aggressive behavior, aggressive cognition, aggressive affect, and cardiovascular arousal, and to decreases in helping behavior (Anderson & Bushman, 2001). Causality and possible psychological mechanisms, however, remain heavily debated. To investigate the neuronal correlates of social behavior in virtual environments, we measured brain activity by functional magnetic resonance imaging (fMRI) (14 subjects; male, age 18-26; 3 T, multi-echo EPI) during gaming (Ego-shooter Tactical Ops). The course of the game was recorded and submitted to a detailed content analysis (5 main / 27 sub-categories; 0.1 sec resolution). Formal testing, interviews, gaming content, and the physiological parameters show behavior comparable to everyday computer gaming during the scans. Even recurrent episodes of exceeding (virtual) violence emerged. Multi-echo imaging with single-shot distortion compensation and spatio-temporal over-sampling for respiration suppression allowed for a sufficient artifact control. Whereas sensory and motor related changes activated the according cortices, contents from the rating were associated with distributed frontal brain activity. Using refined social-psychological constructs, natural behavior in virtual environments can be investigated by functional neuroimaging. VR is perceived similar to operational environments and may allow the study of similar neuronal patterns as occurring in operational environments. Measures of brain activity during natural behavior outside the laboratory, in contrast, are quite limited with respect to spatial resolution and specificity.

1 Introduction

To understand brain functions in complex environments, either virtual reality (VR) has to be adapted to functional brain imaging or brain functions need to be investigated in natural environments. We want to address both issues. In first place however we will present a setup that allows a rather sophisticated investigation of behavior in VR. Video games pose several advantages for such investigations. First, the simulation technique is rather advanced and adapted to as natural perception as possible. Second, the players are very well trained and, thus, act in those environments naturally.

2 An fMRI setup to study virtual social behavior

MR scanners are sensitive to the introduction of metal part. The high static magnetic fields would move magnetic parts und disallow the use of conventional batteries. Cables carry radio frequency (RF) noise to the receiver coil and impair image acquisition. Electronic devices using RF such as fast digital logics create such a noise as well. Moreover, cables and moveable parts pose a security problem to the volunteer because the might get heated by the RF deposition or gradient switching-induced currents or get accelerated in the static magnetic field and harm the subject. This has to be taken into account when designing a VR environment that allows for auditory and visual perception as well as providing sensitive interfaces. Moreover, the physiology and behavior of the volunteers have to be recorded.

For stimulus presentation, we used a video projector that targeted a screen inside the bore of the scanner. The projection was led into the device via a metallic mirror and the subjects could see the screen via a mirror mounted on the RF head coil in front of the eyes. Auditory stimuli were conveyed via custom head phones. We modified conventional dynamic transducer by removing the magnet and placing the membranes with the coils into noise canceling ear muffs. Shielded HF cable led from isolated transformers at the Faraday cage to the head set. The static magnetic field inside the scanner allowed for close to normal operation with high stereophonic quality despite the missing magnets (compare Baumgart et al., 1998). An additional light presentation was realized by directing a laser beam from outside the measurement room on the projection scream (see setup in Fig. 1).

193

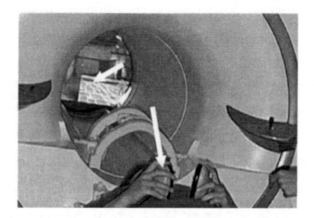

Figure 1: Setup within the MR scanner. Subjects wore MR compatible headphones and could see the projected ame by means of a mirror. They were supine the bore of the magnet.

A modified trackball served for navigation and other interfacing. The internal electronic of a standard optical trackball was removed and connected to the trackball by a 1.20 m long isolated cable. The extracted electronics were kept outside the magnet and distance from the head receiver coil to minimize interferences. These electronics were isolated by an optical USB cable connection which was connected to the PC outside of the Faraday cage which hosted the computer game (Fig. 2). Additionally, optically operating buttons were placed for a reaction time task at the left hand (see Fig. 3).

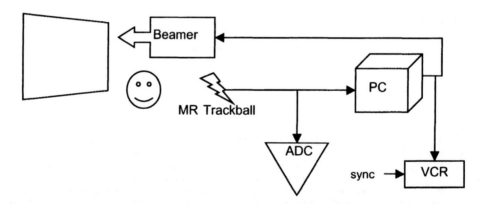

Figure 2: Setup of the gaming equipment. The subjects control via a MR compatible trackball the egoshooter running on the game PC. These actions are recorded via an analog digital converter (ADC). The video of the game is projected into the MR scanner and recorded simultaneously by means of a video cassette recorder (VCR).

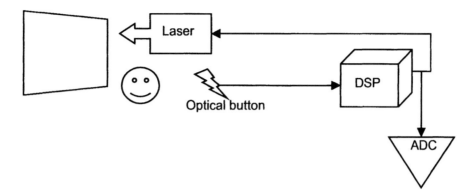

Figure 3: Setup of the distractor loop. A digital signal processor (DSP) output drove a laser diode and a logic channel which was recorded by the ADC device. The DSP was programmed to wait for a Poisson-distributed interval (average duration = 100 s) and then to switch on the laser beam until the optical button was pressed. The reaction required by the subject served as a measure of how much attentional resource was absorbed by the game before the light point in the periphery of the display was detected and led to a reaction.

Physiological variables were recorded to monitor the vegetative state of the subject. The heart rate, pulse amplitude and peripheral oxygenation were measured by a pulseoxymeter with an optical sensor that allowed placing the device outside of the Faraday cage. Skin conductivity was measured at the left food with a device which was connected with an optical serial interface to the recording PC. The latter received as well information from the ADC device via a universal serial bus (USB) interface. The synchronization trigger from the MR scanner was connected via this device as well (Fig. 4).

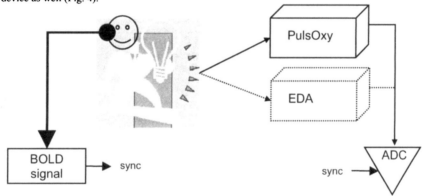

Figure 4: Peripheral physiology was recorded by means of pulseoxymetrie (PulsOxy) and skin conductance (electro dermal activity, EDA). Both devices were connected to the ADC which also recorded the reactions from the distraction task (see Fig. 2) and the synchronization trigger from the MR scanner which recorded in parallel the BOLD signal.

For the data handling three different data sources needed to be integrated: first, the images recorded by the MR scanner, i.e., anatomical and functional images; second, behavioral and physiological data that were input into the ADC or the serial interface of the recording PC; third, the time course of the video game. The latter one was recorded on a VCR (Super-VHS) with the sound and video display. Synchronization of the data sources relied on two different mechanisms. First, a trigger was emitted with the start of every volume acquisition by the MR device and recorded with the ADC as well as on the left audio track of the VCR. Second, the sound from the game was not only recorded by the VCR (mono on the right track) but also converted from analog to digital by the ADC (stereo, 12 bit resolution, 16.7 kHz sampling rate). This indeed provided better synchronization information between the digitized videos and the ADC as compared to the MR trigger. The latter, however, was used to realign the functional MRI signals to the ADC signals (Fig. 5).

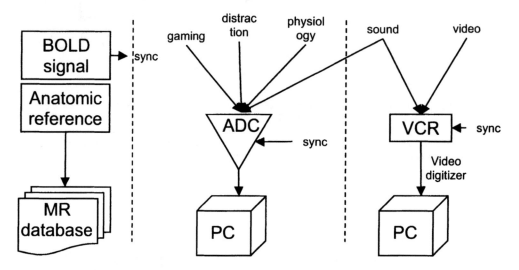

Figure 5: Three different data streams were acquired. The MR scanner sends data directly to its internal database. The time points of data acquisition can be obtained by an extern sync output. Behavior as concerns the gaming device and the distraction task, physiological data from pulseoxymetrie and EDA, and sound track from the game were recorded together with the sync signal by means of a multifunctional ADC on a recording PC using custom software. The VCR recorded video and sound of the game together with the sync signal. It was digitized afterwards into MPEG2 format.

Methods and Subjects

Thirteen male gamers were recruited from local video game and compute shops. They were between 18 and 6 years old and showed no contraindications against MR imaging. Main inclusion criteria were that they played at least 5 h per week for the last year and had experience with ego-shooters (Fig. 6).

Figure 6: To achieve motivation, social reinforcement was applied (in addition to the financial compensation). Likewise, all over recording times of 2 h were possible without loss of interest in involvement of the subjects

Image acquisition was performed at 3 T (TRIO Magnetom, Siemens, Erlangen, Germany) with a multiecho EPI sequence. The sequence acquired 3 echoes at TE = 23, 47, and 79 ms allowing for dynamic undistortion (Weiskopf et al., 2005) and dephasing compensation (Mathiak et al., 2004). Whole brain coverage was achieved with 24 interleaved slices (repetition time TR = 2.25 sec) with spatio-temporal-oversampling reconstruction resulting in an apparent TR of 1.13 sec after spatial filtering. For overlay, anatomical data were acquired of each participant before the functional sessions (T1-weigthed 3D-MPRAGE, 256x224x160 matrix with 1mm isotropic voxels).

Data analysis was conducted using SPM2 (normalization into MNI space of functional and anatomical data, smoothing with 12 mm FWHM kernel; general linear model with phases convoluted with hemodynamic response function as independent variables; random effect model for group analysis corrected for multiple testing across the entire brain volume) and a region-of-interest (ROI) approach. ROIs were defined a-priori based on anatomical coordinates.

After the experiment has ended, the recorded video with sound track was digitized to MPEG2-videos. It was split into single files for each game block (about 12 min each). These videos were content analyzed according to a coding scheme that represents all possible activities within the gaming environment. For this, three trained coder watched the clips on a frame by frame basis and coded events and periods reflecting relevant behavior with a 0.1 s time resolution. The precise strategy and classifications are described elsewhere (Weber et al., in preparation; Fig. 7). The content analysis of the video game play reached an inter-rater reliability of up to 85% and can be considered as highly reliable. External validation criteria, such as hitting the track-ball button (as an indicator for shooting) or "red" in the video (as an indicator for blood or killing within the video game), confirmed the high reliability of the content analysis. The timing data were transferred into the time frame of the ADC data and from there into the time frame of the MR scanner as reflected by the sync signal. Finally, specific events or phases can be chosen to determine a design for the functional imaging analysis with SPM2 (http://www.fil.ion.ucl.ac.uk/spm/). A relatively simple example is illustrated in Fig. 8.

Figure 7: Analysis of each video frame revealed different game phases. The left panel shows rather neutral exploration of the environments whereas the right frame is characterized by violent interaction and potential harm.

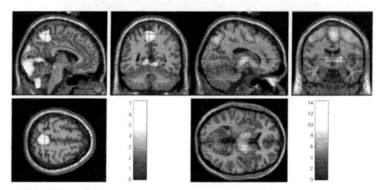

Figure 8: Projections on three planes of activation maps as obtained from stimulus-driven (left panel) and content-driven events (right panel). The stimulation parameters as extracted from the video by physical parameters (brightness, change, …) activated mainly occipital areas whereas content analysis reflects frontal signal changes in the first place.

Emotions interfere with cognitions and vica versa. Thus, to study the neuronal basis of complex behavior the cognitive-affective networks have to be considered as well. Candidates are the anterior cingulate (ACC) and the amygdala. Interactions at the level of cognitive and affective subdivision of the ACC (Bush et al. 2000) as well as the amygala were studied which is of importance because antisocial behavior has been associated with dysfunctions in this network (Veit et al. 2002) and repetitive training as observed in dedicated gamers might affects the functional balance of these areas.

Despite a general improvement of the understanding of human behavior, we might be interested in achieving specific knowledge of video game effects. We particularly might identify risky materials, i.e., phases of the game which may be particular dangerous to elicit aggressive behavior even outside of the game context. Or we might identify subjects which are particularly at risk, i.e., which a prone to high interference of the gaming with social functioning. For both cases alternative strategies have to be developed. That might be in finding comparable exciting but less harmful game content, which induces pro-social behavior. Finally, by the delineation of the affected brain structure, we are closer to find strategies against pathologically aggressive behavior.

3 Transfer to operational environments

It can be easily seen that fMRI or other functional imaging is rather limited in operational environments. Electroencephalography (EEG) and – with certain limitations – functional near-infrared spectroscopy (fNIRS) can be used. In cooperation with the DaimlerCrysler AG-company we started performing measurements of EEG during driving in normal traffic (for further details see Mathiak et al., 2005; Kincses, 2005a, 2005b). Here we use test tones to determine cognitive states and levels of distraction. The elaborate analysis, however, of distributed networks contributing to complex actions cannot be obtained in those setting. Accordingly we need a combination of VR with advanced functional imaging and more simple brain and behavioral during complex actions in the environments of interest. The later mostly serve to verify the validity of the hypothesized underlying brain functions.

4 Conclusion

Modern VR simulations can be used within functional imaging devices such as fMRI. Specific modified equipment and measurement strategies allow for reliable measurements during mildly restrained virtual behavior. Reports of the subjects reflect that mostly the brain imaging device can be ignored and the involvement in the VR becomes very high. Since the behavior should be as unrestrained as possible, classical experimental designs are not suitable. The behavior can be analyzed post-hoc and used as independent variable for the analysis. Moreover, the brain signal can be used as independent variable to predict behavior (or complex visual scenes as in Bartels & Zeki, 2004). VR environments obtain after some training a high validity and can be assumed to achieve a level of *presence* even in the advent of concurrent brain measures.

Acknowledgements

We acknowledge support from the TL Foundation, the DaimlerChrysler AG funded by the DARPA AugCog program, support from DFG (SFB550, Th812/1-1), and support from the Annenberg School for Communication at the University of Southern California (Annenberg Studies on Computer Games). Very special thanks to all the dedicated game users that contributed (Fig. 6) and – as always – to Maike Borutta for here outstanding technical support. Special thanks to Kibum Park and Alfred Aguilar for the codings, and Silke Anders, Nils Birbaumer, Ute Ritterfeld, Ralf Veith, Peter Vorderer, Nikolaus Weiskopf for helpful and valuable discussions. We also want to thank Wolfgang Grodd and Uwe Klose for their help with MR and Ute Ritterfeld and Katharina Behr for their input regarding the content analyses.

References

Anderson, C. A., & Bushman, B. J. (2001). Effects of violent video games on aggressive behavior, aggressive cognition, aggressive affect, physiological arousal, and prosocial behavior: A meta-analytic review of the scientific literature. Psychol. Sci., 12, 353.

Bartels, A., S.Zeki, S. (2004). Funtional brain mapping during free viewing of natural scenes. Hum. Brain Mapp., 21, 75.

Baumgart, F and Kaulisch, T and Tempelmann, C and Gaschler-Markefski, B and Tegeler, C and Schindler, F and Stiller, D and Scheich, H. (1998). Electrodynamic headphones and woofers for application in magnetic resonance imaging scanners. Medical Physics, 25, 2068-2070.

Kincses, W. E. (2005a). Neuronal correlates of Driving. Proceedings of the HCI International 2005. Lawrence Erlbaum Associates, Mahwah, NJ.

Kincses W. E. (2005b). Neural Correlates and Cognitive States of Driving. Proceedings of the HCI International 2005. Lawrence Erlbaum Associates, Mahwah, NJ.

Mathiak, K., Hertrich, I., Grodd, W., & Ackermann, H. (2004). Discrimination of temporal information at the cerebellum: functional magnetic resonance imaging of nonverbal auditory memory. NeuroImage, 21, 154.

Mathiak, K., Hertrich, I., Swirszcz, K., Zvyagintsev, M., Lutzenberger, W., & Ackermann, H. (2005). Preattentive crossmodal information processing under distraction and fatigue. Proceedings of the HCI International 2005. Lawrence Erlbaum Associates, Mahwah, NJ.

Veit, R., Flor, H., Erb, M., Hermann, C., Lotze, M., Grodd, W., & Birbaumer, N. (2002). Brain circuits involved in emotional learning in antisocial behavior and social phobia in humans. Neurosci. Lett., 328, 233.

Weiskopf, N., Klose, U., Birbaumer, N. & Mathiak, K. (2005). Single-shot compensation of image distortions and BOLD contrast optimization using multi-echo EPI for real-time fMRI. NeuroImage, 22, 1068–1079.

Section 2
Cognitive State Sensors

Chapter 5

Functional Near Infrared Technologies for Assessing Cognitive Function

Near-infrared brain imaging and augmented cognition:
A brief overview of advancements, challenges and future developments

Monica Fabiani and Gabriele Gratton

University of Illinois at Urbana-Champaign
Beckman Institute, 405 N. Mathews Ave., Urbana, IL 61822
mfabiani@uiuc.edu; grattong@uiuc.edu

Abstract

This talk will provide a brief overview of recent advancements in the field of near-infrared brain imaging in reference to its application to augmenting human cognitive function. Near-infrared brain imaging is the newest of a series of non-invasive methods for studying human brain function (Gratton & Fabiani, 1998; 2001; Gratton et al., 2003a, b; Villringer & Chance, 1998; Frostig, 2002). It offers the exciting possibility of combining neuronal and hemodynamic measures of brain changes in response to cognitive demands with excellent spatial and temporal resolution. It can also be easily combined with more established brain imaging methods and other physiological sensors (e.g, eye-tracking), thus affording a more extensive and integrated picture of cognitive function. These properties make it an ideal tool for augmenting cognition, especially if current challenges to field applications can be successfully addressed, including signal-to-noise ratio and increased portability

References

Frostig R. (Ed.) (2002). *In Vivo Optical Imaging of the Central Nervous System*, CRC Press.

Gratton, G. & Fabiani, M. (2001). Shedding light on brain function: The event-related optical signal. *Trends in Cognitive Science, 5(8)*, 357-363.

Gratton, G., Fabiani M., Elbert, T., & Rockstroh, B. (2003a). Seeing right through you: Applications of optical imaging to the study of the human brain. *Psychophysiology, 40(4)*, 487-491.

Gratton, G., Fabiani M., Elbert, T., & Rockstroh, B. (Eds.) (2003b). *Optical Imaging*. Special issue of *Psychophysiology*, 40(4), 487-571.

Villringer, A., & Chance, B. (1997). Non-invasive optical spectroscopy and imaging of human brain function. Trends in Neuroscience, 20(10), 435-442.

Non-invasive measurement of functional optical brain changes

Gabriele Gratton & Monica Fabiani

University of Illinois at Urbana-Champaign
Beckman Institute, 405 N. Mathews Ave., Urbana, IL 61822
grattong@uiuc.edu

Abstract

Non-invasive optical methods are used to measures changes in optical properties of brain tissue that are related to its activity. These changes include variations in tissue transparency due to neuronal function (fast signal, Gratton et al., 1995) and subsequent hemodynamic effects due to oxy- and deoxy-hemoglobin concentration changes (slow signal, Hoshi & Tamura, 1993). Given the high absorption at other wavelengths, optical changes are best measured using near-infrared (NIR) light. At these wavelengths, head tissues are highly scattering and the propagation of light is best described using diffusion approximations. In this presentation we will review issues related to the measurement of functional changes of optical brain parameters as they pertain to their practical application in laboratory and field studies. These issues include: (a) advantages and disadvantages of different approaches to the measurement of optical signals (including continuous and various types of time-resolved methods); (b) separation of different optical signals; (c) appropriate quantification of the signal; (d) spatial (2D and 3D) reconstruction of the effects and superimposition on brain anatomy; (e) elimination or reduction of artifacts; (f) signal-to-noise analysis and single-trial classification issues; (g) issues of integration with other techniques.

References

Gratton, G., Corballis, P. M., Cho, E., Fabiani, M., & Hood, D. (1995a). Shades of gray matter: Noninvasive optical images of human brain responses during visual stimulation. Psychophysiology, 32, 505-509.

Hoshi, Y., & Tamura, M. (1993). Dynamic multichannel near-infrared optical imaging of human brain activity. Journal of Applied Physiology, 75, 1842-1846.

A New Approach to fNIR: the Optical Tomographic Imaging Spectrometer

J. Hunter Downs, III
Traci Downs
William Robinson
Erin Nishimura
J Patrick Stautzenberger

Archinoetics, LLC
1050 Bishop St, #240
Honolulu, HI 96813
hunter@archinoetics.com

Abstract

The optical tomographic imaging spectrometer (OTIS) is a functional near-infrared imaging system that can detect the minute changes in neural oxygenation levels that correspond to cognitive activity. The uniqueness of the system is derived from several factors. The hardware was thoughtfully designed to eliminate the noise and NIR light power considerations specific to an fNIR system. The state-of-the-art sensor designs overcome the concerns of non-invasive fNIR imaging and does not limit imaging to areas with exposed skin. The OTIS software converts the reliable raw data gathered by the hardware to oxygenation levels. Algorithms examine short-term relative changes in the oxygenation levels to detect neural activations in the cortical area underlying the OTIS sensors.

1   The Optical Tomographic Imaging Spectrometer

The optical tomographic imaging spectrometer (OTIS) has been developed and tested as a unique functional near-infrared imaging system. Over the past twenty years, a number of technologies have been developed to explore the living human brain. However, of these technologies, only functional near-infrared (fNIR) imaging is capable of imaging the brain using a portable system that can be used in an environment with other electrical equipment.

The OTIS system uses continuous wave technology, recording the changes in the output intensity of continuous input NIR light. The differential signals between the detected data from each of three wavelengths determine the oxy- and deoxy-hemoglobin levels in the tissue underlying the system sensors. Because of its ability to detect minute changes in oxygenation levels that correspond to cognitive activity in the monitored neural region, the OTIS system has an application as a detector of regional neuronal activity.

1.1   OTIS Hardware Overview

There are several factors that make OTIS unique as an fNIR system. The hardware includes a set of multiplexed filters for each of the three wavelengths on each detecting channel. These filters were designed to eliminate noise specific to the noise captured by the silicon photo diodes used by the sensors. The sensor LEDs were custom-designed to provide the stable output power necessary for accurate oxygenation measurements.

1.2   OTIS Sensors

Perhaps the most distinctive feature of the OTIS system is its sensors. The primary concern with using non-invasive fNIR imaging systems to monitor neural signals is that light must be delivered directly to the scalp. Any gap or physical interference between the emitting or detecting optics and the scalp can cause loss of signal or an obliteration of the fNIR signal due to noise. NovaSol's fNIR imaging system designs are state-of-the-art in overcoming this concern. The unique sensor design for the NovaSol fNIR systems allow for reliable neural imaging, even through hair.

The NovaSol sensors are composed of one emitter and multiple detectors. The emitters and detectors contain the electronics and optics, with an opaque urethane encasement that has probe fingers extending from the surface. The central probe finger has an optically clear aperture at the tip, directly above the optic component, that delivers the NIR light to and from the sensor optics. By making contact between the central probe finger and the skin, the optics are effectively coupled to the scalp. A simple, short brushing motion while placing the sensor allows the fingers to penetrate hair, ensuring non-invasive scalp contact.

Older sensor prototypes consisted of optical fibers directly coupling the optics and associated hardware with the scalp. The sensor development deviated from this design to eliminate the attenuation of the signal that occurred at each coupling of differing materials. The fiber added an additional coupling interface, greatly decreasing the signal strength.

Several sensor configurations, with variations in the number of detectors per sensor and the shape of the whole sensor, have been created with these probe designs.

1.3   Processing of the Optical Signal

The reliable NIR signal supplied by the OTIS hardware provides the data required to detect rises in tissue oxygenation levels that indicate neuronal activity in the monitored region. The OTIS software bundle processes the raw data from the OTIS hardware to calculate the oxygenation signal and uses detection algorithms to relate rises in this oxygenation signal to activations in the neural region underlying the system sensors.

1.3.1   Filtering

Noise reduction is accomplished through digital filtering in software, in addition to the preliminary filtering done to the analog signal in the hardware stage. The filter set includes filters to eliminate noise caused by physiological signals, included the heartbeat and breathing.

1.3.2   Oxygenation Calculations

Oxygenation levels are calculated using the Beer-Lambert equation. This calculation relates the changes in intensity for each of the three wavelengths of light used to changes in the levels of oxy- and deoxy-hemoglobin in the tissue.

1.3.3   Cognitive Activity Detection

The calculated oxygenation level is used to detect cognitive activity through rises in brain tissue oxygenation that follows such activity. Long-term drift in the oxygenation signal is removed by examining short-term changes in the signal to reveal increases and decreases in the oxygenation level. Algorithms then determine activations through the short-term rises. These algorithms consider the time and amount that the signal rises over a dynamically-determined threshold.

Since the fNIR technology of OTIS only monitors oxygenation levels of the tissue lying directly beneath the system sensors, the system is used to detect regional neural activity. Therefore, the identification and location of the cortical area used for the desired cognitive activity must be known in order to detect and assess the activity. The OTIS sensors must be placed over the cortical location activated by the cognitive activity under examination.

2   Conclusion

The unique design of the OTIS system overcomes the concerns associated with other existing fNIR imaging systems. The portable, EMI-resistant design makes it an ideal system to detect cognitive activity in nearly any environment. As design development continues, the applications and flexibility of the system will expand.

Hemodynamic Response Estimation During Cognitive Activity Using fNIR Spectroscopy

Meltem Izzetoglu[1], Shoko Nioka[2], Scott Bunce[3], Kurtulus Izzetoglu[1], Banu Onaral[1], Britton Chance[2]

[1] Drexel University, School of Biomedical Engineering, Science and Health Systems
[2] University of Pennsylvania, Department of Biochemistry and Biophysics
[3] Drexel University, School of Medicine

Philadelphia, PA 19104

meltem@cbis.ece.drexel.edu

SUMMARY

Near infrared spectroscopy (NIRs) enables measurement of the hemodynamic changes during functional brain activation. Using the modified Beer-Lambert law and measurements performed at two different wavelengths within the near infrared light range (between 700-900nm) and different times, the relative changes in the concentrations of deoxy-Hb and oxy-Hb can be obtained. These measurements led to the assessment of several types of brain functions such as, motor and visual activation, auditory stimulation and performance of cognitive tasks [1-3]. Modeling the hemodynamic changes is important for several reasons for example these models may lead to better statistical maps, or to the possibility of performing simulations with the model or more importantly they can allow the possibility to give a physiological interpretation of the model parameters. There have been several studies in modeling the hemodynamic response in fMRI using FIR filters, statistical methods such as Bayesian modeling, or mathematical model fitting to Gaussian, Poisson or γ type functions [4-9]. In this work, we implemented the latest approach in functional near infrared (fNIR) spectroscopy (fNIRs) and assume that each evoked hemodynamic response can be modeled by a γ type function as given in the figure 1. Once the evoked hemodynamic responses are estimated, features such as the maximum amplitude or the peak value of oxygenation activation and time to peak or response time can be extracted from the model which can further be used to quantify cognitive state of the subjects.

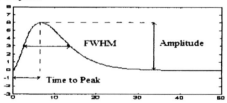

Figure 1: A typical γ function

There have been on-going studies in problem solving of graded difficulties (anagram solution). These studies, using both block and event-related anagram protocols, revealed that a wearable NIRs measurement of metabolic activation and blood flow can be valuable in educational aid [10]. In event related anagram study subjects are presented an anagram for 1 sec. and given 15 sec. to solve it until the next presentation. This procedure allows the hemodynamic response to fully evolve which has been shown in the literature to take 10-12sec. period [8-9]. Event related (ER) studies provided insights as how to model the hemodynamic response and used widely in the assessment of cognitive activation in different regions of the brain for different task loads. However, in such studies, the protocol time is long and they do not reflect real world situations. In block anagram study, subjects are shown as many anagrams as they can solve within 1 min periods. Whenever subjects solved the anagram they press a certain button which results in immediate presentation of the next anagram. Since most of the time subjects solve the anagram within 2-5 sec. period, the hemodynamic responses overlap in time which present challenges to data analysis. Until now, in block anagram studies, it was not possible to evaluate the subject's response times or brain activation for single anagram presentation within a block for graded difficulty analysis.

In this paper, we present a novel single trial hemodynamic response estimation algorithm in a block anagram solution study. We assume that the oxygenation data that is measured by fNIRs is formed by a linear model [8,9] where each hemodynamic response to N single trials or stimuli is added together to form the total oxygenation data;

$$Oxy = \sum_{i=1}^{N} hf_i \qquad (1)$$

where hf_i is the evoked responses for ith stimulus presented at time t_i which is represented by a γ function with unknown parameters as follows $hf_i = A_i t_i^{\alpha_i} e^{\beta_i t_i}$. The unknown parameters (A_i, α_i, β_i) for all the N number of hemodynamic responses are found by minimizing the mean square error between the total oxygenation data, Oxy and the model by using a least mean squares curve fitting technique.

$$\varepsilon = \min_{A,\alpha,\beta}(Oxy - \sum_{i=1}^{N} hf_i)^2 \qquad (2)$$

All calculations are applied to the data gathered from the left hemisphere of the prefrontal cortex (exact location is as shown in figure 2(a)). In block anagram study participated by 14 subjects, the averaged recorded response times, the extracted rise times (min) and the maximum amplitudes from the estimated evoked hemodynamic responses with respect to the 3, 4 and 5 letter anagram sets are presented in figure 2(b). It can be clearly seen that the estimated rise times follow the same pattern as the true response times of the subjects having a correlation of R=0.94 as presented in the scatter plot of the rise time versus response time in figure 2(c). Also the estimated maximum amplitudes are correlated with the true response times (R=0.73) as given in figure 2(d). The rise times and the maximum amplitude values increase as the difficulty level of the anagram solution increases meaning that subjects needed more time and more oxygen to solve difficult anagrams.

Acknowledgments: This work has been sponsored in part by funds from the Defense Advanced Research Projects Agency (DARPA) Augmented Cognition Program, the Office of Naval Research (ONR) and Homeland Security, under agreement numbers N00014-02-1-0524, N00014-01-1-0986 and N00014-04-1-0119.

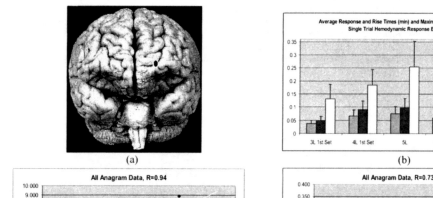

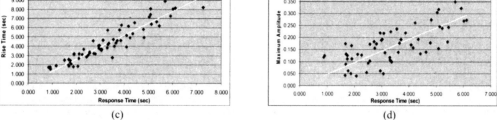

Figure 2: (a) Imaging area of the brain for this study; (b) Subject averages of rise and response times (min) and maximum amplitude; (c) Scatter plot rise time of versus response time averages; (d) Scatter plot of maximum amplitude versus response time averages for all anagram sets of all subjects.

References

[1] Villringer A, Chance B., Non-invasive optical spectroscopy and imaging of human brain function. *Trends in Neuroscience, 20*, 435-442, 1997.

[2] Izzetoglu K, Bunce S, Onaral B., Pourrezaei K., Chance B., "Functional Optical Brain Imaging Using NIR during Cognitive Tasks," *International Journal of Human-Computer Interaction. Special Issue on Augmented Cognition.* 17(2), 211-227, 2004.

[3] Suto T., Ito M., Uehara T., Ida I., Fukuda M., Mikuni M., Temporal Characteristics of Cerebral Blood Volume Change in Motor and Somatosensory Cortices Revealed by Multichannel Near Infrared Spectroscopy, *International Congress Series*, 1232, 383-388, 2002.

[4] Kruggel F., Von Cramon D.Y., Modeling the Hemodynamic Response in Single Trial Functional MRI Experiments, *Magnetic Resonance In Imagigng*, 42, 787-797, 1999.

[5] Ari N., Yen Y.F., Extraction of the Hemodynamic Response in Randomized Event-Related Functional MRI, *Proc. of 23rd Annual EMBS Conference*, 612-615, 2001.

[6] Gossl C., Fahrmeir L., Auer D.P., Bayesian Modeling of the Hemodynamic Response Function in BOLD fMRI, *NeuroImage*, 14, 140-148, 2001.

[7] Goutte C., Nielsen F.A., Hansen L.K., Modeling the Haemodynamic Response in fMRI Using Smooth FIR Filters, *IEEE Trans. On Medical Imaging*, 19:12, 1188-1201, 2000.

[8] Miezin, F.M., Maccotta L., Ollinger J.M., Petersen S.E., Buckner R.L., Characterizing the Hemodynamic Response: Effects of Presentation Rate, Sampling Procedure, and the Possibility of Ordering Brain Activity Based on Based on Relative Timing, *NeuroImage*, 11, 735-759, 2000.

[9] Izzetoglu M., Bunce S., Onaral B, Single Trial Hemodynamic Response Estimation in Event Related fNIR Spectroscopy, *Proc. of OSA*, 2004.

[10] Chance B., Nioka S., Sadi S., Li C., Oxygenation and blood concentration changes in human subject prefrontal activation by anagram solutions, *Adv Exp Med Biol.*;510:397-401, 2003.

When the rules keep changing: Using the fast optical signal to elucidate different brain preparatory states.

Kathy Low, Echo Leaver, Jason Agran, Monica Fabiani, & Gabriele Gratton

University of Illinois at Urbana-Champaign
Beckman Institute, 405 N. Mathews Ave., Urbana, IL 61822
lowka@uiuc.edu

Abstract

Flexible adaptation to changing task demands is essential for successful behavior in a complex environment. Yet, even in relatively simple situations, there are often costs (e.g., slower reaction times or increased errors) associated with having to switch between the rules used in different tasks (Monsell, 2003). These switch costs can be reduced, however, if information about the upcoming task is provided in advance of each trial, presumably due to the engagement of appropriate preparatory processes. To further investigate these preparatory processes, we conducted a series of studies using the event-related optical signal (EROS). EROS is a functional brain imaging tool that relies on frequency-domain measurements of changes in the optical properties of brain tissue (Gratton & Fabiani, 2001). It possesses both good spatial (sub-cm) and temporal (ms) resolution and is therefore ideal for investigating the brain responses that occur following a change in task instruction. To create a task-switching environment, we used a cueing paradigm in which a pre-cue provided specific instructions about the rules to be used to process the upcoming imperative stimulus. For example, in one experiment the pre-cue instructed the participant to attend to either the auditory or the visual modality, with the cued modality varying randomly from trial to trial. The pre-cue was followed, approximately two seconds later, by a compound imperative stimulus that contained an auditory and a visual component. The task was to respond based on the information in the cued modality and ignore the information from the other modality. In separate experiments, we investigated several types of stimulus dimensions to be attended (auditory vs. visual modality, location vs. meaning of a word, local vs. global focus of attention, or left vs. right visual field attention) and the type of response to be emitted (vocal vs. manual). The EROS data were sorted based on whether the current trial contained the same task demand as the previous trial (no-switch) or a different task demand (switch). During the preparatory interval, we found areas of brain activation that were common across a number of different switch cues, presumably reflecting general, task-switching operations that are engaged when the current rule is different from a previous one. Moreover, the EROS data also revealed several task-specific switch effects. That is, brain regions that were active only when the cue dictated a switch to a particular stimulus dimension, but not when switching to one of the other dimensions. Together, these data indicate that optical measures of brain activity can be useful in identifying task appropriate preparatory states. Knowledge about these preparatory states could, in turn, be exploited to augment cognitive processing of upcoming stimuli and reduce the likelihood of human error.

References

Gratton, G. & Fabiani, M. (2001). Shedding light on brain function: The event-related optical signal. *Trends in Cognitive Science, 5(8)*, 357-363.

Monsell, S. (2003). Task switching. *Trends in Cognitive Science, 4(3)*, 134-140.

Telescopic imaging of PFC hemodynamic and metabolic activities by remote sensing NIR imaging

B. Chance, Cindy Wang and Zhongyao Zhao

University of Pennsylvania
250 Anatomy Chemistry Building, Philadelphia, PA 19104
chance@mail.med.upenn.edu

Abstract

A device for remote sensing of the activity of the prefrontal cortex (PFC) at dual wavelengths is appropriate to detecting hemodynamic and metabolic activity related to normal and abnormal brain function. Two systems are proposed to operate remotely (2 meters), both of which will provide a television type of image of hemodynamic activity of the PFC: 1) floodlighting of the PFC with ICCD detection and television display, and 2) flying spot illumination of selected points on the PFC with television type of display plus the ability to focus on a region of activity with automatic tracking of the spot position. Both systems appear feasible and worthy of detailed laboratory and field tests.

The system is designed to operate at a distance of 2 meters from the subject, to be portable (shoulder mounted) and provide a TV type of readout for the operator. The device can monitor PFC function.

1 Introduction

The current technology of contact sensing of PFC signals has been developed by many laboratories following our initial discoveries of pulsed illumination of models of muscles and brain leading to a number of researches on sensorimotor visual and prefrontal imaging of functional activity of the brain. The principal signal that has been measured by most is the optical analogy of the BOLD signal, namely increases of blood flow following functional activation of appropriate areas of the brain, particularly sensorimotor and visual areas. A search for the neuronal activation as measured by the historical increase of transmission of axons during the passage of electrical signal is in progress. However, we have focused on functional changes on stimulation of the PFC such as increased blood volume, a hemodynamic signal (by analogy with BOLD) or a novel signal, a local decrease of hemoglobin saturation due to functional activity, a metabolic signal. These two signals are both activated by the breakdown of ATP to ADP and phosphate, but stimulate quite different physiological functions: 1) increased blood flow due to the release of adenosine into the capillary bed vascular bed, and 2) the release of ADP to the localized mitochondria. Thus, one signal is distant and delocalized (1), and the other is highly localized (2). Current studies of prefrontal activation in problem solving clearly show the expected spatial difference between these two signals [1]. Blood flow can be activated "downstream;" while mitochondria are activated locally and thus are a more readily interpretable functional signal.

2 Rationale

EEG and EP have a high bandwidth but require not only contact detection but also a significant interval of preparation of the electrode contact even with the "hairnet" type of electrode system. Thermography, while readily adapted to remote sensing, gives images of variable blood flow, which have localization possibilities but have a requirement for a stable thermal environment together with the requirement for thermal diffusion of the signals. The optical method, generating metabolic signals due to oxygen extraction from the capillary bed by mitochondrial activity, has the possibility of a fast, non-invasive, remote sensible signal and is the basis for the proposed design.

The following demonstrates that theoretically and experimentally, photon migration at a distance from the object gives an oblique illumination.

2.1 Feasibility Demonstrations – Monte Carlo Simulation - Theory

In order to provide an initial platform for experimental studies, we have set up a Monte Carlo simulation in time-domain of remote detected reflectance. Sufficient photons (10^7) were injected in order to achieve an accurate slope. These simulation results strongly support the hypothesis that remote sensing of diffusive photons technology seem feasible. In fact, simulation of contact TRS and remote TRS does give the same log slope to within 5%. Ab initio, we do not expect to be able to have high resolution imaging, but nevertheless, 1 embedded object (1 cm^3) which has a much higher absorption coefficient than background can be detected. (Fig. 1)

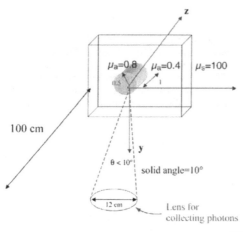

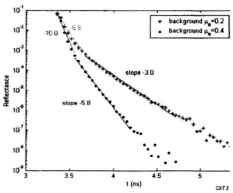

Fig. 1A. Model of Monte Carlo simulation of remote time-resolved sensing. The distance between subject and source/detector is 1m. Lens has a solid detection angle of ~10°. Semi-infinite medium with μa = 0.4 cm^{-1}, μs=10 cm^{-1} with an small object embedded (1cm× 0.5 cm) of μa = 0.8 cm^{-1} and same μs.

Fig. 1B Monte Carlo simulation of photon migration tracks with two models with background μa of 0.2 and 0.4. A small absorber is put in the center of each model, respectively with μa of 0.4 and 0.8. The early slope of the reflectance curve (-6.6 or –10) is bigger than the tail slope (-3 or –5.8), which may provide a clue for detection of the embedded object. The μas calculated from tail slope (0.19 or 0.36 cm^{-1}) fit the theoretical value very well, suggesting the remote time-domain system can detect tissue absorption coefficient very accurately.

2.2 Analytical solution for the reflectance signals [2]

$$R(\rho,t) = (4\pi Dc)^{-3/2} z_0 t^{-5/2} \exp(-\mu_a ct) \exp(-\frac{\rho^2 + z_0^2}{4Dct})$$

$z_0 = [(1-g)\mu_s]^{-1}$, g is isotropic factor of scattering.

$D = \{3[\mu_a + (1-g)\mu_s]\}^{-1}$, diffusion coefficient.

c is light speed, and ρ is distance between detector and source.

For remote sensing, we add a time delay τ, to account for the light traveling in the air (τ=L/c, L is the remote distance), assuming the air doesn't absorb or scatter the light, the reflectance will be:

$$R'(\rho,t) = (4\pi Dc)^{-3/2} z_0 (t+\tau)^{-5/2} \exp[-\mu_a c(t+\tau)] \exp[-\frac{\rho^2 + z_0^2}{4Dc(t+\tau)}]$$

The asymptotic terminal slope at large t:

$$\lim_{t \to \infty} \frac{d}{dt} \ln R'(\rho, t) = -\mu_a c$$

Thus theoretically we can measure the absorption coefficient μ_a from the terminal slope of reflectance in the remote case, just as in the contact case. For finite-slab geometry, the derivation is similar and this conclusion still holds. Depth is proportional to the average photon pathlength, or photon travel time, L=ct. c is light speed in that medium. For depth discrimination, we need to time-gate photons. Late photons, i.e. which have long travel time, correspond to a bigger depth. To increase depth detection, we need to increase input energy and let photons have enough intensity to decay until late time. We have done experiments to measure depth sensitivity.

3   Experimental feasibility studies: Oblique illumination by TRS

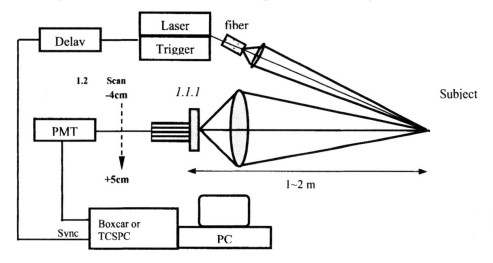

Fig.2 Illustration of current apparatus of remote time-resolved system. Detector can be moved laterally and located usually 1~2 cm off-axis to minimize the photon counts of specular reflection.

We have made a simple version of remote TRS by modifying our currently available contact TRS for breast imaging. The illustration is shown in Fig. 2. The laser source is 1-2m away and the detector is 2.5m away from the subject. This system is composed of pulsed laser diode (FWHM=300 ps) sold by PicoQuant. We used wavelengths of 690 and 780 nm. The detector is a Hamamatsu GaAs PMT (H7422-P50) and time-correlated single photon counting system (TCSPC) which has TAC and computer display of count rate. In this case the optical system is X-Y scanned and the magnification is varied to alter source detector separations.

Specular reflections have a much higher intensity than multiple-scattered light. The time-resolved system has the advantage of discriminating the two by their different photon arrival times. Besides that, we tuned our optics: the source light is at a 45° angle; our detector is put off axis so there is a 2 or 3 cm distance between illuminated spot and detecting spot. We may use 1 or 2 aperture stops to block the specular reflection.

3.1   Continuing experimental studies

We have made many tests using this test system (1-channel, 1-wavelength) on tissue-like phantoms, ICG solutions and human arm cuff. The Patterson, Chance, Wilson equation applies to determination of μ_a remotely from a model containing intralipid at a μ_s of 10 cm⁻¹; variation of the absorption constant by serial additions of India ink giving the four values of μ_a which are calculated to be proportional to the model values. Since the model is homogeneous, the log slope is constant over a wide range of times.

3.2   Accuracy of μ_a Measured Remotely

Quantitative tests of the absorption factor in several models of different μ_a are shown in Fig. 3. The control study with fibers in contact gave the value (cm^{-1}) on the abscissa of Fig. 3B. In the case of remote sensing, the data shows the photon arrivals at the PMT with a maximum delay of 5 or 8 ns. Starting at 1.5 ns thereafter, the traces show a strong peak similar to incident pulse for 1 or 2 ns, which is very likely due to the specular reflection at the illumination spot or shallow traveling photons. Then the peak decays quickly, followed by diffusion photons with 2 - 5 ns of migration time. The photon counts decay logarithmically until drops 3 decades (1 to 10^{-3}) with different decay rate for each model. The decay rate essentially determines medium absorption coefficient. The slopes are approximately constant in selective intervals between approximately 3 and 8 ns. The first three values agreed well with the values measured by contact TRS except for small convolution with the excitation pulse or with specular reflection. The μ_a can be determined remotely.

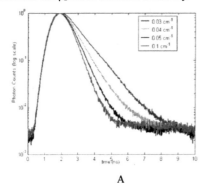
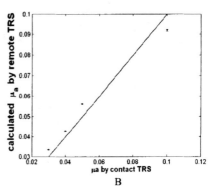

A   B

Figs. 3 A) Remote detected time-resolved curve (normalized) of backscattered light from tissue-like phantoms with $\square_s' = 10$. The values of their absorption coefficient (μ_a) are shown in the small box in A) and also shown in B) x-axis. B) Calculated μ_a (cm^{-1}) from the tail slopes of these curves for 0.3, 0.4, 0.5 cm^{-1} calculated according to Patterson-Chance-Wilson theory (25), are in agreement with the μ_a measured by contact TRS. This indicates the remote TRS system able to measure unknown absorption coefficient of a distant tissue. Error markers in B) measure the standard deviation (s.d) of this linear regression between 4.3 and 7.6 ns and check deviation from Patterson-Chance-Wilson theory. The signal to noise ratio is: 10^4 (peak count)/42 (s.d) =240, using unnormalized data of phantom of μ_a =0.03 cm^{-1}.

The foregoing is intended to demonstrate that not only can photon migration be detected at a distance but also that quantitative determination of tissue optical properties is possible. The following demonstrates how this can be applied to a portable, remote sensing device in which the illumination is parallel to the line of sight.

3.3   Tradeoff between space and time resolution

The spatial resolution obtained by MRI imaging of the BOLD hemodynamic response is impressive and voxels of a few mm in size or better can be obtained. However, the integration time for such images is usually 15-20 minutes and no possibility of TV readout from the PFC (in which additional experimental difficulties for MRI are present) together with confinement in the MRI magnet rule out for this approach to rapid remote imaging of the PFC by a portable device. Here we propose an NIR optical system which is fast and has resolution sufficient to present at least 10 separate voxels covering the PFC and appropriate to detecting hemodynamic and metabolic responses to neuronal function.

4   System Design

4.1   Principle

Oblique illumination of the laser of the forehead has been successfully demonstrated in both the reflectance and absorbance modes but is a geometry which is inconvenient for a portable system. Co-axial illumination in several geometries is here proposed. The system can produce images of the PFC response at a single spot or a raster or spiral scan of large and small areas respectively. The use of this device in monitoring the brain function of an unfettered individual, civilian or military, are described.

4.2   Technical Considerations

Method 1) A TV type of display of PFC signals with a 30 Hz repetition rate and a high sensitivity to hemodynamic and metabolic changes can be accomplished by the flying spot technique, i.e. 100 milliwatt (mW) flying spot illumination of the region of interest, for example 10 voxels (as a TV display) with the ability to provide spiral scan focused on particular voxels of interest.

Method 2) Floodlight of the PFC with a power density of at least 20 mWs per sq cm or a total power of 200 mW with a gated ICCD and data processing of the output to give a TV display of a two wavelength hemodynamic image of the PFC.

The flying spot approach is preferred since all the energy of the light can be concentrated in small areas of interest and in the raster scan; tracking of the focal region by a spiral scan is also possible. However, both methods should be tried, the advantage of the floodlight is the intensity is low enough so that safety considerations are not an issue (i.e. roughly 20 mW/sq cm).

4.3   Feasibility

We have investigated in detail oblique illumination of model systems containing a deep target, for example cardiogreen, and have shown in the following demonstrations that the photon migration signals from a deep target can readily be detected by an appropriate telescopic and area detection PMT system. This system is eminently practical for a fixed system in which the subject can be positioned at the focal point of the optical system and the following slides confirm this hypothesis.

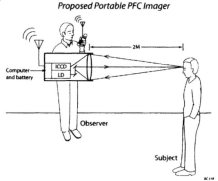

Fig. 4. Proposed portable PFC scanner. The elements are described in the text for method 1 (see above). The device is expected to be able to be battery operated, if necessary, and to give a display of PFC activity at a distance of at least 2 meters, assuming 100 mW of excitation intensity is available. The following describes the properties of the apparatus.

4.4   Geometries

A number of geometries are possible (See Fig. 1). The simplest geometry is a micro-mirror in front of the detector and a projected beam with a lens to an area on the forehead from which there will be early reflection and later photon migration at the selected spot.A single lens with a short distance to the laser source and a long distance to the object is a feasible one.

4.5   Specular reflections

There are, however, some modifications. 1) The use of polarizers. In this case the specular reflected beam should be highly polarized and the polarizers would be normal to the light flux and attenuate the specular flux and admitting the diffusive beam which has been scattered and depolarized. 2) The use of fluorescence, of ICG for example. A) Fluorescence would be delayed so the ability to eliminate the specular reflection on a time resolution basis would be impaired. B) In this case the light passing through a secondary filter would be oblique and a collimator would be required for effective operation of the barrier filters.

4.6   Preferred embodiment

The preferred portable PFC scanner is shown in Fig. 4. It consists of a shoulder strap assembly of the detector, the light source, the CCD together with necessary microcircuit gated integrators, with transmission with digitization with transmission with the necessary gated transmission and reconstruction computation to send the TV picture to a remote operator handheld detector display. The excitation beam is directed towards the preferred target and the LD sends a polarized light beam with moveable mirrors 1 and 2 to give a raster scan of the subject's forehead. The PMT is gated to eliminate overload in the first few nanoseconds and thereafter receives TRS signals of increasing depth from the PFC; repetitive illumination at 50 MHz allows integration of information. Two wavelengths are used and the two scans are subtracted and appropriately processed to bring out the active voxels and the display is transmitted to a handheld display and to any other databank that is desired. The interval of accumulation is fast enough (10 Hz) to display metabolic and hemodynamic responses. Daylight filters are used. Operation in a building illuminated by appropriate LED sodium or mercury arc sources is suitable.

4.7   Apparatus 1

A separate detection system is needed for each of the 10 voxels and for the 2 wavelengths, i.e. 20 channels. While this would be very cumbersome using single photon counting techniques of the SCA, TAC configuration, it is proposed instead to use gated integrators or "boxcar" detectors, which can be made in a mini-circuit configuration with a 3 nsec integration gate starting 1 nsec after the specular reflection signal is received. Under these conditions, assuming that the specular reflection is 100 times the migration signals of 10^4/sec, the count rate of 10^6 counts/sec could readily be tolerated by the PMT without the need for "crowbar" protection of the PMT during the receipt of the specular signal, especially as attenuated by crossed polaroids.

4.8   Apparatus 1 description

The apparatus consists of 100 mW dual wavelength laser diode illuminator operating at 760 and 850 nm. a galvanometer mirror controlled by CPU, one in the X direction and the other in the Y direction (preferably a MEMS mirror), a projecting lens, presumably f-1 and 20 cm with a working distance of 2 meters. The focal point of the specular reflection will be at the laser mirror which serves to minimize the specular reflections. Since this device is essentially a single spot scanner, detection may readily be made by a single large cathode PMT, i.e. 2 cm in diameter or a rosette of small PMTs. In either case, the output of the PMT will be detected at 1 nsec for safety considerations and at 3 nsec for migrating photons, dual wavelength. The output of the PMT will be digitized and computer programs will select out the two wavelength signals, make an appropriate difference ratio of the two wavelengths, and present a raster scan image of the 10 points selected of the PFC. A television type display of localized PFC activity may be identified in a television 30 Hz type of display.

4.9 Minimizing detector overload

The PMT generally can have a fast recovery (a nanosecond if desired here). However, blanking circuits (already published, RSI) which will desensitize the PMT when the specular reflection is about to be received and will establish high sensitivity for the migratory photons time scaled for a 1 nsec input pulse. Recovery 0.5 nsec after the pulse is received seems feasible. This would certainly be appropriate for the television type of scanning. 1) The PMT high voltage might be dropped when the specular reflection comes back and might be gradually increased so that the gain of the PMT would be high as the deep reflections arrive. 2) The polarized filter should be used. 3) The refraction or reflection telescopes can both be considered depending upon the aperture required. 4) Since we focus on the photons that emerge from the site of illumination, the signal-to-noise ratio might be somewhat better than with oblique illumination.

4.10 Apparatus 2

An alternate electronic system uses floodlight illumination of the PFC, ICCD detection and presentation of the image of the floodlit forehead in terms of a two wavelengths display of hemodynamics and metabolism. This system is feasible, but not preferred.

4.11 Sensitivity calculation

The attenuation of signal due to a dual wavelength system and a 10-pixel image at a frame rate of 10 Hz is approximately a factor of 100. Furthermore, time gating of the signal with a 3 nsec gate could cost a net loss of emergent photons of a factor of 10^3. A dual wavelength image at a repetition rate of 10 Hz from a 10 pixel image of the PFC for a signal to noise ratio of 30 requires a count rate of 10^3/second, suggesting that a count rate of 10^6/sec is required of migrating photons from 1 cm depth for 10 pixels at a rep rate of 10 Hz. Typical count rates from breast tissue with inefficient light guide coupling are on the order of 10^3 per sec. Thus, the more efficient optical coupling of the telescope plus an increase of laser power from 5 mWs to 100 mWs is estimated to be an increase of count rate of 200-fold over the 100 counts/sec readily obtained with the present fiber optic configuration (2×10^5/sec).

4.12 Signal separation

The separation of the illumination signal from the migration signal can be achieved by time resolution since specular reflections are returned without delay while photon migraiton signals have the delay characteristic of photon diffusion into the brain cortex and return tehrefrom. Thus time resolution is the method of choise for separation the undesirable reflectance signal from the photon migraition signal.

Another method of separation is by polarization of the incident light and protecting the detector with a cross-polarizer. A similar technique can be used with a CCD. Also, fluorescence excitation and emission can be resolved tby appropriate primary and secondary filters. If a dichroic filter is used, a collimator may be used for its effective use with appropriate aperture stops. However, the most efficient way is that used in radar ranging where the excitation is in a pulse packet, short compared to the signals of interest in radar reflectance from distant objects. In NIR optical imaging, diffuse reflection from deeper objects can be time resolved as in radar ranging, essentially the same problem with light as with radio waves, the well known LIDAR principle.

4.13 Safety considerations

At 7×10^{-7} joules/sq cm, safe illumination of the cornea, 100 mWs would allow 6 microseconds safe illumination. Since the cornea is essentially bloodless, it gives a much bigger reflectance signal than does the skin covering the forehead and accidental illumination of the cornea will give an estimated 10-fold increase in the specular reflection observed in the first nsec of the return signals. This factor of 10 increase of the specular count rate is to be detected by a 1 nsec time gate for integrating the specular reflection count, with a 6 microsec time constant and a comparator detecting a the expected 10-fold increase of the specular signal in going from forehead to cornea. A diode "crowbar" across the laser diode illuminator, would turn off the illuminating laser until for the interval of corneal illumination and provide an appropriate warning signal LED warning signal to the operator.

4.14 Scan type: TV vs spiral

In order to simulate a television display, a raster type of scan is feasible in which the averaging time is currently estimated to be 30 seconds. If however a particular pixel in the scan is of interest, i.e. responsive to a particular type of questioning, then a spiral laser scan would be initiated in which tracking of the particular voxel is feasible. The spiral scan gives a 2 axis error signal appropriate to the servo controlled automatic steering of the device of a particular voxel.

4.15 Scan speed

The scan speed can be fast, i.e. 30 Hz-50 Hz-100 Hz, but the integration time to get a clear image may be 10 seconds. The integration time should be short enough to follow changes of PFC activation; current experimentation suggests 10 seconds or possibly shorter. Thus, a fast raster scan at the television rate together with signal storage of up to 10-20 seconds is desirable.

4.16 Suggested uses of the device

The purpose of the device is to detect brain function in actual field conditions where the PFC function is to be monitored, for example, in any training situation, driving, flying, where operation involves PFC activation and where detection of fatigue and loss of attention might be paramount. At the same time, there may be particular functional activities in which the brain function is to be monitored, particularly for attention, awareness, problem solving ability under field conditions where immediate non-contact coupling to brain function is mandatory and where awareness and bulkiness/wiring of the contact devices is dispensed with.

It is expected that the device will be particularly useful when the pixel or pixels involved in a certain type of attention or problem solving are to be intermittently or continuously monitored throughout an attention-demanding mission.

4.17 Expected delivery of such an apparatus

Layouts of the optics have been made and will shortly be mounted firmly on Newport tables with laser diodes of the appropriate energies for remote sensing. The electronics currently used are SCA, TAC which is unsuitable for the high count rates which are expected with the higher power lasers. Thus, construction of the boxcar detector for the specular reflection (very high count rates for safety protection and the lower count rates for signal detection). It is estimated that the system can be assembled on a Newport table in 6 months, and in 12 months progress is expected towards a portable breadboard unit with a deliverable unit for field testing in 24 months with existing student help; with experienced engineering help plus collaboration of NIM, Incorporated the delivery time can be reduced to 18 months. No component is unavailable and delivery times of components are expected to be short. Project Director B. Chance, Project Engineer Z. Zhao, Graduate Students X. Wang, X. Ye.

References

Chance, B. and Nioka, S. (in preparation). Functional Imaging of PFC Activity.
Patterson, M.S., Chance, B. and Wilson, B.C. (1989) Time Resolved Reflectance and Transmittance for the Noninvasive Measurement of Tissue Optical Properties. J. Appl. Optics 28:2331-2336.

Section 2
Cognitive State Sensors

Chapter 6

EEG for Functional Operator State Assessment

The invisible electrode – zero prep time, ultra low capacitive sensing

Robert Matthews   *Neil J. McDonald*   *Igor Fridman*   *Paul Hervieux*   *Tom Nielsen*
robm@quasarusa.com   neil@quasarusa.com   igor@quasarusa.com   paul@quasarusa.com   tom@quasarusa.com

Quantum Applied Science and Research
5764 Pacific Center Blvd., #107,
San Diego, CA, 92121
(858) 373-0321

Abstract

The principle technical difficulty in measuring bioelectric signals from the body, such as electroencephalogram (EEG) and electrocardiogram (ECG), lies in establishing good, stable electrical contact to the skin. Traditionally, measurements of human bioelectric activity use resistive contact electrodes, the most widely used of which are 'paste-on' (or wet) electrodes. However, the use of wet electrodes is a highly invasive process as some preparation of the skin is necessary in order for the electrode either to adhere to the skin for any length of time or to make adequate electrical contact to the skin. This is uncomfortable for the subject and can lead to considerable irritation of the skin over time, an issue of particular concern in measurements of EEG signals, which typically require an array of electrodes positioned about the head.

Despite over 40 years of investigation, including the development of several alternative electrode technologies, no reliable method for making electrical contact to the skin that does not require some modification of its outer layer has been developed. For example, Ag-AgCl dry electrodes, NASICON ceramic electrodes, and saline solution electrodes do not require any skin preparation, but for each electrode the subject experiences skin irritation over extended periods, and there are various issues that cause the performance of the sensors to degrade over time. Alternatively, insulated electrodes that use capacitive coupling to measure the potential changes on the skin have in the past, for noise considerations, used exotic materials to generate a high capacitive coupling (~1 nF) to the skin. The intrinsic noise of insulated electrodes is adequate for bioelectric measurements, but these high capacitance sensors also exhibit long-term compatibility issues with the skin and are sensitive to motion artifact signals due to the electrode's high sensitivity to relative motion between the skin and the electrode itself.

As a result of advances in semiconductor processing techniques and through the use of innovative circuit designs, QUASAR has developed a new class of insulated bioelectrodes (IBEs) that can measure the electric potential at a point in free space. This has made it possible to make measurements of human bioelectric signals without a resistive connection and with modest capacitive coupling to the source of interest. These electrodes are genuinely non-invasive in that they require no skin preparation, have no long-term compatibility issues with the skin, and can measure human bioelectric activity at the microvolt level through clothing while remaining largely immune to motion artifact signals.

This paper will present measurements of bioelectric activity made using QUASAR's IBEs, and corresponding data measured using conventional wet electrodes will also be presented for comparison. The presentation will include through-clothing measurements of bioelectric signals, the rejection of motion artifact signals, non-invasive (i.e. no skin preparation) EEG measurements of alpha-rhythm signals, and the noise levels observed using both types of sensors.

In a series of tests conducted on unprepared skin, it was observed that both the IBEs and conventional wet electrodes had similar noise levels. This noise level was higher than the expected noise level, which had been predicted based upon the intrinsic noise characteristics of the sensors. The fact that both sensors suffered from this higher noise level suggested a common mechanism, which was later identified as skin noise.

It has been reported in the literature that one of the fundamental noise sources for any bioelectric measurement made on unprepared skin is epidermal artifact noise. This noise is due to potentials developed in the skin itself that are indistinguishable from the bioelectric signal of interest. There exist techniques that can reduce this noise level by as much as a factor of 5, but they involve modification of the skin's outer layer either by abrasion or chemical absorption of conducting fluid. These methods are not comfortable for the subject and may be difficult to perform on subjects with especially sensitive skin, such as neonates, burn victims, or the elderly.

In addition to QUASAR's IBE sensors, this paper will also discuss a new free-space electrode that is designed to be insensitive to epidermal artifact noise, and thus is capable of bioelectric measurements at the microvolt level in the absence of any skin preparation. The new device exhibits significantly less capacitive coupling to the source of interest than the current generation of QUASAR IBEs, without the increase in intrinsic sensor noise that would accompany a reduction in electrode capacitance.

1   Introduction

Over the past 4 years QUASAR has developed a new class of electric potential sensor that can measure the electric potential at a point in free space, in which individual sensors have demonstrated sensitivities at 60 Hz of $5 \mu V/m/\sqrt{Hz}$ (Krupka, Matthews, Say & Hibbs, 2001), representing an improvement of approximately two orders of magnitude in sensitivity from the previous state of the art. This sensitivity is sufficient for the measurement of bioelectric signals without any resistive electrical connection and with only negligible capacitive coupling to the source of interest.

The QUASAR sensor has been used for the measurement of electrocardiogram (ECG) and electroencephalogram (EEG), physiological measurements that have been identified as useful metrics for the development of a cognitive state gauge by the Augmented Cognition (AugCog) program (initiated by the Defence Advanced Research Projects Agency, or DARPA). See St. John, Morrison & Schmorrow (2004) for a discussion of other psychophysiological measures used to identify changes in human cognitive activity during task performance. This ambitious research program has as its aim the maximization of human cognitive capabilities through long-term monitoring of bioelectric signals in diverse settings. It is a requirement of sensors that they be unobtrusive and that their presence not affect an individual's performance. However, the conventional electrodes used for electrophysiological measurements are not suited for long-term measurements of bioelectric signals because they frequently result in skin irritation and discomfort for the subject. This is described in more detail in Section 2.

Capacitive electrodes per se for bioelectric measurements are not new. Detection of human body bioelectric signals using purely capacitive sensors (i.e. no ohmic contact to the body) was first reported in 1968 (Richardson, 1968) and patented in 1970 (Richardson & Lopez, 1970). However, prior technology required high capacitive coupling (~1 nF) to the skin. These high capacitances were generated through the use of exotic materials that suffered from biocompatibility issues and resulted in skin irritation. Furthermore, susceptibility to displacement of these electrodes makes them unsuitable for applications in which even a moderate level of aerobic activity is required.

This paper presents performance data for a new class of insulated bioelectrode (IBE) suitable for use in long-term monitoring of bioelectric signals.

2   Conventional Resistive Contact Electrodes

The principal technical difficulty in measuring bioelectric signals from the body lies in establishing good, stable electrical contact to the skin. Before placing a 'paste-on' (or wet) resistive electrode on a patient, the skin is usually cleaned with alcohol to remove sweat and skin oils that would otherwise prevent the adhesive paste from holding the electrode in contact with the patient's skin. However, it has been known since the Apollo space program that long-term wear of electrodes that require resistive (ohmic) contact can result in major skin irritation, to the extent that NASA has to test astronauts for such reactions (NASA, 2005). While a number of improved gels and connection techniques have been introduced in the last 30 years, resistive contact electrodes are still far from ideal.

For example, if a resistive-contacting electrode pulls away from the skin, or if its gel dries out, then it can no longer collect a signal with the required fidelity. Such problems are commonplace in clinical applications of electrophysiology, and are particularly marked for neonates because, unlike most adult patients, they are usually active during an ECG procedure. Many ECG monitors now have impedance-sensing circuits to determine when an electrode has lost contact. While automatic detection of a loss of contact is useful, it does not replace the missing data, nor enable the system to resume operation.

A further problem is the degradation in performance experienced by the electrodes themselves. Leakage current from the electrode into the amplifier used to make the measurement leads to polarization of the electrode. The almost exclusive use of Ag-AgCl electrodes in ECG measurements is probably due to their very low polarization on current flow, an important consideration for early ECG amplifiers (Klinger, Booth & Schoenberg, 1979), as they do not possess a noticeably better noise performance than other electrodes (Huigen, Peper & Grimbergen, 2002). Although the input current of modern amplifiers is considerably smaller, polarization remains a limitation to the use of NASICON electrodes in long-term measurements (Gondran, Siebert, Yacoub & Novakov, 1996).

More importantly, measurements of electrophysiological signals are degraded in high-impact conditions and states of high physical activity. This is due to the presence of relative motion artifact signals. This degradation is usually a combination of increased sensor noise (caused, for example, by increased contact resistance), direct contamination by motion artifacts (e.g. motion of the sensor in the ambient field of the environment), and modulation of the signal due to relative motion to the subject. The following section presents measurements demonstrating an improved rejection of motion artifact signals using the QUASAR IBEs.

3 Capacitive Measurements of Bioelectric Signals

3.1 Principles of Operation

Measurement of bioelectric signals using electrodes without ohmic contact with the skin requires the use of capacitive coupling. The general circuit architecture for capacitive coupling is shown in Fig. 1. A charge Q flows onto the sensing electrode C_s that is given by the product of C_s and the ambient electric potential at C_s.

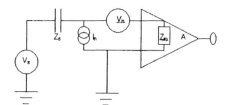

Figure 1: Circuit for a capacitive electrode connected to an amplifier. V_s is the potential at the electrode, Z_s is the sensing electrode, V_n and I_n are the voltage and current noise respectively of the amplifier, and Z_{in} is the amplifier input impedance.

The circuit in Fig. 1 is basically that of an impedance divider. For the ultra high input resistance of presently available amplifiers ($R_{in} = 10^{15}$ Ω), the input impedance of the amplifier, Z_{in}, is dominated by the input capacitance at frequencies where $\omega R_{in} C_{in}$, $\omega R_{in} C_s \gg 1$, and the system has an essentially flat-band response given by $V_{in} = V_s C_s /(C_s + C_{in})$. In most applications it is easy to arrange for $C_s \geq C_{in}$ so that the equation becomes $V_{in} \approx V_s$, thereby yielding a near ideal measurement of the electric potential.

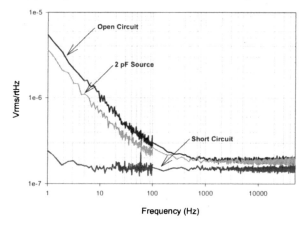

Figure 2: Noise spectrum referred to its input for a QUASAR bioelectrode.
Open Circuit: $C_s = 0$.
2 pF Source: $C_s = 2$ pF.
Short Circuit: Sensor short circuit

The amplifier input current noise acts on the very high impedance of the capacitive electrode, producing an effective input voltage noise that dominates at low frequency. The intrinsic sensitivity (referred to the sensor input) of the current generation of QUASAR bioelectrode technology is shown in Fig. 2. The noise for a shorted input (lowest curve) reflects the input voltage noise of the amplifier. A 1/f dependence for the voltage spectral density is visible in the behaviour of the two upper curves below 100 Hz. For this reason, capacitive bioelectrodes have traditionally used exotic materials to generate a high capacitive coupling (~1 nF) to the skin.

In comparison, QUASAR IBEs use a low dielectric material whose function is to resistively isolate the electrode from the subject. This material exhibits none of the biocompatibility issues of those exotic materials used in traditional high capacitance IBEs, and allows the sensor to measure bioelectric signals when separated from the skin by several layers of fabric. This level of performance has been made possible as a result of recent improvements in transistor technology. State-of-the-art (SoA) devices now have input current noise, $I_n = 0.1$ fA/$\sqrt{Hz}$, which provides an effective input voltage noise of order 2 μV/$\sqrt{Hz}$ at 10 Hz (low enough for off-body EEG). For larger capacitance to the source, or at higher frequency, the effect of the current noise is reduced to even lower levels and the sensor approaches its voltage noise, which is of order 20 nV/$\sqrt{Hz}$ (see Fig. 2).

Two generations of IBEs developed at QUASAR are shown in the photographs of Fig. 3. The first generation is the larger, square sensor (1"x1") in Fig. 3, left (IBEv1). These were used for measurements of human bioelectric signals without touching the body through a cotton T-shirt (reported by Matthews, Krupka, Say & Hibbs (2001)). Under a DARPA Phase I SBIR program, this technology was further developed into the much smaller circular sensor in Fig. 3, right (IBEv2).

 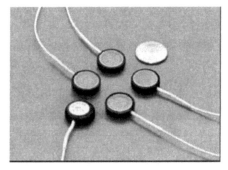

Figure 3: Photographs of QUASAR IBEs.
Left: QUASAR IBEv1 sensor (square sensor) and IBEv2 sensor with a dime for scale.
Right: QUASAR IBEv2 sensors with a US penny included for scale.

3.2   Measurements of Bioelectric Signals Using IBEs

3.2.1   Through-Clothing ECG Measurements

A clinical trial of measuring ECG through clothing, involving 40 healthy people of both sexes and a range of sizes, was conducted at the Walter Reed Army Institute of Research (WRAIR) in the summer of 2002 (Lee, Pearce, Hibbs, Matthews & Morrisette, 2004). All measurements were made through cotton T-shirts that were not modified or prepared in any way. Three IBEv1 sensors were positioned over the T-shirts using a low-cost elastic strap. A typical output for all three sensors obtained when sitting is shown in Fig. 4, in which the top trace is a paste electrode connected to a conventional ECG system. The bottom three lines (tagged by circles, squares and triangles) are the QUASAR electrodes referenced to two other sensors and operating through the T-shirt. The data have been filtered using a 30 Hz low-pass filter, but are otherwise unprocessed.

The correlation between ECG measured by "gold-standard" skin-contacting paste electrodes and the new capacitive sensors through a T-shirt (i.e. without skin contact) was greater than 99% when averaged over all 40 subjects in the trial. Fig. 5 shows correlation between the three QUASAR electrodes and the skin-contacting electrode.

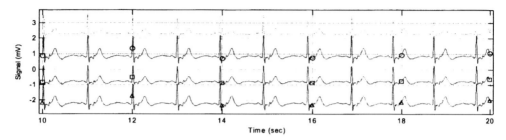

Figure 4: Clinical trial data comparing a conventional skin-contacting electrode (top) with QUASAR electrodes operating through a T-shirt (circle, square, triangle). Subject was sitting.

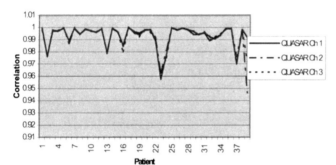

Figure 5: Correlation between each of the three QUASAR electrodes operating through clothing and a skin-contacting paste electrode for all 40 subjects in the clinical test at WRAIR

3.2.2 Effect of High Physical Stress on System Noise

A comparison of the noise performance was made between IBEv2 sensors and a conventional skin-contacting wet sensor. In order to minimize artifacts from other bioelectric signals (EMG, ECG), the IBEv2 measurements were made with two sensors in the center of the forehead. The skin was not prepared in any way and the sensors were held against the subject using a simple elastic strap. To provide a baseline comparison, data were also collected in a subsequent experiment from a pair of conventional wet electrodes attached with their usual adhesive backing. For the wet electrodes, the skin was prepared with an alcohol wipe, but not abraded. In each case the signals from each electrode were differenced using an analog circuit with a very high common mode rejection ratio (CMRR) and subsequently processed using high and low pass digital filters.

Fig. 6 shows time-series and frequency data taken from a 2-minute experiment for a subject sitting in a standard office chair. There is very little difference between the QUASAR sensor and conventional wet electrodes. Around 10 Hz, the wet electrode appears to perform slightly better than the QUASAR sensor, but the wet electrodes exhibit intermittent transient noise spikes. The level of noise for the two sensors is similar in magnitude, suggesting a common mechanism.

The critical issue for many practical applications is the sensor noise level during periods of physical stress on the subject, such as when the subject is moving. Accordingly, data were taken with the subject nodding his head, walking, and running in place. As expected, significant motion artifact was measured from the wet electrode in all modes of motion. This is exemplified by the plot in Fig. 7, left, in which the noise level of the wet electrode has increased to over 150 µV, more than 8 times higher than with the subject at rest. In contrast, the two QUASAR sensors exhibit very little motion artifact for all levels of physical activity, with a noise level almost identical to the level collected while the subject was sitting still. This can be seen in Fig. 7, right. This is an exceptional result, as it has been long believed that significant degradation in signal quality is always experienced when measuring EEG data in motion.

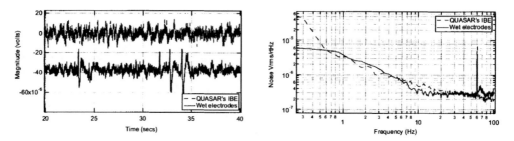

Figure 6: *Left:* Time domain comparison for subject at rest (1-50Hz Bandwidth).
Right: Voltage spectral density for subject at rest

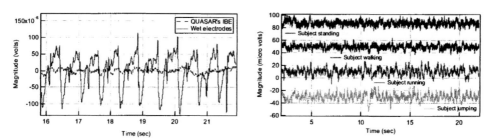

Figure 7: *Left:* Time series comparison between wet electrode and IBEv2 sensor for the subject running in place.
Right: Time series for all levels of physical activity measured by IBEv2 electrodes.
All data filtered using 1-50 Hz bandwidth.

3.2.3 Effect of High Physical Stress on ECG Signals

The effect of physical activity on a bioelectric measurement was determined in a measurement of ECG by IBEv2 electrodes placed on the chest. The electrodes were held in place by an elastic strap. ECG data were recorded while the subject was still and again while the subject was running in place. These measurements were repeated with the ECG recorded through a layer of cotton. The results are presented in Fig. 8, which shows plots for a subject still (lower) and running (upper). Both graphs show excellent ECG signal fidelity with the subject stationary and only slight degradation in signal quality for the subject running in place.

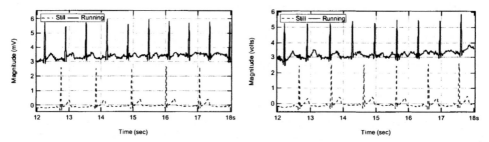

Figure 8: *Left:* Time series comparison still/running in place for IBEv2 sensors placed on the skin.
Right: Time series comparison still/running in place for IBEv2 measurement through cotton.
All data filtered using a 0.05-50 Hz bandpass IIR digital Bessel filter.

3.2.4   Clinical Test of IBEv2 Sensors

The ECG measurements presented in Fig. 8, right demonstrated the fidelity of QUASAR's IBEv1 electrodes in through-clothing measurements of ECG. Fig. 9, right presents data for a clinical demonstration of the IBEv2 sensors placed in cotton pockets of small elastic belts. For the purposes of comparison, wet electrodes were also installed as close as possible to the QUASAR electrodes. ECG data for both sets of electrodes were collected at 1200 Hz. Fig. 9 left shows an ECG chest strap used to position the sensors for measurements on a subject in motion.

The data collected in the clinical tests showed excellent signal fidelity in both lead configurations. A cardiologist who analysed the reported data (Scott, 2004) said that the QUASAR electrodes were "at least equivalent, and often superior, to the standard wet electrodes. The morphologies of the individual components (P, QRS and T waves) were easily recognized and very similar with the two sets of electrodes." In discussing the noise, it was concluded that "despite the noise, the tracings were of sufficient quality for diagnostic use, and similar to some I have seen in clinical settings. In the majority of the recordings the noise level using the QUASAR electrodes was subjectively less than that using the wet electrodes."

 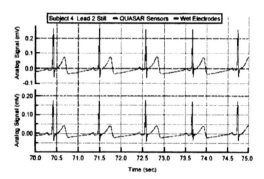

Figure 9: *Left:* ECG chest strap used to position IBEv2 sensors on the chest.
Right: Clinical data recorded with the sensors in the lead-2 configuration.
Upper trace – QUASAR sensors   *Lower trace –* wet electrodes

Further tests were conducted to evaluate the electrodes under moderate-to-high levels of physical activity. For the purposes of comparison, a Polar S180i heart rate monitor system was installed immediately below the QUASAR electrodes on the midline of the chest. Data were collected from three test subjects as they moved through an obstacle course that involved the following physical activities: running in place, stepping up and down from a small block, crawling along the ground on hands and knees, pulling on an elastic rope, lifting a 20-lb weight above the head, climbing up and down a step ladder, and climbing over a 3-ft-high obstacle. Averaged over all of the data collected for the 3 test subjects, the QUASAR electrodes achieved a 99.8% correct classification of the heart rate. This corresponded to 23 points incorrectly classified or missing in 10,417 points. By contrast, the Polar system achieved a 94.5% correct classification of the heart rate.

3.2.5   Comparison of IBEv2 and Wet Electrode for Measurement of EEG

A comparison of EEG signal fidelity for IBEv2 sensors and a conventional skin-contacting wet sensor was conducted using a single IBEv2 sensor and a conventional wet electrode placed side-by-side, with the IBEv2 sensor held in place against the forehead using an elastic strap. The skin under the wet electrode was prepared with an alcohol wipe (with no abrasion), while no skin preparation was used for the IBEv2 sensor. The onset of alpha activity is shown in Fig. 10. In this plot there is near perfect agreement between the sensors (whose traces are shown slightly offset to illustrate the temporal correlation between the two sensors).

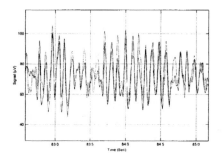

Figure 10: Alpha wave activity observed using QUASAR IBEv2 sensor and conventional resistive contact electrodes. Electrodes were in contact with the top of the head. The two curves are offset to illustrate the temporal correlation between the two sensors. *Blue (darker) trace:* skin-contacting wet electrode. *Red (lighter) trace:* QUASAR IBEv2 sensor.

4   Integrated Bioelectrode Systems

To date QUASAR has developed 6 integrated bioelectric systems based upon the IBEv2 sensors. Fig. 11 presents photographs of five other systems. In the top right photograph of the Figure, IBEv2 sensors have been discretely integrated into a chair for non-contact measurements of ECG. The top center photograph shows a subject lying on a stretcher that similarly has integrated IBEv2 sensors. In both ECG systems, ECG measurements are taken through at least two layers of material (the subject's clothing and the material covering the sensors). In each photograph, there is a real-time display of the ECG measurements (filtered for R-R timing) on the computer screen. The red markers on the plots correspond to the R-wave peak determined using a simple peak algorithm.

The top right photograph is a photograph of a pull-on, adjustable shirt with integrated sensors that is suitable for quick, prep-free measurements of ECG, giving good R-R determination even under high levels of physical motion.

The bottom left photograph shows four sensors mounted in an ordinary baseball cap. The display EEG signal is shown on the laptop screen in the foreground of the photograph. No skin preparation was performed on the subject for this measurement. The bottom right photograph in the Figure shows IBEv2 sensors integrated into a pair of glasses for the measurement of EOG. Data taken with the EOG glasses is shown in Fig. 12.

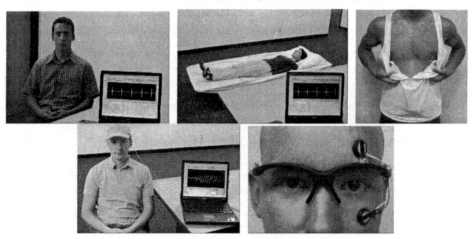

Figure 11: Photographs of integrated ECG systems.
Top Left: ECG chair system. *Top Center:* ECG stretcher. *Top Right:* ECG shirt.
Bottom Left: EEG cap system. *Bottom Right:* EOG glasses.

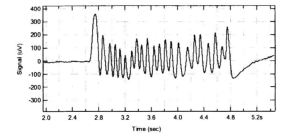

Figure 12: Data taken with EOG glasses

5   Elimination of Skin Noise

It was remarked in Section 3.2.2 that the noise levels shown in Fig. 6 for the QUASAR IBEv2 sensors and conventional wet electrodes were similar in magnitude. This suggests a common mechanism, which can be identified as residual epidermal artifact signal. The epidermal artifact signal is the potential developed in the skin itself (Edelberg, 1973). This skin signal is the primary source of movement-related (motion) artifact even when the subject is still, although contribution to the observed noise can be reduced by careful preparation of the skin. This usually involves abrasion of the skin (Tam & Webster, 1977) and the application of a conducting gel to improve the electrical contact between the electrode and the skin.

Fig. 13 presents data from EEG measurements taken using resistive contact electrodes. The 'no prep' curve (lower) is an example of noise in an EEG measurement using a conventional resistive EEG electrode in the absence of any skin preparation. The noise in the 'no prep' curve is more than an order of magnitude greater than the expected noise that can be explained by the contact impedance. For example, a typical skin-electrode contact resistance of 100 kΩ would possess Nyquist noise with an amplitude equal to $0.4\mu V/\sqrt{Hz}$ (corresponding to approximately $1.2\mu V$ peak-to-peak (p-p) in a 100 Hz signal bandwidth). The 'prep' curve (upper) shows results obtained using a thorough preparation of the scalp involving abrasion combined with the application of a conducting gel. A good surface preparation typically results in a reduction of the surface resistance to ~5 kΩ. This behaviour is in agreement with the results of Huigen *et al.* (2002), which show a clear relationship between electrode noise and the electrode-skin impedance, even though the resulting noise level is still considerably larger than that expected for the electrode-skin contact resistance.

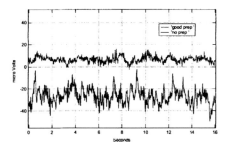

Figure 13: Output of resistive contact EEG electrode. No surface preparation of the scalp was involved for 'no prep' (lower) curve. The surface preparation for the 'good prep' (upper) curve involved abrading the skin and the application of a conducting gel.

QUASAR has developed a new sensor design that is intrinsically insensitive to epidermal artifact noise. The new sensor, which is currently under construction, represents a radical new direction in capacitive electrode technology. Most significantly for this application, it is able to discriminate between skin noise and the bioelectric signal of interest, allowing the sensor to measure EEG signals with a noise level determined by the sensor electronics. More generally, it will allow the operation of capacitive electrodes with significantly lower noise levels.

6   Concluding Remarks

The deployment of bioelectric sensors in settings beyond a clinical environment requires the use of sensors that are non-invasive, unobtrusive, and easy to use. In this paper we presented high fidelity bioelectric signal data collected using QUASAR's IBEv2 technology in the absence of any skin preparation, in which the sensors achieved noise levels similar to those obtained using conventional resistive contact technology. In a clinical trial of measuring ECG through clothing there was excellent agreement between QUASAR's IBEv1 sensors and "gold standard" resistive contact electrodes, with a correlation of greater than 99% when averaged over 40 subjects. Independent analysis of ECG data by a board certified cardiologist from a clinical test using QUASAR's IBEv2 found that the morphologies of individual components in the ECG trace were readily recognizable with a signal fidelity that was often superior to that observed using standard wet electrodes. Furthermore, the QUASAR IBEs exhibited significantly less sensitivity to motion artifact signals than resistive contact electrodes, and bioelectric signals were successfully measured with sensors incorporated into items of clothing including shirts, elastic belts, and glasses.

The QUASAR IBEv2 sensors are state-of-the-art bioelectric sensors that are suitable for most applications involving measurements of ECG or EEG. However, extremely low-noise measurements of bioelectric signals, such as the measurement of event-related potentials (ERPs) require the removal of the epidermal artifact noise, which at present is only possible by aggressively preparing the skin. This limits such measurements to clinical environments and renders long-term monitoring impossible.

The next generation of QUASAR sensors will remove the need for any skin preparation in low-noise bioelectric measurements. The new QUASAR sensors will not only be capable of ultra-low noise, non-contact, through-clothing measurements of bioelectric signals, but they will also be insensitive to epidermal artifact noise. The new design will immediately find uses in any application requiring long-term, high fidelity monitoring of bioelectric signals, such as DARPA's Augmented Cognition program. The combination of long-term monitoring capability, absence of biocompatibility issues, and ease of use will finally allow the real-world deployment of a system to monitor cognitive performance.

References

Edelberg, R. (1973). Local response of the skin to deformation, *J. Appl Physiol, 34*, 334.

Gondran, C., Siebert E., Yacoub, S. & Novakov, E., (1996) Noise of surface bio-potential electrodes based on NASICON ceramic and Ag-AgCl, *Med. & Biol. Eng. & Comput., 34,* 460-466.

Huigen, E., Peper, A., & Grimbergen, C.A. (2002). Investigation into the origin of the noise of surface electrodes, *Medical and Biological Engineering and Computing,* 40, 332-338.

Klinger, D.R., Booth, H.E., & Schoenberg, A.A. (1979). Effects of dc bias currents on ECG electrodes, *Med. Instrum.*, 13, 257-258.

Krupka, M.A., Matthews, R., Say, C., & Hibbs, A.D. (2001). Development and Test of Free-space Electric Field Sensors with Microvolt Sensitivity. Presented at the 2001 Meeting of the MSS Specialty Group, Johns Hopkins Univ., October 23, 2001. Published in the proceedings.

Lee, J.M., Pearce, F., Hibbs, A.D., Matthews, R., & Morrisette, C. (2004). Evaluation of a Capacitively-Coupled, Non-Contact (through-clothing) Electrode for EEG Monitoring and Life Science Detection for the Objective Force Warfighter. Presented at the RTO HFM Symposium on "Combat Casualty Care in Ground Based Tactical Situations: Trauma Technology and Emergency Medical Procedures," St. Pete Beach, USA, August 16-18, 2004.

Matthews, R., Krupka, M.A., Say, C., & Hibbs, A.D. (2001). Capacitively Coupled Non Contact Sensors for Measurement of Human Bioelectric Signals. Presented at Advanced Technology Applications for Combat Casualty Care (ATACCC), Ft. Walton Beach Fl., September 9-14.

NASA website (2005). Experiment Description for: Clinical Aspects of Crew Health (AP001). Retrieved February 23, 2005, from http://lsda.jsc.nasa.gov/scripts/cf/exp_descrp_pop_up.cfm?exp_id=AP001

Richardson, P.C. (1968). The Insulated Electrode: A Pasteless Electrocardiographic Technique. *Proc. Ann. Conf. Eng. Med. Biol.*, 9:15.7.

Scott, W.A., Professor of Pediatrics/Cardiology, University of Texas. Private communication, 2004.

St. John, M., Kobus, D.A., Morrison, J.G., & Schmorrow, D. (2004) Overview of the DARPA Augmented Cognition Technical Integration Experiment. *International Journal of Human-Computer Interaction,* 17, 131-149

Tam, H.W., & Webster, J.G. (1977). Minimizing Electrode Artifact by Skin Abrasion. *IEEE Transactions on Biomedical Engineering,* BME-24, 134-139.

Artifact Detection and Correction for Operator Functional State Estimation

Chris A. Russell

Air Force Research Laboratory
Human Effectiveness Directorate
2255 H Street
Wright-Patterson AFB, OH
45433
christopher.russell@wpafb.af.mil

Ping He

Department of Biomedical,
Industrial, and Human Factors
Engineering
Wright State University
Dayton, OH 45435
phe@cs.wright.edu

Glenn F. Wilson

Air Force Research Laboratory
Human Effectiveness Directorate
2255 H Street
Wright-Patterson AFB, OH
45433
glenn.wilson@wpafb.af.mil

Abstract

Determining operator functional state is a critical component of adaptively aiding closed human-in-the-loop systems. The input measures or features used to define the operator functional state (OFS) model must be free of artifacts to ensure accurate classification of operator functional state. Electroencephalography (EEG) is a major component of the OFS model. Since the EEG magnitude is small relative to other psychophysiological measures it is most susceptible to contamination or artifact. Eye blinks and body movements are the usual culprit of artifact in EEG. This paper discusses an adaptive filter algorithm for removal of both horizontal and vertical eye movements including blinks. Muscle activation is also a major source of EEG contamination. A discussion of a promising muscle artifact removal using independent component analysis is presented. Additional artifacts can be detected but not removed using a simple statistical technique which is presented here.

1 Introduction

Operator functional state (OFS) estimation is an important component of a closed human-in-the-loop automated system. The estimates of the operator state provided to the system for adaptive automation must be as accurate as possible to ensure system reliability and operator acceptance. Possible deterrents to accurate estimates are artifacts in the psychophysiological signals used to classify the operator state.

Filtering is the most obvious choice for removal of artifact. But in many cases, specifically muscle and eye artifacts, filtering the signals using bandpass filters removes the information in the psychophysiological signals that is of interest to the investigator as well as the artifact. Other methods must be considered to remove the artifact and retain the critical information in the signal.

Many artifact correction algorithms are available and exhibit differing levels of success in the removal and detection of artifacts. Both spatial (frequency domain) and temporal (time domain) algorithms exist and both have advantages and disadvantages. Adaptive filters and regression methods are examples of algorithms implemented in the temporal domain while independent component analysis, wavelet techniques, and fast Fourier methods are examples of spatial domain algorithms.

Much of the research has focused on post processing of contaminated signals. This is unacceptable for real-time implementation in a closed loop system. The algorithms must be able to remove or detect and label the artifacts in near real-time to enable accurate on-time estimation of the operator's functional state. Several algorithms present themselves as better candidates for real-time applications. Adaptive filters are used successfully at the Air Force Research Laboratory (AFRL) for the removal of horizontal and vertical eye movement in EEG.

Muscle artifacts pose a unique problem since there are multiple sources all around the skull unlike the eye artifact which comes from two localized sources. Algorithms for removing muscle artifacts from EEG have been pursued with limited success. Independent component analysis (ICA) and principle component analysis (PCA) techniques have been investigated and results suggest while these methods can extract sources of muscle artifact, however, real-time implementation of these algorithms may be difficult.

2 Methods

2.1 Eye Artifact Correction of EEG

Two methods for correcting eye blink and eye movement artifact were compared. The first technique is based on frequency-domain methods (Gasser, Sroka, and Möcks, 1985). This regression method uses a frequency dependent scaling factor to remove the artifact attributed to the EOG signal. The EOG artifact in the contaminated EEG is removed using

$$EEG(\omega) = EEG_{artifact}(\omega) + K(\omega)EOG(\omega) \qquad (1)$$

where $K(\omega)$ is the scaling factor and ω denoted frequency. Note the scaling factor is frequency dependent which indicates each frequency component has a different scale factor. The regression requires training to determine the coefficients of the EOG transfer function K.

The frequency dependent scaling factor represents an improvement over a constant scaling factor. A constant scaling factor removes the same power from the contaminated EEG signal regardless of frequency. The artifact produced in the EEG signal is frequency dependent. More power is in the theta band (4-8 Hz) than is represented in the alpha band (8-12 Hz).

The second approach is a time-domain method based on adaptive filtering (Widrow and Stearns, 1985). Figure 1 shows the two channel adaptive noise canceller with two reference inputs (vertical, VEOG, and horizontal, HEOG, EOG). This method does not require calibration and can artifacts can be remove online and in real-time (He, Wilson, and Russell, 2004)

The primary input to the noise canceller is the contaminated EEG signal and can be modeled as EEG signal plus noise (VEOG and HEOG). The adaptive algorithm used by this model is the recursive least squares algorithm. The recursive least squares algorithm was chosen for its superior stability and fast convergence.

232

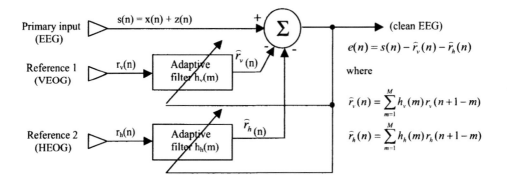

Figure 1. Block diagram of EOG noise canceller using adaptive filtering

The regression algorithm and the adaptive filter algorithms were compared using EEG artificially contaminated with VEOG and HEOG. The simulated noisy EEG signals were generated using a model as shown in Figure 2. An autoregressive (AR) model using the Burg method (Kay, 1988) was used to generate simulated EEG. This signal was summed with a generated EOG signal using an infinite impulse response (IIR) filter.

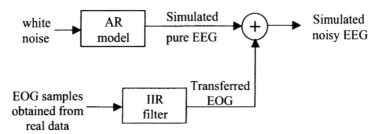

Figure 2. Block diagram of EEG and EOG simulation for verifying correction algorithms.

2.2 EMG Artifact Removal for EEG

The removal of muscle artifact (EMG) from the EEG signal is not as straight forward as the removal of ocular artifacts. Unlike the eye artifact, the EMG artifact does not come from a single source. The artifact caused by muscle has many sources distributed over the scalp and can be caused by small muscle activations producing muscle tension as well as larger muscle movements such as head movements.

The main difficulty in removing EMG noise is due to it's the widespread locations of the generating sources. An EMG signal recorded at any specific site, e.g. F7, is a superposition of action potentials produced by many motor units after traveling to the measurement site. Due the natural neural regulation, these motor units are activated alternately rather than in exact synchronization. In other words, the EMG signal is associated with a group of spatially-distributed sources rather than a single equivalent source. An important consequence of this model is the low coherence between a pair of EMG signals recorded at two different locations. This explains why the EMG noise can't be removed using either the regression method or the adaptive filtering method which require a high degree of correlation between the noise at the recording site and the noise at the reference site.

Two popular methods, PCA and ICA, have been shown to be effective in removing EOG noise (Lagerlund, Sharbrough, and Busacker, 1997, Jung, Makeig, Humphreis, Lee, McKeown, Iragui, and Sejnowski, 2000). An investigation of these two methods for removal of EMG noise from EEG was conducted. A ten second sample of 19 channels of EEG containing visible EMG noise was selected for the evaluation. The PCA and ICA analysis was conducted on this contaminated sample. The ICA or PCA components contributing to the EMG contamination were visually selected for removal from spatial and time domain plots of the components. Using the remained components, the EEG signal was recreated using a reconstruction or spatial filter (Lagerlund, Sharbrough, and Busacker, 1997). Figure 3 shows a block diagram of the EMG removal process.

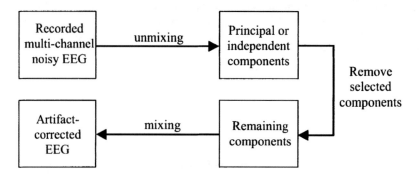

Figure 3. Block diagram of EMG artifact removal using ICA or PCA.

2.3 Artifact Detection in EEG Using Sample Statistics

A method for detecting other types of artifacts in the EEG using simple statistics is being investigated. A sample set of EEG was recorded and the means and standard deviations of each EEG channel were computed. As new data are collected these means and standard deviations were used to detect artifacts. The artifacts that can be detected are the result of muscle activity, movement and other spikes in the signal as well as railed amplifiers. The principle being used with this method is the assumption of Gaussian distribution of the data. The data within three standard deviations of the mean account for 99.7 percent of the data collected. The assumption is that data outside this range are considered outliers and are associated with artifact.

A sample or training set was processed to extract the features, e.g., power of F7 alpha band, and averaged over a one second interval. The means and standard deviations of the averaged features were computed as baseline statistics for subsequent data. The values of the subsequent or new data which lie outside the range of three standard deviations from the mean were tagged as artifact. The data tagged as artifact were visually inspected for accuracy.

3 Results and Discussion

3.1 Eye Artifact Correction of EEG

The ocular artifact correction algorithms were compared using the simulated EEG signal contaminated with EOG noise. Analysis was conducted in both time domain and frequency domain. For the time domain analysis, the total squared error between each true EEG waveform and the corrected EEG waveform, either by using the regression method or the adaptive filter method, is first calculated. The results from 20 simulated waveforms are then averaged to produce the mean square error (MSE). For the adaptive filter method, the results (Table 1) are shown for filter length $M = 1$, forgetting factor $\lambda = 1$ and for $M = 3$ and $\lambda = 0.999$. The adaptive filter method produces the best results with a 63 percent reduction in MSE over the regression method.

Table 1. Time Domain Comparison			
	M	λ	MSE
Regression Method	-	-	8.16
Adaptive Filter	1	1	8.50
Method	3	0.999	3.01

For the frequency domain analysis, the mean spectral difference (MSD) between the true EEG and the corrected EEG are computed for three frequency bands: theta band (3.5 – 7.5 Hz), alpha band (7.5 – 12.5 Hz), and beta band (12.5 – 19.5 Hz). The results are shown in Table 2. The adaptive filter method shows a 66 percent reduction in error for the theta frequency band, a 74 percent reduction for the alpha band, and a 90 percent reduction for the beta band. An example of the signal processing results using the adaptive filter method is shown in Figure 4.

Table 2. Frequency Domain Comparison					
	M	λ	MSE		
			Theta	Alpha	Beta
Regression Method	-	-	0.4588	0.1126	0.0362
Adaptive Filter	1	1	0.4404	0.1000	0.0293
Method	3	0.999	0.1539	0.0297	0.0037

The adaptive filter method is superior to the regression method for the following reasons.
- The method does not need prior calibration.
- The method is extremely stable. The only two parameters, M and λ, can be chosen within wide ranges (M: 1 – 15, λ: 0.99 – 1.0) without compromising the performance of the method.
- The method is fast converging and is suitable for real-time application
- The method is inherently adaptable to the changing measurement condition, e.g. the change of the contact impedance between the tissue and the electrode, while maintaining its effectiveness in removing EOG noise.

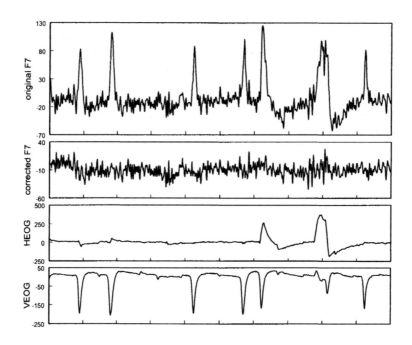

Figure 4. Demonstration of EOG removal using adaptive filtering.

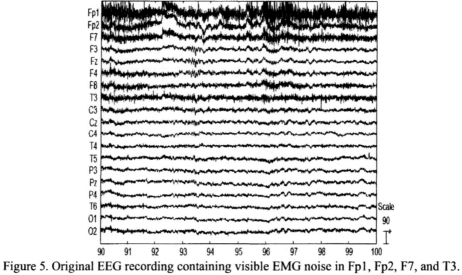

Figure 5. Original EEG recording containing visible EMG noise in Fp1, Fp2, F7, and T3.

3.2 EMG Artifact Removal for EEG

Figure 5 shows the original EEG signal containing visible EMG contamination. The EEG channels containing the most significant contamination are Fp1, Fp2, F7, and T3. The results of PCA are reported in Figures 6 and 7. Figure 6(a) shows the spatial map of the 19 principal components, and Figure 6(b) is a plot of the first 10 principal components. Figure 7 is the reconstructed EEG waveforms for select channels. The spatial filter is constructed from the remaining principal components after removing components 1, 5, and 7 which contain the signal components of eye and muscle artifact.

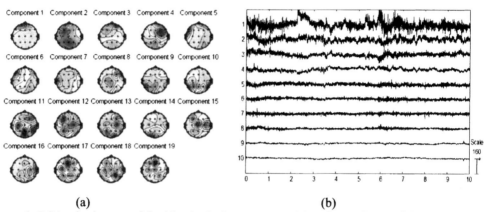

(a)   (b)

Figure 6. PCA - Scalp map of the 19 principal components (a) and waveforms of the first 10 components (b).

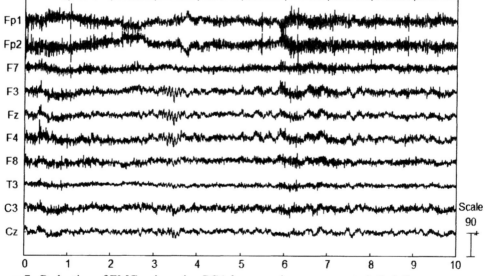

Figure 7. Reduction of EMG noise using PCA by removing components 1, 5, & 7.

The results of ICA are displayed in Figures 8 and 9. Figure 8(a) shows the spatial map of the 19 independent components, and Figure 8(b) is a plot of the components. Figure 9 is the reconstructed EEG waveforms for select channels. The spatial filter is constructed from the remaining components after removing components 1, 5, and 9 which contain the signal components the artifact.

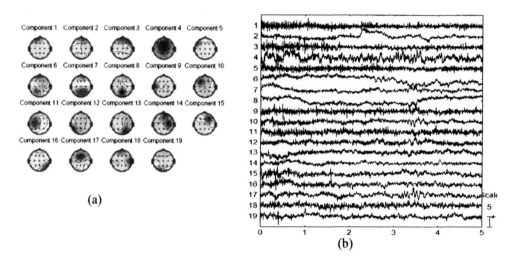

Figure 8. ICA - Scalp map of the 19 independent components (a) and the actual waveforms (b).

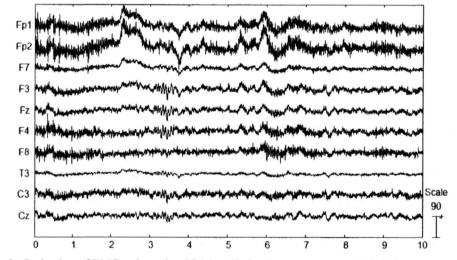

Figure 9. Reduction of EMG noise using ICA by eliminating components 1, 5, & 9.

Both ICA and PCA require offline process and a high degree of human involvement in selecting the components for removal. Both methods reduce EMG but cannot completely remove it.

Additionally, both methods will remove components of the EEG signal as well as the artifact. More research is needed to develop a method for automatic online removal of EMG noise.

3.3 Artifact Detection in EEG Using Sample Statistics

Figure 9 displays the output of the sample statistic artifact detection for electrode F7. The one second power average for each frequency band are displayed in the left column, the raw signal is at the top of the display, and the plots in the right column are the filtered raw EEG for each frequency band. The one second power average plots are bound by a dashed line indicated three standard deviations from the mean for each frequency band. The data outside this range are considered artifact. The raw signal indicates a spike in the signal possibly due to movement followed by muscle activity. The one second power average at time 127 for the beta and gamma bands are outside the range and tagged as artifact. The corresponding filtered signals are also tagged for that one second of data.

This technique is presented for consideration and a thorough evaluation must be conducted to determine the utility of this method. Using the statistics of the signal is more advantageous than simply determining an arbitrary threshold since psychophysiological signals may tend to drift over time and can vary from day to day. This technique also has the advantage of simplicity. Artifacts are determined for each feature based on their statistics, therefore, only the contaminated frequency bands are tagged. Other frequency bands are available for use in classification. A disadvantage to this technique is the necessity for calibration, i.e., data must be recorded to determine the statistics.

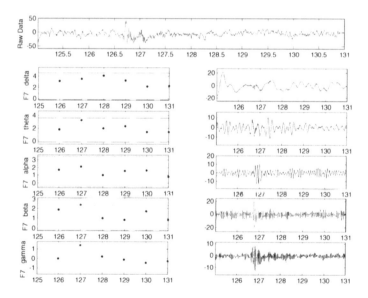

Figure 9. Sample artifact detection using means and standard deviations.

239

4 References

Gasser, T., Sroka, L., & Möcks, J. (1985). The transfer of EOG activity into the EEG for eyes open and closed. *Electroencephalography and Clinical Neurophysiology*, 61, 181-193.

He, P., Wilson, G., & Russell, C. (2004). Removal of ocular artifacts from electro-encephalogram by adaptive filtering. *Medical & Biological Engineering & Computing*, 42, 407-412.

Jung, T. P., Makeig, S., Humphries, C., Lee, T. W., McKeown, M. J., Iragui, V., & Sejnowski, T. J. (2000). Removing electroencephalographic artifacts buy blind source separation. *Psychophysiology*, 37, 163-178.

Kay, S. (1988). Modern spectral estimation: theory and application. Prentice-Hall, Englewood Cliffs, NJ.

Lagerlund, T. D., Sharbrough, F. W., & Busacker, N. E. (1997). Spatial filtering of multichannel electroencephalographic recordings through principal component analysis by singular value decomposition. *Journal of Clinical Neurophysiology*, 14, 73-82.

Widrow, B. and Stearns, S. D. (1985). Adaptive signal processing, Prentice-Hall, Englewood Cliffs, NJ.

On-line Correction of Artifacts in Cognitive State Gauges

Don M. Tucker, Ph.D.
Pieter Poolman, Ph.D.
Phan Luu, Ph.D.

Electrical Geodesics, Inc.
1600 Millrace Drive, Suite 307
Eugene, OR 97405 USA
dtucker@egi.com

Abstract

On-line assessment of operator brain state requires measurement technologies that can analyze the brain signal effectively in operational environments. A major source of artifact or noise is occular artifact, including eye movements and blinks. In collaboration with the Boeing Augmented Cognition team and the University of Oregon Neuroinformatics Center we have applied Independent Components Analysis to separate artifactual from cephalic sources. We illustrate this separation with 256-channel EEG and the high-performance computing cluster (ICONIC Grid) at the NIC.

1. Signal cleaning

The signal cleaning toolbox provides the user with a set of tools related to ICA decomposition. These tools include sequential and parallel versions of the FastICA algorithm and a sequential version of the Infomax algorithm and we are in the process of creating a parallel version of the Infomax algorithm. The ICA algorithms and preprocessing algorithms are called through a user configurable software toolbox. To configure the toolbox, the user edits a set of forms (text files) used to describe the desired preprocessing and processing algorithms as well as options for presenting or storing results from the process.

The FastICA algorithm has been implemented as a distributed MPI (message passing interface) program on the Neuroinformatics Center neuronic cluster and as a shared-memory MPI program on the NIC p655 cluster.

2. Independent Component Analysis (ICA) Toolbox

Independent component analysis (ICA) is a mathematical technique[1] that combines aspects of both linear algebra and statistics to represent EEG data as a linear mixture of a set of statistically independent streams, called the independent components (IC).

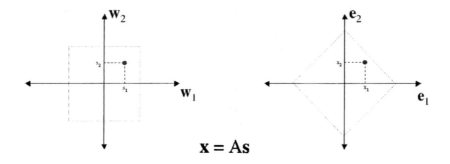

$$\mathbf{x} = \mathbf{As}$$

Distribution of probability is more Gaussian
for coordinate vector (x_1, x_2) than (s_1, s_2).

Figure 1. ICA as a Change of Basis

(x_1, x_2) is coordinate vector for basis $\{\mathbf{e}_1, \mathbf{e}_2\}$: $\mathbf{x} = x_1\mathbf{e}_1 + x_2\mathbf{e}_2$
(s_1, s_2) is coordinate vector for basis $\{\mathbf{w}_1, \mathbf{w}_2\}$: $\mathbf{x} = s_1\mathbf{w}_1 + s_2\mathbf{w}_2$

3. Application of ICA to Blink Removal

Eyeblinks, which occur naturally during the course of EEG data collection, are a source of serious contamination to the recorded EEG and so must be cleanly removed. To systematically study ICA required a framework that could receive pre-processed (cleaned of bad channels, saturated observations, filtered to remove line noise, etc...) EEG data in EGI .raw format, that could apply ICA, and that could extract activity which appeared blink like, as specified by the user. Furthermore, it necessitated the creation of a blink-free baseline, rich in cortical activity, to which simulated blinks could be added and subsequently removed via ICA. By knowing a priori the nature of the blinks and the baseline, we could effectively determine the extent to which the various ICA algorithms were capable of cleaning the data, i.e. removing blinks.

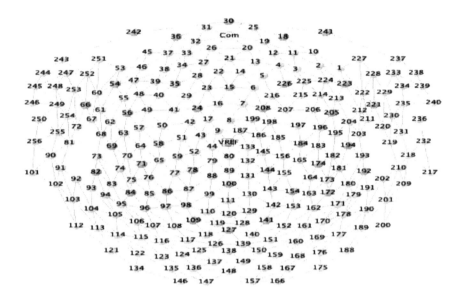

Figure 2. Geodesic Sensor Net, Down-Sampled Channels Highlighted

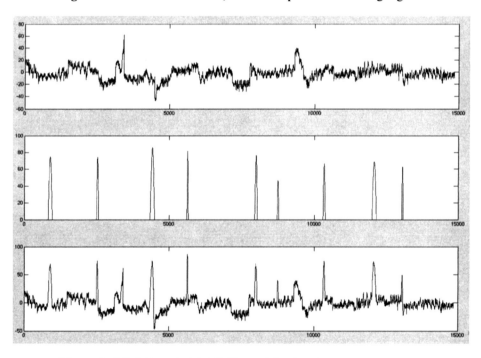

Figure 3. Blink Type # 6, EEG Channel # 1 (15000 samples ~ 60 sec)

243

From Top to Bottom:

- Clean, Blink-Free, Data

- Blink Stream

- Clean, Blink-Free, Data + Blink Template * Blink Stream

The overriding objective in creating these data sets was to develop a suite of data with which we could test the ability of different ICA algorithms to remove blinks from EEG recordings. To this end, we first constructed the "blink-free" EEG data set, rich in cortical activity yet free of blink contamination. This was the baseline or gold-standard to which filtered EEG, after blink removal, would be subsequently compared. We then constructed seven distinct sets of blinks, which were then added to the "blink-free" EEG, resulting in seven sets of "blinky" EEG, each with a unique blink morphology. Now, in subsequent studies, some or all of the following comparisons can be made:

- Compare filtered data to the "blink-free" data;

- Compare extracted blinks to the simulated, inserted blinks;

- Compare corresponding projections of extracted blink activations to the blink template.

This method has produced good separation of both blinks and eye movements from the ongoing EEG data. Results so far show that, once a subject's eye artifacts are characterized with the 256-channel EEG array, and an accurate ICA decomposition is obtained, we can then successfully extract the brain signal with a simple algorithm on a workstation computer with a few milliseconds of computing time.

eXecutive Load Index (XLI): spatial-frequency EEG tracks moment-to-moment changes in high-order attentional resources

D.R. DuRousseau and M. A. Mannucci

Human Bionics LLC.
190 N. 21st Street, Suite 300
Purcellville, VA 20132
don@humanbionics.com

Abstract

Investigators have implanted electrodes in the brains of monkeys to communicate cortical signals that move a cursor or operate a robotic arm hundreds of miles away [1]. Recent human experiments have shown it possible to use brainwaves to control computers and flight simulators, signifying that fully interoperable human-machine systems (HMS) are not far off [2]. Thus, by measuring real-time changes in the brain, investigators have demonstrated brain-actuated devices with narrowly defined control capabilities [3,4]. It is now possible to augment closed-loop systems by manipulating the timing and modality of communications based on a few physiological indicators [5]. By manipulating the timing of an operator's communications, it is possible to improve task performance; however, it's not yet clear if other types of mitigation are available to achieve higher levels of overall system function. Over the next decade our understanding of the cognitive component of the human-system will improve, as will the synergy between the human mind and the systems networked with it. To be commercially viable, the next generation of complex HMS must deliver a seamless interface between the human, with constantly changing operating efficiencies, and a networked system of information processing nodes, each expecting constant efficiency and non-varying inputs and outputs.

From years of research mapping the anatomical and functional networks of the brain, we have made great strides in understanding the dynamics of the mind in terms of its varying capacity, situational awareness, and cognitive efficiency [6,7]. Each day, often for extended periods, humans exist as a system-of-systems in which we operate and are operated upon by the devices and people around us. At times, our cognitive "mind" is capable of operating with several systems simultaneously, although, when drowsy or highly tasked, may become unable to stay focused on more than one item at a time. Thus, to work efficiently next generation HMS must understand how to cope with fluctuations in the human-system by uniquely reacting to them in a manner sufficient to maintain optimal performance. This paper describes current investigations into the design of real-time tools for measuring brain patterns capable of tracking a person's cognitive load that may be capable of detecting an overload state during complex information processing and decision-making tasks. Such tools are needed to help researchers and developers identify and track problems arising from the variability of the cognitive-system and to design improved network architectures able to mitigate deleterious fluctuations in the human-system's performance.

Thus, we theorize that cognitive networks in the brain are responsible for maintaining one's arousal level and controlling the allocation of cognitive resources as we attend, memorize, process new information, and make decisions. We understand that several brain regions participate in higher-order cognitive processing and their cortical networks have been categorized anatomically and functionally using several imaging modalities [8,9,10]. Dynamic spatial-frequency (SF) models of cognitive networks have proved useful in expanding our understanding of the interrelationships among brain regions responsible for specific cognitive tasks, like recognizing a face, moving a mouse, or, working out a difficult math problem [11,12]. Additionally, recent practical hardware and sensor improvements now make real-world use of cognition monitoring devices a feasible means of achieving a sympathetic interface to information systems. Thus, if our efforts are successful, in the near future researchers and developers would have neurometric tools capable of achieving an ubiquitous system-of-systems interface, which would be essential for widespread acceptance of HMS in entertainment, education, and relaxation markets.

1. Background

Human Bionics has built upon a wealth of *a priori* knowledge in the development of an eXecutive Load Index (XLI) and suite of neural network based classification algorithms that utilize SF patterns in the EEG to track changes in coordinated brain activity during complex decision-making activities (Figure 1). Our SF method is intended to provide a neurometric that corresponds to the demands placed on an individual's limited cognitive capacity. In calculating a workload metric capable of differentiating between task conditions, we constructed a neural-classification algorithm (NCA) that uses windowed XLI gauge values from up to eight miniature sensors overlying the anterior cingulate, parahippocampus, and superior parietal cortices. We believe the NCA, which must be trained for each subject, identifies SF patterns in an individual's EEG that mirror changes in an executive communication network corresponding to how hard the brain is working over periods of 2-seconds (Figure 2).

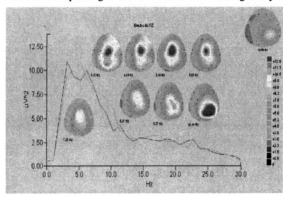

Figure 1

A head down view of several Spatial-Frequency maps displaying oscillatory patterns between 3 to 11 Hz (Red for increasing and Blue for decreasing power in uV^2). XLI map transformations were applied in real-time to differentiate peak activity at sites overlying key cortical areas responsible for coordinating higher-order cognitive network processes. The difference patterns among a few sensors were used to feed algorithms that index cognitive network changes over a 2-second period, allowing the classification of an individual's changing workload state.

Figure 2, describes the operation of the XLI algorithm used to classify taskload dependent SF patterns active in cortical sub-compartments involved in high-order cognitive processing. Our XLI model hypothesizes that the executive control of attentional and memory resources is carried out through oscillatory coupling of interdependent thalamic and cortical networks, where steady (~2-seconds) rhythmic interactions are visible in the EEG at frequencies below 30Hz. The XLI operates by computing the Fast Fourier Transform (FFT) power spectra of 2-second samples of continuous EEG from 8 channels located at F3, F4, FCZ, C3, C4, CPZ, P3, and P4. Ten 2-second FFT sweeps are buffered and the average and standard deviation for each frequency (between 2 and 15 Hz) is held in memory. Upon sampling the next 2-second period and computing its FFT, the algorithm measures peak frequency differences in Delta, Theta, and Alpha bands between the current FFT and the 20-second average at each point on the periodograms (Figure 2).

Figure 2

Spectral amplitudes are derived from electroencephalograph (EEG) signals typically recorded with a bandpass of 0.05 to 100 Hz and sampled at over 256 Hz. Data are filtered from 2 to 30 Hz and power (in uV^2) is computed using FFT. Comparison of peak frequency differences in Delta, Theta, and Alpha band power between a 20-second running baseline and a contiguous 2-second sample provide the basis for the XLI analysis.

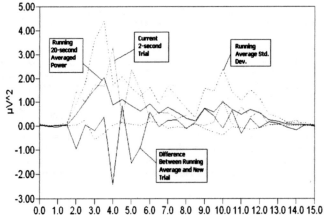

When a new 2-second period is sampled, the prior period replaces the first FFT in the running average, which is recalculated for comparison with the next 2-second sweep. In each new sweep, the XLI algorithm finds the peak frequencies and compares the difference between the running average power and the current power in each channel. The mean and standard deviations of the averaged power signature are used to weight peak differences lying outside two standard deviations for that frequency, sampled over 1 Hz bins. A standardized "eXecutive Load Index" is derived by the algorithm using the equation $[\Sigma(\delta+\theta)/\alpha]$ and passed as input to neural network pattern classifiers to correlate cognitive EEG changes to single-trial measures of taskload. The XLI and NCA were used on data recorded during a complex cognitive task performed within a virtual environment (VE) simulation having two distinct taskload conditions that were alternated over 1.5-minute intervals for a total session time of 12-minutes (Figure 3).

2. Methods

Brain potentials from fourteen male and female undergraduate students using 32 scalp EEG electrodes located according to the modified 10-20 electrode system [13] as well as vertical and horizontal EOG and ECG channels were recorded using Biosemi's Active Two amplifier system and purpose made software created by The University of Western Florida's Institute of Human-Machine Cognition for data collection and closed-loop system mitigation based on arousal, engagement, and XLI gauges. Data were recorded continuously at 256 Hz with a bandpass from 0.05 to 100 Hz and subsequently bandpass filtered from 2 to 30 Hz prior to further processing and analysis. The closed-loop system was developed for Honeywell Laboratory (Minneapolis, MN) for work under DARPA's Augmented Cognition Program. The XLI operated as a real-time component of the data collection and processing system and was used to differentiate task difficulty and detect cognitive task switching events.

Each subject performed an experiment composed of alternated Primary and Secondary taskload components repeated under 2 different conditions; Mitigated and Unmitigated (2 runs per condition). During the Mitigated condition, gauges tracking arousal level, task engagement, and executive workload were monitored continuously and used to trigger an augmented communications scheduler when pre-defined gauge thresholds were met. In the Unmitigated condition, gauge data were recorded but communications to the subject, which included task assignments, were not interrupted or rescheduled. For each condition, the XLI gauge outputs were averaged over the Primary and Secondary taskload intervals and compared to test the gauge's ability to differentiate SF power in the EEG by task difficulty level. To normalize cross-subject differences in XLI results, we calculated the absolute differences between the XLI values from the Primary vs. Secondary taskload periods across all fourteen subjects (Figure 4).

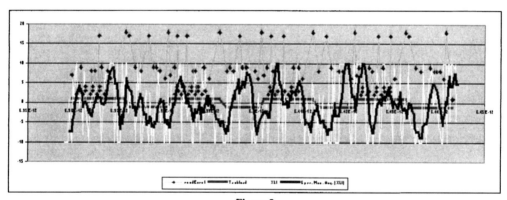

Figure 3
The four graphs represent the taskload level during the experimental session (Red Line), task and communication events experienced by the subject (Light Blue Line), the XLI values (Yellow Line), and the 6-sample Moving Average of the XLI for ease of visualization of XLI values at workload transitions (Black Line). The XLI was averaged over the Primary and Secondary workload intervals and used to differentiate the two conditions. When combined with the NCA, the XLI was used to detect single transition events between the high and low taskload intervals.

3. Discussion

The averaged XLI outputs were examined from 12-minute long Mitigated and Unmitigated sessions producing fifty-two (52) experimental trials (14 subjects unmitigated, 12 mitigated, 2 sessions each). Our goal was to measure the effectiveness of the XLI gauge over each 12-minute trial at differentially classifying the Primary and Secondary workload intervals. The averaged XLI gauge values indicated differences between the Primary (High) and Secondary (Low) workload conditions in thirteen of fourteen subjects under both conditions. However, when the absolute differences between conditions were measured, the Unmitigated case showed significant differences between the Primary and Secondary workload intervals in eight of fourteen subjects (64%) from the XLI gauge values averaged over 6-minutes per condition ($p=0.05$, $n=28$). Whereas, the averaged XLI output was not able to demonstrate a significant difference in taskload under the Mitigated condition (not shown). This result makes perfect sense given that the mitigation strategy performed as predicted and was effective at redistributing the taskload from the Primary to Secondary task periods. In this case, by holding off the delivery of low priority information until the XLI and other gauges indicated sufficient cognitive capacity to uptake and comprehend a message (i.e., during the Secondary task period), Honeywell's scheduler was effective at improving Primary task performance by as much as 500% across both conditions. However, the average improvement in Primary task performance for all fourteen subjects during both high and low workload conditions averaged 26%.

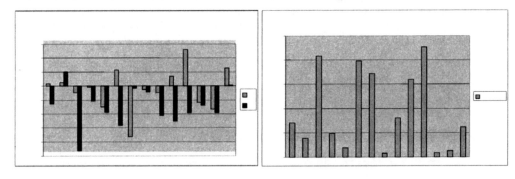

Figure 4a   **Figure 4b**

The graph on the Left (4a) represents the average XLI gauge output for each subject during the Primary (A) and Secondary (B) workload intervals. The task was a continuous performance cognitive battery carried out in a VE simulator at Carnegie Mellon University. The graph on the Right (4b) shows the absolute differences of the XLI averages, used to normalize the data with respect to the polarity of the XLI gauge averages across subjects. Here, significance was measured at an absolute difference $>/ = 0.05$.

Figure 4 shows XLI gauge data averaged over the Primary (high-load) and Secondary (low-load) intervals from all 14 subjects while performing a cognitive task in a VE that consisted of hard and easy workload periods. During the task, subjects were virtually located on a rooftop overlooking buildings on each side and had to identify friendly and enemy soldiers in the structures. Under the high-load conditions (four-1½ minute periods per trial), subjects also had to count friendly soldiers, shoot enemy soldiers, and maintain a count of how many bullets were used. During both high and low conditions, subjects received directions where to look for targets via auditory communications. For the low-load condition, subjects did not see any new targets (enemy or friendly) and had to report their current counts of friendlies, enemies, and ammunition as well as listen for messages with their call sign. The significance of the absolute differences were analyzed using ANOVA to determine the XLI's ability to correctly differentiate Primary from Secondary cognitive task conditions during the complex multimodal VE simulation.

In addition to the XLI gauge analysis, we examined the effectiveness of a simple prototype feed-forward neural network classifier at distinguishing XLI gauge patterns timed to a single discreet-time event when the subject switched from Primary to Secondary task and back. Here, we report results using a real-time neuroclassifier tuned to each individual to identify cognitive task shedding and reacquisition events across subjects. In the past, classification through linear deterministic tools was possible as a first approximation; however, today there are more appropriate non-linear methodologies available for feature selection and pattern recognition. Thus, our approach has been to develop a real-time neural network architecture to manage the highly stochastic EEG phenomena. For this case, our NCA module was configured as a probabilistic feed-forward network to classify the XLI gauge outputs into three classes: 0, which meant no change; 1, which meant a transition to a higher workload state; and -1, which meant a transition to a lower workload state. In real-time operation, the output of the NCA would be returned to the main system architecture to influence mitigation strategies rather than just the XLI values alone.

Training of the NCA, using a 40-second window centered about a critical event (i.e., the high-low workload transition event), was required for each subject and condition prior to classification. In all cases, the first half of the XLI gauge data was used for training and the remaining half for classification. Figure 5 shows the results from one session, where the first half of the data (3 events) was used to create pattern templates and train the NCA to classify workload transition events in the remaining data. In the figure below, the bottom half shows the events generated from the closed-loop system that marked the workload transitions from high to low and visa versa. The graph on top shows that the classifications during the training period were all correct (Left side without box) when compared with the event marks below. Results of the NCA classifications can be seen in the boxed area on the Right of the top graph. Here, based on only 3 prior events, the NCA correctly identified (within 1 XLI sample) the occurrence of 3 transition events, missed 2 events, and added 2 events that didn't exist. Surprisingly, these are quite good results given such a small training set of examples, indicating that individual XLI gauge values about the transition events are repeatable and stable over 2-second intervals.

Preliminary tests of fourteen male and female subjects using our NCA algorithm to classify XLI gauge outputs demonstrated moment-to-moment within-subject performance as high as 72% (mean=64%, std dev=6%), under both Mitigated and Unmitigated conditions over 52 experimental sessions (Figures 6 & 7). The goal of our investigation was to test a real-time application of the XLI as input to our NCA algorithms for neurometric classifications. Here, we report our results using discreet XLI gauge values as input to a neuroclassifier for identifying task switching in the brain patterns of a subject performing a complex continuous performance task in a virtual environment over multiple sessions lasting for 12-minutes.

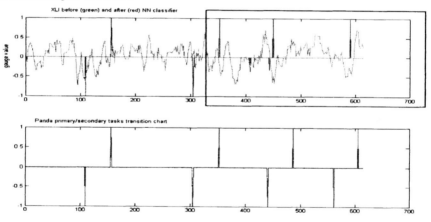

Figure 5
The top figure shows XLI gauge outputs over a 12-minute trial where workload conditions varied between 2 difficulty levels (high & low). The period of each difficulty level was 1-1/2 minutes long and there were 4 periods of each level throughout all trials (total of 6-minutes of data for each workload condition). The bottom figure shows the transition event marks generated by the VE task system. Figures 6 and 7 provide the results from our investigation using a first-stage prototype XLI/NCA to identify an individual's point of cognitive transition from a high workload state to a lower state, or, the other way around.

These results, measured during normal test performance (unmitigated) and when communications were deferred (mitigated), demonstrate that repetitive SF patterns exist in the brain that are associated with higher-order cognitive processes during performance of a routine operational task. Thus, through the application of a moment-to-moment cognitive index that measures a small band of power differences from a few EEG sensors and a novel feed-forward neural network, real-time SF patterns can now be used for identifying cognitive changes in the human-system. By improving the accuracy and sensitivity of these methods, new augmented human-machine systems will soon become prevalent throughout our society.

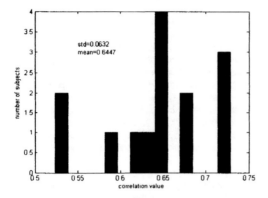

Figure 6
Shows workload classification results using the NCA during an <u>Unmitigated</u> workload session from 14 subjects. Note all levels above 50%

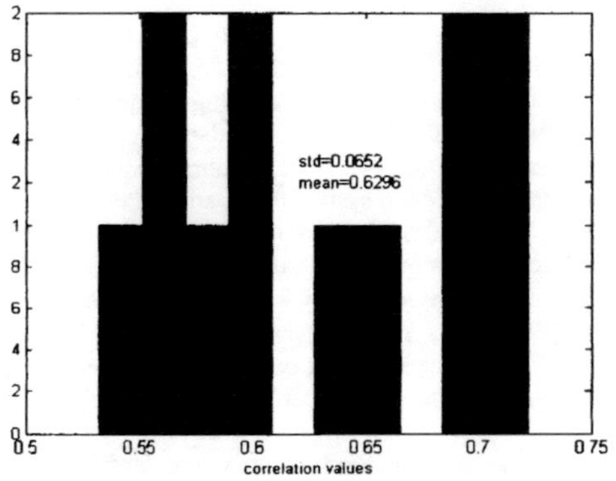

Figure 7
Shows NCA classification results from 12 subjects when communications were <u>Mitigated</u> based on real-time gauge values.

4. Conclusion

Neuroimaging research has expanded our understanding of the brain's cognitive neurodynamics in terms of the variability in information processing, situational awareness, and cognitive capacity. Functional imaging studies have identified several foci for higher-order cognitive processing in the brain and associated cortical networks have been categorized anatomically and functionally using several imaging modalities. Additionally, SF models of cognitive networks have proved useful in expanding our understanding of the interrelationships among brain regions responsible for specific cognitive tasks. From our study of the XLI gauge and NCA, we implemented and tested real-time tools for measuring brain patterns capable of tracking changes in a person's cognitive workload state over extended periods. Additionally, with the NCA, the method worked across all subjects and was able to learn subject specific patterns in XLI gauge outputs sufficient to achieve classification accuracies greater than or equal to 70% across conditions. Training the NCA required very little data to achieve a modest classification success, thus, with more data for longer training periods, improved methods could be implemented to significantly enhance the classification performance of our XLI/NCA technology.

For the XLI alone, given that our analysis used values summed over each trial rather than examining the moment-to-moment gauge changes, we were only able to report the gauge's ability to classify the Primary and Secondary workload conditions over the entire trial. Therefore, we were unable to measure the effectiveness of the gauge at classifying workload over short periods of 1 to 2-seconds. However, that limitation was overcome by the use of the NCA, which successfully identified SF gauge patterns that mirrored individual changes in the executive communication network corresponding to how hard the brain works over periods of 2-seconds. These results, although preliminary, are promising in that they indicate that our XLI/NCA method is capable of distinguishing between high and low levels of cognitive workload over extended periods, while performing complex operationally relevant tasks.

Acknowledgement

Human Bionics recognizes DARPA, Honeywell Laboratories, and particularly, the Institute of Human Machine Cognition and Carnegie Mellon University in support of this R&D effort. This work was supported by contract number W31P4QCR223 through the Information Processing Technology Office and the U.S. Army. This acknowledgement should in no way reflect the position or policy of the U.S. Government and no official endorsement is inferred.

References

[1] Vedantam, S., Report: Monkeys' Thoughts Move Cursor. The Washington Post, March 14, 2002.

[2] McKenna, P. McMillan, G., (2001) Researchers harness brain power to control jets. Report published online by USAF Armstrong Laboratory's Alternative Control Technology Laboratory, Wright-Patterson AFB, OH.

[3] Makeig, S. Enghoff, S. Jung, T. and Sejnowski, T. (2000) A natural basis for efficient brain-actuated control. *IEEE Trans. on Rehab. Eng.* 8(2):208-11

[4] Rick Weiss. (2004) Mind Over Matter: Brain Waves Guide a Cursor's Path; Biomedical Engineers Create Devices That Turn Thoughts Into Action and Could Help the Paralyzed Move Their Limbs. The Washington Post. Washington, D.C.: Dec 13, 2004. pg. A.08

[5] Dorneich M., & Whitlow S., (2004) Improving Warfighter Information Intake under Stress - *Augmented Cognition* Concept Validation Experiment (CVE) Analysis Report for the Honeywell Team, *DARPA AugCog CVE Phase II Analysis Report,* Contract Number: DAAD16-03-C-0054, 8, June 2004

[6] Levy R., Goldman-Rakic P.S., (2000) Segregation of working memory functions within the dorsolateral prefrontal cortex. *Exp. Brain Res.* 133(1):23-32

[7] Gevins, A.S., Cutillo, B., DuRousseau, D.R., Le j., et al. (1994) High-resolution evoked potential technology for imaging neural networks of cognition. In: Thatcher, R.W., et al. (Eds.) *Functional Neuroimaging: Technical Foundations,* Academic Press, Inc. : Orlando, pp. 223-31

[8] Foong, J. et. al.,(1997) Executive function in multiple sclerosis: The role of frontal lobe pathology. *Brain* 120:15-26

[9] Jantzen K.J., Fuchs A., Mayville J.M., Deecke L., Kelso J.A., (2001) Neuromagnetic activity in alpha and beta bands reflect learning-induced increases in coordinative stability. *Clin. Neurophysiology* 112(9):1685-97

[10] Rowe J., Friston K., Frackowiak R., Passingham R., (2002) Attention to action: specific modulation of corticocortical interactions in humans. *Neuroimaging* 17(2):988

[11] Compte A., Brunel N., Goldman-Rakic P.S., Wang X.J., (2000) Synaptic mechanisms and network dynamics underlying spatial working memory in a cortical network model. *Cereb. Cortex* 10(9):910-23

[12] Kremper A., Schanze T., Eckhorn R., (2002) Classification of neural signals by a generalized correlation classifier based on radial basis functions. *Jour. of Neuroscience Methods* 116(2002):179-87

[13] American EEG Society, (1991) Guidelines for standard electrode position nomenclature. *J. Clin. Neurophysiol.* 8(1991):200-2

Cortical Activity of Soldiers During Shooting as a Function of Varied Task Demand

Scott E. Kerick

US Army Research Laboratory
Human Research and Engineering Directorate
Aberdeen Proving Ground, MD 21005
skerick@arl.army.mil

Abstract

The purpose of this research was to investigate dynamic cortical processes of soldiers (17 US Marines, 1 US Army Ranger; age 19.8 ± 4.2) during simulated shooting scenarios as a function of task demand. Task demand was varied among three two-level factors: task load (single, dual), decision load (no-decision, choice-decision), and target exposure time (short [2-4 s], long [4-6 s]). Target shooting scenarios were simulated using the Dismounted Infantryman Survivability and Lethality Testbed (DISALT), which controlled target presentation and exposure times and recorded weapon aim point data. Electroencephalographic (EEG) recordings were acquired continuously from 36 channels during all scenarios and single-trial data from each scenario were epoched offline into 11-s windows time-locked to the onsets of targets (-5 s to +6 s). Event-related spectral perturbation (ERSP) was derived and peak amplitude and latency measures were analyzed in the theta frequency band (4-7 Hz) for this preliminary report. The results revealed that different oscillatory EEG patterns were sensitive to different shooting task events and task load variations. Specifically, two new findings were revealed in the present study with respect to existing marksmanship research: (1) an early theta peak occurred approximately one second after the onset of targets that was sensitive to decision and task load demands but not to target exposure time demands and (2) a late theta peak occurred approximately at the time of trigger-pull that was sensitive to decision load and target exposure time demands but not to task load demands.

1   Introduction

A major objective of the US Army's Future Force Warrior program is to enhance situational awareness of soldiers by promoting technology to communicate real-time battlefield information. More than ever before, soldiers are required to perform in dynamic, complex, and stressful environments and they are faced with increasing task demands such as maintaining situational awareness, monitoring communications, and discriminating among enemy and friendly targets. Given a limited capacity for human information processing and the increasing demands imposed on soldiers to consolidate information from multiple sources, a need exists to better understand how increased cognitive load affects soldier shooting performance. The question is not whether imposing increased task demands consumes attentional resources or whether they may interfere with certain aspects of soldier performance, depending on the nature and complexity of the tasks and the present situation (Scribner, 2002; Scribner & Harper, 2001), but how the brain functions to accommodate multi-task demands. For example, how does the brain organize and integrate sensory information, cognitive operations, and motor planning during the performance of multiple complex tasks?

Although this question has not been previously addressed in the shooting performance literature, the cortical dynamics of skilled and unskilled marksmen have been investigated using real-time electroencephalographic (EEG) recordings. One consistent observation from this research is greater left temporal alpha (8-13 Hz) power in higher-skilled marksmen over the few seconds preceding the trigger pull (Hatfield, Landers, & Ray, 1984; Haufler, Spalding, Santa Maria, & Hatfield, 2000; Kerick, Douglass, & Hatfield, 2004; Kerick et al., 2001). In general, increases in alpha power reflect synchronous oscillations of underlying cortical neurons and is associated with inhibitory or resting brain states (Pfurtscheller, 1992; Pfurtscheller & Lopes da Silva, 1999). Thus, the robust finding of alpha synchronization in more highly skilled marksmen provides evidence in support of an efficiency hypothesis (Hatfield & Hillman, 2001) purporting more highly organized sensorimotor integration and less cognitive effort as skill levels advance from controlled to automatic.

However, previous research has not examined cortical dynamics of soldiers performing secondary tasks while shooting. Further, previous EEG research on marksmen has been limited primarily to the analysis of changes in alpha power during the few seconds preceding the trigger-pull using competition-style shooting paradigms in which the shooter engages a stationary target following a self-paced preparatory aiming period. A more realistic paradigm for investigating the cortical dynamics of soldiers is to involve an externally-paced shooting task in which the shooter must react to targets appearing in the environment, correctly identify them as enemy or friendly, and make a decision whether to aim and fire the weapon or disengage the target. In such a paradigm, event-related changes in the theta frequency band (4-7 Hz) are of particular interest because theta oscillations reflect the encoding of sensory information and they typically increase with attentional demand, task difficulty, and cognitive load (Klimesch, 1996; 1999). Accordingly, this study was designed to investigate cortical dynamics of soldiers during reactive target shooting scenarios with varied task demands. New signal processing techniques were employed to minimize movement artifacts and extract event-related theta perturbations associated with encoding target stimuli and planning responses (shoot or don't shoot). Independent component analysis (ICA) was first applied to separate artifact sources from brain sources of single-trial EEG epochs time-locked to the onsets of targets (Jung et al., 2000). Event-related spectral perturbation (ERSP; Makeig 1993; Makeig, Debner, Onton, & Delorme, 2004) was then computed from the ICA-filtered epochs to quantify the dynamic changes in EEG spectra. It was predicted that the demands of shooting would be reflected by increased 'cortical effort' as indicated by theta power increases exhibiting higher amplitude peaks and longer peak latencies during higher task demand conditions. The results of this research will contribute to our understanding of the functional organization of the brain and may have implications for human systems engineers in the design of closed-loop brain-computer interface systems that support Future Force Warrior objectives of the US Army (e.g., Augmented Cognition). Potentially, neural networks can be trained based on distinct input features exhibited by ERSP and ICA in response to changes in the functional states of the operator and used for adaptive aiding (e.g., see Wilson & Russell, 2004).

2   Methods

2.1   Participants

The participants were 18 male soldiers (17 US Marines, 1 US Army Ranger; age 19.8 ± 4.2). Each read and signed an informed consent form approved by the Human Use Committee of the US Army Research Laboratory, Human Research and Engineering Directorate, and provided demographic information. The participants also completed color vision and visual acuity tests prior to experimental testing (Titmus II, Petersburg, VA) ensuring that all had color vision and 20/30 or better visual acuity.

2.2   Instrumentation

The Dismounted Infantryman Survivability and Lethality Testbed (NAVAIR Training Systems Division, Orlando, FL; NAVAIR TSD ORL) simulated an outdoor shooting range at Aberdeen Proving Ground (i.e., M-Range), controlled the presentation of pop-up targets (target type, location, time of onset, and exposure time), and recorded weapon aim point data and shot results from a demilitarized M16A2 rifle equipped with an infrared emitting diode and collimator lens (see Figure 1). Weapon recoil was simulated using an electromechanical recoil subsystem and weapon sounds were simulated via digital audio surround-sound. A digital terminal board (DT9835) transmitted event markers (type of target, onset and offset times of targets, trigger-pulls, target hits and misses, onset and offset times of secondary task, and onset times of secondary task responses) via parallel cable connection and binary coding from the DISALT system to the EEG recording system. Continuous EEG recordings were acquired (Neuroscan, ver. 4.3; El Paso, TX) during all scenarios from 36 standard scalp locations referenced to the left ear electrode (A1) with FPz as ground and re-referenced offline using an averaged-ears derivation (A1, A2). Electrode impedances were maintained below 10 KOhms. Raw signals were sampled at 500 Hz and amplified 20,000 times. Eye movements were monitored by electro-oculogram (EOG) using bipolar montages attached superior and inferior to the right eye (VEOG) and both orbital fossa (HEOG).

Figure 1. A soldier shooting with the DISALT system while wearing EEG cap.

2.3 Procedures

The soldiers were provided an overview of the study including an explanation of EEG recording procedures and then allowed shooting practice with the DISALT system during the morning hours (0900-1100) on the day of experimental testing. During this period, they completed the scale development phase of the Subjective Workload Assessment Technique (SWAT; Reid & Nygren, 1988), zeroed the weapon, and practiced shooting (four blocks of 36 trials). Following practice and a 1-hr break (1100-1200), they completed eight different shooting task scenarios and an arithmetic-only task (i.e., non-shooting scenario) while EEG was continuously recorded. For the shooting scenarios, task demand was manipulated among three two-level factors: task load (single, dual), decision load (no-decision, choice-decision), and target exposure time (short [2-4 s], long [4-6 s]). Each shooting scenario consisted of 36 targets that were distributed randomly across each of 18 range locations on two occurrences ([left, center, right] x [50, 100, 150, 200, 250, 300 m]) and exposed at variable intervals (10 s ± 2 s). For the task load manipulation, single-task scenarios required the soldiers to shoot with no secondary task, whereas dual-task scenarios required them to shoot while simultaneously listening to and solving arithmetic problems (n = 30) that were presented verbally via digital audio player at variable intervals (2 s ± 1 s) before the onsets of all but six targets, which served as catch-trials. The secondary arithmetic task consisted of addition problems involving two double-digit numbers with a carry operation on the ones column always required (e.g., 54 + 17). Instructions were to verbally answer the problems as quickly and accurately as possible while maintaining optimal shooting performance. For the decision load manipulation, no-decision scenarios consisted of all enemy targets (n = 36), whereas choice-decision scenarios consisted of both enemy (n = 18) and friendly (n = 18) targets. Either an enemy (brown) or friendly (olive) target was exposed on any given trial with equal probability. Instructions were to correctly identify targets and to shoot enemy but not friendly targets. For the target exposure time manipulation, targets remained exposed for variable durations depending on the target exposure time factor (short [2-4 s], long [4-6 s]). For the arithmetic-only task, the soldiers were required only to verbally answer arithmetic problems (n = 30) as quickly and accurately as possible. Subjective reports of workload (SWAT) and stress (Subjective Ratings of Events, SRE) were recorded pre-experimentally and immediately following the completion of each shooting scenario and arithmetic-only task.

2.4 EEG Signal Processing and Data Reduction

Continuous EEG recordings were epoched into 11-s windows with reference to target onset events (-5000 ms before to +6000 ms after target onsets [0 ms]) of all trials for each subject in each shooting task scenario. The epoched data were digitally band pass filtered (.1 to 30 Hz) and then visually inspected for artifacts (single-trial epochs with voltages exceeding ±100 μV were rejected). Independent component analysis (ICA) was applied to all accepted data epochs to derive spatial filters by blind source separation of the EEG signals into temporally independent and spatially fixed components (Jung et al., 2000; 2001; EEGLAB toolbox available at http://www.sccn.ucsd.edu). Independent components (ICs) resembling eye-blink or muscle artifact were removed and the non-artifactual components were then back-projected onto the scalp electrodes by multiplying the input data by the inverse of the spatial filter coefficients. When applicable, missing-channel data were interpolated for all accepted trials of ICA-filtered data using thin-plate smoothing splines (Matlab *tpaps* function). Event-related spectral perturbation (ERSP; Makeig, 1993; Makeig et al., 2002; 2004) was derived in the theta (4-7 Hz) frequency band. Log-spectral estimates were obtained using multi-taper decomposition comprising discrete prolate spheroidal sequences with temporal resolution of 44.98 ms and frequency resolution of 0.37 Hz, yielding a power-time curve with 200 spectral estimates from -3976 ms to +4976 ms. Early (0-1500 ms) and late (1500-6000 ms) peak amplitude and latency values were then derived from the trial-averaged ERSP data for each electrode of each subject in each condition. Data from nine electrode sites (Fz, C3, Cz, C4, P3, Pz, P4, T7, and T8) were subjected to statistical analysis.

2.5   Design and Analysis

The nine different sequences of the nine scenarios were counterbalanced for order and a nine period by nine treatment crossover design was used to analyze primary task performance (decision accuracy, shooting accuracy, and shooting response time), secondary task performance (arithmetic accuracy and arithmetic response time), and subjective ratings (workload and stress) with sequence and order included in the model (analyses of subjective and performance data are not reported here). For ERSP analyses, an additional electrode site factor (9) was included and data from the arithmetic-only condition were not analyzed because targets were not presented during the arithmetic-only task and, therefore, did not have event markers in common with the shooting scenarios (this will be done in a separate report by epoching around the offset times of arithmetic problems in the arithmetic-only condition and in dual-task scenarios).

3   Results

The present paradigm allows both exploratory analysis of stimulus-related cortical dynamics and confirmatory analysis of response-related cortical dynamics by virtue of the multiple event-markers recorded. Precedence from the previous research on marksmen during self-paced shooting at stationary targets is to examine spectral changes time-locked to the trigger-pull. However, because this study is the first to implement a reactive shooting task using pop-up targets, exploratory analyses were conducted on ERSP data time-locked to the onsets of targets rather than to the trigger-pull. Further, observation of the data revealed that activity in the theta band was most responsive to target stimuli and therefore was of exploratory interest. Across all shooting scenarios, theta ERSP exhibited two distinct peaks. An early stimulus-related peak (Grand Mean = 1126 ± 29 ms) occurred after the onset of targets that was maximal over the mid-parietal region and a later response-related peak (Grand Mean = 3014 ± 73 ms) occurred coincident with the trigger-pull (Grand Mean = 3014 ± 50 ms) that was maximal over bilateral temporal regions. Statistical analyses were conducted on both early and late peak amplitude and latency values using mixed linear models with significance levels at α = .05 (SPSS ver. 12.0). Follow-up contrasts of significant main effects and interactions were made by least significant difference tests (LSD). Figure 2 provides topographic maps at the times of early and late theta peaks and the averaged ERSP waveforms time-locked to target onsets by task demand main effects.

3.1   ERSP Theta Early Peak

3.1.1   Amplitude

The main effect of site, $F(8, 832) = 8.73$, $p < .01$, and the Site x Decision Load interaction, $F(8, 832) = 2.45$, $p = .01$, were significant for early peak theta amplitude (dB). Follow-up analyses of the interaction revealed that peak amplitudes were higher in the left (P3) and mid (Pz) parietal sites in the choice-decision vs. no-choice decision load scenarios (P3: $M = 2.44$ vs. 2.13, $SE = .37$; Pz: $M = 2.48$ vs. 2.20, $SE = .37$, respectively; see Figure 2A).

3.1.2 Latency

A significant main effect of site, F(8, 832) = 6.82, p < .01, and the Site x Task Load interaction, $F(8, 832) = 2.06$, $p = .04$, were significant for early peak theta latency (ms). Follow-up analyses of the interaction revealed that peak latencies were longer in the right central (C4) and temporal (T8) sites in the dual- versus single-task load scenarios (C4: $M = 1149$ vs. 1065, $SE = 39$; T8: $M = 1196$ vs. 1080, $SE = 39$, respectively; see Figure 2B). All other main effects and interactions were nonsignificant.

3.2   ERSP Theta Late Peak

3.2.1   Amplitude

Significant main effects of decision load, $F(1, 83) = 28.09$, $p < .01$, and site, $F(8, 832) = 35.01$, $p < .01$, were observed for late peak theta amplitude (dB). For the decision load main effect, peak amplitude was higher in the no-decision vs. choice-decision ($M = 4.02$ vs. 2.72, $SE = .61$, respectively) load scenarios (see Figure 2A). For the site main effect, peak amplitudes were higher in the temporal sites (T7, T8) vs. all other sites (Fz, C3, Cz, C4, P3, Pz, P4) and higher in the right (T8) vs. left (T7) temporal site ($M = 4.28$ vs. $M = 4.10$, $SE = .60$, respectively). All other main effects and interactions were nonsignificant.

3.2.2   Latency

The main effects of decision load, $F(1, 83) = 9.61$, $p < .01$, and target exposure time, F(1, 83) = 119.92, p < .01, were significant for late peak theta latency (ms). Latencies were shorter in the no-decision vs. choice-decision load scenarios ($M = 2968$ vs. 3127, $SE = 56$, respectively; see Figure 2A) and in the short vs. long target exposure time scenarios ($M = 2767$ vs. 3328, $SE = 56$, respectively; see Figure 2C). The Decision Load x Target Exposure Time, $F(1, 83) = 6.13$, $p = .02$, interaction was also significant. Latencies were shorter in no-decision vs. choice-decision load scenarios ($M = 2624$ vs. 2910, $SE = 67$, respectively) with short target exposure times but no differences were observed with long target exposure times ($M = 3313$ vs. 3344, $SE = 67$, respectively). All other main effects and interactions were nonsignificant.

A. Decision Load

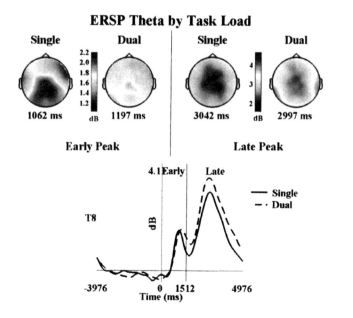

ERSP Theta by Decision Load

No-Decision Choice-Decision No-Decision Choice-Decision

2.2
2.0
1.8
1.6
1.4
1.2

1107 ms   dB   1017 ms   2997 ms   dB   3042 ms

Early Peak   **Late Peak**

4.1 Early   Late

—— No-Decision
—·· Choice-Decision

PZ   dB

-3976   0   1512   4976
       Time (ms)

B. Task Load

ERSP Theta by Task Load

Single   Dual   Single   Dual

2.2
2.0
1.8
1.6
1.4
1.2

1062 ms   dB   1197 ms   3042 ms   dB   2997 ms

Early Peak   **Late Peak**

4.1 Early   Late

—— Single
—·· Dual

T8   dB

-3976   0   1512   4976
       Time (ms)

257

C. Target Exposure Time

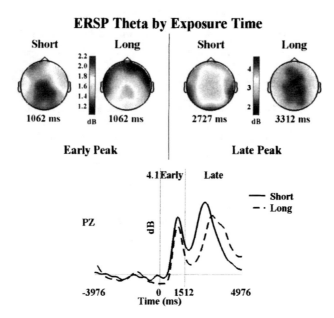

Figure 2. Topographic maps at the times of early (upper left panels) and late (upper right panels) ERSP theta peaks and ERSP theta waveforms time-locked to target onsets (lower panels) illustrating differences among main effects for (A) Decision Load, (B) Task Load, and (C) Target Exposure Time. Note: ERSP plots consist of data from single electrode sites (labelled on the left), where amplitude (dB) or latency (ms) differences were maximal, averaged across subjects, trials, and shooting scenarios. Topographic maps are interpolated from 36 electrode sites. Note also the same amplitude scales within topographic maps of early and late peaks but different amplitude scales between topographic maps of early and late peaks.

4   Discussion

The main objective of this study was to examine differences in dynamic cortical processes of soldiers performing in reactive target shooting scenarios as a function of task load variations. No previous research examining the cortical dynamics of marksmen has employed shooting tasks that require the shooter to search for and identify targets in the environment, to make a decision whether or not to shoot based on visual discrimination of targets, or to perform secondary cognitive tasks while engaged in the shooting task. Time-locking the EEG to target onset events and also providing trigger-pull event markers allowed for exploratory analysis of stimulus-related cortical dynamics which have not been previously investigated, as well as confirmatory analysis of previously reported response-related cortical dynamics (not reported here).

Two new findings were revealed in the present study with respect to existing marksmanship research. First, an early theta peak occurred approximately one second after the onset of targets that was sensitive to decision and task load demands but not to target exposure time demands. Specifically, its amplitude was higher for choice-decision load scenarios in the mid and left parietal regions with no differences in latency, whereas the latency was longer for dual-task load scenarios in the right central and temporal regions with no differences in amplitude. Neither amplitude nor latency of the early theta peak differed between target exposure time scenarios. Second, a late theta peak occurred approximately at the time of trigger-pull that was sensitive to decision load and target exposure time demands but not to task load demands. Specifically, its amplitude was higher in the temporal regions and its latency was globally shorter for no-decision load demands and for short target exposure time demands.

Based on the finding that the amplitude of the early theta peak was higher for the more complex choice-decision load scenarios in the parietal region it appears that theta oscillations in that region at that time are involved in sensorimotor integration related to encoding information about the target stimuli (i.e., encoding specific stimulus features to discriminate among enemy and friendly targets and make a decision whether or not to shoot). Based on the finding that the latency of the early theta peak was longer for the more difficult dual-task load scenarios in the right central and temporal regions it appears that theta oscillations in those regions at that time are influenced by processing of the secondary task at the moment targets appear and imposes a delay on stimulus encoding. Based on the finding that neither amplitude nor latency varied by target exposure time demands it appears that the early theta peak is not related to response timing mechanisms. However, based on the findings that the late theta peak latency was longer for the dual-task demand scenarios and for the long target exposure time scenarios, and that for both main effects the longer latencies coincided with the longer shooting response times, it appears that the late theta peak reflects response timing mechanisms which are influenced by secondary task load demands and by the temporal demands of the primary task. Such interpretations are consistent with those offered by other researchers that the functional significance of increased theta oscillations is in modulating the frequency of action potentials for encoding sensory information and reflects basic mechanisms of information transmission in the cortex (Klimesch, 1996; Lopes da Silva, 1991). According to Lopes da Silva (1991), a change of state in a neuronal network is reflected by a transition from random activity to an oscillatory mode.

In summary, the early and late theta peaks differed temporally and spatially and were sensitive to different stimulus and response events of the shooting task and to different task load variations. Therefore, it can be inferred that different sources of theta oscillations reflect different cortical functions during reactive shooting tasks involving secondary task demands. These data will be further analyzed in a subsequent report to verify by source localization whether the early and late theta peaks arise from different cortical structures and by regression onto shooting performance data to verify whether a functional relation exists between either amplitude or latency and shooting response times or accuracy. Future research is also needed to determine whether deriving ERSP theta amplitude and latency in real-time provides salient input for classification of functional states using a neural networks approach.

References

Hatfield, B. D. & Hillman, C. H. (2001). The psychophysiology of sport: A mechanistic understanding of the psychology of superior performance. In R. S. Singer, H. A. Hausenblaus, & Janelle, C. M. (Eds.), *Handbook of Research on Sport Psychology (2nd Ed.)*, (pp. 362-388). New York: John Wiley.

Hatfield, B. D., Landers, D. M., & Ray, W. J. (1984). Cognitive processes during self-paced motor performance: An electroencephalographic profile of skilled marksmen. *Journal of Sport Psychology 6*, 42-59.

Haufler, A. J., Spalding, T. W., Santa Maria, D. L., & Hatfield, B. D. (2000). Neurocognitive activity during a self-paced visuospatial task: Comparative EEG profiles in marksmen and novice shooters. *Biological Psychology, 53*, 131-160.

Jung, T. P., Makeig, S., Humphries, C., Lee, T. W., McKeown, M. J., Iragui, V., & Sejnowski, T. J. (2000). Removing electroencephalographic artifacts by blind source separation. *Psychophysiology, 37*, 163-178.

Jung, T. P., Makeig, S., Westerfield, M., Townshend, J., Courchesne, E., & Sejnowski, T. J. (2001). Analysis and visualization of single-trial event-related potentials. *Human Brain Mapping, 14*, 166-185.

Kerick, S. E., Douglass, L. W., & Hatfield, B. D. (2004). Neurocognitive adaptations associated with marksmanship training. *Medicine and Science in Sports and Exercise, 36*, 118-129.

Kerick, S. E., McDowell, K., Hung, T. M., Santa Maria, D. L., Spalding, T. W., & Hatfield, B. D. (2001). The role of the left temporal region under the cognitive motor demands of shooting in skilled marksmen. *Biological Psychology 58*, 263-277.

Klimesch, W. (1996). Memory processes, brain oscillations, and EEG synchronization. *International Journal of Psychophysiology, 24*, 61-100.

Klimesch, W. (1999). EEG alpha and theta oscillations reflect cognitive and memory performance: A review and analysis. *Brain Research Reviews, 29*, 169-195.

Lopes de Silva, F. H. (1991). Neural mechanisms underlying brain waves: From neural membranes to networks. *Electroencephalography and Clinical Neurophysiology, 79*, 81-93.

Makeig, S. (1993). Auditory event-related dynamics of the EEG spectrum and effects of exposure to tones. *Electroencephalography and Clinical Neurophysiology, 86*, 283-293.

Makeig, S., Debener, S., Onton, J., & Delorme, A. (2004). Mining event-related brain dynamics. *Trends in Cognitive Science, 8*, 204-210.

Makeig, S., Westerfield, M., Jung, T. P., Enghoff, S., Townsend, J., Courchesne, E. (2002). Dynamic brain sources of visual evoked responses. *Science, 295*, 690-694.

Pfurtscheller, G. (1992). Event-related synchronization (ERS): An electrophysiological correlate of cortical areas at rest. *Electroencephalography and Clinical Neurophysiology 83*, 62-69.

Pfurtscheller, G. & Lopes da Silva, F. H. L. (1999). Event-related EEG/MEG synchronization and desynchronization: basic principles. *Clinical Neurophysiology, 110*, 1842-1857.

Reid, G. B. & Nygren, T. E. (1988). The subjective workload assessment technique: A scaling procedure for measuring mental workload. In P. A. Hancock & N. Meshkati (Eds.), *Human Mental Workload* (pp. 185-218). Holland: Elsevier Science Publishers.

Scribner, D. R. (2002). The effect of cognitive load and target characteristics on soldier shooting performance and friendly targets engaged. *(Technical Report ARL-TR-2838)*. Aberdeen Proving Ground, MD: U.S. Army Research Laboratory.

Scribner, D. R. & Harper, W. H. (2001). The effects of mental workload: Soldier shooting and secondary cognitive task performance. *(Technical Report ARL-TR-2525)*. Aberdeen Proving Ground, MD: U.S. Army Research Laboratory.

Wilson, G. F. & Russell, C. A. (2004). Real-time assessment mental workload using psychophysiological measures and artificial neural networks. *Human Factors, 45*, 635-643.

Author Note

This research was partially funded by the Postdoctoral Research Associateship Program of the National Research Council, National Academy of Sciences, 500 5th Street NW, Washington, DC 20001.

Section 2
Cognitive State Sensors

Chapter 7

Psycho-Physiological Sensor Technologies

Psycho-Physiological Sensor Techniques: An Overview

Robert Matthews

Quantum Applied Science and
Research
5764 Pacific Center Blvd., #107,
San Diego, CA, 92121
robm@quasarusa.com

Neil J. McDonald

Quantum Applied Science and
Research
5764 Pacific Center Blvd., #107,
San Diego, CA, 92121
neil@quasarusa.com

Leonard J. Trejo

Pacific Development and
Technology, LLC
1130 Independence Ave. Ste B
Mountain View, CA 94043
ltrejo@pacdel.com

Abstract

Under the auspices of the Defense Advanced Research Projects Agency (DARPA), the Augmented Cognition (AugCog) program is striving to realize the unambiguous determination of the cognitive state via physiological measurements. The basis of this program is the assumption that as computing power continues to increase, the exchange of information between computers and their human users is fundamentally limited by human information processing capabilities, particularly when the users are fatigued or placed under stressful conditions such as those present in warfare command environments. Knowledge of the cognitive state will enable the development of a human-computer interface that adapts to optimize the processing of information by the user.

At present there is no direct measure of a subject's cognitive state. However, to infer the cognitive state it is possible to use psycho-physiological techniques, in which changes in physiological signals that are affected by the cognitive state (e.g. electroencephalographic signals, variations in the heart rate or blood flow in the brain) are measured using a suite of bioelectric and biophysical sensors, and then processed using sophisticated algorithms based upon theoretical descriptions of the relationship between the cognitive state and the relevant physiological signals.

This paper will provide an overview of current psycho-physiological related research, quoting recent results from groups working in the AugCog program and elsewhere. The latest advancements in sensor technologies will be reviewed in an effort to identify those physiological sensors that are most useful in determining the cognitive state, with particular attention paid to the quality of information provided by the sensor and design elements such as the degree of comfort for the human test subject, sensor power requirements, ease of use, and cost. The compatibility of each sensor type with other technologies will also be assessed, and a psycho-physiological integrated sensor suite that incorporates those sensors identified as being best suited to the determination of the cognitive state will be proposed.

1 Introduction

The AugCog program is an ambitious research program initiated by DARPA with the aim of maximizing human cognitive capabilities. Specifically, it is part of a larger thrust at DARPA to address the complex problems facing the military. The 1992 DoD Key Technologies Plan (Director of Defense Research and Engineering, 1992) listed development of advanced human-system interfaces as one of eleven key DoD technologies and stated that, "Future systems... will adapt to the performance and physiological state of the human" (pp. 11-12). The plan set 2000 as a target date for the development of operational "vigilance monitors," and projected "unobtrusive real-time measures" to become available by 2005. This prediction has come true, with several real-time measures being commercially available.

The AugCog program is a multidisciplinary approach that encompasses measurements of physiological signals, the development of sophisticated algorithms relating the cognitive state to electrical activity in the brain, and an understanding of the nature of dynamic models for augmented cognition. In many situations, the exchange of information between computers and their human users is fundamentally limited by human information processing capabilities. In fact, most of the problems in human performance arise when there is a mismatch between the demands of the task and the internal states of the human.

The cognitive performance of an individual as a function of workload can be represented by the curve in Figure 1. At low workloads, subjects tend to lack vigilance and become inattentive, whereas at high workloads their information processing capabilities are exceeded. In today's complex systems, it is primarily at high workloads where human errors due to cognitive failures occur.

Unlike other cognitive science projects, the aim of the AugCog program is the transition of unobtrusive, real-time measures of the cognitive state from the lab to a variety of hostile environments. Knowledge of the cognitive state then enables the manipulation of information flow in a human-system interface such that the user's cognitive performance is maintained at the peak of the curve in Figure 1. This has become possible as a result of enormous advances in biomedical signal processing and cognitive neurosciences in the past decade. It is now possible to conceive revolutionary new human-computer interfaces, in which the mental state of the user is included and the flow of information is adjusted in order to optimize the processing of information by the user.

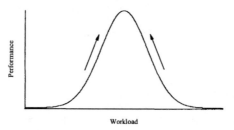

Figure 1: Cognitive performance as function of workload.

2 Cognitive State Detection

In this section we briefly describe how the cognitive state is determined from physiological measurements. There is at present no direct measure of a subject's cognitive state. Indeed, the term *cognitive state* lacks a single accepted definition. It is a combination of the cognitive workload, which describes the allocation of cognitive resources relative to the demands of a particular task (i.e. underloaded/overloaded), the level of arousal (sleepy, drowsy, alert, excited) of a subject, and the subject's degree of vigilance (attentive/inattentive). It is not, however, the purpose of the AugCog program to seek a fundamental definition of the cognitive state, but rather an operational or working description of the cognitive state.

To aid in operational matching of cognitive resources and task demands, the AugCog program is developing *cognitive gauges,* which relate physiological measurements and the cognitive state of an individual through sensors and computational algorithms. Cognitive gauges are made possible by systematic relationships between nervous system activity and the dimensions of cognitive state, and are validated with experimental data and statistical inferences. With reliable cognitive gauges, it is possible to develop real-time advisory systems that detect and prevent impending cognitive dysfunction and serve to maximize an individual's cognitive performance. Detection of the cognitive state involves a sensor suite in which each sensor group measures an objective physiological parameter that indicates an aspect of cognitive function, arousal, or stress. For example, the electroencephalogram or EEG is a spatio-temporal record of brain activity that arises from summed electrical potentials of neurons, primarily in the neocortex. EEG is usually spectrally decomposed into frequency bands, which can be linked to regional sources of brain activity.

However, little is known about the efficacy of cognitive gauges in physically demanding settings. AugCog requires that each cognitive gauge be unobtrusive, which rules out a number of sensitive measurement techniques because of their reliance upon the use of large-scale technologies (e.g. magnetic resonance imaging (MRI), magneto-encephalography (MEG), and high-density, whole-head electroencephalogram (EEG)).

3 Determination of the Cognitive State

3.1 Physiological Manifestation of the Cognitive State

The cognitive state reflects electrical activity in the brain. Cognitive gauges either measure brain activity directly or relate the cognitive state to various measures derived from activity in the central and peripheral nervous systems. The release of hormones by the endocrine system affects and is affected by brain activity. Similarly, the sensory-somatic nervous system, which is mainly concerned with the senses of the body, mediates between the external environment and an individual's cognitive state, Alternatively, the autonomic nervous system maintains homeostasis in the functioning of many organs in the body, such as the lungs and heart.

Although the precise origin of brain activity and the functions that it serves are not currently known, the electric potentials due to activity in neural tissue in the cerebral cortex can be measured directly by the insertion of microelectrodes into brain tissue. This process is highly invasive and therefore is inappropriate for use as a cognitive gauge in the AugCog program. Noninvasive measurements of the neural activity are possible through EEG and MEG measurements. EEG measures the brain's electric field at the scalp, which is spatially diffused by the insulating skull material. MEG is a complementary measurement of the magnetic field due to currents flowing in the brain. Both techniques are sensitive to the activity of a large number of neurons, providing a spatio-temporal record of brain activity that is indicative of the cognitive state.

Brain activity is also reflected in regional cerebral blood flow, which increases in proportion to the metabolic demands of active neurons. Positron emission tomography (PET), function magnetic resonance imagery (fMRI), and function near infrared (fNIR) measurements all monitor the flow of blood and help to infer the cognitive state.

General somatic activity (GSA), characterized by minute motions of the human body, is a consequence of activity in the sensory somatic nervous system. This can be measured by the pressure applied to a mouse (particularly suited for those activities that involve extended periods of computer use), body movement, and body posture.

Activation of the autonomic nervous system can produce organ-specific responses, such that the response in the heart may be different from that of skin conductance. There have been several measures relating the cognitive state to the cardiovascular system. These are primarily based upon the electrocardiogram (ECG) and include heart rate, heart period variability, high frequency QRS ECG, and respiratory sinus arrhythmia (RSA). Additional techniques for determining autonomic arousal include blood oxygenation, pupillometry, and electrodermal activity (EDA).

The sensor technologies involved in each cognitive gauge is described in more detail in Section 3.2 below.

3.2 Psycho-Physiological Sensor Technologies for the Measurement of the Cognitive State

The following technologies are listed according to the types of physiological response they measure. The most fundamental are measurements of brain activity (the Central Nervous System). Other measurements, relating to processes controlled by the Peripheral System, are secondary effects. Several measurement technologies are grouped together under the cardiovascular system.

3.2.1 Central Nervous System

3.2.1.1 Microelectrode Measurements of Brain Activity
Comments: Highly invasive, risk of infection, degradation of sensitivity with time

Microelectrodes have a diameter approximately the width of a human hair. The observed voltages vary greatly over distances of the order of the diameter of a neuron, demonstrating that they can measure the electric potential due to a single neuron. It has been demonstrated that microelectrodes have enabled monkeys to manipulate robot arms (Taylor, Helms Tillery & Schwartz, 2002). The neuroprosthesis described in Mojarradi, *et al.*, 2003, is an extension of this technology that can collect information from thousands of neurons when implanted in the brain.

The spatial resolution of microelectrodes is only possible through surgically implanting the electrodes into brain tissue, carrying the risk of infection. Neural scarring around the electrodes can also build up over time, resulting in a loss of signal. (The loss of signal is not, however, as pronounced for the Mojarradi neuroprosthesis.) Uncertainty about the effects of long-term use of currently available microelctrode technology makes them unsuitable for use in the AugCog program.

3.2.1.2 Electroencephalography (EEG)
Comments: Large body of data supporting links to cognitive state detection, measurements can be noninvasive

The strength and distribution of currents (and therefore potential) in the scalp reflects the intensity and position of activity in the neural tissue. Measurements of EEG are traditionally made through the application of 'paste-on' electrodes, which make resistive contact to the skin. In order to prevent contamination of the signal by epidermal artifact noise, it is common practice to prepare the scalp by removing the outer skin layer before the application of electrodes. However, the electrodes often experience biocompatibility issues in long-term monitoring, resulting in skin irritation for the subject, which can be exacerbated through preparation of the scalp. QUASAR has developed a sensor that can make high-fidelity measurements of EEG without a resistive connection to the subject, in which case there exists no biocompatibility issue. (Matthews, McDonald, Fridman, Hervieux & Nielsen, 2005).

The principal spectral components of EEG are divided into the following signal bands: delta (0-4 Hz), theta (4-8 Hz), alpha (8-12 Hz), beta (above 12 Hz), and gamma (above 40 Hz). Many studies have related the spectral components of EEG to alertness, cognitive functions, and the overall capacity of the brain to operate within its usual limits. The link between EEG and arousal is well enough accepted that some studies simply define vigilance using EEG and electrooculographic (EOG) criteria (Fruhstorfer & Bergstrom, 1969).

EEG spectral data have been linked with accurate alertness estimation as measured using a visual tracking task (Jung, Makeig, Stillwell & Harm, 2001). Interestingly, these experiments also showed that performance decrements appeared as waves of (usually) intermittent detection lapses 4 minutes or less and containing characteristic 15 – 20s cycles. This supports the assertion that objective measures of human cognitive fatigue or alertness and the impact of this cognitive state on error rates requires a pseudo-continuous measurement.

Recent studies at NASA (Trejo, *et al.*, 2004) demonstrate the utility of EEG for assessing cognitive fatigue, in which the trend across subjects was explained by an alpha-theta model, with both alpha and theta signals being greater in fatigued than in normal states. The utility of EEG is such that most groups that form part of the AugCog program use EEG as the basis for their sensor suite.

3.2.1.3 Event-related Potential (ERP) Measures
Comments: Low signal amplitude, increased burden upon mental resources

Event-related potentials are a transient series of EEG voltage oscillations that occur in response to a discrete event. The event may be an external stimulus or an internal mental event, such as recognizing an omitted stimulus. Unlike EEG, ERP is broken down into a series of time rather than frequency domains because cells in different locations of the brain become active at different times after a stimulus. The main behavioral impact of cognitive fatigue is a slowing of mental processes, as response times trend to become significantly and progressively higher over time. There appears to be little consensus as to what the different components measure, but is seems that the early components represent the delivery of sensory inputs while the latter correspond to some form of comparison with internal models. Although EEG has less temporal resolution than the ERP, both are susceptible to the same artifacts Kramer (1991).

The ERP components with well-known relationships to cognitive states can be summarized as follows:

- The N1 (or N100) component is correlated with the accuracy of detection in vigilance tasks. However, the N1 amplitude may be very small and is usually not well suited to on-line estimation because extensive signal averaging is required to measure it.

- The N2 (or N200) component discriminates between levels of single- and dual-task demands. It correlates with monitoring demands (Kramer, 1991) and with response-selection processes. As is the case with the N1 component, the amplitude of the N2 component requires some signal averaging for reliable measurements.

- The mismatch negativity (MNN) component has not received much attention related to alertness or workload, but it does reflect the operation of auditory perception and memory. The amplitude of MMN, which is measured as a difference wave between frequent and rare ERPs in the 100-200 ms range, is mainly related to changes in the properties of an auditory stimulus. However MMN is also sensitive to shifts in attention and operator workload, being larger in amplitude for attended than unattended stimuli or for low workload than high workload conditions when the eliciting stimuli are irrelevant to the task (Trejo, Ryan-Jones & Kramer, 1995). The amplitude and spatial extent of MMN appear to decrease under conditions of low alertness (Sallinen & Lyttinen, 1997). An advantage of MMN its very short refractory period, which allows for frequent estimation of its amplitude.

- The P3 (or P300) component reflects the cognitive processes of evaluating a stimulus or an event within the context of a task. In particular, P300s are related to mental workload (Trejo, Kramer, & Arnold, 1995). Of all the ERP components studied, the P300 seems to be the most suitable for on-line estimation of cognitive state, requiring averages of only 2 to 10 single trials for adequate estimation (Trejo, Kramer & Arnold, 1995). P300s elicited by stimuli associated with a task (primary task stimuli) increase in amplitude along with increasing task difficulty. This effect is explained by a matching increase in mental resources. P300s elicited by irrelevant stimuli (probes) or low priority tasks (secondary tasks) decrease in amplitude as primary task workload increases. This is explained as the result of increased allocation of limited resources to the primary task as workload increases.

- The error-related negativity (ERN) is relatively new compared to other ERP components discussed here and less is known about its applicability to on-line use. The ERN is related to self-awareness of having committed an error in a cognitive or perceptual task, and so may be useful in the analysis of task events and performance. The ERN is a fronto-central negative peak with a latency of about 100 ms relative to the error response. (Gehring, Goss, Coles, Meyer & Donchin, 1993).

The critical practical question in measuring ERP is the signal amplitude. In clinical studies it is possible to increase the ERP SNR at will by simple repeated presentation of the stimulus and signal averaging. Humphrey and Kramer, 1994, examined how much ERP data was needed to discriminate between levels of mental workload in complex, real-world tasks. They showed that 90% correct discrimination could be achieved with 1 to 11 seconds of ERP data. However, a concern regarding ERP for AugCog is the increased mental taxation involved in continuous use of the driving stimulus needed to evoke the response. There has been extensive new work on taking ERP measurements from a single event (known as single-trial ERP), and a measurement of this kind can be taken very quickly without undue disturbance of the subject (Tang, et al., 2003). At present, the research evidence strongly supports the use of P300 as a possible dynamic measure of mental workload.

3.2.1.4 Magnetoencephalography (MEG)
Comments: Requires the use of magnetic shielding and dedicated equipment

Magnetoencephalography is a noninvasive technique for imaging neural activity. An array of SQUID-based magnetic field sensors is placed above the scalp to measure the magnetic fields due to currents flowing throughout the brain. It is regularly used to assess how the brain recovers sensory, cognitive, and language functions after brain injury or stroke, and has been used to test cognitive models of visual attention (Downing, Lui & Kanwisher, 2001).

The source of the currents responsible for EEG are the same as those for MEG and therefore the spectral components relating to the cognitive state are the alpha and theta bands.

Functional imaging of MEG utilizes high temporal resolution to generate real-time brain scans and is used to study how the brain reacts to various stimuli. This is analogous to the ERP signals that were discussed in Section 3.2.1.3. Although the spatial resolution of MEG is higher than that for EEG, the use of SQUID sensors for MEG measurements (necessary because the signal amplitude is typically less than 1 pT) requires the use of cryogens and that measurements be made in a magnetically shielded room. These are considerable limitations for the technology with respect to the Augmented Cognition program.

3.2.1.5 Positron Emission Tomography (PET)
Comments: Requires the use of large-scale, dedicated equipment

Positron emission tomography uses trace amounts of radioactive isotopes to tag specific molecules by emitting positrons. Isotopes of oxygen are transported efficiently to active sites in the brain through blood flow, and are therefore suited for cognitive function measurements. The emission of positrons during radioactive decay of the isotope is then detected by a scanning device.

The isotopes are administered either intravenously or as an inhaled gas. Changes in a subject's mental activities are reflected in changes in the blood flow and metabolism of glucose in the brain. However, PET technology requires the use of specialized hardware that prohibits the development of a practical, operational "vigilance monitor" of the kind required from the AugCog program.

3.2.1.6 Function Magnetic Resonance Imaging (fMRI)
Comments: Requires the use of large-scale, dedicated equipment

Functional magnetic resonance imaging is an extension of well-established MRI technology that can be used to map changes in brain hemodynamics that correspond to mental operations. The technique is based upon the different magnetic susceptibilities of the iron in oxygenated and deoxygenated blood hemoglobin. Oxygenated blood is diamagnetic and possesses a small magnetic susceptibility, while deoxygenation of hemoglobin produces deoxyhemoglobin, which is a significantly more paramagnetic species of iron. Blood Oxygenation Level Dependent (BOLD) measurements measure local variations in the T2* (free induction decay) relaxation time caused by variations in the local concentrations of deoxygenated blood.

A rapidly emerging body of literature documents corresponding findings between fMRI and conventional electrophysiological techniques to localize specific functions of the human brain (George, et al., 1995). Resolutions of less than 1 mm are possible using fMRI, which is better than PET. Studies of increased application of stimuli have been done to test the effects on BOLD fMRI acquisition. Rees et al., 1997, found that as the rate of word presentation to subjects increased, the increase in BOLD response was non-linear, suggesting that the BOLD response was temporally limited.

As is the case with PET, the fMRI technology requires the use of specialized hardware that prohibits the development of a practical, operational "vigilance monitor" of the kind required from the AugCog program

3.2.1.7 Functional Near-Infrared Imaging (fNIR)

Comments: Developing technology, promising noninvasive technique, low dynamic range

fNIR is a relatively novel technology based upon the notion that the optical properties of tissue (including absorption and scattering) change when the tissue is active. Two types of signals can be recorded: fast scattering signals, presumably due to neuronal activity (Gratton, Goodman-Wood & Fabiani, 2001), and slower absorption signals, related to changes in the concentration of oxy- and deoxy-hemoglobin (Villringer & Chance, 1997). However, fNIR lacks the spatial resolution of fMRI and cannot accurately measure deep brain activity.

The fast fNIR signal is measured as an "event-related optical signal" (EROS), which is analogous to the ERP or functional MEG signal. The spatial localization of fast and slow fNIR measurements both correspond to the BOLD fMRI signal (Gratton *et al.*, 1997), and the latency in the EROS signal corresponds in latency to ERP measures (De Soto, Fabiani, Geary & Gratton, 2001). The latency in the slow (hemodynamic) signal roughly corresponds to that for the BOLD fMRI response (see Toronov *et al.*, 2003).

The major limitation of optical methods (both fast and slow signals) is their penetration (max: approximately 3 cm from head surface), which makes it impossible to measure brain structures such as the hippocampus or the thalamus, especially if they are surrounded by light-reflecting white matter. However, the vast majority of the cortical surface is accessible to the measurements. The technology is relatively simple and portable, and may serve a sort of portable, very rough equivalent of fMRI, which may supplement or substitute for some EEG measures.

3.2.1.8 Endocrine System

Comments: Invasive measurement technique, costly to implement

The endocrine system is a group of glands that are responsible for regulating a stable internal environment in the body by controlling the release of hormones. The endocrine system's response to an autonomic arousal state, such as stress, includes "the inhibition of "vegetative" functions while activating energy metabolism, body defenses, blood flow to the skeletal muscles, and a sharpening of the senses" (Bourne & Yaroush, 2003). However, the only way to measure the presence of these hormones is through the analysis of a sample of blood. This would be invasive and therefore entirely unsuitable for long term monitoring of an individual.

3.2.2 Cardiovascular System

3.2.2.1 Heart Rate (HR)

Comments: Measurements simple and reliable, insufficient data to draw link to cognitive state

The heart rate is measured as the interbeat interval (IBI), defined as the time interval between successive R-spikes of the cardiac QRS complex. Electrocardiogram measurements have been used in attempts to measure the cognitive state of pilots in a number of studies finding the broadest use over other measures through the late 1980s (Wilson & Fisher, 1990). Although activity of the autonomic nervous system is related to the cognitive state, the results linking HR to mental workload remain unclear and researchers in the field of brain-machine interfaces generally seem to have abandoned HR. The two underlying reasons are basically that a) HR is affected by physical exertion, and b) it is not clearly linked to the underlying function of the nervous system. However, results from the Air Force Research Laboratory indicates that with adequate normalization, HR data can be very reproducible between for a subject performing the same task on different day (Wilson, 2004).

3.2.2.2 Heart Rate Variability (HRV)

Comments: Measurements simple and reliable, insufficient data to draw link to cognitive state

More recently, variations in R-wave timing in the ECG have been used as a measure of the cognitive state. The underlying paradigm for the relationship between HRV and performance is the Polyvagal model (Porges, Doussard-Roosevelt, Portales & Greenspan, 1996), which views control of heart rate as an example of a negative feedback inhibitory process that allows for the interruption of ongoing behavior and re-deployment of resources to currently relevant tasks. Lower heart rate (but greater HRV) is thought to be a metabolic reaction to conditions perceived of as being of low environmental demand. Several authors (Scerbo, *et al.*, 2001) claim that heart rate (HR) increases and heart rate variability (HRV) decreases as workload increases (Wilson, 1992).

3.2.2.3 High Frequency QRS Electrocardiogram

Comments: Measurements simple, work is in its early stages but results are promising

Recent technology from NASA's Johnson Space Center analyzes high frequency components of the QRS complex (Schlegel, 2005). These signals appear to be indicative of ischemia or infarction and can be tracked in real time. The HFQRS recording system requires ordinary ECG equipment and additional software for the computations. Researchers at the Institute for Human and Machine Cognition in Pensacola are testing the utility of HFQRS for real-time indications of stress or mental overload. Initial findings suggest increases in HFQRS root-mean-square amplitude in response to increasing stress or workload (Raj, personal communication). Workload and stress were controlled by having subjects perform a fast-paced virtual urban combat simulation at different levels of difficulty.

3.2.2.4 Blood Oxygenation
Comments: Inexpensive and noninvasive measure of cognitive state

The oxygen saturation of the blood changes as a result of glucose metabolism, and therefore the color or reflectivity of blood for light in the near infrared band is affected. Blood oxygen is consumed by task-sensitive brain regions during periods of high workload. Using infrared emitter-detector pairs, these changes in reflectance can be tracked in real time and correlated with cognitive factors. The pulse oximeter is an inexpensive, commercially available device for measuring blood oxygen saturation and can be installed in a finger cuff. Increases in the amplitude of the "pulse wave" have been linked to increasing stress or difficulty of cognitive tasks (Dishman, *et al.*, 2000).

3.2.2.5 Respiratory Sinus Arrhythmia (RSA)
Comments: Measurements simple and reliable

Changes in the vagus (the 10th cranial nerve) cause fluctuations in the heart rate during respiration (Frazier, Strauss & Steinhauer, 2004). RSA is calculated from the minimum and maximum heart rates at points of inhalation and exhalation (Task Force, 1996). RSA is thought to be a measure of emotional arousal because it is mainly affected by the vagal fibers that begin in medullary nucleus ambiguous and end in the sino-atrial nerve of the heart. The Frazier *et al.* study observed a decrease in RSA correlated with emotional response.

3.2.3 Peripheral Systems

3.2.3.1 Electrodermal Activity (EDA)
Comments: Susceptible to too many factors to be a reliable measure of cognitive state

Electrodermal activity (EDA, also known as galvanic skin response, or GSR) measures the change in skin conductivity due to sympathetic neuronal activity. EDA is a sensitive psychophysiological index of changes in autonomic sympathetic arousal related to cognitive and emotional states, and is commonly used as a quantitative measure of stress. Recent studies have related brain regions implicated in cognition, attention, and emotion to peripheral EDA responses (Critchley, 2002). Fear, anger, startle response, orienting response, and sexual feelings are all among the emotions that may produce similar GSR responses.

Because EDA measurements are easy to perform, it is one of the few parameters that has been measured in physically demanding settings. EDA is easily measured with pairs of electrodes placed on the hands, and inexpensive electrical equipment. However, the wide range of influences on sweating and conductivity of the skin make the EDA measure somewhat less useful than other measures in applied settings.

3.2.3.2 Electro-Oculogram (EOG)
Comments: Simple, more reliable than EEG, but not clearly linked to cognition

The EOG arises from a standing electric potential difference between the cornea and the retina. When the eye moves or blinks this electric dipole becomes a source of changing electric fields and associated volume currents. With pairs of electrodes above and below or left and right of the eye, the vertical and horizontal motions of the eye (saccades) can be recorded at low fidelity. Blink rate and amplitudes are also derived from the EOG. Compared to EEG measurements, EOG is more simple and reliable. However, its linkage to cognition, like HRV, is less direct than EEG, relying on the changes in eye fixations and blink patterns that individuals produce during workload changes.

3.2.3.3 Puplliometry
Comments: Susceptible to too many factors to be a reliable measure of cognitive state

Beatty (1982) and others have collected vast amounts of experimental data on pupil size as a function of cognitive or emotional factors. Pupil size tends to increase in proportion to mental workload in a variety of tasks. As a rule, the more difficult a task is, the larger the average pupil size will be in a given individual, indicating a link between workload and arousal of the sympathetic nervous system. Pupil size is also relatively easy to measure, even somewhat unobtrusively, with the aid of video cameras and image processing software. For these reasons, pupillometry has gained a foothold in the workload assessment arena. However, pupils also dilate for several other reasons, the greatest influence being the level of ambient or direct illumination of the retina. Additionally, the visual, auditory, and psychological complexity of active workplaces all produce effects on pupil size that are not related to current task workload. At least one recent study examined the relation between pupil size and workload, and found no relationship between pupil size and task difficulty (Schultheis & Jameson, 2004). Nevertheless, with adequate control of lighting and other environmental factors, pupil size may afford a rough index of workload or difficulty in some task situations.

3.2.3.4 Mouse Pressure
Comments: Good measure of cognitive state, limited to applications involving computer use

A mouse designed for the purpose of detecting pressure applied to the mouse during a mouse click is described in Vick & Ikehara, 2003. The pressure applied to a computer mouse is regarded as a potentially useful measure of perceptual-motor performance, and therefore as a cognitive load measure. A mouse pressure measurement provides two measures: the latency of response to an external stimulus, and the delay in the mouse button release (St. John, Kobus, Morrison and Schmorrow, 2004).

3.2.3.5 Body Posture
Comments: Limited to applications involving computer use

Body posture is a physical appearance measure that is most conveniently measured for a seated subject. Posture stability is affected by almost all of the sensory systems, including vision, proprioception, vestibular sensation and audition. In a series of tests in which sensors were integrated in a chair to measure a subject's head movements and back bracing response, Balaban *et al.*, 2004, observed that head movement correlated to the number of targets on a screen. A lack of vigilance was characterized by a smaller-than-expected response. The back bracing response indicates a change in workload. Active back bracing indicated a large workload. Low bracing occurred when the participant perceived a lessoning of the workload. However, in wider applications, gross physical motion and vibration limit the usefulness of body posture measurements.

4   Concluding Remarks

In this paper we have briefly described physiological measurements of parameters that are related to the cognitive state of an individual. Cognitive performance is affected by the cognitive state, and can range between low vigilance at one extreme, to cognitive/perceptual overload at the other. DARPA's AugCog program was conceived with the aim of determining the cognitive state as part of a human-machine interface in order to alter the flow of information to the user and thereby maximize the individual's cognitive performance.

The benefit of physiological methods is that they provide a direct, objective, and continuous approach to assessing or predicting cognitive problems or failure before performance is measurably impaired. However, only a subset of the physiological measurements described in this paper is suitable for use in everyday situations, such as moving about within a building, or even in restricted environments like a cockpit. Our discussion has focused upon measurements that are the most direct indicators of brain function underlying cognition (i.e. EEG, ERPs). Some methods like fNIR or pupil size are either low in dynamic bandwidth or are more distantly connected to brain events underlying cognitive processes than EEG or ERPs. Other methods such as MEG, PET, and fMRI require the use of specialized and unwieldy hardware that is entirely unsuited to use outside of either a laboratory or clinical environment.

The trade-offs among all the methods involve several factors. These include: sensitivity to changing task or internal conditions, diagnosticity (or specificity) of effects for separable cognitive states, validity of relationships to cognitive states, reliability (or reproducibility) of measurements, and usability, which includes cost, ease of use, training, and complexity of data analysis. We have ranked each technique on a 3-point scale, with 3 being the highest, and presented our findings in Table 1. Those techniques that are unsuitable for the AugCog program have been assigned a usability of zero. The table shows that there tends to be a trade-off between usability and sensitivity. The most sensitive methods, EEG and ERP are the least useable.

Method	Sensitivity	Diagnosticity	Validity	Reliability	Usability
Microelectrodes	3	3	2	3	0
EEG	3	3	2	2	2
ERP	3	3	3	1	1
MEG	3	3	2	2	0
PET	2	2	2	2	0
fMRI	3	3	2	2	0
fNIR	2	2	2	2	2
Endocrine	2	1	1	2	0
HR/HRV	2	1	2	3	3
HFQRS	2	1	2	3	3
Blood Oxygen	2	1	1	3	3
RSA	2	1	1	2	2
EDA	1	1	1	3	3
EOG	1	1	2	3	3
Pupillometry	1	1	2	3	3
Mouse pressure	2	1	2	2	2
Body posture	1	1	1	1	1

Table 1: Summary of physiological techniques used in the determination of cognitive state.

References

Balaban, C.D., Cohn, J., Redfern, M.S., Prinkey, J., Stripling, R., & Hoffer, M. (2004). Postural Control as a Probe for Cognitive State: Exploiting Human Information Processing to Enhance Performance. *International Journal of Human-Computer Interaction*, **17**, 275-286.

Beatty, J. (1982). Phasic not tonic pupillary responses vary with auditory vigilance performance. *Psychophysiology*, **19**, 167-172.

Bourne, L.E. Jr., & Yaroush, R.A. (2003). Stress and Cognition: A Cognitive Psychological Perspective. National Aeronautics and Space Administration, Final Report: Grant Number NAG2-1561, February 1, 2003.

Critchley, H.D. (2002). Electrodermal responses: what happens in the brain. *Neuroscientist*, **2**, 132.

DeSoto, M.C., Fabiani, M., Geary, D.C., & Gratton, G. (2001). When in doubt, do it both ways: Brain evidence of the simultaneous activation of conflicting responses in a spatial Stroop task. *J. Cog. Neurosci.*, **13**, 523-536.

Director of Defense Research and Engineering, U.S. Dept. of Defense. (1992). DoD Key Technologies Plan. July, 1992.

Dishman R.K, Nakamura Y., Garcia M.E., Thompson R.W., Dunn A..L, & Blair S.N. (2000). Heart rate variability, trait anxiety, and perceived stress among physically fit men and women. *Int. J. Psychophysiol.*, **37**,121-133.

Downing, P., Lui, J., & Kanwisher, N. (2001). Testing cognitive models of visual attention with fMRI and MEG. *Neuropsychologia*, **39**, 1329-1342.

Frazier, T.W., Strauss, M.E., & Steinhauer, S.R. (2004). Respiratory sinus arrhythmia as an index of emotional response in young adults. *Psychophysiology*, 41, 75-83.

Fruhstorfer H., & Bergstrom, R. M. (1969). Human vigilance and auditory evoked responses. *Electroencephalogr. Clin. Neurophysiol.*, **27**, 346-355.

Gehring, W.J., Goss, B., Coles, M.G.H., Meyer, D.E., & Donchin, E. (1993). A neural system for error detection and compensation. *Psychological Science*, **4**, 385-390.

George, J.S., Aine, C.J., Mosher, J.C., Schmidt, M.D., Ranken, D.M., Schlitt, H.A., Wood, C.C., Lewine, J.D., Sanders, J.A., & Belliveau, J.W. (1995). Mapping Function in the Human Brain with Magnetoencephalography, Anatomical Magnetic Resonance Imaging, and Functional Magnetic Resonance Imaging. *J. Clin. Neurophysiol.*, **12**, 406-429.

Gratton, G., Fabiani, M., Corballis, P.M., Hood, D.C., Goodman-Wood, M.R., Hirsch, J., Kim, K., Friedman, D., & Gratton, E. (1997). Fast and localized event-related optical signals (EROS) in the human occipital cortex: Comparison with the visual evoked potentials and functional MRI. *NeuroImage*, **6**, 168-180.

Gratton, G., Goodman-Wood, M.R., & Fabiani, M. (2001). Comparison of neuronal and hemodynamic measures of the brain response to visual stimulation: an optical imaging study. *Human Brain Mapping*, **13**, 13-25.

Humphrey, D.G. & A. Kramer. (1994). Toward a Psychophysiological Assessment of Dynamic Changes in Mental Workload. *Human Factors*, **36**, 3-26.

Jung, T-P, Makeig, S., Stillwell, D., & Harm, D. (2001). Accessing cognitive state from physiological data. Proceedings of the Bioastronautics Investigators' Workshop, The Universities Space Research Association', Division of Space Life Sciences, January 17–19, Galveston, TX, http://www.dsls.usra.edu/meetings/bio2001/pdf/contents.pdf.

Kandel, E.R., Schwartz, J.H. & Jessell, T.M. (2000) Principles of Neural Science (4th ed.). New York: McGraw-Hill.

Kramer, A.F. (1991). Physiological metrics of mental workload: A review of recent progress. In D. Damos (Ed.), *Multiple Task Performance*. London: Taylor & Francis.

Matthews, R., McDonald, N.J., Fridman, I., Hervieux, P., & Nielsen, T. (2005). The invisible electrode – zero prep time, ultra low capacitive sensing. To be presented at HCI International Conference, Las Vegas, NV, July 22-27, 2005.

Mojarradi, M., Binkley, D., Blalock, B., Andersen, R., Ulshoefer, N., Johnson, T., & Del Castillo, L. (2003). A Miniaturized Neuroprosthesis Suitable for Implantation Into the Brain. *IEEE Transactions on Neural Systems and Rehabilitation Engineering*, 11, 38-42.

Porges, S.W., Doussard-Roosevelt, J.A., Portales, A.L., & Greenspan, S.I. (1996). Infant Regulation of the Vagal "Brake" Predicts Child Behavior Problems: a Psychobiological Model of Social Behaviour. *Dev. Psychobiol.*, 29, 697-712.

Raj, A. (2004). Personal communication, December 16, 2004.

Rees, G., Howseman, A., Josephs, O., Frith, C.D., Friston, K.J., Frackowiak, R.S.J., & Turner, R. (1997). Characterizing the relationship between BOLD contrast and regional cerebral blood flow measurements by varying the stimulus presentation rate. *Neuroimage*, 6, 270-278.

Sallinen, M., & Lyytinen, H. (1997). Mismatch negativity during objective and subjective sleepiness. *Psychophysiology*, 34, 694-702.

Scerbo, M.W., Freeman, F.G, Mikulka, P.J., Parasuraman, R., Di Nocero, F., & Prinzel III, L.J. (2001). The Efficacy of Psychophysiological Measures for Implementing Adaptive Technology. National Aeronautics and Space Administration, NASA Tech. Report TP-2001-211018.

Schlegel, T.T. (2005). Real-Time, High-Frequency QRS Electrocardiograph. Retrieved March 1, 2005, from http://www.nasatech.com/Briefs/July03/MSC23154.html.

Schultheis, H., & Jameson, J. (2004). Assessing cognitive load in adaptive hypermedia systems: Physiological and behavioral methods. In W. Neijdl, P. De Bra (Eds.), *Adaptive hypermedia and adaptive web-based systems: Proceedings of AH 2004, Lecture Notes in Computer Science*. Berlin: Springer-Verlag.

St. John, M., Kobus, D.A., Morrison, J.G., & Schmorrow, D. (2004). Overview of the DARPA Augmented Cognition Integration Experiment. *International Jounral of Human-Computer Interaction*, 17, 131-149.

Task Force of the European Society of Cardiology and the North American Society of Pacing and Electrophysiology. (1996). Heart rate variability: Standards of measurement, physiological interpretation and clinical use. *Circulation*, 93, 1043-1065.

Tang, A.C., McKinney, C.J., Sutherland, M.T., Parra, L., Gerson, H., & Sajda, P. (2003). Extraction of single-trial ERPs from frontal and visual cortex from video game play despite continuous free eye movement. ISBET Conference, University of New Mexico, 2003.

Taylor, D.M., Helms Tillery, S.I., & Schwartz, A.B. (2002). Direct Cortical Control of 3D Neuroprosthetic Devices, *Science*, 296, 1829-1832.

Toronov, V., Waler, S., Gupta, R., Choi, J.H., Gratton, E., Hueber, D., & Webb, A. (2002). The roles of changes in deoxyhemoglobin concentration and regional cerebral blood volume in the fMRI BOLD signal. *Neuroimage*, 19, 1521-1531.

Trejo, L.J., Kochavi, R., Kubitz, K., Montgomery, L., Rosipal, R., & Matthews, B. (2004). Measures and models for estimating and predicting cognitive fatigue. 44th Annual Meeting of the Society for Psychophysiological Research, Santa Fe, NM, October, 2004.

Trejo, L.J., Kramer, A.F., & Arnold, J. (1995). Event-related potentials as indices of display-monitoring performance. *Biological Psychology*, 40, 33-71.

Trejo, L.J., Ryan-Jones, D., & Kramer, A.F. (1995). Attentional modulation of the pitch-change mismatch negativity elicited by frequency differences between binaurally presented tone bursts. *Psychophysiology*, 32, 319-328.

Vick, R.M., & Ikehara, C. (2003). Methodological Issues of Real Time Acquisition from Multiple Sources of Physiological Data. *Proc. of the Hawaii International Conference on System Sciences*, Kona, Hawaii, 2003.

Villringer, A. & Chance, B. (1997). Noninvasive optical spectroscopy and imaging of human brain function. *Trends in Neuroscience*, 20, 435-442.

Wilson, G.F. (1992). Applied use of cardiac and respiration measures: Practical considerations and precautions. *Biological Psychology*, 34, 163.

Wilson, G.F. (2004). Presentation at AugCog Program Review Meeting, January 6, 2004.

Wilson, G. & Fisher, F. (1990). The use of multiple physiological measures to determine flight segment in F4 pilots. In *Proceedings of the IEEE 1990 National Aerospace and Electronics Conference*, Dayton, Ohio, 1990.

A Suite of Physiological Sensors for Assessing Cognitive States

Curtis S. Ikehara

Information and Computer Sciences
University of Hawaii at Manoa
1680 East-West Road, POST 317
Honolulu, HI 86822
cikehara@hawaii.edu

Martha E. Crosby

Information and Computer Sciences
University of Hawaii at Manoa
1680 East-West Road, POST 317
Honolulu, HI 86822
crosby@hawaii.edu

David N. Chin

Information and Computer Sciences
University of Hawaii at Manoa
1680 East-West Road, POST 317
Honolulu, HI 86822
chin@hawaii.edu

Abstract

This study uses a suite of physiological sensors to assess cognitive states. The suite of sensors include eye tracking, skin conductivity, heart rate, peripheral temperature and the pressures applied to a computer mouse. Volunteers connected to the suite of sensors performed 48 tasks involving multiple moving targets and a secondary audio memory task. Volunteers performed a game like task which was structured such that there were three levels of task difficulty. The levels included a memory task, increasing the interaction required per target and increasing the number of targets. Results of physiological measures were similar to task performance measures. The task performance measure was more sensitive to the memory level of difficulty, but the perceptual-motor measure from the computer mouse sensors were sensitive to the order of sets of presentation. Eye tracker measures can significantly distinguish between the different numbers of moving targets. The suite of physiological sensors to

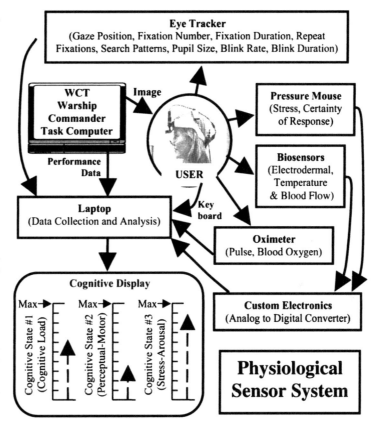

Figure 1: A schematic diagram an eye tracking system as part of a suite of physiological sensors for assessing cognitive states.

assess cognitive states is shown to be a viable alternative to task performance measures. Also, in some cases, these physiological measures can be more sensitive than task performance measures of cognitive states.

1 Introduction

The accurate assessment of cognitive states is essential to identifying and defining cognitive processes and testing models of how cognitive processes operate and interact. Our laboratory is working on combining a suite of physiological sensors to assess cognitive load in real time (see Figure 1). The laboratory assessment of cognitive states have a prescribe methodology, cognitive load is explicated in the following paragraphs, but the need for cognitive assessment in a less structured environment would improve the external validity and utility of the cognitive assessment measures.

There are three methods to assess the cognitive load of a user doing a multitask activity (Charlton 2002). The first method is by using performance measures. Measures of both primary and secondary task performance can be used to assess the cognitive load, but because tasks often differ, every performance measure needs to be customized to the task. Although the primary task measure is important when assessing user performance, care must be taken with a secondary task. A secondary task may co-vary in a non-linear way with the primary task and produce equivocal results for both measures.

The second method is by using rating scales. These are subjective measures usually taken after the activity is completed, although if interruptions do not degrade task performance, ratings can be taken between or during the tasks of an activity. Subjective measures are essential to assessing the user's perceptions of the task and can provide information of the cognitive state of the user, but since most rating scales are taken after the activity, real-time cognitive state assessment is not possible.

The third method is by using physiological measures. Unobtrusive physiological measures are measured in the same way regardless of task, are objective and can be taken in real-time without impacting user performance or user ratings.

Each of the three methods of cognitive assessment adds to the understanding of the user's cognitive state. For real-time measures of cognitive state, only performance and physiological measures are possible, although it is sometimes difficult to incorporate real-time performance measures into an existing computer task since it may involve a significant effort to modify the task program assuming the source code of the program is available.

Table 1 shows eye-tracking as having the potential to contribute many different cognitive state measures. Eye-tracking contributes to evaluation of cognitive load by illuminating the focus of attention. For example, Gilmour and his colleagues at Boeing have shown that for nearly all air to ground search conditions, the observer wastes more than 40 percent of his time in useless search activity during the period after the target has become available (Snyder 1970). Several studies have reported that search times are related to the density or complexity of the background (Erikson 1952; Bloomfield 1972; Monk and Brown 1975; Drury et al. 1978).

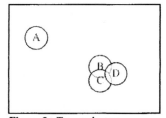

Figure 2: Target clusters.

Fixations are a good measure of task difficulty for fixed images since the number of fixations are greatest during problem solving (Nakano 1971), but extracting measures from eye tracking data when the targets are moving and the target locations are unknown is problematic. Without knowing the location of the targets, the cluster in Figure 2 of the three targets B, C & D, could result in a measure indicating a relatively long fixation on a single target. Clustering of targets would reduce the fixation count, increase fixation duration and reduce the amount of eye movement needed to view the targets.

With moving targets, not knowing the location of the targets further complicates the determination of what is a fixation, especially when targets cross paths (see Figure 3). In Figure 3, since the observer is free to change fixation to the other target, it is not clear whether the number of fixations is one or two. With a large number of moving targets, the formation of transient clusters due to motion and path crossing would make obtaining an accurate fixation count problematic.

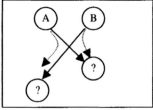

Figure 3: Targets crossing paths.

An indirect method of measuring the number of fixations and in turn measuring problem difficulty would be to measure the amount of eye movement occurring during a fixed time interval. Given the time interval of a task, the increased number and duration of fixations would reduce the amount of time available for eye movement. More difficult problems would have less eye movement due to the increased number and duration of fixations.

Unobtrusive passive physiological sensors can provide measures that are relatively easy to collect across all tasks. Table 1 shows the physiological measures collected by our laboratory.

Table 1: Physiological Measures and Potential Cognitive States

Physiological Measures	Secondary Measures	Potential Cognitive Measures
Eye Position Tracking	Gaze Position, Fixation Number, Fixation Duration, Repeat Fixations, Search Patterns	Difficulty, Attention, Stress, Relaxation Problem Solving, Successful Learner, Higher Level of Reading Skill (Andreassi 1995), (Sheldon 2001)
Pupil Size	Blink Rate, Blink Duration	Fatigue, Difficulty, Strong Emotion, Interest, Mental Activity - Effort, Familiar Recall, Positive / Negative Attitudes, Information Processing Speed (Andreassi 1995)
Skin Conductivity	Tonic and Phasic Changes	Arousal (Andreassi 1995)
Peripheral Temperature		Negative Affect (Decrease), Relaxation (Increase) (Andreassi 1995)
Relative Blood Flow	Heart Rate and Beat to Beat Heart Flow Change	Stress, Emotion Intensity (Andreassi 1995)
Blood Oxygen		General physical status
Mouse Pressure Sensors		Stress, Certainty of Response

Table 2: Positive and Negative Factors of Physiological Sensors

Measures	Positive	Negatives
Eye Tracking	Able to determine several cognitive states during task	Does not provide information when there are no visual targets. Rest periods go unmonitored.
Pupil Size	Able to determine several cognitive states during task	Affected by sudden changes of image intensity
Mouse	Provides information primarily during movement and clicking	Does not provide information during rest periods
Heart Rate	Provides continuous information	Slow rise and decline in heart rate relative to the trigger event
Skin Conductivity	Provides continuous information	Affected by user movements and slow rise and decline in rate relative to the trigger event
Temperature	Provides continuous information	Very slow rise and decline in temperature change relative to the trigger event

Each physiological measure has its positive and negative factors as shown in Table 2. Combining the different measures can fill in temporal gaps in measurement and reaffirm newer measurement methods.

In our present research we are investigating ways to instantaneously assess cognitive load and cognitive capability (Chin et al. 2002). The direction of our investigation involves the use of physiological measures (e.g., eye movements, pupil dilation, pulse, etc.). Our eventual goal is to create effective ways to augment cognition in real time and manage cognitive load for real world tasks.

2   Methods

2.1   Subjects

Eight volunteers, 5 males and 3 females, ranging in age from 22 to 47 years (M = 30.1 yrs, SD = 8.6) participated in the study. Each volunteer participated in up to four sessions of experiments over the course of three days. Results reported in this paper are from one of the four sessions of experiments. Each session used a different set of physiological sensors, but was otherwise similarly designed with each volunteer performing 48 experimental tasks. Our team alternated with another team collecting eye tracking data allowing us to collect eye tracking data from four volunteers. A detailed description of the subjects, experimental methods and the task are found in DARPA Augmented Cognition Technical Integration Experiment, Technical Report 1905 (St. John & Kobus, 2003).

2.2   Session Description

The session being reported involved three different groups of researchers, each using a different set of physiological sensors that were simultaneously attached to a volunteer during the session. The Drexel University research group used a prototype near infrared sensor system, the Advanced Brain Monitoring, Inc. research group used electroencephalographic measures and our research group used eye tracking and a suite of physiological sensors described in the following section. The results reported in this paper are from the data collected by our group.

2.3   Eye Tracker and Physiological Sensors

2.3.1   Eye Tracking and Pupil Size

An Applied Science Laboratories Model 501 eye tracking mobile system was used to obtain gaze position and pupil size. The subject wore the head mounted tracking system. Calibration of the subject's eyes to a calibration chart shown on the screen was performed before starting each block of 12 experimental tests.

2.3.2   Custom Physiological Sensor System

A custom-designed electrically isolated physiological sensor system was used to obtain galvanic skin conductivity (GSR), peripheral temperature, relative blood flow, pulse-oximeter and the pressures applied to a computer mouse. GSR, temperature, blood flow and pulse-oximeter were attached to areas on the left foot so as not to interfere with any keyboard activity. The infrared blood flow detector was attached to the big toe. Attached to the next pair of toes were the GSR electrodes. A peripheral temperature sensor was attached to the next available toe and the pulse-oximeter was attached to the small toe. Attached to the ankle was a general body temperature sensor and a third temperature sensor was place approximately 15 centimeters away from the ankle to obtain the ambient temperature. A computer mouse equipped with pressure sensors on the body and on the buttons of the mouse detected pressures applied during task performance.

2.4   Test Bed Task

The Warship Commander Task (WCT), developed by Pacific Science & Engineering Group, Inc., was the test bed task used during the session. The volunteers performed the task on an IBM-compatible PC with a 17 inch screen, 1024 x 768 pixel color and sound system. The computer mouse was replaced with the pressure sensor equipped mouse.

The WCT computer program was designed to record a number of performance characteristics. There were more than 15 performance measures including the WCT score, response times for different interactions, and a count of the errors and their type. Only the task performance score was used for the analysis of results.

2.5 Task

The WCT tests had three levels of difficulty. First, the number of targets displayed could be 6, 12, 18 or 24 jet fighters approaching a battle ship. Increasing the number of targets would increase the perceptual, cognitive and motor difficulty of the task (i.e., targets level). Second, there were three different types of jet fighter targets each requiring a different amount of interaction to achieve the maximum score. Two levels of motor and cognitive difficulty were introduced by having two different combinations of targets, one with more interaction required than the other (i.e., interact level). Third, a recall task was either present or absent. The memory recall task presented audio information and sometime later the volunteer would be asked a question about the information which required the volunteer to look at a four choice menu on the left of the screen and input a keyboard response (i.e., memory level).

2.6 Procedure

2.6.1 Pre-session Practice

Volunteers were given instructions on how to achieve the highest scores, which entailed identifying the target and taking the appropriate action based on its identified characteristics. Volunteers had varying levels of practice with the WCT test ranging from 2 hours to over 100 hours. Most people can easily achieve a high level of competency with the task after an hour of practice.

2.6.2 Pre-session Questionnaire

A pre-session questionnaire contained questions to collect information on a volunteer's demographics, normal hours of greatest alertness/fatigue, present levels of alertness/fatigue, amount of previous experience and motivation for participating in the pilot study.

2.6.3 Session Activity

Each volunteer was tested individually and performed 48 WCT tests. The 48 WCT test were broken up into four blocks containing 12 WCT tests (see Table 3, shaded blocks). The numbers in the shaded rectangles represent the number of jet fighter targets presented. A block presented to the volunteer was three sets of 4 WCT tests with a short rest period of a minute between sets (e.g., 6, 18, 12, 24, rest, 6, 18, 12, 24, rest, 6, 18, 12, 24). The WCT computer program could only present the number of targets in the order shown. A block could have one of four conditions: less target interaction with no memory test, more target interaction with no memory test, less target interaction with memory test and more target interaction with memory test. Blocks were presented to the volunteer in random order with one to two minute breaks between them.

Table 3: Task Presentation Conditions

		Less Interaction					More Interaction			
	Set #1	6	18	12	24		6	18	12	24
Memory (No)	Set #2	6	18	12	24		6	18	12	24
	Set #3	6	18	12	24		6	18	12	24
	Set #1	6	18	12	24		6	18	12	24
Memory (Yes)	Set #2	6	18	12	24		6	18	12	24
	Set #3	6	18	12	24		6	18	12	24

2.6.4 *Post Session Questionnaire*

A post session questionnaire was given to assess how the volunteer perceived their state of mind during the experiment. Each participant was asked to indicate level of alertness, interest and stress felt during the session, how difficult the session seemed, how challenging it was, whether the duration of the session seemed about right or was too long and to add comments about the task.

3   Results

Results are from performance measures, eye tracking, physiological measures obtained from the pressure sensitive computer mouse, GSR and pulse-oximeter. The performance measure analyzed was the percent of possible correct actions to complete a task. The average amount of eye movement per second during the middle 50% of the WCT task was used to generate the eye tracking measure. The middle 50% was chosen since it would be minimally impacted by task starting and ending artifacts. Physiological measures derived using an algorithm described in Ikehara & Crosby (2005) produced perceptual-motor load and cognitive load measures derived from the pressure sensitive mouse. The arousal-stress measure was derived from the skin conductivity multiplied by heart rate from the pulse-oximeter. Multiplication is used so that the measure reinforces each other when they agree and cancel each other out when they disagree. The resulting measurements are individually standardized giving a percent of perceptual-motor load, cognitive load and arousal-stress.

A repeated measures multivariate analysis used the volunteer's score (i.e., performance data), eye movement (i.e., eye tracking), perceptual-motor load, cognitive load and arousal-stress measures (i.e., physiological data). The analysis was a repeated measures 2 x 2 x 3 x 4 with two memory conditions (with/without), two levels of interaction (high/low), three sets of four WCT tasks (Set #1, Set #2 & Set #3) and four levels of targets per WCT tasks (6, 12, 18 & 24 targets).

Repeated measures multivariate results for the physiological sensor are from eight volunteers. Eye tracking measures are from three volunteers because eye tracking data from one volunteer was over one standard deviation from the other three volunteers.

Results are shown below in Tables 4, 5, 6, 7 and 8. Main effects are listed. Appended to the main effect list are interactions that showed a significant result for any of the five measures. Table 9 is a summary of all the significance levels across the five measures. Interactions that are not significant for any of the measures are not shown. Table 4 shows that using the performance measure (i.e., subject score) it was possible to significantly detect the difference of most of the independent variables. Using each of the physiological measures in Tables 5 to 8, it was possible to significantly detect differences in one or more of the independent variables.

Table 4: Performance Measure from the Volunteer's Task Score

	df	Mean Square	F	Significance Level
Memory (M)	1	703.92	13.68	**0.01**
Interact (I)	1	7390.74	48.91	**0.00**
Set (S)	2	93.82	2.5	0.10
Targets (T)	3	2027.98	19.22	**0.00**
M x I	1	161.89	8.32	**0.02**
M x T	3	52.37	3.56	**0.03**
I x T	3	546.65	6.19	**0.00**
S x T	6	50.33	1.00	0.43
M x I x S x T	6	28.23	2.47	**0.04**

Bold significance level denotes p < 0.05

Table 5: Eye Tracking from the Total Eye Movement During the Middle 50% of a Task

	df	Mean Square	F	Significance Level
Memory (M)	1	28241.95	0.74	0.48
Interact (I)	1	2.82	0.00	0.97
Set (S)	2	2593.98	0.78	0.52
Targets (T)	3	45623.98	16.29	**0.00**
M x I	1	1.01	0.00	0.99
M x T	3	8090.42	6.08	**0.03**
I x T	3	6963.63	1.96	0.22
S x T	6	1566.43	0.53	0.77
M x I x S x T	6	3031.08	2.59	0.08

Bold significance level denotes p < 0.05

Table 6: Perceptual–Motor Load from the Mouse Sensors

	df	Mean Square	F	Significance Level
Memory (M)	1	168.69	0.44	0.53
Interact (I)	1	3226.95	37.73	**0.00**
Set (S)	2	612.37	4.48	**0.03**
Targets (T)	3	53441.86	416.18	**0.00**
M x I	1	93.72	0.76	0.41
M x T	3	4.84	0.10	0.96
I x T	3	303.60	4.59	**0.01**
S x T	6	190.88	2.52	**0.04**
M x I x S x T	6	38.28	0.64	0.70

Bold significance level denotes p < 0.05

Table 7: Cognitive Load from the Mouse Sensors

	df	Mean Square	F	Significance Level
Memory (M)	1	215.10	1.02	0.34
Interact (I)	1	2281.60	13.70	**0.01**
Set (S)	2	481.23	3.02	0.08
Targets (T)	3	57563.49	602.88	**0.00**
M x I	1	87.82	0.47	0.51
M x T	3	5.52	0.13	0.94
I x T	3	337.27	3.85	**0.02**
S x T	6	127.02	1.65	0.15
M x I x S x T	6	31.94	0.63	0.70

Bold significance level denotes p < 0.05

Table 8. Arousal–Stress Measure from Skin Conductivity and Heart Rate

	df	Mean Square	F	Significance Level
Memory (M)	1	0.00	0.00	0.99
Interact (I)	1	0.26	5.65	**0.05**
Set (S)	2	70.07	0.92	0.42
Targets (T)	3	10.60	1.59	0.22
M x I	1	0.00	0.00	1.00
M x T	3	4.53	0.67	0.58
I x T	3	25.61	1.36	0.28
S x T	6	47.78	0.95	0.47
M x I x S x T	6	22.90	1.14	0.36

Bold significance level denotes p < 0.05

Table 9. Summary of Significance Levels

	Score	Eye Track	Percept - Motor	Cog	Arousal - Stress
Memory (M)	**0.01**	0.48	0.53	0.34	0.99
Interact (I)	**0.00**	0.97	**0.00**	**0.01**	**0.05**
Set (S)	0.10	0.52	**0.03**	0.08	0.42
Targets (T)	**0.00**	**0.00**	**0.00**	**0.00**	0.22
M x I	**0.02**	0.99	0.41	0.51	1.00
M x T	**0.03**	**0.03**	0.96	0.94	0.58
I x T	**0.00**	0.22	**0.01**	**0.02**	0.28
S x T	0.43	0.77	**0.04**	0.15	0.47
M x I x S x T	**0.04**	0.08	0.70	0.70	0.36

Bold significance level denotes p < 0.05

4 Discussion

When comparing significance levels in Table 9 of the performance measures (i.e., subject score) to the physiological sensors, the main effect of the four variables (i.e., Memory, Interaction, Set & Targets) could be detected by various combinations of the performance and physiological measures. Issues relating to these results are discussed below.

Using the eye movement measure, it was possible to significantly detect the increasing number of targets as shown in Figure 4, Table 5 and Table 9. It is likely that to optimize performance, volunteers would look for naturally formed clusters of targets which would reduce the amount of eye and mouse movement needed to complete the task. The premise that eye movements would be inhibited by more fixations of longer duration caused by clustering of the targets is supported by the decline in eye movement with an increasing number of targets.

Also, using the eye movement measure it was possible to detect a significant interaction between the memory test and target number as shown in Figure 5, Table 5 and Table 9. The memory test required the volunteer to listen to a question, look at a menu on the left side of the screen and then enter the appropriate answer on the keyboard. The eye tracker detected more eye movements when the number of targets were low (e.g., 6 targets). Some volunteers noted in the post-questionnaire that they only spent time on the memory task when they were not busy on the primary task. So when the target number was low, the volunteer would look more often at the memory test menu increasing the amount of eye movement when there was a memory test.

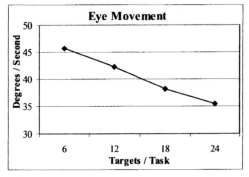

Figure 4: As the target number increases eye movement decreases.

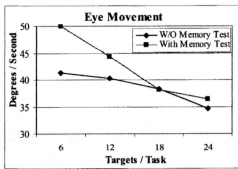

Figure 5: The interaction between target number and eye movement shows that with fewer targets the memory task increased eye movement.

Eye movement as measured by the eye tracker can be used to significantly distinguish between the different numbers of moving targets of unknown location, but a more thorough analysis is needed when there is a secondary task that can also increase eye movement. These eye movement measure results supports using this measure with more dynamic visual scenes where task performance information is not available in real time.

Of the main effects of the four independent variables, the perceptual-motor load measure would not be expected to detect the memory task variable. This is because the perceptual-motor load measure is derived from the pressures applied to the computer mouse, primarily during clicking. It is unlikely that a volunteer entering an input on the keyboard for the memory test would be simultaneously clicking on the computer mouse.

Using the perceptual-motor load measure it was possible to significantly detect the effect of set order and interaction of set order with the number of targets (see Tables 6 and 9). Figure 6 shows the perceptual-motor load increasing for each subsequent task that the volunteer does. Figure 7 shows the interaction between the order of the sets and number of targets. The lower target number tasks show the most increase in perceptual-motor load over subsequent sets, while the high target number task shows the least amount of increase due to a ceiling effect, where the range of increase is limited by the maximum of the range. Both effects were likely due to fatigue during performance of the tasks. The performance measure was not able to significantly detect the difference between the sets. A likely reason is because of a ceiling effect caused by the high mean score of the volunteers (mean = 93.7 %) and a small variance (standard deviation = 4.6).

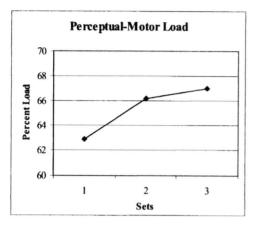

Figure 6: As volunteers progress through each set of four tasks the perceptual-motor load increases.

As with the perceptual-motor measure, minimal mouse clicking during the memory task causes the cognitive load measure not to be able to detect the presence of a memory task (see Tables 7 and 9). As previously discussed, the detection of the sets by the perceptual-motor measure was likely caused by volunteer fatigue while progressing through the sets. The cognitive load measure is not as sensitive to fatigue as the perceptual-motor measure, so that measure does not detect a significant difference between the sets. The arousal-stress measure was only able to detect interaction level.

In summary, results indicate that except for the memory task, the physiological sensors were able to significantly detect the three other main effects. Also, where the performance measure failed to detect the significant difference between sets, the perceptual-motor measure was able to fill that deficiency. Each physiological measure has its advantages and disadvantages. The benefit of using a suite of physiological sensors is that the overlapping advantages of the different sensors can produce reliable measures of cognitive states during task performance.

Acknowledgements

This research was supported in part by Defense Advanced Research Projects Agency grant no. NBCH1020004 and the Office of Naval Research grant no. N00014970578. Also, for technological integration experiment support thanks go to: BMH Associates, Inc., CACI, Pacific Science & Engineering Group, Inc., Space and Naval Warfare Systems Center (San Diego), Science Advisory Panel, Sonalysts, Inc, Strategic Analysis and TLK, Inc.

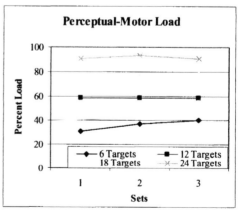

Figure 7: The increase is highest when there are fewer targets. With more targets and a higher load there is a ceiling effect reducing the increase.

281

References

Andreassi, J.L. (1995). Psychophysiology: Human Behavior and Physiological Response (3rd Ed.). New Jersey: Lawrence Erlbaum.

Bloomfield, J. (1972). Visual search in complex fields: Size differences between target disc and surrounding discs. *Human Factors* 14 (2), 139-148.

Charleton, S. G. (2002). Measurement of cognitive states in testing and evaluation, *Handbook of Human Factors and Evaluation*, S. G. Charleton and T. C. O'Brien (Eds.), 97-126.

Chin, D., Crosby, M., Iding, M., Ikehara, C., Gersch, W., and Stelovsky, J. (2002). Real-time assessment of cognitive load for adaptive augmentation of cognition. *DARPA grant proposal*, Award Number NBCH1020004.

Drury, C., Clement, M. and Clement, R. (1978). The effect of area, density, and number of background characters on visual search. *Human Factors* 20 (5), 597-602.

Eriksen, C. (1952). Object location in a complex perceptual field. *Journal of Experimental Psychology* 45 (3), 124-132.

Ikehara, C. S. & Crosby, M. E. (2005). Assessing Cognitive Load with Physiological Sensors, Proceedings of the Hawaii International Conference on System Sciences, Kona, Hawaii, January 2005.

Monk, T. and Brown, B. (1975). the effect of target surround density on visual search performance. *Human Factors* 17 (4), 356-360.

Nakano, A. (1971). Eye movements in relation to mental activity of problem solving. *Psychologia: An International Journal of Psychology in the Orient*, 14, 200-207.

Sheldon, E. (2001). *Virtual agent interactions*, PhD Thesis, Orlando: University of Central Florida.

Snyder, H. (1970). Dynamic visual search patterns. *1970 Spring Vision Symposium*.

St. John, M. & Kobus, D. A. (2003). DARPA Augmented Cognition Technical Integration Experiment (TIE), Technical Report 1905, Retrieved January 23, 2005, from http://www-tadmus.spawar.navy.mil/tr1905.pdf

Use of near-infrared methods to augment cognition

Gabriele Gratton, Kathy Low, & Monica Fabiani

University of Illinois at Urbana-Champaign
Beckman Institute, 405 N. Mathews Ave., Urbana, IL 61822
grattong@uiuc.edu

Abstract

The use of physiological measures to augment cognition requires the use of sensors providing rapid, sensitive, and articulated information about brain and other physiological functions. Furthermore, these measures need to be applicable in a variety of different environmental conditions. Near-infrared (NIR) methods (Gratton & Fabiani, 2001) can be used to study changes in optical parameters of the human cortex that accompany brain activity. They can be particularly useful to augmenting cognition, because of their ability to separate several different types of brain responses, both spatially and temporally. Two types of phenomena can be measured: fast signals (Gratton et al., 1995), related to neuronal activity, and slow signals (Hoshi & Tamura, 1993), related to hemodynamic changes. The slow signals have high reliability, a good signal-to-noise ratio, and can be measured with relatively simple instrumentation. Their main drawback is that feedback about brain activity can only be obtained with some delay (several seconds), which may limit their applications. The fast signals have a lower signal-to-noise ratio but can in principle produce feedback about brain function with very short delays (milliseconds), albeit at a probabilistic level. We will discuss various principles of application of NIR methods in paradigms relevant to the augmented cognition program, as well as some examples from studies conducted in our lab.

References

Gratton, G., Corballis, P. M., Cho, E., Fabiani, M., & Hood, D. (1995a). Shades of gray matter: Noninvasive optical images of human brain responses during visual stimulation. Psychophysiology, 32, 505-509.

Gratton, G. & Fabiani, M. (2001). Shedding light on brain function: The event-related optical signal. Trends in Cognitive Science, 5(8), 357-363.

Hoshi, Y., & Tamura, M. (1993). Dynamic multichannel near-infrared optical imaging of human brain activity. Journal of Applied Physiology, 75, 1842-1846.

Detection of Human Physiological State Change
Using Fisher's Linear Discriminant

Li Yu, Adam Hoover

Eric Muth

Department of Electrical and Computer Engineering
Clemson University
Clemson, SC 29634-0915
{liy,ahoover}@clemson.edu

Department of Psychology
Clemson University
Clemson, SC 29634-1355
muth@clemson.edu

Abstract

This paper describes a new approach to detect physiological state change in human beings through a computer embedded statistical algorithm. Some physiological signals are noisy and therefore changes of such signals are difficult to detect. Also different people generate signals with different characteristics, which makes a simple threshold solution impossible. Fisher's linear discriminant is a widely used statistical algorithm to handle such kind of problems. In this paper we have applied Fisher's discriminant to the detection of physiological state change with some inspiring results. We are designing more structured testing plans to generate data from more subjects and further validate our method[1].

Key Words: Fisher's Discriminant, Heart Rate Variability, Physiological State Change Detection

1 Introduction

Certain physiological signals fluctuate in a manner that makes it difficult to detect "real change" as opposed to meaningless fluctuations. For example, suppose one wishes to detect changes in heartrate. The simplest approach is to categorize heartrate as high or low, and use a threshold to discriminate. However, this approach does not take into account variance due to the individual being monitored. For example, the threshold would need to be adjusted depending on the age of the subject. In addition, a simple threshold cannot discriminate between a subject switching from light to vigorous exercise, as compared to switching from rest to wakefulness. Even if multiple thresholds are used, when does a change across one of the thresholds indicate a meaningful change of physiological state? It could be a temporary increase in a signal that in a moment will dissipate.

We apply Fisher's Linear Discriminant (FLD) to the problem of state change detection of physiological signals. We define a state using the mean and variance of a signal over an arbitrary (but minimum required) amount of time. We define a state change as a significant difference in state between two arbitrarily long (but minimum required) consecutive time periods. There are many kinds of mathematical operators and statistical processes to assess and detect the change of data populations, such as Bayesian parameterization, FLD, Nearest Neighbor Rule (NNR), C-means clustering, etc. The FLD is one of the simplest methods. Its only assumption is that there is no intra-group correlation, and it only requires calculations of means and variances. Thus, the implementation is simple, lending itself to real-time calculation. The FLD minimizes within group variance while maximizing between group differences to find the best split between two groups of data. Based on the output criterion, one can use a threshold or rule set to make decisions on the issue of state change.

We apply our method to physiological arousal. Human arousal is an index measured by heartrate variability (HRV) in the form of inter-beat-interval (IBI) data (Hoover and Muth 2004). The high-frequency component of HRV, due to respiratory sinus arrhythmia (RSA), is well known and many studies have validated the use of various RSA measures as indices of PNS activity. We have collected HRV data taken from a number of human subjects under various tasks such as vigilance, doing math, and playing a computer game. The goal of our methods is to detect the

[1] We gratefully acknowledge the support of grant #N000140210347 from DARPA through the Office of Naval Research. This work was also supported by a subcontract from DARPA through Honeywell Corporation.

284

change in psychological tasks, with the assumption that it causes a state change in physiological arousal because arousal is dependent on the level of task difficulty (Anderson 1994; Hembree, 1988; Robazzo, Bortoli and Nougier, 1998; Sokoll and Mynatt, 1984). Preliminary results on two subjects undergoing a series of different tasks are quite encouraging. We are currently designing a more comprehensive study to explore more subjects by this approach.

2 Methods

Fisher's Linear Discriminant Function is calculated as,

$$FLD = \frac{(\mu_1 - \mu_2)^2}{\sigma_1^2 + \sigma_2^2}$$

where μ_1 and μ_2 are the mean values of the two sampled groups respectively, σ_1 and σ_2 are the standard deviations of the two sampled groups respectively. For our purposes, if data $x_1 \ldots_t$ is divided into two groups $x_1 \ldots_k$ and $x_{k+1} \ldots_t$, and $\mu_1, \mu_2, \sigma_1, \sigma_2$ are respectively,

$$\mu_1 = \frac{\sum_{i=1}^{k} x_i}{k} \qquad \mu_2 = \frac{\sum_{i=k+1}^{t} x_i}{t-k} \qquad \sigma_1 = \sqrt{\frac{\sum_{i=1}^{k}(x_i - \mu_1)^2}{k}} \qquad \sigma_2 = \sqrt{\frac{\sum_{i=k+1}^{t}(x_i - \mu_2)^2}{t-k}}$$

The procedure for computing the Fisher's discriminant in realtime is as follows:

Let x_i, i=0...t be the data, index t is the current or newest data;

1. decide a minimum window size N;
2. when t≥2N, start a set of Fisher's value computations;
3. at each time t, separate the current data into two sequential sets with split at k (k<t) such that N≤ k≤t-N, set one is $x_0..x_k$ and set two is $x_{k+1}...x_t$;
4. at each time t, compute Fisher's value, FLD_k for all possible splits of k;
5. at each time t, take the maximum ($Max_k(FLD_k)$) among all possible Fisher's values FLD_k as the Fisher's value at that time instant and also record the split position k that the maximum Fisher's value resides at.

2.1 Simulated Data

To validate the effectiveness of Fisher's discriminant function, we have first designed and implemented a simulation scheme.

We use two sets of Gaussian random data to simulate IBI data coming out from humans undergoing a state change (switching from one psychological task to another). The two sets of data have the same or different means and standard deviations with one at the end of the other to form a time sequence. The data are all normally distributed. Figure 1 is an example of such a data pair.

Using statistical methods such as Fisher's discriminant function in an approach described above, we can tell where the split between the two sets of data is, in other words, where the physiological state change occurs in terms of IBI data. We calculate the Fisher's value as if we are getting data out one at a time, and the computed curve looks like what is shown in Figure 2. There is always a peak near the split point.

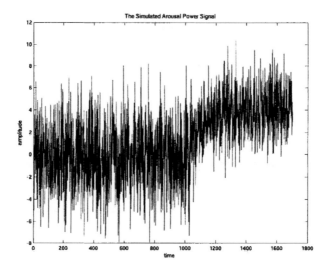

Figure 1: An example of simulated data. The first set starts from time 1 till time 1000 with mean 0.0 and standard deviation 3.0, and the second set starts from time 1201 till time 1700 with mean 4.0 and standard deviation 2.0. There is a transition from time 1001 till time 1200. The transition is formed by taking the weighted average of data from both Gaussian groups.

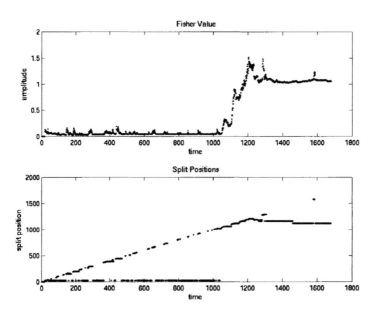

Figure 2: The plots of Fisher's discriminant function and split position. In the upper plot, there is a big overshoot between the transition period 1001 and 1200 for the two sets of simulated data shown in Figure 1, and in the lower plot, the split positions are quite stable after the transition period for the same simulated data.

For any two sets of data with other combinations of means and standard deviations, the curve is more or less the same. The overshoot can be used to detect the state change in a data series like the IBI.

We have also devised another kind of simulation scheme in which a Gaussian series with a certain mean and standard deviation is generated and separated into two groups by a certain value. We test this scheme based on the fact that the real data to be processed may probably fit the category of a single Gaussian distribution rather than two. The real data are actually two sets of data, each forming part of one Gaussian distribution. The simulation results are shown in Figure 3. It can be seen from the plots that there is always an overshoot near the split of the two groups of data using Fisher's function.

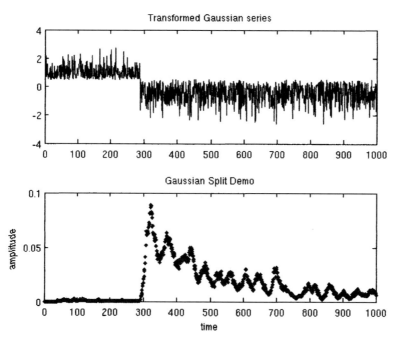

Figure 3: The separated two groups of data from one Gaussian series and split by Fisher's discriminant.

All the simulation results indicate that Fisher's Linear Discriminant is a feasible solution to the problem of detecting real physiological state change.

2.2 Real Data

We have completed multi-task psychological tests on a couple of subjects to obtain the IBI data for our detection. The tasks range from games, vigilance, easy and hard math which signify different degrees of psychological significance. Every subject will first go through a baseline calibration period where a physiologically neutral line is to be determined, then he or she will go through a series of tasks mentioned above. A typical arousal meter output of a subject is shown in Figure 4.

287

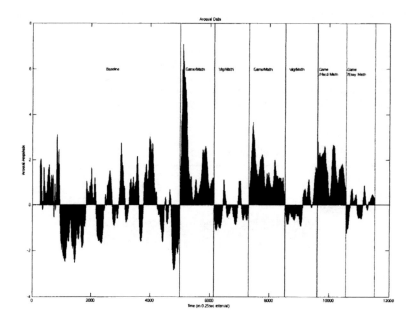

Figure 4: A typical arousal taken from a subject undergoing a series of tasks.

The arousal has been calculated by taking a 256 point FFT of IBI data from subjects at a sampling rate of 4 samples/second. So the arousal data are correlated in a 256-point window, which could pose a somewhat hindrance to Fisher's discriminant because of Fisher's assumption of statistical independence between samples.

The histograms of real data are similar to Gaussian, but have too much noise and correlation in them, which makes the detection much more difficult than the simulated Gaussian data. A typical histogram of such data is shown in Figure 5. It can be seen from the figure that the two groups of data together form a somewhat Gaussian distribution, each representing a particular part of it, one more negative and the other one more positive, with noise and correlation built into it. The correlation comes from the FFT transform mentioned above, while the noise is a common phenomenon in real world data. Both these drawbacks are hard to remove from the data and that is the real difficulty in processing the real data compared with the simulated data.

288

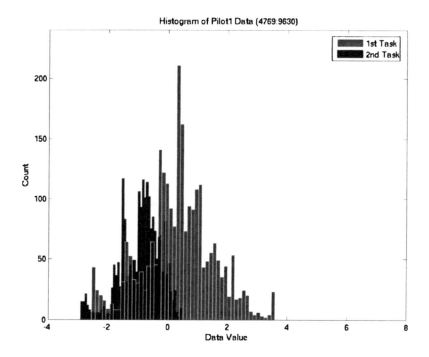

Figure 5: Histogram of arousal data from a subject undergoing a typical pair of tasks

3 Results

During the implementation of applying Fisher's Linear Discriminant to the real data collected from human subjects, we encounter the problem that the FLD output of real data does not show similarity with our simulation counterparts due to the inherent noise and correlation in the data. However after repeated testing and research by varying the minimum window size for each of the two presumed groups, results pretty similar to the simulated data emerge. Successful detection results for a typical pair of tasks are given in Figure 6. In the figure, the top 3 subplots are the original arousal data from a subject, the Fisher's value and the split positions detected by the Fisher's discriminant respectively for window size 300, while the lower 3 subplots are those for window size 800. It is evident that for window size 300, there are multiple peaks with the global peak meaningless while for 800 the global peak is almost at the split position with some other smaller local peaks. The results of other task pairs are more or less similar.

It should be noted that the success of detection depends greatly on the minimum window size selected. The smaller the window, the worse the detection result; the bigger, the better. Based on the current data at hand, the detection will always succeed if the window is large enough. This has proposed the question of time resolution in this research, that is, how small is the time interval that we can detect state change? This is also part of the whole question to be answered both physiologically and computationally in this study, and will be addressed in the ongoing research when data of more subjects are available.

By and large, all the current data available have proved and validated the successful application of Fisher's function in the detection of physiological state change.

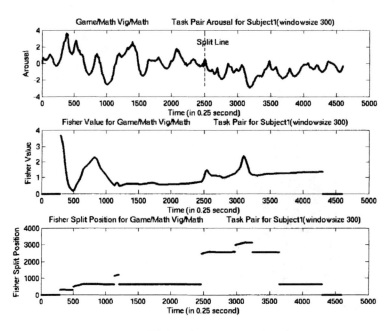

(a) Window size at 300, poor detection

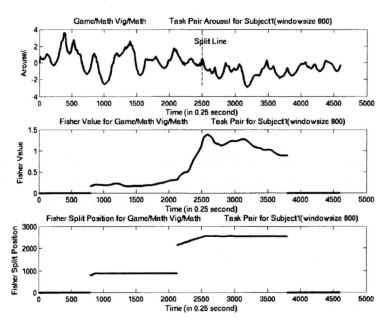

(b) Window size at 800, good detection

Figure 6: Detection of state change in a task pair by Fisher's discriminant with different minimum window sizes (a) 300 (b) 800.

4 Discussion and Conclusions

A new approach to detect state change in physiological signals has been proposed and successfully implemented on the IBI data from a number of human subjects undergoing tasks with different psychological significance. The Fisher's discriminant algorithm only requires the computation of mean and variance, is simple to implement and can be used in a real-time computer embedded system. The tests based on a couple of subjects are successful and promising, and more tests are needed to help us gain a deeper and more thorough understanding of underlying mechanisms in human physiological systems.

Besides testing more subjects, further research will also concentrate on the optimal selection of minimum window size, incorporating the Fisher's value and split positions into one rule set for monitoring and triggering detection.

This approach can be potentially applied to a wide range of scenarios such as training, stressful repetitive work (e.g. air-traffic control, long-distance driving etc.) and military operations. This kind of detection problem is very similar to finding the split points between any pair of four seasons in one year if you have a temperature variation curve in a certain year as shown in Figure 7. And it could also be applied to the detection of flood or drought based on the data of local rainfall or water levels, or the detection of individual obesity based on the data of daily weight measurements. There are potentially numerous applications.

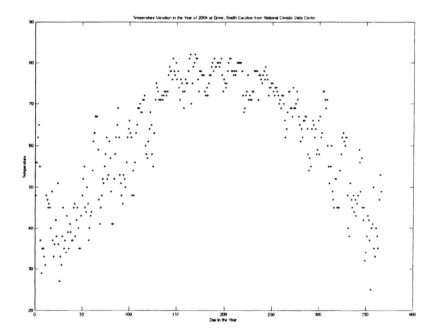

Figure 7: Temperature variation in the year 2004 at Greer, South Carolina. (Data courtesy from National Climatic Data Center at http://www1.ncdc.noaa.gov, a website freely accessible by the general public)

References

Anderson K.J. (1994). Impulsivity, caffeine, and task difficulty: A within-subjects test of the Yerkes-Dodson law. *Personality and Individual Differences*, 16, 813-829.

Hembree R. (1988). Correlates, causes, effects, and treatment of test anxiety. *Review of Educational Research*, 58, 47-77.

Hoover A. and Muth E. (2004). A Real-Time Index of Vagal Activity, *International Journal of Human-Computer Interaction*, Vol.17, No.2, 197-209.

Robazza C., Bortoli L., and Nougier V. (1998). Physiological arousal and performance in elite archers: A field study. *European Psychologist*, 3, 263-270.

Roscoe A.H. (1992). Assessing pilot workload: Why measure heart rate, HRV and respiration? *Biological Psychology*, 34, 259-287.

Schalkoff R.J. (1992). Pattern recognition: statistical, structural, and neural approaches. New York: J.Wiley.

Sokoll G.R. and Mynatt C.R. (1984). Arousal and free throw shooting. *Paper presented at the meeting of the Midwestern Psychological Association*.

Facial Temperature as a Measure of Operator State

Hans J.A. Veltman
Wouter K. Vos

TNO Human Factors
Soesterberg, The Netherlands
hans.veltman@tno.nl
wouter.vos@tno.nl

Abstract

We conducted an experiment to explore the relation between facial temperature and mental effort. Participants had to perform mentally demanding tasks while their face was captured with an infrared camera. The temperature of the nose decreased during these tasks and increased during the successive rest blocks. Other parts of the face did not change due to mental effort. The advantage of this workload measure is that it can provide objective and real-time information about mental effort of operators without attaching sensors to an operator. This measure can be used to measure the workload of operators in relatively stable environmental conditions.

1   Introduction

The need for operator functional state assessment is prompted by a growing worldwide concern with the consequences of performance break down by operators in safety-critical task environments. Operators more often have to deal with complex systems with an increasing level of automation. Despite (or because of) increasing automation, the human operator has an increasingly central role in the execution of tasks, in many cases made more difficult by increased mental workload. Although people can adapt efficiently to changing task demands, a high mental workload does have negative effects such as an increased likelihood of human error.

There are many different techniques to measure mental workload. These techniques can be categorised into performance measures, subjective measures and physiological measures. These different types of measures do not provide the same information (Veltman & Jansen, 2003). Performance is often difficult to measure in applied situations and when it can be measured, it often does not provide adequate information. Operators have the ability to adapt to changing task demands by investing more effort and therefore, an adequate level of performance can often be maintained at the cost of high workload. If the workload becomes too high, the performance often decreases dramatically. It is important to have information about the state of an operator before the level of performance decreases. Subjective workload measures provide more information about the workload but these measures are also difficult to obtain in applied situations. Finally, physiological measures mainly reflect the amount of mental effort that an operator has to invest in order to perform the task adequately. They can provide continuous and objective information about the state of an operator. This is necessary if one wants to prevent a decrease of performance (Hockey, 2003).

An important disadvantage of physiological measures is that most often electrodes or other sensors have to be attached to the person, which restricts the use of these measures in many applied settings. The measurement of facial skin temperature by means of an infrared camera might not have this practical limitation. There are some indications that the face temperature, especially the temperature at the nose, decreases when mental workload increases (e.g. Genno et al., 1997).

In this paper we describe an experiment in which the applicability of facial temperature for the assessment of mental workload is further explored. In this experiment participants had to perform mentally demanding tasks during which the facial temperature was measured with an infrared camera. We explored if the facial temperature changes due to mental effort. Moreover, we explored the most sensitive locations on the face and the sensitivity to different levels of taskload.

This experiment is part of a research program in which the possibility for adaptive automation is investigated. Adaptive automation is a concept in which the level of automation is adapted to a specific situation. The state of the operator can provide relevant information for adaptive automation such as a high workload of the operator. If the mental workload of an operator is too high, the overall performance might increase if the taskload is reduced. This can be accomplished for example by taking over some tasks from the operator, present some tasks to another operator, or wait to present less relevant information until the workload of the operator is reduced.

Facial temperature seems to be a promising element in adaptive automation concept, because it might provide objective information about the workload of an operator and it can be obtained relatively easily.

2 Method

2.1 Participants

The experiment has been performed on eight voluntary participants, six males and two females, their ages ranging from 23 to 41.

2.2 Task

The experiment consisted of seven blocks of three minutes each, following immediately after one another (see Figure 1). The first, third, fifth and seventh blocks were rest blocks. The participants were asked to relax with their eyes open during these blocks. During the other blocks a cognitive demanding task was presented in order to increase the workload of the participant. We used the auditive version of the Continuous Memory-Task (CMT) that has been shown to be a highly cognitive demanding task in earlier experiments (e.g. Veltman & Gaillard, 1998). The participants had to remember two or four target letters (A-B or A-B-X-Y) and had to press the left button on a response box whenever one of the target letters was presented for the first time. Every time a target letter was presented for the second time, the right button on the response box had to be pressed and the participant had to restart counting. Direct feedback was given when the participant had to press the right button. The word "okay" was heard after a correct response and the word "nope" was heard when the right button was incorrectly pressed or when no reaction was provided after a target letter appeared for the second time. The participant had to restart counting after feedback was provided. This means that he also had to restart counting when a target letter was presented once and the right button was pressed. A "ding" sound was played to alert the participant that a block ended and a new block started. The CMT is mentally demanding because the participants have to compare each letter with the letters from the memory set and more importantly, they have to use their working memory continuously. In the remainder of the article CMTs with two and four target letters will be indicated respectively with CMT2 and CMT4.

Figure 1: Scheme of the experimental conditions. Each block lasted three minutes. CMT2 is the Continuous Memory-Task with two target letters and CMT4 is the Continuous Memory-Task with four target letters.

2.3 Procedures and Apparatus

During the experiment the participant was comfortably seated on a chair in front of a table. A chin rest was used in order to stabilise the head. An infrared camera was positioned 1 meter from the participants. The infrared images were saved on a PC memory card. A two-button response box was positioned on the table in front of the participant. The left button had to be pressed when the participant heard a target letter for the first time and the right button when the participant heard the target letter for the second time. Two loudspeakers were used to present the CMT. Together with the response box they were connected to a PC that controlled the experiment.

We used a FLIR SC 2000 infrared camera that was able to take temperature pictures with an accuracy of 0.07 °C (14 bit). The resolution of the camera was 320 x 240 pixels. The camera took pictures with an interval time of 5 sec, resulting in 36 pictures of the face in each block.

Before the experiment the participant was trained in the CMT: one minute training for the CMT2 and one minute of training for the CMT4. The training was primarily meant to inform the participant about the goal of the task. The participants were told that they had to make fast responses and had to avoid errors.

The task was meant to induce a high mental effort only. The performance on the task was not relevant and therefore, this will not be presented.

Since it was expected that the nose temperature would be the most interesting area, four points on the nose were selected manually in the first picture of each participant using Matlab 6.5.1. Another 13 points were selected on the rest of the face (see Table 1 and Figure 2).

Although a chin rest was used to stabilise the participant's head, they were not able to keep their head at the same position throughout the experiment. Therefore, a rectangle around the nose was selected in the first image. This rectangle was correlated in the X and Y-axis with a larger rectangle in the consecutive pictures (20 pixels larger at each side of the original rectangle). Based on this two-dimensional correlation, the points in consecutive pictures were shifted. The adjusted positions were visually checked and it appeared that all selected points remained in place relative to the head.

Table 1: Measurement locations on the face.

NR.	Location
1	Middle forehead
2	Left side forehead
3	Right side forehead
4	Upper inside left eye
5	Lower inside left eye
6	Outside left eye
7	Upper inside right eye
8	Lower inside right eye
9	Outside right eye
10	Nose bridge
11	Left side nose
12	Right side nose
13	Nose tip
14	Left cheekbone
15	Right cheekbone
16	Between nose and upper lip
17	Between lower lip and chin

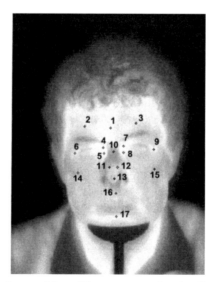

Figure 2: Infrared image of one participant. The measurement locations listed in Table 1 are indicated.

2.4 Data Analysis

In each block we had 36 temperature values for each location. The temperature of each location consisted of an average in a circular area with a diameter of 5 pixels (≈ 7 mm). Within each block, we fitted straight lines (least squares method) through the data points for each location. Measurements outside ± 3 times the standard deviation range were removed. This resulted in 17 (locations) x 7 (blocks) regression lines for each participant.

We used an ANOVA repeated measurement analysis to analyse whether differences in temperature were significant. The following comparisons were made:

- differences between the seven experimental blocks (one factor with seven levels);
- differences between the three task blocks (one factor with three levels);
- differences between the four rest blocks (one factor with four levels);
- differences between the three locations at the nose that appeared relevant in the previous analysis. For this analysis we calculated the average value of the three rest blocks and the average values of the three task blocks and tested whether the temperature changes were different at the three locations.

Differences were further explored with post-hoc analysis (Tukey HSD).

3 Results

Figure 3 presents an example of the data for one participant. The temperature at the forehead and the left side of the nose is plotted for each camera frame. The regression lines are also plotted in this figure. The slopes of these regression lines were used for the statistical analysis. This picture shows that the temperature at the nose decreased during task blocks and increased during the following rest block. The temperature at the forehead is almost stable for this participant. The temperature of the nose changed substantially during a rest and a task block.

The change of the nose temperature started almost directly after the start of a block. This indicates that it is a fast reacting measure. The data of the other seven participants showed similar patterns, but the average range was lower than the data in Figure 3.

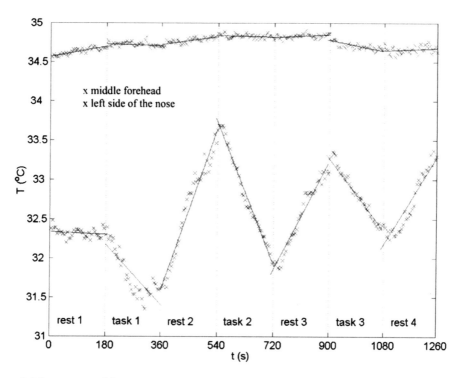

Figure 3: Measurements of the temperature around two locations on the face (forehead and nose) of one participant. Straight lines were fitted through the data with the least squares method for each block.

Figure 4 presents the average temperature slopes for all locations. The rest blocks are presented separately from the task blocks in this figure. The strongest differences between the blocks were found at the left side of the nose [$(F_{(6,42)}=12.25$, $p<0.001)$], at the right side of the nose [$(F_{(6,42)}=10.81$, $p<0.001)$] and at the nose tip [$F_{(6,42)}=9.57$, $p<0.001$]. During the task blocks, the temperature decreased substantially for these locations and during the rest blocks the temperature increased. Post-hoc analysis revealed that the temperature slopes during all three task blocks differed significantly from all four rest blocks for the three locations at the nose (location 11, 12 and 13).

The temperature decreases during the three task blocks was significantly different for the left side of the nose [$F_{(4,14)}=6.6$, $p<0.01$] and for the nose tip [$F_{(4,14)}=4.86$, $p<0.05$]. Post-hoc analysis revealed that the temperature decreased more during CMT4 (second task) than during the first CMT2 task.

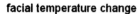

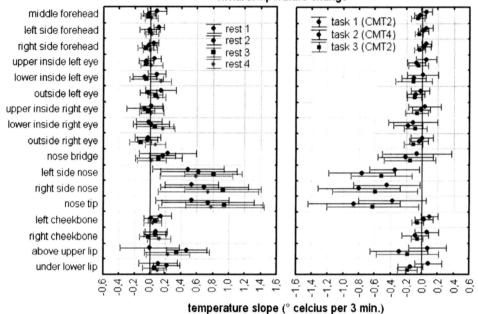

Figure 4: Average temperature change for each measurement location. The left figure shows the temperature changes during rest blocks and the right figure shows the temperature changes during task blocks.

Figure 4 also shows a rather small decrease in temperature at the upper lip (location 16) during task2 and task3 and an increase during rest2, rest3 and rest4. Statistical analysis revealed that only the temperature during task2 was different from rest2. No other statistical effects were found for this location.

Some other significant effects were found, but these effects were very small and were not systematic. Therefore, they are not described here. The smallest temperature changes were found at the forehead. These differences were far from significant.

Figure 5 presents the temperature slopes at the tip of the nose for all eight participants. This figure shows that seven participants had a systematic decrease in nose temperature during task blocks and an increase during rest blocks. Only one participant did not show a systematic difference between the rest and task blocks. There is a large variation between the participants. Some participants only show a small decrease in temperature while other participants show a substantial difference.

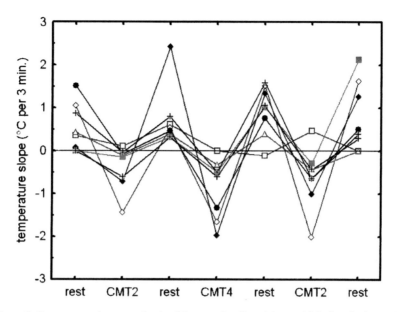

Figure 5: Temperature changes at the tip of the nose for all participants (eight lines in the graph)

4 Discussion

From the results it can be concluded that the skin temperature of the nose is correlated with mental effort. In the case of a mental task the nose temperature drops and when the participant rests the temperature rises again. The task and rest blocks in the present experiment lasted three minutes and the data showed that during these blocks, the change of temperature did not reach a plateau. This can be seen in Figure 3. The other participants showed the same pattern. It is expected that when the task and rest blocks last longer, a plateau will be reached. The position of this plateau might be related to the magnitude of the mental load. Further experiments should provide an answer to the question how long a change in temperature due to mental effort lasts and how long it takes to get back to baseline.

We found a small difference between the nose temperature during the four-target CMT (task2) and the temperature during the two-letter CMTs (task1 and task3). The temperature decrease during task2 (four-letter CMT) was stronger than during task1 (the first two-letter CMT) for the left side of the nose and the nose tip. A smaller and not significant difference was found between the four-letter CMT and the second two-letter CMT. This indicates that there is a relation, albeit not a strong one, between the amount of mental effort invested and the decrease in nose temperature. It should be noted that a higher task demand does not necessarily result in an increased mental effort. It is possible that some participants already did their utmost best during the two-target CMT task. In that case, the more demanding four-target CMT task would not increase their effort investment anymore. The results of further experiments should provide information about the sensitivity of this measure to different levels of mental effort.

Information about the mental effort must be extracted from the change in temperature and can not be extracted from the absolute temperature. Thus, a decrease in nose temperature indicates that a person is investing mental effort and an increase indicates that he is recovering from mental effort. Mental effort is not the only factor that affects the nose temperature. Another important factor is the environmental temperature. If the environmental temperature decreases, it is very likely that the temperature of the face will also decrease. This might make the nose temperature less applicable in applied settings in which the temperature of the environment changes fast.

A possible method to correct for changes in environmental temperature is to calculate the nose temperature relative to the forehead temperature. The temperature of the forehead was not at all affected by mental effort. If the temperature in the environment changes, it is likely that this affects both the forehead and the nose. If the difference between the temperature at the nose and forehead is used, facial temperature as an index of mental effort will be less affected by changes of the environment. However, it is reasonable to assume that changes in environmental temperature will affect the nose temperature stronger than the temperature of the forehead. Therefore, a model that describes the relation between changes in environmental temperature and temperature changes at the nose and forehead is required for a valid correction.

An infrared camera will often capture fixed elements in the environment (e.g., a wall behind the operator). The temperature of the environment is most often available and this can also be used to correct for changes in face temperature due to environmental changes.

The decrease in nose temperature is most probably due to a dilation of the veins in the nose. This causes a reduction of blood flow and as a consequence the nose temperature will adapt faster to the environmental temperature. The diameter of the veins in the nose is mainly regulated by the sympathetic part of the autonomic nervous system (Widdicombe, 1993; Lung, 1995). Mental effort causes a reduced para-sympathetic activity and an increased sympathetic activity. This is the reason why most physiological measures, such as cardiovascular measures are sensitive to mental effort.

An increased sympathetic activity is probably not the only cause of the decrease in temperature. Mental effort often results in increased ventilation (Wientjes, 1993; Veltman & Gaillard, 1998). More relatively cold air might flow through the nose during task performance compared to rests and therefore, the nose temperature drops. The data of the present experiment indicate that respiration might be involved in the present results. The heads of the participants were fixed with a chin rest, which forces them to breathe through the nose. This causes airflow around the nose. We found a small difference in temperature on the measurement location between the lip and the nose. Although these differences were not statistically significant, it indicates that respiration might play a role. Further experiments must clarify what the exact mechanisms of the temperature change are. However, for the application of this measure it is less relevant because both mechanisms result in a temperature decrease during mental tasks.

It is reasonable to assume that the degree of temperature decrease due to mental effort is related to the environmental temperature. If the difference between the body and the environmental temperature is large, then reduced blood flow and/or increased ventilation will cause a large decrease in nose temperature. When there is almost no difference between the body and the environmental temperature, then it is not likely that mental effort will affect the nose temperature. In a hot environment, the relation between mental effort and nose temperature is not clear. Sweat will become an important intervening factor because it will cool the face, especially the forehead. How this will interact with the effect of mental effort is not clear.

The advantage of using the face temperature compared to other physiological measures such as heart rate and heart rate variability is that no electrodes have to be attached. It is likely that the temperature changes are related to the effects found in cardiovascular measures, because cardiovascular measures are also sensitive to changes of the sympathetic nervous system.

Despite the relative small temperature changes, the effects were rather robust. Seven participants showed systematic changes between rest and task blocks. Only one participant did not show differences between rest and task blocks. This is found more often with physiological measures. Therefore, it is important to have information about the physiological reactions of a particular operator before physiological measures (including face temperature) can be used in applied settings.

There are several applications for workload measures. One of the applications is the evaluation of interfaces with regard to differences in mental effort investment. The nose temperature can be used for this application. Another application is adaptive automation for which it is very important to have highly reliable information about the effort investment. Operators will never accept that a system will take over tasks or delay less relevant tasks based on incorrect measurements. The reliability of nose temperature is too low to be used in adaptive automation. The reliability, with which effort can be measured, can be increased if the nose temperature is combined with other measures. Preferably this should be measures that do not require sensors to be attached to the operator or wireless sensors that do not hinder the operator in performing tasks. Examples of such measures are wireless heart rate sensors and eye point of gaze measures. Eye point of gaze can be measures with remote cameras, which can be combined with infrared cameras. The applicability of eye point of gaze measures will be tested within the present research programme.

In the next phase of the project we will get more experimental data about the nose temperature, especially about the applicability during continuous tasks. In the present experiment the difference between low mental effort (rest blocks) and high mental effort (task blocks) was very clear. In most applied settings the operator is performing more continuous tasks in which the differences in mental effort are smaller. We will conduct an experiment to see how the nose temperature can be used is these situations. Furthermore, we will start to develop a method for automatic detection of the nose and forehead form infrared pictures. Face tracking algorithms already exist for pictures taken by visual cameras. The question is if they would also work with pictures taken by an infrared camera. An automatic tracking system is necessary to get real-time information from the operator, which is an important step for the use of adaptive automation.

References

Genno, H., Ishikawa, K., Kanbara, O., Kikumoto, M., Fujiwara, Y., & Suzuki, R. et al. (1997). Using facial skin temperature to objectively evaluate sensations. *International Journal of Industrial Ergonomics*, 19, 161-171.

Hockey, G. R. J. (2003). Operator Functional State as a Framework for the Assessment of Performance Degradation. In G. R .J. Hockey, A. W. K. Gaillard, & A. Burov (Eds.), *Operator Functional State: The assessment and Prediction of Human Performance Degradation in Complex Tasks* (pp. 8-23). Amsterdam: IOS Press, NATO Science series.

Lung, M. A. (1995). The role of the autonomic nerves in the control of nasal circulation. *Biological Signals*, 4, 179-185.

Veltman, J. A., & Gaillard, A. W. K. (1998). Physiological workload reactions to increasing levels of task difficulty. *Ergonomics*, 5, 656-669.

Veltman, J. A., & Jansen, C. (2003). Differentiation of mental effort measures: consequences for adaptive automation. In G. R. J. Hockey, A. W. K. Gaillard, & O. Burov (Eds.), *Operator Functional State: The Assessment and Prediction of Human Performance Degradation in Complex Tasks* (pp. 249-259). Amsterdam: IOS Press.

Widdicombe, J. (1993). The airway vasculature. *Experimental Physiology*, 78, 433-452.

Wientjes, C. J. E. (1993). Psychological influence upon breathing: Situational and dispositional aspects. Thesis. University of Brabant.

Robust Feature Extraction and Classification of EEG Spectra for Real-time Classification of Cognitive State

John Wallerius
Leonard J. Trejo
Pacific Development and Technology, LLC
Mountain View, CA
jwallerius@pacdel.com

Robert Matthews
Quantum Applied Science and Research
San Diego, CA
robm@quasarusa.com

Roman Rosipal
Austrian Research Institute For Artificial Intelligence
Vienna, Austria
roman@oefai.at

John A. Caldwell
Brooks Air Force Base
San Antonio, TX
John.Caldwell@brooks.af.mil

Abstract

We developed an algorithm to extract and combine EEG spectral features, which effectively classifies cognitive states and is robust in the presence of sensor noise. The algorithm uses a partial-least squares (PLS) algorithm to decompose multi-sensor EEG spectra into a small set of components. These components are chosen such that they are linearly orthogonal to each other and maximize the covariance between the EEG input variables and discrete output variables, such as different cognitive states. A second stage of the algorithm uses robust cross-validation methods to select the optimal number of components for classification. The algorithm can process practically unlimited input channels and spectral resolutions. No a priori information about the spatial or spectral distributions of the sources is required. A final stage of the algorithm uses robust cross-validation methods to reduce the set of electrodes to the minimum set that does not sacrifice classification accuracy. We tested the algorithm with simulated EEG data in which mental fatigue was represented by increases frontal theta and occipital alpha band power. We synthesized EEG from bilateral pairs of frontal theta sources and occipital alpha sources generated by second-order autoregressive processes. We then excited the sources with white noise and mixed the source signals into a 19-channel sensor array (10-20 system) with the three-sphere head model of the BESA Dipole Simulator. We generated synthetic EEG for 60 2-second long epochs. Separate EEG series represented the alert and fatigued states, between which alpha and theta amplitudes differed on average by a factor of two. We then corrupted the data with broadband white noise to yield signal-to-noise ratios (SNR) between 10 dB and -15 dB. We used half of the segments for training and cross-validation of the classifier and the other half for testing. Over this range of SNRs, classifier performance degraded smoothly, with test proportions correct (TPC) of 94%, 95%, 96%, 97%, 84%, and 53% for SNRs of 10 dB, 5 dB, 0 dB, -5 dB, -10 dB, and -15 dB, respectively. We will discuss the practical implications of this algorithm for real-time state classification and an off-line application to EEG data taken from pilots who performed cognitive and flight tests over a 37-hour period of extended wakefulness.

1 Introduction

Laboratory and operational research have shown that real-time classification of cognitive states, such as mental fatigue, may be performed using physiological measures, such as EEG recordings, as predictors or correlates of those states (St. John, Kobus, & Morrison, 2003). The purpose of this study was to code and test an algorithm for real-time classification of fatigue states, identify subsets of EEG electrodes that provide adequate performance in predicting cognitive state, and quantify the effect of reduced signal-to-noise ratio (SNR) on the accuracy of the method. We sought to address five specific objectives: a) design an EEG-based computational system for classification of cognitive states, b) simulate and test the effects of reduced electrode density on the accuracy of classification, c) simulate and test the effects of reduced signal-to-noise ratio (SNR) on the accuracy of classification, d) test the effects of reduced electrode density and SNR in human EEGs from a fatigue-inducing task, e) estimate the range of useful electrode densities and SNRs for robust EEG-based classification of fatigue states.

2 Design of the EEG-based classifier

2.1 Selection of EEG features

Previous studies have reported changes in the EEG spectrum as alertness declines. For example, the proportion of low frequency EEG waves, such as theta and alpha rhythms, may increase while higher frequency waves, such as beta rhythms may decrease. In one study, as alertness fell and error rates rose in a vigilance task, researchers found progressive increases in EEG power at frequencies centered near 4 and 14 Hz (Makeig & Inlow, 1993). Thus, the relative power of theta, alpha, and other EEG rhythms may serve to indicate the level of fatigue that subjects experience. However, the EEG spectral changes that relate to cognitive fatigue, in the absence of alertness decrements are unclear because most experiments have used vigilance-like paradigms to examine fatigue or short experimental sessions which induce high workload. The patterns of change in EEG that follow the onset of cognitive fatigue appear to be complex and subject-dependent (Gevins et al., 1990). For these reasons, we designed our algorithm to work with high-resolution estimates of the multi-channel EEG frequency spectrum. Multi-electrode power spectra of individual EEG segments served as the EEG features for training and testing the classifier.

2.2 The PLS algorithm

Recently, a new algorithm derived from machine learning and statistical pattern recognition has proven to be effective for feature extraction and classification of EEG data (Rosipal & Trejo, 2001; Rosipal, Trejo, & Matthews, 2003). The method relies on a core algorithm of partial least squares (PLS). Similar to principal components regression (PCR), PLS is a method based on the projection of input (explanatory) variables to the latent variables (components). However, in contrast to PCR, PLS creates the components by modeling the relationship between input and output variables while maintaining most of the information in the input variables. PLS is useful in situations such as EEG analysis, where the number of explanatory variables exceeds the number of observations and/or a high level of multi-collinearity among those variables is assumed. The PLS method has been paired with support vector classifiers and linear classifiers for robust classification (Rosipal, et al., 2003). Combining PLS with computational kernels (KPLS) allows for extraction of a larger number of PLS components than other methods. This may be useful for isolating unique sources of variance in EEG data, much like independent components analysis (ICA, Bell & Sejnowski, 1995). Unlike ICA, KPLS methods lead to unambiguous results that lend themselves to automatic interpretation. Both linear and non-linear kernels may be applied, the latter allowing for classification of complex multivariate distributions of data. In the experiments documented in this paper, we used a linear kernel. Successful applications of this method have been documented in real-time processing of human EEG for brain-computer systems (Trejo, et al., 2003) and modeling and measuring cognitive fatigue (Trejo et al., 2004).

We regard the power spectral density of single EEG epochs of C channels and F spectral lines as a vector $\mathbf{x}$ (M), with $M = C \times F$ dimensions. Each $\mathbf{x}_i$ ($i = 1, 2, \ldots, n$) is a row vector of a matrix of explanatory variables, $\mathbf{X}$ ($n \times M$), with M variables and n observations. The n observations are the power spectral densities or PSDs of single EEG epochs from two classes, (e.g., experimental conditions, alert/fatigued, etc.). We regard the class membership of each EEG epoch as a matrix $\mathbf{Y}$ ($n \times 1$) in which Class 1 is assigned the value 1 and Class 2 is assigned the value -1. PLS models the relationship between the explanatory variables and class membership by decomposing $\mathbf{X}$ and $\mathbf{Y}$ into the form

$$\mathbf{X} = \mathbf{TP}^T + \mathbf{F}$$
$$\mathbf{Y} = \mathbf{UQ}^T + \mathbf{G}$$

where the $\mathbf{T}$ and $\mathbf{U}$ are ($n \times p$) matrices of p extracted score vectors (components or latent vectors), the ($N \times p$) matrix $\mathbf{P}$ and the ($M \times p$) matrix $\mathbf{Q}$ are matrices of loadings, and the ($n \times N$) matrix $\mathbf{F}$ and the ($n \times M$) matrix $\mathbf{G}$ are matrices of residuals. PLS seeks to maximize the covariance between the components of the explanatory variables and class membership. We use a method known as the nonlinear iterative partial least squares algorithm (NIPALS), which finds weight vectors $\mathbf{w}$, $\mathbf{c}$ such that

$$max_{|r| = |s| = 1}[cov(\mathbf{Xr}, \mathbf{Ys})]^2 = [cov(\mathbf{Xw}, \mathbf{Yc})]^2 = [cov(\mathbf{t}, \mathbf{u})]^2$$

where $cov(\mathbf{t}, \mathbf{u}) = \mathbf{t}^T\mathbf{u}/n$ denotes the sample covariance between the score vectors $\mathbf{t}$ and $\mathbf{u}$. Application of the weight vectors to normalized data produces component scores that serve as inputs to a classifier. We have tested both discretized linear regression (DLR) and support vector classifiers (SVC), and found that in most cases, DLR produces results that are nearly equal to SVC. For this study we rely exclusively on the DLR method.

2.3 High-level design of the classifier

The main blocks of the classifier development procedure are data preprocessing, classifier construction, electrode reduction, and SNR reduction (Figure 1). Each block includes several components or steps, as shown in Table 1.

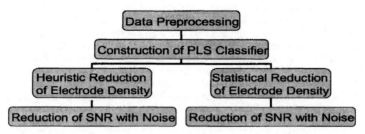

Figure 1: Block diagram of classifier development plan.

2.4 Simulation of EEG and classifier performance tests

We developed a novel system for simulation of EEG sources using autoregressive models (Anderson, Stolz, & Shamsunder, 1995; Figure 2A). This system served to create simulated EEG for testing the classifier with known inputs and SNR level. We simulated a 4-dipole EEG system as a summation of four independent autoregressive processes (BESA Dipole Simulator, MEGIS Software GmbH, Gräfelfing, Germany). The dipoles were situated symmetrically in frontal cortex and parietal cortex. The frontal pair of dipoles approximated sources of EEG theta rhythms, while the parietal pair approximated sources of alpha rhythms. We used the dipole simulator to run a forward solution through the BESA default 3-sphere head model to obtain the source-electrode coupling matrix between these dipole sources and a standard 10-20 system electrode montage (19 electrodes). Using this matrix and exciting the sources with white noise we generated pseudo EEG with properties suitable for testing the effects of noise and electrode density. Using a public-domain autoregressive model estimation package (Hayes, 1999) we solved for the best fitting AR models from the source model output (prior to spatial mixing) and verified that we could recover the model coefficients.

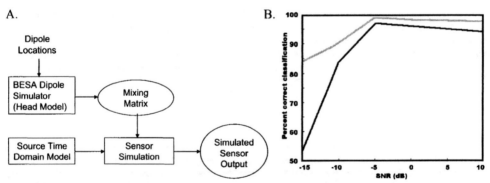

Figure 2: A. System for simulation of multi-channel EEG sources. B. KPLS classifier accuracies for simulated EEG data as a function of SNR. Green line is for training set, blue line is for test set.

We also estimated EEG and fatigue effects and chose reasonable parameters for mimicking the effects of cognitive fatigue (Trejo, et al., 2004) on the pseudo EEG signals. Fatigue was represented by a 50% increase in parietal alpha source amplitudes and a 25% increase in frontal theta source amplitudes over baseline conditions.

We adapted our PLS classifier code to run in Matlab (The Mathworks, Natick, MA) with visualization tools to allow us to inspect the latent variables (components) that are used to classify fatigue periods with reduced electrode density and noise effects. The generator and classifier components which had previously been tested individually are integrated in a single Matlab software system. We also made some adjustments to the code to facilitate efficiently searching the parameter space of SNRs and sensor subsets. We then determined the optimal classifier structure and parameters for the pseudo EEG signals. We performed several informal tests, which verified that our simulation: a) generates realistic EEG data for non-fatigued and fatigued states, b) derives a PLS classification model from a training subset of the simulated data, c) validates the classification model with additional simulated data.

Table 1. Procedure for development of KPLS classifiers. A. Data preprocessing. B. Construction of the PLS classifier. C. Heuristic reduction of electrode density. D. Statistical reduction of electrode density. E. SNR reduction.

A. Data Preprocessing

1 Visual inspection of data
2 Rejection of high artifact / bad data segments
3 EOG artifact removal
4 Bandpass filtering and downsampling
5 Epoch segmentation (10-s segments)
6 Frequency spectrum estimation
7 Binary epoch categorization (normal = 1, fatigued = -1) according to time on task.

B. Construction of the PLS Classifier

1 Partition epochs into training and test sets.
  a Randomly partition training set into equal-sized estimation and validation sets.
  b Use the estimation set to compute a linear PLS model of the binary epoch categories and record test proportion correct (TPC) for the validation set.
  c Repeat steps a. and b. for model orders (1, 2, ..., 10 PLS components) and record TPC.
  d From the 10 PLS models of step c., choose the PLS model order that optimizes the TPC.
2 Compute final PLS model with the fixed order of step 1d and apply to the test set to obtainTPC.

C. Heuristic Reduction of Electrode Density

1 Choose electrode regions that optimize electrode density and ease of placement.
  a Select electrodes for two density levels for each region
    i An oversampled level with 2-4 electrodes per region
    ii A critically sampled level with 1-2 electrodes per region
  b Define special sets of frontal electrodes (Fp1, Fpz, Fp2, with and without Fz).
2 Construct a PLS classifier for each electrode set and record the TPCs for training and test sets.

D. Statistical Reduction of Electrode Density

1 Start with the PLS model for the full set of electrodes.
  a Order the electrodes by the 75th percentile of the weights for each PLS component.
  b For the first PLS component select the electrode set of the model that maximizes TPC.
  c Deflate the input matrix by removing the influence of the first component.
  d Re-compute the PLS model with only the electrodes selected in b. and the deflated input matrix and select the electrode set that maximizes TPC for the second component.
  e Repeat b. – c. for up to 10 PLS components (Example: 1st component: electrodes 1 5 7 12 ; 2nd component: electrodes 3 2 5 4 6 7 etc.
  f Retain the electrode set that is the union of the sets for the 10 components (fully automated) or,
  g Retain the electrode set that "makes sense" after inspection of the components (supervised).
2 Choose two electrode densities based on minimizing PLS components and maximizing TPC values.
  a The oversampled set keeps TPC near the level of the full set on the curve of decreasing TPC vs. electrode density.
  b The critically sampled set keeps TPC near the 80% level on the curve of TPC vs. electrode density.

E. SNR Reduction

1 For each electrode set, add Gaussian noise to reduce SNR in EEG segments in the test set over six levels (dB): -3, -6, -9, -12, -15, -18.
2 Add noise to the test set data for each model, without recomputing the model. Then re-compute the noise-corrupted test-set TPCs. This simulates a classifier which is fixed and is subject to operational noise contamination. Repeat this for all the models developed on noise-free data : the complete set, the heuristically reduced sets, and the statistically reduced sets.

Finally, we formally tested the effects of electrode density and noise on classification accuracy of fatigue states represented by the pseudo EEG signals (Figure 2B). The signals consisted of 60 segments of 2-s length, split into training and test sets of 30 segments each. Noise was manipulated as a factor with six discrete SNR levels ranging from +10 to -15 dB. The dependent measure was the test percentage correct (TPC) for equal-sized test sets of EEG segments from the fatigue and baseline conditions. TPC ranged from 98% to 99% for training data between SNRs of +10 and -5 dB, then declined to 84% at -15 dB. TPC ranged from 94% to 97% for test data between SNRs of +10 and -5 dB, then declined to 53% at -15 dB. As expected, TPC fell to chance accuracy (50%) for low SNR levels. In our experience, human EEG signals from operational settings fall in the higher range of these SNRs (above zero), suggesting that the algorithm will perform satisfactorily in the presence of recording noise.

3   Test classifier testing system with human EEG.

Dr. John Caldwell of Brooks Air Force Base kindly obtained official authorization to provide EEG data for two subjects from a study of the effects of fatigue over a period of 37 hours of sustained wakefulness in pilots (Caldwell, et al., 2003). This study concluded that:

"Over the past 30 years, fatigue has contributed to a number of Air Force mishaps. Resource cutbacks combined with increased operational tempos, sustained operations, and night fighting could exacerbate the problem. Extended wakefulness and circadian factors can be especially problematic in military aviation where mission demands sometimes necessitate flights as long as 17-44 hours. To effectively counter fatigue in such operations, the effects of this threat must be objectively measured and understood. This study assessed F -117 A pilots during a 37-hour period of continuous wakefulness. Although none of the pilots crashed, substantial decrements were found in flight skills, reaction times, mood, and brain activation as early as after the 26th hour of continuous wakefulness. The greatest flight degradations occurred after 27-33 hours awake, even though many pilots believed their worst performance was earlier. The decrements found in this relatively-benign test environment may be more serious under operational conditions unless personnel anticipate the most dangerous times and administer valid fatigue countermeasures."

We acquired 1.6 gigabytes of EEG and EOG data for two subjects from that study and visually assessed levels of EOG artifact and data integrity using graphical displays. We also developed a strategy for labeling the data sets for training and testing of the classifier. We also received confirmation that the data we received were from two subjects for whom there was psychometric evidence of significant fatigue over time, as indicated by the Profile of Mood States and the Visual Analog Mood Scales. We developed classifiers for these data by assuming that the chronologically earlier data represents relatively non-fatigued states, and the data near the end of the 37 hour testing period represents highly fatigued states. We selected an early test period as non-fatigued and a later test period as fatigued. Respectively, these times were 2100 on day 1 versus 1900 on day 2 for subject 212, and 0400 on day 2 versus 1900 on day 2 for subject 220.

3.1   Heuristic reduction of electrode density

We performed combined testing of the signal-to-noise reduction and heuristic electrode reduction tests for the EEG data from the Brooks AFB fatigue study. Using the methods that we developed for simulated EEG data, we trained PLS classifiers for three heuristically chosen electrode set sizes: 21 (full montage), 12, and 4 (Table 2). Our heuristic was to preserve the midline electrodes, reduce electrode densities evenly and symmetrically at off-midline sites.

Table 2. Electrodes selected by heuristic method for SNR tests.

Set Size	Electrodes
21	Fp1, Fpz, Fp2, F7, F3, Fz, F4, F8, T3, C3, Cz, C4, T4, T5, P3, Pz, P4, T6, O1, Oz, O2
12	F3, Fz, F4, C3, Cz, C4, P3, Pz, P4, O1, Oz, O2
4	Fz, Cz, Pz, Oz

The EEG segments for training and testing the classifiers were of duration 1 s, and were drawn from the early (non-fatigued) and late (fatigued) test sessions described above. From periods of 1-minute-long eyes-open, resting EEG recordings in each session, we obtained 60 contiguous epochs, of which 30 each were randomly chosen and set aside for training and testing. For subject 212, the earliest session was 2300 hrs on Day 1 and the latest session was 1900 hrs on Day 2. For subject 220, the earliest session was 0400 hrs on Day 2 and the latest session was 1900 hrs on Day 2. After training and testing the classifiers with data to which no noise was added, we added broadband white noise to the EEG in a range of SNRs from -3 to -18 dB. The classifiers were trained once for each electrode set size and tested at each noise level. This train-and-test procedure was repeated three times for each electrode set size.

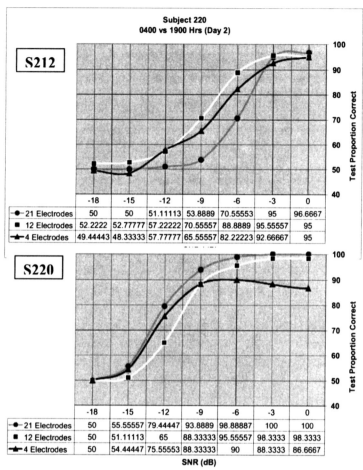

Figure 3: Test accuracy of KPLS classifier for non-fatigued (early) versus fatigued (late) 1-s eyes-open EEG epochs as a function of electrode set size and signal-to-noise (SNR) ration in two subjects (S212, S220). Each point is an average of three replications of training and testing the classifier.

The method for determining the SNR was to first estimate for each epoch, the amplitude of the baseline noise recorded by the EEG amplifiers outside the EEG bandwidth. We estimated the baseline noise power, N, by calculating the average power in the highest part of the recorded EEG band, from 80 to 100 Hz in Watts/Hz. We assumed for purposes of the analysis that the noise power was approximately constant over the 0 to 100 Hz band. We used the average recorded EEG signal power in the band from 0-25 Hz as a signal estimate, which we arbitrarily set to be 0 dB SNR. We did not try to estimate the actual signal to noise ratio; we just estimated the baseline noise power and defined the recorded EEG SNR to be 0 dB. We added additional noise in 3 dB (factors of 2 in power), so that the test SNRs are as follows: S/N (no noise added) ==> 0 dB, S/(N + aN) (add additional noise equal to estimated baseline noise power) ==> -3 dB, etc. The full set of SNRs tested was: 0, -3, -6, -9. -12, -15, -18 dB. As expected, classification accuracy decreased over the range of SNRs (Figure 3).

Overall, we obtained the highest accuracies with 21 electrodes, and lower accuracies with 12 and 4 electrodes, respectively. Accuracies were also higher for Subject 212 than for Subject 220. It is possible this difference derives from a larger interval between the test sessions in Subject 212 than in Subject 220. Interestingly, for Subject 220, the accuracies with 12 and 4 electrodes were both nearly 90% correct at a SNR of -9 dB, only 5% lower than for 21 electrodes. This suggests that with reasonable fatigue effects, out KPLS algorithm may be robust under conditions of severe noise contamination of the EEG, where the added noise is as large as 6 to 7 times the initial background noise level. Since EEG recordings typically have positive SNRs, our method of estimating SNR as a multiple of background noise is conservative, because we assigned the uncorrupted recordings a SNR of 0 dB.

3.2 Frontal electrode sets

We also repeated the preceding tests with two special sets of three or four frontal electrodes, [Fp1 Fpz Fp2] or [Fp1 Fpz Fp2 Fz]. What we had in mind was to select electrodes for which application is unencumbered by scalp hair. This may be beneficial for non-contact E-field sensors and for mounting sensors in headgear. In general, the effects of limiting the set to these four frontal electrodes were positive. Surprisingly, TPCs for the frontal set were slightly higher than for the four midline electrodes, at SNR levels of about -6 dB and below, although the maximal TPCs were lower than for 21 electrodes (Figure 4). In a similar manner, addition of electrode Fz improved high-SNR TPCs but reduced low-SNR TPCs. Thus in the region of the TPC/SNR curve that covers the worst expected levels of noise degradation, the frontal sets performed as well as or better than the midline set. In the other subject (212, not shown), the TPCs for the frontal set between -6 and -9 dB ranged from 82 to 85%. Overall, the results are encouraging for an easily-applied frontal set of electrodes in fatigue classification systems.

3.3 Statistical reduction of electrode density

Our final analyses considered the results of a selection of reduced electrode sets by a statistical learning method. For each subject we performed the steps outlined in Table 1, using a limit of 10 PLS components in each PLS model.

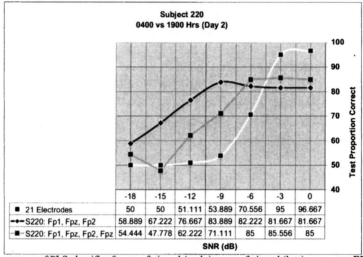

Figure 4: Test accuracy of PLS classifier for non-fatigued (early) versus fatigued (late) eyes-open EEG epochs as a function of electrode set size (21 or Frontal set) and signal-to-noise (SNR) ratio in subject 220. Each point is an average of three replications of training the classifier.

Each model was based on training data, and 10-fold cross validation for determining the optimum number of components. That optimum number, the one that minimized validation set errors, was 4 for each subject. The union of selected electrodes for all four components was then used to build a new PLS classifier using the training set data, as in the heuristic, and frontal sets. For subject 212, that union was 19 (out of 21) electrodes across the four factors. For subject 220, 6 electrodes were in the union set. We summarized the selected electrodes and conditions of each model in Table 3. These served as the *oversampled* sets described in Table 1. We also produced two *critically* sampled sets, on which retained the first four electrodes indicated by their order of importance within the first two PLS factors. The second set retained three such electrodes. We observed that we could produce common sets for both subjects by including electrodes Cz and T5 for both subjects, although each of these electrodes was only indicated in one of each subject's optimal set. Our thinking here was to produce a set that could be used across subjects, retaining the important electrodes from each subject. The two critically sampled sets were [Cz, T5, T6, Oz] or [T5, T6, Oz]. Thus, these electrodes sets met two criteria: they had the greatest sum of PLS weights in the first and second components, and they were indicated in this way for at least one subject.

Table 3. Statistically selected sets of electrodes for PLS classifier in two subjects.

Subject	Optimal/Maximum numbers of Components in PLS Classifier	Statistically Selected Electrode Set (not in order of importance)							Number of Electrodes
212	4 / 10	fp1	fpz	fp2	f7	f3	fz	f4	19
		f8	t3	c3	cz	c4	t4	p3	
		pz	p4	t6	oz	o2			
220	4 / 10	fp1	c4	t5	p4	t6	oz		6

We constructed and tested the classifier as for the heuristic and frontal sets, then corrupted the data with noise and re-tested. The results were not surprising, in that SNR caused the accuracy to degrade smoothly and in roughly the same way as in the other frontal and heuristic simulations (Figure 5). In the figure, the two subjects results are plotted for each of two levels of reduction. The first level is the "optimum" electrode set chosen by our algorithm so as to minimize classification error in cross-validation tests (S 212-19 electrodes, S 220-6 electrodes). The critical levels of reduction are S 212-3, S 212-4, S220-3, S 220-4 for the two subjects and the three- and four-electrode sets respectively. For S 212, the optimal set of 19 electrodes was comparable to the full set of 21 electrodes, and the critical set of three electrodes produced results that were as good as or better than the 12- and 4-electrode sets chosen manually. For S 220, the optimal set of 6 electrodes yielded accuracies comparable to those of the 21-, 12-, and 4-electrode sets chosen manually. Some degradation of TPC occurred for with the four-electrode set for S212 but not for S22. Substantial degradation of TPC occurred for both subjects with the three-electrode set. In S 220, the critical set of 3 electrodes was clearly less successful than any of the manually selected sets. These results are only a coarse indication of what may be possible with automated selection. Nevertheless, as we found with the manual selection methods, automatically selected sets of 4 to 6 electrodes can support a classifier that performs well at SNRs above -6 dB.

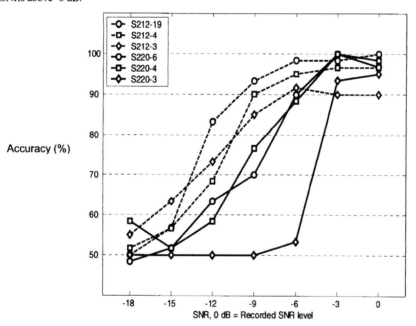

Figure 5: Classification accuracy for statistical electrode selection versus SNR in subjects 212 and 220. The three levels for each subject are the optimal set and the sets which contain the three or four most influential electrodes among both subjects.

3.4   Sample EEG Spectra

Spectra for a single EEG epoch of the four electrodes in the first critical set appear in Figures 6A and 6B. These figures show that the separation of single-trial spectra varies by electrode and subject. In S220 the spectral differences are clear and in the expected direction, but unclear for S212. Accordingly, some epochs are easily classified, while others lead to misclassification. Our KPLS classifier finds the set of components that maximally separates the two classes, leading to the highest TPCs for a given electrode set and subject.

4   Conclusions

Our methods and results successfully assessed the impact of electrode set size and reduced SNR on a real-time EEG-based fatigue classification algorithm. Overall, the results indicate that, our implementation of a PLS classifier functions robustly in the presence of noise. Even for arbitrarily selected sets of as few as four electrodes, with midline or frontal placement, TPCs ranging from 82% to 90% at an SNR of -6 dB. For larger arrays, TPCs ranged up to 99% at the -6 dB SNR level. This is a level at which the power of external noise is three times greater then the EEG signal itself. In laboratory or clinical settings, such a poor SNR would be considered unacceptable and nearly impossible to work with. We expect that operational settings will be noisier than the laboratory, but that SNRs of -6 dB or less should be easily avoided with proper precautions for shielding, grounding, and resistance to motion artifact. We note that the definitions of SNR differ for the simulated and test data. In particular the simulation SNR is known (ratio of AR process signal power to noise power), but for the real EEG data the signal is not known. We arbitrarily set the SNR of the EEG 0 dB and adjusted SNR according to the power of the added noise. In effect, this underestimates the true SNR of the EEG signal, which typically has positive SNR, and makes our results conservative with respect to real-world EEG recordings.

We conclude that robust, real-time classification of EEG patterns associated with fatigue is feasible in noise of up to -6 dB and with as few as four well-placed active electrodes. At present, the results indicate feasibility in a limited context (two subjects) and for unspecific fatigue conditions. Future work should consider extending these results to more subjects and experiment types. In particular, the dissociation of mental or cognitive fatigue from general fatigue and sleepiness must be properly controlled.

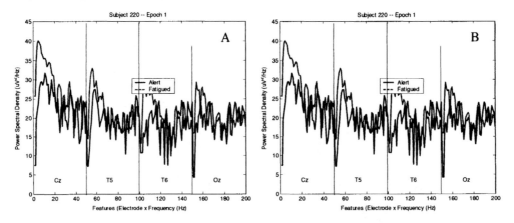

Figure 6: A. Spectra of a single EEG epoch for S212 in alert and fatigued conditions for electrodes in one statistically selected "critical" set. Frequency is numbered in feature space, as input for classification. Frequencies at each electrode range from 0-30 Hz. **B.** Corresponding spectra of a single EEG epoch for S212.

5   References

Anderson, C.W., Stolz, E.A. and Shamsunder, S. (1995). Discriminating mental tasks using EEG represented by AR models. Proceedings of the 1995 IEEE Engineering in Medicine and Biology Annual Conference, September 20--23, Montreal, Canada.

Bell A.J. and Sejnowski T.J. (1995). An information maximisation approach to blind separation and blind deconvolution, Neural Computation, **7**, 1129-1159.

Caldwell J, Caldwell JL, Brown D, Smythe N, Smith J, Mylar J, Mandichak M, Schroeder C. (2003). The Effects of 37 Hours of Continuous Wakefulness on the Physiological Arousal, Cognitive Performance, Self-Reported Mood, and Simulator Flight Performance of F-117A Pilots. AFRL-HE-BR-TR-2003–0086. Brooks City Base, TX: U.S. Air Force Research Laboratory.

Gevins AS, Bressler SL, Cutillo BA, Illes J, Miller JC, Stern J, Jex HR. (1990).10 Effects of prolonged mental work on functional brain topography. Electroenceph. clin. Neurophysiol.. **76**, 339-350.

Hayes, M. H. (1999). Statistical Digital Signal Processing and Modeling. New York, Wiley.

Makeig, S., & Inlow, M. (1993). Lapses in alertness: coherence of fluctuations in performance and the EEG spectrum. Electroencephalogr. Clin. Neurophysiol. **86**, 23-35.

Rosipal R., Trejo L.J. (2001). Kernel Partial Least Squares Regression in Reproducing Kernel Hilbert Space. Journal of Machine Learning Research, **2**, 97-123.

Rosipal R., Trejo L.J., Matthews B. Kernel PLS-SVC for Linear and Nonlinear Classification (2003). In Proceedings of the Twentieth International Conference on Machine Learning (ICML-2003), pp. 640-647, Washington DC.

St. John, M., Kobus, D. A., & Morrison, J. G. (2003, December). DARPA Augmented Cognition Technical Integration Experiment (TIE). Technical Report 1905, SPAWAR Systems Center, San Diego.

Trejo L.J., Kochavi R., Kubitz K, Montgomery L.D., Rosipal R., Matthews B. Measures and Models for Estimating and Predicting Cognitive Fatigue. (2004, October). 44th Society for Psychophysiological Research Annual Meeting (SPR'04), Santa Fe, NM.

Trejo L.J., Wheeler K.R., Jorgensen C.C., Rosipal R., Clanton S.T., Matthews B., Hibbs A.D., Matthews R., Krupka M. (2003). Multimodal Neuroelectric Interface Development. IEEE Transactions on Neural Systems and Rehabilitation Engineering, **11**, 199-204.

Assessing Cognitive Engagement and Cognitive State from Eye Metrics

Sandra P. Marshall

EyeTracking, Inc.
6475 Alvarado Road, Suite 132
San Diego, CA 92120
smarshall@eyetracking.com

Abstract

This paper describes a new approach for estimating cognitive state by using a set of psychophysiological metrics based on eye-tracking data. The metrics are derived from the raw data of pupil size and point-of-gaze information. Each metric requires only 1 second of data for computation, which means that the set can essentially be computed in real time. The set of metrics is described, and three examples of using the set of metrics to determine cognitive state are given. The first example contrasts focused attention with potential overload; the second example compares attention with boredom, and the third example differentiates between fatigue and moderate cognitive effort.

1 Introduction

As today's workplace has become increasingly sophisticated and technologically complex, the demands on the individual operating in that workplace have grown rapidly. Not only is it imperative that the technology and equipment function properly in such an environment, it is also essential that the operator function efficiently and safely. In particular, the cognitive awareness of the operator is key: Is he alert? Is he overloaded? Is he bored? Is he fatigued? If his cognitive state were known at any moment, either the technology itself or the operator's supervisor could raise an alarm and take appropriate action such as relieving the operator from duty or providing additional support.

One way to estimate an operator's cognitive state is unobtrusive monitoring by one or more psychophysiological sensors. Eye-tracking technology offers such an approach. In an ideal environment, the operator would be monitored by a remote camera capable of recording important eye data such as pupil dilation, eye movements, and blinks. Remote eye-tracking cameras with limited capabilities are already available commercially from eye-tracker manufacturers, and more advanced models are expected to be developed as demand for them increases. The eye data they produce would then be processed in real time to provide an estimate of the operator's current state of cognitive awareness.

To go with this new hardware technology, there will be a need for advanced methods to process raw eye data, estimate cognitive state in real time, and provide the estimates in usable form. This paper describes a method that has these three capabilities. The paper has three sections. The first section describes a new set of metrics that are created from the basic raw outputs of eye-tracking systems. The second section briefly presents the results of three studies in which the metrics were applied. The final section summarizes the results and outlines future research.

2 The Metrics

The most basic eye data are recordings taken at a fixed sampling rate. The system we use records data at 4 msec intervals (i.e., the sampling rate is 250 Hz). Each sample consists of three fundamental measurements: the size of the pupil, the vertical location, and the horizontal location for each eye. These are the lowest level measurements that virtually all eye tracking systems provide, at least for one eye. From these three measurements are derived all other metrics of interest, including the well-known measures of fixations and saccades. However, several other metrics can also be computed from the raw eye data, and these prove to be of great value in predicting cognitive state or mental awareness.

Figure 1 illustrates the constructive nature of the metrics. The basic measurements of pupil size, vertical and horizontal location occur every 4 milliseconds. When some sequence of these values becomes available, such as 250 observations in one second, additional metrics can be computed. These metrics are shown in the lower portion of the figure and include blinking, eye movement, the Index of Cognitive Activity (ICA), and divergence. The metrics in Figure 1 can be computed essentially in real time, with only a 1 second delay between actual recording and resulting metric.

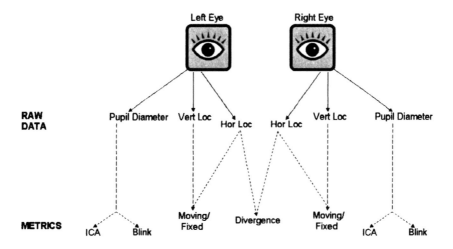

Figure 1. The eye metrics

Each of the metrics shown in Figure 1 provides unique information that can be valuable in predicting cognitive state. Brief descriptions are provided below for the set comprised of:

1. Index of Cognitive Activity
2. Divergence
3. Blinking
4. Movement

2.1 Index of Cognitive Activity (ICA)

The Index of Cognitive Activity is a method for using pupil diameter to estimate cognitive workload (Marshall, 2000; Marshall, 2002). It measures brief bursts of activation that occur in the muscles that govern pupil expansion. These bursts occur during periods of effortful cognitive processing. Calculation of the ICA requires multiple observations. The convention here is to use the number of observations taken in one second (usually 250 or 60) but other time lengths could be chosen as long as they were sufficient for calculation by the wavelets that underlie the ICA derivation. An important property of the wavelet analysis that creates the ICA is that it yields the same values for any subset no matter where the calculation starts. This feature means that it is possible to examine the whole signal or any part of it with no loss of information. Calculation of the index yields a binary vector having the same length as the original pupil signal. The non-zero elements of this vector indicate the times and samples at which unusual pupil activity occurred and hence are indicators of mental effort. The ICA is typically computed as the sum of the binary vector for each second of interest. The average ICA may be then computed across the entire period of

313

observation or each second's ICA may be desired. Both measures may be further transformed by the hyperbolic tangent in order to create output within the range of 0-1.

2.2 Divergence

Eye Divergence is computed by measuring the horizontal location of both left and right eyes and finding the absolute difference between the two. The measurement is taken every 4 msec. When an individual maintains focused attention on a feature, event, or task, the left and right points of gaze are very close together if not overlapping. When attention is lost, these points of gaze tend to separate, as focus shifts from the target to some very distant point. Most of the time, divergence will be measured per second to yield average distance between the left and right points of gaze. It can also be averaged to determine the mean divergence across the entire time period of measurement.

2.3 Blinking

The eye is typically in one of three states: blinking, moving, or fixating. Blinking is defined by eyelid closure, either full or partial. Blinks vary in duration. It is well known that very long blinks are indicative of drowsiness, whereas very short blinks are often indicative of sustained attention. Most blinks last less than 1 second, although longer blinks may be recorded. The calculation of blinks requires multiple observations, and real-time computation necessitates use of a variable time interval that continues until the blink has terminated. Blinking is calculated as the number of observations involved in a blink, with observations taken at a sampling rate of 4 msec. Blinks are not simply zero recordings of pupil size because there are substantial changes in pupil diameter during the initiation and conclusion of each blink. These changes are magnified during drowsiness as the eyes become "heavy," and closure/opening may take extended periods of time. Partial blinks also occur in which the eyelid partially occludes the pupil. For analysis purposes, it is convenient to calculate the number of observations per second that are considered to be part of a blink or partial blink. This value can be transformed by the hyperbolic tangent function to fall within the range of 0-1.

2.4 Moving

When the eye is not engaged in blinking, it is either moving or fixed. Measurement is made of the horizontal and vertical location of each eye every 4 msec. When that location changes from one measurement to the next by more than a specified amount (such as moving more than one pixel width in any direction on some screen displays or such as exceeding a velocity threshold), the eye is said to be moving. The number of observations per second during which movement is detected is recorded. This value can be transformed by the hyperbolic tangent function to fall within the range of 0-1.

3 Cognitive States

The difficulty in estimating a level of cognitive awareness (often called a cognitive state) is that such levels are not discrete entities that can be uniquely and absolutely determined. Rather, there exists a continuum of cognitive awareness ranging from a low end (drowsiness/fatigue) to a high end (cognitive overload). Any number of levels could be identified along this range. For example, we might specify

Drowsy..........Bored..........Distracted...........Attentive........Highly Focused..........Overloaded

as one potential set of states. Others could easily be defined. These states are not necessarily a linear continuum reflected by any single metric.

The cognitive states of interest will be determined by the specifics of any workplace. Two cognitive states that have been frequently studied are drowsiness and cognitive overload. In many sensitive situations, it is desirable to monitor the operator's level of drowsiness (e.g., operators of heavy machinery, truck drivers, air traffic controllers, nuclear power plant operators). Several techniques have been used in this effort. For instance, many systems use a camera to register the closure of the eye. The technique cited most often involves the PERCLOS metric, which

314

measures the extent to which the eyelid covers the pupil over an extended period of time. When the PERCLOS measure reaches a predetermined level—often the proportion of time in one minute that the eye is at least 80% closed—the individual is deemed drowsy (Wierwille, Ellsworth, Wreggit, Fairbanks, & Kirn, 1994). One drawback is that this measure can only sound an alarm after individuals are so drowsy that they cannot keep their eyes open. Safety may already be compromised by that time.

Other physiological measures are available that capture the upper end of cognitive awareness. For example, the Index of Cognitive Activity (ICA) was developed as a method of measuring cognitive workload. The ICA was designed to assess levels of cognitive activity, and many studies using different tasks have validated that the ICA increases for tasks requiring high levels of cognitive activity and decreases for tasks requiring low levels of cognitive activity (Marshall, Pleydell-Pearce, & Dickson, 2003). The Index of Cognitive Activity is based on measurement of changes in pupil dilation. As effortful cognitive processing occurs, the pupil reacts with sharp, abrupt bursts of dilation. These dilations are very small and very frequent. When little processing occurs, they are absent. The ICA algorithm processes the pupil signal, removes the artifacts created by eye blinks, and measures the changes in pupil dilation.

As additional cognitive states are defined and evaluated, it will be necessary to identify characteristics of the operator whenever he or she is deemed to be operating under any particular state. Performance of the operator will always be an important factor in making such a determination. But, performance measures may not always be immediately available. In many instances, some other means of assessing cognitive state is desirable. The eye metrics described here have value in producing cognitive state estimates that are not based on performance but derive only from the simple raw measurements made by eye-tracking systems.

4 Examples of Implementation

To illustrate the use of the metrics, three examples are described below, a driving simulation study, a problem solving study, and a visual search study. The metrics were applied to existing data that had been previously collected during in-house experiments at EyeTracking, Inc. For each study, second-by-second values were computed for each participant for 7 eye metrics: left and right ICA, left and right eye blinks, left and right eye movements, and eye divergence. All were scaled by the hyperbolic tangent function to produce values ranging between 0 and 1.

4.1 Example 1: Driving Simulation Study

The driving simulation program used for the study was a simple driving simulation program, *Driver's Education,* produced by Sierra On-Line, Inc. The program was developed in the late 1990's and is no longer available commercially. The driving simulation program was presented on a 17" monitor at 1024x768 resolution on a Windows 2000 platform. The *Driver's Education* software is a comprehensive multimedia driving course with three-dimensional graphics and realistic motion. In conjunction with the Thrustmaster Grand Prix 1 steering wheel/pedals system, it allows the user to simulate most driving behaviors: scan the road ahead, check mirrors, check speedometer, and use blinkers. For this study, participants completed basic lessons which had them driving through urban and suburban environments, dealing with traffic lights, signs, posted speed limits, schools, and other traffic. Figure 2 shows the basic screen display of the simulation.

The driving simulation study consisted of several episodes of driving and several episodes of driving while doing additional tasks. Four episodes were selected for analysis: two episodes of driving only and two episodes of driving while doing math-related tasks. Each episode lasted between 2-4 minutes. Eleven participants completed all four episodes.

315

Figure 2. Driver's Education screen display

This study should reveal two distinct cognitive states: focused attention (during driving only) and potential overload (with the imposition of the additional tasks while driving). To test whether the observed data corresponds to two distinct states, all seconds of performance were given a binary code, with the seconds during the driving only episodes coded '0' and the seconds during the math added episodes coded '1'. The data from the four episodes were then combined for each participant and analyzed to determine whether the eye metrics could reliably estimate the cognitive state coding. For each individual, all tasks combined resulted in approximately 700 seconds.

Table 1 shows the mean values for the eye metrics for the two conditions averaged across all participants. As Table 1 shows, the mean values for the two conditions are similar for all eye metrics.

Table 1. Mean values of eye metrics across all participants

EYE METRIC	DRIVING ONLY	DRIVING PLUS MATH
ICA Left	.701	.735
ICA Right	.846	.863
Blink Left	.100	.090
Blink Right	.128	.128
Move Left	.485	.495
Move Right	.463	.470
Divergence	.050	.048

Because it is unlikely that decisions in the real world will be made on the basis of a single second, two sets of data were analyzed: the original one-second data as described above and a second set of 5-second averages computed by averaging seconds 1-5, 6-10, etc. As one would expect, the 5-second averages are more stable than the 1-second data.

Both data sets were analyzed in two ways: first with a linear discriminant function and second with a nonlinear back propagation neural network analysis. With only two categories of classification, the discriminant function analysis yielded a single function which was statistically significant in all cases. The neural network was a 3-layer network with 7 inputs (the metrics), 7 hidden units, and 2 output units. Separate analyses using each technique were done for each participant. The results are given in Table 2.

316

Table 2. Classification results for all participants

PARTICIPANT	CLASSIFICATION ACCURACY			
	DISCRIMINANT FUNCTION		NEURAL NETWORK	
	1-SECOND	5-SECOND	1-SECOND	5-SECOND
1	64%	71%	66%	71%
2	70%	83%	70%	74%
3	62%	77%	68%	76%
4	66%	79%	71%	77%
5	59%	68%	63%	67%
6	74%	80%	70%	77%
7	74%	85%	74%	80%
8	69%	73%	69%	71%
9	69%	82%	81%	88%
10	72%	85%	75%	82%
11	72%	81%	75%	81%
Average	68%	79%	71%	77%

These analyses are conservative. All seconds from all episodes are analyzed, including initial start-up seconds for each episode which are generally more cognitively taxing than later periods of time when the participant has adjusted to the demands of the task.

The results are remarkably good; on the basis of only 5 seconds of data, the cognitive state can be correctly predicted more than 75% of the time. And, when limited to a single second, the classification rate drops only five percent, yielding a still useful 70%. Accuracy dips below 60% for only one subject (Participant #5) on one analysis, and even his rates are acceptable for the 5-second average.

4.2 Example 2: Problem Solving Study

The problem solving study consisted of two conditions. In one condition, the participant was asked to solve a set of arithmetic problems which were presented orally at 10 second intervals for 2 minutes. In the other condition, the same participant was asked to sit quietly in a room for 2 minutes. Each participant completed the two conditions twice, performing both conditions one in a dark room and also once in a lighted room. Thirty participants completed all conditions.

The analyses mimicked the analyses of the driving simulation study. Two cognitive states were defined: effortful processing and relaxed/bored. All seconds of data from all four conditions were coded 0 or 1, with both problem solving conditions coded '1' and both non-problem solving conditions coded '0'. Two sets of data were created, the original 1-second data and the averaged 5-second data as described above. And, two analyses were done on each data set: linear discriminant function analysis and neural network classification. As before, the discriminant function analysis yielded a single discriminant function that was statistically significant for all individuals. The neural network had the same structure as that for the driving simulation study, i.e., a 3-layer network with 7 hidden units. Data from each participant were processed separately.

Table 3 shows the mean values for the eye metrics for the two conditions averaged across all participants. As expected, the largest differences are observed in the two ICA metrics although all metrics show differences between the two conditions.

Table 3. Mean values of eye metrics across all participants

EYE METRIC	BLANK SCREEN	PROBLEM SOLVING
ICA Left	.144	.280
ICA Right	.158	.324
Blink Left	.080	.140
Blink Right	.090	.150
Move Left	.621	.550
Move Right	.604	.522
Divergence	.146	.192

Table 4 provides the results for all 31 participants under the two analyses. As with the driving simulation study, the results are strong. The procedures correctly identified the condition in which the second occurred approximately 75% of the time or more.

Table 2. Classification results for all participants

PARTICIPANT	CLASSIFICATION ACCURACY			
	DISCRIMINANT FUNCTION		NEURAL NETWORK	
	1-SECOND	5-SECOND	1-SECOND	5-SECOND
1	72%	83%	74%	83%
2	72%	85%	70%	74%
3	72%	89%	83%	86%
4	61%	70%	67%	67%
5	73%	85%	74%	86%
6	83%	88%	91%	88%
7	72%	79%	78%	88%
8	93%	97%	95%	96%
9	69%	73%	66%	77%
10	82%	94%	89%	91%
11	77%	92%	79%	91%
12	76%	80%	77%	73%
13	61%	64%	58%	67%
14	64%	76%	68%	74%
15	72%	79%	68%	79%
16	72%	84%	79%	91%
17	75%	85%	83%	84%
18	62%	71%	68%	68%
19	81%	91%	87%	90%
20	66%	74%	91%	69%
21	84%	95%	88%	98%
22	70%	81%	79%	74%
23	78%	95%	89%	93%
24	66%	73%	67%	75%
25	77%	89%	81%	79%
26	64%	67%	71%	67%
27	70%	83%	70%	63%
28	81%	87%	91%	92%
29	67%	78%	86%	72%
30	80%	89%	83%	81%
Average	73%	83%	78%	80%

4.3   Example 3: Fatigue during a visual search task

The final example is a visual search task. The task, shown in Figure 3, was a neuropsychological visual search task in which many representations of one letter were displayed in scrambled orientation on the computer screen. The target was the letter in the left-facing orientation ('C', 'E', 'G', or 'S'). A participant responds with key press 'Yes' if target is present and 'No' if target is absent. Each task run presents 104 trials, with some containing a single target and others containing no targets. Each trial presents a random arrangement of the stimuli.

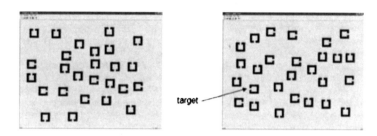

Figure 3. Visual search task

Data on this task were collected repeatedly from a single individual over a six-week period. The participant frequently experienced severe fatigue because she was a young mother with a new baby. At the end of each task run, she provided a subjective estimate of fatigue on a 10-point scale. Her fatigue estimates ranged from a low of '3' to a high of '10'. For analysis purposes here, the four task runs having the lowest fatigue ratings (two rated as '3' and two rated as '4') were contrasted with the five task runs having the highest fatigue ratings (all five were rated '10'). In all, 5167 seconds were recorded for these two sets of tasks.

The discriminant function analysis over these data with two categories (fatigued versus not fatigued) yielded significant results, with correct classification of 76% of all seconds for the 1-second data and correct classification of 84% for the 5-second data. The neural network results were similar. The network accurately classified 81% of the 1-second data and 88% of the 5-second data.

5   Conclusions

These three studies show that the eye metrics are capable of distinguishing among a number of different cognitive states that occur in very different settings. The first study contrasted two states at the upper end of the continuum: focused attention and potential overload. The second study examined two states that fall within the mid-range of the continuum, with comparisons between attention and boredom. The third study captured differences between low levels of cognitive processing and fatigue. The important finding is that the same set of metrics using the same analytic techniques made very successful distinctions in all three studies.

It must be stressed that the method described here produces second-by-second estimates of cognitive state that are made as the individual performs a dynamic task that generally varies along multiple features. The accuracy of the classification will generally improve if the results are aggregated over a longer period of time. This increased accuracy can be seen in the 5-second analyses shown for all three studies. Applying additional constraints should result in greater increases. For instance, if we take a 5 second interval and use the rule that a specific cognitive state will be predicted if at least 4 of the 5 seconds have a common classification, the overall accuracy increases because the occasional lapse is removed.

One expects that most systems designed to estimate an operator's workload level would do so on a much longer time basis than 1 second or even 5 seconds. Presumably, intervals such as 15 seconds or even on a minute-by-minute basis will be most useful. When we extend this procedure described here to these longer time units and apply moderate rules such as having at least 75% of the classifications within each time unit be consistent, accuracy should greatly increase. Larger data sets than the ones available here are required to test the reliability of the longer time intervals. Such testing is a goal of future studies.

In all three studies described above, all possible seconds of performance were analyzed, from the moment the task began until it concluded. Classification could undoubtedly be improved even further by optimizing the times at which comparisons were made. For instance, the driving simulation tasks all begin with the same initial instructions for several seconds, and the problem solving task has intervals of 10 seconds between problems (which is much longer than it takes to solve the simple problems used in the study). Optimization might entail the selection of specific portions of the task for monitoring.

On the other hand, the strength of the procedure is that it is successful even when based on all possible observations. This result is necessary if real-time monitoring is to be achieved. If the technique is to be used as described in the opening paragraphs of this paper, it probably will not be possible to tailor the analysis to fit the task. Rather, the monitoring will begin and the data will be classified. How will the classification rules be determined? In the studies here, all the data were already available and the tasks were designed to be different from each other. In real-world classification, neither of these will be true. Several options exist for initiating real-time evaluation. For example, one could monitor a group of operators prior to the instantiation of real-time monitoring (much as was done in the studies described here) and ask them to perform a series of well-developed and distinguishable jobs that captured the important cognitive states for that situation. From these baseline data, one could then derive a set of weights for each individual from one of the statistical procedures, such as discriminant function analysis. These weights would then be implemented to guide the classification during real-time on-the-job performance.

An unanswered question in this research is whether the eye metrics have a constant relationship to any single identified cognitive state. That is, would we see similar statistical results across individuals and across tasks if we were studying the cognitive state of fatigue? A goal now is to begin evaluation across a number of tasks that elicit the same or similar cognitive states to determine whether common statistical features emerge from them. Such an evaluation would help answer the question and would play an important role in creating norms for real-time monitoring in diverse workplaces.

6   References

Marshall, S. P. (2000). *U.S. Patent No. 6,090,051*. Washington, DC: U.S. Patent & Trademark Office.

Marshall, S. P. (2002). The Index of Cognitive Activity: Measuring cognitive workload. In *Proceedings of the 2002 IEEE 7th Conference on Human Factors and Power Plants* (pp. 7.5 – 7.9). New York: IEEE.

Marshall, S. P., Pleydell-Pearce, C. W., & Dickson, B. T. (2003). Integrating psychophysiological measures of cognitive workload and eye movements to detect strategy shifts. In *Proceedings of the Thirty-Sixth Annual Hawaii International Conference on System Sciences*. Los Alamitos, CA: IEEE.

Wierwille, W.W., Ellsworth, L. A., Wreggit, S. S., Fairbanks, R. J., Kirn, C. L. (1994). *Research on vehicle-based driver status/performance monitoring: development, validation, and refinement of algorithms for detection of driver drowsiness*. National Highway Traffic Safety Administration Final Report: DOT HS 808 247.

Postural measurements seated subjects as gauges of cognitive engagement

Carey D. Balaban, Jarad Prinkey[1], Greg Frank[1] and Mark Redfern

Dept. of Otolaryngology,University of Pittsburgh 107, EEINS , Pittsburgh, PA 15213, U. S. A.
[1]Dept.of Bioengineering,University of Pittsburgh ,763 Benedum Hall, Pittsburgh, PA 15213, U. S. A.
cbalaban@pitt.edu

Abstract

Cognitive engagement is reflected in 'orienting movements' toward events and objects of interest. In seated subjects at a computer monitor or driving a car, these orienting movements can be measured non-invasively and non-obtrusively by pressure-sensitive arrays in the seating surfaces and by ultrasonic or head sensors. This communication reports gauges of postural engagement from that can be measured from seated subjects (1) performing battle space defense tasks at a computer workstation and (2) driving on a highway while performing secondary and tertiary cognitive tasks. These gauges appear to provide signals reflecting global situational awareness and difficulty of workload and have been implemented in real-time for augmented cognition systems.

1 Introduction

Cognitive engagement impacts movements as a direct consequence of the need to orient toward and act relative to events and objects of priority and importance. The orienting movements can therefore be conceptualized as conveying information about (1) the parsing of continuous events into parts for processing and action (review: Zacks and Tversky, 2001) and (2) the relative magnitude of cognitive resource allocation to the event. These movements or static postures are also nonverbal communicative expressions of engagement; for example, we make inferences about whether to interrupt a co-worker on the basis of the content and context of their overt and automatic motor behaviors. The challenge of designing objective methods for decoding this nonverbal language, then, requires development of both sensor arrays and detection algorithms to identify both event parsing and resource allocation.

Humans spend a large proportion of their time seated, particularly in vehicles and the workplace. The detection of postural adjustments and voluntary movements appears to be facilitated by the fact that a seat constrains the ability to move in all directions and provides a stable base for postural control. Since the subject is relatively fixed with respect to the seat surface, the seat can be exploited as a platform (and reference frame) for fixed sensor arrays. The seat reduces the degrees of freedom of body movements for directing gaze, the torso and upper limbs. Hence, seated workplace and vehicular environments are advantageous platforms for using pressure sensor and head tracker technologies to detecting the relationship between postural movements and cognitive engagement. This communication presents evidence that a seat instrumented with an array of pressure sensors and a head sensor can be used to generate gauges related to cognitive engagement in both computer workstation and driving environments.

2 Experimental Methodology

2.1 Computer Workstation Experiments: Warship Commander Task

Fifteen students (18-24 years old) served as subjects in this study. The protocol was reviewed and approved by the Institutional Review Board of the University of Pittsburgh.

An Operator's chair from Lockheed-Martin's Sea Shadow ship was been reupholstered with slipcovers containing Ultrathin 16 X 16 pressure sensor arrays (Vista Medical, FSA pads). The sensors are spaced at 1 inch intervals in a square array. Separate sensor arrays are used to measure the distribution of pressure on the seat bottom and the seat back at a rate of 4.5 Hz. A Flock of Birds (Ascension Technologies) tracker was used to monitor the position of the head and trunk (mid-sternum) with 6 degrees of freedom at 103 Hz. A stand supporting a 15 inch LCD monitor, a keyboard and mouse pad was positioned in front of the subject for presentation of a military air defense simulation, the Warship Commander Task (WCT; St. John et al., 2002).

This task presents a series of nine 75 second trials with waves of friendly, hostile and unidentified planes, which must be tracked, identified (using an 'Identify Friend of Foe' (IFF) interrogation command), warned (if hostile and within the home ship's defensive perimeter) and, if necessary, neutralized. Each new trial is cued by the sound of a bell. The workload was varied by modifying (1) the number of planes in a wave (6, 12, 18 or 24 tracks) and (2) the proportion of unidentified planes ('high yellow' or 'low yellow' conditions); this experiment utilized conditions J, K and L of Warship Commander Version 4.4. The WCT software provides performance statistics, including second-by-second tracking of the number of planes on the screen and the number of tasks pending (e.g., IFF, warning and missile deployment). Posture related data were collected on a PC using custom applications written in LabView (National Instruments) and were analyzed off-line using routines written for MATLAB (MathWorks) and Excel (Microsoft). The WCT task performance data were logged directly to Excel compatible text files (St, John et al., 2002).

2.2   Mercedes-Benz S-Class Platform

Similar instrumentation was used to record postural adjustments of 5 drivers (male DaimlerChrysler employees, 24-33 years old. The postural data were collected Ultrathin 16 X 16 pressure sensor arrays (Vista Medical, FSA pads) on the seat bottom and seat back. The sensors are spaced at 1 inch intervals in a square array and are samples at a rate of 4.5 Hz. The transmitter of a Logitech Ultrasonic Head Tracker (Logitech, Inc.) was mounted on the ceiling above the driver's head and the receiver was mounted on a cap worn by the driver. This tracker monitored the position of the head with 6 degrees of freedom at a 50 Hz sample rate. Data were collected for three experimental protocol sessions from each subject, a *Calibration* session, a *Reference* session and an *AugCog* session (when workload mitigation strategies were employed). The latter sessions each lasted approximately two hours. The primary task in each session was driving a predefined route on German highways. A block design was used for three secondary tasks, interspersed with control periods. The Auditory Workload task required the driver to listen to a story presented by a mail narrator, while ignoring a female newscaster. The Mental Arithmetic task required the subject to count backwards from an arbitrary three digit number in increments of 27. The Special Driving Maneuvers task required the driver to perform specific maneuvers upon receiving commands from the experimenter. The tertiary task consisted of blocks of visual or auditory commands to press a right or left manipulandum for a reaction time task. Since the performance on this task (signal detectability, *d'*)in the *AugCog* session showed 108% improvement during the Auditory Workload task, 103% improvement during Mental Arithmetic task and 72% improvement during Special Driving Maneuvers task, one predicts that postural gauges of cognitive workload show a significant difference between the *Reference* and the *Augcog* sessions during the tertiary task blocks.

The raw data from the seat pads and head sensor are used to calculate ten uncorrelated variables for the analyses:

1. Cosine of seat yaw. Seat yaw as the angle between the seat back and a line between the center of pressure (COP) of the left leg/buttock and the right leg/buttock, with the vertex on the left.
2. Base of support: The distance between the COPs of the left and right leg/buttocks.
3. Seat pressure derivative RMS: The instantaneous standard deviation of the derivative of pressure on at each sensor across the seat pad.
4. Back pressure derivative RMS: The instantaneous standard deviation of the derivative of pressure on at each sensor across the back pad.
5. Global front-back seat pad COP.
6. Global left-right seat pad COP.
7. Global up-down back pad COP.
8. Global left-right back pad COP.
9. Cosine of head yaw.
10. Cosine of head roll.

3 Results

3.1   Warship Commander Task (WCT)

As reported previously (Balaban, Cohn, Redfern, Prinkey, Stripling & Hoffer, 2004), subjects show characteristic postural engagement responses during presentation of waves of tracks in the WCT. Figure 1 shows the relationship between raw scaled (normalized) head position relative to the monitor and the number of tracks on the monitor for one subject during a WCT session. Note that movements of the subjects head relative to the screen often parallel the number of targets. The same relationship is observed for the center of pressure (COP) on the seat, reflecting postural shifts relative to the monitor. This relationship can be quantified as the correlation coefficient between the head position (or seat COP) and the number of tracks on the screen over an 8-12 second period; the correlation coefficient for a standard 10 second period is shown.

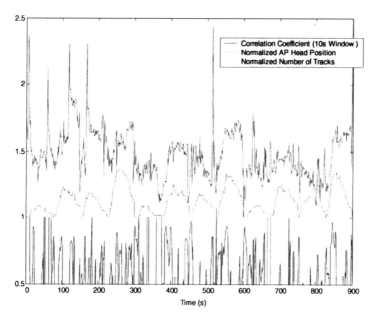

Figure 1: Relationship between head distance to the monitor and the number of tracks during a WCT session. The raw data have been scaled and offset to facilitate comparison of temporal relationships. The number of tracks on the screen (red) has been normalized to a maximum value for 24 tracks and offset by +1. The head position has been normalized to a range of 1 unit and offset for minimal overlap with the trace representing the number of tracks on the screen. Correlation coefficients for 10 second data intervals (above 0.5) are also plotted.

The squared correlation coefficient (coefficient of determination) specifies the percent of the variance that is explained by the linear relationship; a criterion of 50% ($r>0.71$) has been used as a gauge value of cognitive engagement with an 8 second window. Figure 2 shows the slopes of the relationships between head position and the number of tracks and seat COP and the number of tracks for one WCT session in a representative subject. Data points are shown only when the absolute value of the correlation coefficient exceeded 0.71 for the head position (black dots), COP of the left leg/buttock (blue dots) or COP of the right buttock (red circles). These measures of postural and head engagement were partially independent in each subject. An overall *global postural engagement response*, then, was defined as a time when either the head, left seat COP or right seat COP relative to the monitor showed a absolute value of the correlation coefficient greater than 0.71 for a linear relationship to the global number of tracks on the monitor.

An estimate of 'average' population behavior during the WCT was obtained by plotting the number of subjects showing a head or seat engagement response (absolute correlation > 0.71, 8 second window) during each second of the WCT session. These data are shown in Figure 3 for each of 3 waves of 24 tracks in the K (high yellow; high workload) condition. The data are plotted with a scaled urgency score, which increases markedly as unidentified tracks or foes approach the home ship. These plots show three phases to the responses. There is an initial engagement phase, where the majority of subjects show a high linear correlation between head or seat COP movements relative to the monitor and the total number of tracks on the monitor. This finding indicates that the majority of subjects show postural responses reflecting the total number of tracks at the beginning of the wave. However, as the urgency increases (i.e., the subject must react to local tracks converging on the home ship) in the middle of the wave, relatively few subjects show a correlation between postural movements and the global number of tracks on the monitor. Finally, the number of subjects showing global engagement to the number of tracks increases as the urgency decreases. We propose that these relationships reflect transitions between periods of global engagement (or situational awareness to all of the tracks) and local engagement to only the most urgent converging tracks or targets.

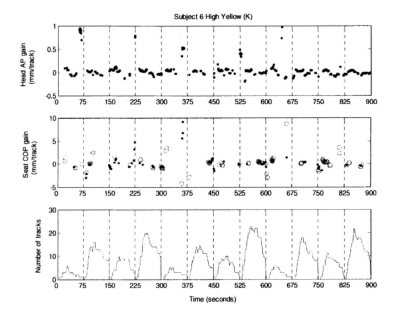

Figure 2. Engagement responses during one WCT task session. The upper two panels plot data only when the value of the squared correlation coefficient exceed 0.50 for the linear relationship between head position and number of tracks (black dots), left seat COP and number of tracks (blue dots) or right seat COP and number of tracks (red circles). The correlation window width was 8 seconds.

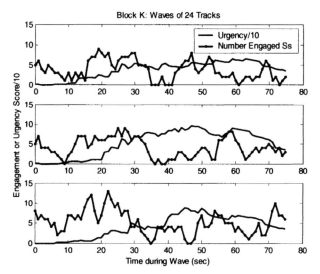

Figure 3. The number of subjects showing a linear correlation (>0.71) over an 8 second window between the number of tracks on the monitor and head position, right seat COP or left seat COP relative to the monitor is plotted for the three waves of 24 tracks in condition K (high yellow) of the WCT. The urgency score from WCT (divided by 10) is also plotted.

This approach was also applied in a Lockheed-Martin Advanced Technology Laboratory AugCog CVE in a simulated Aegis CIC workstation environment (ACIC task). In an experiment on seven subjects, the appearance of a global engagement response (12 second time window) had significant predictive capacity for an impending subject action (hooking, selecting or identifying a track). There was a 0.616 probability of an action within 2 seconds, a 0.766 probability of an action within 3 seconds and a 0.86 probability of an action within 4 seconds. In addition, in a subject of 5 subjects there was a highly significant correlation between the percentage of time with global engagement responses and two performance scores, correct track selection (r^2=0.726, p<0.001) and correct track identification (r^2=0.715, p<0.001). These findings further support the relationship between the global postural engagement metric and task performance.

3.2 Mercedes-Benz S-Class platform

A new cognitive engagement gauge for the tertiary task was constructed from the Calibration and Reference session postural data from all five drivers. An indicator file (binary) was constructed for the tertiary task to indicate blocks of trials with ones and intervals between blocks with zeros. Multiple linear regression analysis was then used to estimate this indicator variable as a linear function of the ten postural variables, first order low pass filtered representations of the ten postural variables (30 second time constant) and a constant term. The resulting gauge was then calculated for each subject across the *Reference*, *Calibration* and *AugCog* sessions. The coefficients for each postural variable are shown in Table 1. The gauge for one representative subject is shown in Figure 4. The calculation of this gauge has been implemented in real-time.

Table 1. Driver Postural Cognitive Engagement Gauge Coefficients

Term (r = 0.41)	Raw signal coefficient	Low pass signal coefficient
Constant	-1.8045	
Cos (Seat yaw)	0.7953	-0.8970
Base of Support	0.0229	-0.0358
Seat pad derivative RMS	-0.0118	-0.9442
Back pad derivative RMS	0.0331	1.2513

Seat COP x	0.0244	-0.0575
Seat COP y	-0.0181	-0.0063
Back COP x	0.0205	0.1704
Back COP y	0.0218	0.0294
Cos (Head yaw)	-0.0151	1.2960
Cos (Head roll)	0.0367	-0.4602

The cumulative distribution of this *driver postural cognitive engagement gauge* across subjects is shown in a full normal probability plot in Figure 5. This plot displays separate distributions for the driving task alone ('No Secondary or Tertiary' task), the tertiary task independent of secondary tasks ('Tertiary'), and different combinations of the secondary and tertiary task conditions. Several features are noteworthy. First, low values of the gauge (<~0.04) are encountered disproportionately in cases where there is no primary or secondary task; 25% of the observations in the absence of a secondary or tertiary task appeared with in this range, as opposed to approximately 2% of the observations when a secondary and tertiary task was present. Secondly, the cumulative distributions of the gauge were shifted toward larger values in the presence of a secondary and/or tertiary task. Thirdly, there was a greater proportion of high gauge values (>0.9) during Special Driving Maneuvers plus a Tertiary task than in either the Audio Workbook plus Tertiary or Mental Arithmetic plus Tertiary task conditions. Since the former condition showed the least improvement in tertiary task performance during the *AugCog* mitigation session, these features of the data are consistent with a positive relationship between the gauge value and cognitive workload.

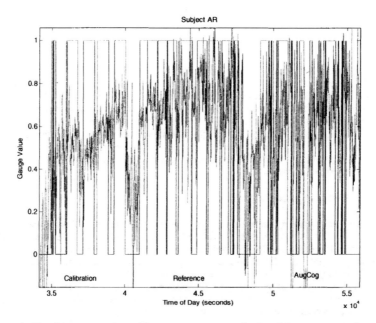

Figure 4. The *driver postural cognitive engagement gauge* is plotted for a representative subject across the three experimental driving sessions. The gauge is plotted in red and the blocks of tertiary tasks are shown in blue. The gauge was estimate (see text) during the Calibration and Reference sessions, then calculated for the AugCog session. Note that the gauge values are reduced during the AugCog session with respect to the Reference session.

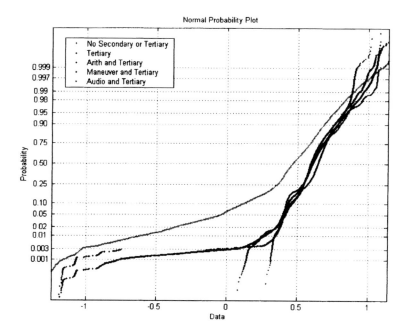

Figure 5. Full normal probability plot of the *driver postural cognitive engagement gauge* for all subjects.

4 Discussion

4.1 Computer Workstation Studies

The results of the WCT studies extend our previous observations that head movements toward the screen are correlated with the appearance of targets at the onset of each cued wave (Balaban et al. 2004). Specifically, the distance from the head to monitor and movements of the left and right seat COP are highly correlated ($r^2 > 0.50$) over an 8-12 second window during the first 30 seconds of waves of tracks. It is suggested that this engagement response represents orientation of the subjects to the global situation on the monitor; hence, it is a signal of global situational awareness. This view was supported by the finding that the cumulative time of engagement responses was correlated with performance in another task requiring global situation awareness, the Aegis CIC task. It also predicted with high fidelity an impending response within 3 seconds (probability: 0.766). This suggests that a high correlation between postural movements toward the monitor and the global number of work items is an indicator of preparation to act on the basis of global situational awareness.

In the most challenging WCT waves (24 tracks with either a low or high number of unidentified tracks), the urgency of action reaches a peak in the middle of the wave (30-60 seconds) because multiple tracks cross the defensive perimeter and converge rapidly on the home ship. Most operators are required to respond rapidly to the simultaneous close proximity of many unidentified and hostile targets. The number of subjects showing a global postural engagement response declined during this time frame, consistent with a switch from global situational awareness to a different orienting strategy. The reappearance of engagement responses was variable between subjects. However, these findings suggest that the disappearance of a postural engagement response to a building workload indicates a shift of strategy from global situational awareness, but it does not indicate directly whether there is a loss of situational awareness or merely a redirection to local feature of the workload or a single task. However, a continuing engagement response to global workload may be promising as a real-time assay of continuing global situational awareness.

327

4.2 Mercedes-Benz S-Class platform

The driver postural engagement gauge has been derived from multiple subjects from behavior during a tertiary reaction time task during driving. Although the gauge values vary across periods of specific secondary and tertiary tasks, higher values of the gauge appear to be associated with difficult workload conditions. For example, the lowest values occur predominantly when there is no secondary or tertiary task, only the primary task of driving. However, because spontaneous driving conditions can vary markedly in difficulty (e.g., with weather and traffic conditions), the primary task only data are expected to include a variety of levels of cognitive workload. The frequency of the highest values of the gauge (e.g., >0.90), on the other hand, were encountered in the secondary plus tertiary task conditions in the order: Mental Arithmetic < Auditory Workbook < Special Driving Maneuvers. This relationship corresponds to the degree of improvement in detectability (d') that was achieved by workload mitigation strategies in the AugCog session , with greater improvement shown during the Mental Arithmetic (103%) and Auditory Workload (108%) tasks than in the Special Driving Maneuvers task (78%). Hence, this driver postural engagement gauge appears be related to global workload during driving and additional task performance. In particular, a low value (less than 0.4-0.5) appears to indicate that workload is low, while higher values (e,g,, greater than 0.8-0.9) would indicate a need to either mitigate workload or avoid additional workload. Intermediate values appear to be associated with significant workload and can be used as one criterion for determining if workload mitigation is warranted.

5 Acknowledgment

This work was supported by DARPA under contract NBCH3030001 and ONR Grant N00014-02-1-0806.

References

Balaban, C.D., Cohn , J.V., Redfern, M.S., Prinkey, J., Stripling, R. & Hoffer, M. (2004). Postural control as a probe for cognitive state: exploiting human information processing to enhance performance, *International Journal of Human-Computer Interaction*, 17, 275-286.

St. John, M., Kobus, D. A., & Morrison, J. G. (2002). *A multi-tasking environment for manipulating and measuring neural correlates of cognitive workload.* Paper presented at the IEEE 7th Conference on Human Factors and Power Plants.

Zacks, J.M. & Tversky, B. (2001) Event structure in perception and conception. *Psychological Bulletin*, 127, 3-21.

Section 2
Cognitive State Sensors

Chapter 8

Wireless, Wearable & Rugged Sensors for Operational Environments

Wearable Cognitive Monitors for Dismounted Warriors

James B. Sampson

US Army Natick Soldier Center

15 Kansas St.

Natick, MA 01760-5020

james.sampson@natick.army.mil

Abstract

The ultimate goal of the Augmented Cognition (AugCog) program is to improve the warfighter's ability to process information while operating under stressful conditions of the battlefield. The mechanism for doing this is to monitor the combatant's psychophysiological state in real time, determine cognitive state, and then influence the information flow so as to maintain or enhance performance. Papers in this session deal with the challenge of taking the biometric indicators of user cognitive states developed from laboratory studies and making them relevant to a user in an operational environment. The research completed to date and reported in this session represents incremental steps toward this objective. Two key questions to be asked for the dismounted warrior are "Can un-tethered wearable and rugged sensor systems be developed for military operations?" and "Are cognitive states of individuals measured in laboratory comparable to cognitive states of individuals moving about in operational environments?" The reliability and durability of sensors, signal-noise ratios, motion artifacts, and simplicity of operation are issues these investigators are pursuing to address the first question. The latter question calls for research on the expected differences in the psychophysiology of individuals in low mobility laboratory experiments to those of the physically more active tasks of missions in the field under stress. Thus, studies in the next phase of the program will have to investigate the more cognitively complicated processes of combat like activities. New classifications and measures of the cognitive dynamics will certainly be needed. Additional sensors may also be required to account for the operator's context in terms of tactical movements and engagement with rapidly changing events. The context monitors will also have to be part of the wearable system that are, at the same time, tough-durable, easy to don and doff and relatively invisible to the operator. These are some of the critical issues that need to be considered for making AugCog operationally viable for a dismounted operator.

1. Introduction

The goal of the Augmented Cognition (AugCog) program is to improve a warfighter's ability to process information while operating under stress (St. John, Kobus, Morrison, & Schmorrow, 2004). The mechanism for doing this is to monitor the warfighter's psychophysiological state in real time, determine his or her cognitive state and then influence information flow through the system to maintain or enhance operator performance. The categories of warfighters and tasks being considered in the AugCog program are diverse across the military services of the United States Army, Navy, Marines and Air Force and each involves a varying mixture of information technologies and display types (visual, auditory, and tactile). In addition, the operational contexts and environments are also varied from the relatively remote and protected command posts to the more vulnerable and environmentally engaging systems such as land, sea, and air based vehicle systems to the dismounted wearable systems of the foot soldier. Systems may also differ in terms of how much time an individual operator may spend working through digital information displays to perform his or her primary tasks. For control room or remotely piloted vehicle operators, for example, the amount of time-on-task with their information displays may be close to 100 percent while for dismounted infantry operators it may be less than 20 percent.

2. Problem Focus

In the present session focus is given to the technical issues related to making the lab based measures and resulting monitoring devices work with mobile users who may have to carry (or wear) the system. Inherent in each of the papers are two questions: "Can an un-tethered wearable and rugged bio-monitoring system be developed for the warrior?" and "Will the cognitive or medical state measures derived in controlled experiments apply to situations where the operator is physically moving while performing mission related tasks?"

Considering the above, the engineering research reported in this session deals with the challenge of taking physiological measures indicative of a system user's cognitive or physiological state, developed largely under fixed or controlled conditions of the lab, and making these relevant to a user operating in dynamic military environments. The research completed to date represents incremental steps toward this objective. The initial concern is with the reliability and durability of sensors, signal-noise ratios, and motion artifacts. Some of the primary sensor-measures tested include the electroencephalogram (EEG), electrocardiogram (ECG), galvanic skin response (GSR), electroooculargram (EOG), respiration, temperature, and general activity (actography). From these come a series of derived measures of inferred psychophysical states such as alertness, mental workload, and various other cognitive and physiological stress states. The challenge of the AugCog program is to enhance warrior performance under stress through the effective use of bio-monitors to essentially change the nature or flow of information to better accommodate the operator's cognitive state.

3. Wearable Sensors and Psycho-Physiological State Measures

In the paper by Muth and associates (2005) a monitoring device for measuring autonomic nervous system (ANS) arousal (aka alertness) of individuals working in operational environments is described. The device with its algorithm has been tested in the context of human physical and psychological performance (movement and cognitive test batteries). The work and product described is a major contribution to biomedical monitoring of individuals with a wearable device that is rugged, simple to use, and provides reliable measures of arousal. The challenge will be to determine if the derived measure translates across a range of activities for various military operators and environments.

For operators where physical movement may not be great and the chance for monotony (vigilance) is high, the measure of arousal level may provide critical information for effectively altering information displays. However, in the case of the foot mobile soldier the arousal measure has not been extensively researched. It is still largely unknown if the measure will be sensitive to the range of ANS arousal experienced by soldiers who are walking or running. Researchers using this monitor face the challenge that there may be qualitatively significant psycho-physiological changes confounded with the motion artifacts that occur when an individual goes from a stationary posture to being ambulatory. One may speculate that the reticular activating system is significantly more innervated in the latter than in the former state. The muscles, the heart, and the brain are certainly more active and the physiological pattern is dynamically different. Questions arise about concomitant changes in the cognitive state of a mobile operator. This is still a vastly unknown area of research and the identified issues are pertinent to nearly everyone conducting experiments on bio-monitoring systems.

DuRousseau's (2005) paper describes experiments run to examine EEG artifacts during cognitive tasks while moving. The author reports on artifacts resulting from eye and heart signals recorded from subjects either seated, walking in place, or completely mobile while performing a cognitive dual-task battery within a simulated communications and target shooting environment. Results include an examination of *post hoc* filtering to remove unwanted artifacts, the identification of residual contamination in the EEG data, and a prescription for new methods of arbitrating the effects of compound artifacts. More research will be needed to determine the degree of filtering needed to obtain essential cognitive state measures for the dismounted operator performing various information processing tasks. The same complexities of the ECG measures are faced with EEG analysis.

In the paper by Mazaeva and associates (2005) experimental trials are described where both EEG and ECG data were collected wirelessly and attempts made to classify cognitive states of attention and cognitive loading in real-time while subjects performed simulated tactical missions in open and wooded areas. Specifically, artifacts related to subject motion included muscle activity, verbal communication and ocular artifacts; whereas artifacts related to the operational environment include instrumental artifacts that create interference with the EEG signal. The authors describe the challenge of mobile data collection with a wireless system where data analysis is done in real time with the filtering of motion artifacts.

In a related paper Whitlow, et al. (2005) describe how the input from the cognitive gauges derived from EEG, ECG and other sensors are used in an adaptive system designed to improve operator performance in a multitask environment. The system works by modifying the mode and timing of communications to the operator based on directly sensed measures of the operator's cognitive state. The experiments described focus on improving working memory performance by reducing divided attention demands. Experimental tasks were based on tasks identified by the U.S. Army as cognitively demanding (working memory of items such as ammunition used) in urban warfare environments. The experiments described involved a suite of biometric 'gauges' that assume to measure different aspects of the performer's cognitive state and are similar to those presented in other papers in this session.

The Stripling et. al (2005) paper describes a low-cost "wireless" PDA-based, 24 channel physiological monitoring system that can be used to explore cognitive states inferred from a wide variety of sensors. The system is capable of supporting user specified combinations of measures such as ECG, respiration, EGG, and EOG. Except for the construction of a low-cost amplifier for EEG signals, the system can be assembled entirely from commercially available technologies. The device uses software written in the LabView programming environment (National Instruments) and all physiological data can be stored locally or downloaded in real-time via wireless communication protocols. The described system should be helpful to many investigators wanting to do basic research on dynamic measures of cognitive functioning with mobile operators.

Buller and associates (2005) describe a unique wearable medical monitoring system that has much to offer the AugCog program in terms of user-centered design. The wearable physiological status monitor (WPSM) system is described as an electronic medical hub of physiologic and medical smart sensors with built-in hub algorithms. The algorithms estimate from various sensor input the medical state of the warfighter in terms of thermal heat balance, hydration, sleep loss, and vital life signs. The system is designed to be simple and easy to wear and rugged enough for field conditions. All sensors in the system are designed to indicate when they are malfunctioning. This allows the system to avoid using false, missing, or erroneous data in providing state measures. More confidence can be given to system reports of operator stress when they are reported. The system is designed to minimize the number and complexity of sensors and connections for ease-of-use by soldiers with limited medical or survival training.

4. User-Operator Applications

In the initial phase of the Army's version of AugCog the operator was conceived of as a frontline rifleman. Tactical vignettes were suggested by the user representatives of the Army's advanced technology program Future Force Warrior (FFW) that involved urban combat. As the program progressed from a sit-down table top demonstration of a first-shooter game to a standup simulation of dismounted rifleman with a head-mounted virtual display and a communication and shooting task, it became apparent the rifleman as operator was not the optimal operator-system for demonstrating an AugCog dismounted capability. In progressing from tabletop to field, more thought was given to who the future operator should be for demonstrating the benefits of the emerging technology.

A relevant criterion for selecting a candidate operator and system is the amount of time the individual spends interacting with the information technology while performing critical tasks. More viable candidates include either a newly conceived robotics NCO (non-commissioned officer) or a concept of the dismounted platoon leader (PL). Both of these positions face the potential information overload in future operational environments. Given the robotics NCO is still being defined and that the current role of dismounted PL is known in great detail, the reasonable choice for initial consideration would be a digitized PL concept. One conception is a dismounted leader who influences action of his squad leaders and rifle teams through digital communication devices and radios. The coordination and control of squad leaders and men would be expected to be done more through multi-display devices like a multi-channel radio-display system. This future dismounted platoon leader, therefore, may be seen as having a variety of communication display options to include a wrist or arm-mounted visual display, a hand-held PDA, head-mounted display and/or assorted non-visual auditory and tactile displays. These will work with an envisioned multi-channel radio system in netted communications across teams and units.

5. Conclusions and Discussion

The papers in this session outline some of the early engineering research issues that must be considered in the exploration of wearable biosensors for enhancing operational performance. The two general questions are whether wireless wearable systems can be engineered to work in militarily relevant environments and if the derivative measures can capture important operator states in a way to influence system performance. These questions are to be considered in light of the tactical missions of a dismounted or free moving warrior. A critical element will be time on task relative to cognitive states and the time to effect change in the system. This may limit application, given current technology, to operator-systems that are not engaged in rapidly changing events. If the decision cycle for action is typically short, relative to the AugCog system cycle for sensing and classifying, then the needed intervention will not happen in time. By addressing some of the applied operational issues early-on, we can identify and direct the research that will best advance our understanding in this emerging area. At the present stage the applicable user-system for the Army is a digitally enabled future dismounted platoon leader. While this operator is still foot mobile he is operating somewhat back from the immediate actions of squad leaders and rifle teams and spending more time working through information displays to communicate with and control tactical units.

With the expected changes in technology, doctrinal changes are also expected (Steeb, Matsumura, Steinberg, Herbert, Kantor, & Bogue, 2004). How small unit operations will look like in ten years is difficult to project since new capabilities are constantly emerging and these interact with tactics, techniques and procedures of the warrior in combat (Connolly, 2004). Thus, the applications of the evolving AugCog capabilities are necessarily moving targets and difficult to address at various stages. The best approach is to use a near-term perspective, working as closely as possible with current capabilities with experienced military personnel and explore slight variations with new technologies that are most likely to be available in the near term.

References

Buller, M.J., Hoyt, R.W., Ames, J. Latzka, J. & Freund, B. (2005) Enhancing Warfighter Readiness through Physiologic Situational Awareness – The Warfighter Physiological Status Monitor – Initial Capability. *Proceedings 1st Augmented Cognition Conference*, 22 – 27 July, Las Vegas, NV.

Connolly, D. (2004) Combined Arms Center: A Catalyst for Change. *Military Training Technology (MTT2)*, 9(5), Oct 27.

DuRousseau, D. R. (2005) Artifact Differences in Seated Vs Mobile EEG Recordings Made During Simulated Military Operations in Urban Terrain (MOUT). *Proceedings 1st Augmented Cognition Conference*, 22 – 27 July, Las Vegas, NV.

Mazaeva, N., Whitlow, S. D., Carciofini, J., Reusser, T., & Rye, J. (2005) Challenges for Honeywell's Mobile AugCog Ensemble. *Proceedings 1st Augmented Cognition Conference*, 22 – 27 July, Las Vegas, NV.

Muth, E, Hoover, A., & Loughry, M. (2005) Developing an Augmented Cognition Sensor for the Operational Environment: The Wearable Arousal Meter. *Proceedings 1st Augmented Cognition Conference*, 22 – 27 July, Las Vegas, NV.

St. John, M., Kobus, D.A., Morrison, J.G., & Schmorrow, D., (2004) Overview of the DARPA Augmented Cognition Technical Integration Experiment. *International Journal of Human Computer Interaction, Special Edition*, 17(2), 131-150.

Steeb, R, Matsumura, J., Steinberg, P., Herbert, T., Kantor, P. & Bogue, P. (2004) *Examining the Army's Future Warrior: Force-on-Force Simulation of Candidate Technologies*. Published by Rand Corp, Santa monica, CA.

Stripling, R., Becker, W., & Cohn, J. (2005) A Low-Cost, "Wireless", Portable, Physiological Monitoring System to Collect Physiological Measures. *Proceedings 1st Augmented Cognition Conference*, 22 – 27 July, Las Vegas, NV.

Whitlow, S.D., Ververs, P.M., Buchwald, B., & Pratt, J. (2005) Scheduling Communications with an Adaptive System Driven by Real-Time Assessment of Cognitive State. *Proceedings 1st Augmented Cognition Conference*, 22 – 27 July, Las Vegas, NV.

Enhancing Warfighter Readiness Through Physiologic Situational Awareness – The Warfighter Physiological Status Monitoring – Initial Capability.

Mark J. Buller
Reed W. Hoyt
John Ames
William Latzka
Beau Freund

United States Army Research Institute of Environmental Medicine
Kansas Street
Natick, MA 01760
mark.buller@na.amedd.army.mil

Abstract

The Warfighter Physiological Status Monitoring – Initial Capability (WPSM-IC) program has developed a soldier wearable system that provides health state information to soldiers, medics and commanders. Thermal casualties are of real concern to the military with over 1800 thermal casualties in 2002. The WPSM-IC system which is comprised of a series of sensors, a health hub, and algorithms, can estimate both thermal and hydration states. The system utilizes a proprietary low power data network, which helps to reduce sensor size and extend battery life. Artifact, and errors are managed by "smart sensors" that "know when they don't know", and a lack of direct ambulatory measures is overcome by an integrated approach of sensors and models designed to fail "gracefully". Methods to provide both thermal and hydration situational awareness to soldiers are outlined, and three retrospective examinations of how physiologic situational awareness could have been used in previous studies are presented. The WPSM-IC system shows great promise in supporting soldiers to effectively maintain their readiness during training and in combat.

1. Introduction

The Warfighter Physiologic Status Monitoring – Initial Capability (WPSM-IC) is a prototype system that provides physiologic and health state situational awareness information to soldiers, medics and commanders. By providing soldiers with timely health state alerts it may be possible to help them maintain their readiness in harsh battlefield environments. Often missions will entail carrying heavy loads, working with limited food and water resources, missing sleep, and combating the elements. These factors combine over time to degrade performance and can pose medical risks.

Providing warfighters with an automated thermal and hydration management system would appear to offer the ability to prevent or slow the number of heat casualties and enable individual soldiers to self maintain their performance and readiness levels.

Heat injury is a concern to the military and has negative effects on both training and operations (Steinman, 1987). The Army reported over 1800 heat injury cases during 2002 ("Heat Related Injuries", 2003), of these nearly one sixth were diagnosed as the more serious condition of "heat stroke". An estimate dollar cost to the military was calculated on earlier heat injury data (1989-1999) at over $10M per year, based on duty days lost, the cost of hospitalization, replacement, and disability (Hoyt et al., 2001). In conjunction with heat injury, dehydration also poses a significant risk to the warfighter, degrading physical and cognitive performance, increasing the risk of heat injury, and can in extreme circumstances lead to death. (Montain, Sawka, & Wenger, 2001; Pandolf, Sawka, & Gonzalez, 1988). Thus, the prevention of heat injury and the maintenance of hydration are critical issues for warfighters, especially in hot environments (IOM, 1994).

For the first time, the U.S. Army has a prototype system that that will monitor physiologic signals from the soldier and through algorithms, estimate a number of critical states. This paper will highlight the current WPSM-IC effort focusing on thermal and hydration state determination and how this information can be used by the warfighter to maintain their readiness during training and combat operations.

2.   The WPSM-IC system

The WPSM-IC system is comprised of a medical hub which hosts a personal area network of physiologic and medical sensors and a number of algorithms. The algorithms, using soldier characteristics and sensor data, estimate the state of the warfighter in the following areas: thermal, hydration, cognitive, life signs, and ballistic impact detection. Each area has four potential states that are color coded. Green represents "normal no action is required"; yellow = "look" or requires attention; red = "look now" or requires attention; and blue = "system fault" or "unknown state". The states enable medics and commanders to apply their own priority based upon context. For each area's state the hub also estimates a confidence level. The states for each medical and physiologic area are based upon input to the algorithms from a number of sensors distributed around a warfighter's body, uniform and equipment and outputs from other models. Figure 1 shows a schematic of the current WPSM-IC sensor system with its physical placement on a warfighter.

Figure 1: The WPSM-IC System. (1. Fluid Intake Monitor, 2. Life Sign Detection System, 3. Sleep Performance Watch, 4. BMIS-T, 5. Medical Hub 6. Core Temperature Ingestible Sensor) "..." Line represents data signal to hub.

The fluid intake monitor (#1 above) measures the amount of fluid consumed from a bladder style canteen. The life sign detection sensor (LSDS) (#2 above), worn around the chest is an integrated system with multiple sensors including: heart rate, respiration rate, body orientation, activity, and skin temperature, and an on-board life signs algorithm. The LSDS also has an integrated ballistic impact detection device which provides alerts when acoustic signals indicate a projectile has impacted the warfighter. The sleep performance watch (#3 above) measures sleep as a consumable quantity and uses an algorithm to equate this to cognitive readiness (Hursh et al., 2004). Information about the state of a soldier can be displayed on a personal digital assistant (PDA) running the biomedical information system - tactical (BMIS-T) software (#4 above) (Telemedicine and Advanced Technology Research Center (TATRC), Ft Detrick, MD). The sensors are connected to the medical hub (#5 above) by a proprietary wireless Radio Frequency (RF) network. The ingestible core temperature sensor (#6 above) (Mini-Mitter Inc., Bend OR) measures core body temperature.

2.1 Operational environment challenges

The WPSM-IC system was developed to overcome a number of key challenges unique to a military operational environment.

Direct measures of both thermal and hydration states are challenging in an ambulatory setting. For hydration state, practical field methods to assess total body water content, or changes to body water compartments in real time are not available. Likewise assessing changes in plasma or urine parameters to determine hydration status is both unreliable and operationally unacceptable. Bioelectric impedance measurements have also consistently failed to demonstrate accurate tracking of water changes (Berneis & Keller, 2000; Koulmann et al., 2000; O'Brien, Young & Sawka, 2002). For thermal state, core temperature ingestible sensor technology (Mini Mitter, Bend OR & Human Technologies Inc., FL) can be a viable substitute for rectal and esophageal probes (O'Brien, Hoyt, Buller, Castellani & Young, 1998), and has been available for field use for almost 10 years. However, in practice this technology is too expensive and too complicated logistically to be used continuously for military settings. Other current thermal state assessment ambulatory techniques are less precise and often work on correlative relationships of skin temperature to core temperature, or employ first-principles thermal strain models (Kraning & Gonzalez, 1997). Obtaining precise inputs to these first principle models can also be challenging. Measuring work rate in the field is difficult (Hoyt et al., 2004). Techniques using load, grade, velocity and terrain factors (Pandolf, Giovoni & Goldman, 1977), foot ground contact time (Hoyt & Weyand, 1996), and pedometry have been suggested, but may be inaccurate as significant changes in total weight (body weight + load) are unaccounted. Changing clothing configurations, such as taking off a shirt, can also impact first principle models. There are also some concerns in using models developed in a laboratory setting over a short time course, for long duration missions in the field. This underscores the importance to know each model's limitations and to use them appropriately in the system.

Few physiologic sensors are well suited for use by soldiers in the field. Most medical research has provided monitors and sensors that rely on wires and adhesives to function and are intended for patients who are supine and stationary. Sensors of this kind have almost unlimited power and processing capability to deal with artifact and error. However, signal artifact and error are exacerbated in the field and can include rapidly changing baselines, signal railing, dropouts, and movement noise. To further compound the problem, classical adhesive methods of attaching electrodes are not acceptable to the foot soldiers (Beidelman et al., 2003).

Another concern is that in a military setting sensors need to be non-invasive, small, consume little power, wearer acceptable, and don't negatively impact mission performance. Size, acceptability and power are all connected. For example: as a device needs more power its size and weight will increase; as size increases acceptability decreases. A soldier's heavy load has been an issue for many years (Harman, Frykman, Pandorf, & LaFiandra, 2002), and continues today, with loads in excess of 90lbs not uncommon (Hoyt et al. 2001). So any additional item that a warfighter is required to carry will always be critically viewed, with an eye to be left behind. If it is small, light, and provides the soldier with a tangible benefit it will more likely be accepted.

2.2 Operational environment solutions

2.2.1 Small size from low power network

One of the main issues in the size, weight, and power conundrum was the choice of the data transport system, and the use of "smart" sensors. Current commercial networking solutions do not fully meet the needs of the dismounted soldier (Shaw, Siegel, Zogbi & Opar, 2004). Wired solutions provide data integrity, high bandwidth, centralized power, and central processing. This network option also allows sensors to almost be no more than electrodes or probes. However, wires transition poorly between clothing layers, and centralizing processing also adds a significant processing burden which in turn requires more power. The WPSM program experimented with a wired network during a field exercise experiment at Ft. Benning, GA (Hoyt, Ft. Benning 1997 reference). The network was successful but at the price of ease of use. Connector failure problems were overcome by co-locating most sensor hardware together, but donning was cumbersome and involved an intricate cable threading process.

Wireless network solutions such as WiFi or IEEE802.11b and Bluetooth® are current commercial standards and both provide generally reliable data transfer with quite high bandwidth, but both are very expensive in terms of power. Even Bluetooth®, which is more power thrifty than IEEE802.11b, causes the battery life of our medical hub to drop from 72 hours to less than 6 hours.

With no viable commercial standard solution available, the WPSM-IC utilized a patented commercial network designed with military operational environments in mind (Mini Mitter Inc., Bend OR). Previous WPSM wireless network efforts proved instructive. Two wireless push protocols were implemented in two experiments in Quantico VA, in 1999 (Hoyt et al., 2001). These two systems were more wearable than the Ft. Benning 1997 wired network, but lost over 34% of data through sensor failure, cross talk and lack of data recovery schemes. An analysis of the network data from these studies suggests a modest performance increase of 13% would have resulted if transmission rates had been increased and a data recovery scheme implemented. These lessons demonstrated that the WPSM-IC network needed to avoid cross talk, and guard against degradation when multiple warfighters congregated or worked in a confined environment

The current WPSM-IC network operates within an unlicensed Radio Frequency (RF) industrial, scientific, and medical (ISM) band. Data are transmitted from sensors to the medical hub in a pseudo-random push transmission scheme (patented). The timing schedule of sensor transmissions is established when sensors are initialized and associated with a hub. Knowing the transmission schedule allows the hub to power up only when it expects a transmission from a sensor. This conserves power consumption in the hub receiver (~0.1% duty cycle) and also guards, to some degree, against cross talk from other sensors. The "push" only transmission scheme has the benefit of removing unneeded sensor receiver circuitry which in a "polled" scheme, is constantly turned on and consuming power. Sensors in the current network sample data every 15 seconds and transmit data on average every 15 seconds (3s to 27s). A data recovery scheme (patent pending) is implemented by transmitting both the current and previous data values in a network packet. The network is designed with an effective range of 18" to avoid cross talk when large numbers of warfighters are in proximity. With the data recovery scheme and minimal range to avoid interference our analysis (unpublished data) suggests
that the network can perform with less than 4% data loss in a group of three outfitted warfighters. It appears that the design of the network has provided reasonable immunity to cross talk and excessive noise.

2.2.2 Coping with artifact by using "smart sensors"

WPSM-IC sensors are designed to identify internal electronic faults, excessive noise, and artifactual readings. For example, the heart rate component of our LSDS is designed to determine the difference of 0 bpm heart rate and heart rate unobtainable as the electrodes are not in contact with a human body. The fluid intake monitor is being designed to differentiate between a sip and a leak in the hydration system. Although the sensors need more processing power to identify poor or incorrect measurements, the sensor circuitry is best suited to make this determination, as compared to any other point in the system. This technique also reduces the bandwidth burden on the network.

2.2.3 Working without direct measures

In the absence of direct measures for hydration state and thermal state, the WPSM-IC system relies on a hybrid of direct physiologic measures and models to provide valid alerts of physiological problems. Soldiers do not necessarily need to know their core temperature to 0.1°C precision, but rather need to know whether he/she is approaching a risky thermal state. When surrogate measures and models are used, it is essential that the measures and models are characterized and their limitations understood. This starts with understanding the quality and error associated with the input data to the models. "Smart sensors" provide both data and some measure of the quality of the information provided, including an indication of missing data and data freshness. Physiologic state can change quickly from "green" to "red." The core temperature of a soldier participating in a hot weather war games at Ft. Polk, Louisiana, increased at a rate of 0.09°C/minute (Hoyt & Friedl, 2004). This suggests that a change from a green state (defined as a core temp. <39.0°C)(actual core temp.= 38.9°C), to a red state (defined as a core temp.

>39.5°C)(actual core temp. = 39.51°C)(see figure 4) would take only 7 minutes, with only 11 minutes more to reach a life threatening state. Thus, WPSM-IC data confidence degrades over time if a parameter is not updated. The temporal decay function ("shelf life") assigned to each parameter depends on how quickly a parameter can change and the effect the change can have in changing a soldier's state. If sensors fail, or model confidence drops too low, the system provides for graceful failure through a cascade of models and algorithms which successively require less data but provides lower-confidence output (Tatbul, Buller, Hoyt, Mullen & Zdonik, 2004). When insufficient information is available the system reports "blue".

2.2.3.1 Thermal State Determination

The best and most reliable method of assessing thermal state in operational environments is direct measurement of core body temperature by using the network-enabled ingestible core temperature sensor. In addition, when heart rate data is available the physiologic strain index (PSI) can be used. The PSI an integrated core temperature/heart rate index that reflects thermal/work strain on a scale of 0 to 10+ (Moran, Shitzer & Pandolf, 1998), may prove useful in assessing acclimation status, guiding heat acclimation routines, and in setting the timing and duration of work/rest cycles. However, the use of a core temperature ingestible sensor is impractical for routine use and so these devices are reserved for use during high thermal stress missions, while encapsulated in nuclear, biological, and chemical protective suits, and/or if use is indicated by other algorithms or medics. When a core temperature ingestible sensor is not being used, the system defaults to a number of lower confidence less-precise methods of assessing thermal state. For example, the system uses first principle models (Kraning & Gonzalez, 1997) to estimate core body temperature. This model takes metabolic rate, environmental conditions, clothing configurations, and biometric data as inputs. Metabolic rate can be derived using a number of techniques (figure 2) which use a number of input parameters such as: heart rate, respiration rate, actigraphy, and geo-location readings in multiple ways with different confidence levels. When it is not possible to use a first principles model to estimate heat strain, a lower confidence skin temperature threshold algorithm is used (Yokota, Moran, Berglund, Stephenson, & Kolka, 2005). This model uses either skin temperature alone or skin temperature, heart rate and body mass index (BMI) to estimate whether there is a "high" probability or not of thermal stress. Figure 2, defines the input parameters for the models, the models that are used by the WPSM-IC system, how the models map input parameters to thermal state, and the relative degree of confidence provided by the models.

2.2.3.2 Hydration State Determination

The current system utilizes either water guidance tables or a first principles thermal model to estimate water loss through baseline requirements and sweat loss. Water intake is measured by the fluid intake monitor. This provides the basis for a fairly simple "leaky bucket" algorithm. A "full bucket" is 100% hydrated. Over time this bucket looses water at a rate determined by the water loss portion of the model, and gains water when indicated by the fluid intake monitor. The current algorithm defines a loss of 2% to 4% of total body weight as a yellow hydration state, and >4% loss as a red hydration state.

3. Physiologic situational awareness in action

The WPSM-IC system has matured to a point where field experiments will examine the benefits or not, of providing thermal and hydration situational awareness to soldiers. Three WPSM studies provide insight into the value of real time physiologic situational awareness.

The first study used a simplified WPSM system in the Smart Sensor Web (SSW) experiments at Ft. Benning, in 2000. The SSW experiment was designed to provide warfighters and commanders with real time situational awareness from a host of sensors including: unmanned aerial vehicle (UAV) video feeds, infra-red pictures from a series of scattered robots, meteorological data from dispersed environmental sensors, and physiologic data measured from the soldiers themselves ("SSW", 2002; Hoyt et al., 2001 "Physio Med Web"). The WPSM system comprised of a core body temperature ingestible sensor, a heart rate strap, and a commercial spread spectrum radio modem.

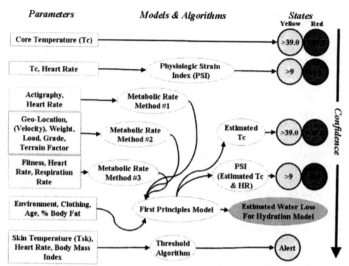

Figure 2: Thermal state. Input parameters, models and state mappings. (Metabolic rate method 1: Moran, Heled & Gonzalez (2004); method 2: Pandolf et al. (1977); method 3: Spurr, Prentice Murgatroyd (1988)

Body core temperature and heart rate were used to calculate PSI. Figure 3 shows the display format for the squad commander and individual soldiers. The left side of the display shows squad member location; the top right shows the group distribution of PSI with color coded state bands; while the bottom right shows a PSI time series for a selected soldier. During this study the yellow and red thermal state thresholds for PSI were set at 8 and 10 respectively. Due to wearable computer failures and information overload the experiment was not able to get information back to the individual soldiers. However, physiologic data were transmitted to, displayed, and recorded at the base control center. On several occasions squad members were able to view live data from other teams engaged in the assault scenario. These squad members were interested in seeing who was "most fatigued", or who was "most able to stand up to the demands of the assault". Subjectively, it appeared that soldiers were interested in using the information as a measure of fitness, or a way to improve training bouts.

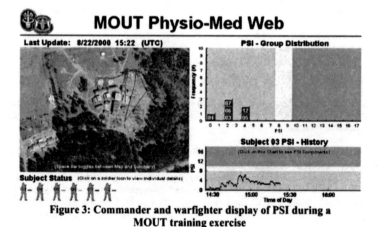

Figure 3: Commander and warfighter display of PSI during a MOUT training exercise

The two other WPSM experiments, although not designed to provide real time physiologic situational awareness, contained full compliments of sensors needed to provide both thermal and hydration states. These studies provide a retrospective look at how our models and health state predictions would have operated. One of these experiments was conducted with U.S. Marines engaged in the marksmanship training at Quantico (Yokota, Berglund, Santee, Buller, & Hoyt, 2004). Marines would daily march out and back to the marksmanship range ~6 Km in total, with pack and rifle, with range training lasting over 8hrs. There was concern from course instructors that students were reaching critical heat strain levels on the march back, especially in hot weather. Figure 4 shows mean group core temperature, and first principles model estimated core temperature for one day of training (ambient temperature = 85°F, RH = 59% , full solar load during most of the day). The measured core temperature and the model estimated core temperature are in reasonable agreement. The Root Mean Square Deviation (RMSD) value (0.23°C) was less than the overall time series mean standard deviation of 0.29°C, and a correlation coefficient (r) of 0.76. The error in agreement is most likely due to the metabolic rate input data which, were estimated by both the Pandolf equation (Pandolf et al. 1977) and from foot ground contact time (Hoyt & Weyand, 1996) methods. Table 1 compares thermal state determinations using: core temperature and estimated core temperature; and PSI and PSI calculated with an estimated core temperature at the end of the return march (1500). Hydration state determination along with estimated water loss and consumed water are also shown.

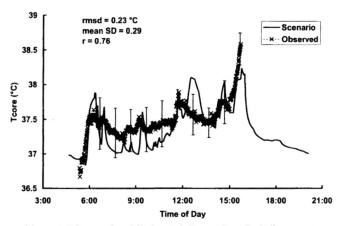

Figure 4: Measured and Estimated Group Core Body Temperature

Table 1: Comparison of Thermal State Determinations from Core Temperature and Estimated Core Temperature

	Mean	SD	N	State Frequencies		
				Green	Yellow	Red
Fluid Consumed (ml)	2280	1140	15			
Estimated Fluid lost (ml)	2480	320	19			
Fluid lost (expressed as a % of body weight)	0.68	0.76	15	15	0	0
Core Temperature (°C)*	38.7	0.3	8	7	1	0
Estimated Core Temperature (°C) *	38.2	0.1	19	19	0	0
PSI	7.8	1.3	5	4	1	0
PSI calculated with Estimated Core Temperature	6.6	1.0	12	12	0	0

*Actual core temperature is significantly different from estimated core temperature (t=4.18, P<0.05). PSI is not significantly different from PSI calculated with an estimated core temperature. (t=1.79, p=0.12).

Although in the return march the group average core temperature exceeds that of the estimated core temperature there is no significant difference in the number of alerts (Mann-Whitney U = 66.5 p = 0.621) between the direct and estimated methods. One subject triggered two yellow state alerts when using a core temperature ingestible sensor with a core temperature of 39.1 and a PSI of 9.0. While the model was able in general to track core temperature during the day, it began to deviate from the population during the high intensity road march. This demonstrates how a non-direct method of assessing state has a lower confidence.

A study conducted at Ft. Polk, LO (Hoyt & Friedl, 2004), demonstrates how PSI would have correctly provided a timely thermal state warning to the soldier and if acted upon would have prevented heat exhaustion. Members of the 509th Infantry Battalion (Airborne) opposing force and West Point cadets were outfitted with WPSM systems. Figure 5, shows the PSI scale for a 509th member and accompanying West Point cadet. Both soldiers were engaged in basically the same activities, wore the same clothing, and consumed similar quantities of water.

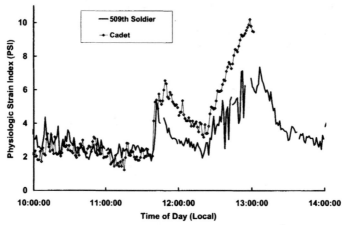

Figure 5: Comparison of Heat Strain Casualty with 509th Soldier

At 1300 the West Point cadet was independently identified as suffering from heat exhaustion. If a PSI based thermal "yellow" state had been set at 9 a "take a look" alert would have been sent at 12:51. Although the work rate and environmental conditions were similar for both candidates, the 509th soldier did not exceed a PSI of 8 and would have received no warning alerts.

4. Conclusion

The means now exist to provide physiologic situational awareness to individual soldiers in the field. This information should help warfighters maintain their hydration state, and reduce the likelihood of heat casualties. As more field data are collected the WPSM-IC system will be able to identify both the normal responses to the rigors of high intensity activity and provide high confidence thermal and hydration alerts.

In conclusion a practical warfighter physiologic monitoring system has been developed by the WPSM-IC program which can help enhance the awareness of soldiers on the ground. A robust data architecture and management strategy is being successfully applied to manage the harsh realities of getting data, understanding data, and providing useful information to warfighters.

5. Disclaimer

6. References

Berneis K, & Keller U. 2000. Bioelectrical impedance analysis during acute changes of extracellular osmolality in man. *Clin. Nutr.* 19:361–366.

Beidelman, B.A., Hoyt, R.W., Pearce, F.J., Sims, N.M., Ditzler, D.T., Ames, J., Speckman, K.L., Blanchard, L.A., Garcia, A., Colquitt, N., Gaffney, W.P., & Freund, B. (2003). User acceptability of design concepts for a life sign detection system. *USARIEM Technical Report.* T04-02.

Harman, E.A., Frykman, P.N., Pandorf, C., & LaFiandra, M.E. (2002) Physical performance benefits of off-loading the soldier. *Proceedings of the Army Science Conference.* pp 2-5, Orlando, FL, 2002.

"Heat Related Injuries" (2003). Heat related injuries, us army, 2002. *MSMR Medical Surveillance Monthly Report.* 9(4), 2-4. from http://amsa/army.mil.

Hoyt, R.W., & Weyand, P.G. (1996). Advances in ambulatory monitoring: Using foot contact time to estimate the metabolic cost of locomotion. In: Marriott BM, Carlson SJ, eds. *Emerging Technologies for Nutrition Research: Potential for Assessing Military Performance Capability.* . (pp. 1–29).Washington, DC: National Academy Press.

Hoyt, R.W., Buller, M.J., Redin, M.J., Poor, M.J., Oliver, S.R., Matthew, W.T., Latzka, W.A., Young, A.J., Redmond, D., Kearns C. (1997). Soldier physiological monitoring - results of Dismounted Battlespace Battle Lab concept experimentation program field study. *USARIEM Technical Report.* T98-6 (AD A332713).

Hoyt, R. W., Buller, M. J., Zdonik, S., Kearns, C., Freund, B., & Obusek, J.F. (2001) Physio-med web: real time monitoring of physiologic strain index (PSI) of soldiers during and urban training operation. In *Blowing Hot and Cold: Protecting Against Climate Extremes.* RTO-MP-076

Hoyt, R.W., Buller, M.J., Delaney, J.P., Stultz, D., Warren, K., Hamlet, M.P., Shantz, D., Matthew, W.T., Tharion, W.J., Smith, P., & Smith, B. (2001). Warfighter physiologic status monitoring (WPSM): energy balance and thermal status during a 10-day cold weather U.S. Marine corps infantry officer course field exercise. *USARIEM Technical Report T-02/02*

Hoyt, R.W., & Friedl, K.E. (2004) Current status of field applications of physiological monitoring for the dismounted soldier. In: National Academies of Science, *Monitoring Metabolic Status – Predicting Decrements in Physiology and Cognitive Performance.* (pp. 247-257). Washington, DC: The National Academies Press.

Hoyt, R.W., Buller, M.J., Santee, W.R., Yokota, M., Weyand, P.G., & Delany, J.P. (2004). Total energy expenditure estimated using foot-ground contact pedometry. *Diabetes Technology & Theraputics.* 6(1), 71-81.

Hursh, S.R., Redmond, D.P., Johnson, M.L., Thorne, D.R. Belenky, G. Balkin T.J., Storm W.F., Miller, J.C., & Eddy, D.R. (2004). Fatigue models for applied research in warfighting. *Aviat Space Environ Med.* 75(3 Suppl.), A44-53.

IOM. 1994. *Fluid Replacement and Heat Stress: Proceedings of a Workshop.* Marriott BM, (ed.) Washington, DC: National Academy Press.

Kraning, K.K., & Gonzalez, R.R. (1997). A mechanistic computer simulation of human work in heat that accounts for physical and physiological effects of clothing, aerobic fitness, and progressive dehydration. *J Therm Biol* 22, 331–342.

Koulmann, N., Jimenez, C., Regal, D., Bolliet, P., Launay, J.C., Savourey, G., & Melin, B. (2000). Use of bioelectrical impedance analysis to estimate body fluid compartments after acute variations of the body hydration level. *Med Sci Sports Exerc.* 32, 857–864.

Montain, S.J., Sawka, M.N., & Wenger, C.B. (2001). Hyponatremia associated with exercise: Risk and pathogenesis. *Exerc Sports Sci Rev* 29, 113–117.

Moran, D.S., Shitzer, A., & Pandolf, K.B. (1998). A physiological strain index to evaluate heat stress. *Am J Physiol* 275, R129-R134.

Moran, D.D., Heled, Y., & Gonzalez, R.R. (2004), Metabolic rate monitoring and energy expenditure prediction using a novel actigraphy method. *Med Sci Monit* 10(11),MT117-120

O'Brien C., Young, A.J., & Sawka, M.N. (2002) Related articles, links bioelectrical impedance to estimate changes in hydration status. *Int J Sports Med.* 23(5), 361-6.

O'Brien, C., Hoyt, R.W., Buller, M.J., Castellani, J.W., & Young, A.J. (1998). Telemetry pill measurement of core temperature in humans during active heating and cooling. *Med Sci Sports Exer* 30,468–472.

Pandolf, K.B., Givoni, B., & Goldman, R.F. (1977). Predicting energy expenditure with loads while standing and walking very slowly. *J Appl Physiol* 43,577–581.

Pandolf, K.B, Sawka, M.N., & Gonzalez, R.R., (Eds.) (1988). *Human Performance Physiology and Environmental Medicine at Environmental Extremes.* Traverse City, MI: Cooper Publishing Group.

Shaw, G.A., Siegel, A.M., Zogbi, G., & Opar, T.P. (2004) Warfighter physiological and environmental monitoring: a study for the U.S. Army Research Institute in Environmental Medicine and the Soldier Systems Center. *MIT Lincoln Laboratory*, Technical Report ESC-TR-2004-077.

"SSW" (2002). Final Report: Smart Sensor Web Project, *Office of the Deputy Under Secretary of Defense for Science and Technology*, November 2002.

Spurr, G.B., Prentice, A.M., & Murgatroyd, P.R. (1988). Energy expenditure from minute-to-minute heart-rate recording: comparison with indirect calorimetry. *Am. J. Clin. Nutr.* 48, 552-559.

Steinman, A.M. (1987). Adverse effects of heat and cold on military operations: History and current solutions. *Military Medicine*, 152, 389-392.

Tatbul, N., Buller, M.J., Hoyt, R.W., Mullen S.P., & Zdonik, S. (2004). Confidence-based Data Management for Personal Area Sensor Networks. *30th International Conference on Very Large Data Bases*: International Workshop on Data Management for Sensor Networks. Toronto.

Yokota, M., Berglund, L.G. , Santee, W. R., Buller, W. R., & Hoyt R. W. (2004) Predicting individual physiological responses during marksmanship field training using an updated SCENARIO-J model. Natick, MA: *USARIEM Technical Report* T04-03.

Yokota, M., Moran, D.S., Berglund, L.G., Stephenson, L.A., & Kolka, M.A. (2005) Noninvasive warning indicator of the "Red Zone" of potential thermal injury and performance impairment: A pilot study. *Proceedings for the 12th International Conference of Environmental Ergonomics.* (In press).

Artifact Differences in Seated Vs Mobile EEG Recordings Made During Simulated Military Operations in Urban Terrain (MOUT)

D.R. DuRousseau

Human Bionics LLC
190 N. 21st Street, Suite 300
Purcellville, VA 20132
don@humanbionics.com

Abstract

The ever increasing complexity of networked human-machine interactions has led to renewed interest in real-time measures of mental effort and stress of soldiers in mobile and dismounted environments. More than a decade of operational workload research has provided cognitive and behavioral metrics that have advanced our laboratory understanding of the human-system (Gevins et al., 1996). Presently, necessity is driving the discovery of human-system metrics for real-time assessments of a war fighter's arousal, vigilance, and attentional capacity. The need for assessment occurs as intelligent systems become capable of monitoring cognition and stress and adapting to variability in the human-machine interface. However, we have yet to transition these metrics from the laboratory into fieldable operational systems that are error-proof and relied-upon. What we lack are the tools to transition from in-the-lab psychophysiological testing to operational on-the-move cognition/stress assessments. In particular, what we require are lightweight, ruggedized, and wireless electronics with ubiquitous sensors and automated systems for de-noising and data analysis.

Considerable efforts have been put forth to develop faster and more reliable EEG artifact identification and removal algorithms over the last 2 decades and today there are several algorithms demonstrating impressive results. Unfortunately, current de-noising methods were designed to deal with artifacts in sedentary recordings under laboratory or clinical conditions. Although, today's filtering algorithms are highly successful at removing simple eye, heart, and 60 Hz artifacts from the EEG, they are not good at identifying and removing compound artifacts that are routinely generated when ambulatory recordings are made during simulated MOUT type operations.

This paper describes research performed to examine the differences in artifact types and severity when EEG's were collected under seated and mobile conditions. Here, we report our artifact difference results from 40-channels of EEG, eye, and heart signals recorded from 26 subjects either seated, walking in place, or completely mobile while performing a complex cognitive dual-task battery within a simulated MOUT environment. Our results include an examination of *post hoc* filtering methods used to remove unwanted artifacts, the identification of residual contaminations in the EEG data, and a prescription for new methods of arbitrating the effects of compound artifacts.

1. Background

To obtain reliable electrophysiology in the field, fast and simple methods are needed for maintaining attachment to and signal quality from several sensors at once. Only then could we start to deal with the many types of artifacts introduced by operational environments. This paper provides an examination of the artifacts present in 40-channels of EEG, EOG, and EMG data recorded under two experimental conditions (one when seated and the other ambulatory), while subject's used an M-16 laser-rifle integrated into two different simulation systems. These data were evaluated to determine the amount and severity of physiological and equipment artifacts specific to each recording environment. Under the first condition, subjects were seated while performing a simulated MOUT training scenario. The second condition necessitated that subjects were mobile while performing a modified MOUT simulation from a rooftop location, which included the use of a body worn virtual environment (VE) system. Results of both evaluations are presented here to discuss issues, solutions, and problems encountered while gathering several different types of physiological signals under highly realistic combat conditions.

345

A concept validation experiment was carried out with 13 subjects at the Carnegie Mellon University (CMU) Mobile Capture Lab as part of a Honeywell team effort sponsored by DARPA's AugCog Program to test a suite of CWA (cognitive workload assessor) gauges measuring EEG, EOG, and ECG data in real-time. In these experiments, subjects were suited with recording hardware, computer, and VR goggles and instructed to move around on a virtual rooftop while maintaining counts of friendly soldiers spotted, enemy soldiers shot, and number of rounds of ammunition used. The purpose of the study was to gather mobile data under out-of-the-lab conditions as well as to test new VE simulation methods. Figure 1 shows a photograph of an experimenter wearing an early version of the mobile system. Luminescent Styrofoam balls were placed on subjects and laser rifle to track body position, responses to movement orders, and shooting accuracy.

Figure 1 shows a subject wearing a 40-channel sensor system with recorder and computer carried on the back i a small pack. The virtual scenario was wirelessl projected onto specialized glasses providing imagery fror a rooftop MOUT simulation. In these test, subjec moved freely to identify, acquire, and shoot enemy targe in surrounding structures that entered their field of viev The lighted balls positioned around the subject and wei used to track body position and targeting accuracy in th virtual environment.

A second concept validation experiment was carried out with 13 subjects at the Institute of Human and Machine Cognition (IHMC) in Pensacola, FL, again, as part of the Honeywell team effort to test real-time gauges in closed-loop cognitive assessment systems. The goal of this study was to test the feasibility of employing mitigation strategies to improve the performance and situational awareness of an operator responsible for navigation, identification of friend or foe, enemy engagements, and netted communications. A central focus of the research was to ameliorate the information overload of a networked soldier by managing the flow of auditory, visual, and tactile communications used to relay command and control data to subjects. These tasking methodologies were used to schedule communications to the most optimal point in time when workload assessment gauges indicated a less taxed cognitive state of the operator. The concept was that when a suite of physiological gauges indicated high workload conditions (i.e., that a cognitive performance limitation had been met), a scheduler would defer all but the highest priority messages. Conversely, in low workload conditions, CWA gauge outputs indicated the availability of ample cognitive resources and therefore would signal an appropriate time for operator interruption.

2. Methods

2.1. Qualitative Evaluation of Artifacts in Seated and Mobile EEG

Artifacts in physiological signals weight heavily on the accuracy of the CWA gauges being developed and tested for use in operational environments. The following section describes the steps we took with both sets of MOUT simulation data to examine artifact similarities and differences under seated and mobile conditions (Figs. 2 & 3), where we:

- Investigated brain, heart, and eye signals collected during seated and walking trials at IHMC;

- Identified artifact types generated under both conditions and provided guidelines for recording mobile physiological signals in operational environments;

- Evaluated CMU mobile data with respect to signal quality, particularly, from sites used in gauge calculations (e.g., Arousal, XLI, Engagement, & P300);

- Examined IHMC vibro-tactoral stimulation data when the device was used as a means of somatosensory communication;

- Researched issues preventing real-time removal of EEG artifacts from mobile data and forwarded solutions to overcome remaining issues as we move ahead to field testing.

The structure of the data displayed in both Figures 2 and 3 are organized with subjects across the top and sensor channels downward. Thirty-five EEG channels are displayed first, followed by eye movement and heart activity channels. Below each set of channels are lists of comments for each dataset followed by a legend describing the artifact color-coding as being Intermittent (Yellow), Bad (Red), Muscle (Light Blue), 60 Hz (Green), and Broadband (Dark Blue). The gauges that were negatively affected by artifacts are shown at the bottom of the figure, where a letter in the block opposite the gauge type indexed a working gauge, a missing letter meant the gauge wasn't working, and a red letter meant one of several sensors used for a particular gauge was not working.

Figure 2 displays a color-coded chart showing 35 EEG, 4 ECG, 1 REF and 1 EOG sensor recorded from 13 subjects performing the IHMC task under seated conditions. Five types of continuous artifacts have been identified that were consistent across both recording conditions. The figure shows that all data from two subjects were lost due to 60 Hz contamination, which caused the largest overall problem. Unfortunately the 60 Hz artifact, which was most likely caused by a poor reference or ground sensor, could have been easily fixed. Here, automated recording procedures to continuously verify signal quality and alarm the operator of poor data quality would have been useful.

Figure 3 shows artifact results from the recordings made at CMU under mobile conditions, where subjects were again using a rifle to shoot enemy targets in the VE while standing, turning, and moving throughout the experiment. As in Figure 2, these data show that the 60 Hz artifact was again a problem, with two subjects being lost. Additionally, broadband noise from an unknown source was responsible for the loss of another subject, which, unfortunately could have been prevented by automated signal checking routines. However, the sensor that suffered the largest problems under mobile conditions was the ECG, where inadequate donning of leadwires led to sway, rubbing, and sensor separation problems. Other than preventable artifacts from better procedures and improved software, the main difference between the data collected under seated and mobile conditions was the amount of muscle activity in the ambulatory recordings, an obvious result.

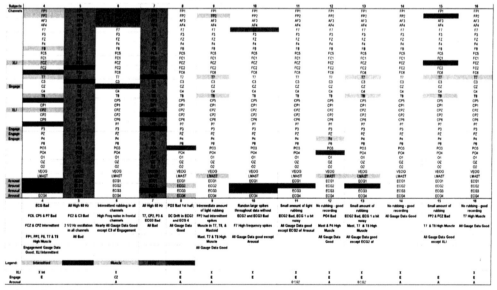

Figure 2: a color-coded map of types and severity of several artifacts measured during seated recordings at IHMC.

347

The results of the signal quality analysis depicted in Figures 2 and 3 from the seated and mobile recording conditions were used to categorize the types and severity of artifacts and improve recording procedures to minimize artifacts in future recordings. With respect to the gauges used to evaluate cognitive workload under realistic task conditions these results provided useful data on the size of the artifact problem. Below are the results for three gauges used to assess cognitive workload during simulated MOUT operations.

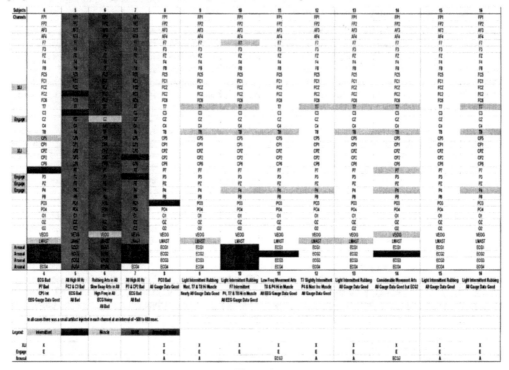

Figure 3

A color-coded map of the types and severity of several artifacts measured during mobile recordings at CMU.

The percentages below represent the amount of acceptable data available to each of the following gauges:

1) XLI gauge (required FCZ and CPZ EEG Channels) had acceptable data in 19 out of 26 subjects (73%);

2) Engagement gauge (required CZ, PZ, P3 and P4 EEG Channels) had acceptable data in 20 out of 26 subjects (77%);

3) Arousal gauge (required 4 ECG Channels) had acceptable data in only 13 out of 26 subjects (50%), however, 4 subjects had good data from 3 of the 4 ECG channels.

Although, the percentages of good data available to each gauge were not very high, significant improvements could be made through more care in positioning sensors and leadwires, implementing data integrity checking procedures, and with improved automated filtering methods. In the next section, we examine a few types of artifacts we encountered.

3. Discussion

3.1 Acceptable vs. Unacceptable Signals

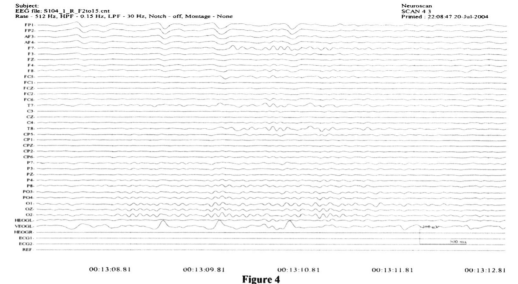

Subject:
EEG file: S104_1_R_F2to15.cnt
Rate - 512 Hz, HPF - 0.15 Hz, LPF - 30 Hz, Notch - off, Montage - None

Neuroscan
SCAN 4 3
Printed : 22:08:47 20-Jul-2004

00:13:08.81   00:13:09.81   00:13:10.81   00:13:11.81   00:13:12.81

Figure 4

Shows "good quality" waveforms bandpass filtered between 0.15 and 30 Hz. The amount of VEOG artifacts in the EEG signals were considered acceptable, owing to the ease of their removal.

Here, we provide a description of steps used to examine multichannel EEG under seated and walking conditions to determine what acceptable signal quality was. Our first task was to identify "Good Quality" signals, where good quality meant that simple bandpass filtering would be sufficient to provide clean data to gauges (Figure 4). Secondly, we compiled a list of artifact types and associated each with a cause. Then we examined the degree to which artifacts contaminated gauge data, and implemented enhanced artifact prevention and removal methods (e.g., sensor and leadwire placements, and real-time filtering techniques). Lastly, we investigated design solutions for the removal of compound artifacts remaining after simple filtering options have failed. Figure 4 provides an example of data that were considered good quality for purposes of this report. EEG and EOG data included for workload assessment were typically bandpass filtered between 2 an 30 Hz prior to use by gauges as shown in the figure. Examples of unacceptable data are presented in Figures 5 and 6, where several artifacts are provided representing contamination from multiple sources within each of the two datasets (Seated vs. Mobile).

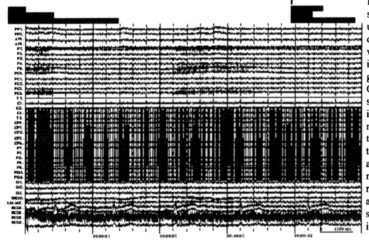

Figure 5 shows multiple sources of contamination in an unfiltered sample of seated data. Here, 60 Hz artifacts are visible in EEG, EOG, & ECG indicating a poor patient ground. Added are the noise in CP2, P3, & ECG3, which were saturated due to input impedance beyond the common mode rejection capabilities of the amplifiers. Artifacts in temporal and occipital sites are also visible due to tensing of neck and jaw muscles. And rubbing artifacts (above 60 Hz), are visible in several EEG signals, most likely due to inappropriately run leadwires.

349

Figure 6 shows multiple artifacts in a sample taken from a recording made during mobile conditions. 60Hz artifacts are again visible in the EEG, EOG, & ECG indicating a cause of a poor patient ground. Signals in FC2 are nearly saturated due to high input impedance, indicating a loose or bad electrode. Signals in channel C3 were saturated and reached the limits of the amplifier; indicated by railing of the signal. Also, a slow wire sway from walking during the task are visible in all channels, primarily, due to the system's less-than-optimal scalp to sensor interface.

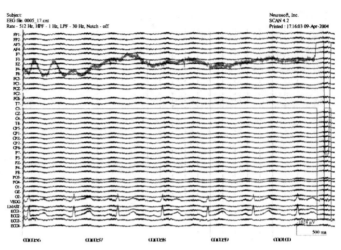

3.2 Common Problems from 60Hz & Broadband Noise

Figures 7a through 7d provide examples of continuous artifacts that were so severe that they prevented the use of entire recordings. These types of contamination could have been prevented if improved sensor placement procedures and automated signal detection software routines were used prior to the recordings.

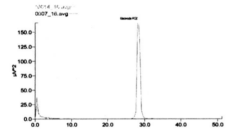

7a) Averaged spectral peaks in 30 Hz contaminated channel with high impedance from period of tactor firing (Red) Vs. clean dataset (Green).

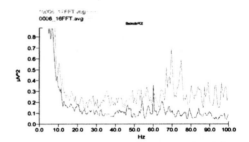

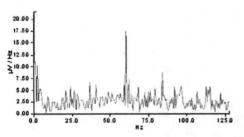

7c) Excessive Broadband & 60 Hz contamination in continuous EEG signals from unknown EMI source. Likely cause is poor ground and reference connections.

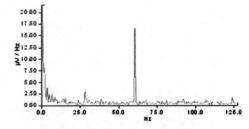

7d) Overwhelming 60 HZ contamination in continuous EEG signals predominantly due to high input impedance at sensors and ground.

The figures above provide a series of spectral plots indicating the levels of 60 Hz and broadband contamination in the EEG data. Figure 7a shows the results from a single subject comparison of the average 30 Hz contamination over brief periods when tactors were used compared to when they weren't. A large signal spike in the Red graph represents the frequency of vibration used in the tactile stimulator, which could be easily removed. The graphs in 7b show the difference in broadband noise between two data samples from the same subject. The noisy (Green) trace completely overwhelms the normal (Red) trace and shows little correlated behavior from 0 to 100 Hz. Figures 7c and 7d provide a measure of the amplitudes of 60 Hz and broadband noise artifacts in single-subject datasets.

3.3 Filtering Solutions

Notch filtering methods have proved effective at limiting 60 Hz contamination from physiological data; however, there are several problems with this method of noise elimination in that it has the potential to allow unwanted high frequency data to alias into the EEG. Given the advanced shielding and high common-mode rejection capabilities of today's modern EEG amplifiers, our preference would be to lower scalp impedances and not use notch filters. For some intermittent artifacts filtering does provide useful solutions. Figure 8 shows compound artifacts from head movement and blinks occurring simultaneously in mobile data recorded at CMU. In these data, neither artifact overwhelmed the other, so source separation wasn't needed and a simple bandpass filter (High-pass = 1 Hz, Low-pass = 15 Hz) was used to remove the movement artifact from the EEG. Notice that blinks and eye movements remained in the filtered data to help visually identify the de-noised quality of the filter's output. These artifacts were removed separately (not shown).

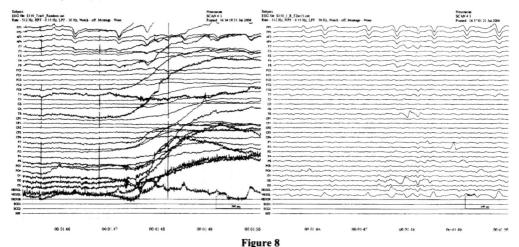

Figure 8
These graphs show artifacts from head movement and blinks (Left) and the effect of applying a simple bandpass filter with HP = 1 Hz and LP = 15 Hz at removing the slow frequency movement artifact (Right). Note the lack of ECG data due to sensor and leadwire problems.

However, there were numerous situations where simple filtering methods were not sufficient to remove severe compound artifacts, leaving badly contaminated and useless data. Although, several methods exist for removing repetitive and non-repetitive signals from mixed sources (e.g., independent and principal components analysis and adaptive linear regression methods) a discussion of these methods would be presented elsewhere. The plots shown in Figure 9 provide examples of severe compound artifacts and the results of simple de-noising methods. From our evaluation of the types and severity of artifacts routinely present when measuring EEG in operational environments, we have identified several continuous and intermittent forms and tested filtering solutions to remove some of these artifacts. Unfortunately, several artifacts remain that require more sophisticated de-noising techniques than tested here.

Thus, several types of artifacts remain without simple filtering solutions. In the following section, we present a brief discussion of the remaining issues for dealing with severe artifacts and recommendations for improved de-noising technologies.

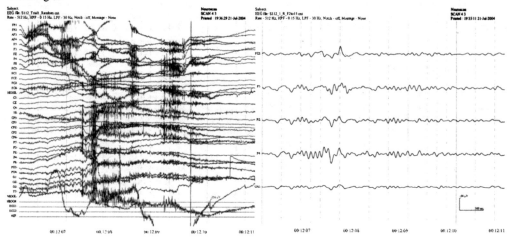

Figure 9

Shows severe contamination in EEG and EOG from large amplitude compound noise caused by movement and wire rubbing (Left). On the Right, is an enlargement of five of the EEG Channels showing noise remaining after filtering of the compound artifact (BP 1 to 15 Hz).

4. Summary and Conclusions

The most severe compound artifacts we encountered involve both movement and rubbing sources, where simple 1 to 30 Hz bandpass filters left considerable data corruption. Application of a narrower 2 – 14 Hz bandpass did improve result somewhat; although the data was not good enough for use by gauges monitoring the cognitive state of an individual. Table 1 provides a list of observations gathered from this study.

Table 1 – Observations from Comparison of Seated vs. Mobile Recordings

1.	Acquiring clean ECG signals during both seated and mobile recordings was the most problematic due to the path of the leadwires allowing them to be easily contacted by limbs, chair arms, and clothing;
2.	When EEG and ECG sensor impedance was high, external contact with the leadwires caused increased triboelectric artifacts from flexing (high frequency), while movement of the wires in the noisy EMI field caused sway artifacts (low frequency);
3.	Surprisingly, very little contamination was evident from heal strike artifacts in the mobile recordings, which were visible as a sway with sharp ending transition in the EEG waveforms;
4.	Unless input impedance was very high, Tactor operation did not interfere with the signal quality of any other channels;
5.	Other than the 3 subjects with overwhelming 60 Hz and Broadband contamination from poor sensor placement, intermittent loss of signal and increased muscle contamination were the predominant artifacts differentially introduced during the Mobile MOUT condition
6.	The incident of rubbing artifacts during the two conditions were the same, but the severity of contamination was greater in the Mobile trials;
7.	Several other types of compound artifacts exist in addition to those presented and they pose as great a problem to signal integrity because simple filtering approaches do not prevent data corruption;
8.	Improved methods for identifying and removing artifacts, or, disqualifying physiological signals must be developed. Existing techniques in spatial filtering and independent component analysis (ICA) have shown promise in this area. What's needed is to adapt these methods for real-time AugCog applications;
9.	With improved sensor interface and application technique, leadwire routing and shielding

	improvements, sensor and system care, and signal quality detection _prior_ to experimental recordings, good quality signals could be available to gauges in the range of 92 to 98% of the time;
10.	With real-time automated artifact removal and integrity detection routines, future fieldable systems would provide robust, self-correcting interface to physiological information from the war fighter in dismounted operations under extreme conditions.

Beyond the observations presented in Table 1, several issues still need further clarification with respect to artifact de-noising from physiological signals used for cognitive gauge assessments. Since filtering removes several artifacts we need to be careful in what we throw away because the artifacts themselves may be highly useful features for state classification. Artifacts that have been removed should be saved for separate feature extraction using adaptive neural networks or similar processing methods. Differential classifications of artifacts generated under varied conditions are needed to provide solutions to minimize future problems in field recordings. And much work remains to identify lingering issues preventing real-time data subset selection, removal of artifacts, and feature weighting across the sets of raw and gauge data from mobile recordings.

Acknowledgement

Human Bionics recognizes DARPA, Honeywell Laboratories, and particularly, the Institute of Human Machine Cognition and Carnegie Mellon University in support of this R&D effort. This work was supported by contract number W31P4QCR223 through the Information Processing Technology Office and the U.S. Army. This acknowledgement should in no way reflect the position or policy of the U.S. Government and no official endorsement is inferred.

Developing an Augmented Cognition Sensor for the Operational Environment: The Wearable Arousal Meter

Eric Muth[1], Adam Hoover[2] and Marty Loughry[3]

1. Department of Psychology
Clemson University
Clemson, S.C. 29634-1355
muth@clemson.edu
2. Department of Electrical and Computer Engineering
Clemson University
Clemson, S.C. 29634-0915
ahoover@clemson.edu
3. UFI Corp.
545 Main St.
Morro Bay, CA, 93442
marty@ufiservingscience.com

Abstract

Clemson University teamed with UFI (Morro Bay, CA) to develop and test an Augmented Cognition (AugCog) sensor for use in operational environments; the Wearable Arousal Meter (WAM). The sensor was first developed by Clemson in the laboratory as a desktop version. The desktop version uses a well-established, non-invasive, cardiovascular index, to objectively assess the parasympathetic activity of the autonomic nervous system, namely, respiratory sinus arrhythmia (RSA). RSA is inversely related to physiological arousal and workload. This sensor is robust in a mobile human but further ruggedization would be required for operational environments. Lessons learned during development of this sensor include: 1) the sensor should be based on sound laboratory science; 2) artifact detection and correction is a must in an operational sensor; 3) data transmission will be limited to the presence of an accessible network (hard wire or satellite) and the power limitations of the device; and 4) it is feasible to incorporate sophisticated high-level program code into a high performance microprocessor platform.

1   Science behind the sensor

When we began our quest to develop an AugCog sensor, we set out with the goal of making the sensor operational in mobile, closed-loop human-system interactions. This meant that the sensor needed to be well based in physiological principles and validated in the literature. It also meant that the sensor needed to be relatively robust at the start of development. For our sensor we chose to use inter-beat-intervals (IBIs; the time between heart beats) as input and RSA as output. This is accomplished by plotting IBIs as a continuous function against time, and deriving the high-frequency (HF) peak between .15-.5 Hz, also known as RSA, using the fast-Fourier transform.

RSA has been validated as an indirect measure of parasympathetic activity (PNS) of the autonomic nervous system (ANS) (Eckberg, 1983; Grossman et al., 1991; Grossman, 1992; Katona & Jih, 1975). When cardiac variability, the majority of which occurs between 0.0-1.0 Hz. (Mezzacappa et al., 1994), is plotted as a continuous function against time, three periodic fluctuations can be observed. These include a low-frequency (LF) peak between .04-.08 Hz., a mid-frequency peak (MF) centered around .1 Hz., and the high-frequency (HF) peak between .15-.5 Hz that we examine.

The ANS regulates internal states by acting as a feed-forward and feedback system from the central nervous system (CNS) to the periphery and from the periphery to the CNS. The PNS is the branch of the ANS that is often associated with homeostatic functions, i.e. returning the organism to "rest". The sympathetic nervous system (SNS) is often associated with "fight or flight", arousing the organism. PNS and SNS interactions are complex in that the PNS and SNS can act independently, coactively or reciprocally (Bernston, Cacioppo & Quigly, 1991). Internal organs such as the heart and stomach respond to PNS and SNS interactions. Typically, during arousal, there is PNS withdrawal and SNS activation, heart rate increases and stomach activity decreases. Our sensor utilizes RSA as an indirect measure of PNS activity and correlate of arousal.

Even though RSA is well described and validated in the literature, little, if any, research utilizes this measure real-time in a way standardized for the individual. Our challenge was to develop and validate an algorithm for deriving a standardized measure of RSA in real-time. Development of this algorithm is described in depth elsewhere (Hoover & Muth, 2004). Briefly, the current version of the Arousal Meter (AM) algorithm reads IBIs from a reliable sensor (e.g. the EZ-IBI unit by UFI Corp, Morro Bay, CA) and plots the IBIs over time. Sixty-four seconds of IBI data are "windowed" and processed using the Fast-Fourier-Transform (FFT -- The mathematical equivalent of passing light through a prism and breaking it into its component colors/frequencies). FFT derived power is plotted across frequencies to determine the HF peak associated with PNS activity (between 9 and 30 cycles per minute). The window moves forward in time 250 msec and the process is repeated. Once two FFT analysis cycles have completed, a running mean and standard deviation of the HF peak are calculated. A standardized "arousal" score is derived as [-(logx– logμ/logσ)] where x = the current HF peak and μ and σ are the running mean and standard deviation of the HF peak on each subsequent FFT cycle. Note that testing has revealed that a minimum of 20 minutes of data are required to establish a stable mean and standard deviation for this standardized arousal calculation to be valid. The HF peak results are available and valid immediately.

Prior to developing the wearable arousal meter (WAM), the results from the desktop algorithm were compared across tasks with varying workload, attention and physiological arousal demands. In general, the results revealed that the AM was sensitive to changes that took place between different types of tasks and that occurred over several minutes. It was not sensitive to changes within a task type or for changes in task that lasted less than several minutes. For example, Figure 1 shows mean arousal from 11 subjects completing a series of 4 first person shooter tasks and then a vigilance task. The duration of the tasks varied by subject with 1A and 2B lasting approximately 16 mins, 1B and 2A lasting approximately 8 mins and the vigilance task lasting approximately 30 mins. There was a significant difference between tasks $F[4/40] = 4.06$, $p < 0.05$. The difference is due to the vigilance task being different from the first person shooter tasks. The first person shooter tasks were not different from one another. Further, when the 30 min vigilance task was broken into three sub-tasks with different target hit rates, there were no significant differences (see Figure 2).

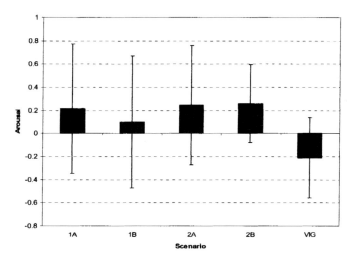

Figure 1: Mean and standard deviation of arousal across 5 different tasks. The first 4 tasks (1A, 1B, 2A, 2B), are all various scenarios in a first person shooter game. The 5th task is a vigilance task.

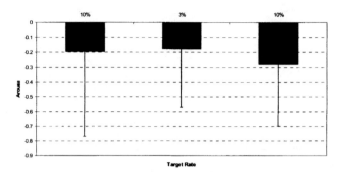

Figure 2: Arousal during 3 different vigilance scenarios with varying target hit rates (either 3% or 10%)

The AM was also tested for feasibility of use in a closed-loop system. By closed loop, we mean a system in which a user is performing a task or series of tasks that cause changes in arousal, arousal is monitored using the AM, the arousal data are fed back to the system, the system then uses the arousal data to somehow modify the task in order to keep arousal in a pre-defined "optimal" zone. While the feasibility experiment did not yield statistically significant results, it did yield meaningful results in that it was possible to run the AM real-time and close-the-loop as described.

After establishing that we had successfully implemented a method well described in the literature and that it had the potential for use in a closed-loop system, we ported the algorithm from the desktop to the WAM. The WAM was then validated against the desktop version for compatibility. This was done by having several subjects wear both the WAM and desktop sensors at the same time. Subjects then completed a series of tasks and body movements as follows:

- 10 minutes rest
- 10 minutes playing a computer game
- repeat the following twice:
  - ➤ 1 minute punching arms in air
  - ➤ 1 minute rest
  - ➤ 1 minute jumping jacks
  - ➤ 1 minute rest
  - ➤ 1 minute running in place
  - ➤ 1 minute rest
  - ➤ 1 minute crunches
  - ➤ 1 minute rest
- 8 minutes rest

Figure 3 is an example of a portion of the recording from the first 10 min rest period. Approximately 3.5 mins of data are shown. The two lines in the center of the figure represent IBIs plotted over time for the desktop (gray line) and the WAM (red-line that is slightly shifted right). The two lines with greater variability represent the standardized arousal measure, the gray line representing the desktop and the green line (slightly right-shifted line) representing the WAM. Note that the slight off-set in the devices is due to slightly shifted start times. The flat line in the beginning represents the 64 seconds to fill the required buffer. Note the excellent correspondence between the desktop and the WAM throughout. This example is typical of the correspondence of the WAM and the desktop from the validation data.

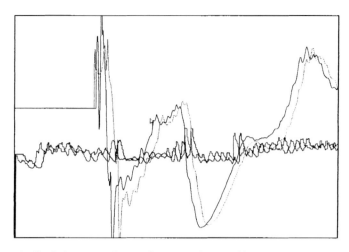

Figure 3: The desktop arousal meter data versus the wearable arousal meter data.

2   Artifact detection: garbage in equals garbage out

There are several unique specifications of the WAM sensor. First, because it is based on heart rate, it will have a high signal to noise ratio prior to any signal processing. This is because the human heart produces one of the strongest electrical signals that can be measured in the body (~20 mV membrane potential). This strong initial signal makes it a good candidate for being measured in a noisy operational environment.

Second, the method for acquiring IBIs has been optimized for a moving human. The person to be monitored is connected to the WAM via 3 electrode leads that are positioned to minimize electrode movement and be in line with the major vector of depolarization of the heart. Two active recording leads are connected, one on the person's right side just below the collar bone and one on the left side just below the left breast. These two electrodes are connected to Fetrodes (field effect transistors) that serve as local amplifiers, amplifying the signal at its source and increasing the signal to noise ratio. The third lead serves as a reference for signal noise reduction and is placed on the subject's right side below the right breast. This sensor configuration has been shown to be more robust to movement artifacts compared to typical silver-sliver chloride electrode configurations and band electrode configurations (Fishel, Muth, Herron, Hoover, & Rand, 2004).

Third, with the WAM, all data processing is completed on board the device and only processed data are transmitted over the wireless electrode. This is atypical of most AugCog sensors in development. Most sensors transmit raw data across either a wired or wireless network. Any errors in transmission cause data losses at rates higher than the transmission error. For example, with our algorithm, 64 secs of data are windowed. If 1 sec of data were lost due to a transmission error, then the algorithm would have to re-fill the data buffer with 64 secs of new data before re-starting to provide processed output. Hence, a 1 sec transmission error would result in at least a 1 min loss of data. This is a directly result of the requirements and assumptions of the types of analysis procedures used, in our case, the FFT. These types of processes and requirements are typical of AugCog sensors and hence, similar data loss multiplication can be expected with other sensors. The WAM avoids this problem completely because all processing takes place on the recording unit. Only processed results are transmitted over a network, which in the case of the WAM is wireless. If 1 sec of processed data is lost due to a transmission error, the system can continue with the next bit of data received as there are no assumptions necessary regarding the continuity of the processed data.

Even in light of these steps to make the WAM robust, without some signal pre-processing, errors in the signal will result in the processing of bad data and erroneous sensor output. During laboratory development it was realized that even a small number of errors in the IBI input (~ 1 per minute) can lead to errors in the output. In a non-moving human, error rates this low can be achieved with the methods just described. Particularly because the WAM uses a unity gain field effect transistor based buffer developed by UFI called a "Fetrode" to "impedance transform" the signal directly at the electrode site. With the Fetrode technology, interference caused by electrode or electrode lead movement, or the environment (e.g. electro-magnetic fields), is minimized. However, in a fully mobile human, muscle potentials (which are stronger than heart potentials, ~30 mV membrane potential), can degrade the signal.

Figure 4 shows an example of how just even a few isolated artifacts can invalidate the arousal calculations. Synchronized IBIs are shown in the upper blue line and arousal is shown on the pink line fluctuating around zero. Note that this plot is from a subject completing the aforementioned tasks and body movements. However, this subject did not complete the 10 minute game task rather they simply rested for the first 20 minutes. Arousal has been magnified by 100 so that it can be displayed in the same graphic.

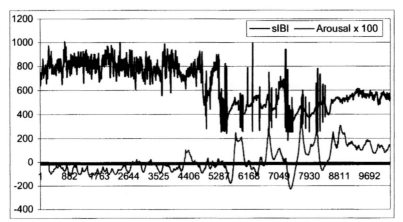

Figure 4: Synchronized IBIs and the corresponding arousal calculations.

During the initial rest period at the beginning of the recording, arousal is generally below zero. During the exercise tasks, arousal looms positive. During the final rest period (last 8 minutes), arousal is slowly decreasing from a high during the exercise back towards zero. Note however, how the errors in the sIBI data during the exercise period cause the arousal measure to err. The faulty sIBI's resemble high variability, which relates to low arousal. This causes the arousal measure during that period to wrongly fall to low levels at some points. This underscores the necessity for IBI error detection and correction. Figure 5 shows a run in which the sIBI data have been manually cleaned of all errors. Note how the large negative dip in arousal that occurred around the 7323 IBI is now minimized. More importantly, note how the final 8 min rest period now looks more like rest than exercise. This is likely due to the errors during exercise affecting the initial calculations during the rest period and leading to errors in the arousal output.

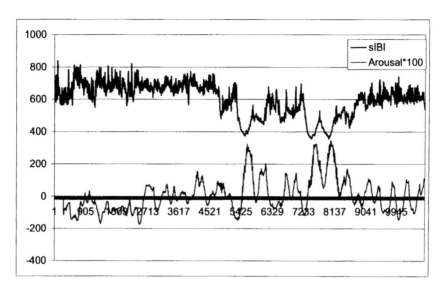

Figure 5: Synchronized IBIs and the corresponding arousal output after all IBI errors have been manually cleaned.

Due to the fact that even minimal artifacts can cause incorrect interpretations in data output, in a fully mobile human, sophisticated error detection and correction needs to be embedded in the sensor suite to handle artifacts caused by muscle movement. The initial version of the WAM incorporated relatively simple error detection and correction and Clemson is currently developing a more sophisticated algorithm that will be incorporated into a future version of the WAM.

3   Limitations: power and network availability

For the WAM, all signal processing and transmission takes place on a battery operated unit that the user is wearing, before it is sent to be integrated with other sensors. This includes deriving the IBI, performing error detection and correction, computing the time series analysis and standardization of the arousal score and then broadcasting the data to an 802.11 network. The fact that all of this is done on a single unit has two advantages. First, the effect of data transmission errors is reduced to single lost data points as described above. Second, band-width is saved because data can be sent at a lower rate required for post-processed data (1-4 Hz) as compared to the higher rate required by pre-processed data (500-1000 Hz). However, all of this processing and transmission requires power. The WAM currently uses six AA alkaline batteries. The WAM will run for approximately 3 hrs on this battery power after fresh batteries have been inserted. The main power limitation is the high current draw of the 802.11 node that is embedded into the unit. If the data were simply stored on the unit, battery life would be extended at least ten-fold. Currently the only alternative would be to go with a different network transmitter such as a proprietary radio frequency transmitter or Blue Tooth type transmitter. Nonetheless, the gains in battery life with these transmitters would still be limited.

The WAM was developed as a wireless sensor in that the user is not tethered to a central computer. For the WAM, we chose to use the 802.11 wireless standard. This lets the device integrate with a common standard into a widely deployed base station network, keeping costs down both in the device and in the base and network infrastructure. In addition, once on a standard network, the data can be transmitted almost anywhere in the world. While 802.11 is a widespread standard, it is not ubiquitous, especially on the battlefield. Hence, data transmission will be limited to the presence of an 802.11 network. Options are available, such as a local on the body network with a storage and download mode. However, this type of system would limit AugCog applications to local on the body closed-loop interventions. Other more ubiquitous networks such as satellite or cellular can and should be investigated. The use of these networks will still be hampered by battery life as described above.

4   Conclusions

Together, a university team and a small business worked to develop a sensor that can be used in closed-loop human-system interactions. The full potential of this device in closed-loop human-systems interactions has yet to be realized. Nonetheless, because of a careful development process which included validation at every step, including validating that the sensor could work in a real-time closed-loop human-system interaction, there is no doubt that the sensor will find utility in applications.

During the development process we learned several lessons that are valuable to all sensor developers. First, a sensor needs to be developed based on well-established physiological or behavioural principles in the literature. If a group is working with a novel physiological or behavioural principle or principles, they must validate that principle or principles before proceeding with sensor development. This validation is necessary, because even well-established principles are often based on very constrained laboratory experiments. These principles are challenged when applied to real-world data. For example, most laboratory measures are validated using aggregate means across subjects. However, for a real-time sensor to be useful, it must be applicable to an individual over time. Hence, even a well-validated principle in the literature must be re-validated when it is applied to real-time, real-world data. If there is no principle to begin with, the situation is much more challenging, as without the bench level work to understand the behavioural and physiological relationships with the sensor, the sensor will be developed in the dark with little understanding of how it is working and hence little understanding of what its likely strengths and weakness are and why they occur.

Second, to make high level interpretations in sensor data, the sensor data must be reliable. This means that they must be free from artifacts. It is not enough that artifacts are minimized and the signal to noise ratio maximized. If one were examining raw data, this would be acceptable, because individual raw data points could be ignored. However, closed-loop AugCog systems utilize processed data that have some meaning. The problem is that the processing can amplify the effects of artifacts yielding false meaning in the processed data. Minimally, a sensor must contain an artifact detection engine. Development of artifact detection and correction algorithms is very difficult, even with a relatively simple and fairly predictable signal such as the cardiac IBI. Hence, it may be the case that the best that can be accomplished with many wearable sensors is an artifact detection system which identifies when the sensor is transmitting good data and when it is transmitting bad data. Minimally, this would allow for the closed-loop system to ignore the sensor data when its meaning was misleading due to bad input data. Artifact correction, while the obvious goal, may be unachievable for a majority of wearable sensors.

Third, even if a sensor shows utility by increasing human performance in a closed-loop type of human-system interaction, its utility will be limited by battery life and transmission capability. Efforts must focus on providing power to these power hungry devices for their potential ever to be realized in military operations.

Finally, our work has demonstrated that it is plausible and indeed possible to take scientific principles from the literature, develop them into real-time processes, make them robust to some level of human movement, embed these processes in a commercial wearable device and achieve real-time closed-loop human-system interaction. Where power is readily available, these types of sensors have merit right now. For example, a real-time measure of alertness in an automobile could be developed and implemented today. No doubt, technological advances will some day make it possible for these sensors to have utility on the battlefield.

Acknowledgements

The majority of this work was funded by a grant from the Defense Advanced Research Projects Agency via the Office of Naval Research (Award# N000140210347) titled "Enhancement of Training and Performance Through Man-Machine Interactions Sensitive to Human Arousal and Task Difficulty".

A portion of the work was funded through a grant from the Defense Advanced Research Projects Agency via a sub-contract from Honeywell Laboratories titled "Cognitive Information Processing Technology Project".

A portion of the work was funded through a grant from the Defense Advanced Research Projects Agency via a sub-contract form Boeing Phantom Works titled "Arousal Meter Gauge Integration, Testing and Upgrades".

References

Berntson, G.G., Cacioppo, J.T. & Quigly, K.S. (1991). Autonomic determinism: The modes of autonomic control, the doctrine of autonomic space, and the laws of autonomic constraint. Psychological Review, 98, 459-487.

Eckberg, D.L. (1983). Human sinus arrhythmia as an index of vagal cardiac outflow. Journal of Applied Physiology, 54, 961-966.

Fishel SR, Muth ER, Herron R, Hoover AW and Rand JR (2004). Accuracy of a low-cost heart rate monitor for research involving inter-beat interval recordings. Psychophysiology 41(supplement), S72.

Grossman, P. (1992). Respiratory and cardiac rhythms as windows to central and autonomic biobehavioral regulation: Selection of window frams, keeping the panes clean and viewing the neural topography. Biological Psychology, 34, 131-161.

Grossman, P., Karemaker, J. & Wieling, W. (1991). Prediction of tonic parasympathetic cardiac control using respiratiory sinus arrhythmia: The need for respiratory control. Psychophysiology, 28, 201-216.

Hoover A and Muth E (2004). A Real-Time Index of Vagal Activity. Journal of Human Computer Interaction, 17 (2), 197-209.

Katona, P.G. & Jih, F. (1975). Respiratory sinus arrhythmia: noninvasive measure of parasympathetic cardiac control. Journal of Applied Physiology, 39, 801-805.

Mezzacappa, E., Kindlon, D. & Earls, F. (1994). The utility of spectral analytic techniques in the study of the autonomic regulation of beat-to-beat heart rate variability. International Journal of Methods in Psychiatric Research, 4, 29-44.

Challenges for Honeywell's Mobile Augmented Cognition Ensemble

Natalia Mazaeva, Jim Carciofini, Jeff Rye, Trent Reusser, and Stephen Whitlow

Honeywell Laboratories
3660 Technology Drive, Minneapolis, MN 55418
natalia.mazaeva@honeywell.com

Abstract

Adaptive human-machine systems derive information about user's cognitive state from a number of physiological and neurophysiological sensors and use it to modify various characteristics of the system to mitigate problems in human-machine interaction and support user performance. Unfortunately, the use of adaptive systems and their sensing technologies has not been explored in real-world operational settings. However, it is important to evaluate adaptive systems and sensing technologies in such environments, specifically as applied to collection and evaluation of physiological sensor data. In the present paper, we describe issues identified while transitioning from laboratory to operational settings and discuss solutions implemented by Honeywell's Augmented Cognition team. Evaluation of the current mobile data collection system suggests that sensing technologies can yield reliable information when used in mobile operational environments.

1   Introduction

Adaptive human-machine systems have been developed in recent years in an attempt to mitigate problems in human-machine interaction and support user performance (Parasuraman, 2003). The purpose of the adaptive systems is to sense or identify aspects of users' behavior, and to change the system adaptively with the goal of assisting the user in successful interaction with the system. Important characteristics of these systems include their ability to sense human state in real time, for example from neurophysiologic measures assumed to be indicative of a state, and use this information as input on the basis of which some characteristic of the system is changed. Thus changes in the user state are used to modify the way information is presented to the user by the system (Hettinger, Branco, Encarnacao, and Bonato, 2003).

One example of such a system, an adaptive closed-loop system, suggests the possibility of changing modes of automation in response to changes in engagement index, which reflects the actual task engagement of the user based on continuous physiological response of the user to the task (e.g., Pope, Bogart, and Bartolome, 1995; Freeman, Mikulka, Prinzel, and Scerbo, 1999). The engagement index is calculated from continuous ongoing electroencephalographic (EEG) signal collected simultaneously while subjects perform tasks. The system allocates tasks to the operator when the index indicates low engagement, and automates tasks when the index indicates high engagement levels.

Another application of the sensor based closed-loop adaptive technology has been realized in the communications scheduler – a system that changes modality and frequency of incoming communications (Dorneich, Whitlow, Ververs, Carciofini, and Creaser, 2004). The system derives information about the user's cognitive state from a number of physiological and neurophysiological sensors, decides how and when messages are presented to the user based on their cognitive state.

As evident from examples above, it is critical that input signal received by sensors from the human is reliable to ensure accurate linking of the sensors to cognitive state identification. Traditionally, adaptive system research and their respective state sensing have been conducted within controlled setting. However, we need to take the additional step of evaluating the sensing technologies in operational settings with their inherent challenges to collecting and processing of physiological sensor data.

The major difficulties associated with collection of EEG data in operational environments stem from factors such as weight of equipment, wireless signal transmission, real-time classification and processing of signals, and most importantly, artifacts associated with both subject motion and the operational environment itself. Specifically, artifacts related to subject motion include high frequency muscle activity, verbal communication and ocular artifacts (EOG) consisting of eye movements and blinks; whereas artifacts related to the operational environment include instrumental artifacts such as electrical noise that create interference with the EEG signal (e.g., Kramer, 1991)

In the present paper, we describe Honeywell's attempt to address issues encountered while applying wireless sensing technologies to collection and processing of EEG and ECG signals in a mobile environment. Specifically, we will discuss problems identified while transitioning from stationary to operational settings and present solutions implemented by the Honeywell Augmented Cognition team.

1.1 Honeywell's Mobile System

Honeywell's Augmented Cognition mobile system uses real-time neurophysiologic measures obtained from EEG and ECG sensors and employs real-time classification algorithms to distinguish cognitive states of a dismounted soldier in various workload conditions in an outdoor field environment. The system also employs augmentation mitigation strategies to modify tasks based on the cognitive states of the soldier through automated communications scheduling and thus decreasing demands on working memory and attention. The following sections illustrate implementation of this system in the field environment.

2   Method

2.1   Data Collection System

EEG data was collected using the ActiveTwo system (Biosemi) where the hardware (fiber connected A-D box and USB receiver) was stored in a backpack on the subject. The hardware system was battery powered and weighed 1.1 kg. The BioSemi active electrode headcap (32 sensors) was placed on the head and connected to the hardware in the backpack via a 140 centimeter electrode cable. ECG data was collected with the Wearable Arousal Meter (UFI) and three fetrodes (field effect transistors) placed on the subject's body and connected to the WAM system stored similarly on the subjects' back. The BioSemi hardware systems was connected to a laptop computer stored in the subject's backpack through a USB2 link while the WAM system was connected to the laptop with RS 232. Remote wireless communication was established between the laptop in the backpack and a base-station PC via an 802.11 connection, a wireless router, and an antenna for wireless access. The latter connection was also possible through the use of Agent Architecture created by the Institute of Human Machine Cognition, which consists of a number of software agents/programs that integrated the data from the sensor hardware and allowed remote communication of these data between the agents in the laptop and the agent-based architecture in the base-station PC. The system is depicted in Figure 1. EEG and ECG data was collected, processes and classified wirelessly in real-time while subjects performed dynamic military scenarios in a wooded area.

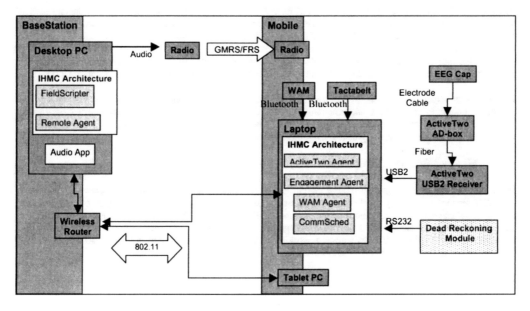

Figure 1: Honeywell's mobile wireless data collection system

2.2   Problems encountered with the wireless mobile system

During several stages of pilot testing of the current data collection system a number of problems have been identified. For instance, variable range of wireless transmission prevented wireless data collection from remote areas in the field. Number of cables on the subject's body restricted mobility. Examination of the EEG signal revealed interference of RF signals with the ongoing EEG activity and artifacts in the form of spikes possibly produced by the movement of the wires on the electrode cap while subjects performed dynamic scenarios. Moreover, we've identified physiological artifacts related to motion and muscle activity as well as ocular activity.

3   Solutions for operation settings

3.1   Identification and removal of artifacts

To improve signal quality and eliminate artifacts that were identified while examining the data, an online algorithm based on adaptive filtering algorithm was implemented. The adaptive linear filter relies on the Widrow Hoff learning rule (Widrow & Hoff, 1960) to derive estimates of the impact of ocular activity at various EEG sites. Once the filter has stabilized, estimates of the ocular activity affecting each channel can simply be subtracted from the signal at each EEG site. This method has found to be robust in real-time detection and removal of artifacts in laboratory settings previously (e.g., Gevins and Smith, 2003; He, Wilson, and Russel, 2004), and does not require preprocessing or calibration.

EEG data was collected at 256 Hz per second and processed in real-time to remove eye blinks using the adaptive linear filter. Information from the vertical ocular sensor (VEOG) was used as the noise reference source and input to the filter. DC drifts were removed using a high-pass filter (0.5Hz cut-off), while spikes identified in the signal were removed using the median filter. A bandpass filter (between 2Hz and 50Hz) was also employed to reject motion induced high frequency components and extract frequency components traditionally associated with cognitive activity.

3.2   Wireless transmission and collection of data

Bluetooth (MSI PC2PC) network was installed to reduce the number of cables on the body and to increase the number of accessible ports for hardware devices. This connection replaced the RS 32 links between the serial devices (i.e., WAM) and the laptop. The laptop was also upgraded with a different wireless card to increase and stabilize the range of wireless transmission. In addition to using the median filter to reduce the impact of artifacts related to movement of the wires, an elastic band was used to secure the wires and reduce their movement.

4   Conclusion

Clearly, there is a need to implement adaptive systems and sensing technologies in operational settings in order to ensure that sensing technologies can be used reliably in mobile environments and utilize the capabilities of adaptive systems fully. As part of transitioning to reliable sensor functioning in mobile settings, Honeywell's Augmented Cognition team sought solutions to the challenges of mobile data collection and processing. Several solutions have been tested and implemented in order to overcome problems associated with the use of the wireless sensing system in the field. Namely, real-time filtering of the EEG data using the adaptive filtering algorithms necessary to also guarantee that the gauged data corresponds to the cognitive state of interest and Bluetooth wireless connection between the sensor hardware and the laptop.

In the future, wireless transmission of the EEG signals form the headcap to the base-station PC will be implemented and supported by the next generation wireless communication standard, WiMAX or 802.16, that supports up to 40 Mbps within 3-10 km WiMAX cells (WiMAX Forum, 2005). It will eliminate the need to store EEG hardware in the backpack and decrease the number of non-wireless connections between the sensors on the body and the hardware.

Acknowledgments

This research was supported by a contract with DARPA and funded through the U.S. Army Natick Soldier Center, under Contract No. DAAD-16-03-C-0054, for which CDR (Sel.) Dylan Schmorrow serves as the Program Manager of the DARPA Augmented Cognition program and Mr. Henry Girolamo is the DARPA Agent. Any opinions, findings, conclusions or recommendations expressed herein are those of the authors and do not necessarily reflect the views of DARPA or the U.S. Army.

References

Dorneich, M., Whitlow, S., Ververs, P., Carciofini, J., and Creaser, J. (2004). Closing the
loop of an adaptive system with cognitive state. Proceeding of the 48th Annual Meeting of the Human Factors and Ergonomics Society. New Orleans, LA.

Kramer, A. (1991). Physiological metrics of mental workload: A review of recent progress. In D. Damos (Ed.), Multiple Task Performance (pp. 279-328). London: Taylor and Francis.

Hettinger, L., Branco, P., Encarnacao, M., and Bonato, P. (2003). Neuroadaptive technologies:
Applying neuroergonomics to the design of advanced interfaces. Theoretical Issues in Ergonomics Science, 4 (1-2), 220-237.

He, P., Wilson, G., and Russell, C. (2004). Removal of ocular artifacts from electro—
encephalogram by adaptive filtering. Medical and Biological Engineering and Computing, 42, 407-412.

Gevins, A., and Smith M. (2003). Neurophysiological measure of cognitive workload
during human-computer interaction. Theoretical Issues in Ergonomics Science, 4 (1-2), 113-132.

Pope, A., Bogart, E., and Bartolome, D. (1995). Biocybernetic sytem evaluated indices of
operator engagement in automated task. Biological Psychology, 40, 187-195.

Freeman, F., Mikulka, P., Prinzel, L., and Scerbo, M. (1999). Evaluation of an adaptive automation system using three EEG indices with a visual tracking task, <u>Biological Psychology,</u> 50, 61-76.

Parasuraman, R. (2003). Neuroergonomics: Research and practice. <u>Theoretical Issues in Ergonomics Science, 4(1-2),</u> 5-20.

WiMax Forum (2005). <u>http://www.wimaxforum.org/about/faq/</u>

Widrow, B. and Hoff, M. (1960). Adaptive switching circuits, in IRE WESCON Convention Record, pp. 96-104.

Scheduling Communications with an Adaptive System Driven by Real-Time Assessment of Cognitive State

Stephen D. Whitlow, Patricia May Ververs

Honeywell Labs
3660 Technology Drive; Minneapolis, MN 55418
Stephen.whitlow@honeywell.com,
trish.ververs@honeywell.com

Ben Buchwald, Jason Pratt

Carnegie Mellon University-Entertainment Tech
Center
700 Pittsburgh Technology Drive
Pittsburgh, PA 15219
bb2@andrew.cmu.edu, pratt@andrew.cmu.edu

Abstract

This paper describes an adaptive system that significantly improved performance in a multi-task environment by scheduling communications based on directly sensed measures of cognitive state. A team of researchers from industry and academia developed an adaptive system that was evaluated in a virtual environment at Carnegie Mellon University's Motion Capture Laboratory. The system took as input neurophysiological and physiological data to status an individual's cognitive state. In conjunction with information regarding the context of the situation, the system adapted the task demands via a communications scheduler to optimize the task load imposed on the user (Dorneich et al, 2004). The participants, acting as lookout scouts, engaged in an identification of a friend or foe task and were responsible to "shooting" the foes. The participants kept a running count of the number of friends and foes as well as the ammunition that they expended during the trial. In addition, verbal messages were played relaying the number of friends and foes that other lookout scouts had found. The participants needed to maintain the count of the number that their teammates had found and ignore the messages from other teams. Information about cognitive state and primary task was used to schedule the delivery of verbal messages during low task load conditions. Findings indicated a 150% performance improvement in the reporting of the number of friends, foes, and ammunition when adapting a communication schedule based on cognitive state as compared to a random scheduler.

1   Experimental Task Environment

The participant was asked to play the part of a military lookout on a virtual rooftop in a simplified urban environment. He or she wore a lightweight, motion-tracked head-mounted display, and was given a motion-tracked M16 rifle prop. The gun prop was visible in the virtual environment, and produced a red laser dot on objects, indicating precisely where the gun was being aimed. In the environment, the participant was surrounded by four buildings, each in one of the cardinal directions of north, south, east, or west. Each building had four columns of evenly spaced windows. The windows of the top four floors on each of the buildings were open, producing a four by four array of windows past which friendly or enemy soldiers could walk.

Computer speakers in the room allowed for simulated radio broadcasts to be heard by the participant. At the beginning of each section of a given trial, a radio message was played instructing the participant to face a particular direction. After that message, groups of friendly and enemy soldiers walked past various windows in the building. There were also radio messages being broadcast periodically, giving numbers of friendly or enemy soldiers spotted by other lookouts. Each message ended with the name of team leader that it is addressed to (e.g., Bravo leader). The participant was instructed to do the following things:

1) Shoot as many enemy soldiers as possible
2) Keep a running count of the number of friendly soldiers seen
3) Keep a running count of the number of enemy soldiers seen
4) Keep a running count of the number of bullets fired
5) Keep a running count of the number of friendly soldiers reported over the radio, only taking into account messages addressed to the Bravo team leader
6) Keep a running count of the number of enemy soldiers reported over the radio, only taking into account messages addressed to the Bravo team leader.

At prescribed times, a radio message was given to "report your status." At this time, the participant reported the running counts that he or she had been keeping. After reporting, the participant's counts were reset to zero and the experiment continued. Each trial was divided into several repeated blocks, each block consisted of two parts. In the first part, groups of soldiers walked past the windows of the building that the participant was facing. The soldiers came in groups, appearing one right after another. Friendly and enemy reports were sent over the radio during this period, in addition to visually identifying and engaging the soldiers in the windows. In the second part of the block, no friendly or enemy soldiers walked past any windows, and the participant was only required to deal with radio messages that he or she received. A "report your status" message was given at the end of each of these parts. Figure 1 depicts the task environment containing 2 enemies (green) and 3 friendlies (tan).

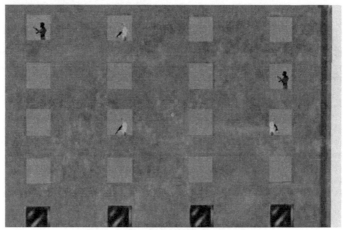

Figure1. The CMU CVE Environment

2   Research Methods

Within this paradigm, the critical comparison is between counting performance for the gauge-driven scheduling condition and the randomly scheduled unmitigated condition. Our primary hypothesis was that gauge-enabled scheduling of incoming radio reports would dramatically reduce catastrophic interference in working memory; thereby, subjects would have dramatically fewer reporting errors with regard to the counts that they had to maintain. The premise of the mitigation strategy is to intelligently schedule incoming radio messages based on the current task load of the subject. Under perfect mitigation, subjects would maintain only their three egocentric counts (total number of friendlies encountered; total number of enemies encountered; total shots fired) during one monitoring block before reporting those counts; at this point, deferred radio messages will be presented during which subjects will maintain two other counts (total of reported number of friendlies reported by Squad A; and total number of enemies reported by Squad A) before reporting them. Perfectly applied mitigation will both reduce occurrence of counting interference and reduce (from five to three) the number of counts to be maintained during the high workload monitoring phase. We would like to note the following assumptions associated with this study:

- Maintaining counts of items is operationally relevant—as suggested by Dr. James Sampson
- Anticipated catastrophic interference is due, in part, to stressing the attention bottleneck and working memory bottleneck.

We would also like to identify some limitations of the current study, including:

- Our protocol is not free-flow as in many operational settings, but is constrained by a block design that affords more experimental control as well as more numerous opportunities to assess the impact of mitigations.

Dismounted soldiers are required to maintain counts of their possession such as ammunition as well as encountered entities such as civilians and combatants. Several factors inherent to dismounted operations conspire to negatively impact soldiers' ability to maintain items in working memory, these include:

- During high-paced operations, soldiers do not have the opportunity to update and rehearse their respective counts which results in failure to maintain an accurate count;
- The inherently stressful nature of dismounted operations consistently disrupts soldiers' capacity to maintain items in working memory; and
- Frequent task switching and communication saturation interferes with maintaining accurate counts

Our basic approach was to extend the application of several derived gauges that use the raw input from EEG and ECG systems to this task-scheduling context. We used a rule-based logic that reasoned about the current state, as well as the direction and rate of change of three gauges (Arousal Meter, Executive Load Index, and the Engagement Index) in order to determine whether incoming messages should be deferred until a later time when the message in question would be less likely to interfere with ongoing task requirements.

Engagement Index. The Engagement gauge indicates level of engagement based on EEG activity. Anything above zero indicates higher then normal engagement and anything below zero indicates lower then normal engagement. The engagement index is measured using the power in three separate frequency bands: theta (4-8Hz), alpha (8-13Hz), and beta 13-22Hz. Frequency analysis is performed using the fast Fourier transform. The band powers are computed from the raw EEG data every X seconds in a window of Y seconds. The average power in each band is further averaged over the relevant electrodes (X,X,Z). Consistent with Freeman et al.'s work, EEG data is recorded from sites Cz, Pz, P3, and P4 with a ground site midway between Fpz and Fz. The Engagement Index (beta/ (alpha + theta)) is calculated from a running average of powers for different EEG frequency bands (Prinzel, et al., 1999).

Arousal Gauge. The Arousal gauge indicates level of arousal based on IBI activity (Hoover & Muth, 2004). Anything above zero indicates higher then normal arousal and anything below zero indicates lower then normal arousal. A three lead ECG is used to detect R-spikes and derive ms resolution IBIs that are then re-sampled at 4 Hz. A fast Fourier transform (FFT) is computed for 16 s, 32 s, or 64 s worth of IBIs. A sliding window is established such that a new FFT is computed every .25 s. When the FFT is computed, the high frequency peak (max power between 9 and 30 cycles per minute) is identified and the power at that peak, termed respiratory sinus arrhythmia (RSA), is stored. Once one minute's worth of FFT results are stored, the arousal meter begins to generate a standardized arousal which is computed every 0.25 s using a z- log-normal score standardization and the running mean and standard deviation of the RSA values.

XLI Gauge. The XLI gauge is a measure of executive load or comprehension, positive values indicate increasing load and negative values indicate decreasing load (DuRousseau, 2004, 2004b). It operates by measuring power in the EEG at frontal (FCZ) and central midline (CPZ) sites. The algorithm uses a weighted ratio of delta + theta/alpha bands calculated during a moving 2-second window. The current reading is compared to the previous 20-second running average to determine if the executive load is increasing, decreasing or staying the same. The basis of the XLI gauge operation is that it reflects the oscillatory activation in a precise cortical network active among spatially disjoint brain compartments. The XLI algorithm measures changes in spatial-frequency (SF) patterns in a manner similar to comparing EP waves using topographic maps and density plots to visualize areas of the brain at rest and during tasks involving attention and working memory (Breakspear and Terry, 2002; Gundel and Wilson, 1992).

Participants
We recruited subjects including students and staff from the Carnegie Mellon University community. All subjects from this pool were not necessarily naïve to the dynamics of the dismounted soldier simulation and the control input devices. They could not have had any alcoholic beverages or sedating medications (for example, cold and flu medications) for at least 12 hours prior to participation. Subjects could be from any race or gender provided they met the above criteria. Pregnant subjects were acceptable.

Experimental Design
This experiment consisted of 2 blocks that each contained 4 pairs of primary-secondary tasks. Each primary-secondary task pair was independent of the other 3 pairs in each block.

Primary Task Period:
- Monitor building for enemies and friendlies, shoot the enemies

- Monitor radio communications
- Maintain cumulative counts of:
  A. number of friendly soldiers seen
  B. number of enemy soldiers seen
  C. number of bullets fired
  D. number of friendly soldiers reported over the radio, only taking into account messages addressed to the Bravo team leader
  E. number of enemy soldiers reported over the radio, only taking into account messages addressed to the Bravo team leader
- Report counts at the end of primary task period

Secondary Task Period:
- Monitor radio communications
- Maintain cumulative counts of:
  A. number of friendly soldiers reported over the radio, only taking into account messages addressed to the Bravo team leader
  B. number of enemy soldiers reported over the radio, only taking into account messages addressed to the Bravo team leader
- Report counts at the end of secondary task period

We compared subject performance under Augmentation ON condition (gauge based scheduling) to a Random scheduling condition. With Augmentation ON gauge values were used to determine if the cognitive state of the participant was overloaded to a point that the system should defer radio messages during the primary task period. Under the Random scheduling condition, radio messages were randomly presented within the primary-secondary task pair period. The assumption underlying the mitigation strategy was to defer radio messages during the primary task load, based on gauge triggers, which would dramatically reduce workload and the working memory interference. The deferred messages would then be presented during the secondary task period when workload and working memory requirements were expected to be less.

The main independent variable was Mitigation ON vs. Mitigation OFF. Dependent variables measured were:

1. Count performance measured by a function of the discrepancy between reported and actual count.
2. Shooting accuracy (# shots fired/ enemies hit)
3. Discrimination between friends and foes
4. # of messages that are deferred during each phase of experiment
5. Neurophysiological and physiological changes measured by:
    a. Engagement Index (EEG based).
    b. eXecutive Load Index (EEG based).
    c. Autonomic Arousal (ECG based).
6. Post-trial subjective report of cognitive workload, including NASA TLX subscales

Table 1. Table Experimental design: 2 (mitigation) x 2 (block) within subject design counterbalanced for order

Group	Block 1				Block 2			
A	Mitigation ON	Mitigation ON	Mitigation ON	Mitigation ON	Mitigation OFF (Random)	Mitigation OFF (Random)	Mitigation OFF (Random)	Mitigation OFF (Random)
	Primary-Secondary Task Pair A	Primary-Secondary Task Pair B	Primary-Secondary Task Pair C	Primary-Secondary Task Pair D	Primary-Secondary Task Pair A	Primary-Secondary Task Pair B	Primary-Secondary Task Pair C	Primary-Secondary Task Pair D
B	Mitigation OFF (Random)	Mitigation OFF (Random)	Mitigation OFF (Random)	Mitigation OFF (Random)	Mitigation ON	Mitigation ON	Mitigation ON	Mitigation ON
	Primary-Secondary Task Pair A	Primary-Secondary Task Pair B	Primary-Secondary Task Pair B	Primary-Secondary Task Pair B	Primary-Secondary Task Pair A	Primary-Secondary Task Pair B	Primary-Secondary Task Pair C	Primary-Secondary Task Pair D

The primary hypothesis is that gauge-enabled scheduling of incoming radio reports will dramatically reduce the catastrophic interference of managing too many counts; thereby, subjects will have dramatically fewer reporting errors with regards to the counts that they have maintained. This is based on the behavioral data we have collected on a similar task in the same environment where a 300% performance improvement was found.

3   System Design

Each participant was outfitted with a BioSemi ActiveTwo EEG cap with 34 scalp electrodes and 3 ECG electrodes from the UFI EZ-IBI system (2 active, 1 ground). The ActiveTwo and EZ-IBI devices connected to PC workstations via USB port and serial port, respectively. The physiologic data was captured by the PC workstations and transferred between agents as needed. The test operator was positioned to monitor the agent and system displays of the PC workstations and could launch and adjust all agents as necessary. Data logging was performed by all agents locally in binary form and the files were collected and posted *post hoc*.

The role of the Mitigation Agent at CMU was to enable radio message scheduling based on sensed cognitive state of participants. It was configured to identify when participants were experiencing the high workload associated with the multi-tasking load inherent to actively monitoring the building while maintaining multiple counts in working memory. When the Mitigation Agent identified such a state, it communicated a "Mitigation ON" state to the Panda 3d virtual environment which would then defer incoming radio messages to be played during the lower workload the secondary task period; otherwise, if high workload was not identified, the Mitigation Agent communicated a "Mitigation OFF" message to the Panda 3d virtual environment which would allow radio messages to pass through to the participants. The Mitigation Agent assumes it will receive input from Z-norm Engagement, XLI, and Arousal Meter. It uses simplified logic that determines if system is in 1 of 8 possible states:

- All 3 gauges up and running (all have certainty >= 0)
- At least gauges up and running (Arousal and Z-norm have certainty >= 0)
- At least gauges up and running (Arousal and XLI have certainty >= 0)
- At least gauges up and running (XLI and Z-norm have certainty >= 0)
- Only one gauge running (Arousal has certainty >= 0)
- Only one gauge running (Z-norm has certainty >= 0)
- Only one gauge running (XLI has certainty >= 0)
- No gauges are up and running

Thresholds were selected for all gauges to maximize differences between hits (high workload during primary task) and false alarms (high workload during secondary task). Gauge thresholds considered both numerical thresholds as well as recent rate of change (ROC) for each gauge individually. Gauges were considered high if either numerical or ROC threshold met. The gauge would return a Boolean value to turn on mitigation is the rules were satisfied, and turn them off is they were not. The mitigation triggering rule set logic is shown in Figure 2.

```
1. If (2 of the 3 is TRUE):
        • Arousal Meter is > .25 OR increased by at least .35 over last 5 seconds
        • Z-norm is > - 1.5 OR increased by at least .25 over last 5 seconds
        • XLI ROC over 3 samples (~ 6 sec) < 0
        THEN Mitigate ON, ELSE Mitigate OFF
2. IF((Arousal Meter is > .25 OR increased by at least .35 over last 5 seconds) OR (IF (Z-norm is > - 1.5
   OR increased by at least .25 over last 5 seconds)), THEN Mitigate ON, ELSE Mitigate OFF
3. IF((Arousal Meter is > .25 OR increased by at least .35 over last 5 seconds) (XLI ROC over 3 samples (~
   6 sec) < 0), THEN Mitigate ON, ELSE Mitigate OFF
4. IF((XLI ROC over 3 samples (~ 6 sec) < 0) OR (Z-norm is > - 1.5 OR increased by at least .25 over last 5
   seconds)), THEN Mitigate ON, ELSE Mitigate OFF
5. IF (Arousal Meter is > .25 OR increased by at least .35 over last 5 seconds), THEN Mitigate ON, ELSE
   Mitigate OFF
6. IF (Z-norm is > - 1.5 OR increased by at least .25 over last 5 seconds), THEN Mitigate ON, ELSE
   Mitigate OFF
7. IF (XLI ROC over 3 samples (~ 6 sec) < 0), THEN Mitigate ON, ELSE Mitigate OFF
8. Mitigate OFF
```

Figure 2. Mitigation trigger rule set logic for the CMU CVE.

4   Results

We ran 2 (condition: mitigated, random) x 2 (task: primary, secondary) ANOVAs looking for main effects and interaction for gauge measures and reported count accuracy; significant interactions were followed up with comparisons. In addition, we ran t-tests to compare effect of condition (mitigated, random) on shooting performance (hit rate) and perceived workload (NASA TLX), as well comparing primary and secondary task for XLI. Accordingly, all reported p values are for either the ANOVA or t-test for the respective measure. We also looked at % of reported counts correct following primary task period as our metric of success for the CMU CVE. We calculated performance improvement as follows:
- (Mitigated % correct – Random % correct) / Random % correct
- Then calculated the average improvement over 10 participants

Reported Count Accuracy. This measure captured the absolute value of the discrepancy (error) between counts reported by the participant and the actual counts. The implications is that the greater the discrepancy reflected overall poorer counting performance. As expected, we found significantly more errors in primary task compared to secondary task across both Mitigation conditions—since participants were required to maintain at least 3 counts while performing a coincident building monitoring task.

To evaluate the impact of our gauge-based mitigation we compared Mitigated and Random condition performance during the critical primary task period, as shown in Figure 3. As expected, primary-mitigated condition showed marginally significantly ($p < .009$) less count errors than primary-random condition (see Table 2 and Figure 3). This comparison is related to the performance improvement metric since it reflects the relative benefit of mitigated counting performance over random counting performance.

Results			
Analysis	df	F-value	p-value
Condition	1	9.35	0.01400
Task	1	55.59	0.00004
Condition*Task	1	9.83	0.01202

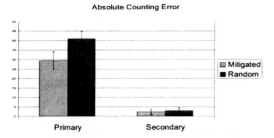

Figure 3. Absolute counting error

Table 2. Comparisons.

Follow-ups				
Comparison	t	df	p-value	Sig
Mitigated-Primary - Mitigated-Secondary	5.79	9	0.00026342	Yes
Mitigated-Primary - RandomPrimary	-3.3	9	0.00919264	Marginal
Mitigated-Primary - RandomSecondary	5.16	9	0.00059401	Yes
Mitigated-Secondary - RandomPrimary	-8.72	9	0.0000110	Yes

Mitigated-Secondary - RandomSecondary	-0.53	9	0.60792697	No
RandomPrimary - RandomSecondary	8.11	9	0.000020	Yes
Alpha = 0.05/6 = 0.008				

Identifying and shooting enemies (hit rate). For this measure we wanted to confirm that our mitigation strategy at a minimum "did no harm" to performance on this ancillary task. Hit rate was calculated by dividing the number of enemy hits by the number of enemies appearing in the monitored building. There was effectively no difference between the mitigation conditions ($p < .48$) (see Figure 4).

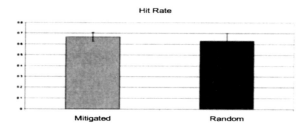

Figure 4. Hit Rate.

Correct Counts (Performance Improvement Metric). Finally, we looked at percentage of reported counts correct following primary task period as our metric of success for the CMU CVE. We calculated performance improvement as follows: (Mitigated % correct – Random % correct) / Random % correct. We found a 155% average performance improvement for 10 participants, as shown in Figure 5. The difference was marginally significant ($p=0.075$). See Table 3 for individual performance.

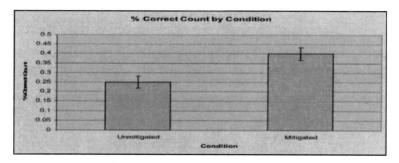

Figure 5. % Correct Count by Condition.

Table 3. Primary Task, percentage of counts correct.

Participant	Random	Mitigated	% improvement
105	16.7%	58.3%	250%
106	8.3%	50.0%	500%
107	50.0%	66.7%	33%
108	16.7%	8.3%	-50%
109	8.3%	33.3%	300%
110	25.0%	41.7%	67%
111	58.3%	41.7%	-29%
112	33.3%	16.7%	-50%
113	8.3%	50.0%	500%
114	25.0%	33.3%	33%
	Average % Improvement		**155%**

Subjective Workload (NASA TLX). We analyzed this to confirm that our mitigation strategy did not negatively impact participants' perceived workload, as measured by NASA TLX subscales. Participants reported a marginally significant ($p < .06$) lower Mental Workload (see Figure 6) for mitigated compared to random conditions. For all other measures (except Physical demand), participants reported a numerically lower workload for mitigated compared to random conditions, as shown in Figure 6.

Comparison	t-value	df	p-value
Mental demand	-2.0147	14	0.06355

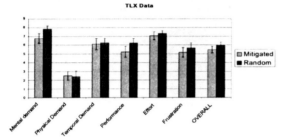

Figure 6. Workload scales for the CMU CVE subjects.

Gauge State Comparisons. We analyzed individual gauge results to evaluate individual gauge response to different conditions (Mitigated, Random) and tasks (primary, secondary) as well as to confirm our expectations developed during the gauge validation studies; accordingly, we would expect differences in gauge response to:
- Task condition: primary to be higher than secondary
- Critical comparison between Mitigated primary (higher) vs. Mitigated secondary– to confirm our thresholds to differentiate between these

Engagement Index Numerical Threshold. We found that there was a numerical difference ($p < .16$) between primary and secondary task in expected direction (see Figure 7) – higher for primary as well as a numerical difference ($p < .16$) between mitigated and random condition in expected direction—lower in. There was a large numerical (0.8) difference between Mitigated-primary and Mitigated-secondary which provided some confirmation regarding our selected threshold; in fact,7 of 10 participants showed this difference.

Results	Df	F	p-value
Condition	1	2.32	0.16
Task	1	2.3	0.16
Condition*Task	1	1.93	0.2

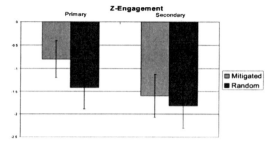

Figure 7. ANOVA for Z-Engagement Gauge.

Engagement Index Rate of Change (ROC). We found a numerical difference (p < .13) between primary and secondary Task in expected direction (see Figure 8) – higher for primary and also numerical difference between mitigated-primary and mitigated-secondary – thus providing some confirmation of our threshold selection; again, 7 of 10 participants showed this difference.

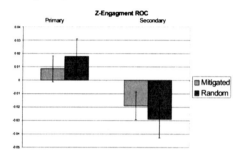

Results	df	F	p-value
Condition	1	0.002	0.97
Task	1	2.75	0.13
Condition*Task	1	3.796	0.083

Figure 8. Z-Engagement ROC.

Arousal Meter. Due to sensor interface issues the ECG signal quality was sufficient for analysis for only five of the subjects. Given the small sample size we looked at numerical trends. First, we found that primary-mitigated to be more arousing than secondary-mitigated - providing confirmation of our specific threshold selection. Second, as expected, we found primary task arousal was higher than secondary task across both mitigation conditions - providing further confirmation our threshold selection. Finally, as expected, we found that the random condition to be numerically more arousing than mitigated for both primary and secondary task conditions--confirming the positive impact of our mitigation strategy. See Figure 9 below.

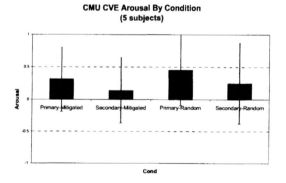

Figure 9. Arousal Meter by condition.

375

XLI. We evaluated differences between primary and secondary task period under the random condition and found a numerical difference that was trending toward significance in a paired sample t-test comparing means ($p < .12$); from this we conclude that the same level of difference did not show up under mitigation condition since the manipulation effectively levels workload to reduce difference between primary and secondary task period—unlike in the random condition.

5   Conclusion

We used a gauge-based mitigation strategy to produce a 155% performance improvement over an unmitigated performance baseline. We confirmed the mitigation triggers by identifying numerical trends for all gauges that were consistent with the expectations and thresholds we established during gauge validation studies intended to differentiate primary and secondary tasks within this task environment. Moreover, our mitigation strategy produced significantly lower perceived mental workload when compared to the unmitigated condition.

Acknowledgments

This research was supported by a contract with DARPA and funded through the U.S. Army Natick Soldier Center, under Contract No. DAAD-16-03-C-0054, for which CDR (Sel.) Dylan Schmorrow serves as the Program Manager of the DARPA Augmented Cognition program and Mr. Henry Girolamo is the DARPA Agent. Any opinions, findings, conclusions or recommendations expressed herein are those of the authors and do not necessarily reflect the views of DARPA or the U.S. Army.

References

Breakspear M,, Terry JR (2002). Topographic Organisation of Nonlinear Interdependence in Multichannel Human EEG. *NeuroImage*, 16, 822 – 835.

Dorneich, M., Whitlow, S., Ververs, P.M., Carciofini, J., and Creaser, J. (2004) "Closing the Loop of an Adaptive System with Cognitive State," to appear in the *Proceedings of the Human Factors and Ergonomics Society Conference 2004*, New Orleans, September.

DuRousseau, D.R., (2004). Spatial-Frequency Patterns of Cognition, *The AUGCOG Quarterly* 1(3):10.

DuRousseau, D.R., (2004b) *Multimodal Cognitive Assessment System*. Final Technical Report, DARPA, DAAH01-03-C-R232.

Gundel A, Wilson GF (1992) Topographical changes in the ongoingEEG related to the difficulty of mental tasks. Brain Topography 5: 17-25.

Hoover, A. & Muth, E.(2004). A Real-Time Index of Vagal Activity. *International Journal of Human-Computer Interaction*.

Prinzel, III, L.J., Hadley, G., Freeman, F.G., & Mikulka, P.J. (1999). Behavioral, subjective, and psychophysiological correlates of various schedules of short-cycle automation. In M. Scerbo & K. Krahl (Eds.), *Automation technology and human performance: Current research and trends*. Mahwah, NJ: Lawrence Erlbaum Associates.

A Low-Cost, "Wireless", Portable, Physiological Monitoring System to Collect Physiological Measures During Operational Team Tasks

Roy Stripling, Ph.D.
U.S. Naval Research Laboratory
4555 Overlook Ave., S.W.
Washington, DC 20375
strpling@itd.nrl.navy.mil

William Becker, Ph.D.
U.S. Naval Postgraduate School
700 Dyer Rd.
Monterey, CA 93943-5001
wjbecker@nps.edu

LT Joseph Cohn, USN MSC
U.S. Naval Research Laboratory
4555 Overlook Ave., S.W.
Washington, DC 20375
cohn@itd.nrl.navy.mil

Abstract

Functional teams exhibit dynamic patters of behavior in the pursuit of their collective goals. In general, the outcome of these patterns of behavior can be subjectively ranked or scored based on a number of performance criteria; and for most applications this remains the preferred method of evaluating team performance and efficacy. In the military domain, almost all operational tasks are team-based. For this reason alone, it is meritorious to explore the potential for objective metrics to quantify the effectiveness of team behaviors. Generating objective measures would arguably enable more accurate evaluation of the team's capabilities, enhance predictability of real-world performance, and, within the training domain, improve the instructor's ability to identify areas in need of improvement (as well as plan the optimal course of instructional intervention). Because of this potential value, numerous efforts have been made to objectively measure team performance along behavioral or cognitive dimensions. However, most such efforts have failed owing to the inherent ambiguity, diversity, and complexity of team-based tasks. Those that have succeeded generally fall within domains of team tasks where team members are restricted to limited, rule-based courses of action (such as airline pilot / co-pilot teams).

Physiological measures are an area receiving more recent attention as a possible source for objective team-based metrics. Deriving team-based metrics from physiological measures has several potential benefits. First, these measures are inherently quantitative. Second, they can be collected in real-time, contributing to continuous evaluation of team performance. Third, they provide equivalent information from each and every member on the team. For example, one can collect heart rates, respiration rates, and electroencephalographic data from everyone on the team. Work in this area has, however, been limited primarily due to three major issues. First, as described above assessing team performance is made difficult by the tasks' inherent ambiguity, diversity, and complexity. Second, thus far basic correlations between team performance and physiological measures from members of the team have not been firmly established. Third, there is a lack of portable, robust, and affordable technologies with which to collect physiological data from all members of a team (which itself limits progress on the second major issues).

To ameliorate the third of these major issues and thereby create opportunities to address the second, we describe here a low-cost "wireless" PDA-based, 16 channel physiological monitoring system. This system includes 16 single-ended EEG channels or 8 differential channels capable of supporting user specified combinations of such measures as ECG, respiration, EGG, and EOG. Except for the construction of a low-cost amplifier for EEG signals, this system is assembled entirely from commercially available technologies, resulting in a total price below $5,000 (U.S.) per unit. Our devices are based on an iPAQ h5550 Pocket PC equipped with a dual-slot PC card expansion pack (Hewlett-Packard). We use one 16 bit PCMCIA PC cards (model 6062E, National Instruments), and run the device using software written in the LabView programming environment (National Instruments). All physiological data can be stored locally and/or downloaded in real-time via 802.11b wireless communication protocols. Optional future configurations may include expansion to 32 EEG channels. Future design modifications will also include the addition of user controlled electronic switching between single-ended and differential recording modes.

1 Introduction

The focus of most if not all present day Augmented Cognition efforts is to enhance the productivity of the individual. However, many of the tasks that could benefit the most from such technologies are team-oriented tasks. Even after individual productivity and capability is enhanced through the application of Augmented Cognition approaches, many tasks will remain team-tasks for several reasons. First, because some of the gains realized through Augmented Cognition systems will arise from the ability to dynamically re-distribute task load among different team members. Second, because tasks requiring problem-solving will continue to benefit from applying multiple (human) minds to the problem. Third, because many Augmented Cognition technologies and approaches may enable teams to multiply their performance enhancements across all of the team members for a collective beneficial impact that far exceeds that of the individual benefit.

Despite these motivations for applying Augmented Cognition approaches to team-tasks, numerous technical and scientific challenges must be surmounted before we will be truly able to collect the real-time, physiological and neurophysiological data necessary to support closed-loop systems. These challenges are not unique to team-task environments, but are multiplied by the need to collect high-quality data from all members of the team. These challenges include: i. overcoming signal artefacts (electromagnetic interference, vibration, muscle contraction, eye movement, foot steps, perspiration, and environmental extremes (including heat, cold, and dirt/dust)); ii. developing of sophisticated software (fast, efficient, and effective data processing and state inference algorithms); and iii. developing of accessible and reliable hardware solutions. The latter most challenge is the focus of this paper. For, while the two former challenges must be met before functional Augmented Cognition systems can be said to exist, surmounting just those challenges will only represent an academic success. Augmented Cognition systems will not achieve mainstream acceptance, nor will they be available for team applications, until these systems are available as low-cost, light weight, compact devices.

To gauge the current potential for hardware solutions to this challenge, we are developing a complete, lightweight (>1 kg), wearable, wireless physiological monitoring system using as much COTS (consumer off-the-shelf) technology as possible. The target system is capable of supporting most, if not all of the physiological signals needed for an Augmented Cognition system, including EEG (electroencephalogram), ECG (electrocardiogram), and respiration measures from its wearer. In addition, our goal is to support simultaneous data collection from multiple (up to four) individuals, enabling the compilation of neurophysiological data from multiple individuals acting independently or as a team. With this approach, a more powerful mobile computer (such as a typical laptop computer) functioning as a LAN (local area network) server could also act as the central processor of a distributed, team-oriented Augmented Cognition system.

While our approach seeks to field a viable neurophysiological tool at minimal cost, the current system will lack several valued features possessed by state of the art systems currently available on the market. Such features include high-density electrode arrays for EEG data collection, greater sampling rates (up to and exceeding 1kHz per channel), and active electrodes for minimizing artefact signals. As such, we do not propose this system to be an across the board substitute for high-end neurophysiological systems. Instead, we suggest that there are three specific advantages for devices such as that described in this paper. First, low-cost systems will enable researchers in this field to begin exploring in earnest team-oriented applications of Augmented Cognition technologies. Second, lower-cost systems will lower the financial barrier to entry for conducting neuro-cognitive research and hopefully increase the number of new researchers entering this field. Third, the low-cost, wearable, and minimal electrode profiles may encourage efforts to explore more rugged operational settings. Although these systems are not so inexpensive as to be disposable, more than half of their cost arises from licensing fees associated with the software programming environment. Of the remaining components, the most fragile (the PocketPC) can be replaced for under $700, making studies of neurophysiological processes in operational environments a more palatable undertaking.

2 Technical Approach

Development of the PocketPC-based physiological sensor is being pursued along three paths: COTS (commercial off-the-shelf) hardware selection, software development, and construction of a low-cost EEG amplifier. The latter most path was deemed necessary because a stand-alone, COTS, low-cost, light-weight amplifier suitable for neurophysiological signals could not be identified.

2.1 COTS hardware

The goal of this project was to develop highly functional, wearable physiological monitors that exceed performance capabilities of complete mobile COTS systems at below market prices (excluding labor). To accomplish this goal, where possible we acquired components from COTS sources. Table 1 lists the components identified and incorporated into our system at the time of writing. These costs include one time charges such as the LabView full development system, LabView PDA module, and licence fees for downloading executables on up to 10 PocketPCs. Researchers already in possession of any or all of these items would incur a much lower development cost. We continue to search for lower-cost alternatives in the hope of reducing the cost of the final system further still.

On a per unit basis, the most costly component we included in our system was the EEG Cap (Neuroscan Quick Cap, Compumedics-Neuroscan). Many alternative EEG electrode placement systems exist and many may be both lower cost and better suited to the useful application of these monitors. This is one area we continue to explore for lower-cost alternatives.

Table 1: COTS (commercial off the shelf) components utilize in the NEURAL-Pack Physiological Monitor.

Item	Unit Costs (US dollars)	Quantity	Total Cost (US dollars)
PocketPC (HP-h5550)	$ 650	4	$ 2,600
SD memory card (512mb)	$ 200	4	$ 800
Dual-Slot Expansion Pack	$ 200	4	$ 800
DAQ card (6062E)	$ 1,295	4	$ 5,180
LabView Full Development System	$ 1,995	1	$ 1,995
LV PDA module	$ 995	1	$ 995
LV PDA licenses (10)	$ 495	1	$ 495
DAQ cables	$ 50	4	$ 200
EEG Cap	$ 1,000	4	$ 4,000
Amplifier Components	$ 140	4	$ 560
Grand Total	$ 7,020		$ 17,625
Cost per unit			$ 4,406

The one component we were not successful finding a suitable low-cost COTS version of was the signal amplifier. Many systems are available on the market, however, all suitable systems were either integrated into a complete physiological monitoring system, or were simply too expensive to be of practical value for this effort.

2.2 Software Development

Development of the software driving our PocketPC-based physiological monitor is being carried out in the LabView graphical programming environment (National Instruments, Inc.). While alternative development environments could be used, LabView offers a range of specific advantages that motivated this choice. First, National Instruments manufactures a broad range of PCMCIA format data acquisition cards compatible with the PocketPC. Second, the professional development package includes numerous drivers and sub-routines designed for data acquisition applications. Third, National Instruments offers an add-on module to LabView that compiles executables specifically for PocketPC or PDA devices.

The software written for the PocketPC drives data acquisition, TCP/IP data transmission, and saving of data to the onboard SD-RAM memory card (Figures 1 and 2). LabView is a graphical development environment whose development system comes with numerous sub-routines (called sub-virtual instruments, or sub-VIs) for data acquisition, and other functions such as TCP/IP communications, file saving, and error handling. The code displayed in Figure 2 creates a TCP listener, identifies the PocketPC's TCP/IP address, establishes a link with the laptop client and displays the laptop's TCP/IP address. Once communication is established, the PocketPC begins continuous data collection at 256 Hz per channel. The number of samples entered by the user on front panel field determines the length of time data is collected before it is transmitted to the laptop and saved to the onboard SD memory card. In this example, the user has selected 512 samples, so data is transmitted and saved every 2 seconds. Because of size limitations on TCP/IP packet sizes, this code transmits the data in 2048 sample packets as soon as it becomes available (4 packets every two seconds, in this example).

The software written for the Laptop receives TCP/IP data transmissions sent from up to four PocketPCs, presents the data from one PocketPC at a time on-screen (the selection is toggle controlled by the user), and saves data from all four to the hard-drive (Figures 3 and 4). The code displayed in Figure 4 opens a TCP/IP connection with up to four PocketPCs with the TCP/IP addresses entered by the user into the front panel. Once the connection is established, this application waits for transmissions sent by each of the PocketPCs. When received, the data is displayed on the respective charts on the front panel, and saved to file. Future versions of this application will conduct online spectral analyses to aid in identifying undesired artefacts.

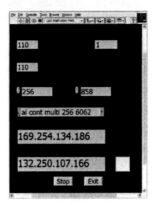

Figure 1: Image of LabView front panel running on PocketPC

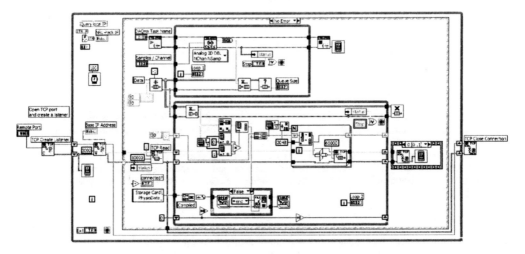

Figure 2: Image of LabView code running on PocketPC

2.3   Amplifier Construction

The amplifier is designed around Linear Technology's LT 1167 precision instrumentation amplifier and the LT 1112 precision picoamp operational amplifier (Figure 5). The LT 1167 was selected because of its low power

380

requirements, wide range of usable supply voltages (±2.3 to ±15), requirement for only one external resistor to set the amplifier gain, and very low internal voltage noise. Also the inputs for this amplifier are ESD protected up to 13kV (human body). The circuit configuration meets IEC 1000-4-2 specification. This chip set is ideal for battery operated medical equipment and in this design only uses ±3 VDC and 1.7mA when in operation. To meet Medical Devise Safety requirements (ISO 9000) this amplifier can only be used in a battery operated configuration.

The circuit design is based on a reference in the specification manual from the manufacture (Figure 5). The first half of the LT 1112 creates a ground for the common mode signal and in conjunction with capacitor C1 maintains the stability of the patient ground. The LT 1167 has a very high CMRR (Common Mode Rejection Ratio) that helps to ensure that the desired differential signal is amplified and attenuates the unwanted common mode signals.

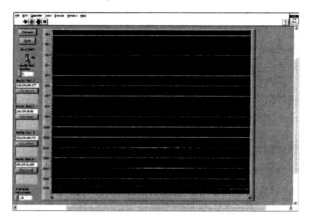

Figure 3: Image of LabView front panel running on Laptop

The second half of the LT 1112 in parallel with capacitor C3 and resistor R7 forms the output amplifier with a lowpass filter to cut off frequencies above 1kHz and give a final output response of 1V/mV. This same circuit is duplicated 16 times for the full instrument. The amplifier is connected to the patient through a 40-pin header to the electrode cap and to the PocketPC by a 68-pin micro-plug.

The first generation amplifier was hand wired using standard discrete components and 8-pin PDIPs making the electronics case relatively large (5 x 7 x 1.25 inches). The second generation will use a manufactured circuit board and surface mount technology that will reduce the size by more then half. Four 3-volt lithium coin power cells connected in a series/parallel circuit with center-tap power the entire circuit providing power for 16 hours of continuous operation. Presently the cells are not rechargeable but a recharging circuit will be installed in the second-generation amplifier.

3   Results

The selected PocketPC outfitted with a dual-slot PCMCIA adaptor sleeve and two PCMCIA data acquisition cards offers the potential for supporting A/D conversion of up to 32 single-ended channels or 16 differential channels. With selection of the appropriate electrode sensors and up to 100 fold software configured amplification, this configuration alone would be capable of supporting a broad range of standard physiological measures including ECG (electrocardiogram), EOG (electro-oculogram), respiration measures, etc. With the addition of a battery-powered signal amplifier this arrangement could theoretically support either multiple physiological measures plus 16 channels of EEG (electroencephalogram) data collection, or without the physiological measures, up to 32 EEG channels. Practical limitations in both software and hardware capabilities, however, have prevented achievement of this ambitious goal. Instead, iterative development efforts have resulted in a system capable of digitizing analog data from 16 single-ended channels at sampling rates of 256 Hz from each channel. These may be configured independently, so theoretically any combination of EEG and physiological measures that require 16 A/D channels or fewer can be supported. As EEG signals are arguably the most challenging physiological signals to collect, we have focused our efforts at developing a 16 channel single-ended EEG system.

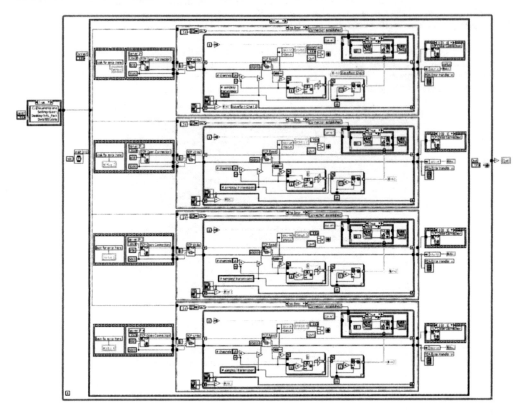

Figure 4. Image of LabView code running on Laptop

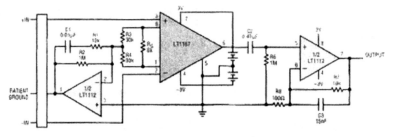

Figure 5. Basic single amplifier circuit diagram

3.1.1 Software Limitations

As described in section 2.2, development of the software needed to acquire, save, and wirelessly transmit the digitized data is being carried out in the LabView graphical programming environment. The reasons for this

382

selection are also listed in section 2.2. Here we focus on the consequences of this decision. These are chiefly, limitations in the drivers for the PocketPC, and limitations in screen display capabilities.

The PDA module and associated drivers for the PocketPC, offered by National Instruments, is version 1.0. Consequently, there remain unsupported capabilities, and undocumented requirements in coding that can severely restrict system performance. For example, current drivers only recognize the first slot in the dual-slot PCMCIA adaptor sleeve. This limits the available number of A/D channels to those that can be obtained from a single PCMCIA card. For the PCMCIA cards we have been working with, this limits us to16 single-ended, or 8 differential channels. At the time of this writing National Instruments had not expressed any intentions to expand this capability.

It was also found that the onscreen displays were not capable of plotting data from more than eight channels (although the device can acquire, digitize, save, and/or transmit the full set of sixteen channels). Furthermore, displaying even one channel of data on screen was found to severely limit performance of the device, resulting in data throughput crashes in under eight minutes when sampling rates were between 75 and 150 Hz per channel (data collected and saved prior to the error remain intact). Data throughput crashes continue to be a problem with this system, however, through a process of trial and error experimentation, we are now able to achieve stable, consistent performance for 90 consecutive minutes while operating at 256 Hz per channel. One of the restrictions necessary to achieve this performance is that the data must be acquired in two second intervals before sending it to file and/or broadcasting it via TCP/IP. Acquiring data in one second intervals results in data throughput crashes after approximately 45 minutes of operation. While longer acquisition intervals would produce longer system operational times, it would also conflict with the need of Augmented Cognition systems for processing the data in near real-time. Research requiring longer recording sessions, and not requiring near real-time data processing would, however, benefit from longer acquisition intervals. The linear relationship between acquisition interval and data throughput crash suggests that the source of this problem should be identifiable and it is our hope that this problem will be resolved completely in the near future.

3.1.2 Hardware limitations

Beyond the limitations posed by the software, the hardware components, particularly the PocketPC itself, also impose certain restrictions on system performance. Indeed, some of the software limitations may actually be partially the result of the physical data throughput limitations of the PocketPC. Beyond this possibility, the chief limitation of the PocketPC is its battery life. Both the PocketPC and the PCMCIA dual-slot sleeve contain rechargeable lithium-ion batteries, however, when acquiring, saving on-board, and wirelessly transmitting data from 16 channels at 256 Hz per channel, we find that the battery life span is approximately 2 hours. Additional operational life could be achieved by constructing a 5V, 2000 mA battery pack and connecting to the adaptor sleeve via the AC/DC adaptor input socket.

4   Discussion

The goal of this project is to develop highly functional, wearable physiological monitors that exceed performance capabilities of complete mobile COTS systems at below market prices. The system described in this paper represents a step along that progression. At the time of writing, the current operational system can support collection of EEG signals from up to 16 channels at 256 Hz per channel. The system saves the data to an on-board 512 MB SD memory card and can simultaneously transmits the data at two second intervals by way of an 802.11b wireless connection. We have successfully acquired, saved, and transmitted data from four PocketPCs to a single laptop computer for 90 minutes. The unit-cost when constructing four systems, including one time licensing fees for the LabView development environment and PDA module, is approximately $4,400 (US). At the time of writing, however, two major limitations remain to be overcome. First, after running approximately 90 minutes, the system experiences a 'data throughput' error and crashes (however, data collected and saved up to that point remain intact). Second, the battery life (when fully charged) is limited to approximately 2 hours of operational time. The second problem may be resolved by adding another battery pack to the configuration, however, solutions for the first problem are still to be determined.

When these two chief limitations of this system are overcome, this device will occupy a unique position in the mobile neurophysiological sensing market. Commercial alternatives include the g.Mobilab portable biosignal acquisition and analysis system (g.tec, inc.; www.gtec.at/products/g.MOBIlab/gMOBIlab.htm) and the series of wearable physiological monitors offered by Cleveland Medical Devices, Inc. (http://www.clevemed.com/index.html). The g.Mobilab supports eight channel of input (2 EEG, 2 EOG, 2 ECG/EMG channels and 2 additional analog inputs) and also operates on an HP 5550 PocketPC. Data is acquired at 256 Hz per channel, and the manufacturer reports a battery life of up to 170 hours. The Cleveland Medical Device systems offer 8 to 11 channels of real-time physiological data collection and wireless transmission, however, none come configured for more than 4 channels of EEG. Sampling rates per channel range from 128 to 960 Hz (configurable). The manufacturer reports battery life ranging from 12 to 16 hours depending on the specific model. Cleveland Medical Devices also offers drivers for MatLab and LabView enabling more flexible configuration control by the researcher and making it the closest match to the system described in this paper.

The ability to receive streams of neurophysiological data from multiple individuals on a single laptop creates opportunities for team-oriented Augmented Cognition research that has not previously been explored. The system described here can collect data from up to four individuals, however, it is theoretically possible to network far greater numbers. Potential benefits from such a capability would include adaptive re-distribution of task responsibilities, scheduling, or communications to under-loaded team members. Low-cost, networked neurophysiological systems such as these may also enable a new line of research into the psychophysiology of teams and team performance. It is conceivable, for example, that experienced, effective teams may be those that compliment each other not only behaviorally, but in the spectral, and/or spatio-temporal patterns of neural activity as recorded via EEG. One testable hypothesis would be that effective team members working together in the same environment and thus exposed to the same external cues and events may be characterized by 'shared' event related potentials (ERPs) or event related desynchronizations (ERDs) of similar timing and power. Conversely, teams that are not as well functioning may be characterized by asynchronous ERPs and ERDs.

The work presented in this paper indicates that the future of wearable neurophysiological sensor devices is promising. PocketPCs and PDAs commercially available now appear to be just sufficiently powerful to support the development of low-cost, wearable, and wirelessly networked neurophysiological monitoring for research purposes. In the near future, wearable computing advances will place the full computational power and processing capabilities of standard laptop and desktop systems in a package small enough and affordable enough to support even more capable and robust sensor systems. These advances will lead to more aggressive neurophysiological field studies, explorations into the neurophysiology of a number of previously unexplored operational environments, and possibly novel and innovative insights into the social-neurobiology of team interactions.

Section 2
Cognitive State Sensors

Chapter 9

Transforming Sensors into Cognitive State Gauges

Transforming Sensors into Cognitive State Gauges

Richard A. Barker

The Boeing Company
P.O. Box 3707 MC 45-85
Seattle, WA 98124
richard.a.barker@boeing.com

Abstract

The problems associated with sensors and cognitive state gauges seem to grow exponentially as we seek to transform them from a laboratory phenomenon into a deployable system. Solutions to the problems dictate the types of mitigations that can be employed and the operator interface software architecture. Some of the problem areas include:

Sensor Noise: Sensors can have good sensitivity and accuracy, and still be too noisy to be used in an applied system. For example, a sensor may correctly show high workload 90% of the time, and yet still oscillate between high (where it stays 90% of the time) and low often enough to confuse an augmentation manager trying to select the correct mitigation.

Types of Cognitive State Gauges: Cognitive state gauges can be classified into part-task gauges and bottleneck gauges. Part-task gauges measure the operator's cognitive state when he is performing a subset of his overall task. The associated mitigations must then be oriented to reducing the workload for that task, but can only do so after the operator has become engaged in that task. Bottleneck gauges measure the cognitive state of the operator's information process resources without respect to a particular task. Bottleneck gauges may, therefore be better suited to a "workload avoidance" strategy, since the need for mitigation can be measure before the operator starts a task. Part-task gauges may be better suite for persistent tasks, rather than episodic tasks. The operator would work on a persistent task for an extended period of time (perhaps his entire shift), and would need workload mitigations during only part of that time.

Response Speed: A cognitive gauge may be validate and reliable, but how fast the gauge must respond is determined by the application in which it will be used. Some tasks may be driven by the need for rapid decision making and shifting among tasks. Other tasks may require more deliberate, in-depth, problem solving and decision making.

These and other issues illustrate the problems with transforming system requirements into requirements for cognitive state gauges. There is no "best" gauge, but rather a "best fit for an application."

Keywords: augmented cognition; bottleneck gauge; cognitive state; cognitive gauge; mitigation

1   Appropriate Acceptance Criteria for Cognitive Gauges

Sometimes, I think the human brain was designed by committee. Wouldn't it be nice if each cognitive bottleneck we wanted to measure had one characteristic sensor signature so that we would know unambiguously that we were measuring what we thought we were measuring. All we would need would be our Acme handy-dandy workload meter, and we could measure unambiguously the cognitive state of anyone on however many bottlenecks we wanted to. Unfortunately, that only works for Commander Data (See Figure 1).

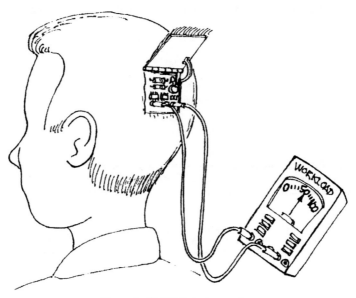

Figure 1: CMDR Data Approach

If we can be sure of anything, though, we can be sure that it's not that simple for the rest of humanity. In fact, It is difficult to even specify what it takes to be a good gauge. This is, I think, a key issue that should guide the development of gauges.

Andrew Belyavin's paper today addresses some of these issues. We must certainly satisfy the requirements for face validity, reproducibility, and sensitivity that he discusses. But, while these criteria are necessary, are they sufficient?

Sometimes laboratory research and operational needs clash in unexpected ways. One way that they clash is that, in the laboratory, the researcher can select which tasks are used to define and validate gauges. In an operational system, however, the tasks are defined by the needs of the warfighter and may not fit neatly into some task classification scheme. A cognitive gauge that works for a laboratory spatial task may not work for an operational task that is *prima facie* a spatial task.

2 Requirements for Multiple Sensors and Cognitive Gauges

Unless you're CMDR Data, there is no one physiological indicator that will give us what we want. Cortical activity corresponding to a cognitive bottleneck is spread over the cortex (e.g., Cabeza & Nyberg, 2002). Therefore, we're looking for activity in more than one cortical area, probably being sensed by more than one sensor and sensor type. For example, Figure 2 through Figure 5 shows relative cortical activity for 4 different bottlenecks.

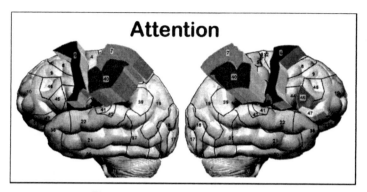

Figure 2: Cortical Locations for Attention

Figure 3: Cortical Locations for Working Memory

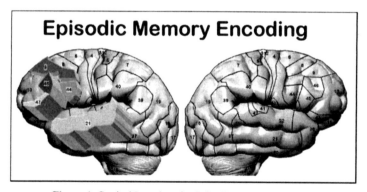

Figure 4: Cortical Locations for Episodic Memory Encoding

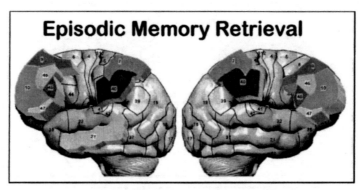

Figure 5: Cortical Locations for Episodic Memory Retrieval

Augmented cognition is one area where redundancy is good! Several of the papers in this session address this issue explicitly.

Sandra Marshall will present a system that helps to unify all of the outputs from different sensors. This is a key area for those of you who have yet to attempt it. Those who have attempted it will understand the importance of data fusion in a way that enables replaying experiments. This is a real "experiment multiplier" in that each replay can use different algorithms for transforming sensors into gauges.

We also have some papers from AFRL. They are doing very important work on cognitive state classifiers for transforming sensors into gauges. Their classifiers use EEG, ECG, EOG, and other sensors and fuse them into cognitive gauges. The work they are going to be presenting in this session is a quantitative comparison of different classifiers. In their other paper they analyze what features of the sensors are contributing to the cognitive gauges. The techniques they present are at least as important as the data they are presenting, and the data they are presenting is very important for almost anyone working in this field. The analytic techniques can be used by all of us in the field.

3   Detection of Cognitive State or Closed Loop Control?

Once we detect cognitive state, how are we going to use it? More importantly, does the use to which we are going to put cognitive state influence the methods we use to detect it?

We have 2 papers on using statistical process control as a method for detecting cognitive state. One paper, by Jerry Tollar, talks about how the Boeing team has been using statistical process control methods in their work. They have been using it as a method for both detecting cognitive state and to control a closed loop system.

4   Diagnosticity of Gauges

Detecting global cognitive workload is useful, but our real objective is to classify cognitive workload in a multidimensional space, where each dimension corresponds to a cognitive bottleneck. Chris Berka and others are presenting their work distinguishing spatial and verbal processing in working memory using EEG. While many people can detect when verbal and/or spatial information is being processed, detecting spatial and verbal processing in working memory is key. It is in working memory where one of the more severe bottlenecks on information processing occurs. They have gotten a very accuracy rate using an artificial neural net that remains stable from week to week and from person to person – very important characteristics for a fieldable system.

5   References

Cabeza, R., & Nyberg, L. (2002). Seeing the forest through the trees: The cross-function approach to imaging cognition. In A. Zani (Ed.), *The Cognitive Electrophysiology of Mind and Brain*: Academic Press Inc.

Feature Saliency Analysis for Operator State Estimation

Chris A. Russell

Air Force Research Laboratory
Human Effectiveness Directorate
2255 H Street
Wright-Patterson AFB, OH 45433
christopher.russell@wpafb.af.mil

Glenn F. Wilson

Air Force Research Laboratory
Human Effectiveness Directorate
2255 H Street
Wright-Patterson AFB, OH 45433
glenn.wilson@wpafb.af.mil

Abstract

Determining operator functional state is a critical component of adaptively aided systems. If the aiding or automation is not provided when it is required by the operator system performance may not improve. Accurate knowledge of the functional cognitive status of the human operator is required for the aiding to be most effective. To determine operator functional state (OFS), we must decide which measured features from the operator best distinguishes between multiple levels of cognitive activity. A battery of 43 psychophysiological signals was collected while operators performed a complex task with two levels of cognitive demand. Three feature reduction methods, principal component analysis, weight-based partial derivative method, and a weight-based signal-to-noise ratio were applied, and the results were used as inputs to an artificial neural network for training and classification. Classification accuracies up to 90 percent were achieved and the number of input features required was reduced by up to 84 percent.

1 Introduction

The estimation of OFS is a critical component of adaptively aided systems so that appropriate aiding is presented at the correct time. If the state of the operator is not correctly assessed then the aiding may be presented at the wrong time or not at all. One approach is to monitor the psychophysiological state of the operator in real-time in order to assess their ability to meet the current cognitive demands of the task. Several components must be present to determine OFS, including real-time measurement and processing of psychophysiological signals and a robust, accurate classification algorithm. Regardless of the classifier algorithm, the many signals that can be collected from the operator must be processed into measures which provide the most accurate classification of OFS. Real-time classification requires the calculation of these measures in real-time. The requirement of real-time processing consequently may reduce the number of measures available to the classification algorithm since computational requirements limit the amount of data that can be processed in the allotted time.

Important considerations of any classification algorithm are the type and number of input features. Identifying the proper features used by the classifier model is one of many barriers to real-time classification of operator functional state. Most classification algorithms can adapt to practically any number of input features, but having many inputs leads to the "curse of dimensionality." As the number of inputs increases, the number of examples required to estimate the free parameters in the classification model also increases. Features that provide no or little information for classification increase the burden on the classifier and on the operator. In addition to the "curse of dimensionality," the inclusion of these unnecessary inputs increases the computational burden on the systems providing data collection and reduction. In real-time systems, any unnecessary delays due to increased processing requirements in unacceptable. Additionally, the sensors are normally physically located on the operator. Too many sensors can have a negative impact on operator acceptance of this emerging technology.

The nature of the input features is also important. The features must provide information to the classifier that makes possible the identification of the patterns of the different levels of OFS. Some features provide no information and are essentially noise. The adage "Garbage in, garbage out" is especially true in the case of classifying operator functional state in complex operational systems.

2 Methods

Three methods of feature selection or saliency analysis were evaluated. Principal component analysis (PCA), a weight-based partial derivative method, and a weight-based signal-to-noise method were compared using psychophysiological measures collected in a complex laboratory task. Eight EEG (F7, Fz, Pz, C4, T4, T5, O1 and O2 of the 10-20 system) and three peripheral measures (heart, respiration and eye activity) were collected. The EEG data were submitted to a FFT and the results were divided into the standard five bands, delta (1-4 Hz), theta (4-8 Hz), alpha (8-12 Hz), beta (13-25 Hz) and gamma (25-42 Hz). Input features selected by each method were presented to a multilayer perceptron artificial neural network (ANN, Haykin, 1999) to determine effects of feature reduction on classification accuracy on two levels of cognitive workload. The NASA Multi-Attribute Test Battery was used to provide the cognitive task.

PCA (Jolliffe, 1986; Flury, 1988) is a useful technique for multivariate analysis that can transform correlated variables into uncorrelated variables, determine linear combinations that have the maximum range of variability, and reduce the numbers of input data. PCA projects the input data on a plane in the direction of each of the eigenvectors determined by the eigenvalues of the characteristic polynomial of the data covariance matrix. The eigenvalues are ordered by size of the magnitude from the largest to the smallest and become the principle components of the input covariance matrix.

The largest principal component is the eigenvalue that accounts for the largest variance of the covariance matrix. Therefore, the first (largest) principal component is the projection in the direction in which the variance of the projection is maximized. Selecting only the largest eigenvalues and their associated eigenvectors reduces the input space. The number of eigenvalues used in this study was those eigenvalues that explain more than 80% of the cumulative variance of the covariance matrix. The input features to the classifier are therefore linear combinations of the derived features weighted by the scalar components of the eigenvector of the more important principal components.

The partial derivative saliency measure (Ruck, Rogers, and Kabrisky, 1990) calculates the partial derivative of each layer of an ANN and rank orders the input features. In essence, this method provides an input-output relationship between the network output layer and the input features. The partial derivative method is possible because although the activation functions of the multilayer perceptron ANN are nonlinear, they are differentiable.

Feature saliency is based on the concept that a fully trained ANN contains all the information for describing the relative importance or saliency of each of the input features. Feature reduction was accomplished by an iterative approach. An ANN was trained using all the input features, the saliency was computed, and the features were rank ordered based on the saliency measure. The least salient feature was removed from the input matrix and the network was retrained using the reduced feature set. This sequence was repeated until only one feature remained. The minimum data set was the smallest data set with the highest classification accuracy on the test data.

The signal-to-noise ratio saliency measure (Bauer, Alsing, and Greene, 2001) compares the input layer weights of the multilayer perceptron for an individual input feature to the weights of an injected white noise feature. Theoretically, because the noise is a random feature, the measure will be significantly larger than zero for salient features and close to or less than zero for insignificant or nonsalient features. As with the partial derivative saliency method this method uses a fully trained ANN to determine the saliency of each of the input features.

The weights of each feature were evaluated with respect to a random noise variable injected into the ANN. The features with high signal-to-noise ratio provide more information for classifying OFS than those with lower signal-to-noise ratios. The features with a negative signal-to-noise ratio provide little or no information for pattern classification since the signal-to-noise ratio of the injected random noise is zero. The feature reduction was accomplished using the same iterative approach as used in the partial derivative saliency measure method.

3 Results and Discussion

The classification accuracy using the reduced feature set for each of the three methods was similar regardless of feature reduction type (Table 1). Table 2 displays the number of features (principle component in the case of PCA) used for each of the feature saliency methods. Each method presents different advantages and disadvantages as techniques for feature reduction. The PCA method will reduce the feature space for the classification algorithm but does not reduce the input space or the number of signals that must be collected. The partial derivative technique does not reduce the feature space by as much as the other two methods; however, it does provide a true input-output relationship for each feature. The signal-to-noise ratio method reduces both the input and feature spaces but requires a noise signal to be injected into the classifier.

Table 1: Classification Accuracy by Subject and Feature Selection Method			
Subject	PCA	Partial Derivative	SNR
1	93.5	89.7	86.5
2	95.0	95.7	97.3
3	98.0	96.0	96.5
4	71.0	71.1	78.7
5	85.2	92.6	89.5
Mean	88.5	89.0	89.7

Table 2: Number of Input Features Remaining	
Feature Selection Method	Number of Features
PCA	8
Partial Derivative	22
SNR	7

The saliency values of each of the significant features were averaged across subjects and rank ordered from highest to lowest saliency. The electrode site locations of the salient features for each of the feature saliency methods are presented in Figure 1. The partial derivative method used frequency band measures from five of the eight EEG electrode sites in addition to eye blink activity and interbreath interval. The signal-to-noise ratio method used frequency band measures from five of the eight electrode sites were used along with interbreath interval. Using the partial derivative saliency method would allow the use of fewer electrodes. The additional measure of interblink interval does not add to the cost of instrumenting for classification of operator functional state since the eye signals are also recorded for removal of eye movement activity from the EEG signals.

The EEG frequency band most often selected as salient was the beta band followed by the delta and gamma bands. These three bands represent the two extremes of the frequency spectra, specifically 1 Hz to 4 Hz and 13 to 42 HZ. The cognitive psychology literature typically associates alpha and theta EEG bands with changes in cognitive load. Both of these variables were selected with the least frequency by all feature selection methods used in this study (Figure 2).

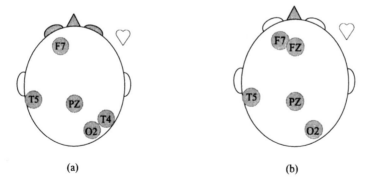

(a)   (b)

Figure 1. Salient feature selected by (a) partial derivative saliency method, (b) signal-to-noise saliency method.

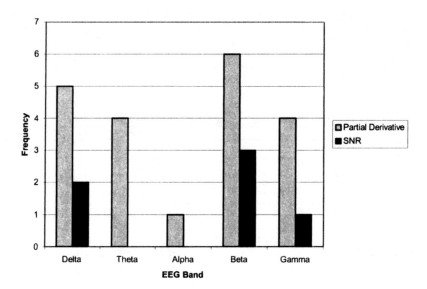

Figure 2. Frequency of EEG band selection associated with feature selection method.

Feature reduction can be accomplished, resulting in a significant decrease in the dimensionality of the ANN model. This reduction not only decreases the amount of training data required by the model but also reduces the training time for the multilayer perceptron. For this study, the selection of a feature reduction method has implications for classification accuracy.

This study indicates that PCA is a good choice of input feature selection for a number of reasons. One advantage of using PCA for input feature reduction is its computational efficiency. Both the weight-based partial derivative method and the signal-to-noise ratio method require the training of multiple ANNs to determine feature saliency and removal of nonsalient features. PCA does not require an ANN for selecting salient features: input feature selection is based on those weighted linear combinations of derived measures that account for the majority of the variance in the covariance matrix.

An additional advantage of PCA is the number of eigenvalues that explain 80% of the variance is robust. Eight weighted linear combinations of the power variable input space were required to explain 80% of the variance of the covariance matrix. This number varied by less than one eigenvalue across subjects. The partial derivative method used an average 22 input features and varied by as much as six features across subjects. The SNR method required an average of seven features and that number varied by as much as eight features across subjects.

The very factors that create advantages for the PCA approach produce a number of disadvantages. One disadvantage to PCA is that this method does not directly reduce the quantity of data that must be collected. Both the weight-based partial derivative method and the signal-to-noise ratio method will reduce the number of electrodes requiring instrumentation. Another disadvantage to PCA is there is no direct relationship between the cognitive state and the frequency bands of each individual electrode site since each component is a linear combination of all inputs.

The partial derivative technique does not reduce the feature space by as much as the other two methods; however, it does provide a true input-output relationship for each feature. This input-output relationship provides a unique advantage over the signal-to-noise ratio method if the interactions between input features provide significant improvements to classification accuracy. The higher order interactions of the ANN classifier model are defined in the hidden layer of the ANN. The signal-to-noise ratio method uses only the input layer weights in its calculations.

Both the signal-to-noise ratio method and the partial derivative method reduce both the input and feature spaces (PCA reduces only the feature space). This fact alone provides a good argument for either of these techniques. Applying operator functional state estimation outside the confines of the laboratory requires the designer to reduce the footprint of the supporting equipment and sensors. Reducing the number of electrodes reduces the size of the supporting electronics as well as reducing the computational complexity of deriving measures from more sensors.

4 References

Bauer, K. W., Alsing, S. G. & Greene, K. A. (2000). Feature screening using signal-to-noise ratios. *Neurocomputing*, 31, 29-44.

Flury, B. (1988). Common principal components and related multivariate models. New York: John Wiley & Sons.

Haykin, S. (1999). Neural networks: A comprehensive foundation. Upper Saddle River, NJ: Prentice Hall.

Jolliffe, I. T. (1986). Principal component analysis. New York: Springer-Verlag.

Ruck, D., Rogers, S. & Kabrisky, M. (1990). Feature selection using a multilayer perceptron. *Journal of Neural Network Computing*, 2(2), 40-48.

Comparing Classifiers for Real Time Estimation of Cognitive Workload

Chris A. Russell[1], Glenn F. Wilson[2], Mateen M. Rizki[3], Timothy S. Webb[4], and Steven C. Gustafson[5]

Abstract

Psychophysiological measures and three types of pattern classification algorithms were used to determine if an optimal algorithm could be found to evaluate how well higher levels of cognitive activity, such as executive function, spatial and verbal working memory and global workload, could be classified. The three pattern classification algorithms were artificial neural networks (ANN), discriminant analysis (DA), and support vector machines (SVM). ANNs, specifically the multilayer perceptron, are compared to discriminant functions and support vector machines. The discriminant functions investigated were linear discriminant analysis, quadratic discriminant analysis, and logistic regression for pattern classification. The SVMs evaluated are linear support vector machines, polynomial support vector machines, and radial basis function support vector machines. In each case the multilayer perceptron outperformed the other classifiers. An uninhabited combat air vehicle (UCAV) simulator was used in which subjects were responsible for monitoring and selecting targets for four vehicles.

1 Introduction

Classification or estimation of operator functional state has numerous applications in the fields of human factors engineering, training and test and evaluation. For example, knowledge of pilot state in an advanced fighter aircraft could be used to increase system efficiency and effectiveness by utilizing this information as real-time guidance for an adaptive control system. In-flight operator state is merely one concern of USAF researchers. Uninhabited Air Vehicle (UAV) and UCAV operators may experience performance degradation during mission segments with high cognitive load. In addition, an understanding of operator workload could support the development and evaluation of human-computer interfaces by providing metrics for operator state. Accurate and reliable assessment of operator functional state is key to successful implementation of adaptive automation, design evaluation, and operational test and evaluation.

The data collected from human studies, especially physiological data, can be voluminous requiring methods for collecting and storing these data. Human studies usually generate

---

[1] Air Force Research Laboratory, Human Effectiveness Directorate, 2255 H Street, Wright-Patterson AFB, OH 45433, christopher.russell@wpafb.af.mil

[2] Air Force Research Laboratory, Human Effectiveness Directorate, 2255 H Street, Wright-Patterson AFB, OH 45433, glenn.wilson@wpafb.af.mil

[3] Dept. of Computer Science and Engineering, Wright State University, Dayton, OH 45435, mrizki@cs.wright.edu

[4] Air Force Institute of Technology, Department of Mathematics and Statistics, 2950 Hobson Way, Wright-Patterson AFB, OH 45433, timothy.webb@afit.edu

[5] Air Force Institute of Technology, AFIT/ENG, Wright-Patterson AFB, OH 45433, steven.gustafson@afit.edu

megabytes or even gigabytes of data from each subject. Therefore it is necessary to develop classifiers that can operate on very large data sets and still learn appropriate models quickly for real-time applications. Classical statistical inference is based on three fundamental assumptions (Casella and Berger, 2002). First, data can be modeled by a set of linear functions. Unfortunately, modern problems are often high-dimensional, and the underlying mapping is usually not very smooth for human studies. Under these conditions linear paradigms need a large number of terms. Also, high dimensionality of the input space implies a large number of independent variables, which leads to "the curse of dimensionality" (Gershenfeld, 1999). Second, the underlying joint probability density is assumed to be Gaussian (i.e., normal), which may not be the case for real data; the data may be far from a normal distribution. Finally, due to the second assumption, the usual induction paradigm for parameter estimation is the maximum likelihood method, which reduces to the minimization of a sum of squared error cost function in most engineering problems. Because of the high dimensionality and the possibly misleading assumption of normality, it may be desirable to replace the maximum likelihood estimator with a different induction algorithm, such as an artificial neural network (ANN).

ANNs have potential advantages that make them attractive for use as classifiers of operator cognitive state. An approach to classification using ANNs seems appropriate, since humans, like neural networks, are adaptive, complex, nonlinear systems, and they have the ability to generalize. Because of the inherent nonlinearity and the complex interactions among the features of cognitive activity during dynamic multiple task situations, accurate cognitive workload classification is difficult. In addition, the relationships between physiological variables and performance in complex, highly dynamic tasks are not well understood; therefore, the relevant features for classification are not known. In particular, the feature probability density functions are mostly unknown, and thus distribution free classification must be performed. Consequently, adaptive neural networks are an attractive choice for classifying cognitive workload in complex real-world situations.

Techniques such as linear discriminant analysis have been used for decades. However, as discussed previously, most real world human cognitive and performance problems are not Gaussian in nature, and linear techniques may not provide adequate results. Other algorithms, such as SVMs developed in the 1970's (Vapnik, 1999), have emerged as alternatives to ANNs and DA. As for ANNs, the model classes are not restricted to linear input-output maps. The parameters are data-driven so as to match the model capacity to the data complexity. SVMs are an attractive alternative to the ANN since after a kernel transformation the data is linearly separable.

2 Methods

Classifiers were developed using 38 psychophysiological features collected in an Uninhabited Combat Air Vehicle (UCAV) simulation. A vehicle health task (VHT) similar to an n-back memory task and an operator vehicle interface (OVI) task were conducted as single task experiments to evaluate the classifiers and investigate the feasibility of multiple cognitive gauges from the same physiological measures. A gauge is a classifier model for a particular cognitive function, i.e., spatial working memory. Trials for four conditions (low VHT, high VHT, low OVI, and high OVI) were conducted and three trials of each condition were presented to each of the operators. The cognitive gauges were spatial working memory, verbal working memory, executive function, global workload, spatial versus verbal working memory, and two task related gauges, VHT task and OVI task.

The cognitive gauges were derived from various combinations of the conditions. The spatial working memory consisted of three levels (no spatial working memory consisted of the low and high VHT conditions, low spatial working memory consisted of the low OVI condition, and high spatial working memory consisted of the high OVI condition). The verbal working memory gauge consisted of no verbal working memory, low verbal working memory, and high verbal working memory (no verbal working memory consisted of the low and high OVI tasks, low verbal working memory consisted of the low VHT trials, and the high verbal working memory class consisted of the high VHT trials).

A gauge to determine the classification of spatial or verbal working memory was also examined. This gauge consisted of two classes, verbal working memory and spatial working memory. The verbal working memory class consisted of the low and high VHT trials. The spatial working memory class consisted of the low and high OVI trials.

The executive function consisted of three classes; low, medium, and high executive function. The low executive function class consisted of the low VHT and low OVI trials, the medium executive function class consisted of the high OVI trials, and the high VHT trials provided the data for the high executive function class. The global workload gauge is a measure of overall workload and consists of two levels, low and high global workload. The low global workload class consisted of the low VHT and low OVI trials, and the high global workload class consisted of the high VHT and high OVI trials.

The OVI and VHT gauges were based solely on the respective trials and task conditions. The low VHT class consisted of the low VHT trials, and the high VHT class consisted of the high VHT trials. The OVI trials were separated into the cruise component and the SAR image processing component. The cruise component is the portion of the trial when the operator is not processing a SAR image and mainly consisted of the ingress to target portion of the trial. The OVI class condition consisted of three classes, cruise, low SAR, and high SAR.

The ANNs used in this study are multilayer feedforward perceptrons with backpropagation training (Haykin, 1999; Duda, Hart, and Stork, 2001). These are the most commonly used ANNs for pattern recognition (Widrow and Lehr, 1990; Lippmann, 1987). Sigmoidal activation functions were used in each of the three layers of the ANN; input, hidden and output layers. Adaptive learning rates and momentum were applied to the learning algorithm to improve backpropagation learning speed and classifier accuracy.

The DA algorithms used in this study are linear discriminant analysis (LDA), quadratic discriminant analysis (QDA), and logistic discriminant analysis (LogDA). LDA performs dimensionality reduction while preserving as much of the class information as possible by maximizing the ratio of between class variance to within class variance (Bishop, 1995; Ripley, 1996; Duda, Hart, and Stork, 2001). QDA extends LDA by including squared and cross products as well as linear functions of the predictor variables or features (Ripley, 1996). LogDA or logistic regression analysis is a well known technique for classification and uses linear classification after a logarithmic transformation.

Kernel based learning algorithms, such as SVMs, are basically comprised of two parts, a general learning machine and a problem specific kernel function (Vapnik, 1995; Burges, 1998). The general learning machine produces an optimal separating hyperplane using quadratic optimization after transformation by a kernel function. Three kernels were used for the classifiers using SVMs; linear, polynomial, and radial basis functions.

Classifiers for the seven cognitive gauges were trained using each of the pattern classification algorithms. Each classifier was trained and tested using the same training and test sets allowing direct comparisons of the classifiers. The direct comparisons conducted were classification accuracy and error rate. Pairwise comparisons were also conducted using a binomial selection procedure (Ripley, 1996). Each test sample can be represented by an independent, identical Bernoulli trial with only two possible outcomes, correct or incorrect. A series of these random Bernoulli trials has a binomial distribution. By comparing the number of successful or unsuccessful trials, comparisons of competing classification algorithms can be made.

3 Results and Discussion

The multilayer perceptron was the baseline algorithm in this study. The DA and SVMs were compared to the ANNs using the same training and test data. Classification accuracy using linear, quadratic and logistic discriminant analysis were compared to the results found using a multilayer perceptron with backpropagation training (Figure 1). Classification accuracies across all algorithms were similar, but the artificial neural network tended to outperform the others. Pairwise comparisons using the multinomial selection procedure were conducted and the wins for each classifier were collapsed across cognitive gauge classification.

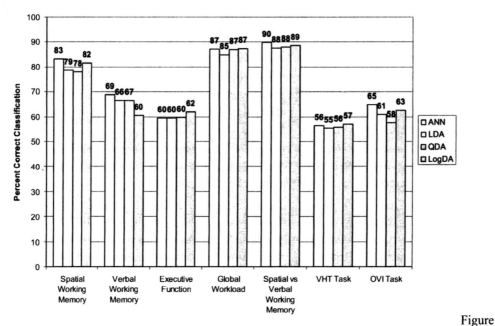

Figure
1. Classification accuracy for the artificial neural network was better as compared to discriminant techniques for most cognitive gauges.

The multilayer perceptron performed better in each case. Figure 2 shows the ANN win percentages against each of the DA classifiers. The worst performer was LDA which lost to the ANN in 80 percent of the models. QDA did better with the ANNs winning 68 percent of the trials. The best discriminant performer against the ANN was logDA which only lost 58 percent of the time.

Comparisons between SVMs and ANNs were also accomplished. The linear SVM is a special case of the polynomial SVM. The order of the polynomial SVM must be determined *a priori* and an evaluation of polynomial SVMs with orders of 1, 2, 3, 4, 5, and 6 was conducted. The best classification accuracy was with a polynomial of order one; however, the linear SVM was already in consideration. The next best order for the polynomial kernel is 3rd order. Additionally, the spread of the radial basis function kernel must be determined *a priori* and was determined in the same manner as the order of the polynomial kernel. Classification accuracy using spreads of 0.01, 0.05, 0.1, 0.25, 0.5 and 1.0 were evaluated and the best spread for the radial basis function kernel was 0.05.

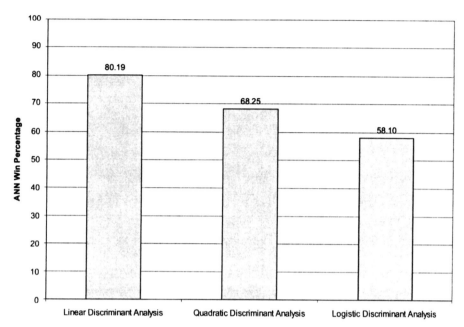

Figure 2. ANN classification win percentage over DA techniques across all trials and cognitive gauges.

The results using the SVMs were compared to the results obtained using the multilayer perceptron. The classification for each algorithm is shown in Figure 3 for each cognitive gauge. As with the results using the discriminant functions, SVMs have comparable classification accuracy to the ANNs. However support vector machines, particularly those using linear and radial basis function kernels, perform almost as well as the artificial neural networks. The 3rd order polynomial SVM did not perform as well as the other SVMs. This was expected since a 1st order polynomial was considered a better choice as a polynomial model.

These results were also compared pairwise and the wins were collapsed across the cognitive gauge classification. The multilayer perceptron performed better in each case. Figure 4 shows the ANN win percentages against each of the SVM classifiers. The worst performer was the 3rd order polynomial SVM, which lost to the ANN in 76 percent of the trials. The radial basis function and the linear SVM classifiers performed about the same, with the ANNs outperforming these algorithms in about 59 percent of the trials.

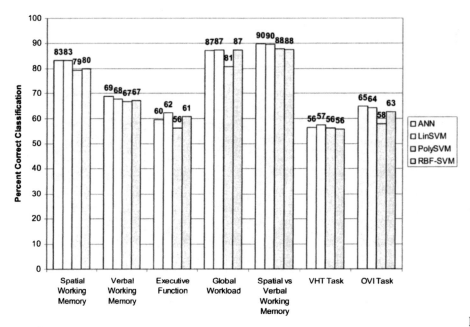

Figure
3. Classification accuracy for comparing results using SVMs and ANNs for each of the cognitive gauges.

Results indicate the multilayer perceptron performed better overall as compared to both the discriminant functions and the SVMs. However, SVMs with a properly selected kernel function provide similar classification and prove to be a good alternative to ANNs. LDA and QDA do not perform as well due to the nonlinear nature of the input space. In fact, the more nonlinear the discriminant function, the better the classifier performs indicating the necessity of nonlinear classifiers for estimating operator cognitive state.

The binomial selection process alone should not be used as the determining factor for evaluation of the strength of classifiers. For example, classifier A shows significant improvement of classifier B using the binomial selection process. However classifier A has a classification accuracy of 85% and classifier B has an accuracy of 82%. Classifier A has a mere 3% improvement over classifier B. Algorithm complexity and ease of use as well as selectivity (classifier accuracy) and specificity (class posterior probabilities) must be considered when determining the best classifier to use in applications.

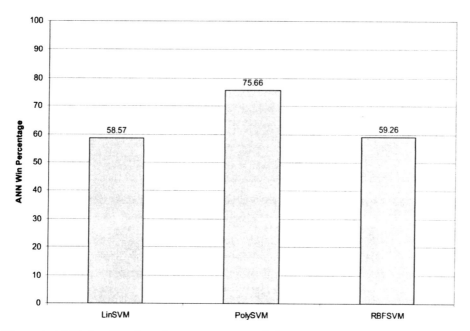

Figure 4. ANN classification win percentage over SVMs across all trials and cognitive gauges.

Other issues should be considered when selecting the classifier. The developer's familiarity with an algorithm will have a role in how well a classifier performs. For example, the ANN will always train to the data with near perfect results. The classifier over learns the data and will not generalize to new data samples. If the developer does not realize this and does not implement techniques such as early stopping using validation data, the ANN will not perform as well as other algorithms that are properly trained.

Algorithm complexity should be a consideration. Occam's razor says the simplest solution is the best solution. More complex algorithms have more parameters that must be selected. Increase in the number of parameters means that more training data must be collected to build a good model of cognitive workload.

4 References

Bishop, C. M. (1995). Neural networks for pattern recognition. New York: Oxford University Press.

Burges, C. C. J. (1998). A tutorial on support vector machines for pattern recognition. *Data Mining and Knowledge Discovery*, 2, 121-167.

Casella, G. & Berger, R. L. (2002). Statistical inference. Pacific Grove, CA: Duxbury.

Duda, R. O., Hart, P. E. & Stork, D. G. (2001). Pattern classification. New York: John Wiley & Sons, 2001.

Fukunaga, K. (1990). Introduction to statistical pattern recognition (2nd ed.). San Diego, CA: Academic Press.

Gershenfeld, N. (1999). The nature of mathematical modeling. Cambridge, Great Britain: Cambridge University Press.

Haykin, S. (1999). Neural networks: A comprehensive foundation. Upper Saddle River, NJ: Prentice Hall.

Lippmann, R. P. (1987). An introduction to computing with neural nets. *IEEE Acoustics, Speech, and Signal Processing Magazine*, 4(2), 4-22 .

Ripley, B.D. (1996). Pattern recognition and neural networks. Cambridge, Great Britain: University Press.

Vapnik, V. N. (1999). The nature of statistical learning theory (2nd ed.). New York: Springer-Verlag.

Widrow, B. & Lehr, M. A. (1990). 30 years of adaptive neural networks: Perceptron, madaline, and backpropagation. *Proceedings of the IEEE*, 78(9), 1415-1442.

EEG Indices Distinguish Spatial and Verbal Working Memory Processing: Implications for Real-Time Monitoring in a Closed-Loop Tactical Tomahawk Weapons Simulation

Chris Berka[1], Daniel J. Levendowski[1], Gene Davis[1], Michelle N. Lumicao[1], Caitlin K. Ramsey[1], Kay Stanney[2], Leah Reeves[2], Patrice D. Tremoulet[3], Susan Harkness Regli[3]

Abstract

This effort focused on developing EEG-derived indicators of verbal versus spatial working memory load. A wireless EEG headset acquired data during execution of both simple and complex tasks associated with a Tactical Tomahawk Weapons Control System (TTWCS). The results established the feasibility of characterizing EEG correlates specific to verbal and spatial working memory. The next goal is to leverage these real-time working memory indices as a feedback loop to direct closed-loop human-system interaction. Specifically, if the preliminary EEG indices derived in this study, in combination with other physiological or behavioral inputs, are shown to relate to the degree of working memory overload in the TTWCS or similar tasks, they could provide a valuable contribution to real-time adaptive aiding of human-system interaction.

1 Introduction

Working memory overload is one of the key contributors to operator errors during complex tasks. The capacity of human working memory has been defined as the number of items that can be held in conscious attention for use in a specific task or for later long-term storage (Baddeley & Logie, 1999). The constraints of working memory are particularly relevant during skill acquisition where working memory capacity is frequently exceeded. Traditional models (c.f. Baddeley & Hitch, 1974) characterize working memory as having two separate and relatively autonomous subsystems: verbal (i.e., phonological loop) and spatial (i.e., visuo-spatial sketchpad), but contemporary models suggest further disassociation (potentially based on sensory modality) (c.f. Miyake & Shah, 1999). Recent investigations suggest that working memory capacity can be enhanced by utilizing verbal, spatial, or alternative sensory modalities in a complementary manner (Wickens, 2002).

The development of an effective method for monitoring working memory load and delineating the verbal and spatial components could greatly enhance the speed and efficiency of human-system interaction. A real-time monitor could identify periods of spatial and/or verbal working memory overload and provide adaptive aiding, such as switching from verbal or spatial presentation formats, to meet operator requirements (Schmorrow, Stanney, Wilson, & Young, 2005). However, frequent switching between one modality or task and another may incur a "cost" due to stimulus competition, ambiguity, or distractions (Baddeley, 2003; Matlin, 1998). Thus, it is essential to develop an understanding of how best to leverage the multimodal capacity of working memory without incurring such costs.

As a first step towards developing a real-time neurophysiological working memory index that could trigger such adaptive aiding, a wireless electroencephalographic (EEG) system was used to acquire data during working memory tasks of varying complexity. In the first phase of the study, participants performed three simple working memory tasks: one spatial and two verbal, designed in accordance with methods previously reported (Proffitt, 2003). The simple verbal and spatial tasks used elements of a Tactical Tomahawk Weapons Control System (TTWCS) so that they could be integrated into a more operationally relevant simulated TTWCS environment that would serve as the testbed for the second portion of the study. The rationale for this design was that if a set of EEG parameters were identified that distinguished between the simple verbal and spatial tasks, these parameters could then be evaluated in

---

[1] Advanced Brain Monitoring, Inc., 2850 Pio Pico Drive, Suite A, Carlsbad, CA USA 92008, chris@b-alert.com
[2] Department of Industrial Engineering and Management Systems, University of Central Florida, 4000 Central Florida Blvd., Orlando, FL USA 32816-2450
[3] Lockheed-Martin Advanced Technology Laboratories, 3 Executive Campus, 6th Floor, Cherry Hill, NJ USA 08002

the more complex TTWCS testbed. The initial goal of the first phase was to select a set of EEG variables optimized for discrimination of verbal from spatial processing in the simple tasks (Proffitt, 2003).

The second phase of the study was designed to evaluate the utility of the EEG variables in a complex TTWCS simulation task and to determine whether additional EEG parameters were required to provide accurate delineation of the spatial and verbal processing within the more complex task environment. The selected EEG variables would then serve as inputs to a neural net that used a combination of physiological parameters to identify relevant cognitive state changes while operators performed the TTWCS simulation tasks. The outputs of the neural net controlled the timing and introduction of multimodal augmentation strategies designed to optimize the distribution of perception and cortical processing within the TTWCS environment to take advantage of the totality of human capacity for multimodal communication. These studies represent the initial steps in building the foundation for designing a robust Command and Control (C^2) system that can adapt interaction techniques to meet specific user conditions.

2 Methods

The current study was performed in two phases. During the first phase, EEG data was acquired while participants performed simple working memory tasks. In the second phase, participants performed a complex working memory task, playing the role of a Tactical Strike Coordinator interacting with a simulated TTWCS environment.

2.1 Participants

For the Simple Working Memory Task experiment, 12 Lockheed Martin employees were studied. The participants were provided verbal instructions but no hands-on training prior to the start of each task. The tasks were presented in the following order: Missile ID, Mental Addition and Missile Location. Data from two participants were excluded from analyses. One participant was dropped due excessive EMG (muscle) artifact in the mono-polar channels due to the mastoid reference electrode. The second participant was dropped due to an inability to determine the start and end of each task due to missing time synchronization data.

Of the ten Lockheed Martin employees who participated in the Complex Working Memory Task study, data from seven participants were included for analyses. Two participants were dropped due to excessive EMG artifact in the monopolar channels due to the mastoid reference electrode. The third participant was dropped due to an inability to determine the start and end of each task due to missing synchronization data. Participants were provided a fixed timeframe for training on the TTWCS tasks in advance of their testing session.

2.2 Simple Working Memory Tasks

The three simple working memory tasks were designed to replicate experiments performed by Proffitt (2003). The tasks were designed to tap into both verbal and spatial working memory, such that they would provide differentiable EEG signatures.

2.2.1 Verbal Task – "Missile ID"

In this auditory recognition task, participants were presented with synthesized speech listing a set of two Missile Identifiers (ID) (e.g. 56U, 15P). After the set of Missile ID's were presented, the computer cued the user by saying the word "listen" and then "spoke" a series of Missile IDs. Participants responded via a keyboard as to whether or not a spoken Missile ID matched one of the sets previously presented.

2.2.2 Verbal Task – "Mental Addition"

In this computation task, participants were presented with a display containing a single number at the center of the display and instructed to respond to the number. A new number was presented each time the participant responded. Participants were instructed to add a series of numbers until prompted to report the total.

2.2.3 Spatial Task – "Missile Location"

In this "grid-task", participants were presented with a 5 x 5 grid that contained from three to five missiles. The display was shown for a brief interval and then removed. After approximately 40 seconds, the grid reappeared with a subset of the missiles shown. Participants were instructed to indicate the locations of missing missiles. Participants had approximately 4 seconds to indicate locations of missing missiles before the grid disappeared.

2.3 Complex Working Memory Tasks

In the Complex tasks the operator performed the role of a Tactical Strike Coordinator interacting with a simulated TTWCS environment. It is the job of the Tactical Strike Coordinator to assign missile strikes to specific targets, and to monitor and reassign missiles to "emergent targets" as these events occur. The scenarios used to select EEG correlates included performing assessment of missile coverage zones, referred to as the "Location Task", and retargeting missiles based on emerging targets of higher priority, referred to as the "Retarget Task".

2.3.1 Spatial Task – "Location"

The location task (Figure 1) was separated into three parts: encoding, rehearsal, and recall. During the "encoding" period, participants were given 15-seconds to study the location and 10-minute coverage zone (i.e., circular region around each missile shown in Figure 1 below) of each missile and the associated targets. During the 45-second "rehearsal" period, the coverage zone circles were removed, however participants were provided the opportunity to continue memorizing the initial information that was previously displayed. During the "recall" period, participants were provided 30-seconds to identify targets that were/were not covered by any missile's coverage zones. Since most participants completed their "recall" responses well within the allotted time, only the first 12 seconds of the recall period were analyzed.

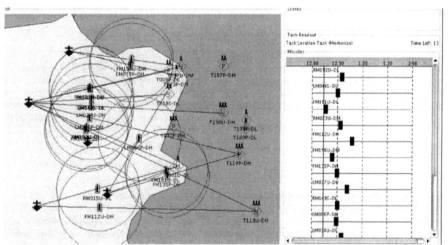

Figure 1: Screen presentation during location task

2.3.2 Verbal Task – "Retarget"

The "Retarget" task (Figure 2) required participants to retarget missiles based on higher priority emergent targets. Participants were provided 10 minutes to reallocate missile coverage to as many emergent targets as possible, while maintaining coverage on as many high and medium default targets as possible. There were four rules for retargeting missiles: a) missile warhead types must match target warhead types, b) the number of missiles servicing a target must match the number of missiles required by that target, c) highest priority emergent targets should be retargeted first, d) only Loiter or Retarget (L/R) missiles may be retargeted. The information available for retargeting was presented in the text below each missile and target. Emergent targets were colored red and their appearance and locations were randomly assigned throughout the task. The Task Readout, to the right of the missile-target map, provided information to the participant helpful in determining a retargeting strategy. This information included the amount of time from missile-target intercept and closest missile to emergent target.

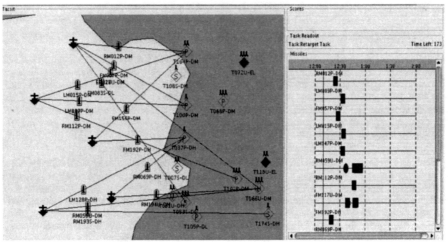

Figure 2: Screen presentation during retarget task

2.4   Data Acquisition

2.4.1   Tactical Tomahawk Weapons Simulation (TTWCS) Test bed

The TTWCS simulation test bed ran on an Intel Pentium 4- 3.0 GHz (800MHz FSB - 1MB Cache) computer with 1 GB (2 64-bit wide DDR data channel) memory, a 60G 7200rpm high performance drive, and an ATI Radeon Mobility Integrated 256MB DDR Video, with Accelerated OpenGL and 8x ultraAGP. The operating system was Red Hat/Fedora Linux with customized 2.6.9 kernel (Fedora Core 2 with latest updates). The visual interface was presented on a 17" WSXGA+ (1680x1050) screen. All user responses were with standard keyboard and 2-button mouse. The audio was presented via 6 channel desktop speakers based on AC'97 2.2 virtual 6-channel output onboard sound.

2.4.2   EEG Acquisition

A modified wireless EEG headset acquired data from 9 monopolar sites referenced to linked mastoids (F3, F4, Fz, C3, C4, Cz, P3, P4, POz) and 2 bipolar sites (Cz-POz, Fz-POz) (see Berka 2004 for basic EEG headset details). The EEG transceiver unit was interfaced to a laptop computer (Pentium 2.4 MHz with 512 RAM and a Windows XP operating system) that operated the EEG acquisition software used to generate an EEG record and provide time synchronization data to the TTWCS test bed via a TCP-IP network protocol.

2.5   Data Reduction

For the Simple Working Memory Tasks, approximately 30 one-second epochs of EEG data per task were extracted from the three task conditions (i.e., Missile ID, Mental Addition, and Missile Location). For the Complex Working Memory Tasks, seventy one-second epochs were extracted for the training data set from each of two segments: the encoding segment of the "location" task and the retargeting segment of the "retarget" task. The balance of data from the tasks was used for testing discriminant function models (see section 2.5).

Processing of the EEG included identification and decontamination or rejection of artifacts, including eye blinks, spikes, saturation, excursions, and EMG, using previously published procedures (Berka 2004). The EEG power spectra for each 1 Hz bin between 3 – 40 Hz were computed for each channel, generating a total of 418 EEG variables from 11 channels for each one-second epoch.

One of the long-term goals of the project is to optimize the selection of sensor locations, number of channels and number of EEG variables to provide accurate classifications while minimizing the amount of data acquired. In order to assess a variety of headset configurations, a total of eight EEG data sets were created. The "Mid-line Bipolar" data set included only the 1 Hz bins from FzPOz and CzPOz for the Simple and Complex Working Memory Tasks, respectively. The "Mid-line All" data sets included the bi-polar sites plus the mono-polar sites Fz, Cz and POz. The "Lateral" data sets included the 1 Hz bins from the mono-polar sites F3, F4, C3, C4, P3 and P4. The "Mid - Lat" data sets included all bi-polar and mono-polar data from all 11 channels.

2.6 Variable Selection and Model Development

Three- and two-class discriminant function models were used to classify data from the Simple and Complex Working Memory Tasks, respectively. Step-wise analysis was used to select predictive variables for each of the eight data sets. The discriminant function models generated with variables and coefficients derived from the Simple (Complex) Working Memory Task for each data set were applied to the Complex (Simple) Task data to evaluate the influence of variable selection. Individualized discriminant function models were derived to assess the benefit of fitting coefficients to the unique EEG patterns of each participant.

3 Results

3.1 Classification Models During the Simple Working Memory Task

The results from four discrimant function models applied to the Simple Working Memory Tasks are presented in Table 1. The Mid-line Bipolar model provided surprisingly good classification for the Missile ID and Missile Location tasks (using just two channels of data and eleven variables), but, due to poor classification during the Mental Addition task, reported an overall classification accuracy of only 50%.

The Lateral Model provided better classification in the Missile ID task compared to the Mid-line All with a similar number of channels and variables, however the overall classification accuracies were similar. The Lateral Model required an additional three sensor sites to achieve the better performance in the Missile ID task. The Mid-Lat Model was clearly superior, although nine sensors, 11 channels and 43 variables were required to achieve these results. Table 2 presents the number of variables selected from each of the sensor sites for the Mid - Lat Model. The rank order of the top seven variables (partial R^2 values > 0.03) was C4 37Hz, P4 11Hz, P3 13Hz, CzPOz 31Hz, Cz 32Hz, CzPOz 18Hz, and C4 26Hz.

Table 1: Classification Accuracy of Three-Condition Simple Task Model

Model	# Sensors	# Channels	# Variables	Percent of epochs correctly classified			Class Accuracy
				Verbal - Missile ID	Verbal-Mental Addition	Spatial - Missile Location	
Mid-line Bipolar	3	2	11	60.4	40.6	61.5	50.8
Mid-line All	3	5	22	65.7	60.7	67.4	64.5
Lateral	6	6	26	75.2	60.3	65.1	66.9
Mid – Lat	9	11	43	77.8	76.4	76.6	77.0

Table 2: Mid – Lat Simple Model: Number of Variables Selected by Sensor Site

| Site | Monopolar | | | Bipolar | |
	Left	Right	Midline		
Frontal	1	5	0	FzPOz	5
Central	6	5	3	CzPOz	6
Parietal	2	6	3*	B-Alert Class	1
Totals	9	16	6	12	

* The midline sensor is parietal occipital (POz)

3.2 Classification Models During the Complex Working Memory Task

The classification distribution trends across the four models for the Complex Tasks were similar to the Simple Task with the Mid-Lat sites providing optimal classification accuracy. For this reason, only results from the Mid-Lat Model are reported. Interestingly, the number of variables selected to discriminate the Complex Tasks was significantly reduced compared to the Simple task from 43 to 11 (Table 3 and 5). The rank order of variables with a partial R^2 values > 0.03 were P4 27Hz, P4 36Hz, and Fz 12Hz. Consistent with the Simple Task, the largest number of variables was selected from the lateral right region (Table 3). Table 4 presents the classification accuracy of the Two-Class Mid-Lat model with variables selected from the Complex Task (second row of data) applied to the three segments of the Location task and two segments of the Retarget task.

Table 3: Mid – Lat Complex Model: Number of Variables Selected by Sensor Site

| Site | Monopolar | | | Bipolar | |
	Left	Right	Midline		
Frontal	1	1	0	FzPOz	0
Central	1	2	2	CzPOz	1
Parietal	0	2	1*	B-Alert Class	0
Totals	2	5	3	1	

* The midline sensor is parietal occipital (POz)

3.3 Influence of Variable Selection on Classification Models

Due to the substantial difference in the number of variables selected for the Simple vs. Complex Tasks, a Three-Class Simple Task Model using the Complex Task variables and a Two-Class Complex Task Model using the Simple Task variables was developed. The result in Table 4 show that the classification accuracies of the two models applied to the Complex Task data were relatively similar. However, the results in Table 5 show that the Complex task variables are unable to discriminate the Simple Tasks better than chance.

Table 4: Results for Complex Task Models Using Variables from Simple vs. Complex Tasks

Model	Variable Type	Percent of epochs classified Verbal (V) or Spatial (S)									
		Spatial Task - Location						Verbal Task - Retarget			
		Encoding		Rehearsal		Recall		Encoding		Retarget	
		S	V	S	V	S	V	S	V	S	V
Mid – Lat	Simple	65.8	34.2	62.6	37.4	51.6	48.4	48.0	52.0	37.6	62.4
	Complex	67.1	32.9	58.9	41.1	50.0	50.0	43.0	57.0	33.9	66.1

Table 5: Results for Simple Task Models Using Variables from Simple vs. Complex Tasks

| Model | Variable Model | # Variables | Percent of epochs correctly classified | | | Class Accuracy |
			Verbal - Missile ID	Verbal- Mental Addition	Spatial - Missile Location	
Mid and Lat- All	Simple	43	77.8	76.4	76.6	77.0
Mid and Lat- All	Complex	11	53.5	55.0	39.5	49.5

To determine if the large difference in the number of simple versus complex variables required to discriminate between the tasks was caused by the additional task in the Simple Model, an alternative Two-class model was evaluated to compare only the verbal ("Missile ID") and spatial ("Missile Location") sub-tests of the Simple Tasks. This model required 34 EEG variables to accurate classify the two states, still significantly more variables than the 11 required for the Complex Task.

3.4 Fitting Model Coefficients to Accommodate Individual Differences in EEG

Table 6 presents the classification accuracies for individual participants using the Two-Class Model with the Simple Task variables and group discriminant function coefficients. These results demonstrate wide variability in the classification accuracies. Table 7 presents findings from Two-Class models using the same variables, with discriminant function coefficients fitted to the individual's EEG patterns.

Table 6: Classification Accuracy Two-Class Model Using Group Coefficients

Participant Number	Percent of epochs classified Verbal (V) or Spatial (S)									
	Spatial Tasks						Verbal Task			
	Encoding		Rehearsal		Recall		Encoding		Retarget	
	S-E	V-R	S-E	V-R	S-E	V-R	S-E	V-R	S-E	V-R
551	48.7	51.3	34.7	65.3	22.5	77.5	65.9	34.1	54.2	45.8
552	91.2	8.8	77.0	23.0	71.8	28.2	77.8	22.2	70.8	29.2
553	38.9	61.1	55.6	44.4	29.6	70.4	13.3	86.7	16.0	84.0
554	63.7	36.3	65.3	34.7	58.3	41.7	46.2	53.8	40.8	49.2
555	73.6	26.4	73.4	26.6	57.5	42.5	41.7	58.3	24.5	75.5
556	70.9	29.1	57.4	42.6	64.1	35.9	30.8	69.2	11.8	88.2
557	74.6	25.4	74.1	25.9	51.4	48.6	53.1	46.9	37.1	62.9
Mean	65.9	34.1	62.5	37.5	50.7	49.3	47.0	53.0	36.4	63.6

Table 7: Classification Accuracy Two-Class Mid-Lat Model Using Individualized Coefficients

Participant Number	Percent of epochs classified Verbal (V) or Spatial (S)									
	Spatial Task - Location						Verbal Task - Retarget			
	Encoding		Rehearsal		Recall		Encoding		Retarget	
	S-E	V-R	S-E	V-R	S-E	V-R	S-E	V-R	S-E	V-R
551	94.1	5.9	89.3	10.7	92.5	7.5	27.3	72.7	17.7	82.3
552	77.5	22.5	52.3	46.7	43.6	56.4	30.6	69.4	32.9	67.1
553	93.7	6.3	88.9	11.1	88.9	11.1	63.3	36.7	21.7	78.3
554	79.0	21.0	58.9	41.1	47.2	52.8	20.5	79.5	21.9	78.1
555	89.3	10.7	79.7	20.3	70.0	30.0	47.2	52.8	12.4	87.6
556	95.3	4.7	77.0	23.0	76.9	23.1	46.1	53.9	11.3	88.7
557	88.7	11.3	87.0	13.0	32.4	67.6	25.0	75.0	15.1	84.9
Mean	88.2	11.8	76.2	23.7	64.5	35.5	37.1	62.9	19.0	81.0

4 Conclusions

This paper presents one model approach to selecting EEG parameters that can be implemented in systems designed to monitor and interpret real-time cognitive state changes. Previous work by the investigative team (Berka, 2004, Berka, 2005) validated this approach to development of an EEG-based closed loop system where EEG correlates of mental workload were first established and validated in a series of simple tasks. The EEG-workload measures were subsequently validated in a complex Aegis simulation environment and were used successfully to control the pacing of stimulus presentation to optimize performance.

These data establish the feasibility of characterizing EEG correlates specific to verbal and spatial working memory in both simple and complex task environments. The variables derived from this analysis can be computed in real-time to provide a second-by-second assessment of verbal and spatial processing. The investigators acknowledge that the terms "verbal" and "spatial" apply only within the context of the tasks and conditions designed for the TTWCS simulation testbed. Although the experimental design team attempted to create task conditions that

411

required predominantly verbal or spatial processing to provide data for selection of EEG parameters, the purity of these conditions, particularly in the complex tasks is questionable.

The difference in the number of EEG variables selected to discriminate Simple Task in comparison to the number required for the Complex Task (i.e., 43 vs. 11) suggests that the two complex tasks were more distinctive as reflected in their EEG characteristics than the three simple tasks, so fewer variables were required. An alternative 2-class model was evaluated to compare only the verbal ("Missile ID") and spatial ("Missile Location") sub-tests of the Simple Tasks. This model required 34 EEG variables to accurate classify the two states, still significantly more variables than the 11 required for the Complex Task. It is also possible that the Simple Task may have required more variables for accurate classification due to the limited attention required to perform the sub-tasks, resulting in greater variability in EEG activity within and between individuals.

Data from the complex tasks reveal that the patterns of EEG associated with "location" were distinctive from those observed during "retargeting". It is possible that the complex tasks were more engaging than the simple tasks and elicited overall a greater percentage of focused time-on-task. Interestingly, the EEG variables selected for the Simple Task model provided similar classification accuracy when applied to the Complex Task, but those selected for the Complex Task were not effective in discriminating the simple sub-tasks. This finding suggests that the applicability of variables selected for one task may or may not be universally applied to other tasks and should be investigated as a component of model development.

The decision on how to accommodate individual differences is relevant to all aspects of model building in the design of closed-loop systems. Although the fitting of the discriminant function coefficients to each individual demonstrated a marked improvement on the consistency of correct classifications across participants the "classification accuracies" of the various models should be interpreted with some caution. Distinctive patterns of EEG classifications may actually reflect differences in strategic approach to the task demands or differential allocation of verbal and spatial memory. More detailed investigations of the relationship of task performance and EEG classifications on a second-by-second basis are required to better understand these associations. In addition, post-test interviews should be conducted to determine whether participants were employing unique strategies for completing the tasks.

If these preliminary data are replicated and the EEG indices, in combination with other physiological or behavioral inputs, are shown to relate to the degree of working memory overload in the TTWCS or similar tasks, they could provide a valuable contribution to real-time adaptive aiding of human-system interaction. The goal of the present model approach was to provide inputs to a neural-net based system that would employ augmentation strategies designed to take advantage of the totality of human capacity for multimodal communication. More specifically, the system involves adaptive multimodal mediation and attention alerting mechanisms, which incorporate multiple display strategies to invoke alternate sensory modalities given a TTWCS user's current cognitive state as measured by real-time biophysical data. Based on output from the physiological sensors, as well as an understanding of current system state (i.e., which verbal and spatial tasks are currently being performed and their relative loading), alternate modality display strategies can be employed (i.e., modality-based task switching/augmenting). Any such aiding must be implemented judiciously, as any gains realized could be tempered if the costs for modality switching are high.

Acknowledgements

This work was supported by the DARPA program, "Improving Warfighter Information Intake Under Stress" in which Advanced Brain Monitoring is a sub-contractor to Lockheed-Martin Advanced Technology Laboratories.

References

Baddeley, A. (2003). Working memory: Looking back and looking forward. *Nature Reviews: Neuroscience*, **4**, 829-839.

Baddeley, A.D., & Hitch, G.J. (1974). Working memory. In G. A. Bower (Ed.), *Recent advances in learning and motivation* (Vol. 8, pp. 47-90). New York: Academic Press.

Baddeley A., & Logie, R.H. (1999). Working memory: The multiple-component model. In A. Miyake & P. Shah (Eds.), *Models of working memory* (pp.28-61). New York: Cambridge University Press.

Berka, C. (2004). Berka C, Levendowski DJ, Olmstead RE, et al., "Real-time Analysis of EEG Indices of Alertness, Cognition and Memory with a Wireless EEG Headset", International Journal of Human-Computer Interaction, 17, 151-170.

Berka, C. (2005). Evaluation of an EEG-Workload Model in an Aegis Simulation Environment. Proceedings of the SPIE Defense and Security Symposium, Biomonitoring for Physiological and Cognitive Performance During Military Operations, In Press.

Matlin, M. W. (1998). *Cognition* (4th edition). Fort Worth, TX: Hartcourt Brace College Publishers.

Miyake, A., & Shah, P. (Eds.). (1999). *Models of working memory: Mechanisms of active maintenance and executive control*. New York: Cambridge University Press.

Schmorrow, D., Stanney, K.M., Wilson, G., & Young, P. (2005, in press). Augmented cognition in human-system interaction. In G. Salvendy (Ed.), *Handbook of human factors and ergonomics* (3rd edition). New York: John Wiley.

Wickens, C.D. (2002). Multiple resources and performance prediction. *Theoretical Issues in Ergonomics Science*, 3(2), 159 -177.

Proffitt, 2003 (unpublished manuscript)

Proffitt, 2004 (Powerpoint slides, and informal communication)

Statistical Process Control as a Triggering Mechanism for Augmented Cognition Mitigations

Jerry R. Tollar

The Boeing Company
P.O. Box 3707 MC 45-85
Seattle, WA 98124
jerry.r.tollar@boeing.com

Abstract

Statistical Process Control (SPC) has been widely used in manufacturing and quality assurance because of its capability to distinguish between normal variation and "special cause" effects. Control Charting, as it is commonly called, has evolved a wide variety of rule sets for identifying special causes in practice, and some of them may prove effective in detecting important cognitive state changes in data streams from neuro-physiological sensors. We wanted to examine the practical use of SPC for recognizing cognitive events, so we selected fNIR (functional Near InfraRed) sensors, which provide neuro-physiological signals of significant interest to our team.

Neuro-physiological sensor signals are rather prone to problems of noisy signals and changes in signal strength. Signal strength may vary day to day due to slight variations is sensor placement or may "drift" within an experimental session due to factors other than the cognitive events of interest. Fortunately, SPC has been refined for reasonably small populations, and it can: 1) reduce data stream noise, and 2) improve the "curve tracking" ability, to enhance the detection of signals of interest. We will discuss some of these techniques, and present test data using them."

Keywords: augmented cognition; control chart; statistical process control; triggering mechanism.

1   Introduction

Statistical Process Control plays a vital role in quality control in manufacturing and service industries. SPC consists of methods for understanding, monitoring, and improving process performance over time. SPC has been widely used because of its capability to distinguish between normal variation and "special cause" effects.

This paper will first give a brief overview of the fundamentals of SPC control charting, and will then discuss why SPC might serve as a triggering mechanism for an augmented cognition workstation. The last section will analyze data from the Boeing team's September 2004 concept validation experiment and evaluate how well SPC worked as a potential triggering mechanism.

2   Control Charts

The foundations of SPC were first described in Walter A. Shewhart's 1931 classic book, *Economic Control of Quality of Manufactured Product*. He was concerned that statistical theory should serve the needs of industry, and the control chart has proven to be a lasting and highly effective tool throughout the modern world. SPC has been proven the world over to be a simple and effective means of understanding data from real-world processes.

The primary advantage of control charts is that they reliably separate potential signals from the probable noise that is common in all types of data. Statistical variation is normal, and the power of the control chart is its ability to distinguish between "special cause" variation and "common cause" variation. Common cause variation is caused by unknown factors, resulting in a steady but random distribution around the average. On the other hand, special cause variation is a shift in output caused by a specific factor, resulting in a non-random distribution or trend.

Figure 1 is a stereotype of a control chart. The horizontal axis is time, and the line running through the center is the control line for fNIR data (an exponential moving average). The horizontal lines are control limits – three above and three below the average line. They are 1, 2 and 3 standard deviations from the moving average. The Boeing team could treat the operator's cognitive state as a process that needs to be kept in control by SPC.

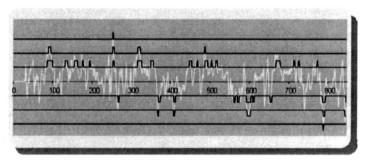

Figure 1: Example SPC Chart of fNIR Data

The "blips" in a control line denote the detection of a special cause based on the SPC rules. Many types of special causes have been studied, and rules have been formulated for when to intervene in a process. The blips in the lines of Figure 1 are based on useful, but fairly elementary rules:

- One point outside the 3-sigma control limits
- Two out of three points that are on the same side of the control line, both at a distance exceeding 2 sigmas from the control line
- Four out of five points that are on the same side of the control line, four at a distance exceeding 1 sigma from the control line.

An informative, highly readable site for SPC and control charts has been provided by iSixSigma at http://www.isixsigma.com/st/control_charts/. The site highlights a number of rules useful in manufacturing and service industries.

3 Why SPC May Serve as a Triggering Mechanism for Augmented Cognition Mitigations

The hallmark of augmented cognition systems is the desire for closed-loop feedback between the human and the system's user interface. Here is a summary of the steps: 1) the human interacts with an application interface, 2) sensors monitor the cognitive processes of the human's brain, 3) a cognitive state assessor "gauges" the human's cognitive state from the sensor signals, 4) an augmentation manager uses the gauges to determine if and when a "mitigation" or change of the user interface would help the human, and 5) the user interface changes, adapting to the human's information processing needs. When the change in the user interface is well-timed and unobtrusive, the operator's overall performance and satisfaction is improved by the mitigation.

3.1 Problem Context of Augmented Cognition Project

The Boeing Team needed to overcome "challenges" within its closed-loop system. First, the previous version of the cognitive state assessor sometimes failed to discriminate between tasks even though the cognitive state assessor's neural net was trained for the task. Moreover, the cognitive state assessor appeared to be slow. It would recognize the task 20 or 30 seconds after task intensity peaked, much too late for the critical event.

Second, the fNIR sensor signal characteristics were not well-understood at the time. The fNIR sensor technology was new to the Boeing Team's sensor suite. Neuro-physiological sensor signals are rather prone to problems of noisy signals and changes in signal strength. Signal strength may vary day to day due to slight variations is sensor placement, or the signal may "drift" within an experimental session due to factors other than the cognitive events of interest. The team did not know how noisy the fNIR signal was, but knew that "signal drift" had occurred in a previous session.

And third, human operators were disturbed by transitions when certain mitigations were removed from the user interface. The change seemed too abrupt to them, and they said it took a lot of time to re-acquire context. They felt that more graduated step-down transitions would help.

3.2 Rationale for Trying SPC Control Charts

The "challenges" in the Boeing team's closed-loop system were the stimuli for brainstorming to discover an alternate way for classifying tasks and producing cognitive gauges. The fNIR sensor technology was most interesting as a starting point, because "activations" in local cortical functions often result in a fast signal ramp-up. This made fNIR signals potentially amenable to statistical process control, a method widely used in manufacturing and service industries. It seemed feasible that SPC might bring relief to the three "challenges" noted in the previous sub-section.

3.2.1 Potential Ability to Discriminate Between Tasks

A major task of the user interface was primarily and heavily verbal. The operator's workload for this "verbal task" started quickly and was sustained in duration. Wernicke's area, a part of the human brain located in the left temporal lobe posterior to the primary auditory complex, is known to be particularly involved in the understanding and comprehension of spoken language. It is a semantic processing area most often associated with language comprehension, or processing of incoming language, whether it be written or spoken. Thus, the signals coming from fNIR sensors placed over Wernicke's area would readily discriminate between the primarily verbal task and the other tasks. Perhaps this ability would at least partially address one of the Boeing team's problems, namely, the inability of cognitive state assessor to identify verbal tasks even though it was trained for it.

The use of SPC offered another immediate benefit too. fNIR sensors provide a steady stream of fast signal values, and the control chart is so sensitive that it responds to even brief, intense periods of verbal workload. Thus, unlike the previous cognitive state assessor, SPC is capable of picking up fast changes. It would no longer take 20 or 30 seconds for the cognitive state assessor to recognize a peak in task intensity.

3.2.2 Recognized Ability to Separate Potential Signals from Probable Noise

Neuro-physiological sensor signals are rather prone to problems of noisy signals and changes in signal strength, and the fNIR sensor technology was new to the Boeing Team's sensor suite. The fNIR data were not stationary. Slight variations in pressure on the sensor against the operator's head affected signal strength significantly, and the signal strength drifted over time.

Fortunately, the primary advantage of control charts is that they reliably separate potential signals from the probable noise common in all types of data. SPC has been refined for reasonably small populations, and it can reduce data stream noise and improve the "curve tracking" ability to enhance the detection of signals of interest. NovaSol's OTIS-2 system provided a steady stream of fNIR data that allowed the signal to be normalized and pre-processed for signal noise and drift. For example, the use of an exponentially weighted moving average for the control line would address drift.

3.2.3 Potential Trigger for Mitigation Shifts – Upward & Downward

One of the perhaps unspoken goals of augmented cognition is to optimise human performance by keeping operators "in the groove." People intuitively understand this, and the phenomenon is often called the Yerkes-Dodson Law. Figure 2 depicts the Yerkes-Dodson curve which predicts an inverted U-shaped function between arousal and performance. Both the high and low end of the curve denotes degraded performance.

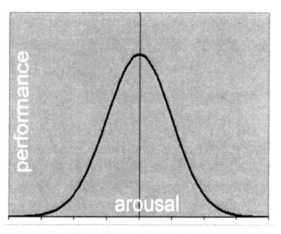

Figure 2: Yerkes-Dodson Law

A familiar form of SPC control chart exhibits similar symmetry, where characteristic values above and below control limits are seen as probable defects resulting from "special causes." Figure 3 exhibits a notional blending of the two ideas, where the operator's cognitive state is considered a process to be kept in control.

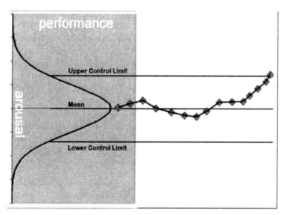

Figure 3: Yerkes-Dodson Curve and SPC Control Chart

The third challenge noted in the sub-section above was that human operators may be disturbed by transitions when mitigations occur. Operator's felt that more graduated step-down transitions would help. If the operator's the Boeing team considered the operator's cognitive state as a process to be kept in control, it was conjectured that mitigations might be incrementally added or removed on the basis of signals as values rose above or fell below the control limits. If the sensor signals are well-behaved, perhaps mitigations could be graduated or delivered incrementally in levels to help operators to keep "in the groove."

Figure 4 shows hypothetical gauge readings derived from control charts. Upper control limit readings could call for the adding a mitigation level, and lower control limits could call for a mitigation level to be removed. This is all very speculative and still un-supported in practice at this time.

417

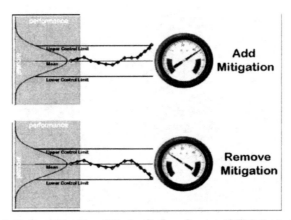

Figure 4: Notional Blending of Yerkes-Dodson Curve and SPC Control Chart

4   Analysis of SPC on fNIR Sensor Signals

The Boeing team explored the use of SPC for controlling closed-loop feedback in its augmented cognition system by deciding to could treat the operator's cognitive state as a process that needs to be kept in control by SPC. Control charting was initiated when the Boeing team collected data during its September 2004 Concept Validation Experiment (CVE). The fNIR signal data was analysed later to determine how well SPC would work as a gauge for identifying cognitive tasks and triggering mitigations.

The Boeing team sensor suite included both fNIR and pupilometry. The fNIR system was an OTIS-2 system from Archinoetics LLB (formerly NovaSol), and the pupilometry system was provided by EyeTracking, Inc. While this paper restricts itself to fNIR sensor data, the conference proceedings also contain Rich Barker's paper entitled Statistical Process Control as a Mathematical Model for Closed Loop Stability, which includes results from using SPC with pupilometry data. This section discusses the pre-processing, the control charts used, and how the control charts were analysed as triggering mechanisms for controlling the augmented cognition workstation.

4.1   Pre-Processing fNIR Signals

The minimum update rate for the system is 30 Hz, but the OTIS-2 system filters and encodes the data received from the sensors into data packets that are deliver to SPC at 1 Hz. The fNIR data received by SPC is not stationary. Slight variations in pressure between the sensors and the operator's head made a large difference in signal strength, and the signal strength drifts over time. An example of un-processed fNIR data that was collected during a trial that lasted 850 seconds is shown in Figure 5.

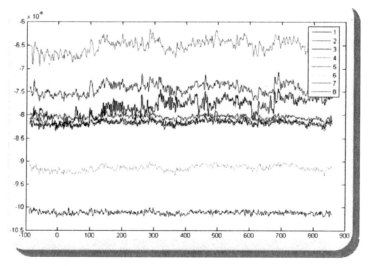

Figure 5: Eight Channels of Un-Processed fNIR Data

Each channel's data was averaged using an exponentially weighted moving average (EWMA). (Note: See Rich Barker's paper entitled Statistical Process Control as a Mathematical Model for Closed Loop Stability in the conference proceedings for a discussion of the EWMA formula.) EWMA acts as a first-order filter, removing any slow drift (up or down) in the data), and it also reacts to changes in the fNIR data faster than the arithmetic mean.

The sensor data was also pre-processed to normalize the mean and standard deviation of the data. The mean was normalized to zero, and the standard deviation was normalized to 1.0 for a 20-second moving window. SDn was calculated using the last 30 data points. (Note: Again, see Rich Barker's paper for a discussion of the formula.) The effect of this was to emphasize rapid changes and filter out changes longer than 20 data points. The results of the pre-processing algorithms were then passed to the SPC charting logic.

4.2   Basic SPC Rules in Place at the September 2004 Concept Validation Experiment

Upper and lower control limits were set at ± 1 sigma, ± 2 sigmas, and ± 3 sigmas. A simple set of pre-defined rules were use to determine when the data begins to change from its nominal state. Upper and lower control limit rules are triggered when:

- One point outside the 3-sigma control limits
- Two out of three points that are on the same side of the control line, both at a distance exceeding 2 sigmas from the control line
- Four out of five points that are on the same side of the control line, four at a distance exceeding 1 sigma from the centerline.

4.3   Evaluation of SPC for Triggering Mitigations for Verbal Tasks

The concept validation experiment of September 2004 had three main tasks: a vehicle health task, a tactical situation task, and a targeting window task. The first task was largely verbal, and the other two tasks were mainly spatial, manipulating visual displays.

Figure 6 shows the results of an 850-second experiment. The line labelled Verbal Task designates instances when vehicle health tasks were delivered to the user interface. Each instance is in the form of a question and response.

419

Varying formats are used, but the dialog for each VHT message has three parts: 1) A logical rule which the operator must read and understand, 2) A few items of variable data which the operator must evaluate against the logical rule, and 3) A choice among alternative actions, only one of which is correct. Operators must hold a logical rule and items of variable factual data in verbal working memory while reasoning and selecting from the alternative choices.

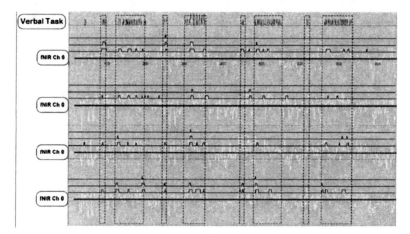

Figure 6: SPC Chart of fNIR Channels During Verbal Task Run

The verbal questions come singly, in pairs, and in "slams" of a dozen questions with a short, fixed interval between them.

Four channels of fNIR data are shown in the figure. Each channel's label is vertically center with the normalized EWMA, and the normalized fNIR signal is in yellow. Three lines appear above each normalized EWMA line for the +1, +2 and +3 standard deviation control limits. The control limit lines also exhibit blips on them when the SPC rules indicates that a "special cause" has been detected.

The SPC rules have been formulated to detect when to intervene in a process. That is the point where it often makes sense to trigger a mitigation! It should be noted that the majority of detections by the SPC rules occur soon after the onset of verbal tasks, and that subsequent blips often continue to appear during slams. False positives, instances where blips appear outside of verbal tasks, occur relatively infrequently. And they may not be false either, because many cognitive tasks have verbal components.

Overall, using SPC as a triggering mechanism for mitigations looks promising. This SPC control chart of fNIR channels from the September 2004 concept validation experiment shows a high correlation with instances of the high-intensity verbal task and few instances where false positives might be indicated.

References

Deming, W. E. (1994), The new economics: For industry, government, education (2nd edition). Cambridge, Massachusetts: Massachusetts Institute of Technology.

Montgomery, D. C. (1996), Introduction to statistical quality control (3rd edition), New York: John Wiley & Sons.

Ryan, T. P. (1989), Statistical methods for quality improvement. New York: John Wiley and Sons.

Yerkes, R.M., & Dodson, J.D. (1908). The relation of strength of stimulus to rapidity of habit formation. Journal of Comparative Neurological Psychology, 18, 459-482.

From Disparate Sensors to a Unified Gauge: Bringing Them All Together

James J. Weatherhead and Sandra P. Marshall

EyeTracking, Inc.
6475 Alvarado Road, Suite 132
San Diego, CA. 92120

Abstract

A fundamental goal of DARPA's Augmented Cognition (AugCog) Program was the unification of multiple sensor channels into one or more gauges of cognitive state. This admirable goal glosses over the enormous underlying problem of how these many different sensors that collect data at different rates, in different metrics, and with different numbers of data streams can pool their results in real time. Moreover, such a pooling of information must be coherent and synchronized, with the capability of selecting specific pieces and forwarding them on to other appropriate software partners. This paper describes one method for solving the data integration problem using the Cognitive Workload Assessment Dashboard (CWAD). One of the CWAD's most important functions is its ability to store all the sensor and simulator data in a meaningful way. A key element in making the data accessible is a common time stamp for all inputs and outputs. The CWAD contains a unique system for synchronizing the data, allowing not only selected extraction but also a fully synchronized playback of the session. This paper describes the CWAD implementation in one component of the AugCog Program.

1 Introduction

The desire for cognitively aware systems, i.e., systems that blend human and machine, is growing rapidly. The ever increasing abundance of computing power of today's age has allowed a large and varied number of scientific disciplines to achieve new heights within their field. Analytical algorithms that once took days to complete their work now take seconds. Tasks that once were thought only achievable by humans now are performed by machines. However, there are many positions that still require human input. But in all likelihood, even these tasks undoubtedly have been made easier by computers. Given that we cannot replace the human in some cases, it is important to ask what more can be done with computers, and if we do not replace the operators, can we increase their performance significantly by altering the way they work with computers?

This very question is one that a number of scientific teams have been trying to tackle around the globe. Any cognitively aware system (at the top level), must contain at least three components: a human, a machine, and a communication medium. The common approach that has been adopted is to try to bring the human and machine closer together. This objective is achieved by sharing information about the user with the machine (and teaching the machine what this means and how to respond) and teaching the machine how best to communicate back to the user, thus distributing and sharing information in the most efficient and meaningful way.

We began working on this problem several years ago. In an internal research and development effort, EyeTracking, Inc. created the Cognitive Workload Assessment Dashboard (CWAD) in order to link for ourselves multiple channels of eye data with other physiological measures. We have been involved in several projects researching and prototyping cognitively aware systems, and the results of our research have been put to use in the Defense Advanced Research Program Agency's Augmented Cognition (AugCog) Program. The AugCog Program provided a forum in which to implement the CWAD and test it with a number of different physiological sensors created by researchers from a number of academic, government, and corporate sites. We initially offered the CWAD for use in the AugCog Program at the Preliminary Technical Integration Experiment (TIE) in early 2003 and subsequently employed it in demonstration mode at the full TIE later that same year.

EyeTracking, Inc. then took its expertise in sensor integration and applied it to lead the integration of all components of a cognitively aware system for Boeing's Uninhabited Airborne Vehicle (UAV) control station project. The project brought together a wide variety of researchers from disciplines including neurology, biology, psychology, computer science and electronic engineering, each bringing unique ideas and knowledge from their respective field. While this breadth of knowledge was seen as the strongest asset (and undoubtedly necessary) to a project of this magnitude, in terms of building the system proper, it was in fact one of the biggest challenges.

2 Identification of the problem

Early in the project, a conceptual architecture was developed by the team (Figure 1). This conceptual design illustrates the major sub-components of the system, and the basic data flow around the system and its sub-components. Reminiscent of the more common closed loop circular diagram, this overview goes deeper, demonstrating the numerous physiological sensors and neural network that would transform the sensor's readings into actionable commands for the control station HMI. It should be appreciated that this is an oversimplified view, and that each subcomponent is generally complex. In some instances, each unit shown here may be made of several subcomponents, encompassing their own unique architecture. The design and architecture of each subcomponent was left to their owner, hence this discussion only pertains to the interconnectivity between these subcomponents.

**System Architecture
(Conceptual)**

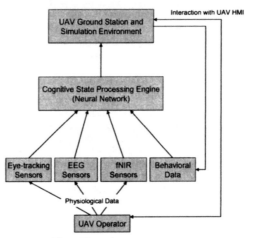

Figure 1

For the ease of describing this 'closed loop' system, we shall say that data starts with the operator. Physiological data is read from the operator via a plethora of sensors. These sensors are fed into a Cognitive State Processing Engine (in this case the CSPE is implemented as a neural net). Data from the control station HMI is also fed into the neural net (labeled here as Behavioral Data). The CSPE runs complex algorithms over these numerous channels of data and then outputs commands to the control station HMI. The commands may change the way data is presented to the operator, and/or may change levels of automation to aid the operator in their goals. Any changes that the CSPE makes to the HMI (in either the display of data or levels of automation) will in turn have an effect on the operator. Thus, the data cycle starts over.

As in any development project, a full understanding of the issues and an identification of the problem(s) must first be made so that a full and suitable plan can be realized. Once this requirements document is complete, the integration project's progress plan can then be implemented.

Only after an initial pass at the requirements document did the research team see the true scale of the issues in front of them. The system that was to be ultimately implemented consisted of several key components. A description of those components follows here.

422

2.1 Sensor Suite

Sensor Suite is a generic term that has been given to the set of 'sensors' that will provide information about the user/operator to the computer. Boeing's portion of the DARPA AugCog Program had a number of sensors that included: functional near infra-red (fNIR), electroencephlagram (EEG), eye-tracking, electrooculography (EOG), and electrocardiogram (EKG).

Characteristics:

- Each sensor may give out only one signal, or multiple signals.
- Each sensor may put out signals at different frequencies than any other sensor.
- If a sensor generates multiple signals, those signals may be generated at different frequencies.
- Signals emitted by the sensor may or may not have been processed before transmittal.

An example of a sensor that demonstrates multiple signals is the sensor developed by EyeTracking, Inc. (selected for discussion here because of the authors' knowledge of the system). This sensor emitted the raw eye-tracking pupil diameter and point-of-gaze data as well as a processed measure of cognitive workload (the Index of Cognitive Activity or ICA) based on the pupil. The six raw data signals (x, y position and pupil data for left and right eye) were emitted at 250Hz, while three processed ICA signals (left, right and global) was generated at 1 Hz.

2.2 Cognitive State Processing Engine (CSPE)

This component serves as the hub of the entire system, and hence must have access to all information involved in the human-machine communication process at all times. Its purpose is to combine and interpret the numerous signals generated from both the sensor suite and from the UAV Control Station.

The CSPE combines all of the sensor suite data to provide a global state for the user at any given time. This global state can then be referenced to the ongoing actions in the system to generate an operator state in relation to a given context. The CSPE is able to provide a suggested action to the control station that may, or may not, trigger an action in the system to enable the system to complement the operator. The CSPE should be regarded as the main source of intelligence in the human-machine symbiot.

It should be appreciated that this description gives an over-simplified view of the true system. Its purpose is to outline the data requirements of the system rather than to provide a detailed description of an implementation. Often the CSPE is broken into several sub-components each with their own specific function.

2.3 The Human-Machine Interface / Control Station

The control station itself must be able to send data out to the CSPE. This data should include state information for the scenario (number or tracks on the screen, keyboard and button presses made by the operator etc.). Not only must the control station be able to send state data, it must also be capable of receiving data; usually such data will be commands from the CSPE.

Typically the control station will be capable of some level of automation by itself. The issue that this overall project addresses is the ability for the automation to be controlled. If the automation is set too high for a whole scenario in our task, the job of the operators becomes easy and hence, they may become too relaxed, lose focus, and lose situational awareness. This is obviously an undesirable situation. Should the operators be called on to perform an action or make a decision, they must first reacquire situational awareness, and the delay introduced could be critical. Conversely, if the automation is set too low, the operators are likely to be working too hard and become 'tunneled' into one, or only a subset of tasks that they should be performing.

By allowing the CSPE to control and throttle the level of automation used, we hope to keep the operator at an optimal level of efficiency and state of heightened Situational Awareness.

2.4 Key Requirements

As well as the requirements imposed by the necessities of the sub-components of the system mentioned above, other important characteristics of the system were identified. These characteristics are described below.

2.4.1 Scalability

The number of sensors and channels to be generated by each sensor was an unknown. It was generally accepted that there would likely be changes to algorithms, processing frequencies, and the type of information to be output by all systems throughout the project. The communication protocol must be able to allow for these increased or decreased numbers of channels and be ready to deal with relatively high data bandwidth usage.

2.4.2 Flexibility

The data being sent to the system was expected to change as the project progressed. As well as being able to accept the varied types of data that would be sent around the system (number of channels per sensor, numerical representation of that data, and frequency of each data channel), the system must be able to cope with the fact that these parameters are not stable. Thus, it must be able to reflect format changes gracefully.

2.4.3 Cross-Platform Compatibility

Due to the nature of the AugCog project, sensor teams had already performed much research and development prior to the integration effort beginning. Thus, many systems were being developed on a number of different platforms and in a variety of languages. Communication methods had to be utilized that would bridge the gap between a variety of technologies.

2.4.4 Reliability / Traceability

It was important that data be stored in a manner that enables extensive post fact analysis. The most low-level sensor and control station activity data should be stored so that higher level data and their algorithms could be re-run and analyzed after the fact.

Hand in hand with the system's ability to store data goes the ability for the system to receive data. While the number of low level network protocols is many, it was apparent that there is a trade off to be made between reliability versus network latency. Among the issues that needed to be considered was the fact that algorithms for the CSPE were still being designed. It was deemed more important that all data be passed through the CSPE than to reduce the communication latency to the CSPE by what was expected to be a few milliseconds.

2.4.5 Time Synchronization

Accurate timing information is arguably one of the most important features that the integration software should possess. Certainly at the beginning of the Augmented Cognition project, there was much discussion as to how accurate the timing would need to be. Initial thinking on an acceptable latency ranged from anywhere between one millisecond to one second. Airing on the side of caution, a lower latency was ultimately preferred for the communication medium. The typical thinking prevailed that too much data is better than not enough. If the time accuracy is too high, no harm is done; if it is too slow, data are lost.

2.4.6 Real-Time Re-Distribution of Data

As well as storing data received from the HMI platform and the sensor suite, the system must be able to distribute its data in real time to one or more components connected to the back end. Examples of this include the ability for the CSPE to receive data from the sensor suite for classifying operator cognitive state, and the HMI software receiving eye-tracking data so that it knows where the operator is looking.

2.4.7 Ease of Integration

Working with so many teams means that integration could become very time consuming if ease is not addressed. The goal of the project was not specifically a data integration issue, but rather a data processing issue. One minor complication that had to be addressed was that programmers for each component had differing levels of expertise, and in several different languages. A method was required that would make integration a non-issue and allow the focus of the work to be placed back on the human-machine interaction processes.

2.4.8 Real-time Data Visualization

It was known that there were to be a large number of variables and unknowns in the system once all of the components were connected. Even once the system as a whole was connected, research on each subcomponent would still to be ongoing.

The authors believed that good visualization of the data would play an important role in the project. Without a good visual representation of data within the system, the only real-time cue that the system was working would simply be an HMI change as the automation triggered on and off, with little traceability as to why. Only tedious offline analysis would reveal answers to the state of each subcomponent at differing times in the scenarios.

If a good method of displaying the data could be developed, then it would help with testing the system to show which components are acting/reacting to specific events in the scenario. It would also aid in the explanation of the system and its subcomponents during the oncoming presentations of the system to visiting partners and the customer.

2.4.9 Playback Capability

This capability would capitalize on the data recorded during live sessions. Playing back the data would allow the entire run to be replayed just as the live run happened. This again would be useful for testing subcomponents, post session analysis, as well as useful for demonstrations.

2.5 Summary of Requirements

The requirements for the system were very broad. One explanation for this breadth was due to the early phases of the AuCog project. At the beginning of the AugCog program, researchers were working independently. Primary focus was on the sensor technologies and investigation of the abilities of each method of data collection to contribute to the evaluation of a global cognitive state.

This approach led to a large number of sensors being developed, using a large number of computer based technologies and software programming languages. The functionality that each sensor system contained also differed: some saved to disk their raw data, some stored the results of processed data, and some had the ability to show graphical representations of their data, and so on. This lack of commonality meant that either a set of specifications had to be made so that each entity must implement the functionality of those specifications, or the communication architecture itself must take on a set of functionality that can be utilized by each and every connecting entity. If feasible, the latter option would reduce numerous project management issues and decrease integration time to the closed loop system.

The original conceptual architecture was taken and adapted to include a central hub that would provide interconnectivity for all components of the system. This communication software suite, known as the Cognitive Workload Assessment Dashboard or CWAD would receive and manage all data on the network. The system architecture shown in Figure 2 became the accepted architecture for the UAV project.

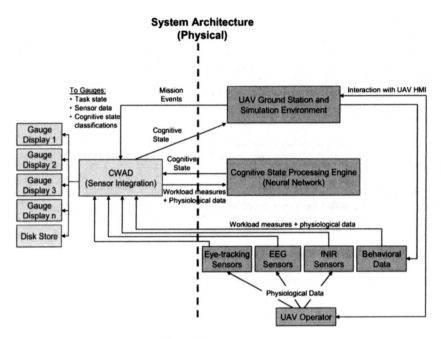

**System Architecture
(Physical)**

Figure 2

3 Solving the Problem

While all of the requirements put forth in the planning stage were taken into consideration, the one that stands out foremost is flexibility. The CWAD software was designed to be used as the main communication system, and to be used from the initial concept of the project through to the latter prototypes. The research aspect drives requirements for data storage, and flexible data formats, while the prototyping aspect drives the need for low data latency and higher bandwidth.

Instead of the common approach (which is to specify a protocol and publish this for the owners of the various subcomponents to implement and abide by) it was decided to make the CWAD a feature-rich system that would manage the complexities of the communication and deal with changes in any aspect of the data flow in the system. The technique that was to be employed by the CWAD was not to take on the role of a data management system, but rather to be the overseer of the given data. By giving the CWAD the ability to be managed by the actual subcomponents, the system became much more flexible and could cope with a greater range of possible changes in the system.

A set of libraries were developed and implemented into a number of software languages. Rather than the subcomponent teams learning a new network protocol and then attempting to implement it, they could integrate these libraries rapidly into their code, and a small number of function calls would then make their technology 'AugCog ready', typically within hours rather than days.

4 Data Flow

As previously discussed, a number of sensors were expected to connect into the Boeing AugCog system. Each one would generate its own unique channel(s) of data in different data formats and at differing frequencies. This characteristic was dealt with by the use of 'data streams' and manipulation of those streams.

4.1 Streams and Elements

Streams contain one or more 'elements' or packets of data that all:
1. originate from the same subcomponent;
2. are connected in some conceptual manner (i.e. all elements of data in the same stream relate to the left eye);
3. refresh at the same rate, and at the same time.

If any of the three rules above are broken, or do not apply for a given set of data, the component should create a separate data stream that contains a set of the conflicting data (where each set of data meets the three constraints above). Each component can make 1 to n streams.

It is up to the sending component (hereafter referred to as a data provider) to request a data stream be made by the CWAD. The request includes a description of all stream elements that are to be contained within the stream (data type, and name). In return, the CWAD will issue a unique identifier for the stream that can be used to reference the stream thereafter.

In addition to the ability for the components to describe their own data to the system, a number of attributes were also made available that would control how the data in those streams should be used by the CWAD. These attributes controlled were:

1. savable state of the stream (This flag was added so that the data provider for any given stream could command the CWAD not to save any data that was in that stream.)
2. distributable state of the stream (This flag was used to control whether other components in the system were able to access data elements contained within it. If the stream was set to be non-distributed, it could only be sent as far as the CWAD system.)

4.2 Network Protocol

The actual network protocol that the system uses is a TCP/IP connection. This protocol addresses a number of requirements that were needed by the system. This protocol is the primary protocol utilized on the Internet, and allows for cross-platform interconnectivity, as well as guaranteed delivery of each and every data packet sent on the network. As with any guaranteed delivery protocol, there exists a performance overhead, and this holds true for TCP/IP. The method involves a series of communications back and forth to ensure that each packet is received, whereas protocols that do not guarantee delivery simply send one packet out and just assume that it is received. The common non-guaranteed delivery UDP protocol was also considered as the main network protocol for the CWAD, and a series of in-house tests were run to assess its reliability versus its speed gain. It was found that the performance difference was negligible when the components were physically connected via a fast network with an intelligent switch. Latency typically only differed in the order of 10-30 ms between the protocols. The trade in 10ms of network latency could not be justified versus guaranteed network delivery.

4.3 Synchronization

One high accuracy timer is used to synchronize all of the data in the system. As soon as any stream receives new data, it receives a time stamp. This stamp is attached to the data and stays with it as it moves around the rest of the system. All data channels are given equal priority within the system, which means that if one stream coming from an EEG sensor contains thirty-two elements (one per EEG channel) and is being updated at 120Hz, it will not receive priority over a 1Hz one value coming from another sensor. Hence, data will move around the system in a timely manner no matter what the attributes of that data may be.

4.4 Disk Storage and Playback

A copy of all data received by the CWAD is written directly to disk. This file can then either be converted to human readable ASCII text for use in analysis programs, or can be replayed back through the system. All of the attributes associated with the system are captured so that the internal configuration of the CWAD will reset to its state during the live data session.

The only thing that differs from the live version is that the data comes from the disk rather than a subcomponent of the system. This functionality was taken further so that during playback, those components that receive data in the system (the neural net for example) could be connected to the CWAD as normal, and would be able to receive the data that it would normally receive in the real sessions.

4.5 Data Re-Distribution

Similar to the method that was employed by data receivers to create and transmit custom data streams though the system, a method was implemented for the back end so that components could receive data from the CWAD.

The components that receive data, termed 'Data Receivers', were also given software libraries that would speed their integration with the system. The libraries incorporated functionality so that interrogation could be carried out on all streams currently available within the CWAD that had their Distributable attribute turned on. Once a list of available streams is obtained, each can be queried for further information such as the human readable name of the stream, the number of data elements in the stream, as well as each element's name and data type.

Data receivers can attach or 'subscribe' to one or more data streams. Similarly, any one data stream can have none or many subscribers. Each time a new sample is delivered to the CWAD for a given data stream, the CWAD time stamps the sample and then checks to see if any subscribers are registered for the stream. If so, the data retains the timestamp and flows straight out of the back end for delivery to that subscribing entity. As this publication/subscription functionality is applicable to real-time applications, the route from data publisher to data receiver is the fastest route through the CWAD system.

In the simplest case, a subscriber can join the network and wait on the CWAD checking for a familiar stream name to appear (this will happen once a provider registers that stream), and then register for the name. While this is the simplest method, it does not take into account whether a stream format is consistent to that which is expected for the incoming data. At the most complex level, once the required stream is registered, the subscriber can interrogate the stream such that any unexpected changes can be flagged or, the subscriber could even reconfigure itself to accept the new data format.

4.6 Data Visualization

Flexibility was also needed for visualizing data in the system as well as for the communication of that data. At the time when data providers are connecting to the network and identifying themselves to the CWAD, the provider may chose to request that a gauge be created for the data streams about to be generated by that provider. Initially, a number of 'stock gauges' were developed that could be requested by the providers, examples of stock gauges are bar graphs, line graphs, pie charts etc. but in actuality, these gauges could be any visual representation whatsoever. Once the provider makes a request for one of the available gauges within the system, a visual representation will be created on the screen on the computer running the CWAD. Attributes can be given to the gauge during the creation of the gauge that will be used to define the behavior of the specific instance of the gauge. The attribute set allowed is specific to the individual implementation of the gauge i.e. the bar graph may expose attributes such as allowing the minimum and maximum values for the X and Y axis to be set.

Upon successful creation of a gauge, a receipt is handed back to the creating data provider. This receipt can now be included in the creation of future data streams, hence linking a specific stream and its elements to the new gauge. Full flexibility is given to the data provider to connect any specific element/s within a stream to any input/s on a gauge. It is also possible to share a gauge amongst differing data streams and their elements for increased flexibility.

It was understood that the stock gauges to be built for the initial release of the CWAD would not be enough to show meaningful representations of all data encompassed in the system. Therefore the gauge support was increased so that new gauges could be built independently. So long as the software code for the new gauge contains a minimum feature set of functions, the separately compiled gauge code can be placed into a gauge library folder on the CWAD computer and can then be requested and used instantly by any data provider that knows how to use the gauge (if the interface has extra requirements from that of the stock gauges). These gauges do not have any constraints as to how they can act. Gauges can be easily built by a programmer and may even include 3-dimensional representations of the data contained within the stream(s).

As the sensor suite developed, more sensors were given gauges, and some sensors even used several gauges to visualize data. This expansion meant that screen space became a commodity. The underlying CWAD publication/subscription was utilized to remedy the problem, and a separate CWAD Viewer application was created. This small and seemingly uncomplex application was set on a number of other computers on the same network as the CWAD. Basically the viewer runs as a data subscriber and interrogates all of the streams on the CWAD server, prompting a user, overseer of the experiment, or commander to select one or more gauges for display. Once selected, the viewer queries the selected streams for their format and gauges they may be connected to. Gauges that are required to complete the operation are downloaded in real-time over the network, a subscription to the stream is made and data begins to flow. The resultant effect is that the graphical display of the CWAD is shared amongst 1 to n computers. Infinite combinations of gauge configuration can be created, allowing for logical arrangements of information for efficient real-time monitoring of components throughout all areas of the system.

5 Conclusion

The CWAD communication system was a strong asset that EyeTracking, Inc. brought to the AugCog Program. At the time of writing this article, a number of sensor technologies have been simultaneously connected via the CWAD, including fNIR, EEG, EOG, and eye-tracking. All of these sensors are known to be some of the more data intensive sensors being used in cognitively aware system's research. Of those numerous sensors that were connected via the CWAD, typical integration times were on the order of one day's work by a intermediate level programmer. This remarkably short integration time was achieved because much of the leg work had been reduced by the pre-written libraries that take care of most of the mundane networking tasks that normally be required.

Integration was further aided by placing a running version of the CWAD on the Internet. As the CWAD's communication protocol runs over the Internet protocol (TCP/IP) the programmers for the various components of the AugCog system could test their code by connecting to the CWAD from their various offices and thus host a virtual test session. This risk reduction exercise enabled testing of a wide number of parameters by all component teams before the actual system was assembled for real in the Boeing laboratory. During this testing phase, data sent to the CWAD was sent back to the component team so that received data could be compared to that sent For those providers that used gauges, a version of the CWAD viewer was provided so that a visualization of the data could be monitored as testing took place.

During trials and live sessions, the CWAD computer was run on a standard Windows 2000 Intel PC. Processor load, memory resources and network bandwidth latency all remained low, even with multiple sensor streaming data. Early in the project, it was not known if computing power would be an issue when all proposed CWAD functionality was added for all streams. The biggest burden on the system is the UI gauge visualizations. This aspect is remedied by the fact that the viewer can be used to offload the work to one or more other processors if required. That said, it has not become necessary to use the offload feature with any system configuration tested to date. Similarly, network bandwidth has not become an issue. Even with multiple high bandwidth sensors connected, bandwidth is minimal when an adequate physical network is used. Typically this would be a 100 Mbps UTP network, although a fiber network could be used to reduce latency somewhat (this is not thought to be necessary). The initial concerns over network latency no longer seem to be an issue, primarily for two reasons. First, the decisions that the CSPE must make are typically in the order of several seconds. If the CSPE tries to move too fast, it may act on anomalies in the data and risk making a critically wrong decision. Second, the CSPE algorithms do not simply act on data at the millisecond level, they generally watch trends of multiple streams of data over one second to five seconds. With these two points in mind, a latency drop of 10-30ms becomes negligible.

The CWAD software has proven to be a robust and stable system tailored to augmented cognition systems in both research and prototype settings. That said, the flexibility required for this application has also made the software applicable to a much wider array of data integration applications.

Construction of appropriate gauges for the control of Augmented Cognition systems

Dr Andy Belyavin

QinetiQ Ltd
A50 Building, Cody Technology Park, Ively Road
Farnborough, Hants GU14 0LX, UK
ajbelyavin@qinetiq.com

Abstract

This paper outlines the development of gauges for the management of operator workload in a closed–loop Augmented Cognition system. Three criteria of sound gauges are outlined: face validity, reproducibility and sensitivity. It is argued in the light of current workload models that the calibration of gauges should be based on relatively simple pure tasks and that more than one task should be employed when multiple workload channels are to be managed. To ensure reproducibility scale–free measures of physiological signals should be employed for state measurement, and a wide range of levels of load should be included in the calibration to ensure sensitivity. These criteria are examined briefly using the experience of QinetiQ and Bristol University in developing the Cognition Monitor.

Keywords: Gauges, workload modeling, Augmented Cognition, Electroencephalogram

1 Introduction

It has long been recognized that for many demanding applications the implementation of systems that are either fully automatic or place the operator in full control are unsatisfactory compromises for some parts of a complex mission. To reduce the need to compromise overall system performance it is argued that the ideal approach is to develop adaptive systems that keep the crew in the loop as much as possible while permitting automatic action when crew performance is likely to be below the expected standard due to high workload or stress such as fatigue (Sheridan 1980, 1987). The interaction between crew and the remainder of such an adaptive system provides an enhancement of overall system performance and is referred to as Augmented Cognition.

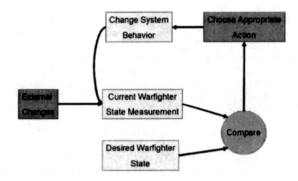

Figure 1: An outline augmented cognition system

At the heart of an adaptive Augmented Cognition system there has to be a methodology for the conversion of information that reflects the physiological or psychological state of the operator(s) into a series of gauges that relate to the key aspects of human performance. The output of these gauges can then be employed in a closed-loop control system to manipulate the way in which the interaction between the crew and the remainder of the system is managed to optimise the overall performance of the system. A sketch of the way such a negative feedback system can operate is displayed in Figure 1.

A large part of the effort during the early stages of the DARPA Augmented Cognition program (Phase I) has been devoted to the development of such measures of crew state. In the majority of systems it has been assumed that the most important gauges should measure aspects of task demand (workload) in the context of human capacity restraints, and there is an implicit assumption that other forms of performance degradation can either be ignored or will be captured as a by product. The aim of the current paper is to review the development of gauges to manage workload in the light of models of how task demand affects performance and to outline the experience of QinetiQ and Bristol University in the development of gauges for the Cognition Monitor.

2 Properties of the gauges

There is a wide range of potential measures of operator activity or state, encompassing physiological signals such as eye pupil size, heart rate and heart rate variability, neuro-physiological signals such as the electroencephalogram (EEG), measures derived from functional near infrared (FNIR) and behavioral measures drawn from crew activity. The aim of selecting particular combinations of measures for use in gauges is that they provide appropriate metrics of crew state that can be used to regulate the operation of the Augmented Cognition system. It is important for the operation of the system both that the gauges correctly recognize when changes in system behaviour are required and that the correct changes are triggered. To meet this goal, the metrics must have three properties:

- Face validity: the set of gauges has to reflect the different crew states in a manner consistent with workload theory
- Reproducibility: the same states should produce a similar set of measures on different occasions.
- Sensitivity: the set of gauges should be able to differentiate the levels of the states

These three requirements reflect the fact that the gauges must measure identifiable states that can lead to degradations of performance in a reproducible manner and they must be capable of detecting those states that are of concern so that correct system behavior is triggered. However, degradation in crew performance can derive from a number of different sources: environmental stressors, such as fatigue or the thermal environment, stress deriving from the mission context such as fear or anxiety, or task demand (Belyavin 1999). It is important that, if the gauges are to be used to trigger interventions, any degradation be identified correctly, so that the appropriate intervention is applied, otherwise it is unlikely that overall performance will be improved. For the purposes of the discussion in this paper it is assumed that the primary purpose of system adaptation is the management of crew workload, since this is likely to be the most rapidly varying external stress.

2.1 Operator workload models and measurement

2.1.1 Workload models

The task load or workload imposed on an operator has long been recognised as a source of performance degradation (Wickens 1984, Aldrich et al 1984). A number of models have been implemented to calculate workload as a state measure, and, in some cases, predict the consequences on performance (Hendy 1997, Farmer 2000). At the root of most approaches lies the concept that performing multiple tasks of a similar nature will impose demands on the operators that are difficult to meet. They will either need to employ the same "resource" for more than one task at once, or they will need to pass information from two sources down the same channel. In both cases, the underlying mechanism of degradation is one of competition or interference between tasks, rather than a consequence of a change in the internal operator state.

The Information Processing/Perceptual Control Theory (IP/PCT) model (Hendy 1997) formally recognizes the need to model task conflict as part of the analysis of task demand, and relates the management of the conflict directly to the performance of tasks, rather than through the medium of a state variable. The Prediction of Operator Performance (POP) model (Farmer et al 1995) employs a stochastic approach to the same problem, and provides an estimate of expected degradation in the face of task contention. For both models, the classical approach of constructing a state measure of workload, derived in some algebraic manner from task characteristics, is rejected in favour of a formal model of task performance based explicitly on task contention. The state concept is replaced by the current level of activity for a number of cognitive processes.

Both models follow Wickens in that they identify input and output modalities (visual/auditory for input or vocal/manual for output) and internal process modes (verbal/spatial) as critical in determining task contention. These general assumptions were confirmed by a meta-analysis of dual task trials designed to investigate sources of contention in task performance (Nicholls et al 2003). For the POP model, it is possible to recover a state measure of workload, and it can be shown that this measure will not be a good predictor of some forms of performance degradation.

2.1.2   Workload measurement

Unlike the formal models of workload, the measurement of workload has treated workload as a state in which a low level will lead to good operator performance whereas a high level will correspond to poor performance, although it is recognized that there is not an exact relationship between the level of workload and performance degradation. Among the best established of the subjective measures of workload is the NASA TLX inventory (Hart and Staveland 1988), which recognizes six dimensions: Mental Demand, Physical Demand, Temporal Demand, Performance, Effort and Frustration. The DRA Workload Scales (Farmer et al 1995) recognize four scales: Input Demand, Central Demand, Output Demand and Time Pressure. In the study conducted by DERA it was found that the TLX ratings were correlated with the DRAWS ratings with strongest associations for TLX Mental Demand and DRAWS central demand, and TLX Physical Demand and DRAWS output demand.

A range of "objective" measures has been used to assess operator workload (O'Donnell and Eggemeier 1986). Most notable among these have been heart rate and heart rate variability, pupil diameter, measures derived from the EEG, and, recently, fNIR. In general these measures have been used to assess the general level of workload by establishing correlations either with performance or subjective ratings such as TLX. Since there is not a one-to-one mapping from the subjective ratings to performance, the distinction between the two approaches is important for applications in Augmented Cognition systems.

2.2   Face validity: gauges based on workload models

If gauges based on objective measures are to be used to manage degradation in operator performance due to task demand in an Augmented Cognition system, they must be able to identify changes in operator activity that will lead to changes in performance. Objective measures such as the EEG are indirect assessments of operator activity and must be calibrated to a set of appropriate standard measurements. Subjective measurements of workload do not necessarily predict performance, so a gold standard measure must be sought by another route. It is argued that validated models of how performance effects of workload arise provide a means to define how a putative set of measures should be calibrated.

2.2.1   Time pressure and task difficulty

Both the POP and IP/PCT models argue that performance degradation is determined by time pressure on any one of a number of channels or resources due to contention between different tasks. If the assumptions of these models are accepted, it will be necessary to detect the presence of time pressure within any of the channels if appropriate mitigating actions are to be undertaken. There are three issues in the identification of suitable calibration strategies. Firstly, it is not clear without further assumptions how time pressure will manifest itself in objective measures such as the EEG. Secondly, it is not simple to allocate task activities precisely to specific channels even with well-understood tasks, since no task with both input and output uses a single channel. Thirdly, individual operators may use different strategies for performing sets of tasks that cause contention, which may lead to different workload and performance effects for different individuals.

In both the POP and IP/PCT models a clear distinction is made between task difficulty and time pressure. An operator is affected by time pressure if they find it difficult to complete one or more tasks within the allotted time. It is assumed that task difficulty is reflected either in the number of processing stages that a task demands or the number of channels that are recruited during task execution. If there is adequate time to complete the task, there will be no time pressure, independent of task difficulty. The level of task difficulty is frequently manipulated by increasing either the number of channels or the number of implicit processing stages involved in task completion, affecting the time required to complete the task. In this way, task difficulty can have a direct effect on time pressure through the need to undertake more work per unit time, although the constructs are strictly independent.

The assumptions of both workload models imply that time pressure should be characterized by more frequent invocation of specific cognitive processes, reflected in the more frequent use of particular cognitive channels. Thus the signature of time pressure should be the more frequent occurrence of a specific pattern in cognitive activity. If a binary variable could be defined that is zero when the process is inactive and one when it is active, time pressure in the associated channel would be reflected in mean values of the variable over a defined time interval close to one, while the absence of time pressure will be characterised by mean values close to zero.

2.2.2 Multiple channels

If it were possible to define a simple task that required the use of a single cognitive channel, and it were possible to define a gauge measure that reflected activity in that channel, there is still a problem in defining the level of task demand in terms of the task characteristics. A procedure that has been tested in workload modeling is to estimate task demand by calculating the time required to execute the task at a given level of accuracy and, using a subject matter expert model of how the task is executed in cognitive terms, derive an estimate of the fractional level of activity. This can be supplemented by subjective estimates of workload in a small trial, combined with confirmatory estimates of task performance. Based on this approach, a procedure for calibrating the gauge would be relatively simple:

- Define a trial with the simple task that provides different levels of task demand
- Derive estimates of the different demand levels using subject matter expert opinion, subjective estimates of workload, and measurement of task performance
- Estimate the relationship between the levels of demand and the gauge using a suitable statistical model

In practice, this idealized situation is very unlikely to arise, as any task involving both input and an operator response will involve the use of more than one channel. If very good estimates of the likely usage of the different channels are available, it is possible to follow the same procedure for multiple channels using a single task, with the rider that it is unlikely that the full range of values will be available for any gauge. If there is some uncertainty about the different ratings, it will be advisable to use more than one task in the calibration, and to derive the calibrations using data from the multiple tasks to provide estimates of the gauges.

2.3 Reproducibility

A key property of any set of gauges is that they should predict a similar degree of demand when presented with the same operator state. There are two effects that can have an impact on reproducibility: the amplitude of the physical signal may not be consistent from one measurement event to the next, and the relationship between state and signal may be unreliable.

If it is anticipated that the amplitude of the physiological signal is unreliable, the gauge must be constructed so as not to depend on the raw amplitude of the signal. For example, the measured signal amplitude from an EEG electrode can vary due to local changes in resistance. To eliminate any dependence on the amplitude of the EEG signal it is necessary to use a scale–independent measure such as the fraction of the spectral power within a particular frequency band. If the relationship between two electrodes is required then it is necessary to use a measure that is independent of the amplitudes of both such as coherence within a frequency band or the ratios of the fractional power within particular bands. In general any measure that is independent of the signal amplitude(s) is likely to have better reproducibility than a measure of raw power.

Given the constraint that a gauge is to be constructed out of amplitude–independent components, it is still necessary to construct gauges that reflect the underlying state in a reproducible manner. If the gauge can be mapped onto a plausible theoretical construct, such as the use of a specific cognitive process, it is more likely that the relationship will be reproducible. When an operator is performing a complex task involving a range of different activities, high task demand can arise from a number of different channels. If the complex task is used to calibrate a particular gauge, it is very likely that the gauge will not be reproducible, since the underlying causes of the high task demand are inconsistent. This strongly indicates that relatively "pure" tasks involving the use of a limited number of cognitive processes should be used for calibration purposes.

2.4 Sensitivity

The requirement that a gauge be sensitive to differences in state implies that it can be tested in circumstances in which changes in the appropriate state can be identified. For complex tasks it is not easy to be sure that a specific state is changing, whereas for relatively pure tasks involving a small number of cognitive processes it is relatively simple to identify states that change with task demand. As part of the calibration procedure, it is important to demonstrate that the selected gauge is sensitive to changes in underlying state, and the most reliable method of ensuring sensitivity is to employ pure tasks as part of the calibration procedure.

2.5 Summary

Combining these requirements leads to the following schema for constructing gauges from physiological signals:

- Attempt to base the set on a widely accepted model of cognitive workload to provide face validity
- Use amplitude–invariant measures as input to the state classifiers as this supports reproducibility
- Calibrate on relatively pure tasks since this provides a clear test of sensitivity

If all these criteria are met we can expect to calibrate our sensors once for an individual and anticipate that the calibration will be stable over prolonged time intervals. We can use fixed thresholds within the system to act as triggers for system intervention, if we make the scale of the difference between different states commensurate for different individuals. We know what our gauges mean and can interpret how our system is working.

3 QinetiQ experience with the EEG

QinetiQ and Bristol University have employed the criteria outlined in 2.5 to develop and test the Cognition Monitor over many years. Three trials in particular bear on the question of gauge construction:

- A trial with the 'pure' gauges task to test calibration reproducibility
- A full test with the Cognitive Cockpit to test calibration transfer from a pure task
- The application of the methodology to data collected by US ARL for the Multi-Attribute Task (MAT) Battery

3.1 Task 1 – Gauge monitoring task

A trial was undertaken to test the reproducibility of calibrations derived for a "pure" gauge monitoring task. The task comprises a multidimensional tracking task, where the individual has to correct for excessive deviation of five gauges by selecting the appropriate gauge and adjusting the offset using either an up or down key. The polarity of some of the gauges is reversed increasing the complexity of the stimulus–response pairing. Performance is scored by measuring the number of gauges at the extreme points of travel.

Six participants were tested in the trial. For the purpose of calibrating the EEG, performance scores were assumed to indicate workload. Although the gauge task is continuous, five levels of task demand were embedded within the sequence by manipulating the frequency of required interventions, enabling discriminations to be made between different levels of task load. EEG was decomposed in the manner described by Pleydell-Pearce et al (2003), and the relationship between each of the EEG measures and gauge task performance (GTP) was investigated using linear regression. For each individual, the EEG measures were ranked on the basis of the strength of the correlation with

GTP. The two EEG measures with the strongest correlation were combined in a single linear regression equation and a predicted GTP calculated. Actual GTP was then ranked from high to low and the success of the predicted GTP in each of these two categories evaluated. Across the six participants, performance in this training data set was successfully classified by the EEG measures 72% of the time.

Participants performed the gauge task on a second occasion, on average 13 weeks after the first time, and the classification accuracy of GTP on the basis of those measures identified during the first visit was assessed. This test of within-participant consistency achieved an accuracy of approximately 71%. The results of this study indicate that it is possible to achieve calibrations for the EEG with a "pure" task that are reliable for long periods within an individual.

3.2 Gauge task used for calibration for use in the Cognitive Cockpit

A closed-loop trial with six participants was undertaken with the Cognitive Cockpit at QinetiQ Farnborough using the Cognition Monitor to control adaptive mitigations in the cockpit interface. A description of the trial is provided in Dickson (2005). The gauge task described in Section 3.1 was used to calibrate EEG measures of executive load in the Cognition Monitor for each of the trial participants individually, imposing the condition that the calibrated predictions were monotonic with the level of imposed task load. As part of the trial, NASA TLX ratings were collected for each of 6 mission segments.

The level of executive load estimated from the EEG measures was examined for the six segments and compared with the subjective TLX ratings. The missions were divided into a relatively lower workload ingress block and a higher workload attack and egress block. Mission difficulty was manipulated by increasing the number of threats in the attack and egress blocks. The mean results for the two blocks by mission difficulty are displayed in Figure 2 for TLX Mental Demand and EEG Executive Load. The three levels of mission difficulty are labeled W1, W2, and W3. The left hand pane displays the TLX Mental Demand ratings and the right hand pane displays the calibrated EEG executive load. There was no evidence for difference in TLX Mental Demand ratings between missions with differing imposed workload, but there was a clear difference between the two mission blocks ($p<0.01$). Similarly there was no overall difference in calibrated executive load between the missions with different imposed workloads, but there was a difference between the mission blocks ($p<0.001$) and there was an interaction between mission blocks and imposed workload level ($p<0.05$).

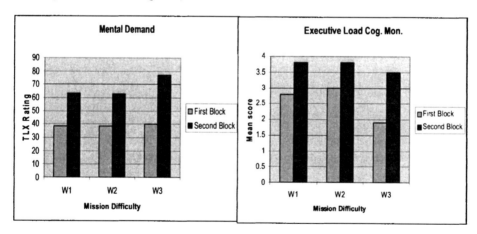

Figure 2: TLX mental demand and executive load ratings

Overall there is similarity between the pattern of values for the EEG executive load ratings and the TLX mean scores, suggesting that they measure similar constructs, although from correlating mean values, there appear to be deviations between the two. There was evidence that participants became fatigued during later trials and this may have a differential effect on the two measures, promoting a different pattern between the mission types in Figure 2.

3.3 MAT Battery analysis

The Multi-Attribute Task Battery (MATB) is a complex task that comprises a number of discrete task components; gauge monitoring, light detection, tracking, resource management, scheduling, and auditory communications tasks. Task load may be manipulated by changing the characteristics of some or all of these tasks, in addition to the number of tasks performed at any one time. The task was developed to reflect the multi-faceted role of aircrew. The data set analyzed here was supplied by Dr. Glenn Wilson of the WPAFB, and comprised EEG and EOG data recorded from eight participants. The participants performed the MATB task for a period of six minutes, at three levels of load, three times a day over a five day period. Thus 45 task repetitions were performed during a single week. MATB was set up with the active components of the gauge monitoring task, light detection task, resource management task and the communications task.

Electro-physiological data were recorded at a rate of 256 Hz from 19 EEG electrodes and two EOG channels. EEG data had previously been corrected for eye movement artifact. EEG was decomposed in the frequency domain to form a number of estimates of spectral density, coherence, cross-power, cross phase and spectral power ratios, with a single time domain class of amplitude measure.

To define a measure of task demand a composite measures of overall MATB performance was calculated, based on response time and accuracy for the task components. A normalized z score was constructed and low performance was defined as $z < 0$ whereas high performance was defined as $z >= 0$. Correlations were calculated for each of the EEG variables against the composite performance score. Predicted performance was then defined as the mean of the predicted scores calculated for each of the identified EEG markers.

Within–session prediction accuracy was assessed, and the EEG was able to predict the correct low workload category 93.7% of the time, and the high workload category with an accuracy of 94.2%. These high success rates are unsurprising as the algorithms were attempting to predict the classification of data they were trained upon. A further measure of classification success was the ability of the techniques to classify the relative load of the different levels of task difficulty. This is indicated by the ability to classify load 2 greater than load 1, load 3 greater than load 1 and load 3 greater than load 2. This metric was termed the internal consistency score (ICS), and for the within–session calibration a 92.65% success rate was achieved.

Between–session classification performance is a better test of gauge construction. The reliability of the EEG gauge based on data drawn from the different days of the trial was then used to classify the load in trials from the other days. The following accuracies were achieved when gauges trained on each of the individual days attempted to classify trial load, on the other trial days.

- Day 1 coefficients – predicting days 2,3,4,5 – 64.7%
- Day 2 coefficients – predicting days 1,3,4,5 – 68.7%
- Day 3 coefficients – predicting days 1,2,4,5 –70.8%
- Day 4 coefficients – predicting days 1,2,3,5 – 71.1%
- Day 5 coefficients – predicting days 1,2,3,4 – 71.3%

The differences between predictive success are not substantial for the different days, but there is an indication that the calibration based on Day 5 data, measured when the participants were more highly trained, provides better classification of the workload on earlier trials than the calibration derived when participants were less experienced.

4   Discussion and conclusions

It is argued in this paper that the calibration of gauges for use in Augmented Cognition systems should be based on "pure" tasks since it is possible to assess what cognitive processes may be required to perform the task in a rigorous manner using established models of cognitive workload. If individual cognitive processes can be identified, then it is possible in principle to derive a generic calibration – at least for a particular crew member – that is applicable whenever the specific cognitive process is involved in a complex task. In this way a robust system for measuring state can be constructed that will be resilient to different system contexts.

The principle of constructing a gauge on a relatively simple task and applying it to a complex context has been tested in a limited form in the trial with the Cognitive Cockpit with some success. In addition, it was established in a separate trial that the calibration based on the simple task was stable over time. It will require further rigorous testing to ensure that the principle can be made to work in general, and that it is possible to construct group calibrations that are both stable in time and applicable to generic contexts.

The problems of using complex tasks with poor definition of the relevant cognitive processes to calibrate EEG markers for the same complex task are highlighted by the analysis of MATB. Since it was not possible to identify specific cognitive processes and calibrate separate gauges for these channels, it was found that there was some lack of reproducibility in the calibrated gauge. If the strategy of calibrating the state measure on the full system task is to be adopted, care must be taken to calibrate individuals when they have fully mastered the task.

Acknowledgements

I would like to thank Dr Glenn Wilson and Chris Russell of WPAFB for supplying the MAT battery data set.

References

Aldrich T. B., Craddock W., McCracken J. H. (1984). A computer analysis to predict crew workload during LHX scout-attack missions. *Volume 1 (MDA903-81-C-0504/ASI479-054-I-84(B))*. *Fort Rucker, AL, USA*: United States Army Research Institute Field Unit.

Belyavin, A. J. (1999). Modeling the effect of stress on human behavior. In *Proceedings of the 8th conference on Computer Generated Forces and Behavioral Representation* (May), 481–487.

Comstock, J. R., Arnegard, R. J. (1992). The Multi-Attribute Task Battery for Human Operator Workload and Strategic Behavior Research. *NASA Technical Memorandum*. 104174

Farmer E. W., Belyavin A. J., Jordan C. S., Bunting A. J. (1995). *Predictive workload assessment: Final Report* DRA/AS/MMI/CR95100/1

Farmer, E. W. (2000). *Predicting operator workload and performance in future systems.* Journal of Defence Science, 5(1), pp F4–F6.

Hart S. G., Staveland L. E. (1988) Development of NASA TLX (Task Load Index): Results of empirical and theoretical research. In Hancock P. A. and Meshkati N. (Eds.), *Human Mental Workload.*, Amsterdam. North-Holland

Hendy K. C., Farrell P. S. E. (1997) Implementing a model of human information processing in a task network environment DCIEM No 97-R-71.

Nicholls, A. P., Farmer, E. W., Peachey, R. I., & Belyavin, A. J. (2003). Dual task interference: Using Hierarchical Cluster Analysis to identify underlying cognitive mechanisms. In *Proceedings of the 47th Human Factors and Ergonomics Meeting.* Denver, Colorado.

O'Donnell R. D., Eggemeier F. T. (1986). Workload assessment methodology. In *Handbook of Perception and Human Performance.* Vol II, 42. John Wiley

Pleydell-Pearce, C. W., Dickson, B. T., Whitecross, B. T. (2003). Multivariate analysis of EEG: predicting cognition basis of frequency decomposition, inter-electrode correlation, coherence, cross phase and cross power. In R. H. Sprague (Ed.) *Proceedings of the Thirty-Sixth Annual Hawaii International Conference on Systems Sciences. IEEE Computer Society.* The Printing House, USA.

Sheridan, T. B. (1980). Computer Control and Human Alienation. *Technology Review,* 83, 61–70.

Sheridan, T. B. (1987). Supervisory control. In *Handbook of Human Factors* (pp. 1243–1268). New York: John Wiley & Sons.

Wickens C. D. (1984) *Engineering psychology and human performance.* Columbus. Ohio. Charles E Merrill..

Section 3
Cognitive State Sensors

Chapter 10

Fundamentals of Augmented Cognition

Session Overview: Foundations of Augmented Cognition

Amy A. Kruse, Dylan D. Schmorrow

Defense Advanced Research Projects Agency
3701 North Fairfax Drive
Arlington, VA 22203
akruse@darpa.mil; dschmorrow@darpa.mil

Abstract

The field of Augmented Cognition has emerged, in part, as the result of a substantial investment from the Department of Defense (DoD) in the tools and technologies to enable the design of revolutionary human-computer interactions. The goal of Augmented Cognition research is to create revolutionary human-computer interactions that capitalize on recent advances in the fields of neuroscience, cognitive science and computer science. The research area of Augmented Cognition that exists today was influenced in part by early investments DoD-funded programs in Biocybernetics, Learning Strategies, the Pilot's Associate program, as well as investments from industry extending into the late 20th century. A review of this foundational history will serve as an introduction to the research papers presented in this session. The papers featured in this session highlight the Augmented Cognition work currently being carried out under DARPA's Improving Warfighter Information Intake under Stress program. As with any developing research area, there are often projects on the vanguard that serve as both examples and trailblazers to the larger research community. These presentations detail the research prototypes from the Improving Warfighter Information Intake under Stress program's seminal teams. This session will also feature a discussion of the overall approach and strategy for the design of these first Augmented Cognition systems. These research teams have taken a bold leap into the future by investing time, resources and personnel in creating physical manifestations of Augmented Cognition systems.

1 Theoretical Foundations of Augmented Cognition

Since the 1960's and the dawning of the Computer Age, visionary researchers like J.C.R. Licklider imagined a "Man-Computer Symbiosis" in which the computer and human user would not be separate entities, but one optimally functioning and symbiotic whole.

> *"The hope is that, in not too many years, human brains and computing machines will be coupled together very tightly, and that the resulting partnership will think as no human brain has ever thought and process data in a way not approached by the information-handling machines we know today." (Licklider, 1960)*

As detailed in two reviews, (Forsythe et al, in press; Schmorrow & Kruse, 2004) over the past several decades there have been numerous attempts -- the majority of which were sponsored by defense funding -- to create symbiotic computational systems. Some of these programs included efforts in Biocybernetics, Learning Strategies and the Pilot's Associate program. Although the initial ideas emerged in the mid-20th century, it has taken considerable time for both the computer sciences and the psychological/neurosciences to reach a point where this symbiosis is now possible. At each step in the evolution of the *field* of Augmented Cognition there have been countless research efforts that have constructed the body of knowledge that the current field now rests on. This has included work on human-computer interfaces, adaptive systems and cognitive modeling.

We assert that Augmented Cognition can be distinguished from its predecessors by the focus on the real-time cognitive state of the user, as assessed through modern neuroscientific tools. Licklider (prophetically) used the terms *"human brains and computing machines"* and we believe that this is the critical component for true symbiosis. The focus on neuro-cognitive assessment of the operator's state does not exclude the use of behavioral, contextual or other state inferences. Yet because those measurements are just that – inferences – their value lies more in what they can add to the assessment provided by the neurophysiological sensors. The direct measurement of cognitive activity, coupled with advanced computational systems is what enables Augmented Cognition to exist today. The current field of Augmented Cognition is both the philosophical and technical realization of Licklider's vision.

This review, and the ensuing conference, represent the first time that the international scientific and technical communities have come together in a public forum to assess the state of the field. Many would agree that this field is in its infancy and that our first systems will look like awkward prototypes in the near future. However, it is enormously important to widen the research community through these scientific interchanges so that we may begin to make the revolutionary leaps to the next generation of Augmented Cognition systems.

2 Critical Components of an Augmented Cognition System

At its core, an Augmented Cognition system is a "closed loop" in which the cognitive state of the operator is detected in real time with a resulting compensatory adaptation in the computational system, as appropriate. This is clearly an oversimplification, but captures the essence. The following section will discuss the critical components of an Augmented Cognition closed-loop system at the same general level. It is not an all-inclusive list, but focuses on the essential elements. All of the systems described in this session will be composed of the same fundamental components but will differ in their applications to the specific domains and platforms. From a platform-independent perspective, the following are necessary components.

2.1 Real-Time Sensors

If the detection of cognitive state is fundamental to an Augmented Cognition system, then it follows that the tools to assess this state in real time would be enabling capabilities. Through the investments in the 'Decade of the Brain' and the efforts of the neuroscience community we now have access to these tools and techniques. An excellent review of the sensors available can be found in the NATO report on Operator Functional State Assessment (NATO HFM-056/TG-008; Wilson). To date, researchers have focused on electroencephalography (EEG), functional near-infrared imaging (fNIR) and a host of physiological measurements – including electro-oculography (EOG), heart rate, galvanic skin response and pupillometry (Muth, in press; Forsythe, in press). Taken together, these techniques comprise a 'sensor suite' that provides robustness for detecting cognitive functioning across a variety of conditions.

Investments in sensor and algorithm development over the past several years have had an important impact on two critical aspects of the sensors. First, the emphasis on making Augmented Cognition systems for genuine operational applications has necessitated a re-design of many sensors to make them usable in non-laboratory conditions. This has included developments to reduce the impact of noise from the environment and make the sensors more 'reasonable' for operational settings by reducing the profile or number of channels needed. Second, the need for real-time assessment of cognitive activity has likewise demanded that the scientific community develop novel signal processing algorithms and techniques. In typical laboratory settings, there is no need for on-line processing since most experiments are a succession of multiple trials of the same stimulus. The single-trial environment has demanded increased signal-to-noise ratios and processing schemes that can run in a feed-forward manner with minimal training. Although additional research and development is needed on sensors for operational environment, the teams presenting here will discuss their successes in detecting cognitive state in real-time. In addition, several sessions at this conference will discuss sensor-based approaches for achieving these challenging goals.

2.2 State Classification

It is one thing to be able to assess cognitive functioning in real time; however, the true challenge comes from making sense of these signals in the context of the specific environment and task. Cognitive state classification is the next critical step in the development of an Augmented Cognition system. At present, a number of general methods are utilized by research teams:

1) Neural net classifiers
2) Statistical Process Control algorithms
3) Rule-based classifiers
4) Classification based on Linear Discriminant Analysis of signal features.

442

These techniques will be discussed in detail during other sessions within this conference. Each of these approaches takes information from the sensors and attempts to make an on-line characterization about the cognitive functioning of the user at that moment. During the classification step, information from the task, environment or other operator behaviors is combined with the cognitive state information to calculate a 'decision' or classification of the relevant state. Through recent efforts, research teams have had significant success in classifying the cognitive state of the user in the context of the specific task environment – these include classifications of verbal and spatial load during demanding command-and-control and driving tasks. Other teams have had success tracking the attentional state of the user during engaging field-based navigation scenarios. It has proved difficult to classify task-independent measures of cognitive state, but this may be possible as real-time signal processing improves and our understanding of the neural signatures expands.

2.3 Adaptive System

The adaptive system is the element that "closes the loop" in an Augmented Cognition system. It is independent from the cognitive classification piece in that the adaptive component does not to know which sensor, behavior or event caused the change in state – it merely receives pre-classified inputs in order to generate its responses within the interface or information delivery scheme. Adaptive systems have been in development for a number of years and this field has benefited from the research that has previously been done. Early adaptive systems such as Pilot's Associate were based on the models of the user's behavior and the operator's state as inferred through behaviors and actions on the system. The adaptive user interface community is now a substantial research community, and the field of Augmented Cognition stands to gain much from their findings and techniques. However, we also maintain that the adaptive systems developed through the research efforts described here will be markedly different from those already in existence. The level of granularity provided by using information about the actual cognitive state of the user will allow researchers to develop precisely-targeted adaptations which not previously possible. For example, if the classification system can yield details about the specific channels overloaded during a task (visual or spatial working memory, attention) this is considerably more instructive than a general measure of workload. If combined with details about the task and behaviors, the resulting adaptations can be very specific for the task and state. Several of the teams have developed adaptations or "mitigation strategies" that can target the information channels currently overloaded and change the interface presentation scheme (i.e. from visual to auditory) based on that information. Without direct sensing of the cognitive state, these elegant adaptations could not be implemented.

2.4 Integrated Architecture

Although it may seem evident, the integrated architecture to pull all of these components together is itself a critical constituent of the Augmented Cognition system. Because disparate research efforts and techniques comprise each of the pieces described above, a solid systems-engineering plan is needed to ensure that the system will function smoothly. In combination with the task environment, the integrated architecture *is* the Augmented Cognition system instantiated in hardware and software. Again, additional sessions during this conference will discuss the details of these systems and their development. The research teams in this session will discuss their particular systems engineering designs, including agent-based approaches. Once again, the platform or task influences how these elements are integrated, as an office-based command and control setting is substantially different from a mobile soldier-based system. As these systems move closer to transition to the targeted military applications, the systems engineering requirements will increase substantially.

3 Highlights of the "Foundations of Augmented Cognition"

This session details the first major efforts to develop full-scale Augmented Cognition systems through the support of the Department of Defense. These efforts did not start out in the advanced state in which they will be discussed. The majority of the work described here was conducted over a three year period during which the fundamental work on cognitive state detection and classification was completed and the creation of the first "closed-loop" prototypes was begun. The technical work for this program is still ongoing and is projected to last through FY06.

3.1 Overview of the DARPA Augmented Cognition Technical Integration Experiment

The paper presented by St. John et al (2005) in this session details the Technical Integration Experiment conducted at the end of the Phase 1 program effort. (St. John et al., 2004, 2003) This experiment served a crucial function for the program as it brought together all of the sensors and teams from Phase 1 for one experiment, on the same testbed, in the same location using a common pool of subjects. The teams used a complex cognitive task (Warship Commander) which presented a varying level of workload to the subjects. It allowed cognitive gauge developers to test their algorithms on a more operationally relevant testbed, a first for many researchers, and permitted the correlation between sensor/gauge activity and specific task load features. The results of this experiment informed the research teams that moved forward into Phase 2 as to what might be expected from specific sensors and algorithms.

3.2 Performance Augmentation through Cognitive Enhancement (PACE)

The paper presented by Thomas et al (2005) in this session details Lockheed Martin Advanced Technology Laboratories' (LMATL) participation in Phase 2 of the program effort. LMATL has been investigating methods for alleviating the working memory load on the Tactical Tomahawk workstation operator. The Tactical Tomahawk represents a new job function and the expected workload is high. Using their PACE architecture, the LMATL team has successful developed a closed loop prototype to reduce overload using a combination of neural and physiological measures.

3.3 Building Honeywell's Adaptive System for the AugCog Program

The paper presented by Ververs et al (2005) in this session details Honeywell Laboratories' participation in Phase 2 of the program effort. Honeywell was faced with a very challenging operational environment – the dismounted soldier – in which to develop a closed loop prototype. They were able to successfully record neurophysiological signals from mobile operators in a field setting and adapt incoming communications based on the attentional state of the soldier. This is a significant development for the eventual deployment of these technologies to the Future Force Warrior program.

3.4 The Boeing Team Fundamentals of Augmented Cognition

The paper presented by Barker and Edwards (2005) in this session details the efforts of Boeing Phantom Works to combat multiple information processing challenges presented in the combat UAV environment. As with other teams, the combat UAV operator is faced with numerous task-based challenges that increase the workload far beyond what one operator might face currently. The Boeing team's goal was, through the use of a closed loop prototype, to demonstrate that one operator could successfully handle three UAV vehicles with no decrement in performance. The team will report its achievement of this goal and strategies for developing Augmented Cognition systems in other testbeds.

3.5 DaimlerChrysler's Closed Loop Integrated Prototype: Current Status and Outlook

The paper presented by Kincses (2005) in this session details DaimlerChrysler's participation in the program effort. DaimlerChrysler tackled the information processing challenges inherent in a mobile driving command and control environment. Utilizing a CAN-BUS architecture already present in their vehicle platform, the DaimlerChrysler team developed a sophisticated classification scheme based on both the cognitive state of the driver and driver behaviors in the car. The result was an extremely accurate classifier across multiple driving conditions which allowed the team to adapt the information flow within the vehicle to suit the cognitive and driving load of the operator.

3.6 The Cognitive Cockpit – a Testbed for Augmented Cognition

The paper presented by Dickson (2005) in this session details Qinetiq's participation in the program effort. The Qinetiq team has been working within a fast-mover jet simulation environment to develop a closed loop "Cognitive

Cockpit." The Qinetiq work involves an advanced 'systems engineering' design which contains modules for the assessment of workload, decision support and an adaptive interface manager integrated with the simulation environment. The result is a seamless cockpit prototype system that is currently being tested for its ability to alleviate pilot task-load within challenging ground to air attack scenarios.

4 Final Thoughts

This session has discussed the historical perspective on Augmented Cognition systems, their critical components, and first efforts to design closed loop systems for military applications. Upon review of the technical obstacles that the field has overcome, there has been substantial progress over the past three years to achieve the results detailed in this session. Much work is still required to develop robust systems capable of being fielded in operational settings. However, these initial prototypes combined with the participation of the larger scientific community will result in even greater developments over the next few years. Augmented Cognition is truly on a course to achieve Licklider's vision

5 References

Barker, R. & Edwards, R. (2005). The Boeing Team Fundamentals of Augmented Cognition. To appear in *Proceedings of the First InternationalConference on Augmented Cognition*. Mahwah, NJ: Lawrence Erlbaum Associates.

Dickson, B. T. (2005). The Cognitive Cockpit – a test-bed for Augmented Cognition. To appear in *Proceedings of the First InternationalConference on Augmented Cognition*. Mahwah, NJ: Lawrence Erlbaum Associates.

Forsythe, C., Kruse, A. & Schmorrow, D. (in press). Augmented Cognition. In C. Forsythe, M.L. Bernard & T.E. Goldsmith (Eds.) *Cognitive Systems: Human Cognitive Models in Systems Design*. Mahwah, NJ: Lawrence Erlbaum Associates

Kincses, W. E. (2005). DaimlerChrysler's Closed Loop Integrated Prototype: Current Status and Outlook. To appear in *Proceedings of the First InternationalConference on Augmented Cognition*. Mahwah, NJ: Lawrence Erlbaum Associates.

Licklider, J. C. R. (1960). Man-Computer Symbiosis. *IRE Transactions on Human Factors in Electronics, 1*, 4-11.

Muth, E., Schmorrow, D., Hoover, A., & Kruse, A. (in press). "Augmented Cognition": Aiding the Soldier in High and Low Workload Environments through Closed-Loop Human-Machine Interactions. In *Studies in Military Psychology; Section 2: Physiological and Mental Dimensions of Warfare; The Human Dimension of Warfare: The intersection between Humans and Technology*. Westport, CT: Greenwood Publishing Group

Schmorrow, D. D., and Kruse, A. A. (2004). Augmented Cognition. In W.S. Bainbridge (Ed.), *Berkshire Encyclopedia of Human-Computer Interaction* (pp. 54-59). Great Barrington, MA: Berkshire Publishing Group.

St. John, M., Kobus, D. A., Morrison, J. G. & Schmorrow, D. (2005). Overview of the DARPA Augmented Cognition Technical Integration Experiment. To appear in *Proceedings of the First InternationalConference on Augmented Cognition*. Mahwah, NJ: Lawrence Erlbaum Associates.

St. John, M., Kobus, D. A., Morrison, J. G. & Schmorrow, D. (2004). Overview of the DARPA Augmented Cognition Technical Integration Experiment. *International Journal of Human-Computer Interaction, 17*(2), 131-150.

St John, M., Kobus, D. A. & Morrison, J. G. (2003). DARPA Augmented Cognition Technical Integration Experiment (TIE). Technical Report 1905, SPAWARS Systems Center, San Diego, CA.

Thomas, M., Tremoulet, P. & Morizio, N. (2005). Performance Augmentation through Cognitive Enhancement (PACE) To appear in *Proceedings of the First InternationalConference on Augmented Cognition*. Mahwah, NJ: Lawrence Erlbaum Associates.

Ververs, P. M., Whitlow, S. D., Dorneich, M. C. & Mathan, S. (2005) Building Honeywell's Adaptive System for the Augmented Cognition Program. To appear in *Proceedings of the First InternationalConference on Augmented Cognition*. Mahwah, NJ: Lawrence Erlbaum Associates.

Wilson, G. F., & Schlegel, R. E. (Eds.) (2004). *Operator Functional State Assessment*, NATO RTO Publication RTO-TR-HFM-104. Neuilly sur Seine, France: NATO Research and Technology Organization.

Overview of the DARPA Augmented Cognition Technical Integration Experiment

Mark St. John, David A. Kobus,
Pacific Science & Engineering Group
9180 Brown Deer Road
San Diego, CA 92121
stjohn@pacific-science.com,
dakobus@pacific-science.com

Jeffrey G. Morrison, &
Space and Naval Warfare System Center
53560 Hull Street, Bldg. A33, Rm. 1405
San Diego, CA 92152
jmorriso@spawar.navy.mil

Dylan D. Schmorrow
Defense Advance Research Projects Agency
3701 North Fairfax Drive
Arlington, VA 22203-1714
dschmorrow@darpa.mil

Abstract

The DARPA Augmented Cognition program is developing innovative technologies that will transform human-machine interaction by making information systems sensitive to the capabilities and limitations of the human component of the human-machine system. By taking better advantage of individual human capabilities, and being sensitive to human limitations, it is expected that overall system performance can be improved by an order of magnitude. There have been many recent advances in the field of Cognitive Science toward understanding human decision-making, and the Augmented Cognition program is taking advantage of them in working toward this potential. The technologies developed over the last decade in measuring brain activity and various facets of cognition are serving as the basis for managing the way information is presented to the human operators of complex systems. The Augmented Cognition program is building demonstrable, quantifiable augmentations to human cognitive ability in realistic operational environments. Towards, this goal, the first phase of the Augmented Cognition program was to empirically assess the utility and validity of various psychophysiological measures in dynamically identifying changes in human cognitive activity as decision-makers engaged in cognitive tasks. This report is the culmination of Phase I of the program – *Measuring Cognitive State*. It describes the empirical results of a Technical Integration Experiment (TIE) involving the evaluation of 20 psychophysiologically derived measures (cognitive state gauges) that were developed under Phase I. The gauges came from 11 different research groups, and were developed with a variety of theories and scientific backgrounds. The TIE brought these disparate approaches to assessing cognitive state together to be assessed with a common test protocol using a relatively complex cognitive task that was derived from the real world decision-making requirements seen with tactical decision-makers. This task was developed specifically to meet the needs of assessing these very different gauges with necessary empirical controls, yet still maintain the essential character of those tasks from a cognitive perspective as would be found in an operational command and control environment. Eleven of the gauges successfully identified changes in cognitive activity during the task. This report also describes the integration of individual gauge technologies into suites of gauges to simultaneously measure multiple cognitive indices, and the issues created with sensor technology integration in developing next-generation cognitive state gauges. Additionally, the gauge developers rated the ability of their sensors to integrate with other sensors as fairly high, and most developers reported no problems integrating multiple sensors onto participants. This report summarizes the results from the TIE, and examines the prospects for, and issues that must yet be addressed for, the successful transition of these cognitive state gauges to field-able military person-machine systems in Phase II of the Augmented Cognition program, and beyond.

1   Technical Integration Experiment

The DARPA Augmented Cognition program is developing technologies capable of extending, by an order of magnitude, the information management capacity of war fighters. This will entail selecting from the myriad of theories and sensor technologies related to the measurement of human cognition developed over the last decade, and marrying them with the many advances in automation and information management. For example, a future C^4I (Command, Control, Computers, Communications, and Intelligence) system may assign a task to the specific operator having the most unused cognitive capacity, or it may filter information or select the mode or style of its presentation based on a particular operator's available capacity to receive information visually, verbally, or through some other sensory modality.

The primary objective for the first phase of the Augmented Cognition program, *Measuring Cognitive State,* was to empirically assess the utility and validity of various psychophysiological measures to dynamically identify changes in human cognitive activity during task performance, and explore potential integration and application issues that would need to be addressed during later phases of the program. Here, we summarize the results of a Technical Integration Experiment (TIE) that provided the culmination for Phase I. This TIE brought together 20 psychophysiological measures (cognitive state gauges) from 11 different research organizations. These measures were demonstrated and assessed in a common test environment that had the complexity and demand characteristics comparable to those seen by a tactical command decision maker.

The gauges used a wide range of sensor technologies, and they were based on very different, yet sometimes overlapping, theoretical approaches. The sensor technologies included functional Near Infra-Red imaging (fNIR), continuous and event-related electrical encephalography (EEG/ERP), eye tracking and pupil dilation, mouse pressure, body posture, heart rate, and galvanic skin response (GSR). Each of the gauges that was evaluated in the study, the type of sensor it used, and the research organization that developed the gauge are listed in Table 1.

Table 1. Summary of Technical Integration Experiment Findings

| | | | | Task Load Factors | | | |
| | | | | Number of Tracks per Wave (6,12,18,24) | Track Difficulty (Hi/Lo) | Secondary Verbal Task (On/Off) | Consistency (% of Participants) |
Gauge	Sensor Type	Research Group	Team				
fNIR							
fNIR (left)	Blood Oxygenation	DrexelU	2	●	○	○	75
fNIR (right)	Blood Oxygenation	DrexelU	2	●	○	○	63
EEG-Continuous							
Percent High Vigilance	EEG	ABM	2	●	○	○	63
Probability Low Vigilance	EEG	ABM	2	●	○	○	75
Executive Load	EEG	QinetiQ/UBristol	3	●	◑	○	100
EEG-ERP							
Motor Effort	ERP-IFF	EGI	1	○	○	○	0
Auditory Effort	ERP-Engage Sound	EGI	1	○	◑	○	0
Loss Perception	ERN-Error Sounds	Sarnoff/Columbia	4	○	○	●	50
Occular-Frontal Source	ERP-Comms	UNewMexico	4	●	○	○	100
Synched Anterior-Posterior	ERP-Comms	UNewMexico	4	○	○	●	100
Visual Source	ERP-Comms	UNewMexico	4	○	○	○	100
Arousal							
Arousal Meter	Inter-Heart Beat Interval	Clemson U	1	○	○	○	0
Arousal	GSR	UHawaii	2	○	○	○	0
Arousal	GSR	AnthroTronix	4	○	○	○	17
Physiological							
Head-Monitor Coupling	Head Posture	UPitt/NRL	1	◑	○	○	43
Head Bracing	Body Posture	UPitt/NRL	1	○	○	○	14
Back Bracing	Body Posture	UPitt/NRL	1	○	○	○	14
Perceptual/Motor Load	Mouse clicks	UHawaii	4	●	●	○	100
Cognitive Difficulty	Mouse pressure	UHawaii	4	●	●	○	100
Index of Cognitive Activity	Pupil dilation	SDSU	floating	◑	○	●	57

Note: Black circles denote statistically significant effects ($p < .05$); half circles denote "marginal" statistical effects ($p < .1$); and White circles denote nonsignificant effects. The final column lists the percentage of participants showing a moderate or high correlation ($r > .3$) between gauge values and the Number of Tracks per Wave. See text for details.

447

The TIE was not the first attempt to combine multiple psychophysiological technologies and measure cognitive activity during a complex task. For example, Fournier, Wilson, and Swain (1999) used a complex personal-computer-based flight simulation to manipulate user workload while measuring cognitive activity using EEG, heart rate, and eye blinks. Smith, Gevins, Brown, Karnik, and Du (2001) used the same task while measuring EEG, and Van Orden, Limbert, Makeig, and Jung 2001 used a mock air warfare target identification and memory task while measuring eye blinks, fixation durations, and mean pupil diameter. However, the TIE, which required coordinating 11 research groups during simultaneous data collection for 20 gauges, was a major undertaking, and the first attempt to bring so many sensor technologies together at the same time.

For the TIE, the 20 cognitive state gauges were assigned to one of four data collection teams to create suites of gauges that could simultaneously monitor participants as they performed the task. This arrangement was done to: 1) assess compatibility issues among the different gauge technologies, 2) allow the direct comparison of results using the different gauges within a team as they assessed the cognitive state changes of the same participants at the same time, yet 3) allow the use of similar sensor technologies, across teams, that would otherwise compete for access to the same physical locations on test participants.

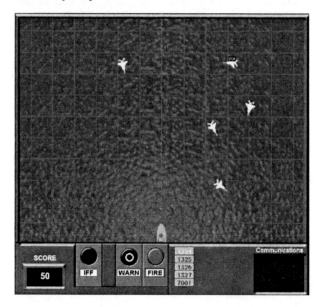

Figure 1. Screen shot of the Airspace Monitoring task in Warship Commander Task

The TIE successfully demonstrated the ability to combine multiple sensors and collect real-time data in a ecologically valid command and control-type decision-making task – which are key requirements of the Augmented Cognition program for transition into Phase II. A key attribute of the TIE was the use of a common experimental test task, under as comparable test conditions as possible across participants and teams. The Warship Commander Task (WCT, St. John, Kobus & Morrison, 2002, see Figure 1) was designed as a basic analog to a Navy air warfare task. This task was based on previous mock air warfare tasks (Ballas, Heitmeyer, & Perez, 1992; Van Orden, Limbert, Makeig, & Jung, 2001), though the pace is faster and the task is more complex in the WCT.

The task was developed to be: 1) suitable for use with undergraduate participants, 2) suitable for stimulating as many aspects of cognition as was feasible, and 3) representative of the complex decision-making environments faced by operational warfighters in tactical command centers. Users performed in a series of 15 minute scenarios during which they monitored a varying number of aircraft (tracks) on a display. They evaluated the tracks and determined if and when it was appropriate to warn them, and if necessary, engage them on the basis of explicit rules of engagement. The task was designed to manipulate a variety of aspects of cognitive activity for the different types of gauges to measure simultaneously, including perception, motor activity, memory, attention, and perceived task load in a semi-realistic command and control-type task.

Figure 2 provides a conceptual illustration of the changing workload demands during the WCT task as perceived by the participants. The pie wedges indicate the proportion of users' activity devoted to each of six dimensions of workload, as defined by the NASA Task Load Index (TLX, NASA-Ames, n.d.; Hart & Staveland, 1988). The left pie chart indicates that during low task load periods of the task, activity on all workload dimensions is low, and users primarily observe and scan the task display. The right pie chart indicates that during high task load periods of the task, temporal and mental demands are high, while other dimensions of workload such as physical demands and frustration remain low, and users have very little time to simply observe the display (The pie wedge proportions are based on previous pilot work). The task, however, did not attempt to explicitly manipulate wakefulness/arousal or physical workload, which has implications for the expected diagnosticity of gauges designed to measure those aspects of cognition.

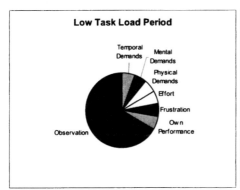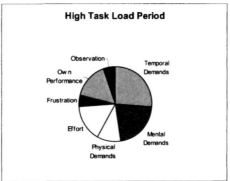

Figure 2. Conceptual illustration of changing workload demands during the WCT task

Cognitive activity, or task load, was varied through three experimental manipulations during the experiment: 1) Number of Tracks per Wave, which varied from 6 to 24 tracks present on the display during each of 12 waves during the course of each scenario, 2) Track Difficulty, which varied between scenarios according to the proportion of potential threat tracks appearing within every wave (High-67% vs. Low-33%) – which required more actions and decisions than other tracks and were thus more complex, and 3) presence or absence of a concurrent secondary auditory/verbal memory task called the Ship Status Task (On or Off) which competed with the primary airspace monitoring task for attentional resources (additional details available at St. John, Kobus, Morrison, & Schmorrow, 2004).

Table 1 summarizes the overall findings of the experiment. For each aspect of task load, a filled black circle in a column indicates that the gauge was statistically sensitive to changes in that specific task load factor ($p < .05$ according to an analysis of variance). A half-filled black circle indicates that a gauge was "*marginally*" sensitive to changes in that task load factor ($p < .10$). Given the limitations of sample size and the complexity of the multi-apparatus data collection sessions, as well as the experimental nature of many of the gauges, we felt that reporting these marginal effects was important. At this early stage of development, subtle changes in technology or procedure may dramatically impact the effectiveness of the gauges. An open circle indicates that a gauge was not sensitive to changes in that task load factor.

The final column of Table 1, "Consistency," is an indicator of the consistency with which a gauge detected changes in task load *across* participants. Consistency was measured by first computing the correlation between gauge values and the Number of Tracks per Wave for each scenario for every participant. Then, the mean correlation was computed for each participant. This correlation provided a measure of gauge sensitivity for each participant. Only the Number of Tracks per Wave factor was examined in this analysis since this factor varied from very low task load to very high task load, and many gauges were able to detect changes in it. Finally, the percentage of participants that showed at least a moderately sized mean correlation was computed. A moderately sized mean correlation was defined to be greater than 0.30. These percentages are list in the final column of Table 1.

While some gauges were consistently sensitive for each participant (e.g. Hawaii's mouse-based gauges and QinetiQ's EEG-based gauge), the majority of gauges were sensitive for some participants but not others. It will be important, in future development of these gauges, to determine the sources of variability and attempt to control them.

2 Discussion of Results

Eleven of the gauges successfully correlated with changes in one or more of the task load factors. Two additional gauges showed specific promise for being diagnostic in detecting changes in task load and warrant further development. Since many of the gauges were very early prototypes that were previously unproven, these results are extremely encouraging. In drawing conclusions from these results, it is important to understand several points. First, positive results indicate that a gauge was successful at detecting changes in the factors that were manipulated in the task. It is likely that these gauges will be similarly successful in tasks that have similar attributes and that are measured under comparable environmental conditions. Specifically, tasks that can be characterized as predominantly involving detection, identification, and memory recall (such as computer-based, fast-paced, command and control-type tasks) and that are presented under similar environmental conditions (such as noise, lighting, and time of day), are likely to show comparable results. These gauges may be successful in other types of tasks, as well.

Second, negative results do not necessarily indicate a "failure." The assessment performed during the TIE involved one task, one narrowly defined context, and a relatively small sample size. The data collection environment might have been too noisy for the gauge, or the small sample size might not have contained sufficient statistical power to reveal the sensitivity of a gauge. Furthermore, due to the rapid development of some gauges, the TIE may have been the first attempt to use them on tasks that differed from those used during their development. There also may have been significant individual differences among participants that require the optimization of various sensor technologies and gauge processing algorithms. The assessment of such issues was well beyond the scope of the TIE, or this paper. Consequently, both positive results, and especially, negative results should be interpreted with healthy skepticism.

More importantly, a gauge might be sensitive to aspects of cognition, but not to the specific cognitive task factors that were manipulated by the WCT. For example, in the WCT, the consequences of error are not severe. Further, participants had limited time to acclimate to the myriad combinations of sensors required for the current state of development of some gauges. As a result, it is reasonable to hypothesize that a gauge that measured the stress induced by severe performance anxiety might not react in the WCT, or be sensitive under the necessary test conditions of the TIE. In sum, conclusions from these results must be viewed within the context of the TIE test conditions and the test task; generalization to other tasks and other situations must be drawn with care.

As a class of gauges, the "arousal" gauges stood out for their inability to detect changes in any of the three task load factors. Since arousal gauges are perhaps the best understood of the gauges used during the TIE, their inability to detect changes in cognitive activity during the WCT is somewhat surprising. These results suggest that there may have been a mismatch between the cognitive states measured by these gauges and the cognitive states elicited by the task, or simply that the gauges themselves were insensitive. As noted above, the WCT does not explicitly manipulate stress, arousal, or physical activity other than in terms of mouse and eye movements. Several of the gauge developers suggested that the introduction of stronger negative consequences for errors committed during the task might have produced more measurable stress changes. For example, game score deductions and loud audio error alerts might have created more stress, especially during high task load periods of the task. It may also be the case that well-practiced command and control-type tasks simply do not evoke strong stress responses, and arousal gauges may not be appropriate for measuring changes in workload in such tasks. However, under operational conditions, the negative consequences of errors can be profound, and changes in stress levels may be important to detect. Therefore, we do not recommend eliminating this class of cognitive state gauges at this time.

In either case, the ultimate success of arousal-type gauges will depend on their ability to predict changes in participant performance, rather than changes in arousal, *per se*. It is well known that highly trained operators, such as pilots, can be highly aroused or stressed, for example while landing on an aircraft carrier, with little or no change in their level of arousal, or operational performance (e.g. Berkan, 2000; Menza, 2002). It may be that arousal gauges are better suited for monitoring novices during training and noting how changes in arousal effect human learning. These issues are complex, the research literature is large and varied, and there appear to be many factors that influence the impact of stress on operational performance. More research is required in this area to better understand the relationships between task load, stress, and performance outcomes in different types of command and control tasks and different levels of expertise and motivation.

Another class of gauges, the ERP gauges, showed mixed results: some were effective, while others were not. The development and use of ERP gauges is somewhat problematic in that the user's task must be well understood to identify appropriate task events to measure. It is also necessary to have some means of determining when these events occur during the task. The WCT provided this information to each gauge, but gauges may not have this luxury in real tasks. If these problems can be addressed, then this class of gauges has the potential to measure specific cognitive processing occurring during a task.

The continuous EEG, fNIR, and ICA gauges, on the other hand, all showed substantial promise for detecting changes in workload. For the TIE, they measured average cognitive activity throughout each wave, but it appears quite possible that they could also measure changes in cognitive activity at much finer time scales. Although the EEG gauges, as a group, measured global cognitive functions, such as attention and executive load, there is support for the idea that EEG measures could also be tailored for more specific cognitive processes (Pleydell-Pearce, Whitecross, & Dickson, 2003)

3 Implications

In addition to evaluating the effectiveness of each gauge individually, the TIE evaluated the practical issue of the ability to combine the sensor hardware into useable suites. The gauge "teams" were arranged so that each contained a mix of compatible technologies, although specific assignments were somewhat arbitrary. Overall, all developers rated the ease of integration as fairly high, and most developers reported no problems integrating sensors onto participants. For example, the gauges from Clemson University (arousal) and the University of Pittsburgh/National Research Laboratory (head and body posture) were designed to compliment any other gauge during the TIE. The most common difficulty arose from the lack of headspace available for multiple sensors and the time required to attach and verify their placement. The development of integrated headgear for multiple sensors should be able to address these concerns. The introduction of wireless technology for transmitting sensor data to computers is also highly promising.

From the user's perspective, the TIE experience highlighted the need to make the gauge hardware comfortable, mobile, and convenient enough to gain user acceptance and to become practical for military applications. War fighters cannot be constrained by bulky, uncomfortable equipment that is difficult or tedious to use. Usability is going to be a critical factor in the successful development of augmented cognition systems in relatively stationary command and control center environments, and especially in more mobile environments. Applications where the performer is relatively mobile, such as vehicle operators and soldiers, will be orders of magnitude more daunting in their challenges. Many of the gauge/hardware systems are promising in these regards, but this issue will only increase in its importance as the Augmented Cognition program moves forward to more applied settings.

Another practical concern is the need to understand and address potential sources of electro-magnetic frequency (EMF) interference, both between sensors, various bio-amplifiers and with environmental factors. Several sources of physical and electro-magnetic interference were identified and resolved prior to the TIE data collection event. Other interference was noted on an intermittent basis in the test facility, with no clear source or technical resolution. As we look to the application of these technologies to military environments, it is almost certain that additional sources will appear - operational environments are often noisy and filled with electrical-magnetic interference from many sources. Again, though many improvements in filtering or adapting to this interference have been made, this issue will only grow in importance as augmented cognition becomes a reality.

In sum, the TIE results point to the great potential for a number of psychophysiological gauges to sensitively and consistently detect changes in cognitive state (activity) during relatively complex command and control-type tasks and to their practical integration into an effective sensor suite. Phase I of the Augmented Cognition program has

451

achieved its goal of providing a solid foundation for the demonstration of augmented cognition systems. The primary objective of the TIE was to demonstrate the successful integration of multiple psychophysiological gauges to detect changes in cognitive states in real-time. The goal for Phase II will be to take these gauges and incorporate them into systems for demonstrating the real-time manipulation of cognitive states as the basis for augmenting cognition.

References

Ballas, J. A., Heitmeyer, C. L., & Perez, M. A. (1992). *Direct manipulation and intermittent automation in advanced cockpits*. Technical Report NRL/FR/5534--92-9375. Naval Research Laboratory, Washington, D. C.

Berkan, M. M. (2000). Performance decrement under psychological stress. Human Performance in Extreme Environments, 5, 92-97.

Fournier, L. R., Wilson, G. F., & Swain, C. R. (1999). Electrophysiological, behavioral, and subjective indexes of workload when performing multiple tasks: manipulations of task difficulty and training. *International Journal of Psychophysiology, 31,* 129-145.

Hart, S. G., & Staveland, L. E. (1988). Development of a multi-dimensional workload rating scale: Results of empirical and theoretical research. In P. A. Hancock & N. Meshkati (Eds.), *Human Mental Workload.* Amsterdam, The Netherlands: Elsevier.

Menza, Lt. M.D. (2002, March). The pucker factor. *Approach*. Retrieved June 27, 2003, from http://www.safetycenter.navy.mil/media/approach/issues/mar02/pucker.htm

NASA-Ames (no date). Task Load Index [TLX] Version 1.0, Users' Manual. Available at http://iac.dtic.mil/hsiac/Products.htm#TLX

Pleydell-Pearce, C.W., Whitecross, S.E., & Dickson, B.T. (2003). Multivariate Analysis of EEG: Predicting cognition on the basis of frequency decomposition, inter-electrode correlation, coherence, cross phase, and cross power. *Proceedings of the 36th Annual Hawaii International Conference on System Sciences.*

Smith, M. E., Gevins, A., Brown, H., Karnik, A., &Du, R. (2001). Monitoring task loading with multivariate EEG measures during complex forms of human-computer interaction. *Human Factors, 43,* 366-380.

St. John, M., Kobus, D. A., & Morrison, J. G. (2002). A multi-tasking environment for manipulating and measuring neural correlates of cognitive workload. In *Proceedings of the 2002 IEEE 7th Conference on Human Factors and Power Plants.* New York, NY: IEEE. pp 7.10 – 7.14.

St. John, M., Kobus, D. A., Morrison, J. G., & Schmorrow, D. (2004). Overview of the DARPA Augmented Cognition technical integration experiment. *International Journal of Human-Computer Interaction, 17,* 131-149.

Van Orden, K. F., Limbert, W., Makeig, S., & Jung, T. (2001). Eye activity correlates of workload during a visuospatial memory task. *Human Factors, 43,* 111-121.

Performance Augmentation through Cognitive Enhancement (PACE)

Nick Morizio, Michael Thomas and Patrice D. Tremoulet

Lockheed Martin Advanced Technology Laboratories
3 Executive Campus
Cherry Hill, NJ 08002
{nmorizio, mthomas, ptremoul}@atl.lmco.com

Abstract

Performance Augmentation through Cognitive Enhancement (PACE) is a domain and application-neutral framework for managing user tasks according to context, including an assessment of the user's cognitive state. This assessment allows the use of presentation mechanisms that are beneficial during certain cognitive states but detrimental during others. For example, it is useful to break a complex task down to smaller pieces during high-stress periods, but this may introduce redundancy, which is undesirable during normal-stress situations.

While the data used by PACE to determine user context has thus far come solely from physiological and neurological sensors, many other possibilities are supported. The PACE framework has also been designed to handle a wide variety of mitigation strategies, that is, context-based manipulations of the user interface designed to maximize operator effectiveness. To date, the strategies tested have involved either deferring the presentation of information and tasks or altering the modality in which they are presented during periods of high workload, high stress, or working memory (WM) overload and delivering them later, at more convenient points for the operator.

The PACE architecture developed under the Improving Warfighter Information Intake Under Stress program provides a powerful, flexible framework that can help support the next generation in computing: making interfaces truly personal, by supporting not only customization through user preferences, but also adaptation based upon workload, cognitive state, and the user's environment.

1 Introduction

As part of the DARPA program Improving Warfighter Information Intake Under Stress, Lockheed Martin Advanced Technology Laboratories (LM ATL) has developed a domain and application-neutral framework for managing user tasks according to context, including an assessment of the user's cognitive state. This framework, called Performance Augmentation through Cognitive Enhancement (PACE), has been used with multiple domains and applications, including Aegis-based command and control operator tasks and Tactical Tomahawk missile monitoring and retargeting. Within these domains, it has been used to perform multiple mitigation strategies for managing undesirable cognitive states, e.g. verbal working memory (WM) overload. Cognitive states were measured by a set of physiological and neurological sensors worn by test subjects while they performed operationally relevant tasks. This data was fused by a Cognitive State Assessment component that produced conglomerate cognitive state gauges which triggered the activation and deactivation of the mitigation strategies used.

While the data used by PACE to determine cognitive state has come solely from physiological and neurological sensors, many other possibilities are supported. A variety of mitigation strategies are also supported by the architecture.

453

2   Background

2.1   Augmented Cognition

PACE was developed under the auspices of the DARPA program Improving Warfighter Information Under Stress. The goal of this program, formerly known as Augmented Cognition, or AugCog, is to optimize performance of combat command and control operators by using neuro-physiological sensors to control the behavior of human computer interfaces, e.g. by tailoring information presentation and task assignments to best suit the currently available cognitive resources of operators. More specifically, if an operator is engaged in a task that has recruited nearly all visio-spatial reasoning resources and critical, task-relevant information arrives, the system may elect to present this information verbally, over an audio channel, to facilitate rapid assimilation.

Augmented Cognition represents the cutting edge in adaptive interfaces, going a step beyond traditional advanced HCI techniques to use neuro-physiological sensors to enable adaptation based upon not only the environment and tasks at hand, but also a real-time assessment of operators' current cognitive capacities.

2.2   Sensors and Cognitive States

As a part of the Augmented Cognition effort, LM ATL has experimented with sensors which provide a variety of neuro-physiological measures including pupil dilation, galvanic skin response (GSR), heart rate variability (HRV), body/head positioning and three different types of brain activation measures: continuous electroencephalography (EEG), event related potential electroencephalography (ERP), and blood oxygenation levels.

One of the objectives of an early series of experiments was developing a sensor suite which could be used to produce reliable and accurate cognitive state 'gauges', based upon multiple neuro-physiological measures. In some cases multiple devices that produce the same sort of physiological data were tested. The selection of sensors included in an integrated suite was based upon a combination of three factors: correlation between sensor data and performance demands, ability to operate the sensing device simultaneously with other sensors with high correlations to performance demands, and the physical discomfort experienced by users due to wearing the sensing device. Other considerations relevant to the Improving Warfighter Information Under Stress program, but not yet used to down-select sensors include portability and robustness of sensing devices. Sensor down-selection was facilitated by the PACE architecture's ability to support plug-and-play of sensing devices, making it possible to quickly and easily add and remove different physiological sensing devices.

Currently, LM ATL's sensor suite consists of a wireless continuous EEG sensor, EKG and GSR sensors, and an off-head binocular eyetracker that logs pupil diameter in addition to point of gaze. These sensors generate EEG, HRV, GSR and pupilometry data that are used to compute values for the following gauges: verbal WM, spatial WM, cognitive workload, arousal.

To date, LM ATL has focused primarily upon using sensor data to estimate the level of utilization of verbal and spatial working memory relative to an individual's capacity, as well as estimating cognitive workload; we have also experimented with individual-sensor based estimates of arousal, e.g. heart rate based arousal and EEG based arousal.

2.3   Mitigation Strategies

The types of adaptations, or behaviors which the system may use to increase the operator's effectiveness include automating or offloading tasks as stress and workload levels approach critical values, changing of the modality in which existing and/or new information is presented, and reinforcing information by representing it in multiple modalities (aural, spatial, verbal). The overall goal of the mitigation strategies is to increase overall performance, while maintaining perceived information load at moderate levels and providing the latitude to increase task difficulty (c.f., Johnstone, 1980).

LM ATL has implemented and pilot tested three mitigation strategies: pacing, intelligent sequencing, and modality switching. The first two have been formally tested through laboratory-based "concept validation experiments" (CVE's). Modality switching will be further explored in future experiments.

2.3.1 Intelligent Sequencing

This strategy ensures that secondary task activities are scheduled to support maximum operator performance by timing the presentation of verbal and spatial secondary tasks when gauges indicate that the operator is not overloaded verbally or spatially, respectively.

So for example, as operators perform primary tasks which are a mix of both spatial and verbal tasks (as they are in most operational environments) and secondary tasks arrive, PACE classifies them as mostly verbal or mostly spatial. Meanwhile, spatial and verbal working memory gauges indicate when the subject's verbal and/or spatial memory stores are taxed by the primary task. Based on secondary task classifications and the gauge readings, PACE schedules the secondary tasks so that they interrupt primary tasks on an optimum schedule. For example, if gauges indicated that an operator's spatial working memory store is taxed, only verbal secondary tasks will be presented until memory taxation has stabilized.

2.3.2 Pacing

This strategy involves directing tasks according to an operator's workload and arousal levels, as determined by arousal and cognitive workload gauges. Research has indicated that the timing of an interruption relative to a user's current task load can affect the user's ability to cope with the interruption (Czerwinski et al., 2000; McFarlane, 2002; Monk et al., 2002). The Pacing mitigation builds upon the Intelligent Sequencing mitigation, a) by adding the identification of appropriate "cognitive breaks" to the total information determining when tasks should be presented, and b) by allowing primary tasks to be decomposed so tasks and subtasks can be optimally scheduled at a higher granularity than just primary vs. secondary tasks.

For example, if arousal and workload gauges indicate that the operator's task load is above maximum threshold, pending tasks are queued for delivery rather than presented immediately. The pending tasks are delivered when the gauges indicate a break in cognitive activity. If too many secondary tasks are competing for delivery, the primary task will be decomposed into subtasks so that cognitive breaks (during which secondary tasks can be delivered) occur more frequently. Overall throughput of memory-taxing tasks are optimized.

2.3.3 Modality-based Task Switching

This strategy involves developing alternate display strategies to invoke specific sensory modalities. Based on output from sensory gauges (e.g., verbal and spatial WM), as well as an understanding of current system state (i.e., which verbal and spatial tasks are currently being performed and their relative loading, c.f., *task intelligence*), display strategies that invoke modalities with spare capacity and/or which are best suited for the information to be communicated will be employed. Table 1 summarizes the various sensory modality display options that will be investigated. For example, while spatial information (e.g., location of a threat) is generally best presented via visual imagery (e.g., target on radar screen), it could be presented as sound localization (e.g., auditory signal at a given location) or as tactile cues (e.g., vibration of a sensor in a tactile vest).

Table 1: Modality-based task switching schemes.

Modality Switching Options	Task Type	
	Verbal	**Spatial**
Visual	Text Instructions or System Status Displayed on Screen (e.g., target ID, type, speed)	Graphics Displayed on Screen (e.g., threat on radar screen)
Auditory	Speech Instructions or Coded Auditory Cues (e.g., auditory warnings)	Sound Localization via Headphones (e.g., audio cue to left ear or to right ear depending on location of target threats)
Haptic	N/A	Vibrations via Tactile Vest (e.g., vibractor tactor in sensory location depending on location of target threats)

455

3 System

3.1 Architecture

A key goal of the design of PACE was to a produce a highly extensible, domain-neutral framework. The resulting architecture provides the flexibility necessary to interchange many different mitigation strategies in response to changes in user cognitive state (or other contextual or environmental factors which PACE is tracking) as well as to integrate with several different applications. To apply PACE to a new domain, it is only necessary to extend a few of the framework components of PACE, such as the Environment Director or Active Task Manager. To incorporate additional mitigation strategies, the management components such as the System Interface Director or Adaptive Workload Director can be extended.

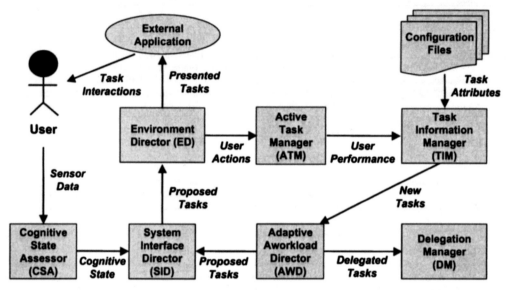

Figure 1: ??

The PACE architecture is divided into seven components each responsible for different aspects of user task management:

1. Active Task Manager – Tracks context associated with a task actively being performed by the user
2. Adaptive Workload Director – Manages all pending tasks
3. Cognitive State Assessor – Analyzes the user's cognitive state based on appropriate cognitive state gauges, computed from real-time sensor data
4. Delegation Manager – Responsible for reassigning tasks to other users or software entities
5. Environment Director – Selects appropriate interaction modalities for proposed tasks, interacts with the external user application
6. System Interface Director – Approves or rejects tasking based on current cognitive state
7. Task Information Manager – Manipulates tasks through decomposition and configuration of presentation options. Also tracks task performance.

3.2 Cognitive State Representation/Use

To this point, experiments with PACE have involved gauging the following cognitive states: verbal working memory, cognitive workload, and arousal. Spatial working memory, executive function, and attention gauges are planned and/or in development. The specific set of cognitive states included is highly configurable within PACE.

456

Only the Cognitive State Assessor needs to be aware of the specific sensors and their relationship to the cognitive gauges used for a particular experiment. Once an assessment is made, it is sent by the CSA to the rest of PACE which uses the assessment to trigger the appropriate mitigations.

3.3 Task Selection in PACE

Tasks in PACE represent goals that need to be achieved by the user. Typically, a task is a straight-forward sequence of actions that must be coordinated between the user and the system. PACE also represents more complex tasks, such as abstract tasks that may have many potential decompositions. In either case, the task itself must specify the complete set of ways that it can be executed.

Task selection refers to the set of processing that determines which tasks to execute at which times and which particular plan to use when executing the task. Task selection is a multi-stage process within PACE. In the first stage tasks are inserted either by the application or by the task generator in the Task Information Manager. These tasks may be provided via messages from another system, or triggered via the completion of other tasks by the user.

If new tasks are not directly executable, the second stage decomposes the task into a forest of subtasks where each tree represents one possible completion of the inserted task and edges in the trees represent temporal dependencies between the subtasks. Through decomposition, a task is able to represent alternative execution strategies. As a trivial example, consider a user that needs to click two buttons (say A and B) in any order. In this case the abstract parent task would decompose into two trees, one which has button A presented first and B second, and the other which has button B presented first followed by A. This sort of decomposition can play a key role in task mitigation, allowing PACE to select the decomposition trees which provide a better fit with the user's and the application's current capabilities. It also allows PACE to delay the execution of a task until an appropriate time.

These tasks and dependency trees are managed by the Task Accomplishment Strategy Manager (TASM) which maintains a queue sorted by task priority. The priority for a task can be determined based upon the application, but defaults to a function based on insertion time and the urgency of the task as defined in the task description. Periodically, the TASM proposes new tasks which have both the highest priorities and resolved dependencies. This process is the third stage of task selection. If the task is rejected at any point after this phase, the task returns to the TASM queue and will be proposed again, up until the deadline specified for the task. Task rejection occurs when there is no way to execute the task, either because there is no empty slot to display the task in the application or because a mitigation strategy was activated.

In the fourth stage, tasks are examined by the environment manager and compared to the latest cognitive workload assessment to make sure the task will not overload the user. The parameters available from the assessment which can be used to make this judgment correspond to the cognitive state gauges built out of neuro-physiological data. Individual parameters may or may not be enabled depending upon available sensors or depending upon the type of tasks. Each task comes with a set of thresholds for each parameter representing either the minimum or maximum acceptable values for that task. After checking to make sure a proposed task will not overload the user, the task moves on to the fifth stage, which is modality selection.

3.4 Modality Selection in PACE

Each PACE task is defined with preferences towards certain modalities. For example, an alert task may be best delivered as text in a window, but if that panel is not available or well-supported by the application, then the alert can be delivered as speech. In this case, we say that the task prefers the panel modality over the speech modality.

In addition, an application interface in PACE specifies which modalities it is capable of, and what type of task quality can be expected for each of those modalities. The application also specifies the number of slots available in each modality. For example, an alert window may have space for only five alerts.

During the last phase of task selection, the System Interface Director examines the available modalities and the proposed task. The task is rejected if no slots are available in any modality that can present the task. Otherwise the SID accepts the task and designates it for the modality with the greatest utility, defined as a combination of the

457

task's preference, the application's modality quality, and the user's current available cognitive capacity with respect to the type of task. If the task passes this stage, it is handed to the Active Task Manager and will be presented to the user immediately.

3.5 Application Domains

The PACE architecture has already been successfully applied to several different domains and applications. These include a surrogate interface representing the Aegis Weapon System Identification Supervisor (IDS) console, a prototype Tactical Tomahawk Weapon Control System (TTWCS) interface for strike monitoring and retargeting, and a number of experimental research-driven applications intended for proof of concept of various mitigation strategies.

4   Future Work

4.1   Delegation

Delegation is a key capability supported by PACE which is not fully developed. Delegation within PACE can take two forms – human-peer delegation and autonomous execution. In human-peer delegation, the Delegation Manager within PACE will react to a user's inability to accomplish a task (due to cognitive state or other limitations) by identifying another user who is capable of accomplishing the task and forwarding the task to that individual. In autonomous execution, software agents can be dispatched to take care of certain tasks for a user. While this is useful in overload situations, it is not desirable for agents to always handle these tasks. Humans may be able to perform the task more effectively than any form of automation, and automation may lead to a loss of operator situational awareness, increasing the risk of suboptimal decision-making after stress levels subside. Moreover, humans may develop strategies to defeat the triggering of automation if they believe that a loss of control makes them less effective.

4.2   Additional Task Selection Strategies

4.2.1   Chunking

While chunking is generally the grouping of information based soley upon semantic content (Miller, 1956), such groupings may also be encouraged via modality differences. In a recent study, Nelson and Bolia (2003) demonstrated that spatial audio displays can enhance auditory-cue identification by approximately 50%, as well as speed reaction time. In this work they placed call signs at different auditory spatial locations (i.e., each call sign sender was assigned a specific spatial location). This is a form of spatialized auditory-verbal grouping that may be employed by PACE in future experiments.

Chunking may also be done according to criticality. For instance, in a high stress and/or high workload situation, the method of delivering alerts to the user could be modified to help focus the user's attention upon the most critical set(s) of information. That is, *knowing a user is currently overloaded* (based on cognitive workload assessment), the system could stop using ordinary methods (modality and information format) to inform the user of new alerts. Those alerts that concern things which are outside a predefined critical context can be queued, e.g., in a small alert window.[1] Meanwhile, the system will immediately inform the user of any and all new high-priority alerts related to the critical task context through alternative interruption strategies that facilitate chunking, such as sound localization or vibrations from a tactile vest.

---

1 All unacknowledged alerts should be visible from this window, so that the user can elect to attend to a 'non-critical' alert if (s)he chooses. There should be some sort of visual distinction between high and low priority alerts, as well as between alerts related to tracks in the critical region and those related to tracks outside this region. The user should be able to sort the alerts according to differing criteria.

4.3 Other Sources of Cognitive State

While the data used by PACE to determine cognitive state come solely from physiological and neurological sensors, many other possibilities are supported. One alternate mechanism would be to have the application system help determine the user's cognitive state, by monitoring the number of tasks being simultaneously performed, the amount of mouse movement the user is making, or any number of other metrics. These values can be sent to the Cognitive State Assessor and taken into account when computing cognitive state gauge values.

The PACE architecture developed under the Improving Warfighter Information Intake Under Stress program provides a powerful, flexible framework that can help support the next generation in computing: making interfaces truly personal, by supporting not only customization through user preferences, but also adaptation based upon workload, cognitive state, and the user's environment.

References

Czerwinski, M., Cutrell, E., and Horvitz, E. Instant Messaging: Effects of Relevance and Time, In S. Turner, P. Turner (Eds), *People and Computers XIV: Proceedings of HCI 2000*, Vol. 2, September 2000, British Computer Society, p. 71-76.

Johnstone, A H. (1980). Nyholm Lecture: Chemical education research: Facts, findings, and consequences. *Chemical Society Reviews*, 9(3), 363-380.

Miller, G.A. (1956). The magical number seven plus or minus two: Some limits on our capacity for processing information. *Psychological Review*, 63, 81-97.

McFarlane, D.C. (2002) "Comparison of Four Primary Methods for Coordinating the Interruption of People in Human-Computer Interaction," Human-Computer Interaction, 17(3), Laurence Erlbaum Associates, Mahwah, New Jersey.

Monk, C., Boehm-Davis, D., & Trafton, J. G. (2002). The Attenional Costs of Interrupting Task Performance at Various Stages. In the *Proceedings of the 2002 Human Factors and Ergonomics Society Annual Meeting*. Santa Monica, CA: HFES.

Nelson, T., & Bolia, R. (2003). Evaluating the effectiveness of spatial audio displays in a simulated airborne command and control task. *Proceedings of the Human Factors and Ergonomics Society 47th Annual Meeting* (pp. 202-206). Denver, CO, October 13-17.

Proffitt, D. (2003). University of Virginia, results reported during the DARPA Augmented Cognition Phase 2 Kickoff and Technical Working Meeting, Redmond, WA. July 9-11, 2003.

Sulzen, J. (2001). *Modality Based Working Memory*. School of Education, Stanford University. Retrieved, February 5 2003, from http://ldt.stanford.edu/~jsulzen/james-sulzen-portfolio/classes/PSY205/modality-project/paper/modality-expt-paper.PDF

Building Honeywell's Adaptive System for the Augmented Cognition Program

Patricia May Ververs, Stephen D. Whitlow, Michael C. Dorneich, Santosh Mathan

Honeywell Laboratories
Minneapolis, MN
trish.ververs@honeywell.com, stephen.whitlow@honeywell.com,
michael.dorneich@honeywell.com, santosh.mathan@honeywell.com

Abstract

The Honeywell AugCog team has developed a closed loop integrated prototype to address the performance advantages of a neurophysiologically driven system. The team has run a series of evaluations to validate the effectiveness of using cognitive state assessment to trigger performance mitigation strategies in an effort to improve overall system performance. This paper reviews the process for creating the closed loop integrated prototype, defining the military-relevant tasks used to test the system, and assessing the system in four separate evaluations. In parallel with the evaluations, the system was evaluated for feasibility on a mobile, dismounted soldier. The demonstration of this feasibility test is discussed.

1 Introduction

The Honeywell team is focused on reducing the limitations imposed on cognitive resources used for information processing in a highly dynamic, information rich environment. Of particular interest to Honeywell is attention, or attention as a *bottleneck* in processing. *Attention* can be broadly defined as a mechanism for allocating cognitive and perceptual resources across controlled processes (Anderson, 1995). In order to perform tasks effectively, one must have the capacity to direct attention to task relevant events and maintain a level of alertness. One must also be able to narrow (focus) or broaden (divide) one's field of attention appropriately depending on the demands of a task. Attention can be stimulated by external events (e.g., responding to an aural warning) as well as being thought of as a state where the level of awareness can also be maintained consciously, as a controlled top-down process.

The Honeywell team is working with the U.S. Army's Future Force Warrior (FFW) program to build a system with mission relevance to the dismounted soldier. The appropriate allocation of attention is key to managing multiple tasks, focusing on the most important ones, and maintaining situation awareness. Appropriate allocation of attention is important to FFW because it directly affects two cornerstone technology thrusts: netted communications and collaborative situation awareness. See Ververs, Whitlow, Dorneich, Mathan, and Sampson (2005) for a perspective on how AugCog technology could be applied to the next-generation warfighter.

This paper outlines the development of a closed loop integrated prototype system driven by neuro-physiological and physiological state information. Described herein are the military tasks used to evaluate the effectiveness of this system, both the cognitive state classification techniques and the mitigation strategies for improving the overall system performance. The findings from four evaluations, two conducted in a desktop environment, one in an immersive virtual environment, and the most recent evaluation conducted both on a desktop and in a field test. A feasibility demonstration conducted in an Army relevant mission, known as a movement to contact, is also discussed.

2 Hardware Sensor System

The sensor suite evolved in response to the effectiveness of cognitive status assessment techniques and to meet the requirements of the various evaluations. The laboratory system consisted of a BioSemi 40 channel Active Two™ to measure EEG and EOG (electro-oculogram), a Cardiax PC-ECG (electrocardiogram) device to measure heart rate and interbeat interval (IBI), an IScan pupilometry device to measure pupil dilation, and a ProComp multi-physiologic measure device used to acquire electrodermal response (EDR) and electromyogram (EMG) for muscle

460

activity. The laboratory navigational cueing task used a Tactile Situation Awareness System (TSAS) belt to provide the vibrotactile cues. See Dorneich, et al. (2004) for a full description of the system used in the laboratory. In the fourth evaluation and demonstration, the system consisted of the BioSemi EEG system, UFI's EZ-IBI unit. The hardware ensemble included a Fujitsu TabletPC and Anthrotronix' vibrotactile belt for tactile alerting cues.

3 Cognitive State Assessment

Researchers at the Institute for Human and Machine Cognition (IHMC) developed an agent-based architecture to provide architecture component developers with a simple interface to the system. See Figure 1. The modular architecture enables the components of cognitive state assessment such as hardware sensors (e.g., EEG, ECG) and software algorithms to be integrated and tested. Mission context is built from event recordings within the virtual environments and most recently from field sensor devices. Information regarding context is made available to the decision making elements (i.e., Augmentation Manager) through the agent architecture.

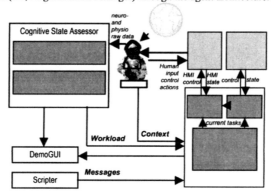

Figure 1. Current System Architecture

Several Honeywell AugCog team members have focused on the algorithms to determine cognitive state. One method used to access an individual's overall cognitive state has been to develop "gauges" to measure different aspects such as stress level, amount of executive function processing, degree of attentional resources engaged, etc. The gauges used in the first three of the evaluations are defined below.

3.1 Engagement Index

Engagement Index is an indicator of alertness. It uses a ratio of EEG power bands, beta/(alpha + theta). Freeman, Mikulka, Prinzel, and Scerbo (1999) have shown the Engagement Index to be a valid measure of an operator's task engagement in vigilance type tasks.

3.2 Arousal Meter

Arousal Meter (AM) is a real-time cardiac-based measure derived from interbeat intervals (IBIs) to status the activity of the autonomic nervous system (ANS). The heart responds to changes in the parasympathetic nervous system (PNS), which is responsible for returning an individual to a resting state. The AM, as developed by Clemson University researchers, uses the PNS subcomponent of ANS to status an individual's level of arousal (Hoover & Muth, 2004).

3.3 Stress Gauge

The Stress Gauge was developed by IHMC researchers and is a composite gauge that has used various inputs including heart rate, pupil diameter and microvolt cardiac QRS waveform root mean square (RMS) amplitude (HFQRS) to determine stress during individual trials (Raj, et al., 2004).

3.4 P300 Gauge

The P300 gauge measures the strength of EEG evoked responses following an alert tone. Once calibrated the detector is optimized to differentiate the EEG response evoked by the alert tones from that activity evoked by a frequent auditory stimulus of no significance. The idea behind the P300 is to equate the strength of the evoked potentials with the availability of attentional resources to process a message following an auditory tone. This gauge was developed by researchers at Columbia University and City College of New York (Parra, Alvino, Tang, Pearlmutter, Yeung, Osman, & Sajda, 2002; Sajda, Gerson, & Parra, 2003)

3.5 Executive Load Index

The Executive Load Index (XLI) Gauge is a measure of executive load or comprehension, where positive values indicate increasing load and negative values indicate decreasing load. It operates by measuring power in the EEG at frontal (FCZ) and central midline (CPZ) sites and was developed by Human Bionics (DuRousseau, 2004, 2004b).

For the fourth evaluation, the Honeywell team used a real-time neural net classification approach to status the cognitive state of the participants. This approach used EEG data as the sole classification input for cognitive state. The classifier used seven EEG channels (CZ, P3, P4, PZ, O2, PO4, F7) and five frequency bands: 4-8Hz (theta), 8-12Hz (alpha), 12-16Hz (low beta), 16-30Hz (high beta), and 30-44Hz (gamma) to form the features for the cognitive state assessment. The neural network classification used three different techniques to estimate cognitive state: Gaussian-Mixture-Model (GMM), K-Nearest-Neighbor (KNN), and a nonparametric Kernel Density Estimate (KDE) (See Erdogmus, Adami, Pavel, Lan, Mathan, Whitlow, Dorneich, 2005 for more information). Cognitive state was classified into high and low cognitive workload states and was based on the agreement from two of the three models. In the event that there was no majority agreement the KDE estimation was used.

4 Task Definition

All four evaluations focused on Army relevant tasks that included a combination of communications, hostile engagements, and navigation tasks. The importance of these tasks was determined from interviews of subject matter experts situated in the FFW program. Of primary importance is the fact that the tasks are Army relevant to immediately enable the transfer of performance benefits to well-known challenges. The tasks also inherently stress the real challenges that must be overcome in order to deploy this technology. Evaluating the effectiveness of the system requires that the warfighter is mobile and frequently communicating to maintain situation awareness from the information being sent via the netted communications.

5 Building the Mitigation Strategies

A four-stage process was used for building effective mitigation strategies. First, the cognitive state assessment technique(s) was defined and built. Two "gauge" approaches to classification of cognitive state either applied preset calculations on the parameters of the signals (e.g., Engagement Index) or used a more dynamic, trained neural network classification. Next, the cognitive state assessment techniques were validated against the tasks to be used in the evaluation. Third, the mitigation strategies for alleviating the cognitive bottlenecks that negatively affect performance were created. The final stage of building the mitigation strategy was to create a ruleset for triggering the mitigation strategies based on the gauge responses. This ruleset included the automation etiquette for turning it on and maybe more importantly turning it off. The Honeywell team developed four mitigation strategies that are described below.

5.1 Communications Scheduler

The Communications Scheduler planned and presented messages to the soldier based on the cognitive state profile as defined by the gauges, the message characteristics (principally message priority), and the current context (tasks). Messaging techniques included drawing attention to higher priority items with the additional alerting tones or text messages or deferring lower priority messages until a later time or to a TabletPC device for later review.

5.2 Tactile Navigation Cueing System

Tactile Navigation Cueing System guided the soldier via tactile cues in the intended correct direction of a target location. This transformed a normally cognitively intense navigation task in an unfamiliar environment into a task that is reactionary in nature to the stimuli.

5.3 Task Offloading Negotiation Agent

Task Offloading Negotiation Agent (e.g., Medevac Agent) reduced lengthy verbal communications exchanges by automatically preloading known information from netted communications into the forms.

5.4 Mixed Initiative Target Identification Agent

The Mixed Initiative Target Identification Agent provided assistance in locating potential targets in a visual search space.

6 System Validation

Four different evaluations were conducted to determine the effectiveness of the Honeywell AugCog system. The first was conducted in a desktop environment where the participants navigated through a virtual Military Operations in Urban Terrain (MOUT) environment. See Figure 2. Additional tasks included identifying foes and shooting them and monitoring and responding to communications. Gauges measured the cognitive responses to task load and availability of attentional resources to comprehend a message, triggering the communications scheduler appropriately (e.g., repeat message, defer message). A noteworthy finding came from the incidences in which the participant failed to acknowledge the message, and the gauges indicated that the lack of attentional resources to comprehend the messages. The scheduler correctly repeated the message 72% of the time, with correctness being verified by lack of an acknowledgement. The availability of the scheduler resulted in significant improvements in situation awareness of message content in the high workload scenarios as compared the same condition when no mitigation strategy was used. In addition, approximately 24% of the time the scheduler repeated a message even if the participant acknowledged its receipt. Though this could be the result of an improper state classification, it also suggests the possibility that gauges were indicating the failure to truly comprehend the message even if the participant said, "Acknowledge." That is, saying "acknowledge" to a message as an automatic response even though it wasn't fully processed and understood would rightfully result in the repetition of the message. See Dorneich et al. (2004) for more details on the experimental conditions and findings.

The findings from the first evaluation indicated that the greatest benefit of the mitigation occurred at the extreme ends of the task load space. Therefore, this was capitalized on for the second evaluation in the desktop environment where long duration and clear task load differences were built into the scenarios. The tasks were similar to the previous evaluation. Several more mitigation strategies were developed. In addition to the communications scheduler, other mitigations tested were the tactile navigation cueing system, the negotiation agent for offloading components of a highly proceduralized task (i.e., Calling for a Medevac), and the mixed initiative target identification agent for enhancing the search in a vigilance task. The mitigations were triggered when the gauges indicated a suboptimal cognitive state. Findings revealed significant improvements in performance in the relevant applicable task with the availability of the automation provided by the mitigation agents. For instance, the navigation cueing reduced the number of enemy encounters by over 300% thereby increasing survivability. Message comprehension and overall situation awareness was improved by over 100% in the conditions when the communications scheduler was available. See Dorneich, et al. (2005) for more details.

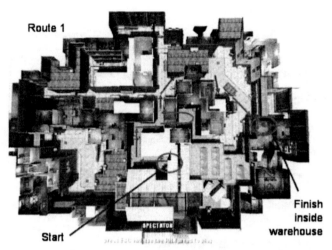

Figure 2. Top-down view of MOUT environment

The third evaluation provided the opportunity to understand the cognitive state gauge assessments in a mobile environment. The scenarios took place in a motion capture laboratory at Carnegie Mellon University. The participants stood in an 8 x 12 foot space with a motion-tracked M16 rifle prop and were tasked with visually scanning their environment for friends and foes, and shooting the foes when they appeared in the windows of buildings. See Figure 3. Participants were responsible for keeping track of the number of friends, foes and ammunition expended and to report out at various times during the scenario. In addition participants were responsible for listening for messages from team members verbally communicating their own enemy and friendly encounters among distractor messages and keeping a running count of the totals. Gauges were used to detect workload periods when the participant was maintaining the counts in working memory. When a high workload state was detected during these periods, the gauge-driven message scheduler delayed incoming messages reducing the task load on working memory. This condition was compared to a random scheduler of messages. Findings indicated over a 150% improvement on the working memory tasks as well as a reduction in the subjective mental workload measures in the mitigated condition over a random message scheduler. See Whitlow et al. (2005) for more details.

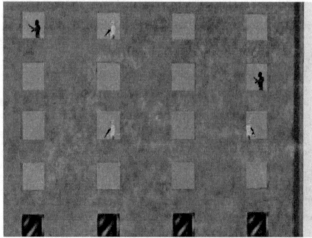

Figure 3. Face of building with 2 enemy targets (green) and 3 friendlies (tan)

In the fourth evaluation, we employed a neural network classification approach to cognitive state assessment and tested the system in a combination desktop/field study. This evaluation tested the communications scheduler and the mixed initiative target identification as mitigation strategies. For the first half of the evaluation, the six participants were seated conducting three tasks on a TabletPC. The tasks consisted of 1) monitoring radio communication reports on the number of enemies, friendlies, and civilians encountered and maintaining the running total, 2) monitoring radio communications on the movement of three squads and directing their movements in a bounded overwatch, and 3) monitoring visual presentations of static images for signs of enemy targets and report the position of the target. There were two task load conditions, a low task load in which the participant received a limited number of communications and visual images, and a high task load condition where the rate of communications increased. The actual rate was predetermined during the training session by taxing the participants to the point where performance began to drop off. Actual rates varied from 6/min to 34/min. In the second half of the evaluation, the participants performed the first two communication tasks while in an upright and mobile position while scanning the woods in the outdoors for targets, that is, concealed "snipers."

Figure 4. Visual search task of MOUT on TabletPC. Highlighted box shows target id agent

The evaluation consisted of two training sessions. The first allowed the participants to gain familiarity with the tasks and system and the second determined the highest event rate to present the verbal messages before performance began to breakdown. This high event rate was developed to cognitively challenge the participant and was later used as the event rate for the high task load condition. Five training blocks were used to calibrate when the high workload conditions were met and to train the neural net classifier. The subsequent experimental and mobility trial scripts were generated using the lowest event rate and the highest event rate just before performance breakdown to generate the low and high task load scenarios. The participants completed five experimental sessions: three seated (low task load/unmitigated, high task load/unmitigated, and high task load/mitigated) and two mobile outdoors (low task load/unmitigated and high task load/unmitigated).

Findings from this evaluation indicated that performance on the radio count accuracy and mission monitoring queries during the *mitigated* high task load condition was equivalent to the performance in the unmitigated low task load condition. The communications scheduler (mitigation) offloaded the radio count messages to the visual modality by sending them to the TabletPC thereby allowing the participant to monitor messages during a later lower workload period. Overall there was a 94% improvement in the mission monitoring task and 36% improvement in the radio count recall task when the communications scheduler was available in the high workload trial as compared to the no mitigation high task load condition. Subjective workload levels in the mitigated conditions mirrored those of the low task load conditions and were statistically lower than in the unmitigated condition.

For the visual search task on the TabletPC (see Figure 4), participants had an average of 87% correct identification of targets in the low task load condition. Their performance dropped to 46% in the high workload but rebounded to 61% in the mitigated condition with the assistance of the target identification agent to identify potential targets. The availability of the agent resulted in a 40% average improvement in performance on the visual search task.

The mobility trials were conducted as a means to assess the feasibility to collect and clean the EEG data in real-time in the field. Hence, not all experimental manipulations needed to be evaluated. Findings indicated that performance in both the low and high task load conditions were lower in the mobile trials indicating that the addition of mobility in an actual environment made performing the mission monitoring and radio count recall tasks more cognitively challenging. Performance declined more in the high task load condition (composite metric: 58% accuracy than the low task load condition (composite metric: 66% accuracy) indicating an effective task load manipulation. An additional test of the classification method was also conducted in the field immediately prior to this evaluation. Findings indicate the ability to correctly classify cognitive state data in a mobile environment. Participants' cognitive states were classified during the execution of three tasks: navigation, visual search, and a combined navigation and communication task. By using the first third of the data to train the classifier and the second third for testing, the neural net classification accuracy was consistently above 90%. See Table 1. For more information on the real-time signal processing and classification methods see Mathan, et al. (2005).

Table 1. Probability of correctly classifying tasks. The diagonal represents the correct classification accuracy.

		Classified as...		
		Navigate	Search	Nav & Comm
True class	Navigate	0.959	0.019	0.000
	Search	0.003	0.981	0.047
	Nav & Comm	0.038	0.000	0.953

The Honeywell AugCog team also demonstrated the system feasibility in a full mission scenario lead by an "augcogified" platoon leader. This mission included a 2-person team traversing a field undercover while communicating via radios with squad leaders during a movement to objective exercise. Cognitive state was assessed during the mission, which included ambushes from opposing forces armed with laser guns. The team also had laser guns to engage the enemy. The simulation verified the feasibility of collecting EEG and IBI data in the field as well as the AugCog system being integrated with a battle dress uniform. The fully mobile system was contained in a backpack worn by the platoon leader and wirelessly transmitted data about his real-time cognitive state to the base station.

7 Conclusions and Summary

The initial findings from the evaluations described above indicate that neurophysiological and physiological data can used be to assess the cognitive state of an individual and drive mitigations to enhance performance. Studies conducted in desktop environments as well as in the field enabled the sensor suite to determine when workload was high and the performance could be enhanced by scheduling the presentation of communication messages, redirecting information to the underutilized tactile modality, offloading procedures and tasks to the automation, and utilize mixed-initiative automation assistance.

The first evaluation validated the ability to create a closed loop system and trigger a mitigation strategy solely based on the input from physiological and neurophysiological sensors. The system assessed the participant's level of comprehension of verbal messages. Findings indicated a 72% effectiveness in detecting when the participant failed to process the incoming message, resulting in a repeat of the unattended messagse. This was found even in the cases where the participant's provided an automatic acknowledgement of the truly unattended message.

In the second evaluation, during high workload periods the Communications Scheduler changed how messages were delivered, by escalating and highlighting high priority messages, and deferring low priority messages to the tablet PC. By doing this at the appropriate time as determined by gauge outputs, this resulted in 100% improvement in message comprehension and 125% improvement in overall situation awareness. These findings were enabled without a subsequent increase in subjective workload to process the communications in multiple modalities as delivered from different sources.

The third evaluation further investigated potential performance improvements of a communications scheduler due to the U.S. Army's concern over the increased information processing requirements resulting from netted communications delivered to dismounted soldiers. In a motion capture laboratory environment, performance on a working memory task was enhanced by over 150% when the scheduler was used to moderate the peaks in workload periods by delaying the arrival of new tasks if a high workload state was detected.

The fourth evaluation set out to evaluate a different approach to cognitive state classification as well as expanding the types of mitigation strategies employed. The primary goal was to move to investigate the system effectiveness with more realistic Army tasks completed in both a seated command center environment as well as a mobile environment outside the laboratory. The evaluation determined that the system could detect a high workload state and trigger the mitigation strategies, including the communication scheduler and target identification agent. These mitigations were able to offload the cognitive demands of the task to show an improvement in performance. The findings indicated a 94% improvement on the mission monitoring task, 36% improvement on the radio count recall task, and 40% improvement on the visual search task. The findings from this evaluation and demonstration determined that the cognitive state assessment from EEG could be collected in real-time on a truly mobile individual outside of the laboratory.

The mitigation strategies and tools developed and implemented for these evaluations are promising solutions to known problems as identified by the Future Force Warrior program and we will continue to explore their utility as we move to the field. However, challenges remain to fully implement a real-time neurophysiologically driven cognitive state assessment system. Real-time cognitive state classifications are difficult due to the complex nature of human physiology. Using the gauge approach limited the ruleset to a static threshold based solution that was applied universally to all participants. Using the neural net approach required specific training for the detection of the desired state. Both approaches are limited by the nonstationary nature of the EEG signals, thereby requiring the retraining of the neural net or recalibration of the ruleset thresholds. We continue to explore developments in sensor technology, such as functional imaging that uses near infrared light to detect changes in blood oxygenation during brain activity, as well as other approaches. Further challenges include the real-time detection of artifacts in data collected on a mobile dismounted soldier. Once reliable, artifact-free data can be collected and characterized, additional work will need to be done to improve the processing efficiency of the algorithms and to miniaturize the solutions in order to deploy these systems with the Future Force Warrior.

Acknowledgments

This research was supported by a contract with DARPA and funded through the U.S. Army Natick Soldier Center, under Contract No. DAAD-16-03-C-0054, for which CDR (Sel.) Dylan Schmorrow serves as the Program Manager of the DARPA Augmented Cognition program and Mr. Henry Girolamo is the DARPA Agent. Any opinions, findings, conclusions or recommendations expressed herein are those of the authors and do not necessarily reflect the views of DARPA or the U.S. Army.

References

Anderson, J.R. (1995). *Cognitive Psychology and its Implications* (2nd Ed.). New York: Freeman.

Dorneich, M. C., Mathan, S., Creaser, J., Whitlow, S., & Ververs, P. M. (2005). Enabling Improved Performance through a Closed-Loop Adaptive System Driven by Real-Time Assessment of Cognitive State. To appear in the *Proceedings of the 1st International Conference on Augmented Cognition.* Mahwah, NJ: Lawrence Erlbaum Associates.

Dorneich, M., Whitlow, S., Ververs, P. M., Mathan, S., Raj, A., Muth, E., Hoover, A. DuRousseau, D., Parra, L., & Sajda, P. (2004). *DARPA improving warfighter information intake under stress – Augmented Cognition: Concept validation experiment analysis report for the Honeywell Team.* DARPA technical report (contract no. DAAD16-03-C-0054).

DuRousseau, D.R., (2004). Spatial-Frequency Patterns of Cognition, *The AUGCOG Quarterly*, 1(3), p. 10.

DuRousseau, D.R., (2004b) *Multimodal Cognitive Assessment System*. Final Technical Report, DARPA, DAAH01-03-C-R232.

Erdogmus, D., Adami, A., Pavel, M., Lan, T., Mathan, S., Whitlow, S., & Dorneich, M. (2005). Cognitive state estimation based on EEG for Augmented Cognition. To appear in the Proceedings of the *2nd International IEEE EMBS Conference on Neural Engineering*. Piscataway, NJ: IEEE.

Freeman, F.G., Mikulka, P.J., Prinzel, L.J., & Scerbo, M.W. (1999). Evaluation of an adaptive automation system using three EEG indices with a visual tracking system. *Biological Psychology, 50*, 61-76.

Hoover, A. & Muth, E. (2004). A Real-Time Index of Vagal Activity. *International Journal of Human-Computer Interaction, 17*(2), 197-210.

Mathan, S., Mazaeva, N., Whitlow, S., Adami, A., Erdogmus, Lan, T., & Pavel, M. (2005). Sensor-based cognitive state assessment in a mobile environment. To appear in the *Proceedings of the 1ˢᵗ International Conference on Augmented Cognition*. Mahwah, NJ: Lawrence Erlbaum Associates.

Parra, L., Alvino, C., Tang, A., Pearlmutter, B., Yeung, N., Osman, A., & Sajda, P. (2002). Linear Spatial Integration for Single-Trial Detection in Encephalography, *NeuroImage, 7*(1).

Raj A. K., Bradshaw, J., Carff, R. W., Johnson, M., and Kulkarni, S. (2004, January). An agent based approach for Aug Cog integration and interaction. In *Proceedings of the Augmented Cognition- Improving Warfighter Information Intake Under Stress, Scientific Investigators Meeting (January 6-8, 2004)*, Arlington, VA: DARPA.

Sajda, P. Gerson, A. & Parra, L. (2003). Spatial Signatures of Visual Object Recognition Events Learned from Single-trial Analysis of EEG, Engineering in Medicine and Biology Society, 2003. *Proceedings of the 25th Annual International Conference of the IEEE*, 3, pp. 2087 – 2090.

Ververs, P. M., Whitlow, S. D., Dorneich, M. C., Mathan, S., & Sampson, J. B. (2005). Augcogifying the Army's Future Warfighter. To appear in the *Proceedings of the 1ˢᵗ International Conference on Augmented Cognition*. Mahwah, NJ: Lawrence Erlbaum Associates.

Whitlow, S., Ververs, P. M., Buchwald, B., & Pratt, J. (2005). Scheduling communications with an adaptive system driven by real-time assessment of cognitive state. To appear in the *Proceedings of the 1ˢᵗ International Conference on Augmented Cognition*. Mahwah, NJ: Lawrence Erlbaum Associates.

The Boeing Team Fundamentals of Augmented Cognition

Richard A. Barker

The Boeing Company
P.O. Box 3707 MC 45-85
Seattle, WA 98124
richard.a.barker@boeing.com

Richard E. Edwards

The Boeing Company
P.O. Box 3707 MC 45-85
Seattle, WA 98124
richard.e.edwards@boeing.com

Abstract

This report discusses how the Boeing AugCog Team approached Augmented Cognition (AugCog) in a closed-loop simulation of an unmanned air vehicle (UAV) system. The Boeing AugCog system was comprised of the UAV simulation and interface, a sensor suite of multiple physiological sensors of the operator's cognitive state, cognitive state assessor (using an artificial neural net and Statistical Process Control software), data collection software called CWAD™, and an augmentation manager which dynamically alters the user interface in response to changes in the UAV operator's cognitive state.

Phase 2 consisted of alternating periods of technology development and technology assessment. Development of specific system components was conducted remotely at each team member site. This presented a number of challenging problems for the Boeing Team. For example, Boeing was responsible for development of the augmentation manager and cognitive bottleneck mitigations, but did not have the sensors and cognitive state classifiers to test the effectiveness of the mitigation concepts. To work around this problem, we tested the mitigation software by assuming the presence of a perfect sensor that always detected cognitive overload at an appropriate point in the mission scenario. Pilot studies were then conducted to assess operator performance under mitigated and non-mitigated conditions. The mitigation concepts were iteratively refined until the desired level of performance enhancement was achieved.

Concept validation experiments (CVE) were conducted bi-monthly to test the effectiveness of the mitigation concepts for each cognitive bottleneck in a closed-loop setting. All CVEs were conducted at the Integrated Technology Development Laboratory in Seattle, Washington. Separate experiments were constructed and conducted for each of five cognitive information processing bottlenecks: 1. Working Memory, 2. Executive Function (our assigned primary bottleneck), 3. Sensory Input, 4. Attention, and 5. Response Generation. During experimental trials, the Boeing AugCog team collected and analyzed data for aided (closed-loop) and un-aided conditions with four trained UAV operators. The results of experimental trials demonstrate that aiding from closed-loop augmented cognition improves UAV operator performance significantly.

Keywords: augmented cognition; cognitive bottleneck; cognitive state

1   Workstation Displays

Software was developed to support operator actions required for weapon-target pairing, selecting weapon aim points, and dealing with vehicle health problems. These actions were performed using the Tactical Situation Display, a Targeting Window, and a Vehicle Health Window. The Tactical Situation Display was always presented on the left workstation monitor. The Targeting Window was always presented on the right workstation monitor. The Vehicle Health Window could be presented on either the left or the right workstation monitor. (Figure 1)

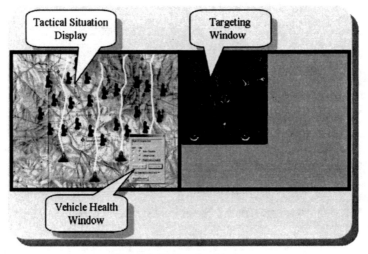

Figure 1: Workstation Displays

The Tactical Situation Display presented a north-up map of the gaming area with icon overlays to represent the relative location of air vehicles and targets as well as changes in the state of air vehicles and targets. The Tactical Situation Display was used to capture Synthetic Aperture Radar (SAR) images of targets and to establish weapon-target pairings. The Targeting window is used to display SAR imagery, and for placing a Designated Mean Point of Impact (DMPI). Once an air vehicle has completed a SAR image capture, the image is automatically displayed in the Targeting window whenever the operator left-clicks on the associated target. When a vehicle has one or more vehicle health alerts activated, its outline turns red on the Tactical Situation Display. The alert is acknowledged by double-clicking on the vehicle icon to display a dialog box for the queued alert.

2   Sensor Suite

Multiple physiological sensors were used to determine the operator functional state. Single sensors provide limited information concerning the overall cognitive involvement in the multifaceted requirements of performing the complex tasks presented by military systems. These demands tax many aspects of the cognitive abilities of operators, often simultaneously. In order to gain the best perspective concerning these demands, sensors that sample activity in several physiological systems are necessary. Combining the input from a suite of sensors should provide a more complete picture of the cognitive demands of the task. The initial sensor suite was composed of the following sensors: pupil size, EEG, ECG, EOG, EMG, and functional near infrared (fNIR).

Initial physiological sensors for electroencephalogram (EEG), electrocardiogram (ECG), and electrooculogram (EOG) were evaluated to reveal possible overlap of information with other candidate sensors and also to eliminate any gaps in the information required to provide the high levels of accuracy of operator state estimation. Prior to Phase 2 the AFRL had developed NuWAM (a neural net cognitive state assessor), and the AFRL made continuous quality improvements to NuWAM during Phase 2. They re-evaluated the EEG sensor configuration to determine the optimal number of sensors and the optimal sensor sites for the UAV tasks. Saliency analysis was conducted on the 20 channels of EEG to determine which electrode sites provided the best information.

Quantum Applied Science and Research, Inc. (QUASAR) continued their development work from Phase 1 on their advanced dry electrode sensors. Their ECG sensor (integrated in a cloth chest strap worn over clothing) and EOG sensor were used in Phase 2.

EyeTracking, Inc. (ETI) provided the sensors for binocular eye tracking and measurement of pupil dilation. ETI's sensor technology applies the Index of Cognitive Activity (ICA) to determine the level of cognitive effort, the amount of sustained attention, and the affective response of the workstation operator. Abrupt changes in pupil diameter indicate current levels of mental effort put forth by the user. At the same time, the point of gaze metric highlights specific elements causing the difficulty.

NovaSol (now Archinoetics) provided the functional Near InfraRed (fNIR) technology, which was used to provide a diagnostic measure for verbal and spatial working memory. The fNIR technology was used to non-invasively gather neural imaging data in the form of oxygenation levels and blood volume in brain tissue. Unique cognitive tasks were identified by associating neuronal activity with the distinct changes in blood oxygenation data detected via the NovaSol fNIR sensors. The fNIR imaging technology allows the system to detect minute hemodynamic response changes due to cognitive activity through a set of fNIR sensors placed on the scalp of the subject.

3 Cognitive State Assessor
3.1 Artificial Neural Net

Our baseline approach to the development of the Cognitive State Assessor was to use an artificial neural network to classify the functional state of the operator. The artificial neural network is a nonlinear classifier that is particularly appropriate with human physiological data that is known to be nonlinear. Thus, our initial approach to the development of the Cognitive State Assessor incorporated the multiple, non-redundant physiological sensors and the neural network to fuse that sensor information into a set of gauges that indicated operator cognitive workload.

3.2 Statistical Process Control

Statistical Process Control (SPC) has been widely used in manufacturing and quality assurance because of its capability to distinguish between normal variation and "special cause" effects. Control Charting, as it is commonly called, has evolved a wide variety of rule sets for identifying special causes in practice, and some of them may prove effective in detecting important cognitive state changes in data streams from neuro-physiological sensors. We wanted to examine the practical use of SPC for recognizing cognitive events, so we selected two neuro-physiological signal of significant interest to our team: 1) fNIR sensors, and 2) pupilometry.

4 Mission Scenarios

All mission scenarios focused on operator tasking within the ingress and attack phases for a suppression of enemy air defenses mission. The scenarios assumed day one of a major theater war with a known threat database and an established single integrated prioritized target list. The operator's task was to control one, two, or three strike packages, where a strike package consisted of four UAV's. The air vehicles were autonomous with regard to flight path control and to management of on-board systems. Therefore, the operators did not remotely pilot the air vehicles with inceptors, but rather acted as a mission manager by sending high-level commands to the air vehicles via datalink.

The scenarios included a variety of targets that required specific types of weapons to ensure their destruction. Therefore, the weapon load varied among the air vehicles within each strike package. When performing weapon-target pairings, the operator had to make sure that the air vehicle selected for the attack carried the correct type of weapon to destroy the designated target and that the proper quantity of weapons were available.

5 Mitigation Software

Mitigation software was developed to assist the operator when various types of cognitive bottlenecks were encountered. Table 1 shows the relationship between specific mitigation concepts and the cognitive bottlenecks that they supported in the concept validation experiments.

Table 1: Mitigation Software for Each Cognitive Bottleneck

	Cognitive Bottleneck				
Mitigation	Attention	Working Memory	Executive Function	Sensory Input	Response Generation
Map Declutter	X		X	X	
Earcons	X		X	X	
Time-Critical Target Voice	X			X	
Bookmark	X		X	X	
Unified Bookmark	X		X	X	
Alert Manager		X		X	
UAV Menu Tokens				X	X
Vehicle Health Tones				X	

5.1 Map Declutter

The declutter mitigation places a "fog layer" over the map, thereby partially obscuring the map and the vehicle flight path lines. This makes the details of interest (e.g., air vehicles and targets) much easier to see. The map declutter mitigation is shown in Figure 2.

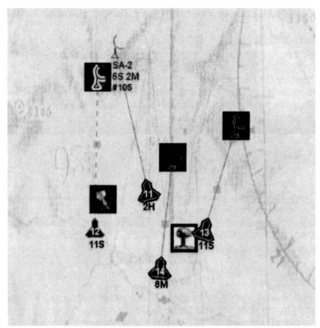

Figure 2: Map Declutter

5.2 Earcons

Earcons are brief distinctive sounds that are typically associated with specific mission events. For example, a camera shutter "click" sound is presented to signal initiation of a Capture SAR command. A bugle call sound is presented to signal initiation of a Direct Attack command. A combination of the camera shutter and the bugle call is used to signal initiation of a SAR/Attack command. These sounds provide feedback to the operator that the intended attack script has been selected from the targeting drop down menu. Otherwise, with just the pull down menu, many operators make wrong menu selections by inadvertently moving the mouse.

5.3 Time-Critical Target Voice

The time-critical voice mitigation is a computer-generated female voice that speaks the words "Tiger leader, we've detected a time-critical target." whenever a time-critical target is activated. This mitigation provides an auditory warning that directs the operator's attention to the Tactical Situation Display to search for the time-critical target.

5.4 Bookmark

The bookmark mitigation provides the operator with some assistance in determining the most expedient "next" action on the Tactical Situation Display. Each strike package has an icon traveling with it at the center of mass of the four air vehicles. The circular center section of the icon is color-coded red, yellow, or green depending on the relative urgency of the most urgent vehicle for each of the strike packages. Four sections of a diamond surrounding the circle are color-coded red, yellow, or green depending on the urgency of the next action for the aircraft to which it points. This allows the user to pick the most urgent strike package by its red circle, and the most urgent vehicle in the package by which pointer around the circle is red. An example of the Bookmark mitigation is shown in Figure 3.

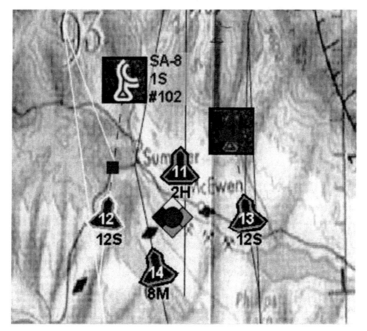

Figure 3: Bookmark

5.5 Unified Bookmark

The unified bookmark expands on the basic bookmark by including in its logic the additional considerations of Vehicle Health Task alerts. It places a red square on the next recommended action. An example of the unified bookmark is shown in Figure 4.

473

Figure 4: Unified Bookmark

5.6   Alert Manager

The Alert Manager is a dialog box that collects all of the currently active vehicle alerts into a single list, sorted by priority. Priority is determined on the basis of time criticality, with emergencies (i.e., engine fires) always on top of the list. Icons appear to the left of each list entry to denote the alert status: A blue hour glass means that the alert is in work (temporarily unable to accept operator input), and a red exclamation point indicates that operator input is required. When the operator selects an item in the list, the associated dialog is opened adjacent to the Alert Manager dialog. An example of the Alert Manager is show in Figure 5.

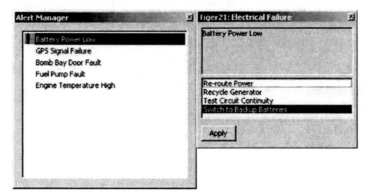

Figure 5: Alert Manager

5.7 UAV Menu Tokens

The UAV Menu Tokens mitigation placed two "tokens" on each air vehicle icon. The tokens were small squares. One token was red and one was green. The red token corresponded to the Direct Attack option on the drop down menu and the green token corresponded to the SAR/Attack option. With this mitigation, a Direct Attack could be initiated by clicking on the red token and a SAR/Attack could be initiated by clicking on the green token. After clicking a token, a target still needed to be selected by clicking on it. Disadvantages to this mitigation were that the tokens could occlude information on the Tactical Situation Display; in particular, when the UAV's got close to one another then the tokens could overlap. An example of the UAV Menu Token mitigation is shown in Figure 6.

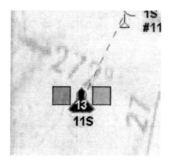

Figure 6: UAV Tokens

5.8 Vehicle Health Tones

The Vehicle Health Tones mitigation presented a unique computer-generated tone for each strike package whenever a vehicle health problem was detected. For example, if an electrical failure occurred on Tiger-21, the tone designated for strike package number 2 would sound. This mitigation provides an auditory warning that directed the operator's attention from the current task to a specific strike package on the Tactical Situation Display or to the Alert Manager. It was possible for a trained operator to remember a sequence of tones and therefore keep some idea of the order in which a series of health problems occurred.

6 Experimental Design

The experimental design was a two-group within-subjects design. There was one independent variable, Mitigation, which had two levels: Available and Not Available. The presentation order for the levels of Mitigation was counterbalanced over the four subjects.

Table 2: Experimental Conditions and Metrics

Bottleneck	Key Technical Idea	Metric	Cognitive Gauge Trigger
Working Memory	Maximize working memory processes via an autonomous intelligent interruption and negotiation strategy	Number of VHT messages successfully resolved during a high workload bottleneck event	Statistical Process Control model using fNIR and pupilometry
Executive Function	Maximize executive functioning and facilitate memory enhancement via an automatic cued retrieval strategy	Number of targets successfully attacked (i.e., bombs on target)	NuWAM Tactical Situation Display Gauge
Sensory Input	Exploit multiple sensory channels via an autonomous information delivery strategy to multiple modalities	Number of times the operator correctly handled coincident time-critical targets and engine fires	Statistical Process Control model using fNIR and pupilometry
Attention	Enhance attention management via a directed attention and autonomous task delegation strategy	Number of time critical targets successfully attacked when the operator had 5 seconds to detect, identify, and initiate an attack against the time critical target	NuWAM Tactical Situation Display Gauge

Response Generation	Maximize response generation and facilitate speed and accuracy of responses via a simpler task-response mapping	Accuracy with which the operator can select the correct attack type under conditions of high time pressure	Statistical Process Control model using fNIR and pupilometry

7 Results

The results are shown in Table 3. Table 3 shows the raw data for each metric for each bottleneck described in Table 2.

Table 3: Raw Results

Metric Results	Working Memory	Executive Function	Sensory Input	Attention	Response Generation
Augmented Condition	34	70	17	15	19
Non-Augmented Condition	5	29	6	2	1
Percent Improvement	680%	241%	283%	750%	1900%

Figure 7 compares the results for all 5 bottlenecks to the program goals for each bottleneck. The goals were exceeded for each cognitive bottleneck.

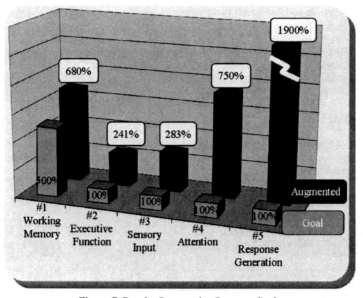

Figure 7: Results Compared to Program Goals

DaimlerChrysler's Closed Loop Integrated Prototype: Current Status and Outlook

Wilhelm E. Kincses

DaimlerChrysler AG, Research & Technology
096/T728 – REI/AI, 70546 Stuttgart, Germany
wilhelm.kincses@daimlerchrysler.com

Abstract

Driving involves a broad variety of human cognitive resources. Unfortunately, these resources are limited, and parallel processing of different pieces of information can easily lead to cognitive overload, with potentially fatal consequences. Therefore, it is our final goal to develop a driver-adaptive vehicle information system capable of defusing critical situations. The current talk presents DaimlerChrysler's closed loop integrated prototype (CLIP) which was developed within the scope of our participation in Phase 2 of the DARPA/IPTO IWIIUS (AugCog) project. This system is able to detect the respective cognitive bottlenecks and helps to avoid any loss of information due to a more effective and efficient use of the driver's limited cognitive resources.

A prerequisite for driver-adaptable vehicle control is real-time assessment of the driver's current cognitive state. Our approach consists in recognizing driver's information processing related mainly to two cognitive bottlenecks: sensory input and working memory. For this purpose, we are using neurophysiological as well as contextual data, since both of them provide measures which constitute correlates of cognitive states.

While current technology offers a wide variety of environmental factors that can be monitored, it is still challenging to obtain reliable neurophysiological measurements in operational environments. Since cognitive phenomena like workload originate within the brain, it is most desirable to record physiological parameters directly from there (instead of using indirect evidence like, e.g., heart rate or electrodermal activity). Currently, scalp electroencephalography (EEG) is the most robust and best documented brain imaging method which, in addition, directly registers electrophysiological activity (as opposed to secondary phenomena like blood-vessel dynamics), features very high temporal and sufficient spatial resolution and – most importantly – is easily realizable with portable equipment.

In preliminary studies we have demonstrated the viability of our concept. Three sets of field experiments have been performed in which the drivers were engaged in secondary tasks inducing certain cognitive states while scalp EEG, EOG and EMG signals were recorded. In addition vehicle data were recorded and video cameras captured the current traffic and environmental situations. After each experimental session the recorded data was used to further develop and refine the cognitive state models and real-time detection algorithms. As soon as the classifiers and algorithms had reached stage of maturity, another experimental block – the actual concept validation experiment – was designed in which we demonstrated the robust and rela-time detection of the driver's cognitive state. At the same time we also incorporated our AugCog mitigation strategies in order to improve driving and mission related performance due to an effective and efficient use of limited cognitive resources.

The sensor suit currently in use comprises devices for EEG recording, seat posture recognition, audio monitoring of vehicle interior, and logging of driver-behavioral variables. Together with vehicle data relating to assessment of driving and traffic situations and context, the output of this sensor suite is fed into the cognitive-state classifiers. These in turn interpret the current driver condition in terms of sensory and cognitive processing on the basis of the previously developed cognitive models.

We also implement a trigger logic that uses the classifiers' interpretations for flexible cognitive-task manipulation in response to the estimated driver's workload. This step closes the measurement–interpretation–manipulation loop, thereby completing the framework of our Closed-Loop Integrated Prototype.

Secondary mission tasks addressing the above-mentioned cognitive bottlenecks were designed. Based on these tasks, dedicated mitigation strategies were impleneted for each bottleneck.

The validation of our AugCog system shows that with activated system drivers were able to use their limited cognitive resources more effectively, resulting in faster and more accurate mission performance. Our results show that it is not only possible to reliably discriminate different cognitive states on the basis of neuronal signals but also that such discrimination can very well be used to control mitigation actions in human-machine interaction under closed loop conditions. In addition, such a system significantly improves the cognitive and behavioral performance of human operators.

A key aspect of our approach is the fact that we are not using any simulated environment. Instead, all experiments are performed under real traffic conditions. Our experimental platform, i.e. our CLIP is a Mercedes-Benz S-Class vehicle. We believe that the specific strength of running a field experiment under real driving conditions lies in the high validity and practical relevance of the collected data.

This work was partly funded by DARPA under contract NBCH3030001.

The Cognitive Cockpit – a test-bed for Augmented Cognition

Blair T Dickson

QinetiQ Ltd
Centre for Human Sciences
Cody Technology Park
Ively Road
Farnborough
Hampshire
GU14 0LX
btdickson@qinetiq.com

Abstract

The Cognitive Cockpit was developed to support the vision of *a 'cockpit which allows the pilot to concentrate skills towards the critical mission event at the appropriate time to the appropriate level'*. This test-bed for principles of Augmented Cognition has been developed through the exploitation of both well-established and emerging technologies that contribute to our understanding of how operators interact with complex systems in naturalistic environments. As core components, models of the operator and the system are employed to understand the nature of interactions. These models are implemented within real-time sub-systems, which identify current operator status and predict operator intent. All the sub-systems are implemented as agents within the system architecture, with further agents employed for the management of the tasks and interfaces within the cockpit environment. The development of the agents has been driven by the requirement to produce a trustworthy, adaptive system that reacts, on a second-to-second basis, to operator requirements by directing attention to mission critical events. A Tasking Interface Manager (TIM) has been developed to enable the real-time management of operator load through the scheduling of tasks, adapting interfaces, and employing a framework for Adaptive-Dynamic Function Allocation (A-DFA). In this paper the concepts that underpin the functional design of the Cognitive Cockpit will be presented alongside data recorded during a simulated air-to-ground strike mission.

1   Introduction

The Cognitive Cockpit (COGPIT) has been developed to enable automated decision support to be investigated in complex military environments with the aim of maintaining the overall efficiency of the platform. COGPIT is focused on the military fast-jet environment, and is designed to support the vision of a *'a cockpit which allows the pilot to concentrate his skills towards the relevant critical mission event, at the appropriate time, and to the appropriate level'* (Taylor et al, 2000). This has resulted in the development of a system designed to provide trustworthy automation, which is sensitive to the ongoing events in the theatre of operation, whilst looking to support the overall mission objectives. These objectives are being addressed through the ongoing analysis of information derived from aircraft sensors and from the pilot. This information is used to provide decision support in the form of a series of planned tasks that are either automated or presented to the pilot depending on the current and projected future task loads. As a result this system is sensitive to the operational context and the ability of the pilot to perform the recommended tasks within the available time.

The cockpit is based on a number of enabling technologies. These are the on-line analysis of the physiological, behavioural and subjective status of the pilot; a Knowledge Based System that enables on-line decision support; and principles of cognitive engineering for the adaptive automation of the task and the display of information.

2   Cognitive Cockpit functional components

The Cognitive Cockpit has been developed as a real-time system, enabling the functional status of the pilot to be monitored, and interpreted within the ongoing dynamic context of the mission. This has enabled a framework for the dynamic allocation of function to be implemented, with the overall aim of optimising overall platform efficiency. To achieve this three functional components have been developed and integrated into a realistic simulation environment, which serves as a test-bed for concepts designed to augment cognition.

2.1   Synthetic Environment

A synthetic environment has been developed to demonstrate the utility of adaptive automation, the monitoring of ongoing cognition, and decision support technologies. This test-bed integrates the primary functional components of the COGPIT as software agents, operating in a synthetic environment with realistic cockpit interfaces and within a representative mission scenario. The simulation environment is a real-time system that enables both the mission and the environment to be simulated, and allows a number of aiding options to be examined in selected mission phases.

A scenario generation engine uses real-world digital terrain data to build an accurate representation of the world, into which air and ground entities may be added. Behaviour of entities can be scripted and this dictates the operation of weapons and sensors. The piloted aircraft is controlled by a high-fidelity model based on the F16. The head-down display (HDD) is constructed to display five re-configurable panels on a touch-screen monitor, and the visible out-the-window display is built from high resolution, rendered digital elevation terrain data, which supports the placement of discrete entities as necessary. This projected display also enables the representation of missile trails and explosions as a result of events in the mission. A simulated head up display (HUD) is overlaid on the display, enabling flight, weapons system, waypoint and other information to be displayed in addition to feed-forward and feedback messages associated with automated system operation.

2.2   Cognition Monitor

The Cognition Monitor (CogMon) has been designed to provide an on-line analysis of the operational status of the pilot. The primary function of the CogMon is to provide estimations of the cognitive–affective status of the pilot in near-real time. These estimations include different aspects of workload and arousal. Overall, this system provides information about the objective and subjective state of the pilot within a mission context, and has been configured to enable both on-line and post-hoc operation.

Inferences about pilot state are derived from four principal sources: behavioural measures (interactions with cockpit controls), physiological measures, subjective measures and contextual information (Pleydell-Pearce et al, 2000). CogMon makes use of all four of these classes of information to identify, as accurately as possible, the status of the pilot. These estimations, which are encapsulated within high-level state descriptors such as levels of stress, alertness and workload, are then provided to the Tasking Interface Manager and may form part of the rationale for a change to the level of automation or method of information presentation.

2.3   Decision support system

A decision support system (DSS) has been developed with the aim of presenting plans to support the current intent of the pilot. Currently the DSS takes two forms: it supports the use of the defensive aids suite (DAS) in one mode, and supports mission-critical functions associated with the timely delivery of air-to-ground munitions in another. The DSS is built upon a knowledge base derived from structured interviews with subject matter experts (SMEs), tactical manuals and the rules of engagement. The DSS monitors the status of the platform sensors for the presence of trigger events. These are pre-coded events that indicate the likely presence of a threat, based on a SAM site entering 'tracking' mode for example. Each plan developed will be based on the sequence of tasks most likely to defeat the current threat, and comprises a task sequence with associated timings. Plans are then forwarded to the TIM system.

2.4 Tasking Interface Manager (TIM)

The operation of the TIM system has been documented elsewhere (Diethe et al, 2004). In essence this system is designed to implement the framework for adaptive automation through the dynamic allocation of function between the automated systems and the pilot, manage the cockpit interfaces, and ensure support for the overall mission goals, tasks and timelines. Primary inputs to the TIM system are from the DSS and the CogMon, enabling pilot load to be accounted for during the allocation of tasks. The adaptive automation framework is known as PACT (Pilot Authorization and Control of Tasks) (Bonner et al, 2000). Six levels of automation are implemented within this framework, based upon the multi-level structure of Sheridan and VerPlanck (1978), as cited in Parasuraman et al (2000). In the implementation of the PACT system, each task is considered prior to the mission, and default, maximum, and minimum levels of automation allocated. These tasks are drawn from the database that is employed in the DSS system. Consequently, when a plan is developed by the DSS, each task within the plan has an associated PACT level. The decision on the automation level of each task is based on an assessment of the status of the pilot, identified by the Cognition Monitor. Following this assessment, either each task in the plan is performed automatically and feedback given to the pilot, or advice is provided through feed-forward text panels in the HUD and HDD.

3 Testing the concept – A closed-loop trial in the Cognitive Cockpit

3.1 Trial design

Six fast-jet pilots with experience of using laser-guided weapons served as participants in this trial. Each participant attended the Cognitive Cockpit facility on two separate occasions; the first to familiarize themselves with the dynamics and performance characteristics of the simulation environment, and with the operation of the automated systems, and the second to reinforce the training from the previous session and to fly the specified mission on six occasions. Three missions were constructed, each with differing levels of threat. Threat was manipulated by inserting two types of simulated surface-to-air missile. These were 'heat seeking' infra-red missiles (SA-14), and radar guidance missiles (SA-6 and SA-8), which could be implemented either within a known static air-defence system, or as portable 'pop-up' threats. Each participant flew each mission variant twice, with and without the aiding systems. The outline for the mission is represented in Figure 1, with a description of each mission segment detailed in the following text.

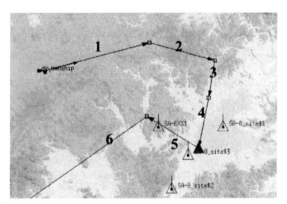

Figure 1: This figure shows the outline of the low threat mission variant.
Flight segment numbers are described in text

The scenario is based on a singleton fast-jet mission with an Air-to-Ground role. The portion of the mission under examination is based on the pilot flying between six waypoints; each segment between waypoints places different demands on the pilot. These are outlined as follows:

Segment 1 – Starts at 3000 ft, pilot performs initial checks and descends to approx. 1000 ft.

Segment 2 – Descends further to 250 ft and performs low-level 'terrain following' flight path.

Segment 3 – More low-level flight (250 ft) and prepares for the Initial Point (IP) to target run. This preparation includes some weapon aiming checks, such as activating the targeting systems, locating the target and locking the guidance system onto the target.

Segment 4 – IP to target run at 150–300 ft, at 450 kts. At 4.1 Nm from target the pilot must apply full power, and at 3.9 Nm must perform a 3G pull to 30° at which point the weapon is released. Within 5 seconds the pilot must activate the laser to guide the weapon to the target, and maintain visual contact with the target through the HDD monitor.

Segment 5 – This segment is flown between 150 and 300 ft, pilots perform post-target switches, such as housing the laser guidance pod, and perform a verbal Battle Damage Assessment.

Segment 6 – During this segment the pilot resumes the egress phase of the mission by following waypoint at approx. 300 ft, 450 kts.

Three levels of workload were manipulated by changing the level of threat by introducing SAM sites during flight segments 4, 5 and 6. Figure 2 shows a typical trace of executive workload, estimated using physiological and behavioural measures, with the vertical lines indicating where a load threshold was crossed and levels of automation increased.

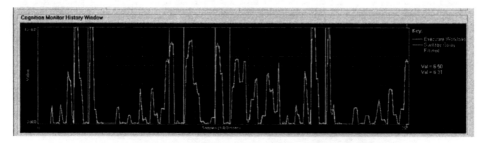

Figure 2: Time history of executive load derived from a combination
of tracking behavior and EEG measures. Vertical lines indicate the periods during
which automation levels were increased

3.2   Results

The trial was designed to answer a number of questions. These included an evaluation of the utility of the Cognition Monitor system for the real-time assessment of cognitive status. Figure 3 shows the average executive load recorded during each of the phases of the mission. These data are averaged across all subjects, across both manual and assisted runs. This produced a significant main effect of mission segment ($p < 0.05$), and demonstrates that the within-mission manipulation of task difficulty was successfully identified by the Cognition Monitor.

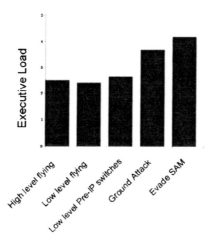

Figure 3: Executive load recorded in each of the flying segments

Executive workload was also examined for the three different workload (threat) scenarios. The plot shown in Figure 4 demonstrates the generalized increase in workload during the second half of the mission, reflecting the effect shown in Figure 3. This plot also indicates that there was no clear effect of the increased levels of threat on the executive workload. An analysis of NASA Task Load indeX (TLX) ratings indicated that overall demand was also insensitive to the threat manipulation employed in this trial; however, subjective 'effort' did increase with the increase threat instance (P<0.025).

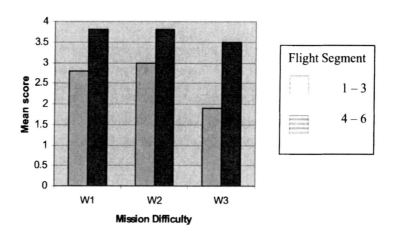

Figure 4: Averaged executive load in each of the threat variants
(W1,W2,W3) for the first and second halves of the mission

483

4 Discussion

The output of the Cognition Monitor system appears to be reflecting the cognitive demands of the task, yet is relatively insensitive to levels of difficulty in the flying task alone. This is evident in the lack of any differences between the estimates of load derived during mission phases 1 and 2. This observation, together with the apparent sensitivity of load estimates to the differing task activities in segments 3, 4 and 5, leads us to suggest that the measures that form the executive load gauge reflect higher-order processes associated with complex manoeuvers and threat avoidance, rather than psychomotor load *per se*. These measures were also supported by subjective workload elements in the NASA TLX scales, where similar patterns of self-rated workload were observed to those identified by the Cognition Monitor.

The observation that the three different mission scenarios did not significantly affect our physiological metric of workload leads us to suggest that the threat manipulations were unsuccessful in manipulating load. This is contradicted by the TLX measure of effort, which rated the high-threat scenario as more demanding that the medium-threat scenario. This finding is also supported by observations of pilot behavior where intense activity was noted during attempts to defeat missile threats. Each threat instance was, however, relatively short in duration (~ 10 sec) and consequently may have exerted relatively little influence on the load observed over the much larger blocks of time examined in the analysis reported here. It would therefore be entirely appropriate to re-examine this data set with an increase in temporal resolution of the analysis window.

A large number of behavioral and performance measures were explored during the trial. One such measure was the number of instances the airframe was buffeted, indicating a hit by an incoming missile. During the runs where the closed-loop system was implemented, the platform was hit half as many times as during the manual runs, although the relatively low number of occurrences of this event precluded a formal statistical analysis.

As a general observation this trial was successful in an initial test of a closed-loop system. Stability of system operation was achieved, and the operation of the PACT system was successfully implemented, where increases in cognitive load successfully triggered increases in automation. Platform survival also appears to be improved by the implementation of the augmentation systems, although this will be confirmed only through an increase in the number of manned trials.

As a final point it is important to recognize that we encouraged pilots to use initiative and their practised skills during these missions. This had the effect, in some cases, of reducing the number of observations of certain types of behavior we hoped to examine. This is typical during a study where relatively little experimental control was exerted over pilot behavior, but is also an important feature of trials that seek to increase the ecological validity of the techniques under examination. If these technologies are to be successfully transferred to application environments it is necessary to move away explicitly from highly controlled laboratory studies to those that enable highly trained individuals to perform a task in a naturalistic fashion.

References

Bonner, M.C., Taylor, R.M., Fletcher, K., & Miller, C. (2000). Adaptive Automation and decision aiding in the military fast jet domain. In *Proceedings of the Conference on Human Performance, Situation Awareness and Automation: User-Centered Design for the New Millennium.Place, publisher*

Diethe, T., Dickson, B.T., Schmorrow, D., & Raley, C. (2004). Toward an Augmented Cockpit. *Proceedings of the Conference on Human Performance, Situation Awareness and Automation II, Daytona Beach, FL, (2) 65–69, March 2004.*

Parasuraman, R., Sheridan, T.B.& Wickens, C.D. (2000) A model for types and levels of human interaction with automation. *IEEE Transactions on Systems, Man, and Cybernetics – Part A Systems and Humans*, 30(3) May 2000.

Pleydell-Pearce, C.W., Whitecross, S., & Dickson, B.T. (2003). Multivariate analysis of EEG: Predicting cognition on the basis of frequency decomposition, inter-electrode correlation, coherence, cross phase and cross power. *Proceedings – 36th Hawaii International Conference on System Sciences (HICSS)*.

Pleydell-Pearce, K., Dickson, B., & Whitecross, S. (2000). Cognition Monitor: A system for real time pilot state assessment. In P.T. McCabe, M.A. Hanson and S.A. Robertson (eds.), *Contemporary Ergonomics 2000*. Taylor and Francis, London. pp65–69.

Sheridan, T.B. & VerPlanck, W.L. (1978). "Human and Computer Control of Undersea teleoperator," *MIT Man-Machine Systems Laboratory, Cambridge, MA, Tech. Rep number?*

Taylor, R.M., Howells, H., & Watson, D. (2000). The cognitive cockpit: operational requirement and technical challenge. In P.T. McCabe, M.A. Hanson, and S.A. Robertson (eds.), *Contemporary Ergonomics 2000*, 55–59. London: Taylor and Francis.

Section 3
Cognitive State Sensors

Chapter 11

Context Modeling for Augmented Cognition

"Look who's talking": Audio Monitoring and Awareness of Social Context

Paul Heisterkamp, Julien Bourgeois, Jürgen Freudenberger, Klaus Linhard, Xavier Lacot*, Guillaume Masson*

DaimlerChrysler AG, Research and Technology, Dialog Systems
Wilhelm-Runge-Str. 11, D-89081 Ulm, Germany
* also: E.N.S.S.A.T, Ecole Nationale Supérieure des Sciences Appliquées et de Technologie
6, rue de Kerampont, F-22305 Lannion, France
Paul.Heisterkamp@DaimlerChrysler.com

Abstract

In very many contexts, people not only interact with technical systems and the external physical world. People interact with people. Much of this interaction goes through the auditive channel. We present and discuss the outlines and the first steps in the technical realization of a complex suite of systems that monitor the auditive channel to gather input for contextual workload assessment. We apply the suite to monitor the human-originating acoustic events inside a passenger car.

First, we sketch the acoustic environment inside a vehicle. We discuss the microphone sensor equipment and some of the signal processing techniques employed in order to separate audio sources, needed to find out who of the people in the car is talking at any given moment.

We then discuss the nature of the auditory events, explaining why we chose to first concentrate on a set of so-called 'para-linguistic' events, non-speech sounds generated by the human articulatory apparatus. We outline how speech recognition technology was expanded to handle the wide frequency range needed. We proceed to show how we overcome the inherent uncertainty in ascribing certain categories to a set of events from a certain source over a certain time.

We show how the technologies mentioned above are integrated in a car and how, using the vehicle's data bus, a communication is established with a contextual workload assessment module that also receives input from a set of other data sources.

We conclude by outlining ongoing and further work on audio monitoring, highlighting a few application scenarios.

1 Introduction: Cognitive load and driver distraction

For a number of years now, the public has been aware of the fact that drivers of road vehicles do not always concentrate on the driving task as it would be necessary. This problem generally goes under the name of 'Driver Distraction', and it is assumed that the growing number and complexity of electronic on-board systems increases this problem (e.g. ElBoghdady 2000). As short reflection reveals, however, to view the perceived lack of concentration on the part of the driver purely, as it is sometime done, as a consequence of the presence and complexity of the electronic devices falls somewhat short. The problem space addressed here is encompassing many more items.

1.1 Cognitive Overload

The difficulty of the individual driving situation, the *demand*, needs to be seen in relation to the skills, the experience, the alertness etc. of the driver (the *resources*). Only if the resources do not or barely meet the demand, additional demand such as created by the operation of devices in parallel to driving a vehicle does this operation become critical and create what is called a distraction. We can also say that this causes "overload" on the part of the resources, and as far as these resources are cognitive, we can speak of 'cognitive overload'.

Obviously, it is desirable to avoid cognitive overload for the driver while driving the vehicle. Many design decisions and many automation activities on the part of the vehicle manufacturers already serve that goal, in generally reducing demand to make the task of driving itself and the related activities as little demanding as possible. However, not all of these activities are honored in the marketplace. Not all of the customers want to be patronized by what is sometimes called 'the nanny car', as, some say, this takes out much of the fun of driving.

This kind of acceptance, or rather reluctance to accept, of systems designed to make driving and related activities as simple and easy as possible, needs to be reckoned with. One has to find the balance between the car creating cognitive overload, and taking care of too many things.

1.2   Cognitive Underload

A second aspect of the car taking care of too many things is one that goes unnoticed more often than not, and that is the aspect that driving a vehicle in present-day traffic can be quite boring. With the number and functionalities of the highly sophisticated driver assistance systems now available in at least premium class vehicles, the driver does not have much to attend to: from automatic gear shifts to radar-based adaptive cruise control to lane departure warning, a host of little helpers takes over routine tasks. The driver is freed of executing these task and left with little to nothing to do. The cognitive demand of driving the car is approaching zero. We can call this state 'cognitive underload': People's minds start wondering, people engage in other activities etc. and – worst of all – people fall asleep.

If in such a state of the driver's mind, in cognitive underload, a sudden and unpredicted situational demand comes up, the human cognitive systems can not react as quickly as might be necessary or even possible had it maintained a certain vigilance. Regrettably, if this kind of lack of attention contributes to an accident happening, these accidents tend to have rather severe consequences, as, for example, the driver sometimes does not break at all. These phenomena then are related in the press quite often as 'human failure'.

So, as necessary as it is to design systems that avoid cognitive overload, it is also necessary to keep track of the cognitive underload problem. One way to reconcile these two opposing demands is to try and endow the vehicle with a system that first of all measures and assesses the cognitive demand for the driver, and to then take those measures that are adequate for a respective situation.

2   Cognitive Workload Assessment in the vehicle

The first thing needed to assess the workload on the driver in a vehicle is a suite of sensors that capture information on the current situation. State-of-the-art passenger vehicles are already equipped with a number of sensors, in their first role used for controlling the car and delivering date for the different driver assistance systems. These data are, again on state-of-the-art vehicles, available via a data bus. To give just a few examples: longitudinal, lateral and skid angle acceleration is measured by an accelerometer. The primary role of this measurement is to provide input for the electronic stability program (ESP) system that ensures the car does not break out in curves. Distance and speed of a vehicle in front of one's own is measured by a Doppler radar. The primary role is to adjust longitudinal acceleration or deceleration in what is known as 'adaptive cruise control'.

Also, data on the driver behavior are available. Gas and brake pedal position, movement and speed and force of activation are measured, both for the direct engine or brake control, but also, for example, to activate the 'Break Assistant' (BAS) system that executes a maximum deceleration to compensate for the known problem of drivers not hitting the brake pedal as hard as they could in an emergency braking situation.

Both these lists are considerably longer, but for the present purposes it is sufficient to point out that there are a large number and variety of sensors around already in the off-the-assembly-line version of modern vehicles to give, as a secondary role, input to the Cognitive Workload Assessment (CWA). The CWA module, implemented in the experimental vehicle by Sandia National Labs, in its turn listens on the vehicle data bus and collects those data that are useful for the assessment task. (cf. Dixon et al 2005).

As stated above, the actual load is dependent on both situational demand and resources available. The CWA captures part of the situational demand through inferences and classification made over the vehicle sensor data. The data that allow inferences on the state of the driver are also used in the assessment (e.g. pedal movement, steering behavior). The overall system in the vehicle encompasses also brain activity measuring and classification methods.

In this paper, however, we concentrate on the one area of influence on the driver that the other sensors so far can not capture. People are not always alone in their cars. They have passengers and co-drivers with whom they interact, and the other people on board also interact with the driver and with each other. As this interaction mostly takes place by voice, it is natural to use voice sensors, i.e. microphones, as the primary equipment to capture this social interaction aspect of the driving situation.

Again, here, these sensors can have multiple uses. The foremost use is to act as an input medium to in-vehicle spoken dialog, interaction with the vehicle's system as part of the Human-Machine Interface. Starting from the existing serial version of the speech control system, we developed the suite of sensors and the necessary signal processing, described in more detail in the next section, that is needed to actually be able, in the first place, to capture the human-originating sound events inside a vehicle with such a quality that allows for the application of further automatic methods like speech recognition technology to get a grasp on the actual interaction events in the vehicle. In the following, we further expand on these technologies and their preliminary results.

3   Audio environment and acoustic signal processing in the vehicle

Above, we have shown why we need sensors to assess the workload created by a situation, and also why we deem it necessary to include sensors that allow for taking into account the social situation inside the vehicle (or in any other multi-party situation). We now discuss the exact nature of the sensors we employ for that specific area, namely, a technology suite that includes microphones, signal processing techniques, and speech recognition. We also sketch the output side that goes along with this technology suite.

The reason that we have a complete cycle from input to output is based in the fact that the primary sensors, the microphones, have multiple uses in the vehicle. They serve, as we are about to detail below, as part of the sensor suite for situation assessment. At the same time, and sometimes in parallel, they also serve other purposes. They are used to enhance interior communication, and they serve as input sensors for Spoken Dialog Systems (SDS) in the car. To use one single piece of hardware equipment in several roles, each creating perceived customer benefit, adds considerably to the commercial viability of introducing this equipment into the product: first on the cost side, as only the accumulated benefits warrants the effort of designing and procuring and assembling the hardware, and – eventually – also on the revenues side, when the customer is willing to honor these benefits.

DaimlerChrysler was the first vehicle manufacturer to introduce a speech recognition command and control system in their serial production vehicles in 1996 (cf. Heisterkamp 2001a). Over time, this system has been further developed. In research, we have for some time now what is called 'conversational dialog system', i.e. systems accepting fluently spoken natural language input. To achieve this, one has to cope with a number of difficulties that are peculiar to the acoustic environment within a car.

3.1   Acoustic problems in the vehicle

The car is considered as a 'hostile acoustic environment'. The acoustic properties of the interior of a car are influenced by a number of factors. First, there is the glazing. The windows of a car are hard surfaces that reflect sound waves directly. They are positioned in a variety of angles, thus creating a complex echo space. The car itself emits noise from the engine, the wheels and the wind noises on the body. These noises also cause the glazing to vibrate, such that the reflections are modified through these vibrations. All of these noises are quasi-stationary, however, i.e. they change more or less monotonously and slowly as a function of engine revolutions and speed. As the car moves along the road, passing vehicles add non-stationary noise. Further causes of for non-stationary noise are turbulent airflow from electric fans, uneven surfaces (like, e.g. pot-holes or gravel), and rain (or even hail) hitting the car with drops of different weight and impact velocity in a chaotic manner.

It is clear that under these circumstances the treatment of noise and, in particular, its suppression, are of paramount importance for the functioning of speech recognition. While stationary noise can in some way be estimated and predicted in signal processing, the non-stationary noise events pose severe difficulties.

One way to tackle this problem is to try and concentrate the primary sensors, the microphones, such that mostly only the desire signal source, the speaker, is picked up, and that other signal sources are disregarded. A close-talking microphone integrated in a headset would greatly help achieving this. However, it is out of the question for civilian drivers to require them to wear a headset to communicate with the car. Rather, the microphone needs to be placed the vehicle such that it best approaches the characteristics of a headset mike.

For a single speaker, such as the driver in a vehicle, so-called microphone arrays are used. A microphone array is a combination of more than one microphone where signal processing, in particular the comparison of signal times, allows for a directed pick-up of the sound waves, so-called beam forming. Microphone arrays are quite good in their directional resolution, i.e. in determining the angle from which a signal comes, and they serve their purpose very well. What is difficult for this technology, though, is to get a grip on depth resolution. For example, when a person sitting on the rear seat behind the driver talks while the microphone is active, it can not distinguish this from the driver, talking at the same time, and the speech recognition running on this signal will most probably end in an undesired result. This is known as the 'crosstalk-problem'.

3.2 Signal processing

To overcome the crosstalk problem, one needs to separate not only noise from speech, but also one needs to differentiate between different sources of speech signals. In an experimental vehicle set-up, we chose to place one microphone over each seat in a passenger car. We selected a flat and small kind of microphone that can be integrated in the headliner. It picks up sound signals mainly from underneath, but also has a directive quality. We could place these microphones in such a way that they are directed towards the seat position of the people on board the vehicle. (cf. Figure 1)

Figure 1: Microphone in housing for in-vehicle use; the arrow indicates the pick-up direction. Scale in centimetres

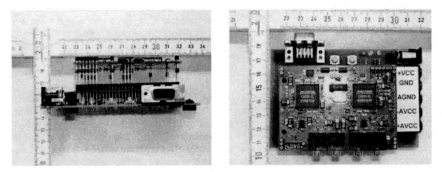

Figure 2: Digital Signal Processor (DSP) set-up, including power supply and interfaces. Scale in centimetres

Figure 3: Loudspeaker arrangement for individual audio in the experimental vehicle

The signals these four microphones pick up can be brought together in a processing device and be compared in a signal processing computing device. In the experimental set-up, this device was specifically designed for the purpose, and comprises two specialized digital signal processors (DSPs) (cf. Figure 2).

Through the application of signal processing algorithms on this device, we are now able not only to pick up speech sounds from each seat separately, but also can determine, e.g. by measuring and comparing energy levels and signal phases, from which seat position one particular sound originates (Bourgeois 2003). We have thus created a 'speaking zone' for each seat.

The sound signals produced in each 'speaking zone' now can also be transferred to other seat positions to help

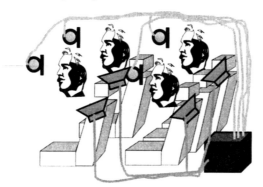

Figure 4: Schematic rendering of the set-up of the four audio input and four audio output channels in the vehicle

reducing communication difficulties between the occupants of the car. Regrettably, it can often be observed that drivers, while driving, turn their head towards the rear passengers such that they can hear the driver better. With the arrangements of microphone and the source separation, now, as a sort of by-product of the set-up, we can reduce the perceived need for this behavior. We can take the signal picked up at one position and forward it to loudspeakers located such that their sound is directed towards an individual hearer on one of the seat positions (cf. Linhard and Freudenberger 2004). Currently, we are employing ordinary speakers (cf. Figure 3), but are also investigating the use of ultrasound loudspeakers directing a sharp beam of ultrasound towards a hearer that is modulated with the original signal. The air then demodulates this signal, and it becomes audible only in the area of the beam. This then creates a 'listening zone' for each seat.

Figure 4 gives an outline of the overall layout of the audio input and output system in the experimental vehicle, the speaking zones and the listening zones.

All eight (input plus output) channels can be active at the same time. Therefore, over and above the noise reduction need in the vehicle anyway, there is also the need to reduce, and if possible eliminate, the echo effects of the loudspeaker output to the respective microphones. This task is also performed by the DSPs. Figure 5 depicts the general process. The sound played out by the loudspeaker (top left) reaches the microphones in two forms. One is the direct

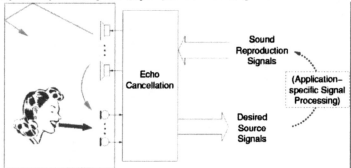

Figure 5: Schematic rendering of the echo cancellation. The green arrows indicate the direct and indirect echos.

sound traveling through the air. The other form is the refracted sound, as it results from the original sound waves hitting surfaces and being reflected. These signals travel longer than the direct form, and in the reflection they are also distorted and lower in energy than the direct form. The echo cancellation algorithm now has to estimate both forms of echo. For this, it uses a mathematical model of the acoustic properties of the room the signal travel through, including this room's reflections. This model, applied to the original output signal, now enables the algorithm to subtract the estimated echo from the sum input signal resulting of the combination of the echo on the one hand and the 'payload' speech signal that is desired to be as clean from the echo as possible. In pauses of the payload signal, such as they occur e.g. during lip closure prior to burst sounds (plosives) (e.g. [b], [p]), the algorithm can adapt the model and re-estimate the echo noise.

Some of the echo, however, will in practice still escape the echo cancellation process. As the sum signal picked up at each microphone is played back through other loudspeakers, the echo that still comes through may cause very annoying feedback noise. We therefore introduced a feedback suppression routine that detects cyclic repetitions in the signal. The frequencies in which this occurs are then immediately reduced in energy. The loss in signal quality is virtually imperceptible.

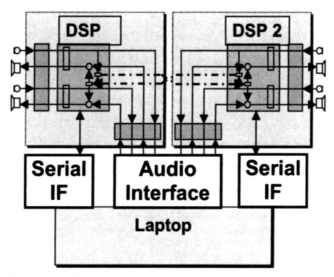

Figure 6: Architecture of DSP device as connected to laptop computer

All the signal processing has to take place in real time, creating considerable computational demand. We found it necessary to distribute the load over two DSPs. The DSPs are connected to each other, to allow for cross-channel operation of the different signal processing methods. They are connected to a laptop computer on which the further processing runs. The computer is equipped with a 4-channel I/O audio card. The DSPs receive from the computer whatever audio needs to be played out for the respective set of speakers as analog data. They output to the computer the pre-processed signals from each of the microphone channels, also in analog form. To program the DSPs and to control their operations, they also have a serial interface that is connected to the computer as well. The whole architecture is shown in Figure 6.

The final output from the microphones through the different stages of signal processing now is largely free of noise and echo, and, in addition, it is also known which channels are active. This output is now taken as the input for one speech recognizer for each channel.

4 Speech recognition for audio monitoring

When one thinks of monitoring the acoustic social interactions inside a vehicle with the purpose of deducing information as to the situational demand and ultimately as to the cognitive workload of the driver, there is a host of possibilities as to which clues actually yield such information. Some of these clues may be straightforward and comparatively easy to detect, others may be more subtle. In any case they have in common one general obstacle, and that is that the data that would allow one to classify these clues tend to be very rare and difficult to obtain. To cover the full range of spoken interaction inside the car, one would have to record, segment, transcribe and classify an enormous amount of speech material. This is beyond the scope of a research project. We therefore must show the validity of the approach, i.e. the value of audio monitoring as part of a sensor suite for CWA, using material that is easier to come by. Also, to do controlled experimentation and verification, we need means to reproduce certain states and series of events in the vehicle. Under these considerations, we set out on an axiomatic approach.

4.1 Recognition scenarios

We identified two extreme positions of cognitive underload and overload, respectively. A severe underload condition occurs, for example, in the following scenario:
You are driving in the night, on a straight and empty highway, with three other people on board. The others need rest. Being polite, you turn the radio off. They fall asleep - and some snore.

It is known for traffic accident research that the fact of passengers sleeping in the car significantly increases the risk of the driver also falling asleep – even if he or she is not all that tired him- or herself. The presence of sleeping people also often inhibits a driver who feels he or she is getting drowsy to take common counter-measures: turn the music volume up, tune to more rhythmic music, turn the heating down, open the window etc. We now assume that if we can detect the acoustic evidence of someone snoring, that this would be a valuable input to the CWA.

The second scenario is an overload example:

You are driving alone in your car, with your baby or infant in the rear. The baby starts crying.

Baby crying gets on your nerves – it is intended to be that way. If a baby starts crying in the car, many people follow the urge to immediately do something about the crying – soothing the baby, start fishing for fallen pacifiers in the rear, etc. We now assume that if we can detect baby crying in the car, that this would also be a valuable input to CWA.

4.2 Paralinguistic sounds and recognizer training data

The sounds presented in the two scenarios above have a number of practical advantages. One of them is that they are so-called para-linguistic sound, i.e. sounds emanated by the human vocal apparatus that are not speech. Other examples of paralinguistic sounds are coughs, laughs, but also pause fillers often represented graphically as 'uh' or 'emm'. They all have in common that they are not dependent of the language a person speak, but seem to be universal for mankind. We therefore could avoid a number of practical difficulties by recording several hundred hours of

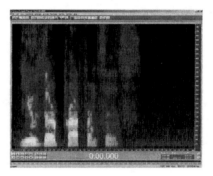

Figure 7: Energy distribution in a sample of normal speech (male): Scale from 0 to 5.500 Hz

both paralinguistic sound categories in Bangalore, India, with the help of the DaimlerChrysler Research and Technology Centre there. The colleagues in Bangalore also edited and labeled the data.

Another advantage of having the paralinguistic sound scenarios is that paralinguistic sounds are (mostly) produced without a conscious effort. This means in recording we do not need to account for Heisenberg effects influencing the quality of the recording, a phenomenon that often distorts recordings of emotional speech.

There is, however, a downside to the use of paralinguistic sounds for speech recognizers. To the best of our knowledge, no one as attempted recognizing baby crying or snoring before. We therefore had to first analyze the data and found that the energy distribution in both types differs considerably from normal speech.

Figure 7 shows the energy distribution in speech of a normal male speaker. The areas with higher energy are brighter. More energy by and large means more information, i.e. the areas with high energy and their position in the signal are of considerable importance in modeling the speech sound for the recognizer. It can be seen that in the spectrum there are bands of harmonics that contain high energy. In standard speech recognizers based on so-called 'Hidden Markov Models' (HMMs), one of the first processing steps is to look at a window of 10 milli-seconds in the speech signal and to apply vector quantization to the energy density in the different areas of the signal. The vectors

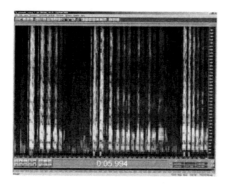

Figure 8: Energy distribution in a sample of baby crying: Scale from 0 to 8.000 Hz

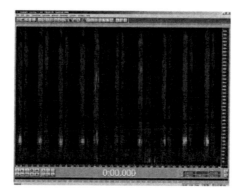

Figure 9: Energy distribution in a sample of adult snoring (male): Scale from 0 to 8.000 Hz

are discrete and not analog. The modeling therefore has to be biased towards the frequency areas which contribute most to the recognition – those with much information and, mostly, with much energy.

If we now look at Figure 8 (mind the slightly different frequency scale), we see that here high energy is almost evenly distributed across the spectral range. Physiologically, this indicates that there is little to no modulation in the articulatory tract. Technically, this means that the standard vector quantization does not work properly, as the bias towards certain spectral regions does not capture the full range of the energy distribution.

A sample of the snoring sound (Figure 9) shows yet another picture. Areas with high energy are distributed in bands, as one would expect from the fact that the sound is modulated by obtrusions in the air-flow from the lungs. However, the vocal cords are not involved. Therefore, the high energy patches are located in quite different areas from both the normal speech and the baby crying. Again, the standard vector quantization falls short to capture properly the characteristics of this signal.

4.3   Recognition results

So, in order to be able to apply the same speech recognizer to detect and classify all these three types of acoustic events – speech, crying and snoring – we had to modify some parts of the DaimlerChrysler speech recognizer. With the modified recognizer in place, we trained the recognizer models for the paralinguistic sound classes. Basically, they are considered just as words would be in a word spotting task. We applied a (pseudo-) grammar, viz. both words may repeat themselves endlessly. In the actual recognition, the utterance length is limited by the breathing pauses the speaker makes. In tests, using an artificial head to play test data to the microphone sensor suite in the vehicle (as describes above in section 2), we achieve a recognition accuracy between those three classes of over

497

90%. This accuracy, achieved n. b. with a standard technology recognizer, is quite sufficient for the monitoring purposes, as we will show in the next section.

5   Audio monitoring integration

The recognition accuracy is sufficient as we still have to do one more processing step anyway, namely to integrate the audio monitoring results over time. This is necessary as isolated events, e.g. a yell from the driver, may correctly lead to the recognition of a crying event, for example. It is not helpful to send such spurious events to the CWA. We therefore implemented a software module, called the Audio Monitoring Collector, that receives the recognition results of the audio monitoring recognizer channels and matches them against a trigger database. Only if a series of

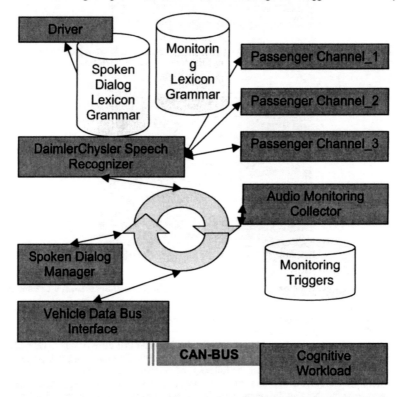

Figure 10: Schematic lay-out of the Audio Monitoring system integrated in the Spoken Dialog architecture via the 'Hub' system, and connected to CWA via the vehicle data bus (CAN)

events of one type from one particular channel persists over a threshold time the trigger fires and the event is sent to CWA. The event message also contains the time of the onset of the event and the duration before the message was dispatched. The trigger database can be configured for the individual channels and events, and it also mildly adapts by lowering thresholds for events that fired the trigger recently.

The recognizer and the collector are an integral part of the generic Spoken Dialog architecture in the vehicle (cf. Figure 10). The architecture is based on the DARPA Communicator hub system, originally developed at MIT. We added a generic interface to the vehicle data bus, that now also handles messages to and from the CWA. Through this interface, we also inform CWA of the different states of the spoken dialog with the driver, such as whether and when the driver or the system is speaking, what is the topic of the dialog etc. From CWA, the dialog receives information it can and does use to adapt the dialog flow to the current and estimated future workload (mitigation).

6   Conclusion

In setting up the system as described above in a real car, and in conducting test rides with it, we have shown that it is technically possible to monitor the social interaction in a vehicle as part of a more global sensor suite for cognitive workload assessment. We have outlined the difficulties we had to overcome, and the technical solutions for this. The work here now can continue in several directions: for one, given one finds sufficient training material, one can try to capture emotion and affect in the social interaction (cf. Andre et al. (Eds.) 2004). One can also expand the coverage of the monitoring recognizers to spot swear words and other indicators of arousal. Ultimately, the technical set-up permits the installation of a spoken dialog interface that can be operated from each seat without the use of an activation device. The dialog system, constantly monitoring what is being said, will know when it is called upon and when it can be of help.

Acknowledgements

The research presented in this paper is partly funded by DARPA under the program 'Improving Warfighter Information Intake Under Stress' (AugCog).

References

André, Elisabeth; Dybkjaer, Laila; Minker, Wolfgang; Heisterkamp, Paul (Eds.) (2004): Affective Dialogue Systems. Tutorial and Research Workshop, ADS 2004. Kloster Irsee, Germany, June 14-16, 2004. Proceedings. Berlin, Heidelberg, New York: Springer. (Lecture Notes in Artificial Intelligence. LNAI 3068).

Bourgeois, Julien (2003): A Clustering Approach to On-line Audio Source Separation. In: Proceedings of the European Conference on Speech and Language Processing (EUROSPEECH), Geneva (Switzerland), 2003.

Bourgeois, Julien; Linhard, Klaus (2004): Frequency-Domain Multichannel Signal Enhancement: Minimum Variance vs. Minimum Correlation. In: Proceedings of the European Signal Processing Conference (EUSIPCO), Vienna (Austria), 2004.

Dixon, Kevin; Lippitt, Carl; Forsythe, Chris (2005): Modeling Human Context Recognition of Driving Situations. In this volume.

ElBoghdady, Dina (2000): Feds fear high-tech car gear. Detnews.com, 23 January 2000.

Heisterkamp, Paul (2001a): Linguatronic: Product-Level Speech System for Mercedes-Benz Cars. In: Proceedings of Human Language Technology (HLT2001), San Diego, Ca.

Heisterkamp, Paul (2001b): Driving forces in the development of speech technology: Safety. Invited talk for a panel discussion at the IEEE Workshop on Automatic Speech Recognition and Understanding (ASRU 2001), Madonna di Campiglio, Italy, December 2001.

Heisterkamp, Paul; Kincses, Wilhelm-Emil (2003): Don't talk to the Driver! Situated Dialog in the vehicle. Invited talk at European ACL 2003 Dialog Workshop, Budapest, Hungary, April 2003.

Lathoud, Guillaume; Bourgeois, Julien; Freudenberger, Jürgen (2005): Multi channel speech enhancement in cars: explicit vs. implicit adaptation control. In: Proceedings of the Workshop for Hands-Free Speech Communication and Microphone Arrays (HSCMA), Pscataway, March 2005.

Linhard, Klaus; Freudenberger, Jürgen (2004): Passenger In-Car Communication Enhancement. In: Proceedings of the European Signal Processing Conference (EUSIPCO), Vienna (Austria), 2004.

Linhard, Klaus; Heisterkamp, Paul (2003): Acoustics aspects of in-vehicle spoken dialogue. Invited talk at DAGA 03, Aachen, Germany, March 2003.

Exploring Human Cognition
by Spectral Decomposition of a Markov Random Field

Monte F. Hancock, Jr.
Principal Staff Scientist
Essex Corporation
Melbourne, Florida USA
hancock@essexcorp.com

John C. Day
Principal Staff Scientist
Essex Corporation
Melbourne, Florida USA
day@essexcorp.com

Abstract

Human cognition is a complex process whose underlying mechanisms are poorly understood. From a behaviorist perspective, the human reasoner can be modeled phenomenologically as a state machine that emits tokens in a manner conditioned by its (known) goals and inputs, and its (hidden) cognitive states.

An "instrumented environment" has been established for collecting a comprehensive set of cognitive artifacts created by human experts as they work. We present an approach based upon Markov Random Fields (MRF's) for modeling the observable phenomenology, and decomposing the resulting model to obtain information about the hidden states of the reasoner.

Markov Random Fields (sometimes referred to as Boltzmann Machines [1]) are regressions that estimate binary system state variables from binary observables (e.g., condition true/false, present/absent). MRF's represent domain knowledge as a weight matrix $W=\{w_{ij}\}$ which quantifies the joint-likelihoods that binary state variables s_i and s_j are simultaneously "true" in a given context. We use this matrix to impose a "soft" relation on the concept space which can be mined to understand the underlying process.

Regarding W as a low-resolution image, we apply a machine vision technique (non-negative matrix factorization, NMF) to partition W into a basis of concepts associated in a known way with (hidden) cognitive states. (It is hypothesized that this will allow the detection of "mixed states" which indicate that a transition from one hidden system state to another is underway.)

Associated with W is an energy function. We show how this function naturally gives rise to a spectral decomposition which expresses the contribution of sets of "concepts" to the (hidden) system state. Three energy decomposition theorems are proven. A physical analogue of the energy function is given.

Acknowledgements: This work has been funded by contract MDA904-03-C-0406 to General Dynamics Advanced Information Systems from the Advanced Research and Development Activity, Novel Intelligence from Massive Data Program.

1 Background

The data domain for this work is "document analysis". In this domain, human experts are assigned analytic tasks; they access open-source information by computer to collect and analyze a variety of information products, then document their conclusions in reports which are forwarded to decision makers.

To generate data for this project, analysts work in the instrumented environment, an experimental monitoring system. As analysts work, instrumentation collects time- stamped work-events and text created while the working expert is logged onto the environment. The collected artifacts include internet browser queries/responses, cut-and-paste operations and buffers, data base queries/ results, and other symbolic exchanges which constitute a kind of "dialog" between the computer and the working analyst.

Use of the instrumented environment differs from conventional knowledge acquisition methods in that it is passive; rather than conducting knowledge acquisition interviews and manually constructing knowledge-based simulants (active), the instrumentation collects symbolic cognitive artifacts by passively "watching" human experts perform a suite of selected tasks. From these observations, knowledge and process models may be inferred.

Analysts perform a variety of analytic tasks. They access data sources, apply tools, spend varying lengths of time on different work segments, and produce products. As analysts work, the instrumented environment logs the activities, inputs, and outputs associated with the specified task. The inputs and outputs are represented as blocks of text; these are tokenized into terms, and aggregated into "concepts".

1.1 Representing the Analytic Process

The analytic process may be represented as sets or sequences of observables generated by a human expert during the execution of an analytic task.

We define five states to represent the human expert's cognitive state during the work process:
- **Define**: receive and understand tasking, initial planning and collection of tools and resources anticipated, formulate initial hypotheses
- **Search**: use web browsers and other search tools for gathering information to generate or support hypotheses
- **Marshal**: assemble evidence to prove or disprove hypotheses
- **Judge**: accept or reject hypotheses
- **Present**: prepare findings for presentation

These states were selected because it was hypothesized that the cognitive process would consist of a repeating cycle of DEFINE-SEARCH-MARHSAL-PRESENT actions. These actions correspond to the human expert performing numerous searches over the Web, retrieving and then editing ("cut, paste, modify") small portions of the retrieved text into other documents.

A long range goal of this work is to develop the ability to robustly infer and explain the (hidden) cognitive states comprising the analytic process. This is necessary to solve the overarching problem: help analyst's perform their tasks by anticipating and pro-actively servicing analytic needs, optimizing the analytic process, overcoming human biases, and mitigating blind-spots.

2   Theoretical Treatment

2.1   The human analyst as a state machine
In the formal computational sense, cognition occurs as a "side effect" of the human expert's transitions through a sequence of the five cognitive states.

The expert's cognitive state is "hidden": it cannot be directly observed, but must be inferred from the concepts emitted by the working analyst and collected by the instrumented environment. A cognitive model of the human expert is naturally a "state-transition" structure, which we assume can be adequately represented by formal state-machine models such as Turing Machines, BBN's, etc., in accordance with Church's Thesis. As this model executes, it ingests token streams from multiple sources, and emits a token stream ("output" produced by the human expert).

Let the "knowledge container" (i.e., the human analyst) be called **A**.

A is both a consumer and producer of "knowledge". For the purposes of this discussion, knowledge will be defined as a hierarchically arranged token string (e.g., a coherent set of rules, or a coherent list of facts).

Taking a behaviorist approach, we do not attempt to look inside the analyst **A**, but model her as a state machine, each state transition, λ_{ij}, being called a "behavior". All states are explicit, in the sense that behaviors are, by definition, observable. The fact that states are observable does not necessarily mean that we know what they all are, and we might want to allow that we do not always know **A**'s state.

The diagram below depicts the analyst as a state machine (Figure 1). Here:

The alphabet B is a finite set of symbols: $B = \{\beta_i\}$
The set S is a finite set of states: $S = \{S_i\}$
The connection matrix, $M = \{p_{ij}(\beta_k)\}$, where $p_{ij}(\beta$, is an array giving the probabilities that the machine will transition from state S_i to state S_j when scanning the input symbol β_k.

The initial state is S_0; state transitions are labeled λ_{ij}, where S_i is the source state, and S_j is the destination state of the transition.

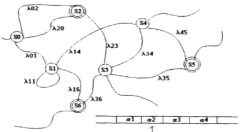

Figure 1: The Analyst as a non-deterministic FSM

Each state in the network can emit symbols. This symbol stream consists of concepts that serve as the basis for reasoning in the domain.

A finite state machine model of a process can be constructed manually or semi-automatically using domain knowledge and/or analytic techniques (e.g., Bayes' Theorem); this is computationally demanding for non-trivial problem spaces. In the case that p_{ij} are all 0 or 1, a conventional memory-less finite state automaton is obtained.

A simple modeling experiment (with C5.0) to predict the analyst's next state from currently observable concepts gives an overall blind-test accuracy of 52% (this is a relatively good result, since ambiguity in the feature space makes the maximum possible accuracy 64%).

2.2 Terms and Definitions

Consider a symbol system representing a cognitive process in some domain. Suppose that in this system, work is performed by organizing symbols into distinguished subsets, that is, by partitioning the set of symbols in some particular way. In a given domain some partitions will be meaningful, while others will not. Symbol sets constituting the members of a partition will be called distinguished sets. Distinguished sets need not be disjoint or exhaustive.

A partition naturally imposes a reflexive and symmetric relation ~ on the set of symbols where r~s iff there exists some distinguished set C having both r and s as elements[1].

2.3 A "crisp" notion of concept space

Let r and s be symbols in a domain D, and C be some distinguished set with r, s ∈ C. Then r~s, and r and s will be said to be consistent with respect to C. If there is no distinguished set C having both r and s as elements, r and s will be said to be inconsistent.

As seen above, it is helpful to associate names with distinguished sets. A distinguished set along with its name will be referred to here as a concept. The set of all concepts in a domain is its concept space **A**.

2.4 A "fuzzier" notion of concept space

If we insist on exact matches in reasoning with concepts as above, the resulting inferencing scheme will be brittle. MRF's provide a natural remedy by replacing the crisp relation r~s with a real-valued symmetric association weight matrix $W = \{W_{rs}\}$. The association weight W_{rs} will have a positive value when r and s usually occur together (are usually consistent), and will have a negative value when s and r usually do not occur together (are usually inconsistent). More precisely, when both p(s|r) and p(r|s) are large, W_{rs} will be positive; when both p(not s|r) and p(not r|s) are large, W_{rs} will be negative. Under this new definition, concepts are arbitrary sets of symbols whose elements generally have pairwise positive association weights, but might also include some element pairs having negative association weights. In keeping with the notion of crisp concepts above, it is customary to require W to be symmetric, so that $W_{rs} = W_{sr}$.[2]

Unlike a simple cooccurrence matrix, W captures the relative strength of the associations of state variables in the context of the values assumed by all other state variables: it retains "context".

2.5 The Problem Domain

We prepared from the instrumented environment collect a feature set from the data of 1140 keywords (out of 3500 words total) which have the highest "mutual entropy" with respect to each annotation. The words have been slightly normalized: forced to lower case and endings removed with the Porter Stemmer. Using these keywords we built a "co-occurrence matrix" by incrementing the co-occurrence counts of word pairs when they occur in the same annotation. This is roughly equivalent to calculating the joint probability distribution of the words given the annotation instances.

There are meta-data in the instrumented environment Annotations Table which reflect the human expert's own comments about the process state. These constitute the "ground-truth" for supervised modeling.

We then computed the principle components of this rather sparse matrix for the purpose of compressing the features into small but hopefully semantically rich clusters. Each annotation is thus coded and projected into this semantic space.

A simple modeling experiment (with C5.0) using observable concept sets to estimate the analyst's current state gives an overall blind-test accuracy of 51% (as above, ambiguity makes the maximum possible performance 64%.)

---

[1] If the distinguished sets are pairwise disjoint, ~ will be an equivalence relation having the distinguished sets as its equivalence classes.

[2] This condition makes practical sense, and is required to insure that learning association weights from tagged data by simulated annealing converges to stable solutions, rather than being trapped in limit cycles.

2.6   Constructing Markov Random Fields from the concept data

As described above, a MRF is a graph consisting of a finite number of fully interconnected vertices (called "units"). The edge from unit s_i to s_j has an associated real weight w_{ij}, where it is required that $s_{ii} = 0$ for all i, and $w_{ij} = w_{ji}$. Specifically, there are M units s_1, s_2, ..., s_M, and a real, symmetric matrix of interconnection weights $W = \{w_{ij}\}$, i, j = 1,..., M, having zeros on the major diagonal.[3] The units are binary; a unit is said to be active or inactive as its value is 1 or 0, respectively.

Certain units can be designated as inputs, and others as outputs. Units that are neither input nor output units are called processing units.

Inferencing in a MRF proceeds according to the following update rule:
1.   assign the (binary) values of the input units ("input clamping")
2.   randomly select a non-input unit s_i. Sum the connection weights of the active units connected to s_i, and make s_i active if the sum exceeds a pre-selected threshold value t_i, else make it inactive.
3.   Repeat step 2 until the unit values stop changing
4.   Read off the (binary) values of the output units

In this scheme, positive weights between units are excitatory, and negative weights inhibitory: an active unit tends to activate units with which is has a positive connection weight, and deactivate units with which it has a negative connection weight.

Formal mathematical analysis of the inferencing procedure above is facilitated by the introduction of an energy function (due to John Hopfield): Let V be a vector of binary inputs. The energy of the MRF at $V = (b_1, b_2, ..., b_L)$ is:

$$E_W(V) = -(1/2) \sum_{i \neq j} s_i s_j w_{ij} + \sum_{i=1}^{M} s_i t_i \qquad (1)$$

With this definition of network energy, Hopfield's updating rule is equivalent to the following: randomly select an unclamped unit; if toggling its state will lower the network energy, do so. Repeat until no single state change can lower the energy, then read the states of the output units.

The values of the input nodes are fixed, but the changes in the states of the processing and output units under updating cause changes in the energy of the network. This suggests that a supervised reinforcement learning algorithm ("simulated annealing", first applied to Hopfield nets by Hinton and Sejnowski) can be used to train the network to associate desired outputs with given inputs:
1.   assign the (binary) values of the input units ("input clamping")
2.   assign the desired (binary) values of the output units ("output clamping")
3.   randomly select a processing unit s_j. Compute the change in energy that will occur if this unit's state is toggled:

$$\Delta_j E = (\sum_{i=1}^{M} s_i w_{ij}) - t_j \qquad (2)$$

4.   Set the state of s_i to active with probability $p(\Delta_j E) = 1/(1+e^{-\Delta j E/T})$, else to inactive.
5.   Repeat step 3 for many epochs, allowing the network to relax to a local energy minimum
6.   Once an energy minimum has been reached, step around the network, incrementing the weights between pairs of units that are simultaneously active (encourage future simultaneity!), and decrementing the weights between pairs having just one active unit (discourage future simultaneity!)

To avoid saturation of the weights, an additional updating sequence having both the processing and output units unclamped is often included (the machine is allowed to "hallucinate", and associations occurring during these "fantasies" have their weights reduced.)

Here T is a parameter (the <u>temperature</u>) that is gradually reduced as training proceeds. At low temperatures, the weight matrix is easily modified; as the temperature is reduced, the weights "lock" into their final values.

Simulated annealing develops a weight matrix that gives local minima in energy when the input pattern is "consistent" with the output pattern: the lower the energy, the stronger the consistency. The interconnect weights, then, codify constraints on outputs the machine is likely to produce when the inputs units are clamped and updating is applied.

2.7   Using the MRF to estimate (hidden) analyst cognitive states
We now map our treatment explicitly to the Program domain.

Let $C = \{c_i\}$ be a concept space over **D**. That is, each value c_i represents a name and distinguished set of domain concepts.

---

[3] Symmetry of the weight matrix is natural and customary where the input pattern is regarded as a binary image, and is necessary to insure that training results in convergence rather than cycling.

Real-world domains are rarely static; to perform inferencing, a mechanism for tracking states in the concept space C is required. We define the truth-value function $s(c_i)$ to have the value "1" when concept c_i is present, and the value "0" when concept c_i is absent. A tabulation of the states of the concept space C at any instant gives the **concept state** of the domain at that instant. In this sense, the bit array $[s(c_i(X_1, X_2, \ldots, X_n))]$, which we will denote simply as $<s_i>$ when no confusion can arise, is the instantaneous concept state vector for D. It is a string of 0's and 1's indicating the current presence/absence of particular concepts.

2.8 How does information arise?
The concept state vector is the raw material needed to support cognitive processes (e.g., planning, prediction, assessment). By itself, though, the concept state vector is just a list of "facts". This does not qualify as information in the sense we intend.

Information is not the merely the presence or absence of facts; it is the consistent presence or absence of groups of facts together [2]. Consistency here is measured with respect to some set of cognitive states in the domain (e.g., "marshalling", "analysis is complete").

A structure that can be used to model instantaneous information is the co-occurrence matrix. A co- occurrence matrix L has entries $[\lambda_{ij}] = [s_i \wedge s_j]$. The set of all such matrices is the concept state space, Λ. The ij entry of L is 1 if the concepts ci and cj are both present, otherwise it is 0. The state history of the domain D can be described as the sequence of co-occurrence matrices, L_K in the concept state space. For each instantaneous concept state there is a co-occurrence matrix showing which concepts were simultaneously present at that time.

2.9 Embedding discrete concepts into a continuous space
In what has been described so far, the domain D has been described as a sequence of discrete objects (the matrices L_K), which are themselves built from the (binary) presence or absence of discrete concepts (the c_i).

In the case that there is a stationary linear state transition transform, this approach gives rise to Markov models of the domain. Other approaches include Bayesian belief nets (BBN's, [3]), dynamic probabilistic relational models (DPRM's, [4]), and Markov Random Fields [5].

Formally, there is no reason that the functions s_i must assume discrete values. Well-known extensions (e.g., fuzzy logic [6]) have shown the utility of generalized truth values. In our approach, though, we introduce continuous values not by extending the definition of s_i, but by means of a weight matrix W having entries $[w_{ij}]$. Each weight quantifies the association between concepts in the domain D. The connection weights w_{ij} are allowed to assume any real values. In keeping with the recent work of Hinton and Sejnowski cited above[7], we interpret the weights as follows:

When the concepts c_i and c_j tend to be present together (i.e, when s_i is 1, it is observed that s_j tends also to have value 1), then w_{ij} will be positive. When concept c_i being present tends to imply that concept c_j is absent, (i.e, $s_i = 1$ usually means that $s_j = 0$), then w_{ij} will be negative. Uncorrelated s_i and s_j are indicated by $w_{ij} = 0$. The relative magnitude of w_{ij} represents the strength of the positive/negative "association" between the corresponding concepts c_i and c_j.

By analogy with physical systems, these notions give rise to an expression for "energy" in the concept space according to the following formula:

$$E(<s_i>) = -(1/2)\sum_{i \neq j} s_i s_j w_{ij} + \sum_{i=1}^{M} s_i t_i \qquad (3)$$

This formula associates with each set of observed concepts real-valued energy. Refer to the Appendix for a treatment of a physical analogy.

2.10 What Does This Theory Tell Us?
The joint consistencies of concepts and their relevance to cognitive states can be learned directly from instrumented environment data. Given examples of cognitive states for a human expert, simulated annealing [8] can be used to set the weights w_{ij} so that the sets of concepts corresponding to target cognitive states will have locally minimum energies. Thus, the energy function, which can be computed for any set of concepts, can be used to objectively and numerically estimate the (hidden) cognitive state.

Further, tracking the cognitive process by means of the concept state space matrices L_K amounts to following trajectories through the concept state space, Λ. Known cognitive states will be at energy minima in this space; the energy of the current cognitive state gives information about the maturity of the hypothesis, and suggests in which direction the human's analysis may be expected to proceed (i.e., toward which energy minimum it appears to be progressing).

As epochs pass, concept truth values are updated in accordance with new evidence. The concept state space matrix L_K tracks these changes, and a trajectory through Λ is generated. Its initial state, L_0, will generally be a high-energy configuration, because virtually any concept is plausible, and inconsistencies will be present. As reasoning proceeds, inconsistencies (simultaneous presence of conflicting concepts, contributing positive energy) are eliminated, reducing the total energy. In the limit, a maximally consistent set of concepts will survive; this occurs at a relative minimum of the energy function, which is a known cognitive state.

The following graphic (Figure 2) shows the reinforcement cycle during which the expert's cognitive state, reflected in instrumented environment artifacts, is transformed through concept state space to a location on the energy surface. This surface gives an estimate of the (hidden) cognitive state, which can be used in the system to triage information services.

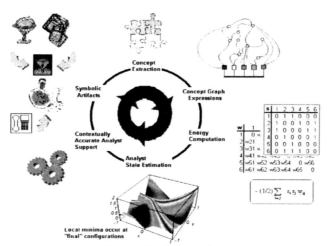

Figure 2: Reinforcement Cycle

In summary, this approach gives a mathematical framework for modeling the human expert's state specifically suited to making estimations of state in the system. It provides objective numeric estimates of the maturity of the expert's hypotheses (as differences in energy between the current and likely goal states), gives a numeric measure of "consistency" of the expert's current beliefs about concepts, and gives a representation scheme for harmonic decomposition of the expert's cognitive states. As will now be seen, harmonic analysis allows the phenomenology corresponding to the human expert's (hidden) states to be decomposed into a set of primitive minimal components; each component is a cluster of related concepts relevant to particular cognitive states.

2.11 Class-wise Decomposition of the energy function

Let a classification problem have M classes, and let the features space consist of N-element binary vectors (e.g., $V = (b_1, b_2, ..., b_N)$ where $b_i = 0$ or1). The feature vectors along with their class names can be thought of as indicating the presence/absence of N concepts in a domain, and so giving the current concept state and its ground-truth:

$$[name, distinguished\ set] = [class, \{s(c_1)=b_1, s(c_2)=b_2, ..., s(c_N)=b_N\}] = [class, <s_i>]$$

A single MRF can be trained to learn these associations. Another way to learn these associations is to build a MRF separately for each class 1, 2, ..., M, and then adjudicate the M outputs (e.g., selecting as "winner" the class machine having the lowest energy, that is, whose class best conforms to the constraints specified by V)[4].

MRF's are <u>commensurable</u> when they have the same topology (their graphs are isomorphic, and they have the same assignment of input, processing, and output units).

An MRF is entirely specified when it topology and weight matrix are given. Given a fixed topology, decomposition of the weight matrix as a sum of commensurable matrices amounts to a decomposition of the MRF W as a <u>sum of machines.</u>

$$W = \sum_1^K c_k W_k \qquad (4)$$

Many such decompositions are possible. Those for which the a_k are all non-negative allow the sum to be viewed as a segmentation of W into positive components. <u>Non-negative matrix factorization</u> (NMF [14]) determines non-negative coefficients a_k and

---

[4] This was done on the 16-feature, 6-class data set described below; the classification accuracies of the single all-class machine and the adjudication of 6 single-class machines differed by a small fraction of a percent (the adjudicated machine being fractionally better).

505

commensurable matrices W_k so that $W = \sum_1^K a_k W_k$ [10].

What can be said about the non-negative sum of commensurable machines,

$W = \sum_1^K a_k W_k$, where the a_k are non-negative?

Let the real weight matrix W for a MRF be:

$w_{11} \ w_{12} \ \ldots \ w_{1N}$

$w_{21} \ w_{22} \ \ldots \ w_{2N}$

$w_{N1} \ w_{N2} \ \ldots \ w_{NN}$

Let W have NMF representation $W = \sum_1^K a_k W_k$

Here a_k are the real reconstruction coefficients for W, and W_k are the real NxN NMF basis elements:

$w_{k11} \ w_{k12} \ \ldots \ w_{k1N}$

$w_{k21} \ w_{k22} \ \ldots \ w_{k2N}$

$w_{kN1} \ w_{kN2} \ \ldots \ w_{kNN}$

Then $W = \sum_1^K a_k W_k$ may be written as the matrix:

$$\sum_1^K a_k w_{k11} \quad \sum_1^K a_k w_{k12} \quad \cdots \quad \sum_1^K a_k w_{k1N}$$
$$\sum_1^K a_k w_{kN1} \quad \sum_1^K a_k w_{kN2} \quad \cdots \quad \sum_1^K a_k w_{kNN}$$

Now, let $\mathbf{V} = (b_1, b_2, \ldots, b_N)$ be a binary vector
(so that $b_i = 0, 1$).

The Boltzmann energy of $\mathbf{V}$ wrt the weight matrix Wi is: $\quad E_W(V) = -\frac{1}{2} \sum_{i \neq j} b_i b_j w_{ij} + \sum_{i=1}^N b_i t_i$ (5)

$$= -\frac{1}{2} \sum_{i \neq j} b_i b_j (\sum_1^K a_k w_{kij}) + \sum_{i=1}^N b_i t_i \quad (6)$$

which can be rewritten as:

$$= [\sum_1^K a_k E_{W_k}(V)] + [\sum_{i=1}^N b_i t_i (1 - \sum_{i=1}^N a_k)] \ (7)$$

When W is a <u>convex combination</u> of the W_k, we have

$1 - \sum_{i=1}^N a_k = 0$, so the second term in (7) is zero. Thus:

Theorem 1: The energy of a convex sum of machines is the convex sum of their energies.
We call a machine homogeneous if all the thresholds t_i are zero. In this case, the second term in (7) will be zero. Thus:

Theorem 2: The energy of a sum of homogeneous machines is the sum of their energies.

Consider now the simplex spanned by the W_k and the origin. This is the set of <u>simplicial sums</u> $\sum_1^K c_k W_k$ having $\sum_1^K c_k \leq$ 1, $0 \leq c_k$.

When the c_k sum to 1 (a convex combination), we are in the face of the simplex opposite the origin. Elsewhere in the simplex, we have $1 - \sum_{i=1}^N c_k > 0$, so by (7) above the function $-E_W(V)$ is convex in the a_k on the simplex. In particular, $E_w(V)$ assumes its minimum on the simplex at one of the vertices {origin, W_1, W_2, ..., W_K}.

Therefore, given a simplicial mixture W of the W_k, there will always be some k so that the minimum over the simplex of $E_w(V) = E_{wk}(V)$. Thus:

Theorem 3: **For a given input vector V, the minimum energy of a simplicial sum of machines occurs at one of the machines.** Averaging machines will not produce more "confident" machines; the average machine will never be more certain than all of its constituents.

3 Decomposing an MRF

3.1 Data Preparation

An analyst in the instrumented environment works in "sessions", which are (usually) contiguous blocks of time during which they perform their customary analyses. Sessions typically last 1 or 2 hours, but occasionally span an entire 24-hour period.

They begin by logging in, and are prompted to identify the type of task they intend to perform. These task types correspond to the five cognitive states: Define, Search, Marshal, Judge, and Report described above. This was not always done in an unambiguous way, so some sessions could not be assigned ground-truth values.

As the analyst works, the instrumented environment collects and annotates the symbolic artifacts (URL's accessed, database searches, web searches, cut-and-paste-buffers, etc.) with times and the task type.

Using six months of instrumented data from multiple analysts, nearly 900 annotated sessions were collected. The term space (set of all "words") for the domain consists of over 3,000 entries. The concept space (distinguished sets) consists of 1,140 concepts[5]. Creation of the concept space from the term space was performed using a combination of automated and manual semantic mapping methods.

With the concept definitions in hand, sessions were tagged as representing one of the five cognitive states.
Sessions that clearly did not consist of analytic work were tagged class 6, "other". Sessions that were not tagged because of time constraints were assigned class 7, "unknown". A class 7 session is actually one of types 1 – 6, but it is not known which. This introduces some ambiguity into the modeling process, since vectors that have the same features occur in different classes. Class 3, for example, consists entirely of vectors which also occur in other classes. We left these ambiguities in the data set for this modeling experiment as a test of robustness.

To evaluate relative ability of the various concepts to distinguish cognitive states, a state identifying k-means classifier was constructed for each concept, and six class precisions were computed for each feature corresponding to the five cognitive states, and the "other" state. The "unknown" state was not analyzed for discrimination.

To winnow the feature set, we selected a few features having the highest precisions for each of the various classes. Three "good" features were selected for each of the six classes except for class 5, which had only one "good" feature.

This gives a feature set having 16 binary features indicating the presence/absence of concepts during a session having known ground-truth (cognitive state).
Seeing a cognitive state's concept sets

For simplicity of presentation, we describe the analysis of class one, which presents only three different concept sets in the training data. These "class 1 archetypes" are:

Feature:	1	2	3	4	5	6	7	8	9	10	11	12	13	14	15	16
A_1	0	0	0	0	0	0	0	0	1	0	0	0	0	0	0	0
A_2	0	0	0	0	0	0	1	1	0	0	0	0	0	0	0	0
A_3	0	0	1	1	0	0	0	1	0	0	0	0	0	0	0	0

The data set was segmented into six sets by class (e.g., set 1 contained only class 1 vectors, set 2 only class 2 vectors, etc.; class 7 was not modeled). Including the original unsegmented file, this gave seven sets for modeling.

An MRF was trained for each of these seven sets. Each had 16 input units, 1 bias unit (always clamped to "1" to allow the machine to process a zero input vector), 30 processing units, and six output units (one for each cognitive state). This gives machines having 53 units. Each MRF represents its constraints in a 53-by-53 real, symmetric interconnection matrix.

---

[5] Each concept is a (usually small) named collection of terms that correspond to some atomic idea in the domain. The terms were stemmed using the Porter Stemmer before being aggregated into concepts.

To make decomposition into positive components possible, we split each matrix in the standard way[6] into a difference of two non-negative matrices, $M^+ - M^-$.

Viewing the seven M^+ 53-by-53 weight matrices as low-resolution images (Figure 3), we applied an NMF algorithm [15] to each to obtain 53 real, non-negative basis matrices (each one 53-by-53); the reconstruction coefficients are stored in a separate matrix, H. This was also done for each of the seven M^- matrices [14].

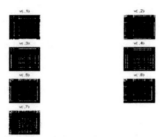

Figure 3: Class 1 - 7 M^+ Weight Matrices

Viewing a weight matrix as an image having rows and columns, it is possible to see the constraints that each basis matrix has captured about a cognitive state. The w_{ij} entry is positive when the concepts c_i and c_j co-occur, negative when they do not co-occur, and approximately zero when they are about as likely to co-occur as not.

Below is an image thumbnail of one of the 53 M^+ basis matrices for the original unsegmented file:

1's M^- matrix. Dark circles represent the weights for their row/col by their radii (Figure 4). Very light circles represent weights that are essentially zero. Patterns known to exist in the class 1 archetypes are clearly indicated in various M^+ matrices; patterns known not to exist in class 1 are clearly indicated in a M^- matrix.

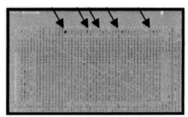

Figure 4: Positive Basis Element #40: "Concept 4 has Multiple Interactions in Cognitive State 1"

4   Future Work

As shown in the example above, these weights designate concept sets that correspond to constraints present in the feature vectors for a class. We believe that using these class-specific constraints, it might be possible to look inside a session and observe transitions in progress. These would show up as "mixtures" of class-specific artifacts.

There are several points in the energy-based state tracking process at which additional analysis may be applied. Probably the most natural is in the concept state space, Λ. By virtue of its construction, Λ is an abelian monoid under component-wise disjunction [9]. This mathematical structure is sufficient for the realization of Λ as a direct product of monoids that can be chosen at our discretion:

$$\Lambda = \mu_1 \otimes \mu_2 \otimes ... \otimes \mu_N \qquad (8)$$

The most useful choice of the μ_i would be sets of primitive concepts strongly correlated with distinct cognitive states. Such a representation allows the consistent, meaningful decomposition of the human expert's current state into a set of representative components. This can be accomplished, for example, by regarding the concept state matrix W as a binary image, and applying non-negative matrix factorization (NMF; [10]). Other harmonic techniques are the wavelet multi-resolution decomposition

---

[6] M^+ has all the non-negative entries of M, with zeros at the negative entries.
M^- has minus the negative entries of M, with zero at the non-negative entries.
Then both M^+ and M^- have non-negative entries, $M = M^+ - M^-$.

(WMRD, e.g. Haar wavelets [11]), and independent component analysis (ICA; [12]). However, these do not segment the image, but deconvolve it.

Analysis of instrumented environment data will show which of these substates correspond to complex analytic behaviors and the corresponding cognitive states (Figure 5). This connection provides a basis for assessing the expert's current state (e.g., "search"), determining when transitions between (hidden) cognitive states is occurring or about to occur, and estimating the maturity (=consistency) of the current hypothesis. The following diagram depicts such a decomposition.

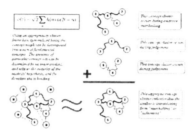

Figure 5: Analysis of Instrumented Environment Data

Having harmonic coefficients will allow us to compactly characterize cognitive states as "sums of archetypes", making the construction of a state recognizer efficient and reliable. This might also support "analysis of the analysis" for process optimization.

5   Conclusions

A human expert reasoning in a complex domain can be modeled formally as a state machine. Consistent associations of sets of concepts with (hidden) cognitive states can be inferred by supervised learning MRF's. In particular, MRF's can be used to estimate the (hidden) cognitive states and detect mixed states.

MRF's can be decomposed into sums of commensurable machines. Viewing the weight matrix of an MRF as a low resolution image, the NMF can be used to segment the machine into a non-negative sum of machines [16]. Facts about state-specific relationships among domain concepts can be derived in an intuitive way from the positive components of a trained MRF when these are viewed as low-resolution images.

Associated with an MRF is an energy function. When the MRF is homogeneous, the NMF segmentation factors the energy function so that the energy of the sum of commensurable machines is the sum of the machine energies. Energy minima of sums occur at summands.

References

[1] Welling, M., Boltzmann Machine, California Institute of Technology
[2] Delmater, R., and Hancock, M., Data Mining Explained, Digital Press, 2001, p. 138
[3] Heckerman, D., "A Tutorial on Learning With Bayesian Networks", Microsoft Technical Report MSR-TR-95-06, 1995
[4] Sanghai, S., Domingos, P., and Weld, D., "Dynamic Probabilistic Relational Models", OSU Colloquia in Computer Science, 2003
[5] Arbib, M., Brains, Machines, and Mathematics, Springer-Verlag, second edition, 1987, p. 105 - 109
[6] Online tutorial: http://www.seattlerobotics.org/encoder/mar98/fuz/flindex.html
[7] Hinton, G., Sejnowki, T., and Ackly, D., "A Learning Boltzmann Machine", Cognitive Science, 1984, volume 9, p. 147 – 169
[8] Geman, S., and Geman, D., "Stochastic Relaxation, Gibbs distributions, and the Bayesian restoration of images", IEEE Transactions on Pattern Analysis and Machine Intelligence, 1984, volume 6, p. 721 – 741
[9] Jacobson, N., Basic Algebra I, W. H. Freeman, 1974, p. 35
[10] Daniel, D., and Seung, H., "Algorithms for Non-negative Matrix Factorization"
[11] Strang, G., "Wavelets", American Scientist, Vol. 82, 1994
[12] Hyyarinen, A., "What is Independent Component Analysis", 2003
[13] Rudin, W., Real and Complex Analysis, McGraw-Hill, 1974, p. 90
[14] Donoho, D., Stodden, V., When Does Non-Negative Matrix Factorization Give a Correct Decomposition into Parts?, Stanford University
[15] Lee, D., Seung, H., Algorithms for Non-negative Matrix Factorization, MIT
[16] Lee, D., Seung, H., Learning the parts of objects by non-negative matrix factorization, Nature, 401, 788-791.

Models and Model Biases for Automatically Learning Task Switching Behavior

Hamilton Link, Terran Lane
University of New Mexico[1]

Joseph P. Magliano
Northern Illinois University[2]

Abstract

Machine learning techniques have been applied to several kinds of human data including speech recognition and goal or user identification. When learning on such data, it is important to use models that are not strongly biased against properties of the data, or the variable assignments learned may be largely incorrect. We are working with data sources for user interface event data and examining the applicability of dynamic Bayesian networks (DBNs) to context tracking. Specifically, we identify the value and transition points of a hidden task variable; this problem is known as segmentation. Our data is drawn from command line interaction collected in a real work setting and window event traces taken during a controlled behavioral study. We have applied discrete time hidden Markov models (HMMs) and DBNs to these data sets, but these methods are fundamentally Markovian and, as a result, cannot correctly learn the properties of hidden variables with nongeometrically distributed dwell times. We believe that using semi-Markov models may better capture some underlying structure and allow for better segmentation. In this paper, we describe the experimental protocols performed, examine the bias of typical DBNs and HMMs towards geometric variable dwell times, and assess the validity of this assumption. We discuss the issues of applying semi-Markov DBNs to the available data.

1 Introduction

One strategy to improve a person's performance on computer-aided tasks is to enable the computer to recognize and respond to patterns in the user's actions. As part of this strategy, we have been researching the automatic identification of task switching from computer user logs. By building systems that detect these switches and develop likely task sets, we hope to automatically identify contexts in user interaction. We would then be able to attach semantic information to these tasks and associate responses with them, at which point reinforcement learning could be used to map the best responses from details about the state of individual users. A learning system would be more flexible than a system tailored to a particular user, while hopefully attaining comparable performance.

One problem of interest in modeling task switching is accurately capturing the probability density function describing how long a person will work on a particular task before moving on. In our experience, hidden Markov models (HMMs) can be effectively used to learn attributes of users through interface data. For example, trained HMMs are sufficient for reasonable user identification performance (Lane, 1999). Conceivably, similar models could be used to segment a data stream accurately as well, to identify a user's work on multiple tasks, but in practice it appears that accurate segmentation is hindered by the strong Markov assumption of such models. In our experiments, automatic segmentation of user data into tasks produces poor results without semi-Markov support.

We have been working with data collected at Purdue and data from a new study conducted at Northern Illinois University (NIU). The NIU data provided an empirical behavioral study of context switching in the presence of multiple information gathering tasks. This data is an application event log and includes some information about each event—for example, web page requests contain their associated URLs. Not all information from the original sessions is available—in particular document and web site content—and not all available information was used by the model. Task labels and segment boundaries are not present in the bulk of the data, so unsupervised learning was used, with one labeled data set available for validation testing. The data gathered at Purdue University is UNIX user command line interactions, collected for previous research on user modeling applied to computer security (Lane and Brodley, 1999; Lane, 2000, etc.). The Purdue data provides more general computer usage data and captures information about unconstrained user task execution patterns, in contrast to the controlled NIU study. No file content was captured. The data is sanitized and is completely unlabeled with respect to the user tasks; shell session boundaries are present in the data but were not originally used for segmentation.

---

1 Dept of Computer Science, Albuquerque, NM 87131 {hamlink,terran}@cs.unm.edu

2 Dept of Psychology, DeKalb, IL 60115, tj0jpm1@wpo.cso.niu.edu

This work involved the design, collection, and preparation of user interface data from NIU, and the initial efforts to automatically segment the NIU and Purdue data with HMMs and dynamic Bayesian networks (DBNs). In our initial estimation, it was hypothesized that Markov analysis would be limited by an inability to model the structured procession of a person from one task to another, and our preliminary results point to the need to more explicitly capture the temporal properties of the data. In future work we hope to leverage the domain expertise of behavioral psychology in automated learning systems.

2 Background

We base our user modeling work on the rich history of statistical user and behavioral modeling. Simple and highly popular statistical models of user behaviors are *atemporal* models such as Latent Semantic Analysis (LSA) (Foltz et al., 1998; Landauer et al., 1997) and other models derived from the "bag of words" assumption. While LSA models have proven to be surprisingly effective in clustering and classifying text documents, they explicitly do not capture temporal interactions among behaviors—precisely the class of effect in which we are interested for the goal segmentation problem. Many authors (including ourselves) have turned to statistically more complex temporal models such as Markov chains (Davison and Hirsh, 1998), hidden Markov models (Lane, 2000), or dynamic Bayesian networks (Albrecht et al.,1998; Zukerman et al., 1999). Of the three, Markov chains make the strongest assumptions about the data, assuming that all statistical information is captured in observable data and that there are no latent or "hidden" variables in the system. Such models are not a good fit for task recognition problems (such as ours) where the existence of one or more hidden variables (such as goals, emotional state, or other mental states) is assumed. User goals are typically not represented directly in observable data, but instead must be inferred from observation and knowledge of temporal dynamics.

HMMs allow a single hidden variable, allowing us to represent a single layer of structure in a goal or task. The state of the hidden variable at a given time can be inferred through the dynamic programming based forward-backward algorithm, while the model parameters can be estimated via the Baum-Welch algorithm, a special case of the Expectation Maximization (EM) algorithm (Rabiner and Juang, 1993). Often, however, we are interested in systems with multiple hidden variables or hierarchically related hidden variables such as goal/subgoal or, in a Goals/Operators/Methods/Selection (GOMS) context, the current stage of a method or operator. HMMs have been extended to hierarchical and layered models for user models (Nguyen et al., 2003; Oliver et al., 2002), but a more general approach to multivariate statistical temporal systems is the DBN model. Bayesian networks arose in the artificial intelligence community in the late 1980's (Pearl, 1988) and have since been widely applied to temporal-process learning and inference problems (Murphy, 2002). Such models are more computationally complex to handle, but possess inference and learning algorithms closely related to the forwardbackward and EM algorithms (Huang and Darwiche, 1994).

A more fundamental difficulty, plaguing both HMMs and DBNs, is that they have a built-in bias toward geometrically distributed state dwell times. This bias arises from discrete-time sampling and an assumption of stationarity—that is, both HMMs and DBNs assume that $Pr[state_t|state_{t-1}]$ is independent of t (ie, $Pr[state_t|state_{t-1}] = Pr[state_{t'}|state_{t'-1}]$ for all t and t') As a result, the process is memory-less and the probability of remaining in a [Fig] single state for k time steps decreases geometrically in k, in contrast to many real-world temporal processes, where the probability of remaining in a particular state is a non-geometric distribution. The probability of staying in some state for k more time steps might depend on how long the process has already been in that state. For example, if a user is doing a task that takes roughly an hour to complete, then the probability that the user completes the task at any moment depends on how long the user has been doing the task. Similarly, factors like boredom thresholds can influence how long a user stays focused on a single task.

The machine learning and human computer interaction (HCI) communities have addressed problems of this nature using a variety of machine learning (ML) techniques to identify the goal or plan of the person under observation or assess a user's cognitive load (Albrecht et al., 1998; Nguyen et al., 2003; Müller et al., 2001, etc.). We are currently using dynamic Bayesian networks (DBNs) as our learning mechanism and ultimately intend to use HCI models to structure the learning process. Comparable work in the field (Oliver et al., 2002; Gurer et al., 1995; Müller et al., 2001, etc.) has indicated that DBNs can be used to accurately infer some attributes of the work environment and the mental state of users under observation, but this work makes limited use of cognitive structure described in HCI and behavioral psychology. Particular models of interest for segmentation of the user's activities are GOMS (Card et al., 1983), Don Norman's seven stage model (Norman, 1986), and Barnard's human information processor (Barnard, 1991).

3 The NIU User Data

The goal of the NIU study was to study task switching and assess human performance in several contexts. The study was conducted to provide problem solving protocols as participants solved complex and naturalistic problems that required web searches and writing documents. These problems were designed to be open-ended. That is, there were a variety of ways a participant could solve the problem. These protocols provided a basis for analyzing their goal structure, which could in turn guide machine learning. We drew upon causal theories to develop an experimental approach to gathering computer use episodes and segmenting them according to goal structures. We created open ended problems for individuals to solve using the web and Microsoft applications (e.g., Word) that would take about an hour to complete. The problems were presented in the context of hypothetical email exchanges with a relative or friend. There was a clearly stated goal provided to the participants at the outset of a session. However, the problem could change as new emails were received. In addition to the primary problem, another task was given to a participant from a different fictive person. This task was social in nature and entirely unrelated to the primary task (e.g., choose a restaurant or a movie). The purpose of this secondary task was to assess when participants would choose to solve it and provide data for domain shifts between two completely unrelated goals.

First message (immediately) Hey, I'm completely swamped at work today. I don't have any time to work on this week's picks. I'm putting my trust in you. Well, sort of!!!! Could you put together a list of potential players that is within the point cap? If you can get it done before lunch, I can take a look at it over my lunch break. Thanks, Jeff.	*Task:* Go to espn Baseball Challenge website. Create a new name and password. Using the rules of the website, select 9 offensive players and pitching staff for todays team. Each player has a point value. The basic rule is that you need to select these players so that the sum of their salaries does not exceed 50 Million. Do NOT, however, enter the names of these players into the game site!!! Instead, Put the names and values of the players in a microsoft word document. Build a brief report in Microsoft Word that justifies your choices, writing a short blurb about each player (or pitching staff) and why you have selected him for todays game. Use any kinds of info that you want here to justify your choices. Assume that your friend wants to review your choices immediately. Send him the Word file as soon as it is complete. Assume that your friend is likely to dispute your decisions because he especially likes players from the Cubs. Write your justification after you send him the Word file.
Second message (after 20 minutes) Hey, Want to see a movie tonight. What's showing in the Naiperville, Aurora, St.	*Task:* Perform an internet search to find movie options in these areas. Create a MS word document listing potential movies and short description if possible. Then create a list of the theaters and movie times for these areas. Send this document as an attachment. Pick a movie that you would like to see and argue for that option in the body of the email.

Figure 1: Fantasy Baseball Protocol

Complex goal directed activities involve a hierarchy of goals, subgoals, and actions (Newtson, 1973; Trabasso et al., 1989; Zacks et al., 2001, etc.). Furthermore, the order in which an individual addresses the goals and subgoals of a problem is often highly systematic (Newell and Simon, 1972). Those subgoals that provide the biggest obstacle in meeting the highest order goal are typically accomplished first. To complicate matters, especially in terms of machine based recognition of goal driven activities, many goals are nested or are accomplished concurrently (Trabasso et al., 1989; Zacks et al., 2001), and the extent to which a person is able to construct a well sequenced set of subgoals varies with domain expertise. Complex goal directed behaviors are sequences of hierarchically related goal episodes (Newtson, 1973; Trabasso et al., 1989; Zacks et al., 2001). The causal network model (Trabasso et al., 1989) of story structure provided the basis for understanding the goal episodes that occurred in the sessions. According to the causal network model, story events can be classier by how they fit into an episodic structure, which consist of a set of story unit categories (Stein and Glenn, 1979). These categories include settings, events, goals, attempts, outcomes, and reactions, but for the present study we disregarded settings and reactions. Events are experienced by a character, but are not the direct result of their action (since they received unsolicited email on a fixed schedule).

We developed two scenarios that would vary in familiarity to a participant: constructing a fantasy baseball lineup (Figure 1) and finding medical information for a sick relative. The email exchanges for both problems are presented to the participant along with information regarding the timing of the reception of the fictive emails throughout the session. All of the participants were experts in fantasy sports, which involve creating hypothetical sports teams based on real players. A team's performance is based on the performance of their players in any given week. There are a high number of websites that provide support for fantasy sports. For this problem, participants were tasked to construct a team for the current week's matchup, which required going to websites that provide information on individual players.

1. Find a diagnosis
    Search symptoms
    Find a symptom search engine
    Find a new search site—nonspecific (1)
    Find a new search site (2)
    Find a new search site (3)
    Find a diagnosis website—nonspecific
    Search new site for symptoms (4)
        Register for symptom website
2. Find out about MDS
    Prognosis info
    Info about types
    Find doctors
    Find hospitals
        Search for hospital site, nonspecific
    Search specific site for MDS
    Search new site for MDS

3. Find doctors and hospitals outside Chicago
    Find doctors
    Find hospitals
4. Find food
5. Sending document
    Locate saved file
6. Close down formerly relevant windows and irrelevant pop-up windows
7. Prepare document
    Extract info from internet: cut and paste to Word file
    Edit document: typing in info and/or editing pasted text
8. Troubleshoot computer problems
9. Open PDF file
    Put PDF link in word file
10. Killing time

Figure 2: Goal structure of a participant engaged in the Health Scenario, as described by the analyst.

The medical information problem involved searching for a diagnosis for a sick relative, and is structurally similar to the fantasy baseball protocol, but involves more interruptions and a more complex search task. The disease chosen, myelodysplastic syndromes, was relatively obscure and would take several attempts to find relevant information. This problem started with an email from a uncle whose wife is exhibiting several unusual symptoms. He is waiting for a diagnosis from his doctor, but wants the participant to find one on the web. After approximately 10 minutes, an email is received with the diagnosis. The task then shifted to finding information regarding the prognosis and treatment options. With respect to treatment options, a restriction is initially placed for local treatment centers, but is eventually opened up to national treatment centers.

We collected three sets of data from each participant for each of these problems. The first was an event log of the computer activities, captured as XML objects. Participants were also instructed to think aloud as they solved the problems and to report whatever thoughts came to mind as they worked through the problems. There is a considerable amount of evidence that suggests that thinking aloud does not change the nature of complex problem solving behavior (Ericsson and Simon, 1994). The third set of data was a continuous record of the information presented on the computer monitor throughout the session. Both the think-aloud protocols and the visual record of the monitor were recorded on DVD. A running clock was used to synchronize all three sets of data. For the user-modeling work in this paper, we have focused on the most easily accessible data: the user interface event log.

In the behavioral analysis, each event recorded was categorized as an event, goal, attempt, or outcome. The goal hierarchy used is shown in Figure 2. We designed the protocols used to allow identification of the major components of the goal episodes a priori. Specifically, the email messages from the relative and friends were considered events that would initiate the goals and sub goals built into the problem and we knew exactly when these would occur in the event log. Furthermore, participants were instructed to send emails to the relative/friends when the explicitly stated goals were completed, which enabled us to identify outcome events in the log. The monitor recordings and think aloud protocols were used to make these classifications as well. For example, when a participant received the initial message in the health scenario, he might say "I need to now find a diagnosis for my relative." As such, we classified the next action as an attempt to achieve this goal. Participants often explicitly stated subgoals, such as finding a symptoms search engine. As such, we would code the next action in the event log as a subgoal attempt. We kept track of and recorded the hierarchical relationships between goals and subgoals. If participants encountered a failed attempt to achieve an explicit goal, they often articulated that failure. As such, these events in the log were coded as a failed outcome. We considered an event a successful outcome of a subgoal if the think-aloud protocols or monitor recordings indicated that it was completed.

4 The Purdue User Data

The Purdue user data was drawn from studies of user modeling applied to computer security—specifically, intrusion and anomaly detection. These studies aimed to differentiate computer users purely by their command line behavior. The data set is UNIX command line session traces from eight graduate students in the Purdue ECE department gathered in the late 1990s. The amount of data available varies among the users from just over 15,000 tokens to well over 100,000 tokens, depending on their work rates and when each user entered and left the study. Because of computational constraints and for testing uniformity, we employed a subset of 10,000 tokens from each user, representing approximately four months of computer usage. All users employed the UNIX shell tcsh and worked in the Xwindows environment. During the original studies, we sanitized the data for privacy by removing file and directory names, user names, email addresses, and web addresses. Command names, switches, and shell metacharacters were preserved, as were the count of arguments to each command. No attempt was made to correct typos in command names or switches, as we hypothesized that patterns of typos might, themselves, be revealing about user identity. Altogether, after sanitization, we identified 2360 unique tokens. An example of raw and sanitized data is given in Figure 3.

```
                              **SOF**     # start of session
                                 cd       # command name
          # Start session        <1>      # one "file name" argument
          cd ~/private/docs       ls       # next command
          ls -laF | more          -laF     # command switches
          cat foo.txt bar.txt \    |       # shell metacharacter
          zorch.txt > somewhere
          exit                    more
          # End session           cat
                                  <3>      # three "file" arguments
                                  <1>
                                  exit
                              **EOF**      # end of session
                 (a)                           (b)
```

Figure 3 Example of Purdue data, pre- and post-sanitization. Example data is synthetic to preserve privacy, but is close in spirit to the original data. (a) Pre-sanitized data. (b) Post-sanitized data. The comment strings (# ...) are purely for clarification and are not present in the actual data streams see by the modeling software.

These data were gathered under uncontrolled circumstances as the users worked normally, and thus represent a spectrum of tasks and goals. All users were informed and consented to the data collection, but we found no evidence that the presence of data gathering affected user behaviors. The command monitoring and recording utility was a built-in capacity of the tcsh shell and imposed no noticeable latency or other overhead that would interfere with user work. We did not attempt to label any data manually, as the original study employed unsupervised learning.

In our prior work with the Purdue data, we showed that users could be distinguished purely through their behavioral patterns with reasonably high reliability (60 - 80% accuracy). Both the instance-based learner and hidden Markov model employed automatically detected natural and semantically meaningful "clusters" of behaviors in the data.

For example, we identified clusters corresponding to text editing, email, and program development. However, segmenting the data into "tasks" was a side effect and not the primary intent. To the degree that the Markov assumption did not interfere with user identification, it was not inappropriate, so we did not examine the dwelltime distribution in detail in that work.

5 Properties of UI Data

It has been noted in domains as varied as economic modeling, semiconductor process design (Ge and Smyth, 2000), control theory (Puterman, 1994), and speech recognition (Russell and Moore, 1985; Ratnayake et al., 1992; Rabiner and Juang, 1993) that many interesting time series processes are significantly nonMarkovian. Specifically, it is often the case that (hidden) state dwell times violate the geometric distribution that naturally arises from a discrete time Markov chain. We are not, however, aware of an examination of this issue for task dwell times in user modeling contexts.

We observe statistical properties in the available data inconsistent with the model bias of Markov HMMs and DBNs. By their iterative nature, hidden variables in such models can be expected to have a geometric distribution of dwell times with a shape parameter corresponding to the fixed likelihood of staying in the same state from one step to the next. Graphing the distribution of session and command lengths for a user against MLE-fit geometric distributions, we see session length likelihoods climb to a sharp peak of a few tokens or commands before trailing off (Figures 4a and 4b). This peak is consistent throughout the Purdue users.

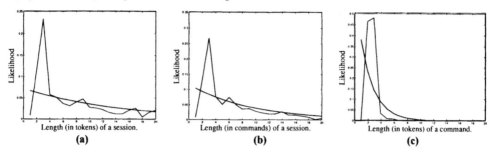

| (a) | (b) | (c) |

Figure 4: Comparison of geometric distributions fit to Purdue user 1's behavior. The natural behaviors in (a) and (b) climb to an early peak before falling. The shell-constrained behavior in (c) shows that short commands with one or two arguments are the norm, with next to no tail on the distribution. (axes differ)

In contrast to the task directed length of sessions, the distribution of *command* lengths in tokens is a function of the UNIX environment. In the original study command boundaries were not used, however the distribution of command lengths does not substantially change the shapes of the curves in Figures 4a and 4b. Bearing in mind that tuples of sanitized tokens were replaced by $<1>$, $<2>$, etc. we see in Figure 4c that short commands of 13 tokens are all fairly common, and that extremely long commands are almost never seen.

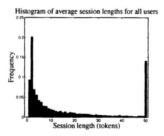

Figure 5: Mean session length distributions across users. The peak seen in User 1 is well represented in the averaged distributions. Zero-length sessions are not included in the graph. The right-most bar is all sessions >50.

Figure 5 shows the average distribution of session lengths (in tokens) across all users in the Purdue study. Zero length sessions have been omitted from the data for this analysis, because they do not represent any particular task information because the windows environment is typically configured to create some number of terminals when a user logs in. In fact the prevalence of zero length sessions had the most variance between users, with the nine users having between five and almost one thousand empty sessions. The graph of "used" session lengths and their average frequency across users shows a robust peak in length and a smooth decline in longer sessions. Looking at the sessions in more detail, we observed a common practice of spawning off one or two independent tasks and exiting; the alternative seems to show users performing more lengthy tasks entirely within the command line. These two different behaviors and the frequency with which an individual user generates zero length sessions could be captured with a mixture model. Such a probability distribution would represent the three individual distributions (zero length sessions, single task sessions, and indefinite sessions) and their relative likelihood. Unfortunately, although this may accurately model terminal window sessions, this mixture model can not be captured by any geometric distribution.

Session boundaries in the Purdue data were available and the quantity of data lends itself to such analysis, but we would expect similar results to hold in the NIU data. Even the simplest task in a graphical user interface will require window selection followed by a spike of mouse or keyboard manipulation within a window, analogous to the creation and brief use of short terminal sessions.

6 ML Methods and Results

In the NIU data, window events are logged as XML objects. We extracted feature values from these sequences, and trained HMMs and DBNs on the resulting sequences of variable values. For the HMMs, we had an unlabeled hidden variable with two to five possible values and used event type as the observable. The DBN used a hidden task variable with eleven possible states and two observables corresponding to the event's application and the event type (Figure 6a). We intended to have a fourth observable mapped to an LSA-chosen cluster for document related event types, but the technical details have not yet been addressed.

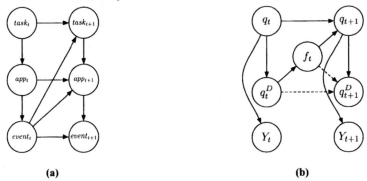

(a)   (b)

Figure 6: (a) The DBN structure trained on the NIU data set. App and event were observed for most time steps, task was an unsupervised hidden variable. (b) The DBN structure of a semi-Markov from HMM (Murphy, 2002) we plan to apply to the Purdue data set. Based on those results, a hierarchical semi-Markov HMM may be used on the NIU data.

The HMMs were trained using the Bayes Net Toolbox (Murphy, 2004), and the DBNs using our own junction tree and EM software (Huang and Darwiche, 1994). As mentioned, this was unsupervised learning using a single validation set. The trained model was used to segment the labeled data, and we chose the heuristically best permutation to map between the abstract task variable values and the semantically laden analyst's labels (Figure 2). The resulting labels were added as annotations to a validation data set (Figure 7a).

We applied the DBN in Figure 6a to the NIU data, but the system proved unable to segment the data into task episodes analogous to those identified by the analyst (Figure 7a). Given the scale and distribution of the specified episodes, this is consistent with our operational hypothesis. If, due to a model bias, the system imposed unnecessary task switches, the system would be forced to conclude that states joined by a highly likely transition exhibited a common, and likely incorrect, pattern of behavior. No clear distinction between those values of the hidden variable would be learned, and the system might segment a new session as we see in our results: a few states, capturing very high mutual transition probabilities, randomly dividing up the entire session (Figure 7c). Relatively common short bursts of activity, such as we see in the Purdue data, would be underpredicted and could lead to a similar result, with transitions being missed and a single state being ascribed to multiple phenomena.

To weakly corroborate this hypothesis, we returned to earlier results from a separate experiment with an HMM—distinct from the DBN that was used to process the NIU data—that had a comparable problem, with the system becoming unable to learn states that had particularly distinct behaviors. An HMM trained to predict event types in the NIU data performed no better than uniform prediction of the most prevalent type identifier. Upon investigation, we found the system bouncing back and forth erratically between the two highest potential states, which were associated with nearly identical probability distributions. This was in an early attempt to predict event classes, but upon reflection the model bias was one reason that the patterns in the data were not learned.

Labelling biases also present a difficulty when applying automated learning methods to partially annotated data. In sparse data, automated learning systems will tend to drift towards whatever patterns are in the data. In unsupervised experiments in particular, these patterns may be somewhat orthogonal to the semantically laden labels an analyst may give to data points in a validation set. Ultimately, an analyst has a lot of prior knowledge about the world, the manner in which people typically solve problems, how information that may be available in the world relates to problems being solved, etc. Take, for example, the descriptive name "register for symptom web site",

516

which semantically captures features about the world and how they relate to the data stream. Even in a large DBN, it would be difficult to represent that the Internet is filled with web sites that are typically comprised of many pages of interrelated material, that some of this material is medical in nature and describes the features an individual may be likely to exhibit when they have succumbed to particular pathogens, and that such information may be accessible only to seekers who have provided identifying information.

7 Conclusions

We have data gathered for two human usage studies, and there are many useful methods in machine learning that we would like to apply to this data. The methods that we originally considered using have a strong bias to interpret data as Markovian and learn properties consistent with that world view. Because the data appears to have several temporal properties that are inconsistent with the model used, we believe that segmentation is not possible with any degree of accuracy using these models unaltered. The direct approach to relaxing the Markov requirement on the data is to use semi-Markov variable structures (Figure 6b), but naïvely running semi-Markov models with a generic DBN is highly inefficient. With linear programming and other methods, it is possible to algorithmically exploit the deterministic nature of the added variables in a semi-Markov DBN. It may also be appropriate to use approximate learning and inference methods. At the moment we are considering these options to making learning and inference tractable while preserving the ability to learn nonMarkov properties and segment the data more accurately.

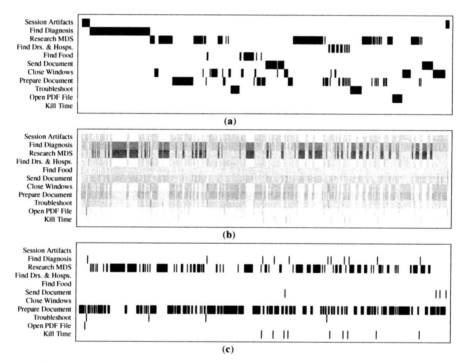

Figure 7: Task variable values over time, depicting analyst labels, inferred likelihoods, and maximum likelihood labels. **(a)** Labels assigned by an analyst to events taken for one user in the Health Scenario. **(b)** Relative likelihood of each task label, after training on unlabeled data. Relative darkness vertically means a label was deemed more likely at that time. **(c)** Labels given to events based on maximum likelihood at each time step.

517

8 Future Work

Adding semi-Markov support to a DBN implementation naïvely can be computationally intensive. Without algorithmically exploiting the counter's deterministic nature, the standard junction tree for a semi-Markov model will have relatively large forward interfaces. For example, the semi-Markov HMM depicted in Figure 6b has a three-dimensional joint probability table as its forward interface, when a Markovian HMM with the same hidden variable has only a single-dimensional potential. In a semi-Markov hierarchical HMM with two levels as might be used to replace the DBN in Figure 6a, the three forward-pointing variables would be replaced by seven, all in the forward interface that determines the structure of the junction tree.

In addition to the computational burden, the addition of hidden variables with potentially high cardinality requires that more information be extracted from the available data, making sparse samples problematic. In both data sets, Purdue and NIU, task labels were not present in the data—forcing us to do unsupervised learning, and further exacerbating the need for data. In future experiments fully annotated data would be desirable, but a balance between the quantity of unsupervised data and annotated data will have to be found. One likely solution is to use richer raw data for events. Document content for the NIU protocols was originally collected for many event types, but was unavailable for this project. We are considering the feasibility of follow-on experiments that preserve such material, allowing us to use LSA or similar techniques for DBN feature extraction.

Although validating our hypotheses on user data is essential, it is difficult to control the necessary variables for an experiment and time consuming to collect enough data for machine learning work. Fortunately, supplementing experimental data with synthetic data is possible and would allow us to study the properties of ML techniques applied to data from a known model. Using generative Markov and semi-Markov variables with the same fundamental variables and similar statistical properties, we can generate representative data sets corresponding to our beliefs about how people work with computers. If the results from small studies on real data and simulated data were analogous, it would justify studies to validate the models experimentally.

A large amount of user study data and exploitable behavioral models could come from HCI projects. We plan to ground our learning methods in these or similar models to add explanatory power. By modeling cognitive processes more explicitly, we hope to gain the ability to assess activities more accurately.

9 Acknowledgements

We thank Chad Lundgren whose editing and formatting assistance with this paper was invaluable. This work was performed in collaboration with Sandia National Laboratories and supported, in part, by the United States Department of Energy under Contract DEAC0494AL85000.

10 References

Albrecht, D. W., Zukerman, I., and Nicholson, A. E. (1998). Bayesian models for keyhole plan recognition in an adventure game. *User Modeling and User Adapted Interaction*, 8(12):5–47.

Barnard, P. (1991). Bridging between basic theories and the artifacts of human computer interaction. In Carroll, J. M., editor, *Designing interaction: psychology at the human computer interface*, pages 103–127. Cambridge University Press, Cambridge.

Card, S., Moran, T., and Newell, A. (1983). *The Psychology of human computer interaction*. Hillsdale, NJ: Erlbaum Associates.

Davison, B. D. and Hirsh, H. (1998). Predicting sequences of user actions. In *Proceedings of the AAAI98/ICML98 Joint Workshop on AI Approaches to Timeseries Analysis*, pages 5–12.

Ericsson, K. A. and Simon, H. A. (1994). Verbal reports as data. *Psychological Review*, 87:215–251.

Foltz, P. W., Kintsch, W., and Landauer, T. K. (1998). The measurement of textual coherence with latent semantic analysis. *Discourse Processes*, 25:285–307.

Ge, X. and Smyth, P. (2000). Deformable Markov model templates for timeseries pattern matching. In *Proceedings of the 2000 ACM SIGKDD*. ACM Press.

Gurer, D., DesJardins, M., and Schlager, M. (1995). Representing a student's learning states and transitions.

Huang, C. and Darwiche, A. (1994). Inference in belief networks: A procedural guide. *International Journal of Approximate Reasoning*, 15(3):225–263.

Landauer, T. K., Laham, D., Rehder, B., and Schreiner, M. E. (1997). How well can passage meaning be derived without using word order? A comparison of latent semantic analysis and humans. In Shafto, M. G. and Langley P., editors, *Proceedings of the 19th annual meeting of the Cognitive Science Society*, pages 412–417, Mawhwah, NJ. Erlbaum.

Lane, T. (1999). Hidden Markov models for human/computer interface modeling. In *Proceedings of the IJCAI99 Workshop on Learning About Users*, pages 35–44.

Lane, T. (2000). *Machine Learning Techniques for the Computer Security Domain of Anomaly Detection*. PhD thesis, Purdue University, W. Lafayette, IN.

Lane, T. and Brodley, C. E. (1999). Temporal sequence learning and data reduction for anomaly detection. *ACM Transactions on Information and System Security*, 2(3):295–331.

Müller, C., GroßmannHutter, B., Jameson, A., Rummer, R., and Wittig, F. (2001). Recognizing time pressure and cognitive load on the basis of speech: An experimental study.

Murphy, K. (2002). *Dynamic Bayesian Networks: Representation, Inference and Learning*. PhD thesis, UC Berkeley, Computer Science Division.

Murphy, K. (2004). Bayes Net Toolbox for Matlab. http://www.cs.ubc.ca/ murphyk/Software/BNT/bnt.html.

Newell, A. and Simon, H. A. (1972). *Human problem solving*. PrenticeHall, Englewood Cliffs, NJ.

Newtson, D. (1973). Attribution and the unit of perception of ongoing behavior. *Journal of Personality and Social Psychology*, 28:28–38.

Nguyen, N., Bui, H., Venkatesh, S., and West, G. (2003). Recognising and monitoring highlevel behaviours in complex spatial environments.

Norman, D. A. (1986). Cognitive engineering. In Norman, D. A. and Draper, S. W., editors, *User centered system design: New perspectives on human computer interaction*. Hillsdale, NJ: Erlbaum Associates.

Oliver, N., Horvitz, E., and Garg, A. (2002). Layered representations for human activity recognition. In *Fourth IEEE Int. Conf. on Multimodal Interfaces*, pages 3–8.

Pearl, J. (1988). *Probabilistic Reasoning in Intelligent Systems: Networks of Plausible Inference*. Morgan Kaufmann.

Puterman, M. L. (1994). *Markov Decision Processes: Discrete Stochastic Dynamic Programming*. John Wiley & Sons, New York.

Rabiner, L. and Juang, B. H. (1993). *Fundamentals of Speech Recognition*. Prentice Hall, Englewood Cliffs, New Jersey.

Ratnayake, N., Savic, M., and Sorensen, J. (1992). Use of semi-Markov models for speaker independent phoneme recognition. In *Proceedings of the 1992 IEEE International Conference on Acoustics, Speech, and Signal Processing (ICASSP92)*, volume 1, pages 565–568. IEEE Press.

Russell, M. and Moore, R. (1985). Explicit modelling of state occupancy in hidden Markov models for automatic speech recognition. In *Proceedings of the 1985 IEEE International Conference on Acoustics, Speech, and Signal Processing (ICASSP92)*, volume 10, pages 5–8. IEEE Press.

Stein, N. L. and Glenn, C. G. (1979). An analysis of story comprehension in elementary school children. In Freedle, R. O., editor, *New directions in discourse processing*, volume 2. Erlbaum, Hilldale, NJ.

Trabasso, T., van den Broek, P., and Suh, S. (1989). Logical necessity and transitivity of causal relations in the representation of stories. *Discourse Processes*, 12:1–25.

Zacks, J. M., Tversky, B., and Iyer, G. (2001). Event structure in perception and conception. *Psychological Bulletin*, 127:3–21.

Zukerman, I., Nicholson, A., and Albrecht, D. (1999). Evaluation methods for learning about users.

Context modeling as an aid to the management of operator state

Dr Andy Belyavin

QinetiQ Ltd
A50 Building, Cody Technology Park, Ively Road
Farnborough, Hants GU14 0LX
ajbelyavin@qinetiq.com

Chris Ryder

QinetiQ Ltd
A50 Building, Cody Technology Park, Ively Road
Farnborough, Hants GU14 0LX
cjryder@qinetiq.com

Abstract

All Augmented Cognition systems have to be based on measurements of operator state to drive the active implementation of appropriate mitigations with a view to maintaining operator performance. The measurement of operator state is necessarily based on observations from the immediate past, but the need for good quality performance management require that future demands on the operator are anticipated rather than measured. The system therefore needs some mechanism that can project future operator state. This paper outlines an approach for projecting future state based on the construction of a user model and the derivation of a simple finite state machine representation of the user model for application in the real time system. It is demonstrated that a plausible user model can be constructed using the Integrated Performance Modelling Environment (IPME), but the simplest approach to forecasting future behavior and workload has limited success. Possible alternative approaches are considered.

Keywords: Workload models, task network models, finite state machines.

1   Introduction

All Augmented Cognition systems have to be based on measurements of operator state to drive the active implementation of appropriate mitigations with a view to maintaining operator performance. The measurement of operator state is necessarily based on observations from the immediate past, but the need for good quality performance management require that future demands on the operator are anticipated rather than measured. The system therefore needs some mechanism that can project future operator state. A natural approach would be to characterise the dynamics of the state measures and to use standard control theory based on first or higher order forecasting to derive an estimate of state a short time in the future. It is argued in this paper that this approach sacrifices a rich source of additional information provided by knowledge of the current context and how the operator should interact with the system.

The construction of a user model is a highly desirable feature of the development of the overall system. If this model is incorporated, it can be exploited as part of the Augmented Cognition system. Short-term forecasts of operator behavior can be based on the embedded user model. Using established models of cognitive workload, future projections of operator workload can be constructed for population means. These estimates can then be combined with measures of current operator state to project estimates of future state. A procedure of this kind has two advantages: if the measured demand on the operator is more than expected the system will react appropriately, compensating for stressed performance in a natural manner; the system will anticipate demand on the operator before it occurs, enabling optimal performance at all times.

The development of an embedded user model that can be executed conveniently in real time is a complex task. To provide an approach that would enable the effectiveness of relatively simple modeling to be tested, a two stage approach is proposed. A detailed user model is constructed so that the general properties of the modeled system can be understood, and the model can be validated against observation. Using the validated user model as a reference, a simplified model can be built that reconstructs the behavior of the full user model and can be embedded in the system. This paper describes the application of this two stage process at QinetiQ to a simplified UAV task. The task is described in Section 2, the task network model describing the interaction between the operator and the task is described in Section 3 and the development of the simplified model is described in Section 4.

2   TaskLite – a UAV management task

2.1   Task outline

TaskLite is a medium fidelity task constructed at QinetiQ, based on the role of a putative Uninhabited Air Vehicle (UAV) operator. The operator's task is to manage a number of assets (UAVs) simultaneously, by tasking each individually to perform a specific role. These roles include searching for a target, taking an image of a target, loitering in an area, engaging an enemy target, or returning to base to replenish fuel, power and weapons. The task requires the strategic deployment of these assets to perform a reconnaissance role, enabling the operator to find and then image each target. Once an image is returned by an asset, the operator must then classify it as friend, foe or neutral. Following classification, the operator may then instruct the UAV to engage the target if it is classed as hostile.

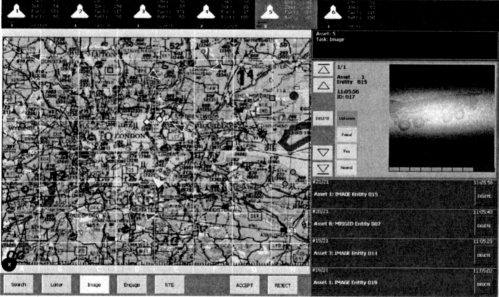

Figure 1: TaskLite display to the operator

521

Figure 1 displays a screenshot of the operator's interface. Along the top is a command task bar which enables the operator to select the UAVs. In addition, it contains information relating to the status of each UAV asset, in terms of fuel level, battery power, number of weapons available, heading and velocity. Below the map display is a tasking bar, which enables the operator to instruct a selected asset to perform a particular operation (i.e. loiter, search, image engage or RTB); in the above display the highlighted UAV is being tasked to IMAGE. Tasking uses a well understood syntactic dialogue with the asset in the form of SUBJECT – VERB – OBJECT. An example of this in operation could be: ASSET 5 (click on UAV symbol) take an image (click IMAGE) of Target 06 (click on target symbol) OK (click ACCEPT). When the asset is within a certain radius an image is returned, for which a classification Identify Friend or Foe (IFF) must then be performed. The target will initially be classified as UNKNOWN. The operator must classify it as FRIEND/FOE/NEUTRAL, using the dialogue in the image window on the right of the screen. Assets may be instructed to engage a target, if the target has been identified as FOE. It is also important that operators ensure that each UAV has enough fuel, power, and weapons to perform the task asked of it. If the fuel levels are low, the operator must instruct the asset to return to base (RTB) in order to refuel and recharge the image pod. Failure to do so will result in the loss of the UAV if fuel runs out, or inability to perform the task if the power or weapons have expired.

2.1.1 Additional workload – Vehicle Health Task (VHT)

In addition to the main UAV task, to provide extra workload, there is a Vehicle Health Task (VHT), which is presented in a window below the UAV task as shown in Figure 1. There are 32 different questions regarding the state of the system as a whole, which are presented as either a straight choice evaluation, a series of conditional statements, or require a calculation, each of which has multiple choice answers. For example:

- Electrical questions: "Switch to the battery with the higher amperage. Batt-A: 4 Amps, Batt-B: 5 Amps",
- Mechanical questions: "Re-cycle Bomb Bay Door ONLY when the Weapon Release Actuator is off: Weapon Release Actuator is ON. Re-cycle Bomb Bay Door, Don't re-cycle",
- Communications questions: "Switch to channel with higher carrier to noise ratio. Channel-A: Carrier 700, Noise 50 Channel-B: Carrier 700, Noise 60. Switch to Channel-A, Switch to Channel-B".

These questions are presented either as a single question which remains on the screen until it is answered or 20s have elapsed, or in a block of three questions, presented at 10s intervals until answered or until 10s has elapsed.

2.1.2 Mitigations

The key operator task is the identification of the images returned by the UAVs and any subsequent instruction to engage targets. The quality of the images is uneven, and a number are not simple to identify. In the event that the operator cannot identify a target, he can request a further image from the UAV from a better position. To provide an adaptive version of the system a number of mitigations are defined that make the operator's task easier during busy periods:

- Returned images are classified using a pseudo Automated Target Detection (ATD) routine. Images are given a suggested classification and an associated confidence rating.
- Images are presented for long enough (e.g. 15 seconds) to enable classification before being overwritten by new images.
- Images are queued and presented on the basis of the ATD classified priority (i.e. hostile targets are presented first).
- An automated monitor for levels of fuel, power and number of weapons onboard each UAV has been implemented. When the UAV is running low a visual queue appears to inform the operator to send the UAV back to base.

The ATD system is not capable of perfect target identification, so that when the operator is less busy he will have better performance than the ATD, but during busy periods he is likely to perform worse, implying that best system performance will be obtained if the mitigations are implemented during busy periods.

3 Development of the full user model

3.1 System states

In developing the user model, the interactions between the operator and Tasklite were viewed as a series of state transitions. Each UAV within the system is in a defined state at a particular stage of its current mission. When it achieves the mission goal, its state changes and operator intervention will be required. For example, a UAV can be tasked to search for possible targets. If it reports the presence of a target, the operator may then request that it provides an image of the target for identification. This interaction provides a further change in state of the system and the cycle continues.

Each of the operator interactions with the system can be identified with a specific task that the operator performs. As a first step in the construction of the model, the pattern of task execution was examined for a small number of runs undertaken as part of the task development. From the preliminary analysis of task performance it was concluded that the overwhelming majority of operator interactions could be grouped together into a small repertoire of behaviors. Each operator behavior triggers a state transition for the system in that, for example, an image is identified, the tasking for a UAV is changed or the VHT is answered. By considering the activities of both UAVs and the operator a small number of system state changes can be identified:

The operator behaviors are:

- Task a UAV to search for targets
- Task a UAV to image a target
- Task a UAV to engage a target
- Task a UAV to return to base
- Identify a target image
- Monitor the display
- Respond to the VHT

The equivalent UAV activities are:

- Loiter
- Search for targets
- Image a target
- Engage a target
- Return to base

A relatively simple abstract model can be defined by linking together the activities of these agents within the system, taking account of the relative priorities of the different operator activities and recording the queues of tasks requiring operator intervention.

3.2 Task network user model

The initial model was implemented in the Integrated Performance Modelling Environment (IPME) (Belyavin and Winkler 2003) as a task network model. The operator behaviors and the UAV activities were modeled as discrete sub-networks composed of a number of atomic tasks. In the overall flow of the model the UAV activities are initiated by operator behaviors; the operator behaviors are determined by the queue of events reported by the UAVs and the relative priorities of the required interventions. The outline model structure is displayed in Figure 2. In the left pane the operator behaviors are highlighted, and in the right pane the UAV activities are identified. In this initial version of the model, mitigations are either always on or always off.

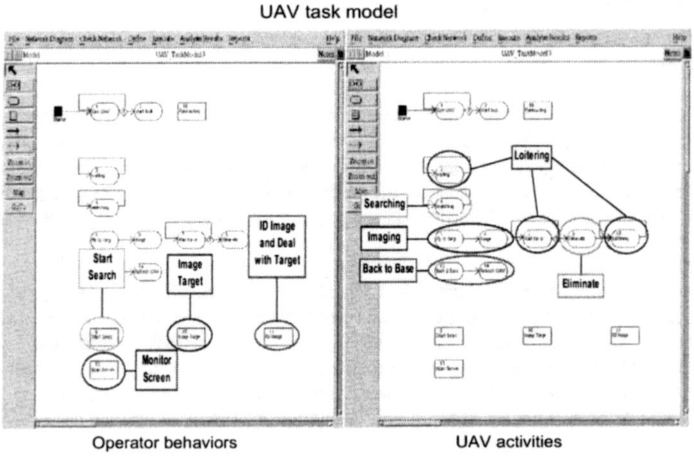

UAV task model

Operator behaviors   UAV activities

Figure 2: Operator behaviors and UAV activities in the IPME task network model

Workload forecasting was undertaken using the POP model (Farmer 2000), which is implemented in IPME as one of the two main task demand models.

3.3   User model validation

The overall user model properties are determined by two key elements: the representation of the times that operator and UAV activities take to execute; the priorities with which operators undertake actions given conflicting queues of demands. The high–level descriptors of these behaviors are the measured times for sequences of activities and the probabilities that particular events follow each other. As a first stage in the development process, a pilot trial was undertaken with the UAV task in which the task was tested in both mitigated and unmitigated forms. Data was then collected on the time taken to undertake a range of activities and the patterns of transition between operator behaviors and comparison was made between the observed and modeled values.

Initial analysis of both observed and simulated times was undertaken to test for differences between mitigated and un-mitigated conditions. No differences were found, so the comparison between observed and simulated times could be made at the gross level. The comparison between the times for operator behaviors is displayed in Table 1.

Table 1: Observed and simulated times for operator behaviors

Behavior	Observed Mean Time (s)	Observed Standard Deviation (s)	Simulated Mean Time (s)	Simulated Standard Deviation (s)
Search	4.45	1.61	4.99	1.76
Image	4.51	1.51	4.74	1.76
Engage	5.00	1.56	4.65	1.56
RTB	3.48	1.63	3.54	2.14
VHT	3.97	1.50	3.79	1.71

The comparison between observed and simulated times is generally good for both mean and standard deviations for all behaviors. Additional comparisons were made for composites of the behaviors by measuring the time taken from receipt of a target image to destruction of a hostile target. A comparison of the two values is displayed in Figure 3.

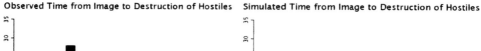

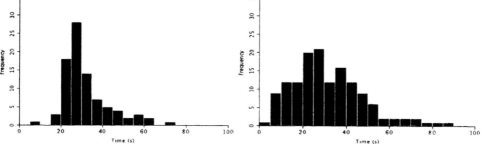

Figure 3: Distributions of observed and simulated time from image to destruction of hostile targets

The means and standard deviations of the two distributions in Figure 3 are very similar, confirming the pattern shown in Table 1. However, it is clear that the distributions are different, with the observed distribution having the larger positive skewness, probably due to small discrepancies in the representation of operator priorities for particular behaviors in the model.

To investigate potential problems with the allocation of operator priorities, a comparison was made between observed and simulated transition probabilities between the five operator behaviors and the results are displayed in Table 2. The columns headed Obs. are the values observed in a trial, while the columns headed Pred. are the values derived from the simulation model.

Table 2: Observed and modeled transition probabilities between successive operator behaviors

		To								
		Search		Image		ID		Engage		RTB	
		Obs	Pred.	Obs.	Pred.	Obs.	Pred.	Obs.	Pred.	Obs.	Pred.
From	Search	0.55	0.42	0.29	0.26	0.12	0.17	0.01	0.01	0.03	0.14
	Image	0.15	0.07	0.24	0.45	0.54	0.38	0.02	0.05	0.04	0.06
	ID	0.32	0.17	0.17	0.30	0.15	0.34	0.25	0.13	0.11	0.06
	Engage	0.26	0.26	0.38	0.19	0.16	0.37	0.04	0.12	0.16	0.07
	RTB	0.23	0.07	0.22	0.07	0.22	0.18	0.09	0.05	0.25	0.32

The match between observed and modeled transition probabilities is quite good, although there are clearly some discrepancies, particularly with respect to transition in the sequence between imaging and engaging. There is some evidence that the initial structure of the operator priorities in the model, based on early trials with a few individuals may need revision to take account of the effects of training. There was evidence within the validation trial that, despite initial training, there was continuing change in subject performance within the trial.

4   Development of the simplified user model

A key component of the augmented cognition version of the UAV task is the short term forecast that will enable the system to adapt to the operator load in the light of future anticipated workload. To support the development of the short term forecast model, the simulation model was adapted so that a simplified representation could be embedded and tested by comparing the results with the full model. It was decided that the short-term forecast should look ahead no more than two behaviors, providing up to a 10 second look-ahead.

To simplify the complex task network model for use in embedded form, it was defined as a simple finite state machine in which the operator behaviors were modeled as state changes, and a series of behavior transition probabilities were constructed using the task network model to supply the data. Five state identifiers were defined:

- One or more UAVs are loitering
- One or more UAVs has detected a target
- One or more images are awaiting identification
- One or more friendly images have been identified
- One or more hostile images have been identified

Labelling the states with a 0 or 1, depending on whether the condition is met, 32 different states of the system were recognised. For each state, a separate probability transition matrix between the operator behaviors was estimated by running the full user model for a long time. The maximum likelihood estimate of the next behavior was then inserted in the simplified model. A sample of the probability transition matrix for a single state is displayed in Table 3.

Table 3: Transition probabilities between behaviors for state 00011[a1] (friend and enemy waiting to be processed)

	SEARCH	IMAGE	ID	ENGAGE
SEARCH	0	0	0	1
IMAGE	0	0	0.25	0.75
ID	0	0.5	0	0.5
ENGAGE	0	0	0.29	0.71

The expected probability of determining the next behavior correctly is derived by summing the weighted probabilities of the corresponding transitions that will occur in the model. The observed probability of predicting the next behavior correctly with the initial set of states is approximately 0.5, consistent with a high incidence of the least well determined state where no events are queued displayed in Table 4. The largest of the maximum likelihood estimates is 0.52 and the smallest 0.38.

Table 4: Transition probabilities between behaviors for state 00000 [a2](General table)

	SEARCH	IMAGE	ID	ENGAGE
SEARCH	0.25	0.38	0.37	0.00
IMAGE	0.12	0.34	0.52	0.02
ID	0.29	0.33	0.26	0.13
ENGAGE	0.18	0.39	0.43	0

This basic form of the simplified model has been implemented successfully in the UAV task, and provides a plausible forecast of workload that is better than current state, since observed and forecast workload are significantly positively related.

5   Discussion

Two stages in the construction of forecasts of workload in augmented cognition systems have been identified: the construction of a full user model to simulate the system and the derivation from the full model of a simplified model for embedding in the system. For the UAV task it has proved possible to construct a task network representation of the overall system that reproduces system characteristics to a reasonable degree, subject to the rider that the operators tested are trained to a consistent level. The derivation of a simplified user model that can be implemented as an embedded model in the system has not proved as tractable although the initial approach shows some promise.

This initial test with the UAV task has demonstrated that it is possible to construct a finite state machine that is a partial representation of operator behavior for this relatively simple case. However, initial estimates of the overall quality of forecast behavior using a single branch for the future forecast suggest a probability of success no better than approximately 0.5 using the simple scheme outlined above. Improvements to the prediction accuracy of this single branch approach are being investigated by examining those cases where the maximum likelihood behavior has a probability less than 0.75. By improving the definition of the current state in these cases and increasing this probability, it may be possible to achieve higher forecasting accuracy for a single branch method of forecasting future workload.

An alternative schema using a probability tree approach is under development, in which all possible branches of the tree up to two behaviors in the future are considered and the expected value of future workload is constructed from all possible routes through the tree. This has the advantage that the finite state machine can be used as an approximation independent of the probabilities of particular events occurring. There is a clear disadvantage when the number of future behaviors is considerable as the complexity of the tree may increase rapidly, making the calculations difficult to derive in real time. Initial development will be undertaken on the UAV task, but the method is to be tested in the Cognitive Cockpit.

References

Belyavin A., Fowles-Winkler A. (2003) Subject variability and the effect of stress In discrete-event simulation. In *the proceedings of 15th European Simulation Symposium* 615–620 Delft. October 2003.

Farmer, E. W. (2000). *Predicting operator workload and performance in future systems.* Journal of Defence Science, 5(1), pp F4–F6.

Automatic Event Structure Parsing for Context Modeling: A Role for Postural Orienting Responses

Carey D. Balaban, Jarad Prinkey, Greg Frank and Mark Redfern

Dept. of Otolaryngology,University of Pittsburgh , EEINS 107 Pittsburgh, PA 15213, U. S. A.

[1] Sandia National Laboratories, P.O. Box 5800, Mailstop 1188, Albuquerque, New Mexico 87185, U. S. A.

E-Mail: cbalaban@pitt.edu

Abstract

Our daily activities depend upon the ability to divide activities and contexts into distinct temporal parts or sequences, both as the actor and the perceiver. In subjects seated at a computer monitor or driving a car, these orienting movements can be measured non-invasively and non-obtrusively by pressure-sensitive arrays in the seating surfaces and by ultrasonic or head sensors. This communication reports detection of postural engagement that can be measured from seated subjects (1) performing battle space defense tasks at a computer workstation and (2) driving on a highway while performing secondary and tertiary cognitive tasks. These gauges appear to provide signals reflecting parsing of contextual information into epochs requiring global versus local situational awareness and contexts requiring imminent action. Since these measures have been implemented in real-time, they can be applied in the future in augmented cognition systems.

1 Introduction

During normal living activities, we are immersed in a stream of activities and contexts. However, we perceive these contexts and activities as being divided into temporal distinct parts or sequences, in a process termed *event structure perception* (Zacks and Tversky, 2001). These perceptual event boundaries can be identified operationally by methods such as subjective identification of meaningful boundaries in video clips or monitoring self-presentation of text on a computer monitor during reading comprehension tasks. This communication presents evidence that measurement of postural orienting responses in seated subjects can demarcate aspects of the parsing of continuous events into parts. Since the measurements are unobtrusive, they provide objective, subject-based framework for real-time contextual cognitive modeling.

The concept that there are automatic physiologic responses associated with orienting to stimuli, parsing information and preparing for action is synthesis of two traditions in experimental psychology and physiology from the mid-twentieth century. The first tradition, summarized by Fraenkel and Gunn (1961), studied directed somatic movements of organisms in relation to stimuli. Directed movements orienting an organism toward or away from a stimulus source were designed as positive (i.e., attractive) or negative (i.e., repellant) forms of 'taxis'. By contrast, they classified movements that varied in rate or intensity with stimulation, but not in orientation, were termed 'kineses.' The second tradition was developed for autonomic responses in the mid-twentieth century by the Pavlovian school in the former Soviet Union. As reviewed by Sokolov (1963), three teleologically distinct classes of responses were distinguished: (1) 'orientation reflexes', (2) 'specific adaptation reflexes' and (3) 'defense reflexes.' Each of these response types was characterized both by distinct temporal patterns of autonomic activity and by distinct adequate stimuli. In the light of clear evidence that autonomic and somatic movements are coordinated (e.g., vergence eye movements: Leigh and Zee 1991; exercise physiology: Waldrop et al. 1996), it is important to regard automatic orienting movements as a synthesis of the concepts from the prior literature. This we may term the performative aspect of orienting responses.

Orienting responses also are also expressive, serving as a form of non-verbal communication of engagement in a task. In effect, we become reasonably adept at reading the 'body language' of orienting responses in others and drawing inferences about immersion in a task. For example, a person listening carefully to a phone conversation assumes a posture and affect that we associate with immersion, while a person listening to music, advertisements and platitudes while 'on hold' assumes a very different pattern of posture and affect.

Postural adjustments reflect a trade-off between the demands of maintaining balance and the demands of generating voluntary movements. Voluntary movements vary in their degree of automaticity. Orienting responses are automatic voluntary movements that direct processing resources toward events of interest. Detecting these movements is facilitated in seated humans by the constraints imposed by the chair and by the reduced degrees of freedom of body movements for directing gaze, the torso and upper limbs. Hence, seated workplace and vehicular environments are advantageous platforms for using pressure sensor and head tracker technologies to detect orienting responses.

2 Experimental Methodology

2.1 Computer Workstation Experiments: Warship Commander Task and Aegis CIC Task

Fifteen students (18-24 years old) served as subjects in this study. The protocol was reviewed and approved by the Institutional Review Board of the University of Pittsburgh.

The Operator's chair from Lockheed-Martin's Sea Shadow ship was reupholstered with slipcovers containing Ultrathin 16 X 16 pressure sensor arrays (Vista Medical, FSA pads). The sensors are spaced at 1 inch intervals in a square array. Separate sensor arrays were used to measure the distribution of pressure on the seat bottom and the seat back at a rate of 4.5 Hz. A Flock of Birds (Ascension Technologies) tracker was used to monitor the position of the head and trunk (mid-sternum) with 6 degrees of freedom at 103 Hz. A stand supporting a 15 inch LCD monitor, a keyboard and mouse pad was positioned in front of the subject for presentation of a military air defense simulation, the Warship Commander Task (WCT; St. John et al., 2002).

This task presents a series of nine 75 second trials with waves of friendly, hostile and unidentified planes, which must be tracked, identified (using an 'Identify Friend of Foe' (IFF) interrogation command), warned (if hostile and within the home ship's defensive perimeter) and, if necessary, neutralized. Each new trial is cued by the sound of a bell. The workload was varied by modifying (1) the number of planes in a wave (6, 12, 18 or 24 tracks) and (2) the proportion of unidentified planes ('high yellow' or 'low yellow' conditions); this experiment utilized conditions J, K and L of Warship Commander Version 4.4. The WCT software provides performance statistics, including second-by-second tracking of the number of planes on the screen and the number of tasks pending (e.g., IFF, warning and missile deployment). Posture related data were collected on a PC using custom applications written in LabView (National Instruments) and were analyzed off-line using routines written for MATLAB (MathWorks) and Excel (Microsoft). The WCT task performance data were logged directly to Excel compatible text files (St. John et al., 2002).

Seven additional subjects participated in an experiment with a modified Aegis CIC workstation task, as part of an AugCog Phase 2 CVE with Lockheed-Martin ATL. The same seat was used, but an ultrasonic head tracker (Logitech, Inc.) was used to sample head position with 6 degrees of freedom at a 50 Hz sample rate. This task was similar to the WCT, but the appearance of waves of targets was uncued.

2.2 Mercedes-Benz S-Class Platform

Similar instrumentation was used to record postural adjustments of 5 drivers (male DaimlerChrysler employees, 24-33 years old) in an experiment described in detail by Bruns et al. (this volume). The postural data were collected Ultrathin 16 X 16 pressure sensor arrays (Vista Medical, FSA pads) on the seat bottom and seat back. The sensors are spaced at 1 inch intervals in a square array and are samples at a rate of 4.5 Hz. The transmitter of a Logitech Ultrasonic Head Tracker (Logitech, Inc.) was mounted on the ceiling above the driver's head and the receiver was mounted on a cap worn by the driver. This tracker monitored the position of the head with 6 degrees of freedom at a 50 Hz sample rate. Data were collected for three experimental protocol sessions from each subject, a *Calibration* session, a *Reference* session and an *AugCog* session (when workload mitigation strategies were employed). The latter sessions each lasted approximately two hours. The primary task in each session was driving a predefined route on German highways. A block design was used for three secondary tasks, interspersed with control periods. The Auditory Workload task required the driver to listen to a story presented by a mail narrator, while ignoring a female newscaster. The Mental Arithmetic task required the subject to count backwards from an arbitrary three digit number in increments of 27. The Special Driving Maneuvers task required the driver to perform specific maneuvers upon receiving commands from the experimenter. The tertiary task consisted of blocks of visual or auditory commands to press a right or left manipulandum for a reaction time task. Since the performance on this task (signal detectability, d')in the *AugCog* session showed 108% improvement during the Auditory Workload task, 103% improvement during Mental Arithmetic task and 72% improvement during Special Driving Maneuvers task (Bruns et al., this volume), one predicts that postural gauges of cognitive workload show a significant difference between the *Reference* and the *Augcog* sessions during the tertiary task blocks.

The raw data from the seat pads and head sensor are used to calculate ten uncorrelated variables for the analyses:

1. Cosine of seat yaw. Seat yaw as the angle between the seat back and a line between the center of pressure (COP) of the left leg/buttock and the right leg/buttock, with the vertex on the left.
2. Base of support: The distance between the COPs of the left and right leg/buttocks.
3. Seat pressure derivative RMS: The instantaneous standard deviation of the derivative of pressure on at each sensor across the seat pad.
4. Back pressure derivative RMS: The instantaneous standard deviation of the derivative of pressure on at each sensor across the back pad.
5. Global front-back seat pad COP.
6. Global left-right seat pad COP.
7. Global up-down back pad COP.
8. Global left-right back pad COP.
9. Cosine of head yaw.
10. Cosine of head roll.

3 Results

3.1 Warship Commander Task (WCT)

During the first half of waves of tracks in the WCT, seated subjects show a characteristic pattern of head and postural adjustments such that movements in the direction of the monitor are correlated linearly with the number of tracks on the screen (Balaban, Cohn, Redfern, Prinkey, Stripling & Hoffer, 2004; Balaban, Prinkey, Frank & Redfern, this volume). These movements are identified by examining the linear correlation between (1) head position, (2) right buttock/leg center of pressure or (3) left buttock/leg center of pressure and the number of tracks on the monitor during a 8 second window. A *global postural engagement response* is defined as any time when the correlation coefficient (r) for any of these calculations is greater than 0.71, which indicates that 50% of the variance is accounted for by the relationship. Figure 1 shows the time of detection of postural engagement responses (black dots along the time axis) in a representative subject. The instantaneous number of tracks on the monitor are plotted the twelve successive waves within each WCT session. These sessions differ in workload because of variations in the proportion of unidentified tracks (yellow icons on the screen) in the waves, which require the subjects to identify the track as 'friend or foe' prior to deciding on an appropriate action. Sessions J and L have a low proportion of unidentified tracks; session K has a high proportion of unidentified tracks. Note that the engagement responses tend to occur (1) during the initial buildup of tracks (first third of the wave) and (2) near the end of a wave with the decline in the number of tracks (last third).

The potential role of *global postural engagement responses* as an index of cognitive event parsing is illustrated in Figure 2, which shows the number of subjects displaying an engagement response during each of the presentations of waves of 12 tracks and waves of 24 tracks in the J (low proportion of unidentified tracks) condition. The distribution of engagement responses in also plotted with respect to a scaled median urgency score, which increases as unidentified tracks or foes cross the perimeter for defensive action and approach the home ship. The increment of the number of tracks during the first 15 seconds of the wave (Figure 1) is accompanied by engagement responses in a majority of subjects. The urgency level is low during this time frame because the tracks are distant from the defensive perimeter. Hence, it is suggested that these postural engagement responses are an orientation response to a specific context and event stream: 'appearance of tracks after the wave onset signal'. Since the response is a movement proportional to the *total number* of tracks on the monitor, it is suggested to be an orientation response indicating global situational awareness.

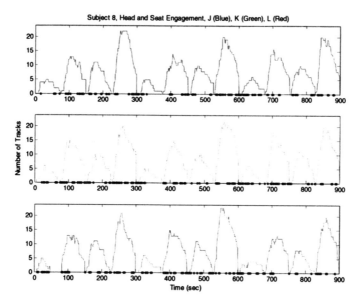

Figure 1. The time of occurrence of global postural engagement responses in the Warship Commander Task is shown for three blocks of trials from one representative subject. The trials varied in the proportion of unidentified tracks. Tasks J and L had a low proportion of unidentified tracks (i.e., high proportion of identified hostile and friendly tracks) at time of appearance on the screen, while Task K had a higher proportion of unidentified tracks and, therefore, a higher workload. The instantaneous number of tracks on the monitor is also displayed for each track.

After the period of global situation engagement across subjects during the first 20-30 seconds of the waves, there was a marked decline in the number of subjects showing global postural engagement responses. This loss of postural orienting to the number of tracks on the monitor appears as tracks cross the defensive perimeter and approach the home ship, which creates a greater urgency to respond. This phase of the wave requires the operator to respond rapidly to converging unidentified or hostile tracks, effectively shifting situational awareness needs from a global to a series of local contexts. Hence, it is suggested that the decrement in subjects showing global postural engagement responses reflects a shift from a global to a reactive local situational awareness strategy. In this sense, it represents a parsing of the event stream into a high workload period requiring urgent action.

531

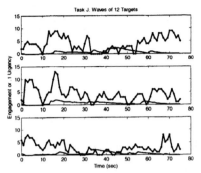
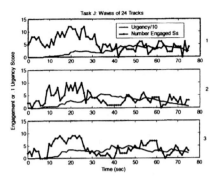

Figure 2. The left and right panels show number of subjects (black line) who displayed a linear correlation (>0.71) over an 8 second window between the number of tracks on the monitor and head position, right seat COP or left seat COP relative to the monitor. The data are plotted for the three waves of 24 tracks in Task J (low proportion of unidentified tracks) for the Warship Commander Task. The urgency score (divided by 10) is also plotted in blue. Note the drop in the number of engaged subjects with increased urgency within 40 seconds of the start of the wave.

After rising from a nadir during high urgency, the number of subjects showing an engagement response tended to remain stable or increase slightly during the final 25 seconds of each wave. This was particularly prominent when urgency fell to zero before end of wave (see left panel of Figure 2). This behavior is consistent with a return to global situational awareness of the number of work items after decline of an urgent workload.

A modified Aegis CIC task offered the opportunity to examine the global postural engagement task in a less structured environment. The postural engagement responses during the Warship Commander Task occurred after a cuing signal. The modified Aegis CIC task, by contrast, requires an operator to select, identify and initiate action toward tracks on a monitor without any cuing. Furthermore, in contrast to the Warship Commander Task where the tracks are a dominant feature of the display, the tracks are embedded in a more complex display for the Aegis CIC task. There was also a marked difference in the interval between the appearances of tracks in the two tasks. For the Warship Commander Task, tracks appear at an average of approximately 2-second intervals, but the interval of track appearance is much longer (approximately 12 seconds) for the Aegis CIC Task. However, the number of tracks in the Aegis CIC task are more complex (including surface vessel, submerged vessels and aircraft) and are greater in number than tracks in the Warship Commander Task.

Figure 3 summarizes data from one of seven subjects performing the Aegis CIC task. Given the difference in the interval of track appearance from the Warship Commander Task, periods of global postural engagement were identified by examining the linear correlation between (1) head position, (2) right buttock/leg center of pressure or (3) left buttock/leg center of pressure and the number of tracks on the monitor during a 12 second window. Again, a *global postural engagement response* is defined by a correlation coefficient (r) for any of these calculations greater than 0.71.

The upper left panel of Figure 3 shows the times of occurrence of global postural engagement responses (red dots) in relation to the normalized number of tracks on the monitor (blue). The periods of engagement coincided almost invariably with a rapid increase in the number tracks, but occurred intermittently during the decline in the number of visible tracks. The former finding is qualitatively similar to the behavior of subjects during the Warship Commander Task. It is perhaps most significant, though, that these engagement responses were clearly precursors to operator actions relative to tracks. The upper right panel shows the cumulative distributions of minimum latencies to an action (hook, identify (ID) or select a track) from the appearance of an engagement response. The lower left panel shows the same data but stratified by type of action. In this subject, there a greater than 0.70 probability of an action within 2 seconds of the appearance of a global postural engagement response. Across all subjects, the median probability of an action within 2 seconds of a global postural engagement response was 0.62, increasing to 0.77 within 3 seconds and 0.84 within 4 seconds of postural engagement. This suggests that global postural engagement responses may be appear as a subject is parsing information about track into a short-latency 'action plan'.

The statistical distribution of the latencies to action after the appearance of a global engagement response may yield further predictive clues about impending actions. The lower right panel of Figure 3 shows an exponential hazard plot of these response latency data. As described in earlier publications (e.g., Balaban, Starcevic & Severs, 1989), a linear segment in this plot represents a time window that has a constant instantaneious probability of the appearance of the subject action. From this perspective, the appearance of a global postural engagement response in the Aegis CIC task appears to mark epochs where there is constant high probability per second of an action.

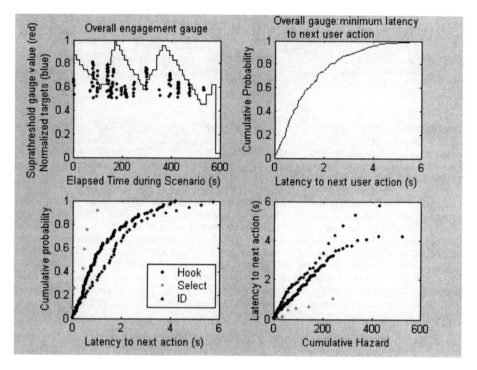

Figure 3. Summary statistics for postural engagement responses of one subject during the Aegis CIC task. The upper left panel shows the occurrence of postural engagement responses (plotted as the value of coefficient of determination, r^2) in relation to the normalized number of tracks on the monitor. Only suprathreshold values (>0.5) are shown. The upper right panel shows the cumulative distribution of the latency to the next user action after the occurrence of a suprathreshold engagement response. The lower left panel shows these latencies stratified by the type of action; the lower right panel shows a cumulative hazard plot of the same data.

3.2 Mercedes-Benz S-class platform

Postural data from five drivers were collected in Mercedes-Benz S-class under normal driving conditions on limited access highways, two lane highways and an urban environment. Driving contexts were logged from a video recording synchronized with the data collection.

One prominent example of context parsing is postural movements that occur prior to manipulations of car controls. Examples of these movements, which represent preparation for action, are shown in Figures 4 and 5. Figure 4 shows an example from a subject while driving on a limited access highway in traffic. The driver was in the right lane being overtaken by traffic. Approximately 5 seconds prior to pressing strongly on the accelerator, the seat pads detected a large preparatory shift in the antero-posterior component of the center of pressure. The center of pressure drifted back to the mean position by the time the accelerator was pressed. The right panel in Figure 4 shows that the posture shift was unrelated to the vehicle acceleration.

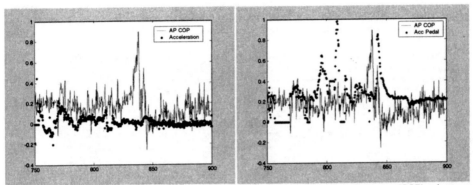

Figure 4. The left panel shows a shift in the antero-posterior (front-back) center of pressure (COP) prior to pressing the accelerator pedal near the time stamp for 850 seconds into the drive. The right panel shows that this shift was unrelated to vehicle acceleration.

Figures 5 and 6 show another example of postural shifts preparatory to operating the accelerator while entering traffic from a ramp onto a limited access highway. In the left panel of Figure 5, the base of support (distance between the centers of pressure of the right and left buttocks/legs) showed a clear relationship to both anticipation and performance of a release and depression of the accelerator. The base of support decreased prior to a release of the accelerator, then increased to a peak prior to depressing the accelerator. A second increase in base of support accompanied the action. The decreased base of support during the initial release of the accelerator was accompanied by a reduction in the RMS movement detected by the seat pads, as well, followed by a small RMS increase that was preparatory to the accelerator movement (Figure 5, right panel). The right panel of Figure 5 also indicates time segments when the coefficient of determination between the back and seat pad RMS (2 second window) exceeded 0.5, which suggests a coordinated whole body movement.

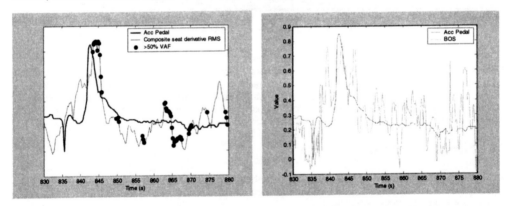

Figure 5. The left panel shows a shift in the base of support prior to releasing pressure (downward shift of accelerator trace) and pressing the accelerator pedal while entering a limited access highway. The right panel shows the simultaneous anticipatory behavior of a composite sum of the RMS of the rate of change of the seat and back sensor pads.

The behavior of seat yaw (torsion of the drivers body relative to the seat and the antero-posterior center of pressure in the seat also showed activity preparatory to this epoch of accelerator manipulation (Figure 6). These actions were temporally correlated with the changes in base of support and RMS, which provide several avenues to detecting a subject's recognition of a context requiring operation of the accelerator. The right panel of Figure 6 also shows that shifts in mediolateral did not show prominent preparatory activity. These unobtrusive measurements, then, may be applicable to recognizing a driver parsing contextual information into epochs requiring action.

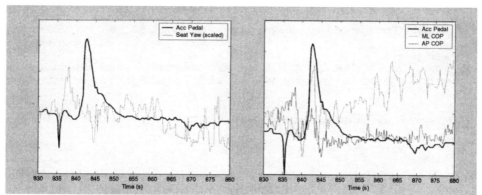

Figure 6. The left panel shows a shift in seat yaw prior to releasing pressure (downward shift of accelerator trace) and pressing the accelerator pedal while entering a limited access highway. The right panel shows the simultaneous anticipatory behavior of the seat center of pressure in the anteroposterior (AP) and mediolateral (ML) dimensions.

4 Discussion

The results of the WCT studies extend our previous observations that head movements toward the screen are correlated with the appearance of targets at the onset of each cued wave (Balaban et al. 2004). Specifically, the distance from the head to monitor and movements of the left and right seat COP are highly correlated ($r^2 > 0.50$) over an 8-12 second window during the first 30 seconds of waves of tracks. Hence, they may be characterized as orienting responses to the workload on the monitor. The time of occurrence of these responses in both the Warship Commander Task and the Aegis CIC task suggest that this response marks transitions between periods when the number of workload tasks is appropriate for maintaining global versus local situational awareness. Furthermore, these orienting responses appear to demarcate periods of preparation for action. Hence, their detection provides insight into a subject's parsing of workload context into a framework for action.

A similar inference may be drawn from detection of preparatory activity to action while driving. Anticipatory postural shifts with respect to specific controls are, in effect, an orienting response related to an impending action. These postural shifts can be detected unobtrusively in real-time with pressure sensors in the seat during driving. Hence, it appears to be feasible to utilize orienting responses of drivers with respect to automobile controls to infer situational awareness, contextual parsing and intent.

5 Acknowledgment

This work was supported by DARPA under contract NBCH3030001 and ONR Grant N00014-02-1-0806.

References

Balaban, C.D., Cohn , J.V., Redfern, M.S., Prinkey, J., Stripling, R. & Hoffer, M. (2004). Postural control as a probe for cognitive state: exploiting human information processing to enhance performance, *International Journal of Human-Computer Interaction*, 17, 275-286.

Balaban, C.D., Starcevic, V.P., Severs, W.B. (1989) Neuropeptide modulation of central vestibular circuits. *Pharmacological Reviews*, 41:53-90.

Fraenkel, G.S. & Gunn, D.L. (1961) The Orientation of Animals: Kineses, Taxes and Compass Reactions. New York: Dover.

Sokolov, Y.N. (1963) Perception and the Conditioned Reflex (Waydenfeld, S.W., translator). New York: MacMillan.

St. John, M., Kobus, D. A., & Morrison, J. G. (2002). A multi-tasking environment for manipulating and measuring neural correlates of cognitive workload. Paper presented at the IEEE 7th Conference on Human Factors and Power Plants.

Zacks, J.M. & Tversky, B. (2001) Event structure in perception and conception. *Psychological Bulletin*, 127, 3-21.

Modeling Human Recognition of Vehicle-Driving Situations As a Supervised Machine Learning Task

Kevin R. Dixon, Carl E. Lippitt, and J. Chris Forsythe
Sandia National Laboratories
Albuquerque, NM 87185 USA
Email: krdixon@sandia.gov clippit@sandia.gov jcforsy@sandia.gov

Abstract

A classification system is developed to identify driving situations from labeled examples of previous occurrences. The purpose of the classifier is to provide physical context to a separate system that mitigates unnecessary distractions, allowing the driver to maintain focus during periods of high difficulty. While watching videos of driving, we asked different users to indicate their perceptions of the current situation. We then trained a classifier to emulate the human recognition of driving situations using the Sandia Cognitive Framework. In unstructured conditions, such as driving in urban areas and the German autobahn, the classifier was able to correctly predict human perceptions of driving situations over 95% of the time. This paper focuses on the learning algorithms used to train the driving-situation classifier. Future work will reduce the human efforts needed to train the system.

1   Introduction

During driving, as in many real-world tasks, humans engage in multitasking such as talking on the telephone, following instructions, and responding to requests. The goal of this research is to develop a system that minimizes the impact of untimely interruptions by providing a physical context to the driving conditions. By mitigating unnecessary tasks during periods of high difficulty, we can improve both driving, by minimizing hazards to safety, and the ability to successfully complete extraneous tasks. For example, talking on a mobile phone increases the likelihood of a traffic accident about 400%, depending on driving difficulty, which is a similar rate to intoxicated driving [1]. A system that, for instance, delayed mobile phone calls during potentially difficult driving situations, such as merging onto a high-speed roadway, could drastically reduce the accident rate, while delaying those important conversations by a short time. The first step toward realizing such a system is the ability to correctly identify potentially difficult driving conditions.

The test vehicle for these experiments is a Mercedes-Benz S-Class sedan, equipped with specialized sensors for this research. The vehicle supplies a wide range of physical data such as speed, turn signals, etc. The posture of the driver is measured by a pressure-sensitive chair and ultrasonic six degree-of-freedom head-tracking system, both developed by the University of Pittsburgh. We collected several hours of data in unstructured driving conditions in both urban areas and on the German autobahn. We also asked humans to label videos of these driving runs according to a list of potential situations. Consequently, the high-level goal of this work is to predict the time-series of human-recognized situations using the various sensors as input. To this end, we used the Sandia Cognitive Framework (SCF) [2] to integrate the information of the driver posture and vehicle state to estimate the current driving situations. The pattern-recognition component of SCF is a type of Nonlinear Dynamical System (NDS). This paper primarily deals with the learning algorithms used to tune the parameters of the NDS to recognize driving situations.

A brief summary of the experimental setup is given in Section 2. In Section 3, we formulate the supervised-learning problem and derive the algorithms in Section 4. We present the results of the learning algorithms on the experimental data in Section 5. In Section 6, we describe related work and give conclusions and lay out future work in Section 7.

2   Methodology

A series of experiments were conducted to ascertain the most useful situations to provide the mitigation strategy with the appropriate driving context. As the goals of the mitigation strategy evolved, its information requirements changed accordingly. For example, during what driving situations would a driver least desire to receive a mobile phone call? Conversely, during what driving situations does a mobile phone call least impact safety? We repeat these questions for each extraneous task that the driver may encounter. In this manner, a list of potentially useful situations was identified and, in later stages, superfluous situations were removed from the list. After several iterations, the remaining situations are assumed to be the minimum set needed to perform the desired classifications and demonstrate an operational performance gain by mitigating extraneous tasks during difficult driving conditions. When the classifier indicates that the driving context is no longer difficult, mitigation ends and the vehicle operates normally. Typically, drivers are unaware that the system proactively intervened on their behalf.

2.1   Data Collection

Five human subjects were instructed to drive on a predefined circuit of German roads measuring about 200 km, ranging from urban streets to the autobahn, and each subject made three runs of the circuit. No modifications were made to the roadways or the ambient driving conditions, such as traffic or road construction. Reference [3] contains more specific information on the experimental setup. Data from the vehicle and driver posture were sampled by our system at a rate of four Hertz. From these data-collection experiments, we obtained a total of almost 24 hours of data (343 946 samples). In addition to the sensor streams, a wide-angle video camera was also used to capture a driver-like perspective out the front of the vehicle, shown in Figure 2.

Figure 1 -A difficult driving condition

Figure 2 - Driving on the German autobahn.

2.2   Generating Ground-Truth Labels

In order to use supervised learning to classify the data according to driving situations, it is first necessary to obtain ground-truth labels. As mentioned earlier, candidate situations were vetted by the information required by the mitigation strategy. After several iterations, the following situations were decided to be the most useful:

> 1) Approaching or Waiting at Intersection
> 2) Leaving Intersection
> 3) Entering On-ramp or High-Speed Roadway
> 4) Being Overtaken
> 5) High Acceleration or Dynamic State of Vehicle
> 6) Approaching Slow-Moving Vehicle

7) Preparing to Change Lanes
8) Changing Lanes

To generate the labels, we created a tool that displays the frontal video (Figure 1 and Figure 2), mentioned in Section 2.1, as well as a set of check-boxes, one for each of the eight candidate driving situation (Figure 3). The human labeler indicates their perception of the current driving situation by checking and unchecking the appropriate box. The tool also allows the user to rewind, pause, save, load, and correct previously labeled time periods. After the user completes the labeling of a video segment, we perform a zero-order hold to associate a label with each input sample. This results in a sequence of binary vectors of "ground truth." In this work, we had two users generate labels for each of the five driver subjects for each of their three circuits, for a total of 15 labeled data sets.

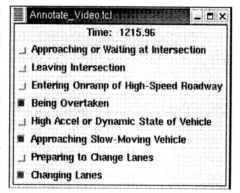

Figure 3 - Part of the GUI presented to a human labeler

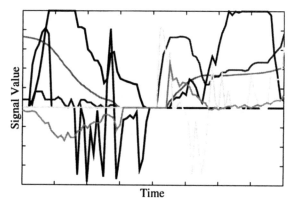

Time

Figure 4 - One example of sensor values that corresponds to "Leaving Intersection," "High Dynamic State," and "Preparing to Change Lanes."

3   Formulation

Formulated as a supervised-learning problem, the goal of our work is to find an "optimal" mapping from a time-series of sensory inputs to a time-series of driving-situation labels. For example, Figure 4 shows one example of sensor inputs that correspond to the driving situations "Leaving Intersection," "High Dynamic State," and "Preparing to Change Lanes." Since the goal of this work is human centric, deriving a classifier that emulates the *human* recognition of driving situations, we used the Sandia Cognitive Framework (SCF). From an engineering perspective, the situation-recognition component of SCF is a type of Nonlinear Dynamical System (NDS). The inputs to the NDS are the processed sensory inputs, described in Section 2. The outputs of the NDS are estimated activation levels of the various situations as a function of time. The estimated situation activations can be considered a trajectory through the state space of the NDS. We use the ordered sequence of labels, described in Section 2.2, as the ground-truth targets for our learning algorithm. From this formulation, the goal of a supervised-learning algorithm is to tune the parameters of the NDS to minimize the error between the estimated situations and the ground-truth situations generated by the human labelers. The type of NDS used in the SCF is a type of Hammerstein-Wiener model, where the linear feedforward- and feedback-parameter gains are unknown, but the nonlinear-function parameters are known.

4   Learning Algorithms

Except in trivial cases, there cannot exist a closed-form optimal solution for the linear parameters of the dynamical equations representing the Hammerstein-Wiener NDS, given a sequence of known sensor inputs and desired output trajectory. Therefore, any learning algorithm will rely on iterative procedures to compute locally optimal estimates. To minimize the error between the ground-truth labels from Section 2.2 and the estimated labels, we pursued two different approaches. The first uses a gradient descent approach and the second uses a Genetic Algorithm (GA) formulation.

According to the human labelers, the majority of the time (52%) during the experiments, none of the target situations were active. The most common driving situation, "Begin Overtaken," occurs 28% of the time, while the rarest situation, "Entering On-ramp or High-Speed Roadway," occurs less than 1% of the time. Given this imbalanced data set, a typical least squares estimation procedure, such as regression, will tend to generate only false negatives because it can achieve 99% accuracy by simply classifying "Not Entering On-ramp." By raising the "punishment" for misclassifying rare situations, a system will be forced to learn the causes of those rare, but important, events. We do this by *weighting* the samples inversely proportional to how frequently they occur, one of several well established approaches to the "rare-event problem" [4]. With such a weighting scheme, incorrectly classifying a time sample as "Entering On-ramp" (false positive) results in the error being weighted by 0.01, whereas missing a classification of "Entering On-ramp" (false negative) results in an error weight of 0.99. This has the effect of minimizing the number of false negatives, which is important in designing a system that mitigates against potentially difficult, though infrequent, driving situations.

In our experiments, we have found that a classifier performs much better when incorporating this biased weighting scheme. This is because it is impossible for a system classifying situations based solely on their relative frequency to achieve better than 50% correct.

We choose our error similar to that of large-margin classification, where learning focuses on finding the most constraining vectors to shatter a training set [5]. We consider that an estimated situation is correct if it is "near" the target label.

4.1   Gradient-Descent Formulation

In this formulation, we minimize the error measure between the target and estimated labels by tuning the parameters according to the gradient of the error. To do this, we need the gradients of the error estimation with respect to the parameters of the dynamical system that describe the SCF, namely the feedback matrix and the feedforward matrix. These gradients were computed using an iterated approach which could lead to unstable updates as the error accumulates. Various researchers in the field of adaptive control have identified sufficient conditions to ensure that gradient-descent update rules yield stable estimates [6]. The selected algorithm and update rules are stable, and consequently guaranteed to converge and terminate at a local minimum, if the step size is decayed in the standard manner [7].

4.2   Genetic Algorithm

We used the DAKOTA optimization package [8] to create a Genetic Algorithm (GA) to find locally optimal solutions for the parameters of the dynamical system. The genome is simply a column-stacked vector of the feedforward and feedback matrices. The fitness criterion for a given parameterization was the error measure between all ground-truth and estimated labels, which is the same as the gradient-descent formulation. Between generations, we keep the best genomes and the stochasticity was handled solely by genome mutations, as we did not allow crossover. The genome mutations were selected by sampling from a Gaussian distribution centered about the keeper genome from the previous generation. The covariance matrix was determined by hand *a priori* to contain reasonable values.

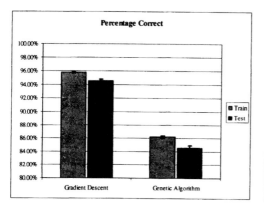

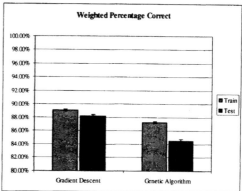

Figure 5 - Results of Gradient Descent and Genetic Algorithm for estimating the correct situation. The error bars indicate the 99% confidence interval.

Figure 6 - Results of Gradient Descent and Genetic Algorithm for estimating the weighted correct situation. The error bars indicate the 99% confidence interval.

5   Results

We divided the driving data into a training set (18.3 hours) and a test set (5.6 hours). Each of the five subjects drove the roadway circuit three times (*cf.* Section II). Each driver had two circuits randomly assigned to the training set and one to the test set. We then ran both the gradient-descent and genetic-algorithm formulations on these data sets. In every performance statistic that we measured, the gradient-descent algorithm outperformed the genetic algorithm. In Figure 5, we show the performance in absolute percentage correct. The gradient-descent algorithm predicted the human recognition of driving situations over 95% of the time on the test set, while the GA managed about 85% correct. As mentioned earlier, we have found the most important statistic in measuring the success of our algorithms to be the weighted percentage correct. In Figure 6, we show the performance in weighted percentage correct. When weighting the classification score, the gradient-descent algorithm predicted human recognition of driving situations about 88% of the time on the test set, while the GA managed about 84% correct. The gradient-descent algorithm is also much more efficient at finding a solution than the GA. On the 18.3 hours driving data in the training set, the gradient descent algorithm typically converged in about one hour of computation time and 101 parameter evaluations. The GA, on the other hand, GA averaged 1611 hours (about 67 days) of computation time and 50 013 evaluations. The relative efficiency of the gradient-descent algorithm is due to the massive amount of problem-specific knowledge incorporated into the algorithm: the gradient of the error. Since this information is not incorporated into the GA, it is not surprising that it used substantially more parameter evaluations to arrive at a locally optimal solution. However, even when the GA was given the solution of the gradient-descent algorithm as its initial genome, the GA was unable to improve performance after several machine-days of computation.

6   Related Work

The identification and extraction of patterns from observed data goes back to ancient times, and pattern recognition is now a necessary capability in many fields. Researchers are still developing novel methods to identify and extract patterns from large amounts of data [9]. Other researchers are interested in identifying contexts, or situations, in response to human behavior to improve system performance, with applications in computing [10] and mobile phones [11]. Incorporating pattern recognition to assist humans in vehicle driving has a relatively long history in robotics, most notably the Navlab project [12]. In order to simplify the process of writing control software, the Navlab project quickly started using supervised-learning algorithms based on observations gained while humans drove the target vehicle [13]. In this paper, we applied learning algorithms to estimate the optimal parameters of a nonlinear dynamical system. Broadly speaking, estimating the parameters of a dynamical system that minimize some cost function is known as optimal control [14]. There has been a substantial amount of work in nonlinear systems, though finding optimal solutions to most nonlinear systems is generally intractable [15]. While we are able to write down the gradient of the system with respect to the tunable parameters and derive stable update equations, this is usually quite cumbersome and time consuming. This is the reason that general stochastic optimization techniques, such as genetic algorithms, are often used to solve for the unknown optimal parameters. While we compared the performance of our gradient-descent algorithm to a standard genetic-algorithm formulation, there is a large number of existing techniques to classify time-series data, including recurrent neural networks [16] and hidden Markov models [17].

7   Conclusions and Future Work

The results of the supervised-learning algorithm on the experimental data are a promising first step toward building a system that can identify potentially difficult driving conditions. A classification rate of over 95% indicates that a fast and efficient gradient-descent formulation, based on the Sandia Cognitive Framework, can accurately predict human recognition of driving situations. Future work will address achieving the same operational performance, but without requiring the tedious human efforts in generating ground-truth labels for the training sets. We hope to use unsupervised learning techniques to classify "interesting" situations from the available raw sensor streams. It is our hope that these self-described situations will then prove useful in aiding similar, learned mitigation strategies.

Acknowledgement

We are grateful to Andreas Bruns, Konrad Hagemann, Paul Heisterkamp, Wilhelm Kincses, and Sven Willmann of DaimlerChrysler for their help and insight. We would like to thank Carey Balaban and Greg Frank of the University of Pittsburgh for their work on the posture-sensing chair. This work was funded by the DARPA/IPTO Augmented Cognition program. Sandia is a multiprogram laboratory operated by Sandia Corporation, a Lockheed Martin Company, for the United States Department of Energy under contract DE-AC04- 94AL85000.

References

[1] D. A. Redelmeier and R. J. Tibshirani, "Association between cellulartelephone calls and motor vehicle collisions," *The New England Journal of Medicine*, vol. 336, no. 7, pp. 453–458, 1997.

[2] C. Forsythe and P. G. Xavier, "Human emulation: Progress toward realistic synthetic human agents," in Proceedings of the 11th Conference on Computer-Generated Forces and Behavior Representation, 2002.

[3] A. Bruns and et al., "EEG- and context-based cognitive-state classifi- cations lead to improved cognitive performance while driving," in print with the First Annual Augmented Cognition International Conference, 2005.

[4] M. Kubat, R. C. Holte, and S. Matwin, "Machine learning for the detection of oil spills in satellite radar images," *Machine Learning*, vol. 30, no. 2-3, pp. 195–215, 1998.

[5] Y. Freund and R. E. Schapire, "Large margin classification using the perceptron algorithm," *Machine Learning*, vol. 3, no. 37, pp. 277–296, 1999.

[6] J.-J. E. Slotine and W. Li, *Applied Nonlinear Control*. Englewood Cliffs, New Jersey: Prentice Hall, 1991.

[7] D. P. Bertsekas, *Nonlinear Programming*. Athena Scientific, 1995.

[8] M. S. Eldred, D. E. Outka, W. J. Bohnhoff, W. R. Witkowski, V. J. Romero, E. R. Ponslet, and K. S. Chen, "Optimization of complex mechanics simulations with object-oriented software design," *Computer Modeling and Simulation in Engineering*, vol. 1, no. 3, 1996.

[9] J. Han, J. Pei, and Y. Yin, "Mining frequent patterns without candidate generation," in *Proceedings of the ACM SIGMOD International Conference on Management of Data*, 2000.

[10] B. N. Schilit, N. Adams, and R. Want, "Context-aware computing applications," in *Proceedings of the Workshop on Mobile Computing Systems and Applications*, 1994.

[11] J. Himberg, K. Korpiaho, H. Mannila, J. Tikanm̈aki, and H. T. Toivonen, "Time series segmentation for context recognition in mobile devices," in *IEEE International Conference on Data Mining*, 2001.

[12] C. Thorpe, M. Hebert, T. Kanade, and S. Shafer, "Vision and navigation for the carnegie-mellon navlab," *IEEE Transactions on Pattern Analysis and Machine Intelligence*, vol. 10, no. 3, pp. 362–373, 1988.

[13] D. Pomerleau, "Efficient training of artificial neural networks for autonomous navigation," *Neural Copmutation*, vol. 3, no. 1, pp. 88–97, 1991.

[14] R. F. Stengel, *Stochastic Optimal Control: Theory and Applications*. Wiley-Interscience, 1986.

[15] H. K. Khalil, *Nonlinear Systems*, 3rd ed. Prentice Hall, 2002.

[16] M. Ḧusken and P. Stagge, "Recurrent neural networks for time series classification," *Neurocomputing*, vol. C, no. 50, pp. 223–235, 2003.

[17] Y. Ephraim and N. Merhav, "Hidden Markov processes," *IEEE Transactions on Information Theory*, vol. 48, no. 6, 2002.

543

Automated Context Modeling through Text Analysis

Travis Bauer

Sandia National Laboratories
MS 1188
Albuquerque, NM 87185-1188
tlbauer@sandia.gov

Abstract

This paper reports on progress made in developing and validating algorithms to automatically derive individualized contextual memory to populate cognitive models of individuals from text they produce and access. In many domains, an expert uses a computer to produce and read documents. In other domains, although we may not have direct access to an individual, the person may express themselves through published documents, such as books or speeches. In either case, we can analyze these documents to discover what key concepts an individual uses and how she relates them to one another. This allows us to keep an up to date evolving model of an individual's knowledge of various domains from which we can get textual documents.

We have an ongoing program for populating cognitive models of individuals. For some applications, these contexts are derived through a process of manual knowledge elicitation whereby a psychologist will interview an individual to elicit the knowledge necessary to build a model of that person. However, in two situations, this is not feasible and automated methods must be developed to populate models. First, in some applications, an individual's model may change over time in response to new experiences and information. In this case, the effort required to keep the models up to date is unreasonable. Second, in some applications it may not be possible to interview the individual.

Our cognitive model framework requires two main types of information: semantic and contextual memory. The semantic memory refers to the key concepts that a person has and with what strength those concepts are related to one another. We use statistical and syntactic analysis methods for discovering what the key concepts are and how related they are to one another.

Although the contextual memory presupposes that a semantic memory has already been computed, it is the methods for deriving and validating contextual memory that is the main focus of this paper. We think of a context as a collection of concepts that tend to occur together. For example, if a group of terms such as "car, tire, engine, road, street sign, and crash" tend to occur together in groups, the "automobile" context would be composed of these concepts. When a sufficient number of these concepts start occurring in the individual's environment, we say that the person is likely to recognize the situation as containing the automobile context.

One aspect involved in determining what constitutes a context is under what circumstances we consider a group of terms to have "occurred" together. In a simple text-based derivation of context, we may split all of the source text into fixed size chunks. However, research (Zwaan and Radvansky, 1998) has demonstrated that people tend to split episodes along six dimensions: time, place, objects/actors, emotions, intention, and causality. Although the dimensions of emotions, intention, and causality can be difficult to determine solely through automated text analysis, the dimensions of time, place, and objects/actors is relatively easy to obtain, especially if the data was collected while the person was actually using the computer so timestamp information can be collected. Using this information, we hope to perform a more psychologically plausible splitting of documents into chunks to acquire more accurate and useful contexts.

In this paper, we report on progress made in automatically constructing context from text collected from individuals and on the methods of validation we plan to apply to determining how well these methods work. Both the construction of contexts and the validation techniques will be partially adapted from information retrieval literature.

Context Modeling for Augmented Cognition

Chris Forsythe

Sandia National Laboratories
Albuquerque, NM 87185 USA
jcforsy@sandia.gov

Abstract

With "Augmented Cognition," the intent is that systems monitor the cognitive state of operators in real-time using physiological and other sensors and adapt in various ways to optimize operator performance. For the success of such systems, it is essential that adaptations be appropriate to ongoing operational contexts. Therefore, an essential element of an Augmented Cognition system is the real-time recognition of operational contexts. This may be accomplished through context modeling where various data inputs provide a basis for predicting operational context(s). The papers in this session describe various approaches for context modeling with particular emphasis placed on the mechanisms by which context models may be acquired through automated techniques.

Where augmented cognition systems adapt to the ongoing cognitive state of an operator based on various physiological and behavioral sensors, the appropriate adaptation will often depend upon the ongoing context. For this reason, a common component for augmented cognition systems will be a "context manager" which utilizes various sensor data, knowledge of the operator and /or information concerning the tasks, equipment, mission, etc. to make inferences concerning the system/operator's current context(s). This component should provide a real-time indication of contexts relevant to given situations. Based on this output, appropriate adaptations may be initiated to mitigate the effects of cognitive load and other states that may hamper human performance. Contextually inappropriate adaptations can have deleterious effects canceling out any gains sought through augmented cognition and potentially leading to overall degradations in performance.

There are various issues vital to real-time context modeling. First, it is necessary to have a mechanism for inferring contexts. Through the DARPA Improving Warfighter Information Intake Under Stress program it has been demonstrated that this may be successfully accomplished using a computational cognitive model. However, using such tools, one must be concerned that the computational cognitive model have certain dynamic properties that enable it to flexibly respond to a broad range of conditions, many of which may be unanticipated by developers.

Second, there must be mechanisms to obtain the knowledge necessary to populate a context model, and for many systems, customize the model to a specific operator. Thus, the broad application of such technologies will require mechanisms to automate, or at least semi-automate, the processes by which the knowledge is obtained to construct or customize the context model.

Third, there must be an understanding of how to implement adaptations to the operator's cognitive state in a contextually appropriate manner. For instance, it has been shown that mitigations that have a beneficial effect in some conditions may have a deleterious effect in others. Finally, it is often necessary to be somewhat opportunistic in utilizing available data sources for the real-time modeling of context, and often there is a reciprocal influence where knowledge of context provides a basis for enhancing the interpretation of sensor data that subsequently enables further recognition of context.

Within the context modeling session, a collection of papers have been combined that illustrate key technical developments in this area that include input sources for modeling context in real-time, approaches to context modeling, mechanisms for training a context model using operational data and system adaptations on the basis of ongoing context.

Section 3
Cognitive State Sensors

Chapter 12

Issues of Trust in Adaptive Systems

Trust in Adaptive Automation:
The Role of Etiquette in Tuning Trust via Analogic and Affective Methods

Christopher A. Miller

Smart Information Flow Technologies
Minneapolis, MN
cmiller@sift.info

Abstract

In this paper, we begin by discussing a definition of trust and settle on one provided by Lee and See (2004). This definition emphasizes the nature of trust as an attitude toward the uncertain future actions of an agent. Some important implications of this definition for adaptive automation systems are discussed including (a) trust is not synonymous with user acceptance; in fact, trust should be tuned to result in accurate usage decisions by operators, (b) trust becomes more important with complex, adaptive automation precisely because it becomes less plausible for a human operator to fully understand what and how the automation will operate in all contents, and (c) as an attitude, trust is produced and affected by methods other than rational cognition. In fact, Lee and See provide a model with three methods of tuning trust—analytic, analogic and affective—with special emphasis on their roles trust for adaptive automation. Of these, we argue that the latter two will be more important in human interaction with adaptive automation than they are with traditional automation. We define and discuss a method of tuning analogic and affective trust: the "etiquette" of human interaction with automation. We provide examples from two recent projects, one involving a laboratory experiment and the other involving human interaction with automation in a realistic full mission simulation, which illustrate trust effects from etiquette-based design and system behavior manipulations.

1   Introduction

1.1   What is "Trust"?

Any discussion of trust, especially as it applies to human-automation interaction, should begin with a definition of terms. Trust is all-too-readily taken as synonymous with user acceptance, but repeated work over the past twenty (Lee and Moray, 1992, 1994; Lee and See, 2004; Parasurman and Riley, 1997) years has shown that both of these concepts are complex and intertwined. Users sometimes use automation that they are suspicious of (perhaps because they do not have the time or workload capacity to do otherwise) and sometimes do not use automation they believe is competent (perhaps because they enjoy doing the job themselves).

Lee and See (2004) in their recent comprehensive review of trust literature in the field of human factors, identify trust as "the attitude that an agent will help achieve an individual's goals in a situation characterized by uncertainty and vulnerability." (Lee and See, 2004, p. 51). This definition is based on a careful analysis of a vast number of studies and experiments of human operators trust-related behaviors and I will use it in the remainder of this paper. It also conveys several subtle distinctions that are important to both the study of trust in human-machine interactions and in designing systems so that *accurate* trust results:

- Trust is an attitude, which means that it is a response to knowledge or belief about world states, but it is not, itself, those beliefs. Even less so is it a decision based on those world states. Many other factors may intervene to produce automation usage or non-usage decisions and, while trust is an important element in those decisions, it is far from the only one.
- Trust is an attitude, which means that, in part, it is affective. Trust and mistrust produce feelings about the agents they are directed at, and we base future trust in part on such feelings.
- Trust is ego-centric in that it is centered on an interpretation about an agent's ability and willingness to help me achieve my goals. In this sense, it is at least potentially separable from basic knowledge about

how and agent works or what it does in an abstract sense, though if I know these things, I can generally infer (at the cost of cognitive effort) whether or not the agent is likely to prove helpful to me in a specific set of circumstances and, hence, whether or not to "trust" it in those circumstances.

- Trust plays its largest role in 'situations characterized by uncertainty and vulnerability". Trust plays a smaller role to the degree that the situation is well understood and predictable. This distinction is critical when discussing trust in highly complex and adaptive automation systems and will be discussed in more depth below.

1.2   Why trust matters (especially) for Adaptive Automation

If trust is an attitude about the future, uncertain behavior of an entity in context, then it is very nearly irrelevant to speak of "trust" in a thoroughly understood and predictable system. I may say that I "trust" that an apple will fall to the ground if I drop it because, after all, there is always some uncertainty in the world, but this is hardly in the same league as saying that I trust that my shares of Company X will increase in value on the stock market. Trust is important precisely to the degree that knowledge and certainty about future behaviors is absent.

When we fail to comprehend all the causal mechanisms associated with an observed behavior, or fail to be able to track them in a timely fashion in context, we are increasingly forced away from an analytic understanding or prediction of an agent's behavior and instead are forced to rely more heavily on trust. As automated systems become ever more complex, their behaviors become ever less predictable and even comprehensible to their users. This is likely to be even more prevalent with adaptive automation systems whose very nature is not to exhibit the same behavior all the time and, therefore, which necessarily introduce another dimension of uncertainty into the human + machine system. Worse, is the fact that we want automation support (and, especially, augmented cognition support) in exactly those situations were our basic human cognitive capabilities are inadequate—either because they are limited in their range of knowledge or in the speed with which we can use them to analyze and decide, or both. Adaptive automation systems should, where possible, behave in a predictable fashion and users should be trained to understand how they will make their behavior decisions (Miller and Hannen, 1999), but this will not be possible to the degree that the automation reduces user workload by removing some elements of decision and execution authority from him or her (Miller and Parasuraman, submitted; Miller, Funk, Goldman, Meisner and Wu, this volume).

Trust matters especially for adaptive automation precisely because it will be largely unreasonable to expect human operators to know (by analytically reasoning through the same cognitive processes as the automation), in all circumstances, whether the adaptive automation is behaving in a correct manner or not. Instead, operators will need to have *accurately tuned* trust as to whether to accept or reject behaviors and recommendations provided by adaptive automation. If we are to assume that the human operator should remain in overall charge of operation with the adaptive automation in a subservient, aiding role, then this puts the designers, and ultimately, the operators, of adaptive automation systems in the position needing to understand how trust is developed and used.

2   Methods for Trust Tuning

Lee and See (2004) identify 3 alternate routes that humans use to develop and tune their trust: analytic, analogic and affective methods. In each case, the method affects or produces the 'attitude about the future, uncertain behavior of an entity' as described above, but the route to achieving that effect is different, making use of different cues, knowledge and different degrees of cognitive processing.

Analytic methods involve a detailed understanding and rational assessment of the mechanisms by which the entity produces its behaviors. Analytic methods assume rational decision making on the basis of what is known about the motivations, interests, capabilities and behaviors of the other party. In other words, I *reason*, taking into account uncertainty, about your likely future behaviors based on what I know about your motives and goals. I may trust you to help me write a paper because I believe that you are motivated to publish good work (and get credit for it).

Analogic methods involve using observable cues to infer broader category membership in a group or context and then to apply trust assumptions for the group as a whole to the newly encountered individual. Analogic trust can also be based on the word or endorsement of intermediaries whom we trust. We trust bankers to behave in responsibly with regards to our money in large part because they are a member of a category that is hired, trained, supervised and watched over to ensure that those behaviors are maintained. When we encounter an individual behind the counter in a bank, we are comfortable handing them our money largely on the basis of this analogic reasoning about their category membership, not because we deeply reason through their motivations with regards to our money.

Affective methods are based strictly on the affect generated by and toward the entity. This method accounts for the frequently recorded and observed finding (e.g., Cialdini, 1993) that we tend to trust those people and devices that please us more than those that do not. While affective trust tuning may be used as a shortcut for detailed reasoning in the analytic or analogic senses, it is nevertheless an effective method of managing our time and attention under conditions of cognitive overload. It is always unpleasant to have someone thwart our goals or plans, and the correlation of unpleasant feelings with thwarted plans is generally stronger than with outcomes that ended up being good for us. Thus, in conditions of uncertainty and especially when there is no time or attentional capacity available to perform more detailed reasoning about the situation, it may well be an effective heuristic to assume that, if something or someone is irritating, it does not have my best interests at heart and is, therefore, not to be trusted.

A interesting and important final element of Lee and See's model is the realization that there is a temporal element to trust building. It takes time to acquire, whether through experience, training or attending to hearsay and the experiences of others, the knowledge required to use either the analogic or analytic methods of trust tuning. When experiencing a new person or a novel system for the first time, with no background knowledge about the agent's motivations, behaviors, or group memberships, the only information a person may have about whether or not to trust it will be affective information. Furthermore, if the affect is negative enough to prompt a strong "do not trust" response, then no further information will be gathered. Analogic trust, in turn, requires less knowledge gathering and cognitive processing than does analytic trust and can, again, serve as a kind of hurdle to further experience. In other words, if a person or system does not do a reasonable job of providing appropriate cues to achieve at least moderate levels of affective and analogic trust, s/he/it may never have a chance to build analytic trust.

3   Applying Trust Tuning Methods to Adaptive Automation

The three approaches to building or tuning trust that Lee and See (2004) lay out can be used to form a coarse taxonomy of approaches to achieving accurate trust in adaptive automation. Furthermore, as noted above, the framework emphasizes the importance of analogic and affective approaches to trust tuning—and, as we will see below, these methods take on particular importance in complex, high criticality systems.

In the case of complex automation, I will generally have a less than complete knowledge of the processes and knowledge by which the automation operates. This is particularly true of automation which is performing millions of computations in a highly time pressured environment, such as much decision support and control automation in high criticality domains. Indeed, one of the prime reasons automation has been incorporated into such systems is because it can either do things that a human simply cannot do, or can do them faster than the human can. Nevertheless, operators are still expected (and generally trained) to understand the basic "motivations" of the automation—that is, the reasons why it was built—and the primary factors on which it bases its conclusions. In this sense, then, analytic trust approaches may not demand a detailed thorough thinking through of the actually computational methods by which automation produces the behaviors it does, but instead an application of general heuristics or presumed intentions.

3.1   Analytic approaches

Simplified analytic approaches provide some method of regularizing, "chunking" and/or constraining (and therefore simplifying) the potential behaviors of the agent. A familiar, if far from simple, example of this is Azimov's "laws" of robotics—a simplified set of heuristics that, at least in Asimov's stories, nevertheless served as a highly abstract view of how his robots actually produced their behaviors. Robots were deemed to be motivated 'never to harm a human or, through inaction, allow harm to come to a human' and this, in turn, served as an element in explaining their behaviors. A human in Asimov's fictional world could "trust" a robot even without completely understanding its thought processes because s/he knew that it would, in turn reason and behave in accordance with that motivation.

A somewhat more practical and current example of simplifying analytic trust tuning is represented by Taylor's (2001) Pilot Authorisation and Control of Tasks (PACT) approach to human-machine interaction whereby the human constricts the set of alternative behaviors available to cockpit aids in an initial "PACT" or contract which coarsely predefines acceptable ranges of automation behavior. Later automation behavior can therefore much more readily be predicted and evaluated against the framework established in the PACT and, thus, analytic trust (positive or negative) is more easily produced even for a very complex and adaptive automation system. Furthermore, the human operator has an enhanced awareness of the system's ability to behavior in accordance with the PACT precisely because the operator provided the behavior constraints in the first place, they were not something s/he had to learn. A similar benefit is sought in our own delegation and Playbook^{TM} approaches to human-automation interaction (Miller, Pelican and Goldman, 2000; Miller, Goldman, Funk, Wu & Pate 2004; Miller, Funk, Goldman, Meisner, Wu, this volume).

3.2   Analogic approaches

Analogic methods of engendering trust for automation may be as simple as the knowledge that the automation was produced by a "trustworthy" company or approved by an agency known to be familiar with the operator's domain. Similarly, user communities and their accumulated impressions and "lore" about how a complex piece of automation behaves can place a very large role in building operator trust or mistrust.

Another way in which automation can tune trust via analogic methods is via elements of the "persona" it assumes—and encourages the operator to perceive in it. If automation uses the professional "jargon" of a highly trained domain (for example, the cadence, syntax and particular vocabulary items of Air Traffic Control), then it encourages operators to assume, via analogic trust reasoning, that it has the competence and range of reasoning abilities of other agents (primarily human air traffic controllers) who use such language. Such an approach might result in over-trust in the system if, in fact, it does not have those capabilities.

Note that the "embodiment" of the system in an animated character (a la Cassell, Sullivan, Prevost and Churchill, 2000) is not required to achieve this effect—simple language usage or the presentation of information in styles unique to a specific domain might be sufficient to trigger analogic trust, whether warranted or not. On the other hand, embodiment and personification open many additional channels (i.e., voice, gesture, movement, appearance, facial expressions, etc.) via which the system's interface can either succeed or fail in conveying group membership and therefore triggering analogic trust mechanisms. Cassell and Bickmore (2004) report a particularly complex, embodied real estate sales agent that manipulates conversational small talk in order to build enough social trust to enable it to introduce personally threatening topics such as income levels and expected family size as a part of its attempts to suggest real estate properties an individual or couple might be interested in.

3.3   Affective approaches

Affective methods involve essentially anything which results in producing positive or negative affect toward the system. While we might like to believe that operators of highly complex equipment in high criticality domains are above being influenced by "interpersonal" annoyances (or, conversely, pleasing layout and design), there is increasing evidence that this is simply not the case (cf. Norman, 2004). Indeed, much of the philosophy behind techniques of Crew Resource Management (e.g., Weiner, Kanki and Helmreich, 1993) are based on understanding and managing these "etiquette" and "style" variables in human-human cockpit interactions.

4   The Importance of Automation Etiquette

As Lee and See (2004) themselves point out, one interesting approach to achieving analogic and affective trust in complex systems is the "etiquette" that they exhibit. By "etiquette", we (Miller, 2002; 2004) mean the largely unwritten codes that define roles and acceptable or unacceptable behaviors or interaction moves of each participant in a common 'social' setting, that is, one between actors which are presumed to be both complex and reasonably understood according to intentional assumptions (Dennett, 1989). Etiquette rules create an informal contract between participants allowing expectations and interpretations to be formed and used about the behavior of others. As such, they can serve an analogic function in human-automation interaction—for example, the use of familiar domain jargon in advice or reporting can serve as a indicator that the agent using it (whether human or automation) falls into the category of "those experienced in the domain" and should, therefore, be accorded the trust reserved for those in that category. Because etiquette behaviors are also intimately tied to what we traditionally interpret as "polite" or "rude" behaviors (cf. Brown and Levinson, 1987), automation etiquette also holds the power to produce affective reactions in users, both positive and negative.

As Nass has pointed out (Reeves and Nass, 1996), humans are highly prone to interpret computer behaviors according to the same scripts or schema that are commonly used for human-human interactions. As discussed above, the sheer complexity of adaptive automation systems provides reason to believe that the role of etiquette in tuning trust via analogic and affective methods may be even larger than it is in less complex systems. We are beginning to see some examples that trust, user acceptance and even human + machine system performance are affected by etiquette variables. We will briefly discuss two of these below.

4.1 Interaction expectations and their role in tuning trust: Rotorcraft Pilot's Associate example

The U.S. Army's Rotorcraft Pilot's Associate (cf. Colucci, 1995), a highly complex adaptive automation system designed to assist the two pilots of an advanced attack-scout helicopter, operated by inferring ongoing and necessary tasks and then adapting information presentations and automation behaviors to best support those tasks. In designing the information management and task allocation aspects of the RPA, we noted (Miller and Hannen, 1999) that rotorcraft crews are trained to communicate with each other about ongoing and future tasks. By some estimates, nearly a third of the crew members' time is spent in crew coordination activities—particularly in maintaining shared awareness and coordination over who is doing what task when. We suspected that for the associate not to have this capability—that is, not to behave in accordance with the expected etiquette for the task domain-- might make it seem less trustworthy (via analogic means). Partly in an effort adhere to the associate metaphor in this domain, to provide an 'associate' that would be a good team player, we attempted to give the RPA some of this capability as well (Miller and Funk, 2001). The RPA cockpit design included an innovative, task-based display to provide the crew with both insight into, as well as some degree of control over, RPA's understanding of their intent. This "Crew Coordination and Task Awareness" display consisted of four small LED buttons located in the upper portion of each pilot's main instrument panel. The buttons report, in textual form, (1) the current inferred high-level mission context, (2) the highest priority current pilot task, (3) the highest priority associate task, and (4) the highest priority copilot task. Pressing these buttons permitted either pilot to override RPA's current inferred tasks and assert new ones (from an automatically scrolled list of higher-level tasks from the overall task network) via a single push button input.

Figure 1 shows the RPA cockpit simulation created for evaluation trials at the Boeing Company in Mesa, Arizona. The location of the Crew Coordination display is circled in Figure 1 and an enlargement and interpretive sketch of the display is provided for clarity.

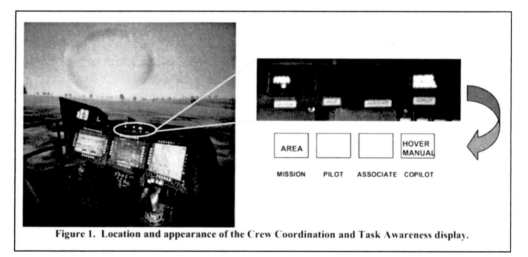

Figure 1. Location and appearance of the Crew Coordination and Task Awareness display.

While pilots were not asked explicitly about their trust in the RPA, pilot's acceptance of the LED Crew Coordination and Task Awareness display was high, as shown in Table 1. In spite of the perceived occasional inaccuracies in RPA's task inferencing mentioned above, and in spite of some pilot complaints about inadequate training in their use, most pilots found the individual task reports of the LED buttons 'Of Considerable Use' or 'Extremely Useful'. This provides some supporting evidence that our inclusion of a capability for the crew to interact directly with the associate's assumptions about active tasks was a capability that pilots welcomed—and that may have served to improve their overall impressions of CIM's capabilities and usefulness.

Table 1. Perceived usefulness of the LED Task Awareness Display (where 4.0='Of Considerable Use' and 5.0='Extremely Useful'.

LED Button for:	Score
Mission Task	4.4
Pilot Task	4.3
Copilot Task	4.3
Associate Task	4.0

The subjective pilot response data obtained from a series of full mission simulation trials (cf. Miller, 1999; Miller and Funk, 2001) suggest that the RPA system we designed and implemented generally met pilot expectations, contributed to perceived pilot effectiveness, reduced workload and gained pilot acceptance. Pilot's clearly felt that RPA provided benefit. Perfection in aiding and tracking pilot intent was not a prerequisite to the levels of acceptance we gained. To say what aspects of RPA design were ultimately responsible for these levels of acceptance is somewhat difficult, but we suspect that two attributes played a significant role. First, RPA was designed to provide high degrees of predictability, even at the cost of flexibility in managing displays in some instances. In some ways, RPA was behaving like a new member of the crew—trying to avoid doing unexpected things and making mistakes until it gained some acceptance. Second, the addition of an effective Crew Coordination and Task Awareness display may have contributed to pilots' willingness to tolerate these occasional 'mistakes' on RPA's part. Again, from the crew's perspective, RPA was adhering to expected crew etiquette and behaving like an acceptable junior member of the crew. Instead of jumping ahead and taking an action when it thought it knew what was going on, RPA generally reported what it thought was happening, followed the pilots' lead, and gave the crew a chance to correct it before it made major mistakes.

4.2 Experimental manipulation of etiquette behaviors: Trust, usage and performance impacts

While the RPA example cited above was for a realistic system tested in a full mission simulation, the results of adapting an interface to support analogic trust tuning are not conclusive. In this section, we summarize the results of a laboratory experiment (Parasuraman and Miller, 2004) in which etiquette variables were directly manipulated and the resulting impact on trust, automation usage decisions and performance we observed. We were interested in whether "good" (i.e., pleasing) etiquette would compensate for poor automation reliability and result in increased usage decisions and whether "poor" (i.e., displeasing) etiquette would wipe out the benefits of good automation reliability.

From the vast range of human etiquette behaviors, we chose to concentrate on a single dimension we label communication style. Communication style referred not to the specific wording of communications (which was held constant across conditions), but to the "interruptiveness" and "impatience" of delivering text messages. This was chosen as an etiquette dimension available to even fairly primitive and non-personified automation interfaces.

We tested 16 participants (both general aviation pilots and non-pilots) on a flight simulation task, the Multi-Attribute Task (MAT) Battery, which has been used extensively in prior high-criticality automation research (Parasuraman & Riley, 1997). The MAT incorporates primary flight (i.e., joystick) maneuvering, fuel management, and engine monitoring/diagnosis task. Participants always performed the first two tasks manually to simulate the busy operator of a high-criticality system. Intelligent automation support, modeled after the Engine Indicator and Crew Alerting System (EICAS) common in modern automated aircraft, was provided for the engine monitoring/diagnosis task. Automation monitored engine parameters, detecting potential engine faults, and advised participants when and what to examine to diagnose the faults. For example one advisory message was: "The Engine Pressure Ratio (EPR) is approaching Yellow Zone. Please check. Also, cross-check Exhaust Gas Temperature (EGT). There is a possible flame out of Engine 1."

"Good" automation etiquette involved a communication style that was "non-interruptive" and "patient" while poor etiquette was the opposite. In the non-interruptive case, advice was provided after a 5-second warning and not at all when the operator was already doing the requested action. Non-interruptive automation was "patient" in that it would not issue a new query until the user had finished the current one. Interruptive/impatient automation, by contrast, provided advice without warning and came on when the user was already querying EICAS . Automation also "impatiently" urged the next query before the user was finished with the current one.

These good and poor communication styles were crossed with the effects of automation reliability. Two levels of reliability were chosen: low, in which automation provided correct advice 60% of the time and high (80%).

We were primarily interested in the effects of etiquette and reliability on users' performance and on their rated trust in the automation. The percentage of correct diagnoses of engine malfunctions in all four conditions are shown on the left side of Figure 2. As expected, user diagnostic performance was significantly (p < .01) better when automation reliability was high (80%) than low (60%). Less obviously, good automation etiquette significantly (p < .05) enhanced diagnostic performance, regardless of automation reliability. Perhaps most interestingly, the effects of automation etiquette were powerful enough to overcome low reliability (p < .05). As the dotted line on the graph on the left side of Figure 2 indicates, performance in the low reliability/good etiquette condition was almost as good as (and not significantly different from) that in the high reliability/poor etiquette condition. These findings on diagnostic performance were mirrored in the results for user trust, shown in the center of Figure 2. High reliability increased trust ratings significantly, but so did good automation etiquette—to an even greater degree. Finally, good communication etiquette seems to have had a consistent effect in reducing perceived user workload in both high and low reliability conditions—as illustrated in the graph on the right side of Figure 2.

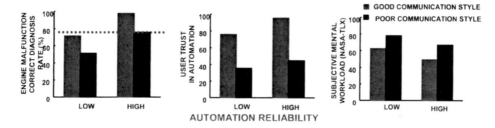

Figure 2. Results of etiquette manipulations in automation diagnostic recommendations presentations on correct performance (left), user trust (center) and subjective workload (right).

A possible objection to these findings is that any interruption might be expected to degrade user performance. Where the findings above due to the "rudeness" of this automation—which interrupted to "nag" (that is, to provide directives that users were working on, thereby effectively telling the operator's to 'hurry up')—or due to the simple interruption itself? To answer this question, Parasurman ran a control group of four participants using interruptions which were non-specific in content—for example, "Maintaining primary flight performance is important, but do not forget to check engine parameters for possible malfunction." These interruptions were varied in their intrusiveness as above—they were either preceded by a warning and not offered if the user was engaged in diagnosis (non-intrusive) or were given with no warning regardless of user activity (intrusive). Under these conditions, intrusiveness had no significant effect on correct diagnosis of engine malfunctions or user trust ratings. Thus, in contrast to the main experiment, less rude, non-specific interruptions were more easily ignored and did not adversely affect user-system performance or trust, implying that the adverse effects on trust and performance observed in the first experiment were the result of the rude "nagging" reminders to do work already being performed rather than the intrusions per se.

These results provide strong, if preliminary, evidence for the impact of automation etiquette on both user performance and trust in using an intelligent fault management system to diagnose engine malfunctions. While we did not collect any measures to indicate whether the trust (and, arguably, the performance) impacts were due to analogic or affective mechanisms, it would seem that affective methods are the most likely. We can speculate that analogic cues in the instructions that were provided would, if anything, have tended to *increase* trust since messages were given in the form and using the language of existing (and, arguably, trusted) cockpit displays and vernacular. On the other hand, interruptions and "nagging" are frustrating and unpleasant and, by the simplified connections which exist at the analogic level, those who do such things are unlikely to have my best interests at heart and are, therefore, not to be trusted or heeded.

These results also clearly show that building reliable automation may not be enough for overall human + machine system efficiency: both user diagnostic performance and trust were lowered by poor automation etiquette even when the reliability of automation advice was high. Furthermore, there is some hint that good etiquette can compensate for low automation reliability since users did better with well behaving automation (good etiquette) even if the automation was unreliable. Some may find this result disturbing, since it suggests that developing robust, sensitive, and accurate algorithms for automation, a challenging task at the best of times, may not be necessary so long as the automation "puts on a nice face" for the user. This would clearly be a violation of the dictum that trust should be tuned to reflect the actual level of performance of the automation, but it might nevertheless be tempting, especially in a commercial products setting. While some "sins" might be papered over in this manner, it was clear that the *best* user performance (and the highest trust) was obtained in the high reliability condition in which the automation also communicated its advice to the user in an etiquette-friendly way. Hence, good automation that behaves well should always be the highest goal. We have shown, however, that automation trust can be tuned to some extent by the etiquette that the automation exhibits and this may prove a useful tool in crafting automation behaviors for optimal use.

5   Conclusions

We have argued above that trust is not identical to user acceptance, but instead should be regarded as a process of tuning the user's attitudes toward automation (and, therefore, willingness to accept it's advice and behaviors) to comply with the actual performance of the automation. As designers of complex, high criticality systems, we must be aware of the factors that influence that trust for better and for worse. The goal is not to get users to accept our systems, but rather to get them to accept and make use of those systems when they will be helpful and to reject them when they will not.

The conceptual framework of Lee and See (2004) provides us with a taxonomy of methods for influencing trust and a model of how they interact. For complex, high criticality automation, and especially for adaptive automation which operates using augmented cognition technologies, there are many reasons to believe that analogic and affective methods of trust tuning will play a conceptually greater role than do more memory, knowledge and experience intensive analytic methods. A promising line of research and design adaptations to tune trust via affective and analogic methods is via the etiquette exhibited by the automation. We have provided two examples, one a laboratory experiment and one a high fidelity simulation of a realistic work environment, of instances in which we believe that the etiquette influenced the trust of operators of advanced automation equipment. Much future work remains to be done to refine our understanding of etiquette and its impact on trust, but this work lays the foundations.

6   Acknowledgements

Work on the Rotorcraft Pilot's Associate was funded by the U.S. Army AATD, contract number DAAJ02-93-C-0008. The author would like to thank Raja Parasuraman for his work in conducting and analyzing the etiquette experiment described above, and John Lee and Katrina See for their comments on earlier drafts of this paper.

7   References

Brown, P. & Levinson, S. (1987). *Politeness: Some Universals in Language Usage.* Cambridge,UK; Cambridge Univ. Press.

Cassell, J. and Bickmore, T. (2004). Negotiated Collusion: Modeling Social Language and its Relationship Effects in Intelligent Agents. *User Modeling and User-Adapted Interaction. 13*(1): 89-132

Cassell, J., Sullivan, J., Prevost, S. and Churchill, E. (2000). *Embodied Conversational Agents*. Cambridge, MA; MIT Press.

Cialdini, R.B. (1993). *Influence: Science and Practice. 3rd Ed.* New York; Harper Collins.

Colucci, F. (1995). Rotorcraft Pilots' Associate update: The Army's largest science and technology program, *Vertiflite*, March/April, 16-20.

Dennet, D. (1989). The Intentional Stance, MIT Press; Cambridge, MA.

Lee, J., & Moray, N. (1992). Trust, control strategies, and allocation of function in human-machine systems. *Ergonomics, 35.* 1243-1270.

Lee, J., & Moray, N. (1994). Trust, self-confidence, and operators' adaptation to automation. *International Journal of Human-Computer Studies, 40,* 153-184.

Lee, J. & See, K. (2004). Trust in Automation: Designing for Appropriate Reliance. *Human Factors, 46* (1), 50-80.

Miller, C. (1999). Bridging the Information Transfer Gap: Measuring Goodness of Information Fit. *Journal of Visual Language and Computation, 10,* 523-558.

Miller, C. A. (Ed.) (2002). *Working Notes of the AAAI Fall Symposium on Etiquette for Human-Computer Work.* Technical Report FS-02-02. Menlo Park, CA: American Association for Artificial Intelligence.

Miller, C. (Ed.), (2004). Human-Computer Etiquette: Managing Expectations with an Intentional Agent. Special section in *Communications of the ACM,* April, 2004.

Miller, C. and Funk, H. (2001). Associates with Etiquette: Meta-Communication to Make Human-Automation Interaction more Natural, Productive and Polite. In *Proceedings of the 8th European Conference on Cognitive Science Approaches to Process Control.* September 24-26, 2001; Munich.

Miller, C., Funk, H., Goldman, R., Meisner, J. and Wu, P. (this volume). Implications of Adaptive vs. Adaptable UIs on Decision Making: Why "Automated Adaptiveness" is Not Always the Right Answer.

Miller, C., Goldman, R., Funk, H., Wu, P. and Pate, B. (2004). A Playbook Approach to Variable Autonomy Control: Application for Control of Multiple, Heterogeneous Unmanned Air Vehicles. In *Proceedings of FORUM 60, the Annual Meeting of the American Helicopter Society.* Baltimore, MD; June 7-10.

Miller, C. & Hannen, M. (1999). The Rotorcraft Pilot's Associate: Design and evaluation of an intelligent user interface for cockpit information management. *Knowledge Based Systems, 12,* 443-456.

Miller, C. & Parasuraman, R. (submitted). Designing for Flexible Interaction Between Humans and Automation: Delegation Interfaces for Supervisory Control. Submitted for publication in *Human Factors.*

Miller, C., Pelican, M. and Goldman, R. (2000). "Tasking" Interfaces for Flexible Interaction with Automation: Keeping the Operator in Control. In *Proceedings of the Conference on Human Interaction with Complex Systems.* Urbana-Champaign, Ill. May.

Norman, D. (2004). *Emotional Design: Why we love (or hate) everyday things.* New York; Basic Books.

Parasuraman, R. and Miller, C. (2004). "Trust and Etiquette in High-Criticality Automated Systems". In C. Miller (Guest Ed.), special section on "Human-Computer Etiquette". *Communications of the ACM. 47*(4), April. 51-55.

Parasuraman, R. & Riley, V. (1997). Humans and automation: Use, misuse, disuse, abuse. *Human Factors, 39,* 230-253.

Reeves, B. and Nass, C. (1996). *The Media Equation.* Cambridge, UK; Cambridge University Press.

Taylor, R.M. (2001). Cognitive Cockpit Systems Engineering: Pilot Authorisation and Control of Tasks. In R. Onken (Ed), CSAPC'01. *Proceedings of the 8th Conferences on Cognitive Sciences Approaches to process Control,* Neubiberg, Germany, September 2001. University of the German Armed Forces, Neubiberg, Germany.

Weiner, E., Kanki, B. and Helmreich, R. (1993). *Cockpit Resource Management.* London; Academic Press.

Automation Etiquette in the Augmented Cognition Context

Santosh Mathan, Michael Dorneich, Stephen Whitlow

Honeywell Laboratories
3660 Technology Dr, Minneapolis, MN 55410
santosh.mathan@honeywell.com

Abstract

Neurophysiologically triggered adaptive assistance offers many advantages over traditional approaches to automation. These systems offer the promise of leveraging the strengths of humans and machines -- augmenting human performance with automation specifically when human abilities fall short of the demands imposed by task environments. However, by delegating critical aspects of complex tasks to autonomous automation components, these systems run the risk of introducing many of the problems observed in traditional human-automation interaction contexts. This paper describes steps taken to mitigate the possibility of introducing these problems in the context of an augmented cognition system to help dismounted soldiers perform taxing cognitive tasks. It describes specific design decisions that address many of the concerns raised by automation researchers, and presents a summary of performance results that support the overall efficacy of implemented measures.

1   Introduction

The pros and cons of automating complex systems have been widely discussed in the literature (e.g. Parasuraman and Miller, 2004; Sarter, Woods and Billings, 1997). Automated systems bring precision and consistency to tasks, relieve operator monotony and fatigue, and contribute to economic efficiency. However, as widely noted, poorly designed automation can impose several undesirable consequences. Automation can relegate the operator to the status of a passive observer – serving to limit situational awareness, and induce cognitive overload when a user may be forced to inherit control from an automated system.

Automation technologies that have emerged under the Augmented Cognition program address several problems associated highly automated systems. They offer the potential to engage the user in a mixed initiative interaction -- leveraging the strengths of both machines and their human operators. Based on real-time assessments of cognitive state, these systems dynamically provide assistance to users when they are likely to be overwhelmed by task demands. However, there are several features of neurophysiologically triggered automation that can have a detrimental impact on performance. First, many neurophysiological indices fluctuate rapidly over short time windows. Triggering automation on the basis of an index with a high degree of inherent non-stationarity can severely disrupt task performance. Second, adaptive assistance can alter the task demand that the controller is subject to. As a consequence, when adaptive assistance is engaged, neurophysiological measures alone may not effectively reflect the overall task demand imposed by the environment. Unless the task context is also assessed and considered, a neurophysiologically triggered adaptive system could potentially return control to the user under circumstances that may be beyond the capability of a user to handle. Third, despite the fact that systems developed under the Augmented Cognition program display high degree of sensitivity to a user's cognitive state, as automated systems they stand to inherit many of the problems commonly observed with highly automated human-in-the-loop systems.

560

This paper describes the considerations that went into the design of an Augmented Cognition system being developed by Honeywell. The system is designed to help soldiers perform effectively under extremely demanding task conditions. We describe specific design decisions that address many of the concerns raised by automation researchers, and present a summary of performance results that support the overall efficacy of implemented measures.

2 Task Context and Mitigation Strategies

Before we describe the design considerations associated with automation in the Honeywell augmented cognition context, we present a brief overview of the task domain.

2.1 Task Context

The Honeywell Augmented Cognition program focuses on the dismounted Future Force Warrior (FFW). A critical element of the FFW program is a reliance on networked communications and high density information exchange. These capabilities are expected to increase situational awareness at every level of the operational hierarchy. It is hoped that an information technology based transformation of the military will facilitate better individual and collaborative decision making at every level. However, effective use of these information sources is constrained by the limitations of the human attentional system. The goal of the Honeywell Augmented Cognition team is to use physiological and neurophysiological sensors to detect states where attentional resources may be inadequate to cope with mission relevant demands. Efforts of the Honeywell team have focused on ways to leverage automation to manage information when human attentional resources may be inadequate for the tasks at hand.

The Honeywell AugCog prototype currently takes the form of a wireless, body-worn system. Information from a variety of sensors, including EEG, EKG, fNIR, are used to provide real time assessments of cognitive state. The Honeywell system is being evaluated in field contexts with mobile users. However, many of the mitigation strategies used to assist users were developed and tested in the context of a virtual environment simulated using the Quake III Team Arena game engine. Much of the discussion about the design of automation will be situated in the context of this virtual environment.

The tasks subjects were asked to perform in the virtual environment imposed extreme demands on attentional resources. Tasks ranged from a long and tedious vigilance task prior to a mission, to the execution of a raid on an enemy compound. During the vigilance task, subjects were asked to maintain sustained attention to detect infrequent targets within images transmitted from a surveillance camera. During raids, subjects had to split their attention among a variety of challenging tasks. These tasks included handling critical communications, navigating, and engaging hostile enemy forces.

2.2 Mitigation Strategies

A variety of mitigation strategies were developed to help users perform each of the tasks mentioned above effectively under extreme conditions. Each of these mitigation strategies leveraged automation to handle aspects of a user's task load when attentional resources were likely to be inadequate. We describe these mitigation strategies below.

2.2.1 Mixed Initiative Search

During the vigilance task subjects were asked to detect infrequent targets in scenes transmitted from a surveillance camera. Maintaining attention over long periods of time in a target deficient environment is a difficult task for subjects. Our pilot studies showed that, on average, alert users were able to detect targets with an accuracy of about 80%. However, following periods ranging from 20 to 40 minutes of sparse targets, performance fell to approximately 40% accuracy.

The system described here was designed to help users during periods of low sustained attention. The system helps users by providing advice on the likely location of targets in a scene. Potential enemies in the scene are tagged with a yellow box (Figure 1). Computer systems trained to detect target stimuli may not perform as well as an alert human. Consequently, they may not be able to completely replace the human operator in operational contexts. However, such a system could aid a human who may not be adequately alert. These systems could play a helpful role if they can be triggered when cognitive state classifiers detect a vigilance decrement. This system was modelled after enabling technology currently under development (Schneiderman & Kanade, 2000). The equipment necessary for such a system would include a display integrated helmet with multi-spectral vision capabilities. Mixed initiative search is a part of the FFW vision. For more information see (US Army, 2003).

Figure 0: *Mixed initiative target identification system tags likely targets with a yellow box*

2.2.2 Communications Scheduling

During scenarios that involved raids on an enemy compound, subjects had to balance demands associated with navigating a difficult course, managing complex communications, and engaging enemy forces. Quite often subjects would encounter situations where concurrent execution of these three tasks would become near impossible. As Ververs et. al. (2005) have noted, battlefield communications are often the most vulnerable task in the chaos of battle. Critical, mission relevant information may go unheeded and overall situational awareness may be compromised in these situations.

An important feature of the Honeywell Augmented Cognition prototype is the Communication Scheduler. The communication scheduler prioritizes information based on mission relevance. The system adapts presentation modality and message sequencing so that only critical mission relevant information is presented to the user. All messages are stored on a Tablet PC that the user carries (see Figure 2). Low and medium priority messages are deferred for later review, high priority messages are presented immediately to the user in audio format. Redundant visual cues summarize the content of high priority messages.

2.2.3 Task Off-Loading

Many mission critical communication transactions follow a well defined format. The transactions associated with a medical evacuation (MEDEVAC) are an example of such a task. The soldier coordinating a MEDEVAC on the field has to accurately relay information about injuries and casualties and communicate precisely where the evacuation can take place. However, the ability

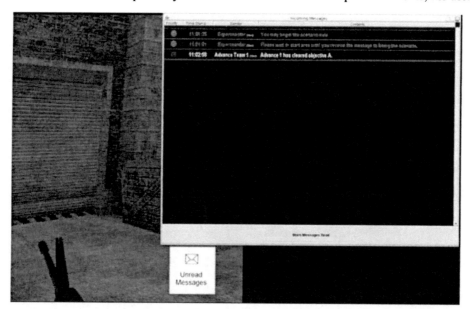

Figure 2: *Icon on the HUD informs that communications scheduler is actively managing messages (left). Deferred messages appear on a Tablet PC, organized by priority (right).*

of a soldier to conduct such a lengthy transaction over voice channel, while in a hostile engagement, may be seriously compromised. Personnel may omit important information or make errors in the information transmitted. Additionally, attention devoted to the MEDEVAC exchange may detract from the performance of other critical tasks.

The MEDEVAC Negotiation Agent was triggered on the basis of task context and cognitive State Profile. If workload was high, and a MEDEVAC had to be coordinated, the MEDEVAC Negotiation Agent was triggered. An icon on the HUD notified the user about the need to coordinate an evacuation using the agent. The user could review the MEDEVAC information on the Tablet PC and transmit information using the interactive form. Figure 3 illustrates a MEDEVAC Negotiation Application. Information available on the FFW Netted Communications network would be automatically entered into the form. The system would present this information to the user for inspection. By off-loading the demands associated with a cognitively demanding and lengthy voice transaction to the MEDEVAC agent, the user is able to direct cognitive resources to other tasks at hand.

2.2.4 Navigation Cues

While in hostile territory, soldiers have to be able to adapt their navigation plans to evolving tactical threats. However, during engagements in hostile territory, the cognitive resources necessary to generate a safe route, while engaging the enemy and handling communications, may simply not be available. To address these concerns, the Honeywell Augmented Cognition prototype incorporated functionality to assist users with navigation tasks. In hostile areas the system would generate navigation plans based on knowledge of the mission's geographical

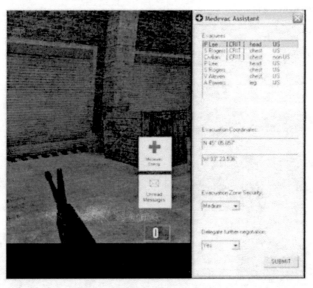

Figure 3: *MEDEVAC icon on HUD informs the user that the MEDEVAC agent is ready to coordinate an evacuation (left). User reviews parameters and engages agents through the negotiation application (right).*

objective and information about enemy locations gathered from the FFW communications networks. The system would provide users with a graphical plan in conjunction with tactile cues to guide them through relatively safe zones. Navigational assistance was invoked when the cognitive state profile indicated workload was high and the subject needed to navigate through an unfamiliar route.

Tactile cues were provided to the user by means of a tactor belt worn around the waist. Tactors were fired to direct subjects toward the bearing of their next waypoint The rate of firing the tactors increased from 1 to 2 to 8 Hz as the subject approached each waypoint. When a waypoint was reached the system provided navigation cues relative to the next waypoint until the subject reached the appropriate location. Tactile cues have been shown to be effective in improving performance of spatial tasks even in the presence of competing secondary workload tasks (Raj, Kass, & Perry, 2000).

3   Automation Etiquette

While the mitigation strategies just described promise to help users perform critical tasks under extreme task contexts, like any complex automated system, they have the potential to hurt task performance in a variety of ways. The system described here was designed on the basis of a close consideration of several problems with human-automation interaction highlighted by Sarter, Woods, and Billings (1997). We describe each of these potential problems and summarize the design features they motivated.

3.1   Uneven Distribution of Workload

As Sarter and colleagues point out, many automated systems actually hinder performance in high workload conditions. Many systems require the user to play the role of a translator or mediator – communicating aspects of the task environment to the system. Operators have to take on the responsibility of explicitly specifying task parameters for the automation to execute. In many cases these demands come during the busiest phases of work.

Automation in the context described here was designed to be invoked and parameterized with minimal involvement from the user. Most of the mitigation strategies described here were triggered based on assessments of cognitive state, and task context. As a result, users received assistance automatically in difficult task contexts -- users were not distracted from the task at hand to configure automation. Parameterization of the automated system was supported by the assumed netted communications infrastructure that is a central component of the FFW program. For example, in the navigation task, likely ambush locations were assumed to be continually assessed using information from human and electronic surveillance assets. Real time access to this information by the mitigation engine would allow the system to come up with route plans without explicit intervention from the user. Similarly, the medical evacuation task could rely on the availability of continually updated health status information over communication networks. As a result, many of the parameters associated with the medical evacuation task could be completed with minimal user intervention.

3.2   Breakdowns in Mode Awareness

Sarter and Woods define mode awareness as the ability of a system user to anticipate the behavior of automated systems. They suggest that breakdowns in mode awareness, so-called automation surprises, can lead to errors of omission in which the operator fails to observe and respond to uncommented or undesirable system behavior.

There are several sources of potential automation surprises in the context described here. First neurophysiological and physiological indices that serve to invoke automation exhibit a great deal of inherent non-stationarity. Triggering mitigations on the basis of signals that vary a great deal over short windows of time can be extremely disruptive for the user. To address this problem, cognitive state classification was based on joint consideration of several indices. Some of the indices employed in our system included EEG, galvanic skin response, heart beat variability and pupilometry. These redundant sources of information combine to provide a more stable indication of cognitive load than any single index would. Our current efforts add additional robustness to cognitive state classification by picking the modal classification output over specified time windows. The tradeoff between mitigation latency and required classification robustness determines the size of the window employed.

Second, once effective mitigation strategies are triggered, they effectively reduce the cognitive load on the user. Consequently, neurophysiological and physiological indices loose their value as indicators of task load. Disengaging mitigations solely on the basis of indices associated with cognitive load can return control to users under very difficult task conditions. To address this issue, mitigations were turned off on the basis of context related information. That is, task related information that is independent of cognitive state assessment. For example, communications scheduling was turned off after the user had indicated that they had caught up on all the deferred messages. Navigation cues were terminated only after a user had arrived at the destination

Third, the system described here provides a range of different types of assistance to users in different tasks contexts. Each of these mitigations assumes control over a certain aspects of a user's task. Unless a user is clearly aware of the status of the adaptive system, the user could encounter a range of automation surprises. To avoid these problems, the system was always explicit concerning the automation mode. For instance, when the communications scheduler was engaged, the user would see an icon on the heads up display (HUD) indicating that messages were being deferred. When system was ready to transact a Medevac, a notification icon would pop up on the HUD; the user would have to explicitly authorize a Medevac based on information populated by the system. Authorization was communicated by clicking a button. Once navigation aiding was turned on, user would feel pulses on a belt that unambiguously conveyed the navigation mode to the user.

3.3   New Coordination Demands

Sarter and colleagues suggest that autonomous automation components effectively become like crew members by taking over aspects of critical tasks. However, unlike good crew members, poorly designed automation may fail to keep users informed about task status. These systems may perform tasks autonomously and silently, but return control to users abruptly when things fail. This serves to raise the coordination requirements and could add to cognitive load.

Elements of the system described here were designed with the assumption that they were fallible. Several mitigations were designed to allow a human to intervene if the system was unable to handle a situation effectively. For example, we recognized that the process of organizing a MEDEVAC may often require more knowledge of the mission context than the MEDEVAC Negotiation Agent might be able to infer from netted data sources. In case further clarifications were required, the MEDEVAC dialog had an option for the user to delegate further negotiation to a platoon member with fewer task demands. The system provided a way to minimize coordination demands by off-loading clarifications to another human in a graceful manner.

3.4   Complacency and Trust in Automation

Sarter and colleagues suggest that complacency induced by automation may be a critical factor in many accidents. They suggest that users may come to relay on automation. Not realizing that these systems, though largely reliable, may be fallible.

Issues of complacency were also a matter of concern in the context described here. By delegating critical tasks such as communications and visual monitoring to an automated system, users stood the risk of missing critical task relevant information. Our approach to reducing possible complacency relied on training to emphasize the fallibility of the system and to provide users with procedures for monitoring the system and resuming control of delegated tasks as soon as practical. For instance, in training associated with the vigilance task, subjects were told that the system was highly fallible, with a performance accuracy of approximately 68%. They were asked to consider the system's assessment, but to remain vigilant. With tasks involving the communications scheduler, the system would only allow high priority messages to be played to users during periods of high workload. However, training emphasized the need to catch up with medium and low priority messages at the earliest possible moment. Results from experiments suggest that these issues of complacency were successfully addressed. Subjects detected targets with an accuracy of over 80%; a level substantially over the 68% base accuracy of the system. Subjects also had a much higher level of situational awareness of the status of the mission relative to conditions where they did not get automatic communications-related assistance from the system.

3.5   Training

Automation of complex tasks often introduces the need for additional training. Besides learning to master the performance of inherently complex tasks, users have to learn about the use of complex automation components to support the execution of these tasks. Sater and colleagues, argue that sophisticated automation components interact with the task environment in complex ways. They argue that training has to occur in the context of use for users to be able to acquire accurate mental models of the system.

All subjects who used the system discussed here received extensive training in the use of automation components in the actual contexts of use. Subjects progressed on to task scenarios only after they were able to successfully demonstrate use of each automation component in training scenarios. Subjects also had to answer a broad range of questions about each automation component and it's interaction with the task environment.

4   Results

Experimental results support the overall efficacy of the automation components incorporated in the system. The Honeywell Augmented cognition system helped subjects perform significantly better on communication, fighting, navigation and vigilance tasks with neurophysiologically triggered automation. Additionally, most subjects felt that the automation components described here actually made task execution easier.

The experiment was conducted with 13 experienced video game players who performed military relevant tasks in the Quake virtual environment. Subjects performed tasks within experimental scenarios that required users to balance communication, fighting and navigation tasks. Subjects performed each scenario with automation and without. Such an experimental design allowed for a within subjects assessment of the effects of neurophysiologically triggered automation on performance. Subjects performing tasks with assistance from the communication scheduler responded to double the number of messages as they did in scenarios without assistance from the system. They performed almost twice as well on assessments of situational awareness in scenarios with automation. Additionally, navigational assistance allowed users to run into a third of the number of ambushes as they did in scenarios without assistance. Subjects also performed significantly better with the help of automation on the vigilance task. Subjects who received neurophysiologically triggered assistance detected 30% more targets than subjects who did not. A between subjects comparison was employed to assess system effectiveness on the vigilance task. The length of time required to induce a vigilance decrement precluded the use of a within subject design in this context.

Subjective assessments of the automation suggest that users found the automation components in the system to be helpful in performing tasks. For instance, 76.9% of the subjects thought it was easier to perform tasks in the mitigated condition. Subjects also found mitigated condition easier for fighting (61.5%), communicating (84.6%) and navigating (76.9%).

5   Conclusion

This paper reviewed several problems with human automation interaction that were relevant to the design of a neurophysiologically triggered adaptive system. Specific strategies to minimize the negative impact of automation were discussed. Experimental evaluations suggest that automation components designed with the considerations specified here were indeed successful in helping users. However, there are several limitations of this work that are important to mention. The automation strategies described here were assessed in a military relevant desktop training environment. It is unclear whether these techniques will remain effective in actual military contexts. Second, it is hard to tell which of the specific automation design strategies implemented here had the most impact on users. The design of the experiment only attests to the relative efficacy of these strategies as a package. Third, the system evaluated here only incorporated support for a small subset of tasks that soldiers are likely to perform. As automation components get added and layered, the overall complexity of the system could grow far beyond the capacity of the strategies described here alone to address.

Acknowledgments

This research was supported by a contract with DARPA and funded through the U.S. Army Natick Soldier Center, under Contract No. DAAD-16-03-C-0054, for which CDR (Sel.) Dylan Schmorrow serves as the Program Manager of the DARPA Augmented Cognition program and Mr. Henry Girolamo is the DARPA Agent. Any opinions, findings, conclusions or recommendations expressed herein are those of the authors and do not necessarily reflect the views of DARPA or the U.S. Army.

References

Dorneich, M.C., Mathan, S., Creaser, J.I., Whitlow, S.D., and Ververs (2005). "Enabling Improved Performance though a Closed-Loop Adaptive System Driven by Real-Time Assessment of Cognitive State." In Proceedings of the 11th International Conference on Human-Computer Interaction (Augmented Cognition International), Las Vegas, Jul 22-27, 2005

Norman DA. The problem with "automation": inappropriate feedback and interaction, not "over-automation". Philosophical Transaction of the Royal Society of London, 1990; B327: 585-593.

Parasuraman, R., & Miller, C. (2004). Trust and etiquette in high-criticality automated systems. Communications of the Association for Computing Machinery, 47(4), 51-55.

Raj, A. K., Kass, S. J., & Perry, J. F. (2000). Vibrotactile displays for improving spatial awareness.Proceedings of the International Ergonomics Association XIV Triennial Congress / 44th Annual Meeting of the Human Factors and Ergonomics Society (1), 181-184.

Sarter, N.B., Woods, D.D., and Billings, C.E. (1997). Automation Surprises. In G. Salvendy (Ed.), Handbook of Human Factors and Ergonomics (2nd edition) (pp. 1926-1943). New York, NY: Wiley.

Schneiderman, H. and Kanade, T. (2000) "A statistical approach to 3d object detection applied to faces and cars". Proceedings of the IEEE Computer Society Conference on Computer Vision and Pattern Recognition (CVPR), pages 746--751, Hilton Head Island, South Carolina, June 2000

US Army (2003), "The Objective Force Warrior: The Art of the Possible...a Vision," Available Online at http://www.ornl.gov/sci/nsd/pdf/OFW_composite_vision.pdf.

Ververs, Whitlow, Dorneich, Mathan, and Sampson (2005) AugCogifying the Army's Future Warfighter. In Proceedings of the 11th International Conference on Human-Computer Interaction (Augmented Cognition International), Las Vegas, Jul 22-27, 2005

Assisted Focus: Heuristic Automation for Guiding Users' Attention Toward Critical Information

Mark St. John, Harvey S. Smallman, & Daniel I. Manes
Pacific Science & Engineering Group
9180 Brown Deer Road
San Diego, CA 92121
stjohn@pacific-science.com
smallman@pacific-science.com
dmanes@pacific-science.com

Abstract

Theaters of operation are busy environments, and displays of tactical situations can quickly become congested and cluttered with military or other symbols. This clutter can distract users from critical information, unnecessarily increase workload, and delay responding. We have developed a concept called Assisted Focus that intelligently augments human attention by reducing clutter and helping users focus on critical information. In the domain of naval air warfare, this concept has required research into three related fields: 1) Algorithms for identifying high and low threat aircraft, 2) methods for directing users' attention toward high threats without reducing overall situation awareness, and 3) the design of human-automation interfaces to help users supervise and interact with realistically imperfect automated systems with sophistication and efficiency. Critically, our research indicates that the algorithms do not need to be perfect in order to significantly help users focus on important threats. The project culminated in an applied experiment to assess the concept within the domain of a realistic naval air defense task. In the experiment, 27 Navy air defense experts used Assisted Focus to identify threats and declutter a complex airspace. Assisted Focus decluttering improved response timeliness to threatening aircraft 25% compared with a baseline display, it was especially beneficial for threats in more peripheral locations, and 93% of the expert warfighters preferred the Assisted Focus display. In related work, we found that Assisted Focus and heuristic automation produced a 23% improvement in search times in a target detection task. Assisted Focus is applicable to a wide range of situation awareness and monitoring tasks.

1 Introduction

Tactical situations can quickly become cluttered. Displays of those situations can become congested and cluttered with military or other symbols. This clutter can serve to distract users from critical information, unnecessarily increase workload, and delay responding. For example, clutter increases search times by increasing the number of objects that must be sifted or searched through to find objects of interest by increasing the search "set size" (e.g. Treisman & Gelade, 1980), and clutter increases the chance for "change blindness," the chronic human inability to detect changes occurring in a scene when attention is, even momentarily, focused elsewhere (Rensink, 2002).
Our goal has been to develop interface concepts that help users manage their attention so that they can concentrate on the most important or threatening aspects of a situation and spend less attention on the less important aspects. Here, we focus on the domain of naval air warfare, where users monitor an airspace for threatening aircraft, though the concepts are applicable to a broad range of situation awareness and monitoring tasks.

A common method for helping warfighters cope with clutter and manage their attention is to "declutter" the display by either removing, or making less salient, the less important aircraft. A number of approaches exist for implementing this method, but their underlying algorithms for determining which aircraft to filter tend to be simple classification rules, such as filter all identified friendly aircraft or filter all unknown aircraft with an altitude over 20,000 feet. Although attractive for their simplicity, these rules often fail to meet the needs of sophisticated users because they do not align with the categories of most interest to these users. Nonetheless, several studies have shown that users appreciate and benefit from even simple filtering methods (Johnson, Liao, & Granada, 2002; Nugent, 1996; Osga and Keating, 1994; Shultz, Nichols, and Curran, 1985; Van Orden, DiVita, & Shim, 1993; Yeh & Wickens, 2001).

Recently, we have begun to investigate the possibility of applying artificial neural network technology to the clutter problem by training an artificial neural network to mimic the threat assessment ratings produced by expert naval personnel and then decluttering the geographical display based on those threat ratings. This approach offers the possibility of decluttering on a much more sophisticated and operationally relevant, albeit complex, criterion. Marshall, Christensen, and McAllister (1996) showed that a simple feed-forward neural network could reliably mimic individual experts' level of interest in different aircraft taken from realistic simulations of naval air warfare. The neural network model reliably classified aircraft that can be safely ignored from aircraft that pose possible problems or potential threats.

Figure 1 shows an example of this approach. A trained neural network sweeps over a geographical display every second, evaluating each aircraft for its level of threat to own ship (the blue aircraft carrier near the center of the display). The network computes threat levels by using a set of aircraft attributes and weightings of those attributes based on extensive interviews with subject matter experts (Liebhaber, Kobus, & Feher, 2002). The resulting level of threat scores are then used to declutter the less threatening aircraft from the display by dimmed their symbols. Using standard navy symbology, potentially threatening aircraft are shown as yellow clover-shapes, and friendly aircraft are shown as blue bullet-shapes. After the decluttering operation, the more threatening aircraft stand out clearly, which helps users focus their attention on the more threatening aircraft easily and effectively.

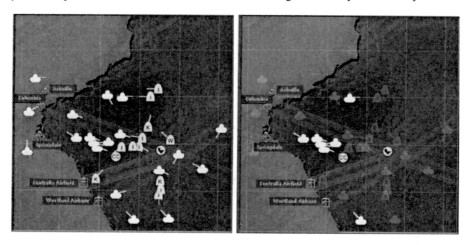

Figure 1: Cluttered geographical display (left) and same display in which less-threatening aircraft have been decluttered by dimming their symbols (right)

Successful implementation of these automation and display concepts requires research into several issues: 1) algorithms for classifying aircraft as high and low threat, 2) interface designs for portraying those classifications, and 3) methods for developing user trust and performance with the system. Here, we briefly review research into each of these three issues.

2   Classification

There are several challenges to producing a useful aircraft threat classification algorithm. One problem is that threat classification can grow very complex, with many variables and high ambiguity, since perfect information is rarely available. Furthermore, even expert decision-makers frequently disagree about the threat level of individual aircraft (Marshall, Christensen, & McAllister, 1996; Morrison, Kelly, & Hutchins, 1996). Consequently, an automated algorithm can never match the threat ratings of every expert. Additionally, well known issues of automation trust, complacency, and confirmation bias (Parasuraman & Riley, 1997) can undermine the effective use of automation and lead to disastrous consequences.

Our automation design philosophy has been to create "mixed initiative systems" that combine "heuristic automation" that is known to be imperfect with engaged, knowledgeable users who use the automation as a guide, or first cut at classification, but who ultimately rely on their own best judgment to make decisions (St. John & Manes, 2002; St. John, Oonk, & Osga, 2000; St. John, Smallman, Manes, Feher, & Morrison, 2005). This philosophy fits well with what is termed "lower levels of automation" (Parasuraman, Sheridan, & Wickens, 2000) and likelihood alarms (Sorkin, Kantowitz, and Kantowitz, 1988).

For example, in a visual search task, St. John and Manes (2002) used heuristic automation in the form of an imperfect target detection tool to make a rough first cut at identifying the likely locations of hidden targets. Users then exploited this information to guide their own searches. This approach led to a 23% improvement in search times, even when the automation was only 70% reliable. In a dual task paradigm, Sorkin, Kantowitz, and Kantowitz (1988) used a "likelihood alarm display" to indicate the likelihood of a signal occurring in the secondary task. Users exploited the likelihood information to decide how carefully to attend to the secondary task. In both studies, knowledgeable users exploited the information provided by imperfect, heuristic automation to guide their attention.

Applying this concept to air warfare, the heuristic automation, in this case a neural network algorithm, evaluates each aircraft and declutters the less threatening ones. There are many relatively clear cases of high threat and low threat for which the algorithm and any trained user would agree. However, there are also a number of more borderline threat cases on which the algorithm and any particular user may disagree. The heuristic automation approach is to allow the automation to provide the first cut, but provide sufficient information to users so that they can tailor the classifications and make final decisions. This approach assists users by helping them to see and focus on the clear threats quickly, while avoiding distraction from clear nonthreats. With Assisted Focusing, situation awareness can be enhanced, and responses speeded, because significant threats are clearly visible.

As with any automated classification tool, there is some danger of confirmation bias. Users might slavishly follow the automation even when it makes classification errors. Specifically, the automation may miss some threatening aircraft and inappropriately leave some nonthreatening aircraft fully visible. The opportunity for these problems is increased by the heuristic nature of heuristic automation. However, the applied experiment described below indicates that the danger of these problems is minimal, at least with the highly trained personnel we tested. One reason why the problems are minimal is that Assisted Focus frees up enough time that was previously devoted to searching the display that users actually have more time to devote both to monitoring threats and to scanning for borderline threats that the algorithm may have miss-classified. In section 4, we demonstrate this finding in a realistic naval air warfare task with expert Navy users.

3 Interface Design

Once an automated threat classification algorithm is developed, there is still the question of how to portray the classifications on the tactical display in order to declutter the less threatening aircraft. A declutter symbology must satisfy two competing demands. First, it must minimize distraction from the fully visible symbols so that users can concentrate on the more threatening aircraft. Second, the declutter symbology must continue to present information about decluttered aircraft so that users can maintain situation awareness about the entire situation. Different symbologies trade off differently against these demands. Replacing standard military symbols with gray dots, for example, should minimize distraction from fully visible symbols, but provide less information about the decluttered aircraft: a gray dot can convey that a aircraft is present at a given location, but it cannot convey heading, affiliation, or platform. At the other extreme, faded, colored military symbols convey as much information as the fully visible symbols, but they are visually much more complex than dots. This complexity, and similarity with the fully visible symbols, may make them difficult to segregate from the fully visible symbols, and thereby increase search times and cognitive workload. These difficulties, if they are substantial, would tend to minimize any potential benefit from the declutter operation.

To investigate this trade-off empirically, we used a visual search task to evaluate six declutter symbologies (see St. John, Feher, & Morrison, 2002, for details). The symbologies were created by manipulating two factors, symbol type and coloring. The symbol types were relatively complex military symbols that coded substantial information about the aircraft, simplified outlines of ships and aircraft that coded an intermediate amount of information, and simple dots that coded minimal information. The simplified outlines were a version of a symbology, called

"Symbicons," that we developed by combining the better attributes of standard military symbols and miniature realistic icons, Smallman, Oonk, St. John & Cowen, 2001.

The coloring factor consisted of using either faded versions of the standard military symbol colors or using gray. To create the faded colored declutter symbols, the brightness of the fully visible colors were reduced from 100% to 30%. To create the gray decluttered symbols, the brightness was reduced to 30%, and the saturation was reduced to 0%. We also investigated the effect of different amounts of declutter, from no declutter (0%) to 25%, 50%, and 75% declutter.

The participants were 52 undergraduate students recruited from local universities. Participants searched for two target symbols among a field of 48 symbols that were a mixture of fully visible, brightly colored military symbols and decluttered symbols. The targets always appeared among the fully visible symbols, and the decluttered symbols merely served as distractors.

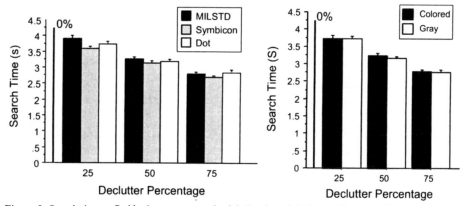

Figure 2: Search time to find both targets on each trial. Graph on left shows effect of declutter symbology on search time for each percentage of declutter. Graph on right shows effect of declutter color on search time for each percentage of declutter.

Each increase in the percentage of declutter produced a significant and linear drop in search time, from no decluttering (0) to having 75% of the symbols on the display decluttered. The intermediately complex symbol outlines produced the least distraction and the fastest search times, but the differences among symbol types were small. Surprisingly, the faded colored symbols produce as little distraction and as fast search times as the gray symbols. Despite the salience of color as a cue in display design, it appears that fading the symbol colors made the decluttered symbols sufficiently different from the fully visible symbols. Participants could easily segregate and ignore the colored decluttered symbols during their searches among the fully visible symbols. Nonetheless, and equally importantly, the faded, colored military symbols remained clearly visible. Therefore, it is likely that users could remain aware of even decluttered aircraft, while at the same time easily focusing on the threatening, brightly colored aircraft.

We concluded that the faded, colored standard military symbols are the best declutter symbols because they are little more distracting than the alternatives while providing much more information. The next step will be to determine how well these declutter symbols help decision-makers manage their attention under more realistic operational conditions.

4 User Trust and Performance

To determine if expert users would actually trust the Assisted Focus concept enough to use it, and if they would realize any performance benefits from it, we evaluated the concept in a realistic, single user simulation of air warfare (see St. John, Smallman, Manes, Feher, & Morrison, 2005, for details).

The participants were 27 US Navy personnel with moderate to very high levels of air warfare experience (3 to 30 years of service in the US Navy with an average of 13 years). The experiment tactical display showed a 170 x 120 nautical mile area reminiscent of the Persian Gulf. In all conditions, users could access a variety of information about an aircraft by clicking on an aircraft's symbol with the mouse and then viewing a set of data that appeared in a window in the lower left corner of the display.

There were three equivalent scenarios, each lasting 15 minutes. During each scenario, aircraft moved about the display at realistic rates: from 95 to 560 nautical miles per hour (10 to 55 pixels per minute or 0.003 to 0.03 degrees of visual angle per second). There were approximately 50 aircraft on the display at all times, with aircraft occasionally entering or exiting the displayed area. Most aircraft appeared benign and non-threatening, behaving like normal commercial airliners, oil platform helicopters, or other light commercial aircraft. At each moment, however, approximately seven aircraft appeared significantly threatening (8 or higher on a 10-point scale), behaving for example like tactical fighter aircraft, moving at high speed, from hostile origins, toward own ship. Approximately 12 additional aircraft appeared potentially threatening, or "borderline" (6 or 7 on a 10-point scale of threat). These aircraft presented a mix of benign and threatening attributes. As aircraft moved about the display, their threat levels changed. In general, the scenario was designed to present a range of aircraft behaviors to keep the participants engaged.

Participants monitored the aircraft and responded to the significantly threatening ones. Participants were instructed that the evaluation part of the task was their own expert judgment. They were also told that the threat algorithm and declutter operation was only an imperfect aid designed to provide a reasonable "first cut" at evaluating threat. These instructions both allowed and encouraged users to judge for themselves which aircraft were significantly threatening. Once a participant deemed an aircraft to be a significant threat, their "rules of engagement" specified a set of responses to execute as the aircraft approached own ship. Participants "notified a superior command element" when significant threats crossed a trip wire at 75 miles of own ship, queried the aircraft at 50 miles, and warned the aircraft at 25 miles. Our hypothesis was that these responses would be made more quickly following the trip wires when Assisted Focus made the threatening aircraft more salient on the display.

Figure 3 illustrates a potentially threatening event from one of the scenarios. An aircraft appeared abruptly from an island off the coast of a hostile country. If the participant believed the aircraft was a significant threat (which it was for all participants), the participant should have immediately queried the aircraft. As the graph shows, the aircraft was queried substantially faster with the Assisted Focus display than with the standard display. While the size of this improvement in response time was not typical, it is an especially powerful example of the benefits of Assisted Focus – for an especially suspect and potentially dangerous aircraft that might have been missed until too late.

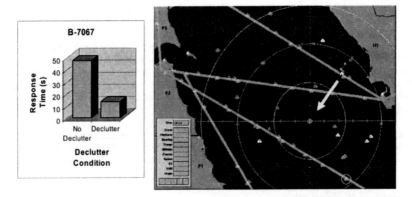

Figure 3: Illustration of the response time benefit of Assisted Focus. The threatening (bright yellow) aircraft approaching own ship from the direction of the arrow is queried sooner with Assisted Focus than without

Overall, Assisted Focus reduced response times by 25% (significant by an analysis of variance). Furthermore, 93% of the participants preferred having the Assisted Focus display to the standard display. These response time benefits obtained in spite of the fact that the participants did not always agree with the automated classifications. While 80% of responses were made to aircraft that the algorithm deemed to be significant threats (8 or higher on a 10 point threat scale), 17% of responses were made to borderline threats (rated as a 6 or 7 on a ten point threat scale), and 3% were made to nonthreats. Hence, the Assisted Focus algorithm worked well, but not perfectly. Nonetheless, participants benefited from the Assisted Focus and preferred it.

Furthermore, the danger from confirmation bias was small. Participants responded to roughly two more borderline threatening aircraft per scenario in the declutter condition than in the baseline condition. When the declutter condition made these aircraft fully visible while decluttering other aircraft, the participants may have been encouraged to interpret them as more threatening than they otherwise would have. Learning to guard against potential bias is important, but we are encouraged by its small size.

In interviews following the experiment, participants claimed that these benefits reduced their workload, relieved the pressure to act and decide quickly, allowed time to concentrate on suspects, and aided situation awareness. Comments included the following: "I didn't have to waste time on low-threat tracks," "I actually had more time to spend scanning the display because I could see where the high threats were," and "Decluttering allowed me to get ahead in my ROE—instead of behind it when mistakes are more likely to happen."

5 Conclusions

In a realistic operational task, Assisted Focus helped users focus on significant threats without having to laboriously search the display. This assistance led to substantial response time benefits and reports of reduced workload and better overall situation awareness. These benefits were demonstrated in the complex tactical situation monitoring task of naval air warfare. This concept of augmented attention with Assisted Focus, however, should apply to a number of other domains in which there are high levels of display clutter, and users much search the display to find, and then re-find, objects of interest.

A key feature of Assisted Focus is use of an imperfect, or heuristic, classification algorithm for identifying objects to declutter. The ability to use an imperfect algorithm dramatically eases the burden on researchers or designers to develop useful algorithms. Traditionally, imperfect algorithms are rightly viewed with skepticism since they naturally lead to misses or false alarms. The insight of Assisted Focus is that imperfect classification is acceptable if the imperfections are understood by users, they use the classification only as a rough first cut to focus their attention, and they remain aware of the situation and have ultimate decision-making control. Here, we showed that these conditions can be met by appropriately instructing trained users in the nature of the Assisted Focus tool and by providing an effective declutter symbology that decluttered the display without removing information.

6 References

Johnson, W. W., Liao, M., & Granada, S. (2002). Effects of symbol brightness cueing on attention during a visual search of a cockpit display of traffic information. In *Proceedings of the Human Factors and Ergonomics Society 46th Annual Meeting* (pp. 1599-1603). Santa Monica, CA: Human Factors and Ergonomics Society.

Liebhaber, M. J., Kobus, D. A., & Feher, B. A. (2002). Studies of U.S. Navy air defense threat assessment: Cues, information order, and impact of conflicting data (technical report SSC-1888). San Diego, CA: Space and Naval Warfare Systems Center.

Marshall, S. P., Christensen, S. E., & McAllister, J. A. (1996). Cognitive differences in tactical decision making. In *Proceedings of the 1996 Command and Control Research and Technology Symposium* (pp. 122-132). Washington, DC: Department of Defense, Command and Control Research Program.

Morrison, J. G., Kelly, R. T., & Hutchins, S. G. (1996). Impact of naturalistic decision support on tactical situation awareness. In *Proceedings of the Human Factors and Ergonomics Society 40th Annual Meeting* (pp. 199–203). Santa Monica, CA: Human Factors and Ergonomics Society.

Nugent, W. A. (1996). Comparison of variable coded symbology to a conventional tactical situation display method. In *Proceedings of the Human Factors and Ergonomics Society 40th Annual Meeting* (pp. 1174-1178). Santa Monica, CA: Human Factors and Ergonomics Society.

Osga, G. & Keating, R. (1994). Usability study of variable coding methods for tactical information display visual filtering (technical report NOSC-2628). San Diego, CA: Naval Command, Control and Ocean Surveillance Center RDT&E Division.

Parasuraman, R. & Riley, V. (1997). Humans and automation: Use, misuse, disuse, abuse. *Human Factors, 39,* 230-253.

Parasuraman, R., Sheridan, T. B., & Wickens, C. D. (2000). A model for types and levels of human interaction with automation. *IEEE Transactions on Systems, Man, and Cybernetics – Part A: Systems and Humans, 30,* 286-297.

Rensink, R. A. (2002). Change detection. *Annual Review of Psychology, 53,* 245-277.

Schultz, E. E., Nichols, D. A., & Curran, P. S. (1985). Decluttering methods for high density computer-generated graphic displays. In *Proceedings of the Human Factors and Ergonomics Society 29th Annual Meeting* (pp. 300-303). Santa Monica, CA: Human Factors and Ergonomics Society.

Sorkin, R. D., Kantowitz, B. H., & and Kantowitz, S. C. (1988). Likelihood alarm displays. *Human Factors, 30,* 445-459.

Smallman, H.S., Oonk, H.M., St. John, M., and Cowen, M.B. (2001) 'Symbicons': advanced symbology for two-dimensional and three dimensional displays (technical report 1850). San Diego, CA:. Space and Naval Warfare System Center

St. John, M., Feher, B. A., & Morrison, J. G. (2002). Evaluating alternative symbologies for decluttering geographical displays (technical report 1890). San Diego, CA: Space and Naval Warfare System Center.

St. John, M. & Manes, D. I. (2002). Making unreliable automation useful. In *Proceedings of the Human Factors and Ergonomics Society 46th Annual Meeting* (pp. 332-336). Santa Monica, CA: Human Factors and Ergonomics Society.

St. John, M., Manes, D. I., Smallman, H. S., Feher, B. A., & Morrison, J. G. (2004). An intelligent threat assessment tool for decluttering naval air defense displays (technical report). San Diego, CA: Space and Naval Warfare Systems Center.

St. John, M., Oonk, H. M., & Osga, G. A. (2000). Designing displays for command and control supervision: Contextualizing alerts and "trust but verify" automation. In *Proceedings of the Human Factors and Ergonomics Society 44th Annual Meeting* (pp. 646-649). Santa Monica, CA: Human Factors and Ergonomics Society.

Treisman, A. M., & Gelade, G. (1980). A feature-integration theory of attention. *Cognitive Psychology, 12,* 97-136.

Van Orden, K. F., DiVita, J. & Shim, M. J. (1993). Redundant user of luminance and flashing with shape and color as highlighting codes in symbolic displays. *Human Factors, 35,* 195-204.

Yeh, M. & Wickens, C. D. (2001). Attentional filtering in the design of electronic map displays: A comparison of color coding, intensity coding, and decluttering techniques. *Human Factors, 43,* 543-562.

Developing Trust of Highly Automated Systems

Thomas M. Spura
Erin E. Accettullo
Lockheed Martin Systems Integration
Owego, NY
Thomas.spura@lmco.com
Erin.Accettullo@lmco.com

Jerry Franke
Vera Zaychik
Lockheed Martin Advanced Technology Laboratory
Cherry Hill, NJ
JFranke@atl.lmco.com
Vera.Zaychik@lmco.com

Abstract

The use of unmanned vehicles today continues to require multiple operators controlling a single vehicle. The user community is looking for methods to increase the vehicle-to-operator ratio, while providing the same or greater levels of capability. This will then allow the force multiplier concept to individual soldiers or platforms. This is a complex problem that requires development in the areas of autonomous behaviors and Human Machine Interface (HMI). As a result, operators are being introduced to new C2 concepts and paradigms. Developing trust in these systems is critical for maximum utilization and system acceptance. This paper looks at the ongoing development of the UCAR Command and Control (C2) interface and the methods being utilized to develop user trust and acceptance.

DARPA's Unmanned Combat Armed Rotorcraft (UCAR) system concept includes the requirement for a single operator to control multiple vehicles. This is one of the first of many new UAV programs that poses this requirement, and also includes the challenge that the operator be one of the cockpit crew of an Apache. The challenges of controlling multiple vehicles by a single operator, who is already in a high workload situation, are great. The solution being developed on the UCAR program utilizes multiple innovative HMI concepts and is applicable for all unmanned systems.

The control problem is amplified on UCAR where the vehicles are considered as force multipliers for combat manned systems. In this case, operators in high workload environments of military engagements currently cannot handle the additional burden with controlling the unmanned vehicles. Recent studies indicate that operators shed the UAV task at these times and focus solely on the local tasks because of the significant increase in workload.

The evolving solution of controlling unmanned vehicles uses a combination of autonomy on the unmanned platforms and the integration of innovative HMI techniques for interfacing with the operator. These techniques mimic the current method of interaction with other manned vehicles, but maintain the recognition that the vehicles are unmanned. Interaction with the unmanned vehicles is at a high level of command and control, releasing the operator from vigilant monitoring of systems.

New human-machine interface (HMI) technology offers the solution to this UAV control problem. Technologies such as multi-modal interfacing and advanced alerting concepts are being studied for use in UAV systems. These capabilities are integrated in the UCAR solution, providing the operator with systems management capabilities.

User trust and acceptance in the system are critical components for new systems to break down usage barriers and hesitance. These aspects are addressed on UCAR by incorporating subject matter expert comments and recommendations into the design of novel HMI components. Development of controls and displays for UCAR operators is guided by accepted human engineering standards for military vehicles, allowing for easy integration into the C2 systems of the future.

The purpose of this paper it to review the challenges that are confronted as the UV designs evolve to single operator / multiple vehicle designs and how those challenges are being met by the HMI design for UCAR. The overall HMI solution for UCAR will be highlighted, with specific attention to its mitigation of operator workload impact of multi-vehicle operations. In addition, preliminary results are presented showing customer trust measurements during the development of UCAR C2 designs.

1 Introduction

An ongoing revolution in the autonomous capabilities of unmanned vehicles is presenting new opportunities and new hazards in the design of human interfaces for their control. Historically, unmanned systems have required a high operator-to-vehicle ratio. Current visions for future unmanned system control call for inverting this ratio, with fewer operators controlling more vehicles. To achieve this vision, unmanned system developers are empowering their vehicles with more autonomous capabilities. With the autonomy comes a new paradigm for interfacing with the vehicles, one where commands are at a higher level than direct control. This new paradigm provides great capability, but also a level of abstraction at the operator level. For the system to operate at it's full potential the operator must have confidence that commands will be correctly interpreted and that the system will respond as expected. In other words, the operator needs to have trust in the system.

As this autonomy revolution takes place, human-machine interfaces (HMI) for system control must react to the new paradigm. However, simply removing functions from human responsibility is not the final solution. In the end, all decisions made in the battlespace must trace to some accountable person, and for unmanned missions, that person is the vehicle's operator. Thus the problem of effective HMI design for these systems is one of dual opposing factors: to allow autonomous functions the freedom needed for their solutions to be effective and to support the operator in attaining a level of comfort and trust that allows the overall manned/unmanned system to be effective. That is, future unmanned system HMI must support increased autonomy and improved oversight simultaneously.

The new paradigm of command and control will place new demands on the operator. The responsibility of operating a remote vehicle that may be capable of targeting, or even executing, targets places stress on the operator. This stress will lead to a level of conservative operation by the operator that may inhibit the full use of system capabilities. As with all systems a certain level of experimentation by the operator is necessary to fully understand system capabilities. To a great extent, this can be accomplished through training. However, for the operator to work fully with the system a certain level of experimentation with the actual system is necessary. This will only occur if there is a level of trust in the system by the operator. With autonomous systems the operator will have a higher level of control, commanding at the task level as opposed to direct control. This abstraction will lead to results that may be unexpected or undesired. The operator needs to trust the system not to react in an undesired manner, or in the case that the results are undesired then the operator needs a mechanism to reverse the command input.

In the subsequent sections, we will explore the issue of trust in autonomous unmanned systems command and control and present several techniques to address the unique problems the domain presents. Our approach to achieving the needed balance comes from combining a robust management-by-exception control strategy with new technologies for low workload oversight and awareness of system activities. The result is an HMI system that effectively demonstrates the goals outlined above.

Section 2 provides a description of the autonomous system developed under the UCAR program, and used in the experiments described. Section 3 provides a description of the experiments along with a discussion of the results. Section 4 provides insight into methods of developing and maintaining trust.

2 Approaches / Relevant Work

Several technological advancements in a few key areas can be applied to the domain of unmanned vehicle command and control to reduce operator workload and improve the interface. To be able to operate multiple unmanned systems, which is the goal of providing autonomous vehicles, the workload per vehicle on the operator must be minimized. In order to reduce the workload, innovative interfacing concepts needed to be developed. This created new interface paradigms for the operators, requiring a shift in operational conduct. For this new interface paradigm to work, the operator(s) must have a certain level of trust that the system will respond as commanded, and that the operator will have the necessary control capability to execute the assigned mission. The following section describes the design strategy and implementation to control multiple autonomous vehicles.

2.1 Operator Interruption

Whether management-by-consent or management-by-exception is employed, the information flows to the operator in bursts, mainly as interruptions to the main task. McFarlane (2002) identified four main interruption management strategies: immediate, mediated, negotiated, and scheduled. Immediate interruption is the typical mechanism employed by most systems today, where all the information is presented to the user as soon as it becomes available. Studies have shown that this strategy adversely affects operator performance on both the primary and the interruption task. McFarlane showed that negotiated interruption strategy is the most effective for those domains where actions are not required immediately. With that strategy, the user is presented with the high-level information about each incoming alert and is left to decide on the order and the timing of handling tasks. A software architecture HAIL (Human Alerting and Interruption Logistics) has been designed and developed to provide support for negotiated interruption combined with powerful context recovery and information assessment mechanisms. This architecture has been successfully applied to improve the navy ship ID supervisor interface. Another effective interruption strategy is a mediated interruption, where some entity, human or software, acts as a mediator between the user and the information, deciding on the timing and order of interruptions based on the type of information and the current user workload. With this approach a concurrent scheduling of several tasks is possible.

Any interruption-driven management has a potential of introducing situation disorientation: the user has to regain the main task context after the interruption has been handled. McFarlane suggests that context recovery, i.e. saving main task context before the interruption and restoring it after the interruption, can reduce context recovery time and effort. Context replay – review of previous contexts – can improve operator situation awareness to improve performance on tasks requiring vigilance-related information. Thus, reduced operator awareness in management-by-exception or management-by-contest control can be mitigated.

2.2 Multi-Modal HMI

Workload is directly related to cross-task interference – the more tasks are scheduled at the same time, the higher the workload, especially if they all use the same channel (usually visual). However, in an environment where the UAV operator is controlling multiple vehicles on top of performing his primary job in the battlefield, parallelizing of tasks is unavoidable. Use of multiple modalities for output has been shown to reduce the workload and improve operator performance (Dixon, Enriquez, Calhoun), which corresponds with the theoretical predictions by Wickens based on the Multiple Resource Theory (Wickens). Multiple Resource Theory argues that the performance of tasks in different modalities (for example, performing one visually/manually and one auditory/verbally) can result in less cognitive interference between the tasks because they use different sets of resources within the cognitive system. Enriquez used a pneumatic tactile alerting system in a driving environment and showed a reduction in response times. Dixon used offloading onto the auditory channel in a UAV control domain and showed reduced interference and workload. Haptic interfaces have been shown to improve situational awareness under certain circumstances as well (Ruff). As long as the tasks can be presented in several different modalities without great loss of information, workload-based modality distribution could be used to manage workload successfully, although general attention saturation is still the limiting factor in how many tasks can be sequenced at the same time.

Another aspect of a control interface with multiple modalities is an ability to provide multi-modal input, i.e. control of the vehicle with means other than a traditional mouse/joystick and keyboard input. Speech-based control has already been incorporated into a few systems (see McMillan, Draper, Williamson). Speech interaction systems have progressed over the years from simple speech recognition to true spoken dialogue (Daniels). The effectiveness of such systems does heavily depend on the capabilities of speech recognition and the noise environment. However, research has shown that speech-based control can improve performance and reduce workload (Nelson).

2.3 The LM UAV Team Approach

The Lockheed Martin UAV HMI team has developed cutting-edge concepts such as natural language interaction, negotiated intervention, tactile alerting, and a management-by-exception approach to command and control of teams of highly automated unmanned vehicles. Some of the key attributes of the HMI design include the use of multiple-modalities to inform and alert the operator, and an effective vocal interface that allows the operator to command teams with natural spoken words. Innovative command and control interfaces that provide adaptive, interactive modes are key to reducing overall operator workload in complex environments.

The use of natural speech patterns to communicate with unmanned vehicles requires significantly less cognitive functioning than requiring the operator to type in commands manually. A highly advanced vocal interface has been implemented to develop this concept. The vehicles may respond to and/or elicit operator response through an auditory channel by which a human-sounding voice speaks through the control headset. This interface mimics the auditory interaction between the operator and manned teammates.

In some cases the vehicles may require to address the operator via a particular modality, even though that modality may be over-saturated. Negotiated intervention makes the operator aware of an impending message (via a display box, aural tone, or tactile stimulation) non-intrusively, allowing him/her to choose when and how to be interrupted by that message.

The use of a tactile vest on the operator implements the use of a third modality to interact with the operator for certain types of messages (such as vehicle location in relation to the controlling element). Tactile alerting frees up the visual and auditory senses to accept more information at once.

3 UAV Experimentation

The UAV team at Lockheed Martin Systems Integration – Owego performed several experiments in which a ground control station and air control station, located on a Longbow Apache, worked collaboratively with a team of simulated vehicles to accomplish several missions. It is especially important to take steps to reduce cognitive workload as much as possible on advanced flight systems such as the Apache Longbow due to the complex nature of the primary flight tasks and potential emergency situations that may arise. Several mission scenarios were lab-tested by a subject matter expert in a simulated cockpit environment. The missions required the operator to control up to four teams of unmanned vehicles at once, while co-piloting the Apache Longbow helicopter. An experimental paradigm was implemented during the demonstrations to measure operator workload, or cognitive resource saturation, at both mission and task levels.

3.1 Participants

The operator of the UCAR system was given the title of UCAR Controlling Element, or UCE. The UCE is the person who is ultimately responsible for the actions of the unmanned systems. Commonly called the operator or controller in current unmanned systems, the use of high degrees of autonomy and decision-making prompted the use of this naming convention. The design concept for the UCAR Command and Control system included multiple instantiations of control capability, and the ability to transfer control from one UCE to another throughout the mission. For maximum control of system operation, the UCE was given an ability to set preferences determining system actions based on mission type, target priority, as well as how operator interfaces perform.

For the purposes of initial development, two UCE designs were developed and tested: the Airborne UCE and the Ground UCE. The Airborne UCE is considered to be the most difficult design for C2 of autonomous systems. It is envisioned that the Airborne UCE will have additional tasks to perform in conjunction with autonomous systems. In the case of the UCAR design, the Airborne UCE is the Longbow Apache Co-pilot/Gunner (CPG).

3.2 Mission Scenarios

Mission scenarios included both Armed Recon and Strike missions. In the recon mission with 25 targets, later increased to 50, the UCE was required to control a single team of four vehicles. In a more complex scenario, the UCE received a new mission plan and was required to split the group into two teams of two vehicles each, flying separate missions. In a third mission scenario, new intelligence was received as the team entered the area of interest, forcing the UCE/UAV team to re-plan for a strike mission.

3.3 Analysis

Because the military customer requires a consolidated technical performance metric for workload assessment, it is necessary to unite a number of aspects of workload and situation awareness as a single, nominal score. The Workload Efficiency Factor (WEF) allows the HMI team to determine adherence to predetermined acceptable levels of operator workload for this system. The WEF is a numeric score derived from a compilation of both qualitative and quantitative workload, situation awareness and technical performance data. Each of the proceeding factors is weighted individually and normalized to a scale of 1-10, with 10 representing the lowest level of workload/highest level of performance and 1 representing an overwhelming level of workload/lowest level of performance. The WEF calculation can be expressed mathematically (for calculations involving any number of variables) as:

$$WEF = \Pi(x_n)/y^n,$$

where x represents the individual ratings and y represents the "acceptable threshold" of UCE workload resulting from unmanned vehicle control. The calculated WEF must be >1 to satisfy the requirement.

Factors that comprise the WEF include a workload score, a situation awareness score, command accuracy metrics (for voice and displays), mission effectiveness (overall performance) metrics, temporal pressure metrics, and an independent physiological workload assessment (blood oxygen saturation) score. A pulse oximeter (BCI 3301 model) was used to collect physiological data non-intrusively through the use of a fingertip sheath sensor.

3.4 Results

In the early stages of experimentation, a single SME was assigned to watch over the UAV system and HMI design development, in order to ensure consistence in inputs to the designers and minimize the personal preference inputs from multiple operators. Thus data collected were minimal and not representative of the overall user population. However, testing in the early stages of system development has indicated positive results as the software develops, and SME comments have helped to define requirements for design. It is expected that the data collected from multiple participants will show a positive trend, demonstrating that performance improves and workload is reduced when controlling the developing UAV systems.

4 Developing and Maintaining Trust

A primary concern of operators with advanced automation is whether the technology will react the way it is expected to. Highly intelligent autonomous vehicles are designed to provide a deterministic response to operator commands and unexpected situations. When situations arise, the vehicles behave in a way the operator can expect a manned teammate to behave, so that with experience operators become comfortable utilizing a more liberal management approach. To accomplish this, the UAV must have the capability to accept a broad range of commands and potential situations and prioritize this information on its own. By mimicking a live teammate as closely as possible, the UAV can establish a confident relationship with the operator.

The concept of natural language interaction allows the operator to interact with the vehicle via verbal communication. Working with a leading speech recognition group, the LM team developed structured command syntax for controlling UAVs, so that the vehicle recognizes a variety of alternative sentence and phrase structures for each possible command or interrogation. For example, to request the most current alert from the system, the UAV will accept "Surface the alert", "Show me the most recent alert", "Select that alert", "Display the latest alert", and any variation of these.

In addition, the vehicles convey status information, respond to operator interrogation, and verify commands via synthetic speech responses. The operator can set preferences for which type of response (aural tone, natural language, or visual display) is received for a variety of communication types received from the vehicle. However, all critical alerts are redundantly displayed and conveyed to the operator by means of natural language interaction.

Key to developing trust in autonomous systems is the ability for the operator to maintain awareness and exert control when needed without interfering with the normal autonomy of the vehicle. While the management-by-exception approach to autonomous vehicle command and control separates the operator from immediate decision-making, the operator is still ultimately responsible for the safety and success of both the vehicle and its mission. The operator must be made comfortable with the amount of information presented about each decision the system makes. Similarly, the operator must be given the ability to correct mistakes by the system in a timely fashion. However, these problems are not insurmountable. With careful operator interface design, both goals can be achieved.

The challenge inherent in providing the operator a sufficient awareness of system progress, state, and decision-making is the effective management of the operator's workload and cognitive resources. Even an autonomous system with control at only the highest levels can produce huge amounts of status information that the operator must navigate. Unless provided help, the operator can easily be overwhelmed by the information flow and miss vital decisions made by the system. The simple alternative is to provide only high-level status information to the operator, but some decisions cannot be adequately explained at a high level.

The solution is to couple the presentation of high-level status information with an interrogation mechanism that provides the ability to request specific details about vehicle operations. In nominal cases, the operator interface presents summary information about the current mission plan, mission progress, and system state that can be absorbed and understood with minimal effort by the operator. This reduces the vigilance workload for the operator while helping the operator maintain a basic awareness about the vehicle's mission execution. When the operator notices an aspect of the situation that bears further exploration, the interrogation interface allows the operator to ask questions about it.

For example, when presented the replan of the vehicle's route, the interrogation interface supports queries about the nature of the replan: What was the reason behind the replan? How will it affect the mission replan? Will all mission goals still be met? How much did the plan change from before?

The other half of the efficient trustworthy control interface is the correction mechanism. Management-by-exception supports operator correction of vehicle decisions. This requires mechanisms for reacting to operator corrections by the vehicle system, but these mechanisms are useless without an effective method for the operator to quickly enter corrective commands to the system. This correction can take multiple forms. Route plans can be modified by shifting waypoints. Threat evasions can be countermanded by selecting the threat and indicating the vehicle should ignore it. Entered re-plans can be rejected, forcing the vehicle to return to the prior plan. These solutions can be used in combinations tailored to the application involved or used individually. When combined with the interrogation mechanism, this provides effective low-workload operator control.

These interface devices can be used a number of ways depending upon the level of trust the operator has built with the system. New users of these systems should be able to set the interface's presentation mechanisms to provide more status information, then reduce the granularity of status information displays as the operator learns which details are necessary for proper oversight. Similarly, preferences can be set to provide a wider variety of corrective operations early in system use, with the operator eventually deciding which corrections should be immediately apparent in the system based on experience. This allows the operator and the vehicle system to grow into a trust relationship over time.

5 Conclusions

We have discussed the implementation of a system that provides high levels of autonomy to unmanned vehicles, and the associated command and control design. The operators need to trust the autonomous systems to react as expected and to be able to conduct the mission predictably.

Current experiments have shown that this capability is productive and can provide the levels of trust necessary for proper operational capability.

Further work is planned to look at how trust can be extended and planned for. These areas include adding preferences selection, incorporating a undo capability, and greater testing of operator acceptability and trust in the system.

6 Acknowledgments

Some of the work described in this paper was funded under the OTA portion of the DARPA Unmanned Combat Armed Rotorcraft (UCAR) program (MDA972-02-9-0011). The authors wish to recognize the following individuals who have made recent contributions to this system:

Draper Laboratory – Brent Appleby, Mark Homer, Leena Sing, Lee Yang

Lockheed Martin Advanced Technology Laboratory - David Cooper, Rich Dickinson, Chris Garrett, Adria Hughes, Steve Jameson, Mike Orr, Brian Satterfield, and Mike Thomas

Lockheed Martin Simulation Training and Support – Angelo Prevete, Ken Stricker, and Brian Vanderlaan

Lockheed Martin Systems Integration – Owego – Rick Crist, Steve DeMarco, Dave Garrison, Carl Herman, Adam Jung, Ateen Khatekhate, John Moody, Donn Powers, Greg Scanlon, Mike Scarangella, Keith Sheppard, Peter Stiles, Sandy Stockdale, Robert Szczerba, and Joel Tleon

UCAR Government Team - Bob Boyd, Marsh Cagle-West, LTC Gerrie Gage, Steve MacWillie, Steve Rast, Randy Scrocca, CW4 Matt Thomas, and Don Woodbury

7 References

1. Calhoun, G.L., Draper, M.H., Ruff, H.A., & Fontejon, J.V. (2002). Utility of a tactile display for cueing faults. Proceedings of the Human Factors and Ergonomics Society 46th Annual Meeting, 2144-2148

2. Daniels, Jody and Hastie, Helen. (2003) "The Pragmatics of Taking a Spoken Language System Out of the Laboratory" Proceedings of the Human Language Technology Conference, HLT-2003, Edmonton, Canada.

3. Dixon, Stephen R. and Wickens, Christopher D. (2003) "Control of Multiple-UAVs: A Workload Analysis" 12th International Symposium on Aviation Psychology, Dayton, OH.

4. Draper, M., Calhoun, G., Ruff, H., Williamson D., & Barry, T. (2003). Manual versus speech input for unmanned aerial vehicle control station operations. Paper presented at the Human Factors and Ergonomics Society 47th Annual Meeting. Santa Monica, CA: Human Factors and Ergonomics Society.

5. Enriquez, M., Afonin, O, Yager, B, MacLean, K (2001). "A Pneumatic Tactile Notification System for the Driving Environment," in Proc. of Workshop on Perceptive User Interfaces (PUI '01), Orlando, FL, 2001.

6. McFarlane, D.C. (2002) "Comparison of four primary methods for coordinating the interruption of people in human-computer interaction," Human-Computer Interaction, 17 (1), 63-139, 2002.

7. McFarlane, D.C. and Latorella, K.A. (2002) "The scope and importance of human interruption in human-computer interaction design," Human-Computer Interaction, 17 (1), 1-61, 2002.

8. McFarlane, Dan (2003). "Engaging Innate Human Cognitive Capabilities to Coordinate Human Interruption in Computer Interaction: The HAIL System," Cognitive Systems: Human Cognitive Models in System Design, Santa Fe, New Mexico, July 2003.

9. McMillan, G. R., Eggleston, R. G., & Anderson, T. R. (1997). Nonconventional controls. In Salvendy, G. (Ed.), Handbook of human factors and ergonomics (2nd ed.). NY, NY: John Wiley & Sons.

10. Navon, D. and Gopher, D. (1979) "On the Economy of the Human Processing Systems" Psychological Review, 86, 254-255.

11. Nelson, W. T., Anderson, T. R., & McMillan, G. R. (2003). Alternative control technology for uninhabited aerial vehicles: human factors considerations. In L. H. Hettinger & M. W. Haas (Eds.). Psychological Issues in the Design of Virtual Environments, Lawrence Erlbaum.

12. Olson, Wesley A. (2001). "Management-By-Consent in Human-Machine Systems: When and Why It Breaks Down", in Human Factors, 43(2), 255-266.

13. Ruff, Heath A., Narayanan, S., and Draper, Mark H. (2002) "Human Interaction with Levels of Automation and Decision-aid Fidelity in the Supervisory Control of Multiple Simulated Unmanned Air Vehicles," Presence: Teleoperators and Virtual Environments, volume 11, issue 4, pp. 335-351, 2002.

14. Sarter, Nadine B. (2001) "Multimodal Communication in Support of Coordinative Functions in Human-Machine Teams", in Journal of Human Performance in Extreme Environments, 5(2), 50-54.

15. Wickens, Christopher D. (2002) "Multiple Resources and Performance Prediction" Theoretical Issues in Ergonomic Science, pp. 1-19.

16. Williamson, D.T., & Barry, T.P. (2000). The design and evaluation of a speech interface for generation of Air Tasking Orders. Proceedings of the Human Factors and Ergonomics Society 44th Annual Meeting, 750-753. Santa Monica, CA: Human Factors and Ergonomics Society.

PACT: Enabling Trust in Adaptive Systems through Contracted Delegation of Authority

Robert M Taylor

Defence Science and Technology Laboratory
Farnborough, Hants, GU14 0LX, UK
rmtaylor@dstl.gov.uk

Abstract

Trust is an important issue for users of automation. Research has indicated that rather like "confidence", trust operates as a psychosocial attitude with inherently variable subjective complexity, and as an intervening varaible between automation reliability and automation use. For the design of adaptable automation and decision support, research effort needs to be directed at constructing and constraining human-automation relationships, interactions and behaviours in manners that naturally engender trust. Methods for delegating authority to automation are needed that manage and control risks of automation in a sensible, regulated and predictable manner, with appropriate safeguards. This paper discusses the PACT framework for delegation of autonomy, and considers the issues arising for engendering and maintaining trust with adaptable automation and adaptive human-computer systems.

1 Understanding Trust

Issues of "trust" have featured strongly in implementation of ideas of dynamic function allocation, automated decision support and adaptable automation for human-computer adaptive systems. This is in response to concerns about human monitoring of unreliable automation and the need for control strategies to mitigate bias from complacency and overtrust, or alternatively, undertrust and disuse (Parasuraman and Byrne, 2004). Early research with aircrew investigated the structure and measurement of trust to help design automation safeguards. A study of twin-crew RAF Tornado aircraft operations, elicited tactical decision making scenarios and aircrew rated them for the importance of factors associated with trust (Taylor, 1988). Demand for trust was associated with perceived risk and the probability of negative consequences, wheras the supply of trust was related to the requirement for judgement and awareness, and uncertainty and doubt in making decisions. Relying on others to make risky decisions calls for a large amount of trust. If the decision requires another person exercising a high degree of awareness and judgement, and there is much uncertainty and doubt in the decision provided, then the actual trust engendered by the decision will be low. A follow-up study investigated the quality of aircrew teamwork and showed that trust was a significant factor in distinguishing between good and poor teamwork performance. Trust was rated at a significantly lower level in single-seat RAF Harrier operations (i.e. human-computer teamwork) than in two-seat RAF Tornado aircaraft tactical operations (i.e. both human computer and human-human teamwork). (Taylor and Selcon, 1993).

Several models of trust have been proposed. Riley (1992) developed a model of the relationships between trust, operator skill level, task complexity, workload and risk, self-confidence and automation reliability. Studies in which workload and reliability were varied, led to refinement of the model to include factors of fatigue and learning about system states (Riley, 1994). Further research has modelled trust as a function of recent performance, and the prescence and magnitude of fault, with subjective trust increasing with automation reliability (Lee and Moray, 1992). The relationship between reliability and trust was confirmed by human monitoring performance measures indirectly measuring trust (May et al, 1993). Recent work has extended modelling to investigate how the dynamics of trust and reliance depend on information sharing (Gao and Lee, 2004), and how trust becomes overtrust through unintended use (Itoh et al, 2004).

Experimental evidence has verified that unexpected automation failure leads to a breakdown of trust, and to difficulty in the recovery of trust with a loss of faith in future teamwork performance. As trust declines, manual intervention increases (Muir,1987; Lerch and Prietula, 1989). Research has showed how when workload is increased, over-trust or complacency develops with automatic systems, and coupled with vigilance problems, this is likely to lead to failure to detect performance deviations and decrements in automation performance (Parasuraman, et al., 1993a). Operator detection of automation failure is substantially degraded with a static allocation fixed over a period of time, favouring dynamic adaptable allocation (Parasuraman, et al.1993b). Manual task reallocation has been proposed as a countermeasure to monitoring inefficiency and complacency since short periods of intermittent manual task reallocation, or cycling between manual and automation control, reduces failures of monitoring (Rouse, 1994). By maintaining manual skill levels, and enhancing situational awareness, manual task re-allocation helps when intervention is needed following automation failure. However, without active involvement, it is difficult to maintain an appropriate dynamic internal model of the important changing relationships needed for regaining manual control following automation (Carmody and Gluckman, 1993). Experimental studies have shown that with competing demands for attention, humans are poor at monitoring automation for occasional malfunctions, exhibiting automation complacency (Parasuraman et al., 1993a). Humans also have poor awareness of adaptable automation failure (Taylor et al, 1995, 1996).

Generally, trust is best considered as an intervening variable between automation reliability and automation use (Parasuraman and Byrne, 2004). For purposes of measuring aiding effectiveness, like "confidence", trust operates as a psychosocial attitude with inherently variable subjective complexity, rather than as a cognitive functional state such as "situation awareness" and "workload". Trust is unlikely to be reliably linked to performance and effectiveness. Furthermore, "trust" is unlikely to be productive of reliable psychometric or concomitant behavioural measures (qualitative or quantitive), with useful sensitivity, discrimination, diagnosticity or predictive power.

2   Enabling Trust

For the design of adaptable automation and automated decision support, research effort is needed to be directed at constructing and constraining human-automation relationships, interactions and behaviours in a manner that naturally engender trust. Methods for delegating authority to automation are needed that manage and control risks of automation in a sensible, regulated and predictable manner, with appropriate safeguards. Safeguards are needed against breakdown or failure in performance to ensure that operator trust in system functioning is maintained at realistically appropriate levels, without adversely affecting situational awareness. The first Law of Adaptive Aiding states that "computers should take tasks, but not give them" (Rouse, 1994). Automatic reallocation of tasks to manual performance seems close to a violation of this law. In particular, variable assistance and allocation could lead to unacceptable unpredictability. So, awareness of the current task allocation strategy is an important factor for system effectiveness. But this may not be easily achieved by seamlessly adaptable aiding. Careful consideration is needed of the proceedures for implementation of dynamic task allocation and re-allocation.

The building of trust between the operator and the computer automation system has been identified as a key issue in increasing the capability for cognitive automation. Trust is built when consistency and correctness is observed in the computer system's decisions and actions. Two important guidelines for building trust have arisen (Reising, 1995):
- *Define the Prime Directives.* These are overall governing rules which bound the behaviour of the aiding system, and yet provide a logical structure for aiding system to act in a rational and reliable manner, avoiding arbitrary behaviour, so that the pilot does not experience any surprises e.g. Asimov's Laws of Robotics.
- *Specify the Levels of Autonomy.* These also bound the behaviour of the aiding system by limiting its decision authority for the performance of specific sub-functions to a set of system configurations specified and set by the operator.

Trust is built on awareness of proven performance. Adaptive strategies for coping with control of complex, dynamic situations, such as automatic unburdening and manual re-allocation, will need careful adaptive logic to ensure their appropriateness. The design of the functional interface needs to ensure that appropriate levels of awareness of the current task allocation are easily maintained. Awareness is needed to avoid task contention, and to ensure that tasks are not overlooked or performed incorrectly.

3   Contracted Delegation of Authority

Under the UK MOD Cognitive Cockpit (CogPit) programme, a framework was created for pilot interaction and delegation of adjustable levels of autonomy, known as the PACT system (Pilot Authorisation and Control of Tasks) (Table 1). PACT spans the range of possible levels of allocation of decision making tasks, or levels of autonomy, between humans and computers (Sheridan and Verplanck, 1978; Parasuraman, Sheridan and Wickens, 2000). The PACT framework is intended to provide trustworthiness in the information and behaviour of adaptable automation

(Taylor, 2001a; Taylor et al, 2001). In addition to the original CogPit work, the PACT system has been had the following control applications:

- Supervisory control of Uninhabited Air Vehicles (UAV) (White, 2002).
- Spatial disorientation safety net (Taylor, Brown and Dickson, 2002).
- Autonomous control of Unattended Cognitive Underwater Vehicles (UCUV) with high levels of adaptive autonomy, using a Bayesian decision system approach (Waters and Taylor, 2004).
- Control of distributed intelligent systems (Taylor, 2003), linked to multi-agent autonomy levels (Barber et al, 2000) and the Extended Control Model (ECOM) (Hollnagel, 2002)
- Control of adaptive automation by psychophysiological state triggers in an Augmented Cockpit (Diethe et al, 2004).

Future manned and uninhabited platforms are likely to have on-board cognitive automation operating with relatively high levels of autonomy or decision authority. Context sensitive technologies or "intelligent" computer software agents (e.g. Bayesian nets) offer the possibility of being able to control, regulate, direct and adapt system behaviour, within constraint boundaries, even in uncertain, novel, and unpredictable situations. The aspiration is to achieve the requisite cognitive agility, precision, reliability and safety of operations with intelligent systems, with the minimum human supervision and human-computer communication.

Table 1. Bonner-Taylor PACT Framework for Pilot Authorisation of Control of Tasks

Primary Modes	Levels	Operational Relationship	Computer Autonomy	Pilot Authority	Adaptation	Information on performance
AUTOMATIC		Automatic	Full	Interrupt	Computer monitored by pilot	On/off Failure warnings Performance only if required.
ASSISTED	4	Direct Support	Advised action unless revoked	Revoking action	Computer backed up by pilot	Feedback on action. Alerts and warnings on failure of action.
	3	In Support	Advice, and if authorised, action	Acceptance of advice and authorising action	Pilot backed up by the computer	Feed-forward advice and feedback on action. Alerts and warnings on failure of authorised action.
	2	Advisory	Advice	Acceptance of advice	Pilot assisted by computer	Feed-forward advice
	1	At Call	Advice only if requested.	Full	Pilot, assisted by computer only when requested.	Feedforward advice, only on request
COMMANDED		Under Command	None	Full	Pilot	None performance is transparent.

The purposes of the PACT framework can be summarised as follows:
- to bound the behaviour of the aiding system,
- to limit its decision authority to the performance of specific sub-functions, and
- to enable a set of system configurations to be specified, set and adjusted by the operator.

The PACT framework provides the necessary and sufficient levels of autonomy for the CogPit Task Manager. PACT is based on the idea of *contractual autonomy*. Using an aircrew term from co-operative air defence, the pilot forms a set of "contracts" with the automation by allocating tasks to PACT modes and levels of automation aiding. The contract defines the specific nature of the operational relationship between the pilot and the computer aiding for co-operative performance of specific sub-functions and tasks. In setting the PACT contract, the operator defines:

- what sub- functions and tasks are aided, when and how
- what level of assistance is provided as primary or default, and when
- what levels of assistance are permissable for anticipateable contingencies, and when
- what are permissable triggers for changing levels of assistance, either contextual or by operator command, and
- what information is provided to the operator, when and how, including status advice, feedforward/feedback couse of action information and saliency

Thus, autonomy is limited by the set of contracts made between the pilot and the computer automation system governing and bounding the performance of tasks to a set of sensible and predictable co-operative behaviours. The contracts are binding agreements for the computer. Only the pilot can set or modify the PACT contracts, or define a priori the contextual circumstances for real-time adaptation PACT changes. Thus, the pilot retains authority and executive control, while delegating responsibility for the performance of tasks in a sensible and predictable manner to the computer. PACT simplifies the number of automation modes required - fully automatic, assisted, commanded - with a further secondary levels nested within the semi-automatic assisted mode, which can be changed adaptably or by pilot command. The PACT framework employs military terminology for categories of support in British Army land forces. This provides realistic operational relationships compatible with military user control schemata. It is a logical, practical set of levels of automation, with progressive operator/pilot authority and computer autonomy supporting situation assessment, decision making and action (Figure 1).

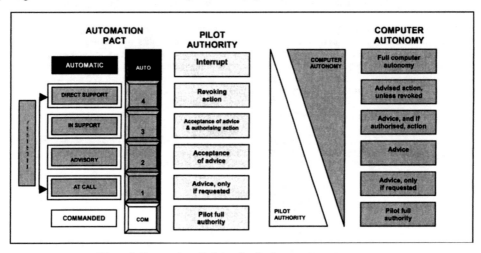

Figure 1. Progression of pilot authority & computer autonomy

Mission functions and tasks, at different levels of abstraction, allocated individually or grouped in related scripts or plays, can be set to these levels in a number of ways:
- pre-set operator preferred defaults,
- operator selection during pre-flight planning,
- changes by the operator during in-flight re-planning (e.g. Direct Voice Input commands), and
- automatic changes according to operator agreed, context-sensitive adaptation rules.

Figure 2 illustrates a set of mission functions and tasks on a timeline in a hypothetical task network.

In the CogPit implementation, the PACT system operates within an adaptive system architecture that couples on-line monitoring of the pilot's functional state and on-line task knowledge management and decision support for context-sensitive aiding, deriving information to mediate the timing, saliency and autonomy of the aiding. Three principle agents with different tasks comprise the CogPit system.
- A *Cognition Monitor* (COGMON) is responsible for monitoring the pilot's physiology and behaviour to provide an estimation of the pilot's functional state.
- A *Situation Assessor Support System* (SASS) is responsible for monitoring the aircraft situation and outside environment, generating advice and recommending courses of action.
- A *Task Interface Manager* (TIM) is responsible for monitoring the mission plan, deciding automation and managing the cockpit interface.

The TIM module provides on-line analysis of higher-order outputs from COGMON and SASS, and other aircraft systems. A central function for this system is maximisation of the goodness of fit between aircraft status, 'pilot-state' and tactical assessments provided by the SASS. These integrative functions enable this system to influence the prioritisation of tasks and, at a logical level, to determine the means by which pilot information is communicated through the TIM and the associated cockpit interfaces. Overall, this system allows pilots to manage their interaction with the cockpit automation, by context-sensitive control over the allocation of tasks to the automated systems.

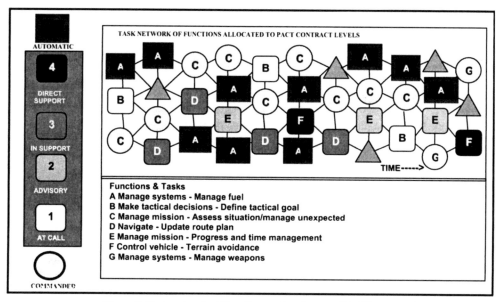

Figure 2. Task network of functions and tasks set to PACT contract levels

The TIM functional architecture comprised modules for goal-plan tracking and for interface, timeline, automation and task management utilising a blackboard for goal-plan tracking information. Details of the TIM functional architecture are provided elsewhere (Bonner, Taylor and Miller, 2000; Taylor, 2001). The idea of a tasking interface exploits the lessons learnt from the US Army's RPA program (Miller, Guerlain and Hannen, 1999). It arose from the need to be able to predict pilot expectations and intentions with reference to embedded knowledge of mission plans and goals. The aim was to provide an adaptive or "tasking" interface that allowed the operators/pilots to pose a task for automation in the same way that they would task another skilled crewmember. It afforded pilots the ability to retain executive control of tasks whilst delegating their execution to the automation. A tasking interface necessitated the development of a cockpit interface that allowed the pilot to change the level of automation in accordance with mission situation, pilot requirements and/or pilot capabilities. It was necessary that both the pilot and the system operated from a shared task model, affording communication of tasking instructions in the form of desired goals, tasks, partial plans or constraints that were in accord with the task structures defined in the shared task model.

Providing flexible or adjustable levels of autonomy for the performance of tasks and functions is a key requirement for implementation of the tasking interface concept. Allowing pilots to choose various levels of interaction for the tasks they are required to conduct can mitigate the problem of unpredictability of automation. TIM utilises the monitoring and analysis of the mission tasks provided by the SASS combined with the pilot state monitoring of the COGMON to afford adaptatation of automation, adaptable information presentation and task and timeline management.

In the CogPit implementation, PACT levels are triggered adaptively, in accordance with PACT contracts, in response to contextual input from COGMON, SASS and TIM mission goal-plan tracking (GPT). The intention is to monitor and manage the variability in performance through a barrier system approach (monitor, detect, correct, reflect performance), and through appropriate cognitive streaming interventions (join, break, divert cognition). TIM feedback and feed-forward control messages are used with appropriate multi-modal intervention saliency (background, hinting, influencing, directing, compelling) developed to reduce cognitive bias with decision support systems. All the tasks in the mission scenario are pre-allocated to *possible* PACT level contracts by the pilot. The individual task PACT levels (defaults and contingencies) were set to mitigate the risks to achievement of the individual task goals. The TIM Task Manager distinguished between pending, active and completed tasks for the current scenario/vignette. Individual tasks progressed from pending, to active and to completed as the scenario progressed.

4 Evaluation

The operation of the PACT framework was successfully demonstrated to the MOD CogPit customer in 2001. It has subsequently been incorporated into interfaces for UAV control and been demonstrated successfully (White, 2002). Diethe et al (2004) reported that a successful DARPA Augmented Cognition programme CogPit closed loop trial was completed in November 2003 during which the stability and performance of the CogPit system were examined under different levels of threat/workload in a realistic deep-strike mission.

Analysis has provided additional sources of evaluation. A risk analysis (Taylor, 2001b) indicated that generic risks of automation are likely to be mitigated by the Assisted PACT levels, as follows:
- PACT 5 Automatic: Automation bias, poor mode awareness and monitoring, surprise, unexpected action, ROE change, out-of-the-loop performance, unpredictability.
- PACT 0 Commanded: Cognitive bias, complexity, pre-occupation, fixation, time pressure, failure to evaluate options, forgetting rules and procedures, breakdown of skill.

The PACT system is designed to support the pilot's cognitive work. The support ranges from providing advice to providing action. The cognitive work required can be represented in terms of a cognitive decision ladder using state flow transition diagrams. Control task analysis (Sanderson et al, 1999) has been used to identify the structure of the cognitive work performed by the pilot and by automation at each PACT level (Taylor, 2001c). Figure 3 provides an example of the control task analysis for PACT Level 3 Assisted-In Support, represented in decision ladder flow terms.

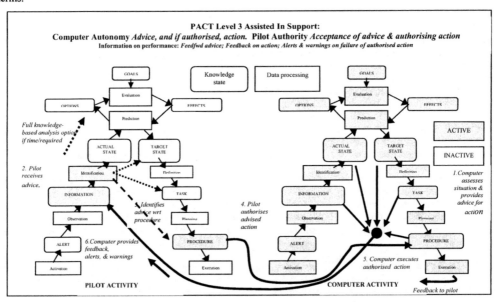

Figure 3 Control task analysis of PACT Level 3 Assisted In Support

Based on control task analyses, estimates of the resultant or residual pilot cognitive load were estimated by the CogPit pilot SME for different degrees of pilot critical involvement. Figure 4 summaries the levels of cognitive work estimated in four decision ladder phases (Perception, Assessment, Decision and Action (PADA)). Workload estimates were provided for PACT levels with immediate acceptance, critical acceptance and independent analysis. The analysis indicated that immediate acceptance of advice, associated with high levels of automation trust, was more likely to occur for Perception, Assessment and Decision phases (i.e. situation assessment, status, goals, options, effects and plans). But immediate authorisation of action was unlikely to occur, without critical appraisal, indicating a basic lack of trust. This may limit the reduction in cognitive load arising from automation of advised action (Direct Support). Concern about the validity of automated action is understandable during early familiarisation and confidence building. Critical appraisal of recommended courses of action probably will continue until the trustworthiness of the system can be established.

Some support for this observation on the untrustworthiness of action automation has been reported in an investigation of multiple UAV control. Ruff et al, (2004) report a study to assess the effects of automation reliability and levels of automation (LOA) on supervisory control of multiple UAVs. The LOA used were Management-by-Consent (MBC) and Management-by-Exception (MBE). These LOA equate to PACT Levels 3 and 4 respectively. Under MBC, the operator had to explicitly agree to suggested actions before they occurred. Under MBE, the system automatically implemented suggested actions after a pre-set time unless the operator objected. Results of two experiments showed that participants reported higher workload and difficulty with MBE. Under MBE conditions, time limits were set on manual intervention before automatic performance of the tasks (re-plans, image prosecution). This manipulation of reliability meant that erroneous action could occur. Participants generally chose to complete the tasks manually before MBE automatic action, indicating a lack confidence or trust in the system reliability.

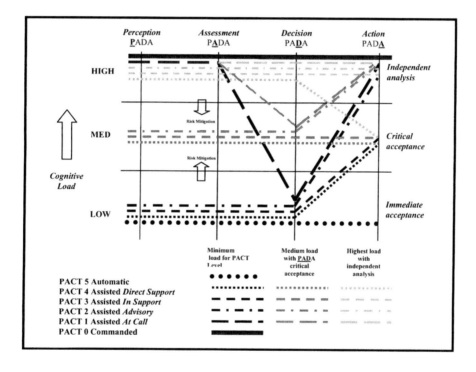

Figure 4. Cognitive load estimates for PACT levels

5   Conclusion

The PACT framework developed for pilot authorisation of control of tasks provides a simplified, practical set of adjustable levels of contractual autonomy capable of engendering trust in automation. This enables the pilot to delegate responsibility for tasks to the computer through a set of contracts that limit autonomy and bound the behaviour of the aiding system, while maintaining the pilot's authority through executive control. Control task analysis, cognitive loading and risk analysis provide useful tools for understanding and modelling the functioning of the PACT system. The PACT framework seems sufficiently robust and useful to be applicable to other systems and environments requiring cognitive control with trustworthy variable levels of autonomy, such as the control of multiple uninhabited vehicles.

References

Barber, K., Goel, A. and Martin C. (2000). Dynamic Adaptive Autonomy for Multi-agent Systems. Journal of Experimental and Theoretical Artificial Intelligence. Vol 12. Part 2, pp 129-148.

Bonner, M., Taylor, R.M. and Miller, C. (2000). Tasking interface manager: Affording pilot control of adaptive automation and aiding. In P.T. McCabe, M.A. Hanson, and S.A. Robertson (Eds.), Contemporary Ergonomics 2000, pp 70-74. London: Taylor and Francis.

Carmody, M.A. and Gluckman, J.P. (1993) Task specific effects of automation and automation failure on performance, workload and situation awareness. Proceeding s of 7th International Symposium on Aviation Psychology, Ohio State University, Vol 1, pp 167-171.

Diethe, T.R., Dickson, B.T., Schmorrow, D. and Raley, C. (2004). Toward an augmented cockpit. In D.A. Vincente, M. Moustapha and P.A. Hancock (Eds). HPSAA II, Vol II, Human Performance, Situation Awareness and Automation: Current Research Trends, pp 65-69. Mawhah, NJ: Erlbuam

Gao, J. and Lee, J.D. (2004). Information sharing, trust and relaince – A dynamic model of multi-operator multi-automation interaction. In D.A. Vincente, M. Moustapha and P.A. Hancock (Eds). HPSAA II, Vol II, Human Performance, Situation Awareness and Automation: Current Research Trends, pp 34-39. Mawhah, NJ: Erlbuam

Hollnagel E. (2002). Cognition as Control: A Pragmatic Approach to the Modelling of Joint Cognitive Systems. Special Issue of IEEE Transactions on Systems, Man and Cybernetics A: Systems and Humans – "Model-Based Cognitive Engineering in Complex Systems (In Press) http://www.ida.liu.se/~eriho/

Itoh, M., Inahashi, H. and Tanaka, K. (2004). Overtrust due to unintended useof automation. In D.A. Vincente, M. Moustapha and P.A. Hancock (Eds). HPSAA II, Vol II, Human Performance, Situation Awareness and Automation: Current Research Trends, pp 11-16. Mawhah, NJ: Erlbuam

Lee, J.D. and Moray, N. (1992). Trust, control strategies and allocation of function in human machine systems. Ergonomics, Vol 35, pp 1243-1270

Lerch, F. and Prietula M. (1989). How do we trust machine advice? In, G. Salvendy and M. Smith (Eds). Designing and Using Human Computer Interfaces and Knowledge-based Systems. Amsterdam, Elsevier

May , P., Molloy,R. and Parasuraman, R. (1993). Effects of automation reliability and failure rate on monitoring performance in a multi-task environment. Proceedings of the 37th Annual Meeting of the Human Factors and Ergonomics Society, Seattle WA.

Miller, C.A., Guerlain, S. and Hannen, M. (1999). The Rotorcraft Pilot's Associate Cockpit Information Manager: Acceptable behaviour from a new crew member. Proceedings of the American Helicopter Society, 55th Annual Forum, Montreal, Quebec, May 25-27, 1999.

Muir, B.M. (1987). Trust between humans and machines and the design of decision aids. International Journal of Man-Machine Studies, Vol 7, pp 527-539

Parasuraman, R and Byrne, E. (2003). Automation and Human Performance in Aviation. In, P.S. Tsang and M.A. Vidulich (Eds). Principles and Practice of Aviation Psychology, Chapter 9, pp 311-356. Mawhah, NJ: Erlbaum.

Parasuraman R., Sheridan T.B, and Wickens C.D. (2000). A model for types and levels of human interaction with automation. IEEE Transactions on Sytsems, Man, and Cybernetics. Part A: Systems and Humans, Vol 30, No 3, pp 286-297. May 2000.

Parasuraman, R., Mouloua, M., Molloy, R. and Hillburn, B. (1993b). Adaptive function allocation reduces the cost of static automation . Proceeding s of 7th International Symposium on Aviation Psychology, Ohio State University, Vol 1, pp 178-181.

Parasuraman, R., Molloy, R. and Singh, I. (1993a). Performance consequences of automation induced complacency. International Journal of Aviation Psychology, Vol 3, (1), 1-23.

Reising J.M. (1995). Must the Human-Electronic Crew Pass the Turing Test? In, The Human Electronic Crew: Can We Trust the Team? WL-TR-96-3039, p 103-108. Wright Patterson AFB, OH., December 1995. Proceedings of the 3rd International Workshop on Human-Computer Teamwork. DRA CHS Report DRA/CHS/HS3/TR95001/02, January 1995

Riley, V. (1992). Modelling the dynamics of pilot interaction with an electronoc crew. In, The Human Electronic Crew: Is the Team Maturing. WL-TR-92-3078. Wright Patterson AFB. OH July, pp103-107.

Riley, V. (1994). A theory of operator reliance on automation. In M. Mouloua and R Parasuraman (Eds). Human Performance in Automated Systems: Current Research and Trends. Hillsdale, New Jersey: Lawrence Erlbaum Associates.

Rouse , W.B. (1994). Twenty years of adaptive aiding: Origins of the concept and lessons learnt. In M. Mouloua and R Parasuraman (Eds). Human Performance in Automated Systems: Current Research and Trends. pp 28-33. Hillsdale, New Jersey: Lawrence Erlbaum Associates.

Ruff, H.A., Calhoun, G.L., Draper M.H., Fontejon J.V. and Guilfoos, B.J. (2004) Exploring automation issues in supervisory control of multiple UAVs. In D.A. Vincente, M. Moustapha and P.A. Hancock (Eds). HPSAA II, Vol II, Human Performance, Situation Awareness and Automation: Current Research Trends, pp 218-222. Mawhah, NJ: Erlbuam

Sanderson P., Naikar N., Lintern G. and Goss S. (1999). Use of cognitve work analysis across the system life cycle: From requirements to decommissioning. Procedings of the Human Factors Society 43rd Annual Meeting, Houston TX. pp 318-322. Santa Monica, HFES

Sheridan, T.B. and Verplank W.L. (1978). Human and computer control of undersea teleoperators. Technical Report. MIT Man-machine Systems Laboratory, Cambridge, MA.

Taylor R.M. (1988) Trust and awareneess in human electronic crew teamwork. In " The Human-Electronic Crew: Can They Work Together? WRDC-TR-89-7008, Wright-Patterson AFB, OH.

Taylor R.M. (2001a). Cognitive Cockpit Systems Engineering: Pilot Authorisation and Control of Tasks. In R. Onken (Ed), CSAPC'01. Proceedings of the 8th Conferences on Cognitive Sciences Approaches to process Control, Neubiberg, Germany, September 2001. University of the German Armed Forces, Neubiberg, Germany.

Taylor, R.M. (2001b). Cognitive Cockpit Control Task Analysis. DERA Memo, DERA/CHS3/6.3/14/7, 07 March 2001.

Taylor, R.M. (2001c). Cognitive Cockpit Risk Analysis. DERA Memo, DERA/CHS3/6.3/14/7, 26 February 2001.

Taylor, R.M. (2003). Cognition and Autonomy in Distributed Intelligent Systems, In, D Harris, V Duffy, M Smith and C Stephanidis Eds. Human Centred Computing: Cognitive, Social and Ergonomic Aspects. Volume 3. Pp330-334. Lawrence Erlbaum Assocites, Mahwah, New Jersey.

Taylor, R.M., Abdi, S., Dru-Drury, R., and Bonner M.C. (2001) 'Cognitive cockpit systems: information requirements analysis for pilot control of cockpit automation', in D Harris (Ed), 'Engineering psychology and cognitive ergonomics Vol. 5, Aerospace and transportation systems, Ch. 10, p81-88. Aldershot: Ashgate.

Taylor, R.M., Brown, L., and Dickson, B.(2002). From safety net to augmented cognition: Using flexible autonomy levels for on-line cognitive assistance and automation. RTO-MP-086 AC/323(HFM-085)TP/42. ISBN 92-837-0028-7. Paper No 27, NATO RTO Human Factors and Medicine Panel, Symposium on Spatial Disorientation in Military Vehicles: Causes, Consequences and Cures. La Coruna, Spain, 15-17 April 2002.

Taylor, R.M. and Selcon, S.J. (1993) Operator and automation capability analysis: Picking the right team. AGARD-CP-520, Paper 20.

Taylor, R.M and Shadrake, R. and Haugh, J. (1995). Trust and adaptation failure: An experimental study of uncooperation awareness. In, The Human Electronic Crew: Can We Trust the Team? Proceedings of the 3rd International Workshop on Human-Computer Teamwork. WL-TR-96-3039. Wright Patterson AFB, OH., December 1995. DRA CHS Report DRA/CHS/HS3/TR95001/02, January 1995, pp 93-98.

Taylor, R.M., Shadrake, R., Haugh, J. and Bunting A. (1996). Situational awareness, trust and compatibility: Using cognitive mapping techniques to investigate the relationships between imprtant cognitve system variables. In Situation Awareness: Limitations and Enhancement in the Aviation Environment, Proceedings of AGARD AMP Symposium, Brussells, Belgium, 24-27 April 1995. AGARD-CP-575, pp 6-1 to 14. NATO, Neuilly sur Seine, CEDEX.

Waters, M. and Taylor, R.M. (2004). A Bayesian Agent Approach to Autonomous Decision Making for an Unattended Cognitive Underwater Vehicle (UCUV). In Proceedings of the Workshop on Uninhabited Military Vehicles (UMVs) - Human Factors of Augmenting the Force, RWS-010- P4, held in Leiden, Netherlands, 10-13 June 2003. NATO RTO Human Factors and Medical Panel 078/Task Group 017, NATO RTO Neully sur Seine.

White, A.D. (2002). The human-machine partnership in UCAV operations. In Proceedings of 17th Bristol UAV Systems Conference, 10-13 April 2002.

The Impact of Operator Trust on Monitoring a Highly Reliable Automated System

Nathan R. Bailey

VMASC/Old Dominion University
Hampton Boulevard, Norfolk VA
NBailey@odu.edu

Mark W. Scerbo

Old Dominion University
Hampton Boulevard, Norfolk VA
MScerbo@odu.edu

Abstract

Technological advances have allowed for widespread implementation of automation in complex systems. However, the increase in both quantity and complexity of advanced automated systems has raised a number of potential concerns including degraded monitoring skills. As such, the present investigation consists of two studies that assessed the impact of system reliability and operator trust on monitoring performance. For both studies, participants monitored a simulated aviation display for failures while operating a manually controlled flight task. In addition, the second experiment assessed an operator's ability to detect a single automation failure over three experimental sessions. Results indicated that realistic levels of system reliability severely impaired an operator's ability to monitor effectively. In addition, results from both studies suggested that operator trust was bolstered as a function of increasing system reliability and that as trust increased, monitoring performance decreased. These results have important implications for operator monitoring in complex systems and suggest that operating highly reliable, highly trusted systems may severely impair an operator's ability to monitor for unanticipated system states.

1 Introduction

Automated systems and computer technology are becoming increasingly prevalent and sophisticated and the need for operators to monitor automated systems for failures or unanticipated states has never been more critical. One negative consequence that may result from increased monitoring demands has been referred to as automation-induced complacency (Parasuraman et al., 1993; Wiener, 1981). Automation-induced complacency refers to the decline in monitoring performance that often follows the shift from performing a task manually to monitoring the automation of that task (Farrell & Lewandowsky, 2000).

A number of previous studies have shown that the reliability of automated systems may influence an operator's ability to monitor effectively (Lee & Moray, 1992; Muir & Moray, 1996; Parasuraman et al., 1993). Lee and Moray demonstrated that both trust and strategies for using automation varied according to its overall reliability. Specifically, highly reliable systems induce trust, which impacts an operator's reliance on automation suggesting that operators are less likely to monitor highly reliable systems. This view is also consistent with the observations of Parasuraman et al. who found that individuals operating highly reliable and consistent automated devices showed poorer overall monitoring performance. By contrast, participants operating automated systems exhibiting lower and/or inconsistent levels of reliability demonstrated better overall monitoring. These findings bolster the assertion that highly reliable systems can engender poor monitoring performance as a result of overreliance or excessive trust in automated devices (Muir, 1989; Parasuraman et al., 1993).

Despite the evidence that system reliability is a fundamental factor impacting monitoring performance, the proportions of system failure used in previous studies on automation-induced complacency have been exaggerated and call into question the external validity of conclusions regarding the impact of reliability on monitoring in real-world systems. In fact, Parasuraman et al. (1993) and Thackray and Touchstone (1989) have suggested the need to conduct research on automation-induced complacency using levels of reliability that approach or exceed 99%. Given that the majority of empirical research on monitoring for automated systems has used high proportions of system failure, it is reasonable to assume that the development of trust described by Muir (1987) and Rempel, Holmes, and Zanna (1985) may be stunted, yielding qualitative differences in how operators interact with and monitor automated systems. The elevated proportions of system failures typically cited as eliciting automation-induced complacency may fail to establish adequate levels of trust because individuals are invariably skeptical of system performance. In addition, although a number of researchers have suggested that highly reliable automated systems lead to poor monitoring performance as a result of overreliance or excessive trust (Lee & Moray, 1992; Muir & Moray, 1996), very few studies have shown a direct relationship between operator trust and monitoring performance; focusing more on fluctuations in trust as a function of changing system reliability or how trust in automation influences operator strategies for using automation.

The purpose of the present set of studies was to examine monitoring performance, as a function of operator trust, in highly reliable automated systems. Consistent with Muir (1987) and Parasuraman et al. (1993), we expected that highly reliable automation would engender higher levels of trust and as a result, subsequent monitoring performance would suffer. Additionally, we anticipated that operator trust would significantly predict monitoring performance with increases in trust associated with degraded monitoring performance.

2 Experiment 1

2.1 Method

2.1.1 Participants

The participants included 32 individuals ranging in age from 20 to 41 years ($M = 25.5$). Twenty-seven of the participants were graduate students from the Old Dominion University Psychology Department. All participants had normal or corrected-to-normal visual acuity.

2.1.2 Experimental Tasks

Participants operated a suite of tasks including a flight task and three different forms of monitoring tasks. The flight task, the operator's primary responsibility, required participants to compensate for disturbances in the attitude of the aircraft using a standard joystick in order to maintain level flight (see Figure 1). Operators were also presented with three concurrent monitoring tasks in the form of a simulated Engine Instrumentation Crew Alerting System (EICAS) display (see Figure 2). The first task required operators to monitor a gauge and detect deviations in any of the six pointers that exceeded a critical value. The second task consisted of monitoring a simulated Mode Control Panel (MCP) to detect any uncommanded changes in the pattern of engaged automation modes. The third task required operators to monitor the digital readout portion of the EICAS display, consisting of four sets of engine parameters representing data from the left and right engines. Only data from the gauge monitoring task will be discussed in the present paper.

For the gauge monitoring task, critical values were represented by two red hatch marks at each end of the circular readouts. Under normal conditions, pointers fluctuated randomly within the normal operating range. Periodically, the gauges would move into the critical zones. Specifically, 10 critical deviations occurred for each 10-min period of operation. Under most circumstances, these critical deviations were accompanied by an amber "Automation System 1" notification that appeared in the upper right hand corner of the EICAS display. Whenever the automation notification was illuminated, the system automatically corrected deviations in the gauge task after four seconds, requiring no action on the part of the operator. Occasionally, a critical deviation occurred that was not unaccompanied by the automation notification. This represented an automation failure for the gauge task. Operators were asked to respond to any deviation in the gauge task that was unaccompanied by the automation notification by pressing the "G" key on the computer keyboard. Following a correct detection, the gauge returned to its normal operating range. If a participant failed to detect a critical deviation within 30 seconds, it was scored as a miss and the pointer reverted back to its normal range. In addition, the reliability of the automated system for detecting deviations in the gauge task was manipulated such that participants in the high reliability condition experienced a 2.0% failure rate while participants under low reliability experienced a 13.0% failure rate.

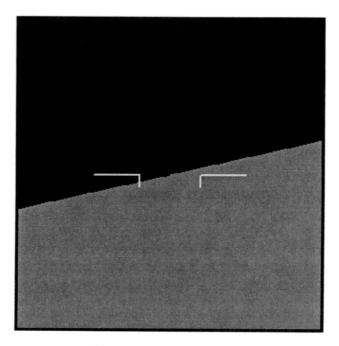

Figure 1: The primary flight task.

2.1.3   Measure of Operator Trust

A 12-item questionnaire was developed to assess operator trust in the automated devices. Each item utilized a 21-point bipolar rating scale. The instrument included four subscales with three items in each scale (e.g., "Indicate how reliable you felt the automated system was at correcting any critical deviations that occurred with the gauge task", and "If you were unable to monitor the automation gauge portion of the display for several minutes, how confident would you be that the automation would detect any problems with the system"?).

Overall ratings on the operator trust questionnaire could range between 12 and 252 with a range of 3 to 63 for any of the four subscales. Internal consistency for the 12-item scale as well as each of the subscales was high. Specifically, the overall reliability for the 12-item scale was $r = .94$ with internal consistency for the trust in the gauge automation subscale at $r = .92$.

2.1.4   Apparatus

Each of the experimental tasks was displayed using a Pentium IV personal computer on separate 17 in monitors. Participants used a standard computer keyboard along with a Microsoft Sidewinder USB joystick. The primary flight task was presented directly in front of the participant at a distance of approximately 20 in. The monitoring task was presented to the participant's left on an adjacent display. This display was angled toward the user at 30° at a distance of 25 in.

2.1.5   Procedure

Each participant was given a set of written instructions along with a brief orientation that used graphical examples of each type of critical deviation. Following the orientation, participants completed a 5-min practice session. After the

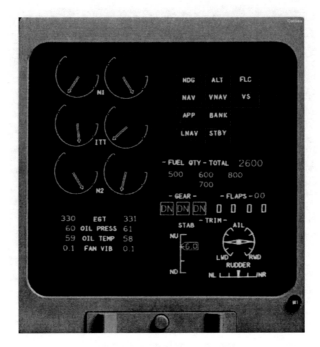

Figure 2: Simulated EICAS display with three monitoring tasks. The display includes the gauge monitoring task (upper left), automation mode monitoring task (upper right), and digital readout (bottom left).

practice session, the participants were asked if they had any questions. They then began the experimental session which lasted approximately 100 min. Immediately following the session, participants completed the measure of operator trust. Participants were then required to return and complete two more experimental trials. After completing the third trial, all participants were debriefed.

2.2   Results

2.2.1   Monitoring Performance

Univariate ANOVAs for system reliability (high or low) were conducted on response time and overall detection performance for finding deviations in the gauge monitoring task. A significant effect for system reliability on response time was found, $F(1, 94) = 9.705$, $p = .002$, $\eta^2 = .094$. Those individuals under high reliability ($M = 15.557$, $SD = 8.996$) demonstrated degraded response time relative to participants in the low reliability condition ($M = 10.486$, $SD = 6.801$). In addition, a main effect for system reliability on detection performance approached significance, $F(1, 94) = 3.677$, $p = .058$, $\eta^2 = .038$, suggesting that participants under high reliability ($M = 72.9\%$, $SD = .385$) had poorer detection performance compared to participants in the low reliability condition ($M = 85.4\%$, $SD = .235$).

2.2.2   Operator Trust and Monitoring Performance

A Univariate ANOVA was conducted for system reliability on operator ratings of trust in the gauge automation and generated a significant effect, $F(1, 94) = 5.405$, $p = .022$, $\eta^2 = .054$. Operators under high reliability reported higher levels of trust ($M = 50.646$, $SD = 7.413$) than operators in the low reliability condition ($M = 46.689$, $SD = 9.175$). In addition, using regression analysis, trust in the gauge automation was found to significantly predict response time, $F(1, 94) = 6.493$, $B = .248$, $SD_B = .097$, $R^2 = .065$, and detection performance, $F(1, 94) = 6.428$, $B = -.010$, $SD_B =$

.004, R^2 = .064. Operators who placed more trust in the gauge automation exhibited degraded detection performance and response times for detecting failures in that system.

2.3 Discussion

Data from Experiment 1 indicated that system reliability influenced the efficiency of operator monitoring. As predicted, operators who monitored a highly reliable system exhibited degraded detection performance and increased response latencies for detecting automation failures compared to individuals who monitored a system with lower reliability.

The impact of system reliability on monitoring performance may be related to operator attentional resources. Malleable Attentional Resources Theory (MART) posits that periods of "underload" or inactivity may degrade operator performance (Young & Stanton, 2003). Specifically, for operators performing tasks with few demands, attentional resources shrink and performance declines as if task demands were high. Consistent with MART, the effects of system reliability on monitoring performance from Experiment 1 may be related to depleted attentional resources. Specifically, participants under high reliability operated a system that demanded few interventions. As a result, their attentional resources became depleted and monitoring performance suffered. By contrast, participants under low reliability were frequently required to make corrections in the gauge. As such, their attentional resources remained intact and their monitoring performance remained high.

In addition, data from Experiment 1 indicated that operators under high reliability reported significantly higher levels of trust in the gauge automation and that those elevated levels of trust predicted lower detection performance and increased response latencies for that task. These data represent some of the first empirical support for the relationship between operator trust and monitoring performance. In general, operator trust varied as a function of system reliability (i.e., when operator trust was high, monitoring performance was poor), supporting the contention by previous researchers that automation-induced complacency is heavily influenced by operator trust (Muir, 1989; Parasuraman et al., 1993).

3 Experiment 2

The second study was specifically designed to assess an operator's ability to detect a single automation failure over several experimental trials. Thackray and Touchstone (1989) suggested that lengthy studies with infrequent failures were necessary to adequately examine automation-induced complacency. Although research by Molloy and Parasuraman (1996) did assess the ability of operators to detect a single critical event, in the present study participants experienced several experimental sessions, some of which included no automation failures. As a result, the reliability of the system was substantially increased, generating a more ecologically valid task structure. Taken together with the findings of the first study, the results from Experiment 2 may represent a more accurate depiction of monitoring performance in complex systems.

3.1 Method

3.1.1 Participants
There were nine participants in Experiment 2 including five men and four women with mean age of M = 22.9 years. Five of the participants were graduate students from the Old Dominion University Psychology Department. All participants had normal or corrected-to-normal visual acuity.

3.1.2 Experimental Tasks

Participants operated a suite of flight tasks similar to those used in Experiment 1. The attitude correction flight task was identical. The monitoring task was also similar, with one critical difference; operators experienced only a single failure across all experimental trials. Each operator received the same instructions used in Experiment 1 and was responsible for monitoring each of the three systems. However, they experienced only one failure in the automation to detect a critical deviation. This deviation occurred in the gauge monitoring task and the timing of the deviation was manipulated across trials, (i.e., occurring for each operator only in the first, second, or third trial). This constituted a 99.7% rate of reliability.

3.1.3 Apparatus and Procedure

The apparatus and procedure were identical to Experiment 1.

3.2 Results

3.2.1 Monitoring Performance

Monitoring performance in Experiment 2 was measured as a function of whether participants detected the single gauge automation failure and their corresponding response time. Performance was poor with two thirds of participants failing to detect the single gauge automation failure yielding a mean response time of $M = 25.496$, $SD = 8.786$.

A comparison of monitoring performance between Experiment 2 and Experiment 1 was also made. Because of large discrepancies in the sample sizes for the monitoring performance data between the experiments, homogeneity of variance tests were conducted on the detection performance and response time data. Although tests did not indicate unequal variances for the response time data, $F(2, 102) = 2.206$, $p = .115$, heterogeneity of variance was present in the detection performance data, $F(2, 102) = 9.260$, $p < .001$. Therefore, a more stringent level of alpha ($\alpha = .01$) was adopted for comparing monitoring data from Experiment 1 and Experiment 2. An ANOVA comparing system reliability from Experiment 1, (i.e., 87.0% for low reliability and 98.0% for high reliability) and system reliability from Experiment 2 (i.e., 99.7% reliability) revealed significant effects for both response time, $F(2, 102) = 14.672$, $\eta^2 = .223$, and detection performance, $F(2, 102) = 9.260$, $\eta^2 = .154$. Participants in Experiment 2 exhibited considerably degraded response times and detection performance compared to either level of reliability used in Experiment 1.

3.2.2 Operator Trust

Ratings of operator trust in the gauge automation from Experiment 2 were compared with ratings from the two reliability levels used in Experiment 1. Similar to Experiment 1, a more stringent level of alpha ($\alpha = .01$) was used for comparisons of operator trust between the experiments. The ANOVA for the three levels of system reliability from both experiments on the level of operator trust in the gauge automation yielded a significant main effect, $F(2, 120) = 20.225$, $\eta^2 = .252$. Operator ratings of trust in Experiment 2 increased by 13.6% over the high and 20.4% over the low reliability system used in Experiment 1.

3.3 Discussion

Data from Experiment 2 showed a precipitous drop in operator monitoring performance for the gauge task compared to performance under both levels of reliability in the first study. Specifically, the 99.7% reliability of the gauge automation in Experiment 2 generated only 33.3% detection rate for the single gauge automation failure. By contrast, data from Experiment 1 indicated 72.9% and 85.4% detection rates for the gauge automation failure for participants under high and low reliability, respectively. Similarly, response latencies increased greatly in Experiment 2. Specifically, mean response times in Experiment 2 increased by 243% and 163% over the high and low reliability conditions from the first experiment. These data suggest that laboratory systems with reliability levels that approach those typically found in the real-world may severely impair an operator's ability to detect unanticipated and/or infrequent system states.

Regarding operator trust, consistent with previous research by Lee and Moray (1992), the comparison of operator trust between Experiments 1 and 2 indicated that system performance is one of the main factors influencing the development of operator trust. Specifically, data from the present set of studies suggest a trend whereby operator trust continues to increase as a function of increasing system reliability.

4 Overall Discussion

Automated systems and computer technology are ubiquitous with applications in domains as diverse as aviation, maritime operations, process control, motor vehicle operation, and information retrieval (Lee & See, 2004). As this trend continues, the need for operators to monitor automated systems for failures or unanticipated states becomes critical. However, results from the present set of studies indicate that realistic levels of system reliability severely impair an operator's ability to monitor effectively. Specifically, data from Experiment 1 and 2 indicate declining operator performance as a function of increasing system reliability with a precipitous drop in monitoring performance for highly reliable systems.

The present set of studies also revealed the direct influence of operator trust on monitoring performance. Specifically, operator trust was bolstered as a function of increasing system reliability. Further, as operator ratings of trust increased, the ability of operators to monitor effectively decreased. This finding indicates a direct and inverse relationship between operator trust and degraded monitoring. Although a number of researchers have argued that monitoring performance in complex systems varies as a function of operator trust in automation, most previous research has failed to show a direct connection. In fact, a recent review of the literature on trust in automation (see for example, Lee & See, 2004) fails to reference any empirical studies that examine monitoring performance in complex systems as a function of operator trust. As such, these data represent some of the first empirical support for a direct connection between operator trust and subsequent monitoring performance.

In addition, despite the declines in monitoring performance indicated by both studies, these findings may still reflect an overly optimistic view of operator monitoring in highly reliable systems. Although three of the nine participants in Experiment 2 did successfully detect the failure, comments from the other participants indicated that they had stopped regularly monitoring the simulated EICAS display. In fact, one operator reported that while he or she "occasionally glanced" at the monitoring tasks in the first and second sessions, he or she did not monitor the systems at all in the third session, focusing exclusively on the primary flight task. Given that most commercial aircraft can travel 3-5 miles in just 30 seconds, it is critical that operators immediately detect any potential problems. However, the present results suggest that in highly reliable systems, monitoring performance may become severely degraded with operators taking up to several minutes to detect deviations.

From a broader perspective, excessive trust due to highly reliable automation may represent an important moderator of monitoring performance in complex systems. One potential method of mitigating the impact of excessive trust may be to optimize attentional resources. Because highly reliable systems induce less than optimal attentional resources (see Young & Stanton, 2002), the use of augmented cognition systems, including biocybernetic systems of adaptive automation (Bailey et al., 2003; Scerbo et al., 2003) may help to improve operator monitoring. Biocybernetic systems are predicated on the notion that psychological states are associated with changes in operator physiology. By measuring the physiological signals that represent these states, it is possible to allocate system functions dynamically, maintaining an optimal level of physiological arousal. Consequently, an operator's attentional and cognitive resources for performing tasks are improved (Gaillard, 1993). Given the degraded attentional resources associated with highly reliable, highly trusted automated systems, augmented cognition systems may represent a viable method for improving operator monitoring.

5 Acknowledgements

This research was supported by the NASA Langley Research Center, Grant No. NGT-1-03019.

6 References

Bailey, N. B., Scerbo, M. W., Freeman, F. G., Mikulka, P. J., & Scott, L. A. (2003). The effects of a brain-based adaptive automation system on situation awareness: The role of complacency-potential. In *Proceedings of the 47th Annual Human Factors and Ergonomics Society Meeting* (pp. 1048-1052), Denver: Human Factors and Ergonomics Society.

Farrell, S., & Lewandowsky, S. (2000). A connectionist model of complacency and adaptive recovery under automation. *Journal of Experimental Psychology: Learning, Memory, & Cognition, 26*, 395-410.

Gaillard, A.K. (1993). Comparing the concepts of mental load and stress. *Ergonomics, 9*, 991-1005.

Lee, J., & Moray, N. (1992). Trust, control strategies and allocation of function in human-machine systems. *Ergonomics, 35*, 1243-1270.

Lee, J. D., & See, K. A. (2004). Trust in automation: Designing for appropriate reliance. *Human Factors, 46*, 50-80.

Molloy, R., & Parasuraman, R. (1996). Monitoring an automated system for a single failure: Vigilance and task complexity effects. *Human Factors, 38*, 311-322.

Muir, B. M. (1987). Trust between humans and machines, and the design of decision aids. *International Journal of Man Machine Studies, 27*, 527-539.

Muir, B. M. (1989). Operators' trust in and use of automatic controllers in a supervisory process control task. Unpublished Doctoral Dissertation, University of Toronto, Canada.

Muir, B. M., & Moray, N. (1996). Trust in automation: II. Experimental studies of trust and human intervention in a process control simulation. *Ergonomics, 39*, 429-460

Parasuraman, R., Molloy, R., & Singh, I. L. (1993). Performance consequences of automation-induced "complacency." *International Journal of Aviation Psychology, 3*, 1-23.

Rempel, J. K., Holmes, J. G., & Zanna, M. P. (1985). Trust in close relationships. *Journal of Personality & Social Psychology, 49*, 95-112.

Scerbo, M. W., Freeman, F. G., & Mikulka, P. J. (2003). A brain-based system for adaptive automation. *Theoretical Issues in Ergonomic Science, 4*, 200-219.

Thackray, R. I., & Touchstone, R. M. (1989). Detection efficiency on an air traffic control monitoring task with and without computer aiding. *Aviation, Space, and Environmental Medicine, 60*, 744-748.

Wiener, E. L. (1981). Complacency: Is the term useful for air safety? In *Proceedings of the 26th Corporate Aviation Safety Seminar* (pp. 116-125). Denver: Flight Safety Foundation, Inc.

Young, M.S., & Stanton, N.A. (2002). Malleable attentional resources theory: A new explanation for the effects of mental underload on performance. *Human Factors, 44*, 365-375.

Human Interaction with Adaptive Automation: Strategies for Trading of Control under Possibility of Over-trust and Complacency

Toshiyuki Inagaki, Hiroshi Furukawa, Makoto Itoh

Department of Risk Engineering, University of Tsukuba
Tsukuba 305-8573 Japan
{inagaki, furukawa, itoh}@risk.tsukuba.ac.jp

Abstract

Function allocation needs to be dynamic and situation-adaptive to support humans appropriately. Machines have thus been given various types of intelligence. It is true, however, humans working with such smart machines often suffer negative consequences of automations. By taking the adaptive cruise control (ACC) system and the lane keeping support (LKS) system as examples of adaptive systems in the real world, this paper investigates human trust in and reliance on adaptive automation, and discusses design of strategies for trading of control under possibility of driver's over-trust in automation.

1   Introduction

The design decision of assigning functions to human and machine is called *function allocation*. In spite of its importance, function allocation is still a kind of art. Various strategies for function allocation have been proposed. Traditional function allocation strategies are classified into three types: (a) comparison allocation, (b) leftover allocation, and (c) economic allocation. These strategies consider "who does what." Such design decisions yield function allocations that are *static*: viz., once a function is allocated to an agent, the agent is responsible for the function at all times. However, the operating environment may change as time goes by, or performance of the human may degrade gradually as a result of psychological or physiological reasons, which suggests that "who does what" design decisions are not sufficient and that "who does what and when" considerations are needed. If the total performance or safety has to be maintained rigorously, it may be wise to reallocate functions between the human and the machine: viz., the resulting function allocation must be dynamic.

A scheme that modifies function allocation dynamically depending on situations is called an *adaptive function allocation*. The adaptive function allocation assumes criteria to determine whether functions have to be reallocated, how, and when. The criteria reflect various factors, such as changes in the operating environment, loads or demands to operators, and performance of operators. The automation that operates under an adaptive function allocation is called *adaptive automation* (Inagaki, 2003; Moray, Inagaki, & Itoh, 2000; Parasuraman, Bhari, Deaton, Morrison, & Barnes, 1992; Rouse, 1988; Scallen & Hancock, 2001; Scerbo, 1996). Adaptive automation works intelligently and reliably. It can sense and analyze situations, decide what must be done, and implement control actions. It may be easy for humans to trust in such a smart machine. However, if human-computer interaction strategies were poorly designed, various types of automation surprises may occur. Based on our experiments with the adaptive cruise control (ACC) system and the lane keeping support (LKS) system, this paper investigates human trust in and reliance on adaptive automation, and discusses design of strategies for trading of control under possibility of driver's over-trust in automation.

2   ACC and LKS Systems in Advanced Automobile

The adaptive cruise control (ACC) system and the lane keeping support (LKS) system may be regarded as real-world examples of adaptive systems. The ACC system is a partial automation for longitudinal control, designed to reduce the driver's workload by freeing the driver from frequent acceleration and deceleration. It controls the *host vehicle* so that it can follow a vehicle ahead (*the target vehicle*) at a driver-specified distance by controlling the engine and/or power train and potentially the brake. When the ACC system detects the deceleration of the target vehicle, it slows down the host vehicle at some deceleration rate. As long as the deceleration of the target vehicle stays within a certain range (say, not greater than 0.2G), the ACC system can control the speed of the host vehicle perfectly and no rear-end collision into the target vehicle occurs. However, when the target vehicle makes a rapid deceleration at a high rate (e.g., 0.4G), the ordinary brake by the ACC system may not powerful enough to avoid a collision into the target vehicle. How to prepare for such cases is one of significant design issues.

The LKS system is a partial automation for lateral control, designed to reduce the driver's workload to keep the host vehicle within its driving lane. Recognizing the lateral position of the host vehicle in the lane, the LKS system assists the driver's steering control to keep the car center of the lane with the use of the power steering system.

3   Two Key Concepts Needed for Systematic Investigations

This section gives a brief summary of two key concepts, *trading of control* and *levels of automation*, that are needed for investigating human interactions with adaptive systems in a systematic manner.

3.1   Trading of Control

Trading of control (or, *trading of authority*) refers to the human-computer collaboration in which either one of the human or the computer is responsible for a function, and an active agent changes alternately from time to time. Trading of control is essential for adaptive automation, because adaptive automation needs to modify function allocation between humans and machines dynamically in response to changes in situations, human workload, or performance. A scheme to implement trading of control is called an *automation invocation strategy*. There are some types of automation invocation strategies (Inagaki, 2003). Among them, this paper discusses the following two classes that are important for transportation systems.

   Critical-event strategies: Automation invocation strategies of this class change function allocations when specific events (called, *critical events*) occur in the human-machine system. It is assumed that human workload may become unacceptably high when the critical events occur. If the critical events did not occur during the system operation, allocation of functions would not be altered.

   Measurement-based strategies: Automation invocation strategies of this class adjust function allocation dynamically by evaluating moment-to-moment workload or total system performance. It is necessary to develop custom tailored algorithms if the system is to be compatible with individual operators. Individual differences in human operator capabilities will also influence the response to multiple task demands.

3.2   Levels of Automation

In describing various types of human-computer interactions, the notion of the *level of automation* (LOA) is useful. Table 1 gives an expanded version in which a new LOA comes between levels 6 and 7 in the original list by Sheridan (1992). The added level, called the level 6.5, has been firstly introduced in (Inagaki, Itoh, & Moray, 1997) with two-fold objectives: (1) to avoid automation surprises that may be induced by automatic actions and (2) to implement actions that are indispensable to assure systems safety in emergency.

Table 1: Scales of Levels of Automation (expanded version)

----
   1. The computer offers no assistance; human must do it all.
   2. The computer offers a complete set of action alternatives, and
   3.   narrows the selection down to a few, or
   4.   suggests one, and
   5.   executes that suggestion if the human approves, or
   6.   allows the human a restricted time to veto before automatic execution, or
   6.5   executes automatically upon telling the human what it is going to do, or
   7.   executes automatically, then necessarily informs humans,
   8.   informs him after execution only if he asks,
   9.   informs him after execution if it, the computer, decides to.
   10. The computer decides everything and acts autonomously, ignoring the human.
----

4   Four Dimensions of Trust

Lee & Moray (1992) distinguished four dimensions of *trust*: (1) *foundation* that represents the "fundamental assumption of natural and social order that makes the other levels of trust possible," (2) *performance* that rests on the "expectation of consistent, stable, and desirable performance or behavior," (3) *process* that depends on "an understanding of the underlying qualities or characteristics that govern behavior," and (4) *purpose* that rests on the "underlying motives or intents."

It would not be hard for a human operator to recognize that the designer's purpose or intention in introducing adaptive automation lies in regulating operator workload at some optimal level. Respecting the second and the third dimensions may not be straightforward: Human's understanding of the automation invocation algorithm may be imperfect if the algorithm is sophisticated or complicated. When the human failed to be certain of the second and the third dimension of trust, she/he would fail to establish trust in the adaptive automation. Human's distrust or mistrust in automation can cause inappropriate use of automation, as has been pointed out by Parasuraman and Riley (1996).

5   Trust in and Reliance on Automation: Experiment 1

This paper discusses human trust in adaptive systems from the viewpoint of *trading of control* and the *levels of automation*. The first two experiments deals with human-automation interactions and trading of control in emergency, and the third experiment discusses those in peacetime. Now, here comes Experiment 1.

5.1   Strategy for Trading of Control in Emergency

Consider a low-speed range ACC system with automatic stopping functionality that controls the host vehicle so that it can follow a target vehicle at a pre-specified distance in a heavy or jammed traffic. The specified distance can be maintained successfully as long as the acceleration/deceleration rate of the target vehicle stays within a certain range. When the target vehicle makes a rapid deceleration to which the ACC's ordinary brake is not strong enough to avoid a crash, and if the ACC system was the only agent to cope with the situation, a rear-end collision may occur eventually. Let us consider the following scheme for trading control from the ACC system to the driver.

  Scheme 1: Upon detecting a rapid deceleration of the target vehicle, the ACC gives an *emergency-braking alert* that tells the driver to hit the brake her/himself hard enough to avoid a collision.

5.2   Participants and Performance Measures

The purpose of Experiment 1 is to investigate drivers' trust in and reliance on the emergency-braking alert. Twenty students participated in the experiment. The subjects were divided into two groups: (1) Group 1 of ten people and (2) Group 2 of ten. Each subject received 17 trials, each of which lasted about 3 minutes. During Trials 1 to 3, no emergency-braking alert functionality was available: Subjects had to judge themselves whether she/he has to take over control from the ACC system and when. During Trials 4 to 10, a correct emergency-braking alert was issued immediately when the target vehicle decelerated rapidly. In Trial 11, an emergency-braking alert was missed when the target vehicle made a rapid deceleration. In Trial 12, subjects in Group 1 received a correct emergency-braking alert, while alert was missed again for subjects in Group 2 upon the target vehicle's rapid deceleration. During Trials 13 to 17, correct alerts were given to all the subjects in either group.

The following data were recorded as performance measures. (1) Response time; viz., time elapsed before the subject initiated to apply the brake, (2) number of unnecessary brakes applied when no rapid deceleration was made by the target vehicle, (3) subjective ratings of trust in emergency-braking alerts, (4) subjective ratings of usefulness of emergency-braking alerts, and (5) what subjects said during and between trials.

5.3   Results

The experiment was conducted with a fixed-based driving simulator. Mean response times of the subjects of the two groups are shown in Figure 1.

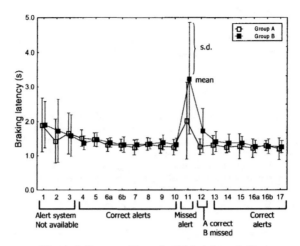

Figure 1: Response Times for Trials 1 through 17

(1) Significant differences were found among the mean response times of 20 subjects for the cases of Trials 3, 10, and 11 ($F(2, 57) = 11.19$, $p < .0001$); see, Figure 2(a). While receiving correct emergency-braking alerts consecutively and repetitively, subjects became reliant on the alerts, and tended to judge based on the alerts whether an emergency brake was necessary or not. When a correct alert was missed at Trial 11, subjects showed significantly late responses to the rapid deceleration of the target vehicle.

(2) Subjects in Group 2 experienced two consecutive missed alerts at Trials 11 and 12. Significant differences were found among response times of subjects for Trials 3, 10, 11, and 12 ($F(3, 36) = 8.86$, $p < .0002$); see, Figure 2(b). Among those response times, the response time for Trial 11 differed significantly from others, and no significant differences between response time for Trial 12 and either one of those for Trials 3 and 10, which implies that, when subject experienced a missed alert, they became vigilant for a while and tried to judge the situation to determine whether an emergency brake was in need.

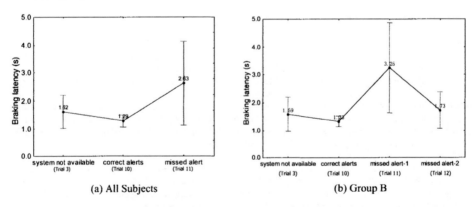

(a) All Subjects   (b) Group B

Figure 2: Effects of Missed Alert(s) on Response Time

(3) A McNemar test found a significant difference between the numbers of unnecessary brakes before and after the missed alert at Trial 11, which suggests that subject behaviors differ between before and after experiencing the missed alert. The same situations were placed symmetrically around Trial 11: viz., each pair of trials (8, 14), (9, 13), and (10, 12) are essentially the same.

(4) Protocol data indicate that subjects' policies in use of alerts changed after experiencing the missed alert. Among 27 protocol data collected before the missed alert trial, 15 protocol data tell that many subjects used emergency-braking alerts to evaluate the situation. After experiencing the missed alert, the number of alert-reliant protocol data was reduced to just 6 out of 28. On the other hand, 20 out of 28 protocol data suggest that subjects tried to evaluate circumstances themselves and used alerts as supplements or confirmation to their judgements.

(5) As described in section 4, *trust* has four dimensions. The authors believe that these dimensions should be distinguished in the subjective rating of trust. Since the first and the fourth dimensions are trivial for emergency-braking alerts, participants of Experiment 1 were requested to evaluate the second and the third dimensions of trust in 7-point scales. They were asked: (Q1): "To what extent do you figure out why an emergency-braking alert was issued (or why it was missed) when the target vehicle made a rapid deceleration?" (1: not at all, 4: partly, 7: completely). (Q2): "Were the alerts given at the right time?" (1: too late, 4: exactly when I expected, 7: too early). Q1 deals with the third dimension of trust (i.e., *process* that depends on "an understanding of the underlying qualities or characteristics that govern behavior"), and Q2 the second dimension (i.e., *performance* that rests on the "expectation of consistent, stable, and desirable performance or behavior"). As for the subjective ratings of usefulness of emergency-braking alerts, the participants were asked: (Q3): "To what extent do you think is this warning system useful?" and requested to answer in the 7-point scale (1: not at all, 4: not sure, 7: very useful).

Figure 3(a) depicts means of the ratings of subjects in Groups A, and Figure 3(b) those of subjects in Group B. All the ratings for Q1 through Q3 behave in a similar manner. However, the rated values for Q2 are smaller than those for Q1 or Q3, which suggests that the subjects were not satisfied with the timings of the emergency-braking alerts.

It is also observed in Figures 3(a) and 3(b) that the subjective ratings of trust in the automated alert recovered quickly, which suggests that it may not be wise to assume that drivers shall become vigilant and continues to be so once they experienced a missed alert. The effects of missed alerts do not remain long and subjects may start relying on the automated alerts again. This implies the need for technological backup measures for ensuring safety of the automobile.

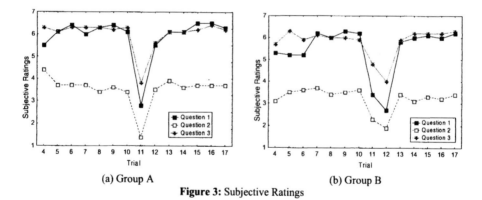

(a) Group A   (b) Group B

Figure 3: Subjective Ratings

6   Trading of Control under Driver's Over-trust in Automation: Experiment 2

The second experiment via a computer simulation approach investigated efficacy of strategies for trading of control in emergency under possibility of driver's over-trust in automation. The experiment tries to investigate the need for technological backup measures for ensuring automobile safety.

6.1   Driver's Psychological States and Their Transitions

When the target vehicle slows down at a certain deceleration rate not greater than 0.2G, the ACC system of the host vehicle begins to decelerate 0.5s after the initiation of the target vehicle's deceleration, in which the ACC system applies the brake at the rate of 0.2G. It would be natural for the driver to trust in the ACC system if she/he observed that the ACC system behaves correctly and appropriately. Sometimes the driver may place excessive trust in the automation. In such cases, the driver may fail to allocate her/his own attention to the driving environment, and may pay attention inappropriately to some non-driving tasks (e.g., use of a mobile phone, manipulation of on-board audio systems). Five psychological states are distinguished for the driver based on a model by (Hashimoto, 1984) as shown in Figure 4.

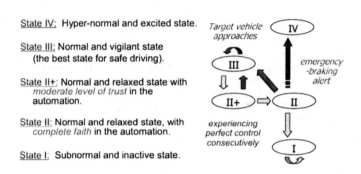

State IV: Hyper-normal and excited state.

State III: Normal and vigilant state
(the best state for safe driving).

State II+: Normal and relaxed state with
moderate level of trust in the
automation.

State II: Normal and relaxed state, with
complete faith in the automation.

State I: Subnormal and inactive state.

Figure 4: Driver's Psychological States and Their Transitions

It is assumed that, when the driver starts driving, her/his psychological state is positioned initially at State III (the best state for safe driving), and that the state changes dynamically as time passes by; see, Figure 5. The following two types of events are distinguished for cases in which the ACC system can perform its longitudinal control with the ordinary brake: (1) *Type-1 event*, in which the target vehicle decelerates at a certain rate and the ACC system performs its longitudinal control perfectly so that it may satisfy the driver completely, and (2) *Type-2 event*, in which the target vehicle decelerates at a rate within 0.2G. Though the ACC system can cope with the situation within the ordinary brake, the driver feels alarm because the host car comes close to the target vehicle.

The contributions of the events to the driver's psychological state transitions are defined as follows: If the driver experiences a certain number (say, 10 or 20) of Type-1 events consecutively, the driver's psychological state changes one step downward (e.g., from III to II+). When the driver experiences a Type-2 event, her/his psychological state changes one step upward (e.g., from II+ to III). State I is the absorbing state: Once the driver enters into State I, she/he never comes out of the state even if a Type-2 event occurs.

6.2 Designs Alternatives for Trading of Control in Emergency

Suppose the ACC system recognizes that the deceleration rate of the target vehicle is much greater than 0.2G, the maximum deceleration rate to which the ACC system can cope with the ordinary automatic brake. In addition to Scheme 1 that was discussed in section 3, the following design alternatives may be feasible:

Scheme 2: Upon recognition of a rapid deceleration of the target vehicle, the ACC system gives an emergency-braking alert, and if the driver does not respond within a pre-specified time (2s in this paper), it applies an automatic emergency brake that decelerates at the rate of 0.4G.

Scheme 3: Upon recognition of a rapid deceleration of the target vehicle, the ACC system applies its automatic emergency brake to implement the deceleration rate of 0.4G simultaneously when it issues an emergency-braking alert.

Scheme 4: Upon recognition of a rapid deceleration of the target vehicle, the ACC system firstly applies its automatic emergency brake with the deceleration rate of 0.4G. Then, the ACC system tells the driver that it applied an emergency brake some seconds ago.

610

In term of the levels of automation, it can be said that the LOA = 4 for Scheme 1, LOA = 6 for Scheme 2, LOA = 6.5 for Scheme 3, and LOA = 7 for Scheme 4. Note that, from the viewpoint of swiftness of the emergency brake, Schemes 3 and 4 are indifferent. However, in case of Scheme 4, the driver may fail to recognize what is going on, when the ACC system applies its emergency brake. Scheme 3 is designed to avoid delay in the automatic emergency brake as well as to avoid automation surprises. Scheme 4 thus shall not be investigated further in this paper.

6.3 Model for Driver's Response to the Alert

When the driver hits the brake hard enough, she/he can make a deceleration at the rate of 0.5G. The driver's response time to the emergency-braking alert varies depending on the psychological state at that time moment:

(1) If the driver was in State I when the alert was set off, he/she does not respond to the alert at all. (2) If the driver was in State II, she/he stays in the state with probability 0.8, and hits the brake pedal in T2 seconds. With probability 0.2, the driver state changes to State IV. (3) If the driver was either in State II+ or in III, she/he applies the emergency brake her/himself either in T2+ or in T3 seconds, respectively. (4) In State IV, the driver panics and fails to take any meaningful actions to attain car safety.

T2, T2+, and T3 are random variables with different means. In the Monte Carlo simulations described in the following section, we assumed: T2 is uniformly distributed over the interval [2.7s, 3.3s], with the mean 3.0s, T2+ is uniformly distributed over [1.8s, 2.2s], with the mean 2.0s, and T3 is uniformly distributed over [1.35s, 1.65s], with the mean 1.5s.

6.4 Monte Carlo Simulations

The experiment has a 2 x 3 x 3 x 2 factorial design, mapping onto (*Headway Distance*) x (*Level of Automation*) x (*Event-Mixture Ratio*) x (*Driver's Psychological State Transition Condition*). Two levels, 80m and 50m, were distinguished for the *Headway Distance* between the host and the target vehicles. Three *Levels of Automation*, LOA-4, LOA-6, and LOA-6.5, were distinguished as the design alternatives for cases of a rapid deceleration of the target vehicle, corresponding to Schemes 1 through 3, respectively. The *Event-Mixture Ratio* denotes the proportion of the number of Type-2 events to the total sum of the numbers of Type-1 and type-2 events before the target vehicle makes a rapid deceleration. The Event-Mixture Ratio was set at 0%, 10%, or 20%. Two cases, 10 and 20, were investigated for the *Driver's Psychological State Transition Condition*: In cases of the former, if the driver experiences 10 Type-1 events consecutively, her/his psychological state changes one step downward. In the latter, 20 consecutive Type-1 events make a downward state transition.

The *Trip Length* was measured in terms of the number of the total sum of the Type-1 and Type-2 events before the target vehicle makes a rapid deceleration. In the present study, the Trip Length was fixed at 100; see Figure 5.

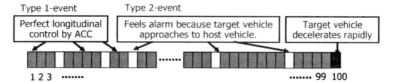

Figure 5: Events in a Trip

For each combination of the above conditions, 5000 Monte Carlo runs were performed with a discrete-event simulation software, WinCrew (Micro Analysis and Design, Inc.), under the following assumptions: The target vehicle makes a rapid deceleration at the rate of 0.4G at some time point while it has been running at the speed of 100km/hr. At the time of the rapid deceleration of the target vehicle, the host vehicle has also been running at 100km/hr, following the target vehicle. When detecting a deceleration of the target vehicle, the ACC system firstly applies its ordinary brake. One second after initiation of the ordinary brake, the ACC system recognizes that the deceleration of the target vehicle cannot be handled by the ordinary brake, and performs safety control action by applying one of Schemes 1 through 3.

6.5 Simulation Results

Table 2 shows the number of accidents (collisions into the target vehicle) during 5000 Monte Carlo runs and the distribution of the driver's psychological states when the trip length was 100.

Table 2: Computational results of automobile safety at the rapid deceleration of the target vehicle

LOA	Psychological state transition condition	Event-mixture ratio	# Runs ended in collision		Driver's psychological state at the emergency-braking alert				
			Headway distance: 80m	Headway distance: 50m	State I	State II	State II+	State III	State IV
4	10	0%	5000	5000	5000	0	0	0	0
		10%	1300	1478	1232	264	1003	2433	68
		20%	28	41	24	19	415	4538	4
	20	0%	4203	4786	3999	797	0	0	204
		10%	9	19	1	16	455	4520	8
		20%	0	0	0	0	34	4966	
6	10	0%	0	5000	5000	0	0	0	0
		10%	0	1280	1218	282	987	2451	62
		20%	0	36	27	32	415	4517	9
	20	0%	0	4192	3981	808	0	0	211
		10%	0	7	3	18	432	4543	4
		20%	0	0	0	0	31	4969	0
6.5	10	0	0	0	5000	0	0	0	0
		10%	0	0	1307	259	984	2375	75
		20%	0	0	22	27	426	4519	6

Let us take a look at the results when the headway distance was 80m. Comparing LOA-4, LOA-6, and LOA-6.5 for the cases where the Driver's Psychological State Transition Condition was set at 10, the number of accidents under LOA-4 was siginificantly larger than either one of those under LOA-6 and LOA-6.5, especially when the Event-Mixture Ratio is small. The result may be interpreted as follows: When the driving is peaceful in which Type-2 events seldom occur and the ACC system continues to be successful in its longitudinal control, the driver is likely to trust in and rely on the ACC system and his/her vigilance may degrade (see, Table 2 for the Driver's Psychological State at the time of the rapid deceleration). If the target vehicle makes a rapid deceleration and if an emergency-braking alert is given when the driver is in such a state with degraded vigilance, the driver may fail to cope with the circumstance in a timely manner.

What happened, when the headway distance was reduced to 50m? It is seen in Table 2 that, in case of LOA-6.5, the number of accidents remains unchanged (naught), irrespective of values of the Event-Mixture Ratio. However, in case of LOA-6, accidents occurred. Moreover, the increase in the number of accidents under LOA-6 is drastic when the headway distance is reduced from 80m to 50m. Why was LOA-6 not effective? The reason lies in the *time-delay* introduced for enabling the driver to initiate her/his emergency brake in that time period (e.g., 2s in the current simulation study). Note here that a discussion such as, "Is it better to make the time-delay shorter (say, to 1s)?" may not always be meaningful. For instance, ask yourself whether 1s is appropriate and long enough for the human to catch an alert, understand its implication, and implement a proper safety control action.

From the viewpoint of trust in and reliance on automation, an interesting observation may be made by comparing the number of accidents for the following two settings: (1) LOA-6 where the Driver's Psychological State Transition Condition was set at 10 and (2) LOA-4 where the Driver's Psychological State Transition Condition was set at 20. In the latter case, the number of accidents was 19, while 1280 accidents occurred in the former case. When the driver is responsible in applying the emergency brake at all times and on every occation, the driver may stay vigilant than in cases in which the driver can expect that the automation takes safety control actions when necessary. If this is the case, automobile safety may be degraded by a design decision to adopt a high LOA for safety control in emergency situations. However, that does not necesarily mean that theLOA has to be kept low. There are two pieces of evidence: First, even under LOA-6, the number of accidents can be reduced, if the Driver's Psychological State Transition Condition may be improved from 10 to 20 by some means, such as by adopting a better design of human interface: The number of accidents was reduced drastically from 1280 to 7 (see, Table 2). Second, the difference between the number of accidents (0) under LOA-6.5 and the number of accidents (19) under LOA-4 is statistically significant.

7 Automation Surprises in the Real World: Experiment 3

With various cars equipped with ACC and LKS systems, the authors conducted Experiment 3 to investigate human interactions with adaptive automation in peacetime. During the experiment, the authors did not read the owner's manuals thouroughly, which is usually the case for an ordinary car driver. While driving by letting the ACC and the LKS systems at work on expressways in Metropolitan Tokyo area, one of the authors (Inagaki) *simulated* various types of drivers, such as active/relaxed/lazy drivers. One day, when he was simulating a highly strained driver who holds the steering wheel tightly, he heard a voice message of the computer on the host vehicle. However, he failed to catch the exact meaning of it. Some time later, he saw a mild curve ahead, and expected the LKS system would steer the wheel appropriately. However, the author's trust was inappropriate: He was surprised to see the host vehicle went straight, passing across the lane boundary. By that time point, the computer had already traded the authority of lateral control from the LKS system to the driver. The computer had determined, by monitoring moment-to-moment steering torque of the driver, that he was not active in driving, and had decided it appropriate to return the steering task to the driver, according to a measurement-based strategy for trading of control in adaptive automation. The problem in this case lies in the fact that the trading of control was executed at a high LOA, by *assuming* (without confirming) that the driver should accept the trading and understand the need of it.

How can the computer communicate with the driver in such a circumstance? Several design alternatives with different LOA are possible: (1) The computer tells the driver, "You seem to be bored," where the LOA is set at 4. (2) The computer says, "Shall I let you drive yourself?" where the LOA is set at 5. (3) The computer gives a stronger message, "I will hand over control to you in a few seconds," where the LOA is positioned at 6. (4) The computer tells the driver, "I have just handed over control to you," where the LOA is set at 7. (5) The computer trades control from the LKS system to the driver silently. The LOA of this design is set at 8 or higher. It was the design alternative (3) that the author experienced in Experiment 3.

8 Concluding Remarks

This paper has discussed some issues of human trust in and reliance on adaptive automation by taking the ACC and the LKS systems as examples. Adaptive automation offers flexible human-machine interactions in a situation-dependent and context-specific manner. This very flexibility, however, may bring various inconveniences or undesired effects. This paper has shown that human-automation interaction design and its effects on humans can be investigated in a unified manner with two key concepts, *trading of control* and *levels of automation*. As the above examples in Experiments 1 through 3 illustrate, *value* or efficacy of LOA varies depending on situational context, such as time-criticality in the given circumstance. In other words, alothough trading of control at such a high LOA may be needed in emergency, trading of authority at such a high LOA in peacetime may lead to possible loss of situation awareness or automation surprises. If the LOA was chosen inappropriately, some undesirable result may come out. In designing human-machine systems, it is important to predict how the design may affect humans and change their behaviors. The following three approaches are available for selecting an appropriate LOA: (a) cognitive experiment, (b) theoretical analysis, and (c) computer simulation. For details with illustrative examples, refer to (Inagaki, 2005).For further discussions on the design of human-automation interactions, refer to (Inagaki & Stahre, 2004).

Since 2004 July, the first author has been conducting a 3-year research project, entitled "Situation and intention recognition for risk finding and avoidance: Human-centered technology for transportation safety," with the research budget (5.3 Million US Dollars) supported by the MEXT of Japanese Government. One of the aims is to develop adaptive automation for automobile with critical-event and measurement-based automation invocation strategies with the aid of real-time technologies for sensing traffic environment as well as driver's psychological states, such as inappropriate trust in automation (see, Figure 6). Research results shall be reported in the near future.

References

Inagaki, T. (2003). Adaptive automation: Sharing and Trading of Control. In E. Hollnagel (Ed.) *Handbook of Cognitive Task Design* (pp. 147-169). LEA.

Inagaki, T. (2005). Design of human-machine interactions for enhancing comfort and safety. *Proceedings of the AAET*, 23-39.

Inagaki, T., & Stahre, J. (2004). Human supervision and control in engineering and music: Similarities, dissimilarities, and their implications. *Proceedings of the IEEE*, 92(4), 589-600.

Inagaki, T., Moray, N., & Itoh, M. (1997). Trust and time-criticality: Their effects on the situation-adaptive autonomy. *Proceedings of the AIR-IHAS Symposium*, 93-103.

Lee, J.D. & Moray, N. (1992). Trust, control strategies and allocation of function in human machine systems. *Ergonomics*, 35(10), 1243-1270.

Moray, N., Inagaki, T., & Itoh, M. (2000). Adaptive automation, trust, and self-confidence in fault management of time-critical tasks. *Journal of Experimental Psychology: Applied*, 6(1), 44-58.

Parasuraman, R., & Riley, V. (1997). Humans and automation: Use, misuse, disuse, abuse. *Human Factors*, 39(2), 230-253.

Parasuraman, R., Bhari, T., Deaton, J.E., Morrison, J.G., & Barnes, M. (1992). *Theory and design of adaptive automation in aviation systems* (NAWCADWAR-92033-60). Naval Air Development Center Aircraft Division.

Rouse, W.B. (1988). Adaptive aiding for human/computer control. *Human Factors*, 30(4), 431-443.

Scallen, S.F. & Hancock, P.A.(2001). Implementing adaptive function allocation. *International Journal of Aviation Psychology*, 11(2), 197-221.

Scerbo, M. W. (1996). Theoretical perspectives on adaptive automation. In R. Parasuraman & M. Mouloua (Eds.). *Automation and human performance* (pp.37-63). LEA.

Sheridan, T. B. (1992). *Telerobotics, automation, and human supervisory control*. MIT Press.

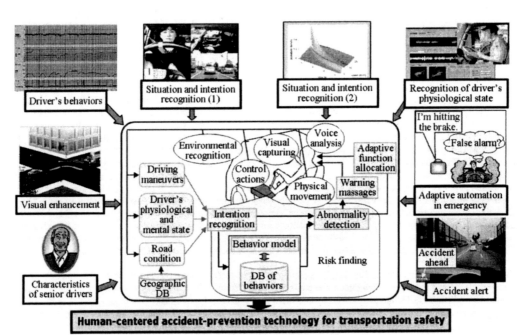

Figure 6: Situation and Intention Recognition for Risk Finding and Avoidance

Section 3
Cognitive State Sensors

Chapter 13

Closed Loop Systems – Stability and Predictability

Closed Loop Systems – Stability and Predictability.

Blair T. Dickson

QinetiQ, Centre for Human Sciences,
Cody Technology Park, Farnborough,
HANTS,UK.
btdickson@qinetiq.com

Abstract

Systems which aim to benefit from the technologies developed in the emerging Augmenting Cognition (AugCog) field employ a closed-loop architecture to optimize interactions between the operator and the system. In the terms of Augmented Cognition the closed-loop nature of the system reflects the real-time monitoring of the status of the operator, and the estimation of the ability to cope with current and possibly future task demands. Knowledge of the status of the operator enables strategies to be employed which seek to focus the skills of the operator on time-critical task performance. Implicit in the operation of such as system is knowledge of the current tasks, goals and consequently intents of the operator. These enable strategies designed to mitigate excessive workload to be targeted in a context sensitive fashion to support the overall efficiency of the man-machine platform. A number of techniques may be employed to mitigate the excessive workload of operators. Broadly these fall into the categories of: attentional management, adaptive interfaces, adaptive automation, decision support, task scheduling and context sensitive support of responses. Critical components of any system which is designed to dynamically change the manner in which control interactions are employed are those issues of operator expectancy and trust. The primary mechanism for maintaining levels of trust in system operation is the development of a shared mental model between the operator and the system. A common knowledge base, or understanding of the principles of operation of the system by the operator, supports the development and consolidation of the mental model, and leads to increased trust in the system which will be reflected in high levels of predictability. The aim of implementing such a system is that the operator will be able to rely upon the systems as they may rely upon other team members. Indeed in common with many systems with multiple operators, well designed AugCog applications enable the distribution of tasks and consequently workload. This is effective team working, and enables a single operator to perform the tasks of many.

Augmentation techniques must also seek to maintain operators in a feed-forward position, supporting strategic and tactical aspects of decision making. These aspects of cognition are traditionally recognized as those that are performed more effectively by trained operators than intelligent systems. We support this observation and suggest that the most effective system configurations are designed to support the operators' requirements as opposed to the operator supporting the deficiencies in the system technology. For this reason the monitoring of operator state drives system configuration. Dynamic interfaces seek to provide essential, timely information in a modality that is most easily assimilated given the current intent of the operator. Similarly the queuing of tasks enables prioritization of information to the operator, ensuring that time critical information is presented before less critical information and that information doesn't appear at inappropriate times. Strategies for keeping the operator apprised of system function are a key element in fostering trust. For this reason, strategies for keeping the operator 'in-the-loop' need careful consideration and are likely to be based on the judicious use of feedback information, based on saliency.

One of the determinants of a successful system configuration is the ability of the operator to predict changes in system operation and configuration. Failure to ensure this will lead to distrust in the adaptive nature of the system and consequently poor overall system efficiency. Similarly it is essential for the stability of the overall system for the reduction in support to be predictable. Too short a period in the augmented configuration may lead to instability in system operation leading to an increase in operator workload owing to the cognitive cost of task-switching. Too long a period with the system augmentations implemented may lead to the reduction of situation knowledge and again reduced platform effectiveness.

This session will address issues fundamental to the successful implementation of AugCog systems, including system stability, the predictability of automation/interface switching and keeping the operator 'in-the-loop' through the use of feed-forward and feedback information flow in dynamic intelligent (cognitive) control systems.

1 Introduction

In the context of the Augmented Cognition (AugCog) programme, a closed-loop system is one in which the operator is viewed as a critical component in the operation of the whole platform. Unlike more traditional approaches to designing systems, where the abilities of the operator are considered only during the initial system configuration phases, in AugCog systems, the state of the operator is viewed as a continually changing element which dictates the overall effectiveness of the platform. As such the ongoing status of the operator can be viewed as a system input and the configuration of the system dynamically changed to reflect the changing abilities of the operator. The concept of dynamically adapting the role of operators in complex systems is not a novel one (Rouse; 1977, Prinzel et al; 2001). In military systems now and for the foreseeable future the operator is viewed as a critical system component either due to the immaturity of automation technology, or because the rules of engagement dictate that a person must have overall sanction over offensive actions. In either case it is important that the abilities of the operator are employed in the most efficient manner, deriving support for a system based technologies to provide the correct level of information detail, at the correct time. This has led to the development of a vision for a pilot of an AugCog Cognitive Cockpit: *'cockpit which allows the pilot to concentrate skills towards the critical mission event at the appropriate time to the appropriate level'* (Taylor et al; 2000). This seeks to ensure that the correct balance of skill and knowledge based tasks are divided as appropriate between the system and the operator.

A number of techniques are available to support the dynamic distribution of tasks between systems and operator. These in summary are: attentional management, adaptive interfaces, adaptive automation, decision support, task scheduling and context sensitive support of responses. Each of these techniques can be employed to reduce the load on the operator, the key to their successful implementation is how changes in the task are achieved. The unexpected change in interface configuration during high tempo, high workload mission phases is likely to increase load, add to task uncertainty, reduce confidence in system operation and foster a mistrust in the adaptive systems. Clearly the predictability of system operation is a determinant of operator trust, A second factor is the stability of operation. The nature of cognition is one of a very dynamic process. If the gauges developed to identify operator state truly reflect the ongoing processing load within the cortex, then the output of these gauges are likely to be characterised by a large number of transients. To switch a series of mitigations on, on the basis of this raw un-filtered data would be unwise as this would lead to gross system instability. This situation would be further exacerbated if the output of the gauge was also used to 'turn off' mitigations. For these reasons it is an important factor in the design of closed loop systems to carefully consider the mechanism for transitioning between augmented and un-augmented states.

2 Predictability of dynamic system operation.

A key element to the successful implementation of any new system is the acceptance of the user community. This is particularly true of a system which seeks to redefine the relationship between operator and system, through the dynamic reallocation of roles and responsibilities. The acceptance of the distribution of work between the operator and system will be greatly enhanced as the predictability of system operation increases. In an AugCog system changes are made to the task load on the basis of an assessment of operator status. Operators may not be aware of their task load especially during high-tempo operational phases, and consequently a system-initiated change to the way a task is performed may not be predictable. It is therefore important to ensure that changes in roles are as unsurprising as possible. One possible mechanism for achieving this is development of a thorough understanding of system operation, based on a common 'mental model'.

The development of shared mental models is an important characteristic of understanding the manner of operator of dynamic systems. In the model for system operation currently under test in the Cognitive Cockpit, the strategy for augmenting cognition is based upon the provision of context sensitive decision support. The manner of implementation of decision support is subject to an adaptive automation frame work based on the Pilot Authorisation and Control of Tasks (PACT) (Bonner et al, 2000). During the population of the decision support system knowledge base, individual pilots are asked to provide an assessment of the acceptable levels of automation associated with each task. The inclusion of the end user in the design of the operation of system automation has the effect of establishing early in the development cycle the role of the automated systems as a pilot support tool. The knowledge encoded in the decision support system is derived from pilots, and consequently, this should be unsurprising when advice is given in response to particular events. As an additional element of the knowledge gathering exercise, operators also dictate the acceptable changes in levels of operation, and those events that trigger these changes. Thus operators not only have an initial input into the design of the system, but also have a core understanding of the operation of the system. This model of system operation defines a mechanism which enables decision support to be tailored to individuals, in terms of both their preferences and their abilities.

A further factor in the development of system trust is one of keeping the operator in the loop. Where the operation of the system is unseen, actions will be performed without the direct involvement of the operator. For the maintenance of situation awareness, it is highly desirable for the operator to be made aware of actions that have been performed either fully automatically, or following authorisation. This may be achieved by the judicious use of feedback information and may exploit the many possible interfaces implemented within the platform. Examples include verbal information presented in the visual or auditory modality, or non-verbal information presented in intelligent icons, implemented within visual displays. It is important to recognise that the presentation of feedback itself also forms part of the operators' task, and as such the manner in which this information is presented should attract the same level of importance as the primary tasks. Considerations should include modality of presentation as well as the scheduling of feedback.

3   Building stability in dynamic systems

Augmented cognition systems monitor the cognitive–affective status of the operator, and derive an estimation of the ability of that individual to cope with current task demands. If an assessment is made that these demands are too great, then mitigations are likely to be implemented which seek to reduce current workload levels. A key objective in the implementation of an augmented cognition, closed loop system is to stabilize system operation. The possibly exists that as workload increases, the mitigations implemented change the characteristics of the ongoing task. This leads to a reduction in load, which is once again sensed by the physiological monitoring systems, and may result in the withdrawal of mitigation strategies; with the consequence that operator load is once again increased. The potential therefore exists for a serious instability in overall system operation. There are a number of strategies that may be employed to reduce the chances of mitigation state flipping during system operation. These are based on a consideration of appropriate mechanisms for the removal of augmentation strategies, and require the consideration of more than operator status alone.

The simplest method that can be used to switch from augmented to un-augmented modes would be based on time alone. The configuration of the system would revert to a default condition a fixed period of time after the implementation of a mitigation strategy. The duration of this period should be derived from a consideration of the nature of the task, and the workload reduction strategy employed. Modeling AugCog system operation is likely to provide the most reliable estimation of appropriate mitigation time constants. A second factor which requires careful consideration are the consequences of the system operating in the augmented mode for too long a period of time. Once again this period is likely to depend on the ongoing operational environment, and the nature of the task. The negative effects of prolonged augmentation are associated with the loss of situational awareness and the associated reduction in overall platform performance. This is likely to occur as systems are less able to dynamically assess complex operational environments than trained operator. Further issues such as the deskilling of operators as a result of over-reliance on automated systems may be viewed as longer term consequences of prolonged augmented cognition system operation.

A second approach to stabilizing system performance involves a consideration of the context of the ongoing operations, and operator behavior. If operator state is used to trigger task augmentation, the task context may be employed alongside operator state to revert system configuration to un-augmented operation. The tracking of behavior by monitoring control interactions can be used to infer the tasks that are currently being performed. A thorough understanding of the nature of tasks, (such as the duration, imposed task load and nature of the load [i.e. verbal or spatial]) and the likely consequences of performing tasks simultaneously, can inform the strategy for implementation withdrawal. The cessation of a certain task, or sequence of tasks, can be identified and used to switch augmentation modes.

A further approach which may be employed to stabilize system performance is based on a combination of task tracking and operator state monitoring. This approach relies on the accurate prediction of future task load, based on the identification of operator intent and a consideration of the current operator status. This methodology for system stabilization employs a pro-active system architecture, mitigating predicted load through the pre-emptive application of mitigation strategies. This modus of operation, whilst still maintaining the reactive properties of the current AugCog systems seeks to activity avoid periods of extreme load.

4 Summary

In summary this session seeks to address the issues of stability and predictability of operation in closed-loop systems, through a consideration of those factors which underpin the successful development of a shared mental model. Furthermore the impact of dynamic system features on expectancy should be carefully considered and should drive the strategy for withdrawal of mitigation support. The use of feed-forward and feed-back messaging enables the operator to remain appraised of system function, whilst being able to concentrate skills on strategic aspects of the role of the system, such as tactical decision making.

As a final, yet important point, the current ability of each and every operator must be considered during all phases of the development cycle, and during the evaluation of these technologies. AugCog systems are novel, and as such no expert operators exist. Subject groups will continue to develop their skill bases in operating these systems, and levels of trust, expectancy and consequently overall system efficiency are likely to develop as levels of system maturity increase.

5 References

Bonner, M.C., Taylor, R.M., Fletcher, K., & Miller, C. (2000). Adaptive Automation and decision aiding in the military fast jet domain. In *Proceedings of the Conference on Human Performance, Situation Awareness and Automation: User-Centered Design for the New Millenium.*

Prinzel, L.J., Pope, A.T., Freeman, F.G., Scerbo, M.W., & Mikulka,P.J. Empirical Analysis of EEG and ERPs for Psychophysiological Adaptive Task Allocation. *NASA/TM-2001-211016.* June 2001.

Rouse,W.B. (1977). Human-computer interaction in multitask situations. *IEEE Transactions Systems, Man, and Cybernetics,* SMC-7, 293-300.

Taylor, R.M., Howells, H., & Watson, D. (2000). The cognitive cockpit: operational requirement and technical challenge. In P.T. McCabe, M.A. Hanson, and S.A. Robertson (eds.), *Contemporary Ergonomics 2000,* 55-59. London: Taylor and Francis.

Enabling Improved Performance though a Closed-Loop Adaptive System Driven by Real-Time Assessment of Cognitive State

Michael C. Dorneich, Santosh Mathan, Janet I. Creaser, Stephen D. Whitlow, Patricia May Ververs

Honeywell Laboratories
3660 Technology Drive; Minneapolis, MN 55418
michael.dorneich@honeywell.com

Abstract

This paper describes an evaluation of an adaptive system that significantly improved joint human-automation performance by "closing the loop" via utilization of a real-time, directly-sensed measure of cognitive state of the human operator. It is expected that in a highly networked environment the sheer magnitude of communication traffic amid multiple tasks could overwhelm the individual soldier. Key cognitive bottlenecks constrain information flow and the performance of decision-making, especially under stress. The Honeywell team, sponsored by the DARPA Augmented Cognition Program, is focusing primarily on varieties of the Attention Bottleneck: tonic arousal required for vigilance and divided attention across multiple tasks. Breakdowns in attention lead to multiple problems: failure to notice an event in the environment, failure to distribute attention across a space, failure to switch attention to highest priority information, or failure to monitor events over a sustained period of time. The appropriate allocation of attention is important to U.S. Army's Future Force Warrior (FFW) program because it directly affects two cornerstone technology thrusts within the FFW program: netted communications and collaborative situation awareness. Honeywell has developed a Closed Loop Integrated Prototype (CLIP) for application to FFW. The CLIP exploits real-time neurophysiological and physiological measurements of the human operator in order to create a cognitive state profile, which is used to trigger mitgation strategies to improve human-automation joint performance. The performance improvements of four mitigation strategies were studied in a Concept Validation Experiment (CVE): task management, modality management, task offloading, and task sharing. Findings indicate that performance can be improved by 100% or more by driving the mitigation strategies with knowledge about the operator's cognitive state.

1   Introduction

Honeywell is addressing the Attention Bottleneck in joint human-machine system performance. The Concept Validation Experiment (CVE) discussed in this paper is part of an on-going research (Dorneich et al 2003; Dorneich et al, 2004) effort to assess the viability of significantly improving joint human-automation performance by "closing the loop" via utilization of a real-time, directly-sensed measure of cognitive state of the human operator.

The U.S. Army is currently defining the roles of the 2010-era Future Force Warrior (FFW). The FFW program seeks to push information exchange requirements to the lowest levels and posits that with enhanced capabilities a squad can cover the battlefield in the same way that a platoon now does. Among other capabilities, the application of a full range of netted communications and collaborative situational awareness will afford the FFW unparalleled knowledge and expand the effect of the Future Force three dimensionally. Task analysis interviews with existing military operations identified factors that negatively impact communications efficacy. In the first few minutes of any intense mission, radio communications are a suboptimal method of communications because everybody is intensely focused on their tasks at hand. In one famous raid, for example, the commander did not hear the radio communications informing him that the plan had changed until he was physically grabbed by the ground force commander and given this critical information. The commander responded by radioing his own troops, who also did not respond. The implications of these kinds of situations are many, but, first and foremost, mission critical information must be reliably communicated.

Adaptive automation can either provide adaptive aiding, which makes a certain component of a task simpler, or can provide adaptive task allocation, which shifts an entire task from a larger multitask context to automation (Parasuraman, Mouloua, & Hilburn, 1999). Currently, adaptive systems derive their inferences about the cognitive state of the operator from mental models, performance on the task, or from external factors related directly to the task environment (Wickens & Hollands, 2000). For example, Scott (1999) used the projected time until impact as an external condition to infer that the pilot's attention was incapacitated, at which point the system would perform the "fly up" evasive maneuver to avoid a ground collision. In that case, the automation took over control of the system (i.e., the aircraft) from the pilot. Adaptive mitigation strategies include task management, optimizing information presentation via modality management, task sharing, and task loading. For instance, in task management, mitigation strategies might include intelligent interruption to improve limited working memory, attention management to improve focus during complex tasks, or cued memory retrieval to improve situational awareness and context recovery. For example, an air traffic controller might be presented with decision aids for conflict detection and resolution by the automated system when it detects a rapid increase in traffic density or complexity (Hilburn, Jorna, Byrne & Parasuraman, 1997). Modality management mitigation strategies might include utilizing available resources (i.e., audio, visual) to increase information throughput. Task offloading and task sharing to automation are also mitigation strategies to reduce workload. Ultimately, the goals of adaptive automation are similar to those of automation in general, such as avoiding "operator out of the loop" conflicts or mistrust in the automation.

An approach was adopted that considers the joint human-computer system when identifying bottlenecks to improve system performance (Dorneich et al, 2003). Key cognitive bottlenecks constrain information flow and the performance of decision-making, especially under stress. From an information-processing perspective, there is a limited amount of resources that can be applied to processing incoming information due to cognitive bottlenecks (Broadbent, 1958; Treisman, 1964; Kahneman, 1973; Pashler, 1994). The DARPA Augmented Cognition program identified four key cognitive challenges related to different components of information processing: 1) the sensory input bottleneck, 2) the attention bottleneck, 3) the working memory bottleneck, and 4) the executive function bottleneck (Raley, et al., in press). The Honeywell team, sponsored by the DARPA Augmented Cognition Program, is focusing primarily on the Attention Bottleneck, though the other bottlenecks are addressed in the studies described herein. There are many varieties of attention that need to be considered to optimize their distribution (Parasuraman & Davies, 1984): executive attention, divided attention, focused attention (both selective visual attention and selective auditory attention), and sustained attention. Breakdowns in attention lead to multiple problems: failure to notice events in the environment, failure to distribute attention across space, failure to switch attention to highest priority information, or failure to monitor events over a sustained period of time. The appropriate allocation of attention is important to FFW because it directly affects two cornerstone technology thrusts of the program: netted communications and collaborative situation awareness (Blackwell, 2003).

Attention (see Figure 1) can be broadly defined as a mechanism for allocating cognitive and perceptual resources across controlled processes (Anderson, 1995). In order to perform effectively in military environments one must have the capacity to direct attention to task relevant events in a dynamic environment (*alertness*). Additionally, one must be able to narrow or

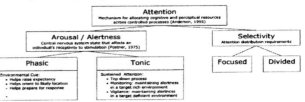

Figure 0. Simplified breakdown of Attention

broaden one's field of attention appropriately depending on the demands of a task (*selectivity*). Attention can be stimulated by external events (*phasic attention*) -- e.g. reacting to gunshots, or a loud aural warning. Alertness can also be maintained consciously, as a controlled top-down process (*tonic attention*). Examples of tonic attention include remaining vigilant while screening baggage at a security checkpoint, or looking for insurgents from surveillance positions over the span of hours. While phasic attention is mostly instinctive and automatic, tonic attention requires active effort on the part of a person. A vast body of literature attests to the difficulty of maintaining tonic attention over prolonged periods of time (e.g. Cabon et al, 1993, Colquhoun, 1985). The CVE explored the use of gauges to detect and drive mitigations during periods when tonic attention levels may be inadequate. We did so in the context of a vigilance task to be described later. Selectivity is another dimension of attention that is critical for task performance. Warfighters have to be able to distribute their attention over information sources effectively in order to accomplish various tasks. Attention has to be highly *focused* in many task contexts. Examples include a bomb disposal expert tuning out distractions to carry out intricate procedures associated witth deactivating an incineary device, or a sniper taking aim at a target. However, many tasks require attention to be *divided* across a diverse range of information sources. This is particulary true in today's information centric warfare environment where the warfighter has to attend to potentially hostile events around him or her while maintaining communications and interacting with a range of information devices. An emphasis of the research reported here was a focus on performance under conditions where limited attentional resources have to be distributed widely in order to perform effectively. Several of the experimental scenarios to be discussed later explored the efficacy of gauge driven mitigations under divided attention demands.

This paper describes an evaluation of the Augmented Cognition Adaptive system built by the Honeywell team. The adaptive system, described in section 2, is driven by real-time non-invasive neuro-physiological and physiological state detection techniques (known in this paper as "cognitive gauges") for determining cognitive workload and comprehension. Subject performed tasks in a virtual environment that represents dismounted soldier combat operations. This virtual environment allowed us to create, and test, a set of cognitive gauges. Cognitive workload was (broadly) defined as the amount of mental effort needed to perform satisfactorily on a task. Comprehension was defined as having sufficient cognitive resources to understand the information at the moment it is presented. Outputs of the cognitive gauges were used to drive an adaptive cognitive assistance system for dismounted combat operations. The experiment, described in section 3, looked at subject performance under a variety of attention states, and aided by a variety of mitigations strategies employed by the adaptive system. Results of the experiment is presented in section 4. With the aid of our proposed adaptive system we hope to increase the soldier's situation awareness, survivability, performance, and information intake by improving their ability to comprehend and act on available information. The results of this experiment will benefit soldiers by creating a system that alters task presentation based on an analysis of that soldier's cognitive state.

2   System Description

2.1   Architecture

The system architecture of the Honeywell Closed Loop Integrated Prototype (CLIP). (see Figure 0) is made up of the following components:

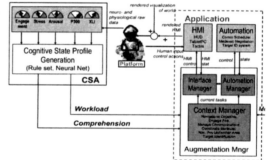

Figure 0. The CLIP architecture

- *Cognitive State Assessor* (CSA) combines measures of cognitive state to produces the cognitive state profile (CSP).
- *Human-Machine Interface*, where the human interacts with the system via a TabletPC, tactor belt, and radio.
- *Automation*, where tasks can be partially or wholly automated.
- *Augmentation Manager* (AM), adapts the work environment to optimize joint human-automation cognitive abilities for specific domain tasks, has:
  - *Interface Manager*, responsible for realizing a dynamic interaction design in the HMI
  - *Automation Manager*, responsible for the level and type of automation
  - *Context Manager*, responsible for tracking tasks, goals, and performance
- *Virtual Environment* (not shown) is a simulated approximation of the real world.
- *Experimenter's Interface* give the experimenter both insight and control over events within the system.

2.2   Cognitive State Assessor

2.2.1   Cognitive Guages

The Honeywell team has developed a comprehensive suite "cognitive gauges" for the CSA:.

- *Engagement Index*. The engagement index is a ratio of EEG power bands (beta/(alpha + theta)), and is a measurement of how cognitively engaged a person is in a task, or their level of alertness (Freeman et al., 1999). Adaptive systems have used this index to successfully control an automation system for tracking performance and a vigilance task (Freeman et al., 1999; Mikulka, Scerbo, & Freeman, 2002; Pope, Bogart, & Bartolome, 1995).The engagement index reflects the selection and focus on some aspect at the expense of the other competing demands, thus it is a measure of focused attention. Low levels of engagement indicate that the subject is not actively engaged with some aspect of the environment.

- *Arousal Meter*. Clemson University's Arousal Meter (Hoover & Muth, 2003) derives autonomic arousal from the cardiac inter-beat interval (IBI), derived from the Electrocardiogram (ECG) at one ms accuracy. The gauge has 3 levels (low, medium high). Increases in score are associated with increased autonomic arousal.
- *Stress Gauge*. The Institute of Human and Machine Cognition developed a composite Stress Gauge (Raj et al., 2003; Kass et al., 2003). The gauge uses a weighted average of the four inputs (Video Pupilometry (VOG), High Frequency Electrocardiogram (HFQRS ECG), Electrodermal Response (EDR), and Electromyogram (EMG) from the left trapezius muscle to detect the subject's response to changes in cognitive load. The gauge detects the cognitive stress related to managing multiple competing tasks on a moment-to-moment basis.
- *P300 Novelty Detector*. The P300-novelty detector gauge (Sjada, Gerson, & Parra, 2003), spatially integrates signals from sensors distributed across the scalp, to discriminate between task-relevant and task irrelevant responses. Mitigation strategies are based on the assumption that the presence of a strong evoked response indicates that subjects have sufficient attentional resources to process the incoming message..
- *Executive Load Index*. Human Bionics developed the eXecutive Load Index (XLI) (DuRousseau, 2004) operates by measuring power in the EEG at frontal (FCZ) and central midline (CPZ) sites. The index was designed to measure real-time changes in cognitive load related to the processing of messages.

2.2.2   Cognitive State Profile

The CSA outputs a Cognitive State Profile (CSP) comprised of two decision state variables: *workload* and *comprehension*. The CSP drives the mitigations of the Augmentation Manager. Currently a simple set of rules is used to derive the assessment of Workload and Comprehension, although work is underway to define this step with neural networks, in order to better account for individual subject differences (see companion paper in these proceedings: Mathan, Mazaeva, & Whitlow, 2005). For this CVE, *workload* was considered high is any of the three gauges, Engagement, Arousal, or Stress, registered high. This was done to allow for the fact that on any given subject, only a subset of the gauges may be able to discriminate differing levels of workload, based on individual differences in subject's cognitive response to the scenario workload manipulation. Likewise, in order to bias *comprehension* towards false positives, both the P300 Novelty Detector Gauge and the XLI gauge had to be high (i.e. that the subject *could* comprehend the message) for the *comprehension* variable to be set to "True".

2.3   Augmentation Manager

2.3.1   Communications Scheduler

Honeywell developed the Communications Scheduler to mitigate divided attention tasks via task-based management and modality-appropriate information presentation strategies.. The Communications Scheduler schedules and presents messages to the soldier based on the cognitive state profile (derived from the gauges), the message characteristics (principally priority), and the current context (tasks). Based on these inputs, the Communications Scheduler can pass through messages immediately, defer and schedule non-relevant or lower priority messages, escalate higher priority messages that were not attended to, divert attention to incoming higher-priority messages, change the modality of message presentation, or delete expired or obsolete messages. Messages are characterized by priority, depending on how critical they are: high (mission-critical and time-critical), medium (mission-critical only), and low (not critical). The Communications Scheduler determined the initial message presentation based on a user's current *workload*. After the first presentation of a message to the user (in audio modality), the Communications Scheduler determined whether to take further action on a message depending on the CSA's assessment of *comprehension*. A more detailed description of the Communications Scheduler can be found in the companion paper in these proceedings (Dorneich et al 2005).

2.3.2   Tactile Navigation Cueing System

The Tactile Situation Awareness System (TSAS) provides navigation cueing during mitigated trials, and consists of a 12 pairs of tractors, representing the cardinal positions of the clock (12 o'clock centered on the umbilicus), worn about the upper abdomen. Operationally, pulses from tectors "tug" the subjects in direction to go. The system was invoked when the CSP indicated *workload* was high and the subject needed to navigate through an unfamiliar route.

However, turning the system off as soon as *workload* fell below some threshold would leave users disoriented in an unfamiliar area. Thus the navigation mitigation persists until users get to the safe destination.

2.3.3 Medevac Negotiation Agent

The evacuation of injured personnel is a crucial warfighter function. The task is lengthy and requires a substantial amount of information to be communicated accurately. Performance on this task may suffer under high workload conditions. Personnel may omit important information or make errors in the information transmitted. Additionally, attention devoted to coordinating the medevac exchange may detract from the performance of other critical tasks. The Medevac Negotiation Agent was triggered on if *workload* was high and a medevac had to be coordinated. A medevac icon on the HUD notified the user about the need to coordinate an evacuation using the medevac agent. Information available on the FFW Netted Communications network would be automatically filled in, and the system would present this information about casualties to the platoon leader for inspection. The platoon leader could review the medevac information on the Tablet PC and transmit information using the interactive form. The system also provided the option to delegate subsequent medevac negotiation to team members facing lower workload demands.. The medevac agent only contained the most critical information needed for a Medevac. A more detailed information exchange might allow for safer and more efficient Medevac operations. Additionally, engaging in Medevac communications may contribute to better situational awareness of a team's status. It is for these reasons that the augmentation is only invoked when the subject's workload is so high and the subject's performance so inadequate, that the differences associated with automated negotiation are acceptable in terms of overall performance.

2.3.4 Mixed-Initiative Identification System

Military personnel sometimes have to maintain high levels of sustained attention in environments where target stimuli may be infrequent and hard to detect. Research suggests that performance on these tasks deteriorates considerably over time. This task cannot be easily automated, since automated systems trained to detect target stimuli in a field may not perform as well as an alert human. However, these systems could play a helpful role if they can be triggered when gauges detect a vigilance decrement. Subjects in the CVE lookined for targets in a series of surveillance photos. If Engagement was low, the system assists in identifying targets by drawing a yellow box around any detected targets. This system was modeled after enabling technology (Schneiderman & Kanade, 2004) currently under development. Such a system of mixed initiative search with intelligent assistance is part of the FFW vision (US Army, 2003).Due to a non-acceptable frequency of errors within the system, the subject used system output for advice, but must continue to scan. In addition, other issues such as over-reliance on automation and the human's generally poor ability to passively supervise automated processes preclude automating the process entirely.

2.4 Virtual Environment

The operational environment was a desktop first-person virtual environment (VE) that simulates Mobile Operations in an Urban Environment . The VE consisted of a city of narrow streets, surrounded by two and three story buildings. The environment had an industrial appearance. The visual complexity of the environment contributed to the subjects' workload. The subject was faced with some number of enemy forces, presented both at street level and above. The enemy forces had attacked with varying levels of success (depending on the workload and difficulty settings). The subject performed all tasks in the environment using a combination of keyboard and mouse controls. The controls allowed the subject to look around the world, to move (walking forward or backward, sidestepping left or right, jumping, and crouching), to shoot their weapon (an approximation of an M16) and to manage messages.

3 Experimental Method

Can the gauges detect the cognitive states of interest consistently enough to drive the mitigations? If the mitigations are driven correctly, do we see performance improvements on the communications? The goal for the CVE was to improve performance on mitigated tasks, with no performance decrement to concurrent tasks, and with no negative effect on overall workload. Three scenarios were developed for the CVE, where each scenario looked at different states of attention, tasks, and mitigations. The CVE focused on two types of attention: Tonic arousal required for vigilance (Scenario 3) and divided attention across multiple tasks (Scenarios 1 and 2).

3.1 Operational Scenarios

3.1.1 Scenario 1: Divided Attention

Scenario 1 focused on three critical task elements: "Navigate to Objective," "Engage Foes" and "Manage Communications" The subject ,a platoon leader, led his or her platoon along a known route through a hostile urban environment to the objective, while being careful to engage enemy soldiers. Subjects also received incoming communications throughout, some requiring an overt response. Subjects received status and mission updates, requests for information, and reports; these communications are a primary source of their situation awareness. Radio communications volume was extremely high. The scenario only included two or three high priority messages (hold at certain locations

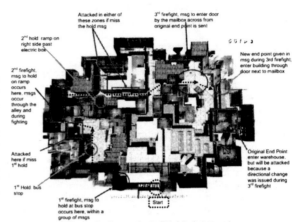

Figure 0. Scenario 1: Divided Attention

for a specified amount of time, or objective location had changed). Failure to heed these high priority messages caused the subject to encounter an ambush. Figure 0 details the route, the points in the scenario where the high priority messages occurred, and the potential ambush locations, if should the subject fails to heed the messages. Scenario 1 was designed to put subjects into distinct periods or extreme high and extreme low workload periods. Validation of the ability of the gauges to distinguish between these periods is vital. When they held at a location for up to three minutes, a low task load situation, it was expected that the gauges would register low workload. When subjects engaged in a firefight, the gauges were expected to register cognitive states associated with high workload. Scenario 1 was principally designed to test the performance improvements derived from the Communications Scheduler. Thus the high workload times included a high volume of communications traffic to the subject, just at the time when their workload was high due to being targeted by foes. Metrics are defined for each task in Table 1.

Table 1. Task metric definitions for Scenario 1

Metric	Definition	Task
Message comprehension	Message comprehension is inferred from compliance with the orders or correct responses from direct quieires in messages.	Manage Communications
Situation Awareness	Score on 4-8 questions asked after each scenario to ascertain if they could recall mission-critical information relayed through the communications.	Manage Communications
Run Time	Time subjects took to complete a known (trained) route, while avoiding hits, was a measure of their effectiveness to navigate to the objective.	Navigate to Objective
Hits Taken	Number of hits by OPFOR (Opposing Forces) on subject.	Engage foes
Hits on OPFOR	Number of hits by subject on OPFOR.	Engage foes
Shot Accuracy	The percentage of shots fired that hit OPFOR	Engage foes

3.1.2 Scenario 2: Divided Attention

The subject traversed the same initial route as in Scenario 1. However, upon reaching the objective area, the subject was informed that the enemy had set a trap. The subject needed to abandon the objective, get back to the safe zone, and avoid the route the just taken to the objective. To return to the safe zone, the subject had to navigate though unfamiliar parts of the city in order to avoid ambushes. The task load stems from having to mentally convert an exocentric 2D representation of an unfamiliar area into an ego centric representation and reasoning

Figure 0. Scenario 2: Map and MEDEVAC

with this newly formed representation. The participants were provided with an updated map on their Tablet PC showing potential ambush zones. Simultaneously, the participants got a request to coordinate a MEDEVAC immediately. This is a quite lengthy and communications- and information-intensive procedure. The map and the information requirements of the MEDEVAC procedure are illustrated in Figure 0. Both the Medevac communications procedure and the navigation to the safe zone task had to be accomplished simultaneously and under extreme time pressure. Additionally the subject had to engage any foes encountered. Metrics are defined for each task in Table 2.

Table 2. Task metric definitions for Scenario 2

Metric	Definition	Task
Time to Safe Zone	Time to navigate from the warehouse to the safe zone	Nav. thru Unfamiliar Area
Ambushes Encountered	Number of ambushes encountered..	Nav. thru Unfamiliar Area
Medevac Questions Answered	Number of medevac-related questions answered correctly	Coordinate Medevac
Time to complete Medevac	Time to complete the medevac negotiation process	Coordinate Medevac
Hits Taken	Number of hits by OPFOR on subject.	Engage foes
Hits on OPFOR	Number of hits by subject on OPFOR.	Engage foes

3.1.3   Scenario 3: Sustained Attention

In Scenario 3, the subject was the leader of a recon unit, and was responsible for identifying any targets (enemy soldiers) from reconnaissance photos received (updates every 2 seconds), via his or her Tablet PC, from external surveillance cameras. The experimental protocol for this scenario was a classic vigilance paradigm. The scenario lasted approximately 30 minutes. The first five minute session had target occurring at a rate of 14%, and served as the measure of baseline performance. This was followed by a 20-minute session with a very low target occurrence rate (3%), to produce a vigilance decrement. The final five minute session had a target occurrence rate identical to the first five minute session. For Scenario 3, the metric of interest is:

• *Target Identification.* The accuracy to which subjects identified targets was taken as a measure of performance. Specifically, the accuracy of stage 1 (baseline) was compared with performance in Stage 3.

3.2   Mitigation Strategies Applied

There are four broad categories of possible mitigations in an Augmented Cognition system: 1) task/Information management, 2) modality management, 3) task offloading, and 4) task sharing. The multiple scenarios of the IHMC CVE provided Honeywell with the opportunity to explore a wide range of possible mitigation strategies. All four categories of mitigation strategies are addressed by the Augmentation Manager. Table 3 summarizes the Mitigation Agents that realize classes of mitigation strategies..

Table 3. Summary of mitigation agents that realize classes of mitigation strategies.

Mitigation Strategy	Mitigation Agent	Task Mitigated	Scenario
Task Scheduling	Communications Scheduler	Manage Communications	1
Task Offloading	Medevac Negotiation	Coordinate Medevac	2
Task Sharing	Mixed-Initiative Target ID	Target Identification	3
Modality Mgmt	Communications Scheduler	Manage Communications	1
Modality Mgmt	Tactile Navigation Cueing	Navigate thru Unfamiliar Area	2

While the mitigations described here have the potential for boosting performance when human cognitive resources may be limited, they could have detrimental effects if left on at all times. Gauge driven mitigation allows these mitigations to be activated when the benefits are likely to outweigh the costs. See the companion paper (Mathan, Dorneich, & Whitlow, 2005) in the proceedings for a thorough discussion of the role of automation etiquette in the design of effective cognitive-state-driven mitigation strategies.

3.3 Experimental Design

The CVE consisted of three two-factor experiments. Each experiment compared performance under gauge driven mitigation with performance without mitigation. For scenarios 1 and 2 this evaluation was a within subjects design, as each subject saw both scenarios in both the mitigated and unmitigated cases. Half of the subject saw Scenario 3 with mitigation, and the half without. Thus for scenario 3 the experiment is a between subjects design.

Hypothesis. In general, the hypothesis for this experiment is as follows: the mitigations will improve performance on the tasks they are mitigating without decrementing other concurrent tasks.

Participants. 14 males (M_{age} = 25.4 years, avg. education = 15 years) participated in this experiment. To reduce the effect of learning for this experiment, participants were chosen who rated their skill level (avg = 3.4 of 5-scale, avg playing time = 5.7 hours/week) at first person shooter games as average to above average. .

Independent Variables. 1) Mitigation (on/off), and 2) Scenario (three, which vary by attention type).

Dependant Variables in addition to metrics defined for each scenario (e.g. see Table 1),subjective workload (via NASA TLX) and subject preferences were assessed.

4 Results

4.1 Scenario 1 Task Metrics

This scenario focused on divided attention bottleneck in multi-tasking, and consisted of the subject performing three tasks. The mitigation strategy employed in this scenario was the Communications Scheduler. The goal was to improve performance on the Manage COmmunciations task while not decrementing performance on the Navigate to Objective and Engage Foes task.

4.1.1 Manage Communications Task Metrics

Message comprehension. Subjects in the unmitigated condition correctly responded to 57 of 143 possible messages (39.9%); . in the mitigated condition they correctly responded to 114 of 143 messages (79.7%). The mitigated condition shows a significant (p<0.0001) performance increase of 100% (see Figure 5).

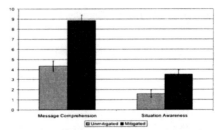

Figure 0. Manage Communications task

Situation Awareness. Subjects in the unmitigated condition correctly responded to 22 of 84 situation awareness questions (26.2%). Subjects in the mitigated condition correctly responded to 49 of 84 situation awareness questions (58.9%). The mitigated condition shows a significant (p=0.009) performance increase of 125% (see Figure 5).

4.1.2 Navigate to Objective Task Metrics

Run Time. The total run time for the mitigated condition (M = 965 sec) was significantly longer than the run time for the unmitigated condition (U = 469 sec), t(12) = -8.29, p < 0.001. This result is in the expected direction. In the mitigated condition, more participants heard the hold messages, which increased the time mission completion time.

4.1.3 Engage Foes Task Metrics

Hits Taken. There was no significant performance change for hits taken while engaging foes (p = 0.29).

Hits on OPFOR. There was a significant difference between the unmitigated (M = 518 sec) versus the mitigated (M = 468) for how many times the subject was able to shoot opposing forces, t(12) = 2.93, p = 0.013. Participants shot the opposing force significantly fewer times during the mitigated condition. This result is expected because in the mitigated condition subjects comprehended the "hold" message more often and thus avoided ambushes.

Shooting Accuracy. No significant performance change for shooting accuracy while engaging foes (p = 0.06). .

4.2 Scenario 2 Task Metrics

This scenario focused on divided attention bottleneck in multi-tasking. The mitigation strategies employed in this scenario were the Tactile Navigation Cueing System and the Medevac Negotiation Tool. These mitigations were expected to improve performance on the Navigate Through Unfamiliar Area task and Coordinate Medevacn task, respectively, all while not degrading performance on the Engage Foes task.

4.2.1 Navigate Through Unfamiliar Area Task Metrics
Time to Safe Zone. The total time it took a subject to reach the safe zone was greater for unmitigated subjects than mitigated subjects. The data is illustrated in [create new figure?]. The average performance improvement was 20%, however this difference was not significant (p = 0.30). [need numbers]

Ambushes Encountered. Subjects in the unmitigated case were almost four times as likely to navigate into an ambush than subjects in the mitigated case. Unmitigated subjects (N=12) ran into 19 ambushes, while the mitigated subjects (N=12) ran into 5 ambushes. The difference was significant (p<0.003). The mitigation resulted in a 380% performance improvement.

4.2.2 Coordinate Medevac Task Metrics

Medevac Questions Answered. Subjects in the unmitigated case answered 50 questions correctly out of a possible 98 (51% correct). Subjects in the mitigated case answered 98 of 98 questions correctly (100% correct). Thus the mitigation was able to significantly (p=0.004) increase performance by 95%, as shown in Figure 6.

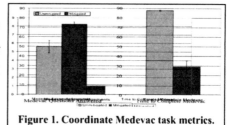

Figure 1. Coordinate Medevac task metrics.

Time to Complete Medevac. Subjects where able to complete the medevac task significantly faster in the mitigated condition (p < 0.001), resulting in a 303% performance improvement, as shown in Figure 6.

4.2.3 Engage Foes Task Metrics

Hits Taken and Hits on OPFOR.,Participants took fewer hits from the OPFOR during the mitigated (M = 102) than the unmitigated (130) condition, t(12) = 2.42, p = 0.03. Additionally, participants had fewer hits on OPFOR in the mitigated (M = 22) versus the unmitigated (M = 27) condition, t(12) = 2.47, p = 0.029. Participants in the mitigated condition potentially encountered fewer ambushes because their ability to navigate the safe route was improved , resulting in participants seeing fewer enemy forces and thus, receiving and inflicting fewer hits.

4.3 Scenario 3 Task Metrics

Target Identification. Baseline performance (stage 1, no mitigation) was 65.8%. Stage Stage 3 mitigation was set at an accuracy rating of 68%. Unmitigated subjects in Stage 3 had an accuracy of 66.2%. Thus, on average, the experiment was not able to produce the decremented human performance one like to see in a vigilance experiment. Nonetheless, subjects in the mitigated condition performed at an accuracy of 85%, much better than the human (66.2%) or automation (68%) accuracy alone. The 30% performance improvement was significant (p=0.022).

4.4 Subjective Workload and Preferences

None of the scales nor the total overall workload were significantly different for either Scenario 1 or 2. Overall 76.9% of the subjects thought it was easier to perform tasks in the mitigated condition. Subjects also found mitigated condition easier for fighting (61.5%), communicating (84.6%) and navigating (76.9%).

5   Acknowledgements

This paper was supported by a contract with DARPA and the U.S. Army Natick Soldier Center, DAAD-16-03-C-0054, for which LCDR Dylan Schmorrow serves as the Program Manager of the DARPA Augmented Cognition program and Mr. Henry Girolamo is the DARPA Agent. The opinions expressed herein are those of the authors and do not necessarily reflect the views of DARPA or the U.S. Army Natick Soldier Center.

The authors would like to acknowledge the contributions of Mr. Jim Carciofini, Mr. Trent Reusser, and Mr. Jeff Rye, and Ms. Danni Bayn of Honeywell. Additional thanks to the Institute for Human and Machine Cognition, lead by Dr. Anil Raj, for hosting the experiment: Mr. Roger Carff, Mr. Matt Johnson, and Mr. Jeremy Higgins .

References

Anderson, JR (1995). Cognitive Psychology and its Implications (2nd Ed.). New York: Freeman

Blackwell, C. (2003), "Objective Force Warrior: Advanced Technology Demonstration", *Natick Soldier Center*, presentation to Honeywell at the NATICK IPR, October 23.

Broadbent, D. E. (1958). *Perception and Communication*. New York: Pergamon.

Cabon, Ph., Coblentz, A., Mollard, R. & Fouillot, J.P. (1993) 'Human Vigilance in Railway and Long-Haul Flight Operation', Ergonomics, 36(9): 1019-1033.

Colquhoun, W.P. (1985) 'Hours of Work at Sea: Watchkeeping Schedules, Circadian Rhythms and Efficiency', Ergonomics, 28(4): 637-653.

Dorneich, M.C., Whitlow, S.D., Mathan, S., Carciofini, J., and Ververs, P.M. (2005). "The Communications Scheduler: A Task Scheduling Mitigation For A Closed Loop Adaptive System." *Proceedings of the 11th International Conference on Human-Computer Interaction (Augmented Cognition International)*, Las Vegas, Jul 22-27, 2005

Dorneich, M., Whitlow, S., Ververs, P.M., Carciofini, J., and Creaser, J. (2004) "Closing the Loop of an Adaptive System with Cognitive State," *Proceedings of the Human Factors and Ergonomics Society Conference 2004*, New Orleans, September.

Dorneich, Michael C., Whitlow, Stephen D., Ververs, Patricia May, Rogers, William H. (2003). "Mitigating Cognitive Bottlenecks via an Augmented Cognition Adaptive System," *Proceedings of the 2003 IEEE International Conference on Systems, Man, and Cybernetics*, Washington DC, October 5-8, 2003.

DuRousseau, D.R., (2004). *Multimodal Cognitive Assessment System*. Final Technical Report, DARPA, DAAH01-03-C-R232.

Freeman, F.G., Mikulka, P.J., Prinzel, L.J., & Scerbo, M.W. (1999). Evaluation of an adaptive automation system using three EEG indices with a visual tracking system. *Biological Psychology, 50*, 61-76.

Hoover, A. & Muth, E.(in press). A Real-Time Index of Vagal Activity. *International Journal of Human-Computer Interaction*.

Kahneman, D. (1973). *Attention and Effort*. Englewood Cliffs, NJ: Prentice-Hall.

Kass, S J, Doyle, M Raj, AK, Andrasik, F, & Higgins, J (2003, April). Intelligent adaptive automation for safer work environments. In J.C. Wallace & G. Chen, *Occupational health and safety: Encompassing personality, emotion, teams, and automation*. Symposium conducted at the Society for Industrial and Organizational Psychology 18th Annual Conference, Orlando, FL.

Mathan, S., Mazaeva, N., & Whitlow, S. (2005). Sensor-based cognitive state assessment in a mobile environment. *Proceedings of the 11th International Conference on Human-Computer Interaction (Augmented Cognition International)*, Las Vegas, Jul 22-27, 2005.

Mathan, S., Dorneich, M., & Whitlow, S. (2005). Automation Etiquette in the Augmented Cognition Context. *Proceedings of the 11th International Conference on Human-Computer Interaction (Augmented Cognition International)*, Las Vegas, Jul 22-27, 2005.

Mikulka, P. J., Scerbo, M. W., & Freeman, F. G. (2002). Effects of a Biocybernetic System on Vigilance Performance. *Human Factors, 44*(4), 654-664.

Parasuraman, R. & Davies, D. R. (1984). *Varieties of Attention*. New York: Academic Press.

Pashler, H. (1994) Dual-task interference in simple tasks: data and theory. *Psychological Bulletin, 116*, 220-244.

Pope, A.T., Bogart, E.H., & Bartolome, D.S (1995). Biocybernetic system validates index of operator engagement in automated task. *Biological Psychology, 40*, 187-195.

Prinzel, III, L.J., Hadley, G., Freeman, F.G., & Mikulka, P.J. (1999). Behavioral, subjective, and psychophysiological correlates of various schedules of short-cycle automation. In M. Scerbo & K. Krahl (Eds.), *Automation technology and human performance: Current research and trends*. Mahwah, NJ: Lawrence Erlbaum Associates.

Raj A.K., Perry, J.F., Abraham, L.J., Rupert A.H. (2003) Tactile interfaces for decision making support under high workload conditions, *Aerospace Medical Association 74th Annual Scientific Meeting*, San Antonio, TX..

Raj A.K., Kass, S.J., and Perry, J.F., (2000). "Vibrotactile displays for improving spatial awareness". In the *Proceedings of the Human Factors and Ergonomics Society 44th Annual Meeting*. Santa Monica, CA: HFES.

Raley, C., Stripling, R., Schmorrow, D., Patrey, J., & Kruse, A. (2004). Augmented Cognition Overview: Improving Information Intake Under Stress. In the *Proceedings of the Human Factors and Ergonomics Society 48th Annual Meeting*. Santa Monica, CA: HFES.

Schneiderman, H. and Kanade, T. (2004) "Object Recognition Using Statistical Modeling". http://www.ri.cmu.edu/projects/project_320.html.

Sajda, P., Gerson, A., & Parra, L. (2003). Spatial Signatures of Visual Object Recognition Events Learned from Single-trial Analysis of EEG, *IEEE Engineering in Medicine and Biology Annual Meeting*, Cancun, Mexico.

Treisman, A.M. (1964). Verbal cues, language, and meaning in selective attention. *American Journal of Psychology, 77*, 206-219.

US Army (2003), "*The Objective Force Warrior: The Art of the Possible...a Vision*," Available Online at http://www.ornl.gov/sci/nsd/pdf/OFW_composite_vision.pdf.

HCI International 2005
The Future of Augmentation Managers

Tom Diethe

QinetiQ Ltd.
Farnborough, UK
tdiethe@QinetiQ.com

Abstract

In every Augmented Cognition (AugCog) application, where a closed-loop architecture is implemented, there is a requirement for a system that is responsible for the implementation of the strategies designed to mitigate the effects of excessive workload. The manner by which these mitigation strategies are implemented is likely to be platform specific, however several generic principles may be defined which enable the outputs of operator state gauges to be interpreted and acted upon.

In the simplest configuration an Augmentation Manager is designed to implement a single or number of mitigation strategies based on a single, high-level descriptor of operator state. In the initial implementation of the Tasking Interface Manager (TIM) in our Cognitive Cockpit (CogPit) the output of a single gauge, executive workload, was passed through a linear filter and a simple thresholding algorithm employed to trigger the implementation of a number of mitigation strategies. Whilst this showed some benefits during our first set of closed-loop trials, it is likely that this approach oversimplifies the problem. In particular, this approach did not take account of the complex relationship between mitigation strategies and state. In addition, any system that employs this, or similar techniques, will be purely reactive rather than proactive. Here we propose a multifaceted approach, in which the outputs of multiple gauges, operator state forecasting and context modeling together with a prediction of the effects of mitigation strategies on operator status will be used to drive the targeted application of a number of mitigation strategies. Achieving this will require a major engineering effort and carefully constructed empirical research.

Furthermore it is apparent that in order to make AugCog systems truly intelligent and adaptive, they will need to be able to monitor their own performance and adapt as necessary. Future systems will also need to be adaptive to the context in which they are operating, enabling the appropriate responses to unfamiliar contextual cues to be made. These occurrences should be stored in order to improve system operation under similar conditions should they arise again. This machine learning element to AugCog systems is probably the most underdeveloped area, but potentially could provide huge benefits to their long-term operational functionality.

Keywords: Augmentation Manager, closed-loop system, adaptive automation, Cognitive Cockpit, mitigation strategies

1   Introduction

The DARPA program "Improving Warfighter Information Intake Under Stress" has entered Phase III, and a number of closed-loop Augmented Cognition systems have been developed to operate on various platforms. During the first two phases, much emphasis was placed on the development of accurate gauges designed to estimate cognitive affective state, whilst Augmentation Managers remained relatively immature. The evolution of these systems has required this, as any Augmentation Manager will need accurate data on its input side in order to effectively implement mitigation strategies. The diverse platforms being used by contributors to the program have ensured that a variety of mitigation strategies have been used, and necessarily these are targeted at the problems specific to the platforms on which they are being implemented. These mitigation strategies have had varying degrees of success, mostly specific to the problems presented by the platform upon which they are being implemented (Raley et al. 2004).

The goal of an Augmentation Manager is to optimize the combined performance of the human and platform through the management of excessive workload by targeted automated assistance. Generally speaking, this has led to the development of gauges that detect what has been defined as excessive workload in any given modality (or global/executive workload). On detecting this state strategies are implemented that aim to mitigate against that excessive workload whilst maintaining system effectiveness. These strategies can take the form of decision support (e.g. provision of advice or modifications to the interface), automated assistance, task scheduling, or modality switching. There is an implicit assumption being made that it would be inappropriate for the mitigation strategy being employed to be "on" all the time, as this will cause unwanted effects such as loss of Situational Awareness (SA), deskilling, and inappropriate support during periods of relative inactivity (Endsley 1995).

We have focused on two different platforms, a known as the Cognitive Cockpit (CogPit), and "TaskLite". CogPit has been developed to enable automated decision support to be investigated in a complex military environment with the aim of maintaining the overall efficiency of the platform (Taylor et al 2000). CogPit is focused on the military fast-jet environment, and is designed to support the vision of 'A cockpit that allows the pilot to concentrate his skills towards the relevant critical mission event, at the appropriate time, and to the appropriate level'. The aim was to develop a platform designed to provide trustworthy automation, which selects and interprets information derived from aircraft sensors and from the pilot. This information is used to provide decision support in the form of a series of planned tasks that are either automated or presented to the pilot, depending on the current and projected task loads. This system is designed to be sensitive to the operational context and the ability of the pilot to perform the recommended tasks within the available time. TaskLite is a medium fidelity task, based on the role of an Uninhabited Air Vehicle (UAV) operator. The operator's task is to manage a number of assets (UAVs) simultaneously, by tasking each individually to perform a specific role. These roles include searching for, imaging, classifying, and, if appropriate, engaging enemy targets, as well as monitoring and maintaining fuel, power and weapons levels. TaskLite was developed to enable rapid testing of augmentation strategies and technologies without requiring the use of trained pilots or the full CogPit facility.

1.1   Description of Cognitive Cockpit methodology

The cockpit is based on a number of enabling technologies. These are the on-line analysis of the physiological, behavioral and subjective status of the pilot; a Knowledge-Based System that enables on-line decision support, and principles of cognitive engineering for the adaptive automation of the task and the display of information to the pilot (Taylor et al. 2000). The Cognitive Cockpit has been designed to be modular at a functional level, enabling the independent development of the core and sub-systems. A number of generic principles are followed, ensuring that the sub-systems may be readily ported to application environments other than a fast-jet cockpit. The main components of the system are shown in Figure 1.

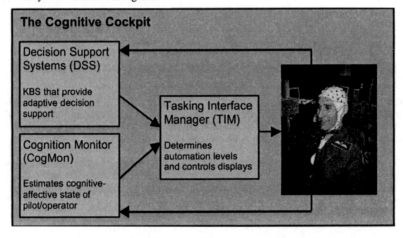

Figure 1: Cognitive Cockpit top-level design, identifying key sub-systems

1.1.1 Tasking Interface Manager/Augmentation Manager

The Tasking Interface Manager (TIM) was developed to allocate pilot and system functions dynamically, and to manage cockpit interfaces, mission tasks and timelines, by interpreting inputs from the Decision Support Systems (DSS) and the Cognition Monitor (CogMon) (Fletcher & Bonner, 2000). These integrative functions enable the TIM to prioritize tasks and to determine the means by which pilot information is communicated. Overall, this system manages the cockpit automation by context-sensitive control over the allocation of tasks to the automated systems and the pilot. The level of automation can be altered in real time in accordance with current mission objectives, pilot requirements and/or pilot capabilities. This capability is realized through the application of a Pilot Authorisation and Control of Tasks (PACT) framework (Bonner, Taylor, Fletcher & Miller, 2000). PACT allows the pilot to form a *contract*, or set of contracts, with the automation by allocating PACT levels on a task-by-task basis. During operation, the TIM monitors the output from the DSS. When a plan is developed the TIM examines the PACT levels of each task within the plan and either performs the task automatically or provides assisted decision support and presents the information in the most appropriate manner. This is derived by examining the pilot status gauges identified by CogMon.

The recognition that a proactive system is more likely to be able to optimize the overall system effectiveness, compared to a reactive system, has led to the development of a second iteration of the system design by the addition of Intent Inferencing and Task Tracking (IITT) based on forecasting of cognitive state (see Figure 2). The 'Augmentation Manager' is therefore the combination of the extant TIM technologies and the new IITT (Diethe 2004).

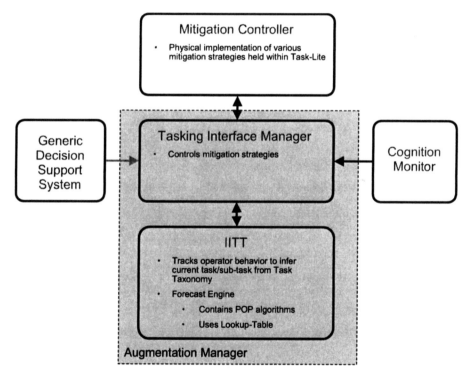

Figure 2: Second iteration of Cognitive Cockpit functional design showing addition of IITT with integrated Prediction of Operator Performance (POP) algorithms

1.1.2 Intent Inferencing/Task Tracking (IITT) System

The IITT serves three main purposes:

- **Task Tracking.** The IITT logs all of the operator's interactions with the interface in real time.
- **Intent Inferencing.** The IITT then uses a "Task Taxonomy" — a database of all known tasks and task sequences — to infer the intent of the operator.
- **Forecasting of Operator State.** The IITT uses a finite state machine represented as a set of probabilities that one system state will follow another – to deduce the most likely future tasks of the operator. These are then used to generate a forecast of cognitive state using Prediction of Operator Performance (POP) algorithms. POP uses a fairly sophisticated model (Farmer, 2000) of human cognition based on the concept of resources. It predicts workload quite accurately and also makes predictions of performance decrement under multiple-task conditions. This forecast is provided to the TIM to enable proactive employment of mitigation strategies.

1.2 Mitigation Strategies

We have approached the problem of augmenting operator performance through a thorough examination of a number of strategies. What follows is a brief discussion of the mitigation strategies that we have taken into consideration.

1.2.1 Task Scheduling

Temporal aspects of task management have long been recognized as playing a major role in operator workload (e.g. Jordan, Farmer & Belyavin, 1995, Hildebrandt & Harrison 2002). We have therefore identified task scheduling according to resource availability as a possible mitigation strategy. This based on the assumption that additional information load at high workload or high tempo sections of the mission is likely to compromise the ability of the operator to perform his/her primary task, such as control of the vehicle (if a pilot), or the maintenance of Situation Awareness. Throughput of information (i.e. warnings, task-related information, and general information) is metered in accordance with available cognitive resources, and as such information of low importance can be postponed during mission-critical events.

1.2.2 Task Queuing and Prioritization

A related mitigation strategy is that of task queuing and prioritization according to saliency, such that higher saliency information is inserted earlier in a queue of messages awaiting presentation than lower saliency information. In addition information of higher saliency is presented in more prominent ways through the use of available interface manipulations, with the intention of controlling attentional focus.

1.2.3 Modality Switching

Modality switching is a potentially powerful mitigation strategy based on a model of human cognition that states that information can be more readily assimilated when parallel non-conflicting input or process channels are employed, and is preferable to loading up a single modality (e.g. Wickens, 1992). Examples of this may be switching from visual alerts to auditory alerts (or *vice versa*).

1.2.4 Adaptive Dynamic Function Allocation (A-DFA)

A-DFA is a form of Adaptive Automation in which a negative feedback loop is formed between the operator and the system, such that the system reacts by increasing automation levels in periods of high workload and *vice versa*. This is based on the assumption that additional task load during high workload sections of the mission is likely to impinge on the primary task(s), as stated before, and increased automation of incoming tasks will enable the operator to concentrate on critical mission events. The shifting of task allocation between the operator and the system must be performed through the use of a structured adaptive automation framework, e.g. the PACT framework mentioned above. The PACT framework (figure 2), is a reduced, practical set of levels, with clear engineering and interface consequences; it is derived from the ten levels of automation for human–computer decision making proposed by Sheridan and VerPlanck (1978), with notable similarities with the levels of control and automation proposed by Endsley and Kiris (1995).

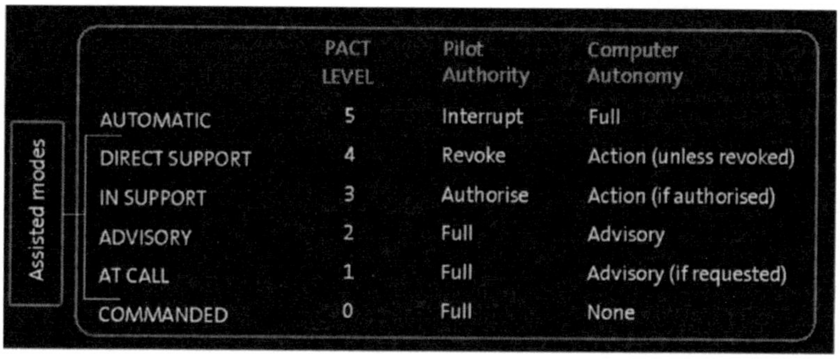

		PACT LEVEL	Pilot Authority	Computer Autonomy
	AUTOMATIC	5	Interrupt	Full
	DIRECT SUPPORT	4	Revoke	Action (unless revoked)
Assisted modes	IN SUPPORT	3	Authorise	Action (if authorised)
	ADVISORY	2	Full	Advisory
	AT CALL	1	Full	Advisory (if requested)
	COMMANDED	0	Full	None

Figure 3: PACT framework for A-DFA

1.2.5   Display Enhancements

There is another class of mitigation strategy where targeted enhancements to the display are made during pertinent periods of a mission. It is important that these mitigation strategies are not simply advanced HCI, in which case there would be no instance in which they would not be helpful. For example, in our TaskLite, the UAV icons and information along the command bar reveal Chicken and Joker fuel levels in order to give an indication of remaining fuel for the given task at the expense of height and bearing information. During periods of relative inactivity, the operator would be able to calculate these levels themselves, and would need the height and bearing information for maintenance of SA, and as such it would be inappropriate to have this mitigation strategy in place all of the time. Mitigation strategies in this class are difficult to design and their efficacy is also difficult to assess.

1.3   System Stability

Once mitigation strategies for the platform in question have been developed (based on an identification of where the problems for that platform lie), and a gauge (or gauges) have been chosen to drive the mitigation strategies, the link between them must be made. In the first instance, the simplest mechanism might be to employ a threshold algorithm that switches mitigation strategies on when workload is high, and switches them off when it is low (using the output of a single gauge, e.g. executive workload). However, from observation of large quantities of data from our psychophysiological sensor suite and resulting higher order gauge output it is apparent that there frequent transient shifts in cognitive state. A simple threshold algorithm would therefore result in many switches occurring, resulting in an unstable system and undesirable consequences. We have implemented strategies to counteract these transient shifts in cognitive state. Namely, these are the use of a digital filter to smooth out some of the finer fluctuations whilst retaining the broader peaks and troughs, and the use of a "refractory period" – a period of time in which the system will not perform a switch (e.g. from mitigated to unmitigated or vice versa) in order to prevent "yo-yoing" between states (Diethe, Dickson, Schmorrow & Raley, 2004). These are used in combination with a simple threshold-based algorithm to implement the required switches in automated states. Although as yet these techniques have yet to be empirically tested, anecdotal evidence suggests that they are having the desired effects.

1.3.1   Digital Filter

The general aim when using some index of psychophysiological state as a trigger for adaptation (whether related to A-DFA or to information presentation) is to determine the point at which the workload is sufficiently high or low to initiate changes. The TIM system employs a low-pass filter to ensure that transient peaks and troughs do not cause rapid switching in the system, along with a simple threshold-based algorithm. The smoothing filter currently used by the TIM is the Savitzky–Golay (1964). These time domain filters remove high frequency fluctuations while still preserving the true amplitudes and widths of the features. Each data value is replaced by a linear combination of itself and a number of nearby neighbors. The filtered value after each iteration is then passed to the thresholding algorithm.

1.3.2   Refractory Period

Each mitigation strategy is assumed to have an effect on the state being measured. Thus if the state met the criterion for being classified as "high", state transition would occur, as a result of which the individual's state might decrease to a level classified as "low", which would trigger a further change of state. Such a cyclic effect would clearly be undesirable and lead to a highly unstable system; hence the refractory period is introduced to determine a minimum time interval between state transitions.

This, due to the prior smoothing of the data, ensures that state transitions occur only when data are consistently above the upper or below the lower threshold. It also has the effect of frequency-limiting the state transitions to reduce the switch rate. Given that the data may exhibit highly transient properties, this ensures that "spikes" in the data are filtered out. An example of the effect of this algorithm with state transitions shown as vertical lines can be seen in Figure 4.

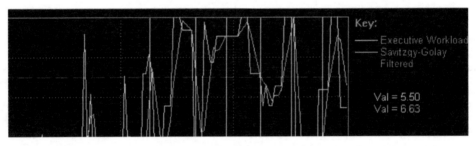

Figure 4: Filtered output from the CogMon with state transitions shown as vertical white lines and the upper threshold shown as a horizontal dashed red line (lower threshold not pictured).

1.4   State Forecasting

One criticism of the approach to controlling mitigation strategies outlined thus far is that it is purely reactive, and that ideally the Augmentation Manager should attempt to prevent excessive workload, rather than detecting that it has occurred before implementing mitigation strategies, as well as being able to predict that an intervention may or may not be necessary contrary to evidence from state gauges. As such some form of state forecasting is required to supplement the state prediction based on physiological behavioral and contextual measures. We have implemented an initial attempt at workload forecasting in the version of the Augmentation Manager that is implemented in our TaskLite (Diethe 2004, Ryder 2004).

Operator state forecasting is driven by the tracking of current behavioral activity, and as such is dependent on a platform which facilitates the tracking of behavior and an inferencing engine capable of interpreting the behavioral outputs provided by the platform. In any given platform, the operator has to perform a series of tasks, each of which can be decomposed into single interactions with an interface; these control interactions we refer to as atoms. In order to build an effective inferencing engine, a full break-down of all of the possible tasks that can be performed into their atomic elements needs to be performed. Once the task of the operator has been fully modeled, a finite state machine can be generated with probabilities that one or more system states (in the form of a tree) will follow another.

During execution, the inferencing engine tracks all of the interactions with the interface, and attempts to discern the current task being performed on the basis of the last few atoms. Once an ongoing task has been inferred, the next stage of the process is to consult the finite state machine to find the most probable next task, or if the prediction probability is too low a range of tasks may be used. This task (or tasks) is then broken up into atoms, and characteristics of each of the atoms in this series are used to generate a profile of future workload. This profile is then passed on to the TIM. The simplest version of this process is represented diagrammatically as shown below:

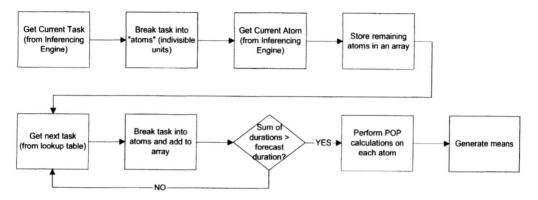

Figure 5: Flow diagram for forecasting procedure

2   Future directions

Thus far the techniques described will allow for the stable control of one or more mitigation strategies on the basis of a gauge output along with an estimation of the future workload for a certain period of time. It is always worth remembering that forecast accuracy will diminish as its duration increases, and on some platforms this may be critical to its utility. Notwithstanding this, the level of sophistication achievable through these methods is sufficient to drive a simple closed-loop Augmented Cognition system with predictable and measurable behavior. The challenge for the future is to go beyond this, to produce systems that take a much more considered view of how to optimize performance. The focus so far has been on strategies intended to mitigate against excessive workload under the assumption that this will prevent degradation of performance. However the dynamic interactions between various aspects of workload (or cognitive activity in general) and performance are necessarily much more complicated. In fact, the constant chase for reduction in workload may even be a red herring, as high workload can be maintained so long as it is not in combination with loss of SA or dangerously high levels of stress or anxiety. As a result there might be situations where the state gauge is indicating very high levels of workload, but performance is at an optimal level. These require extra care, as a choice to implement a mitigation strategy may actually harm performance, whereas a choice not to implement a strategy may result in catastrophic failures or errors. There are, however, ways in which these problems may be addressed, and these are described next.

2.1   Intelligent Control of Mitigation Strategies

Firstly, it is clear that when trying to augment the cognition of an operator, the best possible estimation of current cognitive state is needed. This means that all relevant aspects of cognition need to be considered. In most cases, this will mean producing gauges for executive workload, visual workload (verbal and spatial), auditory workload, psychomotor workload, stress, arousal, and attentional engagement. To achieve this requires a large quantity of empirical research, especially if the gauges are to work on multiple platforms. An advanced model of human cognition is then required in order to interpret the output of these gauges in a meaningful way. Significant testing and evaluation is required at this stage to ensure that the interpretations of multi-modal cognitive function mirror the true cognitive states of the pilot. Alongside this, contextual and environmental conditions should be taken into account, as these can have a significant bearing on the operator's ability to function. In addition, the state forecasting procedure described above should ideally be applied in multiple modalities as well, in order to ensure correspondence with state estimations. All of this information should then be fused within switching algorithms that control mitigation strategies. Assuming that multiple mitigation strategies exist also, a complex mapping between gauges and mitigation strategies will be required. An essential ingredient to the success of this will be modeling of the consequences of each mitigation strategy, and using this model to generate new predictions of future state that can be compared with the existing forecast. One problem that may occur when controlling multiple mitigation strategies using multiple gauges is that triggering on and off different strategies at different times may cause the coherence of the system to diminish. Thus the use of various mitigation strategies in various combinations should also be modeled, and subjective data must be considered an important source of system acceptance, which may well come down to its 'style' as much as for any performance gains to be had.

2.2   Machine-Learning

During the development of our CogMon, it soon became apparent that accurate prediction of state required bespoke profiles of each operator to be built up due not only to significant physiological differences but also the cognitive strategies they employed (Pleydell-Pearce, Dickson & Whitecross, 2000). These differences must equally apply to the effects of introducing various mitigation strategies, and as such the system should be able to adapt to the operator. In effect, the system should learn the consequences of its own actions. For example if a given mitigation strategy was implemented in a given situation and had the effect of increasing workload or adversely affecting some other dimension of cognitive state (e.g. stress) then system could learn to either choose a different mitigation strategy or to not perform any action at all. This would require the ability to learn the important contextual parameters of that situation in order for it to recognize that situation in the future.

The method through which this is implemented is in some ways unimportant, as the end result will be the same, but it is worth mentioning that there are several different technologies that are mature enough to perform the task. An agent based approach (e.g. Barber, Goel & Martin 2000) may offer the simplest way of building smaller learning "units" with data mining capabilities into the system. Other machine-learning techniques that may be considered are standard unsupervised learning methods such as Bayesian statistics, hidden Markov models, Monte Carlo methods. Evolutionary methods such as genetic algorithms may also provide ways of refining models and algorithms.

2.3   Augmenting Team Cognition?

Current team structures are based on fixed function allocations where crew members and technologies have a responsibility for tasks that are defined under their specific role. Dynamic allocation of function as employed by the TIM system between a human operator and automated systems could be expanded to involve allocations between team members as well. Such systems would aim to ensure an optimal workload/performance balance across an entire team (e.g. Scerbo 1996). This clearly adds another level of complexity, but can be conceived more readily if thought of as a resource allocation problem, with each of the operators adding to the pool of resources their particular expertise. Although immature, programs of work are already in place that are investigating team function allocation (Prinzell III, 2003, Harris et al, 2004).

3   Summary

In summary, we are now at a stage in the evolution of Augmented Cognition system design where we can build and test fully closed-loop systems, and the benefits of these are already becoming apparent. Also evident, however, is the need to increase the sophistication of the Augmentation Managers that are controlling the interventions we are employing. In the first instance, this involves ensuring the stability of the switching technology through the use of digital filters and refractory periods, and forecasting of operator state better inform when interventions should be performed. Beyond this, a more detailed understanding of the nature of changes in workload and performance, associated with dynamic changes in task characteristics, is required so that the outputs of multiple gauges can be employed to control multiple mitigation strategies. This engenders a high degree of complexity in the system that will mean increased difficulty with empirical testing. The more gauges, augmentation strategies, and control mechanisms that are in place, the more permutations of possible interventions that need to be considered. For completeness each of these should be tested in isolation, as testing an entire suite at once does not guarantee maximum benefits and makes it very difficult to locate shortcomings. In systems destined for the operations, it also makes it very difficult to accurately predict the behavior of the system in every situation, so certification will be difficult to achieve. In the end the biggest problems for complex systems of this type may be that the cognitive models used to drive the intelligent switching behavior are not mature enough, or even that the sensor technology is too immature, leading to inaccurate or inconsistent gauge output. One possible way, already being explored by DARPA, to counteract some of these problems is to introduce these technologies as training aids. This would allow both acceptance testing and system refinement with a constant source of low-cost participants for studies in a low-risk setting. The other major advance for Augmentation Managers, required for them to be truly "intelligent", will be their ability to learn and adapt to new situations and to the strategies employed by each individual. The potential benefits from this are huge, and must outweigh the potential pitfalls associated with such technology.

4 References

Barber, K.S., Goel, A., & Martin, C. E. (2000). Dynamic adaptive autonomy in multi-agent systems. *The Journal of Experimental and Theoretical Artificial Intelligence, Special Issue on Autonomy Control Software, 12,* 129-157.

Bonner M.C., Taylor R.M., Fletcher K., and Miller C. (2000). Adaptive automation and decision aiding in the military fast jet domain. *Proceedings of the Conference on Human Performance, Situation Awareness and Automation: User-Centred Design for the New Millennium, Savannah, GA, 15-19 October, 2000*

Diethe, T. (2004). Function specification of an Intent Inferencing/Task Task Taxonomy (IITT) system. *QinetiQ Technical Report: QINETIQ/KI/CHS/TR042455*

Diethe, T., Dickson, B.T., Schmorrow, D., & Raley, C. (2004). Toward an augmented cockpit. In D.A. Vicenzi, M. Mouloua, P.A. Hancock (Eds.) *Proceedings of the 2nd Conference on Human Performance, Situation Awareness and Automation, Daytona Beach, FL, (2) 65-69*. Mahwah, NJ: Lawrence Erlbaum Associates.

Endsley, M.R. (1995). Towards a theory of situation awareness. *Human Factors, 37(1),* 32-64.

Endsley, M.R., and Kiris, E.O. (1995). The out-of-the-loop performance problem and level of control in automation. *Human Factors, 37(2),* 381-394.

Farmer, E. (2000). Predicting operator workload and performance in future systems. *Journal of Defence Science, 5(1),* pp F4-F6.

Fletcher, K., & Bonner, M.C. (2000). Technical specification of the Tasking Interface Manager. *DERA Technical Report: DERA/AS/SID/SPEC000453*

Harris, S.L., Bowyer, S.J., Cooper, M.J., Purdy, C.S., Melia, A., & Diethe, T. (2004). Model of an adaptive crew concept. *QinetiQ Technical Report: QINETIQ/KI/CHS/TR040749*

Hildebrandt, M., & Harrison, M.D. (2002). The temporal dimension of Dynamic Function Allocation. In S. Bagnara, S. Pozzi, A. Rizzo & P. Wright (Eds.) *Proceedings of the 11th European Conference on Cognitive Ergonomics, 283-292*. Rome: Consiglio delle Ricerche.

Jordan, C.S., Farmer, E.W., & Belyavin, A.J. (1995). The DRA workload scales: a validated workload assessment technique. *Proceedings from the 8th International Symposium on Aviation Psychology, 2,* pp 1013-1018.

Pleydell-Pearce, K., Dickson, B., Whitecross, S. (2000). Cognition Monitor: A system for real time pilot state assessment. In P. T. McCabe, M. A. Hanson and S. A. Robertson (Eds.). *Contemporary Ergonomics 2000* (pp 65-69). Taylor and Francis, London.

Prinzel III, L.J. (2003). Team-centred perspective for adaptive automation design. *NASA Technical Report: NASA/TM-2003-212154*

Raley, C., Stripling, R., Schmorrow, D., Patrey, J., & Kruse, A. (2004). Augmented Cognition overview: improving information intake under stress. *Proceedings of the Human Factors and Ergonomics Society 48th Annual Meeting.* New Orleans, LA.

Ryder, C. (2004). TaskLite model description. *QinetiQ Letter Report: QINETIQ/KI/CHS/LR042407*

Savitzky, A., and Golay, M.J.E. (1964). Smoothing and differentiation of data by simplified least square procedures. *Analytical Chemistry, 36, 1627-1639.*

Scerbo, M.W. (1996). Theoretical perspectives on adaptive automation. In R. Parasuraman & M. Mouloua (Eds.). *Automation and Human Performance: Theory and Applications (pp 37-64).* Mahwah, NJ: Lawrence Erlbaum Associates.

Sheridan, T.B., and VerPlanck, W.L. (1978). Human and Computer Control of Undersea Tele-operators. *Technical Report. MIT Man Machine Laboratory*, Cambridge MA.

Taylor, R.M., Bonner, M.C., Dickson, B.T., Howells, H., Miller, C., Milton, N., Pleydell-Pearce, C., Shadbolt, N., Tennison, J., & Whitecross. S. (2000). Cognitive Cockpit engineering: coupling functional state assessment, task knowledge management and decision support for context sensitive aiding. *TTCP Technical Panel 7 Human Factors in Aircraft Environments, 2000 Workshop on Cognitive Engineering, Dayton, Ohio*

Wickens, C.D. (1992). *Engineering Psychology and Human Performance.* New York: Harper Collins.

Cybernetics of Augmented Cognition as an Alternative to Information Processing

Thomas J. Smith

School of Kinesiology
University of Minnesota
Minneapolis, MN, USA
Smith293@umn.edu

Robert A. Henning

Psychology Department
University of Connecticut
Storrs, CT, USA
Robert.Henning@UConn.edu

Abstract

Proposed system implementations for augmented cognition incorporate some features of closed-loop control but development efforts to date have largely been guided by open-loop information processing models. A social cybernetic model is presented as a means to re-conceptualize augmented cognition, and to provide more specific research and design guidance for developing these systems whose central purpose is to augment human control over cognition. The premise of this paper is that social cybernetics offers a comprehensive conceptual and empirical framework to address current theoretical shortcomings and guide both research and development of augmented cognition systems. We believe that no other model, theory or paradigm of human-machine performance currently extant offers comparable scientific, methodological and heuristic power for dealing with the complexities and needs of implementing augmented cognition systems.

1   Overview

The current approach being advocated for the development of augmented cognition systems is to assess the cognitive state of the individual in real time, and then to use that information to somehow modify and/or mediate cognition for performance gain (Raley et al., 2004). Directly analogous to how present-day computing machine systems function, the information processing model guiding this effort attributes performance decrements to resource limitations that can occur in one or more of the following sectors; input, cognitive activity or output (Wickens & Hollands, 1999). A more fine-grained analysis for augmented cognition has identified key "cognitive bottlenecks" that are assumed to be a consequence of resource limitations in each of these three sectors (Raley et al., 2004). After identifying psychophysiological response measures associated with the state of each of these bottlenecks, the proposed strategy for implementing augmented cognition systems is to somehow control information flow to the input sector as the means of avoiding conditions of either overload or underload in the system that would degrade cognitive performance.

The basic scientific problem with information processing models that have come to dominate cognitive science in the past four decades is that they are non-refutable. One can hardly deny that when operating complex systems, the human has to process information, be aware of situations, and deal with perceptual and cognitive workloads. The question is, does this conceptual framework enhance understanding and technical implementation of systems to augment cognition?

As described in more detail below, the cybernetic perspective presented here differs in a number of fundamental scientific respects from an information processing model. Behavioral cybernetics (T.J. Smith & Smith, 1987a, 1987b) emphasizes active control of information as sensory feedback via motor-sensory behavior, with motor-sensory behavior mediating both perception and cognition. In contrast, information processing models treat information as a fixed commodity presented on the input side of the processing system, ignoring the need for specific design factors to promote human motor control over this information as a source of sensory feedback. This failure to include motor-sensory control is a direct consequence of using an information processing model, where overt/motor behavior is viewed as a system output with no central role in the organization and control over information as feedback, nor over the control of subsequent behavior through the reciprocal effects of motor control on psychophysiological state.

Lastly, the information processing model provides no guidance on how to develop hybrid automated systems for augmented cognition. The present reliance on the information processing model condones the implementation of augmented cognition systems that are fully automated even though the literature is replete with examples of "clumsy automation" that result in system instabilities and loss of control episodes due to the lack of an appropriate level of supervisory control by the human operator (Woods, Sarter & Billings, 1997; Sheridan, 1997; 2002). In contrast, the social-cybernetic model of human-computer interaction advocates the human factors design of hybrid automation systems with supervisory control (T.J. Smith, Henning, & Smith 1995) for augmented cognition, as described below.

In summary, this report advocates behavioral cybernetics as an alternative and more powerful scientific paradigm for dealing with the challenges of augmented cognition. Conceptual principles of this framework are introduced first, followed by a synopsis of relevant empirical approaches for analyzing human-system performance based on these principles. Implications of this paradigm for realizing the goals of augmented cognition also are addressed.

2   Conceptual Principles

2.1   Cybernetics of Cognition

Cognition in behavioral cybernetic terms emerges as a consequence of active motor control over sensory feedback, a concept that has been delineated from both a behavioral perspective (Smith, 1972; T.J. Smith and Smith, 1987b), and more recently from a neurobiological perspective (Jackson and Decety, 2004). Ecological psychology (Gibson, 1979) shares this idea of an intimate link between action and perception, but behavioral cybernetics places greater emphasis on self-regulatory (as opposed to ecological environmental) control of perception, mediated by kinesiological mechanisms. The cybernetic view of cognitive behavior rejects the notions that environmental stimuli or external information sources can independently drive or control cognition, or that cognition can occur separately from motor behavior. Attention and perception, treated as independent processes under the information processing paradigm (Raley et al., 2004) also are understood in the cybernetic model as interdependent and integrated through motor-sensory control. Seemingly radical in the context of present-day cognitive science, the concept of motor-sensory integration is not new to psychology, having been introduced by functionalists William James and John Dewey over 100 years ago.

Cognition—perception and knowing—manifests itself through a variety of behavioral phenomena, denoted by terms such as thinking, problem-solving, understanding, insight, planning, situation awareness, mental workload, and so forth. In cybernetic terms, what these different manifestations of cognition have in common is predictive activity, which to us represents the essence of cognition. In other words, for effective guidance of behavior you have to be able to predict the sensory and perceptual consequences of your actions (Blakemore, Frith, and Wolpert, 2001; Hawkins and Blakeslee, 2004). Based on empirical studies of brain function, neuroscientists now believe that such predictive guidance is based on what is termed a forward model, in which memory of sensory feedback from past action (the predictive model) is referenced against real time sensory feedback from current action (perception), and the model updated (learning) based on any detected discrepancy. Recently, neurophysiological evidence has been reported documenting neuronal populations in the medial frontal cortex whose cognitive control functioning specifically subserves learned error prediction (Brown and Braver, 2005).

From this perspective, the **first cybernetic concept** is that cognitive demands arise from the need to control future behavior, a process termed feedforward control, anticipatory control, or projective tracking. Unlike compensatory control in which the human operator reacts to sensory feedback about system errors, feedforward control involves the cognitive projection of past memories and associations of feedback control dynamics to anticipate future events and the behavioral requirements for their control, so that control actions can occur to prevent behavioral errors from occurring when these events transpire.

A **second, related cybernetic concept** is that effective cognitive performance depends upon effective motor behavioral control of sensory feedback. Cognitive demands as they are commonly experienced can be understood as resulting from challenges to such feedback control caused by complex sensory environments (control limitations), lack of learning (poor understanding), lack of skill (poor training), and/or poor human factors design. In every case, the cognitive consequence is a compromised ability for projective guidance of behavior.

Particularly dramatic manifestations of compromised sensory feedback control emerge under conditions of displaced sensory feedback from one's own behavior. First documented in the 19th century, these effects were systematically delineated in a sustained program of research by K.U. Smith and colleagues (K.U. Smith, 1962, 1972; K.U. Smith and Smith, 1962; T.J. Smith and Smith, 1987a).Under both spatially displaced and temporally delayed sensory feedback, guidance of behavior is dramatically and immediately degraded (multi-fold reductions in accuracy and increases in performance times), accompanied by loss of body image and increased psychophysiological stress. Some learning is observed under spatially displaced feedback, but under delayed feedback learning is highly limited and labile. The common theme is that confronting an operator with unfamiliar displacements in sensory feedback from movement behavior severely compromises the cognitive ability of that operator to predict perceptual consequences of behavior.

This leads to a **third key cybernetic concept**, that of context—that is, design—specificity in cognitive performance (T.J. Smith, Henning, and Smith, 1994). The essence of this idea is that the preponderance of observed variability in such performance is attributable, not to inherent biological factors or learning ability, but to design of the task. In other words, performance cannot be evaluated outside of its context --- generalized models of performance have little scientific validity. Firm empirical evidence for this conclusion was first compiled almost a century ago from differential learning studies of problem-solving—that is, explicit cognitive—tasks. The authors cited above review other evidence from research on psychomotor performance, such as causes of industrial accidents, effects of displaced sensory feedback, and social tracking, that also supports this conclusion.

Conceptual principles of cognitive cybernetics summarized above carry a number of implications for a behavioral cybernetic approach to augmented cognition, and these can be characterized as a series of general and specific premises. The **first general premise** is that overt motor behavior—action—and the behavior of neuronal populations subserving cognition, collectively represent a closed-loop system mediated by respective functional mechanisms that act as both cause and effect of each other. In this regard, self-regulation of cognition to support both goal realization and learning is recognized as integral to the development and education of children (Snow, Corno, & Jackson, 1996).

The **second general premise**, derived from the first, is that control systems principles and methods can be applied to cognition generally (Hommel, Ridderinkhof, & Theeuwes, 2002), and more particularly to the analysis of degradation in cognitive performance observed under high stress and/or complex interactive environments. The latter concept is aligned with the perspective of Sheridan (2004), who views driver distraction—resulting in a reduction in attention to the driving task---as a control systems problem, and applies control theory to its analysis. In the context of high stress and/or complex interactive environments in battlefield situations, a major focus of augmented cognition research (Raley et al., 2004), we thus assume that compromised cognitive performance of the military operator under high/complex task demand conditions also represents a manifestation of distraction, where we define distraction as a battlefield process or condition that interferes with the attention of the military operator to the task at hand, thereby perturbing behavioral control of the operator (Sheridan, 2004). From this perspective, factors such as operator drowsiness, fatigue, and/or stress are not viewed as sources of distraction per se, but rather as factors that compromise brain activation and thereby alter parameters of cognitive task control such as gain, delay, and variability (IBID).

This leads to the **first specific premise** drawn from the cybernetic principles enunciated above, which is that the key objective of an augmented cognitive system should be to support projective behavior on the part of the operator. This premise assumes that the critical consequence of perturbed behavioral control under distraction (possibly exacerbated by altered brain activation) is a compromised ability to predict the sensory and perceptual consequences of one's action, and therefore that intervention by the system should be to mitigate these effects. The authors are unaware of any real time intervention studies that support this premise. However, in a training study of troubleshooting, Schaafstal and colleagues (2000) clearly show that a modification of the design of a training intervention program, to more effectively and rigorously support predictive cognitive behavior on the part of student technicians in detecting and diagnosing the effects of system malfunctions, resulted in a two-fold improvement in troubleshooting performance.

The **second specific premise**, an extension of the first, is that maintaining or enhancing behavioral control of sensory feedback should represent the key operational goal of augmented intervention for purposes of supporting projective cognitive behavior. The problem is that displacements in sensory feedback, as noted above, severely compromise behavioral control, yet are inherent to contemporary real-time augmented cognition systems, as researchers in this area acknowledge (Raley et al., 2004). Major challenges confronting designers of such systems therefore include: (1) mitigating displaced feedback effects, thereby removing what is arguably the major design obstacle to realization of system goals; (2) delineating which modes, levels and combinations of sensory feedback can be most effectively controlled by operators, and which therefore represent the most attractive target for augmentation; (3) delineating factors related to sources of attentional distraction and also to reduced brain activation that most prominently compromise operator control of sensory feedback; and (4) developing effective real-time methods for characterizing cognitive control and augmenting this control.

Finally, the **third specific premise** is that operator performance with an augmented cognition interface will be context specific, as is the case for performance interaction with any other design artifact. This premise leads to predictions that operator performance and learning with such interfaces will feature: (1) a broad spectrum of individual differences; (2) outcomes and patterns of variability that may not be generalizable from one group of operators to the next; (3) no predictable relationship between the distribution of trainee proficiency scores at the outset of training with that at the completion of training; and (4) emergence of new modes and levels of variability with the introduction of any modifications to the interface design.

In summary, from a behavioral cybernetic perspective, the human factors design challenge for augmented cognition systems is to promote and enhance operator capabilities for self-regulation of projective cognitive behavior, through application of closed-loop intervention strategies that will complement cognitive behavioral efforts by the operator to generate and control sensory feedback in the face of factors that both promote attentional distraction and compromise brain activation. Remaining sections of this report introduce concepts and empirical methods of social cybernetics as a scientific framework for addressing this challenge.

2.2   Social Cybernetics

The basic goal of augmented cognition is to establish a human-computer closed-loop system (Dorneich et al., 2004; Raley et al., 2004). Such a system is modeled on the team, with what Salas and Fiore (2004) term collective team cognition transfigured to what Raley and colleagues (2004) term overall system IQ. There is nothing new about the idea that human cognition has distinct if not predominant social concomitants (Webb and Palincsar, 1996; K.U. Smith, 1974; T.J. Smith and Smith, 1988b)—social interaction probably has contributed to every major evolutionary advance in human cognitive ability since the dawn of the species. What is novel about augmented cognition is the idea that a computer can serve as an effective social surrogate in supporting human cognition.

Our scientific approach to the analysis of system interactive performance involving multiple actors (human-human or human-computer) is grounded in social cybernetics. Social cybernetics is founded on the broader field of behavioral cybernetics (Draper, Kaber & Usher, 1998; T.J. Smith & Smith, 1987a; K.U. Smith & Smith, 1966) which, as noted above, assumes that human behavior is controlled as a closed-loop or cybernetic process. The cybernetic nature of behavior becomes obvious during social interaction between two or more individuals. This is because each individual in a social context must control the sensory feedback generated not only by his/her own behavioral movements and functioning, but also the sensory feedback created by interacting with one or more social participants. The study of interpersonal and group reciprocal sensory feedback and sensory feedback control relationships represents the focus of social cybernetics, directed towards delineating the closed-loop behavioral-physiological manifestations and properties of social interaction (T.J. Smith, Henning & Smith, 1994, 1995; Smith & Smith, 1987a, 1988a; K.U. Smith, 1972, 1974). Social cybernetic theory extends to team contexts a feedback theory of individual movement integration, and thereby assumes that team activity, like all other biological activities, is feedback regulated (Adolph, 1982).

The term social tracking is used to describe the feedback-controlled process by which an individual follows or tracks a social target. Social interaction is conceived as a dynamic linking of the motorsensory behavior of two or more people in a social tracking relationship. In cybernetic terms, the behavioral activities of one person in a social context affect behavioral-physiological changes in others, whose behaviors in turn feedback influence the ongoing behavior and physiology of the first. These effects arise as a consequence of control by each participant of sensory feedback generated by the other social partners. A more detailed example of two-person social tracking is as follows: the movements of one individual generate stimuli that are controlled as sensory feedback by the movements of the social partner, whose motor-sensory control actions generate compliant sensory stimuli back to the first person, who in turn controls this sensory feedback by further movement, and so on. This idea is illustrated in Figure 1.

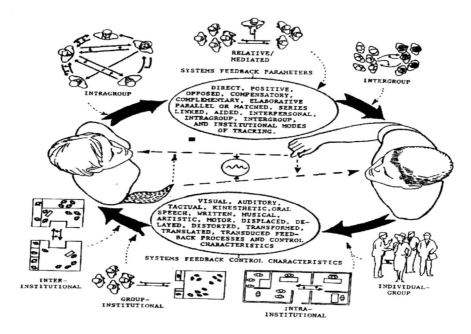

BEHAVIORAL CYBERNETICS OF SOCIAL TRACKING

Figure 1. Behavioral cybernetic model of social interaction. Motor behavior of one partner generates sensory feedback that is sensed and controlled by the other. Motor behavior required for this control generates sensory feedback that, in turn, is sensed and controlled by the first partner. In this manner the two partners establish a social tracking yoke that involves mutual exchange and control of sensory feedback.

Social tracking typically requires each participant to control multiple motor, sensory and cognitive modalities (vision, speech, writing, etc.) and transformations (displacements, delays, etc.) of sensory feedback. The social partners thus become dynamically yoked or interlocked behaviorally and physiologically, as a result of mutual body movement tracking and control of each other's sensory feedback. Through such interlocks, the participants in a social group begin to operate as an integrated system, with definite systems feedback parameters and feedback control characteristics (Fig. 1).

In particular, social cybernetic systems can involve many different types of social tracking modes (direct pursuit, compensatory, parallel, series-linked, etc.) in a variety of different social interactive contexts (interpersonal, group and/or institutional social systems). By definition, mutual control in social interaction to some extent compromises self regulation by the individual but there are also potential gains in behavioral control capabilities, such as improved stability, that could surpass the regulatory abilities of individual participants, thus motivating social behavior.

In summary, social cybernetic theory as outlined above rests upon three basic assumptions: (1) the theory is applicable to conceptual and experimental analysis of all modes and dimensions of human social behavior and interaction; (2) both individual and social behavior are differentially specialized in relation to the organizational and environmental design features in which social interaction occurs; and (3) social human factors dominate all aspects of the human condition, dictating not only the course and level of human development, but specialization of the processes of learning, performance, schooling, aging, organizational design and management, work, and all aspects of machine-related behavior. From this perspective, social cybernetic concepts of interpersonal, group, and institutional behavior create a new comprehensive framework for interpreting human social interaction with machines and technology, including particularly automated systems intended to augment human cognition.

3 Empirical Approaches

3.1 Experimental Analysis of the Social Cybernetics of Human-System Interaction

In order to empirically test the validity of some of the social cybernetic concepts, the feedback properties of social interaction have represented a major focus of behavioral cybernetic research. The experimental approach has involved use of real time, human-computer hybrid systems methods for studying social tracking. During the course of a sustained program of cybernetic research on social tracking behavior, both computer-mediated and television techniques have been used to provide real time feedback indications of error and accuracy in movement correspondence between two or more partners in a variety of different social tracking tasks (Kao & Smith, 1971; Rothe, 1973; Sauter & Smith, 1971; K.U. Smith, 1972, 1974; K.U. Smith & Arndt, 1969; K.U. Smith & Henry, 1967; K.U. Smith, Wargo, Jones, & Smith, 1963; T.J. Smith & Smith, 1987a, 1988a; Ting et al., 1972).

The use of a hybrid automated system to mediate social tracking represents a distinct advance in experimental social behavioral research. Through computer control of social feedback and social tracking conditions, the interactive computer is used to both vary and evaluate the role of interface design-related sensory feedback factors in determining variability in social tracking performance, and thereby enables the social dynamics of both interpersonal and human-machine interactions to be experimentally analyzed in a systematic manner.

This type of social systems measurement had never been achieved previously. It demonstrated that computerized methods could be applied successfully to systematic and precise experimental analysis of the integrative interactions in interpersonal and group behavior, in which the synchronism and dynamic properties of motorsensory exchanges among two or more people are measured and their differences or similarities computed in real time. It may be argued that computer-mediated social tracking methods represent an experimental analog of modern hybrid industrial systems, in which the interactive control characteristics of different modes of interpersonal, group, and human-system social performance are analyzed for a given system design configuration.

3.2 Findings from Studies in Social Cybernetics

Findings derived from computer-mediated social tracking research are reviewed in more detail elsewhere (K.U. Smith, 1972, 1974; T.J. Smith, Henning & Smith, 1994, 1995; T.J. Smith & Smith, 1987a, 1988a). Major results of direct relevance to understanding the social cybernetic basis of human interaction with hybrid automated systems, in which system control is shared between the human operator and computer, are summarized as follows:

- *Design Factors and Variability in Social Tracking*
  As with individual performance, the most variability and specialization in social tracking performance is not due to a learning effect, but rather to the effect of the human factors design of the tracking task and interface (Ting, Smith, & Smith, 1972). Design factors of significance that may be specified include: sensory feedback control parameters and conditions; mix of tracking modalities employed; temporal and spatial properties of sensory feedback; and the level and pattern of interpersonal, group, institutional, and/or human-system social relationships.

- *Sensory Feedback Modality and Social Tracking Skill*
  Relative to visual-manual social tracking, accuracy and proficiency are greater for tactile, kinesthetic, and auditory social tracking (Rothe, 1973; Cherry & Sayers, 1956). Social tracking based on nonverbal sensory feedback often is more effective than verbal tracking in promoting social learning and communication (K.U. Smith & Smith, 1973). For all sensory modalities, the accuracy of interactive social tracking is comparable to that of individual tracking except under feedback delay conditions, when the latter is superior to the former at all delay levels. Mutual social interaction entailing reciprocal exchange of sensory feedback (series- or parallel-linked tracking) is more effective than purely imitative tracking.

- *Learning of Social Tracking Skills*

Even with provision of real time feedback of tracking performance, social learning of specific social tracking tasks is highly variable, relatively limited and inconsistent, and unstable (Sauter & Smith, 1971; Kao & Smith, 1971, 1977).

- *Effects of Sensory Feedback Perturbations on Social Tracking*
  Real time temporal delays or spatial displacements in sensory feedback severely degrade the accuracy of social tracking performance (Probasco, 1969; Rothe, 1973; K.U. Smith, 1974; Kao & Smith, 1977; Yates, 1963), just as they do for individual tracking performance. For example, a feedback delay of 0.2 seconds decreases social tracking accuracy by 50 percent. Therefore, as with individual behavior, the integrity of social behavior also relies upon the ability of each partner to effectively control the temporal and spatial qualities of sensory feedback generated during the social tracking process.

- *Social Tracking in Group Interaction*
  Because of the introduction of additional sources of sensory feedback to control, the demands and complexity of social tracking in groups involving more than two people rapidly escalate (K.U. Smith, 1974; K.U. Smith & Smith, 1973).

- *Social Cybernetics of Cognitive Behavior and Communication*
  Interactive social communication, with its pronounced motorsensory feedback control demands, is central to all modes of cognitive behavior and represents the principal determinant of effective learning of cognitive skills (K.U. Smith, 1974; T.J. Smith & Smith, 1988b).

4   The Social Cybernetics of Human-Computer Interaction

The research described above suggests a number of parallels between interpersonal social interaction and human interaction with automated systems. Various observers have noted (Nickerson, 1986) that full understanding of human interaction with automated computerized systems has been hampered by the lack of a comprehensive theoretical framework. To provide such a framework, this section introduces a social cybernetic theory of human-computer interaction (HCI).

The basic assumption upon which this concept is founded is that human interaction with automated systems embodies in many significant respects the reciprocal communicative and performance interchanges which characterize social interaction among humans. The computer represents the first tool in history with the potential, through variable programming, for adaptive, integrated interaction with the human operator across the interface. That is, a computer can be used to generate, sense, and communicate different patterns, levels, and modalities of social feedback with human users in a compliant, interactive manner. Indeed, through choice or necessity depending on the circumstances, today's computer often serves as a surrogate social partner for the operator in both occupational and non-occupational settings.

From this perspective, we believe that hybrid automated systems embody new patterns of human-machine social dynamics and social human factors design not present in earlier machine systems. We therefore suggest that a new social science of interactive human-computer systems is required, based on a theoretical-experimental understanding of HCI as a social cybernetic process (T.J. Smith, Henning & Smith, 1994, 1995; K.U. Smith & Smith, 1988; T.J. Smith & Smith, 1987a, 1988a).

The cybernetic model assumes that when execution of a task is shared between a computer and a human operator, the interaction takes on many of the fundamental characteristics of interpersonal social interaction. For example, if the primary task of a human operator is to prevent damage to a computer-controlled robot by limiting its behavior, the operator functions in much the same way as a supervisor monitoring the performance of an untrained worker executing a hazardous task. That is, the robot may be able to function most of the time without control input from the human operator, but the operator is largely responsible for preventing the robot from causing damage to itself.

A series of important generalizations may be drawn regarding the evaluation of the nature of variability in hybrid system performance using a social cybernetic model. First, individual, social, and organizational behaviors all are organized as closed-loop or cybernetic processes, in which sensory feedback from design factors in the performance environment is actively controlled on an ongoing basis by behavioral mechanisms of sensory feedback control. Consequently, design factors have a pervasive influence on social and organizational behavior, just as they do on individual behavior. Thirdly, guidance of behavior via sensory feedback control invariably suffers whenever sensory feedback from design factors is inadequate, incomplete, or perturbed in some fashion. This observation provides a social cybernetic basis for understanding performance problems encountered during human interaction with computerized systems because, (1) spatial and temporal perturbations in sensory feedback are inherent to the human-machine interface; and (2) interactive social tracking across the interface is compromised due to the deficient capabilities of the computer for both detecting and controlling sensory feedback in relation to the human partner, and lack of design features that would permit humans to track this sensory feedback control if it existed.

5   Social-Relational Tracking

Social cybernetic theory identifies social-relational tracking as a critical component for maintaining social coordination in joint control situations. In human-human social interaction, social-relational tracking involves one or more participants monitoring the status or quality of the ongoing interaction as an important means of determining when and if feedback-controlled coordinate adjustments in task behaviors are needed. Coordinate adjustments in task behavior could occur in any number of forms such as changes in the pace participants perform a joint task or changes in the specific task roles of the participants. It should be noted that social-relational tracking focuses on the success or failure of the relationship process itself, and not on the output or productivity of the social system, and so it is distinct from tracking system responses to control inputs. "Mode error" (Woods et al., 1997) represent one example of a breakdown in social relational tracking in hybrid automated systems. This occurs when operators of an automated system lose track of which mode of automated operation the system currently is functioning in. The inability of the machine system to exhibit even rudimentary social-relational tracking under these circumstances, and thus having no capability to alert the operator, compounds the problem of mode errors by making detection and recovery more difficult. If augmented cognition systems similarly lack the capacity for social-relational tracking, all of this burden must fall on the human operator.

Therefore, some means of incorporating a hybrid form of social-relational tracking represents a key challenge when designing human-machine systems that are to function at the level of proficiency of human-human teams. In the case of augmented cognition systems, hybrid automation that does not include computerized social-relational tracking would be self defeating because full responsibility for monitoring the success of the human-machine relationship would place additional cognitive demands on the human operator.

6   Summary and Conclusions

The premise of this paper is that social cybernetics offers a comprehensive conceptual and empirical framework to address current theoretical shortcomings and guide both research and development of augmented cognition systems. We believe that no other model, theory or paradigm of human-machine performance currently extant offers comparable scientific, methodological and heuristic power for dealing with the complexities and needs of implementing augmented cognition systems. A series of premises and implications regarding augmented cognition based on this cybernetic framework were outlined in Section 2.1, above. We conclude by listing a series of research imperatives for implementing effective augmented cognition systems we believe this analytic approach points to.

- 'Accessing the cognitive state' of the operator (Raley et al., 2004) clearly represents an essential first step if that state is to be augmented. Rather than exploring alternative interpretations of this term, we emphasize that operator/human cognition is a control process whose 'state' essentially boils down to the degree to which the operator is engaging and executing projective behavior for effective control of task demands. The assumption of the information processing model that cognitive performance decrements are linked to a linear sequence of cognitive 'bottlenecks' (IBID) provides a misleading representation of this control process. In a closed-loop system, performance decrements at any particular stage rapidly propagate throughout the entire system, and a dissective approach to system diagnosis does not adequately deal with its integrative properties. Furthermore, the psychophysiological measures used for purposes of system diagnosis, such as EEG, ECG, GSR and so forth, essentially 'access' brain and CNS activation. These measures perhaps may be considered useful for inferring attentional awareness levels, but for purposes of inferring cognitive functioning, their necessity is uncertain (their validity as a probe of cognition remains to be established), and they certainly are not sufficient. In contrast, use of forcing functions has long been recognized in both engineering and physiology as a valid, effective, and well-documented method for probing the functional state and capabilities of a control system. We therefore recommend monitoring transient cognitive responses by operators to intermittently imposed cognitive forcing functions, based on social cybernetic principles and methods, as a first research imperative for achieving the aims of augmented cognition.

- Enhancement of performance represents the ultimate goal of augmented cognition. To achieve this goal, automated interventions featuring various strategies have been proposed such as intelligent interruption, cued memory retrieval, multi-modal input, and autonomous task delegation (IBID). We assume that to actually realize enhanced performance, any intervention strategy that may be deployed should support one fundamental goal—enhancing the capabilities and scope for projective behavior—feedforward cognitive control---on the part of the operator. This is the augmentation baseline against which all intervention strategies should be designed and validated. We therefore recommend that predictive modeling of task performance in different contexts, and use of such models to devise intervention strategies that will support operator projection of task outcomes, constitutes a second imperative for research in this area.
- Third, since both the task environments and the cognitive functioning of the operator will be highly context specific, the third research imperative is to investigate the nature and extent of context specificity of operator performance during operator-system interaction.
- Fourth, since full automation of augmented cognition systems risks instability and loss of control, the fourth research imperative is to incorporate social cybernetic principles and methods in the development of hybrid automated systems to assure that humans retain control over these tools to augment cognition.

7   References

Adolph, E.F. (1982). Physiological integrations in action. *The Physiologist*, 25(2), Supplement, 1-67.

Blakemore, S-J., Frith, C.D., and Wolpert, D.M. (2001). The cerebellum is involved in predicting the sensory consequences of action. *Neuroreport*, 12(9), 1879-1884.

Bopp, M.I. (1997). Team Self-Management of Performance Feedback. Masters Thesis. Storrs, CT: University of Connecticut.

Brown, J.W., and Braver, T.S. (2005). Learned predictions of error likelihood in the anterior cingulate cortex. *Science*, 307(5712), 1118-1121.

Dorneich, M., Whitlow, S., Ververs, P.M., Carciofini, M., and Creaser, J. (2004). Closing the loop of an adaptive system with cognitive state. In *Proceedings of the Human Factors and Ergonomics Society 48th Annual Meeting* (pp. 590-594). Santa Monica, CA: HFES.

Draper, J.V., Kaber, D.B., and Usher, J.M. (1998). Telepresence, *Human Factors*, 40(3), 354-375.

Gibson, J.J. (1979). The ecological approach to visual perception. Boston: Houghton Mifflin.

Hawkins, J., and Blakeslee, S. (2004). On intelligence. New York: Henry Holt.

Jackson, P.L., and Decety, J. (2004). Motor cognition: a new paradigm to study self-other interactions. *Current Opinion Neurobiology*, 14, 259-263.

Kao, H.S., and Smith, K.U. (1971). Social feedback: determination of social learning, *Journal of Nervous and Mental Disease*, 152, 289-297.

Kao, H.S., and Smith, K.U. (1977). Delayed visual feedback in inter-operator social tracking performance. *Psychologia - An International Journal of Psychology in the Orient*, 20(1), 20-27.

Nickerson, R.S. (1986). Using computers. The human factors of information systems. Cambridge, MA: MIT Press.

Raley, C., Stripling, R., Kruse, A., Schmorrow, D., and Patrey, J. (2004). Augmented cognition overview: improving information intake under stress. In *Proceedings: 48th Annual Meeting of the Human Factors & Ergonomics Society* (pp. 1150-1154). Santa Monica, CA: HFES.

Rothe, M. (1973). Social Tracking in Children as a Function of Age. Masters Thesis. Madison, WI: University of Wisconsin - Madison.

Salas, E., and Fiore, S.M. (Eds.) (2004). Team cognition: understanding the factors that drive process and performance. Washington, D.C.: APA Press.

Sauter, S.L., and Smith, K.U. (1971). Social feedback: quantitative division of labor in social interactions. *Journal of Cybernetics*, 1(2): 80-93.

Schaafstal, A., Schraagen, J.M., and van Berlo, M. (2000). Cognitive task analysis and innovation of training: the case of structured troubleshooting. *Human Factors*, 42(1), 75-86.

Sheridan, T.B. (1997), Supervisory control. In G. Salvendy (Ed.), *Handbook of human factors* (2nd ed.)(Chapter 39, pp. 1295-1327). New York: Wiley.

Sheridan, T.B., (2002). Humans and automation. *HFES Issues in Human Factors and Ergonomics Series, Vol. 3.* New York: Wiley.

Sheridan, T.B. (2004). Driver distraction from a control theory perspective. *Human Factors*, 46(4), 587-599.

Smith, K.U. (1962). Delayed sensory feedback and behavior. Philadelphia: Saunders.

Smith, K.U. (1972). Cybernetic psychology. In R.N. Singer (Ed.), *The psychomotor domain* (pp.285-348).New York:

Lea and Febiger.

Smith, K.U. (1973). Physiological and sensory feedback of the motor system: neural metabolic integration for energy regulation in behavior. In: J. Maser (Ed.), *Efferent organization and the integration of behavior* (pp. 19-66). New York: Academic Press.

Smith, K.U. (1974). *Industrial social cybernetics.* Madison, WI: University of Wisconsin Behavioral Cybernetics Laboratory.

Smith, K.U., and Arndt, R. (1969). Experimental hybrid computer automation in perceptual-motor and social behavioral research, *Journal of Motor Behavior,* 1(1), 11-27.

Smith, K.U., and Henry, J.P. (1967). Cybernetic foundations for rehabilitation. *American Journal of Physical Medicine,* 46, 379-467.

Smith, K.U., and Smith, M.F. (1966). Cybernetic principles of learning and educational design. New York: Holt, Rinehart and Winston.

Smith, K.U., and Smith, M.F. (1973). Psychology. An introduction to behavior science. Boston: Little, Brown.

Smith, K.U., and Smith, W.M. (1958). The behavior of man. Introduction to psychology. New York: Holt, Rinehart and Winston.

Smith, K.U., and Smith, W.M. (1962). Perception and motion. An analysis of space-structured behavior. Philadelphia: Saunders.

Smith, K.U., Wargo, L., Jones, R., and Smith, W.M. (1963). Delayed and space-displaced sensory feedback and learning. *Perceptual and Motor Skills,* 16, 781-796.

Smith, T.J., Henning, R.A., and Li, Q. (1998). Teleoperation in space - modeling effects of displaced feedback and microgravity on tracking performance. SAE Technical Report #981701. 28th Intl. Conference on Environmental Systems, Danvers, MA, July 13-16.

Smith, T.J., Henning, R.H., and Smith, K.U. (1994). Sources of performance variability. In G. Salvendy and W. Karwowski (Eds.), *Design of work and development of personnel in advanced manufacturing* (Chapter 11, pp. 273-330). New York: Wiley.

Smith, T.J., Henning, R.A., and Smith, K.U. (1995). Performance of hybrid automated systems -a social cybernetic analysis. *International Journal of Human Factors in Manufacturing,* 5(1), 29-51.

Smith, T.J., and Smith, K.U. (1987a). Feedback-control mechanisms of human behavior. In G. Salvendy (Ed.), *Handbook of human factors* (pp. 251-293). New York: Wiley.

Smith, T.J., and Smith, K.U. (1987b). Motor feedback control of human cognition – implications for the cognitive interface. In G. Salvendy, S.L. Sauter and J.J. Hurrell, Jr. (Eds.), *Social, ergonomic and stress aspects of work with computers* (pp. 239-254). Amsterdam: Elsevier.

Smith, T.J., and Smith, K.U. (1988a). The social cybernetics of human interaction with automated systems. In W. Karwowski, H.R. Parsei, and M.R. Wilhelm (Eds.), *Ergonomics of hybrid automated systems I* (pp. 691-711). Amsterdam: Elsevier.

Smith, T.J., and Smith, K.U. (1988b). The cybernetic basis of human behavior and performance. In Greg and Pat Williams (Eds.), *Continuing the Conversation. A Newsletter of Ideas in Cybernetics. Special Issue on Behavioral Cybernetics* (Number 15, pp. 1-28, Winter). Gravel Switch, KY: HortIdeas.

Snow, R.E., Corno, L., and Jackson, D. III (1996). Individual differences in affective and conative functions. In D.C. Berliner and R.C. Calfee (Eds.), *Handbook of Educational Psychology* (pp. 243-310, with specific reference to pp. 274). New York: Simon & Schuster Macmillan.

Ting, T., Smith, M., and Smith, K.U. (1972). Social feedback factors in rehabilitative processes and learning. *American Journal of Physical Medicine,* 51, 86-101.

Webb, N.M., and Palincsar, A.S. (1996). Group processes in the classroom. In D.C. Berliner and R.C. Calfee (Eds.), *Handbook of Educational Psychology* (pp. 841-873). New York: Simon & Schuster Macmillan.

Wickens, C.D., and Hollands, J.G. (1999). Engineering psychology and human performance (3rd ed.). Upper Saddle River, N.J.: Prentice Hall,

Woods, D.D., Sarter, N., and Billings, C. (1997). Automation surprises. In G. Salvendy (Ed.), *Handbook of human factors* (2nd ed.)(Chapter 57, pp. 1926-1943). New York: Wiley.

Yates, A. (1963). Delayed auditory feedback. *Psychological Bulletin,* 60, 213-251.

Implementation of a Closed-Loop Real-Time EEG-Based Drowsiness Detection System:
Effects of Feedback Alarms on Performance in a Driving Simulator

*Chris Berka[1], Daniel J. Levendowski[1], Philip Westbrook[1], Gene Davis[1], Michelle N. Lumicao[1],
Richard E. Olmstead[2], Miodrag Popovic[3], Vladimir T. Zivkovic[1], Caitlin K. Ramsey[1]*

Abstract

The management of fatigue is increasingly considered a serious public health and safety concern because impaired vigilance is believed to be a primary contributor to transportation and industrial accidents. Military operations are particularly vulnerable to the effects of fatigue due to the irregular nature of mission-related schedules and the stress of combat conditions. The ability to monitor levels of alertness in real-time, coupled with feedback to the operator or a third party, could prevent accidents and save lives. The electroencephalogram (EEG) is widely regarded as the physiological "gold standard" for the assessment of alertness. This study explored the feasibility of an integrated approach that combined real-time quantification of EEG indices and audio feedback alarms to assist fourteen healthy participants in overcoming performance deficits on neurocognitive tests and in a driving simulator task during a sleep deprivation session. As expected, sleep deprivation significantly increased drowsiness as measured by B-Alert® EEG classifications and impairments in neurocognitive tests and driving simulator performance. Timely administration of feedback resulted in increased alertness as measured by changes in EEG indices and performance, particularly during driving simulator task. Most participants reported that the feedback alarms were beneficial in helping them maintain alertness. This suggests that a closed-loop EEG-based system combined with intelligent feedback can improve performance and decrease operator errors resulting from fatigue.

1   Introduction

With the growing demands of the global economy for round-the-clock operations, fatigue management is increasingly important, particularly in safety-sensitive environments such as military operations and commercial transportation (Akerstedt, 1995; Coleman & Dement, 1986; Mitler et al., 1988). Safety, efficiency and productivity are all impacted by employee alertness. Fatigue-related accidents and decreased productivity associated with drowsiness are estimated to cost the U.S. over $77 billion each year and $377 billion worldwide (Moore-Ede, 1993). It is estimated that more U.S. freeway fatalities are caused by fatigue than alcohol or drugs, with 10% to 50% of motor vehicle accidents attributed to sleepiness (Findley, Fabrizio, Thommi, & Suratt, 1989; Horne & Reyner, 1995). In addition, over 30 million Americans are believed to suffer from sleep disorders, the majority undiagnosed and untreated, resulting in dangerous levels of daytime drowsiness (Dement & Vaughan, 1999). As more workers are forced into shift work to meet the demands of a 24-hour society, sleep is often sacrificed for other activities. Although automation is replacing manual labor, it can have a deleterious effect if it causes the operator to be disengaged from the controls of machinery. Passive monitoring of automated equipment can increase the difficulty of maintaining vigilance with performance decrements increasing with time-on-task (Parasuraman, Molloy, & Singh, 1993).

Several studies revealed that people are not good judges of their own level of fatigue. The AAA Foundation for Traffic Safety interviewed 467 drivers involved in police-reported crashes whose physical condition at the time of the crash was identified as either "asleep" or "drowsy". While most drivers agreed with the police officer's assessment of the role of drowsiness in their accident, close to 50% reported feeling either "slightly" or "not at all" drowsy just prior to the crash. Similarly, AAA Foundation research found that 50% of people tested during sleep deprivation were unable to predict whether they would fall asleep within the next two minutes (Itoi et al., 1993). The study concluded that a "sleepiness indicator device" should be developed to inform users prior to sleep onset

[1] Advanced Brain Monitoring, Inc., 2850 Pio Pico Drive, Suite A, Carlsbad, CA USA 92008; chris@b-alert.com
[2] VA Greater Los Angeles Healthcare System, 11301 Wilshire Blvd., Los Angeles, CA USA 90073
[3] University of Belgrade, School of Electrical Engineering, Serbia and Montenegro

The integration of physiological monitoring into the man-machine interface offers the possibility of allocating tasks based on real-time assessment of operator status. Real-time monitoring could drive intelligent feedback or facilitate active intervention by the operator or through a third party (man or machine), increasing safety and productivity (Parasuraman et al., 1993; Parasuraman, Mouloua, & Molloy, 1996). The achievement of such a system is particularly relevant for the development of future military technology where the emphasis is increasingly on unmanned vehicles and aircraft, maximizing capacity while limiting the need for additional human resources. This study was designed to investigate the utility of a method for real-time detection of drowsiness with alarms delivered directly to the user.

The EEG is the physiological "gold standard" for the assessment of alertness (Akerstedt & Kecklund, 1991; Akerstedt, Kecklund, & Knutsson, 1991; O'Hanlon & Beatty, 1977) with subtle shifts from vigilance to drowsiness identified (Akerstedt, Hume, Minors, & Waterhouse, 1993; Akerstedt & Kecklund, 1991; Akerstedt et al., 1991) on a second-by-second basis (Makeig & Jung, 1995, 1996). Torsvall and Akerstedt identified EEG patterns predictive of sleep onset (Torsvall & Akerstedt, 1988). In their research, eye closures occurred too late in the behavioral chain of events to be useful in a drowsiness detection system. Similarly, in a study of sleep-deprived professional drivers using a driving simulator, EEG was a reliable indicator of alertness and showed evidence of fatigue prior to deteriorations in driving performance (*Eye-Activity Measures of Fatigue and Napping as a fatigue Countermeasure*, 1999).

The current study used an integrated hardware and software solution (B-Alert), proven effective in acquiring high quality EEG in operational environments (Berka et al., 2004; Berka et al., 2005). The system was previously developed and validated in a series of sleep deprivation studies (Levendowski et al., 2001; Levendowski, Olmstead, Konstantinovic, Berka, & Westbrook, 2000; Mitler et al., 2002). A preliminary time-series analysis was developed to monitor patterns in the B-Alert indices and facilitate the delivery of alarms to a user when repeated episodes of drowsiness were identified.

2   Methods

2.1   EEG Acquisition

A patented wireless EEG sensor headset was used to acquire six channels, including bipolar recordings from Fz-POz and Cz-POz, unipolar recordings from Fz, Cz and POz referenced to linked mastoids, and a bipolar configuration for horizontal and vertical EOG (see Berka, 2004 for details). A miniaturized battery-powered data acquisition system (weight <4 oz) incorporates all of the electronic components required to acquire, digitize and transmit six channels of EEG. The digitized EEG is sent via radio frequency (RF) transmission, to a monitoring device that can be placed up to 20 feet away from the user. The EEG sensor headset requires no scalp preparation and provides a comfortable and secure sensor-scalp interface for 12 to 24 hours of continuous use.

2.2   EEG Quantification

Quantification of the EEG in real-time, referred to as the B-Alert system, uses signal analysis techniques to identify and decontaminate eye blinks, and identify and reject data points contaminated with muscle activity (EMG), amplifier saturation, and/or excursions due to movement artifacts (see Berka, 2004 for details). Following quantification and decontamination, each one-second EEG epoch is classified into one of four states of alertness: "high vigilance", "low vigilance", "relaxed wakefulness", and "sleep onset." These four states were derived using EEG acquired from participants participating in sleep deprivation studies. The high and low vigilance states were modeled by varying the level of task engagement. Relaxed wakefulness is the state induced when participants are instructed to relax with eyes closed and is generally characterized by predominance of EEG in the alpha frequency band (8 – 12 Hz.). Data for the sleep-onset class were obtained using EEG samples acquired just subsequent to sleep onset. A "drowsy" classification is also calculated by adding the "relaxed wakefulness" and "sleep-onset" classifications. All classifications are obtained using a discriminant function analysis derived from a large normative database and fitted to each individual's unique EEG patterns with data from three baseline conditions.

2.3 Participants

Fourteen healthy participants (8 males, 6 females, mean ± S.D. age = 38.1 ± 11.8) participated in the study. All participants were screened for prescription medications, history of neurological, psychiatric and attention deficit disorders, head trauma, excessive consumption of alcohol, caffeine and nicotine, and abnormal sleep patterns and /or quality of sleep. The Epworth Sleepiness Scale was used to screen individuals with excessive daytime drowsiness.

2.4 Protocols

2.4.1 Participant Monitoring Prior to Study

Participants completed Quality-of-Life logs detailing the quantity and quality of their sleep for the four nights prior to the screening and sleep deprivation studies. Those who averaged less than 7.25 hours of time in bed and/or had a standard deviation across the nights of sleep greater than one hour were excluded from the studies. Wrist actigraphs (Ambulatory Monitoring, Inc.) monitored sleep patterns for the 72 hours leading up to the sleep deprivation study. One participant was eliminated from the study because his actigraph data did not match the sleep logs. When questioned, the participant admitted to dozing during the reported waking hours.

2.4.2 Screening & Sleep Deprivation Sessions

Screening sessions were scheduled to begin at approximately 9:00 a.m. and to be completed prior to 2:00 p.m. to avoid the circadian dip in alertness. Upon arrival at the study site, the technician reviewed the Quality-of-Life logs and consented the participant for the study. The sensor headset was applied and the participant completed a 4-hour test battery. Participants who exhibited any significant drowsiness during the screening session were terminated from the study. Following successful completion of the screening session, participants were fitted with and provided instructions for use of the actigraph during the four nights prior to their sleep deprivation session.

For the majority of participants, the sleep-deprivation session was scheduled to begin on the third night after the screening session. Participants were allowed a full night of sleep during the night after their screening session (Night One). On the night prior to their study (Night 2), participants were instructed to go to bed 1.5 hours after their previously determined average bedtime and sleep for 3 hours. They were required to leave a message on the answering machine every one half hour to confirm that they were awake. Participants arrived for their study approximately 14 hours after awakening (i.e., ~5:30 PM, Night 3) and completed two 4-hour testing batteries.

2.4.3 Drowsiness Detection and Feedback Alarms

Auditory feedback alarms were used during the sleep deprivation to notify the participants as they became drowsy. A randomized crossover design was used to assign participants into two groups: Feedback First or Feedback Second. Seven participants received alerting feedback during the first four hours of the study (Feedback First) and no feedback in the second four-hour battery. Alternatively, seven participants received no alerting feedback during the first four hours (Feedback Second) followed by a four-hour feedback battery. During the Feedback session, the B-Alert classifications for each one-second epoch of data were used to trigger the feedback alarms each time relaxed wakefulness or sleep-onset was classified in three consecutive epochs or in 7+ epochs from the previous twenty seconds. Six unique sounds were selected for the alarms. The feedback sounds became more urgent each time the alarm was triggered and the three most urgent alarms were sounded for two seconds. The alarm sequence was reset to the least innocuous alarm after 20 consecutive seconds without drowsiness.

2.5 Task Description

2.5.1 3-Choice Vigilance Task (3C-VT)

The 3C-VT (Figure 1a) incorporates features of the most common measures of sustained attention, such as the Continuous Performance Test (Rosvold, 1956), the Wilkinson Reaction Time Test (Wilkinson & Houghton, 1975; Wilkinson, 1958), and the Psychomotor Vigilance Test (Dinges, Orne, & Orne, 1985; Powell, 1999). The 3C-VT is easy to perform and relatively insensitive to practice or training effects. Validity was established in previous sleep deprivation studies by correlation with behavioral evidence, visually scored observations (eye closures, head nods), subjective sleepiness, a handheld PVT-192 test, and driving simulator performance (Levendowski et al., 2000; Levendowski et al., 2002; Mitler et al., 2002).

The 3C-VT requires participants to discriminate primary (70%) from two secondary (30%) geometric shapes and respond as quickly as possible over a 20-minute test period. Each shape is presented for 200 milliseconds with inter-stimulus intervals (ISI) ranging from 1.5 to 10 seconds. Each shape is approximately 6 cm as presented on a 17" computer monitor. Performance measures include reaction time and percentage of correct responses. A 1.5 second threshold was used to measure errors of omission. A brief train-to-criterion period is provided prior to the start of the testing period to minimize practice effects. The first 5 minutes of the 3C-VT are used as one of the three baseline conditions to fit the B-Alert classifications to an individual's unique EEG.

2.5.2 Eyes Open and Eyes Closed Tasks

In addition to the first 5 minutes of the 3C-VT, the Eyes Open and Eyes Closed tasks are used as baseline conditions to fit the B-Alert classifications to each individuals unique EEG. Each task is 5 minutes in length, and the participant is instructed to respond to a red circle presented at 2-second intervals (Eyes Open), or a tone sounded at 2 second intervals (Eyes Closed).

2.5.3 Paired Associate Learning and Memory Task (PAL)

The PAL evaluates attention, distractibility, and image recognition memory. During the training session, a group of 20 images is presented twice. The testing session randomly presents the 20 training images interspersed with 80 additional images. Participants are instructed to identify the images in the training set (Figure 1b). Each image is presented for 500 milliseconds during training and 200 milliseconds during the testing with a 2-second ISI. Five equivalent image categories were used in the study including animals, food, household goods, sports and travel. A practice session was provided to ensure participants understood the task. In the Standard Recognition PAL (PAL-S), the participant must memorize 20 images and identify the 20 training images amidst 80 previously unseen testing images. For the Interference PAL (PAL-I), a set of 20 new images must be memorized and distinguished from the first set of training images and 60 images previously displayed in the Standard PAL. In Numbers PAL (PAL-N), a number is assigned to each image and participants identify the correct image-number pairs.

2.5.4 Driving Simulator Task

A 45-minute driving simulator scenario (Figure 1c) was developed for a Systems Technology Inc. (STI, Hawthorne, CA) stimulator with the NT-version software. A large database has been developed including EEG and the driving performance measures reported to be significant in assessing sleep deprivation by ABM and previous investigators.

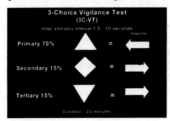 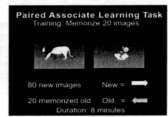 

Figure 1: a. 3-Choice Vigilance Task. b. Paired Associate Learning Task. c. Driving Simulator Task

2.5.5 Verbal Memory Scan (VMS), Verbal Paired Associate Learning (VPA) and Novelty Oddball Task

The VMS is a Sternberg serial probe recognition memory task developed to measure speed and accuracy of verbal working memory (Pineda, Herrera, Kang, & Sandler, 1998). Lists of 5 words (memory sets) are presented, followed by single-word probes (in-set or out-of-set). The Verbal PAL is identical to the image PAL, substituting word pairs for images. The Novelty Oddball task assesses the brain's response to novelty (Courchesne, Hillyard, & Galambos, 1975) with three classes of stimuli: a standard, high-probability event, a low-probability target deviant, and an equally improbable series of unique, unexpected 'novel' events. Data acquired during these tasks will be reported in greater detail in a future publication.

Table 1: Study Protocols for the screening and sleep deprivation sessions

Fully-rested Screening			Sleep Deprivation: Repeated with Feedback on/off		
Paradigm	Time (mins)	Elapsed Time	Paradigm	Time (mins)	Elapsed Time
3C-VT	25	0:25	Eyes Open	5	0:05
Eyes Open	5	0:30	Eyes Closed	5	0:10
Eyes Closed	5	0:35	PAL Standard	10	0:20
Break	10	0:45	PAL Numbers	10	0:30
Verbal Memory	15	1:00	3C-VT	25	0:55
PAL Standard	10	1:10	Break	10	1:05
PAL Interference	10	1:20	Driving Simulator	45	1:50
Break	10	1:30	Break	10	2:00
Verbal PAL	15	1:45	Novelty Oddball	20	2:20
Novelty Oddball	20	2:05	Verbal PAL	15	2:35
PAL Standard	10	2:15	3C-VT	20	2:55
PAL Numbers	10	2:25	Break	10	3:05
Break	10	2:35	Driving Simulator	45	3:50
Driving w/ training	60	3:35	Break	10	4:00

2.6 Data Reduction

The percentage of epochs classified with high vigilance and drowsy was computed for each session. The percent correct and reaction times were computed for each of the 3C-VT, PAL-S, PAL-N, and VMS tests. The B-Alert and performance measures were z-scored across tasks and subjects using comparable data from group of 24 participants during fully rested baseline sessions to compare results across the three conditions: fully rested screening, sleep-deprivation with feedback, and sleep-deprivation with no feedback. These transformations minimized differences across tasks and participants. Z-scores were not computed for the driving simulator task, because there were no comparable data available. Statistical analyses were performed with a repeated measures analysis of variance (ANOVA, SPSS Release 8.0) across the 3 conditions.

Algorithms were developed to derive performance variables for the driving simulator. These performance variables included: the length of time the driver was drifting, veering or outside the lane (drifts and veers), and the combined number of collisions, off-road accidents, and hit pedestrians (accidents). To address variability of driving performance within and between participants particularly during the screening session (possibly due to a learning curve), the driving performance measures obtained for each of the five driving sessions were averaged for each participant. The performance measures were then subtracted from the average to normalize the data prior to analysis. Eleven of the 14 participants had full sets of driving simulator data and were used to compare performance across the feedback and no-feedback conditions. Statistical analyses were performed using a 2 (feedback group) X 4 (driving session) repeated measures ANOVA.

3 Results

3.1 B-Alert EEG Indices

There was a significant increase in the z-scored percentage of drowsy classifications across tasks and participants between screening and no feedback ($F = 7.316$, $p < .02$) and between screening and feedback ($F = 6.391$, $p < .05$) conditions and a marginally significant decrease in drowsiness when the feedback session was compared to no feedback session ($F = 4.225$, $p = .06$) (Figure 2a). A trend towards increasing high vigilance classifications during the feedback session when compared to the no feedback session was also observed (Figure 2b).

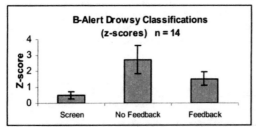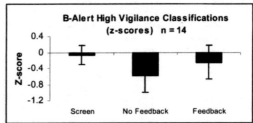

Figure 2: Z-Scored mean ± SE a. drowsy classifications, and b. high vigilance during the three conditions.

To investigate the differential effects of delivering the alerting auditory feedback in the first or second half of the study, the EEG data were computed separately for the two feedback delivery groups. Participants in the Feedback First group evidenced more High Vigilance and less Drowsy classifications during the feedback session (Figure 3b). Once the feedback was turned off, the High Vigilance decreased with a corresponding increase in Drowsy classifications. In contrast, the Feedback Second participants evidenced relatively constant levels of High Vigilance and Drowsy throughout the feedback off and feedback on sessions, suggesting that the feedback prevented the anticipated trend towards increasing drowsiness during the course of the sleep deprivation study (Figure 3c).

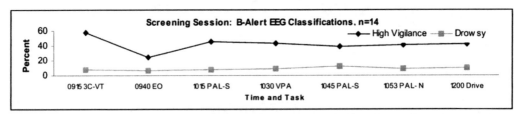

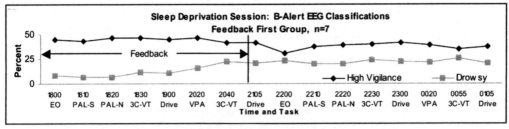

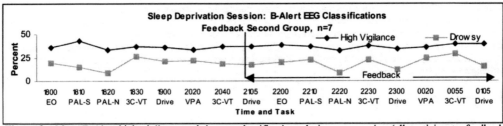

Figure 3: Mean B-Alert high vigilance and drowsy classifications during a. screening (all participants, feedback first and feedback second), b. sleep deprivation feedback first, and c. sleep deprivation feedback second.

The EEG data from the screening session confirm results from previous studies (Levendowski et al., 2001; Levendowski et al., 2000). The 3C-VT, PAL and other memory tests generally elicit high vigilance classification levels of 40 – 70 and drowsy classifications under 10 percent. For fully rested healthy participants, 10 percent drowsy classifications reflect spontaneous bursts of alpha activity during the task post-response periods.

High vigilance reflects the level of task difficulty and the requisite levels of task engagement. A recent study employing a highly demanding naval warship commander simulation task induced levels of B-Alert high vigilance greater than 95 percent for some participants (Berka et al., 2004). In that study, the percentage of epochs classified as high vigilance increased linearly as a function of increasing task workload.

3.2 Performance Measures

A significant increase in the z-scored reaction times across participants and tasks ($F = 5.579$, $p < .05$) was observed in the comparison between the screening session and no-feedback sessions, as a result of sleep deprivation (Figure 4a). Reaction times during the feedback session were not statistically different from the screening session, suggesting that the feedback was effective in preventing the slowing in reaction time related to drowsiness. The z-scored data trended towards increased accuracy during the feedback condition, but the variability prevented statistical significance (Figure 4b).

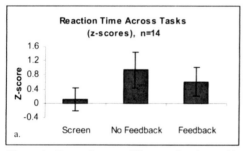
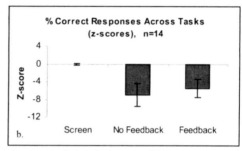

Figure 4: Z-scored mean $\pm$ SE a. reaction time and b. correct responses during the three conditions.

3.3 Driving Performance

There was a significant main effect across the four driving sessions for drifts and veers ($F = 4.259$, $p < .05$) with a significant interaction for feedback group X session ($F = 3.054$, $p < .05$) (Figure 5a). The data trended towards improvements in driving performance as a result of the feedback, but statistical significance was only achieved for the final driving session when feedback was most effective in alerting the participants. The total number of accidents trended towards improvements as the result of feedback, but did not achieve statistical significance. Driving parameters showed greater within- and between-participant variability than other tasks, partially due to the steep learning curve for driving performance.

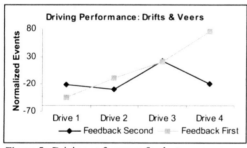
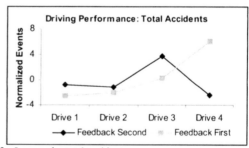

Figure 5: Driving performance for the two groups: a. drifts & veers, b. total accidents.

3.4 Individual Differences

One of the confounding factors in the study was the large variability in the feedback and no-feedback conditions attributed to individual differences in vulnerability to sleep deprivation. Although this factor was identified in the investigators' previous research on sleep deprivation (Levendowski et al., 2000; Mitler et al., 2002), time and budgetary constraints did not allow for studying each participant for two nights of sleep deprivation. It became apparent after the start of the study that quantifying the benefits of feedback was closely linked with how well matched the fatigue levels were during comparable tasks with and without feedback.

3C-VT B-Alert drowsy and reaction time data from two participants are presented in Figure 6 to illustrate these differences. Participant #1 (Figure 6a) was typical of the group most vulnerable to sleep deprivation, whereas participant #8 (Figure 6b) could be classified as a member of the group least vulnerable to sleep deprivation.

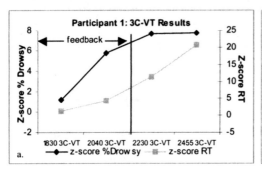 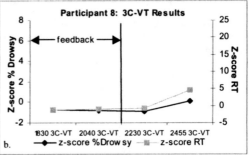

Figure 6: 3C-VT z-scored B-Alert percentage drowsy and reaction time for a. participant 1, and b. participant 8.

4   Conclusions

The feedback alarms, triggered by EEG indices, proved effective in either stabilizing or improving objective measures of performance during extended sleep deprivation. Most participants reported that the feedback alarms were beneficial in helping them maintain alertness.

The high vigilance and drowsy B-Alert parameters clearly evidence the effectiveness of the feedback in assisting participants in maintaining an alert state (see Figures 2, 3, and 4). These data support previous results obtained with the B-Alert indices obtained during studies of fully and partially sleep-deprived healthy participants. The investigators have observed that the relationship between the percentage of epochs classified as High Vigilance and Drowsy can be predictive of future performance degradations in response to increasing levels of fatigue.

The robustness of the feedback alarm effectiveness in preventing performance impairments as a result of sleep deprivation was overshadowed by individual differences in susceptibility to the effects of sleep deprivation. A sub-group of participants never achieved sufficient levels of drowsiness during the sleep deprivation session to evidence changes in the EEG or performance or to fully realize the beneficial effects of the feedback alarms. On the opposite end of the continuum, some participants showed signs of extreme drowsiness very early in the experimental session. Random assignment to the feedback first or no-feedback first group did not ensure that very drowsy participants would receive feedback early enough to significantly impact performance.

The presence of individual differences in susceptibility to sleep deprivation was not unanticipated. A growing number of studies, including several conducted by the investigators, have now shown that individuals differ in their vulnerability to the effects of sleep deprivation (Balkin, 2001; Doran, Van Dongen, & Dinges, 2001); (Baynard et al., 2003; Belenky, Bliese, Wesensten, & Balkin, 2003; Drummond, Salamat, Brown, Dinges, & Gillin, 2003; Leproult et al., 2002). These studies have suggested the possible existence of three groups of participants in the population: those extremely (Stern, 2001) vulnerable to the effects of sleep deprivation, a moderately vulnerable group and a group relatively invulnerable to the effects even after 44 hours of sleep deprivation (Doran et al., 2001); (Stern, 2001). This stratification is a generalization based on small sample sizes and the possibility exists that the level of vulnerability is actually normally distributed across the population.

These differences, however, present a potential confound for any sleep deprivation study designed to assess between-participant changes in performance and alertness over time. An optimal study design would include representation from all three groups, anticipating that the individual differences may override or at minimum interact with circadian and homeostatic effects. In addition, in a drowsiness feedback alarm study, it is likely that the most vulnerable group would evidence earlier habituation to the alarms.

Clearly, these data suggest that a viable drowsiness detection device must have the capability to accommodate individual differences and make accurate predictions regarding performance deterioration as a function of changes in alertness. It is possible that for those individuals who are most vulnerable to the effects of sleep deprivation, audio alarms will not be sufficient countermeasures to drowsiness. The model must be capable of identifying the point at which intervention beyond feedback alarms (e.g. naps, stimulants or third party interventions) are required.

Acknowledgements

This research was supported by NIH NINDS grant R43-NS35387 and NIMH grant R44-MH064344.

References

Akerstedt, T. (1995). Work hours and sleepiness. *Neurophysiol Clin., 25*(6), 367-375. Review.

Akerstedt, T., Hume, K., Minors, D., & Waterhouse, J. (1993). Regulation of sleep and naps on an irregular schedule. *Sleep, 16*(8), 736-743.

Akerstedt, T., & Kecklund, G. (1991). Stability of day and night sleep--a two-year follow-up of EEG parameters in three-shift workers. *Sleep, 14*(6), 507-510.

Akerstedt, T., Kecklund, G., & Knutsson, A. (1991). Manifest sleepiness and the spectral content of the EEG during shift work. *Sleep, 14*(3), 221-225.

Balkin, T. J. (2001, August 21-23). *Sleep Deprivation Research at WRAIR*. Paper presented at the DARPA Workshop, Las Vegas, NV.

Baynard, M. D., Maislin, G., Moest., E. I., Ballas, C., Dinges, D. F., & Van Dongen, H. (2003). Inter-Individual Differences in Psychomotor Vigilance Performance Deficits during Repeated Exposure to Sleep Deprivation. *Sleep, Vol. 26*(Abstract Supplement).

Belenky, G. L., Bliese, P. D., Wesensten, N. J., & Balkin, T. J. (2003). Variation in Sensitivity to Sleep Restriction as a Function of Age Revealed by Growth Modeling Analysis. *Sleep, Vol. 26*(Abstract Supplement).

Berka, C., Levendowski, D. J., Olmstead, R. E., Popovic, M. V., Cvetinovic, M., Petrovic, M. M., Davis, G., Lumicao, M. N., & Westbrook, P. (2004). Real-time Analysis of EEG Indices of Alertness, Cognition and Memory with a Wireless EEG Headset. *International Journal of Human-Computer Interaction, 17*(2), 151-170.

Berka, C., Levendowski, D. J., Ramsey, C. K., Davis, G., Lumicao, M. N., Stanney, K., Reeves, L., Regli, S. H., Tremoulet, P. D., & Stibler, K. (2005). *Evaluation of an EEG-Workload Model in an Aegis Simulation Environment*. Paper presented at the Proceedings of the SPIE Defense and Security Symposium, Biomonitoring for physiological and cognitive performance during military operations.

Coleman, R. M., & Dement, W. C. (1986). Falling asleep at work: a problem for continuous operations. *Sleep research, 15*, 265.

Courchesne, E., Hillyard, S. A., & Galambos, R. (1975). Stimulus novelty, task relevance and the visual evoked potential in man. *Electroencephalogr Clin Neurophysiol, 39*(2), 131-143.

Dement, W. C., & Vaughan, C. C. (1999). *The promise of sleep: A pioneer in sleep medicine explores the vital connection between health, happiness, and a good night's sleep*. New York: Delacorte Press.

Dinges, D., Orne, M., & Orne, E. (1985). Sleep Depth and Other Factors associated with performance upon abrupt awakening. *Sleep research, 14*, 92.

Doran, S. M., Van Dongen, H. P., & Dinges, D. F. (2001). Sustained attention performance during sleep deprivation: evidence of state instability. *Arch Ital Biol, 139*(3), 253-267.

Drummond, S., Salamat, J. S., Brown, G. G., Dinges, D. F., & Gillin, J. C. (2003). Brain Regions Underlying Differential PVT Performance. *SLEEP, Vol. 26*(Abstract Supplement).

Eye-Activity Measures of Fatigue and Napping as a fatigue Countermeasure. (FHWA-RD)(1999). Alexandria, VA: Trucking Research Institute.

Findley, L. J., Fabrizio, M., Thommi, G., & Suratt, P. M. (1989). Severity of sleep apnea and automobile crashes. *N Engl J Med., 320*(13), 868-869.

Horne, J. A., & Reyner, L. A. (1995). *Falling asleep at the wheel*: Report for UK Department of Transport.

Itoi, A., Cilveti, R., Voth, M., Dantz, B., Hyde, P., Gupta, A., & Dement, W. M. D. (1993). *Relationship Between Awareness of Sleepiness and Ability to Predict Sleep Onset.* Stanford, CA: AAA Foundation for Traffic Safety.

Leproult, R., Colecchia, E. F., Berardi, A. M., Stickgold, R., Kosslyn, S. M., & Van Cauter, E. (2002). Individual differences in subjective and objective alertness during sleep deprivation are stable and unrelated. *Am J Physiol Regul Integr Comp Physiol, 284*(2), R280-290.

Levendowski, D. J., Berka, C., Olmstead, R. E., Konstantinovic, Z. R., Davis, G., Lumicao, M. N., & Westbrook, P. (2001). Electroencephalographic indices predict future vulnerability to fatigue induced by sleep deprivation. *Sleep, 24*(Abstract Supplement), A243-A244.

Levendowski, D. J., Olmstead, R. E., Konstantinovic, Z. R., Berka, C., & Westbrook, P. (2000). Detection of Electroencephalographic Indices of Drowsiness in Realtime using a Multi-Level Discriminant Function Analysis. *Sleep, 23*(Abstract Supplement #2), A243-A244.

Levendowski, D. J., Westbrook, P., Berka, C., Popovic, M. V., Ensign, W. Y., Pineda, J. A., Zavora, T. M., Lumicao, M. N., & Zivkovic, V. T. (2002). Event-related potentials during a psychomotor vigilance task in sleep apnea patients and healthy subjects. *Sleep, 25*(Abstract Supplement), A462-A463.

Makeig, S., & Jung, T. P. (1995). Changes in alertness are a principal component of variance in the EEG spectrum. *Neuroreport, 7*(1), 213-216.

Makeig, S., & Jung, T. P. (1996). Tonic, phasic, and transient EEG correlates of auditory awareness in drowsiness. *Brain Res Cogn Brain Res, 4*(1), 15-25.

Mitler, M. M., Carskadon, M. A., Czeisler, C. A., Dement, W. C., Dinges, D. F., & Graeber, R. C. (1988). Catastrophes, sleep, and public policy: consensus report. *Sleep, 11*(1), 100-109.

Mitler, M. M., Westbrook, P., Levendowski, D. J., Ensign, W. Y., Olmstead, R. E., Berka, C., Davis, G., Lumicao, M. N., Cvetinovic, M., & Petrovic, M. M. (2002). Validation of automated EEG quantification of alertness: methods for early identification of individuals most susceptible to sleep deprivation. *Sleep, 25*(Abstract Supplement), A147-A148.

Moore-Ede, M. C. (1993). *The twenty-four-hour society : understanding human limits in a world that never stops.* Reading, Mass.: Addison-Welsey.

O'Hanlon, J. F., & Beatty, J. (1977). *Concurrence of electroencephalographic and performance changes during a simulated radar watch and some implications for the arousal theory of vigilance. pp. 189-201.* New York: Plenum Press.

Parasuraman, R., Molloy, R., & Singh, I. L. (1993). Performance consequences of automation induced "complacency". *International Journal of Aviation Psychology, 3*, 1-23.

Parasuraman, R., Mouloua, M., & Molloy, R. (1996). Effects of adaptive task allocation on monitoring of automated systems. *Hum Factors, 38*(4), 665-679.

Pineda, J. A., Herrera, C., Kang, C., & Sandler, A. (1998). Effects of cigarette smoking and 12-h abstention on working memory during a serial-probe recognition task. *Psychopharmacology (Berl), 139*(4), 311-321.

Powell, J. W. (1999). *PVT-192 and Analysis Software Reference Manual.* Philadelphia: University of Pennsylvania Trustees.

Rosvold, H. E. (1956). A continuous performance test of brain damage. *Journal of consulting psychology, 20*, 343-350.

Stern, Y. (2001, August 21-23 2001). *fMRI Approaches to Enhancing Performance During Sleep Deprivation.* Paper presented at the Continuous Assisted Performance Teaming Workshop, DARPA, Las Vegas, NV.

Torsvall, L., & Akerstedt, T. (1988). Extreme sleepiness: quantification of EOG and spectral EEG parameters. *Int J Neurosci., 38*(3-4), 435-441.

Wilkinson, R., & Houghton, D. (1975). Portable four-choice reaction time test with magnetic tape memory. *Behaviour Research Methods and Instrumentation., 7*, 441-446.

Wilkinson, R. T. (1958). Lack of sleep and performance. *Bulletin of the British Psychological Society, 34*, 5A-6A.

The challenges of designing an intelligent companion

Kempen, M., Viezzer, M., Bisson, P. & Nieuwenhuis, C. H. M.

Delft Co-operation in Intelligent Systems Laboratory – Thales Research and Technology Nederland
P.O. Box 90 – 2600 AB Delft – The Netherlands
{Pascal.Bisson, Manuela.Viezzer, Kees.Nieuwenhuis}@decis.nl; Masja.Kempen@icis.decis.nl

Abstract

In this paper we discuss how the design of an Intelligent Companion constitutes a challenge and a test-bed for computer-based technologies aimed at improving the user's cognitive abilities. We conceive an intelligent companion to be an autonomous cognitive system (ACS) that should be capable of naturally interacting and communicating in real-world environments. It should do so by embodying learning of physically grounded conceptualizations of multimodal perception, decision making, planning and actuation, with the aim of supporting human cognition in both an intelligent and *intelligible* way.

In order to arrive at a proper design of our intelligent companion we start with building and analysing a representative scenario for an intelligent companion within the areas of Education/Entertainment. In the light of such requirement analyses, we discuss three desirable abilities of the companion:
- formation of concepts (concrete as well as abstract concepts, such as emotions and moods);
- reasoning and learning about emotions;
- multimodal communication for a natural (non-disturbing) social interaction.

We argue that these abilities are needed in order to enable the companion to enter a partially grounded linguistic interaction with its user(s). We then present a system-level approach to the design of the cognitive architecture of the companion. This architecture will integrate perceptive (vision, speech, sound, emotions), cognitive (learning, reasoning, knowledge bases) and motor modules.

As a concrete example, we discuss the potential use of an Intelligent companion to assist children in discovering their world. The companion can adapt to a child's cognitive state, first attune to and subsequently challenge the child's cognitive abilities to ever higher levels, in order to support and potentially speed-up the child's development process. The cognitive abilities mentioned above are pre-requisites for the companion to enter a partially grounded linguistic interplay with the child. However, in order to keep up and be ahead of the child we also claim that the system needs to regularly and gradually enlarge and refine its initially built-in conceptual space, and make use of learning mechanisms to evolve and select adequate dialogue schemes.

1   Introduction

The last decade has seen an increasing number of robots aimed at entertainment, opening big challenges in the design of a robot's behaviour, especially concerning multimodal interaction and communication with humans. When robots move out of the manufacturing environment and the research lab, what role or roles will they play? What cognitive and social skills are desirable and/or necessary? We will discuss the design of a robot companion aimed at supporting and eventually enhancing its user's cognitive abilities in both an intelligent and intelligible way.

Human-like robots aimed at companionship are currently very popular in Japan (see for example the Asimo robot, from Honda Motor Inc.) But, should a robot look like a human? One drawback for trying to build a robotic version of a human is that it will constrain the machine to only be able to do what we can. On the other hand, a human-like robot might be socially more acceptable and more likely to be able to interact naturally, as for example with children. This is so because initial contact often relies on anthropomorphic attribution and the novelty effect. When the interaction with robots is repeated and/or long-term, however, problems can arise: the novelty effect wears out quickly, and anthropomorphism might raise false expectations on cognitive and social abilities. An interesting study in this light is one where experimenters investigated whether 10-month-old infants expected adults to talk to a humanoid robot (Aritaa, Hirakia, Kandab & Ishiguro, 2005). The results showed that infants interpret only the interactive humanoid robot as a communicative agent, and that infants characterize non-interactive humanoid robots as objects. A necessary requirement for infants to attribute a mentality to robots is for the human and robot to interact. This suggests that the social interactive ability is more crucial than the humanoid shape to set expectations about the possible internal mental states of the robot. Social competence is especially important for peer-to-peer human-robot interaction, where the robot interacts with the human as a true team member. This contrasts with the abundance of human-robot interaction related work in the literature, which focuses on human robot interaction in a "master-slave" arrangement where the robot takes over jobs from the human, assists the human, or reacts to the human. We refer to (Fong, Nourbakhsh & Dautenhahn, 2003) for a state of the art of socially interactive robots, where the *social* aspect of human-robot interaction is the key.

In terms of control, it seems that there is a continuum of design choices from robots that are under our total direct control via well-designed interfaces (see Shneiderman, 1997) to widely autonomous robots that can be trained (or domesticated, following the dog's metaphor in Dautenhahn, 2004; see also Maes, 1997). We claim that a need for robustness, flexibility and knowledge wideness will put our robot towards the autonomous end. Thus, the robot must be able to learn and develop, and the nature/nurture trade-offs must be examined (what has to be innate versus what has or can be learned, see Sloman & Chappell, 2005).

One example where robots exhibit learning behaviour is from (Breazeal, Hoffman & Lockerd, 2004). They approach human tutelage and collaboration with robots in terms of dialogue and goal-driven joint intention based framework. Another example relating to the grounded acquisition of concepts and communication is research stemming from developmental robotics. Steels' group investigate whether the process of social learning can bootstrap communication. For example, a robot dog was used in experiments to see how language-like communication in robots might be bootstrapped by interaction with a human mediator (Steels & Kaplan, 2001). In a previous set of experiments, the Talking Heads project, Steels and his team demonstrated how agents could self-organize a shared lexicon as a side effect of their interactions. Further, the project studied how a conceptualization can be related to an utterance and how this can result in the self-organization of lexical and ontological constructs that explain meaning and relationships.

From the field of language acquisition, we find the hypothesis that children acquire language through social imitative behaviour. The hypothesis is backed up by recent constructivist views on the problem of language acquisition: the child copies what the caregivers say (see Tomasello, 2002). This is in line with children imitating caregiver's and siblings behaviours, and thus learning how to walk, talk and behave. This social imitative behaviour principle could be applied in the design of intelligent companions.

The organisation of this paper is as follows. We start by presenting the goals that we have in mind for our intelligent companion. Then, in Section 3 we present a short scenario which illustrates the way the intelligent companion and the child could interact. This is followed by a discussion of the desired cognitive abilities our intelligent companion ought to have. In Section 4 we outline of provisional architectural framework for our system. We conclude our paper with a short note on how we envision the assessment of interaction with such an intelligent companion, and also raise some ethical issues which need to be considered.

2 System Goals

The long term motivation for our research is the desire to develop intelligent companions for domestic/education/entertainment purposes that could exhibit some human-like cognitive abilities (e.g. adaptiveness to the interaction context, adaptiveness to the user) and thus gain in acceptance. In the education/entertainment field, a robot companion can be used to trigger the development of a new generation of interactive toys that enhance the cognitive potential of the child, such as a companion baby-doll able to react and behave as a young child, but also able to help/accompany the child in his/her development. Such a companion is an educative toy with a great amount of knowledge and functionalities, able to adapt its behaviour to that of the child. More precisely, we envisage a system with the following high-level abilities:
- Evaluate the knowledge and the emotional state of the child in order to adapt its interaction with the child and its educative "program" (i.e. leveraging assistance to the child development process)
- Help the child to evolve, both at the knowledge level and in the acquisition of her own natural potential.

- Give parents an indirect feedback on the development of the child and her inclination (such a feedback is achieved through a report on the behaviour of the toy and an analysis of the interactions between the toy and the child).

The achievement of such goals involves thinking of intelligent robotic companions as highly complex systems consisting of many different components (many different embodiment modules, such as different modalities, and many different cognition modules, from scene interpreters to dialog managers to learning components) that have to interact with each other in a meaningful way, therefore the issue of the system architecture is really a crucial one to be addressed. We propose to focus on three main topics that involve the integration of different components, namely concept formation, reasoning and learning about emotions, and multimodal communication.

Concept formation: Relying on the fact that the companion will be able to interact with the child assuming different roles (playmate, mentor or pupil, according to the situation and the game being played), we propose as a challenge to investigate and demonstrate the ability to transfer conceptual knowledge both from the companion to the child and from the child to the companion. As a playmate, at the same cognitive level as the child, the companion should be able to learn from the child, both by observing what the child is doing or how the child is reacting (to acquire for example abstract concepts such as moods and emotions), and by means of linguistic interaction. As a coach, at a cognitive level higher than the child, the companion should be able to teach the child (here the interaction will be mainly verbal), to monitor the child's progress and ultimately report to child's parents.

Reasoning and learning about emotions: the ability to deal with emotions should be addressed both at the system level, by providing the companion with the ability to evaluate the emotional atmosphere of its surroundings, and at the level of personal reasoning, by taking into account how the emotional state affects the deliberation process of the companion itself, and by giving the companion the ability to display its emotional state. Awareness of the emotional atmosphere is among the constraints that we envisage for the companion to be able to adapt to the context of interaction and to its playmate. A further challenge to achieve adaptive abilities concerns the coupling of the emotion module with the mechanisms of concept formation (the system gradually enlarges and refines an initially built-in conceptual space) and other learning processes leading to more adequate dialogue schemes. The challenge is in line with the proposal of (Dautenhahn, 2004) to develop *personalised* robot companions, where the robot is both individualized, that is no two robots will be the same, and personalised, that is its individuality reflects the needs and requirements of the environment where it is operating. Personalisation is deemed necessary because of human nature: people have individual needs, likes and dislikes, etc. that a companion would have to adapt to. What is argued is that, rather than relying on an inbuilt fixed repertoire of social behaviours, a robot should be able to learn and adapt to the social 'etiquette' of the people it is living with.

Multimodal communication for natural interaction: our final proposed challenge is the integration of multi-modal knowledge data for language-based interaction. Natural interaction seems to be one of the pitfalls of social robots, as discussed with respect to the BIRON platform, supported by the European Integrated Project Cogniron, where issues for designing robot companions are researched. BIRON functions as a case study (Li, Kleinehagenbrock, Fritsch, Wrede & Sagerer, 2004). The scenario of use is a home-tour where a user is supposed to show BIRON around the home. Functional capabilities constitute individuate and attend to the interacting user, follow the user, learn about new objects. The results of the evaluation show that "a robot has to reach a certain level of verbal competence before it will be accepted as a social communication partner and before its functional capabilities will be perceived as interesting and useful. [...] It turned out that the most interesting features for users were the natural language interface and the person attention behaviour. The more task-oriented function – the following behaviour and the object learning ability – received less positive feedback. [...] The most frequently named dissatisfactions concerned errors with the automatic speech recognition system. Wishes for a more flexible dialog and a more stable system were the only other significant dimensions". Given the availability of different sensor modalities (such as vision, sound and speech) on the robot platform, there should be a focused research effort to integrate the information provided by the different modalities in order to improve both language understanding and language generation.

3 Cognitive Technology in use and desired cognitive abilities

Although primarily targeted to the support and possibly reinforcement of individual cognitive abilities, our ultimate goal through this research is to further advance research on Cognitive Systems by proposing an innovative cognitive architecture enabling us to investigate how emotions can be integrated at the level of cognition in order to first drive autonomous behaviour and second achieve social interaction with the person to whom they are attached to. The development of such "companion" cognitive systems may benefit a broad range of application domains (e.g Healthcare: a nurse companion, Transport: a truck driver companion, etc.) For all these applications it is important to remember that it is necessary for the intelligent companion to alternate between the roles of teacher, playmate, social companion according to the situation at hand.

Since the integration of many cognitive functions represents a major challenge, it is imperative to decrease the total complexity and to state reachable objectives by defining precise scenarios that will describe the functionalities, context of use, and potential cognitive capabilities of the envisaged companion. In this paper we will address the areas of the Education/Entertainment, with a particular attention for robot-child interaction. In the next section we give a vision of an intelligent companion in the role of a cognitive baby-doll. We will illustrate a couple of possible robot-child interactions. In section 3.2 we will list the cognitive abilities the companion needs in order for such interactions to be feasible.

3.1 A Cognitive Baby-Doll Scenario

Robot-child interaction is worth exploring for several reasons. At the level of the overall interaction, children may be more willing to attribute human-like characteristics to the robot even in presence of obvious limitations on the side of the robot, thus judging positively the interaction. As mentioned in the introduction, this is reported for babies as young as 10 months old (Aritaa et al, 2005). For older children (from 2 to 6 years) we think this holds true as well, due to the child's interest for symbolic and imaginative games. In this age-range, some children have a period where they have a strong tendency to attribute life and lifelike qualities to inanimate objects (animism) and/or like to play games where an imagined situation is projected outside onto real objects. Imagine a child playing an imaginative game involving a horse and a tiger. The horse may be in fact the child's chair, a puppet may become a real tiger, and so on. The companion will need to understand that the object the child is referring to is in fact a chair, but for the purpose of the game, now is called a horse.

Certain limitations of the robot could even be used to give rise to 'funny' situations: for example if the robot misunderstands a word or a phrase (the child says "hide and seek" and the robot understands it as "I am sick"), the process of disambiguating might involve some funny language game. What is important here is to ensure that the interaction will continue after the misunderstanding, and that the child's curiosity is maintained alert. Children are easy to fool but they can easily be bored too, so we need an assessment of what kind of repetitive situations are judged boring by children, depending on their age, to find technical ways for preventing or at least limiting them.

Children go through certain developmental stages or phases, the cognitive baby-doll should be able to acknowledge and adapt to these phenomena so that it can interact naturally with the child according to his or her current cognitive level, and perhaps even help the child reach the next stage. Two examples follow:
- Children between 2 and 6 years old display *transductive reasoning*, where children conclude that events that occur next to each other cause each other. This mode of reasoning should be acknowledged by the companion. It should, for example, understand that 'cause' and 'effect' should take place next to each other in time and space, in order for the child to learn causal relationships.
- At the linguistic level, children go to through phases of so called *overextension*, where e.g. at a certain point in time the child's mother is called 'mummy', the child's father 'daddy'. This can be followed by a phase where suddenly *all* adults are referred to as 'mummy'. The companion will need to adapt to these phases of the child's language development, and know that when the child is talking about 'mummy', but looking towards the father, the companion should understand that the child is actually talking to the father and consequently refer to that person as 'daddy', if it is going to enter the conversation.

At the level of the linguistic interaction, the fact that the linguistic interaction with children may be more constrained and less 'open' due to their limited linguistic abilities, can be exploited so that the task to develop a flexible dialogue manager aimed at children may be simpler than the task of developing an analogous module aimed at adults. There are however also some disadvantages that we turn in the context of our research into the following research challenges:

- The need to develop a robust speech recognition system for children's voices;
- The need to have an adequate knowledge of the linguistic abilities at different ages, to adapt the dialogue manager consequently;
- The need to initiate and maintain a (linguistic) two-way interaction with the child;
- The need for the robot to pro-actively select (possibly remotely) information/applications for the sake of achieving its main goal (e.g. for educational purposes, and education select the form of games that can be played with the child robot (from some sort of scribble or linguistic puzzles to story telling and singing) in order to reach some education/entertainment state;
- The need to learn from the child but also the capacity to teach the child. With our intelligent companion we aim towards a two-way communication/interaction behavioural repertoire where the robot can also actually teach the child something new. This means that the companion needs learning skills so that it can broaden its behaviour-base over time, as it interacts with its human companion. The robot learns from the human and the human from the robot companion;
- The need to evolve over the time according to situation and child's progress.

There is a great potential in the kind of games that the companion will play, involving questions and repetitions, and this will allow it to position itself in relation with the child (in the same range of age and in the same range of knowledge). The aim is to elaborate corrective dialogues that will help the child to progressively evolve, in a natural way. For example, we envisage an interactive dialogue game with the companion to figure out the child's age, so that the companion knows how to talk to the child (vocabulary wise) or how he can help move the child to the next stage of linguistic development. Of course, it is important that the natural language processing mechanisms in the robot can be initialised, because the child's utterances will have certain characteristic phonetic, syntactic and semantic properties/forms. Another example involves referring and pointing to objects, such as pointing to objects while saying the object name. This game can be exploited both for the companion to learn new objects and for the companion to teach object names to the child. This requires an ability to point to and to recognise what child is pointing to, and also an ability to guide the child's attention (e.g. by means of joint attention, gaze direction, etc.).

3.2   Desired Cognitive Abilities

In producing a list of challenging and desired cognitive abilities we would like the companion to have, we first take a look at the field of social robotics, to see what abilities are needed for our companion to behave socially. We then continue with some specific requirements for our envisioned cognitive baby-doll.

Social robots are embodied agents that take part in an enlarged society of humans and robots. (Breazeal, 2003) defines four classes of social robots, ranging from socially evocative robots that rely on the human tendency to anthropomorphise, to sociable robots endowed with deep models of social cognition and able to pro-actively engage with humans. The companion that we envisage is situated towards the sociable end, because of the different roles that it will play in the interaction with the child and because of its individuality (the robot will have a personality with internal drives triggering the interaction). According to (Dautenhahn and Billard, 1999) social robots are able to recognise each other and engage in social interactions, they possess histories, and they explicitly communicate with and learn from each other. The following features and capabilities seem to be needed: embodiment, emotion, dialogue, personality, human-oriented perception, user model, social learning, and intentionality. Abilities, that will put our companion in a trend towards a true socially intelligent robot according to (Fong et al, 2003) are:

- *Model formation abilities:* The robot builds a partial model of the situation it is in, distinguishing between other social agents and objects in the environment. It is also aware of certain human interactional structures, such as turn-taking, to smooth the interaction.
- *Learning abilities:* This involves both improving its understanding of the situation by acquiring more objective knowledge, and improving its understanding of its user by acquiring more social knowledge. Learning will be exploited above all to ensure long-term interaction as it allows changes to ensure novelty.

Our goal is not to merely build a social companion, but to build an intelligent social companion. Our companion functions not only as a type of assistant or kind of pet, but also as a playmate and coach. Important in this light is the distinction between 'appearing intelligent' and 'being intelligent': for a playmate or coach it is not enough to appear intelligent, because sometimes the playmate takes the leading role. This pro-activeness requires a certain degree of situation awareness: the companion must be able to characterise a given situation and decide which type of interaction (proposing a game, soothing, being funny, etc.) is adequate.

In addition to the abovementioned required abilities, and in the context of the baby-doll scenario, we propose the following list of default behavioural repertoires for the companion: identify and recognise different people in the family, know about relationships in the family, easily engage in interaction (keep track of nearby persons and recognize when a person is addressing it), identify and recognize rooms in the house, identify the child's mood, identify and recognize toys and the child's preferred objects, learn about objects, recognize and interpret deictic gestures (such as a pointing hand saying this is my cup) to be exploited to learn about objects. Concerning knowledge representation we could use an ontology containing for example: persons, relations among persons, toys, animals above all pets, inanimate objects possibly associated with typical situations of use, rooms and places, property relationships between objects and persons, belong-to relations between objects and places, games, stories and fantastic characters, etc.).

4   Provisional architectural/integrating framework

The successful development and deployment of ACSs as motors for a new generation of cognitive educative toys requires the codification of sound construction rules. From our point of view, such rules should take the form of architectural requirements for cognitive systems, because the design of an architecture presupposes the basic understanding, via a model, of the sort of constructions to be built and the purpose for building them. In general, an architecture defines functional elements, the relationship(s) between the elements and the constraints that apply (Sloman et al., 2005). For a cognitive system, the architecture will have to specify, among other things, how the systems can actively perceive its own goals in relation to its operating environment, how it will acknowledge and evaluate the difference between the present state and the wanted state, how it decides on a plan of action to reach the wanted state and then executes the plan. Without specifying the details, we envisage a need for several functional elements, spanning from perception modules to action modules, such as the ones represented in Figure 1.

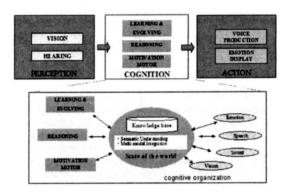

Figure 1: Envisaged functional elements and provisional cognitive organization.

One issue that needs special attention in order to arrive at the desired companion behaviour is the fact that we are dealing with children and adults. Speech recognition systems at present are tuned to adult speech, and child speech has some fundamental differences. Not only do (young) children pronounce words half, differently, etc, they also use different syntactic structures. Speech recognition systems and NLP tools need therefore to be especially trained on child language corpora. This involves, among others, the need to address the following research issues: develop a robust speech recognition system for children's voices; build an adequate knowledge base of the linguistic abilities at different ages, to adapt the dialogue manager consequently; initiate and maintain a (linguistic) two-way interaction with the child.

This last requirement is related to our goal of designing a companion that is able both to learn from the child and to teach the child. With our intelligent companion, in fact, we aim towards a two-way communication/interaction behavioural repertoire where the robot can actually teach the human (in our case child) something new. This means that the companion needs learning skills so that it can broaden its behaviour-base over time, as it interacts with its human companion. Moreover, it needs to evolve over the time according to both the situation and the child's progress. We propose to address the developmental aspect of certain cognitive functions through research focused on concept formation and concept transfer, together with the design of the supporting cognitive architecture. This differs substantially from the more traditional ontology based approaches as it tries to integrate an initial set of innate conceptual structures with both specific and general purpose unsupervised learning mechanisms, thus making the whole cognitive process dynamic. While the concept transfer from the child to the companion actually overlaps with the concept formation problem, the other direction, from the companion to the child, raises new, intriguing research questions. On the one hand, there are issues about adequate monitoring and coaching techniques. On the other hand we have the challenging question concerning the requirements for a cognitive architecture able to support different cognitive roles. With respect to this, we want to investigate the ability to support the coexistence of different knowledge models, such as the playmate knowledge model and the coaching knowledge model, with the related issues of belief maintenance and belief induction in the case of incoherent belief sets.

An exemplary system that simulates the process of categorization in the child is the ROCE system (see http://perso.wanadoo.fr/colette.faucher/ROCE.html). The purpose of this system is to show how information which is stored in episodic memory gradually comes to semantic memory, where an exemplar-based representation has been adopted to model the notion of concept. The core of the ROCE system is its incremental concept formation algorithm. According to studies stemming from developmental psychology, the categorization process starts at the basic level and goes on towards the superordinate and the subordinate levels. Each time an observation is inserted into a category, the algorithm which is used depends upon both the level of the category (basic, superordinate or subordinate level) and the type of the category (biological or functional). The algorithm only focuses on the properties which are predominant in the concerned category. Moreover, whereas at basic and superordinate levels, the classification process is based on a similarity measure, at the superordinate level, it models Nelson's theory according to which the observation functional properties are perceived only indirectly through the events within which they are involved. The input data of this algorithm are observations, i.e. descriptions of real entities, which are observed within their context of occurrence. The concept acquisition process is supervised, insofar as it is possible to provide the user with the name of the category some observations are attached to, as well as links specifying specialization relations between concepts. Thus the language contribution in the elaboration of concepts in the child is simulated. We are planning to explore a similar algorithm for our system.

We propose to follow the approach of (McCarthy, Minsky, Sloman, Gong, Lau, Morgenstern, Mueller, Riecken, Singh, & Singh, 2002) to provide a cognitive organization of the envisaged functional elements within a single architecture. The authors propose an architecture of diversity for commonsense reasoning, an easy task for humans, but a hard problem for artificial systems. Their approach is not to single out one particular artificial intelligence technique, but instead use a multitude of techniques and let the reasoning and representation modules choose the best ones for the particular situation that arises in the commonsense reasoning process. The authors use story understanding to evaluate and scale up their system and believe that the combinatorial explosion in the understanding process is one of the reasons why single mechanism and representation systems have failed thus far. Combinatorial explosion arises in stories due to multiple possible interpretations at all levels of language. What the authors propose is a three-level architecture developed within the Cognition and Affect project (Sloman, 2001). The bottom layer constitutes the reactive processes, the second layer contains the deliberative processes, and the top layer holds the meta-management or the reflective processes. Note that the three layers operate concurrently. Minsky built on this idea and proposed the "Model Six six level architecture", where the layers denote instinctive reactions, learned reactions, deliberative thinking, reflective thinking, self-reflective thinking and self-conscious emotions respectively. The architecture of diversity involves several multiple reasoning and representations schemes from natural language to story-like scripts, transframes, frame-arrays, picture-frames, semantic nets, knowledge-lines, neural nets to micronemes. We refer to (McCarthy et al, 2002) for more detail.

What is important to note is that this architecture of diversity has not been implemented yet – it is a theoretical construct. We think it could be a good inspiration point for our intelligent companion. A provisional architecture for an intelligent sociable companion is represented in Figure 2. The architecture is inspired by the architecture of

BIRON (Li et al. 2004) and the work of (Chella, Infantino & Macaluso, 2004). The interface software provides an interface to the robot's sensors and actuators, and is connected both to the robot hardware and to the game manager module. Besides the hardware controller, the camera and the microphones provide inputs also to the gesture detection system and the sound detection system respectively, which contribute to determine the awake and alertness states of the robot.

If the robot is turned on but asleep, a sudden noise or a voice can wake it up and eventually trigger a state of alertness. Once awake, the robot can become alert for two reasons: either it has detected a person willing to interact with it, or it is itself pro-actively looking for interaction (because of an internal motive, e.g. because it is bored; or because the situation seems to require an intervention, e.g. because the child is crying). So, the first sequence of the potential behaviours of the robot is: asleep, awake, alert. An initial design choice will therefore involve the following: is being asleep the default state of the robot, or does the robot have an internal "biological" rhythm which determines a succession of sleeping periods and waking periods? This second solution would not prevent the robot being suddenly woken up when sleeping (e.g. by a loud noise) and would make the robot more similar to a living creature. On the other hand it will also introduce more complexity on the overall controlling mechanism.

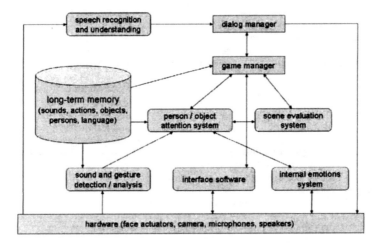

Figure 2: Proposed architecture for cognitive baby doll scenario, based on (Li et al, 2004).

The emotion module describes emotions in terms of discrete categories (e.g. 6 basic emotions: happiness, anger, disgust, fear, sadness, surprise) that can be associated both to internal states (e.g. after a period of inactivity the robot gets bored) and to external situations (e.g. if the child is crying the robot gets sad, and that might trigger a process of cheering up).

When the alertness state is reached, the person/object attention system, with input also from the emotion module, determines whether the situation requires a passive role of listening or an active role of proposing a game to play. A situation requiring a passive role is a situation where a person is individuated who is willing to interact with the robot, so the robot waits for speech and /or gestures from that person to establish the "game" that will be played.

The person/object attention system and the scene evaluation system can vary along a continuum of design choices from a more reactive module to a more deliberative module, depending on how much reasoning capabilities are associated to them. Deliberative processes involve considering possible alternative actions. They include processes such as planning, deciding, and/or reasoning about possible future states, for example in terms of counterfactuals or "what if" questions.

The game manager constitutes the high level controller of the whole architecture: it selects the appropriate game behaviour for the companion, given a certain situation, and then controls the execution of the game selecting the appropriate dialogues to be staged. Some of the "game" behaviours envisaged are: sing, acquire or teach a song, teach a word (correct a mistaken pronunciation), acquire or teach an object, acquire or teach a person or fictive character, acquire or teach a story. In particular, the acquisition game should contemplate a phase during which the robot asks questions about the item to be acquired and store the answers for future use.

In Figure 3 we represent the different states and behaviours available to the companion and individuate the crucial point where the feedback for learning could intervene. Using the terminology proposed in Fong et al. (2003), the robot will be *functionally designed* following an *iterative approach*. The outward appearance of social intelligence will be obtained with an internal design that might not reflect any biologically or psychologically inspired theory. This is so for various reasons: to cope with the limited embodiment of the robot (no navigation, no manipulation, limited gesturing abilities); to be able to focus on certain aspects of the interaction (emotion understanding, emotion displaying, dialogue-based interaction, semantic abilities); to provide engineer-sound solutions (even though these solutions might be unsuitable as psychological models). In iterative design the design process goes through a series of test and re-design cycles. The aim is to address design failures and make improvements based on information from evaluation of use. The reason why we envisage this strategy is because almost no data are available on interaction patterns between young children and robots: developing a workable interaction model is therefore part of the research issues.

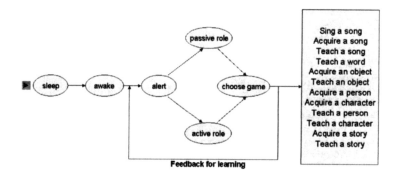

Figure 3: Companion's states and behaviours.

5   Conclusions

In this paper we presented some of the major challenges ahead regarding the design of intelligent companions whose aim would be first to better support/assist and second further advance/develop cognitive abilities of human-being to whom they will be attached to. By promoting a multidisciplinary (psychology, cognitive science, AI, linguistics, etc.) and integrated approach we situate our main research contribution at the level of Human-Robot (also Actor-Agent) Collaborative Interaction and more specifically at the levels of: concept formation, communication and exchange, reasoning and learning about affect (personality traits, emotions, moods) but also multimodal and social interaction. Although applicable to a broad range of application domains (i.e. Healthcare: a nurse companion, Transport: a truck driver companion etc.) we envisaged to first demonstrate and assess our resulting cognitive technology in the context of an educational scenario whose aim will be to speed-up the child development process by stimulating his/her cognitive abilities. The provisional architecture attached to the targeted scenario is considered to be representative of an instance of a more generic architectural scheme for the design of an Intelligent companion seen as an Autonomous Cognitive System.

We are currently at the starting point. On the basis of the baby-doll scenario we uncovered some important cognitive capabilities any robot companion should have, keeping in mind that we want it not only to listen to the child but also the other way around. Future work will be actually building such a cognitive system. Then we will to test it with children and their parents in a real usage context, in order to iteratively modify it on the basis of the outcomes of such experimental studies.

As stated already, although primarily targeted to the support and possibly reinforcement of the individual cognitive abilities of a child, our ultimate goal through this research remains to further advance research on companion Cognitive Systems by proposing a cognitive architecture together with a test-bed platform enabling to investigate how features such as personality traits and affect (emotions, moods) can be integrated at the level of cognition and serve various purpose such as autonomous behaviour and/or social intelligence and social interaction achievement.

Acknowledgements

The research reported here is part of the Interactive Collaborative Information Systems (ICIS) project, supported by the Dutch Ministry of Economic Affairs, grant nr: BSIK03024.

References

Aritaa, A, Hirakia, K, Kandab, T & Ishiguro, H. (2005). Can we talk to robots? Ten-month-old infants expected interactive humanoid robots to be talked to by persons. In *Cognition* xx, 1–9.

Breazeal, C. (2003). Towards sociable robots. In *Robotics and Autonomous Systems* 42, 167-175.

C. Breazeal, G. Hoffman & A. Lockerd (2004). Teaching and Working with Robots as a Collaboration. In *3rd International Joint Conference on Autonomous Agents and Multiagent Systems (AAMAS 2004)*, 19-23 August 2004, New York, NY, USA, pp. 1030-1037.

Chella A., Infantino I., & Macaluso I. (2004). Conceptual spaces and robotic emotions. In Prince, C. G. and Berthouze, L. and Kozima, H. and Bullock, D. and Stojanov, G. and Balkenius, C., (Eds.) *Proceedings Third International Workshop on Epigenetic Robotics: Modeling Cognitive Development in Robotic Systems* (pp. 161-162). Boston, MA, USA.

Dautenhahn, K. (2004), Robots We Like to Live With?! - A Developmental Perspective on a Personalized, Life-Long Robot Companion. Invited paper in *Proc. IEEE Ro-man 2004, 13th IEEE International Workshop on Robot and Human Interactive Communication September 20-22*, 2004 Kurashiki, Okayama Japan, IEEE Press, 17-22.

Dautenhahn, K. & Billard A. (1999). Bringing up robots or the psychology of socially intelligent robots: From theory to implementation. In Proc. Autonomous Agents (Agents '99) Seattle, Washington, USA, 366-367.

Fong, T., Nourbakhsh, I. & Dautenhahn, K. (2003). A Survey of Socially Interactive Robots. In *Robotics and Autonomous Systems* 42, 143-166.

Li, S., Kleinehagenbrock M., Fritsch J., Wrede B. & Sagerer G. (2004). "BIRON, let me show you something": Evaluating the Interaction with a Robot Companion. In Thissen, W., Wieringa, P. Pantic, M. & Ludema, M. (Eds), *Proc. IEEE Int. Conf. on Systems, Man, and Cybernetics, Special Session on Human-Robot Interaction*, (pp. 2827-2834). The Hague, The Netherlands, October 2004. IEEE.

Maes P. (1997). Agenta that reduce work and information overload. In J. M. Bradshaw (Ed.), *Software agents* (pp. 145-164). AAAI Press.

McCarthy, J., Minsky, M., Sloman, A., Gong, L., Lau, T., Morgenstern, L., Mueller, E. T., Riecken, D., Singh, M. & Singh, P. (2002). An architecture of diversity for commonsense reasoning. In *IBM Systems Journal*, 41 (3), 530-539.

Shneiderman B. (1997). Direct manipulation versus agents – Paths to predictable, controllable and comprehensible interfaces. In Bradshaw, J. M. (Ed.) *Software agents* (pp. 97-106). AAAI Press.

Sloman, A. (2001). Varieties of Affect and the CogAff Architecture Schema. AISB01 Convention, 21st – 24th March 2001.

Sloman, A., Chrisley, R. & Scheutz, M. (2005). The Architectural Basis of Affective States and Processes. In Fellous, J.M. and Arbib M. (eds.) *Who needs emotions?* Oxford University Press.

Sloman, A. & Chappell, J. (2005). The altricial-precocial spectrum for robots. Submitted to IJCAI05.

Steels, L. & Kaplan, F. (2001). AIBO's first words. The social learning of language and meaning. In *Evolution of Communication*, 4 (1), 3–32.

Section 4
Cognitive State Sensors

Chapter 14

Stress in the Computing Environment

Operator Performance under Stress

P.A. Hancock , J.L. Szalma, and T. Oron-Gilad

University of Central Florida
3100 Technology Parkway, Suite 337, Orlando FL 32826
phancock@pegasus.cc.ucf.edu; jszalma@mail.ucf.edu; torongil@mail.ucf.edu

Abstract

At the turn of the Twenty First Century, military forces face a fundamental change in the nature of warfare in that they no longer confront an 'obvious' enemy. In recent engagements, it has become progressively more difficult to identify exactly who the opposing forces are. There has also been a comparable change in the site of engagement. Modern conflict occurs in the close quarters of urban landscapes in which the technological support we seek to provide our troops can be negated by local circumstances. Into this new milieu, we are continuously injecting innovative support systems whose functionality is designed to assist our troops but whose complexity can threaten to overwhelm them. All this occurs against a background of threat which ensures that acute and chronic forms of stress are a constant presence. This paper captures some of the theoretical and experimental efforts that have been made by the MURI-OPUS (Operator Performance Under Stress) research program regarding combat stress effects as well as means by which this understanding can be applied to optimize human-machine interaction in crucial military situations and also address the after-effects of stress exposure.

1 Introduction

Recent research on stress (see Hancock & Desmond, 2001) has offered a number of individual taxonomic accounts of performance variation which look at influential effects on either a task by task basis (e.g., memory vs. decision-making, etc.) or on a stress by stress basis (e.g., heat vs. noise) (and see Hockey & Hamilton, 1983). These respective approaches are fundamentally descriptive in nature and provide only a limited framework for unified understanding and subsequent performance prediction. Taxonomic strategies encourage piecemeal efforts toward integrating combinatorial effects of multiple forms of demand on resource capacity. In contrast, the model of Hancock and Warm (1989) (see Figure 1) provides an overarching framework which uses both physiological and behavioral indices and, as a result, deals readily with the question of combined sources of stress. The Hancock and Warm (1989), "extended-U" model can also answer the vital question of under-load or under-stimulation, a form of stress that has received very little attention and yet plays a critical role in military operations in which the absence of information can be the very greatest source of threat. Instantiation of this model allows for the development of a number of tactics to reduce the adverse influence of maladaptive levels of stress, workload, and fatigue before, during, and after engagement. As is evident in the model and as is reflected in many experimental studies the fundamental problem with stress effects is the manner in which incipient failure occurs.

Under extreme stress individuals report a phenomenological 'narrowing' of the range of cues they extract from the surrounding environment. Under these 'dynamically unstable' (Hancock & Warm, 1989) conditions, optimal, compensatory, human-machine response is the goal. One of the concrete problems to achieving this end, is the establishment of a common definition and language by which to assess stress effects. We propose that there are three ways in which to solve this problem. Hancock and Warm previously indicated that stress can be viewed as an influence in three realms; the so-called 'trinity of stress.' Stress effects can be located as values in the environment itself, and are therefore expressed as physical measures, as is the case in setting limits to occupational exposure (see Hancock & Vastimidis, 1998). The second locus of stress effects is in the interaction between such environments and the individuals who are exposed to them. These represent the costs of adaptation, often referred to as 'strain effects'. The final locus, and perhaps most pragmatic way to measure stress effects, is directly through change in performance efficiency of the primary task to hand. Often this is also the easiest approach. However, when prediction of future capacity is paramount, instantaneous changes in primary task performance are not very effective representations. Each of these information sources in the 'trinity of stress' provides windows on the broad-spectrum level of demand imposed by the environment. In the present work, we take this question of measurement one step further. We have engaged in a theoretical search for a common currency to describe the broad-spectrum level of demand. In the base axis shown in Figure 1, these metrics are labelled hypostress and hyperstress and in subsequent work we have broke this axis down into two compounds namely information rate and information structure. The divided dimensions represent the spatial and temporal characteristics of the environment in relation to the capabilities of the exposed individual. As a relational measure, they are theoretically concordant with the Gibsonian notion of an 'affordance' (Gibson, 1979). They represent ratios which express the capability of the individual to respond versus their present level of experienced demand. We use such ratio measures as inputs to both predictive modelling and subsequent mitigation strategies such as adaptive systems (see Hancock & Chignell, 1985). Recently we have further elaborated on temporal characteristics related to information processing and its quantification. Our 'trinity of time' model, now differentiates the axis of time itself (Hancock, Szalma, & Oron-Gilad, 2004). It is to the perception of time to which we now turn.

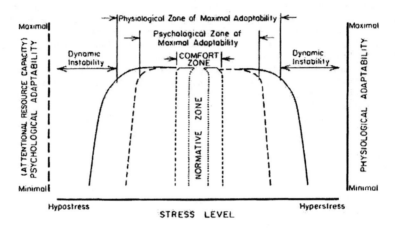

Figure 1 – The Hancock and Warm (1989) model of stress and performance.

2   Time Distortion under Stress

Understanding time distortion under stress is crucial for the operation of complex systems. Time perception influences how operators react to visual, auditory and tactile information. Operators are often required to respond to such information in a timely fashion, and if time estimation is distorted due to stress, the critical information can go unprocessed for a duration which exceeds the time available for solution. In the model describing adaptation under stress, Hancock and Warm (1989) emphasized that the task itself is the proximal source of stress. Individuals cope with high levels of stress by narrowing their attention to specific spatial cues in the environment (Easterbrook, 1959). The temporal domain is also stress-sensitive in a similar manner to the spatial domain and comparable temporal narrowing occurs, resulting in distortions of perceived time (Hancock & Weaver, 2005). Under stress, operators usually consider more pieces of information, but in a more shallow way, and can fixate on one solution and disregard superior alternatives (Hildebrandt & Harrison, 2002). The reduction in attentional capacity with increased stress level results in reduction of information intake and fixation on certain cues by the operator. As stress is increased further, these phenomena become stronger until the operator is fixated on only one source of information (see Hancock & Szalma, 2003). A similar mechanism influences the perception of time. If stress conditions are sufficiently high, such conditions induce time distortion (Block, Zakay, & Hancock, 1999), in which time seems to 'speed up' or 'slow down.'

Under conditions of extreme, life-threatening stress, people often report these distortions of time (Hancock & Weaver, 2005). These distortional experiences are critical since they occur in circumstances where small variations in behavior mean the difference between life and death. Individuals who experience 'time distortion' under stress typically report one of two forms of experience. In the first type of experience, everything appears hazy or as a 'blur' as though events were stacked together almost into one temporal 'moment'. A second pattern has the individual reporting many different events with startling clarity as time appears to 'slow down'. Each of these patterns represents effects of 'time-in-memory.' That is, in these retrospective accounts, individuals are recalling events rather than referring to their immediate experience. In contrast, immediate experience is referred to 'time-in-passing' (see Roeckelein, 2000). The first step toward an understanding of this phenomenon is the linkage between time-in-memory and time-in-passing. A simple approach to understanding this linkage uses the concept of attention and accumulation models of time perception. Accumulation models suggest that individuals 'create' their own time perception by filling 'moments' with a sequence of experiences. A version of the accumulation concept is given in Figure 2. Suppose it takes ten 'events' to create a moment and that those 'events' were fixations of attention to external stimuli. In a normal situation, ten 'events' are recorded and compared to a time base that we each learn as we mature and assimilate the social meaning of time. Given that the perception of an external reference base for time is primarily a learned capacity, when conditions are 'normal' our sense of time thus accords with the social referent, or 'clock' time and we perceived a regular, uninterrupted and 'accurate' sense of duration. Now suppose there is a sudden emergency condition in which the attention of the individual is drawn by the novelty, the intensity, and the complexity of the situation to fixate on stimuli at a rate significantly different from that normally generated. In Figure 2, we can see the ratio between the perceived and actual time can climb from 1:1 to 2:1 to 3:1 and even beyond. This adaptive response may well be useful in searching for critical stimuli that might indicate strategies for successful resolution of the threatening condition. In passing, time for this individual has been 'compressed' such that event registration overwhelms the learned translation between event frequency and clock time. Phenomenologically, time slows in proportion to the increase in event registration. In stress conditions sufficient to induce such time distortion, the increase in event registration must be substantial but this phenomenological "threshold" level yet to be determined by empirical investigation. For such an individual recalling the incident in memory is thus characterized by a recognition of the time distortion itself and a clarity of recall of many events which, since they cannot be reconciled with the learned time translation are reported as being outside the normal run of behavior, i.e., "time seemed to slow down."

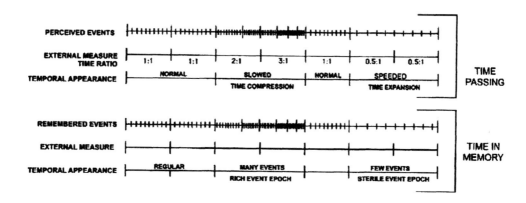

Figure 2 - Time-in-memory and time-in-passing are linked together but not equivalent. When many events occur in the environment, time-in-passing slows. However, when that interval is recalled (i.e., time-in-memory) it appears to be greatly expanded since so many novel and intense stimuli were processed during that interval. The reverse is also true for boring or unstimulating periods (after Hancock & Szalma, 2003).

How do these distortions of time influence augmented cognition and the program directed to improve human capabilities via technological support? Primarily Augcog is directed to improving performance in non-optimal conditions. It is precisely these circumstances in which time is expected to be non-linear. It becomes imperative to distinguish which individual differences distinguish time "dilators" from time "contractors." The former will be able to be "on-loaded" in these critical circumstances, the latter will rely directly on Augcog techniques to "off-load" them as the influences of stress increase. However, these variations in subjective state are exceptionally difficult to capture, especially as what is being distorted (time) is one of the major metrics by which objective performance is captured! Therefore we have to embrace methods which seek quantitative ways to capture uncertainties and one we have recently contributed to is a procedure known as fuzzy signal detection (FSDT). It is to this question of uncertainty measurement to which we now turn.

3   Capturing Uncertainty Via Fuzzy Signal Detection Theory (FSDT)

Dealing with uncertainty by addressing immediate operator needs for information is another major matter of concern for the AugCog program. Fuzzy Signal Detection Theory (FSDT; Hancock, Masalonis, & Parasuraman, 2000; Parasuraman, Masalonis, & Hancock, 2000) captures this inherent uncertainty in real-world stimuli and thus supersedes the limits of traditional SDT in the assessment of real-world detection capacities. Specifically, traditional SDT requires that stimulus categories be fixed and mutually exclusive. However, in domains with complex stimulus events (e.g., baggage screening, security checkpoints, friend-foe identification, etc.) people, objects, or events have properties of both signal and non-signal. FSDT quantifies this uncertainty and permits its incorporation into estimates of ROC curves and their parameters.

A three-pronged approach has been developed for the initial assessment of FSDT. First, psychophysical experiments have evaluated ROC space and the degree to which FSDT meets the assumptions of traditional SDT. Stimuli currently in use include temporal and shape discrimination tasks. Results indicate that FSDT estimates of performance are higher than those computed using traditional SDT (Murphy, Szalma, & Hancock, 2004; see also Masalonis & Parasuraman, 2003). The source of this discrepancy has yet to be determined, but it may be due to the fact that many instances of joint membership in an FSDT analysis (e.g., a response is both a hit and a miss to some degree) would be categorized as an error in traditional SDT (e.g., a miss). However, fuzzy SDT analysis results in an ROC consistent with either an equal variance or unequal variance model, depending on the difficulty of the discrimination (see Murphy et al., 2003). Second, FSDT has been tested using a vigilance task, since sustained attention is an area in which traditional SDT has been extensively applied (e.g., Davies & Parasuraman, 1982; See, Howe, Warm, & Dember, 1995). An initial experiment in which signal probability was shifted showed a corresponding shift in response bias consistent with expectations (Stafford, Szalma, Hancock, & Mouloua, 2003). The third approach has been to evaluate FSDT using laboratory simulations of real-world tasks in threat detection. These include morphing objects such as tanks as well as baggage screening tasks. Theoretical and methodological advances gained from this research are not confined to a particular domain but serve to improve performance of systems across a wide array of applications, including medicine, industrial inspection, threat detection and National priority areas such as detection of deception and criminal/hostile intent and security screening in all facets of Homeland Defense (Hancock & Hart, 2002). In each case the critical step for a valid FSDT analysis is to obtain or derive a valid mapping function that relates one or more physical stimulus dimensions onto a fuzzy stimulus set dimension. As in other areas of psychological measurement, the quantification procedures will only have meaning to the extent that they are isomorphic with actual phenomena. Indeed, it is the derivation of such functions that has proven most challenging in applying these techniques to real-world stimuli. We propose that FSDT can be used to quantify operator state as the crucial input metric to Augcog technologies, especially under stressful conditions. However, even if our measurement problems are resolved, at least to an acceptable degree, we still have the problem of differences between individuals, and it is to this issue which we now turn.

4   Individual Differences Information Intake under Stress

It has now been established that different forms of information processing are linked to specific patterns of emotional and cognitive state (e.g., Hockey & Hamilton, 1983; Matthews, Derryberry & Siegle; 2000; Matthews, Davies, Westerman, & Stammers, 2000). Although it is likely that many of the basic mechanisms of information processing are common across individuals, the coordination of those processes and the context of processing (e.g., expectancies, percepts, etc.) can vary as a function of the characteristics of individuals (trait) and their current cognitive and affective circumstance (state). Thus, visual information is transformed from photic to neural energy in a way that is common across humans (and many animals), but how that information is then organized, and indeed whether the information is processed beyond the sensory level will depend on individual difference factors. Such individual variation is also modified by the context in which the stimuli are presented.

Individual differences clearly affect the way that various people process information and derive decisions. Also, the modality in which the information is presented differentially influences behavioral response, especially when tasks are either learned or subsequently performed in stressful circumstances. Any proposed strategy for augmented cognition, however implemented, will systematically vary in efficacy as a function of these individual acute and chronic pre-dispositions. We are currently examining how individual differences in verbal and spatial working memory capacity predispose individuals to respond to the presence of critical warnings (Hancock, Oron-Gilad, & Helmick-Rich, 2004). We are extending this work to determine if task-based and environment-based sources of stress have systematic effects on performance across the modalities of auditory, graphic, and alpha-numeric presentation in simulated operational conditions. The differentiation of resource-based concepts of working memory remains a controversial issue and it is still debated whether working memory has one unitary pool of resources through which all higher level cognitive activities are supported or if separate pools of resources are dedicated to supporting different processes. Our current experiments focus on the issue of separability of working memory resources of verbal and spatial memory and how such a conceptualized distinction persists in the face of stressful conditions. These effects and their respective change under life threatening stress is critical to military and civilian operations. Augmenting cognition to capture the advantages of such individual variation is clearly critical to future operational success.

If Augmented Cognition technologies do not work for each different individual user, they will eventually fail to provide return on the investment made in their creation. To achieve this goal we have to understand more about variations across individuals especially when they have to perform in aversive and threatening conditions. The diagnostic signals that each individual emit possess some degree of nomothesis, representative of their shared common humanity. However, the more we understand the more we will have to design for each specific person (Hancock, 2003) and there are many sources of variations within one individual, as well as across individuals. Augcog is the natural fruition of concepts such as adaptive human-machine systems (Hancock & Chignell, 1985) but now the theoretical framing is effectively accomplished. Now we have to move to implementation in which all the challenges of real-world operation come to the fore. We have to embrace the complexity of context and individuality, but these are not merely barriers they are significant opportunities to be exploited in the coming years.

5   Discussion and Conclusions

Combat has always an inherently demanding situation which increases stress, heightens arousal, and increases anxiety especially when the outcome is uncertain (and see Jones & Swain, 1995). In any combat environment, maintaining a high level of performance, both of cognitive skills (e.g., decision making tasks) and motor skills (e.g., shooting accurately) is critical and can often mean the difference between life and death. Modern asymmetric warfare is very different from the doctrines of conflict as framed in previous centuries. There has been a comparable change in the site of engagement. No longer expected to occur in wide open spaces, modern conflict happens at close quarters and in shattered urban areas in which the support we seek to provide our troops may perhaps be negated by local conditions. Unlike even guerrilla warfare, today's enemy no longer defends specific locations rather, the enemy now defends a dogma and the battle is not primarily for land or even physical resources per se, it is the battle the mind. However disembodied the mind might appear, it is an emergent property of the brain and brains themselves possess spatial extent. Therefore, modern conflict is hybrid of both previous palimpsestual assemblies of received doctrines, together with this novel expression of conflict incarnation. Since the battle is primarily for the mind, of both public and opposing combatants alike, the perpetuation of stress and anxiety is a crucial operational objective for terrorists in the context of their asymmetric strategy. Into this new milieu, we are continuously injecting innovative technologies whose functionality is designed to assist our troops but whose complexity can threaten to overwhelm them. Advanced technology systems must be recognized by the soldier as a vital piece of their equipment and critical to their survival. Equipment that is cumbersome, unreliable, and ineffective can and will be discarded in the extremes of battle. Consequently, the thought at the forefront of Augcog technology design should not be the technical feasibility of implementation but the utility to the everyday soldier. If it does not work in the stress of battle, if the design is not user-centred, burdening the soldier with advanced technology systems is clearly a disservice.

Acknowledgments

This work was supported in part by the Department of Defense Multidisciplinary University Research Initiative (MURI) program administered by the Army Research Office under Grant DAAD19-01-1-0621, Dr. Elmar Schmeisser, Technical Monitor, P.A. Hancock, Principal Investigator. This work was also facilitated by a DARPA-funded program under Grant NBCH1030012, CMDR Dylan Schmorrow, Technical Monitor, P.A. Hancock, Principal Investigator. This work was also facilitated under Army Research Laboratory through the Micro-Analysis and Design CTA grant DAAD19-01C0065, John Locket, Technical Monitor, P.A. Hancock, Principal Investigator. This work was also facilitated under Army Research Laboratory through the Micro-Analysis and Design CTA grant DAAD19-01C0065, Michael Barnes, Technical Monitor, P.A. Hancock, Principal Investigator. The views expressed in this work are those of the authors and do not necessarily reflect official Army or Department of Defense policy.

References

Block, R.A., Zakay, D., & Hancock, P.A. (1999). Developmental changes in human duration judgments: A meta-analytic review. *Developmental Review, 19*(1), 183-211.

Davies, D.R., & Parasuraman, R. (1982). *The psychology of vigilance.* London: Academic Press.

Easterbrook, J.A., (1959). The effect of emotion on cue utilization and the organization of behavior. *Psychological Review, 66*, 183-201.

Gibson, J. J. (1979). *The ecological approach to visual perception.* Boston, MA: Houghton Mifflin.

Hancock, P.A. (2003). Individuation: Not merely human-centered but person-specific design. *Proceedings of the Human Factors and Ergonomics Society, 47*, 1085-1086.

Hancock, P.A., & Chignell, M.H. (1985). The principle of maximal adaptability in setting stress tolerance standards. In: R. Eberts & C. Eberts (Eds.). *Trends in Ergonomics/Human Factors II.* (pp. 117-125), Amsterdam: North-Holland.

Hancock, P.A., & Desmond, P.A. (Eds.). (2001). *Stress, workload and fatigue.* Lawrence Erlbaum, Mahwah: NJ.

Hancock, P.A., & Hart, S.G. (2002). Defeating terrorism: What can Human Factors/Ergonomics offer? *Ergonomics in Design, 10* (1), 6-16.

Hancock, P.A., Masalonis, A.J., & Parasuraman, R. (2000). On the theory of fuzzy signal detection: Theoretical and practical considerations. *Theoretical Issues in Ergonomic Science, 1*, 207-230.

Hancock, P.A., Oron-Gilad, T. & Helmick-Rich, J. (2004). Warning presentation and retention under varying levels of stress, *Quantifying Human Information Processing* (QHIP) program Report, UCF.

Hancock, P.A., & Szalma, J.L. (2003). Operator stress and display design. *Ergonomics in Design, 11* (2), 13-18.

Hancock, P.A., Szalma, J.L. , & Oron-Gilad, T. (2004). Time, emotion, and the limits to human information processing, *Quantifying Human Information Processing* (QHIP) program Report, UCF.

Hancock, P.A., & Vastimidis, I. (1998). Human occupational and performance limits under stress: The thermal environment as a prototypical example. *Ergonomics, 41*, 1169-1191.

Hancock, P.A., & Warm, J.S. (1989). A dynamic model of stress and sustained attention. *Human Factors, 31*, 519-537.

Hancock, P.A., & Weaver, J.L. (2005). Temporal distortions under extreme stress. *Theoretical Issues in Ergonomic Science, 6* (2), 193-211.

Hildebrandt, M. & Harrison, M.D. (2002). Time-related trade-offs in Dynamic Function Scheduling. In C. Johnson (Ed.). *Proceeding of the 21st European Annual Conference on Human Decision Making and Control* (pp. 89-95). Glasgow, Scotland.

Hockey, R., & Hamilton, P. (1983). The cognitive patterning of stress states. In: Hockey R. (Ed.), *Stress and fatigue in human performance* (pp. 331-360), John Wiley & Sons: Chichester.

Jones, G., & Swain, A. (1995). Predispositions to experience debilitative and facilitative anxiety in elite and nonelite performers. *The Sport Psychologist, 9*, 201-211

Masalonis, A.J., & Parasuraman, R. (2003). Fuzzy signal detection theory: Analysis of human and machine performance in air traffic control, and analytic considerations. *Ergonomics, 46*, 1045-1074.

Matthews, G., Davies, D.R, Westerman, S.J., & Stammers, R.B. (2000). *Human performance: Cognition, stress and individual differences.* Hove, England: Psychology Press/Taylor & Francis.

Matthews, G., Derryberry, D., & Siegle, G.J. (2000). Personality and emotion: cognitive science perspectives. In S.E. Hampson (ed.), *Advances in personality psychology,* vol. 1 (pp. 199- 237), London: Routledge.

Murphy, L., Szalma, J.L., & Hancock, P.A. (2004). Comparison of fuzzy signal detection and traditional signal detection theory: Analysis of duration discrimination of brief light flashes. *Proceedings of the Human Factors and Ergonomics Society*, **48**, 2494-2498.

Parasuraman, R., Masalonis, A.J., & Hancock, P.A. (2000). Fuzzy signal detection theory: Basic postulates and formulas for analyzing human and machine performance. *Human Factors*, **42**, 636-659.

Roeckelein, J.E., (2000). *The concept of time in psychology: A resource book and annotated bibliography*. Wesport, CT: Greenwood Press.

See, J.E., Howe, S.R., Warm, J.S., & Dember, W.N. (1995). Meta-Analysis of the sensitivity decrement in vigilance, *Psychological Bulletin*, **117**, 230-249.

Stafford, S.C., Szalma, J.L., Hancock, P.A., & Mouloua, M. (2003). Application of fuzzy signal detection theory to vigilance: The effect of criterion shifts. *Proceedings of the Human Factors and Ergonomics Society*, **47**, 1678-1682.

Human Temporal Judgment in the Humans-in-Automation Environment

Laurel Allender, Daniel N. Cassenti, and Rene de Pontbriand

U.S. Army Research Laboratory
Human Research and Engineering Directorate
AMSRD-HR-SE
Aberdeen Proving Ground, MD 21005-5425
lallende@arl.army.mil, dcassenti@arl.army.mil, and rdepontb@arl.army.mil

Abstract

Temporal factors are critical to any given task in the humans-in-automation environment. Questions such as, "do I have enough time to accomplish this task?" or "are other tasks going to fail due to the time I need to devote to another activity?" influence the way in which activities are performed or whether those activities will be performed at all. Given the importance of time in the automated systems context, we will explore the intricacies of temporal cognition, then discuss how an understanding of temporal cognition can improve human performance in both human-machine and human-human interaction.

1   Why Focus on Time?

> The team and players have to be completely aware of what's going on for them to be successful. Normally, the quarterback is to blame for this type of mistake (poor performance on a timed play), but I'm not sure all the blame should fall on his shoulders. McNabb had a very up-and-down game, and he was trying to focus on steadying his play. The coaching staff should have realized this and stepped in and taken the clock management portion of the game off his plate. The precious time that was wasted may have ended up deciding this game.
> -- *Eric Allen (February 6, 2005)*

This quote describes the end of the 2005 Super Bowl, when the Philadelphia Eagles fell to the New England Patriots in a football game that may have been decided by time management and temporal cognition. All too often goals succeed or fail because of limited time. The Eagles had the ball and needed to move farther downfield to have a chance of tying or winning the game, but they had only limited time to get there. Unfortunately for the Eagles, time became even more critical when a rushed throw resulted in an interception and the effective end of the game. What if the quarterback, Donovan McNabb, had paid more attention to the clock? Would they have won the game? What happens when someone pays more attention to time? Does that extra attention lost to time hurt other mental processes? Can machines improve temporal performance? Can other team members improve temporal performance? We will explore these questions here.

Sports are an obvious place to start when contemplating the importance of time in human performance. Most sports include clocks that limit the amount of game time. In addition, sports always include discrete performance measures, such as the number of points scored or time of possession. In a very real sense, this sports analogy applies to many aspects of everyday life, and, of interest here, to the military environment. The span of a peacekeeping mission, a defensive operation, or an attack could be days, weeks, months, or longer, but across the entire time course, the success of any single event—a checkpoint encounter, a reconnaissance mission, an enemy engagement—will rely on the split-second timing of decisive, often collaborative, actions. In a checkpoint encounter, the timing of physical actions may be critical—approaching the person entering the checkpoint or interpreting the implications of the person reaching into a pocket. If a reconnaissance mission relies on controlling an unmanned vehicle, the timing of the control actions or noticing an alert icon on a display can mean the difference between success and failure. In an engagement, listening to a message about enemy and friendly locations and then pulling the trigger can have life and death outcomes.

The timing of these actions must be considered in context. Donovan McNabb performed his actions as a part of a well-practiced team. The soldier also performs as part of a team, and increasingly that team includes an automated system. Communication systems, reconnaissance robots, and tactical displays with alerting algorithms and embedded decision aids are becoming ubiquitous on the battlefield. Each of these types of systems operates with specific timing characteristics, independent of the human in the loop. It is that coordination that is critical for the humans-in-automation environment.

2 The Foundations of Temporal Cognition

In order to understand how human-system time coordination and management may be improved, we must understand how people process time. Temporal cognitive research can be found in several, overlapping literatures: psychophysics, neuroscience, cognitive psychology, decision making, and cognitive modelling, as well as in the literature from more applied disciplines such as human factors engineering. In this section, we will sample the literature and describe how people estimate time, what effect training has on temporal-cognitive skill, how length of time affects temporal skill, and what time factors contribute to degraded performance. First, however, we will discuss time perception, a topic that takes us back to the early days of psychology when scientific discussions bordered on philosophical.

2.1 The Perception of Time

The first challenge in talking about time is defining what we all understand to be "the present." It is a conundrum because as soon as we think about the present time, it is in the past. William James (1890) defined the concept of the present in this way: "the prototype of all conceived time is the specious present, the short duration of which we are immediately and incessantly sensible." In other words, "the present" is a cognitive aggregation of near simultaneous perceptions across some small amount of time, just as perception of a photograph is a spatial aggregation of many visual signals. James posited that this "specious present" could be a few seconds, but probably not more than a minute. Attempts to determine more precisely the length of the present require literature on psychophysics and memory.

Fraisse (1984), in his thorough review of the psychophysical literature related to the perception of time, concludes that there are three levels of duration to be considered. At exposures of less than 100 ms, the experience of duration is what he terms "instantaneity" (p. 29, 1984). Between 100 ms and 5 seconds, what he considers the present, duration is experienced essentially directly. Above 5 seconds, beyond the experience of the present, however, duration must be read from memory. Drawing on our everyday experiences, we can certainly agree that saying how long it took to drive from home to the grocery store or from home to the beach (especially if you started in Iowa) without using a clock or a calendar requires recall from memory. It is beyond the scope of this paper to discuss precisely when experiences from the present enter into long term memory, but suffice it to say that there are differences in how time is determined for shorter vs. longer durations. What we do want to take from this brief section is twofold: temporal cognition involves the experience of the relative timing of stimuli or events *and* relies on some aspects of memory.

2.2 Basic Time Findings

2.2.1 Time Estimation

Time estimation refers to the human ability to approximate lengths of time, that is, to estimate duration. Typical methods are either to estimate the length of time of a past event or interval having distinctive start and end markers (e.g., Jones & Boltz, 1989) or to compare two events by selecting the shorter or longer of the two. Another method is to estimate or predict the length of time necessary to perform an activity in the future (see Hollnagel, 2002). It follows that, because temporal cognition relies on memory, it varies depending on the contents of memory, is subject to the effects of various stressors and distracters, and is described differently depending on the measure.

Anecdotes on temporal judgment come from the common experience of having an hour that drags on forever and another hour that flies by like a handful of minutes. An extreme case is of a man who spent 6 months in an underground cave (as reported in Fraisse, 1984). When he estimated the average time between waking and lunch, it was on the order of 5 hours, while the clock time was closer to 10 hours. This effect stems from two factors: what activity (or lack thereof) is filling the time and whether the estimate is provided before the event or after. As we imagine ourselves in the position of the cave dweller, our gut reaction is to say that surely time moved slowly and would be estimated very high; however, for the person in the cave, there were few perceived changes during the interval and so afterward the actual estimate from memory of how much time had passed was understated. There is also a secondary effect of the nature of the activity occurring during the interval being estimated: If it is some engaging activity, and particularly one that is stressed by time—moving the ball downfield; identifying the target before a critical window closes—then afterward the estimate of the duration is short (e.g., Jobidon, Rousseau, & Breton, 2004). Note that this is slightly different from the tradition in the mental workload literature of assessing the subjective experience of time pressure (e.g., the Subjective Workload Assessment Technique (SWAT), Nygren, 1982). The effect here is one of a distortion of the estimation of time itself, rather than the negative subjective or affective aspects of time pressure. Hancock and Weaver (2003) have examined distortion of time perception as a function of some other stressor or arousal factors.

2.2.2 Learning Time

Our temporal cognitive abilities develop over time. This has been shown in neural studies of children and from brain-damaged adults (Szelag, Kanabus, Kolodziejczyk, Kowalska, & Szuchnik, 2004) and, of interest here, through the literature on rhythm entrainment. The particular method used to show entrainment is to present a rhythmic interval, stop the rhythm, and ask the participant to tap when the next beat would have occurred. This procedure is called time production and, as in time estimation, relies on a mental timekeeper. Time production typically results in better performance than time estimation because it involves repeated presentations of intervals, resulting in a greater chance to accurately represent each interval through learning. Rhythms typically help people to perform a task (see Carlson & Cassenti, 2004) and as will be discussed below, non-rhythmic intervals or interruptions that disrupt the rhythm can have a detrimental effect on performance.

2.2.3 Time Granularity

Important to all time studies is a consideration of time granularity. While neurobiological research has examined temporal synchronicity and periodicity in terms of milliseconds, down to the neuronal level (e.g., Mel, Niebur, & Croft, 1997), there has been little information that could help link this research to human-computer or human-human task-level performance, which is the object of discussion here. The field of biophysics is exploring genetic factors related to circadian rhythms (Stelling, Gilles, & Doyle, 2004) in terms of robustness or insensitivity to perturbations of cellular structures. Certainly, the prevalence of automation-paced operations, and of the 24/7 operational tempos sought by military planners, may have dramatic effects upon cell environments. Such changes could moderate evolutionary responses to perturbations in unknown ways. Fortunately, such research dealing with biological complexity is experiencing a growth spurt, as new technologies such as biotechnology make tractable measurement more reliable and accessible. Again, links to task level performance are only now being considered, and thus will not be discussed further here; however, such links represent an area of rich opportunity.

Tasks occur within different types of time granularities. For example, typing one sentence may occur in seconds, but typing an entire paper takes days, weeks, or months, depending on the scope. Kutar, Britton, and Nehaniv (2001) suggest that time granularity is an important concept for all human-computer interaction researchers. Whereas making a double mouse click takes milliseconds, the turnaround in getting a response to an email message, especially in international communications where time zones are a factor, could take hours or days. Estimating longer intervals is much more difficult than estimating shorter intervals. This effect is described by Weber's function (see Killeen & Weiss, 1987), which states that more variability is added to estimates as time goes on. As intervals increase from minutes to hours to days, the ability to estimate time decreases, thus, expectations of precise deadlines for tasks with larger time granularity may be very inaccurate (Kutar, Britton, & Nehaniv, 2001).

2.3 Temporal Factors that Have Implications for Performance

Improving performance sometimes requires avoiding factors that can make performance worse. Three detrimental temporal factors described here are varied intervals, delays, and interruptions. Most tasks benefit from an uninterrupted flow of events to achieve maximal performance (see Carlson, 1997). For example, the perspective of ecological perception (Gibson, 1979) suggests that the ambient flow of changes and constants in the environment provides information on what actions to perform, linking flow of events to quality of performance. Each of these temporal factors has the potential to disrupt the ambient flow of events and therefore hurt performance.

A special case of interruptions, perhaps, is multi-task performance—two or more tasks performed essentially at the same time. The difference between multi-task performance and an interruption is that "interruption" implies that the main task and the interrupting task will be performed sequentially and "multi-task performance" implies the two or more tasks will be performed more or less simultaneously. Task-switching will probably occur several times and at as "micro" a level as the tasks can accommodate. The balancing of multiple tasks has special implications for temporal cognition. These factors are presented below.

2.3.1   Varied Intervals

As discussed above, rhythmicity enhances performance by providing a constant time to allow people to precisely plan and time their actions in synchronization with events. Carlson and Cassenti (2004) found that performance is hurt when varied intervals are used instead of rhythmic intervals. In an event counting task, participants were much less likely to report the correct number of events when the timing of the events was varied (though the average counting time was the same in both conditions). Confidence data suggested that participants were also more aware of errors they made in the varied interval condition, leading Carlson and Cassenti (2004) to conclude that varied intervals hurt performance because of their disruption to a natural flow of events. The disruption of timing prevented the synchronization of the count with the appearance of the new item, leading to noticeable errors.

2.3.2   Delays

A delay does not involve a secondary task but is simply a waiting time between one event and the next in a sequential activity. Cassenti (2004) found that on trials in which participants had to wait approximately 3 seconds between to-be-counted events, there was a decline in performance compared to trials in which the wait between events was only 600 ms. Delays do not seem to cause memory loss for task-relevant information; instead, performance most likely declines because of a strategy used to prevent memory loss (i.e., rehearsal).

2.3.3   Interruptions

Another way of disrupting the flow of events is to include interruptions of a secondary task on a primary task, "surprises," so to speak. We have all experienced interruptions—a phone ringing while in the middle of a conversation or an email alert signal while typing a document. Edwards and Gronlund (1998) found that interruptions hurt primary task performance because of the loss of information from the primary task. When interrupted, people need to remind themselves about what they were doing and what information they had before resuming the primary activity again. The news is not all bad on interruptions. Trafton, Altmann, Brock, and Mintz (2003) found that if participants had a warning before an interruption occurred, they could use memory strategies to overcome some (but not all) of the decrement in performance.

2.3.4   Multi-task Performance

There are no surprises in multi-task performance. You know from the beginning that performance of two (or more) tasks is required at the same time. This invokes the need for managing attentional resources and overloads in working memory for a sort of "self-interruption," over and above the loss of task information due to interruption per se. Brown (1997) found that estimating time while performing a spatial task, decreases performance relative to performance of both tasks alone. Brown (1998), in a review of the nature of divided attention between timing and other secondary tasks, concluded that divided attention hurts performance for a wide range of tasks, but the effect is not symmetrical. Whereas performance of a motor task (pursuit rotor), a visual search task, and higher level cognitive task (mental arithmetic) all hurt performance on the time estimation task, only cognitive task performance decreased due to performance of the time estimation task. Performance on the motor and visual search tasks was not affected. Seeking an explanation, he examined both multiple resource theory (Wickens, 2002) and working memory theory (e.g., Baddeley, 1986). Both theories assume a limited capacity to handle a finite number of tasks at once, but ascribe the bottlenecks differently. The channel limits of multiple resource theory could explain the effect of the time estimation task on the other tasks, but not the asymmetry. Working memory limitations and the associated attentional requirement, however, do adequately explain it. Brown (1997) concludes that temporal estimation tasks, just like high level cognitive tasks, require attentional and working memory resources.

2.4 Time Mechanisms and Models

People have observable ability to estimate time to a reasonable degree of accuracy; however the cognitive processes that lead to time estimation are not directly observable. As discussed above, we are only now developing biotechnological tools to aid in a finer-resolution understanding of internal synchronicity and periodicity. Pending those developments, models have been developed to explain the time estimation phenomenon. Some of the models are described below.

2.4.1 Mental Timekeepers

A current common explanation for time estimation involves a mental timekeeper that has mechanisms that approximate time. The two most prominent models are the internal clock and collections of mental oscillators working in concert with one another (Palmer, 1997). Both of these models imply that temporal cognition is processed and transferred from designated areas of the brain that directly inform the person about lengths of time.

The internal clock model suggests that people have a mechanism that acts just like a stopwatch. When measuring an interval, the clock is initiated at the start of the interval and outputs a time when the interval is over (e.g., Essens & Povel, 1985). The difference between an internal clock and a stopwatch is that the internal clock is not as precise and contains variability in its estimate, first measured by Wing and Kristofferson (1974).

A second model of the mental timekeeper is currently more widely accepted than the internal clock. The mechanism consists of a set of oscillators, with each oscillator possessing a certain frequency. Different combinations of oscillators provide different time intervals. For instance, if there were a 30 ms oscillator and a 70 ms oscillator, two cycles of each could measure a 200 ms interval. Variability in time estimates from this mechanism derive from variability in the timing of each cycle of each oscillator, the variability caused by approximating intervals that are not adequately handled by the frequency of the oscillators (see Jones, 1990), and also possibly from variability caused by stress-induced changes in neurotransmissions (as discussed in McCrone, 1997).

McCrone (1997) goes on to discuss a third view based on neuronal-level evidence from sensory processing: Rather than timekeeping being a separate readout from this internal clock, it is represented widely throughout the brain. He concludes with a final, mediating view that, while temporal information may be present throughout the brain, focused control of or access to "clock" information is located in a single area, the basal ganglia. Clearly research results are continuing to develop; however, for now we can conclude that, whatever the precise mechanism, the brain can handle time internally, without looking at the clock on the wall.

2.4.2 Behavioral Representation

Computational methods have been applied as research tools in cognitive psychology, and recently temporal cognition has been added to the list of processes modeled. Taatgen, van Rijn and Anderson (2004) proposed an augmentation of the Atomic Components of Thought-Rational (ACT-R) architecture based on a timekeeper type of representation. The module fit data from several experiments reasonably well and could be mapped logically onto brain activation patterns found in temporal cognition. Still, the timekeeper concept implies a greater role for implicit cognition than may be justified. According to that concept and the instantiation of it in ACT-R, the timekeeper provides direct access to time estimates in the form of a type of number value in time estimation studies or a timed response in time production studies. In other words, time estimation is represented as effortless; however, earlier we discussed Brown's finding (1997) that time estimation in and of itself requires attentional resources. This evidence that temporal cognition is not a purely implicit task would seem to require a different computation

An alternative to the strict interpretation of the timekeeper theory, and one that addresses the matter of attention, is presented by Cassenti (2004). Cassenti (2004) suggests that the timekeeper combines with inner speech to help estimate time when the durations are on the order of up to several seconds. When judging such intervals, people may use inner speech to pronounce words with a certain numbers of syllables. For example, in Carslon and Cassenti (2004), participants counted items presented one at a time on a computer screen. The items were either presented with a varied interval duration or a rhythmic duration. Cassenti (2004) suggests that when the items are presented with a rhythm, participants will use inner speech to reproduce the current number. This type of internal strategy is essentially the same as that observed in children playing hide-and-go-seek, when the seeker counts "one Mississippi, two Mississippi..." to mark the amount of time for hiding before calling out "ready or not, here I come." This conceptualization of temporal cognition has also been instantiated in an ACT-R model (Cassenti, 2004), without the addition of a specialized ACT-R module (as in Taatgen et al., 2004). Instead, inner speech was represented using ACT-R's output buffer and accounts for both correct performance and errors obtained under both the rhythmic and varied conditions. While neither account is definitive, this recent work highlights the utility of the cognitive modeling approach in exploring the mechanisms underlying temporal cognition.

3 Temporal Cognition at Work

We have talked about some common, everyday experiences that draw upon temporal cognition and we have covered some of the basic psychology underlying it. Now we must look at temporal cognition at work, that is, in the humans-in-automation environment. Recently, Hildebrandt and Rantanen (2004) hosted a panel on time design—the examination of the theories and methods applicable to the design of the temporal features and characteristics of a system. In their introduction, they nicely outline the temporal design space. It includes four dimensions of time: time as it is expressed in the system interface, through user behavior, in task requirements, and in the environment. The collective view of the panel is that time is not a hardware/software feature to be designed around, but rather, a property of the overall system that must be calibrated and orchestrated across the dimensions of the design space. In this section we will briefly discuss each of the four dimensions.

3.1 Temporal Aspects of the Interface

Today's system interface is driven by hardware and software that are built to operate with quartz clock precision, an unrelenting input-process-output cycle that is not necessarily in synch with any of the other dimensions. Early on, when the computer processing speeds were vastly slower than today, human computer interaction researchers spent considerable effort measuring system feedback or response times and urging, along with the hardware and software developers, that the times be as short as possible. It became clear, as discussed in O'Donnell and Draper (1996), that faster was not necessarily better, that a too-fast response negatively affected user performance. Also, similar to the findings from Carlson and Cassenti's task (2004), variability in system response times can cause lower performance, increased user frustration, and even changes in the strategies used to complete a task (O'Donnell and Draper, 1996). In other words, the timing characteristics of a system become yet another feature that the user must adapt to, and not always with the end result of optimized performance.

3.2 Of the User

Most of the key temporal aspects of the user were delineated in the earlier sections of this paper: time perception and estimation; reliance on memory and attention; changes with learning; differences at different time granularities; and some of the vulnerabilities due to interruptions and multi-tasking. It is worth adding a couple more items to the list. For one, there are individual differences both in the ability to estimate time and in the subjective reaction to time as a stressor. Individuals also vary as to their response styles (different from the pure motor aspects of reaction time), so that some simply respond more quickly when a response or decision is needed. All of these sorts of individual factors can affect performance.

3.3 The Task

The temporal aspects of the task have been studied fairly extensively, but, as it turns out, not yet exhaustively. It is well known that most tasks have required end points and that knowing when that end point is affects human performance. The granularity of the time marker available—milliseconds or seconds to days or weeks—is important because time estimation accuracy changes, that is, becomes more variable as the interval lengthens (i.e., Weber's law). Executing a set number of keystrokes within a few seconds is easier to estimate than the successful control of an unmanned system to a rendezvous point. Human computer interaction (HCI) researchers must be distinctly aware of the time granularity in order to properly evaluate\overall human-system performance. Further, the pace of the task, or of the actions required to perform the task, can influence the strategy, overall performance levels, and the actual perception of the time available to complete that task (Jobidon et al., 2004). In other words, if you have to perform faster, you are likely to think that there is actually less (clock) time available.

3.4   And Temporal Aspects of the Environment

The environment is where the system, the user, and the task come together. This is where the interactions abound—all of the issues of multi-tasking, of scheduled and non-scheduled interruptions, of the differential interference of temporal and non-temporal tasks. Brown's finding (1998) that performing a time estimation task negatively impacted the performance of a variety of other tasks has profound implications for the very sort of environment in which users, soldiers, and workers are finding themselves—multi-tasking environments where time management is one of the tasks. Temporal cognition, then, becomes key to successful performance and Hildebrandt and Harrison (2003) argue for using time as a key element in dynamic function allocation. In the following section, we offer some ideas about how to use what we know about temporal cognition as well as the rest of the humans-in-automation design space to improve overall performance.

4   How to Improve the Humans-in-automation Environment

The discussion of improving the humans-in-automation environment includes sections on human-system interactions and also on human-human interactions for a fuller representation of what the expected situations will be.

4.1   Human-system Interactions

4.1.1   Feedback

Skillful temporal estimation improves with feedback (e.g., Philbin & Seidenstadt, 1983). Kladopoulos, Brown, Hemmes, & Cabeza de Vaca (1998) found that when feedback was withheld from participants performing a temporal production task, participants were less accurate and more variable in the timing of their responses over trials than when feedback was provided. Feedback may have the same effect as multiple presentation of the same interval in a rhythm. In both cases, participants may develop an estimate of time and refine that estimate over multiple presentations of the interval. No feedback results in an uncorrected estimate of time and these missed opportunities at correction allow the mental timekeeper to drift over trials.

4.1.2   Time Markers

No matter what the precise form of the mental timekeeper, the timekeeper is best when it has multiple presentations of the same duration. These multiple presentations allow the time keeper to correct its own initial estimates of interval duration, but also allow it to correct for the added variability that grows over time. The timekeeper also needs a set of distinct markers to distinguish the start and end of an interval. Computers may be designed to provide sounds or visual displays to help pace users through any given task. These markers must be distinct enough to properly mark time, but not attention-grabbing enough to divert attention away from the primary task and therefore create a small interruption. The purpose of the time markers is to allow a more reliable method to self-evaluate retrospective judgments of time passed and prospective judgments of time remaining from the initial pool of available time. In the case of relatively short timeframes, a constant rhythm would help to make these judgments less variable and more accurate by providing feedback to the mental timekeeper. In the cases where an overt marker is not reasonable, a "governor" on the system response time so that it is rhythmic or uniform would also support human performance.

4.1.3   Cognitive Effort

The markers would also help by offloading the amount of cognitive effort used to estimate time given the conclusion that time estimation, that is, time management requires workload capacity and attentional resources (Brown, 1997; 1998). Thus, although the resource channels used for the primary task and the time estimation task should be deconflicted to the extent possible in the human-in-automations environment, consideration of working memory and cognitive effort should be primary. Ways of reducing cognitive effort such as computer-issued time markers could do the estimating for the user. Automated time markers would allow the user to concentrate on their primary task rather than relying on divided attention.

4.1.4   Time Granularity

Cassenti (2004) found that rhythms improve performance when they are shorter in duration (i.e., in the range of hundreds of milliseconds), but when increased to three seconds in a counting tasks, the rhythmic intervals hurt performance. It is important to keep the time granularity of a task in mind when deciding to use temporal factors to improve performance. In a counting task, it is relatively simple to assign new numbers to new events, so three second waits between events is too long to benefit from rhythmicity. So, different tasks require different durations between temporal markers. When designers implement temporal markers into a system, they must be aware of about how long the task typically takes and adjust the length of the intervals accordingly. Nittono, Tsuda, Akai, and Nakajima (2000) found that background musical tempo affects the performance of a perceptual-motor task. When tempo is quick, people tend to increase the pace of their actions, but also make more errors. The pace of actions slows with fewer errors when the tempo is slow. The study shows that paced event markers do affect performance. If a task is under time pressure, it would be beneficial to increase the pace of the event markers to quicken the pace, even if that causes more errors. If time pressure is less of an issue, then the pace of the event markers may be slowed to allow for greater accuracy.

4.1.5   Minimizing Detrimental Factors

The final way in which designers should think about temporal cognition is to reduce the number of detrimental factors that hurt temporal performance. To minimize varied timing, designers must be aware of the delay in their own systems. If the computer is subject to over-riding tasks that would derail the precise timing necessary to create rhythmic intervals of markers, then one of the primary roles of the markers would be removed. Temporal markers are only important if they can be presented at regular intervals. The same applies to the inclusion of delays. If the processing speed of the system is slow and causes random delays in the primary task (for example, most have experienced slow connection speed to the internet and the inevitable frustration that occurs if they are used to fast connection speeds), then the designer cannot hope to include improvements in the temporal design of their system. An interruption is typically a greater disruption to a primary task than a delay because with a delay information is still present, whereas interruptions remove attention from the task-relevant information. Interruptions occur all the time and some (e.g., telephone ringing) come from sources external to the computer support. However, there are also interruptions that the computer support causes directly. Interruptions such as the sound and taskbar icon that typically accompany the arrival of new email could be minimized when a user engages a tempo-control system and elects to minimize interruptions.

4.2   Human-human Interaction

Just as machines can support performance, teams may also be organized more efficiently to increase performance as well. The use of temporal factors is already observable in team behavior. A fundamental practice of basic training in the armed forced is marching as a drill sergeant shouts out commands such as, "left-left-left-right-left" to a rhythm while the soldiers coordinate their own marching and synchronize with one another. In this section, we will discuss how teams might be improved by applying temporal cognitive concepts.

4.2.1   The Conductor

Consider a situation in which computer support fails or does not possess a program to issue temporal markers. Without the proper organization a team lacking proper computer support would depend on each individual to estimate their own time. Not only would each individual's time estimate be error prone, but each would also presumably be performing another task. The coupling of their primary task and the secondary time estimation task would decrease their primary task performance (Brown, 1998). The obvious solution to this problem is to have one person as the designated temporal marker producer. This conductor would have no other task but the temporal task, thus freeing the conductor from all divided attention. This is analogous to the conductor of an orchestra, whose only task is to synchronize the band members by providing rhythms. Musical conductors are never asked to play their own instrument. It is understood that conductors would be too distracted to both operate an instrument and provide a rhythm.

4.2.2 Training Task-granularity

To maximize performance, the conductor would need to be trained to recognize which tasks should have certain time granularities. For instance, a rowing captain should know how long each stroke takes and not repeat "row" in the hundreds of milliseconds range, but should instead speak at a pace of about once every two seconds. The conductor should provide the appropriate time intervals to best pace the task. Slower tempos should be reserved for those tasks that can be performed more slowly and fast tempos should be reserved for tasks with fast approaching deadlines.

4.2.3 Minimize Distractions

In order for teams to benefit from a conductor, each team member would have to avoid the pitfalls of the detrimental temporal factors. For one, each member would have to pay attention to each temporal marker, whether it be visual (e.g., see a hand gesture) or auditory (e.g., hear the command, "row"). If too many temporal markers are missed, the intervals will appear to be varied intervals. Also, members should be mindful of distractions, which may cause delays or interruptions. It is essential that each member pay attention to the full task—the assigned task and the markers. Just like a symphony, one misplayed step in a coordinated action can turn a success into a failure.

5   The Outlook for Temporal Cognition in the Humans-in-automation Environment

Temporal cognition is still being explored at some very basic levels; however, a substantial base has been laid down so that it can be considered explicitly in the design of the humans-in-automation environment. First, we know that attending to and managing the temporal aspects of a task require attention and cognitive effort. This effort should be minimized or managed in a way that supports the individual's need for temporal information (how much time is there left before the decision must be made to shoot or not to shoot?). We also know that rhythmic or easily predicted times promote better performance. This is true in the case of feedback from a computer that a key has been pressed or with respect to the control of a remote vehicle or the interpretation of a video image as to the location of an enemy. Fortunately, emerging research tools from sister disciplines such as biotechnology will help explore underlying temporal regulation facets and processes, assisting humans in keeping up with technology-driven changes in their environment. For soldiers and those operating in this emerging high-technology environment, temporal cognition is key to the humans-in-automation environment.

References

Baddeley, A.D. (1986). *Working memory*. New York: Oxford University Press.

Baddeley, A.D., Thomson, N., Buchanan, M. (1975). Word length and the structure of short-term memory. *Journal of Verbal Learning and Verbal Behavior, 14,* 575-589.

Brown, S.W. (1997). Attentional resources in timing: Interference effects in concurrent temporal and nontemporal working memory tasks. *Perception & Psychophysics, 59,* 1118-1140.

Brown, S.W. (1998). Automaticity versus time sharing in timing and tracking dual-task performance. *Psychological Research, 61,* 71-81.

Carlson, R. A. (1997). *Experienced cognition*. Mahwah, NJ: Lawrence Erlbaum Associates.

Carlson, R.A. & Cassenti (2004). Intentional control of event counting. *Journal of Experimental Psychology: Learning, Memory, & Cognition, 30,* 1235-1251.

Cassenti, D.N. (2004). When more time hurts performance: A temporal analysis of errors in event counting. *Unpublished Dissertation.* The Pennsylvania State University.

Edwards, M.B. & Gronlund, S.D. (1998). Task interruption and its effects on memory. *Memory, 6,* 665-687.

Essens, P.J. & Povel, D.J (1985). Metrical and non-metrical representation of temporal patterns. *Perception & Psychophysics, 40,* 69-73.

Fraisse, P. (1984). Perception and estimation of time. *Annual Review of Psychology*, 35, 1-36.

Gibson, J.J. (1979). *The ecological approach to visual perception*. Boston: Houghton Mifflin.

Hancock, P.A. & Weaver, J. L. (2003). On time distortion under stress. *Theoretical Issues in Ergonomic Science, in press.*

Hildebrandt, M. & Harrison, M. (2003). Putting time (back) into dynamic function allocation. In the *Proceedings of the Human Factors and Ergonomics Society 47th Annual Meeting.*

Hildebrandt, M. & Rantanen, E. (2004). Time design. In the *Proceedings of the Human Factors and Ergonomics Society 48th Annual Meeting*, pp 703-707.

Hollnagel, E. (2002). Time and time again. *Theoretical Issues in Ergonomic Science, 3*, 143-158

James, W. (1890/1950). *Principles of psychology*. New York: Dover.

Jobidon, M., Rousseau, R., & Breton, R. (2004). Time in the control of a dynamic environment. In the *Proceedings of the Human Factors and Ergonomics Society 48th Annual Meeting*, pp 557-561.

Jones, M.R. (1990). Musical events and model of musical time. In R.A. Block (Ed.) *Cognitive models of psychological time*, pp. 207-240. Hillsdale, NJ: Erlbaum.

Jones, M.R. & Boltz, M. (1989). Dynamic attending and responses to time. *Psychological-Review*, 96, 459-491.

Killeen, P.R. & Weiss, N.A. (1987). Optimal timing and the Weber function. *Psychological-Review*, 94, 455-468.

Kladopoulos, C.N., Brown, B.L., Hemmes, N.S., & Cabeza de Vaca, S. (1998). The start-stop procedure: Estimation of temporal intervals by human subjects. *Perception and Psychophysics, 60*, 438-450.

Kutar, M., Britton, C., & Nehaniv, C. (2001). Specifying multiple time granularities in interactive systems. In P. Palanque & F. Paternó (Eds.) *Proceedings of Design, Specification and Verification of Interactive Systems Workshop*, 49-61.

McCrone, J. (1997). When a second lasts forever. *The New Scientist*, 1 November 1997.

Mel, B.W., Niebur, E., & Croft, D.W. (1997). How neurons may respond to temporal structure in their inputs. In: J.M. Bower (Ed) *Computational Neuroscience: Trends in Research*. New York: Plenum Press, p. 135-140.

Nygren, T.E. (1982). Conjoint Measurement and conjoint scaling: A user's guide. Wright-Patterson Aerospace Medical Research Laboratory Technical Report. AFAMRL-TR-82-22.

Nittono, Tsuda, Akai, and Nakajima (2000). Tempo of background sound and performance speed. *Perceptual and Motor Skills, 90*, 1122.

O'Donnell, P. & Draper, S. (1996). Temporal aspects of usability: How machine delays change user strategies. SIGCHI Bulletin, 28:2, April 1996. <URL = https://homepages.cwi.nl/~steven/sigchi/bulletin/1996.2/Steve-Draper.html>.

Palmer, C. (1997). Music performance. *Annual Review of Psychology, 44*, 115-138.

Philbin, T. & Seidenstadt, R.M. (1983). Feedback and time perception. *Perceptual and Motor Skills*, 57, 308.

Stelling, J., Gilles, E.D., & Doyle, F.J. III (2004). Robustness properties of circadian clock architectures. *Proceedings of the National Academy of Sciences (Biophysics), 101, 36*, 13210-13215.

Szelag, E., Kanabus, M., Kolodziejczyk, I., Kowalska, J., & Szuchnik, J. (2004). Individual differences in temporal information processing in humans. *Acta Neurobiologiae Experimentalis, 64*, 349-366.

Taatgen, N.A., van Rijn, H. & Anderson, J.R. (2004). Time perception: Beyond simple interval estimation. *Proceedings of the 6th international conference on cognitive modeling*, 296-301. Mahwah, NJ: Erlbaum.

Trafton, J.G., Altmann, E.M., Brock, D.P., & Mintz, F.E. (2003). Preparing to resume an interrupted task: Effects of prospective goal encoding and retrospective rehearsal. *International Journal of Human Computer Studies, 58*, 583-603.

Wickens (2002). Multiple resources and performance prediction. *Theoretical Issues in Ergonomic Science, 3*, 159-177.

Wing, A.M. & Kristofferson, A.B. (1974). Response delays and the timing of discrete motor responses. *Perception and Psychophysics, 14*, 5-12.

Concentration – An Instrument to Augment Cognition

Anthony W.K. Gaillard

TNO Human Factors
PO box 23, 3769 ZG Soesterberg, the Netherlands
gaillard@tm.tno.nl

Abstract

Working in demanding situations that are complex and unpredictable, involving intensive information processing under time pressure, pose special requirements for the cognitive abilities of operators. Measures, support, and training aimed at improving the processing capacity of the operator, mostly focus on the cognitive aspects. However, the magnitude of this capacity and its efficient use, are determined also by other factors as well: e.g., emotion, motivation, activation, and effort. It is argued that these factors may be as important to augment cognition, in particular in situations where fatigue, stress, conflicts, etc., play an important role. This paper presents a framework in which the neglected concept *concentration* plays a key role. This concept has been chosen because it incorporates the processes (attention, energy mobilization, and motivation) that are crucial in cognitive functioning, in particular under demanding conditions. The framework can be used to elucidate the interplay between these processes and to evaluate the influence of inhibiting factors on cognitive processing, such as fatigue and stress.

1 Where is capacity limited?

The human brain has an enormous capacity to process information. We only have to observe a pianist or a surgeon, to realize that a human being is capable to perform very complex tasks at a high rate and apparently without too much effort. Also in our daily life we manage to perform a lot of tasks, such as reading and car driving, without realizing how complex these tasks are. Of course, it could be argued that these tasks are not difficult because they are very well trained and largely automated. Therefore, these tasks can be executed at a low level of both attention and effort. Consequently, when operators complain about information overload we should first examine the attentional and energetical aspects, not only of the task itself, but also of the work environment in general. How attention demanding is the task, and how stimulating is the environment, thereby generating sufficient energy?

To augment the processing capacity of operators we should not only examine the cognitive aspects of the task. When operators complain about the amount of work, their information capacity may not yet be overloaded. It is quite possible that the available processing capacity is not used efficiently, or the work is not stimulating enough. In several studies in work psychology (e.g., Karasek, 1990) and cognitive load (e.g., Neerincx, 2003), it has been shown that inefficiency and non-productive behavior are not only determined by the amount of work, but also by the way the work is *organized*, such as autonomy, work/rest-schedules, interruptions, information exchange, feedback, task allocation, and by *psychosocial factors*, such as social support, coaching, rewards, perspective, and commitment. Even under regular working conditions these factors may lead to feelings of strain, to absenteeism, and in the long run to burnout and cardiovascular diseases. It is conceivable that under conditions of threat, information overload and intensive communication, the chances of negative effects on cognitive readiness and health are even higher.

To augment cognition it is therefore worthwhile to examine not only the cognitive factors that determine the efficiency and effectiveness of information processing, but also to investigate how operators use their resources, which depend on energetical (arousal, activation, and effort), motivational and emotional factors. In the current paper a framework is presented in which the functional state of the operator is determined by the factors at three levels: cognitive, energetical, and emotional. The combination and coordination between the factors on these levels determine the quality of the behavior oriented towards the goals to be achieved.

The framework may be used to improve the organizational and psychosocial determinants of the work environment and to develop measures that enable operators to make optimal use of their resources. The framework may also be used to develop IT support systems that do not only focus on the cognitive state of the operator but also take emotional and energetical aspects into account.

2 Attention, effort, and motivation

The functioning of operators in dynamic and demanding tasks is determined by both positive (skills and experience, but also motivation and satisfaction), and negative factors (information overload, time pressure, fatigue, and stress). Although the interplay between these factors is quite complicated, a few core elements appear to play a critical role in each of these factors. These elements can be divided in three aspects:

- *Attentional* processes referring to the cognitive aspects of the functioning of the operator
- *Energetical processes* (effort, arousal) referring to the biological, physiological, and hormonal state of the operator
- *Emotional processes* referring to feelings, motivation, and attitudes determined by psychosocial determinants

On the basis of this categorization a model is developed that can be used to evaluate how operators mobilize, coordinate, and unite their resources in an attempt to meet the task demands. In this way bottlenecks in their functioning can be detected and measures to improve their performance can be developed.

The model is based on the neglected concept *concentration*. Concentration is a dynamic mechanism, which mobilizes and coordinates our resources in order to develop and sustain goal directed behavior. Concentration encompasses the three aspects: with increased concentration, our attention is directed towards the relevant aspects of the task; energy is mobilized to bring our brain and body in a state that is appropriate to execute the task. Emotions provide the motivation, and thus the intent to do the task and to realize the goal. You always need a goal to concentrate on, it is impossible to concentrate on 'nothing'. Thus, motives steer the concentration process. The present framework can also be used to describe the most important factors that inhibit the functioning of operators. Distraction, fatigue, and stress appear to effectuate their negative influence by degrading the concentration process.

Figure 1 gives a schematic illustration of the concentration model. It consists of three layers: a cognitive environment which involves the steering aspects of our behavior, a psychosocial environment which determines largely our motivation and energy mobilization, whereas task performance takes place between these layers at an executive level. The primary process consists of translating the task demands into performance results. To be able to do this, we need a task set containing the knowledge and actual information necessary to perform the task; but also the global goal to be accomplished, the way of working, and a planning scheme to realize the goal within a specified time frame. At the steering level, cognitive control continuously determines how our attention is distributed over the different aspects of the task. This type of attention is purposeful and consciously directed towards performing the task and realizing of the intended goals. This is called *directed attention* to contrast it with other types of focused attention that are not inspired by our goals, but evoked by stimuli and events that attract our attention. Cognitive control is an 'off-line' process that continually evaluates the progress we make, on the basis of feedback from our performance. If needed we may change our way of working by accepting more errors, or mobilizing more energy through effort. We may also revise our planning by asking for help or postponing our deadlines. Ultimately we may decide to give up the intended goal. In every task situation we are constrained with several other goals we have to meet. There is always competition between the intent to do the current task and other motives that beg for attention and alternative actions. Distractions (e.g., reading email), fatigue (e.g., need to rest) and stress (protect own resources) may disrupt our directed attention to the task and undermine goal-directed behavior, causing performance deterioration. This illustrates how closely connected the processing on a cognitive level is with emotional and energetical factors.

2.1 Attention and goal-directed behavior

Concentration is conceptualized as an instrument to attain an optimal state of our brain to be able to apply our resources efficiently. It regulates our attention (directing, filtering, switching of attention, and dividing between subtasks, etc.), and our energy management (preparation before task, distribution of energy over the work period, mobilizing extra energy through mental effort, deactivation, and recovery after work, etc.). The capacity to concentrate plays an important role in the coordination and organization of our activities, in particular in complex task situations. To give a few examples: the coordination between perceptual, cognitive and motor processes (putting a thread through the eye of a needle), between computational processing on an operational level (task execution itself) and cognitive processing on a strategic level (control of own performance and evaluation of the progress made), and between own performance and that of colleagues (cooperation, information exchange, offering help, etc.). Thus, concentration in a broader perspective not only plays a role in the execution of the current task, but also takes into account possible future developments.

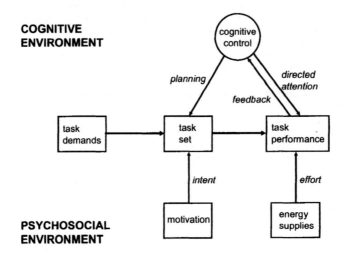

Figure 1: The concentration model, encompassing attention, effort, and motivation.

Since we can only concentrate when we have a goal to concentrate upon, concentration is an important mechanism in goal-directed behavior. When we have a concrete plan on how to realize a particular goal, we are better able to divide our attention between tasks, distribute our energy over the total work period, and to postpone less important subtasks, or neglect interruptions, such as phone calls, emails, etc. The plan has to specify the methods and means to be used, and a time schedule with milestones. The more concrete the plan is, and the more rewarding the goal, the more motivating operators are to mobilize their resources to realize the goals. Only when they are sufficiently motivated, they are able (or willing) to do so, even when there are problems, misunderstandings, changes in plan, and drawbacks which tend to evoke negative emotions and energy.

2.2 Energy mobilization

Under normal circumstances our energetical state is in line with the activities we want to undertake. Energetical mechanisms regulate our brain and body to bring them in an optimal state to process information. The majority of bodily processes are regulated automatically and unconsciously. The execution, and even the planning, of a task prompts energetical mechanisms to adapt our brain and body gradually into an state, which is optimal for efficient processing, and therefore determines the capacity available to execute a task. In most instances we do not have to pay attention to this continuous adaptation, which gives us the opportunity to concentrate on the more interesting aspects of life. Only when our state is far from optimal due to fatigue or strong emotions, do we realize how much cognitive processing is affected by our state. Since these effects are mostly outside our control, we can only attempt to modulate them, adapting to the current demands. When planning our daily activities, we may take into account possible fluctuations in our state, due to fatigue or time-of-day effects. Morning people may prefer to make difficult decisions early in the day, whereas evening types do not mind to do some overtime.

Besides the endogenous effects of our organism (e.g., circadian rhythm) and the influence of environmental factors (e.g., noise, sleep loss, and temperature), two types of task-related energy mobilization may be distinguished: task-induced activation which refers to the stimulating effects of the task or the work environment in general; and internally-guided mental effort which involves voluntary energy mobilization under conditions of cognitive load.

Similarly, Hockey (2003) distinguished two types of energy mobilization, one for routine regulatory activity and one for effort-based control.

2.2.1 Task-induced activation

When we know that a particular task has to be done in the near future, we will prepare for certain activities, not only on a cognitive level, but also on an energetical level. Just thinking about the task to be done, affects the regulation of our state: energetical mechanisms are activated to reach an optimal state. The relation between state and efficiency is dependent on the demands of the task on the one hand and the availability of processing resources on the other. When processing capacity is abundant, a deviation from the optimal state will not result in a reduction of performance efficiency. However, in complex or novel task situations that require all our resources, even a small deviation from the optimal may result in a performance decrement. Thus, for well-trained tasks that do not require many resources the range, in which optimal performance can be obtained, is larger than for tasks for which the amount of resources required approaches the available capacity. We therefore prefer an easy task to a difficult one, when we are tired. In the evening for example, we may be too tired to write a technical paper, but we may be able to read the paper.

2.2.2 Mental effort

When the actual state does not deviate too much from the optimal state, we can attempt to maintain performance at the same level by mobilizing extra energy through mental effort. This "try harder" response can only be maintained for a relatively short period, since the physiological and psychological costs are high, in terms of strain and fatigue. Kahneman (1973) identified effort with the action of maintaining a task activity in focal attention; effort is needed, when lapses in attention immediately result in performance deterioration. This may be the case in the following, apparently quite different situations (see also Gaillard, 1993):

a)   The energetical state is not optimal due to sleep loss or fatigue,
b)   Emotions disrupt the energetical state and the directed attention,
c)   The task is attention demanding because of inconsistent or varying input-output relations, or heavy demands on working memory,
d)   Complex environment requiring divided attention between tasks,
e)   Skills have to be acquired in a new (learning) situation.

These situations have in common that it is difficult to maintain the task set. In a. and b., maintaining the task set is challenged by other goals, such as doing nothing or paying attention to your emotions, problems, and complaints. In c. and d., the task continuously requires attention, because the task set is changing and processing modules are used with limited capacity. In e. a task set has still to be developed.

From the above it can be deducted that there is no direct relation between task difficulty and effort. Increases in task difficulty only result in the mobilization of extra energy through mental effort, when the operator is motivated to do so. This explains why non-specific motivational variables, such as feedback or bonus, have larger effects on indicators of effort than changes in task difficulty. Thus, whether in a particular task situation effort is mobilized, depends not only on the characteristics of the task, but also on the motivation and attitude of the operator (Veltman & Gaillard, 1997).

2.3 The role of emotion

The role of affective processes has been largely neglected in cognitive psychology, human factors, and ergonomics (e.g., Eysenck, 1982). Most theories only distinguish between state and process, dichotomizing between energy (e.g., arousal, activation, and effort) and cognition. Emotions may be regarded as a third layer in between the processing on a cognitive and on a physiological level. Although emotions mostly have a negative connotation in both cognitive psychology and the stress literature, they also have positive influences. Emotions play an important role in motivating people to initiate and maintain task goals in the first place, but they may also interfere with cognitive processing. In particular under time pressure or threatening conditions, the regulation of our emotions is critical for efficient performance. When we are uncertain about our goal, our proficiency to meet the criteria, or about the rewards to be gained, we may have difficulty in maintaining the task set. In particular, under conditions of distraction, fatigue or stress, there will always be competition between other goals and the intent to perform the current task. Since sustaining an effortful state is subjectively aversive and has its costs, it may conflict with other personal goals, such as maintaining well-being and health (see also Hockey, 2003).

Intense negative emotions may reduce performance efficiency in several ways (Gaillard, 2001):

- They continually beg for attention. Since they have "control precedence", they will occupy some processing capacity, leaving less for performing the task.
- They are distracting: operators may have problems following their line of thought and executing their plan.
- In extreme cases, the goal-oriented behavior may be disrupted, and operators may have doubts about their motives, and about the sense and feasibility of the mission.
- They may disrupt the state regulation, which makes it less optimal for task performance due to overreactivity.
- They may be so distracting that they directly interfere with the performance of the task (for example, decision making, and team interactions).
- They may cause psychosomatic complaints, which also demand attention.

Negative emotions evoke energy that in most cases is not functional disrupting the processing of information. Note that the factors which increase our resources are different from the factors that inhibit the proper use of resources (see also Figure 3). In addition, positive and negative energy not necessarily compensate each other. In a poor and unhealthy work environment, people will not work optimally, even when they receive a high salary.

To summarize, emotions affect cognitive processing at least in three ways (see also Figure 1): 1) they determine our goals and consequently direct our plans and behavior; 2) they enable the mobilization of energy; 3) under demanding and threatening conditions, however, they may disrupt the energetical state; and 4) distract our attention and interfere with cognitive processing.

3 Attention and energy

The distinction between directional, content-specific processes (e.g., cognitive control, and attention) and intensive, energizing aspects of our behavior (e.g., energy, arousal, and effort) has a long history in psychology (Kahneman, 1973; see Hockey, Gaillard & Coles, 1986, for a review). As early as 1955 Hebb maintained that "arousal is an energizer, not a guide", indicating that an optimal state of our brain may facilitate information processing, although it does not guide our behavior. It is an engine, not a steering wheel. Energetical mechanisms have to bring our brain in a state that is appropriate to meet the demands of the task. For a long time energy and emotion were not seen as separate processes. The classical one-dimensional concept of arousal combined both types of processes. This is also the case in the well known inverted U-curve, which has yielded controversial interpretations (see section 3.4).

The process of concentration is characterized by a continuous interplay between directing our attention to the task activities and guiding our behavior on the one hand, and the mobilization of sufficient energy to bring and maintain our brain in the appropriate state to process the task information on the other. Under normal conditions the two types of processes converge. Only under extraordinary conditions we experience how difficult it is to remain focused, when our energetical state is not appropriate for the task to be done. If anxiety or anger overwhelms you, or you suffer from sleep loss, it is difficult to keep your attention directed at the task. When planning our daily activities at work and at home, we take into account possible fluctuations in our energetical state.

Figure 2 illustrates that under normal circumstances directing our attention and mobilizing energy are highly correlated. When we are daydreaming we spend very little energy and our attention is not directed to a specific goal, but is wandering without a concrete aim. When we concentrate on a goal, energy is mobilized which enables the direction of attention to the activities at hand. Concentration can vary from passive activities (watching TV), via routine tasks, to complex tasks, which require the full capacity of our brain to meet the task demands. Concentration is assumed to be optimal when both the energy mobilization and attention are intensive and completely directed to the task. This condition is similar to the state of flow (Csikszentmihalyi, 2003): a challenging situation, which makes maximal appeal to the operator to use his/her skills to meet the demands of the task. This is only possible, when the task is inspiring and the goal well defined. The operator should be intrinsically motivated, have self esteem and not be hindered by personal problems or other distracting thoughts.

The diagonal in Figure 2 represents the situations in which there is a balance between attention and energy as is conceptualized by classical one-dimensional activation theories. In these theories it is generally assumed that performance is sub-optimal when the concentration level is too low, given the task demands. Although recent models generally support this view, they differ in the expected effects, when distortions of the concentration process occur. Attention and energy do not converge in all situations. They are independent processes, which are affected by different factors in different ways. For the sake of argument two extreme situations are described that differ in energy mobilization and the direction of attention: stress and fatigue.

3.1 Stress

In the stress literature controversy exists about the effects of high arousal on performance efficiency. At high levels of arousal enhanced energy mobilization no longer improves processing efficiency. In terms of the current framework, operators are no longer able to pay full attention to the task, which impairs effective, goal-directed behavior (lower right corner in Figure 2). When we are very aroused (for example by threat or extreme pleasure), we have a lot of energy while we may not be capable to concentrate on the task, because the energetical state is not optimal and our attention is dispersed. In situations of threat (accident, fire, etc.), we tend to react with basic behavioral patterns, such as the well known 'flight/fight response', which enables a quick and intensive energy mobilization. This state can be advantageous to save your life ('first shoot and then talk'), but it is not appropriate to perform complex tasks and make difficult decisions. In situations where the course of events is unpredictable and the outcomes uncertain, we tend to react with anxiety and to relay on safe, conservative and rigid strategies. With extreme threat people are no longer able to think in a flexible way, which inhibits their problem solving. In panic situations people may adhere to incorrect coping strategies, which may result in more casualties than caused by the threat itself. Also in daily life there are large differences between individuals in their balance between energy and attention. Some persons may have a lot of energy and a large readiness for action (for example, ADHD), while they have problems to focus on a specified goal, maintain a straight line of reasoning, and sustain their attention at the same task for a longer period.

This raises the question which factors determine the difference between the two states characterized by high-energy mobilization with or without directed attention (stress and flow in Figure 2). In both states the attentional range is 'narrowed' and focused in a specific direction. However, under threat this so-called tunnel vision is forced by external events (e.g., fire, aggressive person), whereas with flow the narrowing is a result of concentration on activities needed to realize a specific goal. Thus, under flow attention is voluntarily directed by internal motives and goals. Under threat attention is drawn automatically by external events, which results in rigid thinking, little control over our attention and in primitive behavioral patterns that may not be appropriate. The essential difference between the two states appears to be the type of energy mobilization. With flow, the energy is elicited by the task demands and by mental effort, whereas under threat the energy is evoked by negative emotions, such as anxiety and anger. As was argued in the sections 2.2 and 2.3, this type of energy results in a functional state, which is less appropriate for complex processing of information.

3.2 Fatigue

When operators have to work for a long period, they experience more and more feelings of fatigue, in particular when the task is continuous, repetitive or boring. As a result the working pace slows down and errors are more likely to occur. Operators have problems to remain concentrated on the task, and to keep the focus of their attention on the core activities of the task. Their thoughts may easily drift away from the current task to the environment, to rest and pleasure, or to other competing activities. The magnitude of these effects, however, is not only determined by the amount of fatigue, but also by the operator's motivation. When we are very motivated, due to high rewards or sanctions, we will fight fatigue. We are prepared to invest more effort in the current task and to resist the negative feelings that beg for attention and ask the operator to stop the current activity. As negative emotions, feelings of fatigue have "steering precedence": they are so annoying, so as not to be easily ignored. This implies that less attention, and thus capacity, is available for the processing of task information. Thus, reduced performance may be caused by a depletion of energy resources, or by a loss of motivation resulting in the reduced mobilization of energy. In the latter case, sufficient energy may still be available but the operator is not willing to put more effort in the task.

Even when our energy level is rather low, we may be able to concentrate on a particular activity. Meditation and reflection are examples of concentration at a low energy level. Also creative and associative thinking can be done at low energy levels. However, this type of processing is only possible under optimal conditions. The operator is very sensitive to distraction, such as interruptions and drawbacks in the execution of the task. To be able to perform at low energy levels (e.g., due to sleep loss), the task has to be interesting, the environment stimulating, but stable (no unexpected events). Moreover, it certainly helps to work in a team, to receive regularly feedback on your performance, and high rewards for good results. Finally, one should be inherently motivated to do the job (enjoying the work). Thus working at a low expenditure level is possible, but only for a limited number of activities and under optimal conditions. This explains why in some tasks (monotonous vigilance and continuous reaction tasks) performance deteriorates already within 30 min., whereas in daily life people are able to remain focused for hours: for example, driving in eight hours to your holiday resort after a day's work (not to be recommend though).

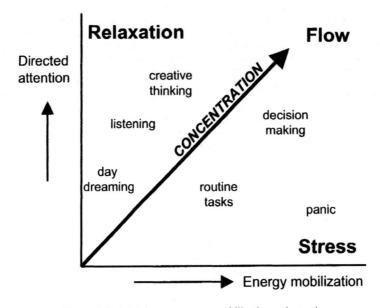

Figure 2: Relation between energy mobilization and attention

3.3 Intensity and specificity

When you concentrate on a difficult task, both attention and energy should not only increase in intensity, but also in *specificity*. Thus, enhanced concentration does *not* mean that the operator pays more attention to everything in the environment and that all parts of the brain are activated. Attention should be directed towards the task and on reaching goals specified in the task set. Similarly, energy mobilization is focused on activating the parts of the brain needed to process the task information.

When energy is mobilized by preparing and exciting the task or via mental effort, only those parts of brain and body that are required by the task demands are activated. For example, the blood flow is directed more intensively to those areas in the brain that are needed to process the information, as can be demonstrated with brain scans (Wilson, 2003). This is in line with the idea that for every task there is an optimal energetical state, which implies that not every type of energy mobilization is appropriate for the performance of a particular task (Hancock, 2003; Hockey et al., 1986; Gaillard, 2001). For example, when you wake up because your house is on fire, you become highly activated within a short period. This results in a state appropriate to flee out of the house, but certainly not for writing a difficult report.

In the present framework effort is regarded as a try-harder response, which attempts to compensate for any deviation from an optimal state, which endangers the proper performance of the task. There are several kinds of possible deviations (see also Figure 2): 1) the energy level is too low due to fatigue; 2) the energetical state may be distorted due too threat or stress; attention may be scattered due to distraction or stress; 3) finally, energy and attention are both insufficient due to poor motivation (you are forced to do a task with great reluctance). In all situations, our effort must prevent us from loosing the task set and letting our attention wander to other activities that originate from alternative goals, and do not contribute to the realization of the current goal.

3.4   Concentration and the inverted U-curve

In human performance research, increases in task demands are assumed to enhance the level of activation, either directly or via enhanced effort or stress. However, studies on the relation between activation and performance efficiency have revealed ambiguous and contradictory results (e.g., Eysenck, 1982). Reliable results are found only at the extremes. Performance is reduced either because the energy level is too low, due to sleep loss or fatigue, or too high, due to anxiety or stress. The relation between the energetical state and performance efficiency is often assumed to be an inverted U-shaped curve (see also Hebb, 1955). So far the U-curve hypothesis has received scant empirical support, and a number of methodological problems have been raised against this type of research (e.g., Neiss, 1988). It has been questioned whether the inverted-U is a correlation or a causal relation. It is quite possible that a reduction in performance efficiency and high activation are affected by a third factor independent from each other. For example, high levels of arousal are often accompanied by intense emotions, which at the same time increase the energy level but also deteriorate attention directed to the task. Events or manipulations, either in the laboratory or in daily-life that enhance the level of activation, may at the same time elicit emotions and distractions, which reduce processing capacity and performance efficiency directly (see also Näätänen, 1973). Secondly, there is no agreement among researchers on how to determine objectively the different levels of activation. A third problem is that (one-dimensional) activation theories do not discriminate between different types of energy mobilization and hardly specify the effects on emotional and cognitive processes, and the consequences this may have for the working behavior of the employee. As a result, arousal theories are not able to explain why under some conditions efficient performance is possible even with high levels of activation, whereas debilitating states that degrade performance may also occur at medium or low levels of activation. It appears that negative emotions reduce performance efficiency, at all levels of activation (Gaillard, 2001).

The concentration model offers a framework in which these problems can be discussed, which may give insight into possible explanations and solutions. It should be notified that the present model and the inverted U-curve are not necessarily contradictory. The current model can be seen as an extension of the U-curve, because this curve can still be drawn on the x-axis of Figure 2. The two models diverge at the medium level where the U-curve bends downwards and performance efficiency is degraded ending in a panic situation. In the current framework, however, concentration can still increase, transcending to the flow state, which results in enhanced efficiency. It is assumed that under optimal conditions, it is possible to mobilize so much energy that operators are able to dedicate all their processing capacity to the task. Thus, you can never have too much energy, as long as it is positive task-related!

This issue also has been discussed in the stress literature. It has been assumed that each form of overreactivity due to a high work load, would elicit too much sympathetic (adrenergic) activity that has aversive effects on well being and health (Frankenhaeuser, 1986). However, it can be argued that overreactivity is not caused by the amount of work, but by negative emotions evoked by detrimental psychosocial determinants in the work environment (Karasek & Theorell, 1990). Moreover, these theories still have to demonstrate why the enhanced physiological activation due to cognitive overload is worse than jogging, which is supposed to be healthy (Gaillard, 2001).

Thus, the nature of the relationship between energy mobilization and performance efficiency appears to depend on the type of energy. It tends to be curvilinear, when evoked by negative emotions and linear when generated by the task or via mental effort. This leads to the subsequent question: which characteristics in the work environment lead to positive and which to negative energy? Research in work psychology on the influence of organizational and psychosocial determinants of performance, well-being, and health, may give the answers. When the models on work stress of various researchers (Frankenheauser, 1986; Hockey, 2003; Karasek & Theorell, 1990; Csikszentmihalyi, 2003) are compared (see Gaillard, 2001) a general picture emerges that may be relevant in the present discussion. In an optimal work environment there should be a balance between positive (e.g., autonomy, rewards, and support) and negative determinants (cognitive load, stress, and fatigue). If the balance is positive, people may work very hard without experiencing stress or adverse health effects. If the balance is negative operators will react with negative emotions, strain, and increased health risks. The positive state (flow in Figure 2) is characterized by efficiency, skillfulness, high control, engagement, and satisfaction; the negative state (stress in Figure 2) is characterized by low control, disengagement, and the underutilization of the operator's skills and cognitive abilities; the work is threatening rather than challenging. The combination of high demands and low control results in strain, low efficiency and negative emotions, whereas high demands and high control lead to energy mobilization, active behavior and efficient performance. Thus, a high level of activation is not always accompanied by a reduction in performance efficiency, which implies that there are different patterns of reactivity to work demands that have different consequences for performance, well-being, and health risks.

4 Positive and negative energy

Under normal conditions there is a balance between the factors generating positive energy (esteem, social contacts, autonomy) and those generating negative energy (cognitive overload, disruptions). When operators have to work under demanding conditions this balance may be distorted. The more demanding the work environment, the more attention should be given to determinants that generate positive energy; that is, the readiness to put effort in the task and to keep attention towards the task goal.

Recent views on stress assume that the effects of work demands are dependent on the balance between positive and negative factors at work, which have been called inhibitors and energizers respectively. Inhibitors refer to the cognitive, emotional and physical demands of the work, whereas energizers refer to the organizational and psychosocial characteristics of the work environment (e.g., autonomy, rewards, social climate, social support, leadership, etc.). Figure 3 illustrates how the level of concentration is determined by the balance between factors that motivate and generate positive energy, and factors that distort energy mobilization and direction attention. Thus, complaints about cognitive overload are not only caused by a shortage of cognitive capacity, but also by a lack of motivation or by distortion of concentration due to distraction, fatigue, and stress.

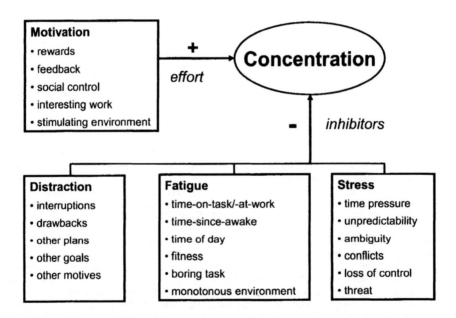

Figure 3: The level of concentration is determined by the balance between factors that motivate and generate positive energy, and factors such as distraction, fatigue, and stress that distort energy mobilization and attention.

5 Epilogue

Measures to augment cognition may be based not only on research of the cognitive aspects, but also on operator's motivation. The present concentration framework is able to incorporate findings coming from different disciplines, such as cognitive psychology, work psychology, and psychophysiology. Combining these findings provides a point of departure for further research on the development of measures, training courses, and IT support tools to augment cognition. Concentration may be enhanced by generating positive energy through coaching, feedback, and social support. At the same time the performance of the operator should be protected from factors that inhibit the concentration process (Gaillard, 2003). In the perspective of the current framework one can think of three related directions to pursue:

1. Increase the resources of the operator by making the task more interesting and the environment more stimulating. Feedback, rewards and teamwork enhance motivation, and therefore intent and cognitive readiness.
2. Find ways in which the available resources can be used more efficiently and effectively by time management and task allocation (between operator and computer, and between team members).
3. Reduce the factors that are distracting, disrupt an optimal state, and inhibit the mobilization of resources.

References

Csikszentmihalyi, M. (2003). *Good business. Leadership, flow, and the making of meaning.* New York, Harper Collins.

Eysenck, M.W. (1982). *Attention and arousal.* Berlin, Springer Verlag.

Frankenhaeuser, M. (1986). A psychobiological framework for research on human stress and coping. In: M.H. Appley & R. Trumball (Eds.), *Dynamics of stress.* New York, Plenum.

Gaillard, A.W.K. (2001). Stress, Workload, and Fatigue as Three Biobehavioral States: A General Overview. In: P.A. Hancock & P.A. Desmond (Eds.), *Stress, Workload, and Fatigue* (pp. 623-639). Mahwah (NJ), Erlbaum.

Gaillard, A.W.K. (2003). Fatigue assessment and performance protection. In: G.R.J. Hockey, A.W.K. Gaillard, & O. Burov (2003). *Operator functional state: The assessment and prediction of human performance degradation in complex tasks* (pp. 24-35). Amsterdam, IOS Press.

Hancock, P.A. (2003). The mitigation of stress, workload, and fatigue in the electronic battlefield In: G.R.J. Hockey, A.W.K. Gaillard, & O. Burov (2003). *Operator functional state: The assessment and prediction of human performance degradation in complex tasks* (pp. 53-64). Amsterdam, IOS Press.

Hockey, G.R.J. (2003). Operator functional state as a framework for the assessment of performance degradation. In: G.R.J. Hockey, A.W.K. Gaillard, & O. Burov (2003). *Operator functional state: The assessment and prediction of human performance degradation in complex tasks* (pp. 8-23). Amsterdam, IOS Press.

Hockey, G.R.J., Coles, M.G.H., & Gaillard, A.W.K. (1986). Energetical issues in research on human information processing. In: G.R.J. Hockey, A.W.K. Gaillard, & M.G.H. Coles (Eds.), *Energetics and human information processing* (pp. 3-21). Dordrecht, M. Nijhoff.

Kahneman, D. (1973). *Attention and effort.* Englewood Cliffs (N.J.), Prentice Hall.

Karasek, R. & Theorell. T (1990). *Healthy work.* New York, Basic Books.

Näätänen, R. (1973). The inverted-U relationship between activation and performance. In: S. Kornblum (Ed.), *Attention and Performance IV* (pp. 155-174). London: Academic Press.

Neerincx, M.A. (2003). Cognitive task load design: model, methods and examples. In: E. Hollnagel (Ed.), *Handbook of Cognitive Task Design.* Mahwah, NJ: Lawrence Erlbaum Associates.

Neiss, R. (1988). Reconceptualizing Arousal: Psychobiological States in Motor Performance. *Psychological Bulletin,* 103, 345-366.

Veltman, J.A. & Gaillard, A.W.K. (1997). Dissociation between task demands and mental effort. *Proceedings of the International Ergonomics Association* (pp. 283-285). Tampere (SF), 29 June - 4 July.

Wilson, (2003). Operator functional state as a framework for the assessment of performance degradation. In: G.R.J. Hockey, A.W.K. Gaillard, & O. Burov (2003). *Operator functional state: The assessment and prediction of human performance degradation in complex tasks* (pp. 8-23). Amsterdam, IOS Press.

Human Robot Teams as Soldier Augmentation in Future Battlefields: An Overview

Michael J. Barnes

U.S. Army Research Laboratory
2520 Healy Ave, Ste 1172
Ft. Huachuca, AZ 85613-7069
michael.j.barnes@us.army.mil

Keryl A. Cosenzo

U.S. Army Research Laboratory
Building 459
Aberdeen Proving Ground, MD 21005-5066
kcosenzo@arl.army.mil

Diane K. Mitchell

U. S. Army Research Laboratory
Building 459
Aberdeen Proving Ground, MD 21005-5066
diane@arl.army.mil

Jessie Y. C. Chen

U.S. Army Research Laboratory
12423 Research Parkway
Orlando, FL 32826
jessie.chen@peostri.army.mil

Abstract

The paper is an overview of a Human Research and Engineering Directorate (HRED), U.S. Army Research Laboratory (ARL) 5-year applied research program addressing human robot interaction (HRI) for both aerial and ground assets in conjunction with the U.S. Army Tank and Automotive Research and Development Engineering Center (TARDEC). The purpose of the program is to understand HRI issues and to develop technologies to improve HRI battlefield performance for Future Combat Systems (FCS). The technologies include adaptive systems to reduce soldier workload in multitasking environments and scalable interfaces for both specialized and common tasking requirements for robotic systems. The paper reviews first year results from modeling, simulations, and laboratory experiments. Based on these results, plans for follow-up research and eventual prototype development and field testing are reviewed as well.

1 Introduction

Modern combat is changing radically. The U.S. and its allies are engaged in a global struggle against forces whose terror tactics negate the advantages of much of U.S. technology. At the same time, the U.S. is developing future systems that will change the nature of the battlefield, especially the use of robotic systems which promise to extend military reach, to increase firepower, and most important, to save soldiers lives. However, the ubiquity of these systems poses serious problems as well as possible solutions (Barnes, Parasuraman, & Cosenzo, 2005). In order to address these issues, HRED, ARL has developed a 5-year applied research program addressing HRI and teaming for Future Combat Systems (FCS) in conjunction with TARDEC.

The purpose of the research program is to understand the nature of these potential problems in terms of how they affect the soldier during combat and to develop software aids and improved interfaces to mitigate these problems. The program is wide ranging because it addresses the impact of multiple robotic systems and their effects on a variety of possible future battlespaces. Future soldiers will not be specialists operating a single piece of equipment; they will have panoply of systems that are either an integral part of their combat duties or which they use occasionally. In either case, misuse or disuse of the system can have lethal consequences (Parasuraman, Sheridan, & Wickens, 2000). The paper will discuss a general research methodology to investigate HRI interfaces and mitigation strategies using HRI modeling and laboratory experimentation as part of an initial winnowing process. Promising concepts will be evaluated and developed further during realistic simulations and early field exercises. Finally, in conjunction with TARDEC, program results will be used to develop prototypes that will be evaluated in large, systems-of-systems field exercises. Also, the paper will highlight some of our first year research results to give the reader a feel for our research methods and our current state of HRI understanding. Specifically, we will detail results related to the modeling of the crew's workload in the mobile combat system controlling unmanned ground vehicles for FCS missions, two HRI simulation experiments manipulating span of control for robotic assets for urban missions, and laboratory experiments investigating multimodal interfaces and adaptive automation to reduce workload.

2 Methodology

A research program needs to be more than a collection of interlocking projects. The projects must have a logical relationship to each other in order to reinforce and to expand evolving results (Barnes & Beevis, 2003). The base of the research pyramid is modeling which captures the multitasking environment and gives preliminary performance estimates (Figure 1). The models also guide empirical research. Laboratory experiments are ideal for hypothesis-driven questions that either are derived from theoretical considerations or are a direct result of issues derived from the modeling. Elegant laboratory experiments are cost effective if they answer simple questions that are generic and influence the course of a general research strategy (e.g., does multimodal information reduce workload for a particular tasking environment?). However, as the questions become less generic and more specific, the results need to be verified in a realistic simulation environment that relates more closely to the actual military situation. A common mistake is to

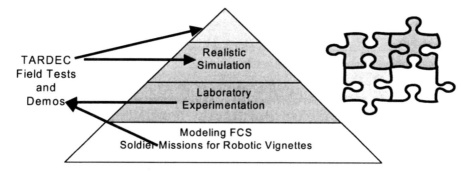

Figure 1: Interrelationships of research components.

rush into a field test or realistic simulations before the relationship of the variables and the underlying theoretical issues are understood fully. The results in such cases can be confusing and contradictory (Barnes & Beevis).

It is best to think of the pyramid in Figure 1 as a winnowing process. As more human related variance is accounted for, complexity can be added to empirical evaluations because more is understood about the soldier's environment. The top of the pyramid represents evaluations during field tests with soldiers interacting with multiple systems while performing realistic night and day missions over an extended operational tempo. These field exercises will be conducted by TARDEC using realistic prototypes; HRED will focus on the soldier performance testing. Not only are these exercises expensive, it is also crucial that individual components be pretested thoroughly to ensure that promising concepts are assessed fairly. It would be a mistake to assume that Figure 1 represents a strictly linear process; the pyramid is part of a spiraling approach in which experimental and simulation results not only reinforce each other (right side of Figure 1) but validate and update early modeling efforts. Similarly, early field tests may uncover problems that can be answered best in simpler simulation experiments as well as provide realistic workload estimates for the evolving modeling efforts (Schipani, 2003). In the first year of the program many of the methods were pursued simultaneously reflecting earlier modeling and experimental work; we will present results in terms of the logical relationships among individual projects rather than follow a strict sequence implied by the pyramid.

3 Modeling: Mounted Combat System Analysis Results

The modeling efforts are the most mature projects of the program. The modeling was conducted using the Improved Performance Research Integration Tool (IMPRINT) modeling environment (Lockett, 1997). IMPRINT models were used to generate crew tasking simulations based on a workload model measuring cognitive, perceptual and motor task loadings during the simulated missions. The workload and performance estimates were then combined with subject matter expertise to evaluate possible HRI problems for realistic missions. The missions were based on vignettes developed from the Future Force Warrior (FFW) and the FCS programs' documentation. IMPRINT models for soldier tasking environments related to the Mitilda small unmanned ground vehicle (SUGV), the Raven unmanned aerial vehicle (UAV), the Shadow UAV, and control of the Armed Reconnaissance Vehicle (ARV) from the Mounted Combat System (MCS) as well as dismounted platoon robotic functions were completed during 2004 and are archived as separate models. HRED is in the process of creating an overall modeling architecture being developed by one of our contractors (Microanalysis & Design, Inc.). The architecture will allow HRI researchers to run multiple models in a common vignette in order to evaluate intra crew workload and soldier interactions among the different systems.

The results from the MCS model are probably the best examples of an analysis of a multitasking environment derived from a FCS vignette (Mitchell & Henthorn, 2005). HRED researchers using the results of IMPRINT tool set and input from ~ 24 subject matter experts analyzed the workload performance of the crew of a MCS Platoon Leader's (PL) Vehicle (PLV) when the crew is required to control an ARV - Reconnaissance, Surveillance, and Target Acquisition (ARV-RSTA) robot. The robotics non-commissioned officer (NCO) is located in the company MCS. When the ARV-RSTA is allocated to a crewmember in the platoon MCS, the risk is that monitoring of the operator control unit (OCU) for long range information and monitoring for line of sight threats will be ignored due to high workload for other task requirements.

Regardless of which MCS PLV crewmember is controlling the ARV, the PL will need the information coming from the ARV sensors. Monitoring the information from the ARV along with his or her other concurrent tasks consistently placed the PL into overload. However, analysis showed that if a dedicated operator is consolidating the information for the PL and therefore reducing the need to constantly monitor the OCU then the PLs workload would be reduced. The gunner is the crewmember with the lowest instances of overload and the crewmember that is a possible candidate to control the ARV. The gunner's role is to scan for targets; a role that is critical to the survivability of the MCS. The scanning role will be more demanding on the MCS than it is for the current tank crew due to the replacement of the fourth crewmember with an autoloader. With the loader gone, the rear security scanning will need to be accomplished by the remaining three crewmembers via cameras. The crews' ability to scan effectively with the cameras will determine how much scanning both the PL and gunner need to accomplish. However, if the gunner's control of the ARV interferes with rear scanning, then the security of the MCS will be compromised.

Our analysis indicated that when gunners had to intervene to control the ARV that their workload was high. The workload estimates were supported by an empirical study that investigated remote driving and mental workload during an actual exercise (Schipani, 2003). This study evaluated operator workload during partially autonomous vehicle operations and found that workload increased significantly during periods when human intervention was necessary. While the average intervention rate was 5%, difficult terrain conditions required higher intervention rates. How much intervention will be required in the future will depend on the final configuration of the autonomous navigation systems (ANS). Until the reliability of the ANS is established, analysis suggested that one of the MCS PLV crew may have to act primarily as a dedicated robotics operator (RO) in order to unload other MCS operators and to control and monitor inputs from the ARV-RSTA. These results must be verified empirically under realistic conditions.

4   Simulations: Span of Control and Robot Teaming Results

The simulation experiments addressed some of the workload problems discussed above. The first experiment (Chen, Durlach, Sloan, & Bowens, 2005) examined workload, span of control, and display factor issues that had been identified in TARDEC sponsored field studies. The experiment simulated a single operator in a FCS like command vehicle performing a target designation task with a UAV asset, a UGV asset, teleoperating the UGV asset, or in the mixed condition controlling all three assets. The experiment was conducted at the U.S. Army Simulation and Training Technology Center (STTC). The operator displays are shown in Figure 2. The simulator was equipped with a steering wheel, and gas and break pedals for control of the teleoperated vehicle. Mechanical buttons on the steering device provided for control of the targeting and weapons systems of the teleoperated vehicle (weapons were not used in this study, however).

In addition to the different robot asset conditions, the display conditions were manipulated across groups. For one group, there was a 250 msec. latency imposed between control inputs and observable responses of the teleoperated UGV. The other group experienced a variable frame rate of 5 to 30 HZ depending on the simulated distance of the operator's vehicle from the robotic asset. There were no significant differences between the two groups. For within group conditions, the teleoperations condition resulted in degraded performance for target designation performance. There was a significant interaction ($p < .01$) between asset condition and whether the operator

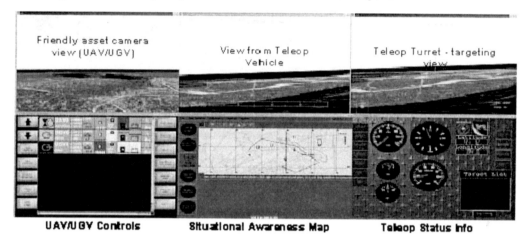

Figure 2: User interface of simulation test bed in the Chin et al (2005) study

controlled a single asset or all three assets. Figure 3 indicates that the mixed assets condition showed pronounced targeting degradation for the UGV and teleoperated UGV when the operator was required to control three assets vice a single asset. In contrast, the UAV performance was the same in both cases, which most likely reflects the fact that the UAV was the most used asset and the operators' preferred strategy was to depend on the UAV for most of their targeting decisions.

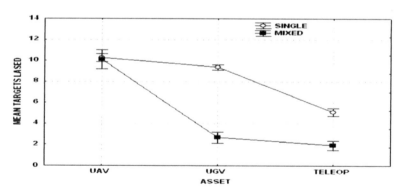

Figure 3: Interaction of asset condition and number of assets controlled.

This result was similar to results from an HRED contract study conducted at the University of Central Florida (UCF). Rehfeld, Jentsch, Curtis, and Fincannon (2005) investigated span of control and human teaming relationship and found marginal utility to adding robotic assets to a two-operator teams performing military target identification tasks in a simulated Iraqi city. The reason for their results was that the two operators tended to concentrate on a

703

single asset in high workload mission segments making the additional robot superfluous. In both cases, adding assets did not improve over all targeting performance but did increase the operator's workload (Chen, Durlach, Sloan, & Bowens, 2005). Similar findings have been reported in actual rescue operations using robotic assets (Murphy, 2004). The above results indicate that under high workload conditions (multiple assets) using robots for targeting is a very demanding task even though the multitasking and mobility factors the operators would experience in actual missions were only represented imperfectly in these studies. Based on these studies and the MCS modeling study, we will examine different mitigation strategies in this year's simulation efforts. Chen and her colleagues will conduct follow-on studies manipulating various crew configurations based on the MCS modeling analysis whereas the UCF researchers will investigate various training, automation, and teaming factors in a parallel effort. Both experimental teams will coordinate their efforts and use more advanced simulation interfaces being developed for STTC.

5   Laboratory Experiments: Multimodal Results and an Adaptive Automation Test Bed

Various interface related experiments were either initiated or completed during the first year. Haas (2005) investigated multimodal interfaces to simulate a soldier receiving targeting information from mobile robots while the soldier maintains situation awareness (SA). The first experiment indicated that when the operator had to search for visual targets with audio cues that signaled the appearance of a target and provided audio positional information (such as an audio cue of "target 15 degrees"); 3-D audio cues produced the shortest visual mean target search times (2.9 – 3.5 seconds). Visual signals with no audio cues produced the longest search times; signals with no audio cues took almost 5 seconds to find at the center of the screen, 6 seconds when the visual target was located halfway across the screen, and almost 7 seconds when located at the edge of the screen (15 degrees field of view). In summary, Haas' first experiment indicated that audio positional cues and 3-D audio in general would be efficacious in helping soldiers maintain SA when evaluating targeting information from robotic systems. Additional work is being pursued this year to investigate the utility of haptic cues as either independent or redundant sources of SA for vehicular operators with robotic assets. Also, HRED researchers are involved in experiments assessing the usefulness of 3-D binocular displays for depth perception requirements for teleoperating small robots.

An important goal of our applied research is to investigate workload reduction technologies for mounted and dismounted applications. In this regard, we are examining adaptive or adaptable aids that unload the operator during high workload, high stress mission segments (Barnes, Parasuraman, & Cosenzo, 2005). The purpose of the aids is to invoke automated interventions during peak workload conditions while allowing operators to maintain optimal SA for important taskings.

Adaptable interfaces allow the soldier to define conditions for automation decisions during mission planning while adaptive interfaces automate tasks as a function of some environmental or behavioral indicator (Parasuraman, Sheridan, & Wickens, 2000). Automation technologies have been applied to basic cognitive task environments and aviation environments. Research to date has not examined the feasibility of adaptive or adaptable automation in an environment in which an operator will control multiple robotic assets from a single control unit. The robotic environment for the future force will be highly complex and there are no existing test-beds to assess automation concepts in a highly controlled manner. Therefore, ARL in collaboration with George Mason University has created a computer task that emulates the essential robotic tasks. The program, Robotic NCO, was based on an existing prototype operator control unit designed by Microanalysis & Design and TARDEC. Figure 4 shows a picture of the displays generated by the software program (Robotic NCO).

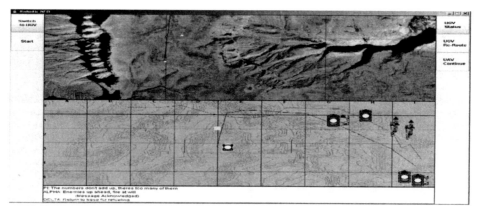

Figure 4: Display showing multitasking environment for initial adaptive automation experiments.

The software represents a multitask environment that includes three main tasks: unmanned ground vehicle (UGV) control, UAV sensor use, and multi-level communications. For the UGV control task, the operator uses a SA map to view the location of the UGV and to plan its path via waypoints. The UGV will stop (i.e., encounters an unknown obstacle) and will require the operator to determine the next course-of-action. The operator will access the UGV view (simulating video images) by clicking on the "Switch to UGV" icon and will see a picture of the UGV obstacle. Some obstacles can be bypassed and others will require re-routing to a safe area (i.e., no enemies in the area). For the UAV task, the operator can access overhead views from a UAV that is flying a predetermined route. The UAV images are updated on a time scale (e.g., 1 per second). Operators are required to identify targets as friend, foe, or neutral by zooming in on suspect locations (i.e., white squares). The targets that are identified are displayed automatically on the SA map. For the communications task, the operator will be prompted at various times for call sign recognition, UGV status update, and location of targets.

Experiments will be designed using Robotic NCO software to assess the utility of adaptive automation and methods of implementation in a robotic environment. The first experiment will manipulate the difficulty level of the tasks described above to determine which tasks contribute to high workload and possible poor performance. From this data we can then determine which tasks may be best to automate and what are the possible invocation mechanisms (Barnes, Parasuraman, & Cosenzo, 2005). In the next series of experiments, we will automate the "high driver" task identified in the first experiment and examine the impact of the automation on the operator's SA and performance. In a separate series of experiments, we will be assessing the utility of various physiological measures for invoking automation.

6 Conclusions

The paper outlined a general methodology that depended on the interrelationships among HRI modeling, laboratory experimentation, and more realistic simulation experiments. In the first year, our modeling efforts and simulation results showed similar trends of possible overload situations and potential crew problems for robotic systems. In particular, SA proved to be a potentially difficult problem both for the ARV (Mitchell & Henthorn, 2005) and smaller UGVs simulated in the UCF study (Rehfeld, Jentsch, Curtis, & Fincannon, 2005). There was evidence in both cases that the number of crewmembers required to maintain SA was greater than anticipated. Murphy (2004) had similar findings when operating robots during real world rescue operations. Chen, Durlach, Sloan, and Bowens (2005) study indicated that their operators had target acquisition problems when operators had to deal with multitasking for mounted missions that included controlling multiple robotic systems. This year simulation efforts will concentrate on understanding these overload problems by investigating various mitigation strategies such as advanced training procedures, extra crew members, better teaming arrangements, and planning aids.

Adaptive automation is a promising technology that may unload operators during peak workload events while allowing them to stay engaged in their most important mission taskings. Laboratory research is ongoing in this area in order to develop adaptive strategies that are useful for mounted RSTA missions; wherein, the operator has information from both aerial and ground unmanned assets. The focus of this year's research is to understand the generic tasking relationships and to test adaptive strategies based on these relationships. As we understand more about these relationships in the laboratory, we will be able to instantiate adaptive algorithms in full mission simulators to evaluate them in more realistic venues.

Two laboratory efforts were discussed whose purpose was to investigate possible interface improvements. Haas' (2005) study indicated that 3-D audio offers a potential augmentation for improving SA for RSTA type applications. This year, Haas will study the possible efficacy of additional haptic cues whereas one of our researchers is leading an effort to investigate 3-D visual information to help operators teleoperate small UGVs by improving depth perception. This year's efforts also include a number of other research tasks directed towards dismounted missions.

An important change in research emphasis will occur in the final three years of the program. Modeling and laboratory research will continue in a supporting role as we learn more about HRI technologies. However, the main thrust will gradually shift to even more realistic simulations and the development of working prototypes of adaptive systems and interface improvements that will be used to validate HRI technologies in large scale, multi-system field exercises.

7 References

Barnes, M.J., & Beevis, D. (2003). Chapter 8: Human system measurements and trade-offs in system design. In Harold R. Booher (Ed.), *Handbook of human systems integration* (pp. 233-259). NY: John Wiley & Sons.

Barnes, M., Parasuraman, R., & Cosenzo, K. (2005). Adaptive automation for robotic military systems. *Human Factors Issues for Uninhabited Military Vehicle*. Manuscript submitted for publication (copy on file with author).

Chen, J.Y.C., Durlach, P. J., Sloan, J., & Bowens, L.D. (2005). *Human robot interaction in a simulated environment.* Manuscript submitted for publication (copy on file with author).

Haas, E. (2005). *Improved visual search times with auditory cues in robotic systems.* Manuscript in preparation.

Lockett, J.F., III. (1997). Task network modeling of human workload coping strategies. In J. Smith, G. Salvendy, & R.J. Koubek (Eds.), *Advances in human factors/ergonomics, 21B. Design of computing systems: Social and ergonomic considerations* (pp. 71-74). San Francisco, CA: Elsevier.

Mitchell, D., & Henthorn, T. J. (2005). *Soldier workload analysis of the MCS platoon.* Aberdeen Proving Ground, MD: U.S. Army Research Laboratory

Murphy, R.R. (2004). Human-robot interaction in rescue robotics [Special issue]. *IEEE Systems, Man and Cybernetics Part C: Applications and Reviews, 34*(2).

Parasuraman, R., Sheridan, T.B., & Wickens, C.D. (2000). A model for types and levels of human interaction with automation. *IEEE Transactions on Systems, Man, and Cybernetics – Part A: Systems and Humans, 30*, 286-297.

Rehfeld, S.A., Jentsch, F.G., Curtis, M., & Fincannon, T. (2005). Collaborative teamwork with unmanned ground vehicles in military missions. *Proceedings of the 11th Annual Human-Computer Interaction International Conference.*

Schipani, S.P. (2003). Evaluation of operator workload, during partially-autonomous vehicle operations. *Proceedings of the Performance Metrics for Intelligent Systems Conference – PerMIS '03.*

Cerebral Hemodynamics and Brain Systems in Vigilance

Joel S. Warm
University of Cincinnati
Department of Psychology
PO BOX 210376
Cincinnati, OH 45221-0376
warmjs@ucbeh.san.uc.edu

Gerald Matthews
University of Cincinnati
Department of Psychology
PO BOX 210376
Cincinnati, OH 45221-0376
matthegd@email.uc.edu

Lloyd Tripp
2215 First Strees, BLDG. 33
Air Force Research Laboratory
Wright Patterson AFB
OH 45433-7947
lloyd.tripp@wpafb.af.mil

P.A. Hancock
University of Central Florida
Department of Psychology
3100 Technology Parkway, suite 337
Orlando FL, 32826
phancock@pegasus.cc.ucf.edu

Abstract

Investigations using positron emission tomography (PET) and functional magnetic imaging procedures (fMRI) have been successful in identifying multiple brain regions that are active in during the performance of sustained attention or vigilance tasks. Included are the nucleus locus coerulus, the brainstem reticular formation, the midbrain tegmentum, the interlaminar region of the thalamus, the cingulate gyrus, and the frontal lobe. Much of this activity also seems to be lateralized to the right cerebral hemisphere. However, as Parasuaman, Warm, & See (1998) have emphasized, the PET and fMRI studies are limited by the failure to correlate brain activity with performance efficiency, perhaps due to the high costs and restrictive environments associated with PET and fMRI use. These costs may be circumvented in studying brain systems in vigilance by employing transcranial Doppler sonography (TCD), a relatively inexpensive and noninvasive neuroimaging procedure that employs ultrasound signals to monitor cerebral blood flow velocity or hemovelocity (CBF) in the middle cerebral arteries which carry 80 per cent of the blood flow to the brain. When an area of the brain becomes metabolically active, as in the performance of mental tasks, by-products of this activity such as CO_2 increase. This results in elevation of blood flow to the region to remove the waste products (Aaslid, 1986). Thus, CBF may be viewed as a metabolic index of information processing (Stroobant & Vingerhoets, 2000). Along this line, our laboratory has carried out an extensive research program in which TCD has been used to examine brain systems in vigilance. A resource model of sustained attention in which changes in blood flow velocity are considered to reflect the availability and utilization of information-processing assets needed to cope with the vigilance task served as the theoretical basis of our work. Three of these studies are described in this report. The studies to be described show that the decline in performance efficiency over time that typifies vigilance performance (the vigilance decrement) is accompanied by a decline in CBF that is keyed to task performance. The time-based reduction in CBF only occurs when observers must actively process the stimuli to be monitored. Blood flow remains stable throughout the testing session when observers are asked to look at the vigilance displays with no work imperative. In addition, we find that the effects of different psychophysical manipulations are also paralleled by changes in blood flow and that these effects are lateralized to the right cerebral hemisphere, confirming PET and fMRI indications of the operation of a right hemispheric system in the functional control of vigilance. Thus, we show that blood flow is greater in the context of capacity demanding successive-type vigilance tasks (absolute judgment tasks in which signals and nonsignals are differentiated by comparing current input against a standard held in working memory) than less demanding simultaneous-type tasks (comparative judgment tasks in which all the information needed to distinguish signals from nonsignals is present in the stimuli themselves and recent memory for the stimulus feature is not required). Moreover, we also show that the temporal decline in CBF can be attenuated by reliable cueing regarding the imminent arrival of critical signals, a procedure which reduces the information-processing demand upon observers, and that the temporal decline in CBF is greater when observers must detect the absence rather than the presence of a stimulus feature, a result consistent with findings in the search literature that detecting the absence of a feature is more capacity demanding than detecting its presence (Quinlan, 2003).

References

Aaslid, R. (1986). Transcranial Doppler examination techniques. In R. Aaslid (Ed.), *Transcranial Doppler sonography* (pp. 39-59). New York: Springer-Veralg.

Parasuraman, R., Warm, J.S., & See, J.E. (1998) *Brain systems in vigilance.* In R. Parasuraman (Ed.), *The attentive brain* (pp. 221-256). Cambridge, MA: MIT Press.

Quinlan, P.T. (2003). Visual feature integration theory. Past, present, future. *Psychological Bulletin,* **129**, 643-673.

Strooband, N., & Vingerhoets, G. (2000). Transcranial Doppler ultrasonography Monitoring of cerebral hemodynamics during performance of cognitive tasks: A Review. *Neuropsychology Review,* **10**, 213-231.

Augmented Reality as a Human Computer Interaction Device for Augmented Cognition

Brian F. Goldiez[1], Ali M. Ahmad[1, 2], Kay M. Stanney[2], Jeffrey W. Dawson[2], and P. A.. Hancock[1, 3]

[1]Institute for Simulation and Training
[2]Department of Industrial Engineering and Management Systems
[3]Department of Psychology

University of Central Florida
4000 Central Florida Blvd.
bgoldiez@ist.ucf.edu, aahmad@mail.ucf.edu, stanney@mail.ucf.edu,
phancock@pegasus.cc.ucf.edu, jeffrey@geomix.com

Abstract

Augmented cognition aims to balance human information processing throughout task execution. It requires the exchange information back and forth between user and system. The objective of augmented cognition is to, in real-time, assess a user's cognitive state and apply augmentation strategies that direct how to best handle pending information dissemination. Such interaction can be characterized as a closed-loop feedback system between user and external task, where the task presents a user with information intelligently to maximize the information management capacity of the entire system. Separately, augmented reality (AR) research has demonstrated approaches for increasing human information processing capabilities by using sensory pathways, to extend working memory in selective tasks. AR technology adds information to the real world using various types of computer displays. AR can facilitate the exchange of information between user and system, and offer additional advantages in terms of mobility and tractability. We propose that sufficient baseline data exist to employ multimodal AR as an input/output device for augmented cognition.

1 Introduction

Augmented cognition is a paradigm shift in interactive computing. The reality is that, with advances in computing technology, human information processing capability is becoming the weak link within human-computer interactive systems. To overcome this limitation, augmented cognition aims to couple traditional electromechanical devices (i.e., mouse, joystick) with human physiological indicators that direct, in real-time, human-system interaction given a user's current cognitive state.

Augmented cognition has the potential of enhancing sensory perception and cognition by exploiting multiple sensory channels for increased input capacity and maintaining multimodal information demands within working memory (WM) capacity. Nevertheless, two technological challenges in augmented cognition are i) the accurate measurement of cognitive state and ii) improving the efficiency and throughput of human computer interfaces so that information can be delivered and maintained through the most efficient means. It is the latter area that is the focus of this paper. Traditional input/output devices, so-called WIMPs (Windows, Icons, Mouse, Pointing devices) have limitations that are further exacerbated when one is mobile or moving (Pew, 2003). The processing time and physical actions these traditional interface devices entail do not efficiently or seamlessly support ubiquitous computing and are likely to fall short if used to support the real-time adaptive interaction envisioned by augmented cognition.

Augmented reality (AR) can address some of these limitations. AR adds computer-generated displays to the real world (Barfield & Caudell, 2001). AR is characterized by a human wearing a display, which may be video, audio, haptic, or any combination of them, see Figure 3. It offers mobile computing capability. AR research has been focused on developing technological foundations in hardware and software, and now prototypical AR equipment is available. It is timely to evaluate this equipment and expand the application domains of AR. Previous AR research has focused mainly on video displays, but some current AR research efforts aim to expand into multimodal inputs and outputs. The availability of prototype AR systems provides the means to better understand the advantages and

709

limitations of this versatile technology. Formative AR research addressing performance shows AR supports improvements in human computer interaction and enhances some aspects of cognitive processing (Goldiez, 2004).

Associated with multimodal delivery of information is the need to better understand the capacities of WM and the most efficient pathways to deliver information. Various models of human information processing exist along with some knowledge on how different types of information are processed in WM. Miller (1956) proposed that WM capacity limits are in the range of 7 ± 2 chunks of information. However, Miller did not comprehensively consider multimodal data being simultaneously delivered to a user. In these cases, WM may process information from several modalities, and questions regarding the independence or interferences between modality-based processing capacities emerge. Understanding how WM behaves when multimodal information is being exchanged is critical in choosing which information to present to a user at a given point in human-system interaction.

Our work has synthesized the results of separate experiments in multimodal and mobile AR tasks. It then proposes a linkage between these efforts, describes usage of existing design guidelines, and suggests additional experiments and considerations to further the development of AR in augmented cognition. It draws connections needed to join two interesting domains: augmented cognition and augmented reality.

2 Background

Clark (2003), Hancock (1997), and Norman (2004) have suggested that various types of equipment can serve as cognitive assistants. A simple example is a wristwatch. Individuals no longer need to keep a mental track of time. The watch serves as an extension of WM for the individual to have access to time. This is a simple example, but more complicated systems have been influenced by use, as have users of those systems. An example is the cellular telephone that no longer requires people to remember telephone numbers or information to convey; a person can be in instant communication with others having the cellular telephone available. Cellular telephones have also adapted to the needs of their users by shrinking in size and incorporating additional, but related, functionality such as text messaging, image capture, and email. These devices have evolved to respond to the needs of the user, but nonetheless represent a form of "augmented cognition," more traditionally known as "external cognition" (Norman, 1993; Preece, Rogers, & Sharp, 2002). With external cognition, external representations (e.g., books, maps, notes, drawings) are leveraged to extend and support an individual's cognitive activities by reducing memory load, offloading computationally intense activities, and/or providing the capacity to annotate or manipulate (e.g., reorder) external representations. Transforming external cognition to augmented cognition is a matter of how the support is realized. With external cognition, for example, one might use a calculator to change the nature of a computational activity from one that is carried out solely in the mind to one that is externally supported. With augmented cognition, a system would detect (via an assessment of cognitive state) that an individual was overburdened by a computational activity and would, in real-time, provide aiding to lessen the burden; there would be no need to access an external support system, such as the calculator. Such automatic detection and aiding could lead to tremendous gains in the human-system dyad.

A properly developed AR system has the potential to serve as a cognitive assistant and an augmented cognition input/output device. Two components are needed to achieve this potential. First, the device must have utility. That is, the device must improve human performance by interpreting, extending and offloading some components of the cognitive process. Secondly, the device must easily accommodate to the physical and behavioral requirements of its user. But by its definition, AR covers a wide swath of technology, some of which might not be appropriate to augmented cognition. So, how does one characterize AR such that it best supports augmented cognition?

Numerous definitions of AR have been suggested, which indicates a broad, yet not always clearly defined, field. For instance, Barfield and Caudell (2001, p. 6) define AR as "*a participant wears a see-through display (or views video of the real world with an opaque HMD) that allows graphics or text to be projected in the real world.*" This definition is typical of much AR work. Dubois and Nigay (2000, p. 165) note that AR is used "to extend the sensory-motor capabilities of computer systems to combine the real and the virtual in order to assist the user in performing a task in a physical setting." They further note that AR information can be presented by the addition of 3D graphics, audio information, or force feedback. Two-dimensional graphics can also be included—simple computer displays, although much of the research has involved the addition of 3D objects. Dubois and Nigay (2000) also mention two characteristics of AR: task focus and the nature of the augmentation, which they use to help define four types of AR systems, which taken together do not exhaust the possible design space for AR.

The above definitions are rather broad making it difficult to determine how particular implementations fulfill the AR mantra. Another classification scheme to better segment different AR implementations may be to consider AR information as one of three types: information that is (1) not part of the environment, (2) integrated into the environment, or (3) is part of the environment but not normally perceived (e.g., infrared images) (Goldiez, 2004). An example of an augmentation that is not part of the environment is navigational information displayed in a heads up display. A physical object overlaid and fused onto (or subtracted from) the environment in a way that makes it indistinguishable from the natural environment is an example of the second type of information. An example of the third type of information is infrared images, which, although part of the environment, are not normally perceived but could be displayed in real time via AR. In considering these three types of AR information vis-à-vis augmented cognition, the display of data that reduce a user's WM load is relevant. If information normally stored in WM is available for display using AR and can be automatically activated via cognitive state indicators, the user may free up mental resources and expand their real-time information processing capabilities. It is important to note that while most AR implementations have focused on visual information, the use of other modalities to display AR information offers an opportunity to expand the utility and effectiveness of AR's role in supporting augmented cognition.

As shown in Figure 1, there are many areas in the information-processing loop (Wickens & Hollands, 2000) where AR-derived augmentation could be helpful. One limiting area in the information-processing loop is WM and its corresponding linkages with perception and response selection. "True" multimodal interaction between humans and systems entails understanding the multimodal capabilities of perception and cognition, both of which are heavily dependent on WM. Several WM models have emerged, (c.f. Miyake & Shah 1999); nevertheless, most of these models are not comprehensive in terms of describing the multimodal components of WM. Several questions arise when attempting to develop a WM model that is multimodally tailored, including 1) are there separable modality-based WM capacities and 2) if these capacities exist, are they independent? Clearly, knowing this information on WM capacity facilitates development of augmented cognition systems, as cognitive state could be characterized, among other dimensions, in terms of each modality's WM capacity. One useful benefit of augmented cognition would be leveraging the whole of modality-based WM capacity in ways that facilitate improved human-system performance.

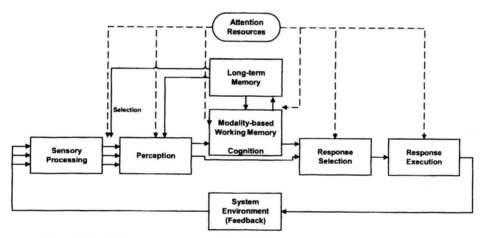

Figure 1: Model of Human Information Processing [adapted from Wickens & Hollands, 2000]

2.1   Working Memory

Working memory (also referred to as short term memory) is where various sensory information is held in temporary storage long enough to process it; whatever is not attended to is lost. Baddeley (1986; 1990; 1992; 1996; 2000; 2001) characterizes WM as having a central executive (see Figure 2), which is said to be aided by two subsidiary closely-related subsystems, the visuo-spatial sketchpad (i.e., holds and manipulates visual images) and the phonological loop (i.e., holds and manipulates speech-based information). This figure includes linkages to long term memory and is consistent with the Haberlandt (1997).

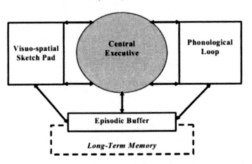

Figure 2: Two Subsystem Model of Working Memory

Baddeley's model illustrates multiple pathways for sensory information to reach WM for further processing. The model also implies caution in providing excessive information in parallel or series across the various sensory channels. Caution is needed with respect to controlling the flow rate of information and minimizing conflict between the sensory channels.

Through the use of this well-established model of WM, a more productive approach for integrating human information processing into the design and use of advanced augmented cognition systems can be achieved. At this juncture, it is important to determine how to leverage technologies, such as AR, to assist augmented cognition systems in distributing information sources across modal sensory or WM systems during real time human-system interaction.

2.2 Design in Augmented Reality and Augmented Cognition

A missing component of many investigations is the subjective "affect" of how well the equipment works with respect to fit, trust, and lack of interference with a user's tasks. In the case of augmented cognition, subjective impacts should include affect from the computer, affect in the design (Norman, 2004), and presence. Norman (2004) discusses three levels of design: visceral, behavioral, and reflective. Visceral design relates to a devices appearance, form, etc. Behavioral design relates to the pleasure and effectiveness of use. These two levels of design are what are commonly used to design many systems. Reflective design reflects how one feels about using a device, including one's self image, personal satisfaction, and trust. Since it is an emerging area of interest, design guidelines for reflective design do not yet exist. However, intuition and common sense tell the astute designer that it must be considered. Reflective design can impact the utility of many systems because it reflects the symbiosis between machine and user. In the case of AR and augmented cognition, one can envision the reflective layer of design where the AR system melds into the background extending ones cognitive abilities in ways that are not necessarily apparent to the user.

Norman's characterization is a good segue into another feature of using an AR system in augmented cognition. That feature is the aspect of presence, which can be defined in part as (ISPR, 2004) *"... even though part or all of an individual's current experience is generated by and/or filtered through human-made technology, part or all of the individual's perception fails to accurately acknowledge the role of the technology in the experience."* The important point is that the equipment generating part of the experience melds into the background and is not distinguishable by the user. The work of Norman and the ISPR are important factors but their measurement remains elusive. Nonetheless, some form of automated feedback from users of AR in an augmented cognition or other context is important to capture. This feedback needs to consider and include information about factors such as user's current WM usage, and a predictive model of how to redirect selective information, to balance the WM load throughout task execution. Monitoring attention and the executive function, shown respectively in Figures 1 and 2, may also be necessary.

2.3   Relevant Experiments in Human Performance

The preceding discussions indicate that the use of AR, in general and multimodal AR in particular, may serve as an excellent interface device for augmented cognition. The following studies support this hypothesis.

Samman (2004) conducted a series of three experiments aiming to develop an expanded WM model that is modality based. The expanded WM model hypothesized separate storage for visual, verbal, spatial, tonal, tactile, and kinesthetic modalities. Visual refers to storage of pictures. Verbal is storage of linguistic information. Spatial is storage of directional or location-based information. Tonal is storage of sounds. Tactile is storage of touches, pressures, or textures. Kinesthetic is the storage of body movement. The experiments were sequentially conducted with each experiment testing a different hypothesis related to WM. The first two experiments were composed from two parts. The first part entailed presenting participants sequentially with items from each modality group, starting from 1 item, to 2 items, and so on, up to 9 items. The second part entailed combining various modality types, i.e., first presenting an item from each modality, second, presenting 2 items from each modality, and so on, up to six items of each modality (for a total of 36 combined items). For each part, the participants were asked to recall the presented items serially, and in addition to that, participants were asked for a free recall of all remembered items. Results from the first experiment indicate separate storage for each modality. Only verbal and spatial modalities approached Miller's "magic number" of 7 ± 2, with capacities of 6.25 letters and 5.58 directional blocks, respectively. Interestingly, multimodal storage capacity was significantly less than the summation of individual modality capacities, which may indicate interference between storage modalities. The second experiment included minor modifications on the experimental procedure, especially in terms of 1) free recall instead of recognition for tonal information, i.e., in the first experiment, participants were asked to recognize played tones, while in the second experiment, participants learned how to play the tones on a computerized scale keys, and used keys to reproduce the tones, 2) marking a figure instead of physically touching for tactile, where in the first experiment participants recalled by pointing to where they were touched, in the second experiment they were given a blank drawing of a human back and front, and were asked to mark and number, where they were touched, and 3) utilizing both serial and free recall. The results of the second experiment match those of the first experiment. Free recall improved performance in verbal, spatial, and tactile modalities. The average recall was 22.11 combined items for the multimodal condition, which is three times what Miller predicts. See Samman, Stanney, Dalton, Ahmad, Bowers, and Sims (2004) for more details.

Samman's (2004) third experiment was designed to test whether interference exists among the various modality resources. This was achieved through a dual task perspective, where a pre-load of information was presented to participants, which they were supposed to recall later. The results revealed no indication of interference between WM modalities. This may suggest that human WM is indeed modality specific. The significance of this work is that utilizing multimodal displays can enhance human performance by tapping different WM resources.

With respect to AR experiments, Tang, Owen, Biocca, and Mou (2003) studied the effectiveness of AR in an assembly task, finding 82% reduction in assembly errors for the AR condition. One finding is that AR reduces the cost of attention switching between paper manuals or manuals displayed on a fixed computer monitor. Reduction in attention switching reduces the cognitive workload for the task performer. Also mentioned is the use of the technique of associating items in memory with spatial locations (as the Greek orator Demosthenes used in his Method of Loci), stating that, "*By spatially relating information to the physical objects and locations in the real world, AR provides a strong leverage of spatial cognition and memory*" (p. 74). One problem, though, that the authors note is that of attention tunneling. Tunneling is the focusing of ones attention on something to the exclusion of attending to other things. In terms of tunneling used by Tang et al. (2003), once an error is made in assembly, users of AR are less likely to correct it than users of traditional reference materials. This is thought to be due to users focusing on a particular display area to the exclusion of other areas, thereby either missing the error or how to correct it. In this context, the term area can encompass the portions of the augmented or real scene, or a particular area of a display, leading to the exclusion of attending to other areas of the display. Fusing augmented with real information can reduce this problem. Methods to mitigate this problem with other forms of augmentation are an open area for research.

Experiments in using AR to support a wayfinding and navigation task were conducted by Goldiez (2004). These cognitive based experiments used a monocular display containing a map of a maze thereby constituting the type of

AR that adds new information to a scene. A mobile AR system, referred to as Battlefield Augmented Reality System (BARS), was used in Goldiez (2004) study. BARS was developed and configured by the Naval Research Laboratory (NRL). BARS is shown in Figure 3. The experiment required participants to traverse a physical maze (shown in Figure 4), obtain a target object and exit the maze, which is a general "search and rescue" task.

Figure 3: Battlefield Augmented Reality System [Front and Side View]

Figure 4: Oblique View of the Maze used in Goldiez (2004)

Goldiez (2004) found that a user directed on-demand AR display improved performance when compared to a continuously on AR display in the search and rescue task scenarios evaluated. Exocentric maps generally resulted in better performance than egocentric maps. It remains to be seen how far this finding can be extended to other tasks or AR equipment. For example, in Goldiez's experiments, participants did not have a habituation period prior to the start of the experiment to get used to the equipment, so they were at the bottom of the learning curve for using the AR display. On the other hand, performance is expected to continue improving as a wayfinding task becomes more difficult and WM becomes saturated. Relative to augmented cognition, however, if the system sensed when a person needed assistance and then offered assistance, the on-demand display scenario may be particularly effective.

Kalawsky and Hill (2000) conducted numerous experiments on cognitive processing in AR. Some of their findings will be noted. They compared an AR display to an online feedback system and found that AR was as effective as the online monitor for the comprehension of abstract symbols, text passages, short-term memory recall, and comprehension and retention. For monitoring a warning display, they found that AR reduced workload relative to having to monitor a side (off-axis) monitor; they noted that this increased situation awareness. Another finding was that there was a need to keep AR information displays relatively simple.

Taken together, the above studies hold promise for leveraging multimodal AR in augmented cognition systems. Such a system would leverage multiple sensory modalities and overlay this information onto a given task context such that WM capacity was enhanced and the cost of attention switching was minimized.

3 Discussion

AR has been shown to facilitate extending WM in a limited and controlled experimental setting. The AR implementations used were unimodal. Other experiments conducted by Samman (2004) demonstrate that human information processing limits can be extended if multimodal inputs are carefully considered. Linking enhanced understanding of multimodal human information processing capabilities, with portable information delivery and acquisition platforms such as AR can facilitate augmented cognition by creating "true" seamless integration between humans and systems thereby facilitating an effective feedback control system that transfers cognitive state into an actionable system inputs to the user.

There are other benefits to implementing AR for augmented cognition. First, AR is not fully immersive, thereby allowing full consideration of the real environment and situation. The appropriate use and type of AR allows us to retain reality and minimize divided attention by using on-demand displays. This is critical in situations where an alarm or visual event may necessitate attending to a situation. Second, AR involves mobile computing technology that can work as an augmented cognition input/output device, processing user inputs and cognitive state and providing appropriate outputs and sensory switching strategies that minimize distraction. Third, AR technology lends itself well to using multimodal information, both for outputs to the user (e.g., visual, auditory), and inputs from the user (e.g., manual, vocal, and cognitive state).

Much work, though, remains to make this conceptual linkage a reality. First, AR systems need to be prototyped incorporating multimodal components. The component implementation(s) must be technologically feasible and contribute to human performance. Interactive design techniques should be employed. The range of tasks where multimodal AR contributes to augmented cognition needs further study. AR systems need to have broad utility to be practical. AR must accommodate a system for measuring cognitive state. This feature might not need to be resident on the AR platform, but a real time linkage between the two is needed along with a transition strategy to sense cognitive overload and take appropriate action. Also, modal switching mechanisms need to be prototyped and evaluated. For example, an approach could be hypothesized that minimizes distraction during an impending visual overload by restricting the visual field of view and replacing the optical flow with spatial audio.

Experimentation merging AR as a tool for enhancing augmented cognition should be cyclical in nature developing prototypical technologies, evaluating their efficacy in a variety of settings, revising the design, etc. Furthermore, enhancing the multimodal capabilities of AR by incorporating auditory and haptic information sources as well as coupling prototype equipment with ability to utilize user's vocal input may enrich the functionalities that AR can offer for augmented cognition. All additional developments on the AR equipment need to be "live" tested using human participants to assess their acceptability and potential in improving human performance.

4 Conclusions

AR has demonstrated its ability to augment cognition at a modest level of functionality and for specific tasks having cognitive components that utilize WM. Prototype systems and guidelines exist and should be used for experimentation to further AR and augmented cognition activities.

In order to successfully couple AR with augmented cognition, several goals need to be accomplished. First, it is essential to understand how best to configure AR to collect augmented cognition inputs in terms of sensor information. This can be accomplished by developing interfaces between the two technologies. Second, human multimodal information processing needs to continue to be investigated in terms of various WM capacities and limitations, processing bottlenecks, and switching and augmentation strategies. The results may be presented as a set of design guidelines that are human-centered, and allow for seamless integration between humans and information sources. Third, strategies accompanied with hardware/software updates need to be investigated in AR to utilize such design guidelines to dynamically affect information flow to the user.

The augmented cognition-AR integration framework proposed in this paper may take augmented cognition one step further by providing a flexible interface device and by utilizing current findings associated with augmented cognition in a more practical real life setting, where the user is mobile. This research may also benefit current AR interfaces, as it may allow such systems to listen to a user's needs and present information in the most effective way such that human information management capacity is optimized.

5   Acknowledgements

This research was facilitated, in part by grants from the Office of Naval Research (N00014-03-1-0677) and the Army Research Office (DAAD 19-01-1-0621). The opinions expressed herein, though, are those of the authors.

6   References

Baddeley, A.D. (1986). *Working memory*. Oxford: Oxford University Press.

Baddeley, A.D. (1990). *Human memory: Theory & practice*. Hove, UK: Lawrence Erlbaum Associates.

Baddeley, A.D. (1992). Working memory. *Science, 255*, 566-569.

Baddeley, A.D. (1996). Exploring the central executive. *Quarterly Journal of Experimental Psychology, 49A*, 5-28.

Baddeley, A.D. (2000). The episodic buffer: A new component of working memory. *Trends in Cognitive Science, 4*, 417-423.

Baddeley, A.D. (2001). Is working memory still working? *American Psychologist, 56*, 849-864.

Barfield, W., & Caudell, T. (2001). Basic concepts in wearable computers and augmented reality. In W. Barfield and T. Caudell (Eds.), *Fundamentals of wearable computers and augmented reality* (pp. 3-26). Mahwah, NJ: Lawrence Erlbaum and Associates.

Clark, A. (2003). *Natural-born cyborgs: Minds, technologies, and the future of human intelligence*. New York, NY: Oxford University Press.

Dubois, E., & Nigay, L. (2000). Augmented reality: which augmentation for which reality? Paper presented at the *Designing Augmented Reality Environments*, Elsinore, Denmark.

Goldiez, B.F. (2004). *Techniques for assessing and improving performance in navigation and wayfinding using mobile augmented reality*. Unpublished doctoral dissertation. University of Central Florida, Orlando, Florida, USA.

Haberlandt, K. (1997). *Cognitive psychology*. Boston, MA: Allyn and Bacon.

Hancock, P. A. (1997). *Essays on the future of human-machine systems*. Eden Prairie, MN. Banta Information Services Group.

International Society for Presence Research (ISPR) website. Retrieved on September 21, 2004 from http://lombardresearch.temple.edu/ispr.

Kalawsky, R., & Hill, K. (2000). *Augmenting the real world: augmented reality and wearable computing*. Retrieved January 23, 2005, from www.avrrc.lboro.ac.uk/VVECC-KevinHill.PDF.

Miller, G.A. (1956). The magical number seven, plus or minus two: some limits on our capacity for processing information. *Psychological Review,.63*, 81-97.

Miyake, A., & Shah, P. (Eds.) (1999). *Models of working memory: mechanisms of active maintenance and executive control.* Cambridge ; New York : Cambridge University Press.

Norman, D.A. (1993). *Things that make us smart.* Reading, MA: Addison-Wesley.

Norman, D.A. (2004). *Emotional design, why we love (or hate) everyday things.* New York, NY. Basic Books.

Pew, R.W. (2003). Evolution of Human-Computer Interaction: From Memex to Bluetooth and Beyond. In J.A. Jacko and A. Sears (Eds.), *Human-Computer Interaction Handbook.* Mhwah, NJ: Lawrence Erlbaum Associates.

Preece, J., Rogers, Y., & Sharp, H. (2002). *Interaction design: Beyond human-computer interaction.* New York: Wiley.

Samman, S.N. (2004). *Multimodal Computing: Maximizing Working Memory Processing.* Unpublished doctoral dissertation. University of Central Florida. Orlando, Florida, USA.

Samman, S.N., Stanney, K.M., Dalton, J., Ahmad, A.M., Bowers, C., and Sims, V. (2004). Multimodal Interaction: Multi-capacity processing Beyond 7 +/-2. *Proceedings of the Human Factors and Ergonomics Society (HFES) 48th Annual Meeting.* New Orleans: Louisiana. September 20-24.

Tang, A., Owen, C., Biocca, F., Mou, W. (April 5-10, 2003). Comparative effectiveness of augmented reality in object assembly. *Proceedings of Association for Computing Machinery Computer Human Interaction* (pp, 73-80). Ft. Lauderdale, FL.

Wickens, C.D., & Hollands, J.G. (2000). *Engineering psychology and human performance* (3rd Edition). Upper Saddle River, NJ: Prentice Hall.

Spatial Orienting of Attention using Augmented Reality

Christian J. Jerome

Consortium Research Fellows Program
at the U.S. Army Research Institute
12423 Research Parkway
Orlando, Florida 32816
christian.jerome2@us.army.mil

Bob G. Witmer

U.S. Army Research Institute
12423 Research Parkway
Orlando, Florida 32816
bob.witmer@us.army.mil

Mustapha Mouloua

University of Central Florida
P.O. Box 160000
Orlando, FL 32816
mouloua@pegasus.cc.ucf.edu

Abstract

The purpose of this research was to empirically examine how well people could locate a visual or auditory cue in a 360° mocked-up urban setting using augmented reality (AR) cues. The speed and accuracy of finding targets were compared for three cueing conditions: Audio cues only, Visual cues only, and Audio and Visual Cues combined. The cues were superimposed on the real world urban setting mock-up in various azimuth and elevation locations. This was accomplished by displaying the visual cues on a video head mounted display (HMD) and audio cues displayed from a two-tier surround sound system including a total of 12 speakers. This augmented reality system was used to accurately place the cues in their correct location in the real environment. The results of this experiment showed that the cues enhanced target acquisition speed and accuracy. Potential applications of this research include automotive, aviation, and military systems used to warn users of possible hazards and to orient their attention to the spatial location in the 360° area surrounding the user.

1   Introduction

Humans are able to take in information from the environment by transforming energy at the sense organs into electro-chemical neural activity. The mechanisms by which each sense modality transforms energy, however, have certain sensory and perceptual capabilities and limitations. For example, humans can hear a wide range of sounds, but are limited by the frequency and intensity that can cause a sensory neuron to fire (Levine, 2000). Clearly, if the sensory neuron does not fire, the stimuli will not be sensed, perceived, or attended to. Being aware of these limitations, we have developed sensory aids that can enhance or amplify the environmental signals so that we may be able to better detect and react to them. These aids include such simple devices as sunglasses to help improve vision by reducing glare and eyeglasses that correct abnormalities in the shape of the eye, smoke and carbon monoxide detectors to warn us of the presence of a fire before our senses can, to telescopes and binoculars that make objects visible that would normally be impossible to see due to their small retinal image

Of course, sensory limitations are not exclusive; attention limitations exist as well. Much research has been done exploring human limitations such as dual-task performance, i.e., the inability to perform two tasks simultaneously when they compete for the same attentional resources (see Wickens, 1976), visual search problems, i.e., the target of interest is surrounded by non-target distracters and does not 'pop out' leading to a longer (serial) search time (see Luck & Hillyard, 1990). On the other hand, a distracter that 'pops out' could pull attention away from the true target, and thus undermine target detection performance. These issues have been studied from different applied goals, e.g., aviation, automotive driving, nuclear power plant operation. The common interest in this topic is that without knowledge about attentional limitations, the design of the interface and controls could inadvertently create a dangerous situation where accidents might happen and people could get hurt.

Previous efforts to overcome these attention limitations include simplifying tasks, reducing the amount of extraneous information that may distract from the stimuli that is of most importance to the task. However, in some ways the task cannot be simplified. Can additional information be added in order to improve the ability to pick out the important information in the environment from the less important/distractions? Augmented reality (AR) is one such tool that may prove to be a valuable attentional aid.

AR provides additional information overlaid upon the real world. The amount of information is increased into the same visual area which can either aid or hinder the user's attentional processes. Information that helps draw the user's attention toward the target increases target detection efficiency by aligning the attentional system with the visual input pathways (Posner, 1980). As humans, we have a bias towards vision as the primary modality for taking in information. However, the information that steers attention does not necessarily need to be visual; it may be from any modality that can provide spatial locations.

1.1 Auditory Cuing of Attention

Providing additional information in order to improve the ability to pick out the important information in the environment from the less important/distractions is thought to be possible using the auditory modality. Exactly how one can determine the spatial location of a target based on its sound is accomplished by the interaural time difference (ITD), i.e., the minute time difference the sound reaches the two separate auditory sensory receivers (the ears).

Auditory cues to auditory targets show a performance improvement for targets on the expected side of the head, supporting the notion that ITDs can be used as a basis for orienting attention (Sach, Hill, & Bailey, 2000). Furthermore, Sach et al. (2000) showed that a centrally located visual cue was successful in orienting attention for subsequent auditory targets, which supports top-down attention control since higher level cognition is required to extract the spatial meaning from the visual cue to the location of the auditory target. Typically, this is referred to as endogenous cueing and is fundamentally different from exogenous cueing, i.e., located at the site of the target, and is a bottom-up attention control. It can also be viewed as a method of priming the sensory neurons and channels associated with the particular spatial location. Spence and Driver (1994) found that endogenous spatial orienting in response to predictive cues can influence localization responses. Ferlazzo, Couyoumdjian, Padovani, and Belardinelli (2002) showed that auditory and visual spatial attention systems are separate, as far as endogenous orienting is concerned. Schmitt, Postma, and DeHaan (2000) found that it is important whether the attention system is activated directly (within a modality) or indirectly (between modalities). Others have found that visual cues affect both visual and auditory localization, but auditory cues only affect auditory localization (see Ward, 1994, and Ward, McDonald, & Lin, 2000). However, Spence and Driver (1997) found that auditory cues affected both visual and auditory target localization, whereas no sign of auditory orienting was found when visual cues were used. As stated previously, the information that steers attention may be from any modality that can provide spatial locations; however, when it is intended to orient attention to a separate modality, there might be costs involved.

1.2 Visual Cuing of Attention

The most obvious and logical method to cue attention to a spatial location is through the visual modality. It is generally accepted that there are two main visual pathways that provide distinct information to humans and primates; the ventral "what" pathway and the dorsal "where" pathway (Niebur & Koch, 1996). Since vision is the primary method of determining the identity and location of a target, then cueing using this modality is highly ecologically valid.

The purpose of a visual cue is to reduce the amount of parallel information and make the important stimuli salient. The 'feature integration theory' explains how vision is broken down into a set of topographic feature maps (Treisman & Gelade, 1980). Within each map, different spatial locations compete for attention, which then feed into a master "saliency map", which codes for conspicuity over the entire visual field (Itti, Koch, & Niebur, 1998). Visual cues that are similar to the targets based on color and location have been shown to improve localization performance (Ansorge & Heumann, 2003). Pratt and McAuliffe (2002) described this as an inclusive rule as opposed to an exclusive rule. In other words, the attention system orients attention to stimuli that shares similar features to the targets, as opposed to a system that does not orient attention to stimuli that does not share similar features. This means that the attention system actively seeks out saliency, or a stimulus that attract attention first from a bottom-up perspective and does not actively ignore from a top-down perspective. This is important from a design standpoint since a warning or alarm should not give many false positives since ignoring is a higher cognitive process and thus requires more resources. If this were to take place, missing a valid warning would more likely occur.

The goal of this research is to explore cues from different modalities in an effort to determine if the simulator used is capable of displaying the cues accurately and if people perceive them where intended, and also to test other factors that may affect the speed and accuracy of detection including cue direction, cue height, and cue location.

2   Methods

2.1   Participants

The participants were 30 university students (15 men and 15 women) attending the University of Central Florida. All of the participants (mean age = 25.17 years) volunteered to participate and were treated in accordance with the principles of ethical treatment of human research participants (American Psychological Association, 1992).

2.2   Materials

A physical mock-up of old world, 2-story buildings (MR MOUT—mixed reality military operations in urban terrain) was used in this research. The building facades (see Figure 1) enclosed a rectangular area of approximately $30m^2$. Lettered labels were affixed to the building and positioned to correspond to potential target locations. In addition to the physical building mock-up, MR MOUT also includes a Canon video-based AR display, a physical mock-up of a simulated infantry weapon, desktop computers for rendering the audio and visual cues, CRT displays allowing the experimenter to monitor the participants' view, and AR software for controlling cue presentation and recording performance. The Intersense IS-900 tracking system was used to track user position and head movement, as well as aiming direction of a simulated weapon. Audio cues were presented through strategically located speakers via a 2-tier (upper and lower) surround sound system. The speakers were visually masked so that their locations were not obvious to participants.

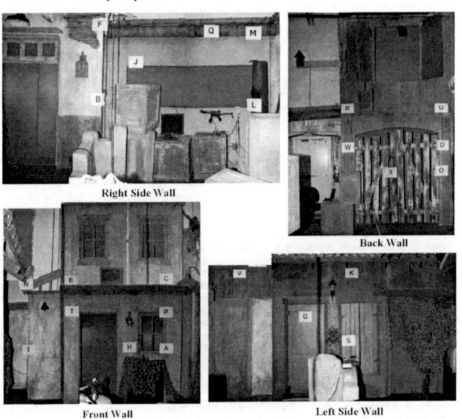

Figure 1: **MR MOUT Building Facades**

2.3   Design and Procedure

For testing purposes the participants were randomly assigned to one of three cue conditions of 10 participants each. Each participant judged the spatial location of audio and/or visual cues located to the participant's front, side, or back. The cues were presented twice in 12 different locations during 24 (2 sets of 12) test trials. The order of the cues was different for the two sets of trials. The cues were presented for a maximum of 6s. on each trial. Participants pulled the trigger on the MR MOUT simulated weapon when they located the target cue, causing the cue to disappear. The time of the trigger pull was recorded as a measure of the speed of acquiring that target. Immediately after pulling the trigger, participants indicated the cue location by calling out the correct lettered location from among 24 potential locations within 10s. of the start of a trial. There were 12 lettered cue locations and 12 lettered distracter locations. If participants were unable to precisely locate the target, they were required to provide their best guess about the target's location. Accuracy was assessed by determining the number of cues correctly located.

A 3 x 3 x 2 (Cue modality x Target direction x Target height) factorial MANOVA design was used to analyze the speed and accuracy of target acquisition. Separate follow-up ANOVAs for speed and accuracy measures were then performed.

3   Results

The MANOVA showed significant differences for each of the independent measures. The primary variable of interest in this study was cue modality. There were significant differences among the 3 levels of cue modality: $F(4, 54) = 13.97$, $p < .001$ indicating that participants acquired fewer audio cues and took longer to acquire them than they did for visual and audio-visual cues (see Figure 2).

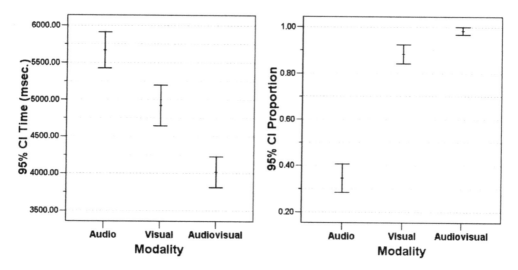

Figure 2: **Speed and Accuracy of Acquiring Targets for Different Cue Modalities**

There was a significant main effect of performance based on the direction of the cue relative to the participant's spatial orientation, $F(4, 24) = 27.28, p < .001$ (see Figure 3) and target height, $F(2, 26) = 21.96, p < .001$ (see Figure 4) indicating that the closer the targets were to the orientation of the participants head in azimuth and elevation, the better performance was for acquiring targets, at least for acquisition time

The univariate results indicate that in general our independent variables had a greater effect on target acquisition speed than on the accuracy of locating targets. Significant speed differences were found for cue modality, $F(2, 27) = 8.22, p < .005$, target direction, $F(2, 54) = 58.49, p < .001$, and target height, $F(1, 27) = 40.55, p < .001$. In contrast, only cue modality had a significant effect on the number of cues accurately located, $F(2, 27) = 120.2, p < .001$

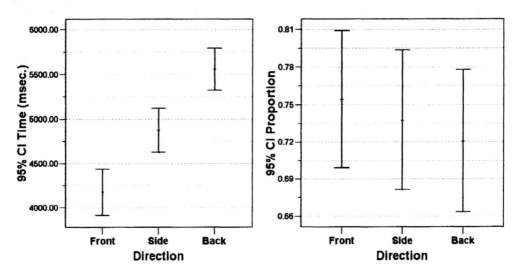

Figure 3: **Effects of Target Direction on Speed and Accuracy of Acquiring Targets**

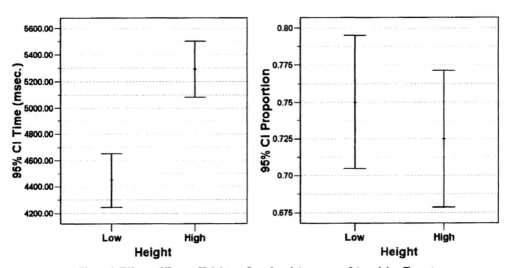

Figure 4: **Effects of Target Height on Speed and Accuracy of Acquiring Targets**

There was a significant interaction between cue modality and target direction, $F(4, 54) = 12.14, p < .001$ indicating that multimodal audio-visual cues are most effective when the target is outside of the user's field of view (see Figure 5). The speed of target acquisition of visually cued targets is as good as for targets cued by audio-visual cues when the target is directly in front of the participant, but not when the targets are located to the participant's side or back.

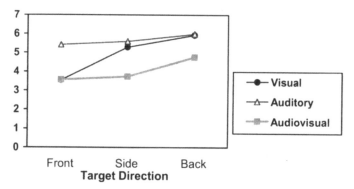

Figure 5: **Modality Effects on Target Acquisition Time for Different Target Directions**

There was also a significant interaction between cue modality and target height, $F(2, 27) = 4.52, p < .05$, for acquisition speed (see Figure 6). The acquisition speed for targets in high positions was about the same as for targets in low positions when targets were visually cued only. In contrast when audio or audio-visual cues are used, the performance for targets in high positions is degraded relative to targets in low positions. This suggests that audio cues help to locate targets in azimuth when targets are at eye level.

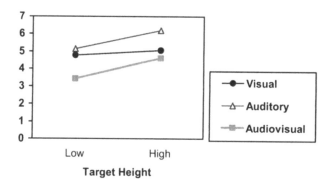

Figure 6: **Modality Effects on Target Acquisition Time for Different Target Heights**

4 Discussion

This research demonstrates the effectiveness of location-based cueing in improving the speed and accuracy of target acquisition. Target acquisition was better than the performance levels that would be expected without these location-based cues. It also shows that using multimodal cueing can further enhance target acquisition beyond what single modality cueing can provide. While visual cues led to more targets being accurately acquired than audio cues, the combination of audio and visual cues together produced even better performance than the visual cues alone. The presence of visual cues in the visual and audio-visual conditions allowed participants to pinpoint the targets more accurately. Still the audio cues enabled participants to orient in the general direction of the target and to narrow down the possible target location within 1 target of the actual target location in both azimuth and elevation. The primary value of including audio cues is to direct participants to targets that are not within the immediate line of sight. However, our participants were not able to use the audio cues very well in differentiating the location of targets placed at different heights. Audio cues and visual cues did not differ significantly in affecting the speed of acquiring targets. In contrast, the audio-visual combination produced significantly faster acquisition times than either cue modality used alone.

A question might arise regarding the practical relevance of speed and accuracy differences found in this study. The difference in target acquisition time among the three modality conditions was only a second or two. However, in combat, where even fractions of a second can mean the difference between becoming a casualty and surviving to fight another day, any gain in target acquisition speed can provide a critical advantage over an adversary. Similarly, increasing the ability of our soldiers to locate an unseen enemy by providing augmented location cues can be critical to their survival. Although we did not use soldiers as our test sample, it would seem reasonable to expect that the advantages provided by location-based cues would help soldiers in acquiring targets as well.

One thing to keep in mind regarding the cues used in this research is that these cues are overlaid onto the existing visual and audio characteristics of the simulated environment using augmented reality technology. That is, the visual and audio aspects of the environment are still present. In the MR MOUT environments there are sounds of weapons firing, choppers flying overhead and other sounds that might be present in a combat zone. In addition all of the visual features of the MR MOUT site provide the context in which the visual cues are embedded. The letters placed in the MOUT environment while providing information about the potential location of the targets also as a distracter because half of the letters do not correspond to any target location. The combat zone sounds, as well as the non-target letters, may distract participants when they are attempting to locate the audio cues.

4.1 Limitations and Advantages of Testing Environment and Procedures

The MR MOUT simulation provided some very real advantages for this research, but like any simulation it also had its limitations. MR MOUT provided a very realistic environment for acquiring targets. Its visual and audio features were very similar to what might be found on some movie sets. MR MOUT allowed participants to search for likely targets within a 360° area completing surrounding the participant. This challenged participants to be fully aware of their surroundings in any direction. Further, targets could appear in high or low positions roughly corresponding to the first and second stories of a building. The video-based HMD provided clear views of the environment and the potential targets. However, the HMD did limit the immediate field of view, thereby restricting views compared to unencumbered views of the environment. Also the HMD used in this research was relatively heavy and difficult to adjust for smaller head sizes.

A sophisticated tracking system smoothly tracked head movement providing for realistic interaction with the visual environment. The tether that was used with the HMD partially restricted free movement, particularly for participants who tended to move their heads without shifting the position of their feet. This may have slowed or prevented some participants from finding targets that were located to their rear.

From the standpoint of sound quality and sound localization, the MR MOUT environment was far from perfect. The tightly enclosed environment with carpeted concrete floors and a high ceiling produced significant sound reflections. This occurred despite using pink noise, an audio cue that is relatively easy to localize. On the other hand, the sound reflections produced by this environment are probably similar to what might occur in a similar real world environment. Although the audio cues were all set to the same volume, one speaker located inside a metal cylinder enclosure seemed to noticeably muffle the sound resulting in lower perceived volume for one audio cue position.

The simulated weapon used in this research made the task more realistic despite the obvious artificiality provided by the lettered targets. Its primary benefit though was enabling the accurate collection of acquisition time data. Participants were instructed to pull the trigger immediately upon acquiring a target. In a few trials, though the target acquisition time was not recorded even though the participant had attempted to pull the trigger. It may be in these instances the participants failed to fully engage the trigger.

Augmented reality shows great promise as a potential hazard detection aid in transportation and the military. Baby carriages and children could wear transponders which send the signal to the receiver equipped in a vehicle that can warn the driver when approaching them if they are in their immediate pathway. The U.S. military can benefit from such a device by providing information about enemy combatant locations collected from unmanned reconnaissance or remote sensors. The information given the automobile driver or the soldier can increase the likelihood of detecting a hazard, or providing advance notice that may provide more time to the individual to react to the situation, increasing the chances of successful countermeasures in time.

Along with the benefits, however, the costs must also be considered. A cueing system that provides hazard location information using different modalities or combinations of modalities may aid or hinder performance. The methods of information delivery thus should be explored and determined prior to system use in order to avoid a poor design, causing more problems than benefits. Although driven by the technology, engaging in user testing in order to make the design more user centered is desirable. The increased safety alone warrants effort into these issues; however, increasing user satisfaction during tasks is important as well. The current work is similar to previous studies, but goes beyond what they have done by looking at a 360° view of the world, as well as looking at an applied military type task in an interactive augmented reality system. The results of such testing could potentially impact the development of an augmented reality system used to warn against potential hazards in the 360° environment around the individual.

This research demonstrated that multimodal cueing in the form of audio-visual cues can be an effective aid to target acquisition in a 360° built up environment. Additional research needs to be performed to determine if other types of multimodal cueing are similarly effective. For example, the use of a haptic plus visual cue combination that employs a haptic vest in conjunction with head and body tracking could be investigated. The realism of the testing scenarios could also be increased by using virtual enemy and friendly combatants as potential targets. MR MOUT is an augmented reality training system that has the capability to provide realistic targets and scenarios. Other research questions might include: (1) how specific must the cues be in indicating the precise location of the target? (2) what types of target cues are most effective in slow or fast tempo scenarios? These are only a few of the many research issues that could be investigated that could provide valuable information related to cued target acquisition.

References

American Psychological Association (1992). Ethical principles of psychologists and code of conduct. *American Psychologist, 47,* 1597-1611.

Ansorge, U. & Heumann, M. (2003). Top-down contingencies in peripheral cuing: The roles of color and location. *Journal of Experimental Psychology / Human Perception & Performance, 29*(5), 937-949.

Ferlazzo, F., Padovani, T., Couyoumdjian, A., Olivetti, M. (2002). Head-centered meridian effect on auditory spatial attention orienting. *The Quarterly Journal of Experimental Psychology, 55A,* 937-963.

Itti, Koch, C., & Niebur, E., (1998). A model of saliency-based visual attention for rapid scene analysis. *IEEE Transactions on Pattern Analysis and Machine Intelligence, 20*(11). 1254 – 1259.

Levine, M. (2000). *Fundamentals of Sensation and Perception,* 3rd Edition. New York: Oxford University Press Inc.

Luck, S. J., & Hillyard, S. A. (1990). Electrophysiological evidence for parallel and serial processing during visual search. *Perception and Psychophysics,* 48, 603-617.

Niebur, E. & Koch, C. (1996). Control of selective visual attention: Modeling the "where" pathway. In Touretzky, D. S., Mozer, M. C., and Hasselmo, M. E., editors, *Advances in Neural Information Processing Systems,* volume 8, pages 802–808. Cambridge, MA: MIT Press.

Pratt, J. & McAuliffe, J. (2002). Determining whether attentional control settings are inclusive or exclusive. *Perception & Psychophisics, 64*(8), 1361-1370.

Sach, A. J., Hill, N. I., & Bailey, P. J., (2000). Auditory spatial attention using interaural time differences. *Journal of Experimental Psychology: Human Perception and Performance, 26*(2), 717-729.

Schmitt, M., Postma, A., & DeHaan, E. (2000). Interactions between exogenous auditory and visual spatial attention. *The Quarterly Journal of Experimental Psychology, 53A*(1), 105-130.

Spence, C. J., & Driver, J. (1994). Covert spatial orienting in audition: Exogenous and endogenous mechanisms. Journal of Experimental Psychology: *Human Perception and Performance, 20*, 555-574.

Spence, C., & Driver, J. (1997). Audiovisual links in exogenous covert spatial orienting. *Perception & Psychophysics, 59*, 1-22.

Treisman, A., & Gelade, G. (1980). A feature-integration theory of attention. *Cognitive Psychology, 12*, 97-136.

Ward, L. M. (1994). Supramodal and modality-specific mechanisms for stimulus-driven shifts of auditory and visual attention. *Canadian Journal of Experimental Psychology, 48*, 242-259.

Ward, L. M., McDonald, J. J., & Lin, D. (2000). On asymmetries in crossmodal spatial attention orienting. *Perception and Psychophysics, 62*, 1258–1264.

Wickens, C. D. (1976). The effect of divided attention on information processing in manual tracking. *Journal of Experimental Psychology: Human Perception and Performance, 2*, 1-13.

Section 4
Cognitive State Sensors

Chapter 15

Human-Machine Symbiosis – Biological Interfaces

Human-Machine Symbiosis Overview

Kevin Warwick and Mark Gasson

Department of Cybernetics
University of Reading, Reading, RG6 6AY, United Kingdom
k.warwick@reading.ac.uk, m.n.gasson@reading.ac.uk

Abstract

The interface between humans and technology is a rapidly changing field. In particular as technological methods have improved dramatically so interaction has become possible that could only be speculated about even a decade earlier. This interaction can though take on a wide range of forms. Indeed standard buttons and dials with televisual feedback are perhaps a common example. But now virtual reality systems, wearable computers and most of all, implant technology are throwing up a completely new concept, namely a symbiosis of human and machine. No longer is it sensible simply to consider how a human interacts with a machine, but rather how the human-machine symbiotic combination interacts with the outside world. In this paper we take a look at some of the recent approaches, putting implant technology in context. We also consider some specific practical examples which may well alter the way we look at this symbiosis in the future.

The main area of interest as far as symbiotic studies are concerned is clearly the use of implant technology, particularly where a connection is made between technology and the human brain and/or nervous system. Often pilot tests and experimentation has been carried out apriori to investigate the eventual possibilities before human subjects are themselves involved. Some of the more pertinent animal studies are discussed briefly here. The paper however concentrates on human experimentation, in particular that carried out by the authors themselves, firstly to indicate what possibilities exist as of now with available technology, but perhaps more importantly to also show what might be possible with such technology in the future and how this may well have extensive social effects.

The driving force behind the integration of technology with humans on a neural level has historically been to restore lost functionality in individuals who have suffered neurological trauma such as spinal cord damage, or who suffer from a debilitating disease such as lateral amyotrophic sclerosis. Very few would argue against the development of implants to enable such people to control their environment, or some aspect of their own body functions. Indeed this technology in the short term has applications for amelioration of symptoms for the physically impaired, such as alternative senses being bestowed on a blind or deaf individual. However the issue becomes distinctly more complex when it is proposed that such technology be used on those with no medical need, but instead who wish to enhance and augment their own bodies, particularly in terms of their mental attributes. These issues are discussed here in the light of practical experimental test results and their ethical consequences.

1 Introduction

Much research is presently being carried out in which biological signals of some form are measured, are acted upon by some appropriate signal processing technique and are then employed either to control a device or as an input to some feedback mechanism (e.g. Penny et.al, 2000). In the vast majority of cases the signals are measured externally to the body, thereby creating all sorts of measurement, communication and noise problems. Whatever technique is employed, in such cases loses and errors occur due to signal attenuation in the body.

Many problems also arise when attempting to translate electrical energy from the computer to the electronic signals necessary for stimulation within the human body. For example, when only external stimulation is employed then it is extremely difficult, if not impossible, to select unique sensory receptor channels, due to the general nature of the stimulation.

Insofaras human studies are concerned, virtual reality and, more recently, wearable computer techniquesprovide one route for creating a human-machine link. In the last few years items such as shoes and glasses have been augmented with microprocessors, although of most interest is perhaps research in which a miniature computer screen was fitted onto an otherwise standard pair of glasses in order to give the wearer a remote visual experience in which additional information about an external scene could be relayed (Mann, 1997). Despite being positioned adjacent to the human body, and even though indications such as stress and alertness can be witnessed, wearable computers and virtual reality systems require signal conversion to take place to interface the human sensory receptors with technology. Of much more interest, if we are considering some form of symbiosis is clearly the case in which a direct link is formed between technology and the nervous system.

Perhaps it is worthwhile considering briefly a number of related animal studies, as these can be something of a pointer for what is to come. In one of these the extracted brain of a lamprey was used to control the movement of a small wheeled robot to which it was attached (Reger et.el, 2000). The basis for this procedure was the fact that the lamprey, which is a small eal like fish, exhibits an innate response to light playing on the surface of water. The lamprey tries to align its body with respect to the light source. When connected into the robot body, this response was made use of by surrounding the robot with a ring of lights. As different lights were switched on and off, so the robot moved around its corral, trying to align itself appropriately.

Meanwhile in studies involving rats, a group were taught to pull a lever in order to receive a suitable reward. Electrodes were then chronically implanted into the rats' brains such that when each rat thought about pulling the lever, but before any actual physical movement occurred, so the reward was proffered. Over a period of a few days, four of the six rats involved in the experiment learned that they did not in fact need to initiate any action in order to obtain a reward; merely thinking about it was sufficient (Chapin, 2004).

The most ubiquitous sensory neural prosthesis in humans is by far the cochlea implant (see Finn & LoPresti eds 2003 for a good overview). Here the destruction of inner ear hair cells and the related degeneration of auditory nerve fibres results in sensorineural hearing loss. The prosthesis is designed to elicit patterns of neural activity via an array of electrodes implanted into the patient's cochlea, the result being to mimic the workings of a normal ear over a range of frequencies. It is claimed that some current devices restore up to approximately 80% of normal hearing, although for most recipients it is sufficient that they can communicate without the need for any form of lip reading. The success of cochlea implantation is related to the ratio of stimulation channels to active sensor channels in a fully functioning ear. Recent devices consist of up to 32 channels, whilst the human ear utilises upwards of 30,000 fibres on the auditory nerve.

Studies looking into the integration of technology with the human central nervous system have varied from diagnostic to the amelioration of symptoms (e.g. Yu et.al. 2001). However in the last few years some of the most widely reported research involving human subjects is that based on the development of an artificial retina (Rizzo et.al. 2001). Here small arrays have been successfully attached to a functioning optic nerve. With direct stimulation of the nerve it has been possible for the, otherwise blind, individual recipient to perceive simple shapes and letters. The difficulties with restoring sight are though several orders of magnitude greater than those of the cochlea implant simply because the retina contains millions of photodetectors that need to be artificially replicated. An alternative is to bypass the optic nerve altogether and use cortical surface or intracortical stimulation to generate phosphenes (Dobelle, 2000). Unfortunately rapid progress in this area has been severely hampered by a general lack of understanding of brain functionality, hence impressive results are still awaited.

Electronic neural stimulation has proved to be extremely successful in other areas though, including applications such as the treatment of Parkinson's disease symptoms and assistance for those who have suffered a stroke. The most relevant to this study is possibly he use of a brain implant, which enables a brainstem stroke victim to control the movement of a cursor on a computer screen (Kennedy et.al, 2004). Functional magnetic resonance imaging of the subject's brain was initially made. The subject was asked to think about moving his hand and the output of the fMRI scanner was used to localise where activity was most pronounced. A hollow glass electrode cone containing two gold wires (Neurotrophic Electrode) was then implanted into the motor cortex, this being positioned in the area of maximum-recorded activity.

Subsequently, with the electrode in place, when the patient thought about moving his hand the output from the electrode was amplified and transmitted by a radio link to a computer where the signals were translated into control signals to bring about movement of the cursor. Over a period of time the subject successfully learnt to move the cursor around by thinking about different movements. The Neurotrophic Electrode uses tropic factors to encourage nerve growth in the brain. During the period that the implant was in place, no rejection of the implant was observed; indeed the neurons grew into the electrode allowing stable long-term recordings.

Sensate prosthetics can also use a neural interface, whereby a measure of sensation is restored using signals from small tactile transducers distributed within an artificial limb. These can be employed to stimulate the sensory axons remaining in the user's stump which are naturally associated with a sensation. This more closely replicates stimuli in the original sensory modality, rather than forming a type of feedback using neural pathways not normally associated with the information being fed back. As a result the user can employ lower level reflexes that exist within the Central nervous system, making control of the prosthesis more subconscious.

Functional Electrical Stimulation (FES) can also be directed towards motor units to bring about muscular excitation, thereby enabling the controlled movement of limbs. FES has been shown to be successful for artificial hand grasping and release and for standing and walking in quadriplegic and paraplegic individuals as well as restoring some basic body functions such as bladder and bowel control. It must be pointed out though that controlling and coordinating concerted muscle movements for complex and generic tasks such as picking up an arbitrary object is proving to be a difficult, if not insurmountable, challenge with this method.

In each of the cases described in which human subjects are involved, the aim is to either bring about some restorative functions when an individual has a physical problem of some kind, e.g. they are blind, or conversely it is to give a new ability to an individual who has very limited abilities of any kind due to a major malfunction in their brain or nervous system. In this paper, whilst monitoring and taking on board the outputs from such research we are however more concerned with the possibility of giving extra capabilities to a human, to enable them to achieve a broader range of skills. Essentially our goal is to augment homosapiens with the assistance of technology. In particular we are focussed on the use of implanted technology to achieve mentally upgraded humans.

2 Augmentation

The interface through which a user interacts with technology provides a distinct layer of separation between what the user wants the machine to do, and what it actually does. This separation imposes a considerable cognitive load upon the user that is directly proportional to the level of difficulty experienced by the user. Manual intention through the use of such as joystick, button or keyboard are widely used, however these traditional interfaces considerably under exploit the processing capabilities of the human user by several orders of magnitude by presenting a bandwidth bottleneck in the link between the human thinking what they want to happen and laboriously pursuing those actions.

Technological development in itself is allowing for the possibility of humans exploiting all sensory modalities in order to receive information from machines to the fullest extent. Currently visual display screens probably most fill this role, although this alone is by no means adequate if the user wishes to fully experience an object in a foreign domain. Augmented Reality (AR) is viewed as something like half way between Virtual Reality and Telepresence since it opens the possibility of immersing the user inside a synthetic environment, on their own terms. However, mainly due to the considerable lack of sensory modalities being exploited, AR has a long way to go in order to supply interactivity rich enough to merge real and virtual domains seamlessly.

The main issue it appears is interfacing human biology with technology. In order to fully exploit all human sensory modalities through natural sensory receptors, a machine would indeed have to be a particularly complex device. One solution is to avoid the sensorimotor bottleneck altogether by interfacing directly with the human nervous system. In doing so it is probably worthwhile first of all considering what might be gained from such an undertaking, in terms of what possibilities exist for human augmentation. In the section which follows the research we have carried out thus far in upgrading a normal human subject will be described. The overall goals of the project are however driven by the desire to achieve improved intellectual abilities for humans, in particular considering some of the distinct advantages that machine intelligence exhibits, and attempting to enable humans to experience some of these advantages at least.

Some of the advantages of machine intelligence are the reasons humans make use of computers in the first place. For example rapid and highly accurate mathematical abilities in terms of number crunching, a high speed, almost infinite, internet knowledge base, and accurate long term memory can all potentially at some time be added to human brain advantages such as resilience, tolerance to ambiguity and an ability to draw together abstract relationships.

We must be clear however that even when functioning to its full potential, the human brain exhibits extremely limited sensing abilities, at least in terms of how we presently understand the functioning of a brain. Humans essentially have 5 senses that we know of, whereas machines offer a view of the world which includes a much wider range, including such as infra-red, ultraviolet and ultrasonic signals. Humans are also limited in that they can only visualise and understand the world around them in terms of a 3 dimensional perception. Meanwhile computers are quite capable of dealing with hundreds of dimensions and conceptualising relationships between these dimensions.

Perhaps the biggest present advantage for machine intelligence over human intelligence is though communication. The human means of communication, getting an electro-chemical signal from one brain to another, is extremely poor, particularly in terms of speed, power and precision, involving conversion both to and from mechanical signals, e.g. pressure waves in speech communication. When one brain communicates with another there is invariably a high error rate due to the serial form of communication combined with the limited agreement on the meaning of ideas that is the basis of human language. In comparison machines can communicate in parallel, around the world with little/no error.

Overall therefore, connecting a human brain, by means of an implant, with a computer network, in the long term opens up the distinct advantages of machine intelligence to the implanted individual. Clearly even the acquisition of only one or two of these abilities would be enough to entice a human to be augmented thus, and certainly is an extremely worthwhile driver for the research.

3   Implant Experimentation

There are two main approaches in the construction of peripheral nerve interfaces, namely extraneural and intraneural. Extraneural, or cuff electrodes, wrap tightly around the nerve fibres, and allow a recording of the sum of the signals occurring within the fibres, (referred to as the Compound Action Potential), in a large region of the nerve trunk, or by a form of crudely selective neural stimulation.

A much more useful nerve interface however is one in which highly selective recording and stimulation of distinct neural signals is enabled, and this characteristic is more suited to intraneural electrodes. Certain types of MicroElectrode Arrays (MEAs) (as shown in Figure 1) contain multiple electrodes which become distributed within the fascicle of the mixed peripheral nerve when inserted into the nerve fibres en block. This provides direct access to nerve fibres from various sense organs or nerve fibres to specific motor units. This device allows for a multichannel nerve interface. The implant experiment described here employed just such a MEA, implanted, during a 2 hour neurosurgical operation, in the median nerve fibres of the left arm of the first named author (KW), acting as a volunteer. There was no medical need for this other than in terms of the investigative experimentation that we wished to carry out.

Before progressing it is worthwhile pointing out that there are other types of MicroElectrode Arrays that can be used for interfacing between the nervous system and technology. For example etched electrode arrays, of which there is quite a variety, actually sit on the outside of the nerve fibres, rather akin to a cuff electrode. The signals obtained are similar to those obtainable via a cuff electrode, i.e. compound signals only can be retrieved, and hence for our purposes this type of array was not selected. To be clear, the type of Microelectrode array employed in our studies thus far consists of an array of spiked electrodes that are inserted into the nerve fibres, rather than being sited adjacent to or in the vicinity of the fibres.

Figure 1: A 100 electrode, 4X4mm MicroElectrode Array, shown on a UK 1 pence piece for scale

Applications for implanted neural prostheses are increasing, especially now that technology has reached a stage that reliable and efficient microscale interfaces can be brought about. In our experiment we were working hand in hand with the Radcliffe Infirmary, Oxford and the National Spinal Injuries Centre at Stoke Manderville Hospital, Aylesbury, UK – part of the aim of the experiments being to assess the usefulness of such an implant, in aiding someone with a spinal injury.

The neural interface obtained via the implant enabled a bi-directional information flow. Hence perceivable stimulation current allowed information to be sent onto my nervous system, while control signals could be decoded from neural activity in the region of the electrodes.(Further details of the implant and techniques involved can be found in Warwick et.al., 2003). A radio frequency bi-directional interface was developed (see Figure 2) to interface between the MEA and a remote computing device. In this way signals could be sent from the nervous system to a computer and also from the computer to be played down onto the nervous system with a signal transmission range of at least 10 metres.

A wide range of experiments were carried out with the implant in place (Gasson et.al. 2002), some of them more concerned with restorative investigations, whilst others involved extensions and upgrades of one kind or another. As an example, with the movement of a finger, neural signals on the nervous system were transmitted to a computer and out to a robot hand as shown in Figure 3. Sensors on the robot hand's fingertips were then employed to pick up signals which were transmitted back onto the nervous system. Whilst wearing a blindfold, in tests KW was not only able to move the robot hand, with his own neural signals, but also he could discern to a high accuracy, how much force the robot hand was applying to an object being gripped.

This experiment was carried out, at one stage, via the internet with KW in Columbia University, New York City, but with the hand in Reading University, in the UK (Warwick et.al., 2004). What it means is that when the nervous system of a human is linked directly with the internet, this effectively becomes an extension of their nervous system. To all intents and purposes the body of that individual does not stop as is usual with the human body, but rather extends as far as the internet takes it. In our case, a human brain was able to directly control a robot hand on a different continent, obtaining feedback from the hand via the same route.

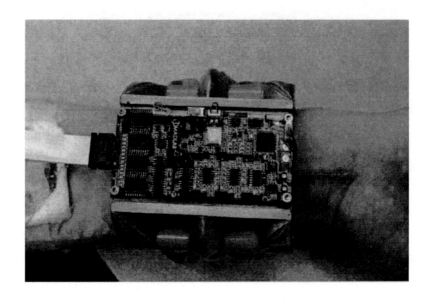

Figure 2: Mobile Interface Module attached to wrist

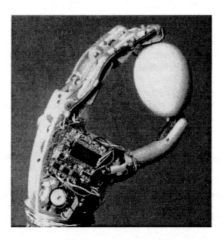

Figure 3: Intelligent anthropomorphic hand prosthesis

Extra sensory input, in the form of signals from ultrasonic sensors, was also investigated as part of the experimentation carried out. In this way KW was able to obtain an accurate bat-like sense of how far objects were away, even whilst wearing a blindfold (Warwick et.al., 2005) The results open up the possibility of senses of different types, for example infra-red or X-Ray also being fed onto the human nervous system and thus into the human brain. What is clear from our one off trial is that it is quite possible for the human brain to cope with new sensations of this type. Just to be clear on this point. It took almost 6 weeks to train the brain to recognise signals of the type shown in Fig.4 being injected onto the nervous system. When the ultrasonic input experiment was subsequently attempted, this was successful after only a few minutes of testing.

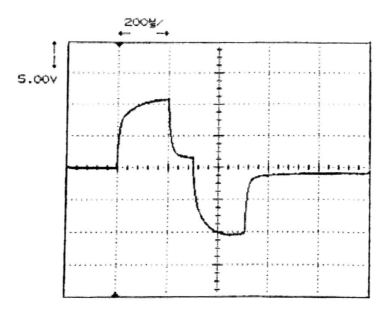

Figure 4: Voltage profile during one bi-phasic stimulation pulse cycle with a constant current of 80μA

The final part of our experimentation occurred when IW (wife of KW), also had electrodes positioned directly into her nervous system. Motor neural signals were transmitted between the two nervous systems to realise a form of radiotelegraphic communication. The next step in this research is undoubtedly to bring about the same sort of communication between two individuals whose brains are both networked in the same way.

4   Conclusion

The interaction of electronic signals with the human brain can cause the brain to operate in a distinctly different way, as is the case when an implant is employed. Such is the situation, for example, with the stimulator implants that are successfully used to counteract, purely electronically, the tremor effects associated with Parkinson's disease.
Rather than merely repairing ineffective body parts, such technology can also be employed to augment the normal functioning of the human brain. Whilst external add-ons, as in the case of a military night sight for example, provide a short term edge, the more permanent implant option has a more direct effect, altering an individual's capabilities, self-opinion and awareness, especially with regard to themselves.

By linking the mental functioning of a human and a machine network, a hybrid identity is created. When the human nervous system is connected directly with technology, this not only affects the nature of an individual's (if they can still be so called) identity, raising questions as to a new meaning for 'I', but also it raises serious questions as to that individual's autonomy. Who are you if your brain/nervous system is part human part machine?
It is clear when we compare the capabilities of machines with those of humans there are obvious differences, this is true both in physical and mental terms. As far as intelligence is concerned, it is apparent that machine intelligence has a variety of advantages over human intelligence. It is these advantages that really express themselves when it comes to linking the human brain directly with a computer network. These advantages then become ways in which a human can be intellectually augmented, providing motivation and reasoning for making the link in the first place.

5 Acknowledgements

The Authors would like to acknowledge the help of Mr. Peter Teddy and Mr. Amjad Shad who performed the neurosurgery described in the applications section, at the Radcliffe Infirmary, Oxford and ensured the medical success achieved thus far. Our gratitude is also extended to NSIC, Stoke Manderville, to the David Tolkien Trust for their support. Ethical approval for our research to proceed was obtained from the Ethics and Research Committee at the University of Reading and, in particular with regard to the involved neurosurgery, was given by the Oxfordshire National Health Trust Board overseeing the Radcliffe Infirmary, Oxford, UK.

This work has been funded in-part by the Institut International de Rechercheen Paraplégie (IRP), Geneva and from financial assistance involving Computer Associates, Tumbleweed Communications and Nortel Networks.

6 References

Chapin, J.K. Using multi-neuron population Recordings for Neural Prosthetics. *Nature Neuroscience*, 7, 452-454, 2004.

Dobelle, W., Artificial vision for the blind by connecting a television camera to the visual cortex, ASAIO J, Vol.46, pp.3-9, 2000.

Finn, W. and LoPresti, P. (eds.), Handbook of Neuroprosthetic methods, CRC Press, 2003.

Gasson,M., Hutt,B., Goodhew, I., Kyberd, P. and Warwick, K; Bi-directional human machine interface via direct neural connection, Proc. IEEE Workshop on Robot and Human Interactive Communication, Berlin, German, pp. 265-270, Sept 2002.

Kennedy, P., Andreasen, D., Ehirim, P., King, B., Kirby, T., Mao, H. and Moore, M., Using human extra-cortical local field potentials to control a switch, *Journal of Neural Engineering*, Vol.1, Issue.2, pp.72-77, 2004.

Mann, S., Wearable Computing: A first step towards personal imaging, *Computer*, Vol. 30, Issue.2, pp. 25-32, 1997.

Penny, W., Roberts, S., Curran, E., and Stokes, M., EEG-based communication: A pattern recognition approach, *IEEE Transactions on Rehabilitation Engineering.*, Vol. 8, Issue.2, pp. 214-215, 2000.

Reger, B., Fleming, K., Sanguineti, V., Simon Alford, S., Mussa-Ivaldi, F., Connecting Brains to Robots: an artificial body for studying computational properties of neural tissues, *Artificial life*, Vol.6, Issue.4, pp.307-324, 2000.

Rizzo, J., Wyatt, J., Humayun, M., DeJuan, E., Liu, W., Chow, A., Eckmiller, R., Zrenner, E., Yagi, T. and Abrams, G., Retinal Prosthesis: An encouraging first decade with major challenges ahead, *Opthalmology*, Vol.108, No.1, 2001.

Warwick, K., Gasson, M., Hutt, B., Goodhew, I., Kyberd, P., Andrews, B., Teddy, P., Shad, A., The application of implant technology for cybernetic systems. *Archives of Neurology*, 60 (10), pp.1369-1373, 2003.

Warwick, K., Gasson, M., Hutt, B., Goodhew, I., Kyberd, P., Schulzrinne, H. and Wu, X., Thought Communication and Control: A First Step Using Radiotelegraphy, *IEE Proceedings on Communications*, Vol.151, No. 3, pp 185-189, 2004.

Warwick, K., Gasson, M., Hutt, B. and Goodhew, I., An Attempt to Extend Human Sensory Capabilities by means of Implant Technology. Proc. IEEE Int. Conference on Systems, Man and Cybernetics, Hawaii, to appear, 2005.

Yu, N., Chen, J., Ju, M.; Closed-Loop Control of Quadriceps/Hamstring activation for FES-Induced Standing-Up Movement of Paraplegics, *Journal of Musculoskeletal Research*, Vol. 5, No.3, pp.173-184, 2001.

Assessment of Invasive Neural Implant Technology

Adam Spiers, Kevin Warwick, Mark Gasson

Department of Cybernetics
University of Reading, Reading, RG6 6AY, United Kingdom
a.j.spiers@reading.ac.uk, k.warwick@reading.ac.uk, m.n.gasson@reading.ac.uk

Abstract

Harnessing signals from the central nervous system is a novel method of providing humans with an extra channel of communication that can be used for the voluntary control of peripheral devices. In chronically paralysed individuals the ability to use brain activity to close even a simple switch would increase quality of life significantly, the command of a computer permitting typing on a virtual keyboard. Alternatively, stimulation of neurons can permit simulation of senses such as touch, facilitating feedback from devices controlled by the brain.

Implantation of devices into the peripheral nervous system (PNS) permits acquisition of useful signals that could, for example, be gathered from the residual limb of an amputee in order to control a prosthetic appendage. Such access to nerves could also bridge the gaps in those with paralysis of limbs via Functional Electrical Stimulation, where movement is artificially incurred by supplying currents to nerves and muscles.

Whereas measuring from the PNS is a very useful technique, an interface that connects a machine directly with the central nervous system (CNS) will permit access to a range of information rich signals not present in the peripheral nerves of patients with Brain-stem stroke and similar conditions. In such cases it is the goal of neural implants to detect activity that illustrates the intention of movement.

The theoretical ideal of harnessing nervous system activity is something that may only be achieved through suitable signal acquisition hardware and software. This paper details work being undertaken to determine and evaluate the importance of numerous characteristics that must be considered prior to warranting a device as safe and worthwhile for implantation into the nervous system. Some evaluation methods are also detailed. These methods will be used to test a large number of different implant types for assessment of suitability.

1   Motivation for Implantable Neuroprosthesis

For individuals with conditions such as Lou Gehrig's diseases, the world is an environment that may be observed and understood but only effected via yes / no eye blinks interpreted by a carer.

Clearly being cognitively intact but with such a low bandwidth of communication is incredibly frustrating. One must answer in binary to questions of food and comfort while remaining unable to easily communicate any of the high level thoughts that make us intelligent and human.

Expanding beyond eye blinks would permit the disabled individual a greater degree of interaction with their environment. Even a simple on/off bell used to call a carer or nurse would permit the patient to make requests rather than being at the mercy of waiting for questions.

In terms of peripheral nervous system devices, the integration of measuring electrodes would permit prosthetic limbs to be controlled via nerve signals in a wholly natural and subconscious way (Warwick et al,. 2003). This goal is a far cry from some current prosthesis, reported so difficult to use that wearers prefer to do without. Other measuring applications include alerting paralyzed individuals to signals that the body is trying to communicate to the brain, such as an indication that their bladder is full and needs emptying via stimulation (Hoffer, 1999).

Stimulating electrodes used in the peripheral nerves affect the control of muscles in order to bridge communication gaps and reclaim control of paralyzed limbs. By stimulating nerves rather than muscles a much lower current is required, reducing the necessity for cumbersome battery packs.

2 Hardware Considerations

It is common knowledge that the human nervous system is a delicate and complex mechanism. As such measurement or stimulation of nerve fibres / brain tissue requires a level of robust precision the must be achieved without compromising considerations of safety.

The design of a neuroprosthesis, like any product, requires manufactures to look to achieve the most effective and efficient method of interfacing with the site while remaining within the constraints of micro-engineering technologies and a sensible budget. The effectiveness of the interface is based upon many variable characteristics dependant on both choice of manufacturing technique and proposed use of the device.

2.1 Site of Implantation

An implantable device, also known as a neuroprosthetic, is designed to either measure from or stimulate the nervous system. Though some implants are designed for use in either the peripheral or central nervous systems, many devices can be used to both. A brief overview of these different implantable areas has been provided in order to identify the characteristics necessary for beneficial respective integration of a device.

2.1.1 CNS – Motor Cortex

The brain is divided into regions that relate to functional and processing abilities. In order to derive a consciously controllable signal from the brain it is necessary to target an area whose neuronal activity can be quantifiably evaluated. The motor cortex, which deals with movement and the intention of movement, is ideal for this purpose, much more so than areas of the brain used for such functions as perception or memory.

Each motor movement executed by an individual causes an increase in firing of a highly specified region of the motor cortex. These spatial regions are distinctly mapped to areas of the body. It is quite easy therefore to target areas such as the arm or hand by simply measuring from the relevant area of the cortex. Numerous previous implant studies have targeted the motor cortex with very promising results (Nicolelis et al., 2003), (Chapin et al., 2004).

Measuring from the Brain is based on the observation of changing voltage potentials of neurons, also known as action potentials. A firing neuron's voltage potential will change from -70mV to +30mV while also inhibiting the voltages of neighbouring neurons. The firing of many neurons at once is referred to as a local field potential and may be detected non-invasively by electroencephalogram (EEG) electrodes placed on the patients scalp. However this method of detection results in imprecise, low-resolution mapping with slow response times for voluntary activity detection (Donoghue, 2002).

The most effective region for measurement of the motor area is the 2mm deep layer of grey matter approximate to the skull. Beneath this area lies white matter, where measurements are impaired by the fatty insulating substance, mylin.

The motor cortex itself consists of six separate layered sheets, each of 300µm depth. It contains six types of neuronal cell. Though all cells and layers will yield some results the most suitable layers for interaction are the Ganglionic and Pyramidal, the latter of which act as motor cortex activity outputs.

Signals emerge from the central (sub-thalamic) area of the brain and propagate outwards towards the skull, becoming more task specific as they disperse into the relevant cortices. Within the cortex neuronal activity can be observed to occur in progressing waves over time and in several locations at once via the Granular / Stellate neuronal cell interconnections that exist between neuronal colonies.

2.1.2   Peripheral Nervous System

The Peripheral Nervous system encompasses the many nerves fibres that permit the Central Nervous System to communicate with sensory receptors (via afferent fibres) and motor effectors (via efferent fibres) located throughout the body.

The major nerves trunks of the body (the median, sacral etc.) are all composed of numerous nerve fascicles, which in turn are composed of many nerve fibres (the axons of individual cells). A single nerve fascicle may contain many tens of thousands of nerve fibres. Diameters of nerve fibres are typically between 2-20µm. As a nerve fibre constitutes the Axon of a neuron it is this region of the nerve trunk that must be targeted for most effective measurement from the nervous system. Coatings of epineurium, perineurium and endoneurium tissue, which separate and insulate nerves and axons from one another, do not make good places from which to measure cellular activity.

Nerves in the Peripheral Nervous system respond with identical all-or-nothing binary action potentials as neurons in the brain.

2.2   Features of a Neuroprosthetic

The aim of a neuroprosthesis is to interface with the nervous system of an organism. In order for a neuroprosthesis to serve a useful purpose in rehabilitation engineering it must be designed to remain in the body for an extended period of time (chronic implantation).

By dictionary definition a prosthesis is "an artificial body part" ("Oxford English Dictionary", 2004) and as such should be expected to exist in harmony with the rest of the body. Where acute electrodes are designed for temporary use in anesthetised animals, chronic electrodes must address the many concerns associated with long term bodily reactions and concerns in mobile environments. These concerns and the features designed to approach them will be described as follows:

2.2.1   Accurate targeting

As previously mentioned it is possible to measure massed brain activity using non-invasive scalp mounted EEG electrodes. In a similar way extra-neural electrode devices exist (Jensen et al., 2001) for measurement of the compound action potential (CAP) of the collective group of nerves that make up a nerve trunk. Though both devices have been used to provide quantifiable results, the generalised nature of data collection means their resolution is very low.

By refining hardware resolution it should be possible to achieve a greater distinction of signals. In the case of measurement this will permit many more 'degrees of freedom' for implantee control, facilitating a higher level of interaction and improved quality of communication. In order to increase resolution one must look to target the action potentials of individual nerves or neurons as mentioned in Section 2.1.

The areas most targettable for implantation in the brain are the Pyramidal and Ganglionic layers, each of which are only 300um deep. From an insertion point of view this proves a challenge in terms of the precise measurement. One is prone to either not reaching the correct layer or bypassing it by inserting the electrode slightly deeper than intended.

In acute measurement from the brain, such as in deep brain stimulation surgery, the patient is held statically in place while electrodes are gradually lowered into the lower regions of the brain via precise, micrometer style tools.
While this is method yields good results it is very time consuming and requires large quantities of very specialised and expensive equipment.

Problems of targeting also arise in the peripheral nervous system, where the numerous insulating sheets need to be bypassed in order to target nerve axons. Again these are very small and may easily be bypassed.

Use of a shaft electrode improves the likelihood of interfacing with the desired region of a nerve trunk or cortical area. By condensing multiple bi-directional electrode sites onto a single needle-like shaft it is possible to target several layers of tissue at once. In an example shaft device (Figure 1) the 16 electrodes are spaced approximately 60µm apart resulting in an overall measuring area spanning 1.6 mm. This implies that an electrode of this sort placed in the motor cortex may result in up to 5 electrodes in a relevant 300µm deep layer whilst almost guaranteeing that *some* of the electrodes will reside proximate to either the Ganglionic or Pyramidal layers.

Figure 1: Shaft Electrode

This same access to relevant areas also applies when implanting into the nerve trunks of the peripheral nervous system. While some electrodes will undoubtedly reside in insulating tissue there is a higher likelihood that some of the sites will make contact with the nerve fibres themselves.

2.2.2 Multiple Recording sites

An increased number of recording sites over the small area of a single implant permits collection of signals from neighbouring neurons and nerve fibres. As brain signals are not determined by individual neuronal activity it is often necessary to observe the inhibition of neighbouring neurons (see section 2.1.1).

There are numerous methods of arranging electrodes around a mechanical structure, each of which contributes to a different purpose. Whereas Shaft Electrodes (Figure 1) identify the vertical propagation of signals, Array devices give access to lateral neuronal interaction, permitting the pinpointing of neural activity and observation of several colonies. This does however result in more difficult positioning of the recording tip. Though investigations are being carried out into the fabrication of an Array of Shaft Electrodes (WIMS, 2003) (permitting lateral inhabitation measurement at different cortical layers), it had not been possible to find a source from which these devices may be acquired.

Multi-site Electrode Arrays originated from the practice of manually placing numerous electrodes into a particular location in the brain. By micro-fabrication techniques the distances between active sites can be reduced and made uniform.

The arrangement of nerve fibres in the PNS means that measurement from fixed depths in a linear function will simply result in repetition of results. In order to measure from multiple nerves using a multi-site electrode it is necessary to arrange electrodes at varying depths. This is possible either with a shaft electrode or a slanted Electrode Array (Figure 2) with different length shafts.

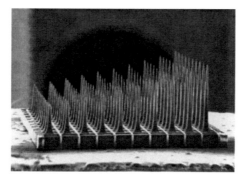

Figure 2: Slant Electrode Array

2.2.3   Stability

The adhesion of a chronic neuroprosthesis to its implant site is a cause of great concern. Movement of less than 300µm will move an electrode to a different layer of the motor cortex. With a distance of only 20µm between synapses it will take an even smaller deflection of 0.1µm to divert measurement away from a particular neuron. Micro-motion of this sort can take on many fundamentally unpreventable guises as the heartbeat-driven pulsing of the brain or simply everyday movement of a patient.

Numerous methods have been identified in order to improve the adherence of an implant to its site. Polymer devices, which are available in many different forms, such as shaft electrodes (Fraunhofer IBMT, 2003), flex to fit the contours of brain, thereby integrating better with the implant site. Penetrating tissue with a flexible device does lead to problems of buckling, though methods do exist for temporary stiffening to assist with such complications (Kipke, 2002).

Holes and cavities are sometimes machined into implants with the intention of anchoring implants via biological symbiosis i.e. the growth of tissue into these anchorage points.

2.2.4   Tissue Trauma

As neuroprosthesis deal with some of the most delicate areas of the body it is important that damage to tissue is kept to a minimum both during and after implantation.

Implantation of 'traditional' needle-like Microneurographic electrodes simply involved pushing devices into tissue and removing them once the acute measurements had been taken. If a device is suitably stiff and thin it may even be possible to enter the brain via the pia and dura (covering membranes that are often removed during surgery). Where chronic insertion is concerned, the motion of stiff electrodes can have a cutting effect that destroys tissue in a similar way to a lobotomy.

Though flexible devices adhere better to an implant site they suffer from penetration difficulties. Similar problems occur for large surface area devices. In the case of the electrode array shown in Figure 2 it is necessary to use a high velocity insertion device to minimize bruising of tissue.

It is important that devices do not cut into tissue in the brain or peripheral nervous system. Not only is this damaging firsthand but also forms scar tissue, making adhesion difficult and micro motion a greater concern. Many devices possess rounded edges and tips, designed to separate cells on implantation rather than destroying them during penetration.

2.2.5 Biocompatibility

An incompatible relationship between device and environment can result in poor results or a harmful biological reaction.

Chemicals within the brain possess a high corrosive property that can be damaging to electrodes of particular material types. Corrosion may also be encouraged if electrodes are to be used for stimulation. Normally the most notable effect of exposure to cerebral fluids is the degradation of device impedance. By measuring the impedance of a variety of devices following exposure to such corrosive tissue it is hoped that the relative susceptibility to long-term degradation (which is very detrimental to chronic implant suitability) can be determined.

Another stimulation concern is the generation of Chlorine, reactive Oxygen or Hydrogen via the process known as *gassing*. These chemicals lead to toxification, acidification or highly reactive hydrogenated compounds respectively. Gassing will inevitably damage or kill tissue surrounding the electrode site. Effects can be reduced by use of lower currents, low pulse frequencies and non-polarizing electrodes with larger surface areas (to reduce local current flow). Note however that reducing current will reduce stimulation effectiveness while larger surface areas will lead to loss of neuronal pinpointing.

Generally the body defends itself from potentially hazardous alien objects via a process known as encapsulation. During encapsulation a layer of Neuroglia cells form over an unrecognized object via a process known as glisos. This is similar to the attachment of white blood cells to a flesh wound.

By containing the object within the neuroglia capsule impedance of the device is increased significantly and electrode interaction with adjacent tissue suffers. Neuroglia in its normal and non-defensive form acts as insulator and mechanical support for neurons. A benefit of capsule formation therefore is improved implant stability.

Recently researchers of brain interfaces have looked to harnessing the beneficial stability properties of encapsulation while simultaneously preventing glisos at the implant site. Use of bioactive coatings (Kipke, 2002) and nerve tropics from the patients own body (Kennedy et al., 2000) encourage neuron growth onto interface sites, giving a better contact than simply placing electrodes adjacent to the targeted area. In cases where neurons are not grown onto interface sites it is important that there is a good electrochemistry relation between an electrode and cerebral fluid.

2.2.6 Thermodynamic Properties

During the passage of current it is characteristic for materials to undergo an increase in temperature. In the brain a temperature increase of only 1°C is all that is needed to damage tissue. It is therefore highly important to establish that using a neuroprosthetic for stimulation purposes will not cause tissue damage via unexpected thermodynamic response.

2.2.7 Electrical Characteristics

In the environment of the central nervous system, where signals are fleeting, small and sporadic, it is essential that the electrical properties of a measuring device permits the optimum acquisition and transferal of activity with minimal degradation and loss of signals. Similarly, in the case of stimulation it is important to monitor the precise amounts of currents applied to the nervous system while also maintaining confidence regarding the consistency of delivered waveform. There are many electrical properties that must be investigated in order to evaluate electrode performance.

2.2.7.1 Impedance

Impedance plays a major role in both stimulation and measurement. The impedance of a device may vary during its lifetime, especially if it is being used for stimulation. When an implant becomes encapsulated by biofilm its impedance increases and less tangible results are taken from a more generalized area. This reduced precision is thought to be due to distribution of the signal through the capsule (also accounting for signal degradation).

2.2.7.2 Harmonic Distortion

After passing through non-linear electronic circuitry it is possible that a signal will acquire unwanted harmonics, or additional signals that occur at integer multiples of the input's centre frequency. These additional higher frequencies can distort the original content of a signal significantly. With Neural prosthesis already subject to large amounts of noise (due to small initial signal and adjacent activity within the nervous system) it is essential that additional confusion from such sources as harmonic distortion are kept to a minimum.

Harmonics of a system may be determined by use of Fast Fourier Transfer (FFT) techniques (Electrical Engineering Training Series). With knowledge of each implants harmonics it will hopefully be possible to use signal processing methods to remove the unwanted harmonics and access the nerve signals.

2.2.7.3 D.C. Current Leakage

D.C current leakage can cause physical shocking sensations when applied to the nervous system. In an article on DBS for chronic pain suppression Young observes this effect from fractured electrodes (Young, 1997). In this case the electrodes ceased to serve their pain suppressing function.

Current leakage is defined as the quantity of current that flows to ground via a person when that person comes into contact with a particular electrical device. If this exceeds 0.5mA a shock is felt. It is important that the current leakage of a neuroprosthesis is significantly lower than this figure or it will have little use as an assistive medical device.

2.2.7.4 Charge Injection

A result of stray capacitance, charge injection occurs when an anomalous quantity of charge is injected from mismatched charges over transformer elements of a circuit. The effect of this would be an increase in measured charge that did not stem from patient activity or was not desired for stimulation purposes.

It is reminded that the nervous system is highly precise and delicate and that damage from over stimulation will destroy cells.

2.2.7.5 Hardware Signal Limitations

The bi-directional measurement or stimulation signals that pass through a neuroprosthesis are of a very low magnitude. Signals in of such order may be regarded as 'delicate' in that they are highly susceptible to noise factors and may be easily absorbed (in whole or part) by elements of parasitic capacitance within an implant. Absorption of a signal in this way will result in severely reduced functionality of an implant.

In order to overcome signal loss in this way it is essential that useable frequencies (that disregard absorbing elements) are determined prior to implantation. A practical method of determining this involves passing signals at different frequencies through a neuroprosthesis immersed in saline. Signals that can be measured through the saline indicated useable frequencies.

A similar experiment will also use saline in order to determine the maximum current that may be passed through a device. As implantable components are only expected to work within the limits of nervous system it is not unexpected for excessive current to cause irreversible damage to electrode areas.

3 Conclusion

Invasive interfacing of the nervous system is a practice that carries a very low margin of error, both in terms of useful results and patient safety. In order to reduce the risks of dangerous or inefficient implantation it is necessary to possess a high degree of confidence regarding the hardware being used for the interface.

Failure to identify features such as parasitic capacitance prior to implantation can lead to confusion and wasted time as researchers attempt to determine why certain stimulation frequencies are not achieving the expected effect.

Acquisition of hardware data will allow researchers to select implants with confidence and design their experiments with consideration of device limitations. The research described in this paper constitutes the first step in compilation of a concise reference of implantable technology data. Armed with such a reference it is hoped that future projects worldwide will be able to skip initial hardware testing and divert their resources more directly towards post-implantation testing.

At the time of writing, a large variety of neuroprosthesis have been acquired by the project team for determination of device characteristics.

4   Acknowledgements

The Authors would like to acknowledge the help of Dr Peter Harris of the Centre for Advanced Microscopy, University of Reading and Dr. R.T.Gladwell of the School of Animal and Microbiological Sciences (AMS), University of Reading.
This work has been funded in-part by the Institut International de Rechercheen Paraplégie (IRP).

5   References

Chapin, J.K. (2004). Using multi-neuron population Recordings for Neural Prosthetics. *Nature Neuroscience*, 7, 452-454.

Donoghue, J.P. (2002). Connecting cortex to machines: recent advances in brain interfaces. *Nature Neuroscience supplement*, 5, 1085-1088.

Hoffer, A. (1999). New *Technology will help control paralyzed limbs*. Simon Fraser University News http://www.sfu.ca/mediapr/sfnews/1999/Jan7/hoffer.html, 14 (1).

Jensen, W., Lawewnxw, S.M., Riso, R.R., Sinkjaer, T., (2001). Effect of initial joint position on nerve cuff recordings of muscle afferents in rabbits. *IEEE Transactions on Neural Systems and Rehabilitation Engineering*, 9 (3), 256-273.

Kennedy, P. Bakay, R., Moore, M., Adams, K., Goldwaith, J., (2000). Direct control of a computer from the human Central Nervous System. IEEE Transactions on Rehabilitation Engineering, 8 (2), 198-202.

Kipke, D.R. (2002), Advanced Neural Implants and Control.

Nicolelis, M.A.L., Dimitrov, D., Carmena, J.M., Crist, R., Gary, L., Jerald D.K., Wise S.P., (2003). Chronic, multisite multielectrode recordings on macaque monkeys. PNAS, 2003. 100(19), 11041-11046.

Warwick, K., Gasson, M., Hutt, B., Goodhew, I., Kyberd, P., Andrews, B., Teddy, P., Shad, A., (2003). *The application of implant technology for cybernetic systems*. Archives of Neurology, 60 (10), 1369-1373.

Young, R.F. (1997). Deep-Brain Stimulation for the Treatment of Chronic Pain. Current Science, 1, 182-191. *Compact Oxford English Dictionary*. Oxford University Press

Catalogue on Available Flexible, Light-weighted Microelectrodes (2001), Fraunhofer Institut Biomedizinische Technik http://www.ibmt.fhg.de/gruppe_l/ibmt_neuro_index_e.html

WIMS ERC Annual Report (2003), Center for Wireless Integrated Microsystems (WIMS). http://www.wimserc.org/

Harmonic Distortion, Electrical Engineering Training Series, Integrated Publishing. Retrieved December 2004 from http://www.tpub.com/neets/book23/101a.htm.

Recording Options for Brain-Computer Interfaces

Gerwin Schalk

Wadsworth Center, NY State Dept of Health
Empire State Plaza, Albany, NY 12201
schalk@wadsworth.org

Jonathan R. Wolpaw

Wadsworth Center, NY State Dept of Health
Empire State Plaza, Albany, NY 12201
wolpaw@wadsworth.org

Abstract

Millions of people in the United States and throughout the world have degenerative diseases that impair the neural pathways that control muscles. The most severely affected patients lose all voluntary muscle control, and thus lose the ability to communicate. Brain-computer interfaces (BCIs) might allow such patients to communicate by creating a new communication channel – directly from the brain to an output device – that does not depend on peripheral nerves and muscles. Recent studies have shown that non-muscular communication is possible and that it might serve useful functions. This paper reviews the different signal recording options that have been explored in these studies.

1   Introduction

Many disorders can disrupt the channels through which the brain communicates with and controls its environment. Amyotrophic lateral sclerosis, brainstem stroke, brain or spinal cord injury, cerebral palsy, muscular dystrophies, multiple sclerosis, and other degenerative diseases impair the neural pathways that control muscles or impair the muscles themselves. Those most severely affected may lose all voluntary muscle control, including eye movements and respiration, and may become completely locked in to their bodies, unable to communicate in any way. Modern life-support technology can allow most individuals, even those who are locked-in, to live long lives, and thus the personal, social, and economic burdens of their disabilities can be prolonged and severe. In the absence of methods for repairing the damage done by these disorders, there are three options for restoring function.

The first is to increase the capabilities of remaining pathways. Muscles that remain under voluntary control can substitute for paralyzed muscles. This substitution is frequently awkward and limited, but can still be useful. Patients largely paralyzed by massive brainstem lesions can often use eye movements, detected by various means, to answer questions, give simple commands, or even operate a word processing program; and severely dysarthric patients can use hand movements to produce synthetic speech (Damper, Burnett, Gray, Straus, & Symes, 1987; LaCourse & Hludik, Jr., 1990; LaCourse et al., 1990; Chen et al., 1999; Kubota et al., 2000).

The second option is to restore function by detouring around breaks in the neural pathways that control muscles. In patients with spinal cord injury, electromyographic activity from muscles above the level of the lesion can control direct electrical stimulation of paralyzed muscles, and thereby restore useful movement. One current implementation of this functional electrical stimulation technology can restore hand-grasp function to patients with cervical spinal cord injuries (Hoffer et al., 1996; Kilgore et al., 1997; Ferguson, Polando, Kobetic, Triolo, & Marsolais, 1999).

The final option for restoring function to those with motor impairments is to provide the brain with a new, non-muscular communication and control channel, a direct brain-computer interface (BCI) for conveying messages and commands to the external world (see Fig. 1). A variety of methods for monitoring brain activity exist, and could in principle provide the basis for a BCI. In addition to electroencephalography (EEG) and more invasive electrophysiological methods such as electrocorticography (ECoG) and recordings from individual neurons within the brain, these methods include magnetoencephalography (MEG), positron emission tomography (PET), functional magnetic resonance imaging (fMRI), and optical imaging. However, MEG, PET, fMRI, and optical imaging are still technically demanding and expensive. Furthermore, PET, fMRI, and optical imaging, which depend on blood flow, detect relatively slow changes in brain function and thus do not readily lend themselves to rapid communication. Non-invasive and invasive electrophysiological methods (i.e., EEG, ECoG, and single-neuron recordings) are the only methods that use relatively simple and inexpensive equipment, can capture fast changes in brain activity, and can function in most environments. Thus, these methods, which are reviewed in the following sections and illustrated in Fig. 2, are at present the primary methods that offer the practical possibility of a new non-muscular communication and control channel – a clinically useful brain-computer interface.

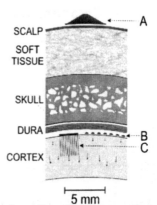

Fig. 1. Basic design and operation of a BCI system. Signals from the brain are acquired by electrodes on the scalp, the cortical surface, or from within the cortex and processed to extract specific signal features (e.g., time- or freqency-domain amplitudes) that reflect the user's intent. These features are translated into commands that operate a device (e.g., a simple word processing program or a neuroprosthesis).

Fig. 2. Signal acquisition methodologies currently used for BCI systems. Signals from the brain are acquired by electrodes on the scalp (A), the cortical surface (B), or from within the cortex (C). Scalp (EEG) electrodes are noninvasive, but are relatively far from the minute signals generated by the brain. Cortical surface (ECoG) electrodes are much closer to the brain and do not require actual penetration of the brain. Microelectrodes that penetrate the brain are closest to the brain signals, but are more invasive, and may face greater issues of long-term stability.

2 Electroencephalography (EEG)

Electroencephalographic activity (EEG) (Vidal, 1977; Sutter, 1992; Elbert, Rockstroh, Lutzenberger, & Birbaumer, 1980; Farwell & Donchin, 1988; Wolpaw, McFarland, Neat, & Forneris, 1991; Pfurtscheller, Flotzinger, & Kalcher, 1993) is recorded from the scalp. This recording method is convenient, safe, and inexpensive, and has recently been shown to support accurate two-dimensional point-to-point movement control (Wolpaw & McFarland, 2004). At the same time, these signals have relatively low spatial resolution (Freeman, Holmes, Burke, & Vanhatalo, 2003; Srinivasan, Nunez, & Silberstein, 1998) and can thus capture only gross changes in brain function. In consequence, users tend to require substantial training to achieve accurate 1D or 2D device control (about 20 and 50 30-min training sessions, respectively) and the maximum information transfer rate that can be attained is currently limited to about 25 bits/min (i.e., 4-5 characters/min). Furthermore, EEG recordings are susceptible to artifacts such as electromyographic (EMG) signals caused by muscle contractions. These problems have impeded the translation of this safe and relatively inexpensive technology into clinical devices.

3 Single-Neuron Recording

Single-neuron activity is recorded within the brain (Georgopoulos, Schwartz, & Kettner, 1986; Kennedy & Bakay, 1998; Laubach & Wessberg, 2000; Taylor, Tillery, & Schwartz, 2002). These signals have higher spatial resolution and have been shown in monkeys to provide control signals with multiple degrees of freedom and high information transfer rates (Taylor et al., 2002). However, BCIs that depend on electrodes within the cortex face substantial problems in achieving and maintaining stable long-term recordings. The small, high-impedance recording sites make penetrating electrodes susceptible to signal degradation due to encapsulation (Shain W et al., 2003). Also, small displacements of the tiny penetrating electrodes can move the recording sites away from individual neurons. These issues require resolution before wide-spread human applications of this method will be practical or possible.

4 Electrocorticography (ECoG)

An intermediate BCI methodology, using electrocorticographic activity (ECoG) recorded from the cortical surface, could be a powerful and practical alternative to these two extremes. ECoG has higher spatial resolution than EEG (i.e., millimeters vs. centimeters (Schmidt, 1980; Ikeda & Shibasaki, 1992; Heetderks & Schmidt, 1995; Levine et al., 1999; Levine et al., 2000; Freeman et al., 2003), broader bandwidth (i.e., 0-200 Hz vs. 0-40 Hz), higher amplitude (i.e., 50-100 μV maximum vs. 10-20 μV), and far less vulnerability to artifacts such as EMG (Srinivasan et al., 1998; Freeman et al., 2003). At the same time, because ECoG is recorded by electrode arrays on the surface of the brain and thus does not require electrodes that penetrate into cortex, it entails less clinical risk and is likely to have greater long-term stability than single-neuron recording. Using ECoG signals, a recent study demonstrated that users can acquire one-dimensional control over a computer cursor (Leuthardt, Schalk, Wolpaw J.R., Ojemann, & Moran, 2004) within a few minutes, much faster than has been reported for EEG (Wolpaw & McFarland, 2004).

5   Conclusions

In summary, studies over the past decade have demonstrated that brain-computer interface technology is possible and indicate that even the current generation of BCIs might serve useful functions. At the same time, while theoretically impressive, most present systems have remained merely laboratory demonstrations. A major reason for this situation is that EEG and single-neuron recording, which are the recording methodologies used in the vast majority of studies, still face significant issues of capability, practicality, and/or safety. ECoG-based BCI systems may prove a practical and powerful alternative recording methodology that could create a direct path towards a BCI system with major clinical implications.

Reference List

Chen, Y. L., Tang, F. T., Chang, W. H., Wong, M. K., Shih, Y. Y., & Kuo, T. S. (1999). The new design of an infrared-controlled human-computer interface for the disabled. *IEEE Trans.Rehabil.Eng, 7*, 474-481.

Damper, R. I., Burnett, J. W., Gray, P. W., Straus, L. P., & Symes, R. A. (1987). Hand-held text-to-speech device for the non-vocal disabled. *J.Biomed.Eng, 9*, 332-340.

Elbert, T., Rockstroh, B., Lutzenberger, W., & Birbaumer, N. (1980). Biofeedback of slow cortical potentials. I. *Electroencephalogr.Clin.Neurophysiol., 48*, 293-301.

Farwell, L. A. & Donchin, E. (1988). Talking off the top of your head: toward a mental prosthesis utilizing event-related brain potentials. *Electroencephalogr.Clin.Neurophysiol., 70*, 510-523.

Ferguson, K. A., Polando, G., Kobetic, R., Triolo, R. J., & Marsolais, E. B. (1999). Walking with a hybrid orthosis system. *Spinal Cord., 37*, 800-804.

Freeman, W. J., Holmes, M. D., Burke, B. C., & Vanhatalo, S. (2003). Spatial spectra of scalp EEG and EMG from awake humans. *Clin Neurophysiol, 114*, 1053-1068.

Georgopoulos, A. P., Schwartz, A. B., & Kettner, R. E. (1986). Neuronal population coding of movement direction. *Science, 233*, 1416-1419.

Heetderks, W. J. & Schmidt, E. M. (1995). Chronic, Multiple Unit recording of Neural Activity with Micromachined Silicon Electrodes. In A.Lang (Ed.), *Proceedings of RESNA 95 Annual Conference* (Arlington: RESNA Press.

Hoffer, J. A., Stein, R. B., Haugland, M. K., Sinkjaer, T., Durfee, W. K., Schwartz, A. B. et al. (1996). Neural signals for command control and feedback in functional neuromuscular stimulation: a review. *J.Rehabil.Res.Dev., 33*, 145-157.

Ikeda, A. & Shibasaki, H. (1992). Invasive recording of movement-related cortical potentials in humans. *Journal of Clinical Neurophysiology, 9*, 509-520.

Kennedy, P. R. & Bakay, R. A. (1998). Restoration of neural output from a paralyzed patient by a direct brain connection. *Neuroreport, 9*, 1707-1711.

Kilgore, K. L., Peckham, P. H., Keith, M. W., Thrope, G. B., Wuolle, K. S., Bryden, A. M. et al. (1997). An implanted upper-extremity neuroprothesis: follow-up of five patients. *The Journal of Bone and Joint Surgery, 79-A*, 533-541.

Kubota, M., Sakakihara, Y., Uchiyama, Y., Nara, A., Nagata, T., Nitta, H. et al. (2000). New ocular movement detector system as a communication tool in ventilator-assisted Werdnig-Hoffmann disease. *Developmental and Medical Child Neurology, 42*, 61-64.

LaCourse, J. R. & Hludik, F. C., Jr. (1990). An eye movement communication-control system for the disabled. *IEEE Trans.Biomed.Eng, 37*, 1215-1220.

Laubach, M. & Wessberg, J. (2000). Cortical ensemble activity increasingly predicts behavior outcomes during learning of a motor task. *Nature, 405*, 567-571.

Leuthardt, E. C., Schalk, G., Wolpaw J.R., Ojemann, J. G., & Moran, D. (2004). A Brain-Computer Interface Using Electrocorticographic Signals in Humans. *Journal of Neural Engineering, 1*, 63-71.

Levine, S. P., Huggins, J. E., BeMent, S. L., Kushwaha, R. K., Schuh, L. A., Passaro, E. A. et al. (1999). Identification of electrocorticogram patterns as the basis for a direct brain interface. *Journal of Clinical Neurophysiology, 16*, 439-447.

Levine, S. P., Huggins, J. E., BeMent, S. L., Kushwaha, R. K., Schuh, L. A., Rohde, M. M. et al. (2000). A direct brain interface based on event-related potentials. *IEEE Trans.Rehabil.Eng, 8*, 180-185.

Pfurtscheller, G., Flotzinger, D., & Kalcher, J. (1993). Brain-computer interface - a new communication device for handicapped persons. *Journal of Microcomputer Applications, 16*, 293-299.

Schmidt, E. M. (1980). Single neuron recording from motor cortex as a possible source of signals for control of external devices. *Ann.Biomed.Eng, 8*, 339-349.

Shain W, Spataro L, Dilgen J, Haverstick K, Retterer S, Isaacson M et al. (2003). Controlling cellular reactive responses around neural prosthetic devices using peripheral and local intervention strategies. *IEEE Trans Neural Syst Rehabil Eng, 11*, 186-188.

Srinivasan, R., Nunez, P. L., & Silberstein, R. B. (1998). Spatial filtering and neocortical dynamics: estimates of EEG coherence. *IEEE Trans.Biomed.Eng, 45*, 814-826.

Sutter, E. E. (1992). The brain response interface: communication through visually-induced electrical brain responses. *Journal of Microcomputer Applications, 15*, 31-45.

Taylor, D. M., Tillery, S. I., & Schwartz, A. B. (2002). Direct cortical control of 3D neuroprosthetic devices. *Science, 296*, 1829-1832.

Vidal, J. J. (1977). Real-time detection of brain events in EEG.

Wolpaw, J. R. & McFarland, D. J. (2004). Control of a two-dimensional movement signal by a noninvasive brain-computer interface in humans. *Proceedings of the National Academy of Sciences of the United States of America, 101*, 17849-17854.

Wolpaw, J. R., McFarland, D. J., Neat, G. W., & Forneris, C. A. (1991). An EEG-based brain-computer interface for cursor control. *Electroencephalogr.Clin.Neurophysiol., 78*, 252-259.

First Investigations of an fNIR-Based Brain-Computer Interface

Evan D. Rapoport
Archinoetics
1050 Bishop Street #240
Honolulu, HI 96813
evan@archinoetics.com

Traci H. Downs
Archinoetics
1050 Bishop Street #240
Honolulu, HI 96813
traci@archinoetics.com

Dennis R. Proffitt
University of Virginia
Box 400400
Charlottesville, VA
22904
drp@virginia.edu

J. Hunter Downs, III
Archinoetics
1050 Bishop Street #240
Honolulu, HI 96813
hunter@archinoetics.com

Abstract

The brain-computer interface (BCI) is clearly not a ubiquitous technology, yet it has tremendous potential to improve the lives of disabled individuals who cannot communicate or control their environments through traditional channels. This experiment is our first investigation into providing means for patients with locked-in syndrome to manipulate computer displays through a direct brain interface. Using functional near-infrared imaging, specifically the Optical Tomographic Imaging Spectrometer (OTIS), we have shown that this new technology can allow a person to communicate through brain signals alone.

1 Introduction

Research into brain-computer interfaces has increased tremendously in recent years (Wolpaw et al., 2000). While the novelty of such systems has made them popular topics for discussion even outside the scientific community ("In the Lab, Monkeys Use Brain Waves", 2000), their utility has not been widespread. For most people, the low bit rate communication associated with direct brain control renders these devices relatively useless as a tool in everyday life. However, for people who are severely disabled, BCI's have been successful by providing a communication channel, albeit a slow one, where one did not previously exist. BCI users have successfully manipulated computer interfaces and machines for a variety of tasks (Kennedy, Bakay, Moore, Adams & Goldwaithe, 2000, Birbaumer et al, 2000, Birbaumer et al, 1999, Wolpaw, McFarland, Neat, Forneris, 1991).

1.1 Locked-in Syndrome

Perhaps the people most in need of a robust brain-computer interface are those with a condition referred to as locked-in syndrome. Individuals lose all ability to speak, move, or externalize any indications that they have conscious control over their bodies ("NINDS Locked-In Syndrome", 2005, "Locked-In Syndrome", 2005). However, all other brain functions remain intact, allowing the person to think, reason, and perceive stimuli in their external environment. The life expectancy has improved recently and can be at least several years. Although locked-in syndrome can occur from slow degenerative diseases, such as amyotrophic lateral sclerosis (ALS), it can also result almost instantly as a result of traumatic brain injury or a brain-stem stroke. While some locked-in patients retain a small amount of motor control allowing them to blink an eye, many do not; it is these people for whom BCI's are most necessary.

1.2 Functional Brain Imaging for Command and Control

A variety of brain imaging devices have been used in BCI's, including electroencephalography (EEG) (Birbaumer et al, 2000, Birbaumer et al, 1999) and surgically-implanted transmitters placed directly in the cortex (Kennedy et al, 2000). Some work has even been done using functional magnetic resonance imaging (fMRI) (Weiskopf et al, 2004). Each of these approaches has had success in allowing people to communicate information. Birbaumer trained patients to control their slow cortical potentials with EEG, enabling them to select items such as words or letters. Kennedy's invasive device allowed two patients to spell words or move a cursor. These are just a few examples of successful BCI's.

1.2.1 Limitations of Current Technologies for Locked-in Patients

A number of brain imaging devices have allowed healthy subjects in laboratory settings to control computers using brain signals; however, a new set of problems arise when a BCI subject is a locked-in patient.

Some systems require a great deal of training before users can control them successfully. This process can take several months and hundreds of sessions (Birbaumer et al 2000, Kennedy et al 2000). This can be frustrating for the patients (Spinney, 2003) and their families, and is made worse by the fact that the patients cannot progress through the standard methods of providing feedback while learning this new task.

One of the most difficult aspects of working with locked-in patients as compared to healthy patients is the physical placement of the brain imaging device on their heads. Because they cannot respond if something is uncomfortable, it is likely a very stressful event for them. Therefore, the easier a device is to put on, the better. Additionally, patients have ventilator tubes, eyeglasses, etc. that make donning the brain imaging device much more complicated than with a healthy subject.

1.2.2 Functional Near-Infrared Imaging (fNIR)

Functional near-infrared (fNIR) imaging is a relatively new technology that measures changes in the relative concentrations of oxy-hemoglobin and deoxy-hemoglobin in the brain (Chance, Cope, Gratton, Ramaunujam & Tromberg, 1998) Like fMRI, it can be used to infer what mental activity someone is doing by monitoring the appropriate brain area, thus little or no training is required to use an fNIR system.

Briefly, fNIR works as follows (Nishimura, Stautzenberger, Robinson, Downs & Downs, in press): Near-infrared light is emitted and passes through the skin, skull, and some of the cortical layers of the brain. Some of this light follows a banana shaped curve and is reflected back out, reaching the photodetectors. The amount of light detected is effected by the amount of blood and oxygen (indicators of neural activity) in the area of the brain being looked at. Therefore, by positioning the sensor over, for example, Broca's area, the system can detect whether or not the person is doing a language task commonly associated with that region.

2 Experiment

This experiment was designed to determine the feasibility of an fNIR-based BCI with the end goal to apply what was learned to enable locked-in patients to use it as a communication tool. Based on our internal research and experience with fNIR, we designed the experiment with OTIS being used as an on/off switch, such that the subject could choose one object or another by performing one of two different tasks.

2.1 Method

2.1.1 Participants

As this was a first investigation, this experiment was done with one healthy, right-handed female subject. We began with a healthy subject, as opposed to a locked-in one, for a number of logistical reasons, including the added stress for the patients and their families that could have been caused by performing a preliminary study.

2.1.2 Hardware

2.1.2.1 The Optical Tomographic Imaging Spectrometer (OTIS)

The OTIS system (see Figure 1 below), developed by Honolulu-based Archinoetics, LLC, is an fNIR system that consists of a small box and an attached set of sensors that are placed on the head (Nishimura et al, in press). The sensor consists of an emitter (an LED for three wavelengths of light) and up to eight detachable photodetectors surrounding it with about three centimeters separation. The entire sensor is about an inch thick and about two to five inches in length and width, depending on how many detectors are attached. It is placed directly on the head, secured using an elastic headband, and covered with dark cloth to block external light from affecting the sensitive detectors.

OTIS transmits packets over Ethernet at a rate of 33Hz. These time-stamped packets contain raw light values at the different detectors as well as additional information that can be controlled by the software interface, as will be discussed later.

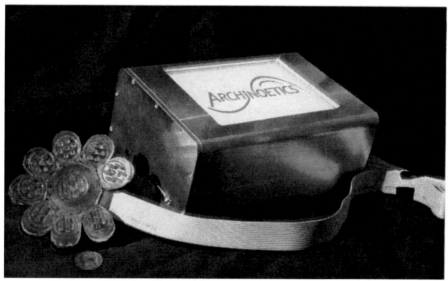

Figure 1: The Optical Tomographic Imaging Spectrometer (OTIS) developed by Archinoetics. The sensor suite here is transparent for viewing purposes; however, the actual sensors are covered with an opaque black dye.

Good, direct contact with the skin is required to get a strong signal from the brain. The unique design of the OTIS sensors allows it to sit firmly on the head when secured with a strap while also penetrating through the hair (see Figure 2 below). Each emitter consists of a central "finger" from which the LED projects near-infrared light. The surrounding fingers do not contain any electronics, and are provided for support and to allow the sensor to be pushed through the hair to make good scalp contact. The detectors have a similar design as the emitter, though their central fingers detect the near-infrared light instead of emitting it.

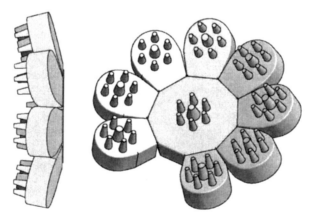

Figure 2: The unique design of the OTIS sensors allows it to get strong brain signals, even through thick hair. This figure shows an OTIS sensor with a central emitter module and eight surrounding detectors.

751

For this study, a single sensor was used since only one brain area was being monitored. The sensor consisted of an emitter and two photodetectors. Although only one photodetector was needed for this study, a second gave the sensor symmetry and a wider base, allowing it to balance on the head when held on by the strap (see Figure 3 below).

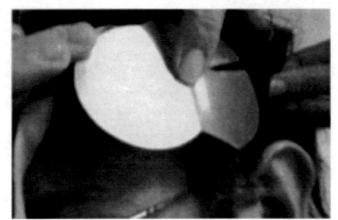

Figure 3: The back of an OTIS sensor as it looks before being secured on the head by an elastic strap.

2.1.2.2 Computer

The OTIS system must be connected to an Ethernet switch for data to be received by a computer. Currently, the software required to connect to OTIS will work on any reasonably fast Windows-based machine with the Microsoft .NET framework. For this experiment, we used a laptop computer with a second LCD monitor attached to allow all software interfaces to be displayed at the same time.

2.1.3 *Software*

The values from OTIS were received in Ethernet packets by one of two different programs developed by Archinoetics. Usage of these applications will be discussed in greater detail in the procedure section. The Data Receiver displays the raw light values for each channel of data (i.e. each detector). This information also appeared on the LCD display on the actual OTIS box. The other program, PuckLink, allowed the user to select from several algorithms and filters to processes the data. The resulting signal was graphed in real-time. Additionally, another program was written for this experiment, which will also be discussed in the procedure section.

2.1.4 *Signal Processing*

PuckLink software provided real-time signal processing by using an adaptive statistical area algorithm. Data was also filtered to remove noise caused by the heart beat and other relatively low frequency patterns. Because the detectors were internally shielded, they were resistant to external electrical noise that would otherwise add unpredictable variations in the signal.

2.1.4.1 Adaptive Statistical Area

The data passed into the adaptive statistical area algorithm represented an oxygenation value. The algorithm took this data stream and maintained a running average of it over a specified time window (a value that could be modified in the user interface). It then computed the area between the actual value curve and the running average for an instantaneous statistical area value. A running average of this value was also maintained. A threshold, which could also be set in the user interface, was established above the running average. To compute a final activation value, values between those two lines were scaled from zero to one; values below the range were negative and values above the range were greater than one. Activations were defined as values greater than one, whereas the theoretical baseline was zero. The threshold could be raised to prevent false positives, or lowered to make the system more sensitive and ensure activations were not missed. For this algorithm, the BCI designer must determine what would be appropriate and most beneficial for the goal of the program.

2.1.4.2 Determining Activations for a Block of Time

Because this research is in its early stages, robust algorithms are still being developed for determining whether a block of time represents a period of activity or of rest in the brain area being looked at. Therefore, this determination was made by experimenters experienced in interpreting OTIS data as they watched the graphs of the oxygenation and adaptive statistical area calculations. Parallel research into appropriate real-time signal processing algorithms to use for this purpose is being conducted internally; however, it was not the focus of this study.

2.1.5 Design

The goal in this study was to allow the subject to communicate with the experimenter through brain signals alone. Specifically, the subject was to communicate a random integer between 1 and 16 using only brain signals. To do so, the list of available numbers was subdivided into two halves with one half being selected by activating a brain area monitored by OTIS and the other half by not activating. The subject had to identify which half contained the number being communicated and then either activate or suppress the activation to select that half. Each activation or suppression halved the available numbers. Consequently, four subdivisions were required to communicate the number from the original list of sixteen (2^4 = 16).

2.1.5.1 The Dorsolateral Prefrontal Cortex

The design of this study was such that it could be easily performed on a locked-in patient in a future iteration. The locked-in patients we visited had all been previously diagnosed with ALS. Therefore, when choosing a brain area to place the sensor over, we selected an area, the left dorsolateral prefrontal cortex (DLPFC), that was not believed to have atrophied due to the disease (such as the motor cortex). This area is activated by a number of tasks, particularly those requiring working memory and sequence learning (Nathaniel & Frith, 2002, Robertson, Tormos, Maeda & Pascual-Leone, 2001, Pascual-Leone, Wassermann, Grafman, Hallett, 1996).

2.1.5.2 Tasks

We chose two tasks for this experiment based on preliminary research using fNIR and fMRI: one task that reliably activated the DLPFC and another that did not. For the first task, which activated the DLPFC, the subject was asked to perform a covert counting task. The specific instructions were to count as quickly as possible while still pronouncing each number in his or her head. For the second task, which did not activate this area, the subject was instructed to repeat a simple word in his or her head over and over, such as, "la, la, la…". This was referred to as the "rest task".

2.1.5.3 Target Number Generation

The numbers to be communicated were created randomly before the experiment began. Ten numbers were selected and were not revealed to the experimenter until after the session was complete.

2.1.5.4 Environment

The testing environment for this experiment was a dark, quiet room illuminated only by the LCD screens. The subject sat in a standard desk chair and was asked to find a comfortable position to sit in that could be maintained for the duration of the experiment.

2.1.6 Procedure

2.1.6.1 Preliminary Setup

Before testing began, the subject was explained the set of tasks and the procedure for positioning the OTIS sensor on the head. The next step was positioning the sensor over the targeted brain area. The sensor was then secured to the head using a standard cotton and elastic headband, making sure that the emitter and detector were both making good contact with the scalp and that no hair was interfering. This was verified using the Data Receiver, which displayed the amount of light reaching each detector. Each detector had a known dark noise and maximum value, creating a range for which a subset existed that represented a strong signal not influenced by external light. To minimize the effect of ambient light from the LCD, the brightness was turned down and a dark cloth was draped over the subject's head. Although these steps were not necessary for a successful session, they added an extra degree of control.

Once the sensor was secured on the head, the experimenter verified a good signal from the appropriate brain area by asking the subject to perform each of the tasks. The experimenter watched the graphs in PuckLink and noted the detection values. If the signal matched the known pattern for the tasks, then the experiment could begin. Otherwise, the sensor was re-positioned and the process repeated.

The subject is also allowed to watch these graphs to provide some performance feedback. Testing performed internally during preliminary research for this study revealed that subjects were not found it harder to focus on the tasks and became easily bored when they were not allowed to watch the graphs.

2.1.6.2 Baseline

Before the various tasks could be performed, baseline levels were established for the signal processing algorithms. The subject was instructed to perform the rest task while the experimenter monitored the graphs. This continued until the signal stabilized, indicating that the algorithm had established a good baseline and that the subject was not activating the DLPFC. This lasted a minimum of one minute for each run (multiple runs were required for the experiment to provide the subject with a break).

2.1.6.3 Number Communication

After the baseline period concluded, the experimenter displayed a custom application with a user interface that allowed for the list of numbers to be selected and distributed. This was available for both the subject and experimenter to see. Prior to the experiment beginning, the subject was instructed that the list on the left could be selected by performing the rest task, whereas the list on the right could be selected by performing the counting task. Thus the subject's goal was to enable the experimenter to choose the correct list each time by performing the specified task, which could be determined by the experimenter through watching the graphs.

The experimenter then signaled the subject to begin the selection process. After a twenty second block, the experimenter informed the subject to rest. If the subject's intentions were clear, the appropriate list on the user interface was selected by the experimenter by clicking the mouse. The subject, watching this selection, informed the experimenter if the selection was incorrect. Otherwise, it was assumed to be correct and the process continued. Between each block, a rest period was used to allow the signal to stabilize, just like during the baseline period (though significantly shorter). However, if the experimenter was not sure of the subject's response, a list would not be selected on the interface and the subject was asked to repeat the same response. This continued until a decision was made by the experimenter.

A written log was kept throughout the study of the experimenter's interpretations and the resulting numbers communicated were written down. Additionally, the PuckLink software kept a detailed log of the OTIS data. By clicking on various buttons in the interface, the experimenter was able to timestamp the beginnings and ends of the blocks for later analysis.

2.2 Results

There were eleven numbers to be communicated in this experiment. Each number selection consisted of four list selections, allowing for a perfect score to be defined as 44 out of 44. As incorrect selections resulted in the subject stopping the experimenter for that number, 44 selections would only be made if they were all correct (except the last one). Thus two scores were given for the session: the one just discussed and another to provide an overall indication of the BCI's ability to communicate useful information (how many numbers were correctly selected). The probability of selecting the correct list by chance (the null hypothesis) was 50%, thus the probability of selecting the correct number over the four selections by chance was 6.25%.

In this study, 27 of 32 selections were correctly interpreted by the experimenter resulting in 6 of 11 numbers being correctly communicated. This represented percentage scores of 84.4% and 54.5%, respectively. A chi square test on the selection score resulted in the rejection of the null hypothesis with a confidence greater than 99.9%.

3   Discussion

This paper is a report on preliminary research that we have conducted with the end-goal of applying our findings to provide locked-in patients with a communication channel. The patients we visited over the past several months had little or no success with other types of brain-imaging devices. Their families had been searching, without success, for any way to allow their loved one to express his or her thoughts. It is our belief that fNIR can and will provide them with this opportunity in the very near future.

3.1   Issues with the Present Study

3.1.1   Signal Processing

In designing this study, it was decided that until our signal processing algorithms could evaluate blocks of data with the same success rate as a trained experimenter, they would not be used to automatically control the selection of lists for the subject. The user interfaces built for allowing locked-in patients to communicate in future iterations of this study were designed assuming that data blocks would be evaluated with a high degree of accuracy, thus an "undo" action was timely through this BCI. Therefore, it would be more effective to request the patient repeat his or her response to reach near certainty than it would be to proceed quickly with lower confidence. Adding control from a second brain area should allow for an easy undo command that should remove this shortcoming of the system.

3.1.2   Physical Movement

We discovered in doing this research that purposeful, repetitive movement (particularly of the fingers) sometimes resulted in OTIS detecting activations when the counting task was not being performed, presumably due to activity in the nearby motor cortex. In the present study, this was controlled by asking the subject to remain motionless during the experiment. Being that locked-in patients were the target population, we knew that movement would not be a problem in the next phase of this study. Although imagined movement could also activate the motor cortex and interfere with our results, it was believed that the paralysis of patients who had been locked-in for at least several months caused their motor cortex to atrophy and prevent it from being a suitable area for use in a BCI (Moore, 2004).

3.1.3   Confirming Selections

One part of this study that obviously could not be used when testing a locked-in patient was the measure for evaluating whether the experimenter's interpretation was correct. In this study, the subject provided that feedback verbally, though of course a locked-in patient could not do that. The solution would be to allow someone else besides the experimenter and the patient to see the numbers and serve that role.

3.2   Future Work

The subject was given numbers that required four independent selections (rather than simply a binary number sequence representing which item to select) because of our end goal: enabling locked-in patients to communicate useful information. Transitioning from sixteen numbers to 26 letters would only require one more selection by the subject. Given a dictionary of common words, and restricting people to it, once one letter is chosen, only a small subset of the alphabet would be possible for the second letter. Using such strategies, words could be spelled through this process-of-elimination method with fewer selections than would be expected. Internal research is currently being done with several locked-in patients to implement this BCI application.

4 Conclusion

The results indicate that OTIS could be a viable option for detecting changes in brain activity and using that information to control a user interface. Future work will focus on improving signal processing both for real-time algorithms and for evaluating blocks of time. Additionally, we continue to focus on developing user interfaces that can be easily controlled even with the low bit rate communication inherent to BCI's.

Though brain-computer interfaces have been receiving more attention in recent years from the academic community and beyond, they have yet to permeate the homes of the families who need them most. Locked-in patients, despite having no means of communication, have fully functional minds and emotions, as well as families who desperately wish for them to "speak" again. Anything that can be done to provide them with a reliable way of doing so would dramatically improve their quality of life.

References

Birbaumer, N., Ghanayim, N., Hinterberger, T., Iversen, I., Kotchoubey, B., Kübler, A., Perelmouter, J., Taub, E., Birbaumer, N., & Flor, H. (1999). A Spelling Device for the Completely Paralyzed Patients. *Nature*, vol. 398, 297-298.

Birbaumer, N., Ghanayim, N., Hinterberger, T., Perelmouter, J., Taub, E., Iversen, I., Kotchoubey, B., Neumann, N., & Flor, H. (2000). The Thought Translation Device (TTD) for the Completely Paralyzed Patients. *IEEE Transactions on Rehabilitation Engineering*, Vol. 8, No. 2, 190-193.

Chance, B., Cope, M., Gratton, E., Ramanujam, N., & Tromberg, B. (1998). Phase Measurement of Light Absorption and Scatter in Human Tissues. *Review of Scientific Instruments*, 689, 3457-3481.

CNN.com Health (2000). In the Lab, Monkey Use Brain Waves to Move a Robotic Arm. Retrived February 22, 2005 from http://archives.cnn.com/2000/HEALTH/11/15/monkey.brain/.

Kennedy, P.R., Bakay, R.A.E., Moore, M.M., Adams, K., & Goldwaithe, J. (2000). Direct Control of a Computer from the Human Central Nervous System. *IEEE Transactions on Rehabilitation Engineering*, Vol. 8, No. 2, 198-202.

M.M. Moore (personal communication, c. October 28, 2004) stated that her experience with ALS patients has shown little or no indication that the motor cortex would be a viable option for controlling a BCI after a patient has been locked-in for several months.

Nathaniel-James, D.A., Frith, C.D. (2002). The Role of the Dorsolateral Prefrontal Cortex: Evidence from the Effects of Contextual Constraint in a Sentence Completion Task. *Neuroimage*, 16(4), 1094-1102.

National Institute of Neurological Disorders and Stroke (2005). NINDS Locked-In Syndrome Information Page. Retrieved February 1, 2005 from www.ninds.nih.gov/disorders/lockedinsyndrome/lockedinsyndrome.htm.

Nishimura, E.M., Stautzenberger, J.P., Robinson, W.J., Downs, T.H., Downs, III, J.H. (in press). A New Approach to fNIR: The Optical Tomographic Imaging Spectrometer (OTIS). *HCI International 2005*.

Pascual-Leone, A., Wassermann, E.M., Grafman, J., Hallett, M. (1996). The Role of the Dorsolateral Prefrontal Cortex in Implicit Procedural Learning. *Experimental Brain Research*. 107(3), 479-85.

Robertson, E.M., Tormos, J.M., Maeda, F., Pascual-Leone, A. (2001). The Role of the Dorsolateral Prefrontal Cortex During Sequence Learning is Specific for Spatial Information. *Cerebral Cortex*, 11(7), 628-35.

Smith, E., Delargy, M. (2005). Clinical Review: Locked-In Syndrome. Retrieved February 22, 2005 from http://bmj.bmjjournals.com/cgi/content/full/330/7488/406.

Spinney, L. (2003). Hear My Voice. *New Scientist*, February 22, 2003, 36-39.

Weiskopf, N., Mathiak, K., Bock, S.W., Scharnowski, F., Veit, R., Grodd, W., Goebel, R., Birbaumer, N. (2004). Principles of a Brain-Computer Interface (BCI) Based on Real-Time Functional Magnetic Resonance Imaging (fMRI). *IEEE Transactions on Biomedical Engineering*, Vol. 51, No. 6, 966-970.

Wolpaw, J.R., Birbaumer, N., Heetderks, J.W., McFarland, D.J., Peckham, P.H., Schalk, G., Donchin, E., Quatrano, L.A., Robinson, C.J., & Vaughan, T.M. (2000). Brain-Computer Interface Technology: A Review of the First International Meeting. *IEEE Transactions on Rehabilitation Engineering*, Vol. 8, No. 2, 164-173.

Wolpaw, J.R., McFarland, D.J., Neat, G.W., Forneris, C.A. (1991). An EEG-Based Brain-Computer Interface for Cursor Control. *Electroenceph clin Neurophysiol*, 78, 252-259.

Using hypermedia to support communication in Alzheimer's disease: The CIRCA project

Arlene J. Astell[1], Maggie Ellis[1]   Normal Alm[2], Richard Dye[2]   Gary Gowans[3], Jim Campbell[3]

[1]University of St. Andrews,   [2]University of Dundee,   [3]Duncan of Jordanstone College of Art
School of Psychology   Division of Applied Computing   Division of Graphic Design,
South Street, St. Andrews   Faculty of Engineering and   Design School, Dundee
Fife, Scotland, KY16 9JU   Physical Sciences   Scotland, DD 4HN,
                              Dundee, Scotland, DD1 4HN

aja3@st-and.ac.uk   nalm@computing.dundee.ac.uk   g.m.gowans@dundee.ac.uk
mpe2@st-and.ac.uk   rdye@computing.dundee.ac.uk   j.campbell@dundee.ac.uk

Abstract

The ability of people with dementia to participate in social interactions is progressively undermined by the illness. This is due in part to memory and speech production difficulties but also to changes in behaviour that can deter potential interaction partners. Therapeutically, reminiscence is a popular medium in dementia care for promoting conversation and positive interactions. By tapping into relatively intact long-term memory, reminiscence encourages people with dementia to recall and share memories from their past. Computer technology provides the opportunity to present a large multimedia database of stimuli to use in reminiscence. CIRCA is an interactive hypermedia-based system designed for people with dementia to facilitate and support conversation and social interactions. The system contains a large database of media presented via a touch screen to act as a prompt for interactions between people with dementia and caregivers. We evaluated the usefulness of CIRCA by comparing it with one-to-one reminiscence sessions conducted using traditional reminiscence stimuli. Unlike traditional reminiscence CIRCA gives people with dementia the opportunity to direct interactions and make choices. This enables them to regain their status as interaction partners.

1 Introduction

Alzheimer's disease (AD) is a progressive neurodegenerative disorder that causes brain cell death. Damage is primarily located in the medial temporal lobes, including the hippocampus, with later extension to the prefrontal cortex (Braak & Braak, 1991.) AD is the most common cause of dementia, accounting for approximately 50% of cases and the biggest single risk factor for developing it is age (Alzheimer's Society, 2003). It results in global impairment in functioning and affects all aspects of the person's life through the progressive deterioration of cognitive, behavioural and social abilities.

The earliest sign of AD is a noticeable short-term memory problem. People find they forget what they have just said or done, that they repeat themselves and their actions and leave tasks unfinished (Boden, 1998). Retrieving the names of familiar people becomes notably more and more difficult and people increasingly speak about experiences from their early life, rather than recent events. As the disease progresses speech is retained but contents become less and less comprehensible to conversation partners (Appell, Kertesz & Fisman 1982). This is due in part to problems that people with AD have with planning the contents of speech (Weingartner et al., 1989), which results in reduced information. In addition there are changes in the types of words used by people with AD. Specifically there is a reduction in nouns and verbs whilst adverbs and pronouns increase (Astell & Ridout, 2001). This has the effect of producing more 'empty' spontaneous speech, comprising phrases such as 'you know', 'thingy', etc. Additionally, people with AD have progressive difficulty initiating conversations and other social interactions (Ripich, Vertes, Whitehouse, Fulton, & Ekelman, 1991).

By contrast some aspects of conversation behaviour, such as turn-taking and nonverbal reinforcers (e.g. nodding and smiling; Astell & Ellis, 2003; Astell & Ridout, 2001), are relatively well preserved in AD. However, the increasing failure of speech as a communication tool typically overshadows these retained skills and little effort is made to make the most of these. The resulting communication impairment has consequently been described as '*the* most devastating consequence of dementia, to both caregivers and patients' (Azuma & Bayles, 1997, 58, emphasis added).

1.1 Social Interaction

However, humans are social beings. We live in social groups and spend most of our waking hours interacting with others. Human infants are born equipped to respond to human faces (Valenza, Simion, Macchi-Cassia & Umilta, 1996) and can mimic simple facial activities, such as sticking out the tongue (Meltzoff & Moore, 1983). Such behaviours suggest an innate predisposition to communicate and interact with others. Whilst serving obvious survival functions (Lester, 1978), it is clear that this predisposition has a much bigger role in the fundamental experience of what it is to be human. Through everyday interactions humans make friends and lifelong partnerships, become socialised and gain comfort and support in times of need. In addition, we define ourselves in relation to others and understand our place in the world through our interactions with others. Indeed, our senses of identity and of self worth are primarily formed through these interactions (Beattie, 1983).

For most people the majority of day-to-day interactions are speech based. Speech develops in a fairly similar way throughout the world's languages, with progression through a set of stages including cooing before babbling before single words and so on (Harley, 2001). However, the development of speech requires more than mere exposure to the native language. It requires interaction with others (Sachs, Bard & Johnson, 1981).

As the staple of most human social interaction conversation has two clear roles. The first of these, termed transactional (Cheapen, 1988) or ideational (Halliday, 1978) is goal-directed. For instance the goal of a conversation may be to receive a cup of tea, find out the results of the big match or be given a lift to somewhere. The second, referred to as interactional (Cheapen, 1988) or interpersonal (Halliday, 1978) is person- rather than message-focused. This relates to the effects of conversation "on the inner lives of the participants, ranging from immediate enjoyment of a social interaction …to longer term effects on self esteem and social relationships" (Todman & Alm, 2003). This clearly illustrates the critical importance for humans of participating in social interaction. Unfortunately, communication and other difficulties can lead to people with dementia being treated as somehow less than human (Kitwood, 1990) and excluded from participation in social interactions.

Finding ways to promote communication in people with dementia is thus vitally important for a number of reasons. First is to tackle the dehumanizing treatment of people with dementia and restore their status as participants in the social world. Second is that caring for someone with dementia can be frustrating and upsetting. When communication fails, caregivers are left to infer intention and meaning from behaviour and this can have negative consequences, such as believing incorrectly that someone is deliberately being difficult. Third, there is a progressive and uneven breakdown in communicative abilities in dementia. Whilst short-term memory impairments make various aspects of conversation difficult and frustrating, activities that do not require keeping a conversation topic active, for instance looking at and commenting on photographs, can provide a structure for meaningful interactions.

Given the uneven nature of the decline in communicative functions, interventions must be targeted at the relatively intact functions (Azuma & Bayles, 1997). Reminiscence provides a way for people with dementia to maximise their retained abilities, in this case long-term memory. Although their short-term memory is impaired, the long-term memory of a person with dementia is often still functioning even at the latter stages of the disease. Within long-term memory in AD older memories are better preserved than newer ones. This temporal gradient can be explained by the Multiple Trace Theory of memory (Nadel & Moscovitch, 1997). This theory holds that memories are stored as extensive distributed networks or traces in the brain. The initial laying down of a memory makes a trace involving the hippocampus in interaction with the neocortex. Each time a memory is retrieved, or reactivated, an additional trace is made in the brain reflecting the changed environment that exists each time. As such "newly acquired traces will be particularly vulnerable whereas older memories, which are multiply represented, will be able to withstand the loss of more hippocampal tissue" (Kopelman & Kapur, 2001, p125). Therefore, memories made longer ago are likely to have many more traces than recent memories and should be easier to retrieve (Kopelman & Kapur, 2001).

This provides an explanation for the success of reminiscence as an intervention for improving social interaction in people with dementia. Taking advantage of the tendency of older people, both with and without dementia, to recall well-rehearsed and well remembered previous experiences or events of personal and emotional significance (Snowden, Griffiths & Neary, 1994; Woods, 1999) reminiscence can prompt retrieval of such memories to serve as topics of conversation. By guiding and supporting the person to take advantage of long term memory they can be helped to take a more active part in conversations (Baines, Saxby & Ehlert, 1987; Finnema, Dröes, Ribbe & Van Tilburg, 1999).

Reminiscence typically employs materials such as archive newsreels and photographs, music and artefacts to stimulate interaction, either in groups or on a one-to-one basis. Music is particularly successful at generating a positive response, even in severely impaired patients (Gaebler & Hemsley, 1991; Norberg, Melin & Asplund, 1986). This fits well with the wide-ranging goals of the reminiscence experience identified by Woods (1999) that include 'increased communication and socialization, and providing pleasure and entertainment' (p324).

In addition to its benefits for the person with dementia, participation in reminiscence activities has been shown to have a positive outcome for caregivers who take part. Baines et al. (1987) found that staff that ran a reminiscence group as part of a research project continued with the group after the end of the project and reported more positive attitudes towards the people they cared for. In addition, the staff felt that the people with dementia enjoyed the group. Jackson (1991) suggested that improving the relationship between caregivers and people with dementia is an important and appropriate aim of dementia care. The provision of a positive interaction, at whatever level a person with dementia understands it, can be considered a successful intervention (Woods, 1999). Thus reminiscence is not only a tool to stimulate interaction, but also contributes to improved quality of life for the person with dementia and their family.

However, traditional reminiscence activities involve careful prior planning and gathering of material, which is time consuming for busy relatives and caregivers. In addition, the process of conducting these sessions can take a great deal of effort on their part. As such, whilst reminiscence sessions can be pleasurable and empowering for the person with dementia, the experience for the caregiver is often far from a relaxed natural interaction.

1.2 Technological intervention

To this end technology offers immense potential for overcoming some of the barriers to communication faced by people with dementia. The benefits of assistive technology for non-speaking people and groups with other impairments that affect communication are well established (Todman, Alm, Elder & File 1994). Conversational modelling has been used to direct predictive speech production systems to follow the flow of naturally occurring conversation (Alm, Newell & Arnott, 1997). In addition, prestored conversation has been found to increase the speaking rate and communicational impact of augmentative and alternative communication for non-speaking people (Alm & Arnott, 1998).

To assist people with dementia, however, the problem is not in the production of speech but in the retrieval of information to speak about. Finding a way to circumvent the memory and initiation deficits of people with AD presents a huge challenge. However, hypermedia offers a possible solution to this problem. Two features of hypermedia make it particularly suitable for use with people with dementia. First is its inherent flexibility. Viewers have freedom to move between interconnected but individual items as they choose. This should be of great benefit for people with memory loss in that it does not put any penalty on 'losing the place' (McKerlie & Preece, 1992). Whatever place the user is in is the right place to be. Exploring and 'getting lost' are actively encouraged as strategies to enjoy experiencing the material. Second is the opportunity to link items from a range of media in a dynamic way. The seamless inclusion of text, photographs, graphics, sound recordings and film recordings makes an inviting and lively contrast to the way reminiscence materials are typically presented. A hypermedia-based reminiscence system has the potential to be both an interactive and a shared experience for people with dementia and their caregivers.

The way that hypermedia operates fits well with the Multiple Trace Theory of memory (Nadel & Moscovitch, 1997). Multiple traces facilitate retrieval by providing more ways to access memories. The specific problems faced by people with dementia make accessing and calling up memories at will extremely difficult. Hypermedia provides a way to present multiple prompts from a range of stimuli to tap into these multiple access routes to their stored memories.

As such we have developed a hypermedia conversation support system based on reminiscence that can be used with the help of a caregiver. The system uses a touch-screen as these have previously been shown to be acceptable and accessible to people with dementia (Ritchie, Allard, Huppert, et al.; 1993). One obvious advantage of a computer based system over using traditional materials is the bringing together of the various media into one easily accessible system. However, we also want to explore ways in which the hypermedia could in some way mimic the way memory is used in conversations.

2 Method

2.1 Participants

Eighteen people with dementia, thirteen women, who met the NINCDS-ADRDA criteria for probable Alzheimer's Disease (McKhann et al., 1983) were recruited from a number of day care and residential facilities. The participants were randomly assigned to one of two groups – CIRCA or traditional reminiscence (TRAD). There was no significant difference between the two groups in age (CIRCA mean 82.0 years (range 65-95), TRAD mean, 81.88 years (71-89) or Mini-Mental State Examination scores (MMSE; Folstein, Folstein & McHugh, 1975) (CIRCA mean 14.88 (range 2-22), TRAD mean 16 (5-23).

A two-stage consent procedure was used. First letters were sent out to relatives of people with dementia in the partner organisations informing them of the study and asking if they were agreeable to the study team approaching their relative to take part. On receipt of this consent, individuals with dementia were approached and asked if they would like to take part in the study. They were asked to give written consent where possible and if they could no longer write, verbal consent was obtained and witnessed by a neutral third party. All participants were free to leave at any time.

Twelve caregivers were also recruited. Each person with dementia was paired with a caregiver for the study sessions. Of the 12 caregivers, five took part in CIRCA sessions only, three took part in TRAD sessions only and four took part in both CIRCA and TRAD sessions.

2.2 Materials

CIRCA. An Apple G4 laptop was used to run CIRCA and it was presented through a 20-inch touch-screen monitor. SONY SRS-T77 speakers were used to output stored speech and music. CIRCA was viewed at a resolution of 1280 x 1024 pixels. Macromedia Director 8.5 was the authoring software for CIRCA and the following additional software were used to provide the content: Adobe Photoshop 6.0; Adobe Illustrator 9.0.1; Adobe Premier 6.0; QTVR Authoring Studio 1.0; SoundEdit 16 version 2; Infini-D 4.5. The CIRCA database contained 113 items comprising 53 photographs, 13 video clips and 23 pieces of music or songs. The average size of photographic content in CIRCA was 800 x 600 pixels. The average length of videoclips was 180 seconds. Songs and pieces of music varied in length, ranging between 30 seconds and 210 seconds. The stimuli were mainly drawn from the1930's to the 1960's and were presented in a simple visual format (see Figure 1).

Figure 1. Screen shot of CIRCA – radio themes

Material in CIRCA was organised into three themes and into three media types. In the interface each theme was associated with a colour, and when a theme was selected, the hue of the background and of all the buttons changed to reflect the hue of the selected theme. Muted colours were chosen for the interface. (see Figure 1).

TRAD. Caregivers were asked to choose their own stimuli for these sessions based on materials they used to run reminiscence sessions in the normal course of their work.

Mini Mental State Examination (MMSE; Folstein, Folstein & McHugh, 1975). This is a standardized measure of global cognitive function for older adults.

A Sony Mini DV Digital Handycam and tripod were used to record the dyads of caregiver and person with dementia.

A checklist was also used to record dyadic activity, with sections for recording the activities of the caregiver and the person with dementia separately. The checklist comprised the following categories:

Person with dementia – choosing with prompt and choosing without prompt

Caregiver – Prompting and conversation maintenance

Dyad– memory, humour, laughter and movement to music

The coding categories employed were used for both types of session and were designed to capture the nature of the interaction between the dyad and their use of reminiscence materials.

2.3 Tasks

Each pair participated in one 20-minute session.

CIRCA sessions. These were one-to-one sessions using CIRCA as the basis for an interaction between a person with dementia and a caregiver.

TRAD sessions. These were one-to-one sessions using typical reminiscence stimuli as the basis for an interaction between a person with dementia and a caregiver.

2.4 Procedure

At the start of each session the study was explained to the caregiver and the person with dementia together and both parties were asked if they had any questions regarding their possible participation. Reiteration of consent to record the sessions was sought at this time. For both groups the MMSE was then conducted with the person with dementia.

Figure 2. Flowchart demonstrating an example decision tree in CIRCA.

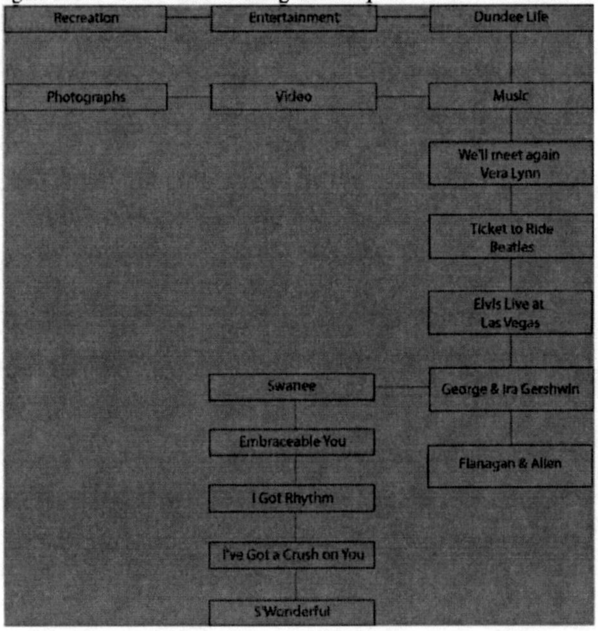

During this time the caregivers in the CIRCA group practised using CIRCA. The caregivers in the TRAD group planned the reminiscence session. All sessions were conducted in a designated unoccupied room within each of the participating care facilities. Each room was set out to allow the caregiver and the person with dementia to sit side by side at a table.

CIRCA sessions. Each pair sat side-by-side in front of the touch screen. Each pair was shown how to start CIRCA and was then left to use it together.

The first screen has a 'start' message. Viewers are then offered a choice of three categories: Entertainment, Recreation, Dundee life. Viewers make a choice by touching the relevant category name on the screen (Figure 2). The next selection is to choose from video, photographs or music.

TRAD sessions. Each pair sat together in positions they chose. Some pairs sat side-by-side and others sat face-to-face. Care staff used the materials they had chosen to facilitate the sessions. Typically the staff members showed the people with dementia photographs or artefacts from the past and used these to generate conversation.

All sessions were video recorded. The video recorder was set up on the tripod in such a position to film both participants at all times. In addition, a member of the research team sat in and observed all sessions. The observer sat behind each pair of participants, out of their view, and kept a tally of items on the checklist. Online coding involved noting down any pieces of information/memories produced by the dyad. The amount of times each event occurred during the session was coded from the videos.

The participants in the CIRCA sessions were interviewed at the end. The participants with dementia were interviewed immediately after the session ended and the caregivers were interviewed after that.

2.5 Coding responses

All reminiscence sessions were observed online and again from videotape. The coding categories and their operational definitions are explained below

Table 1 Coding categories

Person with dementia	Caregiver	Dyad
Choosing with prompt - amount of times the person with dementia chose what they wanted to talk about/see/listen to in response to being offered a choice of stimuli by the caregiver	**Prompting** - prompts given by the caregiver to the person with dementia to make a choice about what he/she wishes to talk about. For example, during TRAD sessions the caregiver might ask, "What would you like to talk about?" During a CIRCA session, the caregiver might ask "Would you like to look at photographs, music or video?"	**Memory** – the number of pieces of information or memories provided by each member of the dyad about themselves or elements of their past. For example: "I had two brothers", "I lived in Smith Street", "It cost tuppence to ride the tram."
Choosing without prompt - amount of times the person with dementia chose what they wanted to talk about/see/listen to in the absence of being offered a choice of stimuli by the caregiver	**Conversation maintenance activities** - contributions from the caregiver classified as serving to maintain the conversation. For instance the caregiver might ask the person with dementia a question such as "did you enjoy going to the pictures when you were younger?"	**Humour** - Instances of the use of humour by either member of the dyad
		Laughter - Instances of laughter within the dyad
		Movement to music - Instances of movement to music such as hand-clapping, swaying, tapping feet, by either member of the dyad

3 Results

To evaluate the efficacy of CIRCA as a conversation support for people with dementia we compared both parties in the dyads on a number of measures. In CIRCA sessions the people with dementia were offered a choice more often by the caregivers ($U = 1.50$, $p<0.001$) and subsequently made more choices ($U = 36.00$, $p<0.001$; Table 2) than in the TRAD sessions. By contrast, in TRAD sessions a lot more of the caregiver role was conversation maintenance activity, such as asking direct questions to the people with dementia ($U = 5.00$, $p=0.01$).

763

Table 2 Prompting and choosing in CIRCA and TRAD sessions

Person with dementia		
Mean (SD) range	CIRCA (n=9)	TRAD (n=9)
Choosing with prompt	6.1 (4.2) 1 – 12	0.33 (0.7) 0 - 2
Choosing without prompt	0.33 (1) 0 - 3	0
Caregiver		
Prompting choice	10 (4.9) 3 - 18	0.77 (1.6) 0 - 5
Maintenance	12.1 (8) 4 - 29	48.1 (28.1) 14 - 98

Examination of the interaction indicators reveals some similarities between the sessions. For instance shared laughter did not differ in occurrence across all sessions for both people with dementia and caregivers, suggesting that this is a product of the one-to-one situation (Table 3). Conversely CIRCA sessions were marked by a large amount of time spent listening to music (35%) whilst there was no music played in any of the TRAD sessions. One person with dementia did sing during a TRAD session but overall, there was significantly more singing together ($z=2.93$, $p<0.01$) and trend of more moving to music ($z=1.83$, $p<0.10$) in CIRCA sessions. Caregivers sang ($z=2.52$, $p<0.05$) and moved to music more ($z=2.19$, $p<0.05$) in CIRCA sessions. There was also far more reporting of personal information by people with dementia in the TRAD sessions ($U = 3.00$, $p=0.001$). On examination the nature of the personal memories retrieved differed between the TRAD and CIRCA sessions. In TRAD sessions the personal information tended to be short responses to direct questions from the caregivers about events from the past. In CIRCA sessions, participants retrieved detailed personal memories in response to the pictures, sounds and videos on the screen.

Table 3 Interaction indicators in CIRCA and TRAD sessions

Person with dementia		
Mean (sd) range	CIRCA n=9	TRAD n=9
Personal memory	12.44 (8) 6 -31	58.2 (21.2) 13 - 84
Humour	1.88 (1.53) 0 – 4	4.1 (3.8) 0 -9
Laughter	8.77 (7.5) 2 - 26	10.1 (8.3) 1 - 29
Moving to music	0.4 (1) 0 -3	0
Singing	1.3 (1.6) 0 -5	0.2 (0.4) 0 -1
Caregiver		
Personal memory	2.6 (5.2) 0 - 16	1.6 (2.6) 0 - 8
Humour	1.5 (1) 0 - 3	1.9 (2.8) 0 –8
Laughter	7 (4.3) 2 - 16	12.7 (10.3) 2 -33
Moving to music	0.4 (0.5) 0 – 1	0
Singing	0.7 (1.3) 0 - 4	0.1 (0.3) 0 - 1

3.1 Evaluation of CIRCA
Person with dementia - general comments
Two people spontaneously commented several times that they were enjoying using CIRCA. One person said, " It takes you back and refreshes your memory." Whilst another commented 'this covers everything', 'Good thing, this', 'It's good to remember things', and 'That's entertainment!' One person said she certainly would like to use CIRCA again. She said she thoroughly enjoyed it and found it very interesting and 'something new'. Another person commented that she enjoyed using the system herself.

Caregivers
When asked whether they thought the session was worthwhile for themselves as caregivers, all of the participants said yes. They gave a wide variety of reasons for this. One caregiver said she enjoyed "using the technology and being able to use the system." Another liked "having more subjects to talk about" and another enjoyed "seeing the client's reactions to CIRCA." Other factors were "getting the chance to spend one-to-one time" and "finding out about things she (the caregiver) didn't know of before (about the client)". One further caregiver said that she "likes

reminiscence and having a one-to-one. It's amazing what the client can remember." Another caregiver commented that she believed "her and the client actually achieved something as the client remembered a lot".

When asked whether they thought the session was worthwhile for the clients, again all said yes. Among the reasons given for this were that CIRCA "gives the client a chance to talk about what he/she wants." Others commented that the "focus of attention was on the client" and that CIRCA "helps the client to remember." One caregiver enjoyed the "one-to-one time" and another "especially enjoyed the singing." One reason given was that "the client was interested in using CIRCA" whilst another thought the most important reason was that "the client used the system herself". One final comment was that it was a "good session & very positive".

4 Discussion

We compared a computer-based reminiscence activity (CIRCA) with a traditional reminiscence activity (TRAD) for people with dementia. As expected people with dementia responded well to the touch screen and were not fazed by the computer. Indeed, although the system was developed with the expectation that a caregiver would touch the screen, we found that people with dementia were willing and able to touch the screen if they received appropriate prompting and encouragement.

The reminiscence activities each lasted twenty minutes and were carried out as one-to-one sessions between a person with AD and a caregiver. However, the nature of the two types of reminiscence sessions differed considerably. For example in CIRCA sessions music was very popular, with 35% of the total time being spent listening to music. However, music did not figure at all in the traditional reminiscence sessions. This is surprising given the power of music to prompt not just direct reminiscences but also a range of interaction alternatives to conversation, such as singing and moving together. Some dyads clapped together or tapped out a beat along with a tune and there was a lot of singing and laughing when music was played. These findings support previous research that music facilitates meaningful and enjoyable communication (Gaebler & Hemsley, 1991).

There was also a difference in the structure of the CIRCA and TRAD reminiscence sessions. In the traditional sessions people with dementia were largely passive, receiving and responding to questions from the caregiver. Caregivers hardly ever offered choices to the people with dementia, much less, did the people with dementia make choices in these sessions. It was rare for a person with dementia to initiate a topic of the conversation in a traditional reminiscence session and as such these sessions were largely one-way traffic from the caregiver to the person with dementia.

In contrast the CIRCA sessions were more equal activities with both the people with dementia and the caregivers contributing to the direction the sessions took. In part this was due to caregivers offering and prompting the people with dementia to make choices about the selections offered by CIRCA. The CIRCA system appears to provide the caregiver with a natural method of offering choices to the person with dementia. For example, within CIRCA sessions, people with dementia were offered a choice of subjects/media and were prompted to choose by means of the caregiver reiterating the options. For example, the caregiver might ask, "Would you like to choose photographs, video or music?" Conversely, in traditional sessions, if the person with dementia was offered a choice of reminiscence stimuli/subjects at all, the options were largely open-ended. For example, the caregiver may ask, "What do you want to talk about?" We have previously shown that prompting people with dementia with specific key words specific is particularly helpful (Astell, et al., 2002). Conversely, open-ended options put more pressure on the person with dementia to suggest a topic. As such CIRCA provides both a framework for caregivers to offer a choice and for people with dementia to make choices that is much more easily accessible than traditional reminiscence methods.

In addition, traditional sessions seemed less conversational and more question and answer marathons. Although this approach elicits previously unknown information about the person with dementia it also puts the onus on the caregiver to keep the conversation going. One caregiver commented that she preferred to use CIRCA as she felt under much less pressure to keep thinking of things to say. These findings suggest that in traditional reminiscence caregivers keeping asking questions as a way to keep the session going. This in turn puts the person with dementia under pressure to answer these questions. Consequently, traditional sessions are less of a shared experience and the relative disparity in status between the person with dementia and caregiver is preserved.

By contrast, in CIRCA sessions the interaction is kept going by the caregiver offering the person with dementia choices. These sessions, unlike the traditional sessions, appear to be genuinely shared experiences with both parties exploring and discovering the system together. In part this may be due to the computer functioning as an object of joint attention. This allows attention to shift from each individual to a shared focus on the 'third party', in this case the touch screen. Joint attention functions to provide a bridge between the interlocutors, giving them a concrete item on which to concentrate (Tomasello & Farrar, 1986). The concreteness facilitates interaction by providing a shared point of discussion and removing the need for one or other partner to keep initiating conversation topics and.

Interactions in the CIRCA sessions also appeared to benefit from the flexibility of hypermedia. By compiling a large database of reminiscence stimuli, there is plenty of scope for finding items that will prompt reminiscence in any given individual. It is not necessary to use personally relevant stimuli for people with dementia as reminiscences can be successfully prompted by generic stimuli (Astell, et al., 2002). A second benefit of hypermedia is novelty. Given the infinite possibilities of journeying around the system, it is possible for each session to be completely different. This is particularly important for care staff that could become bored using the same stimuli time after time. Novelty also seems to be one of the vital ingredients in changing the status of the dyads during CIRCA sessions.

In essence CIRCA appears to restore the status of people with AD to that of equals with their interaction partners. This effect most likely arises from the novelty of the CIRCA experience, in that neither of the partners in the interaction have advance knowledge of what stimuli will be available. As such, people with dementia can make choices and influence the direction of sessions as much as their non-demented communication partners. Additionally, caregivers benefit from not having to remember certain 'routes' around the system, which would place additional burden on them.

CIRCA provides a way to prompt reminiscences that does not require a lot of preparation on the caregiver's part. This is beneficial for hard-pressed care staff that may not have time to gather together and plan reminiscence items. In addition, CIRCA sessions are positively viewed by both interaction partners, with caregivers particularly enjoying finding out more about their clients. This is very important for restoring the personhood of people with dementia, as they become viewed once again as people with lives and histories and stories to tell (Morton, 1999). Further reinforcement of personhood comes from the finding that CIRCA prompts reminiscence and recollection of memories that caregivers have not heard before This also has the effect of further increasing caregiver knowledge about the people they are caring for as does the finding that caregivers have been pleasantly surprised at the ability of people with dementia to use the CIRCA system.

CIRCA was designed to be immediate and not rely on short-term memory. When using CIRCA it does not matter what has previously been spoken about. Rather, whatever is on at the time is the prompt for interaction, whether that is a photograph, a video clip or a song. CIRCA takes advantage of the retention of older, well-established memories, as predicted by the multiple trace theory of memory (Nadel & Moscovitch, 1997) and provides ways to access these memories. As such CIRCA functions in a similar way to more traditional reminiscence methods, such as scrapbooks and audiotapes, that act not only as memory aids, but also as supports for communication. In essence they replace the person's own lost ability to deal with immediate memories (such as what they said five minutes ago), while encouraging them to employ their still effective long-term memory (such as what happened forty years ago).

Our plans are to follow this work by devising ways in which a person with dementia could take control of navigating through such a system. This may facilitate even more meaningful interactions with relatives and caregivers. Finding a way for a person with dementia to independently interact with a helpful computer system could significantly improve the quality of their lives and that of their caregivers. Systems could then be developed to meet a whole range of needs in dementia, such as providing cognitive stimulation and entertainment, and other types of computer-based support in activities of daily living. Whilst none of these interventions could bring about an improvement in the dementia itself, they would empower people with dementia, by enabling them to maximise their remaining skills in new and creative ways to stay engaged with the social world.

References

Alm, N., Newell, A., & Arnott, J. L. (1997). Lessons from applying conversation modelling to augmentative and alternative communication. *Proceedings of the 12th CSUN conference on technology and persons with disabilities.* 18-22 March 1997. Los Angeles. Proceedings on disk --- Publisher : Rapidtext, Newport Beach, California.

Alm, N., & Arnott, J. L. (1998). Computer-assisted conversation for non-vocal people using pre-stored texts. IEEE Transactions on Systems, Man, and Cybernetics, Vol 28 Part C No 3, pp. 318-328.

Appell, J., Kertesz, A., & Fisman, M. (1982). A study of language functioning in Alzheimer patients. *Brain and Language, 17*, 73-91.

Astell, A. J., & Ellis, M.P. (In Press) Communication in severe dementia: a case study. To appear in Special Issue of the Journal of Child and Infant Development.

Astell, A. J., Ellis, M. P, Campbell, J. Alm, N. Dye, R, & Gowans, G. (2002). Be Specific: Tapping Residual Conversation Skills in AD Using Prompts. *British Neuropsychological Society*, Autumn meeting.

Astell, A. J. & Ridout, N. (2001). Verbal communication in dementia of the Alzheimer type: why does it fail? *Proceedings of the 17th Alzheimer's Disease International Conference*, 182.

Åström S, Nilsson M, Norberg A, Sandman PO, & Winblad B. (1991) Staff burnout in dementia care - relations to empathy and attitudes. *International Journal of Nursing Studies*, 28, 1, 65-75.

Azuma, T., & Bayles, J. A. (1997). Memory impairments underlying language difficulties in dementia. *Topics in Language Disorders*, 18, 58-71.

Baines, S., Saxby, P., & Ehlert, K. (1987). Reality orientation and reminiscence therapy: a controlled cross-over study of elderly confused people. *British Journal of Psychiatry*, 151, 222-231.

Beattie, G. W. (1978). Floor apportionment and gaze in conversational dyads. *British Journal of Social and Clinical Psychology*, 17(1), 7-15.

Boden, C. (1998). *Who will I be when I die?* Sydney: Harper Collins.

Finnema, E., Dröes, R-M, Ribbe, M., & Van Tilburg, W. (1999). The effects of emotion-oriented approaches in the care for persons suffering from dementia: a review of the literature. *International Journal of Geriatric Psychiatry*, 15, 141-161.

Gaebler, H. C., & Hemsley, D. R. (1991). The assessment and short-term manipulation of affect in the severely demented. *Behavioural Psychotherapy*, 19, 145-156.

Harley, T. A. (2001). *The psychology of language*, 2nd edition. Hove: psychology press.

Jackson, A. (1991). To reminisce or not to reminisce. *Irish Journal of Psychological Medicine*, 8, 147-148.

Kitwood, T. (1990). The dialectics of dementia: with particular reference to Alzheimer's disease. *Ageing and Society*, 10, 177-196.

Kopelman, M. D., & Kapur, N. (2001). The loss of episodic memories in retrograde amnesia: Single case and group studies. In A. Baddeley, M. Conway, M. & J. Aggleton (Eds.) *Episodic memory: New directions in research*. London, Oxford University Press.

Lester, B. (1978). The organization of crying in the neonate. *Journal of Paediatric Psychology*, 3(3) 122-130.

McKerlie, D and Preece, J. (1992) The hypermedia effect: more than just the sum of its parts, in *Proceedings of the St.Petersburg HCI Conference*, pp 115-127.

Morton, (1999). *Person-centred approaches to dementia care*. Oxford: Speechmark.

Nadel, L & Moscovitch, M. (1997). Memory consolidation, retrograde amnesia and the hippocampal complex. *Current opinion in neurobiology*, 7(2), 217-227.

Norberg, A., Melin, E., & Asplund, K. (1986). Reactions to music, touch and object presentation in the final stage of dementia: an exploratory study. *International Journal of Nursing Studies*, 23, 315-323.

Ripich, D. N., Vertes, D., Whitehouse, P., Fulton, S., & Ekelman, B. (1991). Turn-taking and speech act patterns in the discourse of senile dementia of the Alzheimer's type patients. *Brain and Language*, 40, 330-343.

Ritchie, K., Allard, M., Huppert, F., et al. (1993). Computerized cognitive examination of the elderly (ECO): the development of a neuropsychological examination for clinic and population use. International Journal of Geriatric Psychiatry, 8(11), 899-914.

Snowden, J., Griffiths, H., & Neary, D. (1994). Semantic dementia: Autobiographical contribution to preservation of meaning. *Cognitive Neuropsychology*, 11, 265-288.

Todman, J., & Alm, N. (2003). Modelling conversational pragmatics.

Todman, J., Alm, N., Elder, L, & File, P. (1994). Computer-aided conversation: A prototype system for nonspeaking people with physical disabilities. *Applied Psycholinguistics*, 15(1), 45-73.

Tomasello, M., & Farrar, M. J. (1986). Joint attention and early language. *Child Development*, 57(6), 1454-1463.

Weingartner, H., Grafman, J., Boutelle, W., Kaye, W., & Martin, P.R. (1983). Forms of memory failure. *Science, 221*, 380-382.

Woods, R. T. (1999). Psychological therapies in dementia. In R. T. Woods (Ed.) *Psychological Problems of Ageing*. Chichester: Wiley.

Information Frames in Augmented Reality Mobile User Interfaces

Charles Owen, Frank Biocca[†], Arthur Tang[†*], Fan Xiao*, Weimin Mou[§], Lynette Lim[†]*

†Media Interface and Network Design Laboratories
*Media and Entertainment Technologies Laboratory
Michigan State University, East Lansing, U.S.A.
E-mail: {cbowen, biocca, tangkwo1, xiaofan, lynette}@msu.edu

§Institute of Psychology
Chinese Academy of Sciences,
Beijing, China
E-mail: mouw@psych.ac.cn

Abstract

Mobile augmented reality display technologies allow information fields within any area around the user. The addition of tracking technologies affords many options for placement of these fields relative to the environment, objects, and the body, with many possible ramifications for efficient conveyance of information and potential exploitation of spatial memory. This paper describes a set of information fields around the body, with emphasis on both psychological relevance and physical realisability.

Keywords: Augmented Reality, Spatial Framework, Information Placement, Reference Frames, Mobile Interface

1   Introduction

Wearable computers and see-through head-mounted displays allow for spatial placement of information around the body of a mobile user. With head and body tracking, these displays can present information fields far larger than any display screen with 3D spatial placement anywhere within the visual field even in the presence of large head motion. Many options exist for the stabilization and placement of this data. Data can be display in a way that it appears to be stable relative to the environment, part of the body, the hands, or the head. As these mobile *augmented reality* (AR) systems become practical, it is important to recognize that a wide variety of potential information fields exist around the body and the use of these fields can and will impact user perception, distraction, memory, and usability. This paper describes options for information frames around a mobile user and examines the consequences of informat frame choices.

Billinghurst describes three important ways to display information around a user: head-stabilized displays, wherein information is fixed relative to the user's viewpoint; body-stabilized displays, wherein information is fixed relative to body position; and world-stabilized displays, wherein information is fixed relative to real-world locations [1]. This taxonomy has been popularized in other work as well [2, 3]. The idea of a body-stabilized display can be traced back to Reichlen, who proposed the concept primarily as a method for expanding display resolution through the use of tracking and head-mounted displays (HMD) [4]. We expand on the idea of a body-stabilized display by examining the potential near and distant stabilization alternatives including fields around the arms, legs, and hands. We also examine how composite frames are constructed by coupling elements of more than one frame together. This concept is a generalization of the idea of reduced degrees of freedom presented by Billinghurst [1]. We also discuss how filtering can be applied in the coupling process so as to environmentally stabilize displays, decreasing the apparent *swimming* of information objects due to body motion and noise.

In examining the variety of possible information frames around the body, we have attempted to be complete in providing both an existing literature examination and extensions to additional options. We focus particularly on frames that can be implemented with reasonable tracking technologies, though more advanced frames are also discussed. We also examine the human-computer interface ramifications of these varying frames from the standpoint of human spatial updating and memory, which are critically important issues in future interface designs. The concepts presented in this paper form a set of guidelines for mobile AR systems incorporating advanced tracking technologies and providing users with more flexible information spaces around the body. The concepts of coupling and filtering provide tools for the management of many of the undesirable properties of these environments.

1.1 Use of Proprioceptive Cues in the Mapping of Information Fields to Information Frames

Consider how virtual information can be placed in useable arrays on various locations of the body, some of the automatic neurological processing allocated to tracking the spatial reference frames can be leveraged. Object locations in space are tracked primarily through exteroceptors. Most often the visual system is used unless an object is in direct contact with the surfaces of the body. However, the brain also tracks the disposition and location of the limbs through proprioceptive receptors [5]. These receptors monitor body state through various muscles, skin, and other proprioceptive sensors. By more tightly coupling computer tracking to physio-psychological spatial reference systems, AR interface designers can potentially make use of additional information on body and limb locations to locate, retrieve, and manipulate information. This can take several forms including the mapping of information to body locations. As examples, tools may be arrayed over the length of the arm, or full body gestures may be employed for information manipulation through configurations of muscle groups such as the torso, arms, and limbs. Natural and design gestures can be used to control or manipulate a virtual object or to issue commands.

2 Level of analysis of AR Information Frameworks

The goal of AR systems is to integrate virtual information objects smoothly and seamlessly within a user's space. Information objects in virtual space are integrated within the physical space so as to be visible or audible to a user moving within the integrated space. Information presentation in AR systems can therefore be conceptualized as involving the analysis of an interaction between three levels of spatial analysis:

1. The moving body of the user (the sensorimotor system).
2. The computer's tracking of parts of the user's moving body relative to some reference frames in a 3D physical environment.
3. The organization of information relative to the moving body and the environment.

We will call the interaction of these three levels of analysis AR information frameworks. The primary level of analysis is the moving body of the user, or, more specifically, the user's perception of their moving body and its location in space. We refer to this level as the *Physio-psychological level*. Thinking of the body as a sensorimotor system that perceives space is critical in AR design because the goal of an AR system is the creation of perceptual illusions such as a stable virtual object on a physical table, a stable virtual sign affixed to a real building, or a virtual tool inside the user's physical hand. So in some ways user's body and user's perception of that body is another part of the integration of virtual and physical spaces. This level therefore deals with sensorimotor spatial reference frames, or how the mind uses different reference frames while acting, moving, and perceiving the interaction of the body with the objects in physical space [6, 7].

In creating AR information spaces, AR and VR systems interpose themselves between the natural coupling of the sensory and motor systems [8]. AR and VR mediate the natural linkage between motion and perception that is fundamental to the body's interaction with the environment [9, 10]. As a result, perceptual illusions can be created in VR systems by systematically maintaining the stability of an object in space while the user's head moves. This is why the second level of analysis and implementation – *body tracking* - is so fundamental to the illusion. It is the key to coupling motor actions to perception of space. An analysis and specification of information frames useable in a mobile AR system is the focus of this paper.

The third level of analysis with this AR information framework is the *interface level*. Here the user perceives information fields and the design issues focus on how to organize and array the virtual objects within the various information frames to optimize user performance within the system.

3 Tracking in mobile AR systems

Due to advances in tracking and computing technologies, mobile AR systems have become more and more practical in recent years. Mobile AR integrates AR and mobile computing to provide an augmented environment in which the user is free to locomote. In a mobile AR system, wearable computers are worn on the user's body and information was provided to the user whenever necessary. The Touring Machine developed at Columbia University is a typical mobile AR system [11]. It includes a laptop with 3D graphic accelerator, an optical see-through HMD, orientation and position trackers and a handheld computer. While the user navigates, it provides annotations of buildings through the HMD and detailed information of buildings by the handheld computer. A variety of other mobile AR systems have been developed in recent years [12-14].

Tracking is a crucial problem in mobile AR systems. To overlay virtual objects in the user's view and create the illusion that the virtual objects really exist in the real environment, tracking of the head relative to the world coordinate system must be done very accurately. In AR, there are several traditional coordinate frames that are related to tracking transmitters, markers, and environmental points. A world coordinate frame is often chosen as the central reference frame. The calibration requirements for different frames are described by Tuceryan [15].

A wide variety of tracking technologies exist for mobile AR systems, including magnetic source-based tracking, vision-based tracking, and inertial-based tracking. Discussion of specific tracking technologies is beyond the scope of this paper. Instead, tracking will be modelled as a *point-wise measurement of 6 degrees of freedom* (position and orientation) relative to a world coordinate system. Clearly, alternative non point-wise tracking technologies, such as the ShapeTape [16], exist, but these systems can be modelled using the point-wide model and sets of tracking points for the purpose of this paper.

4   Reference Frames

A *reference frame* is defined as a coordinate system with an origin and orientation. There are a wide variety of reference frames in any mobile AR system. Other than the world coordinate frame, any tracking source induces a local reference frame related to the coordinate system emitted by the tracking system. Likewise, tracker markers emit frames relative to the tracking system source. Vision-based systems emit frames relative to objects in the environment such as fiducial images. However, the focus of this paper is on *information reference frames*, which are reference frames that are useful for placement of information in an AR-based user interface. Most physically induced reference frames, such as tracker marker frames, can be transformed to information reference frames through a rigid geometric transformation, though non-rigid transformations will also be discussed.

The definition of a reference frame requires an origin and orientation of the frame. Reference frames are necessarily relative, often relative to the world coordinate frame. The origin of a reference frame (t_x, t_y, t_z) is a point in the world coordinate system and the orientation is commonly defined with at least two normalized vectors that specify the direction of the coordinate axis in the frame, for example $R_x = (X_x, X_y, X_z)$ and $R_y = (Y_x, Y_y, Y_z)$ may define the direction of the x and y axis of the frame. Given two axis vectors, the third can be computed using a cross product: $R_z = R_x \times R_y$. The three vectors Rx, Ry, Rz represent the rotation matrix elements for the frame. A rotation matrix is a three by three orthonormal matrix with these vectors as the columns and serves to rotate a vector from the frame orientation of the corresponding world coordinate frame orientation.

Given this definition, a transformation matrix T_F can be defined that transforms objects in the defined frame to the world coordinate frame for display:

$$T_F = \begin{bmatrix} X_x & Y_x & Z_x & t_x \\ X_y & Y_y & Z_y & t_y \\ X_z & Y_z & Z_z & t_z \\ 0 & 0 & 0 & 1 \end{bmatrix}$$

In current mobile AR systems, information is often presented world-stabilized: information is fixed to real world locations in the *world coordinate frame* and the data view varies as the user changes viewpoint orientation and position. Thus, user's viewpoint position and orientation are tracked and the transformation from the world frame to the user's view frame is computed. In such a system, the only information frame is the world coordinate frame.

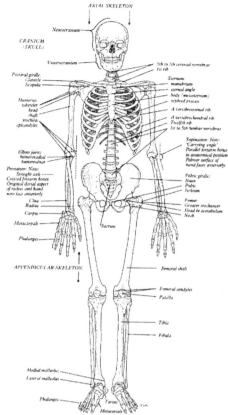

Figure 1: Human skeletal structure [17]

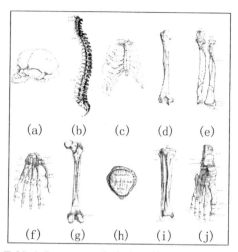

Table 1: Bone groups for body stabilized displays (a) Skull (b) Vertebral Column (c) Sternum and costal cartilages (d) Humerus (e) Forearm group (f) Hand (g) Femur (h) Patella (i) Leg (j) Foot [17].

Some AR systems also present information attached to moving objects. In such AR systems, the position and orientation of objects also needs to be tracked. Systems with object tracking and information placement relative to objects have *object coordinate frames.*

In all cases, graphical objects that are to be displayed will be subjected to transformations amongst various frames until they are in the view frame for each eye and subject to rendering. The focus of this paper is the frame in which the geometric objects are placed from a user-centric viewpoint.

Other well-known options of displaying information include head-stabilized and body-stabilized frames. These methods display information at a fixed position relative to human body. Billinghurst describes a body-stabilized conferencing space and examines the advantages of such a model [1]. These frames have clear application, but are hardly the only possible body-stabilized frames. There are many applications that can utilize frames relative to the hands, arms, or other extremities. To examine the possible body-stabilized frames, it is best to examine the skeletal structure of the human body, exploring the major bone groups that can be used to define useful information frames.

Figure 1 is an illustration of the skeletal structure of the human body. The skeleton represents the rigid structure of the moving elements of the human body and, as such, provides a clear set of possible tracking references that can define information frames relative to the human body. Since direct attachment to bones for tracking purposes is

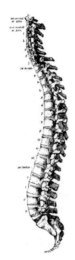

Figure 2: The vertebral column [17].

Tracking source	Comments
World coordinate frame	General reference frame. World stabilized content.
Objects	Object information frames. Manipulation of objects.
Skull	Head-stabilized display.
Cervical vertebrae	Upper-body stabilized display.
Lumbar vertebrae	Lower-body stabilized display.
Sternum and costal cartilages	Torso tracking, but with breath motion.
Humerus	Upper arm tracking. Sleeve-based tools.
Forearm group	Lower arm tracking. Arm-placed tools.
Femur	Upper leg tracking. Pocket area tools.
Patella	Knowledge of knee placement.
Lower leg group	Lower-placement tools. Walking and motion tracking.
Foot	Motion capture and stealth applications.
Metacarpus	General hand manipulations.
Phalanges	Fine grain knowledge of hand geometry.

Table 2: Summary of information frame tracking sources

generally not practical, the attachment is more likely to be to the epidermis (surface of the skin). But proper placement allows the epidural attachment to be a good approximation of the underlying bone tracking.

A human adult skeleton has 206 bones. Clearly, many of these are not useful from an information frame point of view (such as bones in the inner ear) or are redundant (such as the dual function of the ulna and radius or the set of bones in the rib cage). Other bones may have very limited utility in mobile AR environments (such as the bones in the feet). Table 1 describes ten bone groups useful for definition of AR frames. Some of these groups define frames directly (such as the skull); others define sets of frames (such as the vertebral column).

The *skull* is traditionally tracked in AR systems, typically through tracking of a head-mounted display that is assumed to be fixed relative to the head. The skull defines a head-stabilized frame.

The *vertebral column*, illustrated in Figure 2, is commonly tracked in body-stabilized applications. The Vertebral Column is composed of 7 cervical vertebrae, 12 dorsal vertebrae, 5 lumbar vertebrae, sacrum and coccyx. It is situated in the median line of the back of upper torso. The 7 cervical vertebrae, which form the neck, are not well suited for tracking because of the deformation of muscles around the neck. The 12 dorsal vertebrae and 5 lumbar vertebrae are more suitable for tracking. The vertebral column clear emits a variety of tracking points, each with unique characteristics. Tracking of the upper back (dorsal area) will create frame that follow the body through bending motion, whereas tracking in the lumbar area tends to create predominately upright frames in mobile AR environments.

The *sternum and costal cartilages* provide an alternative tracking of the human torso. The disadvantage of sternum tracking is displacement due to breathing.

The *humerus and forearm group* can be tracked as a model for arm motion. Due to the coupling of these systems, tracking of one group in association with torso tracking can yield a good approximation of the other group. Humans traditionally utilize the arm as a tool storage location, often for watches, but also for data entry devices and other physical tools.

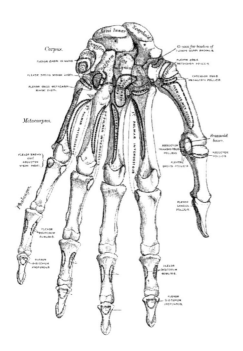

World Frame	Lumbar Frame	Comments
Y	Z, t_x, t_y, t_z	Body stabilized display with gravity. Objects do not tilt with the body.
Y, t_y	Z, t_x, t_z	Stays fixed in height as user moves up and down.
Y, t_y	Z, t_x, t_z	Stays fixed in height as user moves up and down.
Y, Z	t_x, t_y, t_z	Information moves with the body, but does not rotate with the body.

Table 3: World frame to lumbar frame coupling

Figure 3: The human hand [17]

Likewise, the *femur, patella, and lower leg* provide indications of the overall leg motion. As in the arms, the legs are common placement locations for tools and traditionally used as places for pockets. A body-stabilized display based on lower lumbar vertebrae tracking may turn quite independently of the legs. Since many applications of body-stabilized displays assume tool placement around the waist, leg tracking may be of great utility.

Table 2 summarizes tracking sources for near and far information frames, indicating example applications for each tracking source/information frame.

The *human foot* has a large set of bones that can move independently including the ankle and the toes. However, these elements are often confined within shoes. Other than specialized motion capture applications, applications requiring stealth input using the toes, or interpolation of motion over distances, it is unlikely that extensive tracking of the foot would be of great utility in mobile AR environments.

The *human hand* is a complex device with many bones. Figure 3 is an illustration of the bones of the human hand. The most basic configuration of tracking would emit 15 frames for a hand, fourteen for the phalanges, the bones of the fingers, and one for the metacarpus. A variety of glove technologies exist that can provide this level of tracking for the hand. Simple object manipulation can often be accomplished with only metacarpus tracking.

5 Calibration, Coupling, and Filtering

This section discusses the issue of calibration of information frames and introduces the concepts of coupling, blended couplings, filtering, and filtered couplings. Calibration is the mathematic process of computing a transformation from the tracked frame to an appropriate information frame. Coupling is the combination of elements of more than one information frame to form a new frame with desirable properties.

The variety of potential tracker placements in Table 3 imply a basic set of information frames, each requiring only a simple rigid-body transformation to convert from a tracker marker space to a useful information space. The computation of this rigid-body transformation is referred to as *calibration*. Tracing the lower lumbar vertebrae yields a frame that moves identically with the body. However, this frame, if defined by the tracking system alone, may not be oriented as desired. Typically, a body-stabilized display is expected to have a y-axis parallel to the body central axis and a z axis pointing in the forward direction (or reverse, as conventions do vary). We have also adopted the convention that the origin of a body stabilized display should be a point on the floor at the base of the body central axis. In our systems, this calibration is done by requiring a user to stand on the world coordinate system origin and face in the WCF z direction. Then the appropriate transformation is simply the inverse of the transformation from marker space to world space for the tracked frame: T_{Ftow}^{-1}. It is common that these transformations can be computed by simply providing an example of the correct orientation and computing the appropriate transform to multiply the frame by to achieve that orientation (solving for the unknown transformation). In other cases the transformation may be supplied manually based on knowledge of the tracker mounting characteristics.

In addition, presented spaces can be coupled together to create additional, useful information spaces. Billinghurst describes a one degree-of-freedom body-stabilized display [1]. This display was implemented using a simple orientation tracker (likely something as simple as a compass) such that information appears to inhabit a cylinder around the body. We have implemented this functionality using a six degree-of-freedom tracker attached to the lower back so as to track the lumbar vertebrae. The frame is then *coupled* with the world coordinate frame to ensure the up direction of the frame remains up relative to the world. We refer to this concept as a frame with *gravity*, in that down is always down for the frame.

The concept of *coupling* is the combination of two or more frames to create a new information frame. A coupling of the lumbar tracking frame with the world coordinate frame utilizes the origin and z direction vector for the lumbar frame and the y direction vector for the world coordinate frame (0, 1, 0). Note that it is debatable as to if this is a one degree-of-freedom frame or a four degree-of-freedom frame, in that the frame is effectively tracking the body, though that tracking can be eliminated mathematically if necessary.

The introduction of coupling creates a variety of possible new information frames. A relatively large set can be defined just for the coupling of torso tracking and the world coordinate frame, as described in Table 2. Clearly, additional combinations are possible, though of questionable utility.

Similar couplings are possible between skull tracking and the world coordinate frame. There are several applications using head stabilized display. Most of them are Head-up Displays (HUD) for aircraft and military uses [18, 19]. In AR systems, important information such as alert messages may be displayed in this way, though the screen real estate in a head-stabilized display is very limited and there is potential for distraction.

The most basic head-stabilized display is trivial to implement since the transformation from the head information frame to the world coordinate frame is the inverse of the transformation from the world coordinate frame to the head information frame, so simple head-stabilized displays require no tracking. However, a head-stabilized display with gravity may be advantageous in many applications in that text would remain upright relative to the environmental frame of reference users are adapted to.

A *blended coupling* is a coupling between frames that is an interpolation between two frames. Unlike the rigid couplings, a blended coupling creates a flexible information frame, wherein the orientation of the frame is dependent up the position. As an example, information that is to be arrayed around the arm is not actually attached to the bone, but rather to the surface of the skin a non-rigid substance. While an offset from the skeletal model is effective in describing the surface of the skin over parts of the arm, it is not effective around the elbow joint in that that portion of the skin is affected by the motion of both bones. A relative arm location variable *a* can be defined and the arm divided into ranges such that the humerus and forearm groups are used over the appropriate ranges. In the transition area, the frame to be used is an interpolation of the two supplied frames. This interpolation can be easily implemented using cubic blending functions.

Filtering is the modification of a frame though the application of a filter to the terms of the frame. A true body-stabilized display based on tracking of the lumbar vertebrae will move up and down as the user walks due to the natural motion of the spine. This motion causes the body-stabilized display to bounce up and down. One method to eliminate this problem is to fix the height of the frame to ground elevation, effectively forcing the height to the world coordinate frame height. However, this is not always practical, particularly if the user may be moving through varying elevations. The alternative is the filter the t_y value of the body-stabilized display so as to decrease the short-term motion, while retaining longer-duration behaviours. This is a low-pass filtering of the t_y value over time and can be accomplished using a variety of feed-forward and feedback filter designs. However, real-time applications tend to limit the applicability of feed-forward filter solutions due to the potential increase in latency. We have had the best results with simple first and second order feedback filters. The most basic of these is:

$$y_t = (1-a)x_t + ay_{t-1}$$

This filter is in standard recursive filter notation [20]. The parameter a varies the amount of filtering, valid values are in the range [0, 1), with larger values contributing to greater filtering over time. It is also relatively easy to make this filter adaptive by detecting large changes in t_y and decreasing a temporally so as to allow large changes, while filtering smaller changes.

A *filtered coupling* is a coupling between frames wherein filtering is applied to terms during the coupling process. Mathematically, any filtering is a filtered coupling with the world coordinate frame if we assume frames are defined relative to that frame. However, there are many reasons to filter the couplings.

One of the proposed information frames in Table 3 is a body stabilized display that does not rotate with the body. Effectively, this is an information cylinder that moves with the body, but retains a fixed orientation. This would be analogous to a cylinder that follows the user around, but does not turn. An advantage of this information frame is that the user has a 360° range of information and tools arrayed around the body. The user can turn to look at additional information without the cylinder turning the information away, a common criticism of body-stabilized displays. But, over time it may be best that the display adopt an orientation relative to the body, slowly turning to that orientation as the user changes orientation for longer durations. This coupling can be accomplished by filtering the lumbar Y direction relative to the world Y direction over time using a low-pass filter.

6 Psychological Implications of Integrating Physio-psychological Reference Frames into Information Spaces

Just as AR systems track the location of the body in space and simulate the location of objects, a great deal of the brain is allocated to tracking the location of body and objects in space especially in the planning of motor actions [21]. Two neurological systems can be identified, one that can roughly be said to track "where" objects are in space, and another that can loosely be said to classify "what" the objects are [22]. This information is processed via various spatial reference frames that are used to guide the interaction of the moving body with the space [6, 7]. Current AR systems map information primarily to three information frames: world or scene frames, object frames, and head centered frames.

Psychologically all the spatial reference frames are interlinked. In motor control, one can see a hierarchy or heterarchy depending on the task at hand [23]. Information received on what are largely 2D sensory systems is coordinated with the movement of the body to construct a 3D model of egocentric space immediately around the body. It is assumed that egocentric space, organized around the body and especially the head, may be the more primitive space that is most accessible [24]. It appears that objects and the larger world spaces are derived via some mechanism for quickly transforming a location in one coordinate system to another. This use of the different coordinator systems and their transformation is important in guiding and planning motion thru the space. For example, when reaching for an object the location of the object in the head and eye centered reference frame must be transformed into torso, arm, and head coordinate frames to guide hand actions. Humans are largely unaware of the complex processing occurring in every simple reaching motion, although a great deal of complex online transformations are occurring that are so automatic and outside of consciousness.

The goal of the classification of information frames in a proposal for expanded tracking systems for AR is to track and make better use of the user's automatic, effortless processing of the spatial reference frames used in detecting and manipulating objects and information in the physical world. At the interface level, the information fields proposed above can be selectively mapped to difference physio-psychological spatial reference frames to yield better task performance within various applications.

7 Conclusion

The design of AR interfaces prompt a significant human factors challenge of mapping different metaphors, information, and functions of computer usage into a volumetric computing environment so as to maximize information bandwidth and reduce a user's attentional and cognitive load. This paper introduces a taxonomy of different reference frames that can be useful for placing information objects in a mobile computing environment from the physiological perspective. It also lays the mathematical foundations for calibration, coupling and filtering motion of different reference frames. Psychological implications of using different reference frames to organize and hold information objects for a mobile user are also discussed.

8 Acknowledgement

This material is based upon work supported by the National Science Foundation under Grants No. 00-8274, 02-22831. Any opinions, findings, and conclusions or recommendations expressed in this material are those of the author(s) and do not necessarily reflect the views of the National Science Foundation.

References

1. Billinghurst, M., Bowskill, J., Jessop, M., and Morphett, J. A wearable spatial conferencing space, in Proc. 2nd Int. Symp. on Wearable Computers, 1998.
2. Reitmayr, G. and Schmalstieg, D. A Wearable 3D Augmented Reality Workspace, in Proceedings of 5th International Symposium on Wearable Computers (ISCW 2001), , . 2001, Zurich, Switzerland.
3. Biocca, F., Tang, A., and Lamas, D. Evolution of the mobile infosphere: Iterative design of a high-information bandwidth, mobile augmented reality interface., in Euroimage 2001: International Conference on Augmented, Virtual Environments and 3D Imaging, 2001, Mykonos, Greece.
4. Reichlen, B. SparcChair: One Hundred Million Pixel Display, in Proceedings IEEE VRAIS '93, 1993, Seattle WA: IEEE Press.
5. Paillard, J., Motor and Representational Framing of Space, in Bran and Space, Paillard, J., Editor. 1991, Oxford Science Publications: Cambridge, MA.
6. Bryant, D.J., A Spatial Representation System in Humans. Psycholoquy, 1992. 3(16).
7. Previc, F.H., The Neuropsychology of 3-D space. Psychological Bulletin, 1998. 124(2): p. 123-164.
8. Biocca, F., The cyborg's dilemma: progressive embodiment in virtual environments. Journal of Computer-Mediated Communication, 1997. 3(2): http://www.ascusc.org/jcmc/vol3/issue2/biocca2.html
9. Gibson, J.J., The ecological approach to visual perception. 1979, Boston: Houghton-Mifflin.
10. Gibson, J.J., The senses considered as perceptual systems. 1966, Boston: Houghton-Mifflin.
11. Feiner, S., MacIntyre, B., Hollerer, T., and Webster, T. A Touring Machine: Prototyping 3D Mobile Augmented Reality Systems for Exploring the Urban Environment, in Proc, 1st Int'l Symp. Wearable Computers. (ISWC' 97), 1997, Cambridge, MA.
12. Thomas, B., Close, B., Donoghue, J., Squires, J., Bondi, P.d., Morris, M., and Piekarski, W. ARQuake: An Outdoor/Indoor Augmented Reality First Person Application, in Proc. 4th Int'l Symp. Wearable Computers. (ISWC 2000), 2000, Atlanta, GA.
13. Reitmayr, G. and Schmalstieg, D. Mobile Collaborative Augmented Reality, in Proc. ISAR 2001, 2001, New York, NY.
14. Piekarsky, W., Gunther, B., and Thomas, B. Integrating Virtual and Augmented Realities in an Outdoor Application, in Proc. 2nd Int'l Workshop Augmented Reality. (IWAR '99), 1999, San Francisco, CA.
15. Tuceryan, M., Greer, D.S., Whitaker, R.T., Breen, D.E., Crampton, C., Rose, E., and Ahlers, K.H., Calibration Requirements and Procedures for a Monitor-Based Augmented Reality System. Transactions on Visualization and Computer Graphics, 1995. 1(3): p. 255--273.

16. Balakrishnan, R., Fitzmaurice, G.W., and Kurtenbach, G. Digital Tape Drawing, in ACM Symposium on User Interface Software and Technology, UIST99, 1999, Asheville, NC.

17. Gray, H., Bannister, L.H., Berry, M.M., and Williams, P.L., Gray's Anatomy: The Anatomical Basis of Medicine and Surgery. 1995: Churchill Livingstone. 2092.

18. Inzuka, Y., Osumi, Y., and Shinkai, K. Visibility of head up display for automobiles, in 35th Annual Meeting of the Human Factor Society, 1991.

19. Dopping-Hepenstal, L.L., Head-up displays: The integrity of flight information. IEEE Proceedings Part F, Communication, Radar and Signal Processing, 1981. 128(7): p. 440-442.

20. McClellan, J.H., Schafer, R.W., and Yoder, M.A., DSP First: A Multimedia Approach. 1998, Upper Saddle River, NJ: Prentice Hall. 523.

21. Graziano, M.S. and Gross, C., The Representation of Extrapersonal Space: A Possible Role for Bimodal Visual-tacille Neurons, in The Cognitive Neurosciences, Grazzaniga, M., Editor. 1995, M.I.T. Press: Cambridge, MA. p. 1021-1034.

22. Underleider, L.G. and Haxby, J.V., "What and "Where" is the Human Brain. Current Opinion in Neurobiology, 1994. 4: p. 157-165.

23. Redding, G.M. and Wallace, B., Adaptive Spatial Alignment. 1997, Hillsdale, NJ: Lawrence Erlbaum.

24. Grush, R., Self, World, and Space: The Meaning and Mechanisms of Ego- and Allocentric Spatial Representation. Brain and Mind, 2000. 1: p. 59-92.

777

Toward Collective Intelligence

Alex Pentland

MIT Media Laboratory
Room E15-387, 20 Ames St., Cambridge, MA 02139
Pentland@media.mit.edu

Abstract

I argue that we should think of humans as having a collective mind as well as an individual mind; the collective mind operates through unconscious social signaling and in many situations is as powerful a determinant of behavior as the conscious, individual mind. By incorporating this unconscious social signaling into the control of our communications systems we can achieve great advances in group coordination.

1   Tribal Mind

What would it be like to be part of a collective intelligence but still with an individual consciousness? Well for starters, you might expect to see the collective mind `take over' from time to time, directly guiding the individual minds. In humans, the behavior of angry mobs and frightened crowds seem to qualify as examples of a `collective mind' in action, with non-linguistic channels of communication usurping the individual capacity for rational behavior.

But as powerful as this sort of group compulsion can be, it is usually regarded as simply a failure of individual rationality, a primitive behavioral safety net for the tribe in times of great stress. Surely this tribal mind doesn't operate in normal day-to-day behavior --- or does it? If we imagined that human behavior was in substantial part due to a collective tribal mind, you would expect that non-linguistic social signaling --- the type that drives mob behavior --- would be predictive of even the most rational and important human interactions. Analogous with the waggle dance of the honeybee, there ought to be non-linguistic signals that accurately predict important behavioral outcomes.

And that is exactly what I have found. Together with my research group I have built a computer system that objectively measures a set of non-linguistic social signals, such as engagement, mirroring, activity, and stress, by looking at `tone of voice' over one minute time periods (Pentland, 2004).   Although people are largely unconscious of this type of behavior, other researchers (Jaffe et al 2001, Chartrand and Bargh, 1999, France, 2000, Kagen and Snidman 2004) have shown that similar measurements are predictive of infant language development, judgments of empathy, attitude, and even personality development in children.

Using our `social perception' machine, and without listening to words or looking at gestures, we can accurately predict (Human Dynamics Technical Notes, http://hd.media.mit.edu):

- Who will exchange business cards at a meeting
- Which couples will exchange phone numbers at a bar
- Whether or not you think a person would be a good member of your workgroup
- Whether or not you thought a conversation was interesting

- Whether you will be able to negotiate a good deal with a particular opponent
- Whether or not you will feel the negotiation was honest and fair
- Whether or not you are a `connector' within your workgroup

Our accuracy averages almost 90% after excluding cases where we don't have enough `signal' to make a decision. The prediction is made from only a few minutes of `listening in', without taking into account any personal characteristics (sex, attractiveness, etc) or any objective context (e.g., the facts or what was actually said). The experiments were typically 100 subjects each, with subjects 25-35 years of age, and roughly one-third of the subjects were female.

Decision accuracy is calculated by comparing a linear combination of the social signaling features (described later) with the behavioral outcome. With a three-class linear decision (yes, not enough information, no) the yes/no accuracy is typically around 90% with 20% not enough information. The accuracy is typically around 80% with a two class linear decision rule, where we make a decision for every case. More generally, linear predictors based on the measured social signals typically have a correlation of r=0.65, ranging from r=0.40 to r=0.90.

What is surprising about these results is that these decisions are some of the most important interactions a human has: finding a mate, getting a job, negotiating a salary, finding your place in your social network. These are activities for which we prepare intellectually and strategically for decades. And yet the largely unconscious social signaling that occurs at the start of the interaction appears to be more predictive than either the contextual facts (is he attractive? is she experienced?) or the linguistic structure (e.g., strategy chosen, arguments employed, etc.).

So what is going on here? One might speculate that the social signaling we are measuring evolved as a method of establishing tribal hierarchy and cohesion. On this view the tribal mind would function as unconscious collective discussion about relationships and resources, risks and rewards, and would interact with the conscious individual minds by filtering ideas by their value relative to the tribe.

The view that we are part collective mind, part individual mind makes a certain evolutionary sense. Imagine a tribe of 150 individuals on the African veldt. Each day the adults go out gathering and hunting, and in the evening return to sit around a central clearing where they recount the events and observations of the day, and discuss what to do tomorrow. During the group discussion social signaling...reflecting the power hierarchy as well as individual desires...accompanies the new information and the collective social signaling communicates to each individual what the group thinks about it. At the end this social discussion collective decisions have been made, and the required individual behaviors are enforced by the iron hand of social pressure.

2   Modern Mind
Psychology has firmly established that the same sort of social processes are still at work in the modern world (Brown 1986, Dunbar 1998, Provine 2001). Instead of talking about the tribal spirit we now speak of the corporate spirit, and instead of dominance displays we have office politics. Unfortunately socially mediated decision-making has some serious problems, include group polarization (`the risky shift'), `groupthink', and several other types of irrational group behaviors that consistently undermine group decision-making. To have an effective work group

779

you have to think of humans as having a collective social mind in addition to their personal mind.

However just knowing about the problems with group decision making hasn't helped communications technology very much, because to improve group functioning you have to be able to monitor the social process and provide real-time interventions. Human experts can do that...they are called facilitators or mediators....but to date machines have been blind to the social signals that are such an important part of human group function.

Technology has so far focused either on the isolated individual, or has treated the person as a clueless extra wandering in a computer-controlled environment. Researchers seem to have forgotten that people are social animals, and that the quality of their lives is defined by their roles in human organizations. The result is that current communications technology doesn't feel very good. Buzzing pagers, ringing cell phones, and barrages of e-mails are leashes that keep people tethered to their job, and people worry that we are being assimilated into some sort of unhappy Borg Collective.

Our ability to have computers measure social signals and establish social context has the potential to completely change this situation. The measurements tap into the social signaling in face-to-face discussions, and anticipate outcome by use of statistical regularities. For instance, in salary negotiation we found that it was important for the lower-status individual to establish that they are a 'team player' by mirroring, indicating that they are empathetic, while in a potential dating situation the key variable was the female's level of activity, indicating interest. In our data there are patterns of signaling that seem to reliably lead to these desired states, allowing us to gently guide the conversation to a happy ending by providing timely feedback. Similarly, the ability to measure social variables like interest and trust can enable more productive discussions, while the ability to measure social competition can reduce problems like groupthink and polarization.

Quantification of social context may have its largest impact on distance interactions, by propagating social context over physical distance in order to better integrate distant participants. One can imagine harnessing the computing and sensing power of today's nearly ubiquitous wearable computing devices (aka mobile telephones) to provide a 'social intelligence' that improves the functioning of distributed work groups.

To achieve these goals we are building systems that support people's social and organizational roles instead of designing technology for the individual as an isolated entity. Such 'socially aware' communications should enable new types of organizations that are not only more efficient, but also leave the individual with a better balance between their formal, informal, and personal lives. There may be no escape from being assimilated into the Borg Collective, but we can still try to design the Collective to be a human place to live.

3   Social Signals

Malcolm Gladwell's popular book, *Blink*, describes the surprising power of "thin-slicing," defined as "the ability of our unconscious to find patterns in situations and people based on very narrow 'slices' of experience" (2005, P. 23).   Gladwell writes, "Snap judgments and first impressions matter as much as they do because there are…lots of situations where careful attention to the details of a very thin slice, even for no more than a second or two, can tell us an awful lot" (P. 47).

Gladwell's observations reflect decades of research in social psychology, and the term "thin slice" comes from a frequently cited study by Nalani Ambady and Robert Rosenthal (1993). Ambady and Rosenthal had female college students evaluate 30-second, silent video clips of college instructors teaching a class, and found remarkably high correlations between those evaluations and the end-of-semester ratings of those same instructors by their respective students. This result was replicated with high school teachers and using smaller "slices" of video (down to just six seconds for each teacher!).

3.1   Thin Slices

Thin slices of behavioral data have been shown to be predictive of a diverse set of consequences, including therapist competency ratings (Blanck, Rosenthal, Vannicelli, & Lee, 1986), personalities of strangers (Borkenau, Mauer, Riemann, Spinath, & Angleitner, 2004), and even courtroom judges' expectations for trial outcomes (Ambady, Bernieri, & Richeson, 2000; Ambady & Rosenthal, 1992).

Across a wide range of studies, Ambady and Rosenthal (1992) found that observations lasting less than five minutes in duration predicted their criterion for accuracy with an average effect size of $r = .39$. This effect size corresponds to 70% accuracy in a binary decision task (Rosenthal & Rubin, 1982). It is astounding that observation of such a "thin slice" of behavior can predict important behavioral outcomes such as professional competence, criminal conviction, and divorce, when the predicted outcome is sometimes months or years in the future. The key to success lies is in understanding social signalling

3.2   Measuring Social Signals

Animals communicate and negotiate their position within a social hierarchy in many ways, including dominance displays, relative positioning, and access to resources. Humans add to that a wide variety of cultural mechanisms such as clothing, seating arrangements, and name-dropping (Dunbar,1998).   Most of these culture-specific social communications are conscious and are often manipulated.

Nonlinguistic vocal signaling is a particularly familiar part of human behavior. For instance, we speak of someone `taking charge' of a conversation, and in such a case this person might be described as `driving the conversation' or `setting the tone' of the conversation.  Such dominance of the conversational dynamics is popularly associated with higher social status or a leadership role. Similarly, some people seem skilled at establishing a `friendly' interaction. The ability to set conversational tone in this manner is popularly associated with good social skills, and is typical of skilled salespeople to social `connectors' (Gladwell 2000).

In many such situation non-linguistic social signals (e.g., body language, facial expressions, and tone of voice) are as important as linguistic content in predicting behavioral outcomes (Ambady & Rosenthal, 1992; Nass & Brave, 2004). Indeed, there is a serious case that such vocal signaling originally evolved as grooming and dominance displays, and continue to exist today as a complement to human language (Dunbar, 1998; Provine, 2001).

The machine understanding community has studied human communication at many time scales --- e.g., phonemes, words, phrases, dialogs --- and both semantic structure and prosodic structure have been analyzed. However the sort of longer-term, multi-utterance structure associated with social signaling has received relatively little attention (Handel 1989, IEEE Face and Gesture 1997-2004). It is this relatively novel region of minute-scale prodosic and gestural textural structure that my research group is now exploring. Below, I review briefly four general types of social signaling that have been operationalized in our studies.

3.2.1 Mathematical Method

To quantify these social signals I have developed measures for four types of vocal social signaling, which were designated activity level, engagement, stress, and mirroring (Pentland, 2004). These four measures were extrapolated from a broad reading of the voice analysis and social science literature, and we are now working to establish their general validity. To date they have been used to successfully predict outcomes in dating, friendship, and business preferences with accuracy comparable to that of human experts in analogous situations. I have also developed ``motion energy' equivalents for face and hand gesture, and experiments using these visual features are now underway.

Calculation of the activity measure begins by using a two-level HMM to segment the speech stream of each person into voiced and non-voiced segments, and then group the voiced segments into speaking vs. non-speaking (Basu, 2002). Conversational activity level is measured by the z-scored percentage of speaking time.

Engagement is measured by the z-scored influence each person has on the other's turn-taking. When two people are interacting, their individual turn-taking dynamics influences each other and can be modeled as a Markov process [2]. By quantifying the influence each participant has on the other we obtain a measure of their engagement...popularly speaking, were they driving the conversation? To measure these influences we model their individual turn-taking by an Hidden Markov Model (HMM) and measure the coupling of these two dynamic systems to estimate the influence each has on the others' turn-taking dynamics (Choudhury, 2003). Our method is similar to the classic method of Jaffe et al. (2002), who found that engagement between infant-mother dyads is predictive of language development. In our formulation we use a simpler parameterization that permits the direction of influence to be calculated and permits analysis of conversations involving many participants.

Stress is measured by the variation in prosodic emphasis (Handel, 1989). For each voiced segment we extract the mean energy, frequency of the fundamental format, and the spectral entropy. Averaging over longer time periods provides estimates of the mean-scaled standard deviation of the formant frequency and spectral entropy. The z-scored sum of these standard deviations is taken as a measure speaker stress; such stress can be either purposeful (e.g., prosodic emphasis) or unintentional (e.g., physiological stress caused by discomfort). Similar

782

measures of vocal stress have been used to detect deception and also to predict introversion/extroversion in children's personality development.

Mirroring behavior, in which the prosody of one participant is 'mirrored' by the other, is considered a signal of empathy and has been shown to positively influence the outcome of a negotiation and other interpersonal interactions (Chartrand and Bargh, 1999). In our experiments the distribution of utterance length is often bimodal. Sentences and sentence fragments typically occurred at several-second and longer time scales. At time scales less than one second there are short interjections (e.g., 'uh-huh'), but also back-and-forth exchanges typically consisting of single words (e.g., 'OK?', 'OK!', 'done?', 'yup.'). The z-scored frequency of these short utterance exchanges is taken as a measure of mirroring. In our data these short utterance exchanges were also periods of tension release.

4 Collective Mind

These social signaling measurements have been incorporated into the development of three new 'socially aware' communications tools, one based on a badge-like platform (Laibowitz and Paradiso, 2003), one based on the Sharp PDA (Madan, Caneel, and Pentland, 2004), and one based on the Nokia 6600 mobile telephone (Eagle and Pentland, 2004). In each system the basic element of social context is the identity of people in the users' immediate presence. This can be determined by several methods, including Bluetooth-based proximity detection, infra-red (IR) or radio-frequency (RF) tags, and vocal analysis. To this basic context can be added audio feature analysis, sensors for head movement, body motion, and even galvanic skin response (GSR). These sensing capabilities provide a quantitative measure of social context for the user's immediate, face-to-face situation. The systems use these social context measurements to provide users with multi-level 'social awareness' feedback and enable a 'socially intelligent' communications network. The result is a lightweight, unobtrusive wearable system that can identify face-to-face interactions, capture collective social information, extract meaningful group descriptors, and transmit the group context to remote group members.

When a face-to-face interaction is detected (e.g., the combination of proximity and conversational turn-taking), a group context is defined that consists of the identities of the participants, the social features described above, and the audio (and potentially visual) information stream. A 'social gateway' is then created that contains the group context information and which enables pre-approved members of the social or work group to access the on-going conversation and group context information. The social gateway uses real-time machine learning methods to identify relevant group context changes. Any distance-separated user can then access these group context changes.

5 Some Examples

Simple uses of social context are to provide people with feedback on their own interactions. Did you sound forceful during a negotiation? Did you sound interested when you were talking to your spouse? Did you sound like a good team member during the teleconference? Such feedback can potentially head off many unnecessary problems.

The same sort of analysis can also be useful for robots and voice interfaces. While word selection and dialog strategy are very important to achieve a successful human-machine interaction, our experiments and those of others show that social signaling may be even more important.

Here are some more sophisticated and networked examples of communication being mediated by use of social context:

5.1 What Was The Name?
Everyone has had the experience of going to a meeting, discovering interesting people, and then afterwards loosing their business card or forgetting their name. Using the electronic name badge platform (Laibowitz and Paradiso 2004), we have found that we can use audio analysis and body motion to accurately predict when two people will exchange business cards (Gips and Pentland, 2005). One application of this capability is to have the badge keep track of all the interactions where the you acted as if you were likely to exchange business cards, and email you their names and particulars at the end of the day.

5.2 Social Capital
Social capital is the ability to leverage your social network: knowing who knows what, and who to talk to get things done. It is perhaps the central social skill for any entrepreneurial effort, yet for most people such networking is difficult. We have therefore created a set of tools that can help you build your social capital.

One example is Serendipity, a system implemented on Bluetooth-enabled mobile phones (Eagle and Pentland, 2004). This system is built on BlueAware, an application that scans for other Bluetooth devices in the user's local proximity. When a new device is discovered nearby, Serendipity automatically sends a message to a central gateway server with the discovered device's ID. If a "match" is found, then a customized picture message is sent back to each user, introducing them to each other.

The real power of this system is that it can be used to create, verify, and better characterize relationships in on-line 'social network' systems such as Friendster or Orkut. If two people hang out together outside of work, then they are probably social friends. If they only meet at work, or never meet at all, then they likely have a very different type of relationship. Characterization of the relationship can be refined even further by analysis of the social signaling happening during phone calls between the two people. The phone extracts the social signaling features as a background process, so that it can provide feedback to the user about how they sounded, but also to build a profile of the interactions you have had with the other person.

5.3 Being In The Loop
A major problem with distributed workgroups is keeping yourself 'in the loop'. Socially mediated communications, such as our GroupMedia system (Madan, Caneel, and Pentland 2004), can potentially help with this problem by 'patching in' people to important conversations. The social gateway can patch out-of-office group members into on-going face-to-face conversations depending on variables such as measured interest levels, directionality of information flow, and group membership.

When a potentially interesting conversation is detected, the system can notify the distant group member. Upon receiving such a notification a distance-separated group member can either subscribe to this information and begin to receive the raw audio signal plus annotations of the social context, or they can choose to be notified by the system only in case of especially interesting comments, or they can store the audio signal with social annotations for later review.

For instance, if most of your workgroup has gathered together, and the information flow is from your boss, and the interest level is high, it might be wise to either patch in to the audio and keep track of the measured level of group interest for each participant's comments. In such an example the group context information and the 'linking in' notification provided by the gateway can increase both the group cohesion and the distanced individuals' understanding of the raw audio signal.

5.4 Group Dynamics

In my Digital Anthropology seminar every student was outfitted with a GroupMedia system, to enable analysis of group interaction (Eagle and Pentland 2003, Sung et al 2004). Real-time displays of participant interaction could be generated and publicly displayed to reflect the roles and dyadic relationships within a class. An example is the figure below, which shows that the professor (s9) is the dominant member of the seminar while his advisees (s2, s7, s8) concede the floor to him with relatively high probability - indicative of his influence.

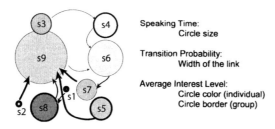

Depiction of the professor and student dynamics during the Digital Anthropology seminar.

This type of analysis can help develop a deeper discussion. Comments that give rise to wide variations in individual reaction can cause the discussion to focus on the reason for disparate opinions, and controversial topics can be retrieved for further analysis and debate. Opinions and comments can also be clustered using 'collaborative filtering' to display groupings of opinion, allowing within-group and between-group debate.

6 Conclusion

This paper describes a framework, measurement method, and examples of communications infrastructure that allow propagation of quantitative social context over distance, thus potentially improving social and group awareness. Using this automated method we have demonstrated predictive power much greater than that suggested by earlier psychology studies, and addressed new areas of human interaction including negotiation, social network formation, and personal attraction.

When considering the personal and societal effects of socially aware communications systems, it is interesting to recall McLuhan's dictum, "the medium is the message"--- that is, the way in which a new technology changes our lifestyle is more important than the information it conveys. By designing systems that are aware of human social signaling, and which adapt themselves to human social context, we may be able to remove the medium's message, and replace it with the traditional messaging of face-to-face communication. Just as computers are disappearing into clothing and walls, the 'otherness' of communications technology may disappear as well.

Acknowledgements: I would like to thank my collaborators, Joost Bonsen, Prof. Jared Curhan, Prof. David Lazar, Dr. Carl Marci, Dr. M. C. Martin, Prof. Joe Paradiso, and my current and former graduate students Sumit Basu, Ron Caneel, Tanzeem Choudhury, Wen Dong, Nathan Eagle, Jon Gips, Anmol Madan, and Mike Sung, for all the hard work and creativity they have added to this project. Thanks also to Prof. Deb Roy, Prof. Judith Donath, and Dr. Tracy Heibeck for insightful comments and feedback. Some sections of this paper are drawn from articles that have appeared in www.edge.org, Int'l Conf. On Developmental Learning 2004, and IEEE Computer March 2005, and are used with permission.

References

Ambady, N., Bernieri, F. J., & Richeson, J. A.(2000). Toward a histology of social behavior: Judgmental accuracy from thin slices of the behavioral stream. *Advances in Experimental Social Psychology, 32,* 201-271.

Ambady, N., & Rosenthal, R. (1992). Thin slices of expressive behavior as predictors of interpersonal consequences: A meta-analysis. *Psychological Bulletin, 111(2),* 256-274.

Basu, B., (2002) Conversational Scene Analysis, doctoral thesis, Dept. of Electrical Engineering and Computer Science, MIT. 2002. Advisor: A. Pentland

Borkenau, P., Mauer, N., Riemann, R., Spinath, F. M., & Angleitner, A. (2004). Thin slices of behavior as cues of personality and intelligence. *Journal of Personality and Social Psychology, 86,* 599-614.

Blanck, P. D., Rosenthal, R., Vannicelli, M.,& Lee, T. D.(1986). Therapists' tone of voice: Descriptive, psychometric, interactional and competence analyses. *Journal of Social and Clinical Psychology, 4,* 154-178.

Brown, R. (1986) Group Polarization, in Social Psychology (2d Edition), New York: Free Press

Chartrand, T., and Bargh, J., (1999) The Chameleon Effect: The Perception-Behavior Link and Social Interaction, J. Personality and Social Psychology, Vo. 76, No. 6, 893-910

Choudhury, T., "Sensing and Modeling Human Networks." PhD thesis, Dept of MAS, MIT, 2003. Advisor: A. Pentland, http://hd.media.mit.edu,

Dunbar, R. (1998) Grooming, Gossip, and the Evolution of Language, Harvard U. Press

Eagle, N., Pentland, A. (2003) Social Network Computing, Ubicomp 2003, Springer-Verlag Lecture Notes in Computer Science,, No. 2864, pp. 289-296, http://hd.media.mit.edu

Eagle, N., Pentland, A. (2004) Social Serendipity: Proximity Sensing and Cueing, MIT Media Lab Tech Note 580, http://hd.media.mit.edu, May 2004

France, D.J., (2000) Acoustical Properties of Speech as Indicators of Depression and Suicidal Risk, IEEE Trans. Biomedical Eng., July, pp. 829-837

Gips, J., Pentland, A. (2005) No More Business Cards: Automatic Measurement of Interest At Conferences, MIT Media Lab Tech Note 586, http://hd.media.mit.edu, Feb 2005

Gladwell, M. (2000). *The Tipping Point: How little things can make a big difference.* New York: Little Brown.

Gladwell, M. (2005). *Blink.* New York: Little, Brown and Company.

Handel, Stephen, (1989) Listening: an introduction to the perception of auditory events, Cambridge: MIT Press

Human Dynamics Technical Notes, See http://hd.media.mit.edu for the most recent papers and technical notes by authors Madan, Caneel, Eagle, Choudhury, Gips, Sung, Dong, and Pentland

Jaffe, J., Beebe, B., Feldstein, S., Crown, C. L., & Jasnow, M. (2001). Rhythms of dialogue in early infancy. Monographs of the Society for Research in Child Development, 66(2), No. 264.

Kagen, J. Snidman, N., (2004) The Long Shadow of Temperment, Belknap Press

Laibowitz, M., and Paradiso, J. (2004) The Uberbadge Project, http://www.media.mit.edu/resenv/projects.html

Madan, A., Caneel, R., Pentland, A., (2004) GroupMedia: Distributed Multimodal Interfaces, IEEE Int'l Conference on Multimedia Interfaces, State College, PA, 2004. http://hd.media.mit.edu

Nass, C., and Brave, S. (2004) Voice Activated: How People Are Wired for Speech and How Computers Will Speak with Us, MIT Press

Pentland, A. (2004) Social Dynamics: Signals and Behavior, Int'l Conf. On Developmental Learning, Salk Institute, San Diego, Oct. 20-22. http://hd.media.mit.edu

Provine, R., (2001) Laughter, Penguin Press

Rosenthal, R., & Rubin, D. B. (1982). A simple, general purpose display of magnitude of experimental effects. *Journal of Educational Psychology, 74*, 166-169.

Russell, J. A., (2003) Core Affect and the Psychological Construction of Emotion, *Psychological Review*, Vol. 110, No. 1, pp. 145-172

Sung, M, Gips, J., Eagle, E., Madan, A., Caneel, R., DeVaul, R., Bonsen, J., Pentland A. (2004) M-Learning Applications for Classroom Settings, to appear J. Computer Augmented Learning, 2005, also MIT Media Lab Tech Note 576, http://hd.media.mit.edu

Section 4
Cognitive State Sensors

Chapter 16

Adaptive User Interfaces

UCAV Operator Workload Issues Using Adaptive Aiding Systems

Bryan A. Calkin
Air Force Research Laboratory, Human Effectiveness Directorate
2255 H Street, Wright-Patterson AFB, OH 45433
bryan.calkin@wpafb.af.mil

Abstract

In order to successfully supervise unmanned combat aerial vehicles (UCAV), while coordinating with other air and ground assets, operators will require significant training, a supportive interface and a high degree of situational awareness (SA). Even with these elements, task saturation and individual cognitive limitations can drive an operator into a state of mental overload, resulting in degraded performance and even mission failure. Real-time adaptive aiding (also adaptive automation) can assist the operator in multiple ways during critical periods throughout the mission. It can also enhance the performance capability of the operators and result in one operator doing the work of many. However, the adaptive automation and decision support tools must be appropriately applied in order to be more helpful than disruptive. The aiding must be reliable, predictable, timely, relevant, and seamlessly introduced and removed. Additionally, predetermined contracts with the adaptive automation can address user-specific, mission-specific and context-specific needs. This paper provides a test perspective from the operator's point of view to workload issues involved in supervising highly-autonomous air vehicles and some of the adaptive aiding strategies that could be used to help manage them. The role of Operator Functional State (OFS) assessment using psychophysiological measures will also be discussed.

1 Introduction

Complex systems and complex interfaces are becoming more prevalent as technology advances. Consequently, military operators are being asked to manage large volumes of information as networked computers increase access to other users, databases and real-time imagery. This influx of technology has, in some ways, outpaced the capabilities of the human component of the system. Automation has been designed into some systems to ease the operator's workload. This automation is meant to complete a portion of the assigned tasks during periods of high cognitive demand. However, the automation can bring with it a new set of problems and additional workload, if improperly introduced (Prinzel, 2003).

Parasuraman, Sheridan and Wickens (2000) define the term "automation" as "a device or system that accomplishes (partially or fully) a function that was previously, or conceivably could be, carried out (partially or fully) by a human operator." They also suggested 10 levels of automation of decision and action selection (Table 1), varying from fully manual to fully automatic. At the low end of the automation scale, humans are responsible for all of the decision-making. At the high end, the computer is fully responsible while the human simply monitors. Operators can incorporate these levels of automation into their operations to help them manage periods of overload, but those periods and levels must be defined and clearly understood.

HIGH	10. The computer decides everything, acts autonomously, ignoring the human.
	9. informs the human only if it, the computer, decides to
	8. informs the human only if asked, or
	7. executes automatically, then necessarily informs the human, and
	6. allows the human a restricted time to veto before automatic execution, or
	5. executes that suggestion if the human approves, or
	4. suggests one alternative
	3. narrows the selection down to a few, or
	2. The computer offers a complete set of decision/action alternatives, or
LOW	1. The computer offers no assistance: human must take all decisions and actions.

Table 1. Levels of Automation of Decision and Action Selection

791

Human operators demonstrate different levels of proficiency and exhibit different thresholds at which they become mentally overloaded. The speed and accuracy with which they perform can affect whether or not an overload condition develops. Adaptively aiding an operator during these overload conditions, based upon the situation and context, can be accomplished in real-time using psychophysiological measures. Wilson (2003) stated "psychophysiological measures can provide insight into the basic responses and functional state of operators" (p. 127). He added that the measures can be used to determine if the brain is able to efficiently process the relevant data from a task (p. 127). Having this insight into OFS in real time would be an effective tool for operators and mission managers before, during and after combat situations. For example, if operators knew a time-critical task was impending and they were already task-saturated, adaptive aiding could be invoked, even anticipated. This could result in alleviating some of the operator's workload and stress. Additionally, routes and missions could be designed to account for situations and tasks that have a high mental demand. Furthermore, it could give supervisors useful metrics on the patterns and levels of performance of their operators.

2 Background of UCAV

The Defense Advanced Research Projects Agency (DARPA) has set out to demonstrate the feasibility of a networked system of unmanned combat aerial vehicles. Combining the past efforts and lessons learned from the DARPA/USAF and DARPA/USN UCAV programs, operational assessment of the vehicles is scheduled to begin in 2007. After its projected fielding in 2010, the primary mission of the UCAV will be the suppression of enemy air defenses (SEAD). Other functions will include intelligence, surveillance, and reconnaissance (ISR), electronic attack, enforcement of no-fly zones and precision strike. These applications will reduce the number of occasions in which piloted aircraft and ground forces are subjected to hostile weaponry.

Mission requirements will dictate the scope of tasks involved for controlling each vehicle, but the UCAV is seen as the evolution of the Predator and Global Hawk unmanned aerial vehicles (UAV), which are used primarily for real-time ISR. Multiple operators are used to control the UAV platforms. The current plan for the UCAV is to have one operator supervise one or more vehicles. Each vehicle's sensors, communications, and attack management protocols will be directly controlled though an interface. This Common Operating System (COS) will manage the resources of the air vehicles and provide the network connections to other manned and unmanned assets (Francis, 2005).

3 Role of the Operator

This complex system has been developed on the premise that operators will be required to simultaneously deal with multiple tasks as they manage these vehicles. Some task examples include capturing and downloading Synthetic Aperture Radar (SAR) images, analyzing those images and discriminating between valid and invalid targets. They will assign munitions to the valid targets, then power on, arm and clear the weapons for release. Additionally, they will have to determine which vehicle should engage a target, evaluate the time-criticality of a target, and decide which type of munition to use. They will also coordinate with other manned and unmanned assets to accomplish ISR, battle management, battle damage assessment, and other integrated functions. This complex interface is shown in Figure 1. Cognitive overload can develop during various points in the mission. Although the UCAV is a highly-autonomous aircraft, operators may shed or postpone tasks during overload conditions and concentrate on one task. They may or may not be aware of their overload condition and may be unable to determine the type, duration and amount of aiding they require. They may not even be aware that a task has been overlooked. The operator's performance can degrade slowly as tasks accumulate, so that additional effort is slowly applied, up to the point of performance breakdown. Consequently, a higher probability of error can occur during critical junctures in the mission.

Figure 1. Illustration of UCAV workstation.

Another use of adaptive aiding is to enhance the operator's performance. This would multiply the efforts of the operator and provide a higher threshold of performance throughout the mission. If an operator is responsible for one or more vehicles, the aiding could be employed in different ways. For example, SAR image analysis will require a significant amount of visual processing. An accurate OFS classification could identify this and employ adaptive aiding to introduce new information through a separate, less-saturated modality. Furthermore, the aiding could help to keep the operator engaged during periods of inattention (Byrne & Parasuraman, 1996). Long ingress periods, or other dull portions of the mission may contribute to operator complacency. The adaptive aiding could stimulate the operator at appropriate intervals using a simple interactive task, or an auditory or tactile cue. In whichever application it's used, well-designed adaptive aiding may help the operator to organize and prioritize the information and increase their "brain on task" resources to improve decision-effectiveness. By appropriately automating a portion of the tasks, and off-loading oversaturated modalities, the operator can focus on the more critical aspects of the mission. This would allow the operator to remain at an optimal level of performance across the spectrum of encountered workload.

4 Operator Functional State (OFS) Estimation

Wilson (2003) referred to OFS as the operator's ability to meet the momentary demands of the task at hand. The key elements of accurate OFS assessment are psychophysiological measures, momentary mission requirements, performance measures and SA measures. These elements provide a comprehensive set of inputs to the classifier, which takes those inputs and derives a calculated measurement of the OFS. Since psychophysical signals are always present and can be collected without interrupting the operator, they afford the real-time measurement aspect necessary for online classification.

Although various classifiers could be used, artificial neural networks (ANN) using these key inputs have shown the levels of accuracy and reliability needed to provide real-time classification (Wilson & Russell, 2003; Wilson & Russell, 2004). Using ANNs in complex tasks, like the ones involving UCAV operations, to classify OFS has been successfully demonstrated (Russell & Wilson, 1998; Wilson & Russell, 2004). Adaptive aiding using real-time assessment can benefit UCAV operators as they encounter situations where time-critical decisions are required.

This comprehensive foundation of classifier inputs is essential to developing a dependable aiding system that operators will accept and consistently use. Operators need to trust the aiding in different situations and across the scope of the aiding. If operators are aided at inopportune times, or in inappropriate ways, which increases the momentary workload, they will disable, or distrust the aiding (Parasuraman & Riley, 1997).

5 Types of aiding

Engineers usually design alarm systems into piloted aircraft to monitor the landing gear, fuel level, altitude, etc., but then assume that the operator is capable of effectively managing resources regardless of situational complexity. The operator's cognitive state is often left out of the equation of system design. As was previously mentioned, operators function at different levels. Even as some operators become task saturated, they may not be overwhelmed mentally. Their proficiency level may allow them to handle that amount of work. Another operator may have a similar degree of workload, with a similar amount of training and experience and yet be unable to cope with anything more than the task at hand. This is where user- or context-specific aiding could be employed.

Additionally, pre-determined contracts can be established so that operators can predict and rely upon the upcoming type, degree, and duration of aiding. These contracts can also be role-, mission-, and user-specific. They can be set to the preferences of the operator and pre-established to consider the context of the situation. Operators can agree beforehand which tasks they are willing to off-load during high workload conditions. They could also predetermine the degree of aiding that would be applied across the range of the automation described back in Table 1.

The concept of adaptive aiding suggests that the operator's workload and SA will be in a state of continuous flux. As the workload state of the operator approaches the point of performance breakdown, aiding will be invoked to make the subsequent workload more manageable, allowing the operator to regain optimal levels of SA and performance. At that time, the aiding would be withdrawn and the cycle would repeat as required during each mission.

An additional concern in the employment of adaptive aiding is the seamlessness of its introduction and removal. For example, as operators enter a situation that requires aiding, they may be fully engaged in a separate time-critical task. The sudden, forced introduction of an automation may create a distraction and, consequently, reduce performance. Furthermore, as aiding is activated at the onset of a task, it should remain active until the end of that task, even though the physiological signals may have dropped down to normal levels. This would prevent the automation from oscillating during the task. If the automation were to alternate off and on, it could create more confusion and cognitive workload. Even though the elevated physiological signals may have subsided, the mental effort required to regain SA in the middle of a task could produce frustration and lead to errors. A well-developed mission management model could be aware of the current task and identify the need to continue aiding.

Other types of aiding that could be useful include "earcons", "tokens", voice-controlled activations of simple tasks, and a prioritization algorithm for attack/alert sequences. Earcons are redundant cues (beeps, bells, etc.) that signify a task has been selected. When rapidly mouse-clicking, operators could benefit from a distinctive audible signal that verified they selected the appropriate action. For example, pull-down menus with multiple options, such as "Direct Attack", "Capture SAR" and "SAR/Attack", can be inadvertently applied by clicking on an action, and then leaving the menu too quickly. On most mouse-driven graphical user interfaces (GUIs), the cursor will select the option that was highlighted on the second, or upward click. In any GUI that uses mouse clicks as its primary method of interaction, time-sensitive multitasking may suffer from this type of error.

Tokens are buttons that can initiate a set of procedures. These tokens act like one-click icons, or a macro function on your computer desktop. For example, one button adjacent to a vehicle may activate a "Capture SAR" sequence between a vehicle and a target of interest. A second adjacent button may initiate a "SAR/Direct Attack" sequence. Instead of having to pull down menus and search for the appropriate action response, the token would be pre-configured to accomplish common tasks.

The improved capability of voice recognition software could allow for simple commands to be inputted without the use of a mouse or a keyboard. Speaking simple words and phrases like "Apply" or "Capture SAR" could serve as an efficient input method for frequently-used tasks. However, caution must be used so that the proper vehicle is affected. For example, if a vehicle is being attended to by an operator, and another vehicle requires a separate action, the operator must ensure the appropriate one is being directed. An additional issue to consider is that speaking could possibly interfere with the collection of psychophysiological data (Byrne & Parasuraman, 1996).

Prioritizing time-critical tasks could be a predetermined function of the mission requirements aspect of the OFS assessment model. Algorithms could be set in place to create a higher priority for a certain type of target. For example, when two (or more) competing tasks become time-critical, the algorithm could list the tasks based on time-sensitivity, criticality or some other contextually-relevant factor. The operator could then sequentially accomplish these tasks. Other types of aiding strategies have been tested and may be useful to different operators in different situations.

6 Conclusions

High workload conditions like those encountered by UCAV operators, or air battle managers, or air traffic controllers may push the cognitive limits of even the most experienced operators. SA and performance may deteriorate at unexpected times, due to fatigue, complacency, task saturation, frustration and other factors. Supporting or augmenting those capabilities through reliable, predictable and appropriate aiding strategies can make the UCAV operator more successful across the diverse range of missions and roles. Using adaptive aiding strategies based on accurate OFS assessment can minimize errors while using complex systems. Furthermore, operators can tailor the level, type and occasions in which adaptive automations are employed. They can integrate it into their concept of operations and specify how and when to best use it.

Some of these aiding strategies could also be designed directly into the interface. High workload conditions may not be needed to warrant their implementation. For certain occasions, one operator may prefer management by consent, which means a suggested action would not be implemented until the operator agreed. Another operator may prefer management by exception, which means the automation would be applied unless the operator declined. For example, if a pop-up target appeared on the screen, a new route to engage or avoid it could be suggested by the system. The context of the mission and/or situation would also help to determine the most appropriate action.

Day-to-day variability of an operator's physiology, motion-induced artifact, operator acceptance of the sensors and other factors can complicate the process of accurately measuring OFS. ANNs can use real-time filtration of motion and signal noise artifact to improve classification. Also, multiple ANN output gauges can be developed to reflect "global workload" or some specific aspect of cognitive functioning, such as "verbal" or "spatial" workload. More OFS assessment research needs to be conducted using complex interfaces, realistic scenarios and actual operators.

Another complicated workload issue involves the distinct workload effect created by continuously switching between the aided and unaided interface. Operators may experience cumulative fatigue and frustration after repeatedly dealing with an interface that converts from an aided view back to an unaided one. The effort expended by operators in reacquiring SA and task context, and possibly having to change to a new management strategy, may ultimately outweigh the benefit of increased performance on a single task.

As these complex systems are studied further and task analyses are completed, these questions can be addressed and operators can begin to train on, trust and use adaptive aiding to its fullest extent. Operators must learn how to accept and effectively use this capability in order to remain at an optimal level of performance. Also, operator training should depict the adaptive aiding as a system partner in the mission. Furthermore, since missions may last for hours, the sensors used to capture the physiological signals should not exceedingly burden the operator's freedom of movement or level of comfort. System designers should consider miniaturized, wireless and off-the-body sensors where feasible. The bottom line is to ensure the operator knows what aiding to expect, knows when to expect it, and trusts that it will in fact be beneficial.

7 References

Byrne, E. A., & Parasuraman, R. (1996). "Psychophysiology and Adaptive Automation," *Biological Psychology*, 42, 249-268.

Francis, M. S. (2005). Overview section. Retrieved Feb 21, 2005, from http://www.darpa.mil/j-ucas/index.htm

Parasuraman, R. & Riley, V. (1997). "Humans and Automation: Use, Misuse, Disuse, Abuse," *Human Factors*, 39(2): 230-253.

Parasuraman, R., Sheridan T., & Wickens C. (2000). "A Model for Types and Levels of Human Interaction with Automation," *IEEE Transactions on Systems, Man, and Cybernetics – Part A: Systems and Humans*, 30(3): 286-297.

Prinzel III, L. J. (2003). "Team-Centered Perspective for Adaptive Automation Design," Technical Memorandum, NASA-TM-2003-212154, Langley Research Center, Hampton VA, 23681-2199

Russell, Chris A. and Glenn F. Wilson (1998). "Air Traffic Controller Functional State Classification Using Neural Networks," in *Smart Engineering Systems: Neural Networks, Fuzzy Logic and Evolutionary Programming* Eds. Cihan H. Dagli, Metin Akay, Anna L. Buczak, Okan Ersoy, and Benito Fernandez, New York: ASME Press, 8: 649-654.

Wilson, G. F. & Russell, C. A. (2004). Real-Time Assessment of Mental Workload Using Psychophysiological Measures and Artificial Neural Networks, *Human Factors*, 45, 635-643.

Wilson, G. F. & Russell, C. A. (2003). Operator Functional State Classification Using Multiple Psychophysiological Features in an Air Traffic Control Task, *Human Factors*, 45, 381-289.

Wilson, G. F. (2003). Adaptive aiding implemented by psychophysiologically determined operator functional state. RTO HFM Symposium on The Role of Humans in Intelligent and Automated Systems, Warsaw, Poland, RTO-MP-088, pp. 18-1 to 18-7. Neuilly sur Seine: NATO/RTO.

Adaptive User Interfaces: Examination of Adaptation Costs in User Performance

Avi Parush

Department of Psychology, Carleton University
Ottawa, Ontario, Canada
Avi_parush@carleton.ca

Yael Auerbach

Industrial Management and Engineering
Israel Institute of Technology, Haifa, Israel
yael_au@techunix.technion.ac.il

Abstract

Adaptive user interfaces are assumed to support and enhance user performance by changing aspects in the user interface in order to better fit the dynamic contexts of interactions. Two user interface adaptation strategies were examined in this study: A fully automatic adaptation and a mixed initiative, user-controlled adaptation. The primary distinction between the strategies is the extent of user involvement and control over the adaptation. In addition, these two strategies were crossed with two levels of adaptation timing: immediate and delayed. These variables were examined using a simulation of an online help desk with varying degrees of workload, which were the trigger for the adaptation. Findings indicate that given the choice, most participants with the user-controlled strategy preferred not to approve user interface changes throughout the experiment despite the varying levels of workload. In addition, there was a significant increase in response time following the adaptation, however, this performance cost decreased significantly as a function of time. The findings are discussed in terms of the impact of involving the user in the decision whether to have an adaptation and aspects of choice behavior. In addition, implicit learning to interact with adaptive user interfaces is discussed.

1   Introduction

Work with interactive systems is highly dynamic. Users change constantly according to their evolving experience and knowledge. In addition, the circumstances in which users interact with systems change often. In contrast to this highly dynamic context, most current user interfaces are fixed. While such fixed interfaces have many advantages, they present a challenge to the users: The need to adapt themselves to the fixed interface when there are changes in their own needs, capabilities, or when there are changes in the interaction circumstances. This requires the user to invest time and resources which in turn may degrade the effectiveness and efficiency of the interaction (Benyon, 1993).

One of the directions that were used to deal with this problem is the development of adaptive systems (Debevc, Meyer, Donlagic & Svecko, 1996). The basic idea of adaptive user interfaces is to have aspects of the interactive system adapt to the changing user and situation (Norcio & Stanley, 1989; Rothrock, Koubek, Fuchs, Hass & Salvendy, 2002). Specifically, elements in the user interface adapt to the changing situations, as opposed to changes in the system itself often referred to as automation. The underlying assumption is that adaptive user interfaces can enhance and facilitate user's performance with the interactive system, particularly in high workload situations (e.g., Trevellyan & Browne, 1987; Gong & Salvendy, 1995; Debvec et.al., 1996; Granic & Galvinic, 2000; Benyon, Crear & Wilkinson, 2001). In this paper we examine this assumption by exploring user performance when interacting with a dynamic adaptive user interface.

There are several basic parameters that characterize the variety of approaches to designing and implementing adaptive user interfaces (Hettinger, Benyon, 1993; Granic & Galvinic, 2000; Rothrock et.al., 2002). These include the adaptation strategy, the trigger for adaptation, the frequency of adaptation, the timing of adaptation, and others. Adaptation strategy is characterized by the level of automation in the adaptation, ranging from a fully manual adaptation (e.g., customization), through a mixed-initiative and user-controlled adaptation, to a fully automatic adaptation. The distinguishing human factor between these various strategies is the extent of user involvement or control over the adaptation.

In the fully manual adaptation, the user initiates and executes the adaptation by deciding whether, when, and how to change the user interface. While giving the user full control over any changes, having to make the decision about the adaptation, consider what to change, and then to actually execute the change, all add another task and increase workload (Theis, 1994; Akoumianakis & Stephanidis, 2001; Keeble & Macredie, 2002). In contract, a system with a fully automatic adaptation strategy initiates and executes the adaptation of the user interface without any involvement of the user. This adaptation strategy is often based on the continuous construction and update of a user model (e.g., changes in the user's experience or workload) or a context model (e.g., changes in the mission, increasing or decreasing system demands, changing environments, etc.). Based on such models, the system initiates changes in the user interface in order to adapt to the dynamic changes and provide a better fit for the user and the task (Greenberg & Witten, 1985; Thies, 1994; Gong & Salvendy, 1995; Rothrock et.al., 2002). This adaptation strategy eliminates the potentially increased workload involved with the fully manual adaptation. However, the exclusion of the user from the decision and execution cycle introduces other problems including poor mental models of the system, degraded situation awareness, and skill loss (e.g., Tattersall & Morgan, 1997; Debevc et.al., 1996; Kaber & Riley, 1999). An adaptation strategy that seems to be a reasonable compromise between the fully manual and fully automatic adaptation is the mixed-initiative, user-controlled adaptation. In this strategy, the system is tasked with monitoring and updating the potential adaptation triggers (e.g., user factors or contextual factors) and recommending or initiating an adaptation. The decision of whether or when to execute the adaptation is left to the user (Debevc et.al. 1996). This strategy follows the human-machine dynamic task allocation rationale (Tattersall & Morgan, 1997) and eliminates some of the additional workload associated with monitoring the situation and making a decision whether an adaptation is appropriate, yet keeping the users in the control loop by letting them approve the adaptation (Gong & Salvendy, 1995; Miah, Kargeorgeou,& Knot, 1997; Miah & Alty, 2000). However, there are some empirical findings indicating that this strategy may distract users from their main task and add the task of approving or rejecting the adaptation recommendation (e.g., Devec, Meyer, & Sveco, 1997; Hudlicka & McNeese, 2002). Taken together, there seem to be inconclusive empirical evidence regarding the relative advantageous and limitations of the various adaptation strategies. Moreover, few studies examined closely what the specific user performance costs are when interacting with an adaptive interface.

Another parameter of adaptation that may have an impact on user performance is the appropriate timing of the adaptation (Rothrock et.al., 2002). The question is when should adaptation take place relative to the occurrence of the adaptation trigger? On the one hand, adaptation should take place immediately to create a clear contingency between the trigger and the user interface change and provide an immediate solution to the changing context. On the other hand, a delay in the adaptation relative to the trigger may serve as an "early warning" that an adaptation is about to take place and the user can prepare for it and perhaps improve performance (Arbuthnott & Woodward, 2002). There is very little empirical research that addressed this question.

In this paper we asked whether user interface adaptation might be associated with a cost in user performance. In view of the inconclusive empirical evidence regarding the relative advantageous of the different adaptation strategies we explored the differences in the impact of adaptation strategy, a fully automatic adaptation as compared to a user-controlled adaptation. If the determining factor is the extent of user involvement and control over the adaptation, it was expected to find less performance cost with user-controlled adaptation. In addition, we examined whether delayed timing of the adaptation may provide the user with an early cue for the adaptation and thus possibly reduce adaptation costs.

2 Method

2.1 Participants

The study included 96 participants, divided into four experimental groups (detailed participant assignment is presented in the Procedure section). All participants were students of the Israel Institute of Technology, within the age range of 18 to 30 with normal or corrected vision and no known colour perception deficiencies. All were familiar with MS Widows and had experience working with a mouse and keyboard.

2.2 Experimental System

A simulation of a web-based help desk was used to address the research questions. There were three primary elements in the simulated system: the online continuous arrival of requests from hypothetical customers, the display of the requests, and the handling of those requests according to prescribed priorities and handling solutions. The simulation lasted 15 minutes and included varying system loads in terms of the amount of arriving requests, with periodic variations from high to low. The introduction of requests to the participant was done within cycles of 90-110 seconds. For a high system load, there was a rate of a request each 2-5 seconds within the cycle, and for a low system load, the rate was a request each 30-33 seconds. Rising or falling system load was used as the trigger for adaptation. Based on many pilot studies it was determined that the threshold criterion for adaptation was when the number of waiting requests in the queue was above or below three. The simulation always started with a low load.

Simulated requests were retrieved from a data base consisting of 180 pre-defined requests which were divided equally into six different topics. The six request topics were divided into three groups according to their handling, and consequently handling priority. The participant was required to search and select the request with highest handling priority. In addition, within a given handling priority, the participant was required to select and handle first the request that has been waiting the longest time.

Two user interfaces were designed (see figure 1): the high-load interface was a list-based screen displaying the queue of requests, and the low-load interface was a form-based screen displaying the request details and the handling options. The design of the displays in relation to the varying system loads is based on the findings of Parush (2004). The form displays included the following elements: a numerical counter of waiting requests at the top of the screen, an area with the details of the request, and three request handling buttons at the bottom of the screen. The list displays also included the numerical counter of waiting requests at the top of the screen, and a list of the requests waiting to be handled. The list was ordered by arrival time of the request into the system, with the most recent one displayed at the bottom of the list.

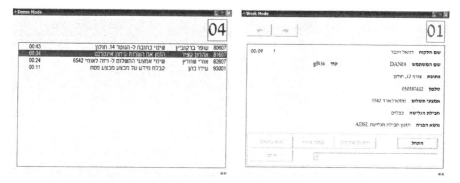

Figure 1: The two screens (in Hebrew) used in the study: The form-based interface (right) and the list-based interface (left)

2.3 Study Design

The study design was a between-participant factorial design with two variables. The first variable, adaptation strategy, was examined by having two between-participant conditions: 1. Fully automatic adaptation: adaptation was linked to system load, and the screen changed from form-based display to list-based display whenever the amount of arriving requests reached the criterion threshold, and changed back to form-based whenever the system load dropped to the criterion threshold; 2. User-controlled adaptation: adaptation was linked to varying system loads by displaying a recommendation to approve an adaptation; i.e., change of screen background colour and the display of a button to approve the adaptation recommendation. The user had the choice to approve or not approve the change. The second variable, adaptation timing, was examined by having two between-participant conditions: 1. Immediate:

adaptation occurred immediately when the system load reached the criterion threshold or immediately when the participant approved the adaptation. 2. Delayed: adaptation, whether automatic or after user approval, was delayed briefly before taking place. The specific delay was contingent upon having two more requests added to the queue before adaptation took place.

2.4 Procedure

Participants were assigned randomly to one of the four experimental conditions and were given different instructions for each of the conditions. There were 25 participants in the immediate fully automatic adaptation condition, 25 participants in the immediate user-controlled adaptation, 22 in the delayed fully automatic adaptation, and 24 in the delayed user-controlled adaptation condition. In all conditions participants received an explanation about the adaptation of the UI and the reasons for it. In order to increase participants' motivation to handle the requests according to the prescribed priorities there was a payoff scheme linking their correct performance and the amount of monetary remuneration for participating in the study.

3 Results

3.1 Executed Adaptations

In order to examine the differences between fully automatic adaptation and user-controlled adaptation, the number of adaptations that actually took place was examined. The results are presented in figure 2.

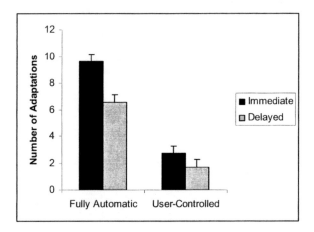

Figure 2: The number of executed adaptations for the four experimental conditions

It can be seen that there were more adaptations in the fully automatic conditions as compared to the user-controlled conditions. In general, there was an average of one to two adaptations in the user-controlled conditions, which reflects a tendency of the participants to reject the recommended adaptation. In addition, more adaptations occurred in the immediate fully automatic conditions as compared to the delayed fully automatic condition. Finally, there were no significant differences between the two user-controlled conditions, the immediate and delayed.

In order to closely examine the behaviour pattern of participants in the user-controlled adaptation conditions, the rate of participants rejecting or approving the recommended adaptation was computed. The findings are presented in figure 3. In general, it was found that participants in these conditions rejected any adaptation recommendation beyond two adaptations for the rest of the duration of the experimental session. Specifically, the majority of participants in the immediate user-controlled condition approved only one adaptation; i.e., from a form-based interface to a list-based interface and then remained with this interface for the rest of the experiment. Only 32% approved an additional adaptation, from the list-based back to the form based interface. Note that 8% of the participants rejected any recommendation and remained with the initial form-based interface for the entire duration of the experiment. A more striking pattern was found with participants in the delayed user-controlled condition. About 41% of the participant rejected all adaptation recommendations and remained with the initial form-based interface for the entire duration of the experiment. About 33% approved only the first adaptation, from the initial forma-based interface to the list-based interface for the remainder of the experiment, and finally, 25% approved the second adaptation recommendation from the list-based to the form-based interface and remained with it for the rest of the experiment. Overall, this pattern reflects that, given the choice, there was a strong tendency to reject most of the system's recommendations to execute an adaptation of the user interface.

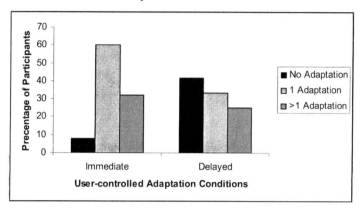

Figure 3: Rate of participants rejecting or approving adaptations in the user-controlled conditions

3.2   Adaptation Cost

One of the measures that were used to assess the possible cost of adaptation was the response time to a waiting request. This measure was calculated from the time the handling of a given request was completed to the time the next correct request was selected for handling. In order to assess the cost of adaptation, the differences between the response times before an adaptation and the respective response times after the adaptation were computed. Since adaptation did not take place in many of the trials with the user-controlled adaptation, the following analysis was performed only for the first adaptation with both strategies. The mean response time differences for the first adaptation are shown in figure 4.

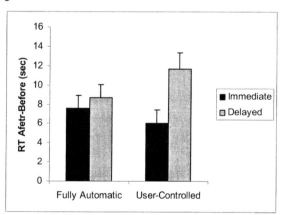

Figure 4: Mean response time differences, before and after the first adaptation, for the two fully automatic adaptation conditions, immediate and delayed.

In general, response times after the first adaptation, for all conditions, were significantly longer than the response times before the adaptation. Mean response time differences were significantly longer for the delayed adaptation conditions as compared with the immediate adaptation. A simple main effects test was performed to examine whether there were any differences between the fully automatic and the user controlled conditions with respect to the response time differences. It was found that within the fully automatic condition, there were no significant differences between the immediate and delayed conditions. However, for the user-controlled conditions, response time differences were significantly longer for the delayed conditions as compared with the immediate conditions.

The above analysis refers only to the first adaptation that took place. The question remains whether such cost of adaptation remained like this throughout the experiment. In order to assess the response time cost of adaptation as a function of time, these computations were made for all the adaptations, only for the fully automatic adaptation conditions. The mean response time differences for both immediate and delayed fully automatic adaptations are shown in figure 5. This figure shows only six adaptation points since this was the average number of adaptations that took place in the delayed fully automatic conditions.

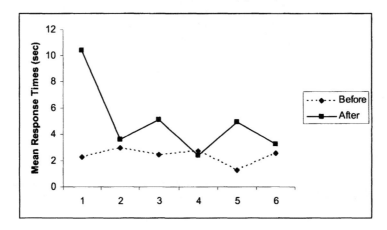

Figure 5: Mean response time differences for fully automatic adaptation conditions as a function of all adaptation points

The response time cost for the first adaptation was significantly longer than the response time cost for the rest of the adaptations that took place in the experiment. A closer examination of the response time costs of all the other adaptations reveals an interesting pattern: response time costs were lower when the adaptation occurred in the transition from a high system load to a low system load (points 2, 4, and 6 in figure 5) as compared to the higher time cost associated with the transitions from a low to a high load (points 1, 3, and 5 in figure 5).

Another indication for the decreasing cost of adaptation as a function of time was reflected in the mean number of requests waiting to be handled. This measure was computed for each minute in the experiment and is presented in figure 6. It can be seen that there were five high peaks and five low troughs in the number of waiting requests reflecting fluctuations in participants' workload associated with the corresponding transitions in system load. However, there was a significant decreasing trend in the number of waiting requests both for the high peaks and the low troughs.

Overall, the pattern of decreasing response times and number of waiting requests reflects a rather rapid decrease in the cost of the adaptation. The fluctuations in this trend were associated with the varying system loads.

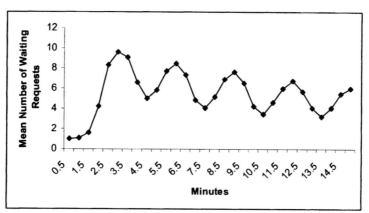

Figure 6: Mean number of waiting requests as a function of each minute in the experiment

4   Discussion

4.1   Summary and account for the findings

The findings of this study show that very few user-controlled adaptations took place. In most cases, no more than the first two changes were approved in contrast to an average of about nine automatic adaptations that took place throughout the experimental session. This pattern was particularly apparent with the delayed user-controlled adaptation with about 40% of the participants not approving even a single adaptation. In other words, most participants with the user-controlled adaptation strategy actually interacted with a fixed user interface despite the very dynamic context of increasing and decreasing workload and with the option to change the interface. Overall, this finding implies that, given the choice, users preferred to continue interacting with the same user interface rather than approve its change.

This pattern seems counter to the expectation that a user-controlled adaptation would be advantageous because the user is given some control and is not excluded from the adaptation decision. One possible explanation is that approving the user interface adaptation is perceived as an additional task (Debevc et.al., 1996). Participants may have felt pressure to simply handle requests as fast and as accurate as they could and did not want to bother with interface changes. Two aspects of choice behaviour can account for this trend. According to the status-quo bias theory (Samuelson & Zeckhauser, 1988), users will tend to remain in their present state rather than change it despite the fact that the change maybe more beneficial for them. The theory also predicts that once users have made the choice to remain in their present state they will tend to keep that choice even in subsequent choice situations. This is supported by the findings of this study where participants in the user-controlled adaptations did not approve more than two adaptations, and most of them approved less than that. The other aspect of choice behaviour is the tendency to adopt a strategy that optimizes present and local performance rather than explore and adopt another strategy that maybe difficult at first but can produce longer term, global benefits (e.g., Herrnstein, Loewenstein, Prelec, & Vaughan 1993; Yehiam, Erev, & Parush, 2004). In the present study, participants with the user-controlled adaptation may have perceived they can still cope and handle the increasing workload without the adaptation of the user interface. Consequently they chose to keep this interface and not "risk" changing it even though the change may have been more beneficial for them in the longer run.

Adaptation itself had a cost in terms of slowing user response after the adaptation. A close examination of the first adaptation in the experimental session showed that response time cost was significantly higher with the user-controlled adaptation strategy. In particular, the longest response time following adaptation was found with the delayed user-controlled adaptation. Overall, this finding implies that the change in user interface, whether fully automatic or user-controlled, may cause a slower response immediately following this change.

The cost in response time, while apparent throughout the entire experimental session, was reduced in the course of time. A dramatic drop in the response time cost after the first adaptation was found with the fully automatic adaptations. Another indication for a progressively drop in the adaptation cost was reflected in the decreasing number of requests waiting to be handled. It was shown that even for the high workload points, there was significant drop in the number of waiting requests, implying a drop in workload as the session progressed. These patterns imply that users may have learnt in a relatively short time the contingency between an increase or decrease in workload (manifested in the number of waiting requests) and the impending change in the user interface. By learning this contingency they were able to prepare for it, which, in turn, decreased the performance cost associated with the change. Taken together, the findings here suggest that user interface adaptation has a short term cost, but users may rapidly learn how to deal with it.

Finally, the magnitude of response time cost was associated with the direction of the workload change. A longer response time cost was found when adaptation took place upon an increase in workload. Conversely, a shorter response time cost was found when adaptation took place upon a decrease in workload. In other words, adaptation cost, and particularly its magnitude, is dependent upon the contextual aspects of the adaptation. Adaptation that takes place and workload is still perceived as high may surprise the user and perhaps cause some dissonance in the perception of the role of the adaptation. This in turn can magnify the performance degradation. In contrast, adaptation that takes place and workload is decreasing may not cause a surprise and a dissonance, and while performance degradation is still apparent, it is significantly smaller.

Delayed adaptation did not produce the hypothesized result; it was actually associated with poorer performance. The possibility that participants have learnt the contingency between workload and adaptation may imply that there was no need for an additional "early warning" or cueing that an adaptation is about to take place. Moreover, the "early warning" seems to have had an adverse impact because of the possible surprise and a dissonance in the understanding of the adaptation.

4.2 Practical implications and research agenda

The findings of this study bring some good news and some bad news regarding adaptive user interfaces. The good news are that users can rapidly learn to work with an adaptive user interface and drastically decrease the potential performance costs associated with the adaptation. The bad news are that there is a temporary performance cost and that users need a learning period to work with user interface adaptation. There are two main practical implications that need further investigations. Users of systems with adaptive user interfaces should be given a training period before actually using the system. This is particularly important in life-critical, mission-critical systems. In other words, users should use the "real system" only when they have achieved a performance level that have minimized or eliminated the potential adaptation costs. The other implication is the critical need for users to understand and experience the potential benefits of the adaptation. In view of the possibility that there is a need to influence user's choice of interaction strategy: making the user realize the longer-term implications of the adaptive interface and then introduce them into the choice situation. In order to achieve that, users may be given at the initial phase of working with the system a fully automatic adaptation system to help them learn the longer-term potential benefits of interactive with an adaptive interface. Once users are at the point of having a good understanding of the adaptation and its impacts, then user-controlled adaptation can be introduced. Such an approach introduces a hybrid adaptive user interface: a phase dependent mix between fully automatic and user-controlled adaptation.

Much research was directed at the development of effective user models to drive user interface adaptation. Little research addressed the issue of effective adaptation as a function of contextual changes. When the adaptation triggers are changes in the users themselves, the users are more often than not aware of these changes and are thus implicitly involved in the adaptation even with a fully automatic adaptation strategy. In contrast, when adaptation triggers are contextual changes, the user is usually less aware of them and consequently less involved and less expects the adaptation. The above discussion places emphasis on the user acquiring the understanding of the adaptation and its benefits. This emphasis suggests that much more research is thus needed to examine adaptive user interfaces, contextual factors, and users' awareness and understanding of these changes.

References

Akoumianakis, D., & Stephanidis, C. (2001). USE–IT: A tool for lexical design assistance. In C. Stephanidis (Ed.), *User interface for all: Concepts, methods and tools.* Mahwah, NJ: Lawrence Erlbaum associates.

Arbuthnott, K. D., & Woodward, T. S., (2002). The influence of cue task association and location on switch cost and alternation switch cost. *Canadian Journal of Experimental Psychology,* 56(1), 18-29.

Benyon, D. (1993). Adaptive system: a solution to usability problems. *User Modeling a User Adapted Interface,* 3(1), 65-87.

Benyon, D., Crerar, A. & Wilkinson, S. (2001). Individual Differences and inclusive design. In C. Stephanidis (Ed.), *User interface for all: Concepts, methods and tools.* Mahwah, NJ: Lawrence Erlbaum.

Debevc, M., Meyer, B., Donlagic, D. & Sveco, R. (1996). Design an evaluation of an adaptive icon toolbar. *User Modeling and User - Adaptive Interaction,* 6, 1-21.

Debevc, M., Meyer, B., & Sveco, R. (1997). An adaptive list of documents on the World Wide Web. *Proceedings of the 1997 International Conference on Intelligent User Interfaces,* Orlando, FL, pp. 209-211.

Gong, Q., & Salvendy, G. (1995). An approach to the design of a skill adaptive interface. *International Journal of Human- Computer Interaction,* 7(4), 365-383.

Granic, A., & Glavinic, V. (2000). Functionality specification for the adaptive user interface. *Proceedings of the IEEE,* 1, 123-126.

Greenberg, S., & Witten, I. H.(1985) . Adaptive personalized interface – A question of validity. *Behavior and Information Technology,* 4(1). 31-45.

Herrnstein, R. J., Loewenstein, G. F., Prelec, D. & Vaughan, W. (1993). Utility maximization and melioration: Internalities in individual choice. *Journal of Behavioral Decision Making,* 6, 149-185.

Hettinger, L. J., Cress, J. D., Brikman, B. J., & Hass M. W. (1996). Adaptive interface for advance airborne crew stations. *Proceedings of the IEEE 1996 virtual reality annual Symposium,* 188-192.

Hudlicka, E., & McNeese M. D. (2002). Assessment of user affective and belief states for interface adaptation: Application to an air force pilot task. *User Modeling and User Adapted Interaction,* 12, 1-47.

Kaber, D. B., & Riely, J. M. (1999). Adaptive automation of a dynamic control task based on secondary task workload measurement. *International Journal of Cognitive Ergonomics,* 3, 169-188.

Keeble, R.J., & Macredie, R. D. (2000). Assistant agents for the World Wide Web intelligent interface design challenges. *Interacting With Computers,* 12, 357-381.

Miah, T., Karageorgou, M. & Knott, R. P. (1997). Adaptive toolbars: An architectural overview. *Proceeding of 3rd ERCIM Workshop on User Interface for All,* 157-163.

Miah, T., & Alty, J. L. (2000). Vanishing windows- a technique for adaptive window management. *Interacting With Computers,* 12, 337-355.

Norcio, A. F., & Stanley, J. (1989). Adaptive human-computer interfaces: A literature survey and perspective. *IEEE Transaction on Systems, Man, and Cybernetics,* 19(2), 399-408.

Parush, A. (2004). An empirical evaluation of textual display configurations for supervisory tasks. *Behavior and Information Technology,* 23 (4), 225-235.

Rothrock, L., Koubek, R., Fuchs, F., Haas, M., & Salvendy, G. (2002). Review and reappraisal of adaptive interface: Toward biologically - Inspired paradigms. *Theoretical Issues in Ergonomics Science,* 3(1), 47-84.

Samuelson, W., & Zeckhauser, R. (1988). Status quo bias in decision making. *Journal of Risk and Uncertainty,* 1, 7–59.

Tattersall, A. J., & Morgan, C. A. (1997). The function and effectiveness of dynamic task allocation. In D. Harris (Ed.), *Engineering Psychology and Cognitive Ergonomics, Vol. 2: Job design and product design* (pp. 247-255). Burlington, TV: Ashgate Publication.

Thies, M. A. (1994). Adaptive user interface. *IFIP- transactions-A- computer- science and technology,* 52(A), 196-202.

Trevellyan, R., & Browne, D. P. (1987). A self - regulating adaptive system. *Proceedings of CHI + GI,* 103-107.

Yechiam, E. Erev, I. & Parush, A. (2004). Easy first steps and their implication to the use of a mouse-based and a script-based strategy. *Journal of Experimental Psychology: Applied,* 10 (2), 89-96.

Augmented Cognition: An Approach to Increasing Universal Benefit from Information Technology

Melody Y. Ivory, Andrew P. Martin, Rodrick Megraw, Beverly Slabosky

University of Washington
Box 352840, Seattle, WA 98195-2840, USA
[myivory, am1982, remegraw, beverlys]@u.washington.edu

Abstract

Complex information technology (IT) is available to a wide audience; however, it is not designed typically to serve the needs of users from diverse populations. As part of the Universal Benefit from Information Technology Research Program, we have examined the design of IT in several areas. Our aim is to identify and mitigate cognitive breakdowns that can occur in the design or the use of IT. We provide an overview of our efforts to augment cognition in the areas of web site design, user assistance, and graphical image translation. For each project, we describe our approach to augmenting cognition, the project's current status, and findings that are relevant to other researchers who are interested in these areas. We think that a focus on resolving cognition breakdowns, as well as usability and accessibility issues, can produce solutions that enhance cognition and benefits from IT.

1   Introduction

The rapid expansion of the Internet, the ubiquity of home and office computers, and the proliferation of mobile computing devices (i.e., small, portable computers) in the last decade has made complex information technology (IT) available to a wide audience. Despite its prevalence, IT has not served the needs of users from diverse populations (e.g., people who have visual or learning impairments) in an equitable and effective manner (Shneiderman, 2000; Stephanidis, Akoumianakis, Sfyrakis, & Paramythis, 1998). Typically, underperforming IT, such as a web site or software application, is attributed to usability or accessibility issues; thus, improvement efforts aim to make it easier for users to use technology. We assert that, in many cases, we can also attribute ineffective IT to breakdowns in cognition (i.e., comprehension, perception, etc.) on the designers' or developers' end (e.g., not understanding how to produce beneficial web sites), the users' end (e.g., not understanding how to complete a task on a web site), or both. By not recognizing that IT deficiencies can result from cognition breakdowns, we miss opportunities to enhance cognition and possibly benefits from IT. Augmented cognition, the identification and mitigation of specific bottlenecks, is one way to address breakdowns that can occur in the design or use of IT.

In this paper, we provide an overview of research efforts that we are undertaking as part of the Universal Benefit from Information Technology (u b i t) Research Program at the University of Washington. The u b i t program aims to enable all people, regardless of their abilities, backgrounds, or experiences, to benefit from IT in all areas of their lives. To achieve this aim, currently, we are exploring ways to augment cognition in the areas of web site design, user assistance (i.e., help systems), and graphical image translation. We are conducting empirical studies and developing solutions to assist: (1) web site designers with developing and evaluating sites, (2) users with locating information to resolve IT problems, and (3) tactile graphics specialists with creating tactile forms of graphical images (i.e., representations of bar charts, line graphs, and illustrations that can be touched and read by people who are blind). We discuss these three efforts in this paper. For each project, we describe our approach to augmenting cognition, the project's current status, and findings that are relevant to other researchers who are interested in these areas.

2   Web Site Design and Evaluation

The Web enables broad dissemination of information and services; however, many web sites are inadequate with respect to usability or accessibility (Forrester Research, 1999; Jackson-Sanborn, Odess-Harnish, & Warren, 2002). This problem arises due to the site's designer being unfamiliar with effective design practices or inexperienced with producing effective sites. In our examination of the work practices of 169 web practitioners, we discovered that 96 percent of them always or sometimes used guidelines to inform their work during the site's design or evaluation process (Ivory, 2003). We conducted four experiments to understand the problems that designers face when they attempt to follow prescriptive guidance. Results of the first three studies demonstrated that novice and professional designers find it difficult to apply guidelines during both the design and evaluation processes and that training is required to make them accessible, especially to novice designers (Chevalier & Ivory, 2003b; Ivory, 2003). The fourth study examined the effect of a short training program in applying usability criteria and recommendations on the evaluation and improvement of a web site (Chevalier & Ivory, 2003a). Although training novice designers to apply guidelines improved their ability to do so, it did not help them to produce sites that did not have problems.

The four studies revealed two key cognitive breakdowns: (1) understanding and applying guidelines to the design of a site and (2) understanding and applying guidelines to the evaluation of a site. To address the design bottleneck, we are developing a tool to generate effective web sites, specifically for classroom based courses (as opposed to distance based courses) that are offered within university settings (Heines, Ivory, Borner, & Gehringer, 2003). To address the evaluation bottleneck, we developed an automated evaluation methodology to help designers to conform to effective design patterns (Ivory & Hearst, 2002). We discuss these two efforts in the remainder of this section.

2.1 Automated Web Site Evaluation

We developed an automated evaluation methodology and software prototype to assist designers with determining whether or not their designs employ practices that are used on effective web sites. As the assessment basis, we used 157 highly accurate, quantitative page- and site-level measures; we identified the measures from an extensive survey of the web design literature and relevant empirical studies. The measures assessed many aspects of web interfaces, including the amount of text on a page, color usage, and consistency. We used these measures (computed for thousands of interfaces) along with ratings from Internet professionals (writers, designers, editors, etc.) to build statistical models or profiles of high quality (usable, accessible, etc.) web interfaces. In essence, the models predicted whether an evaluated page or site is consistent with interfaces that were rated highly, moderately, or poorly by the Internet professionals.

We deployed the models via two applications: (1) a web-based service that enables designers to submit an automated evaluation request for their site and (2) a desktop application that enables designers to explore the results of their site's automated evaluation. Figure 1 depicts the application for viewing evaluation results; the application shows results from various models and rationales for predictions. Rationales can help designers to improve their understanding of how to apply design guidance and help to mitigate cognitive breakdowns that occur during the design and evaluation processes. A small user study demonstrated that novice designers could use the prototype to modify designs and that users preferred pages and sites modified based on the statistical models over the original versions (Ivory, 2003); differences were significant in both cases.

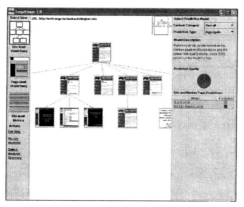

Figure 1: Application for Viewing the Results of an Automated Web Site Evaluation

807

There are several novel aspects to this methodology: (1) the use of quantitative measures, (2) the use of empirical data to develop guidelines, (3) the use of profiles as a comparison basis, and (4) the ability to build models that are context-sensitive (i.e., based on the content category of the site or the functional type or size of a page). The context-sensitive aspect is critical, because it enables us to adapt assessments to the context in which interfaces are used.

2.2 Automated Web Site Generation

After extensive work on automated evaluation, we shifted our focus to the web site design process. To narrow the scope of our work, we focused on the design of sites that instructors use as resources for their classroom based courses. First, we developed a systematic heuristic evaluation methodology (i.e., an expert reviews a site to see if it meets predetermined criteria) (Nielsen, 1994), relevant heuristics, and a toolkit to facilitate evaluations (Ivory, 2004). We then applied the approach to 79 sites that we selected randomly for courses offered at the University of Washington. We found that most sites were of average quality. Sites did well on aspects like visual layout, usability, and performance, but poorly on aspects like accessibility, content organization, and navigation structure. We found that the development environment that an instructor used to create a site has inherent problems (usability or accessibility). Furthermore, even with development tools, instructors needed to expend considerable effort and possess some design and technical expertise to produce sites.

To mitigate the cognitive burden that is inherent in creating effective course sites, we are developing an intelligent site generation system. Our vision is that the instructor will use a simple text based mechanism, which does not require HTML authoring expertise, to maintain the site's content. Unlike existing content management, portal, or site generation systems, our envisioned site generator would use machine learning algorithms to automatically map the text based content into an effective page layout and integrate it into the site. Such functionality would also support the automated transformation of documents (PDF, Word, etc.) into web pages, such that the pages are consistent with the rest of the site and do not require the use of plug-ins. The system would be responsible for all aspects of a site's design (e.g., navigation, layout, consistency, etc.), except for content creation; the instructor would be responsible solely for creating content for a site. The system will also provide the instructor with guidance on content organization or presentation; the feedback mechanism will build upon our earlier automated evaluation approach. Finally, the system will enable the instructor to safely control some design aspects, such as color schemes, font sets, headers, logos, etc., so that a site conforms to effective design patterns, despite these changes. Automated layout and feedback mechanisms can help instructors to improve their understanding of how to apply design guidance and help to mitigate cognitive breakdowns that occur during the design process.

As a first step toward realizing this vision, we created a prototype site generator and have used it to provide web sites for undergraduate- and graduate-level courses. Figure 2 depicts a site created with the generator prototype. The current prototype does not support text based content entry, perform automated content mapping to layouts, or provide feedback to instructors. We developed it mainly to explore ways in which to conform to the heuristics that we used to evaluate sites. For instance, the prototype organizes content in a logical manner, maintains consistent navigation, displays resources like readings in a contextual manner, enhances accessibility, and helps students to stay informed of classroom activities. Informal feedback from students has been promising.

We are currently completing the first phase of this study in which instructors and students complete online questionnaires about their experiences with course web sites. During the second study phase, we will conduct focus groups to gather more insight into course web site experiences and formal feedback on the prototype generator.

3   User Assistance

Users need help systems (e.g., manuals, online content, and wizards) to support their use of complex IT, but often they find these systems difficult to use. We conducted a study to assess the current state of help systems and to identify ways in which to improve them (Martin, Ivory, Megraw, & Slabosky, 2005). We administered an online questionnaire to a diverse population of 107 IT users. The 45-item questionnaire probed users' perceptions and use of help systems that are within PC applications, web sites, and mobile computing devices. Unlike prior studies, our study examined help efficacy in the broadest sense (i.e., across applications, computing environments, and user populations). We identified three groups of users who differed based on their preferred help-seeking strategies and satisfaction with help. We showed that previously documented help system issues, such as incomprehensible

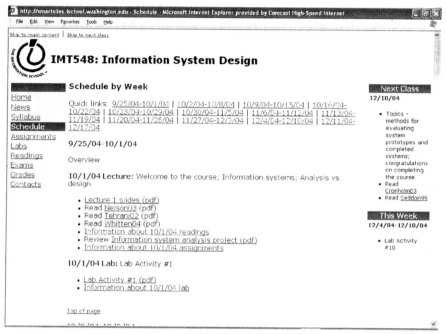

Figure 2: Screen Snapshot of a Site Created with the Site Generator Prototype

terminology and inadequate content, still exist. Furthermore, these problems seemed to be more pronounced in newer technology like web sites and mobile computing devices than in PC applications. Due to inadequate language, content, and search functionality, over two-thirds of users seemed to turn to the Web to resolve IT problems.

The study revealed two key cognitive breakdowns: (1) understanding help content and (2) understanding the right type of help resource to use. To address the first bottleneck, we are developing a help portal system to aggregate various help resources that are available on the Web and organize them in an intelligent manner so that users can have access to multiple perspectives on the same problems. This solution will also streamline help seeking via search engines. To address the second bottleneck, we propose the development of a help matching system to connect users with the right type of help resource in an intelligent and unobtrusive manner. We discuss these two efforts in the remainder of this section.

3.1 Help Portal System

Our study showed that users are increasingly turning to the Web (search engines in particular), when they need assistance with IT problems. Simply put, the help content that users find most useful and understandable does not always exist in a structured help system. We borrow the term "annotation" to indicate a link from a structured help source to an unstructured help source (e.g., discussion group posting, knowledge base article, or FAQ) which contains useful and relevant content (Marshall, 1998). We wish to take the content from structured help systems and annotate it with links to useful, unstructured help content (Figure 3). By doing so, users' information retrieval processes can be streamlined and simplified. As annotations become part of the structured help system itself, a user can query and locate this information just as they would locate structured help content. Annotations may also implicitly bridge the jargon barrier by automatically associating structured, jargon-heavy help content with unstructured help content that is expressed with everyday language and mitigate cognitive breakdowns that occur during the help-seeking process.

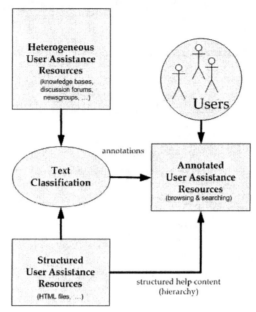

Figure 3: Architecture of the Proposed Help Portal System

We are developing text corpora to examine the use of existing or the development of novel text classification algorithms to annotate structured help with unstructured content. Our corpora include structured and unstructured help content for three email applications that are used within three different computing domains (PCs, Web, and mobile computing devices); it consists of help content taken from Web based (newsgroups, vendors' knowledge bases, and discussion forums) and Windows help sources. Once we complete the corpus, we will conduct text classification experiments. By comparing results across structured help systems and computing domains, we intend to determine how the volume and hierarchical nature of structured help content affects the success of our text classification techniques. Another aim is to use the experiments to possibly provide guidance to help content authors and architects who may wish to support automated retrieval of their unstructured help content.

We view the annotation system as part of a Web based user assistance portal. To explore our portal concept, we constructed a prototype to demonstrate its interface and to describe its envisioned functioning (Figure 4). We are soliciting feedback from help users. Preliminary, informal feedback confirms that users prefer a help system which integrates unstructured sources to one which does not. We will continue to develop the portal prototype and evaluate it during our future focus groups with help users, help developers, and technical support specialists.

3.2 Help Matching System

Our study revealed that users expect help to follow a continuum based on their experience levels (i.e., novice versus expert) or help-seeking preferences (e.g., use tutorials or search). Users suggested the need to support three levels of proficiency (beginner, intermediate, and advanced) by: (1) offering different types of help resources and (2) organizing help content by proficiency level. Even if this continuum is supported, connecting users with the right type of help resource in an intelligent and unobtrusive manner is a major research problem. In future work, we will conduct additional studies to determine appropriate help resources for a wide variety of situations and use results as input to machine learning techniques so that we can build a model to predict useful help resources for a specific context. We will also investigate ways in which to deploy the model. A wizard, which users can invoke as needed, may be a desirable solution; the wizard would ask users a few questions and then recommend the appropriate resource for the context. The help matching system can help to mitigate cognitive breakdowns that occur during the help-seeking process, in particular for less experienced users.

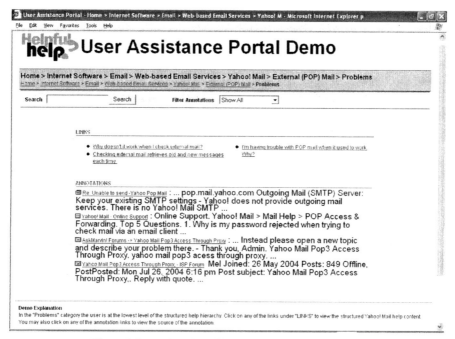

Figure 4: Screen Snapshot of the Proposed Help Portal System

4 Graphical Image Translation

Historically, blind students have been denied access to important graphical materials which their sighted counterparts can access. It is easier and less costly to simply leave images out of a textbook than to have a tactile version (i.e., raised image that is designed to be read by the fingers) created by a tactile graphics specialist. During the manual translation of a graphical image into a tactile form, a tactile graphics specialist, by hand, makes a line drawing or sketch of the graphical image, either on paper or with a drawing package like Visio or Corel Draw. There is an editing program, the BrlGraphEditor (Batusic & Urban, 2002), which supports drawing 2D graphics and printing them on certain Braille printers; however, there is no automated processing of graphics. When completed, these handmade drawings can be used as a master for producing tactile graphics.

Through the use of an on-line questionnaire (administered to 51 people) and 16 in-depth observation sessions, we examined how tactile graphics specialists produce tactile graphics for blind students and how they use software applications during the production process. We found that tactile graphics specialists shy away from using applications, even though their production methods are labor and time intensive. Half of the participants did not use a computer at all and those who did, relied upon simple drawing functionality, such as Microsoft Word's drawing tools, as opposed to fully featured drawing systems like Adobe Photoshop or Illustrator. Participants considered fully featured drawing systems to be cumbersome, non-intuitive, and overkill for their needs. Nonetheless, these tools have valuable image processing algorithms, which could help them to streamline their work practices. For instance, image-editing programs can make automated transformations to simplify the image, such as detecting edges, filtering out noise, thickening lines, or removing unimportant elements.

The study revealed a key cognitive breakdown—understanding how to carry out translation tasks in fully featured drawing software. To address this bottleneck, we are developing the Tactile Graphics Assistant to assist tactile graphics specialists with translating graphical images. We discuss this effort in the remainder of this section.

811

4.1   Tactile Graphics Assistant

We are developing software to support the work practices of tactile graphics specialists. The software will be a plug-in for fully featured drawing packages like Adobe Photoshop or Illustrator. Figure 5 depicts the automated, step-by-step process for translating a graphical image into a tactile graphic. Once the tactile graphics specialist loads a digital image, the software will identify the type of the image (e.g., bar chart, line graph, or illustration). We need to classify images so that similar images can be tactilized using the same image processing steps. Example image processing steps include: segmentation into text and non-textual elements, edge detection, color reduction, resolution reduction, label placement, and so on. After applying the appropriate image processing techniques, the software will assemble graphical and textual image components into an effective layout. Determining an ideal layout could be an iterative process between producing a layout and evaluating it in an automated manner. Tactile graphics specialist can use the final layout as a master for producing a tactile form.

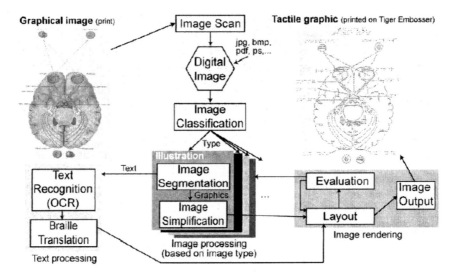

Figure 5: Process for Converting a Graphical Image into a Tactile Graphic

We have used machine learning techniques to develop a highly accurate model for classifying images into three classes: bar charts, line graphs, and diagrams. We have also developed image processing algorithms which can extract text from the image and convert it to ASCII, remove colors, and rescale images. In future work, we will explore the use of machine learning techniques to identify the appropriate image processing steps to apply to an image based on its predicted type. The envisioned automated translation system can help to mitigate cognitive breakdowns that occur during the use of image-editing software to create tactile graphics.

5   Conclusions

During the use or design of IT, there can be cognitive breakdowns on the designers' end, the users' end, or both. We have identified cognitive breakdowns that can occur within three areas: web site design, user assistance, and graphical image translation.

- In web site design, designers have difficulty with understanding and applying guidelines to the design or evaluation of a site. We have developed automated evaluation and web site generation tools to help designers to mitigate these problems.

- In user assistance, users have difficulty with understanding help content or identifying the right type of help resource to use. We are developing a help portal system and a help matching system to help users to mitigate these problems.
- In graphical image translation, tactile graphics specialists have difficulty with understanding how to carry out translation tasks in fully featured drawing software. We are developing an automated translation system to help tactile graphics specialists to mitigate this problem.

By focusing on cognitive breakdowns in addition to usability and accessibility issues, we think that our efforts will produce solutions that enhance cognition and benefits from IT.

Acknowledgments

This research was funded by grants from the National Science Foundation (IIS-0414385l, IIS-0415273, and IIS-0335143) and supplementary funds from the University of Washington's Information School and Equal Opportunity Office. The image processing and classification work was completed by other members of the Tactile Graphics Project: Professors Richard Ladner and Rajesh Rao, Sangyun Hahn, Matt Renzelmann, Satria Krisnandi, Mahalakshmi Ramasamy, Jacob Christensen, Andy Jaya, Terri Moore, Eileen Hash, and Jack Hebert. We thank the participants who completed our various studies. We also thank the anonymous reviewers.

References

Batusic, M., & Urban, F. (2002). *Preparing tactile graphics for traditional braille printers with brlgrapheditor.* Paper presented at the Computers Helping People with Special Needs 8th International Conference, ICCHP 2002, Linz, Austria.

Chevalier, A., & Ivory, M. Y. (2003a, June 22-27). *Can novice designers apply usability criteria and recommendations to make web sites easier to use?* Paper presented at the Proceedings of the 10th International Conference on Human-Computer Interaction, Crete, Greece.

Chevalier, A., & Ivory, M. Y. (2003b). Web site designs: Influences of designer's experience and design constraints. *International Journal of Human-Computer Studies, 58*(1), 57-87.

Forrester Research. (1999). *Why most web sites fail.* Unpublished manuscript.

Heines, J., Ivory, M. Y., Borner, K., & Gehringer, E. F. (2003, February 19-23). *Panel on the development, maintenance, and use of course web sites.* Paper presented at the Proceedings of the Technical Symposium on Computer Science Education (SIGCSE), Reno, NV.

Ivory, M. Y. (2003). *Automated web site evaluation: Researchers' and practitioners' perspectives* (Vol. 4). Dordrecht, The Netherlands: Kluwer Academic Publishers.

Ivory, M. Y. (2004). *SmartSites toolkit for evaluating course web sites* (Technical Report No. IS-TR-2004-12-02). Seattle, WA: Information School, University of Washington.

Ivory, M. Y., & Hearst, M. A. (2002). Improving web site design. *IEEE Internet Computing, 6*(2), 56-63.

Jackson-Sanborn, E., Odess-Harnish, K., & Warren, N. (2002). *Website accessibility: A study of ADA compliance* (Technical Report No. TR-2001-05): University of North Carolina - Chapel Hill, School of Information and Library Science.

Marshall, C. C. (1998). *Toward an ecology of hypertext annotation.* Palo Alto, CA: Xerox Palo Alto Research Center.

Martin, A. P., Ivory, M. Y., Megraw, R., & Slabosky, B. (2005, July 22-27). *How helpful is help? Use of and satisfaction with user assistance.* Paper presented at the 3rd International Conference on Universal Access in Human-Computer Interaction, Las Vegas, NV.

Nielsen, J. (1994). Heuristic evaluation. In J. Nielsen & R. L. Mack (Eds.), *Usability inspection methods* (pp. 25-62). New York: Wiley.

Shneiderman, B. (2000). Universal usability. *Communications of the ACM, 43*(5), 84--91.

Stephanidis, C., Akoumianakis, D., Sfyrakis, M., & Paramythis, A. (1998, October). *Universal accessibility in HCI: Process-oriented design guidelines and tool requirements.* Paper presented at the Proceedings of the 4th ERCIM Workshop on User Interfaces for All, Stockholm, Sweden.

Improving Recovery from Multi-task Interruptions Using an Intelligent Change Awareness Tool

Harvey S. Smallman & Mark St. John
Pacific Science & Engineering Group
9180 Brown Deer Road
San Diego, CA 92121
smallman@pacific-science.com
stjohn@pacific-science.com

Abstract

Detecting changes to complex monitoring tasks is important for situation awareness, yet it is surprisingly difficult. Interruptions due to multi-tasking or other distractions exacerbate this problem. Current display technologies do not provide much support for detecting and identifying significant situation changes and therefore do not provide much support for the recovery of situation awareness following interruptions. Instead, situation displays typically only represent the current situation, which forces users to rely on their own ability to extract changes by cognitively integrating events over time. Sustained situation awareness can be greatly improved by augmenting users' abilities with automated situation change detection. However, the design of the presentation of change information turns out to be crucial and hinges on the issue of how to alert and inform the user effectively without distracting from other important on-going tasks. Here, we review two empirical studies that compare different design approaches to this challenge including visual alerts and an intuitive Instant Replay tool. The experiments use a naval warfare task in which users monitor a busy airspace. The results, however, strongly favor a new set of display concepts that we developed called CHEX (Change History Explicit). CHEX augments the human attentional system with a set of intelligent change detectors whose output is logged in a re-configurable table format that is linked back to the situation display. CHEX is extremely effective both for maintaining situation awareness when monitoring a situation as well as when recovering situation awareness following an interruption.

1 Introduction

A wide range of operations tasks, from airspace management to industrial plant control and disaster relief, involve monitoring a situation display over time and addressing issues that arise. This monitoring process involves detecting significant changes and then responding to them in a timely manner. Current situation display technologies often fail to support change detection and change awareness by representing only the current state of the evolving situation. By representing only the current state, users are required to rely on their own abilities to extract changes by remembering and integrating states of the situation across time.

A burgeoning cognitive science literature is documenting the difficulty of unassisted change detection. Humans are unable to spot significant changes to simple scenes. This poor performance obtains both when users are momentarily distracted by glancing away from the display, and even when they are actively looking at the display, but are attending elsewhere. This change blindness is surprisingly severe and has been documented in many contexts (see Rensink, 2002 for a review), including air warfare (DiVita, Obermayer, Nugent, & Linville, 2004; Smallman & St. John, 2003) and is only just being recognized as a significant problem for human computer interaction (Varakin et al., 2004).

Longer distractions, due to multi-tasking for example, can exacerbate change blindness because memory for prior situation states will decay and the situation will evolve over the course of the distraction, making integration and change detection more difficult. Our contention is that recovering from interruptions and detecting significant situation changes requires computer-assisted augmentation because unassisted recovery and change detection is so poor.

Designing the augmentation involves addressing at least two separate issues. First is the issue of detecting significant changes to information on the display, such as changes in aircraft kinematics. The second issue is the re-presentation of those changes to users in a useful, effective fashion. We do not address the detection issue in this paper, although one straight-forward method would be to develop change detection agents, or "sentinels," that would monitor the displayed information. The criteria for defining an information change that was important enough to bring to a user's attention could be set through consideration of the task domain and through user interviews (for example, "I only care when aircraft altitude drops below 10,000 feet"). More sophisticated criteria could be based on changes in the assessed threat levels of aircraft (see St. John, Smallman, & Manes, 2005 for an example). Some difficulties arise concerning the rate of changes and sizes of changes that should elicit an alert. The human attentional system, of course, possesses transient detectors in abundance for detecting changes to a scene. Rather than define changes based on task analysis, artificial change sentinels could imitate the well-characterized spatio-temporal filtering properties of the Magno-cellular retino-cortical pathways that are thought to subserve visual transient detection in the brain (e.g., Shapley, 1990). It is intriguing to reflect on the findings of the Change Blindness literature which show that the central bottlenecks in human attentional processing lead to the loss of so much of the signals of these transient detectors. In this sense, the goal of our research has been to augment the attentional system by usefully maintaining signals from artificial transient detectors and re-presenting their signals for later exploitation by the user.

The second issue, how to re-present the change information, is the focus here. We review two experiment that first document poor change detection and interruption recovery using conventional displays and then go on to show that augmenting the user's natural abilities with automatic change detection can lead to significantly better performance (Smallman & St. John, 2003; St. John, Smallman & Manes, 2005). In a simplified version of naval air warfare, undergraduate participants monitored a busy airspace to detect significant changes to aircraft such as course, speed, and electronic emissions. Response times and percent correct detections were measured. Augmented change detection, however, is not sufficient for good performance. The design of the re-presentation method is crucial. A poorly designed augmentation may provide no value or even degrade performance below baseline.

2 Experiment One

In the first experiment, a baseline display that showed only the current situation was compared with three methods for automatically detecting and displaying aircraft change information (see Smallman & St. John, 2003, for details). In the first alternative, changes were logged into a static text table next of the map display. The chronologically sorted table listed the time and nature of each significant change. The second alternative also included the table, but it added red "circle alerts" around each changed aircraft on the map. One circle for each change listed in the table. The third alternative also included the table, but the table was linked to the map so that selecting a change entry in the table would highlight the changed aircraft on the map, and vice versa. This table could also be sorted by type of change and type by aircraft as well as chronologically, at the discretion of the user in order to facilitate different tasks. This interactive, linked table of automatically detected changes was called the CHEX tool (Change History EXplicit) because it explicitly and automatically identified changes to the situation rather than leaving users to their own abilities. Aircraft moved around the display at realistic rates over time, occasionally changing their behavior in ways that could be threatening to "own ship", the blue dot at the of the display in Figure 1. The participants' task was to detect and identify the most threatening changes as quickly as possible. They had to report those changes both while monitoring the display and when returning after a minute processing a secondary, mental arithmetic task. Aircraft density was also varied between subjects as high (40 aircraft) and low (13 aircraft).

Our hypothesis was that the CHEX table would 1) facilitate change detection and interruption recovery relative to the baseline display, and 2) it would do so more effectively than either the chronological table or the alert circles. We predicted that the problem with alert circle was that they could quickly clutter the map with distracting and relatively uninformative alerts (all alerts look the same). The CHEX tool, on the other hand, provided a less distracting solution that did not clutter the map, but was effectively linked to it, and that provided more descriptive and better organized information.

Participants in the baseline condition, who used comparable display tools to those available to operators in Navy Combat Information Centers, correctly identified only 34% of critical aircraft changes. Interestingly, these same participants exhibited an over-confidence in their ability to do the task since their confidence ratings dropped a full

26% after having performed the task. This overconfidence in unaided change detection replicates and extends the meta-cognitive underestimation of change detection ability found by Levin et al. (2000).

Augmenting the display by adding an explicit log of automatically detected changes improved performance. However, the static, separated table provided no alerts, and it left to the user the difficult problems of aggregating changes and correlating information in the table to the appropriate symbols on the map. The addition of change alert circles to the map did not further improve performance.

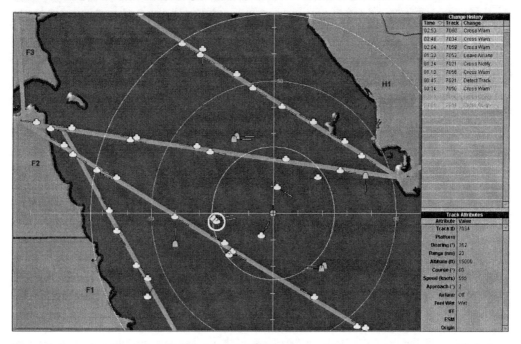

Figure 1. Screenshot of map and CHEX table (upper right) from experiment two (experiment one display was similar). Each row of the table describes one significant change to an aircraft (time, aircraft identification number, and change type). Selecting a row causes the aircraft to be selected on the map and its detailed kinematics information to be displayed in the Track Attribute list (lower right).

CHEX did away with the need for alerts on the map by dynamically linking the output of the (now flexibly sortable) Change History Table. Critical aircraft could be quickly found in the Table, selecting one automatically highlighted each of its changes, and the automatic linking to the map removed the need to search the map to find the location of the relevant aircraft. In the dense display condition, these benefits resulted in an 80% improvement in change identification speed compared with the baseline condition and a 40% improvement in speed over alert circles. Further, whereas participants lowered their confidence in performing the task after using the baseline tools, participants raised their confidence after using the CHEX tools. These benefits were found both for monitoring and recovering changes and across display densities, although there was an interaction between density and tool-type, with CHEX providing impressively density-independent support for change awareness.

Experiment one demonstrated that this serious deficit in change detection and identification can be dramatically improved by augmenting the display with intelligent change awareness tools, but the degree of improvement differed depending on the specific design of the tools.

3 Experiment Two

The first experiment demonstrated that one apparently sensible alternative to CHEX, alert circles, turned out to provide substantially inferior support, especially for longer interruptions. Another intuitively appealing alternative to CHEX is Instant Replay. With Instant Replay, users could replay interrupted periods at high speed to quickly perceive changes. Instant Replay's appeal seems to rest on its familiarity and realistic re-presentation of the temporal sequence of the interrupted situation. We have been developing a theory about user preference for realistic displays and HCI that is maintained in the face of poor performance called *Naïve Realism* (Smallman & St. John, 2005). Naïve Realism predicts that the preference for temporal realism in the Instant Replay tool is misguided.

In a naval air warfare task similar to the one used in Experiment one, we compared two versions of replay against three alternative tools, including CHEX and a baseline condition similar to the baseline used in experiment one (See St. John, Smallman, & Manes, 2005a, for details). Following interruptions of variable length, participants could replay the interrupted period of the simulated task. The map "re-wound" to the beginning of the interruption and then played forward at 20x speed until it caught up with current time. Participants could initiate a replay whenever they desired. The results again supported CHEX over the alternative designs.

In its basic form, Instant Replay simply replays the interrupted sequence and changes remain implicit. With this form of Instant Replay, changes remain difficult to detect, as difficult, in fact, as during real-time monitoring. In both the baseline condition and in the basic Instant Replay conditions, only 60% of the significant changes were detected and reported by participants.

Augmenting the display by adding explicit detection again helped significantly. However, providing the explicit change detection information within the context of a replay tool proved substantially inferior to providing that information in the form of the CHEX tool. Change detection approached 80% using the augmented Instant Replay tool, but participants achieved 100% detection using CHEX. In the augmented replay condition, changes to aircraft were marked on the map as they occurred during the replay sequence. They were removed when the map returned to real-time speed at the end of the replay sequence in order to avoid cluttering the map during real-time monitoring. Again participants could initiate a replay whenever they chose. In the CHEX condition, changes were again logged in an interactive table that was linked to the map.

Relying solely on the Baseline display to detect changes is insupportable, in spite of the commonly held belief that changes are easy to detect. Rather, the Baseline display produced high miss rates, high error rates, and generally slow response times. Replay offered little support for detecting or identifying changes over the Baseline condition. In fact, the Basic Replay tool may even be worse than nothing since it can add a delay as it replays the temporal sequence. The CHEX tool does not offer the same level of realism as the Instant Replay tools, but it offers superior functionality: it provides easily accessible information for both the detection and identification of changes and without cluttering the map.

4 Conclusions

Change detection is difficult enough that augmenting human cognition with automatic detection of significant events is warranted, in spite of the potential reliability and trust issues that accompany automation (but see St. John, Smallman, & Manes, 2005b; St. John, Smallman, Manes, Feher, & Morrison, in press). The method for presenting change detection information, however, can have profound implications for the ultimate success of augmented tool. Furthermore, intuitions about the effectiveness of simple alerts on a map or Instant Replay turn out, on close inspection, to be misguided. The explicit representation of situation changes within an interactive table display proved to be the best design.

5 References

DiVita, J., Obermayer, R., Nugent, W., & Linville, J. M. (2004). Verification of the change blindness phenomenon while managing critical events on a combat information display. *Human Factors, 46,* 205-218.

Levin, D.T., Momen, N., Drivdahl, S.B., & Simons, D.J. (2000) The metacognitive error of overestimating change-detection ability. *Visual Cognition, 7,* 397-412.

Rensink, R. A. (2002). Change detection. *Annual Review of Psychology, 53,* 245-277.

Shapley, R. (1990). Visual sensitivity and parallel retinocortical channels. *Annual Review of Psychology, 41,* 635-658.

Smallman, H. S. & St. John, M. (2003). CHEX (Change History EXplicit): New HCI concepts for change awareness. In *Proceedings of the 46th Annual Meeting of the Human Factors and Ergonomics Society* (pp. 528-532). Santa Monica, CA: Human Factors and Ergonomics Society.

Smallman, H. S. & St. John, M. (2005). Naïve Realism: Misplaced faith in the utility of realistic displays. *Ergonomics in Design,* submitted.

St. John, M., Smallman, H. S., & Manes, D. I. (2005a). Recovery from interruptions to a dynamic monitoring task: The beguiling utility of instant replay. Paper submitted for presentation to the *49th Annual Meeting of the Human Factors and Ergonomics Society,* Orlando, FL, Sept 26-30, 2005.

St. John, M., Smallman, H. S., & Manes, D. I. (2005b). Assisted focus: Heuristic automation for guiding users' attention toward critical information. This meeting.

St. John, M., Smallman, H. S., Manes, D. I., Feher, B. A., & Morrison, J. G. (2005). Heuristic automation for decluttering tactical displays. *Human Factors,* in press.

Varakin, D. A., Levin, D. T., & Fidler, R. (2004). Unseen and unaware: implications of recent research on failures of visual awareness for human-computer interface design. *Human-Computer Interaction, 19,* 389-422.

Augmenting Knowledge Flow and Comprehension in Command and Control (C²) Environments

Ronald A. Moore

Pacific Science & Engineering Group
9180 Brown Deer Road, San Diego, CA 92121
ramoore@pacific-science.com

Abstract

The effective, efficient, and timely exchange of information in military command and control (C²) environments is vital to military operations in the 21st century. The amount and complexity of information that is made available to C² decision makers and their support staffs by modern information technology presents unprecedented opportunities and challenges. However, despite decades of research and development in the domains of military C² and decision making, three major human-related problems have consistently prevented true innovation. These problems are 1) information overload, 2) misplaced emphasis on information analysis and conscious thought, and 3) inappropriate focus on individuals/small teams versus larger groups within an environment.

This paper describes an innovative approach to improving information processing and augmented communication which we refer to as *Intelligent Aided Communication (iaC)*. The iaC concept holds much promise for meeting the current C² challenges and affording new opportunities for improving military information exchange and decision making. The iaC concept and associated prototype address common C² problems and lay the groundwork for more sophisticated, larger-scale efforts to develop an intelligent command center that is aware of, and better able to support its human symbiots in a complex C² environment.

1   Introduction and Background

To be effective, modern military command and control (C²) decision making requires an enormous amount of coordinated communication and collaboration among the various forces, services, and coalition partners involved in military operations. Unfortunately, many of the information technologies intended to assist users in acquiring, filtering, managing, and integrating information have instead added to their information burden. A prime reason that these technologies are unable to assist decision makers is that they are not able to understand human users' knowledge[1] needs and requirements. Consequently, much of the information these technologies provide users is not tailored to meet the requirements of the tasks being performed. The sheer volume of information – much of it irrelevant to the decision or task at hand – slows or prohibits the process of information assimilation, understanding, and decision making.

Although many important incremental improvements have been made in command and control and decision making theories, practices, and supporting technologies over the past several decades, true innovation has been lacking. Three major human-related problems exist that preclude revolutionary progress in military command and control:

- *Information overload.* The first major problem is information overload. Military command and control is constantly changing and is far more complex than it has ever been. Military command and control personnel – from the most junior equipment operator to the most senior commander – now have more data and information available to them, more options to consider, and more decisions to make. Often, the decision maker is faced with simply too much information, and too many choices to consider. The sheer volume of available information – much of it irrelevant to the decision or action at hand – slows the process of information assimilation and response. Dynamic environments, where information changes faster than people can process it, greatly exacerbate this problem. In effect, by giving warfighters access to so much information, we turn experts

---

[1] Knowledge, for purposes of this report, is domain-specific information, facts, principles, and theories.

into novices, we make true situation awareness nearly impossible, and we force decision making to be an exercise in laborious, detailed analysis rather than a intuitive, reflexive, near-instant action.

- *Misplaced emphasis on information analysis and conscious thought.* The second major problem is that there are troubling gaps in our understanding of human cognition. Many of the recent advances in theory, practice, and technology have for the most part approached information acquisition and analysis, and decision making as conscious actions taken on the part of the decision maker. Even those theories that acknowledge near-instant decision making describe it as a sequential, usually conscious process whereby information is sought and gathered, patterns are recognized, and correct courses of action selected. However, real life is filled with examples of people quickly – even instantly – making complex decisions and taking actions without conscious thought or due consideration. True masters in an area of expertise often rely almost exclusively on "instinct" and intuition – there is really little or no analysis or decision making occurring in the traditional sense. For example, master martial artists do not consciously consider their own or their opponent's moves – they simply act and react. Seasoned fighter pilots in the heat of battle do not plan each and every command given to the aircraft, or update/revise their situation awareness, or consider their options – instead, they "feel" their aircraft almost as if it was a second skin and seemingly act without thinking. These actions occur long before new information can be gathered, patterns recognized, or conscious decisions made. Clearly, in such cases, "cognitive leaps" are occurring based on the decision maker's already-existing understanding of a situation and surrounding environment.

- *Inappropriate focus on individuals/small teams versus larger groups within an environment.* The third major problem is that much of the past research and development to improve information acquisition and analysis, decision making, and course of action selection has been focused on improving the lot of the individual operator or decision maker, or on small teams, without due consideration to the larger organizations or systems that they work within. Incremental improvements have been made to individuals' workstations, displays, decision aids, as well as training and procedures, however, little research and development has focused on the larger system as a whole, i.e., the symbiotic relationship between the systems and the humans who use them. As a result, information bottlenecks occur between humans, and between humans and their supporting systems. For example, some users have access to information which they must share with others – but they have no efficient, effective way of doing so. Some users require information from others (human or system) to perform their tasks or make decisions – but have no reliable way of getting it. Sometimes, critical information is available somewhere in the system, but users either don't know it exists or can't access it.

To address these and related problems, new concepts, theories, and technologies must be developed. For example, if we can "intelligently" automate the pre-processing and organization of the information flow to decision makers, it would reduce some of their cognitive load and allow decision makers to focus on relevant and timely information. If we can substantially improve multimodal representations of the information and knowledge space, it could improve a warfighter's perception to the point where understanding and decisions become almost instantaneous. And, if we could re-focus command and control and cognitive decision making research at a higher, holistic level, it should allow us to address system-wide information bottlenecks and greatly reduce decision making error.

Pacific Science & Engineering Group (PSE) has recently completed Phase I of a Defense Advanced Research Projects Agency (DARPA)-sponsored Small Business Innovation Research (SBIR) project titled *Knowledge Flow in Command and Control*. The overall objective of this SBIR effort is to design and develop intelligent, aware command center concepts and technologies based on the symbiotic relationship[2] between human users and their technologies. Phase I demonstrated the feasibility of this objective and laid the foundation for continued development in Phase II.

1.1 The iaC Concept

During Phase I of the SBIR research and development effort, PSE successfully developed a *Knowledge Flow in Command and Control* approach based on lessons learned from previous analyses and research findings in the cognitive, social, organizational and engineering sciences that pertain to information and information exchange

---

[2] In this context, a symbiotic relationship is a cooperative relationship between technologies and their human users that facilitates meeting users' information needs and enhances their decision making.

issues. An initial system architecture was developed to eliminate limitations in existing information- and knowledge-management technologies, infrastructure, and tools. The resulting concept – which we refer to as *Intelligent Aided Communication (iaC)* – and the prototype iaC tool will intelligently facilitate distributed, coordinated information exchange by military decision makers and their support staffs. The iaC tool approaches the information exchange issue from a human- and information-centric perspective, rather then from an individual, task-based, independent point of view.

Information managed by the iaC interface is filtered, sorted, and parsed to iaC users through intelligent and aware agents[3], based on established user profiles, user-defined preferences, and patterned knowledge-based behaviors that are recorded by the system. In a "behind the scenes" manner, agents use a database that contains user profiles to filter, sort, and parse incoming and outgoing information to tailor it to the needs and interests of each system user. Agents thus manage the communication flow between people and the systems they are using. In addition, agents also control information presentation and visualization according to user and situation needs. Innovative visualization interfaces are used within the iaC knowledge space and provide the ability for users to intuitively monitor the flow of information across multiple tasks and missions. iaC agents take on the task of receiving, filtering, integrating, and displaying information in a form that makes it useful for users.

1.2 Characteristics of Modern Command and Control

The design and development of the iaC prototype tool was based on a thorough understanding of C^2 characteristics and requirements. Lessons learned from previous analyses and studies related to C^2 environments were leveraged to support the respective strengths of humans and machines while at the same time engineering the weaknesses out of the system.

Specifically, the following characteristics of C^2 environments were considered:
- *Requirements for communication:* Effective C^2 requires rapid communication and collaboration among U.S. (and coalition) forces. Command centers serve as communication "hubs" within a vast network of distributed commands and individuals.
- *Nature of the communication:* Information exchange in C^2 environments is synchronous and asynchronous; tactical, operational, and strategic.
- *Nature of the participants:* The C^2 information exchange process often involves the participation of individuals and groups that are both co-located, and distributed geographically, organizationally, and/or functionally. Often participants have diverse backgrounds / expertise, are focused on a variety of different tasks, with different goals, and are working under different schedules. Additionally, modern U.S. and coalition C^2 information exchange entails groups who are culturally diverse, have different standard operating procedures, use terminology that differs in application or meaning, and use technologies / infrastructures that are not compatible with other commands / organizations.
- *Nature of the work:* The C^2 environment involves high-stakes, time-compressed decision making, dynamically changing information landscapes, and reliance on stove-piped communication, collaboration, situation awareness, and decision support tools.
- *Nature of the policies and procedures:* Due to the speed with which technologies and policies change, there is often a lack of complete and updated policies, procedures, business rules, and doctrine that guide efficient and effective information exchange.

The C^2 environment characteristics outlined above often exhibit significant information exchange issues and obstacles. Frequently, these issues require that decision-making be based on ambiguous or conflicting information, and result in difficulties identifying related, similar, or redundant information exchanges. Many times it is difficult for decision makers to remain aware of current or emerging communication that is of critical importance; relate spatially- and non-spatially relevant information; and identify overlapping information sources.

Secondary effects from C^2-related information exchange issues also result. For example, individuals and groups operating in a modern Navy C^2 environment reported that they struggle with deciding who needs to know information, what information to share, what format to the provide the information in, if others can receive the

---

[3] Agents are specially designed software applications that act autonomously on behalf of a particular person.

information (i.e., their capabilities, whether bandwidth issues might prevent the exchange of information), what others need to receive, how to alert others to information with high importance, what information can be trusted, and what are others' areas of expertise are (Pester-DeWan, Moore, & Morrison, 2003).

1.3   Factors Influencing Information Exchange

In order to properly consider and address the information exchange issues described in the preceding section, a systematic review of the technical and scientific literature was conducted. The goal of this review was to identify factors that influence the means by which information moves through human–machine systems. Given the breadth of this topic, the literature review covered fields as varied as information and network systems, and cognitive, organizational, and social psychology. The sources for this literature were likewise diverse and included scientific journals, military lessons learned reports, research and development reviews of communication and collaborative methodologies, and technological reviews. Findings from the literature review are summarized below in terms of their relevance to the design and development of the iaC concept and prototype tool.

1.3.1   Context-based knowledge

Context-Based Knowledge, a term developed specifically for this project, is used to highlight the fact that various factors simultaneously impact information exchange among individuals and groups. Although the research areas cited below have been individually reported in the literature, they have been pooled together here to better show how they affect information exchange. Because the research areas conceptually overlap with one another and affect information exchange in mutually dependent ways, the authors propose that it is important to consider all of these areas when addressing the underlying dynamics of information exchange in C^2 decision making.

1.3.1.1   Transactive memory / knowledge of expertise

Transactive memory can be defined as the combination of "the knowledge possessed by individual group members with a shared awareness of who knows what." (Moreland, 1999, p. 5). It includes an understanding by team members of the knowledge distribution within the team (i.e., other team members' expertise, knowledge and resources), so that all members know with whom to share information and from whom to request needed information. Transactive memory systems combine individual knowledge with shared and explicit knowledge about what other group members know.

Wegner (1986) describes the notion of a transactive memory system in a group environment as a system whose individual members rely on external cues from their environment to trigger their memory. Evidence for transactive memory systems has been found in studies of workgroups (Moreland, Argote, & Krishnan, 1996) in which undergraduate student groups that were trained together outperformed groups that were trained individually (Liang, Moreland, & Argote, 1995). Additional research has shown a decrease in performance (Argote, Insko, Yovetich, & Romero, 1995) and work quality (Thompson & Valley, 1997) in workgroups with high membership change or turnover. These results lend support to the transactive memory framework that argues new group members have not had the opportunity to develop transactive memory and as a result, will perform sub-optimally until they develop it.

One way in which a transactive memory system is developed within a group is through previous interaction. When members develop a history of working together through repeated communication, they learn about each others' preferences and work performance—their expertise, previous information products, preferred information sources, etc. (Moreland, 1999). Knowing and understanding the expertise and experience of the provider of information has an important impact on how the recipient learns to trust and, therefore use the information (Smallman, Heacox, & Oonk, 2004). Without such a prior working relationship, which is common in many modern C^2 environments, it is important to consider how to convey expertise and work experience.

1.3.1.2   Situation context / efficient and effective information exchange

The understanding of expertise, knowledge distribution, and team member experience (transactive memory) is a facet of a larger aspect of understanding, referred to as "situation context." Research has demonstrated the importance of displaying situation context – the overall mission and other team member tasks, and information requirements – to information producers in order to facilitate effective and efficient knowledge and information exchange (Oonk, Moore, & Morrison, 2004). Studies of situation context define information exchange as *effective* if producers are creating information and other team members are accessing all the information that they need to perform their tasks. Information exchange is defined as *efficient* if producers are *not* creating information that is extraneous to other team member's tasks. In other words, information exchange should result in the right information, and only the right information, being shared with the right people (Oonk, Moore, & Morrison, 2004).

Situation context allows information producers to understand the information that other team members believe they need to perform their tasks, as well to recognize and share information that is relevant to other team members, where those members may not even be aware they need that information. A failure to understand situation context often results in information producers sharing information that others do not need and not sharing information they do need. At the same time, the lack of such understanding on the part of the individuals who need the information for their tasks prevents them from asking the right questions of the right people to get that information. The importance of a shared understanding of tasks, expertise, team roles and responsibilities, and information requirements of individual team members has also been highlighted in the shared mental model and transactive memory literatures (e.g., Mathieu et al., 2000, Moreland, 1999) and in studies of military supervisory systems and policies (e.g., Shattuck & Woods, 1997).

An important aspect of situation context is information related to *changes to the context*, i.e., changes or new information related to the mission, group member tasking, and group member information requirements. Previous studies indicate that situation context tools should be graphical representations whenever possible and they should include change alerting functionality.

1.3.1.3 Context-Aware Computing

Context-aware computing is a concept, originating within the computing community, which prescribes that the "design of computing artifacts must take into account how people draw on and evolve social contexts to make the artifacts understandable, useful, and meaningful" (Mouran & Dourish, 2001, p. 89). The basic notion underlying the context-aware computing is that technology be "aware" of the user's context and use this awareness to support the user. Context-aware functions include: (a) presenting the user with information and services appropriate to the context, (b) executing services (e.g., commands or system reconfigurations) for the user based on context, and (c) capturing, tagging and attaching context information for later use (Dey, Abowd, & Salber, 2001).

Within the computing literature, the definition of context focuses on the physical environment within which the technology and user are embedded, including geographical location, user or group identity, and the status of physical objects and spaces (Dey et al., 2001). For the purposes of this project, we extend this definition to include other physical characteristics (such as network status or bandwidth access) and situation context variables – user's mission, tasks, and information requirements, and the expertise, tasks, and requirements of other team members – that we have identified as important in previous studies of information exchange environments (see Situation Context section above; Oonk, Moore, & Morrison, 2004). In other words, context also includes aspects of the user's mission and tasks, as well as the information and human resources necessary to support them in the mission.

1.3.1.4 Information visualization

Problems in information exchange arise when users are unable to extract the important information that is available to them from a larger body of information. This problem is often exacerbated by the nature of the teams exchanging the information, which are characterized by functionally distributed and multi-echelon staffs, working under dynamic conditions. Interestingly, this problem is often at least partly caused by new collaborative technologies designed to support information exchange, which allow the transfer of large amounts of information quickly and easily to almost any location. However, this information is often not the right information for the user (see Situation Context section above) and if it is, it is often presented in a difficult-to-extract, difficult-to-use format. Additionally, "stove-piped" communication, collaboration, and knowledge management systems typically in use today are not "aware" of one another and of the needs of the humans using them. Therefore, these technologies cannot merge relevant information into a meaningful representation for the information recipient.

Information visualization is the application of graphical aids and displays to support human cognitive processes and perception in dealing with various forms of information; in other words, "using vision to think" (Bertin, 1977, as cited in Card, Mackinlay, & Schneiderman, 1999). These visual tools provide a formality to information exchange and bring together the numerous and diverse relevant data sets.

Advances in computer and information display technologies have led to a boom in this multi-disciplinary endeavour, with the introduction of thousands of visualizations that apply our understanding of human factors, cognitive psychology, human perception, and computer science (see Card et al., 1999; Tufte, 1990 for some good examples). It is important to review and assess these emerging information displays, understand their attributes, and determine which is best for which purpose, which will vary across users, tasks, and operational contexts.

1.3.1.5 Trust

With the rapid development of collaborative technologies, group members' skills, knowledge, and expertise can be shared and coalesced to create a transactive memory system; however, information exchange is often predicated on the group's ability to trust one another and the source of the information. Simon's concept of bounded rationality (as cited in Simon, 1982) may help to explain a part of the trust issue. Bounded rationality states that no single individual can ever know all of the options in a particular situation; given that no single individual can possess identical knowledge banks. These knowledge banks are an outgrowth of experience; however, persons with vast amounts of experience (also called expertise) know more than they can actually communicate (Polanyi, 1966). Therefore, this human limitation leads to people doubting if they can trust information from another that is not fully expressed or justified. The trust issue is confounded when the recipient of the information has never before interacted with the source of the information, be it human or machine.

The transfer of trust in expertise can be achieved through a concept that is emerging from situational learning research – the idea that learning is constituted through the sharing of purposeful, patterned activity (Lave & Wenger, 1991). Trust in others' expertise and the information they distribute therefore can be learned and developed through interaction in a social unity that shares a stake in a common situation. It can be seen as a practical capability for the recognition of expertise in group related tasks. The emphasis is not on the behaviors that each individual in the group display; instead, it is on the behaviors that each in the relationship or social exchange framework display. Social Exchange theory allows researchers to examine the social exchanges and relationships between individuals involved in large groups or macro-structures (Emerson, 1981). Likewise, it allowed for the examination of "why" people are involved in an exchange instead of "what kind" of exchange they were involved in. Most social exchange models share the following basic assumptions: (a) social behavior is a series of exchanges; (b) individuals attempt to maximize their rewards and minimize their costs; and (c) when individuals receive rewards from others, they feel obligated to respond.

2   Developing the iaC Tool

The overall objective of this SBIR project is to design and develop intelligent, aware command center concepts and technologies that are based on the symbiotic relationship[4] between human users and their technologies. Phase I work included establishing the conceptual foundation for the *Intelligent Aided Communication (iaC)* tool and delineating its primary features. The iaC tool is designed specifically to aid C^2 decision makers in filtering, integrating, interpreting, and acting upon the types of information typically found in operational settings (i.e., complex, dynamic, ambiguous, conflicting, uncertain). Prototype iaC storyboards and notional functionality were developed and revised in Phase I of the project.

2.1   Conceptual Foundation

The findings of the reviewed technical literature and the previous lessons learned served as a foundation for the design of the iaC features and attributes necessary to support intelligent aided communication. Specifically, the iaC is designed to:

---

[4] In this context, a symbiotic relationship is a cooperative relationship between technologies and their human users that facilitates meeting users' information needs and enhances their decision making.

- Help users rapidly develop shared mental models to improve situation awareness among distributed users,
- Allow users working in distributed information spaces to coordinate and collaborate with one another,
- Promote collaborative discussion and analysis of uncertain, ambiguous, or conflicting data,
- Reduce the cognitive load and allow users to focus on relevant, timely information and knowledge,
- Represent knowledge assets, along with their attributes and availability, and
- Filter, organize, and simplify complex information sets without taking away available resources and capabilities required by the warfighter.

The iaC tool is a shell application running on top of the host operating system (e.g., MS Windows) that fully integrates and re-faces existing applications. It is intended to function as a means for all information exchange activities required by a command center and its support groups. The integrative information space within the iaC tool features an information-centric design, support for transactive memory, understanding of situation-context, facilitation of trust among group members and information provided, and alerting capabilities.

Underlying the behavior and display of information for every iaC user is an intelligent "personal agent" that is aware of the user's personal information – their tasking, expertise, experience, current display capabilities, and information preferences, etc. The agent manages the information presented to a user "behind the scenes" by filtering, sorting, and parsing information according to the data contained in evolving user "profiles" These profiles are based on probabilistic needs based on patterned work behaviors, personal information, and user-specified information preferences. The agent is further context-aware in that it understands the overall mission and tasking (situation context) and expertise (transactive memory) of other iaC users within the organization via a shared database. The database is used by the iaC system and all user agents to store, use, and disseminate information about users of the tool. With this awareness, the agent automatically manages communication flow between users and systems, as well as information presentation and visualization. iaC agents will be designed to *adapt information flow and presentation* based on users' changing needs, settings, technologies, priorities. It is important to note that a fundamental iaC design principal is that human users will *always* be able to control, modify, and/or override iaC agents, and they will always be able to modify their personal profile as they deem necessary.

Innovative interfaces will be used within the iaC knowledge space to provide the ability for users to intuitively visualize large data sets, and monitor the flow of filtered information and knowledge. As with all aspects of the iaC tool, specific visualizations used at any point in time are determined by the user's personal agent, based on the situation, user tasking, display capabilities, and user personal preference.

2.2 iaC Features and Attributes

The iaC interface consists of a set of integrated features for collecting, filtering, parsing, and displaying information that can be used collaboratively by team members as they perform their tasks. Each of these features was developed to specifically overcome the information exchange issues identified in the literature review. Each of the interfaces described below are housed within an iaC "container" or shell application, which provides a consistent look and feel throughout the system. The iaC interface also provides access to underlying information agents and integrated functions, help systems, and software tools.

2.2.1 Communication module

All communications will be routed through and displayed by the iaC-managed communications interface. The iaC Communication Module interface will use a full complement of communication media. The iaC Communication Module features:
- *A graphical display* (common operational picture, maps, satellite images, drawings, whiteboards, etc.), tailored to the situation and user.
- *Collaboration/communication capabilities* (e-mail, chat, shared documents, voice/audio etc.), also tailored to the situation and user. The optimal or user-preferred format of the information is presented. In Figure 1, the information is translated to the user from speech to text.
- *Situation information,* to support knowledge of situation context including keywords and information urgency.

2.2.2 Agent communications manager

The iaC Agent Communications Manager interface allows the user to see the expertise, capabilities, and connection speeds of other potential collaborators (selected by the personal agent), and can initiate communication with them. Information displayed by the iaC Agent Communications Manager interface includes:

- *Collaborator personal information.* This includes such information as name, rank, organization, and expertise. Expertise is presented in terms of how relevant it is to the user's current problem/situation/needs. Using this information, the user can decide with whom they should share information, from whom they should request information, and whom they should involve in a collaborative session.
- *Collaborator availability and connection speed.* This allows the user to see how easy it to contact other collaborators. Connection speed gives the user an indication of what sort of information others can receive (in terms of file size). Messages sent to collaborators with low connection speeds may be "stripped" of information, potentially reducing their usefulness. Note that iaC users may have to take both personal information and availability into account when deciding with whom to collaborate. In time-critical situations, they may choose to contact an available individual with lower expertise rather than wait for a more expert individual to become available.
- *Collaborator capabilities.* This includes information about what types of display capabilities other system users have access to, including Chat, VTC, Whiteboard, audio/voice, etc.

2.2.3 iaC Module

The multi-purpose iaC Module includes features such as:

- An *Agent Asset Manager,* which allows the user to create and annotate graphics using other images, maps, weather information, etc.
- *Agent Communication Manager,* described in the previous section.
- *Information annotation,* using "pushpins" to symbolize relevant available information. By selecting a geographical location on the map, the user can initiate a download of other information about that location that can be shared with others. This capability supports efficient and effective information exchange by allowing the user to have access to and quickly share *relevant* information about the situation with others.

2.2.4 Dynamic Adaptable Display

One of the major features envisioned for the iaC tool interface is a dynamic display space that adapts to the changing information focus of each user. Figure 1 shows how the display of the iaC will be able to adapt to changing situation and user requirements. Users (or their agents) select which visualization is most appropriate for the situation and their own preferences. In this example, display options include a map, a tree diagram, and a fisheye view.

- The map is a familiar representation to users, allowing them to view information that is optimally presented geospatially. Locations on the map can be linked to amplifying information, highlighted using overlays, etc. The specific map view (topographic, roads, etc) can be changed according to the needs of a particular situation.
- Tree diagrams are an innovative way of supporting information management, displaying inter-related information and putting it into context. Branches of the tree represent categories of information. The location and relative size of the text for a give item of information within the tree structure indicates its current relevance to the user and its relationship with other information items within the tree.
- In the fisheye view, pointing to particular element will enlarge that element so that it can be more easily viewed. At the same time, the other elements in the display space will become smaller so that more room is available for the element of interest. These de-emphasized elements do not disappear from the display, however, but remain visible in case the user desires to inspect one of them. In that case, pointing to that element will cause it to become the dominant screen object. This fisheye approach to viewing display contents is a good compromise of the competing needs to remain aware of all display content while also being able to view an element of interest in more detail.

Each of these views is augmented by alerting and highlighting, which can be used to draw a user's attention to new or changed information relevant to their current tasking.

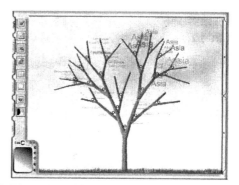

Figure 1. Various situation-relevant iaC visualizations.

3   Summary

The iaC tool represents a new way of receiving, processing, organizing, and displaying information under demanding operational conditions. To be effective, it will require ongoing development of advanced features designed to facilitate the transmission, filtering, organization, display, and use of complex, dynamic, and probabilistic information from multiple sources. Phase I of this SBIR established that a human-centric approach to designing an intelligent aware communication aid can improve information exchange among distributed groups of military C^2 decision makers. Accurate, real-time information exchange is a key requirement for effective C^2. Phase I identified the required iaC functionality, features, and user requirements, evaluated off-the-shelf supporting technologies and infrastructures, and developed metrics to measure the operational impact of the iaC tool. The Phase I effort established these requirements to pave the way for the development of a fully functional iaC tool in Phase II of the SBIR project.

The proposed iaC tool will greatly benefit the military by improving communication, coordination, and collaboration among C^2 decision makers, and thereby dramatically improve shared situational awareness and decision-making. However, fast-paced, dynamic information management and decision-making environments characterized by complex and uncertain data, high-stakes decisions, and distributed, decentralized command structures are not unique to the military. Many government and private organizations operate in similar environments on a daily basis and therefore could be expected to benefit from this technology.

References

Argote, L., Insko, C. A., Yoveovetich, & Romero, A .A. (1995). Group learning curves: The effects of turnover and task complexity on group performance. *Journal of Applied Social Psychology, 25*, 512–529.

Card, S. K., Mackinlay, J. D., & Schneiderman (1999). *Readings in information visualization: Using vision to think.* San Francisco, CA: Morgan Kaufmann Publishers, Inc.

Dey, A. K., Abowd, G. D., & Salber, D. (2001). A conceptual framework and a toolkit for supporting the rapid prototyping of context-aware applications. *Human-Computer Interaction, 16,* 97–166.

Emerson, R. (1981). Social exchange. In M. Rosenberg & R. Turner (Eds.), *Social psychology: Sociological perspectives* (pp. 30–65). New York: Basic Books.

Lave, J., & Wegner, E. (1991). *Situated Learning: Legitimate Peripheral Participation.* Cambridge, UK: Cambridge University Press.

Liang, D. W., Moreland, R. L., & Argote, L. (1995). Group versus individual training and group performance: The mediating role of transactive memory. *Personality and Social Psychology Bulletin, 21(4),* 384–393.

Mathieu, J. E., Goodwin, G. F., Heffner, T. S., Salas, E., & Cannon-Bowers, J. A. (2000). The influence of shared mental models on team process and performance. *Journal of Applied Psychology, 85,* 273–283.

Moreland, R. (1999). Transactive memory: learning who knows what in work groups and organizations. In L. L. Thompson, J. M. Levine, & D. M. Messick (Eds.), *Shared cognition in organizations: The management of knowledge* (pp. 3–31). Mahwah, N.J: Lawrence Erlbaum Associates.

Moreland, R. L., Argote, L., & Krishnan, R. (1996). Socially Shared Cognition at Work: Transactive Memory and Group Performance. In J. L. Nye & A. M. Brower (Eds.), *What's Social about Social Cognition?*: 285–309. Thousand Oaks, CA: Sage.

Mouran, T. P., & Dourish, P. (2001). Introduction to this special issue on context-aware computing. *Human-Computer Interaction, 16,* 87–95.

Oonk, H. M., Moore, R. A., & Morrison, J. M. (2004). Communication of context in multi-echelon information exchange environments. In *Proceedings of the 2004 Command and Control Research and Technology Symposium.* San Diego, CA.

Pester-DeWan, J., Moore, R. A., & Morrison, J. G. (2003). *Knowledge Engineering for Command and Control Transformation at United States European Command (USEUCOM).* (Technical Report). San Diego, CA: Pacific Science & Engineering Group.

Polanyi, M. (1966). *The Tacit Dimension.* Garden City, NY: Doubleday.

Shattuck, L. G., & Woods, D. D. (1997). Communication of intent in distributed supervisory control systems. *Proceedings of the Human Factors and Ergonomics Society 41st Annual Meeting.* Santa Monica, CA: Human Factors Ergonomic Society.

Simon, H. (1982). Theories of Bounded Rationality. In, Herbert Simon (Ed.), *Models of Bounded Rationality. Behavioral Economics and Business Organization* (Vol. 2, pp. 408–423). Cambridge: MIT Press.

Smallman, H. S., Heacox, N. J., & Oonk, H. M. (in press). *Shared Intelligence VisualizEr (SIEVE): Web-based threat assessment for cross cultural intelligence analysis.* (Technical Report). San Diego, CA: Pacific Science & Engineering Group.

Thompson, T. A., & Valley, K. L. (1997). *Changing Formal and Informal Structure to Enhance Organizational Knowledge.* Working Paper 98-060. Harvard University, Cambridge, MA.

Tufte, E. R. (1990). *Envisioning information.* Cheshire, CT: Graphics Press.

Wegner, D. M. (1986). Transactive memory: A contemporary analysis of the group mind. In B. Mullen & G. R. Goethals (Eds.), *Theories of group behavior* (pp. 185–208). New York: Springer-Verlag.

828

Windows as a Second Language:
An Overview of the Jargon Project

Tim Paek

Microsoft Research
One Microsoft Way
Redmond, WA 98052 USA
timpaek@microsoft.com

Raman Chandrasekar

Microsoft Corporation
One Microsoft Way
Redmond, WA 98052 USA
ramanc@microsoft.com

Abstract

When novice users request help from expert sources, they often have difficulty articulating the nature of their problems due to their unfamiliarity with the appropriate technical terms or "jargon." Consequently, in searching for help documents, they struggle to identify query terms that are likely to retrieve the relevant documents. There is often a vocabulary mismatch between the technical language of help documents or experts and the common language of novice users. At Microsoft, as part of the Jargon Project, we are exploring methods to characterize the use of jargon by different communities of users, and techniques to exploit user models based on jargon-related features to provide customized and more appropriate help and support. This paper provides an overview of the Project and presents preliminary results demonstrating the feasibility of distinguishing types of users by their jargon use.

1   Introduction

When novice users request help from expert sources, they often have difficulty articulating the nature of their problems. This is in part due to their unfamiliarity with the appropriate technical terms, or "*jargon*," to describe the constructs and relations of their situation as well as the proper usage for those terms as utilized in a community of experts. For example, consider the following newsgroup posting: "*How do I run the Internet on Windows XP?*" Experts examining the question would surmise that the person was most likely a novice user with respect to the Internet by both the nature of the topic, which is accessing the Internet, and the atypical use of the technical term "*Internet*," in that "*Internet*" is not commonly used in a direct object relation for the verb "*run*." In other words, the question does not conform to the canonical usage, and as such, reveals that the user is most likely not a member of the "language community" of people who know enough about the topic to use that term in a "conventional" or a socially agreed upon manner (Clark, 1996).

In information retrieval, researchers have noted that a vocabulary mismatch often prevails in failed searches between the query expression and the language of document authors (Furnas et al., 1987; Deerwester et al., 1990). This mismatch is particularly evident in the retrieval of help documents, which are written in precise technical terms, and novice queries, which often employ general terms for concepts users may not fully understood. At Microsoft, as part of the Jargon Project, we are exploring methods to characterize the use of jargon by different communities of users, and techniques to exploit user models based on jargon-related features to provide customized and more appropriate help and support. In this paper, we provide an overview of the Project, describing the motivation, objectives, and tools we are using to characterize jargon usage. We also present preliminary findings that demonstrate the feasibility of distinguishing types of users, corresponding approximately to their level of expertise, by their use of jargon.

2   Jargon Project

The Jargon Project is aimed at understanding how jargon usage affects the retrieval of help documents by users at different levels of expertise. The ultimate goal of the Project is to facilitate more personalized retrieval of help documents based on user models inferred from a history of natural language queries. The user models are geared towards inferring level of expertise in various domain topics by examining the linguistic structures pertaining to the use of specific jargon terms. We now describe how the jargon terms were selected and a canonical corpus created to help build such user models.

2.1   Data Collection

In order to create baselines for comparison, we needed to establish a canonical corpus of expert language to which natural language queries could be evaluated. The corpus collected consisted of the text extracted from all help documents written for Windows XP Help and Support Center (HSC) and Office XP products (Word, PowerPoint, Excel, Outlook). The text exceeded 100 MB and covered over 9800 individual help documents. In addition, we extracted text from the body chapters of 21 Microsoft Press books relating to the operation of Windows and Office XP. This text exceeded 23 MB and included over 500 chapters.

The canon, which constitutes a "gold standard" set of jargon terms, was compiled from the corpus using the following definition. Jargon comprises any term that either:

1) Appears in a glossary (e.g., Windows HSC glossary, or the glossary of any of the Microsoft Press books), or
2) Appears in an index of a technical publication.

The general intuition behind this definition is that if a technical term is important enough that authors of technical publications feel necessary to identify it in the index or glossary (so that users may find all references to it or its definition), then that term is considered jargon. Note that for the index, we used only the first-level headings, mostly to limit the number of jargon terms. Overall, we collected roughly 11,000 glossary and index terms, though many of these terms were redundant due to the use of abbreviation or plurals. Observing that some of the terms were abbreviated, we also parsed them to extract any parenthetical acronyms, such as DOS or ODBC. We amassed about 700 acronyms in all.

To obtain natural language queries, we collected over 1.5 GB of postings to 20 "microsoft.public" newsgroups pertaining to Windows and Office XP help and support. We selected particular newsgroups that were known to be geared towards novice users, as opposed to just expert users. In these newsgroups, most of the replies to queries typically come from Microsoft employees monitoring the postings, or MVPs (Most Valued Professionals), non-Microsoft employees officially recognized by the company as experts who have significantly contributed to helping others resolve their software problems. MVPs typically host their own websites with Frequently Asked Questions (FAQ) about particular Microsoft software, and often refer novice users to canned responses they have on their websites.

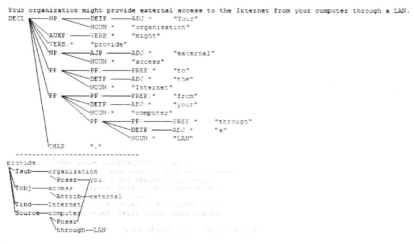

Figure 1: Parse tree and logical form for a sentence containing "Internet" in a help document.

2.2 Characterizing User Groups

Given our interest in learning user models of expertise level based on jargon features, we needed to identify groups of users that corresponded roughly to level of expertise. We divided postings from the user population into the following contrastive groups:

- Experts vs. Non-experts: where Experts were defined as users whose emails ended in either "microsoft.com" or "mvp.org"
- First-in-thread vs. Not-first-in-thread: where First-in-thread contained only postings that started off a thread of replies.
- Queries vs. Solutions: where Queries represented postings that were heuristically selected by key phrases such as "how do you," and Solutions represented replies selected by key phrases such as "Have you tried," that often pointed to known solutions in KB articles and MVP FAQ lists.

We selected the above groups since in reading the postings to the newsgroups we selected, we found that novice users tended to start off query threads, and expert users tended to respond to them. In comparing jargon usage across different types of users, we considered combinations of these contrastive groups, as discussed in section 3. Note that in extracting the content of postings, only new content was considered; that is, we removed all inclusions and indirect quotations, demarcated typically by the line prefix ">".

2.3 Generating Natural Language Features

In order to generate features relating to the usage of jargon, we first obtained natural language parses of all the sentences in the canonical corpus of expert language as well as those in the newsgroup corpus. The parser we used was Microsoft NLPWin (Heidorn, 1999), which not only generates a syntactic tree but also constructs a logical form of the sentence from the tree, representing predicate-argument relations in a semantic graph. Figure 1 displays the parse tree and logical form for a sentence in a Windows XP help document containing the jargon term *"Internet"* in the typical indirect object role for the head "provide." Once the logical form is generated, we extract triples of the structure <object1, *relation*, object2>, where the relations are represented labels on the arcs in the logical form, as shown in the bottom half of Figure 1 (e.g., <provide, **Tind**, Internet>, <provide, Tobj, access>, <provide, Source, computer>).

831

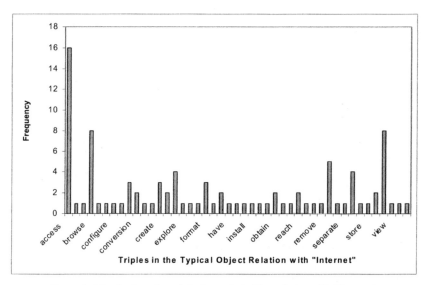

Figure 2: Distribution of words in the typical object relation with "*Internet*"

Given all triples containing a jargon term in either of the two object roles, features can be derived based on the relationship of a given token triple to a distribution of tokens. Consider again the example sentence "*How do I run the Internet on Windows XP?*" The triple for this sentence for the jargon term "*Internet*" in the typical object relation would be < run, **Tobj**, Internet>. We can generate features based on how this triple compares to a distribution of triples as found in canonical text with the same relation and second object. Figure 2 displays a distribution of triples with <*, **Tobj**, Internet> for the Windows XP HSC corpus where the word "*access*" is the most frequently occurring first object of that triple (e.g., "*access the Internet*"). In this case, we would generate a feature specifying that <run, **Tobj**, Internet> did not appear in the Windows XP HSC corpus, and hence, contained zero mass. If it did appear, we would have as our feature its mass with respect to the other triples.

3   Variation of Jargon and Function Word Use by User Community

While it may seem intuitive that users at different levels of expertise might employ jargon in contrasting ways, we sought to empirically verify the feasibility of distinguishing user expertise level from jargon-related features. The most obvious distinction that we expect between user populations is that novice users would be much less likely to employ jargon as frequently as expert users, presumably because they are not as familiar with these terms. To verify this, we examined the contrastive groups of newsgroup postings specified in section 2.2, as well as combinations thereof, to see if any of the groups utilized more jargon terms than the rest. In this section, we present preliminary results demonstrating a clear difference in the percentage of jargon use between groups, suggesting at least one point of distinction.

	Non-Expert & First In Thread	Non-Expert & Query	Expert & Not First In Thread	Expert	Non-Expert	First In Thread	Not First In Thread	Query	Solution
jargon / words	2.12%	1.73%	4.32%	3.37%	2.21%	1.80%	1.49%	1.73%	3.57%
words / postings	84.07	117.54	99.65	133.82	103.30	105.39	114.10	117.54	91.69

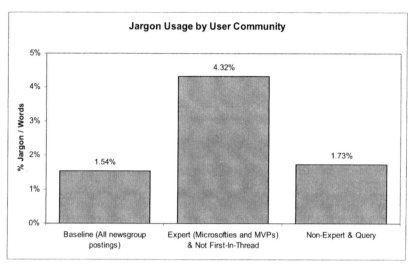

Figure 3: Incidence of jargon use by user community

jargon / postings	1.79	2.03	4.31	4.51	2.28	1.90	1.70	2.03	3.27

Table 1: Jargon usage by different types of newsgroup postings

Table 1 displays a comparison of the jargon usage between the three contrastive groups as well as three combinations of those groups. The first two columns, "Non-Expert & First-in-thread," and "Non-Expert & Query," were meant to capture what seemed to be typical behavior of novices; that is, they tend to be non-MVP, non-Microsoft employees who post a query or comment that starts a newsgroup thread. "Expert & Not First-in-thread" was meant to capture the behavior of experts, typically MVPs or Microsoft employees, who were not starting threads but answering them. To generate the jargon percentages shown in the table, we simply counted the number of times a jargon term in the canon appeared in the various groups of newsgroup postings and normalized that by the total number of words in that group. Every group had well over 2 million words. We also compared the number of words in the postings to the number of postings, as well as the number of jargon terms in the postings to the number of postings.

With respect to the contrastive groups, "Experts" use more jargon than "Non-experts" and postings heuristically defined as "Queries" used less jargon than "Solutions." The later is true despite the fact that many solutions simply referred users to other help resources such as MVP FAQ lists and KB articles. Interestingly, the "First-in-thread" versus "Not-First-In-Thread" group did not display much of a difference; this may be due to the fact that often people who reply to a newsgroup posting simply request more information from the sender.

The key finding in these results relates to the difference between the two groups meant to capture novice behavior (i.e., "Non-Expert & First-in-thread," and "Non-Expert & Query") and the exemplary expert group (i.e., "Expert & Not First-in-thread"). The exemplary expert group was at least twice as great as the novice groups. To verify the significance of the greatest difference, we conducted a chi-square test of homogeneity between "Non-Expert & Query" and "Expert & Not First-in-thread" and found the difference to be significant (χ^2 (1, $N > 14650993$) = 86804.08, $p < .001$). While the difference in percentages may seem small, it must be examined in relation to the baseline, or the percentage of jargon terms to word tokens in the entire newsgroup corpus, as shown in Figure 3. While there is hardly any difference between the baseline and "Non-Expert & Query," the difference between the baseline and "Expert & Not First-in-thread" is even greater than the aforementioned significant difference.

Aside from the percentage of jargon usage, it is interesting to note that "Non-expert & Query" postings had more words per postings than "Expert & Not First-in-thread" group. This suggests quite intuitively that while novices used more words in their queries than experts, experts used more jargon terms. Furthermore, while "Expert" postings employed more words per postings than Non-expert postings, "Solution" had fewer words than "Query," implying that experts have succinct responses to problems. One combination group which we have yet to evaluate against the baseline, is "Expert & Solution," which given the results for "Solution" and "Expert & Not First-in-Thread" is likely to have the highest percentage of jargon usage.

3.1 Function/Stop Words

Since researchers began to apply statistical techniques to infer author attribution of historical texts such as the Federalist Papers (Mostellar and Wallace, 1964), one of the most predictive stylistic discriminators (characteristics of style which remain invariant within a corpus of works for a particular author but which varies from author to author), has been the use of function words: words that serve a grammatical purpose but have no meaning by themselves (e.g., "and", "or", "of"). As features used with various machine learning techniques, function and stop words have been shown to successfully classify authors by gender (Koppel et al., 2002). So, we sought to test if function and stop words, like jargon terms, might be predictive in distinguishing between users of different level of expertise. Using the same set of 467 function and stop words from Koppel et al. (2002), we computed the percentage of function and stop words in the different types of newsgroup postings and obtained interesting results.

"Non-Expert & First-in-thread" (28.2%) differed very little from "Expert & Not First-in-thread" (29.4%), implying that function and stop words by themselves may not adequately distinguish between novices and experts; however, the postings for "Non-Expert & Query" differed significantly from both (21.4%). Here it is important to point out that all "Non-Expert & Query" postings were also "Query" postings and vice versa; there were no query postings by experts. The difference we observed in the percentage of function and stop words between "Query" (21.4%) and "Solution" postings (38.5%) is high. This difference makes sense upon a closer analysis of the distribution of function and stop words. "Solution" postings contained much more of the "you"-related function words (e.g., "you", "your") than "Query" postings. Specifically, 8.8% of all "Solution" function words were "you"-related, compared to only 1.7% of all "Query" postings. Reading the actual messages, the primary reason for this is due to the fact that in advising users to take particular actions, "Solution" postings almost always directly address the users (e.g., "you need to reboot your system").

4 Future Directions and Conclusion

In this paper, we presented an overview of the Jargon Project at Microsoft aimed at understanding how jargon usage affects the retrieval of help documents by users at different levels of expertise. We also described preliminary results demonstrating the feasibility of distinguishing types of users by their jargon usage. In particular, we found that novice users in newsgroup postings employ about half as much jargon as experts in general, presumably because they are less familiar with the proper technical terms, even though they use more words per postings. Given that the ultimate goal of the Jargon Project is to facilitate more personalized retrieval of help documents based on user models inferred from a history of natural language queries, our next step is to build classifiers using the jargon-based features, and to use the classifiers to re-rank search results for natural language queries. We also plan to compare difference classes of computer manuals, and build differential user models based on our analyses.

5 Acknowledgements

We thank Robert Ragno for his willing and patient help with text analysis algorithms. We also thank various colleagues in Microsoft and in Microsoft Press who gave us access to the help documents and the text of MS Press books.

References

Clark, H. (1996). *Using language*. Cambridge University Press.

Deerwester, S., Dumais, S. T., Landauer, T. K., Furnas, G. W. and Harshman, R. A. (1990). Indexing by latent semantic analysis. *Journal of the American Society for Information Science*, 41(6), 391-407.

Furnas G. W., Landauer T. K., Gomez L. M., and Dumais S. T. (1987). The vocabulary problem in human-system communication: An analysis and a solution. *Bell Communications Research.*

Heidorn, G. (1999). Intelligent writing assistance. In Dale, R., Moisl, H., and Somers, H. eds., *A Handbook of Natural Language Processing Techniques.* Marcel Dekker.

Koppel, M., Argamon, S. and Shimoni, A. (2002), Automatically categorizing written texts by author gender, *Literary and Linguistic Computing* 17(4): 401-412.

Mosteller, F., and Wallace, D.L. (1964). *Inference and disputed authorship: The Federalist. Reading*, Mass.: Addison-Wesley.

Section 4
Cognitive State Sensors

Chapter 17

Neuroergonomics – Cognitive Human Factors

Neuroergonomics: An Overview of Research and Applications

Raja Parasuraman

George Mason University
Arch Lab, MS 3F5
4400 University Drive
Fairfax, VA 22030-4444
rparasur@gmu.edu

Abstract

This panel features seven presentations that describe the characteristics and scope of neuroergonomics, which can be defined as the study of brain and behavior at work. The presentations involve studies using diverse neuroergonomic methods and application domains. The methods include neuroimaging and physiological techniques such as electro-encephalography (EEG), event-related brain potentials (ERPs), functional magnetic resonance imaging (fMRI), Transcranial Doppler Sonography (TCDS), near infra-red spectroscopy (NIRS), and (in animals and patients with brain disorders) single-unit recordings. The use of virtual environments for detecting impairments in individuals with cognitive and brain disorders is also described. In addition, studies applying molecular genetic methods to the study of individual differences in cognition are presented. Finally, implications for neuroergonomics from nanotechnology and biotechnology are discussed.

1 Introduction

The goal of neuroergonomics is to harness the power of neuroscience to the engineering of human-machine systems for safety and efficiency (Parasuraman, 2003). Neuroergonomics focuses on investigations of the neural bases of mental functions and physical performance in relation to technology, work, leisure, transportation, health care, and other settings in the real world. The two major goals of neuroergonomics are to use knowledge of brain function and human performance to design technologies and work environments for safer and more efficient operation, and to advance understanding of brain function underlying real-world human performance. This panel features seven presentations that describe the characteristics and scope of neuroergonomics.

2 Panel Presenters and Titles

"Neuroergonomics: An Overview of Research and Applications." **Raja Parasuraman**, George Mason University. Fairfax, VA.

"Neurophysiologic Measures for Neuroergonomics," **Alan Gevins & Michael E. Smith**, SFBRI and SAM Technology, San Francisco, CA.

"The Engineering of Motor Learning and Adaptive Control," **Ferdinando Mussa-Ivaldi**, Northwestern University, Chicago, IL.

"Using Electrical Brain Potentials and Cerebral Oximetry to Detect G-Induced Loss of Consciousness," **Lloyd Tripp & Glenn Wilson**, Wright-Patterson Air Force Base, Dayton, OH.

"Virtual Environments for Localizing Cognitive Impairments in Real World Tasks," **Matthew Rizzo & Joan Severson**, University of Iowa School of Medicine, Iowa City, IA.

"Molecular Genetics of Augmented Cognition," **Raja Parasuraman & Pamela Greenwood**, George Mason University, Fairfax, VA

"Neuroergonomics Support from Bio- and Nano-Technologies," **Rene de Pontbriand**, Army Research Laboratory, Aberdeen Proving Ground, Aberdeen, MD.

3   Discussion

The term neuroergonomics was coined seven years ago to depict an emerging, interdisciplinary area of research and practice devoted to understanding the relation between brain function and work. This panel follows two special issues of a journal (Parasuraman, 2003) and a workshop (International Ergonomics Association in 2003) devoted to the topic, whose scope and utility has steadily increased over the years. The results of the studies presented in this panel provide further evidence of the usefulness of the neuroergonomic approach. Accordingly, recent editorial commentary has identified neuroergonomics as a "burning issue" for contemporary ergonomics (Marek & Pokorski, 2004) and there have been popular descriptions of neuroergonomic research (Huff, 2004). The first technical book devoted to the subject will appear shortly (Parasuraman & Rizzo, 2005).

Neuroergonomics involves the intersection of two disciplines that have rarely communicated in the past, neuroscience and ergonomics. That neuroscience did not consider human behavior in complex environments could be attributed to the fact that the neural mechanisms of human cognitive functions have only been identified recently. Neuroscientists are not standing still, however, as witnessed by calls to move neuroscience "beyond the bench" (Editorial, 2002), the rise of a neuroscience of social behavior (Cacciopo, 2002), and the development of neural prosthetics for control of robots, home automation, and other technologies for physically disabled people (Musallam et al., 2004; Mussa-Ivaldi & Miller, 2003). The future is likely to bring even more impressive achievements in research and practice in neuroergonomics.

4   References

Cacioppo, J. T. (2002). (Ed.) *Foundations in social neuroscience.* Cambridge, MA: MIT Press.

Editorial (2002). Taking neuroscience beyond the bench. *Nature Neuroscience*, Supplement, *5*, 1019.

Huff, C. (2004). The baggage screener's brain scan. *Monitor on Psychology, 35(8)*, 34-36.

Marek, T., & Pokorski, J. (2004). *Quo vadis,* ergonomia?—25 years on. *Ergonomia, 26,* 13-18.

Musallam, S., Corneil, B. D., Greger, B., Scherberger, H. & Andersen, R. A. (2004). Cognitive control signals for neural prosthetics. *Science, 305*, 258-262.

Mussa-Ivaldi, F. A., & Miller, L. E. (2003). Brain–machine interfaces: computational demands and clinical needs meet basic neuroscience. *Trends in Neuroscience, 26(6)*, 329-334.

Parasuraman, R. (2003). Neuroergonomics: Research and practice. *Theoretical Issues in Ergonomics Science. 4*, 5-20.

Parasuraman, R., & Rizzo, M. (2005). *Neuroergonomics: The brain at work.* New York: Oxford University Press.

Neurophysiologic Measures for Neuroergonomics

Alan Gevins and Michael E. Smith

San Francisco Brain Research Institute and SAM Technology
425 Bush St, Fifth Floor, San Francisco, CA 94108
alan@sfbri.org

Abstract

It is a compelling, futuristic idea to use brain activation measures as a basis for automated systems to off-load tasks from an individual if he or she was in a state of high cognitive workload, or allocate more tasks to an individual that appeared to have ample reserve processing capacity. Like many great visions, the devil is in the details. In this presentation we describe recent progress in the development of multivariate neurophysiologic metrics of regional cortical brain activation. In a first experiment, EEG recordings were made during a daytime session while 16 well-rested participants performed versions of a PC flight simulator task that were either low, moderate, or high in difficulty. In a second experiment, the same subjects repeatedly performed high difficulty versions of the same task during an all night session with total sleep deprivation. Multivariate EEG metrics of cortical brain activation were computed for frontal brain regions essential for working memory and executive control processes crucial for maintaining situational awareness, central brain regions essential for sensorimotor control, and posterior parietal and occipital regions essential for visuoperceptual processing. During the daytime session each of these regional measures displayed greater activation during the high difficulty task than during the low difficulty task, and degree of cortical activation correlated positively with subjective workload ratings in these well-rested subjects. During the overnight session, cortical activation declined with time-on-task, and the degree of this decline over frontal regions correlated negatively with subjective workload ratings. Since participants were already highly skilled in the task, such changes likely reflect fatigue-related diminishment of frontal executive capability rather than practice effects. In a third experiment, a prototype online system for concurrently deriving regional indices of cortical brain activation was implemented and tested. The prototype was validated empirically by using it to gauge brain function in ten well-rested participants while workload was varied through task difficulty manipulations. The results confirmed that cortical brain activation indices derived in real time significantly differed between relatively easy and relatively difficult tasks. As in the first experiment, frontal cortical activation correlated positively with subjective workload. The current results thus indicate that a decrease in cortical activation in frontal regions may either reflect a decrease in mental workload, or an increase in mental fatigue and a heightened sense of mental stress. This is problematic for the development of brain-adaptive automation systems. Assigning more tasks to an individual in the former case may indeed serve to increase his or her cognitive throughput, but in the latter case it could be quite counterproductive, even resulting in the sort of tragic accident that too often occurs when fatigued personnel are confronted with unexpected increases in task demands. For such neuroadaptive systems to successfully modulate the cognitive task demands placed on an individual in response to momentary variations in the availability of mental resources as indexed by neural activation measures, sophisticated user models with human-like intelligence will need to be developed that take both task demands and the operator's state of alertness into account. It is realistic to expect that a great deal of methodical research will be needed before brain activation measures can actually be put to use for adaptively augmenting the capabilities of mission-critical personnel working at demanding and stressful tasks. Further systematic developments such as future progress with the measurement approach outlined here will undoubtedly eventually lead to an effective and practical means for monitoring transient changes in cognitive brain function that is prerequisite to actual neuroadaptive automation.

1   Introduction

Over the last three decades scientists and engineers have been evaluating neural signals as tools for monitoring cognitive status in operational situations. Measuring brain function in the real world requires signal processing methods capable of extracting signals of cognitive activity from neural signals related to sensory and motor activity, and, importantly, from sometimes subtle artifactual contaminants not generated in the brain. Such methods must operate in a timely and fully-automated fashion if the measures are to be actionable in adaptive automation contexts. That is, unlike laboratory studies where sources of extraneous signals can be controlled, real world activities can involve extensive eye-, head-, and limb-movements and rapidly changing multimodal sensory information. In addition to the artifactual contaminants generated by such activities, many of the actual brain signals that are recorded under such circumstances may have little to do with cognitive function. Presuming that such difficulties can in fact be overcome, conceptual and pragmatic issues remain. One such issue concerns the fact that brain signals (whether neurophysiologic or neurometabolic) generally lack any one-to-one correspondence to abstract psychological constructs, which implies that any interpretation of the functional relevance of a change in a brain signal in an individual must be seen as inherently ambiguous. For example, a reduction in activation in some brain region could reflect neglect of some important task demand, a change in cortical responsivity due to decreased alertness or a change in neurochemistry, or practice-related development of a more efficient functional brain network. Thus, even when meaningful neural signals can be obtained in a particular context, the interpretation of them by an automated system will require sophisticated, human-level intelligence to infer whether a particular neural signal change reflects a change in mental workload or some other factor such as mental fatigue or an alteration in strategy. Although such complexities are inherent to the problem, steady incremental progress is being made by a number of laboratories in the use of brain signals in neuroergonomic research. In this presentation we review work in our lab that is helping to translate laboratory advances in the measurement of cognitive brain function using EEG measures into practical neuroergonomic technologies.

A significant body of evidence suggests that neurophysiologic measures such as the electroencephalogram or EEG can be used to noninvasively detect states of high cognitive load and mental fatigue. For instance, the frontal midline theta (5-7Hz) EEG signal has been shown to increase in high load task conditions. The frontal medial anterior cingulate cortex, a cortical region critical to attention control mechanisms, is the likely origin of this load-sensitive frontal theta signal (Gevins, Smith et al. 1997). Its increase in demanding task conditions is consistent with the fact that neuroimaging studies have found increased activation in this region with difficult tasks (Paus, Koski, Caramanos, & Westbury, 1998). In contrast, increased task load attenuates the alpha (8-12Hz) rhythm (Gevins & Smith, 1999; Gevins, Smith et al., 1998; Gevins, Smith, McEvoy, & Yu, 1997; Gundel & Wilson, 1992). Because of this load-related attenuation, the magnitude of alpha activity during cognitive tasks has been hypothesized to be inversely proportional to the number of cortical neurons recruited into a transient functional network for purposes of task performance (Gevins and Schaffer 1980). Convergent evidence for this view is provided by observations in fMRI studies of more extensive brain activation when task difficulty increases (Carpenter, Just et al. 2000; Garavan, Ross et al. 2000). Such signals have sufficient stability and sensitivity that multivariate combinations of them can be used to reliably distinguish between different degrees of mental effort in individuals (Gevins, Smith et al., 1998; Smith & Gevins, 1997).

In addition to being sensitive to variations in attention and mental effort, the EEG also changes in a predictable fashion as individuals become sleepy and fatigued, or when they experience other forms of transient cognitive impairment. For example, it has long been known that when subjects are resting quietly in a passive state, the EEG shows diffuse increases in signal strength in the theta band and decreases in signal strength in the alpha band (Davis, Davis, Loomis, Harvey, & Hobart, 1937; Gevins, Zeitlin, Ancoli, & Yeager, 1977; Makeig & Jung, 1995; Matousek & Petersen, 1983; Oken & Salinsky, 1992). Such changes can be detected algorithmically. For example, in one study we used neural network based methods to compare task-related EEG features between alert and drowsy states in individual subjects performing an n-back WM task (Gevins & Smith, 1999). Utilizing EEG features in the alpha and theta bands, average test set classification accuracy was 92% (range 84%-100%, average binomial $p < .001$). In another recent study we explicitly compared metrics based on either behavioral performance measures during an n-back task requiring WM and sustained attention, EEG recordings made during task performance and control conditions, or combinations of behavioral and EEG variables, with respect to their relative sensitivity for discriminating conditions of drowsiness associated with sleep loss from alert rested conditions (Smith, McEvoy, & Gevins, 2002). Analyses based on behavior alone did not yield a stable pattern of results when viewed over test intervals. In contrast, analyses that incorporated both behavioral and neurophysiological measures displayed a monotonic increase in discriminability from baseline with increasing amounts of sleep deprivation. Such results indicate that fairly modest amounts of sleep loss can induce neurocognitive changes detectable in individual subjects, and that the sensitivity for detecting such states is significantly improved by the addition of EEG measures to behavioral indices.

We have been systematically extending EEG-based methods for monitoring EEG signals of attention engagement and their changes with fatigue into more complex task domains For example, in one study (Smith, Gevins, Brown, Karnik, & Du, 2001), EEG was recorded while subjects performed NASA's Multi-Attribute Task Battery (MATB, Comstock & Arnegard, 1992). The MATB is a PC-based multi-tasking game that is often used as a laboratory simulation of some activities performed by a multi-engine air transport pilot and as a criterion task in experiments on adaptive automation (e.g. Fournier, Wilson, & Swain, 1999; Molloy & Parasuraman, 1996; Parasuraman, Mouloua, & Molloy, 1996). Subjects performed versions of the MATB task that varied in difficulty. Consistent with the data reviewed above, frontal theta band EEG activity increased with increasing task difficulty, and alpha band activity tended to decrease. To track variations in attention engagement, subject-specific multivariate functions were derived that discriminated between the EEG spectra of data observed in different task load conditions. We refer to a function of this type as an index of "cortical activation", i.e. the degree to which a set of task demands activates the cortex during task performance. Across subjects mean values of the resulting indices were shown to increase monotonically with increasing task difficulty, and to differ significantly between the task versions (Smith, Gevins, et al., 2001). In another exploratory study (Gevins & Smith, 2003), we extended this approach to monitoring brain activation during a variety of other tasks, including searching for information on the internet, performing exercises with a computerized language learning program, and taking a logical reasoning aptitude test. Such results provided encouraging initial evidence that automated analysis procedures applied to EEG measures can provide a modality for measuring attention engagement from subjects performing everyday attention demanding console tasks.

Task related variation in this sort of uni-dimensional "whole brain" metric might reflect a central bottleneck in the cognitive resources available for task performance. However, the applied psychology and ergonomics literature has long posited a relative independence of the resources involved with higher-order executive processes and those involved with perceptual processing and motor activity (Gopher, Brickner, & Navon, 1982). Similarly, topographic differences can be observed in regional patterns of EEG modulation (Gevins et al 1979abc). Such considerations have led us to extend the sort of whole-brain EEG based measure of cortical activation described above to create multidimensional indices that provides information about the relative activation of a local neocortical region (cf. Smith & Gevins, 2005). Herein we describe experimental results that indicate how systematic variations in task load and degree of mental fatigue affect such regional measures of cortical engagement, and describe progress in the development of a prototype online system capable of assessing regional cortical activation in real time.

2 Experiment I: Effects of Task Demands on Regional Cortical Activation

2.1 Methods
Sixteen healthy adults participated. All participation was fully informed and voluntary, and conformed to all institutional and governmental guidelines for the protection of human subjects. These subjects performed a variety of tasks including simple repetitive vigilance tasks, continuous working memory tasks, and versions of the MATB flight simulator that varied in task load. Neurophysiologic activity during this latter task is the focus of the current report. The versions of the MATB employed for this purpose included four concurrently performed subtasks in separate windows on a computer screen. The four subtasks were systems monitoring, resource management, communications, and tracking. The systems monitoring task required the operator to monitor and respond to simulated warning lights and gauges. In the resource management task, fuel levels in two tanks had to be maintained at a certain level. Fuel level could be controlled by pressing keys to turn on and off a series of pumps. The communications task simulated receiving audio messages from air traffic control, and required pressing keys to make frequency adjustments on navigation and communication radios. Distracter stimuli were occasionally presented in the communications task in which frequency instructions for a different aircraft (identified by a different call sign than the operator's own aircraft) were given. The 2D compensatory tracking task simulated manual control of aircraft position using a joystick with first order control characteristics. Manipulating the difficulty of each subtask in the MATB can serve to vary load; Subjects performed low load (LL), medium load LL, and high load (HL) versions of the tasks. Details of these load manipulations are described elsewhere (Smith, Gevins, et al., 2001). Participants performed the MATB while comfortably seated approximately 60cm. from the computer screen, in blocks or test runs of five minutes.

The data were collected in the context of a large study in which each subject participated in a total of seven experimental sessions conducted on different days (see also Experiment II, below). The first session was a practice session in which subjects learned to perform the tasks. In the second session they performed multiple blocks of the MATB task at the LL, ML, and HL levels of difficulty, as described in Smith, Gevins, et al (2001). EEG was continuously recorded from 28 scalp electrodes using a linked-mastoids reference. EOG was recorded from electrodes placed above and below one eye, and at the other canthus of each eye. Physiological signals were band-pass filtered at 0.01 to 100 Hz and sampled at 256 Hz. Automated artifact detection was followed by application of adaptive contaminant removal filters. The data were then visually inspected and data segments containing possible residual artifacts were eliminated from subsequent analyses.

2.2 Results from Experiment I

The effects of the task manipulations on subjective task difficulty, objective task performance, and EEG variables, are described in detail in Smith, Gevins, et al (2001). Briefly, subjects reported progressively higher task demands on the NASA TLX workload scale across the LL, ML, and HL active performance versions of the MATB. Measures of task performance also tended to vary with overt task demands, but most behavioral measures are not directly comparable across task load because of the different mix of tasks included at each level. Reaction times on one task that was included at each level, the monitoring subtask, were found to be significantly slower in the high load task than in the medium or low load tasks. Finally, EEG parameters in the theta and alpha bands were found to vary systematically with increased task demands, with high load task conditions being associated with increased theta band activity and decreased alpha band activity relative to low load tasks.

In our past research using the MATB, uni-dimensional, subject-specific, cognitive load index functions were derived using examples of data from high and low cognitive load conditions from each subject, and then applied to new data from that subject to derive estimates of cortical activation for the brain as a whole (Smith, Gevins, et al., 2001). In the current project three regional metrics were extracted from the data. One metric was derived from data recorded over frontal cortical areas. Since this region of the brain is known to be involved in executive attention control and working memory processes we will refer to this metric as a measure of cortical activation related to "frontal executive" workload. A second metric was derived from data recorded from central and parietal regions. Since these regions are activated by motor control functions, somatosensory feedback, and the coordination of action plans with representations of extra personal space, we will refer to this second metric as a "sensorimotor" workload component. A third metric was derived from electrodes over occipital regions. Since this region includes primary and secondary visual cortices, we will refer to this third metric as representing variation in "visuoperceptual" workload. (Of course, while these labels are convenient for discussion, they are highly oversimplified with regard to describing the actual operations performed by the underlying cortical systems.)

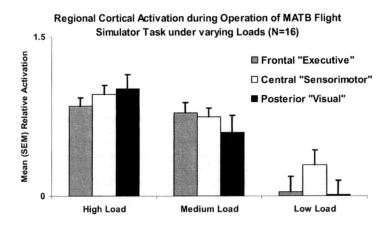

Figure 1. Mean regional cortical activation metrics over frontal, central, and posterior regions of the scalp derived from multivariate combinations of EEG spectral features recorded from N=16 participants performing five minute blocks of high, medium, and low load versions of the MATB task.

To summarize how the three regional cortical activation metrics changed as a result of task manipulations, the means of the regional cortical activation scores were computed across all of the cross validation data segments for each task difficulty level for each subject. **Figure 1** displays the mean and standard error of cortical activation levels across the group of N=16 subjects for the frontal, central, and posterior metrics for the HL, ML, and LL task conditions (for the data presented in both **Figure 1** and **Figure 2** below the metric values were rescaled to range between 0 and 1 across the conditions included in the analysis). In a two-way ANOVA with repeated measures with EEG metric type and task load factors, a highly significant effect of task load was obtained ($F(1,30)=134$, $p < .00001$). For the frontal factor the HL and ML tasks did not significantly differ from one another, but all other pair wise comparisons were significant ($p < .001$ for each). The central metric displayed a more monotonic response to variations in task load, with all pair wise comparisons reaching significance ($p < .02$) for the comparison between the ML and LL conditions, $p < .001$ for all other comparisons). The posterior metric also displayed a monotonic decrease in activation between the HL, ML, and LL task versions ($p < .002$ for all comparisons).

Finally, to examine how variations in regional activation across task versions related to subjective and objective measures of task performance, the workload ratings, monitoring subtask RTs, and EEG activation scores were rank ordered for each participant, and the resulting rankings were correlated using nonparametric statistical methods across the three task load levels and sixteen subjects (using Kendall's tau or τ; Kendall, 1963). Subjective workload ratings were positively correlated with the frontal ($\tau =.67$, $p < .001$), central ($\tau =.58$, p .001), and posterior ($\tau =.74$, $p < .001$) EEG activation scores. Similarly, RTs were also positively correlated with the frontal ($\tau =.53$, $p < .001$), central ($\tau =.54$, p .001), and posterior ($\tau =.64$, $p < .001$) EEG activation scores. That is, across the entire dataset, task blocks that produced a high degree of cortical engagement as inferred from regional EEG activation measures also tended to engender higher subjective workload ratings and to be associated with slowed reaction times.

3 Experiment II: Effects of Sleep Deprivation on Regional Cortical Activation

3.1 Methods

The same 16 subjects that participated in Experiment I also participated in this sleep deprivation experiment. Following the Experiment I test session each subject participated in a total of five additional experimental sessions conducted on different days, each separated by at least one week. Four sessions involved recording from participants before and after they had ingested alcohol, caffeine, antihistamine or placebo. These sessions occurred in a counter-balanced order and the data concerning drug effects are described in a separate report (Gevins, Smith, & McEvoy, 2002). A fifth experimental session was a sleep deprivation manipulation. That session began in the evening and lasted until 0600 the following morning. Data from the sleep deprivation session is the focus of the current report. The alert daytime baseline data was used as a reference point for the data collected in the sleep deprivation session.

In the sleep deprivation session, subjects arrived at the laboratory at approximately 2030, were given a warm-up block of the working memory task and other tasks, and were prepared for the EEG recording. They then participated in five 40-minute recording intervals spaced throughout the night. The first interval occurred on average at 2300, the second at 0030, the third at 0130, the fourth at 0330 and the fifth at 0500. The internal structure of each interval was the same as that during the daytime baseline sessions. Within each interval, participants completed the Karolinska (Akerstedt & Gillberg, 1990) subjective sleepiness rating scale and had their EEG recorded while they performed two blocks each of the easy and difficult versions of the n-back working memory task, and a block of the HL version of the MATB task. The WM task data are the focus of a separate report (Smith, Gevins, et al., 2002).

The subjective measures, the WM task data, and the MATB took approximately 30 minutes to complete during each testing interval. In the periods between these recording blocks, the participants performed other repetitive computer tasks to insure continued wakefulness and to help induce mental fatigue. During periods when the participants were not actively involved in one of the formal tasks they were allowed to read, play video games, or surf the Internet, but no napping was permitted. EEG recording and preliminary analyses followed the same procedures as described in Experiment I.

3.2 Results from Experiment II

Subjective ratings of sleepiness on the Karolinska scale were elevated during the overnight testing session relative to the daytime baseline ($t(15)=4.96$, $p < .001$), and displayed a monotonic increase over the course of the night ($F(4,60)=10.4$, $p < .002$ after Greenhouse-Geisser correction). Subjective ratings of mental workload (using the scaled score version of the NASA TLX rating scale incorporated into the MATB) were generally elevated (on average by approximately 10%) over daytime levels, and tended to increase as the night progressed ($F(4,60)=2.82$, $p < .05$). This elevated subjective workload was accompanied by a modest decline in performance on some subtasks of the MATB. For example, on average subjects responded slower in the monitoring subtask at night than they did during the daytime ($t(15)=2.66$, $p < .02$), with response slowing increasing as the night progressed ($F(4,60)=4.17$, p

< .01). In post hoc comparisons monitoring RTs were not different from the daytime baseline at the 2300 or 0030 test interval, but were significantly slower for the 0130, 0330, and 0530 test intervals (p < .04 for each comparison). Although error rate in the Monitoring subtask was slightly elevated at the latter test intervals, these differences were not significant. Performance in the other subtasks was relatively stable throughout the night. Thus, on the whole, while the sleep deprivation intervention appeared to adversely impact performance ability, subjects nonetheless appeared to continue to make an effort to keep up with the task demands as the night progressed.

During the extended wakefulness session, cortical activation as indexed by the regional EEG cortical activation metrics was observed to change with time-on-task despite task difficulty being held constant and despite the fact that subjects were highly practiced in the task. The changes in these average cortical activation scores during the extended wakefulness session relative to the alert daytime baseline are illustrated in **Figure 2**. The contrast of primary interest here is the comparison with the values representing the average score for the first block of data from the overnight session, where testing on average began around 2300, and with values from the time period within the last four test intervals for each subject when he or she displayed a minimum in total cortical activation (for 15/16 subjects this minima occurred between 0130 and 0530, mean time across subjects was 0400).

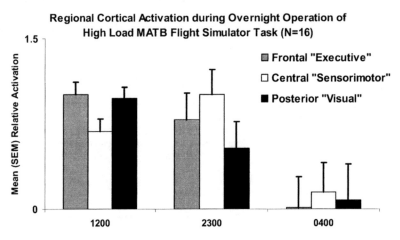

Figure 2. Mean regional cortical activation metrics over frontal, central, and posterior regions of the scalp derived from multivariate combinations of EEG spectral features recorded from N=16 participants performing five minute blocks of the high load MATB task during alert daytime baseline periods, during the first test interval of an all-night recording session, and the average time of the test interval between 0130 and 0530 in which they displayed a cortical activation minima.

These values were included in a two-way ANOVA with repeated measures with the three EEG activation metrics and three test interval values as factors. A significant main effect of test interval was obtained (F(2,30)=16.7, p < .001), but there was no significant effect of cortical activation region by test interval interaction. Simple pair wise comparisons indicated that for the frontal metric there was no significant change in regional activation between the daytime and 2300 test intervals (p=.44), although average cortical activation was lower during the late night period than during either the daytime (p < .004) or 110 PM periods (p < .02). For the central metric there again was no significant difference in regional activation between the daytime and 2300 test intervals (p=.18), or between the daytime and late night test intervals (p = .09), but there was a significant decrease in cortical activation when comparing the late night period with the 2300 interval (p < .001). For the posterior metric there was no significant difference in regional activation between the daytime and 2300 test intervals (p=.20), the daytime and late night test intervals (p = .06), or between the late night period with the 2300 interval (p=.23). The relative insensitivity of the central and posterior metrics to sleep deprivation and sustained performance likely reflects the fact that motor and visuoperceptual demands were held constant across test intervals.

In order to examine how variations in the regional cortical activation scores over test intervals were related to subjective workload, data were averaged over the last three test intervals in the extended wakefulness session and the EEG cortical activation scores were correlated with percent change in subjective workload ratings relative to the alert baseline sessions across the sixteen subjects. Subjective workload ratings were negatively correlated with scores for the frontal region (r=-59, p < .025), positively correlated with scores for the central region (r=.58, p < .025), and not significantly correlated with scores for the posterior region (r=.12, n.s.). This pattern of relations between the EEG cortical activation measures and subjective workload ratings was observed even in partial correlations where variation in subjective sleepiness ratings was controlled for. That is, subjects that displayed relatively low levels of cortical activation over frontal regions during the late night periods also reported an increased sense of task difficulty beyond that which occurred during daytime baseline periods. This suggests that fatigue-related reduction in frontal lobe function increased perceived workload over and above that produced by imposition of a high difficulty task.

4 Experiment III. Real time Monitoring of Regional Cortical Activation

4.1 Methods

Ten healthy adults participated. All participation was fully informed and voluntary, and conformed to all institutional and governmental guidelines for the protection of human subjects. The subjects performed versions of an attentional multiplexing task (AMT) that was designed based on preliminary work by Josyln and Hunt (Joslyn & Hunt, 1998). In the AMT objects descend down a computer screen and the participant must sort them into one of several bins before they reach the bottom based on partial information that is progressively delivered as the stimuli descend. Each object is first presented as a white, open, circle at the top of the screen. As it moves down the screen, an object takes on features that identify it as belonging to 1 of 4 bins by pressing buttons labelled 1, 2, 3, or 4 on a computer keyboard. Bin assignment must occur for each object before it reaches the "time pressure cross" near the center of the screen, or the object will "time-out" and disappear. The bin assignment can be later changed, but at a cost. Once the subject is sure that the bin assignment is correct, the object can be sent to the labelled bin by pressing the spacebar, which is marked "SEND." Each object must be sent to a bin before it reaches the bottom of the screen. Once sent to a bin, the object disappears from the screen and the points earned or lost for that object are presented briefly. Whenever an object is sent to a bin, the score earned is presented briefly. Points are awarded for sending an object to the correct bin and for doing so quickly. Sending an object to the wrong bin or allowing an object to time-out results in penalties. Task difficulty was manipulated by varying event frequency to create Low Difficulty and High Difficulty conditions. In the current study participants were first instructed on task performance using short practice task blocks, and then they performed three critical task blocks of each difficulty level to calibrate and test cortical activation metrics.

The data acquisition electronics used in this experiment consisted of a fast application, EEG headset with nearly dry disposable electrodes and head-mounted, 12-channel, 22-bit wireless EEG amplifiers. Following the three spatial cortical activation metrics utilized in Experiments I & II, for this experiment electrodes were positioned over midline and lateral frontal regions, central regions, and posterior parietal regions. The montage also included electrodes for monitoring eye movement signals (EOG), and ECG and EMG. The system used Bluetooth-based wireless communication from the subject to a PC base-station. The recording system was controlled by a software utility that performed continuous quality control analyses. The data acquisition system also incorporated a real time signal processing module that detected and corrected or eliminated data segments corrupted by artifacts, that concurrently computed regional cortical activation metrics, that displayed the ongoing calculations (updated every four seconds), and that wrote those values to disk permitting post hoc analyses of the results.

4.2 Results from Experiment III

Across the group of subjects, behavioral performance scores on the task were higher in the low difficulty condition than they were in the high difficulty condition (t(9)=5.58, p < .001). Furthermore, subjective workload was lower in the low difficulty task than in the high difficulty task, as measured by the time pressure (t(9)=-7.7, p < .001), mental load (t(9)=-5.00, p < .001), and mental stress (t(9)=-3,2, p < .02) subscales of the NASA TLX subjective workload rating measure. These task difficulty-related differences in performance and subjective measures of mental workload were accompanied by corresponding changes in the spectral composition of the ongoing EEG recorded during task performance. As in the prior studies these included increased spectral power below about 7Hz in the high load task condition (F(1,9)=15.4, p < .004), and decreased spectral power in the alpha band in the more difficult task condition (F(1,9)=7.6, p < .03). Thus, the subjective, behavioral, and EEG results confirm the success of the task manipulation for varying both task difficulty and neural correlates of task load.

To determine whether the online algorithms were capable of reliably detecting these differences, a sample of EEG data collected during the low load and high load AMT were used to automatically calibrate cortical activation measurement functions for frontal, central, and posterior regions, as well as an overall index that summed the outputs from the three regional measures. The resulting functions were then validated by applying them in real time to the data collected during other blocks of the AMT. This resulted in calculation of cortical activation values for each region for each segment of artifact free EEG data, which were displayed concurrently with task performance and saved to disk along with the EEG for subsequent analysis. These values were then averaged within each task difficulty level for each subject, and compared between task difficulty levels in a series of planned comparisons across the group. Cortical activation metric values were found to be significantly higher for the high difficulty version of the AMT than for the low difficulty task version for the frontal (t(9)=-2.3, p < .05), central (t(9)=-3,1, p < .02), posterior (t(9)=-3,5, p < .008), and overall (t(9)=-3,2, p < .02) indices.

5   Discussion and Conclusions

As noted in the introduction, many previous studies have demonstrated that task demands modulate the EEG spectrum. It is also well-known that this modulation displays regional specificity. For example, it is clear that alpha band activity over posterior regions is particularly sensitive to visual stimulation and that increases in motor demands are associated with suppression of alpha and beta band activity over sensorimotor cortex (Arroyo et al., 1993; Jasper & Penfield, 1949; Mulholland, 1995). It is also unambiguous that that when motor and visual confounds are controlled for, increases in cognitive aspects of task difficulty have a significant impact on EEG spectral measures over both frontal and posterior regions of cortex (Gevins & Smith, 2000; Gevins, Smith et al., 1998; Gevins, Smith et al., 1997; Smith, McEvoy, & Gevins, 1999); such task load differences have also been shown to display meaningful regional differences that vary systematically with imposed attentional set (e.g. task requirements to attend to the spatial position of a stimulus versus a verbal label for it).

The results from the current series of studies confirm regional differences in the manner in which the EEG spectrum is modulated by changes in task demands. While each of the regional metrics was on average found to be sensitive to the differences in task demands between the HL and LL task versions, they were differentially impacted by sleep deprivation. In Experiment III we further established that it is also feasible to derive regional estimates of cortical activation from EEG measures in real time.

The observation of regional specificity in task related cortical activation implies that the observed changes in the different regions reflect local sensitivity to different aspects of the task difficulty manipulations. In the current results, manipulations of task difficulty in the MATB and AMT tasks served to increase working memory and decision-making demands, motor control demands, and visuoperceptual demands. Given the well-known functional specialization of the underlying cortical regions, it is tempting and not unreasonable to speculate these different aspects of the task difficulty manipulation differentially and respectively engaged frontal brain regions important for working memory and executive control processes, central regions important for sensorimotor control, and posterior regions important for visuoperceptual processing. However, unambiguous mapping of the type of regional EEG factors observed herein to activation of functionally specialized cortex in response to specific task demands must await future studies in which the cognitive, motor, and visual aspects of tasks are parametrically varied in an independent fashion.

Despite such uncertainties, the changes in the regional measures of EEG activation with sustained task performance and the manner in which those changes relate to both perceptions of mental workload and overt performance ability are of intrinsic interest and provide important information about human brain function under stressful circumstances. In the first set of analyses, regional EEG metrics of cortical activation varied systematically with increases in task difficulty. In Experiment I and Experiment III these differences were positively correlated with subjective mental workload ratings, and are consistent with reports from neuroimaging studies of greater and more extensive cortical activation with increased task demands (Braver et al., 1997; Bunge, Klingberg, Jacobsen, & Gabrieli, 2000). In Experiment II, in response to increases in mental fatigue from sleep deprivation and sustained performance, cortical activation as inferred from the EEG metrics declined, particularly over frontal regions, even though subjective mental workload increased with increased time-on-task. Together the three experiments make it clear that the relationship between activity in the frontal region of the cortex, objective task demands, and perceived mental workload, is not a simple linear one.

It is problematic for the development of brain-adaptive automation systems that perceived mental effort correlated positively with changes in frontal cortical activation in alert individuals, yet correlated negatively with frontal activation when those individuals were sleep deprived. For such neuroadaptive systems to modulate the cognitive task demands placed on an individual in response to momentary variations in the availability of mental resources as indexed by neural activation, sophisticated user models with human-like intelligence will need to be developed that take both task demands and the operator's state of alertness into account.

It is a compelling, futuristic idea to use brain activation measures as a basis for automated systems to off-load tasks from an individual if he or she was in a state of high cognitive workload, or allocate more tasks to an individual that appeared to have ample reserve processing capacity. Like many great visions, the devil is in the details. The current results indicate that a decrease in cortical activation in frontal regions may either reflect a decrease in mental workload, or an increase in mental fatigue and a heightened sense of mental stress. Assigning more tasks to an individual in the former case may indeed serve to increase his or her cognitive throughput. In the latter case it could be quite counterproductive, for instance resulting in the sort of tragic accident that is too often reported to occur when fatigued personnel are confronted with unexpected increases in task demands (Dinges, 1995; Miller, 1996; Rosekind, Gander, & Miller, 1994).

Research using measures of brain function during real world tasks will undoubtedly accelerate understanding of the sources of performance failure under stress. It is realistic, however, to expect that a great deal of methodical research will be needed before such measures can actually be put to use for adaptively augmenting the capabilities of mission-critical personnel working at demanding and stressful tasks. Further systematic developments, for instance such as of the measurement approach outlined here, will eventually lead to an effective and practical means for monitoring transient changes in cognitive brain function that is prerequisite to actual neuroadaptive automation.

6   Acknowledgments

We thank Dr. Linda McEvoy for coordinating the data collection effort, Dr. Robert Du for designing and implementing some of the algorithms used in the analysis, and Cynthia Chan for assistance with data processing. This work was supported by The Air Force Research Laboratory, DARPA and NASA.

7   References

Akerstedt, T., & Gillberg, M. (1990). Subjective and objective sleepiness in the active individual. *International Journal of Neuroscience, 52*, 29-37.

Arroyo, S., Lesser, R. P., Gordon, B., Uematsu, S., Jackson, D., & Webber, R. (1993). Functional significance of the mu rhythm of human cortex: an electrophysiological study with subdural electrodes. *Electroencephalography and Clinical Neurophysiology, 87*, 76-87.

Braver, T. S., Cohen, J. D., Nystrom, L. E., Jonides, J., Smith, E. E., & Noll, D. C. (1997). A parametric study of prefrontal cortex involvement in human working memory. *Neuroimage, 5*, 49-62.

Bunge, S. A., Klingberg, T., Jacobsen, R. B., & Gabrieli, J. D. (2000). A resource model of the neural basis of executive working memory. *Proceedings of the National Academy of Sciences, USA, 97*, 3573-3578.

Comstock, J. R., & Arnegard, R. J. (1992). *The Multi-Attribute Task Battery for Human Operator Workload and Strategic Behavior Research* (No. 104174): NASA Technical Memorandum.

Davis, H., Davis, P. A., Loomis, A. L., Harvey, E. N., & Hobart, G. (1937). Human brain potentials during the onset of sleep. *Journal of Neurophysiology, 1*, 24-37.

Dinges, D. F. (1995). An overview of sleepiness and accidents. *Journal of Sleep Research, 4*(Suppl. 2), 4-14.

Fournier, L. R., Wilson, G. F., & Swain, C. R. (1999). Electrophysiological, behavioral, and subjective indexes of workload when performing multiple tasks: manipulations of task difficulty and training. *International Journal of Psychophysiology, 31*, 129-145.

Gevins, A., & Smith, M. E. (1999). Detecting transient cognitive impairment with EEG pattern recognition methods. *Aviation Space and Environmental Medicine, 70*, 1018-1024.

Gevins, A., & Smith, M. E. (2000). Neurophysiological measures of working memory and individual differences in cognitive ability and cognitive style. *Cerebral Cortex, 10*(9), 829-839.

Gevins, A., & Smith, M. E. (2003). Neurophysiological measures of cognitive workload during human-computer interaction. *Theoretical Issues in Ergonomics Science, 4*, 113-131.

Gevins, A., Smith, M. E., Leong, H., McEvoy, L., Whitfield, S., Du, R., et al. (1998). Monitoring working memory load during computer-based tasks with EEG pattern recognition methods. *Human Factors, 40*, 79-91.

Gevins, A., Smith, M. E., McEvoy, L., & Yu, D. (1997). High-resolution EEG mapping of cortical activation related to working memory: effects of task difficulty, type of processing, and practice. *Cerebral Cortex, 7*, 374-385.

Gevins, A., Smith, M. E., & McEvoy, L. K. (2002). Tracking the cognitive pharmacodynamics of psychoactive substances with combinations of behavioral and neurophysiological measures. *Neuropsychopharmacology, 26*, 27-39.

Gevins, A. S., Zeitlin, G. M., Ancoli, S., & Yeager, C. L. (1977). Computer rejection of EEG artifact. II: Contamination by drowsiness. *Electroencephalography and Clinical Neurophysiology, 42*, 31-42.

Gopher, D., Brickner, M., & Navon, D. (1982). Different difficulty manipulations interact differently with task emphasis: Evidence for multiple resources. *Journal of Experimental Psychology: Human Perception and Performance, 8*, 146-157.

Gundel, A., & Wilson, G. F. (1992). Topographical changes in the ongoing EEG related to the difficulty of mental tasks. *Brain Topography, 5*, 17-25.

Jasper, H. H., & Penfield, W. (1949). Electrocorticograms in man: effect of the voluntary movement upon the electrical activity of the precentral gyrus. *Archives of Psychiatry, Z. Neurology, 183*, 163-174.

Joslyn, S., & Hunt, E. (1998). Evaluating individual differences in response to time-pressure situations. *Journal of Experimental Psychology: Applied, 4*(1), 16-43.

Kendall, M. G. (1963). *Rank Order Correlation Methods (3rd Ed.)*. London: Griffin.

Makeig, S., & Jung, T. P. (1995). Changes in alertness are a principal component of variance in the EEG spectrum. *NeuroReport, 7*, 213-216.

Matousek, M., & Petersen, I. A. (1983). A method for assessing alertness fluctuations in vigilance and the EEG spectrum. *Electroencephalography and Clinical Neurophysiology, 55*, 108-113.

Miller, J. C. (1996). Fit for duty? *Ergonomics in Design*(April), 11-17.

Molloy, R., & Parasuraman, R. (1996). Monitoring an automated system for a single failure: Vigilance and task complexity effects. *Human Factors, 38*, 311-322.

Mulholland, T. (1995). Human EEG, behavioral stillness and biofeedback. *International Journal of Psychology, 19*, 263-279.

Oken, B. S., & Salinsky, M. (1992). Alertness and attention: Basic science and electrophysiologic correlates. *Journal of clinical Neurophysiology, 9*, 480-494.

Parasuraman, R., Mouloua, M., & Molloy, R. (1996). Effects of adaptive task allocation on monitoring of automated systems. *Human Factors, 38*(4), 665-679.

Paus, T., Koski, L., Caramanos, Z., & Westbury, C. (1998). Regional differences in the effects of task difficulty and motor output on blood flow response in the human anterior cingulate cortex: a review of 107 PET activation studies. *Neuroreport, 9*, R37-47.

Rosekind, M. R., Gander, P. H., & Miller, D. L. (1994). Fatigue in operational settings: Examples from aviation environment. *Human Factors, 36*, 327-338.

Smith, M. E. & Gevins, A. (2005). Neurophysiologic monitoring of cognitive brain function for tracking mental workload and fatigue during operation of a PC-based flight simulator. Proceedings of the *SPIE International Symposium on Defense and Security: Symposium on Biomonitoring for Physiological and Cognitive Performance during Military Operations*, Orlando, FL.

Smith, M.E. & Gevins, A. (1997). Monitoring changes in neural indices of working memory load during human-computer interaction. Paper presented at *HCI International, '97*, San Francisco, August.

Smith, M. E., Gevins, A., Brown, H., Karnik, A., & Du, R. (2001). Monitoring task load with multivariate EEG measures during complex forms of human computer interaction. *Human Factors, 43*(3), 366-380.

Smith, M. E., McEvoy, L. K., & Gevins, A. (1999). Neurophysiological indices of strategy development and skill acquisition. *Brain Res Cogn Brain Res, 7*, 389-404.

Smith, M. E., McEvoy, L. K., & Gevins, A. (2002). The impact of moderate sleep loss on neurophysiologic signals during working memory task performance. *Sleep, 25*, 784-794.

The Engineering of Motor Learning and Adaptive Control

Ferdinando A. Mussa-Ivaldi

Department of Biomedical Engineering
Northwestern University
Chicago, IL 60208-3107
sandro@northwestern.edu

Abstract

A distinctive goal of Neural Engineering is to establish bidirectional interactions between nervous system and external devices with a multiplicity of goals. The purpose of this research is twofold: 1) to tap into the power of biological information processing, and 2) to improve the living standards for people suffering from a variety of disorders, ranging from stroke to ALS. To reach either objective it is necessary to understand how the brain learns to carry out complex task and how the biological learning mechanisms can be harnessed and guided toward predefined goals. This is a task which may be defined as the "Engineering of Motor Learning" and which involves addressing two sets of questions:

1) Understanding the biological mechanisms of sensory-motor learning. How does our brain become an expert in mechanics through practice? This question has guided a number of experimental and theoretical studies suggesting that the nervous system organizes a set of building-blocks (or "motor primitives") that are used to control limb movements and to adapt to changing operational conditions.
2) Programming brain-machine interfaces. How can the nervous system be trained to control an artificial device through the bidirectional exchange of information? This question is addressed in different ways by current studies that establish a closed loop interaction in which an artificial device is operated by neural activities and the nervous system receives sensory information from the device.

The biological mechanisms of neural plasticity are essential both to motor learning and to the programming of brain machine interfaces. I will present recent neurobiological, behavioral and computational studies aimed at understanding and gaining access to these mechanisms.

Using Electrical Brain Potentials and Cerebral Oximetry to Detect G-Induced Loss of Consciousness

Lloyd D. Tripp

Air Force Research Lab
Wright-Patterson Air Force Base,
Dayton, OH 45433-7947
lloyd.tripp@wpafb.af.mil

Glenn F. Wilson

Air Force Research Lab
Wright-Patterson Air Force Base,
Dayton, OH 45433-7947
glenn.wilson@wpafb.af.mil

Abstract

G (acceleration) induced Loss of Consciousness (GLOC) remains one of the leading human factors problems in the Air Force today and is responsible for the loss of both aircraft and life. Whinnery, Burton, Boll, and Eddy (1987) have reported that pilots are incapacitated for a total of 24 sec consisting of 12 sec of complete unconsciousness or absolute incapacitation followed by a 12 sec period of subsequent confusion or relative incapacitation. Whinnery and Burton (1988) reported that repeated exposure to GLOC could decrease recovery time of a GLOC event and that GLOC should be incorporated into the centrifuge training curriculum for pilots undergoing high-G centrifuge training. To that end the current study investigated the longitudinal effects of repeated human exposure to GLOC. Electrical brain potentials and cerebral oximetry were used to monitor participants during repeated G exposure and during GLOC episodes. The results provide an example of the application of the neuroergonomic approach to an important practical problem, GLOC.

1   Methods

Ten participants, six males and four females, mean age of 30 years (range 25 to 36 years) experienced GLOC once a week over four weeks on a human centrifuge. Centrifuge facilities located at Wright-Patterson AFB, OH and Brooks City Base TX were used to induce the GLOC events. Participants viewed the task projected onto a screen in front of them and simultaneously performed a mathematical computation task and dual axis compensatory tracking task to emulate flight performance. Thirteen channels of EEG and eye activity were recorded. Additionally, cerebral oximetry using near infrared spectroscopy was employed to measure brain tissue oxygen saturation rSO2

2   Results

The general outline of the allelic association approach to the assessment of individual differences in cognition is as GLOC recovery time for the absolute, relative, and total incapacitation intervals were not significantly different across experimental test days. Performance efficiency of both psychomotor and cognitive tasks deteriorated respectively prior (3.20 to 7.44 sec) to the onset of unconsciousness and did not return to baseline levels until 55.5 sec after the relative incapacitation phase has ended. This study confirms the duration of total incapacitation of 24 sec previously described by Whinnery et al. (1987). The current study had also revealed that the GLOC problem is more serious than they envisioned with an average performance incapacitation time of 87 sec. During an 87 sec period of time a pilot flying at speeds of 500 mph can travel 12.1 miles while not in control of their aircraft. The physiological data have identified seven EEG events that were measured across all subjects. The two most significant were found over widespread scalp sites. They were a 1-2 Hz waveform just prior to G-LOC and a very large, approximately 1 Hz, waveform just prior to regaining consciousness. These changes in EEG which occurred prior to the GLOC event were associated with a drop of rSO2 levels to approximately 90% of the pre-G-LOC baseline levels. Cerebral oxygen saturation levels returned to pre-G-LOC levels within approximately 15 sec following G-LOC while EEG and performance measures took approximately 60 sec

3 Discussion

The application of repeated exposure to GLOC did not discover a longitudinal effect on reducing the time of recovery from a GLOC event. This investigation did, however, illuminate the issue of diminished cognitive and psychomotor performance prior to and following a GLOC episode. These data could be useful to engineers developing automated aircraft recovery systems in terms of when to take control of the aircraft and the amount of time to maintain control of the aircraft following recovery from the unconscious event. The changes in the EEG amplitude and frequency prior to a GLOC event and subsequent recovery of consciousness may prove to be valuable to those currently developing in-flight EEG based GLOC warning/recovery systems. Additionally, cerebral oximetry data may also have utility in predicting a GLOC event. McKinley, Tripp, Bolia, and Roark (in press) developed a model of GLOC using the cerebral oxygen data from this study showing good agreement between the predicted and actual GLOC event with a correlation coefficient of 0.79, best-fit slope of 0.87 with 6 percent error.

4 References

McKinley, R.A., Tripp, L.D., Bolia, S.D., & Roark, M.R. (in press). Modelling the effects of acceleration due to gravity on cerebral oxygen saturation. *Aviation, Space, and Environmental Medicine.*
Whinnery, J., Burton, R.R., Bolls, P.A., & Eddy, D.R. (1987). Characterization of the resulting incapacitation following unexpected +Gz induced loss of consciousness. *Aviation Space and Environmental Medicine, 59,* 631-636
Whinnery, J., & Burton, R. (1988). +Gz-induced loss of consciousness: A case for training exposure to unconsciousness. *Aviation, Space, and Environmental Medicine, 58,* 468-472.

Virtual environments for localizing cognitive impairments in real word tasks

Matthew Rizzo
Division of Neuroergonomics
Department of Neurology
University of Iowa School of Medicine
Iowa City, IA 52242
matthew-rizzo@uiowa.edu

Joan Severson
Digital Artefacts
119 Technology Innovation Center
Oakdale Campus
Iowa City, IA 52242
joan@digitalartefacts.com

Abstract

Our broad goal is to design and develop new tools for evaluating decision-making impairments in cognitively impaired populations and derive tools to help maintain and improve their cognitive functions. Development of a cost-effective and easy-to-maintain system required visual representations that communicate spatial orientation and optical flow in the limited field of view of a standard PC desktop monitor. This approach draws from perceptual psychology, computer graphics, art, and human factors studies of visual cognition. We deviate from traditional simulation approaches, focusing on our assessment needs without assuming visual realism was necessary. The VE tools are designed to provide sufficient pictorial and motion cues relevant for perceiving spatial relationships of objects and user orientation. Similar to high-fidelity driving simulators and in-vehicle navigational systems, the tools utilize motion parallax, optical flow caused by moving objects and the observer, shading, texture gradients, relative size, perspective, occlusion, convergence of parallel lines, and position of objects relative to the horizon. By deviating from a realistic VE design we have an open design space for creative exploration of scenario design and development. Scenario design is guided by cognitive neuroscience (to localize performance errors in specific cognitive domains that are crucial to the real-world task being simulated). We successfully implemented this approach in studies aimed at testing decision making by subjects with cognitive impairments (measured with standardized neuropsychological tests) caused by focal cerebral lesions (defined by magnetic resonance imaging).

Molecular Genetics of Augmented Cognition

Raja Parasuraman

George Mason University
Arch Lab, MS 3F5
4400 University Drive
Fairfax, VA 22030-4444
rparasur@gmu.edu

Pamela Greenwood

George Mason University
Arch Lab, MS 3F5
4400 University Drive
Fairfax, VA 22030-4444
pgreenw1@gmu.edu

Abstract

It is well known that there are large individual differences in most human information processing functions, including attention, memory, decision making, and action selection. In the past, quantification of individual differences in these functions has been dominated by the idiographic approach, as in the development of questionnaire assessments of intelligence and personality. This paper describes a new, complementary approach that capitalizes on the breakthroughs provided by the success of the Human Genome Project. There is a burgeoning research field examining the role of single nucleotide polymorphisms (SNPs) of neurotransmitter genes in different aspects of cognitive function. Different SNPs of genes involved in neurotransmitter synthesis and breakdown and to neurotrophic action sin the brain can be linked to different cognitive functions such as attention and working memory. Specifically, the "gene dose" of a given SNP can be shown to predict, with moderate to high effect sizes, individual differences in different cognitive functions. Recent findings linking SNPs of the DBH and CHNRA genes to specific component cognitive processes are described. These findings provide the beginning basis for a molecular genetics of augmented cognition.

1   Introduction

Neuroergonomics is an emerging field of research and practice that is concerned with the study of brain mechanisms in relation to the use of technology at work and in everyday life (Parasuraman, 2003). The goal is to harness the power of neuroscience to the engineering of human-machine systems for safety and efficiency. Neuroergonomic measurements can be used to change aspects of the human-machine interface or to augment human cognition in real time. The DARPA Augmented Cognition program is similarly aimed at enhancing system effectiveness by managing information presentation and enhancing operator cognitive processing capacity by monitoring neural and physiological signals (St. John, Kobus, Morrison, & Schmorrow, 2004). Both these approaches attempt to use measures of brain function to re-engineer the human-machine interface, not merely in the design phase, but also dynamically during system operations.

In addition to interface design, selection and training of operators to meet system demands represents another approach to enhancing human-machine performance. Individual differences in cognitive functions relevant to operator performance have typically been evaluated in the framework of the idiographic approach, using self-assessments of intelligence and personality. This paper describes a new, complementary approach that capitalizes on the breakthroughs provided by the success of the Human Genome Project. The goal is to increase knowledge of individual variation in cognitive functions by parsing genetic and environmental contributions and to identify specific genes that are related to these functions. By examining the expression of these genes in the brain, genetic and neural information can be combined to yield a more robust understanding of the neural correlates of normal cognitive variation.

Much of what we know about the genetics of cognition has come from twin studies in which identical and fraternal twins are compared to assess the heritability of a trait. This paradigm has been widely employed by behavioral geneticists for over a century and has been used, for example, to show that general intelligence, or *g*, is highly heritable (Plomin & Crabbe, 2000). Recent advances in molecular genetics now provide a different, complementary approach, that of allelic association. A proportion of genes in the human genome show small variations (called alleles) between unrelated individuals in a part of the DNA sequence of base pairs of nucleotides that defines the gene. Such allelic differences between individuals can then be associated with differences in cognitive functions in the same people. The allelic association method has been recently applied to the study of individual differences in cognition in healthy individuals and revealed increasingly compelling evidence of modulation of cognitive task performance by specific neurotransmitter and neurotrophic genes (Egan et al., 2001; Fan, Fossella, Sommer, Wu, & Posner, 2003; Greenwood, Sunderland, Friz, & Parasuraman, 2000; Parasuraman, Greenwood, Kumar, & Fossella, 2005).

2   A Framework for the Study of the Molecular Genetics of Cognition

The general outline of the allelic association approach to the assessment of individual differences in cognition is as follows (for more details, see Greenwood & Parasuraman, 2003; Parasuraman & Greenwood, 2004). The first step involves the identification of candidate genes—genes deemed likely to influence a given cognitive ability or trait due to the functional role of each gene's protein product in the brain. More than 99% of individual DNA sequences in the human genome do not differ between individuals and hence are of limited interest in investigating individual differences in normal cognition. However, a small proportion of DNA base pairs occur in different forms or alleles. Many of these are due to substitution of one nucleotide for another—a SNP—at a particular location within the gene. Other SNPs involve repetitions or deletions of certain base pairs. There are several web sites that list currently known human SNPs, with the databases being updated regularly.

Since SNPs are so numerous (2-3 million), some constraints on a search through the SNP databases is necessary. As a first cut, those SNPs should be selected that are likely to influence neurotransmitter function or to have effects on neurotrophic activity. In the second step, cognitive neuroscience research on the cognitive function in question (e.g., working memory) is examined to identify the brain networks that mediate that function. Third, the neurotransmitter innervation of these networks is identified. Finally, SNPs that influence neurotransmitter or neurotrophic function are identified and associated with the cognitive function.

There are some limitations in the combined allelic association/cognitive neuroscience approach to the genetics of cognition. No component of cognition, no matter how microscopic, is likely to be modified by only one gene, and the interpretation of individual differences in a particular cognitive function will ultimately involve specification of the role of many genes as well as environmental factors (Plomin & Crabbe, 2000). It is also important that SNPs or other candidate genes are chosen in a theory-based manner for their functional significance for cognition, so as to minimize the probability of type I error in finding gene-cognition links.

3   Neurotransmitter Genes, Intelligence, and Working Memory Capacity

Twin studies have shown that Spearman's g, or general fluid intelligence, is highly heritable (Plomin & Crabbe, 2000). This finding leads to two questions that can be addressed with modern neuroimaging and molecular genetic methods. The first question is whether g is associated with genetically based individual differences in brain structure and function. If so, then this would indicate that components of g, or specific cognitive abilities such as working memory, are also likely to be related to variations in brain structure and function controlled by specific genes. With respect to the first issue, individual differences in g are indeed strongly related to differences in the morphological and functional characteristics of prefrontal cortex. By comparing structural MRIs of monozygotic and dizygotic twins, Thompson et al. (2001) showed that gray matter volume within specific regions of cerebral cortex was heritable. Frontal cortex showed the strongest genetic correlation of the four cerebral lobes and estimates of Spearman's g for these participants were most strongly related to frontal gray matter volume. Gray, Chabris, and Braver (2003) also showed that high-g individuals were more accurate in a working memory task and activated areas of prefrontal cortex to a greater degree than low-g individuals.

Thus, there is good evidence for links between genes, general intelligence, and frontal cortex structure and function, Given that working memory is a major component of g, can the specific genes that moderate these links be identified? Recent research suggests a positive answer to this question.

Dopaminergic receptor genes are likely candidates for genetic effects on working memory due to the importance of dopaminergic innervation for prefrontal cortical areas involved in working memory. Dopamine agents have been shown to modulate working memory and prefrontal cortex (PFC) function in monkeys (Sawaguchi & Goldman-Rakic, 1991) and humans (Muller, von Cramon, & Pollmann, 1998). Dopamine plays an important role not only in PFC-mediated processes of working memory, but also in hippocampal inputs to that region (Gurden et al., 2000). Candidate genes include the COMT gene, which is involved in the dopaminergic degradation pathway, (Egan et al., 2001). Another candidate is the DBH gene, which is involved in converting dopamine to norepinephrine in adrenergic vesicles (Cubells et al., 1998). SNPs in the DBH gene, including a G to A substitution at 444, exon 2 (G444A) on chromosome 9q34, have been linked to plasma levels of dopamine beta hydroxylase (DßH), changes in the dopamine to noradrenaline ratio in brain (Cubells & Zabetian, 2004) and to attention deficits in children (Daly et al., 1999). We therefore examined its role in attention and working memory (Parasuraman et al., 2005).

We genotyped a group of healthy adults aged 18-68 years for the G444A polymorphism of the DBH gene and tested them on tasks of working memory and visuospatial attention. DBH is a functional polymorphism, as the A allele is associated with lower plasma and cerebrospinal fluid DßH levels, and the G allele with higher DßH levels (Cubells et al., 1998). The working memory task was a variant of the delayed match to sample paradigm and involved maintaining a representation of up to three spatial locations over a period of three seconds. After a fixation period, participants were shown target circles at one to three locations for 500 ms. Simultaneous with the offset of the dot display, the fixation cross reappeared for a 3 s delay at the end of which a single red test dot appeared alone, either at the same location as one of the target dot(s) (match) or at a different location (non-match). Participants had 2 s to decide whether the test dot location matched one of the target dots.

We first assessed the sensitivity and reliability of the working memory task. As Figure 1 shows, matching accuracy decreased as the number of locations to be maintained in working memory increased, demonstrating the sensitivity of the task to variations in memory load. In follow-up studies with other samples of subjects, we have found this task to provide a well-specified memory load function across a range of memory set sizes and delays. Furthermore, inter- and intra-individual differences on the task are relatively stable. The 2-day test-retest reliability at the highest 3-location memory load was .75, whereas the 6-month stability coefficient was .67.

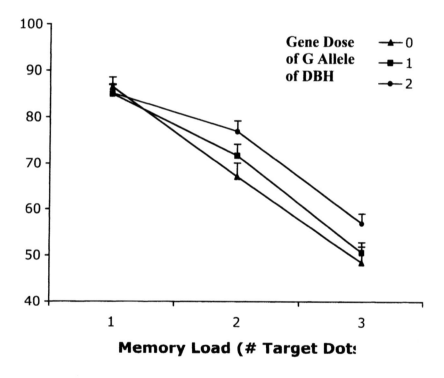

Figure 1. Association between the DBH gene and working memory accuracy at different memory loads.

857

These findings indicate that individual differences on the working memory task were reliable and stable. We then examined to what extent differences between individuals could be associated with the gene dose of the G allele of the DBH gene. Accuracy was equivalent for all three genotypes at the lowest memory load, but increased with higher gene dose of the G allele, particularly for the highest (3 target) load, as confirmed by a simple effects analysis (see Figure 1). Memory accuracy for the GG allele (G gene dose =2) was significantly greater than that for both the AG (G gene dose =1) and AA alleles (G gene dose = 0). The effect size of the G allele on working memory accuracy at the highest memory load was moderate to high, .25. We also administered a visuospatial attention task with little or no working memory component to the same group of participants. Individual differences in performance of this task were not significantly related to allelic variation in the DBH gene. Furthermore, working memory accuracy at the highest memory load was not correlated with performance on the attention task. In sum, these findings point to a substantial association between the DBH gene and working memory performance.

Increasing gene dose of the G allele of the DBH was associated with better working memory performance. This effect was most apparent when the number of target locations to be retained was high. Thus, the association between the DBH gene and working memory was particularly marked under conditions that most taxed the working memory system. Cubells et al. (1998) reported that the G444A polymorphism of the DBH gene influences levels of the DBH enzyme in CSF and there is evidence for high concentrations of DBH-labeled fibers in several prefrontal cortical sites (Gaspar, Berger, Febvret, Vigny, & Henry, 1989). While the precise relationship between the enzymatic activity of DBH and human brain dopamine levels is not known, the association we found between DBH genotype and working memory is consistent with the well-known role of dopaminergic agents in prefrontal cortex and its dopaminergic mediation of working memory (Abi-Dargham et al., 2002).

4 Conclusions

The results of these studies are only the tip of the iceberg. Many other exciting findings on single genes and cognition are emerging at a breakneck pace. The coming decade will witness an explosion of research that may well revolutionize our knowledge of and ability to predict individual differences in various aspects of cognition. To date, neurotransmitter polymorphisms have been linked to individual differences in working memory and attention. In future work, other neurotransmitter polymorphisms need to be examined. Parasuraman et al. (2005) also examined a cholinergic receptor gene, CHRNA4, and found that it was significantly associated with performance on a spatial attention task, but not with the same working memory task that was associated with the DBH gene. These findings are consistent with a double dissociation between the effects of CHRNA4 and DBH on spatial attention and working memory. Also, additional analyses need to be conducted to examine whether the polymorphisms, which have been associated with specific cognitive functions, are inherited together with other DNA loci. Such so-called haplotypes may provide for a better understanding of the functional relationships between genotype and phenotype.

In future research cognitive phenotypes will need to be supplemented by those derived from electrophysiological and neuroimaging measures (e.g., Fan et al., 2003). Positron emission tomography (PET) studies using neurotransmitter ligands may also permit new genetic associations to be discovered. Finally, given progress in understanding the proteins that individual genes code for, functional polymorphisms in neurotransmitter and neurotrophic genes may allow for a "cognitive proteomics", whereby individual differences in cognitive function can be associated with differences in gene-lined protein levels. All these methods will lead to better understanding of sources of individual differences in cognition.

5 References

Abi-Dargham, A., Mawlawi, O., Lombardo, I., Gil, R., Martinez, D., Huang, Y., Hwang, D. R., Keilp, J., Kochan, L., Van Heertum, R., Gorman, J. M., & Laruelle, M. (2002). Prefrontal dopamine D1 receptors and working memory in schizophrenia. *Journal of Neuroscience 22,* 3708-3719.

Cubells, J. F., van Kammen, D. P., Kelley, M. E., Anderson, G. M., O'Connor, D. T., Price, L. H., Malison, R., Rao, P. A., Kobayashi, K., Nagatsu, T., Gelernter, J. (1998). Dopamine beta-hydroxylase: two polymorphisms in linkage disequilibrium at the structural gene DBH associate with biochemical phenotypic variation. *Human Genetics, 102(5),* 533-40.

Cubells, J. F., & Zabetian, C. P. (2004). Human genetics of plasma dopamine ß-hydroxylase activity: Applications to research in psychiatry and neurology. *Psychopharmacology, 174,* 463-476.

Daly, G., Hawi, Z., Fitzerald, M., & Gill, M. (1999). Mapping susceptibility loci in attention deficit hyperactivity disorder: Preferential transmission of parental alleles at DAT1, DBH, and DRD5 to affected children. *Molecular Psychiatry, 4,* 192-196.

Egan, M. F., Goldberg, T. E., Kolachana, B. S., Callicott, J. H., Mazzanti, C. M., Straub, R. E., Goldman, D., & Weinberger, D. R. (2001). Effect of COMT Val108/158 Met genotype on frontal lobe function and risk for schizophrenia. *Proceedings of the National Academy of Sciences U S A, 98(12),* 6917-6922.

Fan, J., Fossella, J. A., Sommer, T., Wu, Y., & Posner, M. I. (2003). Mapping the genetic variation of attention onto brain activity. *Proceedings of the National Academy of Sciences U SA, 100(12),* 7406-7411.

Gaspar, P., Berger, B., Febvret, A., Vigny, A., & Henry, J. P. (1989). Catecholamine innervation of the human cerebral cortex as revealed by comparative immunohistochemistry of tyrosine hydroxylase and dopamine-beta-hydroxylase. *Journal of Comparative Neurology, 279,* 249-271.

Gray, J. R., Chabris, C. F., & Braver, T. S. (2003). Neural mechanisms of general fluid intelligence. *Nature Neuroscience, 6,* 316-322.

Greenwood, P., & Parasuraman, R. (2003). Normal genetic variation, cognition, and aging. *Behavioral and Cognitive Neuroscience Reviews, 2(4),* 278-306.

Greenwood, P. M., Sunderland, T., Friz, J. L., & Parasuraman, R. (2000). Genetics and visual attention: Selective deficits in healthy adult carriers of the varepsilon 4 allele of the apolipoprotein E gene. *Proceedings of the National Academy of Sciences U S A, 97(21),* 11661-11666.

Gurden, H., Takita, M., & Jay, T. M. (2000). Essential role of D1 but not D2 receptors in the NMDA receptor-dependent long-term potentiation at hippocampal-prefrontal cortex synapses in vivo. *Journal of Neuroscience, 20(22),* RC106.

Muller, U., von Cramon, D. Y., & Pollmann, S. (1998). D1- versus D2-receptor modulation of visuospatial working memory in humans. *Journal of Neuroscience, 18(7),* 2720-2728

Parasuraman, R. (2003). Neuroergonomics: Research and practice. *Theoretical Issues in Ergonomics Science. 4,* 5-20.

Parasuraman, R., & Greenwood, P. M. (2004). Molecular genetics of visuospatial attention and working memory. In M. I. Posner (Ed.) *Cognitive neuroscience of attention.* (pp. 245-259). New York: Guilford.

Parasuraman, R., Greenwood, P. M., Kumar, R., & Fosselaa, J. (2005). Beyond herttability: Neurotransmitter genes differentially modulate visuospatial attention and working memory. *Psychological Science, 16(3),* 200-207.

Plomin, R., & Crabbe, J. (2000). DNA. *Psychological Bulletin, 126(6),* 806-828.

Sawaguchi, T., & Goldman-Rakic, P. S. (1991). D1 dopamine receptors in prefrontal cortex: involvement in working memory. *Science, 251*(4996), 947-950.

St. John, M., Kobus, D. A., Morrison, J. G., & Schmorrow, D. (2004). Overview of the DARPA augmented cognition technical integration experiment. *International Journal of Human-Computer Interaction, 17,* 131-149.

Thompson, P. M., Cannon, T. D., Narr, K. L., van Erp, T., Poutanen, P. T., Huttunen, M., Lönnqvist, J., Standertskjöld-Nordenstam, C. G., Kaprio, J., Khaledy, M., Dail, R., Zoumalan, C. I., & Toga, A. W. (2001). Genetic influences on brain structure. *Nature Neuroscience, 4,* 1253-1258.

Neuro-ergonomics Support from Bio- and Nano-technologies

Rene de Pontbriand, Ph.D.

US Army Research Laboratory
rdepontb@arl.army.mil

Abstract

The goal of neuro-ergonomics is a better understanding of the brain's functional structures and activities in relation to work and technology. The objective is to map out the neural systems and the brain's functional structures, activity centers and pathways, allowing diagnostic study of precursor and consequential states. This intimate understanding of the interplay of these structures and phenomena will, it is hoped, lead to development of more intuitive human-system interfaces. These intuitive interfaces---displays, operator control units, driving mechanisms, for example---would require less training, and lead to more consistent and reliable performance by a broader category of individuals than the is the norm today.

Biotechnology, in its elegant sense of biomimetics, provides the means both to study neuronal activities down to the molecular level, in a way that allows us to model the brain's activities with precise depiction, and consequently to study effects of various manipulations upon selected elements of that complex system, in silico (so to speak). This provides the foundation for an in-depth look at possible improvement strategies without the health, political and other concerns that would exist with true human subjects.

Nanotechnology provides the measurement tools needed to actually sense the status and changes in otherwise undetectable structures. In addition, and as importantly, it provides the scale of operations necessary to deliver nutriceuticals and other pharma which might be needed to precisely monitor and modify effects of neurotransmitters, the structure of glial cells, encourage targeted neuro-genesis, and other aspects of neural life, all toward the objective of improved neural health and human performance.

This paper will: discuss each of these three emerging technologies, including a brief review of human-performance related work at the US Army's Institute for Soldier Nanotechnology (MIT) and its Institute for Collaborative Biotechnologies (UCSB, CalTech, MIT); consider the common and differentiating foundation disciplines; assess how the technologies can complement each other; and conclude with a discussion of potential applications to the increasingly automated environment of warfighters, search and rescue personnel, medical practitioners and complex-system managers.

Section 4
Cognitive State Sensors

Chapter 18

Affective Computing

Affect Sensing and Recognition:
State-of-the-Art Overview

Eva Hudlicka

Psychometrix Associates, Inc.
Blacksburg, VA, USA
evahud@earthlink.net

Abstract

The past decade has witnessed an unprecedented growth in user interface and human-computer interaction (HCI) technologies and methods. The synergy of technological and methodological progress on the one hand, and changing user expectations on the other, are contributing to a redefinition of the requirements for effective and desirable human-computer interaction, and are influencing social interaction and collaboration. A key element in these developments is the increasingly important role of affect. The ability to accurately recognize human emotions enhances the effectiveness of HCI and contributes to improved human performance. The ability of machines to manifest behaviours that appear to reflect particular emotions enhances their effectiveness across a range of tasks, including training and education, treatment, and emerging relational agents. In this paper I provide an overview of the state-of-the-art in the emerging technologies supporting emotion sensing and recognition by machines. I also outline several options for the generation of expressive machine 'behaviours', which can be interpreted by humans to reflect particular emotions. I present a framework for describing various approaches to emotion recognition across multiple modalities, including facial expressions, physiological signals, speech, body postures, and gestures and movements. I conclude with a discussion of some fundamental questions regarding the integration of affect in HCI.

Keywords: emotion sensing, emotion recognition, emotion expression, overview.

1 Introduction

The past decade has witnessed an unprecedented growth in user interface and human-computer interaction (HCI) technologies and methods. Immersive virtual environments and the use of multiple modalities to support human-computer interaction are transforming the way we use automated systems and computers, as well how we collaborate with each other. The synergy of technological and methodological progress on the one hand, and changing user expectations on the other, are contributing to a redefinition of the requirements for effective and desirable human-computer interaction, and are influencing social interaction and collaboration.

At the same time, there has been an increasing recognition of the critical role of affect in effective human-computer interaction. Research in the broad area of affective computing[1] (Picard, 1997), along with advances in sensor technologies, pattern recognition methods, and multimedia and VR technologies, is motivating and enabling a range of innovative applications involving affect recognition and expression. Whether we are considering more traditional applications, such as training and tutoring or decision-support systems, or more recent developments, such as various types of relational agents and immersive virtual environments, it is important to take into consideration affective factors. Training and tutoring can be enhanced via adaptive methods if systems capable of detecting student frustration, curiosity or surprise; customer care can be improved if calls are automatically routed to agents capable of handling customer frustration; safety can be improved in various settings, including aviation and medicine, if machines can detect affective states likely to increase error (e.g., frustration and anger, or anxiety).

New developments in HCI technology are making a range of novel human-computer interactions possible, and are narrowing the gap between the human and the machine at the human-machine interface. Machines are increasingly able to sense, or infer, user attributes, and use increasing numbers of available 'modalities' to interact with the user (e.g., virtual reality (VR) technologies used in neuropsychological assessment (Rizzo et al., 2003) and as adjuncts to behavioural treatment of a variety of phobias (e.g., http://www.virtuallybetter.com/) (Zimand et al., 2003)).

---

[1] "computing that relates to, arises from, and deliberately influences emotion"

In this paper I provide an overview of the state-of-the-art in the emerging technologies supporting emotion sensing and recognition by machines, and the generation of expressive machine 'behaviours', which can be interpreted by humans to reflect particular emotions. I first briefly outline the characteristics of emotions relevant for recognition and expression (section 2), and provide a framework for describing the generic requirements for emotion recognition (section 3). I then discuss methods for emotion sensing and recognition across a number of modalities, including facial expression, physiological signals, speech, and various types of movements, in terms of this framework described (section 4). This is followed by brief discussion of the available modalities for producing machine-generated 'behaviours' which can be recognized by humans as reflecting particular emotions (section 5). The paper concludes with a list of questions and issues regarding the integration of affect in HCI (section 6). It should also be noted that this paper does not a critical component of successful use of emotion in HCI: emotion modelling and user modelling.

2   Characteristics of Emotions Relevant for Recognition and Expression

Below I briefly outline some of the characteristics of emotions relevant for machine recognition and expression.

Multiple Modalities Emotions are complex phenomena, manifested across multiple systems within the organism, and multiple expressive modalities (also referred to as data channels). Generally, four categories of emotion manifestations are considered (Ortony, 2001): Cognitive / Interpretive, Somatic / Physiological, Motivational / Behavioral, and Experiential / Subjective. When considering emotion sensing and recognition, whether by humans or machines, two of these categories are most relevant: the *observable behavioural manifestations* (e.g., emotion expressions via facial expressions, posture, gestures and movements, speech attributes (e.g., pitch) and content; and behavioural choices), and the *somatic and physiological manifestations* (e.g., heart rate, galvanic skin response, respiration rate, and blood volume pressure, as well as the less readily accessible chemical sensors).

Characteristic Modalities Associated with Different Emotions and Emotion 'Signatures' Across Modalities While emotions are manifested across all of the modalities outlined above, particular emotions may be more strongly manifested within a specific subset of these modalities. Thus, for example, positive valence is readily detected from facial expressions; arousal correlates with heart rate, etc. This has implications for selecting a particular modality, and associated sensors (for recognition) or output "devices" for expression), to optimize the recognition and expression effectiveness. Identifying characteristic emotion signatures across the distinct modalities can be challenging and represents an active research area. Because of the challenges in emotion recognition and expression, it may be necessary to use multiple modalities in a particular application to ensure the desired level of accuracy of recognition and effectiveness of expression (e.g., see discussion in Hudlicka & McNeese, 2002). Indeed, one of the reasons that machine still lag behind people in recognition of emotion is the typical focus on one modality.

Emotion "Signatures" Over Time Another means of maximizing the use of available emotion 'signals' is to expand our consideration of signal patterns from a static freeze-frame, to a sequence of frames, and the dynamic evolution of these sequences over time. Again, this is the more naturalistic manner in which emotions are recognized (and generated) by humans, and is likely to prove more effective in both machine recognition and expression.

Theoretical and Practical Limits With respect to recognition and expression, all emotions were not created equal. Typically, it is the simpler, so-called 'basic' emotions that are more readily recognized, by humans and by machines. These include fear, anger, joy, sadness, surprise, and disgust. These emotions share more universal triggers (e.g., large approaching objects for fear), and behavioural response patterns, across and within individuals, and even cultures. As we progress from these more 'primitive' emotions to more cognitively complex emotions such as pride, shame, guilt, we begin to see corresponding variability in expression, including the lack of immediately visible or characteristic observable expressive patterns. When considering emotion recognition and expression, it is therefore important to recognize these limits, and, at the same time, recognize that there may be instances where complex pattern recognition methods, using data from multiple modalities, may be more effective in recognizing emotions than humans.

3 Requirements for Affect Sensing, Recognition and Expression

The sensing and recognition of emotions, and the expression of their myriad of manifestations, is a challenging undertaking, requiring the integration of hardware (sensors), mathematical methods for data enhancement and filtering, and pattern recognition and classification, and new and unorthodox output devices for generating machine emotion 'expressions' (e.g., facial expression generators, 3-D avatars capable of gestures and movement). Below we provide a framework for describing the generic requirements for emotion recognition and expression across multiple modalities, by describing the key required components.

Diagnosticity Related to the discussion of unique emotion signatures across modalities, it is important to identify the sources of data most diagnostic of particular emotion, and the associated sensors. Identification of unique patterns of data characterizing a particular emotion, across multiple modalities, either static or dynamic, can greatly increase the chances of successful recognition and expression.

Semantics It is also critical to develop a meaningful vocabulary of 'primitives' associated with each modality; that is, the identification of basic primitive units or tokens within the modality, which then allows an effective encoding of the emotion manifestations. Examples of these units within different modalities are the facial action units comprising the Facial Action Coding System developed by Ekman and Friesen (1978), the 'basic posture units' identified by Mota and Picard (2004) and used to identify boredom and engagement during training, and patterns in speech pitch and tone (Petrushin, 2000).

Emotion Sensors and Input Devices Distinct modalities (e.g., facial expressions, physiological signals, speech, movement) require correspondingly distinct sensors and input devices. These are in turn associated with differences in intrusiveness (e.g., facial EMG electrodes vs. cameras in facial expression recognition), expense, requirements for specialized training, quality of data, and requirements for further mathematical processing for successful feature extraction and classification. All of these must be taken into consideration to ensure successful emotion recognition and expression.

Data Filtering Related to the above, a variety of data filtering and enhancement methods exist, with distinct characteristics, computational requirements, and effectiveness for specific categories of signals. It is important to make an appropriate choice of method(s) at this stage, to ensure the desired recognition performance.

Pattern Recognition Algorithms Once the appropriate features are extracted from the data, the patterns characteristic of particular emotions must be identified. This is usually done via a variety classifier algorithms and methods, which use either static or dynamic data, and some type of supervised, semi-supervised or unsupervised learning algorithms. Depending on the complexity of the problem, hierarchical approaches may be used. Artificial neural networks, multivariate statistical analysis methods and hidden Markov models are among the most popular methods.

Emotion Expression "Output Devices" A number of commercially available options exist for machine 'expression' of emotion, including facial expression on synthetic faces (e.g., PeoplePutty from Haptek.com), 3-D synthetic avatars for gaze, gesture, posture and movements (e.g., DI-Guy from Boston Dynamics). Research is also being conducted in producing emotion in speech (e.g., GRETA (de Rosis et al., 2003), Oudryer, 2003), and novel and experimental devices, such as a child's doll (Paiva, 2003).

The categories above will provide a framework for a more detailed discussion of emotion recognition and expression across the different modalities in the remainder of this paper.

4   Affect Sensing and Recognition

As mentioned above, emotions are manifested in humans across a range of modalities. However, for machine emotion sensing and recognition purposes, the most relevant modalities are the behavioural and physiological / somatic. Below I provide an overview of the technologies and methods available for sensing and recognizing human emotions by machines within these modalities. In particulate, I will focus on the following: facial expressions; physiological signals; gestures and movements; body posture; and speech. While there are other manifestations of emotions that may be used for recognition purposes (e.g., cognitive and behavioural manifestations detected via diagnostic tasks capable of discriminating among emotions), these have not yet been sufficiently developed and will not be addressed here.

4.1   Facial Expressions

Facial expressions are one of the most obvious manifestations of emotion. Extensive literature exists in psychology on emotion expression (Cacioppo et al., 1992; Ekman & Friesen, 2003), and much progress has been made in machine emotion recognition via facial expression (Cohn & Ekman, 2003; Essa & Pentland, 1997; Fasel & Luettin, 2003).

Diagnosticity The so-called 'basic emotions' are reflected in the face via characteristic configurations of the facial 'vocabulary' (e.g., the shape of lips, eyebrows, narrowing of eyes, raising cheeks), controlled by a variety of facial muscles. For example, happiness / joy are characterized by the raising of the lip corners (controlled by the zygomatic major), and the narrowing of the eyelids, causing the characteristic 'crowsfeet' wrinkles, and raising of the upper cheek areas. (Interestingly, frequently seen 'fake' smiles can be identified by lacking some of these components (e.g., the narrowing of the eyelids, crowsfeet), and including other components (e.g., tense broadening of the mouth)). The emotions associated with characteristic patterns are: happiness / joy, sadness / distress, anger, fear, disgust. With the possible exception of happiness, most emotions "cannot be identified by the activity of a single muscle" (Cohn & Ekman, 2003).

It should be noted here that not all researchers subscribe to the view of a one-to-one mapping between basic emotions and unique facial expressions; a view advocated by the so-called 'discrete' emotion theorists, most notably Ekman and Izard. In contrast, the component process theorists postulate that the specific configuration of the facial musculature corresponds to a combination of the results of the cognitive appraisal process and the preparation for action (Kaiser and Scherer, 1998; Ortony and Turner, 1990).

Semantics Emotion expressions have been extensively analyzed by psychologists to identify units of analysis most suitable for effectively describing the variety of observed facial expressions. The most extensively used coding and analysis system is Ekman and Friesen's Facial Action Coding System (FACS) (Ekman & Friesen, 1978). FACS represents an example of a mature semantic vocabulary that is used extensively in machine recognition of emotion, using variety of data (see below). FACS identifies characteristic position of facial muscles in terms of action units (e.g., 'AU6' refers to the muscle controlling the eye shape (orbicularis occuli), AU12 refers to the zygomaticus major, which controls the upward movement of lip corners, and is characteristic of smiling, and AU24 refers to a lip 'tightening' characteristic of anger).

Sensors and Input Devices Two options exist for obtaining the data necessary to recognize emotion from facial expressions: *facial electromyography* (EMG), that is, the direct sensing of the muscle movements via surface or needle electrodes, and *computer vision approaches*, where facial expressions are identified via automatic analysis of face images (Cohn & Ekman, 2003), either enhanced with markers (Kaiser & Wehrle, 1992), such as Vicom™ and Peak Performance™, or markerless, the latter being preferred since they minimize intrusiveness. Of these, the EMG method is the more established, but recently much progress has been made in computer vision analysis. EMG has high temporal resolution but interpretation can be challenging because muscle fibers are intertwined, and it can be difficult to establish that a particular signal corresponds to a specific muscle rather than a combined muscle group. To reduce the intrusiveness of EMG, researchers are experimenting with novel sensors, such as the 'expression glasses' (Picard, 2000), which sense the signals associated with eye muscles and have been used to discriminate between interest and confusion. Whatever the method, the sensors should provide three types of information about the facial 'actions': action type (e.g., which muscle is doing what), intensity, and timing (onset, apex, offset). Both still images and videos are used, the latter being superior.

Data Filtering A number of data filtering and feature extraction methods are available. Cohn and Ekman cite four of the most commonly used for facial image analysis, singly or in combination: difference imaging, principal components analysis (PCA), optical flow, and edge detection (Cohn & Ekman, 2003).

Pattern Recognition Algorithms Once the features are detected, a number of classifier algorithms and methods can be used to identify particular expressions. Typically, the data are divided into a training set, used to train the classifier, and a testing set, used to validate its effectiveness (Cohn & Ekmanm, 2003). The most frequently used methods are artificial neural networks and hidden Markov models, the latter using temporal information (Cohn & Ekmanm, 2003). A variety of classifiers have been explored (see Bartlett et al., 2004 for a review).

Summary Specific success rates for emotion recognition via facial expressions vary, but on the average, researchers report 85% accuracy in recognizing the 'basic' emotions. Valence is most readily assessed, with rates close to 90 %, and facial EMG is effective in discriminating between positive and negative emotion (Cacioppo et al., 1986). Both the EMG and computer vision approaches are expensive, labor intensive and require special training to use effectively.

4.2 Psychophysiological Signals

Emotions are manifested across a variety of physiological systems within the body and assessed via a number of psychophysiological measures (e.g., skin conductance, heart rate, pupil size, respiration, blood volume pressure), which reflect activity of the involuntary autonomic nervous system (ANS), which also influences the arousal level. The ANS is in turn controlled by the neuroendocrine system. While methods are beginning to be used to identify emotions via hormonal and other assays (e.g., cortisol levels in the blood to determine stress levels), and non-invasive imaging methods are also being explored (e.g., fMIR, PET scans), here we focus on the more established physiological signals and their interpretation. A number of assessment methods exist for these signals, varying in intrusiveness, reliability of the obtained data, and diagnosticity with respect to particular emotions.

Diagnosticity To the extent that distinct emotions prepare the organism for distinct behaviour (e.g., approach vs. avoid, and fight vs. flee at the most fundamental level), they ought to be reflected in distinct physiological 'signatures', which could then be used to recognize a particular affective state. A critical theoretical issue here is whether distinct emotions do in fact have distinct, recognizable physiological profiles. This is an area of active research and lively debate, with arguments on both sides (Ekman and Davidson, 1994; Cacioppo et al., 1993). Some data indicate that some of the basic emotions do have unique signatures (LeDoux 1996; Panksepp 1998), and recent work by Picard (Picard et al., 2001) suggests that with sufficient data, appropriate baseline and normalizing procedures, and subsequent pattern recognition algorithms, it is possible to differentiate among a number of emotions with accuracy that begins to match human assessments (81% by machine vs. 80-98% in humans). (Interestingly, Picard and colleagues even departed from the traditional basic emotions typically used in these studies and include differentiation among emotions such as platonic and romantic love, and reverence.) Some psychophysiological signals are most appropriate to assess general levels of arousal, rather than specific emotions (e.g., GSR, heart rate).

Semantics Unlike the well-developed facial coding system discussed above, the 'semantic' units of the psychophysiological channels are typically defined in terms of signal properties and signal processing 'primitives', which are then aggregated into specific features for the particular modality.

Sensors and Input Devices A wide variety of sensors exist for each of the psychophysiological channels outlined above (e.g., Applied Science Laboratories eye tracking system for pupil size; a range of wearable heart-rate monitors), and increasingly, sensors are available combining several modalities (e.g., BodyMediaSenseWear (www.bodymedia.com) captures both GSR and body temperature, Lafayette's DataLab2000 captures GSR, blood volume pressure and heart rate). A variety of experimental sensors are also being developed, for example, under IBM's BlueEyes project, an emotion mouse was developed that senses user's pulse, temperature and GSR (www.almaden.ibm.com/cs/blueeyes/mouse.html), and Picard's group at the MIT Media lab has developed a number of experimental wearable computers, such as earrings, to capture a variety of psychophysiological data (Picard and Healey, 1997). A critical practical issue is the degree of intrusiveness of the sensors (e.g., GSR is best measured via the fingertips, but this clearly interferes with typing and may not be a feasible approach for many tasks.

Data Filtering A detailed discussion of the variety of mathematical filtering and normalizing methods available is beyond the scope of this paper. A good introductory review of these methods can be found in Picard (1997).

Pattern Recognition Algorithms As was the case with facial expression recognition above, a variety of clustering and classification algorithms are appropriate here, and researchers have explored specific combinations for particular data sets (e.g., Picard (1997)). As above, artificial neural networks are used extensively, in either supervised or unsupervised learning modes, as are hidden Markov models for time sequence data.

Summary An advantage of psychophysiological measures is that they reflect autonomic nervous system reactions and are thus difficult to 'fake'. The disadvantage is the difficulty associated with accurate signal detection, and considerable expertise required both for the appropriate use of the sensing apparatus, and the subsequent signal analysis, including collection of baseline data and normalizing procedures. However, recent improvements in wearable and remote, wireless sensors, and promising results indicating the ability to differentiate among basic emotions via complex pattern recognition algorithms, are contributing towards making psychophysiological signals appropriate for emotion recognition. For example, recent work by Picard and colleagues (2000; Vyzas & Picard, 1999) indicates that 80% recognition accuracy can be achieved for several basic emotions using these signals.

4.3 Body Posture

Body posture is not the typical modality used to recognize emotion, but promising work is being done by several research groups (Tan et al., 2001; Mota and Picard, 2004). Both groups used chair-mounted sensors to detect postures and attempt to map them onto unique affective states relevant to learning (e.g., interest, boredom). Tan and colleagues use static, 'intentional' postures (that is, the subjects were asked to assume one of several a priori specific postures), whereas Picard and colleagues use 'naturalistic' postures, as exhibited by children engaged in a computer game. Here we focus on the work of Picard and colleagues (Mota and Picard, 2004), who are exploring the use of chair-mounted sensors to detect both static and dynamic posture patterns reflecting different states of interest and boredom during computer-assisted learning. Mota and Picard explored two means of analyzing body posture data, and mapping them onto emotions. In the 'static' approach, a single 'freeze frame' sensor 'image' was mapped onto a particular state (e.g., leaning forward). In the 'dynamic' approach, a sequence of such static states was used. The latter approach proved to be effective in reliably identifying posture sequences associated with level of interest.

Diagnosticity Diagnosticity of postures has not been explored to the same extent as facial expressions or physiological signal patterns. Some emotion researchers suggest that only emotion intensity can be identified from posture (e.g., Ekman), while others suggest that four emotions and attitudes can be identified (Bull, 1983). Mota and

Picard identified demonstrated that different states of interest can be identified from postures. These states correspond to affective states less differentiated than the basic emotions discussed elsewhere in this paper, reflecting a valence (positive vs. negative) and associated general behavioural tendencies (approach vs. avoid). Nevertheless, discrimination among affective states even at such a high level can be helpful for monitoring and adapting to user behaviour.

Semantics Unlike facial expression of emotion, body posture does not have an associated set of well-defined 'primitives'. Mota and Picard therefore identified several "basic posture units" (2004), which served as the vocabulary for analyzing sensed data, by mapping them onto specific states. Mota and Picard videotaped the subjects and human observers then identified candidates for these 'posture units', such as leaning forward (back) right (left). Sequences of freeze-frame postures were identified to reflect low, medium, or high interest and taking a break.

Sensors and Input Devices Several options exist, including cameras, sensors attached to a chair (e.g., Tekscan), accelerometers, and switches. Camera and computer vision approaches are also possible, but are challenging, due to variability in both the subjects, and the ambient conditions (lighting).

Pattern Recognition Algorithms As is the case with facial expressions and physiological signals, variety options exist here. For classifying the static postures, Mota and Picard used gaussians as input to a feedforward neural network, using unsupervised clustering, and successfully classified 9 postures, with an accuracy of almost 90% for new subjects (subjects whose data were not used during the classifier network training). For classifying the dynamic sequences of postures, they used hidden Markov models, to identify sequences corresponding to interest. The sequences were more idiosyncratic, but the Markov models still achieved an 82% success rate with known subjects and 76% rate with unknown subjects.

Summary Attempts to recognize emotions from posture data are just beginning, but as the initial work of Mota and Picard indicates, this modality may represent a useful and productive addition to the repertoire of methods.

4.4 Gesture and Head and Body Movement

Head and body movement, and gestures, represent powerful means expressing emotions, and consequently a means for emotion recognition, yet these indicators have only recently begun to be explored in affective HCI settings. This is understandable, since their use is limited in traditional desktop computer settings, both in terms of feasibility and utility. However, once we leave the desktop setting and enter the world of wearable sensing devices (possibly sewn into our clothing (e.g., the conductor's jacket (Picard, 2000)), remote sensors for drivers or pilots, and virtual environments and synthetic avatars, these modalities become more relevant.

Diagnosticity In addition to specific established symbolic gestures depicting particular emotions, the overall qualities of movement also reflect distinct emotions. For example, anger tends to be associated with short duration movements, and frequent temp changes; grief is associated with long duration and few tempo changes; and fear with frequent tempo changes and log stops between changes (Camurri et al., 2003).

Semantics Two options exist here for defining gesture and movement semantics. The symbolic option, focuses on well-established symbols and their communicative meanings (e.g., thumbs-up), and a large vocabulary of 'semantic primitives' has been identified by communication researchers, anthropologists, and cross-cultural psychologists. The non-symbolic (non-propositional) option relies on the fact that distinct emotions are often associated with distinct qualities of body movements (e.g., tempo, force), and uses these qualities of movement to infer an emotion (e.g., Camurri et al., 2003), building on the work of Laban, who developed a vocabulary for coding movements in dance.

Sensors and Input Devices Human movement data can be captured via cameras, using either 'natural' or marked images, the latter being enhances by placing markers a specific points of the body to facilitate later movement recognition and analysis. Movement data can also be captured from the sensors placed directly on the body and accelerometers.

Pattern Recognition Algorithms The complexity of movement analysis typically requires layered architectures, which analyze the data (frequently from multiple sources, including images, and sensors attached directly to the body), via a layered series of signal and feature processing algorithms, which map the raw movement data onto the basic emotions via a series of increasingly abstract feature sets (e.g., Camurri et al., 2003).

Summary Emotion recognition from body movement data is still in the early stages but the extensive research in this area, coupled with progress in synthetic avatar technologies, indicates that body movement may become an important modality for emotion recognition.

4.5   Speech

Speech is an important modality for emotion manifestation, and a number of acoustic properties of speech have been used to recognize (and express) affect (Petrushin, 2000; DeRosis et al., 2003, Oudryer, 2003). While the actual content of an utterance can obviously express emotion, here we focus on the generic acoustic qualities of speech instead, such as pitch and rhythm.

Diagnosticity With success rates in averaging around 63%, speech is not the most diagnostic modality with respect to emotion (this is in contrast to $80^{%+}$ range for psychophysiological measures and 85th percentile for facial recognition). Human ability to recognize basic emotions from generic speech qualities such as pitch varies, based on a variety of experimental conditions factors, with different researchers reporting different success rates for human emotion recognition from speech. For example, Dellaert and colleagues (Dellaert et al., 1996) report the following findings: Anger was most readily recognized (48%) happiness is at 44%. Fear had the lowest accuracy of these, at 32%. In contract, Petrushin (2000) reports the following human recognition rates: Happiness 61, Anger 72, Sadness 68,   Fear 49, and Normal 66. Petrushin's data indicate that anger and sadness were more readily recognized by humans than happiness, fear and 'neutral'.

Semantics and Feature Sets   Since different aspects of speech can be used to recognize emotion, there are correspondingly distinct feature sets. However, the literature indicates that pitch is the most important aspect of speech for emotion recognition (Petrushin, 2000), and typical pitch-related feature sets include the mean, standard deviation, min, max and range. Additional features are used relating to rhythm, rate and vocal energy (Petrushin, 2000). 'Smoothed' signals are frequently used to enable the definition of further features (Dellaert et al., 1996).

Sensors and Input Devices   Speech data are relatively easy to capture using standard close-talk microphones, recording speech directly, between 16 and 22 kHz,. Once the features and algorithms are developed, applications frequently use lower-quality data sources, such as telephone speech.

Data Filtering A variety of methods are used in speech analysis, aimed at 'smoothing' the original signals, including linear, quadratic and cubic splines (Delleaert et al., 1996). Such smoothing then often allows the definition of additional productive feature sets as described above.

Pattern Recognition Algorithms As was the case with the analysis of physiological signals, Statistical pattern recognition methods are used to classify speech segments (utterances) into specific emotion categories. A variety of specific clustering and classifier algorithms are available, and researchers often achieve significant improvements in performance by 'tweaking' particular algorithms and experimenting with data filtering methods and specific features used in the classification. Unsupervised, semi-supervised and supervised clustering algorithms have been used.   Examples of specific classifier methods used include Maximum Likelihood Bayes classifier, Kernel Regression, K-nearest neighbors, and neural networks. The effectiveness of specific algorithms appear to be closely linked to the data and feature sets, and at this point no clear 'winners' have been identified (e.g., Delleart and colleagues report the K-nearest neighbor algorithm as the most effective (Delleart et al., 1996), whereas Petrushin reported the best results with neural network 'expert' recognizers, each focusing on a given emotion. Petrushin reports the following success rates: normal 55-75; happiness 60-70; anger 70-80; sadness 75-85; and fear 35-55, with the average accuracy across these basic emotions at 70%. When more abstract classification categories are used, higher rates are achieved. Thus, for example, Petrushin aggregates the basic emotions into two more abstract categories of 'agitation' and 'calm', and was able to achieve rates of 77% for telephone quality speech. Some researchers report a range of success rates as a function of particular algorithms. For example, Oudryer (2003) reports emotion recognition rates of up to 97% using acoustic properties of speech in nonsense (pre-verbal) utterances.

Summary While not the most diagnostic modality, speech nevertheless represents an important source of data for emotion recognition. Progress in emotion recognition from acoustic speech qualities, coupled with progress in natural language understanding, suggests that speech may become an important modality for emotion recognition in the near future.

5   Affect Expression

What does it mean for a computer to 'express emotions', and when might such capabilities be useful in HCI? It should first be made clear that a computer obviously does not feel emotions in any manner resembling human feelings. Rather, there may be particular applications where a synthetic avatar's, agent's, or robot's functionality may be enhanced if it can behave in a way that resembles particular human emotions. For example, a therapeutic agent or a computer tutor may be more effective if it can show appropriate 'empathic' reactions to the users.

As with the emotion sensing above, it is helpful to categorize the modalities available for machine affect 'expression'. These are: facial expression via synthetic animated faces; gestures and body movements and posture for 3-D synthetic avatars; and computer-generated speech. Below I provide several examples of machine 'affect expression' across these modalities, to illustrate the range of emerging options.

Several commercial products are available for generation and display of facial expressions corresponding to distinct emotions. Haptek's PeoplePutty (haptek.com) has been used in a number of research projects. PeoplePutty uses proprietary algorithms for generating a emotional expressions for the basic emotions, using several different faces, both male and female.

Commercial products are also available for generating movements and gesture in 3-D synthetic avatars, for example, Boston Dynamics DI-Guy. Progress is being made in developing mappings of particular emotions into specific movements and gestures, which can then be recognized by human users. A number of efforts are underway in this area (e.g., Marsella and Gratch, 2002), and represent an important synthesis of affect-generation and affect-expression within VR environments, moving in the direction of building realistic and believable synthetic avatars with a broad range of applications.

Progress is also being made in speech generation, both in content reflecting particular emotions (e.g., GRETA (deRosis et al., 2003), and the acoustic qualities of speech, such as pitch and phoneme duration (Oudryer, 2003). Oudryer reports success rates of 57%, using human subjects to recognize computer-generated affect in terms of acoustic speech attributes, using nonsense (pre-verbal) utterances (2003).

A variety of novel methods of affect expression are also being developed. For example, Paiva and colleagues (2003) developed a doll SenToy, which allows users (typically children) to express a particular emotion by manipulating the doll's hands, legs and head. The doll movements are then mapped onto one of 6 basic emotions, which influence the behaviour of a synthetic character in a game.

6   Conclusions

In this paper I provided an overview of methods for sensing and recognizing emotions by machines, including an overview of relevant emotion characteristics and a framework for discussing and comparing alternative methods. I also outlined several options for machine 'affect expression'.

Much progress has been made in emotion recognition, and in many instances computer systems are beginning to approach human success rates in emotion recognition and discrimination for many of the basic emotions, under forced-choice conditions. Progress is also being made in machine 'emotion expression' in terms of several modalities, including gaze, facial expression, and aspects of speech.

The sceptical observer (or user) – and there are many - may ask: Why should consideration of the user's affective factors be necessary? In fact, it may not *always* be necessary, and the degree and form of detecting, and adapting to, the user's affective state, or generating an affective state for an autonomous agent, are likely to vary greatly, depending on the context, as outlined above. There are certainly many situations where user affect is irrelevant, and where it may, and should, be disregarded. Much as consumers did not take to talking cars and elevators, we are unlikely to prefer affective bank machines.

However, there are equally many cases where user affect is critical for the successful completion of a task, for avoiding (often disastrous) errors, for achieving optimal performance, for maintaining reasonable user stress levels, and for enhancing the user experience in general.

Below I list some questions that should be addressed by researchers in affective HCI.

Importance of Affect What are the HCI contexts where affect is critical and must be addressed, when can it safely be ignored, and when might affective considerations interfere with performance? Can we identify features of the situations, and the users, that warrant the investment required to assess, adapt to, model, and express affect? How can we rapidly evaluate the tradeoffs involved?

Selecting Emotions Which emotions must be considered, in which contexts and for which types of users? Are the existing taxonomies of emotions adequate? Or do we need to define more complex cross-products of person-emotion-task features to help answer this question? And what are those features?

Adapting: Who and When Under what circumstances should computers adapt to user affect and when should users be trained so that affect does not play a role? Can we construct tasks and task allocations to eliminate the possibility of affective interference, and thus the need for affective adaptation?

Expressing 'Emotions' What is degree of fidelity required to generate 'convincing' affective behavior in synthetic agents? To generate behavior that will induce an affective response in the user? How does this fidelity vary across the user-emotion-task space?

Measuring Effectiveness How must existing usability criteria be augmented to include affective considerations? How can developments in cognitive systems engineering be used to help design evaluation protocols and metrics?

Plug-and-Play and Emerging Standards How can the affective research community facilitate the development of, and adherence to, standards (e.g., MPEG-4, markup languages, facial expression coding systems and body movement vocabularies). Can these standards be extended to the often confusing and redundant terminology, that is particularly prevalent in models of affect appraisal and emotion architectures? What is the best way to establish web-based libraries of components to facilitate component exchange and system development?

References

Bartlett, M. S., Littlewort, G., & Movellan, J. R. (2004). Machine learning methods for fully automatic recognition of facial expressions and facial actions. *IEEE Conference on Systems, Man, and Cybernetics*, The Hague, The Netherlands

Bull P.E. (1983). *Body movement and interpersonal communication*, Chichester: John Wiley & Sons Ltd.
Cacioppo, J. T., Petty, R. E., Losch, M. E., & Kim, H.-S. (1986). Electromyographic activity over facial muscle regions can differentiate the valence and intensity of affective reactions. *Journal of Personality and Social Psychology, 50*, 260-268.

Cacioppo, J. T., Uchino, B. N., Crites, S. L., Snydersmith, M. A., Smith, G., Berntson, G. G. (1992). Relationship between facial expressiveness and sympathetic activation in emotion: A critical review, with emphasis on modeling underlying mechanisms and individual differences. *Journal of Personality and Social Psychology, 62*, 110-128.

Camurri, A., Lagerlof, I., and Volpe, G. (2003). Recognizing emotion from dance movement. *International Journal of Human-Computer Studies,* 59 (1-2), 213-226.

Cohn, J. and Ekman, P. (2003). Measuring Facial Action by Manual Coding, Facial EMG, and Automatic Facial Image Analysis *Handbook of nonverbal behaviour research methods in the affective sciences*, J.A. Harrigan, R. Rosenthal, & K. Scherer, ed.

Dellaert, F., Polzin, T. & Waibel, A. (1996). Recognizing emotion in speech. In *Proceedings of ICSLP*.

de Rosis, F., Pelachaud, C., Poggi, I., Carofiglio, V., and de Carolis, B. (2003). From Greta's mind to her face. *International Journal of Human-Computer Studies,* 59 (1-2), 81-118.

Ekman, P. (1965). *The differential communication of affect by head and body cues.* Journal of Personality and Social Psychology, 2,726-735

Ekman, P., & Friesen, W. V. (1978). *Facial action coding system.* Palo Alto, CA: Consulting Psychologists Press.

Ekman, P., & Friesen, W. V. (2003). *Unmasking the face: A guide to recognizing emotions from facial cues.* Cambridge, Massachusetts: Malor Books.

Essa, I., & Pentland, A. (1997). Coding, analysis, interpretation and recognition of facial expressions. *IEEE Transactions on Pattern Analysis and Machine Intelligence, 7*, 757- 763.

Fasel, B., and Luettin, J. (2003). Automatic facial expression analysis: A survey. *Pattern Recognition, 36*, 259-275.

Kaiser, S., and Wehrle, T. (1992). Automated coding of facial behavior in human-computer interactions with FACs. *Journal of Nonverbal Behavior, 16*, 67-84.

Hudlicka, E. and McNeese, M. (2002). User's Affective & Belief State: Assessment and GUI Adaptation. *International Journal of User Modeling and User Adapted Interaction,* 12(1), 1-47.

Hudlicka, E. (2003). To Feel or Not To Feel: The Role of Affect in Human-Computer Interaction. *International Journal of Human-Computer Studies,* 59 (1-2), 1-32.

Kaiser, S. Wehrle, T., & Schmidt, S. (1998). Emotional episodes, facial expressions, and reported feelings in human-computer interactions. In A. H. Fischer (Ed.), *Proceedings of the Xth Conference of the International Society for Research on Emotions* (pp. 82-86).Würzburg: ISRE Publications.

Mota, S, and Picard, R.W. (2004). Automated Posture Analysis for detecting Learner's Interest Level. Cambridge, MA: MIT Media Laboratory.

Oudryer, P.Y. (2003). The production and recognition of emotions in speech: features and algorithms. *International Journal of Human-Computer Studies*, 59 (1-2), 157-184.

Paiva, A. (2003). SenToy: An affective sympathetic interface. *International Journal of Human-Computer Studies*, 59 (1-2), 227-235.

Picard, R. W. (1997), *Affective Computing*, MIT Press, Cambridge, MA.

Picard, R. W, and Healey, J., (1997). Affective Wearables, Personal Technologies Vol 1, No. 4 , 231-240.

Picard, R. (2000). Towards Computers that Recognize and respond to User Emotion. *IBM Systems Journal*. 39 (3-4), 705-719.

Petrushin, V. (2000). Emotion recognition in speech signal: Experimental study, development and application. In the Proceedings of the 6th Intl. Conference on Spoken Language Processing (ICSLP).

Rizzo, A.A., Schultheis, M., Kerns, K.A., and Mateer, C. (2003). *Analysis of Assets for Virtual Reality Applications in Neuropsychology.* Neuropsychological Rehabilitation.

Tan H. Z., Slivovsky L. A., and Pentland A., (2001). *A sensing Chair Using Pressure Distribution Sensors*, IEEE/ASME Transactions on Mechatronics 3 (6).

Tekscan (1997). *Tekscan Body Pressure Measurement System User's Manual*. Tekscan Inc., South Boston, MA, USA.

Vyzas, E., and Picard, R. W. (1999). Online and Offline Recognition of Emotion Expression from Physiological Data, Workshop on Emotion-Based Agent Architectures at the Int. Conf. on Autonomous Agents, *Seattle, WA.*

Zimand, E., Anderson, P., Gershon, G., Graap, K., Hodges, L. & Rothbaum, B. (2003). Virtual Reality Therapy: Innovative Treatment for Anxiety Disorders. *Primary Psychiatry*, 9 (7), 51-54.

Voices of Attraction

Anmol Madan, Ron Caneel, Alex "Sandy" Pentland

MIT Media Laboratory
20 Ames Street, Cambridge 02139
{anmol, rcaneel, sandy}@ media.mit.edu

Abstract

Non-linguistic social signals (e.g., `tone of voice') are often as important as linguistic or affective content in predicting behavioral outcomes [1, 12]. This paper describes four automated measures of audio social signaling within the experimental context of speed dating. We find that this approach allows us to make surprisingly accurate predictions about the outcomes of social situations.

1   Introduction

In many situations non-linguistic social signals (body language, facial expression, tone of voice) are as important as linguistic content in predicting behavioural outcome [1, 12].   Tone of voice and prosodic style are among the most powerful of these social signals even though (and perhaps because) people are usually unaware of them [12]. In a wide range of situations (marriage counselling, student performance assessment, jury decisions, etc.) an expert observer can reliably quantify these social signals and with only a few minutes of observation predict about 1/3d of the variance in behavioural outcome (which corresponds to a 70% binary decision accuracy) [1]. In certain areas of human judgement these recorded predictions have been up to 90% accurate [7]. It is astounding that observation of social signals within such a `thin slice' of behaviour can predict important behavioural outcomes (divorce, student grade, criminal conviction, etc.), even when the predicted outcome is sometimes months or years in the future.

Current literature in affective computing tries to link speaking style with specific emotional states. Picard [6,14], Oudeyer [13], and Brezal [3] look for observations of sympathetic or para-sympathetic nervous activity, which also cause variations in the intensity and pitch of voice (e.g. angry versus calm). This technique is well suited for applications that seek to understand the human's internal state, for example, building computers that can measure frustration.

In contrast, we propose that minute-long averages of audio features often used to measure affect (e.g., variation in pitch, intensity, etc) taken together with conversational interaction features (turn-taking, interrupting, making sounds that indicate agreement like 'uh-huh') are more closely related to social signalling theory rather than to an individual's affect. The ability to understand these social signals in an automated fashion allows us to make better predictions about the outcomes of *interactions* – speed-dating, negotiations, trading business cards --- than would affective measurements alone. Our reasoning is that social signals are actively constructed with communicative intent, whereas the connection between individual affect and group behaviour is extremely complex.

2   Measuring Social Signals

Pentland [15] constructed measures for four types of vocal social signalling, designated activity level, engagement, stress, and mirroring. These four measures were extrapolated from a broad reading of the voice analysis and social science literature, and we are now working to establish their general validity

Calculation of the activity measure begins by using a two-level HMM to segment the speech stream of each person into voiced and non-voiced segments, and then group the voiced segments into speaking vs. non-speaking [8,2]. Conversational activity level is measured by the z-scored percentage of speaking time plus the frequency of voiced segments. The activity measure and stress measures are features common in affect recognition literature [6, 13].

Engagement is measured by the z-scored influence each person has on the other's turn taking. When two people are interacting, their individual turn-taking dynamics influence each other and can be modeled as a Markov process [9]. By quantifying the influence each participant has on the other we obtain a measure of their engagement...popularly speaking, were they driving the conversation? To measure these influences we model their individual turn-taking by a Hidden Markov Model (HMM) and measure the coupling of these two dynamic systems to estimate the influence each has on the others' turn-taking dynamics [5]. Our method is similar to the classic method of Jaffe et al. [9], but with a simpler parameterization that permits the direction of influence to be calculated and permits analysis of conversations involving many participants.

Stress is measured by the variation in prosodic emphasis. For each voiced segment we extract the mean energy, frequency of the fundamental format, and the spectral entropy. Averaging over longer time periods provides estimates of the mean-scaled standard deviation of the energy, formant frequency and spectral entropy. The z-scored sum of these standard deviations is taken as a measure speaker stress; such stress can be either purposeful (e.g., prosodic emphasis) or unintentional (e.g., physiological stress caused by discomfort).

Mirroring behaviour, in which the prosody of one participant is `mirrored' by the other, is considered to signal empathy, and has been shown to positively influence the outcome of a negotiation [4]. In our experiments the distribution of utterance length is often bimodal. Sentences and sentence fragments typically occurred at several-second and longer time scales. At time scales less than one second there are short interjections (e.g., `uh-huh'), but also back-and-forth exchanges typically consisting of single words (e.g., `OK?', `OK!', `done?', `yup.'). The z-scored frequency of these short utterance exchanges is taken as a measure of mirroring. In our data these short utterance exchanges were also periods of tension release.

2.1 Signalling Dynamics

These measures of social signalling can be computed on a conventional PDA in real-time, using a one-minute lagging window during which the statistics are accumulated. It is therefore straightforward to investigate how these `social signals' are distributed in conversation. In [10, 16] we analyzed social signalling in 54 hours of two-person negotiations on a minute-by-minute basis. We observed that high numerical values of any one measure typically occur by themselves, e.g., periods with high engagement do not show high stress, etc., so that each participant exhibits four `social display' states, plus a `neutral' relaxed state in which the participant is typically asking emotionally neutral questions or just listening. The fact that these display states were largely unmixed provides evidence that they are measuring separate social displays.

2.2 Attraction Experiment

Speed dating is relatively new way of meeting many potential matches during an evening. Participants interact for five minutes with their `date', at the end of which they decide if they would like to provide contact information to him/her, and then they move onto the next person. A 'match' is found when both singles answer yes, and they are later provided with mutual contact information. Perhaps since speed-dating provides instant-gratification in the dating game, it has recently become the focus of several studies that attempt to predict outcomes based on mate selection theory [11]

Figure 1: Speed dating session in progress. Audio is recorded using Zaurii PDAs.

In this experiment we analyzed 60 five-minute speed-dating sessions. Data was collected from several events held in 2004, and participants were singles from the MIT-Harvard community in the ages of 21-45. In addition to the 'romantically attracted' (provide contact information) question, participants were also asked two other yes/no questions: would they like to stay in touch just as friends, and would they like to stay in touch for a business relationship. These 'stay in touch' questions allowed us to explore whether romantic attraction could be differentiated from other factors.

Speed-dating is of particular experimental interest because there is a clear 'buying' decision at the end of each conversation. Since data was collected from real-life sessions, we were also more immune to the effect of self-entertainment. (i. e. we classify participants' natural intonation, not special 'interested' or 'bored' voices that actors generated).

2.3 Results

The four social signalling measures for both male and female were compared by linear regression to the question responses, and in each case the resulting predictor could account for more than 1/3rd of the variance. For the females responses, for instance, the correlation with the 'attracted' responses were $r=0.66$, $p=0.01$, for the 'friendship' responses $r=0.63$, $p=0.01$, and for the 'business' responses $r=0.7$, $p=0.01$. Corresponding values for the male responses were $r=0.59$, $r=0.62$, and $r=0.57$, each with $p=0.01$.

The engagement measure was the most important individual feature for predicting the 'friendship' and 'business' responses. The mirroring measure was also significantly correlated with female 'friendship' and 'business' ratings, but not with male ratings. The stress measure showed correlation with both participants saying 'yes' or both saying 'no' for the 'attraction' ($r=0.6$,$p=0.01$) and 'friendship' ($r=0.58$,$p=0.01$) questions.

An interesting observation was that for the 'attracted' question female features alone showed far more correlation with both male ($r=0.5$,$p=0.02$) and female ($r=0.48$, $p=0.03$) responses than male features (no significant correlation). In other words, female social signaling is more important in determining a couples 'attracted' response than male signaling. The most predictive individual feature was the female activity measure.

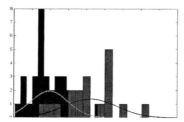

Figure 2: Frequency distribution of female 'attracted' responses (red=yes) vs. predictor value. The cross-validated binary linear decision rule has 72% accuracy

Figure 3 shows a two-class linear classifier for the 'attraction' responses, based on the social signaling measures; this classifier has a cross-validated accuracy of 71% for predicting the 'attracted' response. Feature selection was based on the regression results. The two fitted Gaussians are simply to aid visualization of the distributions' separability.

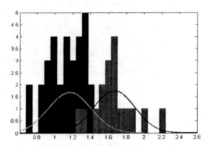

Figure 3: Frequency distribution of female 'business' responses (red=yes) vs. predictor value. The cross-validated three-class linear decision rule produces 83 % accuracy.

Figure 4 illustrates a two-class linear classifier for the 'business' responses, based on the social signaling measures; this classifier has a cross-validated accuracy of 74% for predicting the 'attracted' response. By considering the overlapping region as a third class, we can increase the cross-validation accuracy to 83% for the yes and no response regions. The two fitted Gaussians are simply to aid visualization of the distributions' separability.

It was also observed that the accuracy of the predictions increased when the classifier was trained for a particular person. We believe this is important because people have different speaking styles, which potentially could be captured by our model. We had 8 conversations involving subject J and trained our model only on her data set. For the romantic interest question for subject J, for example, the cross-validation accuracy of our predictions increased to 87.5%.

We also used SVM classifiers to test the separability of the 'yes' and 'no' responses, with linear, polynomial and radial kernels. As seen, the performance of the SVM classifier was only slightly better than a simple linear classifier (with appropriate feature selection). The RBF kernel is more prone to overfitting.

Question	SVM accuracy (Linear kernel)	SVM accuracy (RBF kernel)
Are you interested in this person? (asked of females)	0.71	0.70
Would you like to be friends with this person? (asked of females)	0.76	0.79
Are you interested in maintaining a professional relationship with this person? (asked of females)	0.73	0.71
Are you interested in this person? (asked of males)	0.64	0.62
Would you like to be friends with this person? (asked of males)	0.79	0.82
Are you interested in maintaining a professional relationship with this person? (asked of males)	0.71	0.69

2.4   Real-time Feedback

If our description of social signaling is precise, and the effects are measurable, then a wearable 'social signaling meter' is the next step. We wanted to use such a device to understand how speed daters would react if made more aware of their social signals.

We have built a real-time speech feature calculation engine (SpeedDating Pro) that runs on a Sharp Zaurus Linux PDA. The Zaurus is a wireless, ARM-based handheld with a microphone input and a touch screen interface. The current version of the speech feature code is written in C++, and incorporates all features except the engagement features (or the influence parameters). It also calculates the probability of the other person saying 'yes' based on the coefficients from the model training. Zaurii are time synced using ntpdate (a linux utility) and use UDP broadcasts to communicate with each other. The display can show chances of a second date or the measured social signals, and is updated after every 90 seconds of conversation.

Figure 4 (a) (left): Display of probability of the other person saying YES

Figure 4 (b) (right): Display of measured social signals

We are deploying 20 such Zaurus PDAs running SpeedDating Pro software soon, and look forward to the results. SpeedDating Pro is in some ways similar to the 'Love Detector' software sold by Nemesysco, although we were unable to find any publications documenting its accuracy. We believe that the underlying software for Namesysco is based on vocal stress features, and in our preliminary trials with it, it had ambiguous results.

3   Future Work

We believe that these same social signals may also help us to predict whether people are interested or engaged in a much broader range of situations. In one study that is just beginning to be analyzed, we have people talk to each other on various topics, and then compare the measured signals to their interest rating on a feedback form. Based on the results of the first 8 subjects, which were engaged in 80 conversations of about 3 minutes each, there is a correlation of $r=0.64$, $p=0.1$ between the social signalling and the rated interest level.

Annual US spending on focus groups for qualitative market research is about US$ 1.1 Billion. However, industry experts acknowledge that focus groups are largely lacking in accuracy, since participants often misreport experiences and say what is expected of them, without sharing their insight as a consumer. By measuring these non-verbal signals, the efficacy of focus groups could potentially be improved.

4   Discussion

The social signalling measured in this paper seems to communicate and be involved in mediating social variables such as status, interest, determination, or cooperation, and arise from the interaction of two or more people rather than being a property of a single speaker. Semantics and affect are important in determining what signalling an individual will engage in, but they seem to be fundamentally different types of phenomena. The social signalling measured here seems to be a sort of 'vocal body language' that operates relatively independently of linguistic or affective communication channels, and is strongly predictive of the behavioural outcome of dyadic interactions.

5   References

[1] Ambady, N., & Rosenthal, R. (1992). Thin slices of expressive behaviour as predictors of interpersonal consequences: A meta-analysis. Psychological Bulletin, 111(2), 256-274.

[2] Basu, B., (2002) Conversational Scene Analysis, doctoral thesis, Dept. of Electrical   Engineering and Computer Science, MIT. 2002. Advisor: A. Pentland

[3] Breazal C. (2001), Designing Social Robots, MIT Cambridge Press

[4] Chartrand, T., and Bargh, J., (1999) The Chameleon Effect: The Perception-Behavior Link and Social Interaction, J. Personality and Social Psychology, Vo. 76, No. 6, 893-910

[5] Choudhury, T., and Pentland, A., (2004), NAASCOS, June 27-29, Pittsburgh, PA. PDF available at http://hd.media.mit.edu

[6] Fernandez, R. (2004), A Computational Model for the Automatic Recognition of Affect in Speech, Ph.D. Thesis, MIT Media Arts and Science, February 2004. Advisor: R. Picard

[7] Gladwell Malcolm (2004) , Blink, Little Brown and Co. NY 2004

[8] Handel, Stephen, (1989) Listening: an introduction to the perception of auditory events, Stephen Handel , Cambridge: MIT Press

[9] Jaffe, J., Beebe, B., Feldstein, S., Crown, C. L., & Jasnow, M. (2001). Rhythms of dialogue in early infancy. Monographs of the Society for Research in Child Development, 66(2), No. 264.

[10] Khilnani, R. (2004) Temporal Analysis of Stages in Negotiation, MEng Project, Advisor: A. Pentland.

[11] Kurzban, Weeder, to be published in journal of Evolution and Human Behaviour

[12] Nass, C., and Brave, S. (2004) Voice Activated: How People Are Wired for Speech and How Computers Will Speak with Us, MIT Press

[13] Oudeyer, P-Y. (2003) The production and recognition of emotions in speech: features and algorithms. International Journal of Human Computer Interaction, 59(1-2):157--183.

[14] Picard R. (1997) , Affective Computing, MIT Press

[15] Pentland, A. (2004) Social Dynamics: Signals and Behavior, ICDL, San Diego, CA Oct 20-23, IEEE Press

[16] Pentland, A., Curhan, J., Khilnani, R., Martin, M., Eagle, N., Caneel, R., Madan A (2004) ``Toward a Negotiation Advisor," UIST 04, Oct 24-27, ACM. PDF available at http://hd.media.mit.edu

Developing Computer Agents with Personalities

Sasanka Prabhala and Jennie J. Gallimore

Department of Biomedical, Industrial, and Human Factors Engineering
207 Russ Engineering Center
Wright State University
Dayton OH 45435
[sprabhal; jgalli]@cs.wright.edu

Abstract

Systems with computer agents are collaborative in nature. Therefore, human operators must work with the automated components to complete a task and meet system goals. However, humans have different personalities and it is not unlikely that teams require members to have personalities that enhance the team as a whole. Creating computer agents with personalities may enhance human-machine collaboration and user's trust of the computer system. This paper describes our overall research approach toward modeling computer agents with personalities and investigates the actions, language, and/or behaviors that humans indicate signify personality traits within the Big Five Factor model in the context of collaboration. Participants were asked to rate the personality traits they perceived in a computer game with agents, existing real life team members, and traits they would like in an ideal team member. In each method participants described the actions, language, and/or behaviors that gave them their impressions. Results indicated that participants were able to perceive different personalities and were also able to distinguish specific subtraits. The results are discussed with respect to agent personality modeling.

Keywords

Human-Machine Collaboration, Personality, Big Five Factor Model, Computer Agents, Agent Modeling, Behavior Modifiers

1 Introduction

Many of the systems that surround us such as defense, transportation, medical, and business organizations are complex in nature. Complex systems are characterized by uncertainty, ambiguity, ill-defined goals, dynamically changing conditions, subject to distractions and time pressures. These complex systems generate huge amounts of data that must be fused, integrated, and analyzed by the human operators in the decision making process. The dynamic nature of complex systems and the overwhelming amount of data that must be handled by human operators requires the development of smart systems that can augment human capabilities. Development of computer agents that execute tasks within the overall system creates semi-autonomous systems in which the human-agent collaboration is part of the system dynamic.

Systems with computer agents are collaborative in nature and require the dispersion of system knowledge and awareness among all collaborating agents including human agents. The human operator must work with the automated components to complete a task and meet system goals. There is clearly a need for designing automated systems as collaborative systems that more closely resemble human-to-human collaboration. Questions that need to be addressed include (a) how should the human operator and the automated portion of the human-computer system communicate information, strategies, and commands to each other? and (b) what kinds of interdependencies exist between humans and machines? There is a need for a systematic approach towards development of collaborative human-machine systems. Although it is evident from our everyday experiences that personality is one factor for assigning some types of work to some people and not others, there exists little knowledge for the appropriate use of personality models in the development of computer agents.

Humans have different personalities and it is not unlikely that teams require members to have personalities that enhance the team as a whole. For example, if all members of a team are considered introverts, then it may be difficult for that team to come up with solutions to problems as the members may not express their ideas. Creating computer agents with personalities may enhance the human-machine collaboration because the agent can provide the human participant input and communication in a way that works well with the human's personality. Also, it is possible that creating personalities may enhance the users' trust in the system. Several researchers (Pew & Mavor,

1998; Silverman et al., 2001; Silverman, Cornwell, & O'Brien, 2003; and Wray & Laird, 2003) have noted the need for research and development of computer agents with personalities that behave and act more like humans. Silverman et al., (2001) point out the need to capture personality and individual differences and describe the difficulties in extracting behavior modifiers from the behavioral research. There is clearly a need to customize automated systems and a key research issue of human-machine collaboration is how to effectively incorporate personality into agents and to match agents to humans via personality modeling.

Baker & Salas (1992) noted that collaboration between agents is critical as present day task demands exceed the capabilities of a single individual agent. Hence, there is a need to identify and study the characteristics that facilitate teamwork or collaboration among agents. Studying effective collaboration techniques between humans and machines will lead to better understanding of human constraints for the use of automation tools and also help in understanding the behavior of humans in performing both individual tasks as well as team tasks. Hoc (2001) mentioned that human-machine collaboration in complex systems can be enhanced not only through exclusive improvements in the user interface designs or through the use of expert systems alone, but by addressing the cognitive and social aspects of human-human collaboration to study human-machine collaboration.

According to Lee & Nass (2003) it is essential to influence the human operator's physical presence as well as social presence in the system for effective human-machine collaboration. Social presence is defined as generating feelings of presence in a system where other intelligent agents exist for interaction even though those agents are non-human. They also observed that social presence could be enhanced through similarity-attraction (attracted more to agents that match their personalities than those that do not match with their personalities). The study consisted of matching human personality (extravert v_s introvert) with the computer voice personality by manipulating parameters such as pitch range, speed rate, and fundamental frequency to indicate the computers personality as Extravert or Introvert.

Nass, Steuer, Tauber, & Reeder (1993) observed that social rules guiding human interactions can be applied to human-machine interactions. Their study consisted of matching human personality (extravert v_s introvert) with computer text personality by manipulating parameters such as phrasing of the text displayed by the computer, confidence level expressed by the computer, order or interaction, and name given to the computer to portray an extravert or introvert. Another study conducted by Nass, Moon, Fogg, Reeves, & Dryer (1995) observed that human operator's social presence can be enhanced by changing machine personalities to resemble human personalities using minimal set of cues including assignment of human attitudes, intentions, and motives. Humans responded to these machines in the same way humans respond to other humans who are according to the similarity-attraction theory.

These studies suggest that social rules governing human-human interaction can be applied to human-machine interactions with potential benefits. However, one would argue that most of these studies have a limited scope, focusing on only one aspect of the personality traits with respect to human-machine interaction and have not been attacking the problem with the intention of studying the affects of personality traits on human-machine collaboration. Important questions that need to be answered are (a) is collaboration more effective if the machine agent is given personality and (b) how do we make human-machine communication more similar to human-human communication. In other words how do we model factors such as personality, facial expression, posture, nonverbal cues, direction of gaze, and vocal cues that contribute to the degree of social presence in a face-to-face communication between humans, in human-machine interaction?

2   Research overview

The overall goal of the research is to improve human-machine collaboration through the development of computer agents with personality. In order to move towards this goal three research objectives have been established: (1) to develop agents with personality, (2) validate that humans perceive these agents as having a personality, and (3) to determine if agents with personality will enhance human-machine collaboration.

To meet these objectives three experimental phases are proposed. The purpose of Phase I is to identify actions, language and/or behaviors that human subjects indicate signify personality traits within the Big Five Factor model in the context of collaboration or teamwork. The purpose of Phase II is to create a simulation that models agents with personalities based on the outcome of the first experimental phase and to validate these personalities. The purpose of Phase III is to empirically evaluate human-machine collaboration performance. This paper focuses on Phase I. Phase I consisted of identifying significant actions, language, and/or behaviors that signify personality traits in the context of collaboration or teamwork. Focusing on only one technique for collecting the actions, language, and/or behaviors may result in limited range of output. Therefore, participants were asked to rate personality subtraits using three separate methods described in Section 4.

3   The Big Five Factor personality trait model

According to Webster Dictionary, personality is defined as "the complex of characteristics that distinguishes an individual's behavior and emotional characteristics from one another". The need to classify these tendencies of individuals to behave in a certain way into organized models led to the study of personality theories or personality models. According to Winter & Barengaum (1999) personality models are classified into four classes: motivational models, cognition models, social context models, and trait models. Of these, trait models gained significant importance due to their ability to find small numbers of independent dimensions (factors) or characteristics also known as traits (extraversion v_s introversion) that would account for as much of variation in personality as possible. One of the best known examples of a trait model is the Big Five Factor model. This model uses five factors that are considered central traits to personality. They are: I. Extraversion, II. Agreeableness, III. Conscientiousness, IV. Emotional Stability v_s Neuroticism, and V. Intellect or Openness. To define the central traits more accurately, each central trait is subdivided into six subtraits or facets (Goldberg, 1990). Research conducted by McCrae & Costa (1997), Jang, McCrae, Angleitner, Riemann, & Livesley (1998), and Soldz & Vaillant (1999) also indicated that the Big Five Factors are stable across different domains, cultures and over time. Table 1 lists the central and the subtraits of the Big Five Factor model.

Table 1. The central and subtraits of the Big Five Factor personality model

I. Extraversion	II. Agreeableness	III. Conscientiousness	IV. Emotional Stability v_s Neuroticism	V. Intellect or Openness
Friendliness	Trust	Self-Efficacy	Anxiety	Imagination
Gregariousness	Morality	Orderliness	Anger	Artistic Interests
Assertiveness	Altruism	Dutifulness	Depression	Emotionality
Activity Level	Cooperation	Achievement-Striving	Self-Consciousness	Adventurousness
Excitement-Seeking	Modesty	Self-Discipline	Immoderation	Intellect
Cheerfulness	Sympathy	Cautiousness	Vulnerability	Liberalism

4   Methodology

Thirty six graduate students from Wright State University volunteered to serve as participants. The students were blocked in the following groups: Ideal Team Member (12), Existing Team Member (12), and Computer Team Member (12). The apparatus included 333 MHz Dell personal computer running Windows XP with a 17'' color CRT. The input devices were a standard 101 keyboard and mouse. The experiment took place in an office type environment with dim lighting. The subjects sat in an adjustable office chair, and the keyboard and mouse were placed at comfortable positions determined by each participant. Software included a computer version of the International Personality Item Pool Representation of the NEO PI-RTM™ (IPIP-NEO) questionnaire (Big Five Factor Personality Test) and the computer game Hoyle Casino™ developed by Sierra Attractions Inc. A computer program was created in Java to present the definitions of the Big Five Factor Model to the participants and collect their ratings and descriptions. Each participant was asked to take the IPIP-NEO questionnaire to record their personality type. Each participant was then assigned to one of the three methods to rate and describe their experiences specially focusing on what actions, language, and/or behaviors the Ideal Team Member, Existing Team Member, and the Computer Team Member exhibited that gave them those impressions. Participants in the Computer Team Member method were presented with three computer characters to rate using a balanced Latin Square technique.

4.1   Method I – Ideal Team Member (ITM)

The purpose of this method was to determine what personality traits human subjects think an "ideal" team member should have. Each participant was shown on a computer screen each of the facets of the Big Five Factor model along with its description one at a time in a random order. Participants were asked to rate each facet they think they want in an "ideal" team member on a scale of 1 to 5 (1-extremely important, 2- very important, 3-neutral, 4-not very important, and 5-not at all important). Participants rated a total of 30 facets and were not timed for completion.

4.2   Method II - Existing Team Member (ETM)

The purpose of this method was to ask people to indicate the personality traits they think best describe their "existing" team member. In this method, participants who were teamed in pairs or groups to perform class projects as a normal part of their course work in the Department of Biomedical, Industrial, and Human Factors Engineering were selected. Each participant was shown on a computer screen each facet of the Big Five Factor model along with its description, one at a time in a random order. Participants were asked to rate each facet they think best describes their team member on a scale of 1 to 5 (1-very descriptive, 2-descriptive, 3-neutral descriptive, 4-not very descriptive, and 5-not at all descriptive). Participants rated a total of 30 facets. Participants were also asked to describe their experiences with their team member especially focusing on what language, actions, and/or behaviors the team member exhibited that gave them their impressions and hence the rating.

4.3   Method III – Computer Team Member (CTM)

The purpose of this method was to determine how humans perceive personalities in existing computer media. Participants interacted within a computer game. Each participant played the card game blackjack for 20 minutes using Hoyle Casino™. While playing blackjack there is a dealer as well as three other computer characters at the table. Each computer character's characteristics such as speech, appearance, and attitude in terms of the talkativeness and the animation are already specified in the game. The level of talkativeness can be set using a sliding bar and it was set to its maximum level to provide the highest level of interaction with the agents. During the game, these computer characters interacted with the participant through comments and advice regarding the game. After interacting with the game each participant was shown on a computer screen each facet of the Big Five Factor model along with its description, one at a time in a random order. Participants were asked to indicate if a facet described the character on a scale of 1 to 5 (1-very descriptive, 2-descriptive, 3-neutral, 4-not very descriptive and 5-not at all descriptive). Participants rated a total of 30 facets for each of the 3 computer characters and were not timed for completion. Participant were also asked to describe their experiences with each of the computer characters especially focusing on what language, actions, and/or behaviors the computer characters exhibited that gave them their impressions and hence the rating.

5   Results

5.1   Ratings

The rating results were first analyzed by determining an average score to attribute to the central trait. This was accomplished by determining the average ratings for the five subtraits under each central trait, summing across the subtraits and dividing by 5. The results are plotted in Figure 1. The graph suggests that subjects in the three methods rated the personalities as different. For example, the average rating of Extraversion factor for computer character (CC)1, Ideal team member (ITM), and Existing team member (ETM) indicated that on average subjects thought that Extraversion was descriptive (X=2.083; X=2.055; X=2.361 respectively). CC2 (X=2.958) and CC3 (X=2.875) were rated similarly such that the average was close to neutral. On average subjects rated the Agreeableness factor for ITM (X=1.986) and ETM (X=1.972) as being descriptive compared to CC1 (X=2.625) and CC2 (X=3.263) which were rated as neutral and CC3 (X=3.583) which was rated as not very descriptive. On average subjects rated the Conscientiousness factor for CC3 (X=2.458) and ETM (X=2.138) as being descriptive compared to CC1 (X=2.875) and CC2 (X=2.902) which were rated closer to neutral. Conscientiousness was also a trait participants indicated was important in an ITM (1.638). On average subjects rated the Neuroticism factor for CC2 (X=2.833) and CC3 (X=3.263) as being neutral compared to CC1 (X=3.638), ITM (X=3.611), and ETM (X=3.958) which were rated as being not very descriptive. Not unsurprisingly participants don't want neurotic team members. Similarly, subjects rated the Intellect factor for ITM (X=2.444) as being closer to descriptive compared to CC1 (X=2.833), CC2 (X=3.111), CC3 (X=2.763), and ETM (X=2.833) which were rated closer to neutral.

Although the overall score of each central trait is interesting, the rating of subtraits within each central trait illustrates more clearly the difference in subject perception of personalities in each method and variation in responses. The rating results for the three methods were each analyzed by averaging the data for each of the subtraits corresponding to the central traits across participants in each method. The results are plotted in Figure 2. For example, although CC2 and CC3 are rated similarly in overall central Extraversion factor, subjects rated the Assertiveness facet within Extraversion of CC2 (X=3.833) higher than that of CC2 (X=2.166) which implies that subjects perceived CC2 to be more Assertive compared to CC3. Although CC3 and ITM are rated similarly in Conscientiousness factor, subjects rated the Dutifulness facet within Conscientiousness of CC3 ($\bar{X}$=3.083) higher than that of ITM ($\bar{X}$=1.333) which implies that subjects prefer an ITM to be Dutiful and that CC3 did not express that trait well. Although ITM and ETM are rated similarly in Neuroticism factor, subjects rated the Self-Consciousness facet within Neuroticism of ETM ($\bar{X}$=3.833) higher than that of ITM ($\bar{X}$=2.833) which implies that subjects preferred an ITM to be more Self-Conscious compared to what they have experienced with current ETM. Similarly, although CC2 and ETM are rated similarly in Intellect factor, subjects rated the Intellect sub-facet within Intellect central trait of CC2 ($\bar{X}$=3.083) higher than that of ETM ($\bar{X}$=1.916) which implies that subjects perceived ETM to show more Intellect than CC2. An examination of the graph indicates that most ratings for CC2 fall in the neutral range, indicating that participants did not perceive much personality from this computer character. CC1 and CC3 have quite a bit of variability across the ratings, indicating participants were able to identify certain traits. The graph also indicates that ITM and ETM ratings follow very similar patterns across all traits.

To determine if there were statistically significant differences among methods (CTM, ITM, and ETM) and subtraits, an ANOVA was conducted on the dependent variable ratings. Because the CTM method has three computer characters that were rated by all subjects within that group, we only used one computer character (CC1) in the mixed factor ANOVA, where method is a between-subject variable and subtrait is within-subject. As expected there was a significant interaction $\underline{F}$(58,957)=2.31, (p=0.0001). What we were interested in determining is whether there are differences at each individual subtrait. Table 2 provides the Tukey results indicating which methods resulted in significantly different ratings for the subtraits. Subtraits that had no significant differences are not listed. Methods that have the same letter indicate that the ratings are not different from one another. Eight of the 30 subtraits showed some significant differences among ratings (see Table 2). For all eight subtraits there was no significant difference in rating between the ITM and ETM conditions. Differences were related to CC1.

Table 2. Tukey test results indicating significant differences among methods for each subtrait. Methods with similar letters are not significantly different from one another. Subtraits not listed showed no differences.

Subtraits	Significant Differences		Subtrait	Significant Differences	
Trust	CC1 A		Dutifulness	CC1 A	
	ITM	B		ITM	B
	ETM	B		ETM	B
Self-Efficacy	CC1 A		Cooperation	CC1 A	
	ITM	B		ITM	B
	ETM	B		ETM	B
Morality	CC1 A		Self-Discipline	CC1 A	
	ITM A	B		ITM	B
	ETM	B		ETM A	B
Orderliness	CC1 A		Intellect	CC1 A	
	ITM	B		ITM	B
	ETM A	B		ETM A	B

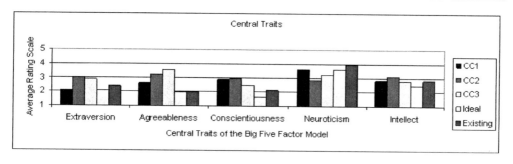

Figure 1. Average rating of the central traits of the Big Five Factor model as perceived by participants in the three methods.

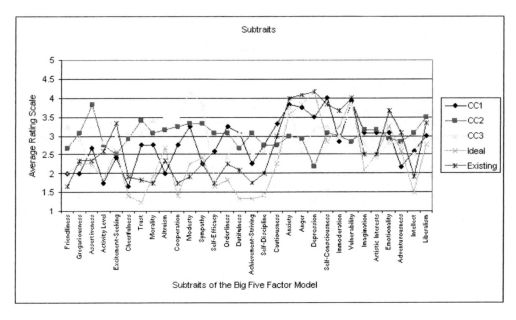

Figure 2. Average ratings of the subtraits of the Big Five Factor model as perceived by participants in the three methods

5.2 Actions, language, and/or behaviors

To determine what influenced subject perceptions, participants were asked to identify actions, language, and/or behaviors that lead them to their impressions. This information is important to develop computer agents with personalities. Table 3 provides a list of subjects' descriptions with respect to each subtrait of the Big Five Factor model. The actions, language, and/or behaviors are represented in (a) **bold** – if they were perceived by participants in all the three methods, (b) ***bold*** – if they were perceived by participants in any of the two methods, and (c) *italics* – if they were perceived by participants in any one method. For example, participants indicated the Friendliness facet with actions such as **smiling a lot**, **speaking amicable**, **showing concern for others**, *cracking jokes*, ***being a good listener***, ***being cooperative***, and *being open to ideas*.

Table 3. Actions, language, and/or behaviors associated with the Big Five Factor model

Central Traits	Subtraits	Actions, Language, and/or Behaviors
Extraversion	Friendliness	**Smiling a lot; Speaking amicably; Showing concern for others;** *Cracking jokes; **Being a good listener; Being cooperative;*** *Being open to ideas*
	Gregariousness	**Enjoying others company;** *Being very talkative; Complimenting others; Sharing thoughts and ideas; Checking to see everything is OK; Having lengthy communication*
	Assertiveness	***Not hesitating while speaking; Being responsible;*** *Being aggressive; Have leadership qualities; Expressing ideas openly*
	Activity Level	***Taking initiative; Being attentive;*** *Displaying active participation; Being ready to make decisions*
	Excitement-Seeking	*Being spontaneous; Taking risks; Being emotionally aroused*
	Cheerfulness	***Being happy; Being optimistic; Being positive;*** *Always smiling*
	Trust	***Being honest;*** *Being self-assured; Showing sympathy; Not being showy; Being committed*

Agreeableness	Morality	**Making better decisions**; *Have logical understanding*; *Showing concern for others*; *Not cheating*
	Altruism	***Encouraging others***; *Displaying non-selfish behavior*; ***Being helpful to others***; *Showing concern for others*
	Cooperation	*Showing appreciation for others*; *Not being a troublemaker*; *Having friendly attitude*; *Encouraging others*; *Understand each others tasks*
	Modesty	*Not insulting others*; ***Not boasting about their wealth***; *Not talking about themselves*; *Being level-headed*; *Being self critic*
	Sympathy	***Showing concern for others***; ***Helping others***; *Understanding others problems*
Conscientiousness	Self-Efficacy	***Having positive attitude***; *Taking risks*; *Being arrogant*; *Hard working*; *Meet deadlines*; *Encouraging oneself*
	Orderliness	**Being well organized**; ***Being consistent***; *Being risk averse*; *Performing well*; *Planning well*
	Dutifulness	***Being responsible***; ***Being sincere in work***; *Following rules*
	Achievement-Striving	**Hard working**; *Planning well*; *Not showing emotions*; *Being ambitious*; *Encouraging others*
	Self-Discipline	***Making good decisions***; ***Doing their own work***; *Do not need prompting*; *Being self-confident*; *Being conservative*
	Cautiousness	**Being risk-averse**; **Being alert**; *Not postponing things*
Emotional Stability v_s Neuroticism	Anxiety	***Doing things in a hurry***; *Showing signs of tension during decision-making*
	Anger	**Being frustrated**; *Passing sarcastic remarks to others*; *Taunting others*; *Using harsh language*; *Being against whatever one does*; *Not responding to others*; *Unwillingness to do work*
	Depression	**Passing negative comments**; ***Being discouraged***
	Self-Consciousness	*Talking negatively*; *Being conscious of oneself*; *Thinking about themselves*; *Being silent in a group*
	Immoderation	***Being greedy***; ***Trying hard to achieve success***; *Taking risks*
	Vulnerability	**Not able to work under pressure**; ***Feeling that they don't have enough time to get the work done***; *Being jealous*
Intellect or Openness	Imagination	**Using creative analogies or phases**; *Using varied response*; *Express ideas clearly*
	Artistic Interests	***Appreciate others***; *Being concern for physical appearance*; *Talking about beautiful things*; *Encouraging others*
	Emotionality	***Being gloomy***; *Being jealous*; *Being upset*; *Speaking emotionally*; *Expressing too much of themselves*
	Adventurousness	**Taking risks**; *Gambling a lot*; *Being innovative*; *Not being afraid of things*
	Intellect	**Having good reasoning abilities**; ***Being aware***; *Making sensible decisions*
	Liberalism	***Having new thoughts***; *Challenging authority*

6 Developing agent personalities

The most important data from Phase I are the actions, language, and/or behaviors associated with perceptions. To develop agents with personalities (Phase II) it is important to determine how to incorporate actions, language, and/or behaviors into a system to provide the perceptions that the computer agents have personality. As a starting point to this phase, we plan to categorize the actions, language, and/or behaviors that are associated within the Big Five Factor personality trait model into two levels for each trait. Two levels will be used because personality traits are measured as a continuum (e.g. Extraversion v_s Introversion). For example, Extraversion factor is indicated by actions such as cracking jokes, being very talkative, speaking amicably, and smiling a lot whereas actions such as complaining about others, being shy, and not being very talkative refers to the Introversion end of the continuum. Providing opposite types of actions, language, and/or behaviors within a trait will provide a possible combination across traits of actions, language, and/or behaviors. As an example, one role of the computer agent may be to draw attention to a task which requires human intervention. Table 4 illustrates different actions, language, and/or behaviors that an Extrovert v_s Introvert computer agent might exhibit. Table 5 illustrates the different actions, language, and/or behaviors for Conscientiousness v_s Conscienceless that may be exhibited when the computer agent takes on the role of collecting, analyzing, and validating information. Similarly, Table 6 illustrates different actions, language, and/or behaviors a Neurotic v_s Emotionally Stable computer agent might exhibit when they take the on the role of providing cues and sending reminders when the user is not paying attention to the tasks. The combination of these traits within a computer agent should lead to the perceptions of different personalities.

Table 4. Example of possible actions, language, and/or behaviors associated with Extravert v_s Introvert computer agent when drawing a user's attention to a task.

Actions, Language, Behaviors	Extrovert	Introvert
Actions	- Taps on the wrist or shoulder of a person to get their attention while simultaneously speaking. - Provides obvious visual indicators	- No physical interaction - Provides simple visual indicators
Language	- Assertive verbal phrases specifying person should pay attention	- No verbal language

Table 5. Examples of possible actions, language, and/or behaviors associated with Conscientiousness v_s Conscienceless computer agent when collecting, analyzing and validating information.

Actions, Language, Behaviors	Conscientiousness	Conscienceless
Actions	- Performs tasks on time - Responds to requests immediately	- Needs prompting by the user to get the task done - Does not respond immediately because it is working on the request but does not tell team member
Language	- Provides verbal statements with confidence level related to the task	- Does not make any confident statements related to the task or is delayed in responding

Table 6. Example of possible actions, language, and/or behaviors associated with Neuroticism v_s Emotional Stability computer agents when providing cues and sending reminders.

Actions, Language, Behaviors	Emotional Stability	Neuroticism
Actions	- Provides gentle reminders - Waits for the operator to complete current task	- Puts information directly over existing task
Language	- Uses positive remarks	- Passing sarcastic remarks or tells users NO – I am busy

After the categorization concepts are developed we plan to develop a computer simulation in which we can test the response of the users to the actions, language, and/or behaviors specified. The purpose of this simulation is to determine if the actions, language, and/or behaviors are perceived by the user interacting with the system as designed. In a simulation we will develop entities with certain defined personality attributes. For example the entity might have an attribute related to friendliness which can be indicated by the level of talkativeness. The simulation would specify an amount of talkativeness that the agent will express throughout the simulation. The characteristics can be triggered by events or time. Assertiveness may be modeled by the level of verbal output and physical interaction through haptic output. Once the collaborative task and entities have been developed, we will investigate human perceptions of the agents to determine if participants perceive the agents as modeled. Our approach is to provide a multimodal form of interaction between the computer agents and the participants. Attributes related to visual and auditory characteristics of the agents (e.g., face, bodies, and accents) could also be included. Trappl &

Petta (1997) describe developing personalities in just these ways. As a starting point we plan to omit the visual attributes to avoid impressions based on stereotypes, but the use of these characteristics will be included in future studies and may provide even stronger impressions of personality.

7   Conclusions

This research addresses the key issue of incorporating personality into agents and describes three methods used to capture actions, language, and/or behaviors that people attribute to different personalities in the context of collaboration or teamwork. Participants were able to describe actions, language and behavior corresponding to personality traits within the Big Five Factor personality model. The next phases of the research are to 1) develop agents with personality for incorporation into a simulation and to evaluate human perceptions of the personalities, 2) determine if agents with personality improve human-machine collaboration.

There are a great deal of issues to explore in the area of affective computing. We have concentrated on the personality aspect of affective computing, but emotions and personality are strongly linked. Participants in this experiment described personalities using terms that would also reflect emotional states. For example, participants used statements such as being happy, being gloomy, being jealous, being upset, speaking emotionally. The central personality trait Emotional Stability v_s Neuroticism is a trait that seems highly linked to emotions. The subtraits in this central trait are anxiety, anger, depression, self-consciousness, immoderation and vulnerability. Cheerfulness under the central trait Extroversion may correspond to emotions of happiness or joy. A similar data collection technique describing emotions could be used to collect perceptions of computer agent emotional states. These states could also be modeled using a simulation approach.

As part of this research we have been evaluating whether the personalities of the participants as measured in the Big Five Factor model are correlated with their perceptions of the computer agents personalities. If correlations exist it may be a way of determining which agent works best with which person. To date, with the limited data of the 12 users rating the computer agents within the casino game, we have not found a correlation. However, this may be due to the lack of personality depth in the computer agents. We have also been examining gender and culture which may influence perceptions of personality (Prabhala & Gallimore, 2005).

Customizing agents with personalities may improve human-machine collaboration by providing a more symbiotic team. Users may develop more trust in the system. Research is progressing in the area of sensing human cognitive and emotional states. The computer agents could adapt as these states are sensed. If the user is cognitively overloaded the computer agent could take over tasks to reduce workload. It is possible that computer agents with personality may also provide a more positive perception of the computer system. Corporations that use web-based sales may have more repeat customers if the users believe they are interacting on a more personal level in order to achieve their desired goal. Additionally, computerized learning systems for students in K-12 could be designed with personalities to match to student's needs thereby increasing student potential. The results from this research may help with the development of more realistic human behavioral models (HMBs) for use in war game simulations, virtual environments, and computer games.

References

1. Baker, D. P., & Salas, E. (1992). Principles for measuring teamwork skills. *Human Factors*, 34, 469-475.
2. Goldberg, L. R. (1990). An alternative "description of personality": The big five factor structure. *Journal of Personality and Social Psychology*, 59 (6), 1216-1229.
3. Hoc, J. (2001). Towards a cognitive approach to human-machine cooperation in dynamic situations. *International Journal of Human-Computer Studies*, 54 (4), 509-540.
4. Jang, K. L., McCrae, R.R., Angleitner, A., Riemann, R., & Livesley, W.J. (1998). Heritability of facet-level traits in a cross-cultural twin sample: Support for a hierarchical model of personality. *Journal of Personality and Social Psychology*, 74, 1556-1565.
5. Lee, K. W., & Nass, C. (2003). Designing social presence of social actors in human computer interaction. *Proceedings of the CHI Conference on Human Factors in Computing Systems*, Ft. Lauderdale, Florida, 289-296.
6. McCrae, R. R., & Costa, P.T., Jr. (1997). Personality trait structure as a human universal. *American Psychologist*, 52, 509-516.

7. Nass, C., Steuer, J., Tauber, E., and Reeder, D. C. (1993). Anthropomorphism, Agency and Ethopoeia: Computers as Social Actors. *Proceedings of the CHI Conference on Human Factors in Computing Systems*, Amsterdam, The Netherlands, 111-112.

8. Nass, C., Moon, Y., Fogg, B.J., Reeves, B., & Dryer, D.C. (1995). Can computer personalities be human personalities? *International Journal of Human Computer Studies*, 43, 223-229.

9. Pew, R. W., & Mavor, A.S. (1998). Modeling human and organizational behavior: Applications to military simulation. Washington, D.C: National Academy Press.

10. Prabhala, S., & Gallimore, J. (Submitted). Perceptions of personality in computer agents: Effects of culture and gender. *Proceedings of the 2005 Human Factors and Ergonomics 49th Annual Meeting*, Orlando, FL, September 26-30.

11. Silverman, B. G., Might, R., Dubois, R., Shin, H., Johns, M., & Weaver, R. (2001). Toward a human behavior models anthology for synthetic agent development. Systems Engineering: University of Pennsylvania.

12. Silverman, B. G., Cornwell, J., & O'Brien, K. (2003). Human Performance Simulation. In J. W. Ness, Ritzer, J.W., and Tepe, V (Eds.), *Metrics and methods in human performance research toward individual and small unit simulation*. Washington, D.C: Human Systems Information Analysis Center.

13. Soldz, S., & Vaillant, G.E. (1999). The big five personality traits and the life course: A 45 year longitudinal study. *Journal of Research in Personality*. Vol. 33, pp. 208-232.

14. Trappl, R., and Petta, P., (eds.), 1997, Creating personalities for synthetic actors: Towards autonomous personality agents. New York: Springer.

15. Winter, D. G., & Barenbaum, N.B. (1999). History of modern personality theory and research. In L. A. Pervin, and O. P. John (Eds.), *Handbook of personality: Theory and research* (pp. 3-27). New York: Guilford Press.

16. Wray, R. E., & Laird, J.E. (2003). Variability in human behavior modeling for military simulations. *Proceedings of the 2003 Conference on Behavior Representation in Modeling and Simulation*, Scottsdale, AZ.

Results from a Field Study: The Need for an Emotional Relationship between the Elderly and their Assistive Technologies

Peggy Wu, Christopher Miller

Smart Information Flow Technologies
211 N 1st St. #300, Minneapolis, MN U.S.A. 55401
{pwu, cmiller} @ sift.info

Abstract

Many of the elder-care giving responsibilities of the aging population fall on the shoulders of adult children. Caregiver burnout is one of the top reasons for the transition from independent living to costly nursing homes or other care-giving facilities. As a result, technologists are rushing to find tools that can assist caregivers and augment the elder's ability to age in place. For a generation whose introduction to technology includes the black and white television, where the metaphors of windows and desktops are not automatically understood, the blitz of PDAs, smart homes, and voice recognition systems can be confusing, intimidating and stressful. Add to this the elder's need for companionship to ward off isolation and dementia, and the design problem transpires to a much bigger challenge than designing around physical limitations of the elderly. The idea of companionship provided by technology is not new. From the simple rule-based Tamagotchi pets to Sony's sophisticated AIBO, there is a large body of evidence that shows that the owners of these robotic pets form genuine and meaningful emotional bonds. However, few solutions combine 'purposeful utility' with entertainment and companionship. Solutions that are 'purposeful' are predictable and seldom achieve the level of autonomy that captures the interest of and engages the user, while solutions with high entertainment value often lack functions that directly aid the elder with daily functions. We describe the details of a smart home field test, and examine focus group discussions that were conducted with both the participants and their adult children caregivers. We identify discussions from focus group transcripts that are related to the user's social behaviours and emotions towards the technology, and changes in the user's interaction with his/her caregiver caused by the introduction of the technology. We describe some aspects of the "personality" the participants projected onto the smart home despite the intended lack of physical character embodiment of the system, and relate this to existing theoretical work from human-human interactions and human learning theory. Finally, we offer insight into how these observations might translate into functional implementations to reap the benefits of assistive technology as a means to both reduce the burden of caregivers and provide companionship.

1   Introduction

As automation becomes more complex and sophisticated, there is increasing evidence that we treat it with the same set of expectations we bring to interactions with other complex, autonomous, social agents (e.g., Reeves and Nass, 1996). There are a number of efforts that are introducing theories from human-human interaction into computer science, including the A.L.I.C.E. chatbot (www.alicebot.org), Carmen's Bright Ideas (Marsella et al., 2003), and KISMIT (Breazeal, 2000). Some are beginning to leverage human-human interaction and emotional intelligence to address social aspects in the health-care domain (Picard, 2001). We describe a smart-home project whose goal is to support independent living for the elderly, and discuss the tendency of users to anthropomorphize technology and its implication for future technology designers.

1.1   Why is Affective Computing Particularly Important to the Elderly?

Affective computing is pertinent to this section of the demographic due to three major reasons. Firstly, emotions play a significant role in cognitive and physical health, which in turn greatly influences an elder's ability to live independently, more so than other age groups. Secondly, elders often lack an understanding of the metaphors used by technology and the technology's intent. Some pedagogic theories may explain tendencies for the personification of technology. Thirdly, irrespective of age, humans who are lacking or desire social interactions tend to actively seek social interactions, even when the interaction is not reciprocal as is the case of inanimate objects.

Due to the potential significant cost savings that can be attained by delaying in-home or nursing home care, technologists are rushing to provide solutions that promote independent living. Everything from ergonomically designed can openers and countertops, to chemical analysis toilets and human washing machines have been proposed or are becoming available. At the high technology end of this spectrum, and showing perhaps the greatest promise, are integrated home monitoring and aiding systems—"smart homes"—that may serve to let the elderly maintain their independence and remain in their legacy homes longer or reduce the need for professional assistance, extending the capabilities of those engaged in it. Smart homes technologies such as Georgia Institute of Technology's Aware Home Research Initiative[1] and Honeywell's Independent LifeStyle Assistant I.L.S.A.™(ILSA) have the primary objective of extending the level and duration of independence of elders. The most common reasons elders are admitted into nursing homes are caregiver burden and the elder's inability to perform Activities of Daily Living (ADLs)[2] and Instrumental Activities of Daily Living (IADLs)[2] (Miller and McFall, Kasper et al., 1990). Solutions for augmenting an elder's cognitive abilities include time and event based reminders and alarms for both the elders and their caregivers, but technologists are beginning to turn their attention to 'preventative' measures. Regular social interaction has been known to maintain emotional health and relationship development between patients and caregivers have been associated with a number of benefits, including improved treatment compliance, improved physiological outcomes, fewer malpractice suits, and more detailed medical histories (Bickmore et al., 2005). Neglect, loneliness, and isolation are often causes for an elder's transition into a nursing home or an independent living facility. The National Elder Abuse Incidence Study (NEAIS), the first study to estimate the national incidence of elder abuse and neglect in the U.S., found that a total of 449,924 elderly persons (adults ages 60 and over) were reported to have experienced some form of abuse or neglect in domestic settings in 1996 (NEAIS, 1996). According to the American Medical Association, thirty-five percent of elder abusers are adult children, and 13 percent are spouses. Elder maltreatment can be intentional or unintentional. Unintentional maltreatment is usually due to caregiver ignorance, inexperience, or inability to provide the necessary care (AMA, 1999). As technology slowly takes on the roles of a caregiver, it must provide more than task-oriented support and address the issues related to the emotion well being of its patients.

Much of the elder generation have lived their childhood and adult life outside of the computer technology sector, and lack a basic understanding of the metaphors used in today's user interfaces. Computer scientists have leveraged anthropomorphism for the development of teaching tools for several decades (Solomon, 1976). In Papert's work with children and the teaching of the Logo programming language[3], a turtle-shaped icon was introduced into the user interface as a means to convey the heading and movement of the "turtle". Papert contends that the children readily projected human characteristics onto the turtle to create an ego-syntonic relationship with it, thus encouraging the Piagetian concept of decentering and eventually bridging the gap between the language of mathematics (used by LOGO) and a means to verbalize it (Rieber, 1994). Although these concepts have traditionally been associated with learning in children, they may also be applicable to "gerogogy" (after Pearson and Wessman[4]). As we will see later, anecdotal evidence suggest that during the learning process, different aspects of a *personality* may be projected onto the technology as a level of abstraction and a means to circumvent the sometimes unnecessary explanations of underlying detail or inexplicable reactions from the technology.

Humans have inherent social needs, regardless of age. The elderly are especially prone to loneliness and depression brought on by Empty Nest Syndrome or neglect, either intentional or not. A means of alleviating the feelings of loneliness is to transfer it to solitude by means of introducing activities or hobbies. As we will explain later in greater detail, companionship through pets has been found to be extremely helpful to elders. The same principles may apply to collectors through the anthropomorphism of things, who will talk to and interact with his/her inanimate objects.

---

[1] See http://www.cc.gatech.edu/fce/ahri/ for more information.
™ Independent LifeStyle Assistant (I.L.S.A.) is a registered trademark of Honeywell.
[2] See the National Center for Health Statistics (http://www.cdc.gov/nchs/) for definitions of ADL and IADL.
[3] See http://el.media.mit.edu/logo-foundation/logo/index.html for more details.
[4] Pearson M, Wessman J (1996). Gerogogy (ger-o-go-gee) in patient education. In Home Healthcare Nurse. 1996 Aug; 14(8):631-6.

1.2 Etiquette, Trust, and User Compliance

In a sophisticated automation system such as a smart home, a large number of interactions between the system and the human user fall into the class of ***directives*** (after Searle[5]). In any directive, the speaker is directing the hearer to perform a task, though we realize that the compelling force of that directive may vary (e.g., command vs. request vs. instruction vs. advice vs. observation) and may come from a variety of sources or motivations (e.g., beseeching, coercing, remonstrating, instructing). There is both theoretical and empirical evidence "etiquette" and "politeness" can impact human performance—specifically, trust, regard and decisions to accept directives. The building of trust and its impact on the human's decision to accept or reject advice has many implications that are the same between human-human and human-machine interactions. Parasuraman and Riley (1997) and Lee and See (2003) provide comprehensive reviews of the impact of trust, perceived reliability, etc. on the acceptance or rejection of advice, as well as the development and tuning of trust. When an agent (human or machine) behaves in a manner which is familiar or pleasing to us, we tend to provide it, through simple affective methods, with attention, trust, and a greater probability of following its directives (cf. Norman, 2004). Parasuraman used the Multiple Aptitude Task (MAT) Battery to experiment with human responses to varying levels of the user interface's "politeness" and accuracy (Parasuraman and Miller, 2004). In one test, he found that human subjects believed the "polite" system to be more accurate, even when it was in fact less accurate. This suggests that etiquette variables can have a profound effect on user compliance, and that designers should use a level of etiquette appropriate to the accuracy of the system, especially when dealing with faulty automation.

1.2.1 Framework for Etiquette

A seminal body of work in the sociological and linguistic study of politeness is the cross-cultural studies and resulting model developed by Brown and Levinson (1978; 1987). Brown and Levinson were interested in cataloging and accounting regular deviations, across languages and cultures, from Grice's (1975) *conversational maxims*. Grice had formulated four "rules" or maxims that characterized efficient conversation. These were:

1. *Maxim of Quality*: Speak truthfully and sincerely
2. *Maxim of Quantity*: Be concise; say neither more nor less than required to convey your message
3. *Maxim of Relevance*: Don't introduce topics at random, follow the conversational "flow"
4. *Maxim of Manner*: Be clear in your statements, avoid ambiguity and obfustication.

Brown and Levinson noted that there is at least one way in which people across cultures and languages regularly depart from the efficient conversation characterized by Grice's Maxims. For example, a caregiver may ask of his/her patient, "Please take your medicine." The use of "please" in that sentence is unnecessary for a truthful, relevant or clear expression of the caregiver's wish and it in fact explicitly violates the Maxim of Quantity since it adds verbiage not required to express the caregiver's propositional intent. Over years of cross linguistic and cross cultural studies, Brown and Levinson collected and catalogued a huge database of such violations of efficient conversation. Their explanation for many of these violations is embodied in their model of politeness. The Brown and Levinson model assumes that social actors are motivated by a set of wants including two important social wants based on the concept of face (Goffman, 1967) or, loosely, the "positive social value a person effectively claims for himself" (cf. Cassell and Bickmore, 2002, p. 6). Face can be "saved" or lost, and it can be threatened or conserved in interactions. Virtually all interactions between social agents involve some degree of threat to the participants' face—what Brown and Levinson call Face Threatening Acts (FTAs). This is especially true for directives, since the speaker is both demanding the attention of the hearer by the act of speaking, as well as directing the hearer to perform a task, which threatens the hearer's autonomy. Brown and Levinson claim that the degree of face threat is a function of the social distance between the speaker and the hearer, a relative power that the hearer has over the speaker, and the ranked imposition of the raw act itself.

---

[5] Searle, J., (1969). *Speech Acts: An Essay in the Philosophy of Language.* (Cambridge, UK.; Cambridge University Press). See also: Searle, J. (1985). *Expression and Meaning: Studies in the Theory of Speech Acts.* (Cambridge, UK.; Cambridge University Press).

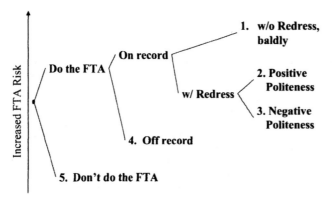

Figure 1. Universal FTA Redress strategies ranked by Brown and Levinson, (1987).

FTAs are potentially disruptive to human-human relationships, and Brown and Levinson offer 5 broad redressive strategies to mitigate the degree of face threat. These are illustrated in Figure 1, and ranked from the most to the least threatening.

The first strategy involves the speaker making the request baldly with no redress. The second strategy, positive politeness, places emphasize on common ground between the speaker and the hearer by invoking in-group identity, by joking and assuming agreement and/or by explicitly offering rewards/promises. The third strategy, negative politeness, strategies focuses on the hearer's negative face needs—independence of action and attention. It minimizes the impact on the hearer by being direct and simple in making the request, offering apologies and deference, minimizing the magnitude of the imposition and/or explicitly incurring a debt. The forth strategy, off record, is the least threatening strategy if the speaker is to conduct the FTA at all. This strategy involves doing the act with a sort of "plausible deniability" by means of innuendo and hints, thus avoiding an overt request or any action from the part of the hearer. The fifth strategy is to avoid the FTA altogether. At some threshold, in some contexts and cultures, it will simply be too threatening for some FTAs to be performed, regardless of the amount of redress offered. Details regarding the strategies used for the prototype system can be found in Miller et al., 2004.

2   The I.L.S.A. System

Haigh et al. (2004) describe the details of the I.L.S.A. system deployment. There are two separate interfaces for the elderly participants and their caregivers. We briefly describe the elder user group and the available interactions with the I.L.S.A. system for this user group below.

2.1   Participant Demographics

The I.L.S.A. system was installed in 11 single occupant homes of elderly participants for a duration of 4 to 6 months. The average age of the participants was 83.42, ranging from 76 - 96. None of the participants had problems associated with dementia, and all were competent in all ADLs. The minimal education level of the group was the completion of high school. As part of the study, each subject was required to identify a family member as a designated caregiver willing to participate in the study.

2.2   Device Suite and User Interaction

2.2.1   Devices
There were slight variations in the set of sensors used in each participant's home due to differences in floor plans and room configurations. The typical sensor suite instrumented a one bedroom apartment, and consisted of one or more passive motion sensors in each zone (living room, dining room, bedroom, kitchen, hallway, and bathroom), a "med caddy" with a contact sensor, and a contact sensor on the entry door. Motion sensors were calibrated to ignore the movement of pets.

2.2.2 User Interface and System Reporting

A Honeywell touch screen WebPAD™ internet appliance was included in each installation for browsing the web-based graphical user interface. Figure 2 shows a screen shot of the ILSA Reminders page. The participant's existing telephone was also used for reporting purposes. The graphic user interface consisted of a set of web pages reporting on the following items:

- Reminders
- Mobility for today and yesterday (amount of motion detected)
- Medication for today and yesterday

Additional controls and information include:
- Controls
  - o Changing System Modes (on, off, away)
  - o User Configuration and Set Up
- List of Caregivers
- Help

Figure 2 Sample Web Page of User Interface (from Haigh et al., 2004)

The participants were free to review the items above at any time. Automatic reminders and alerts were sent to the participant's home telephone. The messages used in the phone system were a combination of a pre-recorded human voice and computerized voice synthesis (for date, time, and numerical information). Participants interacted with the telephone system using the telephone keypad. Depending on the perceived severity of the message, the I.L.S.A. server may also call the caregiver.

3 Focus Group

There were a number of different approaches used in the analysis of sensor and web usage data, as well as cognitive metrics for the subjects and monthly surveys for both the subjects and their caregivers (see ILSA). Below we discuss topics related to ethopoeia (after Nass and Moon[6]), the social needs of the participants, and how ILSA affected the participant's interaction with their caregivers. The focus group was conducted with a subset of the participants and their designated caregiver.

3.1 Focus Group Questions

A user survey designed to assess the politeness of variations of a medication reminder is described in Miller et al. 2003. Discovering the impact on the emotional states and needs of participants was not an explicit goal of the focus group during the time of the study, thus there were little investigator-initiated questions regarding the personality or politeness of the ILSA system. Nonetheless, the flexible, open format of the focus group allowed the subjects to discuss ILSA's "social" characteristics, and allowed the investigators to respond to this concept, and devise questions as the focus group progressed.

A number of open-ended questions acted as guidelines for the discussion. General topics include the subject's perception of the system's impact on safety, independence, caregiver burden, system accuracy and acceptable financial costs of such a system. Note that the initial phrasing of the focus group questions used gender neutral language when referring to ILSA. However, the pre-recorded human voice on the phone system was female.

In the Section 3.2, we select comments regarding the participant's 'social' interaction with ILSA, and how it affected their relationship with others. As the focus group questions were initially categorized using a different scheme, answers do not fall neatly into the two topics we have identified. Regardless, we present a small subset of the focus group questions in the following subsections.

™ WebPAD is a trademark of Honeywell

[6] Nass, C.L., & Moon, Y.(200). Machines and mindlessness: Social responses to computers. Journal of Social Issue, 56(1), 81-103.

3.1.1 Interaction with System

Sample questions asked in the focus group include:
- Did you feel like you were checked on more frequently than you were without ILSA?
- Will you miss ILSA after it is uninstalled from your home?
- What is your comfort level with current ILSA sensor suite?

3.1.2 Social Interaction with Others

Sample questions include:
- How has ILSA affected your relationship with caregivers/elders?
- Did having ILSA mean that people intruded into your life less or more? Was their checking less intrusive? If so, was this a good or bad thing?
- Did you experience a change in behavior as a result of ILSA?

3.2 Focus Group Answers

Below we present some comments from the focus group, as divided into two general topics, (1) the subject's interaction with the system and (2) the impact of the system on the subject's interaction with his/her friends and caregivers.

3.2.1 Interaction with System

"[I checked the ILSA data daily because] I wanted to make sure she was behaving." – Elder subject

In agreement with results from the analysis of webpage usage, we found that most subjects checked their status almost daily, especially reports regarding medication and mobility.

"I tried to find ways to beat the box." – Elder subject

At least two subjects independently invented methods to prevent the system from generating reminders and alerts.

"… I played solitaire." – Elder subject

At least one subject used the WebPAD for games outside of the ILSA interface. There was a general consensus on the need for more features, such as games and email.

"I would be nice if it said a cheerful Good Morning!" – Elder subject
"I would hang up as soon as I heard ILSA's voice." – Elder subject

There were a number of comments regarding the personality and friendliness of the telephone system. In general, subjects found the tone of the telephone voice to be disagreeable, and desired a richer set of interactions.

"It would be nice to get some re-assurance when I'm not feeling well." – Elder subject

Although ILSA had some user modeling capabilities including medication usage and mobility, it did not model external aspects such as hospitalization or illness, factors that dramatically influence the subject's overall well-being. Nonetheless, participants thought it would be useful for the system to offer some of the emotional support usually provided by human caregivers.

"Anything to keep the mind going would be helpful." - Caregiver

One caregiver suggested that the subject's involvement in the field test was helpful as it was a form of cognitive exercise.

3.2.2 Social Interaction with Others

> "My son says I get up a lot at night – [I know] he's looking at it." – Elder subject

Several participants and their caregivers reported using the system as an objective third-party observer and used its reports to discuss issues that the participant would otherwise not disclose to his/her caregiver. For example, during the course of the study, one subject had an abnormal change in mobility, but did not discuss this with his/her caregiver out of fear of 'being a nuisance'. The caregiver initiated a conversation with the subject based on ILSA's 'comments', and later found that the abnormality was due to a new medical problem.

> "I wanted to see what [ILSA] was telling [the caregiver]." – Elder subject

Although participants were curious about the information made available to the caregivers, there was a lesser emphasis on privacy concerns than initially anticipated. This may be because there was minimal personal information displayed on the user interface, and the participants did not see any apparent damaging consequences if unauthorized individuals were to see the subject's mobility and medication patterns. However, subjects were still concerned about the information being disclosed, and examined the caregiver web page with the same amount of frequency as the mobility reports to study the type of information made available to their caregivers.

> "[ILSA] was something for [the subject] to talk about with her friends." - Caregiver

Comments throughout the focus group session suggested that participants enjoyed discussing the ILSA system with their friends and family members. This may be a product of the novelty of the project, or of the feeling of possessing an elite status due to their involvement in the study.

4 Discussion

The inclusion of social characteristics in healthcare technologies is not new. Cassell and Bickmore have been exploring the concept of creating a *working alliance* between a computer agent and its user through the MIT FitTrack project (Bickmore, 2002). Johnson's work on Carmen's Bright IDEAS (Marsella et al., 2003) uses an avatar to teach problem solving to parents of children with leukemia. Lisetti et al. (2003) has taken a different approach by including the detection of emotional states for user modeling in affective interfaces in telehealth and tele-home health care. There was ample anecdotal evidence that subjects projected a personality onto the ILSA system. This is consistent with Reeves and Nass's findings on the tendency for humans to anthropomorphize computer software (Reeves and Nass, 1996). Subjects assigned characteristics to ILSA and interacted with it with the same level of etiquette they perceived was possessed by ILSA. Subjects also referred to ILSA as 'her' and talked about her and 'her thoughts' with their friends and family members, as opposed to 'the system' and 'its reports'. In fact, in one case, the friend of a participant believed ILSA to be a home care nurse until the friend visited the subject's home and saw the system of sensors and the WebPAD. Below we discuss some of the above comments in greater detail, and draw parallels with existing theories.

4.1 Interaction with System

4.1.1 Politeness and Etiquette

In general, the participants regarded the voice of the telephone system as ILSA, and had the notion that some virtual embodiment of ILSA was examining the same user interface to review the subject's status. An analysis of the sensor data revealed that despite the consensus that ILSA phone reminders and alerts were found to be useful, subjects would usual 'hang up on ILSA', and very rarely heard the message in its entirety. This may be due to the subject's lack of understanding about how to use the system, frustrations with the limited interactions available, inappropriate length of message or the perceived impoliteness of the ILSA telephone system (i.e. 'machine-like' tone of voice, lack of a cheerful greeting, and incapable of turn-taking). Comments from the focus group lead us to believe the latter to have the most influence on the subjects' acceptance of the voice system. Subjects were especially opposed to voice synthesis, explaining that it sounded 'cold and insincere'. Some suggestions to resolve this problem include the introduction of small talk, such as starting the voice message with a personalized greeting or asking about the subject's day. Research suggests there may be factors beyond tone that affect a hearer's impression of a synthetic voice (Nass and Lee, 2001), but they were not explored in the focus group due to time constraints.

Subjects also commented on the inappropriateness of phone calls for missed medication reminders. We believe there are two reasons for this. Firstly, almost all the subjects had a highly variable day-to-day schedule, and were often outside the home, thus reminders would be received by answering machines or remained unanswered until the system timed out. However, we believe the same complaint would exist even if the phone reminders were routed to mobile phones due to a second, perhaps more important reason. Although medication compliance was viewed as important, subjects did not view it as possessing enough urgency to warrant a phone call. Subjects in the field test were highly independent elders who relied on their own makeshift systems for medication reminders, such as placing medication in a high traffic area, using pill boxes etcetera. The introduction of a phone reminder may have been viewed as intrusive to their existing lifestyles and threatened their sense of independence. Subjects expressed appreciation for medication reminders when medication was forgotten, but suggested more subtle mechanisms for reminders, such as a light on the top of the med caddy.

Several subjects exhibited behaviors that can be conceived as a byproduct of mistrust. This may be related to the perceived impoliteness of the ILSA telephone system and may support the concepts of trust and etiquette (Parasuraman & Miller, 2004). Without requests from the investigators, most subjects regularly reviewed the reports regarding medication and mobility. Some subjects were vigilant with noting when the said reports were inconsistent with their own activity logs or when the reports were incorrect. Subjects also appeared to take satisfaction in identifying incidents when ILSA 'made a mistake'.

Due to the general dislike of ILSA phone calls, multiple subjects invented ways to prevent the system from generating phone messages. For example, one subject would open and close the med caddy without taking her medication, and then remember on her own to take the medication later. Ironically, the desire to 'beat the box' may have provided the cognitive activities the same subject was seeking when she searched the WebPAD for games.

4.1.2 Increase User-System Interaction Functions and Entertainment Value

While current technology does not allow users to have free form conversations with the system, there was an overwhelming desire for mechanisms with which subjects can provide feedback and personal status. Subjects wanted the capability to verbalize answers to the phone system, but also wanted to provide feedback in other forms. As a simple example, participants suggested the use of physical buttons by the door to indicate when they were leaving the home.

Subjects also reported that the capability for ILSA to provide reassurance would be a valuable function. Subjects did not provide suggestions on specific methods for providing reassurance, nor when it should be provided, but this is an interesting idea that can provide significant value to elders with minimal emotional support from human caregivers.

Most subjects appeared to spend a great deal of time and energy trying to understand the underlying mechanisms of the system, and would sometimes vary their routines to see how ILSA reacted. However, it is conceivable that the process of recruiting subjects for this study favored those who are curious by nature, and that the general public would not react to the system in this way. Subjects found it rewarding when they discovered some of the basic logic in the system's algorithms. Their reactions and comments led us to believe that they enjoyed the mental challenge. This is further supported by the fact that at least one subject used the web pad for its built-in games, and others iterated the desire for card or word games as part of the system.

4.1.3   Interaction with Human Caregivers and Peers

Although the field test was relatively short (4 to 6 months), we believe the subjects' involvement with the study may have increased their social interaction with their human caregivers and peers. Barker (1999) reviews research efforts that have found similar findings in the field of animal therapy research, where a pet can promote the social interaction of the owner with other humans. Further work is needed to investigate whether increased positive socialization will be sustained even after the novelty of the system subsides.

A surprising use of ILSA was as a third party observer who provided 'objective' reports on the subject's actions and status. The caregivers used ILSA's measures as a tool to initiate conversations regarding changes in a subject's level of mobility or number of system reminders generated that may be signs of or may lead to potential problems. This usage model may also be of considerable value to formal caregivers, as the accuracy and amount of information from office visit conversations often rely heavily on the patient's memory, and there lacks automated metrics for comparing a patient's general well being over time.

5   Future Work: Bridging the Gap between Smart Home Capabilities and Social Needs

Elders have difficulties using the typical interaction devices employed in smart home designs (i.e. touch screens, computer interfaces, telephone interfaces, and even panic buttons) because of both physical and psychological barriers. Perhaps more importantly, current smart home systems do not provide the emotional support needed by independent elders. The study revealed that elders would appreciate greater interactivity with a smart home system, especially in entertainment and social aspects, but industry has thus far taken an approach similar to the traditional medical profession – where the patient is observed rather than acts as an active participant in the selection of treatments. Communicating directly with the elder may be the best method to gather information about his/her well being, but voice recognition capabilities and the ability to extract computer-usable information from casual conversations are not readily available.

Within the field of animal-assisted therapy, hundreds of clinical reports show that when animals enter the lives of aged patients with chronic brain syndrome (which follows from either Alzheimer's disease or arteriosclerosis) that the patients smile and laugh more, become less hostile to their caretakers and become more socially communicative. Other studies have shown that in a nursing home or residential care centre, a pet can serve as a catalyst for communication among residents who are withdrawn, and provide opportunities (petting, talking, and walking) for physical and occupational rehabilitation and recreational therapy. More generally, the research literature has established that the physiological health and emotional well-being of the elderly are enhanced by contact with animals (Beck & Katcher, 1996; Center for the Human-Animal Bond). However, many elders live in places that either prohibit pets or are not conducive to animals due to the physical layout of their buildings. In addition, some physiological conditions, such as Alzheimer's disease, may make animal ownership difficult for the individual and unsafe for the animal.

The use of a robotic pet as an addition or alternative to traditional graphical user interfaces may provide the desired increase in emotional support, as well as the means for an elder to communicate with the system. Several projects are underway to develop a robotic pet that can both monitor an elderly user and act as a companion (Center for the Human-Animal Bond; Lavery, 2000; Necoro). A user's communication model with a robotic pet is very similar to that of real pets—and is therefore familiar, effective and comfortable for elders. The challenge to designers will be to create a set of robotic behaviors that can parallel human-pet interactions in the real world while also providing an aid that an elder needs in order to remain independent.

6   Acknowledgement

The authors would like to acknowledge that the work on the ILSA project was made possible by the support of Honeywell and the Advanced Technology Program at the National Institute of Science and Technology, U.S. Department of Commerce under agreement #70NANBOH3020.

References

AMA (American Medical Association), Elder Maltreatment and Neglect: 1999, http://www.medem.com/MedLB/article_detaillb.cfm?article_ID=ZZZE61CTWAC&sub_cat=357.

Barker, Sandra B. (1999). Therapeutic Aspects of the Human-Companion Animal Interaction. Psychiatric Times, Feb 1999 Vol. XVI Issue 2.

Beck, Alan and Katcher, Aaron. (1996). Between Pets and People: The Importance of Animal Companionship (Revised Edition). West Lafayette, IN: Purdue University Press.

Bickmore, T. (2002). "When Etiquette Really Matters: Relational Agents and Behavior Change." Proceedings of the AAAI Fall Symposium on Etiquette for Human-Computer Work, November 15-17, Falmouth, MA.

Bickmore, T., Gruber, A., and Picard, R. (to appear) "Establishing the Computer-Patient Working Alliance in Automated Health Behavior Change Interventions" Patient Education and Counseling.

Breazeal, C. (2000), "Sociable Machines: Expressive Social Exchange Between Humans and Robots". Sc.D. dissertation, Department of Electrical Engineering and Computer Science, MIT.

Brown, P. & Levinson, S. Politeness: Some Universals in Language Usage. Cambridge Univ. Press, UK. 1987.

Center for the Human-Animal Bond, http://www.vet.purdue.edu/chab/. Accessed March, 2005.

Goffman, E. Interaction Ritual: Essays on Face to Face Behavior. Garden City; New York. 1967.

Grice, H.P., Logic and Conversation, In P. Cole and J. Morgan (Eds.), Syntax and Semantics, vol. 3., Speech Acts, Academic Press; New York, 1975, pp. 41-58.

Haigh, Karen Zita, Kiff, Liana M., Myers, Jane, and Krichbaum, Kathleen (2004). "The Independent LifeStyle Assistant[TM] (I.L.S.A.): Deployment Lessons Learned. In The AAAI 2004 Workshop on Fielding Applications of AI, July 25, 2004, San Jose, CA. Pages 11-16.

ILSA (Independent LifeStyle Assistant), http://www.htc.honeywell.com/projects/ilsa/about_introduction.html.

Lavery, Anne (2000). Robo-cat makes purrfect companion. http://news.bbc.co.uk/1/hi/sci/tech/652293.stm. Accessed March 2005.

Lisetti, C.L., Nasoz, F., Lerouge, C., Ozyer, O., and Alvarez, K. (2003). Developing Multimodal Intelligent Affective Interfaces for Tele-Home Health Care. International Journal of Human-Computer Studies Special Issue on Applications of Affective Computing in Human-Computer Interaction, Vol. 59 (1-2):245-255.

Marsella, S., Johnson, W. L., and LaBore, Catherine M. (2003). Interactive Pedagogical Drama for Health Interventions. In *11th International Conference on Artificial Intelligence in Education*, Australia, 2003

Miller, Baila & McFall, Stephanie. Caregiver Burden and Institutionalization, Hospital Use, and Stability of Care. US Department of Health and Human Services. Report, prepared by University of Illinois.

Miller, C., Wu, P., Krichbaum, K., Kiff, L. (2004). "Automated Elder Home Care:Long Term Adaptive Aiding and Support We Can Live With", in proceedings of the AAAI Spring Symposium on Interaction between Humans and Autonomous Systems over Extended Operation, 22-24 March 2004. Stanford, Palo Alto, CA., U.S.A.

Miller, C., Wu, P., Krichbaum, K., Kiff, L. (2003). Etiquette and Effectiveness: How Should a Smart Home Interact? In *Proceedings of the International Conference on Aging, Disability and Independence*, December 4-6; Washington, DC.

Nass C. & Lee, K. M.(2001). Does computer synthesized speech manifest personality? Experimental tests of recognition, similarity-attraction, and consistency attraction. Journal of Experimental Psychology: Applied, vol. 3, pp. 171-181, 2001.

NEAIS (The National Elder Abuse Incidence Study), conducted by the National Center on Elder Abuse at the American Public Human Services Association (formally known as the American Public Welfare Association) and the Maryland-based social science and survey research firm, Westat, 1996.

Necoro, http://www.necoro.com/ Accessed March 2005.

Parasuraman, R. & Miller, C. (2004). "Trust and Etiquette in High-Criticality Automated Systems". In C. Miller (Guest Ed.), special section on "Human-Computer Etiquette". Communications of the ACM. 47(4), Apr. 2004.

Picard, R. W. (2001). "Affective Medicine: Technology with Emotional Intelligence". Chapter in "Future of Health Technology," IOS Press.

Reeves, B. & Nass, C. (1996). The media equation: how people treat computers, television, and the new media like real people and places. Cambridge University Press, Stanford, CA.

Rieber, L.P. (1994). Computers, graphics, and learning. Madison, Wisconsin: Brown & Benchmark.

Solomon, C. J. (1976). Teaching the Computer to Add: An Example of Problem-Solving in an Anthropomorphic Computer Culture, MIT Artificial Intelligence Lab.

Evaluation of Affective Computing Systems from a Dimensional Metaethical Position

Carson Reynolds
MIT Media Laboratory
20 Ames St, Room E15-120F
Cambridge, Massachusetts 02139
carsonr@media.mit.edu

Rosalind W. Picard
MIT Media Laboratory
20 Ames St, Room E15-020G
Cambridge, Massachusetts 02139
picard@media.mit.edu

Abstract

By integrating sensors and algorithms into systems that are adapted to the task of interpreting emotional states, it is possible to enhance our limited ability to perceive and communicate signals related to emotion. Such an augmentation would have many potential beneficial uses in settings such as education, hazardous environments, or social contexts. There are also a number of important ethical considerations that arise with the computer's increasing ability to recognize emotions. This paper will survey existing approaches to computer ethics relevant to affective computing. We will categorize these existing approaches by relating them to different metaethical positions. The goal of this paper is to situate our approach among other approaches in the computer ethics literature and to describe its methodology in a manner that practitioners can readily apply. The result then of this paper is a process for critiquing and improving affective computing systems.

1 Undesirable Scenarios

The film Hotel Rwanda describes historical horrors of a sort that have happened more than once, and thus may happen again, although with new technology and new individuals involved. At several times throughout history, one group has tried to perform "ethnic cleansing," and during many scenes of this film people are asked whether they are Hutu or Tutsi's. In the film, those who admit to being (or are exposed as being) Tutsi are carted off, and eventually a million Tutsi's are brutally murdered. Imagine how much more "efficient" this kind of interrogation process could be if the perpetrators could point a non-contact "lie detector" at each person while questioning them about their race (or other unwelcome beliefs.) Lie detectors typically sense physiological changes associated with increased stress and cognitive load (presuming it is harder to lie than to tell the truth). While honesty is a virtue, and we'd like to see it practised more, it is also possible to imagine cases where a greater virtue might be in conflict with it. If such devices became easy to use in a widespread reliable fashion, they will become easier to misuse as well. It is not hard to imagine an evil dictatorship using such a device routinely, perhaps showing up at your home and pointing it at you, while asking if you agree with their new regime's policies or not, and then proceeding to treat people differently on the basis of their affective responses to such questioning.

Some work in the Media Lab developed wearable devices that could learn to recognize eight emotional states from a person, one of which was anger [Picard et al., 2001]. While that work was just a prototype, suppose that organizations who ran prisons or punishment services were to refine the system to detect a variety of levels of anger. The infliction of pain often gives rise to anger in the recipient of that pain, and such a device could be coupled with a pain infliction device to bring people to a certain level of torment. While we like to think our civilized society would not stoop to such torture, the recent events at the Iraqi prison Abu Ghraib remind us that our systems are subject to fault. It is not hard to imagine efforts to bolster an over - stretched security force with the addition of an electronic interrogator for every suspect, especially when it is believed that the information extracted could save the lives of many others.

A hybridization of affect sensing technology and armed robotics could produce a very menacing device. Suppose a weapons engineer were to take a rifle equipped robot and gate its trigger mechanism such that it only fired at people it didn't know who were expressing anger and fear toward it (some have argued that people fighting on the same side as the most sophisticated technology have no reason to fear it, at least not as much as the enemy has.) Readers may feel that such outlandish devices are the stuff of science fiction and not to be taken seriously. However, with $127 billion in funding going towards future combat systems such as a robot "equipped with a pump - action shotgun system able to recycle itself and fire remotely" [St. Amant, 2004] and with an army researcher suggesting "the lawyers tell me there are no prohibitions against robots making life- or - death decisions" [Ford, 2005] it is appropriate to consider what role ethics ought to play.

These scenarios might seem remote, and merely hypothetical. Meanwhile there are (perhaps more costly) immediate scenarios of emotional paperclips and other software tools that daily annoy people, elicit frustration and even anger, and pose a costly threat to our productivity, health, and even our performance behind the wheel of a car or truck. In general, our research aims to develop guidelines for affective technologies that will reduce risks of harm to people, and address their ethical concerns.

Designers cannot control the ultimate uses of a technology; undesirable uses may occur without any intent whatsoever by a designer. Nonetheless, we think that some designs facilitate certain uses more than others. When exploring possible designs of affective computing systems, it is important to confront the harmful potential applications along with the beneficial. While such considerations may not prevent all harmful uses, they can potentially lessen the likelihood of harmful uses. Thus, we desire to influence the design of technologies that maximize the likelihood for beneficial uses, while minimizing the potential for malicious uses. Toward such a goal, this paper seeks to provide a framework for developing ethical questions about affective computing systems.

2 Questions for Designers

In trying to assess these moral and ethical decisions, it is useful for designers to ask themselves questions that can help gauge the impact of affective technologies on users. For instance:

- "Could a user be emotionally manipulated by a program with the capability to recognize and convey affect?" (this question has also been addressed by Picard and Klein, 2002)
- "Should an affective system try to change the emotional state of a user?"
- "Would a system that allows surveillance of previously invisible affective signals invade privacy?"

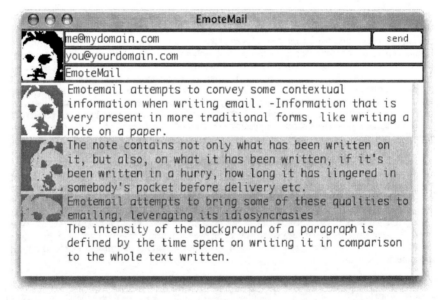

Illustration 1: EmoteMail is an email client that is augmented to convey aspects of the writing context to the recipient. The client captures facial expressions and typing speed and introduces them as design elements. These contextual cues provide extra information that can help the recipient decode the tone of the mail. Moreover, the contextual information is gathered and automatically embedded as the sender composes the email, allowing an additional channel of expression.

In answering questions such as these, a large number of variables come into play. Some philosophies suggest ethical judgements are the result of reasoning and reflection about desires and effects. Still others treat ethical judgements as being the expression of feeling and emotion, having no base in logical or analytical philosophy. What is common

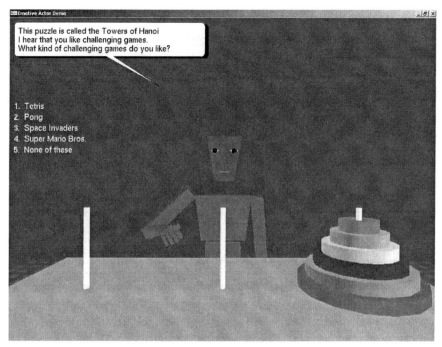

Illustration 2: The learning companion: a relational agent that supports different meta-cognitive strategies to help students overcome frustration. The system makes use of a large number of sensors, facial expression recognition, pressure-sensitive mouse and chair, and skin conductivity sensor. Information from these sensors is fused to achieve effects such as affective mirroring, where the agent subtly mimics certain aspects of the student's affect with the goal of fostering a bond.

among many philosophical approaches is a desire to abstract the core issues away from particulars. We think it is appropriate to take a process of reason and reflection, and will invert this "abstract the core away from particulars" approach and suggest, for the sake of applying ethical theory, the examination of particular dimensions that might bear upon design of ethical affective computing systems.

3   The Social Contextualization of Affective Computing

Before proceeding, we will provide some examples of current affective computing systems. Affective computing is "computing that relates to, arises from, or deliberately influences emotion" [Picard, 1997]. In collaborative efforts, we have developed many prototypes of such systems in a variety of use contexts including the systems pictured in Illustations 1 and 2.

As such systems approach maturity, the use of affective computing technologies by a wider portion of society seems increasingly likely. However, before such systems are widely deployed, it is important to assess what harms and benefits arise from their novel capabilities.

We would like to provide the community with heuristics and design guidelines to help avoid designing affective computing systems that could be exploited in harmful or malicious ways. As a starting point we provide a list of different dimensions that are relevant to affective computing systems. By considering different values for these dimensions, it is our hope that designers can spot different ethical difficulties present in affective computing systems.

3.1   Invasion and Discomfort

Designers make a variety of moral and ethical decisions in the development of an interaction technology. In our initial explorations we explored impacts on comfort and privacy [Reynolds and Picard, 2004]. We presented participants with two hypothetical application contexts (music and news recommendation) that focused on four emotions (joy, anger, sadness, and excitement). In completing surveys regarding these situations we found that participants reported that such systems are invasive of privacy and may also induce feelings of discomfort. Specifically, when asked "Do you think your privacy would be affected ..."and given a choice between "1 -

Completely Invaded" and "7 - Completely Respected" participants reported a mean of 2.6, unless they were given a contract which specified how their affective information could be used, in which case it shifted to 3.4. Likewise, when asked "How comfortable would you feel ..." and given a choice between "7 - Completely Comfortable" and "1 - Completely Uncomfortable" participants also reported a mean of 2.6, in the case that they did not have a contract which precisely specified how their information would be used, and the comfort reported shifted up to 3.5 when a contract was present. These results suggest that individuals may feel threatened by systems that are apparently benign if they do not have information in the form of a contract. Even when a contract is present, their reports average only "neutral", which also leaves room for improvement.

3.2 What is Needed

The use of affective computing technologies by a wider portion of society seems increasingly likely. However, before such systems are widely deployed, it is important to assess what harms and benefits arise from their novel capabilities.

We would like to provide the community with heuristics and design rules to help avoid designing affective computing systems that could be exploited in harmful or malicious ways. However, it is important to avoid ungrounded, ad-hoc rules for the design of systems that deal with information as sensitive as an individual's emotional state. Instead, what is needed is a grounded methodology that designers can use to help assess affective computing systems.

4 Dimensional Metaethics

The methodology we advocate is rooted in metaethical philosophical positions, which are arguments used to justify ethical theory. The field of ethics is divided by Fieser into applied ethics, normative ethics, and metaethics. Applied ethics is the analysis of a domain such as medical, environmental, or computational policy. Normative ethics, in contrast, concerns itself with moral standards that govern statements like "X is right" or "X is wrong." Metaethics concerns itself with the foundation upon which ethical theory and judgements are developed. The affective computing group has explored metaethical positions in computer ethics such as contractualism (e.g., the experiment mentioned above), value ethics, and most recently, a "dimensional metaethical position."

A "dimensional metaethical position" is an evaluation process that expands upon value-sensitive design: "an approach to the design of technology that accounts for human values in a principled and comprehensive manner throughout the design process" [Friedman, 2002]. In contrast, the "dimensional metaethical position" starts by articulating different social dimensions that bear on a system's design and use. Where value-sensitive design articulates "calmness" as a value to consider, a dimensional metaethical position views calmness along the dimension "psychological arousal," recognizing that different applications may interact with many points along the dimension. For example, you might want a workplace technology to facilitate calmness, while you might want your automated exercise advisor to get you angry enough to get back to exercising. It then diverges from value-sensitive design by advocating the exploration of antipodal values along these dimensions. "Power relationship" is another dimension currently being examined. By considering individuals in dominant or submissive roles in power relationships, we seek to understand situations that are viewed as unethical. By considering different points along the dimension of power relationship, we provide a starting point for critique and debate about design and use of affective computing systems. The significance of this approach is primarily that very little work has been done on the dimensions of ethical relevance to affective computing systems.

4.1 Several Dimensions Relevant to Evaluation of Affective Computing Systems

A dimensional metaethical position is centered around different dimensions that can be used to help inform ethical judgements about affective computing systems. The table below presents a listing of various dimensions that have been used as part of an ongoing evaluation of systems that mediate the communication of affect:

Table 1. Several Dimensions Relevant to Evaluation of Systems that Mediate the Communication of Affect (a non-exhaustive list)

Dimension	Examples	Description
Whom	Supervisor, Friends, Nicholas	The individual or individuals who receive the communicated affective message.

Dimension	Examples	Description
What	Telephone, Emotemail, Learning Companion	that acts as a transmitter or receiver for the communicated affective message.
Goal Relationship	Adversarial, Cooperative	The degree of conflict between the goals of the sender and receiver, which can be (but does not have to be) modeled from a game-theoretic perspective.
Power Relationship	Dominant, Submissive, Peer	Role that reflects the ability of either source or destination to alter the political, economic, or social situation of the other.
Genre of Emotion	Valence-Arousal Space, Categories, Emotional Orientation	Model used by the system to describe and encode emotion.
Valence	Positive, Neutral, Negative	Classification of transmitted emotion using an axis with positive or negative poles to describe feeling state.
Demeanor of Recipient	Angry, Sad, Excited	Emotional state of the message destination.
Gender	Female, Male, Intersex	Classification of either message source or destination based on reproductive role.
Ethnicity	Latino, Multi-Ethnic, Asian, Caucasian	Classification of either message source or destination based on racial or cultural identity.
Age	18, Middle-Aged, Mature, Minor	Classification of either message source or destination based on duration of life.
Culture	Rural, Icelandic, Traditional	Cultural context of communication and of either message source or destination.
Risk	Dangerous, Safe, Hazardous, LD50 (lethal dose for 50% of population), LC50 (lethal concentration for 50% of the population)	Potential impact of communication on goals of message source or destination.
Symmetry	Balanced, Skewed	Information or power balance between users of communication system.
Trust	Trustworthy, Deceitful	The degree to which the message source trusts either the destination or the channel.
Designer	Affective Computing Group, Microsoft, GNU, Jussi Angesleva, Employer	Person or organization who created the system that mediates the communication of affect.

Dimension	Examples	Description
Experimenter	Stanley Milgram, Carson Reynolds	The person who conducts an experiment that evaluates the ethical acceptability of communication system.
Time	Now, Ten Years Ago, Tomorrow	When the system that mediates the communication of affect is used.
Informed Consent	None, Compliant with CFR Title 45 Section 46.116	Does message source voluntarily consent to transmission of affective signals?
Security	None, C2, RC5-64, Hardened, Encrypted	Classification of security level of communication system or encoded signal.
Control	None, Partial, Complete	Degree to which message source can control the transmission of affective signals.
Feedback	None, Partial, Complete	Can the message source access the transmitted affective signal?
Transparency	Opaque, Open	Are the workings of the system that mediates the communication of affect visible for inspection, and by whom?
Proximity	Near, Far	Distance between message source and message destination.

The above table presents a non-exhaustive list of many factors that could influence ethical evaluations of systems that mediate the communication of affect.

4.2 An Example Application of Dimensional Metaethics

To make the dimensional metaethical position more concrete, we will now provide an example of its application. Let us begin by choosing an application to evaluate ("what" in the table above). In choosing "what = Emotemail" we specify the artifact which we wish to evaluate.

We then proceed by listing our expectations of how users might interact using Emotemail, providing one value for each dimension and then assessing whether such circumstances would be ethical (through speculation, reasoning, use of survey techniques, and perhaps assessments of actual users)

Dimension	Value
Whom	Friends
What	Emotemail
Goal Relationship	Cooperative
Power Relationship	Peer
Genre of Emotion	Facial Expressions
Valence	Unknown
Demeanor of Recipient	Unknown
Gender	Unknown

Dimension	Value
Ethnicity	Unknown
Age	Over 18
Culture	Unknown
Risk	None
Symmetry	Balanced
Trust	Trustworthy
Designer	Affective Computing Group
Experimenter	Carson Reynolds
Time	Now
Informed Consent	None
Security	None
Control	Complete
Feedback	Complete
Transparency	Open
Proximity	Unknown

The next step of our evaluation is to consider values for dimensions that are extreme or unexpected. For instance, what if the demeanour of the recipient isn't unknown but is "Angry?" (It is known that a neutral face image, perceived by somebody in a negative state, is perceived as more negative, which could facilitate misunderstanding.) Or what if the users of Emotemail are in an adversarial relationship? In exploring these different permutations and making assessments it is possible for the designer to explore potentially unforeseen and potentially unethical difficulties.

5   Comparing Dimensional Metaethics with Other Perspectives

Dimensional metaethics is neither the only nor the first approach at providing a metaethical position for the evaluations of systems. In the sections below we will briefly describe other metaethical positions that have been applied to computer ethics and compare them to dimensional metaethics.

5.1   Value-Sensitive Design

Value-Sensitive Design [Friedman and Kahn, 2002] articulates many dimensions that are relevant to systems that mediate the communication of affect. Value-Sensitive Design (VSD) is "an approach to the design of technology that accounts for human values in a principled and comprehensive manner throughout the design process." It considers Human Welfare, Ownership and Property, Privacy, Freedom From Bias, Universal Usability, Trust, Autonomy, Informed Consent, Accountability, Identity, Calmness, and Environmental Sustainability as values that may be of ethical consequence. Friedman and Nissenbaum applied VSD to evaluation of bias in computer systems [Friedman, 1997]. VSD has been applied by others to problems such as online privacy [Agre ,1997] universal usability [Thomas, 1997], urban planning [Noth, 2000], and browser consent [Friedman et al., 2002]. The Tangible Media Group at the MIT Media Laboratory has considered various ambient displays that support the calmness aspects of VSD in their research on computer-supported cooperative work and architectural space [Wisneski, 1998].

In many ways, a dimensional metaethical position is an extension of value-sensitive design. Both provide a list of criteria which can be used to help structure evaluations and critiques of computing system. The chief difference between Value-Sensitive Design and a dimensional metaethical position is what Kagan refers to as "evaluative focal points" [Kagan, 2000]. Value-Sensitive Design is essentially a virtue ethics that focuses on different values that are of import to the design of computer systems. A dimensional metaethical position instead focuses on dimensions along which the context of use of affective computing systems may vary.

5.2 Disclosive Computer Ethics

Disclosive Computer Ethics [Brey 2000] "is concerned with the moral deciphering of embedded values and norms in computer systems, applications and practices." In contrast to value sensitive design, disclosive computer ethics focuses on justice, autonomy, democracy and privacy. Brey contrasts "mainstream" approaches to computer ethics (which he views as limited) with disclosive computer ethics. Brey sees the disclosive metaethical position as more of a process which is concerned with "disclosing and evaluating the embedded normativity in computer systems."

Our dimensional metaethical position differs from this approach by not focusing on the embedded norms and instead considering the context in which the technology is used and factors that might influence ethical judgements. Put another way, dimensional metaethics is not just artifact-centric, but also is fixated on the environment in which ethical judgements are formed.

Let us make these comparisons more concrete by providing an example ethical analysis of Emotemail. Value-Sensitive Design would ask to consider the virtues of Human Welfare, Ownership and Property, Privacy, Freedom From Bias, Universal Usability, Trust, Autonomy, Informed Consent, Accountability, Identity, Calmness, and Environmental Sustainability in the context of an email system that conveyed emotion. Disclosive computer ethics, on the other hand asks us to examine how a technology embeds various normative judgements. In the case of Emotemail, we would examine how justice, autonomy, democracy, and privacy are embedded and supported by the systems design. The dimensional metaethical position, in contrast would ask us to consider the use of Emotemail while the value of the different dimension vary. Thus we might consider Emotemail's usage when there is and is not a power relationship present between users.

6   Concluding Remarks

While one cannot control the ultimate ways in which an artifact is used, and while we currently know of no designers who deliberately try to make unethical choices in their designs, we recognize that unethical uses may happen. Whether intentional or unintentional, the latter perhaps arising through a simple mismatch in user's goals (e.g. a boss wanting to know which employees aren't happy, while the employees might want to keep their feelings private), there is potential for harm to come about from the use of affective computing technologies. We advocate open consideration up front of such possibilities, and open dialogue about innovative ways to minimize potential misuses.

Granted this attempt at reviewing and extending existing metaethical approaches is only part of a much larger process. Both empirical and theoretical evaluations of affective computing systems are likely to be much more informative than simple model building. It is our hope that the dimensional metaethical position will prove to be of use to others wishing to ethically evaluate systems, and inform design of systems that are more ethical.

References

Agre P.E. and C.A. Mailloux Jr., 1997, Social choice about privacy Intelligent vehicle-highway systems in the United States , in: Human values and the design of computer systems , ed. B. Friedman (Cambridge University Press, Cambridge) p. 289.

Fieser J., 1999, Metaethics, Normative Ethics, and Applied Ethics Contemporary and Historical Readings (Wadsworth Publishing, Belmont, CA)

Ford, P. 2004. Weekly Review for February 22, 2005.
http://harpers.org/WeeklyReview2005-02-22.html

Friedman B and Nissenbaum H. Software agents and user autonomy. In proceedings of the first international conference on autonomous agents, 466-469. 1997. Seattle, Washington.

Friedman B. and P.H. Kahn, Jr., 2002, Human values, ethics, and design , in: Handbook of Human-Computer Interaction , eds. J. Jacko and A. Sears (Lawrence Erlbaum Associates, Mahwah, NJ)

Friedman B, Howe D C, and Felten E W. Informed Consent in the Mozilla Browser Implementing Value Sensitive Design. In proceedings of HICSS 2002, 247-248. 2002. Hawaii, Hawaii.

Kagan, S., 2000, "Evaluative Focal Points", in Hooker, Mason, and Miller, (eds.), pp. 134-55.

Noth M., A. Borning, and P. Waddell, 2000, An extensible, modular architecture for simulating urbandevelopment, transportation, and environmental impacts (UW CSETR 2000-12-01) , http://www.urbansim.org

Picard R.W., 1997, Affective Computing (MIT Press, Cambridge, MA)

Picard, R. W., Vyzas, E., and Healey, J. 2001, Toward Machine Emotional Intelligence: Analysis of Affective Physiological State, IEEE Transactions Pattern Analysis and Machine Intelligence, Vol 23, No. 10, pp. 1175-1191, October 2001.

Picard, R. W. and Klein, J., 2002, Computers that Recognise and Respond to User Emotion: Theoretical and Practical Implications, Interacting with Computers, 14, 2 2002, 141-169.

Reynolds C J and Picard R W. Affective Sensors, Privacy, and Ethical Contracts. Proceedings of 2004 Conference on Human Factors and Computing Systems (CHI 2004). Vienna. 2004, ACM Press.

St. Amant N., 2004, Benning unit tests robot system,
http://www.tradoc.army.mil/pao/TNSarchives/June04/062404.htm

Thomas J.C., 1997, Steps toward universal access within a communications company , in: Human values and the design of computer systems, ed. B. Friedman (Cambridge University Press, Cambridge) p. 289.

Wisneski C, Ishii H, Dahley A, Gorbet M, Brave S, Ullmer B, and Yarin P. Ambient Displays Turning Architectural Space into an Interface between People and Digital Information . Proceedings of International Workshop on Cooperative Buildings (CoBuild '98), 22-32. 1998. Darmstadt, Germany.

Use of a Dynamic Personality Filter in Discrete Event Simulation of Human Behavior Under Stress and Fatigue

Mamadou Seck, Claudia Frydman, Norbert Giambiasi[1]

Université Paul Cézanne – Aix Marseille
Marseille, France
{claudia.frydman, norbert.giambiasi}@univ.u-3mrs.fr

Tuncer I. Ören[2]

University of Ottawa
Ottawa, ON, Canada
oren@site.uottawa.ca

Levent Yilmaz[3]

Auburn University
Auburn, Alabama, USA
yilmale@eng.auburn.edu

1. M&SNet: LSIS (Laboratoire des Sciences de l'Information et des Systèmes)
2. M&SNet: Ottawa Center of the MISS; School of Information Technology and Engineering (SITE)
3. M&SNet: AMSL (The Auburn Modeling and Simulation Laboratory)

Keywords: Augmented cognition, personality filter, stress and fatigue

ABSTRACT

Cognitive complexity is an important factor in decision making in problem solving. As a personality trait, openness is related with cognitive complexity. Hence, dynamic updates of openness corresponding to the changes in its facets can be used to update the values of cognitive complexity which in turn can affect the decision making abilities and performances of humans represented in simulation studies. The possibility to develop dynamic personality filters to affect human performances in human behavior simulation studies are pointed out.

1. Introduction

As a team, our interest in the study of human cognitive abilities is to understand and test the mechanisms of several aspects of cognition to be able to incorporate them in simulation studies. We foresee two types of use: (1) enhance simulation studies and contribute to the advancement of the methodology and technology of cognitive simulation and (2) use cognitive simulation to test hypotheses about human cognition. We also aim that the results should be applicable in human cognition in general and in the study and/or simulation of cognitive abilities relevant to human-computer interaction.

It is well accepted that stress and fatigue are important factors in cognitive abilities. While fatigue deteriorates performance, a minimum level of stress is needed for good performance. Too little and too much stress affects performance negatively. In an article [Seck, Frydman, and Giambiasi, 2004], we reported a discrete event simulation of human behavior under stress and fatigue. Personality representation –based on five traits– for human behavior simulation was reported by Ören and Ghasem-Aghaee [2003]. The dynamics of personality traits each based on six personality facets as well as use of fuzzy logic to represent this personality dynamics is reported by Ghasem-Aghaee and Ören [2003]. In yet another article [Ghasem-Aghaee and Ören, 2004], we elaborated on the effects of cognitive complexity in agent simulation. In that article, the following were reported:

- how cognitive complexity affects problem solving abilities of people,
- how cognitive complexity and one of the personality traits, i.e., "openness" are related,
- how one can recalculate (using fuzzy logic) "openness" as a personality factor, based on its six facets, i.e., fantasy, aesthetics, feelings, actions, ideas, and values, and
- how this personality dynamics can affect the cognitive complexity which affects the problem solving capability of humans.

In this article, we aim to combine the findings of the above cited articles to specify discrete event simulation of human behavior under stress and fatigue taking into account the human personality characteristics. Similar to digital filters in digital photography, personality characteristics can be conceived as a filter –a dynamic personality filter– in human behavior simulation. The dynamic aspect of the personality filter stems from the fact that the fuzzy values of a trait (openness in this application) can be recalculated based on the changes in the values of its six facets. The updated fuzzy value of the corresponding trait (openness inthis application) affects the cognitive complexity of the individual which in turn affects his/her ability to cope with complexity. And this ability may be related with human-computer interaction or any other task.

908

By modifying the values of personality facets and corresponding personality traits, one can use a same mechanism to represent different personalities. This has the following main applications:

(1) Use the same mechanism to represent personalities and taylor the representation to a specific personality. In this way if the personality is represented as a class, different objects can be instantiated to represent different personalities.

(2) For a given personality, the long-term modifications of some of the personality traits can easily be represented as different values of the personality filter. This possibility can be useful to represent and simulate the effects of training on the long-term modifications of personality.

(3) Similarly, for a given personality, the short-term modifications of personality traits can easily be represented as different values of the personality filter. This possibility can represent for example, modifications of openness due to fatigue and/or stress.

Representation of the performance of a person under fatigue and/or stress can systematically be represented by dynamic personality filters. Since this type of conception is rather new, the paradigm offered may represent a good possibility to study and test different hypotheses on the effect of fatigue and stress on performance of different types of people.

In discrete event simulation of human behavior, the study will take into account:

(1) The stress and fatigue factors,

(2) The cognitive complexity of an individual, and

(3) The dynamics of the cognitive complexity by taking into account the updates in the six facets affecting openness trait of personality.

The results will be usable in cognitive discrete simulation in general, in testing hypotheses about human cognitive abilities in general and in human-computer interfaces in particular.

Section 2 of this paper will present theoretical aspects of stress from the physiology and psychology fields. In section 3 and 4, we elaborate on personality knowledge and cognitive complexity. Section 5 introduces the DEVS formalism. Then section 6 presents in more detail our DEVS based cognitive model.

2. Fatigue and Stress

Fatigue and stress are inevitable and they affect performance of humans. While fatigue deteriorates performance, the role of stress is different. For a human to be able to perform well, a stress above a threshold value should exist; similar to have a certain amount of tension on the strings of a violin. Too much tension would result by an inappropriate sound and later to the break of the string; similarly, too much stress may deteriorate the performance and can even be fatal.

2.1 Types of stressors

A synopsis of major types of stressors is given in Table 1 where physical and mental stressors are represented each in two groups:

(1) Physical stressors consist of *environmental stressors* such as heat, cold; vibrations, noise, blast; hypoxia or lack of oxygen and physiological stressors such as lack of sleep, dehydratation, and muscular fatigue.

(2) Mental stressors consist of *cognitive stressors* such as too much or too little information, isolation, and hard judgment and *emotional stressors* such as threats, frustration, and boredom/inactivity/

Table 1: types of stressors.

Physical stressors	Mental stressors
ENVIRONMENTAL	**COGNITIVE**
Heat, cold	Information: too much or too little
Vibrations, noise, blast	Isolation
Hypoxia (lack of oxygen)	Hard judgment ...
PHYSIOLOGICAL	**EMOTIONAL**
Lack of sleep	Threats
Dehydratation	Frustration
Muscular fatigue ...	Boredom/inactivity ...

2.2 Stress-performance relationship

As shown in Figure 1, Stress-performance relationship is represented by an inverted-U model.

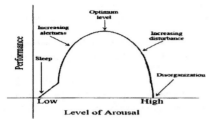

Fig. 1. The inverted-U hypothesis

it states that at low stress, performance is low; it improves until its highest point corresponding to the optimal level of stress or arousal. Then, performance level decreases when stress intensifies, leading to complete disorganization when stress level reaches panic.

3. Personality knowledge

From a simulationist point of view, representation of personality knowledge as well as representation of personality dynamics are summarized by Ören and Ghassem-Aghaee [2003] and by Ghassem-Aghaee and Ören [2003]. "A concise taxonomy of taxonomies of personalities is given at the site of personality project (Personality Project). In contemporary psychology, personality is specified as a function of thirty attributes –each of which called personality facet. The personality facets are clustered in five groups –each called a personality trait [or personality factor]. The five personality factors are also referred to as "the big five" [Costa and McCrae, 1992; Howard, 2000]. The value of each personality factor is determined by the values of its six facets. The five clusters of personality factors are also referred to by letter designation [Acton, 2001; Howard and Howard, 2001a, b]. Acton refers to them as the OCEAN model.

In the OCEAN model, the letters stands for the following meanings: O: Openness, culture, originality, or intellect; C: Conscientiousness, consolidation, or will to achieve; E: Extraversion; A: Agreeableness or accommodation; and N: Need for stability, negative emotionality, or neuroticism" [Ören and Ghassem-Aghaee, 2003].

There is a parallel between representation of a large number of colors and the representation of different personalities. A vector of three elements can represent the values of Red, Green, and Blue, Depending on the resolution to represent each component color, one can have for example 256 values for each color. Similarly, there are five personality traits; hence, a vector of five elements can represent a personality. As explained by Ören and Ghassem-Aghaee [2003], if only two fuzzy values can be distinguished for each element, $2^5 = 64$ types of personalities can be distinguished. This number can become 243 ($= 2^3$) if each trait can be discriminated even having as values low, average, or high.

4. Cognitive complexity

As stated by Streufert and Swezey [1986], persons who are high in cognitive complexity are able to analyze (i.e., differentiate) a situation into many constituent elements, and then explore connections and potential relationships among the elements. Characteristics of high and low-cognitive complexity people are summarized from Streufert and Swezey [1986], in Table 2.

Table 2. Characteristics of high and low cognitive complexity individuals

Characteristics	High cognitive complexity people	Low cognitive complexity people
Information	More open to new information	opposite
Attraction	Attracted to high cognitive complexity people as well as to low cognitive complexity people	Attracted to low cognitive complexity people with similar attitude
Flexibility	More flexible in thinking More fluency of ideas in creativity	opposite
Social influence	Change attitude more easily	More stable in attitudes
Problem solving	Tend to search for more information	opposite
Strategic planning	Greater flexibility in considering alternatives	opposite
Communication	More effective at a communication dependent task	opposite
Creativity	Able to generate more novel ideas	opposite
Leadership	Show leadership	

Figure 2 represents the relationship of level of processed information and situational complexity [Athey 1976]. Based on Athey's work [Athey 1976], Ören [1978] elaborated on the importance of increasing cognitive complexity of an individual to increase his/her effectiveness in coping with complex situations. "However, for each individual there is a critical point beyond which the level of processed information, hence the individual's information processing effectiveness is decreased. After the critical point, an increase in the situational complexity may worsen the individual's ability to cope with complexity, by causing a decrease in his/her level of information processing" [Ghassem-Aghaee and Ören, 2004].

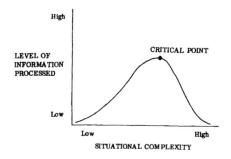

Figure 2: Different levels of information processing of an individual depending on the situational complexity

Figure 3 shows a comparison of information processing curves of two types of individuals, i.e., high and low cognitive complexity individuals "where two important points are shown: First, c_h, the critical point of high cognitive complexity individual is higher than c_l, critical point of low cognitive complexity individual. Thus increasing the cognitive complexity of an individual –within the applicable limits of course– may increase the range of situational complexity within which he/she can perform effectively. Or depending on the task, it may be advisable to assign an individual with cognitive complexity commensurate with the task. Second, for a given situational complexity, the level of information processed by a high cognitive complexity individual i_h is greater than i_l which corresponds to a low cognitive complexity individual" [Ghassem-Aghaee and Ören, 2004]. Taken into consideration the fact that fatigue and stress can affect openness of an individual brings other possibilities for the study of human performance. A given individual may have his/her performance affected by fatigue and stress. The modification of the openness can be represented as an update of the personality filter and can be used in a simulation study.

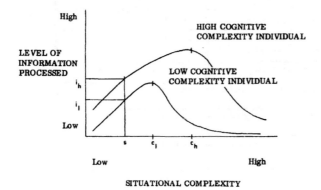

Figure 3: Comparisons of information processing curves of two types of individuals,
i.e., high and low cognitive complexity individuals

4.1 Cognitive complexity and openness

"In Chapter 12 of the handbook of emotional intelligence, it is indicated that "openness has also been associated with other cognitive or quasi-cognitive variables, including moral reasoning, cognitive complexity and wisdom" [McCrae, 2000]. The relationship of cognitive complexity and openness as a personality trait inspires applicability of personality update concept of dynamic personality to cognitive complexity. The personality facets which affect openness are: fantasy, aesthetics, feelings, ideas, and values. The dominant facet, i.e., the one having the largest weighted value determines openness. Any value change in any of the personality facets affecting openness may induce a personality update and change in the value of openness to affect the cognitive complexity of the individual" [Ghassem-Aghaee and Ören, 2004].

5. DEVS (discrete event simulation) formalism

DEVS (Zeigler, B., 1984) is a modular formalism for deterministic and causal systems' modelling. It allows for behavioural description of systems. A DEVS model may contain two kinds of DEVS components: Atomic DEVS and Coupled DEVS.

5.1 Formal specification of an atomic DEVS model

AtomicDEVS = $<S, ta, \delta int, X, \delta ext, Y, \lambda>$
- The time base is continuous and not explicitly mentionned : T= |R
- S represents the set of sequential states: The dynamics consist in an ordered sequence of states in S.
- ta(s) is the lifetime function of a state in the model.
- δint is internal transition function, allowing the system to go from one state to another autonomously.
- Y is the set of outputs of the model.
- X is the set of (external) inputs of the model. They interrupt its autonomous behaviour by the activation of the external transition function δext.
The system's reaction to an external event depends on its current state, the input and the elapsed time.

5.2 Formal specification of a coupled DEVS model

The coupled DEVS formalism describes a discrete events system in terms of a network of coupled components.
   CoupledDEVS = < Xself, Yself, D, {Mi}, {Ii}, {Zi,j}, select >

Self stands for the model itself.
- X*self* is the set of possible inputs of the coupled model.
- Y*self* is the set of possible inputs of the coupled model.
- D is a set of names associated to the model's components, self is not in D.
- {M*i*} is the set of the coupled model's components, with i being in D. These components are either atomic DEVS or coupled DEVS.
- {I*i*} is the set of l'ensemble des influencees of a component. That is what defines the coupling structure.
- {Z*i,j*} defines the model's behaviour, transforming a component's output into another componant's input within the coupled model.
As concurrent components can be coupled, many state transitions can have to occur at the same simulation time. A selection mechanism then becomes necessary, in order to choose which transition is to be executed first. So is the role of the "select" function.

6. DEVS models

This part of the research is being conducted principally in the context of the PIOVRA project (Bruzzone A.G., 2002), aiming at the creation of a new generation of CGFs.
Essentially based on the DEVS formalism (Zeigler, B., 1984), the cognitive model that we are designing consists of various layers, each of which has a specific role, from the detection, appraisal of events and situations, to actions taken by the synthetic agent to attain goals.
External events are sensed by the agents depending on their current attentional focus (yet to be modeled) and are appraised in the appraisal component of the framework. That appraisal is achieved by taking account of the nature of the event, its intensity, the type of mission that is being done and the possible interference of such an event or situation with the outcome of the current task (does it threaten the achievement of a goal?). After that analysis, a mapping with stressor intensities is done. Of course, one can easily accept that different persons, depending on their dominant personality trait, principally Neuroticism, will react differently, and experience varying stressor intensities in such a context. The evaluation of the new stress level is done in the stress component based on the inverted U-hypothesis. Then those stress values are transmitted to the behavioral model and may cause a difference in the agent's behavior. That behavior is in humans materialized by information processing, decision making and a given action on the environment. One may suspect that different persons, depending on their personality may react differently.

6.1 DEVS Appraisal model

Our construct is being designed in such a way to be able to model individuals and groups at the same time, therefore it must be able to represent interactions between individuals like communication or emotional relations (Seck, Frydman, & Giambiasi, 2005).
The appraisal component is where events from the environment are appraised as stressors. The most influential models of psychological stress argue that the stimuli from the environment are appraised as threatening (challenge, harm or loss) or benign and depending on the degree of threat will result or not in a given stress reaction (Lazarus S., 1999). But the appraisal is not only based on the event, it depends also on the goal and maybe subgoals of the current mission. During a mission consisting in reaching a distant point in enemy zone, such a typology of events could be made, in terms of intensity:
-       Suspect noises, blasts, ...
-       Shots, losses in our camps, ...
-       Loss of a team member, wounds, ...
   And those threat intensities result in various stressor intensities. An expert is certainly able to establish a more precise hierarchy of events in a given context, considering goals and subgoals such as success of the mission, necessity of staying alive.
To add more realism to the appraisal phase, adding fuzzy variables representing personality may be an interesting idea. In the five factor personality model, the "Negative Emotionality" trait can be used to represent the

susceptibility to stress for an individual. Then we could have for example three types of soldiers with the following personality types when it comes to appraising events:

- Resilient (Rational, impervious)
- Responsive (Unable to maintain calmness for a long period of time)
- Reactive (Susceptibility to negative emotions

On the following figure, we consider a given set of events to be appraised by an individual in a given context. The individual may be either Resilient (R-), Responsive (R), or Reactive (R+). Negative events X, Y, or Z may occur with varying intensities (1, 2, 3.) or positive events A, B, or C.

The initial state is APPR. When a Responsive (R) individual recieves an event, he/she appraises it normally and generates the appropriate stressor output (when such an individual recieves a x1 event, it goes to state Te1 and a Te!1 value is sent to the stress model.

When a Resilient (R-) individual recieves an event, he/she appraises it as lower threat than it is, (when such an individual recieves a X2, Y2, or Z2 event, it goes to state Te1 and a Te!1 value is sent to the stress model.

When a Reactive (R+) individual recieves an event, he/she appraises it as higher threat than it is, (when such an individual recieves a X2, Y2, or Z2 event, it goes to state Te3 and a Te!3 value is sent to the stress model.

When an individual of any type receives a A, B, or C event, it goes to state Te4, and a Te!4 value is sent to the stress model, which will be interpreted as an element that reduces stress.

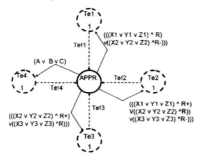

Figure 4: DEVS appraisal model

6.2 DEVS stress model with a personality filter

The understanding of the processes underlying stress has evolved recently, yet leaving numerous uncertainties. We consider a stressor as any event or situation (physical or mental) requiring a change in adaptation or behaviour of an individual (FM-22-51, 1994). It represents a threat to their welfare or survival. It can have positive or negative effects. A stressor implies physiological reflexes that prepare to tackle a situation or to flee it.

An optimal level of stress is useful for task performance. When stress is too low, it implies distraction and tasks are done haphazardly. An intense level of stress may cause poor motor coordination or even choking.

Many models try to link stress to performance. One of the most cited is the Yerkes-Dodson law, also referred to as the inverted-U hypothesis. It states that at low stress, performance is low; it improves until its highest point corresponding to the optimal level of stress or arousal. Then, performance level decreases when stress intensifies, leading to complete disorganization when stress level reaches panic (Lyle E. Bourne, Jr. & Rita A. Yaroush, 2003). Hancock's extended-U model, while focussing on attention, relates stress level to physiological and psychological adaptability (Hancock, P.A., & Warm, J.S. 1989). He defines various zones of the curve corresponding to physical and mental states depending on stress level.

We had the idea of considering discrete states of stress, associating each one with a typical behaviour, and from the physiology and psychology literature, try to model how one goes from one stress state to another.

Based on that, we have proposed a model of stress in the DEVS formalism. It consists in a representation of discrete states of stress. Depending on their nature, external events make the stress level evolve from one state to another. Those events are sorted according to their seriousness or intensity by the appraisal function presented in the previous section. Five discrete stress levels are considered, each representing a given zone of the inverted U.

Here also, personality may play a prominent role in defining the duration in which the individual stays in a non optimal state of stress, the Resilient and the Responsive tend to return to the optimum stress level through coping, while the Reactive, who is more subject to negative emotions, has a harder time coping with such situations. The lifetime function of the DEVS states (x's on figure 6) can thus be calculated as a function of the personality fuzzy variable.

6.3 Modeling fatigue in DEVS

Following [5], transitions of a DEVS model can be conditioned by thresholds of a continuous function. We can model tiredness by a polynomial function of task durations multiplied the task's constant of difficulty. The rate of efficiency of a task in the behavioural model can be conditioned by the computed value of tiredness. The possibility for an agent to start or complete a physically demanding task can also be conditioned by such a function. The values of thresholds and of the constants of difficulty will be defined by experts and refined during validation phases.

6.4 DEVS behavioral model with a personality filter

A behavioural atomic model represents a set of tasks and transitions between tasks that belong to the same sub-mission with a set of distinct sub-goals that permit to appraise events in that given context. When the mission is fulfilled or when another mission is given with a higher priority, the behavioural model and the appraisal models are replaced with the relevant ones. That can be considered as an application of the sequential multi-model concept [Yilmaz, L. and Ören, T.I., 2004].

The current task of the agent is represented by a state. Events that induce a change in behaviour activate transitions that can be conditioned by values of state variables representing stress, tiredness or environmental information…

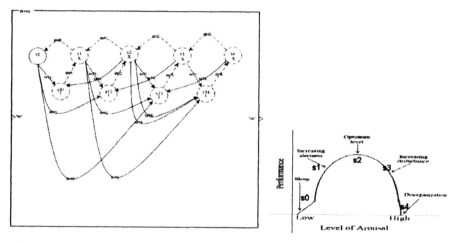

As for the properties of the DEVS formalism, we can represent a reactive behaviour i.e. the agent reacts to external events with constraints expressed by state variables. The decision process is done by actions inside states, decision rules are represented by the transition alternatives. Phases' durations are either set to a static value or defined by a function (of the nature of the task, ability of the performer, stress and fatigue levels…)

Transitions may be internal, subject to the task's duration, or external, when a relevant event occurs.

Stress and fatigue interfere with behaviour through the coupling property of the DEVS formalism, i.e. any time the stress level changes, or any time the fatigue level exceeds a threshold, a message is transmitted to the behavioural model, and if necessary, the behaviour is updated.

The stress states we defined earlier have the following behavioural feature:

S0: Sleep, inactivity
S1: Slow reactions, average precision
S2: Optimal level
S3: Risk taking, bad choices
S4: Bad precision, choking, panic.

915

The coupling property is also very interesting as it allows us to model communication between individuals like orders, or even the possible stress effect of the injury or death of a team member (considered as external events and appraised by the individual). It also may help us fulfil our necessity of aggregation and disaggregation capability.

Stress and fatigue can interfere dynamically on behaviour in performance and decision making (both variables can change within the course of a task). [Seck, M., Frydman, C., and Giambiasi, N. 2005]

We are now considering the effect of personality as performance moderator (does not change within the course of a task). Openness in the five factor personality model is related to cognitive complexity [Ghasem-Aghaee & Ören, 2004]. Based on Athey's work [Athey 1976], Ören elaborated on the importance of increasing cognitive complexity of an individual to increase his/her effectiveness in coping with complex situations [Ören, 1978]. Based on that, we can distinguish on certain tasks, the performance difference between high cognitive complexity people and low cognitive complexity people. A first distinction might be done concerning the time necessary to finish successfully a cognitive task; a second one can be made concerning decision making as high cognitive complexity people are known to be more fluent in idea and more creative thus generally find best solution. To do so, each task of the DEVS atomic behavioural model will contain a variable representing the task's cognitive complexity. Different individuals with different personalities, the openness trait in particular, will have different performances in terms of both time and decision making.

Figure 6 is an example from a DEVS atomic behavioral model. As we stated earlier, states represent tasks, external events or the time advance function determine transitions. An individual whose cognitive complexity is low (ICC == low), by means of an event, is confronted with a mob of armed individuals of average aggressivity. That event triggers a transition to a decision task; the duration of that task is computed by comparing the task's cognitive complexity (TTC) and the individual's cognitive complexity (ICC). Then, she/he chooses to shoot or to negotiate. The decision here depends on the stress level of the individual.

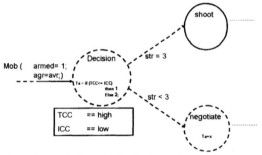

Figure 6: part of a DEVS behavioral model

Conclusion

With all of the above, we see that a dynamic personality filter may be of great help towards the realism of human behaviour simulation, particularly in CGFs. The concept of that filter is a number personality variables representing, at coarse or fine grain, elements of human personality as defined in the five factor model. That filter will be instantiated every time a new agent is to be created. As we noted throughout this presentation, that filter will have effects all over the cognitive process, on the appraisal of events, on the coping with stress and also on the actions of agents. It allows us to obtain more believable models of human behaviour.

Realistic models of human behavior require representation of the effects of fatigue and stress. This part has already been possible for simulationists. See for example, [Seck, Frydman, and Giambiasi, 2004]. Recent advances of the representation of personality traits and relationship of the personality trait openness to fatigue and stress inspires us to represent personality as a dynamically updateable filter that can be used in the simulation studies of human performance. In this article we report the possibilities. We are planning to report simulation results in a sequel article.

REFERENCES

Acton, G,S. (2001). Five-Factor Model: http://www.personalityresearch.org/

Athey, T.H. (1976). Training the Systems Analysts to Solve Complex Real World Problems: In T.C. Willoughby (ed.), Proceedings of the 14th Annual Computer Personnel Research Conference, July 29-30, 1976. The Special Interest Group on Computer Personnel Research (SIGCPR) of the ACM, pp. 103-120.
http://portal.acm.org/citation.cfm?id=811083&jmp=abstract&dl=GUIDE&dl=ACM

Bruzzone A.G., and R.Mosca. 2002. Poly-functional Intelligent Operational Virtual Reality Agents, Technical   Report Eurofinder Proposal, CEPA 11, DIP, University of Genoa, Genoa, Italy.

Costa, P.T., Jr., McCrae, R.R. (1992). NEO PI-R Professional Manual, Odessa, FL: Psychological Assessment Resources, http://psyche.tvu.ac.uk/phdrg/atkins/atws/person/67.html

(FM-22-51, 1994) - Field manual n° 22-51, Leader's manual for combat stress control, Chapter 2: Stress and combat performance. Headquaters, Department of the Army. Washington DC 29 September 1994, www.globalsecurity.org/military/library/policy/army/fm/22-51/22-51_toc.htm .

Ghasem-Aghaee, N. and Ören, T.I. (2003). Towards Fuzzy Agents with Dynamic Personality for Human Behavior Simulation, Proceedings of the 2003 Summer Computer Simulation Conference, Montreal, PQ, Canada, July 20-24, 2003, 3-10.

Ghasem-Aghaee, N. and Ören, T.I. (2004). Effects of Cognitive Complexity in Agent Simulation: Basics, Proceedings of SCSC 2004 - Summer Computer Simulation Conference, July 25-29, 2004, San Jose, CA., pp. 15-19.

Hancock, P.A., & Warm, J.S. (1989). A dynamic model of stress and sustained attention. Human Factors, 31, 519-537.

Howard, P.J. (2000). The Owner's Manual for the Brain, Second Edition, Bard Press, Atlanta, GA, www.bradpress.com

Howard, P.J., Howard, J.M. (2001a). The BIG FIVE Quickstart: An Introduction to the Five-factor Model of Personality for Human Resource Professionals, Center for Applied Cognitive Studies [CentACS], Charlotte, North Carolina, www.centacs.com/quickstart.htm

Howard, P.J., Howard, J.M. (2001b). Owners Manual for the Personality at Work, Bard Press, http://www.bardpress.com/personalityata.htm

Jonathan Gratch. Why you should buy an emotional planner, Proceedings of the Agents'99 Workshop on Emotion-based Agent Architectures (EBAA'99) 1999.

Lazarus S., (1999). "STRESS AND EMOTION, A New Synthesis" Springer Publishing Company, 1999

Lyle E. Bourne, Jr. and Rita A. Yaroush. Stress and cognition: A cognitive psychological perspective NASA. Grant Number NAG2-1561 February 1, 2003

McCrae, R.R. (2000). Emotional Intelligence from the Perspective of the Five-Factor Model of Personality. Ch. 12, in: Handbook of Emotional Intelligence: The Theory and Practice of Development, Evaluation, Education, and Application–at Home, School, and in Workplace (R. Bar-On and J.D.A. Parker, eds.), Jossey-Bass, Wiley.

Ören, T.I. (1978). Rationale for Large Scale Systems Simulation Software based on Cybernetics and General Systems Theories. In: Cybernetics and Modelling and Simulation Large Scale Systems, T.I. Ören (ed.). International Association for Cybernetics, Namur, Belgium, pp. 151-179.

Ören, T.I. and Ghasem-Aghaee, N. (2003). Personality Representation Processable in Fuzzy Logic for Human Behavior Simulation, Proceedings of the 2003 Summer Computer Simulation Conference, Montreal, PQ, Canada, July 20-24, 2003, pp. 11-18.

Yilmaz, L. and Ören, T.I., (2004). Dynamic Model Updating in Simulation with Multimodels: A Taxonomy and a Generic Agent-Based Architecture, Proceedings of SCSC 2004 - Summer Computer Simulation Conference, July 25-29, 2004, San Jose, CA., pp. 3-8.

Personality Project – Taxonomies of Individual Differences: http://pmc.psych.nwu.edu/perproj/readings-taxonomies.html

Seck, M., Frydman, C., and Giambiasi, N. (2004). Using DEVS for Modeling and Simulation of Human Behaviour. Proceeding of AIS 2004, AI Simulation and Planning, October 4-6, 2004, Jeju Island, Korea.

Seck, M., Frydman, C., and Giambiasi, N. (2005). Modeling and Simulation of Military Group Behavior.   2005   Spring Simulation Multiconference (SpringSim'05)

Streufert, S., and Swezey, R.W. (1986). Complexity, managers, and organizations. New York: Academic Press. http://www.css.edu/users/dswenson/web/Cogcompx.htm .

Zeigler, B. Theory of Modeling and Simulation. Krieger Publishing Company. 2nd Edition, 1984.

User Experience Based Adaptation of Information in Mobile Contexts for Mobile Messaging

Timo Saari

Center for Knowledge and Innovation Research,
Helsinki School of Economics and Helsinki
Institute for Information Technology
Tammasaarenkatu 3, 00180, Helsinki, Finland
saari@hkkk.fi

Marko Turpeinen

Helsinki Institute for Information Technology
Tammasaarenkatu 3, 00180, Helsinki, Finland
marko.turpeinen@hiit.fi

Jari Laarni

Center for Knowledge and Innovation Research,
Helsinki School of Economics
Tammasaarenkatu 3, 00180, Finland
laarni@hkkk.fi

Niklas Ravaja

Center for Knowledge and Innovation Research,
Helsinki School of Economics
Tammasaarenkatu 3, 00180, Helsinki, Finland
ravaja@hkkk.fi

Abstract

In this paper, we describe and elaborate an approach to a context sensitive personalization system to systematically facilitate desired user experiences, such as emotional and attentional states and information processing of individual users and groups of users in various contexts. Psychological Customization entails personalization of the way of presenting information (user interface, visual layouts, modalities, structures) per user or user group to create desired transient psychological effects and states, such as emotion, attention, involvement, presence, persuasion and learning. By varying the form of information presented in a mobile device in a contextually intelligent way may be possible to achieve such effects. Theory, conceptual implications, available empirical evidence and an example of an application area in mobile messaging are presented.

1   Introduction

User experience is seen within this article as the transient attentional and emotional states, moods, information processing, learning, persuasion and various other subjective experiences occurring i) just before a user engages with technology, ii) as a result of the user perceiving and processing information mediated by technology during a session of use and iii) immediately after the use of technology within a given task and context of use. User experience is related to action, i.e. the user has an experience and may perform a certain action or decide not to act, i.e. the user has creative and autonomous degrees of freedom for action.

When perceiving external stimuli like information via media and communications technologies users have a feeling of presence. In presence, the mediated information becomes the focused object of perception, while the immediate, external context, including the technological device, fades into the background (Biocca and Levy, 1995; Lombard and Ditton, 1997; Lombard et al, 2000). Empirical studies show that information experienced in presence has real psychological effects on perceivers, such as emotional responses based on the events described or cognitive processing and learning from the events (see Reeves and Nass, 1996).

When using collaborative technology for computer-mediated social interaction, the users experience a state called social presence during which users may, for instance, experience intimacy of interaction or feeling of togetherness in virtual space (Lombard et al, 2000). During social presence users also experience various other types of emotional and cognitive effects, such as interpersonal emotion, emotion based on being successful at the task at hand and learning from shared activities or shared information. Personalization and customization entails the automatic or semi-automatic adaptation of information per user in an intelligent way with information technology (see Riecken, 2000; Turpeinen, 2000). One may also vary the form of information (modality for instance) per user profile, which may systematically produce, amplify, or shade different psychological effects (Saari, 2001; Saari, 2002; Saari, 2003; Saari 2004a; Saari, 2004b; Saari and Turpeinen, 2004a; Saari and Turpeinen 2004b).

2 Psychological Customization

2.1 Mind-Based Media and Communication Technologies

Media- and communication technologies as special cases of information technology may be considered as consisting of three layers (Benkler, 2000). At the bottom is a *physical* layer that includes the physical technological device and the connection channel that is used to transmit communication signals. In the middle is a *code* layer that consists of the protocols and software that make the physical layer run. At the top is a *content* layer that consists of multimodal information. The content layer includes both the substance and the form of multimedia content (Benkler, 2000; Saari, 2001). Substance refers to the core message of the information. Form implies aesthetic and expressive ways of organizing the substance, such as using different modalities and structures of information (Saari, 2001).

With the possibility of real-time customization and adaptation of information for different perceivers it is hypothesized that one may vary the form of information within some limits per the same substance of information. For instance, the same substance can be expressed in different modalities, or with different ways of interaction with the user and technology. This may produce a certain psychological effect in some perceivers; or shade or amplify a certain effect. In Figure 1 the interaction of media and communications technology and the user in context with certain types of tasks is seen as producing transient psychological effects, thereby creating various "archetypal technologies" that systematically facilitate desired user experiences (e.g. Saari, 1998; Saari, 2001). Media and communication technology is divided into the physical, code and content layers. The user is seen as consisting of various different psychological profiles, such as individual differences related to cognitive style, personality, cognitive ability, previous knowledge (mental models related to task) and other differences, such as pre-existing mood. (Ravaja et al, in press; Ravaja et al, 2004; Saari, 2001; Saari, 2003a; Saari, 2003b).

Media- and communication technologies may be called Mind-Based if they simultaneously take into account the interaction of three different key components: i) the individual differences and/or user segment differences of perceptual processing and sense making of different segments of users, ii) the elements and factors inherent in information and technology that may produce psychological effects (physical, code and content layers), and iii) the consequent transient psychological effects emerging based on perception and processing of information at the level of each individual. (Saari and Turpeinen, 2004a).

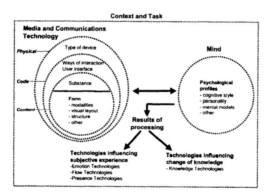

Figure 1. Mind-Based Technologies as a framework for producing psychological effects. Adapted from (Saari, 2001).

This definition can be extended to include both context and task and at least short-term behavioral consequences of a particular user experience. Regarding context and task, a Mind-Based system may alter its functionalities depending on type of task of the user, physical location, social situation or other ad-hoc situational factors that may have psychological impact. As the task of capturing and predicting user´s psychological state in real time is highly complex, one possible realization for capturing user´s psychological state is to have the user linked to a sufficient number of measurement channels of various i) psychophysiological signals (EEG, EMG, GSR, cardiovascular activity, other), ii) eye-based measures (eye blinks, pupil dilation, eye movements) and iii) behavioural measures (response speed, response quality, voice pitch analysis etc.). An index based on these signals then would verify to the system whether a desired psychological effect has been realized.

In more detail, there are several psychophysiological measures that may be used to capture user's emotional and attentional responses (for a review, see Ravaja, 2004a). First, facial EMG measured from the zygomaticus major and corrugator supercilii muscle areas indexes positive and negative emotions, respectively (i.e., the valence of an emotional experience). Electrodermal activity is, in turn, a very sensitive index of emotional arousal. Electroencephalography can be used to measure both emotional valence and attention. That is, (a) greater relative left frontal EEG activity is associated with the processing of positive affects (frontal EEG asymmetry) and (b) a reduction in alpha-wave power characterizes states of increased attention. It may also be advantageous to measure the user's physical activity by motion detectors, given that physical activity may influence both the psychophysiological recordings and affective experiences.

Another approach would be to conduct a large number of user studies on certain tasks and contexts with certain user groups, psychological profiles and content-form variations and measure various psychological effects as objectively as possible. Here, both subjective methods (questionnaires and interviews) and objective measures (psychophysiological measures or eye-based methods) may be used as well interviews (see Ravaja, 2004a). This would constitute a database of design-rules for automatic adaptations of information per user profile to create similar effects in highly similar situations with real applications. Naturally, a hybrid approach would combine both of these methods for capturing and facilitating the user's likely psychological state.

Capturing context and short-term user behavior is a challenge. Computational approach to context utilizes a mass of sensors that detect various states of an environment. AI-based software then massively computes from the signal flow significant events either directly or with the help of some simplifying rules and algorithms. Capturing user behavior in context is easier if the user is using an internet browser to buy an item and clicks the mouse, for instance.

2.2 Psychological Customization Systems

Psychological Customization is one possible operationalization of Mind-Based Technologies in system design. It can be applied to various areas of HCI, such as Augmentation Systems (augmented and contextualized financial news), Notification Systems (alerts that mobilize a suitable amount of attention per task or context of use), Affective Computing (emotionally adapted games), Collaborative Filtering (group-focused information presentation), Persuasive Technology (advertising for persuasion, e-commerce persuasion), Computer Mediated Social Interaction Systems (collaborative work, social content creation templates) and Messaging Systems (emotionally adapted mobile multimedia messaging and email). (Saari and Turpeinen, 2004a)

It can be hypothesized that the selection and manipulation of substance of information takes place through the technologies of the various application areas of Psychological Customization. Underlying the application areas is a basic technology layer for customizing design. This implies that within some limits one may automatically vary the form of information per a certain category of substance of information. The design space for Psychological Customization is formed in the interaction of a particular application area and the possibilities of the technical implementation of automated design variation Initially, Psychological Customization includes modeling of individuals, groups, and communities to create psychological profiles and other profiles based on which customization may be conducted. In addition, a database of design rules is needed to define the desired cognitive and emotional effects for different types of profiles. Once these components are in place, content management technologies can be extended to cover variations of form and substance of information based on psychological profiles and design rules to create the desired psychological effects. (Saari and Turpeinen, 2004a)

In a more detailed account and ideally, Psychological Customization involves i) a given pool of information to be presented to different users within a certain task, ii) a database of desired psychological effects per each user or user segment, such as positive emotion, set by the users themselves or the service provider, iii) a database of user profiles, iv) a database of meta-descriptions of the substance and form of the information to be delivered to users, v) a database of design rules of how the elements of information, such as form, will probably influence the transient psychological states of the different users within given tasks, situations and contexts, vi) sensors and sources of information that assess the psychological state of the user in real-time, such as psychophysiology, eye-tracking, video camera shots, sound, browsing behavior monitoring and other possible sources and vii) an AI-component that monitors the realization of psychological effects and provides the necessary intelligence for the system to function. (e.g. Saari and Turpeinen, 2004a)

At the technically more operationalized level, a Psychological Customization System is a new form of middleware between applications, services, content management systems and databases. It provides an interface for designing desired psychological effects and user experiences for individual users or user groups. The most popular framework for building customized Web-based applications is Java 2 Enterprise Edition. J2EE-based implementation of the Psychological Customization System for Web-based applications is depicted in Figure 2. The basic J2EE three-tiered architecture consisting of databases, application servers, and presentation servers has been extended with three middleware layers: content management layer, customer relationship management layer, and psychological customization layer. The profiles of the users and the communities are available in the profile repository. (see Turpeinen and Saari, 2004)

The Content Management System is used to define and manage the content repositories. This typically is based on metadata descriptions of the content assets. The metadata of the content repositories is matched against the user and community profiles by the Customer Relationship Management (CRM) system. The CRM system includes tools for managing the logic behind content, application and service customization. Rules can be simple matching rules or more complex rule sets. A special case of a rule set are scenarios, which are rule sets involving sequences of the interactions on the Web site. The Customer Relationship Management layer also includes functionality for user and community modeling. This layer can also perform automated customer data analysis, such as user clustering. (e.g. Turpeinen and Saari, 2004)

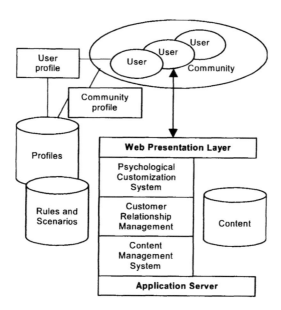

Figure 2. J2EE implementation of the Psychological Customization System [adapted from 49]

The Psychological Customization System layer performs the optimization of the form of the content as selected by the Customer Relationship Management layer. This functionality can be considered similar to the device adaptation by using content transformation rules (for example XSL-T). In the case of the psychological customization, the transformation rules are produced based on the design rules for content presentation variation and the contents of the psychological profile of the user. After this optimization, the content is passed to the Web presentation layer.

One can also add various sensors in the model extracting the state of the environment and users. Technologies such as eye-tracking, video capture of situations and contexts, microphones and psychophysiological recording can be used. Naturally, if these signals can be captured in a non-intrusive manner it would be optimal. Advances in technology related to mobile phones for instance may offer solutions to the capture of psychophysiological signals, for instance.

921

Even though a working prototype of Psychological Customization has not been built yet, several empirical studies support the feasibility of a user-experience driven system that matches the form of information to the psychologically relevant properties and other profile factors of individual users and user groups. Despite this, many complexities, such as the influence of contextual and situational factors on psychological effects remain unknown.

The "design rules" for Psychological Customization of the form of information may be roughly divided into i) contextual rules (what is the psychological impact of a particular task and context, what may be the user´s needs, goals and intentions), ii) social rules (what is the psychological impact of a particular social context, or a particular social interaction) and iii) perceptual and information processing related rules (what is the psychological impact of individual differences on processing various layers of external stimuli, such as mediated information on a mobile phone). Obviously all three levels of rules are needed, but it is very difficult to know what level may be most significant per type of application area or what are the interactions between levels without case specific empirical studies.

In the area of contextual rules for Psychological Customization little cues as to what may be feasible is available. This area is discussed in more detail in the next chapter. In the area of social rules Reeves and Nass (1996) have conducted research revealing that people respond to information technology in ways behaviorally similar to human-human interaction in the areas of flattery, team-building, credibility, persuasion, frustration and a range of other areas. However, in the area of group or community-centric or social interaction focused social rules of Psychological Customization almost no research has been done. This area is indirectly also addressed in the next chapter.

Conversely, in the area of perceptual and information-processing- level rules there is quite a lot of evidence related to the influence of the form of information on psychological effects in conjunction with various individual differences.

For instance, there are individual differences in cognitive processes such as attention, working memory capacity, general intelligence, perceptual-motor skills and language abilities. These individual differences have a considerable effect on computer-based performance and may product sometimes quite large variance in the intensity or type of psychological effects, such as depth of learning, positive emotion, persuasion, presence, social presence and other types of psychological states and effects as well as consequent behavior (e.g. Egan, 1988; Eysenck, 1994; Hampson, 1995; Saari, 2001; Saari, 2002; Saari, 2003a; Saari, 2003b; Saari and Turpeinen, 2004a; Saari and Turpeinen, 2004b; Vecchi et al, 2001).

There is more specific evidence in literature that varying the form of information, such as modality, layouts, background colors, text types, emotionality of the message, audio characteristics, presence of image motion and subliminality creates for instance emotional, cognitive and attentional effects (see Curperfain and Clarke, 1985; Kihlström et al, 1992; Krosnick et al, 1992; Laarni, 2003; Laarni, 2002; Laarni et al, 2002; Lang et al, 1995; Lang et al, 1996; Ravaja et al, in press; Riding and Rayner, 1998). Some of these effects are produced in interaction with individual differences, such as cognitive style, personality, age and gender (Kallinen and Ravaja, in press; Ravaja, 2004b; Ravaja and Kallinen, in press), or pre-existing mood (Ravaja et al, 2004). The role of hardware should not be neglected. A device with a large screen or a portable device with smaller screen with user-changeable covers may also influence the emerging effects (e.g. Laarni and Kojo, 2001). This empiric evidence at least partly validates the possibility for Psychological Customization Systems.

3   Application Areas in Mobile Contexts

3.1   Mobile contexts

The concept of context can be seen as i) psychologically constructed (the user´s interpretation of the present context, or the psychological context "inside the head" including goals, intentions, user experiences of the context as a result of processing external stimuli etc.), ii) socially constructed (context is interaction with other people, such as turntaking in conversations, or context is action in a social sphere or context is shared understanding and shared culture between users enabling dialogue), iii) defined by actions of the user using the resources offered by the situational context (the user actively and autonomously acts to fulfil selected tasks utilizing various situational resources, such as available people, technologies and locations) and iv) objective (measurable factors of context, such as location, temperature, presence of other people, available technologies, psycophysiological signals of users). (see Clark, 1995; Fitzpatrick, 2000; Kulkki, 1996; Kurvinen and Oulasvirta, 2003; Oulasvirta et al, 2003; Rittenbruch, 2002; Saari, 2001; Suchman, 1987; Tamminen et al, 2003)

In the case of mobile contexts an element of mobility is introduced to add more complexity. Mobility adds an element of movement of the user from one location to another carrying a particular technology with him. The user then goes through a series of changes of context that may be quite different from each other. This continuously introduces new factors relevant to behavior and user experience. People also tend to claim personal and group-related space while inside a context. There is also a need to navigate for instance inside an urban environment while moving. Often there are temporal tensions present, such as hurrying to catch a bus. Multitasking is also present as the attention to the mobile device and the social environment fluctuates. (Tammkinen et al, 2003) Mobility also implies potential changes in physical activity (e.g., running to catch a bus) that, in turn, may influence the user's emotional experiences and attentional focus when using a mobile device (see Myrtek et al, 1996; Ravaja, 2004a).

It is the view of the authors that the notion of context has to be taken into account case-by-case or by each particular application area of Psychological Customization. Otherwise it may be that not enough specific information of the psychological impact of situational and contextual factors can be achieved in order to reliably facilitate selected user experiences. However, the basic view of the authors of context and mobile context is mostly psychologically and objectively constructed. This does not imply socially constructed or action-oriented ways of seeing context are irrelevant but rather that they are interpreted psychologically and hence form a subpart of the psychological context. Effect of objective factors is also based on how we interpret them.

Related to ecological validity of psychological effects in real-life conditions there is a clear need to systematically model and classify tasks and contexts, such as social situations and interactions and physical contexts from the point of view of influence to psychological effects. For instance, certain combinations of physical context (bus, café, urban nomadic situations), social interactions (ad-hoc community, close friends, family etc.) and available content and technological resources (mobile phone for messaging, reception of information, taking pictures etc.) may turn to be important facilitators or prohibitors for certain psychological effects to emerge as well as they may act as sources of psychological effects as such.

If this can be done in high-resolution, one may propose Contextually Intelligent Psychological Customization Systems. They would be able to customize i) information presented inside a particular context (media information, advertising, e-commerce sites or mobile messaging, for instance) and ii) user interfaces for software emdedded in various mobile devices, computers and contextually intelligent spaces (such as menus of available location-based services, selecting suitable interaction modalities, etc.) in a psychologically driven context-sensitive way.

3.2 Example: Mobile Multimedia Messaging

Emotional and cognitive effects of information are related to communication within social networks as follows: one may manipulate manually or with automated systems the substance and the form of information. It is obvious that in social interaction the users construct the substance and form of for instance MMS-messages exchanged manually. However, the form of the message may also be varied with automated systems. The information needed to conduct these automated manipulations can be accessed via individual and social modeling and profiling of the users. For instance, a user with an intention to create positive emotion in the other user with the way of presenting his textual message may utilize a background colour suggested by the system predicted to induce a mild positive emotional state in the receiving user based on his profile. The scope of various possibilities for Psychological Customization in mobile contexts is laid out in Table 1, but without the possibility of real-time feedback from psychophysiological signals of the users. As these technologies are not widely available yet, we concenctrate only on non-feedback scenarios.

First, in creating cognitive effects, the system may automatically manipulate the form of the user interface, interaction modality or message received or sent in a manner that is clear and understandable for the receiver based on the rule-database. For instance, typography and screen layout may be optimised for each receiver. Also, when possible, automatic translations from text to audio or vice versa may be sensible depending on environmental conditions and noise. Second, for emotional effects, the system may offer the sender of messages graphical, video, audio or other types of MMS-templates to use in order to communicate a particular emotion of their own to the receiver or create a desired emotion in the receiver. In both cases the system would automatically suggest for the sender of message a possibility to psychologically customize the message for a particular receiver. The user would select a desired effect, such as creating positive emotion in the receiver with a message in which the substance is written in text and the system would present the sender with ready-made and psychologically evaluated templates (consisting of graphics, animation, sounds, videos etc.) that with high probability may create the desired emotion for the receiver with a particular user profile. The sender would type in the text-message in the template, finalize the design and then send the message. The receiver would receive an emotionally optimised message and may then experience the desired emotion. Naturally, if the substance of the message and the form of the message communicate a different emotion, for instance the substance is hostile and the template is joyful, some effects may not be realized.

Table 1. Technological possibilities of Psychological Customization in mobile multimedia messaging.

Layer of Technology	Adaptations of Mobile Multimedia Messaging
1. Physical	-Mobile device: user changeable covers in colours and shapes that facilitate emotion
2. Code *-Windows-type user interface* *-Mouse, pen, speech,*	-The user interface elements (background color, forms, shapes, directions of navigation buttons etc.) may be varied in real-time per page per user in which a certain advertisement is located to create various emotions and ease of perceptual processing -audio channel may be used to create emotional effects (using audio input/output sound, varying pitch, tone, background music, audio effects etc.) -the interaction modalities may be adapted to suit the nature of the task
3. Content	
A. Substance *- Fixed multimedia content created by authors*	-The source of the substance should be taken into account as it may influence trust and processing of information (unknown, familiar, community member, friend, family member, etc.) -Personal relevance of the information to the users can be modeled as it influences involvement -Adding subliminal extra content to create emotion
B. Form Modality *-Multimedia*	-Modality may be matched to cognitive style or pre-existing mood of the receiver to create ease of processing -Background music, audio effects or ringing tones may be used as a separate modality to facilitate desired emotions and moods -Animated text can be used to create more efficient processing of text and also facilitate some emotional effects
Visual presentation	-Emotionally evaluated and positioned layout designs and templates for ads (colors, shapes and textures) may be utilized per type of user segment
Structure -linear/non-linear	-Offering emotionally evaluated and positioned narrative templates for creating emotionally engaging stories

Let's think as an example a person who sits in a café sipping tea and is a bit bored. He needs emotionally uplifting experiences. He may also long for social contact as he knows no one in the café. Simply by playing around with his mobile device that has a happily colored changeable cover makes him feel a bit better. He browses around the user interface and menus of his mobile device as it is a recent acquisition and he is not enough familiar with its functionalities. The user interface has been set to novice- profile that automatically alters the user interface to be as clear and easily understandable for the user as possible. As his cognitive style is a wholist visualizer (see Riding and Rayner, 1998) the system presents an overview of the functions of the interface graphically to aid him grasping the big picture. As he is an outward oriented personality the system uses richly textured and quite a busy graphical interface to match his level of need of stimulation for processing information more efficiently.

He decides to construct a multimedia message and opens the software to do in an emotionally optimized way. The software presents him with several ready-made graphical layouts and designs as well as little background music that are predicted to influence his friend's emotions. He picks a graphical template to create high arousal and negative valence and a similar background music. He decides to write a sarcastic black-humor joke to his friend and smiles while typing it. He sends the message away and gets back to his tea.

His friend in another context logged on to the same customization system receives the joke while walking inside the railway station to catch a train simultaneously talking on his mobile phone. The system notifies him with a vibrating alert as it notices his attention is directed to the conversation and navigating the hallways of the railway station. He ends the call and checks the message. The display of the phone turns bright and adds more contrast to compensate for the bright illumination present in the hallway he is standing. He opens the message. The emotional impact of the joke is intensified by the used visual template and background music and the friend is quite amused. He constructs his own message as a reply, deciding to use a particular emotional template as well as add a highly arousing audio effect to be created while receiving the message and sends it.

The message is received by the user who is drinking tea. As there is some background noise, the level of the audio notification of a new message is automatically increased. The particular audio effect selected by the sender of the message is realized and the tea drinker is somewhat alerted. He quickly looks at the message only to see that it was his friend playing around with the intensive audio effect. He opens the message and is influenced by the layout, laughing at the comments of his friend.

While this use scenario is somewhat pushed to the extreme of possibilities of Psychological Customization, it illustrates some aspects that may add value to the users of such systems in mobile messaging or other application areas in mobile contexts. In the example provided the system offered more opportunities to be playful, have fun and express emotion with mobile messaging. The system also took care of maximum bandwidth of perceptual processing and helped a novice user with his mobile user interface. Also, different modalities were used and adapted to environmental conditions in a smart manner.

4 Conclusion

Too little is known about the cognitive and emotional effects of the manipulation of form and layout factors and about the role of individual differences and situational factors in creating various psychological effects. More basic and applied research is needed. However, in our research for the past four years in varying the form of multimodal information, such as news, entertainment and games presented through color-screen high-capacity mobile phones we have found many emerging commonalities and possible rules for altering designs to reliably facilitate emotions, presence, attention, and learning.

A key issue is a possible way of integrating laboratory research and contextually valid field research. This is elemental as a Psychological Customization system is rule-based and needs explicit representations of users, content, contexts, tasks and other issues. There are two difficulties here to be dealt with: i) an optimal and feasible method to study a real-life case of psychological customization to be able to construct design-rules and ii) technological aspects of psychological customization, such as being able to build a prototype to be tested, or being able to get a feedback signal to verify the realization of a psychological effect and being able to capture and represent task, context and user behavior.

As for the development of the method for studying real-life contextual phenomena and behavior of users in real situations using Psychological Customization, one possibility is the integration of ethnography and mobile psychophysiological research. Ethnographical research richly describes and partly explains the phenomenon while mobile psychophysiology provides predictive information of the influence of various aspects of form of information on attention, cognitive load and emotion. Self-report measures may be also used with questionnaires and interviews. For instance in mobile contexts a small mobile psychophysiological recorder with suitable non-intrusive sensors may be used together with a system of small cameras attached to the mobile device and the user (displaying the user´s face, the user interface actions and social environment). The system would run on shared time code to allow for rating of psychophysiologically significant events to look for responses later on.

The technological challenges of our approach cannot be directly addressed within the limits of this article. It may be said that getting access to the psychophysiological signal of the user may be feasible in the future depending on the application area. Modeling contexts in real-time may be conducted with massively sensor-based software engines that use some ready-made situational models to enhance pattern detection. Modeling task is a bit easier as the interaction of the user and the mobile phone can be tracked. Despite this, if task is seen as user goals, intentions that may be also non-conscious these can only indirectly be inferred and are highly difficult to predict beforehand. Consequently, for the time being Psychological Customization systems may have to rely on ready-made design rules without a specific feedback channel in some selected and narrow application areas within which the prediction of psychological effects is feasible. However, in military applications and perhaps hardcore gaming applications it may be feasible to gain access to psychophysiological signals of users in the near future as these areas are more "mission critical".

Evidently then clear and conclusive hypothesis, best practices for design or other low-level and explicit recommendation on how exactly to best use a Psychological Customization system is beyond the scope of this conceptual paper.

Despite this, the authors argue that the approach to system design presented in this paper may be beneficial in various application areas because: i) it provides a possibility to personalize the form of information that may be more transparent and acceptable to the users than adapting the substance of information as the manipulation of form of information is difficult to recognize by the user, ii) it offers a way of more systematically accessing and controlling transient psychological effects of users of adaptive systems, and consequently influence behavior of individual users and iii) it is potentially compatible with existing and new systems as an add-on or a middleware layer in software with many potential application areas.

The potential drawbacks of the framework include the following: i) it may be costly to build the design-rule databases and actually working real-life systems for creating systematic psychological effects, ii) the rule-databases may have to be adapted also locally and culturally, iii) the method needed to create a rule-database is not easy to use and may be suspect to ecological validity (eye-tracking, behavioral and psychophysiological measures, self-report, field tests are needed to verify laboratory results etc.) and iv) if the system works efficiently it may raise privacy issues, such as the intimacy of a personal psychological user profile (personality, cognitive style, values, other). Also ethical issues related to mind-control or even propaganda may come forward. Some people may also be reluctant to use this kind of a highly adaptive system, exhibiting a need to protect self-determination or so.

Regarding future research, we aim to build, evaluate and field-test prototypes of Psychological Customization in various areas, specifically in mobile, urban ad-hoc contexts and situations but also other areas such as mobile gaming communities. Interesting application areas at the moment are contextually and emotionally sensitive mobile advertising and e-commerce, emotionally adapted games, emotionally loaded mobile multimedia messaging and emotionally adapted mobile "social navigation" systems.

5 REFERENCES

Benkler, Y. (2000) From Consumers to Users: Shifting the Deeper Structures of Regulation. Federal Communications Law Journal 52, 561-63.

Billmann, D. (1998) Representations. In Bechtel, W. and Graham, G. (1998) A companion to cognitive science, 649-659. Blackwell publishers, Malden, MA.

Biocca, F. and Levy, M. (1995) Communication in the age of virtual reality. Lawrence Erlbaum, Hillsdale, NJ.

Cuperfain, R. and Clarke, T. K. (1985) A new perspective on subliminal perception. Journal of Advertising, 14, 36-41.

Clark, H. H. (1996) Using language. Cambridge University Press, Cambridge, UK.

Egan, D. E. (1988). Individual differences in human-computer interaction. In: M. Helander (Ed.), Handbook of Human-Computer Interaction, p. 543 – 568. Elsevier, New York.

Eysenck, M. (1994) Individual Differences: Normal and Abnormal. New Jersey: Erlbaum.

Fitzpatrick, G. (2000) Views on context in the enterprise. In proceedings of Workshop Research Directions in Situated Computing, CHI 2000.

Hampson, S. E. & Colman, A. M. (Eds., 1995) Individual differences and personality. London: Longman.

Kallinen, K., & Ravaja, N. (in press). Emotion-related effects of speech rate and rising vs. falling background music melody during audio news: The moderating influence of personality. Personality and Individual Differences.

Kihlström, J. F., Barnhardt, T. M. and Tataryn, D. J. (1992) Implicit perception. In Bornstein, R. F. and Pittmann, T. S. (eds.) Perception without awareness. Cognitive, clinical and social perspectives, 17-54. Guilford, New York.

Krosnick, J. A. , Betz, A. L., Jussim, J. L. and Lynn, A. R. (1992) Subliminal conditioning of attitudes. Personality and Social Psychology Bulletin, 18, 152-162.

Kulkki, S. (1996) Knowledge creation of multinational corporations. Knowledge creation through action. Doctoral dissertation at Helsinki School of Economics and Business Administration. A-115.

Kurvinen, E. and Oulasvirta, A. (2003) Towards Social Awareness in Ubiquitous Computing: A Turntaking Approach. HIIT Technical Report 2003-1, July 2, 2003

Laarni, J. (2003). Effects of color, font type and font style on user preferences. In C. Stephanidis (Ed.) Adjunct Proceedings of HCI International 2003. (Pp. 31-32). Crete University Press, Heraklion.

Laarni, J. (2002). Searching for optimal methods of presenting dynamic text on different types of screens. In: O.W. Bertelsen, S. Bödker & K. Kuutti (Eds.), Tradition and Transcendence. Proceedings of The Second Nordic Conference on Human-Computer Interaction, October 19-23, 2002, Arhus, Denmark (Pp. 217 – 220).

Laarni, J. & Kojo, I.(2001). Reading financial news from PDA and laptop displays. In: M. J. Smith & G. Salvendy (Eds.) Systems, Social and Internationalization Design Aspects of Human-Computer Interaction. Vol. 2 of Proceedings of HCI International 2001. Lawrence Erlbaum, Hillsdale, NJ. (Pp. 109 – 113.)

Laarni, J., Kojo, I. & Kärkkäinen, L. (2002). Reading and searching information on small display screens. In: D. de Waard, K. Brookhuis, J. Moraal, & A. Toffetti (Eds.), Human Factors in Transportation, Communication, Health, and the Workplace. (Pp. 505 – 516). Shake, Maastricht. (On the occasion of the Human Factors and Ergonomics Society Europe Chapter Annual Meeting in Turin, Italy, November 2001).

Lang, A. (1990) Involuntary attention and physiological arousal evoked by structural features and mild emotion in TV commercials. Communication Research, 17 (3), 275-299.

Lang, A., Dhillon, P. and Dong, Q. (1995) Arousal, emotion and memory for television messages. Journal of Broadcasting and Electronic Media, 38, 1-15.

Lang, A., Newhagen, J. and Reeves. B. (1996) Negative video as structure: Emotion, attention, capacity and memory. Journal of Broadcasting and Electronic Media, 40, 460-477.

Lombard, M. and Ditton, T. (1997) At the heart of it all: The concept of presence. Journal of Computer Mediated Communication, 3 (2).

Lombard, M., Reich, R., Grabe, M. E., Bracken, C. and Ditton, T. (2000) Presence and television: The role of screen size. Human Communication Research, 26(1), 75-98.

Myrtek, M., Scharff, C., Bruegner, G., & Mueller, W. (1996). Physiological, behavioral, and psychological effects associated with television viewing in schoolboys: An exploratory study. Journal of Early Adolescence, 16, 301-323.

Oulasvirta, A., Kurvinen, E. and Kankainen, T. (2003) Understanding contexts by being there: Case studies in bodystorming. Personal and Ubiquitous computing, 2003, Vol 7, pp. 125-134.

Ravaja, N. (2004a). Contributions of psychophysiology to media research: Review and recommendations. Media Psychology, 6, 195-237.

Ravaja, N. (2004b). Effects of a small talking facial image on autonomic activity: The moderating influence of dispositional BIS and BAS sensitivities and emotions. Biological Psychology, 65, 163-183.

Ravaja, N. (in press). Contributions of psychophysiology to media research: Review and recommendations. Media Psychology.

Ravaja, N., & Kallinen, K. (in press). Emotional effects of startling background music during reading news reports: The moderating influence of dispositional BIS and BAS sensitivities. Scandinavian Journal of Psychology.

Ravaja, N., Kallinen, K., Saari, T., & Keltikangas-Järvinen, L. (in press). Suboptimal exposure to facial expressions when viewing video messages from a small screen: Effects on emotion, attention, and memory. Journal of Experimental Psychology: Applied.

Ravaja, N., Saari, T., Kallinen, K., & Laarni, J. (2004). The Role of Mood in the Processing of Media Messages from a Small Screen: Effects on Subjective and Physiological Responses. Manuscript submitted for publication.

Reeves, B. and Nass, C. (1996) The media equation. How people treat computers, television and new media like real people and places. Cambridge University Press, CSLI, Stanford.

Riding, R. J. and Rayner, S. (1998) Cognitive styles and learning strategies. Understanding style differences in learning and behavior. David Fulton Publishers, London.

Riecken, D. (2000) Personalized views on personalization. Communications of the ACM, V. 43, 8, 27-28.

Rittenbruch, M. (2002) Atmosphere: A framework for contextual awareness. International Journal of Human-Computer Interaction, 14(2), pp. 159-180. Rodden, T., Chervest, K. & Davies, N. & Dix, A. (1998). Exploiting context in HCI design for mobile systems. In C. Johnson (Ed.) Proceedings of the First Workshop on Human Computer Interaction with Mobile Devices (pp12-17).

Saari, T. (1998) Knowledge creation and the production of individual autonomy. How news influences subjective reality. Reports from the department of teacher education in Tampere university. A15/1998.

Saari, T. (2001) Mind-Based Media and Communications Technologies. How the Form of Information Influences Felt Meaning. Acta Universitatis Tamperensis 834. Tampere University Press, Tampere 2001.

Saari, T. (2002) Designing Mind-Based Media and Communications Technologies. Proceedings of Presence 2002 Conference, Porto, Portugal.

Saari, T. (2003a) Designing for Psychological Effects. Towards Mind-Based Media and Communications Technologies. In Harris, D., Duffy, V., Smith, M. and Stephanidis, C. (eds.) Human-Centred Computing: Cognitive, Social and Ergonomic Aspects. Volume 3 of the Proceedings of HCI International 2003, pp. 557-561.

Saari, T. (2003b) Mind-Based Media and Communications Technologies. A Framework for producing personalized psychological effects. Proceedings of Human Factors and Ergonomics 2003 -conference. 13.-17.10.2003 Denver, Colorado.

Saari, T. (2004a) Using Mind-Based Technologies to facilitate Positive Emotion and Mood with Media Content. Accepted to proceedings of to Positive Emotion, 2nd European Conference. Italy, July 2004.

Saari, T. (2004b) Facilitating Learning from Online News with Mind-Based Technologies. Accepted to proceedings of EDMedia 2004, Lugano, Switzerland.

Saari, T. and Turpeinen, M. (2004a) Towards Psychological Customization of Information for Individuals and Social Groups. In Karat, J., Blom. J. and Karat. M.-C. (eds.) Personalization of User Experiences for eCommerce, Kluwer.

Saari, T. and Turpeinen, M. (2004b) Psychological customization of information. Applications for personalizing the form of news. Accepted to proceedings of ICA 2004, 27.-31.5. 2004, New Orleans, USA.

Suchman, L. A. (1987) Plans and situated actions. The problem of human machine communication. Cambridge University Press, Cambridge.

Tamminen, S., Oulasvirta, A., Toiskallio, K. and Kankainen, A. (2003) Understanding Mobile Context. Mobile HCI 2003, Udine, Italy. 2003, Springer LNCS, pp. 18-35.

Turpeinen, M. (2000) Customizing news content for individuals and communities. Acta Polytechnica Scandinavica. Mathematics and computing series no. 103. Helsinki University of Technology, Espoo.

Turpeinen, M. and Saari, T. (2004) System Architechture for Psychological Customization of Information. Proceedings of HICSS-37- conference, 5.-8.1. 2004, Hawaii.

Vecchi, T., Phillips, L. H. & Cornoldi, C. (2001). Individual differences in visuo-spatial working memory. In: M. Denis, R. H. Logie, C. Cornoldi, M. de Vega, & J. Engelkamp (Eds.), Imagery, language, and visuo-spatial thinking. Psychology Press, Hove.

Section 5
Cognitive State Sensors

Chapter 19

Augmented Cognition for Training Superiority

Augmented Cognition Technologies Applied to Training: A Roadmap for the Future

Denise Nicholson, Ph.D., Stephanie Lackey, LT Richard Arnold, and Kamaria Scott

NAVAIR Training Systems Division
12350 Research Parkway
Orlando, FL
Denise.Nicholson@navy.mil, Stephanie.Lackey@navy.mil, Richard.Arnold1@navy.mil,
Kamaria.Scott.ctr@navy.mil

Abstract

The military training community strives to meet ever-increasing expectations and challenges. One expectation is to support all levels of training from novice to expert. Supporting the full spectrum of training curricula, from basic knowledge acquisition in a classroom, or e-learning, to on-the-job-training with embedded training and mission rehearsal capabilities, are all of great importance to maintaining a qualified Fleet. Challenges to the delivery of training include: 1) increased operational tempo and mission complexity, 2) the perishable nature of higher order cognitive skills and team behaviours, 3) reduced availability of instructors and role-players, 4) system deployment and usability, and 5) costs including travel and leave costs of personnel while at "brick and mortar" schoolhouses (Lyons, Schmorrow, Cohn, & Lackey, 2002). An approach to addressing theses challenges is to apply advanced artificially intelligent computer assisted instruction technology, also known as Intelligent Tutoring Systems (ITS) (Chipman 2005), throughout the training continuum of delivery mechanisms. We propose that additional improvements could be realized across the training continuum by creating Augmented ITS capabilities that combine state-of-the-art ITS components with augmented cognition technologies, such as cognitive and neurophysiological sensing, and warfighter assessment gauges. This technology is currently being developed under DARPA's Improving Warfighter Information Intake Under Stress Program, also referred to as Augmented Cognition or AugCog. The resulting Augmented ITS will add a new level of capability to blended training solutions.

This paper will provide an introduction and overview into the newly immerging field of Augmented Cognition for Training Superiority and Education. It is being presented as the 1st of seven papers in a session of the 1st International Conference on Augmented Cognition. Throughout the paper we propose approaches for inserting data from cognitive and neurophysiological sensing, and warfighter assessment gauges into the student model, expert model and tutor components of intelligent tutoring systems to create Augmented ITS. Brief summaries of on-going research in this area being conducted by our colleagues are provided for illustration. Notional examples of Augmented ITS applied to both individual and team training are also offered. Finally, we conclude with expected benefits towards addressing the challenges to military training and recommendations for future research.

1 Introduction

The military training community strives to meet ever-increasing expectations and challenges. One expectation is to support all levels of training from novice to expert. Supporting the full spectrum of training curricula (Figure 1), from basic knowledge acquisition in a classroom, or e-learning setting, to on-the-job-training with embedded training and mission rehearsal capabilities, are all of great importance to maintaining a qualified Fleet.

Schoolhouse/Lecture/Lab practice -- e-learning/ADL -- Virtual/M&S -- Live -- Embedded --On-the-job

Figure 1: The continuum of training delivery mechanisms utilized by the military.

Challenges to the delivery of training include: 1) increased operational tempo and mission complexity, 2) the perishable nature of higher order cognitive skills and team behaviours, 3) reduced availability of instructors and role-players, 4) system deployment and usability, and 5) costs including travel and leave costs of personnel while at "brick and mortar" schoolhouses (Lyons, Schmorrow, Cohn, & Lackey, 2002). An approach to addressing theses challenges is to apply advanced artificially intelligent computer assisted instruction technology, also known as Intelligent Tutoring Systems (ITS) (Chipman 2005), throughout the training continuum of delivery mechanisms. We propose that improvements could be realized across the training continuum by combining Augmented Cognition Technologies such as cognitive and neurophysiological sensing, and warfighter assessment gauges with state-of-the-art ITS, providing Augmented ITS capabilities.

The Navy has established several commands to address education and training. The Naval Education and Training Command (NETC) including the Human Performance Center (HPC) and Navy Personnel Development Command (NPDC) focus on the initial portion of the training continuum (individual training), with an innovative initiative called the Integrated Learning Environment (ILE). Fleet Forces Command addresses the other end of the continuum with collective Fleet Training once a sailor or marine is assigned, or deployed, with modeling and simulation (M&S) based, distributed training exercises that can include deployable, embedded and live training platforms. Success in maintaining qualified personnel throughout their careers is contingent on continuous collaboration between these organizations.

The goal of the Integrated Learning Environment (ILE) is to improve and support job performance and mission readiness by providing high quality learning and performance support available anytime and anywhere. This will be accomplished by providing a single system for sailors to view and plan their entire career and advancement process. Sailors will be using elements such as the Navy Knowledge Online website, the Five Vector Model, and an individualized electronic training jacket (ETJ) tool to track their credentials and requirements for advancement (Human Performance Center, 2004).

An essential element of the ILE is migrating relevant areas of its classroom learning to web-based learning combined with modeling and simulation practicum experiences represented by "simulation objects" which could be enhanced with augmented ITS capabilities. This migration is expected to reduce the time and cost to educate and train sailors while increasing operational readiness by:
- Enabling prescriptive learning through reusable learning objects that can be combined "on the fly" to deliver individualized instructional curriculum, referred to as "my course."
- Increasing individual and mission performance by making knowledge available to sailors in the fleet when and where it is needed.
- Providing training according to a "the right sailor with the right skills in the right job at the right time for the right cost" approach.
- Emphasize learner control and responsibility.

In order to maximize the success and payoffs of the entire training continuum including those used by the ILE and M&S based training approaches, it is imperative that we understand and optimize the employment of these delivery systems. Such optimization can only truly be achieved by matching the individual's needs for learning to the time spent in training focused on targeted training objectives. It is this need to understand the individual's state of knowledge, skills and abilities, which we believe can be improved by the addition of augmented cognition technologies in training.

2   Background on Intelligent Tutoring Systems

Scope and Definition

The development of ITS began in the 1970s as an attempt was made to enhance computer based instruction (CBI). . The goal was, and still is, to combine highly individualized CBI with artificial intelligence (AI) methods in an attempt to produce the same kind of flexibility and learning effect that occur when a human tutor and student work together (Siedel and Park, 1994). In meeting this goal, the ITS would have the capability to assess what the student already knows, the curriculum that student needs to learn, and the appropriate instructional strategy and feedback format to be used. These decisions are made continuously and in real time. The flexibility of the system to make these decisions is what distinguishes ITS from CBI. Seidel and Park (1994) cite four features that are intrinsic to ITS that are impossible to include in CBI:

932

1. ITS can generate traditional knowledge, rather that select pre-programmed frames containing knowledge to present to the student.
2. ITS allow both the system and student to initiate instructional activities by applying AI techniques used to develop the computers natural language dialogue capability.
3. ITS can make inferences in interpreting the student's inputs, diagnosing misconceptions and learning needs, and generating instructional presentations.
4. ITS can monitor, evaluate, and improve its own tutoring performance by applying AI techniques commonly used in machine learning.

Components of ITS

By definition, an ITS must contain three essential components: the expert model, the student model, and the tutor (Figure 2). We, and our colleagues in this *Augmented Cognition for Training Superiority and Education* Session, propose various methods to "augment" all three components of ITS. The expert model consists of knowledge of the subject domain and is the basis for comparison of information received from the student model. Although all components of the ITS are integral to its ability to provide instruction, Anderson (1988) considers the expert model to be the backbone to any ITS, as the system could not function without domain knowledge. There are multiple options for encoding information into the expert domain. A black box model involves finding a way of reasoning about the domain that does not require codifying the knowledge that underlies human intelligence. Another option involves developing an expert system and requires the human expert knowledge to be codified and applied to the ITS. A more complex, yet powerful method involves incorporating a cognitive model, which authors such as Anderson (1988) and Zachary (1999) suggest is essential to producing high performance tutoring systems. We propose that representing cognitive state or warfighter assessment data (i.e. working memory load and attention) in the expert models of ITS, particularly when using cognitive models, will provide a deeper level of comparison and diagnosis of the student's knowledge with potential improvement in training effectiveness of the tutoring.

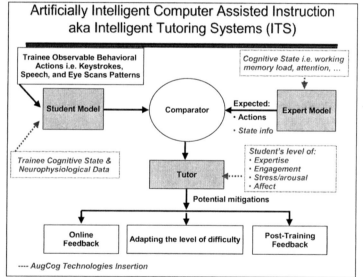

Figure 2: Intelligent Tutoring Systems and proposed "Augmented ITS" approaches.
(Adapted from Cannon-Bowers and Lyons (a.k.a. Nicholson), 1999)

The student model represents the current state of the learners' knowledge. The ITS assesses what the learner already knows during the diagnostic phase. During this phase, the ITS typically uncovers the student's cognitive state from his or her observable behavior (Van Lehn, 1988). There are several diagnostic techniques that are currently being used to assess the student model. The model tracing technique is the easiest and most widely used. This technique assumes that the diagnostic program has access to all of the student's significant mental states. This approach works by "delineating many hundreds of production rules that model curricular chunks of cognitive skill. A learner's acquisition of these chunks and departure from the optimal route is immediately remediated" (Shute and Postka, 1996). There are many other techniques used to construct the diagnostic module that is built in to the student model. They include path finding, condition induction, plan recognition and issue tracing. We propose that combining the trainee's cognitive state and neurophysiological data with the typical observable behavioural data represented in the student model will result in improved diagnosis and improved tutoring results.

The last component of the ITS is the tutor. This component encompasses the knowledge of teaching strategies to be employed during the learning process (Halff, 1988). After a trainee's knowledge has been assessed and compared to the domain knowledge in the expert model, the tutor must then make decisions regarding curriculum and instructional method, in order to bridge the knowledge gap. Although there are many mitigation strategies that could be employed, leading approaches include the delivery of real-time, on-line feedback, adapting the curriculum or scenario and post training debriefing in the form of a summary of performance. We propose that the selection and specific implementation of these strategies can be informed by the data provided through the augmented cognition technologies to effectively map the mitigation strategy to the student. For instance, a novice student may require more explicit, amplifying information in an on-line feedback message where as a more expert trainee could benefit from a simple notification of a mistake. Another area that could benefit from the sensing technologies is maintaining the trainee's interest or level of engagement. A foundational principle of the Augmented Cognition program is that based upon the Yerks and Dodson inverted U curve of human performance, there exists an optimum balance between human performance and a person's level of stress or arousal (Mendl, 1999). One goal of the adaptive human computer interface manipulations used by the Augmented Cognition teams is to maintain the operator at that peak of performance. We propose that the same theory could be used to drive adaptation of a training scenario or level of difficulty of the curriculum, striving to maintain a student at a peak level of learning.

We include a review of our colleagues' specific research efforts exploring several implementations of these concepts for augmenting ITS in section 4 below. However, we will first provide an overview of the augmented cognition technologies that we keep referring to from DARPA's Improving Warfighter Information Intake Under Stress Program, also referred to as Augmented Cognition or AugCog.

3   Augmented Cognition Technologies

New and established neurophysiological sensor technologies, coupled with sophisticated cognitive modeling software, have the potential to provide additional capabilities to traditional ITS approaches employed in existing and emerging human-centric training solutions such as scenario-based training in virtual environments (Lackey, 2004) or web-based training. An impressive collection of cognitive state detection and physiological sensor technologies and tools are currently under investigation by the AugCog researchers (see Table 1). These tools are being investigated to determine which individual technology or combination of technologies contributes to the development of Warfighter Assessment Gauges. Though initially being applied to adaptive human computer interaction (HCI) configurations for operational domains, applying these Warfighter Assessment Gauges to training environments will allow for experimentation to determine the level of contribution each cognitive state detection tool and physiological measure can provide within an augmented ITS.

Cognitive State Detection & Physiological Sensor Technologies	
Cortical blood oxygenation	Interbeat Interval
Cortical blood volume	Respiration Rate
Event related optical signal	Galvanic Skin Response
Neuronal patterns	Posture
Neuronal firing signatures	Eye fixation duration
Frequency of neuronal population firing	Eye gaze location
Sync/Desync of neuronal structures	Pupil dilation
Error related negativity	Mouse Pressure
P300	Rate of task completion
Heart Rate	Error rate

Table 1: Detection and sensor technologies of interest to AugCog (Schmorrow et al., 2005)

The AugCog teams are each working with various combinations of the above sensors to develop Warfighter Assessment Gauges for:

- Attention,
- Verbal Working Memory,
- Spatial Working Memory,
- Memory Encoding,
- Error Detection, and
- Autonomic Activity

The gauges are based upon data output streams from combinations of sensors, simultaneously captured from an individual in real-time. The resulting capabilities have resulted in the ability to address the four cognitive bottlenecks depicted in Figure 3. Sample mitigation strategies follow in Table 2 and are paired with task environments investigated by the four AugCog research teams.

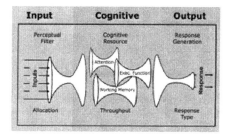

Figure 3: Cognitive bottlenecks (Schmorrow, et al., 2004)

Cognitive Processing Bottlenecks	Sample Task Mitigation Strategy	Task Environment
Attention	Task scheduling strategy based on priority, time and operator readiness	Dismounted Soldier
Working Memory	Alert delivery strategy based on verbal vs. spatial workload of operator	Aegis Command and Control Supervisor
Executive Function	Bookmark cueing strategy based on independent or dependent task context switching	Unmanned Vehicle Controller
Sensory Input	"Sensory shortcut" strategy based on available input channels	Command and Control Vehicle Operator

Table 2: Processing bottlenecks addressed by AugCog

AugCog efforts to mitigate cognitive bottlenecks have provided compelling tools to improve Warfighter performance with computer assisted adaptive HCI in these operational environments. The expectation, based on this success, is that these technologies will also support the improved implementation of computer assisted instruction by improving the effectiveness of training mitigations such as on-line feedback, adaptive scenarios or level of difficulty in curriculum and post training after action review.

4   On-going research in Augmented ITS

This paper is striving to provide an introduction and overview into the newly immerging field of Augmented Cognition for Training Superiority and Education. It is being presented as the 1st of seven papers in a session of the 1st International Conference on Augmented Cognition. The following is a brief summary of each of our colleagues' research approaches towards achieving the Augmented ITS capability described above. The first summary of Campbell and Luu describes an approach for augmenting the Student Model of Figure 2. The remaining researchers provide various approaches or examples of augmenting the Tutor Module of the ITS.

Detection of Slips versus Mistakes (Campbell & Luu, 2005)
Campbell and Luu are performing research, under ONR's Virtual Technologies and Environments (VIRTE) Program, to investigate the feasibility of using neurophysiological indicators, particularly error-related negativity (ERN), when interpreted in conjunction with performance data, to support improved automated diagnosis in a simulation-based ITS. Their concept is that ERN signals have been used to identify when a person recognizes immediately that he or she has made an error. This situation corresponds to a behavioural "slip" in Norma's (1981) discussion of error types, and does not reflect a true mistake indicating an underlying misconception or knowledge gap, which would require the same type or amount of instructional remediation. They propose to first use the ERN response slip data to filter the student performance data set being feed into a student model to increase the amount of variance accounted for by a mathematical model of performance data. Once this is demonstrated, the next step is to evaluate the impact of incorporating this neurophysiological analysis into an augmented ITS in a simulation based training system.

Detection of transition from Novice to Expert (Cohn, Kruse, & Stripling, 2005)
The goal of most training and educational sessions is to move a novice's behaviours and patterns of knowledge and reasoning towards those of the expert's. As discussed previously, typical ITS assess the trainee at the outcome/behavioural level. Another team of scientists under the VIRTE Program are investigating the hypothesis that there are detectable, reliable and repeatable behavioural and neuro-physiological measures that distinguish experts from novices in complex training tasks. They propose the use of continuous collective performance measures, or a "Performance Template", that incorporates both behavioural and neuro-physiological measures. Through the use of these collective measures the goal is to gain the capability to track a trainee, in real time, through the stages from novice to expert using technologies sensitive to the dynamic nature of learning in the brain. If successful, an augmented ITS can be realized in which it will be possible to assess where a trainee is cognitively in the training process and to then move trainees more efficiently and effectively through the training curriculum.

Mitigating Barriers to Sympathetic Teaching and Testing (DuRousseau, 2005)
This effort focuses on the barriers to sympathetic teaching and testing, and addressing those barriers by developing and using a Learning System Developer's Kit (LDSK). DuRousseau describes the process for combining cognitive state assessment algorithms with off-the-shelf computer-adaptive teaching (CAT) software capable of intelligently managing the delivery of instruction. In addition, tools to facilitate the design and test of applications which are capable of manipulating content, timing, style, and form of multimedia stimuli in real-time based on measures of executive control within the human brain are described. A summary of recent work focused on the computer-adaptive testing architecture and analytical methods of the LDSK is included.

Cognition monitoring for pedagogical diagnosis and intervention (Mathan & Dorneich, 2005)
Cognitive tutors, a.k.a. ITS, are used to monitor student behaviour dynamically, and using model tracing techniques, intervene when student behaviours are not consistent with strategies determined to lead to successful task performance. Anderson and Gluck (2001) have noted that ITS have several limitations when compared to live instructors, specifically with respect to the ability to detect cognitive and affective states. However, Mathan and Dorneich propose that the use of neurophysiological sensors to detect cognitive and affective states could be used to effectively augment ITS to enhance the diagnostic process in the Tutor Module, thereby more closely linking student performance and skill level with appropriate automated training interventions. Implications for ITS interventions can be more closely linked to skill maturity level, producing more efficient training.

Physiologial self-regulation training (Pope & Prinzel, 2005)
One challenge of long-duration space flight is cognitive skill maintenance. Pope and Prinzel have developed RESTORE, physiological self-regulation training technologies that can be embedded in engaging/entertaining tasks and adapt the task scenario based on physiological signals derived via EEG to modulate parameters of the task to optimize skill maintenance. These scenario-adapting approaches are proposed to be applied to other operational and training domains, such as submarine operations, to maintain and reinforce cognitive readiness and monitor and maintain effective mood states.

Potential predictors of errors in shooting judgment and cognitive performance by law enforcement personnel during a stressful training scenario. (Meyerhoff, Saviolakis, Burge, Norris, Wollert, Atkins, Speilberger, and Hanson, 2005)
Significant military research has centered on performance during stressful situations. This work adds to this body of literature by investigating the relationship between stress induced elevated arousal levels and performance. The

authors describe a study in which the arousal level of police trainees was assessed after they completed a highly stressful interactive scenario. Although stress induced elevated cortisol levels have been linked to poor performance in a previous study, the results of the current study did not follow that trend. However, the results did suggest that another indicator of arousal, increased heart rate, appears to be linked to performance. The authors then discuss whether these indicators of arousal can be used as possible predictive measures of performance decrement, as well as the development of measure to keep arousal levels balanced during stressful situations. This research could have implications for accounting for levels of stress within an augmented ITS.

5   Aviation Simulation-based Training System Example of Augmented ITS

The following examples, previously mentioned in AugCog International Quarterly (2005), provide insight into the potential impact AugCog technologies through Augmented ITS can have on individual and team training.

Individual Training
The achievements of the AugCog program make it possible to more than imagine, but to speculate what future military training will look like in the next decade. A young Lieutenant seeking additional Naval Flight Officer (NFO) training serves as an illustration. LT Thompson, Air Control Officer (ACO) - Air Wing 5, climbs into his NFO E-2C Hawkeye training simulator to practice tracking and communication skills for an upcoming qualification check. As part of his pre-flight routine, he dons his helmet and safety harness. Integrated into his gear are various cognitive and physiological sensors. From sensor data being fed into the student model and tutor module of the augmented ITS, the NFO simulator automatically alters the scenario on the fly to challenge the ACO's weaknesses and provide extra practice events where needed. LT Thompson's instructor gains deeper understanding of the Lieutenant's intentions, actions, workload, and ultimately his performance in real time from the Warfighter Assessment Gauges being displayed on the instructor's operator station. The increased awareness of the instructor will allow for improved during action feedback and assessment, as well as a more detailed after action review specifically tailored to the Lieutenant's needs. Integration of AugCog technologies into the ITS results in two valuable improvements: a truly individualize training experience for the Lieutenant, and value-added tools that facilitate an instructor's ability to efficiently and effectively enhance individual readiness.

Team Training
An exciting opportunity for the training community is the development of an AugCog sensor suite capable of measuring comprehensive team training performance. This capability involves monitoring team members' mental workload through the use of cognitive and physiological sensors. The ability to capture such data provides an opportunity for an ITS to facilitate training of teamwork skills such as proper communication and back-up behavior between teammates. Team back-up is appropriate when teammates have the responsibility to recognizes one another's workload and step in to cover or reallocate the distribution of tasks when feasible. Additionally, identification of specific scenario events which contribute to the onset of mental overload by the Warfighter Assessment Gauges can provide instructors with much needed insight into team performance.

Extending the example above, LT Thompson, an ACO, returns to his E-2C training simulator for an Air Wing training event that includes members of his immediate team – Combat Information Center Officer (CICO) and Radar Officer (RO) and several F/A-18 Sweep and Strike air assets. During a difficult Dynamic Strike simulation exercise, AugCog sensors track the workload of the three E-2C flight officers. These sensors indicate that LT Thompson's workload level crosses the productive threshold during the middle of the training scenario. At the same time, the sensors indicate that the RO is the best candidate to relieve a portion of the ACO's tasking. Rather than continue to overwhelm his teammate, as part of the established crew contract and his expected teamwork skills, the RO should assume responsibility for monitoring incoming verbal communications and hooking new tracks. If this is not observed by the Team Tutor module, the ITS will provide immediate feedback to the RO notifying him of his teamwork responsibilities and recommend that he provide the needed back-up to his teammate. This allows the ACO to focus on providing outgoing verbal and digital communications to the Sweep and Strike assets until sensors indicate that he is cognitively prepared to recover all of his primary tasking. By incorporating AugCog's cognitive and physiological sensor technology for this training event, individual and team performance is monitored in order to train productive team skills and improve overall team performance.

6   Current capabilities and future directions

The promise of augmented ITS capability will be of no consequence if they fail to deliver effective training defined by a demonstrable improvements in task performance, skill maintenance or other relevant training outcome. To achieve full potential there are several capabilities that must be effectively designed and developed. First, trainee cognitive state must be effectively assessed in real-time. Several efforts currently underway under the DARPA Improving Warfighter Information Intake Under Stress Program are focusing on this problem. One team led by the Boeing Company (Barker, 2004), for example, has developed a closed-loop system for augmenting cognitive performance of Unmanned Aerial Vehicle operators. They have demonstrated real-time measurement of verbal working memory, spatial working memory and executive function using a combination of EEG, pupilometry and functional near infrared (fNIR). Russell (2004) has developed a neural–net based EEG signal processor for this team that can classify EEG activity in real-time, and which feeds into cognitive state classification software that computes relative activity levels of various types of cognitive activity. This team has also developed software that modulates the user interface of the UAV operator, based on cognitive state data, to optimize cognitive processing throughput. In this example, the user interface alternates between relatively textual (verbal) and graphical (spatial) displays to take advantage of relative surpluses in verbal and spatial working memory capacities. They have demonstrated significant improvements in operator task performance under augmented conditions (AugCog International Quarterly, 2005).

Although the Boeing work is an example of adaptive HCI in an operational environment, the application readily transfers to the training domain. The components developed by this team are the same fundamental components that would be required in an augmented ITS: real-time cognitive state measurement and classification, software to modify the interface or task based on cognitive state, and metrics that provide assessment of operator or trainee performance.

Other related efforts are underway under DARPA, ONR and OSD sponsorship to address the usability and ruggedness of the sensor technologies such as non-contact EEG electrodes, wireless electrodes, new optical sensors, and miniaturization of fNIR systems. These initiatives have the potential to provide the components for deployable, maintainable and reliable augmented ITS training systems. However, much research remains for the future, particularly with respect to implementation techniques for augmented ITS as well as training effectiveness evaluations of such systems.

7   Conclusion

We have proposed that improvements could be realized across the training continuum from e-learning to deployed training by combining Augmented Cognition Technologies such as cognitive and neurophysiological sensing, and warfighter assessment gauges with state-of-the-art ITS, providing Augmented ITS capabilities. It is our premise that the realization of these capabilities will be useful for addressing the many challenges currently faced by the military training community.

We predict that the techniques described in this paper will optimize the effectiveness of ITS training systems and be key enablers for providing the individualized and tailored training required to overcome the challenges of increased operational tempo and mission complexity. Such optimization will make the most of every training opportunity, providing the basis for streamlining the training pipeline in an attempt to combat the general high costs of training.

Consequences of increasing operational tempos in today's military environment include reductions in the availability of instructors, teammates or other role-players for training scenarios. For a sailor deployed aboard ship for six months, the non-availability of instructors or teammates could result in significant degradation of perishable higher-order cognitive skills and team behaviours if there were not tools available for skill maintenance. The advent of augmented ITS coupled with realistic synthetic teammates (Schaafstal, 2001) in deployable trainers should provide significant improvement in skill maintenance and refresher training capabilities.

The reduced reliance on live instructors resulting from the implementation of augmented ITS provides new opportunities for cost savings through deployed training. However, deployment of trainers with this technology will place new requirements for the reliability of the system components. The sensor, state monitoring, and adaptive control technologies emerging from the field of augmented cognition must still undergo significant development to become sufficiently rugged, small and reliable before such systems can come into widespread use. Additionally, to capitalize on true cost savings one must consider the return on investment. Certainly there will be cost saving benefits of reduced reliance on live instruction and live teammates, both in terms of manpower and infrastructure. However, if these savings are offset by significant costs related to development, procurement and maintenance of new automated augmented training solutions, the question of their value will remain. Nevertheless, it is encouraging that efforts are being made, by programs such as DARPA's Augmented Cognition, to address issues (i.e., technology maturation and deployability) that are critical to future training capability, so that the time horizon to the widespread availability of affordable and deployable augmented training systems should be significantly shortened.

8   References

Anderson, J.R. (1988). The expert model. In Polson, M.C., and Richardson, J.J. (Eds). Foundations of intelligent tutoring systems. (p. 21-54). Hillsdale, NJ: Laurence Erlbaum Associates, Inc.

Anderson, J. R. & Gluck, K. (2001). What role do cognitive architectures play in intelligent tutoring systems? In D. Klahr & S. M. Carver (Eds) *Cognition & Instruction*, 227-262. Lawrence Erlbaum and Associates.

AugCog International Quarterly (2005). Phase two midterm objectives met: Boeing/AFRL. 2 (1), 6.

Barker, R. (2004). Team mitigation strategies: Boeing. *Presention at Improving Warfighter Information Intake under Stress with Augmented Cognition Principal Investigators and Community Meeting, 5-8 Jan 2004, Orlando, FL.*

Becker, W., Cohn J.C., Lackey S.J., & Allen, R.C. (2004). An empirically driven model for human-centric design. *Paper presented at the Image 2004 Conference*, Scottsdale, AZ.

Campbell, G. & Luu, P. (2005). "Oops I did it again": Using neurophysiological indicators to distinguish slips from mistakes in simulation-based training systems. *Paper presented at the 2005 Augmented Cognition International Conference, Las Vegas, NV*

Cannon-Bowers, J. & Lyons (a.k.a. Nicholson), D. (1999). Advanced Embedded Training (AET). *Presentation at the 1999 Interservice/Industry Training, Simulation and Education Conference, Orlando, FL*

Chipman, S. (2004). Office of Naval Research, Science & Technology, Human Systems. Retrieved February 23, 2005, from http://www.onr.navy.mil/sci_tech/personnel/342/training/cogsci/ai.htm

Cohn, J. V., Kruse, A. & Stripling, R. (2005). Novice to expert training. *Paper presented at the 2005 Augmented Cognition International Conference, Las Vegas, NV*

DeRousseau, D. R. (2005). Will augmented cognition improve training results? *Paper presented at the 2005 Augmented Cognition International Conference, Las Vegas, NV*

Halff, H.H. (1988). Curriculum and instruction in automated tutors. In Polson, M.C., and Richardson, J.J. (Eds). Foundations of intelligent tutoring systems. (p. 79-108). Hillsdale, NJ: Laurence Erlbaum Associates, Inc.

Human Performance Center, Seamless Product Information, Data Exchange and Repository (2004). Retrieved February 23, 2005, from https://www.spider.hpc.navy.mil/index.cfm?RID=WEB_OT_1001348

Just, M., Gonzalez, C., & Schneider, W. (2004). Cognitive, biological and computational analyses of automaticity in complex cognition. Presented at the Augmented Cognition: Improving Warfighter Information Intake Under Stress Scientific Investigators Meeting, Jan 6-8, 2004, Orlando, FL.

Lackey, S.J. (2004). Future applications of AugCog technology. *AugCog International Quarterly.* 2 (1). 9.

Lackey, S.J. (2004) Augmented cognition and the future of training, *Workshop presentation at the 2004 Learning Strategies Conference*, Crystal City, VA.

939

Lyons, D.M., Schmorrow, D., Cohn, J.V., & Lackey, S.J. (2002). Scenario based training with virtual technologies and environments. *Paper presented at the Image 2002 Conference, Scottsdale, AZ.*

Mathan, S. & Dorneich, M. (2005). Augmented tutoring: Improving military training through model tracing and real-time neurophysiological sensing. *Paper presented at the 2005 Augmented Cognition International Conference, Las Vegas, NV*

Mendl, M. (1999). Performing under pressure: Stress and cognitive function. *Applied Animal Behaviour Science, 65,* 221-244.

Meyerhoff, J. L., Saviolakis, G. A., Burge, B., Norris, W., Wollert, T., Atkins, V., Spielberger, C. D., & Hansen, J. H. L. (2005). Potential predictors of errors in shooting judgment and cognitive performance by law enforcement personnel during a stressful training scenario. *Paper presented at the 2005 Augmented Cognition International Conference, Las Vegas, NV*

Norman, D. A. (1981). Categorization of action slips. *Psychological Review, 88,* 1-15.

Pope, A. T. & Prinzel, L. J. III (2005). Recreation embedded state tuning for optimal readiness and effectiveness (RESTORE). *Paper presented at the 2005 Augmented Cognition International Conference, Las Vegas, NV*

Russell, C. A. (2004). UAV: Team state classification methods. *Presentation at Improving Warfighter Information Intake under Stress with Augmented Cognition Principal Investigators and Community Meeting, 5-8 Jan 2004, Orlando, FL.*

Schmorrow, D., Nicholson, D.M., Muller, P., Cohn, J.C., Lackey, S.J., Arnold, R., Patrey, J., & Kruse, A. (2004). Technology and Today's Warfighter: From simulation and training to operational environments. *Panel presentation at the 2004 Human Performance, Situational Awareness, and Automated Technology Conference, Daytona Beach, FL.*

Schaafstal, A., Lyons, D. M. & Reynolds. (2001). Teammates and Trainers: The Fusion of SAF's and ITS's. Proceedings of the 10th Conference on Computer Generated Forces and Behavioral Representation, 225-230.

Seidel, R. J. and Park, O. (1994). An historical perspective and a model for evaluation of intelligent tutoring systems. *Journal of Educational Computing Research,* vol.10(2). P 103-128

Shute, V.J. and Psotka, J. (1996). Intelligent Tutoring Systems: Past, Present, and Future. In D. H. Jones (Ed.), Handbook of research for educational communications and technology: A project of the Association for Educational Communities and Technologies (p570-600). New York, NY: Macmillan Library References

Van Lehn, K. (1988). Student Modeling. . In Polson, M.C., and Richardson, J.J. (Eds). Foundations of intelligent tutoring systems. (p. 55-78). Hillsdale, NJ: Laurence Erlbaum Associates, Inc.

Walwanis Nelson, M.M, Smith, D.G., Owens, J.M., and Bergondy-Wilhelm, M.L., "A common instructor operator station framework: Enhanced usability and instruction capabilities," *Proceedings of the Interservice/Industry Training, Simulation, and Education Conference,* Orlando, FL, 2003.

Zachary, W., Cannon-Bowers, J., Bilazarian, P., Krecker, D., Lardieri, P., Burns, J. (1999). The Advanced Embedded Training Systems (AETS): An Intelligent Embedded Tutoring System for Tactical Team Training. *International Journal of Artificial Intelligence in Education,* vol. 10, p257-277

"Oops, I did it again": Using Neurophysiological Indicators to Distinguish Slips from Mistakes in Simulation-Based Training Systems

Phan Luu

Electrical Geodesics, Inc.
Eugene, OR 97403
PLuu@egi.com

Gwendolyn E. Campbell

NAVAIR Orlando TSD
Orlando, FL 32826-3275
Gwendolyn.Campbell@navy.mil

Abstract

Effective training is based on an accurate diagnosis of a trainee's underlying knowledge, skills and abilities. Current automated and semi-automated diagnostic systems rely on performance data to infer underlying competencies. Unfortunately, performance is an imperfect indicator, and we propose that neurophysiological data may help distinguish two categories of errors, mistakes and slips. In the paper we describe a preliminary study designed to investigate the characteristics of a well-documented neurological indicator, error-related negativity (ERN), during the course of a learning task. We present data that demonstrate that the ERN can distinguish mistakes from slips during the learning of a simple task, and discuss the implications of this for the support of adaptive training in more complex, simulation-based training systems.

1   Introduction

Effective training decisions are often based on a diagnosis of a trainee's underlying knowledge, skills and abilities (KSAs). In simulation-based training systems in particular, instructors use both objective performance data and subjective indicators such as facial expressions and body language to form assessments and decide (a) whether to modify some aspect of the scenario in real-time, (b) when and how to intervene during a scenario with feedback, (c) whether these interventions are being attended to and processed by the trainees, (d) the content and focus of the after action review (or scenario debrief), and (e) the selection of the training activity or exercise that would most appropriately follow the current exercise.

The military's push to support increasingly large and complex distributed, multi-platform simulation-based training exercises, combined with the ever-present need to reduce training costs, is driving this community towards a more automated approach to developing and delivering adaptive training. Most current automated diagnostic systems rely solely on behavioral data such as keystroke, track ball, joystick, and touch screen inputs. Unfortunately, behavioral data acquired through system input devices are an incomplete and imperfect indicator of underlying KSAs and cognitive state.

There are many reasons why KSAs cannot always be derived directly from performance data. One major reason has been identified in the sub-discipline of Human Factors that studies errors (Reason, 1990). Norman (1981) identified two types of performance errors, mistakes and slips. "Mistakes" represent **intentional actions** that happen to be incorrect, and thus may reflect the trainee's underlying understanding of the system, his goal(s), and/or the means that he has at his disposal to accomplish his goals. "Slips", on the other hand, represent **unintentional actions** that happen to be incorrect, and are typically recognized as errors by the trainee almost simultaneously with their performance. An analogous situation in everyday life would be closing a locked car door, and as the motor action of closing the door is executed, the person immediately realizes that the keys are inside the car. The implication of this distinction for training is obvious. Pedagogically, mistakes are likely to represent true training needs, while slips typically will not require the same type and amount of remediation.

Unfortunately, it can be problematic for an automated diagnostic system, relying solely on performance data, to distinguish slips from mistakes in a simulation-based training system. Thus, we have initiated a program of research designed to investigate the capability of neurophysiological measures to support this diagnosis. In fact, there is a strong candidate for just such an indicator. A decade of cognitive neuroscience research has demonstrated that the "oops" response associated with slips elicits the now well-known error-related negativity (ERN). The ERN was first reported in the early 90s by Falkenstein et al. (Falkenstein *et al.*, 1991) and Gehring et al. (Gehring *et al.*, 1993), and is observed as a negative deflection in the ongoing EEG, with a peak negativity approximately 50-150 ms after an erroneous response. The ERN has a mediofrontal distribution, and the associated neural generator has been localized to the anterior cingulate cortex (Dehaene *et al.*, 1994).

Several studies have shown that the ERN is only elicited if subjects is aware or believe that a response is in error (Dehaene et al., 1994; Scheffers & Coles, 2000). This means that the ERN is an ideal marker to distinguish between slips of action (i.e., subjects are aware that an error has been made) and genuine mistakes (i.e., subjects are not aware that a response is in error). Unfortunately, most research on the ERN to date has employed speeded responses tasks with skilled subjects. These tasks are not learning tasks, and thus we know very little about how the ERN emerges and/or develops during the learning process. A study that comes closest to investigating the development of the ERN during learning is the one reported by Holroyd and Coles (Holroyd & Coles, 2002). In their study, subjects had to simply learn the association between two simple stimuli and two responses. The authors reported that as subjects learn, the amplitude of the ERN in response to errors appeared to increase. Unfortunately, the task was too simple and the authors did not provide a clear definition of learning, nor did they show the development of the actual ERN.

Thus, before we can use the ERN as a diagnostic indicator in simulation-based training systems, we must first study the development of the ERN in a complex learning situation, providing both clear criteria for learning as well as investigating the actual development of the ERN as subjects transition between zero knowledge about the task to full knowledge. That is the goal of the present study.

2 Current Study

2.1 Participants

Participants were recruited from the University of Oregon. Subjects were paid $20 for their participation, in addition to the money they earned for their performance on the task. Fifteen subjects participated in the present study. Seven subjects committed enough post-learning errors (> 15 errors) to be included in the analysis.

2.2 Task

The task was a go/nogo learning task. There were 16 stimuli (two digit numbers); eight of the stimuli required a go response and the other eight did not require a response. Moreover, each go stimulus was mapped onto a specific finger such that each stimulus has a specific hand and finger mapping. The subjects' task was to learn which stimulus required a response and then map it to the appropriate finger. Each trial starts with a presentation of the target number. Subjects have 1.5 seconds to respond. A response removed the target number (see Figure 1).

After each response subjects are provided immediately with feedback that stays on the screen until subjects terminate it (max duration is 5 seconds). If the subject makes a response when no response is required or doesn't make a response when it is required, they would see "ErrorNG" or "ErrorGo," respectively. Correct go responses are given different types of feedback, depending on whether the response was appropriately mapped. If the response was a correct "go," but not correct for either hand or finger, subjects saw a "Correct!" feedback. If the response was correct for only response hand, subjects saw a "Correct-H" feedback. If the response was correct for hand and finger, subjects saw a "Correct-F" feedback. These feedback cues guide subjects towards the learning of correct response mappings. After every 100 trials, subjects are asked to provide their best guess of the correct response mappings.

Subjects were informed that they begin the study with zero points, and that each response is associated with either points won or lost: Correct-F = 8, Correct-H = 4, Correct! = 2, ErrorGo = -8, ErrorNG = -8. Subjects were also informed that at the end of the study, they would be compensated at a rate of 0.5 cents/point for the number of points that they earned in the study.

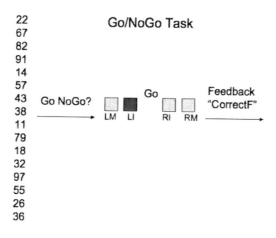

Figure 1. **Schematic of the Go/NoGo task.**

2.3   Measurement

<u>EEG Recording</u>
EEG was recorded from 128 scalp sites using the HydroCel Geodesic Sensor Net (Electrical Geodesics, Inc., Eugene, OR USA). All electrode impedances were brought below 70 KΩ before recording. All channels were referenced to Cz and collected using a 0.1 to 100-Hz band pass. Signals were sampled at 250 samples/second and digitized with a 16-bit A/D converter.

<u>ERP and EEG Analysis</u>The continuous EEG data were filtered with a 30 Hz low pass finite impulse response (FIR) filter with zero phase distortion. The filter was set with a pass band gain of .1dB and stop band gain of 40 dB with a 4 Hz transition band. The continuous EEG was then segmented on the response and sorted according to accuracy and pre-/post-learning trials (see below). The criterion for having learned a mapping for each number is the occurrence of three consecutive Correct-F responses or three consecutive correct response omissions (for no-go numbers). It should be noted that the trial sequence is such that the same number rarely occurs two times in a row.

Next, all trials were corrected for ocular artifacts using the method of Gratton et al. (Gratton *et al.*, 1983). After removal of ocular artifacts, the data were analyzed for additional artifacts. Prior to averaging the data, each trial was re-referenced to the average of all of the sensors at each time point.

2.4   Results

<u>Behavioral Data</u>
Examining the time spent attending to the feedback shows that according to the learning criterion, subjects spent much more time attending to the feedback prior to having learned (mean = 2142 ms, SD = 729 ms) than after having learned the mapping (mean = 650 ms, SD = 208 ms). This difference approached significance, $\underline{F}(1,6) = 5.80$, $\underline{p} < .06$.

Figure 2 shows the response locked data at channel FCz for error responses pre and post learning. The vertical line at time = 0 marks time of the button press. As can be seen, at approximately 50 ms post-response there is a large ERN on those error trials that occur after the learning criterion is reached compared to those error trials that occurred before, $F(1,6) = 5.8$, $p < .06$.

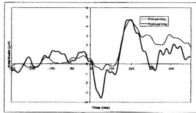

Figure 2. ERN (at channel FCz) associated with error responses before and after achievement of learning criterion.

3 Discussion

In this study, mistakes were defined as error responses prior to learning, and slips were defined as error responses made after participants demonstrated knowledge for correct stimulus-response mapping. The behavioral results showed that we can define a learning criterion that reflects when subjects have acquired knowledge of correct stimulus-response mappings. That is, subjects spent more time examining feedback stimuli prior to meeting the learning criterion than after meeting that criterion. The EEG data show, as expected, that slips can be distinguished from mistakes. In particular, prior to learning, error responses do not elicit a clear ERN. In contrast, error trials committed after subjects have learned the task are associated with a large ERN.

While this is a promising start, for these results to be relevant to the present project, there are still several challenges that must be addressed. The present results are based on averaged data, but to be useful for performance diagnoses, trial-by-trial analyses are required. Therefore, the first challenge is to perform single-trial analysis of the EEG. Single-trial analysis of the EEG involves the extraction of signals of interest, such as the ERN, from the ongoing background EEG, which is much bigger than the signals of interest. One approach to single-trial analysis involves the use of latent variable decomposition, such as ICA (Makeig *et al.*, 1996) and PCA with linear discriminant analysis (Parra *et al.*, 2003). Another approach would involve spectral decomposition of the single-trial EEG data. From the spectrally decomposed data, estimates of the energy within the theta band of the EEG immediately after the response can be used to estimate the amplitude of the ERN because we have shown that the ERN is part of the ongoing theta rhythm (Luu *et al.*, 2004).

Because electrophysiological responses differ between subjects with regards to manifestation (e.g., topography and latency of the ERN) and because we are uncertain how generalizable the results are, both psychologically and electrophysiologically, the second challenge is to study, using single-trial analytic methods, the development of the ERN as a function of learning in single subjects. Overcoming both single-trial and single-subject analytic challenges will allow us to dynamically monitor brain responses during training and to develop training programs that are tailored to the learning needs of each individual.

Beyond the technical challenges associated with identifying the occurrence of an ERN response in a single individual during a single trial, productively applying this technique in a simulation-based training system will present other difficulties. A noteworthy characteristic of simulation-based environments is that they are highly complex and dynamic, and trainees are often required to multitask. In this type of environment, correctly pairing an ERN with the specific action/event that elicited that response will likely pose a significant challenge.

It should also be noted that the addition of EEG measurement and analysis technology to a training system is costly and will not always be necessary. In some domains, performance can be used to distinguish slips from mistakes and careful front-end analyses must be done to determine when and where a physiological indicator is truly necessary. One example of this would be any domain that allows a trainee to undo or redo an action. It seems reasonable to conclude that any error that is immediately followed by a corrective action is more likely to be a slip than a mistake. Finally, as always in the training community, the bottom line will be the impact on the effectiveness and/or efficiency of the training. Researchers and practitioners must be able to show that incorporating this measure into an automated diagnosis system results in a significant improvement to the training system.

Of course, future research may identify other useful applications for the measurement of the ERN response in simulation-based training & assessment systems, as well as other neurophysiological indicators that are capable of contributing to the effectiveness and efficiency of these systems.

4 References

Dehaene, S., Posner, M. I., & Tucker, D. M. (1994). Localization of a neural system for error detection and compensation. *Psychological Science, 5*, 303-305.

Falkenstein, M., Hohnsbein, J., Hoormann, J., & Blanke, L. (1991). Effects of crossmodal divided attention on late erp components. Ii. Error processing in choice reaction tasks. *Electroencephalography and clinical Neurophysiology, 78*, 447-455.

Gehring, W. J., Goss, B., Coles, M. G. H., Meyer, D. E., & Donchin, E. (1993). A neural system for error detection and compensation. *Psychological Science, 4*, 385-390.

Gratton, G., Coles, M. G., & Donchin, E. (1983). A new method for off-line removal of ocular artifact. *Electroencephalography and clinical Neurophysiology, 55*, 468-484.

Holroyd, C. B., & Coles, M. G. H. (2002). The basis of human error processing: Reinforcement learning, dopamine, and the error-related negativity. *Psychological Review, 109*, 679-709.

Luu, P., Tucker, D. M., & Makeig, S. (2004). Frontal midline theta and the error-related negativity: Neurophysiological mechanisms of action regulation. *Clinical Neurophysiology, 115*, 1821-1835.

Makeig, S., Bell, A. J., Jung, T.-P., & Sejnowski, T. J. (1996). Independent component analysis of electroencephalographic data. In D. Touretzky, M. Mozer & M. Hasselmo (Eds.), *Advances in neural information processing systems 8* (pp. 145-151). Cambridge: MIT Press.

Norman, D. A. (1981). Categorization of action slips. *Psychological Review, 88*, 1-15.

Parra, L., Spence, C. D., Gerson, A. D., & Sajda, P. (2003). Response error correction - a demonstration of improved human-machine performance using real-time eeg monitoring. *IEEE Transactions on Neural Systems and Rehabilitation engineering, 11*, 173-177.

Reason, J. (1990). *Human error*. Cambridge University Press.

Scheffers, M. K., & Coles, M. G. H. (2000). Performance monitoring in a confusing world: Error-related brain activity, judgments of response accuraccy, and types of errors. *Journal of Experimental Psychology: Human Perception and Performance, 26*, 141-151.

Investigating the Transition from Novice to Expert in a Virtual Training Environment using Neuro-Cognitive Measures

Joseph V. Cohn

Naval Research Laboratory
4555 Overlook Ave SW
Washington DC 20375
cohn@itd.nrl.navy.mil

Roy Stripling

Naval Research Laboratory
4555 Overlook Ave SW
Washington DC 20375
cohn@itd.nrl.navy.mil

Amy A. Kruse

DARPA
3701 N. Fairfax
Arlington VA, 22309
akruse@darpa.mil

Abstract

Virtual Environment (VE) Training Systems offer a unique solution for a range of training requirements. Yet, while the technologies supporting the advancement of these systems continue to be refined, the degree to which these systems can support training and enhance real world performance is often overlooked or, worse, simply 'assumed'. The critical component which has thus far been lacking is the development of an overarching framework for utilizing the technology solutions offered by VE to enhance training. While this challenge has often times been reduced to one of demonstrating some level of transfer of training from a specific virtual system to a specific real world task, these efforts fail to provide a pedagogy for using these modeling and simulation tools as effective training devices. Of course, it is quite likely that there are many such pedagogies, depending on the types of training requirements that must be satisfied. The current effort reports a first attempt at developing the underlying methodology for determining the types of skills current levels of VE technology can and cannot support, and then developing a training approach, based on current theories, that effectively utilizes these tools. It starts by focusing on what is meant by training cognitive skills, progresses onward to define a viable end-state for such training and concludes by proposing a means for using cognitive based VE training systems to mold an individual's training for maximum efficiency.

1 The Novice to Expert Continuum

1.1 Background

A recurring theme in the Virtual Environment (VE) literature, initially suggested by Roscoe (1982), is that different types of skills have vastly different technical requirements supporting their training. Contrary to what one might expect, there is no guarantee that systems with high fidelity will actually provide better training; as Roscoe (1982) indicated, there is a region of design space within which cost/benefit is optimal; move away from this region, and you run the risk of paying a lot for a system that delivers negative training. Moreover, as Stoffregen, Bardy, Smart and Pagulyan (2003) suggest, there is as yet no tool available for faithfully stimulating the senses virtually the same way they are in the real world setting. So, any skill whose training requires a level of fidelity commensurate with the real world will ultimately fail to be fully supported by VE Simply put, the higher the level of technical fidelity, the more costly the system (Roscoe, 1982). For example, cognitively oriented training, which often features problem solving or decision making as a key training goal, may often times be satisfied by using simple, relatively inexpensive visualization systems (Gopher, Weil & Bareket, 1994; Morris & Tarr, 2002; Munro, Breaux, Patrey, Sheldon, 2002; Stone, 2002; Figueroa, Bischof, Boulanger, Hoover, 2005; Milham, Hale et al, 2004). Yet even limiting the training objectives to cognitive skills does not necessarily assist us in making informed *application* choices. How do we get the training into the system? VEs have been, and for the most part still are, high tech shells within which users may be immersed for what amounts to entertainment purposes. The challenge is to insert training into these shells, so that a given system provides a complete package: a sensorally compelling experience coupled with an instructionally effective one. In other words, we need a pedagogy for injecting training into VEs.

1.2 Defining the Pedagogy

Unlike other training applications, such as the schoolhouse or the field, within a VE information regarding the context, the user's performance, and the relationship between the two is readily available. This presents the possibility of performing real time assessments of performance, and of directly manipulating the environment to propel the trainee along an appropriate training trajectory. In defining this trajectory, we propose treating the behavior of experts as the ideal end state, towards which individual trainees' current behavioral states should move. By identifying different behavioral templates at both the novice and expert level, and by comparing this to the current context within which a trainee is performing, it should be possible to identify early on whether or not the trainee is progressing appropriately and if not, what sorts of remedial feedback should be provided. The challenge then becomes defining what form such an expert end state might take, and determining how to objectively quantify it. A review of human performance literature suggests that the notion of a novice to expert continuum in performance is not new. Dreyfus and Dreyfus (1980) captured this evolution in a five-stage model, defining different transitional stages along the path to expertise the acquisition of cognitive skills. These include:

- *Stage 1:* Novice, learner has no experience in a given domain and performs by explicitly following a set of rules.
- *Stage 2:* Competence, learner has some domain experience, begins to implicitly follow rules, but cannot generalize these rules to a range of domains.
- *Stage 3:* Proficiency, learner can follow rules and generalize to a range of domains.
- *Stage 4:* Expert, learner can generalize rules across domains, generate new rules, yet requires self-monitoring.
- *Stage 5:* Mastery, learner can generate new rules without self monitoring.

In a similar manner, the acquisition of motor skills has also been subjected to a stage-wise decomposition. For example, Fitts (1964) suggested that learners pass through three phases when learning a new motor skill:

- *Stage 1:* Cognitive Phase, where the learner must attempt to understand overall task goal(s).
- *Stage 2:* Associative Phase, where the learner has acquired rudimentary skills associated with the task, and now refines the skill.
- *Stage 3:* Autonomous Phase, where the task can now be performed without fear of interference of other, concurrent tasks.

A feature common to both schema is that, underlying the progression towards expertise is the evolution of task performance from a conscious, deliberated activity that involves explicit concept formation (Ausubel, Novak, & Hanesian, 1978) and that is affected by fear, mistakes and the need for validation (Daley, 1998), to one that is more automatic, involving a constructivist process that uses *active* concept integration driven by contextual factors (e.g., assimilating new information derived from the learning/training context with current knowledge base). In other words, for a wide range of tasks from simple motoric ones to complex cognitive ones, the ability to develop new rules for new situations, without conscious deliberation is a hallmark of the highest level of skill mastery.

Based on this categorical distinction, the hypothesis we propose is that there are detectable, reliable and repeatable measures that distinguish experts from novices in complex training tasks. These measures reflect differences in proficiency in acquiring, processing, synthesizing and acting on information, in the factors that influence this process (e.g., fear versus context), and in their ability to predict potential outcomes of a given behavior (Daley, 1998; Serfaty, Macmillan, Entin, & Entin, 1997; Ericsson and Smith, 1991; Klein & Peio, 1989). We further suggest that these differences are observable at both the behavioral as well as at deeper, neurophysiologic levels. Using the tools of VE and other behavior sensing technologies such as EEG and fMRI to capture performance at the behavioral and neurophysiologic levels, the development of these patterns of behavior could then be represented as Performance *Templates*. The performance template of an expert reflects the ideal; the performance template of a novice indicates the current state. Quantifying differences between these two templates enables the development of a tailored training approach that could bring the one closer to the other. Progress can be monitored over time, with additional training interventions being brought online as needed. The current effort seeks to refine this approach. .

The utility of such templates to assess cognitive load and to cue strategies for reducing these loads when necessary has already been suggested (Balaban, Cohn, et al 2004). The remainder of this paper will follow the development of an ongoing efforts aimed at developing the capability, through neurophysiologic sensing, to be able to characterize whether a trainee is in novice or expert status (correlated with performance), as well as the capability to track a trainee, in real time, through the stages from novice to expert using technologies sensitive to the dynamic nature of learning in the brain, within the framework of developing automaticity for performing a specific task. If these efforts prove successful, an entirely new training approach for VE will be realized, in which it will be possible to correctly assess where a trainee is in the training process and to then move trainees more quickly and effectively through the training process. The challenge then is defining what form an expert end state might take, and determining how to objectively quantify it.

2 Characterizing Trainee State: The Road Towards Automaticity

2.1 Defining Automaticity

We are all familiar with the phenomenon of *automaticity* first qualified as a human performance trait by James (1890), through which routine activities, whether primarily motoric or cognitive in nature, become increasingly less a matter of conscious and deliberate action and more a matter of effortless performance. Examples of this progression range from the simple shift in focus of a young student on memorizing multiplication tables to being able to 'visualize' the result (Koshmider & Ashcraft, 1991), to the more complex development of rapid decision making capabilities in professionals (firefighters, platoon commanders and so forth; Klein, 1998). Most of us, though, would be hard pressed to define the mechanism through which this evolution arises. An easy approach to begin to understand this evolution is to consider what distinguishes automaticity from instances of non-automaticity, as outlined in Table 1.

Table 1: Some Characteristics of Automatic Processes

Automatic processes are:
Fast (Posner & Snyder, 1975)
Effortless (Schneider & Shiffrin, 1977)
Not subject to conscious monitoring (Marcel, 1983)
Not subject to emotion (e.g., fear, mistakes, need for validation (Daley, 1998)

The question of how automaticity precipitates is hotly contested. There is general agreement that automaticity depends in some fashion on attention (Logan, 1988). However, precisely what form this relationship takes is unknown. For the current effort, we consider two competing theories for the development of automaticity, the *modal* view (Kahneman, 1973) and the *instance* view (Logan, 1988). The modal view considers this development process to be a straightforward continual decrease in reliance on attention. As a process becomes more automatic, it becomes increasingly less subject to attentional limitations and is therefore exhibited with increasing speed (i.e. fast and effortless. In the absence of attentional limitations, which suggests an absence of attentional control, the process becomes autonomous.

Significant challenges have been raised against this approach, the most important being concern over the absence of any clear learning mechanism through which automaticity may derive (Logan, 1988). The instance view, proposed to fill this gap, suggests that automaticity precipitates from the development of ever-larger memories of specific instances, such as a set of solutions for a given type of challenge. The progression from deliberate action to automatic one is from an initial dependence on algorithmic processing, essentially rule-based, to a more memory-based processing, relying on an ability to both store and retrieve these memories. This shift in approach is often considered to be a race between these two processes (Palmeri, 1997). As the memory store of instances increases, the balance is tipped towards memory retrieval (Logan, 1988). What is learned is the set of features that comprise a given instance (Logan & Etherton, 1994), what is often called *co-occurrence* (Barsalou, 1990). This learning is, in turn, directed by attention (Logan, Taylor & Etherton, 1996).

There are, of course, difficulties with this approach as well. Dreyfus and Dreyfus (1980), and to some extent Fitts (1964) suggest that high level performers must be able to respond to novel situations; no two instances are ever completely alike. Therefore, in addition to instance comparison, there must be some ability to 'fill in the blocks' by generating novel solutions. As Klein and Wolf (1998) suggest, automaticity must, to some extent, involve the generation of some amount of unique problem solving capability, i.e., it is a constructivist process (Daley, 1998). While the true answer to how automaticity arises is likely not an either/or proposition the work of Logan (1988),

Kahneman (1973), James (1890), Fitts (1964), Dreyfus and Dreyfus (1980) and many others points to a single inescapable conclusion. Namely, that the transition from novice to expert requires an advancement from executing behavior in a deliberate, monitored, and emotional fashion, to performing activities in a manner that is more effortless and natural but which is in no way thoughtless, accidental or haphazard. The outstanding question to which we next turn is how this evolution might be captured using different sensor technologies.

2.2 Neuronal Measures Relating to Automaticity

Within the current effort is a fundamental assumption that any overt behavior must have its neurological correlates (Carlson, 2000). Similarly, the processes through which control of these behaviors derive, such as the process through which automaticity derives, must also have their antecedents in basic neural circuitry. While it is currently impossible to conclusively link a set of neuronal pathways with a given complex, cognitive behavior pattern it is possible to draw some general conclusions based on correlations between the two. For example, it has long been proposed that the process through which both cognitive and motor tasks are learned includes a period of *consolidation* during which time the actual motor process is labile and subject to degradation (Brashers-Krugg, Shadmehr & Bizzi, 1996). Neural imaging studies (Reza & Holcomb, 1997) demonstrated specific evolving changes in the neural substrate that reflect this process. Similarly, cognitive processes, and their movement towards automaticity also have neural antecedents (Kramer & Strayer, 1988; although see Kotchoubey, 2002 for an argument against this from a cognitive perspective). The likely substrate for the development of 'executive' control in both cognitive and motor processes may reside in specific neuronal circuitry (Heyder, Suchan & Daum, 2004; Jenkins, Brooks & Nixon, 1994). These assertions often times gain substantial support through the observation of individuals suffering from neurological impairment (Lang & Bastian 2002).

In considering what neural measures should be used to develop measures of automaticity, three different techniques are explored. These include Electroencephalography, EEG, functional Magnetic Resonance Imaging, fMRI and functional near infrared imaging, fNIR. Each approach brings its own benefits and pitfalls. EEG has a resolution on the order of centimeters, yet can provide a signal within a very short amount of time, measured in the millisecond range. MRI has a much finer spatial resolution, but requires orders of magnitude more time to collect and by itself provides only structural information. Functional MRI, provides data regarding the changes in blood flow and oxygenation that presumably result from changes in neural activity. This technology benefits from the finer spatial resolution of MRI, but the temporal resolution is several orders of magnitude slower than EEG. Both MRI and fMRI require large, expensive, energy demanding, and operator intensive machines. They also place extreme restrictions on the freedom of movement of the experimental participants. fNIR also works by inferring neural activity from blood oxygenation levels, but detects these changes through the differential absorption and reflection of infrared light off of oxygenated and deoxygenated red blood cells. fNIR provides less spatial resolution than fMRI, can only penetrate deep enough to detect changes in cortical blood oxygenation, and generally offers lower temporal resolution than EEG (although recent developments in fNIR processing (Gratton & Fabiani, 2001a & b) suggest that certain fNIR technologies may be able to match the temporal resolution of EEG). We consider the potential applicability of each, in turn, as a tool for providing measures of automaticity.

2.2.1 EEG

Some examples of how EEG technology has been applied to the challenge of monitoring the transformation from novice to expert include:
- Identifying indices of skill level (Deeny, Hillman & Charles, 2003; Ciesielski & French, 1989).
- Identifying indices of skill acquisition (Smith, McEvoy & Gevins, 1999; Kerick, Douglas & Hatfield, 2004).
- Detection of the progression towards automaticity (Gunter & Friederici, 1999).

2.2.2 fMRI

As with EEG, fMRI has been applied to the puzzle of automaticity:
- Identifying indices of skill level (Peres, Van De Moortele, et al, 2000).
- Identifying indices of skill acquisition (Tracy, Flanders, et al, 2003; Debaere, Wenderoth, et al 2004).
- Detection of the progression towards automaticity (Floyer-Lea & Matthews, 2004).

Although this tool has yet to be applied to the problem of detecting the development of automaticity, or other measures of the progression from novice to expert, it has been successfully utilized in the study of a wide range of cognitive tasks, including:

- Execution of motor skills (Okamoto, et al, 2004).
- Wisconsin Card Sorting task (Fallgatter & Strik, 1998).
- Verbal fluency task (Herrmann, Ehlis, & Fallgatter, 2003; Kameyama, et al, 2004).

3 The Novice to Expert Continuum

Having determined that automaticity is both a worthy end state and one that is amenable to measurement using neuroimaging technologies, we must now take a step back and consider how to wrap these factors into a process through which we may enable learning of this trait. Before delving into this, though, it will prove useful to consider what, exactly, is meant by learning. Building on this definition, we will then consider a framework underlying the progression from novice to expert state and, finally, suggest an approach to building a performance template to enable this advancement.

3.1 Operational Definition of Learning

A generally accepted definition of learning, congruent with biological and psychological notions, is a process that modifies overt structures, such as changes in nerve cell responses (Long Term Potentiation and Depression; Huerta & Lisman, 1993; Stanton & Sejnowski, 1989) as well as actual observable behaviours. As Zanone and Kelso (1997) suggest, any theory discussing approaches to enhancing learning must address two points:

- Point 1: Assessing the initial state of the learner
- Point 2: Revealing how this state is modified as a result of the learning interventions

3.2 Detecting the Novice to Expert Evolution

Based on this definition, in conceiving of the movement from novice to expert, we suggest the following steps:

- Step 1: Assess the initial state of the learner
- Step 2: Compare this state to the desired end state
- Step 3: Determine what strategies must be implemented to move the initial state to the desired end state.
- Step 4: Monitor this movement in real time (current state) and correct as necessary

For our purposes, initial state, current state and end state are all different instantiations of an individual's performance templates. Based on the above review of EEG, fMRI, and fNIR one can imagine that such a template would consist of, at the very least, measures derived from the different sensor technologies discussed earlier. These sensor-based metrics would then be coupled with behavioral indicators of performance, which would need to be defined in a context dependent fashion. Each of these elements is discussed in turn, and summarized in Table 2.

Table 2: Notional Performance Template

	Novice	Expert
EEG signature	Changes in power distribution for different frequencies.	
fMRI signature	Changes in type and quantity of cortical structures recruited	
fNIR signature	Changes in type and quantity of cortical structures recruited	
Behavioral indicators	Differential performance on primary and secondary tasks	
Ensemble Assessment	Evolution from effortful processing to automaticity	

3.2.1 Neural Technologies

Current EEG-based research efforts are exploring the neurological activity patterns in a range of tasks including basic working memory, basic motor coordination, visual manipulation, first person shooter-games, target shooting, and cognitive strategy video games. While exploratory in nature, these efforts are building upon past efforts, which have identified such changes as increases in front midline theta (associated with increased attentional focus; Smith et al, 1999), increased central and/or global alpha power (associated with reduced overall cortical activity; Kerick et al, 2004;), sustained increased high frequency alpha activity in regions specifically associated with execution of a task (such as sustained activity over right parietal areas in tasks requiring visuospatial processing; Smith et al 1999), enhanced event related potentials including the N400 and N2 peaks in participants exhibiting task automaticity (Gunter and Friederici, 1999). The current efforts will bring to bear more advanced EEG technologies, including EEG arrays of up to 256 electrodes, and more sophisticated computational approaches, including independent component analysis, discriminant analysis, and artificial neural network processing. It will also systematically explore a wider variety of tasks, including many more complex tasks than have previously been used in such testing.

While in the current effort EEG will be the primary neurotechnology used, fMRI and MRI will also be applied to the problem, in order to gain deeper insight into the region specific activation changes observed during the training progression from novice to expert. fMRI compliments EEG work by offering much greater anatomical specificity. fMRI does not suffer from the blurring of electrical signals that occurs as they move from their sources within the brain through the overlying tissues and fluids to the scalp surface, where EEG electrodes detect them. fMRI avoids this issue because it chiefly detects changes in blood flow and blood oxygenation within the brain that result from changes in local neuronal activity. Due to its lower temporal resolution, researchers exploiting the advantages of fMRI will do so in experiments also utilizing EEG. In addition to fMRI, anatomical MRI (which maps the physical structures of the brain, and overlying tissues) will be utilized to enhance the computational methods used to 'de-blur' the EEG signals and enable probable identification of the original sources of these signals.

MRI technologies (including both fMRI and MRI) are much more expensive to own and operate than EEG, and due to their size and their requirements for subject immobility, these technologies cannot be applied to a number of the tasks being explored. As a result, a new, lower-cost, less restrictive technology for detecting blood flow and changes in blood oxygenation is being explored as well, namely fNIR. The chief limitation of this technology is that light traversing through more than 5-6 cm of tissue is absorbed, thus only blood flow in the superficial levels of the brain can be detected. Fortunately, most conscious cognitive processing is believed to occur at these superficial structures (in the cerebral cortex) and thus fNIR may serve well at the task of detecting when cognitive processing demands shift away from conscious processing to automatic processing.

3.2.2 Behavioral Metrics

While the development of behavioural metrics to capture task performance is somewhat task-specific, it is possible to provide some general principles for bounding this effort. Within the framework of developing response automaticity, it is very likely that attention plays a pivotal role (Logan & Etherton, 1994). As Dreyfus and Dreyfus (1980) and Fitts (1964) suggest one behavioral aspect of automaticity is the ability to perform a task without fear of interference from other tasks. This suggests that the development of automaticity could be monitored, behaviourally, using a classical dual task paradigm (Pashler, 1998). This allows us to propose the following approach to using overt performance as a measure of the novice to expert transition:
- Step 1: Measure, at planned intervals, performance on the actual tasks being trained
- Step 2: Measure performance on a secondary task that is presented randomly throughout the training session, at points that overlap with measurement of performance on the primary tasks.

The expectation is that as trainees progress towards the expert level, they will improve their performance on the specific tasks as well as on their ability to perform a secondary task, without sacrificing performance on either.

4   Conclusion

"Tell me and I forget. Teach me and I remember. Involve me and I learn."
- Benjamin Franklin

In an ideal training environment, the trainee takes an active role in structuring their experience, going beyond passively assimilating information and implementing a process of discovery and metacognitive learning (Alessi & Trollip, 2001; Ford, Smith, Weissbein, Gully & Salas, 1998). The trainee becomes an active participant in structuring their learning. This objective can be only partially realized with most types of training approaches and therein lays the true benefit of using virtual environment systems. At their best, VEs enable individuals to interact with each other and with their environment in real time, developing hypotheses and testing them, and reviewing the consequences of their actions. At their worst, they simply provide an entertaining interlude.

The problem lays in understanding how best to effectively exploit these tools. Recently, a short report in the journal *Nature* (Green & Bavelier, 2003) demonstrated that positive transfer of specific skills from commercially available video games was possible. While this is a critically important result –supporting the notion that simulator fidelity need not be high for effective training to result- it only touches upon the more important question of *how was this training produced?* The Green and Bavelier study, as well as the work of Morris and Tarr (2002) provide *post hoc* indications of where these tools may impact performance. However, they do not provide any indication of the method through which this change is effected. Without both pieces of this puzzle –the how/why as well as the what-the development of truly effective training systems will prove elusive. It is this vacuum that the current work proposes to fill. Working within the structure of VE, it is possible to collect reams of performance data; embedding the types of sensor technologies discussed throughout this paper enables the development of a finer-grained picture of momentary performance. In order to ensure that these disparate sets of information have meaning, though, a pedagogy underlying their utilization, such as the one developed here, must be defined.

The process outlined in this paper has yet to be validated through experimentation. Moreover, while we have suggested a means for monitoring performance through the use of templates, we have not indicated how this performance might be modified should it prove necessary to do so. To do so would require two parallel efforts: one to analyze the different learning strategies through which novices develop expertise. As Frederiksen & White (1989) indicate, there are likely different classes of such strategies into which novices naturally fall. Providing the means for determining from an individual's performance template both what type of strategy they might require for optimal performance and when to provide such a strategy is nontrivial (Donchin, 1989). Nevertheless, indications from other domains suggest that the ideas laid out should prove successful. Frederiksen and White (1989) did show some success in identifying and implementing an optimal training strategy. As well, Zanone and Kelso (1992) demonstrated a unique approach for assessing an individual's initial performance state for a simple coordination task. Using measures of pattern stability, they demonstrated an ability to consolidate novel coordination patterns and to enable the performance of new patterns that were entirely not present in the initial state (Zanone & Kelso, 1997). While the relationship between these overt behaviors to changes in underlying neural states has only been lightly explored (Kelso, Fuchs, et al 1998), such efforts suggest that the current undertaking, although challenging, no longer rest solely in the realm of imagination.

References

Ausubel, D.P., Novak, J.D., & Hanesian, H. (1978). Educational psychology: A cognitive view (2nd ed.). New York: Holt, Rinehart, & Winston.

Alessi, S.M & Trollip, S.R. (2001) Multimedia for Learning (3rd ed.). Boston: Allyn & Bacon.

Balaban, C., Cohn, J.V., Redfern, M., Prinkey, J. & Stripling, R. (2004). Postural Control as a Probe for Cognitive State: Exploiting Human Information Processing To Enhance Performance. *International Journal of Human Computer Interaction*, 17(2), 275-286.

Barsalou, L. W. (1990). On the indistinguishability of exemplar memory and abstraction in category representation. In T. K. Srull & R. S. Wyer (Eds.), *Advances in social cognition* (Vol. 3, pp. 61-88). Hillsdale, NJ: Erlbaum.

Brashers-Krug, T. Shadmehr, R., & Bizzi, E. (1996). Consolidation in human motor memory. *Nature*, 382 (6588): 252-255,

Carlson, N. R. (2000). Physiology of Behavior (7th ed.). Boston: Allyn & Bacon.

Chase, W. G., & Simon, H. A. (1973). The mind's eye in chess. In W. G. Chase (ed.) *Visual Information Processing* (pp. 215-281). New York: Academic Press.

Ciesielski K.T., & French C.N. (1989). Event-related potentials before and after training: chronometry and lateralization of visual N1 and N2. *Biol Psychol*. 28(3):227-38.

Daley, B.J. (1998). Novice to expert: How do professionals learn? In *Proceedings of the 39th Annual Adult Education Research Conference*. University of the Incarnate Word San Antonio, Texas, May 15-16, 1997.

Deeny, S. P., Hillman, C. H., Janelle, C. M., & Hatfield, B. D. (2003). Cortico-cortical communication and superior performance in skilled marksmen: An EEG coherence analysis. *Journal of Sport & Exercise Psychology*, 25 (2): 188-204.

Debaere F., Wenderoth N., Sunaert S., Van Hecke P., & Swinnen S.P. (2004). Changes in brain activation during the acquisition of a new bimanual coodination task. *Neuropsychologia*, 42(7):855-67.

Dreyfus, H.C. & Dreyfus, S.E. (1980). *A five-stage model of the mental activities involved in directed skill acquisition*. ORC 80-2 (F49620-79-C-0063). Bolling, AFB, Washington, DC. Air Force Office of Scientific Research: United States Air Force.

Donchin, E. (1989). The learning strategies project. *Acta Psychologica*, 71: 1-15.

Ericcson, K. A, Krampe, R. T., & Tesch-Römer, C. (1993). The role of deliberate practice in the acquisition of expert performance. *Psychological Review, 100*, 363-406.

Ericsson, K. A., & Smith, J. (1991). Prospects and limits of the empirical study of expertise: An introduction. In K. A. Ericsson & J. Smith (Eds.), *Toward a general theory of expertise: Prospects and limits* (pp. 1-39). Cambridge, England: Cambridge University Press.

Fallgatter AJ, & Strik WK. (1998) Frontal brain activation during the Wisconsin Card Sorting Test assessed with two-channel near-infrared spectroscopy. *Eur Arch Psychiatry Clin Neurosci*. 248(5):245-9.

Figueroa, P., Bischof, W. F., Boulanger, P., & Hoover, H. J. (2005). Efficient comparison of platform alternatives in interactive virtual reality applications. *International Journal of Human-Computer Studies*, 62 (1): 73-103.

Fitts, P.M. (1964). Perceptual motor skills learning. In A.W. Melton (Ed.), *Categories of human learning*. (pp. 243-285). New York Academic Press.

Floyer-Lea A., & Matthews P.M. (2004). Changing brain networks for visuomotor control with increased movement automaticity. *J Neurophysiol*. 92(4):2405-12.

Ford, J., Smith, E., Weissbein, D., Gully, S. & Salas (1998) Relationships of Goal Orientation, Metacognitive Activity and Practice Strategies with Learning Outcomes and Transfer. *Journal of Applied Psychology 83(2)* 218-233.

Frederiksen, J.R. & White, B.Y. (1989). An approach to training based upon principled task decomposition. *Acta Psychologica*, 71:89-146.

Gopher, D., Weil, M., & Bareket, M. (1994) Transfer of skill from a computer game trainer to flight. *Human Factors*, 36 (3): 387-405.

Gratton G., & Fabiani M. (2001a) Shedding light on brain function: the event-related optical signal. *Trends Cogn Sci*. 5(8):357-363.

Gratton G, & Fabiani M. (2001b) The event-related optical signal: a new tool for studying brain function. *Int J Psychophysiol*. 42(2):109-21.

Green, C. S., & Bavelier, D. (2003). Action video game modifies visual selective attention. *Nature* 423 534-537.

Gunter T.C., & Friederici A.D. (1999). Concerning the automaticity of syntactic processing. *Psychophysiology*, 36(1):126-37.

Herrmann M.J., Ehlis A.C., & Fallgatter A.J..2003) Frontal activation during a verbal-fluency task as measured by near-infrared spectroscopy. *Brain Res Bull*. 30;61(1):51-6.

Heyder, K., Suchan, B., & Daum, I. (2004). Cortico-subcortical contributions to executive control. *Acta Psychologica*, 115 (2-3): 271-289.

Huerta, P.T., & Lisman, J. (1993). Heightened synaptic plasticity of hippocampal CAI neurons during a cholinergically induced rhythmic state. *Nature*, 364:723-725.

James, W. (1890). Principles of psychology. New York: Holt.

Jenkins, I. H., Brooks, D. J.; Nixon, P. D.., Frackowiak, R. S. J.; et al (1994). Motor sequence learning: A study with positron emission tomography. *J. Neuroscience*, 14 (6): 3775-3790.

Kahneman, D. (1973). Attention and effort. Englewood Cliffs, NJ: Prentice-Hall.

Kameyama M., Fukuda M., Uehara T., & Mikuni M. (2004) Sex and age dependencies of cerebral blood volume changes during cognitive activation: a multichannel near-infrared spectroscopy study. *Neuroimage*, 22(4):1715-21.

Kelso, J.A.S., Fuchs, A., Lancaster, R., Holroyd, T., Cheyne, D., & Weinberg, H. (1998) Dynamical cortical activity in the human brain reveals motor equivalence. *Nature,* 392: 814-818.

Kerick S.E., Douglass L.W., & Hatfield B.D. (2004). Cerebral cortical adaptations associated with visuomotor practice. *Med Sci Sports Exerc.* 36(1):118-29.

Klein, G. (1998). Sources of power: How people make decisions. Cambridge, MA: MIT Press.

Klein, G. A., & Peio, K. (1989). Use of prediction paradigm to evaluate proficient decision making. *American Journal of Psychology, 102*, 321-331.

Klein, G., & Wolf, S. (1998). The role of leverage points in option generation. *IEEE Transactions on Systems, Man and Cybernetics: Applications and Reviews, 28*(1), 157-160.

Kramer, A. F. & Strayer, David L. (1988). Assessing the development of automatic processing: An application of dual-task and event-related brain potential methodologies. *Biological Psychology*, 26 (1-3): 231-267.

Koshmider, J. W. & Ashcraft, M. H. (1991). The development of children's mental multiplication skills. *Journal of Experimental Child Psychology* 51 (1): 53-89.

Kotchoubey, B. (2002) Do event-related brain potentials reflect mental (cognitive) operations? *Journal of Psychophysiology*, 16 (3): 129-149.

Lang, C. E. & Bastian, A. J. (2002). Cerebellar damage impairs automaticity of a recently practiced movement. *J. Neurophysiology*, 87 (3): 1336-1347.

Logan, G. D. (1988). Toward an instance theory of automatization.*Psychological Review, 95*,492-527.

Logan, G. D., & Etherton, J. L. (1994). What is learned during automatization? The role of attention in constructing an instance. *Journal of Experimental Psychology: Learning, Memory, and Cognition, 20*, 1022-1050.

Logan, G.D., Taylor, S.E. & Etherton, J.L. (1996). Attention in the Acquisition and Expression of Automaticity. *Journal of Experimental Psychology: Learning, Memory, and Cognition.* 22(3): 620-638.

Marcel, A. T. (1983). Conscious and unconscious perception: An approach to the relations between phenomenal experience and perceptual processes. *Cognitive Psychology, 15*,238-300.

Milham, L., Hale, K., Stanney, K., Cohn, J., Darken, R., & Sullivan, J. (2004). When is VE training effective? A framework and two case studies. *Proceedings of the Human Factors and Ergonomics Society 48th Annual Meeting.* New Orleans, LA.

Morris, C.S. & Tarr, R.W. (2002) Templates for selecting PC-based synthetic environments for application to human performance enhancement and training; Virtual Reality, 2002. Proceedings. IEEE 24-28 March p. 109 – 115.

Munro, A., Breaux, R., Patrey, J., & Sheldon, E. (2002). Cognitive aspects of virtual environment design. In K.M. Stanney (Ed), Handbook of virtual environments: Design, implementation, and applications (pp. 415-434). Mahwah: NJ: Lawrence Erlbaum Associates. (pp. 415-434).

Okamoto M., Dan H., Shimizu K., Takeo K., Amita T., Oda I., Konishi I., Sakamoto K., Isobe S., Suzuki T., Kohyama K., & Dan I. (2004) Multimodal assessment of cortical activation during apple peeling by NIRS and fMRI. *Neuroimag,*. 21(4):1275-88.

Palmeri, T. J. (1997) Exemplar Similarity and the Development of Automaticity. *Journal of Experimental Psychology: Learning, Memory, and Cognition.* 23(2): 324-354.

Pashler, H.E. (1998). The psychology of attention. Cambridge MA: MIT Press

Peres M., Van De Moortele P.F., Pierard C., Lehericy S., Satabin P., Le Bihan D., & Guezennec C.Y. (2000). Functional magnetic resonance imaging of mental strategy in a simulated aviation performance task. *Aviat Space Environ Med.* 71(12):1218-31.

Posner, M. I., & Snyder, C. R. R. (1975). Attention and cognitive control. In R. L. Solso (Ed.), *Information processing and cognition: The Loyola symposium* (pp. 55-85). Hillsdale, NJ: Erlbaum.

Romano, D. M., & Brna, P. (2001). Presence and reflection in training: Support for learning to improve quality decision-making skills under time limitations. *CyberPsychology & Behavior*, 4 (2): 265-277.

Roscoe, S. N. (1982) Aviation Psychology. Ames, IA: Iowa State University Press.

Schneider, W., & Shiffrin, R. M. (1977). Controlled and automatic human information processing: 1. Detection, search and attention. *Psychological Review, 84,* 1-66.

Serfaty, D., MacMillan, J., Entin, E. E., & Entin, E. B. (1997). The decision-making expertise of battle commanders. In C. E. Zsambok and G. Klein (Eds.), *Naturalistic Decision Making* (pp. 233-246). Mahweh, NJ: Erlbaum.

Shadmehr, R. &, Holcomb, H. H. (1997). Neural correlates of motor memory consolidation. *Science,* 277 (5327): 821-825.

Smith M.E., McEvoy L.K., & Gevins A. (1999). Neurophysiological indices of strategy development and skill acquisition. *Cogn Brain Res.* 7(3):389-404.

Stanton, P.K., & Sejnowski, T.J. (1989). Associative long term depression in the hippocampus induced by hebbian covariance. *Nature*: 339: 215-217.

Stoffregen, T., Bardy, B. G., Smart, L. J., & Pagulayan, R. (2003) On the nature and evaluation of fidelity in virtual environments. In L.J. Hettinger & M.W. Haas (Eds.), *Virtual and adaptive environments: Applications, implications, and human performance issues.* (pp. 111-128). Mahwah, NJ: Lawrence Erlbaum Associates, Publishers.

Stone, R. (2002). Applications of virtual environments: An overview. In K.M. Stanney (Ed.), *Handbook of virtual environments: Design, implementation, and applications.* (pp. 827-856). Mahwah: NJ: Lawrence Erlbaum Associates.

Tracy J., Flanders A., Madi S., Laskas J., Stoddard E., Pyrros A., Natale P., & DelVecchio N. (2003). Regional brain activation associated with different performance patterns during learning of a complex motor skill. *Cereb Cortex.* 13(9):904-10.

Zanone, P.G., & Kelso, J.A.S. (1992). Evolution of behavioral attractors with learning: nonequilibrium phase transitions. *J. Exp. Psychology: Hum Perc. Perf.* 18(2): 403-421.

Zanone, P.G., & Kelso, J.A.S. (1997). Coordination dynamics of learning and transfer: collective and component levels. *J. Exp. Psychology: Hum. Perc. Perf.* 23(5): 1454-1480.

Will Augmented Cognition Improve Training Results?

D.R. DuRousseau, M.A. Mannucci and J.P. Stanley

Human Bionics LLC
190 N. 21st Street, Suite 300
Purcellville, VA 20132
don@humanbionics.com

Abstract

Behavioral Researchers have expressed considerable interest in assessing an individual's higher order cognitive functions while in the process of learning (Rabinowitz et al., 2001). Much of this focus has been on developing ways to monitor one's level of arousal, attention, and working memory so an intelligent system could account for changes in cognitive state and adapt the flow of information to optimize performance (Trejo, 1986; Gevins and DuRousseau, 1994ab; Gevins and Smith, 2000). Our cause for concern is that without a measure of the cognitive resources a person needs for requisite performance, limitations exist in our ability to close the loop in automated training technologies. Thus, future training systems would be expected to ubiquitously measure physiological changes from a student's normal baselines and respond accordingly when fluctuations in brain and body predict deteriorating performance or diminished ability to process information.

To be successful, next generation training technologies must be able to identify and react to changes in physiological networks that underlie the coordination of cognitive resources during learning and task execution. Additionally, teaching technologies need to adapt their methods of interaction with the student throughout a training session fraught with high variability in arousal level, attentional capacity, and executive workload. By adopting what has been learned from the field of augmented cognition, providing a methodology for assessing patterns from the brain and body, it may be possible for training system designers to create computer-adaptive teaching systems (CATS) that individually fit the presentation of multimedia content by altering information throughput in response to changes in the human-system.

The tuning of information delivery would be done by manipulating the form, timing, and content of an individual's training protocol based on real-time monitoring of physiological networks and feed-forward influence from predictive heuristic algorithms. If done properly, the outcome of such novel improvements in training technologies would deliver efficient and cost effective tools capable of improving acquisition and retention of knowledge (Fletcher, 2001). Of course, CATS would be capable of operation over a wireless distributed network delivering access to augmented learning to nearly everyone, anywhere and anytime.

1. A Cognitive Assessment Method

1.1 Pilot Study

This article provides an overview of recent efforts by Human Bionics to address some of the problems that lie ahead as we begin to design augmented teaching and training systems. In an initial experiment on the campus of Virginia Tech, 32-channels of continuous EEG, eye, and heart activity were recorded from 23 healthy undergraduate students during a two and a half hour computer-based general knowledge assessment task. Subjects were instructed to respond to questions by selecting an answer from a multiple-choice list of answers and selecting a corresponding button with the mouse pointer. When satisfied with their selection, subjects had to press a "Submit answer" button. They were told they could change their selections at any time prior to pressing the Submit answer button, and each question had to be answered to move on. Figure 1 shows a screen from the computer-based test with the question area in the upper half of the screen. Other resource information required to answer questions was presented below as text, pictures, maps, graphs, and audio/video recordings. The media format of the information determined the quadrant where the additional information was presented, so performing the test was much like taking an automated Scholastic Achievement Test (SAT).

To compare brain-activity and test-performance, event marks were sampled with the EEG to time: a) the resting period before each question; b) start of each new question; c) subject's response; and d) system's feedback. Feedback, composed of either the word Correct or Incorrect, plus a number equal to the running percent correct was displayed on the screen one-second after each response. EEG data analysis was performed around the response events in an attempt to identify brain activation patterns correlated with performance, and to evaluate the feasibility of developing real-time methods capable of predicting cognitive-system changes that were detrimental to performance.

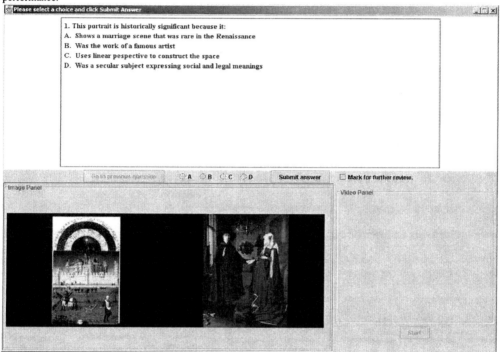

Figure 1

The view above is a screen capture from the general knowledge assessment task used in the study. The picture shows the areas each subject had to navigate to read questions, acquire multimedia information, and submit responses. The test question was displayed in the upper half of the screen and information needed to answer the question was presented in the lower half. The response buttons were located at the top of the shaded region along with the Submit answer button. Depending on media type (e.g., text, graphic, audio, or video) the information was presented either in the area marked Image Panel or Video Panel. If an audio or video recording were used, the subject would press the Start button to begin the recording.

From the general knowledge assessment test, pre-response event-related potentials (ERP) were measured to identify spatial-frequency (SF) pattern differences within a distributed oscillatory neuroelectric network associated with the executive control of attentional and cognitive resources during a complex decision-making task (Gevins et al., 1994). Results indicated that SF pattern characteristics in the pre-response period were predictive of subsequently correct and incorrect answers (DuRousseau et al., 2005). Both the EPs and SF pattern maps measured within-subjects just prior to accurate responses differed dramatically from maps preceding inaccurate responses. Figure 2 shows within-subject averaged spatial power distributions (from 3 to 12 Hz) overlying frontal, central, and parietal cortices one-second before a decision was made. Moment-to-moment calculations of the peak frequency distributions in the EEG were differentially predictive of correct vs. incorrect responses. By examining pre- and post-response EEG from the early and late portions of the same test, peak SF patterns were also identified that were strongly correlated with changes in arousal level, particularly in the 5, 9, and 10 Hz bands (Figure 3). Although, these arousal measures were not significant, predictable improvements in cognitive assessment technologies would drive their acceptance and use in advanced human-system networks. Thus, we are confident that advancements such as these would lead to an improved understanding of cognitive system dynamics during complex decision-making, ultimately leading to the development of fully adaptive training systems capable of teaching more efficiently and at lower cost.

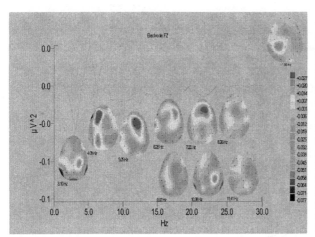

Figure 2

These results show single-subject differences in pre-response averages from accurate and inaccurate trials, where each point on the accurate spectral waveform was subtracted from the corresponding point on the inaccurate waveform. The topographic SF maps identify scalp locations where the peak differences between 3 and 12 Hz were greatest. The GFP colorbar shows the amplitude scale in uV^2, with increased power at Red sites and decreased power at Blue EEG sites. Related to performance, there was significantly increased peak power in anterior-frontal, frontal and fronto-central, and left and right parietal cortices, with decreased peak power in left and right inferior posterior-parietal cortices.

Current efforts at Human Bionics are focused on the design of adaptive neural networks to provide better and faster pattern fitting routines and highly specific physio-classification algorithms. We are working to produce real-time SF algorithms constructed to track spatially distributed neural patterns in cortical networks related to arousal, attention, and the executive control of cognitive resources. If our efforts are successful, we envision a host of human-system interface tools to provide moment-to-moment assessment of, for instance, an individual's stress level, attentional capacity, and cognitive load. Then, CATS technologies would be built with the ability to predict and react to detrimental changes in these levels characteristic of decreased performance (Paulus et al., 2002). For example, in the reactionary case, if a student's arousal level was dropping and attentional indexes indicated malingering or cognitive lapse, a CATS would adapt to the state of the student by pausing the training session and redirecting her attention to a pleasant experience or scene. One way to perturb the student's cognitive systems back to the positive might be through the immediate introduction of a video clip with a favorite artist playing a song or comedian telling jokes. At this time, more work is needed to develop such real-time CATS technologies with predictive cognitive assessment abilities capable of seamlessly interfacing future training system to their human counterparts.

Figure 3

These pre-response group average differences between early and late trials show a much different topography than in Figure 1. In these data, each point on the early spectral waveform was subtracted from the corresponding point on the late waveform. The SF maps identify sites where the peak differences in 3 to 12 Hz amplitudes were greatest. Related to declining arousal, there was significantly increased peak power in focused fronto-central and left parietal cortices and decreased peak power in right anterior-frontal and frontal sites, as well as fronto-central and posterior-parietal cortices.

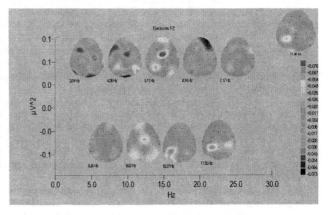

1.2 Further Analysis

Fourteen subjects performed an experiment composed of alternated Primary and Secondary taskload components repeated under 2 different conditions; Mitigated and Unmitigated (2 runs per condition). During the Mitigated condition, gauges tracking arousal level, task engagement, and executive workload were monitored continuously and used to trigger an augmented communications scheduler when pre-defined gauge thresholds were met. In the Unmitigated condition, gauge data were recorded but communications to the subject, which included task assignments, were not interrupted or rescheduled. For each condition, the outputs from an eXecutive Load Index (XLI) gauge (developed by Human Bionics to track changes in workload) were averaged over the Primary and Secondary taskload intervals and compared to test the gauge's ability to differentiate SF power in the EEG by task difficulty level. To normalize cross-subject differences in XLI results, we calculated the absolute differences between the XLI values from the Primary vs. Secondary taskload periods across all fourteen subjects

The averaged XLI outputs were examined from 12-minute long Mitigated and Unmitigated sessions producing fifty-two (52) experimental trials (14 subjects unmitigated, 12 mitigated, 2 sessions each). Our goal was to measure the effectiveness of the XLI gauge over each 12-minute trial at differentially classifying the Primary and Secondary workload intervals. The averaged XLI gauge values indicated differences between the Primary (High) and Secondary (Low) workload conditions in thirteen of fourteen subjects under both conditions. However, when the absolute differences between conditions were measured, the Unmitigated case showed significant differences between the Primary and Secondary workload intervals in eight of fourteen subjects (64%) from the XLI gauge values averaged over 6-minutes per condition (p=0.05, n=28). However, the averaged XLI output was not able to demonstrate a significant difference in taskload under the Mitigated condition (not shown). This result makes perfect sense given that the mitigation strategy performed as predicted and was effective at redistributing the taskload from the Primary to Secondary task periods. In this case, by holding off the delivery of low priority information until the XLI and other gauges indicated sufficient cognitive capacity to uptake and comprehend a message (i.e., during the Secondary task period), Honeywell's scheduler was effective at improving Primary task performance by as much as 500% across both conditions. However, the average improvement in Primary task performance for all fourteen subjects during both high and low workload conditions averaged 26%.

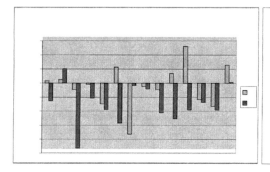 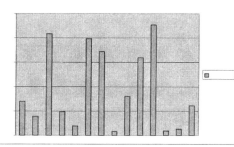

Figure 4a	**Figure 4b**
The bars represents the average XLI gauge output for each subject during the Primary (A) and Secondary (B) workload intervals of a continuous performance cognitive task.	The absolute difference of the XLI averages from 4a. These were used to normalize the data with respect to the polarity of the XLI gauge outputs across subjects. Here, significance was assumed at an absolute difference greater than or equal to 0.05.

Figure 4a shows XLI gauge data averaged over the Primary (high-load) and Secondary (low-load) intervals from all 14 subjects while performing a cognitive task in a VE that consisted of hard and easy workload periods. During the task, subjects were virtually located on a rooftop overlooking buildings on each side and had to identify friendly and enemy soldiers in the structures. Under the high-load conditions (four-1½ minute periods per trial), subjects also had to count friendly soldiers, shoot enemy soldiers, and maintain a count of how many bullets were used. During both high and low conditions, subjects received directions where to look for targets via auditory communications.

For the low-load condition, subjects did not see any new targets (enemy or friendly) and had to report their current counts of friendlies, enemies, and ammunition as well as listen for messages with their call sign. The significance of the absolute differences were analyzed using ANOVA to determine the XLI's ability to correctly differentiate Primary from Secondary cognitive task conditions during the complex multimodal VE simulation.

In addition to the XLI gauge analysis, we examined the effectiveness of a simple prototype feed-forward neural network classifier at distinguishing XLI gauge patterns timed to a single discreet-time event when the subject switched from Primary to Secondary task and back. Here, we report results using a real-time neuroclassifier tuned to each individual to identify cognitive task shedding and reacquisition events across subjects. In the past, classification through linear deterministic tools was possible as a first approximation; however, today there are more appropriate non-linear methodologies available for feature selection and pattern recognition. Thus, our approach has been to develop a real-time neural network architecture to manage the highly stochastic EEG phenomena. For this case, an Neural-Classifier Algorithm (NCA) module was configured as a probabilistic feed-forward network to classify the XLI gauge outputs into three classes: 0, which meant no change; 1, which meant a transition to a higher workload state; and -1, which meant a transition to a lower workload state. In real-time operation, the output of the NCA would be returned to the main system architecture to influence mitigation strategies rather than just the XLI values alone.

Training of the NCA, using a 40-second window centered about a critical event (i.e., the high-low workload transition event), was required for each subject and condition prior to classification. In all cases, the first half of the XLI gauge data was used for training and the remaining half for classification. Figure 5 shows the results from one session, where the first half of the data (3 events) was used to create pattern templates and train the NCA to classify workload transition events in the remaining data. In the figure below, the bottom half shows the events generated from the closed-loop system that marked the workload transitions from high to low and visa versa. The graph on top shows that the classifications during the training period were all correct (Left side without box) when compared with the event marks below. Results of the NCA classifications can be seen in the boxed area on the Right of the top graph. Here, based on only 3 prior events, the NCA correctly identified (within 1 XLI sample) the occurrence of 3 transition events, missed 2 events, and added 2 events that didn't exist. Surprisingly, these are quite good results given such a small training set of examples, indicating that individual XLI gauge values about the transition events are repeatable and stable over 2-second intervals.

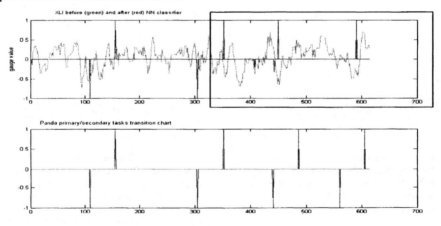

Figure 5
The top figure shows XLI gauge outputs over a 12-minute trial where workload conditions varied between 2 difficulty levels (high & low). The period of each difficulty level was 1-1/2 minutes long and there were 4 periods of each level throughout all trials (total of 6-minutes of data for each workload condition). The bottom figure shows the transition event marks generated by the VE task system.

Preliminary tests of fourteen male and female subjects using our NCA algorithm to classify XLI gauge outputs demonstrated moment-to-moment within-subject performance as high as 72% (mean=64%, std dev=6%), under both Mitigated and Unmitigated conditions over 52 experimental sessions (Figures 6 & 7). The goal of the investigation was to test the real-time application of the XLI as input to the NCA algorithms for neurometric classifications. These preliminary results from our investigation of a prototype gauge that combines the XLI and NCA to identify an individual's point of cognitive transition from a high workload state to a lower state, or, the other way around. These results, measured during normal test performance (unmitigated) and when communications were deferred (mitigated), demonstrate that repetitive SF patterns exist in the brain that are associated with higher-order cognitive processes during performance of a routine operational task. Thus, through the application of a moment-to-moment cognitive index that measures a small band of power differences from a few EEG sensors and a novel feed-forward neural network, real-time SF patterns can now be used for identifying cognitive changes in the human-system. By improving the accuracy and sensitivity of these methods, new augmented human-machine systems will soon become prevalent throughout our society.

2. Applications to Learning

In classroom education as well as in computer-based learning environments, multimedia content is presented to a student in an interactive manner in which the student perceives the information (either correctly or incorrectly) and is tested on the outpouring of that knowledge. In responding, the student formulates a solution and submits an answer through the interaction of higher-order cognitive processes coordinated by an executive-system, which in turn is influenced by the perceptions of the student (i.e., is the question easy or hard, is one's confidence in the answer high or low). Therefore, during complex training and testing tasks, mental processes needed for arousal, attentional focus, working memory and response delivery must be synchronized to achieve optimum performance. Our belief is that the brain coordinates these mental processes through rhythmic oscillations among specialized cortical regions suited to performing a particular perceptual task, or, several networked regions in carrying out complex intended actions (Freeman and Skarda, 1985; Schantze & Eckhorn., 1997).

Preliminary research into the relationships of spatial-frequency dynamics to higher-order cognition, has uncovered signature patterns in the EEG that may be related to one's confidence in an answer prior to submitting it. Given, that such a confidence measure were possible, then interactive learning systems could be developed to manipulate the flow and type of information sent to the student. Such systems would be able to control the timing, content, and style of information presented based on the cognitive state of the student. For instance, if during a math lesson a student's confidence patterns were low, a CATS might present the subject matter in an alternative form (e.g., pictures vs. text), provide hints and background information needed to understand the material better, or completely change the content to a remedial level for review. The benefit of a real-time confidence index would be the creation of new educational tools to enhance human learning potential leading to substantial gains in academic achievement.

Hence, having a real-time indication of student confidence would be a benefit in the development of the next-generation of CATS capable of adapting to a student's varying cognitive state. More important than just designing improved educational technologies would be that access to accurate real-time predictors of declining performance would allow future training system developers to create adaptive human-machine systems capable of measuring when a student was paying attention and when they weren't. Thus, EEG measures may be an important factor in determining how well a student comprehends new information, thereby allowing CATS technologies to react to the student's cognitive-system by manipulating programmed interactions to achieve optimal human-system performance. For instance, a CATS could be developed to respond to a student's cognitive-system in a manner similar to that of a teacher in a classroom. In essence, allowing the software to change a question (making it easier or harder), highlight particular clues or solution paths, or manipulate the style and content of information based on whether the student's confidence in a response was high or low.

A typical classroom analogy would be a teacher reacting to inattentive students by stopping to tell a joke or talking briefly about a completely different topic. For either software or human teacher, the goals would be to present information the student could readily comprehend, direct or redirect the student's attention in the most efficient manner when needed, and augment the uptake of training materials when optimum stress, arousal, and cognitive levels exist. Therefore, with a real-time confidence index in hand, we envision new tools capable of enhancing human learning and comprehension skills, ultimately, leading to substantial gains in academic and corporate achievement at lower cost (Fletcher, 1997).

961

3. Future Directions for Education Research

3.1 Computer-Adaptive Teaching Systems (CATS)

Future CATS technologies would control the presentation of multimedia information content composed of text, pictures, vibrations, and sounds time-locked to the collection of physiological and behavioral responses and system events. The existence of convenient and flexible computerized education and training tools capable of closed-loop cognitive augmentation would accelerate the progress of education research, improve learning results, and reduce educational costs overall. Here, we've described EEG measurement and analysis techniques as a means of advancing our understanding of the human-system and enhancing human learning capacity at the system level. These descriptions lead one to envision Educational Software tools for researchers and training system designers to build reusable and extensible CATS constructed on a distributed Web-based architecture accessible to nearly every student. With such tools, anyone would be able to self-pace the uptake of training materials at times when they were highly vigilant. At other times, CATS might redirect attentional resources such as when drowsy, inattentive, or bored. In these cases, CATS would guide the student's attention back to the task at hand by providing multisensory cues or manipulating the auditory, visual, and tactile content of the materials delivered. Such innovative technologies would only be possible through clever integration of neurometrics, distributed learning, and Web Services technologies to ensure the ethical access and global application of the next generation's augmented training systems.

To get beyond plugging one's self into a computer; several researchers are developing ubiquitous sensor systems capable of instantly measuring brain, eye, heart, and body signals from clothing, furniture, and appliances. With these fast and inexpensive sensors that provide good quality data and leave no residue, a major stumbling block would be cleared in the operational deployment of cognitive assessment technologies. Companies like Human Bionics are developing miniaturized monitoring systems and Web-enabled software tools to integrate measures of stress, arousal, and cognitive load with multimedia information delivery systems used for education and training purposes. Today, training software development companies are beginning to assist researchers and test designers in developing their own institutional CATS as they have finally become the *de facto* standard replacing conventional paper-and-pencil testing methods. Their intention is to make it easier for developers to create easily extensible CATS using software templates rather than having to program everything from scratch. Soon, educators themselves will be designing the most comprehensive set of teaching and testing tools using existing technologies, like the CATSoftware System™ and CAT Builder™ session design and research authoring tools, made by Computer Adaptive Technologies, Chicago, IL, as well as the FastTEST Professional™ and C-Quest™ adaptive session design, administration, and analysis tools from Assessment Systems Corporation, St. Paul, MN. Thus, by seamlessly integrating new cognitive assessment technologies with commercial testing and training design shells, will it be possible to dispense 21st century training technologies that include the human side of the human-computer interface? We believe the answer to this question is yes!

4. Conclusion

The focus of this discussion has been on examining ways to monitor a student's cognitive-system with regard to arousal, stress, and cognitive capacity so intelligent training software could adapt its operation in response to variability in his processing levels. We have examined a suite of real-time algorithms that utilize neurometric cognitive and arousal related methods to derive indices of mental variability that may be used to self-pace learning and enhance student performance. Additionally, we have touched on the topic of CATS development tools that lets educators integrate measures of mental effort with off-the-shelf training design integrated with new human-system augmentation methods. However, limitations exist in our ability to deliver augmented training technologies until we are able to ubiquitously measure changes in physiological networks underlying the coordination of higher-order cognition. The problems associated with entering the education and training markets with new technologies are high, but not insurmountable. Successfully delivering CATS that individually fit the presentation of multimedia content by reacting to individual changes in the human-system is the goal. Optimizing the throughput of training materials over a few hours of a student's continuously varying levels of arousal, stress, motivation, and attentional capabilities is the problem.

Acknowledgement

Human Bionics recognizes DARPA, Virginia Tech, James Madison University, and particularly, Dr. Helen J. Crawford, in support of this R&D effort. This work was supported by contract number W31P4QCR223 through the Information Processing Technology Office and the U.S. Army. This acknowledgement should in no way reflect the position or policy of the U.S. Government and no official endorsement is inferred.

References

DuRousseau, D.R., Stanley, J.P., & Mannucci, M.A. (2005, In Press) Spatial-Frequency Networks of Cognition, *Human-Computer Interface (HCI) International Conference Proceedings on Neural Correlates of Cognitive State*, Lawrence Erlbaum Associates, Inc (LEA)

Fletcher, J.D. (1997) What have we learned about computer based instruction in military training? In R.J. Seidel and P.R. Chatelier (Eds.), Virtual Reality, Training's Future? (pp. 169-177), New York, NY: Plenum Publishing.

Freeman, W.J., & Skarda, C.A., (1985) Spatial EEG patterns, non-linear dynamics and perception: the neo-Sherringtonian view, *Brain Research Review*. Vol. 10 pp. 147-75.

Gevins, A.S., Alexander, J., Cutillo, B., Desmond, J., DuRousseau, D.R., et al. (1994) Imaging the spatiotemporal dynamics of cognition with high resolution evoked potential methods. *Human Brain Mapping*, vol. 1(2), John Wileys: New York, pp 101-16.

Gevins, A.S. & DuRousseau, D.R. (1994a) Biopsychometric signal acquisition and processing system. Final Technical Report, NPRDC N66001-94-C-7012.

Gevins, A.S., & DuRousseau, D.R. (1994b) Spacecrew testing and recording system. Final Report, NASA, NAS 9-19054

Gevins, A. & Smith, M. (2000) Neurophysiological measures of working memory and individual differences in cognitive ability and cognitive style, *Cerebral Cortex, 10(9):829-39.*

Patton and Mussa-Ivaldi (2004) Robot-Assisted Adaptive Training: Custom Force Fields for Teaching Movement Patterns, IEEE *Transactions on Biomedical Engineering* 51(4): 636-646 TBME-00072-2002.R2

Paulus M.P., Hozack N., Frank L., Brown G.G., (2002) Error rate and outcome predictability affect neural activation in prefrontal cortex and anterior cingulate during decision-making. *Neuroimaging* 15(4):836-46

Rabinowitz, M., Blumberg, F.C. and Everson, H. (In Press), The Impact of Media and Technology on Instruction. (Eds) Mahwah, NJ: Lawrence Erlbaum Associates.

Schanze T. and Eckhorn R., (1997) Phase correlation among rhythms present at different frequencies: spectral methods, application to microelectrode recordings from visual cortex and functional implications.

Trejo, L.J., (1986) Brain activity during tactical decision-making: I. Hypotheses and experimental design, HFOSL Tech Note 71-86-6, Navy Personnel Research and Development Center.

Augmented Tutoring: Enhancing Simulation Based Training through Model Tracing and Real-Time Neurophysiological Sensing

Santosh Mathan, Michael Dorneich

Honeywell Laboratories
3660 Technology Dr, Minneapolis, MN 55418
santosh.mathan@honeywell.com

Abstract

Military training simulations provide rich and engaging environments for personnel to develop and maintain mission critical knowledge and skills. However, these systems generally lack the capacity to diagnose student difficulties in real-time and provide automatic, context specific assistance to learners. The lack of pedagogical diagnosis and guidance has several implications for the effectiveness and efficiency of the training process. Without feedback, students can take problem solving paths that deviate far from the solution. Besides introducing inefficiencies, the lack of appropriate guidance can contribute to unproductive floundering and induce confusion and frustration among students. In this paper we describe a set of diagnostic technologies that could provide the basis for inferring both a student's immediate problem solving context and underlying cognitive state. Dynamic guidance within simulation environments based on such a comprehensive assessment of student state could serve to raise both the efficiency and effectiveness of military training simulations.

1   Introduction

Computer based military simulations have emerged as cost effective and engaging mediums for military personnel to develop and maintain operational proficiency. They range in sophistication from embedded training systems incorporated within deployed equipment (e.g. Aegis Combat Training System), to computer games designed with guidance from military domain experts. These systems allow students to practice tasks in environments that bear a high degree of fidelity with real world task contexts. Unfortunately, a key element necessary for achieving the promise of effective ubiquitous training is largely missing in most military training simulations. These systems typically lack the ability to scrutinize student performance and provide context specific guidance to learners in real-time. This could contribute to sub-optimal learning outcomes.

One strategy for assessing student expertise within training simulations is to examine success at accomplishing mission objectives. Unfortunately, global performance outcomes only provide a coarse indication of a student's competence. In order to maximize training effectiveness, it is critical that the training environment assess performance at a finer grain size. In complex task contexts, a broad range of strategies and tactics combine to contribute to the overall performance outcome. For example, consider the case of a unit leader practicing scenarios within a tactical decision simulator. The objective of a particular scenario might be to lead an assault on a building in an urban environment. Besides assessing the ultimate outcome, it may be critical to determine whether the unit leader appropriately coordinated fires — whether a long covered approach selected over a short open route — whether obscuration with smoke was invoked by the student when relevant — whether bounding overwatch maneuvers used when appropriate. Poor performance on one or more of these sub tasks could compromise battle field effectiveness. However, without active monitoring of performance of tactics and strategies that contribute to mission outcomes, many performance lapses may go undetected — potentially denying the trainee an opportunity to acquire and reinforce critical decision making knowledge and skills.

In order to overcome the shortcomings associated with the lack of fine-grained diagnosis and feedback in training simulations, the military has relied on human observers. But such an approach has limitations too. These are highlighted in a study conducted in the context of the Aegis Combat Training System, a ship-based embedded training system (Zachary, Cannon-Bowers, Bilazarian, Krecker, Lardieri, & Burns, 1999). While the ACTS system presents impressive high fidelity simulation capabilities, the lack of functionality to examine student actions in real-time and provide timely remedial feedback, presents numerous difficulties. As Zachary and colleagues have noted, every student in the ACTS environment requires an experienced crew member watching over his or her shoulder to identify difficulties and provide feedback. Unfortunately, these observers are often not training experts. Researchers noted inconsistencies in the quality of help they were able to provide students. Additionally, it was difficult for human observers to analyze and provide feedback on the hundreds of actions that a student might perform in the highly dynamic, simulation environment. Furthermore, the 1:1 student to instructor ratio that Zachary and colleagues noted may be unfeasible in many operational training contexts.

While automated pedagogical diagnosis can raise the efficiency and effectiveness of training simulations, a broad range of technical challenges limit their widespread use. Some of the technical challenges include:

- *Inferring Problem State*: Students interact with dynamic simulation elements in a relatively unconstrained fashion. Learners are capable of employing a broad range of strategies in solving complex tasks. As a consequence, identifying a learner's immediate problem solving state with respect to the problem solving goal is often difficult. The system has to recognize a range of strategies and assess them in terms of their overall effectiveness and efficiency. Additionally, the system must incorporate sufficient knowledge of the domain to provide appropriate context specific feedback.

- *Assessing Evolving Student Knowledge:* In order to optimize the pace at which students progress through a training sequence, it is important that the system be able to assess a student's competence on a potentially large set of knowledge and skill components that may be leveraged to solve problems. Accurate characterization of evolving student knowledge and skill would allow the system to select problems that target specific weaknesses, instead of stepping all students through a canned sequence of problems. Keeping track of a students evolving competence on the broad range of knowledge and skill components necessary to solve problems is a difficult technical challenge

- *Assessing Underlying Cognitive State:* Research suggests that aspects of a student's underlying cognitive state, such as working memory capacity, and attention, have a direct impact on the ability of students to learn. However, assessing these states and tailoring feedback based on these assessments is a difficult challenge and practically never done in the context of computer based learning systems.

This paper presents two powerful technologies — ACT-R based cognitive tutors and non-invasive neurophysiological sensing — that could provide the basis for addressing the technical challenges outlined above. A simulation environment embodying these features would dynamically guide students toward expertise — with feedback based on both an assessment of overt problem solving actions and parameters such as working memory load, attention, and cognitive arousal that have an impact on learning outcomes.

In the following pages we elaborate on each of the technical challenges mentioned above and discuss technical solutions to address each of these problems.

2   Technical Challenges and Promising Solutions

2.1   Tracking Problem State

In complex, scenario-based simulation environments, students have access to a broad range of problem solving actions (operators) that can be combined to transform a problem from some initial state to a goal state through set of intermediate problem states. In complex task domains the number of actions students have at their disposal can be large. By interacting with elements of the simulation environment, these actions can produce a vast set of possible intermediate problem solving states. Certain sequences of problem states may lead to dead ends, other transition paths to the goal state might be inefficient, whereas one or more possible sequence of state might allow learners to get to the goal state efficiently. Navigating unfamiliar problem spaces can be difficult for learners. Dead ends and inefficient paths to the goal can induce frustration and confusion among students; they can also lead to the acquisition of sub-optimal performance strategies.

2.1.1   Model Tracing

A promising approach for boosting the diagnostic capabilities of military training simulations may be found in ACT-R based cognitive tutor technology (Anderson, Corbett, Koedinger, and Pelletier, 1995). ACT-R is a theory of cognition that describes how humans perceive, think and act (Anderson and Lebiere, 1998). ACT-R is instantiated in the form of a programming language whose primitive constructs embody assumptions about human cognition. It has been used to model human performance in a broad array of complex cognitive tasks, ranging from automobile driving (Salvucci, 2001) and tactical decision making (Anderson, Bothell, Byrne, Douglas, Lebiere & Qin, in press) to algebra and computer programming (Anderson et al., 1995). ACT-R's development has been led by psychologist John Anderson and colleagues at Carnegie Mellon University over the course of three decades of research in cognitive psychology and artificial intelligence.

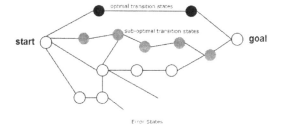

In complex domains, students use a large number of actions or operators to transition from a start state, to a goal state, through a potentially large set of intermediate problem states

Making inferences about where a student might be in a problem space is not a trivial problem.

Figure 0: Challenge of tracking students in complex problem spaces

ACT-R-based cognitive models have had particularly broad impact in the area of computer based tutoring. Through a process called model tracing, cognitive tutors assess student actions as they perform complex cognitive tasks using computers. Model tracing is an instance of a plan recognition algorithm — a class of artificial intelligence programs that deal with the issue of inferring an agent's plans or line of reasoning from its actions (Kautz and Allen, 1986).

The detailed encoding of knowledge embodied in a cognitive model allows the model tracing algorithm to pin-point a student's progress through a problem space. When student actions are consistent with one or more fruitful problem solving strategies, the system remains unobtrusively in the background. However, when student performance is consistent with ineffective or inefficient strategies, the system intervenes with assistance that is tailored to the specific difficulty being faced by a student at a given moment. A student may also ask a tutor for hints that are tailored to a learner's immediate problem solving context.

Model tracing can also serve to simplify after action reviews. Generally, logs generated by simulation environments are difficult for instructors to base instructional feedback on. They commonly express a user's actions in terms of screen coordinates, keystrokes, mouse clicks, and hits taken, to name a few parameters. However, information that is useful from a pedagogical point of view is a trace of a student's thought process in relation to events within simulation scenarios. Since the model tracing process maps a student's actions to cognitive steps expressed in the ACT-R cognitive model, it is possible for the instructor to make inferences about student performance on the basis of a student's thought processes rather than depending on low level actions and events in a simulation alone.

2.2   Assessing Evolving Student Knowledge

Problem solving in complex domains relies on a wide range of knowledge and skill components. Global performance metrics gathered at the end of a task shed little light on a student's evolving competence on the knowledge and skill components that must be leveraged for effective task execution. For example, consider the example of a student trying to master the skill associated with putting a golf ball. The successful execution of task requires competence in areas ranging from the ability to judge the slope of the green, to posture and stroke mechanics. If a student is consistently performing poorly, it is important to be able to identify the specific deficiency that may be contributing to the outcome. This would allow the training process to target particular areas of weakness.

2.2.1 Knowledge Tracing

Many cognitive tutors incorporate functionality known as *knowledge tracing* to optimally pace students through problems and distinguish between slips and errors. Knowledge Tracing relies on Bayesian estimation procedures to estimate a student's strengths and weaknesses relative to the knowledge components in the cognitive model (Corbett and Bhatnagar, 1997). These estimates are dynamically updated as a student is performing tasks. Knowledge tracing estimates are used to pick problems in areas that a student may need most practice on. Additionally, students get estimates of their mastery of various knowledge and skill components via an on-screen bar graph. Unlike many computer based environments, which guide all students through a set sequence of problems, knowledge tracing allows students to work on problems that are appropriate to their competence level. Proficient students can progress quickly to challenging problems, while students who need additional practice get to work on problems that target their particular deficiencies.

2.3 Real world efficacy

ACT-R based cognitive tutors have been rigorously assessed in classroom and laboratory contexts. These systems have been shown to reduce training time by half and increase learning outcomes by a standard deviation or more (Anderson et al., 1993). They have been used to teach concepts ranging from programming to genetics, and represent some of the most broadly used educational systems. ACT-R based tutors for algebra and geometry are in use by 200000 students in over 1800 schools around the country. The US Department of Education has designated cognitive tutors one of 5 exemplary curricula in K-12 mathematics education.

2.4 Assessing Underlying Cognitive State

While replications in numerous domains have shown cognitive tutors to be among the most effective computer based learning platforms, their performance falls short of one-on-one tutoring from highly-skilled human tutors. Researchers have argued that the advantage skilled human tutors have over cognitive tutors may stem from the fine-grain access they have to their student's behaviors. For instance, as Anderson and Gluck (1999) have noted, a human tutor can see frustration on a student's face, hear uncertainty in an utterance, and keep track of how long a student is taking to respond to a problem. Such broad access to a learner's emotive and cognitive state allows the skilled human tutor to display far greater sensitivity and adaptability in the tutorial interaction than a computer based tutor.

While cognitive tutors have conventionally relied on an interpretation of overt behavioral actions to make inferences about students' problem solving progress, it is now feasible to incorporate interpretations of data ranging from eye movements and electroencephalogram (EEG) data to electrocardiogram (EKG) readings to make inferences about a student's cognitive state. Researchers have referred to this expanded diagnostic capacity as *high density sensing* of student state. Computational capabilities on ordinary desktop computers make it possible for the grain-size of tutorial analysis to shift from the 1 to 10 second level of analysis of problem solving actions to the millisecond level of analysis of physiological and neurophysiological states. However, the means by which such a capability can be harnessed to boost tutorial outcomes remains a largely unexplored research area.

The DARPA funded Augmented Cognition program has played a pivotal role in furthering the development of non-invasive physiological and neurophysiological sensing to identify states that could negatively impact human performance. Applications of technology developed under the Augmented Cognition program have primarily focused on ways to aid human task performance. Neurophysiological and physiological sensors are used to detect cognitive bottlenecks and invoke assistance aimed at helping users perform tasks effectively under extremely demanding conditions. Assistance has included strategies such as task offloading, task sharing, modality switching, and task scheduling.

Applications of augmented cognition in the context of the training simulations require a slightly different perspective. The primary objective of using augmented cognition techniques in the context of tutoring applications is not to simplify task performance with adaptive assistance. Rather, neurophysiological sensing could be used optimize instructional efficacy by dynamically tailoring the instructional environment to match the cognitive capacities of a student. AugCog technologies can play a useful role in two distinct phases of the instructional process as detailed below.

2.4.1 Augmented Tutoring

Declarative Instruction

The process of acquiring a novel skill typically begins with a period of declarative instruction when a student learns about the central facts or concepts associated with a domain. In computer based learning environments declarative instruction is typically facilitated through video clips or online textual expositions. Research has shown that unless attention is appropriately directed towards the processing of declarative information, the robustness and accuracy of these facts in memory is likely to be compromised (Anderson, 1993). A poor declarative encoding can impede skill acquisition. Research shows that declarative knowledge of the central concepts in a domain serve to structure early problem solving attempts (Anderson, 1993).

Acquiring a robust and accurate encoding of declarative knowledge requires both sustained attention and active elaboration strategies such as self-explanation. Unfortunately, research indicates that learners have a hard time maintaining attention over long periods of time. For instance, Schooler (in press) observed subjects over the course of a 45 minute reading task. Subjects were interrupted at random and asked if they were still on task. His research revealed that learner's "zoned-out" for close to 20% of the time over reading tasks. Such lapses in attention over the course of reading text and watching video expositions can have a negative impact on the skill acquisition process. Learners may miss critical information that could be of importance in subsequent problem solving efforts.

Research suggests that it may be possible to detect and mitigate low attentional states. Prior efforts aimed at detecting low attention levels using neurophysiological sensors have been fruitful. For instance, Jung, Makeig, Stensmo, and Sejnowski (1997) have shown that EEG spectral power in the 4Hz and 14Hz bands can predict low alertness levels. Additionally, cardiovascular measures such as inter-beat intervals, can serve as an indicator of cognitive arousal. Sympathetic and parasympathetic components of the autonomic nervous system that govern cognitive arousal can be tracked using the Clemson Arousal Meter, a classifier that is an important component of Honeywell's current Augmented Cognition efforts.

Mitigation strategies, triggered by cognitive state classifiers could minimize the negative impact of low attentional states during declarative instruction. For instance, if the low attentional states are detected while a student is reading text online or watching video segments, the system could intervene and step the student through the material with interactive prompts. These prompts could present questions related to concepts just covered and give students the chance to respond using multiple choice responses. Additionally, the system could index text or video segments that may have been encountered by the students during low attentional states. The system could prompt students to revisit these segments at a later time.

Diagnosis during hands-on Practice

Diagnosis based on non invasive neurophysiological and physiological sensing could also play an important role during hands-on practice with simulation platforms. Learning is a working memory intensive process. As Mayer (2001) has suggested, learning involves building connections between incoming materials and existing knowledge. This integration occurs in working memory. Several cognitive theories posit that excessive working memory loads during the learning process can interfere with the acquisition of problem solving schemas (e.g. Sweller, 1998). While the negative consequences of excessive working memory load on learning outcomes are well known, few practical techniques exist to tailor the problem solving process based on a dynamic assessment of a learners working memory capacity.

EEG based neurophysiological sensing could provide a way to assess working memory loads in real-time. Researchers have noted that levels of frontal theta activity increase as a function of working memory load; whereas levels of parietal alpha show a corresponding decrease (Gevins, Smith, McEvoy, Yu, 1997). EEG based classifiers can detect these states and drive mitigations to modulate working memory load within learning environments. Additionally gauges used by the Honeywell AugCog Team to detect high cognitive load (including the EEG based Executive Load Index, and the composite EEG-EKG-pupilometry-based Stress Gauge) could be adapted for use within tutoring environments.

Neurophysiological assessments of working memory may be used to dynamically match working memory demands of a learning environment with a student's working memory capacity. For instance, the level of assistance or scaffolding provided to students following errors during practice could be based on neurophysiological assessments of cognitive state. The grain size of instruction could be dynamically varied based on these assessments (c.f. Anderson et. al., 1995). Students experiencing high levels of cognitive load could be interactively stepped through the series of sub-goals necessary to accomplish a problem solving goal. In contrast, students experiencing lower cognitive load levels could simply be reminded of the overall problem solving goal — maintaining and negotiating the underlying goal stacks in memory could be left to the student. Another strategy that could be used to modulate workload levels might be to adapt the pace of the simulation to the working memory capacity of a student. As a student becomes more proficient at performing tasks, the working memory resources associated with task execution can be expected to diminish.

3   System Implementation

Several elements are necessary for the system to provide feedback based on the joint consideration of cognitive state and problem solving context. First, a simulation environment has to be instrumented to communicate information about simulation objects, events and student actions to the underlying cognitive tutor. Second, an ACT-R cognitive model characterizing the cognitive steps associated with strategies learners are likely to adopt has to be developed for the instructional tasks. Third Information from neurophysiological sensors has to be filtered and classified in terms of cognitive states of interest. Feedback and other mitigations have to be tailored to leverage cognitive state and problem state assessments. Figure 2 shows a possible architecture for an augmented tutor.

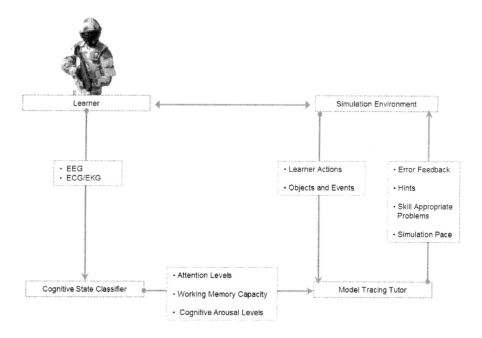

Figure 0: Agmented Tutor Architecture

4 Conclusion

We have described a set of technical solutions — cognitive tutors and neurophysiological sensing — that could combine to improve the efficiency and effectiveness of simulation based training systems. The joint use of these technologies will provide a diagnostic capability of unprecedented scope and will have broad relevance across simulation based training platforms. An instructional system embodying these features could dynamically guide students toward expertise — with feedback based on both an assessment of overt problem solving actions and parameters such as working memory load, attention, and cognitive arousal that have an impact on learning outcomes. These capabilities could combine to transform training simulations from platforms that support unstructured practice to environments that intelligently guide students towards effective and efficient performance.

Acknowledgements

This research was supported by a contract with DARPA and funded through the U.S. Army Natick Soldier Center, under Contract No. DAAD-16-03-C-0054, for which CDR (Sel.) Dylan Schmorrow serves as the Program Manager of the DARPA Augmented Cognition program and Mr. Henry Girolamo is the DARPA Agent. Any opinions, findings, conclusions or recommendations expressed herein are those of the authors and do not necessarily reflect the views of DARPA or the U.S. Army.

References

Anderson, J. R. (1993). Rules of the mind. L. Erlbaum Associates:Hillsdale, N.J.

Anderson, J. R., Corbett, A. T., Koedinger, K., & Pelletier, R. (1995). Cognitive tutors: Lessons learned. The Journal of Learning Sciences, 4, 167-207.

Anderson, J. R., Bothell, D., Byrne M. D., Douglas S., Lebiere, C. & Qin, Y. (in press). An Integrated Theory of the Mind. Psychological Review.

Anderson, J. R., & Gluck, K. (2001). What role do cognitive architectures play in intelligent tutoring systems? In D. Klahr & S. M. Carver (Eds.), Cognition & instruction: twenty-five years of progress (pp. 227–262). Erlbaum.

Anderson, J. R. & Lebiere, C. (1998). The Atomic Components of Thought. Mahwah, NJ: Lawrence Erlbaum Associates.

Bloom, B.S. (1984). The Two Sigma Problem: The Search for Methods of Group Instruction as Effective as One-to-One Tutoring. Educational Researcher, Vol. 13, Nos. 4-6

Byrne, M. D., Kirlik, A., Fleetwood, M. D., Huss, D. G., Kosorukoff, A., Lin, R., & Fick, C. S. (2004, in press). A closed-loop, ACT-R approach to modeling approach and landing with and without synthetic vision system (SVS) technology. To appear in Proceedings of the Human Factors and Ergonomics Society 48th Annual Meeting. Santa Monica, CA: Human Factors and Ergonomics Society.

Corbett, A.T. and Bhatnagar, A. (1997). Student modeling in the ACT Programming Tutor: Adjusting a procedural learning model with declarative knowledge. Proceedings of the Sixth International Conference on User Modeling. New York: Springer-Verlag Wein

Gevins A, Smith ME, McEvoy L, Yu D (1997) High-resolution EEG mapping of cortical activation related to working memory: effects of task difficulty, type of processing, and practice. Cereb Cortex 7:374–385.

Jung, T-P, Makeig, S, Stensmo, M, and Sejnowski, TJ, (1997) "Estimating alertness from the EEG power spectrum," IEEE Transactions on Biomedical Engineering, 44(1), 60-69

Kautz, H. and Allen, J. (1986). Generalized plan recognition. In Proceedings of AAAI-86, pages 32–37

Kulik J.A.& Kulik C.C. " Computer based instruction: What 200 evaluations say?". Centre for Research and Teaching, University of Michigan, 1988

McBreen, B. B, (2003) Close Combat and learning infantry tactics. Retrieved August 8th 2004 from http://www.tecom.usmc.mil/techdiv/ITK/CCM/CCM.htm

Mayer, R.E. (2001). Multimedia Learning, Cambridge: Cambridge University Press

Salvucci, D. D. (2001). Predicting the effects of in-car interface use on driver performance: An integrated model approach. International Journal of Human-Computer Studies, 55,85-107.

Schooler, J.W., Reichle, E.D., & Halpern, D.V. (in press) Zoning-out during reading: Evidence for dissociations between experience and meta-consciousness. In D.T. Levin (Ed), Thinking and Seeing: Visual Metacognition in Adults and Children. MA: MIT Press.

Sweller, J. (1988). Cognitive load during problem solving: Effects of learning. Cognitive Science, 12, 257-285.

United States Marine Core (2004) USMC Family of tactical decision making simulations. Retrieved August 8th 2004 from http://www.onr.navy.mil/sci_tech/special/353_exped/hptetds.asp

Zachary, W., Cannon-Bowers, J., Bilazarian, P., Krecker, D., Lardieri, P., & Burns, J. (1999). The Advanced Embedded Training System (AETS): An intelligent embedded tutoring system for tactical team training. International Journal of Artificial Intelligence in Education, 10, pp. 257-277.

973

Recreation Embedded State Tuning for Optimal Readiness and Effectiveness (RESTORE)

Alan T. Pope & Lawrence J. Prinzel, III

MS 152, D-318
Research & Technology Directorate
NASA Langley Research Center
Hampton, VA 23681
Alan.T.Pope@nasa.gov
Lawrence.J.Prinzel@nasa.gov

Abstract

Physiological self-regulation training is a behavioral medicine intervention that has demonstrated capability to improve psychophysiological coping responses to stressful experiences and to foster optimal behavioral and cognitive performance. Once developed, these psychophysiological skills require regular practice for maintenance. A concomitant benefit of these physiologically monitored practice sessions is the opportunity to track crew psychophysiological responses to the challenges of the practice task in order to detect shifts in adaptability that may foretell performance degradation. Long-duration missions will include crew recreation periods that will afford physiological self-regulation training opportunities. However, to promote adherence to the regimen, the practice experience that occupies their recreation time must be perceived by the crew as engaging and entertaining throughout repeated reinforcement sessions on long-duration missions. NASA biocybernetic technologies and publications have developed a closed-loop concept that involves adjusting or modulating (cybernetic, for governing) a person's task environment based upon a comparison of that person's physiological responses (bio-) with a training or performance criterion. This approach affords the opportunity to deliver physiological self-regulation training in an entertaining and motivating fashion and can also be employed to create a conditioned association between effective performance state and task execution behaviors, while enabling tracking of individuals' psychophysiological status over time in the context of an interactive task challenge. This paper describes the aerospace spin-off technologies in this training application area as well as the current spin-back application of the technologies to long-duration missions - the Recreation Embedded State Tuning for Optimal Readiness and Effectiveness (RESTORE) concept. The RESTORE technology is designed to provide a physiological self-regulation training countermeasure for maintaining and reinforcing cognitive readiness, resilience under psychological stress, and effective mood states in long-duration crews. The technology consists of a system for delivering physiological self-regulation training and for tracking crew central and autonomic nervous system function; the system interface is designed to be experienced as engaging and entertaining throughout repeated training sessions on long-duration missions. Consequently, this self-management technology has threefold capability for recreation, behavioral health problem prophylaxis and remediation, and psychophysiological assay. The RESTORE concept aims to reduce the risk of future manned exploration missions by enhancing the capability of individual crewmembers to self-regulate cognitive states through recreation-embedded training protocols to effectively deal with the psychological toll of long-duration space flight.

1   RESTORE Concept Objective

The RESTORE concept is designed to be a training technology for achieving psychophysiological equanimity under stress that is embedded in individually tailored recreational activities and which will help "restore to health" space travellers in a way that is: (a) intrinsically motivating and rewarding for the space travellers to engage in, (b) targeted to help space travellers enhance their intra-personal skills for dealing with hazardous states of awareness, and (c) provide a physiologically-based monitoring capability for measuring space crew members cognitive state and functioning.

1.1 Relevance to Extended Space Missions

The RESTORE concept uniquely deals with three critical challenges in extended space missions: (1) the problem of deteriorating cognitive skills due to stress, monotony, high workload, and boredom; (2) training and adherence to a physiological self-regulation practice regimen designed to counter the first problem and (3) the need to remotely monitor astronaut psychophysiology and cognitive state for human error vulnerability.

1.1.1 Deteriorating Cognitive and Affective State

Advances in technology have set the stage for long-duration manned missions beyond Earth orbit. A common feature of these expeditions will be extended stays of small groups of humans in space habitats, the success of which depends on the psychological health of those crew members working continuously in confined, isolated, and hazardous environments under changing conditions of boredom and monotony, stress, and high workload.

U.S. astronaut Jerry Linenger, who spent nearly 5 months onboard *Mir*, wrote that he "was astounded at how much I had underestimated the strain of living cut off from the world in an otherworldly environment" (Linenger, 2000, p. 151). Even though he had psychologically prepared for the mission, he experienced profound feelings of confinement and isolation as well as alienation from his crewmates. The Institutes of Medicine (IOM) report, *Safe Passages* (Ball & Evans, 2001), notes that studies of Earth analogues of long-duration spaceflights show an incidence rate of behavioral health problems ranging from 3 – 13 percent per person per year. On-board unobtrusive technologies are needed as astronaut aids for detection and management of these problems. The RESTORE concept aims to reduce the risk of future manned exploration mission by increasing the capability of individual crewmembers to self-regulate cognitive states through recreation-embedded training protocols to restore effectiveness to deal with the significant stress likely to be present during long-duration space flight.

There is compelling evidence to suggest that psychological problems will be encountered by crewmembers of future space explorations. Experience with extended-duration flights longer than 100 days (about 1/10 the anticipated duration of a mission to Mars) have shown that boredom, stress, fatigue, and circadian rhythm and sleep disturbances constitute risk factors that can substantially affect mission safety and reduce human performance effectiveness. The exacting human performance requirements of such missions require that astronauts must be functioning, both cognitively and psychologically, at optimal levels. Otherwise, in the isolated confines of space, mistakes due to lack of alertness or psychological equanimity are compounded as shown by several critical incidences. For example, an exhausted Alexander Lazutkin, a Russian crewmember on Mir 23, mistakenly pulled a vital cable that shut down the Mir's main computer, causing the station to lose power.

The report, "A Strategy for Space Biology and Medical Science" (Space Science Board, National Research Council, 1987), stated that, "it is likely that behavioral and social problems have already occurred during long-term missions and that such problems will be exacerbated as missions become more complex, as mission duration is increased, and as the composition of crews become more heterogeneous. An understanding of the problems and their amelioration is essential if man desires to occupy space for extended periods of time". Santy (1994) further noted that, "thirty years of space flight experience in this country have yielded a gold mine of data and knowledge about the human body and its response to the space environment, but no objective data on the human psyche in that same environment has been produced --- and many scientists consider psychological issues to be the limiting factor in the human exploration of the universe". As Astronaut John Blaha confirms, "Personally, I would not want to repeat a long-duration space flight --- and I didn't have any emergencies. Long-duration orbital duty is much more taxing than the intense, but brief, and heady bursts of five to 14 days orbiting in the space shuttle" (Santy, 1994). A NASA Code W report (NASA Office of Inspections, Administrative Investigations, and Assessments, 1998) stated that there was a "need to improve crew and ground training in recognizing and coping with psychological stressors preflight and during flight". RESTORE is designed to address this specific need through innovative training that will enhance crewmember capabilities to deal with the psychological stressors during extraterrestrial missions.

1.1.2 Training and Adherence

Due to the very real danger and likelihood of psychological and cognitive impairment for space crews on long-duration missions, the *Safe Passages* (2001) IOM report recommended, "developing a technology that will provide an adequate means for assessment of the behavioral health effects of long-duration space flight and that will establish and maintain safe and productive human performance in isolated, confined, and hazardous environments...." The report further recommended that personalized individual training approaches must be incorporated and evaluated as countermeasures based on procedures for evaluation of cognitive state and functioning, that allow for self-assessment and self-management designed within a stress management context which should be combined with biofeedback and relaxation techniques.

A specific recommendation of the *Safe Passages* report was that "the effectiveness of biofeedback and other behavior coping strategies in reducing these patterns [hazardous states of awareness] and their effect on performance should also be explored". Kanas & Manzey (2003) echo these recommendations and stated that, "crewmembers should be taught techniques pertinent to relaxation, meditation, biofeedback, and autogenic training to calm themselves in situations of high workload or tension and to lower anxious arousal by controlling autonomic functions" (p. 149). Such physiological self-regulation techniques have been used with considerable success in teaching astronauts and pilots to control autonomic and cognitive response to stressors under conditions of high workload and stress, and low task engagement and monotony (Cowings & Toscano, 2000; Prinzel, Freeman, & Pope, 2003; Kellar et al., 1993).

The IOM report, *Safe Passages* (Ball & Evans, 2001), noted that, "the role of leisure and recreational activity to combat boredom and maintain fine motor and gross motor skills has not been full evaluated.... relaxation training and leisure activity may promote psychological well-being and improvements in coping strategies, vigilance, and performance" (p. 129). The word "recreation" is derived from the Latin, *recreare*, meaning to "restore to health". The use of recreation time, therefore, can be utilized to help space travellers reinforce coping skills learned during ground-based training to deal with the stressors of long-duration space flight (Figure 1). Furthermore, because space travellers will be required to engage in training for extended periods of time, there is the very likely potential of being disengaged and de-motivated to complete training regimen - a common problem with traditional stress management training.

Figure 1. Leisure activities are an important way for the crew to counter monotony. Training embedded in recreation should be tailored for individual preferences.

RESTORE is tailor designed to be intrinsically rewarding to the individual crewmember to enhance training effectiveness and compliance through embedded recreational activities that would allow a "restoration-to-health" of space travellers psychological and cognitive functioning and reinforce learned coping capabilities for dealing with the stressors of living in outer space. Furthermore, because RESTORE utilizes virtual reality and other digital media to deliver the recreation-embedded training, such training can also be seamlessly embedded within task specific activities through virtual reality training applications. Therefore, the RESTORE training can be reinforced through the delivery of task-specific astronaut tasks which enable conditioning of stress coping responses both to generalized psychological stress, through recreation-embedded training, and to learned targeted responses to specific tasks crew members will have to perform.

1.1.3 Remote Psychophysiological Monitoring

The NASA Research Announcement for Biomedical Research and Countermeasures (NASA Office of Biological and Physical Research, Biomedical Research and Countermeasures Program, 2004) specifically solicited ground-based studies for human health in space "...that will lead to development and validation of predictive tools for the assessment of psychological well-being, cognitive processing, mood, and emotion.... Also of interest are hypothesis-driven ground-based studies that would suggest and evaluate potential proactive techniques or strategies for reducing stress and improving well-being, mood, emotion, and cognitive processing in long-duration crews". The objectives of the RESTORE concept empirically address such research needs through the development of innovative recreation-embedded training technologies for enhancing crew capabilities to deal with the stressors of long-duration space flight. Furthermore, the psychophysiological monitoring capabilities of RESTORE provide ground-based objective techniques and technologies to validly and reliably identify when space travellers are experiencing distress that compromises their performance capabilities in space.

2 Background of RESTORE Concept

On-board unobtrusive technologies are needed as astronaut aids for detection and management of behavioral health problems, for which studies show an incidence rate ranging from 3 – 13 percent per person per year. Physiological self-regulation training is a behavioral medicine intervention that has demonstrated capability to improve psychophysiological coping responses to stressful experiences and to foster optimal behavioral and cognitive performance. Once developed, these psychophysiological skills require regular practice for maintenance. A concomitant benefit of these physiologically monitored practice sessions is the opportunity to track crew psychophysiological responses to the challenges of the practice task in order to detect shifts in adaptability that may foretell performance degradation. Long-duration missions will include crew recreation periods that will afford physiological self-regulation training opportunities. However, to promote adherence to the regimen, the practice experience that occupies their recreation time must be perceived by crew as engaging and entertaining throughout repeated reinforcement sessions on long-duration missions.

NASA biocybernetic technologies and publications have developed a closed-loop concept that involves adjusting or modulating (cybernetic, for governing) a person's task environment based upon a comparison of that person's physiological responses (bio-) with a training or performance criterion. This approach affords the opportunity to deliver physiological self-regulation training in an entertaining and motivating fashion and can also be employed to create a conditioned association between effective performance state and task execution behaviors, while enabling tracking of individuals' psychophysiological status over time in the context of an interactive task challenge.

The closed-loop concept has been implemented at NASA in a number of embodiments: 1) a real-time adaptive automation paradigm (e.g., Prinzel et al., 2000, 2001, 2004); 2) a physiological self-regulation training procedure for improved task engagement (e.g., Prinzel et al., 2002); 3) a neurofeedback system for training attentional state skills in ADHD children (e.g., NASA Office of Aerospace Technology, 2003); 4) a stress counterresponse training method (e.g., Palsson & Pope, 1999); and 5) a training system for enhancing mental state skills for sport performance (Prinzel, et al., NASA LaRC patent case no. LAR-16256). The first of these embodiments is designed to adjust the moment-to-moment human-automation interaction for more effective responsibility allocation, and it was from this research that the second application area emerged. The RESTORE concept technologies derive from the second application area, represented by the remainder of the embodiments, which are designed to train individuals to more readily achieve and maintain effective performance states (attention, mental focus, vigilance under boring conditions, equanimity or sang-froid under stressful conditions.)

2.1 Task Engagement Training

Psychophysiological self-regulation training has shown promise for helping pilots deal with problems associated with the use of automation, such as automation surprises, through the enhancement of cognitive resource management skills. In a NASA study examining the use of psychophysiological self-regulation training in conjunction with adaptive automation of flight management (Prinzel et al., 2002), participants who had received self-regulation training performed significantly better and reported lower mental workload than participants in false feedback and control groups (Figure 2). The physiological self-regulation training procedure used in this study was designed to promote task engagement and attention.

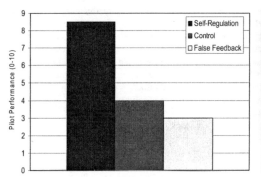

Figure 2. Physiological self-regulation was found to substantially enhance pilot attention and performance.

To promote adherence to a physiological self-regulation practice regimen, the practice experience that occupies their recreation time must be perceived by space crew as engaging and entertaining throughout repeated reinforcement sessions on long-duration missions. NASA has patented and licensed physiological self-regulation training technologies invented by the authors to improve adherence to a physiological self-regulation training regimen by delivering the training through engaging and motivating entertainment technologies (U.S. Patent No. 5377100, 1994; U.S. Patent No. 6450820, 2002; and 6 other invention disclosures). This approach uses physiological signals (e.g., electroencephalogram frequency band ratio) not simply to drive a biofeedback display directly, or periodically modify a task as in other systems, but to continuously modulate parameters (e.g., game character speed and mobility) of a game task in real time while the game task is being performed by other means (e.g., a game controller).

2.2 Neurofeedback Videogame for ADHD

A commercial videogame product based on these inventions (NASA Office of Aerospace Technology, 2003; Kharif, 2004) is currently being employed for the treatment of ADHD in the NASA licensee's network of clinical settings (www.smartbraingames.com). As demonstrated in a study at Eastern Virginia Medical School (EVMS) and in current clinical practice, this recreation-embedded approach has proven effective in overcoming the problems of adherence failure and attrition that plague clinical treatment protocols that require consistent practice. In the EVMS study (Palsson et al., 2001), the video game biofeedback technology produced equivalent results to standard neurofeedback in effects on ADHD problems. Both the video game and standard neurofeedback improved the functioning of children with ADHD substantially in addition to the benefits of medication. However, the video game technology provided advantages over standard neurofeedback treatment in terms of enjoyability for the children and positive parent perception and substantially increased adherence to the practice regimen.

The video game physiological training method is inherently motivating because it blends physiological training into popular entertainment in such subtle ways that the entertainment value is not lost and training is no longer a chore but a treat. The video game or task challenge format motivates trainees to participate in and adhere to the training process through the rewards inherent in mastery of popular video games, and without the demand, monotony or frustration potential of direct concentration on physiological signals. The modulation method employed in this technology explicitly sets up physiological performance criteria, in addition to the usual hand-eye coordination criteria, for success in playing a video game. For producing particular physiological patterns, the player is explicitly rewarded by improved capability and performance in playing the game, that is, production of these physiological patterns is reinforced. In the brainwave study, improved cognitive skills and behavior accompanied these changes. The reinforcement principle involved in this process is known as the Premack prepotent principle, which is stated: "A high probability behavior may be used as the reinforcer for a low probability behavior" (Premack, 1965). A high probability behavior may be understood as that activity in which an individual will engage in the given situation, if unconstrained (for example, video game playing.) The "low probability behavior" in this case is production of target physiological patterns.

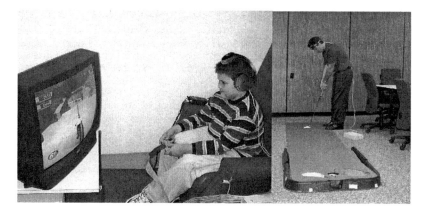

Figure 3. Adaptive physiological training has been shown to enhance the attention of children and adults through a variety of recreational activities, such as playing video games (Left Figure), or playing a sports activity using the ZONE technology (Right Figure).

The RESTORE concept exploits additional NASA inventions to expand the application of the NASA unique physiological modulation technology beyond the licensed product into a variety of recreation devices in order to offer practice choices to trainees (Figure 3).

2.3   Mental State Skill Training for Sports

A third embodiment of the closed-loop modulation concept, the Zeroing Out Negative Effects (ZONE) technology (Prinzel, et al., NASA LaRC patent case no. LAR-16256), integrates physiological self-regulation training into sports practice equipment to enhance mental state skills for sport performance. The ZONE technology allows an athlete to immediately see a physical, mechanical consequence of their physiological responses, and retains a high degree of ecological validity by directly embedding the training in the actual athletic task they will perform. The ZONE technology is designed to help sports performers gain command over "choking," "blocking," and other performance problems that are directly linked to physiological responses of emotions, stress, anxiety, and interfering psychological processes. The ZONE technology is a training method for improving athletes' responses to stress, anxiety and loss of concentration during competition (Figure 3). The technology informs and rewards the trainee for successful attainment of an optimal target state of psychophysiological functioning through real-time changes in the sport practice equipment. These information and reward consequences take various forms, including improved configuration of the task environment (e.g., change of putting surface from moving to still), or improved functioning of a sports implement (e.g., sharper focus of a target sighting device). The technology provides real-time feedback to the athlete about how close their arousal and emotive responses are to an optimal state required to successfully perform the athletic task. The technology makes practicing the optimal mental state and executing the sport movement both part of the same practice challenge by engineering the two tasks into the same practice device.

For the RESTORE concept, the ZONE practice technology is implemented as an extension of the NASA-licensed commercial videogame modulation product. In order to comply with size and weight constraints, changes in the practice task environment in response to the trainee's manipulation of the sports implement are presented in computer graphics rather than in a physical realization. Those personnel who choose to practice their favorite sport in their recreation time will be honing their physiological self-regulation skill at the same time with the ZONE technology.

2.4 Stress Counterresponse Training

A fourth training embodiment of the closed-loop modulation concept, Stress Counterresponse Training (Palsson & Pope, 1999), integrates physiological self-regulation training into the practice of mission-relevant tasks. Stress Counterresponse Training is based upon the concept of Instrument Functionality Feedback (IFF). Using the physiologically-modulated system technology (U.S. Patent No. 6450820, 2002), the method ties the requirement to maintain physiological control to the functionality of a simulator. Instrument Functionality Feedback Training is a concept for training pilots in maintaining the physiological equilibrium suited for optimal cognitive and motor performance under emergency events in an airplane cockpit. Kellar et al. (1993) cite studies to support the assertion that "Reasonable evidence exists to conclude that pilots may lose control of their aircraft as a direct result of reactive stress. The condition in which a high state of physiological arousal is accompanied by a narrowing of the focus of attention can be referred to as autonomous mode behavior. ... A number of studies have produced evidence that this type of training [physiological self-regulation training] effectively reduces arousal which affects operational efficiency in student pilots." IFF training is a conditioning approach for reducing pilot error during demanding or unexpected events in the cockpit by teaching pilots self-regulation of excessive autonomic nervous system (ANS) reactivity during simulated flight tasks. The training method (a) adapts biofeedback methodology to train physiological balance during simulated operation of an airplane, and (b) uses graded impairment of control over the flight task to encourage the pilot to gain mastery over his/her autonomic functions. In Instrument Functionality Feedback Training, pilots are trained to minimize their autonomic deviation from baseline values while, at the same time, operating a flight simulation. This is done by making their skin conductance and hand temperature deviations from baseline impair the functionality of the aircraft controls. Trainees also receive auditory and visual cues about their autonomic deviation, and are instructed to keep these within pre-set limits to retain full control of the aircraft.

Instrument Functionality Feedback, then, means that trainees receive feedback of how their physiology is functioning through changes in the functionality of the equipment that they are operating rather than through graphs and other signal displays, the way physiological self-regulation training is usually presented. Stress Counterresponse Training can be considered a form of counterconditioning (Chance & Lieberman, 2001) in which an adaptive physiological response to a situation or stimulus is learned, through repeated pairing, to replace an original maladaptive, usually arousal, response. The best known form of counterconditioning is systematic desensitization. The approach has much in common with Meichenbaum's (1972) Stress Inoculation Training methodologies that focus on development of cognitive and relaxation coping skills for anxiety reduction, and the use of task-specific stressors that have been shown to improve performance (Meichenbaum & Deffenbacher, 1988).

For the RESTORE concept, Stress Counterresponse Training is implemented by augmenting the NASA-licensed commercial videogame modulation product with autonomic nervous system parameter measurement, or stress measures. In addition to recreational use, the technology permits the use of mission task simulations as games modulated for concurrent physiological self-regulation training. Thus, although the RESTORE concept focuses on a recreational context for the closed-loop modulation method of physiological self-regulation training, the approach is also applicable in a mission-relevant task simulation context.

3   Further Considerations

An important question is addressed in the RESTORE concept because of its theoretical implications for a classically conditioned tie, or association, of state to task stimulus conditions and for its practical implications for transfer of state training to operationally relevant conditions. That question is whether the psychophysiological state fostered by engaging in training with the physiologically modulated technology differs substantially from that fostered by interacting with unmodulated tasks or games. An indication that these states may be very different - and that the difference has important effects - comes from anecdotal reports of users of the videogame biofeedback technology. Parents have reported that, prior to training, their children became stimulated and agitated while absorbed in playing a videogame, but that, after training, videogame play appeared to settle them.

In addition to the self-regulation practice opportunity offered by modulated games and simulations, playing off-the-shelf videogames by themselves has been shown to provide generalizable manual (Rosser et al., 2004) and visual selective attention (Green & Bavelier, 2003) skill benefits.

Heretofore, "smart medical system" projects for space focused on issues such as the diagnosis and treatment of emotional disorders and group conflict resolution; the RESTORE concept is focused on development of training for the self-management of stress-related responses and enhancement of human performance through task- and recreation-embedded physiological self-regulation training.

4　Summary and Conclusions

Conners et al. (1985) make the following recommendations: "Biofeedback employs instrumentation to tell the individual how well he or she is progressing in attempts at self- regulation. It would be desirable to know how these techniques could best be used, either individually or as part of an integrated therapeutic program. A common method of relieving psychological upset is to engage in distracting and rewarding activities. Although there is considerable overlap in the kinds of activities people engage in, there is also considerable individual difference. ... effort should be made to match activity options to the preferences of particular spacecrew. However, in space, activity options will necessarily be limited and space travellers will probably have to learn new distraction/relaxation techniques. We do not yet know how successful such substitutions will be." The RESTORE concept offers an innovative approach to these issues.

RESTORE combines advanced sensors and adaptive technology with recreational activities into a behavioral health problem countermeasure technology with the following characteristics: (a) instrumentation systems for monitoring and archiving space travellers' central and autonomic system function data; (b) instrumentation systems for supporting physiological self-regulation training to maintain and hone cognitive abilities, resilience under psychological stress, and effective mood states; (c) systems interface experienced as engaging and entertaining throughout repeated training sessions on long-duration missions; and (d) systems interface and training protocol that maybe integrated with astronaut task-embedded training.

The RESTORE concept defines a physiological self-regulation training countermeasure technology for maintaining and reinforcing cognitive readiness, resilience under psychological stress, and effective mood states in long-duration crews. The technology consists of a hardware/software system for delivering physiological self-regulation training and for tracking crew central and autonomic nervous system function. RESTORE represents a self-management technology with threefold capability for recreation, behavioral health problem prophylaxis and remediation, and psychophysiological assay (Figure 4).

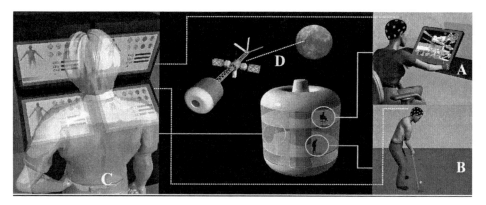

Figure 4. RESTORE is a physiological self-regulation training system that can be embedded in virtual and computer-based training tasks that the space travellers can practice to condition physiological response to actual mission tasks [A] or to virtual reality based or computer simulated recreational activities that allow the crew member to engage in recurrent practice to hone these self-regulation skills in a way that ensure practice regimen adherence [B]. RESTORE also provides for on-board real-time monitoring of physiological and cognitive state under a task challenge [C] that can be transmitted to ground-based mission control [D].

The RESTORE concept builds upon NASA published research as well as NASA patented and commercially licensed technologies to create a physiological self-regulation training countermeasure technology for maintaining and reinforcing cognitive readiness to perform, resilience under psychological stress, and effective mood states.

References

Ball, J.R., & Evans, C.H. (2001). Safe Passages: Astronaut care for exploration missions. National Academy Press: Washington, D.C.

Chance, P. C. & Lieberman, D.A. (2001). Learning and Behavior. Brooks/Cole: Pacific Grove, CA, 107.

Conners, M., Harrison, A., & Akins, F. Living Aloft: Human Requirements for Extended Space-flight. NASA SP-483, 1985.

Cowings, P.S., & Toscano, W.B. (2000). Autogenic-feedback training exercise is superior to promethazine for control of motion sickness syndrome. Journal of Clinical Pharmacology, 40, 1154-1165.

Green, C.S. & Bavelier, D. (2003). Action Video Game Modifies Visual Selective Attention. Nature, v. 423, 29 May 2003, 534-537.

Kanas, N., & Manzey, D. (2003). Space psychology and space psychiatry. El Segundo, CA: Microcosm Press.

Kellar, M.A., Folen, R.A., Cowings, P.A., Toscano, W.B., & Hisert, G.L. (1993). Autogenic Feedback Training Improves Pilot Performance during Emergency Flight Conditions. Flight Safety Digest. Flight Safety Foundation, July 1993.

Kharif, Olga. My Therapist is a Joystick. Business Week. June 14, 2004.

Linenger, J.M. (2000). Off the planet. New York: McGraw-Hill.

Meichenbaum, D. H. (1972). Cognitive modification of test anxious college students. Journal of Consulting and Clinical Psychology, 39, 370-380.

Meichenbaum, D. H., & Deffenbacher, J. L. (1988). Stress inoculation training. The Counseling Psychologist, 16, (1), 69-90.

NASA Office of Aerospace Technology, NASA Headquarters. A Real Attention-Getter, Spinoff 2003, National Aeronautics and Space Administration Headquarters, Washington, DC, NP-2003-08-307-HQ, 12, 13.

NASA Office of Biological and Physical Research, Biomedical Research and Countermeasures Program. Research Opportunities Soliciting Ground-Based Studies for Human Health in Space, NASA Research Announcement NNH04ZUU003N, National Aeronautics and Space Administration Headquarters, Washington, DC, June 4, 2004.

NASA Office of Inspections, Administrative Investigations, and Assessments. Enhancing compatibility of long-duration space flight crews, National Aeronautics and Space Administration Headquarters, Washington, DC, June 19, 1998.

Palsson; O.S., Harris, R.L. & Pope, A.T., (2002). U.S. Patent No. 6450820. Washington, DC: U.S. Patent and Trademark Office.

Palsson, O.S. & Pope, A.T. (1999). Thermal Biofeedback: Clinical Applications and Potential for Pilot Stress Counter-Response Training. American Institute of Aeronautics and Astronautics Paper No. AIAA-99-3501.

Palsson, O.S., Pope, A.T., Ball, J.D., Turner, M.J., Nevin, S., & DeBeus, R. (2001). Neurofeedback videogame ADHD technology: Results of the first concept study. Abstract, Proceedings of the 2001 Association for Applied Psychophysiology and Biofeedback Meeting. March 31, 2001, Raleigh-Durham, NC.

Pope, A.T. & Bogart, E.H., (1994), U.S. Patent No. 5377100. Washington, DC: U.S. Patent and Trademark Office.

Premack, D., (1965). Reinforcement theory. In M. R. Jones (ed.), Nebraska symposium on motivation: 1965. Lincoln: University of Nebraska Press.

Prinzel, L.J., Freeman, F.G., Scerbo, M.W., Mikulka, P.J., & Pope, A.T. (2000). A Closed-Loop System for Examining Psychophysiological Measures for Adaptive Task Allocation. International Journal of Aviation Psychology. 10(4), 393-410.

Prinzel, L.J., Freeman, F.G., Scerbo, M.W., Mikulka, P.J. & Pope, A.T. (2004). Effects of a Psychophysiological System for Adaptive Automation on Performance, Workload, and the Event-Related Potential P300 Component. Human Factors, Vol. 45, No. 4, Human Factors and Ergonomics Society.

Prinzel, L.J., Pope, A.T., & Freeman, F.G., (2001). Application of physiological self-regulation and adaptive task allocation techniques for controlling operator hazardous states of awareness. NASA Technical Memorandum TM-2001-211015.

Prinzel, L.J., Pope, A.T., & Freeman, F.G. (2002). Physiological Self-regulation and Adaptive Automation. International Journal of Aviation Psychology, 12(2), 179-196.

Prinzel, L.J., Pope, A.T., Palsson, O.S., & Turner, M., inventors, NASA LaRC patent case no. LAR-16256.

Rosser, J.C. Jr., Lynch, P.J., Haskamp, L.A., Yalif, A., Gentile, D.A., & Giammaria, L. (2004). Are Video Game Players Better at Laparoscopic Surgery? Paper presented at the Medicine Meets Virtual Reality Conference, Newport Beach, CA, January 2004. Online at: http://www.psychology.iastate.edu/faculty/dgentile/MMVRC_Jan_20_MediaVersion.pdf

Santy, P.A. (1994). Choosing the right stuff: The psychological selection of astronauts and cosmonauts. Westport, Conn: Praeger.

Space Science Board, National Research Council (1987). A strategy for Space Biology and Medical Science for the 1980s and 1990s. Washington, DC: National Academy Press.

983

Potential Predictors of Errors in Shooting Judgment and Cognitive Performance

James L. Meyerhoff, George A. Saviolakis, and Bob Burge
Dept. of Applied Neurobiology, Div. Neuroscience, Walter Reed Army Institute of Research, Silver Spring, MD
20910. emails: james.meyerhoff@na.amedd.army.mil; gsaviolakis@usuhs.mil; bob.burge@na.amedd.army.mil

William Norris, Terry Wollert, and Valerie Atkins
Federal Law Enforcement Training Center (FLETC), Glynco, GA.
wnorris@dhs.gov; terry.wollert@dhs.gov; Valerie.Atkins@dhs.gov

Charles D. Spielberger
Dept. of Psychology, University of South Florida, Tampa, FL
spielber@chuma1.cas.usf.edu

Abstract

Police trainees who were ready to graduate from the Federal Law Enforcement Training Center (FLETC) volunteered, under an IRB-approved protocol, to participate in an exercise designed to evaluate their survivability. In a highly stressful interactive scenario, which included hostage situation arising during a staged domestic dispute, performance was evaluated for a range of responses, including: management of a weapon malfunction, shooting judgment and accuracy, communications, and recall of events. Heart rate, blood pressure and salivary levels of the stress-responsive hormone cortisol, were elevated after the domestic dispute, as were scores on the Spielberger State Anxiety and State Anger Questionnaires (Spielberger, et al. 1983).

Nineteen percent of subjects shot the hostage, a failure rate that falls in the reported range of friendly fire casualties in military combat. Ninety-seven percent failed to meet the criterion of 70% of their rounds hitting the suspect. Many of the students fired blindly, from the minimal cover available. The majority of the students failed to correctly manage a weapon malfunction. During a formal interview after the domestic dispute, only 43% of trainees could accurately describe their shot placement, and only 57% could accurately identify the exact moment when the situation and doctrine first justified the use of lethal force.

In the present study, significant elevations in salivary levels of the stress hormone cortisol were noted after the shooting episode. Of particular interest, subjects achieving passing scores during the two most challenging events had significantly higher heart rates during those episodes. This suggests that differences in arousal or sympathetic/parasympathetic balance could be effective predictors of performance decrement. The Spielberger Trait Anger Scale showed an association with shot placement and performance during the gunfight, as well as with overall performance scores.

1 Introduction.

Soldiers during military operations in urban terrain (MOUT) or in peacekeeping/policing actions are often faced with "Shoot/no shoot" decisions. These decisions must be made in fractions of a second, and have profound consequences. A wrong decision can lead to death of the soldier, death of a comrade, collateral damage or errors in shooting judgment, including a "friendly fire" incident. Authors agree that the most important factor in fratricide is "human inability to cope with the stressors of the battlefield" (Shrader, 1992; Steinweg, 1995).

Civilian Law enforcement officers are often faced with similar challenges and stressors. From 1989 through 1998, 682 police officers were killed in the line of duty in the USA. In addition to interpersonal skills and sound judgment, tactical skills required include vigilance, continuous assessment of threat level, and if threat escalation occurs, a capability for rapid change of tactics to include force, if indicated. Situations associated with high levels of risk include domestic violence investigations, traffic stops and executing search or arrest warrants.

Exposure to environmental stressors stimulates increases in heart rate (HR) and blood pressure (BP) as well as secretion of a variety of hormones. In general, these hormones enable the body to support successful behavioral coping responses [e.g., "fight or flight reaction"] (Cannon, 1929; reviewed in Mason, 1968; Meyerhoff et al., 1988, 1990, 2000). Cortisol is released into the bloodstream from the cortex of the adrenal gland, and is one of several hormones that increase blood glucose as part of the normal response to exertion or psychological stress. Although moderate increases in arousal may enhance performance, extremely high levels may impair it (Yerkes and Dodson, 1908).

In this study, we evaluated the effect of stress on performance of a variety of skills during the course of a realistic and complex scenario, using volunteers recruited from the FLETC officer training courses. Salivary cortisol was measured as the hormonal stress marker, since saliva is easily collected and changes in salivary levels reflect changes in plasma concentrations (Kirschbaum et al., 1992). In addition, we monitored HR and BP throughout and assessed multiple psychometric dimensions before and after completion of the scenario. These indices of arousal in response to the environmental stressor were compared with the scores in individual tasks and with a composite score.

2   Experimental Paradigm.

The FLETC scenario was designed to test the capacity of students to draw on their training and personal resources to "survive" in a novel, rapidly evolving, highly stressful, multi-task paradigm that realistically models lethal force situations often encountered in the line of duty. Under a WRAIR IRB-approved protocol, 90 police trainees who had completed all coursework and were ready to graduate from FLETC were recruited and enrolled in the study after giving their written, informed consent. Students were then instrumented with HR and BP monitoring (Polar and Accutracker) equipment. HR and BP were automatically recorded frequently during the scenario.

One week before the scenario, a pre-baseline sample was taken for salivary cortisol. One and two weeks prior to the scenario and on the day of the scenario, during a baseline period, students were seated and were given the State-Trait Personality Inventory (STPI). This self-report psychometric questionnaire, developed by Spielberger (1983), has been extensively validated and is widely utilized. The STPI State Anxiety asks subjects to endorse feelings ranging from calm, to terror. State Anger scale asks subjects to endorse feelings ranging from mild irritation to fury and rage. Trait Anger is defined as the frequency that state anger is experienced over time. Individuals with high scores for Trait Anger experience a wider range of situations as anger provoking. The Trait Anger Scale includes two subscales: Angry Temperament, a general disposition to frequently experience anger (e.g. "I am a hotheaded person"); and Angry Reaction, a tendency to react with anger in situations that involve frustration and being treated badly (e.g. "it makes me furious when I am criticized in front of others") [Spielberger et al., 1995]. One of the goals in the study was to evaluate the relative contributions of each of these factors. After the psychometric tests were completed, salivary samples were taken for subsequent analysis of salivary levels of cortisol by radioimmunoassay.

After the baseline period, students were studied in a driving paradigm. Students were paired with a confederate of the experimenters (introduced as "another student" who would serve as their duty "partner") and the two were evaluated for routine driving skills. Then the two were evaluated for their performance during an emergency high-speed driving task. After performing first as the driver, while the partner was a passenger, the student was then required to move to the passenger seat and respond to incoming radio communications while the partner drove the test course. The partner (who was actually a driving instructor) rendered a convincing performance of driving erratically, appearing to lose control while taking a turn at high speed, putting the car into a spin and going off the road.

The student then received a radio call from a dispatcher instructing them to proceed with flashing lights and siren to an address to investigate a domestic complaint. Arriving at the address, a sergeant instructed them to enter the house and take a report from a complainant who claimed that his roommate had stolen a large sum of money from him. The student and the partner were given white protective vests and face shields, which they had previously used in interactive firefights with 9mm simunitions (rounds which propel paint capsules). They were also given 9mm semi-automatic pistols modified to fire simunitions, and three 10-round magazines of ammunition. The simunitions rounds were color-coded for the two weapons, so that shooting results could be determined by subsequent visual inspection of the protective vests.

Two instructors, unarmed but also wearing protective gear, were inside the house and served as role players – one as a co-operative "complainant", the other as "suspect". When the student and partner entered the residence, the complainant was seated in the living room and they began taking a report from him. After a few minutes, the "suspect" emerged from a back room. A shouting match between the complainant and the suspect erupted and began to escalate. The partner approached the suspect carelessly and had his holstered 9mm weapon taken away by the suspect. The suspect promptly shot the partner, who staggered and fell, blocking the door that would have provided the only acceptable exit. The suspect took the complainant hostage, holding him to the side for two seconds, thereby presenting himself briefly as a target, before pulling the hostage squarely in front of himself, and backing behind a brick wall. The suspect then shot the complainant with the partner's 9mm weapon, again ducked behind the wall, re-emerging with a shotgun with which he began firing at the student, who had minimal cover available.

The suspect then resumed firing at the student with the 9mm handgun. All shots were exchanged at close quarters, with the suspect and the student within 10 feet of each other. The third round in the student's weapon was a "dud" that failed to fire. This required the student to perform immediate action to clear the malfunction. Many students improperly resorted to inserting a spare magazine. After several minutes of exchanging fire, the suspect terminated this phase of the scenario by falling to the floor, immobile, allowing his weapon to slide a few feet away from him. The student was expected to maintain cover of the downed suspect while recovering the suspect's weapon and making it safe.

The shooting episode was followed by a mock internal affairs (IA) investigation, conducted in a separate room by a role player dressed in civilian clothes. This provided a post-shooting period during which the student was seated at a table and given "official" police forms to fill out related to the IA investigation, and the Spielberger State Anxiety and Anger questionnaires were again administered. After the psychometric tests were completed, the student was questioned about various elements of his or her performance, including shot placement, the regulations covering the use of lethal force, as well as the rationale and the timing of the decision to draw the weapon and fire. One salivary sample was taken at the beginning of the IA interview and another taken 30 minutes later, after subject debriefing at termination of the experiment, to be used for subsequent cortisol analysis.

3   Results.

Students' performances were scored on the following elements of performance: driving, communications, response to weapon malfunction, shooting judgment and accuracy, as well as post-shooting recall. Many of these elements were scored on a pass or fail basis, and the terms "go" or "no go" are customarily used by the instructors at FLETC. For example, if a student shot the hostage, that was rated a "no go".

Serious failures were observed in many of these performance areas. In the emergency driving test, immediately after the partner appeared to lose control of the car, an incoming radio call from the dispatcher was received. Over 60% of students failed to respond to the radio call within the criterion time period of 8 seconds. Fifty-seven percent failed to accurately describe the nature of the call (officer needs assistance) to the partner. Ninety-five percent failed to follow doctrine requiring that they form a plan before any investigation (e.g., who will be contact officer and which one will be cover officer).

During the investigation of the domestic complaint, 19% of students shot the hostage. Moreover, 97% failed to meet the criterion of 70% of their rounds hitting the suspect. Many of the students fired blindly, from the minimal cover available. Trainees were expected to call for backup in high-risk situations and had also been taught that if using their radio while their weapon was in their hand, the weapon should be kept in the dominant hand. Seventy percent failed this element by switching the weapon to the weak hand, in order to operate the radio in the dominant hand. The approved response for coping with the "dud" round that fails to fire is to tap the magazine, rack the slide and re-engage the threat. The majority of the students failed this element, resorting to a variety of methods, all less desirable, to clear the malfunction. During the IA investigation, only 43% of students could accurately describe their shot placement, and only 57% could accurately identify the exact moment when the situation and doctrine first justified the use of lethal force.

Significant elevations in HR and BP were observed during the scenario. Moreover, differences in heart rate (HR) responsivity were observed during two events, between students who achieved passing scores on those events vs. those that failed. Whereas successful students displayed additional HR acceleration while in the passenger role during the "erratic driver" episode [t-test = 3.317, d.f. = 95, p < 0.013] as well as during the gunfight [t-test = 2.429, d.f. = 93, p < 0.017], students who received failing scores on those elements had lower heart rates than successful students.

Salivary cortisol levels were significantly higher after the gunfight, compared to baseline [Dunnett's test, p < 0.013], suggesting increased arousal. Changes in the State Anxiety and State Anger during the scenario were particularly interesting and were significantly elevated after the gunfight, compared to baseline [respective pair wise comparisons by t-tests: t = 15.3, df = 92, p < .001; t = 8.6, df = 88, p < .001]. These two measures did not differ, however, between students receiving passing vs. failing scores on those events.

The Trait Anger score showed an interesting modest association with the shot placement score (r = .27) as well as with overall performance score [r = .32]. This is consistent with the association found between Trait Anger Score and performance during the gunfight [Pearson Chi Square = 2.96, df = 1, p = .086]. An analysis was conducted to determine if the Trait Anger subscales Angry Temperament and Angry Reaction were associated with performance outcomes. While no association was found for Angry Temperament, the data provided evidence of an associative trend between higher Angry Reaction scores and successful performance during the gunfight (Pearson's Chi Square = 3.43, p = 0.064).

4   Discussion.

One definition of stress is the perception that situational demand exceeds resources (Saunders et al., 1996). In the present study, the student relies on the partner as a resource, to perform as a partner, during the investigation of the domestic dispute. Law enforcement officers are routinely assigned to work in pairs and normally expect to rely on their duty partner. The partner demonstrated incompetence, first as a driver and then in failing to protect his weapon during the domestic dispute. Thus, in the FLETC scenario, the shortcomings of the partner constituted a significant deficit in resources and a major stressor for the student.

Stress has been cited as a major factor in errors in shooting judgment (Shrader, 1992; Steinweg, 1995). Friendly fire was responsible for 24% of U.S. forces killed in action (KIA) and 15% of wounded in action (WIA) casualties in Operation Desert Storm (Steinweg, 1995). Thus, the percent of students shooting the hostage (19%) falls in the range of reported incidence of friendly fire casualties in military combat. Errors were prevalent not only in performance, but in post-shooting recall, as over 50% of students could not accurately recall their shot placement. As reported in the instance of friendly fire, psychological stress may be an important factor in the errors in shooting judgment and other elements of performance seen in the present study.

The relationship between arousal and performance has been described as an inverted U-shaped curve (Yerkes and Dodson, 1908). In the present study, significant elevations in salivary cortisol levels were noted after the shooting episode. The administration of exogenous glucocorticoids has an inverted U-shaped effect on memory (Lupien & McEwen, 1997). Low doses increase arousal, whereas high doses are reported to induce hyperarousal, decreased use of relevant cues and impairment of cognitive performance (de Quervain et al., 2000).

Stress-induced increases in endogenous cortisol levels have also been associated with decrement in performance on a memory task (Kirschbaum et al., 1996). In the present study, there were dramatic examples of decrements in many aspects of performance, from threat perception and weapons handling to timeliness in responding to dispatcher. Although the elevated levels of glucocorticoids may have interfered with memory, cortisol levels did not predict overall performance.

Students who performed successfully during particularly challenging events had higher heart rates during those challenges than students that failed those events. This raises several questions, including the possibility that those who failed the scenario were insufficiently reactive, or had a significantly higher degree of vagal tone. Accordingly future studies should include measurement of heart rate variability.

State Anxiety and State Anger were significantly elevated during the scenario, and suggest that subjects had a notable degree of emotional arousal during the scenario. Trait Anger was also an interesting metric with a degree of association with shot placement and performance during the gunfight, as well as with overall performance scores. Two Trait Anger Scale subscales were evaluated, Angry Temperament, a general disposition to frequently experience anger; and Angry Reaction, a tendency to react with anger in situations that involve frustration and being treated badly (Spielberger et al., 1995). The data provided evidence of an associative trend between higher Trait Angry Reaction scores and successful performance during the gunfight.

Stress inoculation training has been shown to improve performance (Saunders et al., 1996). If the stress levels are graduated, confidence and skills may be acquired simultaneously (Keinan, et al, 1996). It is suggested that scenario-based exercises might be introduced earlier in training, with initial stress levels moderate and increased with experience.

5   Conclusions.

FLETC staff recommend that high-stress scenario-based exercises should be introduced earlier and more frequently in training to improve officer capability to adapt to rapidly changing, unpredictable situations. Future studies should include measures of heart rate variability.

6   Acknowledgements.

This work was supported by US Army Medical Research and Materiel Command (MRMC) and the Federal Law Enforcement Training Center (FLETC). Subjects in this study were enrolled after providing written informed consent under an IRB-approved protocol. The authors wish to thank Ms Carolyn Belcher and Ms Tanya Lopez for statistical support, Ms Annabelle Wright and Ms Patricia Stroy for graphics support and SGT Nadia Diaz-Kendall for technical support. Any opinions, findings, and conclusions expressed are those of the authors and do not necessarily reflect the views of U.S. Army.

References

Cannon, W.B. 1929: Bodily changes in pain, hunger, fear and rage. Appleton, New York.

DeQuervain DJ-F, Roozendall, B., Nitsch, R., McGaugh, J. and Hock, C. 2000: Acute cortisone administration impairs retrieval of long-term declarative memory in humans. *Nature Neuroscience*, 3(4), 313-314.

Keinan, G. and Friedland, N. 1996: Training effective performance under stress: queries, dilemmas, and possible solutions. In J.E. Driskell & E. Salas (Eds.) *Stress and human performance*. (pp. 257-277) Mahwah, New Jersey: Lawrence Erlbaum Associates, Inc.

Kirschbaum, C., Wolf, O., May, M., Wippich, W. and Hellhammer, D. 1996: Stress- and treatment-induced elevations of cortisol levels associated with impaired declarative memory in healthy adults. *Life Sciences*, 58(17), 1475-1483.

Lupien, S and McEwen, B. 1997: The acute effects of corticosteroids on cognition: integration of animal and human studies. *Brain Research Reviews* 24,1-27.

Mason, J.W. 1968: Over-all hormonal balance as a key to endocrine organization. *Psychosomatic Medicine* 30: 791-808.

Meyerhoff, J. L., Oleshansky, M.A,, & Mougey, E.H. 1988: Psychological stress increases plasma levels of prolactin, cortisol and POMC-derived peptides in man. *Psychosom. Med.* 50 (3), 295-303.

Meyerhoff, J. L., Oleshansky, M. A., Kalogeras, K. T., Mougey, E. H., Chrousos, G. P., & Granger, L.G. 1990: Responses to emotional stress: possible interactions between circulating factors and anterior pituitary hormone release. In J.C. Porter & D. Jezova, (eds.) *Circulating Regulatory Factors and Neuroendocrine Function: Advances in Experimental Biology and Medicine.* (pp. 91-111). New York:Plenum Press.

Meyerhoff, J.L., M.A. Hebert, K.L. Huhman, E.H. Mougey, M.A.Oleshansky, M. Potegal, G.A. Saviolakis, D.L. Yourick and B.N. Bunnell. 2000: Operational stress and combat stress reaction: neurobiological approaches towards improving assessment of risk and enhancing intervention. In Karl Friedl, Harris Lieberman, Donna Ryan and George Bray (eds). *Countermeasures for Battlefield Stressors; Pennington Nutrition Series*, vol. 10 Baton Rouge: Louisiana University Press.

Saunders, T., Driskell, J.E., Johnston, J.H. and Salas, E. 1996: the effect of stress inoculation training on anxiety and performance. *J. Occup. Health Psychol.* 1(2), 170-186.

Selye, Hans (936: The stress of life. New York: McGraw-Hill Companies.

Shrader, C.R. Friendly fire: the inevitable price. *Parameters.* Autumn 1992, pp. 29-44.

Spielberger, C. D., Jacobs, G. Russell, S., & Crane, R. 1983:. Assessment of anger, the State-Trait Anger Scale. In J.N. Butcher and C.D. Spielberger (Eds.) *Advances in personality assessment* (Vol. 2, pp. 159-187). Hillsdale, NJ: Lawrence Erlbaum Associates.

Spielberger, C. D., Reheiser, E.C., and Sydeman, S.J. 1995: Measuring the experience, expression, and control of anger. In H. Kassinove (Ed.) *Anger disorders: definitions, diagnosis, and treatment.* Washington, D.C.: Taylor & Francis.

Spielberger, C. D., Sydeman, S.J., Owen, A.E., and Marsh, B.J. 1999: Measuring anxiety and anger with the State-Trait Anxiety Inventory (STAI) and the State-Trait Expression Inventory (STAXI). In M.E. Maruish (Ed.) *The use of psychological testing for treatment planning an outcomes assessment (2nd ed.)* Mahwah: Lawrence Erlbaum Associates.

Steinweg, K.K. Dealing realistically with fratricide. *Parameters.* Spring, 1995. pp. 4-29.

Yerkes, R.M. and Dodson, J.D. 1908: The relation of strength of stimulus to rapidity of habit-formation. *J. Comp. Neur. Psychol.* 18, 459-482.

Section 5
Cognitive State Sensors

Chapter 20

Engineering Modular Augmented Systems

Toward Platform Architectures for
Modular Cognitive Cockpits

Mark Austin
Department of Civil and Environmental Engineering, and
Institute for Systems Research,
University of Maryland,
College Park, MD 20742, USA.
E-mail : austin@isr.umd.edu

Abstract

The pathway to practical modular cognitive cockpit design is currently hindered by a lack of appropriate representations for modular end products and processes, coupled with appropriate systems engineering development processes, and so-called "platform architectures" capable of assessing and reconfiguring system-level design options for better modularity. To mitigate this shortcoming, this paper proposes a methodology that combines ideas from systems- and modular object-based engineering development. It identifies bottlenecks that need to be overcome in order for modular cognitive cockpit design platforms to be created.

1   Problem Statement

The common promises of modularity include component-sharing across product lines, shortened delivery times, opportunities for customized products, improved quality and profit margins (Raley & Marshall, 2004). While modularity is something every engineer wants, unfortunately, the details on how to achieve it are often far less clear! As a case in point, modularity in the automobile industry is motivated by the promise of easier assembly and/or pre-testing before final assembly. While variations in the average height/weight of populations in Asia and North America have led to the formulation of parametric models of car frame size, the existence of these models has not led to expanded customer choice in the marketplace. Instead, modularity is largely restricted to piecemeal add-ons (e.g., distinct chunks such as seats, sunroofs and cooling systems) (MacDuffie, 2001). This result can be attributed to economics; suppliers are eager for add-on products to be available to as many types of customers as possible.

The key factor in determining whether or not a product can be easily partitioned into a modular form is its complexity (e.g., no of parts, no of product variants, no of functions, no of technologies, ...etc). As illustrated in Figure 1, there are four types of product architecture: (1) Modular design (one function is allocated to one module); (2) Function distribution (one function is mapped to multiple modules); (3) Function sharing (several functions are allocated to one module); (4) Integrated design (several functions are allocated to several modules). Modular designs maintain

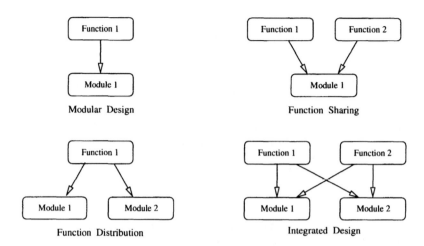

Figure 1: Different types of Product Architecture (Erens and Verhulus, 1997)

a level of simplicity in their implementation by imposing a restriction "one function allocated to one module." A modular architecture has well-defined, standardized, and decoupled interfaces which collectively allow for design changes to be made to one module, without generally requiring a change to other modules. Systems with sophisticated behavior are created through the composition of simple modules. Highly integrated systems, in contrast, have functionality that is shared (or cuts) across modules. Because cause-and-effect relationships can propagate across the entire system, design evaluation procedures must consider all parts of the system and their connectivity. Real systems may assume multiple product architecture types.

As we move from industrial- to information-age systems the problem of creating "modular designs" only becomes more vexing. In the past systems have been viewed from an operations point of view, where information and communication have been regarded as services necessary for the system to operate in pre-defined ways. Nowadays, there is a rapidly evolving trend toward large-scale information-dominated systems (i.e., pilots are empowered with functionality that would not be possible without and ability to gather and work with information), which exploit modular commercial off-the-shelf and communications technologies, have superior performance and reliability, and are derived in response to various types of information drawn from a wide array of sources. A good example is the market for luxury automobiles, where 30% of total cost can be attributed to electronic systems that add "value and services" to the driving experience (e.g., additional comfort; navigation systems). Components that were once purely mechanical – brakes, steering, suspension – are now either electronic or controlled by computers. Indeed, almost anything that makes a modern automobile perform better is likely to involve sensing and fusion of data streams, real-time control, and electronics. To achieve required improvements in system-level performance, control mechanisms often need to affect the behavior of multiple modules, which in turn, requires that modern automobile systems be viewed as integrated architectures. From a control engineering standpoint, this makes modern automobile design more difficult and less modular than before.

As systems become progressively diverse in their functionality, and solutions increasingly reliant on high technology, the challenge in creating good system-level designs will steadily increase unless new approaches are developed. There is now a strong need for methods that can: (1) Identify opportunities for partitioning multidisciplinary systems into modules, (2) Represent and visualize connectivity/communication mechanisms among modules, and (3) Create and evaluate system-level architectures through automated synthesis and formal model-based testing. For our purposes, synthesis of engineering systems is a process whereby provisional and plausible concepts are developed to the point where traditional engineering and design can begin.

2 Separation and Organization of Design Concerns

To keep the complexity of development concerns in check, system-level design methodologies are striving to orthogonalize concerns (i.e., achieve separation of various aspects of design to allow more effective exploration of the space of potential design solutions), improve economics through reuse at all levels of abstraction, and employ formal design representation that enable early predictions of behavior and detection of errors. Validation and verification procedures need to be an integral part of the development process (rather than a postscript to development).

Methodologies for design and evaluation of modules are are guided by the concepts of system coupling, cohesion and module complexity. Coupling is a measure of the interface complexity (or degree of interdependence) between modules. Cohesion is a measure of how well the components of a module are related to one another (put another way, cohesion is a measure of the functional association of the element within an element). Modules should be kept as simple as possible, and hide the details of implementation from the outside environment. Three factors that contribute to module complexity are: (a) its size, (b) the number of internal functions and connections within the modules, and (c) the number of interfaces to the modules. The criteria of coupling and cohesion work together. Generally speaking, modules with components that are well related will have the capability of plugging into loosely coupled systems. Functions in the behavior model will be reorganized so that highly interrelated functions are allocated into common modules. By clustering functions which are highly affected by noises into separate modules, the impact of noises on the total overall system can be minimized. The result is a system architecture that is characterized by a minimized coupling among modules and maximized cohesion within the modules.

A basic principle of architectural design is the need to achieve separation of concerns (e.g., various types of system functionality), by factoring out so-called cross-cutting concerns. As shown in Figure 2, various types of system functionality will then be allocated to separate subsystem/modules. Every module or subsystem fulfills some specific task and is therefore responsible for a particular design concern. Also, "module functionality" concerns can be separated from "connection and communication" among modules. As we move from the system level down to subsystems and components, opportunities for functional modularity increase. Conversely, as we move from components up to the system level, the need for integration and prevalence of crosscutting concerns

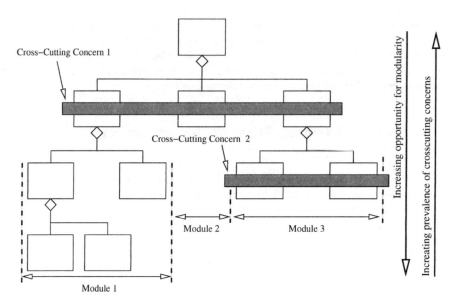

Figure 2: Role of Modularity and Cross-Cutting Concerns in System-Level Design

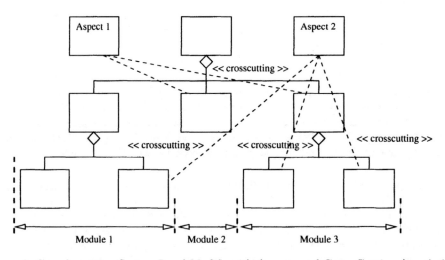

Figure 3: Complementing System-Level Models with Aspects and Cross-Cutting Associations

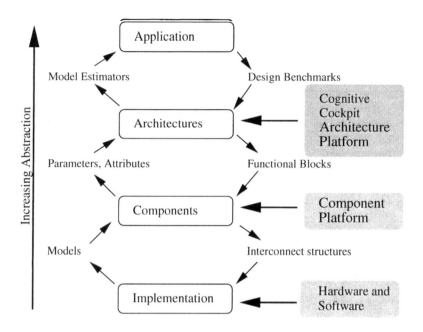

Figure 4: Methodology for Stack of Design Platforms (Adapted from Sangiovanni-Vincentelli, 2003)

increases perhaps to the point where complete factoring of system concerns may not be possible. This leaves concerns that crosscut (or have influence over) several objects (or groups of objects). For example, many non-functional requirements, such as safety and security do not decompose into behavior that maps to a single locus in the system structure.

Researchers are currently investigating the feasibility of modeling cross-cutting concerns in hardware/software systems with aspect hierarchies. The term "aspect" is used for a module that is able to encapsulate (or otherwise treat) a crosscutting concern. In the development of system architectures, all issues associated with the crosscutting concern are housed in an "extra" modular unit, even though it may crosscut mechanical, electronic and software parts. Figure 3 shows how aspects complement the system structure hierarchy. Using UML notation, the << aspect >> stereotype is class used to model units of crosscutting concern. The << crosscut >>; stereotype is an association that links crosscutting relationships (Arnautovic & Kaindl, 2004).

3   Platform-Based Design

The goals of platform-based design are to facilitate development of future product generations that are extensions of original hardware and software investments, thereby allowing for dramatically reduced time to market, while decreasing development and production costs. From a product perspective, product family success is achieved through a set of "well-defined design modules and interfaces" that can be easily customized to a variety of customer requirements. Reuse is a must. From a process perspective, multi-project integration is achieved via a "stable basis of processes and tools" that support networks of cooperation with project partners and suppliers. Design is viewed as a "meeting-in-the-middle process" where successive refinements of specifications meet with abstractions of potential implementations.

Platform-based design procedures are driven by the identification of precisely defined layers where refinement and abstraction processes take place. Abstraction processes need to balance criteria of isolating lower-level details, but letting enough information transpire about lower level abstractions to allow design space exploration (Sangiovanni-Vincentelli, 2003). As shown in Figure 4, a "platform stack" is a stack of layers of abstraction: The upper layer is the abstraction of the layer below so that an application could be developed on the abstraction without referring to the lower layers. The lower layer is the set of rules that allow one to classify a set of components as part of the platform. In general, the components of the platform are partially or completely pre-designed. The upper layer is used to decouple the application from the implementation of the platform. The tools and methods used to map the upper and lower layers of the platform stack are the glue that keeps the platforms together.

4   End-to-End System-Level Development of Modular Cockpits

At a high-level of abstraction, and in the long term, we envision that modular cognitive cockpit design will follow the step-by-step procedure shown in Figure 5. Top-down synthesis of system-level models begins with use cases, and proceeds to fragments of system functionality, expressed as activity and sequence diagrams (i.e., UML-based methods). To minimize the possibility of unforeseen failure we need models of system-level development that will help designers clearly articulate what the system must provide and what must be prevented. Scenarios are partial descriptions of behavior. Positive scenarios specify the intended system behavior. Negative scenarios specify behaviors the system is expected not to exhibit. Several scenarios may be connected to each use case, explaining the different ways the use case may develop in a finer grained description of system functionality. Systems requirements correspond to constraints on system functionality, system interfaces, and non-functional concerns, such as safety and reliability. They are generated, in part, from features in activity and sequence diagrams. For example, sequence diagrams also imply component interfaces needed to support the passing of message between components.

We assume that behavior and structure of cognitive cockpit systems can be modeled as a network of communicating objects/sub-systems. Figure 6 illustrates two key elements of this development style. On the left hand side, object-based systems correspond to networks of communicating objects and systems. They achieve their purpose with modules having having well defined functionality, well defined interfaces for connectivity to other modules and the surrounding environment, and message

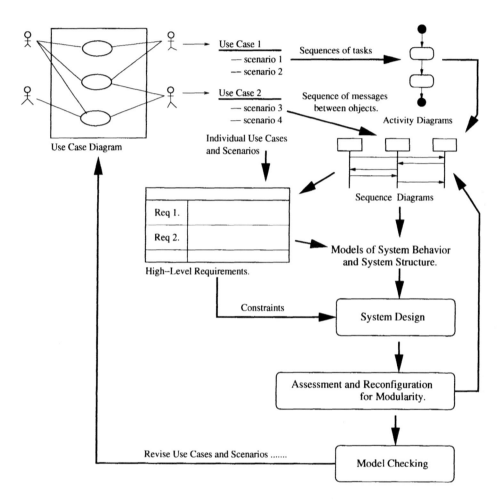

Figure 5: Step-by-step Procedure for Synthesis and Validation of System-Level Cognitive Cockpit Designs (Adapted from Austin, 2004)

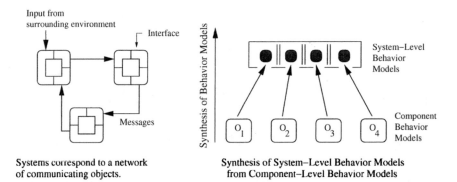

Figure 6: Two Key Elements of Hybrid Object-/Systems- Development. Objects communicate through message passing. Models of System-Level Behavior are synthesized through the parallel composition of Component-Level Behavior Models (Austin, 2004)

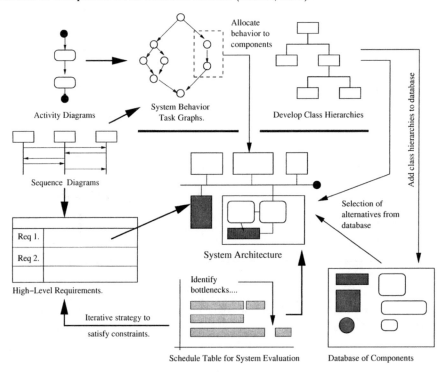

Figure 7: Flowchart for Top-Down/Bottom-Up System Development (Austin, 2004)

passing. The price that we pay for allowing concurrent behavior is the increased complexity in scheduling activities to improve performance, while avoiding failure.

We create system-level design alternatives by mapping fragments of behavior onto the system structure (part of this mapping process is contained in the sequence diagrams mentioned in the previous step), and imposing constraints on performance and operation. In contrast, bottom-up synthesis and verification of executable models of behavior assume that the system can be modeled as a network of interacting finite state machines. Models of architecture-level behavior are obtained through the parallel composition of concurrent processes at the component level. Sophisticated techniques of system analysis and control are justified by the life-critical safety risks and adverse consequences of poor system throughput (Taylor, 2003).

Model checking procedures make sure that the system design: (1) does what it's supposed to do; (2) prevents certain behaviors from occurring; and (3) does not support un-intended behaviors. If any one of these aspects is violated, then we have a "gap" between the intented system and the actual system design. By detecting and validating implied scenarios it is possible to drive the elaboration of scenario-based specifications, behavior models, and system design to a state where there are no more implied scenarios to be validated.

Assessment and reconfiguration for modularity – see Figure 5 – will be guided by analytical procedures, such as design structure matrices (Browning, 1998). Engineers also need to understand the extent to which a system provides functionality beyond what is actually required. Implied scenarios correspond to gaps in the scenario-based specification (Uchitel, 2003). These gaps can occur in two ways: First, perhaps the system architecture does not provide components with a rich enough local view ... i.e., components my not be able to support the required system behavior. A second possibility is that the system architecture may contain feasible of traces of behavior that are not detailed in the scenario specification (i.e., the system architecture might do something that the user is unaware of).

5 Adapting the Platform Concept to Modular Cockpit Development

Once object-based models have been formulated, problem domain concepts are organized into class hierarchies for reuse. Designers need to identify objects and their attributes and functions, establish relationships among the objects, establish the interfaces or each object, implement and test the individual objects, assemble and test the system, and organize classes for reuse (via persistent storage in a database). See Figure 7. These abstractions need to be consistent with the parameter/attribute and functional block divisions located between the architecture- and component-level platforms.

We speculate that in a departure from the goals of regular design platforms (i.e., reduced time to market; increased reuse), initially, platforms for modular system-level cockpit design are likely to focus on assessment of system performance, identification of modular design options (at appropriate levels of abstraction), and associated trade-offs (e.g., diversity of required system performance vs. cost). After these concepts are well understood, then it would seem appropriate to look at opportunities for creating models for multi-disciplinary modularity.

6 References

Arnautovic E. and Kaindl H. (2004). Aspects for Crosscutting Concerns in System Architectures, Proceedings of the 2nd Annual Conference on Systems Engineering Research (CSER), University of Southern California (USC), April 15-16.

Austin M.A. (co-delivered with John Baras) (2004). An Introduction to Information-Centric Systems Engineering, *Tutorial F06*, INCOSE, Toulouse, France.

Browning T.R. (1998). Modeling and Analyzing Cost, Schedule, and Performance in Complex System Product Development, PhD thesis, Massachusetts Institute of Technology, Cambridge, MA.

Erens F. and Verhulst K. (1997). Architectures for Product Families, *Computer Industry*, Vol. 33, pp. 165-177.

Fixxon S.K., The Multiple Faces of Modularity: An Analysis of Product Concepts for Assembled Hardware Products.

MacDuffie J.P. (2001). Modularity and Build-To-Order Pull-Through: Early Trends in the Automotive Industry, Wharton Forum on E-Business, Wharton School, University of Pennsylvania.

Raley C., Marshall L. (2004). Augmented Cognition Modular Design, Class Project, *ENSE 622: System Requirements, Design and Trade-Off Analysis*, Master of Science in Systems Engineering Program, Institute for Systems Research, University of Maryland, College Park.

Sangiovanni-Vincentelli A. (2003). Electronic-System Design in the Automobile Industry, *IEEE Computer Society*, pages 8–18, May-June.

Taylor R.L., et al. (2003). Cognitive Cockpit Engineering: Coupling Functional State Assessment, Task Knowledge Management, and Decision Support for Context-Sensitive Aiding, Chapter 8.

Uchitel S. (2003). Incremental Elaboration of Scenario-Based Specifications and Behavior Models using Implied Scenarios. PhD thesis, Imperial College, London, England.

Modular Design for Augmented Cognition Systems

Colby Raley, Latosha Marshall

University of Maryland
Institute for Systems Research, College Park, MD 20742
cdraley@umd.edu

Abstract

Platform-based design principles can be used to allow faster and more efficient creation of augmented cognition systems from customer specifications. A modular design enables a higher product flexibility and reduction of development time, parallel development of system components, reduction of production time, reduced capital investment in production, reduced material and purchase costs, improved quality, easier service and upgrading, and easier administration of any system (Ericsson, 1999). Platform-based design takes that one step further, and allows systems to be fully developed, designed, validated, and verified while still at high levels of abstraction in the design process. It bridges the gap between custom-designing a new system for each platform instantiation and attempting to merely recycle system components in radically different platforms.

Platform-based design is important in almost any high-technology system, but is particularly relevant to augmented cognition systems for two major reasons:
- Technologies required to implement augmented cognition systems are still being developed, and will continue to improve dramatically over the coming years - for example, sensing technologies will never catch up with the knowledge of architecture design, so a system should be able to accommodate this.
- Augmented cognition systems will continuously be applied to new scenarios and situations; therefore, the systems should be able to adapt to any necessary environment.

This paper will discuss the migration from modular design to platform-based design in augmented cognition systems and will make an argument for the use of platform-based design in all such systems.

1   Platform-Based Design

Both component- and platform-based design capitalize on the principal of modularity. Modularity involves grouping together specific functions into small building blocks of a product to make a complex system more manageable. It helps shorten development, deliver customer-tailored products, and improve profit.

Platform-based design moves one step beyond that, utilizing both top down and bottom up design approaches. In the top down approach, special attention is paid to high level views and ideas of implementation but the entire system is constrained within the framework of these implementations. In the bottom up approach, special consideration is paid to the components and the rest of the system is constrained based upon the component specifications. However, in platform-based design, both views important. A real world example is the Dell Company. Dell uses platform-based design to create customized computers. The implementation of this product is a computing system that fits consumer's needs. The specifications of the system are based upon consumer needs including specific operating system, RAM, storage space, etc. From the Dell example, one can see how platform-based design can be utilized to create a family of products that are tailored to a consumer's specific needs.

The advantages of platform-based design are reuse of design components, reduced design time, and early system verification and validation. In platform-based design, the similarities of the family of products are emphasized in an attempt to reuse key components. For instance with Dell, though the specifications of each system are different, it is known that a hard drive should be a certain size, a monitor is needed, etc... Therefore, when designing a family of products at a general level, engineers can design certain features once and reuse those requirements for all systems, thus eliminating redundant designing and reducing the design time. Additionally, with platform-based design, engineers are able to perform system verification and validation in earlier stages of product design. Using higher levels of abstraction, engineers can ask basic questions such as:

- Will this do what I originally wanted it to do?
- Are all requirements accounted for in the design process?
- Are the system communications in sync with one other?

Many of these questions can be answered by looking at requirements, system architecture, communication flow, and system logic functions before component is actually built.

2   Platform-Based Design for Augmented Cognition Systems

Most of the platforms explored thus far for augmented cognition systems are in the military domain. This is primarily because of DARPA's substantial investment in the Improving Warfighter Information Intake Under Stress program, beginning in 2000. Currently, that program is looking at the platforms of: stationary Command and Control (C2); cockpits (both driving and aviation); and mobile individuals (the dismounted soldier). An additional platform that looks promising and will be explored in the near future is training, particularly in virtual environments. It is assumed that as the enabling technologies develop and advance, that augmented cognition technologies will also be applied to nonmilitary domains, such as office work or traditional learning environments.

In order to highlight the differences and similarities between seemingly disparate platform instantiations, two platforms will be discussed here: driving and learning (in a traditional environment). Driving is a platform to which the technologies have already been applied, and is highly technology driven already. Learning environments have not yet been explored by augmented cognition technologists, and it is a highly "analog" environment, which demands that designers consider creative ways of enabling a closed-loop computational system that includes the students.

In examining these two platforms, it is helpful with a "catalog" of potential system components, including sensors, models, and interfaces. This catalog of components can be used when designing any augmented cognition system and will definitely grow over time. A non-exhaustive list includes:

- Sensors
  - Cognitive
    - Direct Brain Measures: EEG, fNIR
    - Psychophysiological Measures
      - HR, EKG
      - Pulse Ox, GSR
      - Posture
      - Temperature
      - EOG, Pupilometry, Gaze Tracking
  - Environmental
    - Platform Measures
      - Location
      - Internal Conditions
      - Fuel
      - Weapons
    - External Measures
      - Weather
      - Presence of Chemical or Biological Agents
      - Situational Awareness
      - Hostility
      - Obstacles
  - Task
    - Status
- Interfaces
  - Visual
    - Heads up display
    - Traditional display
    - Alert
    - Warning
    - Picture
    - Text
  - Auditory
    - Voice
    - Warning
    - Spatially locatable
  - Tactile
    - Warning
    - Directional cue

Starting with this catalog of components enables the bottom up portion of the design process. Designers starting with a complete version of this list can feel confident that they have exhausted the component research portion of the design process.

Once the design elements have been identified, we must begin implementing constraint-based requirements – this begins the top down portion of the design process. For example, in a driving environment where it expected that a driver would maintain physical contact with the wheel (using her hands), the seat (using her body) and the pedals (using her feet). Therefore, it is reasonable to expect that her hands, body, and feet are the only opportunities for haptic interfaces. Also, knowing that using the pedals requires dexterous action on the part of the feet, one could surmise that haptic interaction might inhibit the manipulation of the pedals and determine that the hands and body are actually the only opportunities for haptic interaction. This constraint would have to be documented and specified in the requirements by the systems engineers to ensure that the human factors engineers would not design in "buzz" alert to be placed on the yet untouched upper arm, for example.

The final step in ensuring top down and bottom up design is the specification of communications between the now constrained components: this allows for the reuse of components between system designs, and the "swapping" in and out of different versions of the components throughout the design cycle.

The best "non-augmented cognition" example of this phenomenon would be the design of seats in a car. As long as systems engineers are aware of the size, space, heft, and interface requirements of each version of the seats, they do not care what kind of seat the mechanical engineers design. The seat is not required for simulation and testing of the car, and the introduction of new seats that are within the specified requirements will not impact the rest of the car.

This is analogous to the integration of updated sensing technologies into an augmented cognition system. While the introduction of a totally new technology would require some redesign of any system, inserting merely updated sensor systems (i.e. a new EEG system) should be as easy as updating the seats in a car. As long as the required inputs (electrical activity of the brain), outputs (electrical activation in μV; localization of activation), and constraints (size, weight, location, etc…) are known, any system that fits or can adapt to the environment can be integrated with minimal impact on the rest of the system.

3   System Architecture and Design

As mentioned in the previous section, these systems are highly temporally dependent, as illustrated in the following diagram.

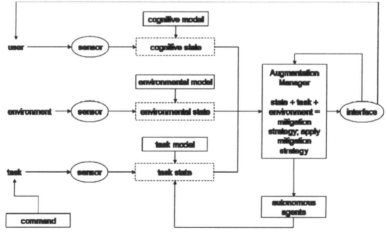

Figure 1: System Level Augmented Cognition Architecture

The arrows in Figure 1 indicate flows of information. The general flow is as follows: commands influence a task; sensors detect activity in the user, environment, and task; sensed and modeled information is combined to create real-time models of the user, environment, and task; state information and interface information is used by the Augmentation Manager to determine an appropriate information-bottleneck mitigation strategy; the Augmentation Manager impacts the interface, which communicates with the user, and autonomous agents if necessary, which complete tasks that the user is to overloaded to complete.

Several things about this architecture illustrate the complexity of augmented cognition systems, but the most obvious is the element of time. Everything in the system is responding in real time, all sensing, modeling, and mitigation in the system is continuous – but a certain flow of information is required in order for the Augmentation Manager to have appropriate information on which to take action. The basic solution for this enigma is that cognitive, environmental, and task state are all continuously updated in real time, but are independent of each other. This way, the Augmentation Manager can use only the current states to evaluate what mitigation strategies are needed to improve information flow. A simplified version of the architecture might look like:

Figure 2: Simplified System Level Augmented Cognition Architecture

The most important aspect of augmented cognition systems is their closed-loop nature – this is how humans are able to be part of the system. As shown in the general system architecture, the only aspects of the system that bring the human in the loop are the real-time sensing and the interface. In this way, the human completes the loop.

By comparing the system-level architecture with instantiations of platform-specific architectures, it is easy to see the importance of including both top-down and bottom-up design methodologies.

The realization of the system architectures is the aspect of the design process where using these principles is most evident – it allows for faster development and design of the required systems. However, platform-based design becomes truly invaluable in requirements generation and the validation and verification processes, as it allows these steps to be accomplished early in the design, rather than at the end, when changes and corrections are extremely expensive to implement.

4   System Verification and Validation

System Verification and Validation is typically done at the end of the design phase of product development. However, when using platform-based design, system verification and validation (V & V) can be completed earlier in the design stage. Tools that assist in doing system V & V are requirements tracing, analysis of communication flow, spatial logic, and temporal logic.

4.1   Requirements Tracing

Requirements traces are useful in ensuring that all requirements have a purpose and all lower level requirements are based upon and congruent with high level requirements. Its purpose is to ensure that the designer takes into account all the general high level requirements when creating his design requirements. When looking at a requirements trace, one needs to make sure that all general high level requirements map to at least one design requirement and vice versa. If this is not achieved, more requirements need to be created so that this criteria is met. Since platform-based design is a combination of bottom up and top down approaches, this sort of requirements trace is necessary to make sure that the two methods align properly with one another.

Another requirements trace that may be useful is tracing requirements to system components. This form of trace ensures that requirements are written to govern the development and interaction with other system components. Again, it is important that every requirement maps to at least one component and vice versa.

4.2 Analysis of Communication Flow

Another useful system V & V tool is analysis of communication flow. This method requires and examination of the system architecture and a determination of what information can flow to and from components. Analysis of communication flow can be done at high and low levels of abstraction. This technique identifies that the information leaving one unit (A) and going into another (B) is compatible with the unit it is entering (B). For example, if unit B is looking for a numeral, and unit A's output is an alpha value, then one can determine that there is already going to be a problem before anything is constructed. For this project, one must look at the communication flow table to ensure that the expected outputs and inputs are appropriate according to the system architecture.

4.3 Spatial Logic

Spatial logic looks at the design area in three dimensions and determines any limitations of the system due to its platform environment. It can also determine if the designer created requirements that take into account the unique features of the platform that may be overlooked when discussing multiple platforms. Constraint-based requirements play a major role in this section. The advantage of using spatial logic is that it answers key questions such as:

- Are platform and interface requirements suitable for each platform?
- Are system components creating an uncomfortable user environment?
- Are there any systems components placed in an unsafe position that inhibit the functions of the platform?
- Did designers adhere to constraint based and design requirements?

Human factor engineers will most likely be involved in this aspect of design and analysis for any augmented cognition system.

4.4 Temporal Logic

Finally, temporal logic takes into account time sequences of the system. Questions that can be answered through the utilization of temporal logic are:
- Is this receiving the data at the correct time?
- What is the sequence of data that needs to happen before the next event can take place?
- What is the critical path of data?
- What are timing constraints for this component to perform its specific task?
- What is the projected reaction time of the system?

Augmented Cognition systems face some unique challenges because they:
- Are primarily controlled be feedback loops (i.e. user-user sensors-user state-Aug Manager-interfaces-user)
- Feature three sub loops running in parallel (i.e. user, environment, and task)
- Feature continuously-streaming data (i.e. user, environment, task, and interfaces are always emitting data)
- Must account for sensing and modeling delays, user reactivity time, and the timing of outside factors

Temporal logic identifies key time issues in complex systems that if otherwise not considered could inhibit system functions.

System V & V for products created by platform-based design can be done earlier than conventional systems to potential save a great deal of money, time, and effort. Key tools for system V &V on complex systems such as augmented cognition platforms would be requirements traces, analysis of communication flow, spatial logic and temporal logic.

5   Conclusion

Augmented Cognition systems will improve human information processing capabilities by providing easily-assimilated information by knowing the human cognitive state and the current tasks being executed. They will enable a huge advance in operator-computer effectiveness by creating cognitive closed-feedback loops between operators and adaptive computer based systems. Additionally, they will have the capability to change information modalities (providing information through a different medium such as aural, spatial, or verbal) or offload tasks – whatever is required to enable and optimally functioning user.

Current Augmented Cognition systems are in the prototype phase, and are totally platform- and task-specific. The components that have been created are not extensible from environment to environment, and the next generation of systems will be largely 'built from scratch.' In envisioning the design cycle for future systems, it was first assumed that a merely modular design would enable the reuse of these sophisticated components. However, it was later realized that aspects of the specific platforms would have to be part of the design process in order for systems to actually be realized: modular design made implementation of augmented cognition designs possible, but platform-based design makes it feasible.

Using the principles of platform-based design will allow Augmented Cognition technologies to be extended from primarily military-based platforms to educational, office-work, and even personal-use environments. This technology has the potential to revolutionize the way and amount of information that humans are able to process, and will become only more important as the amount of and scope of information available continues to increase.

References

Austin, M. (2004, December). Toward Platform Architectures for Modular Cognitive Cockpits. Abstract submitted to Augmented Cognition International, Las Vegas, NV. In review.

Austin, M (2004, September). Class Notes for ENSE 621-623. Not published.

Ericsson, A. and G. Erixon (1999). Controlling Design Variables: Modular Product Platforms. ASME Press, New York, NY.

Raley, C., Stripling, R., Schmorrow, D., Patrey, J., & Kruse, A. (2004, September). Augmented Cognition Overview: Improving Information Intake Under Stress. Proceedings of the Human Factors and Ergonomics Society 48th Annual Meeting, New Orleans, LA. In press.

Rasmussen, J., A. Pejtersen, and L. Goodstein (1994). Cognitive Systems Engineering. John Wiley & Sons, New York, NY. 1994.

Sangiovanni-Vincentelli, A (2002, February). Defining Platform-based Design. EEDesign of EETimes.

Sangiovanni-Vincentelli, A. (2003, May-June). Electronic-System Design in the Automobile Industry. IEEE Micro. pp. 8-18.

Schmorrow, D., Raley, C., Marshall, L. (2004, March). Toward a Cognitive Cockpit. Poster presented at Second Annual Human Performance, Situation Awareness and Automation Technology Conference, Daytona Beach, FL.

Schmorrow, D., Raley, C., et. al.. (2004, March). Toward improved situational awareness through knowledge engineering. Poster presented at Second Annual Human Performance, Situation Awareness and Automation Technology Conference, Daytona Beach, FL.

Modular Software for an Augmented Cognition System

Kenneth O'Neill

Boeing Phantom Works
P.O. Box 3707
Seattle, WA 98124
kenneth.r.o'neill@boeing.com

Abstract

A closed loop Augmented Cognition system can be modeled as five components: an operator, the machine being operated, an operator-machine interface (OMI), physiological sensors, and an augmentation manager. Augmentation manager is the software that provides adaptive control of the OMI. This software must monitor the machine's status, the operator's behavior, and physiological sensor measurements, and compute the optimum presentation of current information and future tasks. The number and complexity of connections between the augmentation manager and the rest of the system seems to require an architecture that is tightly interwoven, without separable modules. However, while the disadvantages should not be overlooked, there are significant advantages to a modular architecture for an augmentation manager.

Boeing developed a modular architecture for the augmentation manager in its Augmented Cognition system. This software runs as a separate process from the sensors, displays, and machines (simulated uninhabited air vehicle control stations) that make up the rest of the system. All interaction is through a fixed set of interface routines. The augmentation manager cannot and does not have knowledge of the OMI displays; instead there is the concept of an "interaction object" that is mapped (not necessarily one-to-one) to features of the display.

Disadvantages found with this approach were mainly associated with the additional effort required. Code had to be added to the OMI and simulated control stations to call the interface routines that send and receive data. A moderately large code module was required to do the translation from augmentation manager's interaction objects into OMI display objects. A major effort was required to create a module that used physiological sensor data to produce cognitive state gauges.

The basic advantage to a modular approach is the capability to connect or disconnect individual modules without impacting the operation of the remainder of the system. In Boeing's Augmented Cognition system, if the augmentation manager itself is completely disconnected the system just reverts to the baseline non-augmented operation. Multiple different sensor suites have been connected to drive closed loop operation, with no changes required to the configuration of the rest of the system. The current augmentation manager's speed and memory requirements have not stressed the available computing resources; but if the software continues to expand, it can be re-hosted on upgraded hardware without affecting the platforms used for the OMI and simulated control stations.

Finally, modular architecture confers a benefit during development, in that individual modules can be developed in parallel. The augmentation manager was developed concurrently with the simulated control stations and with sensor hardware and software. Integration was still required but the effort was minimized because the interfaces had been largely defined up front.

Keywords: augmented cognition; augmentation manager; operator-machine interface; modular architecture; modular software

1   Introduction

The goal of Augmented Cognition is to optimize performance of a human-machine system in the presence of human cognitive limitations. This is accomplished by assessing the operator's cognitive state and adjusting the interface based on that state, in a closed-loop feedback process.

A classical closed loop feedback system is illustrated in figure 1. The external input includes unexpected disturbances. The plant is a device whose output must be kept within specified parameters in spite of those disturbances. The solution is feedback wherein sensors detect the state of the plant and pass that information to a controller. The controller computes commands that will produce the desired output. The commands are then sent along the input path into the plant.

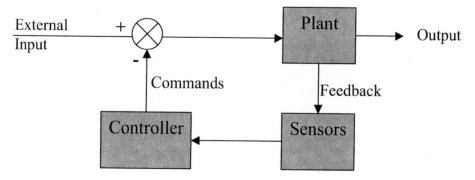

Figure 1: Classical feedback system

The Boeing Company designed and implemented an Augmented Cognition system under Defense Advanced Research Projects Agency' Improving Warfighter Information Intake Under Stress program phase 2. The device studied was a ground control station for a simulated uninhabited air vehicle. The resulting closed loop system, shown in figure 2, resembles the classical feedback system. Note that except for the specific identity of one component ("Aircraft") and its output ("Aircraft State"), this diagram also applies to any Augmented Cognition system.

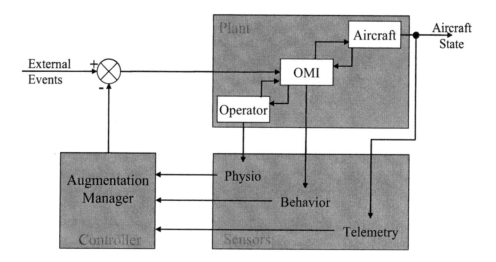

Figure 2: Boeing Augmented Cognition system

In Boeing's system the plant expands into 3 sub-blocks consisting of the human operator, the operator-machine interface (OMI), and the simulated aircraft. There is two-way communication between the operator and the interface, as the operator gathers information and gives commands, and between the interface and the aircraft, as the interface uplinks commands and receives telemetered state information. The output of the plant is the set of aircraft parameters to be controlled, for example heading and speed. The input is external events that have been detected, such as an unexpected shift in wind direction, which need to be displayed to the operator.

The sensor block represents all data collected and passed to the controller. This includes simulated aircraft telemetry although in Boeing's system the data is not generated by an actual sensor. Behavior data is produced by a number of devices, including the mouse and keyboard for operator input, a microphone for voice commands, and a camera for point-of-gaze tracking. These devices can be thought of as behavior sensors since they supply data on the operator's current and potential future activity. The "Physio" label refers to sensors that detect the operator's physiological state. The Boeing system used a number of combinations of EEG, EOG, ECG, pupillometry, and fNIR sensors for physiological data.

The controller is a software module called the augmentation manager. Augmentation manager uses aircraft state data in logic that assesses the current situation with respect to mission goals and plans; uses behavioral data in logic that monitors the operator's current activity and predicts future activity; and uses physiological sensors to assess the operator's cognitive state. All of this information is then used to select the operator-machine interface that will produce the best overall system performance.

Design of the augmentation manager is closely tied to the specific system that is to be controlled. A large quantity of data flows in from the various types of sensors, at various update rates, with widely varying precision and reliability. Interpretation of the data is complex and depends on unique characteristics of its source, either the aircraft, OMI, or a particular operator. Augmentation manager output can only be relevant to the targeted interface. For all these reasons it may seem that the best architecture would be an augmentation manager that is directly integrated with the plant itself. However, there are considerable advantages to an architecture of separable modules, and this is what Boeing selected.

2   Design solution

In Boeing's architecture augmentation manager is a separate module from the remainder of the system. The software runs in its own process, separate from the OMI display software and from any sensor software. All interaction between the augmentation manager and the remainder of the system is through configuration controlled code consisting of three main parts.

The first part is a set of routines collectively called the augmentation manager application program interface (API). The routines in the API are the only avenue to push behavior and aircraft state data in, and to pull OMI data out. The name and argument list (but <u>not</u> the underlying logic) for each routine in the API is set in configuration controlled code that is separate from the augmentation manager. In object oriented programming this type of code is a construct known as an interface. There are many ways of implementing such an interface. Without further detail it is noted that in Boeing's C++ implementation, the interface is a class containing only pure virtual functions, from which the augmentation manager is derived.

The second piece of code to discuss is not a part of augmentation manager at all, but is required for modular operation. This code is called the OMI configuration module. Its function is to translate augmentation manager output into specific modifications to operator displays, because those two types of information are different. The difference is a result of the modular design.

The OMI consists of software and hardware whose purpose is to present information to the operator. The OMI does so through constructs called display objects. Examples of display objects are a blinking icon, a popup text box, or a synthesized voice message, any one of which might be used to warn the operator of a high priority task. In contrast, augmentation manager logic is not concerned with actual displays but rather with the more abstract considerations of cognitive state. The augmentation manager is therefore designed to work with constructs called interaction objects. Interaction objects map to display objects, but not in one-to-one fashion. The display objects above would all map to one interaction object, the warning itself.

The OMI configuration module is the software that performs the mapping. In the example, performing the mapping from warning to icon, text, or voice requires augmentation manager to supply both the interaction object and a property that applies to the object. The property would be "spatial", "verbal text", or "verbal auditory", would be based on operator cognitive state, and would likely apply to all then-current interaction objects. (Note that the feedback loop shown in figure 2 implies that the warning could arise from external events as well as from augmentation manager task prioritization.)

The third and largest block of code necessary to for the modular design is the cognitive state assessor. This code translates the information produced by the sensors into cognitive gauges that can be understood by the augmentation manager. A cognitive gauge is defined as a high level indicator of the operator's cognitive state. Examples are gauges for global workload and for working memory usage. By interpreting sensor data into gauges, the cognitive state assessor divorces the augmentation manager not only from the data format but also from the specific sensors themselves. In a further modularization of the design, the cognitive state assessor is implemented separately from both the augmentation manager and the sensors. It should be noted that creation of a self-contained cognitive state assessor required significant effort, both in design and implementation.

2.1 Advantages

Although the efficiency of an integrated system is lost, the advantages to a modular architecture are numerous. The basic advantage is the ability to swap out individual modules without impacting the operation of the rest of the system.

During 2004 Boeing performed both research experimentation and developmental testing with its Augmented Cognition system. The augmentation manager software was upgraded on a weekly cycle, and was significantly different during each of the quarterly concept-validation experiments. The same was true of other components. The sensor suite that generated data for the first quarterly experiment consisted of EEG, EOG, and ECG leads. For the second experiment the suite included those plus pupillometry; for the third, pupillometry and fNIR; and for the fourth, fNIR alone. Since the fNIR system was under development, different generations of fNIR sensor were used in the different experiments. Conventional EOG sensors were used in one experiment and dry contact electrodes in another.

The cognitive state assessor, whose purpose is only to facilitate the connection between one component and another, is itself a separable module. Two different implementations were used in 2004 experiments, one based on neural nets and one based on statistical process control. A third based on fuzzy logic is planned for 2005.

These changes to software and hardware required no changes to, and had no effect on, the remainder of the system. Modifications could be accomplished in seconds, and were. On numerous occasions the sensor mix was changed between experimental trials, as a particular component failed or quick-look analysis showed anomalous behavior. During one experiment the entire sensor strategy was changed from EEG-heavy to fNIR-heavy when data reduction showed promising results from fNIR. (Boeing's system does not allow for hot-swapping components during experimental trials, however.)

A modular architecture also conveys advantages during system development: individual modules can be developed in parallel by separate teams. Coordination between the teams is accomplished up front and documented in the controlled interface. Teams are then constrained only by the interface, giving considerable freedom to upgrade and make design changes. The OMI and augmentation manager were developed essentially independently within Boeing. A non-Boeing vendor developed each sensor, and one of the cognitive state assessors was developed outside

Boeing as well. Integration of all these disparate pieces was still required, but efforts were minimized by the coordination early in the design cycle.

Another advantage, which has not yet been required, is the ability to physically separate the modules. The augmentation manager software ran on the same computer as the plant (the OMI and simulated aircraft) in 2004 because speed and memory requirements did not stress the available resources. As development continues and the software grows, augmentation manager may be re-hosted on a separate machine with no effect on system operation.

Finally, a significant advantage is the ability to not only swap components but also remove them entirely. Figure 2 illustrates a closed loop system, but when the feedback loop is removed what remains is a fully operational non-augmented cognition system. During 2004 this capability was used to test and develop the simulated baseline control station and aircraft, but if augmented cognition is added to a pre-existing system the capability becomes even more useful. Risk of degrading performance of an existing system is always a barrier to deployment of a new technology, but modular architecture minimizes the risk. If any component of the augmentation feedback fails, or the user decides to opt out, the loop is broken and the system reverts to its baseline operation.

2.2 Disadvantages

There are disadvantages to using a modular architecture. It is inherently inefficient because the addition of interfaces requires both memory and execution time, but these issues can be ameliorated by proper choice of computing platform. The chief disadvantage of a modular architecture was the effort required to implement it.

Implementing the augmentation manager API was a relatively minor task. Some effort was needed to modify the code that called the API, but this integration would have been required for any architecture. Further, since the API was configuration controlled, much of the integration only needed to be done once, where a non-modular architecture would require repetition each time the augmentation manager was upgraded.

Constructing the OMI configuration module was a larger effort. The outputs from augmentation manager and the possible display modifications are both numerous. In addition, the configuration module is closely tied to the implementation of the OMI itself, so it often required updating in lockstep with OMI changes. The OMI configuration module is more closely tied to the specific plant than any other component of the system. While other components (such as the sensors or even much of the augmentation manager) could be used with another plant without modification, the OMI configuration module would need to be written specifically for that system. The lack of re-usability of this moderately sized block of code must be considered a disadvantage of this architecture.

Finally, a major effort was required to create the cognitive state assessor. The initial design was a multilayer perceptron neural net trained with back propagation. The second design used statistical process control with 24 possible rule sets. The planned design for 2005 is a fuzzy logic module that will attempt to distinguish among all possible combinations of 9 different signals, each of which has 4 possible levels. Design and implementation of such a module is plainly a significant task.

3 Conclusion

A closed loop augmented cognition system follows the model of a classical feedback system. This model lends itself to implementation as a modular architecture. Disadvantages of this architecture are chiefly the additional effort required during implementation. Outweighing this are the advantages conferred, including the capability to do concurrent development of system components, the ability to rapidly modify system configuration, and the potential to aid in end-user acceptance of augmented cognition technology.

AMI: An Adaptive Multi-Agent Framework for Augmented Cognition

Matthew Johnson, Shri Kulkarni, Anil Raj, Roger Carff, and Jeffrey M. Bradshaw

Institute for Human and Machine Cognition
40 South Alcaniz, Pensacola, FL 32502
{mjohnson, skulkarni, araj, rcarff, jbradshaw} @ihmc.us

Abstract

In this paper, we report on our efforts to develop an Adaptive Multi-agent Integration framework, called AMI, and its application in the development of Augmented Cognition (AugCog) systems. Key components of AMI include: 1) Pluggable sensor architecture with dynamically substitutable measures for flexible, optimal exploitation of all available *multi-sensory* channels; and 2) An industrial-strength *integration* approach and providing high fault-tolerance and dynamic workload distribution across virtually any implementation platform. AMI leverages IHMC's Knowledgeable Agent-oriented Systems (KAoS) services to provide a flexible and generic approach for linking components. For AugCog systems, these components can be for sensing, detecting cognitive state, and modulating the application itself. AMI has served as the integration platform for a large multi-institutional team led by Honeywell Laboratories under funding from DARPA's Improving Warfighter Information Intake Under Stress program. Multiple sensors and cognitive state measurement gauges have been independently developed by members of the team and straightforwardly integrated for rapid prototype development and evaluation. The flexibility of the AMI framework has been successfully demonstrated by accommodating the myriad of changes to hardware and algorithms required by the research advances achieved throughout the project.

1   Introduction

Research in the field of Augmented Cognition (AugCog) seeks to develop a closed loop human-computer interaction system where the state of the human is measured, analyzed and adapted to automatically to improve performance (Schmorrow & Kruse, 2002). There are many cognitive measures under investigation such as memory, attention, stress, arousal, and workload Schmorrow & Kruse, 2002). From our initial research with the Honeywell team, it is clear that no one indicator will provide a universal solution.   Each cognitive measure provides insight into a particular aspect of the human- computer interaction and its usefulness and accuracy are both highly dependant on both the environment and the domain of application.   There is continuous advancement in sensor technology, cognitive measurement techniques, and adaptation strategies.   Any integration framework has to be flexible enough to cope with the research advances inherent in a new and blossoming research field.   It is clear from these requirements that system integration will be particularly challenging and critical to the viability of an AugCog system.

There are many integration challenges involved in developing an AugCog system.   The highly specialized sensors being employed are often limited to a specific platform by hardware restrictions or driver limitations.   Advanced algorithms and large data sets can be computationally expensive and demanding on limited computational resources.   Both of these considerations indicate the requirement for multiple computational platforms for a complete AugCog system and pose development of such a system as a heterogeneous distributed computing problem.   There are many solutions to these types of problems, such as CORBA (Vinoski, S. 1997), Jini (Venners, 1999), Cougaar (Helsinger, A. & Wright, T. 2005) and the CoABS Grid(Kahn, M., & Cicalese, C. 2001).   The AMI framework does not try to replace these techniques, but instead leverage their capabilities, as discussed in more detail in section 4.   There are two main tasks in distributed heterogeneous computing.   The first is component integration, which involves getting everything to talk the same language.   We address this issue in section 3.   Once everything can talk the same language, the components must find each other.   The discovery process is covered in section 4.   After addressing the distributed computing issues, there still remain several functional and practical challenges of how to configure and manage a complete AugCog System in a way that allows flexible testing and evaluation. One of the goals of the AMI framework was not to provide merely an integration framework, but make use of the framework to provide a set of tools to empower researchers. In section 5 we describe some of the extra features of the AMI framework designed to help a researcher run, test and evaluate a system. Section 6 discusses future work that both addresses areas of weakness and looks toward future advances. Before we address the challenges directly, we will first provide an overview of the history and current state of the AugCog system developed under the DAPRA Improving Warfighter Information Intake Under Stress program.

2   AugCog System Overview

We have developed the AMI framework to support the requirements of the DARPA AugCog project as a member of the Honeywell Team. A basic diagram of the overall architecture is shown in Figure 1. The main components are the Cognitive State Assessor (CSA), the Augmentation Manager, and the application with which the user is interacting. The CSA contains all of the sensors and algorithms for capturing the state of the user. The Augmentation Manager determines how to adjust the application to improve performance based on the inputs of the CSA.

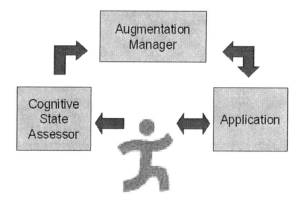

Figure 1 AugCog architecture

The AugCog system we developed was a complex integration challenge, since many components were not ready when we began developing the AMI framework. The cognitive state detection techniques were independently researched and developed by various team members and needed to be easily and seamlessly integrated into the system as they became available. Throughout the project we had several hardware and software additions, as well as frequent software updates to various components. Our current CWA configuration is shown in Figure 2. Most of the hardware provided multiple channels of data. Several of the sensors provide similar data but at different levels of quality. As algorithms were added to interpret cognitive state, they would require certain inputs. The inputs might be from a single device, like Engagement requiring only EEG from Active Two. They may need to be obtained from several devices, like the Gaze algorithm which uses both head and eye movement input. Sometimes different devices provided similar information but had a different quality of data. For example, Arousal needed Inter-beat Interval (IBI) which is available from either Cardiax or Procomp, with Cardiax being the preferred source due to better quality data. AMI also had to handle multiple algorithms requiring the same data. For example, Active Two provided EEG data to Executive Load, Novelty, and Engagement algorithms.

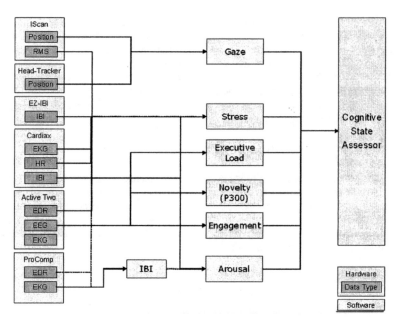

Figure 2 Cognitive Workload Assessor

The AMI framework was not limited to the CWA, but also integrated the Augmentation Manager and the application as well. It has been used to flexibly integrate with both a simulated satellite management application (HCC) provided by NASA (Figure 3A) and a simulated dismounted soldier application developed for the Army Objective Force Warrior (OFW) using the Quake gaming engine (Figure 3B). IHMC extended the normal interfaces of these applications to include haptic devices such as a tongue display shown in Figure 3C.

Figure 3 AugCog Interfaces: A) HCC satellite management B) Quaked based OFW C) haptic tongue display

3   Component Integration Challenge

A core task of the AugCog system is the measurement of physiological data and the processing of that data to detect the cognitive state of the human. The primary goal of AMI is to provide a simple method for data from multiple distributed heterogeneous sensors, algorithms, and interface elements to be collected asynchronously, processed, and used to assist a human. For the Honeywell team, many of the cognitive state detection algorithms were developed independently by researchers whose primary focus is not engineering, but cognitive science. An important design consideration was to minimize the interface burden imposed on those researchers for integrating their component, and free them from the complex issues associated with distributed computing. AMI met this objective by providing a simple and flexible interface that can handle the diverse requirements of different sensors and algorithms. AMI uses agents to handle the complex distributed computing issues, but the framework does not require those wishing to integrate components to have knowledge of agents in order to interface with the framework. Instead developers implement a simple interface, described in section 3.1, which gives the agents the necessary access to the component. The agents themselves can run on any platform that supports Java. Once an interface is created, the component can be plugged into the architecture. Another nice feature of the interface is that it provides a natural method for decoupling the sensor from the algorithm, enabling the reuse of both components separately and minimizing the impact of changes to either component.

The AMI framework provides an architecture, shown in Figure 4, for a *Basic Agent* to handle, and hide, the basic communication and configuration needs of the components being integrated into the AugCog system. The architecture includes four interfaces that allow for separation of different aspects of a component during integration:

1) *Data Provider* (required) provides basic data input and output for sensors or algorithms
2) *Manipulator* (optional) provides data manipulation to incoming or outgoing data
3) *Calibrator* (optional) provides mechanism to start and stop calibration
4) *Graphical User Interface* (optional) provides a graphical window for testing, debugging, and other functions

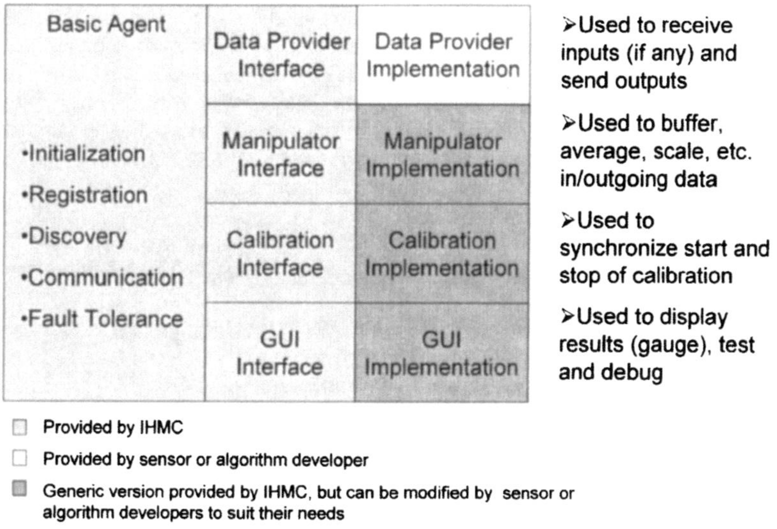

Basic Agent	Data Provider Interface	Data Provider Implementation	➢Used to receive inputs (if any) and send outputs
•Initialization •Registration •Discovery •Communication •Fault Tolerance	Manipulator Interface	Manipulator Implementation	➢Used to buffer, average, scale, etc. in/outgoing data
	Calibration Interface	Calibration Implementation	➢Used to synchronize start and stop of calibration
	GUI Interface	GUI Implementation	➢Used to display results (gauge), test and debug

☐ Provided by IHMC

☐ Provided by sensor or algorithm developer

▨ Generic version provided by IHMC, but can be modified by sensor or algorithm developers to suit their needs

Figure 4 AMI Agent Architecture

3.1 Data provider

Developers are only required to implement a few basic Java methods defined by the AMI *Data Provider* interface. Since many sensor drivers are written in C/C++, the AMI source code also provides a JNI (Java Native Interface) example that can be used to wrap C/C++ code. Developers can protect their source code by providing only the Java class files or a dynamically linked library for C code. They will also need to provide a description of the input requirements and outputs for their component including: initialization parameters (if required), required input including information about format and frequency (if required), expected output in the correct format and a description of what the data means and how to use it in the context of cognitive state assessment. The data format is based on a simple two dimensional array. Each row represents a data sample. Each column is assigned to a specific data channel. Each data sample is also required to have a timestamp generated by the frameworks time synchronization utility discussed in section 5.2, and an assessment of the quality of data as assessed by the developer of the component based on available information. An example of a set of three samples across three channels is depicted in Figure 5.

	Timestamp	Certainty	Channel 1	Channel 2	Channel 3
Sample 1	12345670.0	1.0	33443.5	665.22323	0.3234
Sample 2	12345671.0	1.0	343545.232	2323.3323	-0.2323
Sample 3	12345672.0	1.0	352356.3	1200.0	0.2334

Figure 5 IHMC AugCog Standard Data Format

3.2 Manipulators

To keep the *Data Providers* simple and easy to build, we created the *Manipulator* interface. This interface is used to manipulate the data coming into or going out of a given *Data Provider*. During development we determined several functions that were frequently used to process data either as it was received or before it was sent. For example, we sometimes needed to provide a buffered average of the data to smooth the results. We incorporated many of these generic functions into the AMI framework as *Manipulators*. A *Data Provider* could be nested inside multiple *Manipulators*. For example, data could be scaled and buffered before being sent out. Manipulators can also be applied to single or multiple channels of the data to allow for the flexible application of manipulators to the data. Some Manipulators currently implemented in the AugCog code are:

- *Simple Calibrator* - scales the specified channels by computing the mean and standard deviation for each source channel during the calibration period and then scaling (modified Z-Score using standard deviation) the output of those channels.
- *Output Buffer* - buffers the output allowing the component providing the data to run faster than the component accessing the data.
- *Averager* - outputs the average of the last specified number samples of data.
- *Input Synchronizer* - used to over-sample data or synchronize asynchronous data. It outputs the last value at a specified frequency.
- *Input Buffer* – buffers incoming data to allow the component sending data to run faster than the component receiving data.
- *Input Channel Averager* - modifies the specified channels of the specified agent by replacing the channel data with a rolling average of the last specified number of channel values.
- *Linear Channel Scaler* - scales the specified channel between {-1, +1}
- *Median Channel Filter* - buffers the specified number of values of the specified channels and outputs median value.

3.3 Calibration

Many of the physiological measures used for AugCog systems require calibration. The AMI framework provides some simple calibration techniques, such as linear scaling, but also allows the developer to extend or replace this with their own techniques through a simple interface. The interface adds hooks into their calibration process, and these hooks can then be generically handled by the AMI framework to allow calibration consistent with the domain of application. The calibration can be started or stopped by the graphical interface of the agent controlling the component or by a remote call from a separate component.

3.4 GUI

The *GUI* interface provides a way of separating the visual aspect of the *Data Provider* from the pure non-graphical data retrieval, data manipulation, and data broadcast. The AMI framework provides a basic GUI that provides some testing and debugging features. The basic GUI displays the input and output connections and provides the ability to view the raw inputs and outputs, record the data, and display it on a simple gauge as shown in Figure 6. This GUI can easily be replaced by a more sophisticated one as required, or be disabled to reduce the computational load on the system.

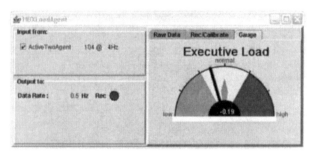

Figure 6 Basic Agent GUI

4   Discovery Challenge

AMI leverages IHMC's Knowledgeable Agent-oriented Systems (KAoS) services (Uszok, Bradshaw, Jeffers, 2004) to provide a flexible and generic approach for linking components. KAoS uses Java Agent Service (Arnold et al., 2002) to abstract the underlying implementation and provides a simplified and consistent interface to several distributed process communication mechanisms including TCP, CORBA (http://www.omg.org), Cougaar (http://www.cougaarsoftware.com), and the CoABS Grid (http://coabs.globalinfotek.com). KAoS also provides an advanced directory service that allows for the publication of not just the agent, but its capabilities as well. This enables component discovery based on capability vice stipulating a specific component name or location. Components with similar capabilities can be seamlessly and dynamically interchanged. The agents that wrap the components hide all the details of registration and querying for needed capabilities. They also handle establishing the required communication. Agents allow data to flow not only between different pieces of code, but also between different platforms. All that is required is a network connection. Agents automate the discovery process required to find necessary data providers. They flexibly handle establishing data exchanges between components and provide robust fault tolerance by handling service interruptions gracefully and attempting to re-establish the connection autonomously. This is critical in AugCog systems where new and untested components are added to the system. These components should not be able to break the entire system, even if an individual component fails completely. Hardware restrictions not withstanding (i.e., a sensor agent is locked to the platform to which the sensor hardware is physically attached), each agent could run on any platform. This allows the agents to harness any available computing power. This flexibility is needed to balance the work load when dealing with algorithms or sensors that were platform restricted or computationally expensive.

5   Tools for running, testing and development

The AMI framework includes a variety of tools for running, testing and evaluating AugCog systems.

5.1   System Management

Although our AugCog system had to handle distributed processes across diverse platforms, the complete system needed to be configured and run by users who may not have an advanced background in distributed computing. It also had to be able to handle changes in overall system configuration, since component availability and requirements frequently changed.   Agent Launcher is a utility provided by the AMI framework to allow for easy system start-up. Running Agent Launcher automatically creates the necessary infrastructure for the AMI framework based on the desired configuration provided by a simple text file. It allows for both local and remote launching of agents and displays the system process ID numbers for each agent to help in debugging. Agent Launcher provides a simple push-button approach to running our AugCog System.

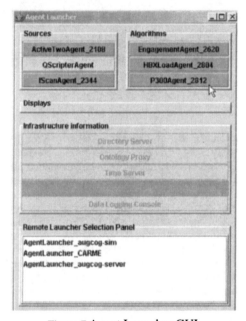

Figure 7 Agent Launcher GUI

5.2   Time Synchronization

Time synchronization is a difficult problem in distributed systems. The AMI framework does not claim to be the panacea for the distributed time synchronization issue, but it does provide a solution that met the accuracy requirements for our particular implementation. By incorporating the time stamping service into the framework, AMI frees the developer from having to worry about time synchronization issues. This also allows the system to be updated to an improved time synchronization technique without affecting the existing code.

To enable data fusion across platforms each data sample must include an associated timestamp synchronized to a global reference. The AMI framework uses a Time Server to provide this reference for distributed agents. A single Time Client runs on each platform and can service many agents. The clients periodically correct their platform time to correlate with the server. The corrections also take into account network delay with an assumption of symmetric network latency. Although this assumption may not be very accurate, it is mitigated by frequent corrections. The corrections, as well as the network latency can be analyzed through tools provided by AMI to verify that the system is working within acceptable margins of error. The resolution of the time returned by AMI framework depends upon the underlying operating system. For example, Microsoft Windows operating systems provide time-of-day accuracy of only 10 milliseconds. However, timestamp intervals can have a much greater resolution if the machine supports a high performance counter. The AMI framework provides this functionality for both Java and C/C++ implementations of the Data Provider. The AMI framework provides a simple utility for obtaining a timestamp for a data sample. However it is up to the developer to ensure the timestamp is obtained as close as possible to the actual sample time.

5.3 Data Logging

The AMI framework includes tools to assist in data collection. A critical component of the AugCog system is to be able to collect data for evaluation of a sensor or algorithm. A generic data logging interface is available to save data in a binary or text format, merge the data into one file and also a means of converting the data to a MATLAB .mat format file. The logging can be controlled locally or managed by the AMI frameworks *Data Logging Agent*. This agent provides an interface with a selectable list of the currently running agents and allows the user to tell all selected agents to start/stop logging. The file name and type can be specified and each session is distinguished by appending a timestamp. Another benefit of allowing the data to be collected by the framework is that it can be used to replace data output of a component when it is not available. This is a very useful testing tool in AugCog systems where specialized sensors are not available, expensive to operate, or require a lot of set up time. We can also use this feature to verify the performance of a component provided by the team members using the sample data they provided. We have also used the component simulation to reconstruct data after an experiment when there had been a failure.

6 Future Work

As we look toward future improvements to the system, we hope to address some of the weaknesses as well as expand the functionality in new ways. Two weaknesses we are addressing are directory service persistence and distribution. The directory service does not currently save it state in a manner that allows for recovery in the event of failure. We are working on a persistence mechanism to allow the directory service to recover its state after failure. Currently components need access to the directory service during initialization. Once connected, the system can run without the directory service. system self configuring by the agents to proactively starting their own data sources. For example, if the stress gauge needs EDR, pupil RMS, and HFQRS, it would take the initiative to launch the required agents. We would also like agents to relocate autonomously to another platform, hardware restrictions not withstanding, to optimize utilization of available computational resources automatically. Currently we can only manually distribute the agents.

Looking further out in the future, if AugCog systems become prevalent, there will be issues of data access privileges. These concerns could be operational driven, for example who needs the information to complete a mission, or they could be privacy driven, for example who is authorized to view the data. By using KAoS as our backbone for AMI, we are afforded access to rich domain and policy services. KAoS Domain Services provide the capability for groups of software components, people, resources, and other entities to be semantically described and structured into organizations of domains and subdomains to facilitate collaboration and external policy administration. KAoS Policy Services allow for the specification, management, conflict resolution, and enforcement of policies within domains. Although not as critical for the current small scale, single-user applications currently being developed, these services could prove to be invaluable in multi-user, cross-organizational applications.

7   Conclusion

AMI has served as the integration platform for a large multi-institutional team led by Honeywell Laboratories under funding from DARPA's Improving Warfighter Information Intake Under Stress program. Multiple sensors and cognitive state measurement gauges have been independently developed by members of the team and straightforwardly integrated for rapid prototype development and evaluation. The flexibility of the AMI framework has been successfully demonstrated by accommodating the myriad of changes to hardware and algorithm required by the research advances achieved throughout the project. The use of agents as a basis for a framework in developing AugCog systems has proven to be an effective means of providing the dynamically adaptable and fault tolerant environment required by the advancing research field of Augmented Cognition. The AMI framework also provides several tools that have proven to be useful in the management, testing and evaluation of AugCog systems.

References

Arnold, G., Bradshaw, J, de hOra, B., Greenwood, D., Griss, M., Levine, D., McCabe, F., Spydell, A., Suguri, H., Ushijima, S. (2002) Java Agent Services Specification. http://www.java-agent.org/

Helsinger, A. & Wright, T. (2005) Cougaar: A Robust Configurable Multi Agent Platform, IEEE Aerospace Conference

Kahn, M., & Cicalese, C. (2001). CoABS Grid Scalability Experiments. O. F. Rana (Ed.), Second International Workshop on Infrastructure for Scalable Multi-Agent Systems at the Fifth International Conference on Autonomous Agents. ACM Press

Schmorrow, D. & Kruse, A. (2002) Tomorrow's human computer interaction from vision to reality: building cognitively aware computational systems. Proceedings of the 38th International Applied Military Psychology Symposium, Amsterdam, Netherlands, 20-24 May.

Uszok, A., Bradshaw, J. M., & Jeffers, R. (2004). KAoS: A policy and domain services framework for grid computing and grid computing and semantic web services. In C. Jensen, S. Poslad, & T. Dimitrakos (Eds.),Trust Management: Second International Conference (iTrust 2004) Proceedings,Oxford, UK, March/April, Lecture Notes in Computer Science 2995, Berlin: Springer, pp. 16-26.

Venners, B. (1999) Objects, the Network, and Jini, http://www.artima.com/jini/jiniology/intro2.html, accessed on 15 FEB 2005

Vinoski, S. (1997) CORBA: Integrating Diverse Applications Within Distributed Heterogeneous Environments. IEEE Communications magazine, Vol. 14, No. 2.

Developing a Human Error Modeling Architecture (HEMA)[1]

Michael E. Fotta
D.N. American
1000 Technology Drive
Fairmont, WV 26554
mike.fotta@dnamerican.com

Michael D. Byrne
Rice University
6100 Main St., MS-25
Houston TX 77005
byrne@rice.edu

Michael S. Luther
Booz Allen Hamilton
8283 Greensboro Dr
McLean, VA 22102
luther_michael@bah.com

Abstract

Although computational cognitive architectures have been applied to the study of human performance for decades no such architecture for modeling human errors exists. We have undertaken the development of a Human Error Modeling Architecture (HEMA), building on the ACT-R cognitive architecture. In developing HEMA we first set the context of what error types HEMA would handle and what overall cognitive performance process (a Framework for Human Performance) was being assumed. We then identified the cognitive functions which were failing and how they are failing when an error occurs. An analysis of these failures, in relation to the Framework, enabled us to specify a set of General Error Mechanisms. Comparison of these mechanisms to existing ACT-R mechanisms identified where ACT-R could be used, where modifications where necessary and where new mechanisms or modules would be need to develop HEMA. A conceptual design for HEMA was then proposed.

1   Introduction

Human error is continually cited as a cause in major disasters and minor mistakes. However, human error can often be traced to a system design which creates situations beyond a human operator's capabilities. Consider, for example, the Defense realm. Given the speed of weapons systems (e.g., supersonic aircraft, missiles) an operator must often filter, process and make decisions at a speed that does not allow for careful consideration of all the information. Human error at such times can lead to serious and even deadly consequences, such as "friendly-fire" incidents. Providing insight into the human error consequences resulting from a particular system design would enable designers to chose between alternative designs and modify a design to reduce error occurrence or enable recovery from human errors. Our research seeks to develop a Human Error Modeling Architecture (HEMA) that provides this insight by simulating errors that operators will experience as a result of a system design.

Efforts at modeling human error to provide predictive power are scarce. Their have, however, been a large number of taxonomic and descriptive efforts to explain human error behavior. Some of the most well known of these are the Generic Error Modeling System (GEMS) approach (Reason, 1991), the stages-of-action model (Norman, 1986) and the Cognitive Reliability and Error Analysis Method (CREAM) (Hollnagel, 1998). However, they are neither mechanistic, which limits their explanatory power, or predictive. Without predictive power these approaches cannot generally be used to determine which of two designs would generate fewer or less serious errors.

In more recent years there have been attempts to predict errors at a more mechanistic level. One example is the work of Byrne and Bovair (1997), which presented a computational account of a class of errors known as postcompletion errors (e.g., leaving a bankcard in an ATM or leaving the original on a photocopier). Another illustrative recent example comes from Anderson, Bothell, Lebiere, and Matessa (1998), which used ACT-R as a model and showed that it was possible to predict both the rate and content of the errors made in a task.

A fundamental problem in modeling human error is that it is the same human perceptual-cognitive-motor system producing all behavior, whether erroneous or not. Thus, to effectively model human error, it will be necessary to have a relatively complete model of the entire perceptual, cognitive, and motor systems. Computational cognitive architectures such as ACT-R, EPIC or SOAR provide such models (see Byrne, 2003a, for a review). Thus, a

---

[1] This research was funded under SBIR Phase I Contract: N00014-03-M-0357 for the Office of Naval Research.

1025

cognitive architecture, properly modified, could serve as a basis for a predictive model of human error (Byrne, 2003b). Our research pursues this approach to develop a Human Error Modeling Architecture (HEMA).

We chose to use ACT-R (Anderson, Bothell, Byrne, Douglass, Lebiere, & Quin, 2004) as the basis for developing HEMA. ACT-R contains mechanisms which can produce "erroneous" behaviors even when the ostensibly "correct" pieces of declarative and procedural knowledge are present in the system. Furthermore, ACT-R has been extensively and successfully applied to model many domains of human performance, has wide acceptance as a computational cognitive architecture, and has been applied to human error modeling in some instances.

However, while ACT-R contains some error modeling mechanisms, it is unlikely to have all the components necessary for comprehensive error modeling. In order to develop HEMA we need to specify the extent to which ACT-R currently provides these mechanisms. Before accomplishing this we obviously had to define the error related mechanisms needed to develop HEMA. In order to define these mechanisms it is necessary to first identify the cognitive functions which can fail and how they can fail when an error occurs. Identification of these failures has to be set in the context of what errors are occurring and what overall cognitive performance process is being followed. Thus, the first two activities undertaken in developing HEMA were to define the cognitive processes that are involved as an operator performs a complex task and define the error types HEMA should handle.

This paper reports on the results of these tasks beginning with the last two and culminating in the development of the Human Error Modeling Architecture.

2 Define the Human Performance Process

In order to define the cognitive processes involved in human error it was necessary to develop a process model of how an operator performs in an environment likely to produce a variety of human errors. This framework, while a synthesis of readings of a variety of papers and texts, and discussions between the authors, owes much to the texts by Anderson (2000), Reason (1990), and Hollnagel (1998), and papers by Endsley (1999) and Leiden, et. al. (2001).

Figure 1 shows the top level diagram for a Framework of Human Performance (FHP) which describes the processes an operator goes through in performing a complex task. In order to reduce the complexity of this figure and supporting process figures the cognitive processes of attention, perception and memory have been combined in an APM Component and used wherever attention is likely to be applied to enable perception.

The process starts with the arrow on the left of the Attention-Perception-Memory (APM) Component. This represents a conscious intent to attend to the situation at hand, e.g., an airline cockpit display, a Combat Information Center. The operator makes a general assessment of the situation via the Understand Problem decision. Current environmental information from the APM and the operator's knowledge is used to quickly make this decision. If there is enough information the process flows to Set Intention(s). If not, the operator will seek further information using the APM and also directly from Memory. There is obviously a time constraint, but this is not shown here. As this process runs the operator is constructing an internal representation of the situation, the Perceived Situation, which is stored in Memory and retrieved and updated as the rest of the processes function.

In order to Set Intention(s) the operator retrieves information on similar past situations from Memory and compares this to the Perceived Situation. If a match is made to a past situation then the intention for this past situation is used. If a direct match is not made then the best match to either a previous situation or rules from a number of similar situations are used to infer an intention. The intention is stored in Memory.

The operator must then *Form Plan(s)* for each intention. Plans are formed on the basis of the extent to which a plan exist in schemas or other knowledge for the current intention. Reason's (1990) scheme of skill-based, rule-based and knowledge-based processing is used to define the type of plan here. The plan is stored in Memory.

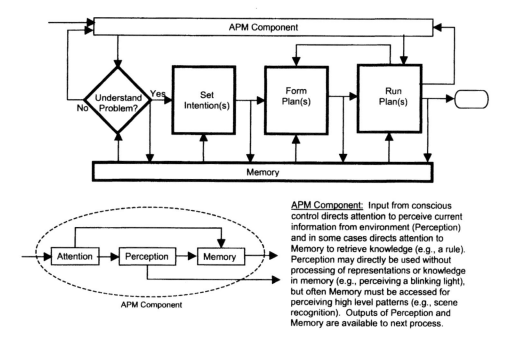

Figure 1. Top level process diagram for a Framework for Human Performance.

Finally, the operator must Run Plan(s). This consists of performing each action in the plan and evaluating the result in relation to the plan and current environmental information, from the APM. The extent of evaluation and modification of plans varies from almost none for skill-based plans to extreme for knowledge-based plans. In the simplest case the next action in the plan will be run or the plan will end. However, results may also indicate that either the intention or plan is no longer appropriate and need modified or dropped. The feedback loops indicate this.

Each process within Figure 1 has been further detailed in its own process description to provide a complete explanation of this framework. This enabled us to use this framework as a basis for identifying the cognitive functions which are involved in producing the error types (see Section 4). Unfortunately space precludes showing this detail here.

3 Define a Sample of Human Error Types

Although our review of the human error literature found many taxonomies and papers on individual errors we came to the conclusion that three approaches formed a broad sample representing the gamut of human error cognitive processing. Each of these approaches are well-described and comprehensive theories of human error and include fairly broad taxonomies. We propose that by identifying mechanisms that cover error types from these taxonomies we form a solid and broad basis for a human error modeling architecture. The three approaches are the Generic Error Modeling System (GEMS) of Reason (1990), Situation Awareness (SA) as put forth by Endsley (1999), and the Cognitive Reliability and Error Analysis Method (CREAM) of Hollnagel (1998). Table 1 shows the error types used in our research from each approach. The taxonomy is shown under the Source column, a high level for the error types in that taxonomy are shown in the Category column, and the Error Types for the Category appear in the last column. Detailed explanations of each error type can be found in the appropriate reference.

Reason's GEMS approach was chosen for a number of reasons: 1) this text is the dominant, most comprehensive descriptive text on human error, 2) it describes a range of performance including skill-based, rule-based and

Table 1. Error types used in study.

Source	Category	Error Types		
GEMS	Skill-based: Inattention	Double-capture slips Reduced intentionality Interference errors	Omissions following interruptions Perceptual confusions	
	Skill-based: Overattention	Omissions Repetitions	Reversals	
	Rule-based: Misapplication of Good Rules	First exceptions Informational overload General rules	Countersigns and non-signs Rule strength	Rigidity Redundancy
	Rule-based: Application of Bad Rules	Lack of Encoding Inaccurate encoding	Protection by specific rules Inelegant rules	Wrong rules Inadvisable rules
	Knowledge-based (KB)	Selectivity Out of sight out of mind Overconfidence	Workspace limitations Confirmation Bias Biased Reviewing	Illusory correlation Causality
	KB: Problems With Complexity	Delayed feed-back Causal series vs. Nets	Thematic vagabonding Processes in time	Encysting
CREAM	Observation (O)	Observation missed		
	O: False Observation	False reaction	False recognition	
	O: Wrong Identification	Mistaken cue	Partial identification	Incorrect identification
	Interpretation (I)	Delayed interpretation	Incorrect prediction	
	I: Faulty Diagnosis	Wrong diagnosis	Incomplete diagnosis	
	I: Wrong Reasoning	Induction error Wrong Priorities	Deduction error	
	I: Decision Error	Decision paralysis	Wrong Decision	Partial Decision
	Planning (P): Inadequate Plan	Incomplete plan	Wrong Plan	
	P: Priority Error	Wrong goal selected		
	Temporary, Person (TP)	Delayed response	Performance variability	Inattention
	TP: Memory Failure	Forgotten	Incorrect recall	Incomplete recall
	TP: Fear	Random actions	Freeze	
	TP: Distraction	Task suspended	Task not completed	Goal forgotten
Situation Awareness	Level 1: Failure to correctly perceive information	Data not available Misperception of data	Data discrimination/detection Failure to monitor or observe data	Memory loss
	Level 2: Failure to correctly integrate or comprehend information	Poor mental model	Use of incorrect mental model Over-reliance on default values	Other
	Level 3: Failure to project future actions or state of the system	Poor mental model	Over-projection of current trends	Other
	General	Habitual schema	Failure to maintain multiple goals	

knowledge-based, and 3) the development and use of plans is a cornerstone of the GEMS approach. Although other authors discuss the use of plans, none go into the detailed description of Reason. Any effort to model human error must account for the development, running, evaluation of and modification of plans.

CREAM is a comprehensive methodology for Human Reliability Analysis (HRA), an extensive area of research to ensure the reliability of complex systems such as aircraft, nuclear reactors, weapons systems, etc. At the highest level CREAM categorizes error types as occurring because of Man (People), Technology or Organizational factors. For our analysis we have used only the People related factors where the cognitive issues and error mechanisms are most likely to occur. The People grouping included one category, Permanent Person Related errors, which we did not include as it went to levels of physical impairments (e.g., deafness, bad eyesight) and individual processing styles (e.g., simultaneous scanning, successive scanning) that is beyond the scope of the first version of HEMA.

Endsley's situation awareness (1999) approach was chosen as: 1) its core tenant is an understanding and projection of an understanding of an entire situation, and 2) the error types form a reasonable taxonomy of perceptual and high-level cognitive processes. The errors which occur when a situation it is not understood is critical to a complete error modeling approach. The perceptual error types are not well covered by Reason, but provide a cross check to the CREAM observational errors. Finally, any error model should be able to account for errors related to complex cognitive processes (e.g., over-projection of current trends) as well as discretized cognitive processing failures.

4 Define Cognitive Functions Involved in Human Error

Our next step was to identify the cause of each error type in terms of a failure within the FHP. We then identified the cognitive function or functions which were involved in this failure. Table 2 shows how this was done for each error type within the Situation Awareness Level 1 category.

Table 2. Identifying the cognitive functions involved in Situation Awareness (SA) Level 1 errors.

Error Type	Location and Cause in FHP	Cognitive Function(s)
Data not available	Not a cognitive process failure.	None
Data hard to discriminate or detect	Occurs in Understand Problem. Could be either a perceptual failure or environment is outside or just at boundaries of perceptual limits.	Perception
Failure to monitor or observe data	Occurs in Understand Problem. Attention failure with perception involved.	Attention: Perception
Misperception of data	Occurs in Understand Problem. If due to influence of prior expectations this is a misperception of the Perceived Situation due to misapplication of declarative knowledge in Memory. If due to distraction this is an Attention failure in APM.	Perception: Memory or Attention.
Memory loss	Occurs in Understand Problem or Set Intentions. No longer in Working Memory, LTM or can not be accessed. Could be that task was shed in Set Intentions process if workload is high.	Memory: Memory loss

An "analysis table" with the type of information as shown in Table 2 was constructed for all error types. Our reasoning was that we would be able to identify commonalities within the cognitive functions by sorting on this column in the table. We hypothesized that these commonalities would enable us to derive a constrained list of General Error Mechanisms. Note that in a few cases (as shown for the "Data not available" error in Table 2) we decided that there was not a plausible cognitive explanation for the error type.

5 Define a set of General Error Mechanisms

Sorting the analysis table on the Cognitive Functions grouped error types which had the same or similar failures. With commonalities in error types now identified and considering the specificity of the error (e.g., if a memory loss then where was it occurring in FHP and/or what other details specified when and how it occurred) we identified the General Error Mechanisms which would have to exist in order to account for all of these error types in our sample.

1029

Table 3 shows a portion of the sorted analysis table. This table shows that by grouping error types by the Cognitive Function Memory with a specific explanation Memory Loss we identified five error types which could be accounted for by the same mechanism - a Decay mechanism. Similar occurrences were found throughout the complete sorted table for all error types, i.e., error types fell into groups by similar Cognitive Functions and explanations. In fact we found it necessary to hypothesize the existence of only 15 error mechanisms, although two did have subsets.

Table 3. Example of identifying a General Error Mechanism for error types with common cognitive functions.

Error Type	FHP Explanation	Cognitive Function	General Mechanism
SA: Memory loss	Loss in Working Memory, LTM, or task was shed in Set Intentions.	Memory: Memory loss	**Decay** -or may be that task was not entered but shed.
SA: Multiple goals	Loss of intention from in Working Memory or a task was shed in Set Intentions.	Memory: Working Memory Loss: Intention	**Decay** (of intention)-or intention was never entered due to task shed.
CREAM: TP-Goal forgotten	Loss of intention from Working Memory (may cause repetition of steps)	Memory: Working Memory Loss: Intention	**Decay** (of intention) -or intention was never entered but shed.
GEMS: Inattention Reduced intentionality	Reduction of strength of intention, or forgetting intention, in Working Memory.	Memory: Working Memory Loss: Intention	**Decay** (of intention) – reduction in strength
CREAM: TP-Loss of orientation	Loss of plan or part of plan from Working Memory.	Memory: Working Memory Loss: Plan	**Decay** (of plan or part of plan)

The hypothesized general mechanisms needed to simulate errors are:
- *Plan developer*: Develops plans given a situation. Can develop incomplete or inappropriate plan.
- *Compare Actions:* Compares expected action to action taken. Can fail due to monitor failure or bias.
- *Monitor*: Performs comparison at certain times to achieve an evaluation, but which can fail to monitor.
- *Attention (for perceptual information):* Allocates attention to perception, but can fail to do so.
- *Bias mechanism*: A bias (strength) which would tend to yield positive comparisons in Compare Action.
- *Rule match*: Matches current information to stored rules. Can fail to retrieve correct rules or apply correct action side of rule. Specific subsets include: General Rules, Rule Bias and Strength of Rule
- *Schema match*: Matches information from perception to entire schema. Can fail to correctly match.
- *Time constraint mechanism:* Places time constraint on various activities, e.g., choosing a rule.
- *Decay:* Reduction of strength of information (e.g., with chunks, rules, intention, plan).
- *Poor Learning (encoding):* Stores incorrect rules, but need to be (somewhat) logically related to learning in previous similar situations. Specific subsets include: Rule conditions not encoded or incorrectly encoded, Rule action incorrect or inefficient, Reduction in rule strength, and Reduction in strength of event schema.
- *Retrieval mechanism*: Retrieves information, but can fail to correctly retrieve.
- *Plan Controller*: Runs plans, but can fail in various ways, e.g., by failing to continue running plans.
- *Perceptual Mechanism:* Inputs perceptual information. Can fail to perceive some information.
- *Association Developer*: Develops associations from memory, but can fail, e.g., by developing narrow association net when deeper one should be developed.
- *Motor Mechanism:* Performs motor execution, but can fail to perform necessary action.

Note that all of these mechanisms are not directly error causing mechanisms. Many are functions that will have to be simulated (e.g., the Plan Developer) to account for the cognitive processes which lead to the error types within the FHP framework. Some mechanisms on the other hand are directly related to errors (e.g., the Bias Mechanism).

6   Compare the General Error Mechanisms to ACT-R mechanisms

Given these necessary error mechanisms we then performed a comparison to ACT-R mechanisms as shown in Table 4. In other words we wanted to identify to what extent ACT-R could account for the General Error Mechanisms. Some of the General Error Mechanisms map straightforwardly to extant ACT-R mechanisms, some will require an extension of existing ACT-R mechanisms and some require new development in HEMA. Table 4 shows both the comparison to ACT-R and the proposed implementation in HEMA.

Table 4. Comparison of General Error Mechanisms to ACT-R

General Error Mechanism	Comparison to ACT-R	Implementation in HEMA
Plan developer	Could use productions and chunks in ACT-R but difficult. AI planning field provides better approaches.	Needs a new mechanism in HEMA - Plan Developer.
Plan Controller Compare Actions Monitor Bias mechanism	Combine into one mechanism for controlling and monitoring plans. Could revise and modify previous ACT-R goal management system, but revision more cumbersome than developing a new mechanism.	Needs a new mechanism in HEMA - Plan Controller which includes Compare Action, Monitor, and Bias Mechanism functions.
Rule Match	Mechanism exists in ACT-R.	Use ACT-R mechanism.
Schema Match	Schemas represented in ACT-R by large hierarchical chunks, but this works poorly and matching is slow.	Modify ACT-R by associating large chunks to develop specific schema structures.
Time Constraint Mechanism	A timing mechanism must be included as ACT-R has no time sense.	Extend ACT-R's Scheduler to include time sense.
Decay	Mechanism exists in ACT-R.	Use ACT-R mechanism.
Attention	Currently too constrained in ACT-R. Include functionality to vary level of attention, spatially or temporally application, etc.	Extend ACT-R mechanism. Start with current attention, but implement as separate module.
Poor Learning (encoding)	This "mechanism" is basically placing poor information into a knowledge representation such as ACT-R's.	Place incorrect information in modified ACT-R knowledge structure, i.e., add schema structure.
Retrieval	Mechanism exists in ACT-R.	Use ACT-R mechanism.
Perceptual Mechanism	Current ACT-R's Perception Module, in general, always correctly perceives the environment.	Modify ACT-R's Perception Module to allow errors.
Association Developer	ACT-R has associations between chunks.	Use ACT-R mechanism and develop poor associations before simulating performance.
Motor Mechanism	ACT-R's Motor Module, in general, always correctly performs the correct action.	Modify ACT-R's Motor Module to allow errors.

7   Design the Human Error Modeling Architecture

Once we identified which General Error Mechanisms ACT-R could handle and where extensions and new mechanisms where needed we designed a conceptual architecture for HEMA. Besides including the ACT-R mechanisms as identified in our analysis this design must also include components to handle all the General Error Mechanisms as described in Table 4 above. Furthermore, HEMA must also handle the processes of the FHP and the error types from which the General Error Mechanisms and hence the HEMA design derive.

Figure 2 presents a conceptual design for HEMA, as a UML component diagram, which meets these criteria. All mechanisms are either specifically shown or can be mapped to existing or modified ACT-R mechanisms (e.g., Rule Match, Decay). Much of HEMA can be accounted for with ACT-R as shown in the shaded area. Note that while generally modules only access buffers, we have included some module to module access via greyed lines to indicate new interactions within HEMA.

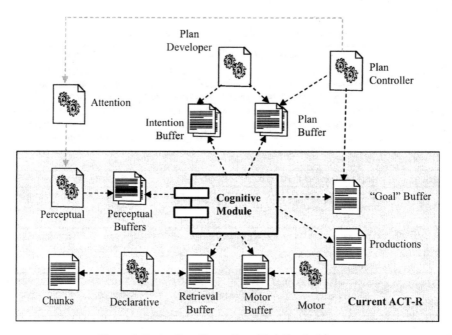

Figure 2. Design for a Human Error Modeling Architecture

The following discussion briefly shows how this component diagram could serve as a design for implementing the FHP (Figure 1):

- *Understand Problem:* The Attention Module directs the Perceptual Module to get information from the environment. The Cognitive Module builds a perceived situation which is placed in the Retrieval Buffer. The Cognitive Module can direct Attention to gather more perceptual information if the operator has time and needs more information.
- *Set Intentions:* Using Productions, Declarative Memory, Chunks and Retrieval Buffer contents the Cognitive Module sets an intention and places this in the Intention Buffer. It is assumed that the Cognitive Module can also estimate the effort needed (from information stored in Chunks) and remove or weigh intentions.
- *Form Plans:* The Plan Developer runs and uses the Cognitive Module to run Productions and Declarative Memory to access Chunks in order to form plans. For skill-based performance a well-rehearsed plan may already be available in Declarative Memory for the current situation. However, for rule or knowledge based performance the Plan Developer will have to run the Cognitive Module repeatedly and form a novel plan. The Plan is placed in the Plan Buffer.
- *Run Plans:* Plan Controller runs plans by retrieving a plan from Plan Buffer and first running the next action in the plan. This action could be: 1) requesting more information from the environment via Attention, Perceptual and Cognition Module to update the Perceived Situation, 2) accessing the Cognitive Module to derive further information from Declarative Memory, or 3) performing a motor activity via Motor Module. If the current action needs an evaluation (and we are assuming information relative to this

need must be placed in the plan) then the Monitor function of Plan Controller requests information on the environment via Attention and Perception to update the Perceived Situation. Plan Controller then activates the Compare Plan mechanism to perform an evaluation to determine current plan status. The Plan Buffer and if necessary the Intention Buffer are updated.

8 Error Types in HEMA

As pointed out above HEMA should be able to explain how to model the error types shown in Table 1. In order to do this we first performed another analysis and sorting of the error types in terms of which error types could be implemented with similar mechanisms in HEMA. This led us to a derivation of a HEMA error taxonomy, which at present has four major categories - Perception, Plan Development, Plan Control and Memory. Each major category has sub-categories of error types and in some cases even a secondary sub-category.

To give some example of the modeling within this error taxonomy we show a few error types in Table 5. The first column in Table 5 gives the name of a HEMA error type, the second column provides a description of how that error could occur using a system implementing the HEMA design, and the third column shows which original error types the HEMA error type can be traced to. Our design includes similarly detailed tables for each of the four major HEMA error categories.

The final step in the conceptual design was the development of sample UML sequence diagrams for each category. These diagrams describe in detail the sequence of actions which would occur in HEMA when a particular error occurs. Sequence diagrams, and supporting information, for each error type will be developed for use in directing the implementation of HEMA.

Table 5. Example of HEMA Error Types and Mapping to original error types.

HEMA Error	Description	Derived From
Plan Development: I. Plan Incorrect: A.Wrong Plan Set	Plan Developer forms wrong plan. The Cognitive module retrieves intention from Intention Buffer then accesses Declarative Memory but retrieves an incorrect schema or rule(s) for the plan. The Plan Developer places this incorrect plan in the Plan Buffer.	CREAM P: Inadequate Plan: Wrong Plan
Plan Control IV. Evaluate Interpretation - Time Delay	Plan Controller runs, but Compare Action runs before Goal Buffer is updated with new information about effect of action. This could be modeled by introducing a delay in the Cognitive Module performing schema/rule matching after action is performed	GEMS KB: Problems with Complexity: Delayed feed-back
Perception: II. Perceptual Attention Failures: C. Wrong Features	Attention directs Perceptual Module to incorrect features in environment (saliency dominates over logic).	GEMS KB: Selectivity
Memory: VI. Memory Loss: B. Decay	This is currently accounted for in ACT-R by activation being reduced on Chunks in declarative knowledge so these Chunks are not retrieved.	CREAM TP: Memory Failure: Forgotten and SA - Level 1: Memory loss

9 Summary

In order to develop a Human Error Modeling Architecture it was first necessary to first develop a complete process flow of human performance - a Framework of Human Performance (FHP). The development of the FHP, along with the identification of a broad sample of human error types enabled the description of cognitive function failures in the context of the FHP. Identifying the commonalities in these failures led to the proposition of fifteen General Error Mechanisms that could account for the error types sampled. Comparison of these mechanisms to the ACT-R architecture demonstrated that ACT-R could account for many of these mechanisms and serve as a basis for HEMA. The proposed HEMA design includes a core ACT-R with extensions (e.g., schema, extended Perceptual Module) and additional modules.

Our final goal is to provide the results of HEMA to system designers for use in assessing the error incidences likely to occur given a proposed system design. To reach this goal HEMA will be the core component on a larger system Human Error Model for Error Tolerant Systems (HEMETS). HEMETS must include capabilities to interface HEMA with system design simulations and provide system designers with some form of "error prediction report" for each system design. Further research will consider the best ways to achieve these goals while implementing the Human Error Modeling Architecture.

References

Anderson, J. R., Bothell, D., Byrne, M. D., Douglass, S., Lebiere, C., & Quin, Y. (2004). An integrated theory of the mind. *Psychological Review*, 111(1036-1060).

Anderson, J.R., Bothell, D., Lebiere, C., and Matessa M. (1998). An Integrated Theory of List Memory. *Journal of Memory and Language*, 38, 341-380.

Byrne, M.D. (2003a). Cognitive Architecture. In J.A. Jacko and A. Sears (Eds.), *The human-computer interaction handbook: Fundamentals, evolving technologies and emerging applications* (pp. 97 - 117). Mahwah: Lawrence Erlbaum Associates.

Byrne, M. D. (2003). A mechanism-based framework for predicting routine procedural errors. In R. Alterman & D. Kirsh (Eds.), *Proceedings of the Twenty-Fifth Annual Conference of the Cognitive Science Society*. Austin, TX: Cognitive Science Society.

Byrne, M.D. and Bovair, S. (1997). A working memory model of a procedural error. *Cognitive Science*, 21, 31-61.

Endsley, M.R. (1999). Situation Awareness and Human Error: Designing to Support Human Performance. *Proceedings of the High Consequence Systems Surety Conference*, Albuquerque.

Hollnagel, E. (1998). Cognitive Reliability and Error Analysis Method. Oxford, UK: Elsevier.

Leiden, K. K., Laughery, R., Keller, J., French, J., Warwick,W. and Wood, S.D. (2001). A Review of Human Performance Models for the Prediction of Human Error. NASA System-Wide Accident Prevention Program. Moffet Field: NASA.

Norman, D.A. (1986). Cognitive Engineering. In D.A. Norman, S.W. Draper (Eds.) *User Centered System Design: New Perspectives on Computer Interaction* (pp. 31 - 61). Hillsdale: Lawrence Erlbaum Associates.

Reason, J.T. (1990). Human Error. New York: Cambridge University Press.

A Modular Architecture for Integrating Cognitive, Communicative, and Situated Behavior

Susann Luperfoy
Stottler Henke Associates, Inc.
48 Grove Street
Somerville, MA 02144-2500 USA
luperfoy@stottlerhenke.com

Bradley Myers
Galactic Village Games, Inc.
174 Littleton Road Suite 3-312
Westford, MA 01886 USA
bmyers@galactic-village.com

Abstract

This paper describes a software architecture for delivering monolog discourse processing and mixed-initiative spoken dialog capabilities as enhancements to or extensions of existing software applications. We show how this architecture enables the integration of cognitive, communicative and situated behavior of synthetic agents, through reliance on an applied theoretical framework for representation and tracking of all available forms of context information. We define the required behaviors that motivated the framework, based on the human model of discourse processing, and the mechanisms through which the information it tracks gets used in service of extended goals for the application system. We then illustrate the runtime integration of modules via their respective API's and the basic steps of the development-time software porting process.

1 Background

The modular software architecture described here is designed to foster straightforward integration of a suite of artificial intelligence (AI) components for spoken dialog technology with non-AI software systems. This architecture has been used to convert game avatars from purely graphical characters into synthetic dialog agents that engage the player in spoken dialog exchanges (Luperfoy, et al., 2003). It was used to augment WIMP (Windows Icons Menus and Pointing) control of a complex military simulation system, to allow spoken dialog interaction between the user and semi-automated forces in the simulation and a disembodied instructional agent (Luperfoy, et al., 1998). We are currently using a version of the architecture to deliver simulated classroom discussions in an Intelligent Tutoring System (ITS) for case-method instruction (Luperfoy, et al., 2004). In that ITS application the context tracking framework is also used in monolog mode as the foundation of a an authoring tool for knowledge acquisition that lets domain experts encode conceptual knowledge as metadata associated with text narratives.

1.1 Motivation

A single interactive application environment can call for a range of dialog and monolog processing capabilities that can be classified, in part, according to who is communicating with whom. (1) **Human-system** dialog lets the user command and control the application using a dialog-enabled user interface, e.g., "Create another enemy tank battalion", or "Zoom in on that bridge", or "Bring up the topological map overlay." (See Walker, et al. 2002 for a comparative evaluation of several human-system dialog interface designs.) (2) **Computer-mediated human-human** dialog lets game players, instructors, operator/controllers, and observers communicate with each other in the context of the shared scenario (Miller, et al. 1996) or decision environment. (3) **Automated analysis** of those human-human dialogs either in real time or retrospectively (Jurafsky, et al., 1997; Glass, et al., 2002) can assist in evaluation of user and/or system performance. These analyses in (3) can also be used to model spontaneous human-human dialog for a fourth category of computational dialog, namely, (4) **computer-generated** synthetic agent dialog. This synthetic dialog exists as part of the situational context of the unfolding scenario to be overheard by the user, but involves no participation of any human speaker. Examples include synthetic dialog between dueling chat bots available on in EMACS or on the worldwide web, or game characters that constitute part of the dynamic context. (5) **Human-DA (Dialog Agent) dialog** lets players engage the dialog-enabled synthetic agents of the simulation world and hear contextually-appropriate DA responses, e.g., "Fifth platoon, decrease speed by two zero miles per hour", "Yes, Sir. Fifth platoon decreasing speed", "First platoon, what is your position?" (Webber et al., 1995; Goldschen, et al. 1998; Nielsen, et al., 2002). (6) **Human-tutor dialog** is a related form of human-machine dialog that provides a personification of the application system or tutor as a disembodied coach (Luperfoy, 1996) or, depending on the application constraints, as a Non-Player Character (NPC) with an overt screen presentation (Graesser, et al., 2002). The dialog-enabled interface can offer 'over-the-shoulder' verbal coaching during the exercise, or it can collect and save observations for After-Action Review when it can engage the user in meta-level dialog about the application, a lesson, a virtual scenario, or the user's performance. In addition to dialog, there can be an application-specific need for monolog processing of natural language speech or text. For example, (7) **monolog analysis** is a component of automated information extraction from text documents or video streaming data. In the output direction, (8) **monolog synthesis** requires turning selections from an internal model or knowledge representation system, into an explanation of its information content to a human reader/listener.

1.2 Overview

In this paper we derive functional requirements dictated by the eight categories of computational discourse processing enumerated above, and present a software architecture that supports those requirements. A central aspect of the architecture is the cognitive framework with its integrated representation of the evolving communicative, cognitive, and situational contexts. This three-level representation is based on a computational theory for both human and synthetic dialog agents that partitions information about surface communicative forms (Mentions), discourse-level abstractions of entities that have been mentioned or can be mentioned in the near future (Discourse-Pegs), and knowledge of the world of reference (KB objects). These three types of information available to the Dialog Agent (DA), have distinct procedures for access, updating, and decay over time. The Belief System, implemented as a Knowledge Base (KB), represents the DA's beliefs about the world under discussion, which may include a model of the DA's dialog partner (e.g., the user model), domain ontologies, rules of sanctioned inference, and a reasoning engine to apply the rules.

The framework was designed to model certain cognitive behaviors exhibited in human dialog interaction that have importance for dialog-enabled applications. These are, the ability to understand an explanation without believing it; the ability to say things that one does not believe to be true; the ability of a listener to correct their flawed interpretation of preceding discourse privately without involving the speaker; the ability to integrate non-communicative events and properties from the evolving situation into the context that informs interpretation and generation of intelligible communicative behavior; the ability to use knowledge about the world and inferential reasoning to construct an internally consistent model of a counterfactual world; and the inability to interpret context-dependent references to concepts that have fallen out of discourse focus due to simple passage of time or due to overwriting by new communicative events that intervene. We show how the framework accounts for these behaviors with data structures and control logic that distinguishes discourse interpretation from assimilation of beliefs, and distinguishes private perception and internal reasoning from joint observations of situational context shared between speaker and hearer. We will illustrate the role of this context representation in the overall software architecture, i.e., how it is accessed, updated, and corrected by the other modules in the architecture.

2 Intelligent Dialog Agents

One way that both human and synthetic agents demonstrate intelligence is through generation and comprehension of communicative contributions and non-communicative actions appropriate to the immediate context state. This skill relies on prior mastery of the grammar (lexicon, syntax, and semantics) for the language[1] in use, awareness of the rules for cooperative discourse in that language community, models of the speakers involved, and knowledge about the domain (world) of reference. A competent Dialog Agent (DA), human or synthetic, uses its unique perspective on the world of reference (arising through perception and reasoning) and its internal goals and plans, to make sense of the communication it receives, and then to decide how to respond to a speaker's intent with an appropriate action, a sufficient verbal response, or both. In this section we decompose dialog competence into three essential capabilities: Context Tracking, Pragmatic Adaptation, and Dialog Management.

---

[1] Here "language" refers to any signal system that generates input and output surface forms according to a lexicon of tokens, with rules of syntax and semantics. Thus, Morse Code, written English, spoken Japanese, human gesture languages, mouse/stylus gesture languages, and output graphical languages that produce animations, appearance of icons, etc. all qualify as languages.

2.1 Context Tracking in Human Discourse

Spontaneous human language contains context-dependent referring expressions, including pronouns, indexical references ("tomorrow", "us", "that room", deictic gestures), elliptical phrases ("No, it doesn't.", "If you want to."), (in)definite noun phrases ("a passenger," "the pilot"), and other forms that receive their semantics in full or in part from the context of their occurrence. (See Appendix for a table of examples.) In order to interpret these **dependent** forms when they occur, humans prepare for upcoming events in advance by mentally tracking the salient elements of the communicative context and the perceptually shared situational context that can **sponsor** the occurrence of subsequent dependent forms. An oversimplified description of the tracking process is that when we hear a new utterance we consult our context representation to find sponsors for any dependent forms in the new utterance, and we add new sponsors to the context representation to prepare for subsequent dependent forms. In human dialog, the speaker's context includes their conceptual model of the state of the listener. That model of the listener guides the speaker's composition of each new dialog utterance. In return, the human listener will often assist the speaker by offering verbal or nonverbal indicators of how well their participation in the dialog is going, e.g., nodding, puzzled facial expression, or vocalized backchanneling ("I see," "uh-huh," "Go on.").

Thus, humans come to the human-machine dialog situation well equipped as listeners and speakers, able to hold up their end of a cooperative mixed-initiative dialog. They also bring valuable expectations of how the DA will behave. Thus, the user is an intelligent resource we can exploit in two ways. First, we use human dialog behavior as a development-time model to guide our design of a cooperative DA, and then at runtime we assume human discourse competence and rely on the behavior of the user as input to DA algorithms for understanding input communications in context and generating appropriate output..

2.2 Pragmatic Adaptation in Human Discourse

Pragmatic adaptation is often required to supplement the literal interpretation of a communicative act in order to arrive at an actionable understanding of the speaker's underlying intent. In human discourse, my understanding of your utterance "Do you know how to open the window?" as a request for specific action, is a complex feat requiring resolution of context-dependent forms, indexical references, indirect speech acts, and more. In this regard, pragmatic adaptation sits at the boundary between communication and action.

However, understanding is only part of the task. Having understood your intent in the context of the current situation, I must still decide on an appropriate action by reasoning about consequences of various actions (or inaction) relative to your intent. For example, I could open the window in silence; answer your yes/no question "Yes, I do," without action; open the window with self narration "Yes of course, I'm opening the window now;" report an execution problem, e.g., tell you that the windows in this building don't open; request a clarification of your intent, "Do you mean this spreadsheet window on my computer here or that glass window to the outdoors;" propose an alternative action "How about if I turn on the air conditioning instead;" or simply ignore the request altogether.

Pragmatic adaptation uses awareness of the immediate situation to supplement a literal interpretation and plan an appropriate response. This skill relies on a prior model of the situational environment, combined with an awareness of the communicative and non-communicative context state. In human communication, such knowledge informs decisions regarding what to say or what action to take, e.g., a pilot communicating with Air Traffic Control, a customer shopping for a sofa, a real estate agent negotiating the sale of a building.

2.3 Dialog Management in Human Discourse

The dialog management skill requires awareness of rules for when to interrupt, when to relinquish the floor to another speaker, how to backchannel (e.g., nodding versus vocalization "Uh-huh"), and how to repair disfluencies. Even in human dialog between two people who are well acquainted, dialog disfluencies occur frequently during normal communicative exchange. Thus, competent speakers of all languages have developed skills for preventing, detecting, diagnosing, and repairing the inevitable disfluencies that arise. Indeed, the dialog repair mode so defined is not an aberration but is as much a part of successful interaction as the primary topic dialog.

McRoy (1996) presents a thorough treatment of dialog repair including prevention of dialog disfluencies. While prevention is essential for any serious work on repair dialog *per se*, we will not address it in this paper. Our description of the repair process (in humans or synthetic DA's) comprises the following subtasks.

1. **Detect**: One of the parties in the dialog must recognize that there is a problem, otherwise the dialog continues, hampered by propagated effects of the miscommunication.
2. **Diagnose**: A determination must be made as to the source of the dialog trouble. This diagnosis can be made unilaterally, or collaboratively by the two conversants. For automated systems, as with human dialog interaction, disruptions to understanding can take place at a number of levels. We have adopted the model of collaborative communication defined by Clark et al. (1997), borrowing their eight levels of presentation acceptance to use as points of potential interpretation failure. In this way we distinguish categories of dialog disfluency based on which component of the dialog agent system indicates difficulty in carrying out its step in the analysis: speech recognition, utterance interpretation, context tracking, or internal elements of pragmatic adaptation (e.g., domain model incompatibility, user model conflict, or ill-formed backend command).
3. **Devise Recovery Plan**: Even given a successful detection and diagnosis, we must query user interface design parameters to determine the preferred method for recovery. For example, upon determining that a user misconception is the cause of an illegally stated question or command, the system has the options of (a) correcting the command without bringing it to the user's attention, (b) correcting the command and reporting back to the user the proper formulation of the command, or (c) reporting the problem without correcting it and suggesting that the user reissue the command using a legal formulation.
4. **Execute Recovery**: For spoken dialog systems this recovery plan must be executed in collaboration with the other dialog agent(s). Since human dialog agents are as unpredictable in repair dialogs as they are in primary dialogs, the DA may have to respond to user input that fails to match behavior prompted for during the repair. For example, the repair prompt "Do you mean this bridge?" calls for a yes/no answer but the user can surprise the DA with "That's not a bridge," or "The assembly area," or "Please repeat."
5. **Close and return to the primary dialog**: Once the dialog trouble has been resolved, both system and user must be brought jointly to the understanding that the next utterance is a return to the primary dialog. Options for achieving this return step include an overt closure statement, or appropriate embellishment of or wrapper around the next utterance to unambiguously associate it with the primary dialog.

3   The Dialog Agent Architecture

The capabilities of an intelligent DA as defined above are implemented using the software architecture shown in Figure 1. In this section we describe the modules for context tracking, pragmatic adaptation, and dialog management as components of this architecture

3.1   Context Tracking

The synthetic DA that interacts with a user can be modeled after a human simultaneous interpreter, translating between speakers of two different languages. In this case, the 'interpreter', i.e., the DA mediates between the human speaking natural language (e.g., English), and the backend software application that communicates only via the functional interface specification of its API. In this role, the DA's job is to make sense of each new human input (statement, query, or command) relative to its own internal representation of the context, then to translate the input into a well-formed command in the language of a backend application. It updates its internal context representation with the input communicative event and issues the backend command. Translating in the other direction, the DA intercepts the backend output response to that API command, converts the output into a context-appropriate natural language output utterance (statement, question, suggestion) for the user, and finally, updates its internal context representation with that output communicative event to prepare for the next user input.

Perception in this model occurs in the case of robotic devices situated in the physical world, or software agents located in a virtual world. For example, in a simulation-based ITS dialog, the context tracker records salient situational events and state properties of the virtual world, as well as communicative information flowing from user to simulation and back. In order to draw useful inferences from these runtime updates received via the simulation API, the DA needs a prior model of the backend simulation/game, its possible states, error conditions, its syntax for well-formed input commands, and its rules of engagement.

Consider the following hypothetical input command to a game-based ITS, "Create one and put it on the other bridge." To interpret this utterance as the user's desire that the DA create a vehicle icon and position it on some bridge other than the bridge in focus, the DA will consult its context representation to find at minimum, entities for the particular vehicle in question, and the spatial layout and location of entities, including at least two bridges, the speaker and the embodied DA itself. In the case of an Intelligent Tutoring System (ITS), to serve as a personified

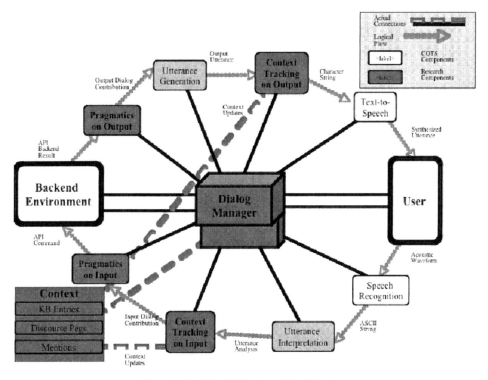

Figure 1: **Modular Dialog Agent Architecture**

tutor, the DA must also have access to the evolving student model, the training objectives, the curriculum model, remediation options, and other knowledge sources required to deliver the desired pedagogical approach.
The combined situational and communicative context is initialized at the start of the session and then all salient communicative and situational events/state changes that occur are recorded by the context tracker during the exercise. A DA that maintains even limited versions of the above forms of contextual information can be construed as having 'beliefs' about the external world, the user, the backend application, and about the dialog itself.

3.1.1 Context State Representation
The three-level representation of context that appears in the lower left corner of Figure 1 is expanded in Figure 2. This representation is based on a computational theory that partitions information about surface communicative forms (Mentions), discourse-level abstractions of entities that have been mentioned or can be mentioned in the near future (Discourse-Pegs), and knowledge of the world of reference (KB objects). These three types of information available to the DA, have distinct procedures for access, updating, and decay over time. The Belief System, implemented as a Knowledge Base (KB), represents the DA's beliefs about the world under discussion, which may include a model of the DA's dialog partner (the user model), ontologies, rules of inference, analogical reasoning engine, and more. In the current implementation we are using our in-house General Representation Inference and Storage Tool (GRIST) populated with an ontology representing the application domain of entities, states, events, and rules of inference.

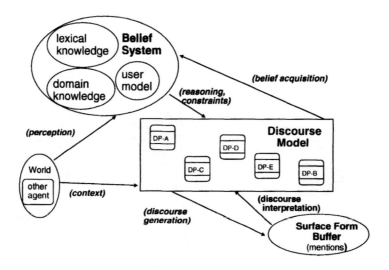

Figure 2: **Cognitive, Communicative, and Situational Context Representation**

The Discourse Model encodes the DA's current understanding of the information content that is currently in shared focus. Content of the discourse model involves systematic uncertainty and is understood by the DA's meta-cognitive awareness to be potentially incomplete or flawed. Information in the Discourse Model is organized around abstract objects called Discourse-Pegs (DPs), that represent the DA's current focus of attention. DPs decay from prominence only when they are ignored by both speakers. When neither DA nor human user mentions a DP for a time it loses its ability to license, or sponsor new dependent forms and is eventually replaced by new DPs for new constructs in focus. There is one Surface Form Buffer for each modality channel (keyboard, speech, joystick, mouse, output graphics, eye tracker, etc.) and its content is supplied by a processor that captures input and output communicative events and interprets them to a level equivalent to first-order predicate logic. Unlike DPs, the objects at the surface level, called Mentions, decay rapidly as a function of time so that linguistic forms, sounds, etc., are soon lost to the context representation while new Mentions replace them. A new Mention can refresh an existing DP or cause a new DP to be introduced into the Discourse Model.

This context representation and updating framework was designed to model cognitive processes exhibited in human dialog interaction: the ability to understand an explanation without believing it, the ability to use knowledge about the world and inferential reasoning to construct an internally consistent model of a counterfactual world, the ability to say things that one does not believe to be true, and the inability to interpret context-dependent references to concepts that have fallen out of discourse focus due to simple passage of time or due to overwriting by new communicative events that intervene. The model enables these behaviors by distinguishing discourse interpretation from assimilation of beliefs, distinguishing private perception and internal reasoning, from joint observations of situational context shared between speaker and hearer as the common ground (Clark, et al. 1989).

Discourse-Pegs can relate to one or more surface Mention. The Mentions involved may be directly related to each other syntactically, or indirectly related semantically through the DP that they share. For example, in "They drove an M1A1 to the assembly area. They were forced to abandon the tank to recover wounded so the vehicle is still there." the Mentions for "tank," "M1A1," and "vehicle" do not show linguistic dependence, but are related semantically through their reference to a common DP. The Context Tracker accesses and updates its own representation of context and affords an interface to remaining components of the DA that access it to reason about appropriate interpretations and their own next actions.

3.2 Dialog Manager

The Dialog Manager (DM) is the facility of the DA that controls the interaction between the human user and all system components that contribute to the user's experience of the dialog, including the Context Tracking and Pragmatic Adaptation modules. The DM orchestrates the firing of modules to process input speech and generate output responses, update context, and translate input requests and queries into well-formed commands in the language of the backend API, then translate backend output into context-appropriate natural language. Global DM property settings control the 'personality' of the DA that the user experiences by determining whether it takes dialog initiative, how it chooses from among repair strategies. Three implementation features that help create DA personality are mixed-initiative dialog, dialog troubleshooting, and backchanneling.

Mixed initiative interaction has long been recognized for its advantages. Instead of long complicated utterances, commands can be spoken in shorter, more natural segments that are easier to both interpret and confirm. If the user's original input is not sufficient for the Utterance-Interpretation subtask, the DA can elicit the missing information, and even suggest an intelligent choice of default values for remaining gaps based on information from the context manager. Mixed initiative interaction requires dialog situation awareness so that, for instance, the DA is responding by voice or prompting the user, the user will be able to interrupt it or "barge in." Likewise, if the DA is in the process of speaking to the user and any other higher priority event takes place, the DA will interrupt its own output and either discard it or record it for later processing. The DA can be designed to cease the initiative when it needs to get the user's attention for any reason—e.g., when it has completed an off-line task that the user had requested, when there is an incoming call, when a new player logs on to the game, etc.

3.3 Pragmatic Adaptation Modules

The DA's unique perspective on the world of reference amounts to its internal goals and plans, its model of the user and its awareness of the immediate state of the backend application. It is through the use of this internal model of perspective that the Pragmatic Adaptation component of our architecture can model the human ability to convert an indirect speech act into the appropriate response based on situational and communicative context.

The pragmatic adaptation module participates in dialog repair by detecting, diagnosing, and remediating dialog disfluencies that occur as a result of some sort of incompatibility between the interpreted input utterance and the DA's model of the backend system. For example, a command to "destroy it" resolved to a non-existent vehicle will result in a detected disfluency leading to a repair dialog. The Pragmatic Adaptation module encodes a set of <condition, response> rules to implement a repertoire of simple strategies to deal with various configurations of dialog disfluency type, user, and backend application. For example, if the DA is stuck on an ambiguity, it can *guess* (randomly select one of the interpretations), *procrastinate* the decision as long as possible, or *request clarification* from the user. If the input is interpretable, but it translates to a command that is impossible to execute or nonsensical in the current communicative and situational context, the DA can report the error and suggest an alternative action, it can try to diagnose the problem and present the user with options for action to remedy the situation, or it can make a unilateral repair and watch for objections from the user.

4 Dialog Enabled Applications

To illustrate the basic procedure for constructing a software system that integrates the DA architecture with existing applications, we integrated the DA architecture with a COTS (Consumer Off The Shelf) game engine to construct an educational game. In this prototype implementation, a human player assumes the role of a Player Character (PC) that engages in spoken dialog with Non-Player Characters (NPC). When the PC is in sufficient proximity with a pair of NPCs engaged in conversation, their voices become audible so that the player overhears them. Salient aspects of their communicative and non-communicative behavior get recorded as part of the dynamic situational context that is used to. The player and DA's were restricted to dialogs about smoking. After the DA interprets a spoken input from the user, "How many people in my staff are smokers?" "How long has Andrew been smoking?" or "Tell Andrew to stop smoking," it translates the interpreted result into a well-formed command to the backend, e.g., an SQL query, or a call to the natural language output system to generate an audible utterance such as "Andrew, stop smoking" that gets conveyed from the XO to the ARA with designation Andrew. In the prototype, ARAs are blindly obedient and robotically self-disciplined so that the XO always conveys the message to the ARAs who always change their value function in response to CO commands and always choose behavior consistent with their value functions.

The prototype leverages our own prior results for knowledge-based DA construction—including the architectural framework, the Context Trackers for Input and Output, our GRIST Knowledge Base system that constitutes the third tier of the Context Representation, the Pragmatics Adaptation modules (for input and output) that form the boundary between communication and action, and the Dialog Manager that controls the overall interaction.

We constructed a temporary grammar and lexicon and defined finite state machines for sentence parsing and generation of output utterances. These are introduced strictly as placeholders for future components to be obtain from sources of mature technology available in the computational linguistics community. In the prototype grammar (enumerated below) personal pronouns are recognized as legal fillers of the <person> slot, and are resolved relative to the Mentions and DPs in the current context. When the prototype ITS game is initialized, Tim and Sally are the only known characters and the player can introduce new characters by asserting their existence in the group, e.g., "George reports to me." The current placeholder lexicon allows the <person> slot to be filled by Tim, Sally, George, or Andrew and the prototype grammar contains these formulas:

                    <smokeQues> = does <personOrPronoun> smoke;
                    <smokeJust> = why does <personOrPronoun> smoke;
                    <follow> = follow me | come here;
                    <stopFollow> = stay here | stop following me | stop there;
                    <age> = how old is <personOrPronoun>;
                    <smokeStart> = when did <personOrPronoun> start smoking;
                    <subjectPronoun> = he | she | they;
                    <objectPronoun> = him | her | them;
                    <smokeCommand> = tell (<person> | <objectPronoun>) to stop smoking;
                    <numberSmokers> = how many smokers are there;
                    <smokingTell> = <personOrPronoun> smokes <quant> (pack | packs) a day;
                    <personCreate> = <person> is in my unit;
                    <quant> = one | two | three;
                    <person> = Tim | George | Andrew | Sally;
                    <personOrPronoun> = <person> | <subjectPronoun>;

We used COTS or open source products to populate remaining components of the architecture, including ASR (automatic speech recognition), TTS (text to speech), and the commercial NeverWinterNights™ game engine. The prototype system runs in distributed client-server mode over a local network of personal computers.

5   References

Clark, Herbert and E. Schaefer. (1987). Collaborating on Contributions to Conversations. Language and Cognitive Processes, pp. 19-41.

Duff, D. and S. LuperFoy (1996) "A Centralized Troubleshooting Mechanism for a Spoken Dialog Interface to a Simulation Application" International Symposium on Spoken Dialog, Philadelphia.

Goldschen, A., Harper, L.D., Anthony, E.R. (1998) The Role of Speech in a Distributed Simulation: The STOW-97 CommandTalk System.

Graesser, A., et al. (2002) "Why-2 Auto Tutor" oral presentation to ONR workshop on Tutorial Discourse.

Grosz, Barbara and Candace Sidner (1985) The Structures of Discourse Structure (Tech. Report). SRI

Luperfoy, S., D. Loehr, D. Duff, K. Miller, F. Reeder, and L. Harper (1998) "An Architecture for Dialog Management, Context Tracking, and Pragmatic Adaptation in Spoken Dialog Systems". In proceedings 36th Annual Meeting of the Association for Computational Linguistics.

Luperfoy, S., E. Domeshek, E. Holman, D. Struck (2003) "An Architecture for Incorporating Spoken Dialog Interaction with Complex Simulations" In proceedings Interservice/Industry Training, Simulation, and Education Conference.

McRoy, S. and G. Hirst (1995) "The Repair of Speech Act Misunderstandings by Abductive Inference" Journal of Computational Linguistics, vol. 21 no. 4.

Miller, K., S. LuperFoy, E. Kim, D. Duff (1995) "Some Effects of Electronic Mediation on Spoken Bilingual Dialog: An Observational Study of Dialog Management for the Interpreting Telephone" Electronic Journal of Communication.

Nielsen, P., Koss, F., Taylor, G., Jones, R. M. (2002) Communication with Intelligent Agents, in proceedings I/ITSEC 2002.

Walker, M., A. Rudnicky, R. Prasad, J. Aberdeen , E. Bratt, J. Garofolo, H. Hastie, A. Le , B. Pellom, A. Potamianos, R. Passonneau, S. Roukos, G. Sanders, S. Seneff, D. Stallard. (2002) "DARPA Communicator Cross-System Results for the 2001 Evaluation" in proceedings, ICSLP (International Conference on Speech and Language Processing).

Webber, B. (1995) Instructing Animated Agents: Viewing Language in Behavioral Terms. *Proc. International Conference on Cooperative Multi-modal Communication*, Eindhoven, Netherlands, May.

Appendix

Phenomenon	Example	Comments
Intra-sentential anaphora	*DA: And so, what about nicotine, is **it** also addictive?*	The pronoun finds its **sponsor** (nicotine) in the current sentence.
	USER: Yes, only 26 people lost **their** lives?	"26 people" sponsors "their"
Inter-sentential anaphora	USER: Yes, **it** is a addictive substance that can hook you quickly.	Here, the pronoun finds its *"sponsor"* (nicotine) in an earlier sentence.
Ellipsis	*DA: Given your definition of illegal, is binge drinking also illegal?* USER: Yes **it is**.	Reconstruction of the elliptical expression yields, "Yes, it is also illegal." The sponsor of "it" is DDT so pronoun resolution yields, "Yes, DDT is also toxic."
	DA: **By whom**?	Main verb is elided
	DA: And **what about** nicotine?	The operator "what about(x) " gets the interpretation APPLY-PROPOSITION-TO(x)
Discourse Deixis	USER: **That** is controversial.	The deictic adverbs, "this," "that," "these," "those," etc. are *sponsored* by something in the prior discourse.
	USER: No it was earlier than **that**.	
	USER: **This** is undecided.	
Totally Dependent Definite Noun Phrase	*DA: Okay. I understand **the term** now.*	This "term" is a second mention of (and sponsored by) a term that was mentioned earlier.
Partially Dependent Definite Noun Phrase	DA: Was it **the purpose**	*The "purpose" is new to the discourse but dependent on a purposeful event mentioned earlier.*
One-Anaphor	DA: Okay, you are telling me about **one** in which nicotine killed someone.	*This new event partially depends on a concept mentioned earlier, the class of deadly events.*
Quantifier as One-Anaphor	USER: I am not aware of **any**?	Quantifiers can behave as one-anaphoric expressions introducing new entities by depending on entities mentioned earlier
	*Does serin gas have **other** uses, unrelated to warfare?*	
Indexical	It has been active since 1982 and **now** has members numbering upwards of---	Functional relationship between indexical expressions, "now," "me," "you," "here," "yesterday," etc. and the situation of the utterance.
	DA: Okay, **you** are telling **me** about one in which serin gas was released into a population.	
	USER: **I** am not aware of any?	

Section 5
Cognitive State Sensors

Chapter 21

Neural Correlates of Cognitive States

Spatial-Frequency Networks of Cognition

D.R. DuRousseau, J.P. Stanley and M.A. Mannucci

Human Bionics LLC.
190 N. 21st Street, Suite 300
Purcellville, VA 20132
don@humanbionics.com

Abstract

Rapid comprehension of new information during decision-making is the most essential factor in efficient and error-free performance of complex operations. Under high stress levels and workloads, or, when inattentive or bored, persons often make costly decision-making errors of which they are typically not aware. Decreased cognitive arousal and limited information processing capacity are important causative factors leading to errors in human performance. The electroencephalogram (EEG) has been used for several decades to study this phenomenon. Here, we propose that rhythmic oscillatory communication networks are active within disparate cortical regions involved in higher order cognitive processing, and that the identification of task specific activation patterns may be used to predict changes in arousal, stress, and cognitive workload. Thus, real-time EEG measures may be able to identify low vs. high quality decision-making prior to an actual response taking place so an operator could be informed with spontaneous feedback to enhance arousal, comprehension, and decision-making performance.

The noninvasive examination of Spatial-Frequency (SF) patterns embedded within the continuous EEG is an area of great interest in medicine and psychophysiology [1,2]. Spectral methods of SF analysis examine the brain's use of oscillatory communication as a means of executive control and are a likely candidate as an index of cognitive functioning in humans [3,4]. The topic of this paper examines the spatial relationship of the banded power in brain waves recorded at the scalp as a predictor of arousal and performance changes during complex higher order cognitive processing in man.

Many researchers have examined the spatial properties of filtered evoked potential components in the delta (2-4 Hz), low theta (4-6 Hz), hi theta (6-8 Hz), low alpha (8-11 Hz), hi alpha (11-13 Hz), sigma (13-15 Hz) and beta (15-30) bands of the EEG and their relationship to attention, working memory, and higher order cognitive processing in humans [5,6,7]. For years researchers have investigated the spatial distribution of evoked potential (EP) waves using topographic maps and current density plots as a means to visualize areas of the brain at rest and during mental tasks involving attention and working memory [8,9]. Some have attempted to quantify the spatiotemporal aspects of cognition using phase and correlation relationships of evoked and non-evoked potentials [10,11]. While others have investigated the use of SF bispectral and bipower techniques to account for nonlinear phase correlated rhythmic oscillations in the human brain [12].

Considerable work in mapping near and distant cortical and subcortical connectivity has laid much of the foundation for our understanding of inter- and intra-cortical communication in the brain [13,14]. Further, the spatial distribution of cognitive processing in the brain has been measured in cerebral blood flow studies with fMRI showing differential metabolic activation in prefrontal and cingulate cortices during attention and decision-making tasks [15,5]. These studies support the long-standing opinion that distributed cortical and subcortical interconnections in the prefrontal cortex, cingulate gyrus, and parahippocampus participate in the executive control of cognitive and attentional resources and in communication among near and distant brain regions [16,17]. Our belief is that this executive control is carried out through oscillatory coupling of interdependent cognitive processes and these rhythmic interactions are visible in the EEG at frequencies from 2 to 30 Hz. In this paper, we report the preliminary results of an ongoing pilot study, which examines the relationship between SF brainwave patterns and changes in arousal and performance of subjects taking a computer-based college level general knowledge assessment test.

Below we report on our first pilot work. In four right-handed adults, the SF patterns measured at the scalp before accurate performance differed from patterns before inaccurate performance. Repeatable spatial power distributions (from 3 to 12 Hz) overlying frontal, central, and parietal cortices one second before a decision was made predicted correct and incorrect responses. Additionally, SF pattern differences were observed to correlate with changes in arousal level, particularly in the 5, 8, and 9 Hz bands. These measurements suggest that brief, spatially distributed oscillatory neural activity patterns in distinct associative, somesthetic-motor, and integrative spatio-motor areas of the human brain may be reliable precursors of arousal and accurate performance during complex tasks [18].

1. Pilot Study

Continuous EEG recordings were made while subjects performed a multimedia computer-based assessment task. With these data, we measured the rapidly changing SF patterns of distributed oscillatory neuroelectric activity associated with the executive control of higher order cognitive processing and decision-making. We found differences in pre-response patterns correlated with subsequently correct and incorrect answers. We also found pre-response pattern differences between early and late portions of the exam. These group differences allowed discrimination of subsequent performance accuracy and indicated a relationship with arousal level. Thus a spatially specific, multiphase rhythmic oscillatory pattern composed of narrow-band frequency bursts at invariant brain areas, may be essential for accurate performance during complex attention and working memory tasks.

Four healthy, right-handed undergraduate male university students were recruited and paid for their participation. They were required to take a 2 1/2 hour computer-based assessment task that was designed to evaluate general knowledge in History and Humanities at the college entry level (Fig 1). Subjects were instructed to evaluate the material until they understood it, then to respond to each question by selecting a single answer from a multiple-choice alphabetical list of answers and pressing a corresponding button with the mouse pointer. They had to decide on a response and press a "Submit answer" button when satisfied with their choice. They were also told they could change their selections at any time prior to pressing the Submit answer button.

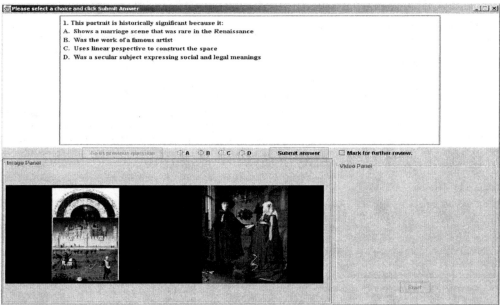

Figure 1
The view above is from the general knowledge assessment task used in the pilot study. The picture shows the program's operational areas each subject had to navigate to read questions, acquire multimedia information, and submit responses. The test question was displayed in the upper half of the screen and information needed to answer the question was presented in the lower half. The response buttons were located at the top of the shaded region along with the Submit answer button. Depending on media type (e.g., text, graphic, audio, or video) the information was presented either in the area marked Image Panel or Video Panel. If an audio or video recording were used, the subject would press the Start button to begin the recording.

Figure 1 provides a screen capture from the computer-based task used in the pilot study showing placement of the question in the upper half of the screen, while other information required to answer each question was presented elsewhere as text, pictures, maps, graphs, and/or audio and video recordings. The media format of the information determined the quadrant where the additional data would be presented. Occasionally, the subject was required to open a separate browser to search for information on the Internet and could minimize the browser window as desired. For each trial, a cue dot was presented to the subject 1-second prior to a new question and feedback indicating a correct or incorrect response and the running average percentage was displayed 1-second after responding.

2. Methods

Brain electrical activity from 29 scalp electrodes located according to the modified 10-20 electrode system [19], vertical and horizontal eye movements, and heart activity were recorded and digitized on the Neuroscan Labs' Synamps and Scan 4.1 recording and analysis system. Data were recorded continuously at 500 Hz with a band-pass from 1 to 55 Hz. The data were subsequently filtered from 2 to 30 Hz prior to further processing and analysis. Two independent judges edited the data for severe artifacts by visual inspection of the brain, eye movement, and heart potential channels and all identified artifacts were eliminated from analysis. Ocular artifact reduction algorithms included in the Scan software were then applied to remove blink and eye movement artifacts from the remaining EEG data [20]. The remaining continuous data were sorted by cue, stimulus onset, response, and feedback events and then epoched into brief segments beginning 1 second before (Pre) and 1 second after (Post) each response.

After manual rejection of contaminated segments and automated artifact removal using the continuous data files, epoched files were generated and visually edited for artifacts to eliminate questionable trials from further signal processing. Next, the epoched files were segmented into two data sets: trials in which the subject's choice matched (accurate) or mismatched (inaccurate) a predetermined answer maintained by the computer-based assessment system. Additionally, to evaluate the potential impact of fatigue and decreasing cognitive arousal over time, accurate and inaccurate trials were sorted according to early and late segments of the task.[1] Finally, all 1-second epochs were spline-fit to 512 points and the Global Field Power (GFP), which is a measure defined as the standard deviation across all channels as a function of time within a sample interval, were generated from the pre-response epochs [21]. These GFP epochs were used to generate power spectral averages with 1 Hz resolution for each accurate and inaccurate condition. The GFP Pre-response averages from each subject were then grouped together to produce the results we report (Fig. 2).

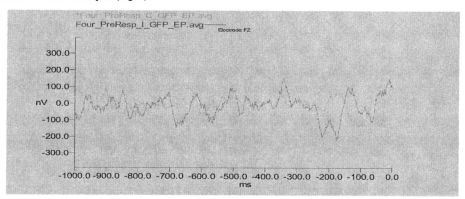

Figure 2

The picture shows group averaged correct (Green) and incorrect (Red) evoked potentials generated from the pre-response period 1-second before a response was made (4 subjects). The average number of trials from the four subjects were 110 accurate and 67 inaccurate. Note the phase differences in the GFP plots occurring -650 ms pre-response and again at -150 ms, just as decisions were finalized.

---

[1] Early segments included trials 1 to 100, which typically covered the first 75 minutes of the exam. The Late segments included trials 101 to 200, covering the second 75-minute period. The average time spent taking the entire exam was 2 1/2 hours and it was taken all at once.

To quantify the spatially distributed oscillatory neural activity patterns visible at the scalp, we examined the pre-response averaged spectral waveforms and plotted the GFP (in uV^2) topographically using the electrode coordinates from a standardized 3-dimensional head model fit to the electrode positioning system. The value at each frequency was color-coded to spatially display the averaged distribution of GFP maxima and minima over the head surface. For the spectral wave at the peak of the response, the power measures were analyzed to determine whether they were significantly different from the averaged GFP of the EEG values taken during eyes-open recordings from each subject. Then, we compared the levels of the oscillatory neural activity patterns under conditions related to performance and arousal.

To perform the analysis of arousal and performance related changes in the EEG, we measured the banded power of all GFP channels in 1 Hz bins from 2 to 30 Hz over brief segments 1-second before the subject responded. From those data we generated pre-response frequency-domain power spectral averages. We compared these results to the same analysis of time-domain averaged evoked potentials from the same time intervals. Statistical comparison of the evoked potential amplitudes made during the pre-response interval starting 1-second before the subject responded did not reveal significant differences between accurate and inaccurate conditions. During this same period, however, well-defined spatially distributed oscillatory patterns differentially related to subsequent accurate and inaccurate performances were discovered (Fig. 3). Additionally, spatial-frequency patterns were discovered that correlated with changes in the arousal level of the subject over relatively long time intervals. However, this latter discovery was based upon results from subjective questionnaires gathered at the end of the assessment task, and therefore, provided a weak means for validation with respect to arousal. We believe that further investigation into the use of SF pattern analysis methods to track changes in arousal is warranted given the preliminary results from our study.

3. Discussion

SF patterns associated with subsequently accurate performance involved 3 to 12 Hz bi-hemispheric oscillations predominantly in the anterior frontal, frontal, central, and parietal cortices. SF patterns associated with subsequently inaccurate performance involved much the same locations as with accurate performance; however, the distributions of peak power were significantly different. Thus prior to responding, clearly visible differences in topographic SF patterns could be seen for subsequently correct and incorrect responses (Fig 4). The map patterns displayed in Figure 3 depict the spatial and frequency distributions of the pre-response GFP for subsequently accurate (Left) and inaccurate (Right) performance.

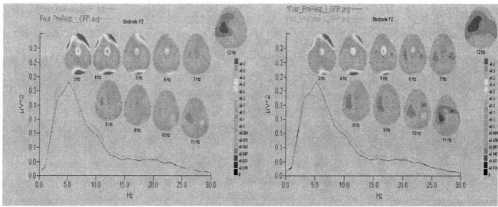

Figure 3
Both pictures show the pre-response accurate (Green) and inaccurate (Red) group averaged GFP spectral waveforms to 30 Hz. The pre-response SF topographic maps are shown for each 1 Hz band from 3 to 12 Hz. The 3-7 Hz maps are in the upper row, 8-11 Hz maps are in the lower row, and the 12 Hz map is located above the color scale. The pre-response accurate (Left) and inaccurate (Right) SF maps show varied spatial distributions in the group averages at each frequency band generated from data taken 1-second before the subject pressed the Submit answer button.

Our SF results are consistent with those obtained earlier using event-related covariance patterns to discriminate between accurate and inaccurate performance in pre-stimulus Continuous Negative Variation averages [22,7]. These results indicated that sites active in predicting subsequent accurate and inaccurate performance were located in the anterior-frontal, midline central, antero-central, left and right antero-parietal, and posterior-parietal cortices. Through the use of pattern subtraction procedures, the SF waveform displayed in Figure 4 shows the difference between the pre-response intervals of accurate and inaccurate performance. The blue and red areas in the maps reflect the greatest peak differences in 3 to 12 Hz oscillations occurring among distinct brain regions. These results are summarized in Table 1.

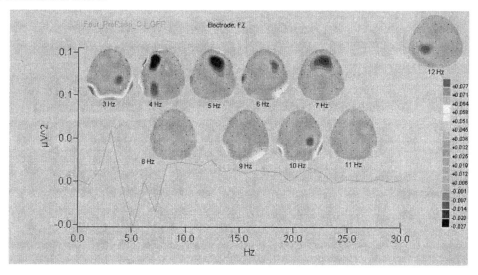

Figure 4

The SF maps show the differences between the pre-response intervals measured 1-second before 4 subjects answered nearly 700 questions. For the GFP waveform above, each point from the incorrect group average was subtracted from the corresponding point on the correct waveform, generating the plot above. The topographic SF maps identify scalp locations where the peak differences in 3 to 12 Hz amplitudes were greatest with increased power in sites colored red and decreased power in blue sites.

The chart below, lists the SF differences between subsequent accurate and inaccurate performance identified in the 1-second interval prior to a subject responding to a complex continuous performance attention and working memory task (from Figure 4):

Table 1 - Spatial-Frequency Changes Prior to Accurate and Inaccurate Performance	
Frequency Band	**Prior to correct responses, the following differences in Global Field Power were found by subtraction of Inaccurate spectra from Accurate.**
3 Hz	1) Increased left and right parietal and posterior-parietal activation with a midline occipital component. 2) Decreased left and right central and inferior antero-parietal foci. 3) Increased broad occipital activation due to an artifact caused by head movement during the task traced to one of the four subjects.
4 Hz	1) Decreased left anterior frontal, frontal, and fronto-central activation foci. 2) Decrease left posterior, central, and anterior-parietal activity. 3) Increased inferior frontal and parietal activation.

1051

5 Hz	1)	Large decreased right midline anterior-frontal, frontal and central foci extending into left inferior frontal and central cortices.
	2)	Decreased inferior centro-parietal and parietal foci.
6 Hz	1)	Decreased right fronto-central foci with left inferior central component.
	2)	Increased right inferior parietal activation.
7 Hz	1)	Large decreased left, right, and midline anterior-frontal, frontal, and fronto-central foci with decreased posterior-parietal component.
8 Hz	1)	Decreased left central and right inferior-central foci extending into left and right inferior-frontal areas.
9 Hz	1)	Decreased left central and right inferior-frontal foci.
	2)	Increased right inferior posterior-parietal activation.
10 Hz	1)	Decreased right central-parietal foci extending to right parietal and inferior-frontal areas.
	2)	Increased left and right inferior parietal and antero-parietal activation.
11 Hz	1)	Decreased right centro-parietal activation extending into left parietal, central, inferior-central and anterior-frontal areas.
12 Hz	1)	Large decreased left centro-parietal activation extending into left inferior central and frontal areas.

4. Conclusion

Although the exact origins of pre-response related SF patterns are unknown, participation by the prefrontal, frontal, premotor, and parietal cortices in higher-order cognitive processing has been demonstrated [16,23]. Preliminary data from our pilot study supports the view that executive functions rely on a distributed cerebral network not restricted to anterior cerebral areas and extends prior research that provides evidence that a fronto-parietal executive control mechanism is involved in working memory and decision-making.

Our analysis of higher-order cognitive communication networks in the brain suggest that decision-making performance in a complex cognitive assessment task involves several brain components: 1) a cognitive component; 2) a perception-based executive control component; and 3) an arousal component. In all cases, SF patterns of oscillatory activity were manifested at invariant activation sites with large differences between accurate and inaccurate group distributions of peak GFP amplitudes in the frequency range from 3 to 12 Hz. These results lead us to believe that it may be possible to identify SF patterns prior to a response taking place and to act upon that information by manipulating the subject matter presented to the operator to enhance overall task performance. If so, this capability could lead to enhanced methods for cognitive assessment, learning, and rehabilitation research.

Our evidence for distributed, coordinated spatial-frequency oscillations during higher-order complex decision-making tasks is consistent with earlier studies of this behavior [24,25]. The involvement of prefrontal and midline frontal sites is consistent with evidence that the dorsolateral prefrontal cortex and anterior cingulate is active during working memory and decision-making [5,15]. The finding of lateralized parieto-central somesthetic-motor sites is consistent with evidence that this area is active in the establishment of a preparatory set for neuronal firing patterns in the cortex prior to motor responses [21,26]. The antero-central integrative motor component is consistent with known activation of premotor and supplementary motor regions in initiating practiced motor responses and establishing new ones [27]. Additionally, the involvements of anterior-parietal and inferior-parietal components are consistent with understood mechanisms of long and short-term memory acquisition and retrieval in humans [28]. Although 3 Hz activity was observed in the occipital cortex during task performance, its presence was not clearly understood. However, after closer examination of the continuous data from each subject, it was found that data from one individual contained sporadic delta activity, which was believed to be caused by movement artifacts generated when the subject pressed his head back against the chair he was seated in.

These results, and similar results that indicate a correlation between SF patterns and working memory, demonstrate that the human brain dynamically communicates among its distributed and specialized subsystems with rhythmic oscillatory patterns in anticipation of decision-making and response selection. These SF patterns have shown site specific increases and decreases in banded spectral power and were effective at predicting accurate and inaccurate performance prior to a response actually taking place. Thus, we theorize that stable pre-response SF patterns may provide an index of a subject's uptake of subject matter or confidence in an answer while a decision was being finalized.

We must caution the results presented here due to small sample size and restriction of female and left handed subjects. Continuation of our pilot study would expand current research and include analysis of heart and eye activity from the same time intervals analyzed in the EEG. Additionally, we would investigate correlations with data provided from post-test behavioral questionnaires and individual SAT scores, as well. Further analysis using higher numbers of EEG channels and digitized electrodes positions would also afford better resolution and cross-validation of SF results with fMRI.

Acknowledgement

Human Bionics acknowledges the financial support of the Defense Advanced Research Projects Agency (DARPA). This work was supported by contract number NBCHC020022 through the Information Processing Technology Office. This acknowledgement should in no way reflect the position or policy of the U.S. Government and no official endorsement is inferred.

References

[1] Cranstoun S.D., Ombao H.C., von Sachs, R., Guo W., Litt B. (2002) Time-Frequency Spectral Estimation of Multichannel EEG Using the Auto-SLEX Method. *IEEE Trans. Biomedical Eng.,* 49(9):988-96

[2] Gobbele R., Waberski T.D., Schmitz S., Sturm W., Buchner H. (2002) Spatial direction of attention enhances right hemispheric event-related gamma-band synchronization in humans. *Neuroscience Letters,* 327(1):57-60

[3] Donald R. DuRousseau: Method and System for Initiating Activity Based on Sensed Electrophysiological Data. Patent Application Number: PCT/US/50509, Patent Date: Dec. 18, 2001

[4] Makeig, S. Enghoff, S. Jung, T. and Sejnowski, T. (2000) A natural basis for efficient brain-actuated control. *IEEE Trans. on Rehab. Eng.* 8(2):208-11

[5] Gevins, A. and Smith, M. (2000) Neurophysiological measures of working memory and individual differences in cognitive ability and cognitive style, *Cerebral Cortex, 10(9):829-39*

[6] Gevins, A.S., Cutillo, B., DuRousseau, D.R., Le j., et al. (1994) High-resolution evoked potential technology for imaging neural networks of cognition. In: Thatcher, R.W., et al. (Eds.) *Functional Neuroimaging: Technical Foundations,* Academic Press, Inc. : Orlando, pp. 223-31

[7] Picton T.W., Bentin S., Berg P., Donchin E., Hillyard S.A., et al. (2000) Guidelilnes for using human event-related potentials to study cognition: recording standards and publication criteria. *Psychophysiology* 37(2):127-52

[8] Breakspear M., and Terry J.R., (2002) Topographic organization of nonlinear interdependencies in multichannel human EEG. *Neuroimaging* 16(3 Pt 1):822-35

[9] Gundel A., and Wilson G.F., (1992) Topographical changes in the ongoing EEG related to the difficulty of mental tasks. *Brain Topography* 5(1):17-25

[10] Gevins, A.S., Alexander, J., Cutillo, B., Desmond, J., DuRousseau, D.R., et al. (1994) Imaging the spatiotemporal dynamics of cognition with high resolution evoked potential methods. *Human Brain Mapping,* vol. 1(2), John Wileys: New York, pp 101-16

[11] Kremper A., Schanze T., Eckhorn R., (2002) Classification of neural signals by a generalized correlation classifier based on radial basis functions. *Jour. of Neuroscience Methods* 116(2002):179-87

[12] Schanze T. and Eckhorn R., (1997) Phase correlation among rhythms present at different frequencies: spectral methods, application to microelectrode recordings from visual cortex and functional implications. *Intern. Jour. of Psychophysiology* 26(1997):171-89.

[13] Compte A., Brunel N., Goldman-Rakic P.S., Wang X.J., (2000) Synaptic mechanisms and network dynamics underlying spatial working memory in a cortical network model. *Cereb. Cortex* 10(9):910-23

[14] Levy R., Goldman-Rakic P.S., (2000) Segregation of working memory functions within the dorsolateral prefrontal cortex. *Exp. Brain Res.* 133(1):23-32

[15] Rowe J., Friston K., Frackowiak R., Passingham R., (2002) Attention to action: specific modulation of corticocortical interactions in humans. *Neuroimaging* 17(2):988

1053

[16] Freeman, W.J., & Skarda, C.A., (1985) Spatial EEG patterns, non-linear dynamics and perception: the neo-Sherringtonian view, *Brain Research Review.* Vol. 10 pp. 147-75

[17] Fuchs, M., Wischmann, H..A., Köhler, Th., Wagner, M., (1996). The local contribution to the field and the noise induced std. dev. as criteria for the iterative refinement of current density reconstruction, *Med. & Biol. Eng. & Computing,* 34(2):249-50

[18] Paulus M.P., Hozack N., Frank L., Brown G.G., (2002) Error rate and outcome predictability affect neural activation in prefrontal cortex and anterior cingulate during decision-making. *Neuroimaging* 15(4):836-46

[19] American EEG Society, (1991) Guidelines for standard electrode position nomenclature. *J. Clin. Neurophysiol.* 8(1991):200-2

[20] Semlitsch H.V., Anderer P., Schuster P., Presslich O., (1986) A solution for reliable and valid reduction of ocular artifacts applied to the P300 ERP. *Psychophysiology* 23(1986):695-703

[21] Lehmann D., and Skrandies W., (1986) Segmentation of EEG potential fields. *Electroencephalography and Clin. Neurophys.* 38(1986):27-32

[22] Gevins A.S., Morgan N.H., Bressler S.L., Cutillo B.A., White R.M., Illes J., et al., (1987) Human Neuroelectric Patterns Predict Performance Accuracy. *Science* 235:(1987):580-85

[23] Fernandez T., Harmony T., Silva-Pereyra J., Fernandez-Bouzas A., et al., (2000) specific EEG frequencies at specific brain areas and performance. *Neuroreport,* 11(2000)2663-68

[24] Jantzen K.J., Fuchs A., Mayville J.M., Deecke L., Kelso J.A., (2001) Neuromagnetic activity in alpha and beta bands reflect learning-induced increases in coordinative stability. *Clin. Neurophysiology* 112(9):1685-97

[25] Gross D.W., and Gotman J., (1999) Correlation of high-frequency oscillations with the sleep-wake cycle and cognitive activity in humans. *Neuroscience* 94(4):1005-18

[26] Smith M., McEvoy L., Gevins A.S., (1999) Neurophysiological indices of strategy development and skill acquisition, *Cognitive Brain Research,* 7(3):389-404

[27] Rogers R.L., Basile L.F.H., Taylor S., Sutherling W.W., Papanicolaou, A.C., (1994) Somatosensory evoked fields and potentials following tibial nerve stimulation. *Neurology* 44(1994):1283-86

[28] Gratton G., and Goodman-Wood M.R., (2001) Comparison of Neuronal and Hemodynamic Measures of the Brain Response to Visual Stimulation: An Optical Imaging Study. *Human Brain Mapping* 13 (2001):13-25

Preattentive crossmodal information processing under distraction and fatigue

Klaus Mathiak

Psychiatry & Psychotherapie, RWTH Aachen
Pauwelsstr. 30, 52074 Aachen, Germany
KMathiak@UKAachen.de

Krystyna Swirszcz

Center for Neurology, University Tübingen
Hoppe-Seyler-Str. 3, 72076 Tübingen
Krystyna.Swirszcz@Med.Uni-Tuebingen.de

Werner Lutzenberger

MEG Center, University Tübingen
Otfried-Müller-Str. 47, 72076 Tübingen
Werner.Lutzenberger@Uni-Tuebingen.de

Ingo Hertrich

Center for Neurology, University Tübingen
Hoppe-Seyler-Str. 3, 72076 Tübingen
Ingo.Hertrich@Uni-Tuebingen.de

Mikhail Zvyagintsev

Psychiatry & Psychotherapie, RWTH Aachen
Pauwelsstr. 30, 52074 Aachen, Germany
MZvyagintsev@UKAachen.de

Hermann Ackermann

Center for Neurology, University Tübingen
Hoppe-Seyler-Str. 3, 72076 Tübingen
Hermann.Ackermann@Uni-Tuebingen.de

Abstract

Stimuli in different sensory modalities are known to interact within the central nervous system. Where spatially divergent stimuli may delay responses, a spatially coherent stimulation design can support correct responses and speed up reaction times. Early connections can be assumed to be rather independent from higher cognitive processes and even fatigue. The current study focused on cross-sensory influences on preattentive auditory processing, which detects change and facilitates orienting reactions and their modulation by distraction and fatigue. We employed an oddball paradigm to assess cortical processing by means of whole-head magnetoencephalography (MEG). While subjects performed distraction tasks of varying difficulty, auditory duration deviants occurred randomly at the left or the right ear preceded (200-400 ms) by oculomotor, visual (static light or flow field), or tactile co-stimulation at either side. Mismatch fields were recorded over both hemispheres. In a second cohort, the MEG paradigm was applied after at least 8, 13, and 18 hours of wakening each. Changes in gaze direction and static visual stimuli cued the most reliable enhancement of deviance detection at the same side, and most strongly at the right auditory cortex. Tactile stimuli at either cheek enhanced the processing of auditory events from the respective ear by the ipsilateral hemisphere. In all three modalities, the lateralized unattended and unpredictive pre-cues acted analogously to shifts in selective attention, but were not reduced by attentional load or sleep suppression. The early cognitive representation of sounds reflects automatic crossmodal interference. We suggest that preattentive crossmodal integration provides a basis for orienting to objects in space and, thus, for selective attention. These networks are largely independent from distraction and fatigue. Crossmodal effects can be expected to reliably enhance perception of information or warning cues even under demanding situations.

1 Auditory perception of space

Crossmodal integration is strongly influenced by the spatial relation of the stimuli in the different sensory modalities. Visual and tactile space encoding relies apparently on rather direct cues, i.e., the position of effect in the sensory organ[1]. In audition, space is not represented at the sensory epithelium in the cochlear. Only indirect information is available on the position of the sound source. We will discuss here the most salient cues used by the central nervous system to extract this information from a sound before addressing the physiological mechanisms underlying crossmodal perception.

---

[1] Certainly for spatial localization also in these modalities, complex calculations are required to transform the information into the correct coordinate system, e.g., depending on gaze direction or position of the limb.

1.1 Cues for space perception

1.1.1 Azimuth encoding

The most commonly investigated cues are interaural level difference (ILD) and interaural time difference (ITD). They code in a rather direct fashion the azimuth angle of a sound, i.e., within the horizontal plane. Or to be more precise, this holds only for frontal sources because fron back cannot be decided on ths information only. Figure 1 illustrates that ITD is created by a higher distance of one ear to the source and ITD by the distance and the acoustic shadow of the head. The brainstem is thought to extract the precise time difference (tenth of μs) by coincidence detectors of signals from both cochlears with different amounts of axonal delays (Peña et al. 2001).

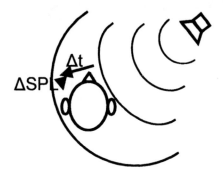

Figure 1: For a sound source from the right, the pressure waves reach the left ear later (30 μs/cm). The greater distance and covering head also lead to reduced amplitude.

1.1.2 Elevation encoding

The encoding of elevation, that is the angle between the sound source direction and the horizontal plane, can be less generally determined. However, the acoustic systems can extract the precise spectral structure of the sound signal which depends upon from which angle the sound reaches the ear. The shape of the pinna is here of high importance but also the head's configuration (hair, hat, nose, ...) and the rest of the body (most importantly the shoulders) also influences the spectral properties of the transfer function (Fig. 2). This information can be measured a coded as the head-related transfer function (HRTF; Algazi et al. 2001). Using this individual information, elevation can be simulated during headphone application of sounds as well.

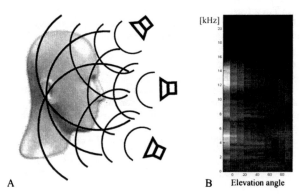

Figure 2: The direction how the sound wave hits the ear influences its frequency transmission properties (A). Each elevation angle shows an individual and specific frequency transfer function (B).

1.1.3 Distance encoding

The least clear is how distance is evaluated during the neuronal processing of sounds. Coleman (1963) showed that one relevant cue the amplitude of the signal. The closer a sound source comes the louder it gets (Fig. 3). However, to estimate an absolute distance, knowledge about the sound source intensity is required. From experience this can be in particularly estimated for voices.

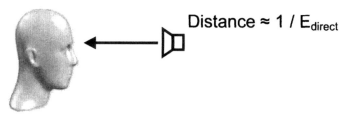

$$\text{Distance} \approx 1 / E_{direct}$$

Figure 3: If the sound energy of an object is known – such as in voices – the distance can be calculated from the received sound energy.

The human auditory system, however, can also estimate distance without explicit information on sound intensity. With knowledge about the acoustical properties of the space, the separation of direct and indirect sound energy can provide information on the distance. As known from the precedence effects, the auditory system separates efficiently direct sound sources from the sounds originating from reflections and arriving later at the listener's ear. The ratio between the received energy from both fractions can be use to estimate the distance independent from the energy of the sound source (Zahorik et al. 2002).

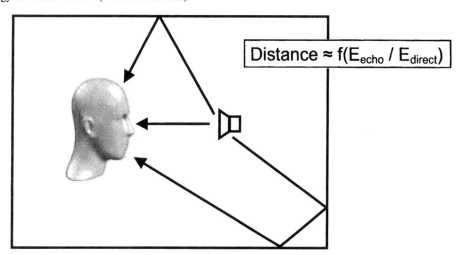

$$\text{Distance} \approx f(E_{echo} / E_{direct})$$

Figure 4: If the sound energy of an object is not known but the geometry of the room then the distance can be calculated from relation of the directly received sound energy and the one from the reflections.

1.2 Simulation of auditory space

The virtual percept of an auditory space can be simulated using headphones. To that means, the auditory cues existing in naturally perceived sounds must be imitated. In particular, the filter function of the ear (HRTF) must be applied depending on the angle of the sound source. Sometimes, it proves to be sufficient to use not individually measured HRTFs but to select a "fitting" one from a given set (see Mathiak et al., 2003). Moreover, the reflections within the room must be imitated. This is very similar to a ray-tracing as applied in computer graphics. Due to the reversibility of rays the sound impression from a reflection can be obtained by reconstructing a virtual sound source that is localized according to the non-reflected beam. From this virtual sound source, the perception characteristics can be evaluated and folded with the dampening effects from the reflections. A perfectly rectangular room – the so called shoebox – makes the calculations rather straight forward (see Fig. 5). Commonly only reflections of low order are calculated.

Almost infinitely more aspects of naturally sounding space simulations can be considered – like in ray-tracing, e.g., dispersion in the air or scattering. More recent systems try o simulate the virtual space percept with free field headphone, which increases the difficulty since the transfer from speaker to ear must be reconsidered newly.

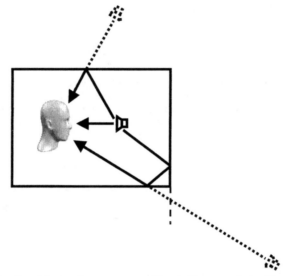

Figure 5: In the "shoebox" model, virtual sources are considered. These are obtained by un-doing the reflections. This virtual source is now used to calculate the sound impression by the according echo component. Additionally, the dampening by the reflections needs to be considered, e.g., a wall and a floor reflection in the case of the right lower source.

1.2.1   Auditory virtual reality

To gain the full versatility of virtual reality, the space simulation must be adapted to dynamic changes in the space. In particular the space must be interactive and if the user moves within the virtual space the entire calculations need to be adjusted with very little latency.

1.3   Automatic encoding of auditory space

One advantage of auditory space simulation or even auditory virtual reality is that it can be applied within functional brain imaging devices. These devices are characterized by tubes or wholes where the brain with the head needs to be placed in. This – in general – does not allow for setting up a natural sound environment. Instead, headphones are used and the sounds are preprocessed according to the before mentioned simulation procedures. Finally even interactive environments are possible. An application for virtual reality – including auditory VR – was presented by Mathiak & Weber (2005).

Early cortical processing of auditory stimuli already reflects spatial processing properties. Using whole-head magnetoencephalography (MEG), we recorded neuromagnetic responses to amplitude variations over both supratemporal planes, without spatial simulation, with HRTF and with full auditory space simulations (Mathiak et al, 2002). In the absence of other cues for distance (direct sound and HRTF condition) amplitude changes elicited enhanced preattentive responses over the right temporal lobe. This indicates a hemispheric lateralization of the 'where' pathway in the human. Lesion studies in monkeys and humans have shown that the rostral part of the right superior temporal cortex contributes to spatial awareness in the visual domain. These data indicate that the distance to a sound source is processed within the adjacent right auditory cortex. This suggestion extends the recent model of a right-hemisphere temporal multisensory matrix that subserves the integration of space-related data across visual and auditory modalities.

2   Selective attention without volition

2.1   Introduction

Spatially directed attention acts across modalities. A 'top-down' influence simultaneously modulates visual, auditory and tactile modalities \cite{eime02}. Moreover, even arbitrary and non-informative stimuli in one modality, e.g., auditory, may involuntarily modulate signal detection in another channel, e.g., visual \cite{mcdo00}. Both exogenous and endogenous mechanisms of response selectivity – subserving the exploration of complex scenes – have been considered to underlie separate neuronal correlates (Frassinetti et al. 2002; McDonald et al. 2001; Macaluso et al. 2001).

Exogenous multi-sensory interactions have been observed in animal experimentation at an early level of neuronal processing. At the auditory pathway, for instance, interactions with several other modalities suggest selective effects: oculomotor input to the inferior colliculus (Groh et al. 2001), audio-visual maps at the superior colliculus \cite{stei98}, and somatosensory evoked responses at the nucleus cochlearis (Young et al. 1995) and the secondary auditory cortex (Schroeder et al.2001). In the human, in contrast, findings on crossmodal interactions at early preattentive stages fail to show spatial selectivity (Foxe et al. 2000).

Endogenous effects of attention can be observed in human studies. Spatial selectivity has been documented only at the level of post-sensory functions. Behavioral effects are well documented (Frassinetti et al. 2002; McDonald et al. 2000). In electrophysiological measures, spatial shifts of selective attention affect sensory decisions \cite{mcdo01} and motor preparation (Kaiser et al. 2000). The auditory evoked component mismatch negativity (MMN) reflects orienting reactions, seen as automatic mechanisms shifting the focus of attention (see Näätänen 1992). This response traces the learning history of relevant auditory contrasts (Cheour et al. 2002). Whole-head MEG measures the neuromagnetic analogue of MMN (MMNm) and focuses selectively on the supratemporal plane (STP) of the left and the right hemisphere (Scherg et al 1986). At this level, dichotic listening yields selective processing of competing stimuli over both ears at the contralateral auditory cortices (Mathiak et al. 2000, 2002).

2.2   Study of early crossmodal interactions

In order to investigate the effect of crossmodal integration on preattentive orienting to objects in space, we studied MMNm in response to lateralized rare events. Subjects performed a cognitive distraction task while unattended crossmodal pre-cues without predictive information were applied at either side. Whole-head MEG recorded the effects of these oculomotor, visual, and tactile pre-cues on preattentive processing of the dichotic sounds over both hemispheres.

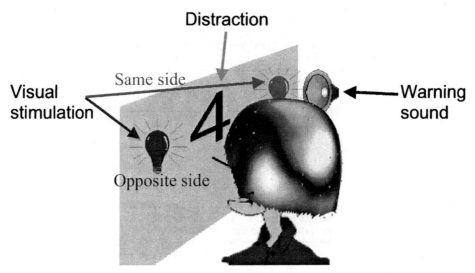

Figure 6: Comic of the experimental setup for the study of crossmodal effects under distraction of voluntary attention. Auditory evoked fields are measured in response to a "warning" sound located at one side. The sound had no behavioral relevant and its side was not known before. The pre-cues, e.g., visual stimuli, were presented randomly either at the same or at the opposite side. None of these stimuli were voluntarily attended. The subjects had a distraction task that was bound to digits displayed in front of the volunteer.

The setup allows for studying the spatially specific effects of unattended pre-cues on pre-attentive processing. Early crossmodal interactions can be found that enhance responses to sounds that were pre-cued at the some side as compared to those pre-cued at the opposite side (Menning et al., 2005; Mathiak et al. 2002). These interactions were most prominent at the right hemisphere conceivably reflecting lateralization of spatial processing. We investigated this interaction with respect to effects of distraction and sleep suppression in order to test how much they rely on attentive processing or up to which level they can be modified.

2.3 Effects of distraction of pre-attentive selective processing

Crossmodal interactions were measured under three different distraction tasks. The lowest demand imposed a simple detection task (ONE), followed by a complex detection task (3-EVEN), and as most involving a continuous performance task (2-BACK). A ceiling effect characterized the relation of mismatch amplitude and cognitive task demand (Fig. 7).

To probe for the influence of attention, interactions between the indicator of complex task (3-EVEN and 2-BACK as compared to ONE) and the lateralization effects were evaluated. Significant effects were revealed only in terms of an enhanced interaction of pre-cue and deviant side under the higher workload conditions: This enhancement yielded significance under the oculomotor condition over both hemispheres (both $p < 0.001$). The visual pre-cues were more effective under the higher workload conditions in the second experiment, over the right hemisphere only ($p < 0.001$).

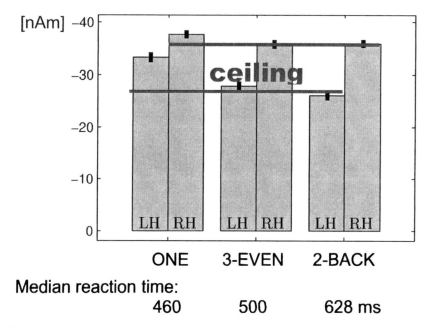

Figure 7: Distraction effects on mismatch response. The simple detection task (ONE), the pattern detection task (3-EVEN), and the demanding continuous performance test (2-BACK) involve increasing demands on the behavioral level, significant increases of the reaction times. The mismatch response in contrast achieves a plateau which may reflect a basic attention independent component.

2.4 Effects of sleep suppression of pre-attentive selective processing

We studied 12 subjects (6 female, 19-36 yrs) during sleep suppression. Subjects filled out a sleeping diary one week prior to the measurements. They were required to sleep for at least 7 h daily and wake up before 9 a.m. during this week. Crossmodal interactions on pre-attentive processing were studied in three equal measurements at 5 p.m., 10 p.m., and 3 a.m., i.e., at least after 8, 13, and 18 h without sleep. Likewise a moderate level of fatigue was obtained.

Analysis followed the established automatized and objective procedures as in the previous experiments. No significant session effect could be observed (Fig. 8). We can conclude that at least at this level, no influence of fatigue on pre-attentive processing and crossmodal interactions can be oserved. It suggests that multisensory and crossmodal integration are robust against vegetative and cognitive interferences.

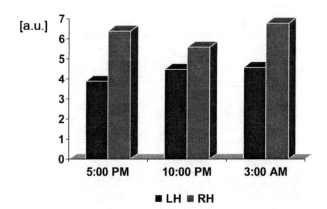

Figure 8: Response amplitudes to the pooled "warning" stimuli that is the mismatch negativity. Right hemispheric responses to the spatial cues are larger. No decrease after sleep suppression can be observed.

3   Discussion and conclusions

Functional imaging – MEG – allows for displaying neuronal processing of auditory space and its interaction with crossmodal stimulation. The relevant auditory cues can be simulated and may elicit a close to natural percept within the imaging devices. The measures reflect automatic processes that prepare attention shift and response preparation. As such, they are relevant for the research of behavior in complex environments. The addressed mechanisms operating without voluntary attention can hardly be studied be means of behavioral measures because they would need to involve directed attention.

It allows a risk-less screening of usually unwanted conditions such as distraction and sleep deprivation. Here we find that these early processing steps and interaction across modality are rather stable. Distraction even tends to stabilize the spatially specific interaction between visual and auditory domain.

The imaged pre-attentive processes do not directly reflect back behavior. A generalization is, thus, only possible to a limited degree. However, basic levels of information processing are addressed which always serve as initial level for higher cognitive processes and the observable behavior. Crossmodal effects can be expected to reliably enhance perception of information or warning cues even under demanding situations.

Acknowledgements

This study was supported by the DFG (SFB550/B1), and the DaimlerChrysler AG funded by the DARPA AugCog program. We acknowledge to Maike Borutta for here reliable technical support.

References

Algazi, V.R., Duda, R.O., Duraiswami, R., Gumerov, N.A., and Tang, Z. (2002) Approximating the head-related transfer function using simple geometric models of the head and torso. J Acoust Soc Am 112:2053-2064.

Cheour M, Martynova O, Näätänen R, Erkkola R, Sillanpää M, Kero P, Raz A, Kaipio M-L, Hiltunen J, Aaltonen O, Savela J, Hämäläinen M (2002) Speech sounds learned by sleeping newborns. Nature 415:599-600.

Coleman, P. D. (1963). An analysis of cues to auditory distance perception in free space. Psychological Bulletin, 60, 302–315.

Groh, J. M., Trause, A. S., Underhill, A. M., Clark, K. R., and Inati, S. (2001). Eye position influences auditory responses in primate inferior colliculus. Neuron, 29, 509–518.

Eimer M, van Velzen J, Driver J (2002) Cross-modal interactions between audition, touch, and vision in endogenous spatial attention: ERP evidence on preparatory states and sensory modulations. J Cogn Neurosci 14:254-271.

Foxe JJ, Morocz IA, Murray MM, Higgins BA, Javitt DC, Schroeder CE (2000) Multisensory Auditory-Somatosensory Interactions in Early Cortical Processing Revealed by High-density Electrical Mapping. Brain Res Cogn Brain Res 10:77-83.

Frassinetti, F., Bolognini, N., Làdavas, E. (2002) Enhancement of visual perception by crossmodal visuo-auditory interaction. Exp Brain Res 147:332-343.

Kaiser J, Lutzenberger W, Preissl H, Ackermann H, Birbaumer N (2000) Right-hemisphere dominance for the processing of sound-source lateralization. J Neurosci 20:6631-6639.

Macaluso E, Frith CD, Driver J (2000) Modulation of Human Visual Cortex by Crossmodal Spatial Attention. Science 289:1206-1208.

Macaluso, E., Frith, C. D., and Driver, J. (2001). Multisensory integration and crossmodal attention effects in the human brain-response. Science, 292, 1791–1792.

Mathiak, K., Hertrich, I., Kincses, W. E., Riecker, A., Lutzenberger, W., and Ackermann, H. (2003). The right supratemporal plane hears distance: Neuromagnetic correlates of virtual reality. NeuroReport, 14, 307–312.

Mathiak, K., Hertrich, I., Kincses, W. E., Rothe, S., Lutzenberger, W., and Ackermann, H. (2002). Pre-attentive shift of covert selective attention in the auditory domain mediated by cross-modal interaction. Society of Neuroscience 32st Annual Meeting, Orlando, FL.

Mathiak, K., Hertrich, I., Lutzenberger, W., and Ackermann, H. (2000). Encoding of temporal speech features (formant transients) during binaural and dichotic stimulus application: a whole-head magnetencephalography study. Cognitive Brain Research, 10, 125–131.

Mathiak, K., Hertrich, I., Lutzenberger, W., and Ackermann, H. (2002). Functional cerebral asymmetries of pitch processing during dichotic stimulus application: a whole-head magnetoencephalography study. Neuropsychologia, 40, 585–593.

Mathiak, K., Weber, R. (2005) fMRI of virtual social behavior: brain signals in virtual reality and operational environments. HCI proceedings, Las Vegas.

McDonald, J. J., Teder-Sälejärvi, W. A., and Hillyard, S. A. (2000). Involuntary orienting to sound improves visual perception. Nature, 407, 906–908.

McDonald, J. J., Teder-Sälejärvi, W. A., and Ward, L. M. (2001). Multisensory integration and crossmodal attention effects in the human brain. Science, 292, 1791–1792.

Menning, H., Ackermann, H., Hertrich, I., Mathiak, K., (2005) Spatial selective attention for peripersonal auditory space is modulated by tactile priming. Experimental Brain Research, in press.

Näätänen, R. (1992). Attention and brain function. Mahwah, NJ: Erlbaum.

Pena, J. L., Viete, S., Funabiki, K., Saberi, K., and Konishi, M. (2001). Cochlear and neural delays for coincidence detection in owls. The Journal of Neuroscience, 21, 9455–9459.

Scherg, M., and Cramon, D. von. (1986). Evoked dipole source potentials of the human auditory cortex. Electroencephalography and Clinical Neurophysiology, 65, 344–360.

Schroeder CE, Lindsley RW, Specht C, Marcovici A, Smiley JF, Javitt DC (2001) Somatosensory Input to Auditory Association Cortex in the Macaque Monkey. J Neurophysiol 85:1322-1327.

Steinschneider M, Reser DH, Fishman YI, Schroeder CE, Arezzo JC (1998) Click train encoding in primary auditory cortex of the awake monkey: Evidence for two mechanims subserving pitch perception. J Acoust Soc Am 104:2935-2955.

Macaluso, E., Frith, C. D., and Driver, J. (2001). Multisensory integration and crossmodal attention effects in the human brain – response. Science, 292, 1791–1792.

Mathiak, K., Hertrich, I., Kincses, W. E., Riecker, A., Lutzenberger, W., and Ackermann, H. (2003). The right supratemporal plane hears distance: Neuromagnetic correlates of virtual reality. NeuroReport, 14, 307–312.

Mathiak, K., Hertrich, I., Lutzenberger, W., and Ackermann, H. (2000). Encoding of temporal speech features (formant transients) during binaural and dichotic stimulus application: a whole-head magnetencephalography study. Cognitive Brain Research, 10, 125–131.

Mathiak, K., Hertrich, I., Lutzenberger, W., and Ackermann, H. (2002). Functional cerebral asymmetries of pitch processing during dichotic stimulus application: a whole-head magnetoencephalography study. Neuropsychologia, 40, 585–593.

McDonald, J. J., Teder-Sälejärvi, W. A., and Hillyard, S. A. (2000). Involuntary orienting to sound improves visual perception. Nature, 407, 906–908.

McDonald, J. J., Teder-Sälejärvi, W. A., and Ward, L. M. (2001). Multisensory integration and crossmodal attention effects in the human brain. Science, 292, 1791–1792.

Näätänen, R. (1992). Attention and brain function. Mahwah, NJ: Erlbaum.

Pena, J. L., Viete, S., Funabiki, K., Saberi, K., and Konishi, M. (2001). Cochlear and neural delays for coincidence detection in owls. The Journal of Neuroscience, 21, 9455–9459.

Scherg, M., and Cramon, D. von. (1986). Evoked dipole source potentials of the human auditory cortex. Electroencephalography and Clinical Neurophysiology, 65, 344–360.

Young, E. D., Nelken, I., and Conley, R. A. (1995). Somatosensory effects on neurons in dorsal cochlear nucleus. Journal of Neurophysiology, 73,743–765.

Zahorik P, Wightman FL (2001) Loudness constancy with varying sound source distance. Nat Neurosci 4:78-83.

EEG- and Context-based Cognitive-State Classifications Lead to Improved Cognitive Performance While Driving

Andreas Bruns[*], Konrad Hagemann, Michael Schrauf, Jens Kohlmorgen[1], Mikio Braun[1], Guido Dornhege[1], Klaus-Robert Müller[1], Chris Forsythe[2], Kevin R. Dixon[2], Carl E. Lippitt[2], Carey D. Balaban[3], Wilhelm E. Kincses

DaimlerChrysler AG, Research & Technology, Machine Perception, 096/T728 – REI/AI, 70546 Stuttgart, Germany
[1] Fraunhofer Institute FIRST, Kekuléstrasse 7, 12489 Berlin, Germany
[2] Sandia National Laboratories, P.O. Box 5800, Mailstop 1188, Albuquerque, NM 87185, U. S. A.
[3] University of Pittsburgh Medical Center, Dept. of Otolaryngology, 146 EEINS, Pittsburgh, PA 15213, U. S. A.
[*] Corresponding author; e-mail: andreas.a.bruns@daimlerchrysler.com

Abstract

In the field of Augmented Cognition (AugCog), the identification of different cognitive states is the primary prerequisite for taking actions that, e.g., relieve the cognitive load currently imposed on the operator of a human-machine interaction system. The ultimate test of such an AugCog system's success, however, is its ability to actually improve the operator's performance through the mitigation actions. Here we present an AugCog system in which cognitive-state discrimination is based mainly on neural correlates of workload, and we demonstrate that with the system implemented on a Mercedes-Benz vehicle, it significantly improves the driver's performance on a number of different tasks.

In our investigation, driving a car serves as an example of an operational environment. Five subjects participated in the study as drivers. All subjects were right-handed, non-smoking men with normal or corrected-to-normal vision, aged 29.4 ± 3.8. Subjects were required to have considerable driving experience and to be used to driving vehicles of the Mercedes-Benz S-Class. The experimental protocol was carried out twice for each subject – once without the AugCog system (*Reference Session*) and once with the system being enabled (*AugCog Session*); each session had an estimated duration of 2 hours. While driving on a pre-defined route on a German highway, the subject performed blocks of three different *secondary tasks*, alternating with baseline blocks (without secondary task). Secondary tasks were meant to induce certain types of cognitive load. In addition to driving and to the secondary tasks, a *tertiary task* was continuously performed. This task was subject to the mitigation measures, i.e., its form of presentation (scheduling vs. no scheduling, visual vs. auditory) was flexibly adapted to the (putatively) current cognitive state. Furthermore, tertiary task performance served the purpose of quantifying the AugCog system's success.

Three types of secondary tasks were presented: (1) Auditory Workload Task (AWT): An audio narration was presented via loudspeakers, with the voice of the (male) narrator superposed by that of a (female) newsreader. The subject was instructed to ignore the latter and to carefully listen to the narration. (2) Mental Arithmetic Task (MAT): The subject had to silently count backward from a 3-digits number in steps of 27 as fast and accurately as possible. (3) Special Driving Maneuvers (SDM): The subject had to perform certain driving maneuvers (e.g., change lanes, overtake or follow other vehicles) which were intended to require increased visual attention to the traffic situation.

In the tertiary task, "left" and "right" messages were randomly (50/50) presented every few seconds and had to be acknowledged by pressing a corresponding (left or right) button as quickly and accurately as possible. During the Reference Session, tertiary stimuli were uniformly presented as voice (i.e. auditory) messages during the AWT and MAT blocks and the pertinent baseline periods, and as arrows (i.e. visual messages) during the SDM blocks and the pertinent baseline periods. During the AugCog Session, two mitigation strategies – scheduling and modality switch – were employed. (A) Scheduling: tertiary stimuli generated during the AWT and MAT blocks were supposed to be suppressed by the mitigation system and to be delayed into the pertinent baseline periods (thereby causing a higher stimulus density during these intervals). (B) Modality switch: tertiary stimuli generated during the SDM blocks were supposed to be switched to the auditory modality (voice messages instead of arrows).

Data acquisition comprised registration of scalp EEG (29 channels), logging the subject's responses and reaction times, and recording vehicle-related data via an extended CAN-bus system. Cognitive workload was assessed on-line by (1) two independent EEG-based classifiers that were trained to detect task-specific neural signal patterns (Fraunhofer Institute FIRST), and (2) a context-based classifier that was trained to detect potentially demanding driving and traffic situations (Sandia National Laboratories; University of Pittsburgh). The subject's reaction times on the tertiary task served for quantifying the AugCog system's success.

EEG-based classifier accuracies were 71.8 ± 14.9 % for the AWT classifier and 78.2 ± 19.0 % for the MAT classifier; the context-based classifiers showed classification accuracies around 95 %. Mean operators' reaction times on the tertiary task decreased, due to the AugCog system, from 730 ms to 700 ms during the AWT, from 759 ms to 663 ms during the MAT, and from 860 ms to 702 ms during the SDM conditions (significant at $p < 0.01$; one-tailed Wilcoxon rank-sum test).

Our results show that it is not only possible to reliably discriminate different cognitive states on the basis of neuronal signals, but that such discrimination can well be used to control mitigation actions in human-machine interaction under closed-loop conditions, and that a system making use of these possibilities is able to significantly improve the cognitive performance of human operators.

This work was partly funded by DARPA under contract NBCH3030001.

Neural Correlates of Simulated Driving:
Auditory Oddball Responses Dependent on Workload

Markus Raabe[1], Roland M. Rutschmann[1], Michael Schrauf[2] Mark W. Greenlee[1]

[1]Department of Psychology, University of Regensburg
Institute of Experimental Psychology, University of Regensburg, Universitaetsstr. 31
93053 Regensburg, Germany
(mark.greenlee@psychologie.uni-regensburg.de)
[2]Institute of Experimental Biological Psychology, University of Düsseldorf, D-40225
Düsseldorf,

Abstract

In the present study, we explored the neural responses to seldom auditory tones presented to volunteers via headphones while they participated in a simulated driving experiment. Experimental manipulation of the workload was achieved in two conditions, one in which the driver set his or her own pace (low workload) and another in which the driver's pace was determined by a lead car (high workload). Using EEG and fMRI, we could determine a significant effect of workload on the amplitude of the P3 component to auditory oddballs. The reduction of P3-amplitude during the high workload conidition was associated with a more focused hemodynamic response in fMRI.

1   Introduction

Ergonomics is the study of human behavior in work settings, with the aim to design work environments optimal with respect to economic and human aspects (Laurig, 1990). Wickens (2002) defines workload in ergonomics as the relation between the (quantitative) demand for resources imposed by a task and the ability to supply those resources by the operator. In this context prototypes of man-machine interfaces should be tested for their usability and experienced workload before rational decisions can be made concerning their production.

As a related discipline, cognitive neuroscience seeks to reveal the neural basis for cognition by employing electroencephalographic (EEG) brain-imaging (PET, fMRI) techniques in human volunteers to determine which cortical regions are involved in psychological tasks. One application of cognitive neuroscience to ergonomics is the study of brain activations during demanding cognitive tasks (Corbetta & Shulman, 2002). For example one can distinguish between different taskloads and states of complexity in verifying the meaning of a sentence by using neuroscientific methods like fMRI (Schmalhofer, Raabe, Friese, Pietruska & Rutschmann, 2004).

This research project intends to combine these two areas - ergonomics and cognitive neuroscience - in a neuroergonomic approach (Baldwin, 2003; Sarter & Sarter, 2003) to study the effect of workload on objective measures of brain activation during simulated driving.

Mental workload assessment techniques in ergonomics are often grouped into three categories: Subjective measures, behavioral measures (primary, dual or multiple task measures) and psychophysiological measures (De Waard, 1996; Wickens, 2000). In our neuroergonomic approach we combine all the three categories. We assess the subjectively experienced effort by a rating scale and introduce a psychophysiological assessed dual task paradigm as an objective measurement of workload. More precisely, methods of electroencephalography (EEG) and functional magnetic resonance imaging (fMRI) are combined to determine the effects of simulated driving on the cortical responses to repetitive auditory stimulation (oddball task). After identifying workload-sensitive brain regions we use EEG and fMRI to gain objective measures of workload, which are not possible with behavioral or subjective measurements alone.

Dual tasks have been developed to test for limited capacity in divided attention paradigms. Compared with single tasks they provide extensive information concerning cross- and intramodal time-sharing capacities (Wickens, 2000). Event-related potentials are averaged event-triggered brain potentials that can be extracted from on-going EEG (Andreassi, 2001). A prominent component of the ERP in cognitive tasks is the P3, which is expressed in a positive wave with a maximum around 300 msec poststimulus. The amplitude of this positive ERP component is dependent on the stimulus and task: seldom occurring stimuli (oddballs) evoke a prominent P3 (Andreassi, 2001) and its amplitude can be modulated by varying levels of workload (Fowler, 1994; Kramer, Trejo & Humphrey, 1995).

Here volunteers perform a primary task (driving under high and low workload demands), while an additional secondary task is performed (listening to intermittent sinusoidal tones of constant frequency). The introduction of seldom, unpredictable oddballs in the form of single tones (one octave above that of the "standards") is expected to evoke a robust positivity in the EEG approximately 300 ms poststimulus. The amplitude of this evoked response is reduced under high, compared to low, workload (Kramer et al., 1995). This neural activity should also evoke hemodynamic responses that can be detected in fMRI (Soltani & Knight, 2000). A further aim of this study is to locate this cortical activity in human cortex and see how workload effects the extent and distribution of the fMRI response.

2 Methods

2.1 Subjects
Overall, 11 participants (females = 5, mean age = 26, SD = 2.7) participated in the EEG experiment at the University of Osnabrück. All of them had normal visual and auditory acuity, had no neurological disturbances and were familiar with playing racecar games. Three of them (n = 3, male, mean age = 27 yrs, SD = 2 yrs) participated in the fMRI experiment. Participants were paid € 22.5 for approximately 3 hours of their time.

2.2 Procedures and material
We adapted a commercially available computer game (DTM-Racedriver, Codemasters, U.K.) for our purposes. The test driver could control the simulated speed and steering of the racecar through appropriate manipulation of the keyboard (response box in fMRI), which was practiced in a warm-up session. During the experiments, volunteers were instructed to "drive" a DTM racecar in separate runs under low (self-paced) and high (pace determined by lead car)

workloads. While driving the volunteers were instructed to attend to frequent and low (80%, 1000 Hz) or seldom and high tones (20%, 2000 Hz), which were presented via headphones and custom-made software in both experiments. In the fMRI experiment one trial consisted of 7 tones. The third position within one trial could either be a frequent or a low tone (see Figure 1 for an illustration of a trial). We presented 96 frequent and 24 seldom trials in a randomized order. For the EEG experiment we presented 40 trials of each condition. To keep the proportion between the frequent and seldom tones constant, the position of the seldom tone within one trial was variable with the restriction that the seldom tones were not allowed to appear consecutively.

Second:	0.5	1.5	2.5	3.5	4.5	5.5	6.5
Tone:	F	F	F/S	F	F	F	F

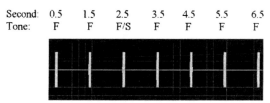

Figure 1: Timing of auditory stimuli presented during a representative trial in the fMRI experiment, F = frequent tone, S = seldom tone (target - one octave higher than F).

In the high workload condition subjects were instructed to obtain a new personal "top score". A lead car, representing the speed recorded in the last lap, was added to provide the participants with feedback about their performance. As a lead car we used a semi-transparent image, since this "ghost" had the advantage of simulating a race without the risk of collisions. In the low workload condition participants were instructed to drive several continuous laps. The lead car was switched off in this condition. In the EEG experiment we inserted an additional control condition. Subjects were instructed to respond to the seldom tones as fast as possible by a button press while maintaining their gaze on a central fixation spot. For assessing the subjectively experienced load of the three experimental conditions, subjects were asked to fill out the Rating Scale Mental Effort (Zijlstra, 1993) after the EEG experiment.

2.2.1 EEG protocol: Acquisition and Analysis

EEG data were collected continuously (sampling rate: 50 Hz; bandpass filtering: 0.1-70 Hz; gain: 1000) with the software SCAN 3.0 (Neuroscan Inc., USA) and a 32-channel Synamps EEG amplifier system Model 5083 (Neuroscan Inc., USA). The impedances of the Ag/AgCl electrodes were below 10,000 ohms. Electroencephalographic activity was recorded from 9 sites (F3, Fz, F4, C3, Cz, C4, P3, Pz, P4) according to the international 10-20 System (Jasper, 1958). They were referenced to both mastoids. During simulated "driving" the volunteers sat in a comfortable examination armchair, seated approximately 80 cm away from a LCD monitor and they were given access to a keyboard.

The data was post-processed and assessed using BrainVision Analyzer (BrainProducts, Germany). Raw data were inspected (maximum allowed voltage step: 50 μV; maximum allowed absolute difference: 200 μV for 200 ms; lowest allowed activity: 0.5 μV for 100 ms) and elements surpassing our thresholds (including a baseline interval of 1000 ms) were excluded from further analysis. In a next step, the EEG data were filtered (bandpass: 0.01-40 Hz; 50 Hz Notch Filter) and segmented by creating overlapping segments lasting from 500 ms pre- until 1000ms poststimulus. Only seldom tones were included in the segmentation. After correcting the baseline (reference interval 500 ms prestimulus until stimulus onset), the ERPs were calculated. Finally we calculated the grand average for each of the three experimental conditions. For statistical analysis we identified the positive maximum of the grand average in the control condition, which appeared 330 ms poststimulus. The mean values of each subject for this time point entered a paired T-test on second level to compare the low with the high workload condition. The subjects' individual P3 peaks of the different conditions were variable. We therefore omitted the latency factor from our approach to significance testing, which was very stringent.

2.2.2 Brain imaging protocol: Acquisition and Analysis

Brain-imaging was performed using functional Magnetic resonance imaging on a 1.5 Tesla Magnetom Sonata scanner (Siemens, Germany) at the Brain-Imaging Center, University of Oldenburg. This scanner is equipped with an echo-planar imaging (EPI) booster for fast gradient switching. Subjects were placed in the scanner and equipped with an MRT compatible LUMI-touch response box (Photon Control Inc., Canada) and headphones (MR Confon, Germany). Their head position was secured in the head coil, and a mirror system was placed on the coil, so that they could see the racing game on a screen mounted on the rear of the scanner bore. High-resolution, sagittal T1-weighted images were acquired at the end of the experiment with the magnetization prepared, rapid acquisition gradient echo sequence (MP-RAGE) to obtain a 3D anatomical scan of the head and brain (TR = 1900 ms; TE = 3.93 ms; 1 mm^3 isotropic voxel size). Functional imaging was performed with T2*-weighted gradient echo-planar imaging (EPI). The time to echo corresponded to TE=50 ms, total scan repetition time was TR=7 sec, the acquisition time was 2 sec, the flip angle corresponded to 90°, and we used a field of view FOV = 192 mm, with a voxel matrix of 3 * 3 mm and a slice thickness of 3 mm, resulting in a 27 mm^3 isotropic voxel size. We acquired volumes with 22 contiguous slices and a distance factor of 10-20 %. The acquisition sequence of the slices was interleaved. The slices were rotated approximately 20° relating to the AC-PC line thus we could image most of the frontal and parietal lobe and superior temporal areas. The stimulation protocol consisted of 122 volumes per run. We used an event-related design and sparse imaging to separate in time the response to the auditory stimuli and minimize the effects of gradient switching noises on the BOLD response (Mueller et al., 2003).

All MR-data were preprocessed and statistically analyzed using SPM2 (Functional Imaging Laboratory, Welcome Department of Imaging Neuroscience, London, UK). Functional images were corrected for acquisition timing, realigned to the first image and coregistered to the structural images. After normalizing all images to the MNI template, functional images were resampled to 2x2x2 mm^3 and spatially smoothed using an isotropic Gaussian kernel with 8 mm FWHM. A general linear model was fitted to the individuals, modelling the standard and the deviant sounds as events. We modelled both events explicitly to avoid mixing up the first scan, which was not task-related, with each of our experimental conditions.

For statistical analysis, a fixed-effects model was applied, treating the different subjects as different experimental sessions. T-contrasts were calculated between the deviant and the standard tone in the high and the low workload condition. Statistical maps were thresholded with T = 3.1 (uncorrected p-value .001) and clusters surpassing a corrected p-value of .05 on cluster level were reported as significantly activated.

3   Results

3.1   Subjectively experienced load

Figure 2 shows the mean values of the Rating Scale Mental Effort (Zijlstra, 1993) in the low and the high workload condition of the EEG Experiment. Both conditions differ significantly from each other as calculated by a paired t-Test (p<0.01, t=4.15). The results support our assumption that there is an increase in the experienced effort and task load from the low to the high workload condition.

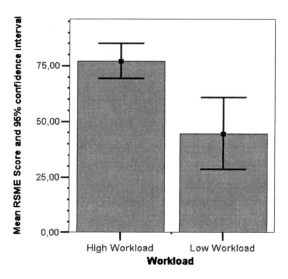

Figure 2: The subjectively experienced workload (mean values of 11 subjects) in the two experimental conditions (high and low workload).

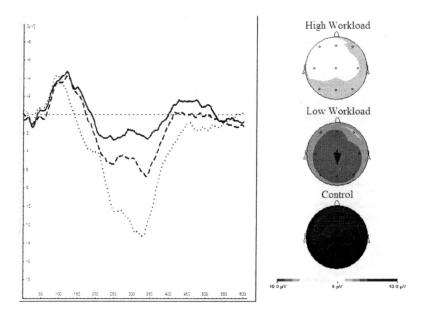

Figure 3: Grand averages and topographic map (300 ms poststimulus, Pz) of the event-related potential recorded in three conditions: high workload (continuous curve), low workload (dashed curve), control condition (dotted curve).

3.2 EEG correlates of (simulated) driving

Figure 3 shows the grand averages for the Pz electrode and a topographic map of their scalp distribution. The continuous and dashed curves illustrate the mean ERP responses to the seldom tones during the high and the low workload condition, while the dotted curve shows the response to the seldom tones in a control condition, in which the subjects were instructed to manually respond after each oddball. The largest P3 amplitude was found at Pz in the control task that demanded a stimulus-dependent response to the oddballs on the part of the subject (dotted curve). For the experimental conditions (passive listening during simulated driving), P3 amplitudes at Pz decreased from the low (dashed curve) to the high (continuous curve) workload condition. The difference between both conditions was significant ($p < 0.05$; T=2.46).

3.3 fMRI correlates of (simulated) driving

The brain activation levels shown in Figure 4 depict an overview of the hemodynamic response to cortical activation under the conditions of low (upper graphs) or high (lower graphs) workloads. Brain activations evoked by seldom auditory stimuli compared with frequent tones are more widely distributed during low workload and this activation is strongly reduced during high workload. Low workload is associated with activation in the right anterior operculum, as well as with modest activation in the right perisylvian region (upper half of figure 4, for more details see figure 5 and Table 1), whereas during high workloads subjects exhibited a focused activation in the right auditory and associated cortex (lower half of figure 4, for more details see figure 6 and Table 1).

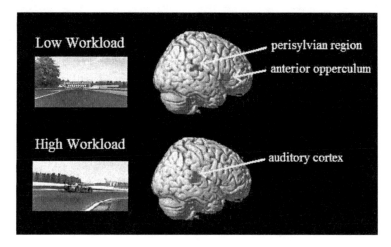

Figure 4: Overview about the two experimental conditions and activated areas in the brain contrasting the deviant sounds against the standards

Table 1: Localization, size, p-values and Z-scores of the activated cluster in the two experimental conditions contrasting the deviant sounds against the standards (abbreviations: R=right; BA = Brodman Area)

Experimental conditions and areas	Size (mm³)	P$_{corr}$	Z-max
Low Workload			
Perisylvian region: R Superior Temporal Gyrus, R Inferior Parietal lobule, R Postcentral gyrus – BA 13, 40, 42(see also figure 5a)	608	< 0.08	4.0
Anterior Operculum: R Inferior Frontal Gyrus, R Insula - BA 13, 45, 47 (see also figure 5b)	1776	<0.001	4.39
High Workload			
Auditory cortex: R Superior Temporal Gyrus, R Postcentral Gyrus, R Middle Temporal Gyrus, R Inferior Parietal Lobule – BA 22, 40, 42 (see also Fig. 6)	2152	< 0.001	4.99

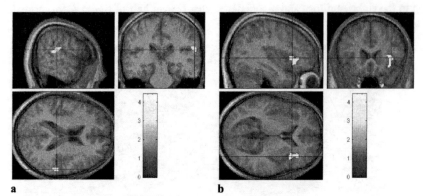

Figure 5: Brain regions activated in the low workload condition contrasting the deviant sounds against the standards: a) right perisylvian area and b) right frontal operculum.

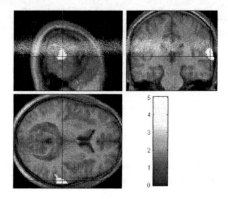

Figure 6: Brain region activated in the high workload condition contrasting the responses to the deviant sounds against those evoked by the standards: right auditory cortex.

4   Discussion

The present results indicate that varying levels of workload during simulated driving have a direct and significant effect on the neural correlates of selective attention: Using an auditory oddball paradigm with seldom target tones, we could derive a P3 component in the ERP, the amplitude of which was reduced by a high workload. Our findings further suggest that workload demands placed on the subjects in a dual-task situation has a significant effect on the fMRI-BOLD response to auditory stimuli. As the P3 component in the time-resolved EEG was the only ERP-component that differed significantly over experimental conditions, we speculate that this response is related to the fMRI findings (which are not time resolved). The differences between responses to the seldom and the frequent tones in the fMRI appear to be the hemodynamic correlate of the measured EEG differences, which were only significant in the P3 component of the ERP to the auditory oddballs.

In our study low workload is associated with a more widespread activation of cortical areas involved in extracting relevant information from the environment. These areas are particularly sensitive to changes in the sensory environment independent from the sensory modality

(Corbetta & Suhlman, 2002; Downar, Crawley, Mikulis & Davis, 2000). The activation in the high workload condition may concentrate on the primary processing of the deviant stimulus (Mueller et al., 2003). With increasing workload the subjects seem to switch to a more efficient neural processing strategy, as evidenced by a more focal response to the stimuli. This phenomenon has already been observed in behavioral (Recarte & Nunes, 2003) and brain imaging (Walter et al., 2001) studies.

It seems quite obvious that time-sharing of separate tasks within the same modality (intramodal) leads to a decrease in performance when the tasks share exactly the same resource (e.g. foveal vision). The efficiency of time-sharing on separate tasks between different modalities (crossmodal) is task-dependent and the experimental findings related to this phenomenon are often mixed (Wickens, 2000). Here we propose a neuroergonomic view of crossmodal interactions between visual, tactile and auditory modalities. Our findings support a limited capacity model of attention, where an increase in the demands of the primary task leads to reduced performance in secondary tasks. Possibly this impaired performance is moderated by a higher level processing resource or by effects of interference between the different sensory modalities. Further research on this question needs to be conducted. Simultaneous acquisition of EEG and fMRI for example could deliver more precise information about the time course of the activations (Mulert et al., 2002) and therefore provide better insight into crossmodal interactions. Combining the three categories of assessment techniques (subjective, behavioral and psychophysiological measures) for assessing a driver's mental workload leads to a better account of the demands of the driving task including its crossmodal aspects. The subjective measurement provides an estimate of validity for our experimental conditions. The high temporal resolution of the EEG, together with the high spatial resolution of the fMRI in a dual task paradigm yields a more precise insight into crossmodal interaction than a behavioral study alone.

5   References

Andreassi, J. (2001). Psychophysiology (4th ed.). Mahwah, NJ: Lawrence Erlbaum Associates.

Baldwin, C. L. (2003). Neuroergonomics of mental workload: New insights from the convergence of brain and behaviour in ergonomics research. *Theoretical Issues in Ergonomics Science*, 4, 132-141.

Corbetta, M., & Shulman, G. L. (2002). Control of goal-directed and stimulus-driven attention in the brain. *Nature Reviews Neuroscience*, 3, 201-215.

De Waard, D. (1996). The measurement of drivers' mental workload. PhD thesis, University of Groningen. Haren, The Netherlands: University of Groningen, Traffic Research Centre.

Downar, J., Crawley, A. P., Mikulis, D. J., & Davis, K. D. (2000). A multimodal cortical network for detecting changes in the sensory environment. *Nature Neuroscience*, 3, 277-283.

Fowler, B. (1994). P300 as a measure of workload during a simulated aircraft landing task. *Human Factors*, 36 (4), 670-683.

Jasper, H. (1958). The ten-twenty electrode system of the international federation. *Electroencephalography and Clinical Neurophysiology*, 43, 397-403.

Kramer, A. F., Trejo, L. J., & Humphrey, D. (1995). Assessment of mental workload with task-irrelevant auditory probes. *Biological Psychology*, 40, 83-100.

Laurig, W. (1990). Grundzüge der Ergonomie. Berlin, Köln: Beuth Verlag GmbH.

Mulert, C., Jäger, L., Pogarell, O., Busfeld, P., Schmitt, R., Juckel, G. & Hegerl, U. (2002). Simultaneous ERP and Event-Related fMRI: Focus on the time course of brain activity in target detection. *Methods and Finding in Experimental and Clinical Pharmacology*, 24 Suppl.D; 17-20

Müller, B. W., Stude, P., Nebel, K., Wiese, H., Ladd, M. E., Forsting, M., & Jueptner, M. (2003). Sparse imaging of the auditory oddball task with functional MRI. *Neuroreport*, 14 (12), 1597-601.

Recarte, M.A. & Nunes, L.M. (2003). Mental Workload While Driving: Effects on Visual Search, Discrimination and Decision Making. *Journal of Experimental Psychology: Applied*, 9(2), 119-137

Sarter, N. & Sarter, M. (2003). Neuroergonomics: opportunities and challenges of merging cognitive neuroscience with cognitive ergonomics. *Theoretical Issues in Ergonomics Science*, 4, 142-150.

Schmalhofer, F., Raabe, M., Friese, U., Pietruska, K., & Rutschmann, R. (2004). Evidence from an fMRI experiment for the minimal encoding and subsequent substantiation of predictive inferences. In: K. Forbus, D. Gentner & T. Regier (Eds.) *Proceedings of the Twenty-Sixth Annual Conference of the Cognitive Science Society*. August 4-7, 2004, Chicago, Illinois, USA, P. 1629

Soltani, M. & Knight, R. T. (2000). Neural Origins of the P300. *Critical Reviews in Neurobiology*, 14(3&4), 199-224.

Walter, H., Vetter, S., Grothe, J., Wunderlich, A., Hahn, S. & Spitzer, M. (2001). The neural correlates of driving. *Neuroreport*, 12 (8), 1763-1767.

Wickens, C. D. (2000). Engineering Psychology and Human Performance (3rd ed.). Upper Saddle River, NJ: Prentice Hall.

Wickens, C. D. (2002). Multiple ressources and performance prediction. *Theoretical Issues in Ergonomics Science*, 3, 159-177.

Zijlstra, F.R.H. (1993). Efficiency in work behavior. A design approach for modern tools. PhD thesis, Delft University of Technology. Delft, The Netherlands: Delft University Press.

Acknowledgements

We thank PD Dr. Kai Christoph Hamborg (University of Osnabrück, Germany) for many helpful discussions on the research project and Prof. Dr. Julius Kuhl (University of Osnabrück, Germany) for providing us his EEG laboratory. MWG and RMR were supported by the German Research Council (SFB 517, C9) and by the Volkswagenstiftung.

An Experiment to Probe Brain Responses to Driver Information Systems

Pieter Poolman
Don Tucker
Phan Luu

Thomas Schnell
Fuat Aktan

Electrical Geodesics, Inc. (EGI)
1600 Millrace Drive Suite 307
Eugene, OR 97403
ppoolman@egi.com
dtucker@egi.com
pluu@egi.com

Operator Performance Laboratory (OPL)
University of Iowa
Iowa City, IA 52242
thomas-schnell@uiowa.edu
faktan@engineering.uiowa.edu

Abstract

Recent advances in sensor, wireless, computing, and Global Positioning System technologies have made sophisticated in-vehicle information systems feasible. However, models of human decision-making processes and information-processing strategies are restricted in their capabilities to assist designers and researchers understand the impact of these futuristic systems, leading to more technology-centered than user-centered solutions. These intelligent operator-vehicle interfaces must be optimized with respect to usability, suitability, safety and user acceptance. Our understanding of human operator behavior is based on psychological principles of behaviorism or rule-based systems. The need is not any longer just to describe behavior, but to understand, model, and predict how intrinsic mechanisms in the human brain give rise to observed behavior. The increasing complexity of the interface requires that we understand and develop computational models for complex human-system interactions to predict operator decision making, distractibility, and situational awareness. Such models, supported by much greater computing power, will revolutionize the transportation scene in terms of efficiency and safety gains.

The paper describes progress made with the setup and application of a modular and mobile, physio-metrology laboratory that can be deployed on many research platforms such as driving simulators, flight simulators, instrumented cars and aircraft. The laboratory provides researchers with the ability to conduct research in an unprecedented manner by collecting a holistic picture of human performance in advanced and complex transportation systems. Furthermore, it enables a completely new insight on the complex interplay amongst stimuli, brain activity, psychophysiological processes, and different engineering designs of human-machine systems, by fusing together cutting-edge insights from both traffic engineering and neuroscience.

An overview is also given of our ongoing research program to discriminate amongst different cognitive states when applied to the study of driver distraction, performance, and safety. Based on the utility of our physio-metrology laboratory, we are investigating the interplay of attention, automaticity, and spatial and verbal working memory on threat detection and performance of drivers engaged in demanding environments. In the end, results from such experiments will define the basis from which driver training and selection, policies, and system design strategies could be evaluated by officials, manufacturers, as well as operators in the transportation sector.

1 Overview of Research Program

Our research effort comprises of two objectives, and focuses on conducting experiments to study driver behavior based on insights from neuropsychology, together with the tools offered by the physio-metrology laboratory. Apart from replicating results from previous driver behavior studies for validation purposes, the emphasis will also fall on demonstrating the utility of the near real-time monitoring capabilities of the laboratory.

1.1 Objective 1

The research calls for the deployment of a suite of measurement systems to collect physiological responses from an operator (driver, pilot, etc.). The modules in the physio-metrology laboratory include multiple psychophysiological measurement systems such as a dense-array EEG (brain wave activity) system, EKG (heart activity), respiration rate, electrodermal activity, eye movement recording, and multi-channel video. Commercial-Off-The-Shelf (COTS) components are used to assemble this fully autonomous laboratory package that can easily be deployed on many research platforms. The focus is on complete data integration and synchronization.

For calibration purposes of a driver behavior model, a simulator provide the ability to place the driver in a fully controlled virtual environment where his/her reactions to external stimuli can be accurately measured in a systematic manner over as long a period as desired. Modern simulation software is based on a sophisticated tile-based database modeling tool that can be used to script driving scenarios in a very short amount of time. A basic simulator is ideal for initial prototyping work and is very low cost to operate. Simulators have a significant role to play in research, particularly in the assessment of driver behavior that would be unsafe to study in a real vehicle surrounded by traffic. With near real-time brain activity measurements of the driver in hand, researchers can exploit the capability of absolute control over the virtual environment inside a simulator to alter task demands on the fly.

1.2   Objective 2

A series of experiments will focus on the distractibility of novice and expert drivers, and to what extent driving performance (as primary task) deteriorates in the face of secondary task loads. In general, both primary and secondary tasks tap verbal and spatial working memory to some degree, and it will be important to account for individual differences by locating these memory areas for each subject. By engaging a driver with a well-routinized primary (spatial) task (e.g. lane tracking), together with a secondary (verbal or spatial) task (e.g. cellular phone conversation), the opportunity exists to study the driver's coping strategies and ability to switch attentional resources by altering the cognitive demand of the primary task, based on the occurrence of specific brain activation patterns.

An important goal is to verify and replicate results from related DARPA (Defense Advanced Research Program Agency) and ONR (Office of Naval Research) research projects relating to automaticity and error-monitoring. In terms of automaticity, it is hypothesized that brain activity in medial frontal areas are strongly engaged in self-monitoring of performance in novices, with a corresponding decrease in working memory available for task performance. In contrast, becoming an expert on the task will require less mental capacity for executive self-monitoring, a decrease in medial frontal activity, and a shift of activity towards posterior brain regions. In terms of error-monitoring, results suggest that dense-array EEG measures can detect the difference between mistakes of ignorance (in which case the operator does not understand the correct response) or slips of performance, in which case the correct action is known but executed wrongly.

Based on modern single-trial analysis methods of brain functioning, and real-time monitoring tools developed by researchers at EGI (funded by DARPA), it will be possible to alter aspects of the driving environment based on the driver's instantaneous mental state. This will enable researchers to regulate the workload a subject driver experiences by tailoring the occurrences and timing of tasks in a traditional dual-task experiment given the individual's current neurophysiology markers. By systematically studying the context-monitoring and learning abilities of different subjects, insights gained from the functional importance of different brain areas engaged by driving, will help uncover human-centered strategies to enhance driver training and selection, transportation policies, and system design strategies. In the end, results from such experiments will define a scientific and robust basis to guide government agencies, manufacturers, and operators towards the betterment of the transportation sector.

2   Traffic Engineering Studies

The most-recent statistics paint the picture of a major national dilemma: In 2001, the estimated economic cost of traffic crashes on the US road network totaled $230.6 billion. Over 6.3 million police-reported motor vehicle crashes occurred on our nation's highways – one every 5 seconds. On average, a person was injured in one of these crashes every 10 seconds, and someone was killed every 12 minutes. Notwithstanding progress with efforts to reduce the number of deaths and serious injuries, we could only manage to decrease the occupant fatality rate per 100,000 population by 1 percent over the last 10 years. Most important, the majority of crashes are caused by some type of driver error, including inattention and distraction, frustration and tunnel vision, fatigue, illegal maneuvers, and driving while under the influence of alcohol or other drugs. The picture for the aviation industry and Air Force is equally bleak: An estimated 95% of all aircraft crashes are attributable to pilot error. Pilots' loss of situational awareness continues to be a major concern.

More than a decade ago, researchers have noted the lack of progress in developing a comprehensive model of driving behavior (Michon, 1985). These include a preoccupation in the highway safety field with accidents and accident-causing behaviors. As a result, it has not been clear whether theories should explain everyday driving, or accident-causing behaviors, or both. Secondly, motivational and risk models, which emerged in the 1960s and 1970s as alternatives to skill-based models of driving, have failed to generate testable hypotheses necessary for developing a body of empirical findings. Importantly, some of these models have been criticized for failing to distinguish between the aggregate (or macroscopic) and individual levels of analysis (Michon, 1989). In addition, the cognitive revolution in psychology has failed to influence driving behavior modeling, not to mention the equally important revolution in research on emotions recently (LeDoux, 1998).

A central shortcoming in most endeavors to enhance the understanding of driver behavior seems to be rooted in the inability to base such behavior as arising from the neural underpinnings of human brain functioning. Like in the past, failing to do so in future will repeatedly lead to numerous incompatible and qualitative models (Poolman, 2002). The impasse in driver behavior modeling is not due to a lack of knowledge on brain functioning and development. An ever-growing body of anatomical, physiological, and developmental facts, supported by neurally-inspired quantitative models of the brain as a key affector of an individual's behavior, is available in the family of neurosciences. This wealth of knowledge amassed in neuroscience lies ready to be tapped by researchers interested in explaining human driver behavior. However, while a neuroscientific framework forms the basis of our driver behavior experiments, we will still draw extensively from results of numerous past driver behavior studies. As will be explained briefly in subsequent sections, these psychological studies have contributed comprehensively to the description of automaticity, dual tasking, and error monitoring in driver behavior.

2.1 Automaticity

Automaticity is characterized as fast, effortless processing, which develops following extended consistent practice (Schneider & Shiffrin, 1977). It is contrasted with controlled processing, which refers to slow, serial, and effortful processing. Their theory suggests that virtually all behaviors include components of controlled and automatic processing, and that the relationship between the various components is constantly changing according to the type and quantity of practice. Summala (1988) discusses the relationship between uncertainty and the development of automaticity in driving, suggesting that novice drivers initially feel a sense of uncertainty in most situations. With practice, skills become automatized and self-confidence replaces uncertainty. In driving, novel or hazardous situations evoke uncertainty, which causes control to shift from automatic to controlled, conscious processing.

Because driving is a time-shared activity, a theory of driving behavior should provide some basis for determining which combinations of skills can and do become automatized with practice. Wickens' (1984) multiple-resource theory may provide a framework for determining the degree of compatibility among various component tasks. He proposed the existence of several different supplies of resources, including the stage of processing (early, late), the modality (auditory, visual), and the processing code (spatial, verbal). Wickens has demonstrated that interference in a dual-task situation will be more likely when the individual tasks draw on the same pool of processing resources.

2.2 Dual tasking

Secondary tasking is a commonly used workload measurement tool which requires a subject, assigned a primary task, to use any spare mental capacity to attend to a secondary task (Gawron, 2000). Measuring workload through primary tasking and other aggregate measures like utilization are important, but the use of secondary task measurements provides a more comprehensive workload analysis (Wickens & Hollands, 2000). Traditional secondary tasking such as tapping and time estimation tasks can be intrusive and introduce an unrealistic artifact during testing (Williges & Wierwille, 1979), however, embedded secondary tasks do not fundamentally change the task or task performance and provide more sensitive measurements.

It was shown that driving experience is related to lane keeping when drivers have to perform a foveal in-car task continuously and rely on peripheral vision only in keeping the car within lane boundaries. A marked change appears in ability to do this successfully between the first 1500 and 50,000 km of driving (Summala et al., 1998). Drivers adapt their behavior to changes in their skills, and improved lane-keeping performance may make them attend increasingly to non-traffic targets such as cellular telephones, other in-car accessories and sight seeing while driving. Early experimental work has shown that perceptual and decision-making tasks are impaired when a driver has to divide attention between the road and a car phone (Brown et al., 1969). Specifically, when drivers attempted to maintain a constant headway to a vehicle ahead and were engaged in a cellular phone conversation, their reactions to headway changes were somewhat delayed.

2.3 Error monitoring by drivers

The contribution of errors to crash causation has been studied extensively. Because drivers commit many errors other than those that precipitate accidents (Brown, 1990), it is clear that accident data alone do not provide appropriate information for the analysis of driving errors. In much the same way that motivational models have shifted the emphasis from accident-precipitating behaviors to all driving behavior, more recent theories have considered errors as a part of normal behavior (Ranney, 1994). This alternative approach advocates studying errors within the larger context of all driving behavior, because they are inevitable in self-regulating systems.

Brown distinguished between factors that influence the production of errors and those that constrain drivers' ability to recover from errors. Because of the potential for catastrophic consequences resulting from driver errors, the highway system has been designed to be tolerant of minor errors, such as deviations from the travel lane. The absence of feedback concerning minor errors in driving can weaken associations between actions and their consequences, which can lead to over-learning of inappropriate behaviors. Drivers' adaptation to error is thus prone to distortion, which may affect the degree to which correct responses can be automatized (Groeger, 1990). Instead, automatic action patterns may include a relatively wide range of both correct and incorrect responses. The inherent variability of human behavior combined with the variability of automatic patterns will inevitably lead to more serious errors. At this point, the driver's ability to recover from error may determine the likelihood of accident. This has led Brown to conclude that factors that influence drivers' ability to recover from errors may be more important to theories of driving behavior than factors that influence error production.

3 Overview of the Physio-Metrology Laboratory

We have developed a modular and mobile, physio-metrology laboratory that can be deployed on many research platforms such as driving simulators, flight simulators, instrumented cars and aircraft. The modules in this physio-metrology laboratory may include a multitude of psychophysiological measurement systems such as a dense-array EEG (brain wave activity) system, blood pressure measurement system, thermal imaging, EKG, respiration rate, electrodermal activity, EMG (muscle activity), eye movement recording, multi-channel video, etc. A data acquisition computer collects and synchronizes data from all these psychophysiological modules, and stores the data for compilation and analysis. Recent research efforts successfully demonstrated that workload and stress can be predicted based on measurements from multiple parameters.

Figure 1. Dense-array (128-channels) EEG demonstration by EGI and OPL researchers in an instrumented vehicle.

As the cornerstone of our laboratory, a dense-array EEG provides the necessary millisecond temporal resolution required to track how the human operator deals with a complex information stream. Dense-array EEG also provides the necessary spatial resolution (about 1cm) required for anatomical analysis of cortical regions that are known to be important to information processing and self regulation, and it holds the promise of portability, which is essential for real world applications. The Geodesic Sensor Net (GSN) from EGI is specifically designed for dense-array (256-, 128-channel) EEG recordings (Figure 1). We have tested the 128-channel EEG in an instrumented vehicle with great success. Typical application time, even for a 256-channel GSN is approximately 15 minutes.

3.1 Psychophysiological signals of cognitive effort, arousal, stress, and affective processes

There are several EEG signals that can be used to index cognitive effort. The most popular is the error-related negativity (ERN), and it is a fairly focal scalp field over the medial frontocentral region. Popular belief is that the ERN indexes either the detection of an error or the detection of response conflict. However, the evidence is more in line with the theory that the ERN is a signal that is generated in response to violations in expected outcomes (Luu & Tucker, 2004). ERN is part of the ongoing EEG theta rhythm (4-7 Hz) generally recorded along the midline, and has been shown to be sensitive to cognitive demand. When task difficulty is increased, either by placing a high demand on working memory or imposing a time limit on performance, theta amplitude is increased (Slobounov et al., 2000), and theta activity has been shown to correlate with information retention in both animals and humans (Mölle et al., 2002).

Another neurophysiological index of cognitive effort is the P3a. The P3a is an event-related potential with a distribution that is maximal positive over the mediofrontal recording sites. Traditionally, the P3a is thought to be responsive to novel events. However, it was recently demonstrated that the amplitude of the P3a to novel events is determined by the effort exerted in a given task (Demiralp, 2001). The researchers showed that the difficulty of a primary task dictates the amplitude of the P3a in response to a novel stimulus: the more difficult the primary task, the larger the P3a is in response to a novel or deviant stimulus. Therefore, a novel or deviant stimulus can be used to periodically probe the cognitive resources of the driver as he/she is performing a task.

The anterior cingulate cortex (ACC) has been shown to contribute to the ERN, theta, and P3a responses. These findings are important in that ACC activity has been shown to be involved in ocular control (Gaymard et al., 1998), and it is also related to pupil responses (Brown et al., 1999). Moreover, the function of the ACC is tightly coupled to the functions of the hypothalamic-pituitary-adrenal (HPA) axis (Diorio et al., 1993), which function is central to the control of responses to stressors, and, therefore, it is also involved in the control of sympathetic and parasympathetic divisions of the autonomic nervous system. These facts allow us to understand how other physiological measures, such as pupilometry, electrocardiogram (EKG), and electrodermal activity (EDA) can be used to index cognitive effort and resources in an integrative, functional framework.

Both fatigue and high demands for cognitive processing are associated with increases in pupil size (Dionisio et al., 2001). However, the distinction between fatigue and cognitive load is not simple using just pupilometry as a measure. This is due to the fact that pupil dilation is controlled by the sympathetic nervous system, and this system responds in a non-specific manner to arousing situations. EDA reflects responses to both emotional and cognitive processes (Stern et al., 2001). Two brain systems are involved in the control of EDA. The first is centered on the ACC, which projects to the thalamus and hypothalamus, and allows for the ACC to directly affect EDA in response to emotional stimuli. The second system is centered on the lateral frontal cortex. It is believed that this system controls EDA during orienting, cognition, and locomotion. Therefore, from this perspective, direct measures of brain activity, as assessed by EEG, requires the ability to spatially resolve the function of at least two cortical areas, which cannot be accomplished with sparse array EEG. A more exclusive index of stress response is one that measures the cardiovascular system. Like EDA, the cardiovascular system responds immediately to stressors. Stressors trigger responses from the sympathetic nervous system and this directly affects the cardiovascular system, which results in increase heart rate and blood pressure.

3.2 Source localization with dense-array EEG

Previous EEG measures for assessing workload or working memory have been derived from scalp recordings. Even dense-array (128- or 256-channel) scalp recordings are inadequate for assessing memory and executive functions because these functions engage limbic regions at the core of the cerebral hemisphere, including medial temporal and memory areas and anterior cingulate executive areas. Instead of scalp gauges, we need brain gauges. Due to the fact that brain sources, such as in the midline cingulate gyrus, may be oriented in complex directions, it is well known that a scalp channel over the left hemisphere may actually reflect brain activity from the midline structures of the right hemisphere. Without source analysis, scalp EEG cannot provide an accurate index to the location, or even the hemispheric lateralization, of brain activity.

To support accurate monitoring of verbal and spatial working memory, a high degree of spatial sampling of scalp EEG is required. As for the time domain, the Nyquist criterion also applies to spatial sampling (Srinivasan et al.,

1998). We have empirically demonstrated for cases that involve large cortical regions, such as stroke, scalp recorded potentials are aliased even with 64-channel recording arrays (Luu et al., 2001). Although it is often claimed that a few electrodes are adequate if one simply wants to determine the difference between two conditions, such as high or low cognitive effort, this is true only if (i) the signal is at a superficial brain structure and thus focal on the scalp, and (ii) one knows beforehand the electrode location at which the signal is maximally different between the two conditions. Because brain geometry varies from subject to subject, and because the distribution of the scalp potential is a function of brain geometry, there is no way to determine a priori where to place the electrode such that the signal that best discriminates between the two conditions is recorded. Therefore, for applications in which measurement of the EEG signals are to be analyzed from individual subjects, such as cognitive effort assessment, adequate sampling of the head surface is required.

4 Neuroscientific Research Framework

Our understanding of human operator behavior is based on psychological principles of behaviorism or rule-based systems. The need is not any longer just to describe behavior, but to understand, model, and predict how intrinsic mechanisms in the human brain give rise to observed behavior. The increasing complexity of the interface requires that we understand and develop computational models for complex human-system interactions to predict operator decision making, distractibility, and situational awareness.

A key aspect of our research is to determine the cognitive state of a driver, based on the monitoring of verbal and spatial working memory, and his/her conflict monitoring or expectancy-violation efforts. As part of the physio-metrology laboratory, we built a platform that will be used for the representation and assessment of multiple physiological signals from a driver. The common representation and assessment of multiple physiological signals will provide converging evidence that will allow for real-time, accurate assessment of the driver's cognitive load. The resulting signals will serve as gauges of cognitive effort that can be output into other systems, such as the event controller of a driving simulator. With the integration of multiple sensor streams (e.g., EKG, eye movements, etc.), it is likely that discriminant functions will not only contribute to gauges that are more accurate, but more reliable (between sessions) and perhaps generalizable (between tasks).

Recent advances in cognitive science and cognitive neuroscience, drawing upon recent developments of non-invasive neuroimaging technologies, have lead to important new insights into brain functions and how to measure them. Traditionally, cognitive functions are regarded as being separable from emotional and motivational processes such as workload stress. However, there is ample evidence that this separation does not capture how mental operations are represented and executed by the brain (Luu & Tucker, 2002). The limbic portion of the frontal lobe, such as the anterior cingulate cortex (ACC), has long been recognized as being important to self-regulation. We now know that the higher cognitive functions are controlled by this structure in relation to current physiological or homeostatic needs, including motivational states (Luu & Tucker, 2004). In demanding driving environments, it is critical to understand the integration of stress and motivation regulation with cognitive control, and how these resources are used to anticipate and meet challenges (both physical and cognitive).

Design and analysis of proposed driver behavior experiments are partially cast within the neuroscientific framework of current EGI studies of operator behavior in military settings, conducted for the US Department of Defense. One of the studies deals with verbal and spatial workload of operators, touching on aspects of automaticity (exhibited by experts) versus slower and effortful control (of novices). A second study addresses the same issues from a learning perspective of the operator, but accentuates the evaluation of errors (slips and mistakes) during the learning process. Both studies incorporate EEG, and are based on a framework of action regulation. These studies are described next.

4.1 Working memory and automaticity

As part of research funded by the US Department of Defense (under DARPA), researchers at EGI have created a real-time analysis engine to compute a gauge of cognitive effort. For this study, researchers used a novel oddball auditory probe as subjects performed a task, called the Warship Commander Task (WCT). The novel oddball auditory probe elicits a P3a response that is sensitive to task difficulty (as was mentioned previously). It was found that as cognitive load increase, the P3a response intensified. After demonstrating the ability to index cognitive work load with the P3a response, researchers from EGI and San Diego State University integrated another EEG measure of cognitive workload (theta power) with pupilometry and reaction time data. The data were analyzed relative to the events that marked when a subject has identified an aircraft as either friend or foe. The independent variable was wave size (i.e. the number of aircraft in a given wave of aircraft). Figure 2 shows the results from these analyses. The time that the subjects required to identify whether an aircraft was a friend or a foe was shown to increase as a function of wave size; the more aircraft that were present the longer the subjects took to identify a particular aircraft (signified by a longer reaction time (RT)) (Figure 2 left). Similarly, the pupilometry measure indicated that as workload increased, so did pupil size (Figure 2 middle). Theta power (4-7 Hz), one second before and one second after the response, showed an increase as cognitive load increased. These results demonstrate convergence between behavioral (RT), autonomic system (pupilometry), and central nervous system (EEG) responses as indices of cognitive workload. However, the EEG response showed one paradoxical effect. As theta power increased with cognitive load, the phase of theta became less aligned to the actual response. This is a potentially important observation in that the alignment of theta phase may indicate the coordination of multiple limbic networks. Therefore, it may be the case that although the subject may have been exerting more cognitive effort (or merely has been more aroused) in the high cognitive load condition, he/she may have been less efficient at processing the information and organizing responses.

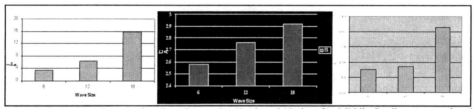

Figure 2. Left: Reaction time as a function of wave size (6, 12, and 18 aircraft). Middle: Pupilometry as a function of wave size. Right: Theta power as a function of wave size.

Closely related to working memory is the issue of automaticity in cognition. In classical reaction time studies, that defined many modern studies of executive control of cognition, researchers (Schneider & Shiffrin, 1977) delineated the characteristics of "automatic" cognition, in which skilled operations (such as monitoring radar tracks) may become sufficiently routinized that they can be carried out successfully without demanding attention and working memory. Over the last decade, EGI researchers have collaborated with Walter Schneider and his associates at the University of Pittsburgh on problems of integrating dense-array EEG with fMRI studies, primarily in sequential (successive) but now in simultaneous (EEG in the magnet) paradigms.

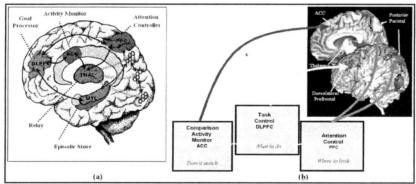

Figure 3. Analysis of cognitive task demands with fMRI measures. (a) Model of executive control with multiple limbic and neocortical contributions to working memory and cognitive effort. (b) fMRI evidence showing anterior cingulate (ACC) contributions to executive monitoring, dorsolateral prefrontal (DLPFC) contributions to maintenance of task goals, and posterior parietal (PCC) contributions to attentional direction within the task space.

In a series of recent studies, Schneider et al. have characterized the brain regions that are engaged in a cognitive task when effortful control is required, and they have shown how practice decreases activation of certain of these executive systems, while not affecting certain task-specific sensory processing regions. This evidence is critically important for understanding the dynamic changes in brain activity that reflect key mechanisms of neurocognitive function in an operational task.

Some of the findings from Schneider's fMRI studies of verbal and spatial working memory, and executive monitoring of task performance, are consistent with the conclusions from EGI's dense-array EEG studies. The importance of the anterior cingulate cortex (ACC), for example, is clear in both lines of work. Furthermore, 128-channel EEG studies on task switching in our laboratory have confirmed roles of dorsolateral prefrontal (DLPFC) and parietal cortex (PPC) that are consistent with the model in Figure 3. Using 256-channel EEG, we have recently demonstrated how the development of expertise activate different brain areas as a subject progresses from being a novice (activation of the anterior cingulate system) to an expert (activation of the posterior cingulate system) during a verbal learning task. During this experiment we could also localize verbal working memory in DLPFC.

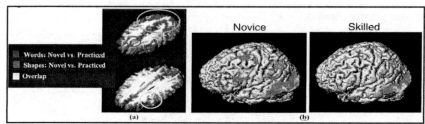

Figure 4. Skilled operators show decreased demands on working memory. (a) Verbal and spatial tasks show both domain general (ACC) and mode specific (verbal = Broca's area; spatial = occipitoparietal) activations when task demands are novel, but decreases when performance becomes practiced. (b) Overall, the domain-specific demands, such as in the visual activation in inferior occipital-temporal areas, remain for skilled performers, but the task goals (DLPFC) and monitoring functions (ACC) decline considerably when the task becomes practiced (b, right).

The dynamic nature of engagement of the executive control system in Schneider's work is particularly important for planning the proposed driver behavior experiments, particularly when it is assumed that there is a working memory system that is relatively fixed and simply engages more or less over time. There appear to be a dramatic decline in working memory demands as a subject shifts from adapting to new or changing demands to performance in routinized or well-automatized circumstances (Figure 4).

4.2 Action regulation and learning

At present, researchers at EGI are conducting another Department of Defense-funded research project designed to assess human learning processes. In particular, the goal is to differentiate, through the use of dense-array EEG, between two types of errors: slips of action and mistakes. The objective is to integrate this information into simulator environments to improve training models. Research results propose that neurophysiological markers of expectancy violations can provide simulation-based training (SBT) systems with the ability to automatically detect and diagnose the state of the trainee with regards to their knowledge, skills, and abilities (KSA). The focus of the mentioned research work is on the evaluation of errors during the learning process, since performance is a measure often used to ascertain learning and establishment of expertise. Although behavioral performance can be used as a measure of learning, performance provides limited diagnostic utility because it is the end result of many cognitive processes and may be non-unique indicators of the underlying learning processes. In the case of errors during learning, there are at least two processes that contribute to performance errors. An error can result from not knowing what to do or it can result from impulsive responses. In the former situation, errors represent inadequate (or faulty) KSA related to the task (i.e. mistakes), and in the later situation, errors represent slips of action, with full KSA related to the task. Therefore, to determine the correct feedback to provide when errors are committed, these two types of errors need to be differentiated.

Human learning can be separated into two complimentary systems centered on limbic circuits. One circuit is made up of the anterior cingulate cortex (ACC) and the medial nuclei of the thalamus (see Figure 5). This circuit is involved in the early stages of learning, being engaged when routine actions and a priori knowledge are no longer appropriate for current demands (Gabriel, 1990). The second circuit is centered on the posterior cingulate cortex and anterior ventral nucleus of the thalamus, and is involved in the later stages of learning, when consolidation of information into long-term memory becomes important. Learning, as controlled by both circuits, is regulated by a simple cognitive phenomenon: violations of expectancy. Traditionally, associative learning was described by Pavlovian conditioning, wherein a stimulus becomes associated with a response through its paring with an unconditioned stimulus. This associative pairing was conceptualized at a very low-level, and assumed that control over a response is passed from one stimulus to another. However, it has been clear for some time that such a conceptualization is inadequate to explain the learning data (Rescorla, 1988), and that a more complex view of learning is required, in which context and expectation play central roles (Nadel et al., 1985). Results have shown that it is unexpected cues that are learned. Cues that are not predictive cannot be conditioned. In essence, it is deviation from expectancy (i.e. the expectancy implicit to the context model) that drives learning.

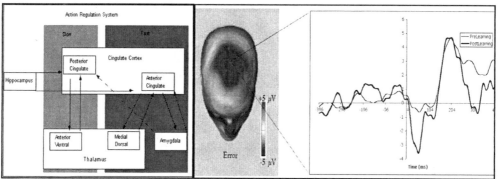

Figure 5. Left: A model of learning systems based upon Gabriel's (Gabriel et al., 1986) model of discriminant learning. Right: Scalp distribution of the ERN. Notice the difference between a mistake of ignorance (pre learning) and a slip of performance (post learning).

When expectancies are violated, such as when an error is committed, a negative deflection in the ongoing EEG is observed over mediofrontal scalp regions (Luu et al., 2000). This negative deflection is referred to as the error-related negativity (ERN) and has been localized to the ACC. An error in these experiments is often defined as slips (i.e. actions or responses that are executed prematurely), and subjects are immediately aware that the response is not the intended action. However, ERN-like activity is observed in other situations when errors are not committed, but when expectancies are violated in other ways. For example, as subjects learn a repeated sequence of stimuli, when a position within that sequence is changed, ERN-like activity is also detected in response to this violation. We have recently shown that dense-array EEG measures can detect the difference between mistakes of ignorance (in which case the operator does not understand the correct response) or slips of performance, in which case the correct action is known but executed wrongly (Figure 5).

Due to the fact that the ERN appears to track expectancy violations, there are two implications for learning. First, the moment of expectancy violation is the moment that the ACC-centered learning system becomes engaged, which means that the underlying learning system is in an optimal learning state. Therefore, tailored feedback under these conditions would predict superior subsequent performance. Second, because it reflects expectancy violations, the ERN can be used to distinguish slips from mistakes; mistakes do not necessary reflect violations of expectancies.

Within our driver behavior experiments, the ERN is used to signal when a driver becomes aware of an expectancy violation in his/her primary task domain, for example when a pedestrian unexpectedly step in front of the vehicle while the driver is moving along a completely untrafficked road. The ERN may also signal to the experimenter when the driver is caught up in his/her secondary task (e.g. a demanding cellular phone conversation), after which the experimenter may then increase the task demand of the primary task (as per the previous example) to investigate the driver's coping strategies (e.g. "not to freeze up") and ability to effectively switch attentional resources.

5   References

Brown, G.G., Kindermann, S.S., Siegle, G.J., Granholm, E., et al. (1999). *Brain activation and pupil response during covert performance of the Stroop Color Word task*. J of Int Neuropsych Soc, (5), 308-319.

Brown, I.D. (1990). *Driver's Margins of Safety considered as Focus for Research on Error*. Ergon, (33), 1307-1314.

Brown, I.D.; Tickner, A.H.; Simmonds, D.C.V. (1969). *Interference between Concurrent Tasks of Driving and Telephoning*. Journal of Applied Psychology, Vol. 53, pp. 419-424.

Demiralp, T., Ademoglu, A., Comerchero, M., & Polich, J. (2001). *Wavelet analysis of P3a and P3b*. Brain Topography, Vol. 13, pp. 251-267.

Dionisio, D. P., Granholm, E., Hillix, W. A., & Perrine, W. F. (2001). *Differentiation of deception using pupillary responses as an index of cognitive processing*. Psychophysiology, Vol. 38, pp. 205-211.

Diorio, D., Viau, V., & Meaney, M. J. (1993). *The role of the medial prefrontal cortex (cingulate gyrus) in the regulation of hypothalamic-pituitary-adrenal responses to stress*. J of Neuroscience, Vol. 13, pp. 3839-3847.

Gabriel, M. (1990). *Functions of anterior and posterior cingulate cortex during avoidance learning in rabbits*. Progress in Brain Research, Vol. 85, pp. 467-483.

Gabriel, M., Sparenborg, S. P., & Stolar, N. (1986). *An executive function of the hippocampus: pathway selection for thalamic neuronal significance code*. In R. L. Isaacson & K. H. Pribram (Eds.), The hippocampus. Plenum.

Gawron, V. J. (2000). *Human Performance Measures Handbook*. Mahwah, NJ:Lawrence Erlbaum Associates.

Gaymard, B., Rivaud, S., Cassarini, J. F., Dubard, T., Rancurel, G., Agid, Y., et al. (1998). *Effects of anterior cingulate cortex lesions on ocular saccades in humans*. Experimental Brain Research, Vol. 120, pp. 173-183.

Groeger, J.A. (1990). *Drivers' Errors in, and out of, Context*. Ergonomics, Vol. 33, pp. 1432-1430.

LeDoux, J.E. (1998). *The Emotional Brain: The Mysterious Underpinnings*. Touchstone, New York.

Luu, P., & Tucker, D. M. (2002). *Self-regulation and the executive functions: electrophysiological clues*. In A. Zani & A. M. Preverbio (Eds.), The cognitive electrophysiology of mind and brain. San Diego: Academic Press.

Luu, P., & Tucker, D. M. (2004). *Self-regulation by the medial frontal cortex: limbic representation of motive setpoints*. In M. Beauregard (Ed.), Consciousness, emotional self-regulation and the brain. John Benjamin.

Luu, P., Flaisch, T., & Tucker, D. M. (2000). *Medial frontal cortex in action monitoring*. Journal of Neuroscience, Vol. 20, pp. 464-469.

Luu, P., Tucker, D. M., Englander, R., Lockfeld, A., Lutsep, H., & Oken, B. (2001). *Localizing acute stroke-related EEG changes: assessing the effects of spatial undersampling*. J of Clinical Neurophys, Vol. 18, pp. 302-317.

Michon, J.A. (1985). *A Critical View of Driver Behaviour Models. What do we know, what should we do?* In: L. Evans, R. Schwing, Eds., Human behavior and traffic safety. New York: Plenum Press.

Michon, J.A. (1989). *Explanatory Pitfalls and Rule-based Driver Models*. Acc Anal and Prev, Vol. 21, pp. 341-353.

Mölle, M., Marshall, L., Fehm, H. L., & Born, J. (2002). *EEG theta synchronization conjoined with alpha desynchronization indicate intentional encoding*. European Journal of Neuroscience, Vol. 15, pp. 923-928.

Nadel, L., Willner, J., & Kurz, E. M. (1985). *Cognitive maps and environmental context*. In P. D. Balsam & A. Tomie (Eds.), Context and learnin (pp. 385-406). Hillsdale: Lawrence Earlbaum Associates.

Poolman, P. *Towards the Extension of the Knowledgebase to Further Understanding and Modelling of Driver Behaviour*. Ph.D. Dissertation, University of Stellenbosch, 2002.

Ranney, T.A. (1994). *Models of Driving Behavior: A Review of Their Evolution*. Acc Anal and Prev, (26), 733-750.

Rescorla, R. A. (1988). *Pavlovian conditioning: it's not what you think it is*. Amer Psychol, Vol. 43, pp. 151-160.

Schneider, W., & Shiffrin, R.M. (1977). *Controlled and autonomic human information processing: I. Detection search and attention*. Psychological Review, Vol. 84, pp. 1-66.

Slobounov, S. M., Fukada, K., Simon, R., Rearick, M., & Ray, W. (2000). *Neurophysiological and behavioral indices of time pressure effects on visuomotor task perfomance*. Cognitive Brain Research, Vol. 9, pp. 287-298.

Srinivasan, R., Tucker, D. M., & Murias, M. (1998). *Estimating the spatial Nyquist of the human EEG*. Behavior Research Methods, Instruments, & Computers, Vol. 30, pp. 8-19.

Stern, R. M., Ray, W. J., & Quigley, K. S. (2001). *Psyhophysiological recording*. New York: Oxford Univ Press.

Summala, H. (1988). *Risk Control is not Risk Adjustment: The Zero-risk Theory of Driver Behavior and its Implications*. Ergonomics, Vol. 31, pp. 491-506.

Summala, H.; Lamble, D.; Laakso, M. (1998). *Driving Experience and Perception of the Lead Car's Braking when Looking at In-car Targets*. Accident Analysis and Prevention, Vol. 30, pp. 401-407.

Wickens, C. D., & Hollands, J. G. (2000). *Engineering Psychology and Human Performance (Third ed.)*. Upper Saddle River, NJ: Prentice-Hall Inc.

Wickens, C.D. (1984). *Processing Resources in Attention*. In: R. Parasuraman, D.R. Davies, Eds., Varieties of attention. Orlando, FL: Academic Press.

Williges, R. C., & Wierwille, W. W. (1979). *Behavioral Measures of Aircrew Mental Workload*. Human Factors, Vol. 21, pp. 549-574.

Section 5
Cognitive State Sensors

Chapter 22

Future Applications of Augmented Cognition

Augmented Cognition for Warfighters; a Beta Test for Future Applications

Henry J. Girolamo

U.S. Army Natick Soldier Systems
Natick, MA 01760
henry.girolamo@us.army.mil

Abstract

For close to thirty-five years, Department of Defense (DoD) Military Platform Program Managers (PMs), Combat Developers and Operators have recognized the need for improved situational awareness, off bore sight targeting, multi-spectral sensor fusion and indirect viewing through integrated information management systems. These systems had historically been focused on fixed wing and rotorcraft applications. Because of revolutionary advancements in miniature displays, sensors, optics, electronics, processor power reduction, weight, and size; man portable systems are being evaluated for dismounted warfighter applications and soldier integration.

 Integrating communication and sensor fused, multi-modality information systems on soldiers is the mandate for the Transformational Future Force Warrior (FFW). Based on past integrated soldier system developments with DARPA, Natick Soldier System human factors specialists and Training and Doctrine Command (TRADOC) combat developer representatives have acknowledged that overloading warfighters with information through technology is easily accomplished. Together they have sought solutions through task and sub-task analysis, as well as minimizing and filtering the information that flows to the warfighter. Examples of these systems relate to combat vehicles, rotorcraft, maintenance, medical, law enforcement and Special Forces. Each user community has their own unique tasks and informational needs. System interfaces and the manner in which information is distributed will be of critical importance to assimilate the users with the transformational information management systems in development.

Overloading the Future Force Warfighter is a major problem and understanding the means by which the information will be disseminated and displayed from the integrated Command, Control, Communications Computers, (C4) systems under development in Future Force Warrior and other transformational warfighter systems requires an intelligent solution. This paper will discuss the soldier of the future who will carry a wide variety of battlefield networked sensors and displays embedded into his equipment, which will data-link to his weapon and to displays on his helmet to information streaming in from the battlefield.

This paper will focus on Augmented Cognition as a new paradigm that assists the information inundated 21st century warfighter with an operational capability that employs an "interactive/intelligent" system "human machine interface", that is correspondingly "normal" and making the "computer–human" interactions fundamentally like "human-human" interactions. The discussion will include descriptions of success stories in the integration and operational evaluation of Augmented Cognition and potential new areas for research that could address other Military Mission Occupational Specialties. This paper will address the importance of working with end users in exploratory research such as experienced warfighters, subject matter experts and the Combat Equipment Developers from TRADOC Schools in support of the Transformational Army's Future Force Warrior (FFW) Advanced Technology Demonstration.

The net-centric Future Force Warrior will operate in squads or sub-squad teams, requiring dedicated wireless communications networks for operation. Each warrior will be equipped with a sensor set, comprised of a helmet integrating day and night cameras and weapon-mounted sight cameras. The images will be displayed on a helmet integrated see-through monocular display, or on a hand held PDA like terminal. Images will also be relayed to other team members for coordination. Future Force Warrior (personal) digital assistants will display navigation and situational awareness pictures, and images received from team members. The radio will be transmitting wideband data communications in addition to the current voice radio. Soldiers will be able to talk naturally with each other, whisper, or chat with text messages during missions. As a novel interface the feasibility and usefulness of a wearable vibrotactile feedback system is also being explored. The Future Force Warrior system has the ability for tying each soldier into tactical local and wide-area networks by an onboard computer that sits at the base of the soldier's back. Troops will also be able to share data with vehicles, aircraft and other individuals.

The Augmented Cognition system supports the Future Force Warrior development by creating a prototype environment with operational utility that employs an "interactive/intelligent" system "human machine interface", mixed-initiative interactions, dialogue and gesturing methodologies. The system is developed to ensure that prioritized information being delivered to the warfighter is in a mode that has utility for human consumption.

The Augmented Cognition program, which involves military agencies, academic institutions, and industry, has generated significant interest from many system and product developers. The development, integration, experimentation and evaluation of Augmented Cognition in Future Force Warrior operational scenarios should lead to many Future Applications of Augmented Cognition such as Homeland Defense, other defense platform integration, law enforcement, security, automotive industry, mass production manufacturing, health care, and professional and industrial occupations where they are all involved in multi-tasking environments and require performance improvement, education or rehabilitation.

1   The Need for Improved Battlefield Situational Awareness

In 1991 under the auspices of the Defense Advanced Research Projects Agency (DARPA) High Definition Systems (HDS) program, miniature displays were being developed as devices that would be integrated into rotorcraft visors to provide sensory information and flight symbology.

This resulted in new programs to develop miniature sensors, optics, and electronics that would be compatible with this new low power, light weight and high resolution display technology. Following these successful developments and evaluations in combat weapon platforms such as combat vehicles and rotorcraft where power was not an issue, new programs commenced where these display and system components would soon be integrated into Head-Mounted Displays (HMD) for dismounted warfighters such as infantry and special operations warfighters, who were evaluating HMDs for hands-free capability when viewing alphanumeric and graphical data.

Man portable system developments using head mounted displays coupled with sensors and cameras required the development of new man portable processors. Components needed to consider space on the body, weight and consume very little power if these man-portable systems were to be objectively evaluated for dismounted warfighter soldier integration and expecting to provide warfighter performance enhancement.

The majority of these systems were developed under the auspices of the DARPA High Definition Systems, Smart Module, and Warfighter Visualization Programs. These systems were developed for the following tactical/operational applications: Combat Vehicles, Rotorcraft, Dismounted, Medical, Maintenance, Special Operations, Reconnaissance Scouts, and Military Police/Law Enforcement applications. Integrating communication and sensor fused multi-modality information management systems on soldiers with HMD systems as the interface called for new graphical User Interfaces (GUI) to the system because a Windows Operating System would not work as a GUI in a mobile environment. Throughout the period from 1991-2001 many variations of systems for the aforementioned warfighter tactical/operational applications were developed. (See Figure 1)

Many operational field experiments were conducted by experts in human factors and research psychologists. Military users were always in the development loop to over see the design through iterations of these experiments so that the Non Recurring Engineering (NRE) design enhancements would capture the feedback of the users. The lessons learned through the years of experiments, operational tests and iterative system enhancements identified critical bottlenecks that pertained specifically to the limitations of the human (warfighter). Human system integration limitations were associated with information dissemination in a multimodality situation especially when the warfighter is mobile; walking, running, climbing and possibly coupled with exertion from the 80 and 120 Lb. load carried by the soldiers. Other stressors to be considered affecting diminished cognition are heat, cold, hunger, fear sleep deprivation etc.

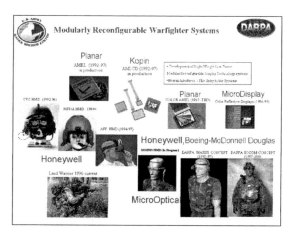

Figure 1: 1991-2001 DARPA –Army Information Management Systems and Component Developments

1.1   The Need for Augmented Cognition

The DARPA Augmented Cognition program, which began in 2001, was viewed by the Army as a means to understand the information management capacity of the human-computer warfighting system by co-developing and demonstrating quantifiable enhancements to human cognitive ability in diverse, stressful, operational environments through experiments that would mitigate information bottlenecks.   This could enable validation of human performance enhancement by understanding the cognitive capacity of a soldier to multitask and process multimodal information.

Future Force Warrior notional concepts seek to create a lightweight, overwhelmingly lethal, fully integrated individual combat system, including weapon, head-to-toe individual protection, netted communications, soldier worn power sources, and enhanced human performance. The program is aimed at providing unsurpassed individual and squad lethality, survivability, communications, and responsiveness. The Human Performance portion of the Future Force Warrior Advanced Technology Demonstration (FFW ATD) was viewed as the ultimate cognitive challenge because the integrated system of systems would be relaying enormous quantities of data and information to soldiers without a full comprehension of a dismounted soldiers capacity for information processing.

The FFW ATD information management system and component subsystems centered on warfighting doctrinal changes that brings about new tactics, techniques and procedures for conducting warfighting operations. The system of systems will consist of Sensors & Communications (C4ISR) Vision: Netted Future Force Warrior small unit/teams with robust team communications, state-of-the-art distributed and fused sensors, organic tactical intelligence collection assets, enhanced situational understanding, embedded training, on-the-move planning, and linkage to other force assets where words like "180-degree, multispectral sensor-fused; see through augmented reality Head-mounted vision system" will allow the warfighter to be one with the environment having 360-degree optimized, intuitive system control. 1.

General Kevin P. Byrnes, Commanding General, TRADOC stated in a recent interview "We talk about a networked force and empowering the force with knowledge—it worked well in Operation Iraqi Freedom at the higher echelons. The picture was much clearer; their ability to communicate data, voice and video was there. At the lower levels, where the fighting was conducted, we're not there yet. We have an awful lot of work to do to push information and intelligence and connectivity to those lower levels, and that's absolutely key to our force. If we're to fight effectively and trade mass for knowledge, we've got to work the lower end, where the fighting occurs." [2.]

Future Force Warrior is expected to provide information management capabilities and the expectations being placed on the ability for human information consumption and understanding is significant. The DARPA-Army information management systems and technologies developed during the 1991-2001 period were designed to enhance warfighter performance and augment their situation awareness but none of these systems were designed with the expectations of the Future Force Warrior system where the cognitive capability of the human may be exceeded. Understanding the capacity of the specific warfighters using the system will be as huge as the Future Force Warrior information management system development itself.

 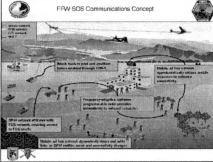

Figure 2: Diagram of the body worn network (Left) connecting the Warfighter (Right) to all echelons space-based and below system of systems (SOS)

1.2 The Relevance of Augmented Cognition

Figure 3: March 2002 Operation Anaconda, Afghanistan: Cognitive Stressors: Heavy Loads, Freezing Temperatures and Sleep Deprivation

Military medical researchers at U.S. Army Research Institute for Environmental Medicine study the effects of sleep deprivation, exertion and the effects of freezing temperatures because these stressors can diminish cognitive functionality in warfighters and can lead to accidents and poor decision making. [3] When the soldiers deployed to Afghanistan in the inaugural battles in the War on Terror, the organic weapon assets were minimal to do the job that was needed to reach the enemy peering out of the cave openings in the mountains. The maps and data used was based on Russian maps from the Russia-Afghanistan war era. The indigenous language turned out to be Russian and the special operations soldiers first to "hit the ground" quickly learned that all their lethality depended on their ability to communicate with support from US Air Force Bombers and AC-130 gunships who's primary missions were close air support, air interdiction and force protection and communications from higher headquarters to provide them with ammunition, food, water, batteries etc. [4] Their cognitive capacities had to be acute because they were on the move in search of Taliban and Al Qaeda. Their combat loads were in excess of 100 Lbs, the cold at night was well below freezing and almost all fire support to eliminate the enemy had to be called in from USAF weapon platforms. The Special Forces would designate the targets with their laser designators and batteries essential for their mission essential electronics had to be air dropped into the theatre of conflict. Many times, they would leave behind clothing and individual equipment articles because of space, weight and power considerations.

In an interview with the U.S. Army 82nd Airborne in October 2002, factors that affected their operations were weight and bulk of mission equipment, electronics and the many different types of batteries that were needed for them.   Although certain items were considered mission essential, if they were not critical they were left behind because of weight, power, bulk and in some instances, performance.   They did without precision Position-Navigation equipment and used commercial-consumer grade products.   They field-stripped their meal rations and took higher fat content components for endurance.   They left behind some warm clothing articles and using their saying which is "travel light-freeze at night" meant that they were huddling with each other to use body warmth as a means to survive.   Leaving behind critical mission equipment, food and clothing articles, compromised, their survivability, situational awareness and because they were susceptible to extreme cold, hunger, exertion, and lacked precision navigation, their cognitive performance might have been compromised as well.   The degree to which extreme cold effects have on human cognition are the subject of a study being completed by the U.S. Army Natick Soldier Systems Center and the U.S. Army Research Institute for Environmental Medicine (USARIEM).

1.3   Augmented Cognition: a Beta Test

The Augmented Cognition program is expected to benefit the warfighter in a digital information abundance environment through synergistic development between DARPA and the Army.   The integration of the Augmented Cognition technology into the Future Force Warrior, both programs simultaneously in development, is a parallel approach where, with the Future Force Warrior Program, Augmented Cognition technology has an optimal platform on which to develop a system that will mitigate informational bottlenecks and maximize the warfighter potential.

The Augmented Cognition system is expected to enhance and maximize the Future Force Warrior's capability to successfully accomplish the functions currently carried out by three or more individuals.   A key objective of the Augmented Cognition program is to foster development of novel and state-of-the-art technologies, in order to experiment with, and understand, the means by which they may be integrated into existing operational systems. Where Future Force Warrior and Augmented Cognition were beginning simultaneously, there was a preconceived opinion by research psychologists and human factors experts that the Future Force Warrior's networked collaborative situational awareness, providing 360° information to 21st century warfighters would severely tax the cognitive capacity of soldiers.   Based on the information management system developments mentioned earlier in this paper, limits of information to assist the warfighter, as opposed to overloading the warfighter were understood fairly well based on experiments executed in controlled environments.   The Future Force Warrior development is a system of systems primarily aimed at netted communications and collaborative situational awareness objectives that, in a tactical environment, could prove to cause cognitive overloaded situations.

 The Augmented Cognition program will accomplish evaluations of cognitive overload and examine mitigation strategies by experimenting with innovative technologies and design principles for human-computer symbiosis. This involves some revolutionary concepts of biomedical and behavioral monitoring interacting with intelligent digital information and control systems. (i.e. gauges consist of arousal (attention) meter, stress gauge, engagement index, executive load gauge, and P300-driven novelty detector and biosensors including: EEG (Electroencephalogram), pupil diameter, eye movements, ECG (Electrocardiogram), and EDR (Electro-Dermal Response)).

Correlating both Augmented Cognition and Future Force Warrior Advanced Technology Demonstration allows for concurrent system design. In using the same processor, bus architecture for carrying the information to the soldiers "human sensors" (eyes, ears, smell, touch etc) the Future Force Warrior development provides the Augmented Cognition team with the opportunity to explore the interaction of cognitive, perceptual, neurological, and digital domains to develop improved performance application concepts for the Future Force Warrior system. The advanced applications will be tailored to operational scenarios and task development in order to experiment and qualify mitigation strategies and demonstrate potential pay-off for first Future Force warrior system operational users. Success will be the improvement in the way the 21st Century warriors interact with computer based systems, advance systems design methodologies, and provide the ability to fundamentally re-engineer military decision making.

1.4   Understanding Stress: Augmenting - Augmented Cognition

There are additional considerations that factor into mitigating information bottlenecks in the Augmented Cognition Program. In the DARPA baseline program, the research is predominantly focused on information understanding and performance. However, there are several additional stressors that can play a role in human cognition. Some of these stressors that decrease cognitive state are mobility, exertion, fatigue, cold, heat, hunger, fear and isolation. The Augmented Cognition team is working on a solution to understand the cognitive capacity of the dismounted soldier. Research conducted with the Medical Research and Material Command and U.S. Army Research Institute for Environmental Medicine (USARIEM) to understand the Warfighter Physiological Status of the soldiers in battlefield situations are part of the Future Force Warrior system development and will provide the capability for assessing the cognitive health of the individual soldiers. The Warfighter Physiological Status Monitor (WPSM) will first determine level of consciousness, and secondly, the level of cognitive workload. Understanding cognitive workload is especially important in environments where there is an abundance of collaborative netted communications and basically too much information going to the soldier through radios, displays, sensors and other information disseminating devices that will inundate the warfighter. The adverse effects of stressors and information overload will be associated to the division of attentional resources among the various components of the task to be performed. Understanding the implications of the overload conditions will be necessary to prioritize relevant tasks through mechanisms such as a message scheduler, a mechanism to filter and disseminate in a proper modality prioritized information to the warfighter. [6.] How will the warfighter handle all the information that is supposed to make them more survivable, lethal and sustainable with the aforementioned stressors factored in to a battlefield situation?

These questions are being addressed under the auspices of new research protocols:
1) "The Effect of Walking Over Irregular Terrain on Soldier Ability to Process Information from the Environment and Communication Displays"
2) "The Effect of Walking with a Load on Soldier Vigilance Performance"

The first new research protocol will investigate to what degree ground soldiers are able to allocate cognitive resources to digital information sources while moving over varied terrain and at the same time remain alert to situations in their immediate environment. Additional objectives are to understand the level of cognitive complexity required to produce a drop off in performance while a soldier is moving over ground with loads, and which presentation modalities might be best suited for information processing while mobile. A choice reaction time task will be used to measure changing capacity to process information from the environment. In addition, a communications task involving two modalities, visual and auditory, will be used to measure ability to monitor and process information from communication channels during mobility. The mobility conditions will be standing, walking, and walking while avoiding obstacles.

The Second new research protocol will investigate whether physical exertion affects concurrent cognitive performance. Participants will perform a vigilance task under a variety of mobility conditions. They will be instructed to detect and respond appropriately to brief signals that are presented at the rate of approximately one per minute. Independent test groups will be exposed to test stimuli (visual, auditory or tactile) from a small, portable device that must be responded to as quickly as possible Each group of test subjects will navigate a walking course of approximately 2.4 km with or without a load. They will be asked to cover this distance as quickly as possible, but at a pace they are comfortable with. The load will be factorially combined with three mobility conditions: standing, walking, and walking while avoiding small obstructions. This experimental design allows us to assess whether walking with a load affects performance on an elemental information processing task, examine potential mechanisms that may underlie any effects, compare potential differences when the critical signal is presented in different sensory modalities and examine whether imposing small obstructions affects performance differentially across sensory modality.

There is much more detail to these protocols but due to the page limitations of this paper, only a cursory summary has been provided. In addition to the two protocols mentioned above, there are three more protocols being prepared for execution within the next year. These are:
3) "Cognitive-Attentional Workload of Soldiers Performing Simulated Combat Tasks While Moving Over Irregular Terrain."
4) "Cognitive Workload Analyses of Infantry Leaders Executing Complex Tactical Missions (field interviews)."
5) "The Effects of Exertion on Soldier Ability to Process Information from the Environment."

The Augmented Cognition team and the Future Force Warrior teams are working together to ensure relevance of Augmented Cognition to Future Force Warrior. In addition to MRMC and USARIEM we have been working with the Human Research Engineering Directorate of the Army Research Labs in at Aberdeen, MD to team and matrix our approach with their research on Motion Artifacts from neurophysiological and physiological sensors to better understand how to deal with motion artifacts.

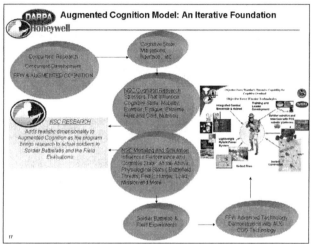

Fig 4: Augmenting - Augmented Cognition, a Concurrent System Design

Performance may be assessed through a number of different experiments to include target acquisition and weapon performance. Army researchers augmenting the program through additional research will look at cognitive processes and complex task performance based on operational scenarios provided by Future Force Warrior. There is also the possibility of designing experiments with both contractor teams in the Joint Rotational Training Center environment. This "real life-like" situation using man-portable communication and information management equipment in this type of experiment might be useful to evaluate how to increase our military's ability to think asymmetrically.

1.5   Conclusion

The Augmented Cognition program experiments thus far show promise for developing a conceptual solution that mitigates informational bottlenecks by classifying cognitive state. Using knowledge of an individual's cognitive state and the development of a communications scheduler protocol to disseminate prioritized communication and information through various modalities to the Warfighter in a hostile environment appears to be an approach that may prove successful in improving overall warfighter performance. With transformational developments expecting more from the individual warfighters, they will continue to be provided with more information technologies that add to their cognitive burden. Throughout the remainder of both Augmented Cognition and Future Force Warrior programs, research and experiments pertaining to the use of physiological, neurophysiological and environmental stress information will continue. The findings of the experiments and evaluation of warfighter performance will advantageously guide developments of information systems to provide future operational capabilities that will maximize the warfighter potential.

1.6 Acknowledgements

This research is funded by DARPA and the U.S. Army Natick Soldier Systems Center supported by a contract through the U.S. Army Natick Soldier Center, under Contract No. DAAD-16-03-C-0054.

The Natick Soldier Systems Center would like to acknowledge CDR (Sel.) Dylan Schmorrow, DARPA Program Manager of Augmented Cognition Program for recognizing the limitations on warfighter performance based on an overloaded human's cognitive state and developing a conceptual solution that mitigates informational bottlenecks. Augmented Cognition technology integrated into the Future Force Warrior system, as an optimal platform, is expected to maximize the warfighter potential.

Dr. Jim Sampson, Natick Soldier Systems Center, Research Psychologist is acknowledged for his guidance, oversight and research in the program and for linking the experimental designs and technology developments directly to user based on tasks analysis and evolving transformational concepts of operation.

Ms. Cynthia Blackwell, Future Force Warrior Human Factors and Training Lead, is acknowledged for ensuring Augmented Cognition is connected to FFW and that the Augmented Cognition team has access to all scenario developments.

The Honeywell Labs extended team, led by Dr. Trish Ververs, Augmented Cognition Principal Investigator, is acknowledged for the high quality teaming approach to successful developments that have the potential to change the manner in which warfighters manage high volumes of information and simultaneously enhance performance.

Any opinions, findings, conclusions or recommendations expressed herein are those of the author's and do not necessarily reflect the views of DARPA or the U.S. Army.

1.7 References

Future Force Warrior (FFW); Introduction: Modernizing the Warrior through Army Transformation. US Army Natick Soldier Systems Center Website; http://www.natick.army.mil/soldier/WSIT/

Farmer, Mark D., Military Training Technology (2004); Interview with General Kevin P. Byrnes, Commanding General, TRADOC; Volume: 9 Issue: 2

Fleming-Michael, Karen (March 2003). Sleepy Soldiers. The Fort Detrick Standard

Major Moores, Drew, Oct-Dec 2002. The 101st Airborne Division Deployable Intelligence Support Element in Operation Enduring Freedom - Air Assault. Military Intelligence Professional Bulletin

Interview with U. S. Army 82nd Airborne. Oct 2002. Soldiers were in Afghanistan (and Anaconda battle) and were at Natick Soldier Center to discuss clothing and individual equipment performance.

Bourne, Jr. Lyle E., and Yaroush Rita A., (Feb 2003) Stress and Cognition: A Cognitive Psychological Perspective. A literature review supported by the National Aeronautics and Space Administration, Grant Number NAG2-1561. Http://psych.colorado.edu/~lbourne/StressCognition.pdf

AugCogifying the Army's Future Warfighter

Patricia May Ververs, Stephen D. Whitlow, Michael C. Dorneich, Santosh Mathan

Honeywell Laboratories
Minneapolis, MN
trish.ververs@honeywell.com,
stephen.whitlow@honeywell.com,
michael.dorneich@honeywell.com,
santosh.mathan@honeywell.com

James B. Sampson

US Army Natick Soldier Center
Natick, MA
james.sampson@us.army.mil

Abstract

The U.S. Army wants to ensure that the Future Force Warrior (FFW) will see first, understand first, act first and finish decisively as the means to tactical success. The Army of the future conceives of small combat units with netted communications enhanced with information from distributed and fused sensors, tactical intelligent assets enabling increased situation assessment, and on-the-move planning (FFW, 2004). The increase in information flow won't come without a cost, however. Information management will be a key aspect of this distributed system. The availability of such technologies as Augmented Cognition (AugCog), will allow the system to be tailored to the situational and cognitive needs of the warfighter. This paper describes an example of AugCog technology applied to a Communications System.

1   The U.S. Army Transformation

The U.S. Department of Defense (DoD) has embarked on a process of change called Transformation to create a highly responsive, networked, joint force capable of making swift decisions at all levels and maintaining overwhelming superiority in any battle space (Parmentola, 2004). In response the U.S. Army is shaping its Future Force to be smaller, lighter, faster, and smarter than its predecessor. The network will be characterized by a network of humans collaborating through a system of C4ISR (command, control, communications, computers, intelligence, surveillance, and reconnaissance) technologies.

Evidence of the Army Transformation could already be seen in the Operations Enduring Freedom and Iraqi Freedom. Some of the most visible and valuable benefits were seen in the speed of operations enabling reduction in the time to plan missions, make decisions, and coordinate and move large groups of soldiers. What was created was a more dynamic and adaptive operation built on the collective capabilities of all the participants. Unprecedented levels of integration took place among the air, naval and land forces. Stone (2003) reports, for instance, that in the middle of Afghanistan special operations soldiers could link with a Navy F-14 or link with a B-52 to pursue a target. "Special Forces on the ground have taken 19th century horse cavalry, combined it with 50-year-old B-52 bombers, and, using modern satellite communications, have produced truly 21st century capability," is a perfect example of the effect of the Transformation (Wolfowitz, 2002).

The wars in Iraq offered an opportunity to examine the direct effect of the Army Transformation. Although there are many similarities between Desert Storm and Operation Iraqi Freedom which were both conducted in the Middle East with similar objectives, to liberate Kuwait or Iraq, with similar successful results, the similarities stop there. In the time between the two wars, the Transformation was enacted, enabling operations to be conducted with substantially fewer troops (540,000 vs. 100,000 ground forces) and resulted in fewer casualties due to fratricide (Keaney & Cohen, 1993; Rumsfeld, 2003; Krepinevich, 2003). The second Gulf War used additional capabilities in Blue Force (friendly force) tracking, GPS technologies and tactical situation displays versus relying heavily on voice transmission. General Tommy Franks, who led the U.S. military operation to liberate Iraq, exploited the effectiveness that small units on the ground can have when supported by airpower. Franks recognized that linking intelligence operations with military operations was a very powerful tool for both intelligence and operational purposes (Barnes, 2003). One of the most highly publicized events, the capture of Saddam Hussein, was enabled due to the rapid intelligence gathering, network structure and high-speed decision making (Stone, 2003).

One of the core capabilities of the Transformation is the availability of netted communications enabling information sharing and real-time collaboration enhancing the kind of situational understanding that drives decisive actions. Just as was seen in Operation Iraq Freedom, FFW will be dependent on the Army command covering more area with fewer warfighters. The Future Force Warrior will have unparalleled connectivity to build situation awareness conceivably right down to the individual soldier. Part of the success will be dependent on the individual warfighter's ability to sort through the vast array of continuous information flow afforded by a full range of netted communications.

2   How Augmented Cognition can Support the Warfighter

As part of the FFW program, the Honeywell Augmented Cognition (AugCog) team is developing warfighting concepts that could substantially increase the combat effectiveness of infantry small combat units. The objective is to enhance human performance and improve survivability through more effective information management. This can only be done if we improve the overall situation awareness from the top of the command down to the adaptable small units and individual soldiers.

The Honeywell AugCog team has developed a set of cognitive measures based on real-time neuro-physiological and physiological measurements of the human operator. The capability to assess cognitive state by determining how resources are being allocated also provides the opportunity to modify the soldier's current task environment by driving adaptive strategies to mitigate information processing bottlenecks. The end result can be the appropriate allocation of attention to the right information at the right time, which is important to FFW because it directly affects two cornerstone technology thrusts within the FFW program: netted communications and collaborative situation awareness.

Early in the program, task analysis interviews concerning existing military operations identified factors that could negatively impact communications efficacy. In one such interview was with a former Army commander from the Vietnam War. The commander reviewed the breakdowns and chaos that can ensue during the execution of a raid. In this famous raid, known as the Son Tay Raid -- a mission to rescue American prisoners of war, the commander recalled the inability to hear and focus on commands once they landed at the site (D. Turner, personal interview, September 19, 2003). Even with the highly rehearsed mission, many things went wrong that day from the breakdown intelligence to the lack of the ability to communicate between teams and the ground force commander. Though the mission was not successful (i.e., the POWs were not rescued), the highly skilled units managed to fight off the enemy without a single causality to their units. The raid highlighted the need to improve the suboptimal method of communications because the soldiers and commanders were intensely focused on the tasks at hand. This need will only intensify for the Future Force Warrior, who will be inundated with information through visual displays and verbal communications, particularly during mission execution. Adding to the challenge is the requirement to cover more ground with fewer troops; these warfighters will be more reliant than ever on the communications and digital information imposed by the additional NLOS (non-line of sight) operations.

3   A new kind of information management

The research described in this paper is aimed at validating the applicability of established non-invasive neuro-physiological and physiological state detection techniques during dismounted soldier combat operations. The system takes as input the soldier's cognitive state and alters information flow and modality of presentation. The incoming information is managed by scheduling the communications to be received by the soldier at the most optimal period, offloading information or task assignments to automation when the soldier is overwhelmed, and providing information in multiple modalities (audio, visual, tactile) to ensure reception and improve comprehension. High task load conditions prompt the automation to defer all but the highest priority information, offload tasks, or change the modality of information presentation; whereas low task load conditions that lead to subsequent cognitive disengagement from the task, indicates an appropriate time for interruption to prompt greater levels of soldier participation in on-going tasks. Without these augmentations the soldier can become overloaded with information having to decide when and where to focus attention among the myriad of high priority communications and high priority tasks. By adapting the soldier's workspace to his or her cognitive state, overall joint human-automation performance can be improved.

4   AugCog Evaluation

The FFW's ability to expand the effect of the Future Force is based on the application of the full range of netted communications and collaborative situational awareness. This paper briefly describes an evaluation conducted to test the AugCog system with Army relevant operational tasks. The evaluation was aimed at collecting the physiological information and using information on the participants' workload level to drive decisions in the communications scheduling.

The evaluation was aimed at mitigating high levels of cognitive workload as induced by information overload due to high task load demands induced by a large number of communicated messages between squad members. This closed loop integrated prototype evaluation integrated a neural-net based classification of cognitive state with an adaptive system designed to maintain levels of performance under increasing task loads. The evaluation involved participants performing three Army-relevant tasks simultaneously and evaluated how effectively a sensor-driven augmentation mitigation strategy addressed increases in cognitive workload. The basic premise was to test the effectiveness of having cognitive state classification drive decisions in the communications flow to the participant. Both the effectiveness of the classification algorithms to detect the user's cognitive state and the augmentation mitigation strategies to moderate high workload states through scheduling communications was evaluated. Six participants from Honeywell Laboratories completed the 3-hour evaluation.

4.1   AugCog System Description

The AugCog System was supplied neurophysiological EEG data from a BioSemi ActiveTwo System with 32 channels. The EEG data were processed to remove artifacts generated from eye blinks using an adaptive linear filter. DC drift was eliminated using a high bandpass filter. The following frequency bands were used to estimate cognitive activity: 4-8Hz (theta), 8-12Hz (alpha), 12-16Hz (low beta), 16-30Hz (high beta), and 30-44Hz (gamma). These five bands, sampled four times per second, formed the features for the cognitive state assessment (CSA). A composite classification scheme consisting of three separate models for estimating cognitive state was employed. The final classification decision was based on an agreement from two of three models, however if there was no majority agreement the output from the nonparametric estimate (i.e., Kernel Density Estimate) was used. See Mathan, Mazaeva, Whitlow, Adami, Erdogmus, Lan, & Pavel (2005) for more information on the classification methodology.

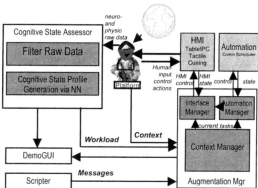

Figure 1. System Architecture

Data were integrated in an agent architecture developed at the Institute of Human and Machine Cognition. See Figure 1. The CSA, described above, determined the cognitive state of the individual and passed the information to the Augmentation Manager via a Cognitive State Profile. The Augmentation Manager contained the rules for adapting the work environment (e.g., displays, automation) to optimize the joint human-automation capability to complete specific tasks. The Communications Scheduler described below is one implementation of the Augmentation Manager.

4.2 The Communications Scheduler

The Communications Scheduler (CS) scheduled and presented messages to the participant based on the cognitive state profile, message characteristics, and the current context (tasks). Based on these inputs, the CS could pass through messages immediately, defer and schedule non-relevant or lower priority messages, escalate higher priority, divert attention to incoming higher-priority messages, change the modality of message presentation, delete expired or obsolete messages, or summarize and filter the content of messages. For this evaluation, the current CS took as input voice and text messages, reasoned over the current tasks and the cognitive state profile, and then scheduled message presentation and modality per task as needed. See Dorneich, Whitlow, Mathan, Carciofini, and Ververs (2005) for a full description of the communications scheduler.

4.3 The Tasks

The participant's cognitive state was assessed while conducting three simultaneous tasks: **target identification**, **mission monitoring**, and **maintain radio counts**. The participant acted in the role of a platoon leader. In the **target identification task**, the participant monitored static images of a MOUT (Military Operations in Urban Terrain) environment on a TabletPC for potential enemy targets. The task load varied by varying the rate of the presentation of images. See Figure 2. Simultaneously, the participant, named Red-6, monitored verbal communications comprised of two types of messages. One message type involved squad movements. In this **mission monitoring task**, communications from three squads moving in bounded overwatch were monitored. One squad (e.g., squad leader, Red-62) moved while the other two squads protected the moving squad. An example communications message included, "Red-62 to Red-6, Squad 2 is in position and ready for overwatch." The participant kept track of the location of three squads as they reported their status. When all three squads reported that they are in position (two squads ready for overwatch and one squad ready to move), the participant ordered the appropriate squad to move forward. The task load was varied by varying the rate of incoming messages. The second type of communication message involved messages between a company commander, the participant and two other platoon leaders. These messages contained reports of the number of civilians, enemies, or friendlies spotted. In this **maintain radio count task**, the participant was responsible for keeping a running total of civilians, enemies, and friendlies reported to him (Red-6), while ignoring the counts reported to the other two platoon leaders (Blue-6 and White-6). An example radio communication was, "Green-66 reporting, 4 civilians spotted, Red-6." The task load varied by varying the rate of incoming messages.

When the Communications Scheduler was activated in the mitigated condition and a high workload state was detected, all the radio count messages were deferred to the TabletPC message box as text messages while the higher priority messages regarding the squad movements were passed through. The participant was able to total the counts during a lower task load period, when the other tasks were completed. Once a high workload state activated the scheduler to redirect the low priority messages, all later messages continued to be redirected until the trial was completed. The automation etiquette and the theory behind the adaptive automation decisions are discussed by Mathan & Dorneich (2005).

Participants completed three experimental trials: low workload trial (E1), high workload with no mitigation (E2), and high workload with the mitigation strategy (E2M). The mitigation strategy (communications scheduler) was triggered by the cognitive state assessment of individual's cognitive state via neurophysiological EEG profile. Performance was measured on percentage of targets correctly identified, correct and timely response to squad movements, and correct reports of enemies, civilians and friendlies spotted by squad members.

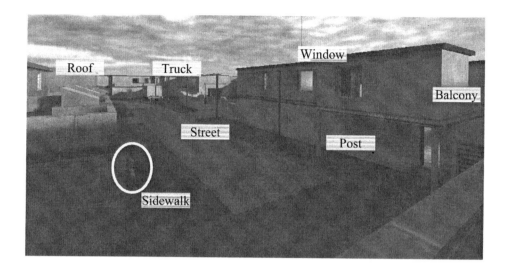

Figure 2. Target Identification Task with target identified in circle

4.4 Preliminary Findings

The main goal was to improve performance in the high task load condition when high cognitive workload states were induced by the increase in radio messages. The communications scheduler assisted the participant by rescheduling tasks to lower task load conditions. All the findings will not be covered here but a brief demonstration of the effectiveness of the cognitive state classifier and its ability to drive the mitigation strategy is described, as well as the overall effectiveness of the communications scheduler to mitigate the high workload effects induced by high task load conditions.

4.4.1 Cognitive State Classification

During the testing prior to the experiment, the system was trained on the first third of the spectral data from each experimental block and tested with the remaining two-thirds of the data; classification accuracy was close to 95%. The classifier correctly indicated a high or low workload state while the participant was performing in the corresponding high or low task load condition. However, the performance of the classifier was limited by the inherent non-stationary aspects of the EEG spectra over large time periods. Findings indicated that the classification performance dropped when the training and testing datasets were from corresponding workload conditions but different experimental blocks. When the classifier was trained and tested in this manner, classification performance ranged from 52% to 77%, with a median accuracy of 68%. Our current efforts focus on the development of technical approaches to address the problems posed by long term non-stationarity of EEG.

4.4.2 Performance data

Performance data are summarized for each of the experimental conditions: E1 – low task load, E2 – high task load, E2M – high task load with mitigation in Figure 3. In all cases in the mitigation trials, the participant's measured workload, as determined by the cognitive state assessor, reached a level to trigger the mitigation. Findings indicate that performance on the radio count accuracy and mission monitoring queries during the mitigated high task load condition (E2M) was equivalent to the performance in the low task load condition (E1). See Figure 3.

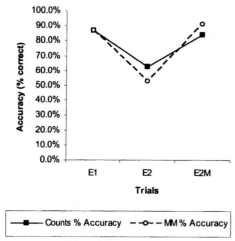

Figure 3. Percent correct (composite score) on Radio Count and Mission Monitoring tasks across three experimental conditions.

Each participants' performance was characterized as a percentage improvement in a mitigated high workload (E2M) as compared to the same condition unmitigated (E2). Overall there was a 94% improvement in the mission monitoring task [$F(1,4) = 31.91$, $p < .01$] and 36% improvement [$F(1,4) = 23.89$, $p < .01$] in the radio count recall task when the communications scheduler was available in the high workload trial as compared to the no mitigation condition. See Table 1. The scheduler offloaded the lower priority radio count messages to the visual modality and sent them to the TabletPC thereby allowing the participant to review the messages during a lower workload period. It is interesting to note that the benefit of the scheduler enhanced performance not only on the radio count task that was offloaded and completed without time pressure, but also the mission monitoring task that could be completed without the competition for resources by the competing task.

Table 1. Percentage performance improvement in mitigated condition vs. unmitigated condition.

Performance Improvement		
Participant	Radio Count Recall	Mission Monitoring
S1	27.24%	63.64%
S2	57.92%	20.54%
S3	18.51%	53.41%
S4	41.40%	110.94%
S5	47.15%	260.00%
S6	23.77%	55.56%
All	36.0%	94.0%

4.4.3 Subjective workload

The participants' perceived workload was measured using Hart and Staveland's NASA TLX scales (Hart and Staveland, 1988) for each of six indices of workload: mental demand, physical demand, temporal demand, performance, effort and frustration. Workload levels in the mitigated conditions (E2M) mirrored those of the low task load conditions (E1) and were statistically improved in the unmitigated condition (E2) for ratings of perceived frustration ($p < 0.05$), and approached significance for temporal demand ($p = 0.10$) and effort ($p = .06$). See Figure 4.

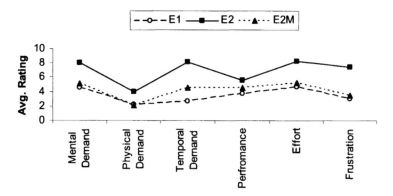

Figure 4. NASA TLX scores in three experimental conditions

5   Conclusions

The findings of the evaluation show promise for applying knowledge of an individual's cognitive state to adapt display devices to improve overall performance. However, several challenges still exist before this technology can be deployed on a dismounted mobile solder. The development of robust sensors that can reliably detect the cognitive status in real-time on a dismounted soldier in a deployed environment is a challenging endevour. However meeting this challenge provides the ability to non-invasively status the cognitive state of an individual to allow a person or system have the right information to make the right decisions and take appropriate actions. We have seen how the Army Transformation accomplished goals with fewer warfighters and resources as it attempts to harness the power of new technologies or old technologies applied in new ways. The key to success for the Future Force Warrior will be better timing, smart integration and rapid exchange of information in combat operations. The way in which the warfighters operate will continue to change. As one example, how an Army medic rapidly assesses the health of his or her distributed platoon will likely change with biosensors. As situations arise assessment of individual warfighter's level of consciousness would not be accomplished face to face manner but remotely. Therefore the ability to remotely status soldiers would better enable the right care and the right time to all warfighters. AugCog technologies could help make this happen.

The full utility of neural classifying cognitive state and its application to netted communications or other applications will continue to be explored in the future. These explorations will continue to research the use of physiological information such as heart rate, interbeat intervals, and respiration to indirectly status cognitive state as well as test the utility of neurophysiological state information generated by such devices as EEG or functional near infrared imagery (fNIR) for a more direct measure of cognitive state. If these technologies are to be deployed into a harsh battlefield environment, the sensors will need to be ruggedized and usefulness in an operational environment fully investigated.

Acknowledgments

This research was supported by a contract with DARPA and funded through the U.S. Army Natick Soldier Center, under Contract No. DAAD-16-03-C-0054, for which CDR (Sel.) Dylan Schmorrow serves as the Program Manager of the DARPA Augmented Cognition program and Mr. Henry Girolamo is the DARPA Agent. Any opinions, findings, conclusions or recommendations expressed herein are those of the authors and do not necessarily reflect the views of DARPA or the U.S. Army.

The Honeywell team would like to acknowledge the efforts of Mr. Jim Carciofini, Mr. Trent Reusser, and Mr. Jeff Rye for the implementation of the mitigation strategies, Ms. Natalia Mazaeva for EEG support and analysis, and Ms. Danni Bayn for work on the pilot studies, scenario definition, and data analyses.

References

Barnes, F. (2003, June 2). "How Tommy Franks won the Iraq war," The Commander, 008(37).

Dorneich, M.C., Whitlow, S.D., Mathan, S., Carciofini, J., and Ververs, P.M. (2005). The communications scheduler: A task scheduling mitigation for a closed loop adaptive system. To appear in the Proceedings of the 1st International Conference on Augmented Cognition, Mahwah, NJ: Lawrence Erlbaum Associates.

Future Force Warrior. (n.d.). Retrieved September 24, 2004, from http://www.natick.army.mil/ffw/content.htm.

Hart, S. G., & Staveland, L. E. (1988). Development of a multi-dimensional workload rating scale: Results of empirical and theoretical research. In P. Hancock & N. Meshkati (Eds.), Human Mental Workload. The Netherlands: Elsevier.

Keaney, T. A. & Cohen, E. A. (1993). Gulf War Air Power Survey: Summary Report. Washington, DC: Government Printing Office.

Krepinevich, A. (2003, October 21). "Operation Iraqi Freedom Outside Perspectives Capitol Hill Hearing Testimony," Available from M. Winslow, On Transformation. Retrieved on January 28, 2005 from http://www.ndu.edu/library/.

Mathan, S. and Dorneich, M. C. (2005). Augmented Tutoring: Improving military training through model tracing and real-time neurophysiological sensing. To appear in the Proceedings of the 1st International Conference on Augmented Cognition, Mahwah, NJ: Lawrence Erlbaum Associates.

Mathan, S., Mazaeva, N., Whitlow, S., Adami, A., Erdogmus, D., Lan, T., & Pavel, M. (2005). Sensor-based cognitive state assessment in a mobile environment. To appear in the Proceedings of the 1st International Conference on Augmented Cognition. Mahwah, NJ: Lawrence Erlbaum Associates.

Parmentola, J. A. (2004). Army transformation: Paradigm-shifting capabilities through biotechnology, The Bridge, 34(3), The National Academy of Engineering of the National Academies. Retrieved January 28, 2005 from http://www.nae.edu/NAE/ bridgecom.nsf.

Rumsfeld, D. (2003, July 9). Lessons Learned from Operation Iraqi Freedom Capitol Hill Hearing Testimony. DefenseLink. Retrieved January 28, 2005 from http://www.defenselink.mil/speeches/2003/sp20030709-secdef0364.html.

Stone, P. (2003). Cebrowski Sketches the Face of Transformation. Retrieved January 28, 2005, from: http://www.defenselink.mil/news/Dec2003/n12292003_200312291.html.

Wolfowitz, P. (2002, April 9). The Imperative for Transformation. Prepared Statement for the Senate Armed Services Committee Hearing On Military Transformation. United States Department of Defense, Washington, DC. Retrieved February 4, 2005 from http://www.defenselink.mil/speeches/2002/s20020409-depsecdef2.html

Augmented Cognition for Fire Emergency Response: An Iterative User Study

Daniel Steingart

UC Berkeley Department of Materials Science
240 Hearst Memorial Mining Building, Berkeley CA 94720
steinda@berkeley.edu

Joel Wilson
Andrew Redfern
Paul Wright

UC Berkeley Department of Mechanical Engineering
2117 Etcheverry Hall, Berkeley CA 94720
jwilson@berkeley.edu
aredfern@kingkong.me.berkeley.edu
pwright@kingkong.me.berkeley.edu

Russell Romero

7201 Santa Rita Pl N.E. Albuquerque, NM 87113
bxc_wolf7@hotmail.com

Lloyd Lim

Open Text Corporation
llim@opentext.com

Abstract

To facilitate the design and development of an advanced IT network specifically for fire crews at the scene of an emergency, the Fire Information and Rescue Equipment (FIRE) team from UC Berkeley conducted interviews at the Chicago Fire Department Training Facility in December of 2002. This review is broken into two main sections. The first details the information gained from the interviews directly, such as department practices and protocols, firefighter experiences with and expectations of communications technology, and fire command operations through the Incident Command System (ICS). The second half discusses how these results, along with other background studies, lead to a design of head of a head mounted display for firefighters as well as an infrastructure to gather and deliver information to firefighters.

1   Introduction

Though communications problems at fireground have existed as long as organized firefighting, the widely reported chaos and confusion at the World Trade Center in New York City on September 11th, 2001 has brought new focus to this critical issue. Shortly after the 11th, Richard Nowakowski, Director of the Office of Emergency Communications of Chicago, began to re-evaluate the emergency response needs of Chicago, with a focus on high-rise disasters. Within a month an order was issued for all buildings over 30 stories to produce detailed digital floor plans, and in March 2002, Mr. Nowakowski assembled a group of academic and commercial research facilities to apply robust information technology and equipment to fireground communications.

We studied the nature of the problem in parallel with investigating solution paths for a head mounted display (HMD) for firefighters since March of 2002. In December 2002, we travelled to Chicago to spend two days learning the practices and protocols of communications at emergency sites. In addition, we informally interviewed Bay Area firefighters many times over the past two years to gain insight regarding the strengths and weaknesses of our systems in development.

The Chicago Fire Department (CFD) uses equipment from a variety of manufacturers including Motorola, and Mine Safety Appliance (MSA). In light of these long standing connections, our role, once the needs of the CFD were determined, was to modify and integrate existing technologies from these companies where applicable. Initial prototypes of such devices are presented in the second half of this paper.

2 Background Studies

The World Trade Center attack cost New York City $33.4 billion in property damage, and over 2800 lives, 350 of whom were firefighters. McKinsey and Co. [1] did a comprehensive analysis of the New York Fire Department's response to the terrorist attacks on September 11th 2001, in terms of management and technology effectiveness. Their study showed a strong coupling of management and technology failures. Many of the losses were related to radio communications breakdowns, creating a chaotic response scene. The fraternal draw for firefighters to save their own, exacerbated by the lack of communication, put many more people in the fire than necessary at a given time with little centralized control. When the command station set up in the bottom of the south tower was destroyed, the ICS broke down completely. While there were many improvements suggested by the study, the overall lesson was that technology must be coupled with strict failsafe and backup protocols to maintain control of a scene.

James Landay's group [2,3] at UC Berkeley has also looked at many issues around graphical user interfaces (GUIs) and firefighters, specifically interfaces for incident commanders.

3 Interviews

3.1 Setup

Roughly 60 firefighters were interviewed in 15 groups according to their current assignments (squad, hazardous materials, truck and engine) at the Chicago Fire Department. Six groups were engines, five were trucks, two were squads and two were haz-mat. The firefighters were interviewed in between training exercises at the academy over two days. It must be noted that over the career of a firefighter, he or she may work in any of aforementioned assignments. In addition, three fire chiefs were interviewed.

Two formats of interviews we conducted. The first consisted of two interviewers (one engineer and one MBA student) with one team. The interviewers started each round with these questions:
  1) Physical location of firehouse
  2) Type of area (Industrial/Commercial/Residential)
  3) Years on the force
  4) What everyday communications/electronic equipment did they use
  5) Comfort with personal computers

The interviews became freeform after these initial questions, and firefighters were encouraged to discuss particular instances where technology failed or excelled, as well periods of technological transition (what new technologies were adopted, how the force dealt with the change, overall feeling of the technological contribution).

The other format was a large roundtable discussion, where three to five groups would sit and have an open discussion on the effectiveness of current technology and what investments should be made in new types of equipment.

3.2   Team Observations

In terms of overall communications technology acceptance/reliance (not need), the groups rank:
1) Hazardous Materials
2) Squad
3) Truck
4) Engine

The hazardous materials (hazmat) teams use a variety of chemical detection apparatus and sometimes cameras to relay information from the scene to a nearby truck. Members are often from scientific backgrounds where use of computers and networking is commonplace. Hazmat teams are the fewest in number, with four teams in Chicago. Hazmat teams are usually called in after evacuations, so they typically, but not always, have the longest amount of time to assess and act on a threat. Of the four groups, the hazmat team requires the most onsite forensic tools to determine the nature and extent of the radioactive, chemical, or biological threat.

The squad is responsible for search and rescue operations first and foremost. While larger in number than hazmat teams, each squad supports multiple truck and engine companies. After the initial truck and engine team arrive at a scene, a squad team may be called in to handle complicated and/or large-scale search and rescue operations (among the firefighters the squad is nicknamed the "Marines" of the CFD.) With such a focus, each squad is equipped with a thermal imaging camera, which was generally described as "invaluable." To quote one firefighter:

> Before we had thermal imaging search and rescue was slow and tedious… in smoke filled rooms we would have to move very slowly, feeling the floor for bodies along the way…. With a thermal imager we can see easily through the smoke, and even a single hand sticking out from under rubble stands out and allowing us to locate the body much faster.

It must be noted that thermal imaging supplements (rather than replaces) traditional sweeping methodologies. The truck team will often assist the squad in rescue operations. The squad is not directly responsible for containing and exterminating fires.

The truck team is the first to respond to fire emergencies. There are six members on each truck, and once arriving at the scene they split up into three groups of two to analyze the situation. While an engine company is automatically called to a fire scene, the truck team, in coordination with the incident commander, determines whether or not more support (including hazmat or squad) is required. The truck also begins evacuation/rescue procedures, and prepares the building for the engine team by breaking obstacles and assuring proper ventilation.

The engine team usually consists of five members; an officer, an engineer, and three firefighters. At the scene the engineer stays behind and manages the controls of the engine, such as routing water from the engine, fire hydrants or standpipes to maintain adequate water pressure in the hose. The officer will either stay with the engineer or join the remaining firefighters on the hose line. Three or four firefighters will hold and steady the line.

The hose line goes in after the initial prep work from the truck team, and the first task is usually to accelerate the ventilation by shooting a stream of water through a breach the truck team creates (through a wall or broken window). This effectively flushes an area of smoke, enhancing evacuation and the survival rate of unconscious victims yet to be located by the truck or squad. The other main duty of the hose line, of course, is to quench the fire.

Overall technology acceptance seems strongly correlated to the funding each team receives. Though we were not provided with budgets, the fire chiefs confirmed that the squad and hazmat teams receive more money per firefighter than the truck and engine teams.

3.3   General Firefighter Safety Measures / Equipment

Except for the hose line, firefighters are always in groups of two on a scene as prescribed by both National Fire Protection Association (NFPA) and Occupation Safety and Health Agency (OSHA) standards. This is largely for the safety of the firefighters, so that a rescue team can be called if one of the firefighters becomes incapacitated. Newer NFPA standards require that for every team of two within a building, there are two firefighters waiting to replace them or rescue them if need be (referred to as "two in, two out").

General safety equipment consists of:

1) Self-Contained Breathing Apparatus (SCBA) setup: a mask and air tank rated for between 30 and 60 minutes (but we were told actual values were 15 to 45 minutes, respectively, because of increased breathing rate). All of the gear must meet National Institute of Science and Technology (NIST) and NFPA flame retardation tests, and new NFPA codes require all masks sold after January 2004 have remaining air indicators.
2) PASS (Personal Safety Alarm): an audible alarm on the firefighter that has two functions. First, it alerts the firefighter to low air pressure in the SCBA tank. Second, after 30 seconds of inactivity (no motion) it sounds an alarm to alert surrounding firefighters that the wearer may be incapacitated.
3) Emergency Band radio: with general and fireground channels. There is also a channel dedicated to use by those who report directly to incident command.
4) Helmet and Turnout gear: heavy, insulated and durable clothing made to be put on quickly when a call arrives and taken off quickly should the fire fighter be set aflame.

3.4 Fire Chief Interviews

We sat down with three Fire Chiefs in Chicago for an informal discussion, to get anecdotal information as well as suggestions regarding an electronic Incident Command System. The chiefs listed the following as the most important pieces of information during an emergency:

1) Proximity to danger
2) Health status
3) Better radios
4) Location and ambient temperature of firefighters.

The fire chiefs were concerned about the validity of the floor plans, particularly with regard to moveable items (desks, cubicle dividers, etc). In order to prevent the firefighters from running into confusion over such objects, they suggested that we not include anything in the floor plan except for elevators, exits, stairwells, standpipes and doorways. Though the firefighters would have to feel their way around office furniture, this is a case where having no information is better than having possibly misleading information.

The fire chiefs also applauded the notion of health monitoring, but gave us warning to possible resistance from the firefighters (this warning proved accurate). Two points at issue were:

1) Firefighters do not want to be pulled from a scene "prematurely" when they feel "ok."
2) Firefighters fear that pension/insurance benefits will suffer if they are found to have pushed themselves beyond reasonable limits after being warned to stop.

If health monitoring is to be implemented in an augmented cognition system for firefighters, these two issues will have to be addressed through a careful combination of technology and policy.

3.5 Overall Lessons Learned

The major differences between the Chicago Fire Department and the Berkeley/San Francisco Fire Departments were fewer high-rise buildings and considerably more training for earthquake awareness in the Bay Area. While strategies for these scenarios varied slightly, OSHA and NFPA regulations and suggestions make the overall protocol and procedure for these fire departments similar.

The most interesting variances between all three departments stem from a trait common to all: limited budgets. Fire departments have little money for technology or equipment that is not mandated by a standards body. The result is trade-offs, not between "wants" and "needs", but between "needs" and "needs more." For example, the Berkeley Fire department gives each firefighter an SCBA mask they can custom fit to their face, but does not have a mobile data terminal (MDT) in each truck, nor does each fire house have a thermal imaging camera. San Francisco, on the other hand, does have an MDT in each truck and a thermal imaging camera in each house, but masks are fixed to

SCBA tanks, and must be shared amongst firefighters (the fit of the masks are worse, giving a leaky seal which leads to a short air duration). In Chicago, firefighters have their own masks but only the squad has thermal imaging cameras. These trade-offs are purely cost driven: there is no engineering advantage to having one item and not the other. Decisions are made based on which items the team feels are most important, but in all cases those who did not have a certain item were envious of those who did.

Firefighters have an almost paradoxical relationship with technology. Though they embrace the ideas of real time tracking and health monitoring, they are wary of the facts that 1) technology, especially IT can be unreliable and 2) prolonged use of such technologies may create dependence. The combination of these downsides would be catastrophic: place a young firefighter (who was never in a fire without an augmented cognition system) in a situation where her head mounted display failed, and the consequences could be awful.

The next section deals with design and prototype implementations for a head mounted display (HMD) for firefighters, an electronic Incident Command System (eICS) and a wireless sensor network to tie them together (SmokeNet). These were designed with the preceding interviews in mind, and the overall lessons learned point to one overarching principle: *limit feature creep at all costs*. Probability of failure and cost are proportional to feature set, and next to the fire itself are the two greatest nemeses to firefighters.

4   Design and Testing of FIRE Components

At first the goal of the fire information and rescue equipment (FIRE) project was to make just a head mounted display (HMD) for firefighters (FireEye). As we began to develop the very first prototypes we found that there was little done to provide information from an external source to the firefighter over digital channels. In response to this we implemented Crossbow Motes running TinyOS to create a mesh network of smoke detectors called SmokeNet. Finally, we realized that this data would also be useful for Incident Commanders after our talks with the fire chiefs; the electronic Incident Command System (eICS) project began to meet that need. Figure 1 provides a simple overview of the system.

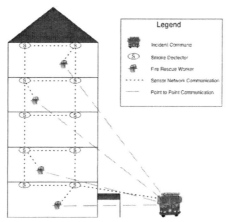

Figure 1 – FIRE project Overview

4.1   FireEye

4.1.1   First Version

Initial prototypes were based on an outside the mask design using a modified Hitachi POMA. Primarily a benchmark for low-cost display quality, no mechanical improvements were made to the HMD itself. The display was mounted on a Mine Safety Appliance (MSA) gas mask with a soft plastic screen along the top of the mask. The position of the display was easily altered for right eye or left eye preference, but due to the top mounting could only be seen by "looking up" (to lower it would impede vision).

The POMA display was a small LCD projected into a holographic prism. The intended effect was to create a display that was somewhat translucent. We found that this actually made the display harder to read due to the natural confusion it caused the eye and the poor quality of the optics used in the prism. Edge distortion and overall barrelling made the display nearly unreadable unless it was perfectly positioned. The firefighters echoed these thoughts during our Chicago interviews.

While we were able to show ambient temperature and location with the display via the Windows CE operating system, wireless TCP/IP (over 802.11b) and the Mica Mote network, most users found the display incredibly fatiguing, though not distracting. Users generally liked the fact that the display could easily be ignored when not in use. From this comment, instead of going towards an augmented reality approach, we kept with an "inconspicuous until necessary" approach in most of the future designs.

Another important lesson learned though fire-fighter interaction on this mask was that "look down" orientation was generally preferred over look up: firefighters spend much time crawling through smoke, and roll their eyes up when moving forward. They expressed concern that this display would inhibit their sight while moving forward, another "inconspicuous display" vs. "augmented reality" debate.

4.1.2   Second Revision

The second prototype was an outside-the-mask design using a MicroOptical (SV-6) display, mated with the internals of a small laptop computer (running Windows XP to simply the prototyping environment). The display was mounted on the side of a MSA SCBA mask, and through software could be modified for left eye or right eye preference. The side mounting allowed for simple vertical placement with no view impediment. The MicroOptical display, though not as capable as the POMA display at differentiating between colors, had a far better resolution and much less distortion around the edges and was much easier to read. It also used a prismatic glass system to route the image to the eye but did not attempt any kind of holographic or translucent effect.

An ABS plastic mount was fabricated with a Statasys 1650 fused deposition modelling (FDM) machine to provide additional mechanical durability outside the mask. We initially decided against placing the display inside of the mask due to 1) the lack of room available for an unmodified display and 2) the added complexity of routing signals and power through the seal of a carefully designed SCBA mask.

Firefighters liked this display much more than the POMA, and could easily read messages displayed in a 16 point font (see map in figure 2) once proper focus was achieved. The lower position preference was also reiterated with this design, with the firefigher essentially wanting to "look down their nose" to see the display. Most criticisms of the display revolved around the external mount, with claims such as "I'd break this thing before I got into the truck." It became clear that future designs would have to be internally mounted, because any amount of ruggedization externally would not offset the fragility of the device the firefighters *perceived*. One firefigher said it best with, "If this mask can protect my face, it can certainly protect the display."

4.1.3   Third Revision

The third prototype was a "through the mask" design (figure 3) using a MicroOptical display disassembled and reconfigured for a direct view. A significant case was fabricated with the FDM, and a custom fit hard plastic display was placed around the new case, maintaining the seal provided by the SCBA. This solution was deemed "rugged enough" by the firefighters, and quite readable, but was a "little too in-your-face."

4.1.4   Fourth Revision

The fourth prototype was a Kopin internal display (figure 4). Using the display's standard direct-view magnifying glass optics, a custom stainless steel mount was fabricated. The entire device was placed on the nose of the mask's rebreather unit, and a clever wiring harness allowed the device to be connected to the real world through the mask's existing oxygen supply lead.

Figure 2 - Basic FireEye GUI

Figure 3 - eICS version 2

While the quality of the Kopin display did not match that of the MicroOptical units, firefighters found this device quite readable, commenting on the seamlessness of the integration. Unfortunately, the wire coupling, while elegant, was too fragile during our tests and broke before we could get full feedback. Prototypes after this did not contain this couple due to its labor intensive construction (time better spent on optimizing optics), but work is currently in progress to manufacture a more durable slip-couple.

4.1.5   FireEye Prototypes (Current Revision)

The two most current prototypes attempt to combine the best features of previous designs, while minimizing their compromises. Important features such as user adjustability and ease of installation from the mask were retained, while overall robustness was increased. The housings holding the electronics were designed using SolidWorks 2003 to be injection moldable for improved manufacturability, and were prototyped on the FDM machine. These housings allowed the FireEye to be mounted level to a nosepiece of a Draeger mask, the brand used by the local Berkeley Fire Department. Mounting to the flexible nosepiece allowed for quick insertion or removal and reduced shock loading.

The two prototypes differed in their optical designs, one being a direct view (figure 5), and the other a see-through HMD using a cube beam splitter (figure 6). The two designs created different user interactions with the FireEye. The direct view acts as a peripheral display, and encourages a graphical user interface (GUI) composed of graphics with filled-in images. It was easily read against a bright background like a fire or flashlight, unlike many see-through designs including this prototype, but required one to momentarily remove their attention from their current task in their external real environment. The see-through design encouraged a GUI with bright lines to outline images, with no color filling, such as GUIs used by military helicopter and jet pilots. An augmented reality GUI makes sense for these pilots, because they need to continually focus attention on flying the aircraft. Moreover, their augmented cognition system, when present, directly helps them to perform their immediate task.

Firefighters, however, generally want to see information about the local incident scene, rather than information immediately relevant to their task. For example, they will frequently want to know locations of the fire, stairways, elevators, standpipes, and other personnel, but do not need to see information from the GUI on how to operate an elevator or other parts of the building at which they are immediately looking. This information may be better shown on a color-coded direct-view FireEye. In the case that a firefighter needs to navigate with the FireEye, a jet pilot-like augmented reality GUI could offer simple arrows directing the user to an escape route or other predetermined locations. It may be safer, however, to see an overview of the scene and a variety of navigation options on a floorplan in case there is a danger unforeseen by the system, or in case the firefighter needs to travel to an unexpected location.

   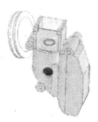

| Figure 3 – | Figure 4 – | Figure 5 – | Figure 6 – |
| Through-Mask design | First in-mask design | Current direct-view mask | See through schematic |

4.2 SmokeNet

4.2.1 SmokeNet Version 1.0

The first SmokeNet was based on TinyOS 0.6.1, using Crossbow Mica wireless sensor nodes, or Motes. The Mica Motes had an Atmel ATMega103L eight bit processor running at 4 MHz with a RFM TR1000 radio and 512 kB of flash memory. Two ADC channels were used to sample a smoke detector (binary) or a thermistor (8 bit sensitivity from 40 F to 120 F). Each Mote would broadcast these values every 10 seconds.

Since each smoke detector node had a static location with a unique Mote identification number, they were used as virtual buoys for firefighters. We made simple digital floorplans indicating the locations of each smoke detector. Each test firefighter would wear an HMD and a Mote attached to a computer running a small Visual Basic program. When they were in broadcast proximity of a SmokeNet node, the floorplan would indicate the general proximity of the firefighter.

A centrally located Mote (base station), which could hear all the other Motes, would aggregate the smoke/temperature data, and provide information to the incident commander via a connected laptop or tablet PC.

While this first attempt at SmokeNet provided a useful test platform for feeding information to FireEye, it had serious drawbacks. First, the single hope network limited the effective range of the Motes. The Mica Motes had an effective range of roughly 50 feet in an open room, 30 feet through doors/walls. Thus, a rough estimation of a 60-foot sphere centered around the base station limited the size of the network.

Second, since the SmokeNet Motes were operating at full radio strength the "buoy" signals to the firefighters would overlap, diminishing the accuracy of the location data. We applied some interpolation algorithms to use the overlapping beacons to increase the accuracy, but we often found that the firefighter Mote would suffer from the "hidden node" problem (two signals reach the Mote simultaneously and create a corrupted packet).

Finally, there was no attempt at energy management on the Motes. They would report at the same rate regardless of fire status. Even with sleep measures in place, the Motes would last not much longer than a month based on a 10 second sleep/wake up cycle.

4.2.2 SmokeNet Version 2.0

The current SmokeNet is based on TinyOS 1.1.7 using CrossBow Mica2 Motes. The CrossBow Mica2 Mote has an improved microprocessor and a superior radio to the Mica Mote. Messages sent from one Mica2 node have up to a 250-foot range. Using this improved platform and improvements to the smoke detectors, like additional temperature sensors and digital output form the smoke detectors we were better able to detect a fire. Custom software was developed for the smoke detector nodes to check the status of the smoke detector every ten seconds if the system was a non-alert state. The node would then send a check-in message containing information about the node's health, such as battery voltage and average room temperature, every five minutes through the multi-hop network to a central building node logger computer. This logger computer is used to track the network's health and see if there were malfunctioning nodes. If a node were to detect a fire, it instantly sent an alter message that put the multi-hop network in an alter state. Once in an alter state, the nodes checked for a fire every five seconds and sent a check in message every two minutes to the logger computer as an alive message if no fire was detected.

The implementation would occur as follows: Once the firefighter teams arrived on scene, the incident commander would wirelessly connect his or her eICS to the network and be able to see the locations of the fire and quickly assess the situation. After the teams have been formed and the firefighters have entered the building, the firefigher's FireEye would show their location and the location of the fire. Location tracking would be done using the smoke detector nodes with a unique Mote identification number and known locations throughout the building as virtual buoys, much like in the first implementation of SmokeNet. The firefighter's Mote would vary radio signal strength in order to fine-tune the firefighter's location to one SmokeNet node. The location of each firefighter would be relayed through the network back to the eICS where the incident commander would monitor progress and make adjustments as needed.

This method worked fairly well and we were able to get room accurate location of the firefighter. For tracking purposes room accuracy is sufficient.

We have started looking at using SmokeNet as way of reducing some of the traffic on the firefighter's voice radios by sending text messages through SmokeNet to individual firefighters and the possibility of automating some of the check-in calls that the incident commander currently does.

4.3 electronic Incident Command System (eICS)

4.3.1 Version 1.0

The first version of the electronic Incident Command System was a simple Visual Basic program with a floor plan of either the Berkeley smoke tower or the second floor of Etcheverry Hall of UC Berkeley. Temperature and smoke data were routed to it via SmokeNet version 1.0, and firefighter location was routed via 802.11b from the firefighter's HMD pack (either the POMA unit or a repackaged laptop computer). The system could only handle one team of firefighters at a time (up to 5), so there was a toggle to choose between teams. This model allowed us to demonstrate the basic functions of SmokeNet v. 1.0 to fire personnel to get feedback.

4.3.2 Version 2.0

We added text messaging support to the second version of the eICS. Interviews revealed that much of the radio chatter consisted of simple "yes/no" questions. In order to free the radios up for more detailed communications, we felt text messages that would pop up on the HMD would be helpful. The eICS contained a set of pre-defined messages such as "Are you ok" "Report Status" and "Wait for order." In addition the incident commander could type custom messages.

Currently, we are adding a toggle switch to the firefighter setup to either provide a yes/no response to the inquiry or if necessary open a radio line to incident command.

5 Conclusion

The firefighters, through testing the various prototypes, were both encouraging and grounding. Often referring to us affectionately as "propeller heads," they always helped us differentiate between life saving technology and "gee-whiz" technology.

Key to all the technology is keeping the information as accurate and brief as possible. For example, 95% room level tracking consistency is much more useful to firefighters than 75% accurate meter resolution. Reliability is the most important feature, and must not be compromised by feature creep. There is also a sweet spot in network reporting. Updates approximately every twenty seconds during a fire were deemed acceptable as long as they were accurate and reliable. Finally, the eICS and the FireEye are only useful to firefighter if they are "crash-proof."

Now that we have achieved this minimum feature set for a tracking/sensor-net system, we are going to implement the system in a trial at the Berkeley fire tower with a truck team. The results of this study will provide a basis for making improvements to the equipment for real world use, including improving durability of the equipment, readability of the display, and what a realistic initial feature set and implementation schedule might look like.

An area that we are only beginning to breach, but is key to the acceptance of any of this technology, is that of allowing the firefighters to customize the displays and networks for the needs of their districts. The final but most daunting problem is implementing this system in such a way that the firefighters can clean and service most of the equipment without what one firefighter called a "tech geek" on-hand.

Acknowledgements

This work has been made possible by funding from the Ford Motor Company and the University of California Center for Information Technology Research in the Interest of Society (UC CITRIS). We give many thanks to the Fire Departments of Chicago, Berkeley and San Francisco for their time, encouragement and feedback.

References

McKinsey and Co. "Increasing FDNY's Preparedness," http://www.nyc.gov/html/fdny/html/mck_report/toc.html, 2002.

Jiang, X., Hong, J., Takayama, L., & Landay, J., "Ubiquitous Computing for Firefighters: Field Studies and Prototypes of Large Displays for Incident Command," *Proc. of CHI 2004*, April 24-29, 2004.

Jiang, X., Chen N. Y., Hong, J. I., Wang, K., Takayama, L., & Landay, J. A., "Siren: Context-aware Computing for Firefighting" in *Proceedings of Pervasive 2004*, April 18-23, 2004.

An Intelligent Deception Verification System

D.R. DuRousseau

Human Bionics LLC
190 N. 21st Street, Suite 300
Purcellville, VA 20132
don@humanbionics.com

Abstract

The devastating events brought about by the destruction of the World Trade Center, war in Iraq, and continuing terrorist actions in Iraq, Afghanistan, and around the globe have changed our world. Today, improved methods of interrogation and detection of deception must be implemented to enhance our operational readiness, make our borders safe, and ensure that our troops have the most accurate field intelligence possible. What's needed in our arsenal are improved methods to uncover those individuals bent on harming American citizens; in other words, an Intelligent Deception Verification System (IDVS). The IDVS provides a miniaturized virtual-reality equipped polygraph used in the field to reliably determine the veracity of suspects and the validity of the information extracted from them (DuRousseau 2004a). Having a robust, rapidly applied, and easy to use digital polygraph that includes computer-aided interrogation (CAI) software would be of huge importance to law enforcement officers and customs agents protecting our cities, embassies, and ports-of-entry, as well as to military forces carrying out covert human intelligence gathering in hostile environments.

1. Background

The psychophysiological detection of deception (PDD) is a procedure routinely used by the U.S. Department of Defense (DoD), various law enforcement agencies, and officers of the court, to determine an individual's truthfulness concerning topics of interest (Dollins, 1997; Lykken, 1981). In theory, the examinee's physiologic reactivity varies with personal relevance to presented stimuli and, more so, with attempts to conceal that relevance from the examiner. In the field of PDD, the variability of electrophysiological activity such as galvanic skin response, heart rate, blood volume, respiratory rate and volume, electroencephalography (EEG), evoked potentials, and even fMRI are visually assessed by a trained examiner (Boucsein, 2001; Spence, 2001; Dollins et al., 1998). Unfortunately, these intuitive assessments, even by the most experienced examiners, are exposed to considerable subjectivity and variability in the accuracy and sensitivity of their interpretation of deceptive behavior. Increased reactivity, defined as a change in response level to some stimuli but not others, is assumed to reflect the personal relevance of an item presented to the examinee. The typical PDD examination is designed to elicit outwardly observable physiologic responses from the examinee to specific questions regarding topics of interest. The physiologic responses are then subsequently scored by one or more methods and interpreted as indicating the truthfulness of the examinee's verbal responses to the questions of interest.

Human Bionics (HB) intends to advance the field of PDD with its IDVS technology; a portable digital polygraph with an automated CAI software system that provides:

A) a time-controlled virtual reality environment (VE) for immersion of the examinee within a 3D multimedia scene, while questions are presented by adaptive software; and

B) an automated user interface and software system for constructing CAI sessions and real-time analysis of the psychophysiological responses of the examinee for detection of deceptive responses.

The existence of a convenient, portable, and fast application digital polygraph with state-of-the-art psychometric analysis tools would accelerate the acceptance and use of PDD. Homeland Security applications for IDVS abound, such as in passenger screening, border interdiction, and interrogation research, where tools to improve deception detection methods and enhance our operational readiness are sorely needed.

Here, we describe the IDVS, which is a method and system for sensing and processing mental and physical signals from the human body through the use of actively attached, passively contacted, and nearby or distant non-contacted sensors that collect information related to the physiological and behavioral activities of an individual or group of individuals for the purpose of determining deceptive intent. Additionally, a preferred embodiment of the IDVS relates to the presentation of questioning materials to an examinee using an immersive multimedia VE while his or her behavioral and/or physiological activities are monitored.

2. Technologies and Methods

Existing PDD methods require rather large and cumbersome analog polygraph devices. Even those systems with somewhat portable digital devices must still use separate and bulky sensing, stimulation, and analysis devices (Farwell & Smith, 2001). In addition to the equipment size problem, the science of PDD continues to rely on the interrogation skills of the examiner and on the examiner's subjective visual interpretation of the polygraph data (Furumitsu, 1999). Unfortunately, there are considerable variations in the accuracy of results across examiners, and human examiners cannot operate as quickly or reliably as automated detection methods (Granhag, 1999; Blackwell, 1996). Further, individuals who are trained to use countermeasures such as tongue biting, toe curling, sphincter flexing, or mental manipulation of numbers can often defeat examiners. Honts, et al., (1994) tested the use of physical and mental countermeasures during a control question test technique and found that both countermeasure methods were equally effective at defeating the polygraph test when administered by human examiners. In one study, fifty percent of examinees defeated examiners, and countermeasures were reported as very difficult to detect.

To be more useful in the future, PDD methods must remove the subjectivity of the human examiner by providing automated detection algorithms that can accurately determine when an examinee is attempting to deceive the examiner or subvert the interrogation by using countermeasures. Although commercial automated software systems for analyzing PDD data and rendering decisions have been developed, studies have found that computer-aided detection are correct only 88 to 91% of the time (Dollins et al., 1999). Thus, a need exists for intelligent automated routines that identify the use of countermeasures and improve the accuracy of deception detection, preferably to 98% or more. In addition, a greater need exists for technologies to disrupt the cognitive-system of an examinee and prevent the use of countermeasures entirely.

The IDVS proposes to address these needs by adding specific muscle sensors to the polygraph suite and through the use of an immersive virtual environment (VE) to disrupt an examinee's ability to coordinate countermeasure activities. To deal with physical countermeasures, the IDVS includes methods to detect toe curls and sphincter flexing from small muscle sensors placed on the ankles and buttocks. Additionally, tongue biting and voice stress patterns are measured from the neck or larynx. The more subtle mental countermeasures are arbitrated using immersive multimedia VE to create unrealistic environments intended to confuse the examinee and prevent overt coordination of higher-order cognitive functions needed to deceive human examiners and existing simple automated computer programs.

2.1 Details of the IDVS

HB is a member in DARPA's Augmented Cognition Program researching human systems integration through the analysis of multimodal neurometrics underlying cognition and stress patterns in humans (DuRousseau, 2002). HB's cognitive and stress assessment methods are derived using highly constrained spatio-temporal EEG analysis, expert-based heart, eye, muscle, voice, electrodermal, and respiration data processing algorithms, and adaptive neural network (ANN) pattern recognition and classification techniques to identify psychophysiological indices of attention and stress (DuRousseau, 2005, Zhang, 2000; Gevins et al., 1995). HB's CEO, Don DuRousseau, has been active in the field of cognitive neuroscience for nearly two decades. He has extensive experience in multimodal neuroimaging, adaptive human-systems interface, operational testing, and fMRI/EEG sensor technologies (DuRousseau, 2004b; Gevins & DuRousseau, 1994a,b).

The IDVS was designed as an automated interrogation and analysis system to rapidly provide digital polygraph data, as well as attention and stress metrics to improve PDD results (Meyerhoff, 2001; Furumitsu, 2000; Dollins, 1999). The IDVS creates a novel or misleading multimedia environment for the examinee, which has been reported to increase the magnitude of physiological reactivity during PDD examinations (Amato-Henderson, 1996). It is our belief that an immersive audio and visual environment would, minimally, place cognitive task demands on the examinee sufficient to disrupt attempts to conceal the use of mental or physical countermeasures (Smith, 1999). Ideally, dealing with the complexity of the VE would actually prevent an examinee's use of countermeasure entirely.

The IDVS uses a wearable ambulatory monitor (WAM) being developed under DARPA funding to record brainwave, eye, heart and/or muscle activity; skin conductance, resistance, and/or impedance; body position, posture, expression, and/or gestured motion; speech; and/or body temperature. The system also has the ability to measure blood flow sensors, as well as stress measurement sensors that process respiration, blood pressure, heart rate and/or other such phenomena. Preferably, the system would be worn in a small back pack or on the utility belt of an officer or security agent. The IDVS delivers all the functionality of existing polygraph devices as well as including functions to process spatio-temporal EEG, eye motion, muscle activity, and voice and motion signals from the larynx (Manfredi, 2000). The IDVS also includes an easy to use application program interface, so an examiner could quickly build and run a CAI program and automatically, or, interactively analyze the data (the IDVS graphically displays ongoing detection results and provides a moment-by-moment rating of subject veracity).

The IDVS provides wireless Web-enabled data transmission capabilities to upload suspect data onto a secure Website for examination by superiors, if needed. The entire system is small enough to be carried on an officer's utility belt and comes with multi-sensor bands to quickly locate electrodes and transducers on or near the examinee in less than three minutes. The current IDVS product specification provides for:

A)   a wearable Web-enabled digital polygraph, which provides a compliment of sensors that can be placed on an individual being examined in under three minutes, including EEG, EOG, EMG, and ECG electrodes; and

B)   an easy-to-use API programming environment for researchers and field examiners to create counterterrorism-based structured interviews and multimedia CAI protocols that automatically or interactively operate the digital polygraph and perform data analysis on brain, eye, heart, muscle, voice, GSR, respiration, and several other PDD related signals.

2.2 Use of an Immersive Virtual Environment

The IDVS works by donning wireless VR glasses, tactile stimulators, and headphones on an individual and locating them into an immersive 3D scene (e.g., the examinee's self is projected into a standard scene or one constructed by the examiner using real crime scene details and evidence). The idea is that the novel environment may, minimally, place cognitive demands on the examinee sufficient to disrupt his or her attempts to conceal the use of mental and physical countermeasures employed to defeat detection attempts. Then, traditional analyses of the EEG and other physiological signals could be performed without confounding problems caused by the use of physical and mental countermeasures (Amato-Henderson, 1996; Furedy, 1996; Cestaro & Dollins, 1994). Thus, the use of immersive multimedia virtual reality for the detection of deceptive and harmful behaviours may lead to vastly improved CAI technologies.

For example, the IDVS might present a 3D graphical scene that places the examinee on a moving rollercoaster as CAI questions are presented by the computer. While CAI takes place, visual, audio, and/or tactile stimuli (i.e., an image of Osama Bin Laden sitting in an adjacent coaster seat) are delivered to the examinee in the form of questions or distractions that limit the examinee's ability to use countermeasures and deceptive test taking tactics. Furthermore, through a simple software toolkit, the IDVS may be configured to present stimuli depicting, for example, one or more images of a crime scene, a weapon used in a crime, an individual involved in a crime (i.e., another participant in the commission of the crime or a victim) or other related items. Lastly, for the actual detection of lying by the examinee in response to questions, sophisticated PDD signal-processing algorithms developed by HB would be applied to measure the examinee's neurometric reactions to the images and sounds presented during interrogation.

In conventional PDD examinations, evoked responses in the EEG are used to determine whether the examinee has previously seen a particular item or not (Farwell & Donchin, 1991). For example, if an item from a murder scene were presented to an examinee who did not commit the murder, that person would not exhibit an expected evoked brain reaction upon viewing the item. However, an examinee that had witnessed the murder scene (presumably because the examinee had been a participant in the crime) would exhibit a pronounced reaction to the item, even though they answered that they did not recognize the object. More subtle deception and voracity detection methods being designed by HB would be included to provide advanced neural network-based detection and classification algorithms as part of the IDVS.

2.3 An Integrated Solution

To deal with the problem of physical countermeasures, the IDVS includes methods for recording and analysing brainwaves as well as eye, heart, and muscle activity. In particular, muscle activity from the ankles (to detect toe curls), buttocks (to detect sphincter flexing), and the neck, tongue, or larynx (to detect tongue biting, as well as to record voice stress patterns) may provide useful details of an examinee's attempts to conceal guilt. Thus, the IDVS may be successful at combating both physical and the more subtle mental countermeasures, such as dividing and counting numbers, imagined pattern manipulation, or other such cognitive workload related processing schemes. Hence, the IDVS has been designed to integrate a wide variety of sensor technologies within a digital polygraph framework that includes computer aided stimulus presentation and automated multimodal signal analysis capabilities.

The IDVS provides an immersive 3D multimedia graphical program that works in concert with system software to synchronize data collection, environment manipulation, and CAI operation. The system uses expert-based signal processing algorithms and adaptive neural network (ANN) classification and adaptation techniques to process multimodal psychometric signals and improve the accuracy of PDD over traditional methods. Our ANN methods utilize measures of power in several EEG frequencies as well as receiving index calculations from several cognitive gauges used to monitor workload, engagement, and stress. Several different types of analyses may be chosen by the examiner, based on established methods, or, for special circumstances where the use of specific crime scene details was required.

A major improvement over conventional PDD methods, IDVS technology would remove the ambiguity of examiner subjectivity by automating the presentation of questions and analysis of psychophysiological signals. The IDVS delivers all the functionality of existing polygraph devices as well as improved functions for processing spatio-temporal EEG, eye motion, muscle activity, and voice and motion signals from the larynx. The integrated system provides an easy to use application program interface, so an examiner could quickly build and run a CAI program and interactively analyze the data (the IDVS would graphically display ongoing detection results and provide a rating of subject veracity). The body-worn polygraph unit itself provides wireless Web-enabled data transmission capabilities to upload suspect data onto a secure Website for examination by superiors, if needed.

Further, by virtually manipulating the visual, auditory and/or tactile environment of the examinee, the system may prevent the successful use of countermeasures to defeat detection. The technology embodied in the IDVS came about through the coupling of cognitive neuroscience, mathematical signal-processing and immersive 3D graphical visualization capabilities. Thus, with the introduction of IDVS technologies, HB has created an inimitable and highly mobile PDD system capable of operation under any conditions.

3. Conclusions

This paper described patent-pending IDVS technologies designed by Human Bionics to combine immersive VE technologies with sophisticated cognitive neurometrics and overcome the limitations in existing lie detection systems. In addition to providing a highly portable polygraph device, the IDVS includes a suite of automated CAI tools for routine examinee interrogations, free from the bias and subjectivity of human examiners. Additionally, to combat the use of physical countermeasures by criminals trained to overcome tradition interrogation methods, the IDVS provides sensors specific to identify covert muscle contractions. To prevent the use of mental countermeasures, the IDVS uses an immersive virtual environment to place examinee's in a totally unfamiliar situation, leading to the inability to coordinate deceptive tactics while questioning was ongoing. The CAI software allows the creation of very specific crime scene re-creations with input capabilities into the VE for several forms of evidence, including, photographs, audio and video recordings, physical items, and documents. As a major component of the IDVS, HB's WAM provides a highly ruggedized and portable system that's easy-to-use, whether in an office or harsh battlefield environment. The near-term goal for our IDVS technologies is to carry out advanced testing under operational conditions to determine the limitations of the system and its suitability for commercial, homeland defense, and military applications. Thus, we have identified several Homeland Security applications for our IDVS technologies for embassy and passenger screening, ports of entry interdiction, and military interrogation, each of which would directly benefit from Human Bionics' technologies, particularly, given the recent problems associated with the mistreatment of detainees at Abu Ghraib and Guantanemo Bay prisons.

References

Amato-Henderson, S.L., (1996) Effects of misinformation on the concealed knowledge test. Dept. of Defense Polygraph Institute, Ft. McClellan, AL: No: 97-R-0001.

Blackwell, N.J., (1996) POLYSCORE: A comparison of accuracy. Dept. of Defense Polygraph Institute, Ft. McClellan, AL: No: 95-R-0001.

Boucsein, W., Schaefer, F., Sokolov, E. N., Schroeder, C., Furedy, J.J., (2001) The color-vision approach to emotional space: Cortical evoked potential data. *Integrative Physiological & Behavioral Science.* Vol 36(2) Apr-Jun 2001, pp. 137 – 153.

Cestaro, V.L., & Dollins, A.B., (1994) An analysis of voice responses for the detection of deception. Dept. of Defense Polygraph Institute, Ft. McClellan, AL: No: 94-R-0001.

Dollins, A.B., Krapohl, D.J., Dutton, D.W., (1999) A comparison of computer programs designed to evaluate psychophysiological detection of deception examinations: Bakeoff 1. Dept. of Defense Polygraph Institute, Ft. McClellan, AL: No: 99-R-0001.

Dollins, A.B., Cestaro, V.L., Pettit, D.J., (1998) Efficacy of repeated psychophysiological detection of deception testing. *Journal of Forenic Sciences.* Vol 43(5) Sep 1998, 1016-1023.

Dollins, A. B., (1997) Psychophysiological Detection of Deception Accuracy Rates Obtained Using the Tests for Espionage and Sabotage: A Replication. Dept. of Defense Polygraph Institute, Ft. McClellan, AL: No: 97-P-0009.

DuRousseau, D.R., (2005) eXecutive Load Index: Spatial Frequency EEG Tracks Moment-to-Moment Changes in High-Order Attentional Resources. *International Journal of Human-Machine Interactions.* In Press, Publication date: July 24, 2005.

Donald R. DuRousseau: Intelligent Deception Verification System. Utility Patent Application Number: 10/736,490, Submitted: Dec. 15, 2003, Pub. No.: US 2004/0143170A1, Pub. Date: Jul. 22, 2004

Donald R. DuRousseau: fMRI Compatible Electrode and Electrode Placement Techniques. Patent Number: 6,708,051, Patent Date: Mar. 16, 2004.

DuRousseau, D.R., (2002) Intelligent Bio-adaptive Warfighter Training System. Defense Advanced Research Projects Agency, BAA-0138, Contract Number: NBCHC020022.

Farwell, L. A. and Smith, S. S. (2001). Using Brain MERMER Testing to Detect Concealed Knowledge Despite Efforts to Conceal Journal of Forensic Sciences 46,1:1-9

Farwell, L. A. and Donchin, E. (1991) The Truth Will Out: Interrogative Polygraphy ("Lie Detection") With Event-Related Brain Potentials. Psychophysiology, 28:531-547.

Furedy, J.J., (1996) The North American polygraph and psychophysiology: Disinterested, uninterested, and interested perspectives. *International Journal of Psychophysiology.* Vol 21(2-3) Feb-Mar 1996, 97-105.

Furumitsu, I., (2000) Laboratory investigations in the psychophysiological detection of deception. *Dissertation Abstracts International: Section B: the Science & Engineering.* Vol 61(1-B), Jul 2000, 583.

Gevins, A.S. & DuRousseau, D.R. (1994a) Biopsychometric signal acquisition and processing system. Final Technical Report, NPRDC N66001-94-C-7012.

Gevins, A.S., & DuRousseau, D.R. (1994b) Spacecrew testing and recording system. Final Report, NASA, NAS 9-19054.

Gevins, A.S., DuRousseau, D.R., Zhang, J., Libove, J., Leong, H., Du, R., Le, J., Smith, M.E, (1995) Towards measurement of brain function in operational environments. *Biological Psychology,* Vol. 6, pp. 22-38.

Granhag, P.A., Stroemwall, L.A., (1999) Repeated interrogations: Stretching the deception detection paradigm. *Expert Evidence.* Vol 7(3), 163-174.

Honts, C.R., Kircher, J.C., (1994) Mental and physical countermeasures reduce the accuracy of polygraph tests. *Journal of Applied Psychology.* Vol 79(2) Apr 1994, 252-259.

Lykken, D.T., (1981). A tremor in the Blood: Uses and Abuses of the Lie Detector. New York: McGraw-Hill.

Manfredi, C., (2000). Adaptive Noise Energy Estimation in Pathological Speech Signals, *IEEE Transactions on Biomedical Engineering.* Vol. 47, No. 11, pp. 1538 – 1542.

Meyerhoff, J.L, Saviolakis, G.A., Koenig, M.L., Yourick, D.L., (2001) Physiological and biochemical measures of stress compared to voice stress analysis using the computer voice stress analyzer (CVSA). Dept. of Defense Polygraph Institute, Ft. McClellan, AL: No: 98-R-0004.

Smith, M., McEvoy, L., Gevins, A.S., (1999) Neurophysiological indices of strategy development and skill acquisition, *Cognitive Brain Research,* 7(3):389-404.

Spence, S. A., Farrow, T.F., Herford, A.E., Wilkinson, L.D., Zheng, Y., Woodruff, P.W., (2001) Behavioural and functional anatomical correlates of deception in humans. *Neuroreport: an International Journal for Rapid Communication of Research in Neuroscience.* Vol 12(13), Sep 2001, 2849-2853.

Zhang, G.P., (2000). Neural Networks for Classification: A Survey, *IEEE Transactions on Systems, Man, and Cybernetics.* Vol. 30, No. 4, pp. 451 – 462.

Augmented Higher Cognition: Enhancing Speech Recognition Through Neural Activity Measures

Erik Viirre M.D. Ph.D.

University of California, San Diego
Departments of Surgery and Cognitive Science
Suite 1A, 9350 Campus Point Dr.
La Jolla CA, 92037
eviirre@ucsd.edu

Tzyy-Ping Jung, Ph.D.

Institute for Neural Computation
University of California, San Diego
San Diego, CA

jung@sccn.ucsd.edu

Abstract

Introduction

The goal of communication is delivery of the content of a message. Vocalizations carry a symbolic representation of a message from the mind of the speaker to a listener (human or machine) who must decode the representation. Machine-based speech interpreters have libraries of sound templates and word and grammar logic systems that assist in decoding a message. Human listeners have some additional information to assist interpretation such as the context of a discourse: the current topic, the time, etc. Though human-to-human discourse is held as the model for delivery of a message from a human to a machine, even speech between humans is sometimes fraught with difficulty in passing meaning. Can we improve sending the meaning of a message from a speaker? In the mind of the speaker is located the content of a message. Can we access information related to that content and help deliver the message? The production of an utterance by the brain of a speaker involves the setting of a message, selection of vocabulary, determination of syntax, fixing of semantics and creation and execution of a motor program that activates the voice. We suggest that it will be possible to assist in content delivery of a message by accessing neural signals related to the production of an utterance.

Since 1980, there have been research studies on electrical signals related to brain events related to speech. Current work on these evoked potentials (EPs) has required averaging. However, new techniques are increasing our ability to resolve signal components that are leading us to the "single shot": pattern detection in a single trial. These techniques include: increasing the number of electrodes and new computational methods such as trained neural networks. Importantly, the technique of Independent Components Analysis (ICA) has lead to the ability to discriminate the many signals that summate to create the potentials recorded at the surface of the scalp. ICA initially requires substantial training of a neural network system based on the signals of a large number of electrical sources. However it can then be used to discriminate features of the data set and match a template of those features to subsequent trials. We propose to use the utterance of a word as a timing signal to match ICA templates to the "pre-potential", the EEG signal preceding the utterance. A statistical match could be made with the pre-potential to candidate meanings for the utterance created by the SR system.

Preliminary Results

We have collected pilot EEG data from a subject who was triggered to utter the sound (tuː) at the presentation a visual cue of the randomly presented words "to" or "too". The subject was instructed to speak the word as soon as they saw it presented. The goal was to review the EEG related to the individual trial utterances and determine if there was a robust set of features related to the utterances that could discriminate the different words. EEG data were collected from 28 scalp electrodes and 1 EOG electrode. Averaged 2-sec EPs related to the visual stimulus presentations under the two different conditions were examined. EPs of two conditions differed largely at site CPz. However, the time courses of single-trials did not always resemble those of the averaged EPs. The trial-to-trial variability made it difficult to distinguish the brain responses under two conditions without further processing. A single-layer perceptron neural network was trained through supervised methods and found a linear boundary between two clusters ('to' and 'too'). The boundary based on the scalp EEG misclassified 19 of a total of 82 trails. ICA was applied to the concatenated 2-sec single-trial epochs from the subject and derived 29 temporally independent components. The averaged activations and scalp topographies of components 6, 8, and 16 suggested that they together account for the large differences in EPs at cite CPz. The trial-to-trial variability was reduced compared to the raw EPs since the contributions of other components were removed from (not projected to) the electrodes. Similarly, the activations of ICA components accounted for the differences in EPs at site T8 were projected and summed at the scalp electrodes. For each trial, the summed projections between 200 and 1400 ms following stimulus presentations at cites CPz and T8 were averaged. The clusters were tighter compared to those from site CPz alone. Again, a single-layer perceptron neural network was trained through supervised methods and found a linear boundary between two clusters ('to' and 'too'). The boundary based on the scalp EEG misclassified 11 of a total of 82 trails (13%).

Conclusions

In this example, ICA separated the distinct brain processes into different independent components and extracted the signals of interest. Mathematically reducing the interference from the other brain or extra-brain activity generators improved the chance to detect small differences in EEG across conditions. It appears feasible that EPs can be used to assist speech recognition.

1 Background

Communication is a central issue in human affairs and interaction with computers. In the management of computerized systems, the ability to provide reliable speech inputs is desirable to improve the accuracy, efficiency and effectiveness of HCI. Because speech is a natural mode of information exchange for people, little training would be necessary if there were a reliable speech recognition (SR) method.

Unfortunately, automated SR is fraught with problems. Many human-computer interface (HCI) projects have the refrain: "wouldn't it be nice if we could use speech input?" The problem is that SR systems make many errors, many of which are not predictable. There are confusable letters (the "E"group) confusable words (words with multiple meanings and spellings: to, right, set…), variable noise environments, variable responses to speakers ("lambs" with good SR and "goats" with poor recognition rates) and differences due to age, gender, sound equipment and more.

A fundamental problem with SR is the fact that automated systems do not take into account "real-world" knowledge. Humans follow a conversation and actually fill in content by anticipating utterances. Good deliverers of speech "set-up" the conversation by describing the topic and use content, prosody and other cues to signal what is coming. SR systems fail to detect these cues, and other sources of information such as the location of a speech interaction. However, in the mind of the utterer is the intended content of the utterance. The processes that result in the production of an utterance have related neural signals: selection of vocabulary, determination of syntax, fixing of semantics and motor plans for articulation. The signals related to these neural activities may serve as a discriminant for an SR system. The EEG signal has also been shown to incorporate signals that code conditions such as neural state (fatigue, stress, etc.) as well as the signals related to the content of the individual word. Neural states such as stress have associated alterations of utterances (speech delivered under stress or noise is sometimes called "Lombard Speech"). By detecting an altered neural state, an EEG based system could signal to the SR system to use a different library for decoding vocalizations.

Note that it is not feasible at this time to envision a complete library of event potentials (EPs) for individual words. What this program is intended to determine is if there is a set of confusables that event potentials could discriminate, given a limited set of options. We then would determine if using the set of EPs would enhance the accuracy of SR in a given set of conditions. For example, we will examine the ability to discriminate "to"," too" and "two". The acoustic based SR system would be expected to discriminate the phoneme tū. The EEG signature would be used to help discriminate which meaning of the phoneme. Thus the EEG system will not act alone, but in concert with other SR technology. It would enhance the reliability of the other technologies.

1.1 Application Areas

SR systems in noisy environments have high error rates. Further, even if the ideal acoustic SR detection system were to be developed there would still be homophones, confusables etc. as well as changes in the user state that affect speech production. This new approach could detect user state and will detect confusables and improve

1123

understanding of the intent of the utterance. It is difficult to quantify the impact of this admittedly blue-sky technology plan for two reasons: we do not know the efficacy of the technology and errors are extremely application specific. We need to define the corpus for test conditions and for field conditions and do some experiments to see if this will work.

Intriguingly this system may be useful for improving human-to-human communication. If we were to develop ways to help signal intended language content, syntax and semantics (say with a visual display system), we could augment the conventional vocal/auditory/body language pathway of communication with this new technique. Mental language would be a means to improve delivery and understanding of discourse.

1.2 Technical Background

1.2.1 Speech Recognition

Success in applications of speech recognition depend on understanding the application that it will be used for (Markowitz J. 2000). There are a variety of application types including: Command and Control, data entry information access and dictation. The Command and Control paradigm for SR in military applications will have certain characteristics. For example, the nature of the utterances that will have to be recognized will often be instructions and responses. Thus the speech flow should be discrete phrases or words, rather than continuous discourse. In fact, with the command communications systems a keyed microphone is commonly used, readily signalling the beginning of an utterance. Further, the vocabulary size will be on the order of 1000 words (though it should be noted that for military aircraft applications, up to 5000 words may be necessary). Thus the language structure for the SR system will likely be a finite-state grammar. Given the ability to train personnel and provide them with specific computer systems, speaker modelling and speaker specificity should be possible in a military C2 speech recognition system. All of these conditions: finite grammar, discrete utterances, keyed microphones, limited vocabulary and speaker training, will all assist in the reduction of errors and in the use of electrophysiologic signals as a method of reducing errors.

1.2.1.1 SR and Errors

There are a variety of errors types that SR systems make. The most common is a substitution error where an inappropriate word is placed for another. Also common are deletion errors where words are missed and insertion errors where words are inserted. Splitting of long words (pandemonium= point and mention) and merging of short ones (lamp and = lampoon) also occur. Logic systems can sometimes deal with these various errors. Where there is ambiguity, such as the phrases "go to sector to (too, two)", grammar rules can be used to make a correct response. However, errors still occur, particularly in noisy environments and where there is grammatical ambiguity such as with homonyms. An important part of determining the ultimate usefulness of electrical potentials augmenting an SR system will be cataloguing error types and frequencies.

The electrophysiologic signal associated with an utterance will be useful in some of these circumstances where errors appear. It will not be possible with current technology to get a distinct library for all phonemes and the intended meaning of the each utterance. Rather, using cues from the utterance, a couple of template matches could be made to the associated evoked potentials with the utterance.

1.2.2 Neural Basis of Speech Production

It is interesting to review the production of speech by humans as an adjunct to machine speech recognition. The central issue of this project is what are the processes involved in the production of speech and are there recognizable brain signals related to production that could be use as SR cues. The answer appears to be yes.

Speech production has several interacting components. There is the "message" that is intended to be delivered. We will not address where this comes from or how it comes about. But from the message, comes the selection of lexicon, the imposition of syntactics and the overall preservation of semantics. Each of these three processes is linked, but they also appear to be distinct systems. Studies of stroke victims suggest that at least in the vast majority of right-handers these processes are left frontal/parietal/temporally located, but are distinct processes, probably separately located. There are also distinct EEG and other functional measures of these processes. Each of these processes is activated in utterances and thus will have potential contributions as cues to an SR system.

1.2.2.1 Lexical Selection

Words have representations in the brain that include their phonetic content, their segments and rules for their combination with other words. Speech production from the word content perspective includes the selection of the word form- articulatory planning and articulatory implementation (motor activation). The final product must: follow rules of pronunciation, have appropriate vocal stress patterns and have appropriate prosodic structure or intonation. The different processing of components of words can be seen in certain aphasics: victims of speech loss from stroke. Some aphasics will produce "words" that have inappropriate phonetic features, but none-the-less are articulated correctly by the voice (Gazzaniga, 1995 Chapter 59). Thus the components of word representations and their activation for speech are potential sources of signals for SR.

1.2.2.2 Syntactics

The processing of syntactics of sentences appears to be another system, most likely located in the Left perisylvian cortext (Broca's area) or near-by structures. The effects of damage to these areas by stroke are well described and suggest that syntax can be damaged in the face of appropriate lexical selection. The syntax of a sentence contains thematic roles, attribution of modification, pronoun referencing and other roles. The activation of brain areas related to these modules in functional imaging studies suggest that there are signals related to syntax that would help with SR. In evoked potential studies, there have been discovered Left anterior negativity signals generated when there are phrase structure violations. For example, when the sentence:
    "The scientist admired Max's of proof of the theorem",
is spoken, a left anterior negative signal appears.
Interestingly, syntactic issues appear to be processed in parallel with lexical selection, apparently on representational ideas ("lemmas") that eventually are phonologically planned. It is seen in subjects such as those with Alzheimer's disease, that syntax can be preserved even in sentences with inappropriate ideas. They speak sentences with good grammar, but have no meaning. This again suggests the modularity of speech production.

1.2.2.3 Semantics

At the highest level of speech production abstractness is the semantics of the utterance. The semantics contribute to the discourse content: topic, focus of attention, novelty of information, temporal order of events, causation etc. Curiously, this high level of abstractness has robust electrical events related to it. In human speech recognition, this was started by Kutas and Hilyard (1980) when they identified the N400, a signal of brain events related to real-time sentence processing. They found that in the right posterior aspect of the brain there were strongly negative signals related to the plausibility of a word in the context of a sentence. "The man spread warm butter over socks" is an example where there would be a strong N400. Since the initial development of the N400 there has been substantial research on refinements of it. (see "The Neurocognition of Language", 1999, chapter 10).

1.3 Single Trial Evoked Potential Analysis

Evoked Potentials (EPs), time series of voltages from the ongoing EEG that are time-locked to a set of similar experimental events, are usually averaged prior to analysis to increase their signal/noise relative to other non-time and phase locked EEG activity and non-neural artifacts. Response averaging ignores the fact that response activity may vary widely between trials in both time course and scalp distribution. This temporal and spatial variability may in fact reflect changes in subject performance or in subject state (possibly linked to changes in attention, arousal, task strategy, or other factors). Thus conventional averaging methods may not be suitable for investigating brain dynamics arising from intermittent changes in subject state and/or from complex interactions between task events. Further, response averaging makes possibly unwarranted assumptions about the relationship between EP features and the dynamics of the ongoing EEG. This proposal proposes to extract event-related brain signals from unaveraged spontaneous EEG data related to speech generation.

Analysis of single event-related trial epochs may potentially reveal more information about event-related brain dynamics than simple response averaging, but faces three signal processing challenges: (1) difficulties in identifying and removing artifacts associated with blinks, eye-movements and muscle noise, which are a serious problem for EEG interpretation and analysis; (2) poor signal-to-noise ratio arising from the fact that non-phase locked background EEG activities often are larger than phase-locked response components; (3) trial-to-trial variability in latencies and amplitudes of both event-related responses and endogenous EEG components. However, interest in analysis of single-trial event-related EEG epochs also comes from the realization that filtering out time- and phase-locked activity (by response averaging) isolates only a small subset of the actual event-related brain dynamics of the EEG signals themselves.

Recently, a set of promising analysis and visualization methods for multichannel single-trial EEG records have been developed that may overcome these problems (Jung et al. 1999, 2000a, 2000b, 2001a, 2001b). The analytical method combines a new signal processing technique, Independent Component Analysis (ICA) and time-frequency analysis (see next section).

1.3.1 Independent Component Analysis

Independent Component Analysis (Comon, 1994) was originally proposed to solve the blind source separation problem, to recover N source signals, $s=\{s_1(t), ..., s_{N(t)}\}$, (e.g., different voice, music, or noise sources) after they are linearly mixed by multiplying by A, an unknown matrix, the vector $x = \{x_{1(t)}, ..., x_{N(t)}\} = As$, while assuming as little as possible about the natures of A or the component signals. Specifically, one tries to recover a version, $u = Wx$, of the original sources, s, identical save for scaling and permutation, by finding a square matrix, W, specifying spatial filters that linearly invert the mixing process. Bell and Sejnowski (1995) proposed a simple neural network algorithm that blindly separates mixtures, x, of independent sources, s, using information maximization (infomax). They showed that maximizing the joint entropy, H(y), of the output of a neural processor minimizes the mutual information among the output components, $y_i = g(u_i)$, where $g(u_i)$ is an invertible bounded nonlinearity and $u = Wx$. Recently, Lee et al. (1999) extended the ability of the infomax algorithm to perform blind source separation on linear mixtures of sources having either sub- or super-Gaussian distributions (for further details, see Bell & Sejnowski, 1995; Lee et al., 1999).

The key assumption used in ICA to solve this problem is that the time courses of activation of the sources (or in other cases the spatial weights) are as statistically independent as possible. The use of temporal independence as a separation criterion for EEG data is a novel approach. Use of ICA for blind source separation of EEG or ECG data is based on two plausible premises: (1) EEG or ECG data recorded at multiple sensors are linear sums of temporally independent components arising from spatially fixed, distinct or overlapping sources. (2) The spatial spread of electric current from sources by volume conduction does not involve significant time delays.

In the past few years, it has been reported that ICA successful separated behaviorally related EP components in an auditory detection task (Makeig et al., 1997) and several complex visual evoked EP data sets (Makeig et al., 1999; 2000; 2002). Jung et al. (1998; 1999; 2000a; 2001a, 2001b) demonstrated that ICA can also be used to remove artifacts from EEG recordings prior to averaging, and to separate distinct brain activities from unaveraged single-trial event-related EEG data. Despite its relatively short history, ICA is rapidly becoming a standard technique in multivariate analysis. For further details regarding ICA assumptions underlying EEG and ECG analysis, see (Makeig et al., 1996, 1997; 1999; Jung et al., 1998, 2000a, 2000b, 2001a, 2001b).

For EEG analysis, the rows of the input matrix, x, are EEG/EP signals recorded at different electrodes and the columns are measurements recorded at different time points. ICA finds an `unmixing' matrix, W, that decomposes or linearly unmixes the multi-channel scalp data into a sum of temporally independent and spatially fixed components, $u = Wx$. The rows of the output data matrix, u, are time courses of activation of the ICA components. The columns of the inverse matrix, W^{-1}, give the relative projection strengths of the respective components at each of the scalp sensors. These scalp weights give the scalp topography of each component, and provide evidence for the components' physiological origins (e.g., eye activity projects mainly to frontal sites). The projection of the ith independent component onto the original data channels is given by the outer product of the ith row of the component activation matrix, u, with the ith column of the inverse unmixing matrix, and is in the original channel locations and units (e.g. μV). Thus brain activities of interest accounted for by single or by multiple components can be obtained by projecting selected ICA component(s) back onto the scalp, $x_0 = W^{-1} u_0$, where u_0 is the matrix, u, of activation waveforms with rows representing activations of irrelevant component activation(s) set to zero.

ICA Decomposition into Independent Components

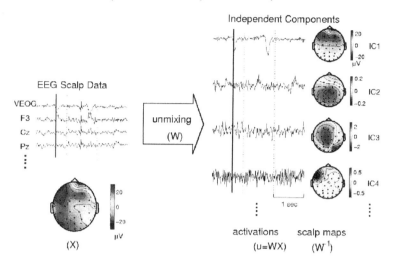

Figure 1. Schematic depiction the decomposition by independent component analysis (ICA) of a 31-channel EEG data set (only 4 channels shown). The left side of the figure shows some of the original data, unmixed or decomposed as the sum of 31 independent components (four shown on the right). Note that ICA separates the contributions of a slow eye blink (top component) and right temporal muscle activity (bottom component) from neural sources (e.g., the two middle maps and time courses). Summing the projected activities of the non-artifactual components (e.g. the two middle components above) will produce artifact 'cleaned' EEG data (in the original μV units) suitable for further processing by any analysis methods (Jung., 2000a,b).

1.3.2 ICA Applied to Unaveraged Single-Trial EPs

In an example, these mathematical tools were used to analyze data from the a visual detection experiment on 28 control subjects plus 22 neurological patients whose EEG data, recorded at 29 scalp and 2 EOG sites, were often heavily contaminated with blink and other eye-movement artifacts. Subjects were asked to respond to a visual stimulus with a button press.

To visualize collections of single-trial EEG records, 'EP image' plots (Jung et al., 1999, 2001b) are useful and often reveal unexpected inter-trial consistencies and variations. 641 single-trial ERP epochs were recorded from an autistic subject, time-locked to onsets of target stimuli (left vertical line). Single-trial event-related EEG epochs recorded at the vertex (Cz) and at a central parietal (Pz) site were plotted as color-coded horizontal traces sorted in order of the subject's reaction time latencies. The EP average of these trials is plotted below the EP image.

ICA, applied to 641 31-channel EEG records, separated out: (1) artifact components arising from blinks or eye movements, whose contributions could be removed from the EEG records by subtracting the component projection from the data; (2) components showing stimulus time-locked potential fluctuations of consistent polarity many or all trials; (3) components showing response-locked activity covarying in latency with subject response times; (4) 'mu-rhythm' components at approximately 10 Hz that decreased in amplitude when the subject responds; (5) other components having prominent alpha band (8-12 Hz) activity, whose inter-trial coherence measuring phase-locking to stimulus onsets, increased significantly after stimulus presentation, even in the absence of any alpha band power increase (middle trace); and (6) other EEG components whose activities were either unaffected by experimental events or were affected in ways not revealed by these measures. This taxonomy could not have been obtained from signal averaging or other conventional frequency-domain approaches.

Better understanding of trial-to-trial changes in brain responses may allow a better understanding of normal human performance in repetitive tasks, and a more detailed study of changes in cognitive dynamics in normal, brain-damaged, diseased, aged, or genetically abnormal individuals. ICA-based analysis also allows investigation of the interaction between phenomena seen in EP records and its origins in the ongoing EEG. Contrary to the common supposition that EPs are brief stereotyped responses elicited by some events and independent of ongoing background EEG activity, many EP features may be generated by ongoing EEG processes.

Decomposition of unaveraged single-trial EEG records allows: (1) removal of pervasive artifacts from single-trial EEG records, making possible analysis of highly contaminated EEG records from clinical populations, (2) identification and segregation of stimulus- and response-locked EEG components, (3) realignment of the time courses of response-locked components to prevent temporal smearing in the average, (4) investigation of temporal and spatial variability between trials, and (5) separation of spatially-overlapping EEG activities that may show a variety of distinct relationships to task events. The ICA-based analysis and visualization tools increase the amount and quality of information in event- or response-related brain signals that can be extracted from event-related EEG data. ICA thus may help researchers to take fuller advantage of what until now has been an only partially-realized strength of event-related paradigms – the ability to examine systematic relationships between single trials within subjects (Jung et al., 2001a, 2001b, Makeig et al, 2002).

1.4 Preliminary Study

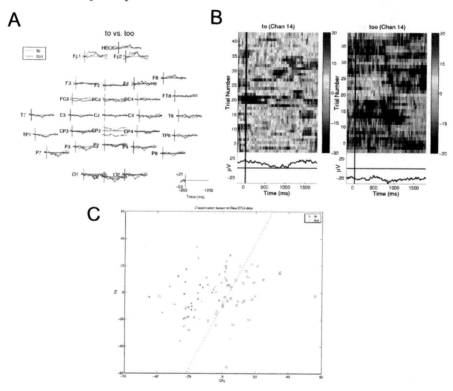

Figure 2. Evoked potentials following the presentation of 'to' and 'too'. (A) Averaged EPs following visual presentations of 'to' (N=43) and 'too' (N=39). (B) EP-image plots of Single-trial EPs recorded at a centro-parietal (CPz) from a normal subject and time-locked to onsets of visual target stimuli (left thin vertical line). (C) A scatter plot of the mean potentials of single-trial EPs at two scalp channels, CPz and T8.

We have collected pilot EEG data from a subject (author EV) who was triggered to utter the sound "tū" based on a visual cue of the randomly presented words "to" or "too". The subject was instructed to speech the word as soon as they saw the word presented. The goal was to review the EEG related to the individual trial utterances and determine if there was a robust set of features related to the utterances that could discriminate the different words. EEG data were collected from 28 scalp electrodes and 1 EOG electrode. Figure 2A (above) shows the averaged 2-sec EPs related to the visual stimulus presentations under the two different conditions. EPs of two conditions differed largely at site CPz. However, the time courses of single-trials did not always resemble those of the averaged EPs (Figure 2B). The trial-to-trial variability made it difficult to distinguish the brain responses under two conditions. To illustrate this difficulty, for each trail, the potentials between 200 and 1400 ms following stimulus presentations at cites CPz and T8 were averaged and plotted in Figure 2C. A single-layer perceptron neural network was trained through supervised methods and found a linear boundary between two clusters ('to' and 'too'). The boundary based on the scalp EEG misclassified 19 of a total of 82 trails.

ICA was applied to the concatenated 2-sec single-trial epochs from the subject and derived 29 temporally independent components. The averaged time courses of activations for each component under two conditions were plotted in Figure 3A. The averaged activations and scalp topographies (not shown) of components 6, 8, and 16 suggested that they together account for the large differences in EPs at cite CPz. The summed projections of the three ICA components to centro-parietal site CPz were plotted in Figure 3B. The trial-to-trial variability was reduced compared to the raw EP-image in Figure 2B since the contributions of other components were removed from (not projected to) the electrodes. Similarly, the activations of ICA components accounted for the differences in EPs at cite T8 were projected and summed at the scalp electrodes. For each trial, the summed projections between 200 and 1400 ms following stimulus presentations at cites CPz and T8 were averaged and plotted in Figure 3C. The clusters were tighter compared to those in Figure 3C. Again, a single-layer perceptron neural network was trained through supervised methods and found a linear boundary between two clusters ('to' and 'too'). The boundary based on the scalp EEG misclassified 11 of a total of 82 trails (13%, compared to 23% in Figure 2C).
In this example, ICA separated the distinct brain processes into different independent components and extracted the signals of interest. Mathematically reducing the interference from the other brain or extra-brain activity generators improved the chance to detect small differences in EEG across conditions.

1.5 Conclusions and Future Research

Our preliminary study suggests that some of the neural activity that precedes a speech utterance may be detectable and have distinguishable components related to the difference meanings of the intended utterance. Further, it appears that the ICA technique can be applied to the individual trials and distinguish with greater than chance reliability which intended meaning the trial was associated with.
Our program is carrying out further research on this technique of distinguishing meaning. We will be examining a variety of word meanings and utterances that are common SR confusables. We are working toward a template matching technique that, based on training data sets, can examine and classify single EP trials in near real-time. We will also work toward development of a limited set lexicon that can be used in an application area such as noisy environments.

1129

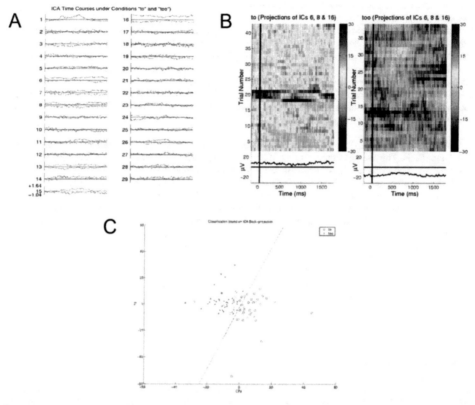

Figure 3: Averaged and single-trial time courses of activations of independent components under two conditions. (A) Averaged activation time course of each independent component following visual presentations of 'to' (red traces) and 'too' (blue traces). (B) Summed projections of the three ICA components to centro-parietal site CPz accounting for large difference in EPs of two conditions ("to", left and "too" right). (C) A scatter plot of the mean summed projections of the ICA components at two scalp channels, CPz and T8.

1.6   References

Bell, A. J., and Sejnowski, T. J. (1995). An information-maximization approach to blind separation and blind deconvolution. Neural Computation, 7(6), 1129-59.

Brown C.M. and Hagoort P. (eds) The Neurocognition of Language (1999) New York, Oxford Press.

Cardoso, J. F., and Laheld, B. H. (1996). Equivariant adaptive source separation. IEEE Transactions on Signal Processing, 44(12), 3017-30.

Cardoso, J.-F., (1998) Multidimensional Independent Component Analysis, Proc. ICASSP 98, 4:1941-4.

Comon, P. (1994). Independent component analysis, a new concept? Signal Processing, 36(3), 287-314.

Ferree, T.C Luu, G.S. Russell and D.M. Tucker (2001) Scalp Electrode Impedance, Infection Risk, and EEG Data Quality. Journal of Clinical Neurophysiology 112/3: 536-544.

Gazzaniga M. (ed.) The Cognitive Neurosciences. (1995) Cambridge Mass, MIT Press.

Jung T-P, Humphries C, Lee T-W, Makeig S, McKeown MJ, Iragui V, Sejnowski TJ. (1998) Extended ICA removes artifacts from electroencephalographic Data, In: Advances in Neural Information Processing Systems, 10:894-900.

Jung, T-P, Makeig, S, Stensmo, M, and Sejnowski, TJ, "Estimating alertness from the EEG power spectrum," IEEE Trans Biomed Eng, 44(1), 60-69, 1997.

Jung T-P, Makeig S, Westerfield M, Townsend J, Courchesne E, and Sejnowski TJ, (1999). Analyzing and visualizing single-trial event-related potentials, In: Advances in Neural Information Processing Systems, 11:118-24.

Jung, T-P, Makeig, S, Humphries C, Lee T-W, McKeown MJ, Iragui V, Sejnowski TJ. (2000a) Removing electroencephalographic artifacts by blind source separation, Psychophysiology, 37:163-78.

Jung, T-P, Makeig, S, Westerfield, M, Townsend, J, Courchesne, E and Sejnowski, TJ. (2000b) Removal of eye activity artifacts from visual event-related potentials in normal and clinical subjects, Clin. Neurophy., 111(10):1745-58.

Jung T-P, Makeig S, McKeown M.J., Bell, A.J. , Lee T-W, and Sejnowski TJ. (2001a) Imaging Brain Dynamics Using Independent Component Analysis , Proceedings of the IEEE 89(7):1107-22.

Jung T-P, Makeig S, Westerfield W, Townsend J, Courchesne E, and Sejnowski TJ (2001b) Analysis and visualization of single-trial event-related potentials, Human Brain Mapping, 14(3):166-85.

Kutas M, Hillyard SA Event-related brain potentials to semantically inappropriate and surprisingly large words Biol Psychol. 1980 Sep;11(2):99-116..

Lee, T.-W., Girolami, M., Sejnowski, T.J. (1999) Independent component analysis using an extended infomax algorithm for mixed sub-Gaussian and super-Gaussian sources, Neural Computation 11(2): 609-633.

Lee, T. W., Girolami, M., Bell, A. J., and Sejnowski, T. J. (2000). A unifying information-theoretic framework for independent component analysis. Computers & Mathematics with Applications, 39(11), 1-21.

Makeig S, Bell AJ, Jung T-P, Sejnowski TJ. (1996) Independent Component Analysis of Electroencephalographic Data, In: Advances in Neural Information Processing Systems 8:145-51.

Makeig S, Jung T-P, Bell AJ, Ghahremani D, Sejnowski TJ. (1997) Blind separation of event-related brain responses into independent components, Proc. Natl. Acad. Sci. USA, 94:10979-84.

Makeig, S., Westerfield, M., Jung, T.-P., Covington, J., Townsend, J., Sejnowski, T. J., and Courchesne, E. (1999) Independent components of the late positive event-related potential in a visual spatial attention task. Journal of Neuroscience, 19(7), 2665-80.

York D.H. et. Al. (1981) Computer extracted Cerebral Activity Preceding Speech. IEEE Transactions on Biomedical Engineering Vol BME-28 (8) p. 593

Augmented Cognition for Bioinformatics Problem Solving

Olga Anna Kuchar and Jorge Reyes-Spindola

Pacific Northwest National Laboratory
P.O. Box 999, 902 Battelle Blvd.
Richland, WA, 99352
{olga.kuchar, jorge.reyes.spindola}@pnl.gov

Michel Benaroch

Center for Creation and Management of Digital
Ventures
Martin J. Whitman School of Management Syracuse
University, Syracuse, NY, 13244
mbenaroc@syr.edu

Abstract

We describe a new computational cognitive model that has been developed for solving complex problems in bioinformatics. This model addresses bottlenecks in the information processing stages inherent in the bioinformatics domain due to the complex nature of both the volume and type of data and the knowledge required in solving these problems. There is a restriction on the amount of mental tasks a bioinformatician can handle when solving biology problems. Bioinformaticians are overwhelmed at the amount of fluctuating knowledge, data, and tools available in solving problems, but to create an intelligent system to aid in this problem-solving task is difficult because of the constant flux of data and knowledge; thus, bioinformatics poses challenges to intelligent systems and a new model needs to be created to handle such problem-solving issues. To create a problem-solving system for such an environment, one needs to consider the scientists and their environment, in order to determine how humans can function in such conditions. This paper describes our experiences in developing a complex cognitive system to aid biologists in knowledge discovery. We describe the problem domain, evolution of our cognitive model, the model itself, how this model relates to current literature, and summarize our ongoing research efforts.

1 Introduction

Bioinformatics is the application of computational techniques to understand and organize the information associated with biological macromolecules. The aims of bioinformatics are three-fold: gather/organize data; analyze data; and interpret data. Gathering and organizing data accomplishes two key criteria. First, it allows researchers to submit new data as it is produced. Secondly, it allows researchers to access existing information. Even though the task of gathering and organizing data is important, the information stored in these databases is relatively useless until analyzed. Thirdly, analyzing data is another aim of bioinformatics; to aid in this data analysis, many tools and resources have been developed that focus on particular aspects of the analysis process. The development of such tools requires extensive knowledge of biology and computational theory. Interpretation of this analysis in a biologically meaningful manner is key to uncovering common principles that apply across many systems, and highlight features that are unique to some. Traditionally, biological studies examined individual systems in detail and compared them with a few systems that are related. When dealing with today's bioinformatics, a need to conduct global analyses of all the available data is prudent. The reader is directed to Luscombe et al. for a detailed description of bioinformatics and its domain (Luscombe, Greenbaum, & Gerstein, 2001).

Over the last several years, advances in biology and the equipment available have resulted in massive amounts of data dealing in the terabytes range; but data does not equate to knowledge – data must be processed and fused in order to discover new knowledge. Humans are very adept at taking heterogeneous data and discovering new knowledge – but even humans cannot easily discover new information in the sea of data that they are facing on a daily basis. For example, biological data is increasing at an unprecedented rate (Reichhardt, 1999). As of February 15, 2004, GenBank contains 44,575,745,176 bases from 40,604,319 reported sequences ("Growth of GenBank", 2005). The amount of information stored in databases around the world continues to grow; for example, the SWISS-PROT database increased its entries by 1% within two weeks to total 170,140 sequence entries ("SWISS-PROT Statistics", 2005). Add to this raw data the knowledge that has been published from scientists' experiments and projects over the years, and a bioinformatician has an enormous quantity and variety of information to harvest and fuse.

The most profitable research in bioinformatics often results from integrating multiple sources of data and placing their pieces of information into context with each other. Unfortunately, it is not always straightforward to access and cross-reference these sources of information because of differences in nomenclature and file formats. The bioinformatician spends most of his/her time in data searching and retrieval, followed by tool applications to find the applicability of these disparate pieces of information to each other and their problem at hand. This is usually a long and tedious process, lasting anywhere between a few days to several months of work (depending on the problem complexity). Thus, the human-processing bottlenecks are: memory (where to go and search for information for its is all distributed); learning (how to relate the disparate pieces of information together and what is the bioinformatician still missing); comprehension (understanding how to use the databases, their outputs, and other tools for their trade); and decision making (what does all this information mean).

So, bioinformatics poses a challenge to intelligent systems because of its constant flux of data and knowledge. There are many different aspects to this problem:

- Solving complex problems involves both implicit and explicit knowledge.
- Not all information is known to the expert at the time of solving the problem. The expert needs to search for additional information in order to solve the higher problem. This is not an easy task based on the massive data that scientists need to sift through.
- Solving these complex problems involves solving sub-problems, each of these being complex in nature and solution.
- Being able to fuse information at a sub-problem level does not necessarily solve the higher problem – knowledge compression must occur as we progress up the hierarchy of solution space.

In our opinion, current cognitive architectures are not robust enough to handle such real environments. In bioinformatics, a scientist is faced with such problems on a daily basis. Many tools have been developed to help the scientists with parts of their task, but no architecture has been designed to help the scientists in knowledge discovery – having the scientist spend more cognitive time in understanding solutions than in digging for information. This paper describes our experiences in developing a new cognitive system to aid biologists in knowledge discovery. The remainder of this paper is divided as follows: description of the problem that we are using for our research; description of the new cognitive model; research issues in implementing this new model; how this new model relates to other cognitive architectures; and a summary of our current research and future work.

2 Bioinformatics

Our observation of bioinformatics cognitive functions centred on the processes involved in analyzing a biological experiment that studied the genetic effects of an endotoxin on mice brains after an induced cerebral stroke. The data output of the experiment was in the form of gene expression data originating from an array of DNA snippets known as a gene chip array. These gene expression experiments quantify the expression levels of individual genes and measure the amount of mRNA or protein products that are produced by a cell. In this particular experiment, the expression levels of mice treated with the endotoxin were compared against a control group treated with a saline solution. DNA samples were then taken at different time points for both groups before and after the induced stroke. It was observed that the mice treated with the endotoxin were able to resist the aftereffects of the cerebral stroke compared to the mice treated with the saline solution. The aim of the experiment is then to investigate what were the genes involved in the stroke resistance and what role did they play in the physiology of the endotoxin-treated mice.

This problem is fairly typical of the high-throughput experimental approach that produces large amounts of results. These experiments provide a "snapshot" of cellular events that can eventually be integrated into a dynamic view of cellular processes through time. In the analysis of this data, gene expression patterns are typically clustered into groups that define different behaviours. At the highest level, biologists want to understand why different types of genes are grouped together and they want to find out what proteins get expressed or what chain reactions of gene-protein-gene they unleash. However, at the most basic level, it is a significant obstacle just to know the identities of proteins in a group, let alone why they are grouped together. This basic protein list is sometimes referred to as the "Molecular Parts List". The process of identifying protein members is referred to as "annotation". Annotation is an important problem for scientists and is the example area that we will use to describe the development of our cognitive model.

To explore the biological meaning of gene expression array experiments, biologists need to address several key sub-problems. First, biologists need to determine what kinds of expression patterns are within their experiment by applying clustering algorithms to the gene expression data. Clustering approaches have been widely applied to this type of analysis (Eisen, Spellman, Brown, & Botstein, 1998; Halkidi, Batistakis, & Vazirgiannis, 2001; Luo, Tang, & Khan, 2003); however, clustering algorithms do not directly provide statistical confidence for clustered expression patterns. As a result, some biologists may triage the genes based on their statistical significance in differential expressions and confirm consistent expression patterns within replicates (Ross et al., 2000; Scherf et al., 2000). Next, biologists may need to determine the aliases of the genes and proteins represented in the gene array. One of the problems is that there are multiple protein names and multiple protein database records that need to be searched and correlated. Once the actors have been identified, other features need to be determined for each one, such as role assignment, domain assignment, superfamily analysis, and ortholog analysis. To determine these, different tools are applied and several databases need to be accessed, such as Similarity Box tool (Sofia, Chen, Hetzler, Reyes-Spindola, & Miller, 2001), and Pfam or SMART databases. Then, key features need to be extracted for each gene. Some of these features are conserved operon patterns, promoter elements, associated transcription factors and metabolic pathway assignments. Once the information is gathered and correlated on each gene, a biologist will start analyzing each group of genes and what are their common attributes. Further data harvesting is required to determine the biochemistry and molecular biology of the involved genes and the associated proteins. This involves both data and literature mining. Finally, a biologist may analyze what network predictions can be made about these genes (Park, 2002). Research in information theory and Bayesian methods assists biologists in this analysis. Overall, biologists must mine and fuse enormous amounts of distributed data by using many different tools to aid them in knowledge discovery. Typically there are tens of thousands of genes in one microarray experiment. The reader is directed to more detailed information ("Dipping into DNA Chips", 1999).

With respect to annotation, proteins fall roughly into three categories: (1) approximately one in ten can be fully and completely annotated with a high level of certainty; (2) a small fraction of proteins are cryptic and cannot be reliably attached to any information; and (3) the majority can be associated with only partial information. Current annotation systems tend to assign misleading names that are partially right but just as equally wrong. The assignments are static and do not provide any basis for evaluating their meaning or quality without repeating multiple bioinformatic searches. Annotation is still an issue for structural data as well, although the biology community has attempted to form a consensus as to what annotation of a structure is currently required.

We worked closely with several biologists and bioinformaticians at Pacific Northwest National Laboratory to determine how they each solve gene expression problems. We interviewed each of our colleagues separately and elicited knowledge from them. During our elicitation process, we created decision trees so that we could capture the expert's strategies and knowledge about how they decompose the problem into manageable parts, and then how they solved those sub-parts. We then transformed our decision trees into verbal protocols. Upon examination of our elicited knowledge, the methodology for solving the gene expression problem can be abstracted to the following:

1. Obtain an initial gene classification by performing clustering.
2. Determine common features of each group.
3. Look for over-represented transcription factors.
4. Search for particular features of transcription factors.
5. Based on these searches, identify important features and gene linkages.
6. Determine particular features of linked genes.

Rather than building an expert system to solve this sub-problem, we built a system that can be used in a dynamic environment and that uses generic methods to solve many different problems. This led us to develop a cognitive model that mimics human higher-level problem solving for challenging problems in complex and dynamic domains.

3   Cognitive Model

A cognitive architecture specifies the underlying structure of an intelligent system that is constant over time and domain. A review of the recent flow of research in cognitive architectures can be found in (Langley & Laird, 2002). Current computational cognitive architectures are not robust enough to handle such dynamic environments as described in section 2. In bioinformatics, a scientist deals with such problems on a daily basis, but solving these problems requires anywhere between a few days to several months. Many tools have been developed to help the scientists with parts of their task, but no architecture has been designed to integrate the cognitive aspects of complex problem solving and associated tools to help the scientists in knowledge discovery *i.e.,* have the scientist spend more cognitive time in understanding solutions than in digging for information. This is the focus of our research work. A human solves problems using the five senses and the brain. The basic functions and operations of the brain remain constant, but the knowledge that is accessed changes based on the problem that is being solved. The brain accesses different knowledge stored within its capacity. In this section, we describe our new cognitive model that is based on this theory. Based on our observations of higher-level problem solving, humans address problem solving using some key knowledge components. We will explain these components in terms of our problem domain.

3.1 Goal Knowledge

Biologists know something about their goal – what it means to have solved the current problem. For example, biologists know that for gene identification, a solution must include a gene name, its chromosome, position within the genome (if available), etc. We define this knowledge as *Goal Knowledge*. Goal knowledge represents a definition of what a solution to a particular problem may contain. For example, a solution to a gene expression problem may contain information about genes, proteins produced, transcription factors, etc. This sort of information can be captured as an entity-relationship diagram that depicts what a person knows about the goal that they are trying to solve. This needs to be in a form that can be expanded by a user (a non-programmer) as new information about solving such a problem is evolving. For example, in future gene expression problems, a biologist may know that the goal of solving this problem would involve a gene's position within the chromosome and they would need to add it to the goal knowledge. As humans, the way we solve certain problems today is probably not the way we solved those problems a decade ago. Since technology is growing and our knowledge and abilities are increasing, our methodologies are also changing and providing such a capability is important. We need to represent such dynamics in an easy fashion so that a user is not dependent on programmers to capture these changes.

3.2 Strategic Knowledge

While solving a problem, biologists have different strategies or heuristics that they follow to get closer to a potential solution. These strategies are not always perfect or lead to the desired solution, but they do recognize a path towards solving the problem. We define this as Strategic Knowledge. Strategic knowledge is knowledge about strategies or problem-solving paths that we have learned during our life. Strategies can differ from person to person, based on their experiences, and some strategies are common. We need the ability to have a user update strategies in our model based on new paths that the user has learned over time. In order to do this, strategic knowledge needs to be separated from domain knowledge so that we can use strategies to a greater advantage. Strategic knowledge can be represented in formal computational logic and, in this form, can provide many advantages. First, strategic knowledge can be added by a user without re-writing all the strategies already in the strategic knowledge base. Secondly, logic engines known as formal theorem provers can provide a consistency check for the knowledge base and can identify any conflicts within it. Thirdly, research in natural languages and translators between English and Computational Logic are being developed (Pease & Murray, 2003). This would ease in having the user update the system and thus keeping it current with new strategies that the biologist may discover work well for their problems.

3.3 Domain Knowledge

Of course, biologists have Domain Knowledge – knowledge about their field (this includes knowledge about tools, location of information, and knowledge deemed to be truthful). One of the key aspects dominant in this field is the understanding of missing knowledge from the Domain Knowledge. This is a key difference from other intelligent models – we assume that not all information is within the cognitive model during the time of processing a solution. As humans, when we do not have all the information in our brains, we know of ways to find the missing information and assimilate it into our working model. We call this New Knowledge – information we find and knowledge created while solving a problem.

Domain knowledge represents the knowledge about a particular problem area(s). Domain knowledge is contained in distributed knowledge bases. The usage of knowledge bases for problem solving has been growing over the past few years. In biology, it is being furthered not just by having vast amounts of data but by progress in ontologies for biology. This work will allow for a richer knowledge representation and manipulation. These techniques are

1135

important for creating an infrastructure that is compatible with computational approaches and is also the key to adding new knowledge into the model.

3.4 From Theory to Model

An overview of the interplay of these types of knowledge is depicted in Figure 1.

Figure 1: A conceptual overview of our cognitive system.

To create a computational cognitive model of Figure 1, we reverted to nature for guidance. A human brain contains billions of neurons, each possessing thousands of synapses connecting to other neurons. When a human is solving a problem, neurons and their pathways are triggered. If we could take a picture of the brain and highlight these neurons and paths that we "used" during problem solving, the result would be a connected graph. No matter what problem a human is solving, the final by-product is a mathematical graph (Weisstein 2002). Thus, the inner-workings of our cognitive model should function on connected graphs. This part of the model has one main requirement – it needs to be generic. Like a human brain, it needs to function no matter what the problem that it is trying to solve. This is represented as the "reasoning engine" denoted in Figure 2.

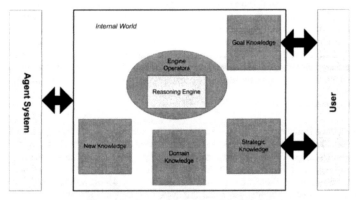

Figure 2: Computational Cognitive Model.

Now, as humans solve problems, there are triggers that fire certain neurons. If we could time-step and take pictures of the brain as it is solving a problem, the connected graph grows in size. In our model, we require "engine operators" to manipulate the graph. These operators are rules for building graphs, such as adding a node or a sub-graph to the current graph. This layer in our model could be viewed as generic to any problem – engine operators function independently of the problem being solved. The knowledge layer of our cognitive model is domain specific. Again, different parts of the brain are triggered for different domain problems, for these areas "store" information about that particular domain or sub-problem. A block-level view of our computational cognitive model is depicted in Figure 2. The different types of knowledge are more computationally complex. Our view of the computations for each of the knowledge is described in the remainder of this section.

New knowledge is knowledge that has been currently "found" to exist; it is not part of our domain knowledge at the time of solving a problem. Humans obtain new knowledge from the senses; since computers do not have senses, a multi-agent system must mine and fuse data to create new knowledge (see left side of Figure 2). The agent tasks are similar to human tasks of finding information and using the correct tools to extract knowledge that is required while solving a complex problem. Ongoing research in problem-solving methods, data mining, and distributed agents will aid in this aspect of the model. One of the key questions is "when does new knowledge become domain knowledge?" Computationally, how does one merge new knowledge into the existing domain knowledge structure? This is similar to humans – you may read words in a book, but to truly understand it, you must fit that information into your current knowledge structure. We are currently developing this aspect of the model.

This section has provided an overview of our new cognitive model. There are some research challenges in developing such a model, and we touch upon some of these challenges in the next section.

4 Research Issues

Even though the cognitive model seeks to address bottlenecks and limitations in cognition, achieving such a model computationally still challenges many fields in computer science. One of the foreseeable hurdles in the generalization of the cognitive model's applicability is the lack of ontologies in most areas of knowledge. Typically, the systematization of science vocabularies and their internal relationships has not been hitherto a widespread activity. As an example, the authors were requested to investigate the applications of the model to problems in atmospheric science. Atmospheric science is another discipline that is currently hampered by problems similar to problems in biology: enormous and constantly-increasing amounts of data supplied by a wide array of instruments. In addition, the constantly changing nature of that field of study makes it difficult to replicate experiments. After interviews with several field practitioners, the authors were faced with the fact that while the physics fundamentals are well-established, there is a lack of meta-knowledge of the sub-field (atmospheric science). The same can be said for non-scientific disciplines such as intelligence analysis. Fortunately, the current drive for the automation of knowledge acquisition and processing is fostering the creation of ontologies. Because an important part of the implementation of our cognitive model is the existence of an adequate ontology, the biological sciences community again provides a significant advance in that matter through the creation of the Gene Ontologies (Gene Ontology Consortium, 2005) which helps establish the knowledge relationships amongst the objects of study.

Another hurdle will be the creation of a flexible and user-programmable reasoning engine. Up to this point, some of the problems of expert systems have been their lack of flexibility and difficulty in reconfiguring the inference engine components. Our reasoning engine is currently programmed using conventional logic programming formalisms. The authors believe that separating strategies, goals and domain knowledge was a step in the right direction. As it was mentioned before, strategies can be expressed in formal computational logic which provides many advantages. However, the point at which the user can input new strategies and rules using natural language is still quite a distance away.

To provide a working proof of the assumptions presented in this paper, the authors have been developing a software prototype of the cognitive model. This software system will take a goal-based description of our problem and through input of user-defined strategies, will attempt to provide a path towards a possible solution. The system, as it stands now, is focused on the solution of the gene expression problem insofar as the domain, goal, and strategic knowledge bases are concerned. Yet the internal workings of the reasoning engine will be knowledge-independent since they are graph-based.

The system is being implemented using a combination of the Java and Prolog languages. Java was chosen for its availability and ease of use as well as the existence of several open-source mathematical graph libraries and graph visualization libraries. The SWI-Prolog implementation ("SWI-Prolog", 2005) was chosen because of its robustness, size, and its inclusion of a well-developed Java-Prolog API library. Calls can be made directly to Prolog queries and if necessary, Prolog can call Java routines as well.

We expect that the cognitive model will help address the problem of mapping and navigating massive information landscapes through the use of our novel reasoning process together with the use of the autonomous agent system. The ultimate purpose of the computational model will be to aid the researcher in solving a complicated question by taking over the onerous tasks of retrieving the appropriate information, processing it and fusing it, thereby freeing the scientist to engage in more analytical work. The question could be raised of whether the authors are building another expert system, however, we believe that the inclusion of the New Knowledge aspect of the model separates this approach from the rest of the pack. By recognizing the fact that all information is dynamic and it should be updated frequently the model takes one step further into injecting intelligence in a well-understood process. The question will linger, though, of at what point the retrieved information becomes part of the established domain knowledge. That is something that will have to be addressed by the appropriate knowledge curators.

This section has provided some insight into the research challenges for building such a cognitive model. Even with these challenges, this research is extending the computer science field. The next section provides a literature overview of how our research relates to the field.

5   Literature Review

There is a great need for intelligent systems in biology. The recent data explosion in this field yielded massive and complex information that requires processing and analyzing it to determine new biological knowledge. As noted in (Altman 2001), certain key features of biological data make intelligent systems critical for their analysis:

- There is a need for robust analysis methods since biological data is normally collected with a relatively low signal-to-noise ratio.
- There is a need for statistical and probabilistic models since bioinformatics is still in its infancy.
- There is a need for complex knowledge representations since we know more about biology in a qualitative rather than in a quantitative sense.
- There is a need for cross-scale data integration methods since the data sources operate at multiple scales that are tightly linked.
- There is a need for data integration methods since biological data is distributed.

Reviewing the research work in biology and intelligent systems, we have seen how intelligence-driven applications have aided biologists, spanning the landscape from extracting weak trends in data to extracting high-quality information-level summary knowledge from scientific databases. The research solutions to some of the challenges facing biology have involved multiple disciplines including data mining, ontologies, knowledge management, mathematics, computer graphics, human-computer interaction, and artificial intelligence. What is missing from this landscape is a model that encapsulates how biologists think and work in their current environment to discover new knowledge. There is a need to support biologists from a cognitive perspective.

The reasoning process underlying our cognitive model relates to Newell's problem-space paradigm of intelligence (Newell 1990). As seen in Figure 3, this paradigm considers problem-solving to be a process that involves two searches – problem search and knowledge search.

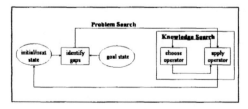

Figure 3: Newell's problem-space paradigm of intelligence.

Problem search assumes the existence of a problem space for the task solved. This implies the availability of: a goal state g that describes what needs to be known at the end of problem solving; an initial state i that describes what is known at the start of problem solving; and a set of operators P that can be used to generate new states in the problem space. As the outer loop in Figure 3 indicates, starting from state i, problem search applies one operator in P at a time to generate a new state in the problem space, with the hope that this state will match (or at least be closer to) the goal state g. Thus, problem search uses operators in P to generate states in the problem space that form a path from state i to state g.

Knowledge search occurs in the inner loop of problem search, as seen in Figure 3. It guides problem search by using search control knowledge to select the operator in P that should be applied to generate the next state on the path from i to g. Knowledge search has usually associated with it a fixed space whose structure (connectivity and how paths through it are described) pre-exists. That is, while problem search occurs in a space that is constructed dynamically, knowledge search typically occurs in a fixed pre-existing structure.

We expand on Newell's paradigm to involve new information that can be gathered from the external world. Thus, we are proposing an extension or elaboration on the knowledge search to include information that can be gathered. Knowledge search still occurs in a pre-existing structure, but it is not fixed.

Research in cognitive models has also been growing. This model relates well to current thoughts on cognitive models of the brain. For example (Wang & Wang, 2002), the authors attempt to develop functional and cognitive models of the brain. The functional model relates to our model in two ways. First, the functional model contains NI-OS (the thinking engine) and NI-App (a set of acquired life applications). Secondly, the functional model has input sensors to gather new knowledge about the external world. As the reader can determine, these functional model items are related to our thinking engine, strategic knowledge, and multi-agent system aspects.

During the last three decades, research on cognitive architectures has been growing and has provided a variety of architectural classes that make different assumptions about the representation, organization, utilization, and acquisition of knowledge. We have surveyed this landscape to find a cognitive architecture that addresses several key features:

- A cognitive architecture that can be updated by a user (non-programmer) to keep the system current in both problem-solving techniques and tools.
- A cognitive architecture that can function in both "fuzzy" defined and dynamic environments.
- A cognitive architecture that is robust to solve many different problems without the need to reprogram the entire system.

Current architectures were not able to handle all of these requirements. This lead us to develop a different cognitive model to base our cognitive architecture upon. For example, Soar (Laird, Newell, & Rosenbloom, 1987) is a cognitive architecture in which all long-term knowledge takes the form of production rules, which are in turn organized in terms of operators associated with problem spaces. Our approach differs – we explicitly define our goal knowledge by describing the structure that our model seeks to produce as an answer. This allows the user to update the goal knowledge (or how should a solution look) and thus allows for expansion of the system as the user finds new ways of solving a problem. Another example is ICARUS (Shapiro & Langley, 1999). ICARUS focuses on reactive execution of existing skills rather than on problem-space search. We feel that the problem-space search is more robust in how humans solve their problems and will lead to a model (and thus architecture) that can be applied to many domains. Another example is ACT-R (Anderson & Lebiere, 1998). ACT-R has two distinct memories: a declarative memory that encodes knowledge about facts and events; and another memory that stores procedural knowledge in the form of production rules. ACT-R mixes strategic and domain knowledge together into its production rules. We feel that strategic knowledge needs to be independent of domain knowledge in order to allow the system to expand as the user discovers new paths for finding a solution to a problem. If domain knowledge is mixed with strategic knowledge, then updating the knowledge in a system is a tedious programming task that involves domain experts and programmers, with usually a massive revamping of code. Overall, intelligent systems have been considered as "black boxes" by users. Users feel that they do not understand how these intelligent systems find answers, and users feel powerless about what information (or intelligence) an intelligent system is using. Biologists have a need for a system that they can manipulate and understand. We want to address this need through the robustness of our cognitive architecture.

6 Conclusions

In this paper, we provided an overview of our new cognitive model to aid biologists in knowledge discovery. Bioinformaticians are overwhelmed at the amount of fluctuating knowledge, data, and tools available in solving their problems. Problem-solving some tasks in biology can take anywhere between a few hours to several months; therefore, there is a need for a system to overcome several computational bottlenecks and thus, augment the biologists performance. We propose a new cognitive model to address this need by providing many key features: the model incorporates any domain since the knowledge bases are separated from the thinking engine; the thinking engine creates a directed graph that gradually grows links between nodes depicting domain objects and relations, but needs no understanding of the domain that it is acting upon; current knowledge is added to the model using a multi-agent system to populate the new knowledge base, thus leading the model to be more dynamic in information content; and new strategies on solving the higher-level problem can be incorporated easily, since domain knowledge and strategic knowledge are not incorporated as one knowledge base within the system.

There are many challenges to information and knowledge processing in biology. In our opinion, current computational cognitive architectures are not robust enough to handle such dynamic environments as biology. Many tools have been developed, but no architecture has been designed to integrate current technology, data, and knowledge into a complex problem-solving environment that can aid biologists in knowledge discovery.

7 Acknowledgements

The research described in this paper was conducted under the LDRD Program at the Pacific Northwest National Laboratory, a multi-program national laboratory operated by Battelle for the U.S. Department of Energy under Contract DE-AC06-76RL01830.

8 References

Altman, R.B. (2001) Challenges for Intelligent Systems in Biology, *IEEE Intelligent Systems*, 16 (6), 14-18.

Anderson, J.R., and Lebiere, C. (1998) The Atomic Components of Thought, Lawrence Erlbaum Associates, Mahwah, NJ.

Eisen, M.B., Spellman, P.T., Brown, P.O, & Botstein, D. (1998) Cluster analysis and display of genome-wide expression patterns, *Proceedings of the National Academy of Sciences*, 95(25), 14863-14868.

Gene Ontology Consortium (2005) Gene Ontology. Retrieved March 4, 2005 from http://www.geneontology.org

Halkidi, M., Batistakis, Y. & Vazirgiannis, M. (2001) Clustering Algorithms and Validity Measures, *Thirteenth International Conference on Scientific and Statistical Database Management*, 3-22.

Laird, J.E., Newell, A., & Rosenbloom, P.S. (1987) Soar: An Architecture for General Intelligence, *Artifical Intelligence*, 33, 1-64.

Langley, P., & Laird, J.E. (2002) Cognitive Architectures: Research Issues and Challenges, (Technical Report) Institute for the Study of Learning and Expertise, Palo Alto, CA.

Luo, F., Tang, K., & Khan, L. (2003) Hierarchical Clustering of Gene Expression Data, *Third IEEE Symposium on BioInformatics and BioEngineering*, 328-335.

Luscombe, N.M., Greenbaum, D., & Gerstein, M. (2001) "What is bioinformatics? An Introduction and Overview", http://bioinfo.mbb.yale.edu/~nick/bioinformatics

Newell, A. (1990) Unified Theories of Cognition, Harvard Press, Boston, MA.

Park, J.H. (2002) Network Biology: Data Mining Biological Networks, *IEEE Intelligent Systems*, 17 (3), 68-70.

Pease, A., & Murray, W. (2003) An English to Logic Translator for Ontology-based Knowledge Representation Languages, *IEEE International Conference on Natural Language Processing and Knowledge Engineering*, 777-783.

Reichhardt, T. (1999) "It's sink or swim as a tidal wave of data approaches." *Nature*, 399 (6736), pp. 517-520.

Ross, D.T., Scherf, U., Eisen, M.B., Perou, C.M., Rees, C., Spellman, P., Iyer, V., Jeffrey, S.S., Van de Rijn, M., Waltham, M., Pergamenschikov, A., Lee, J.C., Lashkari, D., Shalon, D., Myers, T.G., Weinstein, J.N., Botstein, D., & Brown, P.O. (2000) Systematic variation in gene expression patterns in human cancer cell lines, *Nature Genetics*, 24(3), 227-235.

Scherf, U., Ross, D.T., Waltham, M., Smith, L.H., Lee, J.K., Kohn, K.W., Reinhold, W.C., Myers, T.G., Andrews, D.T., Scudiero, D.A., Eisen, M.B., Sausville, E.A., Pommier, Y., Botstein, D., Brown, P.O., & Weinstein, J.N. (2000) A cDNA microarray gene expression database for the molecular pharmacology of cancer, *Nature Genetics*, 24 (3), 236-244.

Shapiro, D., & Langley, P. (1999) Controlling Physical Agents Through Reactive Logic Programming, *Proceedings of the Third international Conference on Autonomous Agents*, 386-387.

Sofia, H., Chen, G., Hetzler, B.G., Reyes-Spindola, J.F., & Miller, N.E. (2001) Radical SAM, a novel protein superfamily linking unresolved steps in familiar biosynthetic pathways with radical mechanisms: functional characterization using new analysis and information visualization methods, *Nucleic Acids Research*, 29(5), 1097-106.

unknown, (1999) Dipping into DNA Chips, NetWatch, *Science*, August 6, 285 (5429), 799d.

unknown, (2005) Growth of GenBank. Retrieved February 28, 2005 from http://www.ncbi.nih.gov/Genbank/genbankstats.html

unknown, (2005) SWI-Prolog. Retrieved March 1, 2005 from http://www.swi-prolog.org

unknown, (2005) SWISS-PROT Protein Knowledgebase Release 46.1 Statistics. Retrieved February 28, 2005 from http://www.expasy.org/sprot/relnotes/relstat.html

Wang, Y., & Wang, Y. (2002) Cognitive Models of the Brain, *IEEE International Conference on Cognitive Informatics*, 259-269.

Weisstein. E.W. (2002) Connected Graph, From MathWorld--A Wolfram Web Resource. Retrieved March 2, 2005 from http://mathworld.wolfram.com/ConnectedGraph.html

Section 5
Cognitive State Sensors

Chapter 23

Augmented Cognition and its Influence on Decision Making

Augmenting Decision Making - What GOES?

Robert M Taylor

Defence Science and Technology Laboratory
Farnborough, Hants, GU14 0LX, UK
rmtaylor@dstl.gov.uk

Abstract

Under the futuristic theme of AUGCOG New Directions, this 1st Augmented Cognition International Conference session focuses on augmented cognition and its influence on decision making. The session features research on the impact of augmented cognition technologies on the field of decision making, in particular the augmentation of decision making by advanced computing and decision systems technologies. The intention is to promote consideration of the relationship between research on augmented cognition and on computational technologies for decision systems. This overview seeks to provide useful perspectives for structuring evaluation in particular with reference to analysis of goals, options and effects for improving status (What GOES?).

1 Problem Solving and Decision Making

The primary challenge to be addressed is how to improve the process of coping with complexity in human decision making and problem solving with the difficulty arising from limitations on human cognition (Simon, 1986). Broadbent (1971), in considering selective attention and motivation, noted that the effects of stress and arousal are on filtering of sources of input information, rather than on central organising of responses (pigeon-holing, categorising). He noted that the Yerkes-Dobson Law states that there is an optimum level of motivation for tasks involving discrimination (choosing between alternative stimuli), the level being lower for harder discriminations i.e. rises in motivation increase the size of response bias in the absence of a stimulus. For decision making, a similiar form of relationship can be proposed between complexity and performance. Poor decision making arises both when simple problems receive low cognitive effort/attentional engagement, and when complexity exceeds cognitive capabilities.

Humans exhibit different forms of bias in human judgement during the decision making process. Silverman (1992) identifies the following sources of error bias:

- Information acquisition - availability, base rate, concreteness, desire for self-fullfilling prophesies, illusion of correlation, selective perception/confirmation.
- Information processing - adjustment, anchoring, conservatism, data solution, expectations, habit, inference, law of small numbers, overconfidence, redundancy, regression effects, representativeness, rule strength.
- Intended output - fact value confusion, illusion of control, wishful thinking.
- Feedback - hindsight, order of effects, reference effect, spurious cues.

Recent evidence has shown that impulsive, sub-optimal decision-making may be linked to decrements in performance of executive control functions of working memory responsible for co-ordination of information in planning and making decisions (Hinson et al., 2003). Increased impulsiveness arises from stress in risky situations, or when considering more information than normal with increasing numbers of options to be evaluated.

Fundamentally, decision making concerns the process of choosing, or "making up one's mind" between alternative choices (*decision* "a conclusion or resolution reached after consideration", Oxford English Dictionary). In purposive behaviour, decisions are choices about what actions should be taken to achieve desired effects. Decision making theory and research typically is about human judgements in uncertainty. In real-world problem solving, decisions often involve inferring causal relationships and perceiving organisation in contexts that are complex, ill-structured and unpredictably variable. The need for decision arises from a causal break in reasoning. Typically, decision is needed when there is incomplete or inadequate knowledge to identify the necessary and sufficient pre-conditions for an observed or intended effect i.e. no observable direct causal link (Hollnagel, 2004). Simon (1986) notes that limits on human rationality, coined "bounded rationality", mean that in order to cope with complexity, people need "to cut problems down to size". He identified that what first is needed is assistance in understanding what brings (and should bring) problems to the head of the agenda (i.e. priority management). Then, when a problem is identified, we need to know how it can be represented in a way that facilitates its solution (i.e. representation design).

Decision tempo and agility are important in military problem solving. Current military doctrine emphasises the manouverist approach. Decision superiority arises from an information and knowledge superiority chain with network enabled capability. Decision superiority is achieved through an effects-based philosophy focusing on attainment of intended outcomes. It involves the meta-cognitive ability to attack an adversary's decision making process by attempting to "get inside" his/her decision cycle, thereby achieving a superior operational tempo (i.e. intent adaptiveness). This includes attaining a needed information and knowledge position, degrading the adversary's position, and defence against attack on one's own position. There is a fine balance between achieving decision superiority and the tempo of decisons. The fulcrum is where just enough decision quality information has been gleaned to enable confident, high-grade decision making. Research on rapid decision making by experts operating under time pressure, show strategies based on sufficing rather than optimising, using rapid recognitional skills (pattern recognition, visualisation or mental simulation, refinment or adaptation of obvious courses of action), rather than exhaustive analytical deliberation. With less time-pressure, such as strategic command and control, expert decision-makers may seek to keep options open to maximise gain from critical discriminating predictive information.

Improved support through computational processing and training is needed to mitigate the pressures of increasing complexity arising from network enabled capability, and from competition on sparce specialist skilled manpower resoures. Rasmussen (1986) has provided a useful model of human problem solving and decision making for adaptive control with complex systems. This distinguishes skill, rule and knowledge-based (SRK) levels of performance. Computational technology can contribute both automation (i.e. unburdening) and decision support to SRK levels (Figure 1). Computational technologies provide reliable error-free performance with simple problems. They can support rational and normative (un-biased) decision-making with optimisation of expected outcome utility. They are limited because the capability for context sensitivity and adaptability is bounded by design. As a general rule, for relatively simple problems, where proceedures are known to work and outcomes are predictable, computer automation can substitute for skill and rule-based behaviours. In contrast, for relatively complex ill-structured problems with unpredictable outcomes, and where existing proceedures need to be refined, or new proceedures need development, human involvement is essential and computers are used to provide decision support rather than to replace knowledge-based behaviour.

Expert human involvement and knowledge-based behaviour provides the capability to respond adaptively to unanticipated and destablising variability in problem context outside the boundaries of the system design. Adaptation is needed for long term stability and control. Adaptive control systems have been developed to deal with the uncertainties of complex systems (e.g. model reference controller, self tuning regulator). Stability can be achieved by regulating performance though a balance of feedforward and feedback control. Anticipation enhances stability by enabling feed-forward control without the prescence of an error resulting from feedback. Humans achieve stable adaptive control with complex problems through using a normative reference model, pre-stored strategies and tactics from past experience, and the ability to observe and self-regulate performance. Flach and Dominguez (2003) describe a cognitive control system that combines components from multiple styles of adaptive control:
- Regulates with respect to error (feedback)
- Anticipates outputs (feedforward)
- Prestored strategies and tactics that have worked in the past (look up table)
- Normative expectations derive from previous experiences (model)
- Capability to observe the system in real time to infer the dynamical properties (observer).

They note that whereas the information processing approach focusses on associating input and output relationships, extracting meaning from information relationships is needed for wise and sensible decisions i.e. structural truths, invariances, constraints, consequences. Human decision-making process can be characterised as a meaning processing system. For meaningful interaction with the environment, there needs to be a set of invariate constraints in the relationships among events in the environment and between human actions and effects. Purposive behaviour must be based on an internal representation of these constraints. This dynamic world model comprises largely tacit knowledge used by experts to discriminate, anticipate, integrate and synchronise with events in their domain of expertise. It reflects internalisation or attunement to the invariant properties or constraints of the task environment. The level of attunement achieved predicts comprehension, problem solving and decision quality. The meaning processing system is adaptive and capable of reorganising and coordinating the flow of information to reflect the constraints and resources within the task environment. Thus, in order to augment meaning processing, computational technologies need to support the flow of information so as to increase the attunement of the individual's dynamic world model. Computational technology for decision support should assist in providing representations that reveal the meaning and structural truths of problems. This will enable people to operate as expert adaptive controllers, and allow the human-computer system to function as a stable adaptive control system (Flach and Dominguez, 2003).

Figure 1: Levels of computer support for decision making (adapted from Rassmussen, 1986)

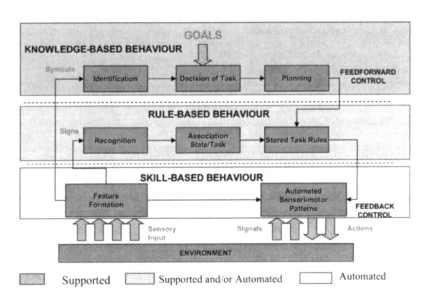

2 Augmented Cognition

Fundementally, augmentation of human intellectual capability is a matter of providing organisation or structure. Organisation holds the prospect of more rapid and better comprehension of problems, gaining useful comprehension in situations previouly too complex, speedier solutions, and finding solutions to problems that seemed insoluble. Englebart (1962) identifies the possibilities for augmentation of intellect arising from use of language (concepts, symbols used to model the world), artifacts (physical objects), methodology (strategies used to organise problem solving) and training (conditioning of skills to be operationally effective). Increased automatic computer processing has been the dominant technology strategy for mitigation of pressures on sparce specialist skilled manpower.

Augmented cognition research focusses on extending a user's abilities via computational technologies that are explicitly designed to address bottlenecks, limitations and biases of cognition. It is well documented humans have limitations in attention, memory, learning, comprehension, sensory bandwidth, visualization abilities, qualitative judgments, serial processing and decision making, with associated cognitive bias. DARPA's aim is to improve warfighter information intake under stress by extending the cognitive, information management capacity and function of the brain by symbiotic human-computer combination (Schmorrow and McBride, 2004). The vision replaces traditional electromechanical interaction devices (e.g.mouse, joystick) with electrophysiological indicators (e.g. EEG, PET,fMRI) for directing human computer interaction. DARPA's envisioned augmented cognition computational system can be defined as follows: "A computational interaction employing a system that monitors the state of the user, through behavioural, psychophysiological and/or neurological data acquired from the user in real time, and adapts or augments the computational interface to significantly improve their performance in real time".

The specific challenge for augmented cognition identified by DARPA is accurately assessing and predicting, from the incoming sensor information, the correct state of the user and having the computer select an appropriate strategy to assist the user at that time. To be successful, it is believed that an augmented cognition system must be capable of identifying at least one cognition bottleneck in real time and be able to alleviate it through a performance enhancing mitigation strategy. These mitigation strategies are conveyed to the user through an adaptive interface. Potential mitigation strategies identified by DARPA include:

- modality switching (between visual, auditory, & haptic),
- intelligent interruption,
- task negotiation and scheduling, and
- assisted context retrieval via book marking.

As a "closed the loop" system, successful augmentation of the user's cognition requires the following activities:

- the user state has to be correctly sensed,
- an appropriate strategy is chosen to alleviate the bottleneck,
- the interface is adapted to carry out the strategy, and
- the resulting sensor information indicates that the aiding has worked.

Augmentation is one coping strategy for dealing with task complexity by adding information and computation (problem solving capability) that is not typically present in the natural work domain (Flach and Dominguez, 2003). Intrinsic cognitive strategies include simplification, applying approximate heuristics or rules of thumb for selective search amongst possibilities, expert judgement intuitive recognition-visualisation-adaptation protocols (recognition-primed decision (RPD)) i.e. satisficing rather than optimising means-ends analysis. These strategies can be used to guide usable, human-centric forms of aiding. Other strategies include aiding by unburdening (delegation to automation), decomposing and delegating problems for collaborating team members, and training. Arising from the field of human reliability and human error (Hollnagel, 2004), ideas for performance shaping and barrier management provide potentially useful coping and mitigation strategies for adaptation (e.g. monitor, detect, reflect, correct).

Recent studies on task interference and working memory provide new understanding of the organisation of cognitive processes, offering a further perspective on strategies for coping with complexity (Macken et al, 2003). Current models of workload assume mental resources comprise separate entities or modules, such as verbal and visual sensory and memory components. Workload arising from interference between concurrent tasks is associated with similarity of content and resultant modular resource competition. In this view, consideration of content and modular resource limitations should guide mitigation through modality switching and intelligent interruption. Research on task interference suggests that effective coping strategies should be based on consideration of the processing operations involved (e.g. seriation, rehearsal, search), rather than task content (visual, verbal). If concurrent activities make use of the same processes interference will be apparent regardless of the modality or content of the task. Consideration of processes, rather than content, should guide possible augmentation mitigation strategies focusing on the organisation of information into potentially concurrent "cognitive streams" (e.g. break, divert, join).

3 Advanced Computing and Decision Systems

Computational technologies have progressed beyond simply providing passive tools for human use. Advanced computing and decision systems seek proactively to help humans cope with complexity in problem solving and decision making by providing intelligent, context sensitive aiding through unburdening (including autonomous decision making), and augmentation (including priority management, problem representation and structuring, and decision support). Levels of computing assistance range from adjustable automatic performance of routine sensing, processing and execution functions for relatively bounded predictable operations, through levels of data and information fusion, situation assessment, to support for information evaluation, knowledge management, mission planning, control and management, and decision making (Antony, 1995). Some degree of coupling or adaptation to the user, or to the context of use, in run time/real time, has been achieved through application of user and task modelling techniques, providing the basis for model-referenced adaptive control. A number of computational technologies applicable to augmenting decision-making and user adaptation can be differentiated:

Decision support system - DSS derive from the information systems community (c 1965). They are interactive information systems that use data and models to help analyse semi-structured problems. DSS include knowledge-oriented expert systems (using proceedural knowledge solvers/models, reasoning logic, and descriptive/environmental knowledge), query and reporting tools, Group DSS, Executive Information Systems, and Business Intelligence (Power, 2003). Traditionally, the DSS focus is on efficiency of information manipulation (robustness, ease of control, simplicity and completeness of relevant detail) rather than human interaction, and thus exhibit little/no integration with modelling the user and context of use.

Artificial Intelligence (AI) - AI methods (KBS, fuzzy systems, neural nets, case-based reasoning, genetic algorithms, and hybrids) can be distinguished in terms of their suitability for performing cognitive tasks autonomously (scheduling, planning, diagnosis, risk analysis, data analysis, monitoring, optimisation, interpretation, classification, control of systems) (Chamberlin, 2004). This provides unburdening of the human operator and enables supervisory control. KBS, case-based reasoning, fuzzy systems, and their hybrids, are best suited to operator decision support. Other relevent AI techniques used for supporting complex problem solving and decision making include agent-oriented software, such as belief-desires-intentions (BDI) based agents, distributed multi-agent systems, and related cognitive systems (systems that know what they are doing). Again, the focus is on efficiency of information processing and unburdening, rather than on human interaction. AI techniques can be used to develop models of user and the context of use. For example, COGNET/iGEN provides a cognitive agent approach used for multi-tasking and for supporting decision making in (Zachary et al, 2000).

Context aware computing/technology – Mobile devices with context sensitivity, focussing on user location, resources, current time (attributes of the environment), and including user tasks and goals (parent and subtasks). User profiles (identity, capabilities, characteristics, preferences, state of user) incorprated include common static parameters (general physical abilities, reading, speaking, writing) and dynamic parameters (current activity, terminal, location, motion-state, orientation) (Tazari, Grimm and Finke, 2003).

Intelligent tutoring systems (ITS) - Student model is updated based on interaction (Nwana, 1991). ITS attempt to mimic the capabilities of the human tutor.

Adaptive user interfaces – A novel solution to usability problems, attempting to exploit the capabilities of interactive systems in order to accommodate a great variety of users and/or functionality. They are intelligent in the degree that the user interface adapts itself and makes communication decisions dynamically at run-time (Benyon 1997).

Intelligent user interfaces – These facilitate a more natural user computer interaction by attempting to immitate human-human communication (Maybury 1999).

Affective computing/gaming – Includes ECG responsive video games, using players physiological status, as governed by level of arousal, to control how the gaming environment reacts to the players prescence, to prevent losing appeal and increase life span (Gilleade and Allanson, 2003).

Adaptive hyper media, semantic web site adaptation – These incorporate individual user modelling (novice, intermediate, expert stereotypes), users behaviour and errors, hypothesis generation of users intentions from limited goal recognition (Brusilovsky and Maybury 2002). K-Web is dynamic status board for facilitating knowledge sharing in command decision making. K-Web uses web based technologies, tools and products to capture, organise dynamically, store and present knowledge in an easily accessible form (Oonk et al, 2002). It is is not a real-time tactical tool, and it does not involve dynamic adaptive user and context of use modelling.

Knowledge management systems (KMS) – KMS employ user profile/ontology models user, including user preferences, skills/competency heuristics, level of activity (very active, active, passive, inactive), level of knowledge sharing (unaware, aware, interested, trial, adopter), and type of activity (knowledge sharing, knowledge creation) (Razmerita et al 2002). CommonKADS methodology for KMS distinguishes the aiding context in terms of model requirements of organisation, tasks, agents, knowledge, and communication prior to implementation. This focuses on the knowledge issues of acquisition, modelling, reuse and maintenance (Schreiber et al., 2000). Additional integration is required to support dynamic agent and task tracking and support adaptation.

Performance monitoring systems (PMS) – PMS provides pseudo-tracking of status information for system supervisory control applications including system process plant changes and operator activities (operator-machine interface, communications, information accessed). PMS uses action recognition templates for pseudo-monitoring, combining information on action pre-conditions, action itself, and post-conditions/outcomes to derive degrees of certainty/confidence of action possibilities. It is useful for managing proceedures to control performance variability and detection of violations, but not for critical functions or operator error detection (Hollnagel and Niwa, 2003).

Cognitive Technologies – A broad category of context sensitive computational technologies developed for intelligent crew assistant systems. Cognitive technologies include the Cockpit Information Manager (CIM) for the US Army's Rotorcraft Pilot's Associate (RPA) - an adaptive interface to planning components of the Cognitive Decision Aiding System, without reference to aircrew status (Miller and Hannen, 1999). CIM provided a forerunner to the Tasking Interfaces, using playbook scripts, developed for multiple UAV mission management ground control stations (Miller et al, 2000). Germany's CASSY and CAMMA programmes have provided systems integration for crew assistant decision systems. The US Navy's on-board Aircrew Decision Support System (ADSS) provided real-time multi-tasking decision and automation support for mission performance optimization, problem prevention and survivability, based on aircraft and aircrew status derived from tracking of observable events (Szczepkowski et al., 2001). DARPA Augmented Cognition has investigated a range of psychophysiological measures to identify changes in human activitity during task performance in real time including fNIR, EEG-Continuous, EEG-ERP, Arousal, Physiological guages (St. John et al, 2004). The UK MOD Cognitive Cockpit programme has developed a systems architecture for adaptive cognitive control of crew tactical decision making and survivability (Taylor et al, 2002). DARPA Augmented Cognition has developed this further with QinetiQ (Diethe et al, 2004). It includes a cognition monitoring system (COGMON) for operator functional state assessment with behavioural/physiological tracking, task load modelling, and intent inferencing. COGMON is coupled with a Tasking Interface Manager (TIM) and decision support, operated through an adaptable crew interface for Pilot Authorisation and Control of Tasks (PACT). TIM/PACT provides adjustable and adaptive levels of task assistance and autonomy, applicable to multiple UAV control. The COGMON/TIM coupling is a forerunner to DARPA Augmentation Manager technology.

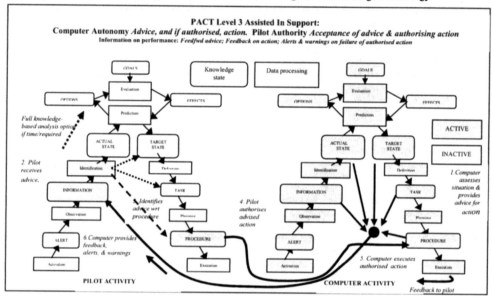

Figure 2: Control task analysis of computer-based decision support

4   Cognitive Systems Engineering

Cognitive Systems Engineering (CSE) provides analytical methods and knowledge gathering tools for decision systems (McNeese and Vidulich, 2002). Relevant modelling frameworks include GOMS (Goals, Operations, Methods, Selection), Soar, ACT-R, EPIC (Executive Process-Interactive Control), CommonKADS, and COGNet (COGnitive NETwork of Tasks), Extended Cognitive Control (ECOM) and Information Processing (IP/PCT). ADSS used COGNet. Relevant CSE methods and tools include Applied Cognitive Task Analysis (ACTA), Critical Decision Method (CDM), PC PACK knowledge acquisition tools, Integrated Performance Modelling Environment (IPME), Ecological Interface Design (EID), Cognitive Work Analysis (CWA), and Control Task Analysis (CTA).

Augmented cognition extends the boundaries of adaptive control to include changes in the user and use context. Useful information can be both directly sensed and model-based. Performance and error data from the user and use context provide information for feedback control but with inherent delays. Anticipatory model-based feed-forward control provides predictiveness and adaptation with incomplete information. Real-time predictive capability is needed for dynamic user task modelling and the development of adaptive systems for augmenting decision making. Some key questions are as follows:

- Why measure users and use contexts?

- Why model users and use context?
- What are the relevant characteristics?
- What types of cognitive processes, functional states, behaviour and contexts of use can be distinguished?
- What modelling techniques should be used to track and maintain user and use context models?
- What are useful adaptation effects?
- What are the advantages/limitations of ontologies in user modelling?
- What are the security and privacy issues?

CTA provides a useful tool for representing and analysing human-computer interactions and functions. A representation of augmentation of decision making by advanced computing and decision systems technologies using the CTA tool is shown in Figure 2. This represents the hypothetical flow of decision tasks and information between the human operator and a computer decision support system providing advice, and if authorised, action (PACT Level 3, Assisted/In Support). The configuration represented assumes immediate acceptance of advice by the human operator, without analysis and critical appraisal. This is a possible interaction that might take place when the solution is relatively obvious, when the information and decision support is trustworthy and reliable, when workload is high, or when there is time pressure to take immediate action. For many complex ill-structured problems, involving uncertain and incomplete information, with unanticipated variation and causal breaks, effective decisions will require expert RPD rapid judgement or, if time, critical appraisal and means-ends analysis of context with respect to goals, options and effects for determining status (i.e. What GOES?). One challenge for augmented cognition is to detect when the human operator is confused and not comprehending meaning and risk, and then provide feedback and effective adaptation. Equally, it would be useful to be able to detect when the required level of human engagement, appreciation and critical appraisal of risk is not provided, and to provide effective feedback and adaptation accordingly. Unless the design of the interface interaction ensures proper human engagement, sensible decision support will be judged as sufficing and accepted without critical appraisal.

5 Decision Aiding Interfaces.

Augmented cognition technologies provide the capability to adapt to the users ongoing tasks, functional state and changing capabilities and needs. These technologies can be expected to have important implications for user interfaces with advanced computing and decision systems. The following is a selection of typical style guide information for the design of computer decision aiding relevant to augmenting decision making by measurement and modelling of user and use context (ASCC, 2004).

User measurement:
Data link - The system should be fully adjustable to user parameters, and personal data-link set-up. However models that have been adapted or modified by the user must be validated using a default system.
Control Loop - If the integration of decision aids places the user out of the "control loop" caution should be used. The decision aids must keep the user informed of any irregularity or uncertainty, thus ensuring smooth and error free change of control between aid and user. If the user takes a less active role in decision-making and control of the aircraft, longer response times may occur in reintegrating the user.

User measurement and user modelling:
Tailoring - The user population (users) characters must be taken into consideration when designing the decision aid interface. This ensures that the communication levels are optimised in all areas. The decision aid must be fully aware as to who is currently in the cockpit. To increase performance, the more tailored the decision aids to the individual, the higher the performance.
Workload Monitoring - If possible the workload of the user must not be increased. The aid must be aware when an increase in cognitive processing by the user is required. The aim is to reduce the amount of difficult information that the user must evaluate. If this means a slight increase in the processing of simple to middle range tasks, the trade off may be significant.
Requirements Adaptation - The decision aid system should be user friendly, beneficial to the user, and present information in an easy to understand, simplistic format, familiar to the users. The systems optimum performance centres on the decision aids "intelligence" to adapt to the user's requirements.
Competence Sensitivity - The decision aid interface system must be flexible enough to take into account varying levels of expertise amongst users. Novices to the system may prefer a rule-based interface, however advanced users may rely on mental imaging techniques.

Use measurement and use modelling:

Decision Information Flow - The decision aid system must use methods that are applicable to the user. The system must also be adaptable to user changes. Throughout the decision process, the user should be informed of progress and method chosen. The user must retain control throughout the entire process. The aid should provide regular feedback to the user on the current stage of processing, and methods used in decision making. This information flow must be minimal, to help reduce information overload, however sufficient information must be given to keep the user in the "control loop". One solution is to have varying levels of information that the user can access, thus allowing the user to choose the level of detail needed.

Information Brokering - The main aim as stated above is to decrease the workload of the user in higher cognitive thought processes. In order to achieve this an "intelligent integrator" must filter out unnecessary information. Focusing on what the user needs and requires is of fundamental importance. Information that is relevant only to the task at hand should be expressed to user, unless there is important information that the user must be made aware of. The integrator should provide the minimum amount of information needed by the user to complete the task. A way of expressing the information in various resolutions and scales of complexity will help. These should have weighting so that important or specific relevant information is highlighted and pointed out to the user.

User measurement, user modelling, use measurement, and use modelling:

Modality Matching - To help with clarity and to maximise performance a different medium of communication with the decision aid may be required for each stage of decision. This could be speech for altering or text in responding. These mediums could adapt and change dependent on internal and external environmental factors. The decision aid whilst monitoring the users actions and vital statistics could interpret the best medium of communication in a given situation. This could include varying degrees of automation relative to the situational awareness of the user.

Intelligent Interaction - The decision aid system must have intelligent interaction between the user and the system. The decision aid should recognise the users needs, goals, abilities, and weaknesses. The aid should have sufficient "intelligence" to interpret poor or confusing queries, adapting and correcting user mistakes. The aid should adapt to the current situation, tailoring the communication method and medium to suit the particular current situation.

6 Conclusions

The following key problems for augmenting decision making, and suggestions for solution, can be identified:

Situation awareness - How do we ensure proper human appreciation of complex situations when highly automated systems are used for gathering, distilling and disseminating information?
- De-bias and tune appreciation of risk and uncertainty
- Ensure attunement and use of critical human judgement

Agenda - How do we ensure the proper prioritisation of problem solving and decision making ?
- De-bias and tune appreciation of critical priorities
- Monitor, model and predict factors driving problem and decision priorities in relation to goals in context of use
- Structure order of information provision and use in accordance with identified priorities

Representation - How do we ensure the proper representation of problems in ways that facilitate their solution?
- Identify the structural truths and invariant constraints in the problem
- Structure the representation to provide clarification of meaning,
- nsure the correct appreciation of meaning is comprehended

Tempo - How do we ensure that proper control of tempo and decision flow is achieved with highly automated systems?
- De-bias and tune control of tempo and decision flow
- Structure and tune speed of information provision

Agility - How can intelligent systems help provide the correct balance of reactive and proactive responding?
- De-bias rigid and innapropriate focusing
- Tune balance of reactive and strategic response

Trust - When can intelligent systems be trusted always to make the correct decisions autonomously?
- Tune correct appreciation of automation effectiveness

Validity - How can we validate the proper and effective integration of advanced automation for critical capabilities?
- Measure decision effectiveness in novel situations
- Evaluate design with reference to capability to provide intended outcomes and effects.

References

ASCC. (2004). Automation Interface Style Guide. Proposed ASCC Advisory Publication 61/116/33, December. Air Standardisation Coordinating Committee, Arlington, VA http://www.airstandards.com.

Antony, R.T. (1995). Principles of Data Fusion Automation. Boston: Artech House.

Benyon, D. (1997). Intelligent Interface Technology. In S Howard, J Hammond, and G Lindgaard Eds. Human Computer Interaction: INTERACT'97 pp 678-679. Chapman Hall

Broadbent, D. (1971). Decision and Stress. London: Academic Press.

Brusilovsky, P. and Maybury, M. (2002). From Adaptive Hypermedia to Adaptive Web. Communications of the ACM45 (5), 31-33.

Chamberlin, M. (2004). Assessment of the potential for application of AI techniques to UCAV mission management, In Proceedings of the Workshop on Uninhabited Military Vehicles (UMVs) - Human Factors of Augmenting the Force, RWS-010- P1, held in Leiden, Netherlands, 10-13 June 2003. NATO RTO Human Factors and Medical Panel 078/Task Group 017, NATO RTO Neully sur Seine.

Diethe, T.R., Dickson, B.T., Schmorrow, D. and Raley, C. (2004). Toward an augmented cockpit. In D.A. Vincente, M. Moustapha and P.A. Hancock (Eds). HPSAA II, Vol II, Human Performance, Situation Awareness and Automation: Current Research Trends, pp 65-69. Mawhah, NJ: Erlbuam

Engelbart, D.C. (1962). Augmenting Human Intellect: A Conceptual Framework. Summary Report AFOSR-3223 under Contract AF 49(638)-1024, SRI Project 3578 for Air Force Office of Scientific Research, Menlo Park, CA: Stanford Research Institute.

Flach, J. and Dominguez, C. (2003). Supporting the Adaptive Human Expert: A Critical Element in the Design of Meaning Processing Systems. In, Virtual and Adaptive Environments: Applications, Implications and Human Performance Issues. Hettinger L.J. and Haas M.W. (Eds). Chapter 20, pp433-460. Mahwah, NJ: Erlbaum.

Gilleade K.M .and Allanson J. (2003). A toolkit for exploring affective interface adaptation in videogames. In, C Stepanidis and J Jacko Eds. Human Computer Interaction: Theory and Practice (Part II). Volume 2. Pp 370-374. Lawrence Erlbaum Assocites, Mahwah, New Jersey.

Hinson J.M., Jameson T.L. and Whitney P. (2003). Impulsive decision making and working memory. Journal of Experimental Psychology: Learning, Memory and Cognition. 29 (2) 298-306.

Hollnagel, E. and Niwa, Y. (2003). Input requirements for a performance monitoring system. In, D Harris, V Duffy, M Smith and C Stephanidis Eds. Human Centred Computing: Cognitive, Social and Ergonomic Aspects. Volume 3. P467-471. Lawrence Erlbaum Assocites, Mahwah, New Jersey.

Hollnagel, E. (2004). Barriers and accident prevention. Aldershot, UK: Ashgate.

Macken, W. J., Tremblay, S., Houghton, R. J., Nicholls, A.P., and Jones, D. M. (2003). Does auditory streaming require attention? Evidence from attentional selectivity in short-term memory. Journal of Experimental Psychology: Human Perception and Performance, 29, 43-51.

Maybury, M. (1999). Intelligent User Interfaces: An Introduction. HCI'99 Tutorial. 8th International Conference on Human-Computer Interaction, HCI '99, Munich, Germany, August 22-27.

McNeese, M. and Vidulich, M. (2002). Cognitive Systems Engineering in Military Aviation Environments: Avoiding Cogminutia Fragmentosa. SOAR-02-01, Human Systems Analysis Centre, Wright Patterson AFB, OH.

Miller, C. and Hannen, M. (1999). "The Rotorcraft Pilot's Associate: Design and evaluation of an intelligent user interface for cockpit information management," Knowledge Based Systems, vol. 12, pp. 443-456.

Miller, C., Pelican, M. and Goldman, R. (2000). "Tasking" Interfaces for Flexible Interaction with Automation: Keeping the Operator in Control. In Proceedings of the Conference on Human Interaction with Complex Systems. Urbana-Champaign, Ill. May.

Oonk, H. M., Rogers, J. H., Moore, R. A., & Morrison, J.G. (2002). "Knowledge Web Concept and Tools: Use, Utility, and Usability During the Global 2001 War Game." Technical Report 1882. SSC San Diego. San Diego, CA.

Nwana, H. (1991). User modelling and user adapted interaction in an intelligent tutoring system. User Modelling and User Adapted Interaction, 1 (1), 1-32.

Power, D.J. (2003). A brief history of Decision Support Systems.DSSResources.COM, World Wide Web, http://DSSResources.COM/history/dsshistory.html, version 2.8, May 31, 2003.

Rasmussen, J., Pejtersen A.M. and Goodstein LP. (1994). Cognitive engineering: Concepts and applications. New York: Wiley

Rassmussen, J. (1986). Information processing and human machine interaction: An approach to cognitive engineering. New York: Holland.

Razmerita. L., Angehrn, A. and Nabeth T. (2002). On the role of User Models and User Modelling in Knowledge Management Systems. In C. Stepanidis and J. Jacko (Eds). Human Computer Interaction, Theory and Practice (Part II). Volume 2. Pp 450-454. Lawrence Erlbaum Associates, Mahwah, New Jersey, 2003.

Schmorrow D., and McBride D. (2004). Introduction. Special Issue on Augmented Cognition. International Journal of Human-Computer Interaction, Vol 17, No 2, pp 127-130.

Schreiber, A. Th., Akkermans, J., Anjewierden, A., De Hoog, R., Shadbolt. N., Van De Velde, W. and Wielinga, B. (1999). Knowledge Engineering and Management: The CommonKADS Methodology. The MIT Press.

Silverman, B.G. (1992). Critiqeing human error. London: Acedemic Press.

Simon, H. (1986). Research Briefings 1986: Report of the Research Briefing Panel on Decsion Making and Problem Sdolving, National Academy Press, Washington DC. Retrieved 31 January 2005 from http://dieoff.org/page163.htm

St. John, M., Kobus, D., Morrison, J.G. and Schmorrow, D. (2004). Overview of the DARPA Augmented Cognition Technical Integration Experiment, International Journal of Human-Computer Interaction, Vol 17, No 2, pp 131-149.

Szczepkowski, M. A., Hicinbothom, J. H., Coury, B. G., Warner, N. W. and Warner, H. D. (2001). Development of an intelligent agent for future tactical aircraft. Proceedings of the 12th International Symposium on Aviation Psychology, Columbus, Ohio, November 2001.

Tazari, M., Grim, M. and Finke, M. (2003). Modelling User Context. In C. Stepanidis and J. Jacko (Eds). Human Computer Interaction, Theory and Practice (Part II). Volume 2. Pp 293-297. Lawrence Erlbaum Associates, Mahwah, New Jersey, 2003.

Taylor, R.M., Bonner, M.C., Dickson, B., Howells, H., Miller, C.A., Milton, N., Pleydell-Pearce, K., Shadbolt, N., Tennison, J. and Whitecross, S. (2002). Cognitive cockpit engineering: Coupling Functional state assessment, task knowledge management, and decision support for context sensitive aiding. In M McNeese and M Vidulich M. Eds., Cognitive Systems Engineering in Military Aviation Environments: Avoiding Cogminutia Fragmentosa. State of the Art Report, SOAR-02-01, Chapter 8, pp253-312. Human Systems Analysis Centre, Wright Patterson AFB, OH.

Zachary, W., Santarelli, T., Ryder, J., and Stokes, J. (2000). Developing a multi-tasking cognitive agent using the COGNET/iGEN integrative architecture. AFRL-HE-WP-TR-2002-0202, December. Human Effectiveness Directorate, Wright Patterson AFB, OH.

Survey of Decision Support Control/Display Concepts: Classification, Lessons Learned, and Application to Unmanned Aerial Vehicle Supervisory Control

Gloria Calhoun[1], Heath Ruff[2], Jeremy Nelson[1], Mark Draper[1]

[1]US Air Force Research Laboratory
AFRL/HECI, 2210 8th Street
WPAFB, OH 45433-7511
gloria.calhoun@wpafb.af.mil
jeremy.nelson@wpafb.af.mil
mark.draper@wpafb.af.mil

General Dynamics[2]
Advanced Information Engineering Services
5200 Springfield Pike, Suite 200
Dayton, OH 45431-1289
heath.ruff@wpafb.af.mil

Abstract

The majority of present day Unmanned Aerial Vehicle (UAV) systems require multiple operators to control a single UAV. Reducing the operator-to-vehicle ratio would reduce life-cycle costs and serve as a force multiplier. Thus, automation technology is under rapid development. The envisioned system involves multiple semi-autonomous UAVs being controlled by a single supervisor. However, it has been documented in studies of manned systems that increasing the use of automation can cause rapid and significant fluctuations in operator workload and can result in loss of operator situation awareness and performance. Decision Support Systems provide a method of supporting the human decision-making process so that the ability of the human to effectively interface with highly automated systems is facilitated.

The Air Force Research Laboratory's Systems Control Interface Branch conducts research addressing human factors challenges associated with UAV operation. To date, research has focused on multi-sensory interfaces and advanced display visualization concepts. However, a new program is underway to design and evaluate situation assessment and decision support interface technologies to maximize flexible, fault-tolerant supervision of multiple intelligent semi-autonomous UAVs by a single operator. As a first step in designing decision support interfaces for UAVs, a survey of controls and displays used in decision support systems was conducted. This paper will describe the methodology used to conduct this survey, classify the major control and display decision support interfaces, and present results that bear on UAV supervisory control.

1 Overview

1.1 UAV Supervisory Control Demands

Though the phrase "unmanned aerial vehicle" (UAV) implies a lack of human interaction, that would in fact be a misnomer as human control of UAVs is vital for mission effectiveness. Until technology matures to the level of complete autonomy able to cope with all unexpected events, there will always be certain attributes only humans can provide. And as UAV platforms and applications continue to proliferate, human factors engineering will become more crucial to fully integrate human and machines. Through advances in automation technology, there is a foreseeable paradigm shift from the current scenario of *multiple operators* controlling a single UAV to a *single operator* supervising multiple semi-autonomous UAVs. This shift will require innovative methods to keep the operator "in the loop" for optimal situation awareness, workload, and decision-making. Operators will be expected to monitor and control multiple UAVs, which require little direct "hands-on" control. The role for humans migrates from more physical to more cognitive tasks, shifting from inner loop variable controls (e.g., flight control surface movements) to more outer loop strategy-oriented decisions (e.g., command UAV to an alternate mission plan). With this new operator role, many interface issues arise that must be properly examined in order to move towards achieving human-automation harmony.

As application of automation technology to the UAV domain increases, the potential for unintentional side effects associated with highly automated systems needs to be considered. These include rapid and significant fluctuations in operator workload, loss of situation awareness, lack of perceived reliability, complacency, skill loss, and operator performance decrements. To truly have a human-centered approach when incorporating automation in UAV applications, it is necessary to employ well designed controls and displays – controls that enable efficient task completion and displays that keep the operator well-informed.

1.2 Decision Support Interfaces for UAVs

Control and display design has always been a key concern for operator station development. However, for future UAV control stations that employ a single operator to control multiple UAVs, the controls and displays need to not only enable conventional operator tasking, but also need to support the operator's quick assessment and judgment of the appropriateness of the automation's actions and the impact of the actions on the overall mission objectives, priorities, etc. Moreover, the number of systems to monitor will increase and it will be a challenge for the operator to maintain situation awareness through long periods of nominal operations interjected with short periods of time-sensitive contingency operations. Thus, the UAV interfaces need to rapidly re-orient the operator and attract the operator's attention to critical information.

To date, the System Control Interface Branch of the Air Force Research Laboratory has focused on multi-sensory interfaces and advanced display visualization concepts in its research targeting human factors challenges associated with teleoperated UAV control. However, a new effort is underway to design and evaluate situation assessment and decision support interface technologies to maximize flexible, fault-tolerant supervision of multiple intelligent semi-autonomous UAVs by a single operator. This new effort will consider the design of the controls and displays associated with decision support systems as well as the knowledge-base software backbone of the decision support. Our interest in the control/display interface is two-fold: 1) it is suspected that there is a dearth of research addressing controls and displays for decision support, and 2) interfaces are key to the operator benefiting from the decision support system – if the interfaces don't support the operator's decision making process, the operator's effective interaction with the intelligent automated systems will be compromised and the benefits of the automation will not be realized. In sum, effective decision support *interfaces* are vital to the UAV operator's ability to make timely, accurate decisions.

1.3 Decision Support Survey

As a first step in designing decision support interfaces, a survey of decision support literature was conducted. One objective of this survey was to determine the degree to which controls and displays are considered in the design of past decision support systems. A second objective was to identify key decision support interface types and develop a classification system to facilitate their description. This paper will include a description of major control and display decision support interfaces addressed in the literature sampled. The methodology used to conduct the survey will be presented, as well as a summary of the results. This will include some lessons learned for designing decision support interfaces for UAV control.

2 Methodology for Decision Support Interface Survey

2.1 Scope

Thirty-four decision support interface research papers were surveyed. Although information continues to be gathered and reviewed, the present paper reports the findings of this initial stage. The sampled literature included journal articles, conference proceedings, book chapters, and technical reports. To be included in the survey, two qualifications had to be met. First, the research had to involve a problem domain in which a human user was challenged by requirements to make quick and accurate decisions. Thus, the documentation did not need to address the UAV domain; there are a multitude of application environments that require decision support, the findings of which are potentially applicable to UAV supervisory control. Second, the research had to involve the evaluation of an interface concept – a decision support control and/or display was manipulated as an independent variable. Thus, this survey focused on empirical evaluations, while steering away from papers that broadly addressed decision support issues, theory, and system design.

2.2   Literature Summary Table

For each document in the survey, notes were made in multiple columns of a table. A sample of this table is shown in Figure 1. Besides bibliographic information, there were seven columns to capture information on the nature of the decision support interface used in the research and the results of the evaluation. The following further describes the content noted in each column:

- *DSI Category*: identifies the general grouping of the decision support interface (DSI) concept (e.g., attentional cue, status display, etc.).
- *DSI Description*: summarizes the purpose or intended benefits of the decision support concept.
- *Intelligent?*: lists the computational functionality of the decision support interface as to whether it is based on artificial intelligence algorithms, knowledge-based systems, modeling efforts, or simple rule-based logic.
- *DSI Control Concepts*: identifies any control concept used to interface to the decision support.
- *DSI Display Concepts*: describes the content and format of information presented to the user in a decision support display.
- *Short Summary of User's Task*: provides an overview of the experimental design, including independent variables and what activities the user performed.
- *Lessons Learned*: summarizes results found with the particular decision support interface.
- *Conclusions*: lists decision support interface findings that may be appropriate for generalizing to other decision support systems.

3   Decision Support Interfaces: Classification

Entries in the table were examined to derive major groupings. First, overall grouping of the decision support interfaces into 'controls' and 'displays' was accomplished. Next, further classification was accomplished based on other factors, such as the degree of computational functionality (intelligence) in implementing the interface and the nature of the support it afforded. The following provides additional detail on each category in the classification scheme developed as a result of this literature survey.

3.1   Decision Support Controls

3.1.1   Control Devices

The majority of controls employed in the decision support interface literature sampled were conventional input devices – mouse and keyboard. The decision maker was presented with displays which he/she controlled by clicking on specific locations on the screen with the mouse or by typing commands. Less often, researchers evaluated decision support control with novel devices. These included voice recognition, touch-screens, and reduced keyboards (Winter, Champigneux, Reising, & Strohal, 1997), and a haptic stick (resistance of the stick depended on the path deviation, Scerbo, 2000). Within the overall conventional versus non-conventional control groupings, each different type of control device constitutes a different category.

3.1.2   Control Method

Besides classifying the control interfaces by device type, it was also deemed useful to classify by the method used to achieve control. For instance, Sycara & Lewis (2002) make the distinction between controls that are 'bottom-up' as opposed to 'top-down'. In bottom-up control, individual elements of the decision problem are specified in detail, and the user then receives updates on each element. In top-down control, an overview of the decision problem is formulated, each element of the problem is then refined, and the user is guided down through

	Reference	DSI Catego ry	DSI Description	Intellig ent?	DSI Control Concepts	DSI Display Concepts	Short Summary of User's Task	Lessons Learned	Conclusions
1	Yeh, M. and Wickens, C.D. (2001). Display signaling in augmented reality: Effects of cue reliability and image realism on attention allocation and trust calibration. Human Factors, 43(3), 355-365.	Attentio nal Cuing	Help the operators detect targets quicker, more reliably	No. (It just moves cuing reticle to target)	None.	Cuing Reticle = four crosshairs (square with tick-marks thru each side) presented conformally over the target to signal its current lateral and vertical location. Cued until target detected or passed. No locator line.	Users piloted a UAV and searched for targets camouflaged in the terrain. Terrain: Low or High Realism Cuing: None or Cued Cue Reliability: 75% or 100%	When image realism is high, subjects rated scene more "compelling" but no change in trust. Higher image realism lead to more conservative actions & more reliance on cue. Higher false alarm rate with cuing (users become risk-takers) With cuing, cognitive attentional tunneling of cue prevents users from catching unexpected (but high priority targets). Tunneling effect (and cue effect) reduced with lower cue reliability.	Cuing aided target detection for expected targets, but drew attention away from unexpected targets. (Cognitive Tunneling) Cognitive tunneling was reduced with lower reliability cuing, BUT cuing benefits (faster RT) also were reduced. Increased Image realism forced higher cue reliance.
2	Crocoll, W.M. and Coury, B.G. (1990). Status or recommendation: Selecting the type of information for decision aiding. Proceedings of the Human Factors Society 34th Annual Meeting, 1524-1528.	status display vs. action recomm endation display	To help the team chief of an air defense task accomplish the tasks of target acquisition, visual identification, and giving the command to engage by presenting relevant info (status or action recommendatio n).	No. (Simple classific ation and rule-based logic	None.	Status Display: Prior to a/c presentation, the sub was provided Friend / Hostile / or Unidentified. Recommendati on Display: Prior to a/c presentation, the sub received Fire / No Fire.	Stinger Team Chief Tasks: Users visually identify a target a/c, use decision rules to make a decision to fire or not fire. Groups: 1-No DA info provided 2-Received only status info 3-Received only recommendation info 4-Received both status and recommendation info	Accuracy > 96% overall. RT faster when decision aid was provided. With lower DA reliability, combined (status and recommendation) and recommendation-only DA resulted in sig. lower accuracy than status-only DA. Low reliability: 95% correct with status-only, 86% with status / recommendation, and 80% with recommendation-only DA Subjective Trust and Tolerance were low with lower reliability DA	DA does impact decision making performance (3 x faster) Strategy in using DA does impact performance. Subs couldn't estimate actual accuracy of the DA Status is best DA especially if there is some chance the DA is wrong.
3	Hughes, S. and Lewis, M. (2002a). Attentive interaction techniques for searching virtual environments. Proceedings of the 46th Annual Meeting of the Human Factors and Ergonomics Society, Sept. 30 - Oct. 04, Baltimore, MD., pp 2159 -2163.	Attentio nal Cuing	DS suggests optimal locations / orientations as the user freely explores a VE.	No. Uses simple rules (a map of ideal gaze vectors)	Mouse used to control virtual being, but automation sometimes controlled gaze vector.	-Attentive camera aligns the user viewpoint with the ideal gaze vector as user moves freely. -Attentive Flashlight shines a spotlight in ideal vector direction At target object	Users explored a VE art gallery with a mouse in order to answer questions about the artwork. 1-No Assistance 2-Attentive Camera 3-Attentive Flashlight	Attentive flashlight technique outperformed the other techniques in all analyzed data.	Continuous feedback is better than intermediate feedback. Providing recommendations independent of user control (not taking actual control) gives the user a greater sense of control (good). Limitation: Direction and distance arrow better than spotlight Limitation: Need user-initiated help, as well as the system forcing views when critical
4	Hughes, S. and Lewis, M. (2002b). Directing attention to open scenes. Proceedings of the 46th Annual Meeting of the Human Factors and Ergonomics Society, Sept. 30 - Oct. 04, Baltimore, MD., pp 1609 - 1612.	Attentio nal Cuing	DS highlights targets in open scenes (photos, videos, etc.)	No. Uses simple rules (knows target location s pre-trial)	None.	Orange circle around target, or highlight target	Users searched digital pictures for target objects. -2 pixel wide orange circle on target -low, medium, high contrast highlighting -20% of images were target-not-present -20% of target-present images had mis-aligned targets	Slower RT with "Target-Absent" images (because users had to search the entire scene to verify) -Higher contrast highlighting leads to shorter RT for target-aligned highlighting, and longer RT when it is mis-aligned with the target (cognitive tunneling) -Annotated targets (circle on target) was approx. same results as strongest highlighting level.	"No Highlight" is better than a high contrast highlighting once reliability drops below some point (around 60%?) Cuing by annotations (arrows, circle, square) is hindered by it's nature of obscuring a portion of the raw data (image background)

Figure 1. Sample from Literature Summary Table.

the hierarchy of information. Another sub-classification is whether the control employed intelligent filters that retrieve, fuse, and manage information during the control process (Scerbo, 2000). In contrast to having intelligent filters, the design can be based on the user changing system constraints, parameters, and/or plans which results in the initialization of processes that change (i.e., control) system states (e.g., Winter et al., 1997). Two opposing control methods make up a final defined control category: closed loop control versus open loop control. With open-loop control, no feedback is sent back to the controller because the control equation is well-accepted. During closed-loop control, feedback is provided to the controller so that any subsequent input brings the system closer to the goal state.

3.2 Decision Support Displays

The sampled decision support interface literature described many type of displays. These were grouped into two major categories: displays that are simply rule-based and ones that are implemented with "intelligent agents".

3.2.1 Rule-based support

Automated warnings, alerts, and *cues* consisting of simple notifications (visual, aural, tactile) when a rule or goal state is broken (or met) constituted one common type of rule-based decision support interface display. These displays can also be multi-modal. *Attentional cuing displays* aid target prosecution and highlight critical information or points of interest through synthetic vision information overlaid conformal on the operator's display or with technologies such as ATC (Automatic Target Cueing) and ATR (Assisted [not Automatic] Target Recognition; Endsley & Kaber, 1999; Hughes & Lewis, 2002a, 2002b). *Status displays* also use a rule-base to present the decision maker with parameters and their associated values. These displays provided system history in the form of graphs, intelligence reports, or logs. Similarly, status displays can offer classification, filtering, or generalizing of situations and systems (Morrison, Kelly, Moore, & Hutchins, 1998; Smith, McCoy, & Layton, 1997). *Action recommendation displays* take status display information and use a rule-base to recommend a course of action to the decision maker (e.g., options, strategies, and/or potential plays), but does not implement the action (Crocoll & Coury, 1990; Woods, 1986). A specific example is a Highway-in-the-Sky display that provides flight guidance (Snow & Reising, 1999). *Adaptive/adaptable systems* are another type of rule-based display. Adaptable systems allow the operator to initiate automated state and mode changes, while adaptive systems allow the operator or automation to initiate the changes depending on the automation management schema. In some designs, a simple rule-base can cause automated mode changes when the operator crosses threshold values in path deviations, EEG index (workload measure based on electroencephalographic signals), or altitude (such as a ground collision avoidance system). Rules could also trigger the system automation to collect, filter, organize, and present information to the operator in anticipation of upcoming decisions. In other designs, the Level of Automation in effect specifies what functions are performed by the operator (Endsley & Kaber, 1999).

3.2.2 Knowledge-based support

Knowledge-based decision support displays are also referred to as intelligent displays. Definitions of "intelligence" varied, but many involved the ability of an entity to achieve goals in complex real world situations. Some definitions specified that the entity must perceive, learn, reason, communicate, and act (McCarthy, 2003). Using intelligent systems to support decision making is most appropriate for well-structured, clearly defined, routine or programmed decision situations (Simon, 1960). Most types of intelligent displays rely on intelligent agents that work in the background to collect, filter, organize, and present information to the decision maker. With an intelligent system, status displays can also provide a feed-forward presentation to support the decision maker in anticipating future system states through formats such as Gantt charts and predictive displays. Intelligent agents can also expand the capabilities of displays in adaptive systems. Other types of knowledge-based decision support displays are plan generators and evaluators. With these displays, operators can a) generate a plan and then have the system evaluate (critique) it, b) tweak a system-generated plan and then have it evaluated, or c) use the intelligent technology to predict future system states by inputting "what if?" scenarios. The generation and evaluation of these plans are usually dependent on optimization algorithms or knowledge-based expert system modeling. It is akin to the operator having a Subject Matter Expert on standby to help with deciding how best to proceed. Examples include support for flying instrument approaches or for air traffic control monitoring activities. Similarly, intelligent adaptive decision support displays rely on intelligent agents to monitor situations and decide what information to present to the user, in what format, and at what time. Some intelligent agents have evolved so readily that they may replace some of the operators in a team decision making environment. To make effective team decisions, the agent-user collaboration, negotiation, and dialogue must work successfully so the agent can provide aggregate, inferential, and decision information to the operator(s) without information overload.

4   Decision Support Interfaces: Lessons Learned

4.1   General Findings

One objective of this survey was to determine the degree to which controls and displays are considered in the design of decision support systems. For the literature sampled, the majority dealt with display interfaces as opposed to control interfaces. This finding is not surprising and could be viewed as reflecting the larger body of research dealing with displays, compared to controls, in overall human factors literature.

Another finding raises a more pressing concern. First, note that all the sampled literature focused on evaluating a control and/or display interface for a decision support system. However, less than a third explained the rationale for the control/display concept chosen to be employed with the particular decision support system. In other words, these reports didn't cite published research or a previous application of the control/display that supported its utility in the decision support system featured in the report. Although this finding may only reflect incomplete documentation by the authors, it is feared that it is indicative of an inadequate prioritization on control/display design. Another possibility is that there are problems inherent in generalizing existing control/display findings to decision support systems. Regardless, this finding supports the need for additional research on how best to apply control/display interfaces for use with decision support systems for UAV operators.

The following sections summarize additional lessons learned gleaned from the literature survey. First, decision support controls will be addressed, followed by decision support displays.

4.2   Decision Support Controls

4.2.1   Control Devices

The literature survey indicated that the use of non-conventional controls, such as speech recognition, touch-screens, and reduced (custom function) keyboards generally improved operator performance and either reduced or had no negative effect on perceived workload for the task environments employed. One example is pilots choosing speech recognition over manual input for making entries into a flight plan decision support system used in a flight test (Winter et al., 1997).

4.2.2   Control Method

There were a few noteworthy results in research that addressed control method, as defined earlier in Section 3.1.2. One study compared alternate control methods for interacting with a system designed to provide decision support on mission selection. In one method, an intelligent agent critiqued any flight path sketched by the operator. In other methods, a planning agent provided flight path choices or the "optimal" choice. The results showed operator performance was worse with the sketchpad control input, as it was not amenable to the operator quickly creating several sketched paths for analysis (Lewis, 1999). Another observation in the literature survey was that in many decision support systems, there are alternate control methods simultaneously available to the operator. For instance, the decision support system may allow the decision maker to choose:
- information search/acquisition method, such as bottom-up (start with detailed query and get updates/notifications) or top-down (global perspective with guides to drill down to detailed information)
- information assessment method (what filters or constraints are employed)
- whether to view recommended alternatives or only the "optimal" solution
- control input device (by sketching, typing, or speaking his/her novel problem solution into the system)

Empirical results showed that these control choices could significantly affect the decision support effectiveness and overall mission performance. For instance, choosing to select a path recommended by the decision support system, as opposed to using a more cumbersome sketch/graphical input method, can bias the operator's decision (Smith et al., 1997). This exemplifies the importance of optimal control interface design for decision support systems.

4.3   Decision Support Displays

4.3.1   Rule-based support

4.3.1.1   Automated warnings, alerts, and cues

Similar to decision support controls, performance with non-conventional decision support displays was generally better than with conventional visual displays. Multi-modal displays were found to improve performance for many tasks, especially when used as a redundant cue to visual information. Examples include haptic (Scerbo, 2000), auditory (Olson & Sarter, 2001), and voice displays/feedback (Winter et al., 1997; Foy & McGuinness, 2000).

Another key finding related to cuing displays is the potential for cognitive tunneling. Cognitive tunneling can occur when the operator becomes focused on the cue being displayed to such an extent that other important objects in the view are not attended. Research has demonstrated that a cue highlighting the location of an expected target can improve detection of that target, while decreasing detection of unexpected/non-cued targets. Manipulation of the reliability of the decision support cue display can impact these detection rates. For instance, reducing the reliability of the cue, in one study, increased detection of unexpected targets, with the tradeoff of decreasing detection of expected targets (Yeh, & Wickens, 2001). Thus, research is needed to explore how cue displays might be designed to minimize the negative effects of cognitive tunneling. At the very least, evaluations are required with the candidate decision support cue and task environment to determine the ideal cue reliability that will result in acceptable detection rates for expected and unexpected targets. Hughes and Lewis (2002a), for example, evaluated the use of a spotlight to show the ideal vector direction to a target in a virtual environment displayed on a monitor, as opposed to automatically changing the operator's viewpoint to be aligned with the ideal gaze vector to the target. The spotlight cueing display was found to improve target designation. These researchers also explored manipulation of cue reliability in a follow-on evaluation with a different task paradigm and found that the cue was beneficial if its reliability was approximately 60% or better (Hughes and Lewis, 2002b). Performance with a cue at a lower reliability level was similar to performance in trials with no cue at all.

Besides reducing the likelihood of detecting unexpected but important information, decision support cues can also unintentionally obscure both expected and unexpected information. For example, cuing by annotations (arrow, circle, or square) can potentially obscure portions of the raw data (image; Hughes & Lewis, 2002b). One technique for minimizing this problem is to implement a rule-based simple heuristic concept whereby cues for high priority information or objects are presented on the top of other layers, while dimming or graying-out lower priority threats, information, or objects. Results from one study suggest that a 25% decrease in time to detect priority threats can be realized with this display concept (St. John, Maines, Smallman, Feher, & Morrison, 2004).

Finally, there is a possibility that operators will over rely on cues displayed from the decision support system, resulting in operator complacency. For example, in an evaluation using the TCAS (Traffic Alert & Collision Avoidance System) decision support system that employs both visual and aural alerts, performance-based situation awareness measures were lower when the TCAS was utilized compared to when it wasn't. In contrast, the subjective ratings of perceived situation awareness were higher when the TCAS was utilized compared to when it wasn't. Analyses indicated that operators waited for the visual/alerts to notify them of problems, rather than continually monitoring the system. Thus, conflicts that were not detected by the TCAS system were more likely to be missed by the operators due to over-reliance on cues from the TCAS (Foy & McGuinness, 2000). .

4.3.1.2   Status and action recommendation displays

In general, the research sampled showed that status and action recommendation type decision support displays were beneficial to operators' performance in the task environments tested. However, the reliability of information displayed begins to have a more consequential impact with recommendation displays. This is because it is more difficult for operators to judge from a recommendation display that underlying data is suspect. Studies have shown that when the decision support system has reduced reliability, operator performance is better with a status display by itself (Crocoll & Coury, 1990; Cummings & Guerlain, in review). Additional findings suggest displays that incorporate graphics are better than text-based displays. It was hypothesized that the graphic displays were rated higher in usability because they support human pattern recognition, or recognition-primed decision-making, through the use of colors, shapes, overlays, and/or coding schemes (Morrison et al., 1998).

4.3.2 Knowledge-based (intelligent) support

Since knowledge-based displays are more difficult and time-consuming to develop and require a well-defined domain, empirical testing is scarce and this was reflected in finding only a few relevant documents in the literature sampled. One evaluation of an advisory tool designed to generate flight paths around weather showed that providing a display to aid operators to manually develop their own plan prior to getting a plan from intelligent agents enabled operators to catch faulty plans better compared to displays that only provided a generated plan (Guerlain, 2000). This result suggests that a cooperative process between the operator and the decision support system would help minimize the occurrence of automation bias.

The nature of the cooperative process can also impact the effectiveness of the decision support system. In a study by Lewis (1999), it was shown that operators could generate plans for distribution to teammates faster with an intelligent planner that provided displays to work the operator through the planning process (providing suggestions, blanks to fill, etc.), than an intelligent planner that instead critiqued an operator generated plan, displaying how good it was according to critical mission measures. These studies also illustrate how display design needs to simultaneously take into account control design as well. In other words, the optimal nature of the operator-system dialogue for capitalizing on the benefits of the decision support system needs to first be determined and then supported by the control/display interface.

Information reliability also plays a role in the effectiveness of decision support system displays. In the study mentioned above (Lewis, 1999), a display with no support was compared to three different types of decision support displays: one that just listed raw data, one with the data classified into columns, and one in which accuracy probability information was also provided. By manipulating the source of errors (raw data, erroneous classification, erroneous probability), the results suggest that presenting the data classified into columns was best, despite the fact that subjective ratings indicated that operators trusted all three types of displays equally. Also, the finding that operators' trust was the same for all display types illustrates the value of presenting key information in the display by which the operator can judge the adequacy of the data and, ultimately, improve the decision making process.

5   Survey Implications on UAV Supervisory Control

Decision support interfaces are key to realizing the benefits of intelligent automated systems that will enable single operator control of multiple UAVs. If the controls and displays fail to support the operator's decision making process, the operator's ability to make timely, accurate decisions will be compromised, as well as the ability to judge the operation of the automated subsystems. Unfortunately, the results of this literature survey suggest that attention to date has focused on development of the knowledge-base for the decision support, rather than the optimal design of the controls and displays used with the decision support system. Thus, there is a need for research that systematically addresses interface issues for various categories of UAV decision support systems. A spiral design/evaluation/refinement process should be utilized whereby candidate control/display approaches are evaluated with numerous UAV related task paradigms.

Some directions for this research are, however, indicated in the surveyed literature. For instance, research with multi-modal interfaces (e.g., speech-based input and visual/aural redundant displays) supports their use with a decision support system. The interaction of control and display design was also illustrated, showing that the desired operator/system dialogue or collaboration must also drive interface design. Other studies demonstrated the value of presenting key information in the display by which the operator can judge the adequacy of the data and, ultimately, improve the decision making process. For UAVs, this would suggest displaying status information for each UAV under supervision, increasing the displayed information by a factor of the number of vehicles. To minimize information overload/retrieval problems, an intelligent system which highlights important information while maintaining the availability of other information may be useful, as well as innovative display approaches that support anticipatory decision-making.

In research evaluating candidate UAV controls and displays, the reliability, bandwidth, and timeliness of information need to be manipulated to determine their effect on the utility of the decision support interface. It is also important that the interface design/evaluation takes into account issues of cognitive tunneling, effective information retrieval, and automation bias. Workload and vigilance demands are also critical. For instance, adaptive controls and displays may be appropriate whereby there is decreasing autonomy in low workload conditions and increasing autonomous functionality during high workload periods. The ability for an intelligent system to vary the level of automation has especially not been explored in terms of the implications on decision support control and display design.

6   References

Crocoll, W.M., & Coury, B.G. (1990). Status or recommendation: Selecting the type of information for decision aiding. *Proceedings of the Human Factors Society 34th Annual Meeting*, pp 1524-1528.

Cummings, M.L., & Guerlain, S. (2005, in review). The decision ladder as an automation planning tool. *Cognition, Technology, and Work*, Springer-Verlag London Ltd.

Endsley, M.R., & Kaber, D.B. (1999). Level of automation effects on performance, situation awareness and workload in a dynamic control task. *Ergonomics*, 42(3), 462-492.

Foy, L., & McGuinness B. (2000). Implications of cockpit automation for crew situational awareness. *Proceedings of the Human Performance and Situation Awareness in Automation Conference*, pp 101-106.

Guerlain, S. (2000). Interactive advisory systems. *Proceedings of the Human Performance and Situation Awareness in Automation Conference*, pp. 166-171.

Hughes, S., & Lewis, M. (2002a). Attentive interaction techniques for searching virtual environments. *Proceedings of the Annual Meeting of the Human Factors and Ergonomics Society*, Baltimore, MD., pp 2159 -2163.

Hughes, S., & Lewis, M. (2002b). Directing attention to open scenes. *Proceedings of the Annual Meeting of the Human Factors and Ergonomics Society*, Baltimore, MD., pp 1609 - 1612.

Lewis, M. (1999) Anticipation delegation, and demonstration: Why talking to agents is hard, in M. Klusch, O. M. Shehory & G. Weiss (Eds.) *Cooperative Information Agents III, Lecture Notes in Computer Science*, vol. 1652, Springer-Verlag Berlin, Heidelberg, pp 365-389.

McCarthy, J. (2003). What is artificial intelligence? Stanford University, web essay, from http://www-fornal.stanford.edu/jmc/whatisai/whatisai.html.

Morrison, J.G., Kelly, J.T., Moore, R.A., & Hutchins, S.G. (1998). Implications of decision making research for decision support and displays, In J. A. Cannon-Bowers & E. Salas (Eds.), *Making decisions under stress: Implications for individual and team training*. Washington, DC: APA Press, pp 375-408.

Olson, W.A., & Sarter, N.B. (2001). Management by consent in human-machine systems: When and why it breaks down. *Human Factors*, (43)2, pp 255-266.

Scerbo, M.W. (2000). Adaptable and adaptive technology in the lab and in the field. *Proceedings of the Human Performance and Situation Awareness in Automation Conference*, pp. 57-62.

Simon, H. A. (1960). *The new science of management decision*, NewYork: Harper & Row.

Smith, P.J., McCoy, C.E., & Layton, C. (1997). Brittleness in the design of cooperative problem-solving systems: The effects on user performance. *IEEE Transactions on Systems, Man, and Cybernetics - Part A: Systems and Humans* (27) 3, May, pp 360-371.

Snow, M.P., & Reising, J.M. (1999). Effect of pathway-in-the-sky and synthetic terrain imagery on situation awareness in a simulated low-level ingress scenario, *Proceedings 4th Annual Symposium on Situational Awareness in the Tactical Air Environment*, Piney Point , pp 198-207.

St. John, M., Maines, D.I., Smallman, H.S., Feher, B.A., & Morrison, J.G. (2004). Heuristic automation for decluttering tactical displays, *Proceedings of the Human Factors and Ergonomics Society 48th Annual Meeting*, HFES: Santa Monica, CA, pp 416-420.

Sycara, K. & Lewis, M. (2002). From data to actionable knowledge and decision. *Proceedings of the Fifth International Conference on Information Fusion*, Annapolis, MD, July 7-11.

Winter, H., Champigneux, G., Reising, J., & Strohal, M. (1997). Intelligent decision aids for human operators. *AGARD Symposium on Future Aerospace Technology in the Service of the Alliance*, AGARD-CP-600, Vol. 2, Palaiseau, France, 14-17 April.

Woods, D.D. (1986). Paradigms for intelligent decision support. In Holnagel, Mancini, and Woods (Eds.). *Intelligent Decision Support in Process Environments*. New York: Springer-Verlag, 153-173.

Yeh, M., & Wickens, C.D. (2001). Display signaling in augmented reality: Effects of cue reliability and image realism on attention allocation and trust calibration. *Human Factors*, 43(3), 355-365.

Enabling Autonomous UAV Co-operation by Onboard Artificial Cognition

Claudia Ertl and Axel Schulte

Munich University of the German Armed Forces
Institute of System Dynamics and Flight Mechanics, 85577 Neubiberg, Germany
[Claudia.Ertl|Axel.Schulte]@unibw-muenchen.de

Abstract

This paper presents an approach, which enables a UAV to act both autonomously and co-operatively dependent on the current situation and the mission goals to be accomplished. The selected method, the so-called 'Cognitive Process', is a theoretical approach to a model of human information processing, which is in turn suitable for the design of artificial autonomous units. It describes, how human-like behaviour can be given to a technical system by gathering information from the environment, interpreting the information, activating goals, planning future actions, and scheduling and executing the planned actions.

An appropriate architecture for the development of cognitive systems is COSA (COgnitive System Architecture), which offers a framework to implement applications according to the theory of the Cognitive Process. This architecture has been used to develop an application example for an autonomously acting UAV, which is capable of understanding a high-level mission order. Based on this one-ship application, a system design concept based on the Cognitive Process is presented for a scenario containing several UAVs. Starting from an examination of co-operation and its implications, namely co-ordination and communication, the knowledge is detailed, which is needed by the Cognitive Process in order to put co-operative behaviour into practice.

1 Introduction

Scenarios in the domain of aviation, that make use of unmanned aerial vehicles (UAVs), range from civil search and rescue missions to military attack missions accomplished by a team of manned and unmanned vehicles. Besides UAVs acting jointly with other UAVs or even human beings, these scenarios include human operators on ground, who might have to supervise several UAVs. In order to keep the workload of the human beings involved on an acceptable level and to enable efficient mission accomplishment, the UAVs have to be capable of accomplishing a mission autonomously and co-operatively.

Autonomous mission accomplishment shall be based on a high-level mission order rather than a pre-compiled detailed flight plan consisting of e.g. waypoints. Moreover, autonomy includes the ability to react to situations which cannot be foreseen by neither the developer nor the operator of the system. The capability to co-operate comprises teamwork with either human or machine partners towards a common objective. Sometimes, this common objective can only be achieved by co-operation, as no single partner has the required capabilities or resources. In other cases, joint action of several partners is not necessary, but facilitates efficient use of given capabilities and resources.

This paper presents an approach to both autonomous and co-operative mission accomplishment, which is based on the so-called 'Cognitive Process', a model of human information processing (section 2). Based on this theory, a framework for the implementation of cognitive systems is described, which has been used to achieve autonomous capabilities (section 3), and which will be used to implement a concept of co-operative behaviour (section 4).

2 The Cognitive Process: A Model of Human Information Processing

The following section describes the Cognitive Process (CP), which serves as a basis for the development of cognitive systems. The blueprint of the CP (section 2.1) can be used several times in order to design miscellaneous capabilities, which together form one cognitive system (section 2.2). Finally, COSA (COgnitive System Architecture) is presented, which facilitates the implementation of a cognitive system according to the theory of the CP (section 2.3).

2.1 The Cognitive Process as a Blueprint for Modelling Cognition

The Cognitive Process (CP) has been developed as a model of human information processing, which is suitable for generating human like behaviour (Putzer & Onken, 2003). As it is compatible with human cognition, and the generated behaviour is driven by goals, which are represented explicitly, the CP is well suited for the development of a cognitive system, which is part of a team consisting of artificial and/or human team mates.

Figure 1 shows the CP consisting of the *body* (inner part) and the *transformers* (outer extremities). The body contains all knowledge, which is available for the Cognitive Process in order to generate behaviour. There are two kinds of knowledge: the '*a-priori knowledge*', which is given to the CP by the developer of an application during the design process and which specifies the behaviour of the CP, and the '*situational knowledge*', which is created at run time by the CP itself by using information from the environment and the a-priori knowledge. The functional units effectively processing knowledge are the above-mentioned transformers, which read input data in mainly one area of the situational knowledge, use a-priori knowledge to process the input data, and write output data to a designated area of the situational knowledge.

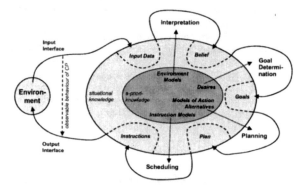

Figure 1: The Cognitive Process

The following steps are performed by the CP in order to finally generate the overall observable behaviour:

- Information about the current state of the environment (*input data*) is acquired via the *input interface*. In this context, the environment includes other objects in the physical world, e.g. another UAV or an obstacle, as well as the underlying vehicle of the CP. Therefore, the input data may for instance contain information about the current autopilot mode or pre-processed sensor information.

- The input data are *interpreted* to obtain an understanding of the external world (*belief*). The interpretation uses environment models, which are concepts of elements and relations that might be part of the environment, to build this internal representation.

- Based on the belief, it is *determined*, which of the desires (potential goals) are to be pursued in the current situation. These abstract desires are instantiated to *active goals* describing the state of the environment, which the CP intends to achieve.

- *Planning* determines the steps, i.e. situation changes, which are necessary to alter the current state of the environment in a way, that the desired state is achieved. For this planning step, models of action alternatives of the CP are used.

- Instruction models are then needed to *schedule* the steps required to execute the plan, resulting in *instructions*.

- These instructions are finally put into effect by the appropriate effectors of the host vehicle. The resulting actions affect the environment, i.e. modify the physical world.

These functional units represent an application-independent inference mechanism, which processes application-specific knowledge. This knowledge-based design approach is of great advantage when implementing the CP: The inference mechanism has to be implemented only once, and can then be used for different applications.

2.2 Different Views on the Cognitive Process

It is desirable to reuse not only the inference mechanism but also knowledge in different applications. For this purpose, a-priori knowledge has to be unitised in so-called 'packages', each of which represents a certain capability.

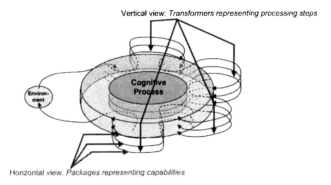

Figure 2: Different views on the Cognitive Process

As indicated in Figure 2, each package (depicted as horizontal layer) implements a capability, which is designed according to the blueprint of the CP. Several packages together form the complete system. They are linked by dedicated joints in the a-priori knowledge and by the use of common situational knowledge. When looking vertically on the packages, a uniform structure of the a-priori knowledge and its order of usage in terms of processing steps according to the transformers of the CP can be recognised.

2.3 The Framework: From Theory To Implementation

An appropriate architecture for the development of cognitive systems is COSA (Cognitive System Architecture), which offers a framework to implement applications according to the theory of the Cognitive Process. It provides an inference mechanism and various means, which make it possible for the developer of an application to use concepts like 'belief', 'goal' and 'plan' rather than a programming paradigm based on a functional decomposition of an application. (Putzer & Onken, 2003)

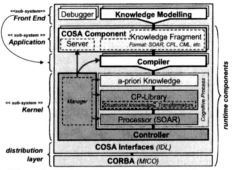

Figure 3: COSA – Cognitive System Architecture

COSA is composed of four building blocks (cf. Figure 3). The **kernel** implements the theory of the CP and does not contain any application-dependent information. Its only task is to generate behaviour from knowledge. In the current implementation, it is based on Soar (Laird, Newell, & Rosenbloom, 1987), which is a general rule-based architecture for developing systems that exhibit intelligent behaviour. The **application** is formed by several COSA-compliant application components, which correspond to packages (see section 2.2). The components provide the a-priori knowledge and may also contain servers with interfaces to the environment or for external calculations. The *front end* provides tools for the developer of an application, which help him to model the knowledge for the application. One part of the front end is CPL (Cognitive Programming Language), which provides a programming support to the implementation of the above-mentioned mental notions 'belief', 'goal', and 'plan' as concepts for knowledge modelling. Finally, the *distribution layer* is responsible for the communication among the modules of COSA. It ensures, that components and modules can run on different computers in a network.

3 Autonomous Mission Accomplishment

A pre-condition for co-operative behaviour of UAVs is the ability to accomplish a mission autonomously. To look into this subject and verify that the CP as underlying theory and COSA as engineering framework are suitable for the design and implementation of an autonomous cognitive system (cf. Ertl, Kriegel & Schulte, 2005), a simulator test bed has been set up, which is capable of simulating a military attack scenario. Besides the UAV, the guidance of which is developed, the scenario consists of a target in a hostile area, which has to be destroyed by the UAV, some SAM sites, and other aircraft. Some of the SAM sites are known a-priori at the beginning of the mission, others occur suddenly during the course of the mission.

3.1 Packages for Autonomous Mission Accomplishment

Two capabilities are the basis for single-ship autonomous mission accomplishment: 'perform safe flight' and 'perform mission'. These are mapped to packages of the CP (see section 2.2) and consecutively to COSA-Components (see section 2.3).

3.1.1 Package 'Perform Safe Flight'

The behaviour of the package 'perform safe flight' is driven by two top-level desires, namely *to follow a given flight plan*, and *to fly safely*. Situation interpretation is based on environment models such as 'own vehicle', 'other vehicle', and 'flight plan', while planning uses action alternatives like an 'evasion script'. Interfaces to the operator and the underlying vehicle including sensors serve as means for interaction with the environment. Figure 4 shows, how these models work together in case an aircraft crosses the flight path of the UAV. Based on input data, which indicate that there is a vehicle at a certain position, an object representing the concept 'other vehicle' is instantiated and becomes part of the belief. As the distance to the own vehicle is dangerously small, an instance of the desire 'avoid traffic' which is a sub-desire of 'flight safety' is instantiated. The UAV now pursues the goal to avoid the newly detected vehicle. The planning step determines that an evasion manoeuvre is suited to achieve this goal. Therefore, an instance of the model parameterised with appropriate autopilot values is created and executed step by step. (see Figure 4)

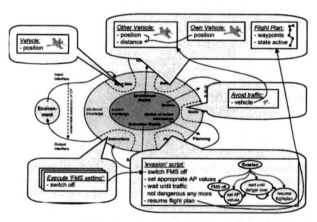

Figure 4: Package 'perform safe flight' considering the occurrence of dangerous traffic as example

3.1.2 Package 'Perform Mission'

As indicated previously, autonomous mission accomplishment is based on a mission order rather than a given flight plan. Therefore, the package 'perform mission' provides the capability to understand a given mission order. This may be incomplete in terms of what command and control (C^2) expects the UAV to do in the consequence of the mission order. Here, domain knowledge on tactical flying has to be deployed. Figure 5 illustrates the behaviour of the package 'perform mission', if the UAV receives a message containing a mission order to destroy the target. This does not explicitly contain the idea of returning to the home-base afterwards, although human operators would bear that in mind as part of their domain specific expertise.

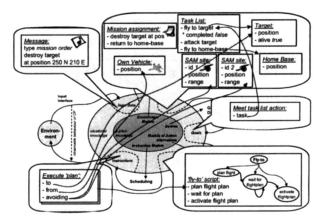

Figure 5: Package 'perform mission' considering the arrival of a new mission order as example

The UAV interprets the received message, which leads to the individual mission assignment. This consists of the elements, which are determined by the UAV as necessary steps having to be completed in order to accomplish the mission and comply with the mission order. In this simplified example, the mission assignment consists of the items 'destroy target' and 'return to home-base'. It is derived from knowledge about the usual course of a mission and possible implications of a mission order. A task list containing the tasks, the UAV should perform in order to accomplish the mission, is closely related to the mission assignment. Here, the tasks are to fly to the target, attack the target, and fly to the home base. As the first task is not completed yet, the desire to meet a not completed task list action, i.e. fly to target, is instantiated. The planning step creates an instance of a script, which is suitable for flying to a certain location. This is executed step by step. In order to execute the planning of a flight plan, the scheduler needs some information about e.g. the current position, the target position, and SAM sites, which are to be avoided. These are gathered from the 'belief' section.

3.2 Results

The packages 'perform safe flight' and 'perform mission' have been implemented using COSA and integrated in the simulation of the scenario. The UAV is enabled to successfully accomplish the mission, i.e. comply with the mission order, by activating the relevant goals and choosing appropriate actions. In unexpected situations, such as another aircraft crossing the flight path of the UAV, or an unknown SAM-site popping up, appropriate actions are taken.
Figure 6 shows the COSA shell, which is used for controlling the COSA kernel and for debugging purposes. Here, the currently bound components can be seen, including the packages 'perform flight' and 'perform mission', which have been described above. In Figure 7, the scenario described at the beginning of chapter 3 is visualised. In the depicted situation, the UAV pursues the active goal to fly to the target (see Figure 8, left), after it has received the mission order to destroy the target. In the following, the goal to avoid traffic becomes active due to a dangerously small distance to another aircraft (see Figure 8, right). As already described in section 3.1.1, the UAV performs an evasion manoeuvre and consecutively resumes normal mission accomplishment.

1169

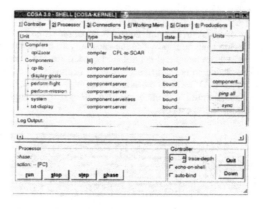

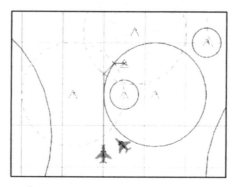

Figure 8: Desires (black) and active goals (magenta) before and during occurrence of traffic conflict

These successful simulator runs prove, that the CP as underlying theory and COSA as engineering framework are suitable for the design and implementation of an autonomous cognitive system.

4   Co-operative Mission Accomplishment

Often, a mission cannot be accomplished by a single UAV, because it does not have the required capabilities and/or sufficient resources. In other cases, a single UAV might have the required capabilities and resources, but working together improves the result or is more efficient. For these reasons, in the following, a team consisting of team mates with various capabilities and varying resources is considered. If all team mates work towards a common objective, they *co-operate* with each other. In most cases, this common objective is related to an externally given mission order. In order to enable co-operation, the team mates have to *co-ordinate* their activities, which requires *communication*. This may be explicit by exchanging messages, or implicit by observing the team mates and trying to reveal their intentions (Strohal & Onken, 1998). Presently, just explicit communication is considered, which requires the use of a common protocol. Co-ordination is organised by an appropriate method, whereas co-operation is mainly driven by goals. (Ertl & Schulte, 2004)

For a demonstration of co-operative behaviour, the scenario from chapter 3 is extended. The number of SAM sites is increased, and besides the UAV, which is capable of destroying the target, UAVs are added, which are capable of destroying SAM sites. In order to accomplish a mission within this scenario, the UAVs have to work together.

4.1 Selection of a Co-ordination Technique

As mentioned above, actions have to be co-ordinated among team mates in order to facilitate efficient co-operation. According to (Jennings, 1996), there are three common co-ordination techniques:

- **Organisational structure**
  Each team mate adopts or is assigned a 'role', which is closely connected to activities that have to be completed when holding this role. I.e. by adopting a role, team mate commit themselves to the associated elements of the common mission assignment.
- **Meta-Level Information Exchange**
  As the name indicates, team mates exchange information on a 'meta-level', i.e. they communicate their intents and commitments rather than detailed sequences of planned actions.
- **Central Planning**
  A central planner, which may for instance be located onboard one dedicated UAV or in a ground control station, co-ordinates the individual mission assignments of all team mates, and provides information on the task allocation to all mates. In contrast to the other methods, the team mates themselves are not involved in the allocation process.

The latter two techniques do not specify, how the allocation is actually done. Nevertheless, a general framework for the co-ordination process with respect to the required communication bandwidth and the flexibility to dynamically (re-)allocate assignments is defined.

Co-ordination using an *organisational structure* needs, if any, little communication for the allocation of responsibilities, but cannot dynamically re-allocate tasks, in case there is a relevant change of the state of the environment or a team mate drops out. Therefore, this method is especially suited for teams consisting of team mates, which differ with respect to their capabilities and cannot dynamically allocate assignments anyway.

Meta-level information exchange requires more communication amongst the team mates, but is more flexible with respect to dynamically (re-)allocating assignments. Unconstrained dynamic re-allocation of assignments can only take place, if the team mates have the same capabilities, i.e. each team mate can in principle commit to each assignment. In such a team, which communicates on an abstract level, resources (e.g. number and type of weapons, fuel, time, distance), opportunities (e.g. a SAM site, which has to be destroyed is located en route of one team mate, while another would have to go a long way round), and conflicts (e.g. a team mate can perform one or another assignment, but not both, because of missing resources) can be considered in order to gain an efficient distribution of assignments.

Co-ordination by the use of a *central planning unit* can take capabilities as well as resources, opportunities, and conflicts into account, provided that it is informed by the team mates. Therefore, a lot of communication amongst the team mates is necessary, which makes this method susceptible to communication loss. If the situation changes significantly, the central planning unit has to re-determine the individual mission assignments for each team mate.

Little communication bandwidth and robustness with respect to loss of communication and team mates is of great importance especially in the military domain. Therefore, central planning is not considered as being an option for the co-ordination of assignments in the UAV domain. Rather, a hybrid approach, which is illustrated in Figure 9, will be used. It shows a team consisting of four members, organised in two groups, whose members have the same capabilities (black/white). All team mates receive the same mission order. As it can be assumed, that all partners interpret it the same way, they will have the same common mission assignment. In a first step, the different capabilities of the groups are used to (statically) allocate the group mission assignments according to an organisational structure. Within the groups, meta-level information exchange will be used to derive the individual mission assignments from the group mission assignment, considering the resources, opportunities, and conflicts of the group members.

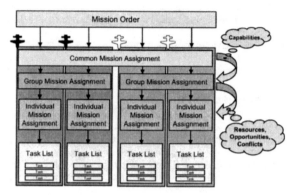

Figure 9: Hybrid approach to task allocation

4.2 Establishing Co-operation by Goals

On top of these considerations on co-ordination, Billings' widely known requirements for human-machine co-operation (Billings, 1997), which have been adapted for machine-machine co-operation (Ertl & Schulte, 2004), lead to the desires driving co-operation (see Figure 10). The top-level desires are to *form a team* and to *achieve the common objective*. In order to form a team, first of all, each mate has to know its team mates with respect to capabilities, intents, commitments, and resources, and the team mates have to support each other, i.e. answer to requests from team mates and send information messages if their own state has changed relevantly. To know the members of the team is a pre-condition for the decision, whether all team mates are involved and none is overburdened.

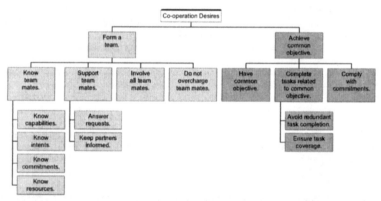

Figure 10: Desires for package 'co-operation'

The sub-desires of 'achieve common objective' are closely related to co-ordination of tasks or assignments, respectively. Once a common objective is available, it has to be ensured, that all tasks are covered and that no task is completed in redundance. If this is not the case, the appropriate desire becomes an active goal, and suitable problem solving mechanisms can be used to determine the necessary steps, which transform the current state (e.g. two team mates are committed to the same task) into the desired state (no redundant task completion).

Usually, messages requesting or providing information, or asking for or offering task completion will be used to achieve desired states related to co-operation. (Ertl & Schulte, 2004)

4.3 Communication as a Pre-Requisite

The communication protocol mentioned above ensures, that all team mates participating in a communication use the same structure for their messages. It does not restrict the content of a message, even though agent communication languages like KQML and the FIPA ACL define performatives such as 'request' or 'inform', which denote the type of a message (Wooldridge, 2002). In the military application domain, the use of protocols like MIDS/Link16 rather than these agent-oriented protocols is conceivable.

Here, communication amongst agents is treated as much as an action leading to a change of the environment as e.g. changing autopilot settings or dropping a weapon. Therefore, the content of a message is generated while planning actions.

4.4 Package for Co-operative Mission Accomplishment

In order to implement co-operative behaviour within the theory of the Cognitive Process, the package 'co-operation', has been designed. Together with the packages 'perform safe flight' and 'perform mission' (see section 3.1), it facilitates co-operative and autonomous mission accomplishment.

Figure 11 shows an example of co-operative behaviour during the phase, in which the members of a team get to know each other. A UAV receives a message requesting its resources. Therefore, the desire to support the team mates, specifically to answer requests, is instantiated. An appropriate action to achieve this goal, is to send a message containing the own resources to the UAV, which requested this information.

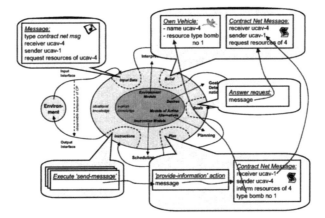

Figure 11: Package 'Co-operation' considering the arrival of a new message from a team mate as example

Currently, the goals to *know team mates* and to *support team mates* (cf. Figure 10) have been implemented, leading to communicative acts requesting and providing information. The information gathered about other team mates by such an information exchange, provides the basis for a model of the other team mates in the belief. This is essential for the task allocation process, which claims to consider capabilities, resources, opportunities, and conflicts of the team mates. The goals related to task allocation will be implemented according to the same methodology as the ones explained above, though leading to messages with different content.

5 Conclusion

In this paper, the Cognitive Process has been presented as a theoretical approach to a model of human information processing, which is in turn suitable for the design of artificial autonomous units. The Cognitive System Architecture COSA offers a framework for the implementation of applications according to the theory of the Cognitive Process. It provides both an inference mechanism which is in the current implementation based on Soar, and various means, which make it possible for the developer of an application to use concepts like 'belief', 'goal', and 'plan' rather than a programming paradigm based on a functional decomposition of an application.

COSA has been used to develop the guidance for a UAV represented as the packages 'perform safe flight' and 'perform mission'. The UAV is supposed to accomplish a military attack mission autonomously, i.e. based on an abstract mission order rather than a detailed flight plan in form of e.g. waypoints. Successful simulator runs have proven that COSA and the Cognitive Process are appropriate for the development of cognitive systems in the flight domain.

A step further is the design of several UAVs accomplishing a mission autonomously and co-operatively. Based on a consideration of co-operation and its implications co-ordination and communication, the design principles of the package 'co-operation' have been illustrated.

Future steps will include the integration and testing of the functionalities presented in this paper in a real-world environment, in which the military attack scenario is mapped to a fire fighting mission consisting of the same core elements. The UCAVs are represented by a model aircraft and a model helicopter guided by a human operator in a mobile ground control station (Ertl, Kriegel & Schulte, 2005).

References

Billings, C. E. (1997). Aviation Automation: The Search for a Human-Centered Approach. Mahwah, NJ: Lawrence Erlbaum Associates.

Ertl, C., & Schulte, A. (2004). System Design Concept for Co-operative and Autonomous Mission Accomplishment of UAVs. In: *Deutscher Luft- und Raumfahrtkongress 2004*, Dresden. 20th-23rd September 2004.

Ertl, C., Kriegel, M., & Schulte, A. (2005). Experimental Set-Up for the Development of Autonomous Capabilities and Operator Assistance in UAV guidance. In: *Proceedings of 6th Conference on Engineering Psychology & Cognitive Ergonomics, in conjunction with HCI International*, Las Vegas, USA, 22nd-27th July 2005.

Jennings, N. R. (1996). Coordination Techniques for Distributed Artificial Intelligence. In: O'Hare, G. M. P., & Jennings, N.R. (Eds.), *Foundations of Distributed Artificial Intelligence* (pp. 187-210). New York: Wiley.

Laird, J. E., Newell, A., Rosenbloom, P. S. (1987). Soar, an architecture for general intelligence. In: *Artif Intell*, 33(1), 1-64.

Putzer, H., & Onken, R. (2003). COSA – A generic cognitive system architecture based on a cognitive model of human behaviour. In: *Cogn Tech Work*, 5, 140-151.

Strohal, M., & Onken, R. (1998). Intent and Error Recognition as part of a knowledge-based cockpit assistant. In: *Proceedings of SPIE – The International Society for Optical Engineering*, Orlando, Florida, 13th-16th April 1998.

Wooldridge, M. (2002). An Introduction to MultiAgent Systems. Chichester: Wiley.

Integration of Human Domains to Enhance Decision Making Through Design, Process, Procedures and Organizational Structure

Jennifer McGovern Narkevicius, PhD

ARINC Engineering Services, LLC
44423 Airport Road, Suite 300, California, MD 20619
jmcgover@arinc.com

Michael Brown

SkillsNET Corporation
310 West Jefferson, Waxahachie, Texas 75165
Michael.Brown@skillsnet.com

Nancy Dolan

Chief of Naval Operations, N-125
FOB #2, Navy Annex, Room 2635, Washington, DC 20370-5120
nancy.dolan@navy.mil

Abstract

Emphasis for automated systems and enhanced cognition has been on the development and implementation of technology, the success of these systems hinges on the successful performance of the humans interacting with these systems to meet the operational capabilities. The System Engineering, Acquisition and PeRsonnel INTegration (SEAPRINT) program provides processes and identifies tool sets that allow successful implementation of decision-making systems that work and are scalable across organizational structures and working within the current structure while it transforms.

1   Introduction

While much of the emphasis on automated systems and enhanced cognition has focused on the development and implementation of technology, the success of these systems hinges on the successful performance of the humans interacting with these systems to meet the operational capabilities. This is true for any organization that wishes to optimize workflow performance. It is particularity relevant to the US Navy and its efforts to revolutionize both warfighting and peacekeeping mission performance. The Navy is focusing on a sailor centered approach which will impact the implementation of automated systems and the utilization of decision-making support. The System Engineering, Acquisition and PeRsonnel INTegration (SEAPRINT) program identifies toolsets, and processes that lay the groundwork that will allow successful implementation of decision-making systems that work and are scalable.

2   Systems Engineering

Thorough systems engineering practice demands meticulous requirements definition detailing the relationship between the human and the system. A number of authors have detailed the systems engineering process (Kossiakoff and Sweet, 2003, Martin, 1997). Human factors is the specialty engineering discipline that specifies the human component of the system. The interaction of systems engineering and human factors results in defining and understanding the needs and requirements of the humans using the system in the operational environment. This inclusion of the human as a major system component is part of the trade space in any system design. However, it is an essential part of the trade space for decision-aiding or automation systems. The challenging part is to understand this portion of the trade space and use it with respect to the remaining trade space to balance "human friendly" solutions for users and program managers.

Figure 1 below illustrates the major phases and iterative nature of the systems engineering process. While this process is quite good and appropriate to the design, development, and deployment of complex systems, systems of systems, and families of systems, the outputs of the process continue to fall below expected performance when deployed and used by representative users and maintainers in operational scenarios. It is clear that human systems integration will bring a great deal of improvement to systems engineering by ensuring the most contentious element of the system is included in the design and development of the system, throughout the systems engineering process.

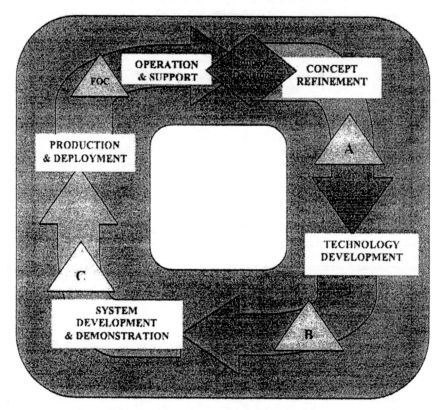

Figure 1. The Systems Engineering Process

3 Human Systems Integration

To define the requirements of humans as a major system component, it is essential to understand the inherent "capacity" of user populations and the operational environment in which they work. A number of authors have detailed Human Systems Integration and to some extent, the place of the human disciplines in systems engineering (Booher, 2003, Chapanis, 1996). This is more than the basic anthropometrics or cognitive capability of the average member of the user population. It requires a detailed description of the target audience of users and maintainers and explicit understanding of the knowledges, skills, and abilities (KSA's) of the people that will be operating (and maintaining) the system as well as other attributes that may impact total system performance. It is also essential to understand the work that will be performed. A number of authors have explored the definition and use of occupational information as well as the effects of organizational structure, business processes, and work structure (Cook, 1996, Kubeck, 1995, Peterson, et al, 1999, Sheridan, 2002, and Wilson and Corlett, 1995 are some examples of these diverse disciplines). These more diverse data must be included in systems engineering and trade space analyses to ensure that the system will perform as envisioned and specified in the operational environment. It is also necessary to address organizational issues. Many automated systems result in overt or covert changes in the organizational structure and business rules of both organizations regardless of size. These organizational changes can affect the work to be performed and must be considered as part of the overarching design. They must be reflected in the information architecture of the system as well, especially if there are automation or decision making support elements.

It is critical to truly understand the work and the context in which it will occur when using the new system. Automated systems have made many technological advances toward being more responsive or appropriate to the humans with whom they interact. However, the effect of the implementation of the system on the work performed by the component humans is not well understood or accounted for in design of those systems. The work the humans perform (including workflow) must be defined. That definition must be utilized locally in the human factors of the design and globally in the overall systems design. In addition, it is important to socialize that definition of the work to be performed into the organization and among the claimants of that work. These business processes, organizational structures, and occupational work must also be factored into the systems thinking and design. This includes the eliciting the information flows necessary to support the automated system and the human decision making processes that will be supported by a system and its performance.

To achieve the goal of successfully integrating humans into the systems engineering of automated systems, especially for decision-making, it is essential to achieve human systems integration, writ large. This requires actually integrating the human domains and applying the products of that successful integration to the design of automated decision-making systems.

4   SEAPRINT

Meaningful integration of the human domains requires more than just inserting human factors into the system design. It requires seeing the system performance as affected by all of the components of that system. All of the human centered domains contribute to the definition, specification, and utilization of the system. The context and predictability measures also contribute greatly to the HSI process. Trade offs must be made inside the human domains and the integration of the domains allows for better, more balanced trade-offs with other specialty engineering disciplines. The effects of manpower, personnel, training, human factors, safety, habitability, survivability, and Environmental Occupational Safety and Health (EOSH) impinge on augmented cognition and decision-making by clarifying the human cognitive, physical, organizational, and personal roles. These domains effect automation and decision-aiding and support by identifying the work to be performed; identifying the target audience; identifying successful and economical training; identifying the optimal design for information architectures, ensuring successful performance of the system not at the expense of the human component.

The SEAPRINT process was developed to achieve this human system integration and insert it appropriately throughout the systems engineering process. The process is not new but rather integrates over 200 established documents generated across the acquisition life cycle. SEAPRINT identifies seven actionable tenants which directly impact programs resulting in enhanced systems engineering through human systems integration. These seven tenants are:

1. Initiate HSI early
2. Identify Issues – Plan Analysis
3. Document/Crosswalk HSI Requirements
4. Make HSI a Factor in Source Selection
5. Execute Integrated Technical Process
6. Conduct Proactive Trade-Offs
7. Conduct HSI Milestone Assessments

and result in programs that have HSI requirements, analysis and issues articulated early and throughout the systems engineering acquisition phases. This ensures that the human domains exercise their full trade space in the same manner the other specialty engineering disciplines do. Figure 2 illustrates these HSI elements within the systems engineering phases.

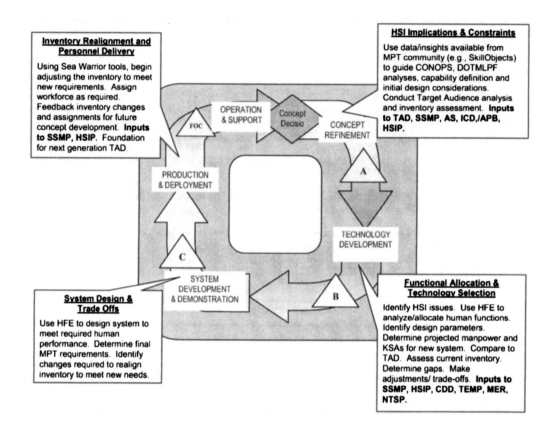

Inventory Realignment and Personnel Delivery

Using Sea Warrior tools, begin adjusting the inventory to meet new requirements. Assign workforce as required. Feedback inventory changes and assignments for future concept development. **Inputs to SSMP, HSIP.** Foundation for next generation TAD.

HSI Implications & Constraints

Use data/insights available from MPT community (e.g., SkillObjects) to guide CONOPS, DOTMLPF analyses, capability definition and initial design considerations. Conduct Target Audience analysis and inventory assessment. **Inputs to TAD, SSMP, AS, ICD,/APB, HSIP.**

System Design & Trade Offs

Use HFE to design system to meet required human performance. Determine final MPT requirements. Identify changes required to realign inventory to meet new needs.

Functional Allocation & Technology Selection

Identify HSI issues. Use HFE to analyze/allocate human functions. Identify design parameters. Determine projected manpower and KSAs for new system. Compare to TAD. Assess current inventory. Determine gaps. Make adjustments/ trade-offs. **Inputs to SSMP, HSIP, CDD, TEMP, MER, NTSP.**

Figure 2. HSI Elements in the Systems Engineering Process

Integration of HSI elements in the systems engineering process and phases allows a number of actions. These include setting realistic systems requirements, identifying future manpower and personnel constraints and evaluation of operator, crew, and maintainer workloads. It is possible to use tools and processes to test alternate system-crew functional allocations assess the work hours required for maintenance and assess performance during extreme conditions. Further processes and tools provide a platform for evaluating performance effects of personnel characteristics and training. Finally, focus of test and evaluation resources can be sharpened.

The SEAPRINT process has direct effect for augmented cognition and decision aiding systems and subsystems. By providing a platform to explore manpower, personnel, and training effects of specific designs and implementations, optimized systems can be explored, designed and fleshed out while requirements and function allocations are still being made. Further, the tools and processes deliver insight to the potential outcomes of requirements from the earliest phases of development. The resulting systems have an apposite mix of an optimized number of human operators or maintainers, an appropriately architected automation capability, are staffed by the optimum selection of people, correctly trained, and organizationally supported to execute suitable business rules.

Case studies performed to date have indicated the potential outcomes of automated systems for crew reduction scenarios. These case studies indicate that careful requirements definition, integration across the human domains, trade offs across the specialty engineering disciplines have the potential to result in more optimized systems that support human users and achieve the necessary capabilities.

References

Booher, H. R. (Ed.). (2003). *Handbook of Human Systems Integration*. Hoboken, NJ: John Wiley & Sons, Inc. A Wiley Interscience.

Chapanis, A. (1996). *Human Factors in Systems Engineering*. Hoboken, NJ: John Wiley & Sons, Inc. Wiley Series in Systems Engineering and Management, Andrew Sage, Series Editor.

Cook, M. A. (1996). *Building Enterprise Information Architectures: Reengineering Information Systems*. Upper Saddle River, NJ: Prentice Hall PTR.

Kossiakoff, A. & Sweet, W. N. (2003). *Systems Engineering Principles and Practice*. Hoboken, HJ. John Wiley & Sons, Inc. Wiley Series in Systems Engineering and Management. Andrew Sage, Series Editor.

Kubeck, L. C. (1995). *Techniques for Business Process Redesign: Tying it all Together*. New York, NY: John Wiley & Sons, Inc. A Wiley-QED Publication.

Martin, J. N. (1997). *Systems Engineering Guidebook: A Process for Developing Systems and Products*. Boca Raton, FL: CRC Press.

Peterson, N. G., Mumford, M. D., Borman, W. C., Jeanneret, P. R., & Fleishman, E. A. (1999). *An Occupational Information System for the 21st Century: The Development of O*NET*. Washington, DC: American Psychological Association.

Sheridan, T. B. (2002). *Humans and Automation: System Design and Research Issues*. John Wiley & Sons, Inc. Wiley Series in Systems Engineering and Management, Andrew Sage, Series Editor. And HFES Issues in Human Factors and Ergonomics Series, Vol. 3, Supervising Editor: David Meister.

Wilson, J. R. & Corlett, E. N. (Eds.). (1995). *Evaluation of Human Work: A Practical Ergonomics Methodology (2nd ed.)*. Philadelphia, PA: Taylor & Francis, Inc.

Implications of Adaptive vs. Adaptable UIs on Decision Making: Why "Automated Adaptiveness" is Not Always the Right Answer

Christopher A. Miller, Harry B. Funk, Robert P. Goldman, John Meisner, Peggy Wu

Smart Information Flow Technologies
Minneapolis, MN
{cmiller, hfunk, rgoldman, jmeisner, pwu@sift.info}

Abstract

In this paper, we begin by contrasting the "adaptive" automation systems prevalent in Augmented Cognition research and development with "adaptable" systems that provide the same range of modifications and behaviors but place control over those adaptations in the hands of the human operator. While there are many obvious reasons to seek adaptive systems, there are also some less well-known consequences to high levels of automation control in human-machine systems. We review the literature on human interaction with high levels of automation in complex and critical work environments to illustrate various problematic effects including loss of situation awareness, poorly tuned trust, skill degradation, unbalanced mental workload, lack of user acceptance and, most importantly, poorer overall human + machine system performance and decision making. We illustrate how these effects may be mitigated by maintaining a more adaptable approach to automation use and suggest an approach to flexible adaptability, our Playbook™ delegation approach, which seeks to provide the best of both approaches.

1   Introduction

Opperman (1994) distinguishes between "adaptive" and "adaptable" systems. In either case, flexibility exists within the system to adapt to changing circumstances, but his distinction centers on who is in charge of that flexibility. For Opperman, an adaptable system is one in which the flexible control of information or system performance automation is resides in the hands of the user; s/he must explicitly command, generally at run time, the changes which ensue. In an adaptive system, by contrast, the flexibility in information or automation behavior is controlled by the system. It is as if Opperman is implying (though not explicitly defining) a kind of "meta-automation" which is present and in control of the degrees of freedom and flexibility in information and performance automation subsystems in an adaptive system, but which is absent (and is replaced by human activities) in an adaptable one. It is unclear whether the Augmented Cognition community consistently uses Opperman's terms or makes his distinction, but it would seem that, in the majority of cases at least, when the phrases "adaptive system", "adaptive user interface" and "adaptive automation" are used in this community, they are used in Opperman's sense of a machine system which controls flexibility in information and performance subsystems, albeit in the service of the human.

Adaptive systems tend to have some distinct advantages over adaptable ones in terms of their impact on human + machine system decision making, and these advantages make them useful in a wide range of military and commercial contexts. By effectively delegating the "meta-automation" control tasks to another agent (that is, off loading them from the human), adaptive systems can frequently achieve greater speed of performance, reduced human workload, more consistency, a greater range of flexibility in behaviors and can require less training time than do human-mediated adaptable systems. On the other hand, by taking the human operator out of that portion of the control "loop", adaptive systems run some risks with regards to decision making that adaptable ones generally do not. Since this community is, perhaps, more familiar with the advantages of adaptive systems than the risks and the complimentary advantages of adaptable approaches, we will concentrate on the risks and disadvantages of adaptive systems below.

2 Disadvantages of Fully Adaptive Systems

Even when automation is fully competent to perform a function without human intervention or monitoring, there may still be reasons to retain human involvement. Increasing the level of automation of a given task and/or giving more tasks to automation, necessarily means decreasing the human role and involvement in those tasks. A wealth of research over the past 20 years points to some distinct disadvantages stemming from reduced human engagement and, by contrast, of advantages to be obtained from maintaining higher levels of human involvement in tasks—a characteristic of adaptable systems. A growing body of research has examined the characteristics of human operator interaction with automation and described the human performance costs that can occur with certain forms of automation (Amalberti, 1999; Bainbridge, 1983; Billings, 1997; Parasuraman & Riley, 1997; Parasuraman, Sheridan, & Wickens, 2000; Rasmussen, 1986; Sarter, Woods, & Billings, 1997; Satchell, 1998; Sheridan, 1992; Wickens, Mavor, Parasuraman, & McGee, 1998; Wiener & Curry, 1980). These performance problems are briefly summarized here.

2.1 Reduced Situation and System Awareness

High levels of automation, particularly of decision-making functions, may reduce the operator's awareness of certain system and environmental dynamics (Endsley & Kiris, 1995; Kaber, Omal, & Endsley, 1999). Humans tend to be less aware of changes in environmental or system states when those changes are under the control of another agent (whether that agent is automation or another human) than when they make the changes themselves (Wickens, 1994). Endsley and Kiris (1995) used an automobile driving decision making task with 5 levels of automation ranging from fully manual to fully autonomous and then asked subjects a series of situation awareness questions. In spite of the fact that there were no distracter tasks and subjects had no responsibilities other than either making driving decisions or monitoring automation in the making of them, results showed situation awareness for the rationale behind decisions was highest in the fully manual condition, intermediate for the intermediate automation levels and lowest for the full automation condition. Studies by Endsley and colleagues, suggest that a moderate level of decision automation providing decision support but leaving the human remains in charge of the final choice of a decision option is optimal for maintaining operator situation awareness.

Mode errors are another example of the impact of automation on the user's awareness of system characteristics (Sarter & Woods, 1994). A mode refers to the setting of a system in which inputs to the system result in outputs specific to that mode but not to other modes. Mode errors can be relatively benign when the number of modes is small and transitions between modes do not occur without operator intervention. For example, in using a remote controller for a TV/VCR, it is commonplace to make mistakes like pressing functions intended to change the TV display while the system is in the VCR mode, or vice versa. When the number of modes is large, however, as in the case of an aircraft flight management system (FMS), the consequences of error can be more significant. Mode errors arise when the operator executes a function that is appropriate for one mode of the automated system but not the mode that the system is currently in (Sarter et al., 1997). Furthermore, in some systems mode transitions can occur autonomously without being immediately commanded by the operator, who may therefore be unaware of the change in mode. If the pilot then makes an input to the FMS which is inappropriate for the current mode, an error can result. Several aviation incidents and accidents have involved this type of error (Billings, 1997; Parasuraman & Byrne, 2002).

2.2 Trust, Complacency, and Over-Reliance

Trust is an important aspect of human interaction with automation (Lee & Moray, 1992, 1994). Operators may not use a well-designed, reliable automated system if they believe it to be untrustworthy. Conversely, they may continue to rely on automation even when it malfunctions and may not monitor it effectively. Both phenomena have been observed (Parasuraman & Riley, 1997). Mistrust of automation, especially automated alerting systems, is widespread in many work settings because of the problem of excessive false or nuisance alarms.

The converse problem of excessive trust or complacency has also been documented. Several studies have shown that humans are not very good at monitoring automation states for occasional malfunctions if their attention is occupied with other manual tasks (Parasuraman, 1993). Parasuraman, Molloy and Singh (1993) showed evidence of increased complacency among users of highly, but not completely, reliable automation in laboratory settings. Metzger and Parasuraman (2001) reported similar findings for experienced air traffic controllers using decision aiding automation. In these studies, users effectively grant a higher level of automation to a system than it was designed to support by virtue of coming to accept automatically the system's recommendations or processed information even though the system sometimes fails. Riley (1994) documents a similar phenomenon, overreliance on automation, by trained pilots. In an experiment where the automation could perform one of a pair of tasks for the operator, but would occasionally fail, almost all of a group of students detected the failure and turned the automation off, while nearly 50% of the pilots failed to do so. While it is impossible to conclude that pilots' increased experience with (albeit, reliable) automation is the cause for this overreliance, it is tempting to do so.

Over-reliance on automation can also be manifest as a bias in reaching decisions. Human decision-makers exhibit a variety of biases in reaching decisions under uncertainty. Many of these biases reflect decision heuristics that people use routinely as a strategy to reduce the cognitive effort involved in solving a problem (Wickens, 1992). Heuristics are generally helpful but their use can cause errors when a particular event or symptom is highly representative of a particular condition and yet is extremely unlikely. Systems that automate decision-making may reinforce the human tendency to use heuristics and result in a susceptibility to "automation bias" (Mosier & Skitka, 1996). Although reliance on automation as a heuristic may be an effective strategy in many cases, over-reliance can lead to errors, as in the case of any decision heuristic. Automation bias may result in omission errors, when the operator fails to notice a problem or take an action because the automation fails to inform the operator to that effect. Commission errors occur when operators follow an automated directive that is inappropriate. Evidence for both types of errors was reported.

2.3 Skill Degradation
If placing automation in a higher, more encompassing role can result in complacency and loss of situation awareness, it is perhaps not surprising that it can also result in skill degradation if allowed to persist over time. The pilots of increasingly automated aircraft feared this effect with regards to psychomotor skills such as aircraft attitude control (Billings, 1997), but it has also been demonstrated to occur for decision making skills (Kaber, et al., 1999). In both cases, the use of an intermediate, lower level of automated assistance proved to alleviate skill degradation assuming the skills had been learned in the first place.

2.4 Unbalanced Mental Workload
Automation can sometimes produce extremes of workload, either too low or too high, That automation can increase workload one of the "ironies of automation", because many automated systems when first introduced are touted as workload-saving moves, and the technical justification for automation often is that it reduces mental workload and hence human error. But this does not always occur. First, if automation is implemented in a "clumsy" manner, e.g., if executing an automated function requires extensive data entry or "reprogramming" by human operators at times when they are very busy, workload reduction may not occur (Wiener, 1988). Second, if engagement of automation requires considerable "cognitive overhead," (Kirlik, 1993), i.e. extensive cognitive evaluation of the benefit of automation versus the cost of performing the task manually, then users may experience greater workload in using the automation. Alternatively, they may decide not to engage automation. (This is, of course, an even greater risk for adaptable automation than for adaptive automation.) Finally, if automation involves a safety-critical task, then pilots may continue to monitor the automation because of its potential unreliability. As Warm, Dember, and Hancock (1996) have shown, enforced monitoring can increase mental workload, even for very simple tasks. Thus any workload benefit due to the allocation of the task to the automation may be offset by the need to monitor the automation.

2.5 Performance Degradation
Most significantly, intermediate levels of human involvement in tasks can produce better overall performance of the human + machine system than either full manual or full automation levels, especially when human and automation roles are well structured and complimentary. In an experiment involving commercial aircraft navigation and route planning, Layton, Smith and McCoy (1994) provided human operators with one of three levels of automation support. In a 'sketching only' condition, (a highly manual approach), operators were required to create route plans using a map-based interface entirely on their own. The system would provide feedback about the feasibility of the human-proposed route in terms such as fuel loading, time of arrival and recommended altitudes. At the other end of the spectrum, a very high level of automation was provided by a system that automatically recommended a 'best' route according to its optimization criteria to the pilot. This 'full automation' mode was capable of providing supporting information about its recommended route plan, and of evaluating suggested alternatives, but only in response to explicit requests from the user. An intermediate level placed the automation in the role of supporting the

user-initiated route planning at a higher level of functionality. The user could ask for a route with specific characteristics (e.g., by way of Denver, avoiding turbulence greater than class 2, etc.) and have the system provide its best route that met such constraints. Each level was cumulative in that the user in the 'full automation' mode could choose to interact in full automation, intermediate or sketching only modes, or could switch between them.

Using this paradigm, Layton et al. found that humans in the intermediate and high automation conditions frequently explored more potential routing alternatives than they did in the highly manual condition; especially when the problem was complex and the range of potential considerations were large. In the sketching only (highly manual) condition, the process of arriving at a route was too difficult for the user to be able to try many alternatives consistently and fully. On the other hand, in the highly automated condition, users tended to accept the first route suggested by the automation without exploring it or its alternatives deeply. Even when they did explore it, the system's recommendation tended to narrow and bias their search. This outcome is similar to that seen in previous studies of complacency and automation bias. Users also tended to check the route provided for obvious mistakes rather than do the work of generating a route on their own to see if the computer's route was similar to the one they would have preferred. Users tended to take more factors into account more fully in their reasoning and the routing options selected were better in the intermediate automation condition. Particularly in trials when the automated route planning capabilities were suboptimal (e.g., because they failed to adequately consider uncertainty in future weather predictions), the intermediate level of automation produced better overall solutions. Layton, et al. suggest this was because users were better aware of the situation and hence, better able to both detect problems in the automation's recommended path, and to explore a range of alternatives quickly.

2.6 Decreased User Acceptance

Finally, one additional reason for preferring automation at the intermediate levels may be operator preference. Our own research and experience has shown that as automation begins to encroach on previously human-held tasks it suffers from a basic sociological problem: human operators want to remain in charge. This is probably particularly true of highly trained and skilled operators of complex, high-criticality systems such as aircraft, military systems, process control and power generation. For example, in developing the Rotorcraft Pilot's Associate (Miller, 1999), we interviewed multiple pilots and designers to develop a consensus list of prioritized goals for a "good" cockpit configuration manager. In spite of offering an advanced, automated aid capable of inferring pilot intent and managing information displays and many cockpit functions to conform to that intent, two of the top three items on the consensus list were "Pilot remains in charge of task allocation" and "Pilot remains in charge of information presented."

Similarly, Vicente (1999) sites examples of human interactions with even such comparatively mundane and low-risk automation as deep fryer timers in fast food restaurants (Bisantz, Cohen, Gravelle, & Wilson, 1996)., illustrating how human operators can become frustrated when they are forced to interact with automation which removes their authority to do their jobs in the best way they see fit. This review goes on to summarize extensive findings by Karasek and Theorell (1990) showing that jobs in which human operators have high psychological demands coupled with low decision latitude (the ability to improvise and exploit one's skills in the performance of a job) lead to higher incidences of heart disease, depression, pill consumption, and exhaustion.

3 A Tradeoff Space for Automation Effects

The above results, taken together, imply that applying sophisticated, adaptive and intelligent automation to manage information flow and equipment behavior for human consumers in complex systems and domains is not a panacea. Users in complex, high consequence domains are very demanding and critical of automation that does not behave according to their standards and expectations, and it has proven difficult to create systems that are correct enough to achieve user acceptance. Worse, as implied above, overly automated systems may well reduce the overall performance of the human + machine system if they do not perform perfectly both because they can reduce the human operator's awareness of the situation, degrade his/her skills and minimize the degree to which s/he has thought about alternate courses of action when the automation fails.

The tradeoff is not a simple two-way relationship between human workload and the degree to which human tasks are given to the automation, as is suggested above. Instead, we have posited a three-way relationship between three factors as illustrated in Figure 1:

1. the *workload* the human operator experiences in interacting with the system to perform tasks—workload that can be devoted to actually ("manually") executing tasks or to monitoring and supervising tasks, or to selecting from various task performance options and issuing instructions to subordinates in various combinations,
2. the overall *competency* of the human + machine system to behave in the right way across a range of circumstances. Competency can also be thought of as "adaptiveness"—which is not to say the absolute range of behaviors that can be achieved by the human + machine system, but rather the range of circumstances which can be correctly "adapted to", in which correct behavior can be achieved.
3. the *unpredictability* of the machine system to the human operator—or the tendency for the system to do things in ways other than expected/desired by the human user (regardless of whether those ways were technically right). Unpredictability can be mitigated through good design, through training and through better (more transparent) user interfaces, but some degree of unpredictability is inherent whenever tasks are delegated and is a necessary consequence of achieving reductions in workload from task delegation.

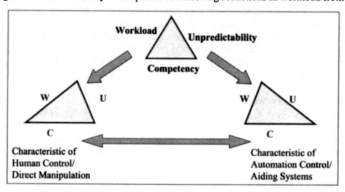

Figure 1. Three-way trade off space and its implications for a spectrum of control alternatives.

As can be seen in Figure 1, human controlled, "direct manipulation" systems (Schneiderman, 1997) are those in which an increased share of the responsibility for achieving a given level of competency has been given to the human. Such systems necessarily entail more workload for the human, but mitigate uncertainty to the human by keeping him or her "in charge" of how things are done. By contrast, automation-controlled systems are those in which the responsibility for achieving the same level of competency has been placed in the hands of automation. Such systems characteristically reduce workload for the human operator, but only at the cost of greater unpredictability. It is possible to achieve a given level of competency through either an expansion in workload or an expansion in unpredictability—or various mixes in between. One implication of this three way relationship is that it is generally impossible to achieve both workload reduction and perfect predictability in any system which must adapt to complex contexts. Another important implication is that the endpoints of full automation vs. full human control describe a spectrum with many possible alternatives in between. Each alternative represents a different mix of task execution and monitoring activities that are the responsibility of the human operator versus that of the system automation. The spectrum of alternatives that result is roughly equivalent to the spectrum of choices that lies between Direct Manipulation (Schneiderman, 1997) and Intelligent Agent interfaces (Maes, 1994).

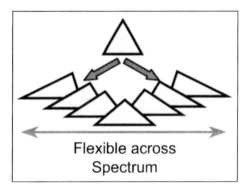

Figure 2. Augmented Cognition technologies (and their like) enable real-time, dynamic flexibility in adapting a design across the spectrum.

At issue, however, is who decides what that mix of automation and human responsibility should be, and when. Traditionally, designers have made that decision statically at design time and called it "function allocation". Increasingly, however, sophisticated world- and operator-sensing technologies (such as Augmented Cognition technologies) are making it possible to adapt that allocation during execution time in a highly dynamic fashion. The effect is as illustrated in Figure 2—the ability to dynamically choose where in the spectrum of human vs. automation task allocation the system should be "designed" at any point in time. The question that this capability raises, however, is where the decision about how to allocate tasks or functions should reside: with the human operator (making the resulting system an "adaptable" one in Opermann's sense) or with automation (making it an "adaptive" one).

In general, the goals of making that decision an "adaptive" one are well understood—decreased workload for human operators, improved human + machine system performance through making better use of both more reliable automation capabilities and more efficient use of human resources. Nevertheless, as we saw above, placing the control of the decision about how to adapt in the hands of automation can have unexpected and undesirable consequences on both human acceptance and overall human + machine system behavior. While retaining that decision making function in the human's hands has drawbacks of its own, we tend to favor more of an "adaptable" approach to interacting with dynamic automation, though the timing and even the scope of that decision can also be adapted as needed by the human operator. Below, we will describe the advantages we anticipate for more adaptable systems, while concluding with a brief description of an interaction architecture we are developing to allow the human operator to control the degree of authority s/he delegates to automation very dynamically and flexibly throughout the context of use.

4   Proposed Effects of Increasing Adaptability

There is reason to believe that adaptable systems do not suffer from the same set of problems as those described above. By keeping the operator in active charge of how much and what kind of automation to use when, we keep him or her more actively "in the loop" and, hence, more aware of more aware of how the system is or should be performing. By maintaining that level of involvement over automation behaviors, we would expect the following effects, each a mitigation or amelioration of the detrimental effects of high levels of automation as described above:

- Increased awareness of the situation and of system performance
- Requiring operators to make decisions about when to use automation, and to instruct automation in what behaviors to exhibit should produce better tuning of trust and better automation reliance decisions
- Allowing the user to perform tasks when needed or desired should keep skills more active and decrease the effects of skill degradation
- Allowing users more control over how much automation to use when will put them in a better position to manage their mental workload and keep it balanced

- If users can make good judgments (or simply better judgments than adaptive automation) about how much automation to use when to best compliment their workload, skills and capabilities, then a more nearly optimized mix of human and automation performance should result, along with the avoidance of performance degradation effects associated with full automation.
- Leaving the user in charge of when and how to use automation will enhance his/her sense of remaining in charge of automation performance, leading not only to a greater degree of acceptance, but also to a sense of being *primarily* responsible for overall task performance—in turn leading to greater attention and concern for the situation and all aspects of system performance.

Of course, adaptable systems have their own set of strengths and weaknesses. While an adaptable system would be expected to provide the benefits described above, it would suffer from increased workload on the part of the human operator and, perhaps, reduced overall task capacity for the human + machine system due to that workload. Indeed, while the human would be expected to have greater awareness of those aspects of the system to which s/he attended in an adaptable system, it might well be the case that fewer aspects/tasks could be attended to overall due to the attentional requirements placed on each task.

In fact, the work of Kirlik (1993) illustrates some of the downsides to adaptable automation. Kirlik developed a UAV simulation in which a human operator was in charge of manually controlling UAVs to have them visit various scoring points while simultaneously flying and navigating his/her own simulated helicopter. The pilot's own helicopter could be flown either manually or by putting it in autopilot mode. In this sense, the piloting of the own ship was an "adaptable" automation task in Opperman's terms. Kirlik performed a Markov Decision Process analysis to determine where decisions to use the autopilot would be optimal given a range of assumptions about (1) how much better or worse the human pilot was at the task than the autopilot, (2) how much time it took to engage and disengage the autopilot, and (3) the degree of inefficiency (as represented by a penalty for non-performance) for not having tasked the UAVs. The results of Kirlik's mathematical analysis showed distinct regions in which deciding to use the auto-pilot would and would not be optimal and, especially, showed that the effects of decision and execution time could eat into the effective performance of automation—implying both that the task of managing automation adds to the human's workload and may make it "more trouble than it's worth" to activate automation, and further implying that if that management task were to be successfully automated (as is the goal for adaptive automation) then there would be a greater likelihood of obtaining value from other forms of task automation. More importantly, in an experiment involving graduate students in the role of pilot, Kirlik found that subjects regularly avoided the least optimal strategies, but were inconsistent in their ability to find the most optimal ones.

In short, adaptable automation is no more a panacea than is adaptive automation. Humans may not always have the time or the expertise to choose the best forms of automation or the best times to use automation. Fortunately, however, we are not required to make hard choice between the adaptive and adaptable alternatives. We must strive to design a proper relationship between human operators and their automation that allows both parties to share responsibility, authority and autonomy over many work behaviors in a safe, efficient and reliable fashion. We are working on developing an approach that, while fundamentally adaptable in nature, nevertheless allows the human to "delegate" large spheres of authority within which adaptive behaviors are welcome and expected. This approach will be outlined in the next section below.

5   An Approach to Integrating Adaptive and Adaptable Systems

We have been exploring an approach to human interaction with complex automation that we call "delegation" because it is patterned on the kinds of interactions that a supervisor can have with an intelligent, trained subordinate. Human task delegation within a team or organizational setting is, overall, an adaptable system, in Opperman's sense, since the human supervisor can choose which tasks to hand to a subordinate, can choose what and how much to tell the subordinate about how (or how not) to perform the subtasks s/he is assigned, can choose how much or how little attention to devote to monitoring, approving, reviewing and correcting task performance, etc. Nevertheless, there can be substantial elements of adaptive system behavior within a delegation interaction in two ways. First, within tasks delegated to the subordinate (in this case, to automation), the subordinate may have full or partial authority to determine how to perform those tasks—including, potentially, what information to report to and even what tasks to ask from the supervisor. Second, there are opportunities for a good subordinate to take initiative within delegated roles to suggest tasks that need to be done and information that needs to be attended to—perhaps even supplying recommendations for who should do those tasks.

In work developing an interface for Uninhabited Combat Air Vehicles (UCAVs), we have explored a method of interacting with automation that attempts to more closely emulate human delegation relationships. In brief, our solution is to allow human operators to interact with advanced automation flexibly at a variety of automation levels and on a task-by-task basis. This allows the operator to smoothly adjust the 'amount' of automation s/he uses and the level at which s/he interacts with the hierarchy of tasks or functions to be performend depending on such variables as time available, workload, criticality of the decision, degree of trust, etc—variables known to influence human willingness and accuracy in automation use (Parasuraman and Riley, 1997].

This does not eliminate the dilemma presented in Figure 1, but it mitigates it by allowing operators to choose various points on the spectrum (as illustrated in Figure 2) for interaction with automation. The fundamental tradeoff between workload and unpredictability remains, but the operator is now put in charge of choosing a point in that tradeoff space. This strategy follows both Rasmussen's (Rasmussen & Goodstein, 1987) and Vicente's (1999) advice that operators should be allowed to 'finish the design' of the system at the time, and in the context, of use. This approach allows the user more control and authority over how and when s/he interacts with automation—and how that automation behaves. Therefore, it should address the desire to remain in charge that operators feel.

The trick, of course, is to design such systems so that they avoid two problems. First, they must make achievable the task of commanding automation to behave as desired without excessive workload. Second, they must ensure that resulting commanded behavior does, in fact, ensure safe and effective overall human + machine system behavior We have created a design metaphor and system architecture that addresses these two concerns. Our approach to enabling, facilitating and ensuring correctness from a delegation interface, we call a Playbook™— because it is based on the metaphor of a sports team's book of approved plays, with appropriate labels for efficient communication and a capability to modify, constrain, delegate and invent new plays as needed and as time permits.

We have written extensively about the Playbook architecture and applications elsewhere (Miller, Pelican and Goldman, 2000; Miller, et al., 2004; Miller and Parasuraman, submitted) and will not repeat that work here. Instead, we will conclude by pointing out three important attributes of delegation systems relevant to the creation of Augmented Cognition systems and the integration of the two:

1. The supervisor in a human-human delegation setting, and the operator of our Playbook™, maintains the overall position of authority in the system. It is not just that subordinates react to their perceptions of his/her intent, but rather they take explicit instructions from him/her. Subordinates may be delegated broad authority to make decisions about how to achieve goals or perform tasks, but this authority must come from the supervisor and not be taken autonomously by the subordinate. The act of delegation/instruction/authorization is, itself, important because it is what keeps the human supervisor "in the loop" about the subordinates activity and authority level. While it costs workload, if the system is well-tuned and the subordinate is competent, then that workload is well spent and results in net payoffs.
2. Even within its delegated sphere of authority, the subordinate does well to keep the supervisor informed about task performance, resource usage and general situation assessment. Beyond simple informing, the delegation system should allow some interaction over these parameters—allowing the supervisor to intervene to correct the subordinate's assumptions or plans on these fronts. The supervisor may choose not to do so, but that should be his/her choice (and again, the making of that choice will serve to enhance awareness, involvement, empowerment and, ideally, performance).
3. There remain substantial opportunities for Augmented Cognition technologies to improve the competency of subordinate systems, and the ability for the human supervisor to maintain awareness of and control over them. Chief among these are methods to improve the communication of plans and recommendations between human and machine systems, to improve negotiation of plans (and plan revisions) so as to take best advantage of the strengths and weaknesses of both human and machine participants, and to provide plans, recommendations and other information when it is needed so as to improve uptake. Note that this last opportunity must be subservient to the first described above—the human supervisor should remain in charge of information presentation. A good subordinate must know when information or plans beyond what the supervisor has requested will be useful and valuable—but

s/he must also know when and how to present them so as not to interrupt the supervisor's important ongoing thoughts and actions.

In short, after years of attempting to design purely adaptive systems, in Opperman's sense, we are skeptical about their utility in high complexity and high criticality domains. Instead, we opt for a more nearly adaptable approach that leaves the decision about when and what kind of automation to be used in the hands of a human operator/supervisor. Nevertheless, Augmented Cognition technologies have an important role to play in both types of systems.

6   Acknowledgements

The authors would like to thank Raja Parasuraman for his help in compiling earlier versions of this paper.

7   References

Amalberti, R. (1999). Automation in aviation: A human factors perspective. In D. Garland, J. Wise & V. Hopkin (Eds.), *Handbook of aviation human factors* (pp. 173-192). Mahwah, NJ: Erlbaum.

Bainbridge, L. (1983). Ironies of automation. *Automatica, 19*, 775-779.

Billings, C. (1997). *Aviation automation: The search for a human-centered approach.* Mahwah, NJ: Erlbaum.

Bisantz, A., Cohen, S., Gravelle, M., & Wilson, K. (1996). "To cook or not to cook: A case study of decision aiding in quick-service restaurant environments," Georgia Institute of Technology, College of Computing, Atlanta, GA, Report No. GIT-CS-96/03.

Endsley, M. & Kiris, E. (1995). The out-of-the-loop performance problem and level of control in automation. *Human Factors, 37.* 381-394.

Kaber, D., Omal, E. & Endsley, M. (1999). Level of automation effects on telerobot performance and human aoperator situation awareness and subjective workload. In M. Mouloua & R. Parasuraman, (Eds.), *Automation technology and human performance: Current research and trends*, (pp. 165-170). Mahwah, NJ: Erlbaum.

Karasek, R. & Theorell, T. (1990). *Healthy work: Stress, productivity, and the reconstruction of working life.* New York: Basic Books.

Kirlik, A. (1993). Modeling strategic behavior in human-automation interaction: Why an 'aid' can (and should) go unused. *Human Factors, 35*, 221-242.

Layton, C., Smith, P., & McCoy, E. (1994). Design of a cooperative problem solving system for enroute flight planning: An empirical evaluation. *Human Factors, 36(1),* 94-119.

Lee, J., & Moray, N. (1992). Trust, control strategies, and allocation of function in human-machine systems. *Ergonomics, 35.* 1243-1270.

Lee, J., & Moray, N. (1994). Trust, self-confidence, and operators' adaptation to automation. *International Journal of Human-Computer Studies, 40,* 153-184.

Maes, P. (1994). Agents that reduce work and information overload. *Communications of the Association for Computing Machinery, 37(7),* 31-40.

Metzger, U. & Parasuraman, R. (2001). The role of the air traffic controller in future air traffic management: An empirical study of active control versus passive monitoring. *Human Factors, 43,* 519-528.

Miller, C. (1999). Bridging the Information Transfer Gap: Measuring Goodness of Information Fit. *Journal of Visual Language and Computation, 10,* 523-558.

Miller, C., Pelican, M. and Goldman, R. (2000). "Tasking" Interfaces for Flexible Interaction with Automation: Keeping the Operator in Control. In *Proceedings of the Conference on Human Interaction with Complex Systems*. Urbana-Champaign, Ill. May.

Miller, C., Goldman, R., Funk, H., Wu, P. and Pate, B. (2004). A Playbook Approach to Variable Autonomy Control: Application for Control of Multiple, Heterogeneous Unmanned Air Vehicles. In *Proceedings of FORUM 60, the Annual Meeting of the American Helicopter Society*. Baltimore, MD; June 7-10.

Miller, C. and Parasuraman, R. (submitted). "Designing for Flexible Human-Automation Interaction: Playbooks for Supervisory Control." Submitted to *Human Factors*.

Mosier, K. & Skitka, L. (1996). Human decision makers and automated decision aids: Made for each other? In R. Parasuraman & M. Mouloua, (Eds.), *Automation and human performance: Theory and applications* (pp. 201-220). Mahwah, NJ: Erlbaum.

Opperman, R., (1994). *Adaptive user support*. Hillsdale, NJ; Erlbaum.

Parasuraman, R. (1993). Effects of adaptive function allocation on human performance. In D. J. Garland and J. A. Wise (Eds.), *Human factors and advanced aviation technologies* (pp. 147-157). Daytona Beach: Embry-Riddle Aeronautical University Press.

Parasuraman, R., & Byrne, E. A. (2002). Automation and human performance in aviation., In P. Tsang & M. Vidulich (Eds.) *Principles and practice of aviation psychology*. Mahwah, NJ: Erlbaum, in press.

Parasuraman, R. Molloy, R. & Singh, I. (1993). Performance consequences of automation-induced 'complacency,' *Int. Jour. of Av. Psych* 3, 1-23.

Parasuraman, R. & Riley, V. (1997). Humans and automation: Use, misuse, disuse, abuse. *Human Factors, 39,* 230-253.

Parasuraman, R., Sheridan, T. & Wickens, C. (2000). A model for types and levels of human interaction with automation. *IEEE Transactions on Systems, Man, and Cybernetics.—Part A: Systems and Humans, 30,* 286-297.

Rasmussen' J. (1986). *Information processing and human-machine interaction*. Amsterdam: North-Holland.

Rasmussen, J. & Goodstein, P. (1987). Decision support in supervisory control of high-risk industrial systems. *Automatica, vol. 23,* pp. 663-671.

Riley, V. (1994). *Human use of automation*. Unpublished doctoral dissertation, University of Minnesota.

Sarter, N. &Woods, D (1994). Pilot interaction with cockpit automation II: An experimental study of pilots' model and awareness of the flight management system. *Int. Jour. of Av. Psych., 4,* 1-28.

Sarter, N., Woods, D. & Billings, C. (1997). Automation surprises. In G. Salvendy, (Ed.) *Handbook of human factors and ergonomics, 2nd ed.* (pp.1926-1943). New York:Wiley.

Satchell, P. (1998). *Innovation and automation*. Aldershot, UK: Ashgate.

Sheridan, T. (1992). *Telerobotics, automation, and supervisory control*. Cambridge, MA: MIT Press.

Shneiderman, B. (1997). Direct manipulation for comprehensible, predictable, and controllable user interfaces. *Proceedings of the ACM International Workshop on Intelligent User Interfaces*, New York pp. 33-39.

Vicente, K. (1999). *Cognitive work analysis: Towards safe, productive, and healthy computer-based work*. Mahwah, NJ; Erlbaum.

Warm, J. S., Dember, W. N., & Hancock, P. A. (1996). Vigilanceand workload in automated systems. In R. Parasuraman & M. Mouloua (Eds.), *Automation and human performance: Theory and applications* (pp. 183–200). Mahwah, NJ: Erlbaum.

Wickens, C.D. (1992). *Engineering psychology and human performance. 2nd ed.* New York: Harper Collins.

Wickens, C. (1994). Designing for situation awareness and trust in automation. In *Proceedings of the IFAC Conference*. Baden-Baden, Germany, pp. 174-179.

Wickens, C., Mavor, A., Parasuraman, R. & McGee, J. (1998). *The future of air traffic control: Human operators and automation*. Washington DC: National Academy Press.

Weiner, E. (1988). Cockpit automation. In E. Weiner & D. Nagel, (Eds.), *Human factors in aviation* (pp.433-461). San Diego: Academic.

Wiener, E., & Curry, R. (1980). Flight-deck automation: Promises and problems. *Ergonomics, 23,* 995-1011.

Supervising UAVs: Improving Operator Performance by Optimizing the Human Factor

Leo van Breda
Chris Jansen
Hans (J.A.) Veltman

TNO Defence, Security and Safety; Business Unit Human Factors
P.O. Box 23, 3769ZG Soesterberg, The Netherlands
leo.vanbreda@tno.nl
chris.jansen@tno.nl
hans.veltman@tno.nl

Abstract

Tele-operated unmanned aerial vehicles (UAVs) have no operators on board and therefore enable extension of the present sensing and communication capabilities in civil and military missions, without unnecessarily endangering personnel or deploying expensive material. One should also realize that tele-operation from a control centre may result in new and formerly unknown human factors problems. For this reason, TNO Human Factors has been performing UAV related research for more than ten years, focusing on the identification of possible human factors problems in UAV supervisory control. The present paper gives an overview of exploratory human-machine interface optimization studies, and reports the findings to overcome specific problems that are inherent in remote camera and platform operation.

Important points of departure for the studies are a UAV mission analysis and an assessed order of possible missions. Since important mission elements involve target location, classification (target acquisition), and damage assessment, operator-in-the-loop performance remains an important issue, depending on the level of automation. We acquired more knowledge on the functioning of the human operator dealing with remote sensing and control, with the emphasis on aspects related to automation and interface aspects. For automation this involved, among others, the distribution of tasks between the human operator and the unmanned system, automation and operator situation awareness, the effects of automation breakdowns and adaptive automation. The interface aspects mainly involved the viewing systems and control devices that form the interface between the remote operator and the UAV. This concerned, among others: image characteristics of the remote sensor, application and consequences of head-slaved viewing systems, effects of time delays, control device configurations and the application of force feedback.

The paper discusses a series of studies that focuses on the negative effects of degraded information, and the possibilities to compensate these effects by innovative human-machine interface designs supporting the operator in charge. Important point of departure was that the improvements did not result in additional claims on the capacity of the up/down data-link. The applied techniques included the use of graphic overlays, ecological interface design, head-coupled control, and the use of prediction techniques. The results show that carefully designed human-machine interfaces are able to partially compensate specific image degradations while operator performance may significantly increase. Further experimental research focused on sub-optimal operator performance when UAV remote information is handled on board a moving platform, e.g., a helicopter. We tested methods to increase spatial awareness. The paper gives an overview of the studies and presents the main results.

1   Introduction

Tele-operated UAVs enable extended sensing and communication operations without an operator on board. In specific situations, UAVs are a prime candidate to (partially) take over specific operations, releasing sophisticated manned aircraft for more important tasks, and reducing the loss rate of manned aircraft in high-risk areas. However, one should realize that tele-operation from a control center, for example in combination with helicopter operations, may result in new and formerly unknown human factors problems. An important point of departure for the studies reviewed in this paper is the fact that a UAV mission particularly involves target location and classification (target acquisition). Important operator tasks in this respect are searching and tracking of targets by remote payload control, in combination with platform supervisory control tasks, possibly in a multiple UAV setting (Van Breda, 1997).

Different levels of automation may be employed in searching and tracking. Eisen and Passenier (1991) mention five levels: direct manual control, indirect manual control, supervisory control with occasional intervention, supervisory control with rare intervention, and finally fully autonomous operation. In general it is assumed that UAVs will for a great deal operate autonomously, particularly when the airframe is moving towards or from the target area. In that case, autonomous operation, or supervisory control with rare interventions is valid, and may enable one operator to monitor a fleet of UAVs.

In the current studies it is assumed that the most important source of information from the remote environment is the image of the on–board camera sensor. This information must be integrated with information on a tactical level (e.g. electronic maps, tactical information, radar images). However, due to data link restrictions, the camera images will be of less quality compared to images perceived directly by the human eye. Combined with the sensory deprivation inherent to a tele–operation situation, it is expected that operator performance degrades.

In UAV control, the following shortcomings are typical:

- No proprioceptive feedback is provided in the controls. In manual control mode, controls will not give feedback on camera or airframe behavior whatsoever;
- No vestibular feedback on airframe attitude. Because the operator is not seated in the vehicle, vestibular information on vehicle behavior (e.g. rotations) is missing;
- No proprioceptive feedback on viewing direction. When the observer is situated on–board a vehicle, proprioceptive information of muscles in neck and eyes provides exact information on the viewing direction. In a tele–operation setting, where visual information is presented on a fixed monitor, this information is missing;
- Limited spatial orientation (gravity). Although this factor may be of minor importance in UAV mission profiles, it is clear that the remote operator lacks this information;
- No direct feedback on control input. When the operator produces an input signal, the result of this action will not directly be available. Delayed feedback may seriously degrade manual control performance, ultimately leading to a go–and–wait strategy (bang–bang control, overshoot) when time delays are considerable;
- No auditory information. Although this information source is probably of minor importance, auditory information could inform on–board operators about certain cues, e.g., on vehicle behavior, on the presence of other airframes, etc.;
- Limited spatial resolution of the camera images. This is a crucial parameter in all camera control tasks (predominantly in detection and identification tasks). Enlarging the limited resolution per degree of visual angle by reducing the field size will also hamper operator performance (see below);
- A limited geometrical field of view (GFOV). A small GFOV may have several consequences. Firstly, the size of the GFOV is directly related to the required camera motion to scan a given area. Secondly, smaller field sizes will hamper the spatial integration of objects in the remote environment, will inhibit building up situational awareness, and may lead to operator disorientation (especially since the sensor slewing will be relatively quickly (Carver, 1987)). Thirdly, if the operator chooses to manually slew the sensor, the workload is expected to increase as the GFOV decreases. And finally, in tracking tasks, the motion of a target relatively to the monitor screen will increase, which will decrease tracking performance (Poulton, 1974);
- A zoomed–in camera image. The limited field size, the limited resolution, and the minimum stand–off distance combined will force the operator to zoom–in on targets. Because a zoomed–in camera image disturbs the normal relation between camera rotation and translational flow in the camera image, this may be an important factor in operator disorientation. Based on the translational flow, the camera rotation will be overestimated;
- Sometimes there are few reference points in the outside world (e.g., above sea surface). Although this is not an element inherent to the tele–operation setting, it is a complicating factor that will further impair the operator in developing a good sense of situational awareness. Furthermore, the operators often have difficulty keeping track of where they have already searched (Carver, 1987; 1988), and where threat areas lie;

- Limited update rate of the sensor image. Lower update rates of a camera sensor image will mainly affect dynamic tasks such as target tracking. Update rates below 10 Hz will decay the perception of the motion of the target, and of the camera and the platform. Very low update rates will lead to a snapshot like presentation of images, without any perceptual information on motion.

Prioritizing the above list, the important bottlenecks in sensor control are the quality of the visual information from the remote environment, and the lacking of (proprioceptive) cues on sensor viewing direction. Lack of auditory and vestibular input is considered of minor importance. Significant consequences for the UAV operator-in-the-loop system are poor tracking performance (resulting in large tracking errors, and losing the target), difficulties in assessing sensor, airframe, and target motions, confusion on the flying direction of the airframe, confusion on the viewing direction of the sensor, disorientation, and degraded situational awareness.

TNO Human Factors has been conducting experiments on UAV control for more than a decade. One of the results is that problems with time delays and low update rates can be compensated for by using a computer-generated grid on or around the camera image (Figure 1).

Figure 1: A simulated camera view of a ship at sea with a computer-generated grid overlay. The grid orientation adheres to the perspective geometry of the camera viewpoint to provide accurate motion feedback on control input.

The grid is a 'virtual' world that is drawn from the viewpoint of the sensor (camera) and moves with the same speed as the UAV, but does not have a time delay between the control input and the system output. The grid can be generated easily in the control station, by directly using the control input signal from the operator. In that case, the movements of the grid are not affected by time delays involved in sending and retrieving the data from the UAV. In addition, the update rate of the grid can be set sufficiently high for adequate motion perception. When there is a time delay in the sensor image, the grid will move sooner than the camera images, which makes control much easier due to direct feedback on the magnitude of the control input. Information presented by such a grid results in improved task performance (Korteling, Van Erp & Kappé, 1997; Van Erp & Van Breda, 1999). Besides control feedback, preview is important as well for adequate supervisory control performance (Van Breda, 1999). Especially when the camera is zoomed in to look at a small part of the environment, preview is very limited. Following a moving object on a road, for example, can be very difficult because the operator does not have information about the direction of the road in front of the object, and therefore, the operator can not anticipate adequately.

Camera sensor images are often presented in combination with digital 2D maps showing the position of the UAV and the part of the earth surface at which the camera is looking (footprint). Such a digital map increases situation awareness of the operator and provides preview, which makes supervisory sensor control easier. However, the problems with time delays and low update rates are not solved with a footprint. These problems can be reduced by using a second footprint that shows the predicted camera position (Van Erp & Kappé, 1998). The principle is similar to the grid that has been described above. From a human factors perspective, a 2D map is not always the optimal solution. A 2D map can be used when global information is required, especially when the map is presented north-up. A 3D map is more optimal for tasks for which local information is required (Wickens & Prevett; 1995). Global information refers to the position of objects relative to other objects in an earth reference frame (e.g., the position of a bridge relative to a city). Local information refers to the position of objects relative to the viewing position of the camera (e.g., a road ahead of a vehicle that has to be tracked).

2   Experiment 1: Support by augmented vision

TNO Human Factors developed a simulation environment in which human factors principles for UAV camera control are demonstrated and in which experimental studies are conducted. In an experiment we used this simulator environment to investigate the benefits of a 3D map with regard to operator performance and mental workload. We constructed a 2D map (oriented north-up) in which feedback about the control input was provided by means of an additional footprint that showed the predicted viewpoint of the camera. Operators were requested to find targets on roads and along wood edges. They could use the map to see to which part of the environment the camera was oriented. In one half of the conditions a 3D map was available together with the 2D map. The 3D map showed identical information, but was presented from the viewpoint of the camera. In some conditions, the quality of the camera images was manipulated by introducing a time delay of 1 second, or by lowering the update rate of the camera images to 3Hz. This had no effect on the 2D and 3D map. Furthermore, in one half of the conditions a secondary task had to be performed. This was done to see whether operators had more spare mental capacity in case the 3D map was available. Several performance measures were distracted and workload was measured with subjective and physiological measures.

The simulator environment consisted of two displays (Figure 2) that were each connected to a graphics computer (SimFusion, Evans & Sutherland). The resolution of each display was 1280 by 1024 pixels. The simulated camera could be manipulated 360 degrees horizontally and 90 degrees vertically by using a joystick. The zoom factor could be adjusted up to 20 times the original size by using a zoom- button on the joystick. The camera images was presented in three modes: 1) normal: low time delay (0.1 s) and image update rate between 10 and 30 Hz (depending on the amount information in the database), 2) low update rate (3 Hz, in combination with the normal time delay) and 3) large time delay (1 s, in combination with the normal update rate).

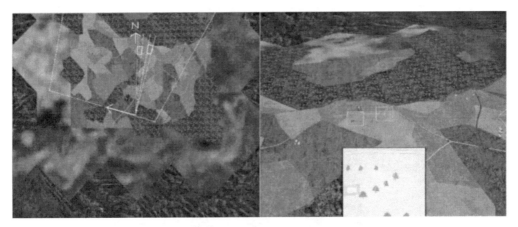

Figure 2: The left panel shows the 2D map with the position of the UAV in the center. The right panel shows the 3D map, which is drawn from the viewpoint of the camera. In both displays, the yellow footprint shows the part of the environment that corresponds to the camera image and the orange footprint shows the predicted position based on operator input signals. The camera image is presented at the bottom of the right display.

The left panel of the simulator displayed a detailed 2D map of the environment, north-up oriented. Apart from terrain information (roads, woods, buildings, etc.) the following relevant information was presented: waypoints and the route of the UAV were shown (yellow line), flight direction of the UAV (orange arrow) and the actual and predicted footprints (see 'Footprint' below for an explanation). The 2D map was a virtual 3D world that was viewed from above at 6500 m altitude. The fixed field of view was 60 degrees (H) by 48 degrees (V) showing an area of 7.2 by 5.6 km. The right panel of the simulator displayed the 3D map, a virtual 3D world presented from the viewpoint of the UAV camera. Observation altitude was 1200 m. The 3D map also had a fixed field of view of 60degrees (H) by 48 degrees (V). Terrain information was better visible in this 3D map because the altitude of the UAV was less than the virtual camera position used for the 2D map. The viewpoint for generating the 3D map depended on the camera control input of the operator, and the angular motion of this viewpoint was identical to the angular motion of the camera. Furthermore, because the vertical viewing angle was most often less than 90 degrees (i.e., when looking straight down), a perspective view of the database environment was obtained. The simulated camera images were presented in the lower center of the right panel and was sized 1/9 of the display (resolution 427 by 341 pixels).

Both the 2D and 3D map had a yellow footprint, showing the section of the map that corresponded to the camera images. The orange footprint provided direct feedback about the control input. In the conditions with low update rates and time delays, the yellow footprint followed the orange footprint. The size of the footprint could be adjusted by the zoom function, providing feedback about the zoom setting. With a zoom factor 1 the size of the footprint on the right display was identical to the size of the camera panel. In conditions in which the 3D maps were not drawn on the right display, the footprints remained visible to provide feedback about the zoom settings.

The primary task was to search and detect military vehicles, while the simulated UAV followed a planned route above a small village that was surrounded by several parcels of wood. The participants had to detect tanks by pressing a button on the joystick. The secondary task was a reaction and memory task. Two white squares were presented above each other. Every 10 s one of these squares became red or blue. A blue square contained a number "1" or "2" indicating the number of time this square had been red before. The participants had to press a 'yes' or 'no' button on the joystick after a blue square appeared, indicating whether the number was correct or not. For an adequate performance, participants had to inspect this display regularly. This task was mentally demanding because changing information had to be kept continuously in working memory. Furthermore, the task was presented on a small display that was placed in the center, below the two simulator displays.

The following performance measures were derived: 1) percentage of areas and roads that were inspected, 2) percentage of vehicles that were detected correctly, 3) performance on the secondary task (percentage correct and reaction times), and 4) camera control: standard deviation of the XY-position of the joystick, and the applied zoom factor. The following workload measures were used: 1) subjective effort rating after each condition (Rating scale Subjective Mental Effort, RSME; Zijlstra, 1993) and 2) physiological measures (heart rate, heart rate variability, respiration and blood pressure). Furthermore, the electroencephalogram (EOG) was measured to distract eye blinks and fast eye movements.

The most important results of the experiment are presented here. The 3D map significantly improved task performance (Figure 3). A larger area was inspected, performance on secondary task improved, indicating that the participants had more spare mental capacity, and the participants reported lower effort.

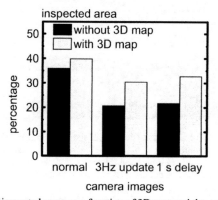

Figure 3: Percentage of inspected areas as a function of 3D map and the quality of camera images.

The positive effect of the 3D map was largest when the quality of the camera images was low. Note that adequate feedback about time delays and low update rates was always available in both the 2D and 3D map display. Without this information, performance would be much worse in the conditions with low update rates and long time delay.

The subjective effort measure showed substantial effects as function of all experimental factors (Figure 4), however, only heart rate showed a small effect as function of the secondary task. This may be due to the lower sensitivity of physiological measures for mental effort. We found such discrepancies between subjective and physiological measures more often (Veltman, Gaillard & Van Breda, 1997) depending of the type of task that is evaluated. Participants most often give higher effort ratings when a task becomes more difficult as a result of a reduced quality of information.

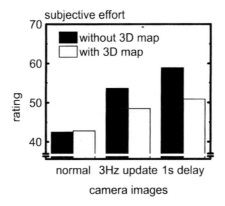

Figure 4: Subjective effort rating as a function of quality of the camera images and 3D map.

Physiological measures most often do not show differences in these situations, because investing more effort most often will not result in better task performance. When an additional task has to be performed, attention has to be divided between more tasks. In these situations, additional effort has to be invested in order to keep an adequate level of performance of the main task. This is reflected in both subjective and physiological measures. The difference between the subjective and physiological measures can be explained along this line of thought. Degrading the quality of the display makes the task more difficult, resulting in higher effort ratings, but does not affect physiological measures because investing more effort will not improve performance. For the secondary task, more effort was required in order to maintain an adequate level of performance on the main task.

3   Experiment 2 and 3: Manned-Unmanned Teaming

Most currently used UAVs are controlled from a (closed) containerized Ground Control Station (GCS). The GCS operators are generally concerned with only one UAV system. Here, the Mission Commander communicates the acquired intelligence to other units. In the last years, experimental setups have been developed in which the information from the UAV sensors is more directly presented to aircraft during a mission. A pioneer project was US Army's AMUST (Airborne Manned Unmanned System Technology; Fayaud 2001). Here, an Apache helicopter and a Hunter UAV form a team, where the UAV may perform all kinds of useful sidekick tasks, such as reconnaissance, laser-designation of targets and acting as decoy. The UAV can be either controlled from a Ground Control Station (GCS) or by the Apache's co-pilot/gunner.

The AMUST concept adds to the complexity of maintaining situation awareness (SA): The operator (the co-pilot / gunner) has to deal with at least three spatial frames of reference, namely the world, the helicopter and the UAV. Our first research questions were: Is such a co-pilot able to built up a SA involving multiple platforms, and how is this SA affected by being aboard of one of the platforms? These questions were investigated in the first study described below.

In the AMUST concept, the co-pilot interacts with the UAV. In future operation concepts, e.g., Network Enabling Capability, 'sensors' and 'shooters' are allocated more dynamically. In a second study, briefly described here as well, we investigated the possible benefits for the pilot in having available a UAV image in the cockpit when performing a simulated Close Air Support mission.

3.1 Multi-platform Situation Awareness

In the first study we focused on the situation awareness of a co-pilot who has to integrate the reference frame of the UAV with that of his own platform (a helicopter). In a simulator experiment, participants serving as UAV operators were seated behind a UAV console situated in a helicopter mock-up. The console displayed an electronic map of the geographical situation of UAV, helicopter and a few other objects using various formats. For each run, the participants watched the movements on the display for a couple of minutes after which their SA was assessed by means of a electronic questionnaire. Questions could be related to an earth-fixed coordinate system (compass) or be orientation-based (relative). Questions addressed the heli, UAV, or a formation of tanks. Our main dependent measurement was angular error, i.e. the difference between the direction indicated by the participant and the real direction. Our main experimental manipulation was the presentation of the outside scenery. On a 180-degree cylinder-shaped screen the visuals could be presented that corresponded with the movements of the helicopter. If presented, we simulated the condition in which the operator was as a co-pilot aboard the helicopter; if not presented the operator was stationed in a Ground Control Station.

The most important result from this study is displayed in Figure 5 (F(5,55)=2.51, p=.04). The Figure depicts for the six different question types the angular error in indicating the location of an object. For most of the questiontypes, the right bars (corresponding with the condition of the UAV operator aboard the helicopter) are much higher than the left bars (corresponding with the situation of an operator in a). This indicates that generally, performance is much worse in conditions in which we simulated that the UAV operator is aboard one of the platforms (except when questions were asked with respect that the operator's platform). Note that the task itself is not different for the stationary and moving operator: just look at the electronic map and built up a spatial SA as good as possible. From this result, we learned that it is indeed more problematic to maintain a multi-platform SA while being aboard a moving platform. Apparently, when the operator is in a situation where one spatial frame of reference is dominant (i.e., while aboard a helicopter), it is very hard to process spatial information from other perspectives.

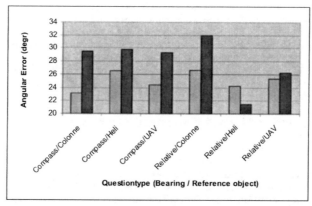

Figure 5: Angular error in indicating an object location for the six question types. The left (blue) bars refer to the situation of an operator in a Ground Control Station; the right (red) bars to the situation of an operator aboard a simulated aircraft.

3.2 Using real-time UAV images while conducting a Close Air Support mission

The above research has shown that performance drops when multiple reference frames need to be integrated while performing a spatial SA task. In our second study, we looked at a more critical situation in which misinterpretation of spatial information directly resulted in failure of a simulated Close Air Support mission. We looked at a situation in which a pilot used a UAV image, presented in the cockpit. As the UAV camera has a viewpoint that differs from that of the pilot, the pilot's interpretation of the spatial layout of the scenery may be prone to error (e.g., if the UAV flies in the opposite direction, the object on the left in the sensor image is actually on the right in the pilot's perspective). The aim of this research was to minimize the chance of such errors by rotating the UAV sensor image such that its orientation is always aligned with the pilot's spatial frame of reference. In our study, four military pilots performed several Close Air Support missions with six different display configurations. The orientation of the electronic map was either North-Up or Heading-Up. The UAV sensor image was either absent, present but non-aligned (i.e., unadjusted image orientation: the image is presented as seen from the UAV viewpoint), or present and aligned with the orientation of the electronic map (adjusted image orientation: the image is presented as if the sensor was placed on the helicopter).

The pilots reported that they generally preferred the Aligned UAV sensor image in combination with a Heading-Up map. This preference was reflected in the their performance, depicted in Figure 6: Targets were identified twice as fast (F(2,6)=16.24, p=0.004). Strikingly, flying performance was also better when an aligned UAV image was used.

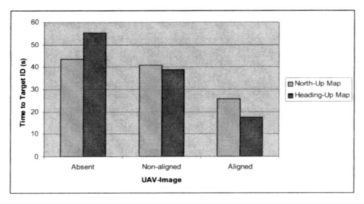

Figure 6: Time to target identifications for the six display configurations.

4   Conclusions

Using a 3D map for UAV sensor control improves task performance for reconnaissance tasks, especially when the sensor image quality is low. But we also found negative effects of the 3D map. Participants inspected more ground surface far away from the UAV and used a higher zoom factor when the 3D map was present. It can be argued that this is a positive effect of the 3D map; making is possible to search at larger distances. However, participants knew that there was insufficient time available for searching the entire area. With the 3D map, some participants had the tendency to search the complete area before they moved the camera to a new location. This might be due to the information about the distance between UAV and the part of the environment where the participant was looking at. The 2D provided much better feedback about this distance than the 3D map. When the participant used mainly the 3D map, it was more difficult to decide when they had to move to a new location. To overcome this problem, better information about the distance must be provided in the 3D map.

We used an additional screen to present the 3D map; the 2D map was always available. Each display representation has its advantages and disadvantages, depending on the type of task. The 2D display provided global information and the 3D display provided local information. Using two different displays is not an optimal situation for a control station, where lack of space is often a problem. A better solution will be to integrate the 2D and 3D display types. This can be achieved by using an adaptable display in which the operator can select the most appropriate presentation, per mission segment, and on his/her demand. By adjusting the viewing point from which the map is drawn, many different map presentations can be achieved. De Vries, Veltman and Van Breda (1999) investigated such an adaptable cockpit display in a high-speed flight task where pilots could select different view points: 2D north up, 2D heading up, 3D egocentric and 3D exocentric representations.

In comparing the (simulated) situations of a stationary UAV operator and a moving UAV operator, we learned that it is very hard to process spatial information from multiple perspectives simultaneously when one perspective is dominant (here, when being aboard a helicopter). In the experiments on presenting UAV images in the cockpit in performing a Close Air Support mission, we tried to facilitate the interpretation of (the non-dominant) UAV image information by aligning that perspective with the dominant perspective of the helicopter. Based on our results and pilot's reports, we conclude that the availability of a UAV sensor image in the cockpit of a fighter aircraft only improves mission performance when its orientation is aligned with the aircraft.

The current paper illustrates that tele-operation from a control center leads to new and formerly unknown human factors issues to be investigated.

References

Fayaud, G. R. (2001). The airborne manned unmanned system. *Unmanned Systems*, 19(4), 16-21.

Carver, E.M. (1987). RPV reconnaissance systems: the image interpreter's task. In: *Proceedings of the sixth International Conference on Remotely Piloted Vehicles* (pp 24.1 – 24.6). Bristol UK: University of Bristol.

Carver, E.M. (1988). *Search of imagery from airborne sensors – implications for selection of sensor and method of changing field of view.* Paper presented at First International Conference of Visual Search. Durham, UK.

De Vries, S.C., Veltman, J.A. & Van Breda, L. (1999). *Use of adaptable displays for fighter aircraft flight support.* Report TNO TM 1999 A011. Soesterberg, The Netherlands: TNO Human Factors.

Eisen, P.S. & Passenier, P.O. (1991). *Technology status of unmanned air vehicles.* Report TNO IZF 1991 A-31. Soesterberg, The Netherlands: TNO Human Factors Research Institute.

Korteling, J.E., van Erp, J.B.F. & Kappé, B. (1997). Visual support for the control of unmanned platforms. In D. Harris (ed.) *Engineering psychology and cognitive ergonomics* (pp. 55-63). Brookfield: Ashgate.

Poulton, E.C. (1974). *Tracking skill and manual control.* New York: Academic.

Van Breda, L (1999). *Anticipating behaviour in supervisory vehicle control.* PhD thesis. Delft, The Netherlands: Delft University Press.

Van Breda, L. (1997). An exploratory study of the man–machine interface for controlling Maritime Unmanned Vehicles. *Proceedings AGARD-CP-591 Subsystem Integration for Tactile Missiles (STIM) and Design and Operation of Unmanned Air Vehicles (DOUAV)*, pp(21:1–21:8). Ankara, Turkey: North Atlantic Treaty Organization.

Van Erp, J.B.F. & Kappé, B. (1998). *Effects of low update rate, time delay, and footprint prediction on camera control from unmanned systems.* Report TM 98 A003. Soesterberg, The Netherlands: TNO Human Factors Research Institute.

Van Erp, J.B.F. & Van Breda, L. (1999). *Human factors issues and advanced interface design in Maritime Unmanned Aerial Vehicles: A project overview 1995-1998.* Report TM-99-A004. Soesterberg, The Netherlands: TNO Human Factors.

Veltman, J.A., Gaillard, A.W.K. & Van Breda, L. (1997). Workload indices: Physiological measures versus subjective ratings. In D. Harris (Ed.), *Engineering Psychology and Cognitive ergonomics. Volume one: Transportation systems.* (pp. 269 275). Aldershot, Ashgate.

Wickens, C.D. & Prevett, T.T. (1995). Exploring the dimensions of egocentricity in aircraft navigation displays. *Journal of Experimental Psychology.* Vol. 1, No. 2, p.110 135.

Zijlstra, F.R.H. (1993). *Efficiency in Work Behaviour: A design approach for modern tools.* Thesis. Technical University of Delft.

Operator in the Loop?
Adaptive decision support for military air missions

Rob J Cottrell,Dave G Dixon
Tom Hope

QinetiQ Farnborough
Hants, GU14 OLX, UK
rjcottrell@qinetiq.com,

Robert M Taylor

Defence Science & Technology Laboratory
DSTL Farnborough
Hants, GU14 OLX, UK

Abstract

Military decision makers must make timely decisions under conditions of high stress. Recent advances in decision support technology offer the opportunity to improve operator decision effectiveness through real-time processing of mission data to offer consistent, timely advice. Applied properly, decision support technology will enable human decision makers to take on roles that no single person could fulfill unaided. To realize this vision, decision support systems must encode the tactical knowledge and reasoning processes of the personnel they support. This development will challenge our traditional concepts of human-computer interaction, posing complex problems that demand novel solutions.

This paper will consider two key issues – 'Adaptation' and 'Critiquing', as they apply to decision support for military air missions involving ground attack or reconnaissance roles. The need for adaptation stems from the dynamic nature of modern military operations, with the ability to tactically re-plan and re-prioritize while maintaining strategic objectives being of paramount importance. The need for critiquing stems from recognition of the fundamental limits applying to decision support systems - even comprehensive systems will be limited by knowledge-base constraints and by variations in the prevailing decision context which can potentially affect decision correctness. Thus there is a need for human operators to augment system behavior, bringing their inevitably greater understanding of the mission – or decision - context to bear. For effective critiquing, the Decision Support System (DSS) needs to be transparent such that an operator can easily discover why particular advice was offered.

These issues will be explored with specific reference to ongoing work in the development of an Adaptive Decision Support System (ADSS) for military air missions.

1 Introduction

Military operators must make timely decisions in stressful circumstances. Even the most proficient and well-trained operators can benefit from technology that supports their decision-making processes. Recent advances in decision support technology promise a revolution in military operations, significantly enhancing the effectiveness of those forces that adopt it. At the same time, decision support systems will challenge our traditional models of human-machine interaction, raising novel issues that demand creative resolution.

1.1 Military need for adaptive decision support

Tactical decision-making is becoming increasingly complex due to restrictive Rules Of Engagement (RoE), dynamically changing environments and enemy tactics, coalition interactions, and embedding of enemy targets and forces within civil or non-combatant populations. Problems are compounded when the decision makers have limited communications, either a dirge or deluge of data from sensors, limited time, and high workload, all of which frequently occur in airborne operations. This may apply equally to operators of inhabited aircraft, and Uninhabited Air-Vehicles (UAVs).

In asymmetric conflicts where political and military success is increasingly reliant on effects based targeting and operations, it is vital that coherence is maintained between strategic intent and tactical necessity. This places additional demands on decision makers. Therefore, where operators of air assets are responsible for tactical decisions, be they aircrew within the asset or remote operators of UAVs, there is a growing need for decision support. Such decision support may be achieved by a variety of techniques from fusion of data and information, better and selective presentation of information, automation or assistance in conducting tasks, cognition aids that enhance situational awareness, and finally technology that takes part directly in the actual decision making. Where operator cognitive ability and/or workload can be gauged, some of these techniques can be modified dynamically, to tailor the information provided, and the level of support and automation employed. Whilst each of these techniques continues to be the subject of research and experimentation, the best gains are to be made from combining them within an Adaptive Decision Support System (ADSS) *[Cottrell & Dixon, 2005]* and it is this innovative research that is the subject of this paper.

2 Challenges in airborne decision support systems

Decision support technology has the potential to hinder, rather than assist operators if not designed and implemented correctly for the particular task and environment. This may be due to poor Human-Machine-Interface (HMI) design, lack of trust in the system, poor engagement with the system (which can result in too much trust), poor choices between task automation and operator-controlled tasks, and in the case of assisted or autonomous decision making, inadequate contextual knowledge. Since in airborne operations the environment and tactical situation can change rapidly, it follows that the DSS must be adaptable to the prevailing circumstances. It is possible to adapt DSS behavior automatically both by gauging operator state, and by assessing the tactical situation/environment. Operator state may be gauged through physiological measurements, inferred through interactions with the DSS (behavioral measures), or a combination. Similarly, the tactical and environmental situation can be gauged from explicit inputs (e.g. threat warning system, aircraft state, flight parameters), inferred from mission phase, or a combination. It is usually also desirable that system behavior can be adapted by the operator, overriding any autonomous behavior.

Thus for a DSS to be effective for airborne operations, an adaptable flexible system is required which strikes the right balance between task automation and operator engagement, and which behaves in a manner transparent and predictable to the operator such that trust is maintained. One way of achieving this is for the ADSS to behave according to a number of configurations, which are programmed with operator involvement pre-mission. The ADSS then switches between configurations during the mission according to operator state and/or tactical situation, or by manual control by the operator.

This is the principal employed within the QinetiQ Cognitive Cockpit (CogPit) *[Taylor, Howells & Watson, 2000]*, which uses 5 configurations - Pilot Authority and Control of Tasks (PACT) levels – which are invoked according to feedback on operator state provided by the Cognition Monitor (CogMon) during mission execution. Each higher PACT level invokes higher levels of task automation and decision making, culminating in full autonomy at the highest level unless the operator overrides the system within a set time. The automation and system behavior within each PACT level, and the rules governing switching between PACT levels, are configured with operator involvement pre-mission by using a Task Interface Manager (TIM) to configure the HMI and system behavior. A pre-cursor to successful implementation of PACT is a process of task modeling and analysis, which establishes the physiological and behavioral traits that typify significant shifts in operator state and/or environment, which can be detected in-mission by the CogMon and associated with the required automation and decision support activities. This task modeling process is usually conducted off-line in representative simulations of the operational task and environment.

As discussed earlier, critiquing has an important role to play in military decision support systems, partly due to the need to verify DSS output against the dynamically changing environment (and therefore potentially an evolving decision context), but also to help maintain good engagement between operator and system and minimize operator bias *[Kahneman, Slovic & Tversky, 1982]*. Critiquing may be employed both by operators critiquing the ADSS, and by the ADSS critiquing operator decisions, depending on situation and workload. Where critiquing of the system by the operator is performed, information provided by the system to justify proposed courses of action or to help in operator evaluation of advice, needs to be presented in a manner that is meaningful and obvious to the operator. Frequently, this means the system needs to translate parametric data into operationally relevant information. Care must be taken to minimize hypothesizing and inferencing by the system in this process, otherwise the critiquing process may have biases that could erroneously support ADSS advice. A hierarchical system of critiquing can be advantageous, such that operators can "drill-down" to obtain more detailed information when time/workload permits, but "headline" information is always available. Where critiquing of the operator by the system is performed, operator decisions or courses of action may be "scored" by the system, but similar challenges exist for passing information back to the operator in terms of expressing these scores in an operationally relevant way and in minimizing system bias.

Selection of tasks to automate is also critical, some tasks are better suited to automation, whilst others are best conducted under operator control. This can also apply to decisions, but with the added complexity that some decisions are sensitive to the context within which they are made - what constitutes a good decision in one context, may be a bad decision under different circumstances. Thus it is usually preferable to automate the more routine tasks or those that are computationally complex but amenable to computer control, in an attempt to reduce operator work-load and provide more time for problem consideration, resolution and decision making. However, as air operations become increasingly complex and dynamic, and with a move to predominantly single-seat operations, this is not always possible and thus there is an emerging requirement for the ADSS to assist directly in the decision making process but in a manner that is sensitive to the prevailing situation and context.

Finally, the ADSS will only be effective if the HMI is intuitive, informative, and sufficiently flexible that it can also be adapted to suit the prevailing situation and context. Research has shown that there is a balance between flexibility and effectiveness, especially within high-workload environments with limited HMI real-estate, such as those encountered in cockpits [Dinadis &Vicente, 1999]. For instance it is usually preferable for display configurations to be constrained rather than fully programmable by the operator (especially during mission), such that at times of pressure, familiarity can assist interactions, rather than operators dealing with a novel HMI design in flight, albeit one of their own creation.

3 Adaptive Decision Support System Design

3.1 Problem context

For inhabited air platforms, context sensitive tactical decision making is particularly problematic due to potentially high workload, limited response time, and risk/threat exposure. For ground-attack roles, this is exacerbated by the move from two-seat to single-seat platforms. With UAVs, operators may face similar decision making challenges, albeit with some different constraints. Operators may be controlling multiple assets with differing endurance, weapons and sensor capabilities, and lack detailed information regarding the platform environs, having to rely on incomplete, remotely sensed data.

3.2 Solution strategy

The ADSS research discussed here seeks to aid operator decision making in the specific context of air-to-ground strike missions, both by inhabited aircraft and by Uninhabited Combat Air-Vehicles (UCAVs). Strike missions have been broken down into phases - each potentially requiring different decision support techniques/strategies. These phases are task allocation, ingress/egress routeing, and target engagement/attack planning. Breaking the problem down in this manner mimics they way in which operators break down the mission planning problem, frequently carrying out the detailed target attack planning and ingress/egress routeing in parallel using different operators, once the high level mission-flow (or tasking) has been established. By using the same type of problem decomposition for computer aided decision support, it is intended that DSS behavior will be similar to skilled operators, facilitating critiquing and fostering trust.

To make the ADSS sensitive to decision context, a process of ADSS configuration is undertaken pre-mission. For the purposes of this research, decision context is considered to be synonymous with mission context, and the configuration is undertaken in a manner that is intended to be intuitive to operators. Pre-mission, operators set desired values for various factors affecting, or pertaining to, the mission. Operators would normally mentally consider these factors during pre-mission planning, and ideally during in-mission tactical re-planning or tactical targeting. This latter is particularly difficult to achieve in practice, especially when it is required to be performed by a single operator.

An example for this type of DSS is the commercial route planning and in-car navigation software that has proven to be highly successful. Typically with such software, "operators" configure the "DSS" prior to their journey, with a set of preferences regarding driving speeds, road categories, time versus distance weighting (quickest route versus shortest route), desire for scenic routes etc. The DSS then attempts to derive the optimum plan that meets these preferences. The more capable systems can apply these preferences during the journey when tactical re-planning is required (e.g. due to road-works, or perhaps more frequently the "operator" attempting to out-smart the "DSS" and becoming "temporarily uncertain of position"). The ADSS operation is similar in principle to this type of vehicle system, but some "preferences" (factors) relate to a single mission phase, while others span multiple phases. Additional areas of sophistication are present, however:

The first is that factors, or preferences, are grouped together under three operationally meaningful categories – survivability, effectiveness and timeliness (S/E/T). In-mission, preferences can be changed at this high-level, without necessarily drilling down to the individual low-level factors.

A second area of sophistication, which seeks to ensure that any such high-level changes are made within the correct context, is that each low-level factor can be set as "tradable" by the operator pre-mission, thus attempting to constrain ADSS behavior in terms of which factors are considered negotiable in trying to formulate an appropriate course of action in-mission in response to dynamic events. This would be analogous to changing the "quickest" versus "shortest" preference in the in-car navigation system while driving, but the system still trying to meet as many of the other driving preferences as possible.

A further area of sophistication compared with the vehicle journey planner, is that ADSS advice is scored against the operator settings and fed-back to the operator. This scoring is performed both at the low-level against each of the individual factors considered by the ADSS, and aggregated at the group S/E/T level.

4 ADSS research implementation

A baseline scenario and architecture has been developed to establish proof-of-principle for the ADSS described above. The baseline scenario involves four UCAVs deployed around a Restricted Operating Zone (ROZ), prosecuting pop-up time sensitive targets. The baseline architecture employs a modular ADSS with separate software modules dealing with task allocation, ingress/egress routeing, and target engagement/attack planning.

4.1 Task allocation

Task allocation considers the allocation of assets to tasks based on task priority, target type, platform, weapons and sensor capabilities, and according to time/distance/fuel constraints. Task allocation within the ADSS is performed by the Task Allocation Module (TAM), which is based on a "hill climbing" Genetic Algorithm (GA). The TAM GA optimizes solutions against a set of constraints which are sensitive to variations in S/E/T, thus providing behavior which is adaptable to decision context through modulation of S/E/T by the operator.

4.2 Ingress/egress routeing

Various techniques for either assisting operators with routeing, or providing auto-routeing solutions, have been developed specifically for military air operations, for both uninhabited and inhabited air platforms. Such techniques offer great potential to missions involving time-sensitive targeting or where airborne re-planning is required. In inhabited aircraft, auto-routeing technology may be required due to time constraints and workload, whereas with UCAV/UAV missions, auto-routeing may be an essential requirement of the platform in addition to any support offered to remote operators to facilitate control of multiple assets.

Whilst these techniques have been shown to be effective within certain constraints, research has shown that for realistic strike missions, rarely does a single routeing technology provide the complete solution. Rather, a "tool-box" of routeing technologies is required where each technology can be applied to parts of the overall problem to which they are best suited, and even then in a form tailored for the particular task or context. Since it is inappropriate for operators to be engaged in the low-level detail of how the routeing technologies function, a mechanism for invoking and controlling them using higher-level knowledge is required. Within the ADSS, this is achieved transparently by the operator as part of the pre-mission configuration of the factors contributing to S/E/T, which influence how the routeing technologies are applied.

Within the ADSS, the routeing technologies have been integrated within a "Mission Computer" (MC). The MC provides a framework for invoking differing routeing algorithms and target engagement plans/tactics, according to S/E/T settings. Within the routeing capabilities are the ability to minimize exposure to specific known threats, minimize generic exposure to unknown threats, avoid designated regions of airspace, and adhere to mandatory waypoints and corridors. The routers employ a variety of technologies, some use Artificial Intelligence (AI) with rules encoded, whilst others are purely cost based. Depending on operator preferences, routeing can be made to include a random element to avoid predictability and repetition.

4.3   Attack planning and target engagement

Specific behaviors associated with attack planning and target engagement are required, and these are accommodated within separate modules to the routers employed for ingress/egress to the target. The technology used to support attack planning and target engagement employs expert tactical knowledge from operators in addition to knowledge of platform, weapons, sensors and target characteristics. The attack planning and target engagement knowledge-base includes rules affecting selection of imaging points, weapons/sensor selection and deployment, weapons release maneuvers and attack profiles for a limited set of weapons/sensor configurations. Although the target engagement and attack planning modules also reside within the MC, they are separate to ingress/egress modules. In principle the MC could reside on inhabited or uninhabited air platforms, whereas the TAM could reside on a ground-station.

4.4   Adaptive behavior according to decision context

The premise is that in both inhabited and uninhabited strike missions, operators will have better understanding of certain mission aspects. These include the political context within which the mission is being conducted, the effects being sought through prosecution of the target(s), the Rules Of Engagement (RoE), the importance of mission success and the impact of failure including collateral damage and late prosecution of targets.

Conversely, it is assumed that there are certain other aspects that can either be assisted or automated through technology. These include calculations involved with time-distance-fuel (airmanship), routeing sensitive to threat exposure, weapons/sensor matching to target, and target engagement itself. Whilst some of these areas are traditionally associated with automation or decision support, the innovation provided by the ADSS is to employ these technologies intelligently within a decision support framework where operators, through their greater understanding of mission (and hence decision) context, can tailor system behavior accordingly. This tailoring is performed through operator knowledge of strategic (level 3) and tactical (level 2) decision contexts as they apply to survivability, effectiveness and timeliness. Factors associated with these three high-level mission aspects are used as control inputs to the decision support system, which modulate system behavior and automation at the operational level – i.e. level 1 *[NATO Code of Best Practice for C2 Assessment, 2002]*.

4.5   Establishing mission (decision) context

Whilst operators would wish for maximum survivability, effectiveness and timeliness, in reality compromises are normally required. For some missions, a compromise might be made between timeliness and survivability (e.g. the most timely solution might be straight-line routeing to the target, but due to ground threats, this might not be survivable). Similarly, if an effective mission is not just destruction of the target, but full conformance with RoE and efficient use of resources, then there may be occasions where operators accept lower effectiveness, and potentially lower survivability, in the interests of prosecuting difficult, high-priority targets on-time. Thus it is clear that S/E/T are inter-related, and that in practice compromises will frequently need to be made between the three.

The detail of these compromises may vary mission-by-mission depending on strategic and tactical objectives. Therefore a system is required that is readily understandable by the operator, for establishing the mission context and influencing system behavior at the detailed level pre-mission, whilst also providing a higher-level interface for in-mission modulation of system behavior.

4.6   Pre-mission configuration of factors affecting S/E/T

The operator establishes the context in which decision support is provided, in terms of a range of factors affecting Survivability, Effectiveness and Timeliness (S/E/T). Pre-mission, the operator will also decide which factors are tradable if (as is likely) compromises need to be made as dynamic events occur and operational (level 1) decision context changes. This approach means that high-level changes to S/E/T in-mission will still result in behaviors that match the mission context, as defined by the operator pre-mission. Thus if the operational decision context changes during the mission (e.g. due to attrition, platform damage, changes in threat environment etc), and the operator is too highly loaded to engage with S/E/T at the low-level, changes can be made at the high-level with knowledge that only those factors that have been set tradable pre-mission will be altered.

4.7   S/E/T modulation during mission execution

Currently, high level changes to S/E/T have to be made manually by the operator, using the interface shown in Figure 2 below. Where a task/routeing solution has been selected by the operator, scores built up from a combination of the lower-level factors influencing S/E/T are reported as percentages against the requested high-level S/E/T settings. However, the intent is to allow autonomous modulation of S/E/T by the system. The paradigm for implementing this functionality will be similar to that employed in the Cognitive Cockpit discussed earlier. Through a process of task modeling and analysis, key operator states and environmental conditions will be associated with required system behaviors. In-mission, the CogMon *[Pleydell-Pearce, Dickson & Whitecross, 2000]* will determine when these states/conditions occur, such that the desired behavior can be invoked. The process of interpretation of the strategic and tactical factors relating to the mission and translation into the low-level S/E/T settings with decisions regarding tradability, are analogous to configuring the Task Interface Manger (TIM). By integrating the Cognition Monitor (CogMon) with the ADSS, it will be possible for the system to modulate the high-level S/E/T settings using an interface similar to the Pilot Authority and Control of Tasks (PACT) system. Thus if platform damage were sustained, or the threat-risk increased due to depletion of countermeasures (chaff and flares), the ADSS could recommend an increase to the high-level survivability setting, and depending on operator state, could automatically invoke this setting if no operator input is received within a set period. Similarly, tasking and routeing solutions offered by the ADSS could be accepted and implemented when operator state monitoring indicates this is appropriate.

4.8   Requested S/E/T versus scored S/E/T

Having set the individual factors contributing to S/E/T, the ADSS calculates potential solutions and scores them both against the individual requested values, and as overall scores against S, E and T. Due to there only being a finite number of options and behaviors within the ADSS, and because S/E/T are inter-related, it is likely that the resulting scores differ from the requested settings. For example, by setting the various factors within S/E/T, an operator may be seeking to achieve maximum survivability, effectiveness and timeliness for a target, but due to fuel constraints, the result is timely (TOT achieved), low survivability (little chance of recovering asset after target prosecution), and of medium effectiveness (asset was not considered to be expendable). However, having been alerted to the high level discrepancy between requested S/E/T and calculated result, the ADSS design allows operators to quickly compare scores for the individual factors within the affected channels (survivability and effectiveness), and thus gives an intuitive way of analyzing (critiquing) tasking and routeing advice offered by the ADSS. The example of insufficient fuel is likely to be fairly easy to determine by other means, but where it is a more complex interaction of factors such as meeting RoE for a particular target within platform weapons/sensor constraints, such a simple way of critiquing ADSS advice is vital.

The design of the ADSS is such that the HMI displays are divorced from other system components, allowing HMI research and development to occur in parallel. However, for engineering and test purposes, simple histogram-based displays and controls are employed for S/E/T with sliders for setting the desired control inputs, and "thermometers" providing the resulting score calculated by the ADSS. The same type of presentation is used at the high level S/E/T settings, and the lower level settings for individual factors contributing to S, E and T. An example of the low-level factors and scores contributing towards "effectiveness" is shown in figure 1 below, similar presentations are used for "survivability" and "timeliness".

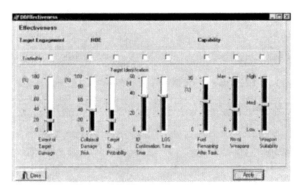

Figure 1 - Factors contributing to "effectiveness"

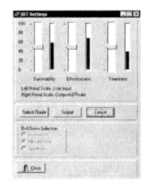

Figure 2: high-level S/E/T HMI

Figure 2 shows an example high-level S/E/T interface designed for engineering and test purposes, in practice polar plots or other representations could be employed:

4.9   Example ADSS output for baseline scenario

In the baseline scenario discussed earlier, the operator is in control of four UCAVs, one loitering at each corner of a Restricted Operating Zone (ROZ), in which ground-based threats and targets are either already present, or expected to appear. New threats or targets appear throughout the mission, posing mission re-planning problems for the operator to solve. A typical re-planning problem involves allocating one or more pop-up targets for engagement by specific UCAVs, based on their time-to-intercept, fuel state, available payload etc. Experiments can be run iteratively with increasing complexity – increasing number of targets, increasing ground threats, targets of differing priority, UCAVs with differing weapons/sensor payloads that need to be matched with particular targets, limiting weapons and fuel etc. Various aspects of these scenarios are sensitive to S/E/T settings (e.g. lower effectiveness may relax the need to conform to ROE and mean that weapons/sensor matching to targets is not required, it may also obviate the need to retain capacity for further targets. Lower survivability means that threat incursion can occur, and ultimately that loss of an asset after target prosecution is acceptable).

In circumstances where scenario complexity is minimized with no threats, low numbers of targets of equal priority and similarly equipped UCAVs, optimal solutions can be generated by skilled personnel. In these cases, time is the principal benefit offered by the system, since it generates all of the appropriate routes automatically. However, the real benefit of the system is its scalability to more complex problems, with more targets and larger groups of UCAVs with heterogeneous payloads. Such a scenario is shown below in figure 3.

In this scenario, UCAVs loiter at each of the four corners of the ROZ. An airfield lies to the West of the ROZ, displayed as a red box. Threats in the ROZ are displayed with their acquisition and engagement ranges, and the fourteen targets are displayed as triangles; the purple targets have higher priority than the blue targets. The UCAVs have different weapon types, each appropriate for different target types, and in this case the number of targets out strips the payload available across the entire group of UCAVs. This type of scenario takes even skilled operators many minutes to solve, and when solutions are critiqued they are often found to be non-optimal in important areas (number of targets destroyed, excessive time-to-target, poor resource balancing, poor conformance to S/E/T and hence mission context). The ADSS makes this problem manageable, by offering a ranked list of four potential solutions in a few seconds. A sample ADSS solution is shown in figure 4 (note – the dots dispersed along the routes are turning points entered by the system to take advantage of terrain screening).

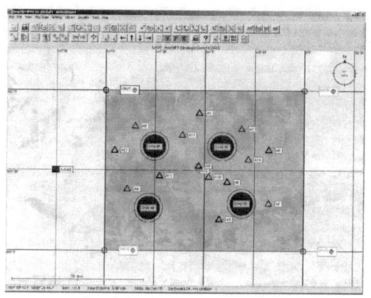

Figure 3: Complex scenario initial condition

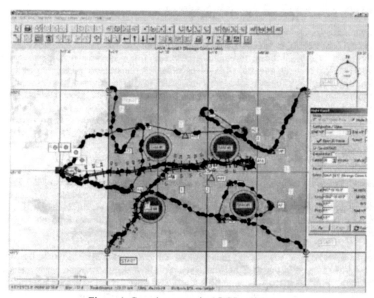

Figure 4: Complex scenario ADSS solution

All of the UCAVs are tasked to return to the airfield after engaging their targets, reflecting system knowledge that this solution will use up their weapons payloads. Notice that the ADSS has advised the operator to avoid engaging the two, lower priority targets in favor of employing their assets against the higher priority targets. The difficulty of this problem is demonstrated by the complexity of its solution. Even if an operator performs the super-human feat of grasping a good solution at a glance, it would take a significant time to specify its details. The ADSS makes practical an otherwise impossible task.

There are many circumstances in which good solutions to these planning problems are counter-intuitive. Research thus far suggests that the natural strategy of human operators when faced with these problems is to segment the ROZ into quadrants, and task the UCAVs to engage targets in their local quadrant. Whilst this works in some circumstances, it is not always the best strategy, and operators can distrust ADSS advice that does not follow this seemingly sensible "rule". A further situation that has confused operators is when the ADSS chooses to swap loiter positions for assets after target prosecution, to minimize return legs and balance fuel. It is therefore vital that operators are able to critique the solutions offered by the ADSS and easily obtain justification for the advice, such that their own biases can be managed effectively *[Kahneman, Slovic & Tversky, 1982]*.

Much of the relevant decision context for this domain has still to be discovered – indeed, it is hoped that systems like this can contribute to that process of discovery. Adaptation is therefore just as important a component of this system as critiquing. In the current ADSS, Adaptation and critiquing systems were designed to work together through the S/E/T process, providing operators with a single 'grammar' through which system operation can be both justified and augmented.

Like critiquing techniques, adaptation techniques must manage the tension between complexity and 'power'. The ADSS described here is complicated, offering a bewildering variety of adaptation options. It is not practical to expect operators, in-mission, to be able to manage the full range of adaptation options in this application. With the help of an experienced military advisors, the research program has attempted to manage this complexity by agglomerating this low level complexity into three, higher level channels – Survivability, Timeliness and Effectiveness. Each channel is essentially a collection of parameters, specific to each suggested course of action generated by the ADSS.

'Survivability' is intended to represent the inverse of the overall 'risk' associated with the mission – solutions with high survivability scores are solutions that present low risk to the assets. The survivability 'score' for a solution is derived by agglomerating the risk to each asset along every part of its route. This risk can be calculated by considering a variety of factors, including exposure to known threats, exposure to un-known threats (generic exposure), threat lethality, actual vs. required fuel and so on. The system defines 'effectiveness' as a combination of the likelihood of successful target prosecution (hitting the target), conformance with RoE, and remaining capacity/resource (better solutions make more efficient use of available assets, leaving greater capacity available to respond to future taskings). 'Timeliness' can also be critical in this domain, where some or all targets may be time-sensitive, requiring engagement either 'as soon as possible', or within some specified time window. Like survivability and effectiveness, timeliness is an agglomerated score. Though the lower-level attributes of each mission solution are available at one level of 'drill-down', operators may not have time to digest those details during complex operations. The value of the agglomeration is that it summarizes those details in a form that can be understood at a glance.

5 Discussion

A unifying characteristic present in many military operational domains is that there may be no 'fact of the matter' that defines an 'optimal' solution. The quality of decision support systems is defined not just by the quality of the algorithms used to derive the outputs, but by the quality of operator interaction with them. This is the defining quality of the domains for which decision support systems, as opposed to traditional computational tools, are appropriate. This characteristic is also the foundation of the requirement for critiquing – perfect solutions are only useful if an operator can trust and understand them at a glance and so respond to them quickly and effectively.

For military air operations, it is unlikely that the relevant decision – or mission – context can ever be fully encoded within a DSS, albeit that the techniques described in this paper for establishing mission context offer significant potential to help bias – or adapt – DSS behavior in the right direction. This means that effective interaction between DSS and operator is vital, such that operators can efficiently bring their advanced appreciation of decision context to bear in augmenting system performance.

This 'gap' between DSS and operators is significant – an instance of the more general distinction between skilled human performance and the behavior of current 'intelligent' machines. DSS research does not just improve decision support technology, but it also provides a practical framework for the study of skilled, human decision-making. In the case of the ADSS described here, the mechanism for adaptation of the system by the operator exercises that quality directly, and renders its results explicit in the detailed behavior of the DSS. Scientists and philosophers have strived for many years to uncover the structure of this elusive capacity of humans. The notion of decision context, in the framework provided by DSS research, offers a practical, systematic way to characterize some of that capacity.

Decision support systems can and do work. The promised revolution is not far away. But DSS research actually promises two revolutions – not one. The first revolution will arise from development of invaluable aids, which will change the face of military (and civilian) operations. The second revolution will follow from the first; as experience with decision support technology grows, so will our ability to characterize the 'gaps' between DSS advice and the decision-making processes of skilled, human operators. Decision support technology will extend the capacity of human operators, but will also improve understanding of those capacities in themselves; a double-benefit well worth pursuing.

6   References

Cottrell, R., Dixon, D. (2005). Adaptive decision support research system status report. QinetiQ S&E/SPI/CR042949/1.0

Taylor, R.M., Howells, H, & Watson, D. (2000). The cognitive cockpit: operational requirement and technical challenge. In P.T. McCabe, M. A. Hanson, & S.A. Robertson (Eds), *Contemporary Ergonomics 2000* (pp55-59). London: Taylor & Francis

Kahneman, D., Slovic, P., & Tversky, A. (1982). Judgment under uncertainty: Heuristics and biases. Cambridge University Press: New York, USA.

Dinadis, N. and Vicente, K. J., (1999). Designing Functional Information for Aircraft System Status Displays. In *International Journal of Aviation Psychology*, Vol. 9 (3): 241-269

NATO Code of Best Practice for C2 Assessment. (2002). www.dodccrp.org.

Pleydell-Pearce, K., Dickson, B. T. & Whitecross S. (2000). Cognition monitor: A system for real time pilot state assessment. In P. T. McCabe, M. A Hanson and S. A Robertson (Eds.), *Contemporary Ergonomics 2000*. London: Taylor and Francis.

Section 5
Cognitive State Sensors

Chapter 24

Team Cognition

Augmented Team Cognition

Nancy J. Cooke

Arizona State University East
Applied Psychology Unit
7001 E. Williams Field Rd., Bldg. 140
Mesa, AZ, 85212
ncooke@asu.edu

and

Cognitive Engineering Research Institute
5810 S. Sossaman
Mesa, AZ 85212

Abstract

Augmented team cognition involves the facilitation of cognitive activities such as decision making, planning, and problem solving at a team level, where team is defined as an interdependent group of specialized individuals working toward a common goal. Several well-known disasters of the 1980s and other more mundane events, which were attributed at least partially to poor cognitive performance by teams, prompted much research and development on team cognition and augmented team cognition. One of the major bottlenecks in implementing augmented cognition for teams is in deciding for any given team, what aspect of team cognition requires augmentation. There are a number of challenges associated with assessment and diagnosis of team cognition including determining the appropriate level of measurement (individual or team level) and the assessment of "shared" cognition in teams that are heterogeneous and clearly not meant to have members who share the "same" cognition. The ultimate challenge is to provide reliable and valid methods for assessing team cognition unobtrusively in an operational setting. Team communication has been targeted by many as a data source on team cognition; however, automation of analysis is required for real-time assessment. Some of our work on automating communication analysis is presented as one solution to real-time assessment of team cognition.

1 Definitions

As work becomes increasingly complex at a cognitive level, the individual human operator or decision maker becomes less capable of working independently. In domains ranging from military command-and-control and civilian emergency management to intelligence analysis and the design of advanced technologies, one finds groups of humans collaboratively engaged in work to accomplish single goals. Unlike work of the industrial age, the individuals do not perform a piece of the task in relative isolation (e.g., assembly lines, typing pools). Rather these groups are actively engaged in information exchange and other joint cognitive activities.

A *team* is a special type of group. Salas, Dickinson, Converse, and Tannenbaum (1992) define a team as "a distinguishable set of two or more people who interact dynamically, interdependently, and adaptively toward a common and valued goal/object/mission, who have each been assigned specific roles or functions to perform, and who have a limited life span of membership" (p. 4). Teams perform military command-and-control operations, civilian emergency operations, surgery in operating rooms, and many other highly cognitive and complex tasks. Teams plan, design, perceive, make decisions, and act as an integrated unit. This team-level thinking is the focus of the relatively new field of *team cognition*. Within this field, concepts and theoretical constructs such as shared mental models, transactive memory, and team situation awareness have come to the forefront. The field resides primarily in the applied disciplines of industrial/organizational psychology, business management, and human factors and has been inspired by theory, methods, and findings from information processing psychology, social psychology, and most recently, ecological psychology.

Augmented team cognition is the application of technology to team training programs or work environments for the purpose of improving cognitive effectiveness at the team level. For instance, intelligent agent technology may be used to simulate teammates in a training environment in order to facilitate team training when other team members are not present. Similarly, decision aids, electronic message boards, and other collaborative technology can enhance team cognition in operational settings. Thus, augmented cognition for teams is similar to augmented cognition for individuals, but the focus is on improving cognition at the team level.

2 Why Augmented Team Cognition?

Even in relatively mundane team-level cognitive activities such as software design and American football, the need for facilitating team cognition has been recognized. Just as individual productivity can improve with cognitive aids and focused training, so can team effectiveness. However, there were a series of disastrous events that crystallized the importance of augmenting team cognition and in turn, spawned a flurry of research and development in the area over the last one and one-half decades.

Indeed much of the research on team cognition was fueled by a handful of disasters in a ten-year period that at least partially pointed to problems with teams performing in highly cognitively demanding arenas. The Three Mile Island accident of 1979 and the Chernobyl accident of 1986 both raised issues about the appropriate response of the operating crews in the control rooms (Gaddy & Wachtel, 1992). On January 28, 1986, faulty decision making at the organizational level resulted in the mistaken and tragic launch of space shuttle Challenger (Vaughan, 1996). Then in July of 1988 the USS Vincennes, a US Naval warship, mistakenly shot down a passenger Iranian airbus (Collyer & Malecki, 1998). This incident, like the others, was tied to a complex web of causes and preexisting system weaknesses (Reason, 1997), but also like the others, was partially attributed to coordination problems in the command-and-control decision making.

Research and development efforts were sparked by these events, but the focus was reinforced by others such as emergency disaster drill failures, inefficiencies in airport security, uncoordinated response to the events of September 11, 2001, and numerous friendly fire incidents, to name a few. Research and development programs such as the US Navy-sponsored TADMUS (Tactical Decision Making Under Stress) program, initiated in response to the USS Vincennes incident, began in the early 90's which also saw the beginnings of the fields of cognitive engineering and naturalistic decision making (Cannon-Bowers & Salas, 1998). This research was motivated primarily by the need to improve team decision making and performance effectiveness in highly complex cognitive environments. The purpose of this work was to augment team cognition.

3 Assessment and Diagnosis Challenges

One of the major bottlenecks in implementing augmented cognition for teams is in deciding for any given team, what aspect of team cognition requires augmentation. Assessment and diagnosis of team cognition is key and indeed is where much of the research in this area has focused. How well is a given team doing on a cognitive task and what specific cognitive strengths or weaknesses underlie the team's performance? For instance, is a team performing poorly because they are cognitively overloaded, have poor team situation awareness, or are biased decision makers? The diagnostic information provides targets for augmented cognition interventions. More so than the augmenting of individual cognition, efforts to augment team cognition have been faced with the challenge of defining, operationalizing, and measuring aspects of team cognition.

The assessment of team cognition is associated with all of the challenges of individual cognitive assessment and more. The "more" includes the appropriate level of measurement (individual or team level) and the assessment of "shared" cognition in teams that are heterogeneous and clearly not meant to have members who share the "same" cognition. In addition, the assessment of team cognition in operational settings necessitates real-time monitoring and measurement at the team level. As is the case for individual cognition, it would not be effective to intrusively test team cognition at multiple points throughout a scenario to determine when workload had reached a limit. The ultimate challenge therefore is to provide reliable and valid methods for assessing team cognition unobtrusively in an operational setting.

3.1 Challenge #1: Defining team cognition

A phenomenological definition of team cognition as cognitive activity in which a team engages was presented in a previous section of this paper. However, in order to measure team cognition, this definition requires some fine tuning. Where does team cognition reside? What are the mechanisms of team cognition?

There are two main ways in which team cognition has been conceptualized. One way is in information processing terms and it is this way that has been adopted by most team cognition researchers. According to this information processing perspective, team cognition resides primarily in the heads of the individual team members viewed as information processing systems. So, the team's cognition can be conceptualized as a collection of the individual information processing units (Figure 1, Panel A). The shared mental models construct arose from this tradition. Individual team members hold knowledge about the task, technology, or team in terms of individual mental models. Original views of shared mental models held that to the extent that the team member mental models were similar, mental models were shared. Shared mental models enabled teams to implicitly coordinate, anticipate behavior, and perform effectively (Cannon-Bowers, Salas, & Converse, 1993). Shared mental models were also proposed to be at the heart of team situation awareness. More recent conceptualizations of shared mental models, team situation awareness, and team cognition, are more flexible in defining the concept of sharing and similarity in order to address weaknesses raised in the second challenge.

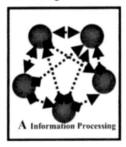 

Figure 1. Depiction of team members represented as spheres and team member interactions as connectors. Panel A represents the emphasis on individual cognition associated with the information processing perspective. Panel B represents the ecological perspective in which emphasis is on team member interactions.

Another perspective on team cognition could be considered ecological in origin. According to this view (Cooke & Gorman, in press), team cognition resides primarily outside of the team members' heads. Team cognition instead resides in the interactions among team members and between team members and the environment (Figure 1, Panel B). Team cognition emerges from these interactions and is different from the cognition associated with any one individual. The interactions *are* team cognition and are therefore observable in team communications and other coordinating behaviors.

In line with other ecological views (Gibson, 1966), this perspective emphasizes the relatively low-dimensional synthetic variable of team coordination in response to ecological events, rather than the local knowledge or processes of individual team members. In this manner, team cognition is characterized as a single organism, ebbing and flowing and adapting itself to novel environmental constraints through the coordination of team perceptual systems.

Though team cognition can to some extent be explained by either the information processing or ecological perspective, there are distinct differences in the foci of the two approaches. The information processing approach focuses on individual cognition and the ecological approach focuses on team interaction. Thus, each approach suggests different measurement methodologies and interventions.

3.2   Challenge #2: Metrics for heterogeneous teams

Teams by the Salas, et al. (1992) definition are heterogeneous. That is they come to the task with different cognitive backgrounds. They have different knowledge, skills, and abilities that pertain to the task. This is clearly the case in an operating room in which the nurse, anesthesiologist, and surgeon bring different knowledge and skills to the table, yet interact to accomplish a single goal. What is team cognition or shared mental models in this context? Here the tenet that shared mental models are similar mental models starts to break down (Figure 2, Panel A). In the case of heterogeneous teams, we might view shared mental models as distributed mental models, where each individual role is assigned to a portion of the knowledge or cognition pertaining to the task. Of course, the real world is even more complicated, as it is likely the case that some knowledge for any two team members overlaps, some is complementary or distributed, some is shared at certain levels of abstraction, some is missing from the team, and some may be more or less accurate. How is team cognition measured in these circumstances?

Recent efforts in the measurement of team cognition have addressed the problem of heterogeneity through the development of heterogeneous knowledge metrics (Cooke, Salas, Kiekel, & Bell, 2004). Knowledge is elicited at the individual level, but is assessed relative to a standard pertaining to a specific team role. A nurse's positional knowledge or mental model can be assessed by comparison to the standard knowledge for a nurse in that setting (Figure 2, Panel B). The nurse's knowledge can also be compared to referent knowledge for other roles to assess the nurse's interpositional knowledge. Thus, these metrics can be used to assess individual and team knowledge and similarity as before, but in addition they can provide measures of positional and interpositional knowledge. Applications of heterogeneous metrics have revealed interesting knowledge profiles of tasks that suggest the degree of specialization required and concomitantly, the degree to which cross training programs would be successful (Cooke, Kiekel, Salas, Stout, Bowers, & Cannon-Bowers, 2003).

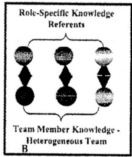

Figure 2. Knowledge measurement for homogeneous teams using a team-level referent (Panel A) or for heterogeneous teams using role-specific referents (Panel B).

3.3 Challenge #3: Collective vs. holistic measurement

Heterogeneous knowledge metrics described in the previous section focus on individual knowledge. Team member knowledge is elicited and assessed at the individual level. A team assessment entails aggregating the individual knowledge values across team members. Thus, the heterogeneous metrics are consistent with the information processing view of team cognition. This aggregation approach, however, is at odds with the ecological perspective that holds that a team is more than the sum of its parts. The ecological perspective espouses a more holistic measurement of team cognition. That is, in order to capture the interactions that are at the heart of team cognition, those interactions need to be reflected in the measures. One way that this can be accomplished is through measurement that is directed at the whole team instead of at individuals. Knowledge, for example, could be elicited at the team level (Cooke, et al., 2004). The assumption here is that the team interactions are reflected in the team-level response.

Another approach to holistic measurement of team cognition is to focus the measurement on the interactions. Thus, team communication has been the target of recent holistic measurement efforts. One can think of team communication as the team analog of individual verbal reports. Though team communication has been seen as an important aspect of or conduit for team cognition (in information processing terms), by the ecological view, team communication is team cognition. That is, team communication is cognitive processing at the team level. In fact, more holistic or ecological views of team cognition hold that a large and important part of team cognition is in team interaction and team communication. According to this view, we do not have to make inferences about team cognition based on team behavior as we do for individuals. Instead, team communication or interaction *is* team cognition and thus provides a unique opportunity to directly observe team cognition (Cooke & Gorman, in press). This identity relationship between communication and team cognitive processing affords the direct observation of team cognition.

In highly cognitive tasks, such as command-and-control, team communication is one form of behavior that is observable. However, team behavior is associated with several other observable indices. Team members can be observed interacting physically or indirectly through computer mediation. Behavioral event logs can then serve as an indication of team cognition. Other aspects of team process have also been proposed as indicators or indices of team cognition. For example, a paper presented in this session on augmented team cognition (Henning, Korbelak, & Smith, 2005), describe the use of social-psychophysiological compliance as a measure of cognitive readiness of the team. Other papers in this session rely on communication to measure various aspects of team cognition (Foltz, 2005; Kiekel, 2005) and as a measure of congruence between team structure and mission requirements (Diedrich, Freeman, Entin, & MacMillan, 2005).

3.4 Challenge #4: Embedded and automated measures for real-time assessment

The identification of the data source, whether it is communications or event records or some biological index is only the first step in assessing and diagnosing team cognition. Data recording is the next step, and a nontrivial one for team communication in which the manual transcription and coding effort required can diminish the value of the data. Research is ongoing to streamline this process using speech recognition tools, custom communication flow recording tools, sequential data analysis, and semantic text analysis. The big challenge in the real-time measurement of team cognition is methodological. How can we extract meaningful patterns from the continuous and rich interaction data and how can we automate this process for real-time assessment that is needed for immediate intervention to augment cognition in an operational setting? Although techniques exist for content and interaction analysis in discourse there are numerous challenges to automating communication analysis. Speech recognition systems are not perfect. Not only is word recognition required, but in addition, speaker and listener identity, as well as the identification of speech segments are important. Often research in this area is hampered by the laborious process of transcription. Methods for content analysis such as latent semantic analysis (Gorman, Foltz, Kiekel, Martin, & Cooke, 2003) further rely on domain-specific language corpora or ontologies. Techniques that focus on content-free interaction patterns show some promise in that they can rapidly record physical data without the cost of transcription or ontologies. More expensive and laborious content approaches might be considered as a second pass once a preliminary physical analysis of communication flow has uncovered interesting patterns.

The next step after patterns in the data are extracted is to map them to very specific, diagnostic aspects of team cognition. These pattern-matching routines are domain-specific and require some objective measure of team performance at the least and of the more specific diagnostic constructs ideally. With this information, it is possible that the pattern matching could proceed automatically using machine learning software that continually extracts patterns and associates them with cognitive team performance—learning over time in an inductive way what each particular pattern means. Real-time augmented team cognition awaits solutions to this set of challenges.
The next section describes a sample of our recent efforts to automate communication analysis for this purpose.

4 Toward Real-Time Assessment of Team Cognition

Over the past five years we have developed, applied, and tested a suite of methods for the automatic analysis of communication data. The methods extract patterns from communication data of two types: 1) flow of speech from one person to another and 2) the content of the communication.

Flow data are recorded in the context of a laboratory UAV (Unmanned Aerial Vehicle) simulation for teams (Cooke & Shope, 2002). In this environment team members press push-to-talk buttons to speak to other team members. Through these button presses, we record (automatically) speaker, listener, and onset and duration of utterance. We have developed four methods to extract patterns in those data (Kiekel, Gorman, & Cooke, 2004). One of the methods called ProNet is based on Pathfinder network analysis of sequential flow patterns. ProNet is used to extract meaningful sequences in the flow data. We have found in the UAV task that ProNet chain lengths is predictive of team performance and qualitative patterns in the resulting networks can provide more diagnostic information. For example, the two ProNet diagrams in Figure 3 are from a single team of three UAV operators interacting in our simulation. Links indicate directional connections between speaker and listener with P, D, and A coding the identities of the three team members (i.e., Payload Operator, DEMPC or Data Exploitation Mission Planning and Communications Operator, and Air Vehicle Operator) and "beg" and "end" coding beginning or end of an utterance. In this example there was a fight between P and D during the fifth mission. The left panel of Figure 3 depicts the pattern of communication that occurred prior to the fight and the panel on the right depicts how the pattern changed after the fight. Notice that prior to the fight the P and D team members displayed a pattern of dyadic interaction which is obviously absent after the fight.

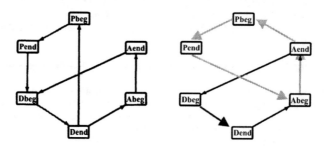

Figure 3. ProNet communication patterns between three team members (A, P, and D) who were beginning ("beg") or ending ("end") a speech event. The left panel is the pattern extracted prior to a fight and the right panel is the pattern extracted after a fight.

ProNet patterns can also provide indications of cliques, interruptions, and monologues and the consistency of ProNet patterns is predictive of team performance. Other methods for flow analysis include a method for extracting patterns relevant to dominance on a team and a method for extracting clusters of speech patterns over time. These methods have been used to predict team performance and have also been indicative of various team states including communication breakdowns, high vs. low workload, and distributed vs. co-located mission environments. This work has benefited from the plethora of criterion measures already available in our UAV laboratory environment. A challenge in applying these methods to the "real world" is in identifying team cognitive state in the absence of these measures or with limited measures so that communication patterns can be mapped onto state.

Our content analysis work has primarily used LSA (Latent Semantic Analysis; Landauer, Foltz, & Laham, 1998) although we are now also exploring the use of keyword techniques and CRA (Centering Resonance Analysis; Corman, Kuhn, McPhee, & Dooley, 2002). These approaches are bottom up approaches to semantic interpretation that are based on statistical properties of the text. Although we have used LSA successfully to identify patterns indicative of effective and ineffective performance, as well as to tag utterances for parts of discourse (e.g., acknowledgement, questions, fact), we have been slowed down in this progress by the need for manual and tedious transcription and coding. This combined with the need for a domain-specific corpus, limits the potential for the near-term application of such content analysis techniques to real-time assessment of team state.

5 Interventions

Although the challenges for augmented team cognition seem to reside primarily in the assessment and diagnosis of team state, there have also been numerous developments by way of cognitive interventions. That is, once we know that a team is in a problematic team state, what do we do to take corrective action? Again, in fast tempo, highly dynamic settings, the successful intervention will occur in real-time or even better, will anticipate the particular team breakdown, and provide just-in-time correction.

Some examples of team-level interventions include synthetic intelligent agents who may play the role of other teammates or enemy forces (for training exercises) or as assistants in operational settings (Scerri, Xu, Liao, Lai, &

Sycara, 2004). In this paper session there are in fact two papers that describe interventions, one on a decision aid to improve team effectiveness for distributed command-and-control teams (Allen, VanEron, O'Hargan, Rached, Jones & Guerlain, 2005) and the other on mitigating memory failures at the team level (Fiore, Jentsch, Salas, & Finkelstein, 2005).

Many interventions for augmenting team cognition have been inspired by the information processing approach. Agents that serve individual cognitive needs and displays that are meant to facilitate "shared mental models" are two such examples. More ecologically inspired interventions might include aids to improve coordination of a team or to guide communication in the most efficient way. Tests of these interventions are not only necessary for valid augmented team cognition, but will also help to provide evidence for or against the different perspectives on team cognition.

6   Conclusions

A number of tools and techniques can be implemented in team environments with the hope of that they will augment team cognition. However, feedback on whether the cognition has been augmented, as well as the ability to tailor the intervention on an as-needed basis rests the ability to assess and diagnose the cognitive state of the team. There are challenges associated with this task and even more associated with real-time assessment and intervention. Progress is being made in this area with research programs dedicated to taking team-level data that occurs as a natural byproduct of team interactions (e.g., communication data) and extracting patterns from the data that are meaningful in terms of team cognition. This approach ultimately requires automated data collection, automated extraction and analysis of patterns, and automated interpretation of those patterns in terms of team cognition.

References

Allen, S., VanEron, K., O'Hargan, K., Rached, T., Jones, C. & Guerlain, S. (2005). Effect of decision support for distributed command and control teams. Paper presented at the First Augmented Cognition International Conference, July 22-28, Las Vegas, NV.

Cannon-Bowers, J.A. & Salas, E. (Eds.) (1998). *Decision Making Under Stress: Implications for Individual and Team Training.* Washington, DC: American Psychological Association.

Cannon-Bowers, J. A., Salas, E., & Converse, S. (1993). Shared mental models in expert team decision making. In J. Castellan Jr. (Ed.), *Current issues in individual and group decision making* (pp. 221-246). Hillsdale, NJ: Erlbaum.

Collyer, S. C. & Malecki, G. S. (1998). Tactical decision making under stress: History and overview. In J.A. Cannon-Bowers & E. Salas (Eds.), *Decision Making Under Stress: Implications for Individual and Team Training* (pp. 3-15). Washington, DC: American Psychological Association.

Cooke, N. C., & Gorman, J. C. (in press). Assessment of team cognition. In the 2nd edition of the *International Encyclopedia of Ergonomics and Human Factors.* Boca Raton, FL: CRC Press.

Cooke, N. J., Kiekel, P.A., Salas, E., Stout, R.J., Bowers, C., Cannon-Bowers, J. (2003). Measuring team knowledge: A window to the cognitive underpinnings of team performance. *Group Dynamics: Theory, Research and Practice, 7,* 179-199.

Cooke, N. J., Salas, E., Kiekel, P. A., & Bell, B. (2004). Advances in measuring team cognition. In E. Salas and S. M. Fiore (Eds.), *Team Cognition: Understanding the Factors that Drive Process and Performance* (pp. 83-106). Washington, DC: American Psychological Association.

Cooke, N. J., & Shope, S. M. (2002). The CERTT-UAV Task: A Synthetic Task Environment to Facilitate Team Research. *Proceedings of the Advanced Simulation Technologies Conference: Military, Government, and Aerospace Simulation Symposium,* pp. 25-30. San Diego, CA: The Society for Modeling and Simulation International.

Corman, S., Kuhn, T., McPhee, R., and K. Dooley (2002), Studying complex discursive systems: Centering resonance analysis of organizational communication. *Human Communication Research*, 28, 157-206.

Diedrich, F. J., Freeman, J., Entin, E. E. & MacMillan, J. (2005). Modeling, measuring, and improving cognition at the team level. Paper presented at the First Augmented Cognition International Conference, July 22-28, Las Vegas, NV.

Fiore, S. M., Jentsch, F., Salas, E, & Finkelstein, N. (2005). Cognition, teams, and augmenting team cognition: Understanding memory failures in distributed human-agent teams. Paper presented at the First Augmented Cognition International Conference, July 22-28, Las Vegas, NV.

Foltz, P. W. (2005). Tools for enhancing team performance through automated modeling of the content of team discourse. Paper presented at the First Augmented Cognition International Conference, July 22-28, Las Vegas, NV.

Gaddy, C. D. & Wachtel, J. A. (1992). Team skills training in nuclear power plant operations. In R. W. Swezey & E. Salas (Eds.), *Teams: Their training and performance* (pp. 379-396). Norwood, NJ: Ablex.

Gibson, J. J., (1966). *The Senses Considered as Perceptual Systems.* Boston, MA: Houghton-Mifflin.

Gorman, J.C., Foltz, P.W., Kiekel, P.A., Martin, M. J., & Cooke, N. J. (2003). Evaluation of latent-semantic analysis-based measures of team communications. *Proceedings of the Human Factors and Ergonomics Society 47th Annual Meeting*, 424-428.

Henning, R., Korbelak, K., & Smith, T. (2005). Social psychophysiology as a gauge of teamwork. Paper presented at the First Augmented Cognition International Conference, July 22-28, Las Vegas, NV.

Kiekel, P. A. (2005). FAUCET: Using communication flow analysis to diagnose team cognition. Paper presented at the First Augmented Cognition International Conference, July 22-28, Las Vegas, NV.

Kiekel, P.A., Gorman, J. C., & Cooke, N. J. (2004). Measuring speech flow of co-located and distributed command and control teams during a communication channel glitch. *Proceedings of the Human Factors and Ergonomics Society 48th Annual Meeting.*

Landauer, T. K., Foltz, P. W., & Laham, D. (1998). An introduction to Latent Semantic Analysis. *Discourse Processes, 25(2 & 3)*, 259-284.

Reason J. (1997). *Managing the Risks of Organizational Accidents*. Brookfield, VT: Ashgate.

Salas, E. Dickinson, T. L., Converse, S. A., & Tannenbaum, S. I. (1992). Toward an understanding of team performance and training. In R. W. Swezey & E. Salas (Eds.), *Teams: Their training and performance* (pp. 3-29). Norwood, NJ: Ablex.

Scerri, P., Xu, Y., Liao, E., Lai, G., and Sycara, K. (2004). Scaling teamwork to very large teams, in *AAMAS'04*.

Vaughan, D. (1996). *The Challenger Launch Decision: Risky Technology, Culture, and Deviance at NASA.* Chicago, IL: The University of Chicago Press.

Acknowledgements

This work was sponsored by the Office of Naval Research (N00014-03-1-0580), the Air Force Office of Scientific Research (FA9550-04-1-0234), and the Air Force Research Laboratory (FA8650-04-6442). The work has benefited from the contributions of Nia Amazeen, Dee Andrews, Christy Caballero, Olena Connor, Janie DeJoode, Pat Fitzgerald, Jamie Gorman, Preston Kiekel, Harry Pedersen, Leah Rowe, Steve Shope, Tom Taylor, and Jennifer Winner.

Modeling, Measuring, and Improving Cognition at the Team Level

Frederick J. Diedrich, Jared Freeman, Elliot E. Entin, Shawn A. Weil, & Jean MacMillan

Aptima®, Inc.
12 Gill St., Suite 1400, Woburn, MA 01801
diedrich@aptima.com, freeman@aptima.com, entin@aptima.com, sweil@aptima.com, macmillj@aptima.com

Abstract

Efforts to augment team cognition, whether through training or application of technology, are a critical aspect of enhancing mission effectiveness. In this manuscript we address this issue through presentation of a method used to model a critical aspect of team cognition, measure it, and manipulate it so as to enhance team performance. In particular, we focus on the fit of organizational structure to mission task structure (congruence). Our review of results from previous studies indicates that our approach can be used to predict team communication patterns and mission effectiveness, and thus, that the methods employed can enable us to augment team cognition by designing organizations that are well matched to mission demands. Moreover, we show that we can also use these techniques to guide adaptation between organizational forms by providing prospective information on the likely effectiveness of alternative organizations, thereby promoting mission effectiveness.

1   Introduction

It is well established that team performance is a critical aspect of mission effectiveness. In turn, team performance is critically dependent on the ability of team members to coordinate through behaviors such as mutual performance monitoring, back-up, proactive communications, and leadership (e.g., Sims, Salas, and Burke, 2004). Hence, efforts to augment team performance, whether through training or application of technology, are a critical aspect of enhancing mission effectiveness.

The challenge in augmenting team performance, however, lies in the complexity of the multiple, interactive processes that shape team cognition. We view team cognition as including the coordinative processes between individuals such as performance monitoring and communications. Yet, as a mediator of team performance, team cognition is not shaped solely by what we might traditionally consider to be cognitive processes that exist inside the head (e.g., decision making, mental models). Team cognition is shaped by the interaction of a group of individuals acting to achieve a common goal, the technologies they use, and the organizational structures in which they are embedded. Team cognition is the outcome of the multiple, interacting processes and constraints that act together, inside and outside of the head, to shape performance.

From this perspective, augmenting team cognition to improve performance becomes something much more than just supporting or increasing knowledge, thoughts or decisions. One can improve team cognition through a variety of strategies that shape not only aspects of individuals, but also their missions, technologies, procedures, and organizations. In particular, in this manuscript we address a method of modeling a critical aspect of team cognition, measuring it, manipulating it, and predicting its effects. We focus on the fit of organizational structure to mission task structure, and show that the modeling method and the empirical evidence that support it enable us to augment team cognition by designing organizations that can communicate efficiently. Moreover, although organizations are often reluctant to change organizational form, we show that we can use these techniques to overcome organizational inertia and guide adaptation between organizational structures, thereby promoting mission effectiveness.

1.1   Organizational Congruence

Different organizations may be better matched to certain missions than others. In other words, organizational effectiveness is partially determined by the congruence between an organization's structural design and an environment or task (e.g., Donaldson, 2001; Van de Ven & Drazin, 1985). If organizational structure is out of "alignment" with the organization's mission, then quality of performance should be reduced, for performance should be affected by whether the right person has the right resources at the right place at the right time.

For instance, two well-studied cases of alternative structures are typically referred to as functional and divisional architectures (e.g., Diedrich et al., 2002; Moon et al., 2000). Typically, in a functional structure (F) a given commander has control over one specialized aspect of a mission such as Strike or ISR, where the specific assets controlled are distributed across the entire area of operations. In contrast, in a divisional structure (D), a given commander has direct control over a single multifunctional platform (ship) with multiple assets (e.g., Strike, Ballistic Missile Defense, *and* ISR) operating within a bounded geographical region. In essence, the divisional structure is akin to a set of large department stores each operating in a bounded region, whereas the functional structure is much like a set of specialized boutiques operating across several regions.

In terms of congruence, one might imagine that a given mission task requires Strike assets and ISR assets for prosecution – for instance, surveillance assets for targeting of strike assets. Note that, in the F case, prosecution of such a task would include coordination between two commanders – the Strike Commander and the ISR Commander. In contrast, in the D case, prosecution of such a task might only include one commander who has both Strike and ISR assets directly under their control. In this F case, there exists a between-commander coordination overhead that does not exist in the D case, for the two commanders must communicate and work together to coordinate their actions. From this perspective, D is more congruent than F with mission demands. The congruence hypothesis holds that the D organization will out perform the F organization due to a reduction in coordination overhead (e.g., communications, workload).

1.2 Modeling Organizational Congruence

Consistent with the congruence hypothesis, results from a variety of studies indicate that modeling techniques can be used to design organizations such that there is a tight congruence or alignment between structures and mission requirements (e.g., Diedrich et al., 2003; Entin et al., 2003; Kleinman et al., 2003). The modeling techniques, based on a variety of approaches (e.g., Carley & Lee, 1998; Handley, Zaidi, & Levis, 1999; Levchuk et al., 2003, 2004), are used first for organizational design and then for simulation to evaluate proposed designs prior to implementation and experimentation. In general, this work requires the identification and formalization of the workflow processes (including communication, command, resource control, information propagation) and constraints of various alternative organizational structures. These organizational structures are then manipulated via optimal and heuristic algorithms, which allow a trade-off between solution optimality and complexity. Finally, the models enable the exploration of algorithms for strategy optimization to specify performance-enhancing rules for task allocation, communication, resource sharing, coordination, synchronization, and decision-making activities that occur at the various levels of the organization. Complete details can be found in Kleinman et al. (2003) and Levchuk et al. (2003, 2004).

In the case presented here, in order to explore our ability to engineer congruence, our basic approach was to define two different organizational structures and then design two missions (scenarios) that exploited these differences. The objective was for the first mission scenario to be "matched" to organization 1 through a high degree of congruence, while also being "mismatched" (i.e., exhibit low congruence) with organization 2. The goal was for the reverse to be true for the second scenario. More specifically, the models enabled generation of the D and F organizational structures, as well as two mission scenarios (d and f) that exploited differences in the organizational structures (Kleinman et al., 2003).

First, employing the models, we manipulated the selection of resource requirements needed to accomplish particular tasks within the scenarios. As noted above, the two organizations differed primarily with regard to the assets that each commander controlled. In the models used to design the scenarios, the degree of predicted (structural) congruence is inversely related to the amount of inter-commander coordination needed to accomplish the mission. Therefore, by adjusting the resource requirements of selected mission tasks it is possible to manipulate the inter-player coordination needed to successfully prosecute selected tasks for a given organization-scenario pairing. We designed tasks within f such that little between-commander coordination would be needed within organization F (tasks primarily required only one resource type that one commander fully controlled), while significant coordination would be needed by organization D (commanders would have to combine their single assets to meet overall task requirements), with the reverse being true for scenario d (tasks primarily required multiple resource types, thus favoring D).

Second, we manipulated congruence through the temporal and geographical distribution of tasks. This manipulation involved mission scenario tasks aimed to exploit the different geographical responsibilities in D versus F. Note that in D players had responsibilities in a defined area, while in F players had responsibilities that covered the entire battle space. Thus, for example, in scenario d we designed "waves" consisting of single functional area attackers (e.g., air) wherein the individual tasks within each wave were distributed geographically. This manipulation imposed a significant load upon one commander (e.g., air warfare) in F in both time and space, but the load was shared among several players in D with little or no coordination needed. In scenario f the attack "waves" consisted of several different functional area tasks where each wave targeted a specific player's area in D. Thus, through this method we could overload a given player in D, but the load would be distributed among several players in F.

1.3 Measuring & Augmenting Organizational Congruence

Based on this modeling approach, below we briefly review two previous studies in order to explore and validate this method of engineering organizational congruence to augment team cognition. In Experiment 1, we used the modeling approach outlined above to manipulate congruence and to identify measures of team cognition correlated with mission effectiveness (for complete details, see Diedrich et al., 2003; Entin et al., 2003). Our empirical strategy was to contrast performance under conditions in which organizational structures and missions were congruent with performance under conditions in which they were incongruent. The models predicted that under incongruent conditions (Df and Fd), there should be more communications as the organizations cope with high between-commander coordination demands. Results indicated that such increases existed, and that moreover, these changes in team cognition were inversely correlated with mission effectiveness. Building on this work, in Experiment 2 we then demonstrated how these models could be used to augment team cognition by guiding organizational adaptation (for complete details, see Weil et al., 2005). In particular, teams were given an opportunity to adapt their structures by moving from Fd to Dd or from Df to Ff. To support this decision, we provided teams with several forms of decision support, including performance forecasts based on models of each organization and anticipated mission demands. Prepared with this information, and facing a state of incongruence, the organizations studied chose to adapt their structures to enhance mission effectiveness. Model-based augmentation of team cognition improved performance.

2 Experiment 1 – Manipulating Congruence

The goal of Experiment 1, therefore, was to manipulate congruence using our model-based approach, thereby demonstrating that we could design organizational fit to augment team performance. In addition, we sought to identify measures of team cognition correlated with mission effectiveness and we focused on communications. More specifically, we predicted that performance would be superior under conditions of congruence (Ff and Dd) as compared to conditions of incongruence (Fd and Df). Due to the basis of the model-based manipulation in between-commander coordination load, we also predicted that there should be enhanced communications volume as the players cope with incongruence by attempting to coordinate.

2.1 Method

Eight teams of six individuals each were organized from 48 officers attending the Naval Postgraduate School. A random half of the teams were assigned to the F organizational structure, while the remaining half were assigned to the D organizational structure throughout all training and testing. Organizational structure was manipulated as a between-subjects factor (F, D), whereas scenario (f, d) was manipulated as a within-subjects factor, counter-balanced across teams, such that each team experienced both congruent (Ff or Dd) and incongruent conditions (Fd or Df).

The simulation was implemented within the Distributed Dynamic Decision-making (DDD) environment. In this case, the organizational structures (F, D) were implemented in the DDD simulation by varying the ownership of the different kinds of assets. For those in F, each commander controlled only one or perhaps two functional areas (e.g., Strike or ISR or Ballistic Missile Defense, etc). In contrast, for those in D, the commanders each controlled almost all areas (Srike, ISR, and Ballistic Missile Defense, etc., in a limited region). As specified above, we varied a number of mission task items in order to design the functional (f) and divisional (d) scenarios including task timings, locations, and resource requirements. The experimental scenarios involved a Joint Forces mission in which the teams were tasked to use a variety of sea, land, and air assets to complete a specified mission that included capturing a command center, two naval bases, two air bases, and a port. Participants began the experiment by completing two hours of DDD "buttonology" training to learn how to control the various assets and how to use the various functions contained in the DDD simulation. A second two-hour session provided training designed to provide experience on the skills necessary to perform either the divisional or functional scenarios. Following training, each team engaged in a two-hour data collection session followed by a second two-hour data collection session on a second day. The DDD simulation environment tracked performance (e.g., proportion of tasks completed) and coders tracked communications to identify frequency of communications and instances of coordination (e.g., "Please help me ....").

2.2 Results & Discussion

Consistent with predictions, teams in congruent conditions outperformed incongruent teams, as demonstrated by the proportion of enemy tasks processed. On average over the entire scenario, misfit teams processed approximately 40% of the required tasks, while the well-fit teams processed 50%, an improvement of 20%. Moreover the congruent conditions out performed the incongruent conditions over time throughout the entire scenario as shown by the frequency of attacks (Figure 1). These results suggested, therefore, that our model-based approach to manipulating congruence was successful in that we were able to augment team performance.

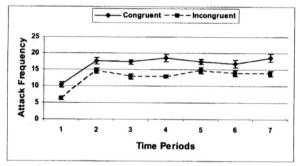

Figure 1: Mean frequency of attacks by 5 minute periods across the F and D structures (see Entin et al., 2003).

Moreover, also consistent with predictions of the computational models, overall communications increased in the incongruent cases and remained high throughout the scenario (Figure 2). In particular, communications specifically about coordination increased throughout the scenarios (Figure 3). Figure 3 shows the mean number of coordination requests ("I need your help") during five-minute periods of the scenarios. Collectively, these results showed that communications load provided an index of team cognition related to fit, and in particular, that communications specifically about coordination increased early in the scenarios in response to incongruence.

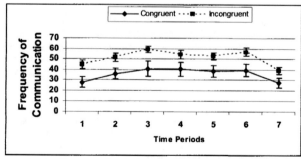

Figure 2: Mean Communication Frequency by 5 minute periods across the F and D structures (see Entin et al., 2003).

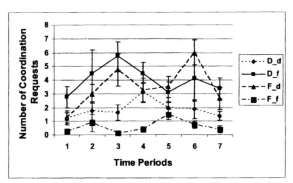

Figure 3: Frequency of requests for coordination by condition (see Entin et al., 2003).

In sum, these data verified our model-based manipulations of organizational congruence. Based on our approach, we were able to model organizations that fit and misfit particular missions, and this manipulation of fit impacted mission effectiveness. These results also showed that as an indicator of team cognition, communications volume and in particular communications about coordination, indexed the state of congruence early in the mission scenarios. Increases in communications volume were inversely correlated with mission effectiveness. As a result, these data indicated that when a match existed between the organizational structure and the mission scenario tasks to be performed, a particular team member within the organization whose job it was to process the task actually possessed the appropriate assets to do so. In contrast, when a mismatch existed, the team member that had to perform the task did not possess all the required resources, thus entailing coordination and communication overhead to amass the correct mix to perform the task. The additional overhead encountered in the mismatched condition likely generated additional workload and workload induced stress that lowered performance (Entin & Serfaty, 1999).

3   Experiment 2 – Guiding Adaptation

Experiment 1 showed that we could model organizational congruence so as to manipulate team cognition and performance. Moreover, the data showed that a measurement of team cognition – communications – could index fit and serve to signal the need for organizational adaptation, for relative to congruent states, high communications signalled incongruence early in the mission scenarios and were inversely correlated with mission effectiveness. Based on these data, in Experiment 2 our goal was to use the model-based approach to augment team cognition by guiding organizational adaptation (for complete details, see Weil et al., 2005).

Previous results indicated that teams are reluctant to alter their organizational structures to increase organizational/environmental congruence (e.g., Hollenbeck et al., 1999). Indeed, Entin et al. (2004) found only limited organizational adaptation despite motivating teams to adapt by providing them with ample data concerning performance deficiencies and justifications for organizational adaptation. Hence, our challenge in this work was to overcome this resistance to change in order to augment team performance by facilitating adaptation from Fd to Dd or from Df to Ff. We employed several strategies to encourage adaptation: First, we provided targeted instruction about the benefits of organizational change in terms of issues such as performance and workload, and we provided extensive training in different alternative organizational structures. Second, as in Entin et al. (2004), we provided a suite of feedback measures, based on proven performance and behavioral indicators of congruence, to the participants at critical times to demonstrate performance decrements in order to emphasize the criticality of organizational adaptation. Third, we presented several fully formed alternative organizational designs to the participants, accompanied by a rationale for their creation, in order to guide adaptation by providing organizations to move toward during the adaptation process. Fourth, along with the organizations, we provided prospective measures based on the modeling techniques in order provide information on likely performance levels and coordination requirements across the various organizations given the upcoming mission requirements. By providing both alternative organizations and model-based prospective measures of likely performance factors, we attempted to overcome inertia against change and guide adaptation such that organizations would improve their fit to the mission requirements. The critical question was whether the organizations would chose to adapt, and if so, whether they adapted so as to increase congruence.

3.1 Method

Twenty-Five officers and non-commissioned officers served as participants for this exercise at the Naval War College in Newport, RI. Participants were organized into four teams of six individuals, with two teams beginning in F and two teams beginning in D. Training began with two hours of DDD "buttonology" in which participants were shown how to use the DDD user interface, followed by training in both the D and F organizations, in a counterbalanced fashion. Following the second training session, the data-collection phase of the study began, such that each team participated in three data-collection sessions. In the first data-collection session, each team performed a mission scenario congruent with their organizational structure (Dd or Ff). In the second data-collection session, each team performed a mission scenario that was incongruent with their organization (Df or Fd). However, participants were not given the opportunity to change their structure; instead they had to stay with the sub-optimal structure so they could experience incongruence (e.g., high communications volume, poor performance, etc.). Following the second data-collection session, after experiencing incongruence, participants were given the opportunity to change their organizational structure. At this time, the teams were given explicit guidance on the merits of organizational change and its expected performance gains. In addition, five model-derived alternative organizational structures were presented to the participants, purported to be from the Joint Task Force Commander "planning shop." The alternative organizations included:

- *Divisional (D)*. Each participant controlled a single platform with multiple functional areas.
- *Divisional/Functional Hybrid (D2)*. Four participants controlled a single platform each with multiple functional areas; two players controlled functional assets across the theater.
- *Regional (R2)*. Theater was divided into two geographic regions. Groups of three participants divided the assets functionally within those two regions.
- *Functional/Divisional Hybrid (F2)*. Four participants controlled functional areas across the theater; two players controlled a single platform each with multiple functional areas.
- *Functional (F)*. Each participant controlled one or two functions across the theatre.

Note that D2, R2, and F2 were engineered so as to be moderate in congruence between D and F with respect to the demands of the f and d scenarios. For example, if facing scenario d, organization D would be the most congruent, then D2, etc. After making their organizational choice, teams experienced a third data collection session in which the mission tasks and requirements were identical to those of the second data-collection session (i.e., incongruent with the team's original structure), but with the organizational structure chosen during the facilitated planning session.

3.2 Results & Discussion

The primary question of interest was whether the organizations changed their structures prior to the third data session, and if so, whether this change was adaptive in nature with respect to congruence. That is, when faced with Fd, did the teams move toward Dd? Similarly, when faced with Df, did the teams move toward Ff?

Results clearly showed that each team made organizational changes when they had the opportunity. The two teams assigned initially to the F organizational structure faced a divisionally oriented mission in the incongruent situation, and they decided to adapt to the D2, which was nearly congruent with d. Similarly, the two D assigned teams faced a functionally oriented mission in the incongruent scenario and fully adapted by selecting an F organizational structure. This change demonstrated maximum adaptation because the F structure was modeled to have the highest congruence with the scenario they were to engage. Note that, while there were not enough teams to make a definitive conclusion, it is possible that the F teams were reluctant to adopt the fully congruent D organization because each team member would have been required to learn a larger variety of assets. In the F organization, each team member has only one or two responsibilities, and thus only needs to learn a few series of actions to interact with the simulator. In contrast, in the D organization, each team member is responsible for a larger variety of assets. Thus, participants starting in F may have been compromising between the modeled advantages for adaptation and the anticipated re-learning required for the transition to the D organization. For the participants starting in D, adaptation may have been more complete because of increased familiarity with all asset types, such that there was less perceived cost for adaptation.

Given these changes, we also explored whether adaptation actually improved mission effectiveness. Each team made organizational changes when they had the opportunity, and the changes made were predicted to be adaptive, based on modeled performance. However, organizational change is only truly adaptive if it leads to improved performance. To explore this issue, we focused on analysis of performance on the subset of tasks that required multiple assets for engagement, for these tasks reflect the crucial areas affected by the congruence manipulation. In a highly incongruent environment, these tasks required multiple team members to coordinate resources within a small window of time, increasing overhead and potentially decreasing performance. In contrast, in a congruent environment, these tasks required single individuals to use multiple assets within their own control. The predicted result was higher performance given the lack of between-player coordination requirements. Thus, with this prediction in mind, given apparent increases in congruence between sessions 2 and 3, we investigated the data to determine if performance improved on tasks requiring multiple assets for engagement. As hypothesized, the percentage of successfully executed tasks requiring multiple assets that were engaged increased by 7% for functional teams and 13.4% for divisional teams following organizational change. Thus, the organizational choices of each team increased the structure/scenario congruence, which decreased the coordination overhead, and allowed teams to engage more targets. Presumably, this effect was greater for the teams originally assigned to the D organizations because their organizational choice (i.e., F) had a higher degree of congruence with the scenario than did the teams originally assigned to the F organizations (i.e., because they chose D2 instead of the more optimal D).

In total, then, results from Experiment 2 showed that when faced with incongruence, and model-based alternative organizations, teams will alter their organizational structures to achieve congruence. These changes were adaptive in nature in that they served to improve performance in a manner proportional to the degree of post-adaptation congruence.

4 Conclusions

Collectively, these findings address a method of modeling team cognition, manipulating it, measuring it, and predicting its effects. Experiment 1 showed that organizational congruence could be engineered by varying coordination load across participants. This study showed that congruent teams outperformed incongruent teams, and these results also demonstrated that we could predict team communication patterns and mission effectiveness such that as required coordination load increased, communications volume increased, and mission effectiveness decreased. Hence, the method explored here enabled us to augment team cognition by designing organizations that can communicate efficiently and accomplish missions effectively due to fit with the mission task environment.

Building on these data, in Experiment 2 we showed that we can also use these techniques to guide adaptation between organizational forms, thereby overcoming resistance to change, and promoting mission effectiveness. Based on instruction concerning alternative organizations and prospective measures of performance in these organizations, teams chose to adapt their organizational structures when faced with incongruence. Improvements in performance following these organizational changes were proportional to the extent of the adaptation toward a congruent state. These data showed that team cognition and team performance could be augmented by employing models to manipulate congruence and by providing prospective information to guide adaptation.

Taken together, these findings provide provocative and important data regarding what it means to augment team cognition and how best to do so. First, these data suggest that efforts to augment team cognition through organizational adaptation must be complex and must promote change along multiple lines. While previous efforts have shown only limited ability to foster organizational adaptation (e.g., Hollenbeck et al., 1999; Entin et al., 2004), the effort outlined here was successful due to the multiple strategies used to guide change. Critical elements included model-based definition of alternative organizations and provision of prospective data on performance. However, participants were also exposed to training in multiple organizations, feedback on performance in incongruent conditions, and general advice on the merits of organizational change. Given previous findings showing reluctance regarding structural organizational adaptation (e.g., Entin et al., 2004), collectively these findings clearly point toward the need for multiple incentives and supports for organizational change. The primary lesson, then, is that much is needed to overcome inertia against organizational change so as to realize the benefits of augmented team cognition.

Second, these data suggest that when augmenting team cognition, it is useful to look beyond the head of the individuals and to the interactions within the organization. By its very nature, teamwork depends on interaction between multiple individuals. The nature of this interaction depends on several factors including the technologies employed, the people involved, and the mission environment. Here we showed how engineering of the mission-organizational fit augmented team performance and cognition as measured by communication patterns. Team cognition depended very much on interaction within a particular task environment. This means, then, that team cognition can be viewed as the outcome of the multiple, interacting processes and constraints that act together, inside and outside of the head, to shape team performance. From the perspective of augmenting cognition, the critical insight is that given the extended nature of cognition, and team cognition in particular, efforts to improve performance should focus not only on the individuals involved, but also on their missions, technologies, procedures, and organizations, for these are the elements that collectively shape team cognition.

Acknowledgements

The research reported here was sponsored by the Office of Naval Research, Contract No. N00014-02-C-0233, under the direction of Gerald Malecki. The opinions expressed are those of the authors and not necessarily those of ONR or the Department of Defense.

References

Carley, K.M. & Lee, J. (1998). Dynamic Organizations: Organizational Adaptation in a Changing Environment. In J. Baum (Ed.), *Advances in Strategic Management, Vol. 15, Disciplinary Roots of Strategic Management Research* (pp. 269-297). JAI Press.

Diedrich, F.J., Entin, E.E., Hutchins, S.G., Hocevar, S.P., Rubineau & MacMillan, J. (2003). When do organizations need to change (Part I)? Coping with incongruence. *Proceedings of the 2003 Command and Control Research and Technology Symposium*, Washington, DC.

Diedrich, F.J, Hocevar, S.P, Entin, E.E., Hutchins, S.G., Kemple, W.G., & Kleinman, D.L. (2002). Adaptive architectures for command and control: Toward an empirical evaluation of organizational congruence and adaptation. *Proceedings of the 2002 Command and Control Research Symposium*, Monterey, CA.

Donaldson, L. (2001). *The Contingency Theory of Organizations*. Thousand Oaks, CA: Sage.

Entin, E. E., Diedrich, F.J., Kleinman, D.L., Kemple, W.G., Hocevar, S.P., Rubineau, B., & MacMillan, J. (2003). When do organizations need to change (Part II)? Incongruence in action. *Proceedings of the 2003 Command and Control Research and Technology Symposium*, Washington, DC.

Entin, E. E., Weil, S.A., Kleinman, D.L., Hutchins, S.G., Hocevar, S.P., Kemple, W.G., Serfaty, D. (2004). Inducing Adaptation in Organizations: Concept and Experiment Design. *Proceedings of the 2004 Command and Control Research and Technology Symposium*, San Diego, CA.

Entin, E.E. and Serfaty, D. (1999). Adaptive team coordination. *Journal of Human Factors*, 41, 321-325.

Handley, H.A, Zaidi, Z.R., & Levis, A.H (1999). The Use of Simulation Models in Model-Driven Experimentation. *Proceedings of the 1999 Command and Control Research and Technology Symposium*.

Hollenbeck, J.R., Ilgen, D.R., Moon, H., Shepard, L., Ellis, A., West, B., Porter, C. (1999). Structural contingency theory and individual differences: Examination of external and internal person-team fit. Paper presented at the *31st SIOP Convention*, Atlanta, GA.

Kleinman, D.L., Levchuk, G.M., Hutchins, S.G., & Kemple, W.G. (2003). Scenario design for the empirical testing of organizational congruence. *Proceedings of the 2003 Command and Control Research and Technology Symposium*, Washington, DC.

Levchuk, G.M., Kleinman, D.L., Ruan, S., & Pattipati, K.R. (2003). Congruence of missions and Organizations: Theory versus data. *Proceedings of the 2003 Command and Control Research and Technology Symposium*, Washington, DC.

Levchuk, G.M., Yu, F., Levchuk, Y., & Pattipati, K.R. (2004). Networks of Decision-Making and Communicating Agents: A New Methodology for Design and Evaluation of Organizational Strategies and Heterarchical Structures. *Proceedings of the 2004 International Command and Control Research and Technology Symposium*, San Diego, CA.

Moon, H., Hollenbeck, J., Ilgen, D., West, B., Ellis, A., Humphrey, S., Porter, A. (2000). Asymmetry in structure movement: Challenges on the road to adaptive organization structures. *Proceedings of the Command and Control Research and Technology Symposium,* Monterey, CA.

Sims, D.E., Salas, E., & Burke, C.S. (2004). Is there a "Big Five" in teamwork? *19th Annual Conference of the Society for Industrial and Organizational Psychology*, Chicago, IL.

Weil, S.A., Levchuk, G., Downs-Martin, S., Diedrich, F.J., Entin, E.E., See, K.E., and Serfaty, D. (2005). Supporting organizational change in command and control: Approaches and metrics, *Proceedings of the 2005 Command and Control Research and Technology Symposium*, Washington, D.C.

Van de Ven, A. H. and Drazin, R. (1985). "The concept of fit in contingency theory." In B.M. Staw and L.L. Cummings (Eds.) *Research in Organizational Behavior, 7,* 333-365.

Social Psychophysiological Compliance as a Gauge of Teamwork

Robert A. Henning

Psychology Department
University of Connecticut
Storrs, CT, USA
Robert.Henning@UConn.edu

Thomas J. Smith

School of Kinesiology
University of Minnesota
Minneapolis, MN, USA
Smith293@umn.edu

Kristopher T. Korbelak

Psychology Department
University of Connecticut
Storrs, CT, USA
Kristopher.Korbelak@UConn.edu

Abstract

Empirical studies verify that team performance is critically dependent on the ability of teams to receive and control sensory feedback. According to social cybernetic theory, cognitive demands unique to teams arise from the need for teams to control future behavior based on limited sensory feedback, a process termed projective control. The key objective of augmented cognition systems for teams should therefore be to extend team capabilities for projective control. Since coordinate motor-sensory control by team members is known to have reciprocal effects on the psychophysiological state of each team member, it is hypothesized that the resulting commonality or behavioral compliance established among team members facilitates social tracking generally and team projective control specifically. Thus, social-psychophysiological compliance (SPC) linked to coordinate motor-sensory control is a candidate measure for assessing a team's capacity for projective control, under ideal or non-ideal conditions of sensory feedback. Efforts to develop SPC measures have focused on coordinate patterns in heart rate, breathing, and electrodermal activity from multiple team members. Laboratory studies reveal that SPC measures are predictive of both present and future team performance on a projective tracking task. Three strategies are proposed for using SPC measures to augment team cognition: (1) as a source of compliant feedback provided directly to teams; (2) as a source of compliant feedback to augmented cognition systems; and (3) as a diagnostic aid for developers of augmented cognition systems.

1   Overview

The central question being addressed in this paper is whether unique measures of social psychophysiological compliance (SPC) represent the most rigorous and meaningful, and therefore the preferred, analytic approach for implementing augmented cognition systems for teams. Current efforts to implement augmented cognition systems for individual operators focus on use of real-time psychophysiological measures that are person-specific (Raley, Stripling, Kruse, Schmorrow & Patrey, 2004). We consider these to be inadequate for assessing aspects of cognition unique to teams due to the added complexities of teams as social systems.

In general, closely corresponding or mirrored social behavior and/or psychophysiological functioning among two or more partners during social interaction is termed yoked behavior. In terms of yoked social behavior, it typically is observed that movement patterns of the social partners correspond in some manner temporally and/or spatially (e.g., Jackson & Decety, 2004; Condon & Ogston, 1966; 1971). In terms of yoked psychophysiological functioning, correspondence in temporal patterning of functional activity such as heart rate or ventilation may be observed. One proposed method of scoring the extent that yoked social behavior affects yoked psychophysiological functioning of social partners is to measure social-psychophysiological compliance (SPC) (T.J. Smith & Smith, 1987a; Henning, Boucsein & Gil, 2001). SPC is defined as the extent to which changes in physiological states among two or more participants in team performance exhibit such close correspondence or reflect mutual influence.

There are two aspects to the background of the SPC approach that can be highlighted. The first pertains to a relative lack of attention to the collective cognitive performance of teams on the part of most current research in the areas of both augmented cognition and psychophysiology. For example, presentations by members of a panel devoted to augmented cognition convened during the 2004 annual meeting of the Human Factors and Ergonomics Society dealt exclusively with augmentation of cognitive performance of individual operators (Dorneich et al., 2004; Fishel et al.,

2004; Raley et al., 2004). Similarly, in a special section of the Winter, 2003 issue of *Human Factors* devoted to psychophysiology in ergonomics (Trimmel, Wright & Backs, 2003), analysis of individuals represents the focus of all 10 reports in the section.

Nevertheless as the second background aspect, when it comes to actual operations in complex environments, particularly those involving integrated human-computer systems, the reality is that in many if not most instances, it is the team rather than the individual operator that represents the key unit of operation (Rothrock, Harvey & Burns, 2005; Salas & Fiore, 2004; Wickens, Gordon & Liu, 1998, pp. 594-600). Three logical scientific consequences have grown out of this reality. The first is the idea of team (as opposed to individual) cognition (Cooke et al., 2000; Entin & Serfaty, 1999; Gutwin & Greenberg, 2004; Langan-Fox, Code & Langfield-Smith, 2000). The second is the idea of replacing social partners by computers, serving as social surrogates, in hybrid human-computer systems operations (T.J. Smith, Henning & Smith, 1995). The third is the idea of using computer intervention to enhance team performance (Sycara & Lewis, 2004).

The foregoing considerations suggest that use of an interactive computer to augment the cognitive performance of team operations has both scientific plausibility and practical significance. Before exploring research strategies that might support this goal, it is appropriate to first introduce prevailing views of the human factors community as to what is meant by the terms 'team' and 'team cognition.' Cooke and colleagues (2000) note that a team is defined as '*a distinguishable set of two or more people who interact dynamically, interdependently, and adaptively toward a common and valued goal/object/mission, who have each been assigned specific roles or functions to perform, and who have a limited life span of membership.*' These authors go on to assert that:

> '*...teams, unlike some groups, have differentiated responsibilities and roles...This division of labor enables teams to tackle tasks that are too complex for any individual. Whereas the team approach is often seen as a solution to cognitively complex tasks, it also introduces an additional layer of cognitive requirements that are associated with the demands of working together effectively with others. Team members must coordinate their activities with other who are working toward the same goal. Team tasks often call for the team to detect and recognize pertinent cues, make decisions, solve problems, remember relevant information, plan, acquire knowledge, and design solutions or products as an integrated unit. Therefore, an understanding of team cognition...is critical to understanding team performance.*'

From the perspective of cognition as a self-regulated, closed-loop control process (T.J. Smith & Henning, 2005, these proceedings), there are a number of key aspects of the foregoing synopsis. One is the idea of the team as a coordinated and integrated unit. A second is the idea that unit integration imposes additional cognitive demands on each team member. A third is the idea of projective planning—that is, feedforward control of unit behavior—as a common feature of team cognition. Surrounding these ideas are a number of basic questions. Does team cognition exhibit the same behavioral attributes as individual cognition, namely closed-loop control, projective behavior (as implied by the above synopsis), and context specificity in performance (IBID)? How is unit coordination and integration of a team actually achieved behaviorally? Is there actually such a thing as collective team cognition, or does team performance simply reflect the separate cognitive contributions of individual team members? If the former is true, is team cognition an additive or multiplicative outcome of the contributions of individual team members? Finally, if computer augmentation of team cognition is to be achieved, how might such an intervention be modeled conceptually, and how might team cognitive performance be monitored as a basis for intervention?

Our answers to these questions are grounded in the social cybernetic model of behavior developed by K.U. Smith and associates (T.J. Smith & Smith, 1987a; 1987b). This model is quite distinct from the information processing model that dominates present-day efforts to implement systems for augmented cognition for individuals (T.J. Smith & Henning, 2005, these proceedings). The social cybernetic model can be used to guide the identification of prospective psychophysiological mechanisms relevant to social-behavioral organization and control, most particularly involving projective control of team behavior. This departs from the convention of focusing solely on instantaneous processing capacity limitations implicated in the information processing model of augmented cognition (Raley et al., 2004).

As noted above, SPC represents the specific social cybernetic paradigm that we believe can be used to account for many of the unit integration, projective tracking, and context specific features of team cognition. As such, our premise is that analysis of SPC among team members also represents the key to effective computer-mediated augmentation of team cognition. Our analysis below of how concepts and methods of SPC might be used to support such intervention is preceded by a review of relevant social cybernetic concepts and related research findings.

2 Cybernetic Principles of Social Interaction

2.1 Self-Regulation of Behavior

Behavioral cybernetic theory—on which concepts of social cybernetics are based—holds that behavior is a self-regulated process, involving highly coordinated, multi-level, closed-loop mechanisms in which motor activity is employed as the primary means for control of behavior (T.J. Smith & Smith, 1987a). Specifically, the theory assumes that because of closed-loop links between action, perception, and physiological functioning, motor behavior integrates perceptual, psychomotor, cognitive, and physiological levels of behavioral organization. Since there is endless variety in the manner in which motor activity can be articulated, through postural, transport and manipulative movements, such integration makes it possible to exert specific projective (or feedforward) control over future behavior.

Our focus in this paper is on the reciprocal links between overt motor behavior and metabolic and physiological functioning (K.U. Smith, 1973), on which concepts and methods of SPC are based. At the level of individual behavior, empirical evidence for skilled physiological tracking dependent on the interroceptor system for feedback control in behavior organization is found in comparisons between elite and novice runners, where the elite athlete exhibits smooth changes in breathing pattern synchronous with footfalls contributing to bioenergetic efficiency, while the novice struggles to establish synchronicity (Bramble & Carrier, 1983). Similar differences in pedaling behaviors occur between novice and elite bicyclists (Jasinskas et al., 1980; Kohl et al., 1981). When professional typists exhibit synchronization between work rhythm, breathing and heart rate, performance improves (Henning & Sauter, 1996) and is associated with subjective ratings of wellbeing (Henning, Sauter & Krieg, 1992). Collectively, this evidence suggests that motor-physiological integration represents a key feature of control of both present and future behavior, and that skill development involves gains in proficiency not only in overt motor skill but also in such integration.

2.2 Social Cybernetics and Social-Psychophysiological Compliance

The study of interpersonal and group reciprocal sensory feedback and sensory feedback control relationships represents the focus of social cybernetics, directed towards delineating the closed-loop behavioral-physiological manifestations and properties of social interaction (T.J. Smith, Henning & Smith, 1995). One unique perspective provided by social cybernetics is that ongoing team behaviors serve in a crucial organizing role for both present and future team behaviors due to the reciprocal effects of motor behavior (i.e., muscle activity) on the psychophysiological states of team members. Motor-sensory control by one individual in response to behavior of the other individual serves as a source of sensory feedback to the other individual that is used to adjust behavior, and vice versa, resulting in a yoked process of motor control during social interaction shown across the centerline of Figure 1. Motor control is assumed to play a central role in the organization of social behaviors through clearly delineated modes of social tracking. Three modes of social tracking have been identified (imitative, parallel-linked, and series-linked) that represent design factors that limit the specific nature and extent of socially yoked behaviors (T.J. Smith & Smith, 1987a). Thus, one can expect specific differences in the nature and extent of psychophysiological changes related to social behaviors in each mode of social tracking due to differences in the reciprocal effects of the associated motor behaviors. Many team tasks will consist of combinations of these modes occurring in sequence or simultaneously depending on task demands. As introduced in the companion paper (T.J. Smith & Henning, 2005, these proceedings), social-relational tracking involves one or more participants monitoring the status or quality of the ongoing interaction as a means of determining when and if some adjustment in social

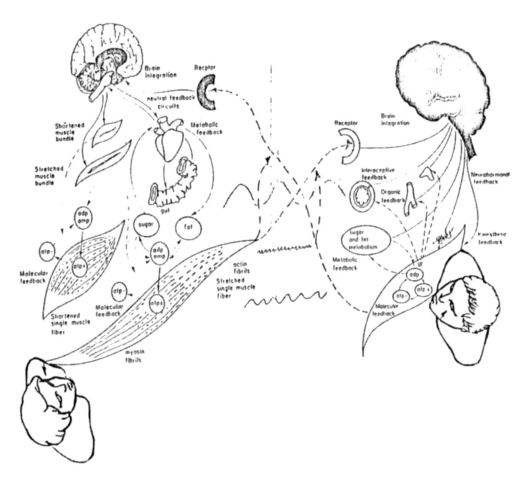

Figure 1. The reciprocal effects of social behavior on psychophysiological state.

behavior is needed to maintain proficiency, such as changing the pace of activity. Social-relational tracking represents a high-order form of tracking of the success or failure of the relationship process itself.

A number of studies report matched psychophysiological changes during social interaction (Boyd & Di Mascio, 1954; Levenson & Gottman, 1983; Levenson & Ruef, 1992; Hatfield, Cacioppo & Rapson, 1994), providing empirical evidence of an association between mutual social tracking and social psychophysiology. No systematic investigation of the effects of the different social tracking modes (imitative, parallel-linked, series-linked) has yet been conducted.

2.3   Measurement of Social-Psychophysiological Compliance (SPC)

One proposed method of assessing social-behavioral organization and control during the parallel mode of social tracking is to score social-psychophysiological compliance (SPC) among team members. SPC is the extent that changes in psychophysiological states among the participants exhibit close correspondence and/or reflect mutual influence. It should be noted that this scoring technique focuses on simultaneous psychophysiological changes among participants. Patterns of heart rate variability, breathing (lung ventilation), and electrodermal activity in two-person teams have been examined with SPC scoring techniques thus far (Henning et al., 2001) but SPC scoring techniques could also be applied to other psychophysiological measures such as EEG.

One method of scoring SPC is to calculate the cross correlation of separate time series data collected from each team member. Cross correlation (at lag =0) scores the extent of proportional matching of simultaneous psychophysiological changes in each team member. For example by not requiring exact matching of hear rate, cross correlation adjusts to individual differences in baseline heart rate as well as to differences in heart rate variability. Autoregressive corrections to each time series before cross correlation analysis may be warranted to isolate the effects of mutual influence (Gottman, 1981) but this also prevents detection of close correspondence in the time series that may also have some relevance to the organization of team behavior, perhaps due to task-induced psychophysiological changes during shared task activities.

A second scoring method is weighted coherence introduced by Porges et al. (1980) for the analysis of psychophysiological signals that naturally vary over a range of frequencies. This scoring method uses cross spectral analysis of psychophysiological changes over a specified frequency range to not only detect proportional matching but also to detect proportional matching that is out of phase. Weighted coherence provides a single summary statistic, varying between 0 and 1 analogous to a squared correlation coefficient. Shared frequencies must represent a high proportion of the total variance in the constituent signals. An adapted method of scoring weighted coherence that is not limited to bivariate analysis has been developed (Bohrer, Henning & Marden, 1992). The weighted coherence scoring technique provides the most accurate estimates of shared variance related to linear changes in coupling (Kettunen & Ravaja, 2000).

Efforts to develop new SPC scoring methods may benefit from the use of artificial neural networks (Wilson & Russell, 2003a; 2003b) to help identify SPC based on ensembles of psychophysiological measures (Raley et al., 2004).

3   Related Research

3.1   Sources of Variability in Team Performance

Relative to scientific investigation into the dynamics of individual cognition, our understanding of the nature and sources of variability in the dynamics of team cognition is very limited. A key complication, with regard to both conceptual and empirical analysis, is that social interaction in a team context introduces the need for each team member to track sensory feedback not only from his/her own behavior but also from other team members, thereby imposing additional multi-tasking tracking demands beyond those required for individual performance.

Of equal importance, our lack of understanding of team performance extends to the level of measurement itself, in that no consensus agreement exists as to how best to evaluate team performance in terms of outcome and/or process measures. Research on team performance generally has focused on performance outcomes rather than on inner coordination processes (McGrath, 1990). Thus, one popular outcome-based approach involves use of game-playing and game theory to provide insight into social interaction (Vogel, 2004; Sosis, 2004). Cooke and colleagues (2000) identify six categories of metrics for assessing team performance (observations, interviews and surveys, process tracing techniques, conceptual methods, task analysis, and decision analysis). All but a subset of about a third of these involve outcome-based measures. This focus may be explained, in part, by the lack of conceptual models and empirical techniques for measurement of the split-second dynamics of team behavior (Bopp, 1997; Brannick et al., 1995; Henning et al., 1997; Zigurs & Kozar, 1994).

As reported in detail elsewhere, behavioral cybernetic research that has been able to incorporate continuous sampling of behavior during team tracking tasks indicates that team tracking performance and team learning are critically influenced by the human factored design of the tracking task (Jentsch et al., 1995; Kao & Smith, 1971, 1977; Rothe, 1973; Sauter & Smith, 1971; K.U. Smith & Smith, 1973; Ting, Smith & Smith, 1972). Based on a meta-analysis of 122 studies, Johnson and colleagues (1981) conclude that, in terms of promoting achievement and productivity, team cooperation is more effective than either interpersonal competition or individual effort.

3.2   Computer-Supported Cooperative Work

A recent line of research focuses on a closely related problem, namely how to compensate for team performance that is degraded by the lack of opportunities for motor-sensory control under conditions of distributed collaboration (Gutwin & Greenberg, 2004). When team members are not at the same location, the research challenge has been to devise ways of compensating for losses in team coordination that inevitably occur. It is significant that the key design features of these compensatory systems re-introduce opportunities for yoked motor-sensory control that have been compromised or eliminated when team members are physically separated. The reported success of this approach provides support for the notion that team cognition for projective control of team behaviors is critically dependent on yoked motor-sensory control. And if enhancing yoked motor-sensory control among team members is able to partially or fully overcome performance losses when team members are physically separated, enhancing yoked motorsensory control opportunities for co-located teams may enhance team performance beyond normal expectations.

3.3   Team Situation Awareness

Research in the area of team situation awareness addresses issues surrounding the measurement of team knowledge that are believed to be a prerequisite to understanding team cognition and its relation to team performance (Cooke, Salas, Cannon-Bowers & Stout, 2000). One of three key characteristics of team situation awareness (SA) is that the team can anticipate future events (Paris, Salas & Cannon-Bowers, 2000; Prince & Salas, 2000), an idea that is complementary to the cybernetic theme of projective control of behavior. However, a single definition of team SA has not yet been accepted, and the proposed assessment tools such as a linear combination of individual SA metrics (Prince & Salas, 2000) are likely to be inadequate.

4   Laboratory Studies of Social Psychophysiological Compliance

4.1   SPC and Tandem Tracking

Based on earlier laboratory research on team control (Wegner & Zeamon, 1956) and social tracking (Sauter & Smith, 1971; Kao & Smith, 1977), SPC was examined during the parallel-linked mode of social tracking by two-person student teams (Henning et al., 2001). Each team member could exert dynamic control over the task system and each team member received immediate visual feedback about how the system responded to the combined team members' control inputs. The computer-based task developed for this experiment involved projective parallel-linked tracking in simulated teleoperation (see Smith et al., 1998), allowing team members to view upcoming task demands and exercise control over the pace the task was performed (e.g., to slow down in order to negotiate an upcoming 90-degree corner while traversing the complex path). Using psychophysiological signals collected from both participants, SPC for each bivariate time series of electrodermal activity, heart rate variability, and breathing pattern was scored using weighted coherence, and also using cross correlation. Each SPC score was then tested separately as a predictor of team performance and coordination.

SPC was found to be predictive of team performance, with the weighted coherence statistic more robust as an index of SPC. Task completion time and team RMS tracking error were predicted by all three SPC measures for electrodermal activity, heart rate variability, and breathing. The clustering of the predictive relationships provided convergent support for a positive linkage between SPC and team performance. The data did not support a counter interpretation that psychophysiological compliance was only the consequence of both team members having the same joystick control actions because no relationship was found between the cross correlation of joystick control actions and SPC. Thus, synchronous joystick control movements by team members did not appear to have induced psychophysiological compliance in this study, nor did they account for the reported predictive relationships between SPC and team performance. Additional analyses using this same set of SPC data were conducted to test for possible relationships with subjective responses. Individual team members had been asked to rate team performance and coordination after each trial in a 7-item survey. Positive linkages between SPC and team member ratings were found (Gil & Henning, 2000).

In a second study of teamwork (Henning & Korbelak, 2005), SPC was studied in relation to loss-of-control episodes under laboratory conditions. Student teams performed the projective tracking task described above but with role specialization in which one team member controlled the horizontal position of the object while the other controlled the vertical position of the object. At a randomly-selected point somewhere between 25% and 75% of total path length, either horizontal and vertical control was swapped between team members, or directional control relationships of the joysticks were reversed (e.g., joystick input that formerly caused the object to move upward now caused the object to move downward), or both horizontal and vertical control was swapped plus directional changes occurred. Higher SPC (cross correlation, lag=0) scores in heart rate variability predicted lower RMS error from path centerline (p < .01) but did not predict collision rate between object and path wall. The results suggest that SPC measures can be used to assess a team's readiness to handle unexpected performance demands.

4.2   Team Biocybernetics

A small pilot study extended biocybernetics (Byrne & Parasuraman, 1996; Freeman et al., 1999; Prinzel et al., 1995) to teams (Henning, Boucsein, Fekieta, Gil, Li & Pratt, 2000). Using a variation of the projective tracking task described above with team members working in the parallel mode of social tracking, task challenge level was dynamically manipulated within experimental trials by changing the size of the inertial object as it moved through the complex path of fixed width. The task became easier when the object was smaller because it was then less likely to collide with the wall of the narrow path. The instantaneous capability of the team was estimated by SPC of heart rate variability. Two conditions were tested; (1) *the matched condition*, where task challenge level was set higher when SPC indicated that the team supposedly could handle increased task demand, and lower when the team could not (the best-case scenario), and (2) *the unmatched condition,* where task challenge level was set higher when SPC indicated that the team could supposedly not handle increased demand, and vice versa (the worst-case scenario). Results indicated that team RMS tracking error from path centerline was lower in the matched condition but a tendency for completion time to increase was also noted, suggesting the need for further study.

5   Use of Social-Psychological Compliance to Augment Team Cognition

The foregoing synopsis indicates that the level of SPC among team members is: (1) predictive of proficiency in team performance, and of team readiness to handle unexpected task demands, in relation to task completion time and tracking error team performance measures; and (2) may have value as an indicator of the potential for improvement in team performance proficiency. These findings support the conclusions that: (1) the level of SPC among team members is predictive of the collective cognitive capabilities of a team; and (2) intervention to promote SPC levels may consequently serve to augment such capabilities. Accordingly, three strategies are proposed for using SPC measures to augment team cognition: (1) as a source of compliant feedback provided directly to teams; (2) as a source of compliant feedback to augmented cognition systems; and (3) as a diagnostic aid for developers of augmented cognition systems.

5.1   As a Source of Feedback to Teams

Alerting teams to more detailed aspects of the nature and extent of social tracking between individual team members may augment projective control over team behaviors, such as team planning efforts to schedule and allocate tasks. Displaying the trajectory of SPC over time in a shared display may be helpful in this regard, and would also promote social-relational tracking. Alternatively, team cognitive performance may benefit by simply alerting teams to periods with high levels of SPC since these periods may represent extraordinary opportunities for teams to engage in projective control or team decision-making. The availability of SPC feedback from what amounts to an aided social tracking system may also reduce the need for ancillary forms of social interaction that are often necessary to maintain a minimal level of social-relational tracking, thus allowing team members to allocate limited cognitive resources to more important tasks.

Providing an indication of SPC levels as compliant feedback to team members during task performance may have an added benefit, namely enhancing the level of SPC itself. That is, providing each team member with a real time indication of their degree of SPC with other team members represents a source of performance feedback that can serve as a training stimulus for improving SPC (availability of augmented performance feedback is widely recognized as one means of enhancing performance proficiency). Based on evidence cited above, such improvement in turn should be accompanied by concomitant improvement in team cognition.

5.2   As a Source of Feedback to AugCog Systems

Augmented cognition systems should be able to adapt to the needs of the team but automated systems typically lack the social skill sophistication necessary to assess team capabilities and then to act accordingly. SPC could be used as a source of compliant feedback to the augmented cognition system by indicating the extent that social tracking among team members occurs in response to actions taken by the augmented cognition system, thereby equipping the system with rudimentary social tracking skills to function like a partner on the team – an identified goal for automated systems (Woods, Sarter & Billings, 1997) that should also improve the overall usability of augmented cognition systems.

Use of SPC as an indicator of social tracking and team capability may also be useful in developing hybrid automation systems for augmented team cognition. As in supervisory control (Sheridan 1997; 2002), control over the specific actions taken by the AugCog system, such as adjusting the amount of feedback information presented to the team, should be shared between the computer system and team members. Dynamic adjustments in the types of control implemented by the AugCog system could be made in relation to the level of SPC but with humans always having control authority.

Lastly, SPC could become part of an ensemble of psychophysiological measures (Raley et al., 2004) in which SPC complements the use of conventional psychophysiological measures collected from individual team members. It may be necessary to combine SPC measures of social tracking with assessments of stress and fatigue of individual team members before augmented cognition systems for teams will become fully effective in addressing the needs of both individual team members and the overall team.

5.3   As a Diagnostic Aid

SPC measures could also be useful as a diagnostic aid during the development of AugCog systems. The extent of SPC could be helpful in the identification of design shortcomings, such as the lack of team involvement when sensory feedback about task-related events demand a coordinated response by team members. SPC as a human factors design tool offers all of the advantages of psychophysiological measures, including continuous monitoring, objective scoring, a scientific basis in known bio-behavioral mechanisms, and a lack of invasiveness (Backs & Boucsein, 1999).

6   References

Adolph, E.F. (1982). Physiological integrations in action. *The Physiologist*, 25(2), Supplement, 1-67.

Backs, R.W., & Boucsein, W. (Eds.) (1999). Engineering Psychophysiology. Mahwah, NJ: Lawrence Erlbaum Associates.

Bohrer, R., Henning, R.A., and Marden, J.I. (1992). A measure of synchronicity in multiple time series analysis. Unpublished manuscript.

Bopp, M.I. (1997). Team Self-Management of Performance Feedback. Masters Thesis. Storrs, CT: University of Connecticut.

Boyd, R.W., and DiMascio, A. (1954). Social behavior and autonomic physiology (a socio-physiologic study). *Journal of Nervous and Mental Disease*, 120, 207-218.

Bramble, D.M., and Carrier, D.R. (1983). Running and breathing in mammals. *Science*, 219, 251-256.

Brannick, M.T., Prince, A., Prince, C., and Salas, E. (1995). The measurement of team process. *Human Factors*, 37(3), 641-651.

Byrne, E. A., and Parasuraman, R. (1996) Psychophysiology and adaptive automation. *Biological Psychology*, 42, 249-268.

Condon, W. S., & Ogston, W. D. (1966) Sound Film Analysis of Normal and Pathological Behavior Patterns. *Journal of Nervous and Mental Disease*, 143(4), p. 338 347.

Condon, W. S., & Ogston, W. D. (1971). Speech and body motion synchrony of the speaker-hearer. In D. L. Horton & J. J. Jenkins (Eds.), *Perception of language*. Columbus, OH: Charles E. Merrill.

Cooke, N.J., Salas, E., Cannon-Bowers, J.A., and Stout, R.J. (2000). Measuring team knowledge. *Human Factors*, 42(1), 151-173.

Dorneich, M., Whitlow, S., Ververs, P.M., Carciofini, M., and Creaser, J. (2004). Closing the loop of an adaptive system with cognitive state. In: *Proceedings of the Human Factors and Ergonomics Society 48th Annual Meeting* (pp. 590-594). Santa Monica, CA: HFES.

Entin, E.E., and Serfaty, D. (1999). Adaptive team coordination. *Human Factors*, 41(2), 312-325.

Fishel, S.R., Owens, J.M., Muth, E.R., Hoover, A.W., and Rand, J.R. (2004). In *Proceedings of the Human Factors and Ergonomics Society 48th Annual Meeting* (pp. 1256-1260). Santa Monica, CA: HFES.

Freeman, F.G., Mikulka, P.J., Prinzel, L.J., & Scerbo, M.W. (1999). Evaluation of an adaptive automation system using three EEG indices with a visual tracking task. *Biological Psychology*, 50, 61-76.

Gil, M.C., and Henning, R.A. (2000). Determinants of perceived teamwork: examination of team performance and social psychophysiology. In *Proceedings of the XIVth Triennial Congress of the International Ergonomics Association and 44th Annual Meeting of the Human Factors and Ergonomics Society* (Vol. 2, pp. 743-746), Santa Monica, CA: HFES.

Gottman, J.M. (1981). *Time Series Analysis*. Cambridge, UK: Cambridge University Press.

Gutwin, C., and Greenburg, S. (2004). The importance of awareness for team cognition in distributed environments. In: E.Sala and S.N. Fiore (Eds.) *Team cognition: Understanding the factors that drive processes and performance*, pp. 177-201, Washington: APA Press.

Hatfield, E., Cacioppo, J.T., and Rapson, R.L. (1994). *Emotional Contagion*. Cambridge, UK: Cambridge University Press.

Henning, R.A., Bopp, M.I., Tucker, K.M., Knoph, R.D., and Ahlgren, J. (1997). Team-managed rest breaks during computer-supported cooperative work. *International Journal of Industrial Ergonomics*, 20, 19-29.

Henning, R.A., Boucsein, W., Fekieta, R.E., Gil, M.O., Li, Q., and Pratt, J.H. (2000). Team biocybernetics based on social psychophysiology. *Proceedings of PIE2000*, July 30th, San Diego, CA.

Henning, R.A., Boucsein, W., and Gil, M.O. (2001). Social-physiological compliance as a determinant of team performance. *International Journal of Psychophysiology*, 40, 221-232.

Henning, R.A., and Korbelak, K.T. (2005). Social-psychophysiological compliance as a predictor of future team performance. Paper submitted to *Pscyhologia*.

Henning, R.A., and Sauter, S.L. (1996). Work-physiological synchronization as a determinant of performance in repetitive computer work. *Biological Psychology*, 42, 269-286.

Henning, R.A., Sauter, S.L., & Krieg, E.F., Jr. (1992). Work rhythm & physiological rhythms in repetitive computer work: Effects of synchronization on well-being. *International Journal on Human Computer Interaction*, 4 (3), 233-243.

Jackson, P.L., and Decety, J. (2004). Motor cognition: a new paradigm to study self-other interactions. *Current Opinion Neurobiology*, 14, 259-263.

Jasinskas, C.L., Wilson, B.A., and Hoare, J. (1980). Entrainment of breathing rate to movement frequency during work at two intensities. *Respiration Physiology*, 42, 199-209.

Jentsch, F.G., Tait, T., Navarro, G., Bowers, C. (1995). Differential effects of feedback as a function of task distribution in teams. In *Proceedings of the 39th Annual Meeting of the Human Factors and Ergonomics Society* (Vol. 2, pp. 1273-1277). Santa Monica, CA: HFES.

Johnson, D.W., Maruyama, G., Johnson, R., and Nelson, D. (1981). Effecs of cooperative, competitive, and individualistic goal structures on achievement: a meta-analysis. *Psychological Bulletin*, 89(1), 47-62.

Kao, H.S., and Smith, K.U. (1971). Social feedback: determination of social learning, *Journal of Nervous and Mental Disease*, 152, 289-297.

Kao, H.S., and Smith, K.U. (1977). Delayed visual feedback in inter-operator social tracking performance. *Psychologia - An International Journal of Psychology in the Orient*, 20(1), 20-27.

Kettunen, J., and Ravaja, N. (2000). A comparison of different time series techniques to analyze phasic coupling: a case study of cardiac and electrodermal activity. *Psychophysiology*, 37(4), 395-408.

Kohl, J., Koller, E.A., and Jager, M. (1981). Relation between pedaling and breathing rhythm. *Eur. J. Appl. Physiol.*, 47,223-237.

Langan-Fox, J., Code, S., and Langfield-Smith, K. (2000). Team mental models: techniques, methods, and analytic approaches. *Human Factors*, 42(2), 242-271.

Levenson, R.W., and Gottman, J.M. (1983) Marital interaction: physiological linkage and affective exchange. *Journal of Personality and Social Psychology*, 45(3), 587-597.

Levenson, R.W., and Ruef, A.M. (1992) Empathy: a physiological substrate. *Journal of Personality and Social Psychology*, 63(2), 234-246.

McGrath, J.E. (1990). Time matters in groups. In J. Galegher, R.E. Kraut, and C. Egido (Eds.), *Intellectual Teamwork*. New Jersey: Lawrence Erlbaum Associates.

Paris, C. R., Salas, E., & Cannon-Bowers, J. A. (2000). Teamwork in multi-person systems: A review and analysis. *Ergonomics, 43*(8), 1052-1075.

Prince, C., & Salas, E. (2000). Team situation awareness, errors, and crew resource management research: Integration for training guidance. In M. R. Endsley & D. J. Garland (Eds.), *Situation Awareness Analysis and Measurement* (pp. 325-347). Mahwah, New JerseyLawrence Erlbaum Associates.

Porges, S.W., Bohrer, R.E., Cheung, M.N., Drasgow, F., McCabe, P.M., and Keren, G. (1980). New time series statistic for detecting rhythmic co-occurrence in the frequency domain: the weighted coherence and its applications to psychophysiological research. *Psychological Bulletin*, 88(3), 580-587.

Prinzel, L.J., Scerbo, M.W., Freeman, F.G., and Mikulka, P.J. (1995). A bio-cybernetic system for adaptive automation. *Proc. of the Human Factors and Ergonomics Society 39th Annual Meeting*, pp. 1365-1369. Santa Monica, CA: Human Factors and Ergonomics Society.

Raley, C., Stripling, R., Kruse, A., Schmorrow, D., and Patrey, J. (2004). Augmented cognition overview: improving information intake under stress. In *Proceedings: 48th Annual Meeting of the Human Factors & Ergonomics Society* (pp. 1150-1154). Santa Monica, CA: HFES.

Rothe, M. (1973). Social Tracking in Children as a Function of Age. Masters Thesis. Madison, WI:University of Wisconsin - Madison.

Rothrock, L., Harvey, C.M., and Burns, J. (2005). A theoretical framework and quantitative architecture to assess team task complexity in dynamic environments. *Theoretical Issues in Ergonomics Science*, 6(2), 157-171.

Salas, E., and Fiore, S.M. (Eds.) (2004). Team cognition: understanding the factors that drive process and performance. Washington, D.C.: APA Press.

Salas, E., Prince, C., Baker, D.P., and Shrestha, L. (1995). Situation awareness in team performance: implications for measurement and training. *Human Factors*, 37(1), 123-136.

Sauter, S.L., and Smith, K.U. (1971). Social feedback: quantitative division of labor in social interactions. *Journal of Cybernetics*, 1(2): 80-93.

Sheridan, T.B. (1997), Supervisory control. In G. Salvendy (Ed.), *Handbook of human factors* (2nd ed.)(Chapter 39, pp. 1295-1327). New York: Wiley.

Sheridan, T.B., (2002). Humans and automation. *HFES Issues in Human Factors and Ergonomics Series, Vol. 3.* New York: Wiley.

Smith, K.U. (1973). Physiological and sensory feedback of the motor system: neural metabolic integration for energy regulation in behavior. In: J. Maser (Ed.), *Efferent organization and the integration of behavior* (pp. 19-66). New York: Academic Press.

Smith, K.U., and Smith, M.F. (1973). Psychology. An introduction to behavior science. Boston:Little, Brown.

Smith, K.U., and Smith, W.M. (1958). The behavior of man. Introduction to psychology. New York: Holt, Rinehart and Winston.

Smith, T.J., and Henning, R.A. (2005). Cybernetics of augmented cognition as an alternative to information processing. In *Proceedings of the 11th International Conference on Human-Computer Interaction*. Mahwah, NJ: Lawrence Earlbaum. To be published.

Smith, T.J., Henning, R.A., and Li, Q. (1998). Teleoperation in space - modeling effects of displaced feedback and microgravity on tracking performance. SAE Technical Report #981701. 28th Intl. Conference on Environmental Systems, Danvers, MA, July 13-16.

Smith, T.J., Henning, R.H., and Smith, K.U. (1994). Sources of performance variability. In G. Salvendy and W. Karwowski (Eds.), *Design of work and development of personnel in advanced manufacturing* (Chapter 11, p. 273-330). New York: Wiley.

Smith, T.J., Henning, R.A., and Smith, K.U. (1995). Performance of hybrid automated systems -a social cybernetic analysis. *International Journal of Human Factors in Manufacturing*, 5(1), 29-51.

Smith, T.J., and Smith, K.U. (1987a). Feedback-control mechanisms of human behavior. In G. Salvendy (Ed.), *Handbook of human factors* (pp. 251-293). New York: Wiley.

Smith, T.J., and Smith, K.U. (1987b).Motor feedback control of human cognition – implications for the cognitive interface. In G. Salvendy, S.L. Sauter and J.J. Hurrell, Jr. (Eds.), *Social, ergonomic and stress aspects of work with computers* (pp. 239-254). Amsterdam: Elsevier.

Sosis, R. (2004). The adaptive value of religious ritual. *American Scientist*, 92(2), 166-172.

Sycara, K., and Lewis, M. (2004). Integrating intelligent agents into human teams. In Salas, E., and Fiore, S.M. (Eds.). *Team cognition: understanding the factors that drive process and performance* (pp. 203-231). Washington, D.C.: APA Press.

Ting, T., Smith, M., and Smith, K.U. (1972). Social feedback factors in rehabilitative processes and learning. *American Journal of Physical Medicine*, 51, 86-101.

Trimmel, M., Wright, N., and Backs, R.W. (2003). Psychophysiology in ergonomics: preface to the special section. *Human Factors*, 45(4), 523-524.

Vogel, G. (2004). The evolution of the golden rule. *Science*, 303(5661), 1128-1131.

Wegner, N., and Zeaman, D. (1956) Team and individual performances on a motor learning task. *The Journal of General Psychology*, 55, 127-142.

Wickens, C.D., Gordon, S.E. & Liu, U. (1998). *An introduction to human factors engineering*. New York: Longman.

Wilson, G.F., and Russell, C.A. (2003a). Real-time assessment of mental workload using psychophysiological measures and artificial neural networks. Human Factors, 45(4), 635- 643.

Wilson, G.F., and Russell, C.A. (2003b). Operator functional state classification using multiple psychophysiological features in an air traffic control task. Human Factors, 45(3), 381-389.

Woods, D.D., Sarter, N., and Billings, C. (1997). Automation surprises. In G. Salvendy (Ed.), *Handbook of human factors* (2nd ed.)(Chapter 57, pp. 1926-1943). New York: Wiley.

Zigurs, I., and Kozar, K.A. (1994). An exploratory study of roles in computer-supported groups. *MIS Quarterly*, 18, 277-297.

Tools for Enhancing Team Performance through Automated Modeling of the Content of Team Discourse

Peter W. Foltz

Pearson Knowledge Technologies
4940 Pearl East Circle, Suite 200
Boulder, CO, 80305
pfoltz@pearsonkt.com

Abstract

Studies of team communication have been hindered by the lack of tools to automatically analyze the large amounts of discourse generated. We describe the use of Latent Semantic Analysis (LSA) for analyzing and coding the content expressed during team discourse. LSA is a fully automatic corpus-based statistical method for extracting and inferring relations of expected contextual usage of words in discourse. It can be used to compute the semantic similarity among units of discourse (e.g., words, sentences, utterances, or entire team dialogues), even when they have no words in common. We describe a range of approaches for applying LSA to analyze the content of team discourse. This work reviews studies that includes measuring overall team performance, determining individual contributions, detecting off-topic language, and the application of automated speech recognition in team tasks. Finally a demonstration of a prototype tool for measuring team communication is shown, along with discussion on future improvements, implications for the design of automated systems for measuring team communications, and suggestions for the development of automated team training systems.

1 Introduction

As technology provides more effective means for teams to interact in complex environments, it becomes increasingly critical to have methods to monitor, assess, and improve team performance. One of the key indicators of team performance is the communication data from teams during their tasks. The communication data provides information about a team's current cognitive and task states, knowledge of individual team members and of the overall team, and amount, types and patterns of information flow among the team members. All of this data can be tied back to measures of team performance in order to develop effective models and tools for characterizing team and individual performance. Yet analyses of verbal interactions have been hindered by the lack of effective tools. Patterns of communication can be measured through statistical analyses of records of which team members talk to which others. However, pattern analyses do not capture that actual content of information being conveyed. The content of the communication among team members provides large amounts of information about team members' thought processes and what they needed to convey to other team members. With appropriate analyses, the content can be tied back to both the team's and each individual's abilities and knowledge. While some methods have been developed which rely on hand coding of verbal interactions, automated analyses of the content through computational linguistic and knowledge representation techniques provide the promise of real-time assessment of teams' and users' states. This paper describes the development of a set of tools for the automatic analysis of the content of verbal communication data in order to assess, monitor, and enhance individual and team performance in training and operational environments.

1.1 Measuring Team Communication

Current manual analysis of team communication shows promising results. For example, Bowers, Jentsch, Salas, and Braun, (1998), found that hand-coding the content of aircrew teams and measuring the frequency of different types of content expressed provided effective characterizations of the team's performance. Nevertheless, the analysis is quite costly. Hand coding for content can take upwards of 28 hours per 1 hour of tape (Emmert, 1989) and can be highly subjective. Since teams can generate large amounts of data in a short time, hand-coding does not permit fast analyses that could be used to monitor or train the teams in near-realtime. Thus, what is required are techniques for automatically analyzing team communications in order to categorize and predict performance.

A number of artificial intelligence, statistical and machine learning techniques have been applied to discourse modelling. These approaches have been used generally for the purpose of improving speech recognition and dialogue systems, although this work is now extending to team research. These methods include decision trees (Core, 1998), statistical model based on current utterance and discourse history (Chu-Carroll, 1998), and hidden Markov models (Stolcke et a.,l 2000,). For example, Stockle et al. were able to predict the tags assigned to discourse within 15% of the accuracy of trained human coders. While the above research has been applied to domains such as tutoring dialogues, techniques have also been used successfully for predicting performance through the analysis of the pattern of flow of information among team members (Graham, Schneider, Bauer, Bessiere, & Gonzalez, 2004; Kiekel, Gorman & Cooke, 2004).

1.2 Latent Semantic Analysis

The ability to produce a team monitoring system of this kind is made possible by machine learning technique called Latent Semantic Analysis (LSA). This technique mimics human understanding of the meaning of natural language. Through training LSA on the language of a domain, LSA can measure the content of free-form verbal interactions among team members. Because it measures and compares the semantic information in these verbal interactions, it can be used to characterize the quality and quantity of information expressed. This permits approaches to determining the semantic content of any utterance made by a team member, as well as to measure the semantic similarity of an entire team's communication to another team.

LSA is a fully automatic corpus-based statistical method for extracting and inferring relations of expected contextual usage of words in discourse (Landauer, Foltz & Laham 1998). In LSA a training text is represented as a matrix, where there is a row for each unique word in the text and the columns represent a text passage or other context. The entries in this matrix are the frequency of the word in the context. There is a preliminary information-theoretic weighting of the entries, followed by singular value decomposition (SVD) of the matrix. The result is a 100-500 dimensional "semantic space", where the original words and passages are represented as vectors. The meaning of a passage is the average of the vector of the words in the passage (Landauer, Laham, Rehder, & Schreiner, 1997). For the analyses described, we use a 300 dimensional semantic space.

More specifically, once the word-by-context matrix is constructed, the word frequency in each cell is converted to its log and divided by the entropy of its row (-sum (p log p)). The effect of this is to "weight each word-type occurrence directly by an estimate of its importance in the passage and inversely by the degree to which knowing that a word occurs provides information about which passage it appeared in." (Landauer et al., 1998) Then SVD is applied in order to decompose the matrix into the product of three other matrices: two of derived orthogonal factor values or the rows and columns respectively and a diagonal scaling matrix. The dimensionality of the solution is reduced by deleting entries from the diagonal matrix, generally the smallest entries are removed first. This dimension reduction has the effect that words, which appear in similar contexts are represented by similar feature vectors. Vectors for individual words, sentences, utterances, and documents can all be compared against each other in this latent, or reduced dimensional space, typically by computing the cosine among the vectors. It should be noted that in the studies described in this paper, LSA represents the primary modelling tool. Nevertheless, other computational linguistic measures which model additional aspects of language can be included into the models.

LSA has been used for a wide range of applications and for simulating knowledge representation and psycholinguistic phenomena. These approaches have included: information retrieval (Deerwester, Dumais, Furnas, Landauer & Harshman, 1990), automated essay scoring (Foltz, Laham, & Landauer, 1998; Landauer, Laham & Foltz, 2001), automated text analysis (Foltz, 1996), and modeling language acquisition (Landauer & Dumais, 1998). In the current context, similar approaches are used for analyzing verbal communication among team members.

1.3 A Team Communications Analysis Toolkit

While LSA on its own has been successful at modelling a number of aspects of language, the overall goal is to apply LSA and related techniques to develop a team communications analysis toolkit. Such a toolkit should be able to take in communication data and output predictions of performance and other useful characterizations of teams. Figure 1 illustrates the process of such a toolkit across what we call the "communications analysis pipeline." In the first part of the pipeline, automated recordings capture all aspects of the communication stream, including both the actual words said and information about who said it to whom and for how long. Spoken communication is converted to text via a trained automated speech recognition system (ASR), while written communication (e.g., person-to-person messages, chatrooms, etc.) can be retained in its original text format. This text is then analyzed using both LSA and other computational linguistic and statistical tools to examine the content, quality, and patterns of communication within a team. The numerical characterizations of those analyses can be incorporated into tools which model performance and characterize it at a cognitive level. For example, the results of the analyses can then be compared to standard metrics of expected performance, as well as models that characterize and predict performance for individuals, teams and tasks. The types of models used could include more cognitively oriented models (e.g., ACT-R, SOAR) or more engineering process-oriented models (e.g., C3TRACE). Finally the modelling output and statistical measures can be incorporated into applications that provide a range of types information about the team performance. These applications can include systems to generate automated debriefings (after action reviews), automated feedback to team members about errors or knowledge gaps, and commander's interfaces that permit commanders to monitor the performance of teams and receive feedback in near-realtime, if there are critical incidents or team problems.

The Communications Analysis Pipeline

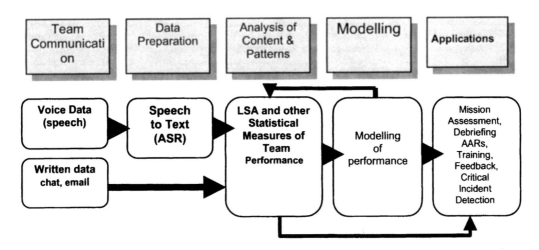

Figure 1. The Communication Analysis Pipeline

This paper will provide an overview of studies which illustrate tools that can be used for content analysis in a team communications analysis toolkit. While an ideal overall communications toolkit involves a range of different technologies, this paper focuses on the use of LSA as well as the application of ASR as input to LSA within such a toolkit. Together these two components illustrate how such a toolkit can successfully operate to make accurate predictions of team performance in near realtime. Finally some demonstrations of an initial prototype of the toolkit will be shown.

2   Automated Content Analyses

To apply LSA in a team communication context, LSA is first automatically trained on a body of text containing knowledge of a domain, for example a set of training manuals. After such training, LSA is able to measure the degree of similarity of meaning of any two units of text. This approach can be used in a variety of ways to measure aspects of performance. For example, one entire transcript can be compared against another by determining the cosine between the vectors for the two transcripts. This would provide a characterization of how much the two teams are talking about the same things in the same way, One can also compute the similarity between any two individual utterances made by team members in order to determine if they said similar things. This can be used as the basis for determining relationships in terms of communication and expressed knowledge among teams as well as among team members.

2.1 Predicting Overall Team Performance

Over a series of studies, LSA-based communications methods have been evaluated favorably in terms of their ability to predict overall team performance. For instance, LSA was successfully able to predict team performance in a simulated UAV task environment based only on communications transcripts (Gorman, Foltz, Cooke, Kiekel, & Martin, 2003; Martin &Foltz, 2004). Using human transcriptions of 67 team missions in a UAV environment, an LSA-based metric was derived to predict the an objective team performance score for each team mission. The Team Performance Score used as the criterion measure is a composite of objective measures including the amount of fuel and film used, the number and type of photographic errors, route deviations, time spent in warning and alarm states, unvisited waypoints and violations in route rules.. The metric was derived using the following method. Given a subset of transcripts, S, with known performance scores, and a transcript, t, with unknown performance score, we can estimate the performance score for t by computing its semantic similarity to each transcript in S. To compute the estimated score for t, we take the average of the performance scores of the 10 closest transcripts in S, weighted by cosines. The results indicated that the LSA estimated performance scores correlated strongly with the actual team performance scores ($r = 0.76$, $p < 0.01$, $r=0.63$, $p<.01$ when correcting for the repeated measure structure). Combining the LSA measures with additional computational linguistic analysis measures, increased the prediction ability very slightly, with correlations increasing an additional 5%. A graph of the LSA-predicted team scores and the actual team scores is shown in Figure 2. The results illustrate that LSA-based methods can successfully measure the overall performance of a team based on their verbal communications. Thus, if one has a set of team transcripts each associated with an objective performance measure, one can then take any new team transcript and predict that team's performance.

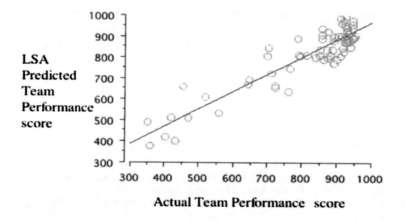

Figure 2. LSA predicted performance compared to actual team performance on transcripts from 67 missions.

A second method for predicting scores uses a proprietary technique based on computing the semantic similarity among a large set of teams using LSA and then determines the best linear ordering of teams from worst to best. The idea of this approach is that if a human read through all the transcripts he/she would be able to make judgements that could order them based on the language perceived in the transcripts. For the 67 UAV transcripts, the method is able to predict team performance moderately well ($r=0.46$, $p<.01$). The advantage of this "no-training" technique is that one can automatically derive a model for predicting performance without the need of initial performance measures or SME input. Instead of training on previous team scores, one just needs to have a set of 50 or more transcripts of the teams and the method will derive predicted scores.

2.2 Predicting types of communication expressed

2.2.1 Tagging type of utterances

While it is important to be able to predict the overall score of a team, it is also important to be able to detrmine what components within the discourse are contributing to the actual performance. Thus, techniques are needed to examine the contributions of individual utterances within a team. Using the same UAV dataset, LSA has used to predict the types of communications said by each team member (see Martin & Foltz, 2004). A corpus of 2916 statements made by team members during the UAV task were hand-coded by at least two experts as to the contextual type of utterance (e.g., *planning, stating a fact, acknowledging, expressing uncertainty, stating an action*). (see Bowers et al., 1998 for the use of this approach to tagging for measuring team discourse). The statements were then coded by an LSA-based agent which was trained on a subset of the corpus and would predict the probability of they type of communication for each statement. Results of LSA's performance, as measured by the agreement with the human coders, were within 5% of the human-human agreement. The frequencies of each type of communication were also predictive of overall team performance. For example, LSA was able to show that the teams that stated more facts and did more acknowledging had better team performance scores, while those teams that had to do more responding and expressed more uncertainty had poorer performance scores. Table 1 shows the correlation of team performance to the frequency of different communications types as predicted by LSA.

Table 1. LSA-based predictions of types of communication correlated to team performance scores

Frequency of Predicted Communication Type	Correlation to team performance score
Acknowledgements	0.34
Facts	0.32
Responses	-0.32
Uncertainty	-.0.46

The results indicate that LSA is able to predict the type of utterances through the statements that team members make. For example, better teams express less uncertainty, but state more facts. As a team member speaks, his or her utterance can be automatically categorized as to the type of utterance can be incorporated into models of team performance. This approach can therefore be adapted to tracking individual and team contexts throughout missions.

2.2.2 Detecting off-topic discussion

Along with being able to tag discussions for types of communications, it's important to be able to detect the amount of relevant (e.g., topic-related) discussion being generated. Lavoie, Psotka, Lochbaum and Krupnick (2004) applied LSA methods of analysis to comments generated from threaded discussions generated on a collaborative learning tool. Army personnel involved in discussions of the US foreign and security policy related to the future of NATO generated over 1600 comments. They then tested two approaches to scoring comments as either "off-topic" or "on-topic". The first method was trained on human ratings of comments to predict the ratings based on their semantic similarity to other comments and was able to classify 92% of the comments correctly. A second method used the "no-training" method described above in section 2.1 and was able to classify 95% of the comments correctly. This approach suggests that such a system is able to accurately monitor discussions and determine when participants are getting off-topic. At appropriate points, a commander or instructor could be notified, or an automated system could identify potential problems in the discussion and intervene when needed.

2.2.3 Additional measures

In addition to the above approaches to measuring performance, a variety of other LSA-based measures have been developed. For example, LSA-based measures of coherence (see Foltz, Kintsch & Landauer, 1998) have shown to be effective ways of measuring the rate of change of topic. For measuring team-based coherence, LSA is used to measure how similar each utterance is to the next utterance within a discourse (or to utterances that happened N utterances previously). This permits a characterization of how well each team member is responding to other team mates and how quickly the team as a whole is moving from one topic to the next. For detecting particular components of discourse, LSA can be trained to search for particular types of knowledge being communicated at particular critical events. These measures can be developed in conjunction with task modelling of the domain in order to determine what types of knowledge and communication should be expected to be expressed at different phases of a mission. As such, this approach illustrates how a toolset can be adapted to different contexts as needed by training the system to measure components of communication that would be critical to specific tasks. Current work is extending this approach to a range of datasets including Naval Command and Control, teams of Air Force F-16 pilots working with AWACS and Army intelligence decision-making during Stability and Support Operations.

2.3   *Automated Speech Recognition and LSA*

For use in the proposed Communication Analysis Pipeline, either in near-real time or in an after action briefing, human typed transcription of the speech to text would not be practical, therefore the speech-to-text transcription must be produced automatically. Output produced by commercial automatic speech recognition (ASR) systems is known to contain numerous errors, even under the best of conditions, and is often unusable under operational conditions. The question is how robust a combination ASR and LSA system can be in a realistic environment, in particular how well an end-to-end LSA-based analysis of mission communications based on ASR input can be expected to correlate with human assessment of the same performance based on manual transcripts. To this end, LSA has been tested operating in conjunction with the analysis of ASR input for portions of a dataset of verbal communication for the UAV team missions studies (Foltz, Laham & Derr, 2003).

In one test, Foltz, et al created simulated ASR errors in the communication stream. They varied the percentage of ASR errors introduced while testing LSA's ability to predict overall team performance based on the level of ASR errors. The results indicated that even with ASR systems degrading word recognition by 40%, LSA's prediction performance degraded less than 10%. In a second study, the audio portion of missions was converted using the Sphinx ASR system. With initial training, the system generated about 40% word error rates. The output of the ASR system was then used to tag individual utterances and these tags were then compared against human performance of tagging. The results indicated that LSA's reliability in tagging was reduced by 17% as compared to tagging typed transcripts.

The results from the two tests indicate that LSA is highly robust to ASR errors. Despite 40% error rates on the incoming text stream, LSA's ability at predicting performance and tags for types of communication is only reduced by 10-17%. This effect is due to the fact that because LSA derives meaning from whole utterances, not from individual words, it is immune to fairly high word level error rates typically found in speech recognition systems. In addition, because verbal interactions in these situations are highly constrained by the actions currently being taken and by the current execution status of the simulation mission plan, and are largely routine, the difficulties of both automatic speech recognition and LSA understanding are greatly reduced. This indicates that ASR can be successfully used in conjunction to LSA to provide accurate performance measurement.

3   A Communications Analysis Toolkit

The results and techniques of the above studies have informed the ongoing development of a set tools that provides a complete prototype system. This system can take speech input, analyze and predict cognitive measures from the communication stream and then provide output with predictions for the appropriate types of feedback to augment cognition. The development of such a communications analysis toolkit permits testing of the different measurement techniques and investigations in how the output can be developed into interfaces that can provide feedback that would be useful to users. Screenshots of an initial prototype are shown in figures 3 and 4. Figure 3 illustrates an interface for reading in communication data from multiple teams and automatically generating predicted scores. This interface could be used by commanders and trainers monitoring multiple teams. Figure 4 shows the integration of graphing tools which permit charting a team's progress over time. The graph shows the overall coherence of the team as well as the predicted score of the team over the 40 minutes that a team is conducting a mission. While still in the initial stages, the prototype illustrates how the tools developed can be easily integrated into a more complete toolset to provide a range of predictions about team performance.

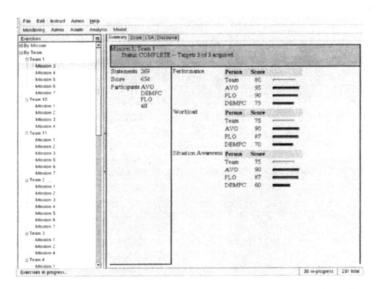

Figure 3. Toolkit Interface to monitor teams across missions with predicted scores. Clicking on any Team brings up additional data and graphs on team performance (see Figure 4).

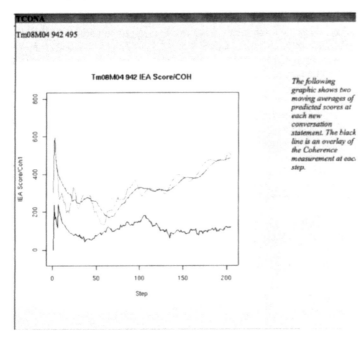

Figure 4. Toolkit interface showing predicted scores and team coherence a team as the team progresses through a mission.

4   Conclusions

Up to now, it has been impossible to systematically parse and evaluate oral communications to identify critical content required by many individual and team-based systems control operations. LSA is promising new technology that has significant potential for training, assessing, and assisting operators. It can process the incoming stream of verbal text and in almost real-time evaluate free-form verbal communication from a variety of sources and match content to stored natural language models. The LSA agent will be integrated with an automated speech recognition system which will permit instant transcription of incoming verbal communication for analysis. This provides a complete communications analysis pipeline, automatically converting team verbal communications to quantifiable performance measures and then analyzing these performance measures within a model of the task domain.

The resulting toolset has the potential for providing near-real-time assessment (within seconds) of individual and team performance including measures of situation awareness, knowledge gaps, workload, as well as predictions of future performance based on analyses of the current context. It can inform cognitive models of team performance as well as provide useful applications. Output from the toolset can be used for tracking teams' behavior and cognitive states, determining when appropriate feedback needs to be given, and for generating automated After Action Reviews. As such this toolset provides great potential to augment the cognition of teams.

Future work can extend this work in a number of ways. First, validation of the existing tools should be performed on novel sets of team communication data in widely different operational venues. For instance, it will be important to examine differences in tasks that are more problem solving oriented as well as tasks that involve coordination. By tying these measures to a wide range of potential measures, (such as situation awareness, critical incidents, team process, performance, knowledge), the toolset can be tested and validated. This will permit characterization of the generalizability of the methods by indicating how well the patterns of communication that are indicative of performance differ across types of tasks, domains and different measures.

Second, additional research and development can focus on novel communication analysis techniques based on additional statistical analysis techniques. For instance, this work can investigate more fully the types of language that are the best indicators of situation awareness and determine its relationship to language. The research can further be combined with existing modelling techniques (e.g., C3TRACE, ACT-R) to track cognition across tasks and permit the analysis techniques to be automatically informed of the appropriate contexts in which to measure performance. It should also be combined with other methods of communication analysis, such as examining patterns of communications.

Finally, a more complete working system can be developed that will permit the integration of a range of tools for measuring team performance.. This toolset can incorporate techniques for:

- Detecting individual and team errors and knowledge gaps
- Predicting cognitive workload
- Predicting overall team performance
- Predicting failures in team process
- Monitoring group dynamics
- Detecting appropriate and inappropriate contexts for the mission

The prototype developed shows that interfaces can developed in was to be used flexibly for monitoring teams (a commander's interface) and providing directed feedback (team member interface). A commander's interface can be used as a tool for observing multiple teams at once and being alerted for critical incident's, inappropriate team contexts, or impending failures. A team member interface can be developed to provide directed feedback within training contexts, such as stopping the simulation, tutoring team members on appropriate actions or knowledge at different points during missions, and generating automated after action reports. Overall, the system will provide a broad ranging toolset for the analysis of verbal communication that should be able to be applied across a number of training and operational contexts.

LSA is promising technology that has significant potential for training, assessing, and assisting operators. It can process the incoming stream of verbal text and in almost real-time evaluate free-form verbal communication from a variety of sources and match content to stored natural language models. The LSA agent will be integrated with an automated speech recognition system which will permit instant transcription of incoming verbal communication for analysis. This provides a complete communications analysis pipeline, automatically converting team verbal communications to quantifiable performance measures and then analyzing these performance measures within a model of the task domain. Therefore, LSA provides powerful tools for processing communication and evaluating teams in both training and operational environments.

5   References

Bowers, C. A., Jentsch, F., Salas, E., & Braun, C. C. (1998). Analyzing communication sequences for team training needs assessment. *Human Factors, 40,* 672-679.

Chu-Carroll, J. (1998) A Statistical Model for Discourse Act Recognition in Dialogue Interactions. Papers from the 1998 AAAI Spring Symposium, J. Chu-Carroll and N. Green, Program Cochairs, *Technical Report SS-98-01. Published by The AAAI Press*, Menlo Park, California (Pp. 12-17).

Core, M. G. (1998) Analyzing and Predicting Patterns of DAMSL Utterance Tags. Papers from the 1998 AAAI Spring Symposium, Jennifer Chu-Carroll and Nancy Green, Program Cochairs, *Technical Report SS-98-01, Published by The AAAI Press*, Menlo Park, California (Pp. 18-24).

Deerwester, S., Dumais, S. T., Furnas, G. W., Landauer, T. K., & Harshman, R. (1990). Indexing By Latent Semantic Analysis. *Journal of the American Society For Information Science, 41,* 391-407.

Emmert, V. J. (1989). Interaction analysis. In P. Emmert and L. L. Barker (Eds.), *Measurement of Communication Behavior*, pp. 218-248. White Plains, NY: Longman, Inc.

Foltz, P. W. (1996). Latent Semantic Analysis for text-based research. *Behavior Research Methods, Instruments and Computers. 28(2)*, 197-202.

Foltz, P. W., Kintsch, W. & Landauer, T. K. (1998). The measurement of textual coherence with Latent Semantic Analysis. *Discourse Processes, 25*(2&3), 285-307.

Foltz, P. W., Laham, R. D. & Derr, M. (2003). Automated Speech Recognition for Modeling Team Performance. In *Proceedings of the 47th Annual Human Factors and Ergonomic Society Meeting.*

Foltz, P. W., Laham, D., & Landauer, T. K. (1999). The Intelligent Essay Assessor: Applications to Educational Technology. *Interactive Multimedia Education Journal of Computer Enhanced Learning. 1,(2).*

Gorman, J. C., Foltz, P. W. Kiekel, P. A., Martin, M. A. & Cooke, N. J. (2003) Evaluation of Latent Semantic Analysis-based measures of communications content. In *Proceedings of the 47th Annual Human Factors and Ergonomic Society Meeting.*

Graham, J.M., Schneider, M., Bauer, A., Bessiere, K., & Gonzalez, C. (2004) Shared Mental Models in Military Command and Control Organizations: Effect of Social Network Distance. *Proceedings of the 47th Annual Meeting of the Human Factors and Ergonomics Society. HFES, California.*

Kiekel, P. A., Gorman, J. C., & Cooke, N. J. (2004). Measuring speech flow of co-located and distributed command and control teams during a communication channel glitch. *Proceedings of the 48th Annual Human Factors and Ergonomics Society Meeting.*

Kiekel, P. A., Cooke, N. J., Foltz, P. W., Gorman, J., & Martin, M. (2002). Some Promising Results of Communication-Based Automatic Measures of Team Cognition. In *Proceedings of the Human Factors and Ergonomics Society 46th Annual Meeting.*

Landauer, T. K, Foltz, P. W., & Laham, D. (1998). Introduction to Latent Semantic Analysis. *Discourse Processes,* *25*, 259-284.

Landauer, T. K, Laham, D., Rehder, B., & Schreiner, M. E., (1997). How well can passage meaning be derived without using word order? A comparison of Latent Semantic Analysis and humans. In M. G. Shafto & P. Langley (Eds.), *Proceedings of the 19th annual meeting of the Cognitive Science Society* (pp. 412-417). Mawhwah, NJ: Erlbaum.

Landauer, T. K., Laham, D. & Foltz, P. W. (2001). Automated essay scoring. *IEEE Intelligent Systems.* September/October

Landauer, T. K. & Dumais, S. T. (1998). A solution to Plato's problem: The Latent Semantic Analysis theory of the acquisition, induction, and representation of knowledge. *Psychological Review, 104,* 11-140

LaVoie, N., Psotka, J., Lochbaum, K. E., & Krupnick, C. (2004). Automated tools for distance learning. Paper presented at the *New Learning Technologies Conference.* February 18-20, 2004. Orlando, FL.

Martin, M. A. & Foltz, P. W. (2004) Automated team discourse annotation and performance prediction using LSA. In *proceedings of the Human Language Technology conference / North American chapter of the Association for Computational Linguistics annual meeting HLT/NAACL 2004.*

Stolcke, A., Ries, K., Coccaro, N., Shriberg, E., Bates, R., Jurafsky, D., Taylor, P., Martin, R., Van Ess-Dykema, C., & Meteer, M. (2000). Dialogue Act Modeling for Automatic Tagging and Recognition of Conversational Speech, *Computational Linguistics, 26(3),* 339-373, 2000.

6   Acknowledgements

This research was completed in collaboration with the researchers at New Mexico State University, Melanie Martin and Ahmed Abdelali, The Arizona State University East CERTT laboratory, Nancy Cooke, Preston Kiekel, Jamie Gorman and Susan Smith, and Pearson Knowledge Technologies, Kyle Habermehl, Noelle Lavoie, Darrell Laham, Mark Rosenstein. and Marcia Derr. This work was supported by Office of Naval Research , Army Research Laboratory, DARPA and the Air Force Research Laboratory.

FAUCET: Using Communication Flow Analysis to Diagnose Team Cognition

Preston A. Kiekel, Ph.D.

Cognitive Engineering Research Institute
5810 S. Sossaman Road Ste. 106, Mesa AZ
pkiekel@yahoo.com

Abstract

This paper discusses FAUCET (Flow Analysis of Utterance Communication Events in Teams) as a methodological toolbox, and illustrates how FAUCET methods can be used to augment team cognition. The methods are applied to team communication flow data, then used to predict outcome of team tasks. These general findings can then be used to diagnose communication behavior for specific teams. Ultimately, with experimental manipulations based on predictive findings, causal relationships between communication behaviors and outcome can be established, allowing for training feedback and other augmentations for team cognition.

1   Introduction

Communication has been a major focus of team research for several decades. It can be used to assess and diagnose team cognition, in much the same way that a think-aloud protocol (Ericsson & Simon, 1993) can be used to assess individual cognition (Kiekel, 2004). Moreover, since individual cognition is frequently used as a basis to form interventions, it follows that team cognition (and hence team communication) can likewise be used for teams (e.g., Kiekel, & Cooke, 2004). For example, analysis of team communication can be used to diagnose team learning shortcomings, so that training interventions are possible.

Communication analysis can focus on either communication content (i.e. the semantic record of what information is passed), or on physical communication data (i.e. the extra-content components of discourse, such as voice frequency, pitch, etc.). One critical form of physical data is the communication flow, i.e. the frequency, timing, direction, and duration of communication events. This paper focuses on a suite of methods developed for communication flow analysis. The suite is called FAUCET (Flow Analysis of Utterance Communication Events in Teams).

This paper will first overview the FAUCET methods. Then they are derived in detail. Third, I apply the methods to predict performance in a dynamic team simulator task. Content specific findings are revealed, which can then be used to design augmented team cognition designs. Finally, the paper concludes by discussing these potential applications of FAUCET to an augmented team cognition design. Context-specific designs will be based on making team-specific diagnoses with FAUCET data.

2   Overview of Methods

FAUCET methods are diverse, addressing many different aspects of team cognition. What they have in common is that they are based on ComLog (communication log; Kiekel, Cooke, Foltz, Gorman, & Martin, 2002; Kiekel, Cooke, Foltz, & Shope, 2001) data. A ComLog is a fine-level event log, recording which team members are speaking, and to whom, at each second, or at a finer level. The ComLog is collected by having team members communicate over headsets. They are required to press buttons when they speak, indicating to whom they will speak. This precise record of who speaks to whom, when, and for how long can be manipulated in a wide variety of ways. These manipulations of communication flow data allow conceptual aspects of team cognition to be assessed with simple, content-independent, physical data.

A number of methods have been developed under the FAUCET umbrella, using both static and sequential data. In each case, it is certainly possible to use the methods in areas other than communication. It is their use with the ComLog that identifies them as FAUCET methods. Sequential methods capitalize on the sequential nature of the ComLog time series by describing team behavior at numerous instances within a session. Static methods describe communication behavior across a larger aggregate of time, such as communication sums for an entire session of team interaction.

Sequential methods discussed in this paper include ProNet (Procedural Networks) and Dominance. Dominance uses cross-correlation functions among team members, with speech quantities as input. It measures how well each team member's speech quantities predict each other team member's speech quantities. ProNet is a tool for identifying outstanding chains of events in a categorical time series. It is essentially lag-sequential analysis, with the Pathfinder algorithm replacing hypothesis tests about node association. In the FAUCET context, ProNet has applied with utterance starts and endings for each team member. This is the most basic unit of communication flow. ProNet can be used to identify specific chains, or to measure the lengths of outstanding chains for each team. The latter is a measure of communication stability.

Static methods include CRP (Communication Required and Passed), and Process Surrogate. CRP measures the extent to which a team member complies with experimenter-defined norms of speech quantities. Process Surrogate describes what communication weighting schemes best account for individual contribution to the team's performance score.

3 Development of Methods

In this section, I derive the four methods to be discussed in this paper. First I consider the two sequential flow methods (ProNet and Dominance), then the two static methods (CRP and Process Surrogate). The methods are discusses in the context of a dynamic three-person environment. Teams controlled a simulator of a ground control station for an Uninhabited Air Vehicle, over several missions. Each team member was randomly assigned to be the pilot, the photographer, or the navigator.

3.1 Sequential-Physical Flow Method: ProNet

With the ComLog data, it is an easy matter to define six nodes of each team member (pilot, P, photographer, F, navigator, N) either beginning or ending a speech act (Pbeg, Fbeg, Nbeg, Pend, Fend, Nend). The second at which a speaker exhibits a "1" is the beginning of a speech act, and the second when the "1" becomes a "0" is the end. This does not, of course, mean that the person actually says anything, but merely that they press the ComLog Push-To-Talk button. For immediate purposes, these will be treated as speech utterances.

For instance, if the Pilot begins speaking, regardless of what she says, this is considered the beginning of a speech act, by the Pilot (Pbeg). The Navigator beginning a speech act is a different event (Nbeg), as is the Pilot's cessation of speech (Pend). A hypothetical turn-taking sequence might consist of the Pilot beginning an utterance, then finishing it, then the Navigator beginning, and the Pilot beginning another utterance before the Navigator finishes (Pbeg→Pend→Nbeg→Pbeg→Nend). These five events would define the Pilot speaking uninterrupted, but then interrupting the Navigator. An infinite number of possible turn-taking sequences is possible, and the goal is to identify prominent patterns.

Among these six nodes, it is interesting to decipher how systematic or orderly the communication is running. This may be equivalent to asking how systematic or orderly an individual's thought processes are. The orderliness in a communication pattern can be defined as the length of prominent chains of events. If for instance, a team only has two events that typically follow one another (e.g., Pbeg followed by Pend), and no other sequences are prominent enough to detect statistically, then it is a rather chaotic communication pattern. On the other hand, if a team exhibits several statistically detectable sequences, some of which are rather long (e.g., 20 events), then this is a highly stable team communication style. It can be likened to an expert in a routine situation, who knows which steps are best to take for a scenario to go efficiently from problem to solution.

The set of observed events for each team-at-mission will be used to define conditional transition probability matrices of various lags. For example, the node sequence Pbeg followed in the next event (i.e. at lag one) by Pend is represented as a cell in the lag 1 transition probability matrix. The event occurs with a probability defined as the number of occurrences for that transition, divided by the sum total of all transitions between Pbeg and all six events. Probabilities, being proximities, can be converted to distances by subtracting them from 1.

These transition probability matrices will be analyzed using ProNet (Cooke, Neville, & Rowe, 1996; Kiekel et al., 2001; Kiekel et al., 2002) to identify typical sequences. The kernel of ProNet is using the Pathfinder algorithm to derive a network, using transitional probability data, rather than other distances (especially distances based on relatedness judgments). FAUCET uses a generalization of the ProNet procedure, which permits statistical detection of all chains within the data set, of arbitrary length. This application of ProNet builds on the lag sequential analysis (Bakeman & Gottman, 1997; Sackett, 1978) to such an extent that it can be said that the routine is a form of lag sequential analysis. The only noteworthy difference is that, rather than relying on probability distributions to converge on the detected sequences, ProNet uses the Pathfinder algorithm. Using this method, we can record chain length for each detectable chain within a team-at-mission, and use summary statistics of chain lengths to predict performance. The generalization of ProNet is discussed in the sequel, with an example.

First, we must realize that the transition probability matrix analyzed by Pathfinder can be of any arbitrary lag. We start with the probabilities that event i follows event j for all i, j. If i and j are immediately adjacent in time, then we have a lag 1 transition probability matrix, because the two events are one step apart. If we add an intervening step, which can be made up of any unknown event, then we have a lag 2 transition probability matrix. If we have k intervening undefined steps, then we have a pair of events that are k+1 steps apart from one another, and so we have a lag k+1 transition probability matrix. Let us use the notation Node-Lag-Node to refer to single links at any lag, and let us denote Node-Node-...-Node as a chain of nodes.

The basic idea of the technique is to take all lag 1 links, identify possible longer chains, build longer chains with lag 2 links, and continue until all supported chains are accounted for. A chain is supported if Pathfinder connects all of its required nodes, at the required lags. At each step, all required subchains must be supported before a new link can be added to any chain. For example, to identify the chain A-B-C-D, we would need to support A-1-B, B-1-C, C-1-D (our lag 1 links), then A-2-C and B-2-D (our lag 2 links), and A-3-D.

In an A-B-C-D example, if any of these subchains is rejected, then A-B-C-D is not retained as a detected chain. For any subchain, if any of the required nodes at the required lags are not connected by the Pathfinder, then the chain is rejected. This means that nodes that are not linked by high co-occurrences do not get built into longer chains. A chain is therefore only as strong as its weakest link. This approach prevents chains from attaining an indefinite length.

An example of how this works follows. Suppose we have a sequence of events that looks like: E-F-E-G-A-B-C-E-G-B-C-D-E-F-A-B-C-D-E-G-A-B-C-B-C-D-E... To the naked eye, there is no obvious pattern, though this is a very simple example. We begin the ProNet routine by identifying the sequentially linked nodes, at lag 1.

Suppose that at lag 1 we identify the following two-link chains:
lag 1: A-1-B; B-1-C; C-1-D; D-1-E; E-1-F; E-1-G

The possible 3 links chains could be any or all of:
A-B-C; B-C-D; C-D-E; D-E-F, but E-F-G is not eligible, because F-1-G is absent. We still have E-F and E-G, which we will keep as 2-link chains in our final set of chains. Chain possibilities are not exhausted, so we get the lag 2 matrix, and rerun Pathfinder. A-B-C will be tested with A-2-C; B-C-D is tested with B-2-D, etc. If any of the required subchains is rejected, then the chain is rejected.

Suppose A-2-C; B-2-D; C-2-E are found, but not D-2-F.
A-B-C (3 link) is supported, because A-1-B; B-1-C; A-2-C (all links) are supported.
B-C-D (3 link) is supported, because B-1-C; C-1-D; B-2-D are supported.
C-D-E (3 link) is supported, because C-1-D; D-1-E; C-2-E are supported.
D-E-F (3 link) is not supported, because D-2-F is absent. In our final list of detected chains, we now have E-F and E-G, plus A-B-C, B-C-D, and C-D-E.

Possible 4 link chains could be A-B-C-D and B-C-D-E, but not C-D-E-F, because the D-E-F sub-chain is absent. Chain possibilities are not exhausted, so we get the next lag (lag 3). A-B-C-D will be tested with A-3-D; B-C-D-E will be tested with B-3-E.

Suppose A-3-D is rejected, but B-3-E is supported. We now add B-C-D-E, because all required sub-chains are supported for this chain. We did not find A-B-C-D, because A-3-D was rejected. We do not need to go on to test A-B-C-D-E, because A-B-C-D is an unsupported sub-chain. There are no more chains to test for. We therefore stop the analysis. Chains supported in this simple example were: E-F; E-G; A-B-C; and B-C-D-E. Notice that we do not count B-C-D or C-D-E, because these sub-chains are absorbed into B-C-D-E. We end up with four distinct chains, with lengths of 2, 2, 3, and 4 respectively.

One variable we can now calculate is the descriptive statistics on the set of chains. If we calculate the mean length of this set of chains, it will be 2.75. On the other hand, if we detected four chains, each with a length of four, the mean would be 4. This higher mean would come from a team for whom longer chains are detectable. Hence, that team's communication pattern is more regular. It is for this reason that chain length descriptive statistics can be used as a measure of communication stability.

In research discussed below, events are defined by each person beginning or ending an utterance. Two other major measures come from the use of ProNet. First, one can determine the detection or absence of certain specific chains, such as interruptions of one person by another. Second, one can return to the original data set, and make frequency counts of how often those specific chains occur.

3.2 Sequential-Physical Flow Method: Dominance

It is useful to define a measure that represents the extent to which each team member's behavior can predict other team members' behavior, in terms of amount of speech. The resulting FAUCET measure for this is called Dominance, of communication quantity. Dominance is juxtaposed against reactivity, the extent to which a team member reacts to other team members' actions. The measure was defined by examining cross correlations (Budescu, 1984; Dillon, Madden, & Kumar, 1983; Wampold, 1984), which are defined as the correlation between one time series (in this case, the time series of one team member's speech quantity), correlated with the lagged values of another time series (i.e. another team member). However, FAUCET expands upon this principle, by calculating the entire cross-correlation functions--i.e. the set of cross-correlations at each lag in the data set--between all pairs of team members.

The FAUCET Dominance measure starts with a time series of mean speech quantity for each team member over some small number of seconds. The actual number of seconds is an average, taken to be either 1/2 the minimum of the three mean speech duration for all team members, or 5 seconds (whichever is longer). This was the set of three time series of speech quantities. The Dominance measures are calculated on these three series.

First, take all pairwise cross-correlation functions between all team members. Square each correlation, so that we have a measure of variance accounted for at each lag. For each team member predicting each other team member, take the weighted average of the squared cross-correlation function, where the weight is the inverse of the lag. One refinement I made to this measure was to only calculate lags up to 20, feeling that 20 events apart was far enough to capture all of the critical predictive information. These pairwise mean squared-cross-correlation functions are measures of the predictability of each team member, to each other team member. These six subscales can be used as separate measures, and can also be refined further, into three dominance scores.

Next, using the mean squared-cross-correlation function for each team member predicting each other team member, take the ratio of each correlation to each other. That is, to determine the dominance of person X over person Y, person X to Y's correlation is divided by the correlation for Y to X, denoted as $CrossR^2XY / CrossR^2YX$. In the current research scenario, calculate the six ratios, for Pilot-Photographer, Pilot-Navigator, Photographer-Pilot, Photographer-Navigator, Navigator-Pilot, Navigator-Photographer.

The scale of these ratios is highly lopsided, going from 0 to 1 if the team member's behavior is reactive, but from 1 to infinity if it is dominant. To rectify this, take the natural log of each ratio. Because of the properties of the correlation coefficient, this ratio will be approximately normally distributed, with a mean of zero and a standard deviation of 1. The proof of this awaits further research.

Now the pairwise ratios are on a commensurate scale, going from negative infinity to 0 when the team member is reactive, and from 0 to infinity when she is dominant. Being commensurate, the pair of log ratios for each team member can now be averaged. This yields a mean score of influence that the team member exhibits over other team members. Hence our dominance score, say for person X, is:

$$X_{dom} = \{\ln(CrossR^2XY / CrossR^2YX) + \ln(CrossR^2XZ / CrossR^2ZX)\} / 2$$

, where $CrossR^2XY$ is the inverse-lag weighted mean of the squared-cross-correlation function from person X to person Y. The final three dominance scores are X_{dom}, Y_{dom}, and Z_{dom}. Only T-1 such dominance scores are independent, where T is the number of team members. The last score is determined by the others. The precise function to find the remaining score from the others awaits further research.

Hence, this results in six subscales of pairwise predictability, plus three measures of dominance, only two of which are independent.

3.3   Static-Physical Flow Method: CRP

When there are known norms of information passage, it is possible to specify how much each team member should ideally speak to each other. FAUCET compares teams to these norms by a set of proportions of Communication Required and Passed (CRP). The specific method is described below.

Start with a separate sum of every second when team member X is talking to team member Y. Let us define the total sum of seconds that X spends speaking to some other person Y as C_{xy}. Create a proportion, relative to all possible seconds.

For each team member, take the proportion of time that that person is either speaking or being spoken to. Take the converse of this proportion, to yield a measure of communicative resources left available to each team member. This is the "Chatter" score. It is normed, based on the idea of leaving available as many cognitive resources as possible, by minimizing speech. The norm is that no speech is the minimum allowable speech quantity.

For the next component, norm the maximum allowable speech quantity. To do this, define the minimal amount of speech each team member must convey to each other team member, in order to complete the mission. For team member X talking to Y, this can be defined by the number of sentence clauses X must convey (or request) from Y (b_{xy}), times the number of events requiring this transmission (U_{xy}), times an arbitrary constant of how long it takes to convey a single clause (k).

Ideally, $C_{xy} = U_{xy}(b_{xy}k)$ for all persons y, x.
Therefore, ideally, for person x, $C_{yx}/C_{zx} = [U_{yx}*(b_{yx}*k)]/[U_{zx}*(b_{zx}*k)]$
Therefore, ideally, $C_{yx}/C_{zx} = [U_{yx}*b_{yx}]/[U_{zx}*b_{zx}]$
Therefore, ideally, $\{C_{yx}/C_{zx}\} / \{[U_{yx}*b_{yx}]/[U_{zx}*b_{zx}]\} = \{C_{yx}*U_{zx}*b_{zx}\}/\{C_{zx}*U_{yx}*b_{yx}\} = 1$

This measure is on a scale ranging from 0-1 for cases when the denominator is larger than the numerator, and from 1 to infinity when the numerator is larger. Suppose we had a number of 500. It is difficult to say if this is an unusually large number. Moreover, if it is a large number, it is difficult to interpret whether the numerator is "too large" or the denominator is "too small," or both. Hence, we are really only concerned with how far the score deviates from the ideal, and not as much with in what way it deviates. The score can be forced to remain between 0 and 1 by inverting ratios greater than 1, and by leaving ratios that are less than 1. Hence, the final score is:

Equation (1) $X_{info} = \min(\{C_{yx}*U_{zx}*b_{zx}\}/\{C_{zx}*U_{yx}*b_{yx}\}, 1 / [\{C_{yx}*U_{zx}*b_{zx}\}/\{C_{zx}*U_{yx}*b_{yx}\}])$

This is the "Information Passing" score.

We wish to identify how close a team member can come to a balance between minimum allowable speech and maximum allowable speech, two separate norms. To do this, for each team member, multiply "Chatter" by "Information Passing" to get a "Communication Required and Passed" (CRP) score. We therefore have three CRP scores.

3.4 Static-Physical Flow Method: Process Surrogate

The FAUCET process surrogate measure is intended to determine what process behaviors would cause a team member to have greater influence over the overall team score, and how these weighted individual contributions would combine to create the overall team score. This measure would therefore serve as a surrogate measure for recording specific team process behaviors, because it only required the ComLog file and the individual performance scores. FAUCET does this by taking the individual performance scores of each team member, and multiplying it by one of the communication process measures. These three weighted individual scores are then aggregated three ways, based on three theories of team interaction.

One measure used is the normalized Dominance score, multiplied by a normalized individual performance score. Since both scores are between 0 and 1, one can think of the extent to which a team member's behavior predicts other team members' behavior as representing a cap on the contribution that team member can make to the overall performance score. Similarly, the second measure used caps individual performance contribution with the CRP score. This way, one can assess the extent to which a team member's norm-appropriate speech quantity determines their influence on the performance score.

Then, given these process-weighted individual performance scores, FAUCET takes three process-based aggregates to estimate the team performance score. To test for an egalitarian team structure, FAUCET uses the arithmetic mean. To test for a team structure where the worst player drags everyone else down, FAUCET uses the minimum. To test for a team structure with elements of both of the other two structures, FAUCET uses the geometric mean. The procedure is serialized below.

Procedure for Process Surrogate:

- 1. Two communication-based functions of individual performance scores are defined. All consist of first proportionalizing the individual performance score. Then said score is:
    - 1a. multiplied by the CRP score defined above,
    - 1b. multiplied by the Dominance score defined above
- 2. Three aggregation schemes are employed on the transformed individual scores yielded by each of these two functions. The sets of individual scores are therefore converted to single scores. A total of six aggregates are taken, in that each of the three functions above were then aggregated by:
    - 2a. arithmetic mean
    - 2b. minimum
    - 2c. geometric mean

4 Predicting Performance for a Team Task

FAUCET methods were developed and tested in a team simulator of an Uninhabited Air Vehicle ground control station. Teams consist of three highly interdependent members: a navigator, a pilot, and a photographer. FAUCET data were used to predict performance and other criteria. Since teams were measured on multiple missions, the predictors were included in a model that corrected for the within-teams factor of trials. Findings were then replicated in a second study. For both studies, the two sequential and two static methods discussed were each predictive of performance.

ProNet chain lengths were measured by the sum of all chain lengths detected by ProNet, for each team (at each mission). Chain lengths were positively predictive of performance for the first study ($F(1, 74) = 3.20, p = .078, R^2 = .04, B = 1.04$) and the second ($F(1, 53) = 6.21, p = .016, R^2 = .11, B = 3.25$). This suggests that teams with greater consistency in communication sequences tend to perform better.

Communication utterance dominance are slightly less straightforward. For the first study, , the pilot predicting the photographer was negatively related to performance, $F(1, 78) = 2.76, p = .100, \eta^2 = .03, B = -5148.16$. For high-performing teams, the photographer's behavior is independent of the pilot's. In the second study, the photographer speaks independently of the pilot, in better performing teams, $F(1, 53) = 3.37$, $p = .072$, $\eta^2 = .06$, $B = -8363.07$. Taken together, the dominance data suggested that the pilot's and photographer's utterances should not be predictive of one another. That is, they should be independent of one another in their utterances.

The two sequential methods discussed were each predictive of performance, elucidating different aspects of team cognition. ProNet indicated a need for well established communication strategies. Dominance findings indicated that the pilot and the photographer should speak independently of one another. The ProNet and Dominance findings, taken together, suggest that that these two team members should have communication strategies that are well established, but which are primarily directed to and/or responding to the other team member, the navigator.

Among the static methods, CRP data revealed that teams do better if the photographer's amount of communication received and passed is closer to the experimenter-specified ideal. This was found in both the first study ($F(1, 54) = 2.88$, $p = .096$, $\eta^2 = .05$, $B = 401.31$), and the second ($F(1, 90) = 3.95$, $p = .050$, $\eta^2 = .04$, $B = 140.34$). Taken with the findings from sequential methods, the implication is that the photographer should have stable communication patterns, which are independent of the pilot, and within experimenter-specified norms.

The process surrogate data address yet another aspect of team cognition. Here we ask which team process behaviors are more influential in determining their performance of the team as a whole. For the first study, process surrogate data revealed that team performance is associated with means of individual performance, weighted by dominance ($F(1, 83) = 8.22$, $p = .005$, $R^2 = .09$, $B = 858.81$). This was replicated in the second study, for both weighting by dominance ($F(1, 94) = 16.98$, $p < .001$, $R^2 = .15$, $B = 1070.52$). Hence more dominant speakers among the team are more influential in team performance--both for better and for worse.

5   Flow-Based Augmentations for Team Cognition

Having identified team behaviors that are associated with high performing teams, the next step is to use these findings to diagnose teams. Given team communication flow data, researchers can predict team outcome, and identify specific shortcomings and strengths of team communication. This is possible because there are so many dimensions of team cognition addressed by FAUCET. It is possible, for example, to use ProNet to identify specific sequences of communication events that are associated with specific team outcomes.

This qualitative technique has been applied to the present experimental context. For example, FAUCET methods have been able to identify team adaptation to a brief outage of communication channel. Some teams create an alternate channel, and some do not seem to notice the glitch. Usage of an alternate channel can be shown with specific ProNet patterns (i.e. increases in adjacent utterances between team members who still have an open channel), and by changes in CRP ratios.

Failure to adapt to the glitch may be a sign of poor team situation awareness, or a related shortcoming in team cognition. Using these highly automatic FAUCET techniques, teams can be monitored for such shortcomings on the fly. This can then be addressed by an intervention, such as just-in-time training or attention filters on team members' screens. It can also be identified during team training or practice sessions, and addressed with special-tailored training. More generally, FAUCET can be used to provide training feedback and other augmentations, either in real time, or after a collaborative session.

Before it is possible to turn descriptions of communication process into design recommendations, it is necessary to establish causal relationships between communication behaviors and outcome. This can be done by taking the general FAUCET findings within a context, and refining them into hypotheses to be experimentally tested. Having then established causal relationships, it is possible to design augmentations. The uncovered causal relationships can be used as feedback in training teams, or as interventions to teams with ineffective communication strategies, or in a host of other approaches to facilitating the team's cognitive processes.

For example, since correlational research in the UAV ground control simulator showed that the pilot and photographer should speak independently of one another, a training routine could be tested to encourage these team members to speak more freely with one another.

Alternatively, even in the absence of causal relationships between communication and performance, the impact of augmentations themselves can be tested with FAUCET measures. This provides a different kind of data than the higher-level performance measures, which primarily focus on an end result at a single point in time.

There are three major steps that are needed before the FAUCET methods are matured. The next step in this research program is to apply the FAUCET measures in a more diagnostic way, and to move toward testing causal hypothetical relationships between communication and performance. Second, an important aspect of this is to test the FAUCET measures on other data sets. This is critical to establish generality for the methods. Finally, in order to facilitate this expansion process, it is necessary to implement FAUCET in a usable stand-alone software package. It is currently available only as a SAS macro, which is difficult for researchers to make practical use of.

However, as these steps are taken, the FAUCET suite will become an increasingly useful tool for team researchers. The idea is to give team researchers an easy way to describe and diagnose the communication behavior of teams. With flow data, this can be done in a very rapid and automatic way, unlike the more time-consuming content methods. Moreover, at least in relatively constrained team environments like the one discussed here, this automated flow analysis can be accomplished without losing the high-level explanatory power that is typical of content measures.

References

Bakeman, R., & Gottman, J. M. (1997). *Observing Interaction: An Introduction to Sequential Analysis* (2nd ed.). Cambridge: Cambridge University Press.

Budescu, D. V. (1984). Tests of lagged dominance in sequential dyadic interaction. *Psychological Bulletin, 96*, 402-414.

Cooke, N. J., Neville, K. J., & Rowe, A. L. (1996). Procedural network representations of sequential data. *Human-Computer Interaction, 11*, 29-68.

Dillon, W. R., Madden, T. J., & Kumar, A. (1983). Analyzing sequential categorical data on dyadic interaction: A latent structure approach. *Psychological Bulletin, 94*, 564-583.

Ericsson, K. A., & Simon, H. A. (1993). Protocol Analysis: Verbal Reports as Data. Cambridge, MS: MIT Press.

Kiekel, P. A. (2004). Developing automatic measures of team cognition using communication data. Doctoral dissertation. Las Cruces, NM: New Mexico State University.

Kiekel, P. A., & Cooke, N. J. (2004). Human factors aspects of team cognition. In R. W. Proctor & K. Vu (Eds.), *Handbook of Human Factors in Web Design* (pp. 90-103). Mahwah, NJ: Lawrence Erlbaum Associates.

Kiekel, P. A., Cooke, N. J., Foltz, P. W., Gorman, J. C., & Martin, M. J. (2002). Some promising results of communication-based automatic measures of team cognition. *Proceedings of the Human Factors and Ergonomics Society 46th Annual Meeting*, 298-302.

Kiekel, P. A., Cooke, N. J., Foltz, P. W., & Shope, S. M. (2001). Automating measurement of team cognition through analysis of communication data. In M. J. Smith, G. Salvendy, D. Harris, and R. J. Koubek (Eds.), *Usability Evaluation and Interface Design* (pp. 1382-1386). Mahwah, NJ: Lawrence Erlbaum Associates.

Sackett, G. P. (1978). Measurement in observational research. In G. P. Sackett (Ed.) *Observing Behavior (Vol. 2): Data Collection and Analysis Methods* (pp. 25-43). Baltimore: University Park Press.

Wampold, B. E. (1984). Tests of dominance in sequential categorical data. *Psychological Bulletin, 96*, 424-429.

The Effect of Command and Control Team Structure on Ability to Quickly and Accurately Retarget Unmanned Vehicles

Stephen C. Allen

Christopher R. Jones

Stephanie Guerlain

Emiror, Inc.
Charlotte, NC
steve@emiror.com

University of Virginia
Charlottesville, VA
crj4f@alumni.virginia.edu

University of Virginia
Charlottesville, VA
guerlain@virginia.edu

Abstract

Teamwork is critical in many command and control problems. A key issue of designing a team structure is to determine the best method of dividing a collaborative task into manageable subtasks to maximize efficiency and accuracy. Thirty-nine college students performed simulated missile retargeting tasks either as members of a three person Supervisor/Subordinate team or as members of a three person Peer team. In the Supervisor team, one team member supervised two subordinate peers each controlling 15 missiles. In the Peer team, all three team members were peers each controlling 10 missiles. The subordinate operators and the peer operators utilized one kind of user interface and the supervisor utilized a different kind of user interface. Team members coordinated through chat messages using a set protocol for retargeting events. Each team ran through the same 20 minute scenario which had five retargeting events. Results showed that mean decision times of the Supervisor/Subordinate teams were significantly slower than the decision times of the Peer teams (134 vs. 89 seconds, p = .039) but the Supervisor/Subordinate teams were significantly more accurate (90% vs. 70% correct, p= .032). These results suggest that allocating the roles of same size teams can significantly influence performance.

1   Introduction

One of The United States Navy's numerous goals is to advance its land attack warfare capabilities, extending its reach far ashore to provide decisive force in conflicts. The Tomahawk cruise missile allows for just such a capability. One of the main benefits of using the Tomahawk over conventional land attack weapons is the fact that the risk to friendly personnel is very low. A missile can be launched and destroy a target while personnel are safely hundreds of miles away.

The current Tomahawk is a "fire and forget" weapon, meaning that once launched there is no way to stop the missile or change its flight path. Because the flight time for a missile can be up to two hours, such weapons are limited to static targets. In order to remedy the situation and to add new capabilities, the Navy has developed a new version of the cruise missile called the Tactical Tomahawk. This missile has the ability to engage in two-way communications with weapon control systems aboard Navy ships. A missile can be retargeted in midair to hit a new target in minutes instead of the hours it would require to launch another missile.

One or more human operators must assume real-time supervisory control over the flight paths of the missile and make tactical decisions during an evolving situation. Early work examined a single operator's ability to monitor and retarget Tactical Tomahawk missiles during simulated strikes. Performance decreased and cognitive workload increased when additional missiles were introduced into the scenarios. In two separate experiments, performance decreased when more than 12 missiles were being monitored simultaneously (Willis, 2001ab; Cummings, 2004). With the potential for hundreds of missiles in flight during a major strike, it is likely that no one operator will be able to manage all of the missiles at once. A solution to this problem is to have multiple operators monitoring and redirecting the missiles. This paper examines ways to design a team management structure and an associated decision support system to handle monitoring and retargeting of Tomahawk missiles in flight, while maintaining situational awareness in a dynamic combat environment.

Two simulated systems, called the Tactical Tomahawk Interface for Monitoring and Retargeting (TTIMR) and the Command Overview (CO), were developed at the University of Virginia in order to test user interface designs and model operator performance during monitoring and retargeting tasks. These systems are being used in a series of human subject experiments to simulate scenarios that require tactical control over a number of missiles in flight.

The study reported on in this paper involves comparing two team structures to address the challenge of cooperative decision making while monitoring and retargeting Tactical Tomahawk missiles. The structures are a Supervisor/Subordinate structure and a Peer structure. The Supervisor/Subordinate structure consists of one supervisor who has responsibility over two subordinate operators. The subordinate operators perform the actual retargeting tasks (each subordinate acts as a peer with control over half of the missiles). In the Peer structure, all three subjects are operators each with control over one third of the missiles. There is no defined supervisor structure beyond any ad hoc collaboration.

2   Background

2.1   Tactical Tomahawk Missile

The Tomahawk is as a long-range cruise missile for striking high value, heavily defended land targets. Launched from Navy surface ships and submarines, Tomahawk cruise missiles are designed to fly at extremely low altitudes at high subsonic speeds, and are piloted over an evasive route by several mission-tailored guidance systems. The Tomahawk missile has a range of over 1,000 miles and a subsonic cruising speed of 550 miles per hour. Each missile supports several different types of warheads. Two basic types used today are the unitary and submunition variants. The Block–III Conventional variant (TLAM–C) contains a 1,000 pound class blast/fragmentary unitary warhead, while the submunition variant (TLAM–D) includes a submunitions dispenser with combined effect bomblets (United States Navy Fact File, 2003). The guidance system of the Tomahawk missile is what gives the weapon its superior accuracy. Because of its long range, lethality, and extreme accuracy, the Tomahawk has become a weapon of choice for the United States Department of Defense.

Conventional tomahawk missiles as just described were first used in an operational context during the 1991 Gulf War. More than 800 Tomahawk land-attack missiles (TLAM) were fired from 35 coalition ships during Operation Iraqi Freedom in 2003. Overall, about 2,000 missiles have been launched in an operational situation.

The Tactical Tomahawk Weapon Control System (TTWCS) will communicate with the newly designed Tactical Tomahawk missiles in flight over a communications network. The missile can send back "health and status" information at pre-designated intervals, and the control system can send a message to the missile, either to redirect it while in-flight to strike any of 15 pre-programmed alternate targets or to redirect it to any GPS-specified target within its fuel range. The Tactical Tomahawk also has an onboard camera system which can be used to assess and send back target battle damage information or perform other surveillance duties. These major new capabilities allow commanders to use an in-flight missile to strike a new, emergent or transitory target within minutes instead of the hours it would take to launch a new missile.

At present, the system and procedures for performing in-flight retargeting have not been well established. There remains a need to study and possibly improve the design of the TTWCS operator (user) interface, taking into account human performance factors and potentially developing various forms of decision support.

2.2   Team Collaboration

Computer Supported Cooperative Work (CSCW) is a broad term that applies to performing collaborative work using computers. "Cooperation occurs when two or more agents work together in a common environment to more effectively reach the maximal union of their goals," (Jones & Jacobs, 2000). These agents must be sharing the same environment, so that their actions or the effects of their actions can be mutually perceived. Each of the agents is pursuing an individual set of local goals. "The agents are working to satisfy some local or mutually agreed upon global goal set or the maximum subset of the union of their local goals. In cooperative systems, each agent has accepted that its goals will best be met by being willing to compromise some of them when they conflict with global goals of the group or local goals of others," (Jones et al.).

Schmidt (as cited in Jones et al, 2000) describes three reasons for cooperation: augmentative, integrative or debative. Augmentative cooperation occurs when an agent can not perform a task alone. Integrative cooperation occurs when a task requires different technique-based specializations. Debative cooperation occurs when the task requires debate among multiple perspectives. In the case of the collaborative Tactical Tomahawk monitoring and retargeting problem, augmentative becomes the main reason for seeking out cooperation. Integrative aspects arise when the Supervisor/Subordinate relationship is used. The supervisor has a different skill set or decision making function that can be brought to bear on the problem. Debative aspects arise when the subjects work together to find the optimal solution. The Peer group may debate more because of the fact that there is no specified leader.

When operators are expected to utilize different skills to perform different sub-goals, they often require different user interfaces to achieve their respective solutions (Bentley, Rodden, Sawyer & Sommerville, 1992). These different interfaces offer different views of the same underlying data. The goal is to provide a layout that maximizes the operator's efficiency in performing each task. However, situational awareness of the entire scenario and overall goal must not be sacrificed to meet the individual operator's sub-goal. The operators are all operating on the same shared entities.

2.3   Previous Tomahawk Research

It is important to look at some previous work done at the University of Virginia on potential Tactical Tomahawk supervisory control interface designs and performance with the prototypes by single operators. Human subject experiments were run to analyze the effectiveness of two evolving versions of the same user interface as well as to provide information about the Tomahawk retargeting task (Willis, 2001ab; Cummings, 2004). The prototype is called the Tactical Tomahawk Interface for Monitoring and Retargeting (TTIMR).

Willis (2001ab) developed the first version of TTIMR, a single-screen user interface that consisted of 1) a geospatial map representation of missiles in flight enroute to targets, 2) a temporal representation of all the missiles (a timebar), and 3) a chat window that prompted users to answer questions about the evolving scenario as a form of secondary tasking and a means to measure human performance with the system. In a human-in-the loop simulation, it was found that as the number of candidate missiles available to redirect to an emergent target increased, time to make a decision on which was the best missile to retarget increased. Accuracy was not affected. Operators were taking the time necessary to make a correct retargeting decision, but these decisions took longer as more candidate missiles needed to be compared. Furthermore, just monitoring a scenario with additional "objects' (20 missiles and targets as compared to 10 missiles and targets) slowed down performance on any secondary (information gathering or reporting) task that required serial search through the map. These findings led to the realization that 1) additional decision support was required during retargeting events, and 2) that there may be an operational limit to the number of missiles and targets that one operator can effectively monitor during a strike.

Cummings (2004) expanded the testbed to be a more interactive, two-screen user interface. This version of TTIMR had a similar map, timebar, and chat, but also had a "decision matrix", specifically designed to aid the operator during a retargeting situation. The matrix lists retargetable missiles down the left column and all targets along the top row, with each of the cells in the matrix showing the relationship between each missile and each target. This facilitates decision making as the users do not need to serially search a map to find missiles and targets, and enables quickly comparing alternative solutions. Cummings performed a similar human-in-the-loop experiment in which the number of retargetable missiles in an operator's control was varied during a set of emerging target scenarios of varying complexity (but the number of non-retargetable missiles was held constant at 16 and the number of targets was held constant). The measurements of that experiment were time and accuracy to retargeting events, time and accuracy to secondary tasking requests, and percent "busy time" during the experiment (known as utilization). It was found that while there was essentially no difference between controlling eight and twelve retargetable missiles in addition to the 16 non-retargetable missiles, an increase to sixteen retargetable missiles produced significantly degraded decision times, increased percent busy times, and increased latency to secondary tasking. The accuracy of the assignments did not suffer as the number of missiles increased. This led to the conclusion that the operators lost situational awareness and their response times suffered as the number of retargetable missiles under control increased above twelve.

1259

3 Experiment Setup

In order to evaluate different team structures, a multi-user version of TTIMR was developed. Several operators can work together, each in control of a subset of the available missiles in a strike. Additionally, a new user interface, the Command Overview, was created for real time interoperation with the multi-user TTIMR. The multi-user Tactical Tomahawk simulation system involves individuals cooperating using a shared information space which reflects the external processes being controlled. Each operator maintains and interacts with a separate view of the shared space. The cooperation between operators is supplemented by direct communication via a text chat box. The Command Overview (CO), which is a focused (target-centric) design and the Tactical Tomahawk Interface for Monitoring and Retargeting (TTIMR) interface, which is a broad (missile-target relationship design), provide the same information in different perspectives to the operators. Depending on the team structure, not all operators will have the same interface yet both types of interfaces implement the ability to simulate a retargeting event without interfering with the activity representation or another operator's view of the data prior to actually implementing the retargeting event. The operators are able to carry out sub-goals and communicate via the chat box to determine which courses of action best meet the final goals of the mission. The shared information space, the different presentations of the data, and the communication among all members enhance the ability for the teams to acquire and maintain situational awareness.

3.1 The Multiuser TTIMR Client

The TTIMR client interface is very similar to the single-user TTIMR (see Figure 1) developed by Cummings (2003). It is composed of two primary subsystems, one for monitoring and one for retargeting. The monitoring capabilities include the map display and time bar, while the retargeting capabilities include the decision matrix and chat box. The main enhancement for the multiplayer version, besides the simulation supporting multiple users, is the concept of missile ownership. Each operator has exclusive control over a portion of the missiles involved in a scenario. Missiles under an operator's control are all displayed together in bold font within a block in the decision matrix. The missiles in control by other operators are dimmed slightly to convey the fact that they are under another operator's control.

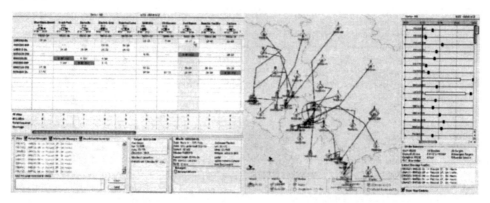

      Retargeting Display   **Monitor Map**

Figure 1. The TTIMR user interface

Similar to the single-user version of TTIMR, in order to retarget a missile, the user selects a combination of a target and a valid missile (e.g., one that meets the target constraints such as required warhead type and within a reachable distance given the remaining fuel) and hits a retarget button. After confirmation by the user, the missile will be retargeted after a simulated communications delay of about four minutes.

3.2 Command Overview

The design for the Command Overview (see Figure 2) was developed as an interface for use by the Commander in the Supervisor/Subordinate team structure. This interface contains the same map and chat box as used in TTIMR. The Decision Matrix is replaced by a more target-centric view of the battlefield. Instead of showing how all missiles relate to all targets simultaneously, the CO displays a "target list". The targets are each displayed in a vertical list, with the number of missiles required vs. the number of missiles assigned. This lets the Commander identify vital statistics for each target and which targets need additional missiles. One target can be selected at a time and a target-specific timebar is displayed, showing all missiles that could hit that target and at what time in a graphical representation. The Commander can then communicate through the chat window with subordinate operators to make a final decision in terms of retargeting.

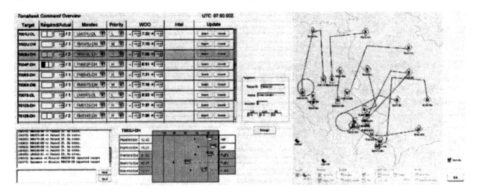

Target-Centric View with Timebar   **Spatio-Temporal Display**

Figure 2. The Command Overview user interface

3.3 Team Structure

In order to operate effectively as a team, a satisfactory team structure is required with coordination strategies depending on team members' roles and authority level.

In a Peer team structure, no operator is ranked above or below any other operator and they all act individually in control of their own subset of missiles. This simulates a command and control structure where each ship is in control of its own missiles. This method is flexible because the operators do not have to be arranged into a team structure beforehand. The interaction is dynamic in that any participant could step up to become a leader, or there could be no leader at all with decisions made by consensus. The downside to this method is that it results in somewhat nebulous responsibilities. No one person is responsible for making a decision.

A more regulated and disciplined team structure exists in the Supervisor/Subordinate model. This model has a more definite chain of command. The supervisor assumes a leadership position over a set of subordinates. The Navy has a hierarchical command structure, and this simulates the situation where a senior commander could play a role supervising and coordinating a set of operators each controlling a subset of missiles. In this model, the role of the subordinate team members is to provide supporting information and to make retargeting suggestions, but the commander makes the final decision as to which missile should be retargeted, and sends an order to the controller of that missile to do so. This differs from the Peer model by placing a better defined responsibility on both the supervisor and the operators but may add more time due to the need for a separate person to verify decision choices and then task a subordinate to execute the final decision made by the supervisor.

4 Design of Experiments

A mixed-factorial experiment was conducted to compare the Peer relationship to the Supervisor/Subordinate relationship for 3-person teams monitoring and controlling a simulated Tactical Tomahawk strike. Team type and role assignments were made in a random fashion. Subjects participated in just one type of team, either as a member of a three person Peer team or as a member of a three person Supervisor/Subordinate team. The Peer team had each of the three operators utilizing their own TTIMR interface in a peer relationship, while the Supervisor/Subordinate team had two of the three operators utilizing their own TTIMR interface, acting as subordinate peers to a third person acting as the Supervisor. The supervisor utilized the Command Overview user interface. The dependent variables are mission accuracy and elapsed time to perform actions. All teams ran through the same 20 minute scenario with 5 retargeting events presented in the same order. Thus, team structure is a between-subjects factor and retargeting event is a within-subjects factor.

4.1 Apparatus

Both the TTIMR and Command Overview interfaces were set up to run on Pentium 3 PCs connected over a network to the simulation server. Each participant's PC was equipped with two 17-inch color monitors arranged side by side along with a standard keyboard and mouse. The display setup for each monitor was 1024x768 pixels with 32 bit color. Total resolution across the two monitors was 2048x768 pixels. During the testing, all user actions were recorded by the software.

4.2 Subjects

The subjects consisted of 36 college students recruited from the University of Virginia. Subjects were monetarily compensated for their involvement. The mean age was 23 with a standard deviation of 6.31 years. The subjects had previously participated in two studies by Jones (2004), one which compared performance with or without the Decision Matrix using the stand alone TTIMR and one which compared performance between using the stand-alone TTIMR vs. the stand-alone Command Overview interface. Subjects were thus trained already on the respective user interfaces, except for on the concept of missile ownership and the protocol for making decisions when on a team.

4.3 Procedure

Six subjects were run at a time. Each subject was handed a note card telling them his/her team assignment and role. They were informed that for the purposes of the simulation they were on different ships and could only communicate through the chat box. The subjects did not know which of the other people in the room were on their team. This was done to make sure all communications occurred through the chat window instead of people talking to teammates directly.

The chat protocol sheet was given to each subject and a brief 5 minute training session was given on how to communicate with teammates and how to reach a consensus in determining who would retarget a particular missile. The protocol was different for the Supervisor-Subordinate vs. Peer team structures. The protocols both started off the same way. Each TTIMR (peer) operator chose the best missile that was controllable by that operator. The operator would then type the Time on Target (TOT) and priority of the missile into the chat box to allow the teammates to see what the best option was for everyone. If the team was formed with the Peer structure, a decision would have to be made among the three peers by consensus. Otherwise, in the Supervisor-Subordinate structure, the Commander would look at the two options from his/her subordinate operators, decide which was the optimal choice and then type back a message authorizing the owning operator to retarget the missile. After the training session, each team ran through a 20 minute practice scenario (which had 5 retargeting events) to allow participants to become familiar with chatting and collectively deciding on a course of action.

After the practice session was complete and any questions were answered, the subjects began the actual test scenario. During the testing session, no talking was permitted. The test scenario lasted about 20 minutes and involved 30 retargetable missiles and five retarget events. Thus, for the Peer structure, each operator controlled 10 missiles and for the Supervisor-Subordinate structure, each of the two subordinate operators controlled 15 missiles. In all situations, all operators could monitor the entire strike albeit through their own user interface.

5 Results

The treatment consisted of twelve teams with three members each. Six were Peer structures and the other six were Supervisor-Subordinate structures.

All statistical tests were performed at a 95% confidence interval (alpha = .05).

5.1 Decision Time

One of the two direct measurements recorded during the experiment was the amount of time it took a team to make a retargeting choice. This was measured as the number of seconds that elapsed from the time the emergent target appeared on the screen to when one of the team members assigned a missile to hit that target. The statistical model that fits this experiment, when examining decision time, is a mixed factorial 5x2 design. The interpretation and general procedures for testing the main effects and the interaction are the same in the mixed factorial as they are in the between-subjects factorial ANOVA. Different error terms, however, are used for the test of the between-subjects main effect and the within-subjects main effect.

In order to meet the assumptions of the ANOVA test, a natural log $(x = \ln(time))$ transformation was applied to the time data to normalize it. The interaction term (Target*Team) was not significant. The mean decision time across all retargeting events was 134 seconds for the Supervisor/Subordinate Team vs. 89 seconds for the Peer teams, showing the Peer teams to be significantly faster $(p = .039)$. The Target variable was also significant $(p < .001)$. Both the Scheffe post hoc test and the Bonferroni test indicated that there are two distinct groupings within the Target variable. Targets 1 and 2 took a similar amount of time to solve, while targets 3, 4 and 5 took a similar but longer time to solve.

An examination of the number of replications is helpful to determine the probabilities of Type I and II errors. Using the 60 trials used in this experiment results in a power level of 0.559, which is less than the commonly desired level of 0.8 (Kuehl, 2000). To reach a power level of 0.8, 106 trials would be needed.

5.2 Accuracy

To measure the accuracy of the teams' decisions, the number of correct answers in the scenario was studied. In this scenario, there were a total of five retargeting events. The scenario was set up so that there was only one "correct" answer for each target according to the training and "rules for retargeting" given to team members. There were viable but suboptimal missile assignments; however, the team was only awarded points for correct answers.

Because of the dichotomous nature (the value is either a 1 for a correct answer or a 0 for a wrong answer) of the accuracy dependant variable, the statistical model used for the accuracy performance was a binomial logistic regression model (Garson, 2004). The likelihood ratio test was run two times. The first time was with the full factorial model, including the main effects and the interaction term. This showed that there was no significance on the interaction effect; however there were not enough degrees of freedom to analyze the main effects. The test was run again on the main effects and both the Team and Target variables showed significance. The Supervisor/Subordinate teams were significantly more likely to get the correct answer (90% correct for the Supervisor/Subordinate teams vs. 70% correct for the Peer teams, $p = .032$).

In summary, the Team structure independent variable was significant for both accuracy and decision time. The decision time increased for the Supervisory/Subordinate structure, a decision took on average 45 seconds longer (134 seconds as opposed to 89 seconds for the Peer structure). However, the accuracy of the decisions increased as well (90% correct as opposed to 70% correct for the Peer structure).

6 Discussion

This experiment was set up to help decide: Given limited human resources (e.g., a set number of people available to assist with a task of high complexity), is it better to divide the workload equally among all the available personnel or to set aside one operator to act in a supervisory role, thereby increasing the workload of each of the remaining operators (see Table 1) but perhaps gaining back the benefit by having an independent review of the situation? The factors that will affect the answer to this question are:

- *The complexity of scenarios* The scenarios will drive the workload requirements of the task.
  - We chose a somewhat complex scenario – 30 retargetable missiles with five different retargeting events, but limited secondary tasking.
- *The maximum workload one operator can handle* Clearly if a scenario is simple enough that a subset of operators can handle the job adequately by dividing the workload evenly, then the difference in performance between the two team structures is likely to be naught.
  - Previous experiments had shown that a single operator becomes overloaded as the number of missiles to control increases beyond 12. Thus, in a peer relationship with 3 operators and 30 missiles, each operator would have 10 missiles to control but in the Supervisor /Subordinate relationship, each operator would have 15 missiles to control, but the Supervisor would be monitoring all the targets independently.
- *The communications overhead imposed by having more operators on a team to communicate with*
  - We held the communications method constant across the two team structures
- *The user interfaces available* Different user interfaces can make a task easier or more difficult given the ease or difficulty with which information can be accessed to support critical decisions.
  - We used a previously tested user interface design and held the user interface design constant across controllers, but gave the supervisor an altered view of the same situation to focus attention on monitoring target status.

The limitations of our experiment are many, thus the results need to be interpreted with caution. We only had 6 teams in each condition. This was due to limited resources, only being able to recruit 36 subjects due to their previous participation in related experiments in our lab. We were only able to test the teams on one, albeit fairly complex scenario with 5 retargeting events. Finally, we were only able to compare the team structure differences for a 3 person team. Given these limitations, the results from this experiment can not be generalized.

Table 1: Theoretical workload differences for controllers in a Peer vs. Supervisor/Subordinate Team relationship, holding number of team members and scenario complexity constant.

Number of People in Team	Number of people acting as controllers (Peer team)	Number acting as controllers (Supervisor/ Subordinate team)	% Workload per controller (Peer Team, assuming load balance)	% Workload per controller (Sup/ sub team, assuming load balance)	Theoretical % increase in workload per controller by making 1 team member a Supervisor
2	2	1	50%	100%	100%
3	3	2	33%	50%	52%
4	4	3	25%	33%	32%
5	5	4	20%	25%	25%

Comparing the Supervisor-Subordinate team structure against the Peer team structure, the lowest decision time favors the Peer team, while higher accuracy favors the Supervisor/Subordinate team. This illustrates a tradeoff between performing a task accurately and performing it quickly. In the Tactical Tomahawk domain it would probably be preferable to sacrifice speed to achieve better accuracy; however in other domains this may not necessarily be true.

The significance of the Target variable for both measures means that the difficulty of making an assignment depended on which target needed a missile. This result was expected because the retargeting events were designed to have differing levels of complexity.

Further data analysis may yield insight into why we found these results. For example, we have yet to analyze the chat communications that led up to each retargeting decision. Such an analysis may lead to some insights as to why some teams were faster or more accurate than others. Not having done this analysis, one speculation is the following: The Peer teams had less missiles to control per operator, thus, as predicted by our prior experiments, as the number of missiles under control increases from 10 to 15, time to make a decision increases. This would explain the increased speed with which the Peer team accomplished the retargeting tasks – because each of the peers could find the "best" missile among their set of missiles faster. Alternatively, in the other team structure, each peer had 15 missiles under control. This would explain the slower time to make a decision for this team, but not the increased accuracy. Previous experiments with single operators showed that accuracy was not affected by the number of missiles (but accuracy was not perfect). Thus, one explanation for the increase in accuracy performance for the Supervisory/Subordinate team is the addition of the Supervisor to the team, who can double-check answers, perhaps using an independent decision making process, particularly given an alternative user interface with which to make the decision.

In regards to the team structure, it appears that the Supervisor may not have an influence on the amount of time it takes to complete a retargeting task but does influence the accuracy with which it is done. This demonstrates that it is not merely the number of people performing a task that matters, but also the type of coordination among the team members. The Peer structure is mostly augmentative and debative because each user performs basically the same tasks as other team members, and deliberation must occur between the subjects. The Supervisor/Subordinate team structure is also augmentative (and perhaps debative), but integrative cooperation is important for successful completion of the objectives. The Commander is provided a different view of the situation and asked to employ a different skill set to help solve the missile retargeting problem. This hypothesis needs to be confirmed with further experimentation, for example, comparing a peer team running a 45 missiles scenario (each operator controlling 15 missiles) to a Supervisory/Subordinate team running a 30 missiles scenario (each operator controlling 15 missiles) and comparing two Supervisory/Subordinate teams of equal size on the same scenarios where the Supervisor has the same or different user interface as the controllers under his/her supervision. Another analysis that could be performed would be a GOMS (Card, Moran, & Newell, 1983) or other human performance model using the data collected through this experiment as inputs into the model. Thus, if we could simulate the team players computationally, we could run many more variations on the theme of this experiment to try to gain more insight into the optimal size and structure of the teams and the human-computer and human-human interactions that would best support rapid decision making.

7 Acknowledgements

This material is based upon work supported by the Naval Surface Warfare Center, Dahlgren Division (NSWC/DD), through a grant from the Office of Naval Research (ONR) Knowledge, Superiority, and Assurance Future Naval Capability (KSA FNC) Program. Any opinions, findings, and conclusions or recommendations expressed in this material are those of the authors and do not necessarily reflect the views of NSWC/DD or ONR.

8 References

Bentley, R., Rodden, T., Saywer, P., Sommerville, I. (1992). An Architecture for Tailoring Cooperative Multi-user Displays, *Proceedings of the 1992 ACM conference on Computer-supported cooperative work*, 187-194.

Card, S. K., Moran, T. P., & Newell, A. (1983). *The Psychology of Human-Computer Interaction.* Hillsdale, NJ: Laurence Erlbaum Associates.

Cummings, Mary (2004). Designing Decision Support Systems for a Revolutionary Command and Control Domains, *University of Virginia Doctoral Dissertation.*

Garson, David (2004). *StatNotes: An Online Textbook,* Retrieved July 7, 2004, from http://www2.chass.ncsu.edu/garson/pa765/statnote.htm

Jones, Chris (2004). A Graphical Multi-User Interface System for the Tactical Tomahawk Missile System, *University of Virginia Undergraduate Thesis.*

Jones, Patricia M., & Jacobs, James L. (2000). Cooperative Problem Solving in Human-Machine Systems: Theory, Models, and Intelligent Associate Systems, *IEEE Transactions on Systems, Man, and Cybernetics – Part C: Applications and Reviews*, 30(4), 397-407.

Kuehl, Robert O. (2000). *Design of Experiments: Statistical Principles of Research Design and Analysis, Second Edition*, Pacific Grove, CA: Brooks/Cole.

Willis, Robert A. (2001a). Tactical Tomahawk Weapon Control System User Interface Analysis and Design, University of Virginia Master's Thesis.

Willis, Robert A. (2001b). Effect of Display Design and Situation Complexity on Operator Performance. *Proceedings of the Human Factors and Ergonomics Society 45th Annual Meeting*, 346-350.

Cognition, Teams, and Augmenting Team Cognition: Understanding Memory Failures in Distributed Human-Agent Teams

Stephen M. Fiore, Florian Jentsch,
and Eduardo Salas
University of Central Florida
3100 Technology Parkway
Orlando, FL 32826
sfiore@ist.ucf.edu, fjentsch@ucf.edu,
esalas@ucf.edu

Neal Finkelstein
U.S. Army Simulation & Training
Technology Center
12423 Research Parkway
Orlando, FL, 32826-3276
neal.finkelstein@us.army.mil

Abstract

Based upon the integration of constructs from organizational and cognitive science we present a theoretical framework for understanding memory function in the context of human-agent teams. To support the development of true *Human Systems Integration*, we use this approach to meld robust concepts in human cognition with human agent team research. Our goal is to illustrate the theoretical and practical importance of these concepts to *team cognition* in general and *augmented cognition* in particular. We discuss this through theory in human memory and memory failures and integrate approaches to illustrate their value to developing research plans for augmenting cognition.

1   Integrating Systems and Humans

From the organizational sciences the field of team research has matured substantially over the latter part of the 20th Century. Similarly, the cognitive sciences have grown tremendously upon a strong theoretical and empirical foundation. Only in the last decade have these two fields begun to more formally interact to produce what is now being called "team cognition" (see Salas & Fiore, 2004). Based upon the integration of constructs from organizational and cognitive science we present a theoretical framework for considering human-agent team functioning. In order to support the development of true *Human Systems Integration*, we use this approach to meld robust concepts from the cognitive sciences with human agent team research. Our goal with this is to illustrate their importance to *team cognition* in general and *augmented cognition* in particular. We first discuss this through the lens of HSI and then narrow our focus to memory and memory failures. Finally we integrate these approaches to illustrate their value to developing research plans for augmenting cognition.

1.1   Human Systems Integration

Over the last decade, the Department of Defense has made increasing use of findings from the cognitive and computational sciences within its "*human-systems integration*" (HSI) program. This is a broad based concept for systems acquisitions programs requiring a level of analysis able to model how tools can support the human in his/her tasks. This includes not only single operators engaged with a given system (Salas & Klein 2001) but also encompasses how teams interact over time and space with distributed technologies (cf. Fiore et al., 2003) and similarly encompasses how intelligent agents are being integrated with modern systems (e.g., McNeese, Salas, & Endsley, 2001; Sycara & Lewis, 2004). HSI doctrine has developed to ensure that both the design of systems and their eventual development are able to fully support the human operator (Clark & Goulder, 2002; Freeman & Paley, 2001; Freeman, Pharmer, Lorenzen, Santoro, & Kieras, 2002; Pharmer, Dunn, & Santarelli, 2001).

Within this context an important development is that of *"human-centered work system design,"* an effort that emerged out of expert systems research in the 1980's and which has since evolved into research in a variety of complex domains (Clancey, 2002) including semiautonomous missions to the moon for NASA (e.g., Clancey, 2004; Sierhuis & Clancey, 2002). This approach applies research and theory to better understand how the human interacts with, and is impacted by, their systems. More specifically, "rather than abstracting human behavior as work processes or tasks... [this models] people's activities comprehensively and chronologically throughout the day (p. 32, Sierhuis & Clancey, 2002). By focusing on how interaction is actually organized and the associated details of such work, this approach takes a broader perspective by considering not only the technologies involved in a task, but also the human operators of these technologies and how they actually use them rather than are believed to use them. As such, this approach is as much anthropological as it is cognitive and engineering – effectively integrating disciplines to create models that appropriately simulate how work really occurs in complex socio-technical systems.

1.2   Overview of Paper

It is this form of human-centered theorizing and design that is foundational to understanding human-agent teams. Specifically, the technology-based characteristics present in such teams have the potential to attenuate the processes and the products occurring during human-agent interaction. For our initial efforts we consider agents broadly, following Fiore et al. and defining them as "ranging from computer-based intelligent decision-support systems with no or minimal anthropomorphism, to highly anthropomorphic machines, such as android robots, robotic animals, and robotic swarms or packs which display group behaviors" (Fiore, Jentsch, Becerra-Fernandez, Salas, & Finkelstein, 2005, p. 1). Understanding the cognitive and social processes emerging within human-agent teams is critical to developing the appropriate tools and techniques for augmenting cognition. To support our efforts we view human-agent teams as a socio-technical system, akin to the way we have viewed distributed teams (see Fiore et al., 2003). While much of the research in Augmented Cognition does emphasize the human, we suggest the research base can be strengthened by more fully exploring human-systems integration and human-centered design. Following this human-centered approach to melding systems with the human (Clancey, 1997; Hoffman, Hayes, & Ford, 2002; Shafto & Hoffman, 2002), we next discuss a theoretical framework that enables us to elucidate a small set of the factors that support human-agent process and are, therefore, targets for augmenting cognition.

2   Understanding Memory Function in Dynamic Environments

Dynamic interaction and distribution over space and time, the rule in technologically-dependant human-agent team environments, forces members to integrate sensory input across differing modalities as they attempt to coordinate their actions. Although the emergence of human-agent teams has led to substantial flexibility in operations, they may also result in undue cognitive load, that is, a workload over and above that experienced in co-located teams (cf. Fiore et al., 2003). When considering that data and interaction can come from, or result from, agent team members, this adds a layer of complexity to coordinative efforts. Specifically, this places additional limits on team members' ability to attend to cues pertinent to their tasks because their teammates are either geographically-dispersed or not human. Although this generally impacts a number of individual and team processes (see Fiore et al., 2003), in this paper we discuss its potential for interacting with the human operator and lead to additional workload that may impact the memory processes of the human team members.

To describe the aforementioned effect we use the general term memory failures. These failures are suggested to occur because the cues normally relied upon by co-located, human-human teams to support memory processes are now attenuated in some way by the human-agent distributed work. This, in turn, may produce faulty coordination leading to poorer performance. In short, we argue that this new interaction environment consisting of humans and agents who are not co-located, may alter and even hinder memory performance for distributed team members. A necessary first step in understanding this phenomenon is the development of a classification of the types and causes of memory failures experienced by team members in these environments. Following earlier work on memory failures (Herrmann, Gruneberg, Fiore, Schooler, & Torres, in press), we suggest that human-agent teamwork would benefit from an investigation of the qualitative nature of the memory failures, as well as the proximal (direct) and distal (indirect) causes of these failures. In Table 1 we describe the broader goals for our approach in understanding memory failures.

Table 1. Goals Associated with Understanding Memory Failures in Human-agent Teams

Broad Goals	Description
Differentiating Memory Failures	Classification of the differing memory failures so as to develop categories that can differentiate lapses in memory
Distinguishing Memory Causes	Development of a taxonomy to parse the differing causes of memory failures in human-agent teamwork
Proposing Guidelines	Specific guidelines able to inform both system designers and operators so as to identify how cognition may be augmented to avoid situations leading to these memory failures

Despite the importance of understanding how memory failures may hinder performance in complex environments there is relatively little research that has been conducted on this phenomenon. As an example of the importance of memory failures, studies of the Aviation Safety Reporting System documented that failures in memory were related to over 10% of the reported errors (Endsley, 1999). Further, these studies found that failures in memory led to problems with decision making, and subsequently could be related to up to 50% of fatal and 35% of non-fatal accidents (Jones & Endsley, 1996). Finally, failures in memory were related to 11% of situation awareness problems as reported in the Aviation Safety Reporting System (Jones and Endsley, 1996). Unfortunately, because no comprehensive system for understanding memory failures is in place, we do not know the nature of these failures.

Some studies have tried to understand memory failures occurring outside the laboratory. For example, classic research on absented-mindedness used diary studies to ascertain the frequency and nature of these memory failures (e.g., Reason & Lucas, 1984). Others have explored how devices such as techniques for reminding can reduce failures (e.g., Beal, 1988). Some have investigated what particular memory improvement methods work in alleviating failures (e.g., Herrmann, Brubaker, Yoder, Sheets, & Tio, 1999; Herrmann, Buschke, & Gall, 1987). Finally, recent studies have investigated "everyday memory failures" by exploring the relation between failure type and cause (Fiore, Schooler, Whiteside, & Herrmann, 1997; Herrmann et al., in press).

It is this latter set of studies on which we base the remainder of this paper. We follow the work of Herrmann and colleagues to highlight its relevance to understanding cognition (and failures in cognition) in human-agent teams. We use this theoretical framework because we argue that a human-centered approach to augmented cognition is necessary to design the technologies that can effectively augment cognition. Specifically, only when we fully understand the limitations of human cognition when operating in human-agent teams, can we understand the particular scaffolds that can support cognition. Research in the area of everyday memory failures uses classification which includes differentiating between lapses that are failures in *prospective* memory or *retrospective* memory. Prospective memory is generally referred to as "memory for the future," or remembering to engage some action at some future time

(Herrmann & Chaffin, 1988). Retrospective memory failures are the more familiar type of failures, that is, failing to recall something that had been previously learned.

Although we do not deny the importance of retrospective memory to cognition and coordination in complex operational environments, for two reasons we focus this paper on prospective memory. First, prospective memory has been studied substantially less than retrospective memory. Recent papers by Kvavilashvili and Ellis (1996) and Ellis and Kvavilashvili (2000) document the changing patterns of interest in this topic within the cognitive sciences (see also Brandimonte, Einstein, & McDaniel, 1996). Second, prospective memory represents an area of cognition where technology may be able to make a significant improvement in functioning through augmented cognition (cf. Herrmann et al., 1999; Herrmann et al., 1987). In sum, although there has been recent interest developing in prospective memory and even some investigations of memory failures in operational environments, we still do not fully understand the nature and causes of these failures. More importantly, in the context of human-agent teams, little if any research has attempted to relate this construct to teams. Fiore et al. (in press) have begun to lay the foundation for understanding this phenomenon in teams and we next discuss their framework.

In sum, memory performance in human-agent teams represents not only an important area of inquiry, but also a rich theoretical area from which to consider how to use agents to augment human cognition. Our approach represents an adaptation of a paradigm developed by Herrmann and colleagues (see Fiore et al., in press; Herrmann et al., in press) for understanding memory failures outside of the laboratory.

3   A Framework for Understanding and Augmenting Memory in Human Agent Teams

Programs in augmented cognition attempt to produce diagnostic methods for understanding cognitive processes for what can be termed a form of dynamic scaffolding of cognition. This generally describes research and development in non-invasive techniques for measuring cortical activation that can be linked to a variety of higher- and lower-level cognitive processes (Schmorrow, 2002; Schmorrow & Kruse, 2002; St. John, Kobus, Morrison, & Schmorrow, 2004). While the short- and medium-term efforts are looking at detection accuracies, in the longer-term, augmented cognition will need to better meld with operationally relevant and time-stressed events. Thus, although these programs are still in their early stages of development, the computational methods and engineering systems will soon be at appropriate levels of sophistication to meet these goals. What we suggest is that, simultaneous to these developments, we must better understand human memory and successful and unsuccessful memory performance in operationally complex environments. These research tracks can develop independently but eventually be integrated when the theories and the technologies are themselves appropriately developed. Towards that end, we present a set of potential research principles for pursuing augmented cognition within human-agent teams. As discussed, for our initial foray into this area we narrowly focus on prospective memory. We argue that intelligent agent technology, given its increasing ubiquity as human-agent teams become more prevalent, represent a viable means for augmenting team cognition. Specifically, given that intelligent agents are being designed as team members and a large body of research is already in progress with respect to cognitive engineering and decision making, it is only prudent that we additionally consider how these agent team members can be used to better augment team cognition.

3.1   Memory Failures Framework

The classification of memory failures and their causes can enable a fuller understanding of team cognition in human-agent teams. Comprehensive programs aimed at augmenting human cognition at the individual and team level must address two primary criteria associated with memory failures. First, research must determine the quantity, and the qualitative nature of, the memory failures. Second, the causes of the memory failures must be determined. Towards this end, the memory failures framework based upon Herrmann and colleagues multi-modal approach to memory (Herrmann, 1996; Herrmann & Parente, 1994) can aid in addressing these criteria. The multimodal framework was developed to account for the multitude of factors associated with memory in everyday environments (i.e., outside the laboratory). Building upon this approach we suggest that understanding effective memory functioning in the complex environments in which human-agent teams operate must encompass physiological and psychological factors that contribute to cognition. Further, we suggest that social and technological factors must be considered within this broader conceptualization.

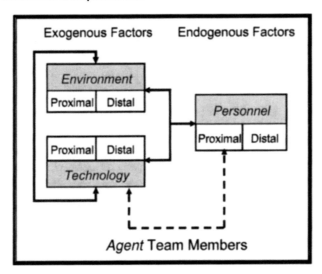

Figure 1. *Memory failures framework for Consideration in Human-agent Teams*

As shown in Figure 1, this framework presents a general classification of memory failure causes. These can arise due to endogenous or exogenous factors and can be either proximally (i.e., directly) related, or distally (i.e., indirectly) related to the memory action (e.g., Herrmann et al., in press). Table 2 shows this distinction and presents a description of each. As shown, causes can emerge due to something internal or external to the human. Exogenous factors influence memory failures due to system or technological problems (e.g., poor understanding of system functioning). Exogenous factors also include the environmental context which could encompass both social (e.g., lack of coordination with agent team member), or natural problems (poor weather). Furthermore, endogenous factors encompass problems arising out of psychological or physiological states, that is, factors internal to the human (e.g., stress). Proximal factors are the causes directly disrupting encoding, retention, or remembering processes while distal factors arise from processes not directly related to memory (e.g., physiological states), but which may hinder memory processing if compromised.

Table 2. Taxonomical Classification of Representative Memory Failures in Human-agent Teams

		Location of Cause	
		Endogenous (Internal)	Exogenous (External)
Directness of Cause	Proximal (Direct)	Cognitive problem arising from task overload	External contextual problem such as cue missing
	Distal (Indirect)	Physiological problems arising due to mission duration	External problem such as social distraction

3.2 Research Guidelines for Augmenting Prospective Memory in Human-agent Teams

As can be see in the Fiore et al. framework presented in Figure 1, what is complicated is that the "agent" team members cut across both the technological and the social components of this framework. As such, research must determine the degree to which there may be additive or even multiplicative effects on coordination that are arising from this socio-technical system. By adopting this framework to the study of memory performance in complex environments we can better understand what failure causes may be amenable to technologically-based interventions. Specifically, it may be feasible to use augmented cognition technology to either mitigate the occurrence, or minimize the effects of, memory failures, both at the individual and team level. Thus, despite gaps in our understanding of prospective memory in complex operational environments utilizing human-agent teams, enough research exists to allow us to introduce preliminary guidelines for consideration in augmented cognition research.

Theories on prospective memory (PM) distinguish between forms of prospective memory that are tied to how it is that the action must be completed. We use these distinctions to suggest how agents can be used to augment these differing forms of prospective memory actions. Specifically, research in prospective memory has noted that the actions associated with prospective memory fall into two major categories (see Ellis, 1988). First are those actions that require precise execution at a given point in time, referred to as "pulses." Other actions can be completed at any time over a wider time frame, referred to as "steps." A related distinguishing characteristic has to do with whether *time* or *events* are considered to be the driving factor in the memory action. This distinction is important because both event-based and time-based prospective memory tasks are diagnosable and similarly represent pertinent targets for augmented cognition. Finally, recent research suggests that matches between ongoing processing and processing required for the prospective memory tasks can be facilitative (see Meier & Graf, 2000, see also Marsh, Hicks, & Hancock, 2000). Following the tenets of Transfer Appropriate Processing theory (see Roediger, Gallo, & Geraci, 2002), this approach notes how synchrony between processes required for the memory task, and the actual operational task processing requirements, can be crucial for performance. This understanding adds an additional level of diagnosticity through which to view human-agent team cognition as it elucidates when memory processes may be compromised. Using the aforementioned theorizing, in Table 3 we present a representative sample of research and development guidelines for augmenting cognition within human-agent teams.

Table 3. Representative Sample of Research Guidelines for Augmenting Cognition Based upon Prospective Memory Theory

Guidelines	Guideline Description
Guideline 1. PM *Pulses*	Agents used to augment successful prospective memory *pulses* must be able to diagnose mission chronology to prompt the human team members appropriately.
Guideline 2. PM *Steps*	Agents used to augment successful prospective memory *steps*, although requiring less rigidity in programming, must be able to monitor task executions over a broader task space to determine when to prompt.
Guideline 3. PM *Events*	Augmenting cognition for prospective memory should map event-based prospective memory tasks onto mission *parameters* so human-agent team members are able to diagnose when *critical events* have occurred and provide reminders appropriately.
Guideline 4. PM *Timing*	Augmenting cognition for prospective memory should map time-based prospective memory tasks onto mission *chronology* so human-agent team members are able to determine when the requisite *time period* has passed and provide reminders appropriately.
Guideline 5. PM *Processing Matches*	Research should determine how agent members of a team can be made aware of the nature of the processing required for a given prospective memory task to determine when memory aids are, or are not, warranted. For example, aids may be warranted when there is a mismatch in processing between the ongoing task and the prospective memory task.

4   Conclusions

In this paper we have presented a preliminary means with which to understand causes of memory failures by illustrating a broad set of categories of causes potentially occurring in human-agent teams. We additionally presented a finer distinction of types of prospective memory tasks. These were used to derive a set of research guidelines for using agents to augment team cognition. Following the general rubric of human-systems integration, and more specifically, human-centered work design, we illustrated how the research base in augmented cognition can be strengthened. We have described how theories on prospective memory emerging out of the cognitive sciences can be used to explore the unique challenges emerging from this new form of organizational structure. Prospective memory presents a rich theoretical and practical area of inquiry in which to explore how agents may be used effectively augment cognition. These are only preliminary guidelines that may be used for research investigating how agents may be designed to target these memory failures so as to attenuate the negative consequences sometimes emerging from human-agent teamwork.

From the perspective of viewing science *strategically*, this approach can be construed of as more of a short- to medium-term effort. In particular, augmented cognition and the research driving us towards that goal is pursuing more of a medium- to long-term emphasis in that the transition from the laboratory to the field is still somewhat in the future. Our argument was that the human-centered approach we have proposed can be pursued simultaneously to other augmented cognition efforts so that convergence can be reached at an earlier date. We bring up this point in our concluding section because there are related efforts to consider from the perspective of long-term research planning. In particular, a growing body of literature is beginning to document the brain regions activated prior to and during prospective memory tasks. For example, West, Herndon, and Ross-Munroe (2000) find that an initial stage of prospective memory known as the noticing component activates the occipital-parietal region. Further, they find that sections of the frontal cortex may be more responsible for a directed search component of this task. Burgess, Quayle, and Frith (2001) varied the intentions associated with a prospective memory task and demonstrated differing areas of activation dependent upon whether the intention was being maintained or was actually realized (see also Burgess & Shallice, 1997; McDaniel et al., 1999). As such, from a programmatic perspective, research and development integrating human-agent teams with augmented cognition can similarly consider this developing literature coming out of cognitive neuroscience.

In sum, there is a growing convergence on our understanding of prospective memory, an important cognitive process supporting the operator in dynamic environments. We argue that it be a strategic target for the efficacious use of agent technology to augment cognition. We suggest that others pursue this approach to human-agent team research so as begin to place bounds around the cognitive and social consequences affecting interaction and team development when work is technology-mediated. More specifically, research must first understand where the problems in team cognition actually are occurring prior to attempting to augment that cognition in human-agent teams.

References

Beal, C. (1988). The development of prospective memory skills. In M. M. Gruneberg & P. E. Morris (Eds.), *Practical aspects of memory: Current research and issues, Vol. 1: Memory in everyday life* (366-370). New York, NY: John Wiley & Sons.

Brandimonte, M.A., Einstein, G.O., & McDaniel, M. A. (Eds.) (1996). *Prospective Memory: Theory and applications*. LEA: Mahwah, NJ.

Burgess, P. W., Quayle, A., & Frith, C. D. (2001). Brain regions involved in prospective memory as determined by positron emission tomography. *Neuropsychologia, 39*, 545-555.

Burgess, P. W. & Shallice, T. (1997). The relationship between prospective memory and retrospective memory: Neuropsychological evidence. In M. A. Conway (Ed.) *Cognitive models of memory* (pp. 247-272). Cambridge, MA: MIT Press.

Clancey, W. J. (1997). The Conceptual Nature of Knowledge, Situations, and Activity. In P. Feltovich, R. Hoffman & K. Ford. (Eds) *Human and Machine Expertise in Context* (pp. 247-291). Menlo Park, CA: The AAAI Press.

Clancey, W. J. (2004). Roles for agent assistants in field science: Understanding personal projects and collaboration. *IEEE Transactions on Systems, Man and Cybernetics, Part C: Applications and Reviews, 34 (2)*, 125-137.

Clancey, W.J. (2002) Simulating activities: Relating motives, deliberation, and attentive coordination. *Cognitive Systems Research 3*, 471-499.

Clark, J. J. & Goulder, R. T. (2002). Human Systems Integration (HSI): Ensuring design and development meet human performance capability early in acquisition process. *PM: JULY-AUGUST*, 88 – 91.

Ellis, J.A. (1988). *Memory for Future Intentions: Investigating Pulses and Steps*. In M. M. Gruneberg, P. E. Morris, and R. N. Sykes (Eds.), *Practical Aspects of Memory: Current Research and Issues, Volume 1: Memory in Everyday Life* (pp. 371 - 376). Chichester: John Wiley & Sons.

Ellis, J. & Kvavilashvili, L. (2000). Prospective memory: Past, present, and future directions. *Applied Cognitive Psychology, 14 (SpecIssue)*, S1-S9.

Endsley, M. R. (1995a). Toward a theory of situation awareness in dynamic systems. *Human Factors. Special Issue: Situation Awareness, 37*, 32-64.

Endsley, M. R. (1999). Situation awareness in aviation systems. In D. J. Garland & J. A. Wise, (Eds.), *Handbook of aviation human factors: Human factors in transportation* (pp. 257-276). Mahwah, NJ: Lawrence Erlbaum Associates, Inc.

Fiore, S. M., Cuevas, H. M., Schooler, J. & Salas, E. (in press). Understanding memory actions and memory failures in complex environments: Implications for distributed team performance. In E. Salas C. A. Bowers, and F. Jentsch (Eds.) *Teams and Technology*. Washington, DC: American Psychological Association.

Fiore, S. M., Jentsch, F., Becerra-Fernandez, I., Salas, E. & Finkelstein, N. (2005). Integrating field data with laboratory training research to improve the understanding of expert human-agent teamwork. In the *IEEE Proceedings of the 38th Hawaii International Conference System Sciences*. Los Alamitos, CA.

Fiore, S. M., Salas, E., Cuevas, H. M. & Bowers, C. A. (2003). Distributed coordination space: Toward a theory of distributed team process and performance. *Theoretical Issues in Ergonomic Science, 4, 3-4*, 340-363.

Fiore, S. M., Schooler, J.W., Whiteside, D., & Herrmann, D.J. (1997). *Perceived contributions of cues and mental states to prospective and retrospective memory failures.* Paper presented at the 2nd Biennial Meeting of the Society for Applied Research in Memory and Cognition - Toronto, Canada.

Freeman, J.T., & Paley, M.J. (2001). A systematic approach to developing system design recommendations. In *Proceedings of the 45th Annual Conference of the Human Factors and Ergonomics Society* (CD-ROM). Minneapolis, MN.

Freeman, J.T, Pharmer, J.A., Lorenzen, C., Santoro, T.P, & Kieras, D. (2002). Complementary Methods of Modeling Team Performance. In *Proceedings of the Command and Control Research and Technology Symposium*, Monterrey, CA.

Herrmann, D. J. (1996). Improving prospective memory. In M. A. Brandimonte, G. O. Einstein, & M. A. McDaniel (Eds.), *Prospective memory: Theory and applications.* Erlbaum: Mahwah, NJ.

Herrmann, D. J., Brubaker, D., Yoder, C., Sheets V., Tio, A. (1999) Devices that remind. In F. T. Durso (Ed.) *Handbook of Applied Cognition* (pp. 377 - 407). Chichester, England: Wiley.

Herrmann, D. J., Buschke, H., & Gall, M. B. (1987). Improving retrieval. *Applied Cognitive Psychology, 1*, 27-33.

Herrmann, D. J. & Chaffin, R. (1988). *Memory in a historical perspective.* New York, NY: Springer Verlag.

Herrmann, D., Gruneberg, M., Fiore, S. M., Schooler, J. W. & Torres, R. (in press). Accuracy of reports of memory failures and of their causes. In L. Nilson and H. Oata (Eds.), *Memory and Society*. Oxford, England: Oxford University Press.

Herrmann, D. J. & Parente, R. (1994). A multi-modal approach to cognitive rehabilitation. *Neuro Rehabilitation, 4*, 133-142.

Hoffman, R., Hayes, P., & Ford, K.M. (2002). The Triples Rule. *IEEE Intelligent Systems, May/June*, 62-65.

Jones, D. G. & Endsley, M. R. (1996). Sources of situation awareness errors in aviation. *Aviation, Space, & Environmental Medicine, 67*, 507-512.

Kvavilashvili, L. & Ellis, J. (2000). Varieties of intention: Some distinctions and classifications. In M. A. Brandimonte, G. O. Einstein, & M. A. McDaniel, (Eds.), *Prospective memory: Theory and applications* (pp. 23 – 51). Erlbaum: Mahwah, NJ.

Marsh, R. L., Hicks, J. L., & Hancock, T. W. (2000). On the interaction of ongoing cognitive activity and the nature of an event-based intention. *Applied Cognitive Psychology, 14 (SpecIssue)*, S29-S41.

McDaniel, M. A. & Einstein, G. O. (2000). Strategic and automatic processes in prospective memory retrieval: A multiprocess framework. *Applied Cognitive Psychology, 14 (SpecIssue)*, S127-S144.

McDaniel, M. A., Glisky, E. L., Rubin, S. R., Guynn, M. J., & Routhieaux, B. C. (1999). Prospective memory: A neuropsychological study. *Neuropsychology, 13*, 103 - 110.

McNeese, M., Salas, E, & Endsley, M. (Editors.) (2001). *New trends in collaborative activities: Understanding system dynamics in complex environments.* Santa Monica, CA: Human Factors and Ergonomics Society.

Meier, B. & Graf, P. (2000). Transfer appropriate processing for prospective memory tests. *Applied Cognitive Psychology, 14 (SpecIssue)*, S11-S27.

Pharmer, J., Dunn, J., & Santarelli, T. (2001). The application of a cognitive modeling tool to provide decision support during Naval operations. In *Proceedings of ASNE Human Systems Integration Symposium, Knowledge Warfare: Making the Human Part of the System* (CD-ROM). Arlington, VA: ASNE.

Reason, J. & Lucas, D. (1984). Absent-mindedness in shops: Its incidence, correlates and consequences. *British Journal of Clinical Psychology, 23,* 121-131.

Roediger, H. L., Gallo, D. A., & Geraci, L. (2002). Processing approaches to cognition: The impetus from the levels-of-processing framework. *Memory, 10,* 319-332.

Salas, E. & Klein, G. (Eds.) (2001). *Linking expertise and naturalistic decision making.* Hillsdale, NJ: Lawrence Earlbaum Associates.

Salas, E., & Fiore, S. M. (Eds.). (2004). *Team Cognition: Understanding the factors that drive process and performance.* Washington, DC: American Psychological Association.

Schmorrow, D. (2002). Augmented Cognition: Building Cognitively Aware Computational Systems. *Presentation at DARPATech Conference,* Anaheim, CA.

Schmorrow, D.D., & Kruse, A.A. (2002). Improving human performance throughadvanced cognitive system technology. In *Proceedings of the Interservice/Industry Training, Simulation and Education Conference* (I/ITSEC'02), Orlando, FL.

Shafto, M., & Hoffman, R. (2002). Human-Centered Computing at NASA. *IEEE Intelligent Systems, September/October,* 10-14.

Sierhuis, M., & Clancey, W.J. (2002). Modeling and Simulating Work Practice: A Method for Work Systems Design. *IEEE Intelligent Systems, September/October,* 32-41

St. John, M., Kobus, D. A., Morrison, J. G., & Schmorrow, D. (2004). Overview of the DARPA Augmented Cognition Technical Integration Experiment. *International Journal of Human-Computer Interaction, 17, 2,* 131-150.

Sycara, K. & Lewis, M. (2004). Integrating Intelligent Agents into Human Teams. In E. Salas and S. M. Fiore (Eds.), *Team Cognition: Understanding the factors driving process and performance* (pp. 203-231). Washington, DC: American Psychological Association.

West, R. Herndon, R. W., & Ross-Munroe, K. (2000). Event-related neural activity associated with prospective remembering. *Applied Cognitive Psychology, 14 (SpecIssue),* S115-S126.

Acknowledgements

Writing this paper was partially supported by Grant Number SBE0350345 from the National Science Foundation and by contract number N61339-04-C-0034 from the United States Army Research, Development, and Engineering Command, to the University of Central Florida as part of the *Collaboration for Advanced Research on Agents and Teams.* The opinions and views of the authors are their own and do not necessarily reflect the opinions of the University of Central Florida, the National Science Foundation, the RDECOM-STTC, the U.S. Army, DOD or the U.S. Government. There is no Government express or implied endorsement of any product discussed herein. All correspondence regarding this paper should be sent to Dr. Stephen Fiore, via email at sfiore@ist.ucf.edu or via regular mail at University of Central Florida, 3100 Technology Parkway, Suite 140, Orlando, FL 32826.

Author Index

Subject Index